FLORIDA

BIOLOGY

JOHNSON • RAVEN

TEACHER EDITION

Teacher Edition WALK-THROUGH

Student Edition CONTENTS IN BRIEF

HOLT, RINEHART AND WINSTON
A Harcourt Education Company
Orlando • **Austin** • New York • San Diego • Toronto • London

PROGRAM OVERVIEW

Holt Biology— Enabling teachers to meet individual needs

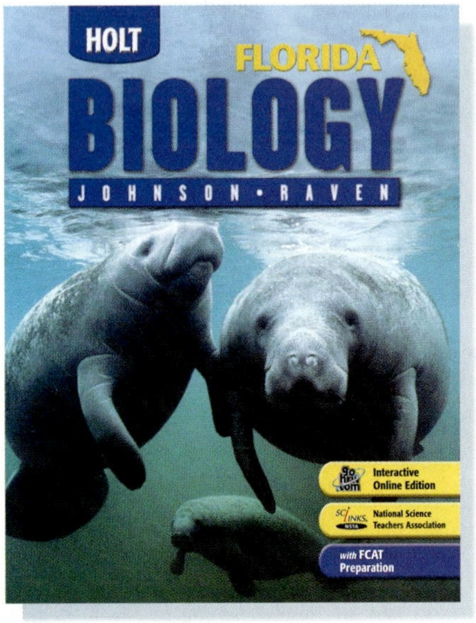

Holt Biology reflects current developments in both biology and the teaching of biology, with a consistent emphasis on complete content. Thorough, accurate explanations and instructions give students a solid foundation in the fundamentals, reducing the likelihood that lower-ability students will get lost or confused. Critical-thinking questions and open-ended activities provide the depth needed to challenge advanced students.

STUDENTS OF DIFFERENT ABILITY LEVELS DEVELOP THE SKILLS THEY NEED FOR SUCCESS

- The *Student Edition* features a clean, easy-to-navigate design.
- Heads and subheads are short and concise, making it easy to preview each chapter, and key vocabulary terms are highlighted at point-of-use to further assist students.
- Activities, labs, worksheets, and ancillaries are labeled by difficulty level (Basic, General, or Advanced) to help you choose the appropriate activities for each of your students. **Chapter Tests** come in two levels, General and Advanced.
- ***Holt Biology Special Needs Activities and Modified Tests with Answer Keys*** presents activities and tests that have been specially modified to reach struggling students.
- **FCAT Prep** allows students to practice their test-taking skills.

A FLEXIBLE LABORATORY PROGRAM BUILDS INQUIRY AND CRITICAL-THINKING SKILLS

- The laboratory program includes in-text labs for each chapter plus additional labs located in the ***Chapter Resource Files.***
- All labs are bench-tested and rated by level of difficulty so you can be sure that they will meet the needs of your students.
- **Quick Labs, Data Labs,** and **Math Labs** are short in-text labs that help you meet your curriculum needs and work within the time constraints of your class schedule.
- Each chapter ends with a two-page **Exploration Lab** or **Skills Practice Lab.**
- **Forensics Labs** in the **Appendix** introduce students to techniques used by modern forensic scientists.

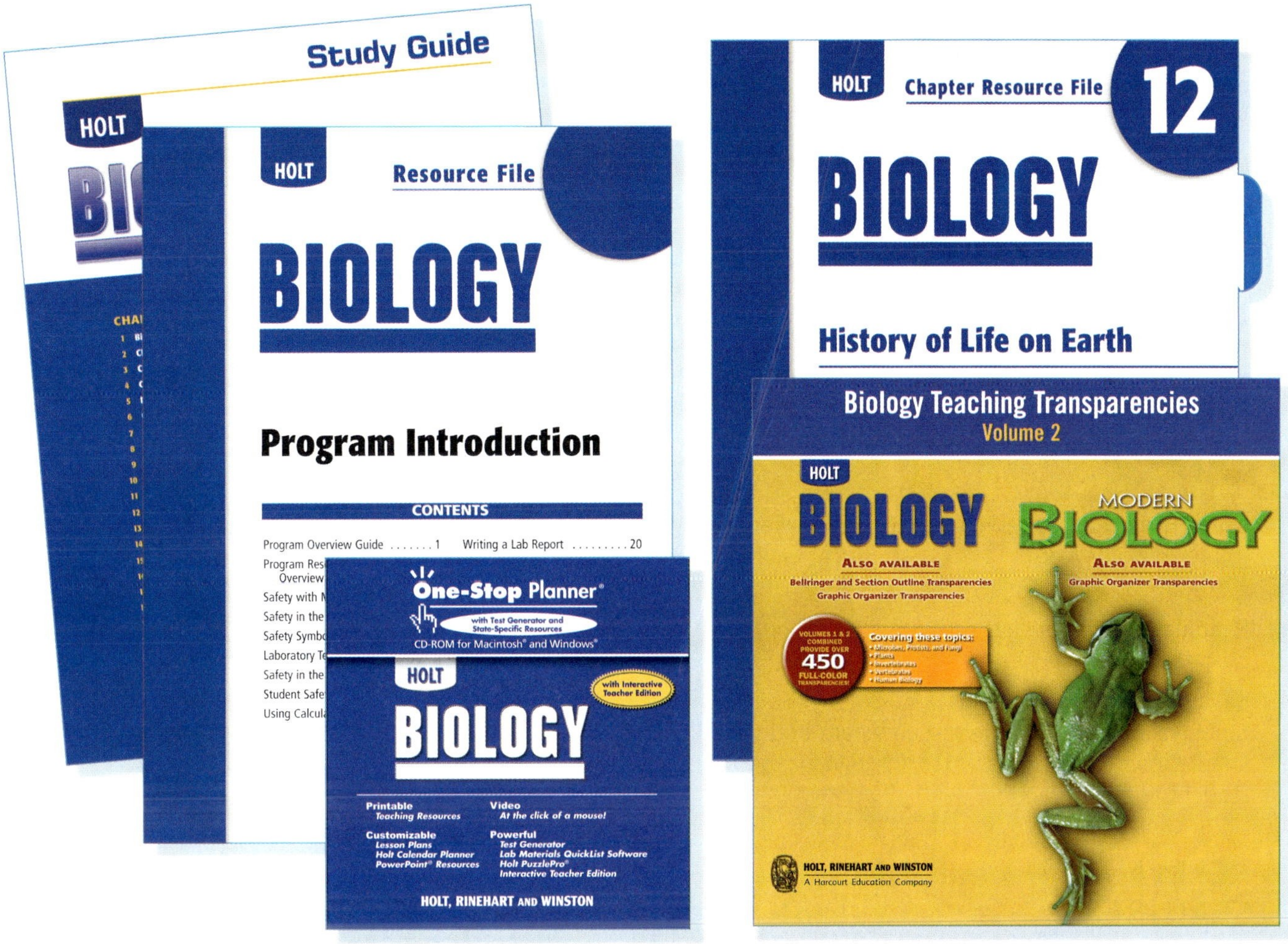

TECHNOLOGY RESOURCES MOTIVATE STUDENTS AND REACH DIFFERENT LEARNING STYLES

- Lighten the load with an interactive *Online Edition* or *CD-ROM Version* of the student text.
- **SciLinks®,** a Web service developed and maintained by the NSTA, contains current and prescreened links that engage students.
- ***Visual Concepts in Biology CD-ROM*** is your guide to **Visual Concepts**—engaging, media-based presentations of core concepts. **Visual Concepts** can be projected for large group instruction or viewed by individuals or groups of students with a computer.
- All the resources you need are on the ***One-Stop Planner® with Test Generator and State-Specific Resources CD-ROM,*** including worksheets, customizable lesson plans, **Holt Calendar Planner, PowerPoint® Resources, Holt PuzzlePro®,** and a powerful test generator.

Student Edition builds skills for success in science

ENGAGING MATERIAL GETS STUDENTS FOCUSED

Sunshine State Standards covered in the chapter are listed on the student page.

Looking Ahead presents a table of contents in brief to guide reading.

Quick Review helps reinforce material covered in previous chapters.

Reading Activity gets students thinking about the chapter to come.

SciLinks and other online resources are tied to chapter content and presented throughout the student text to extend students' learning.

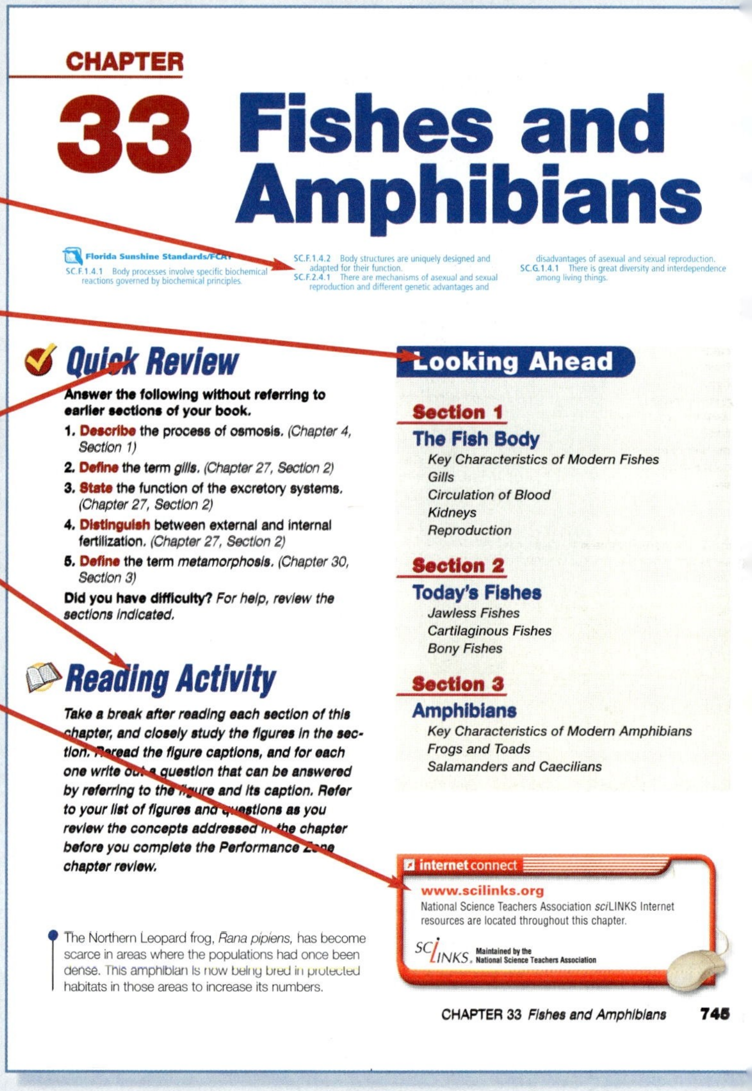

CHAPTER

33 Fishes and Amphibians

Florida Sunshine Standards/FCAT

SC.F.1.4.1 Body processes involve specific biochemical reactions governed by biochemical principles.
SC.F.1.4.2 Body structures are uniquely designed and adapted for their function.
SC.F.2.4.1 There are mechanisms of asexual and sexual reproduction and different genetic advantages and disadvantages of asexual and sexual reproduction.
SC.G.1.4.1 There is great diversity and interdependence among living things.

Quick Review

Answer the following without referring to earlier sections of your book.

1. **Describe** the process of osmosis. *(Chapter 4, Section 1)*
2. **Define** the term *gills*. *(Chapter 27, Section 2)*
3. **State** the function of the excretory systems. *(Chapter 27, Section 2)*
4. **Distinguish** between external and internal fertilization. *(Chapter 27, Section 2)*
5. **Define** the term *metamorphosis*. *(Chapter 30, Section 3)*

Did you have difficulty? *For help, review the sections indicated.*

Reading Activity

Take a break after reading each section of this chapter, and closely study the figures in the section. Reread the figure captions, and for each one write out a question that can be answered by referring to the figure and its caption. Refer to your list of figures and questions as you review the concepts addressed in the chapter before you complete the Performance Zone chapter review.

The Northern Leopard frog, *Rana pipiens*, has become scarce in areas where the populations had once been dense. This amphibian is now being bred in protected habitats in those areas to increase its numbers.

Looking Ahead

Section 1
The Fish Body
Key Characteristics of Modern Fishes
Gills
Circulation of Blood
Kidneys
Reproduction

Section 2
Today's Fishes
Jawless Fishes
Cartilaginous Fishes
Bony Fishes

Section 3
Amphibians
Key Characteristics of Modern Amphibians
Frogs and Toads
Salamanders and Caecilians

internet connect
www.scilinks.org
National Science Teachers Association *sci*LINKS Internet resources are located throughout this chapter.
SCILINKS. Maintained by the National Science Teachers Association

CHAPTER 33 *Fishes and Amphibians* **745**

RELEVANT AND EXCITING FEATURES

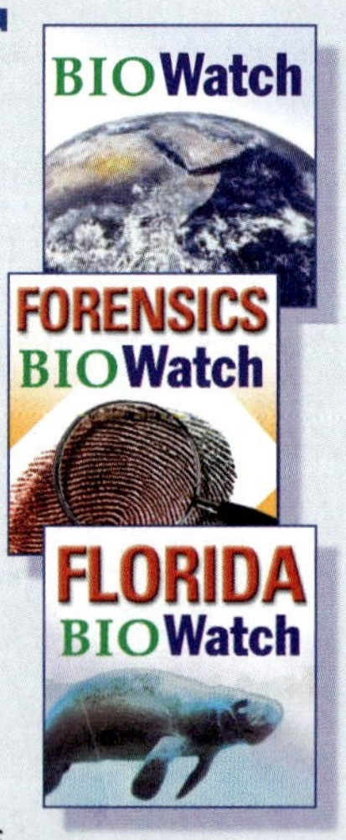

BioWatch features highlight the relationship between biology, ecology, health, and technology issues.

Forensics BioWatch features spark students' interest in the relationship between biology and solving crimes.

Florida BioWatch features focus attention on exciting research being conducted in Florida today.

Biology meets the world in **Real Life** features, explaining biology that students encounter in their daily lives.

Word Origins explore the roots of science language in order to expand students' understanding.

Up Close features provide detailed looks at important organisms.

Exploring Further explores key biological topics in greater depth, giving advanced learners the challenge they need.

Science, Technology, and Society features examine the impact of biology on issues in society.

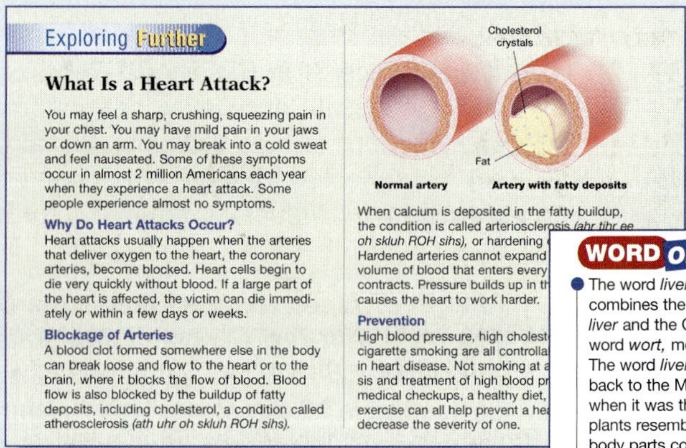

Exploring Further

What Is a Heart Attack?

You may feel a sharp, crushing, squeezing pain in your chest. You may have mild pain in your jaws or down an arm. You may break into a cold sweat and feel nauseated. Some of these symptoms occur in almost 2 million Americans each year when they experience a heart attack. Some people experience almost no symptoms.

Why Do Heart Attacks Occur?
Heart attacks usually happen when the arteries that deliver oxygen to the heart, the coronary arteries, become blocked. Heart cells begin to die very quickly without blood. If a large part of the heart is affected, the victim can die immediately or within a few days or weeks.

Blockage of Arteries
A blood clot formed somewhere else in the body can break loose and flow to the heart or to the brain, where it blocks the flow of blood. Blood flow is also blocked by the buildup of fatty deposits, including cholesterol, a condition called atherosclerosis *(ath uhr oh skluh ROH sihs)*.

When calcium is deposited in the fatty buildup, the condition is called arteriosclerosis *(ahr tihr ee oh skluh ROH sihs)*, or hardening ... Hardened arteries cannot expand ... volume of blood that enters every ... contracts. Pressure builds up in th... causes the heart to work harder.

Prevention
High blood pressure, high cholest... cigarette smoking are all controlla... in heart disease. Not smoking at a... sis and treatment of high blood pr... medical checkups, a healthy diet, ... exercise can all help prevent a hea... decrease the severity of one.

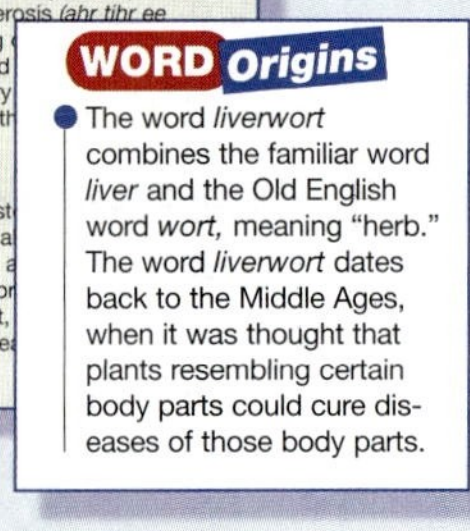

WORD Origins

The word *liverwort* combines the familiar word *liver* and the Old English word *wort*, meaning "herb." The word *liverwort* dates back to the Middle Ages, when it was thought that plants resembling certain body parts could cure diseases of those body parts.

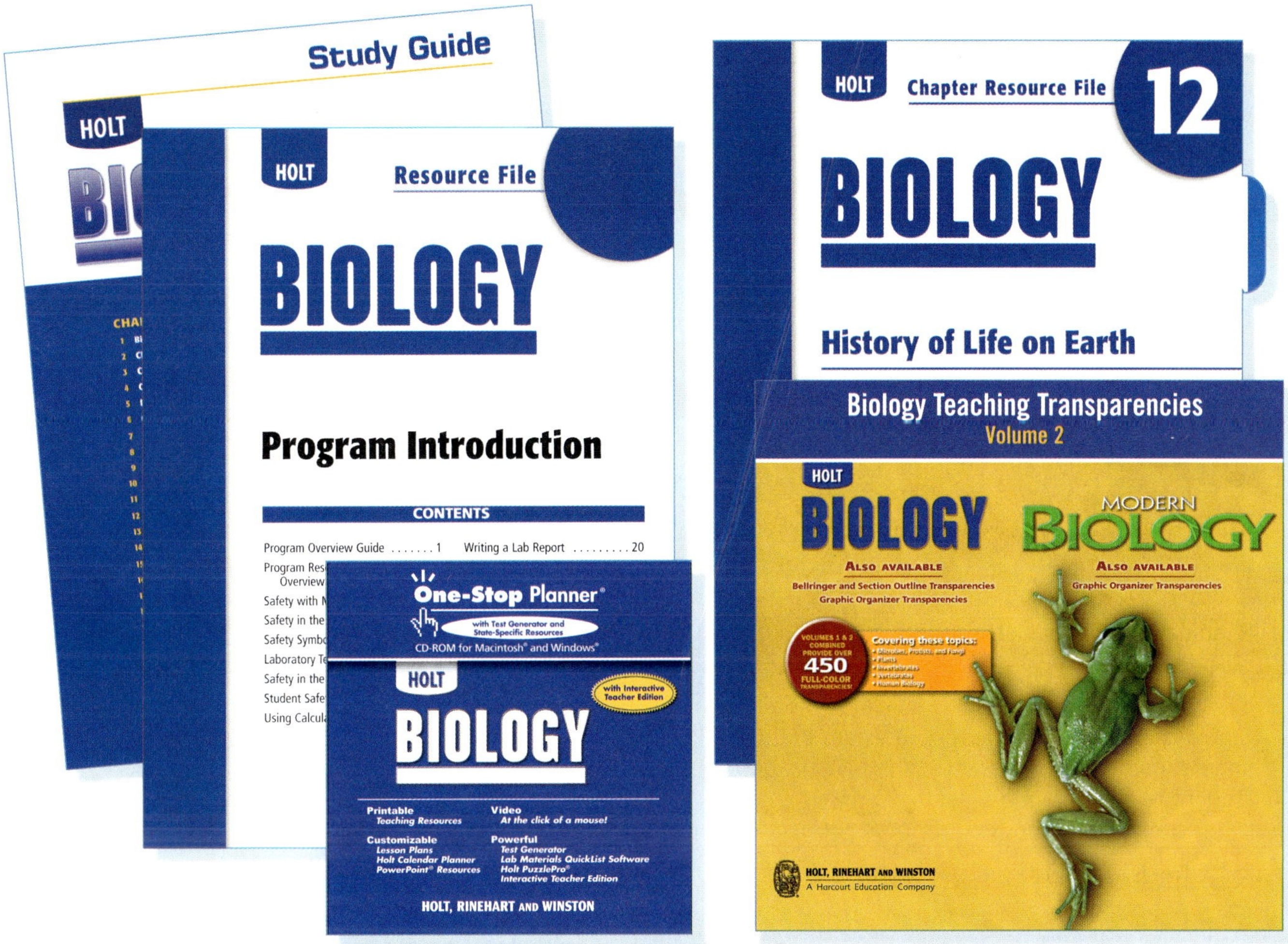

TECHNOLOGY RESOURCES MOTIVATE STUDENTS AND REACH DIFFERENT LEARNING STYLES

- Lighten the load with an interactive *Online Edition* or *CD-ROM Version* of the student text.
- **SciLinks®,** a Web service developed and maintained by the NSTA, contains current and prescreened links that engage students.
- ***Visual Concepts in Biology CD-ROM*** is your guide to **Visual Concepts**—engaging, media-based presentations of core concepts. **Visual Concepts** can be projected for large group instruction or viewed by individuals or groups of students with a computer.
- All the resources you need are on the ***One-Stop Planner® with Test Generator and State-Specific Resources CD-ROM,*** including worksheets, customizable lesson plans, **Holt Calendar Planner, PowerPoint® Resources, Holt PuzzlePro®,** and a powerful test generator.

STUDENT EDITION

Student Edition builds skills for success in science

ENGAGING MATERIAL GETS STUDENTS FOCUSED

Sunshine State Standards covered in the chapter are listed on the student page.

Looking Ahead presents a table of contents in brief to guide reading.

Quick Review helps reinforce material covered in previous chapters.

Reading Activity gets students thinking about the chapter to come.

SciLinks and other online resources are tied to chapter content and presented throughout the student text to extend students' learning.

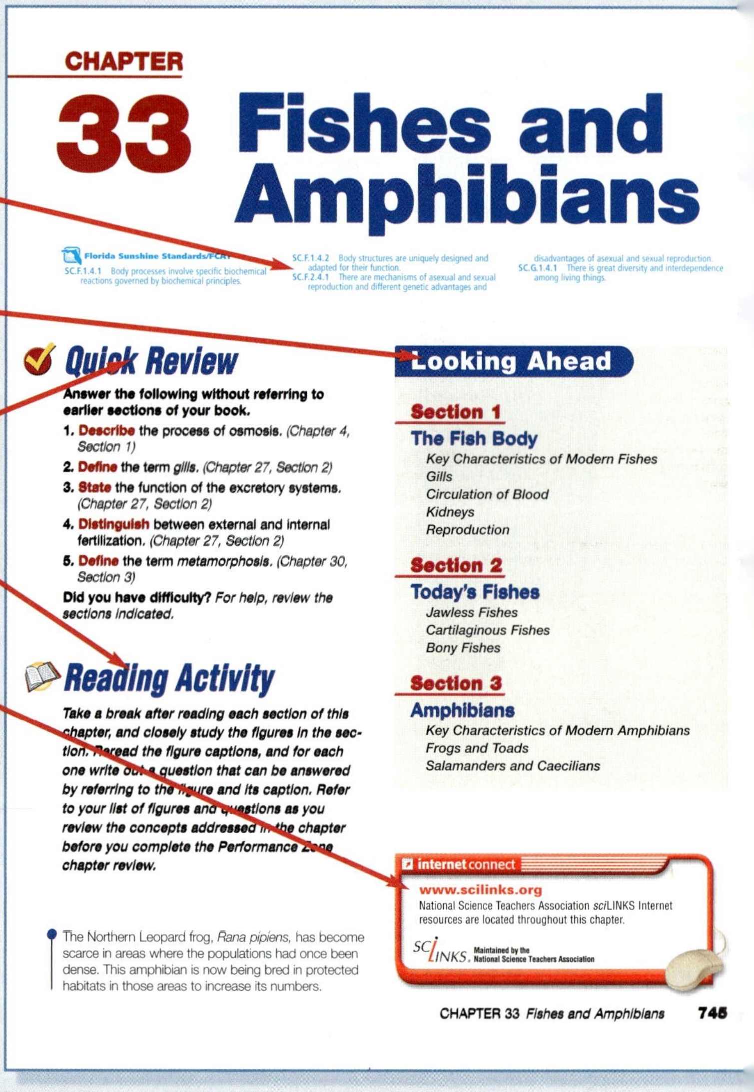

CHAPTER

33 Fishes and Amphibians

Florida Sunshine Standards/FCAT

SC.F.1.4.1 Body processes involve specific biochemical reactions governed by biochemical principles.

SC.F.1.4.2 Body structures are uniquely designed and adapted for their function.

SC.F.2.4.1 There are mechanisms of asexual and sexual reproduction and different genetic advantages and disadvantages of asexual and sexual reproduction.

SC.G.1.4.1 There is great diversity and interdependence among living things.

Quick Review

Answer the following without referring to earlier sections of your book.

1. **Describe** the process of osmosis. *(Chapter 4, Section 1)*
2. **Define** the term *gills. (Chapter 27, Section 2)*
3. **State** the function of the excretory systems. *(Chapter 27, Section 2)*
4. **Distinguish** between external and internal fertilization. *(Chapter 27, Section 2)*
5. **Define** the term *metamorphosis. (Chapter 30, Section 3)*

Did you have difficulty? *For help, review the sections indicated.*

Reading Activity

Take a break after reading each section of this chapter, and closely study the figures in the section. Reread the figure captions, and for each one write out a question that can be answered by referring to the figure and its caption. Refer to your list of figures and questions as you review the concepts addressed in the chapter before you complete the Performance Zone chapter review.

The Northern Leopard frog, *Rana pipiens,* has become scarce in areas where the populations had once been dense. This amphibian is now being bred in protected habitats in those areas to increase its numbers.

Looking Ahead

internet connect

www.scilinks.org

National Science Teachers Association *sci*LINKS Internet resources are located throughout this chapter.

SciLINKS. Maintained by the National Science Teachers Association

CHAPTER 33 *Fishes and Amphibians* **745**

RELEVANT AND EXCITING FEATURES

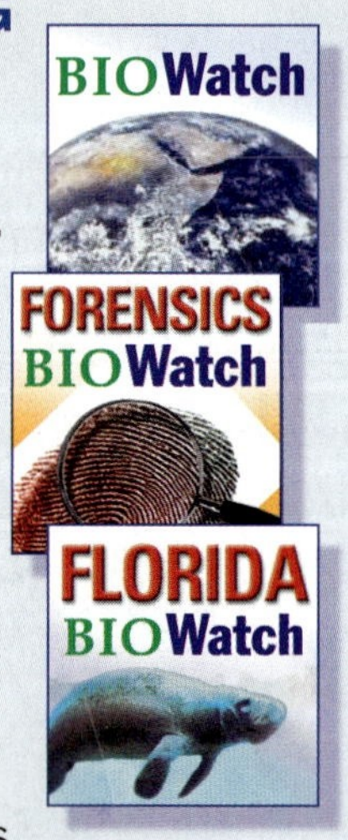

BioWatch features highlight the relationship between biology, ecology, health, and technology issues.

Forensics BioWatch features spark students' interest in the relationship between biology and solving crimes.

Florida BioWatch features focus attention on exciting research being conducted in Florida today.

Biology meets the world in **Real Life** features, explaining biology that students encounter in their daily lives.

Word Origins explore the roots of science language in order to expand students' understanding.

Up Close features provide detailed looks at important organisms.

Exploring Further explores key biological topics in greater depth, giving advanced learners the challenge they need.

Science, Technology, and Society features examine the impact of biology on issues in society.

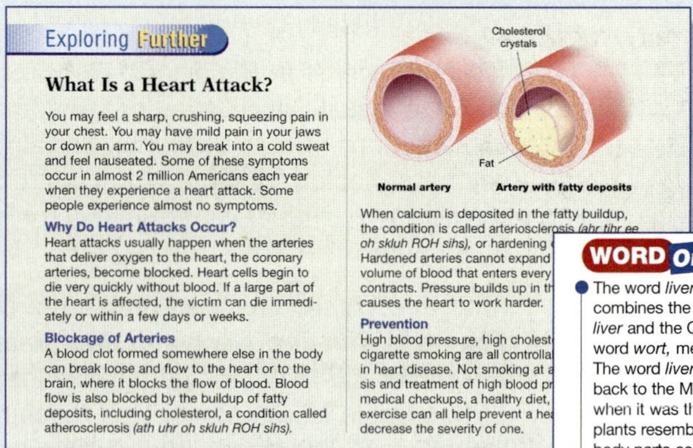

Exploring Further

What Is a Heart Attack?

You may feel a sharp, crushing, squeezing pain in your chest. You may have mild pain in your jaws or down an arm. You may break into a cold sweat and feel nauseated. Some of these symptoms occur in almost 2 million Americans each year when they experience a heart attack. Some people experience almost no symptoms.

Why Do Heart Attacks Occur?
Heart attacks usually happen when the arteries that deliver oxygen to the heart, the coronary arteries, become blocked. Heart cells begin to die very quickly without blood. If a large part of the heart is affected, the victim can die immediately or within a few days or weeks.

Blockage of Arteries
A blood clot formed somewhere else in the body can break loose and flow to the heart or to the brain, where it blocks the flow of blood. Blood flow is also blocked by the buildup of fatty deposits, including cholesterol, a condition called atherosclerosis *(ath uhr oh skluh ROH sihs).*

When calcium is deposited in the fatty buildup, the condition is called arteriosclerosis *(ahr tihr ee oh skluh ROH sihs),* or hardening … Hardened arteries cannot expand … volume of blood that enters every … contracts. Pressure builds up in th… causes the heart to work harder.

Prevention
High blood pressure, high cholest… cigarette smoking are all controlla… in heart disease. Not smoking at a… sis and treatment of high blood pr… medical checkups, a healthy diet, … exercise can all help prevent a he… decrease the severity of one.

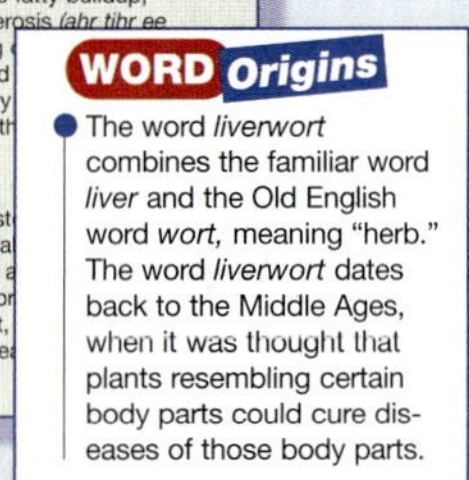

WORD Origins

The word *liverwort* combines the familiar word *liver* and the Old English word *wort,* meaning "herb." The word *liverwort* dates back to the Middle Ages, when it was thought that plants resembling certain body parts could cure diseases of those body parts.

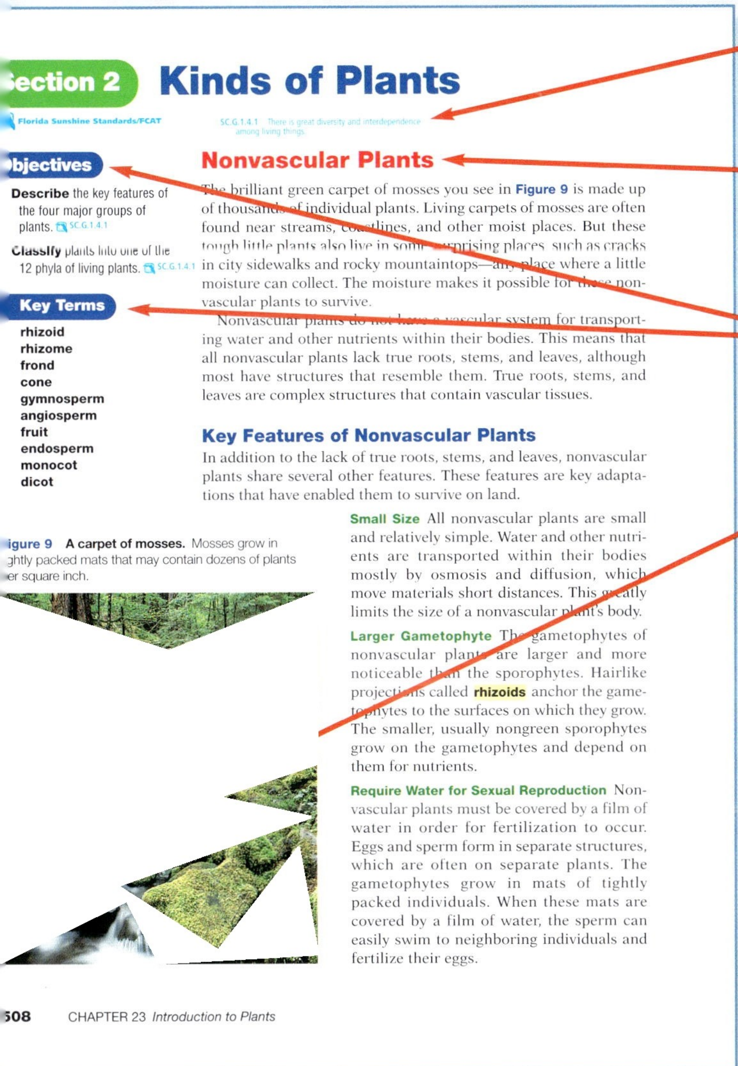

Section 2 **Kinds of Plants**

Florida Sunshine Standards/FCAT SC.G.1.4.1 There is great diversity and interdependence among living things.

Objectives

Describe the key features of the four major groups of plants. SC.G.1.4.1

Classify plants into one of the 12 phyla of living plants. SC.G.1.4.1

Key Terms

rhizoid
rhizome
frond
cone
gymnosperm
angiosperm
fruit
endosperm
monocot
dicot

Nonvascular Plants

The brilliant green carpet of mosses you see in **Figure 9** is made up of thousands of individual plants. Living carpets of mosses are often found near streams, coastlines, and other moist places. But these tough little plants also live in some surprising places, such as cracks in city sidewalks and rocky mountaintops—any place where a little moisture can collect. The moisture makes it possible for these nonvascular plants to survive.

Nonvascular plants do not have a vascular system for transporting water and other nutrients within their bodies. This means that all nonvascular plants lack true roots, stems, and leaves, although most have structures that resemble them. True roots, stems, and leaves are complex structures that contain vascular tissues.

Key Features of Nonvascular Plants

In addition to the lack of true roots, stems, and leaves, nonvascular plants share several other features. These features are key adaptations that have enabled them to survive on land.

Small Size All nonvascular plants are small and relatively simple. Water and other nutrients are transported within their bodies mostly by osmosis and diffusion, which move materials short distances. This greatly limits the size of a nonvascular plant's body.

Larger Gametophyte The gametophytes of nonvascular plants are larger and more noticeable than the sporophytes. Hairlike projections called **rhizoids** anchor the gametophytes to the surfaces on which they grow. The smaller, usually nongreen sporophytes grow on the gametophytes and depend on them for nutrients.

Require Water for Sexual Reproduction Nonvascular plants must be covered by a film of water in order for fertilization to occur. Eggs and sperm form in separate structures, which are often on separate plants. The gametophytes grow in mats of tightly packed individuals. When these mats are covered by a film of water, the sperm can easily swim to neighboring individuals and fertilize their eggs.

Figure 9 **A carpet of mosses.** Mosses grow in tightly packed mats that may contain dozens of plants per square inch.

508 CHAPTER 23 *Introduction to Plants*

A listing of the **Sunshine State Standards** covered in each section help students prepare to learn.

Accessible navigation engages students with outline-style headings and content grouped into small chunks.

Each section begins with a list of Objectives, which are linked to **Sunshine State Standards,** and **Key Terms.**

The illustrations in ***Holt Biology*** help students visualize key concepts. BioGraphic illustrations guide students through multi-step processes linked to in-text explanations.

LABS ARE CORRELATED TO SUNSHINE STATE STADARDS

Quick Labs provide hands-on experiences yet require few materials.

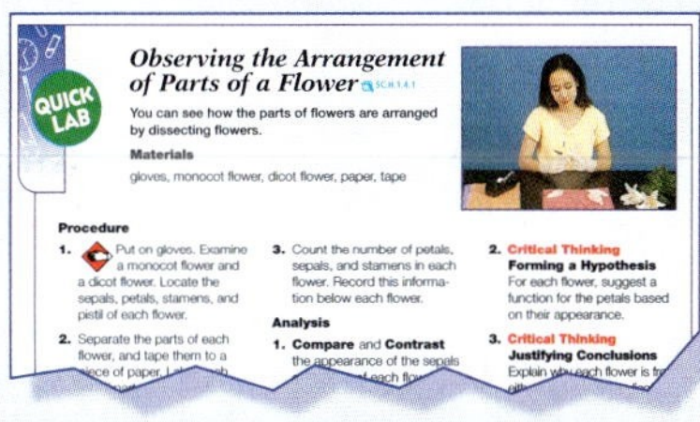

QUICK LAB *Observing the Arrangement of Parts of a Flower* SC.G.1.4.1

You can see how the parts of flowers are arranged by dissecting flowers.

Materials

gloves, monocot flower, dicot flower, paper, tape

Procedure

1. Put on gloves. Examine a monocot flower and a dicot flower. Locate the sepals, petals, stamens, and pistil of each flower.
2. Separate the parts of each flower, and tape them to a piece of paper...
3. Count the number of petals, sepals, and stamens in each flower. Record this information below each flower.

Analysis

1. **Compare** and **Contrast** the appearance of the sepals...
2. **Critical Thinking** **Forming a Hypothesis** For each flower, suggest a function for the petals based on their appearance.
3. **Critical Thinking** **Justifying Conclusions** Explain why each flower is...

Data Labs prompt students to interpret data and draw conclusions.

Math Labs provide practice for real-world math skills as students analyze biological problems.

Exploration Labs allow students to investigate or model a phenomenon and draw conclusions based on their work.

Skills Practice Labs teach lab skills used by biologists.

Forensics Labs allow students to model modern forensic techniques.

REVIEW FOR TEST-READINESS

Study Zone identifies key terms and concepts to help students study for the **Performance Zone Chapter Review.**

Each section and chapter end with **Review** questions that enable students to check their understanding, apply problem-solving skills, and prepare for the FCAT.

FCAT Prep questions that follow Section Review and Chapter Review questions help students review biology concepts in the context of the FCAT format.

Study Tips throughout the chapter help students practice study and test-taking skills.

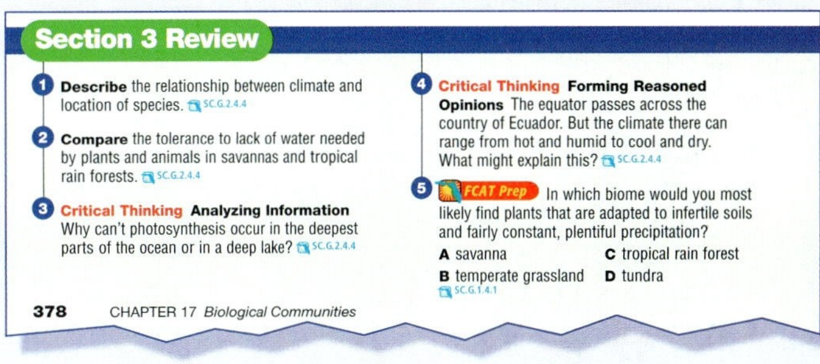

Section 3 Review

1. **Describe** the relationship between climate and location of species. SC.G.2.4.4
2. **Compare** the tolerance to lack of water needed by plants and animals in savannas and tropical rain forests. SC.G.2.4.4
3. **Critical Thinking** **Analyzing Information** Why can't photosynthesis occur in the deepest parts of the ocean or in a deep lake? SC.G.2.4.4
4. **Critical Thinking** **Forming Reasoned Opinions** The equator passes across the country of Ecuador. But the climate there can range from hot and humid to cool and dry. What might explain this? SC.G.2.4.4
5. **FCAT Prep** In which biome would you most likely find plants that are adapted to infertile soils and fairly constant, plentiful precipitation? SC.G.1.4.1
   A savanna
   B temperate grassland
   C tropical rain forest
   D tundra

378 CHAPTER 17 *Biological Communities*

A Teacher Edition that makes planning easy

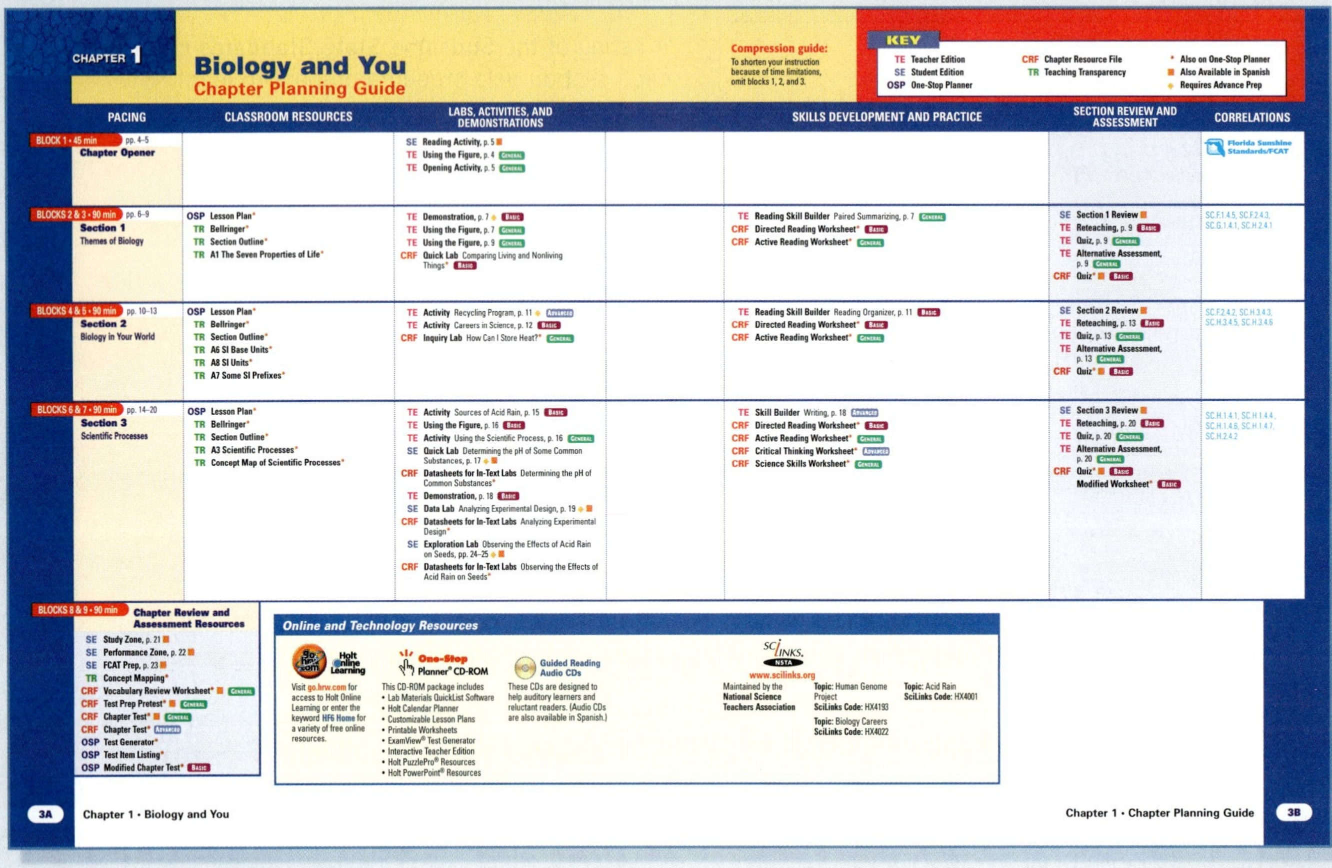

CHAPTER 1 **Biology and You** Chapter Planning Guide

Compression guide: To shorten your instruction because of time limitations, omit blocks 1, 2, and 3.

KEY: TE Teacher Edition; SE Student Edition; OSP One-Stop Planner; CRF Chapter Resource File; TR Teaching Transparency; * Also on One-Stop Planner; ■ Also Available in Spanish; ◆ Requires Advance Prep

PACING	CLASSROOM RESOURCES	LABS, ACTIVITIES, AND DEMONSTRATIONS	SKILLS DEVELOPMENT AND PRACTICE	SECTION REVIEW AND ASSESSMENT	CORRELATIONS
BLOCK 1 • 45 min pp. 4–5 **Chapter Opener**		SE Reading Activity, p. 5 ■ TE Using the Figure, p. 4 GENERAL TE Opening Activity, p. 5 GENERAL			Florida Sunshine Standards/FCAT
BLOCKS 2 & 3 • 90 min pp. 6–9 **Section 1** Themes of Biology	OSP Lesson Plan* TR Bellringer* TR Section Outline* TR A1 The Seven Properties of Life*	TE Demonstration, p. 7 ◆ BASIC TE Using the Figure, p. 7 GENERAL TE Using the Figure, p. 9 GENERAL CRF Quick Lab Comparing Living and Nonliving Things* BASIC	TE Reading Skill Builder Paired Summarizing, p. 7 GENERAL CRF Directed Reading Worksheet* BASIC CRF Active Reading Worksheet* GENERAL	SE Section 1 Review ■ TE Reteaching, p. 9 BASIC TE Quiz, p. 9 GENERAL TE Alternative Assessment, p. 9 GENERAL CRF Quiz* ■ BASIC	SC.F.1.4.5, SC.F.2.4.3, SC.G.1.4.1, SC.H.2.4.1
BLOCKS 4 & 5 • 90 min pp. 10–13 **Section 2** Biology in Your World	OSP Lesson Plan* TR Bellringer* TR Section Outline* TR A6 SI Base Units* TR A8 SI Units* TR A7 Some SI Prefixes*	TE Activity Recycling Program, p. 11 ◆ ADVANCED TE Activity Careers in Science, p. 12 BASIC CRF Inquiry Lab How Can I Store Heat?* GENERAL	TE Reading Skill Builder Reading Organizer, p. 11 BASIC CRF Directed Reading Worksheet* BASIC CRF Active Reading Worksheet* GENERAL	SE Section 2 Review ■ TE Reteaching, p. 13 BASIC TE Quiz, p. 13 GENERAL TE Alternative Assessment, p. 13 GENERAL CRF Quiz* ■ BASIC	SC.F.2.4.2, SC.H.3.4.3, SC.H.3.4.5, SC.H.3.4.6
BLOCKS 6 & 7 • 90 min pp. 14–20 **Section 3** Scientific Processes	OSP Lesson Plan* TR Bellringer* TR Section Outline* TR A3 Scientific Processes* TR Concept Map of Scientific Processes*	TE Activity Sources of Acid Rain, p. 15 BASIC TE Using the Figure, p. 16 BASIC TE Activity Using the Scientific Process, p. 16 GENERAL SE Quick Lab Determining the pH of Some Common Substances, p. 17 ◆ ■ CRF Datasheets for In-Text Labs Determining the pH of Common Substances* TE Demonstration, p. 18 BASIC SE Data Lab Analyzing Experimental Design, p. 19 ◆ ■ CRF Datasheets for In-Text Labs Analyzing Experimental Design* SE Exploration Lab Observing the Effects of Acid Rain on Seeds, pp. 24–25 ◆ ■ CRF Datasheets for In-Text Labs Observing the Effects of Acid Rain on Seeds*	TE Skill Builder Writing, p. 18 ADVANCED CRF Directed Reading Worksheet* BASIC CRF Active Reading Worksheet* GENERAL CRF Critical Thinking Worksheet* ADVANCED CRF Science Skills Worksheet* GENERAL	SE Section 3 Review ■ TE Reteaching, p. 20 BASIC TE Quiz, p. 20 GENERAL TE Alternative Assessment, p. 20 GENERAL CRF Quiz* ■ BASIC Modified Worksheet* BASIC	SC.H.1.4.1, SC.H.1.4.4, SC.H.1.4.6, SC.H.1.4.7, SC.H.2.4.2

BLOCKS 8 & 9 • 90 min **Chapter Review and Assessment Resources**

- SE Study Zone, p. 21 ■
- SE Performance Zone, p. 22 ■
- SE FCAT Prep, p. 23 ■
- TR Concept Mapping*
- CRF Vocabulary Review Worksheet* ■ GENERAL
- CRF Test Prep Pretest* ■ GENERAL
- CRF Chapter Test* ■ GENERAL
- CRF Chapter Test* ADVANCED
- OSP Test Generator*
- OSP Test Item Listing*
- OSP Modified Chapter Test* BASIC

Online and Technology Resources

Holt Online Learning: Visit go.hrw.com for access to Holt Online Learning or enter the keyword HF6 Home for a variety of free online resources.

One-Stop Planner® CD-ROM: This CD-ROM package includes
- Lab Materials QuickList Software
- Holt Calendar Planner
- Customizable Lesson Plans
- Printable Worksheets
- ExamView® Test Generator
- Interactive Teacher Edition
- Holt PuzzlePro® Resources
- Holt PowerPoint® Resources

Guided Reading Audio CDs: These CDs are designed to help auditory learners and reluctant readers. (Audio CDs are also available in Spanish.)

SciLinks NSTA www.scilinks.org Maintained by the National Science Teachers Association
Topic: Human Genome Project SciLinks Code: HX4193
Topic: Biology Careers SciLinks Code: HX4022
Topic: Acid Rain SciLinks Code: HX4001

3A Chapter 1 • Biology and You — Chapter 1 • Chapter Planning Guide 3B

TEACHING RESOURCES DESIGNED FOR CONVENIENCE

The **Chapter Planning Guide** breaks each chapter down into flexible 45-minute blocks and offers a full listing of activities and classroom resources available for each section. Look for guidance on:

- Pacing
- Classroom Resources
- Labs, Activities, and Demonstrations
- Skills Development and Practice
- Section Review and Assessment
- Sunshine State Standards Correlations
- Online and Technology Resources

A **Lesson Cycle** builds structure around every lesson:

- **Focus** uses the objectives listed in the *Student Edition* to focus student attention on the upcoming content.
- **Motivate** uses demonstrations, discussions, and lively activities to get students excited about the material.
- **Teach** presents various teaching techniques including **Skill Builders, Group Activities,** and more.
- Finally, **Close** with quiz questions to ensure students understand the information covered.

ACTIVITIES AND DEMONSTRATIONS FOR EVERY LEARNING LEVEL

Activities are leveled by ability level in the teacher's wrap—**Basic, General,** and **Advanced**—helping you choose the activities appropriate for each student.

Learning styles are addressed throughout—**Interpersonal, Intrapersonal, Auditory, Kinesthetic, Logical, Visual,** and **Verbal**—so you can adapt material to different learning styles. In addition, some are also labeled to identify the activities that help with **Co-op Learning** or **English Language Learners.**

Bellringer activities begin each section with an activity designed to get students thinking. Activities on transparencies get them focused while you attend to administrative duties.

Bellringer
Ask students to think about a snail, a clam, and an octopus. Tell students that these are all mollusks. Ask them to write down some characteristics that these animals have in common. (Students will most likely note the prominent foot, or its modification.)

Opening Activity gets discussion started, while **Group Activity** encourages collaborative effort. **Reading Activity** improves reading skills and comprehension.

Teach

Group Activity — GENERAL

Tissue Types Provide each group of students with a microscope, three microscope slides, and three coverslips. Provide the class with a red onion, a potato, and a stalk of celery. Have students prepare wet-mount slides of the following: red-onion skin (dermal tissue), potato (ground tissue), and celery (dermal, ground, and vascular tissue). Have students draw what they observe in the field of view under the microscope and label the types of tissues they find in each example. Remind them that the field of view should be drawn as a circle.
Co-op Learning   English Language Learners

CREATING RELEVANCE AND UNDERSTANDING

On almost every page you will find exciting features to help ignite class discussion and keep students thinking.

- **Misconception Alert**
- **Real World Connection**
- **Did You Know?**
- **Reading Skill Builder**
- **Demonstration**
- **Skill Builder** for core writing and math skills
- **Attention Grabber**
- **Connection** to other disciplines such as Math, Language Arts, and Social Studies
- **Cultural Awareness**

Trends In... keeps students abreast of the latest thinking in a number of scientific disciplines complementary to biology, including psychology, field biology, and biotechnology.

Integrating Physics and Chemistry ties content on the page to a physics or chemistry principle.

MISCONCEPTION ALERT
The Portuguese Man-of-War Many people think that the Portuguese man-of-war (*Physalia*) is a jellyfish because it is shaped somewhat like one. However, the term jellyfish is used for scyphozoans, a class of cnidarians in which the medusa form dominates the life cycle. *Physalia* is actually a hydrozoan and exists as a colony of medusas and polyps.

REAL WORLD CONNECTION
The cause of multiple sclerosis (MS) is unknown, but many scientists think that it results from the actions of the immune system against the myelin sheaths of nerves of the central nervous system. The myelin sheaths in the brain and spinal cord become sclerotic, or hard, causing poor impulse conductivity. Sometimes the symptoms of MS disappear altogether, but they may return after periods of remission. Current therapies for MS include corticosteroids and beta interferon.

did you know?
Metabolism on the Fly A hummingbird's metabolic rate is about 1,400 calories per kilogram. Compare that to the rate of a mourning dove—about 127 calories per kilogram. Hummingbirds use enormous amounts of energy in part because of their racing wings and hearts. Hummingbirds can move their wings up to 80 times per second, and their heart can beat up to 1,200 beats per minute. Because of their high metabolic rate, hummingbirds feed almost continuously throughout the day.

READING SKILL BUILDER — GENERAL
Reading Organizer Have students create a reading organizer to compare reproduction in bacterial cells and eukaryotic cells. A description of reproduction in eukaryotic cells is given in Sections 2 and 3. LS **Verbal**

INCLUSION STRATEGIES MAKE MATERIAL ACCESSIBLE TO ALL

Written by professionals in the field of special needs education, **Inclusion Strategies** address many different learning exceptionalities in the classroom.

- Learning Disability
- Developmental Delay
- Attention Deficit Disorder
- Gifted and Talented

INCLUSION Strategies
- *Gifted and Talented*
- *Attention Deficit Disorder*

Have the students do a field study of the plants in their schoolyard. The students should diagram or take pictures of different types of plants at school and classify the

Resources that make teaching easier!

CHAPTER RESOURCE FILES

A ***Chapter Resource File*** accompanies each chapter of ***Holt Biology.*** Everything you need to plan and manage your lessons in a convenient timesaving format is included in each chapter book. Also included is a **Program Introduction,** your guide to the resources in each ***Chapter Resource File.***
Each chapter book includes:

Skills Worksheets
- Directed Reading
- Active Reading
- Vocabulary Review
- Science Skills
- Concept Mapping
- Critical Thinking
- Test Prep Pretest

Labs and Activities
- Datasheets for In-Text Labs
- Exploration Labs
- Skills Practice Labs
- Inquiry Labs
- Quick Labs

Assessments
- Quizzes
- Chapter Test (General)
- Chapter Test (Advanced)

Answer Keys
- Lab Notes and Answers
- Answer Keys for Skills Worksheets and Assessments

Teaching Transparency List

One-Stop Planner®
CD-ROM has everything you need!

All the resources for ***Holt Biology*** are here in one place, along with the amazing **Holt Calendar Planner.** See page T10 for more information about this powerful timesaving tool.

ADDITIONAL RESOURCES

These additional resources are designed to help you reinforce and extend lessons.

- ***FCAT Test Preparation Workbook and Answer Key*** provide additional test prep support for students to practice their test-taking skills
- ***Holt Science Laboratory Manager's Professional Reference*** helps you understand risk management and the hazards that can occur in the classroom and lab.
- ***Holt Science Skills Workshop: Reading in the Content Area*** contains exercises that target key reading skills using excerpts from Holt's science textbooks.
- The ***Lesson Plans*** booklet provides pacing, objectives and a suggested teaching outline with time requirements.
- ***Special Needs Activities and Modified Tests with Answer Key*** provides modified worksheets and tests to help meet the needs of your diverse classroom.
- ***Study Guide*** contains review worksheets that reinforce the skills and concepts presented in the *Student Edition.*
- Over 600 **Teaching Transparencies** use graphics directly from the text to enhance classroom presentations.
- ***Lab Manual for Quick Labs, Data Labs, and Math Labs*** and ***Lab Manual for Skills Practice Labs and Exploration Labs*** provide datasheets for the in-text labs and eliminate the need for textbooks in the lab.

SPANISH RESOURCES BRING BIOLOGY TO ENGLISH-LANGUAGE LEARNERS!

- *Student Edition* in Spanish.
- A ***Spanish Glossary*** is right at students' fingertips in the *Student Edition.* It shows the English term, its Spanish equivalent, and a definition in Spanish.
- ***Assessments*** in Spanish include **Section Quizzes** and **General Chapter Tests.**
- ***Guided Reading Audio CD Program*** is a direct read of the student text in Spanish.
- ***Study Guide,*** in Spanish, contains translated review worksheets that reinforce the skills and concepts presented in the *Student Edition.*

Technology that enhances teaching

One-Stop Planner® CD-ROM
with Test Generator

Planning and managing lessons has never been easier than with this convenient, all-in-one CD-ROM. The One-Stop Planner® with Test Generator and State-Specific Resources includes the following features:

PRINTABLE:

- **Teaching Resources**
- **Transparency Masters**

CUSTOMIZABLE:

- **Lesson Plans:** traditional and block-scheduling lesson plans in several word-processing formats.
- **Holt Calendar Planner:** manage your time and resources by the day, week, month, and year.
- **PowerPoint® Resources:** PowerPoint® presentations customized with video resources, images, and standardized test prep questions.

POWERFUL:

- **ExamView® Test Generator:** test items organized by chapter, plus hundreds of editable questions, so you can put together your own tests and quizzes. Plus you can post test questions to Holt Online Assessment for automatic grading.
- **Holt PuzzlePro®:** an easy way to create crossword puzzles and word searches that make learning vocabulary more fun.
- **Interactive Teacher Edition:** the entire teacher text, with links connecting you to the related Teaching Resources.
- **Lab Materials QuickList Software:** helps you create a customized list of the lab materials you need.

GUIDED READING AUDIO CD PROGRAM

This audio program provides a direct read of ***Holt Biology,*** making content more accessible for auditory learners and reluctant readers. Available in English and Spanish.

HOLT BIOLOGY INTERACTIVE TUTOR CD-ROM

A personal tutor guides students through challenging fundamental concepts in biology using animation, real-world application, and interactive problem-solving.

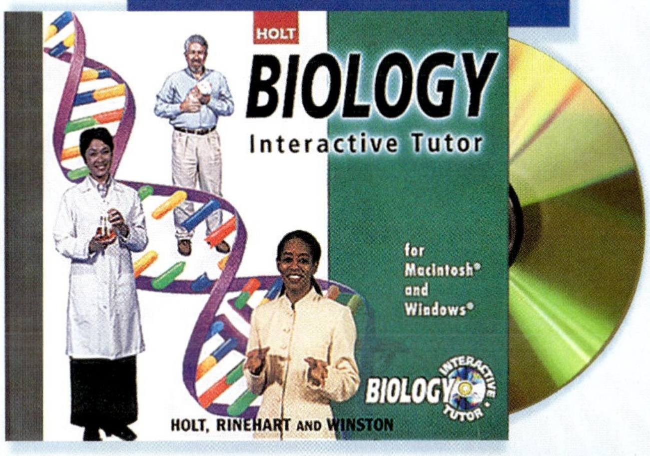

BRAIN FOOD VIDEO QUIZZES

These game-show style quizzes assess students' progress and motivate them to study. Available on DVD or video.

STUDENT EDITION on CD-ROM for Macintosh® and Windows®

Ideal for students who have limited access to the Internet, but who need to lighten the load of textbooks they carry home, the entire *Student Edition* is on one easy-to-navigate CD-ROM, page-for-page.

Online Resources are available anytime, anywhere!

THE *PREMIER ONLINE EDITION* IS PORTABLE, EXPANDABLE, INTERACTIVE, AND YET WEIGHS NOTHING AT ALL

The *Premier Online Edition* of ***Holt Biology*** engages students in ways that were never before possible. And since it's all online, it's available anywhere you or your students connect to the Internet. You'll find interactive exercises and feedback, presentation materials,homework help, and much more, including:

- **Visual Concepts:** media-based visuals that include slides, interactive screens, and short movie clips for student review and teacher presentation
- **Self Quizzes and Tutorials:** objective-based assessments that provide students with immediate feedback and diagnostic responses
- **Interactive Concept Maps:** drag and drop activities for concept review and reinforcement
- **Holt Online Assessment:** allows you to post tests for automatic grading.

Try it for yourself at www.hrw.com/online.

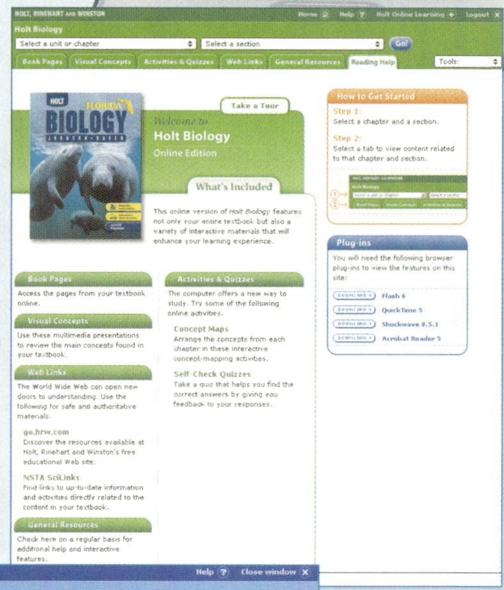

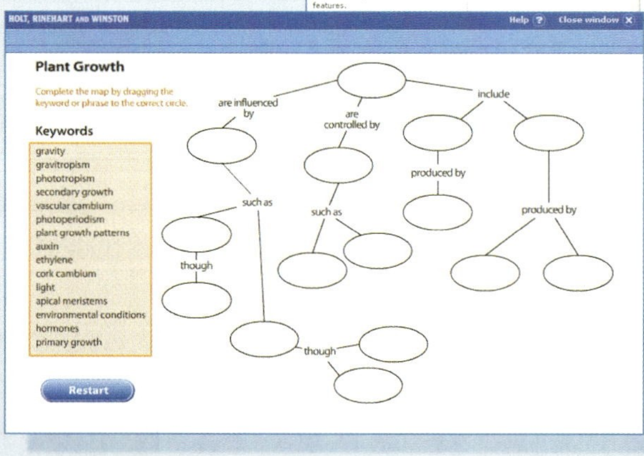

live ink
a new way to read
Research-Proven Reading Help

Improve understanding with **Reading Help** with **Live Ink®**—a scientifically-researched tool to improve students' reading comprehension. **Reading Help** is available for use with the *Premier Online Edition* of ***Holt Biology.***

SCILINKS NSTA

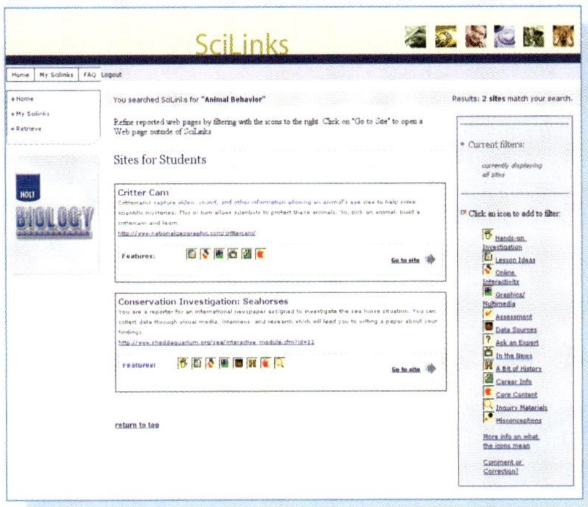

This Web service, developed and maintained by the National Science Teachers Association, contains a collection of prescreened links that include current information and activities that are directly related to chapter topics.

- Saves you valuable time searching for relevant and up-to-date Web sites
- Sites reviewed by science content experts and educators
- **Internet Connect** boxes within each chapter offer opportunities to enrich, enhance, and extend learning

go.hrw.com

Holt's own award-winning Web site includes a variety of resources and reference materials that go along with the text book as well as links to state resources.

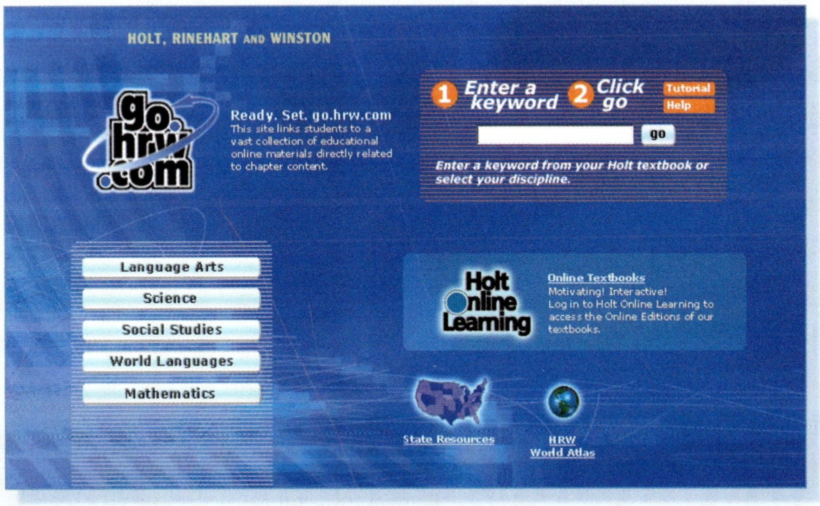

Meeting Individual Needs

Students have a wide range of abilities and learning exceptionalities. These pages show you how ***Holt Biology*** provides resources and strategies to help you tailor your instruction to engage every student in your classroom.

Learning exceptionality	Resources and strategies	
Learning Disabled and Slow Learners Students who have dyslexia or dysgraphia, students reading below grade level, students having difficulty understanding abstract or complex concepts, and slow learners	• Inclusion Strategies labeled ***Learning Disability*** • Activities labeled ***Basic*** • ***Reteaching*** activities	• Activities labeled ***Visual, Kinesthetic,*** or ***Auditory*** • Hands-on activities or projects • Oral presentations instead of written tests or assignments
Developmental Delays Students who are functioning far below grade level because of mental retardation, autism, or brain injury; goals are to learn or retain basic concepts	• Inclusion Strategies labeled ***Developmentally Delayed*** • Activities labeled ***Basic***	• ***Reteaching*** activities • Project-based activities
Attention Deficit Disorders Students experiencing difficulty completing a task that has multiple steps, difficulty handling long assignments, or difficulty concentrating without sensory input from physical activity	• Inclusion Strategies labeled ***Attention Deficit Disorder*** • Activities labeled ***Basic*** • ***Reteaching*** activities • Activities labeled ***Co-op Learning***	• Activities labeled ***Visual, Kinesthetic,*** or ***Auditory*** • Concepts broken into small chunks • Oral presentations instead of written tests or assignments
English as a Second Language Students learning English	• Activities labeled ***English Language Learners*** • Activities labeled ***Basic***	• ***Reteaching*** activities • Activities labeled ***Visual***
Gifted and Talented Students who are performing above grade level and demonstrate aptitude in crosscurricular assignments	• Inclusion Strategies labeled ***Gifted and Talented*** • Activities labeled ***Advanced*** • ***Connection*** activities	• Activities that involve multiple tasks, a strong degree of independence, and student initiative

General Strategies The following strategies can help you modify instruction to help students who struggle with common classroom difficulties.

A student experiencing difficulty with . . .	May benefit if you . . .	
Beginning assignments	• Assign work in small amounts • Have the student use cooperative or paired learning • Provide varied and interesting activities	• Allow choice in assignments or projects • Reinforce participation • Seat the student closer to you
Following directions	• Gain the student's attention before giving directions • Break up the task into small steps • Give written directions rather than oral directions • Use short, simple phrases • Stand near the student when you are giving directions	• Have the student repeat directions to you • Prepare the student for changes in activity • Give visual cues by posting general routines • Reinforce improvement in or approximation of following directions
Keeping track of assignments	• Have the student use folders for assignments • Have the student use assignment notebooks	• Have the student keep a checklist of assignments and highlight assignments when they are turned in
Reading the textbook	• Provide outlines of the textbook content • Reduce the length of required reading • Allow extra time for reading • Have the students read aloud in small groups	• Have the student use peer or mentor readers • Have the student use books on tape or CD • Discuss the content of the textbook in class after reading
Staying on task	• Reduce distracting elements in the classroom • Provide a task-completion checklist • Seat the student near you	• Provide alternative ways to complete assignments, such as oral projects taped with a buddy
Behavioral or social skills	• Model the appropriate behaviors • Establish class rules, and reiterate them often • Reinforce positive behavior • Assign a mentor as a positive role model to the student • Contract with the student for expected behaviors • Reinforce the desired behaviors or any steps toward improvement	• Separate the student from any peer who stimulates the inappropriate behavior • Provide a "cooling off" period before talking with the student • Address academic/instructional problems that may contribute to disruptive behaviors • Include parents in the problem-solving process through conferences, home visits, and frequent communication
Attendance	• Recognize and reinforce attendance by giving incentives or verbal praise • Emphasize the importance of attendance by letting the student know that he or she was missed when he or she was absent	• Encourage the student's desire to be in school by planning activities that are likely to be enjoyable, giving the student a preferred responsibility to be performed in class, and involving the student in extracurricular activities • Schedule problem-solving meeting with parents, faculty, or both
Test-taking skills	• Prepare the student for testing by teaching ways to study in pairs, such as using flashcards, practice tests, and study guides, and by promoting adequate sleep, nourishment, and exercise • During testing, allow the student to respond orally on tape or to respond using a computer; to use	notes; to take breaks; to take the test in another location; to work without time constraints; or to take the test in several short sessions • Decrease visual distraction by improving the visual design of the test through use of larger type, spacing, consistent layout, and shorter sentences

Pacing guide

PACING GUIDE

The following table shows how 45-minute instructional blocks can be organized to meet the needs of your class. The suggested pacing presumes the school year consists of approximately 180 45-minute-long blocks.

The **Basic Course** will expose your students to all fundamental principles of biology and will fulfill the rquirements of the National Science Education Standards.

The pacing of the **Basic Course** assumes that you will cover the 18 *Principles* chapters according to the Compression Guide found in each Chapter Planning Guide. To enhance the depth of the **Basic Course,** one additional 45-minute block has been added to each of the first eighteen chapters to better accommodate the end-of-chapter lab. Each of the five *Introduction* chapters in the **Basic Course** should then be covered at a non-compressed pace.

To implement one of the alternative courses of study shown, you will need to adjust the Chapter Planning Guides that precede each chapter to match these pacing suggestions.

	(23 Chapters) BLOCKS	(29 Chapters) BLOCKS
UNIT 1 PRINCIPLES OF CELL BIOLOGY		
Chapter 1	7	6
Chapter 2	9	5
Chapter 3	7	6
Chapter 4	7	4
Chapter 5	7	5
Chapter 6	9	4
UNIT 2 PRINCIPLES OF GENETICS		
Chapter 7	7	6
Chapter 8	9	7
Chapter 9	8	6
Chapter 10	6	4
Chapter 11	7	7
UNIT 3 PRINCIPLES OF EVOLUTION		
Chapter 12	7	6
Chapter 13	7	5
Chapter 14	5	4
UNIT 4 PRINCIPLES OF ECOLOGY		
Chapter 15	7	4
Chapter 16	7	6
Chapter 17	9	6
Chapter 18	7	6
UNIT 5 EXPLORING DIVERSITY		
Chapter 19	9	6
Chapter 20	–	–
Chapter 21	–	–
Chapter 22	–	–
UNIT 6 EXPLORING PLANTS		
Chapter 23	9	7
Chapter 24	–	–
Chapter 25	–	–
Chapter 26	–	–
UNIT 7 EXPLORING INVERTEBRATES		
Chapter 27	7	6
Chapter 28	–	–
Chapter 29	–	–
Chapter 30	–	–
Chapter 31	–	–
UNIT 8 EXPLORING VERTEBRATES		
Chapter 32	12	7
Chapter 33	–	–
Chapter 34	–	–
Chapter 35	–	–
Chapter 36	–	–
UNIT 9 EXPLORING HUMAN BIOLOGY		
Chapter 37	11	9
Chapter 38	–	7
Chapter 39	–	8
Chapter 40	–	8
Chapter 41	–	10
Chapter 42	–	6
Chapter 43	–	8

(26 Chapters)	(31 Chapters)	(37 Chapters)	(24 Chapters)	(31 Chapters)
BLOCKS	BLOCKS	BLOCKS	BLOCKS	BLOCKS
6	6	3	6	8
6	6	6	10	10
10	6	6	4	10
6	6	6	4	10
9	8	5	9	10
6	6	6	6	10
6	6	6	6	6
9	7	5	8	6
9	8	4	8	8
4	4	3	6	8
8	8	5	6	6
6	6	4	8	6
7	5	3	7	4
6	6	4	8	6
6	4	3	10	4
8	6	4	10	4
6	6	4	10	4
6	4	4	10	4
6	6	5	6	4
–	–	–	4	5
–	–	–	4	5
–	–	–	4	4
9	7	5	8	4
9	–	–	–	–
9	–	–	–	4
9	–	–	–	–
4	6	5	6	4
–	4	4	–	–
–	4	4	–	–
–	6	4	–	–
–	4	4	–	–
4	6	5	6	4
–	6	5	–	–
–	5	4	–	–
–	5	5	–	–
–	5	5	–	–
6	8	8	6	6
–	–	6	–	–
–	–	6	–	–
–	–	6	–	4
–	–	6	–	4
–	–	6	–	4
–	–	6	–	4

Correlation to the Florida Sunshine State Standards

The following list shows the page number correlation of ***Holt Biology*** with the Florida Sunshine State Standards (grades 9–12) for physical science content, Earth science content, life science content, and the nature of science. The benchmarks listed include all of those required for **Biology I.** For further detail, see the Chapter Planning Guide pages before each chapter.

For page numbers in blue, the associated benchmark is printed in the *Student Edition.*
Boldface page numbers indicate labs.

	INSTRUCTION: IN-DEPTH	INSTRUCTION: MENTIONED	ASSESSMENT
Energy			
Standard 1: The student recognizes that energy may be changed in form with varying efficiency. (SC.B.1.4)			
SC.B.1.4.1 The student understands how knowledge of energy is fundamental to all the scientific disciplines (e.g., the energy required for biological processes in living organisms and the energy required for the building, erosion, and rebuilding of the Earth).	6–8, 37, 38–39, 94–96, 444–445	28, 65–66, 74–80, 81–86	37, 42, 44, 45, 68, 96, 103, 615, 756
SC.B.1.4.2 The student understands that there is conservation of mass and energy when matter is transformed.	39, 94–96, 97–103, 104–110, **114–115**	46–47, 64–66, 74–80, 81–86, 90–91	42, 47
SC.B.1.4.7 The student knows that the total amount of usable energy always decreases, even though the total amount of energy is conserved in any transfer.	39, 348–349	345–349, 355	42, 349, 356

	INSTRUCTION: IN-DEPTH	INSTRUCTION: MENTIONED	ASSESSMENT

Processes that Shape the Earth

Standard 1: The student recognizes that processes in the lithosphere, atmosphere, hydrosphere, and the biosphere interact to shape the Earth. (SC.D.1.4)

Benchmark	Instruction: In-Depth	Instruction: Mentioned	Assessment
SC.D.1.4.3 The student knows that changes in Earth's climate, geological activity, and life forms may be traced and compared.	258–263, 264–268, 283–287, 602–603, 618, 622, 629, 633, 642, 651, 664, 665, 693, 700, 714, **716,** 718, 721, 725, 732–734, 737–738, **742–743,** 775, 808–811	603, 619, 642, 729	263, 268, 270, 271, 282, 287, 292, 294, 295, 524, 613, 658, 659, 684, 716, 720, 721, 730, 734, 738, 740, 741, 743, 816, 818
SC.D.1.4.4 The student knows that Earth's systems and organisms are the result of along, continuous change over time.	6–9, 214, 252–257, 272–273, 282, 291–292, 308–310, 362–364, 602, 642–650, 664–669, 714, **716,** 728–729, 733–734, 825–826	627–628, 807	156, 257, 263, 270, 271, 272–273, 282, 292, 294, 295, 310, 313, 334, 364, 429, 636, 658, 669, 716, 730, 740, 750

Standard 2: The student understands the need for protection of natural systems on Earth. (SC.D.2.4)

Benchmark	Instruction: In-Depth	Instruction: Mentioned	Assessment
SC.D.2.4.1 The student understands the interconnectedness of the systems on Earth and the quality of life.	9, 351, 354, 377, 397, **512,** 520, 578, 628, 699	628	354, 356, 357, 381, 494, 512, 522, 524

Processes of Life

Standard 1: The student describes patterns of structure and function in living things. (SC.F.1.4)

Benchmark	Instruction: In-Depth	Instruction: Mentioned	Assessment
SC.F.1.4.1 The student knows that the body processes involve specific biochemical reactions governed by biochemical principles.	7–8, **17,** 28–30, 31–33, 34–37, 38–39, 40–42, **46–47,** 55–60, 62–66, 74–80, **76,** 81–86, 100–103, 104–110, 564, **561, 568–569, 608,** 749, **749, 773, 801, 970–971**	620, 623	9, 17, 20, 42, 44, 46–47, 89, 103, 110, 112, 113, 561, 564, 566, 569, 749, 750, 766, 773, 801, 890, 892, 949, 971

FLORIDA SUNSHINE STATE STANDARDS

Processes of Life, *cont.*	INSTRUCTION: IN-DEPTH	INSTRUCTION: MENTIONED	ASSESSMENT
SC.F.1.4.2 The student knows that body structures are uniquely designed and adapted for their function.	7, 36, 58, 65, 66, 78, 128–129, 422–425, 446, 483, **496–497,** 502–503, **503,** 507, 534–535, 538, 552–559, **557, 558,** 560–561, 574–575, 598–600, 604–608, 618–620, 623, 633, 644–645, **650,** 652, 654–655, 666–669, **667,** 671, 675–677, 681, 695–696, **702, 706–707,** 715, **719,** 719, 731, 747–748, 754–755, **756,** 762–763, 780–781, 784–791, 800–803, **820–821,** 846–849, 850–852, 856–859, 880–882, 885–887, 906–911, 912–915, 924, 927, 944–945, 950–955, 957–958, 975–976, 982–983, 996–998, 999–1000, 1004–1007, **1014–1015**		37, 44, 61, 66, 68, 69, 80, 89, 112, 113, 132, 134, 135, 421, 426, 426, 428, 429, 441, 450, 452, 453, 471, 484, 488, 494, 495, 497, 503, 507, 515, 524, 525, 533, 540, 546, 547, 557, 558, 559, 564, 566, 567, 578, 586, 587, 610, 612, 613, 621, 628, 634, 636, 637, 650, 656, 658, 659, 667, 669, 672, 679, 682, 684, 685, 699, 702, 704, 705, 707, 719, 720, 734, 740, 750, 756, 757, 764, 766, 767, 777, 783, 792, 794, 795, 806, 816, 818, 819, 821, 849, 855, 860, 864, 866, 867, 875, 879, 884, 890, 892, 893, 916, 918, 929, 938, 939, 949, 955, 960, 968, 969, 976, 981, 988, 990, 991, 998, 1003, 1007, 1012, 1013, 1015
SC.F.1.4.3 The student knows that membranes are sites for chemical synthesis and essential energy conversions.	56, 60, 61, 68, 78–80, 86, 88, 112, 60–61, 63–64, 83, 86, 860	608	450, 452, 860, 949
SC.F.1.4.4 The student understands that biological systems obey the same laws of conservation as physical systems.	7–8, 28–30, 38–42, 55, **56,** 74–77, 76, **79, 90–91**	34–36, 55–61, 62–66, 81–86, 608, 756	47, 56, 61, 68, 69, 76, 79, 80, 88, 89, 90–91

Processes of Life, *cont.*	INSTRUCTION: IN-DEPTH	INSTRUCTION: MENTIONED	ASSESSMENT
SC.F.1.4.5 The student knows that complex interactions among the different kinds of molecules in the cell cause distinct cycles of activity governed by proteins.	6, 7, 40–42, **42,** 56–61, 78–80, 81–82, 94–96, 97–103, 104–110, 126–127, **129,** 130–132, **132,** 138–139, 144–145, 148–149, 198–199, **199,** 215–218, **218,** 434, 436–437, 872, 888–889, 926, 931, 933–935, 946	46–47, 62–66, **101**	9, 22, 42, 86, 88, 101, 103, 110, 112, 113, 127, 129, 132, 134, 135, 149, 156, 157, 199, 200, 202, 203, 218,220, 222, 223, 441, 453, 477, 546, 879, 892, 905, 919, 926, 929, 932, 936, 969
SC.F.1.4.6 The student knows that separate parts of the body communicate with each other using electrical and/or chemical signals.	84–86, 858–859, 946–949, **946,** 974, 976, 977–980, 984–988, **986, 992–993,** 1001–1003, **1003**	42, 630–631, 753, 755	86, 88, 110, 859, 939, 946, 949, 955, 968, 976, 981, 986, 988, 990, 991, 993, 998, 1003, 1012, 1013
SC.F.1.4.7 The student knows that organisms respond to internal and external stimuli.	8, **478–479,** 582–584, **584,** 631, **638–639,** 773, **796–797,** 824–829, 833–834, 959–960, **958**	42, 623, **638–639,** 753, 755	22, 112, 182, 479, 584, 586, 587, 631, 639, 658, 777, 797, 836, 838, 839, 855, 892, 955, 958, 960, 969
SC.F.1.4.8 The student knows that cell behavior can be affected by molecules from other parts of the organism or even from other organisms.	42, 55–56, 94–95, 438–439, 580–581, 604–610, 924, 926, 927–929, 930, **932,** 934–935, 936, 961–966, 1005–1006, 1010	60–61, 104–109, 627–628, 758–760	441, 450, 452, 453, 566, 586, 587, 892, 893, 918, 919, 926, 929, 932, 936, 938, 939, 966, 968, 969, 1007, 1010, 1012

	INSTRUCTION: IN-DEPTH	INSTRUCTION: MENTIONED	ASSESSMENT
Standard 2: The student understands the process and importance of genetic diversity. (SC.F.2.4)			
SC.F.2.4.1 The student understands the mechanisms of asexual and sexual reproduction and know the different genetic advantages and disadvantages of asexual and sexual reproduction.	7, 118–120, **136–137,** 146–147, 150–154, **154,** **158–159,** 170–172, **174,** 177–179, 462–463, 484, 486–488, 503–505, 530–533, 537, 540, 541–544, 609–610, 621, 625, 632, 633–634, **811,** 996–998, 999–1003, 1004–1007	506, 621, 625, 631–634, 761, 808–810	124, 135, 149, 151, 154, 156, 157, 165, 169, 174, 176, 182, 184, 185, 294, 463, 477, 484, 488, 494, 495, 507, 515, 533, 540, 544, 546, 547, 610, 621, 628, 634, 636, 679, 750, 764, 794, 811, 816, 838, 998, 1003, 1007, 1012
SC.F.2.4.2 The student knows that every cell contains a "blueprint" coded in DNA molecules that specify how proteins are assembled to regulate cells.	13, 37, 56–58, 62, 118–124, **147,** 194–197, 208–214, **214,** 215–218, **224–225,** 256–257, 435	**136–137,** 826	13, 23, 37, 44,66, 69, 124, 127, 134, 135, 147, 149, 156, 157, 169, 193, 195, 197, 200, 202, 203, 210, 214, 220, 221, 223, 224–225, 245, 270, 271,441, 586, 612, 613
SC.F.2.4.3 The student understands the mechanisms of change (e.g., mutation and natural selection) that lead to adaptations in a species and their ability to survive naturally in changing conditions and to increase species diversity.	7–8, 9, 122–124, **124,** 180–181, 219–220, 259–261	825–826	9, 22, 23, 124, 132, 134, 154, 156, 182, 200, 220, 222, 223, 292, 294, 295, 864, 866, 938

	INSTRUCTION: IN-DEPTH	INSTRUCTION: MENTIONED	ASSESSMENT

How Living Things Interact with Their Environment

Standard 1: The student understands the competitive, interdependent, cyclic nature of living things in the environment. (SC.G.1.4)

	INSTRUCTION: IN-DEPTH	INSTRUCTION: MENTIONED	ASSESSMENT
SC.G.1.4.1 The student knows of the great diversity and interdependence of living things.	9, 132, 200, 258–263, 300–304, 305–306, 320–325, **331,** 340, **342,** 362–364, **363,** 365–367, 367, 371–378, **382–383,** 412–417, 418–421, 422–426, 442, 447–450, 460–461, **466, 468,** 482, 485, 488, 490–492, **491,** 504–505, 508–515, 516–519, **526–527,** 531–533, 572–573, 578, **588–589,** 594–597, **603,** **614–615,** 618–619, 622–623, 624–632, **634,** 642–643, 651–656, 665, 670–671, 673–674, 679–682, 692–695, 700–702, 712–713, 718–719, 724, 731–732, 735–736, 746–750, 751–757, 758–764, 772–777, 778–783, 784–789, 804–806, 1008–1009		9, 20, 22, 132, 134, 165, 263, 268, 271, 287, 294, 295, 304, 310, 313, 325, 331, 344, 342, 349, 356, 363, 364,367, 370, 378, 380, 381, 383, 417, 421, 426, 428, 429, 450, 452, 453, 463, 466, 468, 471, 476, 477, 484, 488, 491, 492, 494, 495, 507, 515, 522, 524, 525, 527, 533, 546, 547, 566, 567, 578, 586, 587, 589, 612, 613, 621, 628, 634, 636, 637, 650, 656, 658, 659, 669, 672, 679, 682, 684, 685, 699, 702, 704, 705, 720, 730, 734, 738, 740, 741, 750, 757, 764, 766, 767, 777, 783, 794, 795, 806, 818, 819, 839, 918, 1010, 1012
SC.G.1.4.2 The student understands how the flow of energy through an ecosystem made up of producers, consumers, and decomposers carries out the processes of life and that some energy dissipates as heat and is not recycled.	94–96, 345–349, 352, 483, **517**	97–103, 104–110, **114–115**	96, 103, 112, 270, 634, 684, 838
SC.G.1.4.3 The student knows that the chemical elements that make up the molecules of living things are combined and recombined in different ways.	95–96, **101,** 107, **114–115,** **548–549,** 579, **581,** 900–905, **920–921**	46–47, 62–66, 101, 60–61, 104–109, 627–628, 758–760	45, 101, 103, 110, 112, 114–115, 354, 356, 357, 495, 549, 581, 584, 586, 893, 905, 916, 918, 919, 921

How Living Things Interact with Their Environment, *cont.*	INSTRUCTION: IN-DEPTH	INSTRUCTION: MENTIONED	ASSESSMENT
Standard 2: The student understands the process and importance of genetic diversity. (SC.G.2.4)			
SC.G.2.4.1 The student knows that layers of energy-rich organic materials have been gradually turned into great coal beds and oil pools (fossil fuels) by the pressure of the overlying earth and that humans burn fossil fuels to release the stored energy as heat and carbon dioxide.	266, 388–389, 392–393		268, 402, 403
SC.G.2.4.2 The student knows that changes in a component of an ecosystem will have unpredictable effects on the entire system but that the components of the system tend to react in a way that will restore the ecosystem to its original condition.	9, 322–325, 326–332, 343–344, **342, 358–359,** 365–370, **695**		332, 334, 335, 342, 359, 492, 495, 684, 704, 795
SC.G.2.4.3 The student understands how genetic variation of offspring contributes to population control in an environment and that natural selection ensures that those who are best adapted to their surroundings survive to reproduce.	9, 279–280, 288–290, **292,** 326–332, **336–337,** 807–816, 825–829	825–826	282, 292, 294, 325, 329, 332, 334, 335, 336–337, 364, 380, 450, 452, 453, 525, 612, 685, 766, 794, 795, 816, 829, 836, 838, 839
SC.G.2.4.4 The student knows that the world ecosystems are shaped by physical factors that limit their productivity.	322–325, 371–378, **374**		325, 335, 357, 374, 378, 380, 381, 476

How Living Things Interact with Their Environment, *cont.*	INSTRUCTION: IN-DEPTH	INSTRUCTION: MENTIONED	ASSESSMENT
SC.G.2.4.5 The student understands that the amount of life any environment can support is limited and that human activities can change the flow of energy and reduce the fertility of the Earth.	103	406–407	23, 103, 325, 334, 356, 380, 402
SC.G.2.4.6 The student knows the ways in which humans today are placing their environmental support systems at risk (e.g., rapid human population growth, environmental degradation, and resource depletion).	**24–25,** 354, 386–389, **389,** 390–391, 394–395, 397, 400, **404–405, 426, 430–431, 678,** 809	325, 406–407, 578, 628, 699	22, 24–25, 244, 245, 344, 389, 395, 400, 402, 403, 405, 426, 431, 477, 637, 678, 704, 818

The Nature of Science

Standard 1: The student uses the scientific processes and habits of mind to solve problems. (SC.H.1.4)

Standard	INSTRUCTION: IN-DEPTH	INSTRUCTION: MENTIONED	ASSESSMENT
SC.H.1.4.1 The student knows that investigations are conducted to explore new phenomena, to check on previous results, to test how well a theory predicts, and to compare different theories.	14–20, **70–71,** 162–165, 166–167, 252–257, **259, 304, 314–315, 454–455,** 456–457, **484, 532, 536, 538, 583, 623, 631, 660–661, 682, 686–687, 768–769, 792,** 830–836, **840–841, 868–869, 886, 894–895**	**24–25, 46–47, 76,** 162–165, **165, 186–187**	20, 22, 24, 47, 70–71, 76, 113, 165, 184, 193, 202, 257, 259, 263, 270, 304, 314–315, 335, 428, 452, 455, 484, 524, 532, 536, 538, 583, 623, 631, 659, 661, 682, 687, 704, 769, 792, 836, 841, 869, 886, 895, 938, 990, 991
SC.H.1.4.2 The student knows that from time to time, major shifts occur in the scientific view of how the world works, but that more often, the changes that take place in the body of scientific knowledge are small modifications of prior knowledge.	19–20, 55, 162, 166–169, 190–193, 434	51, 170–176	23, 156, 165, 184, 203, 270, 428, 441, 567, 938

FLORIDA SUNSHINE STATE STANDARDS

The Nature of Science, *cont.*	INSTRUCTION: IN-DEPTH	INSTRUCTION: MENTIONED	ASSESSMENT
SC.H.1.4.3 The student understands that no matter how well one theory fits observations, a new theory might fit them as well or better, or might fit a wider range of observations, because in science, the testing, revising, and occasional discarding of theories, new and old, never ends and leads to an increasingly better understanding of how things work in the world, but not to absolute truth.	19–20, 55, 162, 166, 169, 282	55, **165,** 216, 282, 737	61, 165, 263, 270, 294, 730, 741
SC.H.1.4.4 The student knows that scientists in any one research group tend to see things alike and that therefore scientific teams are expected to seek out the possible sources of bias in the design of their investigations and in their data analysis.	18–20	162	20, 44, 113, 202, 222, 777, 838
SC.H.1.4.5 The student understands that new ideas in science are limited by the context in which they are conceived, are often rejected by the scientific establishment, sometimes spring from unexpected findings, and usually grow slowly from many contributors.	55, 162, 166–169, 196–197, 216		69, 156, 197, 257, 270

The Nature of Science, *cont.*	INSTRUCTION: IN-DEPTH	INSTRUCTION: MENTIONED	ASSESSMENT
SC.H.1.4.6 The student understands that in the short run, new ideas that do not mesh well with mainstream ideas in science often encounter vigorous criticism and that in the long run, theories are judged by how they fit with other theories, the range of observations they explain, how well they explain observations, and how effective they are in predicting new findings.	14–20, 166–169, 216	**24–25,** 50, 55, **165, 168,** 170–176, **172, 174, 176**	20, 165, 168
SC.H.1.4.7 The student understands the importance of a sense of responsibility, a commitment to peer review, truthful reporting of the methods and outcomes of investigations, and making the public aware of the findings.	14–20	**24–25, 46–47,** 50, **70–71, 90–91, 114–115,** 123, 138–139, **186–187, 204–205, 224–225, 246–247, 272–273, 296–297, 314–315, 336–337, 358–359, 382–383, 404–405,** 406–407, 456–457, 474, **430–431, 454–455, 478–479, 496–497, 526–527, 548–549, 568–569, 588–589, 614–615, 638–639, 660–661,** 688–689, **706–707, 742–743, 768–769, 796–797, 820–821, 840–841, 868–869, 894–895,** 896–897, 903, **920–921,** 936, **940–941,** 953, **970–971,** 981, **992–993,** 1007, **1014–1015**	20, 23, 25, 47, 612
Standard 2: The student understands that most natural events occur in comprehensible, consistent patterns. (SC.H.2.4)			
SC.H.2.4.1 The student knows that scientists assume that the universe is a vast system in which basic rules exist that may range from very simple to extremely complex but that scientists operate on the belief that the rules can be discovered by careful, systemic study.	7–9, 14–20, 162–165, 167–169, 177–182, 300–304, 305–310, 406–407	**24–25,** 50–54, **70–71,** 149, **165,** 170–176	9, 22, 165, 169, 257, 282, 306, 309

FLORIDA SUNSHINE STATE STANDARDS

FLORIDA SUNSHINE STATE STANDARDS

The Nature of Science, *cont.*	INSTRUCTION: IN-DEPTH	INSTRUCTION: MENTIONED	ASSESSMENT
SC.H.2.4.2 The student knows that scientists control conditions in order to obtain evidence, but when that is not possible for practical or ethical reasons, they try to observe a wide range of natural occurrences to discern patterns.	17, **20,** 50–51, 192–193, 277–278, 307–310, 824, **828, 834, 935**	**24–25, 151, 638–639,** 832	20, 22, 169, 257, 282, 306, 309, 639, 794, 828, 829, 834, 839, 935
Standard 3: The student understands that science, technology, and society are interwoven and interdependent. (SC.H.3.4)			
SC.H.3.4.1 The student knows that performance testing is often conducted using small-scale models, computer simulations, or analogous systems to reduce the chance of system failure.	**136–137, 158–159, 172,** 175–176, **176, 186–187, 204–205, 246–247, 253, 257, 280, 296–297, 599, 645, 652, 774, 889, 911, 940–941**	**90–91**	89, 136–137, 158–159, 176, 185, 186–187, 204–205, 246–247, 253, 257, 270, 271, 280, 296–297, 566, 599, 636, 645, 652, 684, 740, 766, 774, 889, 911, 941
SC.H.3.4.2 The student knows that technological problems often create a demand for new scientific knowledge and that new technologies make it possible for scientists to extend their research in a way that advances science.	50–54, **70–71,** 123, 228–**232,** 233–237, 688–689, 862	3, **46–47, 86,** 141, 220, 397, 699, 832, 936, 953, 981	44, 54, 68, 203, 232, 242, 244, 245, 452, 494, 547, 658, 838
SC.H.3.4.3 The student knows that scientists can bring information, insights, and analytical skills to matters of public concern and help people understand the possible causes and effects of events.	36, 10, 12, 109, 123, 138–139, 182, 220, 325, 354, 396–399, 406–407, 445, 456–457, 472–474, 539, 578, 648, 669, 688–689, 699, 729, 832, 853, 862, 864, 878–879, 884, 890, 896–897, 903, 908, 915–916, 936, 953, 981, 987, 1007	3, **24–25,** 50, 141, 634, **1048–1071**	13, 23, 134, 135, 184, 203, 244, 245, 400, 402, 452, 474, 476, 488, 494, 546, 586, 612, 650, 684, 740, 767, 829, 838, 849, 855, 860, 864, 866, 879, 884, 890, 892, 905, 918, 919, 938, 968, 969, 990, 1012, 1013

The Nature of Science, *cont.*	INSTRUCTION: IN-DEPTH	INSTRUCTION: MENTIONED	ASSESSMENT
SC.H.3.4.4 The student knows that funds for science research come from federal government agencies, industry, and private foundations and that this funding often influences the areas of discovery.	406–407, 456–457, 688–689, 902	86, 138–139, 445, 474, 578, 699, 809, 832, 862, 902, 903, 936	905
SC.H.3.4.5 The student knows that the value of a technology may differ for different people and at different times.	11, 123, 234–237, 238–242, 474, 516–522, 688–689, 878, 882–883, 931–932, 981	141, 220, 445, 862, 1007	13, 22, 44, 68, 134, 156, 184, 202, 203, 222, 237, 242, 244, 395, 452, 476, 522, 524, 525, 566, 586, 587, 658, 884, 892, 897, 918, 932, 968, 1012
SC.H.3.4.6 The student knows that scientific knowledge is used by those who engage in design and technology to solve practical problems, taking human values and limitations into account.	11, 36, 109, 123, 236, 456–457, 474, 648, 656, 688–689, 862, 903, 916, 936, 953, 1007	3, 50–54, **46–47,** 139, **1048–1071**	22, 112, 237, 242, 244, 564

Lab Safety

General Safety Guidelines

1. Post laboratory rules in a conspicuous place in the laboratory.
2. Before the class begins an experiment, review safety rules and demonstrate proper procedures.
3. Never permit students to work in your laboratory without your supervision. No unauthorized investigations should ever be conducted, nor should unauthorized materials be in the laboratory.
4. Lock your laboratory (and storeroom) when you are not present.
5. Clearly mark locations of eyewash stations, safety shower, fire extinguishers (ABC tri-class), chemical spill kit, first-aid kit, and fire blanket in the laboratory and storeroom. Check this safety equipment prior to conducting each investigation.
6. Post an evacuation diagram and procedures by every entrance to the laboratory.
7. Provide labeled disposal containers for glass, sharps, and waste chemical reagents.
8. Allow no food or beverages in the laboratory. Caution students to keep their hands away from their face and to wash hands with soap and water before leaving the laboratory.
9. Know the location for the master shut-off for laboratory circuits. Be sure that all outlets have correct polarity and have ground-fault interception. Polarity can be tested with an inexpensive continuity tester available from most electronic hobby shops. All electrical equipment should have 3-prong plugs and 3-wire cords.
10. Follow prescribed procedures for any safety incident, including full documentation. Remind students that any safety incident, no matter how trivial, must be reported directly to you.

Personal Protective Equipment

Chemical goggles: (Meeting ANSI Standard Z87.1) These should be worn with any chemical or chemical solution other than water, when heating substances, using any mechanical device, or observing physical processes that could eject an object.

Face shield: (Meeting ANSI Standard Z87.1) Use in combination with eye goggles when working with corrosives.

Contact lenses: The wearing of contact lenses for cosmetic reasons should be prohibited in the laboratory. If a student must wear contact lenses prescribed by a physician, that student should be instructed to wear eye-cup safety goggles meeting ANSI Standard Z87.1 (similar to swimmer's cup goggles).

Eye-wash station: The device must be capable of delivering a copious, gentle flow of water to both eyes for at least 15 minutes. Portable liquid supply devices are not satisfactory and should not be used. A plumbed-in fixture or a perforated spray head on the end of a hose attached to a plumbed-in outlet is suitable if it is designed for use as an eye-wash fountain and meets ANSI Standard Z358.1. It must be within a 30-second walking distance from any spot in the room.

Safety shower: (Meeting ANSI Standard Z358.1) Location should be within a 30-second walking distance from any spot in the room. Students should be instructed in the use of the safety shower in the event of a fire or chemical splash on their body that cannot simply be washed off.

Gloves: Polyethylene, neoprene rubber, or disposable plastic may be used. Nitrile or butyl rubber gloves are recommended when handling corrosives.

Apron: Rubber-coated cloth or vinyl (nylon-coated) halter is recommended.

Emergency Preparedness

What would you do if a student dropped a liter bottle of concentrated sulfuric acid? RIGHT NOW? Plan how to effectively react BEFORE you need to.

1. Post the phone number of your regional poison control center, fire department, police, ambulance, and hospital on your telephone.
2. Practice fire and evacuation drills. Also have drills on what students MUST do if they are on fire or have chemical contact.
3. Ensure that all personal and other safety equipment is available and tested frequently.
4. Compile an MSDS file for all chemicals. This reference resource should be readily available in case of a spill or other accident.
5. Provide for spill control procedures. Handle only those incidents that you feel comfortable in handling. Situations of greater severity should be handled by trained hazardous materials professionals.
6. Students should never fight fires or handle spills.
7. Be trained in first aid and basic life support (CPR) procedures. Have first-aid kits and spill kits readily available.
8. Fully document ANY INCIDENT that occurs.

Safety With Animals

It is recommended that teachers follow the NABT Position Statement "The Use of Animals in Biology Education" issued by the National Association of Biology Teachers (available at www.nabt.org).

Safety in Handling Preserved Materials

The following practices are recommended when handling or dissecting any preserved specimen:

1. NEVER dissect road-kills or nonpreserved slaughterhouse materials.
2. Wear protective gloves and splash-proof safety goggles at all times when handling preserving fluids and preserved specimens and during dissection.
3. Wear lab aprons. Use of an old shirt or smock under the lab apron is recommended.
4. Conduct dissection activities in a well-ventilated area.
5. Do not allow preservation or body-cavity fluids to contact skin. Fixatives do not distinguish between living or dead tissues.

Biological supply firms use formalin-based fixatives of varying concentrations to initially fix zoological and botanical specimens. WARD'S Natural Science Establishment provides specimens that are freeze-dried and rehydrated in a 10% isopropyl alcohol solution. In these specimens, no other hazardous chemical is present.

Many suppliers provide fixed botanical materials in 50% glycerin.

Reduction of Free Formaldehyde

Currently, federal regulations mandate a permissible exposure level of 0.75 ppm for formaldehyde. Contact your supplier for an MSDS that details the amount of formaldehyde present as well as gas-emitting characteristics for individual specimens.

Prewash specimens (in a loosely covered container) in running tap water for 1–4 hours to dilute the fixative. Formaldehyde may also be chemically bound (thereby reducing danger) by immersing washed specimens in a 0.5–1.0% potassium bisulfate solution overnight or by placing them in 1% phenoxyethanol holding solutions.

Safety With Microbes

Pathogenic (disease-causing) microorganisms are not appropriate investigational tools in the high school laboratory and should never be used.

Safety With Chemicals

Label student reagent containers with the substance's name and hazard class(es) (flammable, reactive, etc.).

Dispose of hazardous waste chemicals according to federal, state, and local regulations. Refer to the

Material Safety Data Sheet for recommended disposal procedures.

Remove all sources of flames, sparks, and heat from the laboratory when any flammable material is being used.

Material Safety Data Sheets

The purpose of a Material Safety Data Sheet (MSDS) is to provide readily accessible information on chemical substances commonly used in the science laboratory or in industry.

The MSDS should be kept on file and referred to BEFORE handling ANY chemical. The MSDS can also be used to instruct students on chemical hazards, to evaluate spill and disposal procedures, and to warn of incompatibility with other chemicals or mixtures.

Storing Chemicals

Never store chemicals alphabetically, as this greatly increases the risk of promoting a violent reaction.

Storage Suggestions:

1. Always lock the storeroom and all its cabinets when not in use.
2. Students should not be allowed in the storeroom and preparation area.
3. Avoid storing chemicals on the floor of the storeroom.
4. Do not store chemicals above eye level or on the top shelf in the storeroom.
5. Be sure shelf assemblies are firmly secured to the walls.
6. Provide anti-roll lips on all shelves.
7. Shelving should be constructed out of wood. Metal cabinets and shelves are easily corroded.
8. Avoid metal, adjustable shelf supports and clips. They can corrode, causing shelves to collapse.
9. Acids, flammables, poisons, and oxidizers should each be stored in their own locking storage cabinet.

Master Materials List

The following lists give the amounts of materials needed for a class of 32 students, divided into 8 groups of 4 students each, to perform all of the laboratory experiments in this book. Full-scale end-of-chapter labs—**Exploration Labs** and **Skills Practice Labs**—are designated by the number of chapter in which they appear. A **Quick Lab** is designated by **QL,** a **Math Lab** by **ML,** and a **Data Lab** by **DL**—followed by the number of the chapter in which the lab appears. **Forensics Labs** are designated by **FOR.** Visit **www.hrw.com/wards** for materials ordering information specific to ***Holt Biology*** (see WARD's item numbers listed below).

Biological Supplies	WARD's Item no.	Quantity per Class	Chapter Number
Alfalfa seeds, pkt.	86R8001	8 seeds	16
Algae, mixed green (wm) slide	91R0560	4 slides	19
Amphioxus—post. to pharynx	92R3014	8	QL31
Animal kingdom survey set, general	67R9911	1	19
Apple and potato slices	Local	8	QL36
Apple, ripe	Local	8	QL26
Archaebacteria, mixture (sm) g(-)	90R0526	4	19
Bacillus megaterium culture	85R1154	1	20
Bacillus subtilis culture	85R0228	1	20
Bacillis thuringiensis culture	85R0203	1	20
Bean seeds, pinto, 1 lb pkg.	86R8014	400 seeds	1
Bean seeds, French	86R8004	48 seeds	25
Beta (beet) root (cs) slides	91R7980	8 slides	QL1

Biological Supplies	WARD's Item no.	Quantity per Class	Chapter Number
Brine shrimp *(Artemia)* eggs, 6 g	87R5102	400 mL	17
Broth, concentrated, beef	Local	8 mL	28
Chameleons, American, 3/pkg.	87R8135	16	34
Clamshell	Local	8	29
Clostridium tetani (sm) g(+) slide	90R2032	4 slides	19
Coleus	86R6800	16	QL26
Coleus stem (cs) slide	91R7831	8 slides	QL25
Corn kernels, dry	86R8080	8	QL11
Crickets *(Acheta domestica)*, 50/pkg.	87R6102	8	16, 33, QL36
Daphnia magna culture, mature	87R5210	1	28, 42
Dianthus—entire leaf (cs) slide	91R7882	8 slides	QL25

Biological Supplies	WARD's Item no.	Quantity per Class	Chapter Number
Diatomaceous earth, jar 30 oz	63R0240	8 oz	QL21
DNA extraction biolab	36R6026	1	QL9
Earthworms *(Lumbricus)*, 10/pkg.	87R4660	8	16
Elodea, 12/pkg.	86R7500	8 sprigs	3, QL5
Fern antheridia and archegonia	91R4865	8	24, QL3
Fruits and vegetables	Local	8	QL23
Frogs, small unsexed, 12/pkg.	87R8217	8 frogs	33
Grapes	Local	8	QL4, QL34
Grass seed, rye, 1 oz pkg.	86R8130	8 pinches	16
Green beans or snow peas	Local	8	15
Hydra culture, brown	87R2020	1	28
Hydra, plain (wm) slide	92R0630	8 slides	27
Isopod (sowbug) culture, 45/pkg.	87R5520	32	16, 30, QL35
Lemna minor (duckweed), 4 oz jar	86R7650	1	24
Lentil seeds	Local	8 seeds	8
Leucosolenia (sponge) (wm) slide	92R0568	8 slides	27
Lilium root (cs) slide	91R7200	8 slides	QL25
Lilium stem (cs) slide	91R7204	8 slides	QL25
Mealworm larvae, 100/pkg.	87R6250	8	16
Micrococcus luteus culture	85V0966	1	20
Micrococcus luteus (sm) g(+) slide	95R0554	4 slides	19
Mitosis, plant (sect) slide	93R2145	8 slides	QL6
Monocot flower and dicot flower	Local	8 of ea.	QL24
Mung bean seed, Oriental, 1 lb pkg.	86R8013	48 seeds	16
Nematode, typical (wm) slide	92R1375	8 slides	27
Obelia medusae (wm) slide	92R0731	8 slides	ML28
Pea seeds	86R8247	8 seeds	8
Pine ovulate cone (ls) qs slide	91R6534	8 slides	QL24
Pine ovule-young archegonium (ls)	91R6540	8	QL24
Pine staminate cone (ls) qs slide	91R6528	8 slides	QL24
Planaria culture, brown	87R2500	1	QL28
Planaria, plain (wm) slide	92R0820	8 slides	27
Plants kept in light and dark	Local	2	QL23
Protists, mixed (wm) slide	92R0005	4 slides	19
Protist (protozoa) survey mixture culture	87R1610	1	21

Biological Supplies	WARD's Item no.	Quantity per Class	Chapter Number
Raisins, dry	Local	8	QL33
Rhizopus sporangium (wm) slide	91W2110	8 slides	QL22
Sea star, 2, 4, 8, 16 cell stages	92R8243	8 slides	43
Sea star, 32, 64 cell	92R8245	8 slides	43
Sea star, blastula	92R8246	8 slides	43
Sea star, early, mid gastrula	92R8249	8 slides	43
Sea star, egg	92R8241	8 slides	43
Sea star, late gastrula	92R8250	8 slides	43
Sea star, zygote	92R8242	8 slides	43
Sea stars *(Asterias)*, 10/pkg.	68R7615	1 pkg.	31
Skin of hairy mammal (sect) slide	93R5023	8 slides	35
Skull, beaver, natural bone	65R6450	1	35
Skull, bullfrog, natural bone	65R5122	1	35
Skull, canine, adult, natural bone	65R5310	1	35
Skull, cat, adult, natural bone	65R5220	1	35
Skull, freshwater dogfish *(Amia)*	65R1410	8	35
Skull, bat, flying fox replica	80R1715	1	35
Skull, snake, natural bone	65R5124	1 skull	35
Skull, turtle, natural bone, w/key	65R5123	1 skull	35
Spirillum volutans culture	85V1140	1	20
Squamous epithelium (sm) slide	93R6003	8 slides	3
Survey set, general, jar mount	67R9904	1	19, 23
Survey set, invertebrate	62R0044	1	27
Sweet corn seed, untreated, 200/pkt.	86R8080	48 seeds	26

Chemicals and Media	WARD's Item no.	Quantity per Class	Chapter Number
Adrenaline, epinephrine solution	250R2060	40 ml	42
Agar (red), 6 bottles/pkg.	88R8902	1 bottle	QL6
Ascorbic acid, L(+) vitamin C	944R4702	10 mL	40
Baking soda, 1 lb pkg.	37R5467	16 oz	1, 5, QL38
DETAIN(tm) protist-slowing agent	37R7950	1	17, 21
Detergent, laundry, 5 different brands	Local	1	2

MASTER MATERIALS LIST

Chemicals and Media	WARD's Item no.	Quantity per Class	Chapter Number
Detergent, liquid household, 14 oz	37R2268	80 drops	QL39
Distilled water, 1 gal	88R7005	4,000 mL	1, 2, 24, QL1, QL5, QL38, QL4, FOR
Ethyl alcohol (ethanol), anhydrous	950R6505	40 mL	QL9
Gelatin, instant	Local	8 pkg.	2
Gelatin solution, 1%, 250 mL	326R8501	40 mL	QL12
Glucose solution, standard, 100 mL	945R6606	80 mL	39
Grape juice, white	Local	1	QL4
Gum arabic solution, 1%, 250 mL	326R8502	24 mL	QL12
Hydrochloric acid, 0.1 M sol.	970R3307	80 mL	30, QL12
Indophenol solution, 0.1%	944R1306	1	40
Knops solution, 500 mL	941R5806	1	24
Lemon juice, 8 oz	37R8542	16 oz	1
Limewater	941-6606	150 mL	22
Liquid soap, antibacterial, 8 oz	18R1533	240 mL	QL1
Lugol's iodine solution, 500 mL	970R4806	120 mL	3
Methyl alcohol (methanol)	950W9205	500 mL	FOR
Methylene blue chloride, 1% aq. sol.	946R1906	80 mL	18, 20
Milk treatment product liquid	Local	64 drops	39
Nail polish, clear, 15 cc	37R1862	1 bottle	QL23
Paints, hobby, 5 colors	Local	8 sets	QL36
Petroleum jelly, 1 oz	15R9832	1 oz	42
Phenolphthalein agar, 800 mL	88R0928	1	4
Plant food, Miracle-Gro, 1.5 lb	20R6020	1 pkg.	24
Pond water, 1 gal	88R7010	1	24, QL28
Preserved specimen conditioner, 500 mL	39W1697	80 mL	1
Rubbing alcohol, 16 oz	950R8506	1 bottle	20, QL1
Silicone culture gum, 25 mL bottle	37R9810	1	28
Sodium carbonate, anhydrous	942R9106	0.7 g	2
Sodium hydroxide 0.1% sol.	970R7807	16 oz	1
Spring water, 1 gal	88R7000	5 gal	33
Sugar (sucrose), granular	946R9608	32 oz	QL4, 22
Vegetable coloring, red	947R3204	8 drops	29, QL29
Vegetable oil	37R9540	160 mL	QL39
Vinegar, white	39R0138	1,200 mL	1, 4, QL1
Yeast, viable baker's, 10 g packet	88R0929	8 pkg.	18, 22, QL7

Laboratory Equipment	WARD's Item no.	Quantity per Class	Chapter Number
Apron, polyethylene, disposable, 100/box	15R1050	32	1, 2, 3, 4, 18, 19, 20, 24, 25, 27, 29, 31, 39, 40, 42, QL1, QL11, QL40, QL42, FOR
Aquarium, 10 gal, 20 × 10 × 12 in.	21R5241	8	33
Aquarium/terrarium, medium, covered	21R2101	8	33
Aquarium/terrarium, large, covered	21R2102	8	QL36
Aquarium/terrarium, small, covered	21R2100	8	16
Beaker, student grade, 50 mL (Griffin)	17R4010	48	2, QL1, FOR
Beaker, student grade, 100 mL (Griffin)	17R4020	8	42, QL33
Beaker, student grade, 150 mL (Griffin)	17R4030	16	2, 20, 26
Beaker, student grade, 250 mL (Griffin)	17R4040	16	1, 4, 22, 24, QL35, QL38, QL39
Beaker, student grade, 400 mL (Griffin)	17R4050	8	3,18, QL17
Beaker, student grade, 600 mL (Griffin)	17R4060		5, 8, 33, QL10, QL42
Blood typing kit, ABO simulated	36R0022	1	QL40
Blood typing kit, simulated, ABO and Rh included	36W0019	8	FOR
Bunsen burner, standard, natural gas	15R0612	8	20
Calculator	27R3055	8	15, 17
Capillary tubes, open-ended	15W2096	16	FOR
Capillary tubes, plunger	15W2155	16	FOR
CBL™ system	Local	8	QL5, QL17, QL18, QL35
CBL probe, temperature	Local	8	QL17, QL18, QL35
CBL probe, dissolved oxygen	Local	8	QL5
Chromatography of simulated plant pigments demo kit	36R6019	1	25
Chromatography paper, strips, 50 per pkg.	15W3708	16	FOR
Chromatography plates, thin layer	17W7650	16	FOR
Clamp, Hoffman, screw compressor, open jaw	15R3910	24	17
Clamp, three finger utility	15W0699	8	FOR

Laboratory Equipment	WARD's Item no.	Quantity per Class	Chapter Number
Corks, size 0, 100/pkg.	15W8360	16	FOR
Corks, size 2, 100/pkg.	15R8362	16	17
Coverslips, 22 mm plastic, 100/box	14R3555	80	3, 18, 42, QL7, QL12, QL21
Dissecting set, basic	14R0773	8	31
Dissecting pins, 1/2 lb./pkg.	14R0200	1 pkg.	31
Dissecting pan, disposable, 10/pkg.	18R3640	1 pkg.	31
Dropping bottle, 15 mL, amber glass	17R6020	8	3, QL12
Dropping bottle, 15 mL, flint glass	17R6018	16	20, 40, FOR
Dropping bottles, polyethylene, hinged close top, 30mL	18W2271	16	FOR
Eyedropper, glass, 12 per pkg.	17W0230	1	FOR
Field guide, Peterson, *First Guide to Wildflowers*	32R0381	8	QL14
Filter paper, medium grade, 9.0 cm	15R2815	8	28
Forceps, dissecting, straight	14R1001	8	3, 20, 21, QL34
Forceps, fine point w/guide pin	14R0541	8	28, QL28
Funnel, glass	17R4300	8	5
Funnel, utility, 60 mm, 2 3/4 in. top	18R1421	8	17
Gauze, wire 6 × 6 in., 12/pkg.	15R0675	8	QL42
Glass bottle with cap, 4 oz.	17W0315	16	FOR
Glass tubing, 30-in. pieces	288-8302	1 pkg.	22
Glass tubing, 25 cm in length, 1/4 in. inside diameter	17W0981	8	FOR
Gloves, heat resistant, one size	15R1095	8	2, QL42, FOR
Gloves, latex, dispos. med., 100/pkg.	15R1071	16	1, 2, 3, 4, 5, 16, 18, 19, 20, 24, 25, 27, 28, 29, 31, 39, 40, 42, QL1, QL9, QL12, QL24, QL40
Grad. cylinder, polypropylene, 10 mL	18R1705	16	42, FOR
Grad. cylinder, polypropylene, 100 mL	18R1730	8	2
Grad. cylinder, polypropylene, 25 mL	18R1710	8	QL12, QL39
Hot plate	15R7999	8	QL17, QL42
Hot water bottle, 2 qt	local	8	17

Laboratory Equipment	WARD's Item no.	Quantity per Class	Chapter Number
Ice bag, 6 in. diameter	local	8	17
Magnifier, dual 3X and 6X	24R1112	32	19, 23, 24, 27, 35, QL24, QL28
Meterstick, wood	15R4065	8	12, 13, 41, QL30
Microscope slide, single concavity, 12/pkg.	14R3510	40 slides	39, 42
Microscope slides, 72/pkg.	14R3500	16 slides	3, 20, 21, 28, QL7, QL12, QL21, QL23
Microscope, compound light	24R2310	8	18, 19, 20, 21, 23, 27, 31, 35, 42, 43, ML28, QL6, QL7, QL12, QL21, QL23, QL24, QL25, QL31
Petri dish, disposable, 25/pkg.	18R7101	40	8, 17, 24, 29, 30, QL34
pH paper strips 1.0-14.0, 50/pkg.	15R3157	56 strips	2, QL1, QL16, QL38
Pipet, graduated polyethylene, 6 in., 500/pkg.	18R2971	40	21, QL21
Pipet, graduated, dropping, 2 mL	17R1706	16	17, 18, 2
Pipets, graduated, polyethylene, 100/pkg.	18R2977	8	QL9
Plant light	20R7500	8	17
Potting soil, growth media, 8 lb bag	20R8306	8 lb	16
Probe, Huber, stainless, 6 in.	14R0958	8	QL36
Probe, stainless steel	14R0950	8	27, 30
Reaction chamber, 15/pkg.	15R9861	8	QL29
Rubber stopper, size 0, black, solid	15R8460	32	17
Rubber stopper, size 0, black, 1 hole	15R8480	16	QL17
Rubber stopper, size 9, black, 2 holes	15R8519	8	22
Safety goggles, general purpose	15R3046	32	1, 2, 3, 4, 18, 19, 20, 24, 25, 27, 29, 31, 39, 40, 42, QL1, QL9, QL12, QL40, QL42, FOR
Sand, coarse, 32 oz container	45R1984	8	16
Sand, fine white Ottawa	942R5806	32 oz	QL17
Scalpel, sterile, No. 22 blade	14R0900	8	26, 29

MASTER MATERIALS LIST

Laboratory Equipment	WARD's Item no.	Quantity per Class	Chapter Number
Scissors, economy, dissection	14R0525	8	6, 7, 9, 11, 13, 21, 28, 30, 31, ML32, QL6, QL7, QL10, QL29 FOR
Specimen dish, 4 1/2 × 2 3/4 in.	17R0550	8	QL28
Specimen dish, 8 1/4 × 3 1/4 in.	17R0560	8	29
Spirometer	14R5070	8	38
Spirometer mouthpieces, 250/pkg.	14R5072	32	38
Spot plate, polystyrene, 12 pack	18W1028	32	FOR
Stage micrometer, ruled, 2 mm scale	94R9910	8	18
Stereomicroscope, beginner	24R4601	8	26, 28, 29, 30, 31
Stir rods, 6 in., glass, 150 × 5 mm, 10 pack	17R6005	8	2, 29, QL9, QL38, QL39, QL42
Stopwatch, digital	15R0512	8	30, 36, 37, 42, QL12, QL13, QL37
Support, rectangular with 4 × 6 in. base	15R0719	8	QL42
Swab applicator, 6 in., 100/pkg.	14R5502	24	20
Teasing Needle	14W0650	8	FOR
Test paper, glucose, 50 strips/pkg.	14R4119	40 strips	39
Test-tube clamp, Stoddard, 6/pkg.	15R0841	8	QL17
Test-tube rack, Nalge, 16 mm	18R4213	8	2, 17, 20, QL9, QL18
Test-tube with rim, Pyrex, 15 × 125 mm	17R0620	32	2, 5, 17, 40, QL5, QL9, QL12, QL17
Thermometer, lab, red alcohol	15R1462	8	2, 22, QL42
Tongs, utility, 8 in.	14R0960	8	QL42
Tray w/cover, polyethylene, 1 gal	18R3650	8	4
Triple beam balance, Ohaus	15R6057	8	15, 2, QL4, QL34, QL38
Tubing, amber latex 3/16 × 1/16 in., 10 ft	15R1133	48	22, QL29
Tubing, vinyl, 1/2 × 1/16 in., 10 ft	18R5084	16 ft	17, QL36
Tubing, vinyl, 3/16 × 1/16 in., 10 ft	18R5081	8 ft	QL29
Vacuum bottle, 500 mL	17-3805	8	22

Miscellaneous	WARD's Item no.	Quantity per Class	Chapter Number
Aluminum foil roll, 12 in. × 25 ft.	15R1009	16 ft.	17
Beads, 12 mm, approx. 100/pkg.	36R4197	8	QL11
Beads, 4 mm, approx. 100/pkg	36R4195	8	QL11
Beads, 8 mm, approx. 100/pkg.	36R4196	8	QL11
Beads, wooden, set of 100	15R0504	200	6, 7
Cardboard	Local	8	QL36
Candy coated chocolates 1 bag	Local	1	FOR
Connector, twistcock LDPE	18R1571	16	QL29
Construction paper, assorted 50/pkg.	15R9841	48	13, 34
Construction paper, black, 50/pkg.	15R9825	8	21
Corrugated box, 10/pkg., 11 1/8 in.	18R1395	8	QL12
Drill for making holes	Local	1	34, QL18
Fabrics, various textures	Local	8 sets	30
Flashlight, 2-D cell	15R3264	8	QL5, QL21
Food coloring, brown	Local	1	FOR
Gas lighter, flat file, 10/pkg.	15R0683	8	20
Graph paper, 5 squares/inch, 100 sheets	15R3835	40 sheets	1, 5, 15, 37, 43, ML15, QL35
Headlight glass, sample	Local	8	FOR
Hole punch, single hole	15W8910	4	FOR
Ice	Local	8 bags	QL35, QL42
Index cards, unlined, 3 × 5 in., 100/pkg	15R9819	32	10, QL6, QL15, QL41
Jar, clear polystyrene, 2 oz, 53 mm	18R1632	8	QL36
Jar, clear polystyrene, 4 oz, 58 mm	18R1633	24	QL4
Jar, 1/2 gal wide-mouth, glass	17R2070	16	34, QL11
Jar, 1 gal wide-mouth, glass	17R2081	16	QL26, FOR
Jar with cap, amber glass, 475 mL	17W0015	1	FOR
Jar, wide-mouth, glass, no cap, 950 mL	17R2060	8	QL16, QL18
Jar cap, white metal, 58 mm	17R2137	24	QL4u
Jar cap, white metal, 89 mm	17R2153	8	QL16, QL18
Jar cap, white metal, 83 mm w/liner	17R2149	16	34

Miscellaneous	WARD's Item no.	Quantity per Class	Chapter Number
Jar cap, 120 mm, white metal w/liner	17R2157	16	QL26
Knife, plastic	250R8128	8	4, QL23
Labels, lab 7/8 × 1 7/8 in.	15R1829	160	QL14
Lamp clamp with reflector	36R4168	8	21, QL18
Light bulb, 150 W, 120 V, clear	36R4173	8	21, QL18
Liver, raw, small pieces	Local	8	QL28
Marker, black lab	15R3083	8	13, 17, FOR
Marking pen, black wax, for glass	15R1155, 15R11578	8	1, 2, 20, 24, 26, 34, QL4, QL23, QL38, QL39
Masking tape,	15R9828	16 ft	10, QL10,
3/4 in. × 60 yds, 3/pkg.			QL14, QL18
Milk, whole	Local	4 L	39, QL16
Mirror, 3 × 4 in.	15R9860	8	35
Pan, aluminum foil, square, 10/pkg.	15R9891	8	ML32, QL29
Paper	Local	8	QL28
Paper bag, 5 in. × 10 1/2 in. × 3 in., 50/pkg.	18R6900	8	QL15
Paper clips, #1, 10 boxes/pkg. clips	15R9815	448	9, 10, 11, FOR
Paper punch, one hole	15R9810	8	21
Paper towel roll, 2 ply, 90 sheets	15R9844	40	1, 4, 20, 26, 42, QL4, QL36
Paper, white	Local	8	21, 43
Pencils, colored, set of 12	15R4690	8 pkg.	10, 12, QL7, QL10
Pencils, 12/box	15R9816	16	43, QL14
Penny or other coin	Local	8	13
Photographs of wildflowers	Local	8	QL14
Pictures of various organisms	Local	8	12
Pipe cleaners, colored, 6 in. 12/pkg.	82R1116	8 pkg.	6, 7
Pipet, dropping, 3 in., glass, 12/pkg.	17R0230	8	3, 28, 29, 39, 42, QL7, QL28, QL29
Plastic wrap, 12 in., 50 ft roll	15R9858	24 ft	2, ML32
Protractor, student, semicircle	14R4507	8	32

Miscellaneous	WARD's Item no.	Quantity per Class	Chapter Number
Pushpins, assorted colors, 100/box	15R0507	200	9, 10, 11, FOR
Rope, clothesline, 5/32 in. × 96 ft	15R3991	30 ft	QL19
Rubber bands, assorted, 1/4 lb	15R9824	16	26
Ruler, clear vinyl, 15/16 × 6 1/8 in.	14R0811	8	5, 32, ML28
Ruler, standard 12 in.	15R4650	8	1, 2, 4, 9, 15, 17, 21, 26, 32 ML32, QL10, QL18, FOR
Screening, fiberglass, 36 × 36 in.	15R0002	8	17
Six-sided die	Local	8	13
Socks, wool and cotton	Local	8	QL35
Soft drink, clear, cold, carbonated	Local	8	QL33
Spoons, plastic, 100/pkg.	15R9800	8	4
Spring hand grip	Local	8	37
Straws, plastic wrapped, 500/pkg. straws	15R9869	160	9, 10, 11, 13, QL38
Straws, plastic, colored	15R9869	8	10
Tags, 1 3/4 × 1 1/8in., 1,000/box	15R1873	80	6, 7
Tape measure, metric/English, 5 ft.	15R2541	8	ML15
Tape, adding machine, 2 1/4 in., 5/pkg.	27R3076	40	12
Tape, transparent w/self-dispenser	15R1959	8	2, 10, 13, 17, 30, QL6, QL7, QL10, QL23, QL24, FOR
Toothpicks, flat, 750/box	15R9864	40 tooth-picks	21, 39, ML32, FOR
Tracing paper, 40 sheets	15R9814	16 sheets	QL10
Water, boiling	Local	8	2
Wax pencil	15W1155	4	FOR
Windowpane glass, sample	Local	8	FOR
Yarn, red, 10 yd skein	15R2544	8	6, 7
Zipper bags, resealable, 4 × 6 in	18R6921	40 bags	1

George Johnson, Ph.D.

Dr. George Johnson is a professor of biology at Washington University in St. Louis, where he has taught general biology and genetics to undergraduates for 30 years. Dr. Johnson is also a professor of genetics at Washington University School of Medicine. He is the author of more than 50 scientific publications and has published several successful college biology texts.

Peter Raven, Ph.D.

Dr. Peter Raven is the director of the Missouri Botanical Garden and a professor of botany at Washington University in St. Louis. Dr. Raven is a MacArthur Fellow, a member of the National Academy of Sciences, a recipient of the National Medal of Science, and the author of 20 other books and several hundred scientific articles.

Printed in the United States of America

ISBN 0-03-074073-8

2 3 4 5 6 7 048 09 08 07 06

Contributing Writers

Ann Bekebrede
Science Writer
Sherborn, Massachusetts

Mary Dylewski
Science Writer
Kassel, Germany

Frances Jenkins
Science Writer
Sunburg, Ohio

David E. LaHart, Ph.D.
Program Director
Energy & Environmental Alliance
Florida State University
Tallahassee, Florida

Matt T. Lee, Ph.D.
Science Writer
Coos Bay, Oregon

Annette Ratliff
Science Writer
Austin, Texas

John A. Solorio
Multi-Technology Lab Facilitator
Austin Independent School District
Austin, Texas

Linda K. Gaul, Ph.D.
Epidemiologist
Texas Department of Health
Austin, Texas

Erik Hahn
Naturalist
Hartley Nature Center
Duluth, Minnesota

Feature Development

Linda K. Gaul, Ph.D.
Epidemiologist
Texas Department of Health
Austin, Texas

Matt T. Lee, Ph.D.
Science Writer
Coos Bay, Oregon

Inclusion Specialists

Joan A. Solorio
Special Education Director
Austin Independent School District
Austin, Texas

Teacher Edition Development

Linda K. Blumenthal
Science Writer
Columbus, Ohio

Alan Eagy
Biology Teacher
The Dalles High School
The Dalles, Oregon

Kevin L. Moore
Science Writer
Denver, Colorado

JoAnne Morgan Mowczko, Ed.D.
Educational Consultant
Gaithersburg, Maryland

Tyson Yager
Science Instructor
Wichita High School East
Wichita, Kansas

Academic Reviewers

David M. Armstrong, Ph.D.
Professor
Environmental, Population, and Organismic Biology
University of Colorado
Boulder, Colorado

Nigel Atkinson, Ph.D.
Associate Professor of Neurobiology
Institute for Neuroscience
The University of Texas
Austin, Texas

Jerry Baskin, Ph.D.
Professor
School of Biological Sciences
University of Kentucky
Lexington, Kentucky

John A. Brockhaus, Ph.D.
Director of Mapping, Charting, and Geodesy Program
Department of Geography and Environmental Engineering
United States Military Academy
West Point, New York

John Caprio, Ph.D.
George C. Kent Professor of Biological Sciences
Louisiana State University
Baton Rouge, Louisiana

Joe W. Crim, Ph.D.
Professor and Head
Department of Cellular Biology
The University of Georgia
Athens, Georgia

Roger J. Cuffey, Ph.D.
Professor of Paleontology
Department of Geosciences
Pennsylvania State University
University Park, Pennsylvania

James Denbow, Ph.D.
Associate Professor
Department of Anthropology
The University of Texas
Austin, Texas

David Futch, Ph.D.
Department of Biology
San Diego State University
San Diego, California

Linda K. Gaul, Ph.D.
Epidemiologist
Texas Department of Health
Austin, Texas

Herbert Grossman, Ph.D.
Associate Professor of Botany and Biology
Department of Environmental Sciences
Pennsylvania State University
University Park, Pennsylvania

William Guggino, Ph.D.
Professor of Physiology
The Johns Hopkins University
School of Medicine
Baltimore, Maryland

David Haig, Ph.D.
Professor
Department of Organismic and Evolutionary Biology
Harvard University
Cambridge, Massachusetts

David R. Hershey, Ph.D.
Education Consultant
Hyattsville, Maryland

David Ho, M.D.
Director and CEO
The Aaron Diamond AIDS Research Center
New York, New York

Joan E. N. Hudson, Ph.D.
Associate Professor
Sam Houston State University
Huntsville, Texas

Leland Lim, M.D., Ph.D.
Year II Resident
Department of Neurology and Neurological Sciences
Stanford University
School of Medicine
Palo Alto, California

Iris F. Litt, M.D.
Marron and Mary Elizabeth Kendrick Professor in Pediatrics
Stanford University
School of Medicine
Palo Alto, California

V. Patteson Lombardi, Ph.D.
Research Assistant Professor Human Biology and Medical Physiology
Department of Biology
University of Oregon
Eugene, Oregon

C. Riley Nelson, Ph.D.
Associate Professor
Department of Integrative Biology
Brigham Young University
Provo, Utah

continued on page 1136

Principles of Cell Biology 2

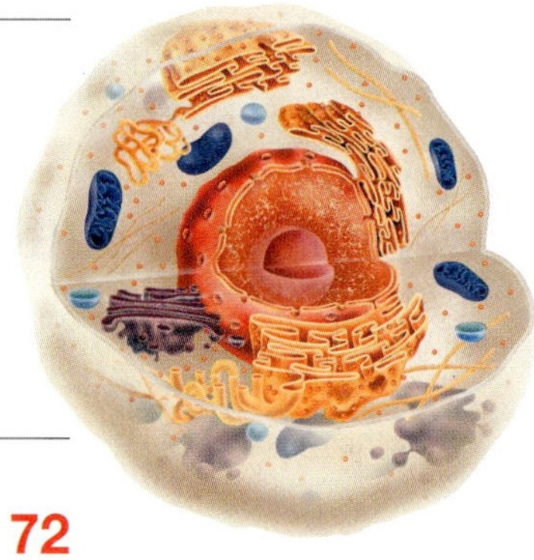

Principles of Evolution 248

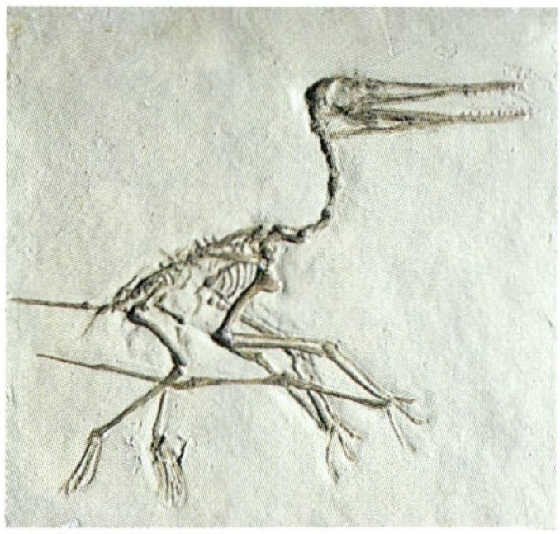

Principles of Ecology 316

Exploring Diversity 408

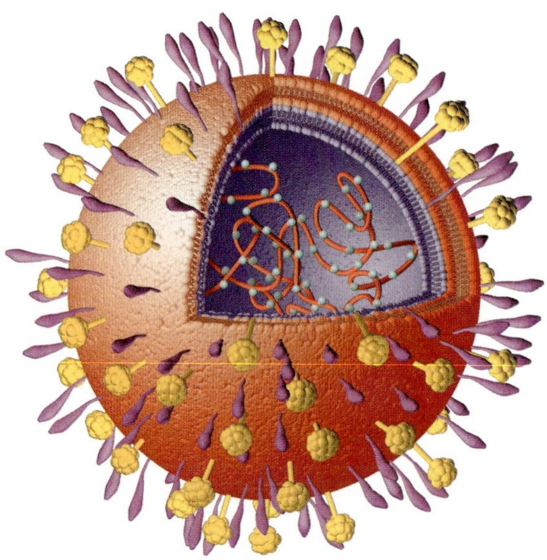

UNIT 6

Exploring Plants 498

Exploring Vertebrates 708

FLORIDA BIOWatch

Florida Biowatch features highlight the role of biology in Florida.

BIOWatch

Biowatch features highlight the impact of biology on health and technology issues.

FORENSICS BIOWatch

Forensics Biowatch features highlight the connection between biology and forensic science.

Up Close

Up Close *features provide detailed looks at important organisms.*

Exploring Further

Exploring Further *features let you explore key biological topics in greater depth.*

SCIENCE • TECHNOLOGY • SOCIETY

Science, Technology, and Society *features examine the impact of new technologies on issues in biology.*

QUICK LAB

***Quick Labs** provide hands-on experience yet require few materials.*

Data Labs allow you to interpret data and draw conclusions.

Math Labs let you practice real-world math skills as you analyze biological problems.

Exploration Labs

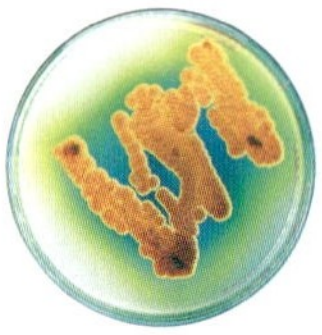

allow you to investigate or model a phenomenon and draw conclusions based on your work.

Skills Practice Labs

teach you lab skills used by biologists.

CHAPTER

CHAPTER

Forensics Labs

allow you to explore the techniques used by forensics scientists in solving crimes and mysteries of history.

How to Use Your Textbook

Your Roadmap for Success with *Holt Biology*

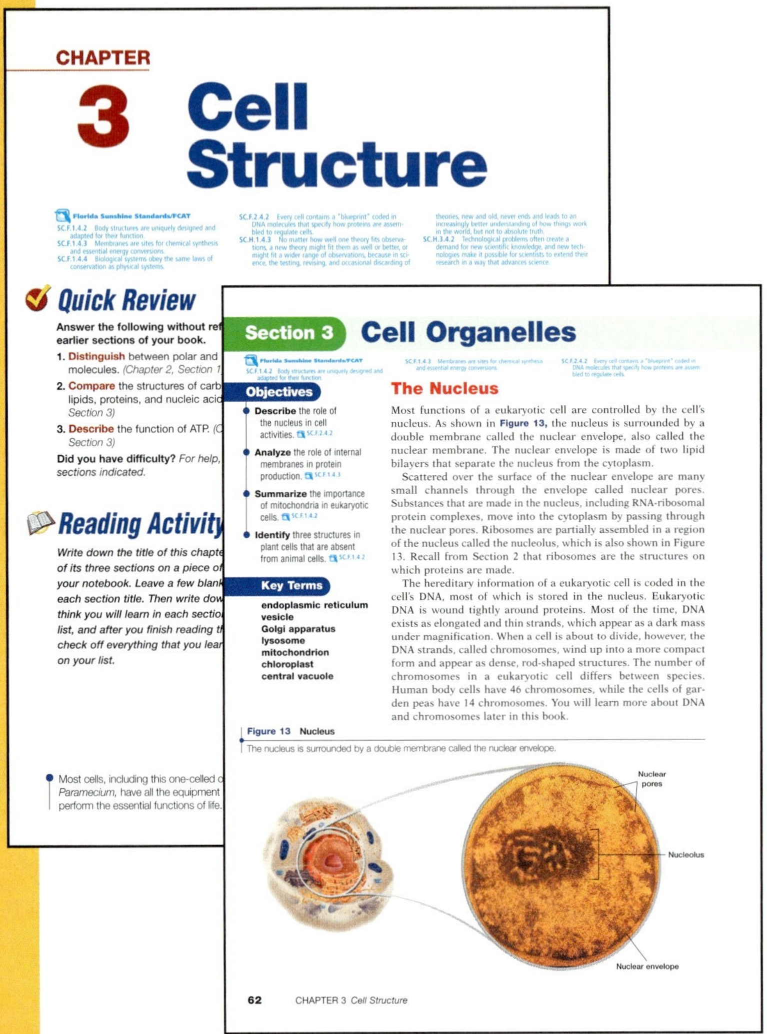

Get Organized

Answer the **Quick Review** questions at the beginning of each chapter to assess your recall of information important to the understanding of new chapter topics. Then complete the **Reading Activity** on the same page to prepare you to read and organize new information.

STUDY TIP Use the **Looking Ahead** outline at the beginning of the chapter to organize your notes on the chapter content in a way that you understand.

Read for Meaning

Read the **Objectives** and the **Florida Sunshine State Standards** at the beginning of each section because they will tell you what you'll need to learn. You should use the objectives and standards listed to help you organize your preparation for the **FCAT. Key Terms** are also listed for each section. Each key term is highlighted in the text. After reading each chapter, turn to the **Chapter Highlights** page and review the **Key Concepts,** which are brief summaries of the chapter's main ideas.You may want to do this even before you read the chapter.

Be Resourceful, Use the Web

Internet Connect boxes in your textbook take you to resources that you can use for science projects, reports, and research papers. Go to **scilinks.org** and type in the **SciLinks code** to get information on a topic.

Visit go.hrw.com

Find resources and reference materials that go with your textbook at **go.hrw.com.** Enter the keyword **HX6 Home** to access the home page for your textbook.

Prepare for Tests

Section Reviews and **Chapter Reviews** test your knowledge of the main points of the chapter. Critical Thinking items challenge you to think about the material in different ways and in greater depth. The **FCAT Prep** questions located in each Section Review and after each Chapter Review help you sharpen your test-taking abilities.

STUDY TIP Reread the Objectives and Chapter Highlights when studying for a test to be sure you know the material.

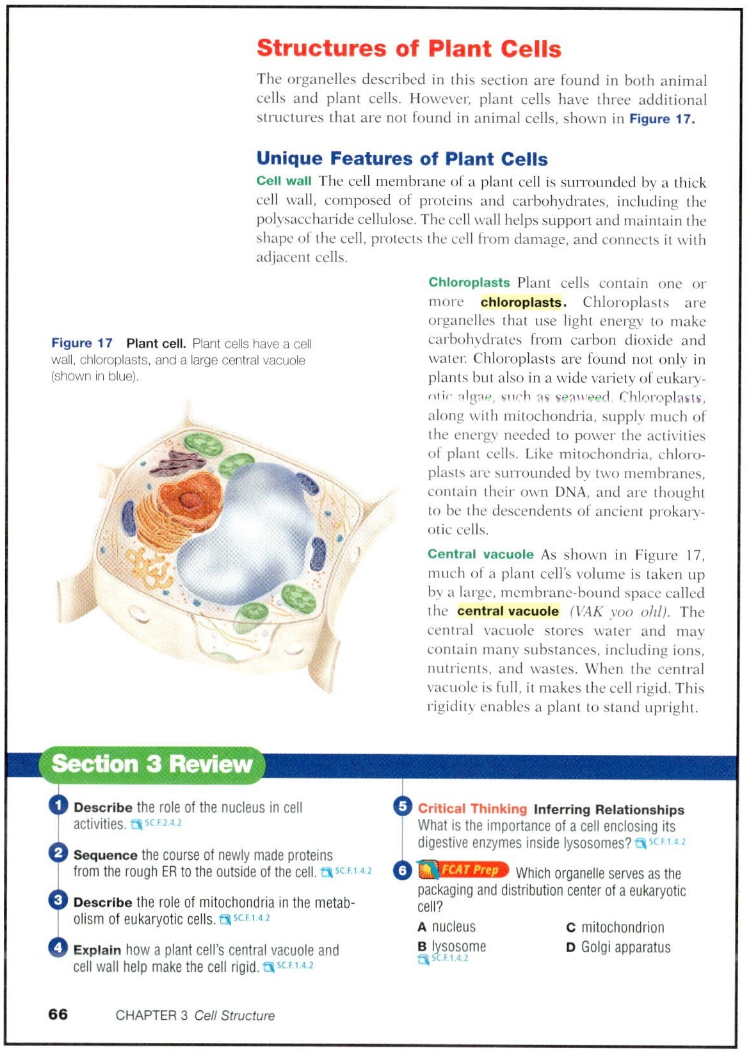

Structures of Plant Cells

The organelles described in this section are found in both animal cells and plant cells. However, plant cells have three additional structures that are not found in animal cells, shown in **Figure 17.**

Unique Features of Plant Cells

Cell wall The cell membrane of a plant cell is surrounded by a thick cell wall, composed of proteins and carbohydrates, including the polysaccharide cellulose. The cell wall helps support and maintain the shape of the cell, protects the cell from damage, and connects it with adjacent cells.

Figure 17 **Plant cell.** Plant cells have a cell wall, chloroplasts, and a large central vacuole (shown in blue).

Chloroplasts Plant cells contain one or more **chloroplasts.** Chloroplasts are organelles that use light energy to make carbohydrates from carbon dioxide and water. Chloroplasts are found not only in plants but also in a wide variety of eukaryotic algae, such as seaweed. Chloroplasts, along with mitochondria, supply much of the energy needed to power the activities of plant cells. Like mitochondria, chloroplasts are surrounded by two membranes, contain their own DNA, and are thought to be the descendents of ancient prokaryotic cells.

Central vacuole As shown in Figure 17, much of a plant cell's volume is taken up by a large, membrane-bound space called the **central vacuole** *(VAK yoo ohl).* The central vacuole stores water and may contain many substances, including ions, nutrients, and wastes. When the central vacuole is full, it makes the cell rigid. This rigidity enables a plant to stand upright.

Section 3 Review

1. **Describe** the role of the nucleus in cell activities. SC.F.2.4.2
2. **Sequence** the course of newly made proteins from the rough ER to the outside of the cell. SC.F.1.4.2
3. **Describe** the role of mitochondria in the metabolism of eukaryotic cells. SC.F.1.4.2
4. **Explain** how a plant cell's central vacuole and cell wall help make the cell rigid. SC.F.1.4.2
5. **Critical Thinking** **Inferring Relationships** What is the importance of a cell enclosing its digestive enzymes inside lysosomes? SC.F.1.4.2
6. **FCAT Prep** Which organelle serves as the packaging and distribution center of a eukaryotic cell?
   - **A** nucleus
   - **B** lysosome SC.F.1.4.2
   - **C** mitochondrion
   - **D** Golgi apparatus

66 CHAPTER 3 *Cell Structure*

Use the Appendix

Your **Reference and Skills** section near the end of the book contains a variety of resources designed to enhance your learning experience. **Reading and Study Skills** provides helpful study aids; **Math and Problem Solving Skills** sharpens your math skills. **Laboratory Skills** summarizes essential safety information and basic laboratory techniques. **Classification in Kingdoms and Domains** organizes living things according to modern principles of classification.

Visit Holt Online Learning

If your teacher gives you a special password to log onto the **Holt Online Learning** site, you'll find your complete textbook on the Web. In addition, you'll find some great learning tools and practice quizzes. You'll be able to see how well you know the material from your textbook.

UNIT 1 Principles of Cell Biology

Chapters

As the runner sprints toward the finish line, 25 trillion red blood cells carry oxygen throughout her body.

Blood: A River of Cells

Yesterday...

In the winter of 1667, a French physician named Jean-Baptiste Denis tried a daring new experiment. After examining a man who exhibited fits of rage, Denis transfused the patient with the blood of a gentle calf. His actions reflected the beliefs of his time—that blood carried the characteristics of the creatures in which it flowed.

Illustration from a 1692 medical textbook

Today...

Biologists know that red blood cells contain a protein called hemoglobin. **Find out how proteins like hemoglobin form their unique shapes.**

In persons with sickle cell anemia, an incorrect form of hemoglobin is made. When this happens, red blood cells become sickle-shaped and cannot adequately perform their job. The shape of the hemoglobin molecule is the key to its role—carrying oxygen throughout the body. **Discover why the cells of your body need oxygen.**

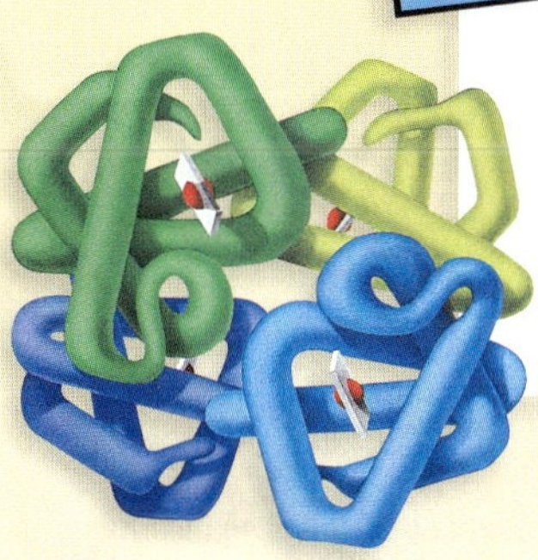

Hemoglobin

Tomorrow...

Researchers are discovering new treatments for sickle cell anemia. New drugs that "turn on" the body's production of normal hemoglobin have shown promise in easing the symptoms of sickle cell anemia.

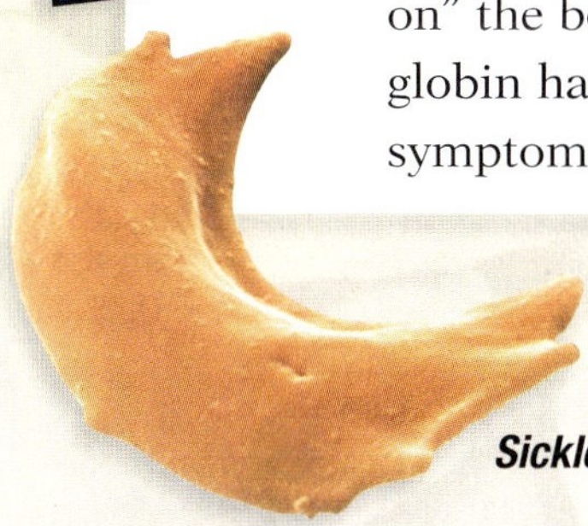

Sickled red blood cell

internet connect

www.scilinks.org
Topic: Sickle Cell Anemia
Keyword: HX4164

SCILINKS. Maintained by the National Science Teachers Association

Section	Resources	Activities and Labs
Chapter Opener		SE **Reading Activity,** p. 5 ■ TE **Using the Figure,** p. 4 GENERAL TE **Opening Activity,** p. 5 GENERAL
& 3 • 90 min pp. 6–9 **Section 1** Themes of Biology	OSP **Lesson Plan*** TR **Bellringer*** TR **Section Outline*** TR **A1 The Seven Properties of Life***	TE **Demonstration,** p. 7 ◆ BASIC TE **Using the Figure,** p. 7 GENERAL TE **Using the Figure,** p. 9 GENERAL CRF **Quick Lab** Comparing Living and Nonliving Things* BASIC
& 5 • 90 min pp. 10–13 **Section 2** Biology in Your World	OSP **Lesson Plan*** TR **Bellringer*** TR **Section Outline*** TR **A6 SI Base Units*** TR **A8 SI Units*** TR **A7 Some SI Prefixes***	TE **Activity** Recycling Program, p. 11 ◆ ADVANCED TE **Activity** Careers in Science, p. 12 BASIC CRF **Inquiry Lab** How Can I Store Heat?* GENERAL
& 7 • 90 min pp. 14–20 **Section 3** Scientific Processes	OSP **Lesson Plan*** TR **Bellringer*** TR **Section Outline*** TR **A3 Scientific Processes*** TR **Concept Map of Scientific Processes***	TE **Activity** Sources of Acid Rain, p. 15 BASIC TE **Using the Figure,** p. 16 BASIC TE **Activity** Using the Scientific Process, p. 16 GENERAL SE **Quick Lab** Determining the pH of Some Common Substances, p. 17 ◆ ■ CRF **Datasheets for In-Text Labs** Determining the pH of Common Substances* TE **Demonstration,** p. 18 BASIC SE **Data Lab** Analyzing Experimental Design, p. 19 ◆ ■ CRF **Datasheets for In-Text Labs** Analyzing Experimental Design* SE **Exploration Lab** Observing the Effects of Acid Rain on Seeds, pp. 24–25 ◆ ■ CRF **Datasheets for In-Text Labs** Observing the Effects of Acid Rain on Seeds*

& 9 • 90 min

Chapter Review and Assessment Resources

- SE **Study Zone,** p. 21 ■
- SE **Performance Zone,** p. 22 ■
- SE **FCAT Prep,** p. 23 ■
- TR **Concept Mapping***
- CRF **Vocabulary Review Worksheet*** ■ GENERAL
- CRF **Test Prep Pretest*** ■ GENERAL
- CRF **Chapter Test*** ■ GENERAL
- CRF **Chapter Test*** ADVANCED
- OSP **Test Generator***
- OSP **Test Item Listing***
- OSP **Modified Chapter Test*** BASIC

go.hrw.com Holt Online Learning

Visit go.hrw.com for access to Holt Online Learning or enter the keyword **HX6 Home** for a variety of free online resources.

One-Stop Planner® CD-ROM

This CD-ROM package includes
- Lab Materials QuickList Software
- Holt Calendar Planner
- Customizable Lesson Plans
- Printable Worksheets
- ExamView® Test Generator
- Interactive Teacher Edition
- Holt PuzzlePro® Resources
- Holt PowerPoint® Resources

Guided Reading Audio CDs

These CDs are designed to help auditory learners and reluctant readers. (Audio CDs are also available in Spanish.)

Compression guide:
To shorten your instruction because of time limitations, omit blocks 1, 2, and 3.

KEY

TE	Teacher Edition	CRF	Chapter Resource File	*	Also on One-Stop Planner
SE	Student Edition	TR	Teaching Transparency	■	Also Available in Spanish
OSP	One-Stop Planner			◆	Requires Advance Prep

SKILLS DEVELOPMENT AND PRACTICE	SECTION REVIEW AND ASSESSMENT	CORRELATIONS
		Florida Sunshine Standards/FCAT
TE Reading Skill Builder Paired Summarizing, p. 7 GENERAL **CRF Directed Reading Worksheet*** BASIC **CRF Active Reading Worksheet*** GENERAL	**SE Section 1 Review** ■ **TE Reteaching,** p. 9 BASIC **TE Quiz,** p. 9 GENERAL **TE Alternative Assessment,** p. 9 GENERAL **CRF Quiz*** ■ BASIC	SC.F.1.4.5, SC.F.2.4.3, SC.G.1.4.1, SC.H.2.4.1
TE Reading Skill Builder Reading Organizer, p. 11 BASIC **CRF Directed Reading Worksheet*** BASIC **CRF Active Reading Worksheet*** GENERAL	**SE Section 2 Review** ■ **TE Reteaching,** p. 13 BASIC **TE Quiz,** p. 13 GENERAL **TE Alternative Assessment,** p. 13 GENERAL **CRF Quiz*** ■ BASIC	SC.F.2.4.2, SC.H.3.4.3, SC.H.3.4.5, SC.H.3.4.6
TE Skill Builder Writing, p. 18 ADVANCED **CRF Directed Reading Worksheet*** BASIC **CRF Active Reading Worksheet*** GENERAL **CRF Critical Thinking Worksheet*** ADVANCED **CRF Science Skills Worksheet*** GENERAL	**SE Section 3 Review** ■ **TE Reteaching,** p. 20 BASIC **TE Quiz,** p. 20 GENERAL **TE Alternative Assessment,** p. 20 GENERAL **CRF Quiz*** ■ BASIC **Modified Worksheet*** BASIC	SC.F.1.4.1, SC.G.2.4.6, SC.H.1.4.1, SC.H.1.4.4., SC.H.1.4.6, SC.H.1.4.7, SC.H.2.4.2

SCILINKS. NSTA
www.scilinks.org
Maintained by the **National Science Teachers Association**

Topic: Human Genome Project
SciLinks Code: HX4193

Topic: Biology Careers
SciLinks Code: HX4022

Topic: Acid Rain
SciLinks Code: HX4001

CHAPTER 1

Overview

Tell the students the purpose of this chapter is to introduce the characteristics shared by all living organisms, including themselves. During their study of this chapter, students will learn how their own lives relate to the study of biology, and how biologists approach the study of living organisms. These themes will recur throughout the book.

Using the Figure

This photo of an orphaned rabbit can be used to discuss the characteristics of living organisms. Have students imagine they are holding and feeding the rabbit. How do they know it is alive? (The rabbit is feeding—metabolism; it is warm—homeostasis; it has characteristics similar to its parents and other rabbits—heredity; it responds to the presence of food and touch; and, over time, growth can be observed.) What does the rabbit need to stay alive? (Nutrients from milk and other foods; water; oxygen in the air.)

Orphaned rabbit

4

Chapter Correlations — Florida Sunshine Standards/FCAT

SC.H.1.4.1 Investigations are conducted to explore new phenomena, to check on previous results, to test how well a theory predicts, and to compare different theories.

SC.H.1.4.4 Scientists in any one research group tend to see things alike and therefore scientific teams are expected to seek out the possible sources of bias in the design of their investigations and in their data analysis.

SC.H.1.4.6 In the short run, new ideas that do not mesh well with mainstream ideas in science often encounter vigorous criticism, and in the long run, theories are judged by how they fit with other theories, the range of observations they explain, how well they explain observations, and how effective they are in predicting new findings.

SC.H.1.4.7 Important to science are a sense of responsibility, a commitment to peer review, truthful reporting of the methods and outcomes of investigations, and making the public aware of the findings.

SC.H.2.4.1 Scientists assume that the universe is a vast system in which basic rules exist that may range from very simple to extremely complex, but scientists operate on the belief that the rules can be discovered by careful, systemic study.

SC.H.2.4.2 Scientists control conditions in order to obtain evidence, but when that is not possible for practical or ethical reasons, they try to observe a wide range of natural occurrences to discern patterns.

SC.H.3.4.3 Scientists can bring information, insights, and analytical skills to matters of public concern and help people understand the possible causes and effects of events.

SC.H.3.4.5 The value of a technology may differ for different people and at different times.

SC.H.3.4.6 Scientific knowledge is used by those who engage in design and technology to solve practical problems, taking human values and limitations into account..

CHAPTER

1 Biology and You

Florida Sunshine Standards/FCAT

SC.G.1.4.1 There is great diversity and interdependence among living things.

SC.H.1.4.1 Investigations are conducted to explore new phenomena, to check on previous results, to test how well a theory predicts, and to compare different theories.

SC.H.1.4.4 Scientists in any one research group tend to see things alike and therefore scientific teams are expected to seek out the possible sources of bias in the design of their investigations and in their data analysis.

SC.H.1.4.6 In the short run, new ideas that do not mesh well with mainstream ideas in science often encounter vigorous criticism, and in the long run, theories are judged by how they fit with other theories, the range of observations they explain, how well they explain observations, and how effective they are in predicting new findings.

SC.H.1.4.7 Important to science are a sense of responsibility, a commitment to peer review, truthful reporting of the methods and outcomes of investigations, and making the public aware of the findings.

SC.H.2.4.1 Scientists assume that the universe is a vast system in which basic rules exist that may range from very simple to extremely complex, but scientists operate on the belief that the rules can be discovered by careful, systemic study.

SC.H.3.4.3 Scientists can bring information, insights, and analytical skills to matters of public concern and help people understand the possible causes and effects of events.

Study TIP

The **Reading Activity** that begins each chapter will help you to comprehend more effectively what you read. **Study Tips** in the margins will help you to organize and review information. **Word Origins** notes in the margins will help you understand how scientific terms are constructed from common root words and suffixes. **Real Life** margin notes link the study of biology to real-world phenomena.

Reading Activity

Before you read this chapter, write the word biology *on your paper. Refer to a dictionary, and write out the definition of the prefix* bio– *and the suffix* –logy. *Think of names of other branches of life science that include the suffix* –logy, *and write them down. Use the* ***Reading and Study Skills*** *appendix at the back of this book to define the prefixes in the words you thought of.*

Just as this orphaned rabbit needs food and water to live and grow, it also depends on oxygen from plants to survive. Our knowledge of biology helps us understand how all life on Earth is interconnected.

Looking Ahead

internet connect

www.scilinks.org

National Science Teachers Association *sci*LINKS Internet resources are located throughout this chapter.

SCILINKS. Maintained by the National Science Teachers Association

Opening Activity — GENERAL

Everyday Biology Write the following statement on the board: *Biology is the study of life.* Under this, write the following list: *Taking care of your pet; Recycling your junk mail; Watering your lawn; Eating breakfast; Taking medicine; Visiting a zoo.* Ask the students: How do all of these activities involve biology? (Answers may vary. Encourage students to recognize the many ways in which biology applies to their everyday activities.)

Reading Activity Answers

The prefix *bio–* refers to life or living organisms and is based on the Greek word *bios*. The suffix *–logy* describes science or the study of a particular subject. The various life sciences listed by the students will vary and may include ecology (the study of living things and their environment), pathology (the study of diseases), or histology (the study of tissues).

Chapter Resource File

- Vocabulary Worksheets
- Concept Mapping

Focus

Overview

Before beginning this section review with your students the objectives listed in the Student Edition. This section describes the seven properties shared by all living organisms at some stage of their life cycles. The first five properties are also major themes in biology, along with evolution and interdependence.

Bellringer

Ask students to list all the characteristics that they have in common with a beetle, then have students share some of their answers. (Answers will vary, but could include eating, growing, having specific body parts, being able to move and respond to the environment, and being able to produce offspring.)

Motivate

Discussion/Question — GENERAL

Without using their textbooks, have students list the characteristics of life. Next, have the students determine if any nonliving object might have any of these characteristics. If so, have the students give a specific example for each (crystals grow, fire requires "food" and gives off heat, viruses reproduce, etc.) and discuss each one. Emphasize that living organisms may not have all seven characteristics at one time (such as the ability to reproduce), but that all organisms have these characteristics at some point in their life.

Section Correlations

SC.F.1.4.5
SC.F.2.4.3
SC.G.1.4.1
SC.H.2.4.1

Section 1 Themes of Biology

Florida Sunshine Standards/FCAT
SC.F.2.4.3 Mechanisms of change (e.g., mutation and natural selection) lead to adaptations in a species and their ability to survive naturally in changing conditions and to increase species diversity.
SC.G.1.4.1 There is great diversity and interdependence among living things.

- **Relate** the seven properties of life to a living organism. SC.G.1.4.1
- **Describe** seven themes that can help you organize what you learn about biology. SC.H.2.4.1
- **Identify** the tiny structures that make up all living organisms. SC.F.1.4.5
- **Differentiate** between reproduction and heredity and between metabolism and homeostasis. SC.F.2.4.3

Key Terms

biology
cell
reproduction
metabolism
homeostasis
gene
heredity
mutation
evolution
species
natural selection
ecology

Characteristics of Living Organisms

You are surrounded by living things, which a scientist calls *organisms*. Many organisms, such as people, plants, and animals, are obvious. Other living things are so small that you cannot see them without a microscope. How do we know if something is alive? What does it mean to be alive?

While most people are capable of distinguishing between living and nonliving, actually defining life can be quite difficult. Perhaps you consider movement, sensitivity, development, and even death as characteristics of living organisms. While present in all living things, these properties are not enough to describe life.

Clouds, for example, move when stimulated by the wind and develop from moisture that is suspended in the atmosphere. Clouds grow and change shapes. Some might view the breakup of clouds as being similar to death. Disorder, however, is not the same as death. Clouds may break up and vanish, but they do not die.

Biology is the study of life. Biologists recognize that all living organisms, such as the cheetahs shown in **Figure 1,** share certain general properties that separate them from nonliving things. As summarized in Figure 1, every living organism is composed of one or more cells, is able to reproduce, and obtains and uses energy to run the processes of life. Living organisms also maintain a constant internal environment and pass on traits to offspring. Responding and adjusting to the environment as well as growing and developing are other characteristics shared by all living organisms.

As you read further, you will have an opportunity to think more about the properties that help define life. Life is characterized by the presence of *all* of these properties at some stage in an organism's life. Remember this fact as you attempt to determine what is living and what is not.

Figure 1 What does it mean to be alive? Life is characterized by the presence of all seven of these properties at some stage in an organism's life.

Properties of Life

- Cellular organization
- Reproduction
- Metabolism
- Homeostasis
- Heredity
- Responsiveness
- Growth and development

6

Chapter Resource File
- Directed Reading BASIC
- Active Reading GENERAL

Transparencies
TR Bellringer
TR A1 The Seven Properties of Life

One-Stop Planner CD-ROM
- Reading Organizers BASIC
- Reading Strategies BASIC

Unifying Themes of Biology

In the study of biology, certain broad themes emerge that both unify living things and help explain biology as a science. The word *science* comes from Latin for "to know." Science is a systematic process of inquiry. As you study the science of biology by reading this textbook, you will repeatedly encounter these themes.

Theme 1 Cellular Structure and Function

All living things are made of one or more cells. **Cells** are highly organized, tiny structures with thin coverings called membranes. A cell is the smallest unit capable of all life functions. The basic structure of cells is the same in all organisms, although some cells are more complex than others. Some organisms have only a single cell, while others are multicellular (composed of many cells). Your body contains more than 100 trillion cells. **Figure 2** shows a single-celled organism called a paramecium.

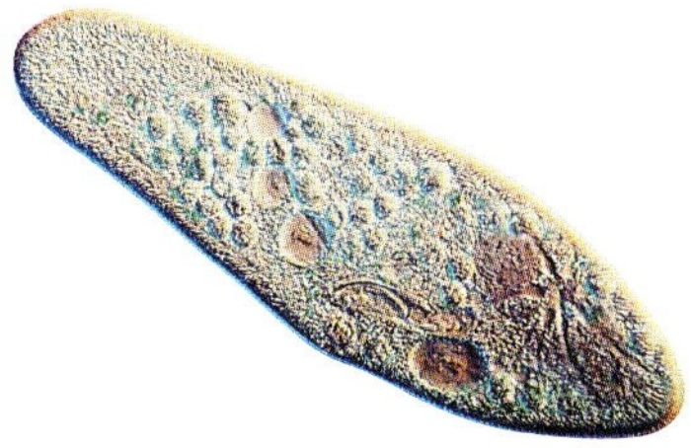

Figure 2
Single-celled paramecium

Theme 2 Reproduction

All living things can reproduce. **Reproduction** is the process by which organisms make more of their own kind from one generation to the next. Some rapidly growing bacteria divide into offspring cells approximately every 15 minutes, and bristlecone pine trees that are 5,000 years old still produce seedlings. Because no organism lives forever, reproduction, as represented in **Figure 3,** is an essential part of living.

Figure 3 Hatchling snakes

Theme 3 Metabolism

Living organisms carry out many different chemical reactions in order to obtain and use energy to run the processes of life. All living things use energy to grow, to move, and to process information. Without energy, life soon stops. **Metabolism** is the sum of all of the chemical reactions carried out in an organism.

Almost all the energy used by living organisms is originally captured from sunlight. Plants, algae, and some bacteria capture this solar energy and use it to make complex molecules in a process called photosynthesis. These molecules then serve as the source of energy, or food, for other organisms. For example, paramecia, such as the one shown in Figure 2, eat bacteria. Humans eat plants or animals that, in turn, have eaten plants. Energy flows from the sun to plants, from these plants to plant-eating organisms, and from plant-eating organisms to meat-eating organisms. The teens shown in **Figure 4** are extracting energy from the food they eat.

Figure 4
Extracting energy from food

Teach

Demonstration — BASIC

Obtain a culture of *Plasmodium polycephalum,* a slime mold. Do not identify it for the students. Instead, have students use their unaided eyes to attempt to determine if it is living or nonliving. Do not let them touch or smell it. Next ask students to think about how they could determine if it is a living organism. (Answers might include observed reproduction, growth, and movement.) Afterwards, explain to the students that this organism is a protist. Show them the protoplasmic movement with a binocular microscope. **LS Visual**

Using the Figure — GENERAL

Figure 3 Explain to the students that reproduction is not needed for the survival of an individual, but that it is necessary for the survival of a species, such as this snake species. Ask students to compare the amount of parental care provided by organisms that have many offspring to the amount provided by organisms that have very few offspring. (Organisms that have large numbers of offspring provide little or no parental care. Usually, few survive. Organisms that have fewer offspring usually provide more parental care. Usually, many of the offspring survive.) Point out that in either case, the offspring that survive allow the continuation of the species, but that extinction can still occur. **LS Logical**

READING SKILL BUILDER — GENERAL

Paired Summarizing Have students read the Unifying Themes of Biology in pairs. As they read each of the seven items, students should discuss what they think each theme means and agree upon a single sentence that summarizes the theme.

did you know?

Metabolism on the Fly A hummingbird's metabolic rate is about 1,400 calories per kilogram. Compare that to the rate of a mourning dove—about 127 calories per kilogram. Hummingbirds use enormous amounts of energy in part because of their racing wings and hearts. Hummingbirds can move their wings up to 80 times per second, and their heart can beat up to 1,200 beats per minute. Because of their high metabolic rate, hummingbirds feed almost continuously throughout the day.

Integrating Physics and Chemistry

Discuss the differences between physical and chemical changes. During physical changes the form of a substance or its properties may change, but its chemical composition remains unchanged. During chemical changes elements recombine to form new chemical compounds with different properties. Have the students classify the following observations as physical or chemical changes: evolution of a gas, mixing oil and water by shaking, color change, temperature increase or decrease when two substances are mixed, crushing a large rock into small pieces, changing ice into water, bioluminescence of marine plankton. Ask them to explain why metabolic processes are chemical rather than physical changes.

Teach, *continued*

Teaching Tip — GENERAL

Maintaining Homeostasis Point out to students that many arctic animals are able to maintain a constant body temperature because they have a thick layer of fat under their skin that keeps much of their body heat from escaping. Ask students for ways in which humans maintain a constant body temperature. (Answers may include sweating, shivering, wearing clothing, and being physically active. If students suggest having body fat, tell them that fat does not insulate humans as it does in many arctic animals.) To illustrate the maintenance of human body temperature, ask a student to volunteer to have his or her body temperature monitored before and immediately after vigorous exercise. Note that human body temperature fluctuates a bit around 37°C (98.7°F) based on the time of day and the amount of activity.

MISCONCEPTION ALERT

Maintaining Homeostasis
Students may believe that the human body temperature remains a constant 37°C under all environmental conditions. When immersed in water, however, the human body loses heat rapidly—up to 26 times faster than in air of the same temperature. The condition that is produced, called hypothermia, becomes serious when body temperature drops below 35°C, as metabolic functions begin to cease.

Figure 5 Harp seal

Theme 4 Homeostasis

All living organisms must maintain a stable internal environment in order to function properly. Organisms respond to changes in their external environment, and their internal processes adjust accordingly. The maintenance of stable internal conditions in spite of changes in the external environment is called **homeostasis** *(hoh mee oh STAY sihs)*. An organism unable to balance its internal conditions with its environmental conditions could become ill and die. Arctic seals, such as the one shown in **Figure 5,** are able to maintain a constant body temperature in spite of their cold environment because of their body shape and thick layer of body fat.

Theme 5 Heredity

All living things are able to pass on traits to their offspring through genes that are passed from parent to offspring each generation. A **gene** is the basic unit of heredity. Genes are coded in a molecule called deoxyribonucleic *(dee AHKS ee rie boh nu klay ik)* acid (DNA) and determine an organism's traits. The passing of traits from parent to offspring is called **heredity.** Heredity is the reason children tend to resemble their parents, as shown in **Figure 6.**

Sometimes damage causes genes to change. A change in the DNA of a gene is called a **mutation.** Most mutations are harmful, but sometimes mutations can help an organism survive. For example, in humans a mutation for the blood protein hemoglobin, which carries oxygen to the body's cells, has both a harmful effect and a positive effect. The harmful effect is that the mutated form of the gene results in sickle cell anemia. Sickle cell anemia is a disease in which the defective form of hemoglobin causes many red blood cells to bend into a sickled—that is, a hooked—shape that reduces the oxygen-carrying capability of the cell. The positive effect is that the mutation produces resistance to malaria, a deadly infectious disease.

Mutations that occur in sex cells (egg and sperm) are passed on to other generations. Mutations that occur in body cells are not passed on, but they can disrupt the control of cell reproduction and result in cancer.

Figure 6 Passing on traits

Section Review Correlations

SC.F.1.4.1: Item 4
SC.F.1.4.5: Item 3
SC.F.2.4.3: Item 6
SC.G.1.4.1: Items 1, 5
SC.H.2.4.1: Item 2

MEDICINE CONNECTION

Proper nutrition is important for maintaining a healthy internal environment. Invite a local clinical nutritionist or dietitian to speak to the class about proper nutrition, dieting, and exercise for teenagers. For several days before the talk, have students prepare a list of the foods they have consumed and a list of questions for the nutritionist. Ask students to compare their lists with the information provided by the speaker.

INCLUSION Strategies

- *Learning Disabled*
- *Attention Deficit Disorder*

Beginning with Chapter 1, students can create a Vocabulary Notebook. Ask students to copy all Key Terms that appear in bold type through out the chapter. The students can then copy the sentence in which the key term appears to help understand the meaning of the word in the context of the chapter. The notebook can be used at any time as a study guide.

Theme 6 Evolution

The great diversity of life on Earth is the result of a long history of change. Change in the inherited characteristics of species over generations is called **evolution.** A **species** is a group of genetically similar organisms that can produce fertile offspring. Individuals in a species are similar, but not identical. Those individuals with genetic traits that better enable them to meet nature's challenges tend to survive and reproduce in greater numbers, causing these favorable traits to become more common. Charles Darwin, the nineteenth-century British naturalist, used the term **natural selection** for the process in which organisms with favorable traits are more likely to survive and reproduce.

Darwin's theory of evolution by natural selection provides a consistent explanation for life's diversity. Most scientists believe that the many different species of animals, plants, and other organisms on Earth today are the result of a long process of evolution. **Figure 7** shows an example of a plant that has flowers modified for attracting insects.

Figure 7 Bee pollinating flower

Theme 7 Interdependence

The organisms in a biological community live and interact with other organisms, as shown in **Figure 8.** A biological community is a group of interacting organisms. **Ecology** is the branch of biology that studies the interactions of organisms with one another and with the nonliving part of their environment. Organisms are dependent on one another and their environment—that is, they are interdependent. Interdependence within biological communities is the result of a long history of evolutionary adjustments. The complex web of interactions in a biological community depends on the proper functioning of all of its members, even those too small to be seen without a microscope.

Figure 8 Owl capturing a rat

Section 1 Review

1. **Identify** the seven properties that all living organisms share. SC.G.1.4.1
2. **Relate** three of the seven major themes of biology to the life of a harp seal. SC.H.2.4.1
3. **Name** the very small, organized structure that is bound by a membrane and that is the basic unit of structure and function in all organisms. SC.F.1.4.5
4. **Define** *homeostasis* and *metabolism,* and describe their differences. SC.F.1.4.1
5. **Critical Thinking Recognizing Verifiable Facts** If you find an object that looks like an organism, how might you determine if your discovery is indeed alive? SC.G.1.4.1
6. *FCAT Prep* What kinds of effects are produced by the mutation that results in sickle cell anemia?
   - **A** only harmful effects
   - **B** only positive effects
   - **C** unimportant effects
   - **D** both harmful and positive effects

   SC.F.2.4.3

Using the Figure — General

Have students look at the image of an owl capturing a rat in **Figure 8.** Tell students to write down as many of the seven themes in biology that are illustrated by this figure. Discuss their answers. (Interdependence—the owl depends on the rat for food; evolution—both the owl and rat have adaptations that have evolved over time; homeostasis—both the owl and rat have coverings to conserve heat; metabolism—the owl has captured the rat for food.) Point out that the other themes are present but not apparent. **LS Visual**

Close

Reteaching — Basic

Have students play a timed game in which one student tries to get his or her team members to guess a particular theme of biology. The student can give clues but cannot use either the actual word or certain designated words. For example, if the theme is heredity, you might designate genetics, traits, and mutation as words that cannot be given as clues.

Quiz — General

1. A change in the DNA of a gene is called a _______. (mutation)
2. The many different species of organisms on Earth today are the result of _______. (evolution)
3. The maintenance of a stable internal environment is called _______. (homeostasis)

Alternative Assessment — General

Seven Themes in Biology Have students cut out magazine photos that depict one or more of the seven themes of biology. Students should glue or tape the photos onto a sheet of paper and then label each photo with the theme(s) that is (are) represented. Students should find enough photos to represent all seven themes.

Answers to Section Review

1. Cellular organization, reproduction, metabolism, homeostasis, heredity, responsiveness, growth and development
2. Answers might include homeostasis, cellular structure and function, or reproduction.
3. the cell
4. Homeostasis is the ability to maintain a stable internal environment in spite of changes in the external environment. Metabolism is the sum total of all the chemical reactions that are carried out in an organism.
5. Answers might include testing for several of the characteristics of living organisms, such as signs of growth, cellular organization, metabolism, or responses to stimuli.
6. **A.** Incorrect. The mutation produces resistance to malaria. **B.** Incorrect. The mutation reduces the oxygen-carrying capability of the cell. **C.** Incorrect. The mutation has an important effect on its carrier. **D.** Correct. The mutation is beneficial with respect to malaria and detrimental to the oxygen-carrying capabilities of a cell.

Focus

Overview

Before beginning this section review with your students the objectives listed in the Student Edition. In this section, students will learn how biologists are currently engaged in some of the most challenging and potentially controversial issues in the world today. These challenges include finding resources to sustain a growing population, fighting deadly diseases, using genetic engineering and cloning responsibly, controlling the spread of AIDS, treating cancer, and limiting the use of bio-warfare.

Bellringer

Ask students to list five issues in the area of biology that they believe are important in the world today. Tally the responses on the board to determine which issues the class feels are most important. Assess what students know about each problem and how scientists are working to solve it.

Motivate

Discussion/ Question — GENERAL

Have students break into groups of four. Within the groups, have students discuss the possible results of human cloning. Students should consider both positive and negative outcomes. Have the class discuss each group's conclusions. (Answers will vary widely. Encourage students to support their positions.)

Co-op Learning

Section Correlations

SC.F.2.4.2
SC.H.3.4.3
SC.H.3.4.5
SC.H.3.4.6

Section 2 Biology in Your World

Florida Sunshine Standards/FCAT

SC.F.2.4.2 Every cell contains a "blueprint" coded in DNA molecules that specify how proteins are assembled to regulate cells.

SC.H.3.4.3 Scientists can bring information, insights, and analytical skills to matters of public concern and help people understand the possible causes and effects of events.

Objectives

- **Evaluate** the impact of scientific research on the environment. SC.H.3.4.3
- **Evaluate** the impact of scientific research on society with respect to increasing food supplies. SC.H.3.4.5
- **Explain** the primary task of the Human Genome Project. SC.H.3.4.6
- **Describe** the contributions of scientists in fighting AIDS and cancer. SC.H.3.4.3
- **Define** the term *gene therapy.* SC.F.2.4.2

Key Terms

genome
HIV
cancer
cystic fibrosis
gene therapy

Solving Real-World Problems

You are unlikely to read a newspaper or magazine today without noticing issues that relate to biology. In this textbook, you will learn about many areas in which biologists are actively working to solve today's problems.

Preserving Our Environment

More than 6 billion people now live on Earth. The increasing human population has had a significant impact on other organisms with which we share this planet. For example, tropical rain forests are home to one-half of the world's species of plants and animals, such as the bird shown in **Figure 9.** The rain forests are being destroyed at the rate of more than one acre every second. At this rate, tropical rain forests—and a million species—may be gone in 30 years. Who knows what potential medicines and foods we are discarding? Like burning a library without reading the books, extinction on this large scale is a tragedy. However, conservation biologists are now exploring ways to achieve a balance between people's growing need for land and the need to preserve the environment.

One of the great achievements of today's biology has been to show the practical benefits of taking better care of our environment. Consider, for example, the fast-food french fry. Because french fries must be formed perfectly to be sold, about one-half of the potatoes used to make french fries were lost as waste. Then a major supplier found a way to use some of the waste: mix it with grain to feed cattle. Leftover potato particles in the potato processing water are also used, serving as a source of methane gas for power plants. Finally, the processing water, rich in nutrients, is used to water and fertilize agricultural crops. The environmental concern that promotes these sensible changes is a major contribution of biology to a better future. Conservation and preservation are now everyday activities of government, industry, and individuals.

Figure 9 Scarlet macaw. The brilliantly colored scarlet macaw, *Ana macao,* lives high in the rain forest canopy in Central America and South America. Its numbers are being reduced by habitat destruction and poaching.

internet connect
www.scilinks.org
Topic: Florida's Environment
Keyword: HX6F06
SciLinks. Maintained by the National Science Teachers Association

10

Chapter Resource File

- Directed Reading BASIC
- Active Reading GENERAL

Transparencies

TR Bellringer
TR A6 SI Base Units
TR A7 Some SI Prefixes
TR A8 SI Units

INCLUSION Strategies

- *Attention Deficit Disorder*
- *Learning Disability*
- *Developmental Delay*

Have students find pictures showing a type of environment. Cut pictures to fit an index card. Use the back of the card as though it were a postcard. Write a "letter" to a friend describing the environment and its plants and animals. Include any problems that could occur if there was a major change to the environment such as fire, flood, or man-made development.

Improving the Food Supply

One of the greatest impacts of modern biology on society in recent years has been the genetic engineering of crop plants. As illustrated in **Figure 10,** biologists have learned how to transfer genes from one kind of plant to another, which changes the hereditary information in its cells. Genetic engineering, shown in Figure 10, has made some crop plants resistant to herbicides (so that weeds can be killed without harming the food crop). Genetic engineering has also made some crop plants poisonous to insect pests, but not humans, and has produced new varieties of crop plants with improved nutritional balance and protein content.

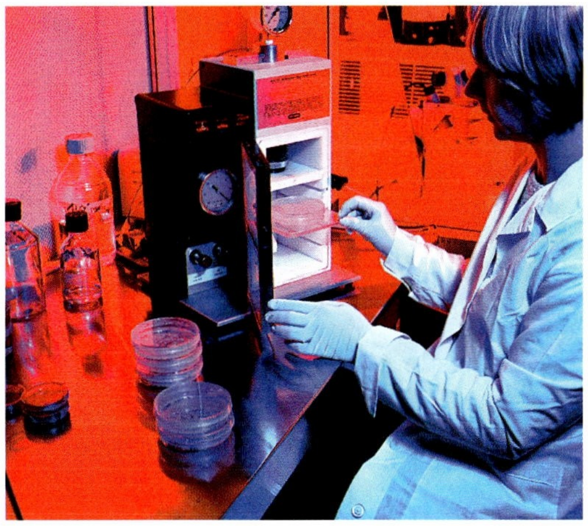

Figure 10 Genetic engineering. Genes for desirable traits can be transferred from one organism to another.

For example, rice, one of the world's most important food crops, lacks iron and vitamin A levels needed for a balanced diet. These deficiencies affect the billions of people worldwide for whom rice is a daily food. The addition of genes from other plants to rice has increased the nutritional value of rice, thus improving the diet of many of the world's people. The long-term safety of genetically engineered food crops is still being studied. These crops may offer great promise, however, of improving the world's food supplies.

Understanding the Human Genome

Another great achievement of today's biology was completed in April of 2003. Government-funded and private research teams from several countries had been racing to complete the sequencing of the human genome, and on that day they announced joint success. The task had been formidable. A **genome** is the complete genetic material contained in an individual. The human genome contains an astonishing 3 billion individual units! At the height of the research, automated gene-sequencing machines were sequencing DNA fragments at a rate of 1,000 units per second, around the clock.

Biologists are now able to read every human gene, providing them with a detailed road map, as you see in **Figure 11,** of human genes. It will be many years before this information is fully analyzed, but it is already proving to be an invaluable tool in medical research.

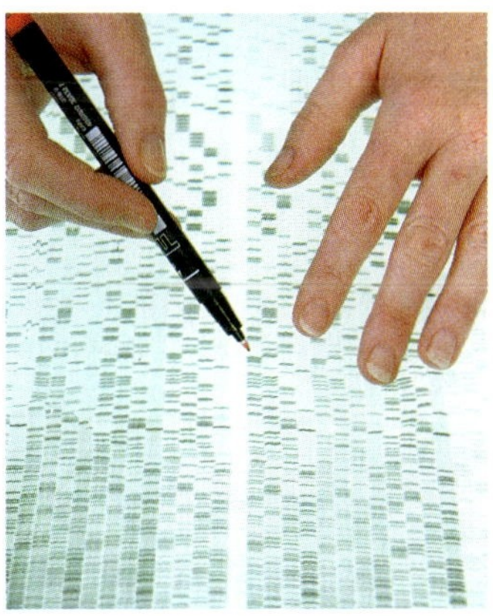

Figure 11 DNA sequencing. The identity of each genetic unit, shown as repeating dark and light bars, has been determined for all human genes found in the human body.

internet connect
www.scilinks.org
Topic: Human Genome Project
Keyword: HX4193
SCILINKS. Maintained by the National Science Teachers Association

Teach

READING SKILL BUILDER — BASIC

Reading Organizer After students read Section 2, have them design a spider map reading organizer to show the main ideas and supporting details of the section. Students should begin by writing the main idea in the center of the diagram and then arrange the details of the section around the center. For more information about reading organizers, see Reading Strategies Worksheets, in Holt BioSources Teaching Resources CD-ROM.

Activity — ADVANCED

Recycling Programs Have students research recycling programs in their community. Invite a speaker from the recycling program to discuss the effectiveness of the program, how the program helps the community financially, and how citizens can get involved in making the program more successful. Ask the students to describe two ways in which the school might contribute to the recycling effort.

One-Stop Planner CD-ROM
- Reading Organizers BASIC
- Reading Strategies BASIC
- Basic Skills Worksheet Words and Roots BASIC

MEDICINE CONNECTION

Flossie Wong-Stall, a prominent AIDS researcher, came to the United States from China in 1965. She earned her Ph.D, and in 1973 began work for the National Institutes of Health, joining the team of Dr. Robert Gallo. In the 1980s, Gallo's team isolated the human immunodeficiency virus (HIV) that causes AIDS. Wong-Staal was the first to make clones of the virus. She discovered how the virus regulates its growth and that of the cells it attacks. She also determined which parts of the virus mutate most frequently.

Attention Grabber

Where in the World Is All the Growth? The human population recently passed 6 billion people and may reach 9 billion within the next 50 years. Highly developed countries, such as the United States, Canada, Japan, and most European countries, have fairly stable populations. Nearly all of the growth is now coming from less developed countries in Asia and Africa. These countries lack the resources to sustain such large populations.

Teach, *continued*

Teaching Tip — BASIC

Politicians and Biology Ask students to suggest what role politicians play or might play in biological research. (Politicians must understand the biology of important diseases, such as AIDS, if they are to make informed decisions about providing care facilities, funding research, and protecting the rights of people with the disease.) Emphasize that most research is supported through grants given by government agencies such as the National Institutes of Health, the National Science Foundation, and the Department of Energy.

Activity — BASIC

Bring several science magazines to class. Have students mark the pages that show photos of scientists. On the board, make a list of the environment in which the scientist is shown working; the type of research he or she is doing; and, if shown, the organism the scientist is working with. Describe the various work environments and areas of study. Point out how biologists work with many different types of organisms.

Teaching Tip — GENERAL

A Biologist's Life Ask students to imagine life as a biologist. Ask them to picture themselves working as biologists. What area would they study? What type of environment would they like to work in? How would they help improve our way of life?

Figure 12 HIV. Individual HIV particles are shown emerging from a white blood cell where they have been assembled.

Fighting Disease

Progress in biology directly affects our lives through medicine, in which scientific advances are curing disease and improving health every day. New technologies have enabled biologists to combat disease in ways scarcely imagined only a few years ago. Among the many diseases that you will study in this text, consider the following.

AIDS

For more than 20 years, biologists have been battling AIDS. AIDS is a fatal disease caused by **HIV,** a virus that attacks and destroys the human immune system. HIV, shown in **Figure 12,** is transmitted by contact with body fluids from an infected person. While biologists have been successful in developing a combination of drugs that slow the progression of AIDS, it has proven very difficult to make a vaccine capable of halting its spread. The problem is that HIV changes as it passes from person to person, altering itself too frequently for any single vaccine to protect many people. This problem soon may be solved. New vaccines now being tested target two or more parts of the virus at the same time. While one part may change, it is very unlikely that two parts will change at the same time in the very same virus particle. For the first time, there is hope of a successful vaccine to control the worldwide outbreak of AIDS.

Cancer

When U.S. President Richard Nixon recruited biologists to join a "War on Cancer" in 1972, we did not know very much about the causes of cancer, although many Americans were dying of it. In the 30 years since then, biologists have learned a lot. **Cancer** is a growth defect in cells, a breakdown of the mechanism that controls cell division.

We now know that many cancers can be largely avoided. To sharply reduce your risk of lung cancer, for example, don't smoke. Many other cancers can be treated successfully when detected early. Colon cancer, for example, develops slowly from intestinal tissue growths called polyps. A simple medical examination enables the detection and removal of the polyps.

Great progress is being made in curing many cancers. More than 25 percent of breast cancers, for example, result from having too many copies of a cell protein that starts cell division. As many as 70 percent of colon and prostate cancers have extra copies of a similar protein. Anticancer drugs that stick to these extra cell proteins, gumming them up so they cannot promote excessive cell division, appear to offer great promise.

internet connect
www.scilinks.org
Topic: Cancer Cells
Keyword: HX4030
SCILINKS. Maintained by the National Science Teachers Association

Emerging Diseases

The past few years have seen the emergence of new diseases not known in the past and the incidence in the United States of diseases from other parts of the world. West Nile virus is one such disease. West Nile virus was not found in the United States until 1999, when

12

Trends in Genome Research

Now that the human genome has been mapped, scientists are hard at work identifying the genes along the DNA strands, as well as the proteins that are encoded by these genes. This research is extending to microbes as well. Eventually, such research may result in new medical treatments and ways of cleaning up the environment, producing energy, and protecting against bioterrorism.

Section Review Correlations

SC.F.2.4.2: Items 4, 6
SC.H.3.4.3: Items 1, 3, 5
SC.H.3.4.5: Item 2

Compression guide:
To shorten your instruction because of time limitations, omit blocks 1, 2, and 3.

KEY

TE Teacher Edition
SE Student Edition
OSP One-Stop Planner
CRF Chapter Resource File
TR Teaching Transparency
* Also on One-Stop Planner
■ Also Available in Spanish
◆ Requires Advance Prep

SKILLS DEVELOPMENT AND PRACTICE	SECTION REVIEW AND ASSESSMENT	CORRELATIONS
		Florida Sunshine Standards/FCAT
CRF Directed Reading Worksheet* BASIC CRF Active Reading Worksheet* GENERAL	SE Section 1 Review ■ TE Reteaching, p. 30 BASIC TE Quiz, p. 30 GENERAL CRF Quiz* ■ BASIC	SC.A.1.4.2, SC.A.1.4.5, SC.A.2.4.2
CRF Directed Reading Worksheet* BASIC CRF Active Reading Worksheet* GENERAL	SE Section 2 Review ■ TE Reteaching, p. 33 BASIC TE Quiz, p. 33 GENERAL TE Alternative Assessment, p. 33 GENERAL CRF Quiz* ■ BASIC	SC.A.1.4.2
TE Reading Skill Builder Paired Reading, p. 35 BASIC TE Reading Skill Builder Reading Organizer, p. 36 GENERAL CRF Directed Reading Worksheet* BASIC CRF Active Reading Worksheet* GENERAL	SE Section 3 Review ■ TE Reteaching, p. 37 BASIC TE Quiz, p. 37 GENERAL TE Alternative Assessment, p. 37 GENERAL CRF Quiz* ■ BASIC	SC.A.1.4.2, SC.B.1.4.1, SC.F.1.4.2, SC.F.2.4.2
CRF Directed Reading Worksheet* BASIC CRF Active Reading Worksheet* GENERAL CRF Critical Thinking Worksheet* ADVANCED CRF Science Skills Worksheet* GENERAL	SE Section 4 Review ■ TE Reteaching, p. 42 BASIC TE Quiz, p. 42 GENERAL CRF Quiz* ■ BASIC Modified Worksheet* BASIC	SC.B.1.4.1, SC.B.1.4.2, SC.B.1.4.7, SC.F.1.4.1, SC.F.1.4.5

SCILINKS. NSTA
www.scilinks.org
Maintained by the **National Science Teachers Association**

Topic: Properties of Water
SciLinks Code: HX4149

Topic: Foods as Fuel
SciLinks Code: HX4086

Topic: Chemical Reactions
SciLinks Code: HX4040

Topic: Enzyme Activity
SciLinks Code: HX4071

Topic: Enzymes
SciLinks Code: HX4072

Overview

Tell students the purpose of this chapter is to demonstrate how chemistry relates to biology. Explain that the term *metabolism* refers to all the chemical reactions that occur in an organism. Inherent to those chemical reactions are atoms, the fundamental units of matter. Atoms join together to form compounds. Much of the body is composed of organic compounds, such as proteins, lipids, and carbohydrates. These compounds are involved in many chemical reactions that help organisms maintain homeostasis.

Using the Figure

There are an estimated 38,000 wild horses on public range lands in the United States, according to the Federal Bureau of Land Management. These horses are actually feral horses (domestic horses returned to the wild), and they compete with native grazing animals and cattle for food and water. Because the horses are more active than cattle, they require much more food. Ask students what they think would happen if the populations of feral horses were to grow unchecked. (Overgrazing would occur, and the natural plant base in the horses' habitat could be decimated.) Tell students that the Wild and Free-Roaming Horse and Burro Act was passed by Congress in 1971 to protect, manage, and control wild horses and burros on public lands.

Assessing Prior Knowledge

Before beginning this chapter, be sure your students can answer the questions in the ***Quick Review*** on the facing page. If they cannot, encourage them to reread the sections indicated.

Stampeding horses

26

Chapter Correlations — Florida Sunshine Standards/FCAT

SC.A.1.4.2 The vast diversity of the properties of materials is primarily due to variations in the forces that hold molecules together.

SC.A.1.4.5 Connections (bonds) form between substances when outer-shell electrons are either transferred or shared between their atoms, changing the properties of substances.

SC.A.2.4.2 There is a difference between an element, a molecule, and a compound.

SC.B.1.4.1 Knowledge of energy is fundamental to all the scientific disciplines (e.g., the energy required for biological processes in living organisms and the energy required for the building, erosion, and rebuilding of the Earth).

SC.B.1.4.2 There is conservation of mass and energy when matter is transformed.

SC.B.1.4.7 The total amount of usable energy always decreases, even though the total amount of energy is conserved in any transfer.

SC.F.1.4.1 Body processes involve specific biochemical reactions governed by biochemical principles.

SC.F.1.4.2 Body structures are uniquely designed and adapted for their function.

SC.F.1.4.5 Complex interactions among the different kinds of molecules in the cell cause distinct cycles of activity governed by proteins.

SC.F.2.4.2 Every cell contains a "blueprint" coded in DNA molecules that specify how proteins are assembled to regulate cells.

CHAPTER

2 Chemistry of Life

Florida Sunshine Standards/FCAT

SC.A.1.4.5 Connections (bonds) form between substances when outer-shell electrons are either transferred or shared between their atoms, changing the properties of substances.

SC.A.2.4.2 There is a difference between an element, a molecule, and a compound.

SC.B.1.4.1 Knowledge of energy is fundamental to all the scientific disciplines (e.g., the energy required for biological processes in living organisms and the energy required for the building, erosion, and rebuilding of the Earth).

SC.B.1.4.2 There is conservation of mass and energy when matter is transformed.

SC.F.1.4.1 Body processes involve specific biochemical reactions governed by biochemical principles.

SC.F.1.4.2 Body structures are uniquely designed and adapted for their function.

SC.F.1.4.5 Complex interactions among the different kinds of molecules in the cell cause distinct cycles of activity governed by proteins.

SC.F.2.4.2 Every cell contains a "blueprint" coded in DNA molecules that specify how proteins are assembled to regulate cells.

Quick Review

Answer the following without referring to earlier sections of your book.

1. **Identify** seven properties of life. *(Chapter 1, Section 1)*
2. **List** seven themes of biology. *(Chapter 1, Section 1)*
3. **Distinguish** between metabolism and homeostasis. *(Chapter 1, Section 1)*

Did you have difficulty? *For help, review the sections indicated.*

Reading Activity

Before you read this chapter, write a short list of all the things you know about the chemistry of organisms. Then write a list of the things that you want to know about the chemistry of organisms. Save your list, and to assess what you have learned, see how many questions you can answer after reading this chapter.

What do these horses have in common with the grass under their feet? They and all other organisms are composed of chemical substances that include water, carbohydrates, proteins, and fats.

Looking Ahead

Section 1
Nature of Matter
Atoms
Chemical Bonding

Section 2
Water and Solutions
Water in Living Things
Aqueous Solutions

Section 3
Chemistry of Cells
Carbon Compounds

Section 4
Energy and Chemical Reactions
Energy for Life Processes
Enzymes

internet connect

www.scilinks.org
National Science Teachers Association *sci*LINKS Internet resources are located throughout this chapter.

SCILINKS. Maintained by the National Science Teachers Association

27

Opening Activity — BASIC

Ask students what the science of chemistry is, and have them suggest how chemistry relates to biology. (Chemistry is the study of the composition and properties of matter. Chemistry is related to biology because all organisms are composed of chemical substances.) Tell students that the properties and reactions of chemical substances are essential to all living things.

Quick Review Answers

1. The seven properties of life are cellular organization, reproduction, metabolism, homeostasis, heredity, responsiveness, and growth and development.
2. Seven themes of biology are cellular structure and function, reproduction, metabolism, homeostasis, heredity, evolution, and interdependence.
3. Metabolism is the sum of all of the chemical reactions carried out in an organism, and homeostasis refers to the maintenance of stable internal conditions in spite of changes in the external environment.

Chapter Resource File

- Vocabulary Worksheets
- Concept Mapping

MISCONCEPTION ALERT

Students may not understand that all living things are composed largely of the same chemical substances even though the organisms may appear to be quite different. Organisms obtain the energy that they need in order to live from carbohydrates, proteins, and fats. Water is also common to all living organisms and is essential for cell function and survival.

Focus

Overview

Before beginning this section, review with your students the objectives listed in the Student Edition. This section introduces students to atoms as the fundamental units of matter. Students learn that atoms can bond to form compounds.

Bellringer

Show students various substances, such as table salt, sugar, water, copper wire, and a piece of aluminum. Ask students to think about which items are composed of elements and which are composed of compounds, and have them write their answers in their notebooks. At the end of this section, explain that salt, sugar, and water are compounds and that copper and aluminum are elements. Emphasize that ions and molecules are important to the functioning of living things.

Motivate

Demonstration — BASIC

To demonstrate the attraction between opposite charges, have students gently rub an inflated balloon through their hair. Then, have students bring the balloon close to their hair without touching their hair. Ask students to describe their observations. Explain that their hair is attracted to the balloon because the negative charges on the balloon are attracted to the positive charges on their hair.

Section Correlations

SC.A.1.4.2
SC.A.1.4.5
SC.A.2.4.2

Section 1 Nature of Matter

Florida Sunshine Standards/FCAT

SC.A.1.4.2 The vast diversity of the properties of materials is primarily due to variations in the forces that hold molecules together.

SC.A.1.4.5 Connections (bonds) form between substances when outer-shell electrons are either transferred or shared between their atoms, changing the properties of substances.

SC.A.2.4.2 There is a difference between an element, a molecule, and a compound.

Objectives

- **Differentiate** between atoms and elements. SC.A.2.4.2
- **Analyze** how compounds are formed. SC.A.1.4.5
- **Distinguish** between covalent bonds, hydrogen bonds, and ionic bonds. SC.A.1.4.2

Key Terms

atom
element
compound
molecule
ion

Atoms

Cooking requires an understanding of how the ingredients in foods interact. A cook's application of chemical principles while preparing recipes affects the flavor and texture of foods. Just as a cook can benefit from a knowledge of basic chemistry, you can better understand principles of biology if you also understand the fundamentals of chemistry. Chemistry will help you learn about biology because organisms, including yourself, are chemical machines.

What does all matter have in common? Matter consists of atoms. An **atom** is the smallest unit of matter that cannot be broken down by chemical means. Atoms are so small and dynamic that their exact structure is difficult to determine. Scientists have developed models, such as the one shown in **Figure 1,** to explain the structure and properties of atoms.

As shown in Figure 1, atoms consist of three kinds of particles: electrons, protons, and neutrons. Protons, shown in red, and neutrons, shown in blue, make up the nucleus, or core, of an atom. The region around the nucleus that electrons may occupy at any time is called the electron cloud, shown as a blue haze around the nucleus. Electrons are negatively charged, so the electron cloud has a negative charge. Protons are positively charged and neutrons have no charge, so the nucleus has a positive charge. Because protons and electrons are oppositely charged, they attract one another. Atoms typically have one electron for each proton, so they have no electrical charge.

Figure 1 Atom. The electron cloud is the region of an atom where electrons are most likely to be found. The nucleus of this atom contains six protons and six neutrons.

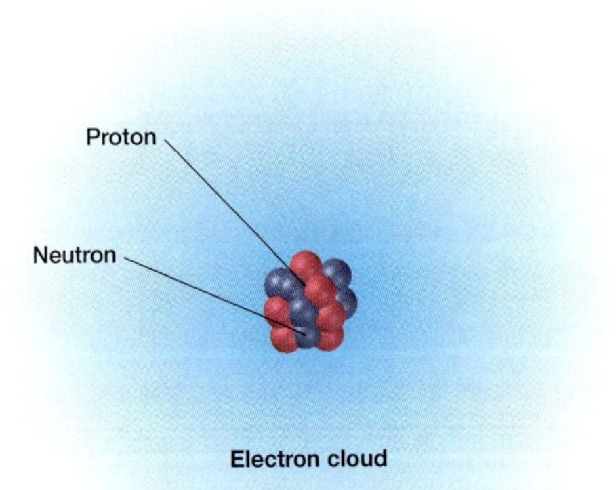

Elements

An **element** is a pure substance made of only one kind of atom. There are more than 100 known elements, and each is represented by a one-, two-, or three-letter symbol. For example, the elements hydrogen, oxygen, and carbon are represented by the symbols H, O, and C, respectively. Elements differ in the number of protons their atoms contain. Atoms of the simplest element, hydrogen, each contain one proton and one electron. In contrast, oxygen atoms contain eight protons and eight electrons. The number of neutrons in an atom is often but not always equal to the number of protons in the atom. Atoms of an element that contain different numbers of neutrons are called isotopes. For example, three common isotopes of carbon, C, are carbon-12, carbon-13, and carbon-14. Each contains six protons, however carbon-13 contains seven neutrons, and carbon-14 contains eight neutrons.

Chapter Resource File

- Directed Reading BASIC
- Active Reading GENERAL

Transparencies

TR Bellringer

One-Stop Planner CD-ROM

- Reading Organizers BASIC
- Reading Strategies BASIC

Chemical Bonding

Atoms can join with other atoms to form stable substances. A force that joins atoms is called a chemical bond. A **compound** is a substance made of the joined atoms of two or more different elements. For example, when sodium atoms, Na, bond with chlorine atoms, Cl, the compound sodium chloride (table salt) forms. Every compound is represented by a chemical formula that identifies the elements in the compound and their proportions. The formula for sodium chloride, NaCl, shows that there is one sodium atom for every chlorine atom in the compound.

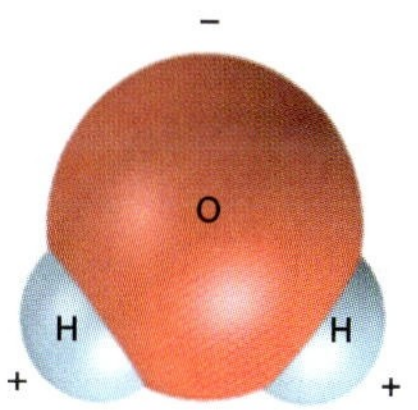

Figure 2 **Water molecule.** Each water molecule is held together by covalent bonds between two hydrogen atoms and one oxygen atom.

Covalent Bonds

Covalent bonds form when two or more atoms share electrons to form a molecule. A **molecule,** such as the water molecule shown in **Figure 2,** is a group of atoms held together by covalent bonds. Like the rivets and welds that connect steel girders in a skyscraper, covalent bonds join the atoms in molecules. Because the number of protons is equal to the number of electrons in a molecule, the molecule has no net electrical charge. Other examples of molecules include carbon dioxide, CO_2, and oxygen gas, O_2.

The arrangement of their electrons determines how atoms bond together. Electrons are grouped into different levels. The levels closest to the nucleus have less energy than the levels farther from the nucleus. Electron levels can hold a limited number of electrons. The outer electron levels of hydrogen and helium can hold up to two electrons. All other atoms, however, have outer electron levels that can hold up to eight electrons. An atom becomes stable when its outer electron level is full. If the outer electron level is not full, an atom will react readily with atoms that can provide electrons to fill its outer level. As Figure 2 shows, water, H_2O, forms when an oxygen atom, which has six outer electrons, combines with two hydrogen atoms, which have one outer electron each.

Hydrogen Bonds

The electrons in a water molecule are shared by oxygen and hydrogen atoms. However, the shared electrons are attracted more strongly by the oxygen nucleus than by the hydrogen nuclei. The water molecule therefore has partially positive and negative ends, or poles. As shown in **Figure 3,** the partially positive end of one water molecule is attracted to the negative end of another water molecule. Molecules with an unequal distribution of electrical charge, such as water molecules, are called polar molecules. This attraction between two water molecules is an example of a hydrogen bond—a weak chemical attraction between polar molecules.

Figure 3 **Hydrogen bonds in water**

Water molecules are attracted to each other by hydrogen bonds.

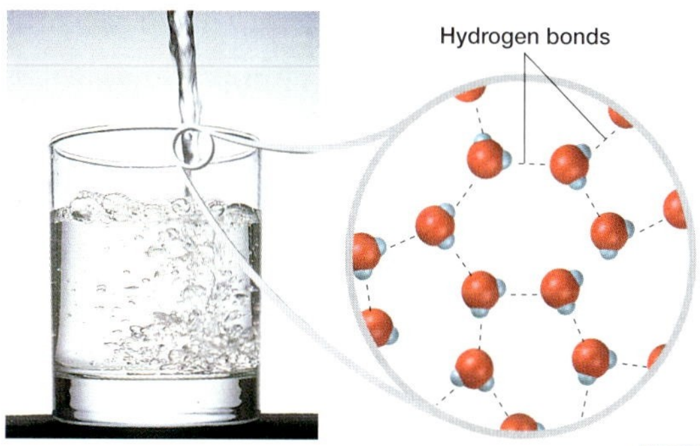

Teach

Teaching Tip — BASIC

Periodic Table of the Elements

Give a brief introduction to the periodic table of the elements. Explain how to use the periodic table to determine the atomic number, the atomic mass, and the numbers of protons, electrons, and neutrons in an atom. For example, the element fluorine, F, has an atomic number of 9, which is equal to both the number of protons and the number of electrons in each atom of fluorine. The atomic mass of fluorine is 18.99, making the number of neutrons about 10 (18.99–9). After presenting a few examples, randomly pick some elements and have the students determine the answers. Elements found in living things include carbon, C, hydrogen, H, oxygen, O, and nitrogen, N. Have students find these elements on the periodic table.

Integrating Physics and Chemistry

Provide the students with a list of 15-20 different substances (e.g., air, water, concrete, an iron nail, the helium inside an inflatable balloon, etc.) that includes a variety of elements, compounds, and mixtures. Have the students work together in pairs or small groups in order to classify the substances. When the students have finished classifying each of the substances as an element, compound, or mixture, check their answers by having them compare and defend their conclusions.

MISCONCEPTION ALERT

Students often do not understand that atoms consist mostly of empty space. If the nucleus of an atom were the size of a typical marble, the first electron level would be about 0.8 km (0.5 mi) away from the nucleus.

MEDICINE CONNECTION

Nuclear medicine uses radioisotopes to treat certain diseases. Isotopes are atoms of the same element that have different numbers of neutrons. A radioisotope is an isotope that has an unstable nucleus and undergoes radioactive decay.

Teach, continued

Real Life

Answer

Humans are endothermic and therefore maintain a fairly constant internal temperature. In response to cold, the body decreases blood flow near the skin's surface. Muscle contractions known as *shivering* may also occur to help generate heat.

Close

Reteaching — BASIC

Assign students to groups of four. Have each group make flashcards listing the key terms from this chapter. Some of the cards should have the definition on one side and the term on the other. Other flashcards should have questions related to this chapter on one side and the answer on the other. Upon completion, have the groups quiz each other.

Quiz — GENERAL

1. Name and describe the three subatomic particles of the atom. (Protons and neutrons make up the nucleus of the atom, and electrons are found in the electron cloud surrounding the nucleus. Protons are positively charged, electrons are negatively charged, and neutrons have no charge.)
2. Name three types of bonds that can form between atoms of two or more elements to form compounds. (covalent bonds, hydrogen bonds, and ionic bonds)

Section Review Correlations

SC.A.1.4.2: Items 4, 5
SC.A.1.4.5: Item 3
SC.A.2.4.2: Items 1, 2, 6

Real Life

Cells contain water and are vulnerable to freezing.

Some species of fish can swim in icy water because their bodies contain chemicals similar to automotive antifreeze.

Analyzing Information
Humans lack these chemicals. Why then can we tolerate short-term exposure to subzero temperatures?

Ionic Bonds

Sometimes atoms or molecules gain or lose electrons. An atom or molecule that has gained or lost one or more electrons is called an **ion** *(IE ahn)*. Ions have an electrical charge because they contain an unequal number of electrons and protons. An atom that has lost electrons is positively charged, whereas an atom that has gained electrons is negatively charged.

Ions of opposite charge may interact to form an ionic bond. For example, an atom of sodium is unstable because it has only one electron in its outer level. Sodium readily gives up this electron to become a stable, positively charged sodium ion, Na^+. An atom of chlorine is also unstable because it has seven electrons in its outer level. Chlorine readily accepts an electron to become a stable, negatively charged chloride ion, Cl^-. The negative charge of a chloride ion is attracted to the positive charge of a sodium ion. Thus, sodium atoms and chlorine atoms readily form an ionic bond to become sodium chloride, as shown in **Figure 4.**

Figure 4 **Ionic bonds in sodium chloride**

Ionic bonds in sodium chloride, NaCl, are formed by the interaction between sodium ions, Na^+, and chloride ions, Cl^-.

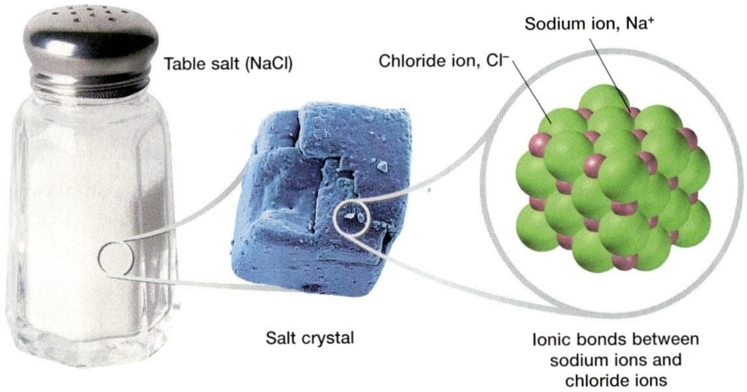

Section 1 Review

1. **Differentiate** between atoms and elements. SC.A.2.4.2
2. **Describe** how an atom differs from a molecule. SC.A.2.4.2
3. **Distinguish** between covalent bonds and ionic bonds. SC.A.1.4.5
4. **Identify** the type of weak bond that forms between water molecules. SC.A.1.4.2
5. **Critical Thinking Recognizing Differences** Explain the difference between polar molecules and nonpolar molecules. Give an example of a polar molecule. SC.A.1.4.2
6. **FCAT Prep** What is sodium chloride an example of?
   A a compound **C** an isotope
   B an element **D** a molecule
   SC.A.2.4.2

Answers to Section Review

1. An atom is the smallest unit of matter that cannot be broken down by chemical means. An element is a substance made of only one kind of atom.
2. An atom is a single unit, whereas a molecule is a group of atoms held together by covalent bonds.
3. Covalent bonds form when two or more atoms share electrons. Ionic bonds form when ions of opposite charge interact with one another.
4. Hydrogen bonds are weak bonds that form between water molecules.
5. Polar molecules have an unequal distribution of electric charge. Nonpolar molecules have an equal distribution of electric charge. An example of a polar molecule is water.
6. **A.** Correct. **B.** Incorrect. Sodium and chlorine do not share electrons as do atoms that form molecules. **C.** Incorrect. Isotopes are atoms of an element that contain different numbers of neutrons. **D.** Incorrect. An ion is an atom that has gained or lost one or more electrons.

Water and Solutions

Section 2

Florida Sunshine Standards/FCAT

SC.A.1.4.2 The vast diversity of the properties of materials is primarily due to variations in the forces that hold molecules together.

Water in Living Things

You may not realize it, but nearly 70 percent of your body is made of water. About two-thirds of the molecules in your body are water molecules. Your body's cells are filled with water, and water is the medium in which most cellular events take place. Your cells are also surrounded by water, and water helps move nutrients and other substances into and out of your cells. What are some of the properties of water that make it such an important substance for life?

Storage of Energy

Water absorbs heat more slowly and retains this energy longer than many other substances do. For example, a pot of boiling water removed from a stove takes a long time to cool down. Many organisms release excess heat through water evaporation. For example, humans cool themselves by sweating. The water vapor lost through the evaporation of sweat carries heat away from the body. In organisms, this ability to control temperature enables cells to maintain a constant internal temperature when the external tempera-ture changes drastically. Water thus helps cells maintain homeostasis.

Cohesion and Adhesion

The hydrogen bonds between water molecules cause the cohesion of liquid water. **Cohesion** *(koh HEE zhuhn)* is an attraction between substances of the same kind. Because of cohesion, water and other liquids form thin films and drops, such as those shown in **Figure 5.** Molecules at the surface of water are linked together by hydrogen bonds like a crowd of people linked by holding hands. This attraction between water molecules causes a condition known as *surface tension*. Surface tension prevents the surface of water from stretching or breaking easily.

Water molecules are also attracted to many other similarly polar substances. **Adhesion** *(ad HEE zhuhn)* is an attraction between different substances. Because of adhesion, some substances get wet. Adhesion powers a process, called capillary action, in which water molecules move upward through a narrow tube, such as the stem of a plant. The attraction of water to the walls of the tube sucks the water up more strongly than gravity pulls it down. Water moves upward through a plant from roots to leaves through a combination of capillary action, cohesion, and other factors.

Objectives

- **Analyze** the properties of water. SC.A.1.4.2
- **Describe** how water dissolves substances. SC.A.1.4.2
- **Distinguish** between acids and bases. SC.A.1.4.2

Key Terms

cohesion
adhesion
solution
acid
base

Figure 5 Cohesion. Because of cohesion, water forms drops like those on this plant.

Chapter Resource File

- Directed Reading BASIC
- Active Reading GENERAL

Transparencies

TR Bellringer

One-Stop Planner CD-ROM

- Reading Organizers BASIC
- Reading Strategies BASIC

Section 2

Focus

Overview

Before beginning this section, review with your students the objectives listed in the Student Edition. This section introduces students to the unique properties of water and discusses why it is an essential compound common to all living things. Students learn about the pH scale and the differences between acids and bases.

Bellringer

Ask students to guess if the following substances are acids or bases, and have them write the reasons for their choices: lemon juice, water, household ammonia, blood, soda, milk, antacid, and shampoo. After taking roll, tell students what each substance is and ask if they can detect any ways in which acids and bases are similar.

Motivate

Demonstration — BASIC

Prior to class, coat a sewing needle with petroleum jelly. Ask students to predict whether you can float the needle on the surface of water in a beaker. The grease keeps the water from wetting the needle; that is, the grease keeps the water molecules from being more attracted to the needle than they are to each other. Using a pair of forceps, carefully lay the needle flat on the surface of the water. The needle will float. Have students observe the results without touching the beaker. Then, push the needle slightly beneath the surface of the water. The needle will sink. Ask students why the needle floated on the water at first but then sank when the water's surface was disturbed. (At first, the surface tension of the water supported the needle. When the surface tension was broken, the needle sank.)

Section Correlations

SC.A.1.4.2

Motivate, continued

Activity — GENERAL

On the day before teaching this section, place one end of a freshly cut stalk of celery in a beaker of water that has a high concentration of red food coloring. Let the celery sit in the water overnight. Have students observe the results. Have each student pull one of the "strings" out of the celery stalk and examine it with a hand lens. Ask students to describe the structure of the "strings." Then, ask students how the water was able to move up through the celery stalk. (Capillary action and other factors enable water to move upward through the celery stalk.)

Teach

Teaching Tip — BASIC

Water Density Hydrogen bonds in ice, unlike those in liquid water, are stable. In ice, the water molecules form a crystalline structure. The molecules in an ice crystal are spaced apart, whereas they are very close together in liquid water. A fixed volume of ice contains fewer water molecules than the same volume of liquid water does. Thus, ice is less dense than water. Fill a drinking glass with water, and put ice cubes in the glass. Ask students why the ice cubes float in the water. (The ice cubes float because they are less dense than the liquid water.)

Section Review Correlations

SC.A.1.4.2: Items 1-5

internet connect
www.scilinks.org
Topic: Properties of Water
Keyword: HX4149
SciLinks. Maintained by the National Science Teachers Association

Aqueous Solutions

Many substances dissolve in water. For example, when you add salt to water, the resulting mixture is a saltwater solution. A **solution** is a mixture in which one or more substances are evenly distributed in another substance. Many important substances in the body have been dissolved in blood or other aqueous fluids. Because these substances can dissolve in water, they can more easily move within and between cells. For example, sugar could not be delivered to your cells if it were not dissolved in water.

Polarity

The polarity of water enables many substances to dissolve in water. Ionic compounds and polar molecules dissolve best in water. When ionic compounds are dissolved in water, the ions become surrounded by polar water molecules. As **Figure 6** shows, ions are attracted to the ends of water molecules with the opposite charge. The resulting solution is a mixture of water molecules and ions. A similar attraction results when polar molecules are dissolved in water. In both cases, the ions or molecules become evenly distributed in the water.

Nonpolar molecules do not dissolve well in water. When nonpolar substances, such as oil, are placed in water, the water molecules are more attracted to each other than to the nonpolar molecules. As a result, the nonpolar molecules are shoved together. This explains why oil forms clumps or beads in water. The inability of nonpolar molecules to dissolve in polar molecules is important to organisms. For example, the shape and function of cell membranes depend on the interaction of polar water with nonpolar membrane molecules.

Figure 6 **Water dissolves ionic compounds**

When sodium chloride, NaCl, is dissolved in water, sodium ions, Na^+, and chloride ions, Cl^-, become surrounded by water molecules, H_2O.

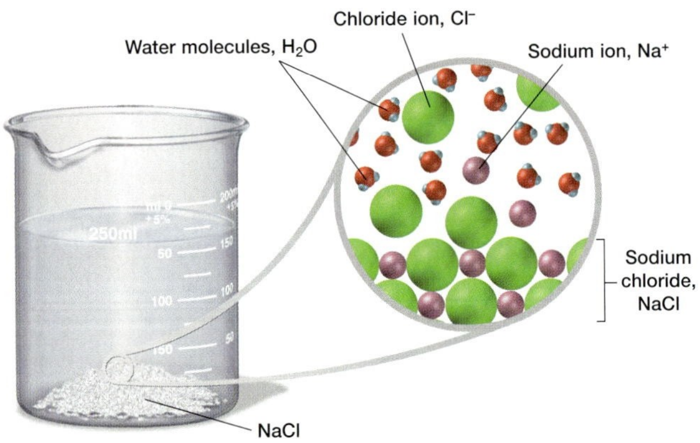

INCLUSION Strategies

- *Attention Deficit Disorder*
- *Learning Disability*

Ask students to use product labels from vinegar products, citrus products, and other household products to identify acids used in the ingredients. Using **Figure 7,** students can classify those acids as strong or weak. Ask students to use product labels from soaps, cleaning agents, and other household products to identify bases used in the ingredients. Using **Figure 7,** students can classify those bases as strong or weak. Students may create two charts showing the products containing acids and bases along with their classification as strong or weak.

Acids and Bases

While the bonds in water molecules are strong, at any given time a tiny fraction of those bonds might break, forming a hydrogen ion, H^+, and a hydroxide ion, OH^-:

$$H_2O \longrightarrow H^+ + OH^-$$

As a result, pure water always has a low concentration of hydrogen ions and hydroxide ions, which are present in equal numbers. Compounds that form hydrogen ions when dissolved in water are called **acids.** When an acid is added to water, the concentration of hydrogen ions in the solution is increased above that of pure water.

In contrast, compounds that reduce the concentration of hydrogen ions in a solution are called **bases.** Many bases form hydroxide ions when dissolved in water. Such bases lower the concentration of hydrogen ions because hydroxide ions react with hydrogen ions to form water molecules.

The pH scale shown in **Figure 7** is based on the concentration of hydrogen ions in solutions. All solutions have a pH value between 0 and 14. Pure water has a pH value of 7. Acidic solutions have pH values below 7, and basic solutions have pH values above 7. Each whole number represents a factor of 10 on the scale. A solution with a pH value of 5, for example, has 10 times as many hydrogen ions as one with a pH value of 6.

Study TIP

Reading Effectively
As you read, you may encounter the terms *alkaline* or *alkalinity.* Basic solutions—whose pH is above 7—are often called *alkaline solutions.* Solutions with pH values below 7 are usually referred to as *acidic solutions.*

Figure 7 **The pH scale.** The pH scale is based on the concentration of hydrogen ions in a solution.

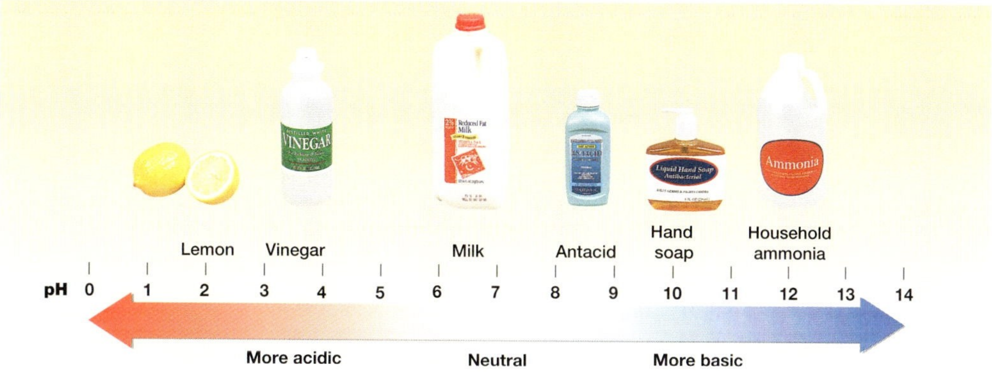

Section 2 Review

1. **Distinguish** between adhesion and cohesion. SC.A.1.4.2
2. **Identify** a substance that would not dissolve well in water. Explain why. SC.A.1.4.2
3. **Differentiate** between acids and bases. SC.A.1.4.2
4. **Critical Thinking** **Inferring Relationships** When salt is added to water, the freezing point of the water decreases. Explain why this occurs. SC.A.1.4.2
5. **FCAT Prep** The pH of solution *A* is 2. The pH of solution *B* is 4. How does the concentration of hydrogen ions in *A* ($[H^+]_A$) compare with the concentration of hydrogen ions in *B* ($[H^+]_B$)?
A $[H^+]_A = 2 \times [H^+]_B$.
B $[H^+]_B = 2 \times [H^+]_A$.
C $[H^+]_A = 100 \times [H^+]_B$.
D $[H^+]_B = 100 \times [H^+]_A$.
SC.A.1.4.2

Answers to Section Review

1. Adhesion is an attraction between different substances. Cohesion is an attraction between substances of the same kind.
2. Oil would not dissolve well in water because oil is nonpolar. Water molecules are polar and are more attracted to each other than they are to the nonpolar oil molecules.
3. Compounds that increase the concentration of hydrogen ions in a solution are acids. Compounds that reduce the concentration of hydrogen ions in a solution are bases.
4. When salt is added to water, the salt dissolves in the water. The dissolved ions interfere with the hydrogen bonds between water molecules, which normally become stable when ice is formed. When the temperature of the solution is lowered, it is more difficult for ice to form after salt has been added to the water. Thus, salt lowers the freezing point of water.
5. **A.** Incorrect. Each consecutive value on the pH scale does not represent a factor of 1. **B.** Incorrect. The concentration of hydrogen ions in solution A is greater that in solution B. **C.** Correct. Each consecutive value on the pH scale represents a factor of 10. **D.** See answer B above.

Close

Reteaching — BASIC

Show students photographs in which the properties of water are illustrated. Use pictures from various magazines. Ask students to identify the property being illustrated in each picture. (Answers may vary. For example, a picture of a tree could be used as an example of capillary action in that water moves through the tree. A picture of a glass filled with water indicates adhesion between water and the glass as well as cohesion of the water molecules.)

Quiz — GENERAL

Decide whether the following statements are true or false. If a statement is false, change the underlined word or words to make the statement true.

1. Nonpolar molecules dissolve very easily in water. (false; Polar)
2. Cohesion and adhesion of water molecules help a plant move water from its roots to its leaves. (true)
3. When a base is added to water, the concentration of hydrogen ions increases. (false, when a base is added to water, the concentration of hydrogen ions decreases)

Alternative Assessment — GENERAL

Immiscible Substances Bring a bottle of vinaigrette salad dressing to class. Ask students to identify the major ingredients in the dressing. (oil and vinegar) Shake the bottle vigorously, and then let the bottle rest during the class period. Ask students to explain why the oil and vinegar eventually separate. (The oil, which is nonpolar, is repelled by the water in the vinegar, which is polar.)

Section 3 Chemistry of Cells

Florida Sunshine Standards/FCAT

SC.B.1.4.1 Knowledge of energy is fundamental to all the scientific disciplines (e.g., the energy required for biological processes in living organisms and the energy required for the building, erosion, and rebuilding of the Earth).

SC.F.1.4.2 Body structures are uniquely designed and adapted for their function.

SC.F.2.4.2 Every cell contains a "blueprint" coded in DNA molecules that specify how proteins are assembled to regulate cells.

Focus

Overview

Before beginning this section, review with your students the objectives listed in the Student Edition. This section introduces students to the kinds of organic compounds that are found in living organisms. Students learn that these compounds are essential for life processes.

Bellringer

Show students several foods, such as cooking oil, sugar, and a piece of meat. Have students write what they think all these substances have in common. (Answers may include that each food contains macromolecules, including lipids, carbohydrates, and proteins, or a combination of molecules.) Emphasize that all of these foods contain the element carbon.

Motivate

Discussion/Question — BASIC

Direct students to look at **Figure 9.** Point out that the saturated fatty acid has no double bonds in its hydrocarbon chain, whereas the unsaturated fatty acid does have double bonds. Ask students what causes the kinks in an unsaturated fatty acid. (the double bonds between carbon atoms in the chain) Ask students why unsaturated fats, such as olive oil, are liquid at room temperature. (In unsaturated fats, the kinks prevent the fat molecules from packing closely together. Thus, unsaturated fats have a low melting point and do not tend to solidify at room temperature.)

Section Correlations

SC.A.1.4.2
SC.B.1.4.1
SC.F.1.4.2
SC.F.2.4.2

Objectives

- **Summarize** the characteristics of organic compounds. SC.A.1.4.2
- **Compare** the structures and function of different types of biomolecules. SC.F.1.4.2
- **Describe** the components of DNA and RNA. SC.F.2.4.2
- **State** the main role of ATP in cells. SC.B.1.4.1

Key Terms

carbohydrate
monosaccharide
lipid
protein
amino acid
nucleic acid
nucleotide
DNA
RNA
ATP

Carbon Compounds

Most matter in your body that is not water is made of organic compounds. Organic compounds contain carbon atoms that are covalently bonded to other elements—typically hydrogen, oxygen, and other carbon atoms. Four principal classes of organic compounds are found in living things: carbohydrates, lipids, proteins, and nucleic acids. Without these compounds, cells could not function.

Carbohydrates

Carbohydrates are organic compounds made of carbon, hydrogen, and oxygen atoms in the proportion of 1:2:1. Carbohydrates are a key source of energy, and they are found in most foods—especially fruits, vegetables, and grains. The building blocks of carbohydrates are single sugars, called **monosaccharides** *(mahn oh SAK uh reyedz)*, such as glucose, $C_6H_{12}O_6$, and fructose. Simple sugars such as glucose are a major source of energy in cells. Disaccharides are double sugars formed when two monosaccharides are joined. For example, sucrose, or common table sugar, consists of both glucose and fructose. Polysaccharides such as starch, shown in **Figure 8,** are chains of three or more monosaccharides. A polysaccharide is an example of a macromolecule, a large molecule made of many smaller molecules.

In organisms, some polysaccharides function as storehouses of the energy contained in sugars. Two polysaccharides that store energy in this way are starch, which is made by plants, and glycogen, which is made by animals. Both starch and glycogen are made of hundreds of linked glucose molecules. Cellulose is a polysaccharide that provides structural support for plants. Humans cannot digest cellulose. Thus, you cannot digest wood, which is mostly cellulose.

Figure 8 Structure of polysaccharides. Starch is a long chain of many linked glucose molecules.

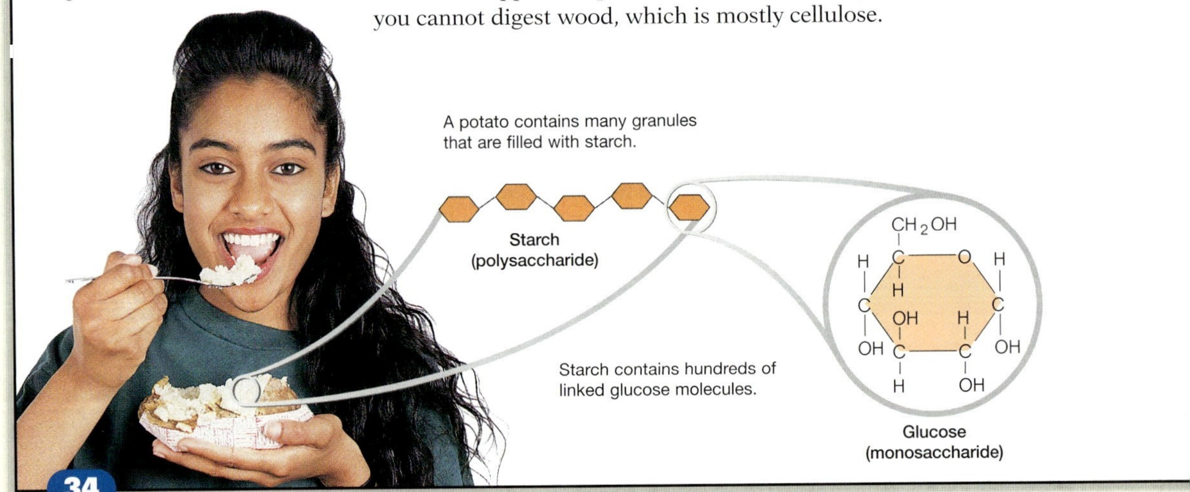

Chapter Resource File

- Directed Reading BASIC
- Active Reading GENERAL

One-Stop Planner CD-ROM

- Reading Organizers BASIC
- Reading Strategies BASIC

Transparencies

TR Bellringer
TR A27 Saturated and Unsaturated Fatty Acids
TR A28 Structure of Nucleic Acids

Figure 13 **Energy and chemical reactions**

Chemical reactions absorb or release energy.

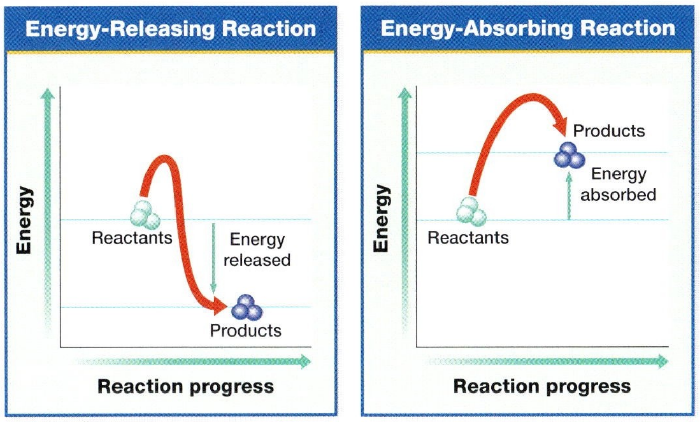

Energy in Chemical Reactions

In chemical reactions, energy is absorbed or released when chemical bonds are broken and new ones are formed. The graphs shown in **Figure 13** compare a chemical reaction that releases energy with a chemical reaction that absorbs energy. The freezing and melting of water are physical, not chemical, changes. Freezing and melting, however, are good examples of how energy is released or absorbed by a system. When water freezes, the process that leads to the formation of ice crystals causes heat energy to be released. When you place water in the freezer to make ice, heat is released from the water as the water freezes. When you remove ice from the freezer, the ice begins to melt. When ice melts, it absorbs heat from the environment. When you hold a piece of ice, your hand gets cold and heat is transferred from your hand to the ice as the ice begins to melt.

Metabolism *(muh TAB uh lihz uhm)* is the term used to describe all of the chemical reactions that occur within an organism. Your cells get most of the energy needed for metabolism from the food you eat. As food is digested, chemical reactions convert the chemical energy in food molecules to forms of energy that can be used by cells.

Activation Energy

The heat from a flame transfers enough energy to ignite the logs in a campfire. The spark from a spark plug causes the gasoline in an automobile engine to ignite. In both cases, energy is needed to start a chemical reaction. The energy needed to start a chemical reaction is called **activation energy**. To better understand activation energy, think of rolling a boulder down a hill. To get the boulder rolling downhill, you must first push it. Activation energy is simply a chemical "push" that starts a chemical reaction. Even in a chemical reaction that releases energy, activation energy must be supplied before the reaction can occur.

internet connect

www.scilinks.org
Topic: Chemical Reactions
Keyword: HX4040

SCILINKS Maintained by the National Science Teachers Association

Teach

Using the Figure — GENERAL

Direct students to look at **Figure 13.** Point out that energy-releasing reactions, represented by the graph on the left, are called *exergonic reactions.* Energy-absorbing reactions, represented by the graph on the right, are called *endergonic reactions.* Ask students to give examples of energy-releasing reactions and energy-absorbing reactions. (Sample answer: The burning of wood gives off heat and light energy. Plants absorb light energy in photosynthesis.)

Teaching Tip — GENERAL

Energy Transformation Have students draw a graphic organizer that is similar to the one at the bottom of this page and that shows an example of energy transformation. Students should list at least two transformations in their diagrams. Ask students to include a paragraph tracing the flow of energy they have charted. Emphasize that energy transformations occur in living things.

MEDICINE CONNECTION

Fever is a response to infection. A high fever is dangerous because critical enzymes are inactivated at high temperatures. Fever exceeding 41°C (105°F) can be fatal. A doctor should investigate fevers accompanied by shaking or chills and fevers that persist for several days.

Graphic Organizer

Use this graphic organizer with **Teaching Tip** on this page.

Close, continued

Demonstration — Basic

When fresh pineapple is added to a mixture containing the protein gelatin, the gelatin will not form a gel. Prior to class, prepare gelatin with fresh pineapple and gelatin without pineapple. Tell students that pineapple contains bromelin, an enzyme that breaks down gelatin. Ask students what may happen if gelatin is prepared by using canned pineapple. (In canned pineapple, the bromelin has been destroyed by a heating process. Thus, the gelatin will form a gel when canned pineapple is used.)

Integrating Physics and Chemistry

Using a periodic table of the elements, show the students where metals and non-metals are located. However, be sure to point out that even though hydrogen is on the "metallic" side of the periodic table, it is a non-metal. Explain that a compound like sodium chloride (NaCl), which is made of the metal Sodium (Na) and the non-metal chlorine (Cl), will form an ionic bond. This is because Na loses an electron and Cl gains an electron. Consequently, the respective positive ion and negative ion are attracted to each other and form an extremely strong ionic bond. In contrast, compounds made of two or more non-metals usually form covalent bonds. In this type of bond electrons are shared. In order to assess the students' understanding of the relationship between an element's bonding and its placement on the periodic table, ask the students to predict whether the following compounds are ionic or covalent: NO_2, MgF_2, H2S04, $CaBr_2$, KI, CO_2, Cl2O, CaO, HNO3, LiF.

Enzymes

Like engines, cells consume fuel because they need energy to function. Just as an engine requires a spark of energy to begin burning gasoline, most biochemical reactions—chemical reactions that occur in cells—require activation energy to begin. The chemical reactions in cells occur quickly and at relatively low temperatures because of the action of many enzymes. **Enzymes** are substances that increase the speed of chemical reactions. Most enzymes are proteins. Enzymes are catalysts *(KAT uh lists)*, which are substances that reduce the activation energy of a chemical reaction. As shown in **Figure 14,** an enzyme increases the speed of a chemical reaction by reducing the activation energy of the reaction.

Figure 14 Enzymes lower activation energy. Enzymes decrease the amount of energy needed to start a chemical reaction. Enzymes do not change the amount of energy contained in either the reactants or the products.

Activation Energy With and Without an Enzyme

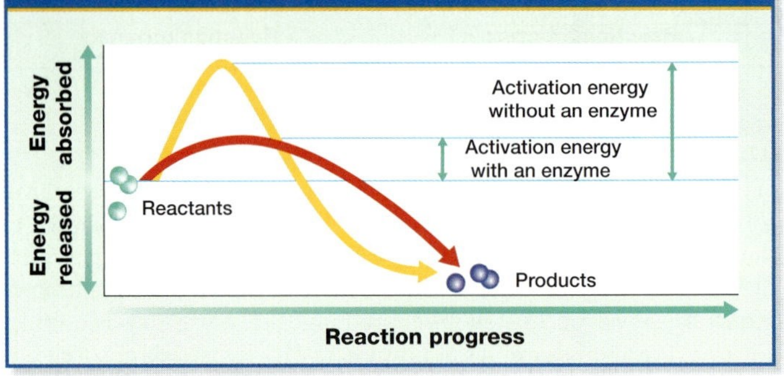

Enzymes help organisms maintain homeostasis. Without enzymes, chemical reactions would not occur quickly enough to sustain life. For example, consider a reaction that takes place in your blood. Blood carries carbon dioxide, CO_2, (a waste product made by cells) to your lungs, where it is eliminated as you breathe out. In the lungs, carbon dioxide reacts with water, H_2O, to form carbonic acid, H_2CO_3, as shown by the following equation:

$$CO_2 + H_2O \xrightleftharpoons{\text{carbonic anhydrase}} H_2CO_3$$

The reverse reaction occurs in your lungs, converting carbonic acid back to carbon dioxide and water. Most enzyme-assisted reactions are reversible, meaning they can proceed in the opposite direction.

Without an enzyme, the reaction that produces carbonic acid is very slow; only about 2000 molecules of carbonic acid are produced in an hour. This rate is not fast enough for your blood to carry away the carbon dioxide released by millions of cells. Fortunately, your blood contains the enzyme carbonic anhydrase *(an HIED rays)*. In the presence of carbonic anhydrase, carbon dioxide and water react to form about 600,000 molecules of carbonic acid per second! The enzyme increases the reaction rate about one million times, enabling your body to eliminate carbon dioxide efficiently.

Study TIP

Reading Effectively
As you read, notice that the names of most enzymes, such as amylase and catalase, end with *-ase*. This will help you identify other enzymes you will encounter in this book.

did you know?

Onions and Crying Onions contain sulfur compounds called *amino acid sulfoxides.* When an onion is cut, peeled, or crushed, enzymes called *allinases* are released from the tissue of the onion. These enzymes convert the amino acid sulfoxides into sulfenic acids, which form a chemical that triggers tear production in the eyes.

INCLUSION Strategies

- *Developmentally Delayed*
- *Attention Deficit Disorder*
- *Learning Disability*

Students can make a poster or create a bulletin board for the classroom about energy. The product should include a definition of "energy," pictures or drawings of forms of energy, and sentences showing examples of how energy can be converted from one form to another. The students can show their understanding of these concepts by presenting their completed product to the class or small group.

Enzyme Specificity

A substance on which an enzyme acts during a chemical reaction is called a **substrate** *(SUHB strayt)*. Enzymes act only on specific substrates. For example, the enzyme amylase *(AM uh lays)* assists in the breakdown of starch to glucose in the following chemical reaction. In this reaction, starch is amylase's substrate.

$$\text{starch} \xrightleftharpoons{\text{amylase}} \text{glucose}$$

The enzyme catalase *(KAT uh lays)* assists in the breakdown of hydrogen peroxide, H_2O_2, a toxin formed in cells. In this case, hydrogen peroxide is broken down to water, H_2O, and oxygen gas, O_2. In this reaction, hydrogen peroxide is catalase's substrate.

$$2H_2O_2 \xrightleftharpoons{\text{catalase}} 2H_2O + O_2$$

An enzyme's shape determines its activity. Typically, an enzyme is a large protein with one or more deep folds on its surface. These folds form pockets called **active sites**. As shown in **Figure 15,** an enzyme's substrate fits into the active site. An enzyme acts only on a specific substrate because only that substrate fits into its active site.

Step 1 When an enzyme first attaches to a substrate during a chemical reaction, the enzyme's shape changes slightly so that the substrate fits more tightly in the enzyme's active site.

Step 2 At an active site, an enzyme and a substrate interact in a way that reduces the activation energy of the reaction, making the substrate more likely to react.

Step 3 The reaction is complete when products have formed. The enzyme is now free to catalyze further reactions.

Real Life

Some laundry detergents contain enzymes.

Enzymes break down carbohydrates or proteins in common stains such as food and blood, making the stains easier to remove. The enzymes can tolerate the high temperatures and alkaline conditions required for cleaning fabrics.

Finding Information

Find out the source for enzymes used in laundry detergents.

Figure 15

BIOgraphic

Enzyme Action

Enzymes assist biochemical reactions by bringing key molecules together.

1. A substrate attaches to an enzyme's active site.
2. The enzyme reduces the activation energy of the reaction.
3. The enzyme is not changed by the reaction.

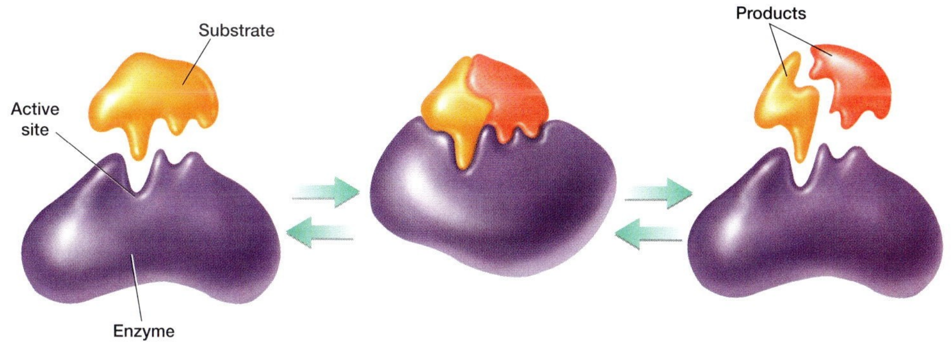

Teaching Tip — ADVANCED

Enzyme Names Enzymes are often named by incorporating the name of an enzyme's substrate with the function that the enzyme performs and then placing the suffix *-ase* at the end. For example, a dehydrogenase is an enzyme that removes hydrogen atoms from a substrate. Adding the suffix *-ase* to the name of the substrate without adding the functional name usually means that the enzyme breaks the substrate down into products. For example, sucrase breaks sucrose down to glucose and fructose. Ask students to suggest the function of the enzymes lactase and lipase. (Lactase breaks lactose down to glucose and galactose. Lipase breaks lipids down to glycerol and fatty acids.) Have the students research the functions of other enzymes.

PHYSICS CONNECTION

X rays, discovered in 1895, were the first means of producing a photographic image of internal body structures, such as bone. However, there are major drawbacks to using X rays. For example, X rays produce only a two-dimensional image of bone, and they do not make soft tissue observable. Exposure to large doses of X rays can cause cancer. Since the discovery of X rays, other imaging techniques have been developed. One is magnetic resonance imaging, or MRI, which uses neither X rays nor high-energy radiation. MRI works by affecting water molecules in the body. The nuclei of the hydrogen atoms in water molecules are randomly oriented. MRI uses powerful magnets to align these nuclei and then uses a brief pulse of radio waves to knock the nuclei out of alignment. The nuclei immediately realign and in the process give out a faint signal, which is picked up by the MRI scanner and translated by a computer into an image. MRI produces images of soft tissue because it consists mainly of water. Bone, on the other hand, contains very little water and is not visualized as clearly with MRI.

Close, *continued*

DATA LAB *Analyzing the Effect of pH on Enzyme Activity*

Skills Acquired
Interpreting graphs, analyzing

Teacher's Notes
Review how to interpret a graph. Tell students that the optimum or greatest activity of each enzyme is the highest point of the curve on the graph.

Answers to Analysis

1. pepsin
2. trypsin
3. trypsin: about 6.6
4. pepsin: about 3.0
5. Pepsin works in the stomach, and functions best at a low pH. Trypsin works in the small intestine, and it functions at a higher pH. Thus, it can be inferred that the stomach is highly acidic and that the small intestine is slightly acidic.

Close

Reteaching — BASIC

Have students make a spider diagram to help them identify the characteristics of enzymes described in Section 4.

Quiz — GENERAL

1. _______ energy is the energy required to start a chemical reaction. (Activation)
2. _______ increase the speed of chemical reactions by reducing the activation energy. (Enzymes)

Section Review Correlations

SC.B.1.4.1: Item 1
SC.B.1.4.2: Item 2
SC.B.1.4.7: Item 2
SC.F.1.4.1: Item 3
SC.F.1.4.5: Items 4, 5, 6

internet connect
www.scilinks.org
Topic: Enzyme Activity
Keyword: HX4071
SCILINKS. Maintained by the National Science Teachers Association

Factors in Enzyme Activity

Any factor that changes the shape of an enzyme can affect the enzyme's activity. For example, enzymes operate most efficiently within a certain range of temperatures. Temperatures outside this range can either break or strengthen some of the enzyme's bonds, changing its shape. Moreover, each enzyme operates best within a certain range of pH values. A pH value outside this range can cause bonds in an enzyme to break, reducing the enzyme's effectiveness.

The enzymes that are active at any one time in a cell determine what happens in that cell. Your body's cells contain many different enzymes, and each enzyme catalyzes a different chemical reaction. Different kinds of cells contain different collections of enzymes. For example, as you read this page, the chemical reactions occurring in nerve cells in your eye are different from the chemical reactions occurring in your red blood cells.

DATA LAB *Analyzing the Effect of pH on Enzyme Activity* SC.F.1.4.5

Background

The graph at right shows the relationship between pH and the activity of two digestive enzymes, pepsin and trypsin. Pepsin works in the stomach, while trypsin works in the small intestine. Use the graph to answer the following questions.

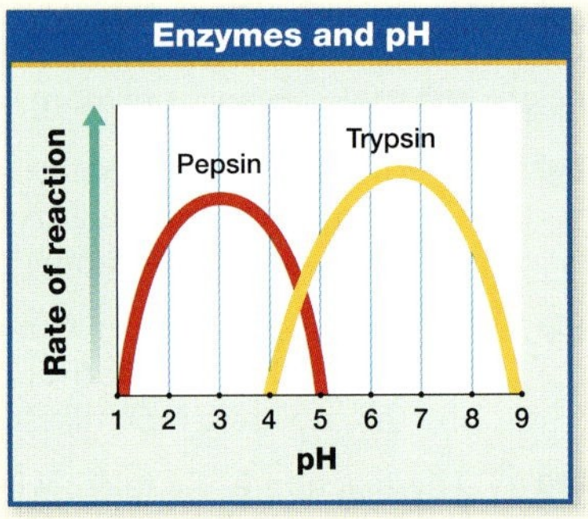

Analysis

1. **Name** the enzyme that works best in highly acidic environments.
2. **Name** the enzyme that works best in less-acidic environments.
3. **Critical Thinking** **Analyzing Data** Identify the pH value at which trypsin works best.
4. **Critical Thinking** **Analyzing Data** Identify the pH value at which pepsin works best.
5. **Critical Thinking** **Inferring Relationships** What does the graph indicate about the relative acidity of the stomach and small intestine?

Section 4 Review

1. **List** three ways that organisms use energy. SC.B.1.4.1
2. **Summarize** how energy is made available by chemical reactions and describe what happens to usable energy. SC.B.1.4.2, SC.B.1.4.7
3. **Explain** how enzymes increase the speed of chemical reactions. SC.F.1.4.1
4. **Describe** how the enzyme amylase affects starch. SC.F.1.4.5
5. **Critical Thinking** **Predicting Outcomes** What effect might a molecule that interferes with the action of carbonic anhydrase have on your body? SC.F.1.4.5
6. *FCAT Prep* Carboxypeptidase is an enzyme that catalyzes reactions in the small intestine. The products of these reactions are amino acids. What are the substrates of carboxypeptidase? SC.F.1.4.5
   A carbohydrates **C** proteins
   B lipids **D** nucleic acids

42

Answers to Section Review

1. Organisms store energy, use energy to power the chemical reactions of metabolism, and use energy to build cell structures.
2. Chemical reactions in the body convert chemical energy in food to other forms of energy that the organism can use. In this conversion, the total amount of usable energy decreases, though the total amount of energy is conserved.
3. Enzymes increase the speed of chemical reactions by lowering the activation energy of the reactions.
4. Amylase assists in the breakdown of starch into glucose.
5. Carbon dioxide would not be broken down. Instead, it would accumulate in the blood and become toxic to the body.
6. **A.** Incorrect. Carbohydrates are made up of sugars. **B.** Incorrect. Lipids are not made up of amino acids. **C.** Correct. Amino acids are the building blocks of proteins. **D.** Incorrect. Nucleic acids are made up of nucleotides.

ZONE

Key Concepts

1 Nature of Matter

- All matter is made of atoms. Atoms consist of electrons, protons, and neutrons.
- Molecules are groups of atoms linked by covalent bonds.
- Hydrogen bonding occurs between polar molecules.
- An ion is a charged atom or molecule. Ions of opposite charge may form an ionic bond.

2 Water and Solutions

- Water, which is essential for life, stores heat efficiently and binds to itself and other substances.
- Water dissolves polar molecules and ionic compounds.
- Acids increase the hydrogen ion concentration of a solution.
- Bases decrease the hydrogen ion concentration of a solution.
- The pH scale measures the strength of acids and bases.

3 Chemistry of Cells

- Organic compounds are found in living things.
- Carbohydrates, such as glucose, are a source of energy and are used as structural materials in organisms.
- Lipids are nonpolar molecules that store energy and are an important part of cell membranes.
- Proteins are chains of amino acids. The sequence of amino acids determines a protein's shape and specific function.
- Nucleic acids store and transmit hereditary information.
- ATP is the main energy currency of cells.

4 Energy and Chemical Reactions

- Chemical reactions absorb or release energy.
- Starting a chemical reaction requires activation energy.
- Enzymes speed up chemical reactions by decreasing the activation energy of the reactions.
- Enzymes bind only certain substrates.
- Factors such as temperature and pH affect enzyme activity.

Key Terms

Section 1

atom (28)
element (28)
compound (29)
molecule (29)
ion (30)

Section 2

cohesion (31)
adhesion (31)
solution (32)
acid (33)
base (33)

Section 3

carbohydrate (34)
monosaccharide (34)
lipid (35)
protein (36)
amino acid (36)
nucleic acid (37)
nucleotide (37)
DNA (37)
RNA (37)
ATP (37)

Section 4

energy (38)
activation energy (39)
enzyme (40)
substrate (41)
active site (41)

Alternative Assessment

Assign students to groups of four. Have each group make a puzzle that emphasizes the key terms and concepts presented in this chapter. Examples of puzzles include cross-word puzzles, word scrambles, and word searches. An answer key must also be included. When the puzzles are complete, have the groups exchange the puzzles. The groups should then solve the puzzles, given to them.

Chapter Resource File

- Science Skills Worksheet GENERAL
- Critical Thinking Worksheet ADVANCED
- Test Prep Pretest GENERAL
- Chapter Test GENERAL

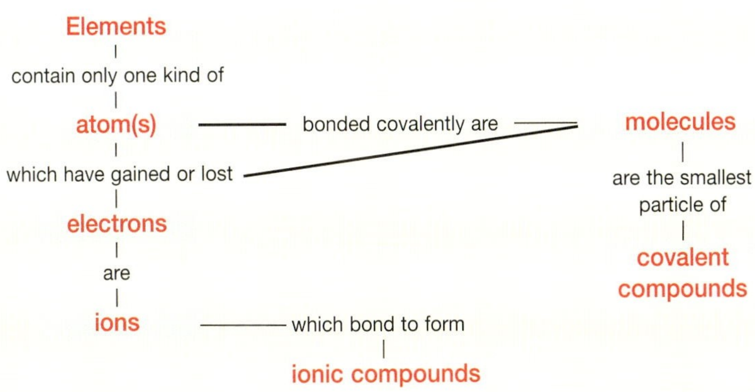

ANSWERS

Understanding Key Ideas

1. d
2. c
3. c
4. d
5. c
6. An enzyme increases the speed of a chemical reaction by reducing the activation energy.
7. Surface tension forms across the surface of the water because of the attraction between individual water molecules. Because of surface tension, the water strider can stand on the water's surface.
8. Because fats store more energy per gram than do carbohydrates, it is more advantageous for the birds to store energy as fat.
9. Lipids are good insulators and help organisms maintain a constant internal temperature. Lipids also store energy efficiently. Carbohydrates, such as glucose, can be stored by the body and released when energy is needed.
10. The answer to the concept map is found at the bottom of the Study Zone page.

Critical Thinking

11. In a covalent bond, atoms share electrons and form molecules. An ionic bond occurs between atoms or molecules that have lost or gained one or more electrons. Ions of opposite charge held together by ionic bonds form ionic compounds.

Alternative Assessment

12. Answers will vary. Some examples of natural additives are carrageen and agar, derived from algae and used as a thickening agent. Salt, sugar, and vinegar are some natural preservatives. Recent concerns over advertising claims have caused the FDA to create guidelines for determining whether a food can be labeled as all natural or not.
13. Answers will vary. Cell fractionation, centrifugation, and electrophoresis are techniques used to separate cells and cell products by size and weight. Separation allows scientists to experiment on one cell type or product at a time.
14. Biochemists study the chemistry of living organisms to increase scientific knowledge and develop ways to apply this knowledge. Most positions require at least a bachelor's degree with courses in biology, chemistry and biochemistry. Many positions require a master's or doctoral degree as well. Employers include colleges, universities, and various biomedical companies. Prospects for career growth are very good. Starting salary will vary by region.

Chapter Review Correlations

SC.A.1.4.2: Items 2, 7
SC.A.2.4.1: Items 1, 11
SC.A.2.4.2: Item 10
SC.B.1.4.1: Items 3, 4, 6
SC.F.1.4.1: Item 5
SC.F.1.4.2: Items 8, 9
SC.H.1.4.4: Item 14
SC.H.3.4.2: Item 13
SC.H.3.4.5: Item 12

Performan

Understanding Key Ideas

1. Most atoms contain one or more
a. neutrons. **c.** protons.
b. electrons. **d.** All of the above
SC.A.2.4.1

2. Water dissolves ionic compounds because water molecules
a. are nonpolar.
b. have a pH value of 14 or greater.
c. have partially charged ends.
d. do not contain atoms.
SC.A.1.4.2

3. In cells, ATP temporarily stores
a. amino acids. **c.** energy.
b. DNA. **d.** lipids.
SC.B.1.4.1

4. Energy needed for metabolism does *not* come from
a. food.
b. lipids.
c. carbohydrates.
d. water.
SC.B.1.4.1

5. Most enzymes are
a. lipids.
b. carbohydrates.
c. proteins.
d. nucleic acids.
SC.F.1.4.1

6. Explain the relationship between an enzyme and the activation energy of the reaction in which the enzyme participates. SC.B.1.4.1

7. Look at the water strider in the photograph below. Using what you have learned about the properties of water, explain how the insect can stand on the water's surface. SC.A.1.4.2

8. **BIOWatch** Ruby-throated hummingbirds migrate 2,000 km every fall. Before migrating, they eat nectar and convert much of the sugar in the nectar to fat. Why is it advantageous for these birds to store energy as fat rather than as glycogen? SC.F.1.4.2

9. Describe how molecules such as carbohydrates and lipids are important in homeostasis. (**Hint:** See Chapter 1, Section 1.) SC.F.1.4.2

10. **Concept Mapping** Make a concept map that illustrates the structure of matter. Include the following key terms in your map: *atom, element, compound, molecule,* and *ion.* SC.A.2.4.2

Critical Thinking

11. Recognizing Differences What are two differences between ionic bonds and covalent bonds? SC.A.2.4.1

Alternative Assessment

12. Evaluating Promotional Claims Analyze the ingredients of various packaged foods. Record the percentage of carbohydrates, fats, and proteins in each food. List any additives that the products contain. Research whether the additives are natural or artificial, and find out why they are added to particular foods. Compare these data to advertising claims about the products you analyzed. SC.H.3.4.5

13. Finding Information Investigate the laboratory techniques of cell fractionation, centrifugation, and electrophoresis. Find out how each technique enables biologists to experiment with cells and to analyze the substances that cells produce. Prepare an oral report and use graphics to interpret and summarize your findings. SC.H.3.4.2

14. Career Connection Biochemist Research the field of biochemistry, and write a report on your findings. Your report should include a job description, the training required, names of employers, growth prospects, and an average starting salary. SC.H.1.4.4

44

Assignment Guide

Section	Questions
1	1, 11
2	2, 7
3	3, 4, 8, 9, 12, 13
4	5, 6

FCAT Prep

For questions 1–6, write your answers on a separate sheet of paper.

1 Animal fats are usually solid, and plant fats are usually (liquid) oils. However, in many animals of the Arctic and Antarctic, animal fats are mostly oils. What adaptive advantage would the storage of body fat as oil instead of as a solid be to animals that live in freezing climates?

A. Plants can never freeze in the Arctic and Antarctic.
B. Oils are soluble in water while fats are not soluble in water.
C. Storing energy in oils as plants do allows animals to photosynthesize.
D. Oils have lower freezing points and give better protection against freezing.

SC.A.1.4.2

2 In an experiment, the rate of an enzyme-catalyzed reaction increased as the substrate concentration increased. Why might the reaction rate increase by only a small amount?

F. The amount of enzyme is limited.
G. The amount of substrate is limited.
H. Enzymes do not affect the rates of reaction.
I. The enzyme stopped catalyzing the reaction.

SC.B.1.4.1

3 Analysis of an unknown substance shows that it has the following characteristics: it contains carbon, hydrogen, and oxygen and is soluble in oil but not in water. What kind of substance is this?

A. carbon dioxide **C.** lipid
B. glucose **D.** polar amino acid

SC.G.1.4.3

Base your answers to questions 4–6 on the diagram below.

MODEL OF BIOLOGICAL MOLECULE

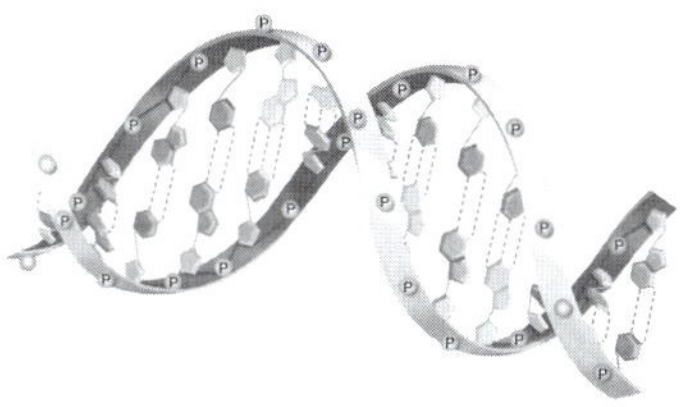

4 What molecule is shown in the diagram?

F. ATP **H.** DNA
G. cellulose **I.** RNA

SC.F.2.4.2

5 What is the main function of this molecule?

A. storing energy
B. making up cell membranes
C. storing hereditary information
D. promoting chemical reactions SC.F.2.4.2

6 How many strands of nucleotides are shown in the diagram?

SC.F.2.4.2

Test TIP

Choose your answer to a question based on both what you already know and any information presented in the question.

45

FCAT Doctor

Question 2 Answer F is the correct choice. When the amount of substrate exceeds the amount of enzyme, there are no additional enzyme molecules to bind the substrate. The enzymes will become available when the other reactions have been completed. Answer G is incorrect because you have increased the amount of substrate, so it is not the limiting factor. Answer H is incorrect because enzymes do speed up chemical reactions. Answer I is incorrect because enzymes can be used over and over again and do not stop catalyzing a reaction unless they are denatured, or there is a lack of substrate.

Question 5 Answer C is the correct choice. DNA is the genetic material that organisms inherit from their parents. Answer A is incorrect because energy is stored by ATP molecules and lipids. Answer B is incorrect because cell membranes are made up of phosphates and lipids. Answer D is incorrect because enzymes promote chemical reactions.

Question 6 The correct answer is 2. DNA molecules consist of two strands of nucleotides held together by hydrogen bonds between the bases. They twist, forming a double helix shape.

Answers

1. D
2. F
3. C
4. H
5. C
6. 2

Test Prep Correlations

SC.A.1.4.2: Item 1
SC.B.1.4.1: Item 2
SC.F.2.4.2: Items 4, 5, 6
SC.G.1.4.3: Item 3

Exploration Lab

Observing Enzyme Detergents

Teacher's Notes

Time Required
45 minutes for Day 1; 30 minutes for Day 2; and 15 minutes for Day 3.

Ratings
EASY → HARD

- Teacher Preparation
- Student Setup
- Concept Level
- Cleanup

Skills Acquired
- Designing Experiments
- Measuring
- Organizing and Analyzing Data

Scientific Methods
In this lab, students will:
- Make Observations
- Analyze the Results
- Draw Conclusions

Materials and Equipment
Materials for this lab can be ordered from WARD'S. See *Master Materials List* at the front of this book for catalog numbers.

Safety Caution
Make sure students wear safety goggles and a lab apron. Caution students to avoid burns by working carefully when heating and pouring boiling water.

Exploration Lab

Observing Enzyme Detergents SC.F.1.4.1

SKILLS
- Using scientific methods
- Measuring volume, mass, and pH

OBJECTIVES
- **Recognize** the function of enzymes in laundry detergents.
- **Relate** temperature and pH to the activity of enzymes.

MATERIALS
- safety goggles and lab apron
- balance
- graduated cylinder
- glass stirring rod
- 150 mL beaker
- 18 g regular instant gelatin or 1.8 g sugar-free instant gelatin
- 0.7 g Na_2CO_3
- tongs or a hot mitt
- 50 mL boiling water
- thermometer
- pH paper
- 6 test tubes
- test-tube rack
- pipet with bulb
- plastic wrap
- tape
- 50 mL beakers (6)

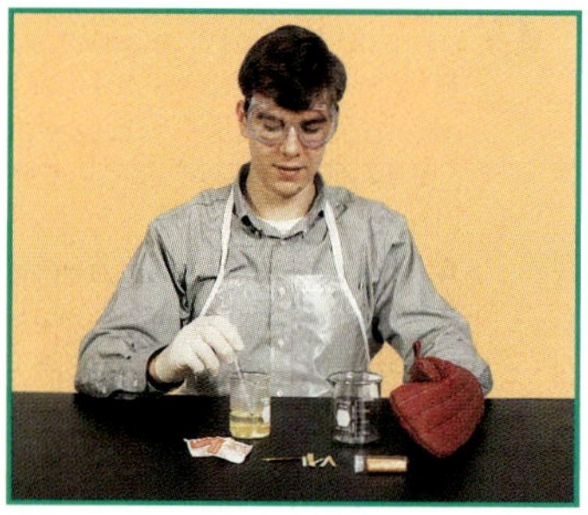

- 50 mL distilled water
- 1 g each of 5 brands of laundry detergent
- wax pencil
- metric ruler

ChemSafety

CAUTION: Always wear safety goggles and a lab apron to protect your eyes and clothing.

CAUTION: Do not touch or taste any chemicals. Know the location of the emergency shower and eyewash station and how to use them. If you get a chemical on your skin or clothing, wash it off at the sink while calling to the teacher. Notify the teacher of a spill. Spills should be cleaned up promptly, according to your teacher's directions.

CAUTION: Glassware is fragile. Notify the teacher of broken glass or cuts. Do not clean up broken glass or spills with broken glass unless the teacher tells you to do so.

Before You Begin

Enzymes are substances that speed up chemical reactions. Each enzyme operates best at a particular **pH** and temperature. Substances on which enzymes act are called **substrates.** Many enzymes are named for their substrates. For example, a **protease** is an enzyme that helps break down proteins. In this lab, you will investigate the effectiveness of laundry detergents that contain enzymes.

1. Write a definition for each boldface term in the paragraph above.
2. Based on the objectives for this lab, write a question you would like to explore about enzyme detergents.

Procedure

PART A: Make a Protein Substrate

1. Put on safety goggles and a lab apron.
2. **CAUTION: Use tongs or a hot mitt to handle heated glassware.** Put 18 g of regular (1.8 g of sugar-free) instant gelatin in a 150 mL beaker. Slowly add 50 mL of boiling water to the beaker, and stir the mixture with a stirring rod. Test and record the pH of this solution.

46

Tips and Tricks
Before starting the lab, ask students why enzymes are added to detergents. (to help break down proteins and other substances from foods that may stain clothing) ask students why detergent enzymes are stable during the hot water cycle. (The enzymes found in commercial laundry soap have been genetically engineered.) The rate of gelatin hydrolysis is slower in instant gelatin that contains sugar than in sugar-free gelatin. Students should bring in labeled samples (3–5 tablespoons) of each of five different laundry detergents. Labels should include active ingredients listed on the container. Students should have a control test tube with 15 drops (1 mL) of water (no detergent) added to the gelatin surface. Require students to present a written procedure for their experiment and a list of all safety precautions before allowing them to gather materials for the lab. When students add the detergent solutions to the gelatin, the mixture will foam. During this reaction, carbon dioxide gas is released. The addition of washing soda (Na_2CO_3) to the gelatin raises the pH of the gelatin from 4 to 8, which is the optimum pH for protease activation in the detergent samples.

Have students use a wax pencil to mark the test tubes at the uppermost level of the cooled gelatin in each tube. They will use this mark to measure the hydrolysis of the gelatin each day. Label the test tubes 1–6. To prepare a 10 percent solution of laundry detergent, students should dissolve 1 g of detergent in 9 mL of distilled water. Have students record the pH for each numbered detergent sample. Students can measure protein hydrolysis after 24 hours by using a wax pencil to draw a second line at the top of the gelatin layer and then measuring the distance (in mm) between the first line and the second line. This indicates the amount of hydrolysis of the protein in the gelatin by the enzymes in the detergent.

3. Very slowly add 0.7 g of Na_2CO_3 to the hot gelatin while stirring. Note any reaction. Test and record the pH of this solution.

4. Place 6 test tubes in a test-tube rack. Pour 5 mL of the gelatin-Na_2CO_3 mixture into each tube. Use a pipet to remove any bubbles from the surface of the mixture in each tube. Cover the tubes tightly with plastic wrap and tape. Cool the tubes, and store them at room temperature until you begin **Part C.** Complete step 12.

PART B: Design an Experiment

5. Work with members of your lab group to explore one of the questions written for step 2 of **Before You Begin.** To explore the question, design an experiment that uses the materials listed for this lab.

You Choose

As you design your experiment, decide the following:

a. what question you will explore
b. what hypothesis you will test
c. what detergent samples you will test
d. what your control will be
e. how much of each solution to use for each test
f. how to determine if protein is breaking down
g. what data to record in your data table

6. Write a procedure for your experiment. Make a list of all the safety precautions you will take. Have your teacher approve your procedure and safety precautions before you begin the experiment.

PART C: Conduct Your Experiment

7. Put on safety goggles and a lab apron.

8. Make a 10 percent solution of each laundry detergent by dissolving 1 g of detergent in 9 mL of distilled water.

9. Set up your experiment. Repeat step 12.

10. Record your data after 24 hours.

PART D: Cleanup and Disposal

11. Dispose of solutions, broken glass, and gelatin in the designated waste containers. Do not pour chemicals down the drain or put lab materials in the trash unless your teacher tells you to do so.

12. Clean up your work area and all lab equipment. Return lab equipment to its proper place. Wash your hands thoroughly before leaving the lab and after finishing all work.

Analyze and Conclude

1. **Analyzing Methods** Suggest a reason for adding Na_2CO_3 to the gelatin solution.
2. **Analyzing Results** Make a bar graph of your data. Plot the amount of gelatin broken down (change in the depth of the gelatin) on the *y*-axis and detergent on the *x*-axis.
3. **Inferring Conclusions** What conclusions did your group infer from the results? Explain.
4. **Further Inquiry** Write a new question about enzyme detergents that could be explored with another investigation.

Do You Know?

Do research in the library or media center to answer these questions:

1. What other household products contain enzymes, and what types of enzymes do they contain?
2. What type of organic compound is broken down by each enzyme that you identified?

Use SciLinks to explore your own questions about products that contain enzymes.

Answers to Before You Begin

1. *enzymes*—molecules (usually proteins) that speed up chemical reactions
2. *pH*—indicates the hydrogen ion concentration of a solution

   substrate—a reactant in a reaction that binds to an enzyme's active site

   protease—an enzyme that aids in the breakdown of proteins
3. Answers will vary. For example: How can you tell if a detergent contains enzymes?

Answers to Procedure

1. Prepare a 10 percent solution of each of five different detergents.
2. Test the pH of each detergent solution with pH paper. Record the pH for each sample in a data table similar to the one below.
3. Add about 15 drops (1 mL) of the first detergent solution to the gelatin surface of the first test tube. Repeat for each of the other samples. Reseal the test tubes and place in a test-tube rack for observation.
4. Add 15 drops of water to the gelatin surface of the sixth test tube.
5. Put the tubes aside for 24 hours at room temperature.
6. Draw another wax pencil line at the top of the gelatin layer. Measure the distance in mm between the first line and the second line. Record the data in a data table.

Sample Data

Detergent	pH	Amount of Protein Broken Down
1	8	7 mm
2	8	0 mm
3	8	9 mm
4	8	0 mm
5	8	6 mm

Answers to Analyze and Conclude

1. The washing soda increases the pH of the gelatin from 4 to 8—the optimum pH for enzyme activity in this reaction.
2. Graphs will vary.
3. Answers will vary. For example, enzymes operate best at a certain temperature and pH.4. Answers will vary. For example: Are enzymes in detergent stable in the presence of bleach?
4. Answers will vary. For example: Are enzymes in detergent stable in the presence of bleach?

Answers to Do You Know?

1. Answers will vary. For example, dairy digestive supplements contain the enzyme lactase.
2. Lactase breaks down lactose, the sugar in milk.

Chapter Opener

- SE **Quick Review,** p. 49 ■
- SE **Reading Activity,** p. 49 ■
- TE **Using the Figure,** p. 48 GENERAL
- TE **Opening Activity,** p. 49 GENERAL

& 3 • 90 min pp. 50–54

Section 1
Looking at Cells

- OSP **Lesson Plan***
- TR **Bellringer***
- TR **Section Outline***
- TR **A10 Metric Units of Length and Equivalents***
- TR **A4 Object Size and Magnifying Power of Microscopes***
- TR **A5 Compound Light Microscope***

- TE **Activity** Magnification, p. 50 GENERAL
- TE **Using the Figure,** p. 52 BASIC
- TE **Activity** Making a Wet Mount, p. 53 ◆ GENERAL
- CRF **Quick Lab** Modeling Cells Surface Area to Volume* BASIC

& 5 • 90 min pp. 55–61

Section 2
Cell Features

- OSP **Lesson Plan***
- TR **Bellringer***
- TR **Section Outline***
- TR **Relationships Between Surface Area and Volume***
- TR **B4 Animal Cells***
- TR **B8 Structure of Lipid Bilayer**
- TR **B10 Membrane Proteins***

- TE **Activity** Bacteria in Yogurt, p. 55 ◆ GENERAL
- TE **Demonstration,** p. 56 GENERAL
- SE **Math Lab** Calculating Surface Area and Volume, p. 56 ■
- CRF **Datasheets for In-Text Labs** Calculating Surface Area and Volume*
- TE **Using the Figure,** p. 58 BASIC
- TE **Activity** Amoeboid Movement, p. 59 GENERAL
- TE **Group Activity** Cell Membrane Model, p. 60 GENERAL
- TE **Using the Figure,** p. 60 GENERAL
- CRF **Skills Practice Lab** Using a Microscope* GENERAL

& 7 • 90 min pp. 62–66

Section 3
Cell Organelles

- OSP **Lesson Plan***
- TR **Bellringer***
- TR **Section Outline***
- TR **B13 Processing of Proteins***
- TR **B12 Mitochondrion***
- TR **B9 Organelles***

- TE **Activity** Cell Structures, p. 62 GENERAL
- TE **Group Activity** Selling Cells, p. 63 GENERAL
- TE **Using the Figure,** p. 63 BASIC
- TE **Using the Figure,** p. 65 ADVANCED
- SE **Skills Practice Lab** Studying Animal Cells and Plant Cells, pp. 70–71 ◆ ■
- CRF **Datasheets for In-Text Labs** Studying Animal Cells and Plant Cells*

& 9 • 90 min

Chapter Review and Assessment Resources

- SE **Study Zone,** p. 67 ■
- SE **Performance Zone,** p. 68 ■
- SE **FCAT Prep,** p. 69 ■
- TR **Concept Mapping***
- CRF **Vocabulary Review Worksheet*** ■ GENERAL
- CRF **Test Prep Pretest*** ■ GENERAL
- CRF **Chapter Test*** ■ GENERAL
- CRF **Chapter Test*** ADVANCED
- OSP **Test Generator***
- OSP **Test Item Listing***
- OSP **Modified Chapter Test*** BASIC

Online and Technology Resources

go.hrw.com Holt Online Learning

Visit go.hrw.com for access to Holt Online Learning or enter the keyword **HX6 Home** for a variety of free online resources.

One-Stop Planner® CD-ROM

This CD-ROM package includes
- Lab Materials QuickList Software
- Holt Calendar Planner
- Customizable Lesson Plans
- Printable Worksheets
- ExamView® Test Generator
- Interactive Teacher Edition
- Holt PuzzlePro® Resources
- Holt PowerPoint® Resources

Guided Reading Audio CDs

These CDs are designed to help auditory learners and reluctant readers. (Audio CDs are also available in Spanish.)

Compression guide:
To shorten your instruction because of time limitations, omit blocks 1, 2, and 3.

KEY

- **TE** Teacher Edition
- **SE** Student Edition
- **OSP** One-Stop Planner
- **CRF** Chapter Resource File
- **TR** Teaching Transparency
- * Also on One-Stop Planner
- ■ Also Available in Spanish
- ◆ Requires Advance Prep

SKILLS DEVELOPMENT AND PRACTICE	SECTION REVIEW AND ASSESSMENT	CORRELATIONS
		Florida Sunshine Standards/FCAT
TE **Skill Builder** Math Skills, p. 51 GENERAL **TE** **Reading Skill Builder** Interactive Reading, p. 51 BASIC **CRF** **Directed Reading Worksheet*** BASIC **CRF** **Active Reading Worksheet*** GENERAL	**SE** **Section 1 Review** ■ **TE** **Reteaching,** p. 54 BASIC **TE** **Quiz,** p. 54 GENERAL **TE** **Alternative Assessment,** p. 54 GENERAL **CRF** **Quiz*** ■ BASIC	SC.H.3.4.2
TE **Skill Builder** Interpreting Visuals, p. 57 GENERAL **TE** **Reading Skill Builder** Reading Organizer, p. 58 BASIC **TE** **Reading Skill Builder** Paired Reading, p. 59 BASIC **CRF** **Directed Reading Worksheet*** BASIC **CRF** **Active Reading Worksheet*** GENERAL	**SE** **Section 2 Review** ■ **TE** **Reteaching,** p. 61 BASIC **TE** **Quiz,** p. 61 GENERAL **TE** **Alternative Assessment,** p. 61 GENERAL **CRF** **Quiz*** ■ BASIC	SC.F.1.4.2, SC.F.1.4.3, SC.F.1.4.4, SC.H.1.4.3
TE **Reading Skill Builder** Reading Hint, p. 64 ADVANCED **CRF** **Directed Reading Worksheet*** BASIC **CRF** **Active Reading Worksheet*** GENERAL **CRF** **Critical Thinking Worksheet*** ADVANCED **CRF** **Science Skills Worksheet*** GENERAL	**SE** **Section 3 Review** ■ **TE** **Reteaching,** p. 66 BASIC **TE** **Quiz,** p. 66 GENERAL **TE** **Alternative Assessment,** p. 66 GENERAL **CRF** **Quiz*** ■ BASIC **Modified Worksheet*** BASIC	SC.F.1.4.2, SC.F.1.4.3, SC.F.2.4.2, SC.H.1.4.1

SCILINKS. NSTA
www.scilinks.org
Maintained by the **National Science Teachers Association**

Topic: Microscopes
SciLinks Code: HX4122

Topic: Cell Features
SciLinks Code: HX4034

Topic: Proteins
SciLinks Code: HX4151

CHAPTER 3

Overview

Tell students that the purpose of this chapter is to describe relationships between structure and function at the cellular and subcellular levels of an organism. Students will learn the common features of cells and the major differences between prokaryotes and eukaryotes. Students will also learn that proper functioning of all cell structures is needed to maintain homeostasis.

Using the Figure

Paramecia, which usually can be found in ponds and slow-moving streams, are classic examples of eukaryotic cells. Like all ciliates, these protozoans have more than one nucleus, which may be difficult to see among the many food vacuoles in the Paramecium in the photo. Cilia, the eyelash-like appendages surrounding the Paramecium, can beat forward or backward, moving the organism. The two contractile vacuoles are clearly visible, characterized by their central space surrounded by several radiating canals. Contractile vacuoles squeeze excess water out of the cell.

Assessing Prior Knowledge

Before beginning this chapter, be sure your students can answer the questions in the ***Quick Review*** on the facing page. If they cannot, encourage them to reread the sections indicated.

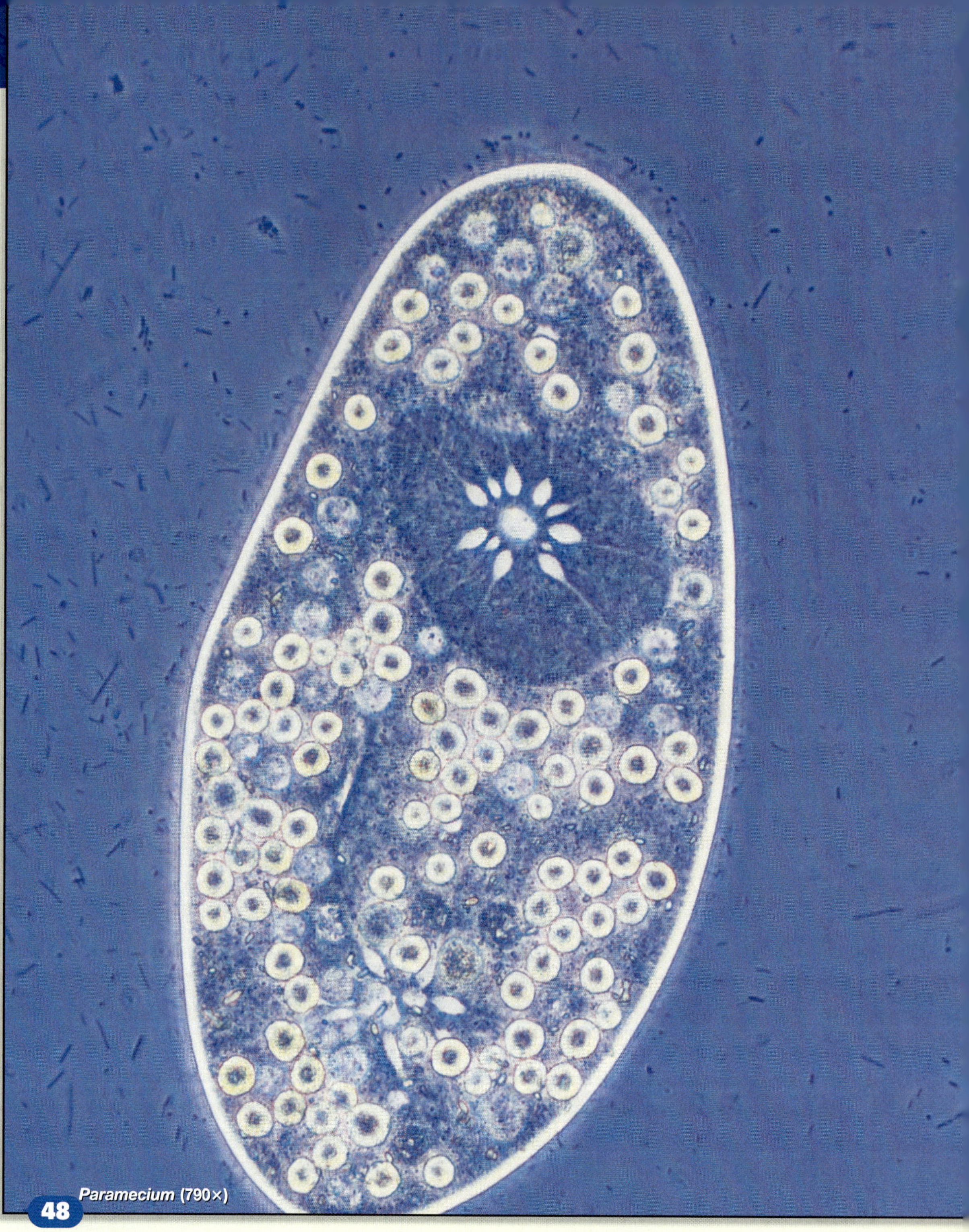

Paramecium (790×)

48

Chapter Correlations — Florida Sunshine Standards/FCAT

SC.F.1.4.2 Body structures are uniquely designed and adapted for their function.

SC.F.1.4.3 Membranes are sites for chemical synthesis and essential energy conversions.

SC.F.1.4.4 Biological systems obey the same laws of conservation as physical systems.

SC.F.2.4.2 Every cell contains a "blueprint" coded in DNA molecules that specify how proteins are assembled to regulate cells.

SC.H.1.4.3 No matter how well one theory fits observations, a new theory might fit them as well or better, or might fit a wider range of observations, because in science, the testing, revising, and occasional discarding of theories, new and old, never ends and leads to an increasingly better understanding of how things work in the world, but not to absolute truth.

SC.H.3.4.2 Technological problems often create a demand for new scientific knowledge, and new technologies make it possible for scientists to extend their research in a way that advances science.

CHAPTER

3 Cell Structure

Florida Sunshine Standards/FCAT

SC.F.1.4.2 Body structures are uniquely designed and adapted for their function.
SC.F.1.4.3 Membranes are sites for chemical synthesis and essential energy conversions.
SC.F.1.4.4 Biological systems obey the same laws of conservation as physical systems.
SC.F.2.4.2 Every cell contains a "blueprint" coded in DNA molecules that specify how proteins are assembled to regulate cells.
SC.H.1.4.3 No matter how well one theory fits observations, a new theory might fit them as well or better, or might fit a wider range of observations, because in science, the testing, revising, and occasional discarding of theories, new and old, never ends and leads to an increasingly better understanding of how things work in the world, but not to absolute truth.
SC.H.3.4.2 Technological problems often create a demand for new scientific knowledge, and new technologies make it possible for scientists to extend their research in a way that advances science.

Quick Review

Answer the following without referring to earlier sections of your book.

1. **Distinguish** between polar and nonpolar molecules. *(Chapter 2, Section 1)*
2. **Compare** the structures of carbohydrates, lipids, proteins, and nucleic acids. *(Chapter 2, Section 3)*
3. **Describe** the function of ATP. *(Chapter 2, Section 3)*

Did you have difficulty? *For help, review the sections indicated.*

Reading Activity

Write down the title of this chapter and the titles of its three sections on a piece of paper or in your notebook. Leave a few blank lines after each section title. Then write down what you think you will learn in each section. Save your list, and after you finish reading this chapter, check off everything that you learned that was on your list.

Looking Ahead

internet connect

www.scilinks.org
National Science Teachers Association *sci*LINKS Internet resources are located throughout this chapter.

SCILINKS. Maintained by the National Science Teachers Association

Most cells, including this one-celled organism *Paramecium,* have all the equipment necessary to perform the essential functions of life.

49

Opening Activity — GENERAL

Cell Structure Using micrographs, show students various cells from a multicellular organism, such as neurons, erythrocytes, and sperm cells. First have students attempt to identify the cells. After the cells have been identified, have students suggest the function of each cell and how the structure of the cell might be involved in that function.

Quick Review Answers

1. Molecules with an unequal distribution of electrical charges are polar molecules. Molecules with an equal distribution of electrical charges are nonpolar molecules.
2. Carbohydrates, lipids, proteins, and nucleic acids are organic compounds containing carbon atoms covalently bonded to hydrogen, oxygen, and other carbon atoms. However, they each differ in their proportions of these atoms, the manner in which their molecules are put together, and whether other types of atoms are present.
3. ATP carries energy in cells.

Chapter Resource File

- Vocabulary Worksheets
- Concept Mapping

Reading Activity Answers

Students may know that cells are the smallest living unit of life and are likely to know that cells contain hereditary material in the nucleus. Some may also know that cells contain organelles that carry out cell functions. However, students are not likely to be able to name these organelles and describe their functions, or distinguish between the features of prokaryotes and eukaryotes.

Focus

Overview

Before beginning this section review with your students the objectives listed in the Student Edition. This section begins by recounting the development of the microscope as a tool for visualizing and studying cells. After a short description of the metric system and the units used in measuring cells, this section describes the characteristics of microscopes and the various types that are used today.

Bellringer

Tell students to examine the SI units in **Table 1.** Then have students answer the following questions:

1. How many meters are there in a kilometer? (1,000 m)
2. How many centimeters are in a meter? (100 cm)
3. What is 20 micrometers in nanometers? (20,000 nm)

Motivate

Activity — GENERAL

Magnification Tell students that microscopes magnify objects, making them appear larger than their actual size. Pair students, and ask each pair to calculate the diameter of a Lincoln penny if it were magnified 270 times. Have students report in meters, in feet (multiply meters by 3.28 ft/m), and in comparison to another object. (A Lincoln penny is 2 cm wide. Magnified 270 times, it would appear 5.4 m wide (17.7 ft), about as wide as a large room.) LS Logical

Section Correlations

SC.H.3.4.2

Section 1 Looking at Cells

Florida Sunshine Standards/FCAT

SC.H.3.4.2 Technological problems often create a demand for new scientific knowledge, and new technologies make it possible for scientists to extend their research in a way that advances science.

- **...ribe** how scientists measure the length of objects. SC.H.3.4.2
- **Relate** magnification and resolution in the use of microscopes. SC.H.3.4.2
- **Analyze** how light microscopes function. SC.H.3.4.2
- **Compare** light microscopes with electron microscopes. SC.H.3.4.2
- **Describe** the scanning tunneling microscope. SC.H.3.4.2

Key Terms

light microscope
electron microscope
magnification
resolution
scanning tunneling microscope

Cells Under the Microscope

Most cells are too small to see with the naked eye; a typical human body cell is many times smaller than a grain of sand. Scientists became aware of cells only after microscopes were invented, in the 1600s. When the English scientist Robert Hooke used a crude microscope to observe a thin slice of cork in 1665, he saw "a lot of little boxes." The boxes reminded him of the small rooms in which monks lived, so he called them cells. Hooke later observed cells in the stems and roots of plants. Ten years later, the Dutch scientist Anton van Leeuwenhoek used a microscope to view water from a pond, and he discovered many living creatures. He named them "animalcules," or tiny animals. Today we know that they were not animals but single-celled organisms.

Measuring Cell Structures

Measurements taken by scientists are expressed in metric units. Scientists throughout the world use the metric system. The official name of the metric system is the International System of Measurements, abbreviated as SI. SI is a decimal system, so all relationships between SI units are based on powers of 10. For example, scientists measure the sizes of objects viewed under a microscope using the SI base unit for length, which is the meter. A meter, which is about 3.28 ft (a little more than a yard), equals 100 centimeters (cm), or 1,000 millimeters (mm). A meter also equals 0.001 kilometer (km). Most SI units have a prefix that indicates the relationship of that unit to a base unit. For example, the symbol "μ" stands for the metric prefix micro. A micrometer (μm) is a unit of linear measurement equal to one-millionth of a meter, or one-thousandth of a millimeter. **Table 1** summarizes the SI units used to measure length.

Table 1 Metric Units of Length and Equivalents

Unit	Prefix	Metric equivalent	Real-life equivalent
Kilometer (km)	*Kilo-*	1,000 m	About two-thirds of a mile
Meter (m)		1 m (SI base unit)	A little more than a yard
Centimeter (cm)	*Centi-*	0.01 m	About half the diameter of a Lincoln penny
Millimeter (mm)	*Milli-*	0.001 m	About the width of a pencil tip
Micrometer (μm)	*Micro-*	0.000001 m	About the length of an average bacterial cell
Nanometer (nm)	*Nano-*	0.000000001 m	About the length of a water molecule

50

Chapter Resource File

- Directed Reading BASIC
- Active Reading GENERAL

One-Stop Planner CD-ROM

- **Reading Organizers** BASIC
- **Reading Strategies** BASIC
- **Basic Skills Worksheet** Microscope Magnification BASIC
- **Problem Solving Worksheet** Operations with Small and Large Numbers GENERAL

Characteristics of Microscopes

Since Robert Hooke first observed cork cells, microscopes have unveiled the details of cell structure. These powerful instruments provide biologists with insight into how cells work—and ultimately how organisms function. Biologists use different microscopes depending on the organisms they wish to study and the questions they want to answer. Two common kinds of microscopes are light microscopes and electron microscopes. In a **light microscope,** light passes through one or more lenses to produce an enlarged image of a specimen. An **electron microscope** forms an image of a specimen using a beam of electrons rather than light.

An image produced by a microscope, such as the one shown in **Figure 1,** is called a micrograph. Many micrographs are labeled with the kind of microscope that produced the image—such as a light micrograph (LM), a transmission electron micrograph (TEM), or a scanning electron micrograph (SEM). Micrographs often are labeled with the magnification value of the image. **Magnification** is the quality of making an image appear larger than its actual size. For example, a magnification value of 200× indicates that the object in the image appears 200 times larger than the object's actual size. **Resolution** is a measure of the clarity of an image. Both high magnification and good resolution are needed to view the details of extremely small objects clearly. As shown in **Figure 2,** electron microscopes have much higher magnifying and resolving powers than light microscopes.

Magnification: 270×

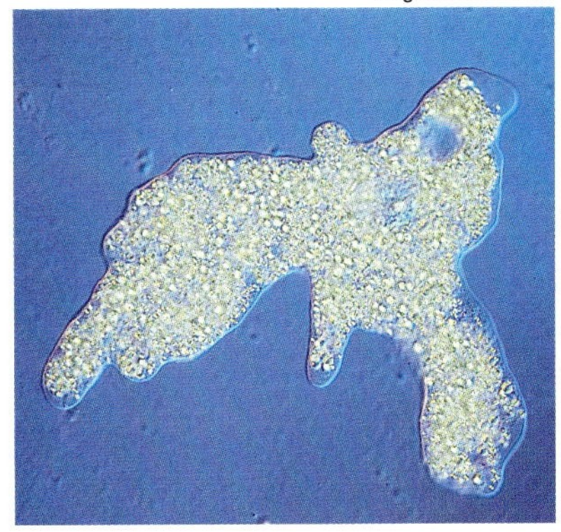

Figure 1 Micrograph. This light micrograph (LM) shows an amoeba.

Figure 2 Magnifying power of microscopes. The scale shows the size range of objects that can be viewed with electron microscopes and light microscopes.

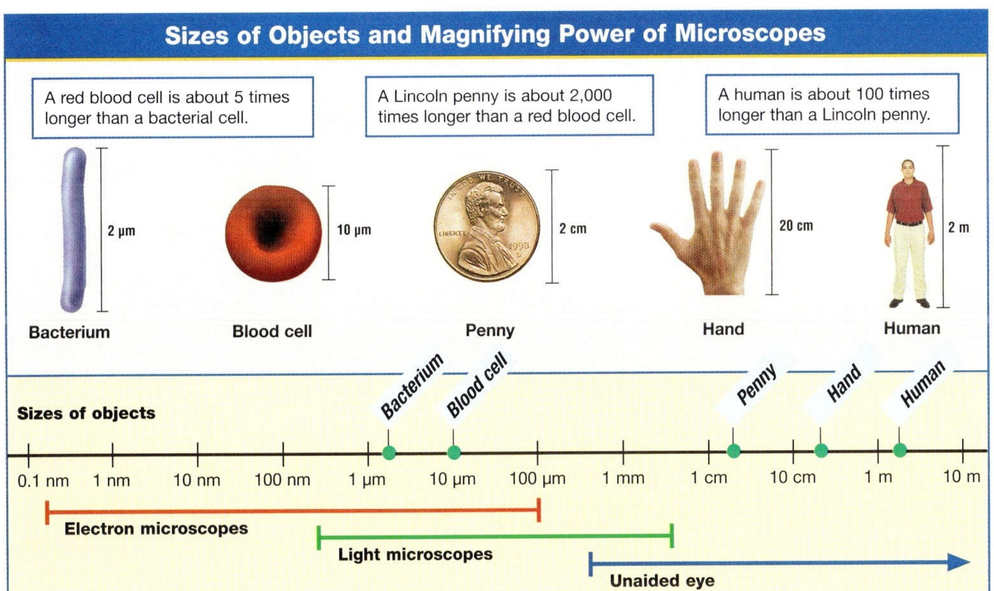

51

Teach

READING SKILL BUILDER — BASIC

Interactive Reading Assign Chapter 3 of the *Holt Biology* Guided Audio CD Program to help students achieve greater success in reading the chapter.

SKILL BUILDER — GENERAL

Math Skills Using yard/meter sticks, show students that the inch is the closest English equivalent to the centimeter, and the yard is the closest English equivalent to the meter. Ask students to measure 5 objects in centimeters and convert that measurement to meters, then to measure 5 objects in inches and convert that measurement to yards. Which system is easier to use? Why? (Students likely had less trouble using the metric system and will likely say that the metric system is easier to use since it is based on equal increments and on multipliers of 10.) You may use this opportunity to have students research and discuss why the U.S. has not totally switched to the metric system. (Tradition is often given as a reason.) **LS Kinesthetic**

Transparencies

- **TR** Bellringer
- **TR A10** Metric Units of Length and Equivalents
- **TR A4** Object Size and Magnifying Power of Microscopes
- **TR A5** Compound Light Microscope

HISTORY CONNECTION

When Hooke observed cork using a microscope, he observed tiny compartments. He gave them the Latin name *cellulae*, meaning "small rooms." That is the origin of the term cell.

did you know?

Standard Measures The International Bureau of Weights and Measures in France provides the standards of the International System of Measurements. The Bureau houses these standard measures, such as a one-kilogram piece of metal and a meter-long metal bar. The current SI system was established in 1960.

Teach, *continued*

Teaching Tip — GENERAL

Electron Microscopy Contact a local college or university and arrange for a field trip that will introduce your students to electron microscopy. Specifically request that the parts of the electron microscope be shown and explained to students and that the students also be shown how a specimen is prepared for viewing and how it is viewed. Also check to see if the institution has any other microscopes normally not available to high schools so that these too can be shown to the students.

LS Visual

Using the Figure — BASIC

The microscopes in your school may look a bit different from the one shown in **Figure 3.** Give a microscope to each pair of students. Demonstrating from the front of the class, point out the parts of your microscopes that correspond to the microscope in the figure. Have students find these parts on their microscopes. Encourage ELL students to make a diagram of the school microscopes that has all of the parts labeled. English Language Learners

internet connect

www.scilinks.org
Topic: Microscopes
Keyword: HX4122

SCILINKS. Maintained by the National Science Teachers Association

Types of Microscopes

Different types of microscopes have different qualities and uses. Microscopes vary in magnification and resolution capabilities, which affect the overall quality of the images they produce. Microscopes also have different limitations. For example, electron microscopes have high magnifying power, but they cannot be used to view living cells. Light microscopes have lower magnifying power, but they can be used to view living cells.

Compound Light Microscope

Light microscopes that use two lenses are called compound light microscopes. In a typical compound light microscope, such as the one shown in **Figure 3,** a light bulb in the base shines light up through the specimen, which is mounted on a glass slide. The objective lens, closest to the specimen, collects the light, which then travels to the ocular *(AHK yoo luhr)* lens, closest to the viewer's eye. Both lenses magnify the image. Thus, a microscope with a 40× objective lens and a 10× ocular lens produces a total magnification of 400×.

Why not add a third lens and magnify even more? This approach does not work because you cannot distinguish between two objects, or "resolve" them, when they are closer together than a few hundred nm. When the objects are this close, the light beams from the two objects start to overlap!

Figure 3 Compound light microscope

In a compound light microscope, a specimen is mounted on a glass slide and is illuminated with a beam of light from below.

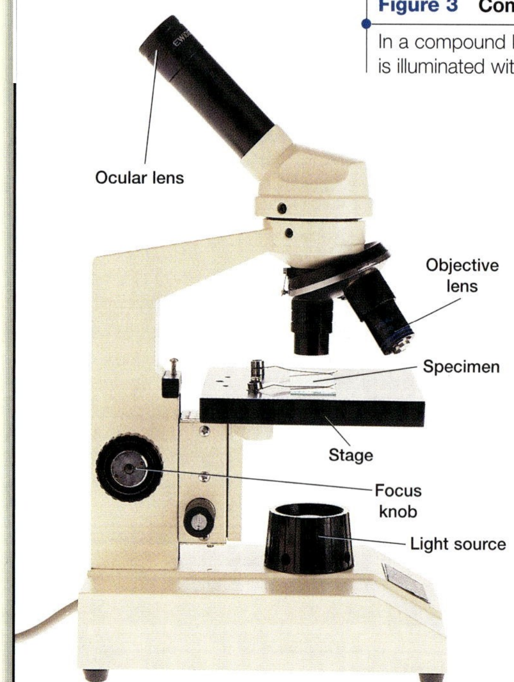

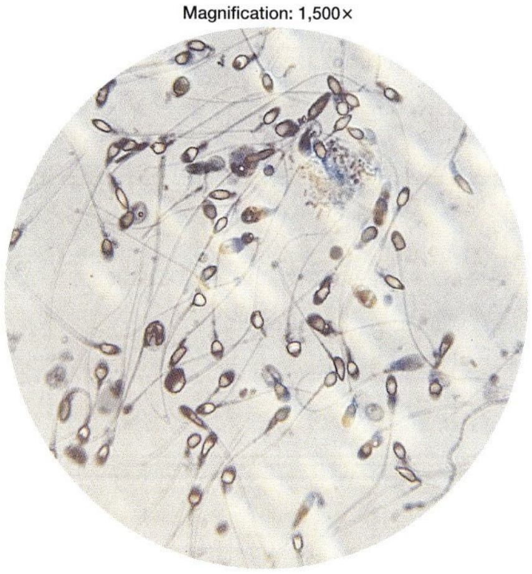

LM of sperm

52

INCLUSION Strategies

• *Gifted and Talented*

Have the student use reference materials (or the Internet) to compare and contrast the light microscope and the electron microscope. The student's work should include diagrams of each microscope with parts labeled, the types of materials that can be studied, and different applications for each type of microscope. Additionally, the student could interview a professional that uses an electron microscope on the job.

Integrating Physics and Chemistry

Remind students that light waves from the specimen are bent as they pass through the ocular and objective lenses of the light microscope. Refraction—the bending of light waves as they pass from one medium to another—results in a "magnified" image of the specimen. To assess the students' understanding of refraction, ask them to identify the media through which the light waves are passing when they use a light microscope. Possible answers include water, air, and glass.

The most powerful compound light microscopes have a total magnification of up to 2,000×, which is sufficient for viewing objects as small as 0.5 µm in diameter. For you to see smaller objects, the wavelength of the light beam must be shorter than the wavelength of visible light. Electron beams have a much shorter wavelength than that of visible light, so electron microscopes are much more powerful than light microscopes.

Electron Microscopes

Electron microscopes can magnify an image up to 200,000×, and they can be used to study very small structures inside cells or on cell surfaces. In electron microscopes, both the electron beam and the specimen must be placed in a vacuum chamber so that the electrons in the beam will not bounce off gas molecules in the air. Because living cells cannot survive in a vacuum, they cannot be viewed using electron microscopes.

Transmission electron microscope In a transmission electron microscope, shown in **Figure 4,** the electron beam is directed at a very thin slice of a specimen stained with metal ions. Some structures in the specimen become more heavily stained than others. The heavily stained parts of the specimen absorb electrons, while those that are lightly stained allow electrons to pass through. The electrons that pass through the specimen strike a fluorescent screen, forming an image on the screen. A transmission electron micrograph (TEM), such as the one of sperm cells shown in Figure 4, can reveal a cell's internal structure in fine detail. TEM images are always in black and white. However, with the help of computers, scientists often add artificial colors to make certain structures more visible.

Real Life

Thirty movies could be stored on a disk the size of a penny.

Using the scanning interferometric apertureless microscope (SIAM) researchers have viewed features that are about four atoms (1 nm) in diameter. The technology could also be used to code information on storage disks.

Applying Information

Would the SIAM likely be more useful in understanding the overall structure of the cell or the structure of biological compounds in the cell?

Figure 4 **Transmission electron microscope**

In a transmission electron microscope, electrons pass through a specimen, forming an image of the specimen on a fluorescent screen.

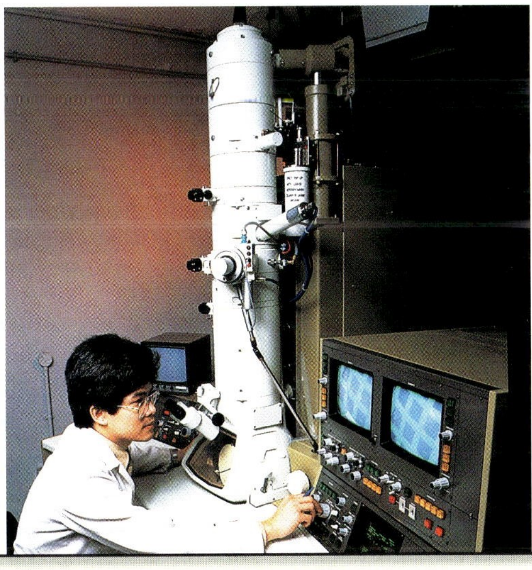

Magnification: 7,730×

TEM of sperm

Real Life

Answer

The SIAM makes molecules visible. Therefore it is useful in studying cells on the molecular level, such as the structure of certain biological compounds like proteins. Other types of microscopes, such as the electron microscope, visualize cells and their organelles. These microscopes are useful in understanding the overall structure of the cell.

Activity — GENERAL

Making a Wet Mount Have students practice preparing wet mounts. CAUTION: Slides and cover slips break easily and have sharp edges. Provide students with glass slides, cover slips, small beakers filled with water, and eyedroppers. Students can begin by viewing cutout letters from newspapers. Challenge students to make a bubble-free slide. Then have students prepare wet mounts using drops of pond water. Emphasize that it will be much easier to find and watch organisms if students make their slides bubble-free.

LS Kinesthetic

SKILL BUILDER — ADVANCED

Writing Skills Electron microscopes are used in areas of research other than biology. For example, they are important tools in Earth science. In this field, electron microscopes have been used to determine the effects of weathering on the microstructural arrangement of the mineral components of rocks, and the effects of chemicals on rock formation and stability. Have students prepare a report that describes uses of electron microscopes in fields other than biology.

LS Verbal

Cultural Awareness

German-born Ernst Ruska and Rheinhold Ruedenberg, working indepedently of each other, invented the transmission electron microscope 1931. In 1933, Ruska built the first electron microscope that was more powerful than the light microscope and, in 1986, was awarded the Nobel Prize in Physics for this achievement.

HISTORY CONNECTION

Hans Janssen, Zacharias Janssen, and Hans Lippershey built the first microscope in the Netherlands between 1590 and 1608. It could magnify objects from 3X to 10X. Anton van Leeuwenhoek made significant improvements to the microscope, developing lenses with magnifications of up to 300X. Van Leeuwenhoek produced about 400 microscopes and used them to study many things, including yeast, muscles, plants, and insects.

Close

Reteaching — BASIC

Ask students which kind of microscope would be most useful to high school students studying the general structure of a leaf. (The light microscope shows good detail for general observations.) Ask students which kind of microscope would be most useful for studying three-dimensional details of a small organism, such as a unicellular protist. (The SEM shows the surfaces of objects in great detail.) Ask them which microscope would be most useful in research on the internal structures of a bacterial cell. (The TEM shows internal detail with great magnification.)

Quiz — GENERAL

1. How many centimeters are in a meter? How do these metric system terms tell you the answer? (There are 100 centimeters in a meter. The prefix centi- means 100, so a centimeter is 1/100th of a meter.)
2. What is the difference between magnification and resolution? (Magnification means to make an image larger, while resolution refers to the sharpness of the image.)

Alternative Assessment — GENERAL

Have students prepare wet mounts of amoebas. Using compound light microscopes, have students bring the amoebas into focus. Using thin centimeter rulers placed on the microscope stages within their fields of view, have students estimate the sizes of the amoebas in SI units and explain how they made their determinations. **LS Kinesthetic**

Section Review Correlations

SC.H.3.4.2: Items 1-6

Figure 5 Scanning electron microscope

In a scanning electron microscope, electrons bounce off a specimen, forming a three-dimensional image of the specimen on a fluorescent screen.

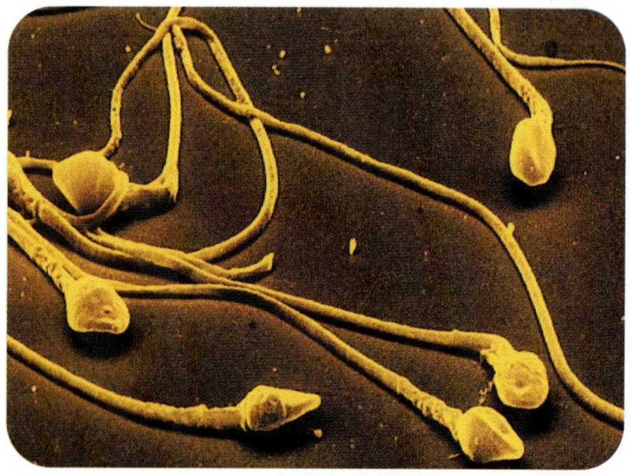

SEM of sperm

Scanning electron microscope In a scanning electron microscope, shown in **Figure 5,** the electron beam is focused on a specimen coated with a very thin layer of metal. The electrons that bounce off the specimen form an image on a fluorescent screen. A scanning electron micrograph (SEM) shows three-dimensional images of cell surfaces, such as the image of sperm cells shown in Figure 5. As with the transmission electron microscope, images produced by the scanning electron microscope are black and white, but often they are artificially colored.

Scanning Tunneling Microscope

New video and computer techniques are increasing the resolution and magnification of microscopes. The **scanning tunneling microscope** uses a needle-like probe to measure differences in voltage caused by electrons that leak, or tunnel, from the surface of the object being viewed. A computer tracks the movement of the probe across the object, enabling objects as small as individual atoms to be viewed. The computer generates a three-dimensional image of the specimen's surface. The scanning tunneling microscope can be used to study living organisms.

Section 1 Review

1. **Describe** the relationship between a meter, a millimeter, and a micrometer. SC.H.3.4.2
2. **Describe** how magnification and resolution affect the appearance of objects viewed under a microscope. SC.H.3.4.2
3. **Compare** the magnifying power of a light microscope with the magnifying power of an electron microscope. SC.H.3.4.2
4. **Critical Thinking Recognizing Differences** Explain why electron microscopes cannot be used to view the structure of living cells. SC.H.3.4.2
5. **Critical Thinking Comparing Functions** Assume that for the purposes of your investigation, you need detailed images of the internal structure of a bacterium. What type of microscope would you select for that that task? Explain your answer. SC.H.3.4.2
6. *FCAT Prep* What did the English scientist Robert Hooke examine with a crude microscope?
   A electrons
   B cork cells
   C individual atoms
   D single-celled organisms
   SC.H.3.4.2

54

Answers to Section Review

1. A millimeter is equal to 0.001 m, or one-thousandth of a meter. A micrometer is equal to 0.000001 m, or one-millionth of a meter.
2. As the magnification increases, the resolution decreases; therefore the clarity of the image is reduced.
3. Light microscopes can magnify an object up to 2,000X. Electron microscopes can magnify an object up to 200,000X.
4. Specimens must be placed in a vacuum as part of the preparation for viewing with electron microscopes; living things cannot survive in a vacuum.
5. You would use a transmission electron microscope because the electron beam passes through the bacterium, revealing the cell's internal structure in great detail.
6. **A.** Incorrect. A crude microscope cannot visualize electrons. **B.** Correct. Hooke studied cork and other plant cells with his crude microscope. **C.** Incorrect. A crude microscope cannot visualize individual atoms. **D.** Incorrect. Leeuwenhoek used a microscope to view single-celled organisms.

Cell Features

Section 2

Florida Sunshine Standards/FCAT

SC.F.1.4.4 Biological systems obey the same laws of conservation as physical systems.

SC.H.1.4.3 No matter how well one theory fits observations, a new theory might fit them as well or better, or might fit a wider range of observations, because in science, the testing, revising, and occasional discarding of theories, new and old, never ends and leads to an increasingly better understanding of how things work in the world, but not to absolute truth.

The Cell Theory

It took scientists more than 150 years to fully appreciate the discoveries of Hooke and Leeuwenhoek. In 1838, the German botanist Mattias Schleiden concluded that cells make up not only the stems and roots but every part of a plant. A year later, the German zoologist Theodor Schwann claimed that animals are also made of cells. In 1858, Rudolph Virchow, a German physician, determined that cells come only from other cells. The observations of Schleiden, Schwann, and Virchow form the **cell theory,** which has three parts:

1. All living things are made of one or more cells.
2. Cells are the basic units of structure and function in organisms.
3. All cells arise from existing cells.

Cell Size

Small cells function more efficiently than large cells. There are about 100 trillion cells in the human body, most ranging from 5 µm to 20 µm in diameter. What is the advantage of having so many tiny cells instead of fewer large ones? All substances that enter or leave a cell must cross that cell's surface. If the cell's surface area–to-volume ratio is too low, substances cannot enter and leave the cell in numbers large enough to meet the cell's needs. Small cells can exchange substances more readily than large cells because small objects have a higher surface area–to-volume ratio than larger objects, as shown in **Table 2.** As a result, substances do not need to travel as far to reach the center of a smaller cell.

Objectives

- **List** the three parts of the cell theory. SC.H.1.4.3
- **Determine** why cells must be relatively small. SC.F.1.4.4
- **Compare** the structure of prokaryotic cells with that of eukaryotic cells. SC.F.1.4.2
- **Describe** the structure of cell membranes. SC.F.1.4.3

Key Terms

cell theory
cell membrane
cytoplasm
cytoskeleton
ribosome
prokaryote
cell wall
flagellum
eukaryote
nucleus
organelle
cilium
phospholipid
lipid bilayer

Table 2 Relationship Between Surface Area and Volume

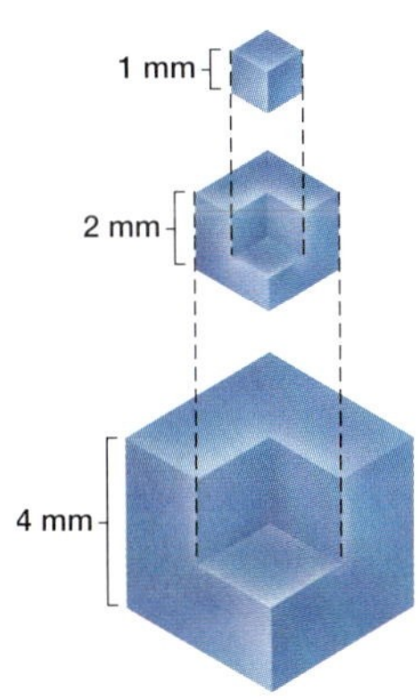

Side length	Surface area	Volume	Surface area/ volume ratio
1 mm	6 mm^2	1 mm^3	6:1
2 mm	24 mm^2	8 mm^3	3:1
4 mm	96 mm^2	64 mm^3	3:2

Section 2

Focus

Overview

Before beginning this section review with your students the objectives listed in the Student Edition. This section introduces students to the cell theory. It continues by discussing the similarities and differences between prokaryotic cells and eukaryotic cells, and closes by describing the cytoskeleton and cell membrane.

Bellringer

Have the following list written on the chalkboard when students enter the room: rocks, grass, soil, water, telephone poles, and grasshoppers. Have students work with a partner to make two new lists: those items on the list that are comprised of cells and those items not comprised of cells. Have them write a rationale for each answer. (Grass, grasshoppers, and telephone poles are comprised of cells. They are either living or were once living. Rocks, soil, and water are not comprised of cells. They are nonliving.) **LS Logical**

Motivate

Activity — GENERAL

Bacteria in Yogurt Smear some yogurt with live cultures on a microscope slide. Mix a drop of water into the smeared yogurt and add a cover slip. Allow students to examine the slide under high power using a compound light microscope. They should see bacteria of the genus *Lactobacillus*. Ask them to compare the size of these cells with the sizes of cells on prepared slides of human and plant tissues. **LS Kinesthetic**

Chapter Resource File

- Directed Reading BASIC
- Active Reading GENERAL
- Data Sheet for Math Lab GENERAL

Transparencies

- TR Bellringer
- TR B2 Relationship Between Surface Area and Volume
- TR B4 Animal Cells
- TR B8 Lipid Bilayers
- TR B10 Membrane Proteins

One-Stop Planner CD-ROM

- Reading Organizers BASIC
- Reading Strategies BASIC
- Basic Skills Worksheets
  Length, Area, and Volume Measurement and Scale Drawings GENERAL
- Problem Solving Worksheet
  Ratios and Proportions GENERAL

Section Correlations

SC.F.1.4.2
SC.F.1.4.3
SC.F.1.4.4
SC.H.1.4.3

Teach

Demonstration — GENERAL

Cut three cubes of varying sizes from a potato. Pour food coloring into three beakers, and place a cube in each beaker. After about 5 min., remove each cube and cut it in half. Have students observe the results. Ask them to determine how the surface area-to-volume ratio of each cube affected the penetration of the food coloring into the cube. (The food coloring traveled the same distance in each cube. However, the smallest cube will be more completely colored than others because it has the greatest surface area-to-volume ratio.) **LS Visual**

MATH LAB: *Calculating Surface Area and Volume*

Skills Acquired
Calculating, analyzing, applying

Teacher's Notes
Bring in cube-shaped game dice of three different sizes. Have students measure the sides of the dice. Then have them calculate the surface area-to-volume ratios of the dice and compare their results.

Answers to Analysis

1. SA = 24 mm^2, V = 8 mm^3, ratio is 3:1
2. SA = 6 mm^2, V = 1 mm^3, ratio is 6:1
3. By being flat, a *Paramecium* spreads its volume over a large area. The surface area-to-volume ratio is increased because there is more surface area per unit volume.

internet connect
www.scilinks.org
Topic: Cell Features
Keyword: HX4034
SciLINKS. Maintained by the National Science Teachers Association

Common Features of Cells

Cells share common structural features, including an outer boundary called the **cell membrane.** The cell membrane encloses the cell and separates the cell interior, called the **cytoplasm** *(SITE oh plaz uhm),* from its surroundings. The cell membrane also regulates what enters and leaves a cell—including gases, nutrients, and wastes. Within the cytoplasm are many structures, often suspended in a system of microscopic fibers called the **cytoskeleton.** Most cells have ribosomes. **Ribosomes** *(RIE buh sohmz)* are the cellular structures on which proteins are made. All cells also have DNA, which provides instructions for making proteins, regulates cellular activities, and enables cells to reproduce. Some specialized cells such as red blood cells, however, later lose their DNA.

MATH LAB: *Calculating Surface Area and Volume* SC.F.1.4.4

Background

You can improve your understanding of the relationship between a cell's surface area and its volume by practicing with the large cube in Table 2.

Magnification: 230×

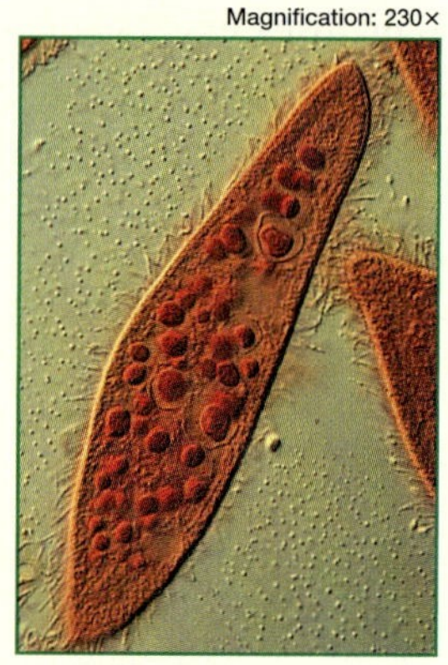

Paramecium (SEM)

1. Find the total surface area of the cube.

- *side length* (l) = 4 mm
- *surface area of one side* = $l \times l = l^2$
- *surface area of one side* (l^2) = 4 mm × 4 mm = 16 mm^2
- *total surface area* = $6 \times l^2$ = 6 × 16 mm^2 = 96 mm^2

2. Calculate the volume of the cube.

- *height* (h) = l = 4 mm
- *volume* = $l^2 \times h$ = 16 mm^2 × 4 mm = 64 mm^3

3. Determine the surface area-to-volume ratio. A ratio compares two numbers by dividing one number by the other. A ratio can be expressed in three ways:

in words	as a fraction	with a colon
x to y	$\frac{x}{y}$	$x{:}y$

For the surface area-to-volume ratio, divide total surface area by volume.

$$\frac{\textit{total surface area}}{\textit{volume}} = \frac{96}{64}$$

Divide both numbers by their greatest common factor:

$$\frac{(96 \div 32)}{(64 \div 32)} = \frac{3}{2}$$

Analysis

1. **Calculate** the surface area–to-volume ratio of the cube with a side length of 2 mm in Table 2.
2. **Calculate** the surface area–to-volume ratio of the cube with a side length of 1 mm in Table 2.
3. **Critical Thinking** **Relating Concepts** How does the flatness of the single-celled *Paramecium* shown above affect the cell's surface area–to-volume ratio?

Integrating Physics and Chemistry

Besides requiring a relatively high surface area-to-volume ratio in order to live, cells also have a total cellular density that permits them to function efficiently in their surrounding environments. Given that cells must have water to survive and are primarily composed of water, ask students if they would expect them to have a density less than, about equal to, or more than that of water. Ask students to defend their responses. If a cell's density is different from that of its nearby habitat, what might account for this density difference?

MISCONCEPTION ALERT

Shape vs. Size Large cells do not necessarily have small surface area-to-volume ratios. Many large cells have shapes that allow them to maintain a large surface area-to-volume ratio. A cell could grow large in one or two dimensions but remain small in others. For example, parts of neurons are very long cylinders with small diameters, whereas epithelial cells are broad and flat. In both cases, the surface area-to-volume ratio is quite large.

Prokaryotes

The smallest and simplest cells are prokaryotes. A **prokaryote** *(proh KAIR ee oht)* is a single-celled organism that lacks a nucleus and other internal compartments. Without separate compartments to isolate materials, prokaryotic cells cannot carry out many specialized functions. Early prokaryotes lived at least 3.5 billion years ago. For nearly 2 billion years, prokaryotes were the only organisms on Earth. They were very simple and small (1–2 µm in diameter). Like their ancestors, modern prokaryotes are also very small (1–15 µm). The familiar prokaryotes that cause infection and cause food to spoil belong to a subset of all prokaryotes that is commonly called *bacteria.*

Study TIP

Reading Effectively
For many words ending in *-um,* the plural is formed by changing the *-um* to *-a.* For example, the plural of *bacterium* is *bacteria,* and the plural of *flagellum* is *flagella.*

Characteristics of Prokaryotes

Prokaryotes can exist in a broad range of environmental conditions. Many prokaryotes, including some bacteria that cause infection in humans, grow and divide very rapidly. Some prokaryotes do not need oxygen to survive. Other prokaryotes cannot survive in the presence of oxygen. Some prokaryotes can even make their own food.

The cytoplasm of a prokaryotic cell includes everything inside the cell membrane. As **Figure 6** shows, a prokaryote's enzymes and ribosomes are free to move around in the cytoplasm because there are no internal structures that divide the cell into compartments. In prokaryotes, the genetic material is a single, circular molecule of DNA. This loop of prokaryotic DNA is often located near the center of the cell, suspended within the cytoplasm.

Prokaryotic cells have a **cell wall** surrounding the cell membrane that provides structure and support. The cells of fungi and plants also have cell walls; only animal cells and some protists lack cell walls. Prokaryotes lack an internal supporting skeleton, so they depend on a strong cell wall to give the cell shape. A prokaryotic cell wall is made of strands of polysaccharides connected by short chains of amino acids. Some prokaryotic cell walls are surrounded by a structure called a capsule, which is also composed of polysaccharides. The capsule enables prokaryotes to cling to almost anything, including teeth, skin, and food.

Many prokaryotes have **flagella** *(fluh JEL uh),* which are long, threadlike structures that protrude from the cell's surface and enable movement. Prokaryotic flagella rotate, propelling the organism through its environment at speeds of up to 20 cell lengths per second. Figure 6 shows a prokaryote with several flagella.

Figure 6 Prokaryotes. Prokaryotic cells have little internal structure. Many also have a capsule and flagella.

Magnification: 61,850×

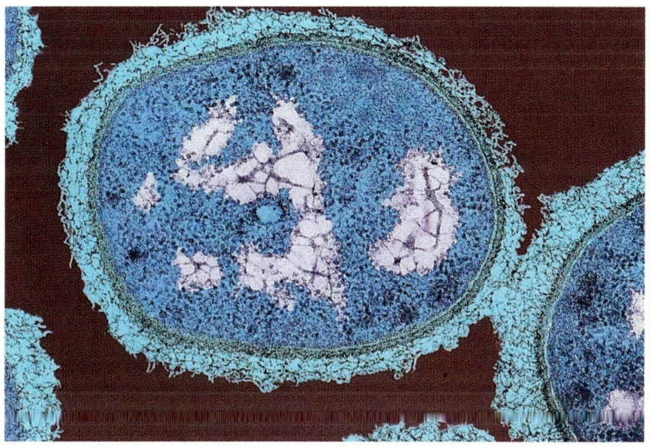

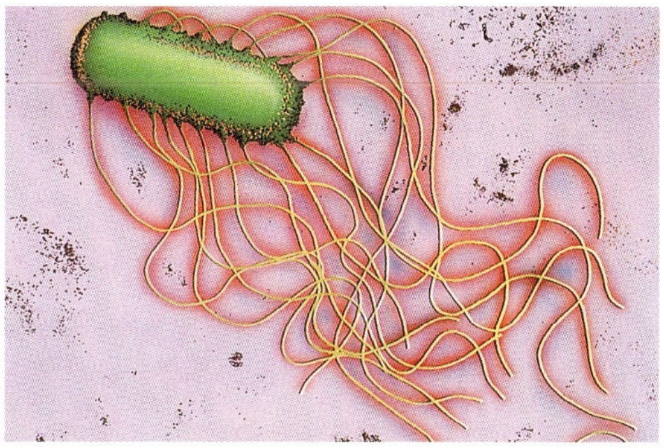

Teaching Tip

Cell Walls Tell students that the cell wall varies among different types of organisms. In plants, the cell wall is composed of cellulose fibers embedded in a polysaccharide and protein matrix. In eubacteria, the cell wall is composed of peptidoglycan, a polymer of sugars cross-linked by a short polypeptide. In fungi, the cell wall is composed of chitin, a polysaccharide.

SKILL BUILDER — GENERAL

Interpreting Visuals Draw students' attention to **Figure 6,** which shows colorized electron micrographs of bacterial cells. Electron micrographs are always black and white. Tell students that any color shown in electron micrographs is usually digitally applied; that is, the colors are "painted on" by computers. These colors are artificial; the living cell is not colored as is shown in the micrograph. Ask students why colorized electron micrographs might be useful. (To highlight structures and "point them out.") Have students determine which structures are being pointed out in **Figure 6.** (Upper micrograph from outside to inside: capsule, cell wall, cell membrane, cytoplasm with ribosomes, DNA. Lower micrograph: the flagella are yellow with a red "halo" and the "body" of the bacterium is green.)

did you know?

Prokaryotes Modern prokaryotes include archaebacteria and eubacteria. They can be found in nearly any environment on Earth, including volcanic vents at the bottom of the ocean and ice in Arctic and Antarctic regions.

Trends in Microbiology

Bacteria Classification This textbook uses the six-kingdom system of classification. However, the research of Carl Woese, Professor of Microbiology at University of Illinois Urbana-Champaign, and his colleagues, suggests that archaebacteria and bacteria are uniquely distinct. In 1980, Woese suggested a three-domain classification. Bacteria and Archaea consist of prokaryotes, while Eukarya consists of all eukaryotes. The numerous kingdoms within each domain correspond to phyla in the six-kingdom system. Today, the three-domain system of classification is accepted and used widely in the field of microbiology.

Teach, *continued*

Using the Figure

Draw students' attention to **Figure 7.** This figure shows most of the organelles that are commonly found in animal cells. Explain that for the sake of clarity only a few of the microtubules and microfilaments have been drawn. In reality, these structures form a dense network (the cytoskeleton) inside the cell. Point out that the nuclear envelope actually consists of two lipid bilayers. **LS Visual**

READING SKILL BUILDER — BASIC

Reading Organizer As students read about prokaryotes and eukaryotes, have them design a chart that identifies and describes all of the differences between prokaryotic and eukaryotic cells. Encourage ELL students to exchange their organizer with native English speakers to check for accuracy. **LS Logical** English Language Learners

Teaching Tip — ADVANCED

Eukaryotic Cells Because all multicellular plants and animals are composed of eukaryotic cells, stress that the eukaryotic cell can be an entity in itself, not just a component of a larger organism. Have students research one-celled eukaryotic organisms, like a yeast or a one-celled protist, and compare them with eukaryotic cells that are part of a multicellular plant or animal. Students should include drawings, and record their findings on a one-page fact sheet. English Language Learners

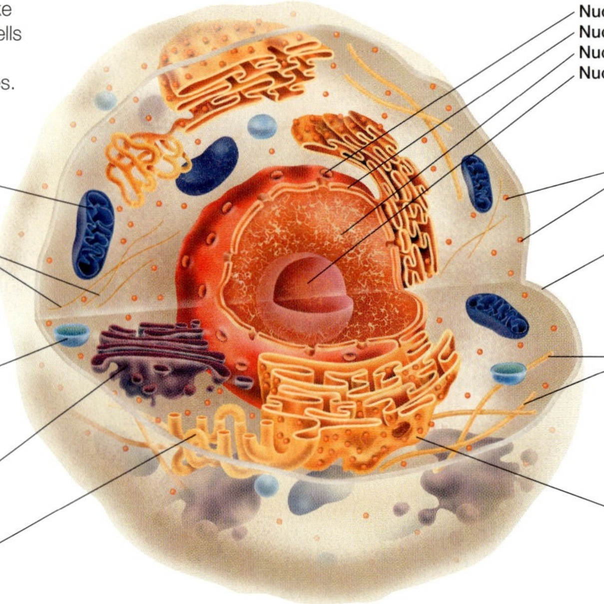

Figure 7 Animal cell. Like all eukaryotic cells, animal cells contain a cell membrane, a nucleus, and other organelles.

Eukaryotic Cells

The first cells with internal compartments were primitive eukaryotic cells, which evolved about 2.5 billion years ago. A **eukaryote** *(yoo KAIR ee oht)* is an organism whose cells have a nucleus. The **nucleus** *(NOO klee uhs)* is an internal compartment that houses the cell's DNA. Other internal compartments, or organelles, enable eukaryotic cells to function in ways different from prokaryotes. An **organelle** is a structure that carries out specific activities in the cell.

The major organelles in an animal cell are shown in **Figure 7.** The cytoplasm includes everything inside the cell membrane but outside the nucleus. A complex system of internal membranes connects some organelles within the cytoplasm. These membranes provide channels that guide the distribution of substances within the cell. The membranes also form envelopes called *vesicles* that move proteins and other molecules from one organelle to another.

Many single-celled eukaryotes use flagella for movement. Short hairlike structures called **cilia** *(SIL ee uh)* protrude from the surface of some eukaryotic cells. Flagella or cilia propel some cells through their environment. In other cells, cilia and flagella move substances across the cell's surface. For example, cilia on cells of the human respiratory system, shown in **Figure 8,** sweep mucus and other debris out of the lungs.

A web of protein fibers, shown in **Figure 9,** makes up the cytoskeleton. The cytoskeleton holds the cell together and keeps the cell's membranes from collapsing. The fluid surrounding the cytoplasm's organelles, internal membranes, and cytoskeleton fibers is called the cytosol.

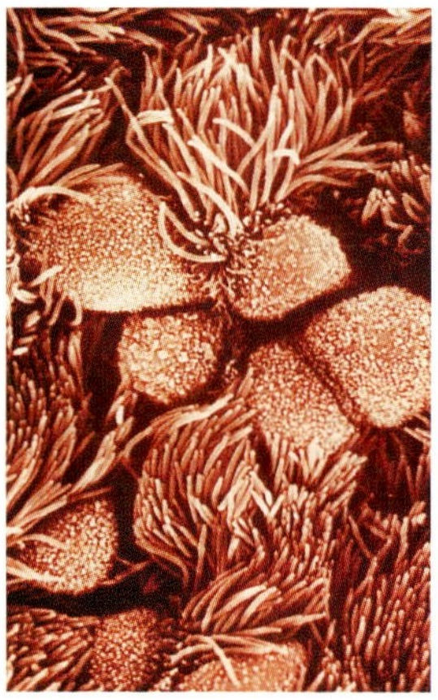

Figure 8 Cilia. Cilia on cells lining the respiratory system remove debris from air passages.

58

Attention Grabber

The Largest Cell Students may be surprised to learn that the largest eukaryotic cell in the world is about the size of a baseball! This cell is the yolk of an ostrich egg.

Career

Cytotechnologist Cytotechnologists stain, mount, and study cells to detect malignancies or other abnormal conditions. They then report all findings to a pathologist. A cytotechnologist must be able to obtain precise results and work under stress during emergency situations. An undergraduate degree in cytotechnology or medical technology is usually needed for entry-level positions. Invite a cytotechnologist from a local hospital to talk to your class. Ask them to discuss recommended high school courses, working conditions, employment opportunities, and salary ranges during their talk.

The Cytoskeleton

The cytoskeleton provides the interior framework of an animal cell, much as your skeleton provides the interior framework of your body. The cytoskeleton is composed of an intricate network of protein fibers anchored to the inside of the plasma membrane. By linking one region to another, they support the shape of the cell, much as steel beams anchor the sides of a building to one another. Other fibers attach the nucleus and other organelles to fixed locations in the cell. Because protein fibers are too small for a light microscope to reveal, biologists visualize the cytoskeleton by attaching fluorescent dyes to antibodies. An antibody is an immune system protein specialized to bind to one particular kind of molecule—in this case—cytoskeleton proteins. When the cell is examined under fluorescent light, the fibers glow because of the fluorescent antibody attached to them.

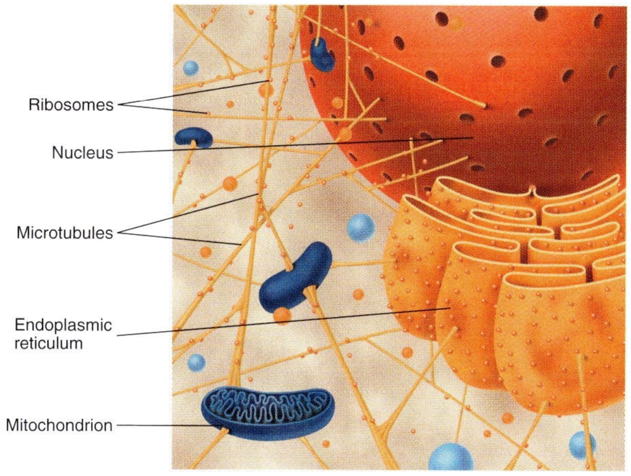

Figure 9 The cytoskeleton. The cytoskeleton's network of protein fibers anchors cells organelles and other components of the cytoplasm.

There are three different kinds of cytoskeleton fibers: (1) long, slender microfilaments made of the protein actin, (2) hollow tubes called *microtubules* made of the protein tubulin, and (3) thick ropes of protein called *intermediate fibers*.

Actin Fibers The actin fibers of the cytoskeleton form a network just beneath the cell surface that is anchored to membrane proteins embedded within the cell membrane. By contracting or expanding, the actin fibers play a major role in determining the shape of animal cells by pulling the plasma membrane in some places and pushing it out in others. If you examine the surface of a protist such as the one shown in **Figure 10,** you will find it alive with motion. Tiny projections extend out from the surface like fingers. Each is a temporary projection of the plasma membrane that shoots out and then retracts.

Microtubules Microtubules within the cytoskeleton act as a highway system for the transportation of information from the nucleus to different parts of the cell. RNA molecules are transported along microtubular "rails" that extend through the interior of the cell like train tracks. The RNA molecules, in complexes with proteins, are attached to so-called motor proteins that chug along microtubules like locomotives on tracks. The motor proteins drag the RNA-protein complexes along with them like freight cars.

Intermediate Fibers The intermediate fibers of the cytoskeleton provide a frame on which ribosomes and enzymes can be confined to particular regions of the cell. The cell can organize complex metabolic activities efficiently by anchoring particular enzymes near one another.

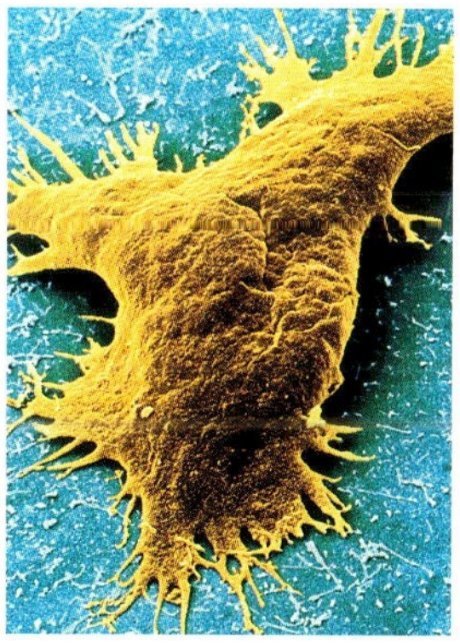

Figure 10 Cytoskeletal projections. The multiple spikes on the surface of this marine amoeba are projections of the cytoskeleton stretching the cell membrane outward.

Teaching Tip

Prokaryotic Cells vs. Eukaryotic Cells Have students make a Graphic Organizer similar to the one on the bottom of this page that compares prokaryotic cells with eukaryotic cells.

Activity — GENERAL

Amoeboid Movement Have students make wet mounts of live amoeba. These protists move and feed by means of pseudopods, or "false feet." Pseudopods are cell extensions formed by actin fibers continuously contracting and expanding, pulling the cell membrane in one place and pushing it out in another, as described on this page. In fact, amoeboid movement is a dramatic example of actin-based cell movement. Although students will not be able to see the actin fibers, they will be able to observe the granular cytoplasm streaming into the projection as it begins to form. Explain to students that their own white blood cells exhibit such movement to catch and ingest foreign invaders such as bacteria. **LS Visual**

READING SKILL BUILDER — BASIC

Paired Reading Assign students to cooperative pairs, and have each student read this page silently. As they read, have students indicate passages they understand with a check mark and passages they do not understand with a question mark. Then have the partners discuss what they did or did not understand. Ask each pair to write a short paragraph summarizing the structure and function of the cytoskeleton. You may wish to pair ELL students with native English speakers. **LS Interpersonal** English Language Learners

Graphic Organizer

Use this graphic organizer with **Teaching Tip** *on this page.*

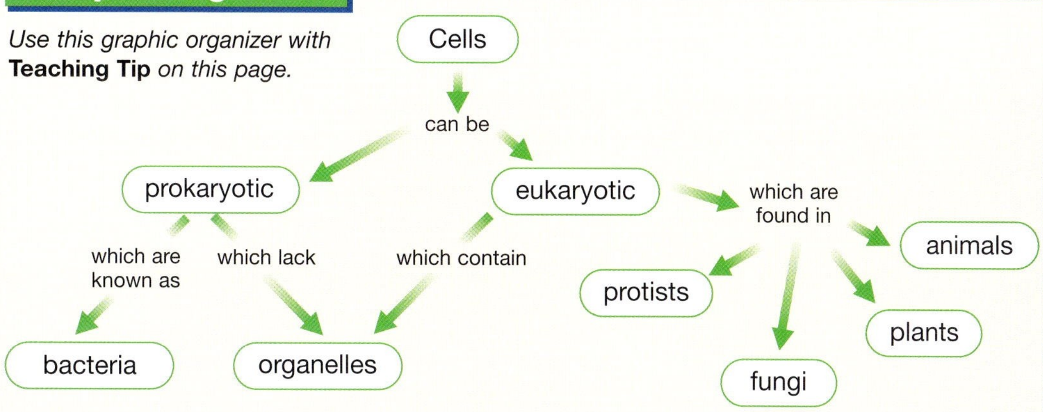

Teach, *continued*

Real Life

Answer

The cells are frozen very quickly, which prevents the formation of large ice crystals that would damage the cells.

Group Activity — GENERAL

Cell Membrane Model Organize the class into groups. Tell each group to make a model of the cell membrane to help visualize its structure. Suggest that students cut off half of one leg of several wooden clothespins. Different colors can be used to represent the hydrophobic (nonpolar) and hydrophilic (polar) ends. Proteins can be represented by buttons, bottle caps, and cardboard tubes. Place completed models around the classroom.
LS Kinesthetic

Using the Figure — GENERAL

Draw students' attention to **Figure 11,** and then have students review the terms *polar* and *nonpolar*. Be certain students understand that the nonpolar interior of the lipid bilayer repels ions and polar substances. Remind them that even though water molecules are polar, they are small enough to move through the lipid bilayer. Make sure students understand that membranes are found in many places in cells, such as around the entire cell, mitochondria, the Golgi apparatus, lysosomes, and the nucleus. Ask students to list cell structures that lack membranes. (ribosomes, microfilaments, microtubules, and cell walls) **LS Visual**

Real Life

Donated blood is frozen in a special process called cryopreservation. Similar methods are used to preserve human eggs, embryos, and blood from the umbilical cord, a rich source of immune-system cells.

Finding Information
Research how cryopreservation methods enable cells to withstand freezing.

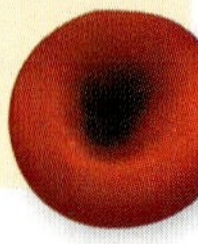

The Cell Membrane

The cytoplasm of a cell is contained by its membrane. Cell membranes are not rigid like an eggshell. Rather, they are fluid like a soap bubble. The fluidity of cell membranes is caused by lipids, which form the foundation of membranes. The lipids form a barrier that separates the inside of the cell from the outside of the cell. This barrier allows only certain substances in the cell's environment to pass through. This selective permeability of the cell membrane determines which substances enter and leave the cell.

The Cell Membrane as a Barrier

The selective permeability of the cell membrane is caused mainly by the way phospholipids interact with water. A **phospholipid** is a lipid made of a phosphate group and two fatty acids. As shown in **Figure 11,** a phospholipid has both a polar "head" and two nonpolar "tails." You may recall that the polar ends of water molecules will form weak bonds with other polar substances. The head of a phospholipid, which contains a phosphate group, is polar and is attracted to water. In contrast, the two fatty acids, or tails, are nonpolar and therefore are repelled by water.

In a cell membrane, the phospholipids are arranged in a double layer called a **lipid bilayer,** as shown in Figure 11. The nonpolar tails of the phospholipids make up the interior of the lipid bilayer. Because water both inside and outside the cell repels the nonpolar tails, they are forced to the inside of the lipid bilayer. Ions and most polar molecules, including sugars and some proteins, are repelled by the nonpolar interior of the lipid bilayer. The lipid bilayer allows lipids and substances that dissolve in lipids to pass through.

Figure 11 **Lipid bilayer**

Cell membranes are made of a double layer of phospholipids, called a lipid bilayer.

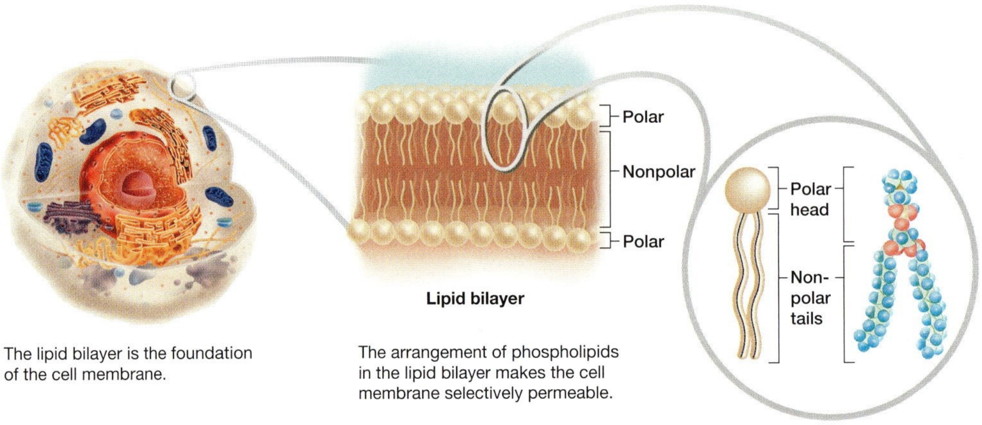

The lipid bilayer is the foundation of the cell membrane.

The arrangement of phospholipids in the lipid bilayer makes the cell membrane selectively permeable.

A phospholipid's "head" is polar, and its two fatty acid "tails" are nonpolar.

60

did you know?

Phospholipids Phospholipids in a lipid bilayer can move about 2 μm in one second. Proteins also move, but because of their greater size, their movement is slower. The membrane remains fluid unless the temperature decreases to the point at which the membrane solidifies. If the phospholipids are rich in unsaturated fatty acids, the membrane will remain fluid at lower temperatures. If the phospholipids are rich in saturated fatty acids, the membrane will be more viscous at any given temperature.

Integrating Physics and Chemistry

Remind students that the cell membrane's selective permeability is the result of how water, ions, and other substances interact with the polar and non-polar portions of the phospholipid bilayer. These differences in polarity also determine how various proteins are embedded in the lipid bilayer.

Section Review Correlations

SC.F.1.4.2: Items 2, 5
SC.F.1.4.3: Item 3
SC.F.1.4.4: Item 1
SC.H.1.4.3: Item 4

Figure 12 Membrane proteins

The cell membrane contains various proteins with specialized functions.

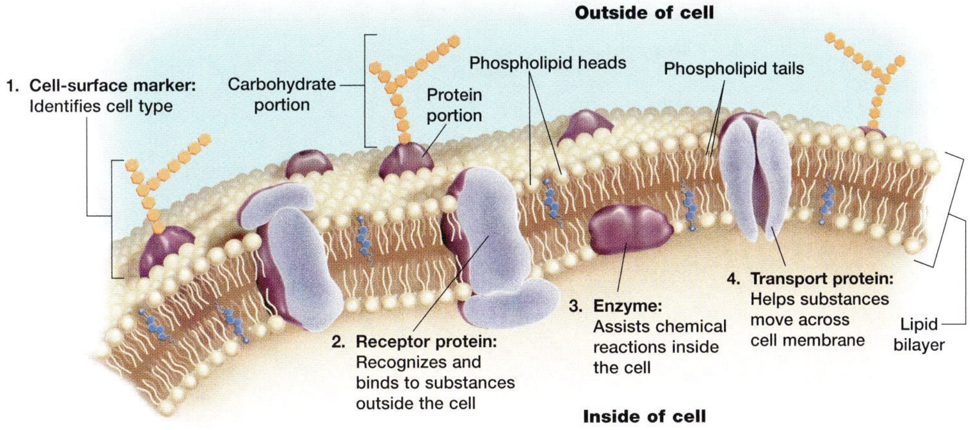

Membrane Proteins

Various proteins are located in the lipid bilayer of a cell membrane. What keeps these proteins within the lipid bilayer? You may recall that proteins are made of amino acids and that some amino acids are polar, while others are nonpolar. The nonpolar part of a membrane protein is attracted to the interior of the lipid bilayer but is repelled by the water on either side of the lipid bilayer. In contrast, the polar parts of the protein are attracted to the water on either side of the lipid bilayer. This attraction helps to hold the protein in the lipid bilayer. The motion and fluidity of phospholipids enable some membrane proteins to move around within the lipid bilayer.

As shown in **Figure 12,** cell membranes contain different types of proteins. Marker proteins attached to a carbohydrate on the cell's surface advertise cell type—such as a liver cell or a heart cell. Receptor proteins bind specific substances, such as signal molecules, outside the cell. Enzymes embedded in the cell membrane are involved in important biochemical reactions in the cell. Transport proteins aid the movement of substances into and out of the cell.

Section 2 Review

1. **Describe** the importance of the surface area–to-volume ratio of a cell. SC.F.1.4.4
2. **Compare** the structure of a eukaryotic cell with that of a prokaryotic cell. SC.F.1.4.2
3. **Critical Thinking Comparing Functions** Describe the functions of two types of cell-membrane proteins. SC.F.1.4.3
4. **Analyze** the three parts of the cell theory and describe two observations of early scientists that support it. SC.H.1.4.3
5. FCAT Prep What would a bacterium that lost its flagella be unable to do?
   - **A** move
   - **B** divide
   - **C** make proteins
   - **D** maintain its shape

   SC.F.1.4.2

Answers to Section Review

1. If the surface area-to-volume ratio of a cell is too small, substances cannot penetrate the cell quickly enough to meet the needs of the cell.
2. Prokaryotic cells lack organelles; eukaryotic cells have organelles, which enable them to carry out unique, specialized functions in organized ways.
3. Answers will vary, but may include marker proteins, which help cells recognize cells of their own type; receptor proteins, which bind to signal molecules; transport proteins, which move substances into and out of the cell; and enzymes, which aid in biochemical reactions.
4. Schleiden, Schwann, and Virchow observed that plants as well as animals are composed of cells, which are the basic units of structure and function in living things and that cells arise only from other cells.
5. **A.** Correct. Flagella are structures of motility. **B.** Incorrect. Flagella are not needed for cell division. **C.** Incorrect. Flagella are not used to make cell proteins. **D.** Incorrect. Flagella play no role in maintaining cell shape.

Close

Reteaching — BASIC

Assign students to two cooperative groups. Have group A represent a prokaryotic cell, and have group B represent a eukaryotic cell. Have each group take turns stating a characteristic that its cell has or something that its cell is able to do. Write these characteristics on the board. Give the other group an opportunity to challenge the answers. Award one point for each correct answer, and deduct one point for each incorrect answer.

Quiz — GENERAL

1. True or false: a spherical large cell has a larger surface area-to-volume ratio than a spherical small cell. (False. The spherical small cell has the larger surface-area-to-volume ratio.)
2. Where is the DNA located in a eukaryotic cell? (in the nucleus)
3. What does the term "selectively permeable" mean in regards to a cell membrane? (Some molecules can pass through the membrane and some cannot.)

Alternative Assessment — GENERAL

Working in pairs, have students construct 3 "cells" out of thick paper, using the measurements given in **Table 2.** Tell students to seal the sides of the cells with cellophane tape, and to leave the top of each cell open. Give the students enough sand to fill each of the cubes, and tell them to use these materials to show visually the surface area-to-volume ratio of the three cells. Ask each pair to explain to you orally how the surface area-to-volume ratio differs among the cells. (Answers may vary depending on the size and shape of the constructed cells.)

Focus

Overview

Before beginning this section review with your students the objectives listed in the Student Edition. This section focuses on the organelles of eukaryotic cells, beginning with the nucleus. Next, this section explores how proteins are packaged and distributed, and the role of mitochondria in the production of ATP. Lastly, the section describes structures found in plant cells that are not found in animal cells.

Bellringer

Provide students with microscope slides of eukaryotic cells. Have the students find the nucleus using compound light microscopes. Have the students give rationales for why they identified the structure they did as the nucleus. LS **Visual**

Motivate

Activity — GENERAL

Cell Structures Using a light microscope, show students a prepared slide of a cheek cell. Then show students electron micrographs of various cell parts and organelles. Have students compare these micrographs with the illustrations in this book. Ask them to attempt to find these structures in the cheek cell. Only the cytoplasm, nucleus, and cell membrane will be observed easily. Remind students that the other structures either are too small to be seen with the light microscope or may need to be stained in a different way to be seen.
LS **Visual**

Section Correlations

SC.F.1.4.2
SC.F.1.4.3
SC.F.2.4.2

Section 3 Cell Organelles

Florida Sunshine Standards/FCAT

SC.F.1.4.2 Body structures are uniquely designed and adapted for their function.

SC.F.1.4.3 Membranes are sites for chemical synthesis and essential energy conversions.

SC.F.2.4.2 Every cell contains a "blueprint" coded in DNA molecules that specify how proteins are assembled to regulate cells.

Objectives

- **Describe** the role of the nucleus in cell activities. SC.F.2.4.2
- **Analyze** the role of internal membranes in protein production. SC.F.1.4.3
- **Summarize** the importance of mitochondria in eukaryotic cells. SC.F.1.4.2
- **Identify** three structures in plant cells that are absent from animal cells. SC.F.1.4.2

Key Terms

endoplasmic reticulum
vesicle
Golgi apparatus
lysosome
mitochondrion
chloroplast
central vacuole

The Nucleus

Most functions of a eukaryotic cell are controlled by the cell's nucleus. As shown in **Figure 13,** the nucleus is surrounded by a double membrane called the nuclear envelope, also called the nuclear membrane. The nuclear envelope is made of two lipid bilayers that separate the nucleus from the cytoplasm.

Scattered over the surface of the nuclear envelope are many small channels through the envelope called nuclear pores. Substances that are made in the nucleus, including RNA-ribosomal protein complexes, move into the cytoplasm by passing through the nuclear pores. Ribosomes are partially assembled in a region of the nucleus called the nucleolus, which is also shown in Figure 13. Recall from Section 2 that ribosomes are the structures on which proteins are made.

The hereditary information of a eukaryotic cell is coded in the cell's DNA, most of which is stored in the nucleus. Eukaryotic DNA is wound tightly around proteins. Most of the time, DNA exists as elongated and thin strands, which appear as a dark mass under magnification. When a cell is about to divide, however, the DNA strands, called chromosomes, wind up into a more compact form and appear as dense, rod-shaped structures. The number of chromosomes in a eukaryotic cell differs between species. Human body cells have 46 chromosomes, while the cells of garden peas have 14 chromosomes. You will learn more about DNA and chromosomes later in this book.

Figure 13 Nucleus

The nucleus is surrounded by a double membrane called the nuclear envelope.

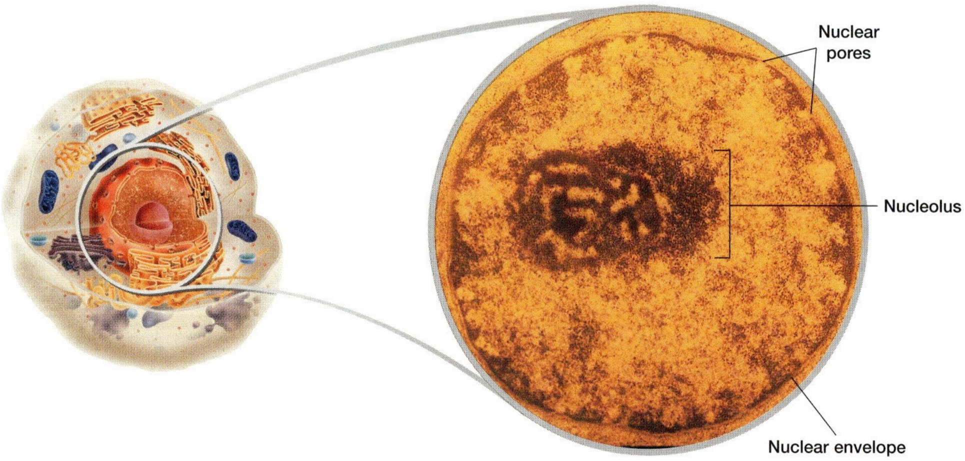

62

Chapter Resource File

- Directed Reading BASIC
- Active Reading GENERAL
- Data Sheet for Quick Lab GENERAL

One-Stop Planner CD-ROM

- Reading Organizers BASIC
- Reading Strategies BASIC
- Supplemental Reading
  The Lives of a Cell ADVANCED

INCLUSION Strategies GENERAL

- *Learning Disability*
- *Attention Deficit Disorder*

Ask students to develop a chart listing the different organelles of a cell. The chart should include the names of each organelle, the functions of each organelle, and three structures found in plant cells that are not found in animal cells. From the chart, the student can design a poster showing all of the cell's organelles.

Ribosomes and the Endoplasmic Reticulum

Unlike prokaryotic cells, eukaryotic cells have a system of internal membranes that play an essential role in the processing of proteins. Cells make proteins on ribosomes. Each ribosome is made of dozens of different proteins as well as RNA. Some of the ribosomes in a eukaryotic cell are suspended in the cytosol, as they are in prokaryotic cells. These "free" ribosomes make proteins that remain inside the cell, such as proteins used to build new organelles.

internet connect
www.scilinks.org
Topic: Proteins
Keyword: HX4151
SCiLINKS. Maintained by the National Science Teachers Association

Production of Proteins

Proteins that are exported from the cell, such as some signal molecules, are made on the ribosomes that lie on the surface of the endoplasmic reticulum, shown in **Figure 14.** The **endoplasmic reticulum** *(ehn doh PLAZ mihk rih TIHK yuh luhm)*, or ER, is an extensive system of internal membranes that move proteins and other substances through the cell. Like the cell membrane, the membranes of the ER are made of a lipid bilayer with embedded proteins.

The part of the ER with attached ribosomes is called rough ER because it has a rough appearance when viewed in the electron microscope. The rough ER helps transport the proteins that are made by its attached ribosomes. As each protein is made, it crosses the ER membrane and enters the ER. The portion of the ER that contains the completed protein then pinches off to form a vesicle. A **vesicle** is a small, membrane-bound sac that transports substances in cells. Because certain proteins are enclosed inside vesicles, these proteins are kept separate from proteins that are produced by free ribosomes in the cytoplasm.

The rest of the ER is called smooth ER because it lacks ribosomes and thus appears smooth when viewed in the electron microscope. The smooth ER performs various functions, such as making lipids and breaking down toxic substances.

Figure 14 Endoplasmic reticulum

The ER moves proteins and other substances within eukaryotic cells.

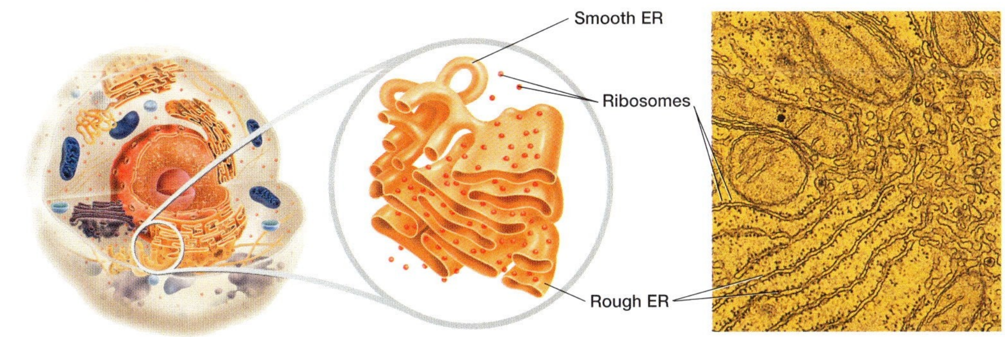

Teach

Group Activity — GENERAL

Selling Cells Have students work in groups of two or three to prepare advertisements for cell organelles. Have each group choose an organelle they want to "sell," and have them prepare print, audio, or video ads. **LS Interpersonal**

Using the Figure — BASIC

Draw students' attention to **Figure 14.** Tell students that the figure shows an increasing level of detail from left to right. On the left, the entire cell is depicted, with the ER circled. To the right of the cell, the ER in the circle is magnified to show a greater level of detail. And to the right of the circle is an electron micrograph of ER. Labels point out the same structures in the diagram as in the ER. However, the diagram shows the ER three-dimensionally, emphasizing its sheet-like nature. The electron micrograph is a thin slice of a cell, so it has a two-dimensional quality, showing a cross-sectional view. **LS Visual**

Transparencies

- TR Bellringer
- TR B13 Processing of Proteins
- TR B12 Mitochondrion
- TR B9 Organelles

MEDICINE CONNECTION

Enzymes found in the abundant smooth ER of the liver help detoxify drugs and environmental pollutants. Detoxification usually involves a series of chemical reactions within the smooth ER of the liver that make the drug more water soluble, allowing it to be excreted in the urine. The ingestion of many drugs, including barbiturates and alcohol, trigger an increase in the amount of smooth ER and accompanying enzymes in liver cells.

Teach, continued

READING SKILL BUILDER — ADVANCED

Reading Hint As students read the section, ask them to list questions they have from the reading. When students have compiled their lists, ask students to share their questions orally with the class. Make a list of questions on the board, and then assign the questions to pairs of students. Ask students to answer the questions by pointing out passages in the book that explain the answers. Have students research answers to the questions unanswered by this book.

Co-op Learning

Teaching Tip — GENERAL

What Am I? Have each student choose one cell part or organelle and write a "What Am I?" essay in which the student pretends to be that structure. The essay should be written in the first person and include the function and structure of that cell part. Encourage students to be creative in their writing styles. Have each student read his or her essay aloud in class. Other students should try to guess the structure or organelle presented in the essay. **LS Verbal**

Co-op Learning

Study TIP

Interpreting Graphics
As you read, use Steps 1–5 in the text, shown in red, to help you follow the same numbered steps shown in Figure 15.

Packaging and Distribution of Proteins

Vesicles that contain newly made proteins move through the cytoplasm from the ER to an organelle called the Golgi apparatus. The **Golgi** *(GOHL jee)* **apparatus** is a set of flattened, membrane-bound sacs that serves as the packaging and distribution center of the cell. Enzymes inside the Golgi apparatus modify the proteins that are received in vesicles from the ER. The modified proteins are then enclosed in new vesicles that bud from the surface of the Golgi apparatus. Other vesicles include **lysosomes** *(LIE seh sohms)*, which are small, spherical organelles that contain the cell's digestive enzymes. The ER, the Golgi apparatus, and lysosomes work together in the production, packaging, and distribution of proteins, as summarized in **Figure 15.**

Step 1 Ribosomes make proteins on the rough ER. The proteins are packaged into vesicles.

Step 2 The vesicles transport the newly made proteins from the rough ER to the Golgi apparatus.

Step 3 In the Golgi apparatus, proteins are processed and then packaged into new vesicles.

Step 4 Many of these vesicles move to the cell membrane and release their contents outside the cell.

Step 5 Other vesicles, including lysosomes, remain within the cytoplasm. Lysosomes digest and recycle the cell's used components by breaking down proteins, nucleic acids, lipids, and carbohydrates.

BIO graphic

Figure 15

Processing of Proteins

Proteins are processed by an internal system of membranes.

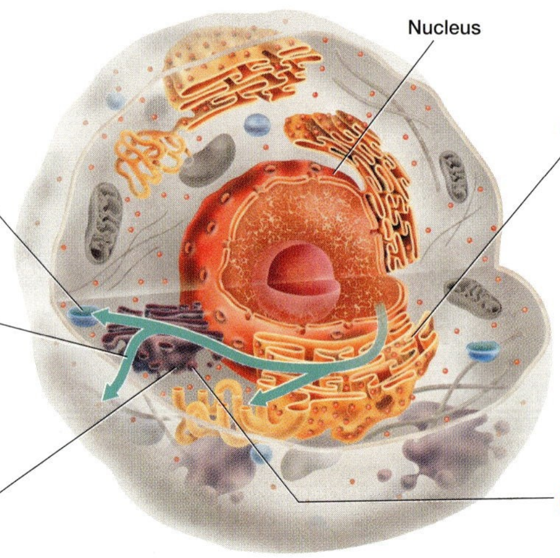

64

REAL WORLD CONNECTION

Several diseases have been attributed to improperly functioning lysosomes. Tay-Sachs disease is caused by the deficiency of a lysosomal enzyme that digests lipids. As a result, cells in the brain become filled with lipids, causing impairment. Pompe's disease results when a lysosomal enzyme that breaks down glycogen is absent, causing liver damage.

HISTORY CONNECTION

In 1945, Belgian-American cytologist Albert Claude began electron microscope studies of the cell and discovered the endoplasmic reticulum and the detailed structure of mitochondria. Along with Christian de Duve and George E. Palade, Claude was awarded the Nobel Prize in Physiology or Medicine in 1974 for discoveries of the structural and functional organization of the cell.

Mitochondria

Nearly all eukaryotic cells contain many **mitochondria** *(miet uh KAHN dree uh),* like the one shown in **Figure 16.** A mitochondrion is an organelle that harvests energy from organic compounds to make ATP, the main energy currency of cells. Although some ATP is made in the cytosol, most of a cell's ATP is made inside mitochondria. Cells that have a high energy requirement, such as a muscle cell, may contain hundreds or thousands of mitochondria. Figure 16 shows that a mitochondrion has two membranes. The outer membrane is smooth. The inner membrane is greatly folded, however, and its surface area is large. The two membranes form two compartments, one inside and one outside the mitochondrion's inner membrane. It is here that the chemical reactions that produce ATP during cell metabolism take place.

WORD Origins

The word *mitochondrion* is from the Greek words *mitos,* meaning "a thread" and *chrondros,* meaning "grain." Knowing this makes it easier to remember that a mitonchondrion is a small, elongated cell organelle.

Mitochondrial DNA

The nucleus is not the only organelle in the cell that contains nucleic acids. Mitochondria also have DNA and ribosomes, and mitochondria make some of their own proteins. However, most mitochondrial proteins are made by free ribosomes in the cytosol. Mitochondrial DNA is independent of nuclear DNA and similar to the circular DNA of prokaryotic cells. This fact supports the widely accepted theory that primitive prokaryotes are the ancestors of mitochondria. You will learn more about the origin of mitochondrial DNA later in this book.

Figure 16 Mitochondrion

In a eukaryotic cell, mitochondria make most of the ATP.

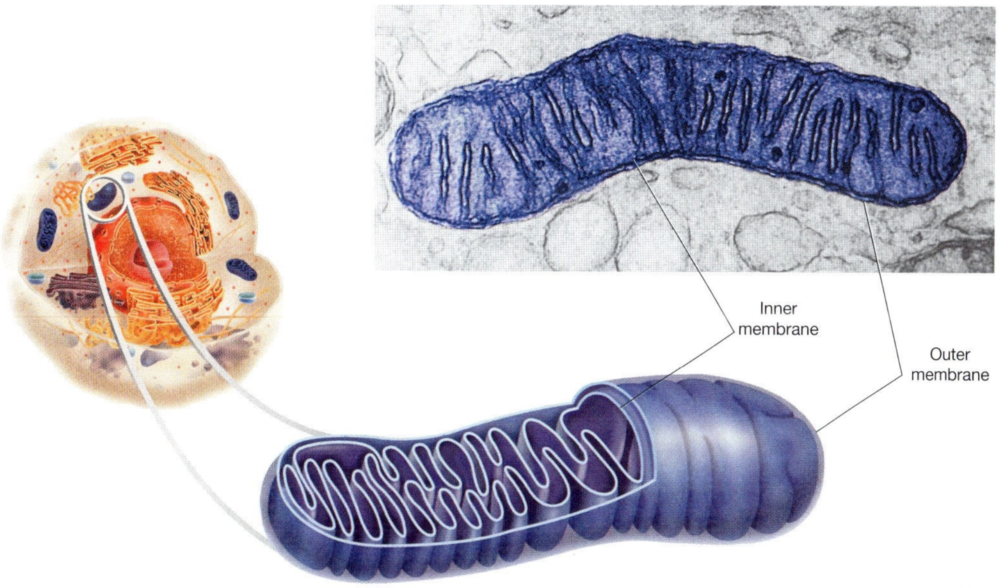

MISCONCEPTION ALERT

Mitochondria in Plant Cells
Make sure students understand that plant cells, as well as animal cells and almost all other eukaryotic cells, contain mitochondria. Many students think that because plants perform photosynthesis, which makes their own food as well as ATP, that they do not contain mitochondria. However, it is important for students to understand that plants use the products of photosynthesis to fuel the manufacture of ATP in the mitochondria. Mitochondrial ATP synthesis takes place in parts of plants that do not undergo photosynthesis, such as the roots, and when it is dark in parts of plants that do undergo photosynthesis, such as the leaves.

Using the Figure —ADVANCED

Point out the infolding of the inner membrane of the mitochondrion. Ask students to suggest a purpose for this infolding of the membrane. (Enzymes for the reactions that take place here are located on these membranes, and the infolding increases the surface area.) Ask students to speculate on how effective a mitochondrion without the folds would be. (With less surface area to carry out its chemical reactions, it might not be able to carry out its series of reactions, or might produce much less ATP.) Next, have students predict what might happen to an organism if the mitochondria in its cells suddenly became only half as efficient. (Life processes that require energy might not be able to take place, might slow down, or might not be able to sustain themselves for as long a time.)

did you know?

Endosymbiosis The theory of endosymbiosis is a widely accepted explanation of the origin of various cell organelles. The theory holds that prokaryotes lived in association with other cells but lost their ability to reproduce independently, becoming organelles of today's eukaryotes. Scientists think that chloroplasts evolved from endosymbiotic photosynthetic bacteria and mitochondria evolved from endosymbiotic aerobic bacteria. This theory is supported by the fact that both chloroplasts and mitochondria have their own DNA and can be inactivated with antibiotics that affect prokaryotes but not eukaryotes.

Close

Reteaching — BASIC

Pass out copies of electron micrographs of cells to students. Ask each student to color in 3 cell organelles of their choosing, each with a different color. Then ask students to name the 3 organelles, identifying them by their color, and to describe the job of those 3 organelles within the cell.

Quiz — GENERAL

1. The mitochondrion has the nickname "the powerhouse of the cell." Explain why this is a good nickname. (The mitochondrion is the organelle [house] in which ATP is synthesized. ATP is the energy currency that powers the cell.)
2. Name two organelles within plant cells that are not present within animal cells. (chloroplasts and central vacuoles)

Alternative Assessment — GENERAL

Have students use food to make a model of a eukaryotic cell. Gelatin (cytoplasm) could be poured into a large, clear cellophane bag (cell membrane). Organelles could be represented by kidney beans (mitochondria), poppy seeds (ribosomes), sections of lasagna noodles (ER and Golgi apparatus), peppercorns (lysosomes), spaghetti noodles (cytoskeleton), and an orange (nucleus). LS **Kinesthetic**

Structures of Plant Cells

The organelles described in this section are found in both animal cells and plant cells. However, plant cells have three additional structures that are not found in animal cells, shown in **Figure 17.**

Unique Features of Plant Cells

Cell wall The cell membrane of a plant cell is surrounded by a thick cell wall, composed of proteins and carbohydrates, including the polysaccharide cellulose. The cell wall helps support and maintain the shape of the cell, protects the cell from damage, and connects it with adjacent cells.

Figure 17 **Plant cell.** Plant cells have a cell wall, chloroplasts, and a large central vacuole (shown in blue).

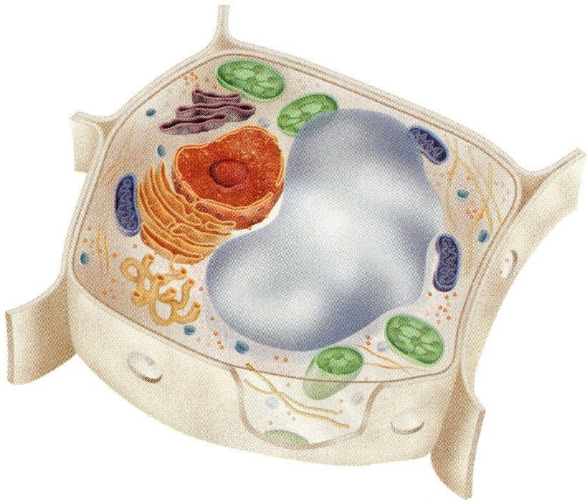

Chloroplasts Plant cells contain one or more **chloroplasts.** Chloroplasts are organelles that use light energy to make carbohydrates from carbon dioxide and water. Chloroplasts are found not only in plants but also in a wide variety of eukaryotic algae, such as seaweed. Chloroplasts, along with mitochondria, supply much of the energy needed to power the activities of plant cells. Like mitochondria, chloroplasts are surrounded by two membranes, contain their own DNA, and are thought to be the descendents of ancient prokaryotic cells.

Central vacuole As shown in Figure 17, much of a plant cell's volume is taken up by a large, membrane-bound space called the **central vacuole** *(VAK yoo ohl).* The central vacuole stores water and may contain many substances, including ions, nutrients, and wastes. When the central vacuole is full, it makes the cell rigid. This rigidity enables a plant to stand upright.

Section 3 Review

1. **Describe** the role of the nucleus in cell activities. SC.F.2.4.2
2. **Sequence** the course of newly made proteins from the rough ER to the outside of the cell. SC.F.1.4.2
3. **Describe** the role of mitochondria in the metabolism of eukaryotic cells. SC.F.1.4.2
4. **Explain** how a plant cell's central vacuole and cell wall help make the cell rigid. SC.F.1.4.2
5. **Critical Thinking** **Inferring Relationships** What is the importance of a cell enclosing its digestive enzymes inside lysosomes? SC.F.1.4.2
6. *FCAT Prep* Which organelle serves as the packaging and distribution center of a eukaryotic cell?
   A nucleus
   B lysosome
   C mitochondrion
   D Golgi apparatus
   SC.F.1.4.2

66

Answers to Section Review

1. The nucleus contains DNA, directs cell activities, and assembles ribosomal proteins and RNA.
2. The proteins, made at the ribosomes attached to the ER, pass through the ER. The proteins are packaged into vesicles and transported to the Golgi apparatus, where they are modified and repackaged. Some of these vesicles release their contents from the cell.
3. Mitochondria produce most of the ATP used for chemical reactions in eukaryotic cells.
4. The cell wall contains cellulose and is thick and somewhat rigid. When the central vacuole is full, it pushes the other cell contents against the cell wall, providing added rigidity.
5. The digestive enzymes of the cell are stored within membranous vesicles so that they do not digest other cell structures.
6. **A.** Incorrect. The nucleus directs the activities of the cell. **B.** Incorrect. The lysosome is a storage vacuole for digestive enzymes. **C.** Incorrect. The mitochondrion generates ATP. **D.** Correct. The Golgi apparatus is the packaging and distribution center of a eukaryotic cell.

Section Review Correlations

SC.F.2.4.2: Item 1
SC.F.1.4.2: Items 2-6

Study ZONE

Key Concepts

1 Looking at Cells

- Microscopes enable biologists to examine the details of cell structure and to understand how organisms function.
- Scientists use the metric system to measure the size of objects.
- Light microscopes have a low magnification and can be used to examine living cells.
- Electron microscopes have a high magnification but cannot be used to examine living cells.
- The scanning tunneling microscope uses a computer to generate a three-dimensional image of an object.

2 Cell Features

- The cell theory has three parts.
- Small cells function more efficiently than large cells because small cells have a higher surface-area-to-volume ratio than large cells.
- All cells have a cell membrane, cytoplasm, ribosomes, and DNA.
- Prokaryotic cells lack internal compartments.
- Eukaryotic cells have a nucleus and other organelles, as well as a cytoskeleton of microscopic protein fibers.
- The lipid bilayer of a cell membrane is made of a double layer of phospholipid molecules.
- Proteins in cell membranes include enzymes, receptor proteins, transport proteins, and cell-surface markers.

3 Cell Organelles

- The nucleus of a eukaryotic cell directs the cell's activities and stores DNA.
- In eukaryotic cells, an internal membrane system produces, packages, and distributes proteins.
- Mitochondria harvest energy from organic compounds to ATP.
- Lysosomes digest and recycle a cell's used components.
- Plant cells have three structures that animal cells lack: a cell wall, chloroplasts, and a central vacuole.

Key Terms

Section 1

light microscope (51)
electron microscope (51)
magnification (51)
resolution (51)
scanning tunneling microscope (54)

Section 2

cell theory (55)
cell membrane (56)
cytoplasm (56)
cytoskeleton (56)
ribosome (56)
prokaryote (57)
cell wall (57)
flagellum (57)
eukaryote (58)
nucleus (58)
organelle (58)
cilium (58)
phospholipid (60)
lipid bilayer (60)

Section 3

endoplasmic reticulum (63)
vesicle (63)
Golgi apparatus (64)
lysosome (64)
mitochondrion (65)
chloroplast (66)
central vacuole (66)

BIOLOGY INTERACTIVE TUTOR **Unit 1**—Use this unit to review the key concepts and terms in this chapter.

Alternative Assessment — GENERAL

Assign students to cooperative pairs. Have each pair devise a test that includes the material covered in this chapter. Each pair must also provide an answer key. Set the parameters for the test. For example, designate the number and type of questions allowed. After the tests are collected, exchange them with the other pairs, and have the pairs take the tests. Let the pairs check their own answers when they have completed the test. Co-op Learning

Answer to Concept Map

The following is one possible answer to Performance Zone item 15.

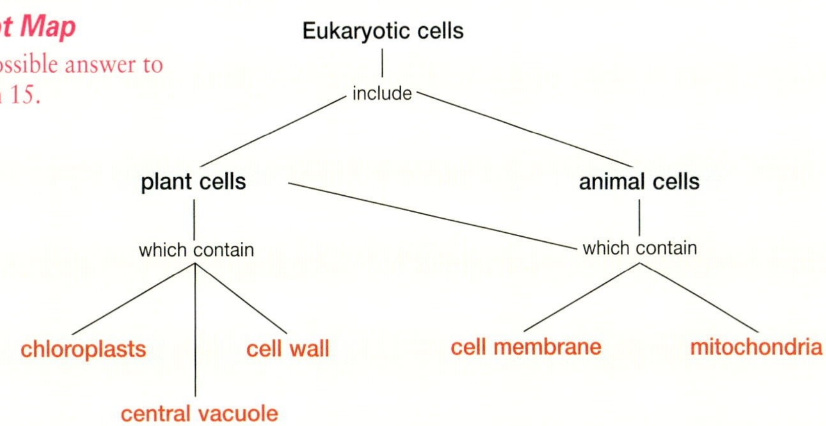

ANSWERS

Understanding Key Ideas

1. c
2. a
3. d
4. d
5. c
6. a
7. The cell membrane regulates what substances enter and leave the cell, thus maintaining the integrity of the cell.
8. A scanning electron microscope produced the image.
9. Because of the presence of hydrogen ions in lysosomes, one can infer that the inside of the lysosome is acidic. Digestive enzymes in a lysosome therefore work best in an acidic environment.
10. One possible answer to the concept map is found at the bottom of the Study Zone page.

Critical Thinking

11. The arrangement of phospholipids in the lipid bilayer influences the permeability of this bilayer because the nonpolar lipid tails repel water (and other polar/water-soluble molecules).
12. Muscles cells use ATP when they contract. Having many mitochondria would help provide the large amount of ATP needed for contraction.
13. Antibiotics can kill bacterial cells, while not harming human cells. Drugs that kill eukaryotic parasites may also harm eukaryotic human cells.

Chapter Review Correlations

SC.A.1.4.2: Item 11
SC.B.1.4.1: Item 12
SC.F.1.4.2: Items 3, 5, 6, 10
SC.F.1.4.3: Items 4, 7, 9, 14
SC.F.1.4.4: Item 2
SC.H.3.4.2: Items 1, 8
SC.H.3.4.5: Item 13

Performance ZONE

Understanding Key Ideas

1. The main advantage of the transmission electron microscope is that it allows the viewer to see
   a. three-dimensional images of cell surfaces.
   b. the organelles of living cells.
   c. a cell's internal structure in fine detail.
   d. the actual colors of a cell's components.
   SC.H.3.4.2
2. The maximum size of a cell is determined by the ratio between the cell's
   a. surface area and volume.
   b. volume and organelles.
   c. organelles and cytoplasm.
   d. cytoplasm and nucleus.
   SC.F.1.4.4
3. Eukaryotic cells differ from prokaryotic cells in that eukaryotic cells
   a. lack organelles.
   b. have DNA but not ribosomes.
   c. are smaller than prokaryotic cells.
   d. have a nucleus.
   SC.F.1.4.2
4. In the cell membrane, the fatty acids of phospholipid molecules
   a. face the cytoplasm.
   b. face the outside of the cell.
   c. are on both sides of the membrane.
   d. are in the interior of the membrane.
   SC.F.1.4.3
5. One function of the Golgi apparatus is to
   a. store DNA.
   b. make carbohydrates.
   c. modify proteins.
   d. digest and recycle the cell's wastes.
   SC.F.1.4.2
6. Structures present in plant cells but not present in animal cells include
   a. chloroplasts and the central vacuole.
   b. mitochondria and the cell wall.
   c. ribosomes and ER.
   d. lysosomes and the Golgi apparatus.
   SC.F.1.4.2
7. Explain how the cell membrane contributes to a cell's ability to maintain homeostasis. SC.F.1.4.3
8. What kind of microscope produced the image of cilia shown below? SC.H.3.4.2

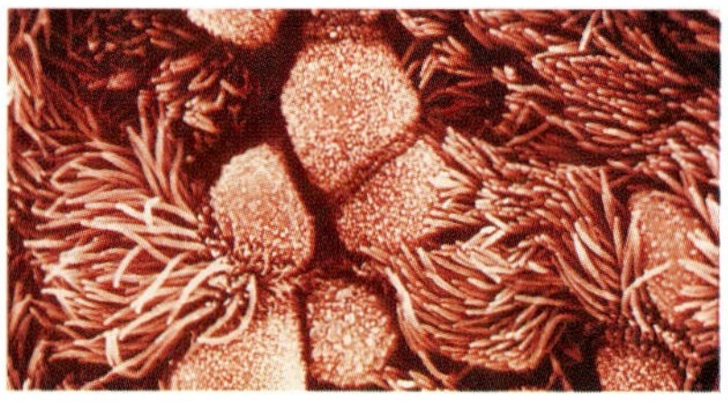

9. Transport proteins in the membrane of a lysosome move hydrogen ions into the lysosome. Use this information to predict whether digestive enzymes in a lysosome work best in a neutral, a basic, or an acidic environment. (**Hint:** See Chapter 2, Section 2.) SC.F.1.4.3
10. **Concept Mapping** Make a concept map that compares plant cells with animal cells. Include the following terms in your concept map: *cell membrane, cell wall, central vacuole, chloroplasts,* and *mitochondria.* SC.F.1.4.2

Critical Thinking

11. **Recognizing Relationships** How does the arrangement of phospholipids influence the permeability of the lipid bilayer? SC.A.1.4.2
12. **Inferring Relationships** Muscle cells have more mitochondria than some other kinds of eukaryotic cells. In what way would having many mitochondria be beneficial to muscle cells? SC.B.1.4.1
13. **Applying Information** Drugs that rid the body of eukaryotic parasites often have more side effects and are harder on the body than drugs that act on bacterial parasites. Suggest a reason for this difference. SC.H.3.4.5

Alternative Assessment

14. **Interactive Tutor Unit 1** Cell Transport and Homeostasis Write a report summarizing the role of the cell membrane in the preservation of body organs donated for transplant. SC.F.1.4.3

Alternative Assessment

14. Answers will vary. Have students access the Internet and other resources for this information. Students may focus on the role of various types of transport proteins in maintaining the organs.

Assignment Guide

Section	Questions
1	1, 2, 8
2	3, 4, 7, 11, 14
3	5, 6, 9, 10, 12, 13

FCAT Prep

For questions 1–6, write your answers on a separate sheet of paper.

1 What structure houses a eukaryotic cell's DNA?
A. cell wall
B. cytoskeleton
C. ER
D. nucleus SC.F.2.4.2

2 Which of the following is a valid explanation as to why small pieces of food cook faster than larger pieces of food?
F. Larger pieces of food heat up faster than smaller pieces of food.
G. Larger pieces of food have fewer enzymes than smaller pieces of food.
H. Smaller pieces of food have a greater volume than larger pieces of food.
I. Smaller pieces of food have a larger surface-area-to-volume ratio than larger pieces.
SC.F.1.4.4

3 What structures are found in both eukaryotic cells and prokaryotic cells?
A. mitochondrion and ER
B. nucleus and cell membrane
C. mitochondrion and ribosome
D. cell membrane and ribosome
SC.F.1.4.2

4 What property of phospholipids makes them ideal for making up the selectively permeable cell membrane?
F. The entire molecule repels small ions.
G. They react readily with water molecules.
H. They form triple layers that insulate the cell.
I. They have a nonpolar region and a polar region.
SC.A.1.4.2

5 READ INQUIRE EXPLAIN Mitochondria are small, elongated organelles that have an outer membrane and a highly folded inner membrane. Unlike other organelles in the cell, mitochondria have their own DNA and ribosomes.

What is the function of the mitochondria? SC.F.1.4.2

6 READ INQUIRE EXPLAIN Microbiologists study the growth, structure, development, and many other characteristics of bacteria and other microorganisms. Some microbiologists also study the action of microorganisms on living and dead tissue. The career requires at least a two-year technical training degree from a community college or technical institution. Many microbiologists have a four-year bachelor's degree plus a master's or doctoral degree.

Part A Determine the tools that would be most useful for a microbiologist. SC.H.3.4.2

Part B Analyze the usefulness that knowledge of larger, multi-celled organisms would have for a microbiologist who specialized in bacteria. SC.H.1.4.5

Test TIP

When possible, use the text in the test to answer other questions. For example, use a multiple-choice answer to "jump start" your thinking about another question.

69

FCAT Doctor

Question 2 Answer I is the correct choice. Small pieces of food have significantly less volume than large pieces of food, yet their surface area is only slightly smaller. Thus, they have a larger surface-area-to-volume ratio. As a result, more surface area is exposed to heat energy while cooking. Answer F is incorrect because the opposite is true. Answer G is incorrect because the opposite is more likely to be true. Answer H is incorrect because the opposite is true.

Question 5 Mitochondria make ATP, the main energy currency of cells, and mitochondria are important for harvesting from organic compounds the energy needed for cell functions.

Question 6

Part A Microbiologists study tiny organisms and require tools that allow them to observe these microorganisms.

Part B In order to understand how microorganisms affect multi-cellular organisms, microbiologists require an understanding of the tissues of these larger organisms.

Answers

1. D
2. I
3. D
4. I
5. The mitochondria make ATP, which is important for energy transfer in living things.
6A. In order to observe microscopic organisms, microbiologists use compound light microscopes and electron microscopes.
6B. Some microbiologists study the action of microorganisms on living and dead tissues found in larger, multi-celled organisms.

Test Prep Correlations

SC.A.1.4.2: Item 4
SC.F.1.4.2: Items 3, 5
SC.F.1.4.4: Item 2
SC.F.2.4.2: Item 1
SC.H.1.4.5: Item 6B
SC.H.3.4.2: Item 6A

Studying Animal Cells and Plant Cells

Teacher's Notes

Time Required One 45-minute class period

Ratings

EASY → HARD

Teacher Preparation
Student Setup
Concept Level
Cleanup

Skills Acquired
- Classifying
- Communicating
- Identifying/Recognizing Patterns
- Inferring
- Interpreting

Scientific Methods

In this lab, students will:
- Make Observations
- Draw Conclusions
- Communicate the Results

Materials and Equipment

Materials for this lab can be ordered from WARD'S. See *Master Materials List* at the front of this book for catalog numbers. Combine all wastes containing Lugol's solution. To this mixture add a few drops of a strong acid, such as 1.0 m sulfuric acid (H_2SO_4), to make the mixture slightly acidic. Then slowly add 0.1 m sodium thiosulfate ($Na_2S_2O_3$) while stirring until the mixture loses its yellowish-orange color. Pour the resulting mixture down the drain.

Skills Practice Lab

Studying Animal Cells and Plant Cells

SC.H.1.4.1

SKILLS
- Using a compound microscope
- Drawing

OBJECTIVES
- **Identify** the structures you can see in animal cells and plant cells.
- **Compare** and **Contrast** the structure of animal cells and plant cells.

MATERIALS
- compound light microscope
- prepared slide of human epithelial cells
- safety goggles
- lab apron
- polyethylene gloves
- sprig of Elodea
- forceps
- microscope slides and coverslips
- dropper bottle of Lugol's iodine solution

Magnification: 530×

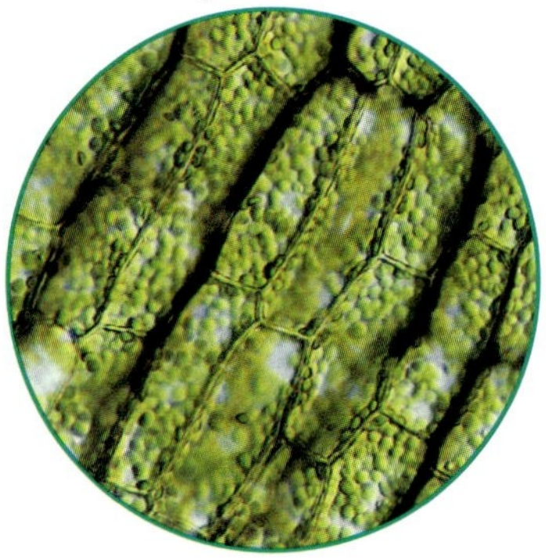

Plant cells

ChemSafety

CAUTION: Always wear safety goggles and a lab apron to protect your eyes and clothing.

CAUTION: Do not touch or taste any chemicals. Know the location of the emergency shower and eyewash station and how to use them. If you get a chemical on your skin or clothing, wash it off at the sink while calling to the teacher. Notify the teacher of a spill. Spills should be cleaned up promptly, according to your teacher's directions.

CAUTION: Glassware is fragile. Notify the teacher of broken glass or cuts. Do not clean up broken glass or spills with broken glass unless the teacher tells you to do so.

Before You Begin

You can see many cell parts with a **light microscope**. In animal cells, the **cytoplasm**, **cell membrane**, **nucleus**, **nucleolus**, and **vacuoles** can be seen. In plant cells, the **cell wall** and **chloroplasts** can also be seen. Stains add color to cell parts and make them more visible with a light microscope. A stain can even make the **endoplasmic reticulum** visible. In this lab, you will use a light microscope to examine animal and plant cells.

1. Write a definition for each boldface term in the paragraph above.
2. Why might a stain be needed to see cell parts under a microscope?
3. Based on the objectives for this lab, write a question you would like to explore about cell structure.

Procedure

PART A: Animal Cells

1. Examine a prepared slide of human epithelial cells under low power with a compound light microscope. Find cells that are separate from each other, and place them in the center of the field of view. Switch to high power, and adjust the diaphragm until you can see the cells more clearly. Identify as many cell parts as you can. *Note: Remember to use only the fine adjustment to focus at high power.*
2. Draw two or three epithelial cells as they look under high power. Label the cell membrane, the cytoplasm, the nuclear envelope, and the nucleus of at least one of the cells. Make a second drawing of these cells as you imagine they might look in the lining of your mouth.

70

Safety Cautions

Review all safety symbols and caution statements with students before beginning the lab. Make sure students wear safety goggles and a lab apron. Caution students that they must wear safety goggles when working with Lugol's iodine solution because iodine can cause severe eye damage. If students are using microscopes with mirrors, caution them not to look directly at the reflected light from the mirror.

Tips and Tricks

Set up labeled containers for the disposal of broken glass and materials stained with Lugol's solution. Place enough materials for one class on a supply table. Allow students to take off their safety goggles when looking through the microscope. Have students use a scale of 1 μm=1 cm when drawing cells. Help students take measurements of cell size.

Part B: Plant Cells

3. Using forceps, carefully remove a small leaf from near the top of an *Elodea* sprig. Place the whole leaf in a drop of water on a slide, and add a cover slip.
4. Observe the leaf under low power. Look for an area of the leaf in which you can see the cells clearly, and move the slide so that this area is in the center of the field of view. Switch to high power, and, if necessary, adjust the diaphragm. Identify as many cell parts as you can.
5. Find an *Elodea* cell in which you can see the chloroplasts clearly. Draw this cell. Label the cell wall, a chloroplast, and any other cell parts that you can see.
6. Notice if the chloroplasts are moving in any of the cells. If you do not see movement, warm the slide in your hand or under a bright lamp for a minute or two. Look for movement of the cell contents again under high power. Such movement is called cytoplasmic streaming.
7. Put on safety goggles, gloves, and a lab apron. Make a wet mount of another *Elodea* leaf, using Lugol's iodine solution instead of water. **CAUTION: Lugol's solution stains skin and clothing. Promptly wash off spills.** Observe these cells under low and high power.
8. Draw a stained *Elodea* cell. Label the cell wall and a chloroplast, as well as the central vacuole, the nucleus, and the cell membrane if they are visible.

Part C: Cleanup and Disposal

9. Dispose of solutions, broken glass, and *Elodea* leaves in the waste containers designated by your teacher. Do not pour chemicals down the drain or put lab materials in the trash unless your teacher tells you to do so.
10. Clean up your work area and all lab equipment. Return lab equipment to its proper place. Wash your hands thoroughly before you leave the lab and after you finish all work.

Analyze and Conclude

1. **Recognizing Patterns** In what observable ways are animal and plant cells similar in structure, and in what observable ways are they different?
2. **Comparing Structures** Compare and contrast the cytoplasm of epithelial cells and *Elodea* cells.
3. **Analyzing Methods** What is the reason for staining *Elodea* cells with iodine?
4. **Inferring Conclusions** Lugol's iodine solution causes the movement of chloroplasts to stop. Explain why.
5. **Inferring Conclusions** If some of the epithelial cells were folded over on themselves but were still transparent, what could you conclude about their thickness?
6. **Further Inquiry** Write a new question about cell structure that could be explored with another investigation.

On the Job

Microscopy is an important tool for biologists who study cell structure. Do research to learn more about how biologists use specialized microscopes to study cell structure. For more about careers, visit **go.hrw.com** and type in the keyword **HX4 Careers.**

Answers to Analyze and Conclude

1. Animal and plant cells both have cytoplasm and a nucleus. Both types of cells are bounded by a cell membrane. Unlike animal cells, plant cells have chloroplasts, a cell wall, and a central vacuole. Plant cells are also more consistent in shape than animal cells.
2. The cytoplasm of epithelial cells appears grainy. The cytoplasm of *Elodea* has observable chloroplasts present, is usually located near the edge of the cell, and typically flows in one direction.
3. The stain makes the nucleus, cell wall, and nucleolus more visible.
4. Lugol's solution is a poison that kills the cell and therefore stops the chloroplasts from moving.
5. A reasonable conclusion is that epithelial cells are extremely thin.
6. Answers will vary. For example: How do other animal and plant cells compare?

Answers to On the Job

Answers will vary. For example, the atomic force microscope is used to study biological surfaces with a resolution of single atoms.

Answers to Before You Begin

1. *light microscope*—a microscope that uses a beam of light passing through one or more lenses; *cytoplasm*—the material inside a cell between the cell membrane and the nuclear membrane; *cell membrane*—a lipid bilayer with embedded proteins that encloses a cell; *nucleus*—an organelle found only in eukaryotic cells that houses the DNA; *nucleolus*—a structure in the nucleus of eukaryotic cells in which ribosomal proteins form; *vacuole*—a membrane-enclosed sac that contains substances, such as nutrients, needed by the cell; *cell wall*—a protective layer outside the cell membrane found in cells of bacteria, plants, fungi, and some protists; *chloroplast*—an organelle found only in the cells of plants and photosynthetic protists that uses light energy to drive the synthesis of organic compounds from carbon dioxide and water; *endoplasmic reticulum*—an extensive system of internal membranes that move proteins and other substances through the cell
2. Because many cell structures are transparent, a stain may be needed to distinguish certain cell parts. Additionally, some stains coat cell parts, such as flagella, making them thicker and easier to see.
3. Answers will vary. For example, how does the size of animal cells compare with the size of plant cells?

Answers to Procedure

Drawings will vary. For the epithelial (animal) cells, make sure students label the cell membrane, cytoplasm, nucleus, nuclear envelope, and any other visible cell parts. For the *Elodea* (plant) cells, make sure students label the cell wall, cell membrane, chloroplasts, central vacuole, nucleus, and any other visible cell parts.

5 min pp. 72–73
Chapter Opener

- SE **Quick Review,** p. 73 ■
- SE **Reading Activity,** p. 73 ■
- TE **Using the Figure,** p. 72 GENERAL
- TE **Opening Activity,** p. 73 GENERAL

& 3 • 90 min pp. 74–80
Section 1
Passive Transport

- OSP **Lesson Plan***
- TR **Bellringer***
- TR **Section Outline***
- TR **B16 Osmosis**
- TR **B17 Hypertonic, Hypotonic, and Isotonic Solutions**
- TR **B19 Ion Channels**
- TR **B18 Facilitated Diffusion**

- TE **Demonstration,** p. 74 BASIC
- TE **Using the Figure,** p. 75
- SE **Quick Lab** Observing Osmosis, p. 76 ■
- CRF **Datasheets for In-Text Labs** Observing Osmosis*
- TE **Demonstration,** p. 77 GENERAL
- TE **Using the Figure,** p. 78
- TE **Using the Figure,** p. 79 GENERAL
- SE **Data Lab** Analyzing the Effect of Electrical Charge on Ion Transport, p. 79 ■
- CRF **Datasheets for In-Text Labs** Analyzing the Effect of Electrical Charge on Ion Transportation*
- CRF **Quick Lab** Demonstrating Diffusion* ◆ BASIC

& 5 • 90 min pp. 81–86
Section 2
Active Transport

- OSP **Lesson Plan***
- TR **Bellringer***
- TR **Section Outline***
- TR **B21 Sodium-Potassium Pump**
- TR **B22 Endocytosis and Exocytosis**
- TR **B20 Changes in Permeability**

- TE **Demonstration,** p. 81 BASIC
- TE **Using the Figure,** p. 82 GENERAL
- TE **Group Activity** Good and Bad Cholesterol, p. 83 ADVANCED
- TE **Using the Figure,** p. 84 GENERAL
- TE **Demonstration,** p. 85 GENERAL
- CRF **Skills Practice Lab** Observing Osmosis in Eggs* ◆ GENERAL
- SE **Exploration Lab** Analyzing the Effect of Cell Size on Diffusion, pp. 90–91 ◆ ■
- CRF **Datasheets for In-Text Labs** Analyzing the Effect of Cell Size on Diffusion* ◆

& 7 • 90 min
Chapter Review and Assessment Resources

- SE **Study Zone,** p. 87 ■
- SE **Performance Zone,** p. 88 ■
- SE **FCAT Test Prep,** p. 89 ■
- TR **Concept Mapping***
- CRF **Vocabulary Review Worksheet*** ■ GENERAL
- CRF **Test Prep Pretest*** ■ GENERAL
- CRF **Chapter Test*** ■ GENERAL
- CRF **Chapter Test*** ADVANCED
- OSP **Test Generator***
- OSP **Test Item Listing***
- OSP **Modified Chapter Test*** BASIC

Online and Technology Reso

Holt Online Learning

Visit go.hrw.com for access to Holt Online Learning or enter the keyword **HX6 Home** for a variety of free online resources.

One-Stop Planner® CD-ROM

This CD-ROM package includes
- Lab Materials QuickList Software
- Holt Calendar Planner
- Customizable Lesson Plans
- Printable Worksheets
- ExamView® Test Generator
- Interactive Teacher Edition
- Holt PuzzlePro® Resources
- Holt PowerPoint® Resources

Guided Reading Audio CDs

These CDs are designed to help auditory learners and reluctant readers. (Audio CDs are also available in Spanish.)

Florida Sunshine Standards/FCAT

TE **Reading Skill Builder** Paired Reading, p. 77 GENERAL **TE** **Reading Skill Builder** Interactive Reading, p. 78 BASIC **CRF** **Directed Reading Worksheet*** BASIC **CRF** **Active Reading Worksheet*** GENERAL	**SE** **Section 1 Review** ■ **TE** **Reteaching,** p. 80 BASIC **TE** **Quiz,** p. 80 GENERAL **TE** **Alternative Assessment,** p. 80 GENERAL **CRF** **Quiz*** ■ BASIC	SC.F.1.4.2, SC.F.1.4.4
TE **Skill Builder** Math Skills, p. 84 ADVANCED **CRF** **Directed Reading Worksheet*** BASIC **CRF** **Active Reading Worksheet*** GENERAL **CRF** **Critical Thinking Worksheet*** ADVANCED **CRF** **Science Skills Worksheet*** GENERAL	**SE** **Section 2 Review** ■ **TE** **Reteaching,** p. 86 BASIC **TE** **Quiz,** p. 86 GENERAL **TE** **Alternative Assessment,** p. 86 GENERAL **CRF** **Quiz*** ■ BASIC **Modified Worksheet*** BASIC	SC.F.1.4.3, SC.F.1.4.4, SC.F.1.4.5, SC.F.1.4.6

Holt Biology Interactive Tutor CD-ROM

This CD-ROM consists of interactive activities that give students a fun way to extend their knowledge of biology concepts.

Unit 1
Cell Transport and Homeostasis

SCILINKS. NSTA
www.scilinks.org

Maintained by the **National Science Teachers Association**

Topic: Water Movement in Cells
SciLinks Code: HX4189

Topic: Ion Channels
SciLinks Code: HX4106

Topic: Receptor Proteins
SciLinks Code: HX4157

Topic: Cell Membrane
SciLinks Code: HX4035

CHAPTER 4

Overview

Tell students that the purpose of this chapter is to learn how substances move into and out of cells. Students will learn two different ways in which this is done: passive transport, which does not require energy, or active transport, which requires energy. They will also learn about channels in membranes that allow charged particles and large particles to move into cells. These processes allow necessary materials to move into cells and wastes to move out.

Using the Figure

Direct students to the Chapter Opener figure. Ask them to identify the major structures in the picture. (They are nerve cells.) Ask them to identify the threadlike structures in the picture. (These are extensions of nerve cells, called dendrites and axons.) Point out to students that cells in the body do not exist independently of each other. They interact with each other in ways that benefit the organism as a whole. This interaction involves different types of communication between cells.

Assessing Prior Knowledge

Before beginning this chapter, be sure that your students can answer the questions in the ***Quick Review*** on the facing page. If they cannot, encourage them to reread the sections indicated.

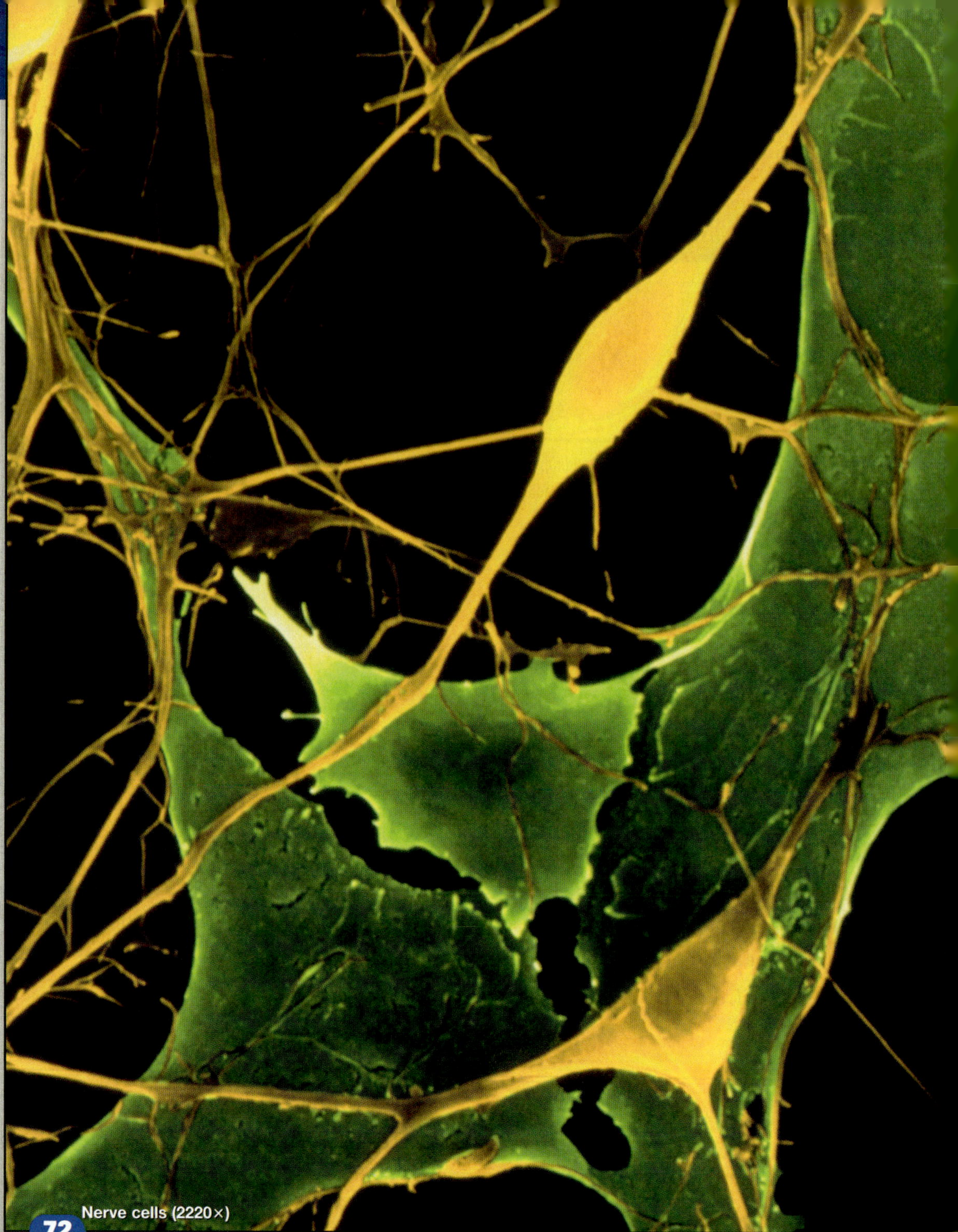

Nerve cells (2220×)

72

Chapter Correlations — Florida Sunshine Standards/FCAT

SC.F.1.4.2 Body structures are uniquely designed and adapted for their function.

SC.F.1.4.3 Membranes are sites for chemical synthesis and essential energy conversions.

SC.F.1.4.4 Biological systems obey the same laws of conservation as physical systems.

SC.F.1.4.5 Complex interactions among the different kinds of molecules in the cell cause distinct cycles of activity governed by proteins.

SC.F.1.4.6 Separate parts of the body communicate with each other using electrical and/or chemical signals.

CHAPTER

4 Cells and Their Environment

Florida Sunshine Standards/FCAT

SC.F.1.4.2 Body structures are uniquely designed and adapted for their function.

SC.F.1.4.3 Membranes are sites for chemical synthesis and essential energy conversions.

SC.F.1.4.4 Biological systems obey the same laws of conservation as physical systems.

SC.F.1.4.5 Complex interactions among the different kinds of molecules in the cell cause distinct cycles of activity governed by proteins.

SC.F.1.4.6 Separate parts of the body communicate with each other using electrical and/or chemical signals.

Quick Review

Answer the following without referring to earlier sections of your book.

1. **Distinguish** *between polar and nonpolar substances. (Chapter 2, Section 1)*
2. **Describe** *the function of ATP in cells. (Chapter 2, Section 3)*
3. **Identify** *different kinds of proteins that compose the cell membrane. (Chapter 3, Section 2)*
4. **Summarize** *the function of vesicles and the Golgi apparatus. (Chapter 3, Section 3)*

Did you have difficulty? *For help, review the sections indicated.*

Reading Activity

Take a break after reading each section of this chapter, and closely study the figures in the section. Reread the figure captions, and, for each one, write out a question that can be answered by referring to the figure and its caption. Refer to your list of figures and questions as you review the concepts addressed in the chapter.

Looking Ahead

internet connect

www.scilinks.org
National Science Teachers Association *sci*LINKS Internet resources are located throughout this chapter.

SCILINKS. Maintained by the National Science Teachers Association

The transfer of information between these nerve cells requires constant movement of substances across their cell membranes.

Chapter Resource File

- Vocabulary Worksheets
- Concept Mapping

Opening Activity —GENERAL

Have students work together in small groups to construct a model of the cell membrane. Provide each group with some small (1- to 2-inch diameter) styrofoam balls and some single-color pipe cleaners. Refer them to the section on cell membrane structure in the previous chapter. Point out the hydrophilic phosphate "heads" and the hydrophobic fatty acid "tails" of the membrane. The molecules of the model membrane should be arranged with the "tails" pointing toward each other and the "heads" pointing away. These "molecules" together represent a cell membrane. Provide students with empty paper towel tubes to represent transport proteins. Provide the students with pipe cleaners of different colors, which they should attach to the "membrane" to represent surface markers.
LS Kinesthetic Co-op Learning

Quick Review Answers

1. Polar substances have opposing characteristics in the same substance, whereas nonpolar substances do not. Polarity may occur because of a relative positive charge in one part of a molecule and a relative negative charge in another part of the same molecule.
2. The role of ATP in cells is to provide energy to build molecules, transport materials, and carry out other cellular activities.
3. The cell membrane contains transport proteins, enzymes, receptor proteins, and surface markers.
4. Vesicles contain materials needed by the cell as well as waste products from the cell. The Golgi apparatus packages materials from the cell into vesicles. These vesicles are used to transport materials out of the cell.

Reading Activity Answers

Answers should show an understanding of the figure information.

Section 1

Focus

Overview

Before beginning this section review with your students the objectives listed in the Student Edition. This section describes how substances move through other substances according to their concentrations in solutions. It also describes how substances move across cell membranes either directly or through channels. Students will learn about diffusion, osmosis, and channel transport.

Bellringer

Ask each student to write three sentences using the word *diffuse* or one of its conjugates. Suggest that they consult a dictionary if they have trouble thinking of how the word is used.

Motivate

Demonstration — Basic

Fill a beaker with lukewarm water. Place two drops of red or blue food coloring in the beaker. As the dye disperses in the water, ask students what they are observing. (Diffusion is one of the processes they are observing.) Explain that diffusion works very quickly over short distances but slowly over long distances. Most of the mixing in this demonstration is caused by slight air currents across the surface of the water as well as temperature differences within the water. This demonstration, therefore, is more a model of convection than an example of actual diffusion.

Section Correlations

SC.F.1.4.2
SC.F.1.4.4

Section 1 Passive Transport

Florida Sunshine Standards/FCAT

SC.F.1.4.2 Body structures are uniquely designed and adapted for their function.

SC.F.1.4.4 Biological systems obey the same laws of conservation as physical systems.

Objectives

- **Relate** concentration gradients, diffusion, and equilibrium. SC.F.1.4.4
- **Predict** the direction of water movement into and out of cells. SC.F.1.4.4
- **Describe** the importance of ion channels in passive transport. SC.F.1.4.2
- **Identify** the role of carrier proteins in facilitated diffusion. SC.F.1.4.2

Key Terms

passive transport
concentration gradient
equilibrium
diffusion
osmosis
hypertonic solution
hypotonic solution
isotonic solution
ion channel
carrier protein
facilitated diffusion

Diffusion

You constantly interact with your environment, whether you are eating or putting on a raincoat to help keep you dry. Your body also responds to external conditions to maintain a stable internal condition. Just as you must respond to your environment to maintain stability, all other organisms and their cells must respond to external conditions to maintain a constant internal condition. Recall that when organisms adjust internally to changing external conditions, they are maintaining homeostasis. One way cells maintain homeostasis is by controlling the movement of substances across their cell membrane. Cells must use energy to transport some substances across the cell membrane. Other substances move across the cell membrane without any use of energy by the cell.

Random Motion and Concentration

Movement across the cell membrane that does not require energy from the cell is called **passive transport.** To understand passive transport, imagine two rooms of equal size separated by a wall with a closed door, as shown in **Figure 1.** Suppose you release several rubber balls into the first room. The balls move randomly, bouncing off the walls, the floor, the ceiling, and each other. Also suppose the balls can bounce forever without slowing down. The balls become evenly distributed throughout the room. What happens when you open the door between the rooms? Some of the balls in the first room bounce through the doorway and into the second room, as shown in Figure 1. You do not have to use energy to make the balls move into the second room. They enter the second room because of their own random motion. Occasionally, a ball will bounce back into the first room. However, most of the balls that pass through the doorway move from the first room, where their concentration is high, to the second room, where their concentration is low. A difference in the concentration of a substance, such as the balls, across a space is called a **concentration gradient**.

As more balls enter the second room, the concentration of balls in the second room increases, while the concentration of balls in the first room decreases. Eventually the concentration of balls in the two rooms will be equal. The balls will still bounce around the rooms, but they will move from the second room to the first room just as often as they move from the first room to the second room. At this point, the system is said to be in equilibrium, as shown in Figure 1. **Equilibrium** *(ee kwih LIHB ree uhm)* is a condition in which the concentration of a substance is equal throughout a space.

Study TIP

Reading Effectively

As you read this chapter, write the objectives for each section on a sheet of paper. Rewrite each objective as a question, and answer these questions as you read the section.

Chapter Resource File

- Directed Reading Basic
- Active Reading General
- Data Sheet for Quick Lab General
- Data Sheet for Data Lab General

One-Stop Planner CD-ROM

- Reading Organizers Basic
- Reading Strategies Basic
- Occupational Applications Pharmacist General

Figure 1 **Models of diffusion**

Because of diffusion, food coloring (blue) will gradually move through uncolored gelatin (yellow), as shown in the beakers below.

1. Randomly bouncing balls are distributed evenly throughout a closed room.
2. If the door to an adjoining room is opened, the balls begin to enter, or diffuse into, that room.
3. At equilibrium, the concentration of balls inside the two rooms will be equal.

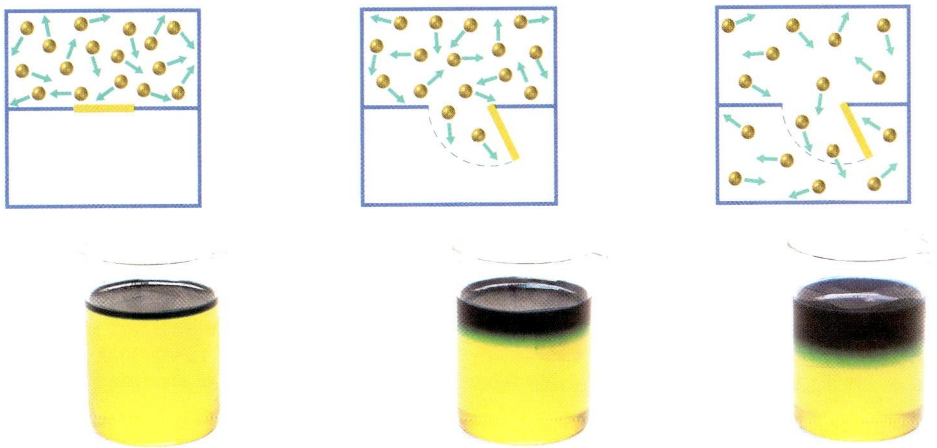

Movement of Substances

Like these imaginary rubber balls, particles of a substance in a solution also move around randomly. If there is a concentration gradient in the solution, the substance will move from an area of high concentration to an area of lower concentration. The movement of a substance from an area of high concentration to an area of lower concentration caused by the random motion of particles of the substance is called **diffusion** *(dih FYOO zhuhn)*. If diffusion is allowed to continue, equilibrium eventually results.

Many substances, such as molecules and ions dissolved in the cytoplasm and in the fluid outside cells, enter or leave cells by diffusing across the cell membrane. Inside the cell, the concentrations of most of these substances are different from their concentrations outside the cell. Thus, for each of these substances a concentration gradient exists across the cell membrane. To diffuse "down" its concentration gradient—from an area of high concentration to an area of lower concentration—a substance must be able to pass through the cell membrane.

The cell membrane is selectively permeable to substances. The nonpolar interior of the lipid bilayer repels ions and most polar molecules. Thus, these substances are prevented from diffusing across the cell membrane. In contrast, molecules that are either very small or nonpolar can diffuse across the cell membrane down their concentration gradient. The diffusion of such molecules across the cell membrane is the simplest type of passive transport.

internet connect

www.scilinks.org
Topic: Water Movement in Cells
Keyword: HX4189

SCILINKS. Maintained by the National Science Teachers Association

Teach

Teaching Tip

Sore Throat Remedy Tell students that a common remedy for a sore throat is to gargle with salt water. Use the concept of osmosis to explain in part how this remedy works. Some of the pain associated with a sore throat is caused by swelling of the throat tissues, which contain water. Salt water has a lower concentration of water than the throat tissues do, so when a person gargles with salt water, water moves by osmosis from the throat tissues into the salt water. The loss of water decreases the swelling and relieves some of the pain. **LS Intrapersonal**

Using the Figure

In **Figure 1**, correlate the model of balls moving between the two rooms with what is happening in the beaker shown below each diagram. The beakers are filled with gelatin. Food coloring moves through the gelatin because of the random motion of food coloring particles. The movement of these particles causes diffusion. Emphasize that the individual particles of food coloring move randomly and independently, causing a net movement of food coloring. The food coloring moves through the gelatin from an area of higher concentration to an area of lower concentration. **LS Visual**

INTERACTIVE BIOLOGY TUTOR

- **Unit 1—Cell Transport and Homeostasis**

This engaging tutorial introduces students to diffusion and passive transport.

Transparencies

- TR Bellringer
- TR B16 Osmosis
- TR B17 Hypertonic, Hypotonic and Isotonic Solutions
- TR B19 Ion Channels
- TR B18 Facilitated Diffusion

MISCONCEPTION ALERT

Some students might think that particles of a substance in solution stop moving when equilibrium is reached. Although at equilibrium there is no concentration gradient (the concentration of the substance is the same throughout the solution), the particles of the substances continue to move randomly and independently. This random movement maintains a continuous equilibrium.

Teach, *continued*

Teaching Tip

Salting Roads Point out to students that during the winter, salt is sprinkled over icy roads. Plants growing along the sides of these roads are often seriously damaged or even killed. The leaves on these plants look similar to the leaves on plants that have been through a very hot, dry summer. The higher-than-normal salt concentration in these plants' external environments causes their cells to lose water through osmosis. Without sufficient water, the plants die.

QUICK LAB — *Observing Osmosis*

Skills Acquired
Observing, analyzing, evaluating

Teacher's Notes
Review how to use a balance to find the mass of objects.

Answers to Analysis

1. Answers will vary. The tap water should cause water to move into the grape. The sugar solution should cause water to move out of the grape. Results with the grape juice may vary depending on the sugar content of the juice.
2. Whether osmosis occurs can be determined by noting an increase or a decrease in the mass of the grapes.
3. Answers will vary. Students who had a clear understanding of osmosis before the experiment probably will not change their thinking.

Osmosis

Water molecules can diffuse through channels in the cell membrane, as shown in **Figure 2.** The diffusion of water through a selectively permeable membrane is called **osmosis** *(ahz MOH sihs).* Like other forms of diffusion, osmosis involves the movement of a substance—water—down its concentration gradient. Osmosis is a type of passive transport.

What causes osmosis? Recall that a solution is a substance dissolved in another substance. In the solutions on either side of the cell membrane, many ions and polar molecules are dissolved in water. When these substances dissolve in water, some water molecules are attracted to them and so are no longer free to move around. If the solutions on either side of the cell membrane have different concentrations of dissolved particles, they will also have different concentrations of "free" water molecules. Then osmosis will occur as free water molecules move into the solution with the lower concentration of free water molecules.

Figure 2 Osmosis

Water diffuses through channels in the cell membrane by osmosis.

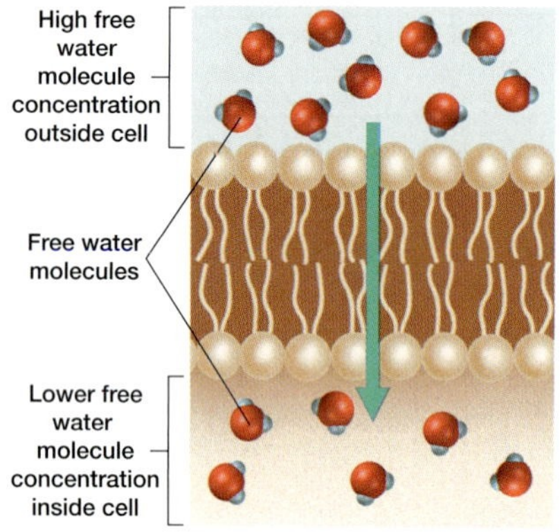

QUICK LAB — *Observing Osmosis* SC.F.1.4.4

You can observe the movement of water into or out of a grape under different conditions.

Materials

3 grapes, 3 small jars with lids, saturated sugar solution, grape juice, tap water, marking pen, paper towel, balance

Procedure

1. Make a data table with four columns (Solution, Original mass, Predicted mass, and Actual mass) and a row for each solution (Sugar solution, Grape juice, and Water).
2. Fill one jar with the sugar solution. Fill a second jar with grape juice. (The grape will be more visible inside the jar if you fill the jar with white grape juice, as shown in the middle jar in the photo above right.) Fill the third jar with tap water. Label each jar according to the solution it contains.
3. Using the balance, find the mass of each grape. Place one grape in each jar, and record the mass of each jar in your data table. Put a lid on each jar.
4. Predict whether the mass of each grape will increase or decrease over time. Explain your predictions.
5. After 24 hours, remove each grape from its jar, and dry it gently with a paper towel. Using the balance, find its mass again. Record your results.
6. Clean up your materials before leaving the lab.

Analysis

1. **Identify** the solutions in which osmosis occurred.
2. **Critical Thinking Evaluating Conclusions** How did you determine whether osmosis occurred in each of the three solutions?
3. **Critical Thinking Evaluating Hypotheses** Did the mass of each grape change as you predicted? Why or why not?

MEDICINE CONNECTION

One of the major functions of the kidneys is to remove wastes from the blood. If a person is in renal (kidney) failure, however, this cannot be done, so wastes remain in the blood. In such cases, hemodialysis is used. Hemodialysis is a process that uses a semipermeable membrane to separate large substances, such as blood cells, from smaller ones, such as urea. Hemodialysis is often referred to as the artificial kidney. Blood from a patient's artery passes through a semipermeable cellophane tube inside a container. Here, the tube is surrounded by a solution that contains varying concentrations of electrolytes and other chemicals. Urea passes through the pores of the membrane and into a wash solution for removal. Blood cells cannot pass through the membrane, so they return to the patient's body through a vein.

Table 1 Hypertonic, Hypotonic, and Isotonic Solutions

If the fluid outside the cell has...	Then outside fluid is...	Water diffuses...	Effect on cell
...lower free water molecule concentration than cytosol	...hypertonic.	...out of cell.  H_2O	Cell shrinks.
...higher free water molecule concentration than cytosol	...hypotonic.	...into cell.  H_2O	Cell swells.
...same free water molecule concentration as cytosol	...isotonic.	...into and out of cell at equal rates. 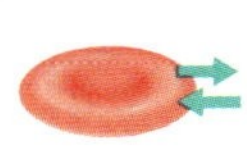 H_2O	Cell stays same size.

The direction of water movement across the cell membrane depends on the relative concentrations of free water molecules in the cytoplasm and in the fluid outside the cell. There are three possibilities for the direction of water movement:

1. **Water moves out.** When water diffuses out of the cell, the cell shrinks. A solution that causes a cell to shrink because of osmosis is called a **hypertonic** *(hie puhr TAHN ihk)* **solution.** If the fluid outside the cell has a higher concentration of dissolved particles than the cytoplasm has, then the outside fluid also has a lower concentration of free water molecules than the cytoplasm.
2. **Water moves in.** When water diffuses into the cell, the cell swells. A solution that causes a cell to swell because of osmosis is called a **hypotonic** *(hie poh TAHN ihk)* **solution.** If the fluid outside the cell has a lower concentration of dissolved particles than the cytoplasm has, then the outside fluid also has a higher concentration of free water molecules than the cytoplasm.
3. **No net water movement.** If the cytoplasm and the fluid outside the cell have the same concentration of free water molecules, water diffuses into and out of the cell at equal rates. This results in no net movement of water across the cell membrane, and the cell stays the same size—a state of equilibrium. A solution that produces no change in cell volume because of osmosis is called an **isotonic** *(ie soh TAHN ihk)* **solution. Table 1** summarizes the effects of hypertonic, hypotonic, and isotonic solutions on cells.

If left unchecked, the swelling caused by a hypotonic solution could cause a cell to burst. Different kinds of cells have different adaptations that deal with this problem. The cells of plants and fungi have rigid cell walls that keep the cells from expanding too much. Some unicellular eukaryotes have contractile vacuoles *(kuhn TRAK tihl VAK yoo ohlz),* which are organelles that collect excess water inside the cell and force the water out of the cell. Animal cells have neither cell walls nor contractile vacuoles. However, many animal cells can avoid swelling caused by osmosis by removing dissolved particles from the cytoplasm. The removal of dissolved particles from a cell increases the concentration of free water molecules inside the cell.

WORD Origins

The words *hypertonic, hypotonic,* and *isotonic* have the same ending, *-tonic,* which is from the Greek *tonos,* meaning "tension." The prefix *hyper–* is from the Greek *hyper,* meaning "over." The prefix *hypo–* is from the Greek *hypo,* meaning "under." The prefix *iso–* is from the Greek *isos,* meaning "same."

77

Integrating Physics and Chemistry

Hypotonic, isotonic, and hypertonic are terms that indicate the relative concentrations of two solutions which are being compared. Distinguish these solution chemistry terms from other relative terms such as dilute and concentrated. Whereas a dilute solution has only a small amount of solute dissolved in it, a concentrated solution has a relatively large amount of solute dissolved in it. For example, even though two sodium chloride solutions may both be considered dilute, whichever one has the higher ion or molecular concentration would be the hypertonic solution while the other one would be the hypotonic solution.

Teaching Tip — Basic

Sticky Drawers Ask students if they have ever noticed drawers in dressers that get sticky when there has been a lot of rain or high humidity. Tell them that on very wet days, wooden drawers in dressers absorb water from the air because of osmosis. This absorption warps the wood, making opening and closing the drawers more difficult. In drier weather, water is lost from the wood, and the drawers return to their original size.

Reading Skill Builder — General

Paired Reading Assign students to cooperative pairs, and have each student read this lesson silently. As they read, have students indicate with a check mark the passages that they understand, and indicate with a question mark those passages that they do not understand. (They should use pencil for their marks.) Have the partners discuss what they did or did not understand.
LS Interpersonal

Demonstration — General

Prepare three potato slices, each approximately 1 cm thick. Place one slice in distilled water, one in tap water, and one in salt water. Ask students to classify the solutions using the terms hypotonic, hypertonic, and isotonic. (Distilled water represents a hypotonic solution; tap water represents an isotonic solution; and salt water represents a hypertonic solution.) Ask students to predict what will happen to each slice. (The slice in the distilled water should swell and become crisp. The slice in the tap water should be unchanged. The slice in the salt water should shrink.)
LS Logical

Teach, *continued*

Teaching Tip

Calcium Channel Blockers
Tell students that calcium channel blockers are drugs that inhibit the flow of calcium ions through channels in the muscle cells of the heart. They decrease heart rate and the force of each heartbeat. They are used to decrease blood pressure, relieve chest pain, and stabilize abnormal heart rhythms.

Using the Figure

Remind students that the cell membrane is selectively permeable to different substances. Only certain ions can pass through each type of ion channel. Only sodium ions, for example, can pass through the ion channel shown in **Figure 3.** They span the entire thickness of the cell membrane. Ions are charged particles, so when one passes through the pore of a channel, it is shielded from the nonpolar interior of the lipid bilayer.

READING SKILL BUILDER — BASIC

Interactive Reading Assign Chapter 4 of the *Holt Biology* Guided Audio CD Program to help students achieve greater success in reading the chapter. LS Auditory

Crossing the Cell Membrane

Recall that most ions and polar molecules cannot pass across the cell membrane because they cannot pass through the nonpolar interior of the lipid bilayer. However, such substances can cross the cell membrane when they are aided by transport proteins. Transport proteins called *channels* provide polar passageways through which ions and polar molecules can move across the cell membrane. Each channel allows only a specific substance to pass through the cell membrane. For example, some channels allow only one type of ion to cross the cell membrane, while others transport a particular kind of sugar or amino acid. This selectivity is one of the most important properties of the cell membrane because it enables a cell to control what enters and leaves.

Diffusion Through Ion Channels

Ions such as sodium, Na^+, potassium, K^+, calcium, Ca^{2+}, and chloride, Cl^-, are involved in many important cell functions. For example, ions are essential to the ability of nerve cells to send electrical signals throughout your body. Muscle cells in your heart could not make your heart beat without the movement of ions between the cells. Although ions cannot diffuse through the nonpolar interior of the lipid bilayer, they can cross the cell membrane by diffusing through ion channels. An **ion channel** is a transport protein with a polar pore through which ions can pass. As **Figure 3** shows, the pore of an ion channel spans the thickness of the cell membrane. Thus, an ion that enters the pore can cross the cell membrane without contacting the nonpolar interior of the lipid bilayer.

The pores of some ion channels are always open. In other ion channels, the pores can be closed by ion channel gates. A model of an ion channel with a gate is shown in Figure 3. Ion channel gates may open or close in response to different kinds of stimuli. These include the stretching of the cell membrane, a change in electrical charge, or the binding of specific molecules to the ion channel. In this way, the stimuli are able to affect the ability of particular ions to cross

Figure 3 Ion channels

Ion channels allow certain ions to pass through the cell membrane.

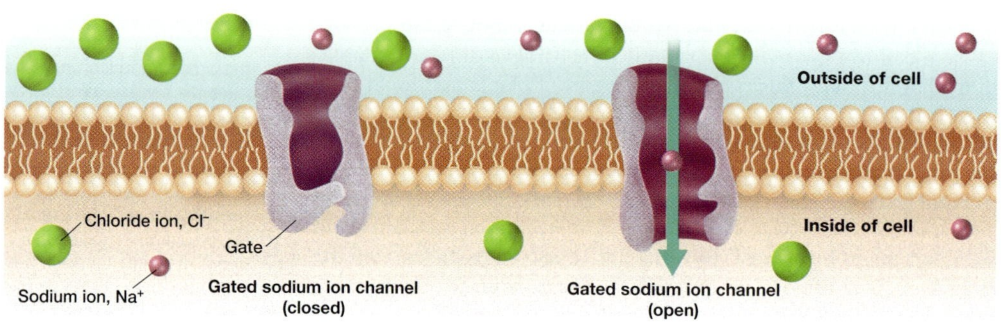

78

REAL WORLD CONNECTION

Everyone has been cautioned about the dangers of dehydration, but too much water can be just as deadly. This condition is known as hyponatremia. Hyponatremia is defined as not having enough sodium in the blood. The main cause is simply drinking too much water while not taking in enough sodium. Sweat contains 2 to 3 grams of salt per liter. Under the right conditions, the body can lose 1 liter of sweat an hour. If a person is exercising heavily and just replacing water and not salt, hyponatremia may develop. The symptoms are similar to those of dehydration: nausea, muscle cramps, disorientation, and poor coordination. Swelling in the brain can induce seizures and even death. Distinguishing dehydration from hyponatremia can be difficult. Typically, hyponatremia is more likely to involve vomiting and a history of recently consuming large amounts of water. Hyponatremia is an urgent medical condition. Drinking sports drinks instead of water to replace lost fluids can help prevent hyponatremia.

the cell membrane. Like the diffusion of small molecules and nonpolar molecules through the lipid bilayer, the diffusion of ions through ion channels is a form of passive transport. No use of energy by the cell is required because the ions move down their concentration gradients.

internet connect
www.scilinks.org
Topic: Ion Channels
Keyword: HX4106
SciLinks Maintained by the National Science Teachers Association

Electrical Charge and Ion Transport

The rate of movement of a substance across the cell membrane is generally determined by the concentration gradient of the substance. The movement of a charged particle, such as an ion, across the cell membrane is also influenced by the particle's positive or negative electrical charge. The inside of a typical cell is negatively charged with respect to the outside of the cell. Opposite charges attract, and like charges repel. Thus, a more positively charged ion located outside the cell is more likely to diffuse into the cell, where the charge is negative. Conversely, a more negatively charged ion located inside the cell is more likely to diffuse out of the cell. The direction of movement caused by an ion's concentration gradient may oppose the direction of movement caused by the ion's electrical charge. Thus, an ion's electrical charge often affects the diffusion of the ion across the cell membrane. This is very important to the functioning of nerve cells in animals.

DATA LAB

Analyzing the Effect of Electrical Charge on Ion Transport SC.F.1.4.4

Background

The electrical charge of an ion affects the diffusion of the ion across the cell membrane. Some ions are more concentrated inside cells, and some ions are more concentrated outside cells. Use the table below to answer the following questions:

Magnification: 13,000×

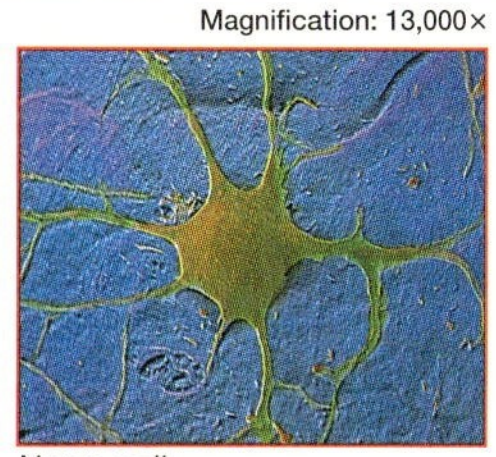

Nerve cell

Ion Charges and Concentration Inside and Outside Cell

Ion	Charge of ion	Concentration of ion outside cell : inside cell
Sodium (Na^+)	Positive	10:1
Potassium (K^+)	Positive	1:20
Calcium (Ca^{2+})	Positive	10,000:1
Chloride (Cl^-)	Negative	12:1

Analysis

1. **Identify** the ion that is more concentrated inside the cell than outside the cell.
2. **Identify** those ions that are more concentrated outside the cell than inside the cell.
3. **Critical Thinking Recognizing Relationships** Do the positive charges of calcium ions and sodium ions make these ions more likely to move into or out of the cell?
4. **Critical Thinking Inferring Relationships** Which ions' electrical charges oppose the direction of movement that is caused by their concentration gradient?

79

Using the Figure — GENERAL

Refer students to **Figure 4** on the next page as you review the process of facilitated diffusion. Compare the carrier proteins in this figure with the ion channels in **Figure 3**. Tell students that when the concentration of a substance in the cell is low, molecules of the substance outside the cell quickly enter the cell through carrier proteins. This process is known as facilitated diffusion. Glucose and amino acids are too large to pass through the cell membrane itself; charged substances also cannot pass through directly.

DATA LAB

Analyzing the Effect of Electrical Charge on Ion Transport

Skills Acquired

Analyzing data, inferring relationships

Teacher's Notes

Review the principles of ions and ionic bonding. Remind students that opposite charges attract one another, and similar charges repel.

Answers to Analysis

1. potassium ion
2. sodium ion, calcium ion, and chloride ion
3. calcium ions and sodium ions are more likely to move into the cell because of their positive charges.
4. Potassium ions move down their concentration gradient out of the cell. This movement, however, is opposed by the ions' attraction to the negatively charged interior of the cell. Chloride ions move down their concentration gradient into the cell, but this movement is opposed by the ions' being repelled by the negatively charged interior of the cell.

Close

Reteaching — Basic

Assign students to cooperative groups of three. Write the following questions on the board.

1. Why are green leafy vegetables sprinkled with water at the supermarket? (This prevents them from wilting due to water loss.)
2. Why is salt sometimes used to preserve foods? (Because a salt solution is hypertonic, the microorganisms present on the foods shrink and die.)
3. Why should you not drink sea water? (Sea water contains high concentrations of solutes. Cells in the body lose water because of osmosis, possibly causing death.)

Quiz — General

True or False:

1. Diffusion occurs as a result of the random motion of substances. (True)
2. Both diffusion and osmosis can occur in an open area or across a cell membrane. (False. Osmosis occurs only across a semipermeable membrane.)
3. The channels through which large and charged particles enter a cell are lined with lipids. (False. The channels are made of proteins.)

Alternative Assessment — General

Ask students to write a paragraph describing how the door on a house is like a cell membrane. Suggest that they think of a screen door in front of the solid door.

Section Review Correlations

SC.F.1.4.2: Items 3, 5

SC.F.1.4.4: Items 1, 2, 4

Facilitated Diffusion

Most cells also have a different kind of transport protein that can bind to a specific substance on one side of the cell membrane, carry the substance across the cell membrane, and release it on the other side. Such proteins are called **carrier proteins.** When carrier proteins are used to transport specific substances—such as amino acids and sugars—down their concentration gradient, that transport is called facilitated diffusion. **Facilitated** *(fah SIHL uh tayt ehd)* **diffusion,** shown in **Figure 4,** is a type of passive transport. It moves substances down their concentration gradient without using the cell's energy.

Step 1 The carrier protein binds a specific molecule on one side of the cell membrane.

Step 2 A change in the shape of the carrier protein exposes the molecule to the other side of the cell membrane.

Step 3 The carrier protein shields the molecule from the interior of the lipid bilayer. The molecule is then released from the carrier protein, which returns to its original shape.

Figure 4

BIO graphic — Facilitated Diffusion

Carrier proteins transport substances down their concentration gradient.

1 A molecule outside the cell binds to a carrier protein on the cell membrane.

2 The carrier protein transports the molecule across the cell membrane.

3 The molecule is released from the carrier protein inside the cell.

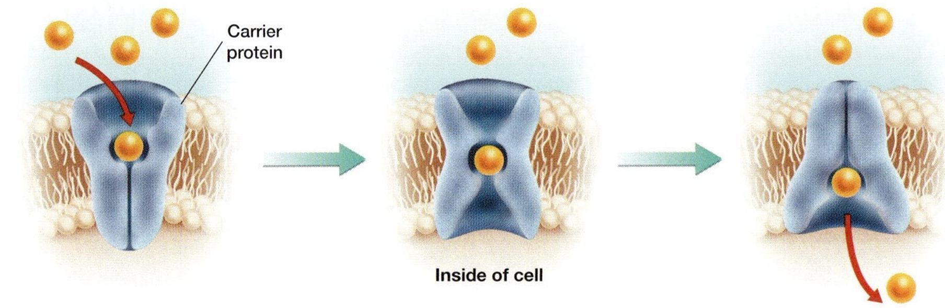

Section 1 Review

1. **Distinguish** between diffusion and equilibrium. SC.F.1.4.4
2. **Describe** how the diffusion of ions across a cell membrane differs from the diffusion of nonpolar molecules across the cell membrane. SC.F.1.4.4
3. **Explain** how some substances cross the cell membrane by facilitated diffusion. SC.F.1.4.2
4. **Critical Thinking Predicting Outcomes** Predict what would happen to a cell that is placed in a hypertonic solution, and explain why this would occur. SC.F.1.4.4
5. **FCAT Prep** Which substance crosses the cell membrane by facilitated diffusion?
   A a sugar
   B water
   C sodium ion
   D chloride ion
   SC.F.1.4.2

80

Answers to Section Review

1. Diffusion is the movement of a substance from an area of high concentration to an area of lower concentration. It is caused by the random motion of particles of the substance. Equilibrium is a condition in which the concentration of a substance is equal throughout a space.
2. Because ions are charged, they cannot pass through the nonpolar interior of the lipid bilayer. Instead, they must pass through ion channels embedded in the lipid bilayer. Nonpolar molecules, however, can pass directly through the lipid bilayer.
3. Facilitated diffusion involves carrier proteins that transport substances across the cell membrane down their concentration gradient.
4. Water would diffuse out of the cell, causing it to shrink. The water would move from an area that has a higher concentration of water (inside the cell) to an area with a lower concentration of water (the solution).
5. **A.** Correct. **B.** Incorrect. Water crosses the membrane by osmosis. **C.** Incorrect. Ions diffuse across the membrane through ion channels. **D.** Incorrect. See answer C.

Active Transport

Section 2

Florida Sunshine Standards/FCAT

SC.F.1.4.3 Membranes are sites for chemical synthesis and essential energy conversions.

SC.F.1.4.5 Complex interactions among the different kinds of molecules in the cell cause distinct cycles of activity governed by proteins.

SC.F.1.4.6 Separate parts of the body communicate with each other using electrical and/or chemical signals.

Objectives

- **Compare** active transport with passive transport. SC.F.1.4.5
- **Describe** the importance of the sodium-potassium pump. SC.F.1.4.5
- **Distinguish** between endocytosis and exocytosis. SC.F.1.4.3
- **Identify** three ways that receptor proteins can change the activity of a cell. SC.F.1.4.6

Key Terms

active transport
sodium-potassium pump
endocytosis
exocytosis
receptor protein
second messenger

Movement Against a Concentration Gradient

Although facilitated diffusion can help move amino acids and sugars across the cell membrane, it can only transport these substances down their concentration gradient. Cells must transport certain amino acids, sugars, and other substances into their cytoplasm from the surrounding fluid. But many of these substances have a low concentration outside cells and a higher concentration inside cells. Their concentration gradients would cause these important substances to move out of the cell rather than into the cell. So, cells also have a way to move some substances against their concentration gradient—from an area of low concentration to an area of higher concentration.

The transport of a substance across the cell membrane against its concentration gradient is called **active transport**. Unlike passive transport, active transport requires the cell to use energy because the substance is being moved against its concentration gradient. Most often, the energy needed for active transport is supplied directly or indirectly by ATP.

Some active-transport processes involve carrier proteins. Like the carrier proteins used in facilitated diffusion, the carrier proteins used in active transport bind to specific substances on one side of the cell membrane and release them on the other side of the cell membrane. But in active transport, the substances bind to carrier proteins where they are low in concentration and are released where they are higher in concentration. Thus, carrier proteins in active transport function as "pumps" that move substances against their concentration gradient. For this reason, these carrier proteins are often called membrane pumps.

Sodium-Potassium Pump

One of the most important membrane pumps in animal cells is a carrier protein called the sodium-potassium pump. In a complete cycle, the **sodium-potassium pump** transports three sodium ions, Na^+, out of a cell and two potassium ions, K^+, into the cell. Sodium ions are usually more concentrated outside the cell than inside the cell, and potassium ions are typically more concentrated inside the cell than outside the cell. Thus, the sodium-potassium pump actively transports both sodium ions and potassium ions against their concentration gradients. The energy needed to power sodium-potassium pumps is supplied by ATP. In some cells, sodium-potassium pumps are so active that they use much of the ATP produced by the cells.

Real Life

Why saltwater frogs aren't in a pickle.

Some frogs have urea—a salty product of metabolism that is usually secreted as urine—in their blood. This makes their bodies nearly as salty as seawater, allowing them to live in saltwater environments.

Finding Information
Find out the species name of a saltwater frog.

Section 2

Focus

Overview

Before beginning this section review with your students the objectives listed in the Student Edition. Tell students that this section describes how substances enter cells when their concentration is higher inside the cells than outside the cells. The section also describes how substances that are too large to pass through carrier proteins move into and out of cells. Membrane receptor proteins, which trigger cellular responses, are also described.

Bellringer

Ask students to write several sentences that describe where hormones are produced, how they reach the cells they stimulate, and how those target cells recognize the hormones. (Hormones are produced in either the hypothalamus of the brain or in endocrine glands located in various parts of the body. Most hormones are released into the blood stream and are transported throughout the body by the circulatory system. Hormones are recognized by and bind to receptor proteins on target cells. This binding triggers a hormonally induced response in the target cells.)

Motivate

Demonstration — Basic

Bring an air pump and an inflatable ball to class. Ask a student to inflate the ball with the pump. Relate this activity to a cell membrane pump. (The air pump uses energy supplied by the student to move air against a pressure gradient.) Tell students that membrane pumps use energy supplied by the cell to move substances against their concentration gradient.

Section Correlations

SC.F.1.4.3
SC.F.1.4.5
SC.F.1.4.6

Chapter Resource File

- Directed Reading Basic
- Active Reading General

One-Stop Planner CD-ROM

- Reading Organizers Basic
- Reading Strategies Basic
- Supplemental Reading Guide
  The Lives of a Cell Advanced

Transparencies

TR Bellringer
TR B21 Sodium-Potassium Pump
TR B22 Endocytosis and Exocytosis
TR B20 Changes in Permeability

Teach

Teaching Tip — GENERAL

Hydrogen Ion Pump During cellular respiration (covered in detail in Chapter 5), cells harvest energy from food molecules. ATP is produced in varying amounts, depending on the active transport of hydrogen ions. In the electron transport chain of cellular respiration, the hydrogen ion pump is used to generate a hydrogen ion concentration gradient that drives the production of ATP. Ask students to suggest where these hydrogen ion pumps are located in animal cells. (Many are located in mitochondria.) Have students construct a Graphic Organizer similar to the one at the bottom of this page that illustrates the function of the sodium-potassium pump.

Using the Figure — GENERAL

Review ATP hydrolysis (the breakdown of ATP into ADP + Pi), which is shown in **Figure 5**. Ask students how and where ATP is formed. (It is formed by the addition of a phosphate group to ADP to make ATP, which occurs during cellular respiration in mitochondria.) Point out to students that the sodium-potassium pump goes through four steps to move three sodium ions out of a cell and two potassium ions into the cell. Remind students that these ions are being transported against their concentration gradients. Cells must use energy (in the form of ATP) to power the sodium-potassium pump.

Figure 5

BIO graphic — Sodium-Potassium Pump

The sodium-potassium pump actively transports sodium ions, Na^+, and potassium ions, K^+, against their concentration gradient.

1. Three sodium ions, Na^+ and a phosphate group (P) from ATP bind to the pump.
2. The pump changes shape, transporting the three sodium ions across the cell membrane.
3. Two potassium ions, K^+, bind to the pump and are transported across the cell membrane.
4. The phosphate group and the two potassium ions are released inside the cell.

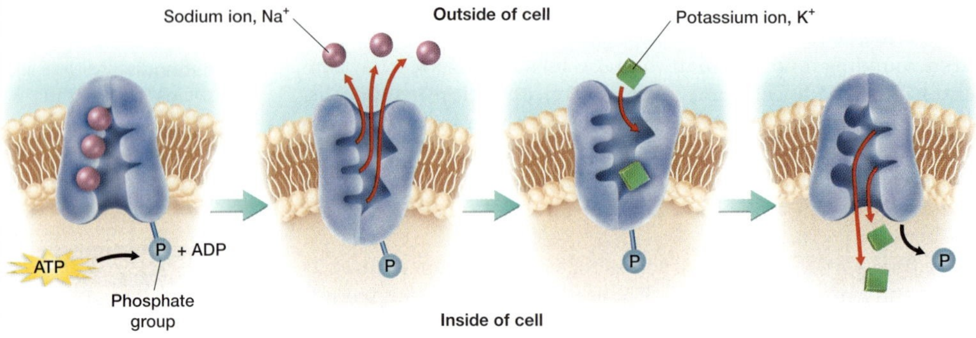

A model of the sodium-potassium pump is shown in **Figure 5.**

Step 1 Three sodium ions inside the cell bind to the sodium-potassium pump. Because energy is needed to move the sodium ions against their concentration gradient, a phosphate group is removed from ATP and also binds to the pump.

Step 2 The pump changes shape, transporting the three sodium ions across the cell membrane and releasing them outside the cell.

Step 3 The pump is now exposed on the surface of the cell. Two potassium ions outside the cell bind to the pump. The phosphate group is released, changing the shape of the pump.

Step 4 The pump is again exposed to the inside of the cell. The two potassium ions are transported across the cell membrane and are released inside the cell.

The sodium-potassium pump is important for two main reasons. First, the pump prevents sodium ions from accumulating in the cell. Sodium ions continuously diffuse into the cell through ion channels embedded in the lipid bilayer of the cell membrane. The increased concentration of sodium ions would then cause water to enter the cell by osmosis, causing the cell to swell or even burst. Second, the sodium-potassium pump helps maintain the concentration gradients of sodium ions and potassium ions across the cell membrane. Many cells use the sodium-ion concentration gradient to help transport other substances, such as glucose, across the cell membrane.

82

Graphic Organizer

Use this graphic organizer with **Teaching Tip** *on this page.*

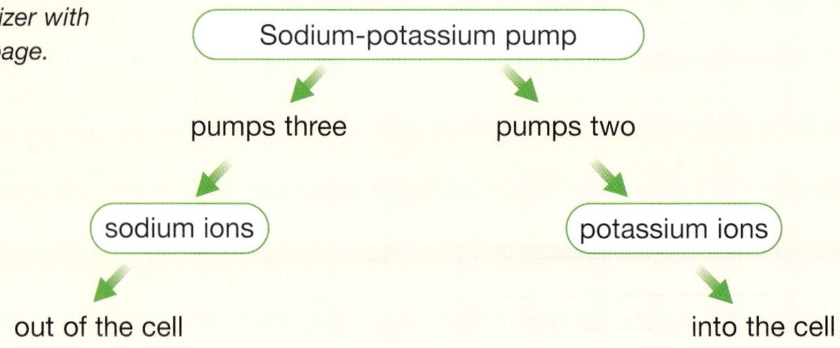

Movement in Vesicles

Many substances, such as proteins and polysaccharides, are too large to be transported by carrier proteins. These substances are moved across the cell membrane by vesicles. The movement of a substance into a cell by a vesicle is called **endocytosis** *(ehn doh sie TOH sihs)*. During endocytosis, the cell membrane forms a pouch around a substance, as shown in **Figure 6.** The pouch then closes up and pinches off from the membrane to form a vesicle. Vesicles formed by endocytosis may fuse with lysosomes or other organelles.

The movement of a substance by a vesicle to the outside of a cell is called **exocytosis** *(ek soh sie TOH sihs),* also shown in Figure 6. During exocytosis, vesicles in the cell fuse with the cell membrane, releasing their contents. Cells use exocytosis to export proteins that are modified by the Golgi apparatus. Nerve cells and cells of various glands, for example, release proteins by exocytosis.

Study TIP

Interpreting Graphics
As you look at Figure 6, notice that during endocytosis, the cell membrane pinches off to become the vesicle membrane. Conversely, during exocytosis, the vesicle membrane becomes part of the cell membrane.

Figure 6 Endocytosis and exocytosis

Vesicles transport substances into and out of cells.

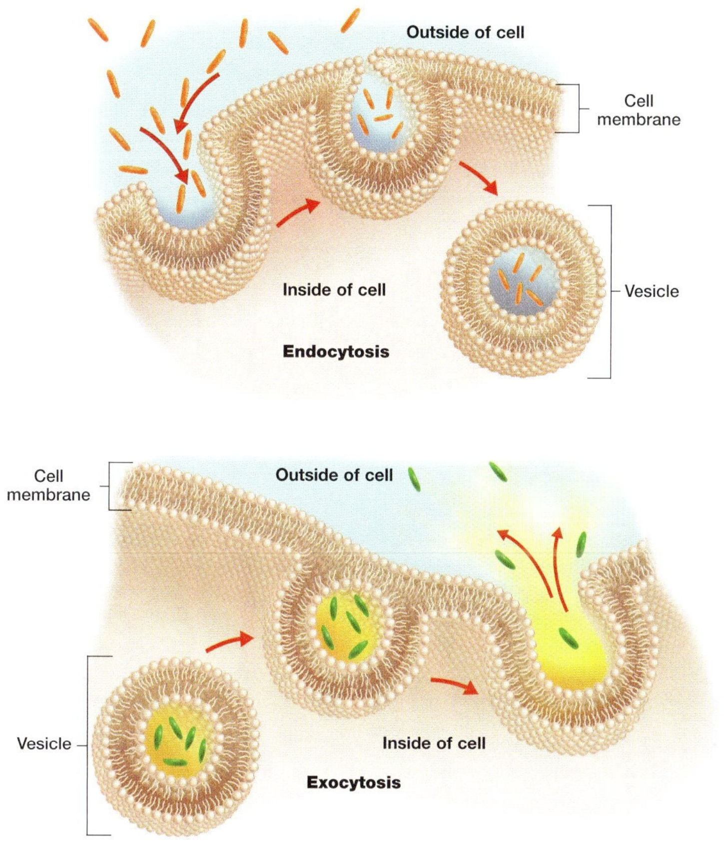

Teaching Tip — BASIC

Endocytosis Tell students that there are actually two types of endocytosis. One is phagocytosis, in which the material brought into the cell is another cell or fragments of organic matter. Phagocytosis is common among unicellular eukaryotes, such as the amoeba, which devour their prey by engulfing them. Ask students if they can think of another example of phagocytosis. (Macrophages engulf invading pathogens.) Another type of endocytosis is pinocytosis, in which the material brought into the cell is a liquid that contains dissolved particles. Maturing human egg cells use pinocytosis to take in dissolved nutrients that are secreted by surrounding cells.

Group Activity — ADVANCED

Good and Bad Cholesterol Tell students that cholesterol is transported into cells by endocytosis. Cholesterol travels in the blood partly as low-density lipoprotein (LDL). LDL binds to receptor sites on the cell membrane, triggering endocytosis. LDL is composed of an outer lipid-protein membrane that surrounds cholesterol molecules. LDL, or "bad cholesterol," contributes to atherosclerosis, in which fatty substances build up on the interior surfaces of artery walls. There is also "good cholesterol," which is essential to the body's functions. For example, the body needs cholesterol to produce lipid-based molecules, such as steroid hormones. Have students work in small groups to research and write a report on the differences between "good" and "bad" cholesterol.

LS Verbal Co-op Learning

TECHNOLOGY CONNECTION

A type of reverse osmosis technology is being used by the U.S. Army to develop lightweight chemical and biological protective clothing. This clothing will provide protection from toxic chemicals and biological agents, including viruses and bacteria. Eventually this clothing may be used by emergency hospital personnel as well as industries that deal with chemicals and pesticides. The new clothing will not only be lightweight, but it will also take up less space, wash easily, be waterproof, and allow moisture to pass through to help cooling.

Attention Grabber

Curare Curare is a poison obtained from a woody vine, *Struchnos toxifer,* that is found in South America. Certain tribes of that continent use curare as an arrow poison. Curare blocks acetylcholine receptors in muscle cells. This change results in paralysis and usually death, since muscles involved in respiration, such as the diaphragm, stop functioning. After World War II, curare was used as a drug, particularly as a muscle relaxant during surgery. The patient was monitored and the lungs were kept working by artificial means. Curare is not an anesthetic.

Teach, *continued*

Teaching Tip

Glucose Transport Tell students that cells transport glucose using a sodium ion concentration gradient. This gradient is established by sodium-potassium pumps. Eventually the sodium ions that moved outside the cell during active transport will reenter the cell through another transport protein down their concentration gradient. When this happens, glucose enters the cell coupled with sodium ions.

Using the Figure — GENERAL

Point out that the receptor protein in **Figure 8** is coupled with an ion channel. The binding of a specific signal molecule to the receptor protein causes the ion channel to open. When the channel is open, sodium ions move through the channel down their concentration gradient. **LS Visual**

SKILL BUILDER — ADVANCED

Math Skills Tell students to consider a person who burns 2,000 Calories per day. Tell them that 180 grams of glucose, when broken down in the process of respiration yields 686 Calories. Also tell students that one pound equals approximately 454 grams. Ask them to calculate how much glucose this person would have to consume in 1 day to meet their energy needs from glucose alone. ([2,000 Cal/day] × [180 g Glu/686 Cal] × [1 lb/454 g] = 1.2 lb Glu/day) **LS Logical**

Figure 7 Action of signal molecules. When you exercise, signal molecules are bound by receptor proteins on your muscle cells, signaling your muscles to contract.

Membrane Receptor Proteins

We are constantly bombarded with information from other people and through television, the Internet, and many other media. To interpret information, we must be able to communicate and to distinguish between important and unimportant information. Similarly, your body's cells must communicate with each other to coordinate your growth, metabolism, and other activities. Cells that do not lie next to each other cannot communicate directly. Instead, some cells release *signal molecules* that carry information to nearby cells and throughout the body. Hormones are one familiar example of signal molecules. Hormones are made in one part of the body and carried in the bloodstream to other parts, where they have their effects.

Cells must also respond to important information and filter out unimportant information. Cells can receive the messages carried by certain signal molecules because the cell membrane contains specialized proteins that bind these signal molecules. Such proteins are called receptor proteins. A **receptor protein** is a protein that binds to a specific signal molecule, enabling the cell to respond to the signal molecule. For example, the muscles of the person exercising in **Figure 7** could not contract without receptor proteins and signal molecules that tell the muscles when to contract and when to relax.

Functions of Receptor Proteins

A signal molecule is bound by a receptor protein that fits that molecule, as shown in **Figure 8.** Most receptor proteins are embedded in the lipid bilayer of the cell membrane. The part of the protein that fits the signal molecule faces the outside of the cell.

The binding of a signal molecule by its complementary receptor protein causes a change in the receiving cell. This change can occur in the following three ways: by causing changes in the permeability of the receiving cell; by triggering the formation of second messengers inside the cell; and by activating enzymes inside the cell.

Figure 8 Changes in permeability

Some receptor proteins are coupled with ion channels.

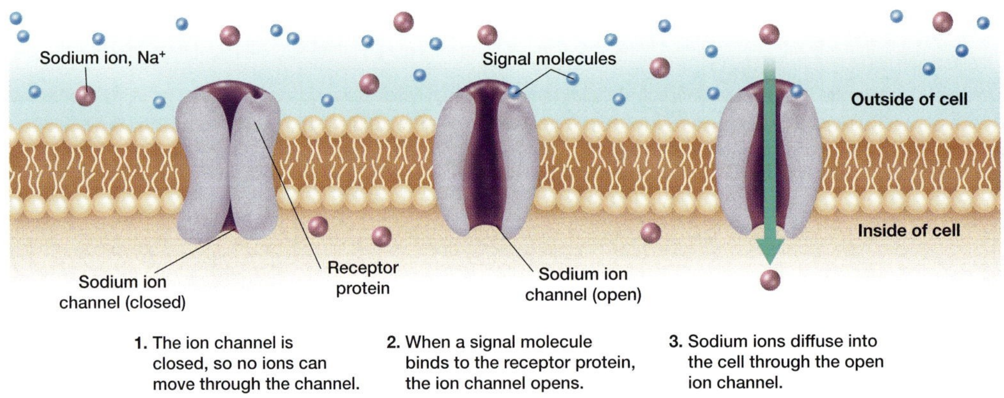

84

Career

A pharmacist not only dispenses medications prescribed by physicians but also provides information or counseling about medications to patients. Pharmacists inform patients about drug interactions, side effects, dosages, and how to take the medication. They also provide information about over-the-counter drugs. The employment conditions of pharmacists vary. For example, a pharmacist may work in a hospital or in a pharmacy. Most employers prefer pharmacists to have a master's degree and sometimes a Ph.D. in pharmacology. To become licensed, a prospective pharmacist must pass a state exam and then serve as an intern under a licensed pharmacist. The demand for pharmacists is expected to increase over the next few years, since the need for medications is increasing with the age of our population. Earnings can be very high, but the hours may be very long.

Figure 9 **Second messengers**

Some receptor proteins trigger the production of second messengers.

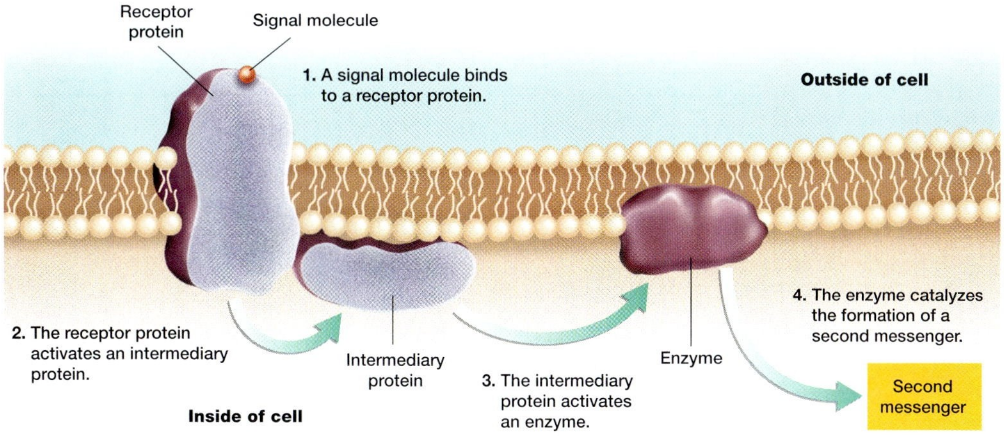

Changes in Permeability The receptor protein may be coupled with an ion channel, as shown in Figure 8. The binding of a signal molecule to the receptor protein causes the ion channel to open, allowing specific ions to cross the cell membrane. This type of receptor protein is especially important in the nervous system.

Second Messengers The receptor protein may cause the formation of a second messenger inside the cell, as shown in **Figure 9.** When it is activated, a **second messenger** acts as a signal molecule in the cytoplasm. The second messenger amplifies the signal of the first messenger—that is, the original signal molecule. Second messengers can change the functioning of a cell in several ways. For example, some second messengers activate enzymes, triggering a series of biochemical reactions in the cell. Other second messengers change the permeability of the cell by opening ion channels in the cell membrane.

Enzyme Action The receptor protein may act as an enzyme. When a signal molecule binds to the receptor protein, the receptor protein may speed up chemical reactions inside the cell. Receptor proteins may also activate other enzymes located inside the cell or in the cell membrane, triggering chemical reactions in the cell. In this way, the signal molecule can cause many changes in the functioning of the receiving cell.

Many drugs affect the binding of signal molecules to receptor proteins. Some drugs, such as the illegal drug heroin, imitate signal molecules by binding to receptor proteins on a receiving cell, altering the function of the cell. Other drugs block or interfere with receptor proteins, preventing signal molecules from binding to the receptor proteins. For example, signal molecules that bind to receptor proteins on heart-muscle cells stimulate the cells, causing the heart rate to

Real Life

Many medicines are drugs that bind to receptor proteins.

Some of these drugs interfere with the receptor's ability to bind to signal molecules.

Finding Information

Research some medicines that bind to receptor proteins.

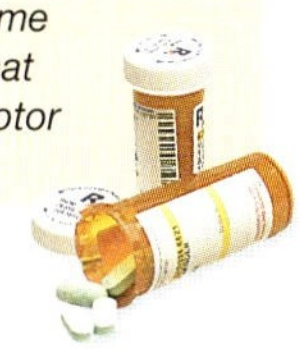

internet connect

www.scilinks.org

Topic: Receptor Proteins

Keyword: HX4157

SCILINKS. Maintained by the National Science Teachers Association

Demonstration — GENERAL

Contact a local pharmaceutical company or the biology department of a local university. Arrange for a speaker to give a presentation to your students on medicines that work by affecting the functions of receptor proteins. Before the presentation, consult with the speaker on the topics to be discussed. Be sure the speaker includes the name of each drug, the receptor it affects, its mode of action, the conditions or diseases it alleviates, and the symptoms of the conditions.

LS Intrapersonal

Real Life

Answer

Examples of medicines that bind to receptor proteins include heart medications (e.g., beta blockers and calcium channel blockers) and narcotics (e.g., morphine and codeine).

BIOWatch

Looking at Membrane Proteins

Teaching Strategies

Tell students that spectroscopy reveals the types of atoms that are in a molecule, the relative number of each type, and how the atoms are connected.

The National High Magnetic Field Laboratory in Tallahassee is a leading center for high field magnet research and development. The laboratory provides high magnetic fields used in a wide range of disciplines.

Discussion

Why is it difficult to determine the structures of cell membrane proteins? (Proteins are very large, usually consisting of more than 100 amino acids. Their three-dimensional structures are complex. Cell membrane proteins are embedded in the sticky surface of the cell.)

increase. Beta blockers, which are drugs prescribed to patients with a rapid heartbeat, bind to some of these receptor proteins. Beta blockers therefore interfere with the binding of signal molecules to the receptor proteins, preventing the heart rate from increasing too rapidly.

FLORIDA BIOWatch

Looking at Membrane Proteins

If you could look closely at the cell surface, you would see a landscape studded with membrane proteins. These proteins act as gateways, enabling the cell to react and respond to the outside world. Membrane proteins receive signals, transport materials, assist chemical reactions, and let one cell recognize another. They are crucial to the life of the cell, yet their structures remain a mystery.

Microscopic Complexity

At the National High Magnetic Field Laboratory at Florida State University in Tallahassee, Florida, Dr. Timothy Cross is studying the molecular structures of key membrane proteins. Cross, a chemist and biophysicist, aims to unravel the mysteries of membrane protein structure. Working with scientists from around the United States, Cross is studying the membrane proteins of *Mycobacterium tuberculosis,* the bacterium responsible for causing tuberculosis (TB).

Because proteins are complex molecules, they are difficult to observe and model. Most proteins consist of more than one hundred amino acids arranged in a precise sequence to form a three-dimensional molecule with a specific shape. Membrane proteins are especially challenging to study because they are embedded in the sticky cell surface.

High-Tech Microscopes

Cross leads a research team of chemists, biologists, physicists, and mathematicians. The team uses nuclear magnetic resonance (NMR) spectroscopy, electron microscopy, x-ray crystallography, and high-resolution mass spectroscopy to identify and classify the cell membrane proteins that make communication possible.

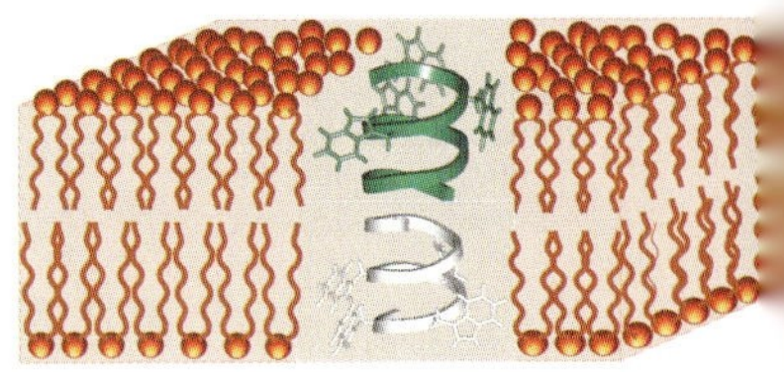

New Drugs to Fight TB

A worldwide disease that is difficult and expensive to treat, tuberculosis killed approximately two million people in 2002. Discovering the structures of *Mycobacterium* membrane proteins is a key step in the development of new drugs to fight the disease. New drugs could fight *Mycobacterium* by disrupting the genes that are responsible for making its membrane proteins. Researchers predict that many new drugs will specifically target the proteins in cell membranes.

Section 2 Review

1. **Distinguish** between passive transport and active transport. SC.F.1.4.5
2. **Describe** how the sodium-potassium pump helps prevent animal cells from bursting. SC.F.1.4.5
3. **Compare** two ways that the binding of a signal molecule to a receptor protein causes a change in the activity of the receiving cell. SC.F.1.4.6
4. **Identify** the terms *endocytosis* and *exocytosis* and distinguish between them. SC.F.1.4.3
5. **Critical Thinking Applying Information** During exercise, potassium ions accumulate in the fluid that surrounds muscle cells. Which cell membrane protein helps muscle cells counteract this tendency? Explain your answer. SC.F.1.4.5
6. FCAT Prep The concentration of molecule *X* is greater inside a cell than outside. If the cell acquires *X* from its surroundings, by what means would *X* cross the cell membrane?
   A exocytosis
   B active transport
   C receptor proteins
   D second messengers
   SC.F.1.4.3

86

Close

Reteaching — Basic

Assign students to cooperative groups of three. Write the following terms on the board: active transport, sodium-potassium pump, endocytosis, exocytosis, and receptor protein. Have the students in each group discuss and write down what they know about each term. Then have them check their answers using the textbook.

Quiz — General

1. Which ion or ions move against a concentration gradient in a sodium-potassium pump? (sodium and potassium)
2. What serves as an "escort" to enable substances to move across a cell membrane by active transport and by endocytosis or exocytosis? (Proteins make up the carriers that escort substances during active transport. A piece of the cell membrane carries substances during endocytosis or exocytosis.)
3. What are the three functions of receptor proteins? (to allow movement of a substance across a cell membrane; to stimulate the formation of a second messenger inside a cell; to act as an enzyme or activate enzymes inside cells)

Alternative Assessment — General

Assign students to cooperative groups of four. Have each group make a crossword puzzle with clues using the terms introduced in this section. Makes copies of all the puzzles, and have each group try to complete the puzzles.

Section Review Correlations

SC.F.1.4.3: Items 4, 6
SC.F.1.4.5: Items 1, 2, 5
SC.F.1.4.6: Item 3

Answers to Section Review

1. *Passive transport*—substances move across membranes down their concentration gradients. *Active transport*—cells use energy to move substances against their concentration gradients.
2. By pumping sodium ions out of the cell, it lowers the cell's concentration of sodium. Thus less water enters the cell by osmosis.
3. When a signal molecule binds to a receptor protein, the receptor protein may cause an ion channel to open. Alternately, the activated receptor protein may activate a second messenger, which may activate enzymes or open ion channels.
4. Endocytosis is the movement of substances into a cell by a vesicle. Exocytosis is the movement of substances out of a cell by a vesicle.
5. The sodium-potassium pump helps counteract this tendency by transporting potassium ions into muscle cells.
6. **A.** Incorrect. Exocytosis moves molecules out of the cell. **B.** Correct. Energy is needed to move molecules against their concentration gradient. **C.** Incorrect. Receptor proteins usually do not actively transport molecules across the membrane. **D.** Incorrect. Second messengers do not transport molecules across the membrane.

Key Concepts

1 Passive Transport

- Passive transport is the movement of substances across the cell membrane without the use of energy by the cell.
- Diffusion is the movement of a substance from an area of high concentration to an area of lower concentration.
- Osmosis is the diffusion of free water molecules across a selectively permeable membrane.
- Ion channels are proteins that have a pore through which ions can cross the cell membrane.
- In facilitated diffusion, a carrier protein transports a substance across the cell membrane down the concentration gradient of the substance.

2 Active Transport

- Active transport is the movement of a substance against the concentration gradient of the substance. Active transport requires cells to use energy.
- In animal cells, the sodium-potassium pump uses energy supplied by ATP to transport sodium ions out of the cell and potassium ions into the cell.
- During endocytosis, substances are moved into a cell by a vesicle that pinches off from the cell membrane.
- During exocytosis, substances inside a vesicle are released from a cell as the vesicle fuses with the cell membrane.
- Communication between cells often involves signal molecules that are bound by receptor proteins on cells.
- A signal molecule that is bound by a receptor protein on a cell can change the activity of the cell in three ways: by enabling specific ions to cross the cell membrane, by causing the formation of a second messenger, or by speeding up chemical reactions inside the cell.

Key Terms

Section 1

passive transport (74)
concentration gradient (74)
equilibrium (74)
diffusion (75)
osmosis (76)
hypertonic solution (77)
hypotonic solution (77)
isotonic solution (77)
ion channel (78)
carrier protein (80)
facilitated diffusion (80)

Section 2

active transport (81)
sodium-potassium pump (81)
endocytosis (83)
exocytosis (83)
receptor protein (84)
second messenger (85)

BIOLOGY INTERACTIVE TUTOR

Unit 1—*Cell Transport and Homeostasis*
Use Topics 1–6 in this unit to review the key concepts and terms in this chapter.

Study ZONE

Alternative Assessment — GENERAL

Have each student list the main types of passive transport mechanisms (diffusion, osmosis, ion channels, and facilitated diffusion) and types of active transport mechanisms (sodium-potassium pump, exocytosis, and endocytosis). Ask students to define each one, state the method of movement (down or against the concentration gradient), and give an example of each. Have them write their answers on paper and make drawings to illustrate the mechanisms. After students have completed their work, assign students to cooperative groups of three. Have the students in each group use the textbook to check their answers and critique their drawings.

Chapter Resource File

- Science Skills Worksheet GENERAL
- Critical Thinking Worksheet ADVANCED
- Test Prep Pretest GENERAL
- Chapter Test GENERAL

Answer to Concept Map

The following is one possible answer to Performance Zone item 15.

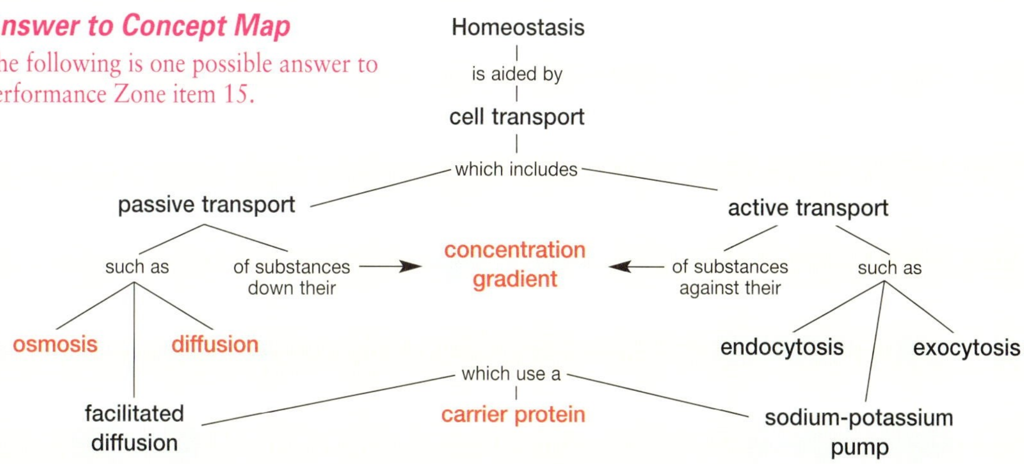

ANSWERS

Understanding Key Ideas

1. d
2. d
3. d
4. b
5. d
6. The solution is hypertonic. The concentration of free water molecules outside the cell was less than the concentration of free water molecules in the cytoplasm.
7. Homeostasis is the maintenance of a constant internal state in a changing environment. The sodium-potassium pump prevents sodium ions from accumulating in cells, which would cause an influx of water and cause cells to burst. The pump also helps transport important substances, such as glucose, into cells.
8. The combined vesicle and lysosome would help because lysosomes contain digestive enzymes that can digest the food.
9. Cell membrane proteins carry out tasks essential for life, such as receiving signals, transporting materials, assisting chemical reactions, and facilitating cell-cell recognition. A change in a membrane protein could destroy its function, leading to cell death.
10. The answer to the concept map is found at the bottom of the Study Zone page.

Critical Thinking

11. Selective permeability can result in certain substances maintaining an equilibrium across the cell membrane, whereas other substances that cannot cross the membrane have different concentrations inside and outside of the cell.

Chapter Review Correlations

SC.F.1.4.3: Items 2, 8, 9, 11, 12

SC.F.1.4.4: Items 1, 6, 13

SC.F.1.4.5: Items 3, 4, 7, 10

SC.F.1.4.6: Item 5

Performance ZONE — CHAPTER REVIEW

Understanding Key Ideas

1. In diffusion, a substance moves
- **a.** against its concentration gradient.
- **b.** from an area of lower concentration to an area of higher concentration.
- **c.** independent of its concentration.
- **d.** down its concentration gradient.

SC.F.1.4.4

2. Substances enter or leave a cell through the
- **a.** cytoplasm.
- **b.** Golgi apparatus.
- **c.** nucleus.
- **d.** cell membrane.

SC.F.1.4.3

3. Facilitated diffusion SC.F.1.4.5
- **a.** is driven by energy from ATP.
- **b.** is a type of active transport.
- **c.** employs receptor proteins.
- **d.** employs carrier proteins.

4. The sodium-potassium pump moves
- **a.** sodium ions into the cell and potassium ions out of the cell.
- **b.** sodium ions out of the cell and potassium ions into the cell.
- **c.** sodium and potassium into the cell.
- **d.** sodium and potassium out of the cell.

SC.F.1.4.5

5. The binding of a signal molecule by a receptor protein can
- **a.** activate a second messenger inside the receiving cell.
- **b.** trigger enzyme activity in the cell.
- **c.** change the permeability of the cell.
- **d.** All of the above

SC.F.1.4.6

6. The drawing below shows a plant cell that has become shriveled after having been placed in a solution. Is the solution most likely hypertonic, hypotonic, or isotonic? Explain your reasoning. SC.F.1.4.4

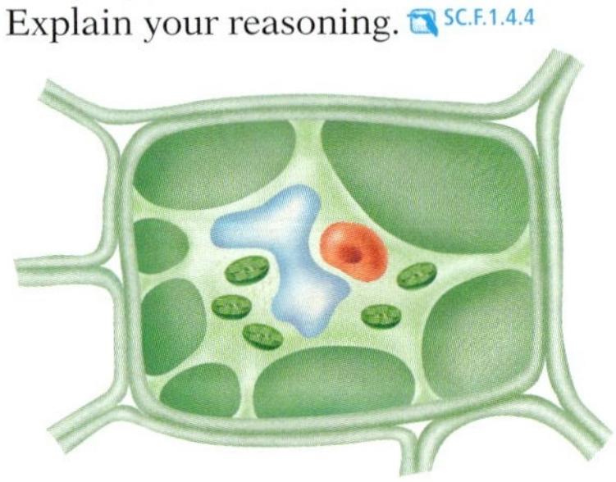

7. Define the term *homeostasis,* and explain how the sodium-potassium pump contributes to homeostasis in an animal. SC.F.1.4.5

8. When a cell takes in a food particle by endocytosis, the vesicle that is formed may fuse with a lysosome. How would fusion with a lysosome help the cell digest the food particle? SC.F.1.4.3

9. **BIOWatch** How could a change in a membrane protein cause the death of a cell? SC.F.1.4.3

10. **Concept Mapping** Make a concept map that shows how cells maintain homeostasis. Include the following terms in your map: *concentration gradient, diffusion, osmosis,* and *carrier protein.* SC.F.1.4.5

Critical Thinking

11. **Applying Information** Identify the property of a cell membrane that allows particles to pass through it. Predict how a cell membrane that allows some types of ions and molecules through but not others would affect equilibrium between the cell and the fluid around it. SC.F.1.4.3

Alternative Assessment

12. **Finding Information** Use library or Internet resources to investigate problems in transport of molecules across the cell membrane. Several human diseases, including cystic fibrosis, hypercholesterolemia, and myasthenia gravis affect the function of specific transport proteins. Describe the symptoms of and treatments for the diseases. Summarize your findings in a written report. SC.F.1.4.3

13. **Interactive Tutor Unit 1** Cell Transport and Homeostasis Write a report summarizing the roles of osmosis and diffusion in the preservation and maintenance of body organs donated for transplants. Why must the organs be preserved in special solutions prior to a transplant? Find out what kinds of substances these solutions contain. SC.F.1.4.4

88

Alternative Assessment

12. The symptoms of cystic fibrosis include persistent diarrhea, bulky stools, frequent wheezing or pneumonia, a chronic cough, and poor growth. The symptoms of hypercholesterolemia include high levels of low-density lipoprotein. The symptoms of myasthenia gravis include muscle weakness in any voluntary muscle and particularly in the muscles of the face. Treatments for each disease involve lessening the symptoms, as none of the diseases has a cure at present.

13. Answers will vary. The solutions in which body organs are preserved must have specific concentrations of solutes to keep the organ tissues functioning prior to transplant.

Assignment Guide

Section	Questions
1	1, 2, 3, 6, 10, 11, 13
2	4, 5, 7, 8, 9, 10, 12

FCAT Prep

For questions 1–6, write your answers on a separate sheet of paper.

1 A cell begins to swell when placed in an unknown solution. What can you conclude about the solution?
A. The solution is hypertonic. **C.** The solution is isotonic.
B. The solution is hypotonic. **D.** The solution is saturated.
SC.F.1.4.4

2 What process allows the cell to dispose of wastes?
F. endocytosis **H.** facilitated diffusion
G. exocytosis **I.** sodium-potassium pumping
SC.F.1.4.2

3 A gelatin block is prepared with a chemical indicator that turns pink in the presence of a base. The block is enclosed in a membrane and placed in a beaker of basic solution. After half an hour, the block begins to turn pink. Why does the gelatin turn pink?
A. The membrane is impermeable to acid, so the gelatin becomes an acid.
B. The membrane is impermeable to base, so the gelatin becomes a base.
C. The acid diffuses down its concentration gradient through the membrane and into the gelatin.
D. The base diffuses down its concentration gradient through the membrane and into the gelatin.
SC.F.1.4.1

4 Why does cooking dried pasta in boiling water make the pasta soft?
F. Heat energy makes the water boil.
G. Heat energy makes the dry pasta melt.
H. Heat energy causes a concentration gradient to form in the boiling water.
I. Heat energy speeds up the diffusion of water down a concentration gradient.
SC.A.1.4.4

5 READ INQUIRE EXPLAIN One type of membrane pump in animal cells is a carrier protein called the sodium-potassium pump, named for the ions that it transports into and out of the cell.

What would happen if a cell's sodium-potassium pump did not function properly? SC.F.1.4.2

6 READ INQUIRE EXPLAIN The diffusion of water through a selectively permeable membrane is called osmosis. Osmosis is a form of diffusion that involves the movement of water down its concentration gradient. When solutions on either side of a membrane have different concentrations of dissolved particles, they also have different concentrations of water molecules that are not interacting with the particles. Water molecules move from the side with a lower concentration of dissolved particles to the side with a higher concentration.

Part A How is osmosis a form of passive transport? SC.F.1.4.1

Part B A student wants to model osmosis by placing a mesh bag in a solution of salt water. The bag fills up with a saltwater solution that is the same concentration as the solution outside of the bag. Why is this **not** a good model for osmosis? SC.H.3.4.1

Test TIP

Pay close attention to words such as *not, only, rather,* and *some* that appear in questions.

89

FCAT Doctor

Question 1 Answer B is the correct choice. In a hypotonic solution, the concentration of the dissolved particles is greater inside the cell. Thus, water diffuses into the cell and the cell swells. Answer A is incorrect because a hypertonic solution would cause the cell to shrink. Answer C is incorrect because an isotonic solution has the same concentration as the inside of the cell, and there would be no concentration gradient. Answer D is incorrect because a saturated solution is very concentrated and would be a hypertonic solution.

Question 3 Answer D is the correct choice. The block turned pink indicating that base diffused across the membrane. Answers A and C are incorrect because there is no mention of acid in the experiment. Answer B is incorrect because the membrane must be permeable to base since it crossed the membrane.

Question 5 The cell would not be able to reduce its sodium ion concentration and would fill with water due to osmosis.

Question 6

Part A Passive transport is transport that does not require energy.

Part B Because the bag is not selectively permeable, the model cannot represent osmosis.

Answers

1. B
2. G
3. D
4. I
5. Sodium ions and water would continuously enter the cell. The cell would thus swell and burst.

6A. Osmosis does not require energy.
6B. The bag is not selectively permeable.

Test Prep Correlations

SC.A.1.4.4: Item 4
SC.F.1.4.1: Items 3, 6A
SC.F.1.4.2: Items 2, 5
SC.F.1.4.4: Item 1
SC.H.3.4.1: Item 6B

Exploration Lab

Analyzing the Effect of Cell Size on Diffusion

Teacher's Notes

Time Required One 45-minute class period.

Ratings

EASY → HARD

TEACHER PREPARATION 2
STUDENT SETUP 3
CONCEPT LEVEL 1
CLEANUP 1

Skills Acquired

- Constructing Models
- Designing Experiments
- Inferring
- Organizing and Analyzing Data

Scientific Methods

In this lab, students will:

- Make Observations
- Analyze the Results
- Draw Conclusions

Materials and Equipment

Materials for this lab can be ordered from WARD'S. See *Master Materials List* at the front of this book for catalog numbers.

Safety Cautions

Make sure that students wear safety goggles and a lab apron.

Phenolphthalein solutions are flammable, and the vapors can explode when mixed with air. Make sure that there are no flames or sources of ignition, such as sparks, when you are using the phenolphthalein solution.

Remind students to read the ChemSafety section before beginning the lab.

Exploration Lab

Analyzing the Effect of Cell Size on Diffusion

SC.F.1.4.4

SKILLS

- Using scientific methods
- Collecting, organizing, and graphing data

OBJECTIVES

- **Relate** the size of a cell to its surface area–to-volume ratio.
- **Predict** how the surface area–to-volume ratio of a cell will affect the diffusion of substances into the cell.

MATERIALS

- safety goggles
- lab apron
- disposable gloves
- block of phenolphthalein agar (3 × 3 × 6 cm)
- plastic knife
- metric ruler
- 250 mL beaker
- 150 mL of vinegar
- plastic spoon
- paper towel

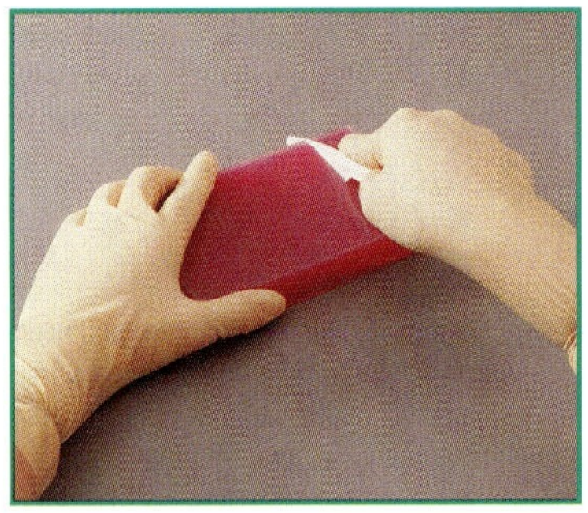

ChemSafety

CAUTION: Always wear safety goggles and a lab apron to protect your eyes and clothing.

CAUTION: Do not touch or taste any chemicals. Know the location of the emergency shower and eyewash station and how to use them. If you get a chemical on your skin or clothing, wash it off at the sink while calling to the teacher. Notify the teacher of a spill. Spills should be cleaned up promptly, according to your teacher's directions.

CAUTION: Glassware is fragile. Notify the teacher of broken glass or cuts. Do not clean up broken glass or spills with broken glass unless the teacher tells you to do so.

Before You Begin

Substances enter and leave a cell in several ways, including by **diffusion.** How efficiently a cell can exchange substances depends on the **surface area–to-volume ratio** (surface area ÷ volume) of the cell. **Surface area** is the size of the outside of an object. **Volume** is the amount of space an object takes up. In this lab, you will investigate how cell size affects the diffusion of substances into a cell. To do this, you will make cell models using agar that contains an indicator. This indicator will change color when an acidic solution diffuses into it.

1. Write a definition for each boldface term in the paragraph above.
2. Based on the objectives for this lab, write a question you would like to explore about cell size and diffusion.

Procedure

PART A: Design an Experiment

1. Work with members of your lab group to explore one of the questions written for step 2 of **Before You Begin.** To explore the question, design an experiment that uses the materials listed for this lab.

You Choose

As you design your experiment, decide the following:

- **a.** what question you will explore
- **b.** what hypothesis you will test
- **c.** how many "cells" (agar cubes) you will have and what sizes they will be
- **d.** how long to leave the "cells" in the vinegar
- **e.** how to determine how far the vinegar diffused into a "cell"
- **f.** how to prevent contamination of agar cubes as you handle them
- **g.** what data to record in your data table

Tips and Tricks

Phenolphthalein agar can be prepared in the laboratory. Add drops of 0.1 M sodium hydroxide solution (prepared by using 4 g NaOH diluted to 1L) to turn the agar red. Pour the mixture into a flat pan to a depth of slightly more than 3 cm. After the agar hardens, cut it into 3 × 3 × 6 cm rectangles. Place enough materials for a class on a supply table. Set up labeled containers for the disposal of solutions, broken glass, and phenolphthalein agar. Before allowing students to gather materials for the lab, have them present a written procedure for their experiment as well as a list of all safety precautions. Make sure that students rinse off their knives between each cutting of different cubes to prevent vinegar from the previous cube from contaminating the next cube. For disposal, wrap the phenolphthalein agar in newspaper, then place the newspaper in the garbage. Dilute the vinegar with water, then pour the vinegar down the sink.

2. Write a procedure for your experiment. Make a list of all the safety precautions you will take. Have your teacher approve your procedure and safety precautions before you begin the experiment.

PART B: Conduct Your Experiment

3. Put on safety goggles, a lab apron, and disposable gloves.

4. Carry out the experiment you designed. Record your observations in your data table.

PART C: Cleanup and Disposal

5. Dispose of solutions, broken glass, and agar in the designated waste containers. Do not pour chemicals down the drain or put lab materials in the trash unless your teacher tells you to do so.

6. Clean up your work area and all lab equipment. Return lab equipment to its proper place. Wash your hands thoroughly before you leave the lab and after you finish all work.

Analyze and Conclude

1. **Summarizing Results** Describe any changes in the appearance of the cubes.
2. **Summarizing Results** Make a graph using your group's data. Plot "Diffusion Distance (mm)" on the vertical axis. Plot "Surface Area–to-Volume Ratio" on the horizontal axis.
3. **Analyzing Results** Using the graph you made in item 2, make a statement about the relationship between the surface area–to-volume ratio and the distance a substance diffuses.
4. **Summarizing Results** Make a graph using your group's data. Plot "Rate of Diffusion (mm/min)" (distance vinegar moved ÷ time) on the vertical axis. Plot "Surface Area–to-Volume Ratio" on the horizontal axis.
5. **Analyzing Results** Using the graph you made in item 4, make a statement about the relationship between the surface area–to-volume ratio and the rate of diffusion of a substance.
6. **Evaluating Methods** In what ways do your agar models simplify or fail to simulate the features of real cells?
7. **Calculating** Calculate the surface area and volume of a cube with a side length of 5 cm. Calculate the surface area and volume of a cube with a side length of 10 cm. Determine the surface area–to-volume ratio of each of these cubes. Which cube has the greater surface area–to-volume ratio?
8. **Evaluating Conclusions** How does the size of a cell affect the diffusion of substances into the cell?
9. **Further Inquiry** Write a new question about cell size and diffusion that could be explored with another investigation.

Do You Know?

Do research in the library or media center to answer these questions:

1. How does cell transport in prokaryotic cells differ from cell transport in eukaryotic cells?
2. Which of the following molecules can diffuse across the cell membrane without the help of a transport protein: water, carbohydrates, lipids, proteins?

Use the following Internet resources to explore your own questions about cell size and cell transport.

internet connect
www.scilinks.org
Topic: Cell Membrane
Keyword: HX4035
SCILINKS. Maintained by the National Science Teachers Association

Answers to Before You Begin

1. *diffusion*—the random movement of particles of a substance from an area of high concentration to an area of lower concentration; *surface area-to-volume ratio*—the ratio of the surface area of an object to its volume; *surface area*—the area of the exterior surface of an object; *volume*—the amount of space an object takes up
2. Answers will vary. For example: How does surface area-to-volume ratio affect the diffusion of a substance into a cell?

Answers to Procedure

Sample Procedure

1. Trim the agar block with a knife to make three cubes with side lengths of 1 cm, 2 cm, and 3 cm. Calculate the surface area, volume, and surface area-to-volume ratio for each cube.
2. Place the three cubes in the beaker, and cover them with vinegar. Using the plastic spoon, turn the cubes frequently.
3. After 10 minutes, use the spoon to remove the agar cubes. Blot the cubes dry using a paper towel.
4. Cut each cube in half. Measure the distance (in mm) the vinegar diffused.
5. Calculate the rate of diffusion (mm/min).

Answers to Analyze and Conclude

1. Cubes should appear light pink near their surfaces.
2. The line plotted on the graph should be horizontal.
3. The diffusion distance is the same regardless of the surface-to-volume ratio.
4. The line plotted on the graph should be horizontal.
5. The rate of diffusion is constant regardless of the surface area-to-volume ratio.
6. Agar models ignore the selective permeability of cell membranes, the role of membrane proteins in facilitated diffusion and active transport, and other mechanisms of cell transport.
7. For a cube with a side length of 5 cm: surface area = 150 cm^2, volume = 125 cm^3; surface-to-volume ratio = 6:5. For a cube with side length of 10 cm: surface area = 600 cm^2; volume = 1,000 cm^3; surface-area-to-volume ratio = 3:5. The small cube has the greater surface-area-to-volume ratio.
8. In a small cell, substances do not have to diffuse as far to reach the center of the cell.
9. Answers will vary. For example, How can the cell membrane be modified to take in materials more efficiently?

Answers to Do You Know?

1. The transport of materials across the cell membrane is essentially the same in prokaryotes and eukaryotes, except that prokaryotes do not exhibit endocytosis or exocytosis.
2. water and lipids

Chapter Opener		**SE Quick Review,** p. 93 ■ **SE Reading Activity,** p. 93 ■ **TE Using the Figure,** p. 92 GENERAL
& 3 • 90 min pp. 94–96 **Section 1** Energy and Living Things	**OSP Lesson Plan*** **TR Bellringer*** **TR Section Outline*** **TR B34 Breakdown of Starch*** **TR B32 ATP Releases Energy***	**TE Activity** Sandwich Energy, p. 94 GENERAL **TE Using the Figure,** p. 95 GENERAL **TE Group Activity** Flow of Energy, p. 96 ADVANCED **CRF Exploration Lab** Plant and Animal Interrelationships* GENERAL **CRF Forensics and Applied Science Experiments** Measuring the Release of Energy from Sucrose* GENERAL
& 5 • 90 min pp. 97–103 **Section 2** Photosynthesis	**OSP Lesson Plan*** **TR Bellringer*** **TR Section Outline*** **TR B52 Photosynthesis*** **TR B25 Absorption Spectra of Photosynthesis Pigments*** **TR B24 Chloroplasts*** **TR B30 Summary of Processes in Light Reactions*** **TR B31 Calvin Cycle***	**TE Demonstration,** p. 97 ◆ **TE Demonstration,** p. 99 **TE Using the Figure,** p. 99 **TE Using the Figure,** p. 100 GENERAL **SE Quick Lab** Identifying a Product of Photosynthesis, p. 101 ◆ ■ **CRF Datasheets for In-Text Labs** Identifying a Product of Photosynthesis* **TE Using the Figure,** p. 102 GENERAL
& 7 • 90 min pp. 104–110 **Section 3** Cellular Respiration	**OSP Lesson Plan*** **TR Bellringer*** **TR Section Outline*** **TR B53 Cellular Respiration*** **TR B37 Glycolysis*** **TR B39 Krebs Cycle*** **TR B54 Electron Transport Chain of Aerobic Respiration***	**TE Using the Figure,** p. 105 GENERAL **TE Using the Figure,** p. 106 **TE Group Activity** Fermentation, p. 108 ADVANCED **TE Demonstration,** p. 109 ◆ **CRF Inquiry Lab** What Conditions Favor Cellular Respiration?* GENERAL **SE Skill Practice Lab** Observing Oxygen Production from Photosynthesis, p. 114 ◆ ■ **CRF Datasheets for In-Text Labs** Observing Oxygen Production from Photosynthesis*

& 9 • 90 min

Chapter Review and Assessment Resources

- **SE Study Zone,** p. 111 ■
- **SE Performance Zone,** p. 112 ■
- **SE FCAT Test Prep,** p. 113 ■
- **TR Concept Mapping***
- **CRF Vocabulary Review Worksheet*** ■ GENERAL
- **CRF Test Prep Pretest*** ■ GENERAL
- **CRF Chapter Test*** ■ GENERAL
- **CRF Chapter Test*** ADVANCED
- **OSP Test Generator***
- **OSP Test Item Listing***
- **OSP Modified Chapter Test*** BASIC

Online and Technology Resources

Holt Online Learning

Visit go.hrw.com for access to Holt Online Learning or enter the keyword **HX6 Home** for a variety of free online resources.

One-Stop Planner® CD-ROM

This CD-ROM package includes

- Lab Materials QuickList Software
- Holt Calendar Planner
- Customizable Lesson Plans
- Printable Worksheets
- ExamView® Test Generator
- Interactive Teacher Edition
- Holt PuzzlePro® Resources
- Holt PowerPoint® Resources

Guided Reading Audio CDs

These CDs are designed to help auditory learners and reluctant readers. (Audio CDs are also available in Spanish.)

Compression guide:
To shorten your instruction because of time limitations, omit blocks 1, 2, and 8.

KEY

TE	Teacher Edition	CRF	Chapter Resource File	*	Also on One-Stop Planner
SE	Student Edition	TR	Teaching Transparency	■	Also Available in Spanish
OSP	One-Stop Planner			◆	Requires Advance Prep

SKILLS DEVELOPMENT AND PRACTICE	SECTION REVIEW AND ASSESSMENT	CORRELATIONS
		Florida Sunshine Standards/FCAT
TE Reading Skill Builder Vocabulary, p. 95 BASIC CRF Directed Reading Worksheet* BASIC CRF Active Reading Worksheet* GENERAL	SE Section 1 Review ■ TE Reteaching, p. 96 BASIC TE Quiz, p. 96 GENERAL TE Alternative Assessment, p. 96 ◆ ADVANCED CRF Quiz* ■ BASIC	SC.B.1.4.1, SC.G.1.4.2, SC.G.1.4.3
CRF Directed Reading Worksheet* BASIC CRF Active Reading Worksheet* GENERAL	SE Section 2 Review ■ TE Reteaching, p. 103 GENERAL TE Quiz, p. 103 GENERAL TE Alternative Assessment, p. 103 ADVANCED CRF Quiz* ■ BASIC	SC.F.1.4.1, SC.G.1.4.2, SC.G.1.4.3, SC.G.2.4.5
TE Reading Skill Builder Anticipation Guide, p. 105 GENERAL CRF Directed Reading Worksheet* BASIC CRF Active Reading Worksheet* GENERAL CRF Critical Thinking Worksheet* ADVANCED CRF Science Skills Worksheet* GENERAL	SE Section 3 Review ■ TE Reteaching, p. 110 BASIC TE Quiz, p. 110 GENERAL CRF Quiz* ■ BASIC Modified Worksheet* BASIC	SC.F.1.4.1, SC.G.1.4.3

Holt Biology Interactive Tutor CD-ROM

This CD-ROM consists of interactive activities that give students a fun way to extend their knowledge of biology concepts.

Unit 2 Photosynthesis

Unit 3 Cellular Respiration

SCILINKS. NSTA

www.scilinks.org

Maintained by the **National Science Teachers Association**

Topic: Light Absorption
SciLinks Code: HX4116

Topic: Factors Affecting Photosynthesis
SciLinks Code: HX4079

Topic: Aerobic Respiration
SciLinks Code: HX4004

Topic: Fermentation
SciLinks Code: HX4080

CHAPTER 5

Overview

Tell the students the purpose of this chapter is to examine how energy is made available to cells to power metabolism. They will learn how energy is captured and stored during photosynthesis in autotrophs, and how that energy becomes usable to both autotrophs and heterotrophs through cellular respiration.

Using the Figure

Ask students to examine the opening photo, then ask them which of the organisms shown obtains energy from the sun. (The tree, grasses, and flowers make their own food using energy from the sun and materials from the environment. The deer gets its energy by eating plants, so indirectly gets its energy from the sun.) Ask students to identify the group of organisms that can use the sun's energy to make organic compounds. (producers, or autotrophs)

Assessing Prior Knowledge

Before beginning this chapter, be sure your students can answer the questions in the ***Quick Review*** on the facing page. If they cannot, encourage them to reread the sections indicated.

Deer eating leaves

Chapter Correlations — **Florida Sunshine Standards/FCAT**

SC.B.1.4.1 Knowledge of energy is fundamental to all the scientific disciplines (e.g., the energy required for biological processes in living organisms and the energy required for the building, erosion, and rebuilding of the Earth).

SC.F.1.4.1 Body processes involve specific biochemical reactions governed by biochemical principles.

SC.G.1.4.2 The flow of energy through an ecosystem made up of producers, consumers, and decomposers carries out the processes of life, and some energy dissipates as heat and is not recycled.

SC.G.1.4.3 The chemical elements that make up the molecules of living things are combined and recombined in different ways.

SC.G.2.4.5 The amount of life any environment can support is limited, and human activities can change the flow of energy and reduce the fertility of the Earth.

CHAPTER

5 Photosynthesis and Cellular Respiration

Florida Sunshine Standards/FCAT

SC.B.1.4.1 Knowledge of energy is fundamental to all the scientific disciplines (e.g., the energy required for biological processes in living organisms and the energy required for the building, erosion, and rebuilding of the Earth).

SC.G.1.4.3 The chemical elements that make up the molecules of living things are combined and recombined in different ways.

SC.G.2.4.5 The amount of life any environment can support is limited, and human activities can change the flow of energy and reduce the fertility of the Earth.

Quick Review

Answer the following without referring to earlier sections of your book.

1. **Describe** different kinds of chemical bonds. *(Chapter 2, Section 1)*
2. **List** the properties of organic compounds. *(Chapter 2, Section 3)*
3. **Distinguish** between mitochondria and chloroplasts. *(Chapter 3, Section 3)*
4. **Differentiate** between passive transport and active transport. *(Chapter 4, Section 2)*

Did you have difficulty? *For help, review the sections indicated.*

Reading Activity

Before you read this chapter, write a short list of all of the things you know about photosynthesis and cellular respiration. Then write a list of the things that you want to know about photosynthesis and cellular respiration. Save your list, and to assess what you have learned, see how many of your own questions you can answer after reading this chapter.

This deer, like many other organisms, depends upon the energy stored in plants for food. Food supplies the energy needed for cellular activities.

Looking Ahead

Section 1
Energy and Living Things
Energy in Living Systems
ATP

Section 2
Photosynthesis
Using the Energy in Sunlight
Stage One: Absorption of Light Energy
Stage Two: Conversion of Light Energy
Stage Three: Storage of Energy

Section 3
Cellular Respiration
Cellular Energy
Stage One: Breakdown of Glucose
Stage Two: Production of ATP
Fermentation in the Absence of Oxygen

internet connect

www.scilinks.org
National Science Teachers Association *sci*LINKS Internet resources are located throughout this chapter.

SCILINKS. Maintained by the National Science Teachers Association

93

Identifying Misconceptions

Many students believe that plants undergo photosynthesis while other organisms undergo respiration. Point out that the reason plants undergo photosynthesis is to produce organic compounds that the plants can then use in respiration. Remind students that plant cells contain both chloroplasts and mitochondria. The mitochondria use the organic compounds to make ATP, which powers cellular activities.

Quick Review Answers

1. Covalent bonds are formed when two or more atoms share electrons to form a molecule. Hydrogen bonds are weaker bonds and form when the positive end of one molecule is attracted to the negative end of a second molecule. Ionic bonds form between positively and negatively charged atoms or molecules that have either lost (1) or gained (2) an electron.
2. Organic compounds contain carbon atoms that are covalently bonded to other elements, especially hydrogen and oxygen.
3. Mitochondria are organelles that use organic compounds to make ATP. Chloroplasts are organelles that use light energy to synthesize organic compounds.
4. In passive transport, materials diffuse freely across a membrane in the direction of their concentration gradient. In active transport, energy is required to move materials across a membrane against the concentration gradient.

Chapter Resource File

- Vocabulary Worksheets
- Concept Mapping

Reading Activity Answers

Answers will vary. In general, students are more familiar with photosynthesis than respiration. They will probably remember the key factors of photosynthesis (sunlight, water, and carbon dioxide) and that plants are able to make their own food through photosynthesis.

Energy and Living Things

Florida Sunshine Standards/FCAT

SC.B.1.4.1 Knowledge of energy is fundamental to all the scientific disciplines (e.g., the energy required for biological processes in living organisms and the energy required for the building, erosion, and rebuilding of the Earth).

SC.G.1.4.3 The chemical elements that make up the molecules of living things are combined and recombined in different ways.

Focus

Overview

Before beginning this section review with your students the objectives listed in the Student Edition. This section is a broad overview of energy flow in natural systems. It also serves as an introduction to the process that converts food into ATP.

Bellringer

Ask students to list as many different forms of energy as they can. (Answers will vary but should include heat, light, chemical energy, mechanical energy, and electrical energy.)

Motivate

Activity — GENERAL

Sandwich Energy Have students trace the energy in a ham and cheese sandwich back to the sun. (Ham → pig → grains → sun. Cheese → milk → cow → grass → sun. Bread → wheat → sun.)

LS Logical

Teach

Teaching Tip — ADVANCED

Deep-Sea Vents Explain that the vast majority of autotrophs are photosynthetic, and include plants, prokaryotes, and algae. The *chemo*autotrophs are prokaryotes, many of which live around deep-sea volcanic vents in complete darkness. Have students research deep-sea vents and some of the organisms that are supported by these unique autotrophs.

Objectives

- **Analyze** the flow of energy through living systems. SC.G.1.4.2
- **Compare** the metabolism of autotrophs with that of heterotrophs. SC.G.1.4.2
- **Describe** the role of ATP in metabolism. SC.G.1.4.3
- **Describe** how energy is released from ATP. SC.B.1.4.1

Key Terms

photosynthesis
autotroph
heterotroph
cellular respiration

Energy in Living Systems

You get energy from the food you eat. Where does the energy in food come from? Directly or indirectly, almost all of the energy in living systems needed for metabolism comes from the sun. **Figure 1** shows how energy flows through living systems. Energy from the sun enters living systems when plants, algae, and certain prokaryotes absorb sunlight. Some of the energy in sunlight is captured and used to make organic compounds. These organic compounds store chemical energy and can serve as food for organisms.

Building Molecules That Store Energy

Metabolism involves either using energy to build molecules or breaking down molecules in which energy is stored. **Photosynthesis** is the process by which light energy is converted to chemical energy. Organisms that use energy from sunlight or from chemical bonds in inorganic substances to make organic compounds are called **autotrophs** *(AWT oh trohfs)*. Most autotrophs, especially plants, are photosynthetic organisms. Some autotrophs, including certain prokaryotes, use chemical energy from inorganic substances to make organic compounds. Prokaryotes found near deep-sea volcanic vents live in perpetual darkness. Sunlight does not reach the bottom of the ocean. These prokaryotes get energy, however, from chemicals flowing out of the vents.

Figure 1 Flow of energy

Energy flows from sunlight or inorganic substances to autotrophs, such as grasses, and then to heterotrophs, such as rabbits and foxes.

Trends in Fuel Production

Food Energy Sometimes food *is* burned for energy. For years, grain has been fermented to make ethanol, an alcohol that can be added to gasoline to boost power and reduce pollution. New research is producing fuels based entirely on renewable sources. These fuels can be made from grains such as corn and wheat, or even from trees and grasses. Biodiesel is a diesel-like fuel made from vegetable oil.

Chapter Resource File

- Directed Reading BASIC
- Active Reading GENERAL

One-Stop Planner CD-ROM

- Reading Organizers BASIC
- Reading Strategies BASIC

Section Correlations

SC.B.1.4.1
SC.G.1.4.2
SC.G.1.4.3

Breaking Down Food for Energy

The chemical energy in organic compounds can be transferred to other organic compounds or to organisms that consume food. Organisms that must get energy from food instead of directly from sunlight or inorganic substances are called **heterotrophs** *(HEHT uhr oh trohfs)*. Heterotrophs, including humans, get energy from food through the process of cellular respiration. **Cellular respiration** is a metabolic process similar to burning fuel. While burning converts almost all of the energy in a fuel to heat, cellular respiration releases much of the energy in food to make ATP. This ATP provides cells with the energy they need to carry out the activities of life.

WORD Origins

The words *autotroph* and *heterotroph* have the same suffix, *-troph,* which is from the Greek word *trophikos,* meaning "to feed." The prefix *auto-* is from the Greek word *autos,* meaning "self," and the prefix *hetero-* is from the Greek word *heteros,* meaning "other."

Transfer of Energy to ATP

The word *burn* is often used to describe how cells get energy from food. Although the overall processes are similar, the "burning" of food in living cells clearly differs from the burning of a log in a campfire. When a log burns, the energy stored in wood is released quickly as heat and light. But in cells, chemical energy stored in food molecules is released gradually in a series of enzyme-assisted chemical reactions. As shown in **Figure 2,** the product of one chemical reaction becomes a reactant in the next reaction. In the breakdown of starch, for example, each reaction releases energy.

When cells break down food molecules, some of the energy in the molecules is released as heat. Much of the remaining energy is stored temporarily in molecules of ATP. Like money, ATP is a portable form of energy "currency" inside cells. ATP delivers energy wherever energy is needed in a cell. The energy released from ATP can be used to power other chemical reactions, such as those that build molecules. In cells, most chemical reactions require less energy than is released from ATP. Therefore, enough energy is released from ATP to drive most of a cell's activities.

Figure 2 Breakdown of starch

Energy is released from starch in a series of enzyme-assisted chemical reactions.

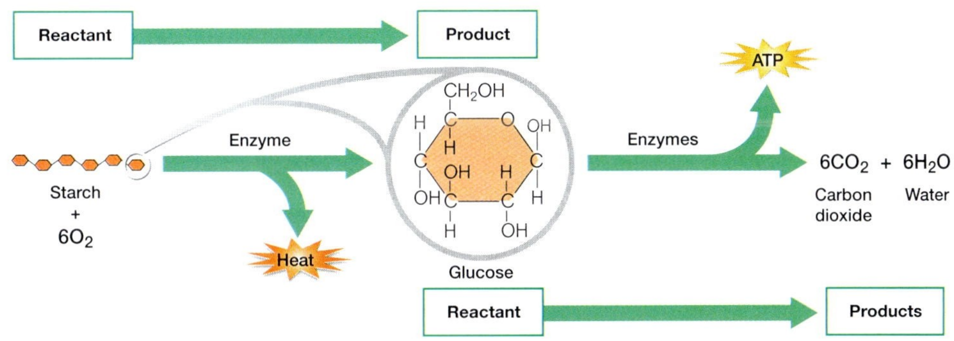

MISCONCEPTION ALERT

Transfer of Energy Because ATP supplies most of the energy that drives metabolism, ATP is sometimes called an energy-rich compound, and the bonds between its phosphate groups are sometimes called "high-energy" bonds. These terms are misleading because they imply that ATP contains an unusually large amount of energy. ATP serves as the cell's energy currency. The bonds between phosphate groups are unstable and therefore break easily. When they break, energy is released that can be use to drive metabolic processes.

READING SKILL BUILDER — BASIC

Vocabulary Have students use a dictionary to define the root words of *autotroph* and *heterotroph*. (auto = self; hetero = other; troph = nutrition or feeding; so an autotroph is literally "self-feeding," and a heterotroph is literally "other-feeding.") **LS Verbal**

Using the Figure — GENERAL

Ask students to identify the autotrophs (grasses) and heterotrophs (rabbits, fox) in **Figure 1.** Next, ask students how the fox is indirectly using the sun to get energy. (The plants convert the energy in sunlight to chemical energy stored in plant tissues. Rabbits eat the plants to get the stored chemical energy. The fox then eats the rabbit. **LS Visual**

Teaching Tip — GENERAL

Autotrophs versus Heterotrophs

Make two columns on the board with the heads *autotroph* and *heterotroph*. Ask each student to name one specific autotroph and one specific heterotroph. Write the student's responses in the columns. If you notice a pattern—such as students naming mostly mammals—bring this to their attention. Tell them other kinds of heterotrophs exist, such as some protists and fungi. Co-op Learning

INTERACTIVE BIOLOGY TUTOR

- **Unit 2 Photosynthesis** This engaging tutorial gives students an overview of photosynthesis.
- **Unit 3 Cellular Respiration** This engaging tutorial reviews the process of cellular respiration.

Transparencies

TR Bellringer
TR B34 Breakdown of Starch
TR B2 ATP Releases Energy

Teach, *continued*

Group Activity — ADVANCED

Flow of Energy Assign students to cooperative groups of three or four. Have each group create a display that shows the transfer of energy through living systems. **LS Interpersonal**

Close

Reteaching — BASIC

Set up a demonstration that includes several organisms, such as a sponge, a worm, a Venus flytrap, and a mushroom. Have students examine the display. Do not let them touch any of the items. Ask students to decide which items represent heterotrophs and which represent autotrophs. **LS Visual**

Quiz — GENERAL

1. Organisms that can make their own food using the energy of the sun are called _______. (autotrophs)
2. Organisms that obtain their energy by eating other organisms are called _______. (heterotrophs)

Alternative Assessment — ADVANCED

Have students build a three-dimensional model of ATP (adenosine triphosphate). Students should use everyday materials, and include the three phosphate groups, the sugar, and the base. Make sure students understand which part of the nucleotide breaks off to give energy for cellular reactions. (the last phosphate group) **LS Kinesthetic**

Section Review Correlations

SC.B.1.4.1: Item 3

SC.G.1.4.2: Items 1, 2, 4, 5, 6

ATP

Recall that ATP (adenosine triphosphate) is a nucleotide with two extra energy-storing phosphate groups. As shown in **Figure 3,** the three phosphate groups in ATP form a chain that branches from a five-carbon sugar called ribose *(RIE bohs).* This phosphate "tail" is unstable because the phosphate groups are negatively charged and therefore repel each other. The phosphate groups store energy like a compressed spring does. This energy is released when the bonds that hold the phosphate groups together are broken.

Breaking the outer phosphate bond requires an input of energy. Much more energy is released, however, than is consumed by the reaction. As shown in Figure 3, the removal of a phosphate group from ATP produces adenosine diphosphate, or ADP. This reaction releases energy in a way that enables cells to use the energy. The following equation summarizes this reaction:

$$H_2O + ATP \rightarrow ADP + P + energy$$

Cells use the energy released by this reaction to power metabolism. In some chemical reactions, two phosphate groups are removed from ATP instead of just one. This tends to make the reaction irreversible because the pair of phosphate groups that is removed is not available for the reverse reaction. Rather, the pair is quickly split into two single phosphate groups.

Figure 3 ATP releases energy

When the outer phosphate group detaches from ATP, energy is released.

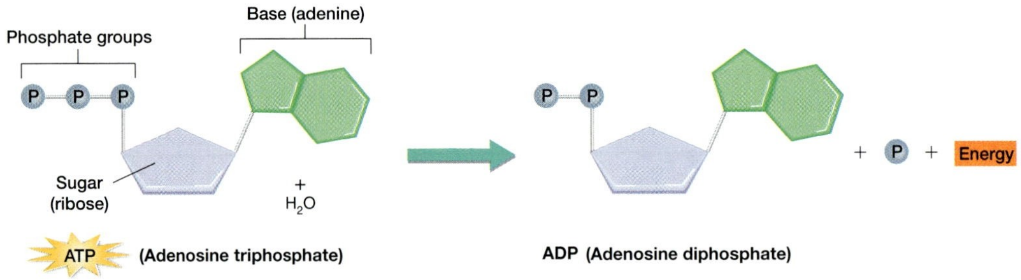

Section 1 Review

1. **Identify** the primary source of energy that flows through most living systems. SC.G.1.4.2
2. **Compare** the metabolism of autotrophs with that of heterotrophs. SC.G.1.4.2
3. **Describe** how energy is released from ATP. SC.B.1.4.1
4. **Critical Thinking Inferring Relationships** How can the energy in the food that a fox eats be traced back to the sun? SC.G.1.4.2
5. **Critical Thinking Analyzing Patterns** Explain how life involves a continuous flow of energy. SC.G.1.4.2
6. **FCAT Prep** A grasshopper obtains energy by eating grass. A snake eats the grasshopper, and a hawk then eats the snake. What is the original source of energy for the hawk?
   A the grass
   B the grasshopper
   C the snake
   D the sun
   SC.G.1.4.2

96

Answers to Section Review

1. the sun
2. Autotrophs use the energy in sunlight or inorganic substances to make organic compounds. Heterotrophs must consume food sources to get energy needed to power their metabolism.
3. When the outer phosphate bond in a molecule of ATP is broken, energy is released.
4. Foxes eat other organisms to get the energy needed for their metabolism. The animals they eat acquired their energy from eating plants. The plants used the energy of the sun to convert compounds into carbohydrates, which power their metabolism. Thus, foxes get their energy indirectly from the sun.
5. The flow of energy between organisms is continuous because energy passes from the sun to autotrophs, then to heterotrophs, and then to other heterotrophs.
6. **A.** Incorrect. Grass produces carbohydrates by using energy from the sun. **B.** Incorrect. The grasshopper obtains energy by eating grass. **C.** Incorrect. The snake obtains energy by eating the grasshopper. **D.** Correct. The original energy source for the hawk is the sun.

Photosynthesis

Section 2

Florida Sunshine Standards/FCAT
SC.G.1.4.2 The flow of energy through an ecosystem made up of producers, consumers, and decomposers carries out the processes of life, and some energy dissipates as heat and is not recycled.
SC.G.2.4.5 The amount of life any environment can support is limited, and human activities can change the flow of energy and reduce the fertility of the Earth.

Using the Energy in Sunlight

When you eat a hamburger, you get energy from the sun indirectly. Plants, such as grass, capture the energy in sunlight. The beef in a hamburger comes from a cow that ate grass. The bun, lettuce, and tomato come from plants. With few exceptions, you end up with plants whenever you trace your food back to its origin. Plants, algae, and some bacteria capture about 1 percent of the energy in the sunlight that reaches Earth and convert it to chemical energy through the process of photosynthesis.

The Stages of Photosynthesis

Photosynthesis is the process that provides energy for almost all life. As **Figure 4** shows, photosynthesis has three stages:

Stage 1 Energy is captured from sunlight.

Stage 2 Light energy is converted to chemical energy, which is temporarily stored in ATP and the energy carrier molecule NADPH.

Stage 3 The chemical energy stored in ATP and NADPH powers the formation of organic compounds, using carbon dioxide, CO_2.

Photosynthesis occurs in the chloroplasts of plant cells and algae and in the cell membrane of certain prokaryotes. Photosynthesis can be summarized by the following equation:

$$\underset{\text{carbon dioxide}}{6CO_2} + \underset{\text{water}}{6H_2O} \xrightarrow{\text{light}} \underset{\text{sugars}}{C_6H_{12}O_6} + \underset{\text{oxygen gas}}{6O_2}$$

This equation, however, does not show how photosynthesis occurs. It merely says that six carbon dioxide molecules, six water molecules, and light are needed to form one six-carbon organic compound and six molecules of oxygen. Plants use the organic compounds they make during photosynthesis to carry out their life processes. For example, some of these sugars are used to form starch, which can be stored in stems or roots. The plant may later break down the starch to make ATP used to power metabolism. All of the proteins, nucleic acids, and other molecules of the cell are assembled from fragments of these sugars.

Figure 4 Photosynthesis

The process of photosynthesis occurs in three stages.

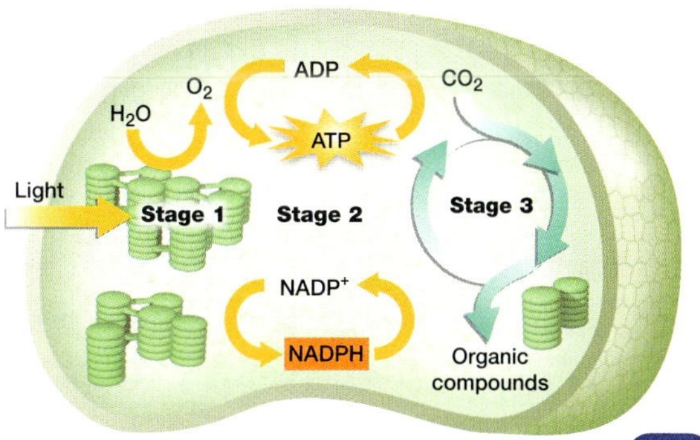

Objectives

- **Summarize** how energy is captured from sunlight in the first stage of photosynthesis. SC.G.1.4.2
- **Analyze** the function of electron transport chains in the second stage of photosynthesis. SC.F.1.4.1
- **Relate** the Calvin cycle to carbon dioxide fixation in the third stage of photosynthesis. SC.F.1.4.1
- **Identify** three environmental factors that affect the rate of photosynthesis. SC.G.2.4.5

Key Terms

pigment
chlorophyll
carotenoid
thylakoid
electron transport chain
NADPH
carbon dioxide fixation
Calvin cycle

Focus

Overview

Before beginning this section review with your students the objectives listed in the Student Edition. In this section, students will learn to describe the major events of the three stages of photosynthesis. These include the capture of energy (stage 1), the conversion of light energy to chemical energy (stage 2), and the formation of organic compounds using stored chemical energy (stage 3).

Bellringer

On the board or overhead, ask students to write down the primary role that sunlight plays in living systems. (Sunlight is the main source of energy in living systems.) Then ask students to define photosynthesis. (Answers should indicate that photosynthesis is the process by which organisms use light energy to make their own food.)

Motivate

Demonstration

Use a scalpel to make thin cross sections of a potato. Use a projection microscope to show the potato section, or have students examine the potato under a compound microscope. Ask them to look for starch granules, which should appear as large, translucent structures inside the potato cells. Explain that much of the mass of the organic compounds plants make during photosynthesis is stored as starch.
LS Visual

Transparencies

- **TR** Bellringer
- **TR B52** Photosynthesis
- **TR B25** Absorption Spectra of Photosynthetic Pigments
- **TR B24** Chloroplasts
- **TR B30** Summary of Processes in Light Reactions
- **TR B31** Calvin Cycle

One-Stop Planner CD-ROM

- **Reading Organizers** BASIC
- **Reading Strategies** BASIC
- **Basic Skills Worksheet** Reading a Thermometer Temperature Conversions BASIC
- **Supplemental Reading Guide** *The Lives of a Cell* ADVANCED

Chapter Resource File

- **Directed Reading** BASIC
- **Active Reading** GENERAL
- **Data Sheet for Quick Lab** GENERAL

Section Correlations

SC.F.1.4.1
SC.G.1.4.2
SC.G.2.4.5

Stage One: Absorption of Light Energy

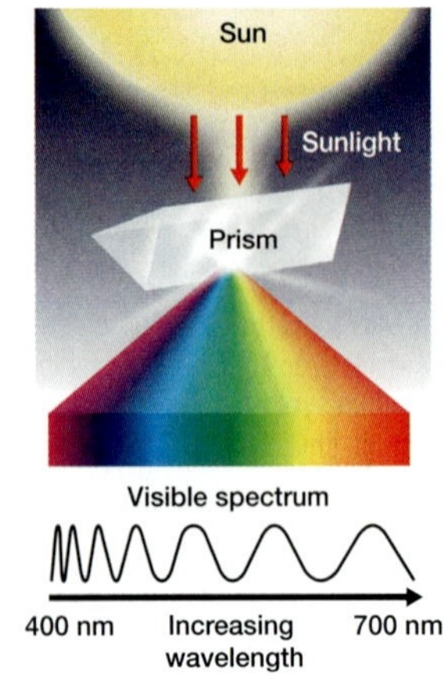

Figure 5 Visible spectrum. Sunlight contains a mixture of all the wavelengths (colors) of visible light. When sunlight passes through a prism, the prism separates the light into different colors.

The chemical reactions that occur in the first and second stages of photosynthesis are sometimes called "light reactions," or light-dependent reactions. Without the absorption of light, these reactions could not occur. Light energy is used to make energy-storing compounds. Light is a form of radiation—energy in the form of waves that travel through space. Different types of radiation, such as light and heat, have different wavelengths (the distance between two consecutive waves). When the sun shines on you, your body is bombarded by many kinds of radiation from the sun. However, you can see only radiation known as visible light. You see wavelengths of visible light as different colors. As shown in **Figure 5,** sunlight contains all the wavelengths of visible light, red through violet.

Pigments

How does a human eye or a leaf absorb light? These structures contain light-absorbing substances called **pigments.** Pigments absorb only certain wavelengths and reflect all the others. **Chlorophyll** *(KLOR uh fihl),* the primary pigment involved in photosynthesis, absorbs mostly blue and red light and reflects green and yellow light. This reflection of green and yellow light makes many plants, especially their leaves, look green. Plants contain two types of chlorophyll, chlorophyll *a* and chlorophyll *b*. Both types of chlorophyll play an important role in plant photosynthesis.

The pigments that produce yellow and orange fall leaf colors, as well as the colors of many fruits, vegetables, and flowers, are called **carotenoids** *(kuh RAH tuh noydz).* Carotenoids absorb wavelengths of light different from those absorbed by chlorophyll, so having both pigments enables plants to absorb more light energy during photosynthesis. The graph in **Figure 6** shows the wavelengths of light absorbed by chlorophyll *a*, chlorophyll *b*, and carotenoids.

Figure 6 Light absorption during photosynthesis. Chlorophylls absorb mostly violet, blue, and red light, while carotenoids absorb mostly blue and green light.

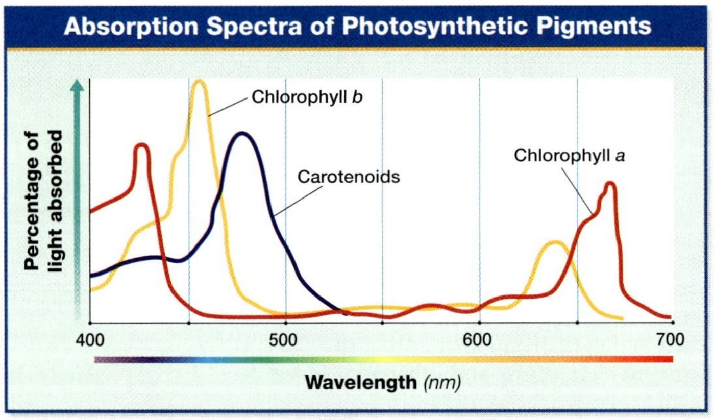

98

Teach

Teaching Tip — ADVANCED

Yellow and orange vegetables are rich in sources of carotenoids. A carotenoid called beta carotene is an important dietary source of vitamin A, which is necessary for proper eyesight, for maintaining the health of membranes, and for tooth and bone development. Have students research and write a report on the effectiveness of beta-carotene as an antioxidant. Also have them compare the effectiveness of food sources of betacarotene with that of other sources, such as dietary supplements. **LS Verbal**

Teaching Tip — GENERAL

Shade-Grown Plants versus Sun-Grown Plants Plants grown in the shade often produce larger leaves than plants grown in full sunlight. Have students propose a hypothesis to explain this. (The larger leaves of shade-grown plants gather more sunlight because they have a larger surface area available to absorb light. This is advantageous because less light reaches the leaves of shade-grown plants.)

did you know?

Spectroscopy The study of specific interactions of light and matter is spectroscopy. The spectrophotometer is the instrument used to accomplish this. Many areas of biology use spectroscopy. The ability of a pigment to absorb various wavelengths of light can be measured using the spectrophotometer. A graph plotting the pigment's light absorption versus wavelength is called an absorption spectrum.

Integrating Physics and Chemistry

Since visible light comprises just one small part of the spectrum of electromagnetic waves, this is an opportune time to go over the full range of the electromagnetic spectrum. Point out that other types of electromagnetic waves, which require no medium in order to travel, include gamma rays, X rays, ultraviolet (UV) rays, infrared waves, microwaves, and radio waves. You might also ask the students to identify which waves have a high frequency and are therefore high energy.

Production of Oxygen

As shown in **Figure 7,** pigments involved in plant photosynthesis are located in the chloroplasts of leaf cells. Clusters of pigments are embedded in the membranes of disk-shaped structures called **thylakoids** *(THIE luh koydz).* When light strikes a thylakoid in a chloroplast, energy is transferred to electrons in chlorophyll. This energy transfer causes the electrons to jump to a higher energy level. Electrons with extra energy are said to be "excited." This is how plants first capture energy from sunlight.

Excited electrons jump from chlorophyll molecules to other nearby molecules in the thylakoid membrane, where the electrons are used to power the second stage of photosynthesis. The excited electrons that leave chlorophyll molecules must be replaced by other electrons. Plants get these replacement electrons from water molecules, H_2O. Water molecules are split by an enzyme inside the thylakoid. When water molecules are split, chlorophyll molecules take the electrons from the hydrogen atoms, H, leaving hydrogen ions, H^+. The remaining oxygen atoms, O, from the disassembled water molecules combine to form oxygen gas, O_2.

internet connect

www.scilinks.org
Topic: Light Absorption
Keyword: HX4116
SCILINKS. Maintained by the National Science Teachers Association

Figure 7 Chloroplast

Pigment molecules are embedded in thylakoid membranes, as are other molecules that participate in photosynthesis.

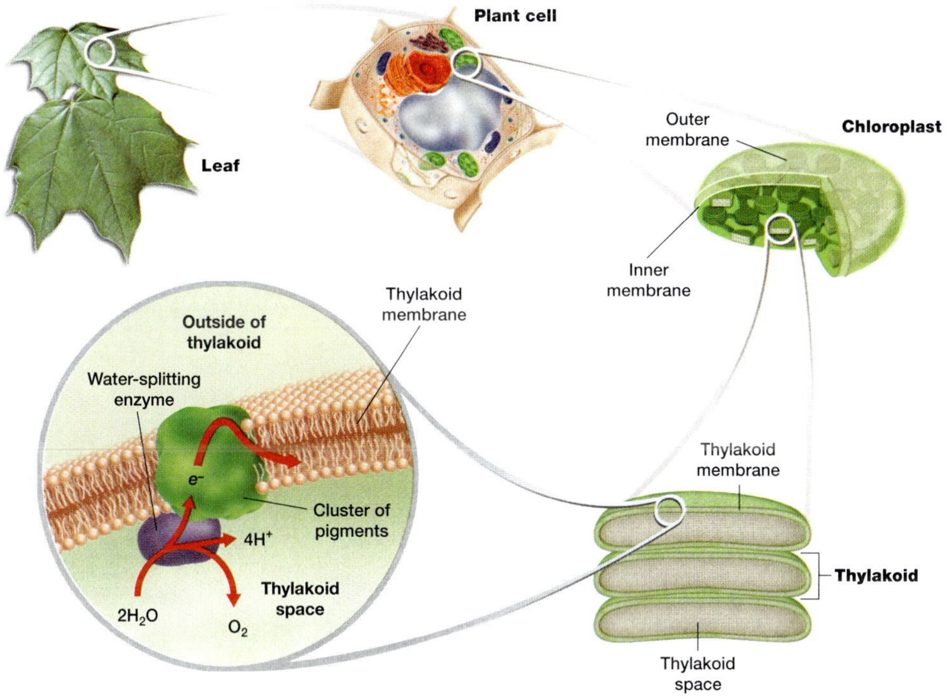

Demonstration

Use small chocolate mints (about 5 cm diameter) with a white center. Make stacks of four or five mints. On one or two of the stacks, cut the top mint in half to expose the center before putting it on the stack. Point out that each stack of mints represents a column of thylakoids. To make the model more realistic, connect one stack to another using strips of paper to represent the membranous connections between thylakoids. Ask students why it is more advantageous for the thylakoids to be in a stack than in a single unit. (Stacks increase the surface area available for light absorption by pigment molecules.) **LS Visual**

Using the Figure

Point out the parts of **Figure 7** that break down the structure of a chloroplast. Note that a leaf is also shown; leaves are the primary sites of photosynthesis in plants because leaf cells contain many chloroplasts. Leaves are generally thin, allowing sunlight to penetrate into the cells. Openings in the leaf surface allow carbon dioxide to enter and oxygen and water vapor to leave. **LS Visual**

MISCONCEPTION ALERT

Why Plants Look Green Many people think that plants are green because plants use green light during photosynthesis. In fact, plants use very little green light, and instead use mainly red and blue light. Tell students that plants look green because they contain chlorophyll, which reflects green and yellow light while absorbing blue and red light.

Attention Grabber

Phytoplankton Some of the most numerous organisms on the planet are phytoplankton. These tiny, floating organisms form the foundation of ecosystems in lakes and oceans. Combined, they may be responsible for 40% of photosynthesis on Earth.

Teach, *continued*

Using the Figure — GENERAL

Figure 8 follows the path of electron transport during the light-dependent reactions of photosynthesis. Guide students through the electron transport chains of photosynthesis by asking the following questions: What is the source of the excited electrons? (chlorophyll molecules) What is the source of some of the replacement electrons? (split water molecules) What type of transport occurs when hydrogen ions are pumped into the thylakoid? (active transport) What type of transport occurs when hydrogen ions move out of the thylakoid? (passive transport) What kind of membrane protein is involved? (carrier protein) Have students record the questions and answers in their notebooks. LS **Visual**

Teaching Tip — GENERAL

Stages of Photosynthesis Help students remember the three stages of photosynthesis by having them make flash cards with a structure or process on one side and a description of its role on the other side. Topics should include pigment molecules, thylakoids, electron transport chains, ATP, and NADPH. LS **Visual**

Study TIP

Interpreting Graphics

Look closely at Figure 8. Electrons are represented by the symbol e^-. The red arrows show the path of excited electrons. Hydrogen ions are represented by the symbol H^+. The blue arrows show the path of hydrogen ions that cross the thylakoid membrane.

Stage Two: Conversion of Light Energy

Excited electrons that leave chlorophyll molecules are used to produce new molecules, including ATP, that temporarily store chemical energy. First an excited electron jumps to a nearby molecule in the thylakoid membrane. Then the electron is passed through a series of molecules along the thylakoid membrane like a ball being passed down a line of people. The series of molecules through which excited electrons are passed along a thylakoid membrane are called **electron transport chains.** Trace the path taken by excited electrons in the electron transport chains shown in **Figure 8.**

Electron Transport Chains

How are electron transport chains used to make molecules that temporarily store energy in the cell? The first electron transport chain shown in Figure 8 lies between the two large green clusters of pigment molecules. This type of electron transport chain contains a protein (the large purple molecule) that acts as a membrane pump. Excited electrons lose some of their energy as they each pass through this protein. The energy lost by the electrons is used to pump hydrogen ions, H^+, into the thylakoid. Recall that hydrogen ions are also produced when water molecules are split inside the thylakoid.

As the process continues, hydrogen ions become more concentrated inside the thylakoid than outside, producing a concentration gradient across the thylakoid membrane. As a result, hydrogen ions have a tendency to diffuse back out of the thylakoid down their

Figure 8 Electron transport chains of photosynthesis

Electron transport chains (represented by the red lines) convert light energy to chemical energy.

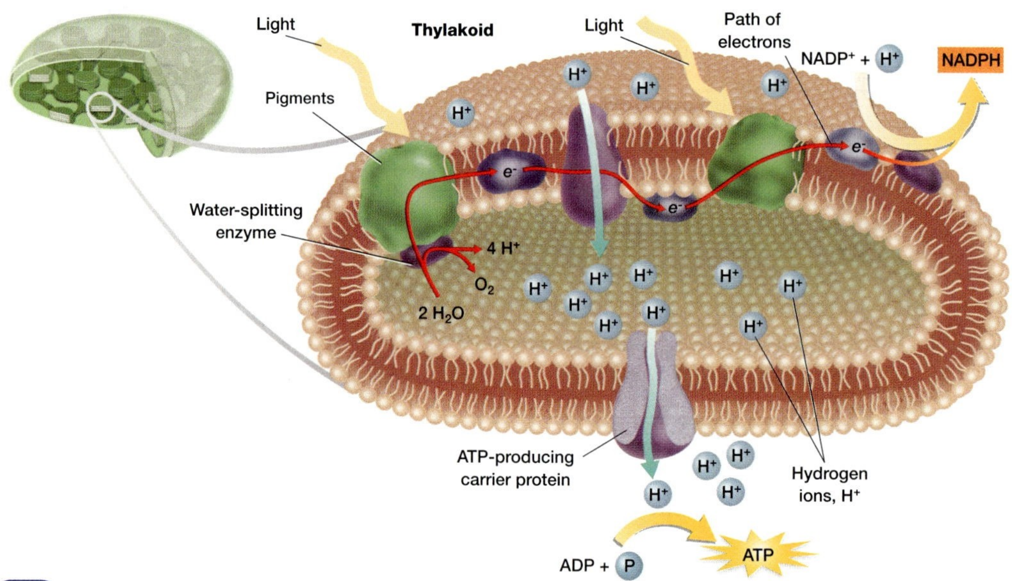

Career

Horticulturist Arrange for a guided tour of a nearby botanical garden with a horticulturist. Before the trip, explain that horticulturists grow plants and trees for use as ornamentals or food. Have students prepare questions that they would like to ask the horticulturist about plants, as well as details of their career.

At the site, ask the horticulturist to point out plants that have different light, temperature, and mineral requirements before answering student questions. LS **Visual**

CHEMISTRY CONNECTION

When an electron in an atom is boosted to a higher energy level, it gains energy. When an electron drops back to a lower energy level, it emits energy. These emissions often take the form of electromagnetic waves that are always specific to the magnitude of the drop in energy. Emissions may be above or below the visible range of the electromagnetic spectrum.

concentration gradient through specialized carrier proteins (illustrated on the lower surface of the thylakoid). These carrier proteins are unusual because they function both as an ion channel and as an enzyme. As hydrogen ions pass through the channel portion of the protein, the protein catalyzes a reaction in which a phosphate group is added to a molecule of ADP, making ATP. Thus, the movement of hydrogen ions across the thylakoid membrane through these proteins provides the energy needed to make ATP, which is used to power the third stage of photosynthesis.

While one electron transport chain provides energy used to make ATP, a second electron transport chain provides energy used to make NADPH. **NADPH** is an electron carrier that provides the high-energy electrons needed to make carbon-hydrogen bonds in the third stage of photosynthesis. The second electron transport chain shown in Figure 8 lies to the right of the second green pigment molecule. In this second chain, excited electrons combine with hydrogen ions as well as an electron acceptor called $NADP^+$, forming NADPH.

The light-dependent reactions of photosynthesis can be summarized as follows. Pigment molecules in the thylakoids of chloroplasts absorb light energy. Electrons in the pigments are excited by light and move through electron transport chains in thylakoid membranes. These electrons are replaced by electrons from water molecules, which are split by an enzyme. Oxygen atoms from water molecules combine to form oxygen gas. Hydrogen ions accumulate inside thylakoids, setting up a concentration gradient that provides the energy to make ATP.

QUICK LAB

Identifying a Product of Photosynthesis SC.G.1.4.3

You can use the following procedure to identify the gas given off by a photosynthetic organism.

Materials

MBL or CBL system with appropriate software, test tube or small glass jar, sprig of *Elodea*, distilled water, cool light source, dissolved oxygen (DO) probe

Procedure

1. Set up an MBL/CBL system to collect and graph data from a dissolved oxygen probe at 30-second intervals for 60 data points. Calibrate the DO probe.
2. Place a sprig of *Elodea* in a test tube or glass jar, and fill the test tube or jar with distilled water.
3. Place the test tube or glass jar under a cool light source, and lower a DO probe into the water. Collect data for 30 minutes.
4. When data collection is complete, view the graph of your data. If possible, print the graph. Otherwise, sketch the graph on paper.

Analysis

1. **Infer** the cause of any change you observed.
2. **Propose** a control for this experiment.
3. **Critical Thinking** **Evaluating Hypotheses** Explain how your data support or do not support the hypothesis that photosynthetic organisms give off oxygen.

QUICK LAB

Identifying a Product of Photosynthesis

Skills Acquired
Inferring, analyzing, evaluating

Teacher's Notes
The *Elodea* should be placed in the test tube cut-side up; the cut should be fresh. Be careful working with glass.

Answers to Analysis

1. Answers will vary. For example, oxygen is produced as water molecules are split in the electron transport chains of *Elodea*.
2. A control could be a cutting of *Elodea* under all of the same conditions except exposure to light.
3. If the probe detects an increase in the oxygen content of the water, then the hypothesis that photosynthetic organisms give off oxygen is supported.

Teaching Tip — Basic

Graphic Organizer Have students make a Graphic Organizer similar to the one at the bottom of this page to illustrate the substances used and produced during each stage of photosynthesis. LS **Visual**

Graphic Organizer

Use this graphic organizer with **Teaching Tip** on this page.

Stages of Photosynthesis		
	Used	Produced
Stage 1	Light, water	Oxygen, hydrogen ions
Stage 2	Electrons, hydrogen ions	ATP, NADPH
Stage 3	ATP, NADPH, carbon dioxide	Organic compounds

Teach, continued

Using the Figure — GENERAL

BIO graphic Have students count the total number of carbon atoms present at each step in the summary of the Calvin cycle, in **Figure 9.** Ask students the following questions: How are ATP and NADPH important to the Calvin cycle? (They supply energy used to form new compounds.) How many carbon dioxide molecules are needed? (3) How many 3-carbon sugars are made? (6) How many of these sugars are actually used to make organic compounds that the plant uses for energy? (1) Emphasize that most of the 3-carbon sugars are "recycled" and used to make the starting 5-carbon compound that begins the cycle again. **LS Visual**

Activity — BASIC

Identifying Variables Have students list three factors that would increase the rate of photosynthesis. Have them identify what stage of photosynthesis each factor would affect. (Examples might include increasing light intensity, which would affect Stage 1; providing more water, which would affect Stage 2; and increasing the carbon dioxide concentration, which would affect Stage 3.)

Stage Three: Storage of Energy

In the first and second stages of photosynthesis, light energy is used to make ATP and NADPH, which temporarily store chemical energy. These stages are therefore considered light-dependent. In the third (final) stage of photosynthesis, however, carbon atoms from carbon dioxide in the atmosphere are used to make organic compounds in which chemical energy is stored. The transfer of carbon dioxide to organic compounds is called **carbon dioxide fixation.** The reactions that "fix" carbon dioxide are sometimes called "dark reactions," or light-independent reactions. Among photosynthetic organisms, there are several ways in which carbon dioxide is fixed.

Calvin Cycle

The most common method of carbon dioxide fixation is the Calvin cycle. The **Calvin cycle** is a series of enzyme-assisted chemical reactions that produces a three-carbon sugar. The Calvin cycle is summarized in **Figure 9.**

Step 1 In carbon dioxide fixation, each molecule of carbon dioxide, CO_2, is added to a five-carbon compound by an enzyme.

Figure 9

Calvin Cycle

The Calvin cycle is a common method of carbon dioxide fixation.

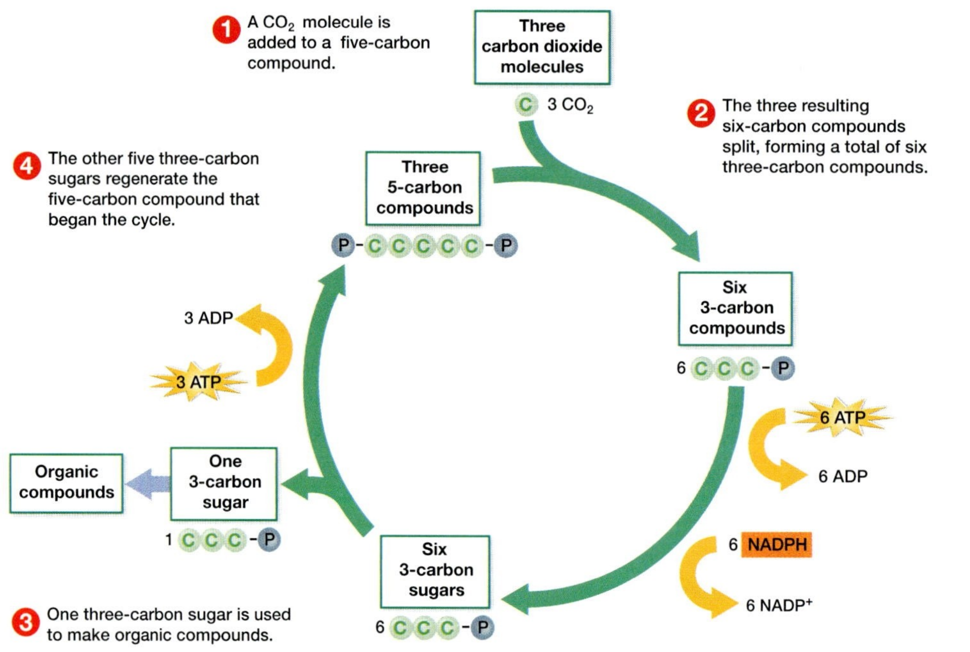

Trends in Plant Physiology

Increasing Carbon Dioxide Researchers around the world are studying plant responses to increasing atmospheric carbon dioxide. In general, more carbon dioxide means that more biomass, or plant tissue, is made during photosynthesis. This may help to counter the effects of rising carbon dioxide levels caused by burning fossil fuels. Other effects appear to be a reduction of nutrients in plant tissues, which may impact the heterotrophs that feed on the plants.

Section Review Correlations

SC.B.1.4.1: Item 6
SC.F.1.4.1: Item 3
SC.G.1.4.2: Item 1
SC.G.1.4.3: Items 2, 4
SC.G.2.4.5: Item 5

Step 2 The resulting six-carbon compound splits into two three-carbon compounds. Phosphate groups from ATP and electrons from NADPH are added to the three-carbon compounds, forming three-carbon sugars.

Step 3 One of the resulting three-carbon sugars is used to make organic compounds—including starch and sucrose—in which energy is stored for later use by the organism.

Step 4 The other three-carbon sugars are used to regenerate the initial five-carbon compound, thereby completing the cycle.

The Calvin cycle is named for Melvin Calvin, the American biochemist who worked out the chemical reactions in the cycle. The reactions are cyclic—they recycle the five-carbon compound needed to begin the cycle again. A total of three carbon dioxide molecules must enter the Calvin cycle to produce each three-carbon sugar that will be used to make other organic compounds. These organic compounds provide the organism with energy for growth and metabolism. The energy used in the Calvin cycle is supplied by ATP and NADPH made during the second stage of photosynthesis.

Real Life

Some houseplants thrive in dim light.

A plant inside a home may receive 100 times less bright light than it would if it were grown outdoors.

Recognizing Patterns

Examine several species of houseplants in a store or nursery. What features are common among house-plants?

Factors that Affect Photosynthesis

Photosynthesis is directly affected by various environmental factors. The most obvious of these factors is light. In general, the rate of photosynthesis increases as light intensity increases until all the pigments are being used. At this saturation point, the reactions of the Calvin cycle cannot proceed any faster. The overall rate of photosynthesis is thus limited by the slowest step, which occurs in the Calvin cycle. The carbon dioxide concentration affects the rate of photosynthesis in a similar manner. Once a certain concentration of carbon dioxide is present, photosynthesis cannot proceed any faster.

Photosynthesis is most efficient within a certain range of temperatures. Like all metabolic processes, photosynthesis involves many enzyme-assisted chemical reactions. Recall that unfavorable temperatures may inactivate certain enzymes.

internet connect

www.scilinks.org
Topic: Factors Affecting Photosynthesis
Keyword: HX4079

SCILINKS. Maintained by the National Science Teachers Association

Section 2 Review

1. **Summarize** how photosynthetic organisms capture the energy in sunlight. SC.G.1.4.2

2. **Compare** the roles of water molecules and hydrogen ions in electron transport chains. SC.G.1.4.3

3. **Describe** the role of the Calvin cycle in the third stage of photosynthesis. SC.F.1.4.1

4. **Critical Thinking Organizing Information** Make a table in which you identify the role of each of the following in photosynthesis: light, water, pigments, ATP, NADPH, and carbon dioxide. SC.G.1.4.3

5. **Critical Thinking Inferring Relationships** What combination of environmental factors affects the rate of photosynthesis? SC.G.2.4.5

6. FCAT Prep During photosynthesis, in what substance do plants store energy?
   A ADP
   B water
   C carbon dioxide
   D 3-carbon sugars
   SC.B.1.4.1

103

Real Life Answer

Most houseplants are small and able to grow in a small amount of soil with relatively low light.

Close

Reteaching — BASIC

Form six cooperative groups of students. Assign one of the following factors to each group: water, chlorophyll, light, carbon dioxide, NADPH, and ATP. Have each group decide on the role their factor plays in photosynthesis. After a designated time, select a person in each group to read the decision of the group to the rest of the class.

LS Interpersonal Co-op Learning

Quiz — GENERAL

1. True or false: Chloroplasts are found within the thylakoid membranes. (False. Thylakoids are the small disk shaped structures found within the chloroplasts.)
2. True or false: Carbon dioxide fixation takes place during the Calvin cycle. (True. The Calvin cycle is the most common method of carbon dioxide fixation.)
3. The electron transport chain converts light energy to chemical energy during the _______ stage of photosynthesis. (second)

Alternative Assessment — ADVANCED

Have students describe the events of the Calvin cycle from the perspective of one of the carbon atoms in a carbon dioxide molecule. Students should include the main events of each of the four steps.

Answers to Section Review

1. Certain pigments within the cells absorb specific wavelengths of light energy.
2. Water molecules are split to provide new electrons for the electron transport chain and hydrogen ions. The excited electrons provide energy used to pump even more hydrogen ions into the thylakoid. The hydrogen ions then diffuse out of the thylakoid in a process that makes ATP.
3. Carbon dioxide is used in the Calvin cycle to produce a 3-carbon sugar that will be used to produce glucose and other organic compounds. Most of the 3-carbon sugars are recycled.
4. Answers will vary, but may be similar to the table on the right.
5. Factors include light intensity, water availability, carbon dioxide concentration, and temperature.
6. **A.** Incorrect. The loss of a phosphate from ATP forms ADP and releases energy. **B.** Incorrect. Water is not directly involved in the Calvin cycle. **C.** Incorrect. Energy is not stored in CO_2. **D.** Correct. Energy stored in 3-carbon sugars is used to make other organic compounds.

Light	Excites electrons
Water	Provides hydrogen ions and replacement electrons
Pigments	Absorb light
ATP and NADPH	Store chemical energy
Carbon dioxide	Used to produce organic compounds

Focus

Overview

Before beginning this section review with your students the objectives listed in the Student Edition. In this section students will learn how organic compounds are broken down into ATP, the energy currency of all cells. Students will learn the basic events of glycolysis and cellular respiration, as well as alternate energy pathways that take place in the absence of oxygen.

Bellringer

Ask students to answer the following questions in their notebooks: How are the products of photosynthesis and respiration related? (The products of photosynthesis are the starting materials for respiration.) What kinds of organisms undergo cellular respiration? (All organisms, including photosynthetic organisms, undergo cellular respiration as long as oxygen is available.)

Motivate

Identifying Preconceptions

Ask students what they know about fermentation. Students may speak of rotting, foul smells, or alcohol production. Explain that fermentation is a way that cells are able to make ATP when in the absence of oxygen. Fermentation can cause foul odors. It also helps produce cheese, yogurt, bread, and wine. Tell them that if they ever exercised to the point of soreness and cramping, they may have forced their muscle cells to undergo fermentation.

Section Correlations

SC.F.1.4.1
SC.G.1.4.3

Section 3 Cellular Respiration

Florida Sunshine Standards/FCAT

SC.F.1.4.1 Body processes involve specific biochemical reactions governed by biochemical principles.

SC.G.1.4.3 The chemical elements that make up the molecules of living things are combined and recombined in different ways.

Objectives

- **Summarize** how glucose is broken down in the first stage of cellular respiration. SC.F.1.4.1
- **Describe** how ATP is made in the second stage of cellular respiration. SC.F.1.4.1
- **Identify** the role of fermentation in the second stage of cellular respiration. SC.F.1.4.1
- **Evaluate** the importance of oxygen in aerobic respiration. SC.G.1.4.3

Key Terms

aerobic
anaerobic
glycolysis
NADH
Krebs cycle
$FADH_2$
fermentation

Cellular Energy

Most of the foods we eat contain usable energy. Much of the energy in a hamburger, for example, is stored in proteins, carbohydrates, and fats. But before you can use that energy, it is transferred to ATP. Like in most organisms, your cells transfer the energy in organic compounds, especially glucose, to ATP through a process called cellular respiration. Oxygen in the air you breathe makes the production of ATP more efficient, although some ATP is made without oxygen. Metabolic processes that require oxygen are called **aerobic** *(ehr OH bihk).* Metabolic processes that do not require oxygen are called **anaerobic** *(AN ehr oh bihk),* meaning "without air."

The Stages of Cellular Respiration

Cellular respiration is the process cells use to harvest the energy in organic compounds, particularly glucose. The breakdown of glucose during cellular respiration can be summarized by the following equation:

$$\underset{\text{glucose}}{C_6H_{12}O_6} + \underset{\text{oxygen gas}}{6O_2} \xrightarrow{\text{enzymes}} \underset{\text{carbon dioxide}}{6CO_2} + \underset{\text{water}}{6H_2O} + \underset{\text{ATP}}{\text{energy}}$$

As **Figure 10** shows, cellular respiration occurs in two stages:

Stage 1 Glucose is converted to pyruvate *(PIE roo vayt),* producing a small amount of ATP and NADH.

Stage 2 When oxygen is present, pyruvate and NADH are used to make a large amount of ATP. This process is called aerobic respiration. Aerobic respiration occurs in the mitochondria of eukaryotic cells and in the cell membrane of prokaryotic cells. When oxygen is not present, pyruvate is converted to either lactate *(LAK tayt)* or ethanol (ethyl alcohol) and carbon dioxide.

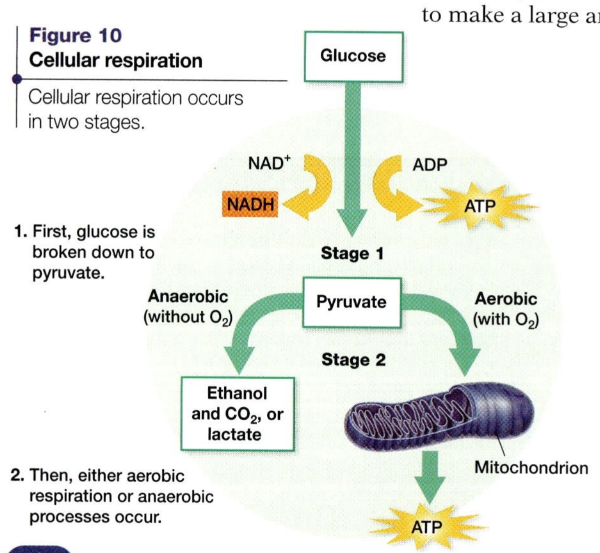

Figure 10 Cellular respiration
Cellular respiration occurs in two stages.

The equation above does not show *how* cellular respiration occurs. It simply shows that the complete enzyme-assisted breakdown of a glucose molecule uses six oxygen molecules and forms six carbon dioxide molecules, six water molecules, and ATP. Aerobic respiration produces most of the ATP made by cells. Intermediate products of aerobic respiration form the organic compounds that help build and maintain cells.

104

Chapter Resource File

- Directed Reading BASIC
- Active Reading GENERAL

One-Stop Planner CD-ROM

- Reading Organizers BASIC
- Reading Strategies BASIC
- Supplemental Reading Guide *The Lives of a Cell* ADVANCED

Transparencies

TR Bellringer
TR B53 Cellular Respiration
TR B37 Glycolysis
TR B39 Krebs Cycle
TR B54 Electron Transport Chain of Aerobic Respiration

Stage One: Breakdown of Glucose

The primary fuel for cellular respiration is glucose, which is formed when carbohydrates such as starch and sucrose are broken down. If too few carbohydrates are available to meet an organism's glucose needs, other molecules, such as fats, can be broken down to make ATP. In fact, one gram of fat contains more energy than two grams of carbohydrates. Proteins and nucleic acids can also be used to make ATP, but they are usually used for building important cell parts.

Glycolysis

In the first stage of cellular respiration, glucose is broken down in the cytoplasm during a process called **glycolysis** *(glie KAHL uh sihs)*. Glycolysis is an enzyme-assisted anaerobic process that breaks down one six-carbon molecule of glucose to two three-carbon pyruvate ions. Recall that a molecule that has lost or gained one or more electrons is called an ion. Pyruvate is the ion of a three-carbon organic acid called pyruvic acid. The pyruvate produced during glycolysis still contains some of the energy that was stored in the glucose molecule.

As glucose is broken down, some of its hydrogen atoms are transferred to an electron acceptor called NAD^+. This forms an electron carrier called **NADH**. For cellular respiration to continue, the electrons carried by NADH are eventually donated to other organic compounds. This recycles NAD^+, making it available to accept more electrons. Glycolysis is summarized in **Figure 11.**

Step 1 In a series of three reactions, phosphate groups from two ATP molecules are transferred to a glucose molecule.

Step 2 In two reactions, the resulting six-carbon compound is broken down to two three-carbon compounds, each with a phosphate group.

Step 3 Two NADH molecules are produced, and one more phosphate group is transferred to each three-carbon compound.

Step 4 In a series of four reactions, each three-carbon compound is converted to a three-carbon pyruvate, producing four ATP molecules in the process.

Glycolysis uses two ATP molecules but produces four ATP molecules, yielding a net gain of two ATP molecules. Glycolysis is followed by another set of reactions that use the energy temporarily stored in NADH to make more ATP.

BIO graphic

Figure 11

Glycolysis

Glycolysis yields a net gain of two ATP molecules.

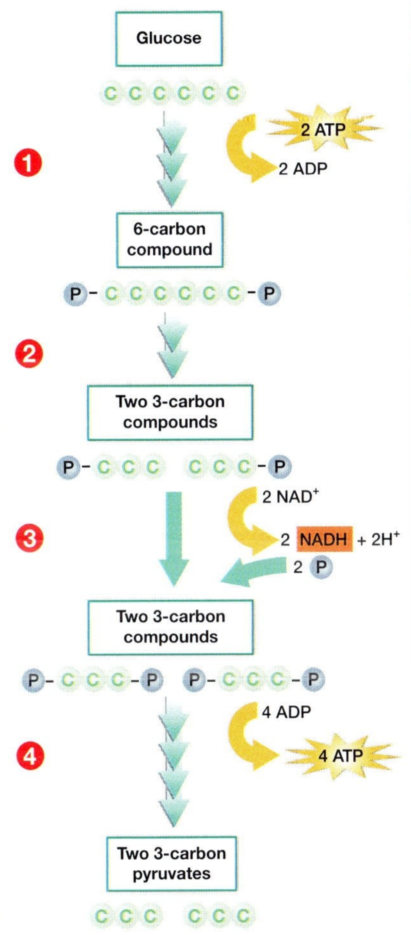

105

Teach

Teaching Tip

Metabolic Pathways Review the term metabolism with students. (Metabolism describes the sum of all chemical reactions within an organism.) Tell students that glycolysis is an example of a metabolic pathway. In glycolysis, the energy stored in glucose is gradually released in a series of enzyme–assisted chemical reactions.

Using the Figure—GENERAL

Have students examine **Figure 11,** then ask them to identify the starting material in glycolysis. (glucose) Work with students to summarize the events that take place in each step. Explain that as glucose is broken down during glycolysis, some of the energy contained in glucose is transferred to the products of glycolysis. Ask students to identify these products. (pyruvate, ATP, and NADH) Explain that pyruvate is the ion of the organic molecule pyruvic acid. **LS Visual**

READING SKILL BUILDER—GENERAL

Anticipation Guide Write the following statements on the board:

1. Organisms do not need oxygen to get energy from organic molecules.
2. It is unlikely that humans will ever run 1,600 m (about 1 mi.) in less than 2 minutes.

Ask students to think about each statement and decide whether they agree or disagree. Students should point to specific passages in the text that support their reasoning. After discussion, ask them if their opinions have changed. **LS Logical**

did you know?

Life Without Oxygen Early life probably used glycolysis to make ATP long before oxygen was present in Earth's atmosphere. According to fossil records, prokaryotes were present on Earth 3.5 billion years ago, but oxygen was not abundant in the atmosphere until around 2.5 billion years ago. Because glycolysis is an anaerobic metabolic pathway that occurs in all cells, glycolysis most likely occurred in early cells.

Teach, *continued*

Teaching Tip

Coenzymes A coenzyme is an organic chemical that is necessary for the action of many enzymes. Ask students why it is important for glucose to be partially broken down to pyruvate. (Unlike glucose, pyruvate is small enough to diffuse across the mitochondrial membranes.) Tell students that when pyruvate enters a mitochondrion and is broken down to a 2-carbon acetyl group, coenzyme A attaches to the acetyl group, forming acetyl-CoA. Coenzyme A enables the acetyl group to enter the Krebs cycle.

Using the Figure

Guide students through the steps of **Figure 12.** Have students count the number of carbon atoms present at each step during the Krebs cycle. Ask them where the Krebs cycle occurs. (in mitochondria) Remind students that for every molecule of glucose that is broken down, two pyruvate ions are produced. Thus the Krebs cycle occurs for each pyruvate. Also tell them that a specific enzyme is involved in each step. Emphasize the role of the Krebs cycle as a precursor of the electron transport chain. **LS Visual**

Stage Two: Production of ATP

When oxygen is present, pyruvate produced during glycolysis enters a mitochondrion and is converted to a two-carbon compound. This reaction produces one carbon dioxide molecule, one NADH molecule, and one two-carbon acetyl *(uh SEET uhl)* group. The acetyl group is attached to a molecule called coenzyme A (CoA), forming a compound called acetyl-CoA *(uh SEET uhl-koh ay).*

Krebs Cycle

Acetyl-CoA enters a series of enzyme-assisted reactions called the **Krebs cycle**, summarized in **Figure 12.** The cycle is named for the biochemist Hans Krebs, who first described the cycle in 1937.

Step 1 Acetyl-CoA combines with a four-carbon compound, forming a six-carbon compound and releasing coenzyme A.

Step 2 Carbon dioxide, CO_2, is released from the six-carbon compound, forming a five-carbon compound. Electrons are transferred to NAD^+, making a molecule of NADH.

Figure 12

BIO graphic

Krebs Cycle

The Krebs cycle produces electron carriers that temporarily store chemical energy.

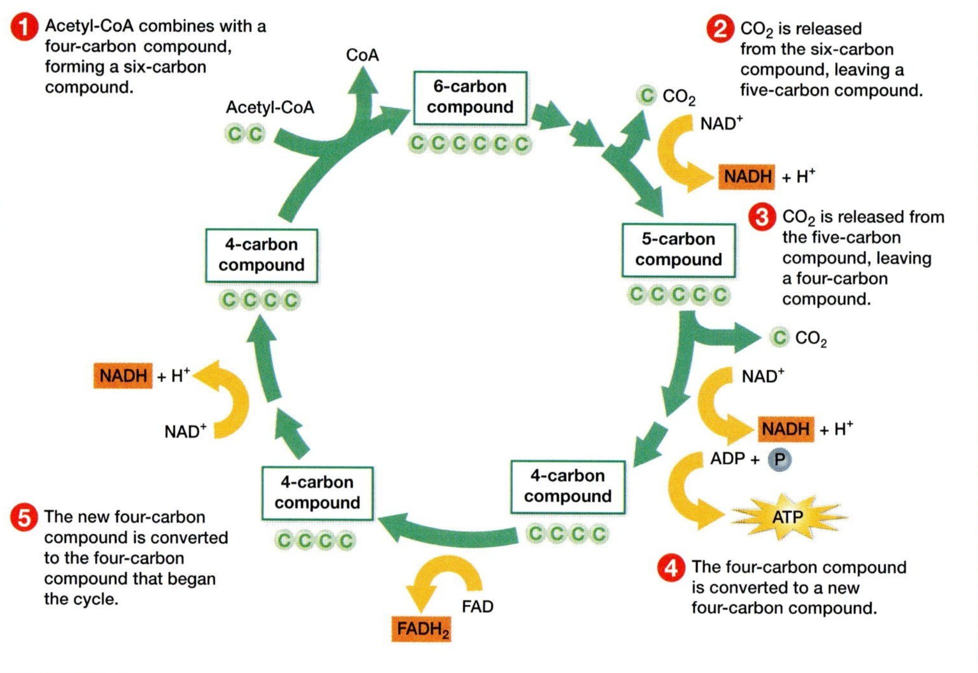

Integrating Physics and Chemistry

Remind students that the end products of photosynthesis and cellular respiration are produced as the result of complex chemical changes that the reactants undergo. In order to review and stress the chemical changes that occur in these essential metabolic processes, have students identify the reactants and products in the overall chemical reactions of photosynthesis and cellular respiration. Ask advanced students to do the same exercise with the Calvin cycle, glycolysis, and the Krebs cycle.

Step 3 Carbon dioxide is released from the five-carbon compound, resulting in a four-carbon compound. A molecule of ATP is made, and a molecule of NADH is also produced.

Step 4 The existing four-carbon compound is converted to a new four-carbon compound. Electrons are transferred to an electron acceptor called FAD, making a molecule of $FADH_2$. **$FADH_2$** is another type of electron carrier.

Step 5 The new four-carbon compound is then converted to the four-carbon compound that began the cycle. Another molecule of NADH is produced.

After the Krebs cycle, NADH and $FADH_2$ now contain much of the energy that was previously stored in glucose and pyruvate. When the Krebs cycle is completed, the four-carbon compound that began the cycle has been recycled, and acetyl-CoA can enter the cycle again.

Electron Transport Chain

In aerobic respiration, electrons donated by NADH and $FADH_2$ pass through an electron transport chain, as shown in **Figure 13.** In eukaryotic cells, the electron transport chain is located in the inner membranes of mitochondria. The energy of these electrons is used to pump hydrogen ions out of the inner mitochondrial compartment. Hydrogen ions accumulate in the outer compartment, producing a concentration gradient across the inner membrane. Hydrogen ions diffuse back into the inner compartment through a carrier protein that adds a phosphate group to ADP, making ATP. At the end of the electron transport chain, hydrogen ions and spent electrons combine with oxygen molecules, O_2, forming water molecules, H_2O.

Figure 13 Electron transport chain of aerobic respiration

In the inner membranes of mitochondria, electron transport chains (represented by the red lines) make ATP.

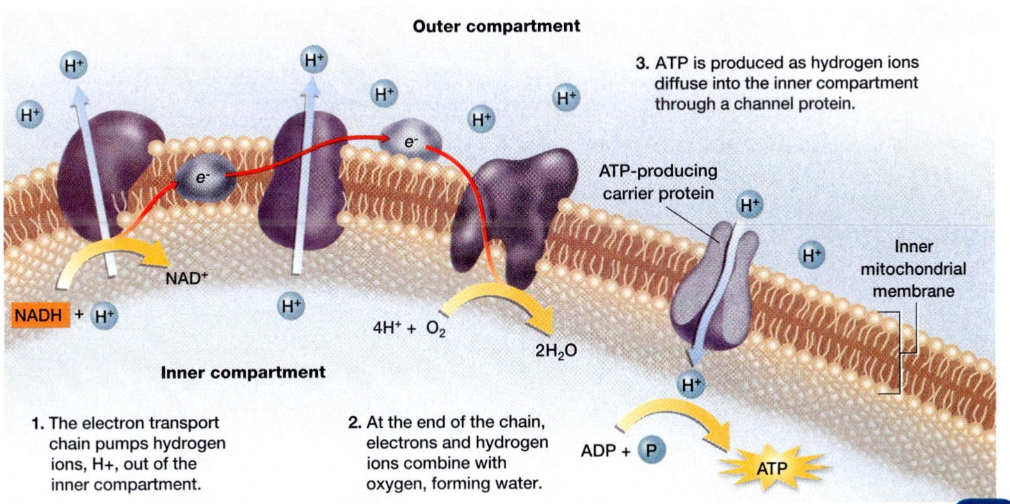

Teaching Tip — BASIC

Electron Transport Ask students how the electron transport chains of photosynthesis and cellular respiration are similar. (In both processes, electrons are passed along an electron transport chain and are picked up by an electron acceptor. The energy of these electrons is used to produce a hydrogen ion concentration gradient, which provides the energy needed to make ATP.)

Using the Figure — GENERAL

In **Figure 13** have students follow the path of electrons, shown by the red arrows, through the electron transport chain. Point out that the energy of these electrons is used to pump hydrogen ions out of the inner compartment. Ask students to identify this type of transport. (active transport) These ions then diffuse back into the inner compartment through the specialized carrier protein (ATP synthase), providing enough energy to make ATP. Ask students to identify this type of transport. (passive transport) Ask students why the folds of the mitochondria are important. (They increase the surface area of the membranes, which allows more ATP to be made.) Ask students to identify the role of oxygen in the electron transport chain. (Oxygen is the final electron acceptor, and water is produced when the spent electrons, hydrogen ions, and oxygen combine.) **LS Visual**

Attention Grabber

ATP Molecules The human body uses about 1 million molecules of ATP per cell per second. There are more than 100 trillion cells in the human body. That's about 1×10^{20}, or 100,000,000,000,000,000,000 ATP molecules used in your body each second!

Cultural Awareness — GENERAL

Hans Adolph Krebs In 1937 Hans Adolph Krebs discovered the details of the Krebs cycle. In 1953 Krebs was awarded the Nobel Prize in physiology or medicine for his discovery. Have students research and prepare a report on the life of Krebs, who had to interrupt his work and leave Nazi Germany prior to World War II because he was Jewish.

Figure 14 **Fermentation.** In cheese making, fungi or prokaryotes added to milk carry out lactic acid fermentation on some of the sugar in the milk.

Respiration in the Absence of Oxygen

What happens when there is not enough oxygen for aerobic respiration to occur? The electron transport chain does not function because oxygen is not available to serve as the final electron acceptor. Electrons are not transferred from NADH, and NAD^+ therefore cannot be recycled. When oxygen is not present, NAD^+ is recycled in another way. Under anaerobic conditions, electrons carried by NADH are transferred to pyruvate produced during glycolysis. This process recycles NAD^+ needed to continue making ATP through glycolysis. The recycling of NAD^+ using an organic hydrogen acceptor is called **fermentation.** Prokaryotes carry out more than a dozen kinds of fermentation, all using some form of organic hydrogen acceptor to recycle NAD^+. Two important types of fermentation are lactic acid fermentation and alcoholic fermentation. Lactic acid fermentation by some prokaryotes and fungi is used in the production of foods such as yogurt and some cheeses, as shown in **Figure 14.**

Lactic Acid Fermentation

In some organisms, a three-carbon pyruvate is converted to a three-carbon lactate through lactic acid fermentation, as shown in **Figure 15.** Lactate is the ion of an organic acid called lactic acid. For example, during vigorous exercise pyruvate in muscles is converted to lactate when muscle cells must operate without enough oxygen. Fermentation enables glycolysis to continue producing ATP in muscles as long as the glucose supply lasts. Blood removes excess lactate from muscles. Lactate can build up in muscle cells if it is not removed quickly enough, sometimes causing muscle soreness.

Figure 15 **Two types of fermentation**

When oxygen is not present, cells recycle NAD+ through fermentation.

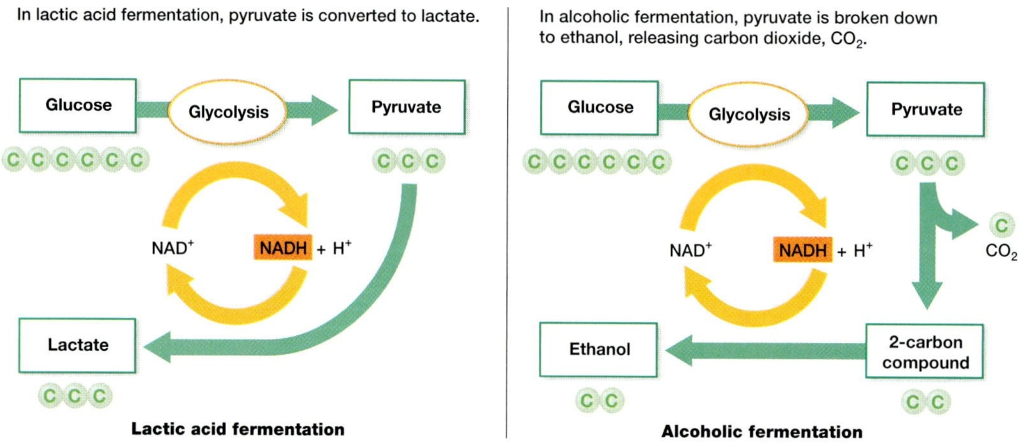

108

Teaching Tip — Basic

Invite one of the physical education teachers or coaches from your school to discuss the physiological effects of exercise on the body. Be certain that the speaker discusses oxygen debt, muscle fatigue, the role of myoglobin in muscles, and the role of lactate in muscle soreness.
LS Intrapersonal

Group Activity — Advanced

Fermentation Have students work in groups of four. Each group should brainstorm a list of foods and beverages that make use of fermentation. Students may wish to use the Web to expand their list. For presentation, students should prepare a poster, using drawings or pictures of fermentation products cut out from magazines. (Answers will vary, but could include various breads, pizza, various cheeses, beer, wine, soy sauce, doughnuts, etc.)
LS Interpersonal Co-op Learning

did you know?

Cyanide shuts down the electron transport chain. Cyanide is a fast-acting poison that blocks the action of the electron transport chain. It exists as hydrogen cyanide gas or cyanide salts used in gold and other metal extractions, electroplating, and metal cleaning. Cyanide enters the body by absorption through the lungs, skin, or gastrointestinal tract. It is highly toxic, and symptoms appear soon after exposure. Ingesting as little as 3 g of cyanide can soon be fatal.

Career

Fitness Trainer Many people use health clubs to exercise or work out. These clubs employ fitness trainers to direct exercise programs for groups or custom-tailor programs for individuals. Have students prepare a report describing the education requirements and responsibilities of a fitness trainer.

Alcoholic Fermentation

In other organisms, the three-carbon pyruvate is broken down to ethanol (ethyl alcohol), a two-carbon compound, through alcoholic fermentation. Carbon dioxide is released during the process. As shown in Figure 15, alcoholic fermentation is a two-step process. First, pyruvate is converted to a two-carbon compound, releasing carbon dioxide. Second, electrons are transferred from a molecule of NADH to the two-carbon compound, producing ethanol. As in lactic acid fermentation, NAD^+ is recycled, and glycolysis can continue to produce ATP.

Alcoholic fermentation by yeast, a fungus, has been used in the preparation of many foods and beverages. Wine and beer contain ethanol made during alcoholic fermentation by yeast. Carbon dioxide released by the yeast causes the rising of bread dough and the carbonation of some alcoholic beverages, such as beer. Ethanol is actually toxic to yeast. At a concentration of about 12 percent ethanol kills yeast. Thus, naturally fermented wine contains about 12 percent ethanol.

internet connect

www.scilinks.org
Topic: Fermentation
Keyword: HX4080

SCiLINKS. Maintained by the National Science Teachers Association

BIOWatch

Muscle Fatigue and Endurance Training

Anyone who runs or exercises for a long period of time soon learns about muscle fatigue. As you continue vigorous exercise, the muscles you are using become fatigued—that is, tired and less able to generate force. The reasons for muscle fatigue are not fully understood, but in most cases the fatigue increases when the production of lactic acid by the exercising muscle increases.

Anaerobic Threshold

Why does an exercising muscle produce lactic acid? A resting muscle obtains most of its energy from aerobic respiration. A continuously exercising muscle, however, soon depletes its available oxygen. At this point, called the *anaerobic threshold,* the exercising muscle begins to obtain the ATP needed anaerobically. In the absence of oxygen, glycolysis extracts the required ATP from glycogen in the muscle. *Glycogen* is a storable form of glucose that acts as an energy reserve. Glycolysis converts the muscle glycogen to pyruvate, which is then fermented to lactic acid.

The ability to perform continuous exercise is limited by the body's stored glycogen. So, physical endurance can increase if glycogen stored in muscles is spared during exercise. Trained athletes such as cyclist Lance Armstrong, shown at right, get a relatively large portion of their energy from aerobic respiration. Thus, their muscle glycogen reserve is depleted more slowly than that in untrained individuals. In fact, the greater the level of physical training, the higher the proportion of energy the body derives from aerobic respiration.

Athletic Endurance

Endurance-trained athletes generally have more muscle mass than untrained people. But it is endurance-trained athletes' high aerobic capacity—rather than their greater muscle mass—that allows these athletes to exercise more before lactic acid production and glycogen depletion cause muscle fatigue.

Lance Armstrong

internet connect

www.scilinks.org
Topic: Anaerobic Threshold
Keyword: HX4192

SCiLINKS. Maintained by the National Science Teachers Association

Demonstration

Thoroughly mix half a packet of yeast with about 400 mL of warm water. Pour the mixture into an Erlenmeyer flask or a beaker. Add a few drops of bromothymol blue indicator and set the flask aside. Ask students what the bubbles indicate. (Carbon dioxide is being released by the yeast.) By the end of the class period, the blue yeast mixture should turn yellow. Ask students why this happens. (The indicator turns yellow in the presence of an acid; when carbon dioxide is produced by the yeast, carbonic acid is formed in the water.) **LS Visual**

BIOWatch

Muscle Fatigue and Endurance Training

Teaching Strategies

Choose a student volunteer who is willing to do a muscle endurance test. Have the student squat with their back against the wall until their thighs are parallel to the floor. They should hold this position until it becomes too uncomfortable. Ask the class to explain why the student could not maintain the position for hours? (The muscles were working continuously until they were fatigued and depleted of oxygen.)

Discussion

Why are endurance athletes able to exercise longer than non-athletes are? (One reason is their muscles are conditioned to obtain ATP through aerobic respiration longer than non-athletes. Aerobic respiration fuels muscles more efficiently than fermentation.)

In what form is glucose stored in muscle tissue? (Glycogen. Glycogen reserves are depleted more slowly in trained athletes than non-athletes.)

Teach, continued

Teaching Tip — BASIC

ATP Production Tell students that up to 36 molecules of ATP can be produced from a single glucose molecule during aerobic respiration. Have students calculate how much more efficient aerobic respiration is than glycolysis, which yields only two ATP molecules. (36 ATP /2 ATP = 18 times more efficient.) LS Logical

Close

Reteaching — BASIC

Assign students to cooperative pairs. Have each pair evaluate the following scenario: Suppose you are an organism that can carry out either aerobic respiration or anaerobic energy pathways. Which one would be more beneficial to you, and why? (If oxygen is present, aerobic respiration is more beneficial because more ATP can be produced.)

Co-op Learning · English Language Learners

Quiz — GENERAL

1. During glycolysis, glucose is broken down to _______. (pyruvate)
2. True or false: When oxygen is not present, a small amount of ATP is still made during the Krebs cycle. (False. The Krebs cycle produces ATP in the presence of oxygen.)

Alternative Assessment — ADVANCED

Have students write a description of the events of the Krebs cycle as if it were a script of a sportscaster giving a play-by-play description. LS Verbal

Figure 16 Effect of oxygen on ATP production

Most ATP is produced during aerobic respiration.

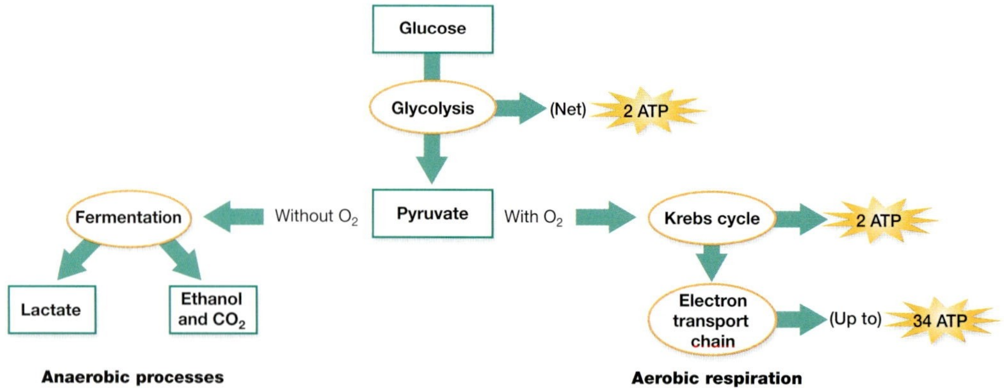

Production of ATP

The total amount of ATP that a cell is able to harvest from each glucose molecule that enters glycolysis depends on the presence or absence of oxygen. As shown in **Figure 16,** cells use energy most efficiently when oxygen is present. In the first stage of cellular respiration, glucose is broken down to pyruvate during glycolysis. Glycolysis is an anaerobic process, and it results in a net gain of two ATP molecules. In the second stage of cellular respiration, the pyruvate passes through either aerobic respiration or (anaerobic) fermentation. When oxygen is present, aerobic respiration occurs. When oxygen is not present, fermentation occurs instead. The NAD^+ that gets recycled during fermentation allows glycolysis to continue producing ATP. Thus, a small amount of ATP is produced even during fermentation. Most of a cell's ATP is made, however, during aerobic respiration. For each molecule of glucose that is broken down, as many as two ATP molecules are made directly during the Krebs cycle, and up to 34 ATP molecules are produced later by the electron transport chain.

Section 3 Review

1. **List** the products of glycolysis. What is the role of each of these products in cellular respiration? SC.G.1.4.3
2. **Summarize** the roles of the Krebs cycle and the electron transport chain during aerobic respiration. SC.F.1.4.1
3. **Describe** the role of fermentation in the second stage of cellular respiration. SC.F.1.4.1
4. **Critical Thinking Comparing Functions** Explain why cellular respiration is more efficient when oxygen is present in cells. SC.G.1.4.3
5. **Critical Thinking Inferring Conclusions** Excess glucose in your blood is stored in your liver as glycogen. How might your body sense when to convert glucose to glycogen and glycogen back to glucose? SC.F.1.4.6
6. **FCAT Prep** When oxygen is present, what process produces most of the ATP made in cellular respiration?
   A glycolysis
   B aerobic respiration
   C alcoholic fermentation
   D lactic acid fermentation
   SC.G.1.4.3

110

Answers to Section Review

1. *Pyruvate:* if oxygen is present, pyruvate will enter the Krebs cycle; if oxygen is absent, pyruvate will undergo fermentation. *NADH:* if oxygen is present, NADH will enter the electron transport chain. *ATP:* temporarily stores energy for cellular processes
2. The Krebs cycle produces electron carriers that donate electrons to the electron transport chain. The electron transport chain produces most of the ATP that is produced in cellular respiration.
3. Fermentation recycles NAD^+, which is needed to continue ATP production in the absence of oxygen.
4. If oxygen is present, aerobic respiration can occur. Aerobic respiration produces much more ATP than anaerobic processes.
5. Sensors in the body monitor the level of glucose in the blood. When the blood glucose level is high, the storage of glycogen is stimulated. When the blood glucose level is low, glucose is released back into the blood.
6. **A.** Incorrect. Glycolysis is an anaerobic process. **B.** Correct. **C.** Incorrect. Alcoholic fermentation is an anaerobic process. **D.** Incorrect. Lactic acid fermentation is an anaerobic process.

Section Review Correlations

SC.F.1.4.1: Items 2, 3
SC.F.1.4.6: Item 5
SC.G.1.4.3: Items 1, 4, 6

Key Concepts

1 Energy and Living Things

- Energy from sunlight flows through living systems, from autotrophs to heterotrophs.
- Photosynthesis and cellular respiration form a cycle because one process uses the products of the other.
- ATP supplies cells with energy needed for metabolism.

2 Photosynthesis

- Photosynthesis has three stages. First, energy is captured from sunlight. Second, energy is temporarily stored in ATP and NADPH. Third, organic compounds are made using ATP, NADPH, and carbon dioxide.
- Pigments absorb light energy during photosynthesis.
- Electrons excited by light travel through electron transport chains, in which ATP and NADPH are produced.
- Through carbon dioxide fixation, often by the Calvin cycle, carbon dioxide in the atmosphere is used to make organic compounds, which store energy.
- Photosynthesis is directly affected by environmental factors such as the intensity of light, the concentration of carbon dioxide, and temperature.

3 Cellular Respiration

- Cellular respiration has two stages. First, glucose is broken down to pyruvate during glycolysis, making some ATP. Second, a large amount of ATP is made during aerobic respiration. When oxygen is not present, NAD^+ is recycled during the anaerobic process of fermentation.
- The Krebs cycle is a series of reactions that produce energy-storing molecules during aerobic respiration.
- During aerobic respiration, large amounts of ATP are made in an electron transport chain.
- When oxygen is not present, fermentation follows glycolysis, regenerating NAD^+ needed for glycolysis to continue.

Key Terms

Section 1

photosynthesis (94)
autotroph (94)
heterotroph (95)
cellular respiration (95)

Section 2

pigment (98)
chlorophyll (98)
carotenoid (98)
thylakoid (99)
electron transport chain (100)
NADPH (101)
carbon dioxide fixation (102)
Calvin cycle (102)

Section 3

aerobic (104)
anaerobic (104)
glycolysis (105)
NADH (105)
Krebs cycle (106)
$FADH_2$ (107)
fermentation (108)

BIOLOGY INTERACTIVE TUTOR **Unit 2, Unit 3**—Use Topics 1–6 in these units to review the key concepts and terms in this chapter.

Alternative Assessment

Have each student write a question on an index card based on the information presented in this chapter. Each student should also record an answer to the question on a second index card. Have students trade question cards with a partner and try to answer their partner's questions. Then have them confer with each other about their answers. Encourage students to refer to the textbook to settle any disagreements.

Chapter Resource File

- Science Skills Worksheet GENERAL
- Critical Thinking Worksheet ADVANCED
- Test Prep Pretest GENERAL
- Chapter Test GENERAL

Answer to Concept Map

The following is one possible answer to Performance Zone item 15.

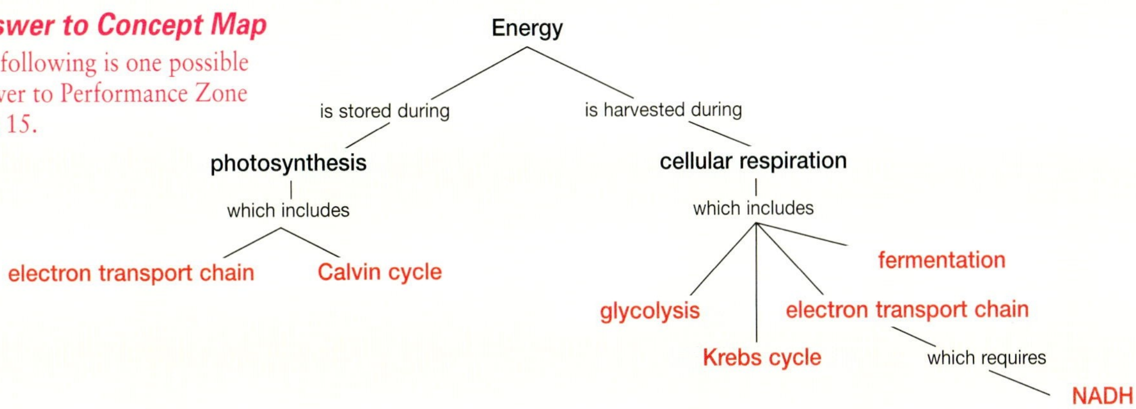

CHAPTER 5 **ZONE**

ANSWERS

Understanding Key Ideas

1. c
2. a
3. d
4. b
5. The structures are thylakoids. During photosynthesis, hydrogen ions are more concentrated inside the thylakoids.
6. a
7. The energy flow begins with sunlight. Autotrophs form organic compounds using energy from sunlight, storing some of this energy in their tissues as carbohydrates. Humans get energy by eating the autotrophs or by eating other heterotrophs that first ate autotrophs.
8. At the anaerobic threshold, the muscle cells have depleted the available oxygen and resort to glycolysis and lactic acid fermentation to obtain ATP. Fatigue and cramping can result.
9. Starch is composed of hundreds of glucose molecules. It is broken down by enzymes, such as amylase, yielding maltose. Maltose consists of two glucose molecules and is broken down by the enzyme maltase.
10. The answer to the concept map is on the bottom of the Study Zone page.

Critical Thinking

11. Humans must obtain thiamine from the foods they eat. Insufficient thiamine in cells may decrease the efficiency of aerobic respiration.
12. More folds in the mitochondrial inner membrane would provide more surface area for cellular respiration, increasing the amount of cellular respiration that can occur.

Alternative Assessment

13. Answers will vary. Bacteria are used to make yogurt, sour cream, and some cheeses. Other cheeses are made with the help of fungi. During bread making, alcoholic fermentation by yeast produces alcohol, which evaporates, and carbon dioxide, which makes the bread rise.

Chapter Review Correlations

SC.F.1.4.1: Item 10
SC.F.1.4.2: Item 5
SC.F.1.4.3: Items 3, 12
SC.F.1.4.5: Items 9, 11
SC.F.1.4.7: Item 8
SC.G.1.4.2: Items 1, 7
SC.G.1.4.3: Items 2, 4, 6
SC.H.3.4.6: Item 13

Performance ZONE

Understanding Key Ideas

1. Energy flows through living systems from?
   a. the sun, to heterotrophs, and then to autotrophs.
   b. autotrophs, to the environment, and then to heterotrophs.
   c. the sun, to autotrophs, and then to heterotrophs.
   d. the environment, to heterotrophs, and then to autotrophs. SC.G.1.4.2
2. The products of photosynthesis that begin cellular respiration are
   a. organic compounds and oxygen.
   b. carbon dioxide and water.
   c. $NADP^+$ and hydrogen.
   d. ATP and water. SC.G.1.4.3
3. The thylakoid membranes of a chloroplast are the sites where
   a. electron transport chains operate.
   b. NADPH and ATP are produced.
   c. pigments are located.
   d. all of the above SC.F.1.4.3
4. The oxygen produced during photosynthesis comes directly from the
   a. splitting of carbon dioxide molecules.
   b. splitting of water molecules.
   c. mitochondrial membranes.
   d. absorption of light. SC.G.1.4.3
5. Study the micrograph of a chloroplast shown below, and identify the structures labeled *X*. During photosynthesis, are hydrogen ions more concentrated in these structures or in the spaces around them? SC.F.1.4.2

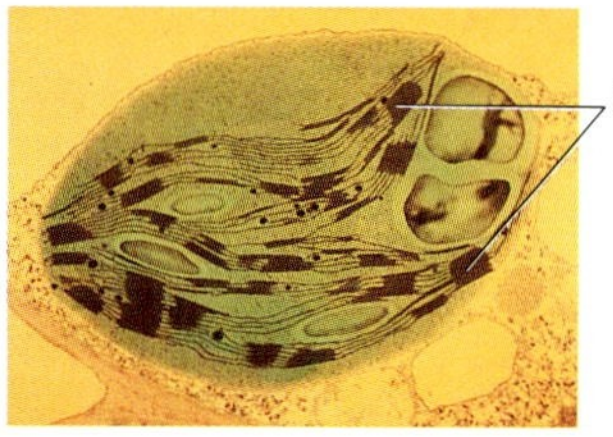

6. Which of the following is the correct pairing of a process and its requirement for oxygen?
   a. glycolysis: no oxygen required
   b. fermentation: oxygen required
   c. Krebs cycle: no oxygen required
   d. none of the above
   SC.G.1.4.3
7. Analyze the flow of energy that enables you to get energy from the food you eat. SC.G.1.4.2
8. **BIOWatch** What change occurs in muscles at the anaerobic threshold? SC.F.1.4.7
9. How is starch broken down to glucose prior to glycolysis? (**Hint:** See Chapter 2, Section 4.) SC.F.1.4.5
10. **Concept Mapping** Make a concept map that shows how photosynthesis and cellular respiration are related. Try to include the following terms in your map: *glycolysis, Krebs cycle, electron transport chain, Calvin cycle, fermentation,* and *NADH.* SC.F.1.4.1

Critical Thinking

11. **Distinguishing Relevant Information** The enzyme that aids in the conversion of pyruvate to acetyl-CoA requires vitamin B1, also called *thiamine.* Thiamine is not made in the human body and must be taken in in foods. How would a deficiency of thiamine in cells affect cellular respiration? SC.F.1.4.5
12. **Inferring Relationships** How might the folding of the inner membrane of mitochondria affect the rate of aerobic respiration? Explain your answer. SC.F.1.4.3

Alternative Assessment

13. **Analyzing Methods** Research several ways that fermentation is used in food preparation. Find out what kinds of microorganisms are used in cultured dairy products, such as yogurt, sour cream, and some cheeses. Research the role of alcoholic fermentation by yeast in bread making. Prepare an oral report to summarize your findings. SC.H.3.4.6

112

Assignment Guide

Section	Questions
1	1, 7
2	3, 4, 6
3	2, 5, 8, 9, 10, 11, 12, 13

FCAT Prep

For questions 1–6, write your answers on a separate sheet of paper.

1. What pigment causes a plant to look green?
   A. carotenoid **C.** NADH
   B. chlorophyll **D.** NAPH
   SC.F.1.4.2

2. A scientist makes the following statement: "If Earth's early atmosphere had been rich in oxygen, photosynthetic organisms would not have been able to evolve." Which of the following statements can be used to argue against this hypothesis?
   F. Earth's present atmosphere is rich in oxygen.
   G. The main component of the atmosphere is nitrogen gas.
   H. Nonphotosynthetic organisms require atmospheric or dissolved oxygen to survive.
   I. Photosynthetic organisms today carry out photosynthesis in the presence of oxygen. SC.H.1.4.4

3. Which of the following is involved during only the anaerobic part of cellular respiration? SC.F.1.4.1
   A. ATP **C.** the Krebs cycle
   B. glycolysis **D.** mitochondria

4. What is the product of the electron transport chain of photosynthesis? SC.F.1.4.5
   F. water **H.** pyruvate
   G. glucose **I.** ATP and NADPH

Base your answers to questions 5 and 6 on the graph below, which shows data on photosynthesis in one type of plant.

EFFECT OF TEMPERATURE ON PHOTOSYNTHESIS

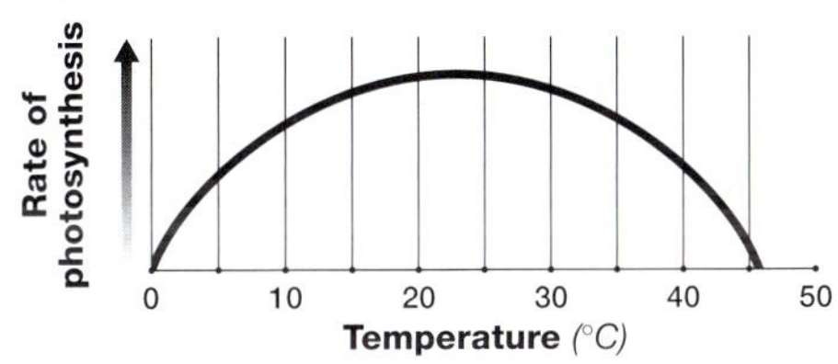

5. Which statement is supported by the data?
   A. Photosynthesis does not occur at 0°C.
   B. The rate of photosynthesis at 40°C is greater than the rate at 20°C.
   C. The optimum temperature for photosynthesis is approximately 46°C.
   D. The rate of photosynthesis increases as temperature increases from 25°C to 30°C.
   SC.H.1.4.1

6. At approximately what temperature (to the nearest 10°C) is the plant producing oxygen at the greatest rate? SC.H.1.4.1

Test TIP

If you come upon a word you do not know, try to identify its prefix, suffix, or root. Sometimes knowing even one part of the word will help you answer the question.

113

FCAT Doctor

Question 2 Answer I is the correct choice. Having an oxygen-rich atmosphere does not affect the existence of modern photosynthetic organisms. Answer F is incorrect because it does not relate photosynthetic organisms to the oxygen content of the atmosphere. Answer G is incorrect because nitrogen does not play a major role in photosynthesis. Answer H is incorrect because the hypothesis does not discuss nonphotosynthetic organisms.

Question 5 Answer A is the correct choice. The graph shows no photosynthesis at 0°C. Answer B is incorrect because the opposite is true. Answer C is incorrect because there is less photosynthesis at 46°C than there is at 35°C. Answer D is incorrect because the opposite is true.

Question 6 The correct answer is 20°C. The graph shows that the optimum temperature for photosynthesis is between 20°C and 25°C. The peak rate temperature to the nearest 10°C is 20°C.

Answers

1. B
2. I
3. B
4. I
5. A
6. 20°C

Test Prep Correlations

SC.F.1.4.1: Item 3
SC.F.1.4.2: Item 1
SC.F.1.4.5: Item 4
SC.H.1.4.1: Items 5, 6
SC.H.1.4.4: Item 2

Skills Practice Lab

Observing Oxygen Production from Photosynthesis

Teacher's Notes

Time Required
One 50-minute class period and about 15 minutes per day for 5 days.

Ratings
EASY → HARD

Teacher Preparation 3
Student Setup 3
Concept Level 2
Cleanup 3

Skills Acquired
- Measuring
- Collecting Data
- Graphing

Scientific Methods
In this lab, students will:
- Make Observations
- Analyze the Results
- Draw Conclusions
- Communicate the Results

Materials and Equipment
Materials for this lab can be ordered from WARD'S. See *Master Materials List* at the front of this book for catalog numbers.

Safety Cautions
Review all safety symbols with students before beginning this lab.

Skills Practice Lab

Observing Oxygen Production from Photosynthesis SC.G.1.4.3

SKILLS
- Measuring
- Collecting Data
- Graphing

OBJECTIVE
- **Measure** amount of oxygen produced by an *Elodea* sprig.

MATERIALS
- 500 mL of 5 percent baking-soda-and-water solution
- 600 mL beaker
- 20 cm long *Elodea* sprigs (2–3)
- glass funnel
- test tube
- metric ruler
- protective gloves

Elodea

ChemSafety

CAUTION: Always wear safety goggles and a lab apron to protect your eyes and clothing.

CAUTION: Glassware is fragile. Notify the teacher of broken glass or cuts. Do not clean up broken glass or spills with broken glass unless the teacher tells you to do so.

CAUTION: Wear disposable polyethlene gloves when handling any plant. Do not eat any part of a plant or plant seed used in the lab. Wash hands thoroughly after handling any part of a plant.

Before You Begin

Plants use **photosynthesis** to produce food. One product of photosynthesis is oxygen. In this activity, you will observe the process of photosynthesis and determine the rate of photosynthesis for *Elodea*.

1. Write a definition for the boldface term above.
2. Create a Data Table like the one below.

DATA TABLE

Amount of Gas Present in the Test Tube		
Days of exposure to light	Total amount of gas present (mm)	Amount of gas produced per day (mm)
0		
1		
2		
3		
4		
5		

Procedure

1. Add 450 mL of baking-soda-and-water solution to a beaker.
2. Put two or three sprigs of *Elodea* in the beaker. The baking soda will provide the *Elodea* with the carbon dioxide it needs for photosynthesis.
3. Place the wide end of the funnel over the *Elodea*. The end of the funnel with the small opening should be pointing up. The *Elodea* and the funnel should be completely under the solution, as shown on the facing page.
4. Fill a test tube with the remaining baking-soda-and-water solution. Place your thumb over the end of the test tube. Turn the test tube upside-down, taking care that no air enters. Hold the opening of the test tube under the solution and place the test tube over the small end of the funnel. Try not to let any solution leak out of the test tube as you do this.

114

Tips and Tricks
Elodea is a common aquarium plant and can be found at some pet stores and most stores that sell aquarium fish. A 5 percent solution of baking soda and water can be made by dissolving 5 g of baking soda in 95 mL of water. You may need to have students practice placing the test tube over the inverted funnel. It may take two or three tries to get the test tube over the funnel stem without letting any air into the tube. First, fill the test tube with the solution. Place your thumb over the opening tightly so no air can get in. Submerge your thumb and the mouth of the test tube underwater. Once the mouth of the test tube is underwater you can release your thumb and maneuver the test tube over the stem of the funnel. Be sure you have the *Elodea* in place under the funnel before you begin.

5. Place the beaker setup in a well-lit area near a lamp or in direct sunlight.
6. Record that there was 0mm gas in the test tube on day 0. (If you were unable to place the test tube without getting air in the tube, measure the height of the column of air in the test tube in millimeters. Record this value for day 0.) In this lab, change in gas volume is indicated by a linear measurement expressed in millimeters.
7. For days 1 through 5, measure the amount of gas in the test tube. Record the measurements in your data table under the heading, "Total amount of gas present (mm)."
8. Calculate the amount of gas produced each day by subtracting the amount of gas present on the previous day from the amount of gas present today. Record these amounts under the heading, "Amount of gas produced per day (mm)."
9. Plot the data from your table on a graph.

Analyze and Conclude

1. **Summarizing Results** Using information from your graph, describe what happened to the amount of gas in the test tube.
2. **Analyzing Data** How much gas was produced in the test tube after day 5?
3. **Drawing Conclusions** Write the equation for photosynthesis. Explain each part of the equation. For example, what ingredients are necessary for photosynthesis to take place? What substances are produced by photosynthesis? What gas is produced that we need in order to live?
4. **Predicting Patterns** What may happen to the oxygen level if an animal, such as a snail, were put in the beaker with the *Elodea* sprig while the *Elodea* sprig was making oxygen?
5. **Further Inquiry** Write a new question about photosynthesis that could be explored with another investigation.

Do You Know?

Do research in the library or media center to answer these questions:

1. What is hydroponic farming?
2. How do coral reefs depend on photosynthesis?

Use the following Internet resources to explore your own questions about photosynthesis.

internet connect
www.scilinks.org
Topic: Photosynthesis
Keyword: HX4136
SciLinks. Maintained by the National Science Teachers Association

Answers to Before You Begin

1. Photosynthesis is the process by which plants use carbon dioxide and the energy from sunlight to make sugar and oxygen.

Answers to Analyze and Conclude

1. Students' graphs should show a gradual increase in the amount of gas in the test tube.
2. Answers will vary according to variables in the classroom, such as the amount of light and temperature.
3. $6CO_2 + 6H_2O + \text{light energy} \rightarrow C_6H_{12}O_6 + 6O_2$. CO_2 is carbon dioxide and comes from the baking soda solution. H_2O is the water in the solution. Light energy comes from the sun. $C_6H_{12}O_6$ is sugar (glucose), and O_2 is oxygen. Photosynthesis produces sugar and oxygen. Plants produce oxygen as a byproduct of photosynthesis; oxygen is the gas that we breathe and cannot live without.
4. Sample answer: The snail would use some of the oxygen generated by the plant. This could decrease the oxygen level.
5. Answers will vary. Example: How would increasing the temperature affect the amount of oxygen produced by the plant?

Answers to Do You Know?

1. Hydroponic farming involves growing fruits and vegetables in nutrient-rich solutions or inert material rather than soil.
2. Autotrophic plankton, which obtain their energy through photosynthesis, are the main food source for many of the animals that inhabit coral reefs.

Chapter Opener		SE **Quick Review,** p. 117 ■ SE **Reading Activity,** p. 117 ■ TE **Using the Figure,** p. 116 GENERAL TE **Opening Activity,** p. 117 GENERAL
& 3 • 90 min pp. 118–124 **Section 1** Chromosomes	OSP **Lesson Plan*** TR **Bellringer*** TR **Section Outline*** TR **B43 Chromosome Structure*** TR **B44 Chromosome Number of Various Organisms*** TR **B45 Karyotype***	TE **Demonstration,** p. 118 GENERAL TE **Demonstration,** p. 119 TE **Using the Figure,** p. 120 GENERAL TE **Activity** Karyotypes, p. 121 GENERAL SE **Quick Lab** Modeling Chromosomal Mutations, p. 124 ■ CRF **Datasheets for In-Text Labs** Modeling Chromosomal Mutations*
& 5 • 90 min pp. 125–127 **Section 2** The Cell Cycle	OSP **Lesson Plan*** TR **Bellringer*** TR **Section Outline***	TE **Demonstration,** p. 125 TE **Group Activity** Cell Cycle Phases, p. 126 BASIC CRF **Exploration Lab** Mitosis* ◆ GENERAL CRF **Skills Practice Lab** Preparing a Root Tip Squash* ◆ GENERAL
& 7 • 90 min pp. 128–132 **Section 3** Mitosis and Cytokinesis	OSP **Lesson Plan*** TR **Bellringer*** TR **Section Outline*** TR **B47 Stages of Mitosis***	TE **Activity** Mitosis, p. 128 GENERAL SE **Math Lab** Calculating the Number of Cells Resulting from Mitosis, p. 129 ■ CRF **Datasheets for In-Text Labs** Calculating the Number of Cells Resulting from Mitosis* TE **Using the Figure,** p. 130 TE **Using the Figure,** p. 131 TE **Group Activity** Performing Mitosis and Cytokinesis, p. 131 GENERAL SE **Quick Lab** Observing Mitosis and Cytokinesis, p. 132 ■ CRF **Datasheets for In-Text Labs** Observing Mitosis and Cytokinesis* SE **Exploration Lab** Modeling Mitosis, pp. 136–137 ■ CRF **Datasheets for In-Text Labs** Modeling Mitosis*

& 9 • 90 min

Chapter Review and Assessment Resources

- SE **Study Zone,** p. 133 ■
- SE **Performance Zone,** p. 134 ■
- SE **FCAT Test Prep,** p. 135 ■
- TR **Concept Mapping***
- CRF **Vocabulary Review Worksheet*** ■ GENERAL
- CRF **Test Prep Pretest*** ■ GENERAL
- CRF **Chapter Test*** ■ GENERAL
- CRF **Chapter Test*** ADVANCED
- OSP **Test Generator***
- OSP **Test Item Listing***
- OSP **Modified Chapter Test*** BASIC

go.hrw.com Holt Online Learning

Visit **go.hrw.com** for access to Holt Online Learning or enter the keyword **HX6 Home** for a variety of free online resources.

One-Stop Planner® CD-ROM

This CD-ROM package includes
- Lab Materials QuickList Software
- Holt Calendar Planner
- Customizable Lesson Plans
- Printable Worksheets
- ExamView® Test Generator
- Interactive Teacher Edition
- Holt PuzzlePro® Resources
- Holt PowerPoint® Resources

Guided Reading Audio CDs

These CDs are designed to help auditory learners and reluctant readers. (Audio CDs are also available in Spanish.)

Compression guide:
To shorten your instruction because of time limitations, omit block 1.

KEY

TE	Teacher Edition	CRF	Chapter Resource File	*	Also on One-Stop Planner
SE	Student Edition	TR	Teaching Transparency	■	Also Available in Spanish
OSP	One-Stop Planner			◆	Requires Advance Prep

SKILLS DEVELOPMENT AND PRACTICE	SECTION REVIEW AND ASSESSMENT	CORRELATIONS
		Florida Sunshine Standards/FCAT
TE **Inclusion Strategies**, p. 118 TE **Reading Skill Builder** Interactive Reading, p. 119 BASIC TE **Reading Skill Builder** Reading Organizer, p. 119 GENERAL TE **Skill Builder** Math, p. 120 GENERAL TE **Skill Builder** Math, p. 122 GENERAL CRF **Directed Reading Worksheet*** BASIC CRF **Active Reading Worksheet*** GENERAL	SE **Section 1 Review** ■ TE **Reteaching**, p. 124 BASIC TE **Quiz**, p. 124 GENERAL TE **Alternative Assessment**, p. 124 GENERAL CRF **Quiz*** ■ BASIC	SC.F.2.4.1, SC.F.2.4.2, SC.F.2.4.3
CRF **Directed Reading Worksheet*** BASIC CRF **Active Reading Worksheet*** GENERAL	SE **Section 2 Review** ■ TE **Reteaching**, p. 127 BASIC TE **Quiz**, p. 127 GENERAL TE **Alternative Assessment**, p. 127 ADVANCED CRF **Quiz*** ■ BASIC	SC.F.1.4.5
TE **Skill Builder** Vocabulary, p. 130 BASIC CRF **Directed Reading Worksheet*** BASIC CRF **Active Reading Worksheet*** GENERAL CRF **Critical Thinking Worksheet*** ADVANCED CRF **Science Skills Worksheet*** GENERAL	SE **Section 3 Review** ■ TE **Reteaching**, p. 132 BASIC TE **Quiz**, p. 132 GENERAL CRF **Quiz*** ■ BASIC **Modified Worksheet*** BASIC	SC.F.1.4.2, SC.F.1.4.5, SC.G.1.4.1, SC.H.3.4.1

Holt Biology Interactive Tutor CD-ROM

This CD-ROM consists of interactive activities that give students a fun way to extend their knowledge of biology concepts.

Unit 4 Cell Reproduction

SCILINKS NSTA
www.scilinks.org
Maintained by the **National Science Teachers Association**

Topic: Chromosomes
SciLinks Code: HX4042

CHAPTER 6

Overview

Tell the students the purpose of this chapter is to examine the structure of chromosomes and the role they play in development and reproduction. Chromosomes and cellular reproduction are further explored by studying the events of the cell cycle. Emphasis is placed on the events of mitosis and cytokinesis.

Using the Figure

Have students look at the opening photograph on this page. Ask students why some of the cells are smaller than others. (They have recently divided and have not grown to full size.) Ask students what allows these few cells to become specialized into lung tissue or heart muscle or skin. (Genes on the chromosomes within each cell contain instructions for building the proteins required for each cell type.) Ask students how each cell receives the proper genes. (Through the process of mitosis, the copied chromosomes are carefully divided so that each new cell receives a complete and correct set of all genes.)

Assessing Prior Knowledge

Before beginning this chapter, be sure your students can answer the questions in the ***Quick Review*** on the facing page. If they cannot, encourage them to reread the sections indicated.

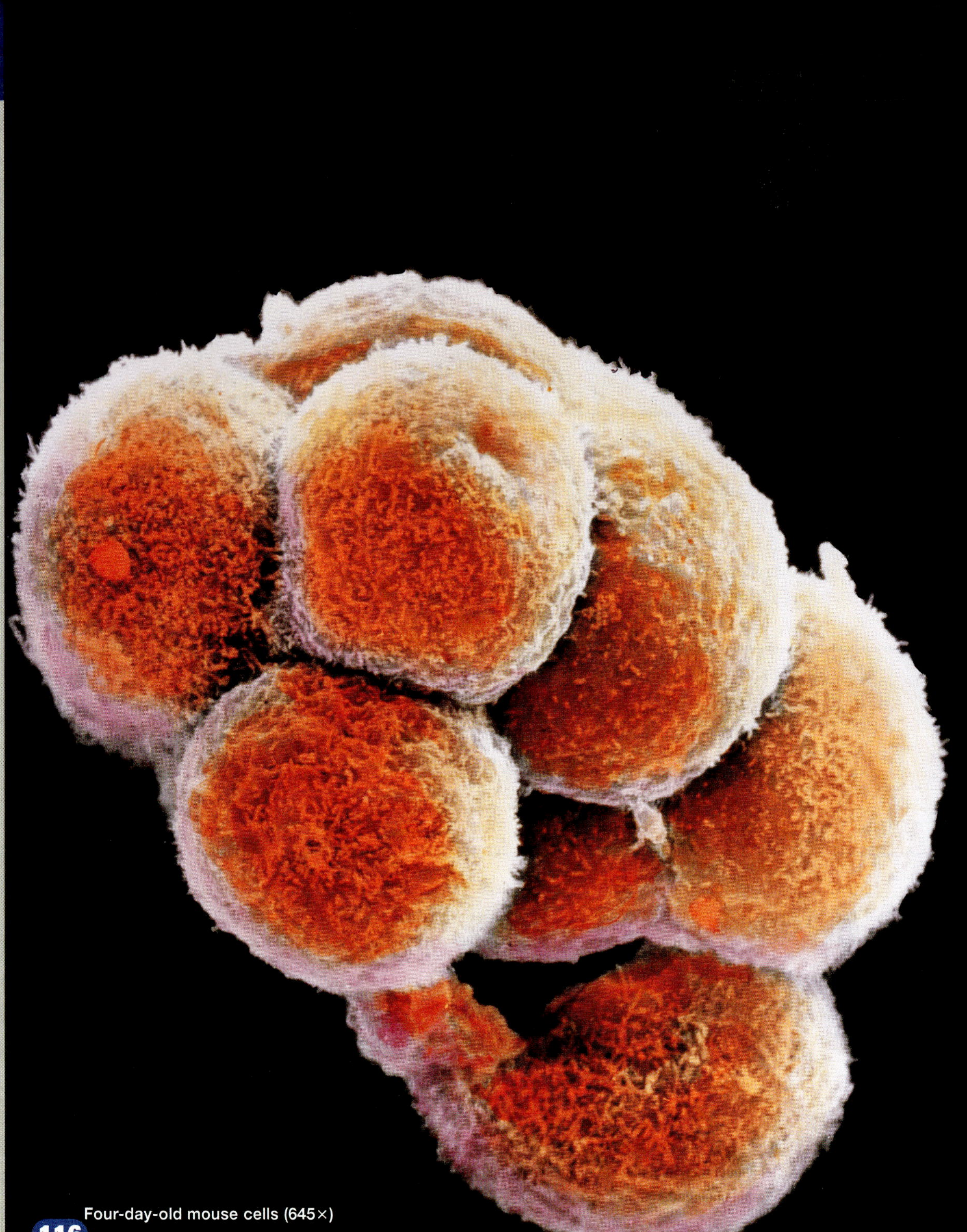

Four-day-old mouse cells (645×)

116

Chapter Correlations — Florida Sunshine Standards/FCAT

SC.F.1.4.2 Body structures are uniquely designed and adapted for their function.

SC.F.1.4.5 Complex interactions among the different kinds of molecules in the cell cause distinct cycles of activity governed by proteins.

SC.F.2.4.1 There are mechanisms of asexual and sexual reproduction and different genetic advantages and disadvantages of asexual and sexual reproduction.

SC.F.2.4.2 Every cell contains a "blueprint" coded in DNA molecules that specify how proteins are assembled to regulate cells.

SC.F.2.4.3 Mechanisms of change (e.g., mutation and natural selection) lead to adaptations in a species and their ability to survive naturally in changing conditions and to increase species diversity.

SC.G.1.4.1 There is great diversity and interdependence among living things.

CHAPTER

6 Chromosomes and Cell Reproduction

Florida Sunshine Standards/FCAT

SC.F.2.4.1 There are mechanisms of asexual and sexual reproduction and different genetic advantages and disadvantages of asexual and sexual reproduction.

SC.F.2.4.2 Every cell contains a "blueprint" coded in DNA molecules that specify how proteins are assembled to regulate cells.

SC.F.2.4.3 Mechanisms of change (e.g., mutation and natural selection) lead to adaptations in a species and their ability to survive naturally in changing conditions and to increase species diversity.

SC.G.1.4.1 There is great diversity and interdependence among living things.

Quick Review

Answer the following without referring to earlier sections of your book.

1. **Define** the term *mutation*. *(Chapter 1, Section 1)*
2. **Describe** the structure of proteins and of DNA. *(Chapter 2, Section 3)*
3. **Summarize** the function and structure of the nucleus and of microtubules. *(Chapter 3, Section 2)*

Did you have difficulty? *For help, review the sections indicated.*

Reading Activity

Copy the following statements in your notebook:

- *Chromosomes from females determine the sex of humans.*
- *Every human cell contains 46 chromosomes.*
- *Healthy cells cannot become cancerous cells.*

Before you read the chapter, write down if you agree with each statement. After you have finished reading the chapter, decide whether you still agree with your first response.

This cluster of cells is smaller than the head of a pin, but over the next 17 days, they will divide repeatedly to form a new mouse. Chromosomes inside each cell carry the instructions for growth and development of an individual.

Looking Ahead

internet connect

www.scilinks.org

National Science Teachers Association *sci*LINKS Internet resources are located throughout this chapter.

SCILINKS. Maintained by the National Science Teachers Association

Opening Activity — GENERAL

Karyotypes Obtain or draw three human karyotypes, one of a female, one of a male, and one of an individual with Down syndrome. Display the karyotypes and ask students to identify the female, the male, and the individual with Down syndrome. (The female will have two large X chromosomes, the male will have one large X chromosome and one small Y chromosome, and the individual with Down syndrome will have three copies of chromosome 21.) **LS Visual**

Quick Review Answers

1. A mutation is a change in the DNA of a gene.
2. Proteins consist of folded chains of amino acids linked by peptide bonds. DNA is a nucleic acid that consists of a double helix of linked nucleotides. Each nucleotide is made of deoxyribose sugar, a base, and a phosphate group.
3. The nucleus is an internal compartment within each eukaryotic cell that houses and protects the DNA. Microtubules are long hollow protein fibers that are part of the cytoskeleton. They also help to move organelles and chromosomes through the cytoplasm.

Chapter Resource File

- Vocabulary Worksheets
- Concept Mapping

Reading Activity Answers

1. Students should disagree. The female always donates an X sex chromosome; the male donates an X or a Y, which determines the sex of the individual.
2. Students should disagree. All somatic cells contain 46 chromosomes. Gametes are haploid and contain 23 chromosomes.
3. Students should disagree. Exposure to mutagens such as radiation, pollution, tobacco smoke, or other chemicals can damage the DNA of cells and cause cancer.

Focus

Overview

Before beginning this section review with your students the objectives listed in the Student Edition. In this section, students will learn that cells must divide in order to grow, replace worn-out cells, or reproduce asexually. It is vitally important that each new cell receives the proper set of chromosomes to function normally. Students will study the structure of chromosomes, the role of chromosomes, and learn that each organism has a characteristic number of chromosomes.

Bellringer

Write the words *haploid, diploid,* and *zygote* on the board. Ask students to write a single sentence that reflects an understanding of each term. (Answers will vary. For example: During fertilization, haploid gametes combine to form a diploid zygote.)

Motivate

Demonstration — GENERAL

Show the class two pictures or photos. One picture should show a person without any visible genetic disorders. The second picture should show a person with recognizable physical abnormalities due to a genetic disorder caused by the presence or absence of a chromosome (such as Klinefelter or Turner syndrome). Discuss with students how a single chromosome contains thousands of genes that code for proteins involved in determining how a person's body develops and functions. LS Visual

Section Correlations

SC.F.2.4.1
SC.F.2.4.2
SC.F.2.4.3

Section 1 Chromosomes

Florida Sunshine Standards/FCAT

SC.F.2.4.2 Every cell contains a "blueprint" coded in DNA molecules that specify how proteins are assembled to regulate cells.

SC.F.2.4.3 Mechanisms of change (e.g., mutation and natural selection) lead to adaptations in a species and their ability to survive naturally in changing conditions and to increase species diversity.

Objectives

- **Identify** four examples of cell division in eukaryotes and one example in prokaryotes. SC.F.2.4.1
- **Differentiate** between a gene, a DNA molecule, a chromosome, and a chromatid. SC.F.2.4.2
- **Differentiate** between homologous chromosomes, autosomes, and sex chromosomes. SC.F.2.4.2
- **Compare** haploid and diploid cells. SC.F.2.4.2
- **Predict** how changes in chromosome number or structure can affect development. SC.F.2.4.3

Key Terms

gamete
binary fission
gene
chromosome
chromatid
centromere
homologous chromosome
diploid
haploid
zygote
autosome
sex chromosome
karyotype

Formation of New Cells by Cell Division

About 2 trillion cells are produced by an adult human body every day. This is about 25 million new cells per second! These new cells are formed when older cells divide. Cell division, also called cell reproduction, occurs in humans and other organisms at different times in their life. In **Figure 1,** the cells of the fawn that is growing and developing and the cells in the wound that is healing are undergoing cell division. The type of cell division differs depending on the organism and why the cell is dividing. For example, bacterial cells undergoing reproduction divide by one type of cell division. Eukaryotic organisms undergoing growth, development, repair, or asexual reproduction divide by a different type of cell division. And the formation of gametes involves yet a third type of cell division. **Gametes** are an organism's reproductive cells, such as sperm or egg cells.

Regardless of the type of cell division that occurs, all of the information stored in the molecule DNA (deoxyribonucleic acid) must be present in each of the resulting cells. Recall from Chapter 3 that DNA stores the information that tells cells which proteins to make and when to make them. This information directs a cell's activities and determines its characteristics. Thus, when a cell divides, the DNA is first copied and then distributed. Each cell ends up with a complete set (copy) of the DNA.

Figure 1 Cell division

The cells of these organisms are undergoing some type of cell division.

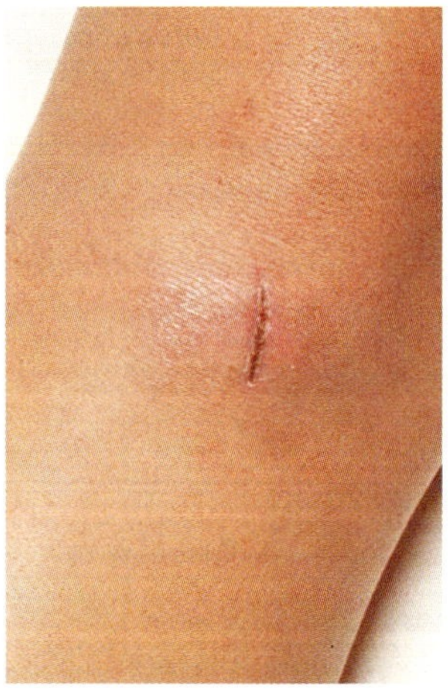

Repair

Growth and development

One-Stop Planner CD-ROM

- **Reading Organizers** BASIC
- **Reading Strategies** BASIC
- **Basic Skills Worksheet** Length, Area and Volume BASIC
- **Occupational Applications** Medical Sonographer GENERAL

INCLUSION Strategies

- *Learning Disability*

Ask students to label three sheets of notebook paper with the titles of each section of the chapter, **Chromosomes, The Cell Cycle,** and **Mitosis and Cytokinesis.** Have students take notes as they read each section. The notes should be divided into three parts, *key terms, notes,* and *questions.* In a small group or with the teacher, students may share what they learned in each section and ask for clarification on their questions.

Prokaryotic Cell Reproduction

A prokaryote's single DNA molecule is circular and is attached to the inner cell membrane. Prokaryotes reproduce by a type of cell division called binary fission. **Binary fission** is a form of asexual reproduction that produces identical offspring. In asexual reproduction, a single parent passes exact copies of all of its DNA to its offspring.

Binary fission occurs in two stages: first, the DNA is copied (so that each new cell will have a copy of the genetic information), and then the cell divides. The prokaryote divides by adding a new cell membrane to a point on the membrane between the two DNA copies. As new material is added, the growing cell membrane pushes inward and the cell is constricted in the middle, like a long balloon being squeezed near the center. A new cell wall forms around the new membrane. Eventually the dividing prokaryote is pinched into two independent cells. Each cell contains one of the circles of DNA and is a complete functioning prokaryote.

Eukaryotic Cell Reproduction

The vast amount of information encoded in DNA is organized into units called genes. A **gene** is a segment of DNA that codes for a protein or RNA molecule. A single molecule of DNA has thousands of genes lined up like train cars. Genes play an important role in determining how a person's body develops and functions. When genes are being used, the DNA is stretched out so that the information it contains can be used to direct the synthesis of proteins.

As a eukaryotic cell prepares to divide, the **chromosomes**—the DNA and the proteins associated with the DNA—become visible, as shown in **Figure 2.** Before the DNA coils up, however, the DNA is copied. The two exact copies of DNA that make up each chromosome are called **chromatids** *(KROH muh tihdz)*. The two chromatids of a chromosome are attached at a point called a **centromere.** The chromatids become separated during cell division and placed into each new cell, ensuring that each new cell will have the same genetic information as the original cell.

Real Life

Escherichia coli **cells can produce 2 million new cells in less than 7 hours.**

A variety of *E. coli* known as O157:H7 is sometimes found in raw or undercooked meat. When such meat is eaten, this bacteria can cause life-threatening intestinal bleeding and kidney failure. Thorough cooking is necessary to destroy the bacteria.

Finding Information

Research outbreaks of E. coli *O157:H7 in your community or state.*

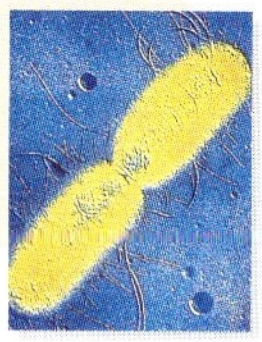

Figure 2 Chromosome structure. A chromosome consists of DNA tightly coiled around proteins. The chromosomes condense as a cell prepares to divide.

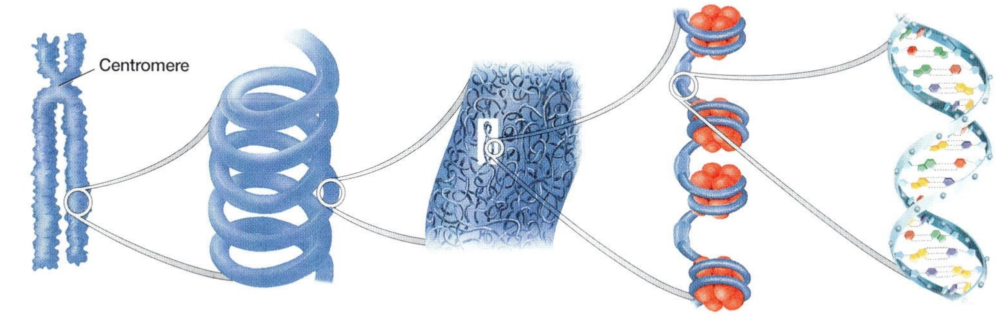

Chapter Resource File

- Directed Reading BASIC
- Active Reading GENERAL
- Data Sheet for Quick Lab GENERAL

Transparencies

TR Bellringer

TR B43 Chromosome Structure

TR B44 Chromosome Number of Various Organisms

TR B45 Karyotype

Teach

READING SKILL BUILDER — BASIC

Interactive Reading Assign Chapter 6 of *Holt Biology* Guided Audio CD Program to help students achieve greater success in reading the chapter.

Real Life

Answer

Answers will vary by community. The strain was first identified in 1982. Between 1992 and 1998, 30 outbreaks each year were reported in various communities, including areas in the Pacific Northwest; Alpine, Wyoming; Milwaukee, Wisconsin; and Indianapolis, Indiana.

Demonstration

Use a microprojector or microscope with a video camera to show the class a cross section of an onion root tip. The root grows because cells in the root tip undergo repeated mitotic divisions. Point out chromosomes that are visible. Tell students that before each root tip cell divides, the chromosomes are replicated and then divided equally into the two resulting cells. Explain that chromosomes consist of genes, which are segments of DNA. **LS Visual**

READING SKILL BUILDER — GENERAL

Writing **Reading Organizer** Have students create a reading organizer to compare reproduction in bacterial cells and eukaryotic cells. A description of reproduction in eukaryotic cells is given in Sections 2 and 3. **LS Verbal**

INTERACTIVE BIOLOGY TUTOR

- **Unit 4—Cell Reproduction: Topics 1–4**

  This engaging tutorial introduces students to principles of chromosome replication and cell division

Teach, *continued*

Teaching Tip — GENERAL

Missing Homologue Ask students to hypothesize what might happen if a human sperm or egg cell did not contain one member of each homologous pair. (The resulting zygote would not have a full set of chromosomes. The zygote might fail to develop. If the zygote did develop, the individual would not be normal because its cells would lack the important information contained in the missing genes.)

SKILL BUILDER — GENERAL

Math Skills Prokaryotic chromosomes are hundreds of times longer than the cell that contains them. For example, if a chromosome in an *E. coli* bacterium were fully extended, it would measure about 1 mm in length. The cell itself is only about 0.002 mm in length. Have students calculate how much longer *E. coli's* chromosome is than the cell itself. (1 mm/0.002 mm = 500 times longer than the cell) **LS Logical**

Using the Figure — GENERAL

Use **Figure 3** to point out how fertilization restores the diploid number of chromosomes. Ask students why sexually reproducing organisms possess an even number of chromosomes in the diploid state. (If students do not realize that $2n$ must be an even number, then ask them to calculate the haploid number in each gamete as if the diploid number in humans were 47 and not 46. It cannot be done without a "half" chromosome.) **LS Visual**

internet connect
www.scilinks.org
Topic: Chromosomes
Keyword: HX4042
SciLinks. Maintained by the National Science Teachers Association

How Chromosome Number and Structure Affect Development

Each human somatic cell (any cell other than a sperm or egg cell) normally has two copies of 23 different chromosomes, for a total of 46 chromosomes. The 23 chromosomes differ in size, shape, and set of genes. Each chromosome contains thousands of genes that play important roles in determining how a person's body develops and functions. For this reason, a complete set of all chromosomes is essential to survival.

Sets of Chromosomes

Each of the 23 pairs of chromosomes consists of two homologous *(hoh MAHL uh gus)* chromosomes, or homologues *(HOH muh logs)*. **Homologous chromosomes** are chromosomes that are similar in size, shape, and genetic content. Each homologue in a pair of homologous chromosomes comes from one of the two parents, as shown in **Figure 3.** Thus, the 46 chromosomes in human somatic cells are actually two sets of 23 chromosomes. One set comes from the mother, and one set comes from the father. A human chromosome is shown in **Figure 4.**

Figure 3 Fertilization

When haploid gametes fuse, they produce a diploid zygote.

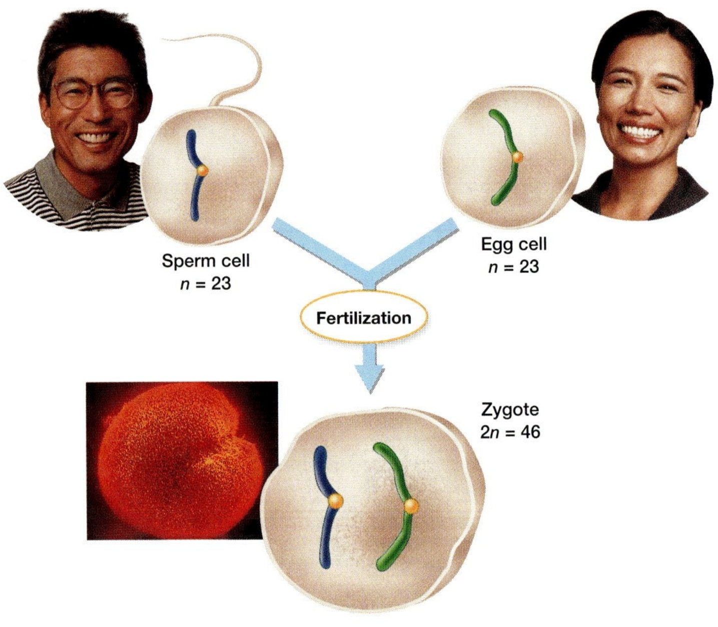

120

Cultural Awareness

Mormons and Gene Mapping
The Human Genome Project suceeded in mapping the entire human genome. The genome consists of over 30,000 genes. Scientists throughout the world have been engaged in this effort. One group has traced genetic markers through three generations of 60 Mormon families living in Utah. Their work with the Mormon families resulted in the mapping of almost 500 genes.

All of the cells in the body, other than gametes, are somatic cells. When a cell, such as a somatic cell, contains two sets of chromosomes, it is said to be **diploid** *(DIHP loyd)*. Unlike somatic cells, human gametes contain only one set of chromosomes (23 total). When a cell, such as a gamete, contains one set of chromosomes, it is said to be **haploid** *(HAP loyd)*. Biologists use the symbol n to represent one set of chromosomes. The haploid number in a human gamete can be written as $n = 23$. The diploid number in a somatic cell can be written as $2n = 46$. The fusion of two haploid gametes—a process called fertilization—forms a diploid zygote, as shown in Figure 3. A **zygote** *(ZY goht)* is a fertilized egg cell, the first cell of a new individual.

As seen in **Table 1,** each organism has a characteristic number of chromosomes. The number of chromosomes in cells is constant within a species. Fruit flies, for example, have only eight chromosomes in each cell. Although most species have different numbers of chromosomes, some species by chance have the same number. For example, potatoes, plums, and chimpanzees all have 48 chromosomes in each cell. Many plants have far more chromosomes. Some ferns have more than 500. A few kinds of organisms—such as the Australian ant *Myrmecia,* the plant *Haplopappus* (a desert relative of the sunflower), and the fungus *Penicillium* (from which the antibiotic penicillin is obtained)—have only one pair of chromosomes.

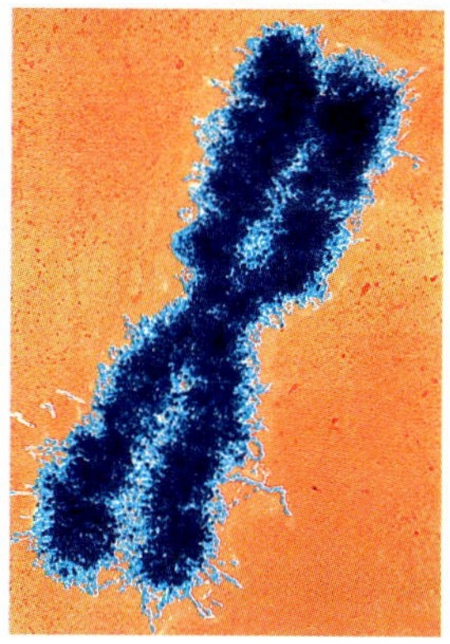

Figure 4 Human chromosome. As many as 500 chromosomes lined up end to end would fit in a 0.2 cm space—about the thickness of a nickel. The chromosome above has replicated and consists of two identical chromatids.

Table 1 Chromosome Number of Various Organisms

Organism	Number ($2n$) of chromosomes
Penicillium	1–4
Saccharomyces (yeast)	16
Mosquito	6
Housefly	12
Garden pea	14
Corn	20
Adder's tongue fern	480–1,020
Frog	26
Human	46
Orangutan	48
Dog	78

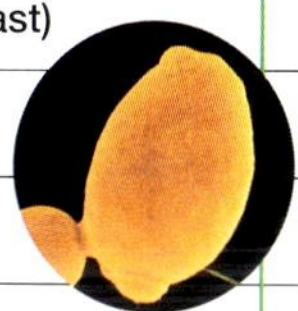

Teaching Tip — BASIC

Prefixes Have students use the prefixes in the words *haploid* and *diploid* to help them remember the meanings of these words. Have them relate the words to the algebraic terms n and $2n$. Haploid (n) represents one set of chromosomes, as seen in gametes. Diploid $(2n)$ represents two sets of chromosomes, as in somatic cells. Tell students that the prefix *hapl-* means "single" and that the prefix *dipl-* means "double." Add that the prefix *poly-* means "many." Then have students predict the nature of a polyploid cell. (It would have multiple sets of chromosomes.) English Language Learners

Activity — GENERAL

Karyotypes Obtain copies of normal human karyotypes from the library or on-line resources. Enlarge them so that each chromosome is no smaller than 2.5 cm (1 in.) in length. Paste the copies of the karyotype on a poster board. Cut out each chromosome and mix up the pieces. Allow the students to sort the chromosomes into pairs using size, length, position of centromere, and banding patterns. Have them compare the karyotypes with each other and with the karyotype shown in **Figure 5.** LS Logical English Language Learners

REAL WORLD CONNECTION

Invite a genetic counselor to speak to the class. Before the speaker attends, have the students research chromosomal abnormalities that result from too few or too many chromosomes. Have the students prepare a list of questions based on their findings. Ask the speaker to bring sample karyotypes of the abnormalities the students researched, if possible. LS Interpersonal

HISTORY CONNECTION

In the 1920s, a biologist reported the diploid number in human cells to be 48. For many years following this report, textbooks described the normal diploid number of human chromosomes as 48. It was not until 1956 that J. H. Tjio and A. Levan showed 46 to be the correct number. These two scientists worked out a procedure to culture white blood cells. Their methods increased the number of cells undergoing mitosis and improved the spreading of the chromosomes, which allowed the scientists to accurately count the chromosomes.

Teach, *continued*

SKILL BUILDER — GENERAL

Math Skills Combined, the 22 somatic chromosomes and one sex chromosome consist of 3,000,000,000 DNA nucleotide pairs. Ask students how many total nucleotides are in a human diploid cell? (6,000,000,000) **LS Logical**

BIOWatch

Prenatal Testing

Teaching Strategies

- Genetic counselors are available before, during, and after each procedure to explain the procedures and results and to offer emotional support.

Discussion

- What risks might be involved in each type of procedure? (infection or injury to the fetus) Tell students that ultrasound techniques are used to help locate the fetus during testing to avoid injury to the fetus.
- Which procedure takes longer to produce results? (amniocentesis) Why? (Amniocentesis requires the cells to be cultured, whereas with CVS the cells come directly from the fetus, which provides a quicker way to examine the cells.)
- What are some of the difficult decisions that must be made after a genetic disorder is diagnosed? (Answers will vary.)

WORD Origins

The word *chromosome* is from the Greek *chroma,* meaning "color," and *soma,* meaning "body." Chromosomes were so named because they absorbed a colored dye that made them more visible under a microscope.

Sex Chromosomes

Of the 23 pairs of chromosomes in human somatic cells, 22 pairs are called autosomes. **Autosomes** are chromosomes that are not directly involved in determining the sex (gender) of an individual. The **sex chromosomes,** one of the 23 pairs of chromosomes in humans, contain genes that will determine the sex of the individual.

In humans and many other organisms, the two sex chromosomes are referred to as the X and Y chromosomes. The genes that cause a fertilized egg to develop into a male are located on the Y chromosome. Thus, any individual with a Y chromosome is male, and any individual without a Y chromosome is female. For example, in human males, the sex chromosomes are made up of one X chromosome and one Y chromosome (XY). The sex chromosomes in human females consist of two X chromosomes (XX). Because a female can donate only an X chromosome to her offspring, the sex of an offspring is determined by the male, who can donate either an X or a Y.

The structure and number of sex chromosomes vary in different organisms. In some insects, such as grasshoppers, there is no Y chromosome—the females are characterized as XX and the males are characterized as XO (the O indicates the absence of a chromosome). In birds, moths, and butterflies, the male has two X chromosomes and the female has only one.

Change in Chromosome Number

Each of an individual's 46 chromosomes has thousands of genes. Because genes play an important role in determining how a person's body develops and functions, the presence of all 46 chromosomes is essential for normal development and function. A person must have the characteristic number of chromosomes in his or her cells. Humans with more than two copies of a chromosome, a condition called trisomy *(TRY soh mee),* will not develop properly. Abnormalities in chromosome number can be detected by analyzing a **karyotype** *(KAR ee uh tiep),* a photo of the chromosomes in a dividing cell that shows the chromosomes arranged by size. **Figure 5** shows a typical karyotype. A portion of a karyotype from an individual with an extra copy of chromosome 21 is also shown in Figure 5. This condition is called Down syndrome, or trisomy 21. Short stature, a round face with

Figure 5 A human karyotype

Karyotypes are used to examine an individual's chromosomes.

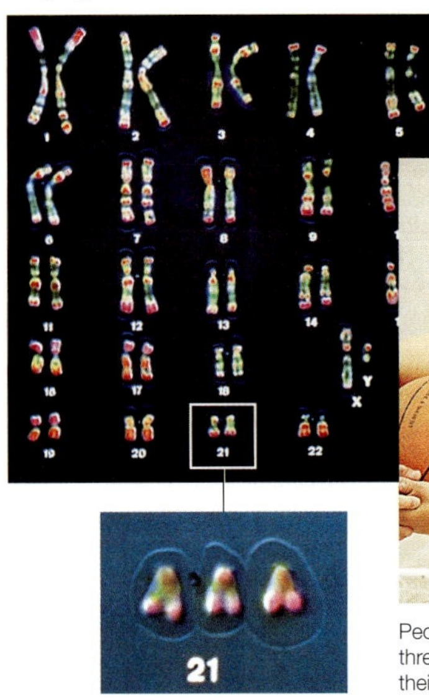

To prepare a karyotype, photographs of the chromosomes are cut out, arranged in pairs from largest to smallest, and numbered.

People with Down syndrome have three copies of chromosome 21 in their karyotype.

Integrating Physics and Chemistry

Tell students the colored dye that makes chromosomes more visible is the result of the reflection of light. Indeed it is the bouncing back of waves to our eyes that makes any object visible. Although there is a tendency to think of reflection as images we can see in a mirror, any object that we can observe with our eyes, whether or not that image is aided with a scientific instrument, is caused by light that bounces back or is reflected from an object. In this instance, the dye increases the amount of reflected light thus making the chromosomes more visible.

Career — GENERAL

Cytogenetic Technologist Cytogenetic technologists aid physicians in diagnosis of genetic disorders by preparing specimens, such as karyotypes. Have students use library references or on-line resources to research the training requirements and duties of a cytogenetic technologist. (Technologist may prepare cells, prepare slides in the correct mitotic stage, stain slides, and prepare accurate karyotypes from photographic prints or computer images.)

upper eyelids that cover the inner corners of the eyes, and varying degrees of mental retardation are characteristics of people with Down syndrome.

In mothers younger than 30, Down syndrome occurs in about 1 in 1,500 births. In mothers 37 years old, the incidence doubles to 1 in 290 births. In mothers over 45, the risk is as high as 1 in 46 births. Older mothers are more likely to have a baby with Down syndrome because all the eggs a female will ever produce are present in her ovaries when she is born, unlike males who produce new sperm throughout adult life. As a female ages, her eggs can accumulate an increasing amount of damage. Because of this risk, a pregnant woman over the age of 35 may be advised to undergo prenatal testing that includes fetal karyotyping.

What events can cause an individual to have an extra copy of a chromosome? When sperm and egg cells form, each chromosome and its homologue separate, an event called disjunction *(dihs JUHNK shuhn)*. If one or more chromosomes fail to separate properly—an event called nondisjunction—one new gamete ends up receiving both chromosomes and the other gamete receives none. Trisomy occurs when the gamete with both chromosomes fuses with a normal gamete during fertilization, resulting in offspring with three copies of that chromosome instead of two. In Down syndrome, nondisjunction involves chromosome 21.

BIOWatch

Prenatal Testing

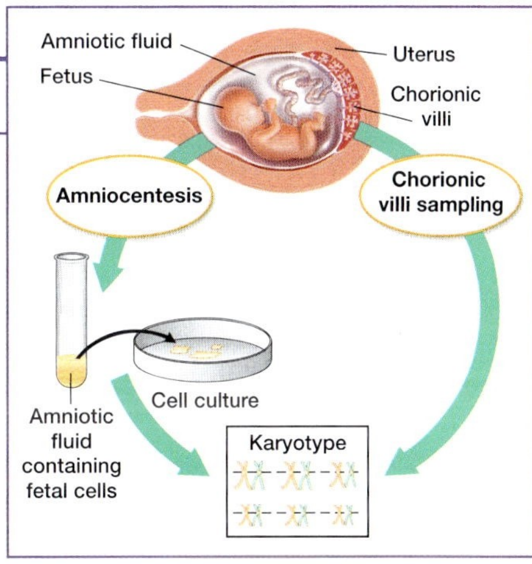

Will our baby be normal? For many expectant parents, prenatal testing can help answer this question. In prenatal testing, the cells of a fetus are tested for normal chromosome number and cell structure by a procedure called fetal karyotyping. Fetal karyotyping allows parents and doctors to view the chromosomes found in the cells of the fetus. The doctor can then check for any abnormalities, such as Down syndrome. There are two ways to obtain fetal cells.

Amniocentesis

In amniocentesis *(am nee oh sehn TEE sihs)*, a needle and syringe are used to remove a small amount of the amniotic fluid that surrounds the fetus. The fluid contains fetal cells. The fetal cells are grown in a laboratory for 1–4 weeks to obtain enough actively dividing fetal cells to make a karyotype, which is then analyzed.

Chorionic Villi Sampling (CVS)

In chorionic villi *(kawr ee AHN ihk VIHL ie)* sampling, a tissue sample is collected from the chorionic villi, fingerlike extensions of the placenta that grow into the mother's uterus. Enough actively dividing cells are obtained to produce a karyotype without having to culture cells. Since the villi have the same genetic makeup as the fetus, the doctor is able to detect abnormalities in the fetal chromosome number.

Teaching Tip — GENERAL

Everyone Needs an X Tell students that most nondisjunctions involving autosomes are lethal, whereas most involving sex chromosomes are not. However, a YO combination in humans *is* lethal, leading biologists to suspect that at least one X chromosome is necessary for development and survival.

QUICK LAB *Modeling Chromosomal Mutations*

Skills Acquired
Modeling, sequencing, interpreting

Teacher's Notes
Cut note card pieces prior to the lab.

Answers to Analysis

Answers will vary based on the type of mutation: *deletion:* cell would be missing a gene, which could prove fatal; *duplication:* cell would have an extra gene, which could prove fatal or result in malfunctioning of the cell; *inversion:* cell may not be able to use gene because it is located in a different area on the chromosome, which could prove fatal; *translocation:* cell may not be able to use the gene because it is located on a different chromosome, which could prove fatal.

Cultural Awareness

Cultural Attitudes Toward Genetic Defects From the Middle Ages through the nineteenth century, most Europeans believed that genetic defects reflected inner corruption. However, many other cultures, including the Celtic people of Europe and Native Americans, thought individuals with such defects had a special insight and a closer connection with nature. These unique members of the community were given responsibility as tribal leaders or healers. Many physically disabled leaders made careful astronomical observations of the sun and moon that helped them advise their communities on the optimum times for planting, harvesting, or migrating to another area.

Close

Reteaching — BASIC

Ask students to differentiate each of the following terms by defining and sketching them: chromosomes, chromatids, DNA, and genes. Students should label their sketches. (chromosomes: coiled DNA and associated proteins; chromatids: identical copies of a given chromosome; DNA: chain of nucleic acids containing genetic code; genes: sequences of DNA that code for specific proteins or RNA.)
LS Visual English Language Learners

Quiz — GENERAL

True or False:

1. At cell division, each chromosome consists of two chromatids attached at the centromere. (True)
2. The normal diploid number for humans is 23. (False)
3. A person with the sex chromosomes XX would be female. (True)

Alternative Assessment — GENERAL

Have each student prepare a complete human karyotype based on **Figure 5.** Each chromosome should be drawn and then cut out. Allow students to choose the sex of their karyotype and determine whether there are mutations, deletions, or extra chromosomes. Have students exchange their diagrams with a partner. Each student must now match homologues and determine the sex of the individual, as well as identify any possible abnormalities.
LS Interpersonal

Section Review Correlations

SC.F.2.4.1: Item 1
SC.F.2.4.2: Items 2, 3, 4
SC.F.2.4.3: Item 5

Change in Chromosome Structure

Changes in an organism's chromosome structure are called *mutations*. Breakage of a chromosome can lead to four types of mutations. In a deletion mutation, a piece of a chromosome breaks off completely. After cell division, the new cell will lack a certain set of genes. In many cases this proves fatal to the zygote. In a duplication mutation, a chromosome fragment attaches to its homologous chromosome, which will then carry two copies of a certain set of genes. A third type of mutation is an inversion mutation, in which the chromosome piece reattaches to the original chromosome but in a reverse orientation. If the piece reattaches to a nonhomologous chromosome, a translocation mutation results.

QUICK LAB

Modeling Chromosomal Mutations SC.F.2.4.3

You can use paper and a pencil to model the ways in which chromosome structure can change.

Materials

14 note-card pieces, pencils, tape

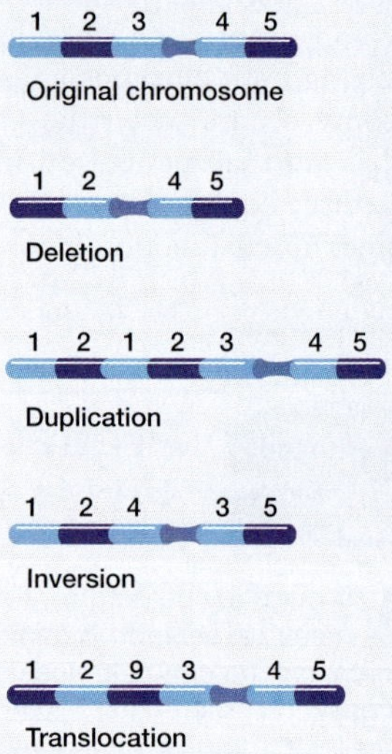

Procedure

1. Write the numbers 1–8 on note-card pieces (one number per piece). Tape the pieces together in numerical order to model a chromosome with eight genes.
2. Use the "chromosome" you made to model the four alterations in chromosome structure discussed on this page and illustrated at right. For example, remove the number 3 and reconnect the remaining chromosome pieces to represent a deletion.
3. Reconstruct the original chromosome before modeling a duplication, an inversion, and a translocation. Use the extra note-card pieces to make the additional numbers you need.

Analysis

Describe how a cell might be affected by each mutation if the cell were to receive a chromosome with that mutation.

Section 1 Review

1. **Summarize** how prokaryotic cells divide by binary fission. SC.F.2.4.1
2. **Identify** the point in a eukaryotic cell cycle at which DNA condenses to form visible chromosomes. SC.F.2.4.2
3. **Summarize** the difference between a haploid cell and a diploid cell. SC.F.2.4.2
4. **Critical Thinking Evaluating Conclusions** Do you agree or disagree that homologous chromosomes are found in gametes. Explain. SC.F.2.4.2
5. **FCAT Prep** How does the karyotype of a person with Down syndrome differ from a normal karyotype?
   A It lacks a chromosome.
   B It has two sex chromosomes.
   C It occurs in XO individuals.
   D It has an extra copy of a chromosome.
   SC.F.2.4.3

124

Answers to Section Review

1. DNA is first copied. Then the cell divides into equal halves by adding new cell membrane between the two DNA copies. The growing cell membrane pushes inward, and the cell constricts to form two new, identical cells. A new cell wall forms around each new membrane.
2. Chromosomes become visible in a eukaryotic cell when the cell prepares to divide (during prophase).
3. A haploid cell *(n)* contains one set of chromosomes. A diploid cell *(2n)* contains two sets of chromosomes.
4. Students should disagree. Homologous chromosomes are pairs of similar chromosomes. Because gametes are haploid *(n)* they contain only one set of chromosomes. Thus, homologous chromosomes are not normally found in gametes.
5. **A.** Incorrect. Lacking a chromosome is usually a fatal abnormality. **B.** Incorrect. All normal individuals have two sex chromosomes. **C.** Incorrect. Down syndrome does not arise from XO chromosomes. **D.** Correct. People with Down syndrome have an extra copy of chromosome 21.

The Cell Cycle

Section 2

Florida Sunshine Standards/FCAT

SC.F.1.4.5 Complex interactions among the different kinds of molecules in the cell cause distinct cycles of activity governed by proteins.

The Life of a Eukaryotic Cell

Cell division in eukaryotic cells is more complex than cell division in bacteria because it involves dividing both the cytoplasm and the chromosomes inside the nucleus. Many internal organelles must be correctly rearranged before the eukaryotic cell can properly divide and form two fully functioning cells.

The Cell Cycle

The life of a eukaryotic cell is traditionally shown as a cycle, as illustrated in **Figure 6.** The **cell cycle** is a repeating sequence of cellular growth and division during the life of an organism. A cell spends 90 percent of its time in the first three phases of the cycle, which are collectively called **interphase.** A cell will enter the last two phases of the cell cycle only if it is about to divide. The five phases of the cell cycle are summarized below:

1. **First growth (G_1) phase.** During the G_1 phase, a cell grows rapidly and carries out its routine functions. For most organisms, this phase occupies the major portion of the cell's life. Cells that are not dividing remain in the G_1 phase. Some somatic cells, such as most muscle and nerve cells, never divide. Therefore, if these cells die, the body cannot replace them.
2. **Synthesis (S) phase.** A cell's DNA is copied during this phase. At the end of this phase, each chromosome consists of two chromatids attached at the centromere.
3. **Second growth (G_2) phase.** In the G_2 phase, preparations are made for the nucleus to divide. Hollow protein fibers called microtubules are rearranged during G_2 in preparation for mitosis.
4. **Mitosis.** The process during cell division in which the nucleus of a cell is divided into two nuclei is called **mitosis** *(mie TOH sihs)*. Each nucleus ends up with the same number and kinds of chromosomes as the original cell.
5. **Cytokinesis.** The process during cell division in which the cytoplasm divides is called **cytokinesis** *(SIET oh kih nee sihs)*.

Mitosis and cytokinesis produce new cells that are identical to the original cells and allow organisms to grow, replace damaged tissues, and, in some organisms, reproduce asexually.

Objectives

- **Identify** the major events that characterize each of the five phases of the cell cycle. SC.F.1.4.5
- **Describe** how the cell cycle is controlled in eukaryotic cells. SC.F.1.4.5
- **Relate** the role of the cell cycle to the onset of cancer. SC.F.1.4.5

Key Terms

cell cycle
interphase
mitosis
cytokinesis
cancer

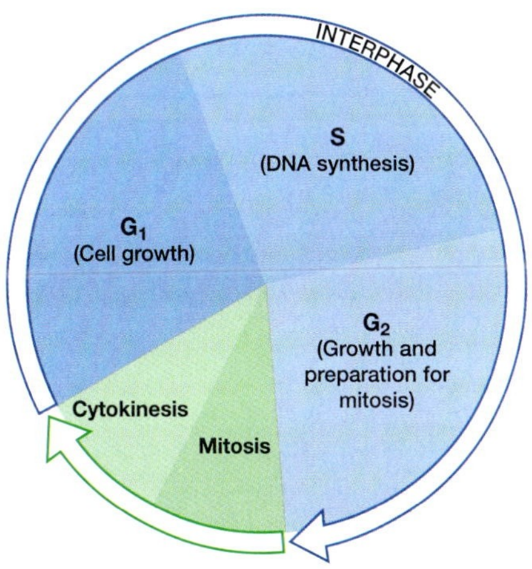

Figure 6 The eukaryotic cell cycle. The cell cycle consists of phases of growth, DNA replication, preparation for cell division, and division of the nucleus and cytoplasm.

125

Section 2

Focus

Overview

Before beginning this section review with your students the objectives listed in the Student Edition. In this section, students will examine the three phases that take place in the life cycle of a cell, which are collectively known as the cell cycle. They will learn that cells spend most of their time in interphase, during which cells grow and DNA is replicated. The cell enters mitosis as it prepares to divide, then divides into two cells during cytokinesis. Students will learn that cancer may result if the controls for the cell cycle break down.

Bellringer

Ask students to finish the following sentence: *A typical eukaryotic cell spends 90 percent of its time in* _______. Write the following choices on the board: mitosis, anaphase, interphase, and cytokinesis. (interphase)

Motivate

Demonstration

Display either a model or a photograph of a human brain. Point out that once the brain is fully formed, most of the nerve cells do not divide, again. These cells remain in the G_1 phase of the cell cycle. Then display either a model or a photograph of a human bone. Point out that red blood cells are produced from cells in the marrow of long bones. An average red blood cell lives for about 120 days. Each second, about 2 million red blood cells are produced by cell division in the bone marrow. Cells in the marrow, unlike those in the brain, continue going through the cell cycle as long as a person lives.

Chapter Resource File
- Directed Reading BASIC
- Active Reading GENERAL

Transparencies

TR Bellringer

One-Stop Planner CD-ROM
- Reading Organizers BASIC
- Reading Strategies BASIC
- Portfolio Project
  Genetics Project GENERAL

Section Correlations

SC.F.1.4.5

Teach

Teaching Tip — GENERAL

Interphase Have students draw a Graphic Organizer like the one below to summarize the events that occur during the phases of interphase. Have them use arrows between the phases, pointing to the right, to indicate the sequence of events.

Teaching Tip — GENERAL

Busy Cells Remind students that during interphase, a cell is not only growing, it is also producing the proteins and carrying out the functions that are characteristic of that type of cell.

Group Activity — BASIC

Writing **Cell Cycle Phases**

Have students read Section 2, then write all the activities and events described during the cell cycle on index cards (one event per card). Draw a large version of **Figure 6** at the front of the room. Then collect the students' cards. Allow one student at a time to come to the front of the room, read a card out loud, and then tape the card to the proper phase of the cell cycle. LS Verbal

Study TIP

Reviewing Information
Learn the stages of interphase by reviewing the steps numbered 1–5 on the previous page. You can see in Figures 6 and 7 that the cell cycle is a repeating series of three steps followed by mitosis and cytokinesis.

Control of the Cell Cycle

If a cell spends 90 percent of its time in interphase, how do cells "know" when to divide? How is the cycle controlled? Just as traffic lights control the flow of traffic, cells have a system that controls the phases of the cell cycle. Cells have a set of "red light–green light" switches that are regulated by feedback information from the cell. The cell cycle has key checkpoints (inspection points) at which feedback signals from the cell can trigger the next phase of the cell cycle (green light). Other feedback signals can delay the next phase to allow for completion of the current phase (yellow or red light).

The cell cycle in eukaryotes is controlled by many proteins. Control occurs at three principal checkpoints, as shown in **Figure 7.**

1. **Cell growth (G_1) checkpoint.** This checkpoint makes the decision of whether the cell will divide. If conditions are favorable for division and the cell is healthy and large enough, certain proteins will stimulate the cell to begin the synthesis (S) phase. During the S phase, the cell will copy its DNA. If conditions are not favorable, cells can typically stop the cell cycle at this checkpoint. The cell cycle will also stop at this checkpoint if the cell needs to pass into a resting period. Certain cells, such as some nerve and muscle cells, remain in this resting period permanently and never divide.
2. **DNA synthesis (G_2) checkpoint.** DNA replication is checked at this point by DNA repair enzymes. If this checkpoint is passed, proteins help to trigger mitosis. The cell begins the many molecular processes that are needed to proceed into mitosis.
3. **Mitosis checkpoint.** This checkpoint triggers the exit from mitosis. It signals the beginning of the G_1 phase, the major growth period of the cell cycle.

Figure 7 Control of the cell cycle. The cell cycle in eukaryotes is controlled at three inspection points, or checkpoints. Many proteins are involved in the control of the cell cycle.

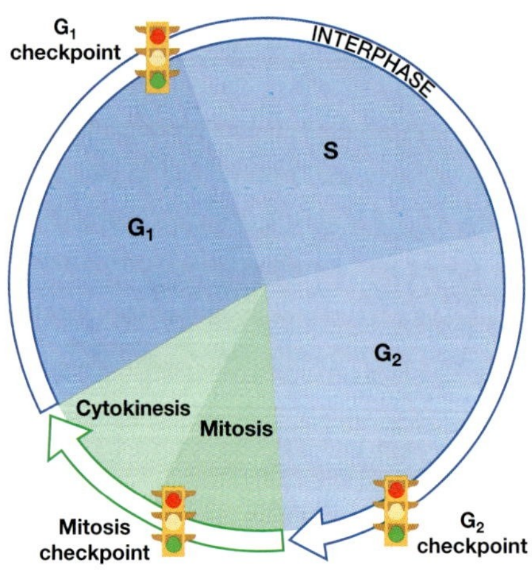

When Control Is Lost: Cancer

Certain genes contain the information necessary to make the proteins that regulate cell growth and division. If one of these genes is mutated, the protein may not function, and regulation of cell growth and division can be disrupted. **Cancer,** the uncontrolled growth of cells, may result. Cancer is essentially a disorder of cell division. Cancer cells do not respond normally to the body's control mechanisms.

Some mutations cause cancer by overproducing growth-promoting molecules, thus speeding up the cell cycle. Others cause cancer by inactivating the control proteins that normally act to slow or stop the cell cycle.

126

Graphic Organizer

Use this graphic organizer with **Teaching Tip: Interphase** on this page.

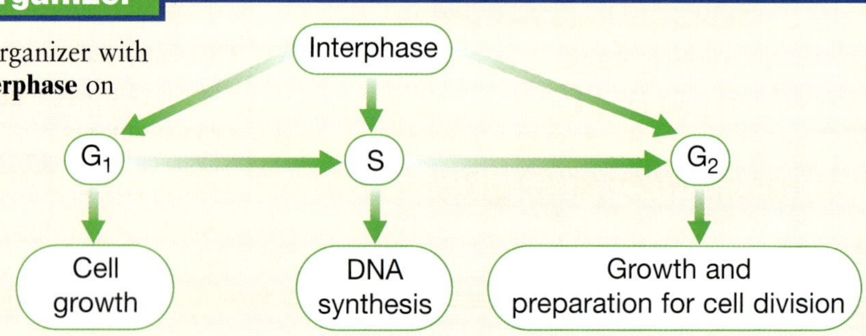

Section Review Correlations

SC.F.1.4.5: Items 1, 2, 4
SC.F.2.4.2: Item 3

Exploring Further

Cancer

Although all cancers are not curable, great progress has been made in cancer research over the last 30 years. We now know that cancer results from damage to a small set of genes that, in normal cells, limits the ability of cells to divide.

What causes this damage? Certain environmental factors appear to be associated with cancer. For example, the incidence of cancer per thousand people is not uniform throughout the United States. Rather, it is higher in cities and in the Mississippi delta, suggesting that pollution and pesticide runoff may contribute to cancer. When pollutants, radiation, and other environmental factors associated with cancer are analyzed, a clear pattern emerges. Most cancer-causing agents are powerful mutagens—that is, they readily damage DNA. The conclusion that cancer is caused by mutation of a cell's DNA is now supported by a very large body of evidence.

How many mutations are required to produce cancer? Research in the last several years indicates that mutation of only a few genes can transform normal cells into cancerous ones. All of these cancer-causing genes are involved with regulating how fast cells grow and divide. How is cell division regulated? As a crude analogy, imagine a car parked on the side of a road. To get it going, you must step on the accelerator and release the brake.

Stepping on the Accelerator

A cell divides when it receives a signal to do so. A "divide" signal is usually in the form of a chemical substance released by another cell. The substance is bound by a protein on the surface of the receiving cell. This binding activates a second protein inside the cell—relaying the signal from the outside of the cell to the inside. Here, a family of proteins then relay the signal inward to the nucleus. One protein molecule passes the signal to the next like a baton in a relay race. The genes for these signal-carrying proteins are called *oncogenes* (*onco* is Greek for "mass" or "tumor."). If oncogenes are changed by mutation to become more active, cancer can result. Like stepping on the accelerator of a car, an increase in the activity of these proteins amplifies the "divide" signal. This causes the cell to divide more often.

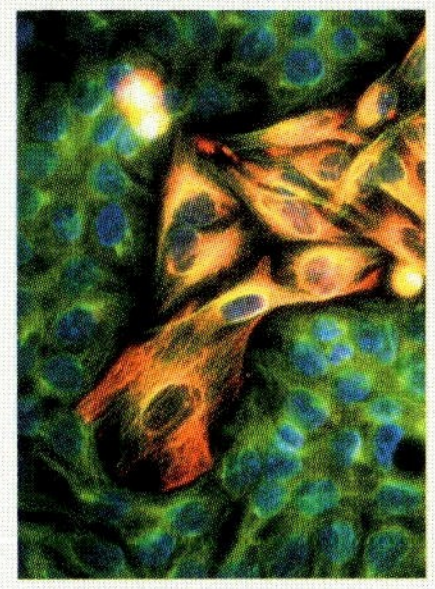

Melanoma cells

Releasing the Brakes

At the nucleus, the divide signal overrides a set of genes that act as "brakes." These braking genes—called *tumor suppressor genes*—prevent cell division from occurring too often. In cancer, these tumor suppressor genes are damaged. Like removing pressure from the brakes of a car increases a car's speed, decreasing the activity of tumor suppressors speeds up cell division.

Cells have three kinds of tumor suppressors, all of which must be disabled before cancer can occur. First, cells have proteins that inhibit DNA replication for limited periods. In cancer cells they are permanently inactivated. Second, cells have error-correcting proteins that detect damage to genes. In most cancers this error-detection has been disabled. Third, cancer cells rebuild the tips of their chromosomes. A little is lost from the ends of chromosomes at each replication, limiting the number of times a normal cell can divide. Adding the deleted material back to the tips removes this limit to a cell's life span.

internet connect
www.scilinks.org
Topic: Cancer Cells
Keyword: HX4030
SCILINKS. Maintained by the National Science Teachers Association

Section 2 Review

1. **Differentiate** between the G_1, G_2, and S phases of the eukaryotic cell cycle. SC.F.1.4.5
2. **Relate** what occurs at each of the three principal checkpoints in the cell cycle. SC.F.1.4.5
3. **Critical Thinking Evaluating Information** Why are individual chromosomes more difficult to see during interphase than during mitosis? SC.F.2.4.2
4. **FCAT Prep** What effect do mutations cause in the cell cycle of typical cancer cells?
   - **A** slower growth
   - **B** a failure in mitosis
   - **C** uncontrolled growth
   - **D** a halt in cell division

   SC.F.1.4.5

127

Answers to Section Review

1. G_1 phase: cell grows rapidly and carries out routine functions; G_2 phase: preparations are made for the nucleus to divide, microtubules are assembled; S phase: DNA is copied
2. Cell growth (G_1) checkpoint: determines if cell is ready to undergo division; DNA synthesis (G_2) checkpoint: DNA repair enzymes check progression; mitosis checkpoint: triggers exit from mitosis at the beginning of the G_1 cycle
3. Chromosomes condense right before a cell divides in mitosis. During most of interphase, chromosomes are not condensed and are harder to see with a microscope.
4. **A.** Incorrect. Cancer is associated with accelerated cell growth. **B.** Incorrect. Cancer does alter the cell cycle, but not by halting mitosis. **C.** Correct. Uncontrolled cell growth is the defining characteristic of cancer. **D.** Incorrect. Cancer cells continue to divide.

Exploring Further

Cancer

Teaching Strategies

- Tell students that most anticancer drugs interfere with the cell cycle of cancer cells. Unfortunately, the drugs also interfere with healthy cells, which is why the patient suffers side effects from the medications.

Discussion

- Where in the cell cycle could scientists target anticancer drugs? (Answers will vary but may mention proteins involved in the three checkpoints, DNA replication, or cytokinesis.)
- What type of environmental factors have been associated with the onset of cancer? (Answers may include diet, UV radiation, hormones, and environmental pollution.)

Close

Reteaching — BASIC

Assign students to groups. Have each group visually interpret a stage in the cell cycle. Have each group display their graphics in the appropriate space in a circle graph that represents the cell cycle.

LS Visual Co-op Learning

Quiz — GENERAL

1. The three phases of the cell cycle are interphase, mitosis, and _______. (cytokinesis)
2. When normal control of the cell cycle fails, _______ may develop. (cancer)

Alternative Assessment — ADVANCED

Have students write and illustrate a description of the cell cycle in the style of a children's book called "A Cell's Life." Be sure students include each phase, including G_1, S, G_2, mitosis, and cytokinesis. LS Verbal

Mitosis and Cytokinesis

Focus

Overview

Before beginning this section review with your students the objectives listed in the Student Edition. In this section, students will examine the processes of mitosis and cytokinesis in more detail. During mitosis, spindle fibers pull the chromatids to opposite ends of the cell, ensuring that each new cell receives the proper assortment of chromosomes. During cytokinesis, the cytoplasm is divided by a new cell membrane or cell wall.

Bellringer

Draw a football field on the board with the players lined up at midfield and the goal posts at each end. Have students write a short paragraph describing how the players, the midfield line, and the goal posts compare to the structures of a cell involved in mitosis and cell division. (The players represent the chromosomes, the midfield line represents the equator of the cell, and the goal posts represent the centrioles at the poles of the cell.)

Motivate

Activity — GENERAL

Mitosis Set up several microscopes with prepared slides of cells at various stages of mitosis, but presented out of order. Have students examine the slides and try to place each slide into the proper order of events. If microscopes and slides are not available, the various stages can be drawn on the board, unlabeled and out of order. English Language Learners

Section Correlations

SC.F.1.4.2
SC.F.1.4.5
SC.G.1.4.1

Florida Sunshine Standards/FCAT

SC.F.1.4.2 Body structures are uniquely designed and adapted for their function.

SC.F.1.4.5 Complex interactions among the different kinds of molecules in the cell cause distinct cycles of activity governed by proteins.

SC.G.1.4.1 There is great diversity and interdependence among living things.

Objectives

- **Describe** the structure and function of the spindle during mitosis. SC.F.1.4.2
- **Summarize** the events of the four stages of mitosis. SC.F.1.4.5
- **Differentiate** cytokinesis in animal and plant cells. SC.G.1.4.1

Key Terms

spindle

Chromatid Separation in Mitosis

Every second about 2 million new red blood cells are produced in your body by cell divisions occurring in the bone marrow. These cells have received the signal to divide. The cells continue past the G_2 phase and enter into the last two phases of the cell cycle—mitosis and cytokinesis. During mitosis the nucleus divides to form two nuclei, each containing a complete set of the cell's chromosomes. During cytokinesis the cytoplasm is divided between the two resulting cells.

During mitosis, the chromatids on each chromosome are physically moved to opposite sides of the dividing cell with the help of the spindle, shown in **Figure 8. Spindles** are cell structures made up of both centrioles and individual microtubule fibers that are involved in moving chromosomes during cell division.

Forming the Spindle

At each of the cell's poles lies a centrosome. The centrosome is an organelle that organizes the assembly of the spindle. In animal cells, a pair of centrioles is found inside each centrosome. As you can see in Figure 8, centrioles are conspicuous. They are not necessary, however, for spindle formation.

Centrioles and spindle fibers are both made of hollow tubes of protein called microtubules. Each spindle fiber is made of an individual microtubule. Each centriole, however, is made of nine triplets of

Figure 8 The spindle

The spindle, made up of centrioles and spindle fibers, helps move chromosomes apart during mitosis.

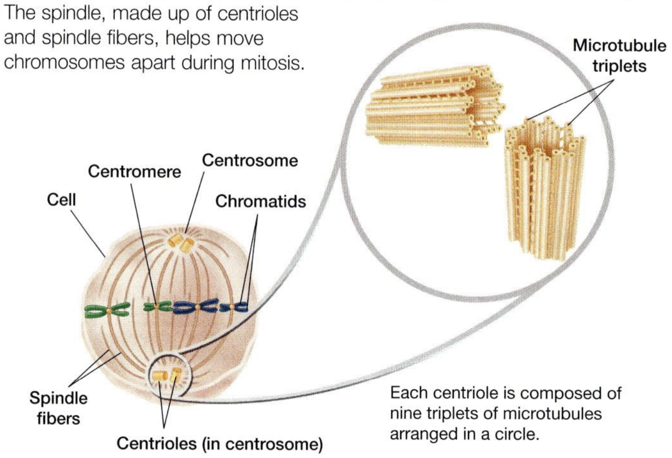

Chapter Resource File

- Directed Reading BASIC
- Active Reading GENERAL
- Data Sheet for Math Lab GENERAL
- Data Sheet for Quick Lab GENERAL

Transparencies

TR Bellringer
TR B47 Stages of Mitosis

One-Stop Planner CD-ROM

- Reading Organizers BASIC
- Reading Strategies BASIC
- Supplemental Reading *The Lives of a Cell*

microtubules arranged in a circle. Unlike animal cells, plant cells do not have centrioles, but they form a spindle that is almost identical to that of an animal cell.

Separation of Chromatids by Attaching Spindle Fibers

Some of the microtubules in the spindle interact with each other. Others attach to a protein structure found on each side of the centromere. The two sets of microtubules extend out toward opposite poles of the cell. Once the microtubules attach to the centromeres and poles, the two chromatids in each chromosome can be separated.

The paired chromatids separate. One of the pair of chromatids will move to one pole of the cell. The second member of the pair will move to the other pole. Once separated, the chromatids move along paths described by microtubules to which they are attached. The chromatids draw closer to the poles of the cell as these microtubules are broken down bit by bit and become shorter.

As soon as the chromatids separate from each other they are called chromosomes. When the chromosomes finally arrive, each pole has one complete set of chromosomes.

MATH LAB

Calculating the Number of Cells Resulting from Mitosis SC.F.1.4.5

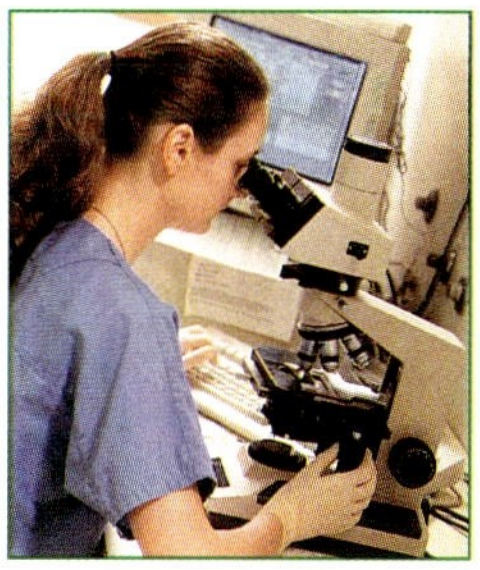

Background

Scientists investigating cancer might need to know the number of cells produced in a certain amount of time. In the human body the rate of mitosis is about 25 million (2.5×10^7) cells produced every second! You can calculate the number of cells produced by mitosis in a given amount of time.

1. **Calculate the number of cells produced by mitosis in the given time.** For example, to find the number of cells produced in 3 minutes, determine how many seconds there are in 3 minutes (since the rate is given in seconds).

$$\frac{60 \text{ seconds}}{1 \text{ minute}} \times 3 \text{ minutes} = 180 \text{ seconds}$$

2. **Multiply the rate of mitosis by the time (in seconds) asked for in the problem (180 seconds).**

$$\frac{2.5 \times 10^7 \text{ cells}}{\text{second}} \times 180 \text{ seconds} = 4.5 \times 10^9 \text{ cells (4,500,000,000 cells)}$$

Analysis

1. **Calculate** the number of cells that would be produced in 1 hour.
2. **Calculate** the number of cells that would be produced in 1 day.
3. **Critical Thinking Predicting Patterns** Identify factors that might increase or decrease the rate of mitosis.

Teach

MATH LAB

Calculating the Number of Cells Resulting from Mitosis

Skills Acquired
Calculating, predicting

Teacher's Notes
Review numbers in scientific notation before starting the lab. Review how to cancel units and set up equal unit proportions. For example, 60 seconds/ 1 minute is an equal proportion.

Analysis Answers

1. 9.0×10^{10} cells = 90,000,000,000 cells
2. 2.16×10^{12} cells
3. Factors that might increase or decrease the rate of mitosis include mutated genes, diet, and exposure to ultraviolet light and tobacco products.

did you know?

Fly Chromosomes The fruit fly *Drosophila melanogaster* has been used to study chromosomal mutations since 1933. The salivary glands of these flies, as in many insects, are composed of cells that do not divide during the larval stage. However, the chromosomes continue to replicate, producing many copies. The copies of each chromosome are closely aligned, resulting in thick chromosomes that are easy to study with a microscope.

Teach, *continued*

SKILL BUILDER — BASIC

Vocabulary To help students master vocabulary, relate the word mitosis to its Greek origin, *mitos*, meaning "thread." Ask students to explain why mitosis is associated with thread. (The hereditary material consists of long, threadlike molecules.)

Using the Figure

BIO graphic Lead students through each stage shown in **Figure 9,** focusing on the behavior of chromosomes. Point out that the various stages are not of equal duration. Using movie film as an analogy, help students avoid the misconception that mitosis "jumps" from stage to stage. Although a movie consists of individual frames of film, the images on the film appear to change continuously. Explain that mitosis progresses in a similar fashion.

Teaching Tip

Prefixes Tell students that the prefixes of the different stages of mitosis describe the order of events or the events themselves. *Pro-* means "earlier than;" *meta-* means "later than" or "after;" *ana-* means "up" or "back" and describes the movement of the chromosomes "up" toward the poles; and *telo-* means "end." English Language Learners

Teaching Tip — GENERAL

Have students hypothesize about how a cell's ATP use changes during mitosis. (The events of mitosis require a lot of additional energy, which is supplied by ATP.)

Mitosis and Cytokinesis

Although mitosis is a continuous process, biologists traditionally divide it into four stages, as shown in **Figure 9.**

Mitosis

Step 1 **Prophase** Chromosomes coil up and become visible during prophase. The nuclear envelope dissolves and a spindle forms.

Step 2 **Metaphase** During metaphase the chromosomes move to the center of the cell and line up along the equator. Spindle fibers link the chromatids of each chromosome to opposite poles.

Step 3 **Anaphase** Centromeres divide during anaphase. The two chromatids (now called chromosomes) move toward opposite poles as the spindle fibers attached to them shorten.

Step 4 **Telophase** A nuclear envelope forms around the chromosomes at each pole. Chromosomes, now at opposite poles,

BIO graphic

Figure 9

Stages of Mitosis

The chromosome copies in the nucleus of a dividing cell are separated into two nuclei.

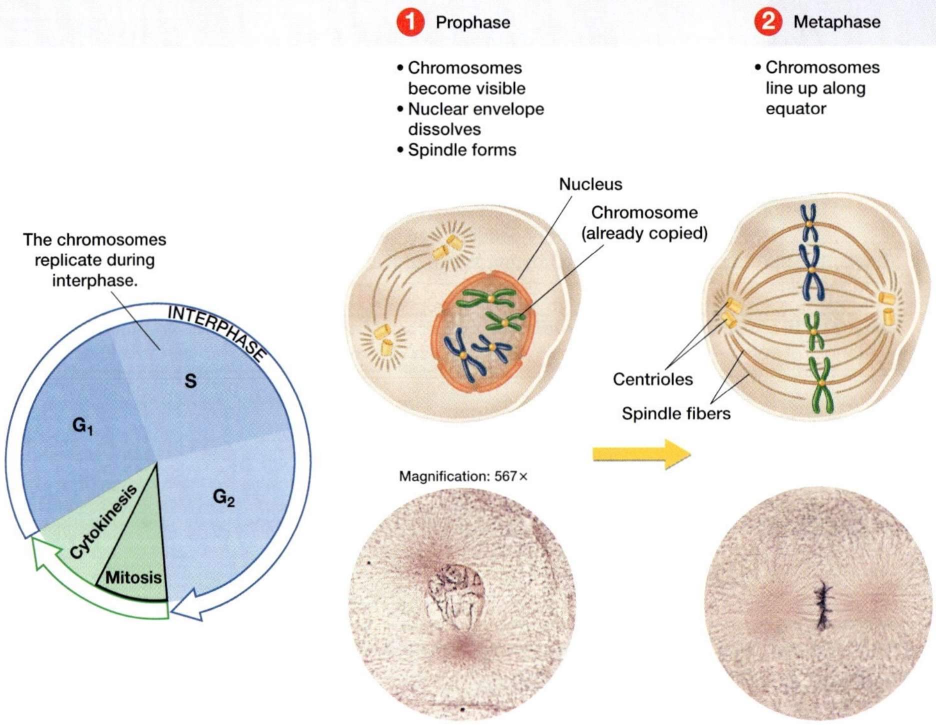

130

MEDICINE CONNECTION

When cultured, cancer cells can divide indefinitely if they are given a continual supply of nutrients. One cell line has been continuously cultured since 1951. The cells in this cell line are called HeLa cells because they were originally from a tumor removed from a woman named Henrietta Lacks. The patient suffered from uterine cervical carcinoma. Cells from this cell line are still used by researchers around the world, especially in research on viruses.

uncoil and the spindle dissolves. The spindle fibers break down and disappear. Mitosis is complete.

Cytokinesis

As mitosis ends, cytokinesis begins. During cytokinesis, the cytoplasm of the cell is divided in half, and the cell membrane grows to enclose each cell, forming two separate cells as a result. The end result of mitosis and cytokinesis is two genetically identical cells where only one cell existed before.

During cytokinesis in animal cells and other cells that lack cell walls, the cell is pinched in half by a belt of protein threads, as shown in **Figure 10.**

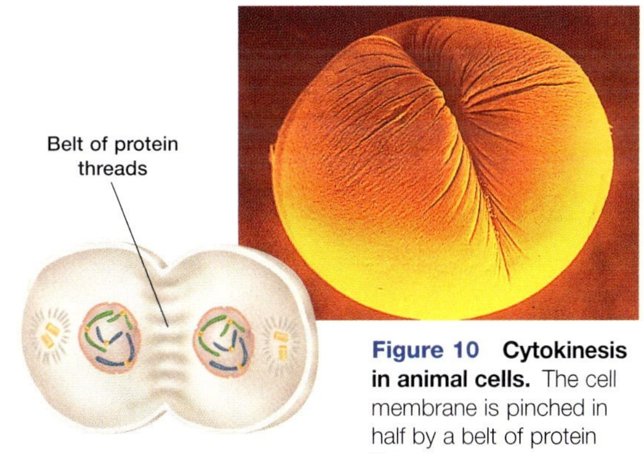

Figure 10 Cytokinesis in animal cells. The cell membrane is pinched in half by a belt of protein threads.

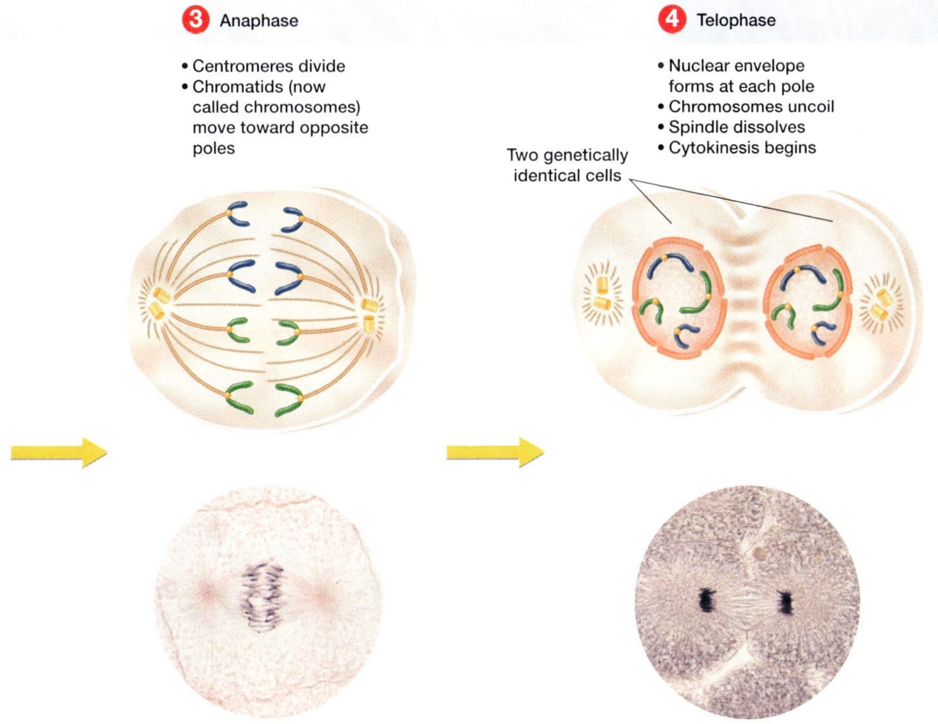

Teaching Tip — GENERAL

Stages of Mitosis Have students draw pictures of each stage of mitosis on four note cards. On one side, have students write the name of the stage. On the opposite side, have them draw an accurate sketch of the stage. They can then use these as flash cards to help each other learn the stages of mitosis.
LS Visual

Using the Figure

Tell students that the belt of protein threads that are labeled in **Figure 10** is a ring of the proteins actin and myosin, the same two proteins that interact to contract muscle cells. The furrowing of the cell membrane occurs perpendicular to the long axis of the spindle. Cytokinesis usually begins in anaphase but is not completed until after the two nuclei have formed.

Group Activity — GENERAL

Performing Mitosis and Cytokinesis Have students work in groups of three or four students. Ask each group to design a "performance" using the entire class as the cast, as well as props, that would demonstrate the processes of mitosis and cytokinesis. Ask groups to present their designs to the class, and have the class decide which "show" is best. If time and materials permit, allow students to stage their mitosis and cytokinesis show.
LS Interpersonal Co-op Learning

MISCONCEPTION ALERT

Mitosis vs. Cytokinesis Students often think that mitosis is the same as cell division. Be sure they understand that mitosis refers strictly to the division of the chromosomes, whereas cytokinesis refers to the division of the cytoplasm. Remind students that cell division is just one of the four events that make up the cell cycle.

Trends in Cell Biology

Turning Off Cancer Currently, researchers at the National Cancer Institute are working to find the proteins that "turn on" or "turn off" gene ECT2, an oncogene that appears to be a critical regulator of cytokinesis. Eventually, researchers may be able to "turn off" the mechanism that causes cancer cells to divide uncontrollably.

Teach, continued

QUICK LAB

Observing Mitosis and Cytokinesis

Skills Acquired
Comparing, inferring

Answers to Analysis

1. Prophase: chromosomes are visible as dark threads; metaphase: chromatids line up along the equator; anaphase: chromosomes appear to pull toward opposite poles; telophase: chromosomes, now at opposite poles, uncoil
2. Answers will vary but should indicate more cells in interphase than in the other stages.
3. Cells spend the majority of their time in interphase.

Close

Reteaching — BASIC

Provide students with yarn to represent chromosomes, nuclear envelopes, and cell membranes; tie tabs to represent centromeres; and string to represent spindle fibers. Have them recreate on their desktops a cell undergoing mitosis.
LS Kinesthetic English Language Learners

Quiz — GENERAL

1. In mitosis, the chromatids move toward opposite poles during ________. (anaphase)
2. During mitosis, the chromosomes line up along the equator during ________. (metaphase)

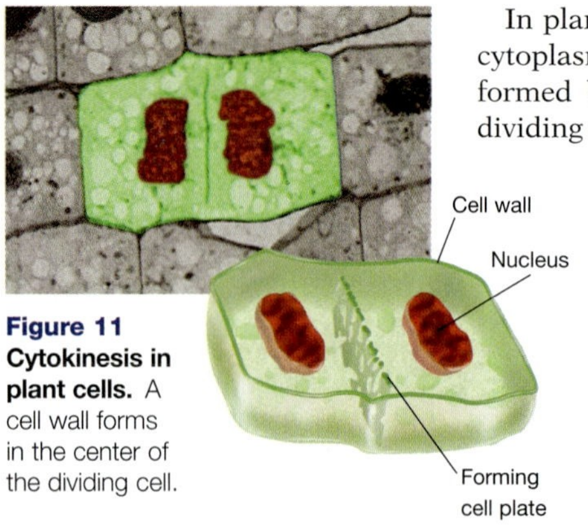

Figure 11 Cytokinesis in plant cells. A cell wall forms in the center of the dividing cell.

In plant cells and other cells that have rigid cell walls, the cytoplasm is divided in a different way. In plant cells, vesicles formed by the Golgi apparatus fuse at the midline of the dividing cell and form a cell plate. A cell plate is a membrane-bound cell wall that forms across the middle of the plant cell. A new cell wall then forms on both sides of the cell plate, as shown in **Figure 11.** When complete, the cell plate separates the plant cell into two new plant cells.

In both animal and plant cells, offspring cells are about equal in size. Each offspring cell receives an identical copy of the original cell's chromosomes. Each offspring cell also receives about one-half of the original cell's cytoplasm and organelles.

QUICK LAB

Observing Mitosis and Cytokinesis

SC.F.1.4.5

You can identify the stages of mitosis and the process of cytokinesis by observing slides of tissues undergoing mitosis using a compound microscope.

Materials

compound microscope, prepared slide of mitosis, paper, pencil

Procedure

1. View a prepared slide of cells undergoing mitosis under low power of a compound microscope.
2. Move the slide until you find a section where different stages of mitosis are visible.
3. Switch to high power. Use the photos in Figure 9 to help you locate and identify cells in interphase and in each stage of mitosis.
4. On a separate piece of paper, sketch an example of each stage. Label each sketch with the following terms where appropriate: *chromosomes, cell membrane, cytoplasm, nucleus, spindle, and cell wall.*
5. Switch to low power, and estimate how many cells are clearly in interphase and how many cells are in one of the stages of mitosis.

Analysis

1. **Describe** the activity of chromosomes in each stage of mitosis.
2. **Compare** the number of cells in interphase with the number of cells in one of the stages of mitosis.
3. **Critical Thinking Inferring Relationships** What does your answer to item 2 indicate about the relative length of interphase?

1. **Describe** the function of the microtubules during anaphase. SC.F.1.4.2
2. **Describe** the events that occur during each of the four stages of mitosis. SC.F.1.4.5
3. **Compare** how cytokinesis occurs in plant cells with how it occurs in animal cells. SC.G.1.4.1
4. **FCAT Prep** Mitosis could not proceed if a mutation interrupted the assembly of what structure? SC.F.2.4.3
   A the cell wall
   B the spindle fibers
   C the cell membrane
   D the nuclear envelope

132

Answers to Section Review

1. The microtubules, attached to the centromeres, shorten and pull the chromatids to opposite poles, similar to a fishing line reeling in a fish.
2. Prophase: chromosomes become visible, nuclear envelope dissolves, spindle forms; metaphase: chromosomes line up at the equator, spindle fibers attach to each chromatid; anaphase: centromeres divide, chromatids move to opposite poles due to shortening spindle fibers; telophase: nuclear envelope forms around the chromatids at each pole, chromosomes uncoil, spindle fibers break down and disappear
3. In plant cells, vesicles formed by the Golgi apparatus fuse at the equator and form the cell plate. A new cell wall forms on both sides of the cell plate. In animal cells, the cell is pinched in half by a belt of protein threads.
4. **A.** Incorrect. The cell wall does not play a key role in mitosis. **B.** Correct. Without the spindle fibers, chromosomes cannot segregate properly into two complete nuclei. **C.** Incorrect. The cell membrane does not play a key role in mitosis. **D.** Incorrect. The nuclear envelope breaks down before mitosis and reforms after the completion of mitosis.

Section Review Correlations

SC.F.1.4.2: Item 1
SC.F.1.4.5: Item 2
SC.F.2.4.3: Item 4
SC.G.1.4.1: Item 3

Study ZONE CHAPTER HIGHLIGHTS

Key Concepts

1 Chromosomes

- Cell division allows organisms to reproduce asexually, grow, replace worn-out or damaged tissues, and form gametes.
- Bacteria reproduce by binary fission.
- Before cell division, DNA coils around proteins and the chromosomes condense. At cell division, each chromosome consists of two chromatids attached at the centromere.
- Each organism has a characteristic number of chromosomes.
- Human somatic cells are diploid, with 23 pairs of homologous chromosomes. Human gametes are haploid, with 23 chromosomes.
- Sex chromosomes carry information that determines an organism's sex.
- Changes in chromosome number or structure can cause abnormal development. Karyotypes are used to examine an individual's chromosomes.

2 The Cell Cycle

- The life of a eukaryotic cell—the cell cycle—includes interphase, mitosis, and cytokinesis.
- Interphase consists of 3 phases: growth, DNA synthesis (replication), and preparation for cell division. A cell about to divide enters the mitosis and cytokinesis phases of the cell cycle.
- The cell cycle is carefully controlled; failure of cellular control can result in cancer.

3 Mitosis and Cytokinesis

- During mitosis, spindle fibers drag the chromatids to opposite poles of the cell. A nuclear envelope forms. Each resulting nucleus contains a copy of the original cell's chromosomes.
- Cytokinesis in animal cells occurs when a belt of protein threads pinches the cell membrane in half. Cytokinesis in plant cells occurs when vesicles from the Golgi apparatus fuse to form a cell plate.

BIOLOGY INTERACTIVE TUTOR **Unit 4—*Cell Reproduction*** Use Topics 1–4 in this unit to review the key concepts and terms in this chapter.

Key Terms

Section 1

gamete (118)
binary fission (119)
gene (119)
chromosome (119)
chromatid (119)
centromere (119)
homologous chromosome (120)
diploid (121)
haploid (121)
zygote (121)
autosome (122)
sex chromosome (122)
karyotype (122)

Section 2

cell cycle (125)
interphase (125)
mitosis (125)
cytokinesis (125)
cancer (126)

Section 3

spindle (128)

Alternative Assessment — GENERAL

Set up a lab practical with various stations. Include microscopes with slides at various stages of mitosis, pictures of chromosomes, and karyotypes for interpretation.

Chapter Resource File

- Science Skills Worksheet GENERAL
- Critical Thinking Worksheet ADVANCED
- Test Prep Pretest GENERAL
- Chapter Test GENERAL

Answer to Concept Map

The following is one possible answer to Performance Zone item 15.

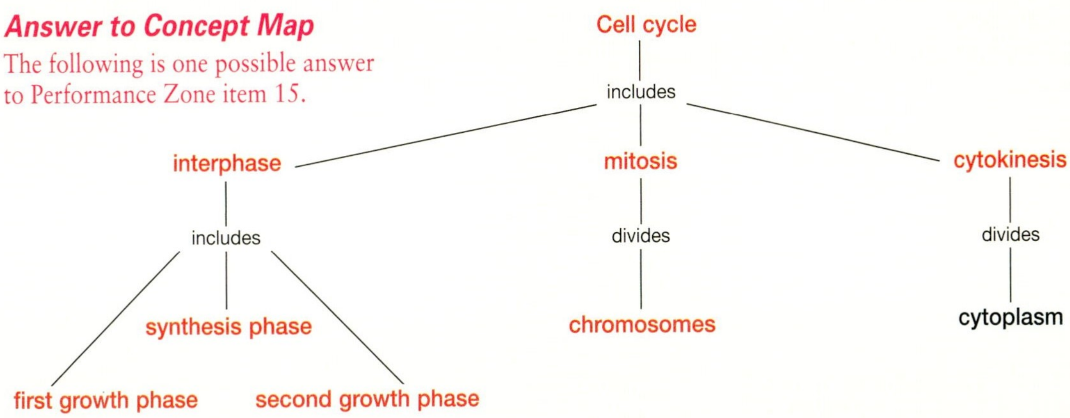

ANSWERS

Understanding Key Ideas

1. b
2. d
3. c
4. a
5. b
6. Answers will vary, but should include organelles such as lysosomes, the Golgi apparatus, the endoplasmic reticulum, ribosomes, and vacuoles.
7. a
8. Normal cells can become cancer cells if a genetic mutation impairs a cell's ability to regulate cell growth and division. The mutation leads to cancer, or uncontrolled cell growth.
9. chromosomal abnormalities
10. The answer to the concept map is found at the bottom of the Study Zone page.

Critical Thinking

11. Both result in new cells with the same number of chromosomes as in the original cell.
12. Most nerve cells are permanently resting in interphase G1. Because they do not undergo mitosis, most damaged nerve cells are not replaced.

Understanding Key Ideas

1. In humans, females have _______ sex chromosomes.
   a. XY **c.** YY
   b. XX **d.** XO
   SC.F.2.4.2
2. The diagram below represents a(n) _______ mutation.
   a. deletion
   b. translocation
   c. inversion
   d. duplication
   SC.F.2.4.3

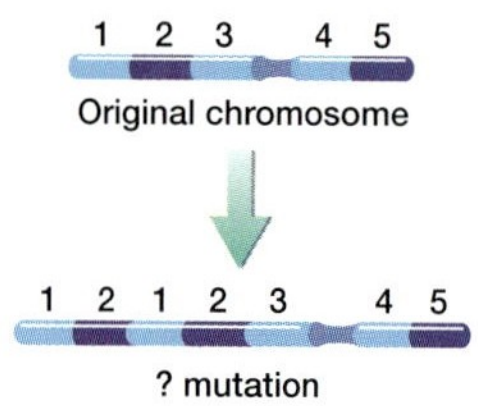

3. When the cell cycle is not controlled, _______ may result.
   a. Down syndrome
   b. binary fission
   c. cancer
   d. a spindle
   SC.F.1.4.5
4. As a result of mitosis, each resulting cell
   a. receives an exact copy of all of the chromosomes present in the original cell.
   b. receives most of the chromosomes from the original cell.
   c. donates a chromosome to the original cell.
   d. receives exactly half the chromosomes from the original cell.
   SC.F.1.4.5
5. During the metaphase stage of mitosis,
   a. the cell membrane folds inward.
   b. chromosomes line up at the cell's equator.
   c. spindle fibers shorten, pulling chromosomes to the poles of the cell.
   d. chromosomes are at opposite ends of the cell.
   SC.F.1.4.5
6. List five organelles that must divide or fragment before the cytoplasm divides. (**Hint:** See Chapter 3, Section 2.) SC.F.1.4.2
7. How does cell division differ between animal and plant cells?
   a. Plant cells do not have centrioles.
   b. Animal cells form a cell plate.
   c. Plant cells are always haploid.
   d. Animal cells do not have centrioles.
   SC.G.1.4.1
8. Summarize how normal cells can become cancer cells. SC.F.2.4.3
9. **BIOWatch** What information, besides chromosome number, can amniocentesis and chorionic villi sampling reveal? SC.H.3.4.5
10. **Concept Mapping** Make a concept map that shows the events in the cell cycle. Try to include the following words in your map: *cell cycle, interphase, synthesis phase, chromosomes, cytokinesis, mitosis, second growth phase,* and *first growth phase.* SC.F.1.4.5

Critical Thinking

11. **Inferring Relationships** Explain the relationships between mitosis in eukaryotic cells and binary fission in prokaryotes. SC.G.1.4.1
12. **Evaluating Conclusions** Damage to the brain or the spinal cord is usually permanent. Use your knowledge of the cell cycle to explain why damaged cells in the brain or spinal cord are not replaced. SC.F.1.4.5

Alternative Assessment

13. **Finding and Communicating Information** Scientists have determined that telomeres (the tips of chromosomes) are shaved down slightly every time a cell divides. When the telomeres reach a certain length, the cell may lose its ability to divide. Find out what scientists have recently uncovered about telomeres and their association with cell division and cancer. Prepare a brief written report to share with your class. SC.H.3.4.3

Alternative Assessment

13. Answers will vary. Cells from large tumors often have unusually short telomeres. Telomerase, an enzyme that catalyzes the lengthening of telomeres, stabilizes telomere length, especially in cancer cells. Researchers are focusing on telomerase as a target for cancer diagnosis and chemotherapy.

Assignment Guide

Section	Questions
1	1, 9, 11
2	2, 3, 8, 10, 11, 12, 13
3	4, 5, 6, 7

Chapter Review Correlations

SC.F.1.4.2: Item 6
SC.F.1.4.5: Items 3, 4, 5, 10, 12
SC.F.2.4.2: Item 1
SC.F.2.4.3: Items 2, 8
SC.G.1.4.1: Items 7, 11
SC.H.3.4.3: Item 13
SC.H.3.4.5: Item 9

FCAT Prep

For questions 1–6, write your answers on a separate sheet of paper.

1. What term describes the asexual reproduction of prokaryotes?
   A. binary fission **C.** disjunction
   B. cytokinesis **D.** mitosis
   SC.F.2.4.1
2. What might happen if cytokinesis were omitted from the cell cycle?
   F. The cell would lose its mitochondria.
   G. The cell would become a cancer cell.
   H. The cell would not divide into two daughter cells.
   I. The daughter cells would not have enough DNA.
   SC.F.1.4.5
3. In what stage of the cell cycle is a cell's DNA copied? SC.F.2.4.2
   A. G_1 **C.** mitosis
   B. G_2 **D.** S
4. What are chromatids?
   F. prokaryotic chromosomes
   G. dense patches of protein within the nucleus
   H. structures that move chromosomes during mitosis
   I. two exact copies of DNA that make up each chromosome
   SC.F.2.4.2
5. READ INQUIRE EXPLAIN A newspaper article describes the concern that more infants with Down syndrome will be born in the United States as more women delay having children.

   What information might account for this statistic? SC.H.3.4.3
6. READ INQUIRE EXPLAIN There are many environmental factors that appear to be associated with cancer, such as pollution and radiation. Cancer occurs when cell division does not respond to the normal signals that regulate the cell cycle. Some cancer-fighting drugs kill cancer cells by interrupting the cell cycle. Examples of such drugs include vincristine and taxol, which prevent the mitosis spindle microtubules from functioning.

   Part A Examine the function that the spindle microtubules play during mitosis. SC.F.1.4.2

   Part B How is cancer a disorder of cell division? SC.F.1.4.5

Test TIP

Test questions may not be arranged in order of increasing difficulty. If you are unable to answer a question, mark it and move on to another question.

FCAT Doctor

Question 2 Answer H is the correct choice. Cytokinesis is cell division. Without cell division, no daughter cells can be produced. Answer F is incorrect because the mitochondria reproduce before cytokinesis; thus, there would be more mitochondria. Answer G is incorrect because cancer cells arise when there is uncontrolled cell growth, which involves cytokinesis. Answer I is incorrect because the cell would have twice the normal amount of DNA but would not divide it between two daughter cells.

Question 5 The eggs a woman has can accrue damages over the years. Later in life, she is more likely to have an egg that has an abnormal number of chromosomes.

Question 6

Part A Spindles are structures made up of both centrioles and individual microtubule fibers that are involved in moving chromosomes during cell division. Both centrioles and microtubule fibers are made up of microtubule proteins.

Part B When cells experience uncontrolled growth, they multiply too quickly and produce clumps of cells called tumors.

Answers

1. A
2. H
3. D
4. I
5. As a woman ages, her eggs accumulate an increasing amount of damage.

6A. Spindle microtubules are involved in moving chromosomes during mitosis.

6B. Cancer is the uncontrolled growth of cells.

Test Prep Correlations

SC.F.1.4.2: Item 6A
SC.F.1.4.5: Items 2, 6B
SC.F.2.4.1: Item 1
SC.F.2.4.2: Items 3, 4
SC.H.3.4.3: Item 5

Exploration Lab

Modeling Mitosis

Teacher's Notes

Time Required 45–55 minutes

Ratings

EASY → HARD

Teacher Preparation 1
Student Setup 2
Concept Level 2
Cleanup 1

Skills Acquired
- Constructing models
- Identifying/Recognizing Patterns
- Inferring
- Predicting

Scientific Methods

In this lab, students will:
- Make Observations
- Ask Questions
- Draw Conclusions

Materials and Equipment

Materials for this lab can be ordered from WARD'S. See *Master Materials List* at the front of this book for catalog numbers.

Safety Cautions

Discuss all safety symbols and caution statements with students.

Tips and Tricks

For each model, students need four pipe cleaners of one color, four pipe cleaners of another color, four wooden beads, 90 cm of yarn, and 16 small white labels. You may want to cut the string before the start of the lab. Review mitosis before starting the lab. If students have difficulty in beginning their models, ask the following questions: Which of these materials would make the best cell membrane? (yarn) Which would make the best spindle fibers? (yarn) Which would make the best chromosomes? (pipe cleaners) Which would make the best centromeres? (beads) Ask students what the cells they make were like before the chromosomes were duplicated. Emphasize that each pair of chromatids represents one chromosome. Most of the supplies for this lab can be saved and used again in Chapter 7.

Exploration Lab

Modeling Mitosis SC.H.3.4.1

SKILLS
- Modeling
- Using scientific methods

OBJECTIVES
- **Describe** the events that occur in each stage of mitosis.
- **Relate** mitosis to genetic continuity.

MATERIALS
- pipe cleaners of at least two different colors
- yarn
- wooden beads
- white labels
- scissors

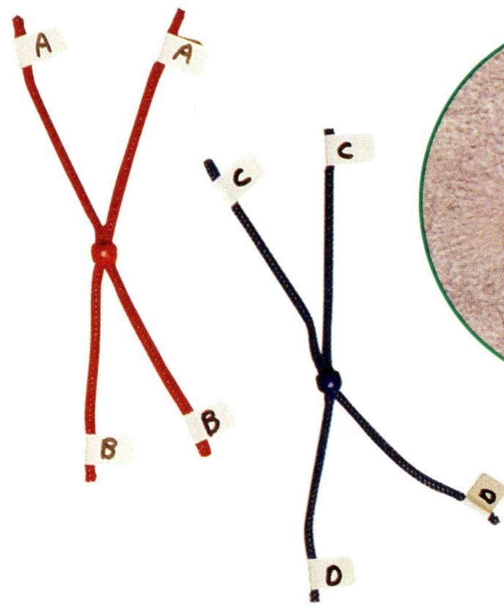

Magnification: 670×

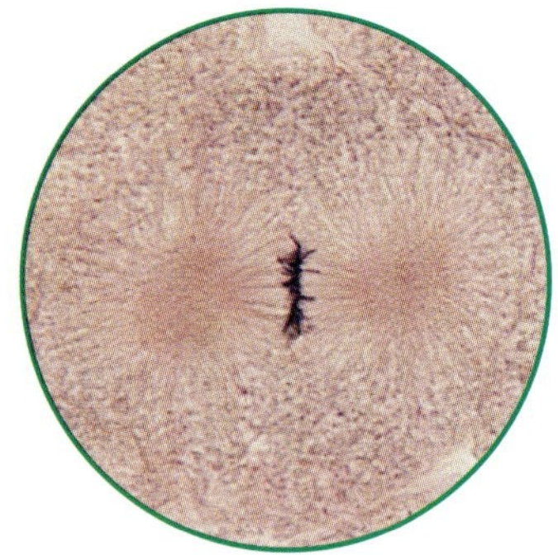

Whitefish cells

Before You Begin

The cell cycle includes all of the phases in the life of a cell. The **cell cycle** is a repeating sequence of cellular growth and division during the life of an organism. Mitosis is one of the phases in the cell cycle. **Mitosis** is the process by which the material in a cell's nucleus is divided during cell reproduction. In this lab, you will build a model that will help you understand the events of mitosis. You can also use the model to demonstrate the effects of **nondisjunction** and **mutations**.

1. Write a definition for each boldface term in the paragraph above and for the following terms: chromatid, centromere, spindle fiber, cytokinesis.
2. Where in the human body do cells undergo mitosis?
3. How does a cell prepare to divide during interphase of the cell cycle?
4. Based on the objectives for this lab, write a question you would like to explore about mitosis.

Procedure

PART A: Design a Model

1. Work with the members of your lab group to design a model of a cell that uses the materials listed for this lab. Be sure your model cell has at least two pairs of chromosomes and is about to undergo mitosis.

You Choose

As you design your model, decide the following:

a. what question you will explore
b. how to construct a cell membrane
c. how to show that your cell is diploid
d. how to show the locations of at least two genes on each chromosome
e. how to show that chromosomes are duplicated before mitosis begins

2. Write out the plan for building your model. Have your teacher approve the plan before you begin building the model.

Answers to Before You Begin

1. *cell cycle*—a repeating five-phase sequence of eukaryotic cellular growth and division; *mitosis*—the process during cell division in which the nucleus of a cell divides into two nuclei, each with the same number and kind of chromosomes; *nondisjunction*—the failure of two chromosomes or chromatids to separate during nuclear division; *mutation*—a change in the DNA of a gene or chromosome; *chromatid*—one of a pair of strands of DNA that make up a chromosome; *centromere*—a section of a chromosome where two chromatids are joined; *spindle fiber*—a structure made of microtubules that helps pull chromatids apart during cell division; *cytokinesis*—the process of dividing the cytoplasm of a cell

3. Build the cell model your group designed. **CAUTION: Sharp or pointed objects can cause injury. Handle scissors carefully. Promptly notify your teacher of any injuries.** Use your model to demonstrate the phases of mitosis. Draw and label each phase you model.

4. Use your model to explore one of the questions written for step 4 of **Before You Begin.** Describe the steps you took to explore the question.

PART B: Test Hypotheses

Answer each of the following questions by writing a hypothesis. Use your model to test each hypothesis, and describe your results.

5. Cytokinesis follows mitosis. How will the size of each new cell that is formed following cytokinesis compare with that of the original cell?

6. Sometimes two chromatids fail to separate during mitosis. How might this failure affect the chromosome number of the two new cells?

7. A mutation is a permanent change in a gene or chromosome. What effect might a mutation in a parent cell have on future generations of cells that result from the parent cell?

PART C: Cleanup and Disposal

8. Dispose of paper and yarn scraps in the designated waste container.

9. Clean up your work area and all lab equipment. Return lab equipment to its proper place. Wash your hands thoroughly before you leave the lab and after you finish all work.

Analyze and Conclude

1. **Analyzing Results** How do the nuclei you made by modeling mitosis compare with the nucleus of the model cell you started with? Explain your result.

2. **Evaluating Methods** How could you modify your model to better illustrate the process of mitosis?

3. **Recognizing Patterns** How does the genetic makeup of the cells that result from mitosis compare with the genetic makeup of the original cell?

4. **Inferring Conclusions** How is mitosis important?

5. **Further Inquiry** Write a new question about mitosis or the cell cycle that could be explored with your model.

Do You Know?

Do research in the library or media center to answer these questions:

1. How often do different cells of the human body undergo mitosis?

2. What are some common chemicals that disrupt the cell cycle?

Use the following Internet resources to explore your own questions about mitosis or the cell cycle.

internet connect
www.scilinks.org
Topic: Cell Cycle
Keyword: HX4033
SCILINKS. Maintained by the National Science Teachers Association

2. everywhere in the human body except in most nerve and skeletal muscle cells

3. During interphase, the cell grows, duplicates its chromosomes and organelles, and assembles microtubules.

4. Answers will vary. For example: How many chromosomes will each new nucleus have after mitosis has occurred?

Answers to Procedure

2. The cell model the students build may vary. For example: The students may use the yarn to represent the cell membrane and spindle fibers, the pipe cleaners to represent the chromosomes, the beads to represent the centromeres, and the labels to indicate the genes on the chromosomes.

3. Student models may vary. See diagram below for an example.

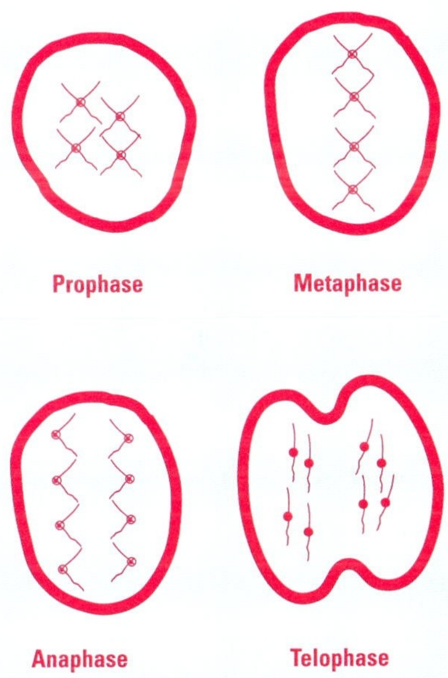

Answers to Analyze and Conclude

1. The nuclei students made should be the same as the nucleus of the original cell except that the chromosomes in the original cell are not replicated until right before mitosis.

2. Answers will vary. Students could mention that the pipe cleaners do not show how the shapes of the chromosomes change as they are pulled apart.

3. The genes found in the cells that result from mitosis are the same as the genes in the original cell. The chromosomes replicate before cell division. One copy of each gene goes to each new cell.

4. Answers will vary. Students should mention that mitosis preserves the chromosome number and genetic makeup of cells.

5. Answers will vary. For example: What happens if the DNA is not replicated before mitosis begins?

Answers to Do You Know?

1. The frequency at which cells in the human body undergo mitosis varies from tissue to tissue. Epidermal cells divide several times a week, whereas most mature skeletal muscle cells and nerve cells never undergo mitosis.

2. Chemicals that disrupt the cell cycle include nicotine and griseofulvin, an antifungal drug.

4. Answers will vary.

5. Each new cell will initially be smaller than the original cell that divides.

6. Nondisjunction of one of the chromosomes will result in one of the new cells having two copies of that chromosome and the other cell having none.

7. All of the cells in the subsequent generations of cells will carry the mutation.

UNDERSTANDING CANCER

Background

Environmental factors overall, including tobacco use, diet, infectious diseases, chemicals, and radiation cause about three-quarters of all cancer cases in the U.S. All of these risk factors are potentially controllable. Some risk factors, however, such as age, sex, and family history of disease cannot be controlled. If a person has one or more risk factors for a type of cancer, it doesn't necessarily mean that they will get the cancer. And cancer does occur in people that have none of the specific cancer's known risk factors.

- Tobacco use, including cigarettes, cigars, chewing tobacco, and snuff, is associated with an increased risk of cancers of the lung, mouth, larynx, bladder, kidney, cervix, esophagus, and pancreas. Smoking alone is estimated to be the cause of one-third of all cancer deaths.
- Unprotected exposure to strong sunlight is related to skin cancer.
- Older age, factors relating to fertility (such as early onset of menses), obesity, being physically inactive, and having a mother or sister who have had breast cancer are risk factors for breast cancer.
- Older age, being African-American, eating a high fat diet, and having a father or brother who have had prostate cancer are risk factors for prostate cancer.

While cancer itself is not an infectious disease, some agents of infectious disease can cause cancer. Examples include the hepatitis B virus, the Epstein-Barr virus, and the human T-cell leukemia/lymphoma virus-1. Infection with the hepatitis B virus is preventable by vaccination.

Teaching Strategies

Tell students that researchers are evaluating the following possible gene therapies for the treatment of cancer:

- Improving the body's natural ability to fight cancer by substituting a normal copy of a gene for one that has undergone mutation and no longer carries out its cancer suppressing function. For example, a healthy copy of a defective gene, such as *p53*, could be used to restore the ability of a defective gene to suppress or block the development of cancer cells.
- Increasing cancer cells' sensitivity to other kinds of treatment, such as by injecting cancer cells with a gene that makes them more sensitive to treatment with an anticancer drug. An effective treatment of this kind would kill only the cells that contain the drug-sensitive gene.
- Enhancing patients' tolerance of a drug by reducing its side effects and thereby enabling patients to take higher doses and/or undergo longer courses of therapy. For example, a multi-drug resistance gene could be inserted into stem cells (cells in the bone marrow that produce blood cells) to make them more resistant to the side effects of the high doses of anticancer drugs.

SCIENC

Do you know what cancer is and what may cause it?

Understanding Cancer

Most healthy cells of the body grow, divide a certain number of times, age, and die. Sometimes, this orderly process is disrupted when cells lose the ability to limit and direct their growth. Cells may divide too often and produce an excess of cells, called a *tumor*. A tumor can be benign or malignant. Benign tumors do not spread to other parts of the body, and can usually be surgically removed. A malignant tumor—cancer—invades and destroys nearby healthy tissues and organs. Cancerous tumors can metastasize, that is, can spread to other parts of the body and form new tumors. There are probably at least 100 different kinds of cancer, with each one affecting different kinds of cells and having different characteristics.

Cancer cells do not respond normally to the chemical signals that regulate the cell cycle. In some cancer cells, the way cell division signals are transmitted to the nucleus is abnormal. In other cancer cells, the entire cell cycle control system may be abnormal. If cancer cells stop dividing, they do so at random points in the cell cycle rather than at the normal checkpoints. When cultured in the laboratory with adequate nutrients, cancer cells can divide indefinitely and are referred to as "immortal." Most mammalian cells grown in culture divide only 20 to 50 times before they stop dividing, age, and die.

How Cancer Begins

Changes in the genes that influence the cell cycle can cause the transformation of a normal cell into a cancer cell. There are two types of these genes. The first type codes for proteins that stimulate cell division. Genes of this type are normally turned off in cells that are not dividing. This type of gene can be converted to an oncogene, that is, a "cancer gene," by mutation. One common oncogene, a gene called *ras*, is present in mutated form in about 30 percent of human cancers and in some forms of leukemia.

The second type of gene associated with cancer is a tumor suppressor gene. Tumor suppressor genes code for proteins that normally restrain cell division. In many cancers, tumor suppressor genes have been inactivated by mutation. An inherited mutation in one copy of a tumor suppressor gene results in higher risk of cancer. However, cancer does not occur unless and until the remaining, healthy copy of the gene is also inactivated by mutation. If a person is born with two normal copies of a tumor suppressor gene, both must be inactivated before cancer can develop.

Almost 50 percent of human cancers are associated with a mutation in the tumor suppressor gene *p53*. These cancers include many breast, colon, lung, prostate, and skin cancers. The protein produced by the *p53* gene normally acts as an emergency brake in the cell cycle. The p53 protein also induces the death of damaged cells. To perform these functions, p53 protein must

Bronchial tumor
Tumor cells have replaced the normal, ciliated tissue in the bronchi.

Lung tumor
A malignant tumor differs from the tissue that surrounds it.

bind to DNA. Mutant p53 protein cannot bind to DNA, thus cell division occurs unchecked.

Mutations that result in cancer, whether they involve oncogenes or tumor suppressor genes, can occur spontaneously. Mutations can also be induced by factors in the environment, such as X-rays and ultraviolet radiation, cigarette smoke, asbestos, and even by the human papilloma virus and the hepatitis B virus.

Surviving Cancer

In the United States, about one of every five deaths—more than half a million each year—are caused by cancer. Only heart disease kills more people. The number of new cancer cases and the number of cancer deaths for every 100,000 persons have decreased over the past 10 years. The death rates for children and adults under the age of 50 have decreased dramatically.

Some kinds of cancer kill more people than others. Lung cancer is the number one cancer killer. Colorectal cancer, breast cancer, and prostate cancer are the next most common. Cancer survival is measured by the percentage of cancer patients who survive a specific number of years. Five-year cancer survival rates for common cancers are shown in the table at the right.

Cancer survival is influenced by the type of cancer and by the stage at which a cancer is diagnosed. For example, the 5-year survival rate for colorectal cancer diagnosed in its first stage is 96 percent. When diagnosed in its fourth and most advanced stage, however, the 5-year survival rate is only 5 percent. Cancer screenings and early detection of cancers, such as those of the breast and colon, greatly increase a person's chances of surviving cancer. ■

Cancer Treatment

Profile

Oncologist

Oncologists are cancer specialists—physicians who care for people who have been diagnosed with cancer.

Job Description

Oncologists may begin to treat a patient soon after he or she is diagnosed with cancer. Oncologists work cooperatively with surgeons, radiologists, and other physicians to devise the best combination of treatments for the particular patient and cancer being treated. Oncologists may manage radiation treatment and/or chemotherapy, as well as other types of cancer therapy.

Science/Math Career Preparation

Biology	Microbiology
Chemistry	Biochemistry
Genetics	Mathematics

Cancer site	5-year survival (percent)
Prostate (males only)	98
Breast (females only)	88
Colorectal	63
Leukemia	47
Brain	32
Lung	15
Pancreas	4
All cancers	64

Analyzing STS Issues

Science and Society

1. Use library resources or the Internet to research factors that increase the risk of cancer. List some of the risk factors and what types of cancer they could lead to. Why are factors in lifestyle or the environment difficult to identify? How can people protect themselves from exposure to known risk factors?

Technology

2. The type, size, stage, and location of a person's cancer determines which cancer treatment is most appropriate for that person. A patient should take an active part in researching and developing an informed treatment plan. Using library resources or the Internet, research and identify some therapies used to treat cancer. What are some advantages and disadvantages to these treatments?

Read Further

1. National Cancer Institute, www.cancer.gov
2. American Cancer Society, www.cancer.org
3. Centers for Disease Control and Prevention, www.cdc.gov

Analyzing STS Issues Answers

1. Students' answers may include the risk factors and cancers identified in the "Background" section of this teacher material. Lifestyle and environmental risk factors are difficult to identify because it is difficult to quantify these risk factors, it takes a long time for their effects to be manifested, and they often occur in combination. People can protect themselves from exposure to known risk factors by not smoking, drinking only moderate amounts of alcohol, eating a healthy low-fat diet, wearing sunscreen, exercising, and avoiding exposure to carcinogenic chemicals and radiation.
2. Answers on the different types of therapies used to treat cancer will vary, but will fall into the general categories of surgery, chemotherapy, and radiation therapy. Students might also mention some new and/or experimental therapies that are being tested for use. The advantages and disadvantages will vary for each specific type of cancer.

UNIT 2 Principles of Genetics

Chapters

A DNA sequencing gel, shown here being examined by a technician, reveals an individual's unique chemical fingerprint.

Hemophilia

Queen Victoria

Yesterday...

Queen Victoria of England carried the gene for hemophilia. "Our poor family," she wrote in her diary, "seems persecuted by this awful disease, the worst I know." One son died of hemophilia and two daughters inherited the gene for the disease. **Why are carriers of hemophilia always female?**

Today...

The blood of normal individuals contains a protein called factor VIII that enables the blood to clot after an injury. But in people with hemophilia, the gene that produces factor VIII is defective and bleeding continues uncontrolled. **Discover how a gene controls the production of a protein.** Injections of factor VIII at the first sign of bleeding allow many hemophiliacs to control bleeding episodes.

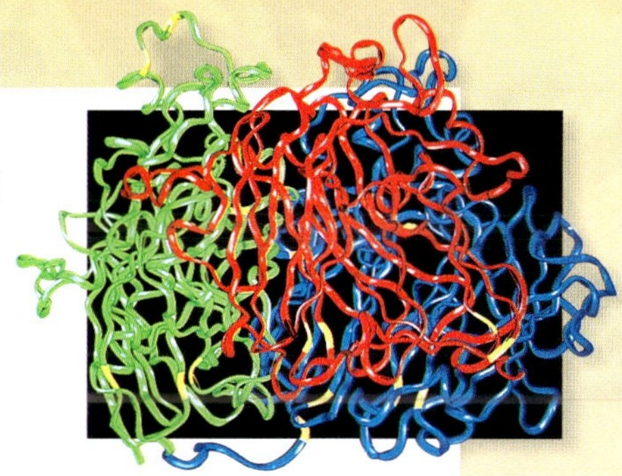

Factor VIII protein

Tomorrow...

Biologists have recently learned to make a genetically engineered form of factor VIII. New research in genetic engineering has yielded female pigs that produce factor VIII in their milk. **What are the safety advantages of genetically-engineered factor VIII?**

internet connect

www.scilinks.org
Topic: Hemophilia
Keyword: HX4097

SCILINKS. Maintained by the National Science Teachers Association

Chapter Opener

SE **Quick Review,** p. 143 ■
SE **Reading Activity,** p. 143 ■
TE **Using the Figure,** p. 142 [GENERAL]
TE **Opening Activity,** p. 143 [GENERAL]

& 3 • 90 min pp. 144–149

Section 1
Meiosis

OSP **Lesson Plan***
TR **Bellringer***
TR **Section Outline***
TR **B49 Stages of Meiosis***
TR **B51 Meiosis in Male and Female Animals***

TE **Demonstration,** p. 144
TE **Group Activity** Modeling Meiosis, p. 145 [GENERAL]
TE **Using the Figure,** p. 146 [GENERAL]
SE **Quick Lab** Modeling Crossing-Over, p. 147 ◆ ■
CRF **Datasheets for In-Text Labs** Modeling Crossing-Over*
TE **Using the Figure,** p. 147 [BASIC]
TE **Using the Figure,** p. 148 [GENERAL]
CRF **Skills Practice Lab** Staining DNA and RNA* ◆ [GENERAL]

& 5 • 90 min pp. 150–154

Section 2
Sexual Reproduction

OSP **Lesson Plan***
TR **Bellringer***
TR **Section Outline***
TR **F13 Haploid Life Cycle***
TR **B48 Diploid Life Cycle***
TR **G5 Alternation of Generations***

TE **Using the Figure,** p. 151 [BASIC]
TE **Activity** Asexual Reproduction in Bacteria, p. 151 [GENERAL]
SE **Quick Lab** Observing Reproduction in Yeast, p. 151 ■
CRF **Datasheets for In-Text Labs** Observing Reproduction in Yeast *
TE **Using the Figure,** p. 152 [GENERAL]
TE **Using the Figure,** p. 153 [GENERAL]
TE **Demonstration,** p. 153 ◆ [GENERAL]
CRF **Exploration Lab** Karyotyping* [GENERAL]
SE **Exploration Lab** Modeling Meiosis, pp. 158–159 ◆ ■
CRF **Datasheets for In-Text Labs** Modeling Meiosis*

& 7 • 90 min

Chapter Review and Assessment Resources

SE **Study Zone,** p. 155 ■
SE **Performance Zone,** p. 156 ■
SE **FCAT Test Prep,** p. 157 ■
TR **Concept Mapping***
CRF **Vocabulary Review Worksheet*** ■ [GENERAL]
CRF **Test Prep Pretest*** ■ [GENERAL]
CRF **Chapter Test*** ■ [GENERAL]
CRF **Chapter Test*** [ADVANCED]
OSP **Test Generator***
OSP **Test Item Listing***
OSP **Modified Chapter Test*** [BASIC]

go.hrw.com Holt Online Learning

Visit **go.hrw.com** for access to Holt Online Learning or enter the keyword **HX6 Home** for a variety of free online resources.

One-Stop Planner® CD-ROM

This CD-ROM package includes
- Lab Materials QuickList Software
- Holt Calendar Planner
- Customizable Lesson Plans
- Printable Worksheets
- ExamView® Test Generator
- Interactive Teacher Edition
- Holt PuzzlePro® Resources
- Holt PowerPoint® Resources

Guided Reading Audio CDs

These CDs are designed to help auditory learners and reluctant readers. (Audio CDs are also available in Spanish.)

Compression guide:
To shorten your instruction because of time limitations, omit block 1.

KEY

TE	Teacher Edition	CRF	Chapter Resource File	*	Also on One-Stop Planner
SE	Student Edition	TR	Teaching Transparency	■	Also Available in Spanish
OSP	One-Stop Planner			◆	Requires Advance Prep

SKILLS DEVELOPMENT AND PRACTICE	SECTION REVIEW AND ASSESSMENT	CORRELATIONS
		Florida Sunshine Standards/FCAT
TE **Reading Skill Builder** Interactive Reading, p. 146 BASIC TE **Skill Builder** Vocabulary, p. 148 BASIC TE **Reading Skill Builder** Discussion, p. 149 ADVANCED CRF **Directed Reading Worksheet*** BASIC CRF **Active Reading Worksheet*** GENERAL	SE **Section 1 Review** ■ TE **Reteaching,** p. 149 BASIC TE **Quiz,** p. 149 GENERAL TE **Alternative Assessment,** p. 149 GENERAL CRF **Quiz*** ■ BASIC	SC.F.1.4.5, SC.F.2.4.1, SC.F.2.4.2
TE **Reading Skill Builder** Reading Organizer, p. 152 BASIC CRF **Directed Reading Worksheet*** BASIC CRF **Active Reading Worksheet*** GENERAL CRF **Critical Thinking Worksheet*** ADVANCED CRF **Science Skills Worksheet*** GENERAL	SE **Section 2 Review** ■ TE **Reteaching,** p. 154 BASIC TE **Quiz,** p. 154 GENERAL TE **Alternative Assessment,** p. 154 GENERAL CRF **Quiz*** ■ BASIC **Modified Worksheet*** BASIC	SC.F.2.4.1, SC.H.3.4.1

Holt Biology Interactive Tutor CD-ROM

This CD-ROM consists of interactive activities that give students a fun way to extend their knowledge of biology concepts.

Unit 4 Cell Reproduction

SCILINKS. NSTA

www.scilinks.org

Maintained by the **National Science Teachers Association**

Topic: Genetic Variation
SciLinks Code: HX4093

Topic: Oogenesis
SciLinks Code: HX4192

Topic: Meiosis
SciLinks Code: HX4120

CHAPTER 7

Overview

Tell students the purpose of this chapter is to understand the process of meiosis and why it is important that reproductive cells have only half the number of chromosomes as other body cells. Students will also learn how sexual reproduction in organisms results in genetic variation. In addition, students will learn how reproduction in some organisms involves only one parent and that the reproductive process is cyclical.

Using the Figure

Each sperm cell surrounding this egg cell is trying to enter it and leave its packet of genetic information in the form of chromosomes containing DNA. Only one will be successful. In order for the newly formed organism to have the same number of chromosomes as its parents, the chromosome number must be halved when the egg and sperm are made. This is accomplished by a process called meiosis. Then, when one sperm successfully fertilizes the egg, the diploid chromosome number is restored.
LS Visual

Assessing Prior Knowledge

Before beginning this chapter, be sure your students can answer the questions in the ***Quick Review*** on the facing page. If they cannot, encourage them to reread the sections indicated.

Chapter Resource File

- Vocabulary Worksheets
- Concept Mapping

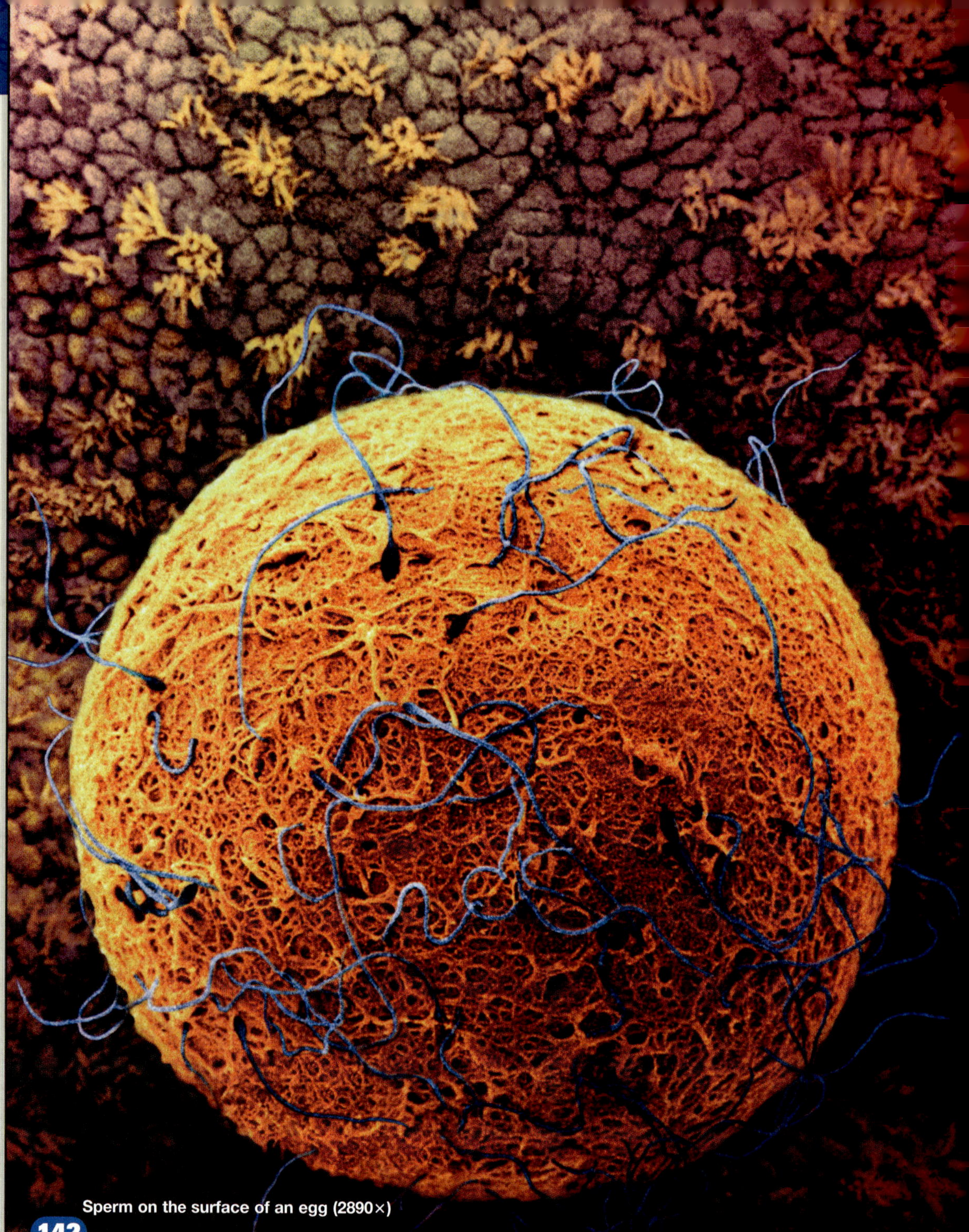

Sperm on the surface of an egg (2890×)

142

Chapter Correlations | **Florida Sunshine Standards/FCAT**

SC.F.1.4.5 Complex interactions among the different kinds of molecules in the cell cause distinct cycles of activity governed by proteins.

SC.F.2.4.1 There are mechanisms of asexual and sexual reproduction and different genetic advantages and disadvantages of asexual and sexual reproduction.

CHAPTER

7 Meiosis and Sexual Reproduction

Florida Sunshine Standards/FCAT

SC.F.1.4.5 Complex interactions among the different kinds of molecules in the cell cause distinct cycles of activity governed by proteins.

SC.F.2.4.1 There are mechanisms of asexual and sexual reproduction and different genetic advantages and disadvantages of asexual and sexual reproduction.

Quick Review

Answer the following without referring to earlier sections of your book.

1. **Define** the terms *evolution* and *natural selection*. *(Chapter 1, Section 1)*
2. **Define** the term *homologous chromosomes*, and identify chromatids. *(Chapter 6, Section 1)*
3. **Differentiate** between haploid cells and diploid cells. *(Chapter 6, Section 1)*
4. **Describe** the structure and function of the spindle. *(Chapter 6, Section 3)*
5. **Summarize** the steps of mitosis. *(Chapter 6, Section 3)*

Did you have difficulty? *For help, review the sections indicated.*

Reading Activity

Take a few moments to study the first two pages in Section 1, including Figure 1. Then on a sheet of paper or in your notebook, answer the following questions:

- *What is the topic of Section 1?*
- *How are meiosis I and meiosis II similar?*
- *How are meiosis I and meiosis II different?*

A special form of cell reproduction produces the egg and sperm cells shown here. When an egg joins with a single sperm cell, genetic instructions from a male and female are combined, and a new individual is formed.

Looking Ahead

Section 1
Meiosis

Formation of Haploid Cells
Meiosis and Genetic Variation
Meiosis and Gamete Formation

Section 2
Sexual Reproduction

Sexual and Asexual Reproduction
Sexual Life Cycles in Eukaryotes

internet connect

www.scilinks.org

National Science Teachers Association *sci*LINKS Internet resources are located throughout this chapter.

SCILINKS. Maintained by the National Science Teachers Association

143

Opening Activity — GENERAL

Multiplying Chromosomes Write the following table on the board:

Organism	Diploid Chromosome Number
Fruit Fly	8
Garden Pea	14
Horse	64

Ask students to calculate the number of chromosomes in each organism through five generations, assuming the chromosome number doubles each generation. (fruit fly 8-16-32-64-128; garden pea 14-28-56-112-224; horse 64-128-256-512-1024) Point out that doubling the chromosome number five times would wreak havoc in the cell. Ask students to propose possible ways this problem could be avoided. (Cells could divide such that each cell only receives one set of chromosomes.) **LS Logical**

Quick Review Answers

1. Evolution is change in genetic makeup of a species over time. Natural selection is the process in which organisms with favorable genes are more likely to survive and reproduce.
2. Homologous chromosomes are similar in shape, size, and genes. Chromatids are copied chromosomes still attached at the centromere.
3. Diploid cells have two sets of chromosomes. Haploid cells have one chromosome set.
4. The spindle is a structure made of microtubules. During mitosis, it helps to organize and separate chromosomes.
5. Prophase—Chromosomes coil, spindle forms, nuclear membrane breaks down. Metaphase—Chromosomes line up on cell equator. Anaphase—Chromatid pairs separate and move to opposite poles. Telophase—chromosomes unwind, nuclear membranes form.

Reading Activity Answers

- formation of haploid cell, how meiosis increases variation, gamete formation
- Both go through similar Prophase, Metaphase, Anaphase, and Telophase steps.
- Only in Meiosis I do homologous chromosomes pair up and crossing-over occurs.

Section 1

Focus

Overview

Before beginning this section review with your students the objectives listed in the Student Edition. This section introduces students to the processes of meiosis, spermatogenesis, and oogenesis. Students will also learn how these processes relate to genetic variation.

Bellringer

Pose this question to your students: "If a *human* sperm and egg each had 46 chromosomes, like other human body cells, how many chromosomes would a fertilized egg have?" (92) Then ask students to explain why increasing the number of chromosomes a human cell has might cause problems. (Answers will vary, but students should note that it would change the genetic makeup of *Homo sapiens*.) **LS Logical**

Motivate

Demonstration

On the board, draw a sperm cell near an egg cell. Draw three homologous chromosome pairs in each cell. Tell students that when the sperm fertilizes the egg, the number of chromosomes in the resulting cell, the new individual, will double. Lead students to the conclusion that the number of chromosomes in sperm and egg cells must be halved, or the number of chromosomes in each subsequent generation will continue to double. Tell students that the process by which the chromosome number is halved when reproductive cells are formed is called meiosis. **LS Visual**

Section Correlations

SC.F.1.4.5
SC.F.2.4.1

Section 1 Meiosis

Florida Sunshine Standards/FCAT

SC.F.1.4.5 Complex interactions among the different kinds of molecules in the cell cause distinct cycles of activity governed by proteins.

SC.F.2.4.1 There are mechanisms of asexual and sexual reproduction and different genetic advantages and disadvantages of asexual and sexual reproduction.

Objectives

- **Summarize** the events that occur during meiosis. SC.F.1.4.5
- **Relate** crossing-over, independent assortment, and random fertilization to genetic variation. SC.F.2.4.1
- **Compare** spermatogenesis and oogenesis. SC.F.1.4.5

Key Terms

meiosis
crossing-over
independent assortment
spermatogenesis
sperm
oogenesis
ovum

Formation of Haploid Cells

Some organisms reproduce by joining gametes to form the first cell of a new individual. The gametes are haploid—they contain one set of chromosomes. Imagine how the chromosome number would increase with each generation if chromosome reduction did not occur!

Meiosis *(meye OH sihs)* is a form of cell division that halves the number of chromosomes when forming specialized reproductive cells, such as gametes or spores. Meiosis involves two divisions of the nucleus—meiosis I and meiosis II.

Before meiosis begins, the DNA in the original cell is replicated. Thus, meiosis starts with homologous chromosomes. Recall that homologous chromosomes are similar in size, shape, and genetic content. The stages of meiosis are summarized in **Figure 1.**

Step 1 **Prophase I** The chromosomes condense, and the nuclear envelope breaks down. Homologous chromosomes pair along their length. **Crossing-over** occurs when portions of a chromatid on one homologous chromosome are broken and exchanged with the corresponding chromatid portions of the other homologous chromosome.

BIO graphic

Figure 1

Stages of Meiosis

Four cells are produced, each with half as much genetic material as the original cell.

1 Prophase I Chromosomes become visible. The nuclear envelope breaks down. Crossing-over occurs.

2 Metaphase I Pairs of homologous chromosomes move to the equator of the cell.

3 Anaphase I Homologous chromosomes move to opposite poles of the cell.

4 Telophase I and cytokinesis Chromosomes gather at the poles of the cell. The cytoplasm divides.

did you know?

Meiosis Timeframe In human males, meiosis and sperm production take about 9 weeks. It's a continuous process that begins at puberty. In females, meiosis and egg production begin before birth. The process stops abruptly, however, and does not begin again until a girl enters puberty. Until a woman reaches menopause, one egg each month resumes meiosis and finishes its development. Therefore, the meiosis of a single egg may take up to 50 years to complete!

MISCONCEPTION ALERT

Identifying Misconceptions Students may think that animals are the only organisms that reproduce sexually. Remind students that many plants, fungi, and algae also reproduce sexually. Point out that a flower, for example, contains both sperm nuclei and egg cells.

Step 2 **Metaphase I** The pairs of homologous chromosomes are moved by the spindle to the equator of the cell. The homologous chromosomes remain together.

Step 3 **Anaphase I** The homologous chromosomes separate. As in mitosis, the chromosomes of each pair are pulled to opposite poles of the cell by the spindle fibers. *But the chromatids do not separate at their centromeres—each chromosome is still composed of two chromatids. The genetic material, however, has recombined.*

Step 4 **Telophase I** Individual chromosomes gather at each of the poles. In most organisms, the cytoplasm divides (cytokinesis), forming two new cells. Both cells or poles contain one chromosome from each pair of homologous chromosomes. *Chromosomes do not replicate between meiosis I and meiosis II.*

Step 5 **Prophase II** A new spindle forms around the chromosomes.

Step 6 **Metaphase II** The chromosomes line up along the equator and are attached at their centromeres to spindle fibers.

Step 7 **Anaphase II** The centromeres divide, and the chromatids (now called chromosomes) move to opposite poles of the cell.

Step 8 **Telophase II** A nuclear envelope forms around each set of chromosomes. The spindle breaks down, and the cell undergoes cytokinesis. The result of meiosis is four haploid cells.

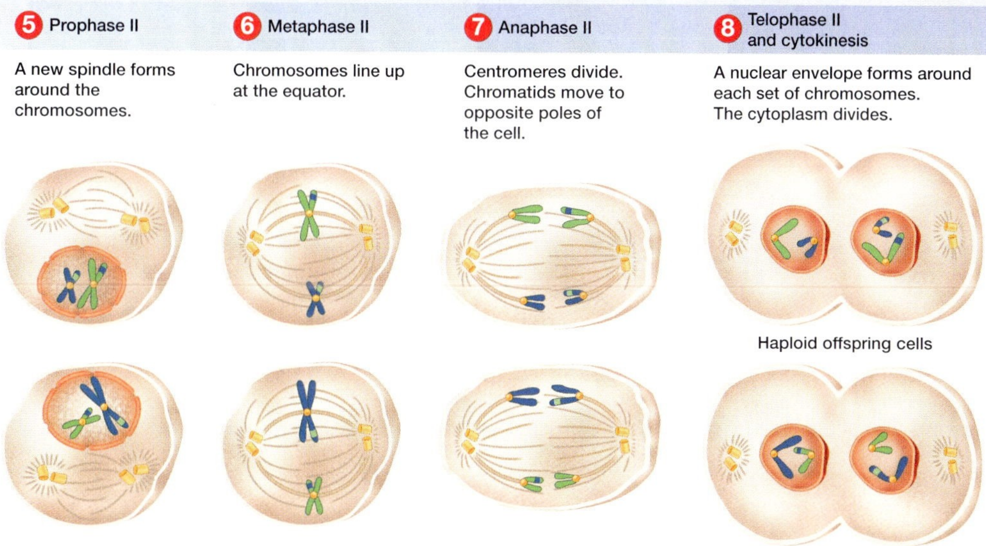

145

Teach

Group Activity — GENERAL

Modeling Meiosis Have students work in groups of four at a table. Give each group 32 beads of one color (yellow, for example) and 32 of a contrasting color (blue, for example). Give each group four 4-inch sections of pipe cleaner to act as centromeres. Have students make two 6-bead yellow strands. Attach with a pipe cleaner. This represents a chromosome with 2 chromatids. Make two 6-bead blue strands and attach together. This represents the homologous chromosome. Make two 10-bead blue strands and attach together, and two 10-bead yellow strands. Have students set the four chromosomes on the table and ask what the diploid number is for this cell. (4) Using these bead strands, lead students through the process of meiosis. Check student progress at each step. Crossing-over can be demonstrated by removing two beads at the ends of non-sister chromatids (having opposite colors) and switching them. When finished, the four resulting gametes should each have one long and one short chromosome, with varying crossover patterns. LS **Kinesthetic**

Transparencies

- **TR** Bellringer
- **TR B49** Stages of Meiosis
- **TR B51** Meiosis in Male and Female Animals

Chapter Resource File

- Directed Reading BASIC
- Active Reading GENERAL
- Data Sheet for Quick Lab GENERAL

One-Stop Planner CD-ROM

- Reading Organizers BASIC
- Reading Strategies BASIC

Trends in Reproductive Biology

Development in Eggs Stanford University researchers recently discovered how sperm-egg contact initiates development in sea urchins. Sea urchin sperm contain an enzyme called nitric oxide synthase that becomes active a few seconds before fertilization. The enzyme quickly causes the production of a large amount of nitric oxide gas, which is injected into the egg once the sperm and egg make contact. The injection of this gas triggers the release of calcium inside the egg. As the calcium levels rise, metabolic changes occur that causes the egg to divide and form into an embryo.

INTERACTIVE TUTOR BIOLOGY

- **Unit 4—Cell Reproduction**

This engaging tutorial helps students to understand how cells divide during the processes of mitosis, meiosis, and gamete formation.

Teach, *continued*

READING SKILL BUILDER — BASIC

Interactive Reading Assign Chapter 7 of the *Holt Biology* Guided Audio CD Program to help students achieve greater success in reading the chapter. LS Visual

Using the Figure — GENERAL

Have students look at the first cell at the top left of **Figure 2.** Ask them how many chromosomes are in the cell. Some will answer "four" while others may say "eight." Tell students that as long as chromatids are connected together (by a centromere), they are considered a single chromosome. As soon as the chromatid pairs separate, they are considered two separate chromosomes. The top cell in **Figure 2** has four chromosomes, the two middle cells each have two chromosomes, and the bottom four cells each have two chromosomes. LS Visual

Teaching Tip — GENERAL

Crossing Over Draw two large chromosomes on the board with colored chalk. Each chromosome should have two chromatids and a centromere. Use different colors along the length of each chromatid to represent genes. Ask volunteers to show crossing-over in two different places by erasing and redrawing the chromosomes with the corresponding gene color. Point out that because of the random orientation of chromosomes at metaphase I and because of crossing-over, children resemble their parents but never look exactly like them. LS Interpersonal

Meiosis and Genetic Variation

internet connect
www.scilinks.org
Topic: Genetic Variation
Keyword: HX4093
SciLinks. Maintained by the National Science Teachers Association

Meiosis is an important process that allows for the rapid generation of new genetic combinations. Three mechanisms make key contributions to this genetic variation: independent assortment, crossing-over, and random fertilization.

Independent Assortment

Most organisms have more than one chromosome. In humans, for example, each gamete receives one chromosome from each of 23 pairs of homologous chromosomes. But, which of the two chromosomes that an offspring receives from each of the 23 pairs is a matter of chance. This random distribution of homologous chromosomes during meiosis is called **independent assortment**. Independent assortment is summarized in **Figure 2.** Each of the 23 pairs of chromosomes segregates (separates) independently. Thus, 2^{23} (about 8 million) gametes with different gene combinations can be produced from one original cell by this mechanism.

Crossing-Over and Random Fertilization

The DNA exchange that occurs during crossing-over adds even more recombination to the independent assortment of chromosomes that occurs later in meiosis. Thus, the number of genetic combinations that can occur among gametes is practically unlimited.

Figure 2 Independent assortment

The same cell is shown twice. Because each pair of homologous chromosomes separates independently, four different gametes can result in each case.

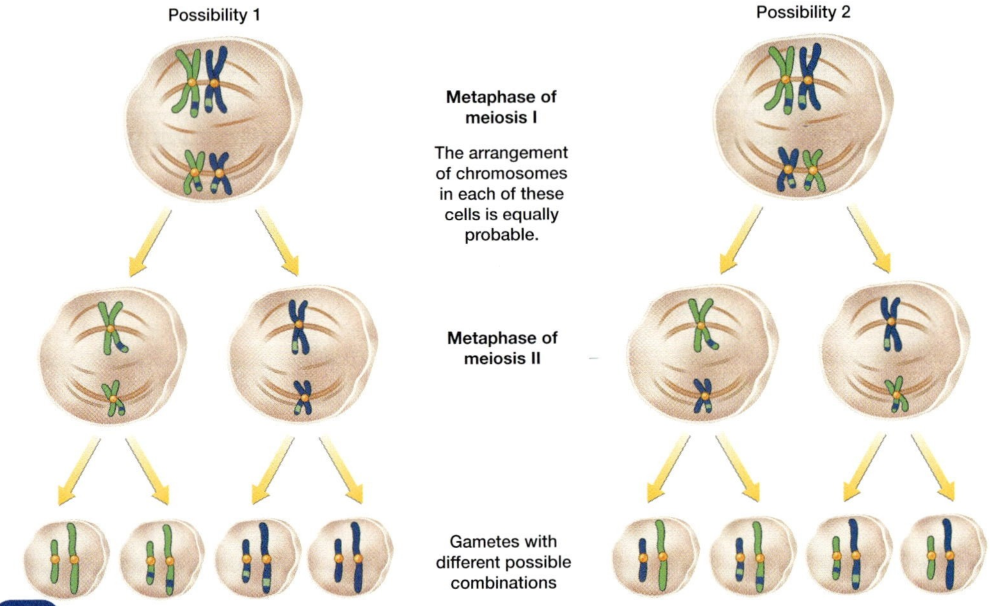

146

MATH CONNECTION

At metaphase I, the orientation of homologous pairs is random: the paternal or maternal homologue can be on either side of the equator. Because of this random orientation, the number of possible chromosome combinations in the gametes is 2^n, where n represents the haploid chromosome number. Therefore, the number of possible chromosome combinations for humans (haploid number = 23) is 2^{23}. Have students use their calculators to determine how many combinations are possible. Have students calculate the possible number of combinations for a mosquito having 6 chromosomes and a dog having 78 chromosomes.

Quick Lab

Modeling Crossing-Over SC.F.2.4.2

You can use paper strips and pencils to model the process of crossing-over.

Materials

4 paper strips, pens or pencils (two colors), scissors, tape

Homologous chromosomes

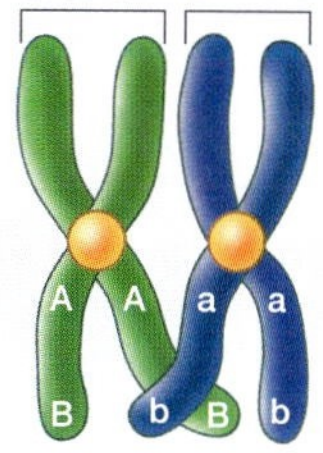

Procedure

1. Using one color, write the letters *A* and *B* on two paper strips. These two strips will represent one of the two homologous chromosomes shown above.
2. Using a second color, write the letters *a* and *b* on two paper strips. These two strips will represent the second homologous chromosome shown above.
3. Use your chromosome models, scissors, and tape to demonstrate crossing-over between the chromatids of two homologous chromosomes.

Analysis

1. **Determine** what the letters *A, B, a,* and *b* represent.
2. **Infer** why the chromosomes you made are homologous.
3. **Compare** the number of different types of chromatids (combinations of *A, B, a,* and *b*) before crossing-over with the number after crossing-over.
4. **Critical Thinking** **Applying Information** How does crossing-over relate to genetic recombination?

Furthermore, the zygote that forms a new individual is created by the random joining of two gametes (each gamete produced independently). Because fertilization of an egg by a sperm is random, the number of possible outcomes is *squared* ($2^{23} \times 2^{23}$ = 64 trillion).

Importance of Genetic Variation

Meiosis and the joining of gametes are essential to evolution. No genetic process generates variation more quickly. In many cases, the pace of evolution appears to increase as the level of genetic variation increases. For example, when domesticated animals such as cattle and sheep are bred for large size, many large animals are produced at first. But as the existing genetic combinations become used up, the ability to obtain larger and larger animals slows down. Further progress must then wait for the formation of new gene combinations.

Racehorse breeding provides another example. Thoroughbred racehorses are all descendants of a small number of individuals, and selection for speed has accomplished all it can with this limited amount of genetic variation. The winning times in major races stopped dramatically improving decades ago.

The pace of evolution is sped up by genetic recombination. The combination of genes from two organisms results in a third type, not identical to either parent. But bear in mind that natural selection does not always favor genetic change. Indeed, many modern organisms are little changed from their ancestors of the distant past. Natural selection may favor existing combinations of genes, slowing the pace of evolution.

WORD Origins

The word *meiosis* is from the Greek word *meioun*, meaning "to make smaller." Knowing this makes it easier to remember that during meiosis, the chromosome number is reduced by half to form haploid gametes.

147

Quick Lab: *Modeling Crossing-Over*

Skills Acquired
Inferring, interpreting, comparing, applying information

Teacher's Notes
Before assigning the Quick Lab, prepare the paper strips for students to use.

Answers to Analysis

1. individual genes
2. The chromosomes are of similar size, shape, and genetic content.
3. Before crossing-over: AB, AB, ab, and ab. After crossing-over: answers will vary (for example, AB, Ab, aB, and ab).
4. Crossing-over causes genetic recombination.

Using the Figure — Basic

Have students look at the diagram of the two homologous chromosomes in the Quick Lab. Point out that the two green strands attached together by a centromere are called *sister* chromatids. Ask students how sister chromatids compare genetically. (They are identical.) The green and blue chromatids are known as *non-sister* chromatids. Ask students what would happen to the genetic makeup of a chromosome if crossing-over occurred between *sister* chromatids. (Since sister chromatids are copies of each other, the genetic makeup would not change.) **LS Visual**

Teaching Tip — General

Distinguishing Between Mitosis and Meiosis Have students make a Graphic Organizer similar to one at the bottom of this page to summarize the difference in chromosomal number between mitosis and meiosis. Ask students to include a brief paragraph that compares the two processes. **LS Logical**

Graphic Organizer

Use this graphic organizer with **Teaching Tip** on this page.

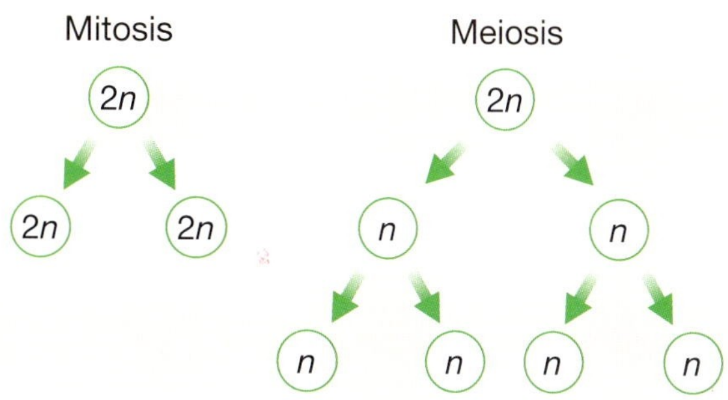

Teach, *continued*

Real Life

Answer

Answers will vary. Some hermaphroditic genera includes *Lates*, which lives throughout the Indo-Pacific region and *Rivulus*, which is found throughout the West Indies and South America.

Teaching Tip — GENERAL

Gamete Development Have students make a Graphic Organizer similar to the one at the bottom of this page that explains the formation of gametes in humans. Be sure they include the terms spermatogenesis, oogenesis, diploid germ cell, sperm, polar bodies, and egg cell.
LS Logical

Using the Figure — GENERAL

Have students examine the immature sperm cells formed after Meiosis I of spermatogenesis in **Figure 3.** Ask students how they compare in size and shape. (They are identical.) Now ask students to compare the two cells formed after Meiosis I of oogenesis. (One cell is larger than the other is.) Ask students why this has happened, and how this helps the egg cell to survive. (The cytoplasm has been divided unequally; this ensures a rich supply of nutrients for the developing egg cell.) Point out to students that while spermatogenesis is continually occurring in human males, oogenesis is suspended after Meiosis I in females. In oogenesis, Meiosis II is not completed until a sperm cell enters the egg. **LS Visual**

Real Life

Some animals are both male and female.

Hermaphrodites *(huhr MAF roh DIETZ)* have both male and female reproductive organs. Most hermaphrodites mate with another individual, while some fertilize themselves.

Other hermaphrodites can reverse their sex each breeding season.

Finding Information
What are some examples of hermaphroditic fish species, and where do they live?

Meiosis and Gamete Formation

The fundamental events of meiosis occur in all sexually reproducing organisms. However, organisms vary in timing and structures associated with gamete formation. Meiosis is the primary event in the formation of gametes—gametogenesis.

Meiosis in Males

The process by which sperm are produced in male animals is called **spermatogenesis** *(spur mat uh JEHN uh sihs)*. Spermatogenesis occurs in the testes (male reproductive organs). As illustrated in **Figure 3,** a diploid cell first increases in size and becomes a large immature cell (germ cell). The large cell then undergoes meiosis I. Two cells are produced, each of which undergoes meiosis II to form a total of four haploid cells. The four cells change in form and develop a tail to become male gametes called **sperm**.

Meiosis in Females

The process by which gametes are produced in female animals is called **oogenesis** *(oh oh JEHN uh sihs)*. Oogenesis, summarized in Figure 3, occurs in the ovaries (female reproductive organs). Notice that during cytokinesis following meiosis I, the cytoplasm divides unequally. One of the resulting cells gets nearly all of the cytoplasm. It is this cell that will ultimately give rise to an egg cell. The other cell is very small and is called a polar body. The polar body may divide again, but its offspring cells will not survive.

Figure 3 Meiosis in male and female animals

Meiosis of a male diploid cell results in four haploid sperm, while meiosis of a female diploid cell results in only one functional haploid egg cell.

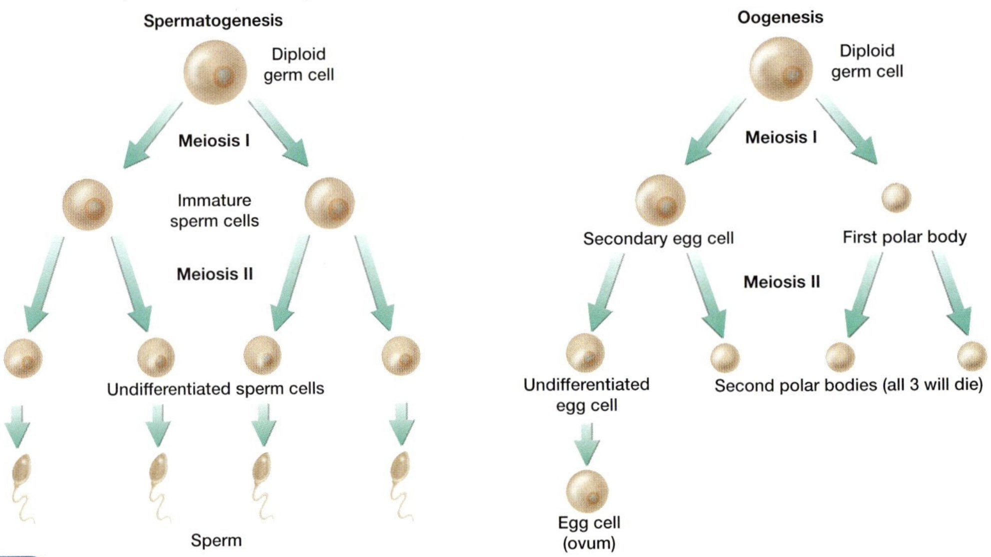

148

Graphic Organizer

Use this graphic organizer with **Teaching Tip** on this page.

Diploid germ cell → Spermatogenesis → Sperm cell

Diploid germ cell → Oogenesis → Egg cell, Polar bodies

Section Review Correlations

SC.F.1.4.5: Items 2, 4
SC.F.2.4.1: Items 1, 3
SC.F.2.4.2: Items 5, 6

The larger cell undergoes meiosis II, and the division of the egg cell during cytokinesis is again unequal. The larger cell develops into a gamete called an **ovum** (plural, ova) or, more commonly, egg. The smaller cell, the second polar body, dies. Because of its larger share of cytoplasm, the mature ovum has a rich storehouse of nutrients. These nutrients nourish the young organism that develops if the ovum is fertilized.

BIOWatch

The Making of an Egg

An intricate series of steps controls the formation of an egg. As oogenesis proceeds, eggs become very different from other cells in the organism. One obvious difference is the size of the egg cells. In many animals, eggs swell to gigantic proportions as they accumulate cellular components. How is this transformation accomplished?

Oogenesis in *Drosophila*

Researchers have examined oogenesis in great detail in the fruit fly, *Drosophila melanogaster.* As in other animals, eggs in *Drosophila* are produced when germ cells divide. Four rapid cell divisions produce 16 cells, which form a cluster known as a germ cell cyst. Bridges called *ring canals* interconnect all cells in the cyst. However, only one of the cells develops into an egg. The other 15 cells become nurse cells, which donate organelles—including mitochondria and parts of the endoplasmic reticulum—to the growing egg. The organelles move through the ring canals by traveling along a network of microtubules. Some scientists believe that this movement reflects an organized sorting process, in which functional organelles collect in the egg and damaged organelles collect in the nurse cells. The nurse cells die as the egg completes its development.

Oogenesis in Other Organisms

A comparable process of cyst formation takes place during oogenesis in the mouse. As in the fruit fly, many cellular components are redistributed among the cells in the cyst. These findings suggest that the early steps in egg formation may be very similar in a wide range of organisms.

Human ovarian follicle

internet connect
www.scilinks.org
Topic: Oogenesis
Keyword: HX4194
SciLINKS. Maintained by the National Science Teachers Association

Section 1 Review

1. **Explain** the significance of meiosis in sexual reproduction. SC.F.2.4.1
2. **Name** the stage of meiosis during which chromatids are separated to opposite poles of the cell. SC.F.1.4.5
3. **Compare** the processes of crossing-over and independent assortment. SC.F.2.4.1
4. **Differentiate** gamete formation in male animals from gamete formation in female animals. SC.F.1.4.5
5. **Critical Thinking Evaluating Information** If one cell in a dog ($2n = 78$) undergoes meiosis and another cell undergoes mitosis, how many chromosomes will each resulting cell contain? SC.F.2.4.2
6. **FCAT Prep** If a cell begins meiosis with two pairs of homologous chromosomes, how many chromatids will be in each cell that is produced at the end of meiosis I?
   A 1 **C** 4
   B 2 **D** 8
   SC.F.2.4.2

Answers to Section Review

1. Meiosis halves the number of chromosomes in gametes so that when two haploid gametes unite to form a zygote, the zygote contains the diploid number of chromosomes.
2. Anaphase II
3. Crossing-over occurs when part of a chromatid is broken and exchanged with the corresponding part on one of the chromatids of its homologous chromosome. Independent assortment is the random distribution of homologous chromosomes during meiosis.
4. In males, a diploid cell forms four haploid cells, which develop tails and become sperm. In females, a diploid cell forms one egg cell and three polar bodies. The polar bodies die.
5. The cells resulting from meiosis will contain 39 chromosomes each. The cells resulting from mitosis will contain 78 chromosomes each.
6. **A.** Incorrect. See answer C. **B.** Incorrect. See answer C. **C.** Correct. There will be 2 chromosomes, with 2 chromatids each. **D.** Incorrect. See answer C.

READING SKILL BUILDER — ADVANCED

Discussion Direct students' attention to the BioWatch feature. Have students brainstorm hypotheses as to how the egg cell might benefit from receiving nurse cell organelles. (Possible hypotheses might be: The egg cell is growing so fast that it cannot make its own organelles fast enough, or only the best organelles are transferred, in a type of mini-selection, etc.) Then ask students how receiving nurse cell mitochondria would help the egg. (The egg cell grows at a tremendous rate, and energy is needed for the numerous cellular activities involved in this rapid growth. Mitochondria are the organelles that release the energy needed for these activities.) Ask students why it is significant to humans that this nurse cell phenomenon was also found to occur in mice. (Mice are physiologically more similar to us than the fruit fly.)

Close

Reteaching — BASIC

Prior to class, make four homologous chromosome pairs of different lengths using different colors of paper. Divide the chromosomes into "genes" with a marker pen. Have students use the chromosomes to model meiosis. Ask students why sexual reproduction provides diversity in organisms. (New organisms have genetic information from two parents.)

Quiz — GENERAL

True or False:

1. Mitosis and meiosis both result in diploid cells. (False; meiosis produces haploid cells.)
2. What type of body cell does the process of meiosis form? (reproductive cells)

Alternative Assessment — GENERAL

Write a humorous story from the viewpoint of a single chromosome going through the stages of meiosis.

Section 2

Focus

Overview

Before beginning this section review with your students the objectives listed in the Student Edition. This section explains the differences between sexual and asexual reproduction, the advantages and disadvantages of each to a species, and how the three major sexual life cycles in eukaryotes function.

Bellringer

Display several numbered pictures of plants and animals (include a picture of a mule, which is sterile). On a piece of scratch paper, have students list whether they think the plant or animal reproduces asexually, sexually, both asexually and sexually, or neither. Discuss answers and reasons for them.
LS Interpersonal

Motivate

Identifying Preconceptions — GENERAL

Ask students to explain the difference between asexual and sexual reproduction. Many students will know that asexual reproduction involves just one parent while sexual reproduction involves two parents. Then ask which type of reproduction is "best" for a species. During discussion, lead students to realize that both types of reproduction have advantages and disadvantages depending on the species involved and the environment they live in.

Section Correlations

SC.F.2.4.1

Section 2 Sexual Reproduction

Florida Sunshine Standards/FCAT

SC.F.2.4.1 There are mechanisms of asexual and sexual reproduction and different genetic advantages and disadvantages of asexual and sexual reproduction.

Objectives

- **Differentiate** between asexual and sexual reproduction. SC.F.2.4.1
- **Identify** three types of asexual reproduction. SC.F.2.4.1
- **Evaluate** the relative genetic and evolutionary advantages and disadvantages of asexual and sexual reproduction. SC.F.2.4.1
- **Differentiate** between the three major sexual life cycles found in eukaryotes. SC.F.2.4.1

Key Terms

asexual reproduction
clone
sexual reproduction
life cycle
fertilization
sporophyte
spore
gametophyte

Sexual and Asexual Reproduction

Some organisms look exactly like their parents and siblings. Others share traits with family members but are not identical to them. Some organisms have two parents, while others have one. The type of reproduction that produces an organism determines how similar the organism is to its parents and siblings. Reproduction, the process of producing offspring, can be asexual or sexual.

In **asexual reproduction** a single parent passes copies of all of its genes to each of its offspring; there is no fusion of haploid cells such as gametes. An individual produced by asexual reproduction is a **clone,** an organism that is genetically identical to its parent. As you have read, prokaryotes reproduce by a type of asexual reproduction called binary fission. Many eukaryotes, as shown in **Figure 4,** also reproduce asexually.

In contrast, in **sexual reproduction** two parents each form reproductive cells that have one-half the number of chromosomes. A diploid mother and father would give rise to haploid gametes, which join to form diploid offspring. Because both parents contribute genetic material, the offspring have traits of both parents but are not exactly like either parent. As shown in **Figure 5,** sexual reproduction, with the formation of haploid cells, occurs in eukaryotic organisms, including humans.

Types of Asexual Reproduction

There are many different types of asexual reproduction. For example, amoebas reproduce by fission, the separation of a parent into two or more individuals of about equal size. Some multicellular eukaryotes undergo fragmentation, a type of reproduction in which the body breaks into several pieces. Some or all of these fragments later develop into complete adults when missing parts are regrown. Other organisms, like the hydra shown in Figure 4, undergo budding, in which new individuals split off from existing ones. The bud may break from the parent and become an independent organism, or it may remain attached to the parent. An attached bud can eventually give rise to a group of many individuals.

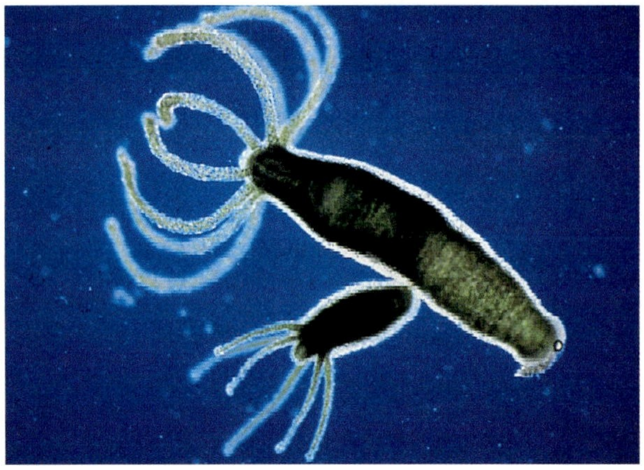

Figure 4 Asexual reproduction. Asexual reproduction creates clones. This hydra is in the process of reproducing asexually. The smaller hydra budding from the parent is genetically identical to the parent.

150

Attention Grabber

Hermaphroditism Students may be surprised to learn that in certain species, a single animal can sexually reproduce alone. For example, in some species of nematodes, sperm are produced and then stored until eggs are produced. This is followed by self-fertilization.

One-Stop Planner CD-ROM

- **Reading Organizers** BASIC
- **Reading Strategies** BASIC
- **Occupational Application Worksheet** Medical Sonographer GENERAL
- **Supplemental Reading Guide** *The Lives of a Cell* ADVANCED

Genetic Diversity

Asexual reproduction is the simplest and most primitive method of reproduction. In a stable environment, asexual reproduction allows organisms to produce many offspring in a short period of time, without using energy to produce gametes or to find a mate. However, the DNA of these organisms varies little between individuals. This may be a disadvantage in a changing environment because a population of organisms may not be able to adapt to a new environment.

On the other hand, sexual reproduction provides a powerful means of quickly making different combinations of genes among individuals. Such genetic diversity is the raw material for evolution.

Figure 5 Sexual reproduction

Sexual reproduction creates genetic diversity. Human gametes contain 23 chromosomes. (Only one chromosome is shown in each gamete below.) After fertilization the resulting zygote has 23 pairs of chromosomes.

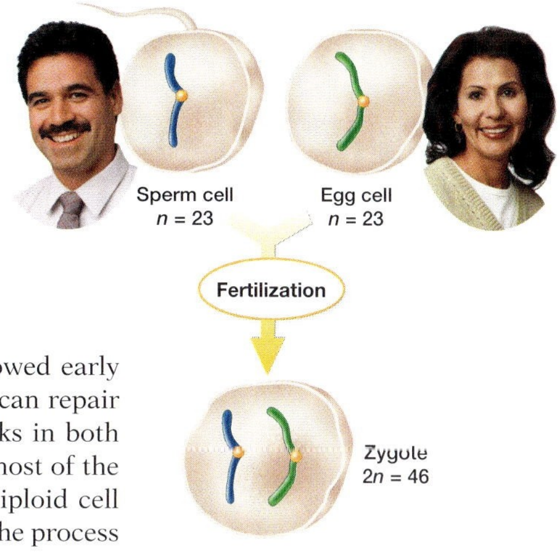

Evolution of Sexual Reproduction

The evolution of sexual reproduction may have allowed early protists to repair their own DNA. Only diploid cells can repair certain kinds of chromosome damage, such as breaks in both strands of DNA. Many modern protists are haploid most of the time, and they reproduce asexually. (They form a diploid cell only in response to stress in the environment.) Thus the process of meiosis and the pairing of homologous chromosomes may have allowed early protistan cells to repair damaged DNA. This hypothesis is further supported by the fact that many enzymes that repair DNA damage are involved in meiosis.

QUICK LAB

Observing Reproduction in Yeast SC.F.2.4.1

Yeast are unicellular organisms that live in liquid or moist environments. You can examine a culture of yeast to observe one of the types of reproduction that yeast can undergo.

Materials

microscope, microscope slides, dropper, culture of yeast

Procedure

1. Make a wet mount of a drop of yeast culture.
2. Observe the yeast with a compound microscope under low power.
3. Look for yeast that appear to be in "pairs."
4. Observe the pairs under high power, and then make drawings of your observations.

Analysis

1. **Infer** the type of reproduction you observed when the yeast appeared to be in pairs.
2. **Identify** the reason for your answer.
3. **Determine,** by referring to your textbook, the name of the type of reproduction you observed.

Teach

Using the Figure — BASIC

Have students examine **Figure 5.** Ask students what the letter "n" represents. (one chromosome set, in this case, 23 chromosomes) Ask students what "2n" represents. (two chromosome sets, in this case, 46 chromosomes)

Activity — GENERAL

Asexual Reproduction in Bacteria Bacteria can reproduce asexually, as fast as once every fifteen minutes. Using this rate, have students calculate the number of bacteria produced from a single bacterium after each hour for 6 hours. (After hour 1, 16 bacteria; hour 2, 256 bacteria; hour 3, 4,096 bacteria; hour 4, 65,536 bacteria; hour 5, 1,048,576 bacteria; hour 6, 16,777,216 bacteria) Tell students that at this rate the earth would quickly be covered with bacteria. Ask students why this doesn't happen. (Ideal growth conditions for bacteria seldom occur in nature for very long.) LS **Logical**

QUICK LAB

Observing Reproduction in Yeast

Skills Acquired
Observing, inferring, locating information

Teacher's Notes
Caution students to use care when working with a microscope, glass slides, and living organisms. Review the steps for making a wet-mount slide.

Answers to Analysis

1. asexual reproduction
2. Only one parent was observed. The new organism split off from the parent.
3. budding

Chapter Resource File

- Directed Reading BASIC
- Active Reading GENERAL
- Data Sheet for Quick Lab GENERAL

Transparencies

TR Bellringer
TR F13 Haploid Life Cycle
TR B48 Diploid Life Cycle
TR G5 Alternation of Generations

MISCONCEPTION ALERT — GENERAL

Reproduction and sex are not the same process. Students often think that reproduction and sex are the same thing. Reproduction is the formation of new individuals, whether of unicellular or multicellular organisms. Sex, on the other hand, is the combining of genetic material from two cells.

Teach, *continued*

READING SKILL BUILDER — BASIC

Reading Organizer Point out to students that they will encounter three reading organizers in this section. Ask students if they can spot the reading organizers provided. (They depict the haploid life cycle, the diploid life cycle, and alternation of generations.) Have them explain why these organizers are appropriate graphic representations of the text. **LS Visual**

Using the Figure — GENERAL

Draw students' attention to **Figure 6.** Use this figure to discuss the steps involved in a haploid life cycle. Be sure students know the difference between a diploid cell and a haploid cell. Have students answer the following questions. What is the diploid stage in the life cycle? (the zygote) What is the result of meiosis? (haploid cells) Why is this type of life cycle called a haploid life cycle? (The haploid cells occupy the major portion of the life cycle.) What event determines when the organism enters into the haploid stage of its life cycle? (meiosis) What event determines when the organism enters into the diploid stage of its life cycle? (fusion of gametes) What process occurs to produce haploid gametes from haploid cells? (mitosis) What event produces haploid cells from the diploid organism? (meiosis)
LS Visual English Language Learners

Study TIP

Interpreting Graphics
After reading this chapter, trace or make a sketch of Figures 6, 7, and 8 without the labels. On separate pieces of paper, write down the labels. Without referring to your book, match the labels with the correct part of your sketch.

Sexual Life Cycles in Eukaryotes

The entire span in the life of an organism from one generation to the next is called a **life cycle.** The life cycles of all sexually reproducing organisms follow a basic pattern of alternation between the diploid and haploid chromosome numbers. The type of sexual life cycle that a eukaryotic organism has depends on the type of cell that undergoes meiosis and on when meiosis occurs. Eukaryotes that undergo sexual reproduction can have one of three types of sexual life cycles: haploid, diploid, or alternation of generations.

Haploid Life Cycle

The haploid life cycle is the simplest of sexual life cycles. In this life cycle, shown in **Figure 6,** haploid cells occupy the major portion of the life cycle. The zygote is the only diploid cell, and it undergoes meiosis immediately after it is formed, creating new haploid cells. The haploid cells give rise to haploid multicellular individuals that produce gametes by mitosis (not meiosis). In a process called fusion, the gametes fuse to produce a diploid zygote, and the cycle continues.

When the diploid zygote undergoes meiosis it provides an opportunity for the cell to correct any genetic damage, as discussed earlier. The damage is repaired during meiosis, when the two homologous chromosomes are lined up side-by-side in preparation for crossing over. Special repair enzymes remove any damaged sections of double stranded DNA, and fill in any gaps. This type of life cycle is found in many protists, as well as in some fungi and algae, such as the unicellular *Chlamydomonas (KLUH mih duh moh nuhs),* shown in Figure 6.

Figure 6 Haploid life cycle. Some organisms, such as *Chlamydomonas,* have haploid cells as a major portion of their life cycle.

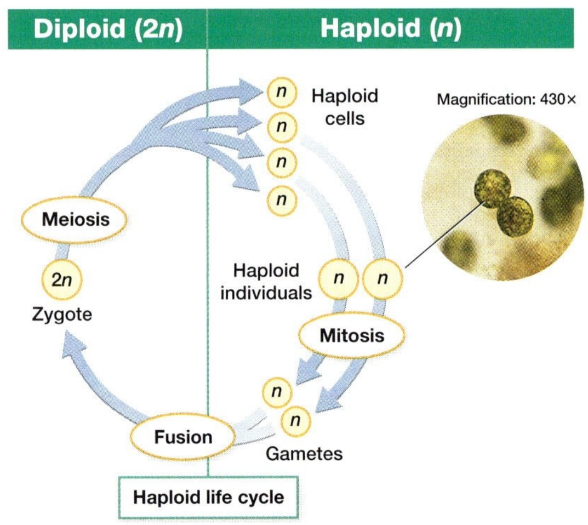

HISTORY CONNECTION

Although Charles Darwin (1809–1882) recognized the importance of genetic variation in natural selection (the mechanism for evolution), he could not explain why offspring resembled but were not identical to their parents. Gregor Mendel (1822–1884) later published his theory of inheritance, which helped explain genetic variation. But Mendel's work was not recognized by biologists until more than 15 years after his death.

Diploid Life Cycle

The outstanding characteristic of the diploid life cycle is that adult individuals are diploid, each individual inheriting chromosomes from two parents. In most animals, including humans, a diploid reproductive cell undergoes meiosis to produce gametes.

As shown in **Figure 7,** the gametes (sperm and egg cells) join in a process called **fertilization,** which results in a diploid zygote. After fertilization, the resulting zygote begins to divide by mitosis. This single diploid cell eventually gives rise to all of the cells of the adult. The cells of the adult are also diploid since they are produced by mitosis.

The diploid individual that develops from the zygote occupies the major portion of the diploid life cycle. The gametes are the only haploid cells in the diploid life cycle; all of the other cells are diploid.

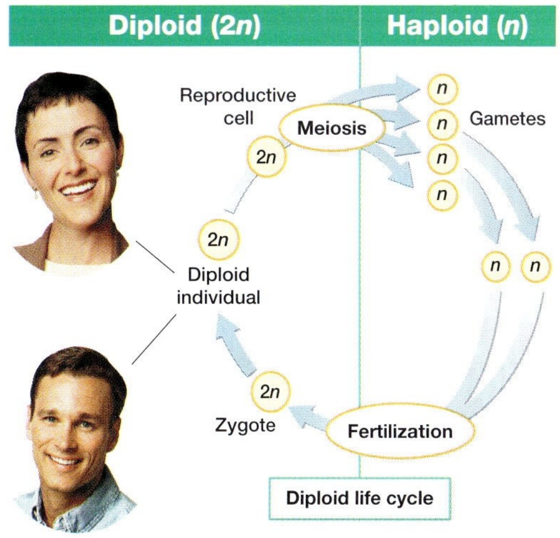

Figure 7 Diploid life cycle. Humans and other organisms have a life cycle dominated by a diploid individual.

Using the Figure — GENERAL

Use **Figure 7** to discuss the steps involved in a diploid life cycle. Ask students to point out the diploid stage in the life cycle. (the zygote, which grows into an individual) Ask students what the result of meiosis is. (haploid cells) Ask students why this type of life cycle is called a diploid life cycle. (The diploid individual occupies the major portion of the life cycle.) English Language Learners
LS Visual

Demonstration — GENERAL

Invite a technician from a fertility clinic to discuss the process of in vitro fertilization. Encourage the speaker to bring equipment or pictures to help students understand the process.

Exploring Further

Cloning by Parthenogenesis

A snake is born to a mother that did not have a mate. Although this may sound impossible, or like some headline in a tabloid magazine, this can actually occur in nature. Parthenogenesis *(pahr thuh noh JEHN uh sihs)* is a type of reproduction in which a new individual develops from an unfertilized egg. Since there is no male that contributes genetic material, the offspring is a clone (genetically identical) of the mother. Clones are usually produced in nature by asexual reproduction. Parthenogenesis, however, is a special form of cloning.

How Does Parthenogenesis Occur?

Parthenogenesis in snakes has usually occurred in older females that have lived many years without male companionship, such as those in a zoo. It is hypothesized that in the mother snake, her own chromosomes are copied in place of the missing father's chromosomes, thereby self-fertilizing her egg. Other scientists think that after a long absence of males, some unknown signal (such as a hormone) triggers the egg to start dividing.

Whiptail lizard

Organisms That Undergo Parthenogenesis

Organisms capable of reproducing by parthenogenesis include dandelions, hawkweeds, and some fishes, lizards, and frogs. Whiptail lizards are all females that lay eggs that hatch without any male contributions. Honeybees also produce male drones by parthenogenesis.

Parthenogenesis is not thought to be possible in mammals. Embryos of mammals that do not have genes from both a female and a male parent do not develop normally. The only natural mammalian clones known are identical twins, which develop when a fertilized egg splits and two individuals develop.

153

Exploring Further — ADVANCED

Cloning by Parthenogenesis

Teaching Strategies

Before students read the article, tell them that parthenogenesis is a type of reproduction in which a new individual develops from an unfertilized egg. Ask students what organisms might exhibit this type of reproduction. (Answers will vary.)

Discussion

Discuss the following questions with students.

- Why do scientists think parthenogenesis is not possible in mammals? (Mammalian embryos that do not have genes from both a female and a male do not develop normally.)
- What do scientists think initiates parthenogenesis? (Some unknown signal, such as a hormone, may trigger the egg to start dividing.)
- What organisms undergo parthenogenesis? (dandelions, hawkweeds, and some fishes, frogs, and lizards)
- What are the only natural mammalian clones? (identical twins)

did you know?

Sexual Variations Parthenogenesis is not the only variation on sexual reproduction that occurs among vertebrates. Among many species of fish, individuals can change their sex. Some begin life as females and then change into males, a phenomenon called protogyny. Others change from males into females, which is known as protandry. Even more unusual, some deep-sea fish are hermaphrodites—both male and female at the same time.

Close

Reteaching — BASIC

Ask students to list an advantage of asexual reproduction (many offspring produced in a short time period) and sexual reproduction (genetic diversity). Have students name the three sexual life cycles and give an example of an organism of each kind. (Haploid life cycle—*Chlamydomonas;* Diploid life cycle—human; Alternation of Generations—moss)

Quiz — GENERAL

1. What does sexual reproduction involve? (the union of gametes from two parents)
2. Does sexual or asexual reproduction allow for the greatest genetic variation in offspring? (sexual reproduction)
3. True or false: humans spend most of their life cycle in the diploid state. (true)

Alternative Assessment — GENERAL

Instruct students to make two columns on a piece of paper. Have students head each column with one of the two types of reproduction. Have students list at least two advantages and two disadvantages of that particular type of reproduction under each heading. On the back of the sheet, have them make three columns, heading each of the columns with one of the three kinds of sexual life cycles. In each column, have students make a simple diagram of the life cycle named. ELL students may wish to label the diagram in English as well as their native language.

Alternation of Generations

Plants, algae, and some protists have a life cycle that regularly alternates between a haploid phase and a diploid phase. As shown in **Figure 8,** in plants, the diploid phase in the life cycle that produces spores is called a **sporophyte** *(SPOH ruh fiet).* Spore-forming cells in the sporophyte undergo meiosis to produce spores. A **spore** is a haploid reproductive cell produced by meiosis that is capable of developing into an adult without fusing with another cell. Thus, unlike a gamete, a spore gives rise to a multicellular individual called a gametophyte *(guh MEET uh fiet)* without joining with another cell.

In the life cycle of a plant, the **gametophyte** is the haploid phase that produces gametes by mitosis. The gametophyte produces gametes that fuse and give rise to the diploid phase. Thus, the sporophyte and gametophyte generations take turns, or alternate, in the life cycle.

Figure 8 Alternation of generations

Some organisms, such as roses, have a life cycle that alternates between diploid and haploid phases.

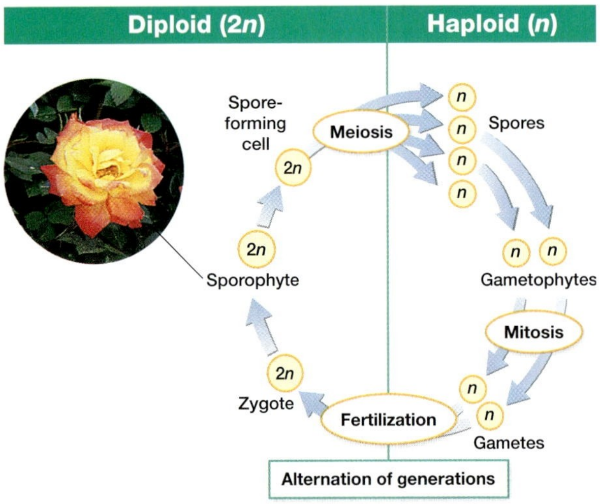

In moss, for example, haploid spores develop in a capsule at the tip of the sporophyte "stalk." When the lid of the capsule pops off, the spores scatter. The spores germinate by mitosis and eventually form sexually mature gametophytes. The male gametophytes release sperm which swim through a film of moisture to the eggs in the female gametophyte. The diploid zygote develops as a sporophyte within the gametophyte and the life cycle continues.

It is important not to lose sight of the basic similarity of all three types of sexual life cycles. All three involve an alternation of haploid and diploid phases. The three types of sexual life cycles differ from each other only in which phases become multicellular.

Section 2 Review

1. **Identify** the type of reproduction that results in offspring that are genetically identical to their parent. SC.F.2.4.1
2. **Describe** two different types of eukaryotic asexual reproduction. SC.F.2.4.1
3. **Compare** the haploid life cycle found in *Chlamydomonas* with a diploid life cycle. SC.F.2.4.1
4. **Summarize** the process of alternation of generations. SC.F.2.4.1
5. **Critical Thinking Evaluating Information** Evaluate the significance of mutations and repair of mutations to the evolution of sexual reproduction. SC.F.2.4.3
6. *FCAT Prep* The amount of genetic variation in offspring is greatest in organisms that reproduce through which of the following processes?
   A sexually through fission
   B sexually through meiosis
   C asexually through mitosis
   D asexually through budding
   SC.F.2.4.1

154

Answers to Section Review

1. asexual
2. *fission*—parent separates into two or more individuals of equal size; *fragmentation*—the body breaks into several pieces; *budding*—new individuals split off from existing ones.
3. The haploid cells of *Chlamydomonas* give rise to haploid multicellular individuals that produce gametes by mitosis. The gametes fuse to form a diploid zygote, which undergoes meiosis, creating new haploid cells. In a diploid life cycle, haploid gametes are produced by meiosis. The gametes join by fertilization to form a diploid zygote.
4. Spore-forming cells in the sporophyte undergo meiosis to produce spores, which develop into gametophytes. The gametophyte produces gametes by mitosis. The gametes fuse and give rise to the sporophyte.
5. Sexual reproduction is believed to have evolved as a way for cells to repair damaged or mutated DNA.
6. **A.** Incorrect. Fission is a type of asexual reproduction. **B.** Correct. **C.** Incorrect. Mitosis produces genetically identical cells. **D.** Incorrect. Budding produces genetically identical individuals.

Section Review Correlations

SC.F.2.4.1: Items 1-4, 6
SC.F.2.4.3: Item 5

ZONE

Key Concepts

1 Meiosis

- Meiosis reduces the number of chromosomes by half to form reproductive cells. When the reproductive cells unite in fertilization, the normal diploid number is restored.
- During meiosis I, homologous chromosomes separate. Crossing-over during prophase I results in the exchange of genetic material between homologous chromosomes.
- During meiosis II, the two chromatids of each chromosome separate. As a result of meiosis, four haploid cells are produced from one diploid cell.
- Independent assortment, crossing-over, and random fertilization contribute to produce genetic variation in sexually reproducing organisms.
- In sexually reproducing eukaryotic organisms, gametes form through the process of spermatogenesis in males and oogenesis in females.

2 Sexual Reproduction

- Asexual reproduction is the formation of offspring from one parent. The offspring are genetically identical to the parent.
- Sexual reproduction is the formation of offspring through the union of gametes. The offspring are genetically different from their parents.
- A disadvantage to asexual reproduction in a changing environment is the lack of genetic diversity among the offspring.
- Sexual reproduction increases variation in the population by making possible genetic recombination.
- Sexual reproduction may have begun as a mechanism to repair damaged DNA.
- Eukaryotic organisms can have one of three kinds of sexual life cycles, depending on the type of cell that undergoes meiosis and on when meiosis occurs.

Key Terms

Section 1

meiosis (144)
crossing-over (144)
independent assortment (146)
spermatogenesis (148)
sperm (148)
oogenesis (148)
ovum (149)

Section 2

asexual reproduction (150)
clone (150)
sexual reproduction (150)
life cycle (152)
fertilization (153)
sporophyte (154)
spore (154)
gametophyte (154)

BIOLOGY INTERACTIVE TUTOR Unit 4—*Cell Reproduction*
Use Topics 5–6 in this unit to review the key concepts and terms in this chapter.

Alternative Assessment — GENERAL

Have students design a concept map that relates the two forms of cell division to the human life cycle. Be sure they include the terms *haploid*, *diploid*, *mitosis*, *meiosis*, and *gamete*.

Chapter Resource File

- Science Skills Worksheet GENERAL
- Critical Thinking Worksheet ADVANCED
- Test Prep Pretest GENERAL
- Chapter Test GENERAL

Answers to Concept Map

The following is one possible answer to Performance Zone item 15.

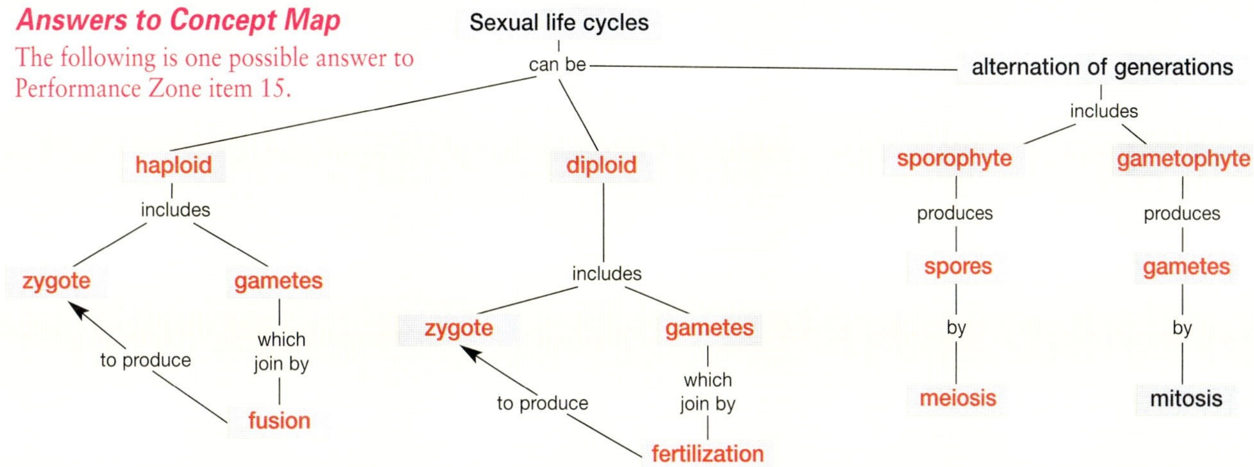

ANSWERS

Understanding Key Ideas

1. a
2. b
3. a
4. b
5. a
6. ABcdE; ABcde; aBCDE; aBCDe
7. Mitosis produces two genetically identical diploid cells. Meiosis produces four haploid cells that are not genetically identical to the original cell. In metaphase I of meiosis, homologous chromosomes pair up along the equator. In metaphase of mitosis, homologous chromosomes line up independently along the equator. Crossing-over occurs in meiosis but not mitosis. The stages of meiosis II are identical to mitosis.
8. Nurse cells donate organelles to a growing egg cell.
9. Proposed hypotheses include copying of egg chromosomes (self-fertilization) and an unknown signal (such as a hormone) triggers eggs to start dividing in the absence of males. The strength of the self-fertilization hypothesis is the fact that parthenogenesis has been observed in female snakes confined in zoos; a weakness of this hypothesis is that parthenogenesis also occurs in some snakes in the wild. A strength of the hormone hypothesis is the fact that we know that many complex reproductive cycles are controlled by hormones; a weakness is that no "parthenogenesis hormone" has ever been isolated. Student answers will vary.

Chapter Review Correlations

SC.D.1.4.4: Item 13
SC.F.1.4.2: Item 8
SC.F.1.4.5: Items 1, 3, 7
SC.F.2.4.1: Items 4, 10
SC.F.2.4.2: Items 2, 5, 6
SC.F.2.4.3: Items 11, 12
SC.H.1.4.2: Item 9
SC.H.1.4.5: Item 14
SC.H.3.4.5: Item 15

Performance ZONE

Understanding Key Ideas

1. Which of the following events occurs during prophase I of meiosis?
   a. crossing-over
   b. duplication of chromatids
   c. reduction in chromosome number
   d. separation of chromatids to opposite poles
   SC.F.1.4.5
2. Homologous pairs of chromosomes move to opposite poles during
   a. prophase I. **c.** metaphase II.
   b. anaphase I. **d.** anaphase II.
   SC.F.2.4.2
3. Spermatogenesis produces
   a. four haploid cells.
   b. four diploid cells.
   c. four polar bodies.
   d. two haploid cells.
   SC.F.1.4.5
4. Sexual reproduction may have originated as a way for cells to
   a. shuffle genetic material.
   b. repair damaged DNA.
   c. produce diploid individuals.
   d. increase their population growth at a maximum rate.
   SC.F.2.4.1
5. In plants, the sporophyte generation produces ______ spores through meiosis.
   a. haploid
   b. triploid
   c. diploid
   d. mutated
   SC.F.2.4.2
6. After crossing-over as shown below, what would the sequence of genes be for each of the chromatids? SC.F.2.4.2

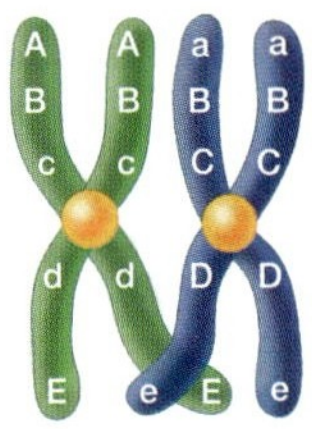

7. Compare and contrast the processes of mitosis and meiosis. (**Hint:** See Chapter 6, Section 4.) SC.F.1.4.5
8. **BIOWatch** What is the function of a nurse cell in egg formation? SC.F.1.4.2
9. **Exploring Further** Review two hypotheses that have been proposed to explain parthenogenesis according to the strengths and weaknesses of those hypotheses. Which hypothesis do you think is more supportable, and why? SC.H.1.4.2
10. **Concept Mapping** Make a concept map that shows the three sexual life cycles in eukaryotic organisms. Include the following words in your map: *meiosis, fusion, gametes, spores, fertilization, zygote, gametophyte, sporophyte, haploid,* and *diploid*. SC.F.2.4.1

Critical Thinking

11. **Evaluating Results** Occasionally homologous chromosomes fail to separate during meiosis I. Using the hypothetical example of an adult organism that has two pairs of chromosomes, describe the chromosomal makeup of the eggs that would result from this error in meiosis. SC.F.2.4.3
12. **Evaluating Results** If normal human sperm fertilized the eggs described above, what would the chromosomal makeup of the resulting zygote be? SC.F.2.4.3
13. **Applying Information** How do independent assortment, crossing-over, and random fertilization affect the rate of evolution? SC.D.1.4.4
14. **Critiquing Hypotheses** A student states that organisms that reproduce asexually are at a disadvantage in a stable environment. If you agree with this hypotheses, name one or more of its strengths. If you disagree, name one or more of its weaknesses. SC.H.1.4.5

Alternative Assessment

15. **Interactive Tutor Unit 4** Cell Reproduction
   Write a report summarizing the effects of various treatments for infertility. Find out how the production of gametes may be affected in some people who are infertile. SC.H.3.4.5

156

10. One possible answer to the concept map is found at the bottom of the Study Zone page.

Critical Thinking

11. If one of the pairs of chromosomes failed to separate, eggs would have either three chromosomes or one chromosome.
12. Zygotes would be diploid.
13. They increase the rate of evolution by increasing the level of genetic variation.
14. Most students should disagree. In a stable environment, asexual reproduction allows organisms to produce offspring without using energy to produce gametes or to find a mate.

Alternative Assessment

15. Answers will vary but might include summaries of one or more of the following types of treatments for infertility: in vitro fertilization (IVF); donor egg IVF; artificial insemination; tubal embryo transfer (TET); gamete intrafallopian transfer (GIFT); and drug treatments (such as clomiphine citrate, pergonal, and metrodin).

Assignment Guide

Section	Questions
1	1, 2, 3, 6, 7, 8, 11, 12, 15
2	4, 5, 9, 10, 14

FCAT Prep

For questions 1–6, write your answers on a separate sheet of paper.

1. How do most multicellular eukaryotes form specialized reproductive cells?
   A. binary fission **C.** meiosis
   B. fragmentation **D.** mitosis
   SC.F.1.4.5
2. How does meiosis differ from mitosis?
   F. Mitosis includes cytokinesis, while meiosis does not.
   G. The DNA is replicated before mitosis but not before meiosis.
   H. Meiosis produces haploid cells; mitosis produces diploid cells.
   I. Mitosis produces gametes; meiosis produces daughter cells.
   SC.F.1.4.5
3. Plants experience alternation of generations with a sporophyte phase and a gametophyte phase. How are sporophytes and gametophytes different in the life cycle of a plant?
   A. Sporophytes are diploid, while gametophytes are haploid.
   B. Sporophytes undergo mitosis, while gametophytes undergo meiosis.
   C. Sporophytes come from spores, while gametophytes come from gametes.
   D. Sporophytes produce diploid cells, while gametophytes produce haploid cells.
   SC.F.2.4.1
4. When are genes exchanged between homologous chromosomes?
   F. during crossing-over **H.** during meiosis
   G. during fertilization **I.** during telophase II
   SC.F.2.4.1

Base your answers to questions 5 and 6 on the diagram below, which shows how an egg and sperm unite to form a zygote with a complete set of genes.

HUMAN REPRODUCTION

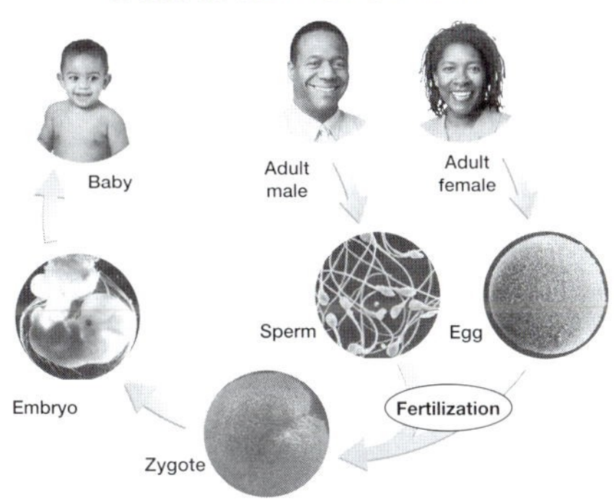

5. Which type of eukaryotic life cycle does this diagram represent?
   A. asexual **C.** diploid
   B. cloning **D.** haploid
   SC.F.2.4.1
6. If the sperm and egg in this diagram each have 23 chromosomes, how many chromosomes does the fertilized egg have? SC.F.2.4.2

Test TIP

When using a diagram to answer questions, carefully study each part of the figure as well as any lines or labels used to indicate parts of the figure.

FCAT Doctor

Question 2 Answer H is the correct choice. Meiosis of a diploid cell produces four haploid cells. Mitosis of a diploid cell produces two diploid cells. Answer F is incorrect because both meiosis and mitosis include cytokinesis. Answer G is incorrect because DNA is replicated before both processes. Answer I is incorrect because meiosis produces gametes, while mitosis produces daughter cells.

Question 5 Answer C is the correct choice. Humans have a diploid life cycle. Answer A is incorrect because asexual reproduction does not involve fusion of haploid cells. Answer B is incorrect because cloning is a form of asexual reproduction and does not involve fusion of haploid cells. Answer D is incorrect because, in a haploid life cycle, the adult is haploid and produces gametes through mitosis.

Question 6 The correct answer is 46. Sperm and egg cells are haploid and each has only 23 chromosomes. When the sperm and egg meet and fertilization occurs, the cell that results is diploid. Each diploid cell in the human body has 23 pairs of homologous chromosomes and thus 46 chromosomes in all.

Answers

1. C
2. H
3. A
4. F
5. C
6. 46

Test Prep Correlations

SC.F.1.4.5: Items 1, 2
SC.F.2.4.1: Items 3, 4, 5
SC.F.2.4.2: Item 6

Exploration Lab

Modeling Meiosis

Teacher's Notes

Time Required One 45-minute class period

Ratings

EASY → HARD

Teacher Preparation 1
Student Setup 1
Concept Level 2
Cleanup 1

Skills Acquired

- Communicating
- Constructing Models
- Identifying/Recognizing Patterns
- Interpreting
- Predicting

Scientific Methods

In this lab, students will:

- Make Observations
- Ask Questions
- Test the Hypothesis
- Analyze the Results
- Draw Conclusions
- Communicate the Results

Materials and Equipment

Materials for this lab can be purchased from a local craft store or ordered from WARD'S. See *Master Materials List* at the front of this book for catalog numbers. For each model, students need 4 pipe cleaners of one color and 4 pipe cleaners of another color, 4 wooden beads, 90 cm of yarn, and 16 small white labels.

Safety Cautions

- Review all safety symbols with students before beginning the lab.

Exploration Lab

Modeling Meiosis SC.H.3.4.1

SKILLS

- Modeling
- Using scientific methods

OBJECTIVES

- **Describe** the events that occur in each stage of the process of meiosis.
- **Relate** the process of meiosis to genetic variation.

MATERIALS

- pipe cleaners of at least two different colors
- yarn
- wooden beads
- white labels
- scissors

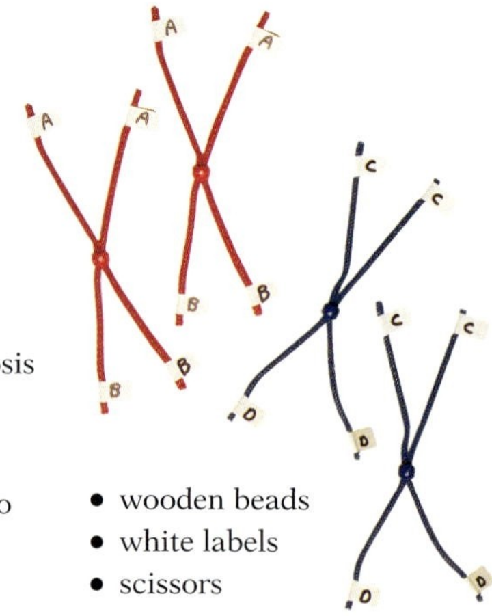

Glass frog with eggs

Before You Begin

Meiosis is the process that results in the production of cells with half the normal number of chromosomes. It occurs in all organisms that undergo **sexual reproduction**. In this lab, you will build a model that will help you understand the events of meiosis. You can also use the model to demonstrate the effects of events such as **crossing-over** to explain results such as **genetic recombination**.

1. Write a definition for each boldface term in the paragraph above and for the following terms: homologous chromosomes, gamete.
2. In what organs in the human body do cells undergo meiosis?
3. During interphase of the cell cycle, how does a cell prepare for dividing?
4. Based on the objectives for this lab, write a question you would like to explore about meiosis.

Procedure

PART A: Design a Model

1. Work with the members of your lab group to design a model of a cell using the materials listed for this lab. Be sure that your model cell has at least two pairs of chromosomes.
2. Write out the plan for building your model. Have your teacher approve the plan before you begin building the model.

> **You Choose**
>
> **As you design your experiment, decide the following:**
>
> **a.** what question you will explore
> **b.** how to construct a cell membrane
> **c.** how to show that your cell is diploid
> **d.** how to show the locations of at least two genes on each chromosome
> **e.** how to show that chromosomes are duplicated before meiosis begins

3. Build the cell model your group designed. **CAUTION: Sharp or pointed objects can cause injury. Handle scissors carefully.** Use your model to demonstrate the phases of meiosis. Draw and label each phase you model.
4. Use your model to explore one of the questions written by your group for step 4 of **Before You Begin.** Describe the steps you took to explore your question.

158

Tips and Tricks

Review the stages of meiosis before beginning the lab. If students have difficulty in building their models, ask the following questions:

1. Which of these materials would make the best cell membrane? (yarn)
2. Which would make the best spindle fibers? (yarn)
3. Which would make the best chromosomes? (pipe cleaners)
4. Which would make the best centromeres? (beads)

Emphasize that each pair of chromatids represents one chromosome. Be sure students show that homologous chromosomes pair during metaphase I of meiosis.

PART B: Test Hypotheses

Answer each of the following questions by writing a hypothesis. Use your model to test each hypothesis, and describe your results.

5. In humans, gametes (eggs and sperm) result from meiosis. Will all gametes produced by one parent be identical?
6. When an egg and a sperm fuse during sexual reproduction, the resulting cell (the first cell of a new organism) is called a zygote. How many copies of each chromosome and each gene will be found in a zygote?
7. Crossing-over frequently occurs between the chromatids of homologous chromosomes during meiosis. Under what circumstances does crossing-over result in new combinations of genes in gametes?
8. Synapsis (the pairing of homologous chromosomes) must occur before crossing-over can take place. How would the outcome of meiosis be different if synapsis did not occur?

PART C: Cleanup and Disposal

9. Dispose of paper and yarn scraps in the designated waste container.
10. Clean up your work area and all lab equipment. Return lab equipment to its proper place. Wash your hands thoroughly before you leave the lab and after finishing all work.

Analyze and Conclude

1. **Analyzing Results** How do the nuclei you made by modeling meiosis compare with the nucleus of the cell you started with? Explain your result.
2. **Recognizing Relationships** How are homologous chromosomes different from chromatids?
3. **Forming Reasoned Opinions** How is synapsis important to the outcome of meiosis? Explain.
4. **Evaluating Methods** How could you modify your model to better illustrate the process of meiosis?
5. **Drawing Conclusions** How are the processes of meiosis similar to those of mitosis? How are they different?
6. **Predicting Outcomes** What would happen to the chromosome number of an organism's offspring if the gametes for sexual reproduction were made by mitosis instead of by meiosis?
7. **Further Inquiry** Write a new question about meiosis or sexual reproduction that could be explored with your model.

Do You Know?

Do research in the library or media center to answer these questions:

1. What types of human abnormalities arise when chromosomes do not separate properly during meiosis?
2. How do chemicals such as nicotine affect meiosis?

Use the following Internet resources to explore your own questions about meiosis and gamete formation.

internet connect
www.scilinks.org
Topic: Meiosis
Keyword: HX4120
SCI LINKS. Maintained by the National Science Teachers Association

Answers to Before You Begin

1. See the Glossary in the Appendix of this textbook for the definitions of the listed terms. Homologous chromosomes are chromosomes that are the same size and shape and carry genes for the same traits. A gamete is a reproductive cell, either a sperm or an egg.
2. ovaries and testes
3. A cell's chromosomes duplicate, and certain organelles replicate.
4. Answers will vary. For example: How many chromosomes will each new nucleus have after meiosis has occurred?

Answers to Procedure

2. Plans will vary. For example, students may use the yarn to represent the cell membrane and spindle fibers, the pipe cleaners to represent the chromosomes, and the beads to represent the centromeres.
3. Drawings will vary.
4. Answers will vary.
5. Hypotheses will vary. Students should find that the gametes will be identical only if the parent is homozygous for every one of its genetic traits.
6. Hypotheses will vary. Students should find two copies of each chromosome and two copies of each gene in a zygote.
7. Hypotheses will vary. Students should find that crossing-over can produce new combinations of genes when an organism has different versions of the genes on the parts that cross over.
8. Hypotheses will vary. Students may find the wrong chromosome number in a gamete if the homologous pairs do not separate properly during anaphase I. Also, there would be less genetic variation.

Answers to Analyze and Conclude

1. The nuclei made by meiosis have half the original chromosome number. Two divisions of the nuclear material occur.
2. Homologous chromosomes are the same size and have genes for the same traits, but they are not identical, as are chromatids.
3. Synapsis ensures that each new cell will get one member of each pair of homologous chromosomes.
4. Answers will vary. Students may suggest using different materials.
5. Mitosis and meiosis are both forms of nuclear division. Both consist of prophase, metaphase, anaphase, and telophase. Mitosis consists of one division and results in two cells that are genetically the same and have the same number of chromosomes as the original cell. Meiosis consists of two divisions and results in four cells that each have half the number of chromosomes as the original cell and are not the same genetically.
6. The chromosome number would be twice that in its parents' cells.
7. Answers will vary. For example: How do mitosis and meiosis compare?

Answers to Do You Know?

1. Answers will vary but might include Down syndrome, Klinefelter syndrome, and Turner syndrome.
2. Chemicals can affect the ability of chromosomes to separate normally during meiosis.

Chapter Opener		SE **Quick Review,** p. 161 ■ SE **Reading Activity,** p. 161 ■ TE **Using the Figure,** p. 160 GENERAL TE **Opening Activity,** p. 160 GENERAL
pp. 162–165 **on 1** rigins of Genetics	OSP **Lesson Plan*** TR **Bellringer*** TR **Section Outline*** TR **C1 Three Steps of Mendel's Experiment***	TE **Demonstration,** p. 162 BASIC TE **Group Activity** Benefits of Peas, p. 163 ADVANCED TE **Using the Figure,** p. 164 GENERAL SE **Math Lab** Calculating Mendel's Ratios, p. 165 ■ CRF **Datasheets for In-Text Labs** Calculating Mendel's Ratios*
& 5 • 90 min pp. 166–169 **Section 2** Mendel's Theory	OSP **Lesson Plan*** TR **Bellringer*** TR **Section Outline*** TR **C5 Monohybrid Cross of Homozygous Plants*** TR **C6 Monohybrid Cross of Heterozygous Plants*** TR **C3 Mendel's Factors***	TE **Using the Figure,** p. 167 SE **Quick Lab** Identifying Dominant or Recessive Traits, p. 168 ■ CRF **Datasheets for In-Text Labs** Identifying Dominant or Recessive Traits* TE **Activity** Graphic Organizer, p. 168 GENERAL TE **Activity** Hairy Knuckles, p. 169 GENERAL CRF **Inquiry Lab** Corn Seed Germination* ◆ GENERAL
& 7 • 90 min pp. 170–176 **Section 3** Studying Heredity	OSP **Lesson Plan*** TR **Bellringer*** TR **Section Outline*** TR **C4 Probability with Two Coins***	TE **Using the Figure,** p. 171 BASIC SE **Data Lab** Analyzing a Test Cross, p. 172 ■ CRF **Datasheets for In-Text Labs** Analyzing a Test Cross* TE **Demonstration,** p. 173 GENERAL TE **Using the Figure,** p. 174 and p. 175 GENERAL SE **Math Lab** Predicting the Results of Crosses Using Probabilities, p. 174 ■ CRF **Datasheets for In-Text Labs** Predicting the Results of Crosses Using Probabilities* SE **Data Lab** Evaluating a Pedigree, p. 176 ■ CRF **Datasheets for In-Text Labs** Evaluating a Pedigree*
& 9 • 90 min pp. 177–182 **Section 4** Complex Patterns of Heredity	OSP **Lesson Plan*** TR **Bellringer*** TR **Section Outline*** TR **C30 Some Human Genetic Disorders***	TE **Demonstration,** p. 178 GENERAL TE **Demonstration,** p. 179 BASIC TE **Using the Figure,** p. 179 GENERAL TE **Activity** Using Punnet Squares, p. 179 GENERAL TE **Demonstration,** p. 179 BASIC TE **Group Activity** Patterns of Heredity, p. 181 GENERAL CRF **Quick Lab** Interpreting Information in a Pedigree* BASIC SE **Skills Practice Lab** Modeling Monohybrid Crosses, p. 186* ■ CRF **Datasheets for In-Text Labs** Modeling Monohybrid Crosses*

0 & 11 • 90 min **Chapter Review and Assessment Resources**

SE **Study Zone,** p. 183 ■
SE **Performance Zone,** p. 184 ■
SE **FCAT Test Prep,** p. 185 ■
TR **Concept Mapping***
CRF **Vocabulary Review Worksheet*** ■ GENERAL
CRF **Test Prep Pretest*** ■ GENERAL
CRF **Chapter Test*** ■ GENERAL
CRF **Chapter Test*** ADVANCED
OSP **Test Generator***
OSP **Test Item Listing***
OSP **Modified Chapter Test*** BASIC

Holt Online Learning

Visit go.hrw.com for access to Holt Online Learning or enter the keyword **HX6 Home** for a variety of free online resources.

One-Stop Planner® CD-ROM

This CD-ROM package includes
- Lab Materials QuickList Software
- Holt Calendar Planner
- Customizable Lesson Plans
- Printable Worksheets
- ExamView® Test Generator
- Interactive Teacher Edition
- Holt PuzzlePro® Resources
- Holt PowerPoint® Resources

Guided Reading Audio CDs

These CDs are designed to help auditory learners and reluctant readers. (Audio CDs are also available in Spanish.)

Compression guide:
To shorten your instruction because of time limitations, omit blocks 1, 2, and 3.

KEY

- TE Teacher Edition
- SE Student Edition
- OSP One-Stop Planner
- CRF Chapter Resource File
- TR Teaching Transparency
- * Also on One-Stop Planner
- ■ Also Available in Spanish
- ◆ Requires Advance Prep

SKILLS DEVELOPMENT AND PRACTICE	SECTION REVIEW AND ASSESSMENT	CORRELATIONS
		Florida Sunshine Standards/FCAT
TE **Skill Builder** Math, p. 164 GENERAL TE **Reading Skill Builder** K-W-L, p. 163 GENERAL CRF **Directed Reading Worksheet*** BASIC CRF **Active Reading Worksheet*** GENERAL	SE **Section 1 Review** ■ TE **Reteaching,** p. 165 BASIC TE **Quiz,** p. 165 GENERAL TE **Alternative Assessment,** p. 165 GENERAL CRF **Quiz*** ■ BASIC	SC.H.1.4.1, SC.H.1.4.2
CRF **Directed Reading Worksheet*** BASIC CRF **Active Reading Worksheet*** GENERAL	SE **Section 2 Review** ■ TE **Reteaching,** p. 169 BASIC TE **Quiz,** p. 169 GENERAL TE **Alternative Assessment,** p. 169 GENERAL CRF **Quiz*** ■ BASIC	SC.F.2.4.1, SC.H.1.4.1, SC.H.2.4.1
TE **Skill Builder** Writing, p. 174 ADVANCED TE **Skill Builder** Vocabulary, p. 175 GENERAL CRF **Directed Reading Worksheet*** BASIC CRF **Active Reading Worksheet*** GENERAL	SE **Section 3 Review** ■ TE **Reteaching,** p. 176 BASIC TE **Quiz,** p. 176 GENERAL TE **Alternative Assessment,** p. 176 ADVANCED CRF **Quiz*** ■ BASIC	SC.F.2.4.1, SC.H.3.4.1
TE **Reading Skill Builder** Discussion, p. 179 GENERAL CRF **Directed Reading Worksheet*** BASIC CRF **Active Reading Worksheet*** GENERAL CRF **Critical Thinking Worksheet*** ADVANCED CRF **Science Skills Worksheet*** GENERAL	SE **Section 4 Review** ■ TE **Reteaching,** p. 182 BASIC TE **Quiz,** p. 182 GENERAL TE **Alternative Assessment,** p. 182 ADVANCED CRF **Quiz*** ■ BASIC **Modified Worksheet*** BASIC	SC.F.2.4.1, SC.F.2.4.2, SC.F.2.4.3, SC.H.3.4.1, SC.H.3.4.3

Holt Biology Interactive Tutor CD-ROM

This CD-ROM consists of interactive activities that give students a fun way to extend their knowledge of biology concepts.

Unit 5 Heredity

SCILINKS. NSTA

www.scilinks.org

Maintained by the **National Science Teachers Association**

Topic: Genetic Counseling
SciLinks Code: HX4090

Topic: Genetic Disorders
SciLinks Code: HX4091

CHAPTER 8

Overview

Tell students the purpose of this chapter is to explain how genetic traits are passed from one generation to another. The passing on of different genetic traits is fundamental to evolutionary change. The genetic variation provided by the possible combinations of parental genes and by crossing-over allows populations to change, and increases the probability that evolution will occur within a given population. Genetic mutations make evolution possible by introducing random changes into the inherited information that is passed on.

Using the Figure

Ask students to look at the show jumper and ask them if they know why racehorses and show jumpers have been selectively bred. (for speed, strength, and size). Thoroughbred horses have been selectively breed since the 1600's. All thoroughbreds can be traced back to three stallions (the Darley Arabian, the Godolphin Barb and the Byerly Turk). With such a limited gene pool, inbreeding becomes a significant problem. The fertility rate is around 67%, which is much lower than other animals, such as cattle (85%). In addition, the performance of the animals is not greatly improving, even though they are bred for speed.

Assessing Prior Knowledge

Before beginning this chapter, be sure your students can answer the questions in the ***Quick Review*** on the facing page. If they cannot, encourage them to reread the sections indicated.

Show jumper

160

Chapter Correlations — Florida Sunshine Standards/FCAT

SC.F.2.4.1 There are mechanisms of asexual and sexual reproduction and different genetic advantages and disadvantages of asexual and sexual reproduction.

SC.F.2.4.2 Every cell contains a "blueprint" coded in DNA molecules that specify how proteins are assembled to regulate cells.

SC.F.2.4.3 Mechanisms of change (e.g., mutation and natural selection) lead to adaptations in a species and their ability to survive naturally in changing conditions and to increase species diversity.

SC.H.1.4.1 Investigations are conducted to explore new phenomena, to check on previous results, to test how well a theory predicts, and to compare different theories.

SC.H.1.4.2 From time to time, major shifts occur in the scientific view of how the world works, but more often the changes that take place in the body of scientific knowledge are small modifications of prior knowledge.

SC.H.2.4.1 Scientists assume that the universe is a vast system in which basic rules exist that may range from very simple to extremely complex, but scientists operate on the belief that the rules can be discovered by careful, systemic study.

SC.H.3.4.1 Performance testing is often conducted using small-scale models, computer simulations, or analogous systems to reduce the chance of system failure.

SC.H.3.4.3 Scientists can bring information, insights, and analytical skills to matters of public concern and help people understand the possible causes and effects of events.

CHAPTER

8 Mendel and Heredity

Florida Sunshine Standards/FCAT

SC.F.2.4.1 There are mechanisms of asexual and sexual reproduction and different genetic advantages and disadvantages of asexual and sexual reproduction.
SC.F.2.4.2 Every cell contains a "blueprint" coded in DNA molecules that specify how proteins are assembled to regulate cells.
SC.F.2.4.3 Mechanisms of change (e.g., mutation and natural selection) lead to adaptations in a species and their ability to survive naturally in changing conditions and to increase species diversity.
SC.H.1.4.1 Investigations are conducted to explore new phenomena, to check on previous results, to test how well a theory predicts, and to compare different theories.
SC.H.1.4.2 From time to time, major shifts occur in the scientific view of how the world works, but more often the changes that take place in the body of scientific knowledge are small modifications of prior knowledge.
SC.H.2.4.1 Scientists assume that the universe is a vast system in which basic rules exist that may range from very simple to extremely complex, but scientists operate on the belief that the rules can be discovered by careful, systemic study.
SC.H.3.4.1 Performance testing is often conducted using small-scale models, computer simulations, or analogous systems to reduce the chance of system failure.
SC.H.3.4.3 Scientists can bring information, insights, and analytical skills to matters of public concern and help people understand the possible causes and effects of events.

Quick Review

Answer the following without referring to earlier sections of your book.

1. **Define** the term *gamete*. *(Chapter 6, Section 1)*
2. **Summarize** the relationship between chromosomes and genes. *(Chapter 6, Section 1)*
3. **Differentiate** between autosomes and sex chromosomes. *(Chapter 6, Section 1)*
4. **Describe** how independent assortment during meiosis contributes to genetic variation. *(Chapter 7, Section 1)*

Did you have difficulty? *For help, review the chapters indicated.*

Reading Activity

Before you read this chapter, write a short list of all the things you know about inheritance. Then write a list of the things that you want to know about inheritance. Save your list, and to assess what you have learned, see how many questions you can answer after reading this chapter.

Horses like this show jumper are bred for certain characteristics, such as speed and agility. Most characteristics are inherited—passed down from parents to offspring.

Looking Ahead

internet connect

www.scilinks.org
National Science Teachers Association *sci*LINKS Internet resources are located throughout this chapter.

SCILINKS. Maintained by the National Science Teachers Association

Opening Activity — GENERAL

Physical Resemblance in Families

Have students volunteer to bring photographs of parents or siblings to class. Ask them to try and match the names of fellow students with each photograph. After the students have matched the photographs, show each picture individually and ask the students whose relative is pictured to stand. Ask students to identify how the two are similar and different. Then ask them what techniques they think scientists might use to find a family connection. Be sensitive to those students who may be adopted.
LS Interpersonal

Quick Review Answers

1. A gamete is a haploid cell that participates in fertilization by fusing with another haploid cell.
2. A gene is a segment of DNA that codes for a protein or RNA molecule, while a chromosome is a structure on which genes are located and which is composed of DNA and associated proteins.
3. A sex chromosome is one of a pair of chromosomes that is involved in determining the sex of an individual, while autosomes are chromosomes not directly involved in determining the sex of an individual.
4. Independent assortment during meiosis contributes to genetic variation because each homologous chromosome pair of an individual segregates (separates) independently of every other pair, the number of combinations is 2^n; where n = the haploid chromosome number.

Chapter Resource File

- Vocabulary Worksheets
- Concept Mapping

Reading Activity Answers

Students may know that phenotypic traits such as hair and eye color are inherited. Some students may be aware that all instructions for the construction of the body are inherited. Students may want to know how these traits are inherited.

Section 1

Focus

Overview

Before beginning this section review with your students the objectives listed in the Student Edition. This section explains Mendel's discoveries in modern terms and explains traits expressed as ratios.

Bellringer

Ask students to list on paper five characteristics that are passed on in families (eye, hair and skin color, height, and so on), and to name one characteristic that may also be inherited but that is also influenced by behavior or environment (muscle size, body weight, having a sun tan, and so on). **LS Intrapersonal**

Motivate

Demonstration — BASIC

Display large pictures of a few flowering plants or bring in real plants. Ask students to come up with a list of traits that could be inherited in plants. Encourage students to think of many different traits, such as flower shape, flower color, flower position on stem, leaf shape, leaf color, pattern of veins, pattern of stem growth, presence of hairs on stems, and inner structure of flower. Ask them if they think the traits are inherited together or separately.
LS Visual

Section 1 The Origins of Genetics

Florida Sunshine Standards/FCAT

SC.H.1.4.1 Investigations are conducted to explore new phenomena, to check on previous results, to test how well a theory predicts, and to compare different theories.

SC.H.1.4.2 From time to time, major shifts occur in the scientific view of how the world works, but more often the changes that take place in the body of scientific knowledge are small modifications of prior knowledge.

Objectives

- **Identify** the investigator whose studies formed the basis of modern genetics. SC.H.1.4.2
- **List** characteristics that make the garden pea a good subject for genetic study. SC.H.1.4.1
- **Summarize** the three major steps of Gregor Mendel's garden pea experiments. SC.H.1.4.1
- **Relate** the ratios that Mendel observed in his crosses to his data. SC.H.1.4.1

Key Terms

heredity
genetics
monohybrid cross
true-breeding
P generation
F_1 generation
F_2 generation

Mendel's Studies of Characters

Many of your characteristics—or *characters*—including the color and shape of your eyes and the texture of your hair resemble those of your parents. The passing of characters from parents to offspring is called **heredity.** From the beginning of recorded history, humans have attempted to alter crop plants and domestic animals to give them traits that are more useful to us. Before DNA and chromosomes were discovered, heredity was one of the greatest mysteries of science.

Mendel's Breeding Experiments

The scientific study of heredity began more than a century ago with the work of an Austrian monk named Gregor Johann Mendel, shown in **Figure 1.** Mendel carried out experiments in which he bred different varieties of the garden pea *Pisum sativum,* shown in **Figure 2** and in **Table 1.** British farmers had performed similar breeding experiments more than 200 years earlier. But Mendel was the first to develop rules that accurately predict patterns of heredity. The patterns that Mendel discovered form the basis of **genetics,** the branch of biology that focuses on heredity.

Mendel's parents were peasants, so he learned much about agriculture. This knowledge became invaluable later in his life. As a young man, Mendel studied theology and was ordained as a priest. Three years after being ordained, he went to the University of Vienna to study science and mathematics. There he learned how to study science through experimentation and how to use mathematics to explain natural phenomena.

Figure 1 Gregor Mendel. Mendel's experiments with garden peas led to our modern understanding of heredity.

Mendel later repeated the experiments of a British farmer, T. A. Knight. Knight had crossed a variety of the garden pea that had purple flowers with a variety that had white flowers. (The term *cross* refers to the mating or breeding of two individuals.) All of the offspring of Knight's crosses had purple flowers. However, when two of the purple-flowered offspring were crossed, their offspring showed both white and purple flowers. The white trait had reappeared in the second generation!

Mendel's experiments differed from Knight's because Mendel counted the number of each kind of offspring and analyzed the data.

162

Chapter Resource File

- Directed Reading BASIC
- Active Reading GENERAL
- Data Sheet for Math Lab GENERAL

Transparencies

TR Bellringer

TR C1 Three Steps of Mendel's Experiment

One-Stop Planner CD-ROM

- Reading Organizers BASIC
- Reading Strategies BASIC
- Problem Solving Worksheet Ratios and Proportions GENERAL

Section Correlations

SC.H.1.4.1
SC.H.1.4.2

Figure 2 **Pollen transfer in Mendel's experiments**

To cross-pollinate flowers of different colors, Mendel first removed the stamens—the pollen-producing structures—from one flower.

Mendel transferred pollen from a second flower to the pistil of the original flower.

Useful Features in Peas

The garden pea is a good subject for studying heredity for several reasons. **Table 1** shows the seven characters that Mendel chose to study.

1. Several characters of the garden pea exist in two clearly different forms. For example, the flower color is either purple or white—there are no intermediate forms. Note that the term *character* is used to mean inherited characteristic, such as flower color. *Trait* refers to a single form of a character—having purple flowers is a trait.
2. The male and female reproductive parts of garden peas are enclosed within the same flower. You can control mating by allowing a flower to fertilize itself (self-fertilization), or you can transfer the pollen to another flower on a different plant (cross-pollination). To cross-pollinate two pea plants, Mendel removed the stamens (the male reproductive organs that produce pollen) from the flower of one plant. As shown in Figure 2, he then dusted the pistil (the female reproductive organ that produces eggs) of that plant with pollen from a different pea plant.
3. The garden pea is small, grows easily, matures quickly, and produces many offspring. Thus, results can be obtained quickly, and there are plenty of subjects to count.

Table 1 The Seven Characters Mendel Studied and Their Contrasting Traits

Flower color	Seed color	Seed shape	Pod color	Pod shape	Flower position	Plant height

163

Teach

READING SKILL BUILDER — GENERAL

K-W-L Before they read this chapter, have each student write a short list of all the things they already **K**now (or think they know) about inheritance. Ask them to contribute their entries to a group list on the board or overhead projector. Then have the students list things they **W**ant to know about inheritance. Have students save their lists for later use in section 4.

Teaching Tip — BASIC

Genetic make-up Bring photos or stuffed toys of animals with different traits. Use these props to emphasize that many genes are involved in giving an animal its overall appearance, and that the genes for most traits have two or more versions. Ask them if they can estimate how many genes animals have in common with each other. For example, chimpanzees and humans share approximately 98% of their genetic makeup. **LS Visual**

Group Activity — ADVANCED

Benefits of Peas Divide the class into small groups. Have each group design newspaper ads that would have attracted someone like Mendel to purchase peas for genetic research. The ads should mention all of the benefits of *Pisum sativum* that make it useful for genetic research. Ask students to use illustrations in their ads. Encourage students to be creative. They may use butcher paper, computer paper, construction paper, and so on. Post the ads on the bulletin board and lead a discussion on the benefits of the garden-pea for genetic research. **English Language Learners**

Trends in Genomics

Cats and Humans Researchers working on the genomes of organisms have found that when it comes to the arrangement of genes on our chromosomes, we are closer to cats than to any other groups studied so far except for primates. Stephen J. O'Brien, a geneticist and chief of the National Cancer Institute's Laboratory of Genomic Diversity, began studying the genetics of the house cat in the 1970's. The Cat Genome Project is a comprehensive genetic analysis of *Felis catus*. The results of this research have proved useful in boosting human AIDS research and have been useful in criminal forensics.

INTERACTIVE BIOLOGY TUTOR

- **Unit 5—Heredity: Introduction**

This engaging tutorial introduces students to principles and practical applications of Mendelian genetics.

Teach, *continued*

Teaching Tip — GENERAL

Hidden traits Ask students if they can tell by looking at the purple pea flowers in **Figure 3** which ones are true-breeding for the purple trait and which ones are not. Point out that you cannot always tell the genetic makeup of an organism by looking at it. Ask students how a cross helps determine if a plant is true-breeding for a trait.

Using the Figure — GENERAL

Point out to students the male and female flower structures illustrated in **Figure 2,** (pistil on the left, stamen on the right). Explain the difference between cross-pollination and self-pollination, and the significance of removing the stamens from the flower on the left. (In cross-pollination, the pollen from one flower is transported to the female structures of a different flower. In self-pollination, the pollen from one flower is transported to the female structures of the same flower. By removing the male stamens of a flower, all offspring will be the result of cross-pollination.)

SKILL BUILDER — GENERAL

Math Skills Ask students to practice reducing ratios to their simplest forms. Survey the class for some numbers to work with. For example, ask how many students own a cat. Have them divide each number (class size; cat owners) by the smallest number (cat owners) and write it as a ratio. If there are 30 students in class and 10 own a cat, the ratio is 30 to 10. Simplified, $\frac{30}{10} = 3$ and $\frac{10}{10} = 1$, the ratio is 3:1. **LS Logical**

Traits Expressed as Simple Ratios

Mendel's initial experiments were monohybrid crosses. A **monohybrid cross** is a cross that involves *one* pair of contrasting traits. For example, crossing a plant with purple flowers and a plant with white flowers is a monohybrid cross. Mendel carried out his experiments in three steps, as summarized in **Figure 3.**

Step 1 Mendel allowed each variety of garden pea to self-pollinate for several generations. This ensured that each variety was **true-breeding** for a particular character; that is, all the offspring would display only one form of the character. For example, a true-breeding purple-flowering plant should produce only plants with purple flowers in subsequent generations.

These true-breeding plants served as the parental generation in Mendel's experiments. The parental generation, or **P generation,** are the first two individuals that are crossed in a breeding experiment.

WORD Origins

The word *filial* is from the Latin *filialis,* meaning "of a son or daughter." Thus F (filial) generations are all those generations that follow a P (parental) generation.

Step 2 Mendel then cross-pollinated two P generation plants that had contrasting traits, such as purple flowers and white flowers. Mendel called the offspring of the P generation the first *filial* generation, or **F_1 generation.** He then examined each F_1 plant and recorded the number of F_1 plants expressing each trait.

Step 3 Finally, Mendel allowed the F_1 generation to self-pollinate. He called the offspring of the F_1 generation plants the second filial generation, or **F_2 generation.** Again, each F_2 plant was characterized and counted.

BIO graphic

Figure 3

Three Steps of Mendel's Experiments

Mendel studied traits in three generations of plants.

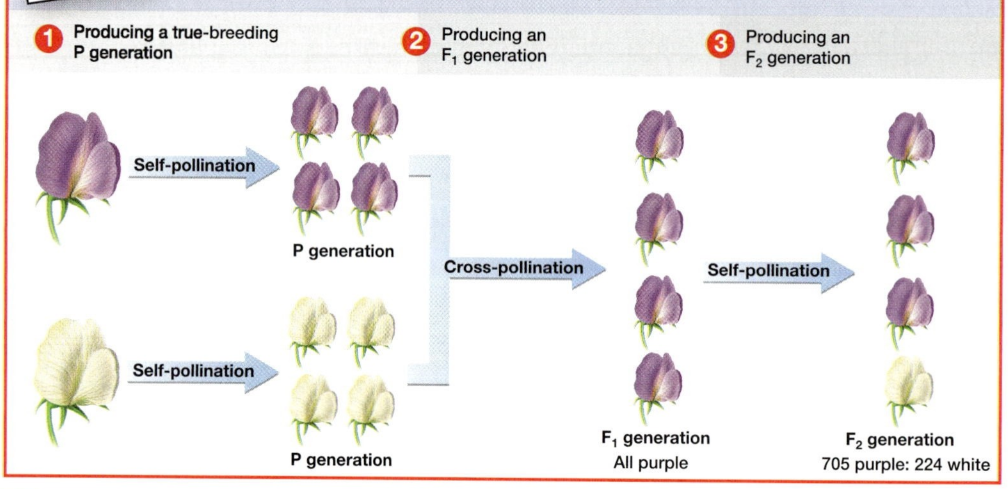

Trends in Genetics

Flies and Worms Many scientists who study genetics use the fruit fly *Drosophila melanogaster* or the roundworm *Caenorhabditis elegans* in their research. These organisms show a variety of traits, are easy to obtain and breed, have short generation time (less than 2 weeks for fruit flies; less than 3 days for roundworms), and produce a large number of offspring. How long would it take to study three generations of humans?

Section Review Correlations

SC.F.2.4.1: Item 6
SC.G.1.4.1: Item 5
SC.H.1.4.1: Items 2, 3, 4
SC.H.1.4.2: Item 1

Mendel's Results

Each of Mendel's F_1 plants showed only one form of the character. The contrasting trait had disappeared! But when the F_1 generation was allowed to self-pollinate, the missing trait *reappeared* in some of the plants in the F_2 generation. When Mendel crossed purple flowers with white flowers, all of the offspring in the F_1 generation had purple flowers. In the F_2 generation, 705 plants had purple flowers and 224 plants had white flowers—a ratio of 705 to 224.

A ratio is a comparison of two numbers and can be written as a fraction ($\frac{705}{224}$) or with a colon (705:224). You can see patterns more easily in data if you reduce a ratio to its simplest form. To do this, divide each term by the smaller of the two terms. This reduces 705:224 to 3.15, which is then rounded to 3:1.

$$\frac{705}{224} \div \frac{224}{224} = 3.15 \text{ (or about 3)}$$

For each of the seven characters Mendel studied, he found the same 3:1 ratio of plants expressing the contrasting traits in the F_2 generation.

MATH LAB: *Calculating Mendel's Ratios* SC.H.1.4.1

Background

You can calculate the ratios Mendel obtained in the F_2 generation for the characters he studied. First copy the partially completed table below on a separate piece of paper.

Analysis

1. **Calculate** the ratio for each contrasting trait. Use colon form.
2. **State** the ratio for each pair of contrasting traits in words and as a fraction.
3. **Critical Thinking Interpreting Results** Do the data confirm a 3:1 ratio in the F_2 generation for each of the characters he studied?

Characters	F_2 generation results		Ratio
Flower color	705 purple	224 white	3.15:1
Seed color	6,022 yellow	2,001 green	
Seed shape	5,474 round	1,850 wrinkled	
Pod color	428 green	152 yellow	
Pod shape	882 round	299 constricted	
Flower position	651 axial	207 top	
Plant height	787 tall	277 dwarf	

Section 1 Review

1. **Describe** the contribution of Mendel to the foundation of modern genetics. SC.H.1.4.2
2. **Describe** why garden-pea plants are good subjects for genetic experiments. SC.H.1.4.1
3. **Summarize** the design of Mendel's pea-plant studies. SC.H.1.4.1
4. **State** the ratio Mendel obtained in each F_2 generation for each of the characters he studied. SC.H.1.4.1
5. **Critical Thinking Evaluating Outcomes** What differences would be expected in experiments with squash plants, which usually do not self-pollinate? SC.G.1.4.1
6. **FCAT Prep** When two true-breeding pea plants that show contrasting forms of a trait are crossed, what characteristic do all of the offspring show? SC.F.2.4.1
   - **A** a different trait
   - **B** one form of the trait
   - **C** both forms of the trait
   - **D** one-fourth of each trait

165

Answers to Section Review

1. Mendel was the first to develop rules that accurately predict patterns of heredity.
2. Garden-pea plants have many characters with two clearly different forms. Mating is easy to control. The plants are small, easy to grow, mature quickly, and produce many offspring.
3. Mendel self-pollinated different varieties of pea flowers for several generations to produce true-breeding P generations. He then cross-pollinated P generation plants with contrasting forms of a character and recorded the number of F_1 plants expressing each trait. He also self-pollinated F_1 plants and recorded the number of F_2 plants expressing each trait.
4. A ratio of about 3:1.
5. Without a true-breeding (self-pollinating) P generation, there would not be a consistent ratio of traits in the F_1 or F_2 generations.
6. **A.** Incorrect. Only parental traits appear in the offspring, unless there is a mutation. **B.** Correct. When one of each contrasting trait is inherited, one trait is masked by the other. **C.** Incorrect. For both traits to appear, both parents must have at least one of the recessive traits. **D.** Incorrect. See answer C.

MATH LAB: *Calculating Mendel's ratios*

Skills Acquired

Interpreting, identifying patterns, organizing data

Teacher's Notes

Provide an example of how to calculate ratios. For example, the ratio of students who own a dog to those who do not. Imagine the ratio is 20 to 5. Explain that the ratio is simplified by dividing each number by the smaller of the two numbers ($\frac{20}{5}$ and $\frac{5}{5}$). The simplest ratio is 4:1. (4 to 1 in words and $\frac{4}{1}$ as a fraction.)

Answers

1. 3.01:1, 2.96:1, 2.82:1, 2.95:1, 3.14:1, 2.84:1
2. Three yellow seeds to one green seed, $\frac{3}{1}$; three round to one wrinkled, $\frac{3}{1}$; three green pods to one yellow pod, $\frac{3}{1}$; three round pods to one constricted pod, $\frac{3}{1}$; three axial to one top, $\frac{3}{1}$; three tall to one dwarf, $\frac{3}{1}$.
3. Yes.

Close

Reteaching — BASIC

Ask students to replicate the three steps in Mendel's experiments as illustrated in **Figure 3,** substituting another trait from **Table 1.**

Quiz — GENERAL

1. Why did Mendel allow the pea plants to self-pollinate for several generations before beginning with his crosses? (To ensure that each variety was true breeding.)
2. Define heredity. (Heredity is the passing of traits from parents to offspring.)

Alternative Assessment — GENERAL

Ask students to select an organism that would be useful for genetics research and list the reasons why. (Examples include fruit flies, roundworms, and garden peas.)

Section 2

Focus

Overview

Before beginning this section review with your students the objectives listed in the Student Edition. This section explains Mendel's discoveries in modern terms and explains the law of segregation and the law of independent assortment.

Bellringer

Tell students that a gardener noticed that some of the flowers on her plants were white. In previous years, the flowers had been purple. Ask students to write down their proposed explanation for this difference. Tell them they will be finding out more on this topic as they read this section. (Instead of buying hybrid seeds from the store, she decided to plant pea seeds from the crop she harvested the previous year. Her plants were the F2 generation, which shows a 3:1 ratio of purple to white flowers.)

Motivate

Identifying Preconceptions — BASIC

Ask students if it is possible for an offspring to have traits different from both of their parents. Some students will respond that this is not possible. Explain that some hidden traits in the parents can combine together and appear in the offspring. An example would be two right-handed parents having a left-handed child.

Section 2 Mendel's Theory

Florida Sunshine Standards/FCAT

SC.H.1.4.1 Investigations are conducted to explore new phenomena, to check on previous results, to test how well a theory predicts, and to compare different theories.

SC.H.2.4.1 Scientists assume that the universe is a vast system in which basic rules exist that may range from very simple to extremely complex, but scientists operate on the belief that the rules can be discovered by careful, systemic study.

Objectives

- **Describe** the four major hypotheses Mendel developed. SC.H.1.4.1
- **Define** the terms *homozygous, heterozygous, genotype,* and *phenotype.* SC.F.2.4.1
- **Compare** Mendel's two laws of heredity. SC.H.2.4.1

Key Terms

allele
dominant
recessive
homozygous
heterozygous
genotype
phenotype
law of segregation
law of independent assortment

A Theory of Heredity

Before Mendel's experiments, many people thought offspring were a *blend* of the traits of their parents. For example, if a tall plant were crossed with a short plant, the offspring would be medium in height. Mendel's results did not support the blending hypothesis. Mendel correctly concluded that each pea has two separate "heritable factors" for each character—one from each parent. As shown in **Figure 4,** when gametes (sperm and egg cells) form, each receives only one of the organism's two factors for each character. When gametes fuse during fertilization, the offspring has two factors for each character, one from each parent.

Mendel's Hypotheses

The four hypotheses Mendel developed were based directly on the results of his experiments. These four hypotheses now make up the Mendelian theory of heredity—the foundation of genetics.

1. *For each inherited character, an individual has two copies of the gene—one from each parent.*
2. *There are alternative versions of genes.* For example, the gene for flower color in peas can exist in a "purple" version

Figure 4 **Mendel's factors**

Each parent has two separate "factors," or genes, for a particular trait.

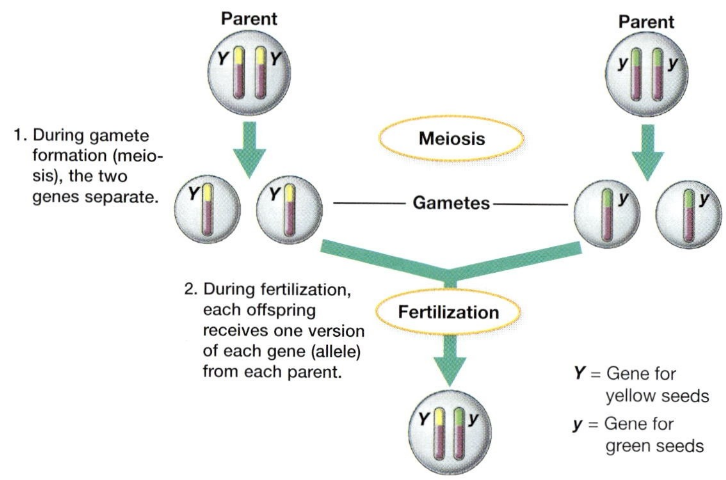

Cultural Awareness

Blood and Inheritance The Greek philosopher Aristotle associated inheritance with blood. He thought the blood carried hereditary information from the body's various structures to the reproductive organs. We know this is not true, but the idea is ingrained in many languages. For example, "blue blood," "blood stock," and "It is in the blood" (English); "Corre en la sangre" (Spanish); "Bon sang ne peut mentir" and "celle est dans le sang" (French); and "Es liegt im Blute" and "von gutem Blut" (German) all associate inheritance with blood.

Section Correlations

SC.F.2.4.1
SC.H.1.4.1
SC.H.2.4.1

or a "white" version. Today the different versions of a gene are called its **alleles.** As shown in Figure 4, an individual receives one allele from each parent. Each allele can be passed on when the individual reproduces.

3. *When two different alleles occur together, one of them may be completely expressed, while the other may have no observable effect on the organism's appearance.* Mendel described the expressed form of the character as **dominant.** The trait that was not expressed when the dominant form of the character was present was described as **recessive.** For every pair of contrasting traits that Mendel studied, the allele for one form of the character was always dominant and the allele for the other form of the character was always recessive. For example, if a plant has both purple and white alleles for flower color but blooms purple flowers, then purple is the dominant form of the character; white is the recessive form. This is shown in **Figure 5.**
4. *When gametes are formed, the alleles for each gene in an individual separate independently of one another. Thus, gametes carry only one allele for each inherited character. When gametes unite during fertilization, each gamete contributes one allele.* As shown in Figure 4, each parent can contribute only one of the alleles because of the way gametes are produced during the process of meiosis.

Mendel's Findings in Modern Terms

Geneticists have developed specific terms and ways of representing an individual's genetic makeup. For example, letters are often used to represent alleles. Dominant alleles are indicated by writing the first letter of the character as a capital letter. For instance, in pea plants, purple flower color is a dominant trait and is written as *P.* Recessive alleles are also indicated by writing the first letter of the dominant trait, but the letter is lowercase. For example, white flower color is recessive and is written as *p.*

If the two alleles of a particular gene present in an individual are the same, the individual is said to be **homozygous** *(hoh moh ZIE guhs)* for that character. For example, a plant with two white flower alleles is homozygous for flower color, as shown in Figure 5. The allele for yellow peas, *Y*, is dominant to the allele for green peas, *y.* A plant with two yellow-pea alleles, *YY,* is homozygous for seed color.

If the alleles of a particular gene present in an individual are different, the individual is **heterozygous** *(heht uhr oh ZIE guhs)* for that character. As shown in Figure 5, a plant with one "purple flower" allele and one "white flower" allele is heterozygous for flower color. A plant with one "yellow pea" allele and one "green pea" allele is heterozygous for seed color.

PP
Purple flowers, homozygous dominant
Pp
Purple flowers, heterozygous

pp
White flowers, homozygous recessive

Figure 5 **Recessive alleles.** Alleles can be present but not expressed. The allele for purple flowers, *P*, is dominant to the recessive allele, *p*.

Teach

Using the Figure

Have students look at **Figure 5.** Point out to the students that the stages of meiosis result in gametes that have only one version of each gene. During meiosis, a cell completes two successive divisions that produce 4 cells, each with a chromosome number that has been reduced by half.

Teaching Tip — GENERAL

Dominant and Recessive Ask students why some traits appear more often than others do. For example, there are more dark haired people than light haired people. To emphasize the point, do a hand count of some contrasting traits such as eye color, tongue curling and free ear lobes. The majority of students will be dominant for a given character. Discuss dominant and recessive and explain that for the dominant trait to appear, only one allele for the character is needed, but for the recessive trait to appear, both alleles for the character must be inherited. However, emphasize that dominant phenotypes are not always more common than recessive phenotypes. If there are very few alleles for a dominant phenotype in a population, it will not occur often.

Teaching Tip — BASIC

Genotype and Phenotype Have students practice using the boldface terms in this section by providing several examples. For example, tell them that the gene for plant height has two versions: T = tall and t = dwarf. Ask students to identify the two alleles for plant height. (T and t) Write Tt, tt, and TT on the board and ask students to identify the genotype and phenotype of each set of alleles. (genotypes—Tt, tt, TT; phenotypes: tall, dwarf, and tall) Ask students to identify whether a plant with TT alleles is homozygous or heterozygous. (homozygous)

Chapter Resource File
- Directed Reading BASIC
- Active Reading GENERAL
- Data Sheet for Quick Lab GENERAL

Transparencies

TR Bellringer
TR C3 Mendel's Factors

One-Stop Planner CD-ROM
- Reading Organizers BASIC
- Reading Strategies BASIC
- Portfolio Project
  Genetics Project GENERAL

Identifying Dominant or Recessive Traits

Skills Acquired
Summarizing, calculating, applying information

Teacher's Notes
Emphasize that dominant phenotypes are not more common than recessive phenotypes. Point out that the expression of some phenotypes (such as freckles) may be influenced by the environment.

Analysis Answers

1. Answers will vary.
2. Answers will vary.
3. The recessive traits. Recessive traits must be homozygous to be expressed.

Teaching Tip — GENERAL

Gene Expression Point out that the environment may influence the expression of some phenotypes (such as freckles). Ask students for other examples (muscle size).

Activity — GENERAL

Graphic Organizer Have students work in pairs to make a graphic organizer to demonstrate the law of independent assortment (see bottom of this page). Ask students to illustrate their graphic organizer with at least one example showing the inheritance of two pairs of contrasting traits. Have them write a brief explanation. Ask student to volunteer to put their examples on the board or overhead projector. English Language Learners

Figure 6 Dominent alleles. In heterozygous individuals, freckles, *F*, is the dominant allele. Similarly, the allele for a cleft chin is dominant to the allele for a chin without a cleft.

In heterozygous individuals, only the dominant allele is expressed; the recessive allele is present but unexpressed. An example of a human trait that is expressed in a heterozygous individual is freckles. Freckles *F*, is a dominant allele. The recessive allele is *f*, no freckles. The recessive allele may be present but not expressed. As shown in **Figure 6,** people who are heterozygous for freckles (*Ff*) will have freckles even though they also have the allele for no freckles, *f*.

The set of alleles that an individual has for a character is called its **genotype** *(JEE noh tiep)*. The physical appearance of a character is called its **phenotype** *(FEE noh tiep)*. Phenotype is determined by which alleles are present. For example, if *Pp* is the genotype of a pea plant, its phenotype is purple flowers. If *pp* is the genotype of a pea plant, its phenotype is white flowers. When considering seed color, if *Yy* is the genotype of a pea plant, its phenotype is yellow seeds. If *yy* is the genotype of a pea plant, its phenotype is green seeds. Note that by convention, the dominant form of the character is written first, followed by the lowercase letter for the recessive form of the character.

QUICK LAB *Identifying Dominant or Recessive Traits* SC.F.2.4.1

You can determine some of the genotypes and all of the phenotypes for human characters that are inherited as simple dominant or recessive traits.

Materials
pencil, paper

Procedure

1. **Make** a table like the one at right. For each character, circle the phenotype that best matches your own phenotype.
2. **Determine** how many students in your class share your phenotype by recording your results in a table on the chalkboard.

Dominant trait	Recessive trait
Cleft chin	No cleft
Dimples	No dimples
Hair above knuckles	Hairless fingers
Freckles	No freckles

Analysis

1. **Summarize** the class results for each character.
2. **Calculate** the class dominant:recessive ratio for each character.
3. **Critical Thinking Applying Information** For which phenotypes in the table can you determine a person's genotype without ever having seen his or her parents? Explain.

168

Graphic Organizer

Use this graphic organizer with **Activity** on this page.

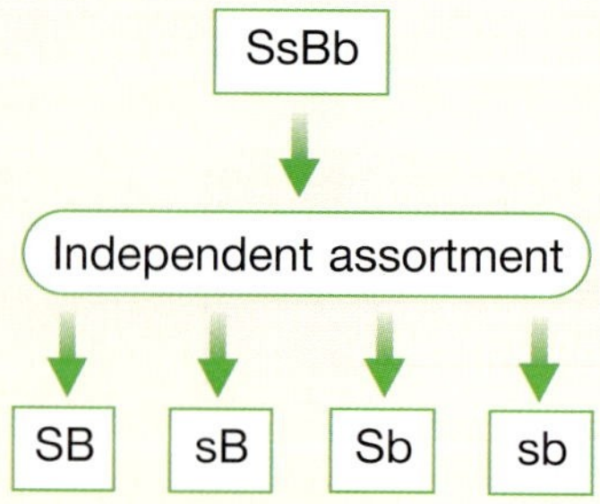

The Laws of Heredity

Mendel's hypotheses brilliantly predicted the results of his crosses and also accounted for the ratios he observed. Similar patterns of heredity have since been observed in countless other organisms. Because of their importance, Mendel's ideas are often referred to as the laws of heredity.

The Law of Segregation

The first law of heredity describes the behavior of chromosomes during meiosis. At this time, homologous chromosomes and then chromatids are separated. The first law, the **law of segregation,** states that the two alleles for a character segregate (separate) when gametes are formed (as shown in Figure 4).

The Law of Independent Assortment

Mendel went on to study whether the inheritance of one character (such as plant height) influenced the inheritance of a different character (such as flower color). To study how different pairs of genes are inherited, Mendel conducted dihybrid crosses. A dihybrid cross is a cross that considers two pairs of contrasting characters. For example, a cross that considers both plant height and flower color is a dihybrid cross.

Mendel found that for the characters he studied, the inheritance of one character did not influence the inheritance of any other character. The **law of independent assortment** states that the alleles of different genes separate independently of one another during gamete formation. For example, the alleles for the height of the plant shown in **Figure 7** separate independently of the alleles for its flower color. We now know that this law applies only to genes that are located on different chromosomes or that are far apart on the same chromosome.

The search for the physical nature of Mendel's "factors" dominated biology for more than half a century after Mendel's work was rediscovered in 1900. We now know that the units of heredity are portions of DNA called *genes,* which are found on the chromosomes that an individual inherits from its parents.

Figure 7 The law of independent assortment. Mendel found that the inheritance of one character, such as plant height, did not influence the inheritance of another character, such as flower color.

Section 2 Review

1. **Differentiate** between alleles and genes. SC.F.2.4.2
2. **Apply** the terms *homozygous, heterozygous, dominant,* or *recessive* to describe plants with the genotypes *PP* and *Pp.* SC.F.2.4.1
3. **Identify** the phenotypes of rabbits with the genotypes *Bb* and *bb*, where *B* = black coat and *b* = brown coat. SC.F.2.4.1
4. **Determine** whether the rabbits in item 3 are heterozygous or homozygous. SC.F.2.4.1
5. **Critical Thinking Critiquing Explanations** Review Mendel's two laws according to their strengths and weaknesses in terms of our modern understanding of meiosis. SC.H.2.4.2
6. **FCAT Prep** If a pea plant is heterozygous for a particular character, how can the alleles that control the character be described?
   A two recessive **C** one dominant, one recessive
   B two dominant **D** three dominant, one recessive
   SC.F.2.4.1

Answers to Section Review

1. Genes are pieces of DNA that code for a particular character. There are alternative versions, or alleles, for each gene.
2. The *Pp* plant is heterozygous; the *PP* plant is homozygous dominant.
3. The *Bb* rabbit has a black coat and the *bb* rabbit has a brown coat.
4. *Bb* = heterozygous; *bb* = homozygous
5. A weakness in the law of independent assortment is that it applies only to genes that are located on different chromosomes or that are far apart on the same chromosome.
6. **A.** Incorrect. A pea plant that has two recessive alleles would be called homozygous. **B.** Incorrect. A pea plant that has two dominant alleles would be called homozygous. **C.** Correct. A pea plant that has one dominant and one recessive allele would be called heterozygous. **D.** Incorrect. Multiple alleles, genes with three or more alleles, would describe a pea plant that has three dominant and one recessive alleles.

Activity — GENERAL

Hairy Knuckles Have students determine whether or not they have hair above their knuckles. Tell students that the presence of hair above the knuckles is caused by a dominant allele, H. Then ask them to identify the genotype of a person who does not have hair above their knuckles (hh). Have students determine under what circumstances a parent without hair above their knuckles can produce a child with hair above their knuckles. (The second parent must have the hair above the knuckles gene.) **LS Intrapersonal**

Close

Reteaching — BASIC

Have students review the steps involved in Mendel's scientific investigations. Then have them apply these methods to Mendel's discoveries about heredity. Ask students to summarize Mendel's hypotheses and predictions.

Quiz — GENERAL

1. How are the genotype of a dominant allele and a recessive allele written? (A dominant allele is represented by a capital letter, and the same letter in lower case represents a recessive allele.)
2. What is the genotype of a purple-flowered pea plant? (PP or Pp)
3. What is a dihybrid cross? (A cross that considers two pairs of contrasting traits.)

Alternative Assessment — GENERAL

Ask students to relate Mendel's four hypotheses to his experimental results. Refer the students to **Figure 3** for Mendel's experimental results.

Section Review Correlations

SC.F.2.4.1: Items 2, 3, 4, 6
SC.F.2.4.2: Items 1, 5

Focus

Overview

Before beginning this section review with your students the objectives listed in the Student Edition. This section explains the use of Punnett squares for predicting outcomes, probability and pedigrees.

Bellringer

Since the dawn of agriculture, people have used selective breeding to improve crops and domestic animals. Modern applications of Mendelian genetics and gene technology have resulted in major changes in crops and animals. Ask students to list on paper some examples of selective breeding in domestic animals or crops. Ask students to explain how they might go about selecting for a particular character.

Motivate

Discussion/ Question — BASIC

Tell students that the basenji is a dog that cannot bark. However, they can make a yodeling type of sound. Basenjis are small dogs with pointed ears, short silky hair and rows of wrinkles on their foreheads. Ask students to hypothesize a genetic explanation for why the basenjis cannot bark. (The ability to bark is a dominant trait in dogs. All basenjis have two recessive genes for this character.) Ask them if they can suggest other traits that have been selected for in dogs or cats.

Section 3 Studying Heredity

Florida Sunshine Standards/FCAT

SC.F.2.4.1 There are mechanisms of asexual and sexual reproduction and different genetic advantages and disadvantages of asexual and sexual reproduction.

SC.H.3.4.1 Performance testing is often conducted using small-scale models, computer simulations, or analogous systems to reduce the chance of system failure.

Objectives

- **Predict** the results of monohybrid genetic crosses by using Punnett squares. SC.F.2.4.1
- **Apply** a test cross to determine the genotype of an organism with a dominant phenotype. SC.F.2.4.1
- **Predict** the results of monohybrid genetic crosses by using probabilities. SC.F.2.4.1
- **Analyze** a simple pedigree. SC.H.3.4.1

Key Terms

Punnett square
test cross
probability
pedigree
sex-linked gene

Punnett Squares

Animal breeders try to breed animals with very specific characteristics. Thus, breeders must be able to predict how often a trait will appear when two animals are crossed (bred). Likewise, horticulturists (plant breeders) need to produce plants with very specific characteristics. One simple way of predicting the expected results (not necessarily the actual results) of the genotypes or phenotypes in a cross is to use a Punnett square.

A **Punnett square** is a diagram that predicts the outcome of a genetic cross by considering all possible combinations of gametes in the cross. Named for its inventor, Reginald Punnett, the simplest Punnett square consists of four boxes inside a square. As shown in **Figure 8,** the possible gametes that one parent can produce are written along the top of the square. The possible gametes that the other parent can produce are written along the left side of the square. Each box inside the square is filled in with two letters obtained by combining the allele along the top of the box with the allele along the side of the box. The letters in the boxes indicate the possible genotypes of the offspring.

One Pair of Contrasting Traits

Punnett squares can be used to predict the outcome of a monohybrid cross (a cross that considers one pair of contrasting traits between two individuals). For example, a Punnett square can be used to predict the outcome of a cross between a pea plant that is homozygous for yellow seed color (*YY*) and a pea plant that is homozygous for green seed color (*yy*). Figure 8 shows that 100 percent of the offspring in this type of cross are expected to be heterozygous (*Yy*), expressing the dominant trait of yellow seed color.

Figure 8 **Monohybrid cross: homozygous plants**

A cross between a pea plant that is homozygous for yellow seeds (*YY*) and a pea plant that is homozygous for green seeds (*yy*) will produce only yellow heterozygous offspring (*Yy*).

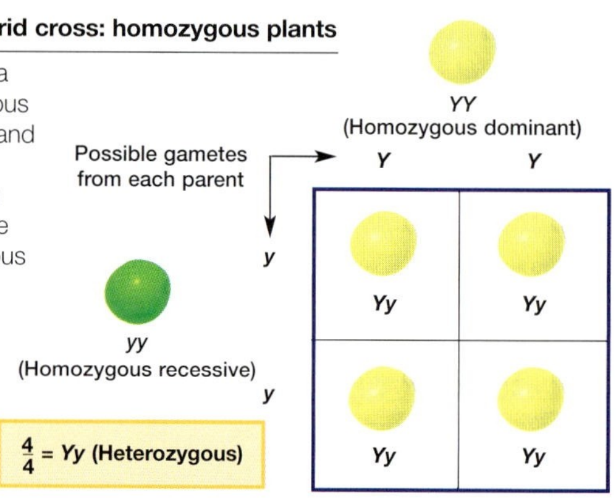

170

did you know?

Point out to students that breeders use Punnett squares to help them select individuals that will be most likely to produce offspring of the phenotype they want.

One-Stop Planner CD-ROM

- **Reading Organizers** BASIC
- **Reading Strategies** BASIC
- **Problem Solving Worksheet** Genetics and Probability GENERAL

Section Correlations

SC.F.2.4.1
SC.H.3.4.1

Figure 9 shows a Punnett square that predicts the results of a monohybrid cross between two pea plants that are both heterozygous (*Yy*) for seed color. One-fourth of the offspring would be expected to have the genotype *YY,* two-fourths (or one-half) would be expected to have the genotype *Yy,* and one-fourth would be expected to have the genotype *yy*. Another way to express this is to say that the genotypic ratio is 1 *YY* : 2 *Yy* : 1 *yy.* Because the *Y* allele is dominant over the *y* allele, three-fourths of the offspring would be yellow, and one-fourth would be green. The phenotypic ratio is 3 yellow : 1 green.

Punnett squares allow direct and simple predictions to be made about the outcomes of genetic crosses. Although animal breeders and horticulturists are not always certain what characteristics will turn up in the offspring, they can use the predictions from Punnett squares to cross individuals that they know will be most likely to produce offspring with the desired phenotypes.

Figure 9 Monohybrid cross: heterozygous plants

Crossing two pea plants that are heterozygous for seed color (*Yy*) will produce offspring in the ratio shown in the Punnett square.

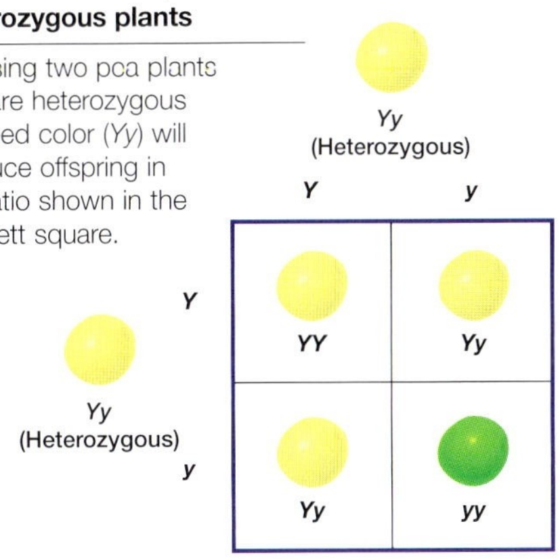

$\frac{1}{4}$ = *YY* (Homozygous dominant)
$\frac{2}{4}$ = *Yy* (Heterozygous)
$\frac{1}{4}$ = *yy* (Homozygous recessive)

Exploring Further

Crosses That Involve Two Characters

Suppose a horticulturist has two characters that she wants to consider when crossing two plants. A cross that involves two pairs of contrasting traits is called a dihybrid cross. For example, she may want to predict the results of a cross between two pea plants that are heterozygous for seed shape (*R* = round, *r* = wrinkled) and seed color (*Y* = yellow, *y* = green).

Determine possible gametes
To use a Punnett square to predict the results of this cross, first consider how the four alleles from either parent (*RrYy*) can combine to form gametes that are either *RY, Ry, rY,* or *ry* (Figure A).

Figure A Gametes

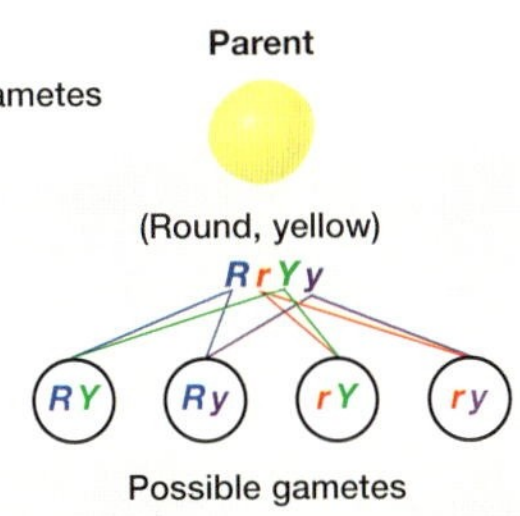

Then write the genotypes of these gametes on the top and left sides of a Punnett square (Figure B).

Complete the Punnett square
On a separate sheet of paper, make a copy of the Punnett square in Figure B, which has been partially filled in with the predicted genotypes. Fill in the remaining genotypes, then do the following:

- **List** all of the possible genotypes that can result.
- **Calculate** the genotypic ratio for this cross.
- **List** all of the possible phenotypes that can result.
- **Calculate** the phenotypic ratio for this cross.

Figure B Punnett square

Possible gametes from each parent (Yellow × Yellow)

	RY	*Ry*	*rY*	*ry*
RY	*RRYY*	*RRYy*	*RrYY*	*RrYy*
Ry	*RRYy*		*RrYy*	
rY	*RrYY*	*RrYy*		
ry	*RrYy*			

171

Chapter Resource File

- Directed Reading BASIC
- Active Reading GENERAL
- Data Sheet for Data Lab GENERAL
- Data Sheet for Math Lab GENERAL
- Data Sheet for Data Lab GENERAL

Transparencies

TR Bellringer
TR C5 Monohybrid Cross of Homozygous Plants
TR C6 Monohybrid Cross of Heterozygous Plants
TR C4 Probability with Two Coins

Teach

Using the Figure — BASIC

Point out to students that **Figure 8** shows a Punnett square used for predicting the outcome of a genetic cross. The genotype of a parent determines the possible alleles that can be found in its gametes. The possible gametes are written along the top and left sides of the square. Review with students how the genotype in each square was obtained. Assign several monohybrid crosses for students to practice.

Teaching Tip — BASIC

Possible gametes Have students practice listing possible gametes when given the genotype of a dihybrid parent. LS **Logical**

Exploring Further

Crosses that Involve Two Traits

Teaching Strategies
- Explain how genotypes are written in dyhibrid crosses. (for example, RrYy, not RYry)

Discussion
- Under what circumstances might a dihybrid cross fail to produce four different kinds of gametes? (if the genes are close together on the same chromosome)

Answers
- Possible genotypes—RRYY, RRYy, RrYY, RrYy, Rryy, RRyy, rrYY, rrYy, rryy
- Genotypic ratio—1 RRYY, 2 RRYy, 1 RRyy, 2 RrYY, 4 RrYy, 2 Rryy, 1 rrYY, 2 rrYy, 1 rryy
- Possible phenotypes—round yellow, round green, wrinkled yellow, wrinkled green
- Phenotypic ratio—9 round, yellow : 3 round, green : 3 wrinkled, yellow : 1 wrinkled, green

Teach, *continued*

Teaching Tip — GENERAL

Test Cross Tell students that they have been presented with one of Mendel's purple-flowering pea plants. Ask them if they can identify the genotype of the plant in regard to flower color. Then have them propose a method for discovering the genotype of the plant. Give the students a clue by asking them what genotype can be determined from the phenotype (homozygous recessive). (The purple flowered pea plant should be crossed with a white flowered pea plant; if any offspring are white, the unknown was heterozygous)

DATA LAB *Analyzing a Test Cross*

Skills Acquired
Analyzing, interpreting, inferring, drawing conclusions, predicting outcomes

Teacher's Notes
Encourage students to recognize the importance of sample size in making conclusions about the genotype of an unknown individual.

Answers to Analysis

1. possible gametes each parent can produce
2. the genotype of each possible kind of offspring
3. The genotypic ratio will be 4 *Pp* : 0 *PP* : 0 *pp*. The phenotypic ratio for the offspring will be 4 purple : 0 white.
4. heterozygous

Determining Unknown Genotypes

Animal breeders, horticulturists, and others involved in breeding organisms often need to know whether an organism with a dominant phenotype is heterozygous or homozygous for a character. How do they determine this? For example, how might a horticulturist determine whether a pea plant with a dominant phenotype, such as yellow seeds, is homozygous (*YY*) or heterozygous (*Yy*)? The horticulturist could perform a test cross. In a **test cross**, an individual whose phenotype is dominant, but whose genotype is not known, is crossed with a homozygous recessive individual.

For example, a plant with yellow seeds but of unknown genotype (*Y?*) is test-crossed with a plant with green seeds (*yy*). If all of the offspring produce yellow seeds, the offspring must be *Yy*. Thus, the genotype of the "unknown" plant must be *YY*. If half of the offspring produce yellow seeds and half produce green seeds, the genotype of the unknown plant must be *Yy*. In reality, if the cross produces even one plant that produces green seeds, the genotype of the unknown parent plant is likely to be heterozygous. After performing a test cross, the horticulturist can continue breeding the original plant with more certainty of its genotype.

DATA LAB *Analyzing a Test Cross* SC.H.3.4.1

Background

You can use a test cross to determine whether a plant with purple flowers is heterozygous (*Pp*) or homozygous dominant (*PP*). On a separate sheet of paper, copy the two Punnett squares shown below, and fill in the boxes in each square.

Is this purple flowering pea plant *Pp* or *PP*?

	P	*p*
p		
p		

Figure A Heterozygous (*Pp*) plant

	P	*P*
p		
p		

Figure B Homozygous (*PP*) plant

Analysis

1. **Determine** what the letters at the top and side of each box represent.
2. **Determine** what the letters in each box represent.
3. **Calculate** the genotypic and phenotypic ratios that would be predicted if the parent of the unknown genotype were homozygous for the character (Figure B).
4. **Critical Thinking Predicting Outcomes** If half of the offspring have white flowers, what is the genotype of the plant with purple flowers?

172

Outcomes of Crosses

Like Punnett squares, probability calculations can be used to predict the results of genetic crosses. **Probability** is the likelihood that a specific event will occur. Probabilities can be expressed in words, as decimals, as percentages, or as fractions. For example, if an event definitely will occur, its probability can be expressed as either 1 out of 1 (in words), 1 (as a decimal numeral), 100 percent (as a percentage), or $\frac{1}{1}$ (as a fraction). If an event definitely will not occur, its probability can be expressed as either 0 out of 0, 0, 0 percent, or $\frac{0}{0}$.

In order to simplify our discussion of probability, we will express probabilities as fractions. Probability can be determined by the following formula:

$$\text{Probability} = \frac{\text{number of one kind of possible outcome}}{\text{total number of all possible outcomes}}$$

Consider the possibility that a coin tossed into the air will land on heads (one possible outcome). The total number of all possible outcomes is two—heads or tails. Thus, the probability that a coin will land on heads is $\frac{1}{2}$, as shown in **Figure 10.**

Probability of a Specific Allele in a Gamete

The same formula can be used to predict the probability of an allele being present in a gamete. If a pea plant has two alleles for seed color, the plant can contribute either allele (yellow or green) to the gamete it produces (the law of independent assortment). For a plant with two alleles for seed color, the total number of possible outcomes is two—green or yellow. The probability that a gamete will carry the allele for green seed color is $\frac{1}{2}$. The probability that a gamete from this plant will carry the allele for yellow seed color is also $\frac{1}{2}$.

Probability of the Outcome of a Cross

Because two parents are involved in a genetic cross, both parents must be considered when calculating the probability of the outcome of a genetic cross. Consider the analogy of two coins being tossed at the same time. The probability of a penny landing on heads is $\frac{1}{2}$, and the probability of a nickel landing on heads is $\frac{1}{2}$. The way one coin falls does not depend on how the other coin falls. Similarly, the allele carried by the gamete from the first parent does not depend on the allele carried by the gamete from the second parent. The outcomes are independent of each other.

To find the probability that a *combination* of two independent events will occur, multiply the separate probabilities of the two events. Thus, the probability that a nickel *and* a penny will both land on heads is

$$\frac{1}{2} \times \frac{1}{2} = \frac{1}{4}$$

Study TIP

Reviewing Information
Because probability is a ratio of a subset of all possible outcomes to all possible outcomes, the value for probability is never greater than 1. When it is less than one, it can be expressed as a fraction or as a percentage of the whole.

Figure 10 Probability of heads or tails. The probability that a tossed coin will land on heads is $\frac{1}{2}$. The probability that a tossed coin will land on tails is $\frac{1}{2}$.

Demonstration — GENERAL

Shuffle a deck of cards. Ask students to determine the probability of drawing an ace from the deck. (Students may suggest $\frac{4}{52}$ or $\frac{1}{13}$.) Ask how they arrived at this conclusion. Deal 13 cards from the top of the deck. Count the number of aces in those 13 cards and compare that number with the students' prediction. If the number varies from the prediction, have the students speculate about the reasons for the difference. **LS Logical**

Teaching Tip — GENERAL

Using Probabilities in Genetic Crosses Point out that the probability of a specific genotype occurring in a cross can be obtained by setting up a Punnett square similar to those in **Figures 8 and 9.** The probability of finding a specific allele in a gamete is written next to the possible allele across the top and along the side.

Teach, *continued*

Using the Figure — GENERAL

Make sure students understand that the probabilities in each square in **Figure 11** were obtained by multiplying the probability at the top of the box by the probability along the side of the box.

MATH LAB *Predicting the Results of Crosses Using Probabilities*

Skills Acquired
Calculating, applying information

Teacher's Notes
Have students set up Punnett squares similar to those in **Figures 8 and 9.** Then ask them to write in the probabilities of finding a specific allele in a gamete.

Answers to Analysis

1. $\frac{1}{4}$
2. $\frac{1}{2}$
3. 1
4. 0

SKILL BUILDER — ADVANCED

Writing Skills Have students develop stories from which a pedigree can be drawn. Encourage them to be creative in thinking of characters and traits that they choose to follow through several generations. To illustrate their pedigree stories, student can add "family portraits." Read some of the stories in class, and have students draw the pedigrees from the information given in each story. LS **Verbal**

Figure 11 **Probability with two coins**

The probability of the results of flipping two coins is easy to compute.

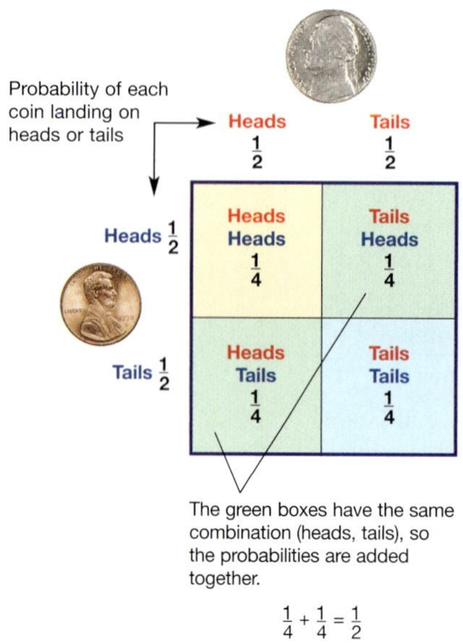

The possible results of tossing a nickel and a penny at the same time and the probability of each outcome are shown in **Figure 11.** Since the combination of heads and tails can occur in two possible ways, those two probabilities are added together.

$$\frac{1}{4} + \frac{1}{4} = \frac{2}{4} \text{ or } \frac{1}{2}$$

Consider the possible results that can occur in a cross between two pea plants that are heterozygous for seed shape (*Rr*). The *R* allele for round seed shape is dominant over the *r* allele for wrinkled seed shape. The probability of each parent carrying gametes with *R* or *r* alleles is $\frac{1}{2}$. The probability of offspring with *RR* alleles is

$$\frac{1}{2} \times \frac{1}{2} = \frac{1}{4}$$

Similarly, the probability of offspring with *rr* alleles is

$$\frac{1}{2} \times \frac{1}{2} = \frac{1}{4}$$

The combination of *Rr* alleles can occur in two possible ways. One parent can contribute the *R* allele, and the second parent the *r* allele, or vice versa. Thus, the probability of offspring with *Rr* alleles is

$$\frac{1}{4} + \frac{1}{4} = \frac{1}{2}$$

MATH LAB *Predicting the Results of Crosses Using Probabilities*

SC.F.2.4.1

Background

In rabbits, the allele *B* for black hair is dominant over the allele *b* for brown hair. You can practice using probabilities to predict the outcome of genetic crosses by completing the genetic problems below. Draw Punnett squares for each problem.

Analysis

1. **Calculate** the probability of homozygous dominant (*BB*) offspring resulting from a cross between two heterozygous (*Bb*) parents.
2. **Calculate** the probability of heterozygous offspring resulting from a cross between a heterozygous parent and a homozygous recessive (*bb*) parent.
3. **Calculate** the probability of heterozygous offspring resulting from a cross between a homozygous dominant parent and a homozygous recessive parent.
4. **Calculate** the probability of homozygous dominant offspring resulting from a cross between a heterozygous parent and a homozygous recessive parent.

174

MISCONCEPTION ALERT

Probabilities Students may think that probabilities in genetic crosses show the definite outcome of a genetic cross. Point out that probabilities are used only to predict the possible outcome of a genetic cross.

Cultural Awareness

Albinism in Hopi Tribes A survey of a Hopi tribe in Arizona found the frequency of albinism to be 1 in 277. In contrast, albinism is very rare or nonexistent in other Native American communities in Arizona and New Mexico. Why is the frequency so high among the Hopi? The Hopi people have always had a high regard for albinos and clan leaders have taken special care to protect them from the harsh desert sun. This type of selection could explain the increase in albinism in the community.

Inheritance of Traits

Imagine that you want to learn about an inherited trait present in your family. How would you find out the chances of passing the trait to your children? Geneticists often prepare a **pedigree,** a family history that shows how a trait is inherited over several generations. Pedigrees are particularly helpful if the trait causes a genetic disorder and the family members want to know if they are carriers or if their children might get the disorder. Carriers are individuals who are heterozygous for an inherited disorder but do not show symptoms of the disorder. Carriers can pass the allele for the disorder to their offspring.

Figure 12 shows an example of a pedigree for a family with albinism. In the genetic disorder albinism, the body is unable to produce an enzyme necessary for the production of melanin. Melanin is a pigment that gives dark color to hair, skin, scales, eyes, and feathers. Without melanin, an organism's surface coloration may be milky white and its eyes may be pink, as shown in Figure 12.

Scientists can determine several pieces of genetic information from a pedigree:

Autosomal or Sex-Linked? If a gene is autosomal, it will appear in both sexes equally. Recall that an autosome is a chromosome other than an X or Y sex chromosome. If a trait is sex-linked, its effects are usually seen only in males. A **sex-linked gene's** allele is located only on the X or Y chromosome. Most sex-linked genes are carried on the X chromosome and are recessive. Because males have only one X chromosome, a male who carries a recessive allele on the X chromosome will exhibit the sex-linked condition.

A female who carries a recessive allele on one X chromosome will not exhibit the condition if there is a dominant allele on her other X chromosome. She will express the recessive condition only if she inherits two recessive alleles. Thus, her chances of inheriting and exhibiting a sex-linked condition are significantly less.

Figure 12 **Albinism pedigree**

Albinism is a genetic disorder transmitted by a recessive allele.

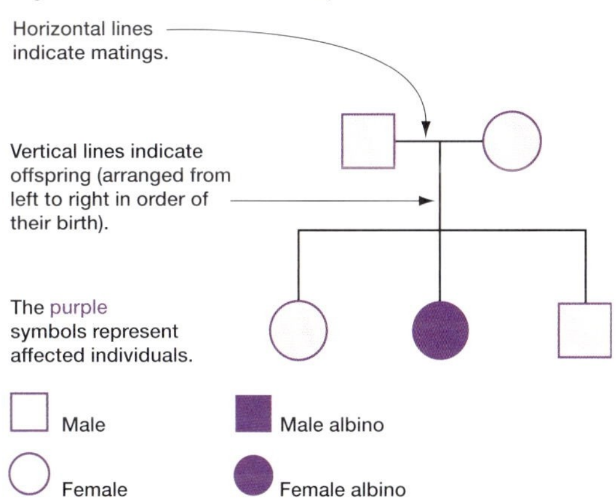

In the wild, albino animals have little chance of survival. They lack the pigments that provide protection from the sun's ultraviolet rays.

Real Life

About 10 percent of Dalmatians are deaf.

Because many purebred dogs are inbred—that is, they have closely related parents—some of them are homozygous for certain recessive disorders.

Finding Information

If you have a purebred dog, find out if that breed is prone to a genetic disorder.

175

Real Life

Answer

Some pedigreed dogs that are prone to genetic diseases include Irish setters (blindness), German shepherds (hip dysplasia) and dachshunds (dwarfism).

SKILL BUILDER — GENERAL

Vocabulary Ask students to differentiate between Punnett squares, probabilities, and pedigrees. (Punnett squares predict the expected outcome of a cross by considering all possible combinations of gametes in a cross. Probabilities predict the mathematical likelihood that a specific event, such as the outcome of a cross, will occur. Pedigrees provide a visual representation of how a trait is inherited over several generations.)

Teaching Tip — BASIC

X-linked Tell the students that some genes are not inherited equally by both sexes. An X-linked gene is usually expressed only in males, and most are recessive. Ask students how a male might inherit an X-linked trait from his mother. (The mother carries the recessive allele on one of her X-chromosomes; the son inherits this chromosome from his mother and a Y from his father.)

Using the Figure — GENERAL

Before teaching students to interpret a pedigree such as that shown in **Figure 12,** introduce the symbols: male (square), female (circle), trait expressed (shaded circle or square), and trait not expressed (circle or square not shaded). Once students are comfortable with the meanings of the symbols, have them interpret the pedigree in **Figure 12.** Tell students the gene for this trait not only results in a deficiency of skin, hair, and eye pigmentation but also causes defects in vision. **LS Visual**

HISTORY CONNECTION

There is a high frequency of hemophilia among members of the royal families throughout Europe. Queen Victoria was a carrier of sex-linked hemophilia. Because members of the European nobility usually married within their own social class, the hemophilia gene was passed via Queen Victoria's daughters to the Russian, German, and Spanish royal families, increasing the frequency of the recessive allele among European nobility.

DATA LAB: Evaluating a Pedigree

Skills Acquired
Analyzing, interpreting, drawing conclusions, applying information

Teacher's Notes
Encourage students to use "If-then" statements to organize their thoughts and interpret the pedigree. Example: If a gene is expressed by an offspring but not by either parent, then the gene must be recessively inherited.

Answers to Analysis

1. autosomal recessive
2. homozygous
3. $\frac{1}{2}$

Close

Reteaching — BASIC

Write the following genotypes on the board: (1) PP, (2) Pp, and (3) pp. Pair each student with a partner. Have students choose two of the genotypes and construct and complete a Punnett square showing the cross. Have them share their results with their partners.

Quiz — GENERAL

1. What is the probability of two parents each carrying a recessive gene for an inherited disease to produce a child that will have that disease? ($\frac{1}{4}$)
2. Explain how the parents of an individual who expresses a recessive gene can both not express the gene. (Parents are both heterozygous dominant.)

Dominant or Recessive? If the gene is autosomal dominant, every individual with the condition will have a parent with the condition. If the condition is recessive, an individual with the condition can have one, two, or neither parent exhibit the condition.

Heterozygous or Homozygous? If individuals with autosomal traits are homozygous dominant or heterozygous, their phenotype will show the dominant allele. If individuals are homozygous recessive, their phenotype will show the recessive allele. Two people who are heterozygous carriers of a recessive mutation will not show the mutation, but they can produce children who are homozygous for the recessive allele.

DATA LAB: Evaluating a Pedigree SC.H.3.4.1

Background

The photo shows a family with an *albino* member. Pedigrees, such as the one below, can be used to track different genetic traits, including albinism. Use the pedigree below to practice interpreting a pedigree.

Albino

Analysis

1. **Interpret** the pedigree to determine whether the trait is sex-linked or *autosomal* and whether the trait is inherited in a dominant or recessive manner.
2. **Determine** whether Female A is homozygous or heterozygous.
3. **Critical Thinking Applying Information** If Female B has children with a homozygous individual, what is the probability that the children will be heterozygous?

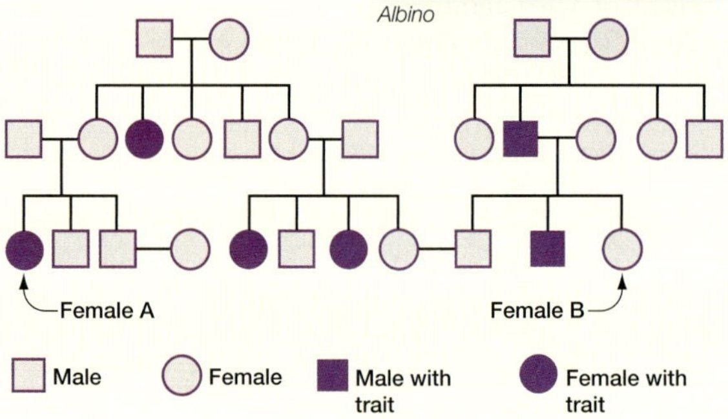

Section 3 Review

1. **Predict** the expected phenotypic and genotypic ratios among the offspring of two individuals who are heterozygous for freckles (*Ff*) by using a Punnett square. SC.H.3.4.1
2. **Summarize** how a test cross can reveal the genotype of a pea plant with round seeds. SC.F.2.4.1
3. **Calculate** the probability that an individual heterozygous for a cleft chin (*Cc*) and an individual homozygous for a chin without a cleft (*cc*) will produce offspring that are homozygous recessive for a chin without a cleft. (*cc*) SC.F.2.4.1
4. **Critical Thinking Analyzing Graphics** When analyzing a pedigree, how can you determine if an individual is a carrier (heterozygous) for the trait being studied? SC.H.3.4.1
5. **FCAT Prep** A cross between two pea plants that produce yellow seeds results in 124 offspring: 93 produce yellow seeds and 31 produce green seeds. What are the likely genotypes of the plants that were crossed?
   A both *Yy*
   B both *YY*
   C both *yy*
   D one *YY*, one *Yy*
   SC.F.2.4.1

176

Answers to Section Review

1. 3 freckles:1 no freckle; 1FF:2Ff:1ff
2. If, after a testcross, all of the offspring have round seeds, the parent of the unknown is likely to be homozygous dominant. If, after a test cross, any of the offspring have wrinkled seeds, the parent with the unknown genotype is likely to be heterozygous.
3. $\frac{1}{2}$
4. An individual is a carrier if any of his or her children express the trait.
5. **A.** Correct. This cross would produce about $\frac{1}{4}$ green seeds. **B.** Incorrect. This cross would not produce any green seeds. **C.** Incorrect. This cross would not produce any yellow seeds. **D.** Incorrect. This cross would not produce any green seeds.

Section Review Correlations

SC.F.2.4.1: Items 2, 3, 5
SC.H.3.4.1: Items 1, 4

Complex Patterns of Heredity

Section 4

Florida Sunshine Standards/FCAT

SC.F.2.4.3 Mechanisms of change (e.g., mutation and natural selection) lead to adaptations in a species and their ability to survive naturally in changing conditions and to increase species diversity.

SC.H.3.4.3 Scientists can bring information, insights, and analytical skills to matters of public concern and help people understand the possible causes and effects of events.

Complex Control of Characters

A horse with red hair mates with a horse with white hair, and their offspring has both red and white hair. How can this be? If characters are controlled by single genes with simple dominant and recessive alleles, the colt's hair should be one color or the other. Not always! Most of the time, characters, such as hair color in horses, display more-complex patterns of heredity than the simple dominant-recessive patterns discussed so far.

Characters Influenced by Several Genes

When several genes influence a character, this is an example of **polygenic inheritance.** The genes for a polygenic character may be scattered along the same chromosome or located on different chromosomes. Determining the effect of any one of these genes is difficult. Due to independent assortment and crossing-over during meiosis, many different combinations appear in offspring. Familiar examples of polygenic characters in humans include eye color, height, weight, and hair and skin color. All of these characters have degrees of intermediate conditions between one extreme and the other, as shown in **Figure 13.**

Intermediate Characters

Recall that in Mendel's pea-plant crosses, one allele was completely dominant over another. In some organisms, however, an individual displays a phenotype that is intermediate between the two parents, a condition known as **incomplete dominance.** For example, when a snapdragon with red flowers is crossed with a snapdragon with white flowers, a snapdragon with pink flowers is produced. Neither the red nor the white allele is completely dominant over the other allele. The flowers appear pink because they have less red pigment than the red flowers. In Caucasians, the child of a straight-haired parent and a curly-haired parent will have wavy hair. Straight and curly hair are homozygous dominant traits. Wavy hair is heterozygous and is intermediate between straight and curly hair.

Figure 13 Polygenic inheritance. Many characters—height, weight, hair color, and skin color—are influenced by many genes.

Objectives

- **Identify** five factors that influence patterns of heredity. SC.F.2.4.1
- **Describe** how mutations can cause genetic disorders. SC.F.2.4.3
- **List** two genetic disorders, and describe their causes and symptoms. SC.F.2.4.2
- **Evaluate** the benefits of genetic counseling. SC.H.3.4.3

Key Terms

polygenic inheritance
incomplete dominance
multiple alleles
codominance

Section 4

Focus

Overview

Before beginning this section review with your students the objectives listed in the Student Edition. Some characters are controlled by several genes, alleles may be equally dominant, or may be influenced by the environment or mutations. This section discusses complex patterns of inheritance such as incomplete dominance, codominance, polygenic traits, mutations and environmental influences

Bellringer

Ask students to study the animals shown in **Figure 16.** Ask them to list possible mechanisms that allow the arctic fox to change its fur color with changing seasons. (The temperature triggers enzymes involved in hormonal responses that influence the genes controlling fur color.)

Motivate

Discussion/Question — GENERAL

Ask students to look at the variations in human traits as show in **Figure 13.** Ask them to propose a mechanism for the inheritance of a trait such as eye color in humans, which can appear as brown, green, blue and gray. (There are at least three genes involved: brown, green and blue; with brown dominant to green and blue, and green dominant to blue.) LS **Visual**

Chapter Resource File

- Directed Reading BASIC
- Active Reading GENERAL

One-Stop Planner CD-ROM

- Reading Organizers BASIC
- Reading Strategies BASIC
- Supplemental Reading Guide The Double Helix ADVANCED

Integrating Physics and Chemistry

Evaluate students' ability to analyze, review, and critique scientific explanations by asking them to identify and describe the limitations of Mendel's understanding of inheritance based on his pea plant experiments. Complete this exercise by comparing and contrasting simple patterns of trait inheritance associated with pea plants with more complex patterns of trait inheritance such as polygenic traits, incomplete dominance, codominance, and multiple alleles.

Transparencies

TR Bellringer
TR C30 Some Human Genetic Disorders

Section Correlations

SC.F.2.4.1
SC.F.2.4.2
SC.F.2.4.3
SC.H.3.4.3

Characters Controlled by Genes with Three or More Alleles

Genes with three or more alleles are said to have **multiple alleles.** For example, in the human population, the ABO blood groups (blood types) are determined by three alleles, I^A, I^B, and *i*. The letters *A* and *B* refer to two carbohydrates on the surface of red blood cells. In the *i* allele, neither carbohydrate is present. The I^A and I^B alleles are both dominant over *i*. But neither I^A nor I^B is dominant over the other. When I^A and I^B are both present they are codominant. Even for characters controlled by genes with multiple alleles, an individual can have only two of the possible alleles for that gene. **Figure 14** shows how combinations of the three different alleles can produce four different blood types—A, B, AB, and O. Notice that a person who inherits two *i* alleles has type O blood.

Characters with Two Forms Displayed at the Same Time

For some characters, two dominant alleles are expressed at the same time. In this case, both forms of the character are displayed, a phenomenon called **codominance.** Codominance is different from incomplete dominance because both traits are displayed.

The situation of human ABO blood groups, as discussed above, is an example of co-dominance. The genotype of a person who has blood type AB is I^AI^B, and neither allele is dominant over the other. Type AB blood cells carry both *A*- and *B*-types of carbohydrate molecules on their surfaces.

Figure 14 Multiple alleles control the ABO blood groups

Different combinations of the three alleles I^A, I^B, and *i* result in four different blood phenotypes, A, AB, B, and O. For example, a person with the alleles I^A and *i* would have blood type A.

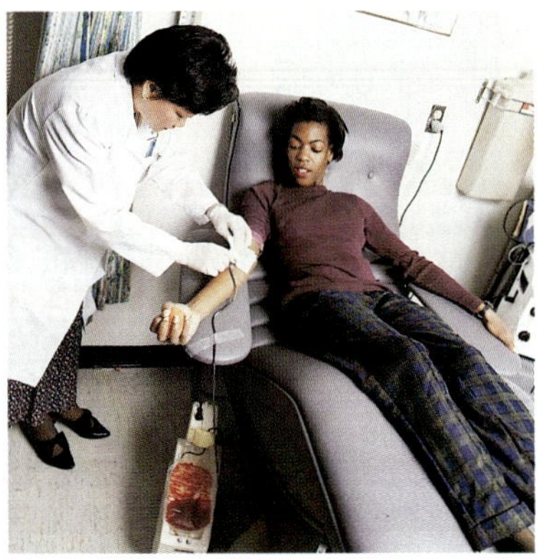

Possible alleles (columns); Possible alleles (rows)

	I^A	I^B	*i*
I^A	I^AI^A	I^AI^B	I^Ai
I^B	I^AI^B	I^BI^B	I^Bi
i	I^Ai	I^Bi	*ii*

Blood types: A | AB | B | O

178

Teach

Teaching Tip — BASIC

Incomplete Dominance Ask students whether a plant breeder could produce only pink flowering snapdragons by crossing pink-flowering snapdragons and white-flowering snapdragons. Lead students to understand that since all pink-flowering snapdragons are heterozygous, mating a pink-flowering snapdragon with a white-flowering one would produce pink-flowering and white-flowering offspring in a ratio of 1:1.

Demonstration — GENERAL

To convey the concept of universal donor and universal acceptor, set up four flasks to represent each blood type, and label them appropriately. The flasks should contain the following: "A" blood (water with red food color); "B" blood (water with blue food color); "AB" blood (water with red and blue food color); and "O" blood (water only). Take an empty beaker and pour "O" blood into it. Show students that pouring "A" blood into the beaker containing "O" blood will "contaminate" the "O" blood (change its color). Demonstrate the possible mixtures and have students derive which blood types are compatible with each other. Point out that A blood and B blood each contain unique carbohydrates that O does not, which is why O is a universal donor, and AB is a universal acceptor. LS Visual

did you know?

Human Inheritance Mendel's work with garden-pea plants showed that the characters he studied are controlled by single genes. In humans, single-factor inheritance has been found in about 600 recessively inherited characters, and in such dominant conditions as Huntington's disease. However, many more conditions are determined by polygenic inheritance, which involves several genes. Such conditions include cleft lip and palate, schizophrenia, hypertension, and diabetes.

Characters Influenced by the Environment

An individual's phenotype often depends on conditions in the environment. In plants, hydrangea *(hie DRAYN juh)* flowers of the same genetic variety range in color from blue to pink, as shown in **Figure 15.** Hydrangea plants in acidic soil bloom blue flowers, while those in neutral to basic soil will bloom pink flowers.

The color of the arctic fox is affected by temperature. During summer, the fox produces enzymes that make pigments. These pigments darken the fox's coat to a reddish brown, as shown in **Figure 16,** enabling the fox to blend in with the summer landscape. During the winter, the pigment-producing genes of the arctic fox do not function because of the cold temperature. As a result, the coat of the fox is white, and the animal blends in with the snowy background.

Figure 15 Environmental influences on flower color. Hydrangea with the same genotype for flower color express different phenotypes depending on the acidity of the soil.

Fur color in Siamese cats is also influenced by temperature. In a Siamese cat, the fur on its ears, nose, paws, and tail is darker than on the rest of its body. The Siamese cat has a genotype that results in dark fur at locations on its body that are cooler than the normal body temperature. Thus, the darkened parts have a lower body temperature than the light parts.

In humans many characters, such as height, are influenced by the environment. For example, height is influenced by nutrition, an internal environmental condition. Exposure to the sun, an external environmental condition, alters the color of the skin. Many aspects of human personality, such as aggressive behavior, are strongly influenced by the environment, although genes appear to play an important role. Because identical twins have identical genes, they are often used to study environmental influences. Because identical twins are genetically identical, any differences between them are attributed to environmental influences.

Figure 16 Environmental influences on fur color

Can the same species of fox look so different? Many arctic mammals, such as the arctic fox, develop white fur during the winter and dark fur during the summer.

READING SKILL BUILDER — GENERAL

Discussion Ask students to look at the girl having a blood test done in **Figure 14.** Lead a brief discussion on blood tests. Tell students that testing for specific materials in the blood can discover many disorders. For example: anemia (too few red blood cells, thus test for red blood cell count), diabetes (inability to break down blood sugar, thus test for blood sugar levels), and high cholesterol (thus test for HDL).

Using the Figure — GENERAL

Have the students study **Figure 14.** Explain how the table shows the possible blood types, and the use of "I" with the subscripts A, B, AB to denote alleles dominant to i.

Activity — GENERAL

Using Punnett Squares Ask students to use a Punnett square to figure out the following problem: If the Mother of a type O child is A, list the mother's genotype and the possible genotypes for the father. (Mother is I^Ai; possible genotypes for the father are I^Ai, I^Bi, ii)

Demonstration — BASIC

Use litmus paper to test a weak acid such as vinegar, and a weak base such as a baking soda solution. Have students note the color change, red to blue for base and blue to pink (red) for acid. Then have students look at the flowers in **Figure 15.** Ask them is they can relate the color of the flowers to the litmus test used for acids and bases. In the case of the *Hydrangea*, the flowers are blue if the soil is acidic. (Litmus is a dye made from organisms called lichens.) Ask student how the *Hydrangea* could be used as a bio-indicator for the acidity of the soil. **LS Visual**

Integrating Physics and Chemistry

Review the pH scale with students and have them design an experiment to test the effects of various pH environments on hydrangea. Be sure to have them identify the experimental hypothesis, which according to the textbook, is that an acidic environment produces blue flowers while a neutral or basic environment produces pink flowers. If time permits, have students conduct the experiment in the laboratory in order to demonstrate how chemistry—pH in this instance—affects the physical expression of a genotype.

MEDICINE CONNECTION

The blood groups A, B, AB, and O all have an identical sugar chain on their cell surface. Type A cells have an additional sugar, type B cells have a different additional sugar, and type AB cells have both additional sugars. Researchers hope to produce type O cells by using enzymes to remove the additional sugars from type A, B, and AB cells. Ask students why this would be an important medical breakthrough. (Any blood type could be converted to type O, which would make it compatible with all other blood types as a universal donor.)

Teach, *continued*

Discussion/ Question — GENERAL

Ask student why most bald people are male. Tell them both males and females can inherit a "baldness" allele. (The male hormone testosterone activates the allele and eventually leads to baldness. Women produce small amounts of testosterone. However, baldness does not occur in females unless they have both alleles for baldness. The presence of only one allele for baldness causes men to become bald.)

Teaching Tip — ADVANCED

Hemoglobin Ask students how a faulty gene can alter a hemoglobin molecule. (Hemoglobin is a protein. A mutation in a hemoglobin gene that results in a change in the amino acid sequence of the gene can alter the structure of the protein and ultimately the protein's function.)

Teaching Tip

Sickle Cell Anemia and Malaria

Explain to students the adaptive value for individuals that have one gene for sickle cell anemia. Ask them how this might explain the higher incidence of the disease in African Americans. (The adaptive value is protection from malaria. Since malaria is more prevalent in Africa, those individuals inheriting one allele for sickle cell anemia are somewhat protected against malaria. This protection retains the gene in the population. People with two alleles for sickle cell anemia would become ill with the disease, lessening the chance of passing on the gene.)

Genetic Disorders

In order for a person to develop and function normally, the proteins encoded by his or her genes must function precisely. Unfortunately, sometimes genes are damaged or are copied incorrectly, resulting in faulty proteins. Changes in genetic material are called mutations. Mutations are rare because cells have efficient systems for correcting errors. But mutations sometimes occur, and they may have harmful effects.

The harmful effects produced by inherited mutations are called genetic disorders. Many mutations are carried by recessive alleles in heterozygous individuals. This means that two phenotypically normal people who are heterozygous carriers of a recessive mutation can produce children who are homozygous for the recessive allele. In such cases, the effects of the mutated allele cannot be avoided. Several human genetic disorders are summarized in **Table 2.**

Sickle Cell Anemia

An example of a recessive genetic disorder is sickle cell anemia, a condition caused by a mutated allele that produces a defective form of the protein hemoglobin. Hemoglobin is found within red blood cells, where it binds oxygen and transports it through the body. In sickle cell anemia, the defective form of hemoglobin causes many red blood cells to bend into a sickle shape, as seen in **Figure 17.** The sickle-shaped cells rupture easily, resulting in less oxygen being carried by the blood. Sickle-shaped cells also tend to get stuck in blood vessels; this can cut off blood supply to an organ.

Figure 17 Sickle cell. One out of 500 African Americans has sickle cell anemia, which is caused by a gene mutation that produces a defective form of hemoglobin.

The recessive allele that causes sickle-shaped red blood cells also helps protect the cells of heterozygous individuals from the effects of malaria. Malaria is a disease caused by a parasitic protozoan that invades red blood cells. The sickled red blood cells of heterozygous individuals cause the death of the parasite. But the individual's normal red blood cells can still transport enough oxygen. Therefore, these people are protected from the effects of malaria that threaten individuals who are homozygous dominant for the hemoglobin gene.

Cystic Fibrosis (CF)

Cystic fibrosis is the most common fatal, hereditary, recessive disorder among Caucasians. One in 25 Caucasian individuals has at least one copy of a defective gene that makes a protein necessary to move chloride into and out of cells. About 1 in 2,500 Caucasian infants in the United States is homozygous for the *cf* allele. The airways of the lungs become clogged with thick mucus, and the ducts of the liver and pancreas become blocked. While treatments can relieve some of the symptoms, there is no known cure.

HISTORY CONNECTION

Working with limited laboratory facilities and a strong determination to fight the disease that was killing their friends and families, two African-American researchers, Dr. Angela Ferguson and Dr. Roland Scott, published a paper on sickle cell anemia in the 1940's—25 years ahead of other researchers. Dr. Scott, known as the "father of sickle cell anemia research," is the founder and former director of Howard University's Center for Sickle Cell Anemia Research. Dr. Ferguson was an associate professor of pediatrics at Howard University.

Hemophilia

Another recessive genetic disorder is hemophilia *(hee moh FIHL ee uh)*, a condition that impairs the blood's ability to clot. Hemophilia is a sex-linked trait. More than a dozen genes code for the proteins involved in blood clotting. A mutation on one of these genes on the X chromosome causes the form of hemophilia called hemophilia A. If the mutation appears on the X chromosome, which a male receives from his mother, he does not have a normal gene on the Y chromosome to compensate. Therefore, he will develop hemophilia.

internet connect
www.scilinks.org
Topic: Genetic Disorders
Keyword: HX4091
SCILINKS. Maintained by the National Science Teachers Association

Huntington's Disease (HD)

Huntington's disease is a genetic disorder caused by a dominant allele located on an autosome. The first symptoms of HD—mild forgetfulness and irritability—appear in victims in their thirties or forties. In time, HD causes loss of muscle control, uncontrollable physical spasms, severe mental illness, and eventually death. Unfortunately, most people who have the HD allele do not know they have the disease until after they have had children. Thus, the disease is unknowingly passed on from one generation to the next.

Table 2 Some Human Genetic Disorders

Disorder	Dominant or Recessive	Symptom	Defect	Frequency Among Human Births
Sickle Cell Anemia	Recessive 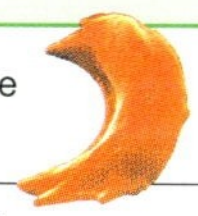	Poor blood circulation	Abnormal hemoglobin molecules	1 in 500 (African Americans)
Hypercholes-terolemia	Dominant	Excessive cholesterol levels in blood, leading to heart disease	Abnormal form of cell surface receptor for cholesterol	1 in 500
Tay-Sachs Disease	Recessive in early childhood	Deterioration of central nervous system; death	Defective form of a brain enzyme	1 in 3,500 (Ashkenazi Jews)
Cystic Fibrosis	Recessive 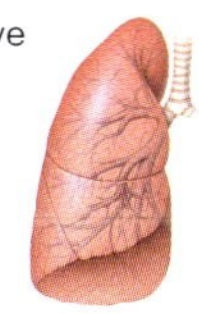	Mucus clogs organs including the lungs, liver, and pancreas; affected individuals usually do not survive to old age	Defective chloride-ion transport protein	1 in 2,500 (Caucasians)
Hemophilia A (Classical)	Sex-linked recessive	Failure of blood to clot	Defective form of a blood-clotting factor	1 in 10,000 (males)
Huntington's Disease	Dominant 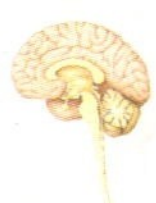	Gradual deterioration of brain tissue in middle age; shortened life expectancy	Inhibitor of brain-cell metabolism is made	1 in 10,000

Group Activity — GENERAL

Patterns of Heredity Pair students and ask each pair to make a table to organize information about patterns of heredity that are more complex than simple dominant-recessive patterns. The students should write the following headings across the top: *Explanation, Example(s)*. Along the sides, students should write the following: *Polygenic Traits, Incomplete dominance, Codominance, Multiple alleles,* and *Environmentally influenced traits.* Have students add information to the table as they review this section. **LS Logical**

READING SKILL BUILDER — BASIC

Reading Organizer Have students make a reading organizer describing the cause and effect of each of the genetic disorders discussed in this section. Students should construct a cause effect graph for each disease. **LS Logical**

REAL WORLD CONNECTION

Women who have PKU often have babies with mental retardation, not because the baby has PKU, but because the mother's body chemistry is altered during pregnancy. These babies cannot be helped with a special diet. However, the mental retardation can be avoided if the mother follows a low-phenylalanine diet before and during pregnancy.

Attention Grabber

Witchcraft or Disease? In the United States, many cases of Huntington's disease can be traced back to two brothers. The two men immigrated to North America from England in the 1600's because of accusations of witchcraft in their family. The family members of these two brothers were apparently persecuted because of their strange behaviors, which are now understood to be symptoms of Huntington's disease. Constant dance-like movements in its victims characterize the disease.

Close

Reteaching — BASIC

K-W-L Tell students to return to their list of things they want to know about inheritance, which they created in Section 1. Have them place check marks next to the questions that they are now able to answer. Students should finish by making a list of what the have **L**earned. Conclude by asking students which questions are still unanswered. Ask if they have new questions.

Quiz — GENERAL

Give an example of a possible trait for each of the following genotypic conditions:

1. Incomplete dominance (flower color in snapdragons or hair shape in humans)
2. Multiple alleles (blood type in humans; roan coat in horses)

Alternative Assessment — ADVANCED

Assign pairs of students a genetic disorder and have the students design an informative brochure about the disorder, similar to brochures found in a doctor's office. Set guidelines on information you expect students to include, such as symptoms, causes, prognosis and support groups. Sample brochures can be obtained from various medical facilities. Have students present their brochures to the class. Co-op Learning

Section Review Correlations

SC.F.1.4.7: Item 2

SC.F.2.4.1: Items 1, 4, 5

SC.F.2.4.3: Items 3, 6

internet connect
www.scilinks.org
Topic: Genetic Counseling
Keyword: HX4090
SCiLINKS. Maintained by the National Science Teachers Association

Treating Genetic Disorders

Most genetic disorders cannot be cured, although progress is being made. A person with a family history of genetic disorders may wish to undergo genetic counseling before becoming a parent. Genetic counseling is a form of medical guidance that informs people about genetic problems that could affect them or their offspring.

In some cases, a genetic disorder can be treated if it is diagnosed early enough. For example, an individual with the genetic disorder phenylketonuria (PKU) lacks an enzyme that converts the amino acid phenylalanine into the amino acid tyrosine. As a result, phenylalanine builds up in the body and causes severe mental retardation. If PKU is diagnosed soon after birth, however, the newborn can be placed on a low-phenylalanine diet. Because this disorder can be easily diagnosed by inexpensive laboratory tests, many states require PKU testing of all newborns.

Gene Therapy

Gene technology may soon allow scientists to correct certain recessive genetic disorders by replacing defective genes with copies of healthy ones, an approach called *gene therapy.* The essential first step in gene therapy is to isolate a copy of the gene. The defective *cf* gene was isolated in 1989. In 1990, a working *cf* gene was successfully transferred into human lung cells growing in tissue culture by attaching the *cf* gene to the DNA of a cold virus. The cold virus—carrying the normal *cf* gene piggyback—easily infects lung cells. The *cf* gene enters the lung cells and begins producing functional CF protein. Thus, the defective cells are "cured" and are able to transport chloride ions across their plasma membranes.

Similar attempts in humans, however, were not successful. Most people have had colds and, as a consequence, have built up a natural immunity to the cold virus. Their lungs therefore reject the cold virus and its *cf* passenger. In the last few years, similar attempts using a different virus to transport the *cf* gene into lung cells have been initiated. This virus, called *AAV,* produces almost no immune response and so seems a much more suitable vehicle for introducing *cf* into cells. Clinical trials are underway, and the outlook is promising.

Section 4 Review

1. **Differentiate** between incomplete dominance and codominance. SC.F.2.4.1
2. **Identify** two examples of characters that are influenced by environmental conditions. SC.F.1.4.7
3. **Summarize** how a genetic disorder can result from a mutation. SC.F.2.4.3
4. **Describe** how males inherit hemophilia. SC.F.2.4.1
5. **Critical Thinking Justifying Conclusions** A nurse states that a person cannot have the blood type ABO. Do you agree or disagree? Explain. SC.F.2.4.1
6. FCAT Prep Which of the following descriptions applies to the mutated allele that causes Huntington's disease? SC.F.2.4.3
   A autosomal and dominant
   B autosomal and recessive
   C sex-linked and dominant
   D sex-linked and recessive

182

Answers to Section Review

1. Incomplete dominance produces phenotypes that are intermediate between two contrasting forms of a character. In codominance, both dominant forms of a gene are displayed at the same time.
2. Answers will vary but may include fur color in Siamese cats and arctic mammals, flower color in hydrangea plants, and height and skin color in humans.
3. A genetic disorder results when a mutation is inherited and the mutation produces harmful effects.
4. The male receives from his mother an X chromosome with a mutated blood-clotting gene.
5. Students should agree. It would require that an individual have three alleles—I^A, I^B, and i.
6. **A.** Correct. Huntington's disease is caused by an autosomal dominant allele. **B.** Incorrect. Sickle cell anemia, Tay-Sachs and cystic fibrosis are examples of autosomal recessive alleles. **C.** Incorrect. Huntington's is dominant but not sex-linked. **D.** Incorrect. Hemophilia A is an example of a sex-linked recessive allele.

Key Concepts

1 The Origins of Genetics

- Gregor Mendel bred varieties of the garden pea in an attempt to understand heredity. Mendel observed that contrasting traits appear in offspring according to simple ratios.
- In Mendel's experiments, only one of the two contrasting forms of a character was expressed in the F_1 generation. The other form reappeared in the F_2 generation in a 3:1 ratio.

2 Mendel's Theory

- Different versions of a gene are called alleles. An individual usually has two alleles for a gene, each inherited from a different parent.
- Individuals with the same two alleles for a gene are homozygous; those with two different alleles for a gene are heterozygous.
- The law of segregation states that the two alleles for a gene separate when gametes are formed. The law of independent assortment states that two or more pairs of alleles separate independently of one another during gamete formation.

3 Studying Heredity

- The results of genetic crosses can be predicted with the use of Punnett squares and probabilities.
- A test cross can be used to determine whether an individual expressing a dominant trait is heterozygous or homozygous.
- A trait's pattern of inheritance within a family can be determined by analyzing a pedigree.

4 Complex Patterns of Heredity

- Characters usually display complex patterns of heredity, such as incomplete dominance, codominance, and multiple alleles.
- Mutations can cause genetic disorders, such as sickle cell anemia, hemophilia, and Huntington's disease.
- Genetic counseling can help patients concerned about a genetic disorder.

BIOLOGY INTERACTIVE TUTOR Unit 5—*Heredity*
Use this unit to review the key concepts and terms in this chapter.

Key Terms

Section 1

heredity (162)
genetics (162)
monohybrid cross (164)
true-breeding (164)
P generation (164)
F_1 generation (164)
F_2 generation (164)

Section 2

allele (167)
dominant (167)
recessive (167)
homozygous (167)
heterozygous (167)
genotype (168)
phenotype (168)
law of segregation (169)
law of independent assortment (169)

Section 3

Punnett square (170)
test cross (172)
probability (173)
pedigree (175)
sex-linked gene (175)

Section 4

polygenic inheritance (177)
incomplete dominance (177)
multiple alleles (178)
codominance (178)

Alternative Assessment — GENERAL

Assign each student one of the following topics: incomplete dominance, codominance, or multiple alleles. Have each student think of a concrete example for teaching their assigned topic to others. Have each student give a short oral report, using their chosen example to explain their topic.

Chapter Resource File

- Science Skills Worksheet GENERAL
- Critical Thinking Worksheet ADVANCED
- Test Prep Pretest GENERAL
- Chapter Test GENERAL

Answer to Concept Map

The following is one possible answer to Performance Zone item 15.

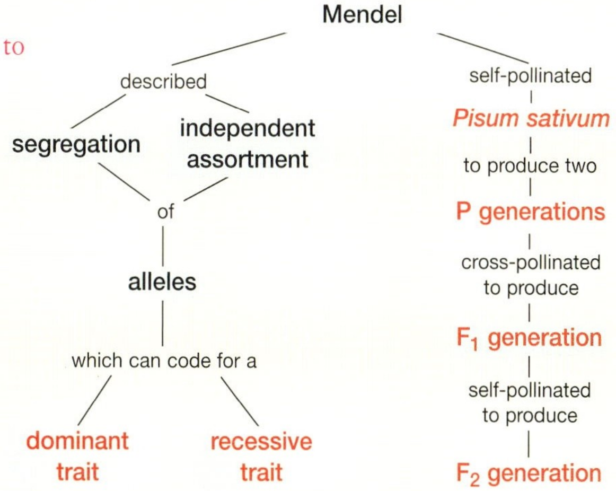

Performance ZONE

CHAPTER REVIEW

ANSWERS

Understanding Key Ideas

1. b
2. b
3. b
4. a
5. c
6. c
7. d
8. A copy of the functional gene is attached to the DNA of a virus. The functional gene gets into the defective cells by "piggybacking" on the virus. Once inside the cells, it produces a functional protein that helps remedy the disease.
9. During meiosis II, the members of each pair of alleles separate when gametes are formed as described in the law of segregation.
10. 1 YYRR : 2 YyRR : 1 yyRR; 3 yellow, round : 1 green, round.
11. One possible answer to the concept map is found at the bottom of the Study Zone page.

Critical Thinking

12. Patterns obtained from large samples are less likely to be distorted by rare events that can occur by chance.
13. Cystic fibrosis is a recessive autosomal disorder. Thus, each parent must have the recessive allele. Chances are increased that his sister is a carrier (heterozygote) for cystic fibrosis.

Chapter Review Correlations

SC.F.2.4.1: Items 3-7, 9-11

SC.H.1.4.1: Item 2

SC.H.1.4.2: Item 1

SC.H.2.4.1: Item 12

SC.H.3.4.3: Item 13

SC.H.3.4.5: Items 8, 14

Understanding Key Ideas

1. The scientist whose studies formed the basis of modern genetics is
a. T. A. Knight. **c.** Louis Pasteur.
b. Gregor Mendel. **d.** Robert Hooke.
SC.H.1.4.2

2. Which of the following is *not* a good reason why *Pisum sativum* makes an excellent subject for genetic study?
a. Many varieties exist.
b. They require cross-pollination.
c. They grow quickly.
d. They demonstrate complete dominance.
SC.H.1.4.1

3. If smooth peas are dominant over wrinkled peas, the allele for smooth peas should be represented as
a. *W.* **b.** *S.* **c.** *w.* **d.** *s.*
SC.F.2.4.1

4. The law of segregation states that pairs of alleles
a. separate when gametes form.
b. separate independently of one another during gamete formation.
c. are always the same.
d. are always different.
SC.F.2.4.1

5. The trait shown below is

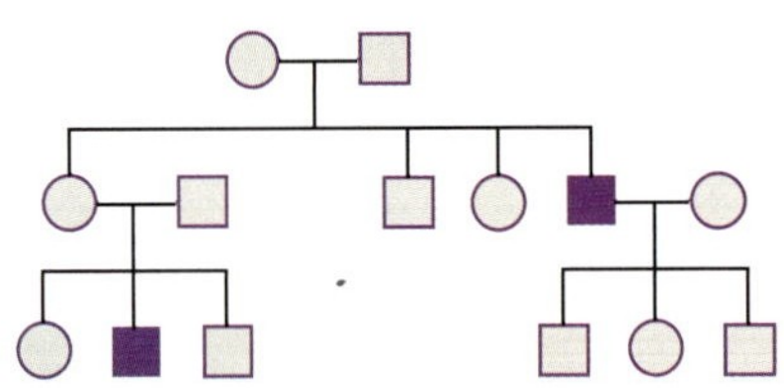

a. sex-linked and dominant.
b. autosomal and dominant.
c. sex-linked and recessive.
d. autosomal and recessive.
SC.F.2.4.1

6. *D*, dimples, is the dominant allele to the recessive allele, *d*, no dimples. The probability of parents with *Dd* and *dd* genotypes having a child with no dimples (*dd*) is
a. $\frac{1}{8}$. **c.** $\frac{1}{2}$.
b. $\frac{1}{4}$. **d.** 1.
SC.F.2.4.1

7. The unknown genotype of an individual with a dominant phenotype can be determined using
a. a ratio. **c.** probability.
b. a dihybrid cross. **d.** a test cross.
SC.F.2.4.1

8. Explain how working genes have been inserted into defective cells during gene therapy. SC.H.3.4.5

9. Relate the events of meiosis to the law of segregation. (**Hint:** See Chapter 7, Section 1.)
SC.F.2.4.1

10. Exploring Further State the genotypic and phenotypic ratios that would result from a cross between two *YyRR* pea plants.
SC.F.2.4.1

11. **Concept Mapping** Make a concept map about Mendel's experiments. Try to include the following words in your map: Pisum sativum, *P generation,* F_1 *generation,* F_2 *generation, dominant trait,* and *recessive trait.* SC.F.2.4.1

Critical Thinking

12. Evaluating Results Mendel based his conclusion about inheritance patterns on experiments involving large numbers of plants. Why do you think the use of large numbers of individuals is advantageous when studying patterns of inheritance?
SC.H.2.4.1

13. Justifying Conclusions A 20-year-old man who has cystic fibrosis has a sister who is planning to have a child. The man encourages his sister to see a genetic counselor. What do you think the man's reasons are for giving such advice?
SC.H.3.4.3

Alternative Assessment

14. Technology and Learning Find out how new technologies have changed plant-breeding methods since Mendel's time. Prepare an oral report to summarize your findings. Or create a display that compares the methods and equipment Mendel might have used with those used by plant breeders today. SC.H.3.4.5

184

Alternative Assessment

14. Reports and displays will vary. Gene technology is now used in plant breeding. Many plant breeders use gene technology equipment to conduct their breeding program.

Assignment Guide

Section	Questions
1	1, 2, 3
2	9, 11, 12, 14
3	5, 6, 7, 10
4	8, 13

FCAT Prep

For questions 1–6, write your answers on a separate sheet of paper.

1 What are the offspring of true-breeding parents called?
A. F_1 generation
B. F_2 generation
C. dominant offspring
D. recessive offspring
SC.F.2.4.1

2 A man and a woman who do not have hemophilia visit a genetic counselor. What tool might the counselor use to determine the risk of the couple having a child with hemophilia?
F. blood test **H.** karyotype
G. DNA fingerprint **I.** pedigree
SC.H.3.4.1

3 What characteristic is described in the statement: The dog's coat is brown? SC.F.2.4.1
A. dominance **C.** pedigree
B. genotype **D.** phenotype

4 Albinism is rare among wild animals, but common among some domesticated species. What factor might account for this difference?
F. Domesticated animals do not eat a variety of foods.
G. Wild animals tend to benefit from recessive traits such as albinism.
H. Parents of wild animals usually reject offspring with recessive traits.
I. Domesticated animals are inbred and more likely to have recessive traits.
SC.F.2.4.1

Base your answers to questions 5 and 6 on the diagram below.
***T* and *t* represent the alleles for the tall and dwarf traits, respectively.**

PEA PLANT CROSS

	?	?
?	*Tt*	*tt*
?	*Tt*	*tt*

5 What genotypic ratio is expected in the offspring of this cross?
A. 1 *Tt* : 1 *tt* **C.** 1 *Tt* : 3 *tt*
B. 3 *Tt* : 1 *tt* **D.** 1 *TT* : 1 *tt*
SC.F.2.4.1

6 If this cross produced 240 offspring, how many of the offspring would be expected to have the dwarf trait? SC.F.2.4.1

Test TIP

Before answering word problems that involve genetics, write the problem down by using letters to symbolize genotypes.

FCAT Doctor

Question 2 Answer I is the correct choice. A pedigree would allow the counselor to determine who in their family had the recessive disease. The counselor could then determine the probability of either parent having the recessive allele. Answer F is incorrect because a blood test would determine only if they had the recessive disease. Answer G is incorrect because a DNA fingerprint is used to identify the source of a sample of DNA and to compare DNA sequences. Answer H is incorrect because a karyotype shows the number and kinds of chromosomes a person has, which does not show hemophilia.

Question 5 Answer A is the correct choice. 2 *Tt* : 2 *tt* reduces to 1 *Tt* : 1 *tt*. Answer B is incorrect because the offspring are 50% *Tt* and 50% *tt*. Answer C is incorrect because to produce any *Tt*, one parent has to have at least 1 *T*, which, using the Punnett square, gives the probability of 50% of the offspring inheriting *T*, not 25%. Answer D is incorrect because to produce any *tt*, each parent must have at least 1 *t*. If the parents are Tt, the ratio would be 1 *TT* : 2 *Tt* : 1 *tt*. If one parent was *Tt* and the other was tt, the ratio would be 2 *Tt* : 2 *tt*.

Question 6 The correct answer is 120. Punnett square predicts 1/2 will be *tt*, or 1/2 of 240, which is 120.

Answers

1. A
2. I
3. D
4. I
5. A
6. 120

Test Prep Correlations

SC.F.2.4.1: Items 1, 3-6
SC.H.3.4.1: Item 2

Skills Practice Lab

Modeling Monohybrid Crosses

Teacher's Notes

Time Required
One 50-minute class period

Ratings
EASY ——→ HARD

TEACHER PREPARATION (1)
STUDENT SETUP (1)
CONCEPT LEVEL (3)
CLEANUP (1)

Skills Acquired
- Predicting outcomes
- Calculating data
- Organizing data
- Analyzing data

Scientific Methods
In this lab, students will:
- Make Observations
- Analyze the Results
- Draw Conclusions
- Communicate the Results

Materials and Equipment
Materials for this lab can be ordered from WARD'S. See *Master Materials List* at the front of this book for catalog numbers.

Safety Cautions
Remind students to avoid tasting the peas and lentils.

Skills Practice Lab

Modeling Monohybrid Crosses SC.H.3.4.1

SKILLS
- Predicting outcomes
- Calculating data
- Organizing data
- Analyzing data

OBJECTIVES
- **Predict** the genotypic and phenotypic ratios of offspring resulting from the random pairing of gametes.
- **Calculate** the genotypic ratio and phenotypic ratio among the offspring of a monohybrid cross.

MATERIALS
- lentils
- green peas
- 2 Petri dishes

Before You Begin

A **monohybrid cross** is a cross that involves one pair of contrasting traits. Different versions of a gene are called **alleles.** When two different alleles are present and one is expressed completely and the other is not, the expressed allele is **dominant** and the unexpressed allele is **recessive.**

1. Write a definition for each boldface term in the paragraph above.
2. Based on the objectives for this lab, write a question you would like to explore about heredity.

Procedure

PART A: Simulating a Monohybrid Cross

1. You will model the random pairing of alleles by choosing lentils and peas from Petri dishes. These dried seeds will represent the alleles for seed color. A green pea will represent *G*, the dominant allele for green seeds, and a lentil will represent *g*, the recessive allele for yellow seeds.
2. Each Petri dish will represent a parent. Label one Petri dish "female gametes" and the other Petri dish "male gametes." Place one green pea and one lentil in the Petri dish labeled "female gametes" and place one green pea and one lentil in the Petri dish labeled "male gametes."
3. Each parent contributes one allele to each offspring. Model a cross between these two parents by choosing a random pairing of the dried seeds from the two Petri dishes. Do this by simultaneously picking one seed from each Petri dish *without looking*. Place the pair of seeds together on the lab table. The pair of seeds represents the genotype of one offspring.
4. Record the genotype of the first offspring in your lab report in a table like Table A shown below.

DATA TABLE A

	Gamete Pairings	
Trial	Offspring genotype	Offspring phenotype
1		
2		
3		
4		
5		
6		
7		
8		
9		
10		

5. Return the seeds to their original dishes and repeat step 3 nine more times. Record the genotype of each offspring in your data table.

186

Answers to Before You Begin

1. A monohybrid cross is a genetic cross that involves one pair of contrasting traits. Alleles are different versions of a gene that code for the same trait. A dominant allele is a version of a gene that is expressed in the presence of other alleles that code for the same trait. A recessive allele is a version of a gene that is not expressed when it is in the presence of other alleles that code for the same trait.
2. Answers may vary.

Procedure

4. Answers will vary.
5. Answers will vary.
6. *GG* = green seeds; *Gg* = green seeds; *gg* = yellow seeds.
8. Answers will vary but genotypic ratios should approximate 1 *GG*:2 *Gg*:1 *gg*. Phenotypic ratios should approximate 3:1, or three individuals with green seeds to one individual with yellow seeds.
9. See answer to Procedure step 8.
10. See answer to Procedure step 8.
11. See answer to Procedure step 8.
12. See answer to Procedure step 8.

6. Based on each offspring's genotype, determine and record each offspring's phenotype.

PART B: Calculating Genotypic and Phenotypic Ratios

7. In your lab report, prepare a data table similar to Table B shown below.

DATA TABLE B

Offspring Ratios		
Genotypes	Total	Genotypic ratios
Homozygous dominant (*GG*)		
Heterozygous (*Gg*)		___ : ___ : ___
Homozygous recessive (*gg*)		
Phenotypes		Phenotypic ratios
Green seeds		
Yellow seeds		_____ : _____

8. Determine the genotypic and phenotypic ratios among the offspring. First count and record the number of homozygous dominant, heterozygous, and homozygous recessive individuals you recorded in Table A. Then record the number of offspring that produce green seeds and the number that produce yellow seeds under "Phenotypes" in your data table.
9. Calculate the genotypic ratio for each genotype using the following equation: phenotypic ratio =

$$\frac{\text{number of offspring with a given genotype}}{\text{total number of offspring}}$$

10. Calculate the phenotypic ratio for each phenotype using the following equation: phenotypic ratio =

$$\frac{\text{number of offspring with a given phenotype}}{\text{total number of offspring}}$$

11. Now pool the data for the whole class, and record the data in your lab report in a second table like Table B.
12. Compare the class's sample with your small sample of 10. Calculate the genotypic and phenotypic ratios for the class data, and record them in your data table.
13. Construct a Punnett square showing the parents and their offspring in your lab report.
14. Clean up your materials before leaving the lab.

Analyze and Conclude

1. **Summarizing Results** What character is being studied in this investigation?
2. **Analyzing Data** What are the genotypes of the parents? Describe the genotypes of both parents using the terms homozygous or heterozygous, or both. Did Table B reflect a classic monohybrid-cross phenotypic ratio of 3:1?
3. **Drawing Conclusions** If a genotypic ratio of 1:2:1 is observed, what must the genotypes of both parents be?
4. **Predicting Patterns** Show what the genotypes of the parents would be if 50 percent of the offspring were green and 50 percent of the offspring were yellow.
5. **Further Inquiry** Construct a Punnett square for the cross of a heterozygous black guinea pig and an unknown guinea pig whose offspring include a recessive white-furred individual. What are the possible genotypes of the unknown parent?

? Do You Know?

Do research in the library or media center to answer these questions:

1. How are hybrids of plants such as orchids produced?
2. What is *hybrid vigor?*

Use the following Internet resources to explore your own questions about genetics.

internet connect
www.scilinks.org
Topic: Genetic Code
Keyword: HX4089
SCILINKS. Maintained by the National Science Teachers Association

187

Answers to Analyze and Conclude

1. The character being investigated is seed, or fruit, color.
2. Both parents are heterozygous green, *Gg*. Each seed represents an allele. The pairs represent the gametes that an offspring will receive. Answers will vary. In a sample size of 10 crosses, no clear ratio may be evident. When combining data from the entire class, the 3:1 ratio should be seen.
3. Both parents must be heterozygous, *Gg*.
4. One parent would be heterozygous, *Gg*, and the other parent would be homozygous recessive, *gg*.
5. Students should draw two Punnett squares. One should show a cross between two heterozygous individuals; the other should show a heterozygous individual crossed with a homozygous recessive individual. If *B* represents the dominant black fur color and *b* stands for white fur color, the unknown parent could be genotypically *Bb* or *bb*.

Answers to Do You Know

1. Normally, hybrids are produced through cross-pollination. However, for many plants, such as orchids, hybrids are often produced using tissue culture techniques.
2. Hybrid vigor is a term that is used to describe hybrids that are more vigorous, disease-resistant, and sometimes more free-blooming than either of the parental species.

Chapter Opener		**SE Quick Review,** p. 189 ■ **SE Reading Activity,** p. 189 ■ **TE Using the Figure,** p. 188 GENERAL **TE Opening Activity,** p. 189 GENERAL
& 3 • 90 min pp. 190–193 **Section 1** Identifying the Genetic Material	**OSP Lesson Plan*** **TR Bellringer*** **TR Section Outline*** **TR C9 Griffith's Discovery of Transformation*** **TR C10 The Hershey-Chase Experiment***	**TE Using the Figure,** p. 191 GENERAL **TE Group Activity** Role Playing, p. 192 ADVANCED
& 5 • 90 min pp. 194–197 **Section 2** The Structure of DNA	**OSP Lesson Plan*** **TR Bellringer*** **TR Section Outline*** **TR C11 DNA Double Helix***	**TE Demonstration,** p. 194 **SE Quick Lab** Observing Properties of DNA, p. 195 ■ **CRF Datasheets for In-Text Labs** Observing Properties of DNA* **TE Using the Figure,** p. 196 BASIC **CRF Exploration Lab** Extracting DNA* GENERAL **CRF Skills Practice** Identifying Genetic Disorders by Karyotyping* GENERAL
& 7 • 90 min pp. 198–200 **Section 3** The Replication of DNA	**OSP Lesson Plan*** **TR Bellringer*** **TR Section Outline*** **TR C12 DNA Replication*** **TR C13 Replication Forks***	**TE Using the Figure,** p. 199 GENERAL **SE Math Lab** Analyzing the Rate of DNA Replication, p. 199 ■ **CRF Datasheets for In-Text Labs** Analyzing the Rate of DNA Replication* **TE Demonstration,** p. 200 **SE Exploration Lab** Modeling DNA Structure, pp. 204–205 ■ **CRF Datasheets for In-Text Labs** Modeling DNA Structure* ◆

& 9 • 90 min

Chapter Review and Assessment Resources

- **SE Study Zone,** p. 201 ■
- **SE Performance Zone,** p. 202 ■
- **SE FCAT Test Prep,** p. 203 ■
- **TR Concept Mapping***
- **CRF Vocabulary Review Worksheet*** ■ GENERAL
- **CRF Test Prep Pretest*** ■ GENERAL
- **CRF Chapter Test*** ■ GENERAL
- **CRF Chapter Test*** ADVANCED
- **OSP Test Generator***
- **OSP Test Item Listing***
- **OSP Modified Chapter Test*** BASIC

Online and Technology R

Holt Online Learning

Visit go.hrw.com for access to Holt Online Learning or enter the keyword **HX6 Home** for a variety of free online resources.

One-Stop Planner® CD-ROM

This CD-ROM package includes
- Lab Materials QuickList Software
- Holt Calendar Planner
- Customizable Lesson Plans
- Printable Worksheets
- ExamView® Test Generator
- Interactive Teacher Edition
- Holt PuzzlePro® Resources
- Holt PowerPoint® Resources

Guided Reading Audio CDs

These CDs are designed to help auditory learners and reluctant readers. (Audio CDs are also available in Spanish.)

Compression guide:
To shorten your instruction because of time limitations, omit blocks 1 and 2.

KEY

TE	Teacher Edition	CRF	Chapter Resource File	*	Also on One-Stop Planner
SE	Student Edition	TR	Teaching Transparency	■	Also Available in Spanish
OSP	One-Stop Planner			◆	Requires Advance Prep

SKILLS DEVELOPMENT AND PRACTICE	SECTION REVIEW AND ASSESSMENT	CORRELATIONS
		Florida Sunshine Standards/FCAT
TE **Reading Skill Builder** Paired Reading, p. 191 GENERAL TE **Reading Skill Builder** Reading Organizer, p. 192 BASIC CRF **Directed Reading Worksheet*** BASIC CRF **Active Reading Worksheet*** GENERAL	SE **Section 1 Review** ■ TE **Reteaching,** p. 193 BASIC TE **Quiz,** p. 193 GENERAL TE **Alternative Assessment,** p. 193 GENERAL CRF **Quiz*** ■ BASIC	SC.H.1.4.2, SC.H.2.4.2
TE **Skill Builder** Interpreting Visuals, p. 195 BASIC CRF **Directed Reading Worksheet*** BASIC CRF **Active Reading Worksheet*** GENERAL	SE **Section 2 Review** ■ TE **Reteaching,** p. 197 BASIC TE **Quiz,** p. 197 GENERAL TE **Alternative Assessment,** p. 197 GENERAL CRF **Quiz*** ■ BASIC	SC.F.2.4.2, SC.H.1.4.5
CRF **Directed Reading Worksheet*** BASIC CRF **Active Reading Worksheet*** GENERAL CRF **Critical Thinking Worksheet*** ADVANCED CRF **Science Skills Worksheet*** GENERAL	SE **Section 3 Review** ■ TE **Reteaching,** p. 200 BASIC TE **Quiz,** p. 200 GENERAL TE **Alternative Assessment,** p. 200 GENERAL CRF **Quiz*** ■ BASIC **Modified Worksheet*** BASIC	SC.F.1.4.5, SC.G.1.4.1, SC.H.3.4.1

Holt Biology Interactive Tutor CD-ROM

This CD-ROM consists of interactive activities that give students a fun way to extend their knowledge of biology concepts.

Unit 6 Gene Expression

SciLinks NSTA
www.scilinks.org
Maintained by the **National Science Teachers Association**

Topic: DNA
SciLinks Code: HX4058

Topic: DNA Replication
SciLinks Code: HX4059

CHAPTER 9

Overview

Tell students the purpose of this chapter is to describe DNA—the hereditary material that directs the development of a new generation of individuals patterned after the parental generation. The chapter continues the story of heredity from the work of Mendel, who elucidated the patterns of inheritance. Section 1 describes how scientists determined that the hereditary material is DNA. Section 2 chronicles the groundbreaking work that led to an understanding of the structure of DNA. Section 3 describes how DNA replicates prior to cell division, ensuring that each new cell has a full complement of genetic material.

Using the Figure

Explain that the computer model of DNA shown in the photo is a representation of the molecular structure of DNA. However, the photo shows only a small fraction of a typical DNA molecule. DNA molecules are so long that short stretches are wound around proteins like thread on spools. The molecule is further coiled and coiled again, forming a manageable package within the cell. Tell students this amazing thought: If just one molecule of DNA could be stretched out and magnified to the size of the molecule in the photo, it would likely stretch from New York City to Disney World in Florida!

Assessing Prior Knowledge

Before beginning this chapter, be sure your students can answer the questions in the ***Quick Review*** on the facing page. If they cannot, encourage them to reread the sections indicated.

Computer model of DNA

188

Chapter Correlations — Florida Sunshine Standards/FCAT

SC.F.1.4.5 Complex interactions among the different kinds of molecules in the cell cause distinct cycles of activity governed by proteins.

SC.F.2.4.2 Every cell contains a "blueprint" coded in DNA molecules that specify how proteins are assembled to regulate cells.

SC.G.1.4.1 There is great diversity and interdependence among living things.

SC.H.1.4.2 From time to time, major shifts occur in the scientific view of how the world works, but more often the changes that take place in the body of scientific knowledge are small modifications of prior knowledge.

SC.H.1.4.5 New ideas in science are limited by the context in which they are conceived, are often rejected by the scientific establishment, sometimes spring from unexpected findings, and usually grow slowly from many contributors.

SC.H.2.4.2 Scientists control conditions in order to obtain evidence, but when that is not possible for practical or ethical reasons, they try to observe a wide range of natural occurrences to discern patterns.

CHAPTER

9 DNA: The Genetic Material

Florida Sunshine Standards/FCAT

SC.F.1.4.5 Complex interactions among the different kinds of molecules in the cell cause distinct cycles of activity governed by proteins.
SC.F.2.4.2 Every cell contains a "blueprint" coded in DNA molecules that specify how proteins are assembled to regulate cells.
SC.G.1.4.1 There is great diversity and interdependence among living things.
SC.H.1.4.2 From time to time, major shifts occur in the scientific view of how the world works, but more often the changes that take place in the body of scientific knowledge are small modifications of prior knowledge.
SC.H.1.4.5 New ideas in science are limited by the context in which they are conceived, are often rejected by the scientific establishment, sometimes spring from unexpected findings, and usually grow slowly from many contributors.
SC.H.2.4.2 Scientists control conditions in order to obtain evidence, but when that is not possible for practical or ethical reasons, they try to observe a wide range of natural occurrences to discern patterns.

Quick Review

Answer the following without referring to earlier sections of your book.

1. **Describe** the structure of chromosomes. *(Chapter 6, Section 1)*
2. **Define** the term *gene*. *(Chapter 6, Section 1)*
3. **Identify** the stage in a cell cycle in which DNA is copied. *(Chapter 6, Section 2)*
4. **Identify** changes in DNA as mutations. *(Chapter 6, Section 3)*
5. **Summarize** Mendel's theory of heredity. *(Chapter 8, Section 2)*

Did you have difficulty? *For help, review the sections indicated.*

Reading Activity

Before you read this chapter, write a short list of all the things you know about DNA. Then write a list of the things that you want to know about DNA. Save your list, and to assess what you have learned, see how many of your own questions you can answer after reading this chapter.

Each human cell contains about 6 billion pairs of nucleotides—one pair for each "step" in the winding staircase of the DNA model shown on the facing page.

Looking Ahead

internet connect

www.scilinks.org
National Science Teachers Association *sci*LINKS Internet resources are located throughout this chapter.

SCILINKS. Maintained by the National Science Teachers Association

189

Opening Activity — GENERAL

Make a stack of books totaling about 10,000 pages. Tell students that the stack of books represents only about one-fiftieth of the information contained in the DNA of every human cell. Correlate this with the amount of information required to code for a human being.

Quick Review Answers

1. A chromosome consists of two replicated strands of DNA tightly coiled around proteins. The two strands, called chromatids, are attached at a point called a centromere.
2. A gene is a segment of DNA that codes for a protein or RNA molecule.
3. A cell's DNA is copied during the synthesis (S) phase.
4. When chromosomes break, the broken pieces can detach completely or can reattach in various ways. Therefore, the chromosome is changed, or mutated.
5. (a) For each inherited trait, an individual has two copies of the gene, one from each parent. (b) There may be alternative versions of genes. (c) When two different alleles occur together, one of them may be completely expressed, while the other may have no observable effect on the organism's appearance. (d) When gametes are formed, the alleles for each gene in an individual separate independently of one another, and when gametes unite during fertilization, each gamete contributes one allele.

Chapter Resource File
- Vocabulary Worksheets
- Concept Mapping

Reading Activity Answers

Students may already know that DNA is the hereditary material and that it is located in the nucleus of a cell. Some may also know that the DNA molecule looks like a twisted ladder, but they are unlikely to recall the stories of how scientists discovered that DNA is the hereditary material and the details of its structure.

Focus

Overview

Before beginning this section review with your students the objectives listed in the Student Edition. Section 1 describes how the work of Griffith and Avery led to the discovery that hereditary material is composed of DNA, and how the work of Hershey and Chase revealed that genes are made of DNA.

Bellringer

Show students photographs of various organisms. Ask them: Does the DNA of all organisms have the same structure? How does your DNA differ from that of other species? (Explain that the basic structure of DNA—a polymer made of nucleotide subunits—is the same in all organisms. However, the sequence of the nucleotides differs. These differences are minor within species and become greater the more distantly related organisms are.)

Motivate

Discussion Question — BASIC

Tell students that Griffith was an English medical officer who was studying the bacterium responsible for a pneumonia epidemic in London in the late 1920s. Ask students what the leading cause of death was then (infectious diseases) and today. (Heart disease and cancer, two non-infectious disease.) Ask them why they think this shift has occurred. (One reason is the discovery of antibiotics.)

Section Correlations

SC.H.1.4.2
SC.H.2.4.2

Section 1 Identifying the Genetic Material

Florida Sunshine Standards/FCAT
SC.H.1.4.2 From time to time, major shifts occur in the scientific view of how the world works, but more often the changes that take place in the body of scientific knowledge are small modifications of prior knowledge.
SC.H.2.4.2 Scientists control conditions in order to obtain evidence, but when that is not possible for practical or ethical reasons, they try to observe a wide range of natural occurrences to discern patterns.

Objectives

- **Relate** Griffith's conclusions to the observations he made during the transformation experiments. SC.H.1.4.2
- **Summarize** the steps involved in Avery's transformation experiments, and state the results. SC.H.1.4.2
- **Evaluate** the results of the Hershey and Chase experiment. SC.H.2.4.2

Key Terms

vaccine
virulent
transformation
bacteriophage

Transformation

Mendel's experiments and results answered the question of why you resemble your parents. You resemble your parents because you have copies of their chromosomes, which contain sets of instructions called genes. But Mendel's work created more questions, such as, What are genes made of? Scientists believed that if they could answer this question they would understand how chromosomes function in heredity.

Griffith's Experiments

In 1928, an experiment completely unrelated to the field of genetics led to an astounding discovery about DNA. Frederick Griffith, a bacteriologist, was trying to prepare a vaccine *(vahk SEEN)* against pneumonia. *Streptococcus pneumoniae* (abbreviated *S. pneumoniae*), is shown in **Figure 1.** *S. pneumoniae* is a prokaryote (of the type commonly called a bacterium) that causes pneumonia. A **vaccine** is a substance that is prepared from killed or weakened disease-causing agents, including certain bacteria. The vaccine is introduced into the body to protect the body against future infections by the disease-causing agent.

Griffith worked with two types, or strains, of *S. pneumoniae,* as shown in **Figure 2.** The first strain is enclosed in a capsule composed of polysaccharides. The capsule protects the bacterium from the body's defense systems. This helps make the microorganism **virulent** *(VIHR yoo luhnt),* or able to cause disease. Because of the capsule, this strain of *S. pneumoniae* grows as smooth-edged (*S*) colonies when grown in a Petri dish. The second strain of *S. pneumoniae* lacks the polysaccharide capsule and does not cause disease. When grown in a Petri dish, the second strain forms rough-edged *(R)* colonies.

Griffith knew that mice infected with the *S* bacteria grew sick and died, while mice infected with the *R* bacteria were not harmed, as shown in Figure 2. To determine whether the capsule on the *S* bacteria was causing the mice to die, Griffith injected the mice with dead *S* bacteria. The mice remained healthy. Griffith then prepared a vaccine of weakened *S* bacteria by raising their temperature to a point at which the bacteria were "heat-killed," meaning that they could no longer reproduce. (The capsule remained on the bacteria).

Figure 1 ***Streptococcus pneumoniae***

Certain types of *S. pneumoniae* bacteria can cause the lung disease pneumonia.

Magnification: 17,250×

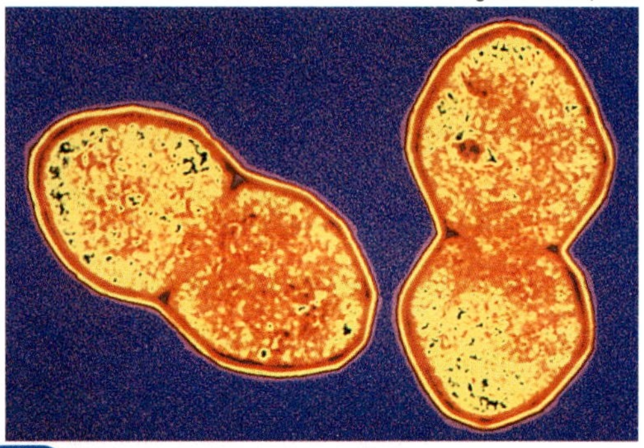

Chapter Resource File
- Directed Reading BASIC
- Active Reading GENERAL

One-Stop Planner CD-ROM
- Reading Organizers BASIC
- Reading Strategies BASIC

Transparencies

TR Bellringer
TR C9 Griffith's Discovery of Transformation
TR C10 The Hershey-Chase Experiment

Figure 2 Griffith's discovery of transformation

Griffith discovered that harmless bacteria could turn virulent when mixed with bacteria that cause disease.

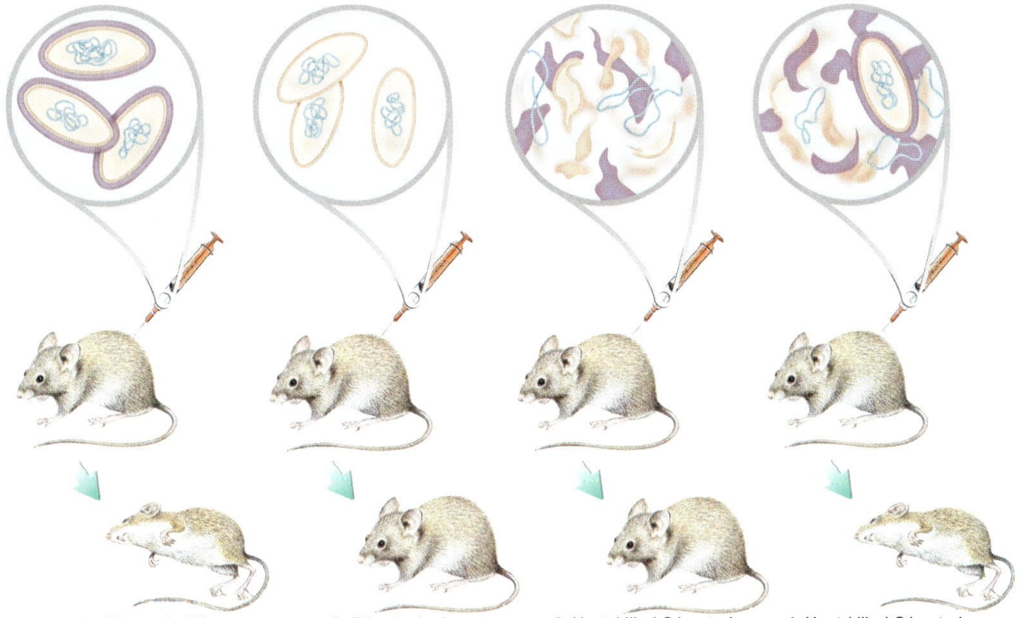

When Griffith injected mice with heat-killed *S* bacteria, the mice still lived. Thus, Griffith knew it was not the capsule on the *S* bacteria that killed the mice. He then mixed the harmless live *R* bacteria with the harmless heat-killed *S* bacteria. Mice injected with this mixture of previously harmless preparations died. When Griffith examined the blood of the dead mice, he found that the live *R* bacteria had acquired capsules. Somehow, the harmless *R* bacteria had changed and became virulent *S* bacteria. Griffith had discovered what is now called transformation. **Transformation** is a change in genotype caused when cells take up foreign genetic material. But the cause of the transformation was not known at the time.

Avery's Experiments

The search for the substance responsible for transformation continued until 1944. Then, an elegant series of experiments showed that the activity of the material responsible for transformation is not affected by protein-destroying enzymes. The activity is stopped, however, by a DNA-destroying enzyme. Thus, almost 100 years after Mendel's experiments, Oswald Avery and his co-workers at the Rockefeller Institute, in New York City, demonstrated that DNA is the material responsible for transformation. DNA contains the instructions for the making of the capsule in the *S* strain of *S. pneumoniae*.

WORD Origins

The word *virulent* is from the Latin *virulentus*, which means "full of poison." Knowing this makes it easier to remember that a microorganism's virulence is its ability to cause disease.

191

Teach

Using the Figure — GENERAL

Use **Figure 2** to review each step of Griffith's experiment. Encourage students to put the explanation into their own words so that they will better understand the results and the implications of the experiment. **LS Visual**

Teaching Tip — GENERAL

The Heat Factor DNA can tolerate temperatures around 90°C without being altered, whereas proteins are denatured around 60°C. Ask students what effect temperature had on Griffith's work. (The DNA of the heat-killed bacteria was not altered and became incorporated into the DNA of the live bacteria.)

READING SKILL BUILDER — GENERAL

Paired Reading Give students self-stick notes, and have them mark areas of difficulty as they silently read Section 1. Have students use a question mark to signify an area that is unclear and a check mark to indicate understanding. Remind students to study the captions, diagrams, and illustrations as they read. After completing the section, have students work with a partner to discuss what they did or did not understand. Have students write their remaining questions on the board for later discussion.

TECHNOLOGY CONNECTION

During his experiments, Griffith unwittingly manipulated genes. Today, the manipulation of genes is known as genetic engineering or recombinant DNA technology. Transformation, in particular, is a common modern-day genetic engineering technique. For example, bacteria can be used to introduce foreign genes into plant cells. Whole plants can then be grown from the altered genetic material, producing plants with desirable traits such as long shelf life or resistance to crop pests.

Integrating Physics and Chemistry

Before students read about the Hershey-Chase experiment, review the term "radioactive." Have students describe the causes of radioactivity. Radioactive elements have unstable nuclei that emit alpha and beta particles or energy as gamma rays. These emissions make it easier for scientists to detect and trace the path of the radioactive elements as they interact with other materials or are transformed during metabolic processes. In addition, point out that although any element can be radioactive, this chemical property is more common as atomic mass increases.

BIOLOGY INTERACTIVE TUTOR

- **Unit 6—Gene Expression**

This engaging tutorial reviews the structure and replication of DNA, as well as the processes of transcription and translation.

Teach, *continued*

READING SKILL BUILDER — BASIC

Reading Organizer To summarize Hershey and Chase's experiment, have students design a Graphic Organizer like the one shown at the bottom of this page. Have them include a paragraph that explains how the experiment showed that DNA is the genetic material. **LS Logical**

Teaching Tip — GENERAL

Heredity and DNA Inform students that many scientists were reluctant to accept the results of experiments that suggested DNA was the hereditary material. Ask students to explain why most scientists initially thought that the hereditary material was protein, not DNA. (Proteins are made from 20 different amino acids, which can be combined in a variety of ways. Additionally, scientists knew that proteins were important to many aspects of cell structure and metabolism. They knew very little about DNA.)

Group Activity — ADVANCED

Role Playing Divide the class into three groups: One group will represent Hershey and Chase, the second group will represent a team of scientists who will question the first group on their results, and the third group will form a committee that will decide if the work is to be published in a recognized scientific journal. The questions of the second group will help the committee make a decision, and the committee will write a summary supporting its decision. **LS Interpersonal**

Section Review Correlations

SC.F.2.4.2: Items 2, 3

SC.H.1.4.1: Items 1, 4, 5, 6

Viral Genes and DNA

Even though Avery's experiments clearly indicated that the genetic material is composed of DNA, many scientists remained skeptical. Scientists knew that proteins were important to many aspects of cell structure and metabolism, so most of them suspected that proteins were the genetic material. They also knew very little about DNA, so they could not imagine how DNA could carry genetic information.

BIO graphic

Figure 3

The Hershey-Chase Experiment

Bacteriophages were used to show that DNA, not protein, is the genetic material of viruses.

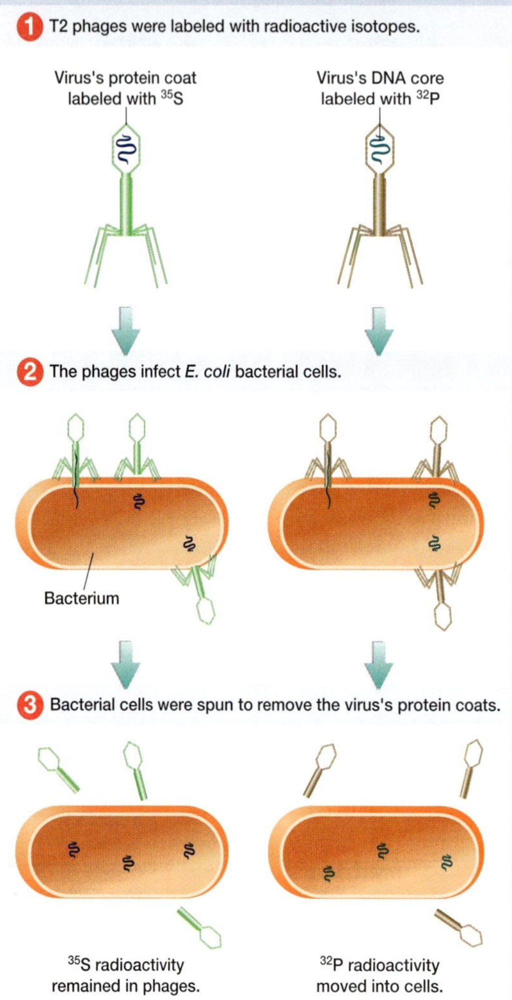

DNA's Role Revealed

In 1952, Alfred Hershey and Martha Chase, scientists at Cold Spring Harbor Laboratory, in New York, performed an experiment that settled the controversy. It was known at that time that viruses, which are much simpler than cells, are composed of DNA or RNA surrounded by a protective protein coat. A **bacteriophage** *(bak TIHR ee uh fayj)*, also referred to as phage *(fayj)*, is a virus that infects bacteria. It was also known that when phages infect bacterial cells, the phages are able to produce more viruses, which are released when the bacterial cells rupture.

What was not known at the time was how the bacteriophage reprograms the bacterial cell to make viruses. Does the phage DNA, the protein, or both issue instructions to the bacteria?

Hershey and Chase used the bacteriophage T2, shown in **Figure 3,** to answer this question. Hershey and Chase knew that the only molecule in the phage that contains phosphorus is its DNA. Likewise, the only phage molecules that contain sulfur are the proteins in its coat. Hershey and Chase used these differences to carry out the experiment shown in Figure 3.

Step 1 Hershey and Chase first grew T2 with *Escherichia coli* (abbreviated *E. coli*) bacteria in a nutrient medium that contained radioactive sulfur (^{35}S). The protein coat of the virus would incorporate the ^{35}S. They grew a second batch of phages with *E. coli* bacteria in a nutrient medium that contained radioactive phosphorus (^{32}P). The radioactive phosphorus would become part of the phages' DNA.

192

Graphic Organizer

Use this graphic organizer with **Reading Skill Builder** on this page.

T2 phage grown in ^{32}P → *E. coli* → Blender → Most ^{32}P found in bacteria → ^{32}P in new phages

T2 phage grown in ^{35}S → *E. coli* → Blender → Most ^{35}S found in phage

Step 2 The ^{35}S-labeled and ^{32}P-labeled phages were used to infect two separate batches of *E. coli* bacteria. Because radioactive elements release particles that can be detected with machines, they can be followed, or traced, in a biological process. Scientists could determine whether it was the DNA, the protein, or both that were being transferred into the bacterial cells to reprogram the bacteria.

Step 3 After a few minutes, the scientists tore the ^{35}S-labeled phages off the surfaces of the bacteria (with the help of a blender). The bacteria infected with the ^{32}P-labeled phage were likewise mixed in a blender. The investigators used a centrifuge to separate the bacteria and phages. The heavier, bacterial cells formed a solid layer at the bottom of the centrifuge tubes. The lighter, viral parts remained in the upper, liquid layer.

Hershey and Chase examined the layers from the ^{35}S-infected bacteria. The scientists found that most of the ^{35}S label was still part of the phage (the upper layer), meaning the protein was not injected into the bacteria. When they examined the layers from the ^{32}P-infected bacteria, the scientists found the ^{32}P label mostly in the layer containing the bacterial cells (the lower layer). The DNA had been injected into the hosts. Moreover, the new generation of phages that was produced by these bacteria also contained radioactive DNA.

Hershey and Chase concluded that the DNA of viruses is injected into the bacterial cells, while most of the viral proteins remain outside. The injected DNA molecules causes the bacterial cells to produce more viral DNA and proteins. This meant that the DNA, rather than proteins, is the hereditary material, at least in viruses.

These important experiments, and many others since, have shown that DNA is the molecule that stores genetic information in living cells. As you will see in the next section, the structure of DNA makes DNA particularly well suited to this function.

Real Life

Many viruses infect humans.

Although the T2 viruses used by Hershey and Chase infect bacteria, many other viruses infect humans. For example, viruses are responsible for causing polio, the common cold, measles, and rabies.

Finding Information

Research other human diseases that are caused by viruses.

Section 1 Review

1. **Summarize** Griffith's transformation experiments. SC.H.1.4.1
2. **Describe** how Avery's experiment supplied evidence that DNA, and not protein, is the genetic material. SC.F.2.4.2
3. **Describe** the contributions of Hershey and Chase to the understanding that DNA is the genetic material. SC.F.2.4.2
4. **Critical Thinking Evaluating Methods** Why did heat kill Griffith's *S* bacteria? SC.H.1.4.1
5. **Critical Thinking Applying Information** What might Hershey and Chase have concluded if they had found ^{32}P and ^{35}S in the bacterial cells? SC.H.1.4.1
6. *FCAT Prep* Who performed the first experiments that correctly identified the molecule that carries genetic information?
   A Oswald Avery
   B Martha Chase
   C Frederick Griffith
   D Alfred Hershey
   SC.H.1.4.1

193

Answers to Section Review

1. Mice were injected with either *S* bacteria, *R* bacteria, heat-killed *S* bacteria, or a mixture of heat-killed *S* bacteria and *R* bacteria. The mice injected with the *S* bacteria or the mixture of heat-killed *S* bacteria and *R* bacteria died; those injected with the *R* or the heat-killed *S* bacteria lived.
2. Avery showed that DNA-destroying enzymes, but not protein-destroying enzymes, prevented transformation.
3. Hershey and Chase showed that the DNA of viruses was injected into bacterial cells and that this DNA caused the bacterial cells to produce more viruses containing DNA.
4. The heat denatured bacterial enzymes and proteins that were necessary for survival.
5. Hershey and Chase would have concluded that both the DNA and the protein coats had entered the bacterial cells, and would have been unable to determine which was the genetic material.
6. **A.** Correct. **B.** Incorrect. See answer D. **C.** Incorrect. Griffith's experiments showed that harmless bacteria could turn virulent when mixed with bacteria that cause disease, but they did not identify DNA as the genetic material. **D.** Incorrect. The experiments of Hershey and Chase confirmed the earlier work of Avery.

Real Life

Answer

Answers will vary. Other human diseases caused by viruses include rubella, mumps, AIDS, influenza, hepatitis B, and infectious mononucleosis. Viruses have also been associated with several types of cancer, such as liver cancer.

Close

Reteaching — Basic

Arrange students into pairs. Provide each pair with materials to make cards on which they diagram each step of Griffith's experiment. Tell students to use **Figure 2** as a guide. After they have finished making the cards, tell them to shuffle the cards and place them face up on their tables in the correct order.

Quiz — General

True or False:

1. Griffith's experiments with *S. pneumoniae* in mice showed that harmless bacteria could turn virulent when mixed with heat-killed bacteria that cause disease. (True)
2. Avery's experiments clearly demonstrated that the genetic material is composed of DNA. (True)
3. The experiments of Hershey and Chase cast doubt on whether DNA was the hereditary material. (False)

Alternative Assessment — General

Tell students to imagine they are news reporters working from 1920 to 1955. They are to choose a scientific discovery from the work of Griffith, Avery, or Hershey and Chase, and write a front-page newspaper, magazine, radio or TV report describing the discovery and its contribution to scientific understanding and to society.

Section 2

Focus

Overview

Before beginning this section review with your students the objectives listed in the Student Edition. In this section, the double-helical, nucleotide-subunit structure of DNA is explained, as well as the contributions of Chargaff, Franklin, and Wilkins in helping Watson and Crick determine that structure. The rules for pairing between nitrogen bases in DNA are also reviewed.

Bellringer

Tell students that DNA is often compared to a ladder. Ask them to look at **Figure 4** to answer the following questions: How is the structure of DNA similar to that of a ladder? How is it different from that of a ladder? (Students have not yet studied the structure of DNA, so answers will be simple. The DNA has "side rails" like a ladder and rungs like a ladder. However, the DNA is twisted, while a ladder is flat, and the rungs of a ladder are single units, whereas those in DNA are made of two units.) **LS Visual**

Motivate

Demonstration

Construct a model of DNA as you describe the structure for students. Use either a model purchased from a biological sciences supply house, or use everyday materials such as candy, paperclips, pipe cleaners, or pop beads. In your materials, use differences in color or structure to represent sugars, phosphates, and each of the four bases. **LS Visual**

Section Correlations

SC.F.2.4.2
SC.H.1.4.5

Section 2 The Structure of DNA

Florida Sunshine Standards/FCAT

SC.F.2.4.2 Every cell contains a "blueprint" coded in DNA molecules that specify how proteins are assembled to regulate cells.

SC.H.1.4.5 New ideas in science are limited by the context in which they are conceived, are often rejected by the scientific establishment, sometimes spring from unexpected findings, and usually grow slowly from many contributors.

Objectives

- **Describe** the three components of a nucleotide. SC.F.2.4.2
- **Develop** a model of the structure of a DNA molecule. SC.F.2.4.2
- **Evaluate** the contributions of Chargaff, Franklin, and Wilkins in helping Watson and Crick determine the double-helical structure of DNA. SC.H.1.4.5
- **Relate** the role of the base-pairing rules to the structure of DNA. SC.F.2.4.2

Key Terms

double helix
nucleotide
deoxyribose
base-pairing rules
complementary base pair

A Winding Staircase

By the early 1950s, most scientists were convinced that genes were made of DNA. They hoped that the mystery of heredity could be solved by understanding the structure of DNA. The research of many scientists led two young researchers at Cambridge University, James Watson and Francis Crick, to piece together a model of the structure of DNA. The discovery of DNA's structure was important because it clarified *how* DNA could serve as the genetic material.

Watson and Crick determined that a DNA molecule is a **double helix**—two strands twisted around each other, like a winding staircase. As shown in **Figure 4,** each strand is made of linked nucleotides *(NOO klee oh tiedz).* **Nucleotides** are the subunits that make up DNA. Each nucleotide is made of three parts: a phosphate group, a five-carbon sugar molecule, and a nitrogen-containing base. Figure 4 shows how these three parts are arranged to form a nucleotide. The five-carbon sugar in DNA nucleotides is called **deoxyribose** *(dee ahk see RIE bohs),* from which DNA gets its full name, deoxyribonucleic acid.

Figure 4 DNA double helix

Watson and Crick's model of DNA is a double helix composed of two nucleotide chains that are twisted around a central axis and held together by hydrogen bonds.

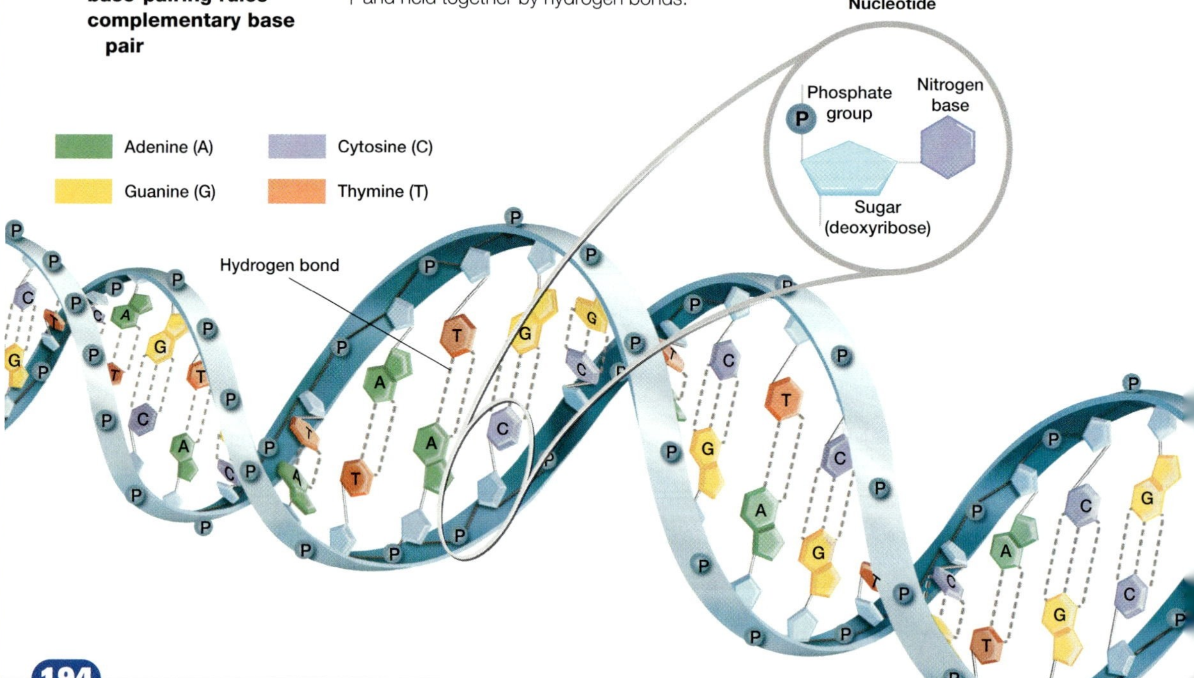

194

Chapter Resource File

- Directed Reading BASIC
- Active Reading GENERAL
- Data Sheet for Quick Lab GENERAL

One-Stop Planner CD-ROM

- Reading Organizers BASIC
- Reading Strategies BASIC
- Supplemental Reading Guide The Double Helix ADVANCED

HISTORY CONNECTION

Tell students that DNA was first isolated in 1869 by a Swiss scientist named Friedrich Miescher. Miescher reported that the substance he isolated from white blood cells was white, sugary, and slightly acidic. Because he found it only in the nuclei of these cells, he called it nuclein.

While the sugar molecule and the phosphate group are the same for each nucleotide in a molecule of DNA, the nitrogen base may be any one of four different kinds. **Figure 5** illustrates the four different nitrogen bases in DNA: adenine *(AD uh neen)*, guanine *(GWAH neen)*, thymine *(THIE meen)*, and cytosine *(SIET oh seen)*. Adenine (A) and guanine (G) are classified as purines *(PYUR eenz)*, nitrogen bases made of two rings of carbon and nitrogen atoms. Thymine (T) and cytosine (C) are classified as pyrimidines *(pih RIHM uh deenz)*, nitrogen bases made of a single ring of carbon and nitrogen atoms.

Note how the DNA shown in Figure 4 resembles a ladder twisted like a spiral staircase. The sugar-phosphate backbones (the blue "ribbons") are similar to the side rails of a ladder. The paired nitrogen bases are similar to the rungs of the ladder. The nitrogen bases face each other. The double helix is held together by weak hydrogen bonds between the pairs of bases.

Figure 5 **Purines and pyrimidines**

The nitrogen base in a nucleotide can be either a bulky, double-ring purine, or a smaller, single-ring pyrimidine.

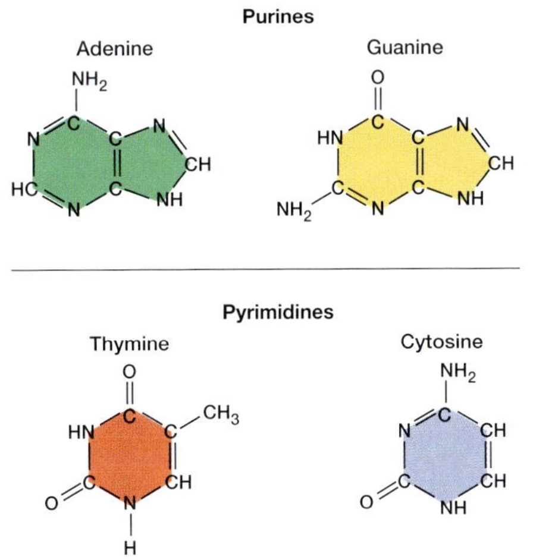

Teach

SKILL BUILDER — **BASIC**

Interpreting Visuals Draw students' attention to the four bases shown in **Figure 5.** Ask students which bases are complementary. (Adenine pairs with thymine; guanine pairs with cytosine.) Ask students to give reasons why adenine does not pair with guanine and thymine does not pair with cytosine, based on their structures. (Students should be able to see that adenine and guanine, the purines, have double-ring structures, and that thymine and cytosine, the pyrimidines, have single-ring structurse. When purines pair with pyrimidines exclusively, the "rungs" of the DNA "ladder" remain a consistent width. If purines paired or pyrimidines paired, rungs that are too wide or too narrow would result.) **LS Visual**

QUICK LAB

Observing Properties of DNA

SC.F.2.4.2

You can extract DNA from onion cells using ethanol and a stirring rod.

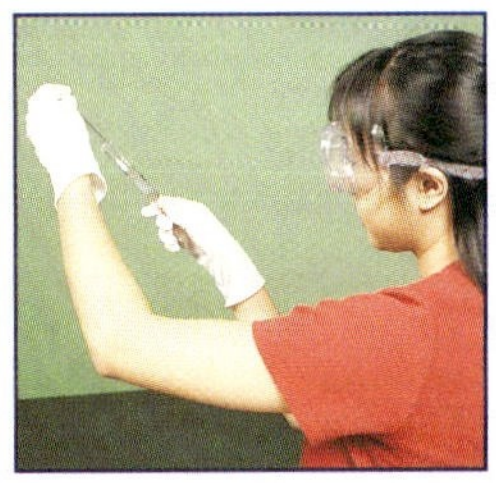

Materials

safety goggles and plastic gloves, 5 mL of onion extract, test tube, 5 mL of ice-cold ethanol, plastic pipet, glass stirring rod, test tube rack

Procedure

1. Place 5 mL of onion extract in a test tube.
2. CAUTION: *Ethanol is flammable. Do not use it near a flame.* Hold the test tube at a 45° angle. Use a pipet to add 5 mL of ice-cold ethanol to the tube one drop at a time. *NOTE: Allow the ethanol to run slowly down the side of the tube so that it forms a distinct layer.*
3. Let the test tube stand for 2–3 minutes.
4. Insert a glass stirring rod into the boundary between the onion extract and ethanol. Gently twirl the stirring rod by rolling the handle between your thumb and finger.
5. Remove the stirring rod from the liquids, and examine any material that has stuck to it. Touch the material to the lip of the test tube, and observe how the material acts as you try to remove it.
6. Clean up your materials and wash your hands before leaving the lab.

Analysis

1. **Describe** any material that stuck to the stirring rod.
2. **Relate** the characteristics of your sample to the structural characteristics of DNA.
3. **Propose** a way to determine if the material on the stirring rod is DNA.

QUICK LAB

Observing Properties of DNA

Skills Acquired
Observing, relating

Teacher's Notes
Order the DNA Extraction Biolab kit from WARD'S. Demonstrate the lab procedure for adding ethanol to the test tube and for using the stirring rod. Tell students not to shake or mix the onion extract-ethanol combination.

Answers to Analysis

1. The material is clear, viscous, and elastic.
2. It sticks to itself and to the test tube, and it forms long strings.
3. Answers will vary. Students might suggest that a laboratory test might confirm whether the material is DNA.

Career

Molecular Biologist Molecular biologists have determined the nucleotide sequence of the human genome through an international cooperative effort called the Human Genome Project. (A genome is all of the DNA in an organism.) They now hope to identify the function of every gene on each human chromosome. The total number of human genes—estimated at 30,000 to 35,000—is much less than previously thought. Ask students to find out the requirements to become a DNA researcher and the opportunities for employment in industry and research—in the U.S. and internationally.

Cultural Awareness

True Ancestors of Modern Japanese The Jomon settled Japan more than 10,000 years ago, and the Yayoi people inhabited Japan at a later date. There is a debate about which group of people are the true ancestors of the modern Japanese. Researchers who have compared the genes of modern Japanese to the genes of people in different parts of Asia have found that the Jomon originated in northern Asia. Data suggests that both the Jomon and Yayoi peoples have contributed genes to the modern Japanese.

Transparencies

TR Bellringer
TR C11 DNA Double Helix

Teach, *continued*

Teaching Tip — GENERAL

The Nobel Prize Ask students to name some prestigious awards. (Answers may include the Oscar or the Heisman Trophy.) Discuss the Nobel Prize. Alfred Nobel (1833–1896), the Swedish inventor of dynamite, provided a $9 million endowment in his will, with the interest to be used as awards for people whose work showed ingenuity and most benefited humanity. Nobel prizes were first awarded in the areas of chemistry, physics, physiology or medicine, and literature. Later, prizes in peace and economics were added. James Watson, Francis Crick, and Maurice Wilkins received the Nobel Prize in 1962 for their work in elucidating the structure of DNA. The Nobel Prize is only given to living individuals, so Rosalind Franklin (who died 4 years earlier) did not receive it.

Using the Figure — BASIC

Use **Figure 8** to show students how the structure of DNA can be simplified. Ask students to follow one of the strands on the upper drawing and compare the nitrogen base sequence to the sequence in the simplified diagram on the bottom. Point out that the sequence in both drawings is the same—they must follow the twist of the double helix. Have students verify that Chargaff's base-pairing rules are indeed followed in each DNA molecule. Ask students to identify the deoxyribose and the phosphate groups in each model. (In the simplified diagram, the horizontal line represents the sugar and phosphate groups.)

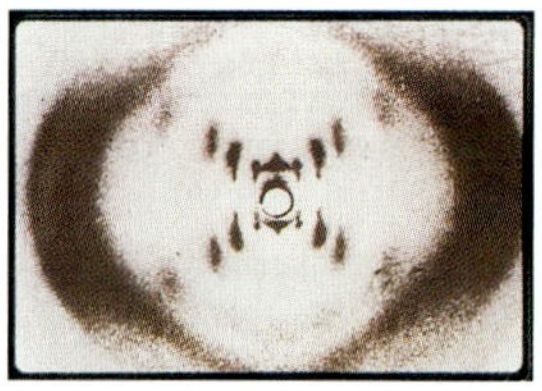

Figure 6 Franklin and her X-ray diffraction photo. The photographs revealed the X pattern characteristic of a helix. Franklin died of cancer when she was 37 years old.

Discovering DNA's Structure

How were Watson and Crick able to determine the double helical structure of DNA? As with most discoveries in science, other scientists provided crucial pieces that helped them solve this puzzle.

Chargaff's Observations

In 1949, Erwin Chargaff, a biochemist working at Columbia University, in New York City, made an interesting observation about DNA. Chargaff's data showed that for each organism he studied, the amount of adenine always equaled the amount of thymine (A=T). Likewise, the amount of guanine always equaled the amount of cytosine (G=C). However, the amount of adenine and thymine and of guanine and cytosine varied between different organisms.

Wilkins and Franklin's Photographs

The significance of Chargaff's data became clear in the 1950s when scientists began using X-ray diffraction to study the structures of molecules. In X-ray diffraction, a beam of X rays is directed at an object. The X rays bounce off the object and are scattered in a pattern onto a piece of film. By analyzing the complex patterns on the film, scientists can determine the structure of the molecule (much like shining a light on an object and then analyzing its shadow).

In the winter of 1952, Maurice Wilkins and Rosalind Franklin, two scientists working at King's College in London, developed high-quality X-ray diffraction photographs of strands of DNA. These photographs, such as the one in **Figure 6,** suggested that the DNA molecule resembled a tightly coiled helix and was composed of two or three chains of nucleotides.

Figure 7 Watson and Crick's model. The double-helical model of DNA takes into account Chargaff's observations and the patterns on Franklin's X-ray diffraction photographs.

Watson and Crick's DNA Model

The three-dimensional structure of the DNA molecule, however, was yet to be discovered. Any model had to take into account both Chargaff's findings and Franklin and Wilkins's X-ray diffraction data. In 1953, Watson and Crick used this information, along with their knowledge of chemical bonding, to come up with a solution. With tin-and-wire models of molecules, they built a model of DNA with the configuration of a double helix, a "spiral staircase" of two strands of nucleotides twisting around a central axis. **Figure 7** shows Watson (left) and Crick next to their tin-and-wire model of DNA.

196

did you know?

Rosalind Franklin Studying the work of their colleagues, Watson and Crick pooled knowledge to determine the structure of DNA. However, Watson has since commented that Rosalind Franklin would have been able to determine the structure of DNA if she and Crick had simply talked with one another for an hour. Although Franklin's X-ray diffraction photograph was crucial to solving the puzzle of the structure of DNA, she never received the credit that Watson and Crick did. She died from cancer at the age of 37, before her accomplishments were fully recognized.

Integrating Physics and Chemistry

Explain to students that diffraction is a wave property that occurs when waves bend as they pass by or around an edge of an object. Wilkins and Franklin produced X-ray diffraction photographs of DNA fibers in order to help them ascertain the structure of DNA.

Section Review Correlations

SC.F.2.4.2: Items 1, 2, 4, 5, 6

SC.H.1.4.5: Item 3

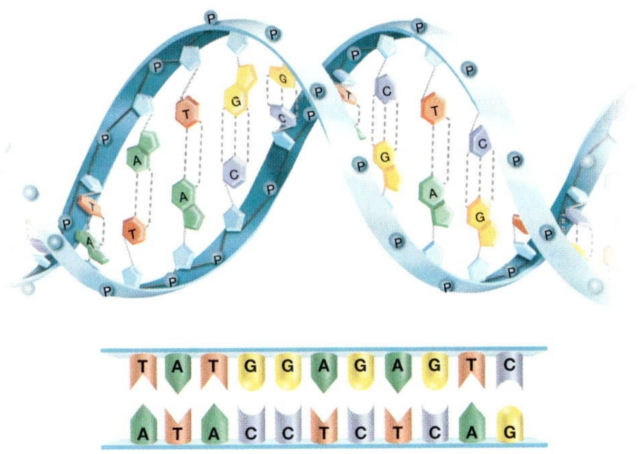

Figure 8 Base-pairing in DNA

The diagram of DNA below the helix makes it easier to visualize the base-pairing that occurs between DNA strands.

Pairing Between Bases

Watson and Crick determined that a purine on one strand of DNA is always paired with a pyrimidine on the opposite strand, as you can see in **Figure 8.** More specifically, an adenine on one strand always pairs with a thymine on the opposite strand, and a guanine on one strand always pairs with a cytosine on the opposite strand. The structure and size of the nitrogen bases allows for only these two paired combinations. These **base-pairing rules** are supported by Chargaff's observations. One easy way to visualize base-pairing is by simplifying the way in which DNA structure is represented, as shown in Figure 8.

Adenine forms two hydrogen bonds with thymine, and cytosine forms three hydrogen bonds with guanine. The hydrogen bonds between the nitrogen bases keep the two strands of DNA together. The strictness of base-pairing results in two strands that contain **complementary base pairs**. That is, the sequence of bases on one strand determines the sequence of bases on the other strand. For example, if the sequence of nitrogen bases on one strand of a DNA molecule is TCGAACT, the sequence of nitrogen bases on the other strand must be AGCTTGA.

Study TIP

Organizing Information

Create a timeline that summarizes the people and events that led to the discovery that DNA is the molecule where genetic information is stored. Start with 1928, and end with 1953.

Section 2 Review

1. **Describe** the three parts of a DNA nucleotide. SC.F.2.4.2
2. **Relate** the base-pairing rules to the structure of DNA. SC.F.2.4.2
3. **Describe** the two pieces of information from other scientists that enabled James Watson and Francis Crick to discover the double-helical structure of DNA. SC.H.1.4.5
4. **Explain** why the two strands of the double helix are described as complementary. SC.F.2.4.2
5. **Critical Thinking Applying Information** Suppose a strand of DNA has the nucleotide sequence CCAGATTG. What is the nucleotide sequence of the complementary strand? SC.F.2.4.2
6. **FCAT Prep** Which pattern shows how bases pair in complementary strands of DNA?
   A A-A and C-C
   B A-C and T-G
   C A-G and T-C
   D A-T and C-G
   SC.F.2.4.2

197

Answers to Section Review

1. The three parts of a nucleotide are a phosphate group, a sugar, and a nitrogen base (which can be adenine, guanine, cytosine, or thymine).
2. Thymine on one strand of DNA base pairs with adenine on the complementary strand. Cytosine on one strand of DNA base pairs with guanine on the complementary strand.
3. Chargaff's data on the ratio of nitrogen bases in organisms, and Franklin and Wilkins's X-ray photos of DNA enabled Watson and Crick to propose the structure of DNA.
4. The two strands are described as complementary (not identical) because each strand provides one member of an obligatory pair.
5. GGTCTAAC
6. **A.** Incorrect. These pairs show bases pairing with themselves, which does not occur.
   B. Incorrect. Although these pairs show purines with pyrimidines, the pairing is with incorrect members of those groups.
   C. Incorrect. These pairs show purines pairing and pyrimidines pairing, which does not occur.
   D. Correct. These pairs show the correct purines pairing with the correct pyrimidines.

Close

Reteaching — BASIC

Give pairs of students index cards with the components of DNA written on them: sugar on one card, phosphate group on another card, and each of the four nitrogen bases on four different cards. Each set of cards should include at least eight sugars, eight phosphates, and two of each of the nitrogen bases. Ask students to arrange the cards in a pattern to demonstrate the structure of DNA. Students should be able to reconstruct a DNA molecule.

Quiz — GENERAL

1. What does the abbreviation DNA stand for? (DNA stands for deoxyribonucleic acid.)
2. Distinguish between purines and pyrimidines. (Purines are nitrogen bases made of two rings of carbon and nitrogen atoms. Pyrimidines are nitrogen bases made of a single ring of carbon and nitrogen atoms.)
3. What was the significance of Wilkins and Franklin's X-ray diffraction photographs regarding DNA structure? (Their photographs suggested that the DNA molecule resembled a tightly coiled helix and was composed of two or three chains of nucleotides.)

Alternative Assessment — GENERAL

Place students in groups of three. Have the groups search magazines, newspapers, or the Internet to collect as many images as they can of DNA, such as the photo used in this chapter's opener. Then ask students to analyze the images and determine whether they correctly depict DNA. Ask each group to write a brief description of how each image shows the molecular structure of DNA and to attach their descriptions to the images. Discuss the groups' descriptions, and display their images and descriptions in the classroom.

The Replication of DNA

Florida Sunshine Standards/FCAT

SC.F.1.4.5 Complex interactions among the different kinds of molecules in the cell cause distinct cycles of activity governed by proteins.

SC.G.1.4.1 here is great diversity and interdependence among living things.

Objectives

- **Summarize** the process of DNA replication. SC.F.1.4.5
- **Describe** how errors are corrected during DNA replication. SC.F.1.4.5
- **Compare** the number of replication forks in prokaryotic and eukaryotic DNA. SC.G.1.4.1

Key Terms

DNA replication
DNA helicase
replication fork
DNA polymerase

Roles of Enzymes in DNA Replication

When the double helix structure of DNA was first discovered, scientists were very excited about the complementary relationship between the sequences of nucleotides. They predicted that the complementary structure was used as a basis to make exact copies of the DNA each time a cell divided. Watson and Crick proposed that one DNA strand serves as a template, or pattern, on which the other strand is built. Within five years of the discovery of DNA's structure, scientists had firm evidence that the complementary strands of the double helix do indeed serve as templates for building new DNA.

The process of making a copy of DNA is called **DNA replication**. DNA replication is summarized in **Figure 9.** Recall from your reading of earlier chapters that DNA replication occurs during the synthesis (S) phase of the cell cycle, before a cell divides.

Step 1 Before DNA replication can begin, the double helix unwinds. This is accomplished by enzymes called DNA helicases. **DNA helicases** open the double helix by breaking the hydrogen bonds that link the complementary nitrogen bases between the two strands.

BIO graphic

Figure 9

DNA Replication

DNA replication results in two identical DNA strands.

1. The two original DNA strands separate.
2. DNA polymerases add complementary nucleotides to each strand.
3. Two DNA molecules form that are identical to the original DNA molecule.

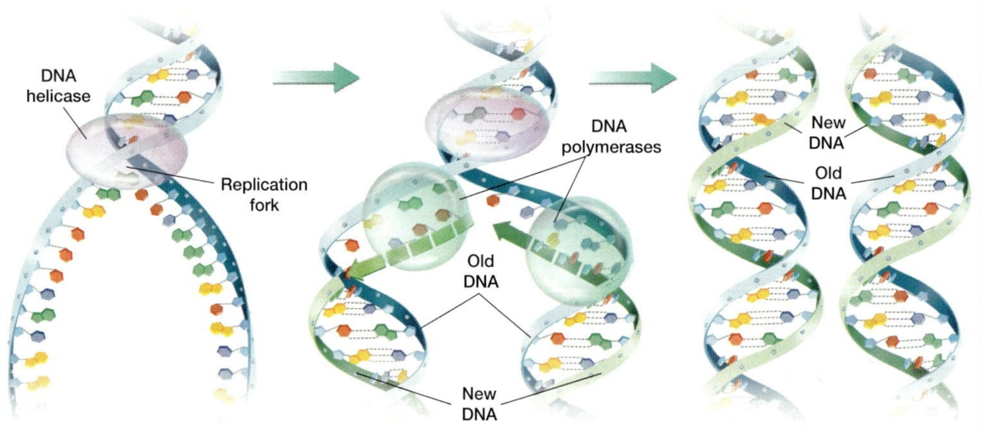

Focus

Overview

Before beginning this section review with your students the objectives listed in the Student Edition. Section 3 first explains how DNA is copied and how errors are corrected. Then, replication forks, and their role in increasing the speed of replication, are described.

Bellringer

DNA is considered to be a relatively stable molecule. Ask students what gives it this stability, even though the hydrogen bonds between the nitrogen bases are easily broken. Have students write their answers, and then ask them to share and compare their ideas. (The bonds connecting the deoxyribose and phosphate groups, which form the backbone of DNA, are covalent bonds. Covalent bonds are strong and keep the strands of nucleotides intact.)

Motivate

Identifying Preconceptions Tell students that before DNA can make copies of itself, it must "unzip" the hydrogen bonds holding the complementary nitrogen bases together. Ask students how they think this happens. (Enzymes catalyze the reactions in which the hydrogen bonds are broken.) Then ask students how they think nucleotides are added to each "unzipped" strand. (Enzymes catalyze the bonding of nucleotides to the "free" DNA strands.) After students recognize that enzymes are needed to catalyze these reactions, introduce the names of the specific enzymes involved.

Section Correlations

SC.F.1.4.5
SC.G.1.4.1

Chapter Resource File

- Directed Reading BASIC
- Active Reading GENERAL
- Data Sheet for Math Lab GENERAL

One-Stop Planner CD-ROM

- Reading Organizers BASIC
- Reading Strategies BASIC

Attention Grabber

The Same Yet Different Genetically, humans and chimps are amazingly similar—their DNA sequences only differ by between 1 and 5 percent. So why don't we see chimps attending classes in our schools or carrying briefcases to their jobs on Wall Street? One major reason is that our brains are different. Results from a study published in the April 12, 2002 issue of the journal *Science* notes that genes are activated differently in the brains of chimps and humans. Additionally, the amount of gene and protein expression is greater in the brains of humans than in the brains of chimps.

Once the two strands are separated, additional proteins attach to each strand, holding them apart and preventing them from assuming their double-helical shape. The areas where the double helix separates are called **replication forks** because of their Y shape, as shown in Figure 9.

Step 2 At the replication fork, enzymes known as **DNA polymerases** move along each of the DNA strands. DNA polymerases add nucleotides to the exposed nitrogen bases, according to the base-pairing rules. As the DNA polymerases move along, two new double helixes are formed.

Step 3 Once DNA polymerases have begun adding nucleotides to a growing double helix, the process continues until all of the DNA has been copied and the polymerases are signaled to detach. This process produces two DNA molecules, each composed of a new and an original strand. The nucleotide sequences in both of these DNA molecules are identical to each other and to the original DNA molecule.

Checking for Errors

In the course of DNA replication, errors sometimes occur and the wrong nucleotide is added to the new strand. An important feature of DNA replication is that DNA polymerases have a "proofreading" role. They can add nucleotides to a growing strand only if the previous nucleotide is correctly paired to its complementary base. In the event of a mismatched nucleotide, the DNA polymerase can backtrack. The DNA polymerase removes the incorrect nucleotide and replaces it with the correct one. This proofreading reduces errors in DNA replication to about one error per 1 billion nucleotides.

MATH LAB: *Analyzing the Rate of DNA Replication* SC.F.1.4.5

Background

Cancer is a disease caused by cells that divide uncontrollably. Scientists studying drugs that prevent cancer often measure the effectiveness of a drug by its effect on DNA replication. During normal DNA replication, nucleotides are added at a rate of about 50 nucleotides per second in mammals and 500 nucleotides per second in bacteria.

Magnification: 83,640×

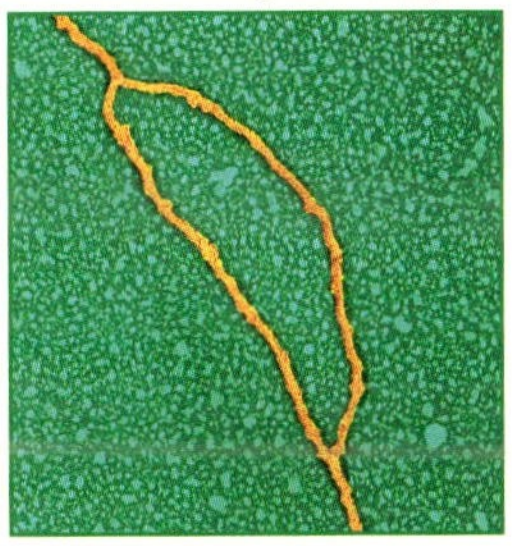

DNA replication forks

Analysis

1. **Calculate** the time it would take a bacterium to add 4,000 nucleotides to one DNA strand undergoing replication.
2. **Calculate** the time it would take a mammalian cell to add 4,000 nucleotides to one DNA strand undergoing replication.
3. **Critical Thinking** **Predicting Outcomes** How would the total time needed to add the 4,000 nucleotides be affected if a drug that inhibits DNA polymerases were present?

199

Teach

Using the Figure — GENERAL

Use the overhead projector to help students identify the replication fork and helicase enzymes shown in Step 1 of **Figure 9.** Ask students to identify the original DNA strands, the new strands, and the location of the DNA polymerases in Step 2. Have them compare the two DNA molecules in Step 3 with the DNA in Step 1 and with each other. (They are identical.) **LS Visual**

Teaching Tip — BASIC

Modeling Replication Draw a sequence of eight nitrogen bases on the board. Have students copy it onto a piece of paper and draw the base sequence for the complementary strand. Then, using **Figure 9** as a model, have students separate the two strands by drawing each sequence of eight bases. Finally, have students "replicate" each strand and compare the two new molecules with the original sequence. **LS Kinesthetic**

MATH LAB: *Analyzing the Rate of DNA Replication*

Skills Acquired
Calculating, predicting

Teacher's Notes
Remind students that rates are ratios that compare quantities of different units, such as miles per hour.

Answers to Analysis

1. 8 seconds
2. 80 seconds
3. The total time would increase

Transparencies

TR Bellringer
TR C12 DNA Replication
TR C13 Replication Forks

Integrating Physics and Chemistry

Remind students that there are three primary types of chemical bonds: ionic, covalent, and metallic. Describe the characteristics of each bond in terms of electron behavior (ionic—electron transfer; covalent—electron sharing; metallic—electron sea) and then review the nature of the hydrogen bond. Ask students to compare and contrast the hydrogen bond with the three primary bond types. Which one of these is the hydrogen bond most like? Have students defend their responses.

Teach, *continued*

Demonstration

To model the separation of DNA strands in the circular chromosomes of bacteria, form a circle with a 12-inch zipper. While holding the ends of the zipper together with one hand, unzip the zipper with the other. Note that DNA "unzipping" and replication begin at one spot but actually proceed in both directions at two replication forks. LS **Visual** English Language Learners

Close

Reteaching — BASIC

Assign students to cooperative groups to build DNA models. Let students choose their own construction materials to represent the components and bonds that make up DNA. Ask students to use their models to demonstrate DNA replication. LS **Kinesthetic** Co-op Learning

Quiz — GENERAL

1. How is the exact replication of DNA ensured? (It is ensured by complementary base pairing and "proofreading" by DNA polymerases.)
2. What are replication forks? (Areas of DNA where the double helix separates prior to replication.)

Alternative Assessment — GENERAL

Have students write two sentences to describe each of the three steps summarizing DNA replication shown in **Figure 9.**

Section Review Correlations

SC.F.1.4.5: Items 1, 3
SC.F.2.4.2: Item 5
SC.F.2.4.3: Items 2, 4

internet connect

www.scilinks.org
Topic: **DNA Replication**
Keyword: **HX4059**
SCILINKS. Maintained by the National Science Teachers Association

The Rate of Replication

Replication does not begin at one end of the DNA molecule and end at the other. The circular DNA molecules found in prokaryotes usually have two replication forks that begin at a single point. The replication forks move away from each other until they meet on the opposite side of the DNA circle, as shown in **Figure 10.**

In eukaryotic cells, each chromosome contains a single, long strand of DNA. The length presents a challenge: The replication of a typical human chromosome with one pair of replication forks spreading from a single point, as occurs in prokaryotes, would take 33 days! To understand how eukaryotes meet this challenge, imagine that your class has to carry 25 boxes to another building. Carrying one box over, returning, carrying the second box, and so on, would be very slow. It would be much faster if everyone in the class picked up a box so that all of the boxes could be carried in one trip. That is similar to replication in eukaryotic cells, as shown in Figure 10. Each human chromosome is replicated in about 100 sections that are 100,000 nucleotides long, each section with its own starting point. With multiple replication forks working in concert, an entire human chromosome can be replicated in about 8 hours.

Figure 10 Replication forks

Prokaryotic and eukaryotic DNA have a different number of replication forks.

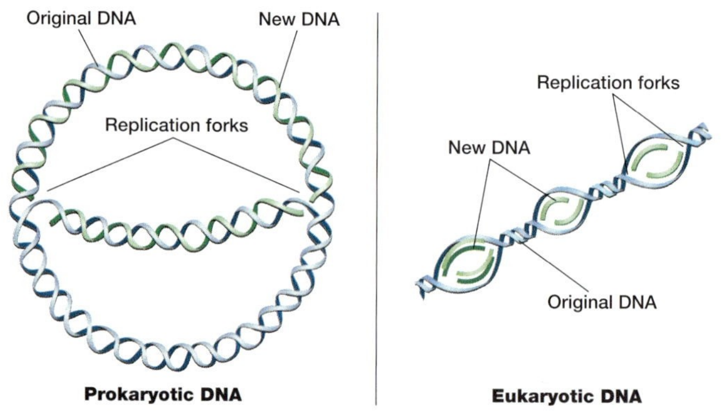

Section 3 Review

1. **Explain** the two roles that enzymes play in DNA replication as is illustrated in Figure 9 in this section. SC.F.1.4.5
2. **Explain** the relationship between DNA polymerases and mutations. SC.F.2.4.3
3. **State** the effect of multiple replication forks on the speed of replication in eukaryotes. SC.F.1.4.5
4. **Critical Thinking** **Evaluating Information** If a mutation occurs during the formation of an egg cell or sperm cell, is that mutation more significant or less significant than a mutation that occurs in a body cell? Explain your answer. SC.F.2.4.3
5. *FCAT Prep* How many DNA strands exist after one molecule of DNA has been replicated?
   A 1 **C** 4
   B 2 **D** 8
   SC.F.2.4.2

Answers to Section Review

1. DNA helicases unwind the double helix by catalyzing the break up of the hydrogen bonds between the nitrogen bases. DNA polymerases enable the addition of complementary nucleotides.
2. DNA polymerases act as "proofreaders" by checking for mismatched nucleotides that can lead to mutations. If a nucleotide is placed incorrectly, polymerases remove the incorrect nucleotide and replace it with the correct nucleotide.
3. Multiple replication forks increase the speed of replication.
4. Mutations in gametes may be inherited and thus are likely to have a more significant effect. It would also depend on the type of mutation and whether the gamete is involved in fertilization. Body cells must usually accumulate multiple mutations before they become harmful.
5. **A.** Incorrect. After replication, 2 double-stranded molecules exist, resulting in a total of 4 strands of DNA. **B.** Incorrect. This is the number of strands before replication. **C.** Correct. See answer A. **D.** Incorrect. This would be the number of strands after two rounds of replication.

Study ZONE

CHAPTER HIGHLIGHTS

Key Concepts

1 Identifying the Genetic Material

- The experiments of Griffith and of Avery yielded results that suggested DNA was the genetic material.
- Hershey and Chase used the bacteriophage T2 and radioactive labels to show that viral genes are made of DNA, not protein.
- DNA stores the information that tells cells which proteins to make and when to make them.

2 The Structure of DNA

- DNA is made of two strands of nucleotides twisted into the form of a double helix.
- Each nucleotide in DNA is made of the sugar deoxyribose, a phosphate group, and one of four nitrogen bases. The four nitrogen bases found in DNA nucleotides are adenine (A), thymine (T), guanine (G), and cytosine (C).
- The two strands of DNA are complementary—each A on one strand pairs with a T on the opposite strand, and each G on one strand pairs with a C on the opposite strand.
- Watson and Crick announced their discovery of the structure of DNA in 1953 with the help of data gathered by Wilkins, Franklin, and Chargaff.

3 The Replication of DNA

- Before a cell divides, it copies its DNA by a process called DNA replication.
- In DNA replication, enzymes work to unwind and separate the double helix and add complementary nucleotides to the exposed strands.
- The result of DNA replication is two exact copies of the cell's original DNA. Each new double helix is composed of one original DNA strand and one new DNA strand.
- DNA polymerase proofreads DNA during its replication so that very few errors occur.

Key Terms

Section 1

vaccine (190)
virulent (190)
transformation (191)
bacteriophage (192)

Section 2

double helix (194)
nucleotide (194)
deoxyribose (194)
base-pairing rules (197)
complementary base pair (197)

Section 3

DNA replication (198)
DNA helicase (198)
replication fork (199)
DNA polymerase (199)

INTERACTIVE BIOLOGY TUTOR

Unit 6—*Gene Expression*
Use Topics 1–2 in this unit to review the key concepts and terms in this chapter.

201

Alternative Assessment — GENERAL

Provide students with three-dimensional objects that represent sugars, phosphate groups, and eight nitrogen bases. Provide a key for what each object represents. Instruct students to construct a DNA model with the objects. Have them demonstrate replication by separating the bases. Instruct them to draw the model they have made on paper and assign names to the bases. Then have them draw the finished product of replication with the sugar, phosphate, and bases labeled.

Chapter Resource File

- Science Skills Worksheet GENERAL
- Critical Thinking Worksheet ADVANCED
- Test Prep Pretest GENERAL
- Chapter Test GENERAL

Answer to Concept Map

The following is one possible answer to Performance Zone item 15.

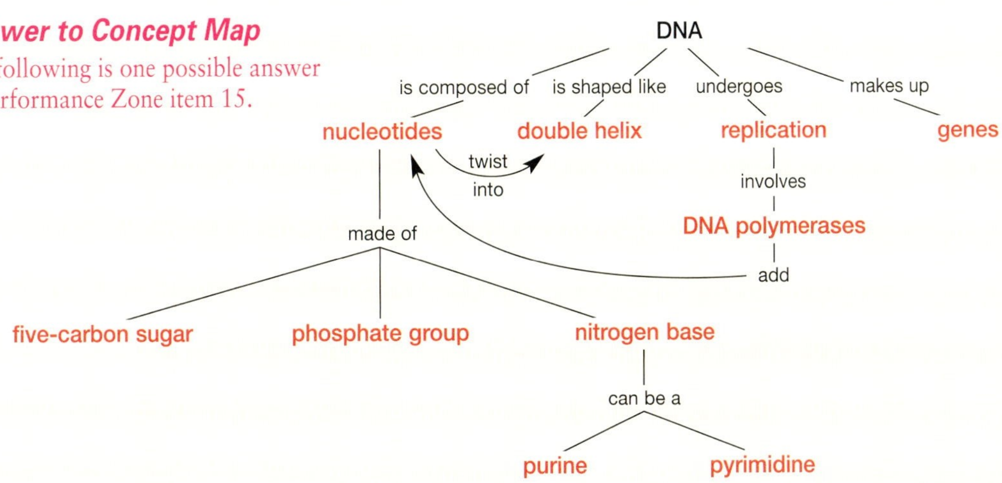

ANSWERS

Understanding Key Ideas

1. a
2. b
3. a
4. b
5. a. about 1:1
   b. A and T, C and G
   c. yes
6. c
7. both
8. DNA polymerases add complementary nucleotides to a DNA strand during replication, and they act as "proofreaders" by checking for mismatched nucleotides. If a nucleotide is placed incorrectly, polymerases remove the incorrect nucleotide and replace it with the correct nucleotide.
9. A gene is a segment of DNA that codes for a protein or RNA molecule. Thousands of genes, which are packaged with proteins and coiled up before cell division, make up a chromosome. Two chromatids make up a chromosome; chromatids are the replicated DNA formed before a cell undergoes cell division.
10. The answer to the concept map is found at the bottom of the Study Zone page.

Critical Thinking

11. Watson and Crick's model of DNA needed to account for the equal amounts of A/T and G/C in DNA molecules, and that the amounts of A/T and G/C varied among organisms. The model had to resemble a tightly coiled helix. Their model accounted for all known facts and could be considered a good representation of the structure of DNA.
12. The process of DNA replication synthesizes new molecules of DNA. It requires DNA helicases for the unwinding of DNA prior to replication. Reducing the helicases would slow down replication.

Alternative Assessment

13. Answers will vary. Techniques may include the Maxam-Gilbert method or the more frequently used Sanger (dideoxy) sequencing method. Fluorescent dyes are now used in the Sanger method to tag the modified nucleotides. Sequencing technology can be used to find mutations that result in genetic disorders.

Chapter Review Correlations

SC.F.1.4.5: Items 4, 8, 12
SC.F.2.4.1: Item 7
SC.F.2.4.2: Items 3, 5, 6, 9, 10
SC.H.1.4.1: Items 1, 2
SC.H.1.4.4: Item 11
SC.H.3.4.5: Item 13

Assignment Guide

Section	Questions
1	1, 2
2	3, 5, 11
3	4, 6, 7, 8, 9, 10, 12, 13

CHAPTER REVIEW ZONE

Understanding Key Ideas

1. In his experiments on *Streptococcus pneumoniae,* Griffith found that
   a. harmless *R* bacteria changed and became virulent.
   b. the *S* bacteria were transformed.
   c. the capsule did not protect the bacterium.
   d. mice injected with the *R* bacteria died.
   SC.H.1.4.1
2. Hershey and Chase showed that
   a. bacteriophages can infect human cells.
   b. DNA controls heredity.
   c. bacteria undergo transformation.
   d. a vaccine for pneumonia could be produced. SC.H.1.4.1
3. James Watson and Francis Crick
   a. built a structural model of DNA.
   b. discovered DNA replication.
   c. used X-ray diffraction.
   d. discovered DNA polymerases.
   SC.F.2.4.2
4. Multiple replication forks along the DNA
   a. correct replication errors.
   b. reduce DNA replication time.
   c. ensure that the new and old DNA strands are complementary.
   d. signal DNA polymerase to stop. SC.F.1.4.5
5. The table below summarizes the percentage of each nitrogen base found in an organism's DNA.

Percentage of Each Nitrogen Base

	A	T	G	C
Human	30.4	30.1	19.6	19.9
Wheat	27.3	27.1	22.7	22.8
E. coli	24.7	23.6	26.0	25.7

   a. What is the ratio of purines to pyrimidines?
   b. Within each organism, which nucleotides are found in similar percentages?
   c. Do the ratio and percentages in (a) and (b) follow Chargaff's rule?
   SC.F.2.4.2
6. If the sequence of nucleotides on one strand of a DNA molecule is GCCATTG, the sequence on the complementary strand is
   a. GGGTAAG.
   b. CCCTAAC.
   c. CGGTAAC.
   d. GCCATTC.
   SC.F.2.4.2
7. Does DNA replication occur immediately before asexual reproduction, before sexual reproduction, or before both? SC.F.2.4.1
8. What are two functions of DNA polymerases during DNA replication? SC.F.1.4.5
9. Differentiate between DNA, genes, chromatids, and chromosomes. (**Hint:** See Chapter 6, Section 1.) SC.F.2.4.2
10. **Concept Mapping** Make a concept map that shows the structure of DNA and how it is copied. Try to include the following words in your concept map: *nucleotides, phosphate group, five-carbon sugar, nitrogen base, purine, pyrimidine, double helix, replication, DNA polymerases,* and *gene.* SC.F.2.4.2

Critical Thinking

11. **Evaluating Models** Explain why you do or do not think Watson and Crick's model of DNA illustrated in **Figure 7** in this chapter is a good representation of the structure of DNA. What existing information about DNA did Watson and Crick's model have to take into account? SC.H.1.4.4
12. **Predicting Results** Identify the process by which new molecules of DNA are synthesized, and predict the effect on this process of reducing available DNA helicases. SC.F.1.4.5

Alternative Assessment

13. **Selecting Technology** Research two methods used to sequence the nucleotides in a gene. Compare and contrast the two methods. Give examples of how this technology might be used in a clinical setting. Prepare a poster to summarize the nucleotide-sequencing methods you researched. SC.H.3.4.5

202

FCAT Prep

For questions 1–6, write your answers on a separate sheet of paper.

1 What is the name of the process that was involved in changing Griffith's *R* bacteria to *S* bacteria?
A. crossing over
B. DNA replication
C. polymerization
D. transformation
SC.F.1.4.5

2 A scientist extracted 4.6 picograms (or 4.6×10^{-12} grams) of DNA from mouse muscle cells. How much DNA could be extracted from the same number of mouse sperm?
F. 10^{-12} picograms
G. 2.3 picograms
H. 4.6 picograms
I. 9.2 picograms
SC.F.2.4.2

3 Which of the following is a component of a DNA nucleotide?
A. double helix
B. phosphate group
C. phospholipid
D. three-carbon sugar
SC.F.2.4.2

4 What molecule did Hershey and Chase's work show was the genetic material of the T2 bacteria?
F. DNA
G. a protein
H. DNA helicase
I. DNA polymerase
SC.H.1.4.2

5 READ INQUIRE EXPLAIN X-rays damage DNA in organisms. Rosalind Franklin died of cancer at an early age.

Analyze how her work with X-ray diffraction might have led to her death. SC.H.3.4.3

6 READ INQUIRE EXPLAIN The polymerase chain reaction (PCR) is a technique used in genetic engineering and applied in procedures such as DNA fingerprinting and genetic screening. This technique allows a small sample of DNA to be copied many times. Genetic analysis, which determines the presence of restriction length polymorphisms, or mutations on specific genes, can be performed with *PCR-amplified DNA* samples.

Part A How could someone known to have relatives with a genetic disease benefit from PCR? SC.H.3.4.5

Part B Infer the meaning of the term "PCR-amplified DNA" in the passage above. SC.H.3.4.2

Test TIP

You can sometimes figure out an answer to a question before you look at the answer choices. After you answer the question in your mind, compare your answer with each answer choice. Choose the answer that most closely matches your own answer.

FCAT Doctor

Question 2 Answer G is the correct choice. Sperm cells are haploid and contain half the amount of genetic material found in body cells. Answer F is incorrect because 10–12 is the power of the units for picograms. Answer H is incorrect because only body cells would have the same amount of genetic material as other body cells, which are diploid. Answer I is incorrect because the amount of genetic material in sperm is half, not double, that of muscle cells.

Question 3 Answer B is the correct choice. A DNA nucleotide consists of a five-carbon sugar, a phosphate group, and a nitrogen base. Answer A is incorrect because many DNA nucleotides make up a molecule shaped like a double helix. Answer C is incorrect because phospholipids are a kind of lipid and not related to nucleic acids such as DNA nucleotides. Answer D is incorrect because DNA contains a five-carbon sugar, ribose, and not a three-carbon sugar.

Question 5 X-rays are a type of radiation that can lead to damage or mutation of DNA. Mutations can lead to cancer.

Question 6

Part A PCR is used to replicate a single DNA molecule, producing many more molecules that can then be analyzed for genetic screening.

Part B Amplification means toincrease the amount.

Answers

1. D
2. G
3. B
4. F
5. X-rays could have caused DNA mutations that led to her cancer.
6A. They could undergo genetic screening that involved PCR techniques in order to determine the likelihood that they have the disease.
6B. PCR-amplified DNA is a sample of DNA that has been replicated many times using PCR.

Test Prep Correlations

SC.F.1.4.5: Item 1
SC.F.2.4.2: Items 2, 3
SC.H.1.4.2: Item 4
SC.H.3.4.2: Item 6B
SC.H.3.4.3: Item 5
SC.H.3.4.5: Item 6A

Exploration Lab

Model DNA Structure

Teacher's Notes

Time Required

One 45-minute class period

Ratings

EASY → HARD

Teacher Preparation (1)
Student Setup (1)
Concept Level (2)
Cleanup (1)

Skills Acquired

- Communicating
- Constructing Models
- Identifying/Recognizing Patterns
- Interpreting
- Predicting

Scientific Methods

In this lab, students will:

- Make Observations
- Ask Questions
- Test the Hypothesis
- Analyze the Results
- Draw Conclusions
- Communicate the Results

Materials and Equipment

Materials for this lab can be ordered from WARD'S. See the *Master Materials List* at the front of this book for catalog numbers. Obtain enough plastic soda straws to make 48 3-cm sections for each group. Students will also need 48 standard-size paper clips and 48 colored pushpins (start each group with 12 red, 12 blue, 12 yellow, and 12 green pushpins).

Exploration Lab

Modeling DNA Structure

SC.H.3.4.1

SKILLS

- Modeling
- Using scientific methods

OBJECTIVES

- **Design** and analyze a model of DNA.
- **Describe** how replication occurs.
- **Predict** the effect of errors during replication.

MATERIALS

- plastic soda straws, 3 cm sections
- metric ruler
- pushpins (red, blue, yellow, and green)
- paper clips

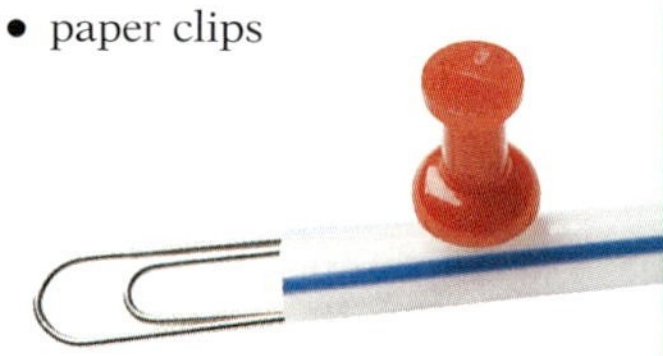

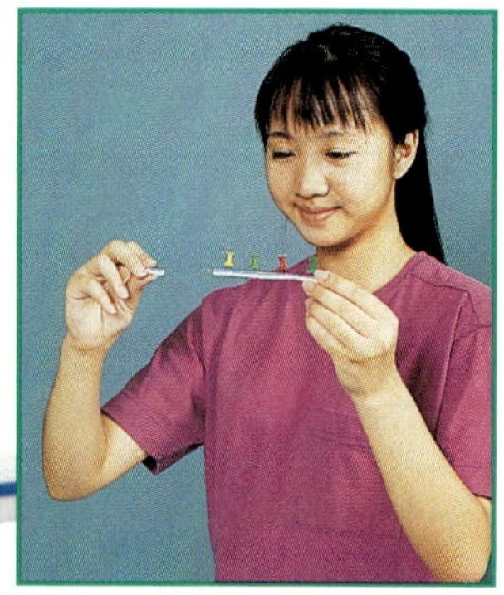

Before You Begin

DNA contains the instructions that cells need in order to make every **protein** required to carry out their activities and to survive. DNA is made of two strands of **nucleotides** twisted around each other in a **double helix**. The two strands are **complementary,** that is, the sequence of bases on one strand determines the sequence of bases on the other strand. The two strands are held together by hydrogen bonds.

In this lab, you will build a model to help you understand the structure of DNA. You can also use the DNA model to illustrate and explore processes such as **replication** and **mutation**.

1. Write a definition for each boldface term in the paragraphs above and for each of the following terms: replication fork, base-pairing rules.
2. Identify the three different components of a nucleotide.
3. Identify the four different nitrogen bases that can be found in DNA nucleotides.
4. Based on the objectives for this lab, write a question you would like to explore about DNA structure.

Procedure

PART A: Design a Model

1. Work with the members of your lab group to design a model of DNA that uses the materials listed for this lab. Be sure that your model has at least 12 nucleotides on each strand.

> **You Choose**
>
> **As you design your model, decide the following:**
>
> **a.** what question you will explore
>
> **b.** how to use the straws, pushpins, and paper clips to represent the three components of a nucleotide
>
> **c.** how to link (bond) the nucleotides together
>
> **d.** in what order you will place the nucleotides on each strand

2. Write out the plan for building your model. Have your teacher approve the plan before you begin building the model.
3. Build the DNA model your group designed. **CAUTION: Sharp or pointed objects may cause injury. Handle pushpins carefully.** Sketch and label the parts of your DNA model.
4. Use your model to explore one of the questions written for step 4 of **Before You Begin.**

204

Safety Cautions

Review all safety symbols with students before beginning the lab. Caution students to be careful of the sharp tips of the pushpins. Account for all materials at the end of the lab.

Tips and Tricks

Let students know they may need more or less of each color as they decide on the sequence of their DNA. If desired, ask students to save the assembled "nucleotides" for use in other DNA modeling labs.

Answers to Before You Begin

1. *DNA*—double-stranded, helical nucleic acid that stores hereditary information; *protein*—an organic molecule composed of linked amino acids; *nucleotides*—subunits of nucleic acids that consist of a nitrogenous base, a sugar, and a phosphate group; *double helix*—a spiral structure characteristic of the DNA molecule; *complementary*—characteristic of nucleic acids in which the sequence of bases on one strand determines the sequence of bases on the other; *replication*—the process by which DNA in a cell is copied; *mutation*—a change in the nucleotide sequence of a gene or chromosome;

PART B: DNA Replication

5. Discuss with your lab group how the model you built for Part A may be used to illustrate the process of replication.
6. Write a question you would like to explore about replication. Use your model to explore the question you wrote. Sketch and label the steps of replication.

PART C: Test Hypothesis

Answer each of the following questions by writing a hypothesis. Use your model to test each hypothesis, and describe your results.

7. Mitosis follows replication. How might the cells produced by mitosis be affected if nucleotides on one DNA strand were incorrectly paired during replication?
8. What would happen if only one strand in a DNA molecule were copied during replication?

PART D: Cleanup and Disposal

9. Dispose of damaged pushpins in the designated waste container.
10. Clean up your work area and all lab equipment. Return lab equipment to its proper place. Wash your hands thoroughly before you leave the lab and after you finish all work.

Analyze and Conclude

1. **Analyzing Results** In your original DNA model, were the two strands identical to each other?
2. **Relating Concepts** How does DNA structure ensure that the two DNA molecules made by replication are the same as the original DNA molecule?
3. **Drawing Conclusions** Did the two DNA molecules you made in step 6 have the same nitrogen-base sequence as your original model DNA molecule?
4. **Inferring Relationships** The order of nitrogen bases on a DNA strand is a code for making proteins. What does this mean has happened to the "code" in one of the DNA molecules you made in step 7?
5. **Predicting Outcomes** What would happen if the DNA in a cell that is about to divide were not replicated?
6. **Inferring Information** What are the advantages of having DNA remain in the nucleus of a cell?
7. **Further Inquiry** Write a new question about DNA that could be explored with your model.

Do You Know?

Do research in the library or media center to answer these questions:

1. Are there any pollutants in the environment that disrupt replication when an organism is exposed to the pollutant?
2. How do DNA molecules differ among various species of animals and plants? How are they similar?

Use the following Internet resources to explore your own questions about DNA.

internet connect
www.scilinks.org
Topic: DNA
Keyword: HX4058
SCILINKS. Maintained by the National Science Teachers Association

205

2. a nitrogen base, a sugar, a phosphate
3. adenine, guanine, cytosine, and thymine
4. Answers will vary. For example: Will the sequence of nucleotides on the two strands of DNA be identical?

Answers to Procedure

2. Answers will vary. Students might use straw segments to model the sugar-phosphate backbone of DNA, different-colored pushpins to represent the nitrogen bases, and paper clips to represent the bonds that hold the nucleotides in a chain.
6. Answers will vary. For example: What happens to the new DNA strand if the wrong nucleotide is added during replication?
7. Hypotheses will vary. The DNA in the new cells that result from cell division will not be identical.
8. Hypotheses will vary. The single strand would not have a strand to pair with and would not be complete.

Answers to Analyze and Conclude

1. No, they were complementary.
2. The DNA structure is such that the two strands are complementary to each other.
3. yes
4. The code has been changed.
5. One of the resulting cells would die because it would not receive any protein-making instructions.
6. The nucleus provides an isolated and protected area for storing genetic information.
7. Answers will vary. For example: How do DNA molecules differ among various species of animals and plants?

replication fork—a Y-shaped point that results when the two strands of a DNA double helix separate so that the DNA molecule can be replicated; *base-pairing rules*—rules stating that cytosine pairs with guanine and adenine pairs with thymine in DNA.

Answers to Do You Know?

1. Answers will vary and may include benzo [a] pyrene and other polyaromatic compounds.
2. The number of chromosomes varies among species, but the relative chromosome number is not an indication of the kingdom to which an organism belongs. Lettuce has 18 chromosomes, while an adder's tongue fern has 1,262 chromosomes. A human has 46 chromosomes, and a sand dollar has 52 chromosomes. The DNA sequences differ between plants and animals. The DNA structure in plants and animals is similar.

Chapter Opener			TE **Using the Figure,** p. 206 GENERAL SE **Quick Review,** p. 207 ■ SE **Reading Activity,** p. 207 ■
& 3 • 90 min pp. 208–214 **Section 1** From Genes to Proteins		OSP **Lesson Plan*** TR **Bellringer*** TR **Section Outline*** TR **C15 Transcription*** TR **C16 Codons in mRNA*** TR **C17 Translation: Assembling Proteins***	TE **Discussion,** p. 209 BASIC TE **Activity** Triplet Spelling, p. 211 GENERAL TE **Using the Figure,** p. 212 GENERAL TE **Group Activity** Retroviruses, p. 213 ADVANCED SE **Quick Lab** Modeling Transcription, p. 210 ■ CRF **Datasheets for In-Text Labs** Modeling Transcription* SE **Data Lab** Decoding the Genetic Code, p. 214 ■ CRF **Datasheets for In-Text Labs** Decoding the Genetic Code* CRF **Exploration Lab** Determining Protein Concentration* GENERAL
& 5 • 90 min pp. 215–220 **Section 2** Gene Regulation and Structure		OSP **Lesson Plan*** TR **Bellringer*** TR **Section Outline*** TR **C22 Controlling Transcription in Eukaryotes*** TR **C26 Major Types of Mutations***	TE **Demonstration,** p. 215 BASIC TE **Demonstration,** p. 216 GENERAL SE **Quick Lab** Modeling Introns and Exons, p. 218 ■ CRF **Datasheets for In-Text Labs** Modeling Introns and Exons* CRF **Skill Practice Lab** Transforming Bacteria with Foreign DNA* GENERAL SE **Exploration Lab** Modeling Protein Synthesis, pp. 224–225 ■ CRF **Datasheet for In-Text Labs** Modeling Protein Synthesis* ◆

& 7 • 90 min

Chapter Review and Assessment Resources

- SE **Study Zone,** p. 221 ■
- SE **Performance Zone,** p. 222 ■
- SE **FCAT Test Prep,** p. 223 ■
- TR **Concept Mapping***
- CRF **Vocabulary Review Worksheet*** ■ GENERAL
- CRF **Test Prep Pretest*** ■ GENERAL
- CRF **Chapter Test*** ■ GENERAL
- CRF **Chapter Test*** ADVANCED
- OSP **Test Generator***
- OSP **Test Item Listing***
- OSP **Modified Chapter Test*** BASIC

Online and Technology Resources

go.hrw.com Holt Online Learning

Visit go.hrw.com for access to Holt Online Learning or enter the keyword **HX6 Home** for a variety of free online resources.

One-Stop Planner® CD-ROM

This CD-ROM package includes
- Lab Materials QuickList Software
- Holt Calendar Planner
- Customizable Lesson Plans
- Printable Worksheets
- ExamView® Test Generator
- Interactive Teacher Edition
- Holt PuzzlePro® Resources
- Holt PowerPoint® Resources

Guided Reading Audio CDs

These CDs are designed to help auditory learners and reluctant readers. (Audio CDs are also available in Spanish.)

Compression guide:
To shorten your instruction because of time limitations, omit blocks 2 and 3.

KEY

- **TE** Teacher Edition
- **SE** Student Edition
- **OSP** One-Stop Planner
- **CRF** Chapter Resource File
- **TR** Teaching Transparency
- * Also on One-Stop Planner
- ■ Also Available in Spanish
- ◆ Requires Advance Prep

SKILLS DEVELOPMENT AND PRACTICE	SECTION REVIEW AND ASSESSMENT	CORRELATIONS
		Florida Sunshine Standards/FCAT
TE Reading Skill Builder Brainstorming, p. 210 BASIC **TE Skill Builder** Math Skills, p. 211 GENERAL **CRF Directed Reading Worksheet*** BASIC **CRF Active Reading Worksheet*** GENERAL	**SE Section 1 Review** ■ **TE Reteaching,** p. 214 BASIC **TE Quiz,** p. 214 GENERAL **TE Alternative Assessment,** p. 214 GENERAL **CRF Quiz*** ■ BASIC	SC.D.1.4.4, SC.F.2.4.2
CRF Directed Reading Worksheet* BASIC **CRF Active Reading Worksheet*** GENERAL **CRF Critical Thinking Worksheet*** ADVANCED **CRF Science Skills Worksheet*** GENERAL	**SE Section 2 Review** ■ **TE Reteaching,** p. 220 BASIC **TE Quiz,** p. 220 GENERAL **TE Alternative Assessment,** p. 220 GENERAL **CRF Quiz*** ■ BASIC **Modified Worksheet*** BASIC	SC.F.1.4.5, SC.F.2.4.2, SC.F.2.4.3

Holt Biology Interactive Tutor CD-ROM

This CD-ROM consists of interactive activities that give students a fun way to extend their knowledge of biology concepts.

Unit 6 Gene Expression

SCILINKS. NSTA

www.scilinks.org

Maintained by the **National Science Teachers Association**

Topic: Genetic Code
SciLinks Code: HX4089

Topic: Genetic Disorders
SciLinks Code: HX4091

CHAPTER 10

Overview

Tell the students that this chapter examines the process of making proteins. This process begins with transcription, during which a mRNA copy is made using a section of DNA as a template. Translation follows transcription. During translation, tRNA segments bring specific amino acids to the ribosome for assembly into protein.

Using the Figure

Have students examine the opening photograph on this page. Ask students to look for evidence of proteins. (Many proteins are structural and are found in both the grass and the firefly beetle. Some proteins are enzymes and are involved in chemical reactions in the grass and firefly, including the firefly's glowing abdomen.) Ask students why only the abdomen glows. (The cells of the abdomen produce the protein that activates the luminescence; the other cells do not.)

Assessing Prior Knowledge

Before beginning this chapter, be sure your students can answer the questions in the ***Quick Review*** on the facing page. If they cannot, encourage them to reread the sections indicated.

Firefly (19×)

206

Chapter Correlations — Florida Sunshine Standards/FCAT

SC.D.1.4.4 Earth's systems and organisms are the result of a long, continuous change over time.

SC.F.1.4.5 Complex interactions among the different kinds of molecules in the cell cause distinct cycles of activity governed by proteins.

SC.F.2.4.2 Every cell contains a "blueprint" coded in DNA molecules that specify how proteins are assembled to regulate cells.

SC.F.2.4.3 Mechanisms of change (e.g., mutation and natural selection) lead to adaptations in a species and their ability to survive naturally in changing conditions and to increase species diversity.

CHAPTER

10 How Proteins Are Made

Florida Sunshine Standards/FCAT

SC.D.1.4.4 Earth's systems and organisms are the result of a long, continuous change over time.
SC.F.1.4.5 Complex interactions among the different kinds of molecules in the cell cause distinct cycles of activity governed by proteins.
SC.F.2.4.2 Every cell contains a "blueprint" coded in DNA molecules that specify how proteins are assembled to regulate cells.
SC.F.2.4.3 Mechanisms of change (e.g., mutation and natural selection) lead to adaptations in a species and their ability to survive naturally in changing conditions and to increase species diversity.

Quick Review

Answer the following without referring to earlier sections of your book.

1. **Summarize** the structure and function of proteins. *(Chapter 2, Section 3)*
2. **Describe** the function of ribosomes. *(Chapter 3, Section 2)*
3. **Differentiate** between DNA and genes. *(Chapter 6, Section 1)*
4. **Describe** the structure and function of DNA. *(Chapter 9, Section 2)*
5. **State** the base-pairing rules. *(Chapter 9, Section 2)*

Did you have difficulty? *For help, review the sections indicated.*

Reading Activity

Before you read this chapter, write a short list of all the things you know about how proteins are made. Then, write a list of the things that you want to know about how proteins are made. Save your list, and to assess what you have learned, see how many of your own questions you can answer after reading this chapter.

Looking Ahead

The firefly shown here gives off light because of a chemical reaction activated by an enzyme (which is a type of protein) made by its cells.

internet connect

www.scilinks.org
National Science Teachers Association *sci*LINKS Internet resources are located throughout this chapter.

SCILINKS. Maintained by the National Science Teachers Association

Identifying Preconceptions

Students may believe that each cell has a unique set of instructions, in the form of DNA, that code for the unique products of that cell. In fact, each cell (with the exception of egg and sperm) contains the entire set of chromosomes, which includes two copies of every gene. Only a small fraction of each chromosome contains genes, and only a small fraction of the genes are converted into proteins in any one cell. For example, eye cells contain the instructions for making hair proteins, but they are not "turned on."

Quick Review Answers

1. Proteins are chains of linked amino acids that have been folded into compact shapes. Some proteins play important roles as enzymes. Other proteins are structural components of cells and tissues.
2. Ribosomes are structures built of proteins and RNA that serve as assembly sites for protein synthesis.
3. DNA is an extremely long, linear molecule containing many genes; genes are relatively short segments of DNA that code for protein or RNA.
4. DNA is made of two strands of complementary nucleotides and contains the four bases adenine, thymine, cytosine, and guanine. DNA contains the sugar deoxyribose and serves to store the complete set of an organism's genetic material.
5. In DNA, adenine (A) pairs with thymine (T), and guanine (G) pairs with cytosine (C).

Chapter Resource File

- Vocabulary Worksheets
- Concept Mapping

Reading Activity Answers

Answers will vary. Students should remember that genes are specific sequences of DNA nucleotides, and that the different genes code for different proteins. They also should remember that proteins are made from chains of amino acids.

From Genes to Proteins

Florida Sunshine Standards/FCAT

SC.D.1.4.4 Earth's systems and organisms are the result of a long, continuous change over time.

SC.F.2.4.2 Every cell contains a "blueprint" coded in DNA molecules that specify how proteins are assembled to regulate cells.

Objectives

- **Compare** the structure of RNA with that of DNA. SC.F.2.4.2
- **Summarize** the process of transcription. SC.F.2.4.2
- **Relate** the role of codons to the sequence of amino acids that results after translation. SC.F.2.4.2
- **Outline** the major steps of translation. SC.F.2.4.2
- **Discuss** the evolutionary significance of the genetic code. SC.D.1.4.4

Key Terms

ribonucleic acid (RNA)
uracil
transcription
translation
gene expression
RNA polymerase
messenger RNA
codon
genetic code
transfer RNA
anticodon
ribosomal RNA

Decoding the Information in DNA

Traits, such as eye color, are determined by proteins that are built according to instructions coded in DNA. Recall that proteins have many functions, including acting as enzymes and cell membrane channels. Proteins, however, are not built directly from DNA. Ribonucleic *(rie boh noo KLAY ihk)* acid is also involved.

Like DNA, **ribonucleic acid (RNA)** is a nucleic acid—a molecule made of nucleotides linked together. RNA differs from DNA in three ways. First, RNA consists of a single strand of nucleotides instead of the two strands found in DNA, as shown in **Figure 1.** Second, RNA nucleotides contain the five-carbon sugar ribose *(RIE bohs)* rather than the sugar deoxyribose, which is found in DNA nucleotides. Ribose contains one more oxygen atom than deoxyribose contains. And third, in addition to the A, G, and C nitrogen bases found in DNA, RNA nucleotides can have a nitrogen base called **uracil** *(YUR uh sihl)*—abbreviated as U. No thymine (T) bases are found in RNA. Like thymine, uracil is complementary to adenine whenever RNA base-pairs with another nucleic acid.

A gene's instructions for making a protein are coded in the sequence of nucleotides in the gene. The instructions for making a protein are transferred from a gene to an RNA molecule in a process called **transcription.** Cells then use two different types of RNA to read the instructions on the RNA molecule and put together the amino acids that make up the protein in a process called **translation.** The entire process by which proteins are made based on the information encoded in DNA is called **gene expression,** or protein synthesis. This process is summarized in Figure 1.

Figure 1 Gene expression

The instructions for building a protein are found in a gene and are "rewritten" to a molecule of RNA during transcription. The RNA is then "deciphered" during translation.

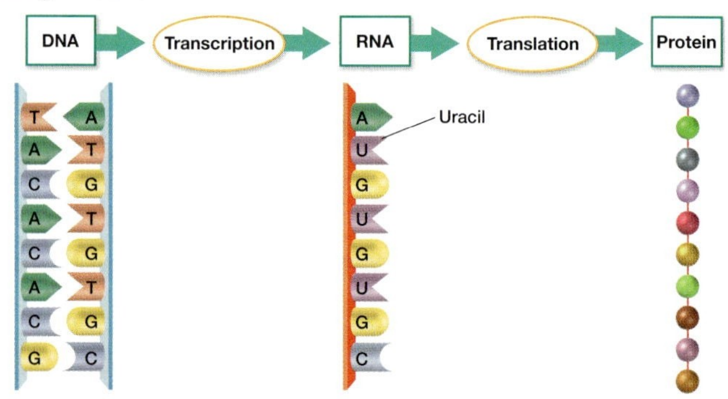

Focus

Overview

Before beginning this section review with your students the objectives listed in the Student Edition. In this section, students will follow the events of transcription, during which information is transferred from a strand of DNA to a complementary strand of RNA. This RNA is then used during translation, the process that builds proteins. Students will learn how the three-base codons of mRNA are used to link specific amino acids in protein synthesis.

Bellringer

Ask students to name some body parts that contain proteins. (hair, skin, nails, internal organs) Point out that chemical structures involved in physiology are also made of proteins—hemoglobin in the blood, insulin that regulates blood glucose levels, and enzymes that regulate all body functions.

Interactive Biology Tutor

- Unit 6—Gene Expression: Topics 3–6

This engaging tutorial introduces students to principles of protein synthesis within the cell.

Transparencies

TR Bellringer
TR C15 Transcription
TR C16 Codons in mRNA
TR C17 Translation: Assembling Proteins

Section Correlations

SC.D.1.4.4
SC.F.2.4.2

Chapter Resource File

- Directed Reading BASIC
- Active Reading GENERAL
- Data Sheet for Quick Lab GENERAL
- Data Sheet for Data Lab GENERAL

One-Stop Planner CD-ROM

- Reading Organizers BASIC
- Reading Strategies BASIC

INCLUSION Strategies

- *Learning Disability*

To help students remember the steps of transcription and translation, have students write each step on an index card. The steps should be numbered on the back of the index card. Students can mix the steps and re-order them from memory. When they have attempted to put the steps in order, they can turn the cards over and check to see if they have them in the correct order. This can be a study guide for an individual student or small group.

Transfer of Information from DNA to RNA

The first step in the making of a protein, transcription, takes the information found in a gene in the DNA and transfers it to a molecule of RNA. **RNA polymerase,** an enzyme that adds and links complementary RNA nucleotides during transcription, is required. **Figure 2** summarizes the steps of transcription.

Step 1 Transcription begins when RNA polymerase binds to the gene's promoter—a specific sequence of DNA that acts as a "start" signal for transcription.

Step 2 RNA polymerase then unwinds and separates the two strands of the double helix, exposing the DNA nucleotides on each strand.

Step 3 RNA polymerase adds and then links complementary RNA nucleotides as it "reads" the gene. RNA polymerase moves along the nucleotides of the DNA strand that has the gene, much like a train moves along on a track. Transcription follows the base-pairing rules for DNA replication except that in RNA, uracil, rather than thymine, pairs with adenine.

As transcription proceeds, the RNA polymerase eventually reaches a "stop" signal in the DNA. This "stop" signal is a sequence of bases that marks the end of each gene in eukaryotes, or the end of a set of genes in prokaryotes.

Real Life

Death cap mushrooms are deadly if eaten.

One of the poisons in death cap mushrooms (*Amanita phalloides*) is taken up by liver cells, where the poison binds to an RNA polymerase. The poison prevents liver cells from making RNA and, thus, from making proteins. Liver failure—and death—can result.

Finding Information

Research other poisons found in Amanita *spp. and determine their methods of action.*

BIOgraphic Figure 2

Transcription: Making RNA

RNA polymerase adds complementary RNA nucleotides as it reads the gene.

1 RNA polymerase binds to the gene's promoter.

2 The two DNA strands unwind and separate.

3 Complementary RNA nucleotides are added.

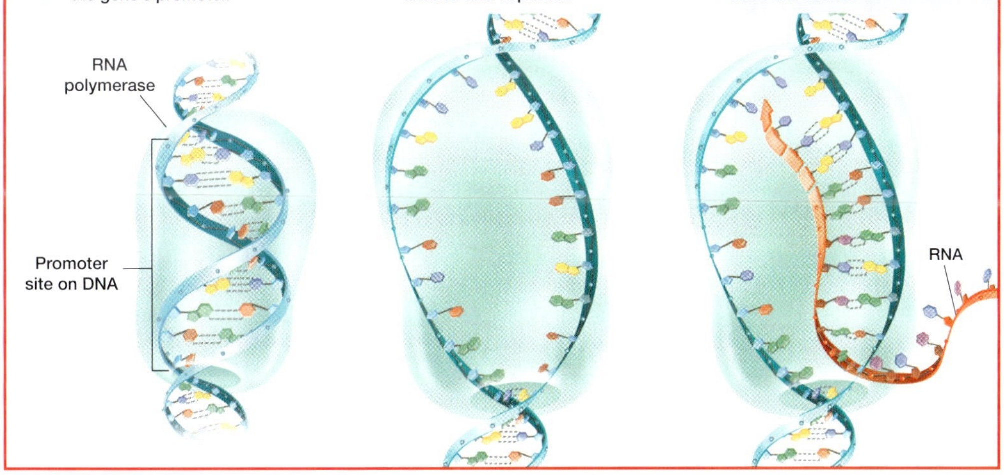

Motivate

Discussion/Question BASIC

Draw the structures of deoxyribose and ribose on the board. Ask students to compare the two sugars. (Students should notice that the structures are almost identical, except deoxyribose has one less oxygen than ribose.) Ask students to explain how this difference is reflected in the names of the two sugars. (Deoxyribose is deoxygenated, relative to ribose.) English Language Learners **LS Visual**

Deoxyribose

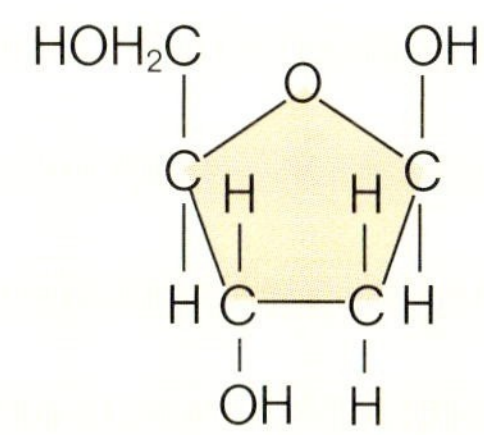

Ribose

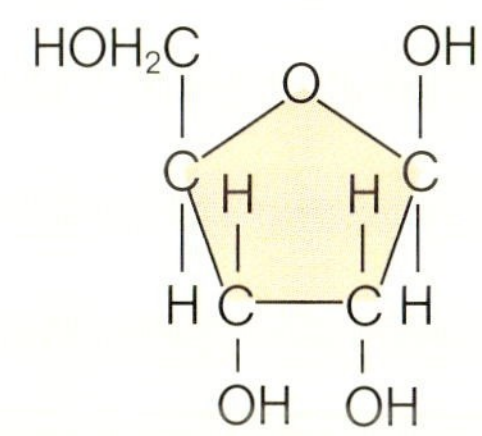

MEDICINE CONNECTION

Some diseases, such as sickle cell anemia, are caused by changes in just one or a few nucleotides of a gene. Contact a local laboratory where genetic testing is performed, or contact an organization such as the March of Dimes. Ask for a speaker to give a presentation about such genetic diseases, including the specific point mutations, the gene and chromosome involved, and the consequences of the mutations.

Teach

Real Life

Species of *Amanita* including fly agaric *(A. muscaria)* and panthercap *(A. pantherina)* contain the poisons ibotenic acid and muscimol. These poisons are related and have a similar effect on the body, although muscimol is generally more potent. These poisons affect the nervous system and may lead to drowsiness, dizziness, delirium, and hyperactivity. Unlike the liver-damaging poison found in *A. phalloides,* these poisons are rarely fatal, but can cause coma and convulsions in children.

Teach, *continued*

Teaching Tip — Basic

Comparing Transcription and Replication Have students make a Graphic Organizer similar to the one at the bottom of this page to illustrate the differences between transcription and DNA replication.
LS Verbal

Reading Skill Builder — Basic

Brainstorming Pair each student with a partner. Have each pair read the first page of the section aloud, sharing the reading equally. Then ask students to answer the following questions:

- What is the main idea in this section? What passages or words led you to this conclusion?
- How will this section be organized? What words or sentences support your conclusion?

LS Auditory English Language Learners

Quick Lab: *Modeling Transcription*

Skills Acquired
Analyzing, predicting

Answers to Analysis

1. Two colors represent the two different molecules.
2. The mRNA sequence would not be the same as the one constructed in the activity.
3. Their second mRNA is different from the first mRNA.

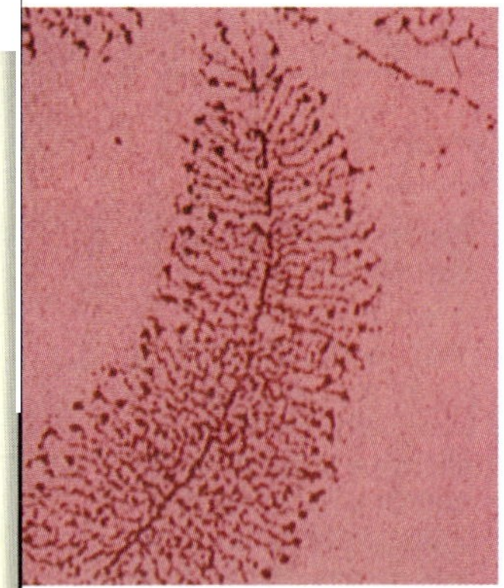

Figure 3 **Multiple copies of RNA.** In eukaryotes, RNA polymerase adds about 60 nucleotides per second. There are typically about 100 RNA polymerase molecules per gene.

When the RNA nucleotides are added during transcription, they are linked together with covalent bonds. As RNA polymerase moves down the strand, a single strand of RNA grows. Behind RNA polymerase, the two strands of DNA close up by forming hydrogen bonds between complementary bases, re-forming the DNA double helix.

Like DNA replication, transcription uses DNA nucleotides as a template for making a new molecule. However, in DNA replication, the new molecule made is DNA. In transcription, the new molecule made is RNA. In addition, in DNA replication, both strands of DNA serve as templates, whereas in transcription, only part of one of the two strands of DNA (a gene) serves as a template.

Transcription in prokaryotic cells occurs in the cytoplasm (because prokaryotic cells have no nucleus); transcription in eukaryotic cells occurs in the nucleus, where the DNA is located. During transcription, many identical RNA molecules are made simultaneously from a single gene, as shown in **Figure 3.** The RNA being made fans out from the gene to give a "feathery" appearance. The long line along the length of the "feather" is the DNA being transcribed. The circles along the length are the RNA polymerase molecules. The "hairs" on the feather are the RNA chains being made.

Quick Lab: *Modeling Transcription* SC.F.2.4.2

You can use paper and pens to model the process of transcription.

Materials

paper, scissors, pens or pencils (two colors), tape

Procedure

1. Cut a sheet of paper into 36 squares, each about 2.5 × 2.5 cm (1 × 1 in.) in size.
2. To make one side of your DNA model, line up 12 squares in a column. Using one color, randomly label each square with one of the following letters: A, C, G, or T. Each square represents a DNA nucleotide. Use tape to keep the squares in a column.
3. To make the second side of your DNA model, line up 12 squares next to the first column. Use the same color you used in step 2 to label each square with the complementary DNA nucleotide. Tape the squares together in a column.
4. Separate the two columns. The remaining 12 squares represent RNA nucleotides. Use a different color to "transcribe" one of the DNA strands.

Analysis

1. **Propose** a reason for using different colors for the DNA and RNA "nucleotides."
2. **Predict** how a change in the sequence of nucleotides in a DNA molecule would affect the mRNA transcribed from the DNA molecule.
3. **Critical Thinking** **Applying Information** Use your model to test your prediction. Describe your results.

210

Graphic Organizer

Use this graphic organizer with **Teaching Tip** on this page.

Transcription	DNA replication
RNA polymerase is used.	DNA polymerase is used.
RNA nucleotides are linked.	DNA nucleotides are linked.
An RNA molecule is made.	A DNA molecule is made.
Only one part of one strand (a gene) is used as a template.	Both DNA strands serve as templates.

The Genetic Code: Three-Nucleotide "Words"

Different types of RNA are made during transcription, depending on the gene being expressed. When a cell needs a particular protein, it is messenger RNA that is made. **Messenger RNA** (mRNA) is a form of RNA that carries the instructions for making a protein from a gene and delivers it to the site of translation. The information is translated from the language of RNA—nucleotides—to the language of proteins—amino acids. The RNA instructions are written as a series of three-nucleotide sequences on the mRNA called **codons** *(KOH dahnz)*. Each codon along the mRNA strand corresponds to an amino acid or signifies a start or stop signal for translation.

In 1961, Marshall Nirenberg, an American biochemist, deciphered the first codon by making artificial mRNA that contained only the base uracil (U). The mRNA was translated into a protein made up entirely of phenylalanine amino-acid subunits. Nirenberg concluded that the codon UUU is the instruction for the amino acid phenylalanine. Later, scientists deciphered the other codons. **Figure 4** shows the **genetic code**—the amino acids and "start" and "stop" signals that are coded for by each of the possible 64 mRNA codons.

internet connect
www.scilinks.org
Topic: Genetic Code
Keyword: HX4089
SciLinks. Maintained by the National Science Teachers Association

Figure 4 Interpreting the genetic code

The amino acid coded for by a specific mRNA codon can be determined by following the three steps below.

1. Find the first base of the mRNA codon along the left side of the table.
2. Follow that row to the right until you are beneath the second base of the codon.
3. Move up or down in that section until you are even, on the right side of the chart, with the third base of the codon.

Codons in mRNA

First base	Second base: U	Second base: C	Second base: A	Second base: G	Third base
U	UUU, UUC – Phenylalanine UUA, UUG – Leucine	UCU, UCC, UCA, UCG – Serine	UAU, UAC – Tyrosine UAA, UAG – Stop	UGU, UGC – Cysteine UGA – Stop UGG – Tryptophan	U C A G
C	CUU, CUC, CUA, CUG – Leucine	CCU, CCC, CCA, CCG – Proline	CAU, CAC – Histidine CAA, CAG – Glutamine	CGU, CGC, CGA, CGG – Arginine	U C A G
A	AUU, AUC, AUA – Isoleucine AUG – Start	ACU, ACC, ACA, ACG – Threonine	AAU, AAC – Asparagine AAA, AAG – Lysine	AGU, AGC – Serine AGA, AGG – Arginine	U C A G
G	GUU, GUC, GUA, GUG – Valine	GCU, GCC, GCA, GCG – Alanine	GAU, GAC – Aspartic Acid GAA, GAG – Glutamic Acid	GGU, GGC, GGA, GGG – Glycine	U C A G

211

Activity — GENERAL

Triplet Spelling Write the four letters *A, E, R* and *T* on the board. Ask students to write as many meaningful three-letter words as they can using any three of the four letters. (Answers may include art, are, rat, tar, tea, eat, ate, ear.) Continue by asking students to form words that use the letters more than once. (New answers may include tee, and tat.) Students should see that many three-letter combinations can be produced. Compare this exercise to the table in **Figure 4.** Ask students how the two compare. (In the exercise, letters are combined to form words, but in the table the letters stand for combinations of nucleotides.) English Language Learners

SKILL BUILDER — GENERAL

Math Skills Tell students they can predict the number of possible combinations if they know how many items there are to choose from and how many items will be in a set. For example, there are 4 possible bases in an mRNA codon, and three bases per codon. That predicts a possible $4 \times 4 \times 4 = 64$ possibilities. Have students confirm this in the table in **Figure 4.** Ask student to predict how many possible codons would exist if there were only 2 bases per codon? ($4 \times 4 = 16$) How many codons would exist if there were only three possible bases, and 3 bases per codon? ($3 \times 3 \times 3 = 27$) LS Logical

READING SKILL BUILDER — BASIC

Interactive Reading Assign Chapter 10 of the *Holt Biology* Guided Audio CD Program to help students achieve greater success in reading the chapter. English Language Learners

REAL WORLD CONNECTION

The prokaryotic ribosome is smaller than the eukaryotic ribosome. Its protein and RNA content are also dissimilar. These differences allow selective antibiotics, such as tetracycline, to bind to the prokaryotic ribosomes and interfere with prokaryotic protein synthesis. These antibiotics can be safely given to humans because they do not affect the ribosomes and protein synthesis of eukaryotic cells.

MISCONCEPTION ALERT

Translation When given an mRNA strand to translate, many students mistakenly use the anticodons, instead of the codons, to determine the amino acid sequence. Point out that the genetic code is based on the codons found on the mRNA and not on the anticodons of tRNA. The codon sequence is the genetic code (DNA language) rewritten or transcribed in RNA language.

Teach, continued

Teaching Tip — GENERAL

Ribozymes A substance called peptidyl transferase catalyzes the formation of a peptide bond that joins the polypeptide chain from the P site to the amino acid at the A site. Evidence suggests that peptidyl transferase is not a protein, as might be expected, but an rRNA molecule that acts as an enzyme. RNA catalysts are known as ribozymes. Ask students what impact this evidence has on previously held beliefs about biological catalysts. (All biological catalysts were once thought to be protein in nature; ribozymes are nucleic acids.)

Using the Figure — GENERAL

BIO graphic Work with students to help them summarize the events that take place during translation shown in **Figure 5.** Point out the complementary nature of codons on the mRNA and the anticodons on the tRNA, as shown in step 1. Refer students to the table in **Figure 4** to make the connection between the amino acid the tRNA is carrying and the codon it binds to. Be sure students understand the function of each of the following: mRNA (carries code for making protein), tRNA (carries specific amino acids to site of translation), and ribosome (coordinates protein assembly). Some students may find this process confusing. Encourage questions. When you are satisfied that they understand the process, ask students to close their books and write a summary of translation. LS **Visual**

RNA's Roles in Translation

Translation takes place in the cytoplasm. Here transfer RNA molecules and ribosomes help in the synthesis of proteins. **Transfer RNA** (tRNA) molecules are single strands of RNA that temporarily carry a specific amino acid on one end. Each tRNA is folded into a compact shape and has an anticodon *(an tee KOH dahn)*. An **anticodon** is a three-nucleotide sequence on a tRNA that is complementary to an mRNA codon. As shown in **Figure 5,** the amino acid that a tRNA molecule carries corresponds to a particular mRNA codon.

Ribosomes, shown in Figure 5, are composed of both proteins and ribosomal RNA (rRNA). **Ribosomal RNA** molecules are RNA molecules that are part of the structure of ribosomes. A cell's cytoplasm contains thousands of ribosomes. Each ribosome temporarily holds one mRNA and two tRNA molecules. Figure 5 summarizes the process of translation:

Step 1 Translation begins when the mRNA leaves the nucleus and enters the cytoplasm. The mRNA, the two ribosomal subunits, and a tRNA carrying the amino acid methionine *(muh THIE uh neen)* together form a functional ribosome. The mRNA "start" codon AUG, which signals the beginning of a protein chain, is oriented in a region of the ribosome called the P site, where the tRNA molecule carrying methionine can bind to the start codon.

Figure 5

BIO graphic **Translation: Assembling Proteins**

Amino acids are assembled from information encoded in mRNA.

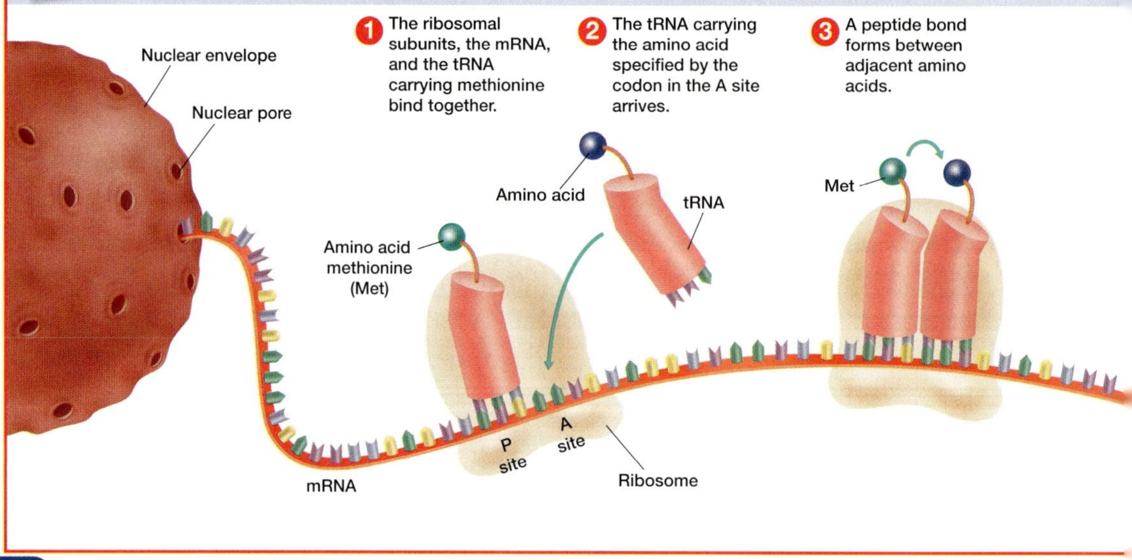

Attention Grabber

In 2002, as the Human Genome Project was nearing completion of its first goals, scientists had sequenced nearly all 3 billion base pairs. These base pairs include about 30,000 genes.

did you know?

E Sites In addition to the A and P sites, researchers have discovered a third tRNA binding site on the ribosome—the E (exit) site. When a tRNA molecule detaches, as shown in Step 4 of **Figure 5,** the tRNA first moves from the P site to the E site on the ribosome. The tRNA is then released from the ribosome.

Step 2 The codon in the area of the ribosome called the A site is ready to receive the next tRNA. A tRNA molecule with the complementary anticodon arrives and binds to the codon. The tRNA is carrying its specific amino acid.

Step 3 Now both the A site and the P site are holding tRNA molecules, each carrying a specific amino acid. Enzymes then help form a peptide bond between the adjacent amino acids.

Step 4 Afterward, the tRNA in the P site detaches, leaves behind its amino acid, and moves away from the ribosome.

Step 5 The tRNA (with its protein chain) in the A site moves over to fill the empty P site. Because the anticodon remains attached to the codon, the tRNA molecule and mRNA molecule move as a unit. As a result, a new codon is present in the A site, ready to receive the next tRNA and its amino acid. An amino acid is carried to the A site by a tRNA and then bonded to the growing protein chain.

Step 6 The tRNA in the P site detaches and leaves its amino acid.

Step 7 Steps 2 through 6 are repeated until a stop codon is reached. A stop codon is one of three codons (UAG, UAA, or UGA) for which there is no tRNA molecule with a complementary anticodon. Because there is no tRNA to fit into the empty A site in the ribosome, protein synthesis stops. The newly made protein is released into the cell.

4 The tRNA in the P site detaches and leaves its amino acid behind.

5 The tRNA in the A site moves to the P site. The tRNA carrying the amino acid specified by the codon in the A site arrives.

6 A peptide bond is formed. The tRNA in the P site detaches and leaves its amino acid behind.

7 The process is repeated until a stop codon is reached. The ribosome complex falls apart. The newly made protein is released.

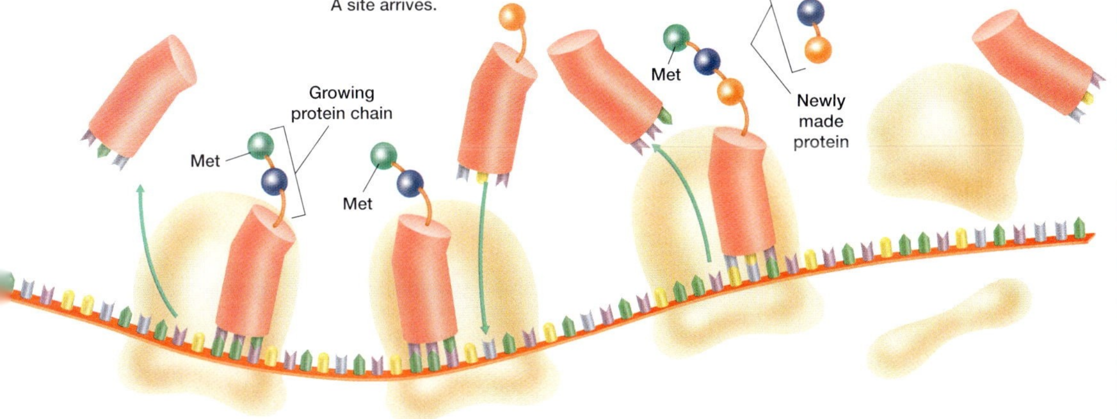

Group Activity — ADVANCED

Retroviruses Retroviruses contain RNA rather than DNA as their nucleic acid, and they use RNA as a template to make DNA. To do this, an enzyme called reverse transcriptase is used. HIV is an example of one such virus. Several drugs have been developed to slow the replication of the virus. Two of these drugs are lamivudine and zidovudine. Have students work in teams of three or four to research how these drugs function, write a brief report on their findings, and make simple models or drawings comparing DNA to RNA. (These drugs inhibit the action of reverse transcriptase. In retroviruses, DNA is made from viral RNA. This DNA is incorporated into the host DNA, which is then transcribed, resulting in the viral mRNA being translated to make specific viral proteins.)
LS Verbal Co-op Learning

DATA LAB *Decoding the Genetic Code*

Skills Acquired
Recognizing patterns, interpreting information

Teacher's Notes
Use **Figure 5** to review the process of translation. Make sure students understand the terms codon and anticodon. Point out that because of space limitations, the stop codon is not included on the mRNA strand that is shown.

Answers to Analysis

1. Serine-arginine-glutamic acid-phenylalanine-serine
2. AGA, GCA, CUU, AAA, AGG
3. AGAGCACTTAAAAGG
4. TCTCGTGAATTTTCC

INCLUSION *Strategies*

- *Attention Deficit Disorder*
- *Learning Disabilities*

Ask students to research the life of Barbara McClintock. The students should find information about her life and her work. Additionally, have students give their opinion of how it would feel and how they would react if they made a major scientific discovery that the scientific world denounced and would not recognize for twenty years. They can report their findings and opinions in a tape-recorded message to the class.

Integrating Physics and Chemistry

Point out that the metabolic processes by which *E. coli* absorb and break down lactose is a form of digestion whereby food—"milk sugar" in this instance—is chemically changed into simpler compounds that our bodies can then use for energy or making other structures needed by cells. Ask students to name the enzyme involved in the chemical digestion of lactose and to identify the portion of its name that indicates it is an enzyme.

Close

Reteaching — BASIC

Write the following on eight different pieces of paper: *transcription, translation, DNA, RNA, mRNA, tRNA, codon,* and *anticodon.* Put these pieces of paper in a small container. Write the question How are they linked? on the board. Have a student pick two pieces of paper from the container. Show the terms to the class and give them a couple of minutes to write down the answer. Return the papers to the container and repeat as long as time will allow.

Quiz — GENERAL

True or False:

1. RNA is similar in structure to DNA, except it contains the sugar uracil rather than thymine. (False. RNA does contain uracil, but uracil is a base.)
2. The anticodon is the complementary sequence of the codon. (True. Codons are mRNA, while anticodons are tRNA.)
3. The codons are the same for most organisms. (True. Genes may differ, but the codons within the genes are the same.)

Alternative Assessment — GENERAL

Have students create a colorful poster that compares and illustrates the functions of mRNA, tRNA, and rRNA.

Section Review Correlations

SC.F.2.4.2: Items 1-6

As the mRNA moves across the ribosome, another ribosome can find the AUG codon on the same mRNA and begin making a second copy of the same protein. In this way many copies of the same protein are made from a single mRNA molecule.

With few exceptions, the genetic code is the same in all organisms. For example, the codon GUC codes for the amino acid valine in bacteria, in eagles, in plants, and in your own cells. For this reason, the genetic code is often described as being nearly universal. It appears that all life-forms have a common evolutionary ancestor with a single genetic code. Some exceptions include the ways cell organelles that contain DNA (such as mitochondria and chloroplasts) and a few microscopic protists read "stop" codons.

DATA LAB

Decoding the Genetic Code SC.F.2.4.2

Background

Keratin is one of the proteins in hair. The gene for keratin is transcribed and translated by certain skin cells. The series of letters below represents the sequence of nucleotides in a portion of an mRNA molecule transcribed from the gene for keratin. This mRNA strand and the genetic code in Figure 4 can be used to determine some of the amino acids in keratin.

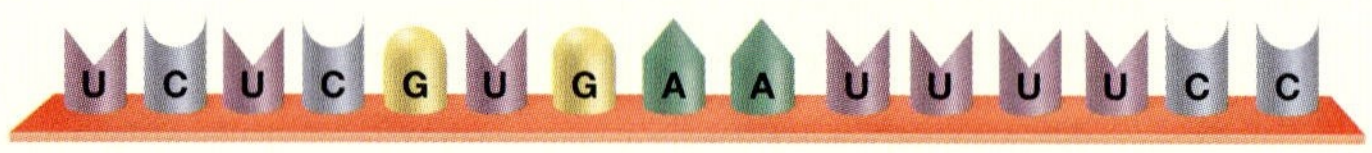

Analysis

1. **Determine** the sequence of amino acids that will result from the translation of the segment of mRNA above.
2. **Determine** the anticodon of each tRNA molecule that will bind to this mRNA segment.
3. **Critical Thinking Recognizing Patterns** Determine the sequence of nucleotides in the segment of DNA from which the mRNA strand above was transcribed.
4. **Critical Thinking Recognizing Patterns** Determine the sequence of nucleotides in the segment of DNA that is complementary to the DNA segment described in item 3.

Section 1 Review

1. **Distinguish** two differences between RNA structure and DNA structure. SC.F.2.4.2
2. **Explain** how RNA is made during transcription. SC.F.2.4.2
3. **Interpret** the genetic code to determine the amino acid coded for by the codon CCU. SC.F.2.4.2
4. **Compare** the roles of the three different types of RNA during translation. SC.F.2.4.2
5. **Critical Thinking Justifying Conclusions** Evaluate the following statement: The term *transcription* is appropriate for describing the production of RNA, and the term *translation* is appropriate for describing the synthesis of proteins. SC.F.2.4.2
6. **FCAT Prep** What is the maximum number of amino acids that could be coded for by a section of mRNA with the sequence GUUCAGAACUGU?
   A 3 **C** 6
   B 4 **D** 12
   SC.F.2.4.2

214

Answers to Section Review

1. RNA is single stranded; DNA is double stranded. RNA contains the sugar ribose; DNA contains the sugar deoxyribose. RNA contains the bases A, G, C, U; DNA contains the bases A, G, C, T.
2. RNA polymerase binds to the promoter, unwinds and separates the DNA strands, then adds and links complementary RNA nucleotides as it "reads" the gene.
3. proline
4. mRNA carries the instructions for making a protein; tRNA temporarily carries a specific amino to the site of translation; rRNA is part of the ribosome.
5. Transcription means "writing out;" instructions on the gene are written out as mRNA. Translation means, "to put into words of a different language;" the instructions for making a protein are translated from the language of nucleic acids to amino acids.
6. **A.** Incorrect. Codons are based on sets of three nucleotides, not four. **B.** Correct. There are four codons, which consist of three nucleotides each. **C.** Incorrect. Codons are not based on two-nucleotide segments. **D.** Incorrect. Codons are not based on a single nucleotide.

Gene Regulation and Structure

Section 2

Florida Sunshine Standards/FCAT

SC.F.1.4.5 Complex interactions among the different kinds of molecules in the cell cause distinct cycles of activity governed by proteins.
SC.F.2.4.3 Mechanisms of change (e.g., mutation and natural selection) lead to adaptations in a species and their ability to survive naturally in changing conditions and to increase species diversity.

Objectives

- **Describe** how the *lac* operon is turned on or off. SC.F.1.4.5
- **Summarize** the role of transcription factors in regulating eukaryotic gene expression. SC.F.1.4.5
- **Describe** how eukaryotic genes are organized. SC.F.2.4.2
- **Evaluate** three ways that point mutations can alter genetic material. SC.F.2.4.3

Key Terms

operator
operon
***lac* operon**
repressor
intron
exon
point mutation

Protein Synthesis in Prokaryotes

Although prokaryotic organisms, such as bacteria, might seem simple because of their small size, prokaryotic cells typically have about 2,000 genes. The human genome, which is the largest genome sequenced to date, has about 30,000 genes. Not all of the genes, however, are transcribed and translated all of the time; this would waste the cell's energy and materials. Both prokaryotic and eukaryotic cells are able to regulate which genes are expressed and which are not, depending on the cell's needs.

An example of gene regulation that is well understood in prokaryotes is found in the bacterium *Escherichia coli.* When you eat or drink a dairy product, the disaccharide lactose ("milk sugar") reaches the intestinal tract and becomes available to the *E. coli* living there. The bacteria can absorb the lactose and break it down for energy or for making other compounds. In *E. coli,* recognizing, consuming, and breaking down lactose into its two components, glucose and galactose, requires three different enzymes, each of which is coded for by a different gene.

As shown in **Figure 6,** the three lactose-metabolizing genes are located next to each other and are controlled by the same promoter site. There is an on-off switch that "turns on" (transcribes and then translates) the three genes when lactose is available and "turns off" the genes when lactose is not available.

Figure 6 Turning prokaryotic genes on and off

The *lac* operon allows a bacterium to build the proteins needed for lactose metabolism only when lactose is present.

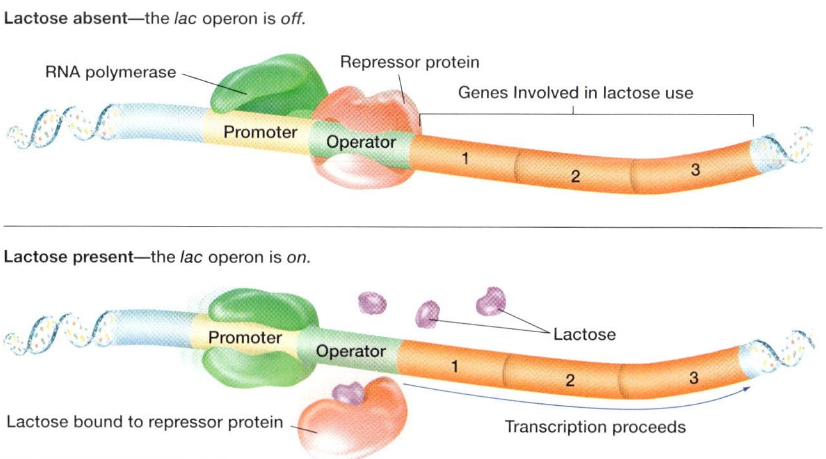

Section 2

Focus

Overview

Before beginning this section review with your students the objectives listed in the Student Edition. In this section, students will learn that cells express only a small percentage of the genes they contain at any given moment. The methods of gene regulation—or how genes are turned "on" and "off" are examined for both prokaryotes and eukaryotes. Students will also learn about the structure of eukaryotic genes, and the ways that mutations can alter the function of genes.

Bellringer

Challenge students by writing the following question on the board: *Why has protein synthesis research been focused on prokaryotes rather than eukaryotes?* Have them answer the question in their notebooks, and then discuss. (Prokaryotes are easier to grow and simpler to study. Like eukaryotes, prokaryotes have DNA, genes, and codons, but prokaryotes have much less DNA, and their DNA is circular rather than linear.) **LS Verbal**

Motivate

Demonstration — Basic

Hold up a copy of the student edition textbook. Point out to students that their textbook is about 1,000 pages long. Ask students to imagine the textbook as 1,000 pages of genetic information found within each of their cells. A typical cell at any given time is only using 3–5 percent of this information, or 30 to 50 pages. Flip through 30–50 pages to illustrate this—about one chapter's worth. English Language Learners

Chapter Resource File

- Directed Reading Basic
- Active Reading General
- Data Sheet for Quick Lab General

Transparencies

TR Bellringer
TR C22 Controlling Transcription in Eukaryotes
TR C26 Major Types of Mutations

One-Stop Planner CD-ROM

- Reading Organizers Basic
- Reading Strategies Basic
- Supplemental Reading Guide *A Feeling for the Organism* Advanced

Section Correlations

SC.F.1.4.5
SC.F.2.4.2
SC.F.2.4.3

Teach

Demonstration — GENERAL

Cut shapes from colored paper to represent the components of the lac operon as shown in **Figure 6.** These shapes should be large enough for students to see from their seats. Secure these shapes to the board and use them to demonstrate the process by which genes are turned on and off in the lac operon. Emphasize the role of feedback systems in gene regulation. **LS Visual** English Language Learners

Exploring Further

Jumping Genes

Teaching Strategies

- Help make transposons relevant to students' lives by pointing out that one type of transposon has been identified as a cause of leukemia. In humans, a piece of chromosome 22 breaks off and binds to chromosome 9.
- Tell students that transposase, an enzyme encoded by transposons, is responsible for transposon. The movement of transposons to different sites of the genome is random and rare.

Discussion

- What effect do transposons have on other genes? (Transposons can inactivate the genes they jump into.)

The piece of DNA that overlaps the promoter site and serves as the on-off switch is called an **operator.** Because of its position, the operator is able to control RNA polymerase's access to the three lactose-metabolizing genes.

In bacteria, a group of genes that code for enzymes involved in the same function, their promoter site, and the operator that controls them all function together as an **operon** *(AHP uhr ahn)*. The operon that controls the metabolism of lactose is called the ***lac* operon** and is shown in Figure 6.

What determines whether the *lac* operon is in the "on" or "off" mode? When there is no lactose in the bacterial cell, a repressor turns the operon off. A **repressor** is a protein that binds to an operator and physically blocks RNA polymerase from binding to a promoter site. The blocking of RNA polymerase consequently stops the transcription of the genes in the operon, as shown in Figure 6.

When lactose is present, the lactose binds to the repressor and changes the shape of the repressor. The change in shape causes the repressor to fall off of the operator, as shown in Figure 6. Now the bacterial cell can begin transcribing the genes that code for the lactose-metabolizing enzymes. By producing the enzymes only when the nutrient is available, the bacterium saves energy.

Exploring Further

Jumping Genes

The spotted and streaked patterns seen in Indian corn result from genes that have moved from one chromosomal location to another. Such genes are called transposons *(trans POH zahns).* When a transposon jumps to a new location, it often inactivates a gene or causes mutations. In Indian corn, some pigment genes are not expressed in some cells because they have been disrupted by jumping genes.

The Discovery of Transposons

In the 1950s, the geneticist Barbara McClintock discovered transposons while studying corn. Most scientists rejected her ideas for more than 20 years. The idea that genes could change locations on the chromosome contradicted the prevailing view that genes and chromosomes are stable parts of the cell. Over time, additional research supported her hypothesis, and her model gradually gained acceptance. In 1983, McClintock received a Nobel Prize for her discoveries involving transposons.

Importance of Transposons

All organisms, including humans, appear to have transposons. Transposons probably play a role in spreading genes for antibiotic resistance among bacteria. Transposons that affect flower color in morning glory flowers have been found. Transposons may also have medical applications, such as helping scientists discover how white blood cells make antibodies and what causes cancer.

Barbara McClintock

Although the movement of transposons is very rare, transposons are important because they can cause mutations and bring together different combinations of genes. The transfer of these mobile genes could be a powerful mechanism in evolution and could help solve certain mysteries about evolution, such as how larger organisms developed from single cells and how new species arise.

REAL WORLD CONNECTION

Many adults lack sufficient lactase enzymes to digest the lactose found naturally in dairy products. This can lead to gas, bloating, and diarrhea after consuming products containing lactose. Lactase enzymes are available as a dietary supplement, and in some products, lactase is added before consumption to break down the lactose sugar into more digestible sugars.

Cultural Awareness

The *lac* Operon Two French scientists, Francois Jacob and Jacques Monod, were responsible for the discovery of operons. They both joined the staff of Institut Pasteur, where they collaborated on genetic research. By 1960, these two men were able to determine how genes influence the making or degrading of certain substances—a system referred to as an operon. They were awarded a Nobel Prize in 1965.

Protein Synthesis in Eukaryotes

Eukaryotic cells contain much more DNA than prokaryotic cells do. Like prokaryotic cells, eukaryotic cells must continually turn certain genes on and off in response to signals from their environment. Operons have not been found often in eukaryotic cells. Instead, genes with related functions are often scattered on different chromosomes.

Because a nuclear envelope physically separates transcription from translation in a eukaryotic cell, more opportunities exist for regulating gene expression. For example, gene regulation can occur before, during, and after transcription. Gene regulation can also occur after mRNA leaves the nucleus or after translation, when the protein is functional.

Study TIP

Organizing Information
Make a table to organize information about the regulation of protein synthesis. Across the top write the headings *Prokaryotes* and *Eukaryotes.* Along the sides write *Protein(s) that regulate(s) the genes* and *Details of regulation.* Add information to the table as you read Section 2.

Controlling the Onset of Transcription

Most gene regulation in eukaryotes controls the onset of transcription—when RNA polymerase binds to a gene. Like prokaryotes, eukaryotic cells use regulatory proteins. But many more proteins are involved in eukaryotes, and the interactions are more complex. These regulatory proteins in eukaryotes are called transcription factors.

As shown in **Figure 7,** transcription factors help arrange RNA polymerases in the correct position on the promoter. A gene can be influenced by many different transcription factors.

An enhancer is a sequence of DNA that can be bound by a transcription factor. Enhancers typically are located thousands of nucleotide bases away from the promoter. A loop in the DNA may bring the enhancer and its attached transcription factor (called an activator) into contact with the transcription factors and RNA polymerase at the promoter. As shown in Figure 7, transcription factors bound to enhancers can activate transcription factors bound to promoters.

Figure 7 **Controlling transcription in eukaryotes**

Transcription factors bind to the enhancer and to the RNA polymerase. The binding activates transcription factors bound to the promoter.

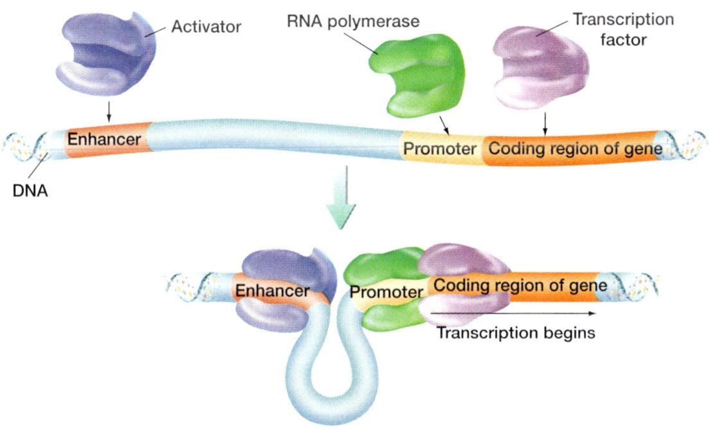

Teaching Tip — GENERAL

Adult versus Fetal Genes Fetal hemoglobin, which differs somewhat from adult hemoglobin, has a higher affinity for oxygen. A short time after birth, fetal hemoglobin is replaced by adult hemoglobin. We do not know what mechanism stops the production of fetal hemoglobin and begins the production of adult hemoglobin. Apparently, the genes for fetal hemoglobin are turned off, and those for adult hemoglobin are turned on just after birth.

Teaching Tip

Unexpressed Genes A typical human cell only expresses 3–5 percent of its genes at any given time. Also, not all mRNA is translated once it is made. The egg cells of many organisms synthesize and store large amounts of mRNA molecules that are translated after fertilization.

Trends in Biotechnology

Designing Proteins Current computing systems can simulate the folding of a protein into its compact form, but it takes 40 months of processor time to run such a simulation. Advances in terascale computing will allow one trillion operations per second. Instead of 40 months, the simulation would take just one day. Such technology would allow scientists to custom design proteins for medical treatments and other applications.

Career

Molecular Geneticist Molecular genetics deals with the molecular nature of genes and their role in the function and development of an organism. Molecular geneticists with a B.S. degree can work as laboratory technicians. With advanced degrees, molecular geneticists can design and supervise research projects. Molecular geneticists work for universities, government agencies, and agricultural, pharmaceutical, and biotechnological firms.

Teach, *continued*

Teaching Tip

Introns and Exons Tell students that two scientists independently found evidence that introns and exons existed. These two scientists, Richard Roberts and Phillip Sharp, shared a Nobel Prize in 1993 for this discovery. We now know that the cutting and pasting of introns and exons is carried out by a complex of proteins and RNA molecules together called snRNPs (pronounced "snurps"), or small ribonucleoproteins.

QUICK LAB *Modeling Introns and Exons*

Skills Acquired
Applying information, predicting outcomes

Teacher's Notes
Review the terms intron and exon with the students. Ask students to explain the graphic before beginning the lab.

Answers to Analysis

1. The strip with the letters "apprialyjoed" represents the introns. The strip with the letters that spell protein represents the exons.
2. Answers will vary. Because the function of a protein is ultimately a result of its amino acid sequence, a protein with additional amino acids will most likely not function.

WORD Origins

The "int" in the word *intron* comes from the "int" in the word *intervening*. The "ex" in the word *exon* comes from the "ex" in the word *expressed*.

Intervening DNA in Eukaryotic Genes

While it is tempting to think of a gene as an unbroken stretch of nucleotides that code for a protein, this simple arrangement is usually found only in prokaryotes. In eukaryotes, many genes are interrupted by **introns** *(IN trahnz)*—long segments of nucleotides that have no coding information. **Exons** *(EK sahnz)* are the portions of a gene that are translated (expressed) into proteins. After a eukaryotic gene is transcribed, the introns in the resulting mRNA are cut out by complex assemblies of RNA and protein called *spliceosomes*. The exons that remain are "stitched" back together by the spliceosome to form a smaller mRNA molecule that is then translated.

Many biologists think this organization of genes adds evolutionary flexibility. Each exon encodes a different part of a protein. By having introns and exons, cells can occasionally shuffle exons between genes and make new genes. The thousands of proteins that occur in human cells appear to have arisen as combinations of only a few thousand exons. Some genes in your cells exist in multiple copies, in clusters of as few as three or as many as several hundred. For example, your cells each contain 12 different hemoglobin genes, all of which arose as duplicates of one ancestral hemoglobin gene.

QUICK LAB *Modeling Introns and Exons*

SC.F.1.4.5

You can use masking tape to represent introns and exons.

Materials

masking tape, pens or pencils (two colors), metric ruler, scissors

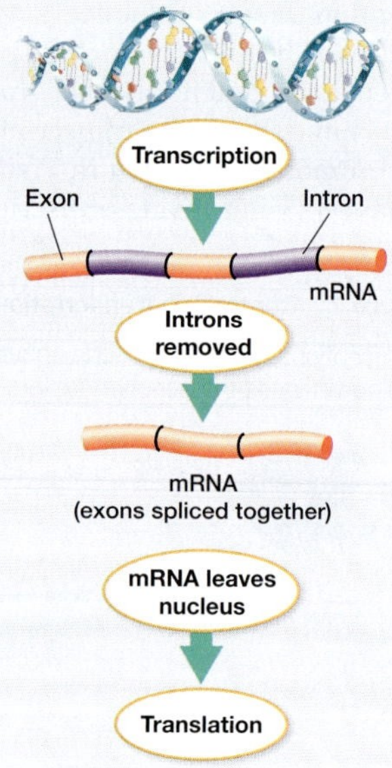

Procedure

1. Place a 15–20 cm strip of masking tape on your desk. The tape represents a gene.
2. Use two colors to write the words *appropriately joined* on the tape exactly as shown in the diagram below. Space the letters so that they take up the entire length of the strip of tape. The segments in one color represent introns; those in the other color represent exons.

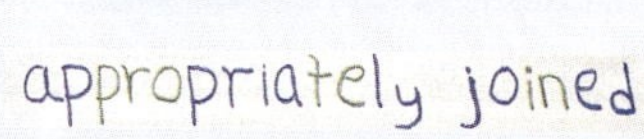

3. Lift the tape. Working from left to right, cut apart the groups of letters written in the same color. Stick the pieces of tape to your desk as you cut them, making two strips according to color and joining the pieces in their original order.

Analysis

1. **Determine** from the resulting two strips which strip is made of "introns" and which is made of "exons."
2. **Critical Thinking Predicting Outcomes** Predict what might happen to a protein if an intron were not removed.

did you know?

Eukaryotic mRNA At one end of the mRNA molecule, a cap consisting of a nucleotide called 7-methylguanylate attaches to the molecule. At the other end of mRNA, a poly A tail, consisting of many adenines, attaches. Scientists think that the cap prevents the mRNA from degradation, making it more stable than prokaryotic DNA. In a eukaryotic cell, mRNA can persist for hours, and sometimes for days or weeks. Compare this to two minutes in a prokaryotic cell. The function of the poly A tail is unknown, although it is thought to aid in the export of mRNA from the nucleus and to prevent degradation in the cytoplasm.

Mutations

Although changes in an organism's hereditary information are relatively rare, they can occur. As you learned in Chapter 6, a change in the DNA of a gene is called a mutation. Mutations in gametes can be passed on to offspring of the affected individual, but mutations in body cells affect only the individual in which they occur.

Mutations that move an entire gene to a new location are called *gene rearrangements.* Changes in a gene's position often disrupt the gene's function because the gene is exposed to new regulatory controls in its new location—like what would happen if you moved to France and couldn't speak French. Two types of gene rearrangements are shown in **Figure 8.** Genes sometimes move as part of a transposon. That is, the genes are carried by the moving transposon like fleas on a dog. Other times, the portion of the chromosome containing a gene may be rearranged during meiosis.

Mutations that change a gene are called *gene alterations.* Gene alterations such as those shown in Figure 8 usually result in the placement of the wrong amino acid during protein assembly. This error will usually disrupt a protein's function. In a **point mutation,** a single nucleotide changes. In an *insertion* mutation, a sizable length of DNA is inserted into a gene. Insertions often result when mobile segments of DNA, called transposons, move randomly from one position to another on chromosomes. Transposons make up 45 percent of the human genome. In a *deletion* mutation, segments of a gene are lost, often during meiosis.

Figure 8 Major types of mutations

The substitution, addition, or removal of one or more nucleotides is called a gene alteration. If the mutation changes the original position of a gene of the chromosome, the gene may not function normally.

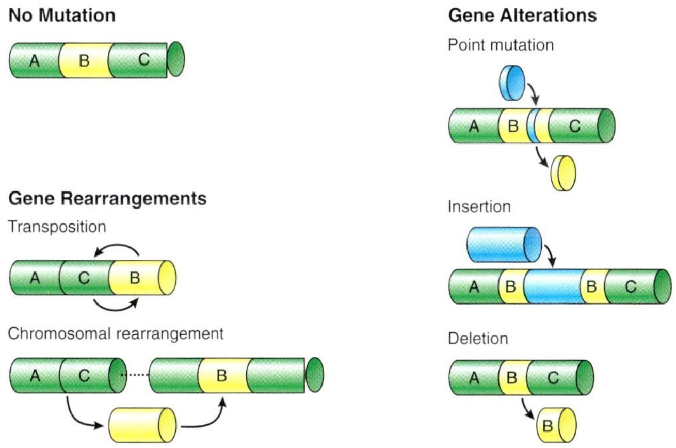

Teaching Tip — ADVANCED

Mutagens Many environmental factors (such as X rays and gamma rays) and chemicals can cause mutations. Have pairs of students research a specific mutagen. Their report should include the type of mutation caused and the ill effects of the mutation on the human body.

Co-op Learning

BIOWatch — GENERAL

Gene Therapy Cured Puppies' Blindness

Teaching Strategies

- Tell students that about 2,000 people in the United States have a genetic disease. Scientists hope to someday be able to use gene therapy on human babies that have inherited two defective genes from their parents.
- Tell students that only one pair of genes was at fault in the puppies' eye disease. With most common health problems, many genes play a role, as does a person's environment and behavior. Gene therapy may become a useful tool to help prevent or treat illness, but it won't become a substitute for good health habits!

Discussion

Have the class research and discuss ethical issues related to gene therapy. Note that while some feel that gene therapy holds great promise for preventing human suffering, some people have concerns about scientists developing technology that can change people's genetic makeup.

did you know?

Mitochondrial DNA Mothers and their offspring have identical mitochondrial DNA because sperm mitochondria are destroyed in the developing zygote. Mutations rarely occur in mitochondrial DNA. Families can be traced by this DNA because it remains essentially unchanged from generation to generation.

HISTORY CONNECTION

In 1986, there was an explosion of a nuclear reactor at the Chernobyl power plant, in the former Soviet Union (now the Ukraine). Scientists predicted that the residents of the surrounding towns would develop high rates of thyroid cancer and leukemia as a result of radiation exposure. Have students read articles about the effects of the explosion and then discuss their findings in brief reports that link the elevated incidence of cancer to radiation damage of DNA.

Because the genetic message is read as a series of triplet nucleotides, insertions and deletions of one or two nucleotides can upset the triplet groupings. Imagine deleting the letter C from the sentence "THE CAT ATE." Keeping the triplet groupings, the message would read "THE ATA TE," which is meaningless. A mutation that causes a gene to be read in the wrong three-nucleotide sequence is called a *frameshift mutation.*

Gene Therapy Cured Puppies' Blindness

Most of the time, viruses mean trouble. The microscopic invaders get into our bodies and make us sick. However, some scientists think that viruses may be useful in treating or preventing certain genetic diseases. In fact, with the help of one altered virus developed by scientists at the University of Florida, puppies that were born blind can now see. The technique that the scientists used to restore the puppies' vision is called *gene therapy.*

A Scientific First

The puppies in this case had a genetic disease that caused their blindness. Defective genes in their retinas could not produce a protein needed to make their eyes work properly.

To correct the defective gene, scientists inserted a normal gene into the genome of the cells in the puppies' eyes. Three months later, the young dogs could find their way through a dimly lit maze. This experiment was the first time that an animal regained a lost function after receiving gene therapy.

Special Transport

To insert the desired genes into the puppies' eye cells, the scientists modified the genetic material of a type of virus that usually causes the common cold. The virus then transported this genetic package to the nucleus of the puppies' retinal cells. Once in the nucleus, the inserted gene could then give the cell the instructions to manufacture the missing protein. This protein helped the brain receive messages from the eye, and let the puppies see for the first time.

internet connect
www.scilinks.org
Topic: Medicine in Florida
Keyword: HX6F10
SCILINKS. Maintained by the National Science Teachers Association

Section 2 Review

1. **Describe** the effect a repressor has on the *lac* operon when lactose is present. SC.F.1.4.5
2. **Explain** the role of transcription factors and enhancers in eukaryotic gene expression. SC.F.1.4.5
3. **Differentiate** between exons and introns. SC.F.2.4.2
4. **Critical Thinking Evaluating Significance** Which type of mutation would have a greater effect on the sequence of amino acids in a protein, a base-pair substitution or a frameshift mutation? Explain your answer. SC.F.2.4.3
5. **FCAT Prep** What term describes a mutation in which one nucleotide in a gene is replaced with a different nucleotide?
   A a deletion **C** a substitution
   B an insertion **D** a frameshift mutation
   SC.F.2.4.3

220

Close

Reteaching — BASIC

Divide the class into groups of three. Assign each group one of the following topics: lac operon, eukaryotic gene expression, and mutations. Have each group prepare a 5-minute presentation that summarizes their topic. Each group should also prepare a four-question quiz. Combine the quizzes, and have the class answer them in writing after the presentations.

Quiz — GENERAL

1. In prokaryotes, gene expression is regulated by _______. (operons)
2. Mutations that change one nucleotide are called _______ mutations. (point)
3. True or false: Introns are the parts of a gene that are translated. (False. Introns are segments that are cut out of a gene.)

Alternative Assessment — GENERAL

Form groups with six to eight students per group. Have students use each other to physically demonstrate the operation of the lac operon and to walk through the control of transcription in eukaryotes. Each student should be able to explain his or her role in the processes.

Section Review Correlations

SC.F.1.4.5: Items 1, 2
SC.F.2.4.2: Item 3
SC.F.2.4.3: Items 4, 5

Answers to Section Review

1. Lactose binds to the repressor, which changes the shape of the repressor and causes the repressor to fall off of the operator. RNA polymerase is able to bind, allowing transcription.
2. Transcription factors are regulatory proteins. Some help to arrange RNA polymerase in the correct position on the promoter, while others, called activators, bind to an enhancer (a segment of DNA). Transcription begins when the activator bound to the enhancer comes in contact with the transcription factor and RNA polymerase at the promoter.
3. Exons are portions of a eukaryotic gene that are translated into proteins. Introns contain no coding information and are cut out before translation occurs.
4. A frameshift mutation causes a disruption in the triplet groupings, which results in the incorrect sequence of amino acids. A base-pair substitution may or may not change the amino acid coded for in the triplet.
5. **A.** Incorrect. A deletion drops a nucleotide. **B.** Incorrect. An insertion adds a nucleotide. **C.** Correct. A substitution changes one nucleotide for another. **D.** Incorrect. A frameshift mutation causes the entire code beyond the mutation to shift.

Key Concepts

1 From Genes to Proteins

- The instructions needed to make proteins are coded in the nucleotides that make up a gene. The instructions are transferred to an mRNA molecule during transcription. The RNA is complementary to the gene, and the RNA nucleotides are put together with the help of RNA polymerase.
- During translation, the mRNA molecule binds to a ribosome, and tRNAs carry amino acids to the ribosome according to the codons on the mRNA. Each codon specifies an amino acid. The amino acids are joined to form a protein.
- The genetic code (codons) used by most organisms to translate mRNA is nearly universal.

2 Gene Regulation and Structure

- Prokaryotic and eukaryotic cells are able to control which genes are expressed and which are not, depending on the cell's needs.
- In prokaryotes, gene expression is regulated by operons. Gene expression is switched off when repressor proteins block RNA polymerase from transcribing a gene.
- In eukaryotes, an enhancer must be activated for a eukaryotic gene to be expressed. Transcription factors initiate transcription by binding to enhancers and to RNA polymerases.
- Many eukaryotic genes are interrupted by segments of DNA that do not code for proteins; these segments are called introns. The segments of DNA that are expressed are called exons. After transcription, the introns are cut out, and the exons are joined. The exons are then translated.
- Mutations are changes in DNA. Gene alterations are mutations that change a gene. These mutations can involve a change in a single nucleotide or an entire gene.

Key Terms

Section 1

ribonucleic acid (RNA) (208)
uracil (208)
transcription (208)
translation (208)
gene expression (208)
RNA polymerase (209)
messenger RNA (211)
codon (211)
genetic code (211)
transfer RNA (212)
anticodon (212)
ribosomal RNA (212)

Section 2

operator (216)
operon (216)
lac operon (216)
repressor (216)
intron (218)
exon (218)
point mutation (219)

BIOLOGY INTERACTIVE TUTOR
Unit 6—*Gene Expression*
Use Topics 3–6 in this unit to review the key concepts and terms in this chapter.

221

Alternative Assessment — GENERAL

Have students work in groups of four. Provide each group with a large poster board. Ask each group to create its own DNA sequence by writing the names of 24 bases on the poster. The first three bases must be TAG, and the last three bases must be either ATT, ATC, or ACT. The bases in-between should be random but when reading the bases as triplets, should not be any of the four triplets indicated above. Have each group determine the mRNA sequence that would form and the protein that would be produced. Next have the group choose a point mutation—substitution, insertion, or deletion—and "mutate" its DNA, and determine how the mutation affects the protein. Co-op Learning

Chapter Resource File

- Science Skills Worksheet GENERAL
- Critical Thinking Worksheet ADVANCED
- Test Prep Pretest GENERAL
- Chapter Test GENERAL

Answer to Concept Map

The following is one possible answer to Performance Zone item 15.

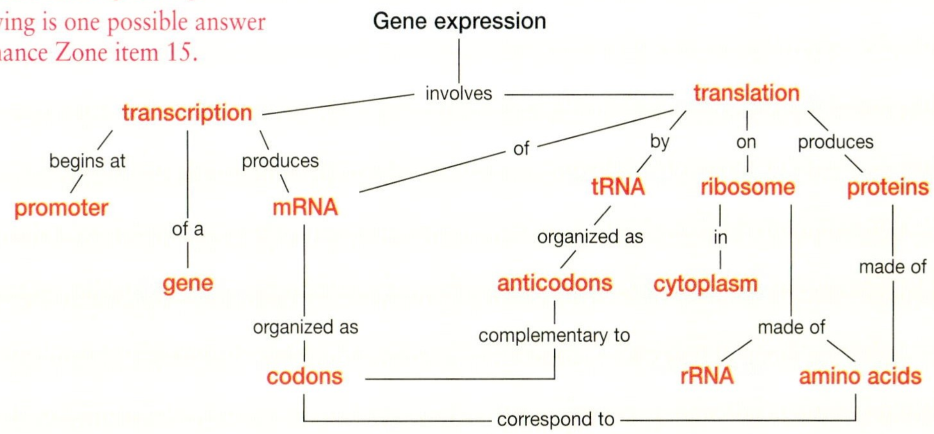

ANSWERS

Understanding Key Ideas

1. d
2. b
3. d
4. c
5. a
6. b
7. Transposons can inactivate a gene or cause a mutation. Exons do not affect genes—they are the gene, or the parts of a gene that are translated.
8. RNA—DNA does not contain uracil.
9. Sample answer: Gene therapy may be less reversible than other forms of treatment.
10. One possible answer to the concept map is found on the bottom of the Study Zone page.

Critical Thinking

11. The results are probably correct. Different types of cells synthesize different proteins, so they could have different mRNA molecules.
12. The classmate is correct. Exons are the portions of a gene that are translated, while introns are not. Introns are cut out of the gene before translation.
13. Chromosomal mutations involve sections of a chromosome (thousands of nucleotides). Point mutations involve changes in one nucleotide. Chromosomal mutations are potentially more severe, but both can lead to nonfunctional proteins that are potentially life threatening.

Alternative Assessments

14. Students' reports will vary. Many antibiotics inhibit bacterial protein synthesis by combining with ribosomal proteins. Erythromycin and chloramphenicol combine with the 50S ribosomal subunit. The tetracyclines, streptomycin, gentamicin, kanamycin, and the nitrofurans combine with the 30S ribosomal subunit. Mupirocin and puromycin inhibit protein synthesis at the tRNA level.

Understanding Key Ideas

1. Anticodons are found on _____ molecules.
- **a.** mRNA
- **b.** DNA
- **c.** rRNA
- **d.** tRNA

SC.F.2.4.2

2. Unlike DNA, RNA contains
- **a.** the sugar deoxyribose.
- **b.** the nitrogen base uracil.
- **c.** a phosphate group.
- **d.** nucleotides.

SC.F.2.4.2

3. A short chain of DNA has the nucleotide sequence ATA CCG. Its complementary mRNA nucleotide sequence is
- **a.** TAT GCC.
- **b.** UAU GCC.
- **c.** TUT GCC.
- **d.** UAU GGC.

SC.F.2.4.2

4. The *lac* operon allows a bacterium to build the proteins needed for lactose metabolism when
- **a.** RNA polymerase is not bound to the promoter.
- **b.** lactose is absent.
- **c.** lactose is present.
- **d.** the repressor is bound to the operator.

SC.F.1.4.5

5. Transcription of lactose-metabolizing genes is blocked when the _____ is bound to the operator.
- **a.** repressor
- **b.** operon
- **c.** inducer
- **d.** enhancer

SC.F.1.4.5

6. In eukaryotes, gene expression can be regulated by SC.F.1.4.5
- **a.** mutations.
- **b.** transcription factors.
- **c.** repressors.
- **d.** operons.

7. **Exploring Further** Compare the way transposons and exons affect genes. SC.F.2.4.2

8. Does the drawing below represent a strand of RNA or a strand of DNA? Explain your answer. SC.F.2.4.2

9. **BIOWatch** How does gene therapy differ from other forms of medical treatment? SC.H.3.4.5

10. **Concept Mapping** Make a concept map that shows the role of RNA in gene expression. Try to include the following words in your map: *transcription, translation, mRNA, tRNA, rRNA, gene, promoter, codons, anticodons, proteins, amino acids, ribosome,* and *cytoplasm.* SC.F.2.4.2

Critical Thinking

11. **Evaluating Results** A molecular biologist isolates mRNA from the brain and from the liver of a mouse and finds that the mRNA molecules are different from each other. Can these results be correct or has the biologist made an error? Explain your answer. SC.H.1.4.4

12. **Evaluating an Argument** A classmate states that damage to exons is very likely to affect the synthesis of a protein, while damage to introns is not. Evaluate that statement. SC.H.1.4.4

13. **Evaluating Significance** Compare and contrast chromosomal mutations with point mutations, and evaluate the significance of each. SC.F.2.4.3

Alternative Assessment

14. **Interactive Tutor Unit 6 Gene Expression** Write a report summarizing how antibiotics inhibit protein synthesis in bacteria. How do some antibiotics interfere with translation? SC.H.3.4.5

Assignment Guide

Section	Questions
1	1, 2, 3, 8, 10, 11, 14
2	4, 5, 6, 7, 12

Chapter Review Correlations

SC.F.1.4.5: Items 4, 5, 6
SC.F.2.4.2: Items 1, 2, 3, 7, 8, 10
SC.F.2.4.3: Item 13
SC.H.1.4.4: Items 11, 12
SC.H.3.4.5: Items 9, 14

FCAT Prep

For questions 1–6, write your answers on a separate sheet of paper.

1. Which of the following shows the correct order of events in producing a protein from a DNA sequence?
   A. exon splicing, transcription, translation
   B. exon splicing, translation, transcription
   C. transcription, exon splicing, translation
   D. translation, transcription, exon splicing
   SC.F.2.4.2

2. What term describes mutations that change one nucleotide in a gene? SC.F.2.4.3
   F. codon mutation **H.** point mutation
   G. operon mutation **I.** repressor protein

3. What process involves making RNA based on the sequence of nucleotides in DNA? SC.F.2.4.2
   A. DNA replication **C.** transcription
   B. gene regulation **D.** translation

4. How does gene replication of the *lac* operon promote homeostasis in intestinal *E. coli* bacteria?
   F. If lactose is present, the genes that code for producing lactose are translated, conserving resources.
   G. If lactose is present, the enzymes that metabolize lactose are not produced, conserving resources.
   H. If lactose is not present, the genes that code for producing lactose are not translated, conserving resources.
   I. If lactose is not present, the enzymes that metabolize lactose are not produced, conserving resources. SC.F.1.4.5

Base your answers to questions 5 and 6 on the diagram below.

TRANSLATION MODEL

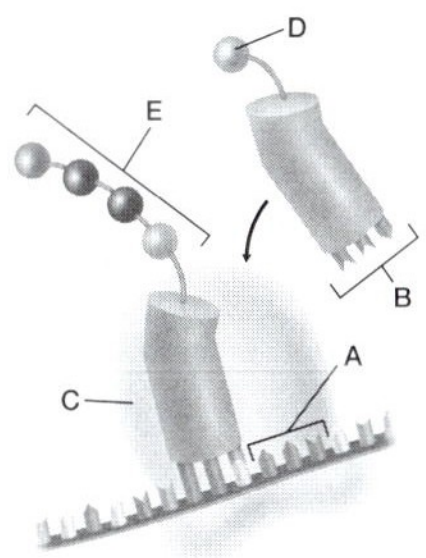

5. What cellular function does this model represent?
   A. DNA replication **C.** transcription
   B. gene regulation **D.** translation SC.F.2.4.2

6. How many codons are visible on the segment of mRNA shown in the diagram? SC.F.2.4.2

Test TIP

Slow, deep breathing may help you relax. If you suffer from test anxiety, focus on your breathing in order to calm down.

FCAT Doctor

Question 4 Answer I is the correct choice. When lactose enters the cell, the lac operon is activated and the necessary enzymes needed to metabolize lactose are produced. If lactose is not present, these enzymes are not produced. Answers F and H are incorrect because the lac operon does not involve producing lactose, it involves producing enzymes that metabolize lactose. Answer G is incorrect because this procedure would mean that the body would not produce the enzymes for metabolizing lactose when they are most needed.

Question 5 Answer D is the correct choice. The information in DNA is translated to make proteins. The diagram shows a piece of mRNA being translated at a ribosome, producing a protein chain. Answer A is incorrect because replication involves making DNA from DNA. Answer B is incorrect because gene regulation involves regulating which genes are translated from DNA according to when the cell needs them. Answer C is incorrect because transcription involves making tRNA from DNA.

Question 6 The correct answer is 5. A codon is a three-nucleotide sequence on mRNA. The mRNA segment depicted in the diagram shows 5 sets of 3 nucleotides.

Answers

1. C
2. H
3. C
4. I
5. D
6. 5

Test Prep Correlations

SC.F.1.4.5: Item 4
SC.F.2.4.2: Items 1, 3, 5, 6
SC.F.2.4.3: Item 2

Exploration Lab

Modeling Protein Synthesis

Teacher's Notes

Time Required

One 45-minute class period

Ratings

EASY → HARD

Teacher Preparation (2)
Student Setup (2)
Concept Level (3)
Cleanup (1)

Skills Acquired

- Communicating
- Constructing Models
- Identifying/Recognizing Patterns
- Interpreting

Scientific Methods

In this lab, students will:

- Make Observations
- Ask Questions
- Draw Conclusions
- Communicate the Results

Materials and Equipment

Materials for this lab can be ordered from WARD'S. See *Master Materials List* at the front of this book for catalog numbers.

Safety Cautions

Discuss all safety symbols and caution statements with students. Caution students to be careful of the sharp tips of the pushpins.

Tips and Tricks

Use materials from Chapter 9 Exploration Lab, if they were saved. Each group will need enough plastic soda straws to make about 40 sections of one color (3 cm each) and 15 sections of a different color. Each group will also need about 40 standard-sized paper clips, 40 colored pushpins, and enough materials to make at least eight tRNA/amino acid cards. Before beginning the lab, use **Figures 2 and 5** of the chapter to review the steps in gene expression.

Exploration Lab

Modeling Protein Synthesis SC.F.2.4.2

SKILLS

- Modeling
- Using scientific methods

OBJECTIVES

- **Compare** and **Contrast** the structure and function of DNA and RNA.
- **Model** protein synthesis.
- **Demonstrate** how a mutation can affect a protein.

MATERIALS

- masking tape
- plastic soda-straw pieces of one color
- plastic soda-straw pieces of a different color
- paper clips
- pushpins of five different colors
- marking pens of the same colors as the pushpins
- 3 × 5 in. note cards
- oval-shaped card
- transparent tape

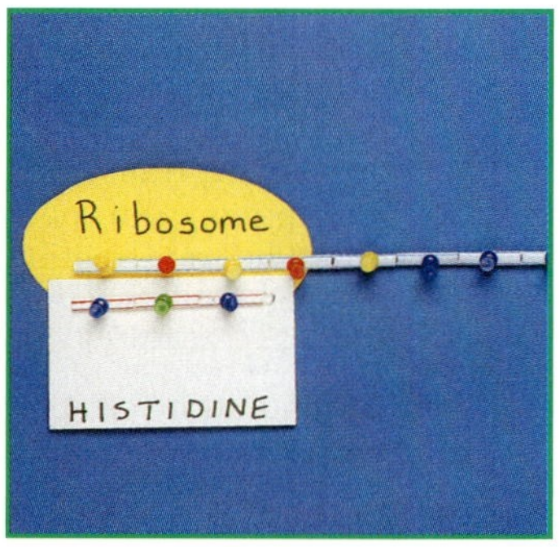

Before You Begin

The nature of a **protein** is determined by the sequence of amino acids in its structure. During **protein synthesis,** the sequence of nitrogen bases in an **mRNA** molecule is used to assemble **amino acids** into a protein chain.

A **mutation** is a change in the nitrogen-base sequence of DNA. Many mutations lead to altered or defective proteins. For example, the genetic blood disorder **sickle cell anemia** is caused by a mutation in the gene for **hemoglobin**.

In this lab, you will build models that will help you understand how protein synthesis occurs. You can also use the models to explore how a mutation affects a protein.

1. Write a definition for each boldface term in the paragraph above and for each of the following terms: transcription, translation, tRNA, ribosome, codon, anticodon.
2. Describe three differences between DNA and RNA.
3. Based on the objectives for this lab, write a question you would like to explore about protein synthesis.

Procedure

Part A: Design a Model

1. Work with the members of your lab group to design models of DNA, RNA, and a cell. Use the materials listed for this lab.

> **You Choose**
>
> **As you design your models, decide the following:**
>
> a. what question you will explore
> b. how to represent DNA nucleotides
> c. how to represent RNA nucleotides
> d. how to represent five different nitrogen bases
> e. how to link (bond) nucleotides together
> f. how to represent tRNA molecules with amino acids
> g. how to represent the locations of DNA and ribosomes

2. Write out the plan for building your models. Have your teacher approve the plan before you begin building the models.
3. Build the models your group designed. **CAUTION: Sharp or pointed objects may cause injury. Handle pushpins carefully.** Start your model of DNA with a strand of nucleotides that has the following sequence of nitrogen bases: TTTGGTCTCCTC.

Account for all materials at the end of the lab, and have students store the pushpins by color. If desired, ask students to save the assembled "nucleotides" for use in Chapter 11 Skills Practice Lab.

Answers to Before You Begin

1. *protein*—an organic molecule made of amino acids; *protein synthesis*—the process by which proteins are made based on the information encoded in a gene; *mRNA*—an RNA copy of a gene that carries the instructions for making a protein from a gene and delivers it to a ribosome for translation; *amino acids*—organic molecules that are the building blocks of proteins; *mutation*—a change in the DNA of a gene or chromosome; *sickle cell anemia*—a condition caused by a mutant allele that produces a defective form of hemoglobin; *hemoglobin*—a protein found in red blood cells that binds with and carries oxygen throughout

PART B: Model Protein Synthesis

4. Use your models and Figure 10-5 on pp. 212–213 to demonstrate how transcription and translation occur. Draw and label the steps of each process.
5. Use your models to explore one of the questions written for step 3 of **Before You Begin.**

PART C: Test Hypothesis

Answer each of the following questions by writing a hypothesis. Use your models to test each hypothesis, and describe your results.

6. The DNA model you built for step 3 represents a portion of a gene for hemoglobin. Sickle cell anemia results from the substitution of an A for the T in the third codon of the nitrogen-base sequence given in step 3. How will this substitution affect a hemoglobin molecule?
7. The addition of a nucleotide to a strand of DNA is a type of mutation called an *insertion.* What happens when an insertion occurs in the first codon in a DNA strand, before the DNA strand is transcribed?

PART D: Cleanup and Disposal

8. Dispose of damaged pushpins in the designated waste container.
9. Clean up your work area and all lab equipment. Return lab equipment to its proper place. Wash your hands thoroughly before you leave the lab and after you finish all work.

Analyze and Conclude

1. **Comparing Structures** How did the nitrogen-base sequence of the mRNA you made compare with that of the DNA it was transcribed from?
2. **Recognizing Relationships** How is the nitrogen-base sequence of a gene related to the structure of a protein?
3. **Recognizing Patterns** What is the relationship between the anticodon of a tRNA and the amino acid the tRNA carries?
4. **Drawing Conclusions** How does a mutation in the gene for a protein affect the protein?
5. **Further Inquiry** Write a new question about protein synthesis that could be explored with your model.

Do You Know?

Do research in the library or media center to answer these questions:

1. What are mutagens, and how do they affect DNA?
2. What are two other genetic disorders that result from mutations?

Use the following Internet resources to explore your own questions about DNA.

internet connect
www.scilinks.org
Topic: Genetic Disorders
Keyword: HX4091
SCILINKS. Maintained by the National Science Teachers Association

225

the body; *transcription*—a stage of gene expression in which the information in DNA is transferred to an RNA molecule; *translation*—a stage of gene expression in which the information in mRNA is used to make a protein; *tRNA*—an RNA molecule that temporarily carries a specific amino acid to a ribosome during translation; *ribosome*—a cytoplasmic organelle on which proteins are made; *codon*—a three-nucleotide sequence in mRNA that encodes an amino acid or signifies a start or stop signal; *anticodon*—a three-nucleotide sequence on tRNA that recognizes a complementary codon on mRNA.

2. RNA is single stranded, and DNA is double stranded. RNA nucleotides contain the sugar ribose and the nitrogen bases uracil, guanine, cytosine, or adenine. DNA nucleotides contain the sugar deoxyribose and the nitrogen bases thymine, guanine, cytosine, or adenine.
3. Answers will vary. For example: How does the cell's DNA determine the proteins made by a cell?

Answers To Procedure

2. Answers may vary. Students might use straw segments to model the sugar-phosphate backbone of DNA, different-colored pushpins to represent the five different nitrogen bases, paper clips to represent the bonds that hold the nucleotides in a chain, note cards to model tRNA molecules with amino acids attached, and oval-shaped cards to represent ribosomes.
4. Answers may vary. Check for students' understanding of each step in gene expression.
5. Students' results will vary. Some students might produce a mutation and demonstrate how it affects protein synthesis. Some mutations cause the wrong amino acid to be inserted in the protein. Other mutations might insert a start or stop codon at an inappropriate place.
6. Hypotheses will vary. The substitution will change the amino acid sequence from *lysine-proline-glutamic acid-glutamic acid* to *lysine-proline-valine-glutamic acid.*
7. Hypotheses will vary. The codon triplets are shifted.

Answers To Analyze and Conclude

1. They are complementary.
2. The nitrogen-base sequence of a gene contains a code that determines the amino-acid sequence of a protein.
3. A tRNA with a particular anticodon always carries the same amino acid.
4. A mutation can change the amino acid sequence of a protein, and thus may change its activity.
5. Example question: What happens if a mutation occurs that changes a codon to a stop codon?

Answers To Do You Know?

1. Mutagens are agents, such as certain chemicals and radiation that cause mutations.
2. Answers will vary but may include cystic fibrosis and Huntington's disease.

Chapter Opener

- TE **Using the Figure,** p. 226 GENERAL
- SE **Quick Review,** p. 227 ■
- SE **Reading Activity,** p. 227 ■
- TE **Opening Activity,** p. 227 GENERAL

& 3 • 90 min pp. 228–232

Section 1

Genetic Engineering

- OSP **Lesson Plan***
- TR **Bellringer***
- TR **Section Outline***
- TR **C34 Genetic Engineering***
- TR **C32 Restriction Enzymes Cut DNA***
- TR **C33 Gel Electrophoresis***

- TE **Activity** Recombinant DNA, p. 229 GENERAL
- TE **Using the Figure,** p. 229
- TE **Using the Figure,** p. 230
- TE **Using the Figure,** p. 231
- SE **Quick Lab** Modeling Gel Electrophoresis, p. 232 ■
- CRF **Datasheets for In-Text Labs** Modeling Gel Electrophoresis*
- SE **Exploration Lab** Modeling Recombinant DNA, pp. 246–247 ■ GENERAL
- CRF **Exploration Lab** Modeling Recombinant DNA, pp. 246–247* ◆ GENERAL
- CRF **Exploration Lab** DNA Whodunnit* GENERAL

& 5 • 90 min pp. 233–237

Section 2

Human Applications of Genetic Engineering

- OSP **Lesson Plan***
- TR **Bellringer***
- TR **Section Outline***
- TR **C35 Genetically Engineered Medicine***
- TR **C38 Making a Genetically Engineered Vaccine***

- TE **Demonstration,** p. 233 BASIC
- TE **Group Activity** Genetic Privacy, p. 234 BASIC
- TE **Group Activity** Genetic Diseases of Different Ethnicities, p. 235 GENERAL
- TE **Group Activity** Smallpox, p. 236 ADVANCED
- CRF **Skills Practice Lab** Transforming Bacteria with a Firefly Gene* GENERAL
- SE **Forensics Labs Appendix** DNA Fingerprinting, p. 1056
- CRF **Exploration Lab** DNA Fingerprinting GENERAL
- **Forensics and Applied Science Experiments** The Murder and the Blood Sample* GENERAL

& 7 • 90 min pp. 238–242

Section 3

Genetic Engineering in Agriculture

- OSP **Lesson Plan***
- TR **Bellringer***
- TR **Section Outline***

- TE **Group Activity** Genetically Engineered Crop Plants, p. 239 GENERAL
- TE **Group Activity** Ads for Gene Products, p. 240 ADVANCED

& 9 • 90 min

Chapter Review and Assessment Resources

- SE **Study Zone,** p. 243 ■
- SE **Performance Zone,** p. 244 ■
- SE **FCAT Test Prep,** p. 245 ■
- TR **Concept Mapping***
- CRF **Vocabulary Review Worksheet*** ■ GENERAL
- CRF **Test Prep Pretest*** ■ GENERAL
- CRF **Chapter Test*** ■ GENERAL
- CRF **Chapter Test*** ADVANCED
- OSP **Test Generator***
- OSP **Test Item Listing***
- OSP **Modified Chapter Test*** BASIC

go.hrw.com Holt Online Learning

Visit go.hrw.com for access to Holt Online Learning or enter the keyword **HX6 Home** for a variety of free online resources.

One-Stop Planner® CD-ROM

This CD-ROM package includes

- Lab Materials QuickList Software
- Holt Calendar Planner
- Customizable Lesson Plans
- Printable Worksheets
- ExamView® Test Generator
- Interactive Teacher Edition
- Holt PuzzlePro® Resources
- Holt PowerPoint® Resources

Guided Reading Audio CDs

These CDs are designed to help auditory learners and reluctant readers. (Audio CDs are also available in Spanish.)

Compression guide:
To shorten your instruction because of time limitations, omit blocks 1, 6, and 7.

KEY

TE	Teacher Edition	CRF	Chapter Resource File	*	Also on One-Stop Planner
SE	Student Edition	TR	Teaching Transparency	■	Also Available in Spanish
OSP	One-Stop Planner			◆	Requires Advance Prep

SKILLS DEVELOPMENT AND PRACTICE	SECTION REVIEW AND ASSESSMENT	CORRELATIONS
		Florida Sunshine Standards/FCAT
TE Skill Builder Vocabulary, p. 230 CRF Directed Reading Worksheet* BASIC CRF Active Reading Worksheet* GENERAL	SE Section 1 Review ■ TE Reteaching, p. 232 BASIC TE Quiz, p. 232 GENERAL CRF Quiz* ■ BASIC	SC.H.3.4.1, SC.H.3.4.2
CRF Directed Reading Worksheet* BASIC CRF Active Reading Worksheet* GENERAL	SE Section 2 Review ■ TE Reteaching, p. 237 BASIC TE Quiz, p. 237 GENERAL TE Alternative Assessment, p. 237 ADVANCED CRF Quiz* ■ BASIC	SC.H.3.4.2, SC.H.3.4.5
TE Reading Skill Builder Brainstorming, p. 240 GENERAL TE Reading Skill Builder Discussion, p. 241 ADVANCED CRF Directed Reading Worksheet* BASIC CRF Active Reading Worksheet* GENERAL CRF Critical Thinking Worksheet* ADVANCED CRF Science Skills Worksheet* GENERAL	SE Section 3 Review ■ TE Reteaching, p. 242 BASIC TE Quiz, p. 242 GENERAL TE Alternative Assessment, p. 242 GENERAL CRF Quiz* ■ BASIC Modified Worksheet* BASIC	SC.H.3.4.5

Holt Biology Interactive Tutor CD-ROM

This CD-ROM consists of interactive activities that give students a fun way to extend their knowledge of biology concepts.

Unit 6 Gene Expression

SCILINKS. NSTA

www.scilinks.org

Maintained by the **National Science Teachers Association**

Topic: Genetic Engineering
SciLinks Code: HX4092

Topic: Cloning
SciLinks Code: HX4047

CHAPTER 11

Overview

Tell students that the purpose of this chapter is to teach them about how a particular gene is isolated and cloned; how genetic engineering can be used to treat and cure certain diseases and identify organisms; how genetic engineering can be used to improve food crops and farm animals; and how to alter farm animals to make medically useful proteins.

Using the Figure

Tools and techniques of gene technology are being applied in a number of different ways, including gene therapy. Many genetic diseases are caused by a defect in a single gene. Scientists have already developed ways of engineering bacteria to produce human proteins that can be ingested or injected into people whose own bodies don't have the genes to make the proteins. Examples include insulin and growth hormone.

Assessing Prior Knowledge

Before beginning this chapter, be sure your students can answer the questions in the ***Quick Review*** on the facing page. If they cannot, encourage them to reread the sections indicated.

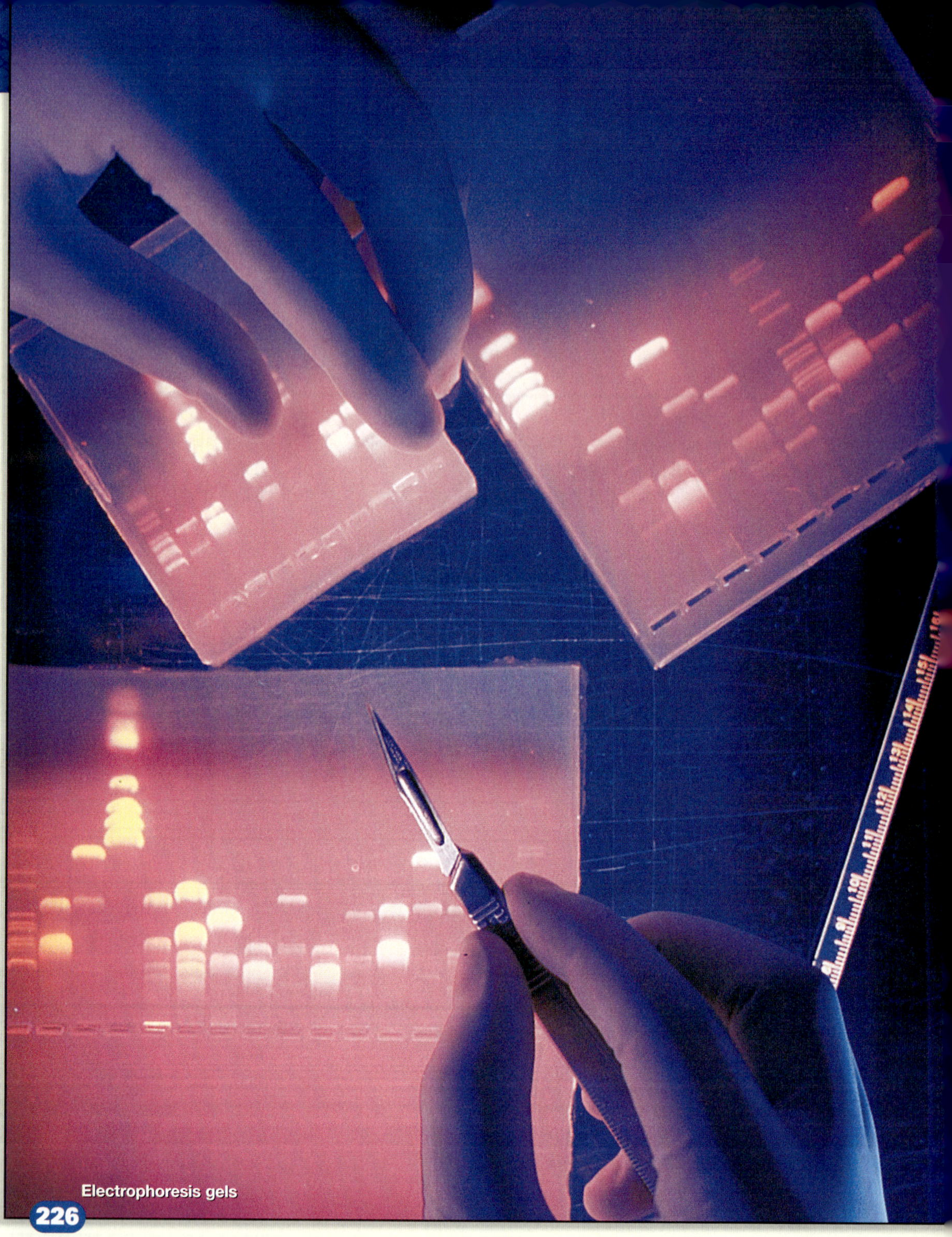

Electrophoresis gels

226

Chapter Correlations | **Florida Sunshine Standards/FCAT**

SC.H.3.4.2 Technological problems often create a demand for new scientific knowledge, and new technologies make it possible for scientists to extend their research in a way that advances science.

SC.H.3.4.5 The value of a technology may differ for different people and at different times.

CHAPTER

11 Gene Technology

Florida Sunshine Standards/FCAT

SC.H.3.4.2 Technological problems often create a demand for new scientific knowledge, and new technologies make it possible for scientists to extend their research in a way that advances science.

SC.H.3.4.5 The value of a technology may differ for different people and at different times.

Quick Review

Answer the following without referring to earlier sections of your book.

1. **Define** the term *gene*. *(Chapter 6, Section 1)*
2. **Describe** the structure of DNA. *(Chapter 9, Section 2)*
3. **State** the base-pairing rules that determine the structure of DNA. *(Chapter 9, Section 1)*
4. **Explain** why the genetic code is said to be universal. *(Chapter 10, Section 1)*

Did you have difficulty? *For help, review the sections indicated.*

Reading Activity

Before you read this chapter, write a short list of all the things you know about gene technology. Then, write a list of the things that you want to know about gene technology. Save your list, and to assess what you have learned, see how many of your own questions you can answer after reading this chapter.

Looking Ahead

Electrophoresis is a technique used in a laboratory that results in the separation of charged particles. DNA is a negatively charged molecule, and is moved by electric current through an electrophoresis gel.

Opening Activity — GENERAL

Gene Technology Have students discuss why gene technology is controversial. (Lead students to suggest that DNA technology allows researchers to produce new variants of life forms with specific characteristics or abilities.) Students should understand that gene technology is a potentially powerful tool for fighting diseases and for understanding organisms. Explain that people are excited about the prospects of using this tool, but concerned about its misuse. LS **Verbal**

Quick Review Answers

1. A gene is a segment of DNA that codes for a protein or RNA molecule.
2. DNA is constructed of nucleotide subunits made up of a phosphate group, a sugar molecule, and one of four nitrogen-containing bases. The nucleotides are linked end-to-end in two strands that are twisted into a double helix.
3. The base-pairing rules state that a purine base in one of the two strands of a DNA molecule always has across from it on the other strand a pyrimidine base. These rules further specify that the purine adenine always pairs with the pyrimidine thymine and the purine guanine always pairs with the pyrimidine cytosine.
4. The genetic code is universal because it is nearly the same in every living thing.

Reading Activity Answers

Instruct students to save their lists for use after they have finished reading the chapter.

Chapter Resource File

- Vocabulary Worksheets
- Concept Mapping

Transparencies

TR Bellringer
TR C34 Genetic Engineering
TR C32 Restriction Enzymes Cut DNA

Focus

Overview

Before beginning this section review with your students the objectives listed in the Student Edition. This section introduces students to the techniques and tools used to combine DNA from two or more different organisms. Students learn how it is determined that different DNAs have been combined and the different uses for the recombinant DNA.

Bellringer

Ask students to imagine the appearance of an organism that has some genetically determined characteristics from two different species. Ask them to draw a picture of their imaginary organism.

Motivate

Discussion — BASIC

Show students a bowl with various types of fruits and vegetables. Ask students to identify their favorite fruits and vegetables. Ask students what characteristics they would change in their favorites, if they could. (Answers may include making the fruit or vegetable sweeter, firmer, easier to bite into, or a different color.) Discuss how scientists are able to manipulate many characteristics of fruits and vegetables using genetic engineering techniques. Once the gene that controls a trait is found and isolated, scientists can try to manipulate it in various ways. **LS Visual**

Section Correlations

SC.H.3.4.2

Section 1 Genetic Engineering

Florida Sunshine Standards/FCAT

SC.H.3.4.2 Technological problems often create a demand for new scientific knowledge, and new technologies make it possible for scientists to extend their research in a way that advances science.

Objectives

- **Describe** four basic steps commonly used in genetic engineering experiments. SC.H.3.4.2
- **Evaluate** how restriction enzymes and the antibiotic tetracycline are used in genetic engineering. SC.H.3.4.2
- **Relate** the role of electrophoresis and probes in identifying a specific gene. SC.H.3.4.2

Key Terms

genetic engineering
recombinant DNA
restriction enzyme
vector
plasmid
gene cloning
electrophoresis
probe

Basic Steps of Genetic Engineering

Not too long ago, using bacteria to produce human insulin and inserting genes into tomatoes and human cells were ideas that existed only in science fiction books and movies. But now, the techniques required to carry out these ideas have been developed and are used daily.

In 1973, Stanley Cohen and Herbert Boyer conducted an experiment that revolutionized genetic studies in biology. They isolated the gene that codes for ribosomal RNA from the DNA of an African clawed frog and then inserted it into the DNA of *Escherichia coli* bacteria, as summarized in **Figure 1.** During transcription, the bacteria produced frog rRNA, thereby becoming the first genetically altered organisms. The process of manipulating genes for practical purposes is called **genetic engineering.** Genetic engineering may involve building **recombinant DNA**—DNA made from two or more different organisms.

The basic steps in genetic engineering can be explored by examining how the human gene for insulin is transferred into bacteria. Insulin is a protein hormone that controls sugar metabolism. Diabetics cannot produce enough insulin, so they must take doses of insulin regularly. Before genetic engineering, insulin was extracted from the pancreases of slaughtered cows and pigs and then purified. Today, the human insulin gene is transferred to bacteria through genetic engineering. Because the genetic code is universal, bacteria can transcribe and translate a human insulin gene using the same code a human cell uses in order to produce human insulin.

Figure 1 Genetic alteration of an organism

Cohen and Boyer produced the first genetically engineered organisms.

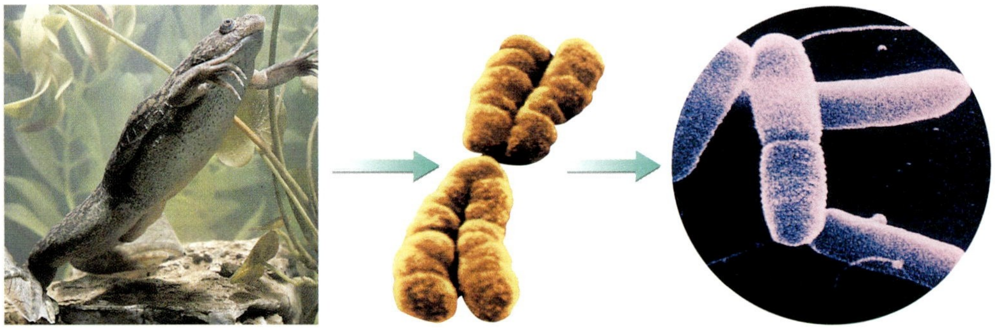

1. Cohen and Boyer used an African clawed frog as their experimental organism.
2. They isolated an rRNA gene from one of its chromosomes.
3. They inserted the gene into bacteria. The bacteria produced frog rRNA.

Chapter Resource File

- Directed Reading BASIC
- Active Reading GENERAL
- Data Sheet for Quick Lab GENERAL

Transparencies

TR Bellringer
TR C33 Gel Electrophoresis

One-Stop Planner CD-ROM

- Reading Organizers BASIC
- Reading Strategies BASIC

Steps in a Genetic Engineering Experiment

Genetic engineering experiments use different approaches, but most share four basic steps, as illustrated in **Figure 2.**

Step 1 Cutting DNA The DNA from the organism containing the gene of interest (in our example, the insulin gene) is cut by restriction enzymes. **Restriction enzymes** are bacterial enzymes that recognize and bind to specific short sequences of DNA, and then cut the DNA between specific nucleotides within the sequences. The DNA from a vector also is cut. A **vector** is an agent that is used to carry the gene of interest into another cell. Commonly used vectors include viruses, yeast, and plasmids. **Plasmids,** shown in Figure 2, are circular DNA molecules that can replicate independently of the main chromosomes of bacteria.

Step 2 Making recombinant DNA The DNA fragments from the organism containing the gene of interest are combined with the DNA fragments from the vector. An enzyme called DNA ligase is added to help bond the ends of DNA fragments together. In our example, human DNA fragments are combined with plasmid DNA fragments. The host cells then take up the recombinant DNA.

Step 3 Cloning In a process called **gene cloning,** many copies of the gene of interest are made each time the host cell reproduces. Recall from your reading that bacteria reproduce by binary fission, producing identical offspring. When a bacterial cell replicates its DNA, its plasmid DNA also replicates.

Step 4 Screening Cells that have received the particular gene of interest are distinguished, or separated, from the cells that did not take up the vector with the gene of interest. The cells can transcribe and translate the gene of interest to make the protein coded for in the gene.

internet connect

www.scilinks.org
Topic: Science and Technology of Florida
Keyword: HX6F14

SciLinks. Maintained by the National Science Teachers Association

BIO graphic Figure 2

Genetic Engineering

Many genetic engineering experiments use one or more of these basic steps.

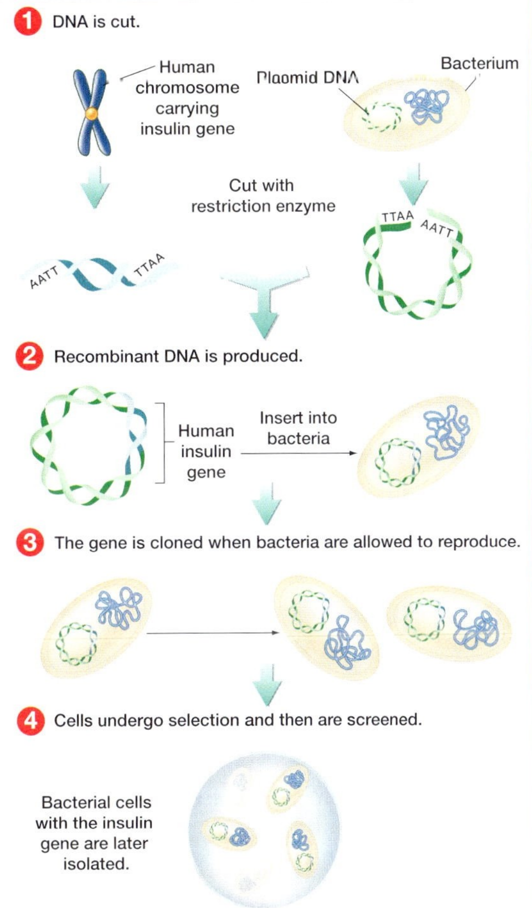

Teach

Using the Figure

Direct students' attention to **Figure 2.** Use this figure to review the steps used in many genetic engineering experiments. In step 4, emphasize that the bacterial cells with the insulin gene will later be isolated. These cells will transcribe and translate the insulin gene just as they would their own gene. Thus, scientists can isolate the insulin from the cells, purify it, and then use it for medical purposes.
LS Visual

READING SKILL BUILDER

Assign Chapter 11 of the *Holt Biology* Guided Audio CD Program to help students achieve greater success in reading the chapter.

Activity — GENERAL

Recombinant DNA Give pairs of students a piece of yarn (to represent a human chromosome), a long pipe cleaner a different color from the yarn (to represent plasmid DNA), and some tape. Ask the students to form a circle with the pipe cleaner, and twist the ends once to secure the circle. Have students use scissors to cut a "gene" from the yarn and to untangle the twisted end of the pipe cleaner. Ask the students what the scissors represent. (restriction enzymes) Tell the students to insert the human gene (piece of yarn) into the opening of the plasmid and secure the ends with tape. Ask the students the following: What does the tape represent? (DNA ligase as it helps bond the DNA pieces together) What does the plasmid represent? (a vector) Why is this recombinant DNA? (The plasmid now contains foreign DNA.) What is the next step in genetic engineering? (The plasmid is inserted into a host cell, typically a bacterium.) Why? (The host cell will reproduce; each time it does, all the genetic material will be copied, including the human gene.)
LS Visual

did you know?

Naming Restriction Enzymes Enzymes are named after the specific bacteria from which they are isolated. For example, the restriction enzyme *EcoRI* is named after the bacterium *Escherichia coli*. The first letter, "E," is the initial letter of the genus name of the organism from which the enzyme was isolated (*Escherichia*). The second and third letters, "co," are usually the first two letters of the species name (*coli*). These three letters are always italicized since they are part of the scientific name. The fourth letter, R, if present, represents the strain of the organism (strain RY 13). The Roman numerals (I) indicate the order of discovery (i.e. first endonuclease isolated in this strain of bacteria). **LS Verbal**

Teach, *continued*

Teaching Tip

Natural Selection and Screening
Discuss with students how screening cells involves events similar to those of natural selection. Only certain organisms will survive in a given environment. In screening, only bacteria with the gene for antibiotic resistance will survive in a culture medium containing an antibiotic. In natural selection, the bacteria with the successful trait survive and produce future generations. **LS Verbal**

Using the Figure

Have students examine **Figure 3.** Reinforce the importance of sticky ends and how sticky ends are complementary. Ask students why the genes for replication and tetracycline resistance must be present in the vector. (replication gene—so that the plasmid can replicate; tetracycline resistance gene—to screen the bacterial cells that take up the recombined vector) There are thousands of different restriction enzymes, and each one recognizes a different DNA sequence. **LS Verbal**

SKILL BUILDER

Vocabulary Tell students that the word clone comes from the Greek *klon* meaning "twig." Pieces of a plant can be cut off and rooted, producing a new and identical plant—a clone. **LS Verbal**

Figure 3 Restriction enzymes cut DNA

The restriction enzyme *Eco*RI recognizes the nucleotide sequence GAATTC and makes its cut between the G and the A.

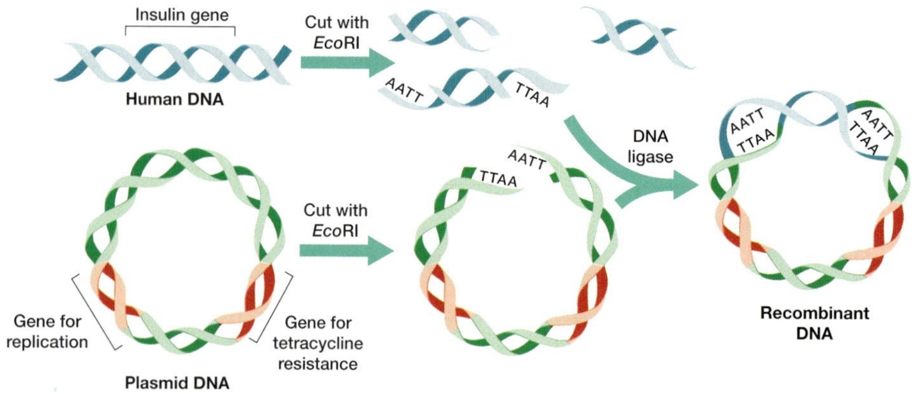

Cutting DNA and Making Recombinant DNA

An example of how restriction enzymes work is shown in **Figure 3.** The enzyme recognizes a specific sequence of DNA. The sequence the enzyme recognizes and the sequence on the complementary DNA strand are palindromes—they read the same backward as they do forward (such as the word *noon*).

The cuts of most restriction enzymes produce pieces of DNA with short single strands on each end that are complementary to each other. The ends are called *sticky ends*. As illustrated in Figure 3, the vectors that are used contain only one nucleotide sequence that the restriction enzyme recognizes. Thus, vectors such as the circular plasmids "open up" with the same sticky ends as those of the cut human DNA. The two DNA molecules bond together by means of complementary base pairing at the sticky ends. The plasmid DNA has both the gene for plasmid DNA replication and the gene that makes the cell carrying the plasmid resistant to the antibiotic tetracycline.

Figure 4 Screening. Only the cells that take up the vectors are resistant to tetracycline and survive when tetracycline is added.

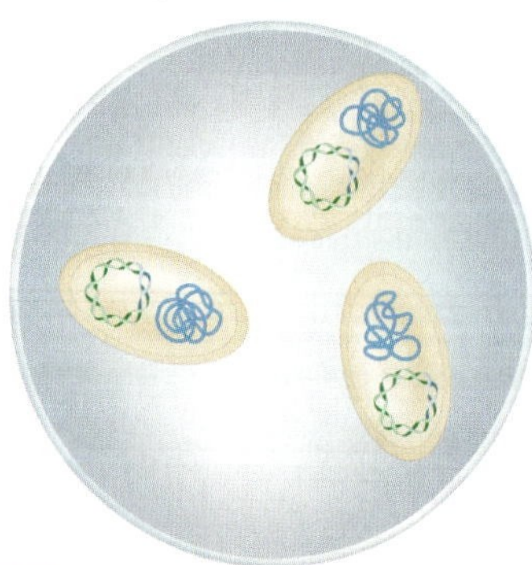

Cloning, Selecting, and Screening Cells

One difficult part in a genetic engineering experiment is finding and isolating the cells that contain the gene of interest. First, the cells that have taken up the plasmid must be identified. The bacterial cells that have taken up the plasmid are identified by growing the bacteria on plates that contain the antibiotic tetracycline. As shown in **Figure 4,** only the cells that have taken up the vectors (which contain the gene for tetracycline resistance) survive when exposed to tetracycline. Each surviving cell makes a copy of the vector every time the cell reproduces. Eventually, each surviving cell forms a colony of genetically identical cells, or clones. Some vectors contain the gene of interest, and some do not.

230

Attention Grabber

Molecular Scissors Restriction enzymes are often described as being like molecular scissors because they cut a specific nucleotide sequence at the first nucleotide. Many teaching aids developed for classroom use actually depict a small pair of scissors snipping the DNA. Of course, teachers will state that these aren't really microscopic scissors. Scientists have recently discovered that one restriction enzyme's structure actually does resemble a tiny pair of scissors! BGLII is a restriction endonuclease that recognizes and cleaves the DNA sequence AGACTC. Using X-ray crystallography, the scientists found that the enzyme's two subunits swing away from each other in a dramatic scissorslike motion. The sliding of the subunits past each other results in cutting of the DNA.

Confirmation of a Cloned Gene

The surviving bacterial colonies are tested for the presence of the gene of interest. One method used to identify a specific gene is a technique called a Southern blot, as summarized in **Figure 5.**

Step 1 In a Southern blot, the DNA from each bacterial clone colony is isolated and cut into fragments by restriction enzymes.

Step 2 The DNA fragments are separated by gel **electrophoresis** *(ee LEK troh fuh REE sis)*, a technique that uses an electric field within a gel to separate molecules by their size. The gel is a rectangular slab of gelatin with a line of little rectangular wells near the top edge. The DNA sample is placed in the pits. Because DNA is negatively charged, it migrates toward the positive pole when the electric field is applied. The DNA fragments move through the gel, with the smallest DNA fragments moving fastest. A pattern of bands is formed. The gel is soaked in a chemical solution that separates the double strands in each DNA fragment into single-stranded DNA fragments.

Step 3 The DNA bands are then transferred (blotted) directly onto a piece of filter paper. The filter paper is moistened with a probe solution. **Probes** are radioactive- or fluorescent-labeled RNA or single-stranded DNA pieces that are complementary to the gene of interest.

Step 4 Only the DNA fragments complementary to the probe will bind with the probe and form visible bands.

WORD Origins

The word *electrophoresis* is from the Latin *electrocus,* meaning "electricity," and the Greek *phoresis,* meaning "to carry." Knowing this makes it easier to remember that electrophoresis uses electricity to separate DNA fragments.

BioGraphic

Figure 5

Southern Blot: Identifying a Gene of Interest

A DNA or RNA probe can be used to identify a cloned gene.

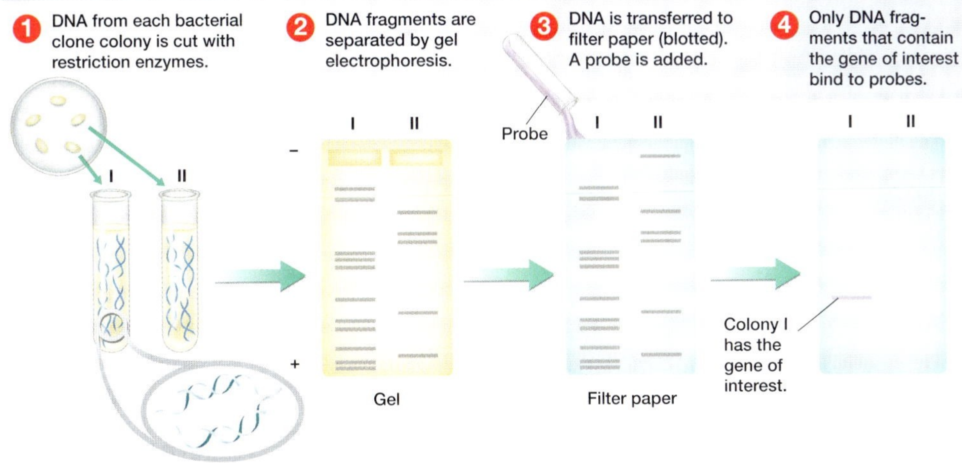

Teaching Tip

Discovery of Restriction Enzymes Tell students that the first restriction endonuclease (restriction enzyme) was purified in 1972 by Hamilton Smith, a molecular biologist who was working with *Haemophilus influenzae* and a bacteriophage that naturally infects *Salmonella.*

Using the Figure

Direct students' attention to **Figure 5.** Remind them that the bacterial colonies in Step 1 were produced from the basic genetic engineering steps described in **Figure 2.** Point out that the fragments on the gel in Step 2 are all of different sizes, with the smallest fragments closest to the positive pole. The transfer or blotting of the DNA fragments that occurs in Step 3 is the reason for the second part of the name of this technique—Southern blot. The first part of the name, Southern, is named after E. M. Southern, who developed the technique. In Step 4, the colony that contains the gene of interest is actually identified. Because the original Petri dish (from Step 1) is stored while the researcher conducts the Southern blot, the researcher can then return to the original Petri dish and conduct further research on the colony containing the gene. **LS Visual**

Integrating Physics and Chemistry

Help students make the connection between electrophoresis and electrolytic behavior of solutions. Inform them that when an electric current is applied to the gel, which is an electrolytic solution, electrolytes or dissolved ions are attracted to the electrodes that are introducing the electric field into the gel. In a Southern blot, the negatively charged DNA molecules migrate toward the positive electrode and as a result are separated from other molecules.

internet connect

www.scilinks.org
Topic: Genetic Engineering
Keyword: HX4092

SCILINKS. Maintained by the National Science Teachers Association

Once the bacterial colonies containing the gene of interest are identified, the researcher can manipulate the genetically engineered bacteria in many different ways. For example, the gene of interest can be isolated so that the researcher has pure DNA to use in genetic studies. The researcher can then study how the gene is controlled. Pure DNA allows the researcher to determine the sequence of nucleotides that make up the gene. By comparing the nucleotide sequence of several different organisms, researchers can study the evolution of a particular gene.

The gene of interest can also be isolated and then transferred to other organisms. The bacterial colonies can be used to produce large quantities of the protein coded for by the gene so that the protein can be studied further or used to make drugs, such as insulin.

QUICK LAB

Modeling Gel Electrophoresis

SC.H.3.4.2

You can use beads to model how DNA fragments are separated in a gel during electrophoresis.

Materials

500 mL beaker, large jar, 3 sets of beads—each set a different size and different color

Procedure

1. Fill a large jar with the largest beads. The filled jar represents a gel.
2. Mix the two smaller beads in the beaker and then pour them slowly on top of the "gel." The two smaller size beads represent DNA fragments of different sizes.
3. Observe the flow of the beads through the "gel." Lightly agitate the jar if the beads do not flow easily.

Analysis

1. **Identify** which beads flowed through the "gel" the fastest.
2. **Relate** the sizes of the beads to the sizes of DNA fragments.
3. **Determine** whether the top or the bottom of the jar represents the side of the gel with the positively charged pole.
4. **Critical Thinking Forming Conclusions** Why do the beads you identified in Analysis question 1 pass through the "gel" more quickly?

Section 1 Review

1. **Apply** the four steps commonly used in genetic engineering experiments to describe the cloning of a human gene. SC.H.3.4.2
2. **Relate** the role of DNA "sticky ends" in the making of recombinant DNA. SC.H.3.4.2
3. **Summarize** how cells are screened in genetic engineering experiments. SC.H.3.4.2
4. **Evaluate** the role of probes in identifying a specific gene. SC.H.3.4.2
5. **Critical Thinking Evaluating Conclusions** A student performing electrophoresis on a DNA sample believes that her smallest DNA fragment is the band nearest the negative pole of the gel. Do you agree with her conclusion? Explain. SC.H.3.4.2
6. FCAT Prep What circular DNA molecules are used in many genetic engineering experiments that are performed in bacteria?
   A phages
   B plasmids
   C probes
   D promoters
   SC.H.3.4.2

QUICK LAB

Modeling Gel Electrophoresis

Skills Acquired

Modeling, relating information, forming conclusions

Teacher's Notes

The large beads should be big enough so that when placed in the jar there are spaces between them. The smallest beads should flow through the spaces of the large beads.

Answers to Analysis

1. the smallest beads
2. The smaller beads represent the smaller DNA fragments.
3. bottom; DNA is negatively charged and will flow to the pole with the opposite charge.
4. The smaller size allows them to flow through the space in between the large beads.

Close

Reteaching — BASIC

Ask students to rewrite each objective as a question. Then ask the students to answer each question. LS Verbal

Quiz — GENERAL

1. What is recombinant DNA? (DNA made from two or more different organisms)
2. What is the role of gel electrophoresis in genetic engineering? (It separates DNA fragments according to their size.)

Section Review Correlations

SC.H.3.4.2: Items 1-6

Answers to Section Review

1. Human and vector DNA are cut and then combined to produce recombinant DNA. The recombinant DNA is inserted into host cells. The host cells reproduce, cloning the human gene. The host cells are screened for the gene.
2. The bases of the sticky ends are complementary, which allows the DNA from the two different organisms to combine.
3. The plasmids contain the gene for tetracycline resistance. Bacteria with the recombined plasmid are grown in the presence of tetracycline. Only bacteria with the plasmid survive.
4. Probes are pieces of single-stranded DNA or RNA complementary to the gene of interest. The DNA fragments on the gel are made single stranded. The probe will bind to any complementary fragments. The genes are identified because the probes are made with radioactive or fluorescent-labeled tags.
5. No, disagree. DNA is negatively charged and opposite charges attract. The smallest fragments will move fastest through the gel.
6. **A.** Incorrect. Phages are viruses. **B.** Correct. **C.** Incorrect. Probes are radioactive or fluorescent labeled RNA or DNA pieces. **D.** Incorrect. Promoters are individual genes.

Human Applications of Genetic Engineering

Section 2

Florida Sunshine Standards/FCAT SC.H.3.4.2 Technological problems often create a demand for new scientific knowledge, and new technologies make it possible for scientists to extend their research in a way that advances science.

- **Summarize** two major goals of the Human Genome Project. SC.H.3.4.2
- **Describe** how drugs produced by genetic engineering are being used. SC.H.3.4.5
- **Summarize** the steps involved in making a genetically engineered vaccine. SC.H.3.4.2
- **Identify** two different uses for DNA fingerprints. SC.H.3.4.5

Key Terms

Human Genome Project
vaccine
DNA fingerprint

The Human Genome Project

In February of 2001, scientists working on the Human Genome Project published a working draft of the human genome sequence. The sequence of an organism's genome is the identification of all base pairs that compose the DNA of the organism. The **Human Genome Project** is a research project that has linked over 20 scientific laboratories in six countries. Teams of scientists, such as those shown in **Figure 6,** cooperated to identify all 3.2 *billion* base pairs of the DNA that makes up the human genome. Scientists were surprised by some of the discoveries they made.

The Geography of the Genome

One of the most surprising things about the human genome is the large amount of DNA that does *not* encode proteins. In fact, only 1 to 1.5 percent of the human genome is DNA that codes for proteins. Each human cell contains about six feet of DNA, but less than 1 inch of that is devoted to exons. Recall that exons are sequences of nucleotides that are transcribed and then translated. Exons are scattered about the human genome in clumps that are not spread evenly among chromosomes. For example, chromosome number 19 is small and is packed with transcribed genes. The much larger chromosomes 4 and 8, by contrast, have few transcribed genes. On most human chromosomes, great stretches of untranscribed DNA fill the chromosomes between scattered clusters of transcribed genes.

The Number of Human Genes

When they examine the complete sequence of the human genome, scientists were surprised at how few genes there actually are. Human cells contain only about 20,000 to 25,000 genes. This is only about double the number of genes in a fruit fly. And it is only about one-fifth of the 120,000 genes that scientists had expected to find. How had scientists made this prediction of the number of human genes, and why was it wrong? When scientists had counted unique human messenger RNA (mRNA) molecules, they had found over 120,000. Each of these different forms of mRNA molecules can, in turn, be translated into a unique protein. So the scientists expected to find as many genes as there are types of mRNA molecules.

Figure 6 Genetic Research. Hundreds of scientists around the world worked to identify the human genome sequence.

Section 2

Focus

Overview

Before beginning this section review with your students the objectives listed in the Student Edition. This section discusses the efforts and discoveries of the Human Genome Project, drugs and vaccines that have been produced using the techniques of genetic engineering, and the techniques and applications of DNA fingerprinting.

Bellringer

Ask students to make a list of characteristics they possess that are determined by genes. (Answers will vary but may include eye color, skin color, enzymes that catalyze different reactions in the body, specific proteins such as hormones and antibodies, height, and intelligence.)

Motivate

Demonstration — Basic

Use an inkpad and white paper to take a fingerprint of each student. Display the prints. Ask the class what is special about each print. (They are all different.) Explain that just as everyone (except identical twins) has different fingerprints, we all have different DNA prints. Inform students that the DNA prints are called DNA fingerprints because they create a pattern similar to the way that the ridges on the fingers leave a pattern in a fingerprint. Display autoradiographs of DNA fingerprints from science journals or magazines.

LS Intrapersonal

English Language Learners

Chapter Resource File

- Directed Reading BASIC
- Active Reading GENERAL

Transparencies

TR Bellringer
TR C35 Genetically Engineered Medicine
TR C38 Making a Genetically Engineered Vaccine

One-Stop Planner CD-ROM

- Reading Organizers BASIC
- Reading Strategies BASIC

Section Correlations

SC.H.3.4.2
SC.H.3.4.5

Teach

Teaching Tip

Junk DNA Tell students that the findings of the Human Genome Project indicate that at least 50 percent of the human genome is made up of repeated sequences of DNA that do not code for proteins. This DNA is often referred to as "junk DNA" because scientists have not been able to determine that it has any function. Recent research suggests that these repeats are involved in reshaping the genome by rearranging it, creating entirely new genes, and modifying and reshuffling existing genes.

Teaching Tip — GENERAL

Graphic Organizer Have students make a Graphic Organizer to outline how genetically engineered drugs are made. LS Visual

Group Activity — BASIC

Genetic Privacy By 2010, scientists predict, $100 will buy a test that effectively identifies genetic markers for a myriad of conditions and diseases. Have students work in groups of five to discuss genetic information and privacy by reacting to questions such as: Should employers have access to genetic information? Should health and life insurance companies? Should school officials? Should anyone except you (or your parents, as a child) know your genetic predispositions and your genetic potential or lack thereof? Moreover, should they know your genetic information even if you choose not to know it? LS Interpersonal Co-op Learning

Genetically Engineered Drugs and Vaccines

Much of the excitement about genetic engineering has focused on its potential uses in our society. The possibilities for the applications of these techniques in medicine and research are endless. Many applications are already commonplace, such as the production of genetically engineered proteins used to treat illnesses and the creation of new vaccines used to combat infections.

Drugs

Many genetic disorders and other human illnesses occur when the body fails to make critical proteins. Juvenile diabetes is such an illness. The body is unable to control levels of sugar in the blood because a critical protein, insulin, cannot be made. These failures can be overcome if the body can be supplied with the protein it lacks. The proteins that regulate the body's functions are typically present in the body in very low amounts. Today hundreds of pharmaceutical companies around the world produce medically important proteins in bacteria using genetic engineering techniques as summarized in **Figure 7.**

Factor VIII, a protein that promotes blood clotting, is an example of a GM medicine (**g**enetically **m**odified; a drug manufactured by genetic engineering). A deficiency in factor VIII leads to one type of hemophilia, an inherited disorder characterized by prolonged bleeding. For a long time, hemophiliacs received blood factors that had been isolated from donated blood. Unfortunately, some of the donated blood was infected with viruses such as HIV and hepatitis B. The viruses were sometimes unknowingly transmitted to people who received blood transfusions. Today, the use of genetically engineered factor VIII eliminates these risks.

Figure 7 Use of genetically engineered medicines. Many medicines, such as medicines used to treat burns, are produced by genetic engineering techniques.

Genetically Engineered Medicines

Product:	Used for treatment of:
• Erythropoetin	Anemia
• Growth factors	Burns, ulcers
• Human growth hormone	Growth defects
• Insulin	Diabetes
• Interferons	Viral infections and cancer
• Taxol	Ovarian cancer

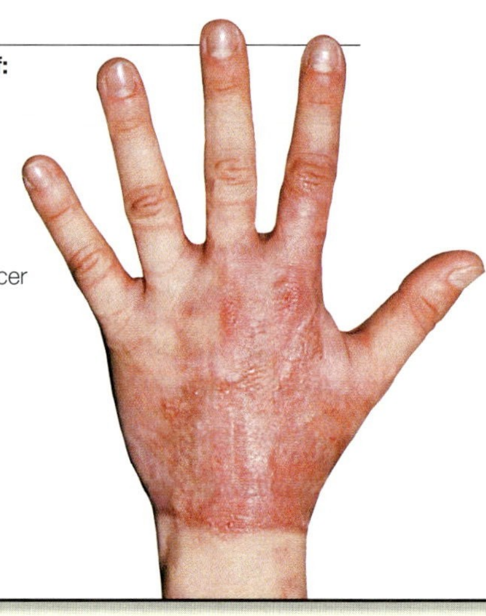

234

Graphic Organizer

Use this graphic organizer with **Teaching Tip: Graphic Organizer** on this page.

Desired gene identified → Gene transferred to bacterial plasmid → Plasmid containing desired gene introduced into bacteria → Bacteria grown in large numbers to produce needed drug

Vaccines

Many viral diseases, such as smallpox and polio, cannot be treated effectively by existing drugs. Instead they are combated by prevention—using vaccines. A **vaccine** is a solution containing all or part of a harmless version of a pathogen (disease-causing microorganism). When a vaccine is injected, the immune system recognizes the pathogen's surface proteins and responds by making defensive proteins called antibodies. In the future, if the same pathogen enters the body, the antibodies are there to combat the pathogen and stop its growth before it can cause disease.

Traditionally, vaccines have been prepared either by killing a specific pathogenic microbe or by making the microbe unable to grow. This ensures that the vaccine itself will not cause the disease. The problem with this approach is that there is a small but real danger that a failure in the process to kill or weaken a pathogen will result in the transmission of the disease to the very patients seeking protection. This danger is one of the reasons why, for example, rabies vaccines are administered only when a person has actually been bitten by an animal suspected of carrying rabies.

Vaccines made by genetic engineering techniques avoid this danger. As illustrated in **Figure 8,** the genes that encode the pathogen's surface proteins can be inserted into the DNA of harmless viruses such as cowpox (Vaccinia). The modified but harmless cowpox virus becomes an effective and safe vaccine, as illustrated in Figure 8. The surfaces of the modified virus display herpes surface proteins in addition to the virus's own surface proteins. When the modified virus is injected into a human body, the body's immune system quickly responds to this challenge. The immune system makes antibodies that attack any virus displaying the herpes surface protein. As a result, the body is thereafter protected against infection by the herpes virus.

Real Life

You might get a vaccine in a banana.

Genetic engineers are putting genes from disease-causing microbes into fruits and vegetables to create vaccines that are inexpensive and easy to take. Clinical trials using different foods, including potatoes, are underway.

Finding Information
What are the most common ways vaccines are now administered?

Figure 8 **Making a genetically engineered vaccine**

A person vaccinated with a genetically engineered vaccine, such as the genital herpes vaccine, will make antibodies against the virus.

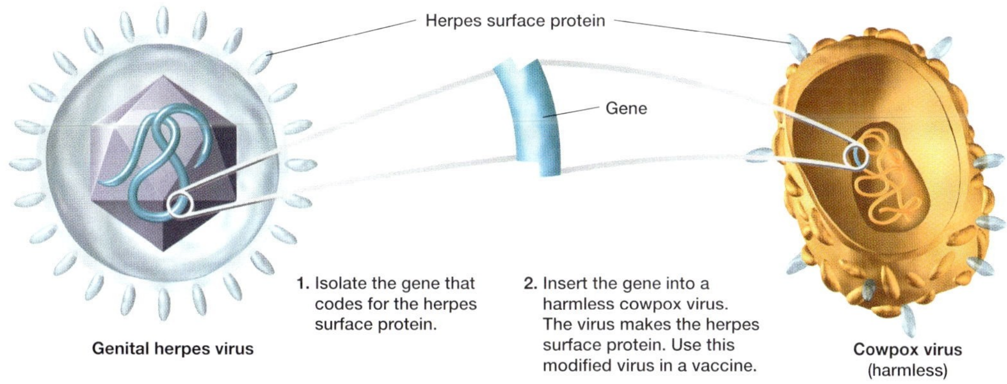

Real Life GENERAL

Answer

Nearly all vaccines are currently administered by injection. Most vaccines are injected into muscle tissue or under the skin.

Group Activity — GENERAL

Genetic Diseases of Different Ethnicities Have students work in small groups to conduct research on diseases that are more common in people of specific ethnic backgrounds. Some examples they could research are sickle cell anemia, thalassemia, Tay-Sachs disease, and tyrosinemia. Have each group prepare an oral report on the disease, including its prevalence in the specific population and the general population, its symptoms and severity, current treatments available, and current genetic engineering research being conducted as part of efforts to prevent, control, and/or treat the disease.
LS Verbal Co-op Learning

Group Activity — ADVANCED

Smallpox Have students work in small groups to conduct library and Internet research on smallpox. Assign one group to research the history of smallpox and its control, including its status as the first vaccine-preventable disease and its eradication in the 1970s. Assign another group to research the only approved smallpox vaccine, which is made from a virus called Vaccinia which is a live "pox"-type virus related to smallpox. Assign another group to research the development of new smallpox vaccines, the testing and approval processes for new vaccines, and the current status of smallpox vaccines. Assign a final group to research the potential use of smallpox as a bioterrorism agent and the public health system's current plans for controlling a possible bioterrorist attack using the smallpox virus.
LS Verbal Co-op Learning

TECHNOLOGY CONNECTION

Gene therapy is a novel approach to treat, cure, or ultimately prevent disease by changing the expression of a person's genes. Gene therapy is in its infancy and current gene therapy is primarily experimental, with most human clinical trials only in the research stages. Gene therapy can be targeted to somatic (body) or germ (egg and sperm) cells. In somatic gene therapy the recipient's genome is changed, but the change is not passed along to the next generation. Thus the gene therapy may have to be repeated in future generations. There are several diseases for which gene therapy is currently being investigated using somatic gene therapy, with mixed results to date. In germline gene therapy, the parents' egg and sperm cells are changed with the goal of passing on the changes to their offspring. Germline gene therapy prevents the trait from being passed on to further generations. Germline gene therapy is not being actively investigated, at least in larger animals and humans, although a lot of discussion is being conducted about its value and desirability.

BIOWatch

Polymerase Chain Reaction (PCR)

Teaching Strategies

- Let students know that PCR is now a very common technique used to obtain sufficient DNA samples not just in criminal cases but for basic research as well.
- When the source of DNA is meager or impure, PCR is a quicker and more selective method than gene cloning. Because PCR is performed completely *in vitro* (in test tubes), cells are not used as with gene cloning. Billions of copes of DNA can be made in a few hours with PCT, whereas gene cloning would take days.

Discussion

- If most proteins (such as enzymes) denature or break down with high heat, how can the DNA polymerase used in PCR withstand the heating cycles? (The DNA polymerase is isolated from bacteria living in hot springs. The DNA polymerase from these bacteria can withstand the heat needed to separate the DNA strands.)
- Under conditions of PCR amplification, DNA replicates about every 5 minutes. Determine how many copies of a DNA fragment will result from 85 minutes of PCR. ([85 min] / [5 min/cycle] = 17 cycles; 2^{17} or 1.31×10^5 copies.)

Forensics Labs Appendix

Turn to pp. 1048–1071 for forensics lab activities and more student information on forensics techniques.

Vaccines for the herpes II virus and for the hepatitis B virus are now being made through genetic engineering. The herpes II virus produces small blisters on the genitals (the external sex organs). The hepatitis B virus causes an inflammation of the liver that can be fatal. A major effort is underway to produce a vaccine that will protect people against malaria, a protozoan-caused disease for which there is currently no effective protection.

FORENSICS BIOWatch: Polymerase Chain Reaction (PCR)

A detective finds a single hair as the only evidence left behind at a crime scene. Will this hair provide enough DNA to analyze? For DNA fingerprinting and many of the genetic engineering uses discussed throughout this chapter, a certain amount of DNA is needed. Sometimes, however, only a very tiny amount of DNA is available.

Today scientists use a technique called the polymerase chain reaction (PCR) to quickly make many copies of selected segments of the available DNA. With PCR, a scientist can produce a billionfold increase in DNA material within a few hours!

Heating and Replication Cycles

In PCR the double-stranded DNA sample to be copied is heated, which separates the strands. The mixture is cooled, and short pieces of artificially made DNA called primers are added. The primers bind to places on the DNA where the copying can begin.

DNA polymerase and free nucleotides are added to the mixture. The DNA polymerase extends the DNA by attaching complementary free nucleotides to the primer. The result is two strands of DNA that are identical to each other and to the original strand. The heating and replication process is repeated over and over again. Every 5 minutes, the sample of DNA doubles again, resulting in many copies of the sample in a short amount of time. Today, scientists use PCR machines, which automatically cycle the reaction temperature.

PCR's Many Uses

PCR can duplicate DNA from as few as 50 white blood cells, which might be found in a nearly invisible speck of blood. PCR is important for diagnosing genetic disorders and for solving crimes. PCR is also used in different types of research and for studying ancient fragments of DNA found in fossils or in preserved material.

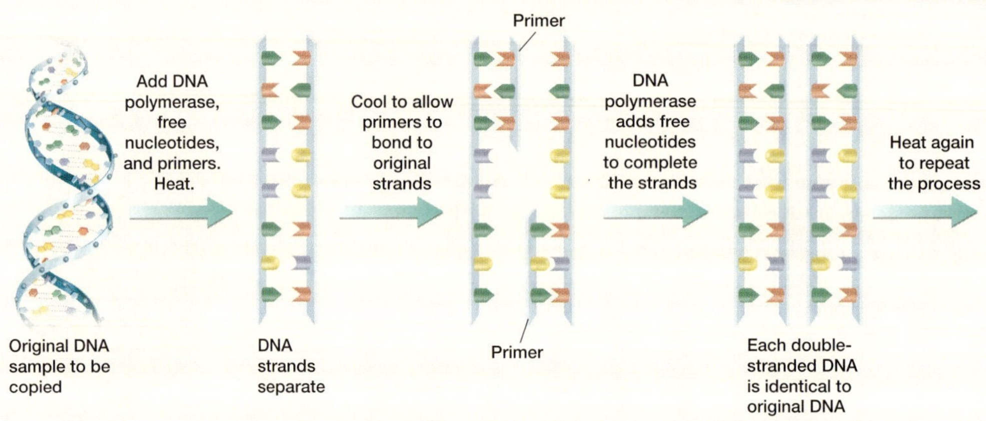

REAL WORLD CONNECTION

In cases of bioterrorist attack such as the anthrax mail incidents, officials need to understand the situation quickly. Early detection and identification of the biological organism and its source are crucial for minimizing the potentially catastrophic human and economic costs. Clues may lie hidden in the weapon itself. Does the bacterium or virus harbor information in its DNA that could lead to its source? Is it resistant or sensitive to vaccines and antibiotics? To find answers to these questions, revolutionary new approaches are being developed. Intensive research on *Bacillus anthracis* and *Yersinia pestis* (causes of anthrax and plague, respectively) has led to a wealth of genetic information and unique technologies for detecting and identifying these organisms. The studies have focused on tracking selected, highly variable, or very specific DNA signatures in the microbes. These analyses may offer clues to the agents' genetic identity, geographic origin, and genetic modification to enhance its resistance to antibiotics and vaccines.

DNA Fingerprinting

Other than identical twins, no two individuals have the same genetic material. Scientists use DNA sequencing gel technology to determine a DNA fragment's nucleotide sequence, as shown in **Figure 9.** Because the places a restriction enzyme can cut depend on the DNA sequence, the lengths of DNA restriction fragments will differ between two individuals. Such DNA fragments of different lengths (polymorphisms) are called restriction fragment length polymorphisms, or RFLPs.

RFLPs can be used to identify individuals and to determine how closely related members of a population are to one another. The Southern blot technique, shown in Figure 5, is used to show an individual's RFLP profile. The result is called a *DNA fingerprint*. A **DNA fingerprint** is a pattern of dark bands on photographic film that is made when an individual's DNA restriction fragments are separated by gel electrophoresis, probed, and then exposed to an X-ray film. Because restriction enzymes cut the DNA from different individuals into DNA fragments of different lengths (RFLPs), each individual (other than identical twins) has a unique pattern of banding, or DNA fingerprint.

The banding patterns from two individuals can be compared to establish whether they are related, such as in a paternity case. Because it can be performed on a sample of DNA found in blood, semen, bone, or hair, DNA fingerprinting is useful in forensics. Forensics is the scientific investigation of the causes of injury and death when criminal activity is suspected. DNA fingerprints are also valuable for identifying the genes that cause genetic disorders, such as Huntington's disease and sickle cell anemia.

Figure 9 DNA sequence. The nucleotide sequence of DNA fragments can be determined using DNA sequencing gel technology.

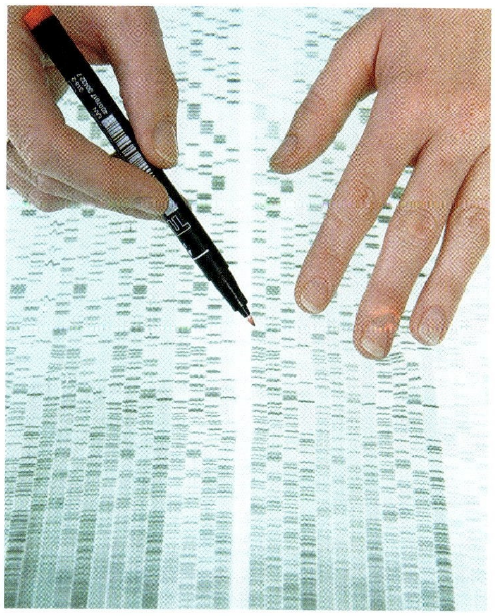

Section 2 Review

1. **Relate** the use of genetic engineering to the treatment of human illnesses such as hemophilia. SC.H.3.4.5
2. **Relate** genetic engineering techniques to the making of vaccines. SC.H.3.4.5
3. **List** two ways in which DNA fingerprinting has been useful to society. SC.H.3.4.5
4. **Critical Thinking Distinguishing Relevant Information** A student states that genetic engineering is "perfectly safe and sound." What safety and ethical issues do you think might arise over the use of genetic engineering? SC.H.3.4.5
5. **FCAT Prep** One medicine made in bacteria using genetic engineering techniques is insulin. What is insulin used to treat?
   - **A** cystic fibrosis
   - **B** diabetes
   - **C** heart attacks
   - **D** smallpox

   SC.H.3.4.6

Close

Reteaching — Basic

Have students write one sentence for each of the Key Terms. Each sentence should demonstrate the meaning of the term as it is defined in the text. LS Verbal

Quiz — General

1. What is the Human Genome Project? (a multinational scientific research project that has elucidated the DNA sequence of the entire human genome)
2. Provide an example of a genetically engineered medicine. (erythropoetin, growth factors, human growth hormone, insulin, interferons, and taxol)
3. What is the polymerase chain reaction, or PCR? (a laboratory technique used to quickly make multiple copies of selected segments of a DNA sample)

Alternative Assessment — Advanced

Instruct students to prepare themselves for a visit to a university or pharmaceutical company that uses genetic engineering techniques. Have them prepare a list of questions about the research conducted at the facility, the staff and financial resources available, the techniques and equipment used, and any applications of their research that are currently in use.

Answers to Section Review

1. Many genetically engineered proteins are used to treat illnesses. For example, factor VIII, a protein that promotes blood clotting, is now made by genetic engineering and sold as a drug to hemophiliacs. Genetic engineers are also attempting to replace defective human genes with healthy ones.
2. Instead of using a killed or weakened pathogen, the genes that code for the proteins found on the surface of the pathogen are inserted into the DNA of harmless bacteria or viruses. People are then vaccinated with the modified virus or bacteria.
3. DNA fingerprinting has been useful in forensics, in paternity suits, and in identifying the genes that cause genetic disorders.
4. Answers may include concerns about who has access to personal genetic information and how they might use that information and whether gene technologies are safe to individuals and to the human population as a whole.
5. **A.** Incorrect. Insulin is not used to treat cystic fibrosis. **B.** Correct. **C.** Incorrect. Insulin is not used to treat heart attacks. **D.** Incorrect. Insulin is not used to treat smallpox.

Section Review Correlations

SC.H.3.4.5: Items 1-4

SC.H.3.4.6: Item 5

Section 3 Genetic Engineering in Agriculture

Focus

Overview

Before beginning this section review with your students the objectives listed in the Student Edition. This section discusses the genetic engineering of crop plants and animals to improve yield and quality and reviews concerns about the safety of this kind of engineering.

Bellringer

Have students write the name of a fruit or vegetable that they don't like to eat and explain why they don't like it. Then ask them to write about ways in which the fruit or vegetable could possibly be changed by genetic engineering so that they would like it.

Motivate

Discussion — GENERAL

The Flavr-Saver™ tomato was the first genetically altered fruit to reach supermarket shelves. It was altered to last longer and taste better than other types of tomatoes. Genetically engineered crops also present an alternative to the use of pesticides. A large percentage of the world's food supply is lost to pests. Pesticide use, though effective, has its disadvantages. Ask students what some of the disadvantages might be. (Over time the pests become resistant to the pesticide; other plant and animal life may be harmed) Ask students to discuss concerns about using genetically engineered crops. (The genetically engineered plant may pass the genes on to close relatives in the wild.) LS Verbal

Section Correlations

SC.H.3.4.5

Florida Sunshine Standards/FCAT

SC.H.3.4.5 The value of a technology may differ for different people and at different times.

Objectives

- **Describe** three ways in which genetic engineering has been used to improve plants. SC.H.3.4.5
- **Summarize** two ways in which genetic engineering techniques have been used to modify farm animals. SC.H.3.4.5
- **Summarize** the cloning of sheep through the use of differentiated cells. SC.H.3.4.5

Key Terms

transgenic animal

Improving Crops

Farmers began primitive genetic breeding by selecting seeds from their best plants, replanting them, and gradually improving the quality of successive generations. In the twentieth century, plant breeders started using the principles of genetics to select plants. Today, genetic engineers can add favorable characteristics to a plant by manipulating the plant's genes, as shown in **Figure 10.**

Genetic engineers can change plants in many ways, including making crop plants more tolerant to drought conditions and creating plants that can adapt to different soils, climates, and environmental stresses.

Genetic engineers have developed crop plants that are resistant to a biodegradable weedkiller called *glyphosate*. This has enabled farmers to apply glyphosate to kill weeds without killing their crops. Because the field does not need to be tilled to control weeds, less topsoil is lost to erosion. Half of the 72 million acres of soybeans planted in the United States in 2000 were genetically modified to be glyphosate resistant.

Scientists have also developed crops that are resistant to insects by inserting a certain gene isolated from soil bacteria into crop plants. This gene makes a protein that injures the gut of chewing insects. Crops that are resistant to insects do not need to be sprayed with pesticides, many of which can harm the environment.

More Nutritious Crops

Genetic engineers have been able, in many instances, to improve the nutritional value of crop plants. For example, in Asia many people use rice as a major source of food, yet rice has low levels of iron and beta carotene, which your body uses to make vitamin A (necessary for vision). As a result, millions suffer from iron deficiency and poor vision. Genetic engineers have added genes to rice from other plants, as shown in **Figure 11,** to overcome this deficiency.

Figure 10 Genetically engineered plants. At least 50 plants have been genetically engineered, including potatoes, soybeans, and corn. The researcher Athanasios Theologis genetically engineered tomatoes to ripen without becoming soft.

238

Chapter Resource File

- Directed Reading BASIC
- Active Reading GENERAL

Transparencies

TR Bellringer

One-Stop Planner CD-ROM

- Reading Organizers BASIC
- Reading Strategies BASIC
- Supplemental Reading Guide *The Double Helix* ADVANCED

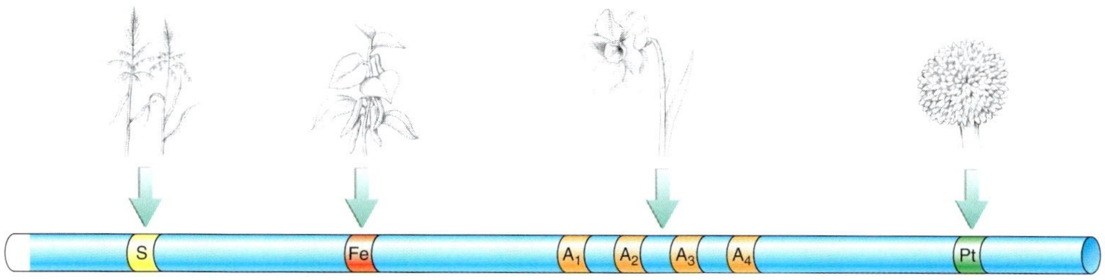

Figure 11 **Rice enriched with iron and vitamin A.** Genetically modified "golden" rice offers the promise of improving the diets of people in rice-consuming countries, where iron and vitamin A deficiencies are a serious problem.

Risks of Genetically Modified Crops

Many people, including influential scientists, have expressed concern that genetically modified crops (GM crops) might turn out to be dangerous. What kind of unforeseen negative effects might "improved" GM crops have?

Potential Problems

Some food crops, such as corn and soybeans, have been genetically rendered resistant to glyphosate, a weed killer that is harmless to humans. Glyphosate, when used on a food crop, will kill the weeds but will not harm the GM crop, thus increasing food crop yields. Some scientists are concerned that the use of GM crops and the subsequent use of glyphosate will eventually lead to glyphosate-resistant weeds. This will leave farmers with few weed-control alternatives.

Some GM crops have genes added to improve nutritional character, as was done in rice. It is important to check that consumers are not allergic to the product of the introduced gene. For this reason, screening of GM crops for causes of allergy problems is now routine.

Are GM Crops Harmful to the Environment?

Will introduced genes pass from GM crops to their wild or weedy relatives? This sort of gene flow happens naturally all the time, so this concern is legitimate. For most crops, no closely related wild plant is around to receive the gene. The GM gene cannot pass to a nonrelative, because crop plants cannot successfully reproduce with unrelated species, any more than a cat can breed with a giraffe. There are wild relatives of corn in Mexico and Guatamala, which frequently exchange genes with corn crops. Scientists are divided about whether it makes any difference if one of the genes is a GM gene.

Might pests become resistant to GM toxins? Pests are becoming resistant to GM toxins just as they have become resistant to the chemical pesticides that are sprayed on crops.

Scientists, the public, and regulatory agencies must work together to evaluate the risks and benefits of GM products.

REAL WORLD CONNECTION

The Green Revolution of the 1960s and 1970s, with its package of improved seeds, farm technology, better irrigation, and chemical fertilizers was highly successful at meeting its primary objective of increasing crop yields and augmenting food supplies. In Asia, where the package was the most widely adopted, food production increased substantially. Yet, despite its success at increasing aggregate food supply, the Green Revolution has not translated into benefits for most of the lower strata of the rural poor. Poor nutrition and poverty are still prevalent, and the distribution of food remains skewed. This has led to the recognition by development agencies of the need to formulate a more equitable and sustainable Green Revolution aimed at improving food security for the poorest people in rural areas. Much of the success of this new approach will depend on its ability to respond to the realities of women farmers, who are the critical people involved in producing, providing, and managing food supplies in the poorest rural households.

Teach

Teaching Tip

Plants of The Green Revolution In the 1950s, Norman Borlaug of the International Maize and Wheat Research Centre, Mexico, crossed short-stem wheat plants with Mexico's best wheat plants. In 1963, Borlaug responded to an urgent request from the government of India to provide them with seeds of his dwarfed, high-yielding plants. The tall native wheat plants of India were growing too high when heavily fertilized with nitrogen and would fall over from the weight of their seed heads. The short wheat plants Borlaug provided, along with fertilizer and irrigation, increased wheat production from 12 million metric tons in 1965 to over 20 million in 1970 and over 37 million in 1982. Since the new wheat plants were broadly adapted, Green Revolutions also took hold in countries sharing similar latitudes, such as Pakistan, Turkey, and Afghanistan. For his contributions, Borlaug was awarded the Nobel Prize.

Group Activity — GENERAL

Genetically Engineered Crop Plants Have students work in small groups to conduct library and Internet research on genetically engineered crop or ornamental plants. Assign different groups to research one of the following kinds of plants: One that has been engineered to enhance its nutritional value, another engineered to make it resistant to an herbicide, another engineered to enhance its resistance to an insect or other pest, and another engineered to enhance its aesthetic value. Have each group prepare a written report of their findings, including a description and picture or drawing of the plant, the methods used to develop it, the cost and time involved in engineering it, the public's acceptance of the plant, and any problems associated with producing or marketing the plant. Co-op Learning

Teach, *continued*

READING SKILL BUILDER — GENERAL

Brainstorming Lead students in a discussion comparing food crops and farm animals produced by selective breeding and by genetic engineering. (A new variety is developed in a shorter period of time using DNA technology instead of selective breeding. You can also insert medically useful human proteins into animals so that they produce the protein in milk.) Emphasize that the development of organisms to enhance certain features has been occurring for many years through selective breeding. Ask students to envision ways in which genetic engineering might not be beneficial. (may lead to a loss of biodiversity or create unforeseen problems) LS Verbal

Group Activity — ADVANCED

Ads for Gene Products Have groups of students choose and then research a genetically engineered product such as human insulin. Ask students to create a full-page advertisement. The ad must be creative, name the product and its use, list side effects, and include a well-thought-out effort to convince the public that the genetically engineered product is safe.
LS Visual Co-op Learning

internet connect
www.scilinks.org
Topic: Cloning
Keyword: HX4047
SciLinks. Maintained by the National Science Teachers Association

Gene Technology in Animal Farming

Farmers have long tried to improve farm animals and crops through traditional breeding and selection programs. In the past, the cow that produced the most milk on a farm may have been mated to male offspring of high producers in hopes that the cow's offspring would also produce a lot of milk. But these traditional processes were slow and inefficient.

Now, many farmers use genetic-engineering techniques to improve or modify farm animals. Some farmers add growth hormone to the diet of cows to increase milk production. Previously, the growth hormone was extracted from the brains of dead cows. But now the cow growth hormone gene is introduced into bacteria. The bacteria produce the hormone so cheaply that it is practical to add it as a supplement to the cows' diet.

By altering the gene responsible for GH production, scientists have stimulated natural GH in pigs, increasing their weight. Though these procedures are still new, they may lead to the creation of new breeds of very large and fast-growing cattle and hogs.

Making Medically Useful Proteins

Another way in which gene technology is used in animal farming is in the addition of human genes to the genes of farm animals in order to get the farm animals to produce human proteins in their milk. This is used especially for complex human proteins that cannot be made by bacteria through gene technology. The human proteins are

Figure 12 Cloning a sheep from mammary cells

In 1997 scientists announced the first successful cloning using differentiated cells—a lamb named Dolly.

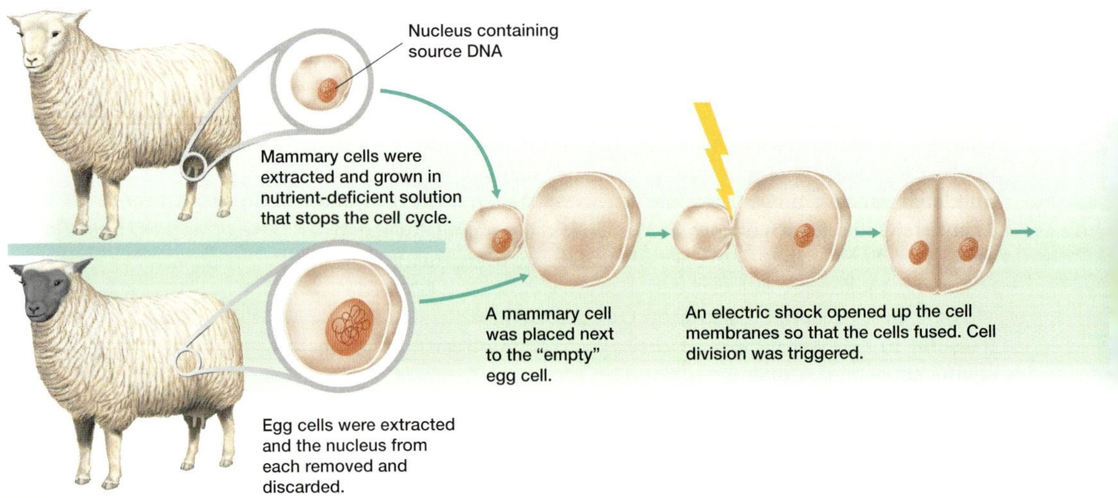

240

TECHNOLOGY CONNECTION

Having already cloned cattle, pigs and goats, researchers at Texas A&M University have moved on to house pets, with the first successful cloning of a common house cat. The fact that the cloned kitten was one of only 87 cloned cat embryos to survive underscores obstacles still remaining in this area of research. In a process similar to that used with Dolly the cloned sheep, the researchers transplanted DNA derived from the nuclei of cumulus cells (near the ova) of a calico cat into the empty (i.e. nucleus removed) egg cell of another cat, then transplanted the embryo into a third cat.

Genetic tests confirm that the kitten is indeed a genetic copy of the original calico cell donor. Interestingly, the kitten does not have the same coloring as the genetic parent, a fact the researchers attribute to the play of dueling X chromosomes and developmental factors outside the control of the nucleic DNA.

extracted from the animals' milk and sold for pharmaceutical purposes. The animals are called **transgenic animals** because they have foreign DNA in their cells.

Most recently, scientists have turned to cloning animals as a way of creating herds of identical animals that can make medically useful proteins. The intact nucleus of an embryonic or fetal cell (whose DNA has been recombined with a human gene) is placed into an egg whose nucleus has been removed. The egg with the new nucleus is then placed into the uterus of a surrogate, or substitute, mother and is allowed to develop.

Cloning From Adult Animals

In 1997, a scientist named Ian Wilmut captured worldwide attention when he announced the first successful cloning using differentiated cells from an adult animal. A differentiated cell is a cell that has become specialized to become a specific type of cell (such as a liver or udder cell). As summarized in **Figure 12,** a lamb was cloned from the nucleus of a mammary cell taken from an adult sheep. Previously, scientists thought that cloning was possible only using embryonic or fetal cells that have not yet differentiated. Scientists thought that differentiated cells could not give rise to an entire organism. Wilmut's experiment proved otherwise.

An electric shock was used to fuse mammary cells from one sheep with egg cells without nuclei from a different sheep. The fused cells divided to form embryos, which were implanted into surrogate mothers. Only one embryo survived the cloning process. Dolly, born on July 5, 1996, was genetically identical to the sheep that provided the mammary cell.

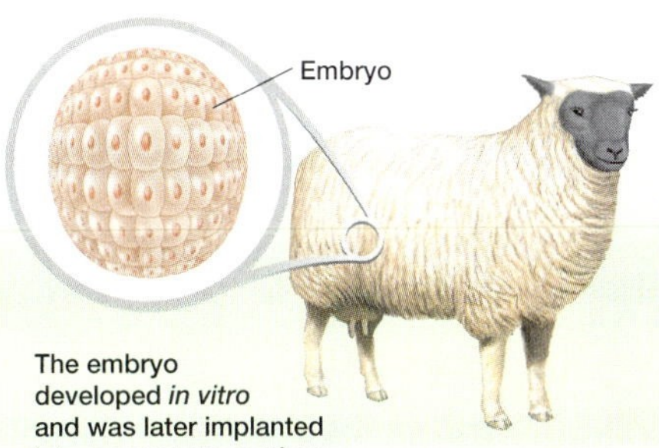

Teaching Tip

Technology Terminology Tell students that the terms *genetic modification* (GM) and biotechnology are often used interchangeably. GM is a special set of technologies that alter the DNA of such living organisms as animals, plants, or bacteria. Biotechnology, a more general term, refers to using natural living organisms or their components. Combining DNA from different organisms is known as recombinant DNA technology, and the resulting organism is said to be "genetically modified," "genetically engineered," or "transgenic."

READING SKILL BUILDER — ADVANCED

Discussion Dolly, the lamb cloned from undifferentiated cells, developed some unforeseen problems as she grew older. Her chromosomes began showing signs of premature aging. In 2002 she was diagnosed with arthritis, a condition of aging that normally develops in sheep older than her 5 years. Dolly was euthanized in 2003. Ask students why the chromosomes could appear abnormally old when Dolly was relatively young. (Dolly was cloned from mammary cells from an adult sheep.) How does this information affect other cloning experiments? (It appears that when organisms are cloned from adult cells, their chromosomes may age prematurely. These findings suggest that Dolly's cellular aging clock was not reset at her own "birth" but that her clock is actually set at her age plus 6 years, the age of the sheep from which she was cloned. This in turn may eventually affect the metabolism of the organism.) **LS Verbal**

did you know?

Genetically Modified Foods Genetically modified (GM) crops are grown in over 40 countries and on 6 continents. In 2000, about 109.2 million acres were planted with transgenic crops, the principal ones being herbicide- and insecticide-resistant soybeans, corn, cotton, and canola. Other crops grown commercially or field tested are a sweet potato resistant to a virus that could decimate most of the African harvest, rice with increased iron and vitamins that may alleviate chronic malnutrition in Asian countries, and a variety of plants able to survive weather extremes. On the horizon are bananas that produce human vaccines against infectious diseases such as hepatitis B, fish that mature more quickly, fruit and nut trees that yield years earlier, and plants that produce new unique plastics.

In 2000, countries that grew 99% of the global transgenic crops were the United States (68%), Argentina (23%), Canada (7%), and China (1%). Although growth is expected to plateau in industrialized countries, it is increasing in developing countries.

Close

Reteaching — BASIC

Ask students to create a table summarizing the ways in which genetic engineering has been used to improve food crops and farm animals and to make medically useful proteins in the milk of farm animals. **LS Verbal**

Quiz — GENERAL

1. Why is glyphosate a valuable weedkiller? (it is biodegradable and crop plants can be genetically engineered to resist its toxic effects)
2. What is an animal called that has foreign DNA in its cells? (a transgenic animal)
3. What is genomic imprinting and what are the implications of this process for cloning? (Genomic imprinting is the process of conditioning DNA to the early stage of development of an embryo. Because cells divide rapidly during the cloning process, there is not enough time for reconditioning of the DNA to occur.)

Alternative Assessment — GENERAL

Tell students to return to the list of things they want to know about gene technology from the previous Active Reading exercise. Have them place check marks next to the questions they can now answer. Students should finish by making a list of what they have learned. Conduct a discussion of the remaining questions that have gone unanswered. **LS Verbal**

Section Review Correlations

SC.H.3.4.2: Items 2, 3
SC.H.3.4.5: Items 1, 5
SC.H.3.4.6: Item 4

Problems With Cloning

Since Dolly's birth in 1996, scientists have successfully cloned animals. Only a few of the cloned offspring survived for long, however. Many become fatally oversized. Others encounter problems in development. For example, three cloned calves were born healthy in March, 2001, only to die a month later of immune system failure.

The Importance of Genomic Imprinting

Technical problems in reproductive cloning lie within a developmental process that conditions eggs and sperm so that the right combination of genes are turned "on" or "off" during early development. When cloned offspring become adults, a different combination of genes is activated. The process of conditioning the DNA during an early stage of development is called genomic imprinting.

In genomic imprinting, chemical changes made to DNA prevent a gene's expression without altering its sequence. Usually, a gene is locked into the "off" position by adding methyl (–CH_3) groups to its cytosine nucleotides, as shown in **Figure 13.** The bulky methyl groups prevent polymerase enzymes from reading the gene, so the gene cannot be transcribed. Later in development, the methyl groups are removed and the gene is reactivated.

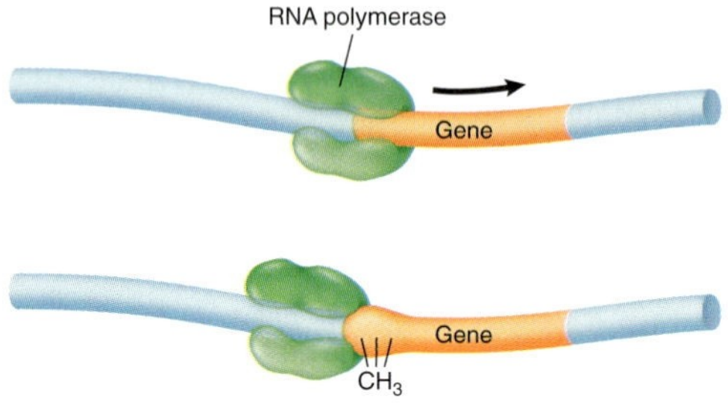

Figure 13 Methylated gene. In one model of genomic imprinting, methyl groups attached to a gene prevent the gene from being expressed.

Why Cloning Fails

Normal vertebrate development depends on precise genomic imprinting. This process, which takes place in adult reproductive tissue, takes months for sperm and years for eggs. Reproductive cloning fails because the reconstituted egg begins to divide within minutes. There is simply not enough time in these few minutes for the reprogramming to process properly. Key genes fail to become properly methylated, and this leads to critical errors in development.

Because of these technical problems, and because of ethical problems, efforts to clone humans are illegal in most countries.

Section 3 Review

1. **List** three ways in which food crops have been improved through genetic engineering. SC.H.3.4.5
2. **Compare** the cloning of sheep through the use of differentiated cells with the cloning of sheep through the use of embryonic cells. SC.H.3.4.2
3. **Critical Thinking Analyzing Methods** In the movie *Jurassic Park*, scientists used DNA to bring back extinct species. How is that different from the creation of cloned sheep using differentiated cells? SC.H.3.4.2
4. **Critical Thinking Forming Reasoned Opinions** List reasons you would or would not be concerned about consuming milk from cows treated with growth hormone. SC.H.3.4.6
5. **FCAT Prep** Genetic engineering can be used to produce rice with high levels of beta-carotene. This technology would help people who suffer from a deficiency in what substance? SC.H.3.4.5
   A complex proteins
   B glyphosate
   C growth hormone
   D vitamin A

Answers to Section Review

1. making food crops more tolerant to drought conditions, different soils, or climates; enhancing their nutritional value; controlling the process by which fruit ripens; making them resistant to the weedkiller glyphosate
2. Cloning with embryonic cells allows an organism to develop from cells that have not undergone specialization. Cloning with differentiated cells requires that the cell be manipulated so that the cell cycle is stopped.
3. The DNA used in Jurassic Park was fragmented. The researchers filled it in with DNA from other species. Thus, the dinosaurs had foreign DNA and were not truly clones.
4. Answers will vary. Students may not be concerned if evidence provided by the company indicates the hormone is safe. Concerns may include that the hormone may contribute to breast and prostate cancer and that cows get more infections and have to be given antibiotics, which could enter the milk.
5. **A.** Incorrect. Beta-carotene would not alleviate a deficiency of complex proteins. **B.** Incorrect. Humans do not require glyphosate. **C.** Incorrect. Beta-carotene would not alleviate a deficiency of growth hormone. **D.** Correct.

Key Concepts

1 Genetic Engineering

- Genetic engineers manipulate DNA for practical purposes.
- Restriction enzymes cleave DNA into fragments that have short sticky ends. Sticky ends allow DNA fragments from different organisms to join together to form recombinant DNA.
- Recombinant DNA is inserted into host cells. The cells are screened to identify cells that have the recombinant DNA. Each time the cells reproduce, the gene of interest is cloned.
- Electrophoresis uses an electric field within a gel to separate DNA fragments by their size.
- Specific genes can be identified with the Southern blot technique.

2 Genetic Engineering in Medicine and Society

- Genetic engineering is used to manufacture human proteins for use as drugs and to make safer and more effective vaccines.
- Some human genetic disorders are being treated with gene therapy.
- DNA fingerprinting is used to identify individuals and determine relationships between individuals.
- The Human Genome Project is an effort to determine the nucleotide sequence of and map the location of every gene on each human chromosome by the year 2003. The sequence of the genomes of many organisms has already been determined.

3 Genetic Engineering in Agriculture

- Crop plants can be genetically engineered to have favorable characteristics, including improved yields and resistance to herbicides and destructive pests.
- Genetically engineered growth hormone increases milk production in dairy cows and weight gain in cattle and hogs.
- Success in cloning animals using differentiated cells was announced in 1997. In addition, transgenic animals can be cloned and used to make proteins that are useful in medicine.

Key Terms

Section 1

genetic engineering (228)
recombinant DNA (228)
restriction enzyme (229)
vector (229)
plasmid (229)
gene cloning (229)
electrophoresis (231)
probe (231)

Section 2

Human Genome Project (233)
vaccine (235)
DNA fingerprint (237)

Section 3

transgenic animal (241)

Study ZONE

Alternative Assessment — ADVANCED

The use of DNA technology presents many ethical and moral issues. Divide your class into two groups. Ask one group to prepare information that supports DNA technology development, including specifics about the benefits of genetic engineering to society. Have the other group prepare information that demonstrates the negative aspects of these technologies. Have the two groups debate their stands on the issue. LS Verbal

Chapter Resource File

- Science Skills Worksheet GENERAL
- Critical Thinking Worksheet ADVANCED
- Test Prep Pretest GENERAL
- Chapter Test GENERAL

Answer to Concept Map

The following is one possible answer to Performance Zone item 15.

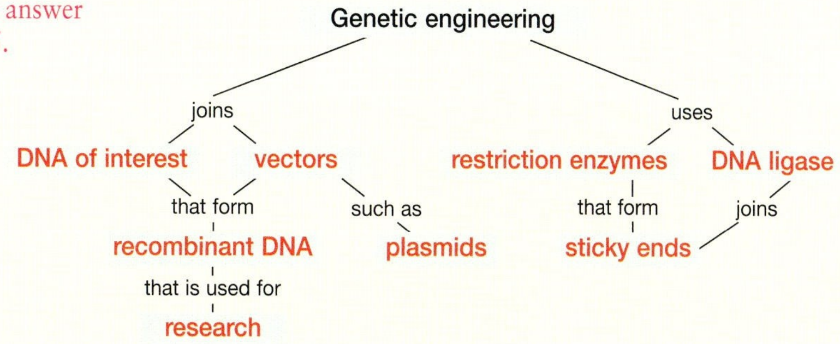

ANSWERS

Understanding Key Ideas

1. a
2. b
3. b
4. c
5. b
6. a
7. d
8. Molecule A was produced through genetic engineering. Plasmid DNA and DNA from a different organism are cut with restriction enzymes and then combined to produce recombinant plasmid DNA. The recombinant plasmid DNA molecules are then inserted into bacterial cells.
9. DNA is extracted from the bone. DNA primers, DNA polymerase, and nucleotides are added. The DNA is heated and then cooled. In a short time the original DNA from the fossilized bone is replicated. The sample is heated again to repeat the process until an adequate sample size is obtained.
10. The answer to the concept map is found at the bottom of the Study Zone page.

Critical Thinking

11. Answers will vary. The regulations were prompted by concerns that genetically engineered organisms might cause disease or have harmful effects on the environment. By limiting their survival outside the laboratory, scientists are preventing the organism from possibly harming people or other organisms and interfering in food chains.

Performance ZONE

Understanding Key Ideas

1. Gel electrophoresis is used to _______ DNA fragments.
   a. separate **c.** cut
   b. join **d.** copy
   SC.H.3.4.2
2. Which of the following human illnesses can be treated using a product of genetic engineering?
   a. malaria **c.** flu
   b. hemophilia **d.** a sinus cold
   SC.H.3.4.5
3. Injecting a healthy copy of a gene into a person who has a defective gene is called
   a. probing.
   b. gene therapy.
   c. PCR.
   d. DNA cloning.
   SC.H.3.4.2
4. The major effort to map and sequence all human genes is called
   a. the RFLP Project.
   b. the PCR Project.
   c. the Human Genome Project.
   d. DNA fingerprinting.
   SC.H.3.4.2
5. A transgenic organism is produced as a result of
   a. hybridization.
   b. recombinant DNA.
   c. mutation.
   d. RFLPs.
   SC.H.3.4.2
6. The process of making recombinant DNA is *least* related to
   a. clones.
   b. DNA fragments.
   c. restriction enzymes.
   d. sticky ends.
   SC.H.3.4.2
7. Genetic engineers can make plants
   a. resistant to insects.
   b. more tolerant to droughts.
   c. that are adapted to different soils.
   d. All of the above
   SC.H.3.4.5
8. Describe how molecule *A* was produced. SC.H.3.4.2

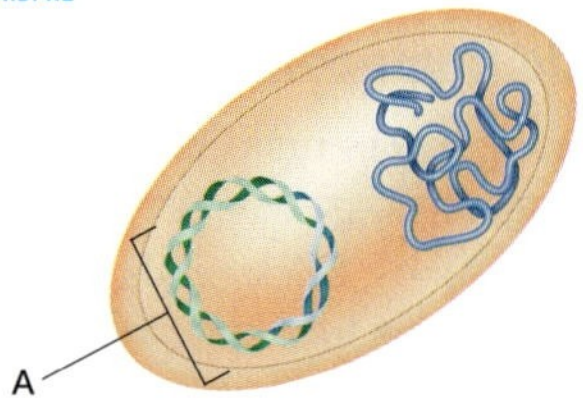

9. **BIOWatch** You have discovered a fossilized bone. How can you use PCR to obtain sufficient DNA for DNA analysis? SC.H.3.4.5
10. **Concept Mapping** Make a concept map about genetic engineering. Try to include the following words in your map: *DNA of interest, vectors, recombinant DNA, plasmids, restriction enzymes, sticky ends,* and *research.* SC.H.3.4.2

Critical Thinking

11. **Forming Reasoned Opinions** In the United States, government regulations require researchers to contain experimental genetically engineered organisms inside a laboratory and to ensure that the organisms could not survive outside the laboratory. Why do you think these strict regulations are necessary? SC.H.3.4.6
12. **Distinguishing Fact from Opinion** A judge presiding over a highly publicized murder trial dismissed the prosecution's request to admit DNA fingerprints as evidence, calling it "unproven." Do you agree with the judge? Explain your answer. SC.H.3.4.3
13. **Distinguishing Relevant Information** Organize and videotape a class debate about the safety questions raised by the potential release of genetically engineered plants, bacteria, and animals into the environment. Use library references and on-line databases to back up your arguments. SC.G.2.4.6

244

12. Students should disagree. A match of DNA fingerprints from two different samples has only two explanations—the prints are from the same person or from identical twins.
13. Answers will vary. For example: Will genetically engineered foods contain new proteins that are allergenic or toxic to some people? Will genetically engineered crops pass their new genes to close relatives in nearby wild areas and create weeds that are very difficult to control?

Chapter Review Correlations

SC.G.2.4.6: Item 13
SC.H.3.4.2: Items 1, 3-6, 8, 10
SC.H.3.4.3: Item 12
SC.H.3.4.5: Items 2, 7, 9
SC.H.3.4.6: Item 11

Assignment Guide

Section	Questions
1	1, 6, 8, 10, 11
2	2, 3, 4, 9, 12
3	5, 7, 13

FCAT Prep

For questions 1–6, write your answers on a separate sheet of paper.

1 What term describes a molecule containing DNA from two different organisms?

A. plasmid
B. probe
C. recombinant DNA
D. RFLP DNA

SC.F.2.4.2

2 Two pieces of DNA have been cut by restriction enzymes. The first piece of DNA has a sticky end with the nucleotide sequence: TGGCCA. What nucleotide sequence must the sticky end of the second piece of DNA have if it is to bond with the sticky end of the first piece of DNA?

F. ACCGGT
G. CTTAAG
H. TCCGGA
I. UGGCCU

SC.H.3.4.2

3 What agent allows genetic engineers to cut DNA at specific sites?

A. DNA ligase
B. DNA polymerase
C. plasmid DNA
D. restriction enzyme

SC.H.3.4.2

4 What technique is used to identify individuals in paternity cases and criminal cases?

F. DNA fingerprinting
G. gene therapy
H. genomic imprinting
I. vaccination

SC.H.3.4.3

5 READ INQUIRE EXPLAIN In natural selection, organisms that are best suited to their environment will reproduce more successfully than others.

Examine how natural selection could be affected by genetic engineering. SC.G.2.4.6

6 READ INQUIRE EXPLAIN The question of awarding patents on genetically engineered organisms arose when a microbiologist named Ananda Chakrabarty filed for a patent on a bacterium capable of digesting the components of crude oil. Chakrabarty identified enzymes that degrade different components of crude oil and added the enzymes to *Pseudomonas* bacteria. His patent request was brought before the U. S. Supreme Court, which ruled in 1980 that human-engineered organisms are patentable under federal law.

Part A What type of genetic engineering did Chakrabarty use to add enzymes to *Pseudomonas* bacteria? SC.H.3.4.2

Part B Briefly describe the process of adding genes to organisms such as bacteria. SC.H.3.4.2

Test TIP

When answering short-response questions, be sure to write in complete sentences.

FCAT Doctor

Question 2 Answer F is the correct choice. T bonds with A, and G bonds with C. Answer G is incorrect because the complementary sequence is GAATTC. Answer H is incorrect because the complementary sequence is AGGCCT. Answer I is incorrect because this is RNA. Its complementary DNA sequence is ACCGGA.

Question 3 Answer D is the correct choice. Restriction enzymes recognize, bind to, and cut specific short sequences in DNA. Answer A is incorrect because DNA ligase is used to bond together fragments of DNA, not cut them. Answer B is incorrect because DNA poly merase is used in the replication of DNA, not in the cutting of DNA. Answer C is incorrect because plasmid DNA are circular forms of DNA and are the molecules that are acted upon by restriction enzymes and DNA ligase.

Question 5 If the genetically modified organisms are fitter than naturally evolved organisms, the genetically modified organisms would be naturally selected.

Question 6

Part A To insert genes for specific enzymes into the bacteria genome, Chakrabarty would have to use recombinant DNA.

Answers

1. C
2. F
3. D
4. F
5. Fit organisms that have naturally evolved might not be able to compete with genetically modified organisms that have newly engineered genes.

6A. He used recombinant DNA.

6B. The DNA sequence that contains the gene is cut from the DNA of one organism using restriction enzymes. The DNA of a second organism is cut with the same restriction enzymes. The DNA sequence from the first organism is added into the second organism's DNA, and DNA ligase is added to help bond the ends of the DNA fragments together.

Test Prep Correlations

SC.F.2.4.2: Item 1
SC.G.2.4.6: Item 5
SC.H.3.4.2: Items 2, 3, 6A, 6B
SC.H.3.4.3: Item 4

Exploration Lab

Modeling Recombinant DNA SC.H.3.4.1

SKILLS
- Modeling
- Comparing

OBJECTIVES
- **Construct** a model that can be used to explore the process of genetic engineering.
- **Describe** how recombinant DNA is made.

MATERIALS
- paper clips (56)
- plastic soda straw pieces (56)
- pushpins (15 red, 15 green, 13 blue, and 13 yellow)

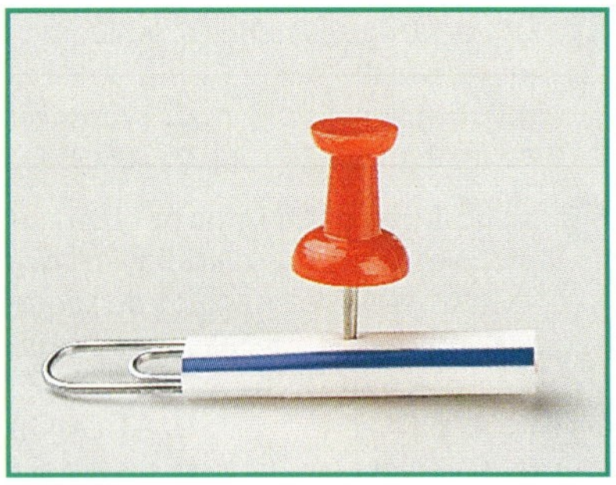

Before You Begin

Genetic engineering is the process of taking a gene from one organism and inserting it into the DNA of another organism. The gene is delivered by a **vector,** such as a virus, or a bacterial **plasmid.**

First, a fragment of a chromosome that contains the gene is isolated by using a **restriction enzyme,** which cuts DNA at a specific nucleotide-base sequence. Some restriction enzymes cut DNA unevenly, producing single-stranded **sticky ends.** The DNA of the vector is cut by the same restriction enzyme. Next, the chromosome fragment is mixed with the cut DNA of the vector. Finally, an enzyme called **DNA ligase** joins the ends of the two types of cut DNA, producing **recombinant DNA.**

In this lab, you will model genetic engineering techniques. You will simulate the making of recombinant DNA that has a human gene inserted into the DNA of a plasmid.

1. Write a definition for each boldface term in the paragraph above and for the term *base-pairing rules.*
2. Based on the objectives for this lab, write a question you would like to explore about the process of genetic engineering.

Procedure

PART A: Model Genetic Engineering

1. Make 56 model nucleotides. To make a nucleotide, insert a pushpin midway along the length of a 3 cm piece of a soda straw. **CAUTION: Handle pushpins carefully. Pointed objects can cause injury.** Push a paper clip into one end of the soda-straw piece until it touches the pushpin.
2. Begin a model of a bacterial plasmid by arranging nucleotides for one DNA strand in the following order: blue, red, green, yellow, red, red, blue, blue, green, red, blue, green, red, blue, blue, green, yellow, and red. Join two adjacent nucleotides by inserting the paper clip end of one into the open end of the other.
3. Using your first DNA strand and the base-pairing rules, build the complementary strand of plasmid DNA. **Note:** *Yellow is complementary to blue, and green is complementary to red.*
4. Complete your model of a circular plasmid by joining the opposite ends of each DNA strand. Make a sketch showing the sequence of bases in your model plasmid. Use the abbreviations B, Y, G, and R for the pushpin colors. Your sketch should be similar to the one at the top of the next page.

246

Exploration Lab

Modeling Recombinant DNA

Teacher's Notes

Time Required 40–50 minutes

Ratings

EASY → HARD

Teacher Preparation 2
Student Setup 1
Concept Level 3
Cleanup 1

Skills Acquired
- Constructing Models
- Identifying/Recognizing Patterns
- Inferring
- Interpreting
- Predicting

Scientific Methods

In this lab, students will:
- Make Observations
- Ask Questions
- Analyze the Results
- Draw Conclusions

Materials and Equipment

Materials for this lab can be ordered from WARD'S. See *Master Materials List* at the front of this book for catalog numbers.

Safety Cautions

Review all safety symbols and caution statements with students before beginning the lab. Caution students to be careful of the sharp tips on the pushpins.

Tips and Tricks

Make sure each group has enough materials before they begin the activity. Use materials from Experimental Design in Chapter 10, if they were saved. Have students work in cooperative groups of four students each. Divide each group into pairs. One two-person team should complete steps 2–4 while the other team completes steps 5 and 6. The entire group should work together to complete steps 7–11. Before beginning the lab, review the structure of DNA and to review the steps of genetic engineering.

Answers to Before You Begin

1. *genetic engineering*—the process of isolating a gene from the DNA of one organism and inserting it into the DNA of another organism

   vector—agent, such as a virus or plasmid, used to carry a DNA fragment into a cell

   plasmid—small circular DNA molecule, usually found in bacteria, that can replicate independently of the main chromosome

   restriction enzyme—bacterial enzyme that cuts DNA at a specific sequence of nucleotides

   sticky ends—single-stranded ends of DNA that are produced when a restriction enzyme cuts DNA

   DNA ligase—enzyme that joins ends of DNA

   recombinant DNA—DNA that contains DNA segments from different organisms

   base-pairing rules—rules stating that in DNA, adenine on one strand always pairs with thymine on the opposite strand and cytosine on one strand always pairs with guanine on the opposite strand
2. Answers will vary. For example: What steps are involved in transferring a gene from one organism to another one?

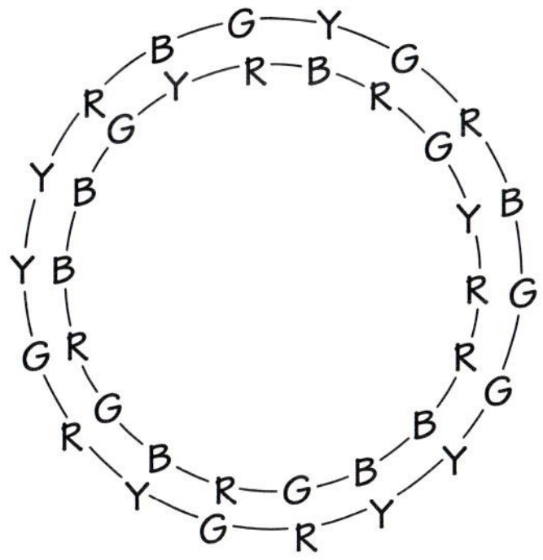

5. Begin a model of a human chromosome fragment made by a restriction enzyme. Place nucleotides for one DNA strand in the following order: BBRRYGGBRY. Build the second DNA strand by arranging the remaining nucleotides in the following order: BRRYGBYYGG.
6. Match the complementary portions of the two strands of DNA you made in step 5. Pair as many base pairs in a row as you can. Make a sketch showing the sequence of bases in your model of a human chromosome fragment.
7. Imagine that the restriction enzyme that cut the human chromosome fragment you made in steps 5 and 6 is moving around your model plasmid until it finds the sequence YRRBBG and its complementary sequence, BGGYYR. This restriction enzyme cuts each sequence between a B and a G. Find such a section in your sketch of your model plasmid's DNA.
8. Simulate the action of the restriction enzyme on the section you identified in step 7. Open both strands of your model plasmid's DNA by pulling apart the adjacent green and blue nucleotides in each strand. Make a sketch of the split plasmid DNA molecule.
9. Move your model human DNA fragment into the break in your model plasmid's DNA molecule. Imagine that a ligase joins the ends of the human and plasmid DNA. Make a sketch of your final model DNA molecule.

PART B: Cleanup and Disposal

10. Dispose of damaged pushpins in the designated waste container.
11. Clean up your work area and all lab equipment. Return lab equipment to its proper place. Wash your hands thoroughly before you leave the lab and after you finish all work.

Analyze and Conclude

1. **Comparing Structures** Compare your models of plasmid DNA and human DNA.
2. **Relating Concepts** What do the sections of four unpaired nucleotides in your model human DNA fragment represent?
3. **Comparing Structures** How did your original model plasmid DNA molecule differ from your final model DNA molecule?
4. **Drawing Conclusions** What does the molecule you made in step 9 represent?
5. **Further Inquiry** Write a new question that could be explored with another investigation.

On the Job

Genetic engineering is used to produce many products that are useful to humans. Do research to discover how genetic engineering is used to better our lives. For more about careers, visit **go.hrw.com** and type in the keyword **HX4 Careers.**

Answers to Analyze and Conclude

1. Both the plasmid DNA model and the human DNA model are double-stranded. The plasmid DNA model is circular and the human DNA model is linear. The human DNA model represents a fragment of a DNA molecule that has been cut with a restriction enzyme and has sticky ends.
2. sticky ends
3. The original model plasmid DNA was smaller (had fewer nucleotides) and did not contain a gene from a human chromosome.
4. recombinant DNA
5. Answers will vary. For example: What happens if the plasmid and human DNA do not have complementary sticky ends?

Answers to On the Job

Answers will vary. Advances have been made in the production of proteins used to treat illnesses, in the development of new vaccines, in the replacement of defective genes, and in the improvement of food crops.

Answers to Procedure

4.

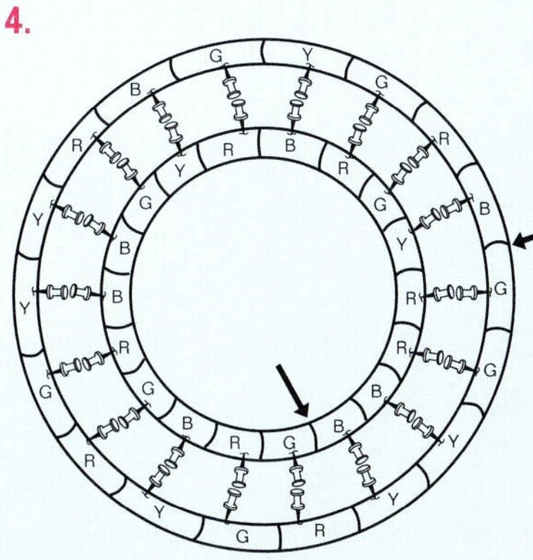

Model of plasmid DNA molecule

6.

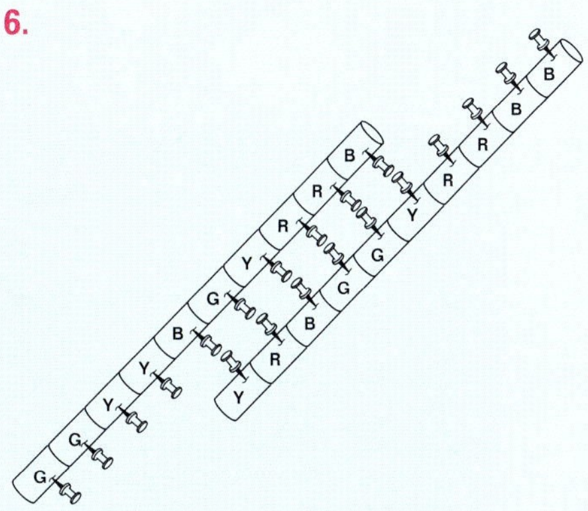

Human donor DNA fragment

7. The sketch should show the split at the nucleotides indicated by the arrows in item 4.

9.

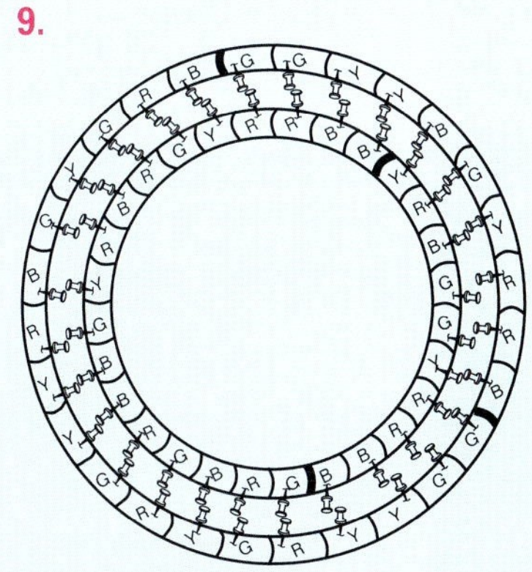

Final recombined DNA molecule

UNIT 3 Principles of Evolution

Chapters

Tropical rain forests contain more than one-half of all the world's animal and plant species, such as this red-eyed tree frog from Central America.

Tropical Rain Forests

Yesterday...

During their travels, early European visitors were fascinated by the variety and number of tropical rain-forest plants and animals so different from those they knew at home. **Discover how early scientists named and classified living things.**

Exploring the Brazilian jungle in 1820

Rain-forest researcher at Madre de Dios, Peru

Today...

Today's explorers are faced with the challenge of classifying living things in lesser-known areas, such as tropical rain forests, that have not yet been studied by scientists. **What organisms are found in tropical rain forests, and what is the impact of human activity on these species?**

Coffee beans

Tomorrow...

To preserve species, new ways of protecting tropical habitats are being explored. For example, modern coffee-growing techniques eliminate native species as fields are planted solely with coffee, a major source of revenue for many tropical countries. But shade coffee plantations grow coffee in the midst of the forest canopy, thus preserving habitat for tropical species.

internet connect

www.scilinks.org
Topic: Rain Forests
Keyword: HX4155

SCILINKS. Maintained by the National Science Teachers Association

Chapter Opener

SE **Quick Review,** p. 251 ■
SE **Reading Activity,** p. 251 ■
TE **Using the Figure,** p. 250 GENERAL
TE **Opening Activity,** p. 251 GENERAL

& 3 • 90 min pp. 252–257

Section 1
How Did Life Begin?

OSP **Lesson Plan***
TR **Bellringer***
TR **Section Outline***
TR **D4 Radioactive Decay***
TR **D5 Miller-Urey Experiment***
TR **D6 Lerman's Bubble Model***

TE **Using the Figure,** p. 252 GENERAL
SE **Quick Lab** Modeling Radioactive Decay, p. 253 ■
CRF **Datasheets for In-Text Labs** Modeling Radioactive Decay*
SE **Quick Lab** Modeling Coacervates, p. 256 ■
CRF **Datasheets for In-Text Labs** Modeling Coacervates*
CRF **Skills Practice Lab** Determining the Age of Artifacts Using C14* GENERAL

& 5 • 90 min pp. 258–263

Section 2
The Evolution of Cellular Life

OSP **Lesson Plan***
TR **Bellringer***
TR **Section Outline***
TR **D8 Evolution of Eukaryotes***

TE **Demonstration,** p. 258 GENERAL
SE **Data Lab** Analyzing Signs of Endosymbiosis, p. 259 ■
CRF **Datasheets for In-Text Labs** Analyzing Signs of Endosymbiosis*
TE **Group Activity** Models of Endosymbionts and Bacteria, p. 260 GENERAL
TE **Activity** Favorite Fossils, p. 261 ADVANCED
TE **Demonstration,** p. 261 GENERAL
TE **Activity** Burgess Shale, p. 262 ADVANCED

& 7 • 90 min pp. 264–268

Section 3
Life Invaded the Land

OSP **Lesson Plan***
TR **Bellringer***
TR **Section Outline***
TR **D7 Ozone Shields the Earth***

TE **Activity** Coevolution of Insects and Plants, p. 266 ADVANCED
TE **Demonstration,** p. 267 GENERAL
TE **Demonstration,** p. 267 BASIC
CRF **Quick Lab** Analyzing Adaptations, Living on Land* ◆ BASIC
SE **Exploration Lab** Making a Timeline of Life on Earth, pp. 272–273 ■
CRF **Datasheets for In-Text Labs** Making a Timeline of Life on Earth*

& 9 • 90 min

Chapter Review and Assessment Resources

SE **Study Zone,** p. 269 ■
SE **Performance Zone,** p. 270 ■
SE **FCAT Test Prep,** p. 271 ■
TR **Concept Mapping***
CRF **Vocabulary Review Worksheet*** ■ GENERAL
CRF **Test Prep Pretest*** ■ GENERAL
CRF **Chapter Test*** ■ GENERAL
CRF **Chapter Test*** ADVANCED
OSP **Test Generator***
OSP **Test Item Listing***
OSP **Modified Chapter Test*** BASIC

go.hrw.com Holt Online Learning

Visit go.hrw.com for access to Holt Online Learning or enter the keyword **HX6 Home** for a variety of free online resources.

One-Stop Planner® CD-ROM

This CD-ROM package includes
- Lab Materials QuickList Software
- Holt Calendar Planner
- Customizable Lesson Plans
- Printable Worksheets
- ExamView® Test Generator
- Interactive Teacher Edition
- Holt PuzzlePro® Resources
- Holt PowerPoint® Resources

Guided Reading Audio CDs

These CDs are designed to help auditory learners and reluctant readers. (Audio CDs are also available in Spanish.)

Compression guide:
To shorten your instruction because of time limitations, omit blocks 1, 2, and 3.

KEY

TE	Teacher Edition	CRF	Chapter Resource File	*	Also on One-Stop Planner
SE	Student Edition	TR	Teaching Transparency	■	Also Available in Spanish
OSP	One-Stop Planner			◆	Requires Advance Prep

SKILLS DEVELOPMENT AND PRACTICE	SECTION REVIEW AND ASSESSMENT	CORRELATIONS
		Florida Sunshine Standards/FCAT
TE **Skill Builder** Writing, p. 253 ADVANCED **CRF** **Directed Reading Worksheet*** BASIC **CRF** **Active Reading Worksheet*** GENERAL	**SE** **Section 1 Review** ■ **TE** **Reteaching,** p. 257 BASIC **TE** **Quiz,** p. 257 GENERAL **TE** **Alternative Assessment,** p. 257 GENERAL **CRF** **Quiz*** ■ BASIC	SC.D.1.4.4, SC.F.2.4.2, SC.H.1.4.1, SC.H.3.4.1
TE **Reading Skill Builder** Reading Organizer, p. 259 BASIC **TE** **Reading Skill Builder** Reading Organizer, p. 259 BASIC **CRF** **Directed Reading Worksheet*** BASIC **CRF** **Active Reading Worksheet*** GENERAL	**SE** **Section 2 Review** ■ **TE** **Reteaching,** p. 263 BASIC **TE** **Quiz,** p. 263 GENERAL **TE** **Alternative Assessment,** p. 263 GENERAL **CRF** **Quiz*** ■ BASIC	SC.D.1.4.3, SC.F.2.4.3, SC.G.1.4.1, SC.H.1.4.1
TE **Skill Builder** Math, p. 265 ADVANCED **CRF** **Directed Reading Worksheet*** BASIC **CRF** **Active Reading Worksheet*** GENERAL **CRF** **Critical Thinking Worksheet*** ADVANCED **CRF** **Science Skills Worksheet*** GENERAL	**SE** **Section 3 Review** ■ **TE** **Reteaching,** p. 268 BASIC **TE** **Quiz,** p. 268 GENERAL **TE** **Alternative Assessment,** p. 268 GENERAL **CRF** **Quiz*** ■ BASIC **Modified Worksheet*** BASIC	SC.D.1.4.3, SC.D.1.4.4, SC.G.2.4.1

SCILINKS. NSTA
www.scilinks.org
Maintained by the **National Science Teachers Association**

Topic: Radioactive Decay
SciLinks Code: HX4154

Topic: Endosymbiosis
SciLinks Code: HX4068

Topic: Extinction
SciLinks Code: HX4078

CHAPTER 12

Overview

Tell students the purpose of this chapter is to explore how scientists estimate the age of Earth and how life may have begun. Students will also learn about how life increased in complexity from its first forms on Earth. Finally, they will learn about the types of environments inhabited by Earth's early organisms.

Using the Figure

Direct students' attention to the figure opening this chapter. Tell them that the history of life on Earth began approximately 3.5 billion years ago. How this occurred has been and will continue to be a topic for inquiry and debate. Scientific evidence indicates that since life's beginning, each organism that has inhabited this planet has been the product of evolution—the change in organisms over time.

Assessing Prior Knowledge

Before beginning this chapter, be sure your students can answer the questions in the ***Quick Review*** on the facing page. If they cannot, encourage them to reread the sections indicated.

Lightning at sea

250

Chapter Correlations — Florida Sunshine Standards/FCAT

SC.D.1.4.3 Changes in Earth's climate, geological activity, and life forms may be traced and compared.

SC.D.1.4.4 Earth's systems and organisms are the result of a long, continuous change over time.

SC.F.2.4.2 Every cell contains a "blueprint" coded in DNA molecules that specify how proteins are assembled to regulate cells.

SC.F.2.4.3 Mechanisms of change (e.g., mutation and natural selection) lead to adaptations in a species and their ability to survive naturally in changing conditions and to increase species diversity.

SC.G.1.4.1 There is great diversity and interdependence among living things.

SC.G.2.4.1 Layers of energy-rich organic materials have been gradually turned into great coal beds and oil pools (fossil fuels) by the pressure of the overlying earth, and humans burn fossil fuels to release the stored energy as heat and carbon dioxide.

SC.H.1.4.1 Investigations are conducted to explore new phenomena, to check on previous results, to test how well a theory predicts, and to compare different theories.

CHAPTER

12 History of Life on Earth

Florida Sunshine Standards/FCAT

SC.D.1.4.3 Changes in Earth's climate, geological activity, and life forms may be traced and compared.
SC.D.1.4.4 Earth's systems and organisms are the result of a long, continuous change over time.
SC.F.2.4.2 Every cell contains a "blueprint" coded in DNA molecules that specify how proteins are assembled to regulate cells.
SC.F.2.4.3 Mechanisms of change (e.g., mutation and natural selection) lead to adaptations in a species and their ability to survive naturally in changing conditions and to increase species diversity.
SC.G.1.4.1 There is great diversity and interdependence among living things.
SC.G.2.4.1 Layers of energy-rich organic materials have been gradually turned into great coal beds and oil pools (fossil fuels) by the pressure of the overlying earth, and humans burn fossil fuels to release the stored energy as heat and carbon dioxide.
SC.H.1.4.1 Investigations are conducted to explore new phenomena, to check on previous results, to test how well a theory predicts, and to compare different theories.

Quick Review

Answer the following without referring to earlier sections of your book.

1. **Compare** the structure of proteins, lipids, and nucleic acids. *(Chapter 2, Section 3)*
2. **Describe** the role of enzymes in catalyzing chemical reactions. *(Chapter 2, Section 4)*
3. **Contrast** prokaryotes and eukaryotes. *(Chapter 3, Section 2)*
4. **Identify** the structure and function of chloroplasts and mitochondria. *(Chapter 3, Section 3)*
5. **Summarize** the role of DNA in heredity. *(Chapter 9, Section 1)*

Did you have difficulty? *For help, review the sections indicated.*

Reading Activity

Before you begin to read this chapter, write down all of the key words for each section of the chapter. Then, write a definition next to each word that you have heard of. As you read the chapter, write definitions next to the words that you did not previously know, and modify as needed your original definitions of words familiar to you.

Billions of years ago, the combination of simple molecules and energy from sources such as lightning may have given rise to the complex organic molecules necessary for life.

Looking Ahead

Section 1
How Did Life Begin?
The Age of Earth
Formation of the Basic Chemicals of Life
Precursors of the First Cells

Section 2
The Evolution of Cellular Life
The Evolution of Prokaryotes
The Evolution of Eukaryotes
Multicellularity
Mass Extinctions

Section 3
Life Invaded the Land
The Ozone Layer
Plants and Fungi on Land
Arthropods
Vertebrates

internet connect

www.scilinks.org
National Science Teachers Association *sci*LINKS Internet resources are located throughout this chapter.

SCILINKS. Maintained by the National Science Teachers Association

251

Opening Activity — GENERAL

To help students visualize the enormity of one billion of anything, ask students if they could physically carry one million dollars. One million dollars would be a stack of $100 bills approximately 1 m in height (10,000 bills). A billion dollars would be a stack of the same bills to a height of the Washington Monument (one million bills). Now relate this image to the first billion years of our planet's existence without any form of life.

Quick Review Answers

1. Proteins are polymers of amino acids. Lipids are mostly or completely nonpolar substances containing either chains or rings of primarily carbon and hydrogen atoms. Nucleic acids are polymers of nucleotides.
2. Enzymes interact with molecules to place strain on a chemical bond so that it breaks easily, or to place substances in close proximity so that a small amount of energy input results in a bond being formed between the substances.
3. Prokaryotes are single-celled organisms that lack membrane-bounded organelles. Eukaryotes are single or multicellular organisms that have membrane-bounded organelles, including a nucleus.
4. Chloroplasts and mitochondria are oval and have a complex system of internal membranes. Chloroplasts carry out all of the reactions of photosynthesis. Mitochondria carry out all of the reactions of cellular respiration.
5. The DNA of an organism makes up its genetic information. When a new organism is formed from two parents, it contains its own genetic information, half of which came from each parent.

Chapter Resource File

- Vocabulary Worksheets
- Concept Mapping

Reading Activity Answers

Student lists will likely indicate some understanding of the Earth's beginnings and the origin of life. Few will likely show an understanding of how the complexity and diversity of life on Earth today could have evolved from the earliest organisms that appeared on Earth.

Focus

Overview

Before beginning this section review with your students the objectives listed in the Student Edition. They will learn about how scientists measure the age of the Earth. They will also learn about several models scientists have developed to explain how the first life forms originated on the early Earth. Finally, they will learn how scientists think the complex chemicals found in living things could have arisen from the chemicals present on Earth at the time.

Bellringer

Ask students to draw a picture of what they think Earth's first life form looked like, and have them label the parts of the organism.

Motivate

Using the Figure — GENERAL

Direct students' attention to **Figure 1.** Ask students to explain why the graph of the rate of decay shown in this figure is not a straight line. (The rate of radioactive decay is based on the half-life of a radioactive material. The amount of material that decays in one half-life is one-half of the undecayed portion of the sample present at the beginning of the time period. With each successive half-life, the total amount of material decayed increases. LS Visual

Section Correlations

SC.D.1.4.4
SC.F.2.4.2
SC.H.1.4.1

Section 1 How Did Life Begin?

Florida Sunshine Standards/FCAT
SC.F.2.4.2 Every cell contains a "blueprint" coded in DNA molecules that specify how proteins are assembled to regulate cells.
SC.H.1.4.1 Investigations are conducted to explore new phenomena, to check on previous results, to test how well a theory predicts, and to compare different theories.

Objectives

- **Summarize** how radioisotopes can be used in determining Earth's age. SC.H.1.4.1
- **Compare** two models that describe how the chemicals of life originated. SC.D.1.4.4
- **Describe** how cellular organization might have begun. SC.D.1.4.4
- **Recognize** the importance that a mechanism for heredity has to the development of life. SC.F.2.4.2

Key Terms

radiometric dating
radioisotope
half-life
microsphere

The Age of Earth

When Earth formed, about 4.5 billion years ago, it was a fiery ball of molten rock. Eventually, the planet's surface cooled and formed a rocky crust. Water vapor in the atmosphere condensed to form vast oceans. Most scientists think life first evolved in these oceans and that the evolution of life occurred over hundreds of millions of years. Evidence that Earth has existed long enough for this evolution to have taken place can be found by measuring the age of rocks found on Earth.

Measuring Earth's Age

Scientists have estimated the age of Earth using a technique called radiometric dating. **Radiometric dating** is the estimation of the age of an object by measuring its content of certain radioactive isotopes *(IE soh tohps)*. An isotope is a form of an element whose atomic mass (the mass of each individual atom) differs from that of other atoms of the same element. Radioactive isotopes, or **radioisotopes,** are unstable isotopes that break down and give off energy in the form of charged particles (radiation). This breakdown, called radioactive decay, results in other isotopes that are smaller and more stable.

For example, certain rocks contain traces of potassium-40, an isotope of the element potassium. As **Figure 1** shows, the decay of potassium-40 produces two other isotopes, argon-40 and calcium-40. The time it takes for one-half of a given amount of a radioisotope to decay is called the radioisotope's **half-life.** By measuring the proportions of certain radioisotopes and their products of decay, scientists can compute how many half-lives have passed since a rock was formed.

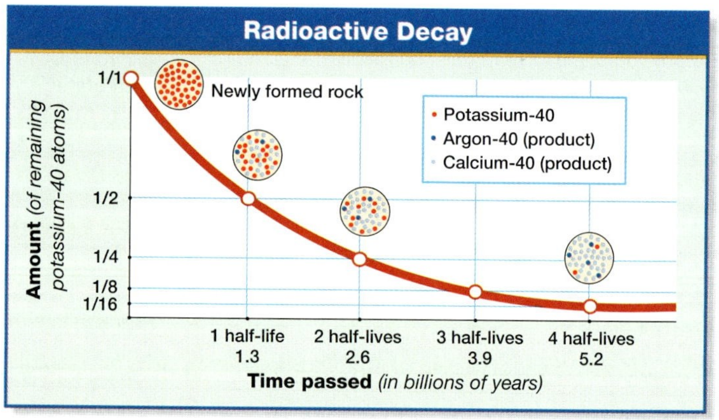

Figure 1 Rate of decay for potassium-40. This graph shows the rate of decay for the radioisotope potassium-40. After one *half-life* has passed, half of the original amount of the radioisotope remains.

Chapter Resource File

- Directed Reading BASIC
- Active Reading GENERAL
- Data Sheet for Quick Lab GENERAL

One-Stop Planner CD-ROM

- Reading Organizers BASIC
- Reading Strategies BASIC
- Basic Skills Worksheet Mass and Density BASIC

INCLUSION Strategies

- Attention Deficit Disorder
- Developmental Delay
- Learning Disability

Ask students to draw illustrations of their interpretation of the Earth's formation 4.5 billion years ago. The illustrations should show their idea of at least four stages of the evolution of the Earth's surface. The stages should reflect information in the opening paragraphs of Section 1. Students should narrate their drawings through a tape-recorded description of the events shown in their drawings.

Formation of the Basic Chemicals of Life

Most scientists think that life on Earth developed through natural chemical and physical processes. It is thought that the path to the development of living things began when molecules of nonliving matter reacted chemically during the first billion years of Earth's history. These chemical reactions produced many different simple, organic molecules. Energized by the sun and volcanic heat, these simple, organic molecules formed more-complex molecules that eventually became the building blocks of the first cells. The hypothesis that many of the organic molecules necessary for life can be made from molecules of nonliving matter has been tested and supported by results of laboratory experiments.

internet connect
www.scilinks.org
Topic: Radioactive Decay
Keyword: HX4154
SCILINKS. Maintained by the National Science Teachers Association

QUICK LAB
Modeling Radioactive Decay SC.H.3.4.1

You can use some dried corn, a box, and a watch to make a model of radioactive decay that will show you how scientists measure the age of objects.

Materials

approximately 100 dry corn kernels per group, cardboard box, clock or watch with a second hand

Procedure

1. On a separate sheet of paper, make a data table like the one below.
2. Assign one member of your team to keep time.
3. Place 100 dry corn kernels into a box.
4. Shake the box gently from side to side for 10 seconds.
5. Keep the box still and remove and count the kernels that "point" to the left side of the box, as shown below. Record in your data table the number of kernels you removed.

6. Repeat steps 4 and 5 until all kernels have been counted and removed.
7. Calculate the number of kernels remaining for each time interval.
8. Make a graph using your group's data. Plot "Total shake time (seconds)" on the *x*-axis. Plot "Number of kernels remaining" on the *y*-axis.

Analysis

1. **Identify** what the removed kernels represent in each step.
2. **Calculate** the half-life of your sample, in seconds, that is represented in this activity.
3. **Calculate** the age of your sample, in years, if each 10-second interval represents 5,700 years.
4. **Evaluate** the ability of this model to demonstrate radioactive decay.

DATA TABLE

Total shake time (seconds)	Number of kernels removed	Number of kernels remaining
10		
20		
30		

Teach

SKILL BUILDER —ADVANCED

Writing Skills Have students research radioactive decay using the Web site in the Internet Connect box. Have students write a report summarizing their findings.

QUICK LAB
Modeling Radioactive Decay

Skills Acquired
Forming a model, inferring, calculating

Teacher's Notes
Have kernels precounted and placed in paper or plastic cups. If time allows, multiple trials (two or three) would provide a better set of data. Make sure all kernels are accounted for; they can be slippery if stepped on.

Analysis Answers:

1. The kernels that are removed with each step represent the molecules that decayed.
2. The half-life of the sample is the number of trials it took to remove 50 kernels multiplied by 10 sec/trial.
3. Age of the sample = (number of trials) × (5,730 years/trial)
4. Answers will vary, but should be supported by the data.

Transparencies

TR Bellringer
TR D4 Radioactive Decay
TR D5 Miller-Urey Experiment
TR D6 Lerman's Bubble Model

Trends in Particle Physics

Trace Analysis A new ultrasensitive trace analysis technique—able to detect single atoms in a large sample—has been developed by Argonne National Laboratory researchers. The technology, called Atom Trap Trace Analysis, or ATTA for short, holds promise for advancing the state of the art in many fields, from groundwater studies to environmental monitoring. One of its first applications may be dating ancient Greenland ice cores. While carbon-14 dating has been used to date ice up to about 50,000 years old, ages of older samples have to be inferred from other evidence, such as their depth. Trace analysis of krypton-81, which lasts 40 times longer than carbon-14, can be used to date samples up to one million years old.

ATTA could be applied to other isotopes for applications in a range of scientific disciplines. One possibility is using calcium-41 as a medical tracer to monitor bone loss from osteoporosis. A patient could ingest a small, non-harmful amount of the isotope, which would be taken up into the bones. Over the following years, ATTA could be used to detect the miniscule amounts of calcium-41 in urine samples, which would signal the loss of calcium from bones.

Teach, *continued*

Real Life — GENERAL

Answers

Scientists are working on technology to eventually examine (by remote telescopes) the planets in and beyond our solar system whose spectral "signatures" indicate the presence of oxygen, water, and carbon dioxide. NASA plans a sample recovery mission to Mars for 2008.

Teaching Tip — GENERAL

Skin Health Remind students that ultraviolet (UV) radiation in sunlight can be very damaging to their health. Sunburn indicates that skin has been damaged by UV radiation. The American Cancer Society reminds us that people who have been seriously sunburned, even once, have an increased risk of developing skin cancer. Remind students that if they are going to be exposed to the sun, they should wear a high SPF sunscreen or protective clothing to reduce UV damage. Explain that the "peeling" that occurs after sunburn is the body's way of getting rid of cells that are too damaged to repair.

Real Life

Is there life on other planets?

Many scientists think life could have arisen on other planets the same way it did on Earth.

Analyzing Information
Find out about research on extraterrestrial life, and compare scientists' predictions about possible life-forms on other planets.

The "Primordial Soup" Model

In the 1920s, the Russian scientist A. I. Oparin and the British scientist J.B.S. Haldane both suggested that the early Earth's oceans contained large amounts of organic molecules. This hypothesis became known as the primordial *(prie MAWR dee uhl)* soup model. Earth's vast oceans were thought to be filled with many different organic molecules, like a soup that is filled with many different vegetables and meats. Oparin and Haldane hypothesized that these molecules formed spontaneously in chemical reactions activated by energy from solar radiation, volcanic eruptions, and lightning.

Oparin, together with the American scientist Harold Urey, and other scientists also proposed that Earth's early atmosphere lacked oxygen. They hypothesized that the early atmosphere was instead rich in nitrogen gas, N_2; hydrogen gas, H_2; and hydrogen-containing gases such as water vapor, H_2O; ammonia, NH_3; and methane, CH_4. They reasoned that electrons in these gases would have been frequently pushed to higher energy levels by light particles from the sun or by electrical energy in lightning. Today, high-energy electrons are quickly soaked up by the oxygen in Earth's atmosphere because oxygen atoms have a great "thirst" for such electrons. But without oxygen, high-energy electrons would have been free to react with hydrogen-rich molecules, forming a variety of organic compounds.

In 1953, the primordial soup model was tested by Stanley Miller, who was then working with Urey. Miller placed the gases that he and Urey proposed had existed on early Earth into a device like the one seen in **Figure 2.** To simulate lightning, he provided electrical sparks. After a few days, Miller found a complex collection of organic molecules in his apparatus. These chemicals included some of life's basic building blocks: amino acids, fatty acids, and other hydrocarbons (molecules made of carbon and hydrogen). These results support the hypothesis that some basic chemicals of life could have formed spontaneously under conditions like those in the experiment.

Figure 2 Miller-Urey experiment. Miller simulated an atmospheric composition that Oparin and other scientists incorrectly hypothesized existed on early Earth. His experiment produced several different organic compounds.

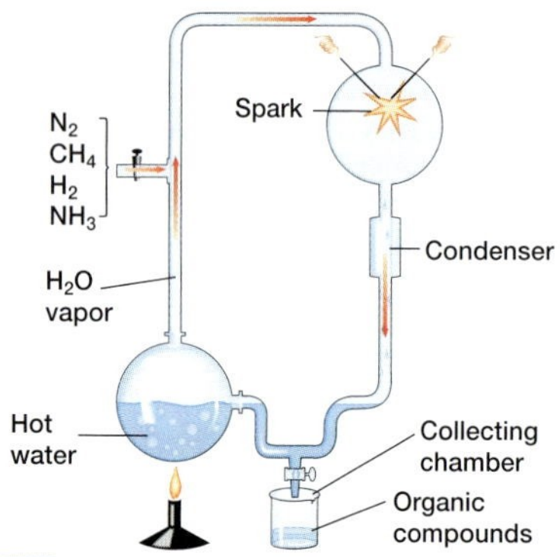

Reevaluating the Miller-Urey Model

Recent discoveries have caused scientists to reevaluate the Miller-Urey experiment. We now know that the reductant molecules used in Miller's experiment could not have existed in abundance on the early Earth. Four billion years ago, Earth did not have a protective layer of ozone gas, O_3. Today ozone protects Earth's surface from most of the sun's damaging ultraviolet radiation. Without ozone, ultraviolet radiation would have destroyed any ammonia and methane present in the atmosphere. When these gases are absent from the Miller-Urey experiment, key biological molecules are not made. This raises a very important question: If the chemicals needed to form life were not in the atmosphere, where did they come from? Some scientists argue that the chemicals were produced within ocean bubbles. Others say that the chemicals arose in deep sea vents. The correct answer has not been determined yet.

did you know?

The Light of Life Experiments similar to Miller and Urey's that used UV light instead of electrical sparks as an energy source also produced amino acids.

Integrating Physics and Chemistry

Remind students that electrons occur in regions around atoms called orbitals which exist within each of several discrete energy levels surrounding the atom's nucleus. Excited electrons pushed into higher energy levels are more reactive, since interactions between the outermost electrons cause chemical bonding. Ask students to identify the number of valence electrons for nitrogen, carbon, and hydrogen (the component elements found in the "primordial soup" of the Miller-Urey experiment), and then relate this information to their bonding pattern.

The Bubble Model

In 1986, the geophysicist Louis Lerman suggested that the key processes that formed the chemicals needed for life took place within bubbles on the ocean's surface. Lerman's hypothesis, also known as the bubble model, is summarized in **Figure 3.**

Step 1 Ammonia, methane, and other gases resulting from the numerous eruptions of undersea volcanoes were trapped in underwater bubbles.

Step 2 Inside the bubbles, the methane and ammonia needed to make amino acids might have been protected from damaging ultraviolet radiation. Chemical reactions would take place much faster in bubbles (where reactants would be concentrated) than in the primordial soup proposed by Oparin and Haldane.

Step 3 Bubbles rose to the surface and burst, releasing simple organic molecules into the air.

Step 4 Carried upward by winds, the simple organic molecules were exposed to ultraviolet radiation and lightning, which provided energy for further reactions.

Step 5 More complex organic molecules that formed by further reactions fell into the ocean with rain, starting another cycle.

Thus, the molecules of life could have appeared more quickly than is accounted for by the primordial soup model alone.

Study TIP

Reading Effectively
Before reading this chapter, write the Objectives for each section on a sheet of paper. Rewrite each Objective as a question, and answer these questions as you read the section.

BIO graphic

Figure 3

Lerman's Bubble Model

Lerman proposed that gases formed simple organic molecules.

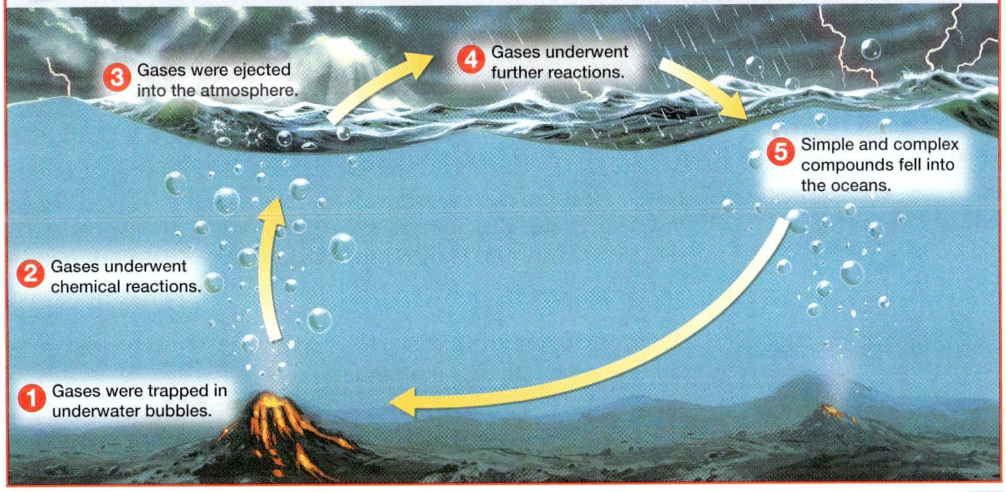

255

ASTRONOMY CONNECTION

In the late 1890's, telescopic studies of Mars spawned theories about vegetative and possibly intelligent life on Mars. These ideas had great popular appeal and, given the poor quality of the data available, possibly even a modicum of scientific credibility. There was no question that the Martian poles were ice-locked and so Mars indeed seemed to possess one of the essential ingredients for life—water. At issue, then as now, was whether this water was ever in liquid form. Even as late as 1962, respected astronomers continued to interpret data gleaned from fifty years of telescopic observation of Mars as supportive of theories of life there. The high-quality images returned by the Mariner flybys showed Mars to be heavily cratered, more akin to the Earth's moon than to the Earth.

READING SKILL BUILDER — GENERAL

Interactive Reading Assign Chapter 12 of the *Holt Biology* Guided Audio CD Program to help students achieve greater success in reading the chapter.

Teaching Tip — GENERAL

Space Science The origin of life on this planet is of interest to scientists in many fields. Investigators in the fields of biology, chemistry, physics, geology, and astrophysics are all working together to learn more about other planets in the quest to learn more about our own. Indeed, whole new fields of study, such as planetary geology, were unheard of a few years ago. Ask students to discuss new fields of study that might arise from exploration of the planets in our solar system. (Answers will vary, but might include planetary mining and refining, and mass transit between planets.)
LS Verbal

Teaching Tip — ADVANCED

Earth's Changing Atmosphere Students may think that the atmosphere is an infinite resource that will always be present as it is today. Challenge this misconception by having students compare the changes in the atmosphere of early Earth with the changes that are occurring in our present atmosphere. Begin by describing how oxygen levels steadily increased in the ancient atmosphere due to the activity of photosynthetic bacteria. Then ask students what kind of gaseous products human activity is producing today. (Carbon dioxide, methane, sulfur dioxide, nitrous oxide, and others) Ask which gas of particular importance is increasing worldwide and which is decreasing worldwide. (Carbon dioxide is increasing, and upper atmospheric ozone is decreasing.)

Teach, *continued*

Teaching Tip — GENERAL

The First Hereditary Molecule
Armed with the knowledge of the function of RNA in protein synthesis, many students may not be able to understand immediately how RNA, which relies on a template provided by DNA, could have developed before DNA. Explain to students that today certain viruses (called retroviruses), including HIV, contain only RNA as their genetic material. Their viral RNA, when released into a host cell, is used as a template to then make DNA.

QUICK LAB

Modeling Coacervates

Skills Acquired
Forming a model, predicting outcomes

Teacher's Notes
Solutions must be fresh and prepared accurately. Always wear safety goggles and lab aprons when mixing chemicals. Tell students to notify you in case of an accident. Supervise the cleanup of all spills.

Analysis Answers:

1. The solution became cloudy.
2. Coacervates are spherical and membrane-bound but have no internal structures.
3. Answers may vary but may include that coacervates might break up.
4. Answers will vary. Students might state that coacervates are discrete structures that resemble cells and that arise in certain chemical environments.

Figure 4 **Proposed stages leading to RNA self-replication and protein synthesis.** Chemical reactions between inorganic molecules formed RNA nucleotides. The nucleotides assembled into RNA macromolecules. These molecules were able to self-replicate and to catalyze the formation of proteins.

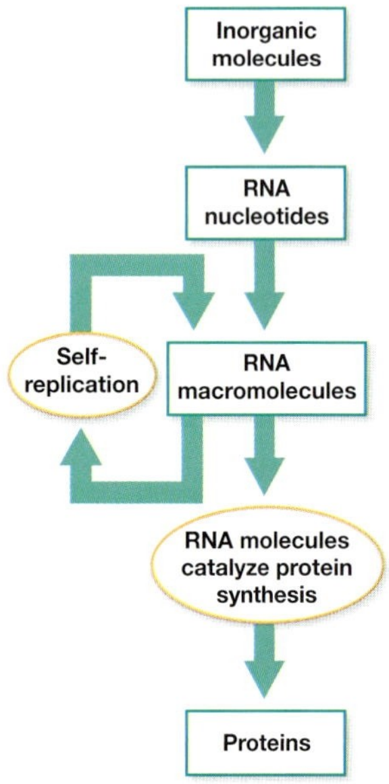

Precursors of the First Cells

Scientists disagree about the details of the process that led to the origin of life. Most scientists, however, accept that under certain conditions, the basic molecules of life could have formed spontaneously through simple chemistry. But there are enormous differences between simple organic molecules and large organic molecules found in living cells. How did amino acids link to form proteins? How did nucleotides form the long chains of DNA that store the instructions for making proteins? In the laboratory, scientists have not been able to make either proteins or DNA form spontaneously in water. However, short chains of RNA, the nucleic acid that helps carry out DNA's instructions, have been made to form on their own in water.

A Possible Role As Catalysts

In the 1980s, American scientists Thomas Cech of the University of Colorado and Sidney Altman of Yale University found that certain RNA molecules can act like enzymes. RNA's three-dimensional structure provides a surface on which chemical reactions can be catalyzed. Messenger RNA acts as an information-storing molecule. As a result of Cech's and Altman's work and other experiments showing that RNA molecules can form spontaneously in water, a simple hypothesis was formed: RNA was the first self-replicating information-storage molecule and it catalyzed the assembly of the first proteins. More important, such a molecule would have been capable of changing from one generation to the next. This hypothesis is illustrated in **Figure 4.**

Microspheres and Coacervates

Observations show that lipids, which make up cell membranes, tend to gather together in water. By shaking up a bottle of oil and vinegar, you can see something similar happen—small spheres of oil form in the vinegar. Certain lipids, when combined with other molecules, can form a tiny droplet whose surface resembles a cell membrane. Similarly, laboratory experiments have shown that, in water, short chains of amino acids can gather into tiny droplets called **microspheres.** Another type of droplet, called a *coacervate (koh AS suhr VAYT),* is composed of molecules of different types, including linked amino acids and sugars.

Scientists think that formation of microspheres might have been the first step toward cellular organization. According to this hypothesis, microspheres formed, persisted for a while, and then dispersed. Over millions of years, those microspheres that could persist longer by incorporating molecules and energy would have become more common than shorter lasting microspheres were. Microspheres could not be considered true cells, however, unless they had the characteristics of living things, including heredity.

did you know?

Ancient Oceans Early oceans were probably similar in salt content to modern freshwater lakes. The salinity of today's oceans is the result of millions of years of water running over rocks, picking up mineral salts, and flowing to the sea.

CHEMISTRY CONNECTION

Carbon atoms can form an almost infinite number of compounds by combining with each other in long chains and in other types of large, complex molecules. They not only form a total of four bonds, they can also join other atoms in single, double, and triple bonds. The structure of silicon, however, is far more rigid. Their differences can be seen by comparing carbon-based plastics to silicon-based ceramics.

Origin of Heredity

Although scientists disagree about the details of the origin of heredity, many agree that double-stranded DNA evolved after RNA and that RNA "enzymes" catalyzed the assembly of the earliest proteins. Many scientists also tentatively accept the hypothesis that some microspheres or similar structures that contained RNA developed a means of transferring their characteristics to offspring. But researchers do not yet understand how DNA, RNA and hereditary mechanisms first developed. Therefore, the subject of how life might have originated naturally and spontaneously remains a subject of intense interest, research, and discussion.

QUICK LAB

Modeling Coacervates SC.H.3.4.1

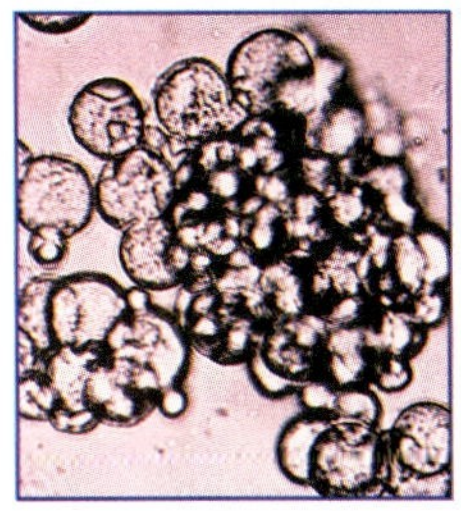

By using simple chemistry, you will see that some properties of coacervates resemble the properties of cells.

Materials

safety goggles, protective gloves and lab apron, graduated cylinder, 1 percent gelatin solution, 1 percent gum arabic solution, test tube, 0.1 M HCl, pipet, microscope slide and coverslip, microscope

Procedure

1. **CAUTION: Hydrochloric acid is corrosive. Put on safety goggles, gloves, and apron. Avoid contact with skin and eyes. Avoid breathing vapors. If any of this solution should spill on you, immediately flush the area with water, and notify your teacher.**
2. Mix 5 mL of a 1 percent gelatin solution with 3 mL of a 1 percent gum arabic solution in a test tube.
3. Add 0.1 M HCl to the gelatin–gum arabic solution one drop at a time until the solution turns cloudy.
4. Prepare a wet mount of the cloudy solution, and examine it under a microscope at high power.
5. Prepare a drawing of the structures that you see. They should resemble the structures in the micrograph above.

Analysis

1. **Describe** what happened to the solutions after the acid was added.
2. **Compare** the appearance of coacervates with that of cells.
3. **Predict** what would happen to the coacervates if a base were added to the solution.
4. **Critical Thinking** **Evaluating Hypotheses** Based on the evidence you obtained, defend the hypothesis that coacervates could have been the basis of life on Earth.

Section 1 Review

1. **Explain** how radioisotopes are used to determine the age of a rock. SC.H.1.4.1
2. **Critique** two scientific models that explain the origin of life. SC.H.2.4.2
3. **Describe** the first step that may have led toward cellular organization. SC.D.1.4.4
4. **Explain** how heredity may have arisen. SC.D.1.4.4
5. FCAT Prep Miller and Urey's model is inconsistent with the finding that Earth's early atmosphere lacked what substance?
   A hydrogen
   B nitrogen
   C ozone
   D water
   SC.H.1.4.5

257

Answers to Section Review

1. The amount of the radioisotope in the rock is compared to the amounts of isotope species that resulted from decay. An older rock would have less of the original radioisotope.
2. In the "primordial soup" model, the early atmosphere contained large amounts of nitrogen and hydrogen-containing gases. Recent discoveries indicate that without ozone, any ammonia and methane would have been destroyed. In the bubble model, ammonia, methane, and other gases from underwater volcanoes were trapped in bubbles in the seas. Chemical reactions took place faster in the bubbles, and the underwater bubbles were protected from UV radiation. This model relies on the assumption that bubbles remained intact for a significant time.
3. The formation of microspheres might have led to cellular organization.
4. Microspheres that contained replicating RNA might have transferred both their structure and their RNA to offspring.
5. **A.** Incorrect. Hydrogen is believed to have been present. **B.** Incorrect. Nitrogen is believed to have been present. **C.** Correct. The Earth's early atmosphere did not have a protective layer of ozone. Without ozone, the ammonia and methane in Miller and Urey's model would have been destroyed by ultraviolet radiation. **D.** Incorrect. Water is believed to have been present.

Close

Reteaching — BASIC

Ask students to draw a rough diagram of the Miller-Urey apparatus. Be sure to have them label all components of the diagram. (Students should include in their sketch an area for holding water, an area where water and gases can combine and be subjected to electrical sparks, an area where any by-products can cool, and an area where the by-products can be collected.)

Quiz — GENERAL

1. What is a "half-life"? (It is the amount of time that it takes for half of an amount of material to break down.)
2. What does "spontaneous" mean, in reference to the origin of life? (It means that life originated from nonliving things.)
3. Every theory about the origin of life has a source of energy for the transformation. What are some of these sources? (Possible sources include ultraviolet or other light energy from the sun, an electrical spark as from lightning, and volcanic eruption.)
4. What molecule is thought to be the first hereditary molecule? (RNA)

Alternative Assessment — GENERAL

Ask students to write an answer to this question: Were the first cells that evolved on Earth prokaryotes or eukaryotes? Have them justify their choice. (The first cells were prokaryotes. At this point in the chapter, students should conclude that, since prokaryotes are single-celled organisms lacking a nucleus, they probably evolved earlier than more complex eukaryotes.)

Section Review Correlations

SC.D.1.4.4: Items 3, 4
SC.H.1.4.1: Item 1
SC.H.1.4.5: Item 5
SC.H.2.4.2: Item 2

Focus

Overview

Before beginning this section review with your students the objectives listed in the Student Edition. Tell students that the purpose of this lesson is to teach them how the first eukaryotic cells and their organelles might have evolved and how multicellular organisms likely evolved. How explosions of new species of organisms occurred in several different periods of Earth's history and how mass extinctions have also played an important role in the diversity of life on Earth are also discussed.

Bellringer

Ask students to write down the names of all the kingdoms of living things, writing them in the order in which they appeared on Earth.

Motivate

Demonstration — GENERAL

Some students may have difficulty understanding the timeline of Earth's history. This analogy should provide some perspective. Draw a large circle on the board to represent a clock. Place an arrow at 12:01 to indicate when Earth was formed. Place additional arrows at 1:15 (the first bacteria appeared), 8:00 (the first eukaryotes), 11:00 (the first life on land), 11:50 (dinosaurs became extinct), and 11:59 (first humans appeared). Students will be able to relate more life-on-Earth "milestones" to a 24-hour day when they complete the Exploration Lab. LS Visual

Section Correlations

SC.D.1.4.3
SC.F.2.4.3
SC.G.1.4.1

Section 2 The Evolution of Cellular Life

Florida Sunshine Standards/FCAT
SC.F.2.4.3 Mechanisms of change (e.g., mutation and natural selection) lead to adaptations in a species and their ability to survive naturally in changing conditions and to increase species diversity.
SC.G.1.4.1 There is great diversity and interdependence among living things.

Objectives

- **Distinguish** between the two groups of prokaryotes. SC.G.1.4.1
- **Describe** the evolution of eukaryotes. SC.F.2.4.3
- **Recognize** an evolutionary advance first seen in protists. SC.D.1.4.3
- **Summarize** how mass extinctions have affected the evolution of life on Earth. SC.D.1.4.3

Key Terms

fossil
cyanobacteria
eubacteria
archaebacteria
endosymbiosis
protist
extinction
mass extinction

The Evolution of Prokaryotes

When did the first organisms form? To find out, scientists study the best evidence of early life that we have, fossils. A **fossil** is the preserved or mineralized remains (bone, tooth, or shell) or imprint of an organism that lived long ago. The oldest known fossils, which are microscopic fossils of prokaryotes, come from rock that is 2.5 billion years old.

Recall that prokaryotes are single-celled organisms that lack internal membrane-bound organelles. Among the first prokaryotes to appear were marine cyanobacteria. **Cyanobacteria** *(SIE an oh bak TIR ee ah)* are photosynthetic prokaryotes. Before cyanobacteria appeared, oxygen gas was scarce on Earth. But as ancient cyanobacteria carried out photosynthesis, they released oxygen gas into Earth's oceans. After hundreds of millions of years, the oxygen produced by cyanobacteria began to escape into the air, as shown in **Figure 5.** Over time, more oxygen was added to the air. Today oxygen gas makes up 21 percent of the Earth's atmosphere.

Two Groups of Prokaryotes

Early in the history of life, two different groups of prokaryotes evolved—eubacteria (which are commonly called *bacteria*) and archaebacteria. Living examples include *Escherichia coli,* a species of eubacteria, and *Sulfolobus,* a group of archaebacteria. **Eubacteria** are prokaryotes that contain a chemical called peptidoglycan *(PEP tih doh GLIE kan)* in their cell walls. Eubacteria include many bacteria that cause disease and decay.

Archaebacteria are prokaryotes that lack peptidoglycan in their cell walls and have unique lipids in their cell membranes. Modern archaebacteria are thought to closely resemble early archaebacteria. Chemical evidence indicates that archaebacteria and eubacteria diverged very early.

Figure 5 Evolutionary timeline. This timeline shows some of the major events that occurred during the evolution of life on Earth.

Age (in Millions of Years Ago)

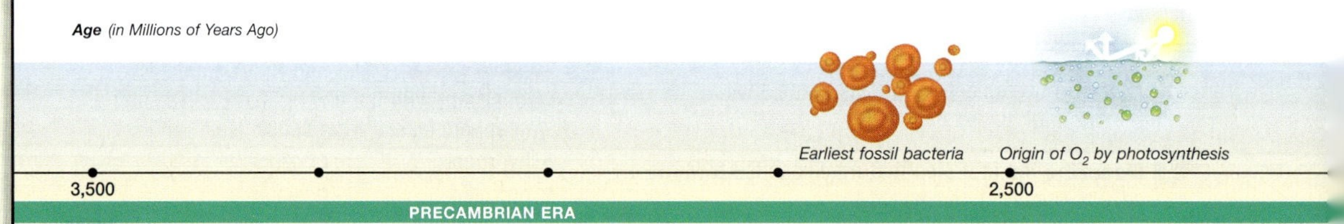

258

Chapter Resource File

- Directed Reading BASIC
- Active Reading GENERAL
- Data Sheet for Data Lab GENERAL

One-Stop Planner CD-ROM

- Reading Organizers BASIC
- Reading Strategies BASIC

Transparencies

TR Bellringer
TR D8 Evolution of Eukaryotes

The Evolution of Eukaryotes

About 1.5 billion years ago, the first eukaryotes appeared. A eukaryotic cell is much larger than a prokaryote is. Eukaryotic cells have a complex system of internal membranes. Eukaryotic DNA is enclosed within a nucleus. Almost all eukaryotes have mitochondria. Chloroplasts, which carry out photosynthesis, are found only in protists and plants. Mitochondria and chloroplasts are the size of prokaryotes, and they contain their own DNA.

The Origins of Mitochondria and Chloroplasts

Most biologists think that mitochondria and chloroplasts originated as described by the theory of **endosymbiosis** that was proposed in 1966 by the American biologist Lynn Margulis. This theory proposes that mitochondria are the descendants of symbiotic, aerobic (oxygen-requiring) eubacteria and chloroplasts are the descendants of symbiotic, photosynthetic eubacteria.

DATA LAB: *Analyzing Signs of Endosymbiosis*

SC.H.1.4.1

Magnification: 6930x

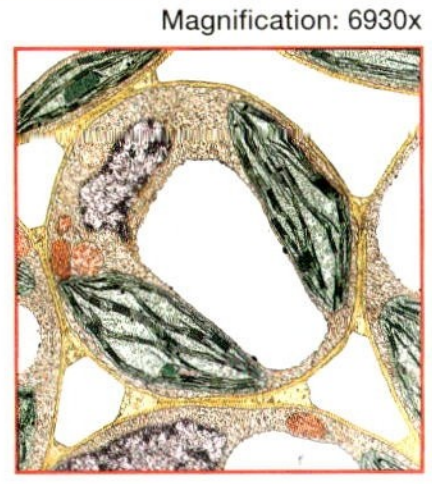

Background

You may recall that mitochondria have their own DNA and produce their own proteins. The data below were collected by scientists studying the proteins produced by mitochondrial DNA. The scientists found that the three-nucleotide sequences (codons) in the nucleus of an organism's cells can code for different amino acids than those coded for in the cell's mitochondria. Examine the data below, and answer the questions that follow.

Analysis

1. **Defend** the theory of endosymbiosis using these data.
2. **Infer** what these data indicate about the evolution of plant cells.
3. **Describe** how these data can be used to support the idea that more than one type of cell evolved early in the history of life.

Amino Acids Made in the Nucleus and Mitochondria

	Amino acids or other instructions coded for in the nucleus	Amino acids or other instructions coded for in mitochondria	
Codon	Plants and mammals	Plants	Mammals
UGA	Stop	Stop	Tryptophan
AGA	Arginine	Arginine	Stop
AUA	Isoleucine	Isoleucine	Methionine
AUU	Isoleucine	Isoleucine	Methionine
CUA	Leucine	Leucine	Leucine

First eukaryotes

1,500

Graphic Organizer

Use this graphic organizer with **Teaching Tip** *on this page.*

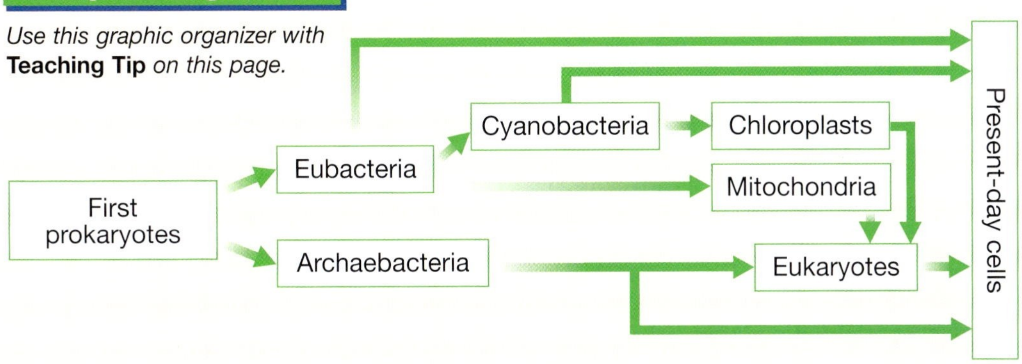

Teach

Teaching Tip

Remembering What Came from What Use a Graphic Organizer similar to the one at the bottom of this page to help students grasp the progression of the complexity of life. Reproduce the diagram, without the eight terms, on the board or overhead. Next, write an alphabetical list of the eight terms beside the graphic organizer. Ask students to insert the terms into their correct rectangles.

DATA LAB: *Analyzing Signs of Endosymbiosis*

Skills Acquired

Analyzing data, relating conclusions

Teacher's Notes

Point out to students that they are looking at a "doubled double" set of data. That is, they are considering the nucleus with respect to mitochondria and plants with respect to mammals.

Answers to Analysis

1. The fact that in mammals the genetic code of nuclear DNA differs from that of mitochondrial DNA suggests that mitochondria have evolved separately from the animal cells that contain them.
2. We can infer that plant mitochondria and plant cells together have been subjected to selective pressures different from those that acted on mammalian cells and on mammalian mitochondria.
3. We can infer that the prokaryotic ancestors of mammalian mitochondria probably diverged very early from the prokaryotic ancestors of plant mitochondria because the genetic codes are quite different in the two types.

Teach, *continued*

Teaching Tip — GENERAL

Endosymbiosis The term endosymbiosis refers to any symbiotic relationship in which one organism lives inside another. Help students understand Margulis's hypothesis for the endosymbiotic origin of mitochondria and chloroplasts by reviewing other examples of endosymbiosis. Examples include wood-digesting bacteria that live within the digestive tracts of termites, bacteria that break down plant cell walls within the digestive tract of cows, and bacteria in the human digestive tract that break down otherwise undigested materials and synthesize vitamin K as a by-product. English Language Learners
LS Verbal

Group Activity — GENERAL

Models of Endosymbionts and Bacteria Have students work in small groups to create models or posters of chloroplasts, mitochondria, non-photosynthetic bacteria cells, and photosynthetic bacterial cells. Students should use reference materials for their models or posters to ensure that they accurately depict these organelles and cells. Tell students that their models or poster pictures should have clearly labeled internal structures and the colors should be realistic (i.e., green for chloroplasts or photosynthetic bacteria). They should also identify the names of the bacteria used for these models or pictures. Display the models and posters where all students can observe them. Co-op Learning

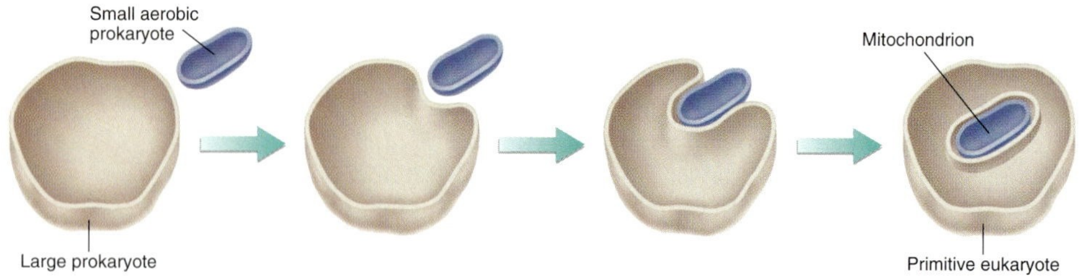

Figure 6 Endosymbiosis

Mitochondria are thought to have evolved from small, aerobic prokaryotes that began to live inside larger prokaryotes.

internet connect
www.scilinks.org
Topic: Endosymbiosis
Keyword: HX4068
SCILINKS. Maintained by the National Science Teachers Association

According to Lynn Margulis's theory of endosymbiosis, bacteria entered large cells either as parasites or as undigested prey as illustrated in **Figure 6.** Instead of being digested, the bacteria began to live inside the host cell, where they performed either cellular respiration (mitochondria) or photosynthesis (chloroplasts). The invading bacteria that became chloroplasts were probably closely related to cyanobacteria. Both mitochondria and chloroplasts have characteristics that are similar to those of bacteria. The following observations support the idea that mitochondria and chloroplasts descended from bacteria:

1. **Size and structure.** Mitochondria are about the same size as most eubacteria, and chloroplasts are the same size as some cyanobacteria. Both mitochondria and chloroplasts are surrounded by two membranes. The smooth outer membrane of mitochondria is thought to be derived from the endoplasmic reticulum of the larger host cell. The inner membrane of mitochondria is folded into many layers, so it looks like the cell membranes of aerobic eubacteria. Inside this membrane are proteins that carry out cellular respiration. Both chloroplasts and cyanobacteria contain thylakoids, structures in which photosynthesis takes place.
2. **Genetic material.** Mitochondria and chloroplasts have circular DNA similar to the chromosomes found in bacteria. Both chloroplasts and mitochondria contain genes that are different from those found in the nucleus of the host cell.
3. **Ribosomes.** Mitochondrial and chloroplast ribosomes have a size and structure similar to the size and structure of bacterial ribosomes.
4. **Reproduction.** Like bacteria, chloroplasts and mitochondria reproduce by simple fission. This replication takes place independently of the cell cycle of the host cell.

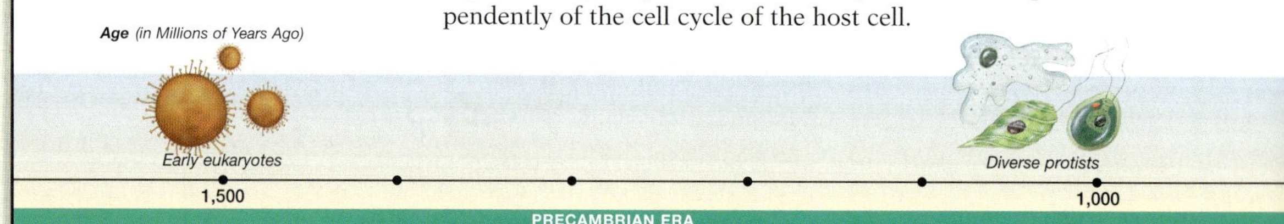

260

MEDICINE CONNECTION

Inside nearly every cell are mitochondria, the tiny structures that provide the cell with energy. Each mitochondrion contains a circular chromosome. Several rare diseases are caused by abnormal genes carried by the chromosome inside a mitochondrion. Mitochondrial genes code primarily for proteins in the electron transport chain and for ATP synthase. Defects in one or more of these proteins thus reduce the amount of ATP the cell can make. One disease, mitochondrial myopathy, causes weakness, exercise intolerance, and muscle deterioration. In addition, at least some cases of diabetes and heart disease, as well as other disorders such as Alzheimer's disease, have been linked to mitochondrial genes.

When an egg is fertilized, only mitochondria from the egg become part of the developing fetus; all mitochondria from the sperm are discarded. Therefore, diseases caused by abnormal mitochondrial genes are transmitted by the mother. A father with abnormal mitochondrial genes can't transmit any such diseases to his children.

Multicellularity

Many biologists group all living things into six broad categories called kingdoms. The two oldest kingdoms, Eubacteria and Archaebacteria, are made up of single-celled prokaryotes. The first eukaryotic kingdom was the kingdom Protista. **Protists** make up a large, varied group that includes both multicellular and unicellular organisms. The other three kingdoms (fungi, plants, and animals) evolved later and also consist of eukaryotes.

The unicellular body plan has been tremendously successful, with unicellular organisms today constituting about half the biomass (the total weight of all living things) on Earth. But a single cell must carry out all of the activities of the organism. Distinct types of cells in one body can have specialized functions. For example, some organisms may have specific cells that help the organism protect itself from predators or disease. Other cells may help the organism resist drying out. Other examples of specialized cells include cells that help a multicellular organism move about in order to find a mate or food. With all these advantages, it is not surprising that multicellularity has arisen independently many times.

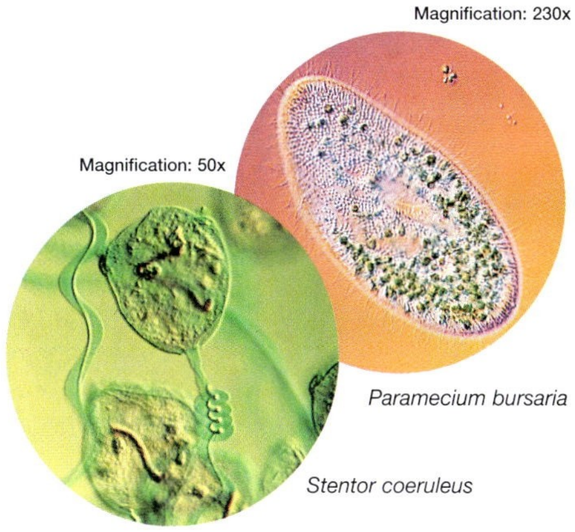

Figure 7 Single-celled protists. Single-celled protists occur in many shapes and can live in many different types of environments, including water and land.

Almost every organism large enough to see with the naked eye is multicellular. Most protists, such as those shown in **Figure 7,** are single celled, but there are many multicellular forms. The development of multicellular organisms of the kingdom Protista marked an important step in the evolution of life on Earth. The oldest known fossils of multicellular organisms were found in 700 million year-old rocks.

Some of the multicellular lines that resulted did not produce diverse groups of organisms. Among those groups of organisms that survive today are plantlike red, green, and brown algae, shown in **Figure 8.** You may know these algae as seaweed. Three of the multicellular groups that evolved from the protists were very successful, producing three separate kingdoms—Fungi, Plantae, and Animalia. Each of these three kingdoms evolved from a protistan ancestor.

Figure 8 Brown algae. Brown algae, called kelps, are multicellular protists that form vast underwater "forests" in some coastal waters.

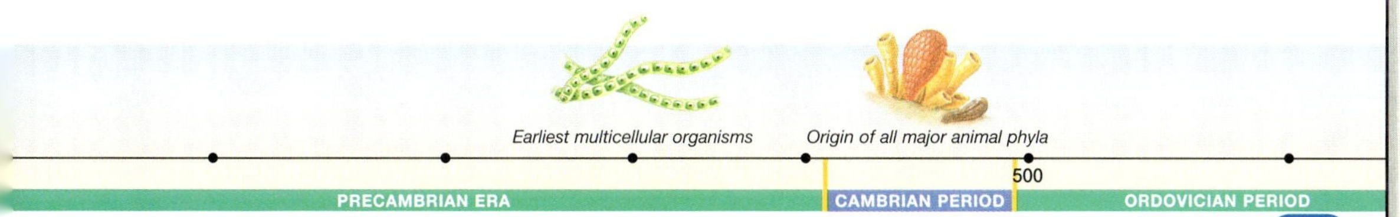

261

OCEANOGRAPHY CONNECTION

Sponges have a variety of sexual and asexual reproductive modes. They are best known for their quite remarkable regenerative powers. This is illustrated by the classic experiment of forcing a piece of sponge through a sieve, which separates the cells of the organism. Within a short period of time the dissociated cells aggregate and reform a multicellular organism. This cannot be achieved by all sponge species.

The ability to regenerate is related to asexual reproduction. A bud or small fragment broken from the parent sponge can generate a new sponge. Some sponges produce specialized asexual reproductive bodies called gemmules. These consist of food-filled archeocytes and amoebocytes. The amoebocytes are capable of giving rise to any other type of cell. Gemmules remain viable for extended periods of time. When suitable conditions are found, a gemmule can grow to form a new sponge that is genetically identical to the parent. Gemmules provide a means of dispersal and are a way of maintaining local distribution and abundance.

Teaching Tip — GENERAL

Multicellularity and Cellular Specialization It is important that students recognize the relationship between multicellularity and the division of labor among different types of cells. Illustrate these concepts with the following analogy. Modern industry uses individual workers' skills and abilities to perform different tasks. Use as an example a local industry (such as an automotive plant, steel mill, canning company, or meat packing plant) to show how more is produced, both in quantity and complexity, by breaking the total task into smaller parts that are accomplished by workers trained to do that specific task. Point out that cells exhibit the same type of departmentalizing.

Activity — ADVANCED

Favorite Fossils Have students pick an extinct species that interests them and write a short report about it. Ask students to address the following questions: When, where, and how did the species live? What evidence do we have that the species lived at all? Why might it have become extinct? Do you see a relationship between this species and a modern species? Have students include with their reports an artist's rendition of their species or a picture of a fossil of the species.
LS Verbal

Demonstration — GENERAL

Bring to class a variety of fossils. Petrified wood, diatomaceous earth, trilobites, casts and molds of small fish, mollusks, and coprolites (petrified feces) usually are inexpensive and easy to obtain. Fossils are also available through biological supply houses or other retail dealers. Assign identification or classification activities using the specimens.
LS Visual

Teach, *continued*

Teaching Tip — GENERAL

Cambrian Explosion Point out to students that the diversity of multicellular organisms expanded rapidly during the period known as the Cambrian explosion. Changing geological and atmospheric conditions during this time period are thought to have increased the number of available habitats for animals. Primitive multicellular organisms could then specialize differently, become more complex, and take advantage of the newly available resources. This first explosion of biological diversity was the foundation from which most complex life existing today evolved.

Activity — ADVANCED

Burgess Shale Have students use library or Internet resources to research the Burgess Shale. Fossil evidence from the shale suggests that animals with a wide variety of body plans existed very early in evolutionary history. Have students search for organisms that would be considered unusual (even bizarre) today, and make simple sketches of them.

Origins of Modern Organisms

Most animal phyla that exist today probably originated during a relatively short time (most estimates range from 10 to 100 million years) during the late Precambrian and early Cambrian periods. This rapid diversification of animals is sometimes known as the "Cambrian explosion." The Cambrian period was a time of great evolutionary expansion, as shown in **Figure 9.** Many unusual marine animals also appeared at this time, animals for which there are no close living relatives. A very rich collection of Cambrian fossils was uncovered in 1909 in a geological formation in Canada called the Burgess Shale. The fossils in the Burgess Shale include those of strange animals that are not like anything alive today.

The Ordovician period, which followed the Cambrian period, lasted from about 505 million to 438 million years ago. During this time, many different animals continued to abound in the seas. Among them were trilobites, marine arthropods that became extinct about 250 million years ago, shown in Figure 9.

Figure 9 Appearance of a Cambrian sea. By studying fossils from the Cambrian period, such as the trilobite fossil below, artists re-create a scene from the shallow seas of the Cambrian period.

262

Trends in Astrophysics

Cause of Mass Extinction Earth's most severe mass extinction, which happened 250 million years ago, may have been triggered by a collision with a comet or asteroid. Over 90 percent of all marine species and 70 percent of land vertebrates perished as a result. Some scientists hypothesize that the collision wasn't directly responsible for the extinction, but rather triggered a series of events, such as massive volcanism and changes in ocean oxygen, sea level, and climate. That in turn led to wholesale loss of species.

Scientists don't know the site of the impact 250 million years ago, when all Earth's land formed a supercontinent called Pangaea. There is strong evidence suggesting the extinction happened very rapidly, on the order of 8,000 to 100,000 years.

Mass Extinctions

The fossil record indicates that a sudden change occurred at the end of the Ordovician period. About 440 million years ago, a large percentage of the organisms on Earth suddenly became extinct. **Extinction** is the death of all members of a species. This was the first of five major mass extinctions that have occurred on Earth. A **mass extinction** is an episode during which large numbers of species become extinct.

Another mass extinction of about the same size happened about 360 million years ago. The third and most devastating of all mass extinctions occurred at the end of the Permian period, about 245 million years ago. About 96 percent of all species of animals living at the time became extinct. About 35 million years later, a fourth, less devastating mass extinction occurred. Although the specific causes of these extinctions are unknown, evidence indicates that worldwide geological and weather changes were likely factors. The fifth mass extinction will be discussed in more detail in a later chapter. It occurred 65 million years ago and brought about the extinction of about two-thirds of all land species, including most of the dinosaurs.

Some scientists think that another mass extinction is occurring today. These scientists reason that this new extinction is taking place because the Earth's ecosystems, especially tropical rain forests, are being destroyed by human activity, as shown in **Figure 10.** The world has already lost half its tropical rain forests. If the current rate of destruction continues, from 22 percent to 47 percent of Earth's plant species will be lost, along with 2,000 of the world's 9,000 species of birds and countless insect species. This would be an astonishing loss of biodiversity.

internet connect
www.scilinks.org
Topic: Extinction
Keyword: HX4078
SCILINKS. Maintained by the National Science Teachers Association

Figure 10 Rain forests are being destroyed at an alarming rate. Although tropical rain forests cover only 7 percent of the Earth's land surface, they contain more than one-half of all the world's animal and plant species.

Section 2 Review

1. **Contrast** the two major groups of prokaryotes. SC.G.1.4.1
2. **Analyze** Margulis's theory of endosymbiosis, citing its strengths and weaknesses. SC.H.1.4.3
3. **Compare** bacteria with eukaryotes. SC.G.1.4.1
4. **Summarize** how multicellularity advanced the evolution of protists. SC.D.1.4.4
5. **Critical Thinking Justify** the argument that today's organisms would not exist if mass extinctions had not occurred. SC.D.1.4.3
6. **FCAT Prep** What is the kingdom that includes both multicellular and unicellular eukaryotes called?
   - **A** Plantae
   - **B** Protista
   - **C** Eubacteria
   - **D** Archaebacteria

   SC.G.1.4.1

Animal diversity abounds; first jawless fishes

First mass extinction

500 — 440

ORDOVICIAN PERIOD | SILURIAN

Close

Reteaching — Basic

Have students write a one-page summary of this section. Ask them to be sure to include a timeline indicating all of the evolutionary advancements they describe.

Quiz — General

1. When did the first life form appear on Earth? (Over 2.5 billion years ago)
2. What cellular organelles are thought to have derived from bacterial cells? (Mitochondria and chloroplasts were probably bacteria that became endosymbionts with pre-eukaryotic organisms.)
3. During what geologic time period did multicellular life forms on Earth significantly increase in diversity? (The Cambrian period)

Alternative Assessment — General

Divide the class into two teams. Have each team write ten true or false questions about the material in this section. A spokesperson for one team should pose the questions to the other team. Give two points for each correct answer. Give students one point for each of their own questions that was answered correctly. Tally the scores, and declare a winning team.

Answers to Section Review

1. Eubacteria contain peptidoglycan in their cell walls. Archaebacteria lack peptidoglycan and their cell membrane lipids differ from eukaryotes.
2. Eukaryotes probably evolved through endosymbiosis. Refer to the text to check students' answers.
3. Unlike eukaryotes, bacteria have no membrane-bounded cellular inclusions; bacterial DNA is in a single loop, unlike the chromosomal DNA of eukaryotes; and bacteria have cell walls, unlike eukaryotic protists.
4. The development of multicellularity in protists allowed for cell specialization.
5. The ancestors of many living species filled niches vacated by organisms that died in mass extinctions.
6. **A.** Incorrect. The organisms in kingdom Plantae are all multicellular eukaryotes. **B.** Correct. Organisms in the kingdom Protista may be unicellular or multicellular eukaryotes. **C.** Incorrect. Eubacteria are single-celled prokaryotes. **D.** Incorrect. Archaebacteria are single-celled prokaryotes.

Section Review Correlations

SC.D.1.4.3: Item 5
SC.D.1.4.4: Item 4
SC.G.1.4.1: Items 1, 3, 6
SC.H.1.4.3: Item 2

Section 3 Life Invaded the Land

Florida Sunshine Standards/FCAT

SC.D.1.4.3 Changes in Earth's climate, geological activity, and life forms may be traced and compared.
SC.G.2.4.1 Layers of energy-rich organic materials have been gradually turned into great coal beds and oil pools (fossil fuels) by the pressure of the overlying earth, and humans burn fossil fuels to release the stored energy as heat and carbon dioxide.

Objectives

- **Relate** the development of ozone to the adaptation of life to the land. SC.D.1.4.3
- **Identify** the first multicellular organisms to live on land. SC.D.1.4.3
- **Name** the first animals to live on land. SC.D.1.4.3
- **Explain** the relationship between plants and fossil fuels. SC.G.2.4.1
- **List** the first vertebrates to leave the oceans. SC.D.1.4.3

Key Terms

mycorrhizae
mutualism
arthropod
vertebrate
continental drift

The Ozone Layer

The sun provides both life-giving light and dangerous ultraviolet radiation. Early in Earth's history, life formed in the seas, where early organisms were protected from ultraviolet radiation. These organisms could not leave the water because ultraviolet radiation made life on dry ground unsafe. What enabled life-forms to leave the protection of the seas and live on the land?

Formation of the Ozone Layer

During the Cambrian period and for millions of years afterward, organisms did not live on the dry, rocky surface of Earth. However, a slow change was taking place. About 2.5 billion years ago, photosynthesis by cyanobacteria began adding oxygen to Earth's atmosphere. As oxygen began to reach the upper atmosphere, the sun's rays caused some of the molecules of oxygen, O_2, to chemically react and form molecules of ozone, O_3. In the upper atmosphere, ozone blocks the ultraviolet radiation of the sun, as shown in **Figure 11.** After millions of years, enough ozone had accumulated to make the Earth's land a safe place to live.

Figure 11 **Ozone shields the Earth**
As ancient cyanobacteria added oxygen to the atmosphere, ozone began to form.

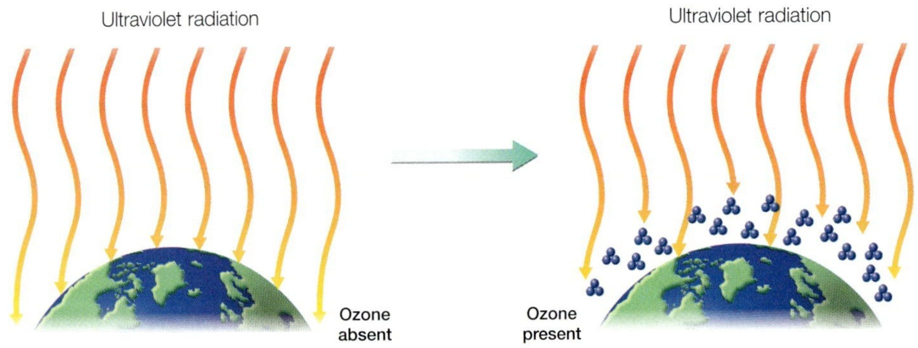

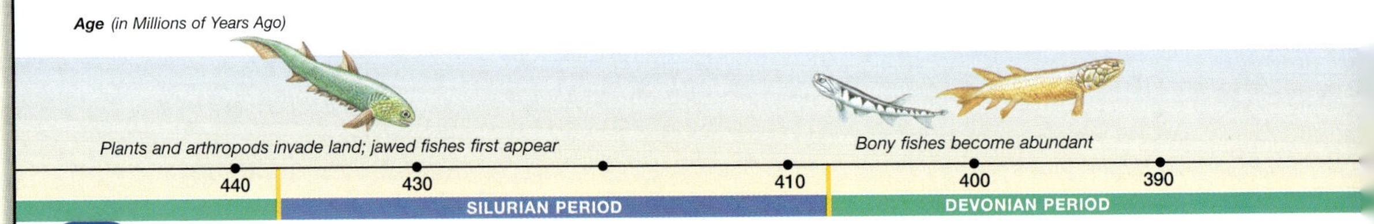

264

Section 3

Focus

Overview

Before beginning this section review with your students the objectives listed in the Student Edition. Tell students that the purpose of this section is to teach them about how the conditions of Earth's atmosphere make it possible for organisms to live on land. They will learn about the challenges of living on land and about the succession of organisms that moved from the oceans onto land.

Bellringer

Have students write down some of the differences they recognize between ocean-dwelling and land-dwelling organisms. (You might suggest that they consider how organisms support themselves and move around, how they obtain food, how they protect themselves from dangers, and how they reproduce.) Ask them to consider all kinds of organisms, not just animals.

Motivate

Discussion/Question — GENERAL

Ask students if ozone is a desirable part of our atmosphere. (Most will answer yes.) Then ask if ozone is always a desirable part of our atmosphere. (At lower elevations, ozone contributes to pollution; at higher elevations, however, ozone absorbs much of the damaging ultraviolet radiation that would otherwise reach Earth where it could harm living things.) LS **Verbal**

Section Correlations

SC.D.1.4.3
SC.G.2.4.1

Chapter Resource File

- Directed Reading BASIC
- Active Reading GENERAL

Transparencies

TR Bellringer
TR D7 Ozone Shields the Earth

One-Stop Planner CD-ROM

- Reading Organizers BASIC
- Reading Strategies BASIC
- Supplemental Reading Guide *The Dinosaur Heresies* ADVANCED

Plants and Fungi on Land

The first multicellular organisms to live on land may have been fungi living together with plants or algae. Such paired organisms were able to live on land because each group possessed a quality needed by the other.

Plants, which likely evolved from photosynthetic protists, could carry out photosynthesis. In photosynthesis, plants use the energy from sunlight to make carbohydrates. Plants cannot, however, harvest needed minerals from bare rock. In contrast, fungi cannot make nutrients from sunlight but can absorb minerals—even from bare rock.

Early plants and fungi formed biological partnerships called mycorrhizae *(MIE koh RIE zee)*, which enabled them to live on the harsh habitat of bare rock. **Mycorrhizae,** which exist today, are symbiotic associations between fungi and the roots of plants, as shown in **Figure 12.** The fungus provides minerals to the plant, and the plant provides nutrients to the fungus. This kind of partnership is called mutualism. **Mutualism** is a relationship between two species in which both species benefit. Plants and fungi began living together on the surface of the land about 430 million years ago.

internet connect

www.scilinks.org
Topic: Discovering Florida's Past
Keyword: HX6F02

SCILINKS. Maintained by the National Science Teachers Association

Fossilized *Cooksonia*

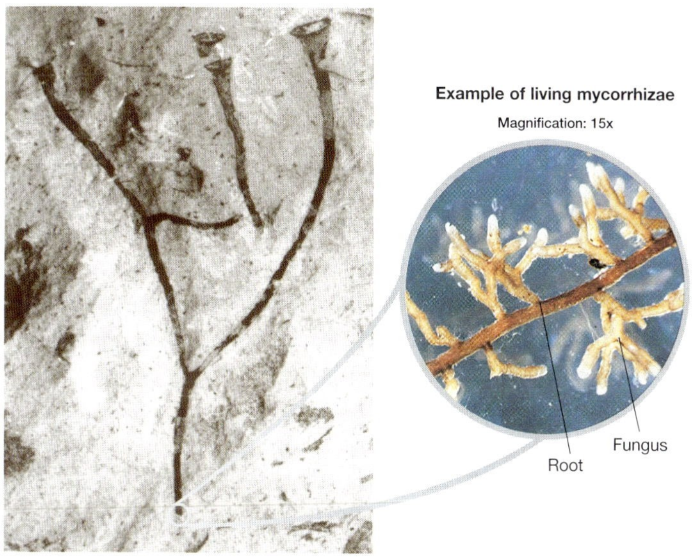

Figure 12 Mycorrhizae formed on the roots of the first plants. This fossil is of *Cooksonia*, the first known vascular plant, which lived 410 million year ago. *Cooksonia* was only a few centimeters tall. *Cooksonia*'s roots formed mycorrhizae similar to the living mycorrhizae shown in color to the left.

265

Teach

Teaching Tip — GENERAL

Hard As a Rock Ask students: what kinds of organisms could survive on bare rock? What would it take to be successful? (Students may suggest an association similar to lichen, in which a fungus and a cyanobacterium or alga coexist in a symbiotic relationship.) Remind students that as life emerged from the oceans, living things were confronted with this kind of problem. **LS Verbal**

SKILL BUILDER — ADVANCED

Math Skills Tell students that an experiment was conducted in which pine seedlings were grown from seed in soil with or without mycorrhizae. After several months of growth, the plants were harvested and dried, and their weights and contents of certain nutrients were measured. The following averaged data were obtained:

Plant Composition

	Without Mycorrhizae	With Mycorrhizae
Dry weight (g)	321	405
Nitrogen (g)	0.85	1.24
Phosphorus (g)	0.07	0.20
Potassium (g)	0.43	0.74

Have students calculate the percent increase in dry weight, nitrogen content, phosphorus content, and potassium content for the plants grown in soil with mycorrhizae, compared with plants grown in soil without mycorrhizae. (Dry weight: 26.2% higher with mycorrhizae; nitrogen content: 45.9% higher; phosphorus content: 185.7% higher; potassium: 72.1% higher) Have students prepare an appropriate graph of these results and write a paragraph of conclusions about them. **LS Logical**

Trends in Paleobotany

Closest Living Relative of First Land Plants

Some 470 million years ago, the first land plants emerged from prehistoric waters, put down roots in soil, and ended up ruling the plant world. But scientists haven't been certain about the family history of those pioneer plants.

By studying gene sequences of common freshwater algae, a team of University of Maryland researchers, funded by the National Science Foundation (NSF), has traced this family tree and identified the closest living relatives of the first land plants. Charales, a group of green algae which survives today in freshwater around the world, was confirmed to be the descendant. Data show that land plants and the Charales both evolved from a common ancestor that was a fairly complex organism. Among its characteristics were branching threads and reproduction with eggs and sperm.

Teach, *continued*

Activity — ADVANCED

Coevolution of Insects and Plants Have students conduct library or Internet research on an example of an insect species and a plant species that have coevolved. Some examples that students could investigate are yucca plants and their moth pollinators, and various orchid species and their bee or wasp pollinators. Have students write a brief report on their findings that includes a description of the structural and functional adaptations of both members of the partnership, the "costs" and "benefits" to both species, and characteristics of the environment in which the two species are found that might influence the partnership.

INCLUSION *Strategies*

- *Learning Disability*
- *Developmental Delay*

Have students create their own resource documents of animal classification. The *Arthropod Guide* and the *Vertebrate Guide* would each contain lists, pictures, descriptions, and approximate time of each animal's appearance on Earth. Students can design a cover for each guide and a page for each type of animal they research. The completed guides may be used as tools for students to exchange with each other and study.

Arthropods

By 100 million years after their first union with fungi, plants had covered the surface of the Earth, forming large forests. These land plants provided a food source for land-dwelling animals. The first animals to successfully invade land from the sea were arthropods. An **arthropod** is a kind of animal with a hard outer skeleton, a segmented body, and paired, jointed limbs. Examples of arthropods include lobsters, crabs, insects, and spiders, like the one in **Figure 13.** Biologists think a type of scorpion was the first arthropod to live on land.

A unique kind of terrestrial arthropod—the insect—evolved from the first land dwellers. Insects have since become the most plentiful and diverse group of animals in Earth's history. The success of the insects is probably connected to their ability to fly. Insects were the first animals to have wings. Flying allowed insects, like the dragonfly shown in **Figure 14,** to efficiently search for food, mates, and nesting sites. It also led to partnerships between insects and flowering plants. In the great coal swamps of the Carboniferous period, organic materials such as the plants also shown in Figure 14 were subjected to pressure from overlying earth. Over millions of years this produced fossil fuels—beds of coal and reservoirs of oil. Humans now burn both oil and coal to release stored energy, in the process, also releasing carbon dioxide.

Figure 13 An arthropod. This marbled spider is a member of the phylum Arthropoda, which includes about 1 billion billion (10^{18}) individuals in about 1.5 million described species.

Figure 14 Swamp 320 million years ago. Forested swamps were dominated by tall, seedless canopy trees and shorter tree ferns. Dragonflies had wingspans of more than 1 m (about 3.25 ft).

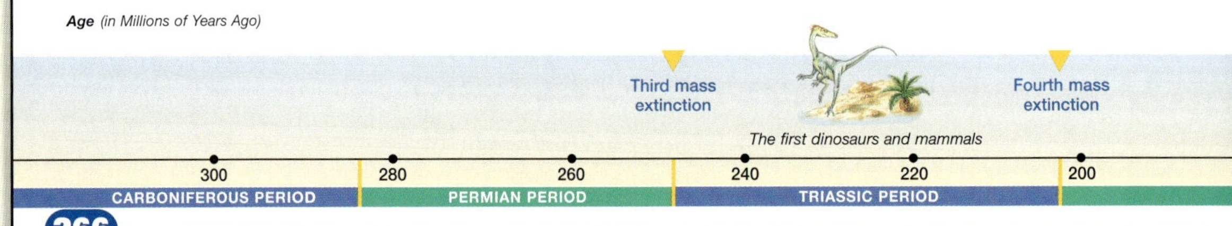

MISCONCEPTION ALERT

Evolution Some students mistakenly view evolution as individuals changing into new forms, rather than populations changing over periods of time longer than the lifetime of an individual organism. An individual organism begins life with a particular set of genes. This organism may take on different characteristics depending on the kind of environment it grows in, but its genes are not modified. When the organism reproduces, it passes half of its genes (in sexual reproduction) to each of its offspring. The organisms that reproduce are those that have a combination of genes that make them well adapted to their environment. The process of natural selection operates on every generation. Over many generations, the genetic composition of the organisms living in the environment might become different from what it was when their ancestors first lived in that environment. It is this change that constitutes evolution.

Vertebrates

A **vertebrate** is an animal with a backbone—vertebrates are the animals most familiar to us. Humans are vertebrates, and almost all other land animals bigger than our fist are vertebrates as well.

Fishes

Figure 15 Fossilized fish skeleton. This fish skeleton clearly shows the backbone, the structure that is characteristic of all vertebrate animals.

According to the fossil record, the first vertebrates were small, jawless fishes that evolved in the oceans about 530 million years ago. Jawed fishes first appeared about 430 million years ago. Jaws enabled fishes to bite and chew their food instead of sucking up their food. As a result, jawed fishes were efficient predators. A fossilized example of a jawed fish is shown in **Figure 15.** Fishes soon came to be among the most abundant animals in the seas, and for hundreds of millions of years the sea is where vertebrates stayed. Fishes are the most successful living vertebrates—they make up more than half of all modern vertebrate species. After nearly 200 million years of living in the sea, fishes have become uniquely adapted for success in water. Major changes had to occur in fish body organization, however, before some descendants of fishes became capable of living on land.

Amphibians

The first vertebrates to inhabit the land did not come out of the sea until 370 million years ago. Those first land vertebrates were early amphibians. Amphibians are smooth-skinned, four-legged animals that today include frogs, toads, and salamanders.

Several structural changes in the bodies of amphibians occurred as they adapted to life on land. Amphibians had moist breathing sacs—lungs—which allowed the animals to absorb oxygen from air. The limbs of amphibians are thought to have derived from the bones of fish fins. The evolution of a strong support system of bones in the region just behind the head made walking possible. This system of bones provided a rigid base for the limbs to work against. Because of their strong, flexible internal skeleton, the bodies of vertebrates can be much larger than those of insects. While amphibians were well adapted to their environment, a new group of animals more suited to a drier environment evolved from them.

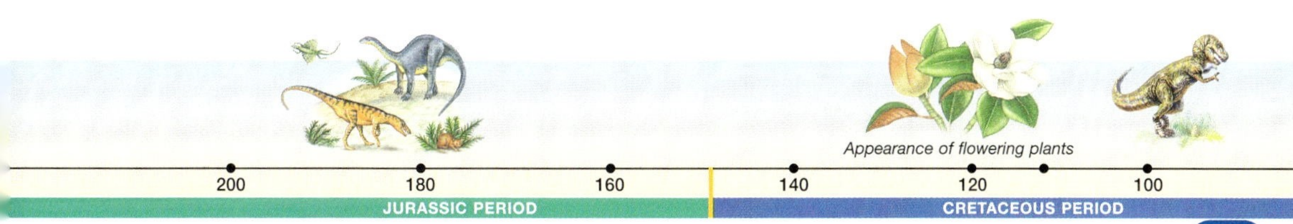

267

did you know?

Mammals of the Sea Mammals that live in the ocean range in size from the furry, five-foot-long sea otter to the hundred foot long blue whale. Like us, these creatures are warm-blooded animals that breathe air and nurse their young. The cetaceans (whales, dolphins, porpoises) descended from a cow-like ancestor that returned to ocean life about 65 million years ago. The streamlined, fishlike whale doesn't resemble its ancestor. With one or two nostrils on the top of its head, a whale can easily breathe at the surface without lifting its head. Its tail is powerful enough to push the whale through the water. The pinnipeds (seals and sea lions) are agile swimmers with tapering bodies and strong flippers, but they still retain many ties to the land. Most return to shore to mate and give birth. The sea otter adapted to life in the ocean over the past four million years. It still resembles the weasel, its relative on land. Sea otters live close to shore; they are not as well equipped for the open ocean as the streamlined, deep-diving seals and whales.

Teaching Tip — BASIC

Clamping Down Advantages

The lamprey, a jawless fish alive today, is a parasite. Without jaws, it can merely attach to a host and try to gain nutrition by sucking. Help students appreciate the importance of jaws by having each name a vertebrate predator. (Answers will vary.) How does each predator catch and eat its prey? For every case, direct the discussion to the important role of the predator's jaws in its successful lifestyle. (If not directly involved in capture, jaws are important in ingesting food.)

Demonstration — GENERAL

Distribute soft gelatin-based candy such as "gummy worms" to the class and challenge small groups of students to create an internal skeleton for some worms and an external skeleton for others, using toothpicks and tape. Caution students not to eat the candy. Have the groups share their progress and results with the class. Ask which type of skeleton requires less material and weighs less. (The internal [endo-] skeleton should be smaller and lighter.)
LS Kinesthetic Co-op Learning

Demonstration — BASIC

Crack open an egg in a Petri dish. In another Petri dish, place an intact egg. Weigh each egg-and-dish combination and record the weight, then store the eggs in a refrigerator, if possible. At the end of 2 days, crack open the intact egg and compare it to the exposed egg. Record the similarities and differences between the two, then reweigh both Petri dishes. Have students explain which egg setup showed the greater difference in weight and why. (The opened egg should weigh less, because of evaporation, and may show signs of decomposition. The other egg was protected by a shell that prevents water loss and should be in good condition.) **LS Visual**

Close

Reteaching — BASIC

Draw a table on the board. Across the top of the table, write these four headings: Evolutionary History, Structural Features, Special Adaptations, and Examples. Down the left side of the table write these seven classes: Jawless fishes, Sharks (cartilaginous fishes), Bony fishes, Amphibians, Reptiles, Birds, and Mammals. Have students take turns filling in the information.

Quiz — GENERAL

1. In what ways did the evolution of flight and an endoskeleton provide advantages for the organisms that possessed these traits? (Both flight and an endoskeleton aid an animal's movement on land.)
2. What are amphibian legs thought to be derived from? (The fins of bony fishes are thought to be the precursors of amphibian legs.)

Alternative Assessment — GENERAL

Tell students to return to the list of things they want to know about how life began. Have them place check marks next to the questions that they are now able to answer. Students should finish by making a list of what they have learned. Ask students the following: Which questions are still unanswered? What new questions do you have? You may wish to assign projects that require students to research their unanswered questions.

Section Review Correlations

SC.D.1.4.3: Items 1-4

SC.G.1.4.1: Item 6

SC.G.2.4.1: Item 5

Figure 16 Reptiles. Reptiles, such as this crocodile, were the largest group of land-dwelling organisms until the end of the Cretaceous period.

Reptiles

Reptiles evolved from amphibian ancestors about 340 million years ago. Modern reptiles include snakes, lizards, turtles, and crocodiles. Reptiles are better suited to dry land than amphibians because reptiles' watertight skin slows the loss of moisture. Reptiles also have a watertight egg, such as the one shown in **Figure 16.** Unlike amphibians, reptiles can lay their eggs on dry land. Amphibians must lay their eggs in water or in very moist soil because their eggs are unable to retain enough water to remain alive.

Mammals and Birds

Birds apparently evolved from feathered dinosaurs during or after the Jurassic period. Therapsids, reptiles with complex teeth and legs positioned beneath their body, gave rise to mammals about the same time dinosaurs evolved, during the Triassic period. Sixty-five million years ago, during the fifth mass extinction, most species disappeared forever. All of the dinosaurs except for the ancestors of birds became extinct. The smaller reptiles, mammals, and birds survived. Although many resources were available to the surviving animals, the world's climate was no longer largely dry. Thus the reptiles' advantages in dry climates were not so important. Birds and mammals then became the dominant vertebrates on land.

Both extinctions and continental drift played important roles in evolution. **Continental drift** is the movement of Earth's land masses over Earth's surface through geologic time. Continental drift resulted in the present-day position of the continents. The movement of continents helps explain why there are a large number of marsupial (pouched) mammal species in both Australia and South America, continents that were once connected.

Section 3 Review

1. **Summarize** why ozone was important in enabling organisms to live on land. SC.D.1.4.3
2. **Name** the first multicellular organisms that colonized land. SC.D.1.4.3
3. **Identify** the first kinds of animals to live on land. SC.D.1.4.3
4. **Describe** the first kinds of vertebrates that inhabited land. SC.D.1.4.3
5. **Critical Thinking Defend** the argument that fossil fuels are *not* a renewable resource. SC.G.2.4.1
6. *FCAT Prep* Mycorrhizae are mutualistic relationships between the roots of plants and what organisms?
   A amphibians **C** fungi
   B cyanobacteria **D** insects
   SC.G.1.4.1

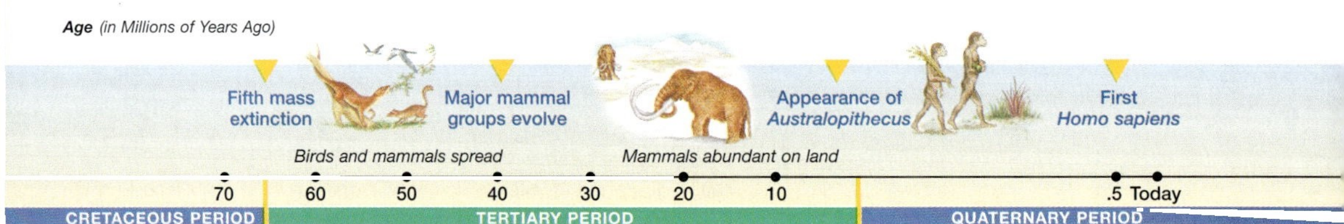

Answers to Section Review

1. Ozone was essential to protect early land organisms from the sun's ultraviolet rays.
2. Plants and fungi (or mycorrhizae) were the first multicellular land-dwelling organisms.
3. Arthropods were the first animals to live on land.
4. The first land-dwelling vertebrates were amphibians, which had lungs, an endoskeleton, and smooth skin.
5. The extremely long period of time required to produce fossil fuels from organic materials makes fossil fuels not renewable in real time.
6. **A.** Incorrect. Amphibians and plant roots do not form a mutualistic relationship. **B.** Incorrect. Cyanobacteria are aquatic organisms that do not form relationships with plant roots. **C.** Correct. Mycorrhizal associations involve plant roots and fungi. **D.** Incorrect. Insects and plant roots may form a relationship, but it is not mycorrhizal.

Study ZONE

Key Concepts

1 How Did Life Begin?

- The Earth formed about 4.5 billion years ago according to evidence obtained by radiometric dating.
- The primordial soup model and the bubble model propose explanations of the origin of the chemicals of life.
- Scientists think RNA formed before DNA or proteins formed.
- Scientists think that the first cells may have developed from microspheres.
- The development of heredity made it possible for organisms to pass traits to subsequent generations.

2 Complex Organisms Developed

- Prokaryotes are the oldest organisms and are divided into two groups, archaebacteria and eubacteria.
- Prokaryotes likely gave rise to eukaryotes through the process of endosymbiosis.
- Mitochondria and chloroplasts are thought to have evolved through endosymbiosis.
- Multicellularity arose many times and resulted in many different groups of multicellular organisms.
- Extinctions influenced the evolution of the species alive today.

3 Life Invaded the Land

- Ancient cyanobacteria produced oxygen, some of which became ozone. Ozone enabled organisms to live on land.
- Plants and fungi formed mycorrhizae and were the first multicellular organisms to live on land.
- Arthropods were the first animals to leave the ocean.
- The first vertebrates to invade dry land were amphibians.
- The extinction of many reptile species enabled birds and mammals to become the dominant vertebrates on land.
- The movement of the continents on the surface of the Earth has contributed to the geographic distribution of some species.

Key Terms

Section 1

radiometric dating (252)
radioisotope (252)
half-life (252)
microsphere (256)

Section 2

fossil (258)
cyanobacteria (258)
eubacteria (258)
archaebacteria (258)
endosymbiosis (259)
protist (261)
extinction (263)
mass extinction (263)

Section 3

mycorrhizae (265)
mutualism (265)
arthropod (266)
vertebrate (267)
continental drift (268)

Study ZONE

Alternative Assessment — GENERAL

Have students prepare a blank bingo card with 25 squares, and label the center square FREE. Write 24 terms or short phrases on the board. Have students write in each term or phrase in random order, filling all of the squares on their bingo cards. Next, have students cross out the appropriate square each time a definition or clue is provided. Students call out "bingo" whenever a horizontal, vertical, or diagonal line is completed.

Chapter Resource File

- Science Skills Worksheet GENERAL
- Critical Thinking Worksheet ADVANCED
- Test Prep Pretest GENERAL
- Chapter Test GENERAL

Answer to Concept Map

The following is one possible answer to Performance Zone item 15.

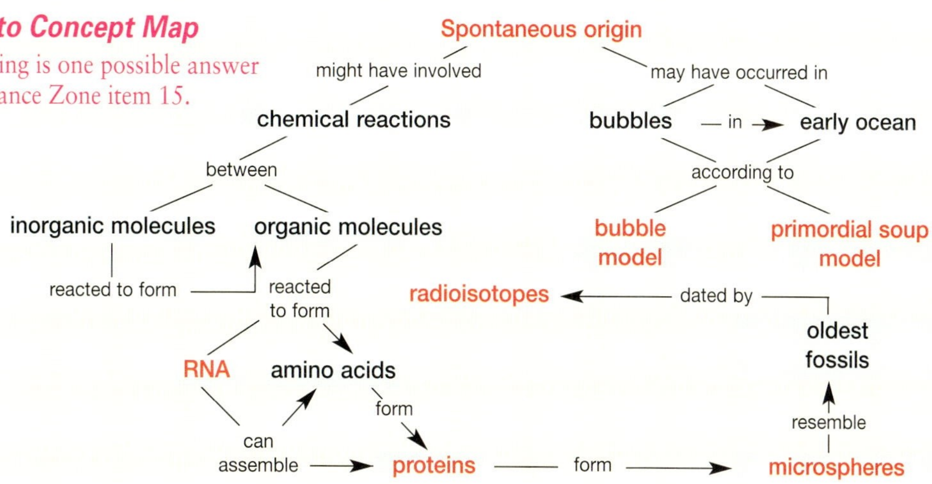

Understanding Key Ideas

1. After three half-lives of a radioisotope have passed, how much of the original radioisotope has decayed?

a. 1/8 **c.** 3/4
b. 1/2 **d.** 7/8

SC.H.1.4.1

2. Unlike the primordial soup model, Lerman's bubble model takes ______ into account.

a. ozone
b. ultraviolet radiation
c. lightning
d. volcanoes

SC.H.3.4.1

3. Cells are different from microspheres because cells

a. contain amino acids.
b. have a two-layer outer boundary.
c. grow by taking in molecules from their surroundings.
d. transfer information through heredity.

SC.F.2.4.2

4. Cell specialization came about as a result of

a. endosymbiosis. **c.** the Cambrian period.
b. archaebacteria. **d.** multicellularity.

SC.D.1.4.4

5. The first multicellular organisms to invade the land were

a. reptiles. **c.** fungi and plants.
b. amphibians. **d.** mammals.

SC.D.1.4.3

6. Describe the evidence that supports the theory of endosymbiosis. SC.H.1.4.3

7. The half-life of carbon-14 is 5,700 years. If a sample originally had 26 g of carbon-14, how much would it contain after 22,800 years? SC.H.1.4.1

8. What is the half-life of the radioisotope represented in the graph? SC.H.1.4.1

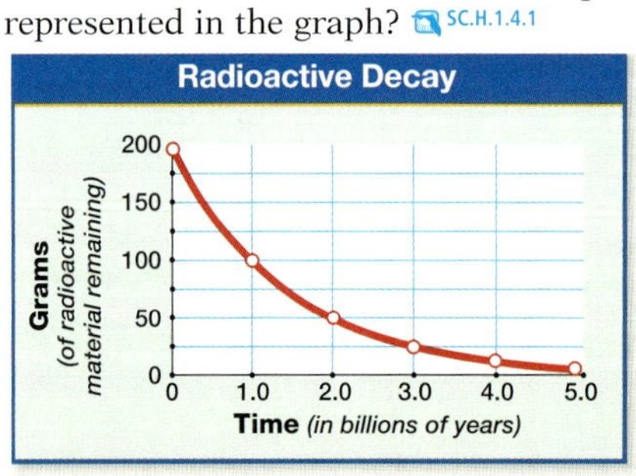

9. Relate the order in which different types of organisms invaded the land to the flow of energy. (**Hint:** See Chapter 5, Section 1.) SC.G.1.4.2

10. **Concept Mapping** Construct a concept map that shows how life might have originated by natural forces. Include the following items in your map: *spontaneous origin, primordial soup model, bubble model, RNA, proteins, microspheres,* and *radioisotopes.* SC.D.1.4.4

Critical Thinking

11. **Evaluating Viewpoints** Several scientists have said that if a large asteroid struck the Earth, the impact could result in a mass extinction. If an asteroid impact did not kill all organisms, would evolution continue or stop? Explain. SC.D.1.4.3

12. **Recognizing Relationships** Propose a hypothesis for the appearance of all animal phyla on Earth within a relatively short period during the late Precambrian and early Cambrian periods. SC.D.1.4.3

Alternative Assessment

13. **Being a Team Member** Work together in groups to design a poster to illustrate the different models that describe how life's chemicals may have originated. Show how the compounds on early Earth would have participated in each of these models. SC.D.1.4.3

14. **Finding and Communicating Information** Thomas Cech and Sidney Altman shared a Nobel prize in 1989 for their work on RNA. Research their work and the rewards associated with winning a Nobel prize. Relate your findings in an oral report. SC.H.1.4.2

15. **Finding and Communicating Information** Use the media center or Internet resources to study scientific hypotheses for the origin of life that are alternatives to the hypotheses proposed by Oparin and Lerman. Analyze either Oparin's or Lerman's hypotheses as presented in your textbook along with one alternative scientific hypotheses that you discover in your research. SC.H.1.4.5

270

CHAPTER 12 ZONE

ANSWERS

Understanding Key Ideas

1. d
2. b
3. d
4. d
5. c
6. The theory of endosymbiosis is supported by the observations that mitochondria reproduce by binary fission, and that they have their own genes that are separate and distinct from nuclear genes.
7. After four half-lives, the sample would have 1.625 g of carbon-14.
8. 1 billion years
9. Producers (plants) colonized the land first, providing food for consumers of plants and, in turn, consumers of the consumers, or predators. In this way, energy was gathered from the sun by land plants and then passed through the food web, from herbivores to carnivores.
10. One possible answer to Performance Zone item 15 is on the Study Zone page.

Critical Thinking

11. Evolution would continue after a mass extinction. In the remaining organisms, there would be traits that would help ensure some survival to the point of reproduction. These traits would be amplified within populations. New niches would then cause new species to evolve.
12. Answers may vary. Students may propose that changes in the sea environment might have increased selection pressure and resulted in rapid evolution and diversification.

Alternative Assessment

13. Answers will vary.
14. Answers will vary.
15. Students' research may identify a variety of scientific hypotheses for the origin of life. Possibilities include the proposed assembly of complex biological molecules on clay or other mineral substrates, hypotheses of the "RNA world" which holds that RNA acted both as a catalyst and as the first hereditary molecule, and so on."

Chapter Review Correlations

SC.D.1.4.3: Items 5, 11, 12, 13
SC.D.1.4.4: Items 4, 10
SC.F.2.4.2: Item 3
SC.G.1.4.2: Item 9
SC.H.1.4.1: Items 1, 7, 8
SC.H.1.4.2: Item 14
SC.H.1.4.3: Item 6
SC.H.1.4.5: Item 15
SC.H.3.4.1: Item 2

Assignment Guide

Section	Questions
1	1, 2, 3, 7, 8, 10, 13, 15
2	4, 6, 11, 12, 14
3	5, 10

FCAT Prep

For questions 1–6, write your answers on a separate sheet of paper.

1. What did the development of heredity allow organisms to store and pass on to their offspring?
   A. energy **C.** radioisotopes
   B. information **D.** UV radiation
   SC.F.2.4.2

2. A cell's surface-area-to-volume ratio limits the size that unicellular organisms can achieve. How can multicellularity solve this problem?
   F. Multicellular organisms tend to have larger cells than unicellular organisms, which makes them stronger.
   G. Communication between the nucleus and other parts of the cell is faster in larger cells than in smaller cells.
   H. As a cell grows in size, its surface-area-to-volume ratio increases, which improves the efficiency of the cell.
   I. An organism with many small cells has a greater surface-area-to-volume ratio than an organism with one large cell.
   SC.D.1.4.4

3. Photosynthesis in what organisms originally formed the oxygen that became ozone in Earth's atmosphere?
   A. archaebacteria **C.** cyanobacteria
   B. arthropods **D.** mycorrhizae
   SC.D.1.4.3

4. In Stanley Miller's 1953 primordial soup model, Miller attempted to simulate the atmospheric composition on early Earth. He combined several gases in a chamber and sent electrical sparks through the mixture. What event on Earth were the sparks intended to simulate?
   F. lightning **H.** an earthquake
   G. a forest fire **I.** a volcanic eruption
   SC.H.3.4.1

5. READ INQUIRE EXPLAIN Unlike prokaryotes, eukaryotes have membrane-bound organelles, such as mitochondria and chloroplasts.

   Identify and describe the event that resulted in the origin of eukaryotes. SC. D.1.4.3

6. READ INQUIRE EXPLAIN Originally, there were no organisms living on dry land. One factor that contributed to this occurrence was that outside of an aquatic environment, organisms could not get all of the nutrients and minerals they needed to survive. Then, a symbiotic relationship developed between plants and fungi, which provided nutrients for the fungi and minerals for the plants. This allowed them to colonize the bare land. Land plants served as food for insects, which then served as food for amphibians and other organisms.

   Part A Categorize the type of relationship that the plants and fungi had. SC.G.1.4.1

   Part B Determine the characteristics that must be present for this kind of relationship. SC.G.1.4.1

Test TIP

Allow a few minutes at the end of the test-taking period to check for mistakes made in marking answers.

271

FCAT Doctor

Question 2 Answer I is the correct choice. Materials for cellular processes must come through the cell surface, and the cell's ability to get enough materials falls as the cell grows in size. Answer F is incorrect because multicellular organisms have a variety of cell sizes, and their size does not necessarily make them stronger. Answer G is incorrect because the opposite tends to be true. Communication between the nucleus and other parts of the cell tends to be slower in larger cells than in smaller cells. Answer H is incorrect because the opposite is true. As a cell grows in size, its surface-area-to-volume ratio decreases.

Question 5 Endosymbiotic theory proposes that mitochondria are the descendants of symbiotic, aerobic eubacteria, and chloroplasts are the descendants of symbiotic, photosynthetic eubacteria.

Question 6

Part A The fungus provided minerals to the plant and the plant provided nutrients to the fungus. This kind of partnership is called mutualism.

Part B Mutualism is a relationship between two species in which both species benefit. Both plants and fungi benefited from their relationship.

Answers

1. B
2. I
3. C
4. F
5. According to the theory of endosymbiosis, bacteria entered large cells and began to live inside host cells. Cyanobacteria likely became chloroplasts and carried out photosynthesis for the host cell. Other bacteria became mitochondria and carried out cellular respiration for the host cell.
6A. The plants and fungi had a mutualistic relationship.
6B. Both species must benefit from a relationship in order for it to be mutualistic.

Test Prep Correlations

SC.D.1.4.3: Items 3, 5
SC.D.1.4.4: Item 2
SC.F.2.4.2: Item 1
SC.G.1.4.1: Items 6A, 6B
SC.H.3.4.1: Item 4

Exploration Lab

Making a Timeline of Life on Earth

Teacher's Notes

Time Required 90 minutes

Ratings

EASY → HARD

Teacher Preparation 2

Student Setup 1

Concept Level 1

Cleanup 1

Skills Required
- Communicating
- Identifying/Recognizing Patterns
- Inferring
- Measuring
- Organizing and Analyzing Data

Scientific Methods

In this lab, students will:
- Make Observations
- Ask Questions
- Draw Conclusions
- Communicate the Results

Materials and Equipment

Materials for this lab can be ordered from WARD'S. See *Master Materials List* at the front of this book for catalog numbers.

Tips and Tricks

Set up stations in various parts of the classroom with specimens, fossils, photographs, or slides of different types of organisms. Provide compound light microscopes if needed. During the first hour (or lab period), have students make their timelines and observe and record data for as many organisms as possible. During the second period, have students complete the investigation.

Exploration Lab

Making a Timeline of Life on Earth

SC.D.1.4.4

SKILLS
- Observing
- Inferring relationships
- Organizing data

OBJECTIVES
- **Compare** and **contrast** the distinguishing characteristics of representative organisms of the six kingdoms.
- **Organize** the appearance of life on Earth in a timeline.

MATERIALS
- 5 m roll of adding-machine tape
- meterstick
- colored pens or pencils
- photographs or drawings of organisms from ancient Earth to present day

Before You Begin

About 4.5 billion years ago, Earth was a ball of molten rock. As the surface cooled, a rocky crust formed and water vapor in the atmosphere condensed to form rain. By 3.9 billion years ago, oceans covered much of the Earth's surface. Rocks formed in these oceans contain **fossils** of bacterial cells that lived about 3.5 billion years ago. The **fossil record** shows a progression of life-forms and contains evidence of many changes in Earth's surface and atmosphere.

In this lab, you will make a **timeline** showing the major events in Earth's history and in the history of life on Earth, such as the evolution of new groups of organisms and the mass extinctions. This timeline can be used to study how living things have changed over time.

1. Write a definition for each boldface term in the paragraphs above.
2. Make a data table similar to the one at right.
3. Based on the objectives for this lab, write a question you would like to explore about the history of life on Earth.

Procedure

PART A: Making a Timeline

1. Make a mark every 20 cm along a 5 m length of adding-machine tape. Label one end of the tape "5 billion years ago" and the other end "Today." Write "20 cm = 200 million years" near the beginning of your timeline.
2. Locate and label a point representing the origin of Earth on your timeline. Use your textbook as a reference. See the timeline at the bottom of Section 2 and Section 3 of this chapter. Also locate and label the 11 periods of the geologic time scale beginning with the Cambrian period.
3. Using your textbook as a reference, mark the following events on your timeline: the first cyanobacteria appear; oxygen enters the atmosphere; the five mass extinctions; the first eukaryotes appear; the first multicellular organisms appear; the first vertebrates appear; the first plants,

DATA TABLE

Organism	Kingdom	Characteristics/adaptation for life on Earth

272

Answers to Before You Begin

1. Definitions: *fossil*—preserved remains or traces of a past life form; *fossil record*—the record of past life-forms found in fossils; *timeline*—a series of events arranged in chronological order.
2. Students should each create a data table as shown.
3. Answers will vary. For example: How did life on Earth change after oxygen entered the atmosphere?

Answers to Procedure

2. The Earth is estimated to be about 4 billion years old. Students should label the 12 divisions of the geologic time scale approximately as follows: *Precambrian time*—>3.5 billion to 540 million; *Cambrian*—540 million to 505 million years ago (mya); *Ordovician*—505 to 438 mya; *Silurian*—438 to 408 mya; *Devonian*—408 to 360 mya; *Carboniferous*—360 to 283 mya; *Permian*—283 to 247 mya; *Triassic*—247 to 203 mya; *Jurassic*—203 to 144 mya; *Cretaceous*—144 to 67 mya; *Tertiary*—67 to 2 mya; and *Quaternary*—1 mya to the present.

fungi, and land animals appear; the first dinosaurs and mammals appear; the first flowering plants appear; the first humans appear.

4. Look at the photographs of organisms provided by your teacher. Identify the major characteristics of each organism. Record your observations in your data table.

5. Lay out your timeline on the floor in your classroom. Place photographs (or drawings) of the organisms you examined on your timeline to show when they appeared on Earth.

6. Fold the timeline at the mark representing 4.8 billion years ago. This leaves 24 segments, each representing 200 million years, in your timeline. Now you can think of each segment as 1 hour in a 24-hour day.

7. When you are finished, walk slowly along your timeline. Note the sequence of events in the history of life on Earth and the relative amount of time between each event.

PART B: Cleanup and Disposal

8. Dispose of paper scraps in the designated waste container.

9. Clean up your work area and all lab equipment. Return lab equipment to its proper place.

Analyze and Conclude

1. Analyzing Information Think of each segment of your timeline as 1 hour in a 24-hour day as you answer each of the following questions:

a. How long has life existed on Earth?

b. For what part of the day did only unicellular life-forms exist?

c. At what time of day did the first plants appear on Earth?

d. At what time of day did mammals appear on Earth?

2. Summarizing Information Identify the major developments in life-forms that have occurred over the last 3.5 billion years.

3. Inferring Relationships How do mass extinctions appear to be related to the appearance of new major groups of organisms?

4. Justifying Conclusions Cyanobacteria are thought to be responsible for adding oxygen to Earth's atmosphere. Use your timeline to justify this conclusion.

5. Calculating Determine the amount of time, as a percentage of the time that life has existed on Earth, that humans *(Homo sapiens)* have existed.

6. Further Inquiry Write a new question about the history of life on Earth that could be explored in another investigation.

On the Job

Timelines are used to organize events in chronological order. Do research to discover how other scientists use timelines in their work. For more about careers, visit **go.hrw.com** and type in the keyword **HX4 Careers.**

3. Cyanobacteria appear—3.5 billion years ago; oxygen enters the atmosphere—2.75 billion years ago; five mass extinctions—440 million years ago (mya), 360 mya, 247 mya, 203 mya, and 65 mya; eukaryotes appear—1.5 billion years ago; multicellular organisms appear—700 mya; vertebrates appear—500 mya; land plants and fungi appear—430 mya; land animals appear—370 mya; dinosaurs and mammals appear—230 mya; flowering plants appear—120 mya; humans appear—500,000 years ago.

4. Answers will vary. See sample data table at the bottom of page 273.

5. Students should place the photograph or specimen at the point on the timeline where that type of specimen appeared. For example, a photograph of a flowering plant should be placed on the timeline at 120 million years ago.

Answers to On the Job

Many types of scientists use timelines in their work. For example, an epidemiologist (a scientist who tracks the emergence and spread of disease) might arrange the known cases of a disease on a timeline to track the events of an outbreak.

Answers to Analyze and Conclude

1. a. Life on Earth has existed for at least 17.5 "hours." **b.** Solely unicellular life forms existed for about 14 "hours." **c.** The first plants appeared at about 9:30 PM. **d.** Mammals appeared at about 11 PM.

2. Life has changed from all prokaryotic, unicellular forms to complex multicellular forms composed of eukaryotic cells. A vast variety of living things have lived on Earth during the last 3.5 billion years. Most life forms that evolved have become extinct.

3. Mass extinctions appear to be followed closely by the evolution of many new types of organisms.

4. Cyanobacteria appeared before the accumulation of oxygen in the atmosphere, and cyanobacteria produce oxygen. Therefore, it is reasonable to conclude that they are responsible for adding oxygen to Earth's atmosphere.

5. Humans have existed for 0.014 percent of the time that life has existed on Earth.

6. Answers will vary. Sample answer: How have Earth's climates changed since Precambrian times?

Sample Data Table

ORGANISM	KINGDOM	CHARACTERISTICS/ADAPTATIONS FOR LIFE ON EARTH
Insect (fossil in amber)	Animal	Segmented body, jointed legs, and wings
Fern (fossil mold in shale)	Plant	Large surface area to capture sunlight; reproductive structures on fronds

Section	Resources	Activities
Chapter Opener		SE **Quick Review**, p. 275 SE **Reading Activity**, p. 275 TE **Using the Figure**, p. 274 GENERAL TE **Opening Activity**, p. 275 GENERAL
& 3 • 90 min pp. 276–282 **Section 1** The Theory of Evolution by Natural Selection	OSP **Lesson Plan*** TR **Bellringer*** TR **Section Outline*** TR **D10 Darwin's Finches***	TE **Demonstration**, p. 276 BASIC TE **Using the Figure**, p. 278 GENERAL TE **Group Activity** Influences on Darwin, p. 278 ADVANCED SE **Quick Lab** Modeling Natural Selection, p. 280 CRF **Datasheets for In-Text Labs** Modeling Natural Selection* TE **Demonstration**, p. 281 BASIC
& 5 • 90 min pp. 283–287 **Section 2** Evidence of Evolution	OSP **Lesson Plan*** TR **Bellringer*** TR **Section Outline*** TR **D13 Evidence of Whale Evolution*** TR **D14 Forelimbs of Vertebrates*** TR **D16 Hemoglobin Comparison***	TE **Using the Figure**, p. 284 GENERAL TE **Demonstration**, p. 285 GENERAL TE **Using the Figure**, p. 286 GENERAL CRF **Quick Lab** Comparing Limb Structure and Function* BASIC CRF **Skills Practice Lab** Melanism in Insects* GENERAL
& 7 • 90 min pp. 288–292 **Section 3** Examples of Evolution	OSP **Lesson Plan*** TR **Bellringer*** TR **Section Outline***	TE **Group Activity** Examples of Evolution, p. 289 ADVANCED TE **Using the Figure**, p. 290 GENERAL TE **Group Activity** Modeling Natural Selection, p. 290 BASIC SE **Math Lab** Analyzing Change in Lizard Populations, p. 292 CRF **Datasheets for In-Text Labs** Analyzing Change in Lizard Populations* SE **Exploration Lab** Modeling Natural Selection, pp. 296–297 CRF **Datasheets for In-Text Labs** Modeling Natural Selection*

& 9 • 90 min

Chapter Review and Assessment Resources

- SE **Study Zone**, p. 293
- SE **Performance Zone**, p. 294
- SE **FCAT Test Prep**, p. 295
- TR **Concept Mapping***
- CRF **Vocabulary Review Worksheet*** GENERAL
- CRF **Test Prep Pretest*** GENERAL
- CRF **Chapter Test*** GENERAL
- CRF **Chapter Test*** ADVANCED
- OSP **Test Generator***
- OSP **Test Item Listing***
- OSP **Modified Chapter Test*** BASIC

go.hrw.com Holt Online Learning

Visit go.hrw.com for access to Holt Online Learning or enter the keyword **HX6 Home** for a variety of free online resources.

One-Stop Planner® CD-ROM

This CD-ROM package includes
- Lab Materials QuickList Software
- Holt Calendar Planner
- Customizable Lesson Plans
- Printable Worksheets
- ExamView® Test Generator
- Interactive Teacher Edition
- Holt PuzzlePro® Resources
- Holt PowerPoint® Resources

Guided Reading Audio CDs

These CDs are designed to help auditory learners and reluctant readers. (Audio CDs are also available in Spanish.)

Compression guide:
To shorten your instruction because of time limitations, omit blocks 1, 6, and 7.

KEY

TE	Teacher Edition	CRF	Chapter Resource File	*	Also on One-Stop Planner
SE	Student Edition	TR	Teaching Transparency	■	Also Available in Spanish
OSP	One-Stop Planner			◆	Requires Advance Prep

SKILLS DEVELOPMENT AND PRACTICE	SECTION REVIEW AND ASSESSMENT	CORRELATIONS
		Florida Sunshine Standards/FCAT
TE **Reading Skill Builder** Interactive Reading, p. 278 BASIC TE **Reading Skill Builder** Discussion, p. 279 GENERAL TE **Skill Builder** Graphing, p. 279 GENERAL TE **Skill Builder** Math Skills, p. 281 ADVANCED TE **Inclusion Strategy,** p. 280 BASIC CRF **Directed Reading Worksheet*** BASIC CRF **Active Reading Worksheet*** GENERAL	SE **Section 1 Review** ■ TE **Reteaching,** p. 282 BASIC TE **Quiz,** p. 282 GENERAL TE **Alternative Assessment,** p. 282 GENERAL CRF **Quiz*** ■ BASIC	SC.D.1.4.4, SC.G.2.4.3, SC.H.2.4.2, SC.H.3.4.1
TE **Integrating Physics and Chemistry,** p. 287 CRF **Directed Reading Worksheet*** BASIC CRF **Active Reading Worksheet*** GENERAL	SE **Section 2 Review** ■ TE **Reteaching,** p. 287 BASIC TE **Quiz,** p. 287 GENERAL TE **Alternative Assessment,** p. 287 ADVANCED CRF **Quiz*** ■ BASIC	SC.D.1.4.3
TE **Integrating Physics and Chemistry,** p. 290 CRF **Directed Reading Worksheet*** BASIC CRF **Active Reading Worksheet*** GENERAL CRF **Critical Thinking Worksheet*** ADVANCED CRF **Science Skills Worksheet*** GENERAL	SE **Section 3 Review** ■ TE **Reteaching,** p. 292 BASIC TE **Quiz,** p. 292 GENERAL TE **Alternative Assessment,** p. 292 GENERAL CRF **Quiz*** ■ BASIC **Modified Worksheet*** BASIC	SC.D.1.4.4, SC.F.2.4.3, SC.G.2.4.3, SC.H.3.4.1

SciLinks NSTA
www.scilinks.org
Maintained by the **National Science Teachers Association**

Topic: Natural Selection
SciLinks Code: HX4128

Topic: Theory of Evolution
SciLinks Code:HX4175

Topic: Fossil Record
SciLinks Code: HX4088

Topic: Evolution
SciLinks Code: HX4074

Topic: Paleontology
SciLinks Code: HX4134

Topic: Species Formation
SciLinks Code: HX4167

Overview

In this chapter, students will study how Darwin developed the theory of evolution by natural selection. They will develop an understanding of the basic elements of the theory and will evaluate supporting evidence from fossils, molecules, anatomy, and embryonic development. Finally, students will examine examples of changes in natural populations, including the formation of new species, that illustrate the process of evolution.

Using the Figure

Have students study this photograph of *Phyllium pulchrifolium* (the name means "beautiful leaf"). Ask students to think of an advantage of looking like a leaf. (Leaf-dwelling insects that resemble leaves may avoid being eaten by predators.) Ask students to predict what would happen to this insect's chances of survival if it were growing in an environment that had only plants with red leaves. (It would not blend in as well, so it would be more likely to be captured by predators.)

Assessing Prior Knowledge

Before beginning this chapter, be sure your students can answer the questions in the ***Quick Review*** on the facing page. If they cannot, encourage them to reread the sections indicated.

Phyllium pulchrifolium (beautiful moving leaf)

274

Chapter Correlations — Florida Sunshine Standards/FCAT

SC.D.1.4.3 Changes in Earth's climate, geological activity, and life forms may be traced and compared.

SC.D.1.4.4 Earth's systems and organisms are the result of a long, continuous change over time.

SC.F.2.4.3 Mechanisms of change (e.g., mutation and natural selection) lead to adaptations in a species and their ability to survive naturally in changing conditions and to increase species diversity.

SC.G.2.4.3 Genetic variation of offspring contributes to population control in an environment, and natural selection ensures that those who are best adapted to their surroundings survive to reproduce.

SC.H.2.4.2 Scientists control conditions in order to obtain evidence, but when that is not possible for practical or ethical reasons, they try to observe a wide range of natural occurrences to discern patterns.

CHAPTER

13 The Theory of Evolution

Florida Sunshine Standards/FCAT

SC.D.1.4.3 Changes in Earth's climate, geological activity, and life forms may be traced and compared.
SC.D.1.4.4 Earth's systems and organisms are the result of a long, continuous change over time.
SC.F.2.4.3 Mechanisms of change (e.g., mutation and natural selection) lead to adaptations in a species and their ability to survive naturally in changing conditions and to increase species diversity.
SC.G.2.4.3 Genetic variation of offspring contributes to population control in an environment, and natural selection ensures that those who are best adapted to their surroundings survive to reproduce.
SC.H.2.4.2 Scientists control conditions in order to obtain evidence, but when that is not possible for practical or ethical reasons, they try to observe a wide range of natural occurrences to discern patterns.

Quick Review

Answer the following without referring to earlier sections of your book.

1. **Describe** the structure of proteins. *(Chapter 2, Section 3)*
2. **Relate** the sequence of nucleotides in DNA to the amino acid sequence in proteins. *(Chapter 10, Section 2)*
3. **Define** genetic mutations. *(Chapter 10, Section 2)*
4. **Describe** gene sequencing. *(Chapter 11, Section 1)*
5. **Summarize** the concept of radiometric dating. *(Chapter 12, Section 1)*

Did you have difficulty? *For help, review the sections indicated.*

Reading Activity

Create a Reader Response Log to record your personal responses to the concepts presented in this chapter. Divide your paper in half. On the left side of the paper, copy a word, phrase, or passage from the text. On the right side, write your reactions, thoughts, or questions about your entries from the text.

The body of the beautiful moving leaf insect closely resembles the leaves on which it lives. Camouflage such as this helps protect animals from predators.

Looking Ahead

internet connect

www.scilinks.org
National Science Teachers Association *sci*LINKS Internet resources are located throughout this chapter.

SCLINKS. Maintained by the National Science Teachers Association

Opening Activity — GENERAL

Living Things Change Ask students what the word *evolution* means. Students should recall that evolution means "change over time." Ask students to brainstorm types of animals or plants that have changed over time.

Quick Review Answers

1. Proteins are made up of chains of amino acids linked together like pearls on a necklace.
2. Each triplet of nucleotides of DNA codes for a particular amino acid or instruction. A specific sequence of nucleotides of DNA codes for only one amino acid sequence.
3. A change in DNA is a mutation. Mutations in gametes can be passed on to the next generation and so are called genetic mutations.
4. Gene sequencing is the determination of the nucleotide sequence of DNA and mapping of the location of all of the genes carried by this sequence of DNA.
5. Radiometric dating is the calculation of the age of an object by measuring the proportions of the radioactive isotopes of certain elements in the object.

Reading Activity Answers

Encourage students to maintain their Reader Response Logs as they read the entire chapter, and work together in pairs or small groups to answer any questions that individual students could not answer.

Chapter Resource File

- Vocabulary Worksheets
- Concept Mapping

Section 1

Focus

Overview

Before beginning this section review with your students the objectives listed in the Student Edition. This lesson describes Darwin's experiences as a naturalist, how his curiosity about the things he saw prompted him to conclude that evolution occurs, and how he developed an explanation for the mechanism of evolution. Students will also learn how Darwin's theory of evolution by natural selection has been modified by other scientists with regard to formation of new species.

Bellringer

Write the following headings on the board: Long life span; Short life span. Have students write a few sentences in which they relate an organism's life span to the potential rate of evolution of a species. (Species with shorter life spans can potentially undergo evolutionary change much faster than species with longer life spans.)

Motivate

Demonstration — BASIC

Show students pictures of several different varieties (breeds) of a commercially important plant or animal. Ask students how these varieties originated. (All originated through selective breeding by humans.) If possible, show students pictures of the wild ancestors of the organisms. Tell students that observations of change in domesticated animals and plants helped people recognize that species can change over time.

Section Correlations

SC.D.1.4.4
SC.G.2.4.3
SC.H.2.4.2

Section 1 The Theory of Evolution by Natural Selection

Florida Sunshine Standards/FCAT

SC.D.1.4.4 Earth's systems and organisms are the result of a long, continuous change over time.
SC.G.2.4.3 Genetic variation of offspring contributes to population control in an environment, and natural selection ensures that those who are best adapted to their surroundings survive to reproduce.

Objectives

- **Identify** several observations that led Darwin to conclude that species evolve. SC.H.2.4.2
- **Relate** the process of natural selection to its outcome. SC.G.2.4.3
- **Summarize** the main points of Darwin's theory of evolution by natural selection as it is stated today. SC.G.2.4.3
- **Contrast** the gradualism and punctuated equilibrium models of evolution. SC.D.1.4.4

Key Terms

population
natural selection
adaptation
reproductive isolation
gradualism
punctuated equilibrium

Darwin Proposed a Mechanism for Evolution

The idea that life evolves may have been first proposed by Lucretius, a Roman philosopher who lived about 2,000 years ago before the modern theory of evolution was proposed. Then, in 1859, the English naturalist Charles Darwin, shown in **Figure 1,** published convincing evidence that species evolve, and he proposed a reasonable mechanism explaining how evolution occurs.

Like all scientific theories, the theory of evolution has developed through decades of scientific observation and experimentation. The modern theory of evolution began to take shape as a result of Darwin's work. Today almost all scientists accept that evolution is the basis for the diversity of life on Earth.

As a youth, Darwin struggled in school. His father was a wealthy doctor who wanted him to become either a doctor or a minister. Not interested in the subjects his father urged him to study, Darwin frequently spent more time outdoors than in class. At the age of 16, Darwin was sent to Edinburgh, Scotland, to study medicine. Repelled by surgery, which at the time was done without anesthetics, Darwin repeatedly skipped lectures to collect biological specimens. In 1827, Darwin's father sent him to Cambridge University, in England, to become a minister. Although he completed a degree in theology, Darwin spent much of his time with friends who were also interested in natural science.

Figure 1 Charles Darwin. Darwin was born in England in 1809 and died in 1882.

In 1831, one of Darwin's professors at Cambridge recommended him for a position as a naturalist on a voyage of HMS *Beagle*. Although the ship had an official naturalist, the *Beagle*'s captain preferred to have someone aboard who was of his own social class. At the age of 22, Darwin set off on a journey that would both change his life and forever change how we think of ourselves. The ship and its route are shown in **Figure 2.**

276

Chapter Resource File

- Directed Reading BASIC
- Active Reading GENERAL
- Data Sheet for Quick Lab GENERAL

One-Stop Planner CD-ROM

- Reading Organizers BASIC
- Reading Strategies BASIC

Transparencies

TR Bellringer
TR D10 Darwin's Finches

Figure 2 The route of HMS *Beagle*. HMS *Beagle* sailed around the world along the route shown on this map. The purpose of the ship's 5-year voyage was to survey the coast of South America.

Science Before Darwin's Voyage

In Darwin's time, most people—including scientists—held the view that each species is a divine creation that exists, unchanging, as it was originally created. But scientists had begun to seek to explain the origins of fossils. Some scientists tried to explain their observations by altering traditional explanations of creation. Others (including Darwin's own grandfather) proposed various mechanisms to explain how living things change over time.

In 1809, the French scientist Jean Baptiste Lamarck (1744–1829) proposed a hypothesis for how organisms change over generations. Lamarck believed that over the lifetime of an individual, physical features increase in size because of use or reduce in size because of disuse. Further, according to Lamarck, these changes are then passed on to offspring. This part of Lamarck's hypothesis is now known to be incorrect. However, Lamarck correctly pointed out that change in species is linked to the "physical conditions of life," referring to an organism's environmental conditions.

Darwin's Observations

During his voyage on the *Beagle*, Darwin found evidence that challenged the traditional belief that species are unchanging. During the voyage, Darwin read Charles Lyell's book *Principles of Geology*. Lyell proposed that the surface of Earth changed slowly over many years. As Darwin visited different places, he also saw things that he thought could be explained only by a process of gradual change. For example, in South America, Darwin found fossils of extinct armadillos. These fossilized animals closely resembled, but were not identical to, the armadillos living in the area.

277

Teach

Teaching Tip

Galapagos Giant Tortoises Tell students that the Galápagos Islands were named for the large tortoises that inhabit this archipelago. When Spanish explorers arrived, they found so many giant tortoises they called the islands "Galápagos," Spanish for tortoise. The scientific name for the tortoises is *Geochelone elephantopus*. Galápagos tortoises can weigh up to 227 kg (499 lb) and measure 150 cm (5 ft) across the carapace (the upper shell). Galápagos tortoises metabolize fat stored in their tissues and, therefore, can survive without food or water for long periods of time. Because of this characteristic, they became a convenient source of fresh meat for early explorers. The explorers caught the tortoises on land and stored them live in ship holds for up to a year. When the Spanish arrived, there were an estimated 250,000 tortoises. Today, approximately 15,000 remain.

Teaching Tip

Law of Use and Disuse Point out to students that Lamarck recognized that the environment played an important role in evolution. He theorized that when an organism uses a part of its body, that part becomes more developed as a result of its use. He reasoned that the modified part is then passed on to the organism's offspring. For example, he might have argued that a bird that eats tough seeds will develop a thicker beak and will in turn have offspring with thicker beaks. Tell students that if Lamarck were correct, the offspring of bodybuilders would be born with enormously developed muscles.

Teaching Tip

Acquired Traits Have students design a simple and hypothetical experiment in which Lamarck's theory of evolution—the inheritance of acquired characteristics—could be tested. They should ask questions, formulate a testable hypothesis, and suggest how they would conduct the experiment. If time permits and the experiment is doable, have students carry out their investigations.

Teach, *continued*

Using the Figure — GENERAL

Have students examine **Figure 3** and describe the features of the finches' bills. Have students describe how each finch's bill is adapted to the bird's diet. (For example, the large ground finch's bill is thick and heavy, which enables these birds to crack open large seeds.) **LS Visual**

READING SKILL BUILDER — BASIC

Interactive Reading Assign Chapter 13 of the *Holt Biology* Guided Audio CD Program to help students achieve greater success in reading the chapter. **LS Auditory**

Group Activity — ADVANCED

Influences on Darwin Assign students to work in small groups to research what scientific data and theories may have influenced Darwin as he began conducting his investigations into the mechanism of evolution. Have one group collect information on George Cuvier, who developed the theory known as catastrophism from his work on fossils. Have another group collect information on James Hutton, who developed the theory known as gradualism from his work on geological formations. Have a final group collect information on Charles Lyell, who developed the theory known as uniformitarianism from his work on geological processes. Have each group prepare a presentation on its findings, including information on each man's work and how Darwin's ideas were influenced by the work. Co-op Learning

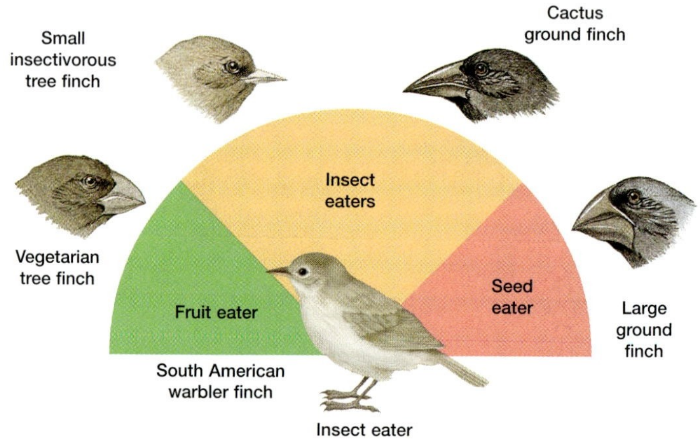

Figure 3 Darwin's finches. Darwin discovered that these finches closely resembled South American finches.

Darwin visited the Galápagos Islands, located about 1,000 km (620 mi) off the coast of Ecuador. Darwin was struck by the fact that many of the plants and animals of the Galápagos Islands resembled those of the nearby coast of South America. Darwin later suggested that the simplest explanation for this was that the ancestors of Galápagos species such as those shown in **Figure 3,** migrated to the islands from South America long ago and changed after they arrived. Darwin later called such a change "descent with modification"—evolution.

When Darwin returned from his voyage at the age of 27, he continued his lifelong study of plants, animals, and geology. However, he did not report his ideas about evolution until many years later. During those years, Darwin studied the data from his voyage. As Darwin studied his data, his confidence that organisms had evolved grew ever stronger. But he was still deeply puzzled about how evolution occurs.

Growth of Populations

The key that unlocked Darwin's thinking about how evolution takes place was an essay written in 1798 by the English economist Thomas Malthus. Malthus wrote that human populations are able to increase faster than the food supply can. Malthus pointed out that unchecked populations grow by geometric progression, as shown in **Figure 4.** Food supplies, however, increase by an arithmetic progression at best, also shown in Figure 4. He suggested that human populations do not grow unchecked because death caused by disease, war, and famine slows population growth.

Figure 4 Geometric and arithmetic progressions. The blue graph line shows uncontrolled population growth, in which the numbers increase by a multiplied constant. The red graph line shows increased food supply, in which the numbers increase by an added constant.

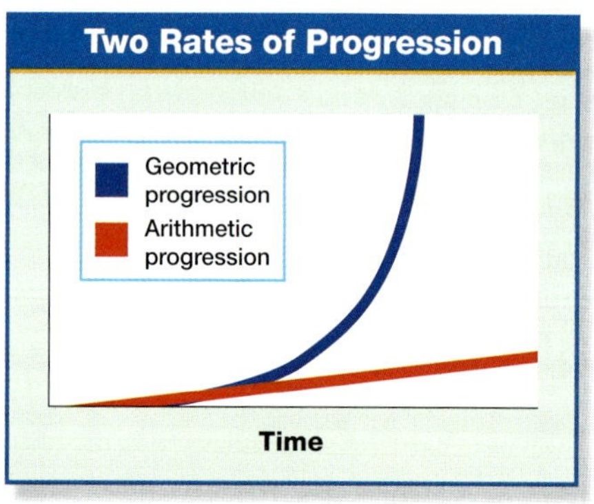

The term *population,* as it is used in biology, does not only refer to the human population. In the study of biology, a **population** consists of all the individuals of a species that live in a specific geographical area and that can interbreed.

278

Attention Grabber

Survival of the Fittest The most famous statement describing Darwin's theory, "survival of the fittest," did not appear in Darwin's original work. The phrase was coined by another biologist, Herbert Spencer, upon learning about Darwin's theory. Darwin liked the phrase and used it to summarize his theory's implications. Unfortunately, others misunderstood and misinterpreted the phrase. Soon the meaning of fittest was distorted to mean "most powerful or worthy." Political leaders and industrialists used it to justify conquest, colonialism, and oppression under the guise of "natural law."

Trends in Field Biology

Role of Naturalist Most scientists today must specialize in one field. However, many of the great scientists of the past were called naturalists—people who studied nature from a variety of different perspectives. Darwin's role on the Beagle was not just to study the native plants and animals encountered at the ship's many stops but to study the geology, climate, and people of those areas as well. In fact, his interest in geology led him to collect many fossils, some high in the Andes Mountains of South America. Such findings helped to stimulate his thinking about how environments and organisms might have changed over time.

Evolution by Natural Selection

Darwin realized that Malthus's hypotheses about human populations apply to all species. Every organism has the potential to produce many offspring during its lifetime. In most cases, however, only a limited number of those offspring survive to reproduce. Considering Malthus's view and his own observations and experience in breeding domestic animals, Darwin made a key association. *Individuals that have physical or behavioral traits that better suit their environment are more likely to survive and will reproduce more successfully than those that do not have such traits.* Darwin called this differential rate of reproduction **natural selection.** In time, the number of individuals that carry favorable characteristics that are also inherited will increase in a population. And thus the nature of the population will change—a process called *evolution.*

Darwin further suggested that organisms differ from place to place because their habitats present different challenges to, and opportunities for, survival and reproduction. Each species has evolved and has accumulated adaptations in response to its particular environment. An **adaptation** is an inherited trait that has become common in a population because the trait provides a selective advantage.

internet connect
www.scilinks.org
Topic: Natural Selection
Keyword: HX4128
SciLinks. Maintained by the National Science Teachers Association

Publication of Darwin's Work

In 1844, Darwin finally wrote down his ideas about evolution and natural selection in an early outline that he showed to only a few scientists he knew and trusted. At about this time, both a newly published book that claimed that evolution occurred, and Lamarck's hypotheses about evolution were harshly criticized. Shrinking from such controversy, Darwin put aside his manuscript.

Darwin decided to publish after he received a letter and essay in June 1858 from the young English naturalist Alfred Russel Wallace (1823–1913), who was in Malaysia at the time. Wallace's essay described a hypothesis of evolution by natural selection! In his letter, he asked if Darwin would help him get the essay published. Darwin's friends arranged for a summary of Darwin's manuscript to be presented with Wallace's paper at a public scientific meeting.

Figure 5 Political cartoon of Charles Darwin. This 1874 cartoon of Darwin with a monkeylike "ancestor" is an example of how some people ridiculed Darwin because of his work.

279

READING SKILL BUILDER — GENERAL

Discussion Tell students that the term "teleology" is the doctrine of design which holds that the phenomena of life can be explained by conscious or purposeful causes directed to definite ends. Tell them that many people refer to adaptations in living things as having evolved "in order to" accomplish some goal. For example, someone might say that "birds evolved wings in order to fly." Lead a class discussion of teleology and why this concept differs from evolution by natural selection. (Natural selection acts on variation already present in a population — it does not induce variation.)

SKILL BUILDER — GENERAL

Graphing Have students draw a graph with numerical data to reinforce the difference between geometric progression and arithmetic progression. Have students draw a graph that depicts the population size of both an animal and its primary food source over a 30-year time period. Have students connect the points on their graphs with a line. Assume that at a particular point in time (time "zero" on the graph), there are 10 animals in the population. The population doubles in size every 5 years. For the graph depicting the food source, tell students that at time "zero" there are 10 units of food in the animal's environment and that the food supply increases by 10 units every 5 years. Ask them which line depicts a geometric progression. (the line depicting population size) Ask them which line depicts an arithmetic progression. (the line depicting units of food) **LS Logical**

REAL WORLD CONNECTION

Antibiotics are powerful drugs that have saved many lives. Penicillin, the first antibiotic discovered, was observed to inhibit bacterial growth on Petri dishes by the scientist Alexander Fleming in 1928. Many other antibiotics have been discovered or synthesized since then. In the last several decades, the effectiveness of antibiotics has become compromised by the evolution of antibiotic-resistant strains of many species. Some diseases must now be treated with a "cocktail" of several different antibiotics to cure patients. Antibiotic resistance has become a problem because of the widespread use of antibiotics, not only for treatment of human diseases, but also for prevention of diseases in humans and domestic animals. With such widespread use, a mutation in a bacterium's DNA that allows it to survive treatment will be perpetuated. Soon, bacteria with these mutations may constitute the majority of the species alive.

Teach, *continued*

Modeling Natural Selection

QUICK LAB

Skills Acquired
Collecting data, interpreting, summarizing, evaluating

Teacher's Notes
Remind students that individuals have two copies of each gene, and most mutations code for recessive traits, meaning that the genes are expressed only if an individual has two copies.

Answers to Analysis

1. the different things that can happen to an organism if it is exposed to change in its environment
2. Most mutations will be passed on because they are harmful only if an individual has two copies.
3. The survivors avoided chance death (a "die" card) and two copies of the mutation, which is "lethal" when expressed.
4. sample answer: does not distinguish between beneficial and harmful mutations; does not distinguish between living and reproducing

Darwin's Theory

Darwin's book *On the Origin of Species by Means of Natural Selection* appeared in November of 1859. Many people were deeply disturbed by Darwin's theory, including the suggestion, made in a later work, that humans are related to apes, as **Figure 5** on the previous page suggests. But Darwin's arguments and evidence that evolution occurs slowly convinced biologists around the world. Darwin's theory of evolution by natural selection is supported by four major points:

1. Inherited variation exists within the genes of every population or species (the result of random mutation and translation errors).
2. In a particular environment, some individuals of a population or species are better suited to survive (as a result of variation) and have more offspring (natural selection).
3. Over time, the traits that make certain individuals of a population able to survive and reproduce tend to spread in that population.
4. There is overwhelming evidence from fossils and many other sources that living species evolved from organisms that are extinct.

QUICK LAB *Modeling Natural Selection* SC.H.3.4.1

By making a simple model of natural selection you can begin to understand how natural selection changes a population.

Materials

paper, pencil, watch or stopwatch

Procedure

1. On a chalkboard or overhead projector, make a data table like the one shown below.
2. Write each of the following words on separate pieces of paper: live, die, reproduce, mutate. Fold each piece of paper in half twice so that you cannot see the words. Shuffle your folded pieces of paper.
3. Exchange two of your pieces of paper with those of a classmate. Make as many exchanges with additional classmates as you can in 30 seconds. Mix your pieces of paper between each exchange you make.
4. Look at your pieces of paper. If you have two pieces that say "die" or two pieces that say "mutate," then sit down. If you do not, then you are a "survivor." Record your results in your class table.
5. If you are a "survivor," record the words you are holding in the data table. Then refold your pieces of paper and repeat steps 2 and 3 two more times with other "survivors."

Analysis

1. **Identify** what the four slips of paper represent.
2. **Describe** what happens to most mutations in this model.
3. **Identify** what factor(s) determined who "survived." Explain.
4. **Evaluate** the shortcomings of this model of natural selection.

DATA TABLE

Student name	Trial 1	Trial 2	Trial 3

280

did you know?

Erasmus Darwin Charles Darwin's grandfather, Erasmus Darwin, wrote about evolution more than 60 years before his grandsons' theory was presented. Erasmus Darwin cited things such as the metamorphosis of insects, the new varieties produced by selective breeding, the variations among similar organisms in different climates, and the similarities of vertebrate structure as evidence that all life was "produced from a similar living filament."

INCLUSION *Strategies*

- *Learning Disability*
- *Gifted and Talented*
- *Attention Deficit Disorder*

Using a box of animal crackers, have the students select several different "animals." For each of the "animals" selected, have the student describe and draw changes that would have to take place for each of these "animals" to evolve to live in a water environment. Also have the student describe and draw changes if these "animals" had to evolve to an environment like that at the South Pole.

Darwin's Ideas Updated

Since the time Darwin's work was published, his hypothesis—that natural selection explains how evolution happens—has been carefully examined by biologists. New discoveries, especially in the area of genetics, have given scientists new insight into how natural selection brings about the evolution of species.

internet connect
www.scilinks.org
Topic: Theory of Evolution
Keyword: HX4175
SCiLINKS. Maintained by the National Science Teachers Association

Change Within Populations

Darwin's key inference was based on the idea that in any population, individuals that are best suited to survive and do well in their environment will produce the most offspring. So, the traits of those individuals will become more common in each new generation.

Scientists now know that genes are responsible for inherited traits. Therefore, certain forms of a trait become more common in a population because more individuals in the population carry the alleles for those forms. In other words, natural selection causes the *frequency* of certain alleles in a population to increase or decrease over time. Mutations and the recombination of alleles that occurs during sexual reproduction provide endless sources of new variations for natural selection to act upon.

Species Formation

The environment differs from place to place. Thus, populations of the same species living in different locations tend to evolve in different directions. **Reproductive isolation** is the condition in which two populations of the same species do not breed with one another because of geographic separation, a difference in mating periods, or other barrier to reproduction. As two isolated populations of the same species become more different over time, they may eventually become unable to breed with one another. Generally, when the individuals of two related populations can no longer breed with one another, the two populations are different species. As shown in **Figure 6,** the Kaibab squirrel, which lives on the North Rim of the Grand Canyon in Arizona, has a black belly and other characteristics that distinguish it from the Abert squirrel. The Abert squirrel, which has a white belly, lives on the South Rim of the Grand Canyon. Because they have been so isolated from one another, they have become different enough that some biologists consider them separate species.

Figure 6 Reproductive isolation in action

These two squirrel populations became isolated from each other about 10,000 years ago, thus preventing their interbreeding.

Kaibab squirrel

Abert squirrel

MATH CONNECTION

In 1908, two scientists named Hardy and Weinberg developed a simple equation that can be used to determine the genotype frequencies in a population and to track their changes from one generation to another. This is known as the "Hardy-Weinberg equilibrium equation." In this equation ($p^2 + 2pq + q^2 = 1$), p is defined as the frequency of the dominant allele and q as the frequency of the recessive allele for a trait controlled by a pair of alleles (A and a). The frequency of one plus the frequency of the other must equal 100%, or $p + q = 1$. The chances of all possible combinations of alleles occurring randomly is therefore $(p + q)^2 = 1$, or more simply, $p^2 + 2pq + q^2 = 1$. In this equation, p^2 is the frequency of homozygous dominant (AA) individuals in a population, 2pq is the frequency of heterozygous (Aa) individuals, and q^2 is the frequency of homozygous recessive (aa) ones.

SKILL BUILDER —ADVANCED

Math Skills Provide students with the information contained in the Math Connection: Hardy Weinberg Equilibrium at the bottom of this page. Tell them to consider the example of a flowering plant in which red color is dominant ("A" allele) and white color is recessive ("a" allele). Have them consider a population of 100 of these plants, of which 16 are white-flowered ("aa" genotype). Have them calculate the frequencies of the A allele (frequency = p) and the a allele (frequency = q). (Since q^2 is the frequency of the aa genotype, and $q^2 = 0.16$ [16/100], then $q = 0.4$ and $p = 0.6$.) Next, tell students that after a period of time there aren't as many white-flowered plants around—only 4 out of 100 plants. Ask them what the new frequencies of the A allele and a allele are. ($q^2 = 0.04$, so $q = 0.2$ and $p = 0.8$.) **LS Logical**

Demonstration —BASIC

Bring to class as many of the following food crops as possible: cabbage, Brussels sprouts, broccoli, cauliflower, kale, and kohlrabi. Try to bring a wild mustard plant or a picture of one. (This can be found in many biology and horticulture texts.) Point out the parts of the plants that are eaten. Tell students that all of these different varieties have been bred from the same species, *Brassica oleracea*. Ask students how these varieties of the same species could look so different. (There is so much variation in the genes of this species.) Is it possible for these varieties to become separate species? (yes) How? (Through isolation, their genetic make-ups may become so different that they can no longer interbreed.)

Exploring Further

Punctuated Equilibrium

Teaching Strategies
Explain that these models are just that: abstractions that describe and help us understand the mechanics of evolutionary change. Emphasize that a good scientist tries to make a model fit his or her research results, not the other way around.

Close

Reteaching — BASIC

Remind students that while Wallace was in Indonesia, he wrote a paper describing his idea about how evolution occurred and sent it to Darwin. Ask students to write a letter that Darwin could have written to Wallace in response. **LS Verbal**

Quiz — GENERAL

1. What is a feature called that provides a selective advantage to a population? (An adaptation)
2. At what level of grouping of organisms does evolution occur? (the population level)

Alternative Assessment — GENERAL

Have students make visual representations of the four major points that support Darwin's theory. For example, for point 1 students could cut out pictures of different individuals of the same species and paste them onto the board. Below the pictures, they could list possible variations in the traits of the species. **LS Kinesthetic**

Section Review Correlations

SC.D.1.4.3: Item 5
SC.D.1.4.4: Item 4
SC.G.2.4.3: Items 2, 3
SC.H.2.4.2: Item 1

internet connect
www.scilinks.org
Topic: Fossil Record
Keyword: HX4088
SCILINKS. Maintained by the National Science Teachers Association

The Tempo of Evolution

For decades, most biologists have understood evolution as a gradual process that occurs continuously. The model of evolution in which gradual change over a long period of time leads to species formation is called **gradualism.** But American biologists Stephen Jay Gould and Niles Eldredge have suggested that successful species may stay unchanged for long periods of time. Gould and Eldredge have hypothesized that major environmental changes in the past have caused evolution to occur in spurts. This model of evolution, in which periods of rapid change in species are separated by periods of little or no change, is called **punctuated equilibrium.**

Exploring Further

Punctuated Equilibrium

How could major environmental changes lead to spurts in evolution? The fossil record shows that drastic environmental changes have occurred very infrequently, separated by periods of time that often last tens of millions of years. Events such as volcanic eruptions, asteroid impacts, and ice ages have been linked to sudden and drastic changes in climate. Such changes have also been linked to the extinction of many groups of organisms. As a result, environments that were once inhabited became empty. This provided opportunities for colonization by species that could quickly adapt to the new conditions through natural selection.

What Fossils Reveal
Despite large gaps, due most likely to poor conditions for fossilization, there is some evidence of both gradualism and punctuated equilibrium in the fossil record. Many groups of organisms appear suddenly in the fossil record. Some of these groups remain virtually unchanged for millions of years, while other groups disappear as suddenly as they appear. Still other groups of organisms appear to change slowly through time, as predicted by the gradualism model of evolution. More study of the fossil record may reveal additional examples of one or both types of evolution.

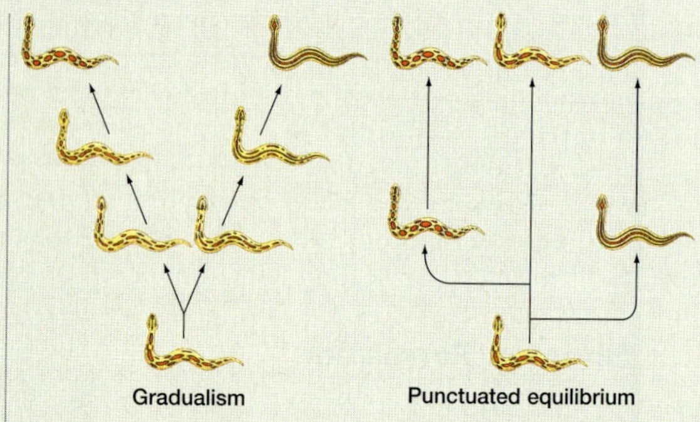

internet connect
www.scilinks.org
Topic: Evolution
Keyword: HX4074
SCILINKS. Maintained by the National Science Teachers Association

Section 1 Review

1. **List** two observations made by Charles Darwin during his 5-year voyage that led him to conclude that living species evolved from extinct species. SC.H.2.4.2
2. **Describe** how natural selection occurs. SC.G.2.4.3
3. **Summarize** the modern theory of evolution by natural selection. SC.G.2.4.3
4. **Compare** the punctuated equilibrium model of evolution with the gradualism model. SC.D.1.4.4
5. **FCAT Prep** What must two populations become in order for speciation to result?
   A extinct
   B interbred
   C one population
   D reproductively isolated
   SC.D.1.4.3

282

Answers to Section Review

1. Darwin found fossils of armadillos that closely resembled living armadillos. He also observed the resemblance between organisms on the Galápagos Islands and those on the nearest coast.
2. Individuals with traits well-suited to their environment are more likely to survive and reproduce than individuals without such traits.
3. Genetic variations in a population enable some organisms to produce more offspring than others. Over time, populations evolve and reflect the survival of organisms with the most advantageous heritable traits.
4. According to the punctuated equilibrium model, evolution occurs in spurts in response to strong environmental pressures. According to gradualism, species evolve gradually over long periods of time.
5. **A.** Incorrect. Extinct populations cannot be acted upon by natural selection. **B.** Incorrect. If two populations are interbreeding, they will not diverge into different species. **C.** Incorrect. A single population of a species will remain a single species unless isolation occurs. **D.** Correct. Isolation prevents interbreeding.

Evidence of Evolution

Section 2

Florida Sunshine Standards/FCAT SC.D.1.4.3 Changes in Earth's climate, geological activity, and life forms may be traced and compared.

The Fossil Record

Have you ever looked at a series of maps that show how a city has grown? Buildings and streets are added, changed, or destroyed as the years pass by. In the same way, fossils of animals show a pattern of development from early ancestors to modern descendants. Fossils offer the most direct evidence that evolution takes place. Recall that a fossil is the preserved or mineralized remains or imprint of an organism that lived long ago. Fossils, therefore, provide an actual record of Earth's past life-forms. Change over time (evolution) can be seen in the fossil record. Fossilized species found in older rocks are different from those found in newer rocks, as you can see in **Figure 7.**

After observing such differences, Darwin predicted that intermediate forms between the great groups of organisms would eventually be found. Since Darwin's time, some of these intermediates have been found, while others have not. For example, fossil intermediaries have been found between fishes and amphibians, between reptiles and birds, and between reptiles and mammals, adding valuable evidence about the fossil history of the vertebrates.

Today, Darwin's theory is almost universally accepted by scientists as the best available explanation for the biological diversity on Earth. Based on a large body of supporting evidence, most scientists agree on the following three major points:

1. Earth is about 4.5 billion years old.
2. Organisms have inhabited Earth for most of its history.
3. All organisms living today share common ancestry with earlier, simpler life-forms.

Objectives

- **Describe** how the fossil record supports evolution. SC.D.1.4.3
- **Summarize** how biological molecules such as proteins and DNA are used as evidence of evolution. SC.D.1.4.3
- **Infer** how comparing the anatomy and development of living species provides evidence of evolution. SC.D.1.4.3

Key Terms

paleontologist
vestigial structure
homologous structure

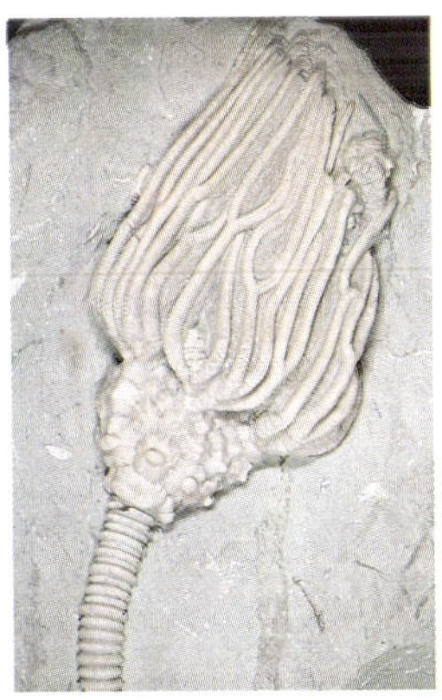

Crinoid

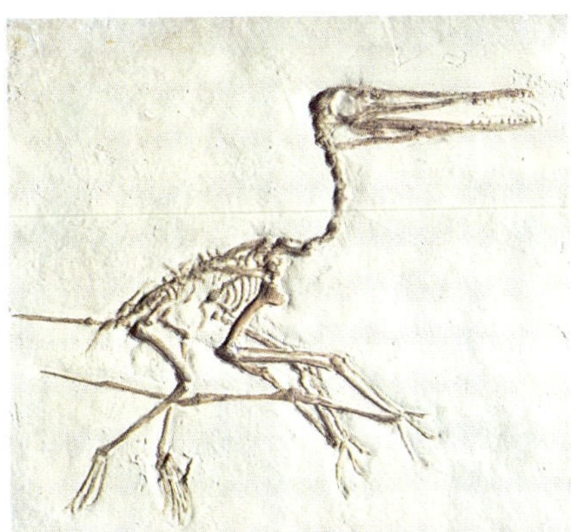

Pterodactyl

Figure 7 Fossils. Fossils of early multicellular life-forms, such as the crinoid, occur in 800-million-year-old rocks found in Indiana. Fossils of the pterodactyl, an extinct reptile, occur in 140- to 210-million-year-old rocks.

Section 2

Focus

Overview

Before beginning this section review with your students the objectives listed in the Student Edition. Tell students that the purpose of this lesson is to teach them about the different types of information scientists have found that provide evidence of evolution. These include the fossil record; molecular similarities in proteins and nucleic acids; anatomical similarities among organisms; and similarities among developmental patterns of organisms.

Bellringer

Bring in several fossils and pass them around to student groups. Ask students to brainstorm what organisms they think the fossils represent. Examples of fossils include fossilized bones, shells, footprints, or leaf prints. You can then start the lesson by discussing what fossils are and how they formed.

Motivate

Discussion/ Question — GENERAL

Read the following passage from Darwin's *On the Origin of Species:* "Of this history [of the world], we possess the last volume alone… Of this volume, only here and there a short chapter has been preserved; and of each page, only here and there a few lines." Ask students to describe what Darwin meant by this passage. (Darwin was referring to the incompleteness of the fossil record unearthed so far.) LS **Verbal**

Chapter Resource File

- Directed Reading BASIC
- Active Reading GENERAL

One-Stop Planner CD-ROM

- Reading Organizers BASIC
- Reading Strategies BASIC
- Supplemental Reading *The Origin of Species* ADVANCED

Transparencies

TR Bellringer
TR D13 Evidence of Whale Evolution
TR D14 Forelimbs of Vertebrates
TR D16 Hemoglobin Comparison

Section Correlations

SC.D.1.4.3

Teach

Teaching Tip

Extinct Organisms Tell students that scientists estimate that 99 percent of all animal and plant species that ever existed are now extinct. Also, a large number of extinct species have been described by paleontologists. For example, in one small area in Wyoming, early Eocene rocks have yielded fossils of more than 50 species of animals.

Using the Figure — GENERAL

Tell students that the first three animals in **Figure 8** are known only from their fossil remains. The fourth animal is a living species. Tell students that it is rare to find a complete skeleton in any one fossil. As more fossils of a species are found, all species' bones may eventually be found. Using their knowledge of anatomy, paleontologists project what these bones will look like. Ask students how the backbone changed relative to the time these animals spent in water. (The backbone became heavier.) What is the advantage of this change? (Whales use up-and-down motions of their bodies to swim. A heavier backbone better supports the muscles used for this motion.) **LS Visual**

Study TIP

Reading Effectively
Read the heading "Formation of Fossils," and ask one or more Who, What, Where, When, Why, or How questions. For example, How are fossils formed? As you read, answer your questions.

Formation of Fossils

The fossil record, and thus the record of the evolution of life, is not complete. Many species have lived in environments where fossils do not form. Most fossils form when organisms and traces of organisms are rapidly buried in fine sediments deposited by water, wind, or volcanic eruptions. The environments that are most likely to cause fossil formation are wet lowlands, slow-moving streams, lakes, shallow seas, and areas near volcanoes that spew out volcanic ash. The chances that organisms living in upland forests, mountains, grasslands, or deserts will die in just the right place to be buried in sediments and fossilized are very low. Even if an organism lives in an environment where fossils can form, the chances are slim that its dead body will be buried in sediment before it decays. For example, it is very likely to be eaten and scattered by scavengers.

Figure 8 Evidence of whale evolution

Whales are thought to have evolved from an ancestral line of four-legged mammals, which are represented here by their fossils and artistic reconstructions showing what scientists think they may have looked like.

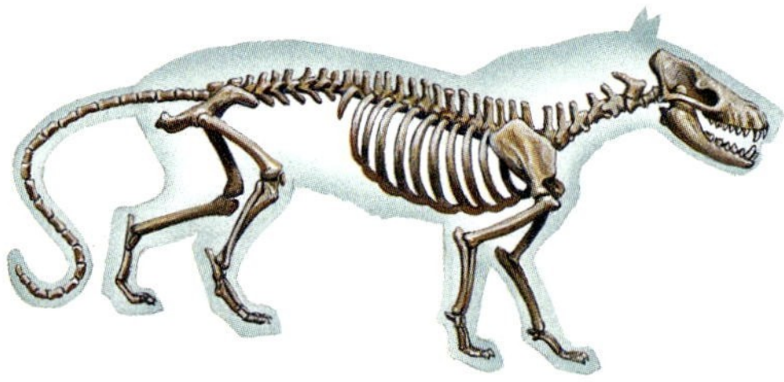

Mesonychids are one hypothesized link between modern whales and certain hoofed mammals. They were about 2 m (6 ft) long. They are thought to have lived about 60 million years ago. Some scientists favor an alternative hypothesis linking whales to other ancestral hooved mammals. These hooved mammals are also ancestral to hippopotamuses or pigs.

Ambulocetus natans apparently walked on land like modern sea lions and swam by flexing its backbone and paddling with its hind limbs (as do modern otters). They were about 3 m (10 ft) long. They existed about 50 million years ago.

CHEMISTRY CONNECTION

Scientists can determine the ages of fossils and other artifacts by measuring the amount of decay of radioactive atoms in the specimen or in the surrounding sediment. Since a radioactive atom is unstable, it will eventually change into a more stable atom. The term *half-life* describes how long it takes for one-half of the radioactive atoms in a sample to decay. For example, the half-life of carbon-14 is 5,730 years. (The most common form of carbon in living things is carbon-12.) When an organism dies, the carbon-14 it has steadily decreases as it decays, causing the ratio of carbon-14 to carbon-12 to decrease over time. If a fossilized organism has one-fourth the carbon-14 to carbon-12 ratio of a living organism, the fossil is 11,460 years old. (This is two half-lives of carbon-14.) Because the half-life of carbon-14 is relatively short, this radioisotope is only used for dating fossils and artifacts less than about 50,000 years old. To date older fossils, scientists use radioactive isotopes with longer half-lives. For example, uranium-235 has a half-life of 704 million years.

Furthermore, the bodies of some organisms decay faster than others do. For example, an animal with a hard exoskeleton (such as a crab) would have a better chance of becoming fossilized than would a soft-bodied organism, such as an earthworm.

Although the fossil record will never be complete, it presents strong evidence that evolution has taken place. When a fossil is discovered, **paleontologists** (scientists who study fossils) analyze the sediments around it. By radiometric dating certain types of rocks and minerals in those sediments, paleontologists can arrange the fossils in order from oldest to youngest. When this is done, orderly patterns of evolution can be seen. Based on existing fossils, **Figure 8** shows an artist's idea of the appearance of three extinct species that might have been ancestral to modern whales. They are arranged in the order that they evolved, based on their fossil's age as determined by radiometric dating.

internet connect

www.scilinks.org
Topic: Paleontology
Keyword: HX4134

SCILINKS Maintained by the National Science Teachers Association

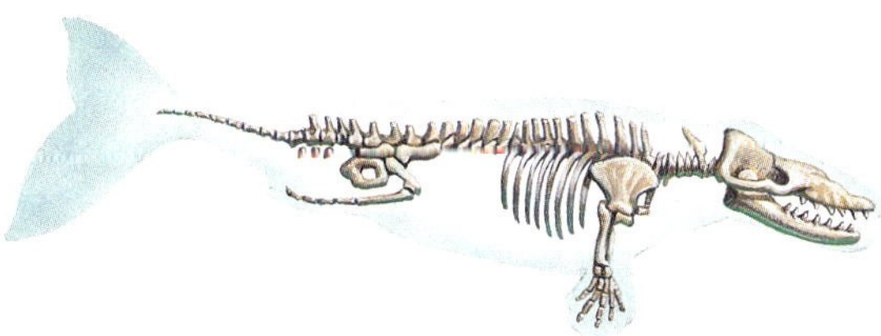

Rodhocetus kasrani, a more recent ancestor of modern whales, probably spent little time on land. Its reduced hind limbs could not have aided in walking or swimming. It is thought to have existed about 40 million years ago.

Modern whales have forelimbs that are flippers and hind limbs that have been reduced to only a few internal functionless hind-limb bones.

Teaching Tip — BASIC

Penguins "Fly" Underwater

Tell students that, although penguins are not able to fly in the air, the movement of their wings under water resembles the motion of the wings of birds that do fly. However, penguins have much heavier bones than birds that fly in the air. Ask students what adaptive advantage this might give penguins. (The hefty bone structure of a penguin's wings is an advantage for moving through water, which is far denser than air. Also, heavier bones are not a disadvantage because of the buoyancy provided by water.)

Demonstration — GENERAL

Have students study samples or pictures of different types of fossils. Ask how these organisms are similar to modern organisms. (Answers will vary. Students should note the similarities in bone and shell structure.) Diatomaceous earth used in aquarium filters provides excellent examples of microfossils for examination. **LS Visual**

MISCONCEPTION ALERT

Vestigial Structures A discussion of vestigial structures begins on the next page. Many people think that structures are labeled as vestigial because they are useless. Many such structures do, in fact, have functions. Structures are labeled vestigial if they are smaller, less functional, or perform a different function from that of homologous structures in ancestral life forms. For example, the yolk sac of a mammal is homologous to the yolk sac of birds and reptiles, but it is considered vestigial because it does not provide nutrients for the growing embryo. However, the mammalian yolk sac does produce blood cells.

Teach, *continued*

Teaching Tip — GENERAL

Making Mutations To show students that the amino acid sequence of a protein is determined by the nucleotide sequence of a gene, put the following sequence of nucleotide bases on the board or overhead: CUU, GUU, CCU, GGC, AGG. Have students look up the amino acids encoded by these triplets. (leucine, valine, proline, glycine, and arginine) Have students take turns substituting one of the other three nitrogen bases for a base in one of the triplets. Have students look up the name of the amino acid encoded by the new triplet. Tell students that the substitutions they made represent mutations. Ask how mutations affect the proteins encoded by DNA. (Mutations change the genetic blueprints for protein production.)

Using the Figure — GENERAL

Direct students' attention to **Figure 10.** Ask them which species listed in this table shares the most recent common ancestor with humans. (gorilla) Ask them if a family tree produced with the amino acid data in **Figure 10** would show the same relationships as a family tree based on nucleotide substitutions. (yes) Why? (A nucleotide sequence determines an amino acid sequence.)

Anatomy and Development

WORD Origins

The word *vestigial* comes from the Latin word *vestigium*, meaning "footprint." *Homologous* is from the Greek word *homologos*, meaning "agreeing."

Comparisons of the anatomy of different types of organisms often reveal basic similarities in body structures even though the structure's functions may differ between organisms. For example, sometimes bones are present in an organism but are reduced in size and either have no use or have a less important function than they do in other, related organisms. Such structures, which are considered to be evidence of an organism's evolutionary past, are called **vestigial** *(vehs TIJ ee uhl)* **structures.** For example, the hind limbs of whales are vestigial structures.

As different groups of vertebrates evolved, their bodies evolved differently. But similarities in bone structure can still be seen, suggesting that all vertebrates share a relatively recent common ancestor. As you can see in **Figure 9,** the forelimbs of the vertebrates shown are composed of the same basic groups of bones. Such structures are referred to as homologous *(hoh MAHL uh guhs).* **Homologous structures** are structures that share a common ancestry. That is, a similar structure in two organisms can be found in the common ancestor of the organisms.

Most scientists believe that the evolutionary history of organisms is also seen in the development of embryos. At some time in their development, all vertebrate embryos have a tail, buds that become limbs, and pharyngeal *(fuh RIN jee uhl)* pouches. The tail remains in most adult vertebrates. Only adult fish and immature amphibians retain pharyngeal pouches (which contain their gills). In humans, the tail disappears during fetal development, and pharyngeal pouches develop into structures in the throat.

Figure 9 Homologous structures

The forelimbs of vertebrates contain the same kinds of bones, which form in the same way during embryological development.

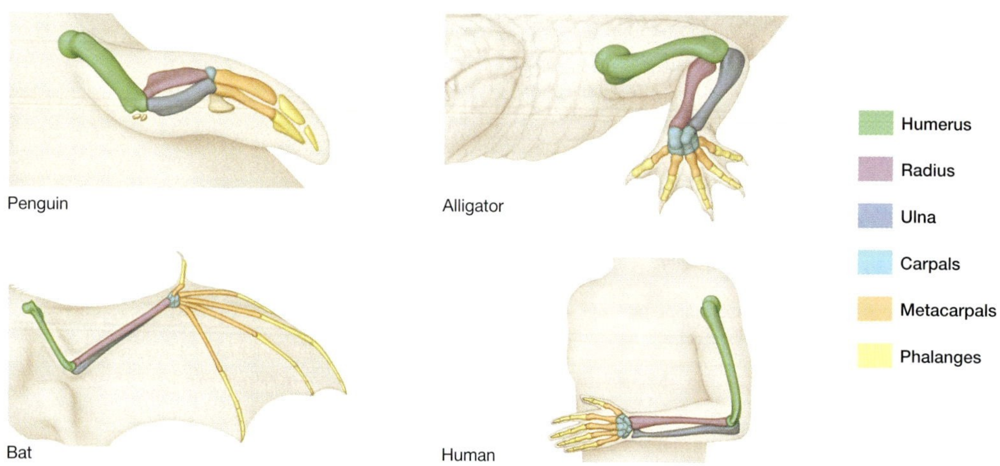

Graphic Organizer

Use this graphic organizer with **Reteaching** on the next page.

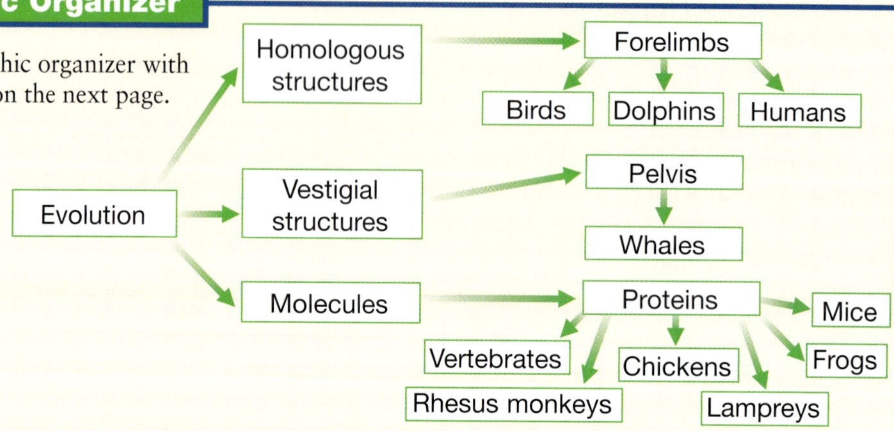

Biological Molecules

The picture of successive change seen in the fossil record allows scientists to make a prediction that can be tested. If species have changed over time as the fossil record indicates, then the genes that determine the species' characteristics should also have changed by mutation and selection. As species evolved, one change after another should have become part of their genetic instructions. Therefore, changes in a gene's nucleotide sequence should build up over time.

Proteins

This prediction was first tested by analyzing the amino acid sequences of similar proteins found in several species. If evolution has taken place, then, in general, species descended from a recent common ancestor should have fewer amino acid differences between their proteins than do species that share a common ancestor in the more distant past.

Comparing the same hemoglobin protein in several species reveals the pattern shown in **Figure 10**. Species that are thought to have shared a common ancestor more recently (for example, humans and gorillas) have few amino acid sequence differences. However, those species that are thought to have shared a common ancestor in the more distant past (such as humans and mice) have many amino acid sequence differences.

Hemoglobin Comparison

Species	Amino Acid Differences from Human Hemoglobin Protein
Gorilla	1
Rhesus monkey	8
Mouse	27
Chicken	45
Frog	67
Lamprey	125

Figure 10 Hemoglobin differences. The more similar organisms' hemoglobin proteins are, the more recent the organisms' common ancestor is likely to have been.

DNA Sequences

This pattern, however, does not hold true for all proteins. A certain protein may evolve more rapidly in some groups than others. Comparisons of proteins, therefore, may not reflect evolutionary relationships supported by the fossil record and other evidence. Evolutionary histories, however, are generally not inferred from any single protein's amino acid sequences. More accurate hypotheses about evolutionary histories are based on large numbers of gene sequences. These evolutionary histories based on DNA sequences tend be very similar to evolutionary histories inferred by biologists based comparative anatomy and evidence from the fossil record.

Section 2 Review

1. **Relate** how the fossil record provides evidence that evolution has occurred. SC.D.1.4.3
2. **State** how comparing the amino acid sequence of a protein can provide evidence that evolution has taken place. SC.D.1.4.3
3. **Describe** how comparing the anatomy of living species provides evidence of evolution. SC.D.1.4.3
4. *FCAT Prep* Which two organisms would likely have the least-similar nucleotide sequences in a given gene?
   A chimpanzee and gorilla
   B gorilla and dog
   C dog and shark
   D shark and butterfly
   SC.G.1.4.1

Close

Reteaching — BASIC

Have students design a graphic organizer that describes at least two kinds of physical traits that can be used to support the theory of evolution by natural selection. **LS Logical**

Quiz — GENERAL

1. What does the presence of hind limb remnants in the whale indicate? (The presence of these bones indicates that whales evolved from a land-dwelling ancestor.)
2. What is a vestigial structure? (bones or other structures that are present in an organism but are reduced in size and either have no use or have a less important function than they do in other, related organisms)

Alternative Assessment — ADVANCED

Tell students that the earliest phylogenetic diagrams were constructed using only evidence of morphological characteristics. Ask students why such diagrams might not reflect true evolutionary relationships. (Similar morphological characteristics might reflect adaptation to a similar environment, rather than a common ancestry.) Have students use library resources to construct phylogenetic diagrams of an organism or organisms of their choosing. Remind them to consider more than morphological characteristics.

Answers to Section Review

1. Answers will vary. The fossil record shows changes in organismal forms that can be traced forward and backward in time along different ancestral lines.
2. Similarities in amino acid sequences suggest that organisms are related, and may indicate the degree of relatedness. Differences in amino acid sequences suggest that mutations have introduced genetic variation into populations of organisms that eventually resulted in evolution.
3. Anatomical similarities between living species, such as similar bones used for similar functions, suggest that these species may have evolved from a common ancestor.
4. **A.** Incorrect. These two animals should have relatively similar nucleotide sequences because they are both primates and thus closely related. **B.** Incorrect. These two animals should have somewhat similar nucleotide sequences because they are both mammals. **C.** Incorrect. These two animals are both vertebrates and should have nucleotide sequences that are more similar than those of a vertebrate (shark) and invertebrate (butterfly). **D.** Correct. The shark is a vertebrate, and the butterfly is an invertebrate. They are the least related and would have the least similar nucleotide sequences.

Section Review Correlations

SC.D.1.4.3: Items 1, 2, 3
SC.G.1.4.1: Item 4

Focus

Overview

Before beginning this section review with your students the objectives listed in the Student Edition. This lesson provides examples of population changes that have resulted from natural selection. It also describes how the process of evolution may result in the formation of new species.

Bellringer

Have student groups brainstorm to identify characteristics of birds that provide information about their diets. (This information could be inferred from beak characteristics. A large, anvil-like beak suggests a diet of large seeds. A long, narrow beak suggests a diet of insects or possibly nectar. A very large claw-like beak suggests a diet of the flesh of animals.)

Motivate

Discussion/Question — GENERAL

Point out that within the phylum Arthropoda, the class Insecta has more species than any other class. In fact, about one-third of all animals are beetles, which constitute just one order of insects. To date, more than 1 million species of insects have been classified, and many scientists estimate that there are probably several million more. Ask students how they think so many different kinds of species of insects could have evolved. (Insects are adapted to many different kinds of environments and rely on many different kinds of food sources.)

Section Correlations

SC.D.1.4.4
SC.F.2.4.3
SC.G.2.4.3

Section 3 Examples of Evolution

Florida Sunshine Standards/FCAT

SC.F.2.4.3 Mechanisms of change (e.g., mutation and natural selection) lead to adaptations in a species and their ability to survive naturally in changing conditions and to increase species diversity.

SC.G.2.4.3 Genetic variation of offspring contributes to population control in an environment, and natural selection ensures that those who are best adapted to their surroundings survive to reproduce.

Objectives

- **Identify** four elements in the process of natural selection. SC.G.2.4.3
- **Describe** how natural selection has affected the bacteria that cause tuberculosis. SC.F.2.4.3
- **Relate** natural selection to the beak size of finches. SC.G.2.4.3
- **Summarize** the process of species formation. SC.D.1.4.4

Key Terms

divergence
speciation
subspecies

Natural Selection at Work

How does evolution occur? The heart of Darwin's theory of evolution is that natural selection is the mechanism that drives evolution. Darwin wrote: "Can we doubt . . . that individuals having any advantage, however slight, over others, would have the best chance of surviving and of procreating their kind? On the other hand, we may feel sure that any variation in the least degree injurious would be rigidly destroyed. This preservation of favorable variations, I call Natural Selection." In his writings, Darwin offered examples of how natural selection has shaped life on Earth. There are now many well-known examples of natural selection in action.

The key lesson scientists have learned about evolution by natural selection is that the *environment* dictates the direction and amount of change. If the environment changes in the future, the set of characteristics that most help an individual reproduce successfully may change. For example, the polar bear's white fur, shown in **Figure 11,** enables it to hunt successfully in its snowy environment. In a warmer environment, having white fur would no longer be an advantage.

Factors in Natural Selection

The process of natural selection is driven by four important points that are true for all real populations:

1. **All populations have genetic variation.** That is, in any population there is an array of individuals that differ slightly from each other in genetic makeup. While this may be obvious in humans, it is also true in species whose members may appear identical, such as a species of bacteria.
2. **The environment presents challenges to successful reproduction.** Naturally, an organism that does not survive to reproduce or whose offspring die before the offspring can reproduce does not pass its genes on to future generations.
3. **Individuals tend to produce more offspring than the environment can support.** Thus individuals of a population often compete with one another to survive.
4. **Individuals that are better able to cope with the challenges presented by their environment tend to leave more offspring than those individuals less suited to the environment do.**

Figure 11 **Polar bear.** Camouflage benefits predators and prey alike.

Chapter Resource File

- Directed Reading BASIC
- Active Reading GENERAL
- Data Sheet for Math Lab GENERAL

Transparencies

TR Bellringer

One-Stop Planner CD-ROM

- Reading Organizers BASIC
- Reading Strategies BASIC

Example of Natural Selection

The lung disease tuberculosis (TB) is usually caused by the bacterium *Mycobacterium tuberculosis,* shown in **Figure 12,** and it kills more adults than any other infectious disease in the world. In the 1950s, two effective antibiotics, isoniazid and rifampin, became available, and they have saved millions of lives. In the late 1980s, however, new strains of *M. tuberculosis* that are largely or completely resistant to isoniazid and rifampin appeared. Rates of TB infection began to skyrocket in many countries, and in 1993 the World Health Organization declared a global TB health emergency.

How did antibiotic-resistant strains of *M. tuberculosis* evolve? A detailed look at a single typical case reveals how: through natural selection. This case is of a 35-year-old man living in Baltimore who was treated with rifampin for an active TB infection. After 10 months, the antibiotics cleared up the infection. Two months later, however, the man was readmitted to the hospital with a severe TB infection, and despite rifampin treatment, he died 10 days later. The strain of *M. tuberculosis* isolated from his body was totally resistant to rifampin.

How had TB bacteria within his body become resistant to rifampin? Doctors compared DNA of the rifampin-resistant bacteria to DNA from samples of normal, rifampin-sensitive *M. tuberculosis.* There seemed to be only one difference: a single base change from cytosine to thymine in a gene called rpoB.

Figure 12 Tuberculosis. TB may be diagnosed from an X-ray of the lungs. TB is caused by *Mycobacterium tuberculosis.*

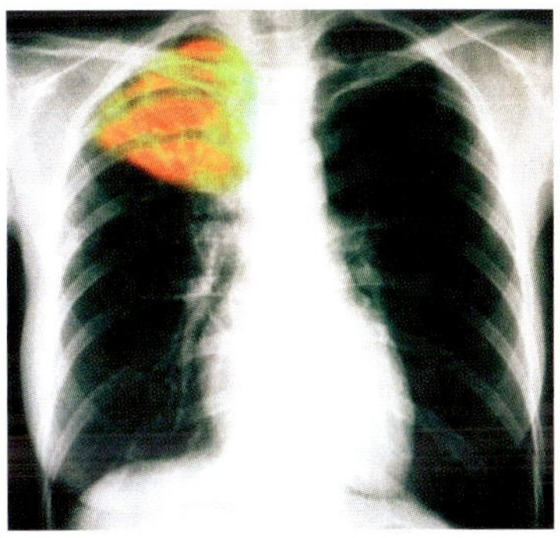

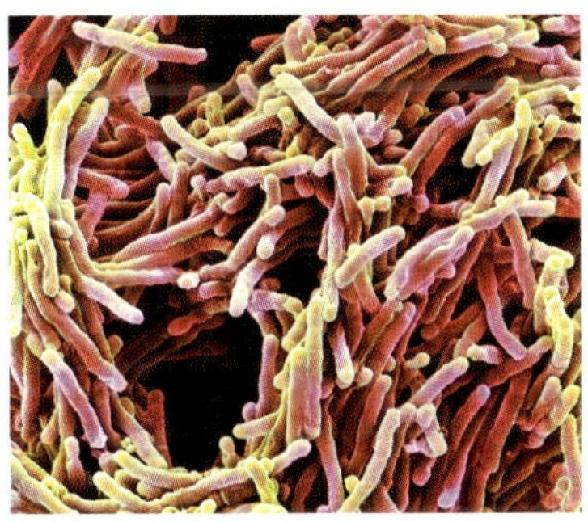

Mycobacterium tuberculosis

Evolution of Antibiotic Resistance

Rifampin acts by binding to *M. tuberculosis* RNA polymerase, preventing transcription and so killing the bacterial cell. The mutation in the polymerase's rpoB gene prevents rifampin from binding to the polymerase. The mutation, however, does not destroy the polymerase's ability to transcribe mRNA. The mutation likely occurred in a single *M. tuberculosis* bacterial cell sometime during the first infection. Because its polymerase function was no longer normal, the mutant bacterium could not divide as rapidly as normal bacteria can, but it still could divide. The antibiotic caused the normal bacterial cells to eventually die. The mutant bacteria continued to grow and reproduce in the antibiotic-containing environment.

Because the total number of *M. tuberculosis* bacteria was reduced drastically by the first antibiotic treatment, the patient's infection had seemed to clear. However, mutant, antibiotic-resistant bacteria survived and continued to grow in his body. The mutant bacteria could reproduce more effectively in the presence of the antibiotic than the normal bacteria could. Therefore, the mutant bacteria became more common in the bacterial population, and they eventually became the predominant type. When the patient became acutely ill again with TB, the *M. tuberculosis* bacterial cells in his lungs were the rifampin-resistant cells. In this way, natural selection led to the evolution of rifampin resistance in *M. tuberculosis.*

Teach

Teaching Tip — GENERAL

Selection Pressure and Rates of Evolution Tell students that the term *selection pressure* refers to the impact of natural selection on a particular trait. If individuals expressing a particular trait all die before they reach reproductive age, then we say that there is very strong selection pressure against that trait. An example is the human disease cystic fibrosis, whose victims have until only recently rarely survived to reproductive age. An allele that has strong selection pressure against it is not likely to persist in a species. However, if the trait is recessive, as it is for cystic fibrosis, individuals who are heterozygous will not be subjected to selection pressure themselves but will pass the allele on to future generations. In this way, harmful alleles can persist for a long time in a species.

Group Activity — ADVANCED

Examples of Evolution Have students work in small groups to research examples of evolution using library, Internet, or other sources. Their examples should show evolutionary change within a human lifespan. Assign one group to work on examples of artificial selection by humans. This group could select examples among domestic and farm animals, or crop plants. Assign a second group to work on examples of selection in bacteria resulting from human use of antibiotics. This group could select examples among pathogenic bacteria. Assign a third group to work on examples of selection in laboratory organisms. This group could select examples of organisms such as *Neurospora, Drosophila melanogaster,* or small rodents. Have each group prepare an illustrated poster displaying their findings. Display the posters for the whole class to examine.

Co-op Learning

INCLUSION Strategies

• *Developmental Delay* • *Learning Disability*

Using the examples of a polar bear and a brown bear, have the students make a chart of the differences between how the two bears live and survive in their environments. The chart should include the habitats of the bears, what kind of food they eat, how their camouflage helps them survive in their habitat, and why each might not survive in the other's environment.

TECHNOLOGY CONNECTION

Natural selection has made insect pests harder to fight. When DDT was first introduced, for example, it was a highly effective insecticide. Over time, DDT became less and less effective; individuals that were resistant to the insecticide survived and produced the next generation. In fact, many populations of insects are now resistant to DDT. Although DDT is now banned in the United States because of its persistent toxicity, farmers have repeatedly had to deal with insect populations that develop resistance to insecticides.

Teach, *continued*

Using the Figure — GENERAL

Direct students' attention to **Figure 13.** Ask students to suggest a reason why beak size returns to normal following each dry year. (The environmental conditions that selected for large beak size were removed.) Ask students what would have to happen for beak size to remain large in the population over many years. (The area would have to experience continually dry weather.) LS **Visual**

Group Activity — BASIC

Modeling Natural Selection
Divide students into small groups and provide each group with one 30-cm (1-ft) square of aluminum foil and 10 each of 2.5-cm (1-in.) squares of aluminum foil and white paper. (Crinkle the foil and flatten it out before cutting the squares.) Have one student in each group spread out the 20 small squares on the larger square of aluminum. Give another student in each group 10 seconds to pick up and remove as many squares as possible, one at a time, from the big sheet. Have each group count and report the number of aluminum squares and paper squares "captured." Keep a tally on the board or overhead. More white paper squares than aluminum squares should be picked up. Thus, the camouflaged aluminum squares have an adaptive survival advantage. Ask: If the little squares could reproduce, which type would be more numerous in the next generation? (small aluminum squares) Why? (More of them were left to reproduce.)
LS **Kinesthetic**

Evolution in Darwin's Finches

Darwin collected 31 specimens of finches from three islands when he visited the Galápagos Islands. In all, he collected 9 distinct species, all very similar to one another except for their bills. Two ground finches with large bills feed on seeds that they crush in their beaks, while two with narrower bills eat insects. One finch is a fruit eater, one picks insects out of cactuses, and yet another creeps up on sea birds and uses its sharp beak to drink their blood.

Darwin suggested that the nine species of Galápagos finches evolved from an original ancestral species. Changes occurred as different populations accumulated adaptations to different food sources. This idea was first tested in 1938 by the naturalist David Lack. He watched the birds closely for five months and found little evidence to support Darwin's hypothesis. Stout-beaked finches and slender-beaked finches were feeding on the same sorts of seeds. A second, far more thorough study was carried out over 25 years beginning in 1973 by Peter and Rosemary Grant of Princeton University. The Grants' study presents a much clearer picture that supports Darwin's interpretation.

It was Lack's misfortune to study the birds during a wet year, when food was plentiful. The size of the beak of the finch is of little importance in such times. Slender and stout beaks both work well to gather the small, soft seeds which were plentiful.

During dry years, however, plants produce few seeds, large or small. During these leaner years, few small, tender seeds were available. The difference between survival and starvation is the ability to eat the larger, tougher seeds that most birds usually pass by. The Grants measured the beaks of many birds every year. They found that after several dry years, the birds that had longer, more-massive beaks had better feeding success and produced more offspring.

When wet seasons returned, birds tended to have smaller beaks again, as shown in **Figure 13.** The numbers of birds with different beak shapes are changed by natural selection in response to the available food supply, just as Darwin had suggested.

Figure 13 **Natural selection in finches**

By relating the environment to beak size, the Grants showed that natural selection influences evolution.

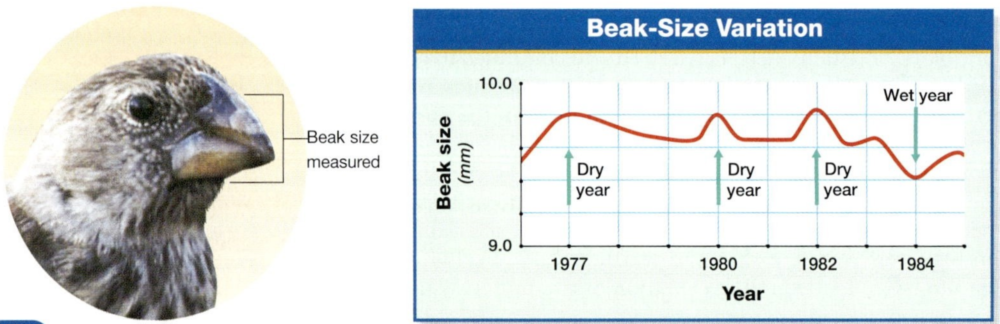

290

MISCONCEPTION ALERT

Selection, Not Induction Many people think that selection pressure induces changes. For example, people think that antibiotics induce mutations in bacteria that make them antiobiotic resistant. Remember, natural selection does not induce changes in the genes of organisms; rather it selects for those genes that are most adaptive. Genes for antibiotic resistance are already present in many bacteria, but organisms expressing these genes do not predominate until they are grown in the presence of antibiotics and the resistance provides a survival advantage.

Formation of New Species

Species formation occurs in stages. Recall that natural selection favors changes that increase reproductive success. Therefore, a species molded by natural selection has an improved "fit" to its environment. The accumulation of differences between groups is called **divergence** *(die VUHR jehns)*. Divergence leads to the formation of new species. Biologists call the process by which new species form **speciation** *(spee see AY shun)*.

Forming Subspecies

Separate populations of a single species often live in several different kinds of environments. In each environment, natural selection acts on the population. Natural selection results in the evolution of offspring that are better adapted to that environment. If their environments differ enough, separate populations of the same species can become very dissimilar. Over time, populations of the same species that differ genetically because of adaptations to different living conditions become what biologists call **subspecies.** The members of newly formed subspecies have taken the first step toward speciation. Eventually, the subspecies may become so different that they can no longer interbreed successfully. Biologists then consider them separate species.

Maintaining New Species

What keeps new species separate? Why are even closely related species usually unable to interbreed? Once subspecies become different enough, a barrier to reproduction, like the one shown in **Figure 14,** usually prevents different groups from breeding with each other.

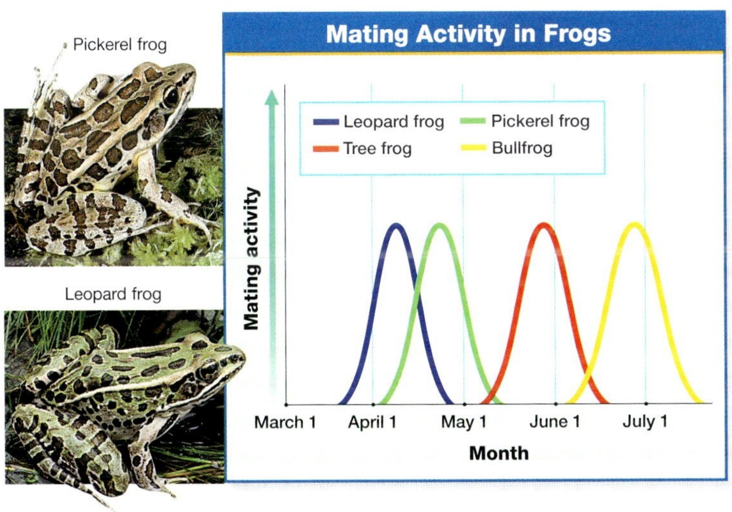

Figure 14 Mating activity in various frogs. Though they appear to be similar, pickerel frogs (*Rana palustris*) and leopard frogs (*Rana pipiens*) are different species. The graph shows that the time of peak mating activity varies between four species of frogs.

Real Life

Why do we find certain people pretty or handsome?

Some evolutionary biologists think that many traits that contribute to a person's attractiveness actually reveal the person's fitness as a mate.

Finding Information

Research and compare hypotheses concerning biological and cultural reasons that people judge others as attractive.

Real Life

Answers will vary. Sample answers might include the hypothesis that body form, especially the degree of left-right symmetry, correlates with genetic fitness. Studies have confirmed that people find symmetry an attractive physical quality.
LS Logical

Teaching Tip — GENERAL

Sequential Diagram of Speciation Have students make a Graphic Organizer that shows each step of speciation in the proper sequence. Have students use the following terms: divergence, isolation, natural selection, new species, and variation.

MATH LAB: *Analyzing Change in Lizard Populations*

Skills Acquired
Interpreting, analyzing, concluding, predicting

Teacher's Notes
Point out to students that the graph shown here is a scatter plot. The data points on such a graph are not connected by lines. Rather, these graphs illustrate the distribution and patterns of the data.

Answers to Analysis

1. The average hind limb length of each population changed in response to differences in the average perch diameter of plants on the different islands.
2. The population could evolve and have longer average hind limbs, or it could go extinct.
3. The experiment illustrates that characteristics of populations can change over time in response to environmental pressures.

Graphic Organizer

Use this graphic organizer with **Teaching Tip** on this page.

Variation → Isolation → Natural selection → Divergence → New species

Close

Reteaching — Basic

Have students suggest why sharks and alligators, which are considered "living fossils," have changed little over millions of years. (Their environments, to which they are well adapted, have remained fairly constant over this time period.)

Quiz — General

1. Divergence can lead to the formation of new _______ and new _______. (subspecies, species)
2. If two similar species of flowering plants bloom at different times, what is the name of the process that keeps these species separate? (reproductive isolation)

Alternative Assessment — General

Have students make a labeled diagram showing the process of natural selection using species of their choice. The chosen species could be one the student makes up. In their diagrams, students should include the four important points about natural selection discussed in this section.

internet connect
www.scilinks.org
Topic: Species Formation
Keyword: HX4167
SciLinks. Maintained by the National Science Teachers Association

There are several types of barriers that may isolate two or more closely related groups. For example, groups may be geographically isolated or may reproduce at different times. Physical differences may also prevent mating, or they may not be attracted to one another for mating. The hybrid offspring may not be fertile or suited to the environment of either parent.

Biologists have seen the stages of speciation in many different organisms. Thus, the way that natural selection leads to the formation of new species has been thoroughly documented. As changes continue to build up over time, living species may become very different from their ancestors and from other species that evolved from the same recent common ancestor, leading to the appearance of new species.

MATH LAB: *Analyzing Change in Lizard Populations*

SC.G.2.4.3

Background

In 1991, Jonathan Losos, an American scientist, measured hind-limb length of lizards from several islands and the average perch diameter of the island plants. The lizards were descended from a common population 20 years earlier, and the islands had different kinds of plants on which the lizards perched. Examine the graph at right and answer the following questions:

Hind-Limb Length Variation

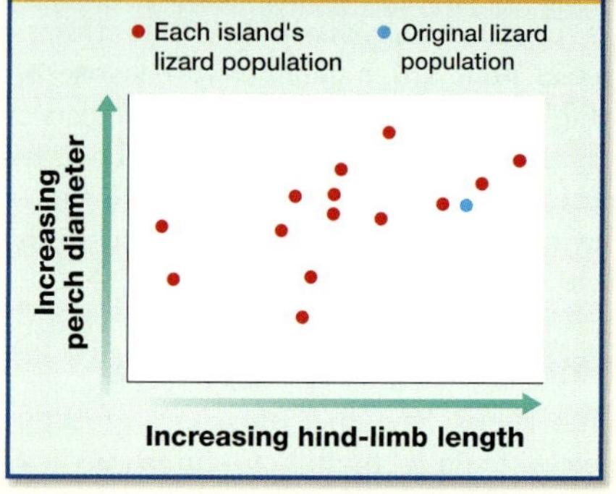

Analysis

1. **Interpreting Graphics** How did the average hind-limb length of each island's lizard population change from that of the original population?
2. **Predict** what would happen to a population of lizards with short hind limbs if they were placed on an island with a larger average perch diameter than from where they came.
3. **Justify** the argument that this experiment supports the theory of evolution by natural selection.

Section 3 Review

1. **List** four elements of natural selectionction. SC.G.2.4.3
2. **Describe** the mechanism that causes population changes in antibiotic-resistant bacteria. SC.F.2.4.3
3. **Identify** what caused the change in the finch's beaks as seen in the Grants' study. SC.G.2.4.3
4. **Describe** how speciation takes place. SC.D.1.4.4
5. **Critical Thinking Evaluating Results** Based on the results of David Lack's study and the Grants' study of finches, what conclusion can you make about the length of time required for evolution of a new species to take place? SC.D.1.4.4
6. **FCAT Prep** The beaks of finches on the Galápagos Islands enlarged over generations in response to what factor?
   A isolation **C** pollution
   B limited food supply **D** rain
   SC.D.1.4.3

292

Section Review Correlations

SC.D.1.4.3: Item 6
SC.D.1.4.4: Items 4, 5
SC.F.2.4.3: Item 2
SC.G.2.4.3: Items 1, 3

Answers to Section Review

1. (1) genetic variation, (2) environmental challenges to reproduction, (3) overproduction of offspring and a struggle for survival and (4) an increase in the number of individuals with characteristics suited to the environment
2. Bacteria have genetic variations that enable some to survive and reproduce in the presence of antibiotics. The non-resistant bacteria die, while the resistant bacteria reproduce.
3. The changes in the finches' beaks were a response to changes in their food sources that were caused by climate changes.
4. As populations of a species spread throughout an environment, they are exposed to varying conditions (environmental pressures). Over time, the separate populations become distinct and split into subspecies, and eventually, separate species.
5. It would take at least several generations and would depend on how long it takes an organism to reach reproductive maturity.
6. **A.** Incorrect. The populations of finches were not isolated from one another. **B.** Correct. Finches with larger beaks were able to eat the larger, tougher seeds. **C.** Incorrect. There was no evidence of pollution. **D.** Incorrect. Rain did not cause beak enlargement (though lack of rain, indirectly, was involved).

ZONE

Key Concepts

1 The Theory of Evolution by Natural Selection

- Charles Darwin concluded that animals on the coast of South America that resembled those on the nearby islands evolved differences after separating from a common ancestor.
- Darwin was influenced by Thomas Malthus, who wrote that populations tend to grow as much as the environment allows.
- Darwin proposed that natural selection favors individuals that are best able to survive and reproduce.
- Under certain conditions, change within a species can lead to new species.
- Gradualism is a process of evolution in which speciation occurs gradually, and punctuated equilibrium is a process in which speciation occurs rapidly between periods of little or no change.

2 Evidence of Evolution

- Evidence of orderly change can be seen when fossils are arranged according to their age.
- Differences in amino acid sequences and DNA sequences are greater between species that are more distantly related than between species that are more closely related.
- Similarities of structures in different vertebrates provide evidence that all vertebrates share a common ancestor.

3 Examples of Evolution

- Individuals that have traits that enable them to survive in a given environment can reproduce and pass those traits to their offspring.
- Experiments show that evolution through natural selection has occurred within populations of antibiotic-resistant bacteria and in Darwin's finches.
- Speciation begins as a population adapts to its environment.
- Reproductive isolation keeps newly forming species from breeding with one another.

Key Terms

Section 1

population (278)
natural selection (279)
adaptation (279)
reproductive isolation (281)
gradualism (282)
punctuated equilibrium (282)

Section 2

paleontologist (285)
vestigial structure (286)
homologous structure (286)

Section 3

divergence (291)
speciation (291)
subspecies (291)

ZONE

Alternative Assessment — GENERAL

Have students list the key terms used in this chapter and a brief definition of each term. Have them construct crossword puzzles using the definitions as the clues.

Chapter Resource File

- Science Skills Worksheet GENERAL
- Critical Thinking Worksheet ADVANCED
- Test Prep Pretest GENERAL
- Chapter Test GENERAL

Answer to Concept Map

The following is one possible answer to Performance Zone item 15 on the next page.

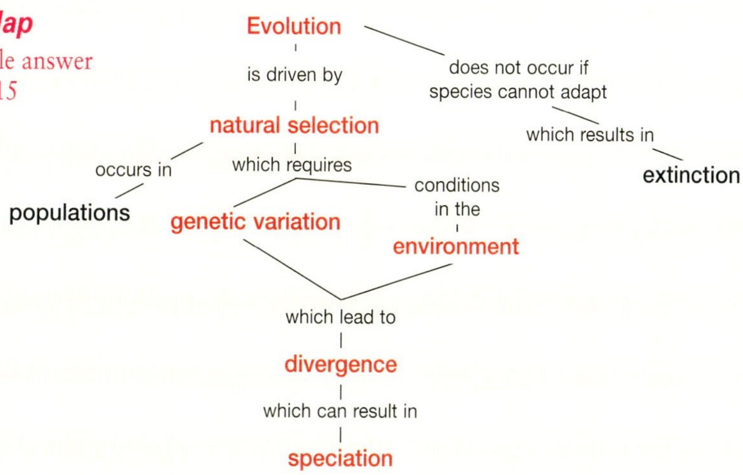

Performance ZONE

CHAPTER 13

ANSWERS

Understanding Key Ideas

1. b
2. b
3. a
4. a
5. a. An adaptation enables an organism to better survive in its environment; natural selection is the process by which populations change in response to their environment. b. Extinction is the death of all members of a species; isolation is the separation of populations of the same species.
c. Populations are groups of individuals of the same species living in the same area; subspecies differ genetically because of adaptations to different conditions.
d. Homologous structures share a common ancestry. Vestigial structures are reduced in size or function. e. Divergence is the accumulation of differences between groups; speciation is the process by which species form.
6. It may indicate that these animals share a common ancestor.
7. No; natural selection operates on genetic variation.
8. The greater the number of nucleotide differences between two species, the more distant their most recent common ancestor may be.
9. Subspecies are populations of the same species that differ genetically because of adaptation. This is the first step toward speciation.
10. Meiosis provides a source of genetic variation upon which natural selection can act.
11. erosion
12. The answer to the concept map is found at the bottom of the Study Zone page.

Chapter Review Correlations

SC.D.1.4.3: Items 5, 8, 11, 12, 16
SC.D.1.4.4: Items 3, 4, 9
SC.F.2.4.1: Item 10
SC.F.2.4.3: Item 2
SC.G.1.4.1: Item 6
SC.G.2.4.3: Items 7, 13
SC.H.1.4.3: Items 1, 14, 15

Performance ZONE CHAPTER REVIEW

Understanding Key Ideas

1. According to the modern theory of evolution,
   a. Lamarck was completely wrong.
   b. random gene mutation is a part of evolution.
   c. punctuated equilibrium has replaced natural selection.
   d. the diversity of life-forms resulted from the inheritance of acquired characteristics. SC.H.1.4.3
2. With respect to the problem of antibiotic-resistant tuberculosis, which entity evolves?
   a. the patient
   b. the bacterium
   c. the antibiotic
   d. None of the above. SC.F.2.4.3
3. What is true about gradualism with respect to punctuated equilibrium?
   a. Each is a model of evolution.
   b. Neither is a model of evolution.
   c. Only gradualism portrays true evolution.
   d. Only punctuated equilibrium portrays true evolution. SC.D.1.4.4
4. The process by which isolated populations of the same species become new species is called
   a. speciation.
   b. reproductive isolation.
   c. genetic variation.
   d. natural seclection. SC.D.1.4.4
5. For each pair of terms, explain the differences in their meanings.
   a. adaptation, natural selection
   b. extinction, reproductive isolation
   c. population, subspecies
   d. homologous, vestigial
   e. divergence, speciation SC.D.1.4.3
6. Adult lobsters and barnacles look very different. The larvae of barnacles and lobsters, however, are practically identical. What does this indicate about the evolutionary history of these organisms? SC.G.1.4.1
7. Could a population of identical organisms undergo natural selection? Why or why not? SC.G.2.4.3
8. Explain the relationship between the number of nucleotide differences between two species and the time since the species shared a common ancestor. SC.D.1.4.3
9. What is a subspecies, and how is formation of a subspecies related to the process of speciation? SC.D.1.4.4
10. How is meiosis beneficial to the evolution of a species by natural selection? (**Hint:** See Chapter 7, Section 1.) SC.F.2.4.1
11. Exploring Further Other than punctuated equilibrium, what naturally occurring phenomena might explain large gaps in the fossil record? SC.D.1.4.3
12. **Concept Mapping** Make a concept map that shows how natural selection leads to speciation. Try to include the following terms in your map: *evolution, natural selection, genetic variation, environment, speciation,* and *divergence.* SC.D.1.4.3

Critical Thinking

13. **Applying Information** If a favorable trait increases the life span of an organism without affecting reproductive success, does evolution occur? SC.G.2.4.3
14. **Evaluating** Analyze Darwin's theory of evolution by natural selection and describe one strength and one weakness. SC.H.1.4.3
15. **Justifying Conclusions** About 40 years after the publication of *On the Origin of Species*, genetics was recognized as a science. Explain how information about genetics might support Darwin's theory of evolution. SC.H.1.4.3

Alternative Assessment

16. **Career Connection** Paleontologist Research the field of paleontology, and write a report on your findings. Your report should include a job description, training required, kinds of employers, growth prospects, and starting salary. SC.D.1.4.3

Critical Thinking

13. No
14. Darwin's theory of evolution by natural selection is supported by the observation of evolution in real time, evidence of change of organisms over time found in the fossil record, and so on. Weaknesses of Darwin's theory may include the gaps in the fossil record, the lack of real-time confirmation of speciation events, and so on.
15. This understanding of genetics supports Darwin's theory by explaining how offspring can have characteristics found in their parents, whether or not such characteristics are expressed.

Alternative Assessment

16. Paleontologists study fossils and other remains of past life. Paleontologists usually have at least an undergraduate degree in zoology and/or geology, including training in chemistry and physics but may require a Ph.D. Most paleontologists are employed by universities, museums, or large oil and construction companies.

Assignment Guide

Section	Questions
1	1, 3, 10, 11, 13, 14
2	2, 6, 8, 15, 16
3	4, 5, 7, 9, 12, 15, 16

FCAT Prep

For questions 1–6, write your answers on a separate sheet of paper.

1. What term describes the process by which a species becomes better suited to its environment?
   A. adaptation **C.** gradualism
   B. equilibrium **D.** natural selection
   SC.D.1.4.4
2. What are anatomical structures that share a common ancestry?
   F. analogous structures
   G. evolutionary structures
   H. homologous structures
   I. vestigial structures
   SC.F.2.4.3
3. In the Grants' study, the effect of climate on the size of the finch's beak provides an example of which of the following?
   A. fossilization **C.** reproductive isolation
   B. natural selection **D.** speciation
   SC.D.1.4.3
4. The woodpecker finch and the warbler finch are different species. Which of the following can you conclude about these two birds?
   F. They cannot interbreed.
   G. The lack a common ancestor.
   H. They lack homologous structures.
   I. They have very different embryos.
   SC.G.1.4.1
5. READ INQUIRE EXPLAIN Generation time is the time from the beginning of an organism's life to the point of reproduction.

   Assess the effect that the time from the beginning of an organism's life to the point of reproduction has on the rate of evolution of a species. SC.D.1.4.4

6. READ INQUIRE EXPLAIN Alfred Russel Wallace was a biologist who collected insects on an 1848 expedition to the Amazon. He also made observations in the Malay Archipelago between 1854 and 1862. Wallace discovered that animals on the western islands of the Malay Archipelago differed sharply from those on the eastern islands.

   Part A Infer what condition might have caused these animals to evolve into different species. SC.D.1.4.3

   Part B Why might species of fish collected on the shores of these different islands **not** show the same differences that the animals collected by Wallace showed? SC.D.1.4.3

Test TIP

For a question about a structure or phenomenon that has a complex name, write down the name and review its meaning before answering the question.

295

FCAT Doctor

Question 3 Answer B is the correct choice. The numbers of the birds with different beak shapes are changed by natural selection in response to the available food supply determined by the climate. Answer A is incorrect because fossilization occurs after an organism's death and does not affect natural selection. Answer C is incorrect because the climate was not preventing two populations from interbreeding. Answer D is incorrect because the Grants studied differences in populations from one year to the next, which is too short a time to show speciation.

Question 5 A shorter period between one generation and the next means that more generations can occur in a specific period of time, and thus evolution can occur faster.

Question 6

Part A Reproductive isolation can occur when two populations of the same species do not breed with one another because of geographic separation, such as by being on different islands. Reproductive isolation can lead to formation of separate species.

Part B There is no separation in the ocean like there is on the islands in the archipelago.

Answers

1. A
2. H
3. B
4. F
5. A shorter generation time causes a faster rate of evolution for an organism.

6A. Reproductive isolation might have caused these animals to evolve into different species.

6B. The fish do not have reproductive isolation, because the waters they breed in are connected.

Test Prep Correlations

SC.D.1.4.3: Items 3, 6A, 6B
SC.D.1.4.4: Items 1, 5
SC.F.2.4.3: Item 2
SC.G.1.4.1: Item 4

Exploration Lab

Modeling Natural Selection

Teacher's Notes

Time Required

One 50-minute class period

Ratings

EASY → HARD

Teacher Preparation 2

Student Setup 2

Concept Level 3

Cleanup 1

Skills Acquired

- Collecting Data
- Constructing Models
- Experimenting
- Organizing and Analyzing Data

Scientific Methods

In this lab, students will:

- Make Observations
- Ask Questions
- Test the Hypothesis
- Draw Conclusions

Safety Cautions

Remind students to be careful with scissors and always cut in a direction away from the face and body.

Exploration Lab

Modeling Natural Selection

SC.H.3.4.1

SKILLS

- Modeling a process
- Inferring relationships

OBJECTIVES

- **Model** the process of selection.
- **Relate** favorable mutations to selection and evolution.

MATERIALS

- scissors
- construction paper
- cellophane tape
- soda straws
- felt-tip marker
- meterstick or tape measure
- penny or other coin
- six-sided die

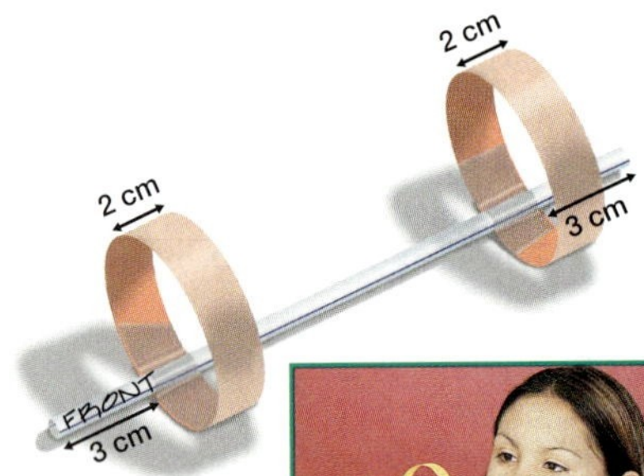

Egyptian origami bird

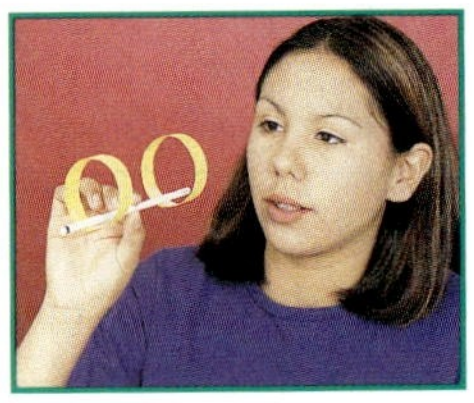

Before You Begin

Natural selection occurs when organisms that have certain **traits** survive to reproduce more than organisms that lack those traits do. A population evolves when individuals with different **genotypes** survive or reproduce at different rates. In this lab, you will model the selection of favorable traits in a new generation by using a paper model of a bird—the fictitious Egyptian origami bird (*Avis papyrus*), which lives in dry regions of North Africa. Assume that only birds that can successfully fly the long distances between water sources will live long enough to breed successfully.

1. Write a definition for each boldface term in the preceding paragraph.
2. Make a data table similar to the one shown below.
3. Based on the objectives for this lab, write a question you would like to explore about the process of selection.

Procedure

PART A: Parental Generation

1. Cut two strips of paper, 2 × 20 cm each. Make a loop with one strip of paper, letting the paper overlap by 1 cm, and tape the loop closed. Repeat for the other strip.

DATA TABLE

Bird	Coin flip (H or T)	Die throw (1–6)	Anterior wing (cm)			Posterior wing (cm)			Average distance flown (m)
			Width	Circum.	Distance from front	Width	Circum.	Distance from back	
Parent	NA	NA	2	19	3	2	19	3	
Generation 1									
Chick 1									
Chick 2									
Chick 3									
Generation 2									
Chick 1									
Chick 2									
Chick 3									

Tips and Tricks

Students will simulate the breeding of several generations of the birds and observe the effect of various phenotypes on the evolutionary success of these animals. The random nature of mutations is demonstrated by randomly changing the anterior and posterior wing position and wing circumference of the birds. Each student group will need one sheet of construction paper, a meter stick or tape measure, a pair of scissors, a straw, cellophane tape, a coin, and a die. To save time, have students create their table before beginning the lab, and have students work in pairs or small groups. You may wish to review and clarify the principles of selection. This investigation reinforces the concept of natural selection as differential reproduction rather than merely differential survival. Have students dispose of their paper scraps and paper birds in the trash.

2. Tape one loop 3 cm from each end of the straw, as shown above. Mark the front end of the bird with a felt-tip marker. This bird represents the parental generation.

3. Test how far your parent bird can fly by releasing it with a gentle overhand pitch. Test the bird twice. Record the bird's average flight distance in your data table.

PART B: First (F_1) Generation

4. Each origami bird lays a clutch of three eggs. Assume that one of the chicks is a clone of the parent. Use the parent to represent this chick in step 6.

5. Make two more chicks. Assume that these chicks have mutations. Follow Steps A–C below for each chick to determine the effects of its mutation.

   Step A Flip a coin to determine which end is affected by a mutation.

   Heads = anterior (front)

   Tails = posterior (back)

   Step B Throw a die to determine how the mutation affects the wing.

 = Wing position moves 1 cm toward the end of the straw.

 = Wing position moves 1 cm toward the middle of the straw.

 = Wing circumference increases by 2 cm.

 = Wing circumference decreases by 2 cm.

 = Wing width increases by 1 cm.

 = Wing width decreases by 1 cm.

   Step C A mutation is lethal if it causes a wing to fall off the straw or a wing with a circumference smaller than that of the straw. If you get a lethal mutation, disregard it and produce another chick.

6. Record the mutations and the wing dimensions of each offspring.

7. Test each bird twice by releasing it with a gentle overhand pitch. Release the birds as uniformly as possible. Record the distance each bird flies. The most successful bird is the one that flies the farthest.

PART C: Subsequent Generations

8. Assume that the most successful bird in the previous generation is the sole parent of the next generation. Repeat steps 4–7 using this bird.

9. Continue to breed, test, and record data for eight more generations.

PART D: Cleanup and Disposal

10. Dispose of paper scraps in the designated waste container.

11. Clean up your work area and all lab equipment. Return lab equipment to its proper place. Wash your hands thoroughly before you leave the lab and after you finish all work.

Analyze and Conclude

1. **Analyzing Results** Did the birds you made by modeling natural selection fly farther than the first bird you made?

2. **Inferring Conclusions** How might this lab help explain the variety of species of Galápagos finches?

3. **Further Inquiry** Write another question about natural selection that could be explored with another investigation.

On the Job

Population biology is the study of populations. Do research to discover how population biologists study evolution. For more about careers, visit **go.hrw.com** and type in the keyword **HX4 Careers.**

Answers to Before You Begin

1. *natural selection*—process by which populations change in response to their environment as individuals better adapted to the environment leave more offspring; *traits*—distinguishing characteristics; *genotype*—the genetic constitution of an organism as indicated by its set of alleles
3. Answers will vary. For example: Are the effects of natural selection obvious in only a few generations?

Answers to Procedure

3. Answers will vary.
6. Answers will vary.
7. Answers will vary.
8. Answers will vary.
9. Answers will vary.

Answers to Analyze and Conclude

1. Most students should answer "yes" as the best-flying birds are selected as the sole parents of the next generation.
2. This lab demonstrates that organisms can change significantly over only a few generations. The lab therefore shows how isolated populations could diverge to the point that they constitute different species, as happened to finches on the Galápagos Islands.
3. Answers will vary. For example: Does natural selection act on one trait at a time, or can selection pressure affect the evolution of several traits at once?

Answers to On the Job

Population biologists study evolution by examining changes in population size over time. Techniques population biologists use to study evolution include population sampling, computer simulations, and various statistical techniques for analyzing data.

Chapter Opener

SE **Quick Review**, p. 299 ■
SE **Reading Activity**, p. 299 ■
TE **Using the Figure**, p. 298 GENERAL
TE **Opening Activity**, p. 299 BASIC

3 • 90 min pp. 300–304
Section 1
Categories of Biological Classification

OSP **Lesson Plan***
TR **Bellringer***
TR **Section Outline***
TR **D29 Biological Hierarchy of Classification***
TR **D30 Classification of a Bee***

TE **Demonstration**, p. 301 ◆ GENERAL
TE **Demonstration**, p. 302
TE **Activity** Comparing Classifications, p. 303 GENERAL
TE **Group Activity** These Are the Names of My Favorite Things, p. 303 ADVANCED
SE **Quick Lab** Using a Field Guide, p. 304 ■
CRF **Datasheets for In-Text Labs** Using a Field Guide* GENERAL
CRF **Exploration Lab** Classification* GENERAL

& 5 • 90 min pp. 305–310
Section 2
How Biologists Classify Organisms

OSP **Lesson Plan***
TR **Bellringer***
TR **Section Outline***
TR **D31 Evolutionary Systematics and Cladistic Taxonomy***

TE **Activity** Classifying Hardware, p. 305 ◆ GENERAL
SE **Data Lab** Analyzing Taxonomy of Mythical Organisms, p. 306 ■
CRF **Datasheets for In-Text Labs** Analyzing Taxonomy of Mythical Organisms*
TE **Using the Figure**, p. 307 GENERAL
TE **Activity** Loss of Biodiversity, p. 307 ADVANCED
SE **Data Lab** Making a Cladogram, p. 309 ■
CRF **Datasheets for In-Text Labs** Making a Cladogram*
SE **Skills Practice Lab** Making a Dichotomous Key, pp. 314–315 ■
CRF **Datasheets for In-Text Labs** Making a Dichotomous Key*

& 7 • 90 min
Chapter Review and Assessment Resources

SE **Study Zone**, p. 311 ■
SE **Performance Zone**, p. 312 ■
SE **FCAT Test Prep**, p. 313 ■
TR **Concept Mapping***
CRF **Vocabulary Review Worksheet*** ■ GENERAL
CRF **Test Prep Pretest*** ■ GENERAL
CRF **Chapter Test*** ■ GENERAL
CRF **Chapter Test*** ADVANCED
OSP **Test Generator***
OSP **Test Item Listing***
OSP **Modified Chapter Test*** BASIC

go.hrw.com Holt Online Learning

Visit go.hrw.com for access to Holt Online Learning or enter the keyword **HX6 Home** for a variety of free online resources.

One-Stop Planner® CD-ROM

This CD-ROM package includes
- Lab Materials QuickList Software
- Holt Calendar Planner
- Customizable Lesson Plans
- Printable Worksheets
- ExamView® Test Generator
- Interactive Teacher Edition
- Holt PuzzlePro® Resources
- Holt PowerPoint® Resources

Guided Reading Audio CDs

These CDs are designed to help auditory learners and reluctant readers. (Audio CDs are also available in Spanish.)

Compression guide:
To shorten your instruction because of time limitations, omit blocks 1, 3, and 6.

KEY

TE	Teacher Edition	CRF	Chapter Resource File	*	Also on One-Stop Planner
SE	Student Edition	TR	Teaching Transparency	■	Also Available in Spanish
OSP	One-Stop Planner			◆	Requires Advance Prep

SKILLS DEVELOPMENT AND PRACTICE	SECTION REVIEW AND ASSESSMENT	CORRELATIONS
		Florida Sunshine Standards/FCAT
TE Reading Skill Builder Interactive Reading, p. 302 BASIC TE Skill Builder Writing, p. 302 ADVANCED TE Reading Skill Builder Brainstorming, p. 301 CRF Directed Reading Worksheet* BASIC CRF Active Reading Worksheet* GENERAL	SE Section 1 Review ■ TE Reteaching, p. 304 BASIC TE Quiz, p. 304 GENERAL TE Alternative Assessment, p. 304 GENERAL CRF Quiz* ■ BASIC	SC.G.1.4.1, SC.H.2.4.1
TE Reading Skill Builder Anticipation Guide, p. 306 GENERAL TE Reading Skill Builder Brainstorming, p. 307 BASIC TE Skill Builder Writing, p. 307 ADVANCED CRF Directed Reading Worksheet* BASIC CRF Active Reading Worksheet* GENERAL CRF Critical Thinking Worksheet* ADVANCED CRF Science Skills Worksheet* GENERAL	SE Section 2 Review ■ TE Reteaching, p. 310 BASIC TE Quiz, p. 310 GENERAL TE Alternative Assessment, p. 310 GENERAL CRF Quiz* ■ BASIC Modified Worksheet* BASIC	SC.D.1.4.4, SC.G.1.4.1, SC.H.2.4.1, SC.H.2.4.2

SciLinks NSTA
www.scilinks.org
Maintained by the **National Science Teachers Association**

Topic: Naming Species
SciLinks Code: HX4127

Topic: Systems of Classification
SciLinks Code: HX4173

Topic: Taxonomy
SciLinks Code: HX4174

Topic: Evolutionary Systematics
SciLinks Code: HX4077

Topic: Classification
SciLinks Code: HX4044

CHAPTER 14

Overview

Tell students that the purpose of this chapter is to explain how living organisms are classified, and to identify the traits biologists use to classify them. After studying this chapter, students will appreciate life's diversity, how it reflects the ongoing process of evolution, and how scientists have developed classification systems in order to understand this diversity and the evolutionary relationships among organisms.

Using the Figure

Direct students' attention to the photo on this page. Ask them to point out similarities and differences between the cactuses and the other plants shown, and to suggest their own methods of classifying them (e.g., according to categories such as size, type, color, and other characteristics). As an extension, have interested students find out the biologic classification of the plants shown.

Assessing Prior Knowledge

Before beginning this chapter, be sure your students can answer the questions in the ***Quick Review*** on the facing page. If they cannot, encourage them to reread the sections indicated.

Desert organisms

298

Chapter Correlations | **Florida Sunshine Standards/FCAT**

SC.D.1.4.4 Earth's systems and organisms are the result of a long, continuous change over time.

SC.G.1.4.1 There is great diversity and interdependence among living things.

SC.H.2.4.1 Scientists assume that the universe is a vast system in which basic rules exist that may range from very simple to extremely complex, but scientists operate on the belief that the rules can be discovered by careful, systemic study.

SC.H.2.4.2 Scientists control conditions in order to obtain evidence, but when that is not possible for practical or ethical reasons, they try to observe a wide range of natural occurrences to discern patterns.

CHAPTER

14 Classification of Organisms

Florida Sunshine Standards/FCAT

SC.D.1.4.4 Earth's systems and organisms are the result of a long, continuous change over time.

SC.G.1.4.1 There is great diversity and interdependence among living things.

SC.H.2.4.1 Scientists assume that the universe is a vast system in which basic rules exist that may range from very simple to extremely complex, but scientists operate on the belief that the rules can be discovered by careful, systemic study.

SC.H.2.4.2 Scientists control conditions in order to obtain evidence, but when that is not possible for practical or ethical reasons, they try to observe a wide range of natural occurrences to discern patterns.

Quick Review

Answer the following without referring to earlier sections of your book.

1. **Define** the term *species*. *(Chapter 13, Section 2)*
2. **Relate** macromolecules to evolutionary history. *(Chapter 13, Section 2)*
3. **Relate** homologous structures to evolutionary relationships. *(Chapter 13, Section 2)*
4. **Summarize** speciation. *(Chapter 13, Section 3)*

Did you have difficulty? *For help, review the sections indicated.*

Reading Activity

Take a break after reading each section of this chapter, and closely study the figures in the section. Reread the figure captions, and, for each one, write out a question that can be answered by referring to the figure and its caption. Refer to your list of figures and questions as you review the concepts addressed in the chapter before you complete the Performance Zone chapter review.

As illustrated by this variety of desert plants, most of Earth's surface—including its oceans—is populated by a great diversity of organisms. Scientists have developed systems of naming and classifying organisms.

Looking Ahead

internet connect

www.scilinks.org

National Science Teachers Association *sci*LINKS Internet resources are located throughout this chapter.

SCILINKS. Maintained by the National Science Teachers Association

Opening Activity — BASIC

Show students two plants that have obvious differences, such as an ivy plant or a fern and a plant in bloom. Ask them to devise a single-word name that could be used to describe both plants. Next have them distinguish the plants from each other by adding a second word that is unique to each plant. Have students share their names and point out similarities and differences in their nomenclature.
LS Visual

Quick Review Answers

1. A species is a group of genetically similar organisms that naturally mate to produce fertile offspring.
2. Organisms that have similar compositions of various macromolecules, such as proteins and nucleic acids, have a more recent common ancestor than do organisms that have less similar compositions of macromolecules. Having a recent common ancestor means that organisms have a similar evolutionary history and are more closely related.
3. Homologous structures have similarities in basic structure that indicate a common ancestry. The homologous structures may be modified and no longer serve the same function.
4. Speciation is the process by which new species are formed. It occurs when populations of the same species living in different locations evolve in different directions, eventually becoming unable to breed with one another.

Reading Activity Answers

Students' answers will vary but should show an understanding of the subject matter of the chapter.

Chapter Resource File

- Vocabulary Worksheets
- Concept Mapping

Focus

Overview

Before beginning this section review with your students the objectives listed in the Student Edition. In this section, students learn Linnaeus's role in developing the modern system of biological classification and the eight levels of classification used by scientists today.

Bellringer

Have students review the chapter and list the name and number of each type of organism illustrated (e.g., cactuses, bees, humans, oaks, etc.). Ask them to suggest reasons why a scientific method of classification is useful to study these organisms. (Classification facilitates: organizing the huge diversity of living things into categories; identifying similarities, differences, and evolutionary relationships; understanding the characteristics of all living things, etc.)

Motivate

Discussion/Question — BASIC

Ask students if they have ever heard of two people who have the same name. Have them consider what kinds of problems could arise from this situation. Ask students how they could differentiate themselves from an individual with the same name. (social security number, address, age, and so on)
LS Interpersonal

Section Correlations

SC.G.1.4.1
SC.H.2.4.1

Categories of Biological Classification

Florida Sunshine Standards/FCAT
SC.G.1.4.1 There is great diversity and interdependence among living things.
SC.H.2.4.1 Scientists assume that the universe is a vast system in which basic rules exist that may range from very simple to extremely complex, but scientists operate on the belief that the rules can be discovered by careful, systemic study.

Objectives

- **Describe** Linnaeus's role in developing the modern system of naming organisms. SC.G.1.4.1
- **Summarize** the scientific system for naming a species. SC.H.2.4.1
- **List** the seven levels of biological classification. SC.G.1.4.1

Key Terms

taxonomy
binomial nomenclature
genus
family
order
class
phylum
kingdom
domain

Taxonomy

Just as it is impossible for postal workers to sort mail bearing only the addressee's first name, it is impossible for biologists to memorize every name of the estimated 10–30 million organisms on Earth. To make sorting mail easier, postal workers sort first by zip code, then by street name and house number. In the same way, biologists group organisms into large categories that in turn are assigned to smaller and more specific categories.

More than 2,000 years ago, the Greek philosopher and naturalist Aristotle grouped plants and animals according to their structural similarities. Later Greeks and Romans grouped plants and animals into basic categories such as oaks, dogs, and horses. Eventually each unit of classification came to be called a *genus (JEE nuhs)* (plural, *genera*), the Latin word for "group." Starting in the Middle Ages, genera were named in Latin. The science of naming and classifying organisms is called **taxonomy** *(tak SAH nuh mee)*.

Until the mid-1700s, biologists named a particular type of organism by adding descriptive phrases to the name of the genus. These phrases sometimes consisted of 12 or more Latin words. They were called polynomials (from *poly*, meaning "many," and *nomen*, meaning "name"). As you can see in **Figure 1,** the polynomial for the European honeybee became very large and awkward. Polynomials were often changed by biologists, so organisms were rarely known to everyone by the same name.

Figure 1 European honeybee. The European honeybee once had a 12-part scientific name.

Polynomial: *Apis pubescens, thorace subgriseo, abdomine fusco, pedibus posticis glabis, untrinque margine ciliatus*

A Simpler System

A simpler system for naming organisms was developed by the Swedish biologist Carl Linnaeus. His ambition was to catalog all the known kinds of organisms. In the 1750s, he wrote several books that used the polynomial system. But Linnaeus also included a two-word Latin name for each species. Linnaeus's two-word system for naming organisms is called **binomial** *(bie NOH mee uhl)* **nomenclature** (from *bi*, meaning "two"). His two-part name for the European honeybee was *Apis mellifera*, the genus name followed by a single descriptive word. Over the past 250 years since Linnaeus first used two-part binomial species names, his approach has been universally adopted. Most of the species he described in 1753 still have the two-part names he gave them.

300

Chapter Resource File
- Directed Reading BASIC
- Active Reading GENERAL
- Data Sheet for Quick Lab GENERAL

Transparencies
TR Bellringer

One-Stop Planner CD-ROM
- Reading Organizers BASIC
- Reading Strategies BASIC
- Basic Skills Worksheet Words and Roots BASIC

Scientific Names

The unique two-part name for a species is now referred to as its *scientific name*. The first word is the genus to which the organism belongs. A **genus** is a taxonomic category containing similar species. Organisms in a genus share important characteristics. For example, the genus *Quercus* is composed of oak trees. The second word in a scientific name identifies one particular kind of organism within the genus, called a species. A species is the basic biological unit in the Linnaean system of classification. **Table 1** lists and describes two species of oaks in the genus *Quercus*.

The first letter of the genus name is always capitalized, and the first letter of the second word is always lowercase. Scientific names are italicized or underlined. After the first use of the full scientific name, the genus name can be abbreviated as a single letter. For example, *Quercus rubra* can be abbreviated *Q. rubra*.

The scientific name of an organism gives biologists a common way of communicating, regardless of their native language. One species may have many common names, and one common name may be used for more than one species. For example, the bird called a robin in Great Britain is *Erithacus rubicula*. The very different bird called a robin in North America is *Turdus migratorius*.

The name given to a species must conform to the rules established by an international commission of scientists. All scientific names must have two Latin words or terms created according to the rules of Latin grammar. Two different types of organisms cannot have the same scientific name. Since all the members of a genus share the genus name, the second word in the name of each member of that genus must be different. For example, only one species of the genus *Homo* can be given the name *sapiens*.

WORD Origins

The word *Quercus* is Latin for "oak." The word *rubra* is Latin for "red," and the word *phellos* is Greek for "cork." (Cork is part of the bark of a tree.)

internet connect

www.scilinks.org
Topic: Naming Species
Keyword: HX4127

SCiLINKS. Maintained by the National Science Teachers Association

Table 1 Two Species of Oak

Common name	Genus	Scientific name	Traits
Red oak 	*Quercus* 	*Quercus rubra* 	Lobed leaves; produces acorns approximately 25 mm (1 in.) long
Willow oak 	*Quercus* 	*Quercus phellos* 	Unlobed leaves; produces acorns approximately 15 mm (0.6 in.) long

301

HISTORY CONNECTION

Over a number of years, Linnaeus observed that certain plants opened and closed their flowers at particular times of the day and that he could deduce the approximate time of day according to which plants had opened or closed their flowers. Arranged in sequence of flowering over the day, they constituted a "floral clock" or *horologium florae*, as Linnaeus called it in his *Philosophia Botanica* (1751).

Various attempts have been made to illustrate the concept, mostly in the form of a clock face decorated with an image of the flower for the time of day.

Teach

Teaching Tip

Binomial Nomenclature Tell students that Linnaeus did not invent binomial nomenclature. The use of two-word names for species, or for types of organisms within a group, occurs in many languages and goes back to ancient times. It is indeed the common practice in vernacular nomenclature. Linnaeus improved these methods of classification by adding consistency and precision to his system. He linked each of the specific everyday names with a descriptive name, which helped to identify the species and limited the application of its two-word name to that one species. During the second half of the 18th century, the adoption by biologists of this consistent two-word nomenclature for species came about because Linnaeus introduced it in comprehensive works, which naturalists soon found indispensable.

READING SKILL BUILDER

Brainstorming Explain to the class that in situations with an overwhelming amount of information, a useful classification system might include both broad and specific categories. Geography provides a good example. A broad category might be North America, followed by the United States, then the Midwest, then Minnesota, then Minneapolis, and so on, until you reach a specific street address. Work with the class to develop a list of other things that are categorized in this way. **LS Verbal**

Demonstration — GENERAL

Show students a picture of *Felis concolor*, and ask them to name the animal. (lion, mountain lion, puma, cougar, panther, or catamount) Emphasize that besides being ambiguous, common names can be misleading. Ask students to give examples of misleading common names. (Some examples include the starfish, sea horse, silverfish, horseshoe crab, and ringworm.) **LS Visual**

Teach, *continued*

READING SKILL BUILDER

Interactive Reading Assign Chapter 14 of the *Holt Biology* Guided Audio CD Program to help students achieve greater success in reading the chapter. **LS Auditory**

SKILL BUILDER — ADVANCED

Writing Skills Have students research how species are named using the Internet or library resources. Have students write a report summarizing their findings. **LS Verbal**

Demonstration

Display photos of the following organisms, and write their scientific names on the board: *Carnegiea gigantea* (giant saguaro cactus), *Nymphaea ordata* (fragrant water lily), *Canis familiaris* (domestic dog), *Viola tricolor* (pansy with three-colored flowers), and *Peromyscus californicus* (a mouse common in California). Before telling them the common names, have students suggest why these organisms were given these particular scientific names. **LS Visual**

Classifying Organisms

Linnaeus worked out a broad system of classification for plants and animals in which an organism's form and structure are the basis for arranging specimens in a collection. The genera and species that he described were later organized into a ranked system of groups that increase in inclusiveness. The different groups into which organisms are classified have expanded since Linnaeus's time and now consist of eight levels, as shown in **Figure 2.**

Similar genera are grouped into a **family.** Similar families are combined into an **order.** Orders with common properties are united in a **class.** Classes with similar characteristics are assigned to a **phylum** *(FIE luhm)*. Similar phyla are collected into a **kingdom.** Similar kingdoms are grouped into **domains.** All living things are grouped into one of three domains. Two domains, Archaea and Bacteria, are each composed of a single kingdom of prokaryotes. The third domain, Eukarya, contains all four kingdoms of eukaryotes.

In order to remember the eight categories of classification in their proper order, it may prove useful to memorize a phrase, such as **D**o **K**indly **P**ay **C**ash **O**r **F**urnish **G**ood **S**ecurity, to remember **D**omain **K**ingdom **P**hylum **C**lass **O**rder **F**amily **G**enus **S**pecies.

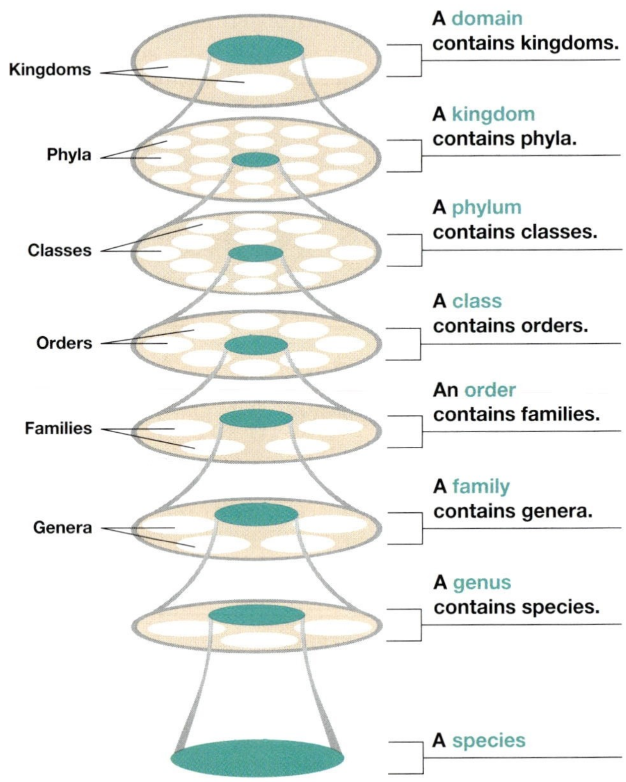

Figure 2 System of classification. Each living thing is assigned to a series of groups, beginning with domain (most inclusive) and ending with species (least inclusive).

internet connect

www.scilinks.org
Topic: Systems of Classification
Keyword: HX4173

SCILINKS. Maintained by the National Science Teachers Association

302

Transparencies

TR D29 Biological Hierarchy of Classification

Attention Grabber

"A rose by any other name . . ." Common names in different cultures can cause confusion. The flowering plant *Mertensia virginica* is known in the United States as the bluebell. In Europe and Asia, however, bluebells are plants from the genus *Endymion*. Elsewhere, certain species of the genera *Campanula*, *Clematis*, and *Polemonium* also are known as bluebells.

Classification of the Honeybee

Scientific names are particularly powerful because they tell you so much about the organism. Each level of classification is based on characteristics shared by all the organisms it contains. For example, consider the classification of the honeybee, shown in **Figure 3.**

The honeybee's scientific name, *Apis mellifera*, indicates that it belongs to the genus *Apis*, which is classified in the family Apidae. Knowing the honeybee's family is Apidae tells you a great deal about the honeybee. All members of the family Apidae are bees that live either alone or in hives, as does *Apis mellifera*.

Knowing the order bees belong in tells you even more. The order to which the honeybee belongs, Hymenoptera, includes ants, bees, and wasps, which usually have two pairs of wings and are likely to be able to sting. At each higher level, the information becomes more general.

At the next higher level of classification, *A. mellifera* belongs to the class Insecta, meaning it is an insect with three major body parts and three pairs of legs. Its phylum, Arthropoda, indicates that *A. mellifera* is an arthropod, an organism with a coelom, segmented body, jointed appendages, and a hard outer skin made of a complex

Real Life

You named *what* after me?

Several plant and animal names honor famous people. *Shakesperia* was the original name of a genus of parasitic wasp (see *The Taming of the Shrew*, Act 2, Scene I), and cartoonist Gary Larson has both a beetle and a biting louse named for him.

Finding Information

Look up the scientific name of your favorite plant or animal, and create a binomial based on your own name.

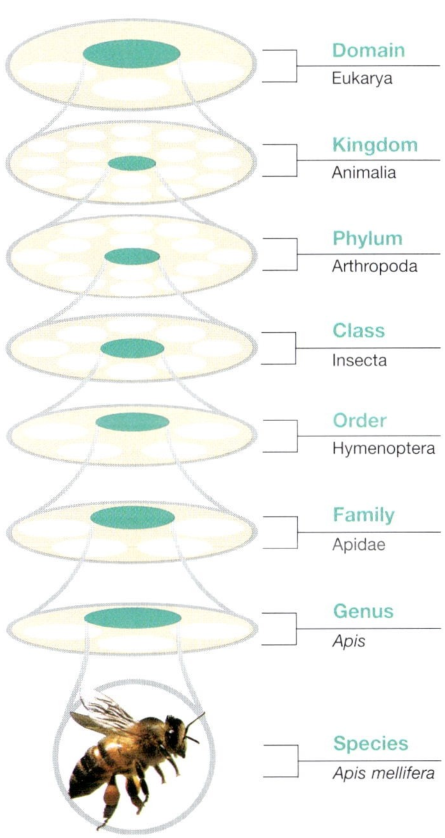

Figure 3 Classification of a bee. The European honeybee is classified in eight different levels of biological classification.

Real Life

Students' names should be Latinized. Point out that Carolus Linnaeus Latinized his own name; it originally was Carol von Linne.

Activity — GENERAL

Comparing Classifications

Ask students to devise a Graphic Organizer similar to the one at the bottom of this page that compares humans, chimpanzees, and gorillas in a hierarchy, beginning with kingdom Animalia. Provide books that help students identify the kingdom, phylum, class, order, family, genus, and species for each organism. **LS Visual**

Group Activity — ADVANCED

These Are the Names of My Favorite Things Have students work in groups to list the common names of their favorite plants and animals. Then have each group research the classification of these organisms and report to the class the scientific name of each organism, as well as its kingdom, phylum, class, order, and family. Make sure that students in each group choose different organisms to research. **LS Intrapersonal**

Co-op Learning

Graphic Organizer

Use this graphic organizer with **Activity** on this page.

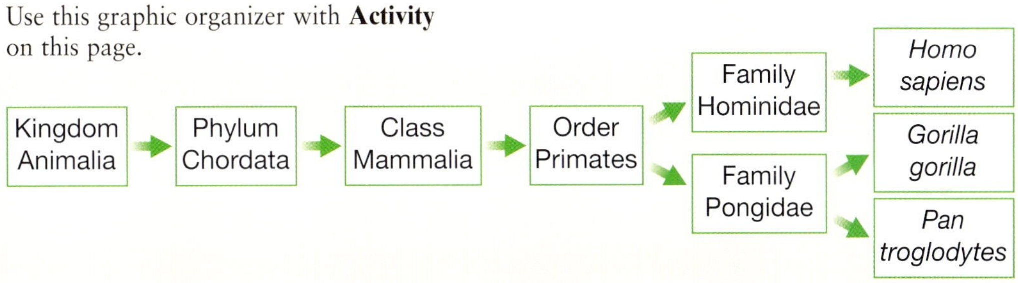

carbohydrate called chitin. Arthropods have been the most successful of all animals. Two-thirds of all the named species on Earth are arthropods.

Its kingdom, Animalia, tells you that *A. mellifera* is a multicellular heterotroph whose cells lack walls. And finally, *A. mellifera* is classified in the domain Eukarya, which is composed of all eukaryotes. Recall that eukaryotes are composed of one or more cells that have membrane-bound organelles and whose DNA is enclosed within a nucleus.

QUICK LAB

Using a Field Guide SC.H.1.4.1

You can use a standard pictoral field guide or a dichotomous key to help you identify species of plants, animals, or other organisms.

Materials

paper and pencil, a plant or animal field guide

Procedure

1. **CAUTION: Wear protective gloves when handling any wild plant. Keep your hands away from your face.** Using a dichotomous key or other field guide, identify several species of plants that share the same phylum and class. Collect specimens only if your teacher tells you to do so.
2. Try to identify two plants of the same genus.
3. In a table like the one on the right, record the scientific name of each specimen.
4. Read the description of each species in the field guide. Determine the set of charateristics that fit each specimen. Write these characteristics in a table like the one below.

Analysis

1. **List** the characteristics shared by two specimens that are in the same genus but are different species.
2. **Describe** how the binomial name of these two species show that they are members of the same genus.
3. **Identify** the key characteristics your field guide uses to tell these two species apart.
4. **Critical Thinking** **Analyzing Data** Based on your observations, are two species from the same genus more similar or less similar than two species from different genera?

Specimen	Genus name	Binomial species name	Identifying characteristics
A			
B			
C			

1. **Explain** how Linnaeus helped develop the modern system of taxonomy. SC.G.1.4.1
2. **List** rules that scientists use to name organisms. SC.H.2.4.1
3. **Sequence** the eight levels of biological classification, beginning with the most specific level. SC.G.1.4.1
4. *FCAT Prep* Which organism is most similar to the sandhill crane, *Grus canadensis*?
   - **A** *Branta canadensis*
   - **B** *Falcipennis canadensis*
   - **C** *Grus americana*
   - **D** *Recurvirostra americana*

   SC.G.1.4.1

304

QUICK LAB

Using a Field Guide

Skills Acquired

Observing, recognizing relationships

Teacher's Notes

It is easiest to identify plants when they are flowering.

Analysis Answers

1. Answers will vary but may include leaf shape.
2. They have the same genus name.
3. Answers will vary.
4. more similar

Close

Reteaching — Basic

Write scientific names on the board. Have students research the meanings of each word and the common name of each organism.

Quiz — General

1. Why is Latin useful for naming organisms? (because it is recognized by people all over the world)
2. Similar families are grouped together in the same _______. (order)

Alternative Assessment — General

Have students determine what is incorrect about each of the following "scientific" names: spotted leopard (common name), *Tabanus b.* (species name missing), *Acer Rubrum* (species name is capitalized).

Answers to Section Review

1. Linnaeus added consistency and precision to the practice of identifying species with two words and of grouping species into broader, more inclusive categories.
2. The first letter of the genus name is capitalized; the first letter of the species name is not. Both names are printed in italics or underlined. The names must be Latinized, and no two species of the same genus can have the same species name.
3. species, genus, family, order, class, phylum, kingdom, domain
4. **A.** Incorrect. This organism is not in the genus *Grus*. **B.** Incorrect. This organism is not in the genus *Grus*. **C.** Correct. **D.** Incorrect. This organism is not in the genus *Grus*.

Section Review Correlations

SC.G.1.4.1: Items 1, 3, 4

SC.H.2.4.1: Item 2

How Biologists Classify Organisms

Section 2

Florida Sunshine Standards/FCAT

SC.G.1.4.1 There is great diversity and interdependence among living things.

SC.H.2.4.1 Scientists assume that the universe is a vast system in which basic rules exist that may range from very simple to extremely complex, but scientists operate on the belief that the rules can be discovered by careful, systemic study.

What Is a Species?

Have you ever wondered how scientists tell one species from another? For example, how can you tell a mushroom that can be eaten from a similar looking mushroom that is poisonous? Scientists traditionally have used differences in appearance and structure to group organisms. Sometimes, however, these differences cannot be used to determine if two organisms are of the same species. For example, all single-celled protists classified as *Paramecium syngens* were once thought to be a single species. These organisms actually represent several protistan species that look very similar but have other characteristics that differ.

In 1942, the biologist Ernst Mayr of Harvard University proposed a biologically based definition of species, which is called the biological species concept. Mayr defined a **biological species** as a group of natural populations that are interbreeding or that could interbreed, and that are reproductively isolated from other such groups.

Reproductive isolation occurs when a barrier separates two or more groups of organisms and prevents them from interbreeding. In nature, however, reproductive barriers between sexually reproducing species are not always complete. Sometimes individuals of different species interbreed and produce offspring called *hybrids*. For example, wolves and dogs are members of separate species in the genus *Canis*. But interbreeding between wolves and dogs produces fertile offspring, such as the hybrid shown in **Figure 4.** Another example of a fertile hybrid is triticale, a hybrid of wheat and rye. It is important to remember that when reproductive barriers between two species are not complete, the two species are closely related.

Objectives

- **List** the characteristics that biologists use to classify organisms. SC.H.2.4.1
- **Summarize** the biological species concept. SC.G.1.4.1
- **Relate** analogous structures to convergent evolution. SC.D.1.4.4
- **Describe** how biologists use cladograms to determine evolutionary histories. SC.H.2.4.2

Key Terms

biological species
phylogeny
convergent evolution
analogous character
cladistics
ancestral character
derived character
cladogram
evolutionary systematics
phylogenetic tree

Figure 4 Dog-wolf hybrid. Wolves and dogs can produce fertile offspring.

Dog (*Canis familiaris*)

Wolf (*Canis lupus*)

Dog-wolf hybrid

Section 2

Focus

Overview

Before beginning this section review with your students the objectives listed in the Student Edition. In this section, students learn the characteristics biologists use to classify organisms, the biological species concept, the relationship between an organism's classification and its evolutionary history, and how biologists use cladograms to study organisms' evolutionary histories.

Bellringer

Have students write down the names of as many different kinds of cats as they can think of. Tell them that all domestic cats belong to the same genus, *Felis*. Ask them to identify which cats they think belong to the same species. (All house cats are in the same species.)

Motivate

Activity — GENERAL

Classifying Hardware Allow students to work with a partner. Give each pair of students a small handful of hardware—nuts, bolts, screws, washers, and so on. Have them group similar pieces into "genera," and give each piece a unique "species" name. Have students share with the class their genus and species names for their various pieces of hardware. LS Verbal English Language Learners

Chapter Resource File

- Active Reading GENERAL
- Data Sheet for Data Lab GENERAL

Transparencies

TR Bellringer
TR D30 Classification of a Bee

One-Stop Planner CD-ROM

- Reading Organizers BASIC
- Reading Strategies BASIC
- Supplemental Reading Guide *The Dinosaur Heresies* ADVANCED

Section Correlations

SC.D.1.4.4
SC.G.1.4.1
SC.H.2.4.1
SC.H.2.4.2

Teach

READING SKILL BUILDER — GENERAL

Anticipation Guide Before reading this section, have students write the answer to the following question in their notebooks: How do scientists determine if similar organisms should be categorized as the same species? Discuss their answers, and have students make any necessary corrections or additions after reading this section.
LS Interpersonal

DATA LAB

Analyzing Taxonomy of Mythical Organisms

Skills Acquired
Applying information, classifying, evaluating

Teacher's Notes
Have students create a field guide to accompany the given information. Ask students to place the mythical creatures into real-life classes and orders and to list the traits they used to categorize each creature.

Analysis Answers

1. Answers will vary.
2. Answers will vary. Students may group the creatures according to size, presence of wings, or other features.
3. The biological species concept cannot be used to classify them without knowledge of their breeding compatibility with other groups.

Figure 5 **New species of tree kangaroo.** *Dendrolagus mbaiso* is a black-and-white tree kangaroo that lives in the mountains of New Guinea.

Evaluating the Biological Species Concept

The biological species concept works well for most members of the kingdom Animalia, in which strong barriers to hybridization usually exist. For example, Asian elephants and African elephants do not interbreed in nature. But the biological species concept fails to describe species that reproduce asexually, such as all species of bacteria and some species of protists, fungi, plants, and even some animals.

Within many groups of organisms, there are no barriers to interbreeding between the species. Many species of plants, some mammals, and many fishes are able to form fertile hybrids with one or a few closely related species. In practice, modern biologists recognize species by studying an organism's features.

Number of Species

The number of species in the world is much greater than the number described. Large numbers of species, such as the kangaroo shown in **Figure 5,** are still being discovered. Only about 1.5 million species have been described to date. Scientists estimate that 5 million to 10 million more species may live in the tropics alone. Since no more than 500,000 tropical species have been named, it is clear that our knowledge of Earth's diversity of species is limited.

DATA LAB

Analyzing Taxonomy of Mythical Organisms SC.H.2.4.2

Background

Classification of organisms often requires grouping organisms based on their characteristics. Use the following list of mythological organisms and their characteristics to complete the analysis.

- **Pegasus** stands 6 ft tall, has a horse's body, a horse's head, four legs, and two wings.
- **Centaur** stands 6 ft tall, has a horse's body with a human torso, a male human head, and four legs.
- **Griffin** stands 4–6 ft tall, has a lion's body, an eagle's head, four legs, two wings, fur on its body, and feathers on its head and wings.
- **Dragon** can grow to several hundred feet, has a snake-like body, from 1–3 reptile-like heads, four legs, scales, and breathes fire.
- **Chimera** stands 6 ft tall, has a goat's body, snake's tail, four legs, a lion's head, fur on its body and head, scales on its tail, and breathes fire.
- **Hydra** is several hundred feet long, has a long body with four legs and a spiked tail, 100 snake heads, scales, and is poisonous.

Analysis

1. **Identify** the characteristics that you think are the most useful for grouping the organisms into separate groups.
2. **Classify** the organisms into at least three groups based on the characteristics you think are most important.
3. **Evaluate** the use of the biological species concept to classify these mythical organisms.

SOCIAL STUDIES CONNECTION

In 1966, Congress passed the Endangered Species Preservation Act, which stipulated that only native animal species be listed as *endangered* and provided limited means for their protection. Under this law, the Departments of Interior, Agriculture, and Defense were to protect the listed species, preserve their habitats, and acquire land for their protection. The Endangered Species Conservation Act of 1969 provided additional protection for species in danger of "worldwide extinction."

The Endangered Species Act of 1973 combined and strengthened the provisions of the 1966 and 1969 legislation and also broke new ground. Some of its provisions include: the requirements that plants and all classes of invertebrates be protected, that all federal agencies undertake species conservation programs and provide matching funds for states with cooperative agreements, and that land be acquired for listed animals and plants.

Evolutionary History

Linnaeus's classification system was based on his observation that organisms have different degrees of similarity. For instance, a tiger resembles a gorilla more closely than either resembles a fish. According to Darwin's views, organisms that are more similar to one another than they are to other organisms have descended from a more recent common ancestor. Therefore, classification based on similarities should reflect an organism's **phylogeny,** that is, its evolutionary history. Inferring evolutionary connections from similarities, however, can be misleading. Not all features—or *characters*—are inherited from a common ancestor. Consider the wings of a bird and the wings of an insect. Both enable flight, but the structures of the two kinds of wings differ. Moreover, fossil evidence indicates that they evolved independently of one another. Through the process called **convergent evolution,** similarities evolve in organisms not closely related to one another, often because the organisms live in similar habitats. Similarities that arise through convergent evolution are called **analogous** *(ah NAHYL uh guhs)* **characters. Figure 6** shows an example of convergent evolution.

internet connect
www.scilinks.org
Topic: Aquariums and Museums of Florida
Keyword: HX6F01
SCILINKS. Maintained by the National Science Teachers Association

Cladistics

Most biologists today analyze evolutionary relationships using cladistics *(kluh DIHS tihks)*. **Cladistics** is a method of analysis that reconstructs phylogenies by inferring relationships based on shared characters. Cladistics can be used to hypothesize the sequence in which different groups of organisms evolved. To do this, cladistics focuses on the nature of the characters in different groups of organisms. With respect to two different groups, a character is defined as an **ancestral character** if it evolved in a common ancestor of both groups. Thus when considering the relationship between birds and mammals, a backbone is an ancestral character. Having feathers, however, is a derived character. A **derived character** evolved in an ancestor of one group but not of the other. Feathers evolved in an ancestor of birds that was not also ancestral to mammals.

internet connect
www.scilinks.org
Topic: Taxonomy
Keyword: HX4174
SCILINKS. Maintained by the National Science Teachers Association

Figure 6 Similar structures in the cactus and spurge families

Although they evolved in different parts of the world, cactuses and spurges look similar.

Organ pipe cactus, *Sternocereus thurber*

Gifboom, *Euphorbia virosa*

307

Using the Figure — GENERAL

Direct students' attention to **Figure 6.** Point out that the organ pipe cactus is found in some of the deserts of the Western Hemisphere, and cactus-like spurges, such as the gifboom, are found in the Eastern Hemisphere. Ask students how the organisms could be so similar yet not be closely related. (The two organisms evolved in similar environments, developing similar adaptations for survival in hot, dry deserts.) LS **Visual**

READING SKILL BUILDER — BASIC

Brainstorming Ask students to brainstorm and derive a list of analogous structures that are found in organisms. (Examples: insect wing/bird wing; platypus bill/duck bill) Have them explain their hypotheses about the adaptive value of each structure. LS **Logical**

SKILL BUILDER — ADVANCED

Writing Skills Have students research taxonomy using the Web site in the Internet Connect box on this page and write a report summarizing their findings. LS **Verbal**

Activity — ADVANCED

Loss of Biodiversity Inform students of the following data regarding the diversity of life in the rain forest: 1,200 different species of beetles have been found on just 19 tree species; 80,000 species of plants have been identified in the Amazon Basin (compared to 1,200 species in New England); Brazil has more terrestrial vertebrates than any other nation. Ask students to write an essay explaining why the South American rain forests are so rich in diversity. Have them share their ideas, then discuss the impact of deforestation on the region and the world. LS **Interpersonal**

MISCONCEPTION ALERT

Teleology Evolution is sometimes mistakenly described in terms of purposefulness—that wings, for example, evolved so an organism could fly. This view of evolution is based on teleology (Greek *telos*, "end"; *logos*, "discourse"), or the doctrine that explains phenomena in terms of ends or final causes. Teleology is based on the proposition that the universe has a design and purpose, and that natural processes are determined by their utility in an overall natural design. Darwin's theory of evolution by natural selection, which has been supported by countless experiments and observations, does not include a design or plan, but states that populations (not individuals) change in response to their environment as individuals better adapted to the environment leave more offspring than those not suited to the environment. It is also incorrect to say that every feature of an organism is present because it has adaptive value. Some traits are neither selected for or against and are neutral in terms of natural selection.

Teach, *continued*

Teaching Tip

Dolphin Evolution Tell students that Aristotle recognized that dolphins and land mammals shared many characteristics, and he classified them with land creatures. This fact was overlooked by biologists for almost 2,000 years. It is now thought that the ancestors of dolphins lived on land and that their legs became modified into flippers as they became aquatic.

Exploring Further

Cladograms

Teaching Strategies
Point out that DNA analysis is changing some long-held beliefs about evolutionary relationships based on anatomical traits alone. For example, the cheetah was thought to be a unique standout in the cat family. DNA studies show, however, that it is closely related to the pumas.

Discussion

- How can DNA analysis alone be used to construct a cladogram? (A similar chart and graph showing evolutionary relationships could be created based on similarities and differences between genes.)
- How can the presence or absence of certain traits be misleading when making judgments about evolutionary relationships? (An organism's phenotype does not necessarily reveal all of the genes that it carries. Some organisms no longer have traits that were found in their ancestors.)

Cladistics is based on the principle that shared *derived* characters provide evidence that two groups are relatively closely related. Shared *ancestral* characters, however, do not. For example, lizards and dogs have limbs, but a whale has no limbs. Having limbs does not provide evidence that dogs and lizards are more closely related than are dogs and whales. Recall that whales are descended from an ancestor that had limbs. Therefore, the presence of limbs is a shared ancestral character of all three groups. However, both dogs and whales have mammary glands—a shared derived character not found in lizards or lizard ancestors. This provides evidence that dogs and whales share a more recent common ancestor than either shares with lizards.

Exploring Further

Cladograms

How many different ways can you organize your possessions? For example, should all your clothes be grouped according to their type or according to color? Biologists sometimes disagree about how to organize groups of organisms.

Why Study Cladograms?
Some biologists use cladograms to study the evolutionary relationships among certain groups of organisms, such as species within a genus or genera within a family. Cladograms show how closely two or more groups are related, based on important characteristics. Cladograms convey comparative information about relationships. Organisms that are grouped more closely on a cladogram share a more recent common ancestor than those farther apart. Because the analysis is comparative, cladistic analysis deliberately includes an organism that is only distantly related to the other organisms. This distantly related organism is called an *out-group.* The out-group serves as a base line for comparisons with the other organisms being evaluated, the in-group.

	Characters		
Organisms	**Vascular tissue**	**Seeds**	**Flowers**
Mosses (out-group)	0	0	0
Pine trees	1	1	0
Flowering plants	1	1	1
Ferns	1	0	0

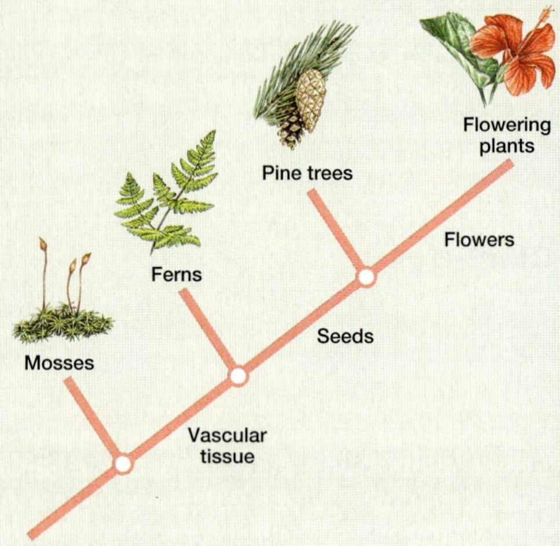

Constructing a Cladogram
This example shows the evolutionary relationships among plants.

1. In the table at the left, the characters in the row for the out-group are marked with a 0. A character not found in the out-group is considered a derived character and is marked with a 1.
2. Starting with a diagonal line, as shown above, the out-group (mosses) is placed on the first branch. Just past this first branch, the most common derived character is listed—vascular tissue. Vascular tissue is a series of tubes and vessels within a plant.
3. Next the second most common derived character—seeds—is listed. Ferns lack seeds and so are placed on the second branch of the cladogram.
4. The third most common derived character—flowers—is then listed. Conifers do not have flowers and so are placed on one branch, and the flowering plants are placed on the other.

MEDICINE CONNECTION

Cytochrome *c* is a protein found in all aerobic organisms. The amino acid sequences of this protein have been determined for a wide variety of species and provide data about the degree of relatedness among species. For example, human and canine cytochrome *c* differ by 13 amino acids, human and rattlesnake cytochrome *c* by 20 amino acids, and human and tuna cytochrome *c* by 31 amino acids. The 104 amino acids of human and chimpanzee cytochrome *c* match exactly.

Career

Museum Curators Natural history museums, zoos, gardens, and aquariums hire curators to maintain and expand exhibits and collections, create new exhibits, conduct research, and provide public programs. There is overlap with the fields of research and education. A bachelor's degree in the sciences may be required, and many curators hold advanced degrees. Encourage students to find out more about the work of curators. Visit a local museum with the class, or invite a curator from a nearby natural history or science museum to speak to the class.

A biologist using cladistics constructs a branching diagram called a **cladogram,** which shows the evolutionary relationships among groups of organisms. Organisms that share derived characters are grouped together on the cladogram. As groups evolve, new derived characters appear on the cladogram that were not present in earlier organisms.

Considering Characters

The great strength of cladistics is objectivity. If a computer is fed the same set of data repeatedly, it will make exactly the same cladogram every time. The disadvantage of cladistics is that the *degree* of difference between organisms is not considered. Cladistic analysis simply indicates that a character does or does not exist.

Cladistic analysis does not take into account variations in the "strength" of a character, such as the size or location of a fin or the effectiveness of a lung. Each character is treated equally. Thus in a cladogram of vertebrate evolution, birds are grouped with reptiles. This accurately reflects their true ancestry—birds evolved from dinosaurs. The immense evolutionary impact of a derived character like feathers, however, is ignored.

Evolutionary success often depends on high-impact events, such as the evolution of feathers. Some modern cladistic studies therefore attempt to weigh the evolutionary significance of the characters being studied.

Study TIP

Organizing Information
The numbered list in the Exploring Further feature tells you how to make a cladogram. Use this list to prepare a cladogram of the organisms listed in the Data Lab.

DATA LAB

Making a Cladogram SC.H.2.4.2

Background

A cladogram is a model that represents a hypothesis about the order in which organisms evolved from a common ancestor. Scientists construct a cladogram by first analyzing characters in a data table. The absence of a vascular system and the absence of seeds is ancestral. Use the data below to construct a cladogram on a separate sheet of paper.

Horsetail

Liverwort

Pine tree

Analysis

1. **Identify** the out-group.
2. **Name** the least common derived trait.
3. **List** the order in which the plants in the table would be placed on a cladogram.

	Characters	
Plants	Seeds	Vascular system
Horsetails	No	Yes
Liverworts	No	No
Pine trees	Yes	Yes

Teaching Tip

Ancestral and Derived Characters Clarify that the designation of ancestral and derived characters is useful only when it is related to a particular group under study. A character that is derived relative to one group may be ancestral for a less inclusive group. For example, the presence of fur is a derived character if one is studying tetrapods (four-footed vertebrates), and it serves to distinguish mammals from their ancestors, the reptiles. However, it is an ancestral character for the group consisting only of mammals, and it is not useful for determining relationships among them.

DATA LAB

Making a Cladogram

Skills Acquired
Interpreting results, applying information

Teacher's Notes
Inform students that millions of years ago, horsetails were more treelike than are modern ones. Pines and flowering trees, which developed later, outcompeted horsetails. Today, horsetails grow primarily in wet areas and are usually less than 1 m (39 in.) tall. Ask students to place the horsetail and liverwort on the cladogram they constructed in the activity on the previous page. (Students should place the horsetail with the ferns and the liverwort with the mosses. Not enough information is given to place them more precisely.)

Analysis Answers

1. Liverworts are the out-group.
2. seeds
3. liverwort, horsetail, pine tree

did you know?

Cladistics The premise behind cladistic analysis is that by examining ancestral and derived characters, diagrams can be drawn that clarify the evolutionary relationships between groups. Branching points (nodes) on the diagram are generated every time a derived character (or group of them) is identified which one group possesses and another does not. The two groups on alternate sides of a node are called sister-groups. By analyzing enough different characters, or traits, eventually a true picture of the family tree can be generated. The goal is to define a group in which all members are descended from a single, common ancestral species, and all descendant species are included. This is called a *monophyletic* group. If all members of a group are not descended from a single common ancestor, the group is termed *polyphyletic*. If the group doesn't contain every descendant of that common ancestor, it is called *paraphyletic*. An example of a paraphyletic group is the reptiles. The Class Reptilia in its traditional sense is a useful concept, but it doesn't contain all the descendants of a common ancestor, because mammals and birds are generally placed in their own classes.

Close

Reteaching — BASIC

Have students work with a partner to devise a table and cladogram for the following animals: eagle, frog, goldfish, dog, and mouse. They should compare at least five different traits. Allow students to rank the animals as they choose.

Quiz — GENERAL

1. The formation of fertile hybrids presents an exception to the concept of biological _______. (species)
2. Closely related organisms share _______ characters. (derived)
3. What is the main disadvantage of cladistics? (It simply indicates that a character does or does not exist, without taking into account variations in the strength, or importance, of a character.)

Alternative Assessment — GENERAL

Have students prepare a graphic organizer to show the relationships of the following shapes: triangle, hexagon, trapezoid, parallelogram, square, pentagon, rectangle, quadrilateral, and polygon. Suggest that they begin by grouping the shapes according to their general structure, and then group them according to their more specific shapes. Ask them to justify the rationale for their graphic organizer.

Transparencies

TR D31 Evolutionary Systematics and Cladistic Taxonomy

Section Review Correlations

SC.D.1.4.4: Items 3, 4, 5
SC.G.1.4.1: Item 2
SC.H.2.4.1: Item 1

internet connect
www.scilinks.org
Topic: Evolutionary Systematics
Keyword: HX4077
SCILINKS. Maintained by the National Science Teachers Association

Evolutionary Systematics

In **evolutionary systematics,** taxonomists give varying degrees of importance to characters and thus produce a subjective analysis of evolutionary relationships. In this type of analysis, evolutionary relationships are displayed in a branching diagram called a **phylogenetic tree.** As you can see in **Figure 7,** evolutionary systematics places birds in an entirely separate class from reptiles, giving more importance to characters like feathers that made powered flight possible. Evolutionary systematics involves the full observational power of the biologist, along with any biases he or she may have. A phylogenetic tree and a cladogram are similar in that each represents a hypothesis of evolutionary history, which must be inferred because it was not observed.

Figure 7 Evolutionary systematics and cladistic taxonomy

Biologists differ in the ways that they classify organisms.

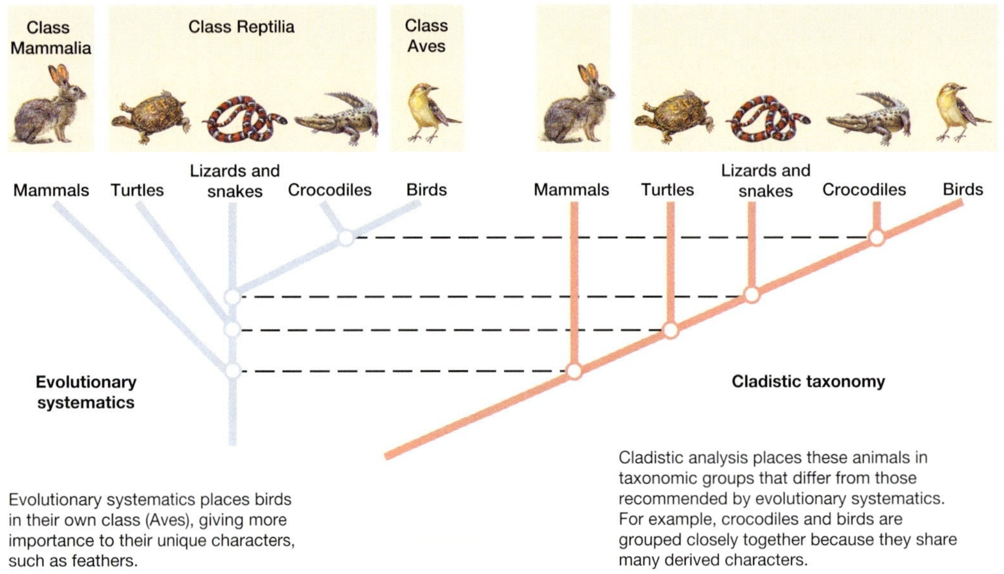

Section 2 Review

1. **Identify** the kinds of information that scientists use when they classify organisms. SC.H.2.4.1
2. **Summarize** what scientists mean by the biological species concept. SC.G.1.4.1
3. **Define** *cladistics,* and list the kind of information cladistic analysis reveals about evolutionary histories. SC.D.1.4.4
4. **Critical Thinking Inferring Relationships** Explain the relationship between convergent evolution and analogous characters. SC.D.1.4.4
5. FCAT Prep What do different species that are placed in the same group on a cladogram have in common?
   A appearance
   B scientific name
   C derived characters
   D analogous characters
   SC.D.1.4.4

310

Answers to Section Review

1. Scientists classify organisms based on appearance and structure, breeding patterns, and, more recently, similarities of DNA sequences.
2. This concept defines a species as a group of naturally interbreeding populations that is reproductively isolated from other such groups.
3. Cladistics is a method of analysis that uses similarities of organisms to infer evolutionary relationships. Cladistics allows scientists to infer the sequence in which different groups likely evolved.
4. In convergent evolution, similar features called analogous characters evolve in relatively unrelated species living in similar environments.
5. **A.** Incorrect. Appearance is only once characteristic used to group organisms. **B.** Incorrect. The scientific name is not used to group organisms. **C.** Correct. **D.** Incorrect. Analogous characters do not indicate evolutionary relationships and so are not used to group organisms.

ZONE

Key Concepts

1 Categories of Biological Classification

- Swedish biologist Carl Linnaeus developed binomial nomenclature, the two-word system of naming organisms.
- Scientific names are written in Latin and give biologists a common way of communicating, regardless of the language they speak.
- The scientific name of an organism consists of its genus name followed by a second name, which identifies its species.
- Each category of classification is based on characteristics that are shared by all the organisms in the category.
- Scientists use an eight-level system to classify organisms.
- The modern system of classification includes the groups domain, kingdom, phylum, class, order, family, genus, and species.

2 How Biologists Classify Organisms

- Biologists usually define species according to their appearance and structure.
- The biological species concept defines species according to their sexual reproductive potential.
- The biological species concept cannot be used to classify asexually reproducing species.
- Similar organisms may have analogous structures that arose through convergent evolution.
- Cladistics focuses on sets of unique characteristics found in a particular group of organisms to reconstruct an evolutionary history.
- Evolutionary systematics is a more subjective method of classification than is cladistics, but evolutionary systematics allows greater evolutionary importance to be placed on certain characters.

Key Terms

Section 1

taxonomy (300)
binomial nomenclature (300)
genus (301)
family (302)
order (302)
class (302)
phylum (302)
kingdom (302)
domain (302)

Section 2

biological species (305)
phylogeny (307)
convergent evolution (307)
analogous character (307)
cladistics (307)
ancestral character (307)
derived character (307)
cladogram (309)
evolutionary systematics (310)
phylogenetic tree (310)

Alternative Assessment — GENERAL

Have students use a field guide to identify trees and other plants in the schoolyard. Using weather-resistant materials, have them make identification labels that include the common and scientific names of the identified organisms. Place the labels on or near the appropriate specimens. **LS Logical**

Chapter Resource File

- **Science Skills Worksheet** GENERAL
- **Critical Thinking Worksheet** ADVANCED
- **Test Prep Pretest** GENERAL
- **Chapter Test** GENERAL

Answer to Concept Map

The following is one of several possible answers to Performance Zone item 15.

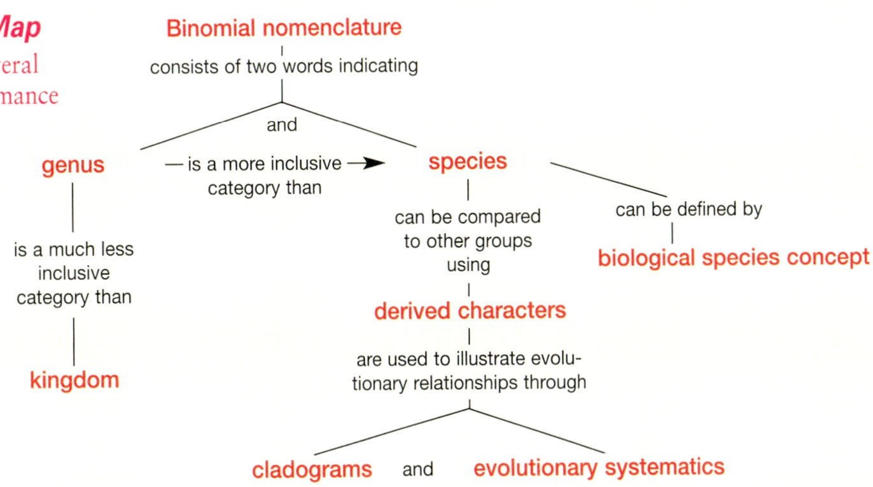

ANSWERS

Understanding Key Ideas

1. a
2. c
3. c
4. All living things are grouped into one of three domains. Each domain, therefore, contains one or more kingdoms of organisms.
5. Yes. According to the biological species concept, the horse and donkey would have fertile offspring if they were the same species.
6. d
7. Bats and birds both have wings. Many bats and birds also migrate seasonally.
8. A cladogram can show the sequence of the development of certain traits over evolutionary time, thereby illustrating evolutionary relationships between groups of organisms.
9. Analogous characters are similar characteristics that evolved separately in two relatively unrelated organisms. Homologous characters are characteristics that may or may not appear similar in two organisms but developed from a common ancestor.
10. One possible answer to the concept map is found at the bottom of the Study Zone page.

Critical Thinking

11. Homologous traits may develop differently in different organisms, but such traits originated in a common ancestor. By arranging related organisms on the basis of the presence or absence of cer-

Performance ZONE

Understanding Key Ideas

1. The system of binomial nomenclature was developed by
a. Linneaus. **c.** Mayr.
b. Aristotle. **d.** Darwin.
SC.G.1.4.1

2. The scientific name for humans is correctly written as
a. Homo sapiens. **c.** *Homo sapiens.*
b. Homo Sapiens. **d.** *Homo Sapiens.*
SC.H.2.4.1

3. A difference between the scientific name of an organism and the classification of that organism is that
a. the scientific name includes the family and class of the organism.
b. the scientific name always contains three words (trinomial nomenclature).
c. the classification includes more categories than the scientific name.
d. classification can vary from place to place.
SC.H.1.4.4

4. Explain the relationship of domains to kingdoms. SC.G.1.4.1

5. The offspring of a donkey *(Equus asinus)* and a horse *(Equus caballus)* is a mule, shown below. The mule is sterile. Is the classification of donkeys and horses into different species justified according to the biological species concept? Explain. SC.G.1.4.1

6. Biologists classify organisms based on
a. their appearance.
b. their structure.
c. their ability to interbreed.
d. All of the above
SC.H.2.4.1

7. What features common to both bats and birds suggest that convergent evolution has occurred? SC.D.1.4.4

8. **Exploring Further** Explain how constructing a cladogram might help scientists understand evolutionary relationships. SC.H.2.4.2

9. Contrast analogous characters with homologous characters. (**Hint:** See Chapter 13, Section 2.) SC.D.1.4.4

10. **Concept Mapping** Construct a concept map that shows how biologists determine the classification of a new species. Use the following terms: *genus, species, binomial nomenclature, kingdom, biological species concept, derived characters, cladogram,* and *evolutionary systematics.* SC.G.1.4.1

Critical Thinking

11. Evaluating Hypotheses Scientists infer that groups of organisms that have homologous traits must be related, so they use cladograms to study relationships. Explain the reasoning that supports this inference. SC.H.2.4.2

12. Evaluating Viewpoints Explain this statement: Diversity is the result of evolution; classification systems are the inventions of humans. SC.G.1.4.1

13. Forming Reasoned Opinions In the laboratory, a scientist studied two identical-looking daisies that belong to the genus *Aster*. The two plants produce fertile hybrids in the laboratory, but they never interbreed in nature because one plant flowers only in the spring and the other only in autumn. Do the plants belong to the same species? Explain. SC.H.1.4.1

312

tain traits, scientists can gain insight into their evolutionary background.

12. The diversity that we observe is a result of ongoing evolutionary processes. Humans have developed classification systems to make sense out of the bewildering diversity.
13. No, the plants are not the same species because they do not *naturally* interbreed in the wild. They are reproductively isolated.

Assignment Guide

Section	Questions
1	1, 2, 3, 4
2	5, 6, 7, 8, 9, 10, 11, 12, 13

Chapter Review Correlations

SC.D.1.4.4: Items 7, 9
SC.G.1.4.1: Items 1, 4, 5, 10, 12
SC.H.1.4.1: Item 13
SC.H.1.4.4: Item 3
SC.H.2.4.1: Items 2, 6
SC.H.2.4.2: Items 8, 11

FCAT Prep

For questions 1–6, write your answers on a separate sheet of paper.

1 For grasshoppers and locusts to be in the same family, what other category must they also share?
A. genus **C.** order
B. group **D.** species
SC.G.1.4.1

2 In which category are two members most closely related?
F. class **H.** genus
G. family **I.** order
SC.G.1.4.1

3 What kind of features can convergent evolution lead to?
A. analogous characters
B. ancestral characters
C. derived characters
D. homologous structures
SC.D.1.4.4

4 What can you infer about the relationship between *Escherichia coli* and *Entamoeba coli* from their names?
F. They are in the same class because the second parts of their scientific names are the same.
G. They are not in the same genus because the first parts of their scientific names are different.
H. They are not in the same order because the first parts of their scientific names are different.
I. They are the same species because the second parts of their scientific names are the same.
SC.G.1.4.1

Base your answers to questions 5 and 6 on the diagram below, which shows biological classifications from kingdom to species.

SEVEN LEVELS OF BIOLOGICAL CLASSIFICATION

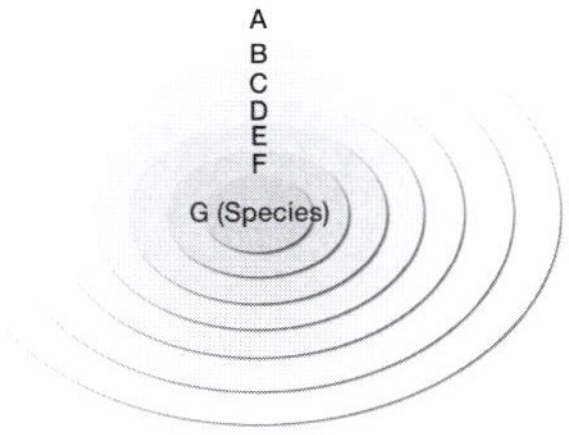

5 What can you infer about different species that belong to the same group at level *D*?
A. They belong to the same group at *C*.
B. They belong to different groups at *C*.
C. They belong to the same group at *E*.
D. They belong to different groups at *E*. SC.G.1.4.1

6 A certain family includes 16 species. A certain order includes 5 times the number of species as that family. A certain class includes twice the number of species as that order. How many species belong to the group at Level *D*? SC.G.1.4.1

Test TIP

Before looking at the answer choices for a question, try to answer the question yourself.

313

FCAT Doctor

Question 3 Answer A is the correct choice. Similarities that arise through convergent evolution are called analogous structures. Answer B is incorrect because an ancestral character appears in organisms that share a common ancestor, which is not necessarily true in convergent evolution. Answer C is incorrect because derived characters evolve in the ancestor of one group and not another, causing those groups to have different characters. Answer D is incorrect because homologous structures arise in organisms that have a common ancestor.

Question 5 Answer A is the correct choice. Members of the same order belong to the same class. Answer B is incorrect because members of the same order cannot belong to different classes. Answer C is incorrect because members of the same order are not necessarily in the same genus. Answer D is incorrect because members of the same order are not necessarily in different genera.

Question 6 The correct answer is 80. The family (Level *E*) includes 16 species. The order (Level *D*) includes 5 × 16, or 80, species. The class (Level *C*) includes 5 × 16 × 2, or 160, species.

Answers

1. C
2. H
3. A
4. G
5. A
6. 80

Test Prep Correlations

SC.D.1.4.4: Item 3
SC.G.1.4.1: Items 1, 2, 4, 5, 6

Skills Practice Lab

Making a Dichotomous Key

Teacher's Notes

Time Required 50 minutes

Ratings

EASY → HARD

- Teacher Preparation: 1
- Student Setup: 2
- Concept Level: 2
- Cleanup: 1

Skills Acquired
- Communicating
- Designing Experiments
- Identifying/Recognizing Patterns
- Organizing and Analyzing Data

Scientific Methods
In this lab, students will:
- Make Observations
- Ask Questions
- Draw Conclusions
- Communicate the Results

Materials and Equipment
Items commonly found in a classroom can be used or the teacher can ask students to bring in an assortment of items from home.

Skills Practice Lab

Making a Dichotomous Key

SC.H.1.4.1

SKILLS
- Identifying and comparing
- Organizing data

OBJECTIVES
- **Identify** objects using dichotomous keys.
- **Design** a dichotomous key for a group of objects.

MATERIALS
- 6 to 10 objects found in the classroom (e.g., shoes, books, writing instruments)
- stick-on labels
- pencil

Before You Begin

One way to identify an unknown organism is to use an **identification key,** which contains the major characteristics of groups of organisms. A **dichotomous key** is an identification key that contains pairs of contrasting descriptions. After each description, a key either directs the user to another pair of descriptions or identifies an object. In this lab, you will design and use a dichotomous key. A dichotomous key can be written for any group of objects.

1. Write a definition for each boldface term in the paragraph above.
2. Based on the objectives for this lab, write a question you would like to explore about making or using a dichotomous key.

Procedure

Part A: Using a Dichotomous Key

1. Use the **Key to Forest Trees** to identify the tree that produced each of the leaves shown here. Identify one leaf at a time. Always start with the first pair of statements (**1***a* and **1***b*). Follow the direction beside the statement that describes the leaf. Proceed through the key until you get to the name of a tree.

Key to Forest Trees

1*a*	Leaf edge has no teeth, waves, or lobes	go to **2**
1*b*	Leaf edge has teeth, waves, or lobes	go to **3**
2*a*	Leaf has a bristle at its tip	**shingle oak**
2*b*	Leaf has no bristle at its tip	go to **4**
3*a*	Leaf edge is toothed	**Lombardy poplar**
3*b*	Leaf edge has waves or lobes	go to **5**
4*a*	Leaf is heart-shaped	**red bud**
4*b*	Leaf is not heart-shaped	**live oak**
5*a*	Leaf edge has lobes	**English oak**
5*b*	Leaf edge has waves	**chestnut oak**

Part B: Design a Dichotomous Key

2. Work with the members of your lab group to design a dichotomous key using the materials listed for this lab.

314

Safety Cautions
Have students wash hands thoroughly before leaving the lab.

Tips and Tricks
Having students make dichotomous keys allows them to practice observing, recording, and organizing data. Lead students through a discussion of how to use the process of elimination when working with a dichotomous key.

You may want to have students complete parts A and B on two different days. Part A will take about 10 minutes to complete. Part B will require about 40 minutes. Encourage students to select objects that can be easily distinguished from one another. Emphasize that a dichotomous key includes pairs of descriptions that lead to the identification of an object.

> **You Choose**
> **As you design your key, decide the following:**
> **a.** what question you will explore
> **b.** what objects your key will identify
> **c.** how you will label personal property
> **d.** what distinguishing characteristics the objects have
> **e.** which characteristics to use in your key
> **f.** how you will organize the data you will need for writing your key

3. Before you begin writing your key, have your teacher approve the objects your group has decided to work with.

4. Using the **Key to Forest Trees** as a guide, write a key for the objects your group selected. Remember, a dichotomous key includes pairs of contrasting descriptions.

5. Use your key to explore one of the questions written for step 2 of **Before You Begin.**

6. After each group has completed step 5, exchange keys and the objects they identify with another group. Use the key you receive to identify the objects. If the key does not work, return it to the group so corrections can be made.

PART C: Cleanup

7. Clean up your work area and all lab equipment. Return lab equipment to its proper place. Wash your hands thoroughly before you leave the lab and after you finish all work.

Analyze and Conclude

1. Drawing Conclusions What tree produced each of the leaves shown in this lab?

2. Forming Hypotheses What other characteristics might be used to identify leaves with a dichotomous key?

3. Analyzing Methods How was the key your group designed dichotomous?

4. Evaluating Results Were you able to use another group's key to identify the objects for which it was written? If not, describe the problems you encountered.

5. Analyzing Methods Does a dichotomous key begin with general descriptions and then proceed to more specific descriptions or vice versa? Explain your answer, giving an example from your key.

6. Further Inquiry Write a new question about making or using keys that could be explored with another investigation.

Do You Know?

Do research in the library or media center to answer these questions:

1. Are the identification keys used in biology always dichotomous?

2. What other types of identification keys do biologists use?

Use the following Internet resources to explore your own questions about the diversity of life on Earth.

internet connect
www.scilinks.org
Topic: Classification
Keyword: HX4044
SCILINKS. Maintained by the National Science Teachers Association

Answers to Do You Know?

1. No, many keys use groups of three or more contrasting descriptions instead of only two.
2. Some keys, such as field guides, use descriptive paragraphs, diagrams, and/or photos for comparison to identify organisms.

Answers to Before You Begin

1. *identification key*—a tool that can be used to identify an unknown organism or object; *dichotomous key*—a key that uses pairs of contrasting descriptive statements to lead to the identification of an organism or some other object
2. Answers will vary. For example: How do you use a dichotomous key?

Answers to Procedure

3. Objects chosen for the dichotomous key should be in a single category but different from each other in appearance.
4. Students' dichotomous keys should include sufficient pairs of contrasting statements to distinguish one object from the next.
5. Answers will vary.
6. Students may find that their keys need refining.

Answers to Analyze and Conclude

1. A. shingle oak; B. Lombardy poplar; C. red bud; D. live oak; E. English oak; and F. chestnut oak
2. Other leaf characteristics might include whether the leaf is compound or simple, whether the leaf is needlelike, the arrangement of leaves on the branches, and the leaf's vein pattern.
3. The key was dichotomous because the descriptive statements were written as pairs of contrasting statements.
4. Answers will vary. Students might have discovered that some of the opposing statements were not contrasting, that some of the objects were not correctly described, or that the key did not include a sufficient number of descriptions to distinguish all the objects from one another.
5. Dichotomous keys proceed from general characteristics to specific characteristics. Examples from keys will vary but should reflect this gradation.
6. Answers will vary. For example: Can the members of any kind of group be identified using a dichotomous key?

Chapters

Fewer than 4,000 gray wolves exist in the lower 48 states.

The Gray Wolf

Yesterday...

Steel trap

The gray wolf once roamed North America from coast to coast. In an effort to protect their livestock, early settlers hunted and killed wolves nearly to the point of extinction. By the mid-1950s, however, few wolves existed in the lower 48 states. **How does the relationship between predator and prey help to maintain balance in an ecosystem?**

Today...

Gray wolves play a vital role in ensuring the diversity and health of ecosystems in which they live. Because they tend to prey upon the old, the sick, and the injured, wolves help to produce stronger populations of deer, elk, and moose. **Discover how competition between species affects a biological community.** In each of the lower 48 states except Minnesota, the gray wolf is listed as endangered.

Biologist David Mech

Tomorrow...

Wolf researcher Diane Boyd

Because the numbers and range of the gray wolf have been increasing, the U.S. Fish and Wildlife Service is reviewing potential changes to the Endangered Species Act protection for gray wolves. Researchers must continue to study wolf populations before a long-term plan to conserve gray wolves can be implemented. **Read to learn why species diversity is important.**

internet connect

www.scilinks.org
Topic: Wolves
Keyword: HX4191

SCILINKS. Maintained by the National Science Teachers Association

Chapter Opener

SE **Quick Review,** p. 319 ■
SE **Reading Activity,** p. 319 ■
TE **Using the Figure,** p. 318 GENERAL
TE **Opening Activity,** p. 319 GENERAL

& 3 • 90 min pp. 320–325

Section 1

How Populations Grow

OSP **Lesson Plan***
TR **Bellringer***
TR **Section Outline***
TR **D32 Exponential Growth Curve***
TR **D33 Logistic Growth***

TE **Activity** Sampling, p. 321 GENERAL
TE **Using the Figure,** p. 322 GENERAL
TE **Using the Figure,** p. 323 GENERAL
TE **Activity** Campus Field Trip, p. 324 GENERAL
CRF **Quick Lab** Using Random Sampling* BASIC
CRF **Skills Practice Lab** Determining Growth Rate* GENERAL

& 5 • 90 min pp. 326–332

Section 2

How Populations Evolve

OSP **Lesson Plan***
TR **Bellringer***
TR **Section Outline***
TR **D20 Two Kinds of Selection***

TE **Group Activity** Genetic Drift: The Founder Effect, p. 328 GENERAL
SE **Quick Lab** Demonstrating the Hardy-Weinberg Principle, p. 329 ■
CRF **Datasheets for In-Text Labs** Demonstrating the Hardy-Weinberg Principle*
TE **Demonstration,** p. 330
TE **Group Activity** Normal Distribution, p. 330 ◆ GENERAL
TE **Using the Figure,** p. 331 BASIC
SE **Math Lab** Building a Normal Distribution Curve, p. 331 ■
CRF **Datasheets for In-Text Labs** Building a Normal Distribution Curve*
SE **Exploration Lab** Observing How Natural Selection Affects a Population, pp. 336–337 ■
CRF **Datasheets for In-Text Labs** Observing How Natural Selection Affects a Population*

& 7 • 90 min

Chapter Review and Assessment Resources

SE **Study Zone,** p. 333 ■
SE **Performance Zone,** p. 334 ■
SE **FCAT Test Prep,** p. 335 ■
TR **Concept Mapping***
CRF **Vocabulary Review Worksheet*** ■ GENERAL
CRF **Test Prep Pretest*** ■ GENERAL
CRF **Chapter Test*** ■ GENERAL
CRF **Chapter Test*** ADVANCED
OSP **Test Generator***
OSP **Test Item Listing***
OSP **Modified Chapter Test*** BASIC

Online and Technology Resources

go.hrw.com Holt Online Learning

Visit go.hrw.com for access to Holt Online Learning or enter the keyword **HX6 Home** for a variety of free online resources.

One-Stop Planner® CD-ROM

This CD-ROM package includes
- Lab Materials QuickList Software
- Holt Calendar Planner
- Customizable Lesson Plans
- Printable Worksheets
- ExamView® Test Generator
- Interactive Teacher Edition
- Holt PuzzlePro® Resources
- Holt PowerPoint® Resources

Guided Reading Audio CDs

These CDs are designed to help auditory learners and reluctant readers. (Audio CDs are also available in Spanish.)

Compression guide:
To shorten your instruction because of time limitations, omit block 1.

KEY

TE	Teacher Edition	CRF	Chapter Resource File	*	Also on One-Stop Planner
SE	Student Edition	TR	Teaching Transparency	■	Also Available in Spanish
OSP	One-Stop Planner			◆	Requires Advance Prep

SKILLS DEVELOPMENT AND PRACTICE	SECTION REVIEW AND ASSESSMENT	CORRELATIONS
		Florida Sunshine Standards/FCAT
TE Skill Builder Interpreting Graphics, p. 321 GENERAL **TE Reading Skill Builder** K-W-L, p. 321 BASIC **TE Skill Builder** Graphing, p. 323 GENERAL **TE Reading Skill Builder** Interactive Reading, p. 323 BASIC **TE Reading Skill Builder** Reading Organizer, p. 324 BASIC **CRF Directed Reading Worksheet*** BASIC **CRF Active Reading Worksheet*** GENERAL	**SE Section 1 Review** ■ **TE Reteaching,** p. 325 BASIC **TE Quiz,** p. 325 GENERAL **TE Alternative Assessment,** p. 325 GENERAL **CRF Quiz*** ■ BASIC	SC.G.1.4.1, SC.G.2.4.2, SC.G.2.4.4
TE Reading Skill Builder Paired Reading, p. 327 BASIC **CRF Directed Reading Worksheet*** BASIC **CRF Active Reading Worksheet*** GENERAL **CRF Critical Thinking Worksheet*** ADVANCED **CRF Science Skills Worksheet*** GENERAL	**SE Section 2 Review** ■ **TE Reteaching,** p. 332 BASIC **TE Quiz,** p. 332 GENERAL **TE Alternative Assessment,** p. 332 GENERAL **CRF Quiz*** ■ BASIC Modified Worksheet* BASIC	SC.G.1.4.1, SC.G.2.4.2, SC.G.2.4.3

Holt Biology Interactive Tutor CD-ROM

This CD-ROM consists of interactive activities that give students a fun way to extend their knowledge of biology concepts.

Unit 7 Ecosystem Dynamics

SciLinks NSTA
www.scilinks.org
Maintained by the **National Science Teachers Association**

Topic: Population Characteristics
SciLinks Code: HX4143

Topic: Population Growth Factors
SciLinks Code: HX4145

Topic: Population Pyramids
SciLinks Code: HX4146

Topic: Hardy-Weinberg Equation
SciLinks Code: HX4095

CHAPTER 15

Overview

Tell students the purpose of this chapter is to clarify the important characteristics of populations—how populations grow and evolve. They will also learn the three key features of populations (size, density, and dispersion), how populations can be modeled, population growth patterns, and the factors that influence how populations change over time.

Using the Figure

Pose the following question to your students: If you heard of an area inhabited by only seven birds, would you consider it to be crowded? (It would depend on the size of the area.) Ask students if they think the puffins shown in the figure appear "crowded." Most students will probably agree that the puffins are crowded *by human standards*. Ask students to investigate the habitats of puffins, and how the puffins can obtain enough food to survive in such "crowded" conditions. (They spend most of their lives at sea, feeding in the ocean, and come ashore only to breed.) Point out that the *number* of organisms in a population may not be as important for survival as the *density* of the organisms in the population (the number of individuals in a given area).

Assessing Prior Knowledge

Before beginning this chapter, be sure your students can answer the questions in the ***Quick Review*** on the facing page. If they cannot, encourage them to reread the sections indicated.

Atlantic puffins

318

Chapter Correlations — Florida Sunshine Standards/FCAT

SC.G.1.4.1 There is great diversity and interdependence among living things.

SC.G.2.4.2 Changes in a component of an ecosystem will have unpredictable effects on the entire system, but the components of the system tend to react in a way that will restore the ecosystem to its original condition.

SC.G.2.4.3 Genetic variation of offspring contributes to population control in an environment, and natural selection ensures that those who are best adapted to their surroundings survive to reproduce.

SC.G.2.4.4 The world ecosystems are shaped by physical factors that limit their productivity.

CHAPTER

15 Populations

Florida Sunshine Standards/FCAT

SC.G.1.4.1 There is great diversity and interdependence among living things.

SC.G.2.4.2 Changes in a component of an ecosystem will have unpredictable effects on the entire system, but the components of the system tend to react in a way that will restore the ecosystem to its original condition.

SC.G.2.4.4 The world ecosystems are shaped by physical factors that limit their productivity.

Quick Review

Answer the following without referring to earlier sections of your book.

1. **Describe** Mendel's laws of inheritance. *(Chapter 8, Section 2)*
2. **Define** *phenotype* and *genotype*. *(Chapter 8, Section 2)*
3. **Define** *probability*. *(Chapter 8, Section 3)*
4. **Evaluate** the significance of mutations. *(Chapter 10, Section 2)*
5. **Define** *natural selection*. *(Chapter 13, Section 1)*

Did you have difficulty? *For help, review the sections indicated.*

Reading Activity

Copy the following statements on a piece of paper or in your notebook, leaving a few blank lines after each statement.

1. *Very small populations are more likely to become extinct than larger populations.*
2. *A single bacterium that divides every 30 minutes will become a population of more than a million in only 10 hours.*
3. *Natural selection acts only on genes themselves, not on phenotypes.*

Before you read the chapter, write down whether you agree or disagree with each statement. Save your responses, and after you have finished reading the chapter, decide whether or not you still agree with your first response.

Individuals of a species that live together form a population. This group of puffins is part of the population of puffins that live on the far northern Atlantic coast of North America.

Looking Ahead

internet connect

www.scilinks.org

National Science Teachers Association *sci*LINKS Internet resources are located throughout this chapter.

SciLINKS. Maintained by the National Science Teachers Association

Opening Activity

Population Density Have the class calculate the population density of their town or city. Ask students what information they need to get started. (the size of the population and the size of the geographic area in question) Supply the needed information, and have students compare their results with data from other nearby towns and from national and international cities. Have them calculate the population density of their state (supply the needed information), and ask them why the calculated population density may be misleading. (Certain areas, such as mountaintops, deserts, or lakes, may be uninhabitable, making the population density seem less than it really is.)

Quick Review Answers

1. The law of segregation states that the two alleles of a trait are separated when gametes are formed. The law of independent assortment states that the alleles of different genes separate independently of one another during the formation of gametes.
2. The genotype is the actual genetic makeup of an organism. The phenotype is the physical appearance of a trait.
3. Probability is the likelihood that an event will occur.
4. Mutations are changes in the nucleotide sequence of DNA. Some have little or no effect on an organism, others are harmful, and some are beneficial, providing the variation upon which natural selection occurs.
5. Natural selection is the process by which organisms with favorable variations reproduce at higher rates than those without such variations.

Chapter Resource File

- Vocabulary Worksheets
- Concept Mapping

Reading Activity Answers

1. agree
2. agree
3. disagree

Section 1

Focus

Overview

Before beginning this section review with your students the objectives listed in the Student Edition. In this section, students learn about populations, their characteristics, and how different populations grow. Students also learn how mathematical models can be used to explain population growth trends and the different growth patterns in real populations.

Bellringer

Tell students to imagine that they have just been hired for a summer job for the month of July only. They are to be paid one cent on the first day of the month, to be doubled each day. They will be paid at the end of each day. Have them take a few minutes—no calculators—to write down estimates as to how much money they would earn on the tenth, twentieth, and final day of July. Ask students to volunteer their estimates. Then calculate the answers with them: *(Day 10 = $5.12; day 20 = $5,242.88; day 31 = $10,737,418.24!)* Relate this growth in earnings to exponential population growth.

Interactive BIOLOGY Tutor

- **Unit 7—Ecosystem Dynamics**
  This engaging tutorial introduces students to the living and non-living components of an ecosystem and how changes occur through time.

Section Correlations

SC.G.1.4.1
SC.G.2.4.2
SC.G.2.4.4

Section 1 How Populations Grow

Florida Sunshine Standards/FCAT

SC.G.2.4.2 Changes in a component of an ecosystem will have unpredictable effects on the entire system, but the components of the system tend to react in a way that will restore the ecosystem to its original condition.

SC.G.2.4.4 The world ecosystems are shaped by physical factors that limit their productivity.

Objectives

- **Distinguish** among the three patterns of dispersion in a population. SC.G.1.4.1
- **Contrast** exponential growth and logistic growth. SC.G.2.4.2
- **Differentiate** *r*-strategists from *K*-strategists. SC.G.2.4.4

Key Terms

population
population size
population density
dispersion
population model
exponential growth curve
carrying capacity
density-dependent factor
logistic model
density-independent factor
***r*-strategist**
***K*-strategist**

What Is a Population?

The people on a city sidewalk shown in **Figure 1** are members of a population. Since 1930, the world's human population has nearly tripled. What causes populations to grow? What determines how fast they grow? What factors can slow their growth?

A **population** consists of all the individuals of a species that live together in one place at one time. This definition allows scientists to use similar terms when speaking of the world's human population, the population of *Escherichia coli* bacteria that live in your intestine, or the population of Devil's Hole pupfish that swim in the tiny pool shown in **Figure 1.**

Every population tends to grow because individuals tend to have multiple offspring over their lifetime. But eventually, limited resources in an environment limit the growth of a population. The statistical study of all populations is called demography *(dih MAH gruh fee)*. Demographers study the composition of a population and try to predict how the size of the population will change.

Figure 1 Populations

A population can be can be widely distributed, as is Earth's human population. Or a population can be confined to a small area, such as the population of Devil's Hole pupfish, which lives in this small pool and nowhere else.

Chapter Resource File

- Directed Reading BASIC
- Active Reading GENERAL

Transparencies

TR Bellringer
TR D32 Exponential Growth Curve
TR D33 Logistic Growth

One-Stop Planner CD-ROM

- Reading Organizers BASIC
- Reading Strategies BASIC
- Problem Solving Worksheet Population Size GENERAL

Three Key Features of Populations

Every population has features that help determine its future. One of the most important features of any population is its size. The number of individuals in a population, or **population size,** can affect the population's ability to survive. Studies have shown that very small populations are among those most likely to become extinct. Random events or natural disturbances, such as a fire or flood, endanger small populations more than they endanger larger populations. Small populations also tend to experience more inbreeding (breeding with relatives) because only relatives are available as mates. Inbreeding produces a more genetically uniform population and is therefore likely to reduce the population's fitness—more individuals will be homozygous for harmful recessive traits. For example, the worldwide cheetah population is very small, and the individuals are almost genetically identical. Many biologists think that a disaster, such as a new disease, could cause their extinction.

A second important feature of a population is its density. **Population density** is the number of individuals that live in a given area. If the individuals of a population are few and are spaced widely apart, they may seldom encounter one another, making reproduction rare.

A third feature of a population is the way the individuals of the population are arranged in space. This feature is called **dispersion.** Three main patterns of dispersion are possible within a population, and each is shown in **Figure 2.** If individuals are randomly spaced, the location of each individual is self-determined or determined by chance. If individuals are evenly spaced, they are located at regular intervals. In a clumped distribution, individuals are bunched together in clusters. Each of these patterns reflects the interactions between the population and its environment.

internet connect

www.scilinks.org
Topic: Population Characteristics
Keyword: HX4143

SCILINKS. Maintained by the National Science Teachers Association

Figure 2 Patterns of dispersion

These are the three possible patterns of dispersion in a population.

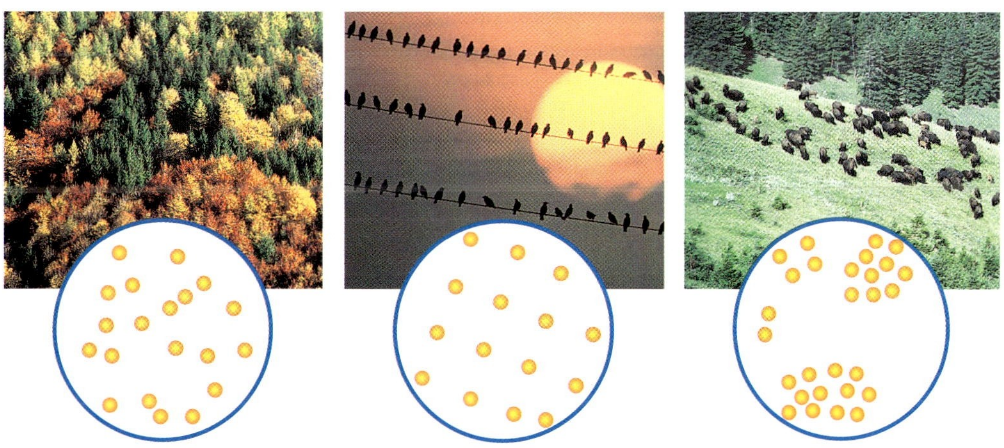

Pine trees in a *random distribution* | Birds in an *even distribution* | Buffalo in a *clumped distribution*

321

Career

Ecologist Ecologists are biologists who specialize in the study of how species interact within a population, with other species, and with the environment. Have students use library or Internet references to find information about the work of ecologists. Have them present their findings to the class.

Motivate

Activity — GENERAL

Sampling Divide the class into small groups. Ask each group to estimate the number of blades of grass in a local athletic field. Since it is impossible to count the blades of grass, ask each group to outline a method of estimating the number of blades of grass. Some groups may suggest a method of population sampling. Ask students how other types of populations could be sampled, such as animals. (trapping, tagging and releasing) What is the value of tagging? (The same animals won't be counted twice.) **LS Interpersonal**

Teach

SKILL BUILDER — GENERAL

Interpreting Graphics Before class, draw a 5 × 5 grid on the board. Tell students each square on the grid represents one acre of land. Using chalk, make a total of 75 dots as follows: In the upper left part of the grid, make the dots in a random distribution; in the upper right part of the grid, make the dots in an even distribution, and in the lower right make the dots in a clumped distribution. Ask students to determine: a) The population size (75); b) the population density (75/25 = 3 organisms/acre); and c) the dispersion pattern of each area (upper left = random; upper right = even; lower right = clumped). **LS Visual**

READING SKILL BUILDER — BASIC

K-W-L Before students read this section, have them write short lists of all the things they already **K**now about populations. Ask them to contribute their entries to a group list on the board or overhead. Then have students list the things they **W**ant to know about populations. Have students save their lists for later use in the Reteaching at the end of Section 1.

Teach, *continued*

Real Life

Answer

Census information is used to obtain federal and state funding for many public projects. Improvements include scanning responses into computers and cooperating with state, local, and tribal governments to correct census maps and address lists.

Teaching Tip

What Values Can *r* Take? Explain to students that the value for *r* (growth rate) may be negative, positive, or 0. For a population in which the birthrate and death rate are the same, *r* equals 0. For populations in which the birthrate exceeds the death rate, *r* is positive, and for those in which the death rate exceeds the birthrate, *r* is negative. Ask students if the *r* value for the United States is positive, negative, or 0. (positive)

Using the Figure — GENERAL

Have students construct a graph similar to the one shown in **Figure 3.** Instruct them to put an "x" at the spot on the graph where they believe human population growth is now. (Globally, the human population is growing exponentially—the "x" should be near the top right of the graph. However, see the Did You Know? in the bottom margin.) **LS Visual**

Real Life

Uncle Sam wants to count you.

The United States census, conducted every 10 years, collects detailed information on the country's population.

Finding Information

Explore Internet resources to find out more about the United States census. Why should every household complete a census form? What steps has the government taken in the past few years to improve the accuracy of the census?

Modeling Population Growth

When demographers try to predict how a population will grow, they make a model of the population. A **population model** is a hypothetical population that attempts to exhibit the key characteristics of a real population. By making a change in the model and observing the outcome, demographers can predict what might occur in a real population. To learn how demographers study a population, consider a simple model of population growth in three stages of complexity.

Growth Rate

A population grows when more individuals are born than die in a given period. So a simple population model describes the rate of population growth as the difference between the birthrate and the death rate. For human populations, birth and death rates are usually expressed as the number of births and deaths per thousand people per year.

Growth Rate and Population Size

When population size is plotted against time on a graph, the population growth curve resembles a J-shaped curve and is called an exponential *(ehks poh NEHN shuhl)* growth curve. An **exponential growth curve** is a curve in which the rate of population growth stays the same, as a result the population size increases steadily.

Figure 3 shows an exponential growth curve. For example, a single bacterial cell that divides every 30 minutes will produce more than 1 million bacteria after only 10 hours. To calculate the number of individuals that will be added to the population as it grows, multiply the size of the current population (N) by the rate of growth (r).

However, populations do not usually grow unchecked. Their growth is limited by predators, disease, and the availability of resources. Eventually, growth slows, and the population may stabilize. The population size that an environment can sustain is called the **carrying capacity** (K).

Figure 3 Exponential growth. This J-shaped curve is characteristic of exponential growth.

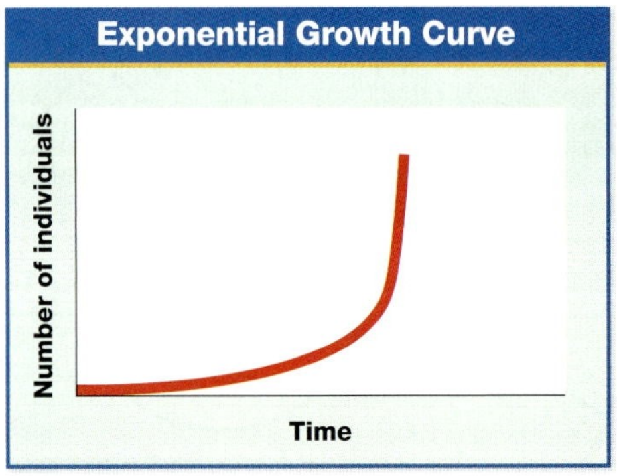

Resources and Population Size

As a population grows, limited resources (that is, resources in short supply) eventually become depleted. When this happens, the growth of the population slows. The population model can be adjusted to account for the effect of limited resources, such as food and water. These resources are called **density-dependent factors** because the rate at which they become depleted depends upon the population density of the population that uses them.

did you know?

Declining Populations Ukraine and Russia are two examples of countries in Europe and Asia that have huge gaps between birth and death rates. Ukraine's population is decreasing by about 0.7% yearly and Russia's population is decreasing by 0.8% yearly. This sharp decrease is not occurring in any other world region besides southern Africa. In the absence of offsetting international migration, the decrease in populations of these countries will continue. In addition, a major cause of the increased number of deaths in Europe is the high proportion of the region's elderly population. Fifteen percent of Europe's population is age 65 or older, compared with 7 percent globally.

The population model that takes into account the declining resources available to populations is called the logistic model of population growth, after the mathematical form of the equation. The **logistic model** is a population model in which exponential growth is limited by a density-dependent factor. The everyday meaning of the word *logistics* refers to the ability to obtain, maintain, and transport materials. In other words, logistics is about solving the day-to-day problems of living. Unlike the simple model, the logistic model assumes that birth and death rates vary with population size. When a population is below carrying capacity, the growth rate is rapid. However, as the population approaches the carrying capacity, death rates begin to rise and birthrates begin to decline. As a result, the rate of growth slows. The population eventually stops growing when the death rate equals the birthrate. In real situations, the population may, for a short time, actually exceed the carrying capacity of its environment. If this happens, deaths will increase and outnumber births until the population falls down to the carrying capacity. Many scientists are concerned that the Earth's human population, which passed 6 billion in 1999, may have exceeded its carrying capacity. A curve that shows logistic growth is illustrated in **Figure 4.**

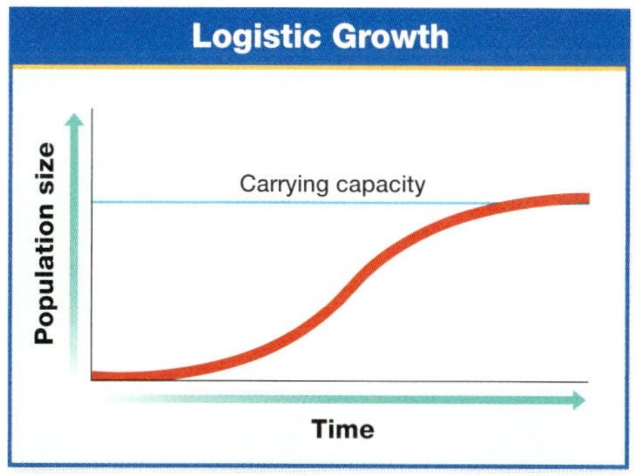

Figure 4 Logistic growth. The curve of logistic growth looks like a stretched-out letter *S*.

The logistic model of population growth, though simple, provides excellent estimates of how populations grow in nature. Competition for food, shelter, mates, and limited resources tends to increase as a population approaches its carrying capacity. The accumulation of wastes also increases. Demographers try to make logistic models based on current population sizes and predict how much a population will increase. **Figure 5** summarizes the three stages of a population model.

Population Growth Models

- **Simple model (part one): calculating the population growth rate**

  r (rate of growth) = birthrate − death rate

  The rate of population growth equals the rate of births minus the rate of deaths.
- **Simple model (part two): exponential growth curve**

  ΔN (change in population, read as "delta N") = rN

  Once r has been determined for a population (part one), the number of individuals that will be added to a population as it grows is equal to the rate of growth multiplied by the number of individuals in the current population (N).
- **More realistic model: logistic model**

  $$\Delta N = rN\frac{(K - N)}{K}$$

  Population size calculations often need to be adjusted by the number of members of the population at carrying capacity (K).

Figure 5 Population model. A population model can be described using three stages.

HISTORY CONNECTION

Tell students that rapid human population growth is a relatively recent phenomenon. About 10,000 years ago, cultivation of food crops began, and populations started to grow more rapidly. By the mid-1600s, advances in hygiene, health care, agriculture, commerce, and technology accelerated population growth. During the period from 1750 to 1900, the population doubled from 800 million to 1.6 billion. From 1900 to 1965, it doubled again to 3.2 billion. (In 2002, the estimated global population was about 6.24 billion.)

Using the Figure — GENERAL

Have students identify the three growth rates of a logistic growth curve, as shown in **Figure 4:**
1. initial slow growth during the establishment of a population;
2. rapid exponential growth; and
3. stabilization around the carrying capacity. **LS Visual**

READING SKILL BUILDER — BASIC

Interactive Reading Assign Chapter 15 of the *Holt Biology* Guided Audio CD Program to help students achieve greater success in reading the chapter. English Language Learners

Teaching Tip — GENERAL

Zero Population Growth Ask students if people in the U.S. would need to stop having children to achieve ZPG. (no) Ask them how birth and death rates should compare to reach ZPG. (They must be equal.) Explain that if each couple has only two children to replace them when they die, the population would reach ZPG. Ask students how many generations could be produced before 25-year-old parents die. (Answers will vary. There would likely be at least 2 more generations before the parents die.) Ask students how long they think the population would continue to grow before it leveled off. (Answers will vary, but assuming the parents are 25 and the life expectancy is around 75, the population would continue to increase for at least another 50 years.) **LS Logical**

SKILL BUILDER — GENERAL

Graphing Contact your local city government for historical population data for your area, town, or city. Display the data on the board or overhead and have students use it to make a line graph of population growth. (Example: Title: "Population of Halsey, Oregon, 1885–2004"; X-axis label: "Years"; Y-axis label: "Number of People.")

Teach, *continued*

READING SKILL BUILDER — BASIC

Reading Organizer Before reading this section, have students prepare a two-column table to compare the features of *r*- and *K*-strategists. Students should fill in their tables as they read. Headings for the two columns are: ***r*-strategists** and ***K*-strategists.** The headings for the six rows are: **Growth pattern** (*r*-: exponential, *K*-: slow); **Population size** (*r*-: temporarily large, *K*-: small); **Environment** (*r*-: unpredictable, *K*-: stable); **Reproductive strategy** (*r*-: early in life when conditions are favorable, *K*-: later in life under most conditions); **Offspring characteristics** (*r*-: many in number, small in size, mature rapidly; *K*-: few in number, large in size, mature slowly); **Parental care** (*r*-: little or none, *K*-: much). LS **Logical**

Activity — GENERAL

Campus Field Trip If possible, take your class on a walk around your school campus. Point out different organisms and ask them whether they are *r*- or *K*-strategists. (Examples: dandelion, *r*-strategist; dog or cat, *K*-strategist) Ask students why they would classify each organism as an *r*- or *K*-strategist. (See Reading Skill Builder answers, above.)

Section Review Correlations

SC.G.1.4.1: Item 1
SC.G.2.4.2: Item 2
SC.G.2.4.4: Items 3, 5
SC.G.2.4.5: Item 4

Growth Patterns in Real Populations

internet connect
www.scilinks.org
Topic: Population Growth Factors
Keyword: HX4145
SciLinks. Maintained by the National Science Teachers Association

Many species of plants and insects reproduce rapidly. Their growth is usually limited not by density-dependent factors but by environmental conditions, also known as **density-independent factors.** Weather and climate are the most important density-independent factors. For example, mosquito populations increase in the summer, while the weather is warm, but decrease in the winter. The growth of many plants and insects is often described by an exponential growth model. The population growth of slower growing organisms, such as bears, elephants, and humans however, is better described by the logistic growth model. Most species have a strategy somewhere between the two models; other species change from one strategy to the other as their environment changes. (Note that the use of the word *strategy* here means "pattern of living." An organism does not consciously plan its strategy.)

Rapidly Growing Populations

Many species, including bacteria, some plants, and many insects like cockroaches and mosquitos, are found in rapidly changing environments. Such species, called ***r*-strategists,** grow exponentially when environmental conditions allow them to reproduce. This strategy results in temporarily large populations. When environmental conditions worsen, the population size drops quickly. In general, *r*-strategists have a short life span. In addition they reproduce early in life and have many offspring each time they reproduce. Their offspring are small, and they mature rapidly with little or no parental care. The cockroaches shown in **Figure 6** are *r*-strategists.

Figure 6 **Different species have different growth patterns**
Cockroaches are *r*-strategists, while whales are *K*-strategists.

Cockroaches

Humpback whale

Attention Grabber

Global Population Most of the world's population growth occurs in less-developed countries. Global population increases by about 83 million annually—99 percent occurring in the less-developed areas of Africa, Asia, Latin America and the Caribbean, and Oceania. The top ten most populated countries in 2001 were: 1) China; 2) India; 3) United States; 4) Indonesia; 5) Brazil; 6) Pakistan; 7) Russia; 8) Bangladesh; 9) Japan; 10) Nigeria. The Population Reference Bureau projects the following top ten most populated countries for the year 2025: 1) China; 2) India; 3) United States; 4) Indonesia; 5) Pakistan; 6) Brazil; 7) Nigeria; 8) Bangladesh; 9) Russia; 10) Mexico.

Slowly Growing Populations

Organisms that grow slowly, such as whales, often have small population sizes. These species are called **K-strategists** because their population density is usually near the carrying capacity (K) of their environment. K-strategists are characterized by a long life span, few young, a slow maturing process, and reproduction late in life. K-strategists often provide extensive care of their young and tend to live in stable environments. Many endangered species, such as tigers, gorillas, and the whale shown in **Figure 6,** are K-strategists.

FLORIDA BIOWatch

Saving Florida's Panthers

In the 1980s, the total population of Florida panthers was about 30 individuals. With such small numbers, scientists knew the population was in danger of completely dying off.

Studying the Problem

People who wanted to save the Florida panthers tried to understand how the population had become endangered. The causes of panther deaths included illegal hunters, car collisions, and diseases. But the major problem was that roads, farms, and cities were overtaking panther habitat. The panthers could no longer travel over large areas in search of food or mates.

When populations are small and a few individuals mate only among themselves, the variety of genes in that population is limited. As a result, harmful traits arising from recessive genes occur more often. The shrinking population of Florida panthers began to show these kinds of problems.

Making a Plan

To help save the Florida panther, people worked to educate hunters, lower speed limits, and build fences around roads. But the biggest help came from other panthers. Texas cougars and Florida panthers are two subspecies of the same species,

Felis concolor. In 1995, Texas cougars were brought to Florida, thus increasing the variety of genes in the population.

Now, about 80 Florida panthers are living in the wild of Southern Florida, and scientists think their population may continue to grow.

internet connect

www.scilinks.org
Topic: Animals of Florida
Keyword: HX6F17
SCILINKS. Maintained by the National Science Teachers Association

Section 1 Review

1. **Identify** the pattern of dispersion of fans attending a basketball game as random, even, or clumped. Explain your answer. SC.G.1.4.1
2. **Differentiate** a logistic growth pattern from an exponential growth pattern. SC.G.2.4.2
3. **Describe** why an r-strategist might be better suited for an unpredictable environment than a K-strategist is. SC.G.2.4.4
4. **Critical Thinking Applying Information** If healthcare improvements lead to a decreased death rate, how must the birth rate change in order to achieve a stable population size? Explain your reasoning. SC.G.2.4.5
5. **FCAT Prep** Which model best describes population growth that is limited by density-independent factors?
   A exponential growth model
   B logistic model
   C stage I model
   D stage III model
   SC.G.2.4.4

Answers to Section Review

1. Fans may be clumped into different sections (for best viewing) but be evenly dispersed within each section (due to their seat placement).
2. Logistic growth begins slowly, increases exponentially, then stabilizes as the population nears the carrying capacity. Exponential growth shows slow growth at first, then increases steadily.
3. If environmental conditions become favorable for even a short time, r-strategists can succeed because they reproduce and mature rapidly. K-strategists require long, favorable periods to mature and reproduce.
4. To achieve a stable population, birth rates must decrease, otherwise the population will grow rapidly.
5. **A.** Correct. Populations controlled by density-independent factors often exhibit exponential growth until environmental conditions worsen and the population size drops quickly. **B.** Incorrect. Populations controlled by density-independent factors typically do not level off. **C.** Incorrect. Stage I is simply a model for determining growth rate. **D.** Incorrect. Stage III is a model for logistic growth.

BIOWatch GENERAL

Saving Florida's Panthers

Teaching Strategies

Tell students that mountain lions, panthers, and cougars are all the same species. Ask students: Why bring in Texas cougars instead of Montana mountain lions? (The introduced animals should be able to survive in the wild in Florida. The climate and other environmental factors of Florida are more similar to those of Texas than of Montana. So, the Texas cougars would be better adapted and more likely to survive and reproduce.)

Close

Reteaching — BASIC

K-W-L Have students return to their lists of things they **W**ant to know about populations. (See the Reading Skill Builder at the beginning of this section.) Have them place check marks next to the questions that they are now able to answer. Students should finish by making a list of what they have **L**earned.

Quiz — GENERAL

True or False:

1. A population consists of individuals of the same species living together in one place at one time. (true)
2. The human population is experiencing logistic growth. (false; it is experiencing exponential growth)

Alternative Assessment — GENERAL

Have students choose a familiar organism and indicate whether it is an r- or K-strategist, and explain why. (Answers should include growth pattern, population size, environment, reproductive strategy, offspring characteristics, and amount of parental care.)

Section 2

Focus

Overview

Before beginning this section review with your students the objectives listed in the Student Edition. This section reviews how populations evolve—the forces that affect allele frequencies, how selection changes populations, and how natural selection changes the distribution of traits in populations.

Bellringer

Tell students to imagine they are in charge of a sheep ranch. The cost of fencing is high, so the student must implement a breeding program that would produce short-legged sheep within the next 20 years. Have students write down how they would meet this goal. (Sheep with the shortest legs would be mated with other short-legged sheep. This is artificial selection. Explain how this is similar to natural selection, except "nature" is not the selector.)

Motivate

Identifying Preconceptions

The gene producing extra fingers (polydactyly) in humans is caused by a dominant allele. Ask students: Is the frequency of polydactyly increasing, decreasing, or staying the same? Discuss. (Some students will believe that polydactyly must be increasing since dominant alleles "overpower" recessive alleles. Explain that a mathematical equation shows that allele frequency will remain the same if no evolutionary forces are acting.)

Section Correlations

SC.G.1.4.1
SC.G.2.4.2
SC.G.2.4.3

Section 2 How Populations Evolve

Florida Sunshine Standards/FCAT

SC.G.2.4.2 Changes in a component of an ecosystem will have unpredictable effects on the entire system, but the components of the system tend to react in a way that will restore the ecosystem to its original condition.
SC.G.2.4.3 Genetic variation of offspring contributes to population control in an environment, and natural selection ensures that those who are best adapted to their surroundings survive to reproduce.

Objectives

- **Summarize** the Hardy-Weinberg principle. SC.G.2.4.3
- **Describe** the five forces that cause genetic change in a population. SC.G.2.4.2
- **Identify** why selection against unfavorable recessive alleles is slow. SC.G.2.4.3
- **Contrast** directional and stabilizing selection. SC.G.2.4.2

Key Terms

Hardy-Weinberg principle
gene flow
nonrandom mating
genetic drift
polygenic trait
normal distribution
directional selection
stabilizing selection

The Change of Population Allele Frequencies

In the more than a century that has passed since Darwin's death, the science of genetics has blossomed. This has allowed biologists to better understand how natural selection changes the proportions of alleles within populations. But before you can understand how populations change in response to evolutionary forces, you need to learn how populations behave in the absence of these forces.

Allele Frequencies

When Mendel's work was rediscovered in 1900, biologists began to study how frequencies of alleles change in a population. Specifically, they wondered if dominant alleles, which are usually more common than recessive alleles, would spontaneously replace recessive alleles within populations.

In 1908, the English mathematician G. H. Hardy and the German physician Wilhelm Weinberg independently demonstrated that dominant alleles do not automatically replace recessive alleles. Using algebra and a simple application of the theories of probability, they showed that the frequency of alleles in a population does not change. Moreover, the ratio of heterozygous individuals to homozygous individuals does not change from generation to generation unless the population is acted on by processes that favor particular alleles. If a dominant allele is lethal, for example, it will not become more common just because it is dominant. Their discovery, called the **Hardy-Weinberg principle,** states that the frequencies of alleles in a population do not change unless evolutionary forces act on the population.

Figure 7 Fruit fly mutation. *Drosophila melanogaster* experiences about one mutation in every 840,000,000,000 base pair replications.

The Hardy-Weinberg Principle

The Hardy-Weinberg principle holds true for any population as long as the population is large enough that its members are not likely to mate with relatives and as long as evolutionary forces are not acting. There are five principle evolutionary forces: *mutation,* illustrated in **Figure 7,** *gene flow, nonrandom mating, genetic drift,* and *natural selection.* These evolutionary forces can cause the ratios of genotypes in a population to differ significantly from those predicted by the Hardy-Weinberg principle. The Hardy-Weinberg principle can be expressed as an equation that can be used to predict genotype frequencies in a population.

326

Chapter Resource File

- Directed Reading BASIC
- Active Reading GENERAL
- Data Sheet for Quick Lab GENERAL
- Data Sheet for Math Lab GENERAL

Transparencies

TR Bellringer
TR D20 Two Kinds of Selection

One-Stop Planner CD-ROM

- Reading Organizers BASIC
- Reading Strategies BASIC
- Problem Solving Worksheet Genetics and Probability GENERAL
- Supplemental Reading Guide *Silent Spring* ADVANCED

Mutation

Although mutation from one allele to another can eventually change allele frequencies, mutation rates in nature are very slow. Most genes mutate only about 1 to 10 times per 100,000 cell divisions, so mutation does not significantly change allele frequencies, except over very long periods of time. Furthermore, not all mutations result in phenotypic changes. Recall that more than one codon can code for the same amino acid. Therefore, some mutations may result in no change in the amino acid coded for in a protein, and other changes in an amino acid that do occur may not affect how the protein works. Mutation is, however, the source of variation and thus makes evolution possible.

internet connect
www.scilinks.org
Topic: Hardy-Weinberg Equation
Keyword: HX4095
SCILINKS. Maintained by the National Science Teachers Association

Exploring Further

Using the Hardy-Weinberg Equation

You can use the Hardy-Weinberg principle to predict genotype frequencies. The Hardy-Weinberg principle is usually stated as an equation.

$$p^2 + 2pq + q^2 = 1$$

p^2	$2pq$	q^2
frequency of individuals that are homozygous for allele *A*	frequency of heterozygous individuals with alleles *A* and *a*	frequency of individuals that are homozygous for allele *a*

By convention, the frequency of the more common of the two alleles is referred to as *p*, and the frequency of the rarer allele is referred to as *q*. A frequency is the proportion of a group that is of one type. The frequency of allele *A* is the proportion of all alleles that are *A* for this gene in the population. Similarly, the frequency of allele *a* is the proportion of alleles that are *a*. The sum of the allele frequencies must always equal 1.

Individuals that are homozygous for allele *A* occur at a frequency of *p* times *p*, or p^2. Individuals that are homozygous for allele *a* occur at the frequency of *q* times *q*, or q^2. Heterozygotes have one copy of *A* and one copy of *a*, but heterozygotes can occur in two ways—*A* from the father and *a* from the mother or *a* from the father and *A* from the mother. Therefore, the frequency of heterozygotes is $2pq$.

Calculating the frequency of cystic fibrosis

How do you calculate the number of people in a crowd, like the one below, who are likely to be carriers of the cystic fibrosis gene?

1. **Calculate the frequency of the recessive allele.** Recall from Chapter 9 that cystic fibrosis is caused by the recessive allele *c*. If q^2, the frequency of recessive homozygotes, is 0.00048, then *q* is $\sqrt{0.00048}$, or 0.022.
2. **Calculate the frequency of the dominant allele *C*.**

   Because
   $$p + q = 1, p = 1 - q.$$
   So
   $$p = 1 - 0.022,$$
   or 0.978.
3. **Determine the frequency of heterozygotes.**
   $$2pq = 2 \times 0.978 \times 0.022 = 0.043$$

This means that 43 of every 1,000 Caucasian North Americans are predicted to carry the cystic fibrosis allele unexpressed (without disease).

Hardy-Weinberg proportions seldom, if ever, occur in nature because at least one of the five causes of evolution is always affecting populations.

Cultural Awareness

Allele Frequencies Allele frequencies in humans often differ among populations. The A, B, and O blood type alleles are a well-studied example. This chart shows the A-B-O frequencies for several populations.

Group	Frequency of A	Frequency of B	Frequency of O
United States (Caucasians)	0.258	0.069	0.673
United States (African-Americans)	0.168	0.131	0.701
China	0.181	0.158	0.662
Peru (Native Americans)	0.073	0.000	0.928

Teach

READING SKILL BUILDER — BASIC

Paired Reading Assign students to cooperative pairs. Have each student copy the blue headings of each section onto a piece of paper. Have them read Section 2 silently. As they read, have them indicate parts they understand with a check mark and parts they do not understand with a question mark. Then have the partners discuss what they did or did not understand.
Co-op Learning

Exploring Further — ADVANCED

Using the Hardy-Weinberg Equation

Teaching Strategies

- Ask students, How do we know that $p + q = 1$? (We are assuming that there are only two alleles for this gene, and they are found in the frequencies of *p* and *q*. Because there are no other possibilities, the sum of the two frequencies must be 1, or 100 percent of the possible alleles. Note also that if both sides of the equation $p + q = 1$ are squared, the equation still holds true, because we treat both sides of the equation in the same manner. The result is $p^2 + 2pq + q^2 = 1$).
- Tell students that the Hardy-Weinberg equation applies to any trait, no matter how many alleles are involved. For example, blood type is controlled by three alleles (A, B, and O). In this case, the Hardy-Weinberg equation becomes $p + q + r = 1$, where *p*, *q*, and *r* represent the frequencies of the three alleles.

Discussion

- Which of the five forces of evolution affect the outcome of the Hardy-Weinberg proportions in human populations? (all of them)
- Which of the factors play the strongest role? (Answers will vary.)

Teach, *continued*

Group Activity — GENERAL

Genetic Drift: The Founder Effect Tell students that in a small population, chance events, such as storms, earthquakes, floods, or fires, can change allele frequencies significantly; they can even result in new populations with allelic frequencies much different from the original population's. Have students imagine a population of 100 birds inhabiting a coastal area. Two alleles, *W* and *w*, affect wing color. *W* is the dominant allele that produces red wings, and *w* is the recessive allele that produces white wings when homozygous. In this population, 96/100 birds have red wings (64 *WW* and 32 *Ww*), and only 4 have white wings (*ww*). A hurricane blows 10 birds 500 miles out to sea, where they land on a small island. Two of the birds have white wings (*ww*) and 8 have red wings (4 *WW*, 4 *Ww*). Have students work in pairs to determine the frequency of the *w* allele in the new population. (Total alleles = 20; *W* = 12/20, or 60%; *w* = 8/20, or 40%. The frequency of allele *w* has suddenly jumped from 20% to 40%! The 10 "founder" birds have colonized a new location and started a new population with radically different allele frequencies. This phenomenon is a form of genetic drift called the "founder effect".)
LS Interpersonal

Figure 8 Nonrandom mating. Female widowbirds prefer to mate with males, such as the one shown, that have long tails over males that have short tails. This increases the proportion of alleles for long tails in the population.

Gene Flow

The movement of individuals from one population to another can cause genetic change. The movement of individuals to or from a population, called migration, creates **gene flow,** the movement of alleles into or out of a population. Gene flow occurs because new individuals (immigrants) add alleles to the population and departing individuals (emigrants) take alleles away.

Nonrandom Mating

Sometimes individuals prefer to mate with others that live nearby or are of their own phenotype, a situation called **nonrandom mating.** Mating with relatives (inbreeding) is a type of nonrandom mating that causes a lower frequency of heterozygotes than would be predicted by the Hardy-Weinberg principle. Inbreeding does not change the frequencies of alleles, but it does increase the proportion of homozygotes in a population. For example, populations of self-fertilizing plants consist mostly of homozygous individuals. Nonrandom mating also results when organisms choose their mates based on certain traits. In animals, females often select males based on their size, color, ability to gather food, or other characteristics, as shown in **Figure 8.**

Genetic Drift

In small populations the frequency of an allele can be greatly changed by a chance event. For example, a fire or landslide can reduce a large population to a few survivors. When an allele is found in only a few individuals, the loss of even one individual from the population can have major effects on the allele's frequency. Because this sort of change in allele frequency appears to occur randomly, as if the frequency were drifting, it is called **genetic drift.** Small populations that are isolated from one another can differ greatly as a result of genetic drift.

The cheetah, shown in **Figure 9,** is a species whose evolution has been seriously affected by genetic drift. Cheetahs have undergone drastic population declines over the last 5,000 years. As a result, the

Figure 9 Cheetahs are endangered. Cheetahs have gone through at least two drastic declines in population size.

Cultural Awareness

Inbreeding Discuss with students how inbreeding can significantly affect small populations. Explain that the entire population of the Old Order Amish of Pennsylvania has about 17,000 people. The Old Order Amish are descended from only a few individuals whose offspring continued to marry within the group. As a result, approximately 12 percent of the Old Order Amish are heterozygous for the recessive allele that causes a syndrome of dwarfism and polydactylism (extra fingers). Since the early 1770s, when the group was founded, some 61 cases of this rare syndrome have been reported, about as many as in the rest of the world.

cheetahs alive today are descendants of only a few individuals, and each cheetah is almost genetically uniform with other members of the population. One consequence of this genetic uniformity is reduced disease resistance—cheetah cubs are more likely to die from disease than are the cubs of lions or leopards. This reduction in genetic diversity of cheetahs may hasten their extinction.

Natural Selection

Natural selection causes deviations from the Hardy-Weinberg proportions by directly changing the frequencies of alleles. The frequency of an allele will increase or decrease, depending on the allele's effects on survival and reproduction. For example, the allele for sickle cell anemia is slowly declining in frequency in the United States because individuals who are homozygous for this allele rarely have children. Heterozygotes are resistant to malaria, a significant health problem in many parts of the world. Heterozygotes, however, do not have an advantage over normal homozygotes as they would have in a malaria area. As a result, homozygotes are selected *against* in the United States, and the frequency of the sickle cell allele decreases. Natural selection is one of the most powerful agents of genetic change.

Study TIP

Compare and Contrast
To compare and contrast the five forces that cause evolution, list each force and describe how it causes populations to evolve. Write the ways in which the five forces are alike and the ways in which they are different.

QUICK LAB

Demonstrating the Hardy-Weinberg Principle SC.G.2.4.3

You can model the allele frequencies in a population with this simple exercise.

Materials

equal numbers of cards marked *A* or *a* to represent the dominant and recessive alleles for a trait, paper bag

Procedure

1. Make a data table like the one below.
2. Work in a group, which will represent a population. Count the individuals in your group, and obtain *that* number of *both A* and *a* cards.
3. Place the cards in a paper bag, and mix them. Have each individual draw two cards, which represent a genotype. Record the genotype and phenotype in your data table.
4. Randomly exchange one "allele" with another individual in your group. Record the resulting genotypes.
5. Repeat Step 4 four more times.

Analysis

1. **Determine** the genotype and phenotype ratios in your group for each trial. Do the ratios vary among the trials?
2. **Hypothesize** what could cause a change in the "genetic makeup" of your group. Test one of your hypotheses.

DATA TABLE

	Trial 1	Trial 2	Trial 3	Trial 4	Trial 5
Genotype					
Phenotype					

Teaching Tip — GENERAL

Huntington's Disease
Huntington's disease causes gradual deterioration of the brain and is caused by a dominant allele. The symptoms usually appear between the ages of 30-50. Have students hypothesize as to whether or not this allele will persist or decline in the population. (Hypotheses will vary. It may persist because by the time a person realizes he (or she) has the disease, he may already have had children to whom he passed the defective allele. Barring a new mutation, the allele may decline because children of a person with Huntington's disease will often know their parent has the disease (since the allele is dominant and is expressed in the parent) and may decide not to have children.) LS **Logical**

QUICK LAB

Demonstrating the Hardy-Weinberg Principle

Skills Acquired
Observing, applying, identifying variables, drawing conclusions

Teacher's Notes
To increase interest, allow groups to choose the recessive and dominant phenotypes (for example, straight or curly hair). If groups have less than five members, students will have to exchange alleles with the same person more than once.

Answers to Analysis

1. Unless all group members are heterozygous or all are homozygous (a rare situation, but more likely in small groups), the ratio of genotypes and phenotypes should remain constant for each trial.
2. A new individual could join the group, or a member could leave (gene flow). Homozygous recessive individuals could be prevented from passing on their alleles (natural selection). Groups may also consider genetic drift and nonrandom mating.

MISCONCEPTION ALERT

Individuals Do Not Evolve A common misconception is that *individuals* evolve, changing to match changes in the environment. Explain that natural selection acts on a *population*, eliminating organisms that are poorly suited to their environment. This causes a shift in the genetic makeup of the *population*. The genetic makeup of individual organisms in the population does not change.

Teach, *continued*

Demonstration

Reinforce the concept that natural selection operates on phenotypes and not genotypes. Show the class two photographs, one of a human population in which differences among phenotypes are obvious, and one in which phenotypic differences are not very apparent. The latter could be one of penguins, which, like humans, form large populations of sexually reproducing individuals. Point out that no two penguins are exactly alike, but every penguin is able to recognize its mate, its offspring, and its nesting-ground neighbors. Ask students what type of variations in the phenotypes might be favored by natural selection. (Answers will vary.)

Group Activity — GENERAL

Normal Distribution Have students work in groups of 2–4. Give each group 50 peanuts (in the shell) or a similar item, and have them measure them in millimeters and graph their results. (If using peanuts, be sure no students are allergic to them.) Have each group share their graphs with the class. Co-op Learning

Teaching Tip — GENERAL

How Populations Evolve Have students create a Graphic Organizer that represents the five forces that cause populations to evolve: mutation, gene flow, nonrandom mating, genetic drift, and natural selection. A completed Graphic Organizer is shown at the bottom of this page. English Language Learners

WORD Origins

The word *hemophilia* is from the Greek words *haima,* meaning "flowing blood," and *philos,* meaning "loving." Knowing this information makes it easier to remember that in the human disease hemophilia, uncontrollable bleeding results from even minor injuries.

Action of Natural Selection on Phenotypes

Natural selection constantly changes populations through actions on individuals within the population. However, natural selection does not act directly on genes. It enables individuals who express favorable traits to reproduce and pass those traits on to their offspring. This means that natural selection acts on phenotypes, not genotypes.

How Selection Acts

Think carefully about how natural selection might operate on a mutant allele. Only characteristics that are expressed can be targets of natural selection. Therefore, selection cannot operate against rare recessive alleles, even if they are unfavorable. Only when the allele becomes common enough that heterozygous individuals come together and produce homozygous offspring does natural selection have an opportunity to act. For example, Alexei Nikolayevich, the only son of the last Tsar of Russia, shown in **Figure 10,** suffered from *hemophilia,* a disease caused by a recessive gene. Without modern medical care, a cut can lead to uncontrollable bleeding and death. This kind of selection would remove a homozygous person from the gene pool. However, such an act of natural selection does not affect heterozygotes, who do not express hemophilia. Therefore, the gene is not eliminated from the population.

Why Genes Persist

To better understand this limitation on natural selection, consider this example. If a recessive allele (*a*) is homozygous in only 1 out of 100 individuals, then 18 out of 100 individuals will be heterozygous (*Aa*) and will therefore carry the allele unexpressed. So natural selection can act on only 1 out of every 19 individuals that carry the allele. As a result, this leaves 18 individuals that maintain the allele in the population. Many human diseases caused by recessive alleles have frequencies similar to this. For example, cystic fibrosis, the most common fatal genetic disorder among Caucasians, produces a thick mucus that clogs the lungs and other organs. About 1 in 25 Caucasians has a copy of the defective gene but shows no symptoms. Homozygous recessive individuals, which include about 1 in 2,500, die from the disease. Genetic conditions are not eliminated by natural selection because very few of the individuals bearing the alleles express the recessive phenotype.

Figure 10 Hemophilia in a family. The last Tsar of Russia's only son, Alexei Nikolayevich (yellow circle), had hemophilia, a blood-clotting disorder that affects males who have a single copy of a recessive gene.

330

Graphic Organizer

Use this graphic organizer with **Teaching Tip** on this page.

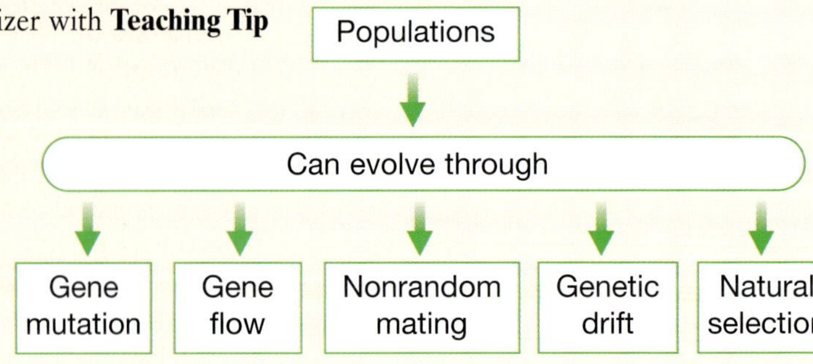

Natural Selection and the Distribution of Traits

Natural selection shapes populations affected by phenotypes that are controlled by one or by a large number of genes. A trait that is influenced by several genes is called a **polygenic** *(pah lee JEHN ihk)* **trait.** Human height and human skin color, for example, are influenced by dozens of genes. Natural selection can change the allele frequencies of many different genes governing a single trait, influencing most strongly those genes that make the greatest contribution to the phenotype. Like following one duck in a flock, it is difficult to keep track of a particular gene. Biologists measure changes in a polygenic trait by measuring each individual in the population. These measurements are then used to calculate the average value of the trait for the population as a whole.

Because genes can have many alleles, polygenic traits tend to exhibit a range of phenotypes clustered around an average value. If you were to plot the height of everyone in your class on a graph, the values would probably form a hill-shaped curve called a **normal distribution,** as illustrated in **Figure 11.**

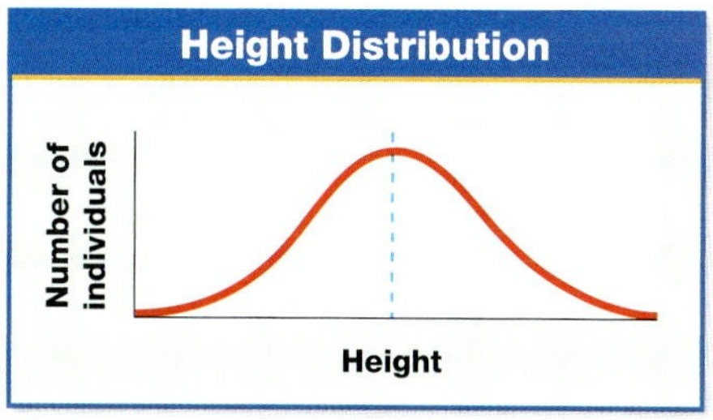

Figure 11 Normal distribution. This hill-shaped curve represents a normal distribution. The blue, dashed line represents the average height for this population.

MATH LAB

Building a Normal Distribution Curve SC.G.1.4.1

Background

You can help your class build a normal distribution curve by measuring the length of your shoes and plotting the data.

Materials

paper, pencil, measuring tape, graph paper

DATA TABLE

Shoe length (centimeters)	Gender

Procedure

1. Prepare a table like the one above.
2. Measure and record the length of one of your shoes to the nearest centimeter. Record your measurement and your gender.
3. Formulate a hypothesis about whether female shoes as a group are longer, shorter, or the same as shoes from males.
4. Determine the number of shoes of each length represented in the class.
5. Make a graph showing the distribution of shoe length in your class. Show the number of students on the y-axis and shoe length on the x-axis. Your graph should resemble the graph in Figure 11.
6. Make a second graph using data only from females.
7. Make a third graph using data only from males.

Analysis

1. **Describe** the shape of the curve that resulted from the graph you made in step 5.
2. **Distinguish** how the distribution curve for shoe length of females differs from the curve for the shoe length of males.
3. **Predict** how the distribution curve that you made in step 5 would change if the data for males were deleted.

Using the Figure — BASIC

Teachers often hear, "Do you grade on a curve?" from students. Have students look at **Figure 11** and ask them why a normal distribution is often called a "bell-shaped curve." Ask them: If a teacher grades "on a curve," what grade (in a traditional A-F grading system) would most students get? ("C" grade) Also ask them how the number of "A" grades would compare to the number of "F" grades. (very similar) Therefore, if 25 students took a test, and the highest grade was 79%, what grade, on a "curve," would that student receive? (an "A") On the same test, if the lowest grade was 70%, what grade would that student receive? (an "F") **LS Visual**

MATH LAB

Building a Normal Distribution Curve

Skills Acquired

Measuring, applying, drawing conclusions, predicting

Teacher's Notes

Emphasize to students that many human characteristics are polygenic, so individuals will have a range of traits. The distribution of that range usually conforms to a normal curve. Have students enter their data into a table for the whole class, as well as one for their gender. This can be done on the board or an overhead to facilitate graphing.

Answers to Analysis

1. Their graphs should resemble Figure 11.
2. Although the curves should appear similar, the peak of the graph of female shoe lengths should be to the left of (lower than) the peak of the graph of male lengths.
3. The curve would match the curve from step 6.

MISCONCEPTION ALERT

Human Heredity Most students transfer what they learn about Mendelian genetics to the inheritance pattern for all traits, including those in humans. Tell students that in humans, very few traits involve only one gene with two alleles. In fact, most traits in humans are the result of polygenic inheritance. In addition to height, other polygenic traits include weight, skin color, and metabolic rate. How genes interact in most polygenic traits is not well understood.

Close

Reteaching — BASIC

Tell students that a certain fictional trait (fuzzy wings) is caused by a recessive allele and prevents birds from flying swiftly. Ask them if they would expect rapid or slow selection against this allele, and why. (Slow; only recessive homozygotes would be selected against.) Ask students if the Hardy-Weinberg principle applies to the frequency of this allele, and why or why not. (No; A recessive, fuzzy-wings phenotype either makes the bird unable to catch food or unable to flee from predators. Therefore, natural selection is acting on this population.)

Quiz — GENERAL

True or False:

1. In nature, evolutionary forces are at work and will cause allele frequencies to shift. (true)
2. Natural selection disrupts Hardy-Weinberg equilibrium. (true)

Alternative Assessment — GENERAL

Tell students a certain population of plants contains only purple or red flowers. Purple (*R*) is dominant to red (*r*). The population has 16% red-flowered plants. Have the students calculate a) the frequency of the red allele (.4); b) the frequency of the purple allele (.6); c) the percentage of flowers that are homozygous dominant (36%); and d) the percentage that are heterozygous (48%). If no evolutionary forces are acting, what would be the frequencies of *R* and *r* after 3 generations? (The frequencies would remain the same.) **LS Logical**

Directional Selection

When selection eliminates one extreme from a range of phenotypes, the alleles promoting this extreme become less common in the population. In one experiment, when fruit flies raised in the dark were exposed to light, some flew toward light and some did not. Only those flies that had the strongest tendency to fly toward light were allowed to reproduce. After 20 generations, the average tendency to fly toward light increased.

In **directional selection,** the frequency of a particular trait moves in one direction in a range. Directional selection is illustrated in the upper panel of **Figure 12.** This type of selection has a role in the evolution of single-gene traits, such as pesticide resistance in insects.

Stabilizing Selection

When selection reduces extremes in a range of phenotypes, the frequencies of the intermediate phenotypes increase. As a result, the population contains fewer individuals that have alleles promoting extreme types.

As you can see in the lower panel of Figure 12, in **stabilizing selection,** the distribution becomes narrower, tending to "stabilize" the average by increasing the proportion of similar individuals. Stabilizing selection is very common in nature.

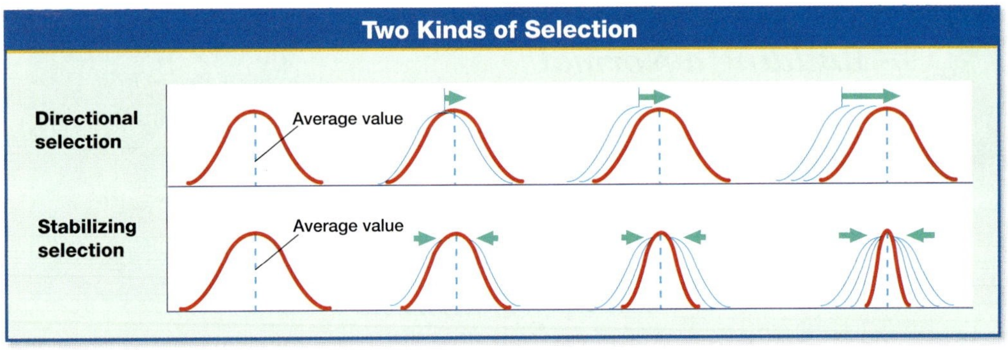

Figure 12 Two kinds of selection on polygenic traits. Directional selection is the change of the average value of a population. Stabilizing selection is the increase of the number of average individuals in a population.

Section 2 Review

1. **Describe** the Hardy-Weinberg principle. SC.G.2.4.3
2. **List** the five forces that cause genetic change in a population. SC.G.2.4.2
3. **Describe** why natural selection against an unfavorable recessive allele is a slow process. SC.G.2.4.3
4. **Critical Thinking Comparing Concepts** Explain how directional selection and stabilizing selection differ, and whether they tend to increase or decrease diversity. SC.G.2.4.2
5. **Critical Thinking Justifying Conclusions** Individuals who are heterozygous for the allele for sickle cell anemia are resistant to malaria. Explain the effects of natural selection on the frequency of the sickle cell allele in an area where malaria occurs. SC.G.2.4.3
6. **FCAT Prep** Which evolutionary force decreases the genetic diversity of a population by increasing the proportion of similar individuals?
   A gene flow **C** directional selection
   B mutation **D** stabilizing selection
   SC.G.2.4.2

332

Answers to Section Review

1. For large populations with no evolutionary forces, allelic frequencies stay the same.
2. mutation, gene flow, nonrandom mating, genetic drift, natural selection
3. Natural selection works on phenotypes. Only homozygous recessive individuals would be selected against; heterozygotes would maintain the allele in the population.
4. Directional selection: One extreme phenotype is favored, shifting the normal distribution toward that trait (increasing diversity). Stabilizing selection: The average phenotype is favored, resulting in more individuals with the "average" value (decreasing diversity).
5. Natural selection favors those heterozygous for sickle cell anemia, as it confers resistance to malaria and prevents deaths. Sickle cell anemia allelic frequency is higher in areas where malaria is common.
6. **A.** Incorrect. Gene flow increases diversity. **B.** Incorrect. Mutation increases diversity. **C.** Incorrect. Directional selection increases diversity. **D.** Correct. Stabilizing selection increases the number of similar organisms.

Section Review Correlations

SC.G.2.4.2: tems 2, 4, 6

SC.G.2.4.3: Items 1, 3, 5

Study ZONE

Key Concepts

1 How Populations Grow

- A population consists of all the individuals of a species that live together in one place at one time.
- A population's future survival is determined by its size, density, and dispersion.
- Though a population's growth is limited by factors such as predation and availability of resources, a population can grow rapidly and may eventually stabilize at a size that the environment can sustain.
- Some populations grow quickly in response to density-independent factors, and other populations grow more slowly and their size is controlled by density-dependent factors.

2 How Populations Evolve

- The Hardy-Weinberg principle states that the frequencies of alleles and genotypes remain constant in populations in which evolutionary forces are absent.
- Allele frequencies in a population can change if evolutionary forces, such as mutation, migration, nonrandom mating, genetic drift, and natural selection, act on the population.
- Natural selection acts only on phenotype, not on genotype.
- Natural selection reduces the frequency of a harmful recessive allele slowly; very few individuals are homozygous recessive, so very few express the allele.
- The range of phenotypes that are controlled by polygenic traits result in a normal distribution when plotted on a graph.
- Directional selection results in the range of phenotypes shifting toward one extreme.
- Stabilizing selection results in the range of phenotypes narrowing.

Key Terms

Section 1

population (320)
population size (321)
population density (321)
dispersion (321)
population model (322)
exponential growth curve (322)
carrying capacity (322)
density-dependent factor (322)
logistic model (323)
density-independent factor (324)
r-strategist (324)
K-strategist (325)

Section 2

Hardy-Weinberg principle (326)
gene flow (328)
nonrandom mating (328)
genetic drift (328)
polygenic trait (331)
normal distribution (331)
directional selection (332)
stabilizing selection (332)

BIOLOGY INTERACTIVE TUTOR

Unit 7—*Ecosystem Dynamics*
Use Topic 2 in this unit to review the key concepts and terms in this chapter.

Study ZONE

Alternative Assessment

Have each student create a fictitious trait controlled by a recessive allele. If 1 in every 20,000 people express the trait, what is the frequency of heterozygotes? (q^2 = 1/20,000 = 0.00005; $q = \sqrt{0.00005} = 0.007$. Since $p + q = 1$, $p = 1 - q$, or $1 - 0.007 = 0.993$. So the heterozygote, $2pq = 2(.993)(.007) = 0.014$.) Have students describe evolutionary factors at play that may affect the Hardy-Weinberg prediction. (Answers will vary, but may include mutation, gene flow, nonrandom mating, genetic drift, or natural selection.)

Chapter Resource File

- Science Skills Worksheet GENERAL
- Critical Thinking Worksheet ADVANCED
- Test Prep Pretest GENERAL
- Chapter Test GENERAL

Answer to Concept Map

The following is one possible answer to Performance Zone item 15.

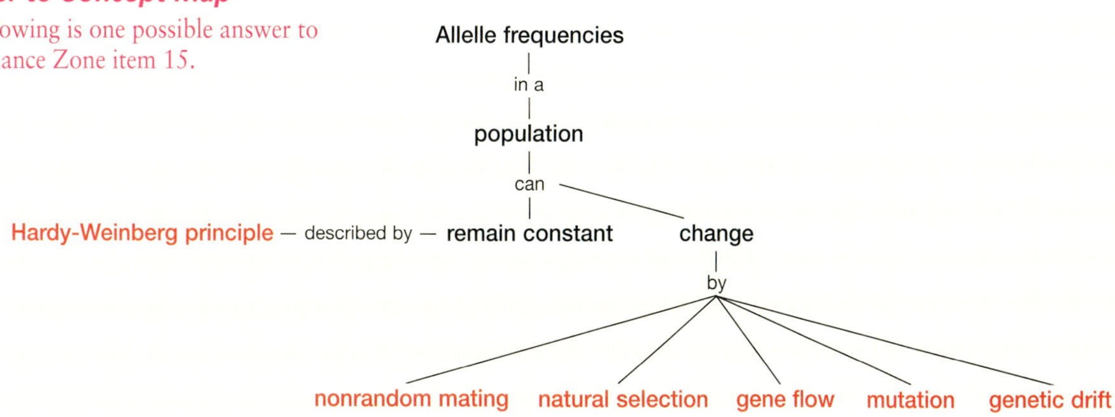

ANSWERS

Understanding Key Ideas

1. b
2. d
3. b
4. a
5. The carrying capacity is about 1600 individuals, based on the way the logistic curve stabilizes, then fluctuates around this value.
6. No. For evolution to occur, there must be a change in allele frequencies.
7. If the environment becomes favorable toward individuals with an extreme trait, those individuals may survive to reproduce while others without the trait do not. Thus, the frequency of the extreme trait increases, which is directional selection.
8. The biggest remaining threat to the panthers has been loss of habitat due to human activities.
9. The frequency of the dominant allele is 0.9.
10. One possible answer to the concept map is found at the bottom of the Study Zone page.

Critical Thinking

11. They are more likely to be *r*-strategists, which grow and reproduce quickly, making them better suited to colonize vacant habitats.
12. The exponential growth model is less realistic because it does not account for the carrying capacity that eventually limits the growth of the population.
13. Recessive disorders are more likely to show up in pure breeds.

Understanding Key Ideas

1. A colony of bacteria that has a limited food supply likely will undergo _____ growth.
   a. exponential **c.** natural
   b. logistic **d.** random
   SC.G.2.4.5
2. According to the Hardy-Weinberg principle, allele frequencies in randomly mating populations without selection
   a. change when birth rate exceeds death rate.
   b. increase and then decrease.
   c. decrease and then increase.
   d. do not change.
   SC.G.2.4.3
3. Which of the following is *not* a cause of genetic change?
   a. genetic drift
   b. random mating
   c. natural selection
   d. mutation
   SC.G.2.4.3
4. Why is it unlikely that natural selection will quickly reduce the frequency of hemophilia?
   a. Natural selection acts only on recessive homozygotes.
   b. Hemophilia is not a genetic disorder.
   c. The frequency of recessive homozygotes is too great.
   d. Dominant homozygotes can have affected children.
   SC.G.2.4.3
5. Biologists introduced pheasants onto an island in Washington State in the 1930s. Using the data shown below, determine the island's carrying capacity. SC.G.2.4.5

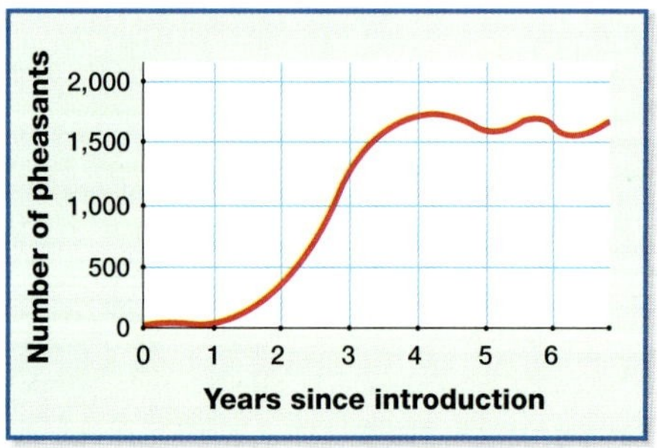

6. Can evolution *not* be associated with genetic change? SC.D.1.4.4
7. Relate the concept of directional selection to extreme traits. SC.G.2.4.2
8. **BIOWatch** Why did the population of Florida panthers continue to decline after being protected from hunters? SC.G.2.4.2
9. **Exploring Further** The frequency of homozygous recessive albino rats in a population is 0.01. Calculate the expected frequency of the dominant allele in this population. SC.G.2.4.3
10. **Concept Mapping** Draw a concept map that shows how the forces of genetic change cause evolution. Try to include the following terms in your map: *Hardy-Weinberg principle, genetic drift, nonrandom mating, natural selection, mutation,* and *gene flow.* SC.G.2.4.3

Critical Thinking

11. **Applying Information** After a forest fire, certain plants quickly recolonize the burned area. Are these plants more likely to be *r*-strategists or *K*-strategists? Explain your answer. SC.G.2.4.2
12. **Evaluating Models** Is a population growth model based on exponential growth more or less realistic than a logistic population model? Explain your answer. SC.G.2.4.2
13. **Applying Information** Why might purebred dogs and cats be subject to more inherited disorders than are mixed breeds? SC.G.2.4.3

Alternative Assessment

14. **Identifying Variables** Formulate a hypothesis about human population growth. Then use library or Internet resources to find estimates of the current rate of human population growth and forecasts for future growth. Predict trends from the data and communicate your conclusions in the form of a report to your class. SC.G.2.4.2

Alternative Assessment

14. Hypotheses will vary. The rate of growth of the human population in 2000 was 1.4%. This yields an annual increase of about 85 million people. Because of the large and increasing population size, the number of people added to the global population will remain high for several decades, even as growth rates continue to decline.

Assignment Guide

Section	Questions
1	1, 5, 8, 11, 12
2	2, 3, 4, 6, 7, 13

Chapter Review Correlations

SC.D.1.4.4: Item 6
SC.G.2.4.2: Items 7, 8, 11, 12, 14
SC.G.2.4.3: Items 2, 3, 4, 9, 10, 13
SC.G.2.4.5: Items 1, 5

FCAT Prep

For questions 1–6, write your answers on a separate sheet of paper.

1 A scientist hypothesizes that natural selection causes human populations to evolve. How could the bubonic plague in the Middle Ages be used to justify this hypothesis?
A. Such diseases have little effect on the human population.
B. People who die from the plague would have relatives who survive and pass on their genes.
C. Populations of the bacterium that causes the plague have evolved due to a lack of resistance in humans.
D. People resistant to disease would survive the plague and pass on their genes for disease resistance.
SC.G.2.4.3

2 In a population model, what is a limited resource known as?
F. density-dependent factor **H.** logistic model
G. density-independent factor **I.** polygenic trait
SC.G.2.4.4

3 What could happen if a rancher maintains a population of livestock that exceeds the carrying capacity of the ranch?
A. The livestock population would evolve to need fewer resources.
B. The increased population would deplete more resources.
C. Resources on the ranch would increase naturally to compensate.
D. There could be food and water shortages or rapid spread of diseases.
SC.G.2.4.4

4 Which of the following results in a population whose individuals have extreme traits?
F. directional selection **H.** random mating
G. exponential growth **I.** stabilizing selection
SC.G.2.4.2

Base your answers to questions 5 and 6 on the chart below.

DISTRIBUTION OF BODY COLORS IN BARK BEETLES

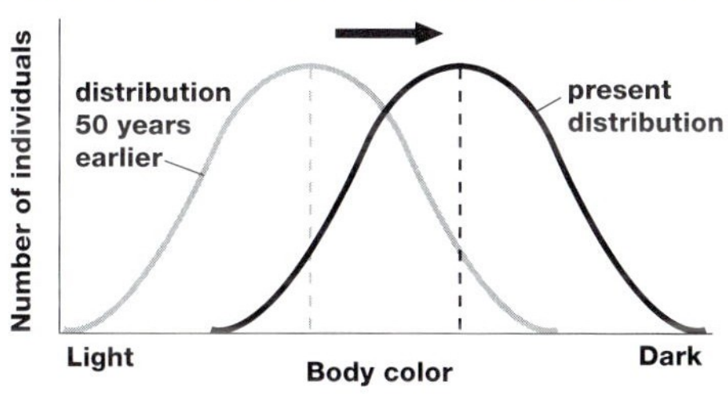

5 Which evolutionary force is represented by the chart?
A. gene flow **C.** mutation
B. directional selection **D.** stabilizing selection
SC.G.2.4.2

6 The researchers assigned numerical values to the body color of individuals; 1 was very light, 100 was very dark. In the first year of the study, the average body color for the population was valued at 40. The average body color 50 years later was 1.5 times darker. What was the value of the average body color for the population in the 50th year? SC.H.1.4.1

Test TIP

When using a graph to answer a question, make sure you know what variables are represented on the x- and y-axes before answering the question.

335

FCAT Doctor

Question 1 Answer D is the correct choice. Natural selection will cause the population to evolve disease resistance. Answer A is incorrect because only individuals who survive to reproduce can pass on their genes. Answer B is incorrect because the plague and other devastating diseases have a great affect on human populations when they spread. Answer C is incorrect because changes in the population of bacteria that cause the plague cannot be used to justify natural selection of humans.

Question 5 Answer B is the correct choice. Directional selection causes a trait to move toward one extreme. Answer A is incorrect because gene flow changes would not graphically form a hill-shaped curve. Answer C is incorrect because mutation would not result in a hill-shaped curve, but, rather, more abrupt changes. Answer D is incorrect because stabilizing selection would cause the body to be more intermediate between light and dark.

Question 6 The correct answer is 60. The value of the average body color 50 years later was 1.5×40, or 60.

Answers

1. D
2. F
3. D
4. F
5. B
6. 60

Test Prep Correlations

SC.G.2.4.2: Items 4, 5
SC.G.2.4.3: Item 1
SC.G.2.4.4: Items 2, 3
SC.H.1.4.1: Item 6

Exploration Lab

Observing How Natural Selection Affects a Population

SC.G.2.4.3

SKILLS
- Using scientific methods
- Collecting, graphing, and analyzing data

OBJECTIVES
- **Measure** and collect data for a trait in a population.
- **Graph** a frequency distribution curve of your data.
- **Analyze** your data by determining its mean, median, mode, and range.
- **Predict** how natural selection can affect the variation in a population.

MATERIALS
- metric ruler
- graph paper (optional)
- green beans or snow peas
- calculator
- balance

Before You Begin

Natural selection can occur when there is **variation** in a **population**. You can analyze the variation in certain traits of a population by determining the mean, median, mode, and range of the data collected on several individuals. The **mean** is the sum of all data values divided by the number of values. The **median** is the midpoint in a series of values. The **mode** is the most frequently occurring value. The **range** is the difference between the largest and smallest values. The variation in a characteristic can be visualized with a **frequency distribution curve**. Two kinds of natural selection—**stabilizing selection** and **directional selection**—can influence the frequency and distribution of traits in a population. This changes the shape of a frequency distribution curve. In this lab, you will investigate variation in fruits and seeds.

1. Write a definition for each boldface term in the paragraph above.
2. Based on the objectives for this lab, write a question you would like to explore about variation in green beans or snow peas.

Procedure

PART A: Design an Experiment

1. Work with the members of your lab group to explore one of the questions written for step 2 of **Before You Begin.** To explore the question, design an experiment that uses the materials listed for this lab.

> **You Choose**
>
> **As you design your experiment, decide the following:**
>
> **a.** what question you will explore
> **b.** what hypothesis you will test
> **c.** which trait (length, color, weight, etc.) you will measure
> **d.** how you will measure the trait
> **e.** how many members of the population you will measure (keep in mind that the more data you gather, the more revealing your frequency distribution curve will be)
> **f.** what data you will record in your data table

2. Write a procedure for your experiment. Make a list of all the safety precautions you will take. Have your teacher approve your procedure and safety precautions before you begin the experiment.
3. Conduct your experiment.

Exploration Lab

Observing How Natural Selection Affects a Population

Teacher's Notes

Time Required
Two 40-minute periods

Ratings
EASY → HARD

Teacher Preparation: 1
Student Setup: 1
Concept Level: 2
Cleanup: 1

Skills Acquired
- Collecting Data
- Designing Experiments
- Experimenting
- Organizing and Analyzing Data

Scientific Methods
In this lab, students will:
- Ask Questions
- Test the Hypothesis
- Analyze the Results
- Draw Conclusions

Materials and Equipment
Materials for this lab can be ordered from WARD'S. See the *Master Materials List* at the front of this book for catalog numbers. Provide enough metric rulers so each lab group has one. Standard graph paper will be sufficient for the graph. The balance should be sensitive enough to measure 0.1 g. Green beans or snow peas can be obtained from a local grocery store.

Safety Cautions
Review all safety symbols with students before beginning the lab. Check to make sure no students are allergic to the food item used.

Tips and Tricks
- Encourage students to read the lab and complete the Before You Begin section before coming to class on the day of the lab.
- During the first lab period, have students make the measurements and record their data. During the second period, have students graph their data and answer the questions.
- Pooling the data will increase the size of the populations studied, thus making students' statistical analyses more valid. To compile class data, record each group's data on the board and have students that measured a similar trait pool the data.
- Make sure students choose a sufficient number of beans or peas to measure so that their data will be more meaningful. Students might choose to measure length of beans or peas, number of seeds in beans or peas, etc.

PART B: Cleanup and Disposal

4. Dispose of seeds in the designated waste containers. Do not put lab materials in the trash unless your teacher tells you to do so.

5. Clean up your work area and all lab equipment. Return lab equipment to its proper place. Wash your hands thoroughly before you leave the lab and after you finish all work.

Analyze and Conclude

1. **Summarizing Results** Make a frequency distribution curve of your data. Plot the trait you measured on the x-axis (horizontal axis) and the number of times that trait occurred in your population on the y-axis (vertical axis).
2. **Calculating** Determine the mean, median, mode, and range of the data for the trait you studied.
3. **Analyzing Results** How does the mean differ from the mode in your population?
4. **Drawing Conclusions** What type of selection appears to have produced the type of variation observed in your experiment?
5. **Evaluating Data** The graph below shows the distribution of wing length in a population of birds on an island. Notice that the mean and the mode are quite different. Is the mean always useful in describing traits in a population? Explain.

Distribution of Wing Length

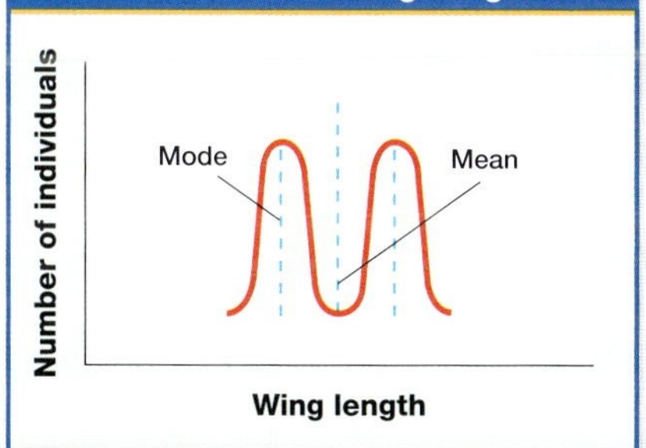

6. **Forming Hypotheses** What type of selection (stabilizing or directional) would be indicated if the mean of a trait you measured shifted, over time, to the right of a frequency distribution graph?
7. **Further Inquiry** Write a new question about variation in populations that could be explored in another investigation.

Do You Know?

Do research in the library or media center to answer these questions:

1. What factors have contributed to the rise in bacterial resistance to antibiotic drugs?
2. How do farmers use directional selection to improve domestic plants and animals?

Use the following Internet resources to explore your own questions about how populations change.

internet connect
www.scilinks.org
Topic: Populations
Keyword: HX4147
SCILINKS. Maintained by the National Science Teachers Association

Answers to Before You Begin

1. *natural selection*—the process by which populations change in response to their environment, as better adapted individuals leave more offspring; *variation*—a difference in traits among individuals in a population; *population*—a group of individuals that belong to the same species, live in the same area, and breed with others in the group; *mean*—a mathematical average; *median*—the midpoint in a series of values; *mode*—a value that occurs most frequently; *range*—the difference between the largest and smallest values; *frequency distribution curve*—a type of graph that maps the number of occurrences of each variation of a trait; *stabilizing selection*—a type of selection that increases the proportion of individuals with the average trait; *directional selection*—a type of selection that increases one extreme form of a trait over another form.
2. Answers will vary. For example: How much variation is there in the length of green beans?

Sample Procedure

1. Measure the lengths or the weights of 20 green beans or snow peas.
2. Organize the measurement values from the lowest to the highest value, and round to the nearest whole number. Beneath each value, write the number of green beans of that length. On the graph paper, make a graph showing the distribution curve.
3. Calculate the mean, median, and range for each type of measurement.

6. directional selection
7. For example: How does the size of a population of snow peas that has been grown in hot conditions compare to a population of snow peas grown in normal weather conditions?

Answers to Do You Know?

1. Students should cite the use of antibiotics as a selective force in which the antibiotic eliminates the susceptible bacteria but leaves those bacteria that are resistant.
2. Farmers may select for faster growth, higher seed production, or higher milk production.

Sample Data Graph

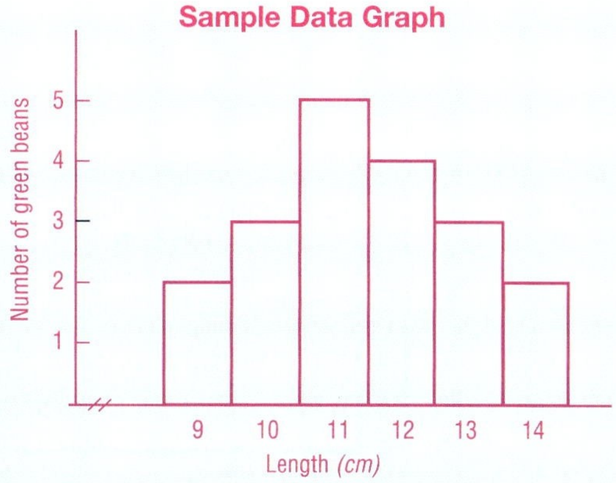

Answers to Analyze and Conclude

1. Most will approximate a normal distribution.
2. Answers will vary.
3. Most means will be similar to modes.
4. Most will show stabilizing selection.
5. No. In the example given, the two modes in the population are very different from the mean. Thus, the mean is not always useful in describing traits in a population.

CHAPTER 16

Ecosystems
Chapter Planning Guide

PACING	CLASSROOM RESOURCES	LABS, ACTIVITIES, AND DEMONSTRATIONS
BLOCK 1 • 45 min pp. 338–339 **Chapter Opener**		**SE** **Quick Review,** p. 339 ■ **SE** **Reading Activity,** p. 339 ■ **TE** **Using the Figure,** p. 338 **TE** **Opening Activity,** p. 339
BLOCKS 2 & 3 • 90 min pp. 340–344 **Section 1** What Is an Ecosystem?	**OSP** Lesson Plan* **TR** Bellringer* **TR** Section Outline* **TR** E25 Ecological Succession at Glacier Bay*	**TE** **Activity** Species Count, p. 340 GENERAL **TE** **Group Activity** Biodiversity, p. 341 ◆ GENERAL **SE** **Quick Lab** Evaluating Biodiversity, p. 342 ◆ ■ **CRF** **Datasheets for In-Text Labs** Evaluating Biodiversity* **TE** **Demonstration,** p. 343 ◆ BASIC **TE** **Using the Figure,** p. 343 GENERAL **SE** **Quick Lab** Modeling Succession, p. 344 ◆ ■ **CRF** **Datasheets for In-Text Labs** Modeling Succession*
BLOCKS 4 & 5 • 90 min pp. 345–349 **Section 2** Energy Flow in Ecosystems	**OSP** Lesson Plan* **TR** Bellringer* **TR** Section Outline* **TR** E4 Trophic Levels* **TR** E5 Food Chain in an Antarctic Ecosystem* **TR** E6 Food Web in an Antarctic Ecosystem* **TR** E8 Energy Transfer Through Trophic Levels* **TR** E11 Energy Efficiency in Food Consumption*	**TE** **Demonstration,** p. 345 GENERAL **TE** **Using the Figure,** p. 346 BASIC **TE** **Group Activity** Food Webs & Biodiversity, p. 347 ◆ GENERAL **TE** **Activity** Energy in Trophic Levels, p. 349 GENERAL **CRF** **Quicklab** Making a Food Web* BASIC
BLOCKS 6 & 7 • 90 min pp. 350–354 **Section 3** Cycling of Materials in Ecosystems	**OSP** Lesson Plan* **TR** Bellringer* **TR** Section Outline* **TR** E12 Water Cycle* **TR** E13 Carbon Cycle* **TR** E14 Nitrogen Cycle*	**TE** **Using the Figure,** p. 351 GENERAL **TE** **Using the Figure,** p. 352 BASIC **TE** **Demonstration,** p. 353 BASIC **TE** **Using the Figure,** p. 353 GENERAL **SE** **Exploration Lab** Modeling Ecosystem Change Over Time, pp. 358–359 ■ **CRF** **Datasheets for In-Text Labs** Modeling Ecosystem Change Over Time* ◆ **CRF** **Skills Practice Lab** Mapping Abiotic Factors*

BLOCKS 8 & 9 • 90 min

Chapter Review and Assessment Resources

- **SE** **Study Zone,** p. 355 ■
- **SE** **Performance Zone,** p. 356 ■
- **SE** **FCAT Test Prep,** p. 357 ■
- **TR** Concept Mapping*
- **CRF** Vocabulary Review Worksheet* ■ GENERAL
- **CRF** Test Prep Pretest* ■ GENERAL
- **CRF** Chapter Test* ■ GENERAL
- **CRF** Chapter Test* ADVANCED
- **OSP** Test Generator*
- **OSP** Test Item Listing*
- **OSP** Modified Chapter Test* BASIC

Online and Technology Resources

Holt Online Learning

Visit go.hrw.com for access to Holt Online Learning or enter the keyword **HX6 Home** for a variety of free online resources.

One-Stop Planner® CD-ROM

This CD-ROM package includes
- Lab Materials QuickList Software
- Holt Calendar Planner
- Customizable Lesson Plans
- Printable Worksheets
- ExamView® Test Generator
- Interactive Teacher Edition
- Holt PuzzlePro® Resources
- Holt PowerPoint® Resources

Guided Reading Audio CDs

These CDs are designed to help auditory learners and reluctant readers. (Audio CDs are also available in Spanish.)

Compression guide:
To shorten your instruction because of time limitations, omit blocks 1, 3, and 8.

KEY

TE Teacher Edition
SE Student Edition
OSP One-Stop Planner
CRF Chapter Resource File
TR Teaching Transparency
* Also on One-Stop Planner
■ Also Available in Spanish
◆ Requires Advance Prep

SKILLS DEVELOPMENT AND PRACTICE	SECTION REVIEW AND ASSESSMENT	CORRELATIONS
		Florida Sunshine Standards/FCAT
TE **Reading Skill Builder** Reading Hint, p. 341 BASIC CRF **Directed Reading Worksheet*** BASIC CRF **Active Reading Worksheet*** GENERAL	SE **Section 1 Review** ■ TE **Reteaching,** p. 344 BASIC TE **Quiz,** p. 344 GENERAL CRF **Quiz*** ■ BASIC	SC.G.1.4.1, SC.G.2.4.2
TE **Reading Skill Builder** Interactive Reading, p. 346 BASIC TE **Skill Builder** Interpreting Visuals, p. 348 BASIC CRF **Directed Reading Worksheet*** BASIC CRF **Active Reading Worksheet*** GENERAL	SE **Section 2 Review** ■ TE **Reteaching,** p. 349 BASIC TE **Quiz,** p. 349 GENERAL TE **Alternative Assessment,** p. 349 ADVANCED CRF **Quiz*** ■ BASIC	SC.B.1.4.7, SC.G.1.4.2
TE **Reading Skill Builder** Paired Reading, p. 351 BASIC CRF **Directed Reading Worksheet*** BASIC CRF **Active Reading Worksheet*** GENERAL CRF **Critical Thinking Worksheet*** ADVANCED CRF **Science Skills Worksheet*** GENERAL	SE **Section 3 Review** ■ TE **Reteaching,** p. 354 BASIC TE **Quiz,** p. 354 GENERAL TE **Alternative Assessment,** p. 354 ADVANCED CRF **Quiz*** ■ BASIC **Modified Worksheet*** BASIC	SC.D.2.4.1, SC.G.1.4.2, SC.G.2.4.2

Holt Biology Interactive Tutor CD-ROM

This CD-ROM consists of interactive activities that give students a fun way to extend their knowledge of biology concepts.

Unit 7 Ecosystem Dynamics

SciLINKS. NSTA
www.scilinks.org

Maintained by the **National Science Teachers Association**

Topic: Biodiversity
SciLinks Code: HX4020

Topic: Food Chains
SciLinks Code: HX4085

Topic: Energy Pyramids
SciLinks Code: HX4069

Topic: Water Cycle
SciLinks Code: HX4188

Topic: Carbon Cycle
SciLinks Code: HX4031

Topic: Sustainable Agriculture
SciLinks Code: HX4170

Topic: Ecosystems
SciLinks Code: HX4066

CHAPTER 16

Overview

Tell students the purpose of this chapter is to study the interactions between organisms and their environment. Students will learn what ecosystems are and how energy flows through ecosystems. Then students will learn how materials cycle through ecosystems.

Using the Figure

Have students study this picture and then ask, "If all these fish are of the same species, what is the name for this group of fish?" (a population) Next, ask them to notice the orange and green living coral in the picture. Ask, "If we include these organisms with the fish, what kind of group is it considered to be now?" (a community) Ask students if these are the only things in the picture, and if not, what else is present? (water, rocks, and so on) Tell students this is a new kind of grouping—one that includes nonliving things along with the living organisms. What kind of grouping is this? (an ecosystem)

Assessing Prior Knowledge

Before beginning this chapter, be sure your students can answer the questions in the ***Quick Review*** on the facing page. If they cannot, encourage them to re-read the section indicated.

Chapter Resource File

- **Vocabulary Worksheets**
- **Concept Mapping**

Coral-reef ecosystem

338

Chapter Correlations | **Florida Sunshine Standards/FCAT**

SC.B.1.4.7 The total amount of usable energy always decreases, even though the total amount of energy is conserved in any transfer.

SC.D.2.4.1 The systems on Earth and the quality of life are interconnected.

SC.G.1.4.1 There is great diversity and interdependence among living things.

SC.G.1.4.2 The flow of energy through an ecosystem made up of producers, consumers, and decomposers carries out the processes of life, and some energy dissipates as heat and is not recycled.

SC.G.1.4.3 The chemical elements that make up the molecules of living things are combined and recombined in different ways.

SC.G.2.4.2 Changes in a component of an ecosystem will have unpredictable effects on the entire system, but the components of the system tend to react in a way that will restore the ecosystem to its original condition.

CHAPTER

16 Ecosystems

Florida Sunshine Standards/FCAT

SC.D.2.4.1 The systems on Earth and the quality of life are interconnected.
SC.G.1.4.2 The flow of energy through an ecosystem made up of producers, consumers, and decomposers carries out the processes of life, and some energy dissipates as heat and is not recycled.
SC.G.1.4.3 The chemical elements that make up the molecules of living things are combined and recombined in different ways.
SC.G.2.4.2 Changes in a component of an ecosystem will have unpredictable effects on the entire system, but the components of the system tend to react in a way that will restore the ecosystem to its original condition.
SC.B.1.4.7 The total amount of usable energy always decreases, even though the total amount of energy is conserved in any transfer.

Quick Review

Answer the following without referring to earlier sections of your book.

1. **Contrast** autotrophs with heterotrophs. *(Chapter 5, Section 1)*
2. **Summarize** the process of photosynthesis. *(Chapter 5, Section 2)*
3. **Describe** the process of cellular respiration. *(Chapter 5, Section 3)*
4. **Compare** the energy flow in photosynthesis with the energy flow in cellular respiration. *(Chapter 5, Sections 2 and 3)*

Did you have difficulty? *For help, review the sections indicated.*

Reading Activity

Copy the following statements on a piece of paper or in your notebook, leaving a few blank lines after each.

1. In an ecosystem, more energy is stored in plants than in consumers.

2. The extinction of one species in an ecosystem can have an impact on all other species.

Before you read the chapter, write down whether you agree or disagree with each statement. After you have finished reading the chapter, decide whether or not you still agree with your first response.

Materials and energy cycle continuously through the components of this coral reef. The complex relationship of organisms and their physical environment makes up an ecological system, or *ecosystem*.

Looking Ahead

Section 1
What Is an Ecosystem?
Interactions of Organisms and Their Environment
Diverse Communities in Ecosystems
Change of Ecosystems over Time

Section 2
Energy Flow in Ecosystems
Movement of Energy Through Ecosystems
Loss of Energy in a Food Chain

Section 3
Cycling of Materials in Ecosystems
Biogeochemical Cycles
The Water Cycle
The Carbon Cycle
The Phosphorus and Nitrogen Cycles

internet connect

www.scilinks.org
National Science Teachers Association *sci*LINKS Internet resources are located throughout this chapter.

SCILINKS. Maintained by the National Science Teachers Association

339

Opening Activity

Ecological Spheres Tell students that scientists have developed sealed glass spheres in which a self-sustaining ecosystem exists. Although there is no input of materials, ask students whether there is an input of anything else into the ecosystem. (Yes; there is a constant input of energy in the form of light.)

Identifying Misconceptions

Students may believe that an ecosystem must be an area that is delineated by specific boundaries—like the walls of an aquarium or the edge of a pond. Point out that the limits of an ecosystem are defined by the observer. Thus, an ecosystem could be in a jar of pond water, or it could be the entire pond itself.

Quick Review Answers

1. Autotrophs are able to make food by using available energy and materials; heterotrophs must obtain their energy from the food produced by autotrophs.
2. In photosynthesis, organisms use light energy to rearrange the atoms in carbon dioxide and water to make carbohydrates. Oxygen is given off to the environment or used by the organism itself.
3. In cellular respiration, a carbohydrate is combined with oxygen, rearranging its atoms into carbon dioxide and water. Some of the energy released during this process is available for use by the cell.
4. In photosynthesis, an input of energy is required, and the energy is stored in carbohydrate molecules. In cellular respiration, this stored energy is released for the cell to use.

Reading Activity Answers

1. Agree; as consumers use the plants for food, some of the energy is lost to the environment.
2. Agree; some species, called keystone species, may be critical links in the food web.

Focus

Overview

Before beginning this section review with your students the objectives listed in the Student Edition. This section introduces students to the differences between communities and ecosystems. Students will also explore the concept of biodiversity, and how one species may gradually be replaced by another.

Bellringer

Point to your aquarium (or show a picture of an aquarium with a variety of organisms—different species of fish, snails, plants, etc.) and ask your students to list all of the organisms they see. Tell them that when they finish, they are to make a list of all the factors that affect the survival of the organisms in the aquarium. (They should include such things as water, food, temperature range, light, pH, oxygen, and so on.)

Motivate

Activity — GENERAL

Species Count If possible, take the students on a walk-around tour of your school. Point out different species of plants and animals that you observe. Have one student keep a count of the number of species identified on the campus. When finished, mention that biologists sampling a tropical rain forest in Ecuador obtained samples suggesting there are as many as 24,000 different insect species alone per acre.

Section Correlations

SC.G.1.4.1
SC.G.2.4.2

Section 1 What Is an Ecosystem?

Florida Sunshine Standards/FCAT

SC.G.1.4.1 There is great diversity and interdependence among living things.

SC.G.2.4.2 Changes in a component of an ecosystem will have unpredictable effects on the entire system, but the components of the system tend to react in a way that will restore the ecosystem to its original condition.

Objectives

- **Distinguish** an ecosystem from a community. SC.G.1.4.1
- **Describe** the diversity of a representative ecosystem. SC.G.1.4.1
- **Sequence** the process of succession. SC.G.2.4.2

Key Terms

ecology
habitat
community
ecosystem
abiotic factor
biotic factor
biodiversity
pioneer species
succession
primary succession
secondary succession

Interactions of Organisms and Their Environment

It is easy to think of the environment as being around but not part of us—something we always use, sometimes enjoy, and sometimes damage. But in fact, we are part of the environment along with all of Earth's other organisms. All of Earth's inhabitants are interwoven in a complex web of relationships, such as the one illustrated in **Figure 1.** To understand how the interactions of the parts can affect a whole system, think about how a computer operates. Removing one circuit from a computer can change or limit the interactions of the computer's many components in ways that influence the computer's overall operation. In a similar way, removing one species from our environment can have many consequences, not all of them easily predictable.

In 1866, the German biologist Ernst Haeckel gave a name to the study of how organisms fit into their environment. He called this study *ecology,* which comes from the Greek words *oikos,* meaning "house," or "place where one lives," and *logos,* meaning "study of." **Ecology** is the study of the interactions of living organisms with one another and with their physical environment (soil, water, climate, and so on). The place where a particular population of a species lives is its **habitat.** The many different species that live together in a habitat are called a **community.** An **ecosystem,** or ecological system, consists of a community and all the physical aspects of its habitat, such as the soil, water, and weather. The physical aspects of a habitat are called **abiotic** *(ay bie AHT ihk)* **factors,** and the organisms in a habitat are called **biotic factors.**

Figure 1 Organisms interact within an ecosystem. Organisms within an ecosystem continually change and adjust. This plant species is dependent on the bat for its reproduction, and the bat uses part of the flower for food.

Cultural Awareness

Isolated Ecosystem The Yanomami are a tribe living in remote rain-forest jungles of Venezuela and Brazil. The Yanomami in Venezuela have had very little contact with outsiders and are one of the few cultures remaining in the world in which people are still an integral part of an intact natural ecosystem. The Yanomami and their land in Venezuela are protected as an international biosphere reserve. Outsiders must get written permission to visit, but even these few visits have given the Yanomami a taste of the outside world. Many observers question how long the Yanomami hunting and gathering culture can last.

Diverse Communities in Ecosystems

The variety of organisms, their genetic differences, and the communities and ecosystems in which they occur is termed **biodiversity**. Consider a pine forest in the southeastern United States, such as the one shown in **Figure 2.** If you could fence in a square kilometer (0.4 mi^2) of this forest and then collect every organism, what would you expect to get? Which of the six kingdoms of organisms would be represented in your collection?

Figure 2 Pine forest. Pine forests like this one are common in the southeastern United States.

Ecosystem Inhabitants

Large animals in the forest might include a bear or a white-tailed deer. The woods also contain smaller mammals—raccoons, foxes, squirrels, rabbits, and chipmunks. Snakes and toads often remain hidden among the leaves. Many birds can be found, including hawks, warblers, and sparrows. If the square kilometer included a lake, you might find catfish, bass, perch, a variety of turtles, and perhaps an alligator.

There are pine trees, a variety of smaller trees, and shrubs. Beneath the trees, grasses and many kinds of flowers grow on the forest floor.

The soil contains an immense number of worms. Hidden under the bark of trees and beneath the leaves covering the ground are many different species of insects and spiders, such as those shown in **Figure 3.**

Many of the life-forms in the soil and water of a pine forest are too small to be seen without a microscope. Protists, which include algae and related microscopic eukaryotes, thrive in water. There may be billions of bacteria in a handful of soil.

internet connect
www.scilinks.org
Topic: Biodiversity
Keyword: HX4020
SciLinks. Maintained by the National Science Teachers Association

Figure 3 Forest spider and insect

The jumping spider is found in sunny, dry parts of the forest. The larvae of the stag beetle live in and eat decaying wood and bark.

Jumping spider

Male stag beetle

Teach

Group Activity — GENERAL

Biodiversity Before class, find an area at your school with a good amount of biodiversity. Organize the students into groups, and have each group create a chart with the following headings: Name, Diagram, Approximate Average Size, and Approximate Number. Take your students to the area and have them use their charts to make an inventory of the organisms found there. Students may not know the names of the plants or animals, but they can make a small diagram of each organism, and for the time being, make up a fictitious name for each organism they draw. Have them save this information to use for a food web in the next section.

READING SKILL BUILDER — BASIC

Reading Hint Point out to students the importance of studying the figures in each section. Strongly encourage them to stop after each section and read the captions for each figure. Prepare questions that correspond to each figure. For example, in reference to **Figure 1**, you might ask students to hypothesize as to what would happen if the plant species in the photograph were eradicated from the area.

Interactive Biology Tutor

- Unit 7—Ecosystem Dynamics

This engaging tutorial introduces students to Ecosystem Dynamics.

Career

Wildlife Biologists are hired by federal, state, and sometimes local government agencies to study and manage wildlife. Common game and nongame species, and sometimes rare and endangered species are studied and/or managed by wildlife biologists. Knowledge of each species' role in the ecosystem and its relation to other species is critical. Starting salary varies with agency and geographic region.

Chapter Resource File

- Directed Reading BASIC
- Active Reading GENERAL
- Datasheet for Quick Lab GENERAL

One-Stop Planner CD-ROM

- Reading Organizers BASIC
- Reading Strategies BASIC

Transparencies

TR Bellringer

TR E25 Ecological Succession at Glacier Bay

Teach, continued

Teaching Tip — GENERAL

Detritivores Bacteria and fungi belong to a group of organisms known as detritivores, which live on detritus, the dead remains of plants and animals. Ask students to think of other organisms that would be detritivores. (Maggots, grubs, and earthworms would be examples.)

QUICK LAB *Evaluating Biodiversity*

Skills Acquired
Observing, drawing conclusions

Teacher's Notes

Have each student use an area of equal dimensions, at least 50 × 50 m, to make the task more manageable and to allow comparisons between different ecosystems. Encourage students to take their time; they will observe much more if they do.

Answers to Analysis

1. Answers will vary.
2. Answers will vary. Example: 3 robins/29 total organisms = 0.103 = 10%
3. Answers will vary.
4. Answers will vary. In general, the abiotic factors in an ecosystem provide organisms (biotic factors) with a physical place to live, energy, nutrients, and water. The organisms alter and recycle some of these abiotic factors, changing the landscape in the process.

Figure 4 Forest fungi

These fungi digest plants and other materials they find in the forest.

Mushrooms are often found on moist forest floors.

Shelf fungi grow on and digest trees.

You might find many kinds of fungi growing on fallen trees and spreading as fine threads through the decaying material on the forest floor, as illustrated in **Figure 4.** Other fungi are found on the surface of trees or rocks as lichens. Lichens are associations between fungi and algae or cyanobacteria.

If you were to remove every organism from your square kilometer, the nonliving surroundings that remain make up the abiotic factor. This would include the minerals, organic compounds, water, wind that blows over the Earth, rain, and sunlight.

Ecosystem Boundaries

The physical boundaries of an ecosystem are not always obvious, and they depend on how the ecosystem is being studied. For example, a scientist might consider a single rotting log on the forest floor to be an ecosystem if he or she is interested only in the fungi and insects living in the log. Often individual fields, forests, or lakes are studied as isolated ecosystems. Of course, no location is ever totally isolated. Even oceanic islands get occasional migrant visitors, such as birds blown off course.

QUICK LAB *Evaluating Biodiversity* SC.G.1.4.1

By making simple observations, you can draw some conclusions about biodiversity in an ecosystem.

Materials

note pad, pencil

Procedure

1. **CAUTION: Do not approach or touch any wild animals. Do not disturb plants.** Prepare a list of biotic and abiotic factors that you observe around your home or in a nearby park.

Analysis

1. **Identify** the habitat and community that you observed.
2. **Calculate** the number of different species as a percentage of the total number of organisms that you saw.
3. **Rank** the importance of biotic factors within the ecosystem you observed.
4. **Infer** what the relationships are between biotic factors and abiotic factors in the observed ecosystem.

342

Attention Grabber

Introduced Species Introducing species to new areas can wreak havoc on an ecosystem. In the 1870s Lord George Bennet founded the small coastal town of Bandon, Oregon. Bennet is also the one believed by many to have imported a plant called gorse (*Ulex europaeus*) from his native Ireland. Gorse spreads out rapidly, leaving a dead, dry center. Its nasty thorns make it difficult to remove, and its naturally high oil content makes it very flammable and causes it to reach high temperatures when it burns. Having no natural enemies in Oregon, gorse spread quickly and could not be controlled. On Sept. 26, 1936, a spark from a nearby fire ignited a gorse thicket which eventually spread to burn 287,000 acres, destroying all but 16 of Bandon's nearly 500 buildings and killing 14 people. In 1994 the gorse spider mite was released near Bandon to help control this noxious weed.

Change of Ecosystems over Time

When a volcano forms a new island, a glacier recedes and exposes bare rock, or a fire burns all of the vegetation in an area, a new habitat is created. This change sets off a process of colonization and ecosystem development. The first organisms to live in a new habitat where soil is present tend to be small, fast-growing plants, called **pioneer species.** They may make the ground more hospitable for other species. Later waves of plant immigrants may then outcompete and replace the pioneer species.

Succession

A somewhat regular progression of species replacement is called **succession.** Succession that occurs where life has not existed before is called **primary succession.** Succession that occurs in areas where there has been previous growth, such as in abandoned fields or forest clearings, is called **secondary succession.** It was once thought that the stages of succession were predictable and that succession always led to the same final community of organisms within any particular ecosystem. Ecologists now recognize that initial conditions and chance play roles in the process of succession. For example, if two species are in competition, a sudden change in the climate may favor the success of one species over the other. For this reason, no two successions are alike.

Figure 5 Glacier Bay

A receding glacier makes primary succession possible.

Recently exposed land has few nutrients.

Alders, grasses, and shrubs later take over from pioneer plants.

As the amount of soil increases, spruce and hemlock trees become plentiful.

Glacier Bay: an Example of Succession

A good example of primary succession is a receding glacier because land is continually being exposed as the face of the glacier moves back. The glacier that composes much of the head of Glacier Bay, Alaska, has receded some 100 km (62 mi) over the last 200 years. **Figure 5** shows the kinds of changes that have taken place as time passed.

The most recently exposed areas are piles of rock and gravel that lack the usable nitrogen essential to plant and animal life. The seeds and spores of pioneer species are carried in by the wind. These include lichens, mosses, fireweed, willows, cottonwood, and *Dryas*, a sturdy plant with clumps about 30 cm (1 ft) across. At first all of these plants grow close to the ground, severely stunted by mineral deficiency, but *Dryas* eventually crowds out the other plants.

After about 10 years, alder seeds blown in from distant sites take root. Alder roots have nitrogen-fixing nodules, so they are able to grow more rapidly than *Dryas*. Dead leaves and fallen branches from the alder trees add more usable nitrogen to the soil. The added nitrogen allows willows and cottonwoods to invade and grow with vigor. After about 30 years, dense thickets of alder, willow, and cottonwood shade and eventually kill the *Dryas*.

Demonstration — BASIC

Bring to class a rock covered with lichens or mosses. Ask students why these organisms are known as "pioneer species." (Like pioneer settlers, they are the first to inhabit a new area.) Point out that lichen can absorb nutrients from the bare rock.

Teaching Tip — GENERAL

Succession Have each student draw an illustrated timeline representing the succession that occurs after a forest fire has burned all of the vegetation in an area. Starting at time zero, the land should be barren, followed by the appearance of grasses and weeds, then small bushes, and finally trees. **LS Visual**

Using the Figure — GENERAL

Have students examine the top photo in **Figure 5.** Ask them to think of other events that might occur that would lead to primary succession. (Landslides and volcanism would be examples.) Tell students that primary succession can be a very slow process. Scientists estimate that the primary succession from sand dunes to the beech-maple forest along the shores of Lake Michigan took about 1,000 years. In contrast, secondary succession may take less than 100 years. Ask students why the hemlock and spruce trees don't come before the grasses and shrubs. (The grasses and shrubs often provide a microhabitat that makes it possible for the seeds of the trees to survive.)

did you know?

Mount St. Helens After the eruption of Mount St. Helens on May 18, 1980, succession was aided by an unlikely source—pocket gophers. Pocket gophers that survived the blast dug miles of tunnels under the barren pumice fields. The digging helped mix the soil, which improved its quality and ability to hold water. When a gopher emerged, the mound they made would catch seeds blowing across the moon-like landscape, helping plants to get started. In another surprise to scientists, the gophers helped to save amphibians, which used the tunnels as cool underground areas in which to avoid the hot, dry landscape.

About 80 years after the glacier first exposes the land, Sitka spruce invades the thickets. Spruce trees use the nitrogen released by the alders and eventually form a dense forest. The spruce blocks the sunlight from the alders, and the alders then die, just as the *Dryas* did before them. After the spruce forest is established, hemlock trees begin to grow. Hemlocks are very shade tolerant and have a root system that competes well against spruce for soil nitrogen. Hemlock trees soon become dominant in the forest. This community of spruce and hemlock proves to be a very stable ecosystem from the perspective of human time scales, but it is not permanent. As local climates change, this forest ecosystem may change too.

QUICK LAB
Modeling Succession SC.G.2.4.2

You can create a small ecosystem and measure how organisms modify their environment.

Materials

1 qt glass jar with a lid, one-half quart of pasteurized milk, pH strips

Procedure

1. Prepare a table like the one below.
2. Half fill a quart jar with pasteurized milk, and cover the jar loosely with a lid. Measure and record the pH. Place the jar in a 37°C incubator.
3. Check and record the pH of the milk with pH strips every day for seven days. As milk spoils, its pH changes. Different populations of microorganisms become established, alter substances in the milk, and then die off when conditions no longer favor their survival.
4. Record any visible changes in the milk each day.

Analysis

1. **Identify** what happened to the pH of the milk as time passed.
2. **Infer** what the change in pH means about the populations of microorganisms in the milk.
3. **Critical Thinking Evaluating Results** How does this model confirm the model of succession in Glacier Bay?

DATA TABLE

Day	pH	Appearance
1		
2		
3		

Section 1 Review

1. **Identify** what components of an ecosystem are *not* part of a community. SC.G.1.4.1
2. **Relate** how gardening or agriculture affects succession. SC.G.2.4.6
3. **Differentiate** primary succession from secondary succession. SC.G.2.4.2
4. **Critical Thinking Applying Information** Why do some ecosystems remain stable for centuries, while others undergo succession? SC.G.2.4.2
5. *FCAT Prep* What characteristic of alders allows them to grow relatively rapidly in the succession that occurs as a glacier recedes?
   A nitrogen-fixing nodules
   B no need for minerals
   C no roots
   D shade tolerance
   SC.F.1.4.2

Teach, *continued*

QUICK LAB *Modeling Succession*

Skills Acquired
Observing, inferring

Teacher's Notes
Make sure students do not screw the lid down on the jars too tightly because the growing culture will suffocate, or the jar may explode from anaerobically produced gas.

Answers to Analysis

1. The pH dropped as the environment became more acidic.
2. By-products from the microorganisms change the pH. Then new organisms that are better adapted to the changed pH begin to thrive.
3. Like the Glacier Bay model, organisms colonize and slowly change a new environment such that it becomes more suitable for other organisms.

Close

Reteaching BASIC

Show students a picture of a forest ecosystem. Have them describe the events that would take place after a fire.

Quiz GENERAL

1. What is the difference between communities and ecosystems? (An ecosystem is a community and its abiotic factors.)
2. How are the boundaries of an ecosystem set? (in whatever way makes it easier to study)

Answers to Section Review

1. the abiotic environment
2. Because the land is kept in a constant state of disturbance, secondary succession in gardens and farms does not usually get beyond the first stage of fast-growing weeds and grasses.
3. Primary succession occurs on land where there has not been any previous plant growth, such as recently exposed land that has few nutrients. Secondary succession occurs in areas where there has been previous plant growth, such as in abandoned fields or forest clearings.
4. Answers may vary, but students should cite factors such as climate, geology, and human intervention.
5. **A.** Correct. **B.** Incorrect. They do require minerals. **C.** Incorrect. Alders do have roots. **D.** Incorrect. Alders are not shade tolerant.

Section Review Correlations

SC.F.1.4.2: Item 5
SC.G.1.4.1: Item 1
SC.G.2.4.2: Items 3, 4
SC.G.2.4.6: Item 2

Energy Flow in Ecosystems

Section 2

Florida Sunshine Standards/FCAT

SC.G.1.4.2 The flow of energy through an ecosystem made up of producers, consumers, and decomposers carries out the processes of life, and some energy dissipates as heat and is not recycled.

SC.B.1.4.7 The total amount of usable energy always decreases, even though the total amount of energy is conserved in any transfer.

Movement of Energy Through Ecosystems

Everything that organisms do in ecosystems—running, breathing, burrowing, growing—requires energy. The flow of energy is the most important factor that controls what kinds of organisms live in an ecosystem and how many organisms the ecosystem can support. In this section you will learn where organisms get their energy.

Primary Energy Source

Most life on Earth depends on photosynthetic organisms, which capture some of the sun's light energy and store it as chemical energy in organic molecules. These organic compounds are what we call food. The rate at which organic material is produced by photosynthetic organisms in an ecosystem is called **primary productivity.** Primary productivity determines the amount of energy available in an ecosystem. Most organisms in an ecosystem can be thought of as chemical machines driven by the energy captured in photosynthesis. Organisms that first capture energy, the **producers,** include plants, some kinds of bacteria, and algae. Producers make energy-storing molecules. All other organisms in an ecosystem are consumers. **Consumers** are those organisms that consume plants or other organisms to obtain the energy necessary to build their molecules.

Trophic Levels

Ecologists study how energy moves through an ecosystem by assigning organisms in that ecosystem to a specific level, called a **trophic** *(TROHF ihk)* **level,** in a graphic organizer based on the organism's source of energy. Energy moves from one trophic level to another, as illustrated in **Figure 6.**

Objectives

- **Distinguish** between producers and consumers. SC.G.1.4.2
- **Compare** food webs with food chains. SC.G.1.4.2
- **Describe** why food chains are rarely longer than three or four links. SC.G.1.4.2, SC.B.1.4.7

Key Terms

primary productivity
producer
consumer
trophic level
food chain
herbivore
carnivore
omnivore
detritivore
decomposer
food web
energy pyramid
biomass

Figure 6 Trophic levels

The sun is the ultimate source of energy for producers and all consumers.

Section 2

Focus

Overview

Before beginning this section review with your students the objectives listed in the Student Edition. This section discusses how producers and consumers facilitate the flow of energy through ecosystems in the form of food webs and food chains. Students will then discover that a depletion in the amount of available energy limits the number of steps that can occur in a food chain.

Bellringer

Tell students to think about their local area and make a diagram of a food chain that would be typical for your area. Ask them to try to put 6 organisms into the food chain. (It is very difficult to do—ask students why this is so, and lead into a discussion of trophic levels and energy loss.)

Motivate

Demonstration — GENERAL

List the following organisms that can be found in an open field: robin, hawk, snake, frog, grasshopper, mouse, and rabbit. Have students draw arrows to show what eats what in this field ecosystem. Students should see the complexity of even this simple food web in which each predator can take more than one type of prey and each type of prey could be exploited by several different species of predators.

Chapter Resource File

- Directed Reading BASIC
- Active Reading GENERAL

One-Stop Planner CD-ROM

- Reading Organizers BASIC
- Reading Strategies BASIC
- Occupational Application Worksheet Wildlife Biologist

Transparencies

TR Bellringer
TR E4 Trophic Levels
TR E5 Food Chain in an Antarctic Ecosystem
TR E6 Food Web in an Antarctic Ecosystem

Section Correlations

SC.B.1.4.7
SC.G.1.4.2

Teach

Real Life

Answer

These bacteria are chemosynthetic (as opposed to photosynthetic) producers. They are at the bottom of their food chain.

READING SKILL BUILDER — **BASIC**

Interactive Reading Assign Chapter 16 of the *Holt Biology* Guided Audio CD Program to help students achieve greater success in reading the chapter.

Teaching Tip — GENERAL

Trophic Levels Ask a volunteer from the class to describe his or her dinner last night. Then have students describe the trophic level of each different food item in the meal. What was the trophic level of the student as he or she ate each item? (Answers will vary.) **LS Verbal**

Using the Figure — BASIC

Ask students to look at the food chain in **Figure 7.** Ask students if this picture shows a lot of diversity. (no) Ask them what would happen to this food chain if leopard seals ate only cod, killer whales ate only leopard seals, and a disease wiped out all of the cod. (Leopard seals and killer whales would starve.)

Real Life

Not all producers are photosynthetic.

At the bottom of oceans near volcanic vents live bacteria that harvest energy from the reduced sulfur compounds ejected by these volcanic vents.

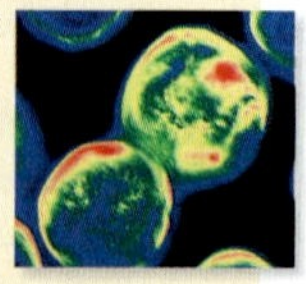

Applying Information
Where in their food chains do these bacteria lie?

First Level The path of energy through the trophic levels of an ecosystem is called a **food chain.** An example is shown in **Figure 7.** The lowest trophic level of any ecosystem is occupied by the producers, such as plants, algae, and bacteria. Producers use the energy of the sun to build energy-rich carbohydrates. Many producers also absorb nitrogen gas and other key substances from the environment and incorporate them into their biological molecules.

Second Level At the second trophic level are **herbivores** *(HUHR beh vohrz),* animals that eat plants or other primary producers. They are the primary consumers. Cows and horses are herbivores, as are caterpillars and some ducks. A herbivore must be able to break down a plant's molecules into usable compounds. However, the ability to digest cellulose is a chemical feat that only a few organisms have evolved. As you will recall, cellulose is a complex carbohydrate found in plants. Most herbivores rely on microorganisms, such as bacteria and protists, in their gut to help digest cellulose. Humans cannot digest cellulose because we lack these particular microorganisms.

Third Level At the third trophic level are secondary consumers, animals that eat other animals. These animals are called **carnivores.** Tigers, wolves, and snakes are carnivores. Some animals, such as bears, are both herbivores and carnivores; they are called **omnivores** *(AHM nih vohrz).* They use the simple sugars and starches stored in plants as food, but they cannot digest cellulose.

In every ecosystem there is a special class of consumers called detritivores, which include worms and fungal and bacterial decomposers. **Detritivores** *(deh TRIH tih vohrz)* are organisms that obtain their energy from the organic wastes and dead bodies that are produced at

Figure 7 **Aquatic food chain**

This food chain shows one path of energy flow in an Antarctic ecosystem.

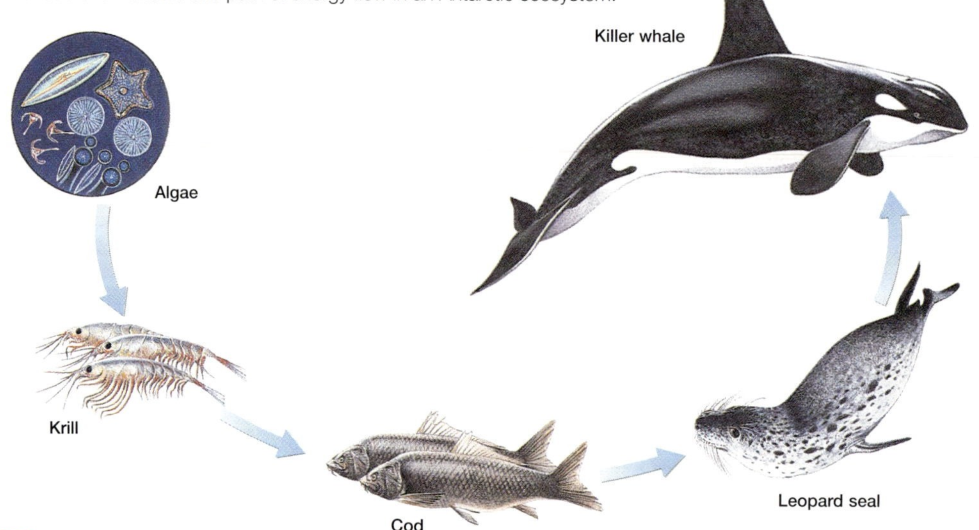

Integrating Physics and Chemistry

Reiterate that digestion involves physical and chemical changes. Whereas physical processes of digestion change large pieces of food into smaller ones primarily by means of chewing and muscle movements of the stomach and the intestines, chemical digestion takes place when various enzymes are added to the ingested food. These enzymes chemically break down food particles into molecules that are small enough to be absorbed and used by an organism's cells.

MISCONCEPTION ALERT

Food Chains vs. Food Webs Emphasize that food chains, while being useful tools for showing the flow of energy through an ecosystem, can be misleading to students because they imply that each organism eats ONLY the organism below it in the food chain. We know, of course, that most organisms eat a wide variety of other organisms, and for this reason the term "food web" gives a more accurate picture of what actually happens in an ecosystem.

all trophic levels. Bacteria and fungi are known as **decomposers** because they cause decay. Decomposition of bodies and wastes releases nutrients back into the environment to be recycled by other organisms.

Many ecosystems contain a fourth trophic level composed of those carnivores that consume other carnivores. They are called tertiary consumers, or top carnivores. A hawk that eats a snake is a tertiary consumer. Very rarely do ecosystems contain more than four trophic levels.

In most ecosystems, energy does not follow simple straight paths because individual animals often feed at several trophic levels. This creates a complicated, interconnected group of food chains called a **food web,** as illustrated in **Figure 8.**

internet connect
www.scilinks.org
Topic: Food Chains and Webs
Keyword: HX4085
SCLINKS. Maintained by the National Science Teachers Association

Figure 8 Aquatic food web

This food web shows a more complete picture of the feeding relationships in an Antarctic ecosystem.

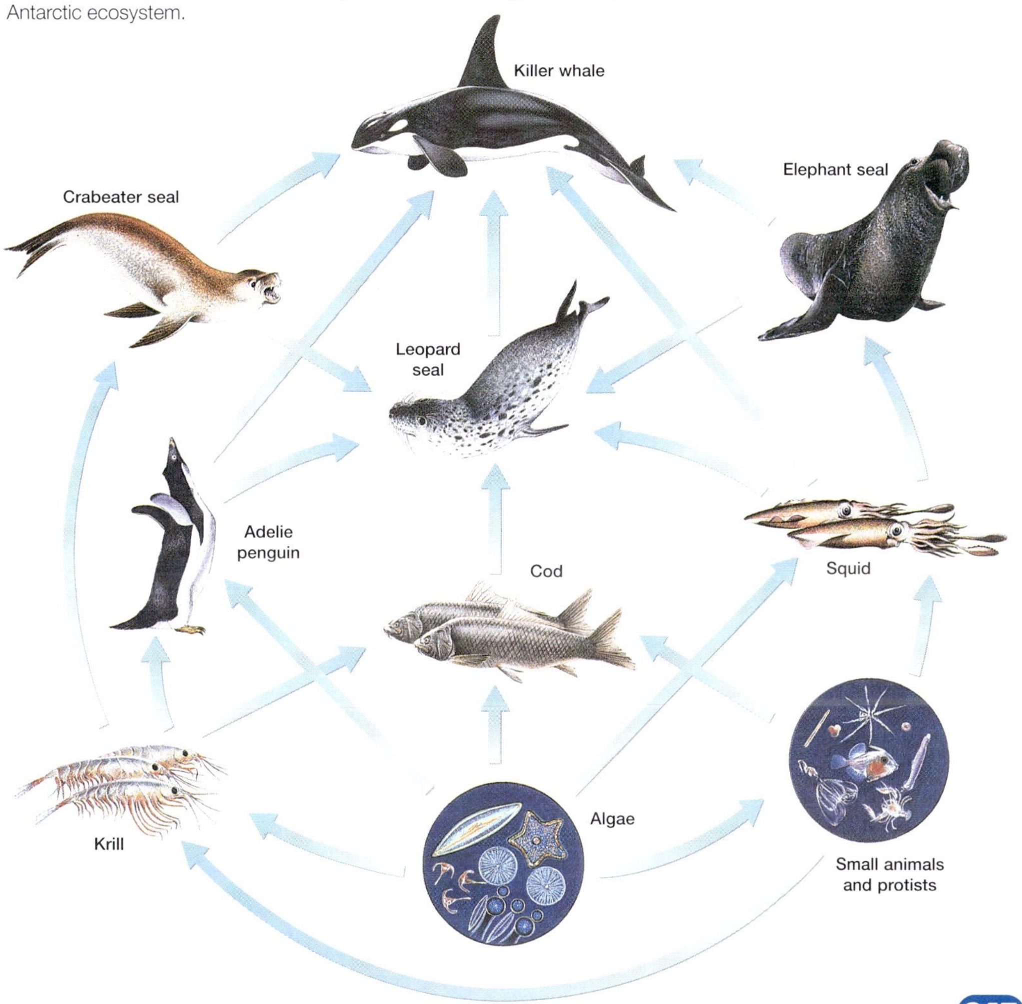

Teaching Tip

Understanding the Flow of Energy Have students draw a Graphic Organizer that summarizes the flow of energy from producers to herbivores, omnivores, carnivores, and detritivores. A completed graphic organizer is show at the bottom of this page.

Group Activity — GENERAL

Food Webs & Biodiversity
Organize students into groups of 4. Obtain sheets of butcher paper (each about 2–3 feet on a side). Have one student in each group begin by writing the name of an organism anywhere on the sheet in large letters. After the name they should write one of these letters: P (producer), C (carnivore), H (herbivore), O (omnivore), and D (detritivore). Moving clockwise, each student writes the name of another organism randomly on the sheet. Each student draws an arrow from this organism to anything that eats it, and an arrow to this organism from organisms it would eat. Continue until the food web gets very messy. Stop the activity and have each group hold up their paper and have the class "vote" on which food web shows the greatest biodiversity. Discuss what would happen to the "winning" food web if just one organism became extinct. (In most cases there would be no effect if there were a lot of biodiversity, but if the extinct organism was the only producer, it could have a profound effect.) Co-op Learning

Graphic Organizer

Use this graphic organizer with **Teaching Tip** *on this page.*

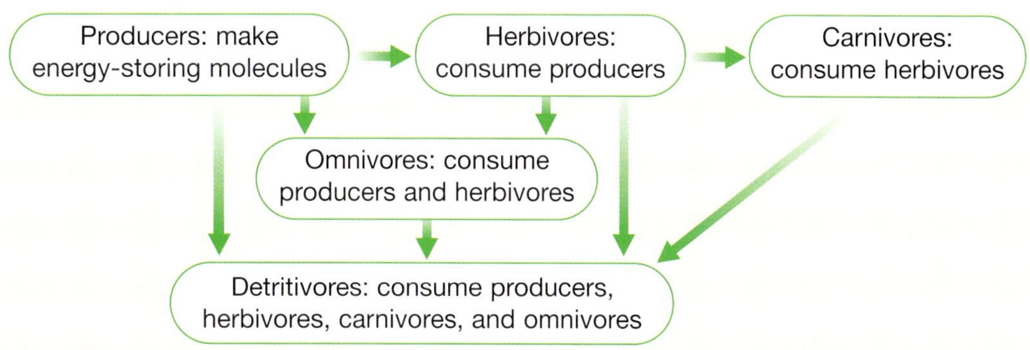

Teach, *continued*

Teaching Tip — GENERAL

Physics Remind students that the first law of thermodynamics states that energy cannot be created or destroyed but only changed in form. Have students give examples of energy changes from one form into another. (Examples: in a light bulb, electrical energy is changed to light energy and heat; batteries change chemical energy into electrical energy) The second law states that energy change between forms is never 100 percent efficient. Ask students to think of an example that shows the inefficiency of energy conversions. (In an automobile engine, gasoline—chemical energy—is changed to kinetic energy to turn the wheels; however, much of the energy is converted to heat, which is unavailable to power the automobile.) LS **Logical**

SKILL BUILDER — BASIC

Interpreting Visuals Have students examine the energy pyramid shown in **Figure 9.** Ask students how many trophic levels there are in the diagram. (4) Have them note that the top trophic level is very small. Ask them how the size of the upper level relates to the fact that there are a small number of trophic levels. (There is not enough energy to support another level of consumers.)

Transparencies

TR E8 Energy Transfer Through Trophic Levels

TR E11 Energy Efficiency in Food Consumption

WORD Origins

The word *ecosystem* is from the Greek words *oikos,* meaning "house," and *systematos,* meaning "to place together." Knowing this information makes it easier to remember that an ecosystem includes a community of living things as well as all physical aspects of its environment.

Loss of Energy in a Food Chain

A deer browsing on leaves is acquiring energy. Potential energy is stored in the chemical bonds within the molecules of the leaves. Some of this energy is transformed to other forms of potential energy, such as fat. Some of it aids the deer in running and breathing, and in fueling cellular processes. But much of the energy is dispersed into the environment as heat.

Energy Transfer

During every transfer of energy within an ecosystem, energy is lost as heat. Although heat can be used to do work (as in a steam engine), it is generally not a useful source of energy in biological systems. Thus, the amount of useful energy available to do work decreases as energy passes through an ecosystem. The loss of useful energy limits the number of trophic levels an ecosystem can support. When a plant harvests energy from sunlight, photosynthesis captures only about 1 percent of the energy available to the leaves. When a herbivore uses plant molecules to make its own molecules, only about 10 percent of the energy in the plant ends up in the herbivore's molecules. And when a carnivore eats the herbivore, about 90 percent of the energy is lost in making carnivore molecules. At each trophic level, the energy stored by the organisms in a level is about one-tenth of that stored by the organisms in the level below.

The Pyramid of Energy

Ecologists often illustrate the flow of energy through ecosystems with an energy pyramid. An **energy pyramid** is a diagram in which each trophic level is represented by a block, and the blocks are stacked on top of one another, with the lowest trophic level on the bottom. The width of each block is determined by the amount of energy stored in the organisms at that trophic level. Because the energy stored by the organisms at each trophic level is about one-tenth the energy stored by the organisms in the level below, the diagram takes the shape of a pyramid, as shown in **Figure 9.**

internet connect
www.scilinks.org
Topic: Energy Pyramids
Keyword: HX4069
SCILINKS. Maintained by the National Science Teachers Association

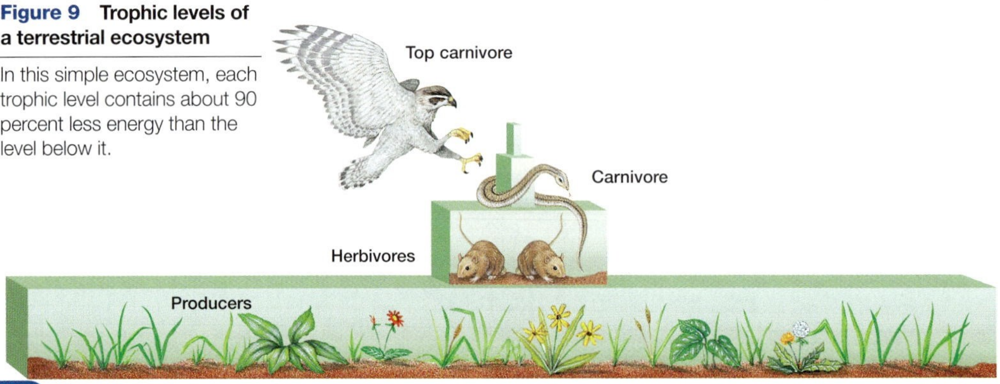

Figure 9 Trophic levels of a terrestrial ecosystem

In this simple ecosystem, each trophic level contains about 90 percent less energy than the level below it.

348

MISCONCEPTION ALERT

Lost Energy It is quite common for teachers and students to refer to energy being "lost." Remind students that the first law of thermodynamics says energy cannot be "lost"—it can only be converted from one form into another. What people mean when they say "lost" is that the energy is made unavailable—usually in the form of heat. For example, when an automobile engine heats up, the heat energizes nearby air molecules as it leaves the engine, but this heat cannot realistically be used to move the automobile.

Limitations of Trophic Levels

Most terrestrial ecosystems involve only three or, on rare instances, four levels. Too much energy is lost at each level to allow more levels. For example, a large human population could not survive by eating lions captured on the Serengeti Plain of Africa because there are too few lions to make this possible. The amount of grass in that ecosystem cannot support enough zebras to maintain a large enough population of lions to feed lion-eating humans. In other words, the number of trophic levels that can be maintained in a community is limited by the dispersal of potential energy.

Humans are omnivores, and unlike lions, we can choose to eat either meat or plants. As illustrated in **Figure 10,** about 10 kg (22 lb) of grain are needed to build about 1 kg (2.2 lb) of human tissue if the grain is directly ingested by a human. If a cow eats the grain and a human eats the cow, then about 100 kg (220 lb) of grain are needed to build about 1 kg (2.2 lb) of human tissue.

Also, the number of individuals in a trophic level may not be an accurate indicator of the amount of energy in that level. Some organisms are much bigger than others and therefore use more energy. Because of this, the number of organisms often does not form a pyramid when one compares different trophic levels. For instance, caterpillars and other insect herbivores greatly outnumber the trees they feed on. To better determine the amount of energy present in trophic levels, ecologists measure biomass. **Biomass** is the dry weight of tissue and other organic matter found in a specific ecosystem. Each higher level on the pyramid contains only 10 percent of the biomass found in the trophic level below it.

Figure 10 **Energy efficiency in food consumption**

Adding a trophic level to a food chain increases the energy demand of consumers by a factor of about 10.

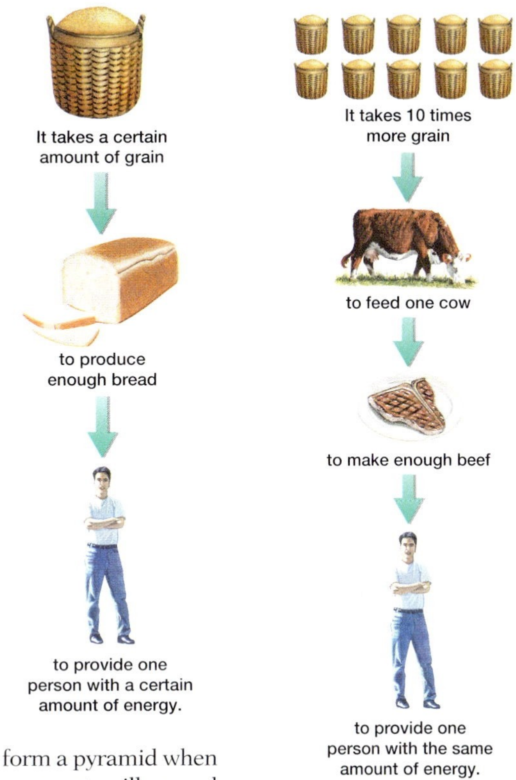

Section 2 Review

1. **Explain** how producers differ from consumers. SC.G.1.4.2
2. **Analyze** the flow of energy through a food chain that contains four tropic levels, one of which is a carnivore. SC.B.1.4.7, SC.G.1.4.2
3. **Construct** a food web, and explain the interactions of the organisms that compose it. SC.G.1.4.2
4. **List** the reasons why food chains do *not* tend to exceed four links. SC.G.1.4.2
5. **Critical Thinking** **Justifying an Argument** Explain why scientists believe that most animals would become extinct if all plants died. SC.G.1.4.1
6. *FCAT Prep* Which series shows a correct path of energy flow in a marine food chain?
   A algae → krill → cod
   B cod → leopard seal → krill
   C krill → cod → algae
   D leopard seal → algae → krill
   SC.G.1.4.2

349

Activity — GENERAL

Energy in Trophic Levels Tell students that if an average of 1,500 kilocalories of light energy per day falls on a square meter of land surface covered by plants, only about 15–30 kilocalories become incorporated into chemical compounds through photosynthesis. How much of this energy could end up in a person who eats these plants? (1.5–3.0 kilocalories) How much of this energy could end up in a person who eats a steak from a steer that ate the plants? (0.15–0.30 kilocalories) **LS Logical**

Close

Reteaching — BASIC

Have students construct a food web and energy pyramid for the schoolyard. Ask students to identify producers, herbivores, omnivores, carnivores, and detritivores. What species would they expect to find if the area had been a natural park.

Quiz — GENERAL

1. What do all consumers rely on for their food? (producers)
2. Give an example of consumers outnumbering the producers that they are feeding upon. (One tree might have thousands of insects feeding on it.)

Alternative Assessment — ADVANCED

Have students draw a food chain, a food web, and an energy pyramid for any ecosystem. They must choose enough organisms for several trophic levels and feeding pathways.

Answers to Section Review

1. Producers use energy (usually from the sun) to assemble food molecules, hence "producing food." Consumers must take in (or consume) these food molecules to obtain their energy.
2. As energy moves through the food chain, about 90 percent is "lost" at each level. So, 1,000 kilocalories in grass eaten by a rabbit would result in only about 100 available to the fox that eats the rabbit. The eagle that eats the fox would only have 10 kilocalories of the original 1,000 kilocalories available.
3. Answers will vary. Students should be able to explain the interactions between all the organisms.
4. Because so much energy is "lost" at each trophic level, it is difficult to exceed four links, or trophic levels.
5. Plants are the producers that make food. Without them, animals would starve.
6. **A.** Correct. Algae is the producer, and must come first in the food chain. **B.** Incorrect. Cod is not a producer. **C.** Incorrect. Algae is the producer. **D.** Incorrect. A leopard seal is not a producer.

Section Review Correlations

SC.B.1.4.7: Item 2
SC.G.1.4.1: Item 5
SC.G.1.4.2: Items 1-4, 6

Focus

Overview

Before beginning this section review with your students the objectives listed in the Student Edition. This section explains how important materials necessary for survival, such as water, carbon, phosphorus, and nitrogen, rotate through natural systems, constantly re-supplying organisms. Students will also learn that bacteria play an important role in the recycling of many elements.

Bellringer

Have students make a list of anything they have used in the past week that had previously been recycled or is made of recycled materials. After they have made their lists, have students volunteer items and discuss what the item is recycled from, if known, and the pluses and minuses (if any) of recycling each item.

Motivate

Discussion — GENERAL

Make the following statement to your class: "Someone once said that if a person dumped a glass of water on the ground, one year later, no matter where they were on Earth, any glass of water they pour would contain at least one molecule of water from the original glass." Ask students if they think this is true, and if so, how could this occur? (Probably not, but it makes for a great discussion about the water cycle!)

Section Correlations

SC.D.2.4.1
SC.G.1.4.2

Section 3 Cycling of Materials in Ecosystems

Florida Sunshine Standards/FCAT

SC.D.2.4.1 The systems on Earth and the quality of life are interconnected.
SC.G.1.4.2 The flow of energy through an ecosystem made up of producers, consumers, and decomposers carries out the processes of life, and some energy dissipates as heat and is not recycled.

Objectives

- **Summarize** the role of plants in the water cycle. SC.D.2.4.1
- **Analyze** the flow of energy through the carbon cycle. SC.G.1.4.2
- **Identify** the role of bacteria in the nitrogen cycle. SC.D.2.4.1

Key Terms

biogeochemical cycle
ground water
transpiration
nitrogen fixation

Biogeochemical Cycles

Humans throw away tons of garbage every year as unwanted, unneeded, and unusable. Nature, however, does not throw anything away. Most energy flows through the Earth's ecosystems from the sun to producers to consumers. The physical parts of the ecosystems, however, cycle constantly. Carbon atoms, for example, are passed from one organism to another in a great circle of use. Producers are eaten by herbivores, herbivores are eaten by carnivores, and carnivores are eaten by top carnivores. Eventually the top carnivores die and decay; their carbon atoms then become part of the soil to feed the producers in a long and complex cycle that reuses this important element. Carbon is not the only element that is constantly recycled in this way. Other recycled elements include many of the inorganic (noncarbon) substances that make up the soil, water, and air, such as nitrogen, sulfur, calcium, and phosphorus.

All materials that cycle through living organisms are important in maintaining the health of ecosystems, but four substances are particularly important: water, carbon, nitrogen, and phosphorus. All organisms require carbon, hydrogen, oxygen, nitrogen, phosphorus, and sulfur in relatively large quantities. They require other elements, such as magnesium, sodium, calcium, and iron, in smaller amounts. Some elements, such as cobalt and manganese, are required in trace amounts.

Figure 11 Trees and the carbon cycle. Approximately 500 million tons of carbon were taken up as a result of forest regrowth in the Northern Hemisphere between 1980 and 1989.

The paths of water, carbon, nitrogen, and phosphorus pass from the nonliving environment to living organisms, such as the trees in **Figure 11,** and then back to the nonliving environment. These paths form closed circles, or cycles, called biogeochemical (*bie oh jee oh KEHM ih kuhl*) cycles. In each **biogeochemical cycle,** a pathway forms when a substance enters living organisms such as trees from the atmosphere, water, or soil; stays for a time in the living organism; then returns to the nonliving environment. Ecologists refer to such substances as cycling within an ecosystem between a living reservoir (an organism that lives in the ecosystem) and a nonliving reservoir. In almost all biogeochemical cycles, there is much less of the substance in the living reservoir than in the nonliving reservoir.

350

Chapter Resource File

- Directed Reading BASIC
- Active Reading GENERAL

One-Stop Planner CD-ROM

- Reading Organizers BASIC
- Reading Strategies BASIC
- Supplemental Reading Guide *Silent Spring* ADVANCED

Transparencies

TR Bellringer
TR Graphic Organizer
TR Concept Map
TR E12 Water Cycle
TR E13 Carbon Cycle
TR E14 Nitrogen Cycle

The Water Cycle

Of all the nonliving components of an ecosystem, water has the greatest influence on the ecosystem's inhabitants. In the nonliving portion of the water cycle, water vapor in the atmosphere condenses and falls to the Earth's surface as rain or snow. Some of this water seeps into the soil and becomes part of the **ground water,** which is water retained beneath the surface of the Earth. Most of the remaining water that falls to the Earth does not remain at the surface. Instead, heated by the sun, it reenters the atmosphere by evaporation. The path of water within an ecosystem is shown in **Figure 12.**

In the living portion of the water cycle, much water is taken up by the roots of plants. After passing through a plant, the water moves into the atmosphere by evaporating from the leaves, a process called **transpiration.** Transpiration is also a sun-driven process. The sun heats the Earth's atmosphere, creating wind currents that draw moisture from the tiny openings in the leaves of plants.

In aquatic ecosystems (lakes, rivers, and oceans), the nonliving portion of the water cycle is the most important. In terrestrial ecosystems, the nonliving and living parts of the water cycle both play important roles. In thickly vegetated ecosystems, such as tropical rain forests, more than 90 percent of the moisture in the ecosystem passes through plants and is transpired from their leaves. In a very real sense, plants in rain forests create their own rain. Moisture travels from plants to the atmosphere and falls back to the Earth as rain.

internet connect
www.scilinks.org
Topic: Water Cycle
Keyword: HX4188
SciLinks. Maintained by the National Science Teachers Association

Figure 12 Water cycle

This diagram shows the major steps in the water cycle.

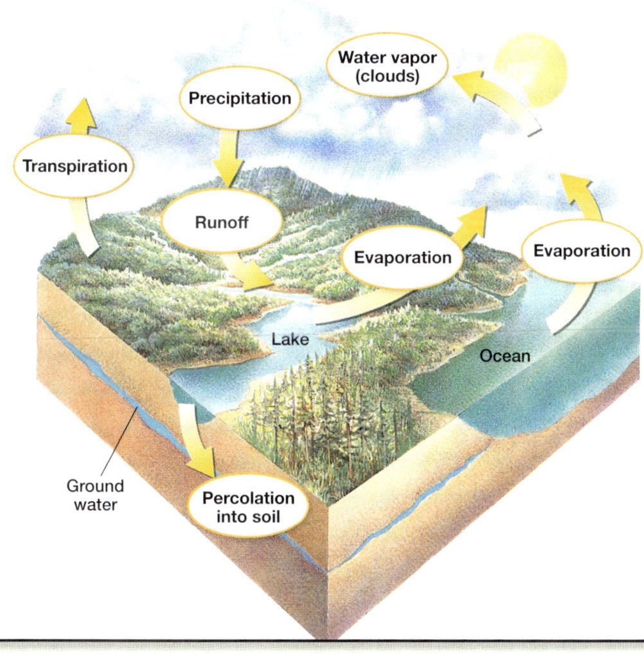

Teach

READING SKILL BUILDER — BASIC

Paired Reading Pair each student with a partner. You may want to pair ELL students with native English speakers. Have each student read about the water cycle silently while making a question mark on a sticky note next to the passages that they find confusing. After they finish reading, ask one student to summarize and the second to add anything omitted. Both readers should then help each other with any passages that either (or both) did not understand. Have them create a list of questions to ask the class. Students should repeat this process for the carbon and nitrogen cycles. LS Interpersonal

Using the Figure — GENERAL

Have each student look at **Figure 12** and notice the two arrows labeled "evaporation" and the one arrow labeled "transpiration." Ask them what is needed for evaporation or transpiration to occur. (an input of energy) Next ask students if this input of energy that causes evaporation and transpiration can be harvested by humans, and if so, how? (As the clouds that hold the evaporated and transpired water move across the land, precipitation may occur, and the water runoff may eventually enter a river. If the river has a hydroelectric dam on it, the water will run through the turbines, spinning them and converting the energy of the water into electrical energy, to be used by man.)

CHEMISTRY CONNECTION

Changes in State of Water Water is a unique liquid because, unlike other liquids, it is not at its densest point when it freezes. Picture a pond in early winter. As the water cools in the pond, its molecules pack in tighter and tighter until the water reaches 4° C, at which point it is at its densest. As the water continues to cool, the molecules now begin moving apart from each other as they swing outward—like a door on a hinge—being attracted to other water molecules and bonding with them. When they reach 0° C they have formed a uniform latticework of water molecules called ice. Because ice is less dense than the warmer water around it, it floats. This property is critical for living organisms, because the floating ice insulates the water beneath it. This prevents the water under the ice from freezing solid, killing the organisms in it.

Teach, *continued*

Teaching Tip — GENERAL

Atmospheric Carbon Dioxide
Have each student make a list of activities that contribute to rising levels of carbon dioxide. (Burning gasoline in a car, propane in a stove, charcoal in a barbecue, etc.) Ask them how electrical power is generated in your area. If fossil fuels are burned, then their use of electricity also contributes (indirectly) to carbon dioxide levels. Have students identify ways to reduce the amount of carbon dioxide they contribute. Afterwards, encourage interested students to use the Internet Connect box to learn more about the carbon cycle. **LS Intrapersonal**

Using the Figure — BASIC

Have students observe the processes of the carbon cycle as depicted in **Figure 13.** Point out that an increase in "activity" in one part of the cycle will affect other parts of the cycle. Ask students which part of the carbon cycle has changed the most drastically over the last 200 years. (the amount of combustion due to the Industrial Revolution) **LS Visual**

internet connect
www.scilinks.org
Topic: Carbon Cycle
Keyword: HX4031
SCILINKS. Maintained by the National Science Teachers Association

The Carbon Cycle

Carbon also cycles between the nonliving environment and living organisms. You can follow the carbon cycle in **Figure 13.** Carbon dioxide in the air or dissolved in water is used by photosynthesizing plants, algae, and bacteria as a raw material to build organic molecules. Carbon atoms may return to the pool of carbon dioxide in the air and water in three ways.

1. **Respiration.** Nearly all living organisms, including plants, engage in cellular respiration. They use oxygen to oxidize organic molecules during cellular respiration, and carbon dioxide is a byproduct of this reaction.
2. **Combustion.** Carbon also returns to the atmosphere through combustion, or burning. The carbon contained in wood may stay there for many years, returning to the atmosphere only when the wood is burned. Sometimes carbon can be locked away beneath the Earth for thousands or even millions of years. The remains of organisms that become buried in sediments may be gradually transformed by heat and pressure into fossil fuels—coal, oil, and natural gas. The carbon is released when the fossil fuels are burned.
3. **Erosion.** Marine organisms use carbon dioxide dissolved in sea water to make calcium carbonate shells. Over millions of years, the shells of the dead organisms form sediments, which form limestone. As the limestone becomes exposed and erodes, the carbon becomes available to other organisms.

Figure 13 Carbon cycle
This diagram shows the major steps of the carbon cycle.

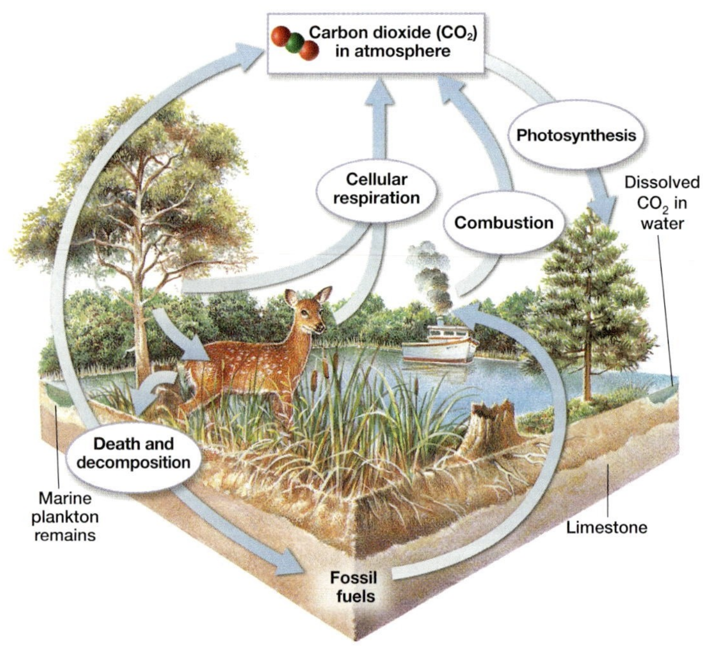

did you know?

Greenhouse Effect As light energy enters a car windshield in the heat of summer, some of the energy is lost, and the high energy, short-wavelength light is changed to lower energy, long-wavelength heat. The longer wavelength cannot escape through the glass, and the inside of the car heats up. This process also occurs in greenhouses and is known as the "greenhouse effect." Many scientists believe that a buildup of carbon dioxide in the atmosphere may act like the glass in a car, absorbing the heat, causing the atmosphere to slowly heat up. This could have disastrous effects. If the greenhouse effect causes an increase in the melting of the polar ice caps, coastal areas could experience serious flooding.

The Phosphorus and Nitrogen Cycles

Organisms need nitrogen and phosphorus to build proteins and nucleic acids. Phosphorus is an essential part of both ATP and DNA. Phosphorus is usually present in soil and rock as calcium phosphate, which dissolves in water to form phosphate ions, PO_4^{3-}. This phosphate is absorbed by the roots of plants and used to build organic molecules. Animals that eat the plants reuse the organic phosphorus.

The atmosphere is about 78 percent nitrogen gas, N_2. However, most organisms are unable to use it in this form. The two nitrogen atoms in a molecule of nitrogen gas are connected by a strong triple covalent bond that is very difficult to break. However, a few bacteria have enzymes that can break it, and they bind nitrogen atoms to hydrogen to form ammonia, NH_3. The process of combining nitrogen with hydrogen to form ammonia is called **nitrogen fixation.** Nitrogen-fixing bacteria live in the soil and are also found within swellings, or nodules, on the roots of beans, alder trees, and a few other kinds of plants.

The nitrogen cycle, diagramed in **Figure 14,** is a complex process with four important stages.

1. **Assimilation** is the absorption and incorporation of nitrogen into organic compounds by plants.
2. **Ammonification** is the production of ammonia by bacteria during the decay of organic matter.
3. **Nitrification** is the production of nitrate from ammonia.
4. **Denitrification** is the conversion of nitrate to nitrogen gas.

Study TIP

Reviewing Information
Using your own words, write four sentences, each one describing one of the four biogeochemical cycles.

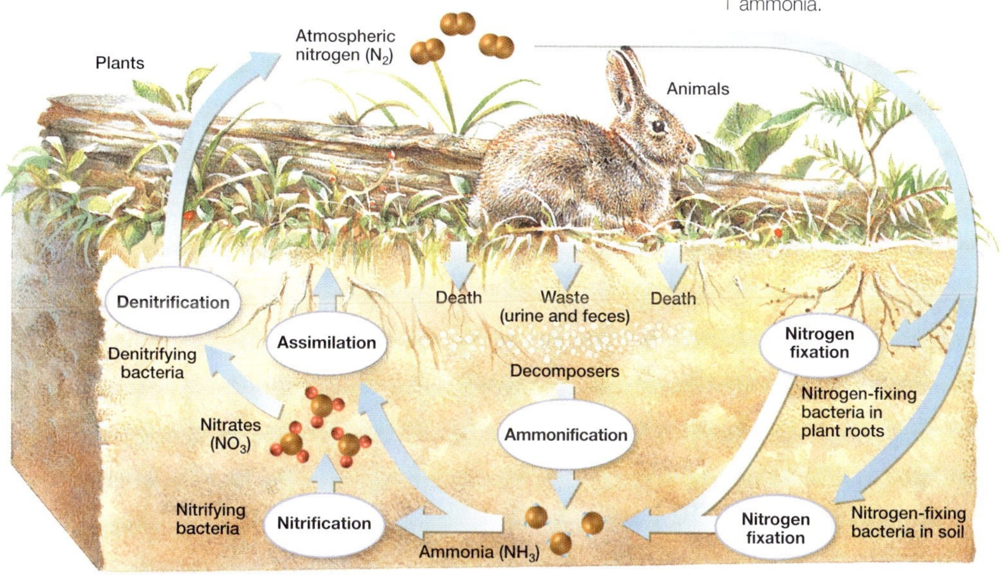

Figure 14 Nitrogen cycle

Bacteria carry out many of the important steps in the nitrogen cycle, including the conversion of atmospheric nitrogen into a usable form, ammonia.

353

Trends in Agriculture

Crop Rotation Farmers often rotate a nonleguminous crop, such as corn, with a leguminous one, such as alfalfa. The alfalfa will fix nitrogen and release some of it into the soil. If a crop of alfalfa is plowed back into the soil, it may add as much as 350 kg (770 lb.) of nitrogen per hectare (2.5 acres) of soil, enough to grow a crop of nonleguminous plants without the need for additional fertilizer.

Demonstration — BASIC

Tell students that beans are leguminous plants, the roots of which have nodules containing nitrogen-fixing bacteria. Then tell them that other such plants include clover, peas, alfalfa, lupines, and locust and alder trees. If any of these plants are available nearby, see if you can dig up an example to show the class any nodules that may be present. LS **Visual**

Teaching Tip — GENERAL

Gene Splicing Only some types of plants have the symbiotic relationship with nitrogen-fixing bacteria that enables them to convert nitrogen gas into a usable form of nitrogen. Scientists are working to isolate and remove the "nitrogen-fixing gene" in hopes they can splice it into other types of plants and make them nitrogen fixers as well. Have students discuss the pros and cons of doing this. (pro—reduce amount of fertilizer needed, tremendous economic impact; this could help feed starving peoples in countries with poor agricultural results; con—no one knows for sure what will happen with a new, genetically-engineered organism)

Using the Figure — GENERAL

Have students study **Figure 14.** Point out the difference between nitrogen fixation (nitrogen gas→ammonia) and nitrification (ammonia→nitrates). Tell students that lightning also changes nitrogen gas to ammonia, but such atmospheric action amounts to less than 10 percent of that carried out by organisms through nitrogen fixation. Finally, have students recognize that denitrification returns nitrates to the atmosphere as nitrogen gas. LS **Logical**

Teach, *continued*

BIOWatch

Nutrient Balance in the Everglades

Discussion

Ask students: How can changing an ecosystem have unpredictable consequences? (All of the organisms in an ecosystem are connected in a complex web of relationships. A change that affects one species or one resource in an ecosystem can have many unforeseen consequences.)

Close

Reteaching — BASIC

Ask students to write a short essay describing what the world would look like without fungi or bacteria.

Quiz — GENERAL

True or False:

1. Of all the abiotic components of an ecosystem, water has the greatest influence on the ecosystem's inhabitants. (true)
2. Carbon is necessary for photosynthesis. (true)

Alternative Assessment — ADVANCED

Have students select a biogeochemical cycle (except the water cycle). Being specific and writing in essay form, they should explain how humans would be affected if their chosen element did NOT cycle.

Section Review Correlations

SC.D.2.4.1: Item 3

SC.G.1.4.2: Items 1, 2

SC.G.1.4.3: Items 4, 5

The growth of plants in ecosystems is often limited by the availability of nitrate and ammonia in the soil. Today most of the ammonia and nitrate that farmers add to soil is produced chemically in factories, rather than by bacterial nitrogen fixation. Genetic engineers are trying to place nitrogen-fixing genes from bacteria into the chromosomes of crop plants. If these attempts are successful, the plants themselves will be able to fix nitrogen, thus eliminating the need for nitrogen-supplying fertilizers. Some farmers adjust their farming methods to increase natural recycling of nitrogen.

FLORIDA BIOWatch: Nutrient Balance in the Everglades

Known as "the river of grass," the Everglade ecosystem forms a broad expanse of marshland covering 4,000 square miles of southern Florida. Through it, water moves slowly from Lake Okeechobee to the Gulf of Mexico and Florida Bay.

Water Management

Water—its quantity and quality—has always shaped the fate of the Everglades. South Florida's early residents struggled with flooding during the rainy season, while crops perished during the dry season. During the 1940s, water management—a system of canals to drain water from wet areas and send it to dry areas—opened the door to wide-scale development. At the same time, levees and gates were built to prevent Lake Okeechobee from overflowing.

A Difficult Balance

Development has disrupted the natural flow of water into and through the Everglades. Water needed for homes and farms is extracted from the Everglades during dry seasons. During wet seasons, rainwater runs through farms and cities and carries chemicals into the Everglades. Nutrients in fertilizers, such as phosphorus and nitrogen, cause excessive growth of algae and disrupt the ecosystem.

Connecting Land to Sea

Today, a massive environmental project aims to restore some of the original water flow through the Everglades. However, this change may create a new threat to coral reefs near Florida's coasts. Recent increases in the flow of water into Florida Bay were followed by coral deaths, and scientists fear that excess nitrogen was to blame. Future management efforts must consider water and nutrients in all parts of the ecosystem.

internet connect

www.scilinks.org
Topic: Florida's Environment
Keyword: HX6F06
SCILINKS. Maintained by the National Science Teachers Association

Section 3 Review

1. **Identify** the role of energy in the part of the water cycle in which plants transfer water to the atmosphere. SC.G.1.4.2
2. **Analyze** the carbon cycle's relationship to the flow of energy. SC.G.1.4.2
3. **Describe** how bacteria participate in the nitrogen cycle. SC.D.2.4.1
4. **Critical Thinking** **Defend** the argument that nutrients can cycle but energy cannot. SC.G.1.4.3
5. **FCAT Prep** Which component of the carbon cycle removes carbon dioxide from the atmosphere?
   A cellular respiration
   B combustion
   C erosion
   D photosynthesis
   SC.G.1.4.3

354

Answers to Section Review

1. The sun's heating causes wind currents, which draws moisture from plant leaves.
2. Carbon and energy both move through ecosystems. Light energy is captured by photosynthesizers and used to make organic molecules using carbon dioxide. Energy flows out of ecosystems mainly as heat during cellular respiration and combustion. These processes also release carbon, but it does not "leave" the ecosystem; photosynthesizers recycle it.
3. Nitrogen-fixing bacteria convert nitrogen gas to ammonia. Nitrifying bacteria convert ammonia to nitrates, also usable by some plants. Denitrifying bacteria turn the nitrates into nitrogen gas.
4. Nutrients can cycle because they are in a form usable by at least some organisms, which keeps the nutrient moving through (but still within) the ecosystem. Most energy, on the other hand, only flows in one direction. By the end of the food chain, nearly all the original energy has been "lost" as unusable heat.
5. **A.** Incorrect. This process puts CO_2 into the atmosphere. **B.** Incorrect. Combustion puts CO_2 into the atmosphere. **C.** Incorrect. Erosion puts CO_2 into the atmosphere. **D.** Correct.

ZONE

Key Concepts

1 What Is an Ecosystem?

- Ecology is the study of how organisms interact with each other and with their environment.
- A community of organisms and their nonliving environment constitute an ecosystem.
- Ecosystems contain diverse organisms.
- Ecosystems change through the process of succession.
- Succession on a newly formed habitat is primary succession.
- Secondary succession occurs on a habitat that has previously supported growth.

2 Energy Flow in Ecosystems

- Energy moves through communities in food chains, passing from photosynthesizers (producers) to herbivores (consumers) to carnivores (consumers), creating a food web.
- Energy transfers between trophic levels transfer only 10 percent of the energy in a trophic level to the next level.
- Most terrestrial communities have only three or four trophic levels because energy transfers between trophic levels are inefficient.

3 Cycling of Materials in Ecosystems

- Minerals and other materials cycle within ecosystems among organisms and between organisms and the physical environment.
- In the water cycle, water falls as precipitation and either evaporates from bodies of water, is stored in ground water, or cycles through plants and then evaporates.
- Carbon enters the living portion of the carbon cycle through photosynthesis. Organisms release carbon through cellular respiration. Carbon trapped in rocks and fossil fuels is released by erosion and burning.
- Bacteria fix atmospheric nitrogen, thus making ammonia available to other organisms.

Key Terms

Section 1

ecology (340)
habitat (340)
community (340)
ecosystem (340)
abiotic factor (340)
biotic factor (340)
biodiversity (341)
pioneer species (343)
succession (343)
primary succession (343)
secondary succession (343)

Section 2

primary productivity (345)
producer (345)
consumer (345)
trophic level (345)
food chain (346)
herbivore (346)
carnivore (346)
omnivore (346)
detritivore (346)
decomposer (347)
food web (347)
energy pyramid (348)
biomass (349)

Section 3

biogeochemical cycle (350)
ground water (351)
transpiration (351)
nitrogen fixation (353)

BIOLOGY INTERACTIVE TUTOR Unit 7—*Ecosystem Dynamics*
Use Topics 1, 3–6 in this unit to review the key concepts and terms in this chapter.

355

Alternative Assessment — GENERAL

Have students choose an ecosystem from anywhere in the world. (For example, students could choose the Costa Rican rain forest, Sonoran Desert, Kalahari Plain, or Florida Everglades.) Have them research and report on the biotic and abiotic components. Tell them to include descriptions of four producers, three herbivores, two small carnivores, one top carnivore, and one detritivore. Have them draw a food chain, food web, and energy pyramid for this ecosystem.

Chapter Resource File

- Science Skills Worksheet GENERAL
- Critical Thinking Worksheet ADVANCED
- Test Prep Pretest GENERAL
- Chapter Test GENERAL

Answer to Concept Map

The following is one of several possible answers to Performance Zone item 15.

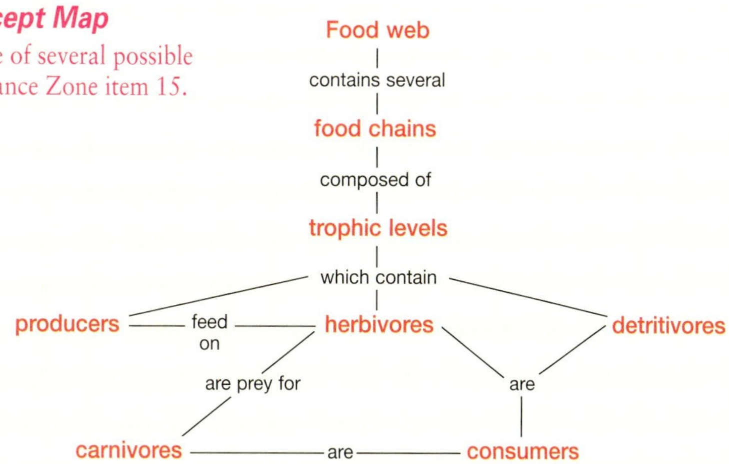

Performance ZONE
CHAPTER REVIEW

Understanding Key Ideas

1. Ecosystems differ from communities in that ecosystems usually contain
a. several climates.
b. several communities.
c. only one habitat.
d. only one food web.
SC.G.1.4.1

2. What critical role is played by fungi and bacteria in any ecosystem?
a. primary production
b. decomposition
c. boundary setting
d. physical weathering
SC.G.1.4.1

3. Which sequence shows the correct order of succession at Glacier Bay, Alaska?
a. alder, *Dryas,* hemlock
b. *Dryas,* hemlock, alder
c. *Dryas,* alder, Sitka spruce
d. mosses, hemlock, Sitka spruce
SC.G.2.4.2

4. Which role is *not* performed by bacteria in the nitrogen cycle?
a. fixing nitrogen
b. changing urea to ammonia
c. turning nitrates into nitrogen gas
d. changing nitrates to ammonia
SC.G.1.4.3

5. How would the food web below be affected if the plants were eliminated?

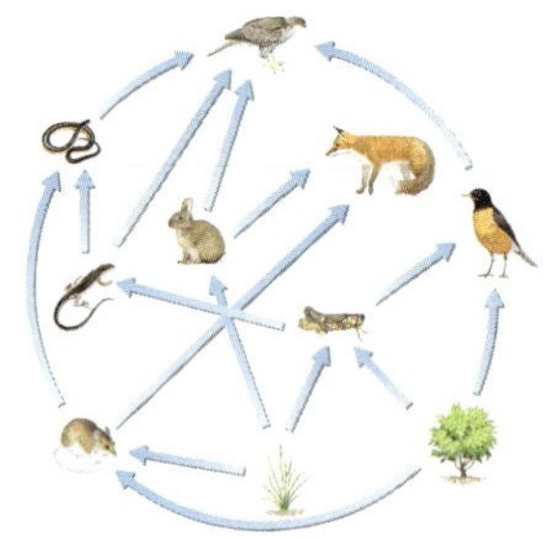

a. Herbivores would become carnivores.
b. The food web would collapse.
c. The herbivores would change trophic levels.
d. Nothing would happen.
SC.G.1.4.2

6. How much energy is available at the third trophic level of an energy pyramid if 1,000 kcal is available in the first level?
a. 1,000 kcal **c.** 10 kcal
b. 100 kcal **d.** 1 kcal
SC.G.1.4.2

7. Humans, raccoons, and bears are omnivores. What adaptive advantage might this feeding strategy provide? SC.G.1.4.2

8. BIOWatch How could increasing the amount of water flow through the Everglades threaten coral reefs? SC.G.2.4.5

9. Relate photosynthesis to the nitrogen cycle. (**Hint:** See Chapter 5, Section 2.) SC.D.2.4.1

10. **Concept Mapping** Make a concept map that describes the flow of energy through an ecosystem. Try to include the following terms in your map: *trophic level, food web, food chain, producer, consumer, carnivore, detritivore,* and *herbivore.* SC.G.1.4.2

Critical Thinking

11. **Inferring Relationships** Analyze the flow of energy between an ecosystem and one of its top carnivores, such as a hawk. SC.G.1.4.2

12. **Applying Information** Is nitrogen cycling or carbon cycling more important to a pioneer species during primary succession? Explain your answer. SC.G.1.4.3

13. **Predicting Results** Describe the probable effects on an ecosystem if all decomposers were to die. SC.G.2.4.2

Alternative Assessment

14. **Identifying Functions** Obtain photocopies of nature paintings by American painters such as John James Audubon or Edward Hicks. Choose three animals, and write a report that compares the animals, the ecosystems in which they live, their roles in biogeochemical cycles, and the trophic level they occupy. SC.G.1.4.1

356

CHAPTER 16 ZONE

ANSWERS

Understanding Key Ideas

1. b
2. b
3. c
4. d
5. b
6. c
7. Greater flexibility in diet allows omnivores to eat whatever food is available.
8. Nitrogen in runoff enters Florida Bay, causing excessive algal growth, which damages marine corals.
9. Photosynthesis enables plants to incorporate nitrogen into proteins, making it available to consumers. When consumers excrete waste or decompose, plants (with the aid of bacteria) may take up the nitrogen again, completing the cycle.
10. One possible answer to the concept map is found at the bottom of the Study Zone page.

Critical Thinking

11. Each trophic level contains about 90% less energy than the level below it. If 1,000 kilocalories of seeds were eaten by a mouse, it would result in about 100 kilocalories for the snake that eats the mouse. And there would only be about 10 kilocalories available for the hawk that eats the snake.
12. Many pioneer organisms, such as lichens, have the ability to fix nitrogen, but all pioneer plants are able to fix carbon during photosynthesis. Therefore, nitrogen cycling is probably more important to a pioneer species during primary succession.
13. Dead organisms and wastes would not decay and nutrients would not be recycled back into the ecosystem.

Alternative Assessment

14. Answers will vary depending on the organisms pictured.

Chapter Review Correlations

SC.D.2.4.1: Item 9
SC.G.1.4.1: Items 1, 2, 14
SC.G.1.4.2: Items 5, 6, 7, 10, 11
SC.G.1.4.3: Items 4, 12
SC.G.2.4.2: Items 3, 13
SC.G.2.4.5: Item 8

Assignment Guide

Section	Questions
1	1, 3, 11, 12, 14
2	2, 5, 6, 7, 10, 13
3	4, 8, 9, 12

FCAT Prep

For questions 1–6, write your answers on a separate sheet of paper.

1. Which of the following situations describes a carnivore and an herbivore?
   A. A horse eats an apple.
   B. A rabbit eats a dandelion.
   C. A mountain lion eats a rabbit.
   D. A fungus breaks down a dead oak tree.
   SC.G.1.4.2
2. What term describes most humans?
   F. carnivore **H.** herbivore
   G. detrivore **I.** omnivore
   SC.G.1.4.2
3. Ecologists once referred to stable ecosystems as a final or climax community. Now most ecologists say that no ecosystem can truly have a final end point. Why have ecologists changed their viewpoint?
   A. Stable ecosystems never change.
   B. Ecosystems are dynamic by nature.
   C. Ecosystems are rarely stable for more than a few years.
   D. Ecologists can change the ecosystems with very little effort.
   SC.G.2.4.4
4. What is the process by which materials pass between the nonliving environment and living organisms?
   F. biogeochemical cycle **H.** food web
   G. energy pyramid **I.** primary succession
   SC.D.2.4.1

Base your answers to questions 5 and 6 on the graph below, which shows the monthly variation in atmospheric carbon dioxide concentration over a deciduous forest.

ATMOSPHERIC CARBON DIOXIDE VARIATION

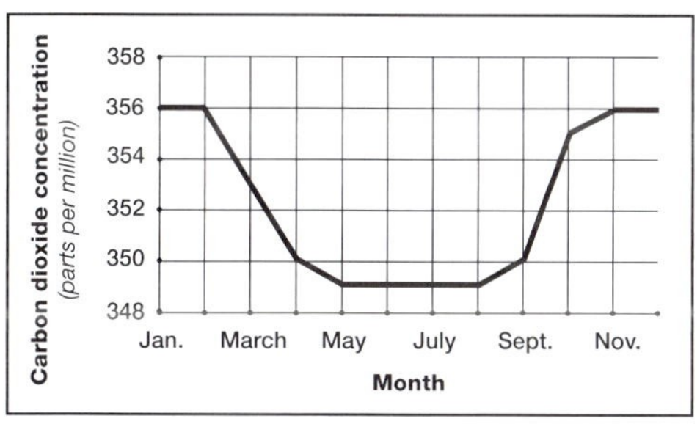

5. During which of the following months is the rate of photosynthesis greatest?
   A. January **C.** May
   B. March **D.** September
   SC.G.1.4.3
6. What is the difference between the carbon dioxide concentration in January and that in April, in parts per million? SC.G.1.4.3

Test TIP

For multiple-choice questions, try to eliminate any answer choices that are obviously incorrect, and then consider the remaining answer choices.

357

FCAT Doctor

Question 3 Answer B is the correct choice. Changes in climate, large-scale disturbances such as fire or volcanic eruption, and even changes in biotic factors, such as an insect pest outbreak, can change the conditions of an ecosystem. Answer A is incorrect because this answer does not support the change in viewpoint. Also, even stable ecosystems experience some change. Answer C is incorrect because ecosystems can be stable for many years. Answer D is incorrect because it generally takes great forces to change an ecosystem, and ecologists do not generally try to change them.

Question 5 Answer C is the correct choice. Because photosynthesis consumes carbon dioxide, the lowest level of carbon dioxide would indicate the greatest rate of photosynthesis. Answer A is incorrect because the carbon dioxide level is highest in January. Answer B is incorrect because March shows a higher carbon dioxide level than May. Answer D is incorrect because September shows a higher carbon dioxide level than May.

Question 6 The correct answer is 6 ppm. The carbon dioxide concentration in January is 356 ppm. The concentration in April is 350 ppm. 356 – 350 = 6.

Answers

1. C
2. I
3. B
4. F
5. C
6. 6

Test Prep Correlations

SC.D.2.4.1: Item 4
SC.G.1.4.2: Items 1, 2
SC.G.1.4.3: Items 5, 6
SC.G.2.4.4: Item 3

Exploration Lab

Modeling Ecosystem Change over Time

Teacher's Notes

Time Required About 20 minutes on day 1 and about 10 minutes each day thereafter over a period of a few weeks.

Ratings

EASY → HARD

- Teacher Preparation 1
- Student Setup 2
- Concept Level 2
- Cleanup 1

Skills Acquired
- Collecting Data
- Constructing Models
- Organizing and Analyzing Data

Scientific Methods

In this lab, students will:
- Make Observations
- Test the Hypothesis
- Analyze the Results

Materials and Equipment

Materials for this lab can be ordered from WARD'S. See *Master Materials List* at the front of this book for catalog numbers. Have students bring clear plastic 2- or 3-L soda bottles from home. Soil can be collected from around the school, brought from home, or purchased from a garden center. You may also be able to find earthworms and crickets in the local environment or at bait shops.

Safety Cautions

Review all safety symbols with students before beginning the lab. Warn students to take care when handling insects and other small animals. Small animals are easily harmed, and some are capable of biting when disturbed.

Exploration Lab

Modeling Ecosystem Change over Time SC.G.2.4.2

SKILLS
- Using scientific methods
- Modeling
- Observing

OBJECTIVES
- **Construct** a model ecosystem.
- **Observe** the interactions of organisms in a model ecosystem.
- **Predict** how the number of each species in a model ecosystem will change over time.
- **Compare** a model ecosystem with a natural ecosystem.

MATERIALS
- coarse sand or pea gravel
- large glass jar with a lid or terrarium
- soil
- pinch of grass seeds
- pinch of clover seeds
- rolled oats
- mung bean seeds
- earthworms
- isopods (pill bugs)
- mealworms (beetle larva)
- crickets

Before You Begin

Organisms in an **ecosystem** interact with each other and with their environment. One of the interactions that occurs among the organisms in an ecosystem is feeding. A **food web** describes the feeding relationships among the organisms in an ecosystem. In this lab, you will model a natural ecosystem by building a **closed ecosystem** in a bottle or a jar. You will then observe the interactions of the organisms in the ecosystem and note any changes that occur over time.

1. Write a definition for each boldface term in the paragraph above and for each of the following terms: producer, decomposer, consumer, herbivore, carnivore, trophic level.
2. Based on the objectives for this lab, write a question you would like to explore about ecosystems.

Procedure

PART A: Building an Ecosystem in a Jar

1. Place 2 in. of sand or pea gravel in the bottom of a large, clean glass jar with a lid. **CAUTION: Glassware is fragile. Notify your teacher promptly of any broken glass or cuts. Do not clean up broken glass or spills with broken glass unless your teacher tells you to do so.** Cover the gravel with 2 in. of soil.
2. Sprinkle the seeds of two or three types of small plants, such as grasses and clovers, on the surface of the soil. Put a lid on the jar, and place it in indirect sunlight. Let the jar remain undisturbed for a week.
3. After one week, place a handful of rolled oats into the jar. Place the mealworms in the oats, and then place the other animals into the jar and replace the lid. Place the lid on the jar loosely to enable air entry.

You Choose

As you design your experiment, decide the following:

- **a.** what question you will explore
- **b.** what hypothesis you will test
- **c.** how you will plant the seeds
- **d.** where you will place the ecosystem for one week so that it remains undisturbed and in indirect sunlight
- **e.** how often you will add water to the ecosystem after the first week
- **f.** how many of each organism you will use
- **g.** what data you will record in your data table

358

Tips and Tricks

Remind students that the ecosystems are dependent on humans for care. They should not be permitted to be overheated or become too cold. If water evaporates from the ecosystem, it should be replenished. Recording the number of organisms may be tricky in some cases and estimates may be required.

Answers to Before You Begin

1. *ecosystem*—an ecological system encompassing a community and its abiotic factors; *food web*—a network of feeding relationships in an ecosystem; *closed ecosystem*—an ecosystem that does not exchange materials outside of itself; *producer*—organisms that first capture energy; *decomposer*—organisms that decompose dead organic material; *consumer*—organisms that consume producers; *herbivore*—organisms that eat plants or other primary producers; *carnivore*—organisms that are secondary consumers; *trophic level*—ecosystem level based on the organism's source of energy.

PART B: Design an Experiment

4. Work with the members of your lab group to explore one of the questions written for step 2 of **Before You Begin.** To explore the question, design an experiment that uses the materials listed for this lab.
5. Write a procedure for your experiment. Make a list of all the safety precautions you will take. Have your teacher approve your procedure and safety precautions before you begin the experiment.
6. Set up your group's experiment. Conduct your experiment for at least 14 days.

PART C: Cleanup and Disposal

7. Dispose of solutions, broken glass, and other materials in the designated waste containers. Do not put lab materials in the trash unless your teacher tells you to do so.
8. Clean up your work area and all lab equipment. Return lab equipment to its proper place. Wash your hands thoroughly before you leave the lab and after you finish all work.

Analyze and Conclude

1. **Summarizing Results** Make graphs showing how the number of individuals of each species in your ecosystem changed over time. Plot time on the x-axis and the number of organisms on the y-axis.
2. **Analyzing Results** How did your results compare with your hypothesis? Explain any differences.
3. **Inferring Conclusions** Construct a food web for the ecosystem you observed.
4. **Recognizing Relationships** Does your model ecosystem resemble a natural ecosystem? Explain.
5. **Analyzing Methods** How might you have built your model ecosystem differently to better represent a natural ecosystem?
6. **Evaluating Methods** Was your model ecosystem truly a "closed ecosystem"? List your model's strengths and weaknesses as a closed ecosystem.
7. **Further Inquiry** Write a new question about ecosystems that you could explore with another investigation.

Do You Know?

Do research in the library or media center to answer these questions:

1. What is Biosphere 2?
2. What problems were encountered by the Biosphere 2 crew during the 1991–1993 project?

Use the following Internet resources to explore your own questions about ecosystems.

internet connect
www.scilinks.org
Topic: Ecosystems
Keyword: HX4066
SciLinks. Maintained by the National Science Teachers Association

2. Answers will vary. For example, What are the effects of continuous exposure to bright light on the ecosystem?

Part B: Sample Procedure

1. Place the jar to the side of a window so that it receives indirect sunlight throughout the day.
2. After one week, add three earthworms, five isopods, three mealworms, and three crickets to the jar. Using a mister, add four squirts of water per square decimeter of soil surface every other day.
3. Record population data every other day for two weeks.

Answer to Analyze and Conclude

1. Answers will vary. Students should make one graph for each species observed or use different colors to indicate each species.
2. Answers will vary.
3. Answer will vary. All plants are producers (primary trophic level); earthworms feed on dead plant material in the soil; crickets feed on plants; mealworms (beetle larvae) feed on plants; isopods (pill bugs) eat wood.
4. Yes and no. Natural ecosystems and the model ecosystem both contain organisms at several trophic levels, have living and nonliving components, and depend on the sun for energy. However, the model ecosystem is less diverse, much younger, and has more definite boundaries than a natural ecosystem.
5. Answers will vary.
6. No. Strengths are that the organisms in the model ecosystem did not leave the ecosystem and that other organisms could not enter from the outside. Weaknesses are that water and air probably had to be added to maintain a healthy ecosystem.
7. Answers will vary. For example: What are the effects of certain abiotic factors, such as temperature, light, and moisture, on the organisms in an ecosystem?

Answers to Do You Know?

1. Biosphere 2 is a large artificial ecosystem near Tucson, Arizona. The 204,000-m^3 glass and steel structure contains seven ecosystems. The structure has been used for research on the biosphere, the sum of all of Earth's ecosystems.
2. Answers will vary. Students should list several examples of problems, including food shortages, oxygen shortages, and population explosions of microorganisms, ants, and cockroaches.

Chapter Opener		SE **Quick Review,** p. 361 ■ SE **Reading Activity,** p. 361 ■ TE **Using the Figure,** p. 360 TE **Opening Activity,** p. 361 GENERAL
45 min pp. 362–364 **Section 1** How Organisms Interact in Communities	OSP **Lesson Plan*** TR **Bellringer***	SE **Data Lab** Predicting How Predation Would Affect a Plant Species, p. 363 ■ CRF **Datasheets for In-Text Labs** Predicting How Predation Would Affect a Plant Species* CRF **Exploration Lab** Life in a Pine Cone* BASIC
& 4 • 90 min pp. 365–370 **Section 2** How Competition Shapes Communities	OSP **Lesson Plan*** TR **Bellringer*** TR **Section Outline*** TR **E22 Warbler Foraging Zones*** TR **E21 Effects of Competition on Two Species of Barnacles***	SE **Data Lab** Predicting Changes in a Realized Niche, p. 367 ■ CRF **Datasheets for In-Text Labs** Predicting Changes in a Realized Niche* TE **Demonstration,** p. 368 TE **Group Activity** Investigating Competition, p. 368 ADVANCED
6 & 7 • 90 min pp. 371–378 **Section 3** Major Biological Communities	OSP **Lesson Plan*** TR **Bellringer*** TR **Section Outline*** TR **E26 Elements of Climate*** TR **E27 Earth's Biomes*** TR **E28 Three Lake Zones***	TE **Activity** Local Climate Graphs, p. 371 ADVANCED TE **Demonstration,** p. 372 GENERAL TE **Using the Figure,** p. 372 BASIC TE **Group Activity** Biome Collages, p. 373 GENERAL TE **Activity** Identifying Biomes, p. 373 GENERAL SE **Quick Lab** Investigating Factors That Influence the Cooling of Earth's Surface, p. 374 ■ CRF **Datasheets for In-Text Labs** Investigating Factors That Influence the Cooling of Earth's Surface * TE **Activity** Viewing Freshwater Communities, p. 376 ADVANCED SE **Skills Practice Lab** Observing How Brine Shrimp Select a Habitat, pp. 382–383 ◆ ■ CRF **Skills Practice Lab** Examining Owl Pellets* GENERAL

& 9 • 90 min
Chapter Review and Assessment Resources

- SE **Study Zone,** p. 379 ■
- SE **Performance Zone,** p. 380 ■
- SE **FCAT Test Prep,** p. 381 ■
- TR **Concept Mapping***
- CRF **Vocabulary Review Worksheet*** ■ GENERAL
- CRF **Test Prep Pretest*** ■ GENERAL
- CRF **Chapter Test*** ■ GENERAL
- CRF **Chapter Test*** ADVANCED
- OSP **Test Generator***
- OSP **Test Item Listing***
- OSP **Modified Chapter Test*** BASIC

Holt Online Learning

Visit go.hrw.com for access to Holt Online Learning or enter the keyword **HX6 Home** for a variety of free online resources.

One-Stop Planner® CD-ROM

This CD-ROM package includes
- Lab Materials QuickList Software
- Holt Calendar Planner
- Customizable Lesson Plans
- Printable Worksheets
- ExamView® Test Generator
- Interactive Teacher Edition
- Holt PuzzlePro® Resources
- Holt PowerPoint® Resources

Guided Reading Audio CDs

These CDs are designed to help auditory learners and reluctant readers. (Audio CDs are also available in Spanish.)

Compression guide:
To shorten your instruction because of time limitations, omit block 1.

KEY

TE Teacher Edition
SE Student Edition
OSP One-Stop Planner
CRF Chapter Resource File
TR Teaching Transparency
* Also on One-Stop Planner
■ Also Available in Spanish
◆ Requires Advance Prep

SKILLS DEVELOPMENT AND PRACTICE	SECTION REVIEW AND ASSESSMENT	CORRELATIONS
		Florida Sunshine Standards/FCAT
TE **Inclusion Strategies,** p. 363 CRF **Directed Reading Worksheet*** BASIC CRF **Active Reading Worksheet*** GENERAL	SE **Section 1 Review** ■ TE **Reteaching,** p. 364 BASIC TE **Quiz,** p. 364 GENERAL CRF **Quiz*** ■ BASIC	SC.D.1.4.4, SC.G.1.4.1
TE **Reading Skill Builder** Paired Reading, p. 366 BASIC TE **Skill Builder** Interpreting Visuals, p. 366 GENERAL TE **Reading Skill Builder** Paired Summarizing, p. 369 BASIC TE **Skill Builder** Interpreting Graphics, p. 369 GENERAL CRF **Directed Reading Worksheet*** BASIC CRF **Active Reading Worksheet*** GENERAL	SE **Section 2 Review** ■ TE **Quiz,** p. 370 GENERAL TE **Alternative Assessment,** p. 370 ADVANCED CRF **Quiz*** ■ BASIC	SC.G.1.4.1, SC.G.2.4.2
TE **Inclusion Strategies,** p. 373 CRF **Directed Reading Worksheet*** BASIC CRF **Active Reading Worksheet*** GENERAL CRF **Critical Thinking Worksheet*** ADVANCED CRF **Science Skills Worksheet*** GENERAL	SE **Section 3 Review** ■ TE **Reteaching,** p. 378 BASIC TE **Quiz,** p. 378 GENERAL CRF **Quiz*** ■ BASIC **Modified Worksheet*** BASIC	SC.G.1.4.1, SC.G.2.4.4

SciLinks NSTA
www.scilinks.org
Maintained by the **National Science Teachers Association**

Topic: Symbiosis
SciLinks Code: HX4171

Topic: Biomes
SciLinks Code: HX4023

Topic: Estuaries
SciLinks Code: HX4073

Topic: Adaptation
SciLinks Code: HX4002

CHAPTER 17

Overview

Tell students the purpose of this chapter is to explain the concept of biological communities: populations of different species living together. Section 1 begins the chapter by describing how organisms interact in communities through predator/prey relationships, parasitism, mutualism, and commensalism. Section 2 describes competition among species and how it can limit the ways in which species use resources. Section 3 lists and describes the seven major terrestrial biomes, which are biological communities that occur over large areas of land. The chapter concludes with a discussion of aquatic communities.

Using the Figure

Ask students to describe the desert. Write their descriptive phrases on the board. Students generally have the preconception that deserts are always warm or hot. The photograph challenges that notion. Deserts are biomes that have 25 centimeters (10 inches) or less of precipitation annually. They are further classified into semi-, true, and extreme deserts, depending on their level of precipitation. Most deserts are in warm climates, but some are in climates so cold that moisture generally freezes, making it unavailable to plants. Therefore, very little vegetation grows. The opening photograph was taken in a true desert, which receives less than 12 centimeters (4.7 inches) of rain per year. The Mojave Desert in southeastern California and southwestern Nevada is an example of such a desert. During the winter, temperatures can dip below the freezing mark, and snow sometimes falls.

Assessing Prior Knowledge

Before beginning this chapter, be sure your students can answer the questions in the ***Quick Review*** on the facing page. If they cannot, encourage them to reread the sections indicated.

Desert biome in winter

360

Chapter Correlations **Florida Sunshine Standards/FCAT**

SC.D.1.4.4 Earth's systems and organisms are the result of a long, continuous change over time.

SC.G.1.4.1 There is great diversity and interdependence among living things.

SC.G.2.4.2 Changes in a component of an ecosystem will have unpredictable effects on the entire system, but the components of the system tend to react in a way that will restore the ecosystem to its original condition.

SC.G.2.4.4 The world ecosystems are shaped by physical factors that limit their productivity.

CHAPTER

17 Biological Communities

Florida Sunshine Standards/FCAT

SC.D.1.4.4 Earth's systems and organisms are the result of a long, continuous change over time.
SC.G.1.4.1 There is great diversity and interdependence among living things.
SC.G.2.4.2 Changes in a component of an ecosystem will have unpredictable effects on the entire system, but the components of the system tend to react in a way that will restore the ecosystem to its original condition.
SC.G.2.4.4 The world ecosystems are shaped by physical factors that limit their productivity.

Quick Review

Answer the following without referring to earlier sections of your book.

1. **Compare** the energy flow in photosynthesis to the energy flow in cellular respiration. *(Chapter 5, Section 1)*
2. **Differentiate** between the terms *habitat, community,* and *ecosystem. (Chapter 16, Section 1)*
3. **Analyze** the relationship between primary productivity and energy flow in ecosystems. *(Chapter 16, Section 2)*

Did you have difficulty? *For help, review the sections indicated.*

Reading Activity

Before you read this chapter, create a list of all the ways that two species in an ecosystem can interact. Then develop a list of all the different types of communities or ecosystems you can think of. Can any of the ecosystems be grouped into larger systems (biomes)? Read the chapter to see how scientists have defined interactions and biomes.

Harsh and unforgiving, the desert is home to plants and animals equipped to thrive in the face of environmental challenges. No other terrestrial biome displays a wider range of extreme conditions.

Looking Ahead

internet connect
www.scilinks.org
National Science Teachers Association *sci*LINKS Internet resources are located throughout this chapter.
SCILINKS. Maintained by the National Science Teachers Association

361

Opening Activity — GENERAL

Ask each student to choose two animals that live in the same community and interact. Prepare a list on the board or overhead with three headings: Both Benefit, One Benefits/One Suffers, One Benefits/One Not Affected. Ask students to place the interaction of each animal pair into the appropriate category. **LS Logical**

Identifying Misconceptions

Students do not often recognize that, while an organism's everyday behavior is shaped by its encounters with other organisms, the genetic basis for its behavior does not change in that organism's lifetime. Traits in a population change over time, as individuals with the genes for those traits survive and reproduce more than individuals without those traits.

Quick Review Answers

1. In photosynthesis, light energy and carbon dioxide are converted to oxygen and chemical energy which is stored in sugars (food). In cellular respiration, the chemical energy stored in food is released as ATP to do work. Oxygen is converted to carbon dioxide in this process.
2. A habitat is the place in which a population lives. The many species that live together in a habitat are called a community. An ecosystem consists of a community and all the physical aspects of its habitat.
3. Primary productivity, the rate at which food is produced through photosynthesis by autotrophs, determines the amount of energy available in an ecosystem. Energy flow is decreased or increased in an ecosystem based on primary productivity.

Chapter Resource File

- Vocabulary Worksheets
- Concept Mapping

Reading Activity Answers

Students will likely mention predator/prey relationships. Students may also be familiar with symbiotic relationships, such as the interaction between flowers and their pollinators.

Section 1

Focus

Overview

Before beginning this section review with your students the objectives listed in the Student Edition. Section 1 describes types of interactions among species. This section discusses how predators and prey coevolve and describes symbiotic relationships.

Bellringer

Show the class a photograph of a tapeworm. Point out adaptations that make it a successful parasite. (Hooks on the head enable the tapeworm to attach to the intestinal wall of its host. The tapeworm's permeable body wall allows absorption of nutrients from the host. The body consists of segments specialized for reproduction.) Tapeworms in humans may grow as long as 6 m (20 ft). Ask students why it is not in the tapeworm's best interest to kill its host. (Killing its host would destroy its home and food supply.)

Motivate

Identifying Preconceptions — GENERAL

Have students read the paragraph entitled "Interactions Among Species." Ask students to then describe coevolution. Students may have difficulty understanding that the environment affects the survival of traits *after* they appear in the population. They often think that the change was the result of the environment acting on organisms to *produce* change. Provide examples of coevolution. LS Verbal

Section Correlations

SC.D.1.4.4
SC.G.1.4.1

Section 1 How Organisms Interact in Communities

Florida Sunshine Standards/FCAT

SC.D.1.4.4 Earth's systems and organisms are the result of a long, continuous change over time.

SC.G.1.4.1 There is great diversity and interdependence among living things.

Objectives

- **Describe** coevolution. SC.D.1.4.4
- **Predict** how coevolution can affect interactions between species. SC.G.1.4.1
- **Identify** the distinguishing features of symbiotic relationships. SC.G.1.4.1

Key Terms

coevolution
predation
parasitism
secondary compound
symbiosis
mutualism
commensalism

Evolution in Communities

What are the most important members of an ecosystem? When you try to answer this question, you soon realize that you cannot view an ecosystem's inhabitants as single organisms, but only as members of a web of interactions.

Interactions Among Species

Some interactions among species are the result of a long evolutionary history in which many of the participants adjust to one another over time. Thus, adaptations appeared in flowering plants that promoted efficient dispersal of their pollen by insects and other animals. In turn, adaptations appeared in pollinators that enabled them to obtain food or other resources from the flowers they pollinate. Natural selection has often led to a close match between the characteristics of the flowers of a plant species and its pollinators, as you can see in **Figure 1.** Back-and-forth evolutionary adjustments between interacting members of a community are called **coevolution.**

Predators and Prey Coevolve

Predation is the act of one organism killing another for food. Familiar examples of predation include lions eating zebras and snakes eating mice. Less familiar, but no less important, examples occur among arthropods. Spiders are exclusively predators, as are centipedes.

Figure 1 **Coevolution.** With its long beak and tongue, the hummingbird is able to reach the nectar deep within this flower.

In **parasitism,** one organism feeds on and usually lives on or in another, typically larger, organism. Parasites do not usually kill their prey (known as the "host"). Rather, they depend on the host for food and a place to live. The host often serves to transmit the parasite's offspring to new hosts. Many parasites (such as lice) feed on the host's outside surface. Among the external parasites that may have fed on you at some time are ticks, mosquitoes, and fleas. More highly specialized parasites like hookworms live entirely within the body of their host.

362

Chapter Resource File

- Directed Reading BASIC
- Active Reading GENERAL
- Data Sheet for Quick Lab GENERAL

Transparencies

TR Bellringer

One-Stop Planner CD-ROM

- Reading Organizers BASIC
- Reading Strategies BASIC

Plant Defenses Against Herbivores

As you might expect, animal prey species have ways to escape, avoid, or fight off predators. But predation is also a problem for plants, which live rooted in the ground. The most obvious way that plants protect themselves from herbivores is with thorns, spines, and prickles. But it is even more common for a plant to contain chemical compounds that discourage herbivores. Virtually all plants contain defensive chemicals called **secondary compounds**. For some plants, secondary compounds are the primary means of defense.

As a rule, each group of plants makes its own special kind of defensive chemical. For example, the mustard plant family produces a characteristic group of chemicals known as mustard oils. These oils give pungent aromas and tastes to such plants as mustard, cabbage, radish, and horseradish. The same tastes that we enjoy signal the presence of chemicals that are toxic to many groups of insects.

How Herbivores Overcome Plant Defenses

Surprisingly, certain herbivores are able to feed on plants that are protected by particular defensive chemicals. For example, the larvae of cabbage butterflies feed almost exclusively on plants of the mustard and caper families. Yet these plants produce mustard oils that are toxic to many groups of insects. How do the butterfly larvae manage to avoid the chemical defenses of the plants? Cabbage butterflies have the ability to break down mustard oils and thus feed on mustards and capers without harm.

Real Life

Leaflets three, let it be. Members of the genus *Toxicodendron*, which includes poison ivy, produce a defensive chemical called urushiol *(OO roo shee awl)*, which causes a severe, itchy rash in some people.

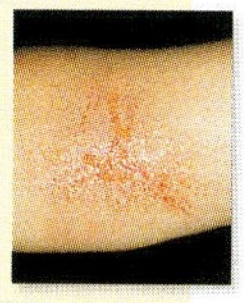

Finding Information
Do research to discover effective treatments for the rash caused by poison ivy.

internet connect
www.scilinks.org
Topic: Florida Parks and Preserves
Keyword: HX6F03
SciLinks. Maintained by the National Science Teachers Association

DATA LAB
Predicting How Predation Would Affect a Plant Species

SC.G.1.4.1

Background

Grazing is the predation of plants by animals. Some plant species, such as *Gilia*, respond to grazing by growing new stems. Consider a field in which a large number of these plants are growing and being eaten by herbivores.

Analysis

1. **Identify** the plant that is likely to produce more seeds?
2. **Explain** how grazing affects this plant species.
3. **Evaluate** the significance to its environment of the plant's regrowth pattern.
4. **Hypothesize** how this plant species might be affected if individual plants did not produce new stems in response to grazing.

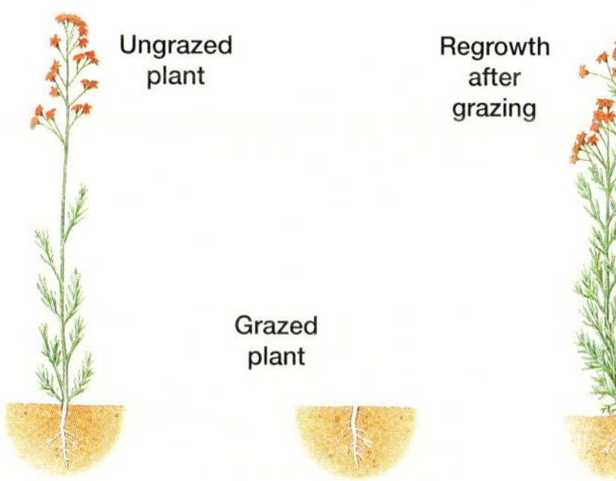

REAL WORLD CONNECTION

As mentioned in the student text, virtually all plants contain defensive chemicals. Many of these defenses can be used to deter pests in a flower or vegetable garden. Some pest repellant plants and the pests they repel are:

- mint, chives, onions and catnip, which repel ants
- basil, which repels mosquitoes
- rosemary, wormwood, onions and chives, which repel slugs

Have students search the Internet for more examples of pest repellant plants.

INCLUSION *Strategies*

• Gifted and Talented • Attention Deficit Disorder

Using reference materials or the Internet, have students create a chart of plants around their home or school that are toxic to humans or pets. Included in this chart should be names and photos or drawings of the plants, descriptions of how each plant affects humans or their pets, and ways to resolve the reactions caused by plant ingestion or contact.

Teach

Real Life
Answer

Calamine lotion, Epsom salts, and bicarbonate of soda may all reduce the severity of a poison ivy rash.

DATA LAB
Predicting How Predation Would Affect a Plant Species

Skills Acquired
Interpreting results, applying information, predicting

Teacher's Notes
Point out that a grazed Gilia plant has more stems than an ungrazed plant.

Answers to Analysis

1. The grazed plant would most likely produce more seeds because it would have more stems and flowers.
2. Because grazing leads to dense regrowth and the production of more flower heads, the grazed plant would produce more offspring.
3. Dense regrowth and the production of more flower heads may allow this plant to spread in its environment and outcompete other plants.
4. If new stems were not produced in response to grazing, the grazed plants would produce few, if any, seeds. Over the years, the plant might become rare or extinct in areas of heavy grazing.

Close

Reteaching — Basic

Have students make a chart to help them remember the different types of species interactions—predation, parasitism, mutualism, and commensalism. Tell them to use the four interactions as headings. Then have them place plus and minus signs in the row below to indicate whether each of the two organisms is helped or harmed by the interaction above. For example, mutualism would have a + / + because both organisms benefit. **LS Visual**

Quiz — General

True or False:

1. Commensalism is a symbiotic relationship in which both participating species benefit. (False. In a mutualism, both partners benefit. In commensalism, one partner benefits and the other is not affected.)
2. A predator/prey relationship in which the predator does not kill the prey is called parasitism. (True)
3. Evolutionary change is the result of the environment acting directly on organisms to produce a change in an organism's traits. (False. The environment affects the survival of traits after their appearance in the population.)

Figure 2 Mutualism. The small green insects on this plant stem are aphids. They are protected by their ant guards.

Symbiotic Species

In **symbiosis** *(sim bie OH sis)*, two or more species live together in a close, long-term association. Symbiotic relationships can be beneficial to both organisms or benefit one organism and leave the other harmed or unaffected. Parasitism, mentioned earlier, is one type of symbiotic relationship that is detrimental to the host organism. While it is relatively easy to determine that an organism in a symbiotic relationship is being helped, it can be difficult to determine that an organism is neither harmed nor helped.

Mutualism

Mutualism is a symbiotic relationship in which both participating species benefit. A well-known instance of mutualism involves ants and aphids, as shown in **Figure 2.** Aphids are small insects that use their piercing mouthparts to suck fluids from the sugar-conducting vessels of plants. They extract a certain amount of the sucrose and other nutrients from this fluid. However, much of the fluid—so-called honeydew—runs out in an altered form through their anus. Certain ants have taken advantage of this fact and "milk" the aphids for the honeydew, which they use as food. The ants, in turn, protect the aphids against insect predators. Thus, both species benefit from the relationship.

Figure 3 Commensalism. The clown fish can survive the stings of the sea anemone, which protects it from predators.

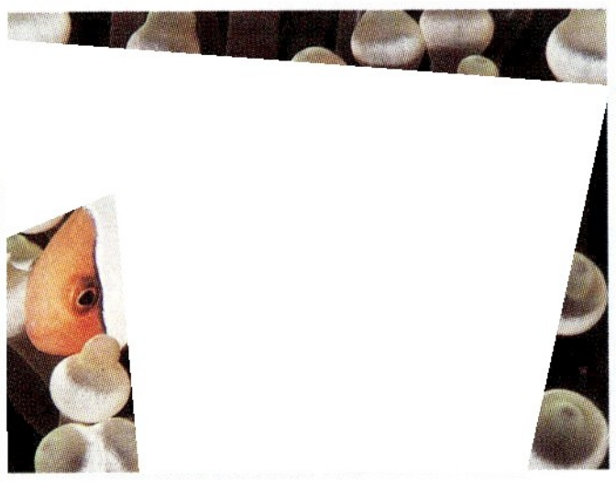

Commensalism

A third form of symbiosis is **commensalism,** a symbiotic relationship in which one species benefits and the other is neither harmed nor helped. Among the best-known examples of commensalism are the relationships between certain small tropical fishes and sea anemones, marine animals that have stinging tentacles. These fishes, such as the clown fish shown in **Figure 3,** have the ability to live among and be protected by the tentacles of the sea anemones, even though these tentacles would quickly paralyze other fishes.

Section 1 Review

1. **Explain** why predator-prey coevolution can be described as an "arms race." SC.D.1.4.4
2. **Critical Thinking Applying Information** Is the relationship between a plant and its pollinator mutualistic? Why or why not. SC.G.1.4.1
3. **Critical Thinking Interpreting Interactions** In a relationship that is an example of commensalism, would the species that is neither helped nor harmed evolve in response to the other species? Defend your answer. SC.G.1.4.1
4. **Critical Thinking Illustrating Principles** In Japan, native honeybees have an effective defense strategy against giant Japanese hornets. Imported European honeybees, however, are unable to defend themselves. Use this example to illustrate the results of natural selection in adaptation. SC.G.2.4.3
5. *FCAT Prep* Which pair of organisms has a mutualistic relationship?
   A aphid and ant
   B clown fish and sea anemone
   C flea and dog
   D lion and zebra
   SC.G.1.4.1

Answers to Section Review

1. In predator-prey coevolution, prey with certain defensive traits (or "weapons") survive predation and reproduce. Predators with traits that allow them to catch prey, survive and reproduce. Thus, traits in predator and prey populations change in response to each other, in an "arms race."
2. Plants and pollinators have a mutualistic relationship. Pollinators carry pollen to plants so the plants are fertilized. Plants provide nectar and pollen to the pollinator as food.
3. No, the unaffected species neither gains nor suffers from the relationship, so there is no selective pressure for coevolution.
4. Native honeybees have been exposed to Japanese hornets for thousands of years, so traits that protect the honeybees have become prevalent in the population. European honeybees have only recently been exposed to the Japanese hornet and do not have the adaptations native honeybees do, so they cannot defend themselves from the hornets.
5. **A.** Correct. Ants use aphid honeydew for food, and in turn protect the aphids from insect predators. **B.** Incorrect. Clown fish and sea anemones have a commensal relationship. **C.** Incorrect. Dogs are hosts to parasitic fleas. **D.** Incorrect. Zebra are prey for predator lions.

Section Review Correlations

SC.D.1.4.4: Item 1
SC.G.1.4.1: Items 2, 3, 5
SC.G.2.4.3: Item 4

Section 2

How Competition Shapes Communities

Florida Sunshine Standards/FCAT

SC.G.1.4.1 There is great diversity and interdependence among living things.

SC.G.2.4.2 Changes in a component of an ecosystem will have unpredictable effects on the entire system, but the components of the system tend to react in a way that will restore the ecosystem to its original condition.

Objectives

- **Describe** the role of competition in shaping the nature of communities. SC.G.1.4.1
- **Distinguish** between fundamental and realized niches. SC.G.1.4.1
- **Describe** how competition affects an ecosystem. SC.G.2.4.2
- **Summarize** the importance of biodiversity. SC.G.1.4.1

Key Terms

competition
niche
fundamental niche
realized niche
competitive exclusion
biodiversity

Common Use of Scarce Resources and Competition

When two species use the same resource, they participate in a biological interaction called **competition.** Resources for which species compete include food, nesting sites, living space, light, mineral nutrients, and water. Competition occurs for resources in short supply. In Africa, for example, lions and hyenas compete for prey. Fierce rivalry between these species can lead to battles that cause injuries to both sides. But most competitive interactions do not involve fighting. In fact, some competing species never encounter one another. They interact only by means of their effects on the abundance of resources.

To understand how competition influences the makeup of communities, you must focus on the day-to-day events within the community. What do organisms eat? Where do they live? The functional role of a particular species in an ecosystem is called its **niche** *(NICH).* A niche is how an organism lives—the "job" it performs within the ecosystem.

A niche may be described in terms of space utilization, food consumption, temperature range, requirements for moisture or mating, and other factors. A niche is not to be confused with a habitat, the place where an organism lives. A habitat is a location; a niche is a pattern of living. **Figure 4** summarizes some aspects of the jaguar's niche in the Central American rain forest.

A niche is often described in terms of how the organism affects energy flow within the ecosystem in which it lives. For example, the niche of a deer that eats a shrub is that of a herbivore. The niches of some organisms overlap. If the resources that these organisms share are in short supply, it is likely that there will be competition between the organisms.

A Jaguar's Niche

- **Diet** Jaguars feed on mammals, fish, and turtles.
- **Reproduction** Jaguars give birth from June to August, during the rainy season.
- **Time of activity** Jaguars hunt by day and by night.

Figure 4 Each organism has its own niche. All of the ways that this jaguar interacts with its environment make up its niche.

Section 2

Focus

Overview

Before beginning this section review with your students the objectives listed in the Student Edition. This section focuses on competition and how it affects populations comprising a community.

Bellringer

Tell students that a niche is very complex and includes all the ways an organism affects and is affected by its environment. Then draw students' attention to **Figure 4.** This figure shows only three aspects of the jaguar's niche. Have students make a list of other aspects of the jaguar's niche.

Motivate

Discussion/Question — GENERAL

Ask students to discuss why competition is usually most intense between closely related organisms. (Closely related organisms are likely to be very similar and therefore are likely to use resources in similar ways.) LS Verbal

Chapter Resource File

- Directed Reading BASIC
- Active Reading GENERAL
- Data Sheet for Data Lab GENERAL

Transparencies

TR Bellringer

One-Stop Planner CD-ROM

- Reading Organizers BASIC
- Reading Strategies BASIC

Section Correlations

SC.G.1.4.1
SC.G.2.4.2

Teach

READING SKILL BUILDER — **BASIC**

Paired Reading Give the students self-stick notes, and have them mark areas of difficulty as they silently read Section 2. Have students use a question mark to signify an area that is unclear and a check mark to indicate understanding. Remind students to study the captions, diagrams, and illustrations as they read. After completing the section, have students work with a partner to discuss what they did or did not understand. Have students write their remaining questions on the board for later discussion.

LS Interpersonal English Language Learners

Teaching Tip

Warblers Tell students that all five warbler species that MacArthur studied belong to the same genus (*Dendroica*). Because closely related species are often competitors, MacArthur was interested in how these very similar species could coexist.

SKILL BUILDER — **GENERAL**

Interpreting Visuals Ecologists use the phrase "resource partitioning" to describe the patterns of resource use in a community. Have students examine **Figure 5,** and then discuss why resource partitioning is an appropriate description of the feeding behavior of the five warbler species. (The warblers partition, or divide up, the insect populations on which they feed by foraging in different parts of the tree.) **LS Visual**

internet connect

www.scilinks.org
Topic: Symbiosis
Keyword: HX4171

SCILINKS. Maintained by the National Science Teachers Association

Size of a Species' Niche

To gain a better understanding of what a niche is, you can look more closely at a particular species. Imagine a Cape May warbler (a small, insect-eating songbird) flying in a forest and landing to search for dinner in a spruce tree. The niche of this bird is influenced by several variables. These variables include the temperature it prefers, the time of year it nests, what it likes to eat, and where on the tree it finds its food. (The Cape May warbler spends its summers almost exclusively in the northeastern United States and Canada. It nests in midsummer, eats small insects, and searches for food high on spruce trees at the tips of the branches.) The entire range of resource opportunities an organism is potentially able to occupy within an ecosystem is its **fundamental niche**.

Dividing Resources Among Species

Now reconsider what the Cape May warbler is doing. It feeds mainly at the very top of the spruce tree even though insects that the warbler could eat are located all over the tree. In other words, Cape May warblers occupy only a portion of their fundamental niche. Why?

Closer study reveals that this surprising behavior is part of a larger pattern of niche restriction. In the late 1950s, the ecologist Robert MacArthur, while a graduate student at Yale University, carried out a classic investigation of niche usage, summarized in **Figure 5.** He studied the feeding habits of five warbler species—the Cape May warbler and four of its potential competitors. MacArthur found that all five species fed on insects in the same spruce trees at the same time. As Figure 5 shows, however, each species concentrated on a different part of the tree. Although all five species of warbler had very similar fundamental niches, they did not use the same resources. In effect,

Figure 5 Niche restriction

Each of these five warbler species feeds on insects in a different portion of the same tree, as indicated by the five colors shown below.

366

REAL WORLD CONNECTION

An analogy may help students understand the difference between habitat and niche. Point out that their houses or apartments and the places they frequent make up their habitats. What they do in their habitats—their patterns of living—are their niches. Ask students to write short descriptions of their habitats and their niches. Have them also write about what would be different if they lived in other habitats (such as rural or urban habitats).

Transparencies

TR E22 Warbler Foraging Zones
TR E21 Effects of Competition on Two Species of Barnacles

they divided the range of resources among them, each taking a different portion. A different color is used to represent the feeding areas of each of the five warbler species shown in Figure 5.

The part of its fundamental niche that a species occupies is called its **realized niche**. Stated in these terms, the realized niche of the Cape May warbler is only a small portion of its fundamental niche. How does this species of warbler benefit from hunting for food in only a portion of the tree? MacArthur suggested that this feeding pattern reduces competition among the five species of warblers. Because each of the five warbler species uses a different set of resources by occupying a different realized niche, the species are not in competition with one another. MacArthur concluded that natural selection has favored a range of preferences and behaviors among the five species that "carve up" the available resources. Most ecologists agree with this conclusion.

Study TIP

Reading Effectively
To better understand the relationship between fundamental and realized niches, draw two circles, one within the other. Label the larger circle "Fundamental niche, entire tree." Label the smaller circle "Realized niche."

DATA LAB

Predicting Changes in a Realized Niche SC.G.1.4.1

Background

Two features of a niche that can be readily measured are the location where the species feeds and the size of its preferred prey. The darkest shade in the center of the graph below indicates the prey size and feeding location most frequently selected by one bird species (called Species A).

Prey Length and Location for Species A

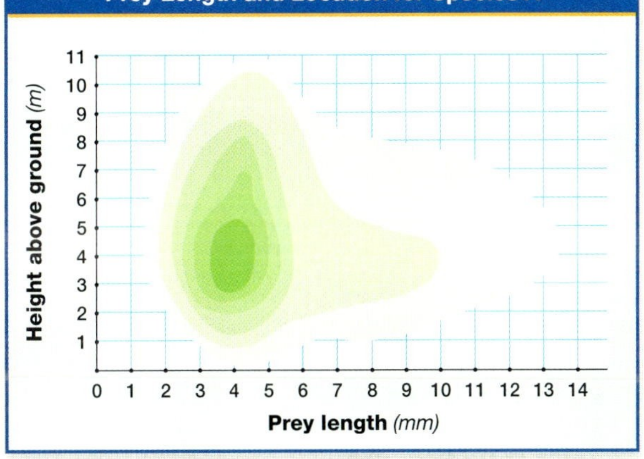

Analysis

1. **State** the range of lengths of Species A's preferred prey.
2. **Identify** the maximum height at which Species A feeds.
3. **Critical Thinking**
Predicting Outcomes
Species B is introduced into Species A's feeding range. Species B has exactly the same feeding preferences, but it hunts at a slightly different time of day. How might this affect Species A?
4. **Interpreting Graphics**
Species C is now introduced into Species A's feeding range. Species C feeds at the same time of day as Species A, but it prefers prey that are between 10 and 13 mm long. How might this affect Species A?
5. **Critical Thinking**
Predicting Outcomes
How would the introduction of a species with exactly the same feeding habits as Species A affect the graph?
6. **Interpreting Graphics** What does the lightest shade at the edge of the contour lines represent?

Teaching Tip

Realized Niches Point out that an organism's realized niche is a subset of its fundamental niche. Thus, the realized niche can be smaller than (or the same size as) the fundamental niche, but never larger.

DATA LAB

Predicting Changes in a Realized Niche

Skills Acquired
Interpreting results, applying information, analyzing information, predicting outcomes

Teacher's Notes
Explain to students that this type of graph is a convenient way to represent three variables: height above the ground, prey length, and frequency of selection of a combination of prey length and feeding height.

Answers to Analysis

1. Most selected prey are approximately 3.5 to 4.5 mm.
2. Maximum feeding height is nearly 11 m.
3. Even though it is feeding at a different time of day, Species B might reduce the prey available to Species A, since it has the same feeding preference.
4. Species C would reduce Species A's realized niche by competing with Species A for large prey. Since Species A prefers smaller species, however, competition from Species C would be minimal.
5. Accept any well-reasoned answer. Sample answer: It might not affect how the graph looks. However, since both species have the same niche, one could be forced to extinction in this area.
6. The lightest shade represents the combination of feeding height and prey length least frequently selected but still exploited by Species A.

did you know?

Bold Coyotes The coyote's broad fundamental niche has allowed it to thrive in areas of human encroachment. For example, coyotes can survive on a variety of animal and plant foods, including most of the food humans throw away. The coyote's brazenness is a behavioral adaptation and therefore part of its niche as well. Its boldness pays off by allowing it access to areas that competitors, such as bobcats, usually avoid.

Teach, *continued*

Demonstration

Competition between two species can result in one being eliminated from the community. Show the class photographs of a starling and a bluebird. Starlings were first introduced into Central Park in New York City in 1890. Today starlings are found throughout the continental United States. In many areas they have bested bluebirds for nesting sites, causing a drastic decline in bluebirds. Make bluebird house plans available to students and select an area to post them. Emphasize the importance of monitoring the houses to keep out non-native competitors like starlings.

Group Activity — Advanced

Investigating Competition Divide students into groups of three. Tell students that each group is to investigate an example of competition. Each group's first task is to find an organism to study. If the weather is warm enough, have students explore the school grounds or areas around their homes to select an organism. Next, have each group determine at least one organism that competes for resources with the organism they selected. Finally, ask groups to gather information about the organisms they selected using the Internet, the library, or local experts. With this information, have them build a story about the interaction between the organisms, focusing on how this interaction affects the organisms' resource use and realized niche. Encourage students to take photos of their "study" organisms or to find images of them in magazines or on the Internet. LS Interpersonal

Competition and Limitations of Resource Use

WORD Origins

The word *niche* is from the Latin word *nidus*, meaning "nest." The place an organism occupies in its environment is part of its niche—its overall functional role.

A very clear case of competition was shown by experiments carried out in the early 1960s by Joseph Connell of the University of California. Connell worked with two species of barnacles that grow on the same rocks along the coast of Scotland. Barnacles are marine animals that are related to crabs, lobsters, and shrimp. Young barnacles attach themselves to rocks and remain attached there for the rest of their lives. As you can see in **Figure 6,** one species, *Chthamalus stellatus,* lives in shallow water, where it is often exposed to air by receding tides. A second species, *Semibalanus balanoides,* lives lower down on the rocks, where it is rarely exposed to the atmosphere.

When Connell removed *Semibalanus* from the deeper zone, *Chthamalus* was easily able to occupy the vacant surfaces. This indicates that it was not intolerance of the deeper environment that prevented *Chthamalus* from becoming established there. *Chthamalus*'s fundamental niche clearly includes the deeper zone. However, when *Semibalanus* was reintroduced, it could always outcompete *Chthamalus* by crowding it off the rocks. In contrast, *Semibalanus* could not survive when placed in the shallow-water habitats where *Chthamalus* normally occurs. *Semibalanus* apparently lacks the adaptations that permit *Chthamalus* to survive long periods of exposure to air. Connell's experiments show that *Chthamalus* occupies only a small portion of its fundamental niche. The rest is unavailable because of competition with *Semibalanus.* As MacArthur suggested, competition can limit how species use resources.

Figure 6 Effects of competition on two species of barnacles

The realized niche of *Chthamalus* is smaller than its fundamental niche because of competition from the faster-growing *Semibalanus*.

Career

Naturalist Parks (local, state, and national) as well as school districts and private nature centers, employ naturalists. Passionate about the workings of living communities, naturalists educate youths and adults about natural history, ecology, and environmental science in some of the most beautiful settings in the world. Strong communications skills and knowledge of ecology and natural history are required.

MISCONCEPTION ALERT

Environment Can Influence Competition
Be sure students do not develop the misconception that the same species of a competing pair will always eliminate the other species. Point out that it is possible to alter the environment so that the outcome is reversed. Thomas Park and his colleagues at the University of Chicago conducted competition experiments on two species of flour beetles (*Tribolium*). The temperature and humidity at which the beetles were raised determined which species was the superior competitor.

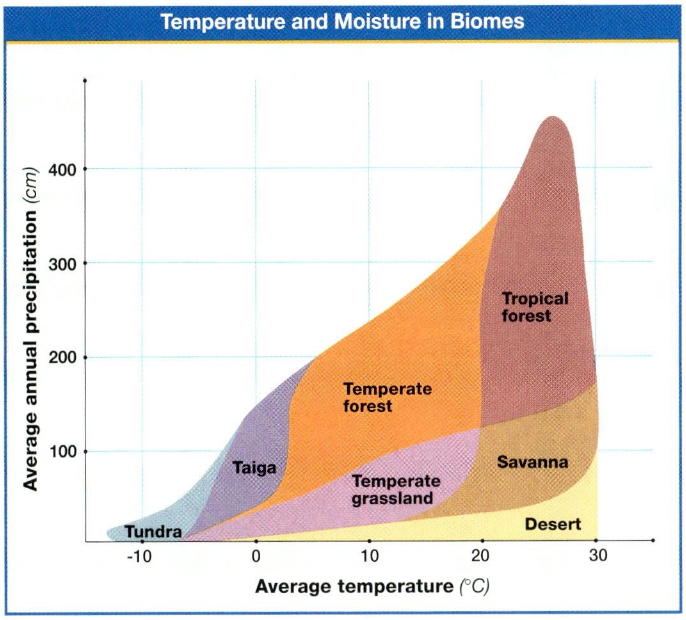

Figure 11 Conditions in biomes. Different biomes have characteristics of temperature and humidity.

Terrestrial Biomes

Tropical Rain Forests Tropical rain forests receive an average of as much as 450 cm (180 in.) of rain per year, with little difference in distribution from season to season. The richest biome in terms of number of species is the tropical rainforest. Tropical rainforests may contain at least half of the Earth's species of terrestrial organisms—more than 2 million species. Tropical rain forests have a high primary productivity even though they exist mainly on quite infertile soils. Most of the nutrients are held within the plants; the soil itself contains few nutrients.

Tropical rain forest in Puerto Rico

Savannas The world's great dry grasslands, called savannas, are found in tropical areas that have relatively low annual precipitation or prolonged annual dry seasons. Annual rainfall is generally 90 to 150 cm (35 to 60 in.) in savannas. There is a wider fluctuation in temperature during the year than in the tropical rain forests, and there is seasonal drought. These factors have led to an open landscape with widely spaced trees. Many of the animals are active only during the rainy season. Huge herds of grazing mammals are found on the savannas of East Africa.

Savanna in East Africa

Group Activity — GENERAL

Biome Collages Place students in groups of three. As students study each biome, have them collect images of plants and animals that are characteristic of that biome. After studying the biomes and collecting the images, have students develop a collage of each biome. Have the group present each collage to the class, explaining why their images are appropriate to each biome. Hang all the collages of each type of biome together on the wall and label. **LS Visual** English Language Learners

Teaching Tip

Tropical Rain Forest Competition Point out that the strongest competition among trees in a rain forest is for light. In the rain forest, about 70 percent of all plant species are trees. Their trunks are usually slender and tall. Their leaves are usually large and dark

Activity — GENERAL

Identifying Biomes Find images of the various biomes on the Internet or in magazines. As you project each image onto a screen, ask students to identify the biome. Then ask students to justify their answers, supporting their identifications with evidence. **LS Visual**

Attention Grabber

Termite Mounds The savannas of Australia and Africa are home to the world's tallest non-human-built structures. These are termite mounds, which are built by Australian and African termites. Termites live about 5 to 6 feet beneath the mounds. Tunnels within the mounds are used as ventilation systems for the termite "living quarters" below. The largest termite mounds are so tall that if the termites that built the mounds were the size of humans, then the mounds would be 180 stories tall!

INCLUSION Strategies

• Learning Disabled *• Attention Deficit Disorder*

Using index cards have students create postcards to send to the class from a biome they have visited. Students should include a description of the characteristics of the biome, the kind of climate there, the kinds of animals there, and locations where this biome exists. On the postcard, students should draw a picture of the biome. **LS Visual**

Taiga in Manitoba, Canada

Tundra in Denali National Park, Alaska

Taiga Cold, wet climates promote the growth of coniferous forests. A great ring of northern forests of coniferous trees, primarily spruce and fir, extends across vast areas of Eurasia and North America. This biome, one of the largest on Earth, is called by its Russian name, taiga *(TIE guh)*. Winters in the taiga are long and cold, and most of the precipitation falls in the summer. Many large mammals, including herbivores such as elk, moose, and deer and carnivores such as wolves, bears, lynxes, and wolverines live in the taiga.

Tundra Between the taiga and the permanent ice surrounding the North Pole is the open, sometimes boggy biome known as the tundra. This enormous biome covers one-fifth of the Earth's land surface. Annual precipitation in the tundra is very low, usually less than 25 cm (10 in.), and water is unavailable for most of the year because it is frozen. The permafrost, or permanent ice, usually exists within 1 m (about 3 ft) of the surface. Foxes, lemmings, owls, and caribou are among the vertebrate inhabitants.

QUICK LAB

Investigating Factors That Influence the Cooling of Earth's Surface SC.G.2.4.4

You can discover how the amount of water in an environment affects the rate at which that environment cools.

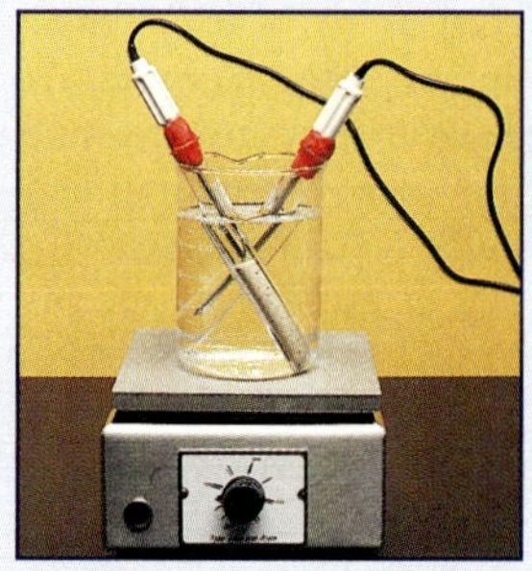

Materials

MBL or CBL system with appropriate software, temperature probes, test tubes, beaker, hot plate, one-holed stoppers, water, sand, test-tube tongs, test-tube rack

Procedure

1. Set up an MBL/CBL system to collect and graph data from each temperature probe at 5-second intervals for 240 data points. Calibrate the probe using stored data.
2. Fill one test tube with water. Fill another test tube halfway with sand.
3. Place a temperature probe in the sand, and suspend another temperature probe at the same depth in the water, using one-holed stoppers to hold each temperature probe in place.
4. Place both test tubes in a beaker of hot water. Heat them to a temperature of about 70ºC. **Caution: Hot water can burn skin.**
5. Using test-tube tongs, remove the test tubes and place them in the test-tube rack. Record the drop in temperature for 20 minutes.

Analysis

1. **Critical Thinking** **Analyzing Results** Did the two test tubes cool at the same rate? Offer an explanation for your observations.
2. **Critical Thinking** **Predicting Outcomes** In which biome—tropical rain forest or desert—would you expect the air temperature to drop most rapidly? Explain your answer.

QUICK LAB

Investigating Factors That Influence the Cooling of Earth's Surface

Skills Acquired

Observing, interpreting results, applying information, analyzing results, predicting outcomes

Teacher's Notes

Preparation tip: Be sure that each lab station has the required materials before the start of class.

Safety notes: Have students wear safety goggles. Remind them to handle the hot test tubes with tongs and to use extreme care. Remind them to be careful of the hot plate at all times. Moisten the ends of the probes to facilitate pushing them through the stoppers.

When test tubes are heated, the sand and water may not reach 70°C at the same time. Students may remove each tube as it reaches 70°C. Remind them that they are interested in observing the rate of cooling, and that the starting temperature is not critical.

Answers to Analysis

1. The sand should cool faster, as water has a greater capacity to store heat and therefore a slower cooling rate.
2. The desert temperature would drop more rapidly, due to the lack of water in the ground and the atmosphere.

Cultural Awareness

The Joshua Tree Show the class a photograph of a Joshua tree, a North American desert plant. This tree was named by early Mormon colonists, who thought it resembled the biblical prophet Joshua waving them toward the Promised Land.

REAL WORLD CONNECTION

Many of today's domesticated plants and animals, including cattle, sheep, horses, wheat, barley, and oats, are descendants of the wild grassland communities of Eurasia.

Deserts Typically, less than 25 cm (10 in.) of precipitation falls annually in the world's desert areas. The scarcity of water is the overriding factor influencing most biological processes in the desert. In desert regions, the vegetation is characteristically sparse. Deserts are most extensive in the interiors of continents. Less than 5 percent of North America is open desert. The amount of water that actually falls on a particular place in a desert can vary greatly, both during a given year and between years.

Desert in Texas

Terrestrial Biomes in Florida

Temperate Grasslands Moderate climates halfway between the equator and the poles promote the growth of rich temperate grasslands called prairies. Temperate grasslands once covered much of the interior of North America. Such grasslands are often highly productive when converted to agriculture. The roots of grasses characteristically penetrate far into the soil, which tends to be deep and fertile. Herds of grazing animals often populate temperate grasslands. In North America, huge herds of bison once inhabited the prairies.

Temperate grasslands in Lake Kissimmee State Park, Florida

Temperate Deciduous Forests Relatively mild climates and plentiful rain promote the growth of forests. Temperate deciduous forests (deciduous trees shed their leaves in the fall) grow in areas with relatively warm summers, cold winters, and annual precipitation that generally ranges from 75 to 250 cm (30 to 100 in.). Temperate deciduous forests cover much of the eastern United States and are home to deer, bears, beavers, raccoons, and other familiar animals. The trees are hardwoods (oak, hickory, and beech).

Temperate deciduous forest in O'Leno State Park, Florida

Temperate Evergreen Forests In other temperate areas, drier weather and different soil conditions favor the growth of evergreens. Large portions of the southeastern and western United States have temperate evergreen forests—extensive areas where pine forests predominate over deciduous forests. Where conditions are even drier, temperate forests give way to areas of dry shrubs, such as in the chaparral areas of coastal California and in the Mediterranean.

Temperate evergreen forest in Paynes Prairie Preserve, Florida

Teach, *continued*

Teaching Tip

Deserts in the United States

Desert covers less than 5 percent of North America. The North American deserts are the Mojave, Sonoran, Great Basin, and Chihuahuan. Have students write a report that contrasts the animals, plants, geography, and climate of these four deserts.

Teaching Tip

Desert Temperatures

Temperatures in the desert undergo drastic changes over a 24-hour period. Daytime temperatures can be 30°C (54°F) higher than nighttime readings. These fluctuations occur due to the lack of heat-retaining moisture in the desert air.

Teaching Tip

Numbers Versus Diversity

A deciduous forest usually has only a few species of trees, but there are numerous individuals of each species. In contrast, a tropical rain forest usually has many species of trees, but each species may be represented by only a few individuals.

Integrating Physics and Chemistry

Just as temperature is one of the most important elements of climate, it is also a critical factor that influences solubility. The solubility of most solids rises with increasing temperature. However, there are exceptions, such as cerium chloride, whose solubility decreases with increasing solvent temperature. Ask students whether or not the solubility of gases goes up or down with increasing solvent temperature. They may be surprised to learn that increasing temperature actually decreases the solvent's ability to retain a dissolved gas.

did you know?

Sunlight and Biomes Hours of sunlight per day range widely through the year in northern biomes. The northernmost areas of taiga receive only 6 to 8 hours of sunlight during the winter but nearly 19 hours during the summer. The extremes are even more exaggerated in the tundra, which receives less energy from the sun than any other biome.

Teach, *continued*

Teaching Tip

Deep-water Detritivores In freshwater and marine communities, there are no photosynthetic organisms in the deep-water zone. Most of its inhabitants are detritivores, including scavenging fish, bacteria, and aquatic worms, which eat the detritus (decaying material) that falls from above.

Teaching Tip

Photosynthetic Organisms When sunlight strikes the water surface, red, orange, and yellow wavelengths of light are absorbed (converted into heat) first. Only blue and green can penetrate deeply. Thus, below a depth of a few meters, only those photosynthetic organisms capable of using these shorter wavelengths of light can survive.

Activity — ADVANCED

Viewing Freshwater Communities Have students collect water samples from a nearby pond or lake. Ask students to collect samples from various areas of the pond, collecting as much plant and animal life as they can see. Be sure to have them collect surface water as well. In the classroom, place the pond samples in an aquarium for viewing. Have students make wet mounts of water samples to see which phytoplankton (microscopic plants) and zooplankton (microscopic animals) live in the freshwater community in your area. Have students choose an organism in the community to research and report on to the class. LS Kinesthetic

Aquatic Communities

At a glance, you might at first think that freshwater and marine communities are separate from terrestrial biomes. Yet large amounts of organic and inorganic material continuously enter both bodies of fresh water and ocean habitats from communities on the land.

Study TIP

Organizing Information Make a concept map that describes the zones of a pond or lake as described at right and in Figure 12. For each zone, include the plant and animal life found there.

Freshwater Communities

Freshwater habitats—lakes, ponds, streams, and rivers—are very limited in area. Lakes cover only about 1.8 percent of the Earth's surface, and rivers and streams cover about 0.3 percent. All freshwater habitats are strongly connected to terrestrial ones, with freshwater marshes and wetlands constituting intermediate habitats. Many kinds of organisms are restricted to freshwater habitats, including plants, fish, and a variety of arthropods, mollusks, and other invertebrates too small to be seen without a microscope.

Ponds and lakes have three zones in which organisms live, as illustrated in **Figure 12.** The **littoral zone** is a shallow zone near the shore. Here, aquatic plants live along with various predatory insects, amphibians, and small fish. The **limnetic zone** refers to the area that is farther away from the shore but close to the surface. It is inhabited by floating algae, zooplankton, and fish. The **profundal zone** is a deep-water zone that is below the limits of effective light penetration. Numerous bacteria and wormlike organisms that eat debris on the lake's bottom live in this zone. The breakdown of this debris releases large amounts of nutrients. Not all freshwater systems are deep enough to include a profundal zone.

Figure 12 Three lake zones. Each region, or zone, of a lake contains characteristic organisms.

376

Trends in Limnology

Following Fish Odors Limnology, which comes from a Greek word meaning lake or marsh, is the study of bodies of fresh water. One area of study for limnologists is the effect of fish odors on the behavior of invertebrates such as the water flea, *Daphnia hyalina.* Limnologists have discovered that fish odors could trigger the daily migrations of these tiny freshwater invertebrates. A chemical nicknamed TMA (trimethylamine), which is a component of fish odor, appears to cue *Daphnia* when predator fish are present. *Daphnia* respond by moving to areas where the fish are not swimming. Limnologists are currently researching such aquatic chemical communication.

ENVIRONMENTAL SCIENCE CONNECTION

The physical link between rivers or streams and their surrounding terrestrial communities is the stream bank, also called the *riparian zone.* Riparian zones often appear as ribbons of vegetation around waterways. These areas provide humans with many services, including filtering pollution and absorbing floodwaters. Riparian zones also provide vital habitat for a broad array of plant and animal species.

Florida's Aquatic Communities

FLORIDA BIOWatch

Florida is full of fresh water and surrounded by salt water. Along with a subtropical climate, these conditions make Florida home to an extraordinary variety of aquatic communities with unique species. These communities include streams, rivers, lakes, swamps, marshes, coastlines and everything in between.

Wetlands

A *wetland* is generally defined as an area of land that is regularly or permanently covered with water and that supports semi-aquatic organisms. Wetlands exist in many forms, including freshwater and saltwater varieties of swamps, marshes, bogs, and flats. About 25 percent of Florida is currently classified as wetlands, although this percentage is half of what is was prior to modern development.

Major freshwater wetland types in Florida include *cypress stands* and other *bottomland*, also called *riverside.* These are swamps that are characterized by specific trees or shrubs. *Pocosins* are a type of bog with soggy, acidic soils and evergreen trees or shrubs.

In contrast, *marshes* are mostly treeless and filled with grasses and low shrubs. Cattail and sawgrass marshes are common in Florida. Deep marshes are called *sloughs. Wet prairies* are a blend of marsh and swamp that may include grasses, palmettos, ferns, and other plants.

Florida's coastal areas include a variety of saltwater wetlands such as *salt marshes,* also called *tidal marshes,* and *mangrove swamps.*

The Everglades is a vast expanse of mixed fresh water and salt water wetlands.

Cypress trees are common in Florida's freshwater swamps.

Estuaries

Florida's coastlines are rimmed with *estuaries,* where fresh water flows into and mixes with salt water from the ocean. These areas include saltwater wetlands as well as *lagoons, bays, harbors,* and *inlets.*

Estuaries are constantly changing. Water alternates between fresh, brackish, and salty, and the water level rises and falls with the tides. Despite these challenging conditions, estuaries provide critical habitat for many species of birds, mammals, reptiles, fishes, shellfish, and other invertebrates.

Estuaries serve important roles by controlling floods, protecting land from storms, filtering out pollutants, and improving water quality. They are important to Florida's economy because they serve as nurseries for many species of sport fish and seafood. Indeed, estuaries are often called "the cradle of the sea."

The Everglades

The vast Everglades region includes several terrestrial and wetland communities, but mainly consists of sawgrass marshes, wet prairies, salt marshes, mangrove swamps, and estuaries.

internet connect

www.scilinks.org
Topic: Florida's Freshwater
Keyword: HX6F07
SCILINKS. Maintained by the National Science Teachers Association

377

BIOWatch GENERAL

Florida's Aquatic Communities

Teaching Strategies

Tell students that many of Florida's aquatic communities are connected by water flow. Inland wetlands and streams are the source of many nutrients that eventually feed estuarine organisms. Wetlands slow down floodwaters and help keep soil from being washed into estuaries or out to sea.

Tell students that estuaries are among the Earth's most productive, yet endangered biological communities. Draining, dredging, and filling for human development have destroyed thousands of hectares of Florida's estuaries.

Activity

Have students find out which estuary areas of Florida have been identified as important by agencies such as the USEPA, NOAA, or Florida DEP. Have students create a map of these locations. (Locations include Charlotte Harbor, Sarasota Bay, Tampa Bay, Indian River Lagoon, and many smaller areas.)

Transparencies

TR E28 Three Lake Zones

did you know?

Wetlands and Avian Botulism The most serious disease of wetland bird species is avian botulism. It is estimated that millions of waterfowl die annually from this disease, which is an international concern.

OCEANOGRAPHY CONNECTION

The National Oceanic and Atmospheric Administration (NOAA) conducts research and gathers data about the oceans, atmosphere, space, and sun. One activity of NOAA is to guide the use of ocean and coastal resources and protect them. Since 1993, NOAA has maintained the world's only permanent undersea research laboratory. Located 4 miles off the Florida coast, the "Aquarius" lab addresses scientific questions pertaining to ocean environments and marine life.

Teach, *continued*

Teaching Tip

Deep-sea Waters Humans first reached the floor of the Mariana Trench in 1960. Two scientists used the bathyscaphe *Trieste* to descend 11 km (7 mi) below the ocean's surface. It descended when water was pumped into air tanks at either end, and it ascended when water was pumped out. The *Trieste* obtained data about the conditions at such depths, and its scientists observed many never-before-seen deep-water organisms.

Close

Reteaching — BASIC

Show students pictures of various biomes. Ask them to identify each and to discuss its physical and biological characteristics, as well as where on the planet such a biome might be found. English Language Learners
LS Visual

Quiz — GENERAL

1. Why do mountains often show the same sequence of ecosystems that is found as one goes north or south from the equator? (Temperature and precipitation decrease as latitude or elevation increases.)
2. Which biome is a transitional area between tropical rain forest and desert? (Savanna)
3. In a small pond or lake, which of the three life zones might not be present and why? (A small pond or lake might be shallow and therefore not have a deep-water profundal zone.)

Section Review Correlations

SC.G.1.4.1: Item 5

SC.G.2.4.4: Items 1-4

Fish in shallow ocean waters in Key Largo, Florida

Diatoms from surface of the open sea

Lanternfish from the ocean depths off of the coast of Florida

Marine Communities in Florida

Nearly three-fourths of the Earth's surface is covered by ocean, which consists of three major kinds of marine communities.

Shallow Ocean Waters The zone of shallow water is small in area, but compared with other parts of the ocean, it is inhabited by large numbers of species. The seashore between high and low tide, called the *intertidal zone,* is home to many species of marine invertebrates. Coral reef communities, the world's most diverse, occur in shallow tropical waters. The world's great fisheries are located in the coastal zones of cooler waters, where nutrients washed out from land support huge numbers of fishes.

Surface of the Open Sea Drifting freely in the upper waters of the ocean is a diverse community of **plankton,** composed of bacteria, algae, fish larvae, and many small invertebrate animals. Fishes, whales, and invertebrates such as jellyfishes feed on plankton. And larger fishes and birds, in turn, feed on some of these animals. Photosynthetic plankton (algae such as diatoms and some bacteria) that form the base of this food chain account for about 40 percent of all the photosynthesis that takes place on Earth. Because light penetrates water only to the depth of about 100 m (328 ft), this rich community is confined to the ocean's surface.

Ocean Depths In the deepest waters of the sea, the marine community lives in total darkness, in deep cold, and under great pressure. Despite what seem like hostile conditions, the deep ocean supports a diverse community of bizarre invertebrates and fishes. This includes great squids and angler fishes that attract prey with projections from their head that emit light. On the ocean floor, at an average depth of more than 3 km (1.9 mi), researchers have found an unexpected abundance of species, a diversity that rivals the tropical rain forest.

Section 3 Review

1. **Describe** the relationship between climate and location of species. SC.G.2.4.4
2. **Compare** the tolerance to lack of water needed by plants and animals in savannas and tropical rain forests. SC.G.2.4.4
3. **Critical Thinking Analyzing Information** Why can't photosynthesis occur in the deepest parts of the ocean or in a deep lake? SC.G.2.4.4
4. **Critical Thinking Forming Reasoned Opinions** The equator passes across the country of Ecuador. But the climate there can range from hot and humid to cool and dry. What might explain this? SC.G.2.4.4
5. *FCAT Prep* In which biome would you most likely find plants that are adapted to infertile soils and fairly constant, plentiful precipitation?
   A savanna
   B temperate grassland
   C tropical rain forest
   D tundra
   SC.G.1.4.1

378

Answers to Section Review

1. The climate of a region determines what species can live in that region. Only those well adapted to the conditions will prevail.
2. Savannas have low annual precipitation or prolonged dry seasons, while tropical rain forests have high rainfall year-round. Therefore, rain forest dwellers need little tolerance to lack of water, while savanna dwellers need high tolerance.
3. Photosynthesis cannot occur in deep water because light is completely absorbed before it reaches these depths.
4. Ecuador has mountains, which create climate conditions different from those associated with latitude alone.
5. **A.** Incorrect. Savannas have low annual precipitation or prolonged dry seasons. **B.** Incorrect. The soil of temperate grasslands tends to be fertile. **C.** Correct. Tropical rain forests have high annual precipitation with infertile soil, since most of the nutrients are held within the plants. **D.** Incorrect. Annual precipitation in the tundra is very low.

Key Concepts

1 How Organisms Interact in Communities

- Species within communities coevolve, making many adjustments to living together.
- In a predator-prey interaction, prey often evolve ways to escape being eaten. Predators evolve ways to overcome the defenses of the prey.
- In mutualism and commensalism, species evolve in ways that benefit one or both parties.

2 How Competition Shapes Communities

- Interactions among species help shape communities.
- Competition occurs when two species use the same limited resource.
- An organism's niche is its way of life. An organism may occupy only a part of its fundamental niche, which is called its realized niche.
- Competition can limit how species use resources.
- Biodiversity tends to promote stability and productivity.

3 Major Biological Communities

- Climate and genes largely determines where species live.
- Temperature and moisture are key factors in determining where biomes occur.
- The seven major biomes are tropical rain forest, desert, savanna, temperate deciduous forest, temperate grassland, taiga, and tundra.
- Freshwater communities have three zones of life—littoral, limnetic, and profundal.
- The three major marine communities are shallow ocean waters, open sea surface, and deep-sea waters.

Key Terms

Section 1

coevolution (362)
predation (362)
parasitism (362)
secondary compound (363)
symbiosis (364)
mutualism (364)
commensalism (364)

Section 2

competition (365)
niche (365)
fundamental niche (366)
realized niche (367)
competitive exclusion (369)
biodiversity (370)

Section 3

climate (371)
biome (372)
littoral zone (376)
limnetic zone (376)
profundal zone (376)
plankton (378)

Alternative Assessment — GENERAL

Have students imagine that a large island has suddenly appeared on the surface of the planet. Have students work together to create climate conditions for the island based on its location on the earth, then predict what biome(s) would be found on the island in the future. Have them discuss some of the species they could expect to find, and how those species might interact. **LS Interpersonal**

Chapter Resource File

- Science Skills GENERAL
- Critical Thinking ADVANCED
- Test Prep Pretest GENERAL
- Chapter Test GENERAL

Answers to Concept Map

The following is one possible answer to Performance Zone item 15.

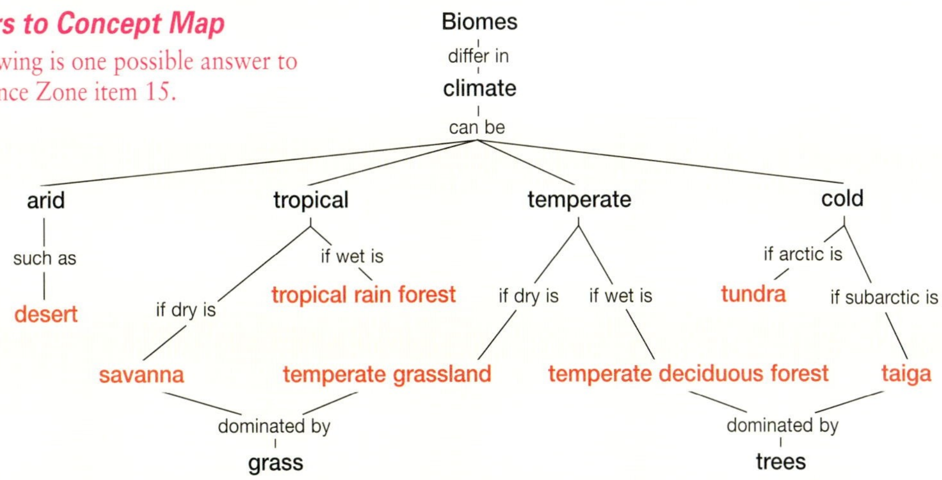

ANSWERS

Understanding Key Ideas

1. a
2. c
3. The grass grows in sunlight and recovers quickly from dry spells and grazing. Energy from the sun is stored in its tissues and passed on to feeding zebra. Zebra are herbivores and live in herds, which offer some protection from predators such as lions. Lions feed primarily on herbivores, capturing a portion of the grazed energy. Lions hunt in small groups, which enables them to capture large prey, such as zebra.
4. d
5. b
6. a
7. Temperatures generally decrease with increasing elevation (height above sea level) and latitude (distance from the equator). In other words, increasing elevation and latitude positively correlate with the transition from tropical to temperate to arctic biomes.
8. Answers may include: freshwater—cypress stands, bottomland hardwood swamps, and pocosins; saltwater—salt marshes and mangrove swamps.
9. The flow of energy through ecosystems ensures that biological communities have the same basic structure. Ecosystems have autotrophs, such as plants. There are heterotrophic herbivores, such as rabbit or deer and heterotrophic carnivores, such as foxes. An abundance of dead organic matter provides for decomposers, such as bacteria.

Chapter Review Correlations

SC.G.1.4.1: Items 2, 11
SC.G.1.4.2: Items 3, 6, 9
SC.G.2.4.2: Item 12
SC.G.2.4.3: Item 1
SC.G.2.4.4: Items 4, 7, 8, 10, 13, 14
SC.G.2.4.5: Item 5

Performance ZONE

Understanding Key Ideas

1. In predator-prey coevolution, if the prey gains a defense to stop predation, then the predator may evolve
a. in a way that enables it to overcome the prey's defense.
b. so that it can parasitize the prey.
c. secondary compounds.
d. into a prey species.
SC.G.2.4.3

2. The principle of competitive exclusion indicates that
a. a niche can be shared by two species if their niches are very similar.
b. niche subdivision may occur.
c. one species will eliminate a competing species if their niches are very similar.
d. competition ends in worldwide elimination of a species.
SC.G.1.4.1

3. Describe the niches of a lion, a zebra, and the grass that grows on the African plain in terms of how each species affects energy flow in the ecosystem. SC.G.1.4.2

4. Which abiotic factor is likely *not* a reason for the desert biome's low primary productivity?
a. extreme temperatures
b. frequent flooding
c. high predation
d. availability of sunlight
SC.G.2.4.4

5. When populations of similar species occupy the same area at the same time, these populations often
a. share all their resources equally.
b. divide their range of resources.
c. compete for resources to the death.
d. look in other areas for different resources.
SC.G.2.4.5

6. Which of the words sets below describes a vulture eating a dead rabbit?
a. heterotroph, scavenger
b. parasite, predator
c. herbivore, mutualism
d. competitor, commensalisms
SC.G.1.4.2

7. Describe how elevation and latitude affect the distribution of biomes. SC.G.2.4.4

8. **BIOWatch** List two types of freshwater wetlands and two types of saltwater wetlands that are common in Florida. SC.G.2.4.4

9. How does the flow of energy through living systems help determine the components of a biological community? (**Hint:** See Chapter 5, Section 1.) SC.G.1.4.2

10. **Concept Mapping** Make a concept map that shows how the biomes can be classified based on precipitation, temperature, and geographical location. Try to include the following terms in your map: *tropical rain forest, savanna, desert, temperate deciduous forest, temperate grassland, taiga,* and *tundra.* SC.G.2.4.4

Critical Thinking

11. **Justifying Conclusions** In Gause's experiments, *Paramecium caudatum* could coexist with *P. bursaria* but not with *P. aurelia.* Predict what would happen if *P. aurelia* and *P. bursaria* were grown together, and justify your conclusions. SC.G.1.4.1

12. **Justifying Conclusions** Newly introduced predators often prove devastating to native animals. Explain why prey are often more vulnerable to introduced predators than to native predators. SC.G.2.4.2

13. **Analyzing Data** Using the data presented in this chapter, explain why many ecologists refer to the tundra as a frozen desert. SC.G.2.4.4

Alternative Assessment

14. **Summarizing Information** Work with a small group of students to develop a map that shows the most prominent terrestrial and aquatic communities within your state, and for coastal states, those immediately offshore as well. Be certain to include any large swamps or wetlands that connect terrestrial and aquatic communities. SC.G.2.4.4

380

10. The answer to the concept map is found at the bottom of the Study Zone page.

Critical Thinking

11. The two could coexist if *P. bursaria* feeds on yeast at the bottom of the tube and *P. aurelia* feeds on bacteria at the top.
12. Native prey have often evolved defense mechanisms to protect them from native predators. These mechanisms may be ineffective against introduced predators.
13. Annual precipitation in deserts is less than 25 cm, which is the same level as in the tundra. Additionally, water in the tundra is unavailable for most of the year because it is frozen.
14. Students' maps should realistically represent your state and its predominant communities.

Assignment Guide

Section	Questions
1	1, 6, 9, 12
2	2, 3, 5, 11
3	4, 7, 8, 10, 13, 14

FCAT Prep

For questions 1–6, write your answers on a separate sheet of paper.

1 Both a spruce tree and a hemlock tree require nitrogen from the soil. What is the interaction between these two species? SC.G.1.4.1

A. commensalism **C.** mutualism
B. competition **D.** succession

2 What term describes the ways in which an organism interacts with its environment?

F. ecosystem **H.** niche
G. habitat **I.** space

SC.D.2.4.1

3 Which of the following is a transition zone between tropical rain forest and desert?

A. savanna
B. taiga
C. temperate deciduous forest
D. tundra

SC.G.2.4.4

4 Why is the open ocean biome considered only slightly more productive than the desert biome?

F. Neither the open ocean nor the desert receive very much rain, which restricts productivity.
G. Light penetrates only the top 100 meters of water in the open ocean, which restricts productivity.
H. The open ocean receives the same amount of light, yet has significantly more water than the desert.
I. Both lack large trees that block sunlight, but the open ocean lacks the soil needed for plants to grow.

SC.G.2.4.4

5 READ INQUIRE EXPLAIN Terrestrial biomes include tropical rainforests, savannas, the taiga, the tundra, deserts, temperate grasslands, temperate deciduous forests, and temperate evergreen forests.

Describe the characteristics of the tundra. SC.G.2.4.4

6 READ INQUIRE EXPLAIN A keystone predator is one who regulates the populations of various competitors in an ecosystem. The *Pisaster* sea star is one example of a keystone predator. Sea stars are fierce predators of marine animals that live in an intertidal community, such as clams and mussels. Mussels can outcompete other species for space on the rocks. Keystone predators reduce the occurrence of competitive exclusion of weak competitors.

Part A Why is the *Pisaster* sea star classified as a keystone predator? SC.G.1.4.2

Part B How does a keystone predator promote biodiversity? SC.G.1.4.1

Test TIP

After you finish writing your answer to a short-response question, proofread it for errors in spelling, grammar, and punctuation.

FCAT Doctor

Question 1 Answer B is the correct choice. The trees compete for the same resource. Answer A is incorrect because commensalism involves one of the organisms benefiting from the relationship. Answer C is incorrect because in mutualism both organisms benefit from the relationship. Answer D is incorrect because succession describes changes in ecosystems.

Question 4 Answer G is the correct choice. Despite the abundance of water in the open ocean, the lack of sunlight beneath 100 meters limits productivity. Answer F is incorrect because a lack of rain would not be a limitation for the oceans' large aquatic environment. Answer H is incorrect because the abundance of water in the ocean does not limit its productivity. Answer I is incorrect because the ocean's primary producers do not require soil.

Question 5 The tundra is located between the taiga and the North Pole.

Question 6

Part A By preying on mussels, sea stars keep the mussel populations too low to drive out other species.

Part B Keystone predators prey on superior competitors, allowing inferior predators to thrive, increasing the variety of living organisms.

Answers

1. B
2. H
3. A
4. G
5. The tundra has cold and long winters, very few trees, and little precipitation.

6A. The *Pisaster* sea star preys on, or eats, mussels that might otherwise outcompete other species.

6B. A keystone predator reduces competition by decreasing the populations of superior competitors.

Test Prep Correlations

SC.D.2.4.1: Item 2
SC.G.1.4.1: Items 1, 6B
SC.G.1.4.2: Item 6A
SC.G.2.4.4: Items 3, 4, 5

Skills Practice Lab

Observing How Brine Shrimp Select a Habitat

Teacher's Notes

Time Required Two lab periods of 40-50 minutes each

Ratings

EASY → HARD

- Teacher Preparation 2
- Student Setup 3
- Concept Level 2
- Cleanup 1

Skills Acquired
- Observing
- Designing experiments
- Collecting data
- Communicating

Scientific Methods

In this lab, students will:
- Make Observations
- Ask Questions
- Test the Hypothesis
- Analyze the Results
- Draw Conclusions

Materials and Equipment

Materials for this lab can be ordered from WARD'S. See *Master Materials List* at the front of this book for catalog numbers.

Safety Cautions
- Make sure students do not use very hot water.
- Students should not put their hands near their face while handling the brine shrimp.
- Tell students to wash their hands thoroughly after concluding the experiment.

Skills Practice Lab

Observing How Brine Shrimp Select a Habitat SC.G.1.4.1

SKILLS
- Using scientific methods
- Collecting, organizing, and graphing data

OBJECTIVES
- **Observe** the behavior of brine shrimp.
- **Assess** the effect of environmental variables on habitat selection by brine shrimp.

MATERIALS
- clear, flexible plastic tubing
- metric ruler
- marking pen
- corks to fit tubing
- brine shrimp culture
- screw clamps
- test tubes with stoppers and test-tube rack
- pipet
- Petri dish
- Detain™ or methyl cellulose
- aluminum foil
- calculator
- fluorescent lamp or grow light
- funnel
- graduated cylinder or beaker
- hot-water bag
- ice bag
- pieces of screen
- tape

Magnification 20×

Brine shrimp

Before You Begin

Different organisms are adapted for life in different **habitats.** For example, **brine shrimp** are small crustaceans that live in salt lakes. Given a choice, organisms select habitats that provide the conditions (e.g., temperature, light, pH, salinity) to which they are adapted. In this lab, you will investigate habitat selection by brine shrimp and determine which environmental conditions they prefer.

1. Write a definition for each boldface term in the paragraph above.
2. Based on the objectives for this lab, write a question you would like to explore about habitat selection by brine shrimp.

Procedure

Part A: Making and Sampling a Test Chamber

1. Divide a piece of plastic tubing into 4 sections by making a mark at 12 cm, 22 cm, and 32 cm from one end. Label the sections *1*, *2*, *3*, and *4*.
2. Place a cork in one end of the tubing. Then transfer 50 mL of brine shrimp culture to the tubing. Place a cork in the open end of the tubing.
3. When you are ready to count shrimp, divide the tubing into four sections by placing a screw clamp at each mark on the tubing. *While someone holds the corks firmly in place,* first tighten the middle clamp and then the outer clamps.
4. Starting at one end, pour the contents of each section into a test tube labeled with the same number. After you empty a section, loosen the adjacent clamp and fill the next test tube.

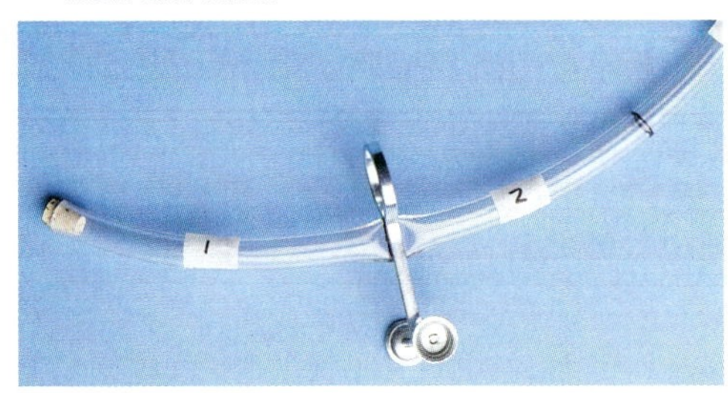

382

Tips and Tricks

Students should begin by observing the movement of brine shrimp through water. Tell students to allow the shrimp enough time to distribute themselves in response to environmental factors. This activity works well with groups of four students. When tubing is divided as directed, it allows for space taken up by stoppers and forms equal quarters. Students may count shrimp by looking in the Petri dish or in the pipet. Students might hold the pipet up to a light for better visibility. The Procedure section provides students with one method for randomly dividing brine shrimp into groups that can be experimentally manipulated in the tubing or in the Petri dishes.

5. Stopper one test tube, and invert it gently to distribute the shrimp. Use a pipet to transfer a 1 mL sample of shrimp culture to a Petri dish. Add a few drops of Detain™ to the sample. Count and record the number of live shrimp.

6. Repeat step 5 three more times for the same test tube. Record the average number of shrimp for this test tube.

7. Repeat steps 5 and 6 for each of the remaining test tubes.

PART B: Design an Experiment

8. Work with the members of your lab group to explore one of the questions written for step 2 of **Before You Begin.** To explore the question, design an experiment that uses the materials listed for this lab.

You Choose

As you design your experiment, decide the following:

a. what question you will explore
b. what hypothesis you will test
c. how to set up your control
d. how to expose the brine shrimp to the conditions you chose
e. how long to expose the brine shrimp to the environmental conditions
f. how you will set up your data table

9. Write a procedure for your group's experiment. Make a list of all the safety precautions you will take. Have your teacher approve your procedure and safety precautions before you begin the experiment.

10. Set up and conduct your group's experiment. Do *not* use water over 70°C, which can burn you. **CAUTION: If you are working with the hot-water bag, handle it carefully. If you are working with a lamp, do not touch the bulb. Light bulbs get very hot and can burn your skin.**

PART C: Cleanup and Disposal

11. Dispose of broken glass in the designated waste container. Put brine shrimp in the designated container. Do not pour chemicals down the drain or put lab materials in the trash unless your teacher tells you to do so.

12. Clean up your work area and all lab equipment. Return lab equipment to its proper place. Wash your hands thoroughly before you leave the lab and after you finish all work.

Analyze and Conclude

1. **Summarizing Results** Make a bar graph of your data. Plot the environmental variable on the x-axis and the number of shrimp on the y-axis.

2. **Analyzing Results** How did the shrimp react to changes in the environment?

3. **Analyzing Methods** Why was a control necessary?

4. **Analyzing Methods** Why was it necessary to take many counts in each test tube (step 6 of Part A)?

5. **Further Inquiry** Write a new question about brine shrimp that could be explored with another investigation.

Do You Know?

Do research in the library or media center to answer these questions:

1. What are some predators of brine shrimp?
2. What is the ideal habitat for one of your favorite animals?

Use the following Internet resources to explore your own questions about habitat selection.

internet connect
www.scilinks.org
Topic: Adaptation
Keyword: HX4002
SciLinks. Maintained by the National Science Teachers Association

Answers to Before You Begin

1. *habitats*—areas where organisms live and to which they are adapted
brine shrimp—small crustaceans that live in salt lakes
2. Answers will vary. For example: Will brine shrimp prefer a warm habitat or a cool habitat?

Sample Procedure

Perform steps 1–4 of part A. Record the number of shrimp and temperature in each test tube, then calculate the average number and temperature. This is the control sample. Then make a new test chamber using the procedure outlined in part A steps 1 and 2. Tape the tubing to the desk top and cover it with aluminum foil. Mark the foil to show the approximate positions of sections 1 and 4. Carefully place a bag of very cold water over section 1 and a bag of very warm water over section 4. After 10 minutes, quickly complete steps 4 and 5 of part A. Immediately read and record the temperature of the solution in the two test tubes. Then count the shrimp in each tube. Make a histogram plotting temperature on the x-axis versus number of shrimp on the y-axis versus. Include a bar for the control sample.

Sample Data Table

Treatment	No. of Shrimp
control	even distribution . . . about 25
warm water	about 35
room temperature	about 17

Answers to Analyze and Conclude

1. Answers will vary.
2. Answers will vary depending on the species of *Artemia* used.
3. The control was necessary to show that the brine shrimp did not inherently prefer one part of the tube.
4. Many counts were taken to make allowances for variations in populations and to provide data for calculating an average.
5. Answers will vary. For example: How do brine shrimp react to water movement?

Answers to Do You Know?

1. Fish are the main predators of brine shrimp.
2. Answers will vary depending on the species of animal chosen.

Chapter Opener		SE **Quick Review,** p. 385 ■ SE **Reading Activity,** p. 385 ■ TE **Using the Figure,** p. 384 GENERAL TE **Opening Activity,** p. 385 GENERAL
& 3 • 90 min pp. 386–389 **Section 1** Global Change	OSP Lesson Plan* TR Bellringer* TR Section Outline* TR E30 Ozone "Hole" Over Antarctica* TR E29 The Greenhouse Effect TR E31 Atmospheric Temperature and Carbon Dioxide Levels* TR E32 Change in Global Temperature*	SE **Quick Lab** Modeling the Greenhouse Effect, p. 389 ◆ ■ CRF **Datasheets for In-Text Labs** Modeling the Greenhouse Effect*
& 5 • 90 min pp. 390–395 **Section 2** Effects on Ecosystems	OSP Lesson Plan* TR Bellringer* TR Section Outline* TR E33 Biological Magnification of DDT* TR E34 World Population Growth Patterns*	TE **Activity** Pollution, p. 390 ◆ GENERAL TE **Group Activity** Ground Water in Your Back Yard, p. 393 GENERAL CRF **Exploration Lab** Effects of Acid Rain* ◆ GENERAL CRF **Inquiry Lab** How Does Pollution Affect a Lake?* ◆ GENERAL **Forensics and Applied Science Experiments** Wetlands Acid Spill* GENERAL
& 7 • 90 min pp. 396–400 **Section 3** Solving Environmental Problems	OSP Lesson Plan* TR Bellringer* TR Section Outline*	TE **Demonstration,** p. 397 ◆ TE **Activity** Technology, p. 398 ADVANCED TE **Using the Figure,** p. 399 BASIC TE **Activity** Making a Commitment, p. 399 ADVANCED SE **Skills Practice Lab** Studying Population Growth, pp. 404–405 ■ GENERAL CRF **Datasheets for In-Text Labs** Studying Population Growth* ◆ GENERAL

& 9 • 90 min

Chapter Review and Assessment Resources

- SE **Study Zone,** p. 401 ■
- SE **Performance Zone,** p. 402 ■
- SE **FCAT Test Prep,** p. 403 ■
- TR Concept Mapping*
- CRF Vocabulary Review Worksheet* ■ GENERAL
- CRF Test Prep Pretest* ■ GENERAL
- CRF Chapter Test* ■ GENERAL
- CRF Chapter Test* ADVANCED
- OSP Test Generator*
- OSP Test Item Listing*
- OSP Modified Chapter Test* BASIC

go.hrw.com Holt Online Learning

Visit go.hrw.com for access to Holt Online Learning or enter the keyword **HX6 Home** for a variety of free online resources.

One-Stop Planner® CD-ROM

This CD-ROM package includes
- Lab Materials QuickList Software
- Holt Calendar Planner
- Customizable Lesson Plans
- Printable Worksheets
- ExamView® Test Generator
- Interactive Teacher Edition
- Holt PuzzlePro® Resources
- Holt PowerPoint® Resources

Guided Reading Audio CDs

These CDs are designed to help auditory learners and reluctant readers. (Audio CDs are also available in Spanish.)

Compression guide:
To shorten your instruction because of time limitations, omit blocks 1, 2, and 8.

KEY

TE Teacher Edition
SE Student Edition
OSP One-Stop Planner
CRF Chapter Resource File
TR Teaching Transparency
* Also on One-Stop Planner
■ Also Available in Spanish
◆ Requires Advance Prep

SKILLS DEVELOPMENT AND PRACTICE	SECTION REVIEW AND ASSESSMENT	CORRELATIONS
		Florida Sunshine Standards/FCAT
CRF Directed Reading Worksheet* BASIC CRF Active Reading Worksheet* GENERAL	SE Section 1 Review ■ TE Reteaching, p. 389 BASIC TE Quiz, p. 389 GENERAL TE Alternative Assessment, p. 389 ADVANCED CRF Quiz* ■ BASIC	SC.G.2.4.1, SC.G.2.4.6
TE Reading Skill Builder Reading Hint, p. 391 BASIC TE Skill Builder Writing, p. 392 ADVANCED TE Skill Builder Math, p. 392 BASIC TE Skill Builder Interpreting Graphics, p. 394 BASIC TE Skill Builder Graphing, p. 394 BASIC CRF Directed Reading Worksheet* BASIC CRF Active Reading Worksheet* GENERAL	SE Section 2 Review ■ TE Reteaching, p. 395 BASIC TE Quiz, p. 395 GENERAL TE Alternative Assessment, p. 395 GENERAL CRF Quiz* ■ BASIC	SC.G.2.4.1, SC.G.2.4.6
CRF Directed Reading Worksheet* BASIC CRF Active Reading Worksheet* GENERAL CRF Critical Thinking Worksheet* ADVANCED CRF Science Skills Worksheet* GENERAL	SE Section 3 Review ■ TE Reteaching, p. 400 BASIC TE Quiz, p. 400 GENERAL TE Alternative Assessment, p. 400 GENERAL CRF Quiz* ■ BASIC Modified Worksheet* BASIC	SC.G.2.4.6, SC.H.3.4.3

Holt Biology Interactive Tutor CD-ROM

This CD-ROM consists of interactive activities that give students a fun way to extend their knowledge of biology concepts.

Unit 7 Ecosystem Dynamics

SCILINKS. NSTA
www.scilinks.org
Maintained by the **National Science Teachers Association**

Topic: Greenhouse Effect
SciLinks Code: HX4094

Topic: Chemical Pollution
SciLinks Code: HX4039

Topic: Aquifers
SciLinks Code: HX4013

Topic: Population Growth
SciLinks Code: HX4144

Topic: Solving Environmental Problems
SciLinks Code: HX4166

CHAPTER 18

Overview

Tell students that the purpose of this chapter is to explain how human activities affect the quality of Earth's environment; specifically how chemical pollution, loss of natural resources, and human population growth affect the environment. They will also learn about world-wide efforts being undertaken to reduce pollution, as well as steps that they can take to improve or preserve their environment.

Using the Figure

Instruct students to read the legend on the opposite page that corresponds to this figure. Tell them humans have depleted numerous kinds of natural resources and have reduced the quality of the environment in many ways, and that the story of the bald eagle is a typical example. Explain that it is also an example of a "success" story, illustrating how understanding a problem, thoughtful planning, and dedicated efforts can reverse environmental damage.

Assessing Prior Knowledge

Before beginning this chapter, be sure your students can answer the questions in the ***Quick Review*** on the facing page. If they cannot, encourage them to reread the sections indicated.

Bald eagle

384

Chapter Correlations **Florida Sunshine Standards/FCAT**

SC.G.2.4.1 Layers of energy-rich organic materials have been gradually turned into great coal beds and oil pools (fossil fuels) by the pressure of the overlying earth, and humans burn fossil fuels to release the stored energy as heat and carbon dioxide.

SC.G.2.4.6 Humans today are placing their environmental support systems at risk in various ways (e.g., rapid human population growth, environmental degradation, and resource depletion).

SC.H.3.4.3 Scientists can bring information, insights, and analytical skills to matters of public concern and help people understand the possible causes and effects of events.

CHAPTER

18 The Environment

Florida Sunshine Standards/FCAT

SC.G.2.4.1 Layers of energy-rich organic materials have been gradually turned into great coal beds and oil pools (fossil fuels) by the pressure of the overlying earth, and humans burn fossil fuels to release the stored energy as heat and carbon dioxide.

SC.G.2.4.6 Humans today are placing their environmental support systems at risk in various ways (e.g., rapid human population growth, environmental degradation, and resource depletion).

Quick Review

Answer the following without referring to earlier sections of your book.

1. **Define** *pH* and describe acid rain. *(Chapter 1, Section 3)*
2. **Identify** the role of the ozone layer in Earth's atmosphere. *(Chapter 12, Section 1)*
3. **Describe** the relationship between the long-term survival of species and the resources on which they depend. *(Chapter 15, Section 1)*
4. **Summarize** the events of the water cycle and the carbon cycle. *(Chapter 16, Section 3)*

Did you have difficulty? *For help, review the sections indicated.*

Reading Activity

Before you read this chapter, write a short list of environmental issues and efforts that you are familiar with. Then, write a list of questions about the environment and environmental issues. Save your list, and to assess what you have learned, see how many of your own questions you can answer after reading this chapter.

Looking Ahead

The bald eagle was a familiar sight in colonial America. By 1963, however, fewer than 1,000 bald eagles remained in the lower 48 states—their numbers diminished by habitat loss, sport hunting, and, in the mid-twentieth century, the pesticide DDT. After efforts to protect and manage our national bird, the bald eagle has made a remarkable comeback.

internet connect

www.scilinks.org

National Science Teachers Association *sci*LINKS Internet resources are located throughout this chapter.

SCILINKS. Maintained by the National Science Teachers Association

385

Opening Activity

Humans and the Environment

So much has been written and said about the negative impact of humans on the environment that students may be "turned off" when the issue is raised. To stimulate their interest, show them a picture that reveals the magnitude of the problem. For example, show them a picture of the Fresh Kills landfill in New York State, and emphasize the enormity of the problem at this site. It covers nearly 20 km^2 (8 mi^2), and it receives 34 million pounds of garbage every day.

Quick Review Answers

1. The term *pH* refers to a measure of the concentration of hydrogen ions in a solution, which indicates its acidity or alkalinity. The term *acid rain* refers to rainfall with an acidic pH, usually by the mixing of water in clouds with certain gaseous pollutants.
2. The ozone layer in the Earth's upper atmosphere absorbs ultraviolet radiation, thus reducing the amount of UV radiation that reaches the Earth's surface.
3. The long-term survival of species depends on the availability of resources in the environment that they can use to meet their needs for food, shelter, or other activities.
4. In the water cycle, water moves between living and nonliving things in the environment: It falls as precipitation and either evaporates from bodies of water, is stored in ground water, or cycles through plants and then evaporates. In the carbon cycle, carbon moves between living and nonliving things in the environment: Carbon enters the living portion of the cycle through photosynthesis. Organisms release carbon through cellular respiration, Carbon trapped in rocks and fossil fuels is released by erosion and burning.

Chapter Resource File

- Vocabulary Worksheets
- Concept Mapping

Reading Activity Answers

Students will likely be familiar with issues such as environmental pollution, global warming, and human population growth. They may have only general knowledge about these issues and wish to learn more details. They also may wish to learn more about local issues in their immediate environment.

Section 1

Focus

Overview

Before beginning this section review with your students the objectives listed in the Student Edition. Section 1 identifies the causes and effects of acid rain and the consequences of atmospheric ozone depletion. It also explains how the burning of fossil fuels alters the atmosphere, and it clarifies the connection between the greenhouse effect and global warming.

Bellringer

Have students brainstorm examples of solid, liquid, and gaseous pollutants in the environment.

Motivate

Discussion/ Question — GENERAL

Ask students why they think acid rain has long-term effects. Explain that the United States, Canada, and many European nations have reduced sulfur emissions in recent years, but that the lakes and forests damaged by acid rain have not recovered as quickly as hoped. (One reason is that acid rain alters soil chemistry, an effect which may take many decades to reverse.)

Transparencies

TR Bellringer

Section Correlations

SC.G.2.4.1
SC.G.2.4.6

Section 1 Global Change

Florida Sunshine Standards/FCAT
SC.G.2.4.1 Layers of energy-rich organic materials have been gradually turned into great coal beds and oil pools (fossil fuels) by the pressure of the overlying earth, and humans burn fossil fuels to release the stored energy as heat and carbon dioxide.

Objectives

- **Recognize** the causes and effects of acid rain. SC.G.2.4.6
- **Evaluate** the long-term consequences of atmospheric ozone depletion. SC.G.2.4.6
- **Explain** how the burning of fossil fuels has changed the atmosphere. SC.G.2.4.1
- **Analyze** the proposed relationship between the greenhouse effect and global warming. SC.G.2.4.6

Key Terms

acid rain
chlorofluorocarbon
global warming
greenhouse effect

The Atmosphere and Ecosystems

You may be surprised to learn that some kinds of human activity can ultimately influence every ecosystem on Earth. Human-induced environmental changes are affecting ecosystems worldwide and may lead to global change.

Acid Rain

Coal-burning power plants send smoke high into the atmosphere through tall smokestacks, often more than 300 m (984 ft) tall. This smoke contains high concentrations of sulfur because the coal that the plant burns is rich in sulfur. The intent of those who designed the power plant was to release the sulfur-rich smoke high into the atmosphere, where winds would disperse and dilute it. Tall smokestacks, first introduced in the mid-1950s, are common in the United States and in Europe.

Scientists have since discovered that the sulfur introduced into the atmosphere by smokestacks can combine with water vapor to produce sulfuric acid. Rain and snow carry the sulfuric acid back to Earth's surface. This acidified precipitation is called **acid rain.** In North America, acid rain is most severe in the northeastern United States and in southeastern Canada, areas that are downwind from coal-burning plants in the Midwest. Recall that the pH of pure water is 7.0. In the northeastern United States, rain and snow have an average pH of about 4.0–4.5. This is over 10 times as acidic as the typical pH values for precipitation in the rest of the United States.

What is the impact of acid rain? Rainwater and some soils are naturally slightly acidic. Research suggests, however, that the acidity added by human activity is having a dramatic effect. In the United States and Canada, thousands of lakes are "dying" as their pH levels fall below 5.0. Forests in the eastern United States and southern Canada are being damaged. The acid pH may be harming symbiotic fungi in tree roots. The trees shown in **Figure 1** may have been affected by acid rain along with other factors.

Figure 1 One effect of acid rain. Acid rain may have played a role in the death of these trees on Mount Mitchell, North Carolina. Scientists suspect that other factors, including infestation by an insect pest, also contributed to the trees' decline.

386

CHEMISTRY CONNECTION

Sulfur released by smokestacks reacts with oxygen in the atmosphere to form sulfur trioxide: $2S + 3O_2 \rightarrow 2SO_3$. In turn, the sulfur trioxide reacts with water vapor in the atmosphere to form sulfuric acid: $SO_3 + H_2O \rightarrow H_2SO_4$. Burning coal also produces sulfur dioxide, which reacts with water to form sulfurous acid: $SO_2 + H_2O \rightarrow H_2SO_3$.

Integrating Physics and Chemistry

As the text indicates, rain in the northeastern United States has a pH of approximately 3.8. If this pH is about 100 times more acidic than precipitation in the rest of the United States, and each whole number on the pH scale represents a 10-fold shift in an acid's concentration, what would be the average pH of typical U.S. rainfall? In order to simplify the problem, round the 3.8 pH to 4.

The Ozone Layer

You may recall from your reading that organisms were able to leave the oceans and colonize the land only after a protective shield of ozone, O_3, had developed in the upper atmosphere. Imagine what would happen if that shield were destroyed. Alarmingly, it appears that this is just what is happening. The ozone layer is being reduced, and human activity may play a large role in its reduction.

The Ozone Hole

In 1985, a researcher in Antarctica noticed that ozone levels in the atmosphere seemed to be as much as 35 percent lower than the average values during the 1960s. Satellite images taken over the South Pole revealed that the ozone concentration was unexpectedly lower over Antarctica than elsewhere in the Earth's atmosphere, as shown in **Figure 2.** It was as if an "ozone eater" were causing a mysterious zone of below-normal concentration, an area that researchers called the ozone hole. Alarmed, scientists examined satellite images taken in previous years. They found that the disintegration of the Earth's ozone shield was evident as far back as 1978. Every year since then, more ozone has disappeared, and the ozone hole has grown larger. Moreover, a smaller hole has appeared over the Arctic.

Because the decrease in ozone allows more ultraviolet radiation to reach the Earth's surface, scientists expect an increased incidence of diseases caused by exposure to ultraviolet radiation. These diseases include skin cancer, cataracts (a disorder in which the lens of the eye becomes cloudy), and cancer of the retina, the light-sensitive part of the eye. In fact, in the United States, the number of cases of malignant melanoma, a potentially lethal form of skin cancer, has almost doubled since 1980.

Real Life

Ozone can be harmful.

In many cities and towns, summer weather reports include warnings of high ozone levels. Ozone at ground level irritates people's nasal passages, throats and lungs and may harm the immune system.

Finding Information

Find out why ground-level ozone is increasing even though the ozone layer in the upper atmosphere is decreasing.

What Is Destroying Ozone?

The major cause of ozone destruction is a class of chemicals called **chlorofluorocarbons** (CFCs). Invented in the 1920s, CFCs were considered extremely stable, supposedly harmless, and a nearly ideal heat exchanger. Throughout the world, CFCs were commonly used as coolants in refrigerators and air conditioners, as aerosol propellants in spray cans, and as foaming agents in the production of plastic-foam cups and containers. Though CFCs were escaping into the atmosphere, at first no one worried.

By 1985, the scientific community had learned that CFCs are the primary cause of the ozone hole. High in the atmosphere, ultraviolet radiation from the sun is able to break the usually stable bonds in CFCs. The resulting free chlorine atoms then enter into a series of reactions that destroy ozone. As a result of this discovery, CFCs have been banned as aerosol propellants in spray cans in the United States. Today many countries limit or ban the use of CFCs.

Figure 2 Ozone "hole" over Antarctica. In this satellite view of the South Pole, the pink area indicates the region with the least amount of ozone.

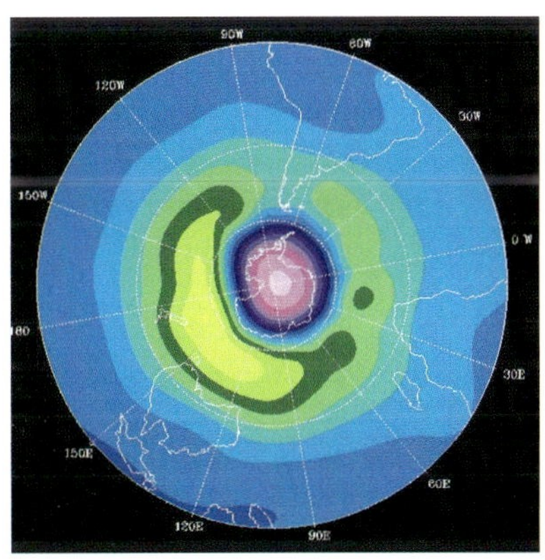

Teach

Teaching Tip — BASIC

UV Damage Tell students that scientists estimate that every 1 percent drop in upper atmosphere ozone concentration will lead to a 6 percent rise in skin cancer. But perhaps most seriously, certain marine algae (the producers in marine ecosystems) are very sensitive to UV levels. Reductions in algal populations would have a severe impact on marine ecosystems. Have students draw a Graphic Organizer similar to the one on the bottom of this page that shows the impact on human health of releasing CFCs into the atmosphere.

Real Life

Answer

Ground-level ozone is increasing because it is produced when sunlight strikes car exhaust, and there are more cars on the road each year. Ground-level ozone rarely makes it to the upper atmosphere, where ozone levels are decreasing, possibly due to the effects of CFCs.

Chapter Resource File

- Directed Reading BASIC
- Active Reading GENERAL
- Data Sheet for Quick Lab GENERAL

INTERACTIVE BIOLOGY TUTOR

- Unit 7—Ecosystem Dynamics
  This engaging tutorial covers ecosystems, population growth, succession, food chains and food webs, and the importance of chemical cycles to the health of ecosystems.

One-Stop Planner CD-ROM

- Reading Organizers BASIC
- Reading Strategies BASIC
- Occupational Applications
  Wildlife Biologist GENERAL

Graphic Organizer

Use this graphic organizer with **Teaching Tip** *on this page.*

CFCs escape into atmosphere → CFCs destroy ozone → More UV radiation reaches Earth's surface → Skin cancer cases increase

Teach, *continued*

QUICK LAB

Modeling the Greenhouse Effect

Skills Acquired
Observing, interpreting results, applying information, predicting outcomes

Teacher's Notes
If an MBL/CBL system is not available, have students record temperature manually every minute for 15 minutes. Use lamps as a heat source if direct sunlight is not available.

Answers to Analysis

1. The jar's glass kept heat from escaping, raising the temperature inside the jar.
2. Like the glass of the jar, carbon dioxide retains heat energy from the sun that would otherwise dissipate.
3. The temperature of the Earth would be much colder.

LS Kinesthetic

Integrating Physics and Chemistry

Remind students that solar energy is transferred to the earth by means of radiation in the form of electromagnetic waves. When the solar energy penetrates the earth's atmosphere, it warms air molecules such as CO_2. Once these molecules have absorbed the heat energy, they can pass the energy on to other molecules by conduction (direct contact) or convection currents (the movement of the heated substance itself).

Global Temperatures

The average global temperature has been steadily increasing for more than a century, particularly since the 1950s. In Earth's long history there have been many such periods of **global warming,** often followed by centuries of cold. Scientists hypothesize that sunspot cycles may contribute to these cyclical changes in global temperature. Many scientists suspect, however, that human activity may be significantly contributing to global warming in modern times.

The Greenhouse Effect

Our planet would be as cold as the moon except for the insulating effects of certain gases—called greenhouse gases—such as water vapor, carbon dioxide (CO_2), methane, and nitrous oxide in the atmosphere. The chemical bonds in carbon dioxide molecules absorb solar energy as heat radiates from Earth. As shown in **Figure 3,** this process, called the **greenhouse effect,** traps heat within the atmosphere in the same way glass traps heat within a greenhouse. There has been a large increase in carbon dioxide in the Earth's atmosphere in recent times. This increase seems to be related to the burning of fossil fuels that has accompanied clearing of forests and urban industrialization.

Is Global Warming Occurring?

Figure 4 shows the average change in global temperature since 1960, and, on the same graph, the average concentration of atmospheric carbon dioxide. The correlation of increasing temperatures with increasing carbon dioxide levels is very close. Therefore many scientists are convinced temperature and carbon dioxide levels are related.

Figure 3 Earth's atmosphere traps heat. Just as the glass panes of a greenhouse retain heat, CO_2 and other greenhouse gases in the atmosphere capture heat radiated from Earth's surface.

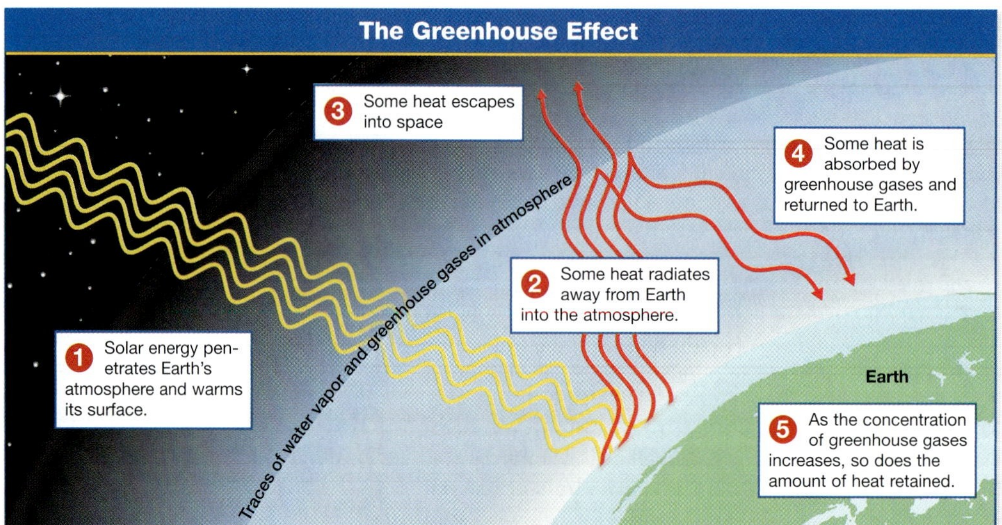

388

METEOROLOGY CONNECTION

If the global warming trend continues, many scientists think average global temperatures will increase at least 2°C (about 4°F) by 2050. This increase would probably be enough to trigger droughts in some inland regions as rainfall patterns change. It would also cause flooding in coastal regions as the polar icecaps melt. Powerful storms may become more frequent worldwide due to the increased energy in the lower atmosphere.

Attention Grabber

Greenhouse Effect on Venus A greenhouse effect also exists on Venus but to a much greater extent than on Earth. Early in their formation, Earth and Venus had similar concentrations of carbon dioxide, but Earth's oceans and plants eliminated much of the gas from the atmosphere. Venus lacks Earth's oceans and plants, so the dense carbon dioxide gas remains, causing surface temperatures to reach 460°C (865°F).

In science, however, correlation does not prove cause and effect. Both global temperature and levels of greenhouse gases may be changing because of other variables that have not been recognized yet. Some countries take seriously the possibility that increasing greenhouse gases play a role in global warming. These countries are attempting to formulate international treaties that place limits on greenhouse-gas emissions. The matter remains controversial, however, and the role of greenhouse gases in global warming is hotly contested.

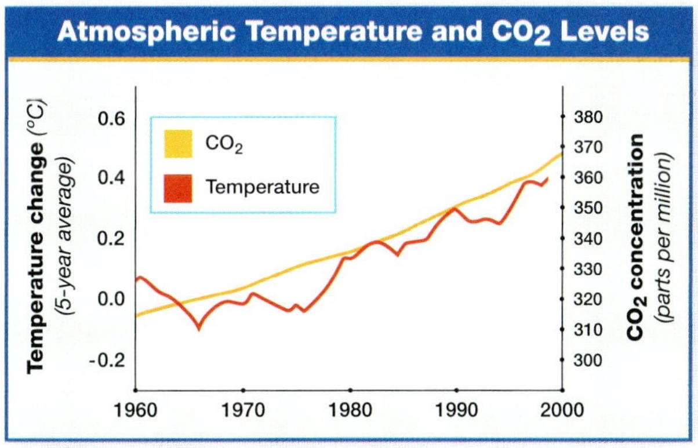

Figure 4 **Global changes.** Atmospheric CO_2 concentration and the average change in global temperature have risen together over the past several years.

QUICK LAB

Modeling the Greenhouse Effect SC.G.2.4.6

You can use a quart jar to explore the greenhouse effect.

Materials

MBL or CBL system with appropriate software, 2 temperature probes, 1qt jar, lid with a 0.5 cm hole in the center, tape, heat source

Procedure

1. Set up an MBL/CBL system to collect data from two temperature probes at 6 second intervals for 150 data points.
2. Insert the end of one probe into the hole in the lid of a quart jar, and tape the probe in place. Place the other probe about 4 in. from the jar and at the same height as the first probe.
3. Place the jar about 30 cm from a heat-radiating source, and begin collecting data.
4. After 5 minutes, turn off (or remove) the heat source. Collect data for another 10 minutes.

Analysis

1. **Propose** an explanation for any differences between the two probes.
2. **Critical Thinking** **Comparing Functions** How does carbon dioxide gas in the atmosphere function in a way similar to the glass jar?
3. **Critical Thinking** **Predicting Outcomes** How would the temperature on Earth be different if there were no carbon dioxide in the atmosphere?

Section 1 Review

1. **Summarize** two harmful effects of acid rain. SC.G.2.4.6
2. **Differentiate** the greenhouse effect from global warming. SC.G.2.4.6
3. **Critical Thinking** **Predicting Outcomes** How might the depletion of the ozone layer affect human health? SC.G.2.4.6
4. **Critical Thinking** **Evaluating Models** Is the greenhouse-gases model adequate to explain all warming of Earth over its existence? Why or why not? SC.G.2.4.6
5. **FCAT Prep** Which activity might reduce the severity of the greenhouse effect?
   A burning vegetation **C** planting more trees
   B cutting trees **D** promoting decomposition
   SC.G.2.4.6

Answers to Section Review

1. Acid rain can weaken or kill trees and reduce the pH of lakes and streams, sometimes to levels deadly to most aquatic life.
2. The greenhouse effect is the trapping of heat within the Earth's atmosphere by gases such as carbon dioxide. Global warming is an increase in the average global temperature of the Earth, which could possibly be caused by the greenhouse effect.
3. Ozone depletion could result in an increased incidence of skin and retinal cancer and cataracts.
4. No. It is likely that other factors, such as sun spots, have intermittently increased the temperature of the Earth.
5. **A.** Incorrect. Burning vegetation releases CO_2 into the air. **B.** Incorrect. Trees remove CO_2 from the air. Cutting trees will contribute to the greenhouse effect. **C.** Correct. Trees remove CO_2 from the air. **D.** Incorrect. Decomposing organisms release CO_2 into the air.

Close

Reteaching — BASIC

Invite a guest speaker to address how your community might be affected by one of the following global changes: acid rain, ozone depletion, or global warming. Have students prepare questions ahead of time.

Quiz — GENERAL

1. Identify one environmental pollutant that can contribute to the formation of acid rain. (sulfur)
2. What is the "ozone hole"? (It is a region over the South Pole where the ozone layer is depleted.)
3. What are two greenhouse gases? (carbon dioxide, methane, nitrous oxide)

Alternative Assessment — ADVANCED

Ask students to write a short essay on ecosystem interdependence and to hypothesize how damage to one ecosystem can affect other ecosystems. (For example, destruction of the rain forest in South America can destroy the wintering grounds of migratory birds, affecting North American ecosystems where the birds spend summers.)

Transparencies

TR E30 Ozone "Hole" over Antarctica
TR E29 The Greenhouse Effect
TR E31 Atmospheric Temperature and Carbon Dioxide Levels
TR E32 Change in Global Temperature

Section Review Correlations

SC.G.2.4.6: Items 1-5

Focus

Overview

Before beginning this section review with your students the objectives listed in the Student Edition. In this section, students will learn how the release of chemical pollutants can harm both living and nonliving parts of the environment, how nonreplaceable resources are depleted, and the potential consequences of uncontrolled population growth.

Bellringer

Ask students to list examples of chemical pollution from industry, agriculture, or even everyday use. Then, write their examples on the blackboard or an overhead transparency and discuss the possible effects of each pollutant with the class.

Motivate

Activity — GENERAL

Display pictures of different forms of environmental pollution, such as water pollution, air pollution, and solid waste. Give students old magazines and have them cut out examples of pollution. Have students identify the causes and effects of each type of pollution. For example, sewage, agricultural runoff, and industrial waste cause water pollution. Polluted water may harm the plants, animals, and humans that use it. LS **Visual**

Section 2 Effects on Ecosystems

Florida Sunshine Standards/FCAT

SC.G.2.4.1 Layers of energy-rich organic materials have been gradually turned into great coal beds and oil pools (fossil fuels) by the pressure of the overlying earth, and humans burn fossil fuels to release the stored energy as heat and carbon dioxide.

Objectives

- **Describe** the effects of chemical pollutants on the environment. SC.G.2.4.6
- **Identify** three nonrenewable resources. SC.G.2.4.1
- **Predict** the potential consequences of uncontrolled population growth. SC.G.2.4.6
- **Contrast** population growth in developing countries with that in industrialized countries. SC.G.2.4.6

Key Terms

biological magnification
aquifer

Effects of Chemical Pollution

Examples of global change such as ozone depletion or global warming occur on a large scale. But what about serious environmental problems that occur in our own backyard? For example, one important urban environmental problem is chemical pollution. Until recently, people assumed that the environment can absorb any amount of pollution. Lake Erie and other large lakes became polluted because of the assumption that they could absorb unlimited amounts of industrial chemicals. Because of this incorrect assumption, pollution has often risen to a serious level.

In a highly publicized example of pollution, a very large oil tanker ran aground off the coast of Alaska in 1989. Oil from the tanker heavily polluted 1,600 km (1,000 mi) of coastline and injured or killed thousands of marine animals, such as the one shown in **Figure 5.** Despite costly and heroic cleanup efforts, damage to local wildlife was extensive. This dramatic example is not an isolated occurrence. Smaller oil spills and leaks that receive little or no publicity account for more than 90 percent of all pollution from oil seepage.

Many of the most disastrous incidents of pollution involve industrial chemicals that are toxic or carcinogenic (cancer-causing). Until recently, there has been relatively little regulation of the manufacture, transportation, storage, and destruction of such chemicals. A particularly clear example of this problem occurred in Basel, Switzerland, in 1986. Firefighters putting out a warehouse fire accidentally washed 27,000 kg (30 tons) of mercury and pesticides that were stored in the warehouse into the Rhine River. These poisons flowed down the Rhine, through Germany and the Netherlands, and into the North Sea, killing fish and other aquatic animals and plants. Today the river is recovering, but its species diversity remains far lower than it was before the disaster.

Figure 5 Oil spill victim. This common scoter was one of thousands of animals injured or killed when an accident released a large volume of oil off the Alaskan coast in 1989.

Agricultural Chemicals

In many countries, modern agriculture introduces large amounts of chemicals into the global ecosystem. These chemicals include pesticides, herbicides, and fertilizers. Industrialized countries, like the United States, now attempt to carefully monitor side effects of these chemicals. Unfortunately, large quantities of many toxic chemicals that are no longer manufactured still circulate in the ecosystem.

390

Chapter Resource File

- Directed Reading BASIC
- Active Reading GENERAL

One-Stop Planner CD-ROM

- Reading Organizers BASIC
- Reading Strategies BASIC
- Supplemental Reading Guide *Silent Spring* ADVANCED

Transparencies

TR Bellringer
TR E33 Biological Magnification of DDT

Section Correlations

SC.G.2.4.1
SC.G.2.4.6

Pesticides For example, molecules of chlorinated hydrocarbons—a class of compounds that includes the pesticides DDT, chlordane, lindane, and dieldrin—break down slowly in the environment. They also accumulate in the fatty tissue of animals. As these molecules pass up through the trophic levels of the food chain, they become increasingly concentrated. This process is called **biological magnification,** illustrated in **Figure 6.** The presence of DDT in birds causes thin, fragile eggshells, which can break during incubation. Because of DDT use, many predatory birds in the United States and elsewhere failed to reproduce, and their numbers dwindled. In 1972, the use of DDT was severely restricted in the United States, and threatened bird populations slowly began to increase. However, chlorinated hydrocarbons are still manufactured in the United States and exported to other countries, where their use continues.

In order for us to meet the needs of an increasingly crowded world, the use of chemicals is necessary. We must learn to use them as intelligently as possible. Doing so will enable us to protect the productive capacity of the Earth. Failure is not a rational option.

internet connect

www.scilinks.org
Topic: Chemical Pollution
Keyword: HX4039

SCILINKS. Maintained by the National Science Teachers Association

Figure 6 **Biological magnification of DDT**

Because DDT accumulates in fatty tissue, DDT concentrations (in parts per million, ppm) increase as this chemical moves up the food chain.

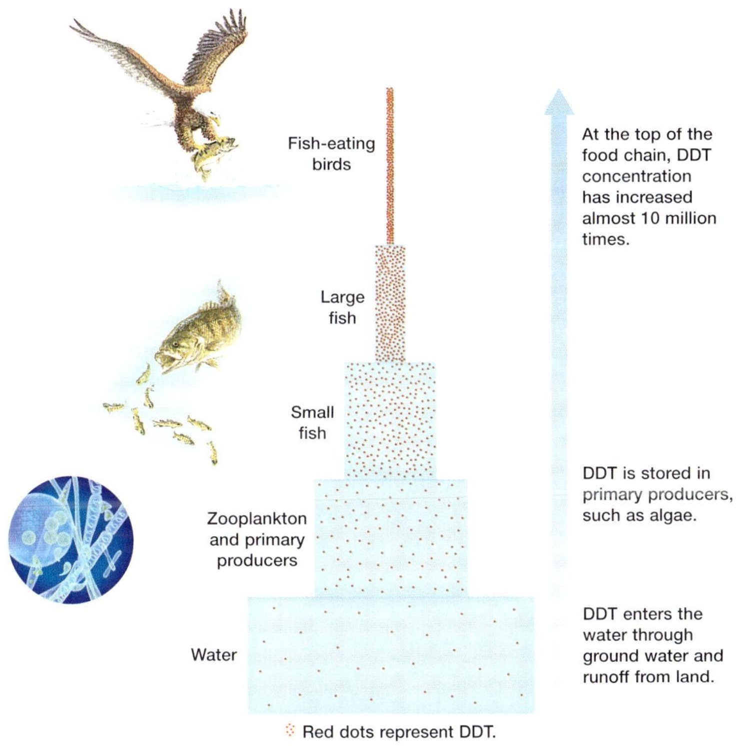

Teach

Teaching Tip — GENERAL

Agriculture and Water Pollution
Ask students what they think is the primary source of water pollution. They may be surprised to learn that agriculture, not industry, has the greatest impact on water quality. Explain that the runoff of silt, pesticides, and fertilizers from farmland is difficult to control because it comes from so many sources rather than a single point, such as a pipe or a smokestack. That is why such pollution is known as *nonpoint-source* pollution.

READING SKILL BUILDER — BASIC

Reading Hint As students read this section, have them write down the main points and briefly describe the impact or importance of each. Tell them to look for section headings to help identify the main points. **LS Logical**

Teaching Tip — GENERAL

Chemical Pollution Have students research chemical pollution using the Web site in the Internet Connect box on this page. Have students write a short report summarizing their findings. **LS Verbal**

Teaching Tip — ADVANCED

Ground-water Contamination
Ask students to suggest how chemicals might pollute the ground water in urban areas. Explain that underground storage tanks, which hold liquid fuels and industrial chemicals below ground, often leak. In the United States, more than one quarter million underground tanks at service stations are thought to be leaking. Disposing of the tanks and contaminated soil is often very costly and difficult. As an extension, have interested students investigate the methods and costs entailed in disposal of ground water contaminated with gas or oil and report their findings to the class.

ENVIRONMENTAL SCIENCE CONNECTION

When rainfall washes nitrogen-containing fertilizers or plant and animal wastes into a body of fresh water, the ecosystem can be seriously disrupted. Large amounts of nitrates and phosphates stimulate explosive algal growth. Only the topmost algal layer receives enough light for photosynthesis, while the lower layers die and are decomposed by bacteria. These bacteria use up large amounts of dissolved oxygen, which slowly suffocates fish and other lake organisms.

Teach, *continued*

SKILL BUILDER — Advanced

Writing Skills Tell students that the Endangered Species Act of 1973 is the most powerful legislation protecting rare and endangered species in the United States. Have students write a report that summarizes the provisions of this law and discusses which species it protects in their area. **LS Verbal**

READING SKILL BUILDER — Basic

Interactive Reading Assign Chapter 18 of the *Holt Biology* Guided Audio CD Program to help students achieve greater success in reading the chapter. **LS Auditory**

English Language Learners

SKILL BUILDER — Basic

Math Skills Inform students that the rate of extinction today is not known because no one knows how many species exist. About 1.6 million species have been described and named. Estimates of the total number of species on Earth range from 5 million to 100 million. Tell students to assume that there are 10 million species on Earth and to calculate how many species would remain if 20 percent of them became extinct in the next 50 years. (8 million) Then ask how many species would remain if the same rate of extinction occurred for the following 50 years. (6.4 million) **LS Logical**

English Language Learners

Study TIP

Organizing Information
Make a table to organize information about *loss of resources.* Across the top, write the headings *Extinction of species, Loss of topsoil,* and *Ground-water pollution and depletion.* Along the sides, write *Causes, Effects,* and *Possible solutions.* Add information to the table as you read this chapter.

Loss of Resources

Among the many ways that ecosystems are being damaged, one problem stands out as potentially more serious than the rest—the consumption or destruction of resources that we cannot replace. Though a polluted stream can be cleaned up, no one can restore an extinct species. Three kinds of nonrenewable resources are being consumed or destroyed at alarming rates: species of living things, topsoil, and ground water.

Extinction of Species

Over the last 50 years, about half of the world's tropical rain forests have been burned to make pasture and farmland or have been cut for timber. Many thousands of square miles more will be destroyed this year. The people responsible, often poor farmers, view the forest lands as a resource to be developed, much as Americans viewed North American forests a century ago.

The problem is that as the rain forests disappear, so do their inhabitants. No one knows how many species are being lost. To find out, scientists carefully catalogue all of the residents of one small segment of forest and then extrapolate their data. That is, scientists use what they know to predict what they don't know. The resulting estimates vary widely, but it is clear Earth is losing many species. Some 10 percent of well-known species teeter on the brink of extinction. Worst-case estimates are that we will lose up to one-fifth of the world's species of plants and animals—about 1 million species—during the next 50 years. An extinction of this size has not occurred in at least 65 million years, since the end of the age of dinosaurs.

Figure 7 Beneficial species. Two potent anticancer drugs have been isolated from the leaves of the rosy periwinkle, a native of Madagascar.

The tragedy of extinction is that as species disappear, so do our chances to learn about them and their possible benefits. This situation is comparable to burning a library before reading the books—we lose forever the knowledge we might have gained. For example, two potent anticancer drugs have been isolated from the rosy periwinkle, shown in **Figure 7,** a flower that occurs naturally only on Madagascar, an island being devastated by deforestation.

internet connect
www.scilinks.org
Topic: Florida Parks and Preserves
Keyword: HX6F03
SCILINKS. Maintained by the National Science Teachers Association

MISCONCEPTION ALERT

Preservation of Species Students may think that the best way to preserve a species is to set aside habitat as a park or reserve. Clarify that "locking away" large tracts of land may not be practical, especially in developing countries with large, rapidly growing populations that depend on the land for subsistence. A better solution may be *extractive reserves,* which preserve habitats while allowing the sustainable harvest of products such as fruits, seeds, and rubber. Local populations need a sustainable income, which depends on keeping the habitat intact for all species.

Loss of Topsoil

The United States is one of the most productive agricultural countries on Earth, largely because of its fertile soils. These soils have accumulated over tens of thousands of years. The Midwestern farm belt sits astride what was once a great prairie. The topsoil of that ecosystem accumulated slowly as the remains of countless animals and plants decayed. By the time humans came to plow the prairie, the topsoil was more than a meter thick.

This rich topsoil cannot be replaced, and it is being lost at a rate of several centimeters each decade. Turning over the soil to eliminate weeds, allowing animals to overgraze ranges and pastures, and practicing poor land management all permit wind and rain to remove more and more of the topsoil. Since 1950, the world has lost one-third of its topsoil, primarily because of human activity.

Ground-Water Pollution and Depletion

A third resource that we cannot replace is ground water. Much ground water is stored within porous rock reservoirs called **aquifers** *(AHK wuh furz)*, as shown in **Figure 8.** Water seeps into aquifers too slowly to replace the large amount of water now being withdrawn. In most areas of the United States, there is relatively little effort to conserve ground water. Consequently, a very large portion of it is wasted on watering lawns, on washing cars, and through leaky and inefficient faucets and toilets. A great deal more ground water is being polluted by irresponsible disposal of chemical wastes. Once pollution enters the ground water, there is no effective way to remove it.

internet connect

www.scilinks.org
Topic: Aquifers
Keyword: HX4013

SCILINKS. Maintained by the National Science Teachers Association

Figure 8 Aquifer

Large amounts of ground water are being removed from many aquifers far faster than natural processes can replenish it.

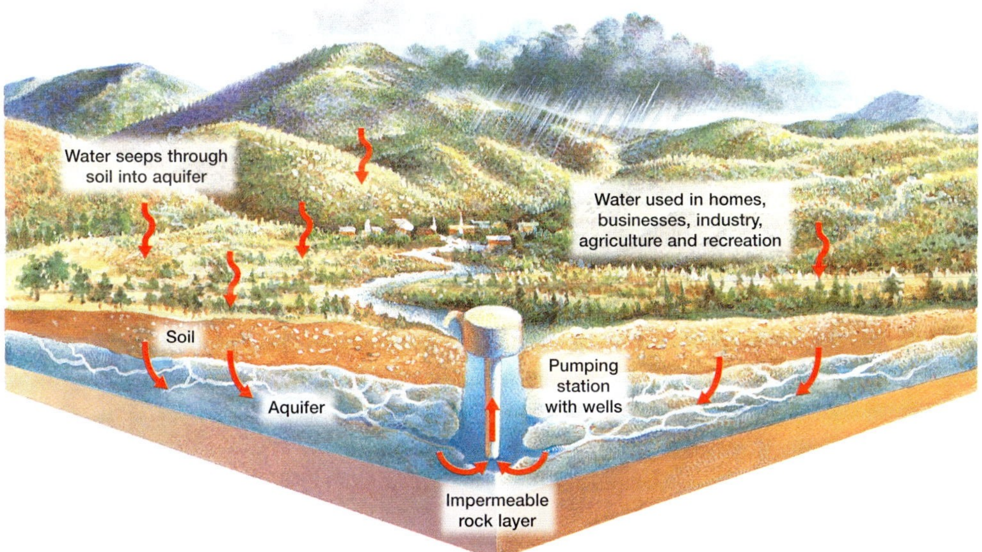

Teaching Tip — GENERAL

Topsoil Point out to students that soil typically has three layers. The top layer (topsoil) contains the most organic matter, or humus. The second layer (subsoil) consists of inorganic particles and minerals that have washed down from the topsoil. The bottom layer consists of loose rock that extends down to the bedrock.

Teaching Tip — ADVANCED

Topsoil in the Mississippi River Basin Approximately 1 million tons of topsoil are lost each year in the Mississippi River Basin. Most of this topsoil washes into the Mississippi River and is carried downstream to the Gulf of Mexico. Have students use library resources to find out more about the loss of topsoil in this area, and ask them to summarize their findings in a newspaper article or TV news bulletin.

Group Activity — GENERAL

Ground Water in Your Back Yard Assign students to work in five groups. Have students in each group conduct research using library, Internet, and/or local resources on one of the following issues in their home city, region, or state: (1) characteristics of the ground water including its locations; (2) the sources and types of pollutants reaching the ground water; (3) quality issues in the ground water; (4) factors depleting the ground water; and (5) possible future threats to ground-water quality and/or quantity (e.g., from projected increases in population size). Have each group prepare an oral presentation on their topic and present it to the class.

Co-op Learning — English Language Learners

did you know?

Sustainability The idea of *sustainability* is key to preventing the loss of our natural resources. When a resource is used sustainably, it is consumed only as fast as it is naturally replenished, and therefore it remains abundant. Some resources, such as fossil fuels, ground water, and topsoil, are replenished on a time scale of thousands or millions of years. The creation of these resources proceeds far too slowly to keep up with today's rapid rate of consumption. Living resources, such as trees, are easier to use sustainably because they replace themselves more quickly.

Teach, *continued*

SKILL BUILDER — Basic

Interpreting Graphics Refer students to **Table 1**. Ask them how much time it took for the world's population to double from 1 billion to 2 billion. (130 years) Ask them how much time it took for the world's population to double from 2 to 4 billion. (44 years) Ask them to project how long it will take for the world's population to double from 4 billion to 8 billion. (47 years) English Language Learners
LS Logical

SKILL BUILDER — Basic

Graphing Have students prepare line graphs illustrating the data presented in **Table 1.** Instruct them to divide the horizontal axis into even time intervals (e.g, 10 years each) from 1800 through 2000 and to label it, and to divide the vertical axis into even number intervals from 0 through 6 (billion). Ask them to draw a curved line connecting the dots on the graph. English Language Learners
LS Logical

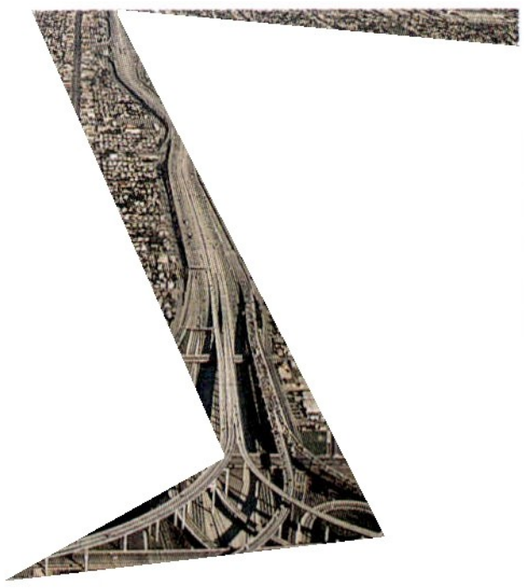

Figure 9 **Population growth.** The world's population has topped 6 billion.

internet connect
www.scilinks.org
Topic: Population Growth
Keyword: HX4144
SciLinks. Maintained by the National Science Teachers Association

Growth of the Human Population

If all of the problems associated with pollution mentioned earlier in this chapter were solved, would the future of our environment be free from danger? Not necessarily. Scientists recognize that we would still need to address a more fundamental problem: the rapid growth of the human population, as shown in **Table 1.**

Humans first reached North America at least 12,000 years ago, by crossing the land bridge between Siberia and Alaska. Humans then spread throughout North America and South America. Ten thousand years ago the continental ice sheets that covered northern Europe and North America receded, and agriculture soon developed. There were only about 5 million people on Earth then. As agriculture produced more dependable sources of food, the human population began to grow. By about 2,000 years ago, there were an estimated 130 million people on Earth. By 1650, the world's population had reached 500 million.

Since then, the average global birthrate has remained near 30 births per 1,000 people per year. However, with the development of technology to ensure better sanitation and improved medical care, the death rate has fallen steadily. In 2002, the estimated death rate was about 9 deaths per 1,000 people per year. The difference between the annual birthrate in 2002 (now estimated to be 21 births per 1,000 people) and death rate results in an annual worldwide increase in the human population of approximately 1.3 percent. This number may seem small, but don't be deceived. The world's population would double in just over 60 years if it continued to grow at this rate!

Table 1 Number of Years to Add 1 Billion People

Human population	Year	Years to add
1 billion	1800	All of human history
2 billion	1930	130
3 billion	1960	30
4 billion	1974	14
5 billion	1987	13
6 billion	1999	12
	Projected	
7 billion	2009	11
8 billion	2021	11

394

did you know?

Population Growth in Different Regions Human population growth rates vary greatly by region. Less-developed nations (excluding China) are growing at approximately 2.4 percent annually and will double in 30 years if no changes in fertility or mortality rates occur. The average annual rate of increase in more-developed nations is 0.5 percent, with a doubling time of 137 years. The regional contrast in age structures is even more striking. The mean fraction of the population under 15 years of age in more-developed countries is 21 percent. In less-developed countries (excluding China) it is 39 percent; in Kenya it is fully 50 percent.

Worldwide Rates of Growth

The world's population exceeded 6 billion in October 1999, and the annual increase is now about 94 million people. About 260,000 people are added to the world population each day, or about 180 every minute. Population growth is fastest in the developing countries of Asia, Africa, and Latin America. It is slowest in the industrialized countries of North America, Europe, Japan, and in New Zealand, and Australia. In industrialized countries like the United States, about one fifth of the population is under 15 years of age. In developing countries like Nigeria, the percentage is typically twice as high, which may lead to explosive population growth in coming decades. **Figure 10** shows the population growth rates of developing and developed countries. The population growth rate in the United States is only 0.8 percent. Most European countries are growing even more slowly, and the populations of Germany and Russia are actually declining. By contrast, as of 1996, Nigeria's population was increasing by about 3.05 percent per year.

The global rate of population growth has been declining. The United Nations projects that the world's population will stabilize at 9.7 billion by the year 2050. As you have read, however, population growth rates are uneven across Earth. Population growth tends to be the highest in countries that can least afford it. Already limited resources are strained further, and natural resources—ground water, land for farming, forests—are ever more quickly depleted or polluted.

No one knows whether the Earth can support six billion people indefinitely, much less the far larger population that lies in our future. Building a sustainable world is the most important task facing humanity's future. The quality of life available to your children in the new century will depend to a large extent on our success.

Figure 10 World population growth patterns. Most of the world population increase since 1950 has been in developing countries.

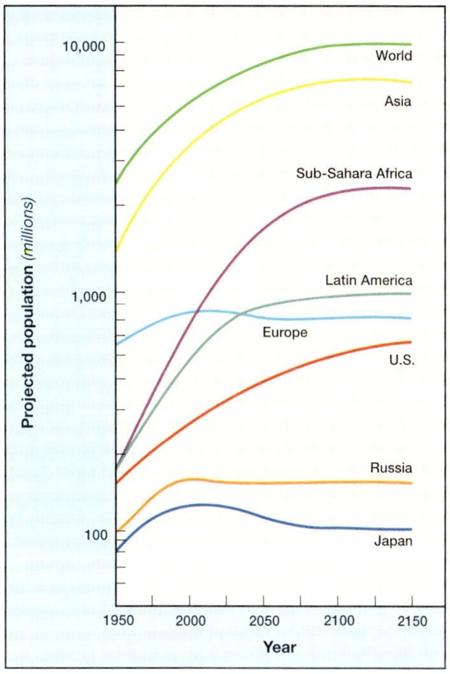

Section 2 Review

1. **Evaluate** how research showing that chlorinated hydrocarbons are an environmental threat may impact people in the United States. SC.G.2.4.6
2. **Summarize** why supplies of soil and ground water are dwindling even though these resources are replenished by natural processes. SC.G.2.4.6
3. **Critical Thinking Inferring Relationships** Describe two instances in which technology has caused the growth rate of the human population to increase. SC.H.3.4.5
4. **Critical Thinking Evaluating Viewpoints** A classmate claims that the growth of the human population will not affect populations of other species. Evaluate this viewpoint. SC.G.2.4.6
5. *FCAT Prep* Which organism would have the highest concentration of a pollutant that enters the water and is absorbed by aquatic primary producers?
   - **A** large fish
   - **B** small fish
   - **C** fish-eating bird
   - **D** zooplankton

   SC.G.2.4.6

Answers to Section Review

1. U.S. companies may stop producing products containing chlorinated hydrocarbons, resulting in the loss of jobs. People may have to avoid eating food from other countries that has been exposed to these chemicals.
2. Soil and ground-water supplies are dwindling because they are being used much faster than they are being replenished.
3. Technological advances in farming have increased the food supply dramatically, and medical advances have resulted in reduced death rates.
4. The growing human population will inevitably use more land, which other organisms need for habitat.
5. **A.** Incorrect. Fish are not the top consumer in this food chain. **B.** Incorrect. See answer A. **C.** Correct. The bird is the top consumer in this food chain and therefore will have the highest concentration of pollutants. **D.** Incorrect. Zooplankton is near the bottom of this food chain.

Close

Reteaching — BASIC

Have students prepare an outline for a movie script about an actual ecological disaster. Students should include the environmental conditions as well as the actual consequences of the disaster. (Accept any well-reasoned scenarios that are based on fact.) **LS Verbal**

Quiz — GENERAL

1. Why is the dumping of hazardous chemicals into the environment a serious concern? (Because these chemicals are toxic to a variety of living things and often circulate in the ecosystem for a long time.)
2. What is the current major cause of biological extinction? (destruction of areas where organisms live)
3. At what rate is the world's population currently increasing? (1.3% per year)

Alternative Assessment — GENERAL

Have students write an essay about which environmental concern will have the greatest impact on their community in the next decade. In their essays, students should justify their choice. (Answers will vary. For example, if the community is located near a beach resort, there would be an increase in skin cancer caused by the loss of ozone in the atmosphere, which could result in fewer people at the beach and loss of economic resources for the community.) **LS Verbal**

Transparencies

TR E34 World Population Growth Patterns

Section Review Correlations

SC.G.2.4.6: Items 1, 2, 4, 5
SC.H.3.4.5: Item 3

Focus

Overview

Before beginning this section review with your students the objectives listed in the Student Edition. In this section, students will learn about successful strategies to reduce pollution nationally and globally. They will also learn the five necessary steps to solve environmental problems, and approaches individuals can adopt to address environmental problems.

Bellringer

Instruct students to write 3–4 sentences describing a global, national, or local environmental problem and any efforts that are being undertaken to correct the problem.

Motivate

Discussion/Question

Have students discuss their answers to this question: Could an organism exist that does not have an impact on its environment in any way? (The answer is no. All organisms require substances from their environment and change the abundance of these substances by using them. Likewise, all organisms produce wastes. Students should see that humans are not alone in exploiting the environment. However, no other organism disrupts ecosystems on the scale that humans do.)

Transparencies

- TR Bellringer
- TR Concept Map

Section 3 Solving Environmental Problems

Florida Sunshine Standards/FCAT

SC.G.2.4.6 Humans today are placing their environmental support systems at risk in various ways (e.g., rapid human population growth, environmental degradation, and resource depletion).

Objectives

- **Describe** two effective approaches that have been taken to reduce pollution in the United States and abroad. SC.H.3.4.3
- **Evaluate** the five major steps necessary to solve environmental problems. SC.H.3.4.3
- **Determine** how individuals can take personal action to help solve environmental problems. SC.G.2.4.6

Key Terms

None

A Worldwide Effort

As you have seen, environmental problems affect all inhabitants of an ecosystem without regard to state or national boundaries. As human activities continue to place severe stresses on ecosystems, worldwide attention must be focused on solving these problems.

One of the most encouraging developments of the early 1990s was the global increase in efforts to reduce pollution. International agreements to stop CFC production are one example. And the release of many dangerous industrial and agricultural chemicals—notably the insecticide DDT and the carcinogens asbestos and dioxin—has been restricted in the United States.

A great deal of progress has also been made in reducing air and water pollution. The number of secondary sewage treatment facilities, which remove chemicals as well as bacteria from sewage, is on the increase. **Figure 11** shows a "scrubber," a device that reduces harmful sulfur emissions from industrial smokestacks. Emissions of sulfur dioxide, carbon monoxide, and soot—three pollutants produced by the burning of coal—have been cut by more than 30 percent in 10 years. However encouraging, this progress represents only a beginning. Serious attempts to address the overall problem of pollution have also brought about more fundamental changes in our society. For example, a reduction of the number of automobiles on the road is encouraged by providing dedicated lanes to cars with several occupants, as shown in **Figure 12.**

Figure 11 Reducing pollution

Waste water is cleaned in several steps. Scrubbers remove many pollutants from factory emissions.

Sewage treatment plant

Smokestacks with scrubbers

Chapter Resource File

- Directed Reading BASIC
- Active Reading GENERAL

One-Stop Planner CD-ROM

- Reading Organizers BASIC
- Reading Strategies BASIC
- Problem Solving Worksheet Population Size GENERAL
- Occupational Applications Worksheet Blood-Bank Technologist GENERAL

Section Correlations

SC.G.2.4.6
SC.H.3.4.3

In the United States

Two effective approaches have been taken to reduce pollution in the United States. The first approach has been to pass laws forbidding it. In the last 30 years, laws have begun to significantly slow the spread of pollution. These laws impose strict standards for what can be released into the environment. For example, all cars are required to have catalytic converters to reduce emissions. Similarly, the Clean Air Act of 1990 requires scrubbers on the smokestacks of power plants. Converters and scrubbers make cars and energy more expensive. The effect is that the consumer pays more to avoid polluting the environment.

A second effective approach to reducing pollution has been to make it more expensive by placing a tax on it. The gasoline tax is a good example of such a tax. To be fully effective, however, a tax must be high enough to reflect the actual cost of the pollution. By adjusting the tax, the government attempts to balance the conflicting demands of environmental safety and economic growth. Such taxes, often imposed on industry in the form of "pollution permits," are becoming increasingly common.

Figure 12 Carpooling. Some cities reserve certain lanes for carpool use during peak travel times.

FLORIDA BIOWatch

A Mercury Mystery in the Everglades

In the 1980s, scientists measuring pollutants in Florida lakes and streams saw a disturbing trend. Mercury concentrations in freshwater wildlife across the state were rising. The situation was worst in the Everglades.

In some of the most pristine areas, where other contaminants were barely detectable, mercury concentrations had risen to dangerous heights in animals. Two freshwater fish species, the warmouth and largemouth bass, contained 3.4 ppm (parts per million) of mercury—almost seven times higher than established safety limits. A mercury concentration of 110 ppm claimed the life of at least one endangered Florida panther.

Finding Pollution Sources

Research programs were begun to identify causes of the problem and possible solutions. Scientists discovered that 98 percent of the mercury comes from the atmosphere, in rainfall. Thus, they looked for sources, such as factories, power plants, and incinerators, that could be adding pollutants to the atmosphere. These facilities often burn coal or waste materials and release mercury vapor into the atmosphere.

Curiously, scientists did not find consistently higher mercury levels in wildlife near the facilities. Scientists worried that sources of mercury might be far away—and hard to regulate—because elemental mercury can be carried by wind and remain in the atmosphere for years. Still, the government began to require local incinerators to keep mercury out of their emissions.

Regulations Helped

Within a decade, mercury levels in bird feathers and fish tissue from the Everglades had dropped by 60 to 70 percent. This was clear evidence that local incinerators had played a large role in the Everglades' mercury problem.

internet connect

www.scilinks.org
Topic: Solving Environmental Problems
Keyword: HX4166

SCILINKS. Maintained by the National Science Teachers Association

Teach

Demonstration

Show the class a catalytic converter or a cutaway diagram of one. Explain that a few grams of platinum, palladium, or rhodium are embedded as pellets in the catalytic converter. The pellets act as catalysts in reactions that break down toxic emissions, which would otherwise pass through the car's exhaust system and into the atmosphere.

BIOWatch

A Mercury Mystery in the Everglades

Teaching Strategy

Tell students that sulfates in runoff from agricultural fertilizers also play a role in the mercury problem. The connection is through aquatic bacteria. When mercury enters a lake or stream from the atmosphere, sulfate-reducing bacteria use sulfates to change elemental (or inorganic) mercury to methylmercury—an organic form that is then passed up aquatic food chains.

Research Activity

Divide the class into small groups, and have each group research the role of mercury and sulfates in aquatic food webs. Ask each group to create a poster, computer presentation, or diorama showing how mercury bioaccumulates in a typical Florida food web.

INCLUSION Strategies

• *Attention Deficit Disorder* • *Learning Disability*

Ask students to design flyers titled "Your Contribution" to distribute to classes in their school. The flyers should include examples and ideas of what students and their families can do to conserve energy and reduce pollution and waste in their community. The ideas and examples should make students and teachers aware of things they can do at school and at home.

Attention Grabber

Pollution Permits The first auction of pollution permits in the United States was held in 1993. Some 150,000 permits were sold for more than $21 million. Some public interest groups also bid in an effort to keep the pollution permits out of the hands of industry. Their efforts amounted to less than 1 percent of bidding.

Teach, *continued*

Real Life

Answers

Pros include that recycling conserves natural resources, saves land from mining and logging, and reduces solid waste production. Cons include that recycling can be expensive and it uses energy resources.

Teaching Tip — GENERAL

History of Pollution Ask students when they think large-scale pollution problems first occurred. Explain that studies have shown that large-scale pollution existed several thousand years ago. Sediment samples from the beds of lakes in Sweden show that lead concentrations were above natural levels 2,600 years ago. Ice cores removed from Greenland's icecap also have higher lead levels, which started at about the same time. What caused this ancient pollution? It was most likely caused by early metal smelting operations in what is now Spain. Have students research sources of early pollutants in Earth's history and report their findings. Co-op Learning

Activity — ADVANCED

Technology Have students use library resources to locate information concerning how an individual, group, or community solved an environmental problem using knowledge of biology and technology.
Co-op Learning | English Language Learners

Real Life

Better the second time around?

Although many communities have recycling programs, some researchers argue that recycling doesn't pay off.

Evaluating Viewpoints

Find out the main arguments for and against recycling, and propose the kinds of waste-management programs that would work in your community.

Solving Environmental Problems

It is easy to get discouraged when considering the world's many serious environmental problems. But do not lose track of the conclusion that emerges from our examination of these environmental problems—each of the world's many problems is solvable. If one looks at how environmental problems have been overcome, a clear pattern emerges.

Five Steps to Success

Viewed simply, there are five components to successfully solving any environmental problem.

1. **Assessment.** The first stage is scientific analysis of the problem, the gathering of information about what is happening. To construct a scientific model of an ecosystem, data must be collected and analyzed. A model makes it possible to describe the current state of the ecosystem. A model would also allow scientists to make predictions about the future of the ecosystem.
2. **Risk analysis.** Using the information obtained by scientific analysis, scientists predict the consequences of different types of environmental intervention. It is also essential to evaluate any negative effects associated with a plan of action.
3. **Public education.** When it is possible to describe alternative courses of action, the public must be informed. This involves explaining the problem in understandable terms, such as at the public meeting shown in **Figure 13,** presenting the alternative actions available, and explaining the probable costs and results of the different choices.
4. **Political action.** The public, through its elected officials, selects and implements a course of action. Individuals can be influential at this stage by exercising their right to vote and by contacting their elected officials.
5. **Follow-through.** The results of any action should be carefully monitored to see if the environmental problem is being solved.

Figure 13 Public participation. Public discussion of environmental problems helps citizens to evaluate alternative courses of action.

398

Integrating Physics and Chemistry

Point out that worldwide population growth is also a problem because energy demands inevitably rise as the population increases. Currently, most of our energy is derived from fossil fuels with lesser amounts provided by nuclear, hydroelectric, geothermal, solar, and wind resources. Have students compare and contrast the economic and environmental impacts of each of these resources. Ask them to identify the energy resource on which they are most dependent and if there is a way for them to reduce this dependency.

Career

Toxicologist Toxicologists protect the public health by studying and identifying toxic substances in the environment. Usually employed by state or federal agencies, toxicologists provide vital information and education to health departments, physicians, legislators, and citizens. Most toxicologists receive a degree in biology, chemistry, or environmental science and then pursue advanced degrees in their chosen field.

Two Success Stories

The development of appropriate solutions to the world's environmental problems often rests partly on the shoulders of politicians, economists, bankers, scientists, and engineers. However, it is important not to lose sight of the key role often played by informed individuals. Two examples serve to illustrate the point.

Figure 14 Marion Stoddart. The recovery of the Nashua River shows that polluted environments can be restored when committed individuals, like Marion Stoddart, work to bring about a change.

The Nashua River Running through the heart of New England, the Nashua River was severely polluted by mills established in Massachusetts in the early 1900s. When Marion Stoddart, shown in **Figure 14,** moved to a town along the river in 1962, she was appalled. Stoddart organized the Nashua River Cleanup Committee. The committee presented bottles of dirty river water to politicians, spoke at town meetings, recruited business people to help finance a waste treatment plant, and began to clean garbage from the Nashua River's banks. This citizen's campaign contributed to the passage of the Massachusetts Clean Water Act of 1966. Industrial dumping into the river is now banned, and the river has largely recovered.

Lake Washington Following World War II, this very large lake east of Seattle became surrounded by a ring of 10 suburbs, each with its own municipal sewage treatment plant. Between 1940 and 1953, these 10 municipal sewage plants discharged their treated outflow into the lake. Safe enough to drink, the outflow was believed to be harmless. Starting in the early 1940s, the combined daily discharge in the lake was 80 million liters (20 million gallons).

In 1954, an ecology professor at the University of Washington in Seattle, W. T. Edmondson, noted that his research students were reporting blue-green algae growing in the lake. Such algae require an abundance of the nutrients nitrogen and phosphorus to grow. Because deep freshwater lakes like Lake Washington usually lack these nutrients, the presence of the algae was surprising. The researchers found that phosphates and nitrates in the sewage had been fertilizing the lake! Edmondson was alarmed and began a campaign in 1956 to educate public officials about the danger: Bacteria decomposing the dead algae would soon deplete the lake's oxygen. This would kill all life in the lake, and it would never recover. After five years, as a direct result of his efforts, joint municipal taxes financed the cleanup of Lake Washington with a massive trunk sewer that rings the lake and carries treated discharge far out into Puget Sound. Today, through the efforts of many people, the lake is healthy, its waters clean and blue, as shown in **Figure 15.**

Figure 15 Lake Washington. Once choked with algae that were nourished by the outflow from sewage treatment plants, Lake Washington is a healthy lake today.

REAL WORLD CONNECTION

Although the 1990 Clean Air Act is a federal law covering the entire country, states do much of the work to carry it out. For example, state air pollution agencies hold hearings on permit applications for a power or chemical plant, or fine companies for violating air pollution limits. The EPA sets limits on the amount of a pollutant allowed in the air anywhere in the United States. The law allows individual states to enforce stronger pollution controls, but not weaker controls than those set for the country. It also recognizes that pollution control often requires knowledge of local industries, geography, housing patterns, etc.

In fact, states have to develop state implementation plans (SIPs). A SIP includes the regulations a state will use to clean up polluted areas. The EPA must approve each SIP. The United States government, through the EPA, assists states by providing scientific data, engineering designs, and money to support clean air programs.

Teaching Tip — GENERAL

Air Pollution and Vehicles Tell students that each of today's cars produces 60 to 80 percent less pollution than cars did in the 1960s. Ask them to suggest reasons for the decrease. (More people are using mass transit; cars now use catalytic converters; and leaded gas has been phased out, resulting in dramatic declines in atmospheric levels of lead, a very toxic chemical.) Despite this progress, in the United States, motor vehicles are still the sources of at least half of all hazardous air pollutants. What went wrong? In part, more people are driving more cars for more miles.

Using the Figure — BASIC

Ask students if they participate in any of the activities illustrated in **Figures 16 and 17**. Have them brainstorm actions they can take on a daily basis to help any aspect of the environment—including rivers, lakes, or their immediate environment. (Possible suggestions include conserving water use; building a compost pile; using reusable plates, cups, and napkins; buying and recycling certain products and materials; and carpooling or riding a bus or a bicycle whenever possible instead of a car; and drying clothes on a line.) Ask interested students to investigate other notable success stories and the people involved, and present their findings to the class in an oral or written report, collage, or bulletin-board display.

Co-op Learning | English Language Learners

Activity — ADVANCED

Making a Commitment Have students consult books and magazines to identify ways they themselves can help the environment. Then have each student carefully consider and select four actions he or she is not already doing and carry out these actions for 2 weeks. At the end of this period, students should write a brief and honest appraisal of their efforts, addressing how consistently they performed their actions, why some actions were easier than others, and any unexpected benefits or impediments they encountered.

LS Kinesthetic

Close

Reteaching — Basic

Have the class organize and implement a school-wide project that addresses one of the environmental concerns discussed in this chapter. Tell students that their project must be organized according to the five-step plan outlined previously in this chapter. **LS Interpersonal**

Quiz — General

1. Name one pollutant that has been dramatically reduced in the past decade. (CFC, DDT, asbestos, dioxin, sulfur dioxide, carbon monoxide, soot)
2. What are the five steps to success in solving environmental problems? (assessment, risk analysis, public education, political action, follow-through)

Alternative Assessment — Advanced

Have students research the practice of selling pollution permits and write a report about what they learn. In their reports, students should present the arguments for and against this practice and evaluate its effectiveness as a method of pollution control.

Figure 16 Conserving energy. By riding a bicycle instead of traveling in a car, bus, or train, this commuter helps save energy.

Your Contribution

You cannot hope to preserve what you do not understand. Humans rely on the Earth's ecosystems for food and all of the other materials our civilization depends on. It has been said that we do not inherit the Earth from our parents but borrow it from our children. Therefore, we must preserve for them a world in which they can live.

Although solving the world's environmental problems will take the efforts of many people, including politicians, economists, and engineers, the issues are largely biological. When all is said and done, your knowledge of ecology is the essential tool that you can contribute to the effort. **Figures 16** and **17** show some simple ways you can participate in solving the problems described in this chapter by conserving energy and reducing pollution and waste. You can save energy by walking, riding a bicycle, or taking public transportation to work or school. Newspapers, aluminum products, glass containers, and many plastic containers can be recycled. A resource- and energy-use inventory of your home can identify additional ways to help the environment. For example, installing inexpensive, low-flow shower heads can reduce shower water use by up to 50 percent.

Figure 17 Your lifestyle affects the environment. Choices that you make in your day-to-day activities can benefit the environment.

internet connect
www.scilinks.org
Topic: Florida's Environment
Keyword: HX6F06
SciLinks. Maintained by the National Science Teachers Association

You Can Help
- Conserve energy by walking, riding a bicycle, or taking public transportation
- Do a resource- and energy-use inventory of your home
- Learn about the environment
- Recycle newspapers, aluminum products, glass, and plastic
- Create rich soil by making your own compost heap from leaves, grass, and fruit peelings

Section 3 Review

1. **Describe** how a tax can reduce pollution. SC.H.3.4.3
2. **Critical Thinking Justifying Conclusions** Of the five steps listed in this chapter for solving environmental problems, which step might be the most difficult to implement? SC.H.3.4.3
3. **Critical Thinking Analyzing Information** At which step in the solution of an environmental problem could you have the greatest influence? Explain your answer. SC.G.2.4.6
4. How does the release of certain nutrients into a lake harm the lake? SC.G.2.4.6
5. **FCAT Prep** What is one good way to conserve natural resources?
   A drive instead of walk
   B throw away cans and jars
   C use high-flow shower heads
   D use public transportation
   SC.G.2.4.6

Answers to Section Review

1. A tax will raise the price of the good, in turn reducing its consumption and production. The money from the tax can also be used to implement pollution controls.
2. Accept any reasonable answer. For example: step 5—follow-through—may be the most difficult to implement because it requires a clear commitment of money and resources as well as cooperation among many parties.
3. Accept any reasonable answer. For example: Individuals can probably have the most influence at step 4—political action—because voting and contacting elected officials provide direct channels of influence.
4. Fertilizers can cause overgrowth of algae. The bacterial decomposition of algae when they die deprives the lake of the oxygen that aquatic organisms need to survive.
5. **A.** Incorrect. Driving instead of walking does not conserve fossil fuels. **B.** Incorrect. Throwing away cans and jars does not conserve metals and glass used in these products. **C.** Incorrect. Using high-flow shower heads does not conserve water. **D.** Correct. Using public transportation conserves fossil fuels.

Section Review Correlations

SC.G.2.4.6: Items 3, 4, 5
SC.H.3.4.3: Items 1, 2

Study ZONE

Key Concepts

1 Global Change

- Acid rain, which is caused by airborne pollutants that lower the pH of rain, has damaged many forests and lakes, especially in the Northeast.
- Destruction of the ozone layer is caused by chlorofluorocarbons (CFCs) and several other manufactured chemicals.
- The greenhouse effect occurs when greenhouse gases, such as carbon dioxide, trap heat within Earth's atmosphere.
- Many scientists think that increased concentrations of CO_2 and other greenhouse gases in the atmosphere have led to global warming.

2 Effects on Ecosystems

- The release of toxic chemicals into the environment can have serious effects, particularly when their concentration is magnified by food chains.
- Three nonreplaceable resources—species of living things, topsoil, and ground water—are being consumed or destroyed at a rapid rate.
- Rapid growth of the human population places serious stress on the Earth's ecosystems.

3 Solving Environmental Problems

- Worldwide efforts to reduce pollution are being made, but they are only part of the solution to the overall pollution problem.
- Taxing products or services that create pollution and creating laws requiring pollution-control devices are two methods that have been used to reduce pollution.
- Each of the world's many environmental problems can be solved if seriously addressed. A combination of scientific investigation and public action can solve many environmental problems.

Key Terms

Section 1

acid rain (386)
chlorofluorocarbon (387)
global warming (388)
greenhouse effect (388)

Section 2

biological magnification (391)
aquifer (393)

Section 3

None

BIOLOGY INTERACTIVE TUTOR

Unit 7—*Ecosystem Dynamics*
Use this unit to review the key concepts and terms in this chapter.

Alternative Assessment

Have the class create a display titled "Human Impact on the Environment" to inform the rest of the school about the severity of the environmental issues raised in this chapter. The display should cover a range of issues, provide specific information, and show possible solutions and how individuals can make a difference. Make arrangements to place the display where it will be viewed by the entire student body.

Co-op Learning | English Language Learners

Chapter Resource File

- Science Skills Worksheet GENERAL
- Critical Thinking Worksheet ADVANCED
- Test Prep Pretest GENERAL
- Chapter Test GENERAL

Answer to Concept Map

The following is one of several possible answers to Performance Zone item 15.

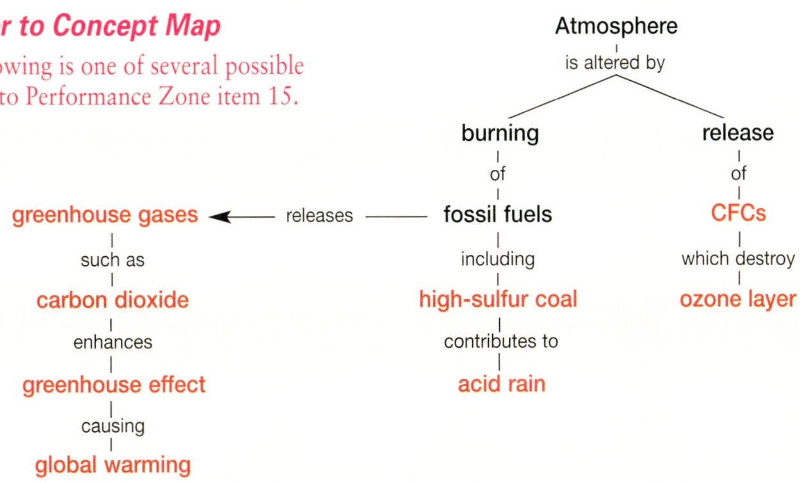

CHAPTER 18 ZONE

ANSWERS

Understanding Key Ideas

1. b
2. d
3. c
4. c
5. b
6. Loss of topsoil diminishes the productivity of agricultural land. Species extinctions may prevent the discovery of new food sources.
7. Through biological magnification, DDT became increasingly concentrated as it moved up through the food chain, with the highest levels found in bald eagles. The chemical reduced the eagles' birth rate causing the population to decline.
8. As factories, power plants, and incinerators burn coal or waste materials, they release elemental mercury into the atmosphere. The mercury enters the water cycle and thus enters the Everglades ecosystem.
9. The loss of topsoil and ground water reduces the resources that plants need to grow. Fewer plants mean less food for consumers, thereby reducing the carrying capacity of Earth.
10. One possible answer to the concept map is found at the bottom of the Study Zone page.

Critical Thinking

11. Increased levels of CO_2 may cause global warming, which disrupts the rainfall patterns that crops rely on.
12. Stopping all pesticide use would result in decreased food production unless alternate forms of pest control are used. With more insects around, the rate of insect-borne disease would increase.

Chapter Review Correlations

SC.G.2.4.1: Items 2, 10
SC.G.2.4.5: Items 6, 9
SC.G.2.4.6: Items 1, 4, 7, 8
SC.H.3.4.3: Items 3, 5, 11-15

Performance ZONE
CHAPTER REVIEW

Understanding Key Ideas

1. Which of the following causes acid rain?
   a. releasing chlorofluorocarbons
   b. burning high-sulfur coal
   c. polluting ground water
   d. scrubbing smokestack emissions
   SC.G.2.4.6
2. The burning of fossil fuels has changed the atmosphere by
   a. increasing the global concentration of ozone in the upper atmosphere.
   b. reducing the amount of CFCs.
   c. producing an ozone hole.
   d. increasing the concentration of carbon dioxide.
   SC.G.2.4.1
3. Which of the following statements about the extinction of species that is now occurring is *not* true?
   a. It is the largest extinction event since the dinosaurs disappeared.
   b. One of its causes is the destruction of tropical rain forests.
   c. It is confined to tropical countries.
   d. Potentially useful species are becoming extinct.
   SC.H.3.4.3
4. Which of these countries has the largest population growth rate?
   a. Germany
   b. United States
   c. Nigeria
   d. Russia
   SC.G.2.4.6
5. Of the five major steps to solving environmental problems, which involves determining the potential outcomes of an environmental plan before it is tried?
   a. assessment
   b. risk analysis
   c. follow-through
   d. political action
   SC.H.3.4.3
6. Explain why the loss of topsoil and the extinction of living species are a threat to increased food production. SC.G.2.4.5
7. Analyze the relationship between biological magnification and the reduction of the bald eagle population by the pesticide DDT. SC.G.2.4.6
8. **BIOWatch** How does mercury pollution get into the Everglades? SC.G.2.4.6
9. Describe how the global decrease in available ground water and the loss of topsoil could affect the carrying capacity of the Earth. (**Hint:** See Chapter 16, Section 1.) SC.G.2.4.5
10. **Concept Mapping** Make a concept map that shows how human activities are disrupting the atmosphere and that describes the effects of these disruptions. Try to use the following terms in your map: *greenhouse effect, carbon dioxide, greenhouse gases, global warming, CFCs, ozone layer, acid rain,* and *high-sulfur coal.* SC.G.2.4.1

Critical Thinking

11. **Evaluating an Argument** Evaluate the statement that rising levels of carbon dioxide in the atmosphere might lead to increased food production. SC.H.3.4.3
12. **Predicting Outcomes** How would stopping all pesticide use likely affect rates of food production and of diseases, such as malaria, that are spread by insects? SC.H.3.4.3

Alternative Assessment

13. **Summarizing Information** Write a report that identifies several alternatives to CFCs and describes the requirements established for CFC replacements. SC.H.3.4.3
14. **Career Connection** Environmental Scientist Use library or Internet resources to research the educational background necessary to become an environmental scientist. Describe the degrees or training that is recommended for this career. Summarize the employment outlook for this field. SC.H.3.4.3
15. **Interactive Tutor Unit 7** Ecosystem Dynamics Write a report summarizing ways that humans can work to reduce the depletion and pollution of ground water. How would more-efficient use and recycling of ground water benefit ecosystems? SC.H.3.4.3

402

Alternative Assessment

13. Alternatives to CFCs include hydrochlorofluorocarbons (HCFCs) and hydrofluorocarbons (HFCs). Alternate CFCs must have: (1) similar boiling points to CFCs, (2) nonflammability and low toxicity, (3) technical and economic viability, (4) chemical stability, and (5) low potential for causing ozone depletion and global warning.
14. Answers will vary. Environmental scientists use scientific techniques to study the natural and human-altered environment. Most positions require a bachelor's degree and often a graduate degree in science or engineering. Recommended high-school courses include as much math and science as possible. The employment outlook for this field is fair. Starting salary will vary by region.
15. Answers will vary. Humans can conserve ground water by planting crop, yard, and garden plants suited to the local climate, as well as using and recycling ground water efficiently.

Assignment Guide

Section	Questions
1	1, 2
2	3, 4, 6, 7, 9, 11, 12
3	5, 8, 10, 13, 14, 15

FCAT Prep

For questions 1–6, write your answers on a separate sheet of paper.

1 What is the term for liquid precipitation with a low pH that results from sulfur emissions reacting with water?

A. acid rain **C.** sulfuric acid
B. greenhouse gas **D.** thermal pollution

SC.G.2.4.6

2 What process involves the warming of Earth by radiated energy trapped by gases such as carbon dioxide, and nitrous oxide?

F. biological magnification **H.** ozone destruction
G. greenhouse effect **I.** thermal pollution

SC.G.2.4.6

3 Tall smokestacks are part of many coal-burning power plants. What is the purpose of these tall smokestacks?

A. to carry pollution back to Earth downwind by precipitation
B. to disperse smoke into the atmosphere and away from people
C. to generate energy by cooling the smoke as it leaves the plants
D. to reduce the amount of pollution released into the atmosphere

SC.G.2.4.1

4 What term describes the increasing concentration of substances in animal tissue toward the top of the food chain?

F. biological magnification **H.** DDT concentration
G. chlorofluorocarbon toxicity **I.** hydrocarbon pollution

SC.G.2.4.6

Base your answers to questions 5 and 6 on the graph below.

GRAIN PRODUCTION

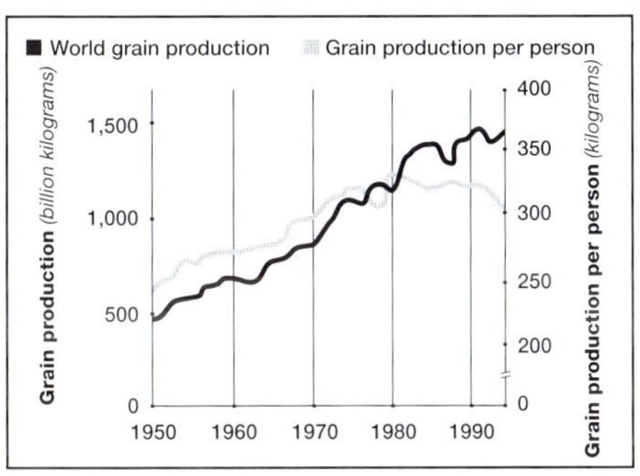

5 How have world grain production and the human population changed since the early 1980s?

A. Both have decreased at the same rate.
B. World grain production has increased; the human population has decreased.
C. Both have decreased, but the world grain production has decreased more rapidly.
D. The human population has increased more rapidly than world grain production.

SC.G.2.4.6

6 To the nearest 500 billion kilograms, by what amount has the world grain production changed between 1950 and 1990? SC.G.2.4.6

Test TIP

If time permits, take short mental breaks to improve your concentration during a test.

403

FCAT Doctor

Question 3 Answer B is the correct choice. The intent for the smokestacks is to release wastes high into the atmosphere, where winds would disperse and dilute them. Answer A is incorrect because this is an unintentional drawback of smokestacks. Answer C is incorrect because cooling the smoke in smokestacks does not generate useful energy for the power plant. Answer D is incorrect because the smokestacks have the opposite effect on the atmosphere.

Question 5 Answer D is the correct choice. Total grain production has increased, but grain production per person has decreased, indicating an increase in human population. Answer A is incorrect because, if this statement were true, the total grain production and the grain production per person would both have decreased and by the same amount. Answer B is incorrect because, if this statement were true, the grain production per person would have risen, not fallen. Answer C is incorrect because, if this statement were true, the world grain production would show a negative, not a positive, slope.

Question 6 The correct answer is 1,000. In 1950, the amount was 500 billion kg. In 1990, the amount was 1,500 billion kg. 1,500 – 500 = 1,000 billion kg.

Answers

1. A
2. G
3. B
4. F
5. D
6. 1,000

Test Prep Correlations

SC.G.2.4.1: Item 3
SC.G.2.4.6: Items 1, 2, 4, 5, 6

Skills Practice Lab

Studying Population Growth

Teacher's Notes

Time Required 20–30 minutes per day for 4 consecutive days

Ratings

EASY ——— HARD

Teacher Preparation 1
Student Setup 2
Concept Level 3
Cleanup 1

Skills Acquired
- Collecting Data
- Inferring
- Measuring
- Organizing and Analyzing Data
- Predicting

Scientific Methods

In this lab, students will:
- Make Observations
- Analyze the Results
- Draw Conclusions

Materials and Equipment

Materials for this lab can be ordered from WARD'S. See the *Master Materials List* at the front of this book for catalog numbers.

Safety Cautions

Caution students to treat all microorganisms as potential pathogens. Remind students to keep their hands away from their faces as they handle the yeast cultures. Remind students to wash their hands after this lab.

Skills Practice Lab

Studying Population Growth SC.G.2.4.6

SKILLS
- Using a microscope
- Collecting, graphing, and analyzing data
- Calculating

OBJECTIVES
- **Observe** the growth and decline of a population of yeast cells.
- **Determine** the carrying capacity of a yeast culture.

MATERIALS
- safety goggles
- lab apron
- yeast culture
- (2) 1 mL pipets
- 2 test tubes
- 1% methylene blue solution
- ruled microscope slide (2 × 2 mm)
- coverslip
- compound microscope

Magnification: 1100×

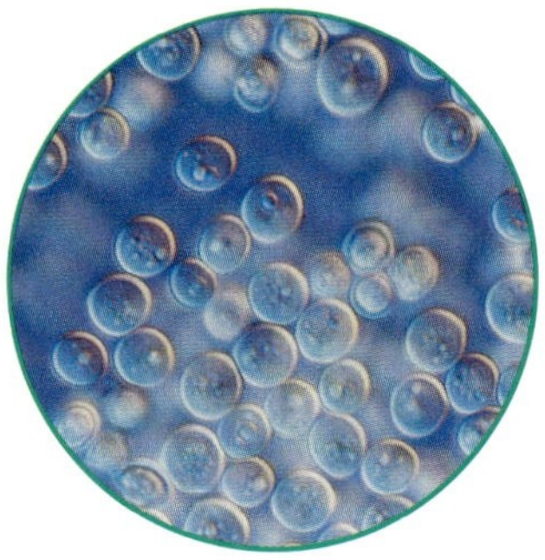

Yeast cells

ChemSafety

CAUTION: Always wear safety goggles and a lab apron to protect your eyes and clothing.

CAUTION: Do not touch or taste any chemicals. Know the location of the emergency shower and eyewash station and how to use them. If you get a chemical on your skin or clothing, wash it off at the sink while calling to the teacher. Notify the teacher of a spill. Spills should be cleaned up promptly, according to your teacher's directions.

CAUTION: Glassware is fragile. Notify the teacher of broken glass or cuts. Do not clean up broken glass or spills with broken glass unless the teacher tells you to do so.

Before You Begin

Recall that population size is controlled by **limiting factors**—environmental resources such as food, water, oxygen, light, and living space. **Population growth** occurs when a population's **birthrate** is greater than its **death rate**. A decline in population size occurs when a population's death rate surpasses its birthrate. In this lab, you will study the concepts of population growth, decline, and carrying capacity by growing and observing yeast.

DATA TABLE

Time (hours)	Number of cells per square: Squares 1–6	Average	Population size (cells/0.1 mL)
0			
24			
48			
72			
96			

1. Write a definition for each boldface term in the previous paragraph.
2. Make a data table similar to the one below at left.
3. Based on the objectives for this lab, write a question about population growth that you would like to explore.

Procedure

PART A: Counting Yeast Cells

1. Put on safety goggles and a lab apron.
2. Transfer 1 mL of a yeast culture to a test tube. Add 2 drops of methylene blue to the tube. **Caution: Methylene blue will stain your skin and clothing.** The methylene blue will remain blue in dead cells but will turn colorless in living cells.

404

Tips and Tricks

Prepare the yeast population by dissolving 1.0 g of yeast and 1.0 g of sugar in 40 mL of warm water. Remove 1 mL of this solution and dilute with 9 mL of water. Mix well and again remove 1 mL of the solution and dilute with 9 mL of water. Fresh yeast may be prepared up to 4 hours before the investigation. If substituting dried yeast or freeze-dried yeast (available from WARD'S), prepare about a week in advance. Keep the yeast in a warm, dark area for the duration of the investigation. Ruled microscope slides can be purchased from WARD'S. Alternatively, make a transparency copy of a piece of graph paper and then cut the transparency into coverslip-size pieces. Students can use these transparency pieces as their coverslips. Use gloves when preparing methylene blue and avoid creating dust while working. Make a 1.0 percent solution by dissolving 1.0 g of methylene blue in 100 mL of distilled water. Have students work in teams of two. Solutions of yeast and of methylene blue may be rinsed down the drain. Wash thoroughly and air dry all glassware.

3. Make a wet mount by placing 0.1 mL (one drop) of the yeast and methylene blue mixture on a ruled microscope slide. Cover the slide with a coverslip.
4. Observe the wet mount under the low power of a compound microscope. Notice the squares on the slide. Then switch to the high power. *Note: Adjust the light so that you can clearly see both stained and unstained cells.* Move the slide so that the top left-hand corner of one square is in the center of your field of view. This will be area 1, as shown in the diagram below.

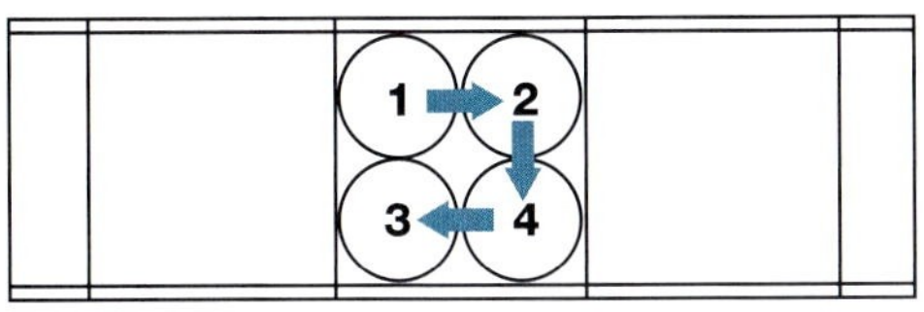

5. Count the live (unstained) cells and the dead (stained) cells in the four corners of a square using the pattern shown in the diagram above. In your data table, record the numbers of live cells and dead cells in the square.
6. Repeat step 5 until you have counted 6 squares on the slide. Complete Part B.
7. Find the total number of live cells in the 6 squares. Divide this total by 6 to find the average number of live cells per square. Record this number in your data table. Repeat this procedure for dead cells.
8. Estimate the population of live yeast cells in 1 mL (the amount in the test tube) by multiplying the average number of cells per square by 2,500. Record this number in your data table. Repeat this procedure for dead cells.
9. Repeat steps 1 through 8 each day for 4 more days.

PART B: Cleanup and Disposal

10. Dispose of solutions and broken glass in the designated waste containers. Do not pour chemicals down the drain or put lab materials in the trash unless your teacher tells you to do so.
11. Clean up your work area and all lab equipment. Return lab equipment to its proper place. Wash your hands thoroughly before you leave the lab and after you finish all work.

Analyze and Conclude

1. **Analyzing Methods** Why were several areas and squares counted and then averaged each day?
2. **Summarizing Results** Graph the changes in the numbers of live yeast cells and dead yeast cells over time. Plot the number of cells in 1 mL of yeast culture on the *y*-axis and the time (in hours) on the *x*-axis.
3. **Inferring Conclusions** What limiting factors probably caused the yeast population to decline?
4. **Further Inquiry** Write a new question about population growth that could be explored in another investigation.

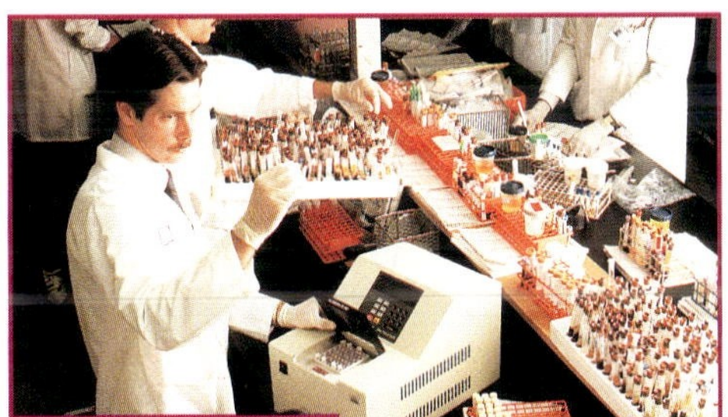

On the Job

Biologists must sometimes estimate the number of cells in a given volume. Do research to discover how a **hemacytometer** is used in medical tests. For more about careers, visit **go.hrw.com** and type in the keyword **HX4 Careers.**

405

Answers to Before You Begin

1. *limiting factors*—resources needed for survival obtained from or provided by the environment; *population growth*—increase in population size; *birthrate*—number of births in a population per unit time; *death rate*—number of deaths in a population per unit time;
3. For example: How does the size of a population with limited resources change over time?

Answers to Procedure

5. Answers will vary.
6. Answers will vary.
7. Answers will vary.
8. Answers will vary.
9. Answers will vary.

Answers to Analyze and Conclude

1. An average was taken to allow for variation within the population.
2. Answers will vary. See the Sample Data Graph below for one example.
3. A lack of food and lack of space limit the yeast cells. They could also be poisoned by their own wastes.
4. Answers will vary. For example: Would the carrying capacity of the yeast's environment expand if the size of the environment increased?

Answers to On The Job

Biologists use hemocytometers to manually count the number of red blood cells in a given volume (usually 1 mm^3) of blood. A low red blood cell count can indicate a variety of medical conditions, including acute blood loss.

Sample Data Graph

Estimate of Yeast Population Growth

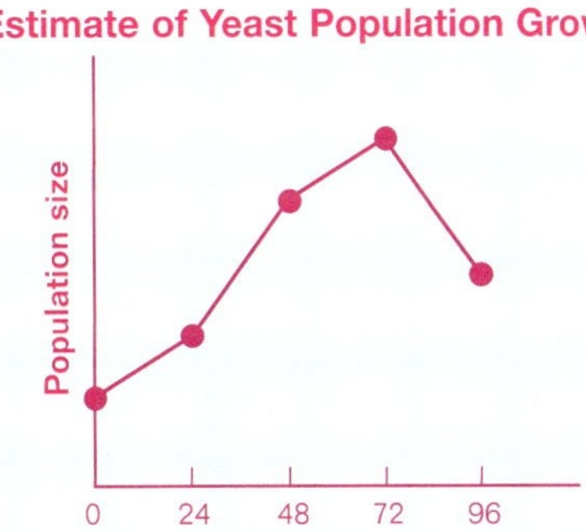

Saving Biodiversity

Background

When it comes to saving biodiversity, both time and money are short, so conservationists are trying to focus their efforts on areas that are richest in biodiversity. Finding these areas is the problem, because biologists know surprisingly little about the Earth's inhabitants. Also, biological exploration cannot keep pace with habitat destruction—particularly in the species-rich tropics—because of a shortage of systematists and funds for systematic research.

While on a bird-watching expedition in South America, Murray Gell-Mann, a Nobel laureate in physics and a keen amateur ornithologist, and Ted Parker, a tropical bird expert at Louisiana State University, came up with a new way to focus conservation efforts. Both men agreed that traditional field research was gathering data too slowly, and they proposed conducting short, intensive surveys to quickly fill some of the gaps in the knowledge of biodiversity. From their conversation came the Rapid Assessment Program described in the feature. As head of the MacArthur Foundation's world environment and resources committee, Gell-Mann was in a position to help secure funding for RAP expeditions, which began in 1990 and are sponsored by Conservation International.

Biologists race against time to save species threatened with extinction.

Saving Biodiversity

How many species of organisms live on Earth? That seems like a simple question, but no one knows the answer. The uncertainty over this basic question illustrates how little scientists actually know about biodiversity.

One thing is certain about biodiversity, though: it is disappearing fast. Biologists need good information about biodiversity in order to choose the actions that will save the most species. To focus conservation efforts, biologists agree that the first priority must be to take an inventory of global biodiversity, determining which species live where.

A New Approach

Beginning in the late 1980s, a group of biologists and conservationists decided a shortcut was needed. They launched the Rapid Assessment Program, or RAP, to speed the study of biodiversity. RAP sends small teams of experts to tropical habitats to quickly determine what kinds of organisms live there. The program focuses on "hot spots," threatened areas that are high in biodiversity and contain large numbers of unique species. The aim is to survey as many of these hot spots as possible and identify areas that need to be protected.

To conduct these surveys, Conservation International, the nonprofit organization that sponsors the program, brings together several of the world's leading tropical biologists. Each expedition also includes several scientists from the country where the team is working. (One of the goals of RAP is to train biologists in tropical countries.) The RAP team uses satellite photos, aerial observation, and discussions with scientists and government officials in tropical countries to choose the locations it will visit.

Instead of taking an exhaustive, time-consuming inventory of all species, the team may focus on a few familiar groups. An inventory of mammals, birds, fishes, flowering plants, butterflies, and reptiles gives a good indication of an area's total biodiversity. A habitat with many species of plants and birds, for example, also probably has many species of bacteria, insects, and other less conspicuous organisms.

Rare treasures
Loss of habitat threatens many tropical rain forest species.

406

Teaching Strategies

- Point out that in the last 200 years, scientists have named only about 1.5 million species. About 15,000 new species are described and named each year.
- Remind students that extinction is a problem even in developed countries. Have students research a species that became extinct in the United States within the last century. Each student should give a short presentation to the class describing where the species lived and what caused it to become extinct.
- Explain that although most newly discovered species belong to poorly known groups like insects, eubacteria, and nematodes, new vertebrates—even new mammal species—turn up frequently. In 1994, for instance, scientists exploring the rain forests of New Guinea discovered a new species of kangaroo.
- Have students identify ways that the high standard of living in the United States and other developed countries affects biodiversity. Suggest that they consider how resource use, pollution, and agriculture might endanger species. Then ask them what they would give up or change to protect biodiversity.

Tropical rain forest
Tropical rain forests are home to many species.

The Rewards of RAP

RAP scientists describe their work as exhausting but exhilarating. "It's what we live for," says Tom Schulenberg, an ornithologist (a biologist who studies birds) and RAP team leader. "We're always scheming to get back to the field." While exploring new places, Schulenberg feels "an incredible sense of excitement, knowing no other biologist has been there and everything you see is being seen for the first time."

How RAP Helps Conserve Biodiversity

Though less exciting than exploring a rain forest, the next stage of RAP is just as important. The scientists return to the United States to analyze their data. They then present a report containing their recommendations to the host country. RAP scientists stress that their role is to provide scientifically sound advice, not to tell tropical countries how to manage their natural resources. They leave all decisions to the individual governments. Using the information from RAP reports, tropical countries can guide their land-use decisions to help preserve biodiversity. ■

Biologist

RAP biologists

Profile

When biologists recognized that traditional field research was gathering data too slowly, the Rapid Assessment Program was born. Biologists help conduct short, intensive surveys to quickly fill some of the gaps in the knowledge of biodiversity.

Job Description

Biologists study all aspects of the biology of living things—anatomy, physiology, behavior, ecology, and evolutionary relationships. Their research may involve lab work, field studies, or a combination of both. Many biologists work for universities, museums, or government agencies.

Job Duties

Biologists who are members of a RAP team apply their knowledge of basic biology in parts of the world that are most important for saving species for future generations. Working in remote areas far from medical care has its dangers. Team members have been laid low by bubonic plague, malaria, and hepatitis. Besides disease, they must also watch out for poisonous snakes, biting insects and spiders, and falling trees.

Science/Math Career Preparation

- Biology
- Zoology
- Botany
- Evolutionary biology
- Mathematics
- Genetics
- Biochemistry

Read Further

1. Conservation International, www.conservation.org
2. Convention on Biological Diversity, www.biodiv.org
3. Center for Applied Biodiversity Science, www.biodiversityscience.org

Analyzing STS Issues

Science and Society

1. **What are the benefits of preserving biodiversity?** About 25 percent of medicines are derived from chemicals made by plants. Research a medicine derived from a tropical plant. What plant was it isolated from? Where does the plant live? What disease or diseases is the medicine used to treat?

2. **What are other countries doing to preserve biodiversity?** The government of many countries throughout the world are participating in the comprehensive protection of the Earth's biological resources. Using library resources or the Internet, research the various measures these governments have taken to catalog and preserve biodiversity and write a report summarizing your findings.

Technology: PCR

3. **Who should get the benefits of biodiversity?** The enzyme that copies DNA during the polymerase chain reaction (PCR) was isolated from an archaebacterium that lives in Yellowstone National Park. Although PCR generates more than $200 million in income each year, the federal government receives no royalties from the use of the enzyme. Research this issue and then write an essay supporting or opposing this statement: Companies that profit from PCR should be required to compensate the government for using a species discovered on federal land.

Analyzing STS Issues Answers

1. Answers will vary. One example is the rosy periwinkle of Madagascar, which provides the anticancer drugs vincristine and vinblastine.
2. Governments of more than 150 countries throughout the world have signed the Convention of Biological Diversity (CBD). Some of the measures they have agreed to implement include:
   - creating national plans for protection of biodiversity
   - identifying ecosystems, species and genomes important for conservation and sustainable use of biodiversity
   - monitoring biodiversity and any factors that might negatively impact it
   - establishing a system of protected areas
3. Answers will vary. Students' positions should be logical and supported by facts.

UNIT 5 Exploring Diversity

Chapters

Surrounded by protective gear, a worker in a biotechnology clean room is isolated from contaminants in the surrounding environment.

Microbes: Unseen Agents of Disease

Yesterday... Surgeons worked without masks, gowns, and gloves? Although this early operation looks strange to us today, Dr. Joseph Lister was actually ahead of his time. In the mid-1860s, Dr. Lister became the first physician to treat patients with an antiseptic during surgery. Lister recognized that spraying an airborne mist of carbolic acid over a patient reduced the likelihood of infections. **Read to learn how bacteria can cause disease.**

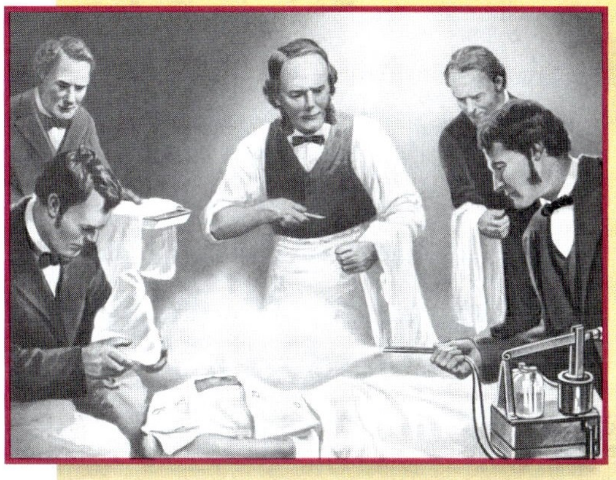

Dr. Joseph Lister and patient in 1867

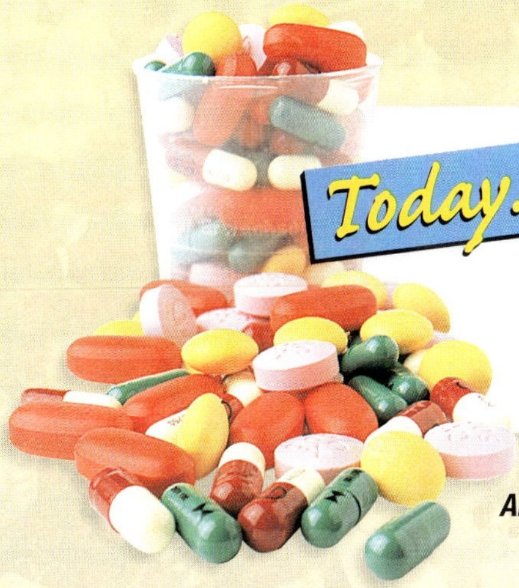

Today... Physicians fight bacteria with a far more sophisticated array of weapons than those used by Dr. Lister and his colleagues. Antibiotic drugs target infections throughout the body. **What is the Gram reaction?**

Antibiotics

Tomorrow... Some kinds of bacteria have become resistant to certain antibiotics. When an antibiotic fails to stop a bacterial infection, a physician will usually prescribe a different antibiotic—and hope that it works. As antibiotics are used with increasing frequency, strains of antibiotic-resistant bacteria (such as those that cause drug-resistant tuberculosis) continue to become more widespread. **Read to discover how antibiotics kill bacteria.**

Tuberculosis bacterium

internet connect

www.scilinks.org
Topic: Bacteria
Keyword: HX4018

SCILINKS. Maintained by the National Science Teachers Association

Pacing	Resources	Activities
Chapter Opener		SE **Quick Review,** p. 411 ■ SE **Reading Activity,** p. 411 ■ TE **Using the Figure,** p. 410 TE **Opening Activity,** p. 411 BASIC
& 3 • 90 min pp. 412–417 **Section 1** Introduction to Kingdoms and Domains	OSP **Lesson Plan*** TR **Bellringer*** TR **Section Outline*** TR **D27 Six Kingdoms*** TR **D28 Kingdom and Domain Characteristics***	TE **Group Activity** Grouping by Complexity, p. 413 GENERAL TE **Demonstration,** p. 414 BASIC TE **Using the Figure,** p. 415 TE **Using the Table,** p. 416 GENERAL CRF **Quick Lab** Using Bacteria to Make Food* ◆ BASIC
& 5 • 90 min pp. 418–421 **Section 2** Advent of Multicellularity	OSP **Lesson Plan*** TR **Bellringer*** TR **Section Outline***	TE **Activity** Distinguishing Organisms, p. 420 ◆ GENERAL CRF **Inquiry Lab** Slowing the Movements of Protozoans* ◆ GENERAL
& 7 • 90 min pp. 422–426 **Section 3** Complex Multicellularity	OSP **Lesson Plan*** TR **Bellringer*** TR **Section Outline***	TE **Demonstration,** p. 422 ◆ ADVANCED TE **Demonstration,** p. 423 GENERAL TE **Group Activity** Diversity of Living Things, p. 424 GENERAL TE **Activity** Animal Characteristics, p. 424 ADVANCED TE **Group Activity** A Zoological Expedition, p. 425 ◆ GENERAL SE **Quick Lab** Modeling True Multicellularity, p. 426 ■ CRF **Datasheets for In-Text Labs** Modeling True Multicellularity* SE **Exploration Lab** Surveying Kingdom Diversity, pp. 430–431 ■ CRF **Datasheets for In-Text Labs** Surveying Kingdom Diversity*

& 9 • 90 min

Chapter Review and Assessment Resources

SE **Study Zone,** p. 427 ■
SE **Performance Zone,** p. 428 ■
SE **FCAT Test Prep,** p. 429 ■
TR **Concept Mapping***
CRF **Vocabulary Review Worksheet*** ■ GENERAL
CRF **Test Prep Pretest*** ■ GENERAL
CRF **Chapter Test*** ■ GENERAL
CRF **Chapter Test*** ADVANCED
OSP **Test Generator***
OSP **Test Item Listing***
OSP **Modified Chapter Test*** BASIC

go.hrw.com Holt Online Learning

Visit go.hrw.com for access to Holt Online Learning or enter the keyword **HX6 Home** for a variety of free online resources.

One-Stop Planner® CD-ROM

This CD-ROM package includes
- Lab Materials QuickList Software
- Holt Calendar Planner
- Customizable Lesson Plans
- Printable Worksheets
- ExamView® Test Generator
- Interactive Teacher Edition
- Holt PuzzlePro® Resources
- Holt PowerPoint® Resources

Guided Reading Audio CDs

These CDs are designed to help auditory learners and reluctant readers. (Audio CDs are also available in Spanish.)

Compression guide:
To shorten your instruction because of time limitations, omit block 1.

KEY

- TE Teacher Edition
- SE Student Edition
- OSP One-Stop Planner
- CRF Chapter Resource File
- TR Teaching Transparency
- * Also on One-Stop Planner
- ■ Also Available in Spanish
- ◆ Requires Advance Prep

SKILLS DEVELOPMENT AND PRACTICE	SECTION REVIEW AND ASSESSMENT	CORRELATIONS
		Florida Sunshine Standards/FCAT
TE **Reading Skill Builder** Interactive Reading, p. 413 BASIC TE **Skill Builder** Math Skills, p. 415 GENERAL CRF **Directed Reading Worksheet*** BASIC CRF **Active Reading Worksheet*** GENERAL	SE **Section 1 Review** ■ TE **Reteaching,** p. 417 BASIC TE **Quiz,** p. 417 GENERAL TE **Alternative Assessment,** p. 417 GENERAL CRF **Quiz*** ■ BASIC	SC.G.1.4.1
TE **Reading Skill Builder** Reading Organizer, p. 419 BASIC TE **Reading Skill Builder** Discussion, p. 420 BASIC CRF **Directed Reading Worksheet*** BASIC CRF **Active Reading Worksheet*** GENERAL	SE **Section 2 Review** ■ TE **Reteaching,** p. 421 BASIC TE **Quiz,** p. 421 GENERAL TE **Alternative Assessment,** p. 421 GENERAL CRF **Quiz*** ■ BASIC	SC.G.1.4.1
TE **Reading Skill Builder** Reading Hint, p. 423 BASIC TE **Skill Builder** Interpreting Visuals, p. 425 GENERAL CRF **Directed Reading Worksheet*** BASIC CRF **Active Reading Worksheet*** GENERAL CRF **Critical Thinking Worksheet*** ADVANCED CRF **Science Skills Worksheet*** GENERAL	SE **Section 2 Review** ■ TE **Reteaching,** p. 426 BASIC TE **Quiz,** p. 426 GENERAL TE **Alternative Assessment,** p. 426 ADVANCED CRF **Quiz*** ■ BASIC **Modified Worksheet*** BASIC	SC.F.1.4.2, SC.G.1.4.1, SC.G.2.4.6

SCILINKS. NSTA
www.scilinks.org
Maintained by the **National Science Teachers Association**

Topic: Domains of Life
SciLinks Code: HX4060

CHAPTER 19

Overview

Tell students that the purpose of this chapter is to introduce the six-kingdom classification system. They will learn about the basic features of the two groups of prokaryotes and the eukaryotes and about how true multicellular organisms are made up of coordinated, communicating cells. Students will also learn about multicellular protists and the basic features of fungi. Finally, they will learn about how the cells of complex multicellular organisms are arranged to form tissues and the basic features of plants and animals.

Using the Figure

Direct students' attention to the picture of the meadow. Tell them that populations of organisms must be able to adapt to changing ecosystems to survive. As environmental conditions throughout the world have changed over time, a fantastic array of organisms has evolved and survives in nearly every environment on Earth.

Assessing Prior Knowledge

Before beginning this chapter, be sure your students can answer the questions in the ***Quick Review*** on the facing page. If they cannot, encourage them to reread the sections indicated.

A meadow

410

Chapter Correlations — **Florida Sunshine Standards/FCAT**

SC.F.1.4.2 Body structures are uniquely designed and adapted for their function.

SC.G.1.4.1 There is great diversity and interdependence among living things.

CHAPTER

19 Introduction to the Kingdoms of Life

Florida Sunshine Standards/FCAT

SC.F.1.4.2 Body structures are uniquely designed and adapted for their function.
SC.G.1.4.1 There is great diversity and interdependence among living things.

Quick Review

Answer the following without referring to earlier sections of your book.

1. **Calculate** the surface area–to–volume ratio of a cell. *(Chapter 3, Section 2)*
2. **Summarize** the characteristics of prokaryotes and eukaryotes. *(Chapter 3, Section 2)*
3. **Differentiate** introns from exons. *(Chapter 10, Section 2)*
4. **Identify** the terms *bacteria* and *archaebacteria* *(Chapter 12, Section 2)* and *kingdom*. *(Chapter 14, Section 2)*
5. **Summarize** the system of classification of organisms. *(Chapter 14, Section 1)*

Did you have difficulty? *For help, review the sections indicated.*

Reading Activity

When you look at a living thing, how do know if it is an animal or some other type of organism? Animals have a particular set of characteristics that distinguish them from other organisms. Develop a list of the characteristics you look for when determining whether a living thing is an animal.

Members of the plant kingdom, such as the trees and herbs shown, have many similar features, including body organization and mode of nutrition. Scientists group organisms according to similarities.

Looking Ahead

internet connect

www.scilinks.org
National Science Teachers Association *sci*LINKS Internet resources are located throughout this chapter.

SCILINKS. Maintained by the National Science Teachers Association

Opening Activity — Basic

Bacteria Homes Find photos of the arctic, a geyser, tundra, sea vents, and a desert. Include micrographs of body structures such as skin and intestinal linings. Ask students what each environment has in common. After discussing possibilities, explain that they are all a home to bacteria. Emphasize that bacteria have evolved to survive in nearly every environment. **LS Visual**

Identifying Misconceptions

Tell students that although organisms may look similar, they do not necessarily belong to the same kingdom. For example, some corals may resemble plants, but they belong to the kingdom Animalia.

Quick Review Answers

1. A cell's surface area-to-volume ratio is its surface area divided by its volume, with both surface area and volume measured in the same units.
2. Prokaryotes are single-celled organisms that lack nuclei and membrane-bound organelles. Eukaryotes are unicellular or multicellular organisms that have nuclei and membrane-bound organelles in their cells.
3. Exons are segments of genes that are translated into proteins. Introns are segments of DNA that do not code for proteins.
4. Bacteria are prokaryotes that have peptidoglycan in their cell walls and have membrane lipids like those of eukaryotes. Archaebacteria are prokaryotes that lack peptidoglycan in their cell walls and have unique lipids in their membranes. A kingdom is a group of organisms that have certain characteristics in common.
5. The current system uses the form and structure of organisms as a basis for classification. The system is composed of ranked groups that increase in inclusiveness.

Chapter Resource File

- Vocabulary Worksheets
- Concept Mapping

Reading Activity Answers

Students' lists will likely include the animal characteristics of obtaining food by ingestion and of being motile. They might also note that most animals have cells organized into tissues, organs, and organ systems; they can reproduce sexually; they have a nervous system that enables them to sense and respond to their environment; and most have a circulatory system that transports oxygen, nutrients, and other needed substances to all the cells of the body.

Focus

Overview

Before beginning this section review with your students the objectives listed in the Student Edition. This section introduces the organization of the six kingdoms of living things and the criteria used to classify organisms. The grouping of the six kingdoms of organisms into three domains and the characteristics of the organisms in each domain will also be discussed.

Bellringer

Ask students to draw a diagram depicting a system for organizing the following office supplies: pencils, pens, paper clips, scissors, typing paper, lined paper, tape, rubber bands, and rulers. (One possible organizational scheme would involve grouping writing instruments, fasteners, papers, and tools into four groups.) English Language Learners
LS Logical

Transparencies

- TR Bellringer
- TR D27 Six Kingdoms

Section 1 Introduction to Kingdoms and Domains

Florida Sunshine Standards/FCAT

SC.G.1.4.1 There is great diversity and interdependence among living things.

Objectives

- **Identify** the characteristics used to classify kingdoms. SC.G.1.4.1
- **Differentiate** bacteria from archaebacteria. SC.G.1.4.1

Key Terms

None

The Six Kingdoms of Life

Biologists have long organized living things into large groups called kingdoms. For example, a hummingbird, an earthworm, an elephant, and a butterfly are all members of the animal kingdom. But what *is* an animal? Given the diversity of living things, it is easy to forget that all living things have much in common. Focusing on a few fundamental characteristics makes it easier to see the relatedness within kingdoms.

Cell Type Organisms are either prokaryotes, which have prokaryotic cells, or eukaryotes, which have eukaryotic cells. Scientists generally recognize two kingdoms of prokaryotes and four kingdoms of eukaryotes.

Cell Walls In four of the six kingdoms organisms have cell walls. In one of the six kingdoms, organisms lack cell walls. In the remaining kingdom, some organisms have cell walls and some do not.

Figure 1 Six kingdoms. Living organisms are divided into six kingdoms and are grouped according to their cell type, complexity, and method for obtaining nutrition.

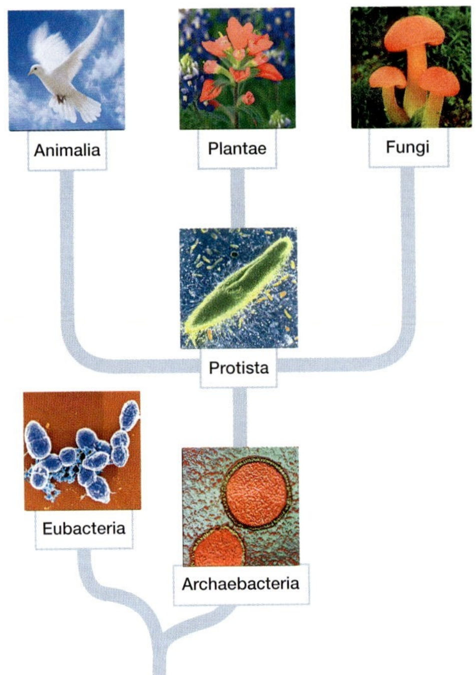

Body Type Organisms are either unicellular or multicellular. Two kingdoms consist only of unicellular organisms. Two other kingdoms have both unicellular and multicellular organisms. The two remaining kingdoms consist entirely of multicellular organisms, many of which have tissues and organs.

Nutrition Many organisms are autotrophs. Autotrophs make nutrients from inorganic materials. Many other organisms are heterotrophs. Heterotrophs get nutrients by consuming other organisms. Three kingdoms consist of both autotrophic and heterotrophic organisms. One kingdom consists of autotrophs. The other two kingdoms have only heterotrophic organisms.

Today, biologists group organisms into six kingdoms, based on their similarities: Eubacteria, Archaebacteria, Protista, Fungi, Plantae, and Animalia. Eubacteria and the Archaebacteria were once grouped in the kingdom Monera, which contained all the prokaryotes. However, data from RNA and DNA sequencing led biologists to divide the monerans into two distinct kingdoms. **Figure 1** shows how biologists think the six kingdoms are related to one another.

Chapter Resource File

- Directed Reading BASIC
- Active Reading GENERAL

One-Stop Planner CD-ROM

- Reading Organizers BASIC
- Reading Strategies BASIC

Attention Grabber

Huge Bacteria The organism *Epulopiscium fishelsoni*, which lives in the intestines of surgeonfish, puzzled scientists. It is over a million times larger than the *E. coli* in our own intestines. Its size led scientists to believe it might be a protist. However, they now classify the organism as a bacterium. A comparison of the organism's DNA with that of other organisms shows that the closest relative is gram-positive bacteria. That means the bacterium *E. fishelsoni* is larger than many eukaryotic cells.

Section Correlations

SC.G.1.4.1

The Three Domains of Life

For many decades, scientists recognized two basic forms of life, prokaryotes and eukaryotes. Then in 1977, the American scientist Carl Woese and his colleagues proposed that some prokaryotes are so fundamentally different from others that they merit their own broad division.

Woese and his team based their proposal on comparisons of ribosomal RNA sequences. The scientists showed that the group of prokaryotes that make up the kingdom Archaebacteria are more closely related to eukaryotes than they are to the other kingdom of prokaryotes, Eubacteria.

In 1996, scientists made the first comparison between complete DNA sequences of an archaebacterium and a bacterium. In recognition of the vast differences between the two groups of prokaryotes, biologists have adopted a classification system that divides all organisms into three superkingdoms, or domains: Bacteria, Archaea, and Eukarya. **Figure 2** shows how the three domains are related.

The domain thought to be the oldest is Bacteria, which is composed of the organisms in the kingdom Eubacteria. Archaea is the second prokaryotic domain and is also composed of a single kingdom, Archaebacteria. A third domain, Eukarya, contains all four of the eukaryotic kingdoms: Animalia (animals), Fungi (fungi), Plantae (plants), and Protista (protists). As you can see in Figure 2, the rRNA sequences of protists indicate that they are a very diverse group.

internet connect
www.scilinks.org
Topic: Domains of Life
Keyword: HX4060
SciLinks Maintained by the National Science Teachers Association

Figure 2 Three domains

This phylogenetic tree, based on rRNA sequences, demonstrates the division of all living things into three broad domains.

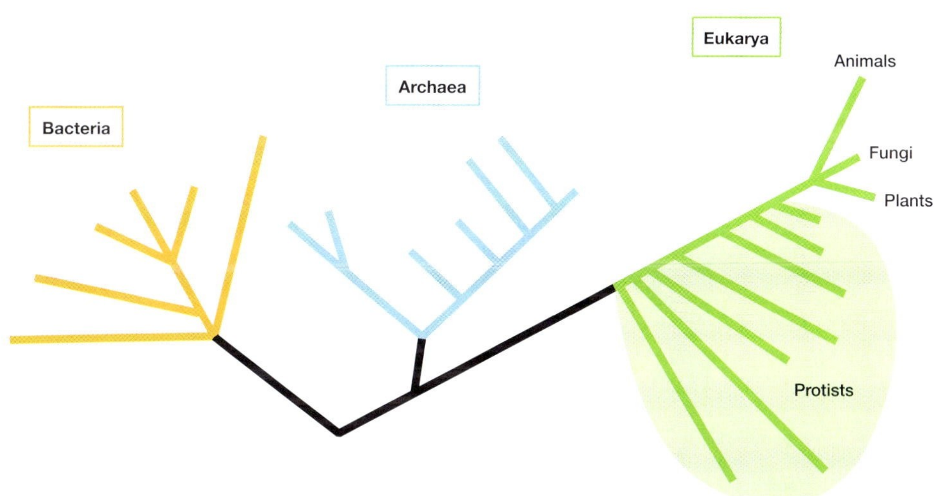

413

Motivate

Discussion — BASIC

Point out to students that we often organize things — such as products in a store, books in a library, and subjects in a school curriculum — into fundamental classes or groups. Work with the class as a whole to develop a list of other things that are organized into classes or groups. After students have created a list, tell them that biologists have organized all life forms into six major groups, or kingdoms. LS Verbal

Teach

READING SKILL BUILDER — BASIC

Interactive Reading Assign Chapter 19 of the *Holt Biology* Guided Audio CD Program to help students achieve greater success in reading the chapter. LS Auditory

Teaching Tip — ADVANCED

Recognizing Relationships Show students pictures of living organisms. Include organisms that are unusual in appearance, such as the slime mold Plasmodium, yeast, moss with brown fruiting bodies, a sea anemone, a shelf fungus, and a sponge. Ask students to determine in what kingdom they would place each organism. LS Logical

Group Activity — GENERAL

Grouping by Complexity Have students work in groups, and give each pair a set of pictures that includes bacteria, protists, plants, fungi, and animals. Have students design a family tree using the pictures. Ask them to justify their answers. Give them additional pictures and ask them to place these on their trees as well. Have students transfer their family trees onto paper using words in place of pictures. Ask them to label each organism as unicellular or multicellular. LS Visual English Language Learners

did you know?

Bacteria Are Used to Mine Minerals The first U.S. patent for using bacteria in mining was issued in 1958 to the Kennecott Copper Corporation. However, bacteria were first used to remove copper from ore by Roman miners over 2,000 years ago. The ancient Romans did not realize bacteria worked as a catalyst, but they recognized that the blue salts ($CuSO_4$ produced by the metabolism of bacteria) in the pools of some mines were copper-containing compounds.

Career

Bacteriologist Many bacteriologists are employed by health departments and medical institutions to assist in protecting the public from potential pathogens. Increasingly, however, bacteriologists are working for industry and agriculture. These scientists use bacteria as "micro factories" to produce medical and agricultural chemicals. A degree in biology, microbiology, or chemistry is required, with advanced studies in microbiology or immunology.

The Domain Bacteria

The domain Bacteria contains a single kingdom, the kingdom Eubacteria. Some scientists call this kingdom Bacteria. The common name for its members is *bacteria*. Bacteria are prokaryotes that have the same kind of lipid in their cell membranes as do eukaryotes. Most bacterial cells vary in size from 0.1 to 15 µm. Bacteria are found in practically every environment on Earth, and they affect humans every day. Bacteria are similar in physical structure, with no internal compartments, and they obtain nutrients in one of many different ways. There are several key characteristics common to all bacteria.

Study TIP

Organizing Information

Draw two overlapping circles. Label one circle "Bacteria" and the other circle "Archaea." In the circle labeled "Bacteria," write down the characteristics of bacteria. Do the same for the circle labeled "Archaea." Finally, in the area where the circles overlap, write down the characteristics that the two kingdoms share.

Characteristics of Bacteria

Cell Wall Bacteria have strong exterior cell walls made of peptidoglycan *(pep tih doh GLY kan)*, a weblike molecule complex made of carbohydrate strands cross-linked by short peptide bridges.

Gene Translation Apparatus Scientists infer evolutionary relationships between organisms based in part on the amino acid sequences in their proteins. The amino acid sequences of the ribosome proteins and RNA polymerases found in bacteria differ from those found in eukaryotes or in archaebacteria. This information, together with data from ribosomal RNA, is used to infer evolutionary relationships of groups within the kingdom Eubacteria.

Kinds of Bacteria

Bacteria are the most abundant organisms on Earth. There are more bacteria living in your mouth than there are mammals living on Earth. Some bacteria cause disease. Other bacteria are used by humans to process foods, such as the bacterium shown in **Figure 3.** Bacteria are used to control agricultural pests, to produce various chemicals, and to perform genetic engineering. Traditionally, bacteria have been classified according to their shape, the nature of their cell wall, and their type of metabolism. Traditional groupings of bacteria may change, however, as we get more information about their DNA and RNA.

Some bacteria obtain energy from inorganic compounds such as hydrogen sulfide, ammonia, and methane. Some bacteria are photosynthetic and are found in ocean and freshwater ecosystems, where they are primary producers. Other bacteria are heterotrophs. Some heterotrophic bacteria are capable of living in the absence of oxygen, while others must have oxygen to live. Heterotrophic bacteria are important decomposers in many ecosystems. They are responsible for the recycling of carbon, nitrogen, and phosphorus.

Figure 3 Useful bacterium. Bacteria, such as *Lactobacillus bulgaricus,* which turns milk into yogurt, can be useful to humans.

Teach, *continued*

Teaching Tip

Downsizing To help students understand the absolute and relative sizes of living things and their parts, draw the table in the Physics Connection below on the board. Define the abbreviations: mm = millimeter, µm = micrometer, nm = nanometer. Tell students that very small multicellular organisms, eukaryotic cells, and many bacteria can be seen with a light microscope. Objects of about 0.1 µm or smaller can be seen with an electron microscope.

Demonstration — BASIC

Show students food products that are made with the help of bacteria such as yogurt and tofu. Ask students why we worry about bacteria if we add them to certain foods. (Most bacteria are harmless or beneficial, but others can cause severe illnesses.) **LS Verbal**

INCLUSION *Strategies*

- ***Learning Disability***
- ***Attention Deficit Disorder***

Have the students list all the students in their class. Have the students divide the students into two groups of similar properties of characteristics. Next, the students should divide those groups into more specific and smaller groups because of similar properties or characteristics. The students should continue to divide these groups until each student is in their own individual group.

PHYSICS CONNECTION

The following chart shows how large several different organisms are in comparison to atoms and molecules:

Approximate size	Organism, cell, or particle
1mm	rotifers, roundworms
0.1mm	protozoa, algae, moulds, onion epidermis cells
0.01mm (10µm)	yeasts, human red blood cells
0.001mm (0.1µm)	bacteria
0.1µm	viruses
0.01µm	human hemoglobin molecules
0.001µm (1nm)	glucose molecules
0.1nm	atoms

The Domain Archaea

The domain Archaea also contains a single kingdom, the kingdom Archaebacteria. Archaebacteria are prokaryotes that seem to have diverged very early from the bacteria. They are more closely related to eukaryotes than to bacteria. Although they are a diverse group, all archaebacteria have certain features in common.

WORD Origins

The word *archaebacteria* is from the Greek words *archaios,* meaning "ancient," and *baktron,* meaning "a staff" (which refers to the rodlike shape of some bacteria).

Characteristics of Archaebacteria

Cell Wall and Membrane The cell walls of archaebacteria do not contain peptidoglycan, as the cell walls of bacteria do. Archaebacteria contain lipids very different from those of bacteria or eukaryotes.

Gene Structure and Translation As with the genes of eukaryotes, the genes of archaebacteria are interrupted by introns. The ribosomal proteins of archaebacteria are very similar to those of eukaryotes and different from those of bacteria.

Kinds of Archaebacteria

The archaebacteria first identified by scientists live in extreme environments, such as hot springs and salty lakes. This led scientists who studied archaebacteria to think that all of these unusual organisms live in very harsh environments. In recent years, however, scientists have learned that archaebacteria are much more common than was first thought. Several "signature sequences" of DNA are common to all archaebacteria. Using these sequences as probes, scientists were surprised to find archaebacteria in ordinary soil and even in seawater. There are three basic kinds of archaebacteria.

Methanogens These archaebacteria obtain energy by combining hydrogen gas, H_2, and carbon dioxide, CO_2, to form methane gas, CH_4. Methanogens live deep in the mud of swamps and are poisoned by even traces of oxygen.

Extremophiles A group of extremophiles called *thermophiles* lives in very hot places—up to 106°C. **Figure 4** shows a thermophilic species. *Halophiles* inhabit very salty lakes that can be three times as salty as seawater. Still other extremophiles live in very acidic places—with a pH below 1—or under enormous pressure—up to 800 atmospheres.

Nonextreme Archaebacteria Nonextreme archaebacteria grow in all the same environments that bacteria do.

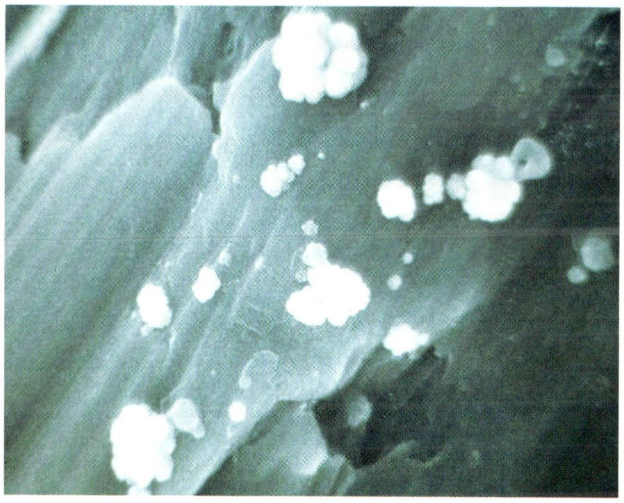

Figure 4 Thermophile. This species of *Acidianus brierleyi* is a thermophilic bacterium and is found near volcanic vents.

did you know?

Salty Bacteria Scientists studying chunks of crystallized salt in Baja California were hoping to find new species of archaebacteria by looking in material that was extremely salty. They were rewarded for their efforts when they found fabulously technicolored microbes ranging from yellow-green to pinkish purple inside the rocks. When the team compared DNA sequences from the bacteria in the rock samples with those available in databases, they discovered that they had collected a wide variety of both archaebacteria and bacteria. And many of these were new to science. Of the 150 bacteria species found, nearly a third had not previously been catalogued. And of the 90 archaebacteria species found, two-thirds had not been catalogued.

SKILL BUILDER — GENERAL

Math Skills Tell students that an average size for a bacterial cell is about 2 µm. Tell them that 1 µm (micrometer) is one-millionth of a meter. Ask them how large an image of an average bacterial cell would be if it were magnified one million times. ([2 µm] × [1,000,000 µm / m] = 2 m) **LS Logical**

Teaching Tip

Sources of Energy Point out to students that all organisms must use energy to carry out their metabolic activities—to grow, move around, sense aspects of their environment, reproduce, and conduct other activities. All organisms get their energy from energy stored in chemical bonds. Heterotrophic organisms get their energy from breaking down the carbon-carbon bonds in sugar and other molecules. Photosynthetic autotrophs, such as plants and cyanobacteria, also get their energy from breaking down carbon-carbon bonds in sugar molecules. However, photosynthetic organisms first build the sugar molecules using energy they harvest from the sun. Chemosynthetic autotrophs are organisms that produce their own food, but do so using chemical energy rather than the sun's energy. They use chemicals like hydrogen sulfide (H_2S) and ammonia (NH_3), and oxidize them by removing electrons from their chemical bonds.

Using the Figure — GENERAL

Direct students' attention to **Figure 4.** Tell them that organisms living in harsh environments, such as the thermophiles shown in this figure, stretch our concept of life-supporting habitats. If life can exist in such places on Earth, ask students to consider the possibility of life on other planets.
LS Intrapersonal

Teach, *continued*

Using the Table — GENERAL

Direct students' attention to **Table 1** on the opposite page. Point out to them the three domains (two prokaryotic, one eukaryotic) and six kingdoms (two prokaryotic, four eukaryotic) of organisms. Ask them to recall the four characteristics used to classify organisms. (cell type, cell structure, body type, nutrition) Ask them to identify the kingdoms of organisms that have cells with cell walls. (bacteria, archaebacteria, protists, fungi, plantae) Ask them to identify the kingdoms of organisms that have multicellular members. (protista, fungi, plantae, animalia) Next ask them to identify the kingdoms of organisms that have autotrophic members. (bacteria, archaebacteria, protista, plantae) Point out that organisms with nearly every possible combination of these four characteristics can be found in one of the six kingdoms. LS **Verbal**

Transparencies

TR D28 Kingdom and Domain Characteristics

The Domain Eukarya

The third domain of life, Eukarya, is made up of four kingdoms: Protista, Fungi, Plantae, and Animalia. Members of the domain Eukarya are eukaryotes, organisms composed of eukaryotic cells. A complex internal structure enabled eukaryotic cells to become larger and, eventually, led to the evolution of multicellular life. While different in many fundamental respects, eukaryotes share several key features.

Characteristics of Eukarya

Highly Organized Cell Interior All eukaryotes have cells with a nucleus and other internal compartments. This allows specialization of functions within a single cell.

Multicellularity True multicellularity, in which the activities of individual cells are coordinated and the cells themselves are in contact, occurs only in eukaryotes.

Sexual Reproduction Although exchange of genes occurs in bacteria, genetic exchange in eukaryotes is a more regular process. Eukaryotes have a life cycle that involves sexual reproduction. In this type of reproduction, meiotic cell division forms haploid gametes, and two gametes unite to form a diploid cell in fertilization. Genetic recombination during meiosis and fertilization causes the offspring of eukaryotes to vary widely, thus providing raw material for evolution.

Kinds of Eukaryotes

A wide variety of eukaryotes are unicellular. Most unicellular eukaryotes are grouped in the kingdom Protista. Protista contains both unicellular and multicellular organisms, many of which are aquatic. The protists are grouped together primarily because they do not fit in any other kingdom of eukaryotes.

Fungi are a group of heterotrophs that are mostly multicellular. Fungi are composed of cells with cell walls of chitin. One group of fungi, the yeasts, is unicellular. Many fungi live on and decompose dead organisms, and many other fungi are parasitic.

Figure 5 Multicellular eukaryotes. This jaguar and the vegetation it lives among are examples of complex multicellular eukaryotes.

HISTORY CONNECTION

Cholera is one of the most feared diseases on earth. In 1855, a wave of cholera ravaged the citizens of some parts of London. Thousands became ill and died before John Snow identified the Broad Street water pump as the single point source of that outbreak. His classical maps showing where people who became sick lived convinced him that the only possible source of the infection was the water pump.

MISCONCEPTION ALERT

What About Viruses? Students may wonder why viruses are not discussed in this lesson. Tell them that viruses do not meet any generally accepted definition of life. They are not made up of cells, they do not carry out metabolic activities, and they cannot reproduce on their own. Thus, they are considered nonliving. However, because they affect living things in many ways, their characteristics and activities are discussed in biology texts.

Section Review Correlations

SC.G.1.4.1: Items 1-5

Plants and animals, such as those in **Figure 5,** are all multicellular organisms. Almost all plants are autotrophs and have cells with cell walls composed of cellulose. All animals are heterotrophs composed of cells that do not have cell walls. Most plants and animals have tissues and organs.

Table 1 below summarizes the major characteristics of the organisms in the six kingdoms and three domains. You will learn more about the structure and diversity of living things as you continue to study biology.

Table 1 Kingdom and Domain Characteristics

Domain	Kingdom	Characteristics				
		Cell type	Cell structure	Body type	Nutrition	Example
Bacteria	Eubacteria	Prokaryotic	Cell wall, peptidoglycan	Unicellular	Autotrophic and heterotrophic	Enterobacteria Spirochetes
Archaea	Archae-bacteria	Prokaryotic	Cell wall, no peptidoglycan	Unicellular	Autotrophic and heterotrophic	Methanogens
Eukarya	Protista	Eukaryotic	Mixed	Unicellular and multicellular	Autotrophic and heterotrophic	Amoebas Euglenas Kelps
Eukarya	Fungi	Eukaryotic	Cell wall, chitin	Unicellular and multicellular	Heterotrophic	Yeasts Mushrooms
Eukarya	Plantae	Eukaryotic	Cell wall, cellulose	Multicellular	Autotrophic	Ferns Pine trees
Eukarya	Animalia	Eukaryotic	No cell wall	Multicellular	Heterotrophic	Birds Earthworms

Section 1 Review

1. **Analyze** the relationship between kingdoms and domains. SC.G.1.4.1
2. **Identify** the characteristics that distinguish the six kingdoms. SC.G.1.4.1
3. **Describe** how kingdom Eubacteria differs from kingdom Archaebacteria. SC.G.1.4.1
4. **Critical Thinking Evaluating Conclusions** Justify the division of prokaryotes into two kingdoms. SC.G.1.4.1
5. **FCAT Prep** Organism *X* is a multicellular, heterotrophic eukaryote whose cells lack cell walls. To which kingdom does organism *X* belong?
   A Animalia
   B Archaebacteria
   C Fungi
   D Plantae
   SC.G.1.4.1

417

Answers to Section Review

1. All living things are classified in one of six kingdoms of organisms. These six kingdoms are then organized into three different domains; two of these parallel the two different kingdoms of prokaryotes, and one is composed of the four kingdoms of eukaryotes.
2. Kingdoms are distinguished by cell type (eukaryote vs. prokaryote); cell structure (presence of a cell wall vs. absence of a cell wall, and cell wall composition); body type (unicellular vs. multicellular); and nutrition (heterotrophic vs. autotrophic).
3. Bacteria are prokaryotes with cell membrane lipids similar to those of eukaryotes. Bacterial ribosomal proteins and RNA polymerase differ from those of archaebacteria and eukaryotes. Archaebacteria are prokaryotes with unique membrane lipids and no peptidoglycan in their cell wall. Archaebacteria have genes with introns.
4. Although superficially similar, bacteria and archaebacteria are very different at the molecular level.
5. **A.** Correct. **B.** Incorrect. Archaebacteria are not eukaryotes. **C.** Incorrect. Fungi have cells with cell walls. **D.** Incorrect. Plants are autotrophic and their cells have cell walls.

Close

Reteaching — BASIC

Assign students to cooperative groups. Have them write down as many different forms of transportation as they can think of. Then ask them to place each mode of transportation into a transportation "kingdom." Have students compare kingdoms from each group. Ask them if some are more general while others are more specific. Tell students that scientists face the same challenge when trying to organize the biological kingdoms.

Co-op Learning | English Language Learners

Quiz — GENERAL

1. Why do many biologists now use a three-domain system for classifying organisms? (because they recognize that bacteria, archaebacteria, and eukaryotes are so different from each other that they should be considered to be different life forms)
2. How is gene translation different in bacteria compared to both eukaryotes and archaebacteria? (Bacterial ribosomal proteins and RNA polymerases differ from eukaryotic and archaebacterial ribosomal proteins and RNA polymerases)
3. In what kinds of environments might you find archaebacteria? (in water near deep-sea vents, in high sulfur-containing environments, and in very salty environments)

Alternative Assessment — GENERAL

Place students in cooperative groups. Give each group a list of 10 organisms. Next to each organism list a kingdom to which it may or may not belong. Have students determine if the kingdom associated with each organism is correct. Students must provide a reason for each decision.

LS Logical | English Language Learners

Focus

Overview

Before beginning this section review with your students the objectives listed in the Student Edition. This section discusses the difference between associations between unicellular organisms and true multicellularity. It describes the complexity of biological organization that can only occur with true multicellularity, and also discusses the characteristics of multicellular protists and of fungi.

Bellringer

Ask students what type of society they would prefer to live in, one in which they must do everything for themselves (grow and gather food, build shelter, etc.), or one in which each person does the job that they do best. Discuss the advantages of having each person do a specialized job. (Less time and energy is spent on basic survival.) Next, discuss the benefits of doing everything yourself. (You wouldn't have to rely on others for your survival.)

LS Intrapersonal

Motivate

Discussion — GENERAL

Ask students to list the advantages of being a multicellular organism and the advantages of being a unicellular organism. Discuss the advantages of each in class. (Multicellular organisms are larger and more complex. Unicellular organisms are smaller and simpler, and they reproduce very rapidly.)

LS Verbal

Section Correlations

SC.G.1.4.1

Section 2 Advent of Multicellularity

Florida Sunshine Standards/FCAT

SC.G.1.4.1 There is great diversity and interdependence among living things.

Objectives

- **Contrast** the terms *colony* and *aggregate.* SC.G.1.4.1
- **List** the characteristics of protists. SC.G.1.4.1
- **List** the characteristics of fungi. SC.G.1.4.1

Key Terms

colonial organism
aggregation
multicellular
differentiation
tissue
organ
organ system
protist
hypha

The Many Forms of Multicellularity

More than half of the biomass on Earth is composed of unicellular organisms—prokaryotes and some eukaryotes. For these organisms, unicellularity has been tremendously successful. However, many other organisms have found success not as individual cells but as members of a coordinated group of cells. Groups of cells that live together can have different levels of cooperation, as shown in **Figure 6.**

Colonies

Occasionally, the cell walls of bacteria adhere to one another. Some bacteria, such as cyanobacteria, form filaments, sheets, or three-dimensional formations of cells. These formations are not considered multicellular if the cells do not communicate and coordinate their activities. Such bacteria may properly be considered colonial. A **colonial organism** is a group of cells that are permanently associated but that do not communicate with one another. A colonial in the kingdom Protista is shown in Figure 6.

Aggregations

An **aggregation** *(a gruh GAY shuhn)* is a temporary collection of cells that come together for a period of time and then separate. For example, a plasmodial slime mold, such as the one shown in Figure 6, is a unicellular organism (a member of the kingdom Protista) that spends most of its life moving about and feeding as single-celled amoebas. When starved, however, these cells aggregate into a large group. This weblike mass produces spores, which are then dispersed to distant locations where there may be more food.

Figure 6 Multicellularity. *Volvox* is a colonial organism. Each is a hollow ball of hundreds or thousands of flagellated cells embedded in a jellylike layer. A plasmodial slime mold is an aggregate organism. Its cells form a large mass temporarily.

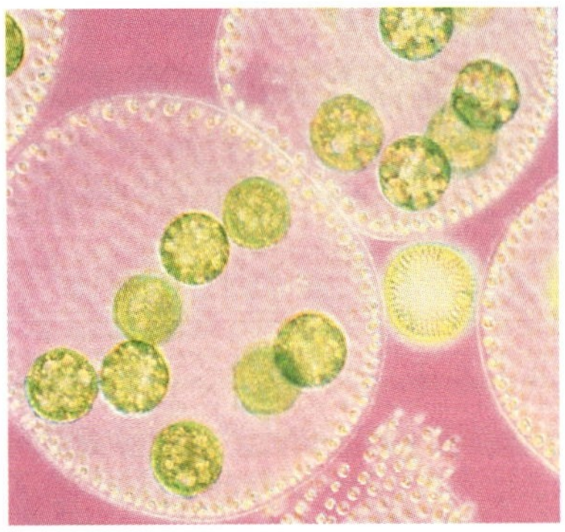

Volvox sp.

Plasmodial slime mold

Chapter Resource File

- Directed Reading BASIC
- Active Reading GENERAL

Transparencies

TR Bellringer

One-Stop Planner CD-ROM

- Reading Organizers BASIC
- Reading Strategies BASIC

True Multicellularity

A **multicellular** organism is an organism composed of many cells that are permanently associated with one another, such as the green alga shown in **Figure 7.** Multicellularity occurs only in eukaryotes. While single cells cannot grow larger than a certain size, multicellular organisms can be large. True multicellularity occurs when individual cells are in contact with each other and when their activities are coordinated.

Multicellularity enables cells to specialize in different functions. With this division of labor, a multicellular organism can have cells that protect it. Other cells help the organism move about, and still others play roles in reproduction and feeding. Cell specialization begins as a new organism develops. For example, as a chicken develops from an egg, new cells form by cell division. These cells grow and undergo **differentiation,** the process by which cells develop a specialized form and function.

Figure 7 **Cell specialization.** This green alga has specialized cells that hold it to the bottom of ocean tidal pools. It has many other specialized cells.

Complex Multicellularity

Plants and animals have complex multicellularity. The specialized cells of most plants and animals are organized into structures called tissues and organs. A **tissue** is a distinct group of cells with similar structure and function. Muscle, for example, is a tissue composed of many muscle cells that work together. Different tissues may be organized into an **organ,** which is a specialized structure with a specific function. An example of an organ is the heart, which is composed of muscle, nerve, and other tissues that work together as a pump. Various organs that carry out a major body function make up an **organ system.** The circulatory system, which is composed of the heart, the blood vessels, and the blood within them, is an example of an organ system. The relationships between tissues, organs, and organ systems are shown in **Figure 8.**

Figure 8 **Complex multicellularity**

Specialized cells form tissue that makes up an organ called the lung. The lungs and other organs constitute an organ system.

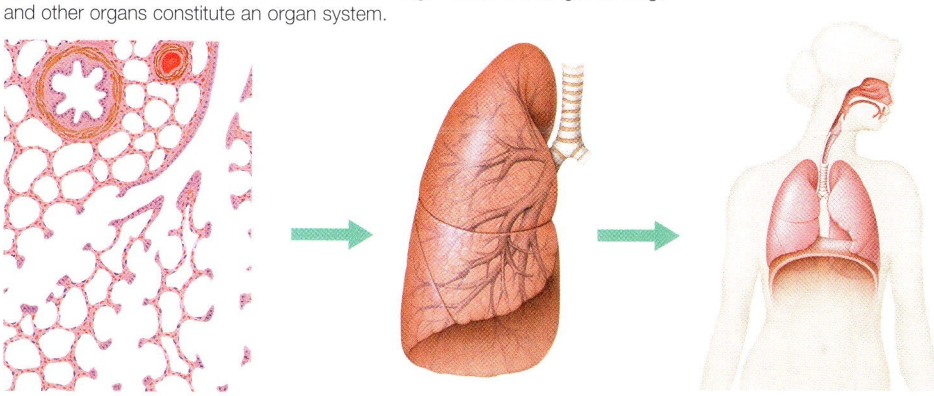

Teach

Teaching Tip — GENERAL

Cell Specialization Ask students to name different kinds of cells that make up the body. (skin cells, muscle cells, bone cells, etc.) Challenge students to explain how all these cells can be different if they all have the same genes. After discussing possibilities, tell them that cells begin to specialize very early in development, shortly after fertilization. As cells develop in different parts of the body, various chemicals are produced that cause the cells to turn genes "on" or "off," which leads to differentiation. **LS Verbal**

READING SKILL BUILDER — BASIC

Reading Organizer To help students who are having difficulty understand the levels of organization within an organism, instruct them to write the following headings across the top of a piece of paper:

Cell, Tissue, Organ, Organ System

Tell students to list at least three examples of each under the headings. Then ask students to share their information with their classmates. **LS Verbal**

REAL WORLD CONNECTION

Your skin cells depend on other body cells for nutrients and oxygen, which is why they cannot live independently of your body. When someone suffers a serious burn, that part of their body no longer has a protective covering of skin cells. Fortunately, scientists can now grow skin for them in a laboratory. Scientists take healthy skin cells, mix them with collagen proteins from cow skin, and suspend them in a nutrient solution. These new cells are applied to the burned skin as a bandage while the damaged skin heals.

Teach, *continued*

READING SKILL BUILDER — BASIC

Discussion Lead a class discussion on the level of complexity of the simplest eukaryotes, the protists and the fungi, compared to the prokaryotes and more complex eukaryotes (the plants and the animals). Help students to understand that single-celled organisms, which include all prokaryotes and many protists, are simple organisms because they don't have multiple cells that can specialize in different activities such as feeding and reproduction. The multicellular protists and the fungi do not show much more specialization. In most cases, most of the body's cells carry out all of the activities of the whole organism. LS Verbal

Activity — GENERAL

Distinguishing Organisms Set up microscope stations at which students can observe different structures of unicellular, colonial, aggregate, and multicellular algae. Have students make drawings of their observations and label the cell parts. LS Kinesthetic

Kingdom Protista

Of the six kingdoms of organisms, the kingdom Protista is the most diverse. Members of the kingdom Protista, **protists,** are defined on the basis of a single characteristic: they are eukaryotes that are not fungi, plants, or animals. Many are unicellular; in fact, all single-celled eukaryotes (except yeasts) are protists. Some protists, such as some kinds of algae, have cell specialization. Protists can vary widely in size, as shown in **Figure 9.** Most are microscopic, but some are as large as trees.

Protists are very diverse in many other ways. While all protists have a cell membrane, some—including algae and slime molds—have strong cell walls. Others—including diatoms and forams—produce shells of glassy silica. Some protists are photosynthetic autotrophs. Other protists are heterotrophs. Many protists move about, using structures such as flagella, cilia, or pseudopods. Protists normally reproduce asexually by means of mitotic cell division. When they are under stress, many protists reproduce sexually. The most ecologically important protists are probably the algae that live in the ocean and form the base of many ocean food chains.

Figure 9 Protists range in size. Protists can be so small that a microscope is required to see them. Some multicellular algae, such as the giant kelp can be as tall as 100 m.

Magnification: 1,500×

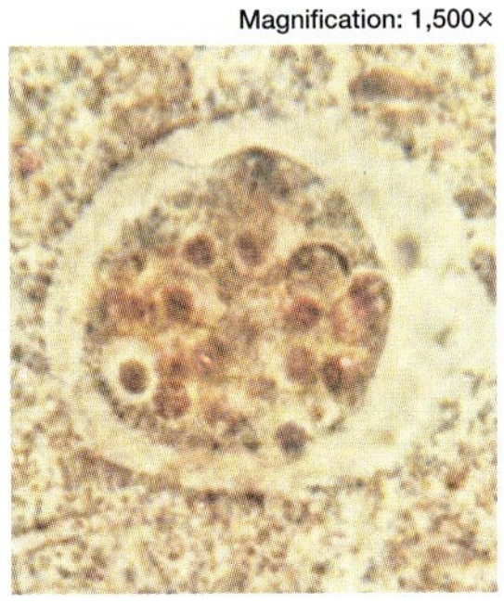

Entamoeba histolytica

Giant kelp

Kinds of Protists

Biologists recognize several general groups of protists. The six groups listed below are based on physical or nutritional characteristics.

Protists That Use Pseudopodia Amoebas are protists that have flexible surfaces with no cell walls or flagella; they move by using extensions of cytoplasm called pseudopodia *(soo doh POH dee uh)*. Forams, by contrast, have porous shells through which long, thin projections of cytoplasm can be extended.

Protists That Use Flagella Many protists, including autotrophs and heterotrophs, move by using flagella. The ciliates, which have large numbers of cilia, are so different from other protists that some biologists place them in a separate kingdom.

Protists with Double Shells Diatoms are photosynthetic protists with unique double shells made of silica, like boxes with lids. Diatoms are part of the plankton and may be found in fresh water or in marine environments.

Photosynthetic Algae Algae are photosynthetic and are distinguished by the kinds of chlorophyll they contain. Many algae are multicellular and reproduce sexually. Algae may be found in marine and freshwater environments.

Funguslike Protists Slime molds and water molds are often confused with fungi because they aggregate in times of stress to form spore-producing bodies. Slime molds are often found in fresh water, in damp soil, and on forest floors.

Spore-Forming Protists Sporozoans are nonmotile unicellular parasites that form spores. Responsible for many significant diseases, including malaria, sporozoans have complex life cycles.

OCEANOGRAPHY CONNECTION

Giant Kelp Giant kelps, *Macrocystis pyrifera*, are marine algae that grow to 30 meters (98.4 feet) tall. Much simpler than true plants, kelp has no roots, stems, leaves, or flowers. They have leaf-like blades that are photosynthetic. The stem-like strips carry sugars from the blades to the rest of the plant. And the root-like holdfasts anchor the organism to rocks. Giant kelps live on rocky substrates from a depth of 6–30 meters and form dense stands called kelp forests. In North America, these range from Alaska to Baja California, where the water stays cool year-round. A forest of *M. pyrifera* creates a reef-like habitat that is home to many other organisms. Snails, abalone, urchins and others graze on the kelp. They, in turn, fall prey to fishes, sea stars and sea otters. Sharks and sea lions hunt the hunters, while crabs feed on the dead.

Kingdom Fungi

Fungi are a very unusual and successful group of organisms. Although most fungi are multicellular, one group, the yeasts, are unicellular. The cell walls of all fungal cells contain chitin, the same tough material found in a crab's shell. The bodies of fungi consist of long strands of cells that are connected end to end and that share cytoplasm. The slender strands of fungi, like those in **Figure 10,** are called **hyphae** *(HIE fee)* (singular, *hypha*). Often, hyphae are packed together to form complex reproductive structures, such as mushrooms. Fungi reproduce by a variety of asexual and sexual methods.

In the past, fungi were classified in the plant kingdom. Like plants, fungi do not move from place to place. In addition, the general appearance of many fungi is similar to that of plants. But, fungi lack the green pigment chlorophyll and the ability to conduct photosynthesis. Like animals, fungi are heterotrophs. Unlike most animals, however, fungi do not ingest their food. Instead, fungi obtain food by secreting digestive enzymes onto whatever they grow on. Many fungi are saprophytes that live on dead organisms. Many other fungi are parasites that live on living organisms and cause diseases that affect plants and animals.

Figure 10 Fungus. The hyphae of fungi look like threads or filaments when they are viewed up close.

Kinds of Fungi

There are three phyla of fungi, and they are distinguished by their type of reproductive structures.

Zygomycetes Zygomycetes *(zie goh MIE seets)* form structures for sexual reproduction called zygosporangia. Zygomycetes include species such as *Rhizopus stolonifer,* common bread mold.

Basidiomycetes Basidiomycetes *(buh sih dee oh MIE seets)* include fungi that make mushrooms. Mushrooms are the sexual reproductive structures produced by basidiomycetes. Basidiomycetes almost always reproduce sexually.

Ascomycetes Ascomycetes *(as koh MIE seets)* form sexual spores in special saclike structures called asci. The sexual reproductive structure formed by ascomycetes often resembles a cuplike structure called an ascocarp.

Real Life

Some wild mushrooms are deadly.

Mushrooms, the fruiting bodies of some fungi, are eaten by many people throughout the world. In the United States, however, up to 100 people die each year from eating poisonous mushrooms. Always consult an expert before eating any wild mushrooms.

Finding Information

Read about a common poisonous species, Amanita *sp.*

Section 2 Review

1. **Differentiate** a colony from an aggregation. SC.G.1.4.1
2. **Describe** the characteristics of the protists. SC.G.1.4.1
3. **Describe** the characteristics of the fungi. SC.G.1.4.1
4. **Evaluate** the argument that insects and fungi are closely related because both have chitin covering their bodies. SC.G.1.4.1
5. *FCAT Prep* What is the process by which cells become specialized in form and function during development?
   A aggregation
   B association
   C coordination
   D differentiation
   SC.F.1.4.2

Real Life

Answer

Amanita species contain extremely dangerous toxins that can destroy a person's liver in one week.

Close

Reteaching — BASIC

Have each student list the differences between multicellular and unicellular organisms. Then have students share and discuss their lists with a partner. Have each pair combine and revise their lists as needed to create one master list. Students should then work together to see how well their list matches the material presented in the text.

LS Interpersonal English Language Learners

Quiz — GENERAL

1. Different cells of a multicellular organism become ________ to carry out different activities. (specialized)
2. List the distinguishing characteristics of fungi. (Fungi are almost all multicellular, heterotrophic, and their cells have cell walls containing chitin.)

Alternative Assessment — GENERAL

Ask students to explain how multicellularity contributed to biodiversity. (Cell specialization made it possible for different structures and different ways of functioning to evolve.) **LS Verbal**

Answers to Section Review

1. A colony consists of a group of cells that are permanently associated but do not communicate with one another. An aggregation consists of a group of cells that come together temporarily.
2. Protists are eukaryotic organisms. Many of them are unicellular; some are autotrophic and some are heterotrophic; and some have cell walls and some do not.
3. Fungi are multicellular (except yeast) eukaryotes that exist primarily as threadlike strands called hyphae. Their cell walls contain chitin. They are heterotrophic, digesting organic matter by secreting enzymes into their environment and then absorbing the nutrients.
4. The presence of chitin is a single, superficial trait and probably arose through convergent evolution.
5. **A.** Incorrect. An aggregation is a cluster of something, such as cells. **B.** Incorrect. An association is a relationship between two things. **C.** Incorrect. Coordination is a manifestation of communication between two entities and can occur in living things. **D.** Correct.

Section Review Correlations

SC.F.1.4.2: Item 5

SC.G.1.4.1: Items 1-4

Section 3 Complex Multicellularity

Florida Sunshine Standards/FCAT

SC.F.1.4.2 Body structures are uniquely designed and adapted for their function.

SC.G.1.4.1 There is great diversity and interdependence among living things.

Objectives

- **List** the levels of cellular organization that occur in plants and animals. SC.F.1.4.2
- **Name** the characteristics of plants. SC.G.1.4.1
- **Identify** the characteristics of animals. SC.G.1.4.1
- **Differentiate** plants from animals. SC.G.1.4.1

Key Terms

vascular tissue
invertebrate
vertebrate

internet connect

www.scilinks.org
Topic: Plants of Florida
Keyword: HX6F16

SCILINKS. Maintained by the National Science Teachers Association

Kingdom Plantae

Plants are complex multicellular autotrophs; they have specialized cells and tissues. Most plants have several different types of cells that are organized into many specialized tissues. For example, **vascular** *(VAS kyoo lur)* **tissue** is made up of specialized cells that play a role in transporting water and dissolved nutrients. Plant cells are different from all other cells in that their cell walls are composed of cellulose, a complex carbohydrate.

Unlike many other organisms, plants cannot move from one place to another. A few groups have motile sperm, but most plants are rooted in the ground. Portable reproductive structures, such as spores and seeds, enable the dispersal of plants.

As autotrophs, plants are the primary producers in most terrestrial food webs. Thus, they provide the nutritional foundation for most terrestrial ecosystems. Plants also release oxygen gas to the atmosphere. They are also very important in the cycling of phosphorus, water, nitrogen, and carbon.

Like the fungi, plants evolved on land and are the dominant organisms on the surface of Earth. Plants cover every part of the terrestrial landscape, except for the extreme polar regions and the highest mountaintops.

Plants are sources of food for humans and other animals. They are also a source of medicines, dyes, cloth, paper, and many other products. As shown in **Figure 11,** plants vary in size from the 1 mm tall duckweed *(Wolffia microscopica)* to the giant sequoia redwood *(Sequoia sempervirens)*, which can grow to over 90 m (296 ft) tall.

Figure 11 Range of plant size. Plants can be as small as duckweed or as large as a redwood.

Duckweed

Redwood tree

Focus

Overview

Before beginning this section review with your students the objectives listed in the Student Edition. This section discusses the characteristics of the complex organisms that are classified in the plant and animal kingdoms. The section also covers the distinguishing characteristics of the main groups of plants and animals.

Bellringer

Ask students to write down the common name of the smallest and largest plants they can think of. Then ask them to write down the common name of the smallest and largest animals they can think of. LS Intrapersonal

Motivate

Demonstration —— Advanced

Show students pictures or overheads of the internal structures of plant and animal cells. Point out the chloroplasts and mitochondria. Remind students that many scientists think that these organelles were originally symbiotic prokaryotes living inside early eukaryotic cells (see the theory of endosymbiosis in Chapter 12). Eventually they became permanent parts of the eukaryotic cell. LS Visual

Section Correlations

SC.F.1.4.2
SC.G.1.4.1

Chapter Resource File

- Directed Reading Basic
- Active Reading General

Transparencies

TR Bellringer
TR Graphic Organizer

One-Stop Planner CD-ROM

- Supplemental Reading *The Andromeda Strain*

Mosses

Ferns

Figure 12 **Seedless plants.** Mosses and ferns are two groups of plants that do not produce seeds.

Kinds of Plants

There are four basic kinds of plants, as shown in **Figure 12** and **Figure 13.** They differ from one another according to the type of vascular tissue and reproductive structures that they have.

Nonvascular Plants Plants without a well-developed system of vascular tissues are called nonvascular plants. These plants are all relatively small. They lack the tissue to transport water and dissolved nutrients. They also lack true roots, stems, and leaves. Mosses, such as the one shown in Figure 12, are the most familiar example of nonvascular plants.

Plants with a well-developed system of vascular tissues are called vascular plants. Their larger, more-complex bodies are organized into roots, stems, and leaves. Most plants are vascular plants. One group of vascular plants, called seedless vascular plants, does not produce seeds.

Seedless Vascular Plants Ferns are the most common and familiar seedless vascular plants. They have roots, stems, and leaves, and their surfaces are coated with a waxy covering that reduces water loss. They reproduce with spores that are resistant to drying. Both haploid and diploid phases occupy significant parts of the life cycle.

Most vascular plants produce seeds. Vascular plants that produce seeds are called seed plants. There are two general types.

Nonflowering Seed Plants Gymnosperms *(JIHM noh spuhrmz)* are vascular plants that reproduce using seeds but do not produce flowers. Gymnosperms include plants that produce seeds in cones, such as pines and spruces. Seeds enable plants to scatter offspring and to survive long periods of harsh environmental conditions, such as drought and extreme temperatures.

Flowering Seed Plants Most plants that produce seeds also produce flowers. Flowering plants are called *angiosperms (AN jee oh spuhrmz)*. Angiosperms, such as roses, grasses, and oaks, produce seeds in fruits. Fruits are structures that enable the dispersal of seeds.

Figure 13 **Seed plants.** Plants that make seeds are either nonflowering, such as the pine trees, or flowering, such as these bluebonnets.

Pine trees

Bluebonnets

Teach

READING SKILL BUILDER — **BASIC**

Reading Hint Before students read further, have them create two lists of as many different kinds of plants and animals as they can think of. Have them write down broad categories rather than individual species. Ask them to compare their lists to the lists in the text. Have students then correct their lists. LS Verbal

Teaching Tip — GENERAL

Organizing Cellular Complexity
Have students create a Graphic Organizer similar to the one at the bottom of this page to illustrate the increasing levels of complexity found in organisms. Ask students to include an example for each level of complexity, if there is one. LS Visual

Demonstration — GENERAL

Show students a specimen or photograph of a plant crown gall (one that was not induced by insects). Point out that the cells of complex, multicellular organisms must be highly regulated by internal mechanisms to prevent unwanted growth and changes in cell function. Changes in the DNA (mutations) of cells can cause this to happen. When cells of a particular tissue in an animal begin to grow out of control, the animal is said to have a tumor, or cancer. Tell students that such a growth in a plant is usually called a gall. LS Visual

Graphic Organizer

Use this graphic organizer with **Teaching Tip** on this page.

Single cell (Example: any bacterium)

Colony/Aggregate (Example: *Volvox*, plasmodial slime mold)

Multicellular (Example: any Fungus, except yeast)

Organs (Example: any plant or animal)

Organ systems (Example: any plant, any animal except a sponge or a cnidarian)

Teach, *continued*

Group Activity — GENERAL

Diversity of Living Things Have students work in pairs to collect pictures or drawings of representatives of the plant and animal kingdoms. Assign each group to work on one of the groups of plants (nonvascular, seedless vascular, non-flowering seed, flowering seed) and animals (sponges and cnidarians, mollusks, worms, arthropods, echinoderms, vertebrates). Encourage them to collect 2–3 examples for each group. Display the pictures and drawings on a poster board. **LS Visual**

Co-op Learning | English Language Learners

Activity — ADVANCED

Writing **Animal Characteristics** Sometimes the specialized or differentiated cells of an animal develop into a unique characteristic. For example, all mammals have hair, but only the male lion has a large, bushy mane. Have students research unusual physical characteristics of animals that are exaggerated versions of ordinary characteristics. Tell students to prepare a multimedia presentation of their research results. (Suggest that students investigate a peacock's feathers, an elephant's trunk, or the "suction pad" toes of various frogs.) **LS Verbal**

Figure 14 A running animal. This impala is demonstrating the ability to avoid predators. Running is the result of several organ systems working well together.

Kingdom Animalia

Animals are complex multicellular heterotrophs. Their cells are mostly diploid, lack a cell wall, and are organized as tissues. In addition, their zygotes develop through several stages. These adaptations have enabled animals to be successful in different habitats. The specialized tissue called *muscle* enables animals to move about readily. As illustrated in **Figure 14,** the ability of animals to move more rapidly and in more complex ways than members of other kingdoms is one of their most interesting characteristics. A remarkable form of movement unique to animals is flight, an ability that is well developed among both insects and vertebrates. Movement enables animals to avoid predators and to look for food and mates.

Most animals reproduce sexually. In animals, cells formed in meiosis function directly as gametes. The haploid cells do not divide by mitosis first, as they do in plants and fungi, but rather fuse directly with one another to form the zygote. The zygote then gradually develops into an adult, going through several developmental stages.

Almost all animals (99 percent) are **invertebrates;** that is, they lack a backbone. Of the more than 1 million living species, only about 42,500 have a backbone; they are referred to as **vertebrates.** The animal kingdom includes about 35 phyla, most of which live in the sea. Far fewer phyla live in fresh water, and fewer still live on land.

internet connect
www.scilinks.org
Topic: Animals of Florida
Keyword: HX6F17
SCILINKS. Maintained by the National Science Teachers Association

Kinds of Animals

Animals are very diverse in form, as shown in **Figure 15.** They can range in size from 0.5 mm (0.02 in.) microscopic mites *(Demodex folliculurum)* that live on your skin to enormous whales, which are vertebrates, and giant squids, which are invertebrates. Blue whales can reach a length of 30 m (100 ft) and weigh up to 220 tons. The many kinds of animals can be grouped by phylum.

Figure 15 Variety of animals. Organisms as small as the mite share basic characteristics with organisms as large as the whale.

Demodex follicularum

Blue whale

424

Integrating Physics and Chemistry

Having the ability to move more rapidly than living organisms of other kingdoms is one of the unique features of the animal kingdom. One way that physicists analyze movement is by calculating speed *(speed = distance/time).* Have students look up the maximum speeds reached by a variety of animals and then calculate how far these animals can move in some limited time period.

INCLUSION Strategies

- *Gifted and Talented*
- *Attention Deficit Disorder*

Have the students do a field study of the plants in their schoolyard. The students should diagram or take pictures of different types of plants at school and classify the diagrams or pictures by similarities of each plant. Additionally, have the student identify the scientific names of each plant to look for similarities of these plants.

Sponges Sponges, like the one shown in **Figure 16,** are the only animals that do not have tissues, but they do have specialized cells.

Cnidarians Cnidarians *(nih DAYR ee uhnz)* are mostly marine animals; they include jellyfish, sea anemones, and corals.

Flatworms Flatworms have flat, ribbonlike bodies. Some are parasitic and live inside the bodies of animals. Others are freeliving and may live in soil or water.

Roundworms Nematodes, or roundworms, are small worms that have long, very slender bodies. Some roundworms are freeliving in soil or water, while others are parasites of animals and plants.

Segmented Worms Annelids, or segmented worms, live in both water and soil and include the familiar earthworm. Bristled marine worms are segmented worms, as are leeches, which can be blood-sucking parasites.

Mollusks Mollusks have a saclike cavity called a coelom that encloses internal organs. Mollusks are very diverse aquatic and terrestrial animals. They include snails, oysters, clams, octopuses, and squids. Most mollusks have a hard external skeleton (a shell).

Arthropods By far the most diverse of all animals, arthropods have an external skeleton. They also have jointed appendages, such as antennae and jaws. These structures enable arthropods to sense their environment and obtain food. Two-thirds of all named species of animals are arthropods, most of them insects. The high rate of reproduction of insects has contributed to their success.

Echinoderms This group of invertebrates includes sea stars, sea urchins, and sand dollars. Many echinoderms *(ee KIE noh duhrmz)* are able to regenerate a lost limb. In fact, some echinoderms can lose a limb to a predator in order to escape that predator.

Invertebrate Chordates Invertebrate chordates are aquatic animals that have much in common with vertebrates, though they do not have a backbone. Some invertebrate chordate are swimmers that resemble fish, while others live attached to a rock or other object.

Vertebrates have an internal skeleton made of bone, a vertebral column (backbone) that surrounds and protects the spinal cord, and a head with a brain contained in a bony skull. Vertebrates include mammals, fish, birds, reptiles, and amphibians.

Study TIP

Reviewing Information
After reading this section, make a list of the questions you have about multicellularity. Exchange your list with a partner, and find answers to the questions.

Figure 16 Sponge. A sponge has specialized cells in its body that enable it to eat and reproduce sexually.

SKILL BUILDER — GENERAL

Interpreting Visuals Use a microprojector or a videomicroscope to show students live *Euglena* specimens. Ask them to determine whether they would place this organism in kingdom Plantae because it contains chlorophyll and is autotrophic, kingdom Animalia because it moves, or kingdom Protista because it is unicellular. Point out to students that research at the molecular level is often needed in order to make these kinds of determinations. (*Euglena* is classified as a Protist.) **LS Visual**

Teaching Tip — GENERAL

Ants and Termites Tell students that arthropods, and especially insects, make up most of the world's animal life. Then tell them some scientists estimate that ants account for 10 percent of the animal biomass in the world, and that termites account for another 10 percent.

Group Activity — GENERAL

Writing **A Zoological Expedition** Take the class on a field trip to a local zoo. Before the trip, explain that different animals have evolved different structures and tissues that help them exist in their natural habitat. While at the zoo, point out such structures on some of the animals that you see. After returning to class, have students write a report on the animals they saw and the unique adaptations that enable those animals to survive in their environment. **LS Visual**

Cultural Awareness

Giant clams live on coral reefs in warm tropical waters of the Pacific Ocean. They have long been a staple in the diet of many Pacific islanders and are a delicacy in many Asian countries. During the past 20 years, however, giant clams have been threatened with extinction due to overfishing. Scientists from Australia, Japan, Indonesia, and several other countries have been working to restore giant clam populations by raising thousands of these mollusks on giant clam "farms" along protected coasts. When farmed clams are large enough to be safe from predators, they are transplanted to coral reefs. As a result of these efforts, giant clams are once again common around many Pacific islands.

Teach, *continued*

QUICK LAB

Modeling True Multicellularity

Skills Acquired
Observing, interpreting results

Teacher's Notes
Make sure the colonial group does not use any communication, such as gestures or grunts.

Answers to Analysis

1. The multicellular group.
2. Communication increased their efficiency.
3. The multicellular group is more advanced because its cells are coordinated.
4. Answers will vary.

Close

Reteaching — BASIC

Assign one of the six groups of animals discussed in the text to each student. Have students create a poster of the features of that group including a species. LS **Visual**

Quiz — GENERAL

1. How are cells organized in plants and animals? (Cells are organized into tissues, tissues form an organ, and organs are organized into organ systems.)
2. What characteristics distinguish members of the four major groups of plants? (vascular tissues, seeds, and flowers)

Section Review Correlations

SC.F.1.4.2: Items 1, 5
SC.G.1.4.1: Items 2, 3, 4

Ecological Roles

Animals fulfill various roles in an ecosystem. Some animals are detritivores, animals that feed on waste and dead tissue. The microscopic mites that live on your body and eat dead skin cells are detritivores. Other animals, such as buffalo that eat grass, are primary consumers. Many animals, such as humans, bears, and lions, are secondary consumers that eat primary consumers. Finally, some animals, such as intestinal worms, act as parasites and may cause disease in other animals.

QUICK LAB

Modeling True Multicellularity SC.G.2.4.6

In order to understand the advantage that true multicellular organisms have over colonial organisms, you will model multicellular and colonial life.

Materials

two 15 ft lengths of rope, several objects in the classroom

Procedure

1. Working as a class, divide into two groups. One group will model a colonial organism, and the other will model a true multicellular organism. Your teacher will loosely tie a rope around each group.
2. One student in each group will receive a set of instructions for collecting objects from around the classroom.
3. As each group carries out its instructions, students modeling the true multicellular organism may talk with one another, but students modeling the colonial organism must remain silent.

Analysis

1. **Identify** which group finished the assigned task first.
2. **Infer** why the first group to finish was able to accomplish its task so quickly.
3. **Choose** which type of organism is more advanced. Explain.
4. **Predict** how the more advanced organism could become more efficient.

Section 3 Review

1. **Identify** the type of cellular organization found only in plants and animals. SC.F.1.4.2
2. **Describe** the identifying characteristics of plants. SC.G.1.4.1
3. **Relate** the characteristics of animals to a dog or a cat. SC.G.1.4.1
4. **Compare** the characteristics of plants and animals. SC.G.1.4.1
5. *FCAT Prep* What are tissues composed of?
   A cells
   B organs
   C organ systems
   D organisms
   SC.F.1.4.2

426

Alternative Assessment — ADVANCED

Dichotomous Key Have students create a dichotomous key that includes 3 plants and 3 animals. LS **Logical**

Answers to Section Review

1. Plants and animals have specialized cells that are organized into different tissues.
2. Plants are multicellular, photosynthetic autotrophs with specialized cells and tissues. They lack the ability to move from place to place.
3. Dogs and cats are typical animals in that they have specialized cells, tissues, and organs, and are heterotrophic and motile. They also have complex organs and organ systems.
4. Both plants and animals are multicellular eukaryotes that have specialized cells that form tissues. Plants are autotrophic, while animals are heterotrophic. Plants lack the ability of animals to move from place to place. Plant cells have a cell wall, but animal cells do not.
5. **A.** Correct. **B.** Incorrect. Tissues are organized into organs, not vice-versa. **C.** Incorrect. Tissues are organized into organs, and organs into organ systems. **D.** Incorrect. Tissues are organized into organs, which are organized into organ systems, which are organized into organisms.

Study ZONE

CHAPTER HIG

Key Concepts

1 Introduction to Kingdoms and Domains

- Members of the six kingdoms are grouped according to their cell type, cell complexity, body type, and nutrition.
- Bacteria are heterotrophic and autotrophic prokaryotes that have peptidoglycan in their cell wall.
- Bacteria are classified according to their nutrition, their cell shape, and the nature of their cell wall.
- Archaebacteria are prokaryotes that have unusual lipids in their cell membrane, have no peptidoglycan in their cell wall, and have introns in their genes.

2 Advent of Multicellularity

- A colonial organism is a group of cells that live together permanently but do not coordinate most cell activity. Aggregations are collections of cells that come together for a limited period of time.
- Only eukaryotes exhibit true multicellularity, which occurs when the activities of the individual cells are coordinated and the cells are in contact with one another.
- Protists include multicellular and unicellular eukaryotes and can be heterotrophs or autotrophs.
- Fungi are eukaryotic, principally multicellular heterotrophs that exist mainly as slender hyphae.

3 Complex Multicellularity

- Specialized cells are organized into structures called tissues, organs, and organ systems. These cells have special functions and coordinate their activities with one another.
- Plants are photosynthetic eukaryotes with tissues. Their cells have cell walls.
- Plants are the primary producers in most terrestrial food webs. They release oxygen gas and aid in resource cycling.
- Animals are multicellular heterotrophs with cells that lack a cell wall, that are organized as tissues, and that are mostly diploid. They reproduce sexually, and their zygotes develop through several stages.

Key Terms

Section 1

None

Section 2

colonial organism (418)
aggregation (418)
multicellular (419)
differentiation (419)
tissue (419)
organ (419)
organ system (419)
protist (420)
hypha (421)

Section 3

vascular tissue (422)
invertebrate (424)
vertebrate (424)

Alternative Assessment — GENERAL

Use the names of the six kingdoms of living things as the categories for a game. Assign students to cooperative groups of four. Have each group develop four questions and answers for each of the six categories. Arrange competitions by having two groups use the questions and answers of a third group. Competitors choose a category and select an answer (already revealed) at random. They must then formulate a question to match the answer. **LS Verbal**

Co-op Learning

Chapter Resource File

- Science Skills Worksheet GENERAL
- Critical Thinking Worksheet ADVANCED
- Test Prep Pretest GENERAL
- Chapter Test GENERAL

Answer to Concept Map

The following is one of several possible answers to Performance Zone item 15.

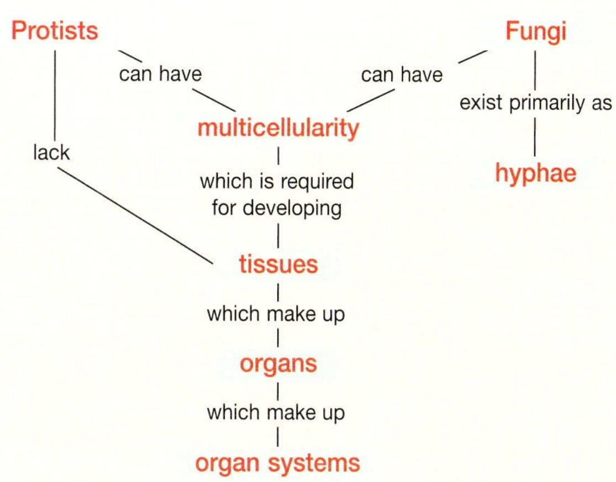

ANSWERS

Understanding Key Ideas

1. b
2. a
3. c
4. a
5. b
6. Accept all reasonable answers, such as the relatively large size of the bird, the presence of complex structures (skin, feathers, beak, etc.), reproduction (chicks), and nutrition (eating insects).
7. The cell walls of bacteria are made of peptidoglycan, the cell walls of fungi are made of chitin, and the cell walls of plants are made of cellulose.
8. The ribosomal RNA sequences, cell membrane lipids, and cell wall compounds of archaebacteria are different from those of other prokaryotes.
9. Protists are eukaryotes that lack the distinguishing characteristics of other eukaryotes, such as tissues, organs, chitin, and septae. Therefore, they are not fungi, plants, or animals.
10. One possible answer to the concept map is found at the bottom of the Study Zone page.

Critical Thinking

11. Multicellular organisms evolved cell specialization, which enabled the development of tissues, organs, and organ systems.

Chapter Review Correlations

SC.F.1.4.2: Items 4, 6, 10, 11
SC.G.1.4.1: Items 1, 2, 3, 5, 7, 9, 13
SC.H.1.4.1: Items 12, 14
SC.H.1.4.2: Item 8

Performance ZONE — CHAPTER REVIEW

Understanding Key Ideas

1. Which of the following characteristics is *not* used to classify organisms into kingdoms?
   a. cell type
   b. photosynthetic pigment
   c. body type
   d. nutrition
   SC.G.1.4.1
2. Eubacteria exhibit all of the following except
   a. nuclei.
   b. strong exterior walls made of peptidoglycan.
   c. a lack of internal compartments.
   d. size less than 15 μ.
   SC.G.1.4.1
3. Archaebacteria differ from eubacteria because archaebacteria
   a. have cell walls made of peptidoglycan.
   b. have cell membranes made of phospholipids.
   c. contain genes interrupted by introns.
   d. cannot live in harsh environments.
   SC.G.1.4.1
4. Tissues occur only in
   a. animals and plants.
   b. animals.
   c. protists and plants.
   d. animals and protists.
   SC.F.1.4.2
5. Animals differ from plants in that animals
   a. have chlorophyll.
   b. have no cell walls.
   c. have eukaryotic cells.
   d. reproduce sexually.
   SC.G.1.4.1
6. What information can you find in the photograph below that indicates that the bird is a complex multicellular organism? SC.F.1.4.2

7. How do cell walls differ among bacteria, fungi, and plants? SC.G.1.4.1
8. What evidence led Carl Woese to propose that archaebacteria should be classified separately from other prokaryotes? SC.H.1.4.2
9. What criteria are used to classify organisms as members of the kingdom Protista? SC.G.1.4.1
10. **Concept Mapping** Construct a concept map that shows the characteristics of protists and fungi. Try to include the following terms in your map: *protists, fungi, multicellularity, tissues, organs, organ systems,* and *hyphae.* SC.F.1.4.2

Critical Thinking

11. **Applying Information** How did the advent of multicellularity enable organisms to grow in complexity? SC.F.1.4.2

Alternative Assessment

12. **Finding Information** Use the media center or Internet resources to find out what kind of cell types and cell organization occurs in different kinds of organisms. What kind of animals are scientists using now to research basic questions about embryo development? What questions are scientists asking, and what information are they finding? Prepare a poster to summarize your findings. SC.H.1.4.1
13. **Comparing Structures** Visit the zoo, and list the scientific names of all the animals you see, or use your library to research 10 organisms. Record the scientific and common names of these organisms. For each organism, identify a trait that led taxonomists to classify the organism in its particular genus or family. SC.G.1.4.1
14. **Career Connection** Zoologist Research the field of zoology, and write a report on your findings. Your report should include a job description, training required, kinds of employers, growth prospects, and a starting salary. SC.H.1.4.1

Alternative Assessment

12. Answers will vary but should include a discussion of multicellularity, cell specialization, tissues, and organs.
13. Answers will vary depending on the organisms that are selected.
14. Answers will vary. Zoologists study animals, including their activities, how they interact with one another and their environment, and how different species are related to one another. Most zoology jobs require a college education. Certain zoology careers require additional training in graduate or professional schools. Employers include universities, zoos, and museums. Starting salary will vary by region.

Assignment Guide

Section	Questions
1	1, 2, 3, 7, 8
2	4, 6, 9, 10, 11
3	5, 12, 13, 14

FCAT Prep

For questions 1–6, write your answers on a separate sheet of paper.

1. In which kind of organism does multicellularity occur? SC.G.1.4.1
   A. aggregates
   B. archaebacteria
   C. eukaryotes
   D. prokaryotes

2. Through what process do cells become specialized? SC.F.1.4.2
   F. aggregation
   G. colonialism
   H. differentiation
   I. vascularization

3. What term is used to describe an organism that has cells that are permanently associated but do not communicate with one another? SC.G.1.4.1
   A. aggregate
   B. colonial
   C. complex
   D. heterotrophic

4. How are the cells of multicellular fungi arranged? SC.G.1.4.1
   F. in sheets
   G. in tissues
   H. in colonies
   I. in filaments

5. READ INQUIRE EXPLAIN Paleontologists think that many different organisms evolved in the history of Earth, but not one kind of cell nor level of complexity completely dominated Earth at any one time.

   What mechanism could explain the diversity of life? SC.D.1.4.4

6. READ INQUIRE EXPLAIN Developmental biologists often study the tissues and cells of simple animals in order to understand more complex animals. For example, they might study embryonic development in fruit flies and echinoderms in order to understand embryonic development of more-complex animals.

   Part A What do all animals, whether simple or more complex, have in common? SC.G.1.4.1

   Part B Evaluate the frequent similarity of the mechanisms of the processes in simple and complex organisms. SC.G.1.4.1

Test TIP

For short-response questions, be sure to answer the prompt as fully as possible. Include supporting details in your response.

429

FCAT Doctor

Question 2 Answer H is the correct choice. Differentiation is the process by which cells develop a specialized form and function. Answer F is incorrect because the cells in aggregations are not specialized and can easily disperse. Answer G is incorrect because the cells in colonies are all similar. Answer I is incorrect because vascularization has to do with the formation of tissues that transport water and nutrients and does not describe all forms of cell differentiation.

Question 5 Because there are so many different environments, the results of natural selection in those environments will vary.

Question 6

Part A An organism with complex multicellularity has specialized cells that are organized into tissues and organs. A heterotroph is an organism that obtains organic food molecules by eating other organisms or their byproducts and that cannot synthesize organic compounds from inorganic materials.

Part B Because both simple and complex organisms have similar needs, they require structures with similar functions.

Answers

1. C
2. H
3. B
4. I
5. Different environments on Earth provide different kinds of natural selection.
6A. All animals are complex multicellular heterotrophs.
6B. There is a correspondence between cell structures and their functions.

Test Prep Correlations

SC.D.1.4.4: Item 5
SC.F.1.4.2: Item 2
SC.G.1.4.1: Items 1, 3, 4, 6A, 6B

Skills Practice Lab

Surveying Kingdom Diversity

Teacher's Notes

Time Required 40–50 minutes

Ratings

EASY → HARD

Teacher Preparation 2

Student Setup 2

Concept Level 2

Cleanup 2

Skills Acquired
- Classifying
- Collecting Data
- Identifying/Recognizing Patterns
- Inferring
- Organizing and Analyzing Data

Scientific Methods

In this lab, students will:
- Make Observations
- Analyze the Results
- Draw Conclusions

Materials and Equipment

Materials for this lab can be ordered from WARD'S. See *Master Materials List* at the front of this book for catalog numbers.

Safety Cautions

Caution students to use care when observing the preserved specimens. The preservative can leak if the jars are tilted. Tell students to wash their hands if they come into contact with the preservative. Have students wear safety goggles and a lab apron. Remind students about microscope procedures when they use oil-immersion lenses. Remind students to wash their hands before they leave the laboratory.

Tips and Tricks

Provide students with preserved, mounted, or dried samples of archaebacteria, an *Anabaena* species, a protozoan, algae, fungi (mold and yeast), plant tissues, and animals. Set up the specimens at different stations and have students circulate among the stations. Have students make their data table and an answer sheet before class. This will make it easier for students to answer the questions, since not all students will start with Station 1. Students should work in groups of two to four and move clockwise from station to station.

Skills Practice Lab

Surveying Kingdom Diversity

SC.G.2.4.6

SKILLS
- Using a microscope
- Comparing

OBJECTIVES
- **Observe** representatives of each of the six kingdoms.
- **Compare** and **contrast** the organisms within a kingdom.
- **Analyze** the similarities and differences among the six kingdoms.

MATERIALS
- specimens from each of the six kingdoms
- compound microscopes
- hand lenses or stereomicroscopes

ChemSafety

CAUTION: Always wear safety goggles and a lab apron to protect your eyes and clothing.

CAUTION: Do not touch or taste any chemicals. Know the location of the emergency shower and eyewash station and how to use them. If you get a chemical on your skin or clothing, wash it off at the sink while calling to the teacher. Notify the teacher of a spill. Spills should be cleaned up promptly, according to your teacher's directions.

CAUTION: Glassware is fragile. Notify the teacher of broken glass or cuts. Do not clean up broken glass or spills with broken glass unless the teacher tells you to do so.

Before You Begin

Many biologists classify living things into six **kingdoms.** The organisms in a kingdom have fundamental characteristics in common. For example, the organisms of two kingdoms are made of **prokaryotic cells,** while the organisms in the other four kingdoms are made of **eukaryotic cells.** Some kingdoms contain only **unicellular** or **colonial** organisms, while others contain only **multicellular** organisms. In this lab, you will examine representatives of six kingdoms of organisms. You will see that each kingdom is distinct from the others.

1. Write a definition for each boldface term in the previous paragraph and for each of the following terms: tissue, organ, organ system, autotroph, heterotroph.
2. Make a data table similar to the one below.
3. Based on the objectives for this lab, write a question you would like to explore about the kingdoms of organisms.

DATA TABLE

Kingdom name	Type of cells	Level of organization	Other characteristics	Examples

Answers to Before You Begin

1. *kingdom*—the largest taxonomic group; *prokaryotic cell*—a cell that has no organized nucleus or organelles; *eukaryotic cell*—a cell that has a nucleus and several different types of organelles; *unicellular organism*—an organism that consists of only one cell; *colonial organism*—a group of cells that are permanently associated but do not communicate with one another; *multicellular organism*—an organism made of many cells that are permanently associated with one another; *tissue*—a group of cells that work together to perform a certain function; *organ*—a specialized structure with a specific function; *organ system*—a group of organs that work together to perform a function; *autotroph*—an organism that obtains energy and makes organic compounds from inorganic materials; *heterotroph*—an organism that obtains energy and organic compounds by eating other organisms
3. Answers will vary. For example: How do organisms in the six kingdoms differ?

Procedure

PART A: Conducting a Survey

1. Put on safety goggles and a lab apron.
2. Visit the station for each kingdom listed below, and examine the specimens there. Answer the questions, and record observations in your data table.
3. **Archaebacteria** Examine the prepared slides.
   a. What does a microscope reveal about the structure of archaebacteria?
   b. How do these organisms get energy for life processes?
4. **Eubacteria** Examine the prepared slides.
   a. What does a microscope reveal about the structure of eubacteria?
   b. Would you consider *Anabaena* to be unicellular or multicellular? Explain.
   c. How does *Anabaena* appear to obtain energy for life processes? Explain.
5. **Protists** Examine the prepared slides.
   a. What does a microscope reveal about the structure of protozoans?
   b. How do protozoans appear to obtain energy for life processes? Explain.
   c. Are the algae unicellular or multicellular? Explain.
   d. How do algae differ from protozoans?
6. **Fungi** Examine the specimens.
   a. Are fungi unicellular or multicellular? Explain.
   b. What does a microscope reveal about the structure of fungi?
   c. How do the fungi appear to obtain energy for life processes? Explain.
7. **Plants** Examine the specimens.
   a. What is the most striking characteristic shared by these plants?
   b. What does a microscope reveal about the structure of plants?
8. **Animals** Examine the specimens.
   a. What is the most striking characteristic shared by these animals?
   b. What is the most striking difference among these animals?

PART B: Cleanup and Disposal

9. Dispose of broken glass and solutions in the designated waste containers. Do not pour chemicals down the drain or put lab materials in the trash unless your teacher tells you to do so.
10. Wash your hands thoroughly before you leave the lab and after you finish all work.

Analyze and Conclude

1. **Summarizing Data** What are the main differences observed among the six kingdoms?
2. **Recognizing Patterns** How does the size of bacterial cells compare with the cell size in the other kingdoms?
3. **Analyzing Methods** How did you determine the cell type for each kingdom?
4. **Inferring Conclusions** Which kingdom exhibits the most diversity?
5. **Further Inquiry** Write a new question about the kingdoms of life that could be explored with another investigation.

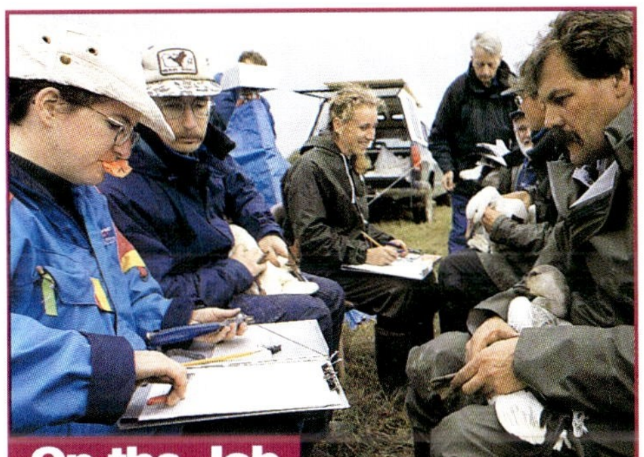

On the Job

A **survey** is a detailed study that is conducted through observation and analysis. Do research to learn about the surveys conducted by a famous biologist, such as Charles Darwin, Carl Linnaeus, or John James Audubon. For more about careers, visit **go.hrw.com** and type in the keyword **HX4 Careers.**

Answers to Procedure

3 a. Archaebacteria are microscopic and prokaryotic.
b. They can be either heterotrophic or autotrophic.

4 a. Eubacteria are microscopic, prokaryotic, and many different shapes.
b. Anabaena appears to be multicellular. It is made of chains of cells.
c. Anabaena appears to be an autotroph that obtains energy through photosynthesis.

5 a. Protozoans are unicellular and much larger than bacteria. They are eukaryotic, meaning that they have a nucleus and several types of organelles.
b. Protozoans appear to be heterotrophic since they are not green and therefore do not contain chlorophyll.
c. Some algae are unicellular, and some are multicellular.
d. Algal cells have cell walls and pigments, such as chlorophyll, which enable them to be photosynthetic autotrophs.

6 a. Some fungi are unicellular, but most are multicellular.
b. Fungi are made of strands of undifferentiated cells. These strands are called hyphae.
c. Fungi appear to be heterotrophic since they are not green and cannot photosynthesize.

7 a. Answers will vary. For example, students may note that all of the plants are green.
b. Plants are made of several types of specialized cells that are arranged in tissues and organs.

8 a. Answers will vary. For example, students may note that all of the animals have an organized structure with specialized parts.
b. Answers will vary. For example, students may describe differences in size, shape, or symmetry.

Answers to Analyze and Conclude

1. Answers will vary. Students should include differences in cell structure, body organization, cell specialization, and method of getting food.
2. Bacterial cells are much smaller than the cells of organisms in the other four kingdoms.
3. The cell type (prokaryotic or eukaryotic) is determined by microscopic inspection for a nucleus and organelles.
4. Protista is the most diverse, although that may not be evident from the specimens examined.
5. Answers will vary. For example: How does the method of reproduction differ among the kingdoms?

Answers to On the Job

Answers will vary. Darwin surveyed a wide variety of living organisms and fossils during his five-year voyage on HMS Beagle. Linnaeus classified and named more than 11,000 species of plants and animals. Audubon surveyed and painted American wildlife, especially birds.

Chapter Opener		SE **Quick Review,** p. 433 ■ SE **Reading Activity,** p. 433 ■ TE **Using the Figure,** p. 432 GENERAL TE **Opening Activity,** p. 433 BASIC
& 3 • 90 min pp. 434–441 **Section 1** Viruses	OSP **Lesson Plan*** TR **Bellringer*** TR **Section Outline*** TR **F9 Viral Replication in Bacteria*** TR **F7 Structures of Adenovirus and Bacteriophage*** TR **F8 Structures of TMV and Influenza Virus*** TR **F10 Important Viral Diseases***	TE **Using the Figure,** p. 435 BASIC TE **Demonstration,** p. 436 ◆ GENERAL TE **Using the Figure,** p. 436 GENERAL TE **Activity** Future Headlines, p. 438 GENERAL TE **Group Activity** AIDS Drug, p. 439 GENERAL CRF **Quicklab** Modeling Viruses* BASIC
& 5 • 90 min pp. 442–450 **Section 2** Bacteria	OSP **Lesson Plan*** TR **Bellringer*** TR **Section Outline*** TR **F2 Three Bacterial Cell Shapes*** TR **F3 Gram Staining*** TR **F4 *Escherichia coli**** TR **F6 Important Bacterial Diseases***	TE **Activity** Organizing Bacteria, p. 444 BASIC TE **Using the Table,** p. 447 GENERAL TE **Demonstration,** p. 447 GENERAL TE **Using the Figure,** p. 448 TE **Group Activity** "Rational" Drug Design, p. 449 ADVANCED TE **Group Activity** Breakthroughs in Science, p. 449 GENERAL SE **Skills Practice Lab** Staining and Observing Bacteria, pp. 454–455 ■ CRF **Datasheets for In-Text Labs** Staining and Observing Bacteria* ◆ GENERAL CRF **Skills Practice** Gram Staining Bacteria* ◆ GENERAL

& 7 • 90 min Chapter Review and Assessment Resources

- SE **Study Zone,** p. 451 ■
- SE **Performance Zone,** p. 452 ■
- SE **FCAT Test Prep,** p. 453 ■
- TR **Concept Mapping***
- CRF **Vocabulary Review Worksheet*** ■ GENERAL
- CRF **Test Prep Pretest*** ■ GENERAL
- CRF **Chapter Test*** ■ GENERAL
- CRF **Chapter Test*** ADVANCED
- OSP **Test Generator***
- OSP **Test Item Listing***
- OSP **Modified Chapter Test*** BASIC

Online and Technology Resources

Holt Online Learning

Visit go.hrw.com for access to Holt Online Learning or enter the keyword **HX6 Home** for a variety of free online resources.

One-Stop Planner® CD-ROM

This CD-ROM package includes
- Lab Materials QuickList Software
- Holt Calendar Planner
- Customizable Lesson Plans
- Printable Worksheets
- ExamView® Test Generator
- Interactive Teacher Edition
- Holt PuzzlePro® Resources
- Holt PowerPoint® Resources

Guided Reading Audio CDs

These CDs are designed to help auditory learners and reluctant readers. (Audio CDs are also available in Spanish.)

Compression guide: To shorten your instruction because of time limitations, omit blocks 1 and 6.

KEY

- **TE** Teacher Edition
- **SE** Student Edition
- **OSP** One-Stop Planner
- **CRF** Chapter Resource File
- **TR** Teaching Transparency
- * Also on One-Stop Planner
- ■ Also Available in Spanish
- ◆ Requires Advance Prep

SKILLS DEVELOPMENT AND PRACTICE	SECTION REVIEW AND ASSESSMENT	CORRELATIONS
		Florida Sunshine Standards/FCAT
TE Reading Skill Builder K-W-L, p. 435 BASIC TE Skill Builder Graphing, p. 436 ADVANCED TE Reading Skill Builder Paired Summarizing, p. 437 BASIC TE Reading Skill Builder Brainstorming, p. 438 BASIC TE Skill Builder Writing, p. 439 ADVANCED TE Reading Skill Builder Interactive Reading, p. 439 BASIC CRF Directed Reading Worksheet* BASIC CRF Active Reading Worksheet* GENERAL	SE Section 1 Review ■ TE Reteaching, p. 441 BASIC TE Quiz, p. 441 GENERAL TE Alternative Assessment, p. 441 GENERAL CRF Quiz* ■ BASIC	SC.F.1.4.5, SC.F.1.4.8, SC.F.2.4.2, SC.H.1.4.2
TE Reading Skill Builder Brainstorming, p. 443 BASIC TE Skill Builder Math, p. 444 GENERAL TE Skill Builder Writing, p. 448 ADVANCED TE Skill Builder Graphing, p. 448 ADVANCED CRF Directed Reading Worksheet* BASIC CRF Active Reading Worksheet* GENERAL CRF Critical Thinking Worksheet* ADVANCED CRF Science Skills Worksheet* GENERAL	SE Section 2 Review ■ TE Reteaching, p. 450 BASIC TE Quiz, p. 450 GENERAL TE Alternative Assessment, p. 450 GENERAL CRF Quiz* ■ BASIC Modified Worksheet* BASIC	SC.B.1.4.1, SC.F.1.4.2, SC.G.1.4.1, SC.H.1.4.1

Holt Biology Interactive Tutor CD-ROM

This CD-ROM consists of interactive activities that give students a fun way to extend their knowledge of biology concepts.

Unit 4 Cell Reproduction

SciLinks NSTA

www.scilinks.org

Maintained by the **National Science Teachers Association**

Topic: Viral Diseases
SciLinks Code: HX4186

Topic: Bacteria
SciLinks Code: HX4018

CHAPTER 20

Overview

Tell students the purpose of this chapter is to learn about the diversity of viruses and bacteria and their impact on humans. By learning about viral structure and reproduction, students will be able to better understand how viruses cause disease. Learning about the structure of bacteria and the variety of ways that they obtain energy will help students to understand how bacteria can live everywhere. Students will also learn how bacteria are beneficial to humans and how they cause disease.

Using the Figure

Tell students that sore throats can be caused by a species of bacterium from the genus *Streptococcus*, the bacterium shown in the photo on this page. Ask students to describe these bacteria. (They are microscopic in size, spherical in shape, and form chains.)

Assessing Prior Knowledge

Before beginning this chapter, be sure your students can answer the questions in the ***Quick Review*** on the facing page. If they cannot, encourage them to reread the sections indicated.

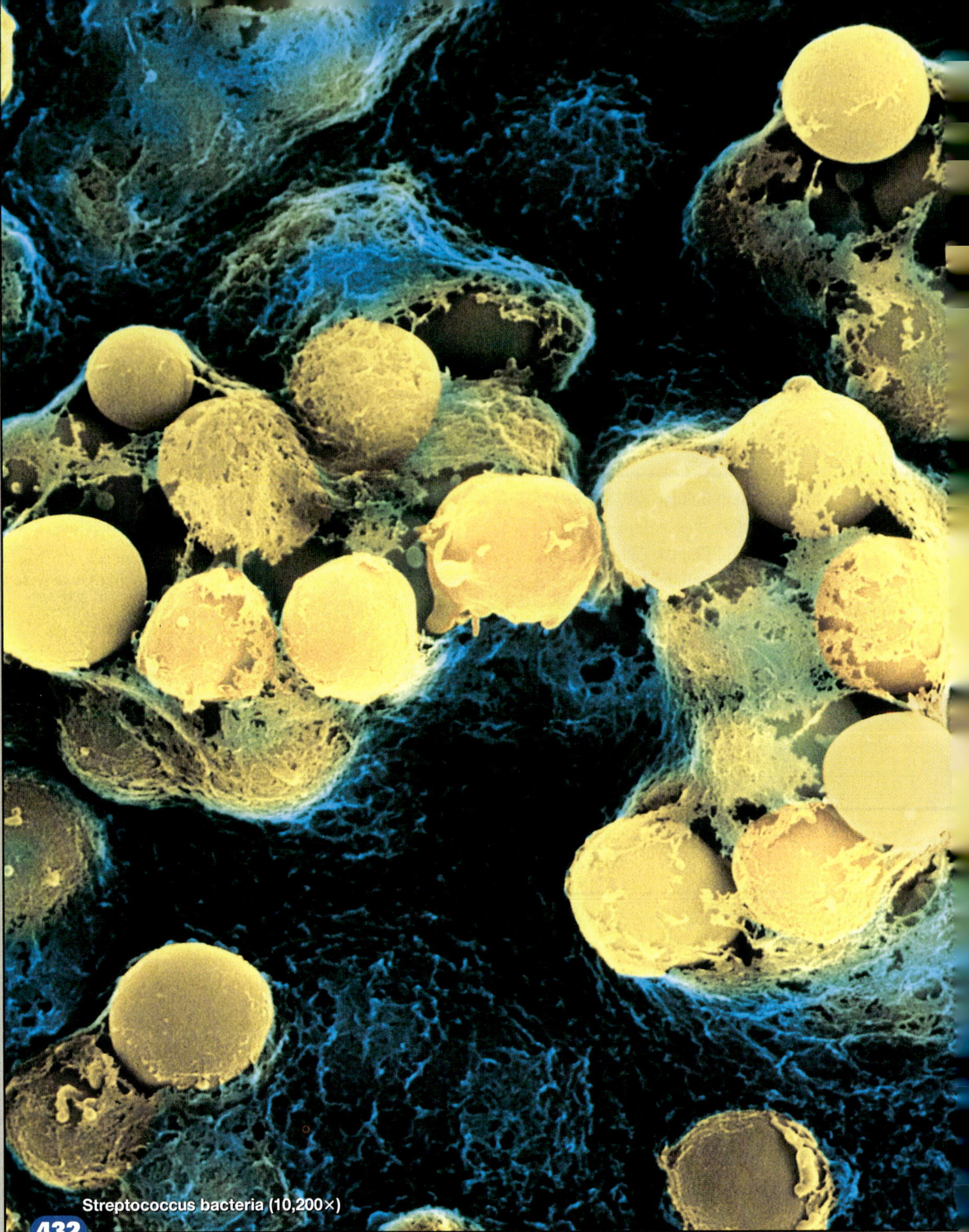

Streptococcus bacteria (10,200×)

432

Chapter Correlations — Florida Sunshine Standards/FCAT

SC.B.1.4.1 Knowledge of energy is fundamental to all the scientific disciplines (e.g., the energy required for biological processes in living organisms and the energy required for the building, erosion, and rebuilding of the Earth).

SC.F.1.4.2 Body structures are uniquely designed and adapted for their function.

SC.F.1.4.5 Complex interactions among the different kinds of molecules in the cell cause distinct cycles of activity governed by proteins.

SC.F.1.4.8 Cell behavior can be affected by molecules from other parts of the organism or even from other organisms.

SC.F.2.4.2 Every cell contains a "blueprint" coded in DNA molecules that specify how proteins are assembled to regulate cells.

SC.G.1.4.1 There is great diversity and interdependence among living things.

SC.H.1.4.2 From time to time, major shifts occur in the scientific view of how the world works, but more often the changes that take place in the body of scientific knowledge are small modifications of prior knowledge.

CHAPTER

20 Viruses and Bacteria

Florida Sunshine Standards/FCAT

SC.B.1.4.1 Knowledge of energy is fundamental to all the scientific disciplines (e.g., the energy required for biological processes in living organisms and the energy required for the building, erosion, and rebuilding of the Earth).
SC.F.1.4.2 Body structures are uniquely designed and adapted for their function.
SC.F.1.4.5 Complex interactions among the different kinds of molecules in the cell cause distinct cycles of activity governed by proteins.
SC.F.1.4.8 Cell behavior can be affected by molecules from other parts of the organism or even from other organisms.
SC.F.2.4.2 Every cell contains a "blueprint" coded in DNA molecules that specify how proteins are assembled to regulate cells.
SC.G.1.4.1 There is great diversity and interdependence among living things.
SC.H.1.4.2 From time to time, major shifts occur in the scientific view of how the world works, but more often the changes that take place in the body of scientific knowledge are small modifications of prior knowledge.

Quick Review

Answer the following without referring to earlier sections of your book.

1. **List** the properties of life. *(Chapter 1, Section 1)*
2. **Define** *prokaryote*. *(Chapter 3, Section 2)*
3. **Describe** a bacteriophage. *(Chapter 9, Section 1)*
4. **Differentiate** DNA from RNA. *(Chapter 10, Section 1)*

Did you have difficulty? *For help, review the sections indicated.*

Reading Activity

Take a break after reading each section of this chapter, and closely study the figures in the section. Reread the figure captions, and, for each one, write out a question that can be answered by referring to the figure and its caption. Refer to your list of figures and questions as you review the concepts addressed in the chapter before you complete the Performance Zone chapter review.

Streptococcus bacteria include a number of strains that can produce a wide range of illnesses. Some, like "strep throat," are easily treated. Others are rare and require immediate medical attention.

Looking Ahead

Section 1

Viruses

Is a Virus Alive?
Viral Structure
Viral Reproduction
How HIV Infects Cells
Viral Diseases

Section 2

Bacteria

Bacterial Structure
Eubacteria and Archaebacteria
Pathogenic Bacteria
Antibiotics
Importance of Bacteria

internet connect

www.scilinks.org
National Science Teachers Association *sci*LINKS Internet resources are located throughout this chapter.

SCILINKS. Maintained by the National Science Teachers Association

Opening Activity — Basic

Location of Bacteria Ask students to make a list of places where they might find bacteria. Tell students that every answer is correct because bacteria live almost everywhere—in the human body and on everything we touch and eat. Emphasize that bacteria are an important part of the environment.

Quick Review Answers

1. Properties of life include cellular structure and function, reproducing, metabolizing to obtain and use energy, maintaining homeostasis, passing on traits through heredity, and changing over time.
2. Prokaryotes are single-celled organisms that lack a nucleus and other internal compartments.
3. A bacteriophage is a virus that infects bacteria. It consists of a nucleic acid surrounded by a protein coat.
4. DNA is double stranded and contains thymine and deoxyribose. RNA is single stranded and contains uracil and ribose.

Chapter Resource File

- Vocabulary Worksheets
- Concept Mapping

Reading Activity Answers

Students should examine the chapter figures closely. A typical question written for Figure 4, for example, might ask for a comparison of the lytic and lysogenic cycles of viral replication.

Section 1

Focus

Overview

Before beginning this section review with your students the objectives listed in the Student Edition. This section introduces students to the structure of viruses and examines how they replicate. The section also examines how viruses cause disease. Viral characteristics are discussed as they apply to HIV.

Bellringer

Distribute newspapers or news magazines to student groups and ask students to find an article about viruses. Have them quickly read the article and write a question about material in the article. Most newspapers will contain at least one article about viruses, for example, articles on HIV or West Nile virus.

Motivate

Discussion/Question — GENERAL

Ask students what tomatoes, white potatoes, and garden peppers have in common. (All three are from the nightshade family, Solanaceae, and they are vulnerable to the same diseases.) Tell students that tobacco is also a member of the nightshade family. A virus that plagues this family, the tobacco mosaic virus (TMV), was first recognized in tobacco. Have students speculate as to why TMV infects only members of this family. (Students may mention specific receptor sites on the plants' cell surfaces.)

Section Correlations

SC.F.1.4.5
SC.F.1.4.8
SC.F.2.4.2
SC.H.1.4.2

Section 1 Viruses

Florida Sunshine Standards/FCAT

SC.F.1.4.5 Complex interactions among the different kinds of molecules in the cell cause distinct cycles of activity governed by proteins.
SC.F.2.4.2 Every cell contains a "blueprint" coded in DNA molecules that specify how proteins are assembled to regulate cells.
SC.H.1.4.2 From time to time, major shifts occur in the scientific view of how the world works, but more often the changes that take place in the body of scientific knowledge are small modifications of prior knowledge.

Objectives

- **Describe** why a virus is not considered a living organism. SC.F.1.4.5
- **Summarize** the discovery of the tobacco mosaic virus. SC.H.1.4.2
- **Describe** the basic structure of a virus. SC.F.2.4.2
- **Summarize** the steps of viral replication. SC.F.1.4.5
- **Explain** how HIV infects immune system cells. SC.F.1.4.8

Key Terms

virus
pathogen
capsid
envelope
glycoprotein
bacteriophage
lytic
provirus
lysogenic
prion
viroid

Is a Virus Alive?

Throughout the book, you have learned about the properties of life. All living things are made of cells, are able to grow and reproduce, and are guided by information stored in their DNA. The smallest organisms that have these properties are prokaryotes. **Viruses** are segments of nucleic acids contained in a protein coat. Viruses are not cells. Viruses are smaller than prokaryotes and range in size from about 20 nm to 250 nm (0.02–0.25 µm) in diameter. (One nanometer is equal to 0.001µm or 0.00000004 in.) Most viruses, such as the Ebola virus shown in **Figure 1,** can be seen only with an electron microscope. Viruses are **pathogens**—agents that cause disease. Viruses replicate by infecting cells and using the cell to make more viruses. Because viruses do not have all the properties of life, biologists do not consider them to be living. Viruses do not grow, do not have homeostasis, and do not metabolize. Because they cause diseases in many organisms, viruses have a major impact on the living world.

Discovery of Viruses

Near the end of the nineteenth century, scientists were trying to find the cause of tobacco mosaic disease, which stunts the growth of tobacco plants. Scientists filtered bacteria from the sap of infected plants. They were surprised to find that the filtered sap could still cause uninfected plants to become infected. The scientists concluded that the pathogen is smaller than a bacterium. The pathogen was called a *virus,* a Latin word meaning "poison."

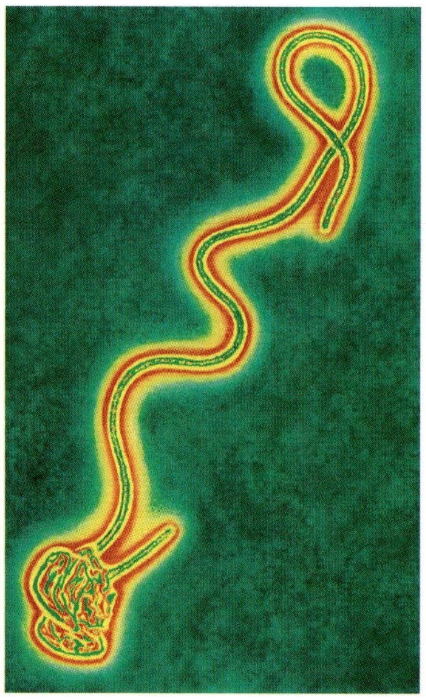

Figure 1 Ebola virus. This virus causes an often-fatal disease and has been recognized only since 1976.

For many years after this discovery, viruses were thought to be tiny cells. In 1935, biologist Wendell Stanley of the Rockefeller Institute purified tobacco mosaic virus (TMV). He crystallized the purified virus. Stanley concluded that TMV is a chemical rather than an organism.

Each particle of TMV is composed of RNA and protein. Scientists were able to separate the RNA from the protein and reassemble the virus so that it could infect plants.

Chapter Resource File

- Directed Reading BASIC
- Active Reading GENERAL

One-Stop Planner CD-ROM

- Reading Organizers BASIC
- Reading Strategies BASIC
- Occupational Applications Worksheet Sanitarian GENERAL
- Supplemental Reading Guide *Microbe Hunters* ADVANCED

Viral Structure

The virus protein coat, or **capsid,** may contain either RNA or DNA, but not both. RNA viruses include the human immunodeficiency virus (HIV), which causes AIDS, influenza viruses, and rabies virus. DNA viruses include those viruses that cause warts, chickenpox, and mononucleosis. Many viruses, such as the influenza virus shown in **Figure 2,** have a membrane, or **envelope,** surrounding the capsid. The envelope helps the virus enter cells. It consists of proteins, lipids, and **glycoproteins** *(glie koh PROH teenz),* which are proteins with attached carbohydrate molecules. Some viruses also contain specific enzymes.

Viruses exist in a variety of shapes. Some viruses, such as the Ebola virus, shown in Figure 1, are long rods that form filaments. Spherical viruses, such as the influenza virus shown in Figure 2, typically are studded with receptors. These receptors help the virus enter cells. A helical virus, like the tobacco mosaic virus shown in Figure 2, is rodlike in appearance, with capsid proteins winding around the core in a spiral. Polyhedral viruses have many sides and are roughly spherical. The capsid of most polyhedral viruses has 20 triangular faces and 12 corners. This odd shape is an efficient one for containing a viral genome. Figure 2 shows the polyhedral shape of an adenovirus, which can cause several different kinds of infections in humans.

Viruses that infect bacteria, called **bacteriophages,** have a complicated structure. A T4 bacteriophage, for example, has a polyhedron capsid attached to a helical tail. A long DNA molecule is coiled within the polyhedron.

Study TIP

Compare and Contrast To compare and contrast the properties of life as defined in Chapter 1 and the properties of viruses, make a two-column list. In one column, write the properties of life. In the other column, write the properties of life that viruses have.

Figure 2 Viral structures

Viruses can have characteristic shapes.

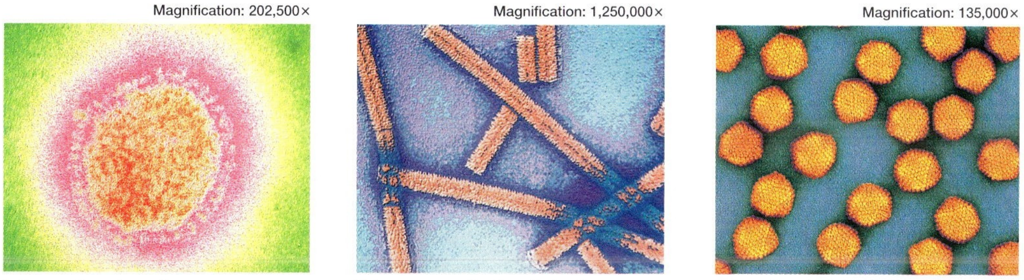

Influenza (enveloped) | Tobacco mosaic virus (helical) | Adenovirus (polyhedral)

CHEMISTRY CONNECTION

Only a few different kinds of proteins are found in a viral capsid. For example, the capsid of TMV contains more than 1,000 protein molecules, but they are all copies of the same protein. This means that one gene can code all of the capsid components.

Integrating Physics and Chemistry

Due to the way in which elements interact and form bonds, viruses have characteristic shapes (helical or polyhedral). In a similar way, water also has a characteristic shape because the hydrogen atoms are arranged with the oxygen atom to form a specific angle, which is the same for all water molecules (i.e., 104.45°). Because oxygen is more electronegative than hydrogen, oxygen has a greater tendency to attract electrons. The unequal sharing of bonding electrons produces water's polar property.

Teach

READING SKILL BUILDER — BASIC

K-W-L Before students read this section, have them write short individual lists of all the things they already **K**now (or have heard) about viruses. When students have finished their lists, ask them to contribute their entries to a class list on the board or overhead. Then have students list things they **W**ant to know about viruses. Have students save their lists for later use in Reteaching at the end of Section 1.

Teaching Tip — GENERAL

Design a Virus Have students create a fictitious virus. They should produce a drawing or a model of their virus and provide information about the origin of the virus, its components, the host it infects, how the virus is transmitted, and the effects of the virus on the host. Have students share their creations with the class.
LS Kinesthetic

Using the Figure — BASIC

Use **Figure 2** to point out two important parts of a virus: the capsid (made of protein) and the nucleic acid (RNA or DNA). Some viruses contain DNA, whereas others contain RNA. Ask students what constitutes the membranous envelope surrounding the capsid of the influenza virus. (proteins, lipids, and glycoproteins from the host cell) Emphasize that different viruses can have very different appearances; some are larger and more complex than others. **LS Visual**

Transparencies

- **TR** Bellringer
- **TR F9** Viral Replication in Bacteria
- **TR F7** Structures of Adenovirus and Bacteriophage
- **TR F8** Structures of TMV and Influenza Virus
- **TR F10** Important Viral Diseases

Teach, continued

Demonstration — GENERAL

Show students labels from tomato plants or ads in nursery catalogs that state that the plants are resistant to TMV. Explain that people who handle tobacco products can carry the virus on their hands and infect plants by touching them. An infected plant will develop light and dark patches on its leaves—the mosaic-like symptoms of the disease. LS Visual

Using the Figure — GENERAL

Point out that some of the capsids in **Figure 3** appear empty. Ask students to explain why. (Those capsids that appear empty have already delivered their DNA to the bacterial cell. Those capsids that appear dark still contain their DNA.) English Language Learners

Teaching Tip — GENERAL

Viral Reproduction Tell students that a single virus that infects a bacterial cell can produce about 100 new viruses in 20 minutes. Have students calculate how many viruses would exist after 1 hour, assuming that there are sufficient cells to support continuous viral replication. (1 million) LS Logical

SKILL BUILDER — ADVANCED

Graphing Have students graph the number of viruses described in the Teaching Tip above. Students should graph the number of viruses present at 0, 20, 40, and 60 minutes. Student graphs should show exponential growth.

Figure 3 Bacteriophage infecting a bacterium. Bacteriophages (pink) first attach to a bacterial cell (blue) and then push their DNA into it. The cell then produces more viruses.

Viral Reproduction

Viruses lack the enzymes necessary for metabolism and have no structures to make protein. Therefore, viruses must rely on living cells (host cells) for replication, as shown in **Figure 3.** Before a virus can replicate, it must first infect a living cell. A plant virus, like TMV, enters a plant cell through tiny tears in the cell wall at points of injury. An animal virus enters its host cell by endocytosis. A bacterial virus, or bacteriophage, punches a hole in the bacterial cell wall and injects its DNA into the cell.

Lytic Cycle

The reproduction of bacterial viruses has been well studied. Inside a cell, some viruses, such as T4—a virus that infects *E. coli*— will set out on one of two paths: the lytic cycle or the lysogenic cycle.

In bacterial viruses, the cycle of viral infection, replication, and cell destruction is called the **lytic** cycle. After the viral genes have entered the cell, they use the host cell to replicate viral genes and to make viral proteins, such as capsid proteins. The proteins are then assembled with the replicated viral genomes to form complete viruses. The host cell is broken open and releases newly made viruses. Though reproduction in a single bacterial virus is illustrated here, these stages are common to infections by some other viruses as well. The lytic cycle is shown in **Figure 4.**

Lysogenic Cycle

During an infection, some viruses stay inside the cells but do not make new viruses. Instead of producing virus particles, the viral gene is inserted into the host chromosome and is called a **provirus.**

Figure 4 Viral replication in bacteria. The bacterial virus, T4, provides a model by which viruses replicate through the lytic cycle or lysogenic cycle.

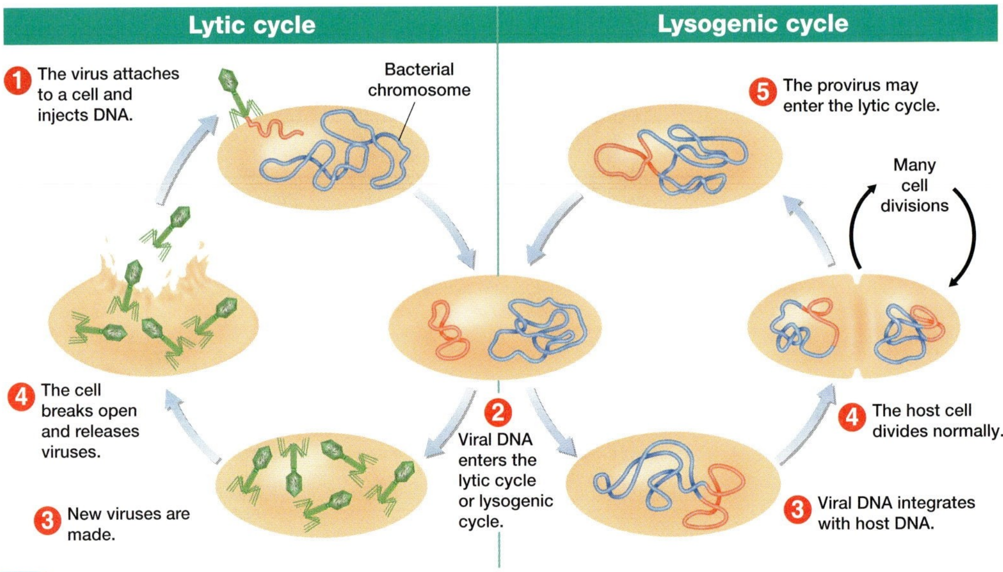

REAL WORLD CONNECTION

Commonly called the flu, influenza has been perhaps the most lethal viral disease in human history. Because the influenza virus is airborne, it is highly contagious. A common flu virus can be lethal to people who have respiratory ailments or compromised immune systems, as well as to elderly people and young children. Sometimes a new type of flu virus is very lethal. For example, in 1918 and 1919, a new type of influenza killed 22 million people—more people than were killed in combat during World War I.

Whenever the cell divides, the provirus also divides, resulting in two infected host cells. In this cycle, called the **lysogenic** *(lie soh JEHN ihk)* cycle, the viral genome replicates without destroying the host cell. This cycle is shown in Figure 4. In some lysogenic viruses, a change in the environment can cause the provirus to begin the lytic cycle. This results in the destruction of the host cell.

In animal cells, viruses can replicate slowly so that the host cell is not destroyed by the virus. For example, the virus that causes cold sores in humans hides deep in the nerves of the face. When the conditions in the body become favorable for the virus, such as when a person is under stress, the virus then begins to cause tissue damage that is seen as a cold sore or fever blister.

Host Cell Specificity

Viruses are often restricted to certain kinds of cells. For example, TMV infects tobacco and related plants, but does not infect animals. Scientists hypothesize that this specificity may be due to the viruses' origin. Viruses may have originated when fragments of host genes escaped or were expelled from cells. The hypothesis that viruses originated from a variety of host cells may explain why there are so many different kinds of viruses. Biologists think there are at least as many kinds of viruses as there are kinds of organisms.

Structure of HIV—an Enveloped Virus

Many viruses that infect only animals, such as the influenza virus shown in Figure 2, have an exterior viral envelope. **Figure 5** shows human immunodeficiency virus (HIV), the virus that causes acquired immune deficiency syndrome (AIDS). Figure 5 illustrates the envelope and other features common to several animal viruses. In many cases, the viral envelope is composed of a lipid bilayer derived from the membrane of the host cell. On the surface of the virus, glycoproteins are embedded within the envelope.

Within the envelope lies the capsid, which in turn encloses the virus's genetic material. In the case of HIV, the genetic material is composed of two molecules of single-stranded RNA. The approximately 9,000 nucleotides of HIV make up nine genes. Three of these genes are common to many different viruses.

Figure 5 **HIV.** HIV infects human white blood cells.

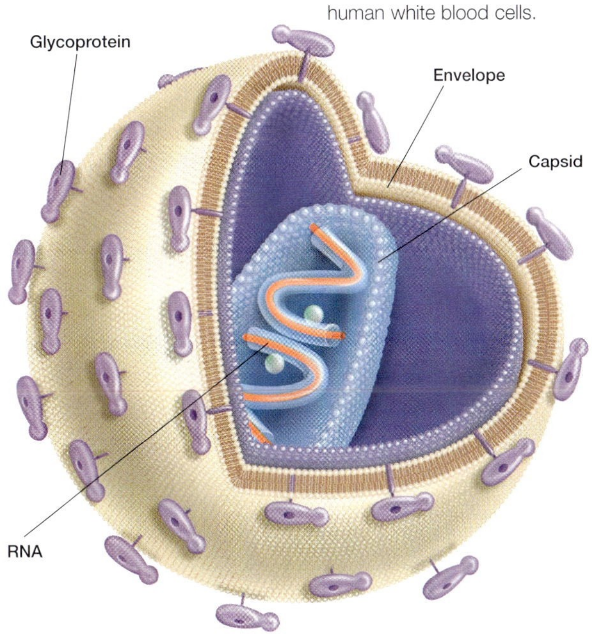

READING SKILL BUILDER — Basic

Paired Summarizing Have pairs of students silently read about the two methods of viral replication. As they read, have them note passages they do not understand. After students finish, ask one member of each pair to summarize the passage, referring to the text as needed. The listener should add anything omitted and continue the discussion by pointing out what he or she did not understand. Partners should work together to clarify what was not clear and to formulate questions to present to the class.

Co-op Learning | English Language Learners

Teaching Tip

County Health Department Ask a county public health official to speak to your class about the kind of work the county public health department does to prevent viral diseases. Be sure to ask the speaker to discuss viral diseases that are of concern in your county (e.g., mosquito-transmitted viral encephalitis, measles, polio, influenza, HIV). Check with your school's guidelines about the kinds of subject areas that are appropriate. Ask the speaker to bring samples of equipment that the county uses in its health work. LS Intrapersonal

Career

Epidemiologist Epidemiologists are employed by hospitals, health departments, universities, and private consulting firms. Their work is often a combination of biology, medicine, and detective-like problem solving. Epidemiologists study human health and disease, investigate and identify new cases, and work to prevent epidemics. A strong background in biology—especially microbiology—statistics, and evolution is vital. Many epidemiologists also have advanced degrees in public health or medicine.

Teach, *continued*

READING SKILL BUILDER — BASIC

Brainstorming Ask students what becomes of the proteins made by the host cell under the direction of the viral genome. (The proteins become the capsids for the new viruses and the enzymes needed for replication.) English Language Learners

Activity — GENERAL

Writing Future Headlines Ask students to write headlines about HIV that they might expect to see in 10 years. Discuss the headlines and the predictions behind them with the class. Then have students write short essays to explain their predictions about AIDS in the future. Post the headlines and essays around the room.
LS Verbal

How HIV Infects Cells

HIV, shown in **Figure 6**, provides a good example of how animal viruses enter cells. HIV entry is a two-step process. First, the virus attaches to the cell at specific sites called receptors. Second, the viral envelope fuses with the cell membrane, opens, and releases the capsid into the host cell.

Attachment

Studding the surface of each HIV are spikes composed of a glycoprotein. This particular glycoprotein precisely fits a human cell surface receptor called CD4. Thus the HIV glycoprotein can bind to any cell whose membrane has CD4 receptors. In humans, immune system cells called lymphocytes and macrophages, as well as certain cells in the brain, possess CD4 receptors.

Entry into Macrophages

HIV cannot enter a cell merely by docking onto a CD4 receptor. Rather, the HIV glycoprotein must also bind to a co-receptor called CCR5. This binding to CCR5 allows the viral capsid to enter the cell. **Figure 7** shows the process of HIV infecting a cell. Because human macrophages possess both CD4 and CCR5 receptors, HIV can enter macrophages. Lymphocytes, which are critical to immune system function, do not have CCR5 receptors. HIV therefore does not enter lymphocytes until later in the course of HIV infection.

Replication

Once inside a cell, the HIV capsid comes apart and releases its contents, which include the viral RNA. Accompanying the RNA is an enzyme called reverse transcriptase. Reverse transcriptase uses the viral RNA as a template for making a DNA version of the viral genome. This process is mistake prone, so it creates many new mutations in the viral genome. The viral DNA molecule then enters the cell nucleus and becomes a part of, or integrates into, the host's DNA. The integrated viral DNA uses the host cell's machinery to direct the production of many copies of the virus. HIV doesn't kill the macrophage by rupturing it; instead, the new viruses are released from the cell by budding. The new virus particle's envelope is thus derived from the cell membrane.

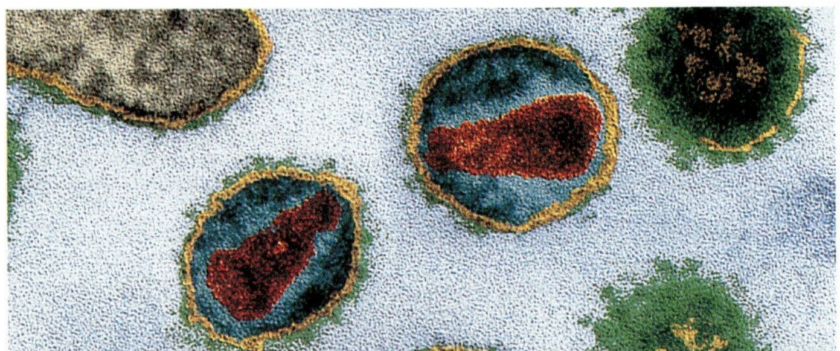

Figure 6 HIV. The spherical structure of HIV is visible in this transmission electron micrograph of individual virus particles.

MISCONCEPTION ALERT

HIV Infection and AIDS Students may not realize that an individual who has been infected with HIV can pass the virus to others even though he or she shows no symptoms of AIDS. During the long latency period between infection and the appearance of AIDS, typically 8 to 10 years, the virus multiplies in the body but does not kill the cells. Eventually, however, the tissues of the lymphatic system cease to function, and AIDS develops.

Integrating Physics and Chemistry

Have students identify and describe the strengths and weaknesses of the hypothesis concerning the origin of viral specificity. Ask them if they can think of a way this hypothesis might be tested.

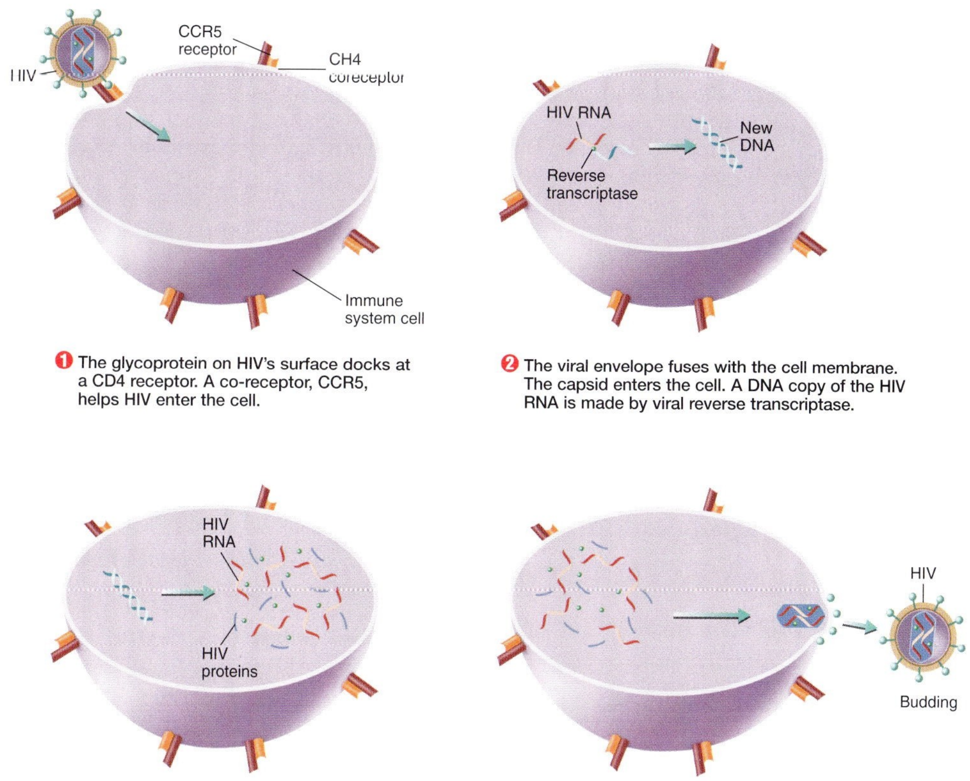

1 The glycoprotein on HIV's surface docks at a CD4 receptor. A co-receptor, CCR5, helps HIV enter the cell.

2 The viral envelope fuses with the cell membrane. The capsid enters the cell. A DNA copy of the HIV RNA is made by viral reverse transcriptase.

3 The viral DNA integrates into the host genome and directs synthesis of HIV proteins and HIV RNA.

4 New HIV particles exit macrophages by budding. HIV particles exit T cells by budding or bursting through the membrane.

Figure 7 HIV infection. HIV docks at specific receptors on cell membranes. The virus is reproduced by the infected cell.

AIDS

For years after the initial infection, HIV continues to replicate (and mutate). Eventually and by chance, HIV's surface glycoproteins change such that they now recognize a new co-receptor. This co-receptor is found on the subset of lymphocytes called T cells.

Unlike its activity in macrophages, HIV reproduces in T cells and then destroys them. This increases the number of virus particles in the blood, which then infect other T cells, widening the circle of cell death. It is this destruction of the body's T cells that blocks the body's immune response and signals the onset of AIDS. AIDS is a disease in which a person's immune system is unable to defend against infections that do not normally occur in healthy individuals.

Usually, HIV-infected people do not develop AIDS symptoms until years after infection. As a result, an HIV-infected individual can feel healthy and still spread the virus to others. HIV is not passed from an infected person to a healthy one through casual contact. It is transmitted in body fluids (such as semen or vaginal fluid) through sexual contact and in blood through the sharing of nonsterile needles. It is also transmitted to infants during pregnancy or through breast milk.

SKILL BUILDER — ADVANCED

Writing Skills The first cases of AIDS were reported in the early 1980s. Have students obtain statistics on the spread of AIDS from its discovery to the present time. Ask students to present the information in brief reports, including a bar graph that compares the number of AIDS-related deaths over the past few decades. **LS Verbal**

Group Activity — GENERAL

AIDS Drug Ask student groups to design a hypothetical drug that will disable the infection cycle of HIV. Students' drugs can target any step within the viral replication cycle, such as preventing the virus from attaching to white blood cells, inhibiting endocytosis of the virus, or destroying reverse transcriptase. Ask students to make diagrams of the HIV infection cycle that show their drug at work. Remind students that prevention is the best way to fight AIDS. English Language Learners

READING SKILL BUILDER — BASIC

Interactive Reading Assign Chapter 20 of the *Holt Biology* Guided Audio CD Program to help students achieve greater success in reading the chapter.

Attention Grabber

Immunity to AIDS A small number of individuals have a natural resistance to HIV. This resistance seems to be due to their having a defective cell surface receptor for HIV. Researchers are trying to make use of this mutation to develop a vaccine for AIDS.

Teach, *continued*

Teaching Tip

Eradicating Smallpox Tell students that smallpox is the first disease to be eradicated globally. Then tell students that the World Health Organization was able to wipe out the disease through a highly effective worldwide vaccination strategy. The last known smallpox case outside of the lab was reported in 1977 in Somalia. Although officially declared eradicated in 1980, samples of the smallpox virus are stored in the United States and in Russia. Controversy exists over whether to destroy these samples or maintain them for future study. Have a class discussion about this controversy, and ask students for their views.

Teaching Tip — ADVANCED

Prions Disease-causing prions have the same amino acid sequence as the normal prion protein, but the disease-causing prions are folded in a way that changes their function and physical characteristics. Have students research the shape of a disease-causing prion and that of its normal protein and make diagrams showing the differences. **LS Visual**

Viral Diseases

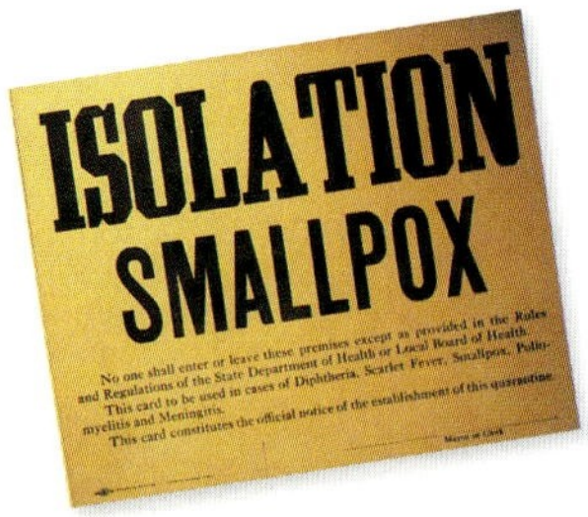

Diseases caused by viruses have been known and feared for thousands of years. Perhaps the most lethal virus in human history is the influenza virus. Commonly known as the flu, influenza is characterized by chills, fever, and muscular aches. The virus infects cells of the upper respiratory tract. There the viruses replicate and spread to new cells. About 22 million Americans and Europeans died of flu during an 18-month period in 1918–1919. **Table 1** lists some familiar viral diseases.

Certain viruses can also cause some types of cancer. Recall that cancer is a condition in which cells reproduce uncontrollably as a result of the failure of mechanisms that control cell growth and division. Viruses associated with human cancers include hepatitis B (liver cancer), Epstein-Barr virus (Burkitt's lymphoma), and human papilloma virus (cervical cancer).

Table 1 Important Viral Diseases

Disease	Description of illness	How the disease is transmitted
AIDS	Immune system failure	Sexual contact, contaminated blood, or contaminated needles
Common cold	Sinus congestion, muscle aches, cough, fever	Inhalation, direct contact
Ebola	High fever, uncontrollable bleeding	Body fluids
Hepatitis A	Flulike symptoms, swollen liver, yellow skin, painful joints	Contaminated blood, food, or water
Hepatitis B	Flulike symptoms, swollen liver, yellow skin, painful joints; can cause liver cancer	Sexual contact, contaminated blood, or contaminated needles
Influenza (flu)	Fever, chills, fatigue, cough, sore throat, muscle aches, weakness, headache	Inhalation
Mumps	Painful swelling in salivary glands	Inhalation
Polio 	Fever, headache, stiff neck, possible paralysis	Contaminated food or water
Rabies	Mental depression, fever, restlessness, difficulty swallowing, paralysis, convulsions; fatal	Bite of infected animal
SARS (Severe acute respiratory syndrome)	High fever, headache, dry cough; can be fatal	Inhalation, direct contact
Smallpox 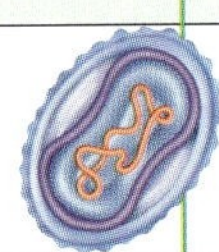	Blisters, lesions, fever, malaise, blindness, disfiguring scars; often fatal	Inhalation
Yellow fever	Fever, weakness, yellow skin; often fatal	Bite of infected mosquito

Cultural Awareness

Accidental Spread of Viral Diseases

Isolated communities are at particular risk of epidemics when outsiders visit. Diseases spread by travelers can wipe out a small isolated community. Spanish conqueror Hernán Cortés was aided in his conquest of the Aztecs by a smallpox epidemic that struck those Native Americans. Prior to the arrival of the Europeans, smallpox had been unknown in the New World. The disease killed millions of Native Americans.

Today, the Yanomamo tribe of Brazil and Venezuela is being decimated because of the onslaught of malaria, influenza, measles, and chicken pox brought by miners in search of gold.

Emerging Viruses

Newly recognized viruses or viruses that have reappeared or spread to new areas are called emerging viruses. These new pathogens are dangerous to public health. People become infected when they have contact with the normal hosts of these viruses.

In 1999, a mosquito-borne virus called West Nile virus began to spread across the United States. West Nile virus probably was brought from overseas to America by an infected bird. While it is an emerging virus in North America, West Nile virus is common in Africa, eastern Europe, and western Asia. People who are infected with the virus from mosquito bites typically experience mild flulike symptoms. In some people, particularly the elderly, inflammation of the brain may occur, which can be fatal.

First detected in the southwestern United States, hantavirus is spread in rodent droppings and can cause a lethal illness in humans. At least 38 percent of its human victims die.

internet connect
www.scilinks.org
Topic: Viral Diseases
Keyword: HX4186
SCILINKS. Maintained by the National Science Teachers Association

Prions and Viroids

In 1982, the American scientist Stanley Prusiner, of Stanford University, described a new class of pathogens that he called **prions** *(PREE awnz)*. Prions are composed of proteins but have no nucleic acid. A disease-causing prion is folded into a shape that does not allow the prion to function. Contact with a prion will cause a normal version of the protein to misfold, too. In this way the misfolding spreads like a chain reaction.

Prions were first linked to a brain disease in sheep called scrapie. Later, brain diseases such as mad cow disease, displayed by the cow shown in **Figure 8,** and Creutzfeldt-Jakob disease were also associated with prions. Eating meat containing the disease-causing prion can cause infection.

A **viroid** *(VEER oid)* is a single strand of RNA that has no capsid. Viroids are important infectious disease agents in plants. Viroids have affected economically important plants such as cucumbers, potatoes, avocados, and oranges.

Figure 8 Infected cow. This cow, which is unable to stand and walk, is showing signs of mad cow disease.

Section 1 Review

1. **Compare** the properties of viruses with the properties of cells. SC.F.1.4.2
2. **Describe** Stanley's experiment with the tobacco mosaic virus. SC.H.1.4.2
3. **Name** the parts of a virus. SC.F.2.4.2
4. **List** the steps by which viruses replicate. SC.F.1.4.5
5. **Describe** how HIV causes AIDS. SC.F.1.4.8
6. **Critical Thinking Evaluate** the argument that emerging viruses are new viruses. SC.H.1.4.2
7. FCAT Prep Which of the following is a characteristic of both cells and viruses?
   A growth
   B metabolism
   C homeostasis
   D nucleic acids
   SC.F.1.4.5

441

Close

Reteaching — BASIC

K-W-L Have students return to their lists of things they **W**ant to know about viruses that they created when they began this section. Have them check off the questions that they can now answer. Students should then make a list of what they have **L**earned. Conclude by asking:

- Which questions are still unanswered?
- What new questions do you have?

You may wish to have students research their unanswered questions. LS Verbal

Quiz — GENERAL

1. What is the genetic material of HIV? (HIV is composed of two molecules of single-stranded RNA.)
2. What is a bacteriophage? (A bacteriophage is a virus that infects bacteria.)
3. What is the general structure of a virus? (A virus has a nucleic acid core surrounded by a protein coat.)

Alternative Assessment — GENERAL

Writing Have students write a story telling how an HIV virus infects a cell. LS Verbal

Answers to Section Review

1. Cells metabolize, grow, maintain homeostasis, and reproduce. Viruses reproduce, but not without the help of a host cell.
2. Stanley purified the tobacco mosaic virus, found that it was crystalline, and then used the crystal to infect healthy tobacco plants.
3. Viruses consist of a nucleic acid core, a protein capsid, and may be surrounded by a membranous envelope (proteins, lipids, and glycoproteins).
4. Viruses enter the host cell. This is followed by replication of the viral genome, which is then used to build viral proteins. The proteins and nucleic acids are assembled into new viruses, which cause the cell to burst, releasing new viruses.
5. HIV replicates in macrophages over a period of years. Eventually, the HIV surface glycoproteins change and they attack T cells, killing them. This destroys the body's immune system.
6. Emerging viruses have been recorded only recently as causing diseases. However, these viruses may not be new; they simply may have gone unnoticed or unrecorded.
7. **A.** Incorrect. Viruses do not grow. **B.** Incorrect. Viruses do not metabolize. **C.** Incorrect. Viruses do not maintain homeostasis. **D.** Correct. Viruses have nucleic acids.

Section Review Correlations

SC.F.1.4.2: Item 1
SC.F.1.4.5: Items 4, 7
SC.F.1.4.8: Item 5
SC.F.2.4.2: Item 3
SC.H.1.4.2: Items 2, 6

Section 2

Focus

Overview

Before beginning this section review with your students the objectives listed in the Student Edition. This section introduces students to Eubacteria and Archaebacteria, their structure, and their methods of obtaining energy. Their ability to cause disease and their importance to humans are also discussed. The section ends with a discussion of antibiotic resistance in bacteria.

Bellringer

Have students make individual lists of as many beneficial uses of bacteria as they can think of.

Motivate

Identifying Preconceptions — GENERAL

Ask students which of the following diseases are caused by viruses and which by bacteria: measles, AIDS, tuberculosis, syphilis, influenza, chicken pox, botulism, bubonic plague, polio, mumps, Lyme disease, strep throat, and the common cold. (Measles, AIDS, influenza, chicken pox, polio, mumps, and colds are caused by viruses; the other diseases are caused by bacteria.) Ask whether antibiotics can cure all these diseases and, if not, why. Explain that although antibiotics can cure many bacterial infections, they are not effective at curing viral infections.

Section 2 Bacteria

Florida Sunshine Standards/FCAT

SC.B.1.4.1 Knowledge of energy is fundamental to all the scientific disciplines (e.g., the energy required for biological processes in living organisms and the energy required for the building, erosion, and rebuilding of the Earth).

SC.F.1.4.2 Body structures are uniquely designed and adapted for their function.

SC.G.1.4.1 There is great diversity and interdependence among living things.

Objectives

- **List** seven differences between bacteria and eukaryotic cells. SC.G.1.4.1
- **Describe** three different ways bacteria can obtain energy. SC.B.1.4.1
- **Describe** the external and internal structure of *Escherichia coli*. SC.F.1.4.2
- **Distinguish** two ways that bacteria cause disease. SC.G.1.4.1
- **Identify** three ways that bacteria benefit humans. SC.G.1.4.1

Key Terms

pilus
bacillus
coccus
spirillum
capsule
antibiotic
endospore
conjugation
anaerobic
aerobic
toxin

Bacterial Structure

The prokaryotes referred to in this chapter as bacteria include the organisms that compose the kingdom Eubacteria (Domain Bacteria) and the organisms that compose the kingdom Archaebacteria (Domain Archaea). Bacteria differ from eukaryotes in at least seven ways.

1. **Internal compartmentalization.** Bacteria are prokaryotes. Unlike eukaryotes, prokaryotes lack a cell nucleus. Many species of bacteria have no internal compartments or membrane systems.
2. **Cell size.** Most bacterial cells range in size between 1μ and 5μ, while eukaryotic cells tend to range between 10μ and 100μ. There are, however, some very large bacteria—up to 750μ—and some very small eukaryotic cells.
3. **Multicellularity.** All bacteria are single cells. Some bacteria may stick together or may form strands. These formations, however, are not truly multicellular, and the activities of individual cells are not specialized.
4. **Chromosomes.** Bacterial chromosomes consist of a single circular piece of DNA. Eukaryotic chromosomes are linear pieces of DNA that are associated with proteins.
5. **Reproduction.** Bacteria reproduce by binary fission, a process in which one cell pinches into two cells. In eukaryotes, however, microtubules pull chromosomes to opposite poles of the cell during mitosis. Afterward, the cytoplasm of the eukaryotic cell divides in half, forming two cells.
6. **Flagella.** Bacterial flagella are simple structures composed of a single fiber of protein that spins like a corkscrew to move the cell. Eukaryotic flagella are more-complex structures made of microtubules that whip back and forth rather than spin. Some bacteria also have shorter, thicker outgrowths called **pili** *(PIHL ee)* (singular, pilus), shown in **Figure 9.** Pili enable bacteria to attach to surfaces or to other cells.
7. **Metabolic diversity.** Bacteria have many metabolic abilities that eukaryotes lack. For example, bacteria perform several different kinds of anaerobic and aerobic processes, while eukaryotes are mostly aerobic organisms.

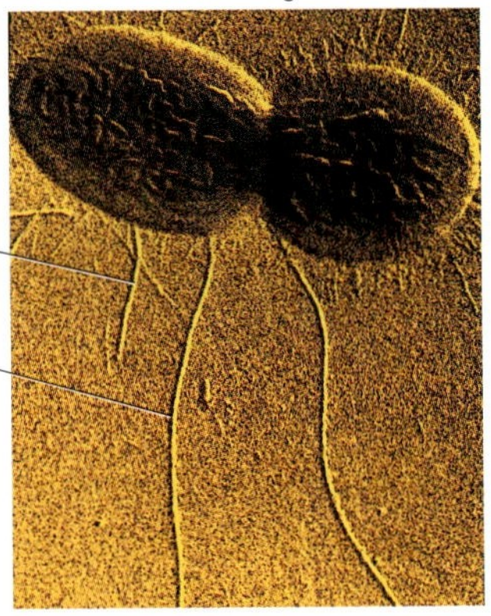

Figure 9 Flagella and pili. Bacteria have flagella that provide them with movement and pili that enable adherence to surfaces.

Chapter Resource File

- Directed Reading BASIC
- Active Reading GENERAL

One-Stop Planner CD-ROM

- Reading Organizers BASIC
- Reading Strategies BASIC
- Supplemental Reading Guide *The Andromeda Strain* ADVANCED

Section Correlations

SC.B.1.4.1
SC.F.1.4.2
SC.G.1.4.1

Bacterial Cell Shapes

A bacterial cell is usually one of three basic shapes, as shown in **Figure 10:** **bacillus** *(buh SIHL uhs),* a rod-shaped cell; **coccus** *(KAHK us),* a round-shaped cell; or **spirillum** *(spy RIHL uhm),* a spiral cell. A few kinds of bacteria aggregate into strands. Species that form filaments are indicated by the prefix *strepto-,* while species that form clusters are indicated by the prefix *staphylo-.*

Members of the kingdom Eubacteria have a cell wall made of peptidoglycan, a network of polysaccharide molecules linked together with chains of amino acids. Outside the cell wall and membrane, many bacteria have a gel-like layer called a **capsule.** Members of the kingdom Archaebacteria often lack cell walls.

Cell walls Eubacteria can have two types of cell walls, distinguished by a dye staining technique called the Gram stain. One group is called Gram-negative, and the other Gram-positive.

Gram staining is important in medicine because the two groups of eubacteria differ in their susceptibility to different antibiotics. **Antibiotics** are chemicals that interfere with life processes in bacteria. Thus, Gram staining can help determine which antibiotic would be most useful in fighting an infection.

Endospores Some bacteria form thick-walled **endospores** *(EHN doh spohrz)* around their chromosomes and a small bit of cytoplasm when they are exposed to harsh conditions. These conditions can be the depletion of nutrients, a drought, or high temperatures. Endospores can survive environmental stress and may germinate years after they were formed, releasing new, active bacteria.

Pili Pili enable bacteria to adhere to the surface of sources of nutrition, such as your skin. Some kinds of pili enable bacteria to exchange genetic material through a process called conjugation. **Conjugation** *(kahn juh GAY shuhn)* is a process in which two organisms exchange genetic material. In prokaryotes, pili from one bacterium adhere to a second bacterium, and genetic material is transferred from the first bacterium to the second bacterium. Conjugation enables bacteria to spread genes within a population.

Study TIP

Reviewing Information

Prepare flash cards for each of the **Key Terms** in this chapter. On each card, write the term on one side and its definition on the other side. Use the cards to review meanings of the Key Terms.

internet connect

www.scilinks.org
Topic: Bacteria
Keyword: HX4018

SCILINKS. Maintained by the National Science Teachers Association

Figure 10 Bacterial shapes. Bacteria are usually one of three shapes.

Magnification: 117,300×

Magnification: 2,295×

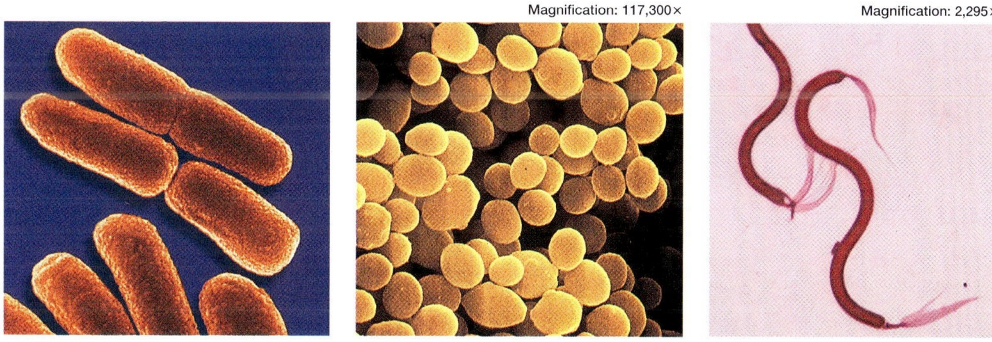

Bacillus (rod-shaped) *E. coli*

Coccus (round-shaped) *Micrococcus luteus*

Spirillum (spiral-shaped) *Spirillum volutans*

Teach

READING SKILL BUILDER — BASIC

Brainstorming Ask students to discuss whether the following statement is true or false: Bacteria are successful in part because they have cellular structures that enable them to live in a wide variety of environments. Students should give examples supporting their viewpoints.

Teaching Tip — GENERAL

Movement and Locomotion

Explain that bacteria have several mechanisms of locomotion, including sliding over slimy surfaces, twisting through fluids, and propelling themselves with flagella. However, not all bacteria move. Bacteria that are capable of movement are called motile bacteria. Those that cannot move on their own are called nonmotile bacteria. Have students discuss how nonmotile bacteria might move from place to place. (Possible answers are transport by other organisms, through the air, through water.)

Transparencies

- **TR** Bellringer
- **TR F2** Three Bacterial Cell Shapes
- **TR F3** Gram Staining
- **TR F4** *Escherichia coli*
- **TR F6** Important Bacterial Diseases

Attention Grabber

Bacterial Flagella Flagella are composed of proteins. A flagellum is connected to a protein shaft, which passes through the capsule of the bacteria. The shaft attaches to a double set of rings in the cell wall, much like rings of ball bearings. When the inner ring turns, the flagellum rotates. The outer ring is fixed to the cell wall and does not move. Interested students may want to make a model of a bacterial flagellum.

Teach, *continued*

SKILL BUILDER — GENERAL

Math Skills Explain to students that, in theory, each colony on a Petri dish arose from a single bacterium and that bacteriologists use this understanding to count the number of bacteria in a sample. Give students the following problem. A 1-mL sample of bacterial culture was diluted 1,000 times in water, and then 0.1 mL was added to a Petri dish. Twenty-four hours later, 38 colonies were counted. Ask students to calculate how many bacteria were in the original sample. (38 × 1,000 × 10 = 380,000 bacteria per mL) Then ask students to calculate how many bacteria would have been in the sample if the dilution factor had been only 1 to 250. (38 × 250 × 10 = 95,000 bacteria per mL) LS Logical

Teaching Tip — GENERAL

Nitrogen-Fixing Bacteria Ask students what would happen if all nitrogen-fixing bacteria were eliminated. (The amount of nitrogen available to plants would fall, plant growth would be reduced, and the amount of organic compounds available at higher trophic levels would decline.)

Activity — BASIC

Organizing Bacteria Have students make a table classifying common bacteria by the way they obtain energy. Tables should divide bacteria into three groups: photosynthetic, chemoautotrophic, and heterotrophic. LS Logical

Real Life

Big, big bacteria
In 1999, scientists announced the discovery of the largest bacteria ever discovered. *Thiomargarita namibiensis* was found off the coast of Namibia. More than 100 times larger than the previously known largest bacterium, *T. namibiensis* is 0.5 mm wide.

Obtaining Energy

Over 4,000 species of bacteria have been named, and probably many more haven't yet been discovered. Bacteria occur in the widest possible range of habitats and play key ecological roles in nearly all of them. As you may recall from an earlier chapter, bacteria thrive in hot springs, frigid arctic seas, and groundwater. They are even found at high pressures in the deep sea and inside solid rock.

Bacteria can be classified in several different ways. Classifying bacteria by the different ways in which they obtain energy, for example, gives a good general sense of the great diversity among bacteria.

Bacteria can also be classified according to their phylogenetic relationships. By comparing the sequence of their ribosomal RNA, scientists have determined that there are at least 12 phyla of eubacteria and four phyla of archaebacteria.

Photosynthesizers

A significant fraction of the world's photosynthesis is carried out by bacteria. Photosynthetic bacteria can be classified into four major groups based on the photosynthetic pigments they contain: purple nonsulfur bacteria, green sulfur bacteria, purple sulfur bacteria, and cyanobacteria. Green sulfur bacteria and purple sulfur bacteria grow in **anaerobic** (oxygen-free) environments. They cannot use water as a source of electrons for photosynthesis and instead use sulfur compounds, such as hydrogen sulfide, H_2S. Purple nonsulfur bacteria use organic compounds, such as acids and carbohydrates, as a source of electrons for photosynthesis.

Figure 11 Photosynthetic bacterium. *Anabaena* is a photosynthetic cyanobacterium in which individual cells adhere in filaments. The two large orange-colored cells are encased in a structure where nitrogen fixation occurs.

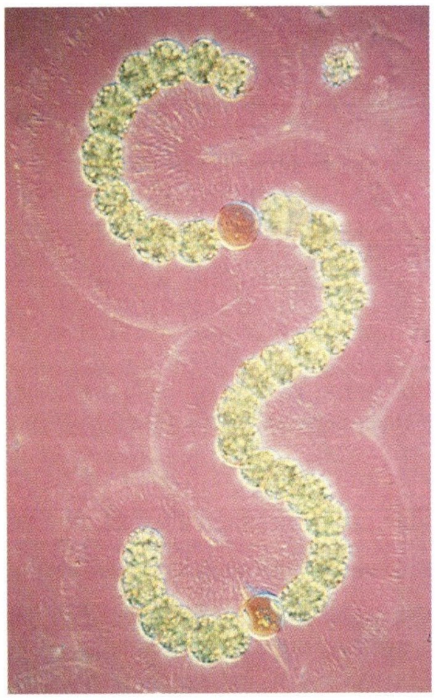

Of particular importance are the cyanobacteria, which often clump together in large mats of filaments. Recall that cyanobacteria are thought to have made the Earth's oxygen atmosphere. Each filament is a chain of cells encased in a continuous jellylike capsule. Many cyanobacteria, such as species of *Anabaena,* shown in **Figure 11,** are capable of fixing nitrogen.

Chemoautotrophs

Bacteria called chemoautotrophs *(KEE moh AW toh trohfs)* obtain energy by removing electrons from inorganic molecules such as ammonia, NH_3, and hydrogen sulfide, H_2S, or from organic molecules such as methane, CH_4. In the presence of one of these hydrogen-rich chemicals, chemoautotrophic bacteria can manufacture all their own amino acids and proteins. Chemoautotrophic bacteria that live in the soil, such as *Nitrosomonas* and *Nitrobacter,* are of great importance to the environment and to agriculture. They have an important role in the nitrogen cycle called nitrification. Nitrification, as you may recall from an earlier chapter, is the process in which bacteria oxidize ammonia into nitrate. Nitrate is the form of nitrogen most commonly used by plants.

444

Cultural Awareness

Making Yogurt Yogurt was first made by nomadic Middle Eastern tribes. The hot desert sun, aided by the gentle rocking of camel transportation, provided the perfect environment for making yogurt. Yogurt is created from milk that has been curdled by two types of bacteria: *Lactobacillus bulgaricus* and *Streptococcus thermophilus.*

Heterotrophs

Most bacteria are heterotrophs. Together with fungi, heterotrophic bacteria are the principal decomposers of the living world; they break down the bodies of dead organisms and make the nutrients available to other organisms. Many are **aerobic,** that is, they live in the presence of oxygen. Some other bacteria can live without oxygen.

Other activities of heterotrophic bacteria may be helpful or harmful to humans. For example, more than half of our antibiotics are produced by several species of *Streptomyces,* a filamentous bacterium found in soil. On the other hand, one species of *Staphylococcus* can secrete a poison into food. This poison causes nausea, diarrhea, and vomiting in people who eat the *Staphylococcus*-contaminated food.

Species of the symbiotic bacteria *Rhizobium* are by far the most important of all nitrogen-fixing organisms. *Rhizobium* species are heterotrophic bacteria that usually live within lumps on the roots of legumes (plants such as soybeans, beans, peas, peanuts, alfalfa, and clover), as shown in **Figure 12.** Farmers take advantage of *Rhizobium*'s nitrogen-fixing abilities when they "rotate" their crops every few years and grow legumes, which replenish the soil with nitrogen-containing compounds.

Figure 12 Nitrogen-fixing bacteria. The bacteria inside the lumps on these soybean roots contain *Rhizobium,* a nitrogen-fixing bacteria.

FORENSICS BIOWatch

Bioterror Detectives

When a criminal commits a crime, investigators look for fingerprints at the crime scene. But what happens when the criminal is a bioterrorist and the weapon is a deadly bacterium?

The new field of forensic genomics offers investigators a new way to track bacteria to their source. Forensic genomics uses genome sequencing to identify a specific bacterial type—a subset of a species—called a *strain.* Just as a fingerprint can belong to only one individual, the genomic sequence of each different strain is unique.

Anthrax is a lethal bacterium that produces tough spores. When anthrax spores were sent through the mail in the fall of 2001, five people died. Investigators wanted to know the source of the bacterium. They thought it might have been stolen from a research laboratory. Because labs work with specific strains, a bacterium could possibly be linked to one—and only one—lab.

Scientists sequenced the genome of the anthrax strain from the first victim. Then they compared it to the genome sequence of a known strain. In this case, researchers identified more than 60 genetic markers that differed between the two strains. With time, genome sequencing of other anthrax strains will create a "library" of genome markers, increasing the speed and success rate of future identification attempts.

Teaching Tip — Basic

Classifying Bacteria Have students draw a table like the Graphic Organizer at the bottom of this page to classify bacteria by the way they obtain energy. Tables should divide bactria into three groups: photosynthetic, chemoautotrophic, and heterotrophic.

BIOWatch

Bioterror Detectives

Teaching Strategies

Explain to students that forensic genomics can help researchers track natural pathogens such as tuberculosis, pneumonia, cholera, and *E.coli.* Forensic geno-mics can also be used to distinguish wild strains from lab strains, as in the case of West Nile virus.

Discussion

A contagious bacterial infection spreads throughout a hospital. How could forensic genomics help scientists identify the source of the outbreak? (By comparing genome markers in the bacterial strains found in the patients, forensic genomics can help scientists find the point of origin of the outbreak.)

Forensics Labs Appendix

Turn to pp. 1048–1071 for forensics lab activities and more student information on forensics techniques.

Graphic Organizer

Use this graphic organizer with **Teaching Tip** *on this page.*

Type of bacteria	Energy source	Examples
Photosynthetic	sunlight	cyanobacteria, green sulfur bacteria, purple sulfur bacteria
Chemoautotrophic	electrons in inorganic molecules	*Nitrosomonas, Nitrobacter*
Heterotrophic	organic material from other organisms	Streptomyces, Rhizobium

Up Close

Escherichia coli

Teaching Strategies
Tell students that in 1993, four children died from infection with a strain of *E. coli* called O157:H7 found in undercooked hamburger meat, and more than 600 cases of food poisoning were reported in northwestern states. The symptoms of *E. coli* O157:H7 food poisoning can include bloody diarrhea and kidney failure. Severe cases can cause permanent damage to organs or even death. The Centers for Disease Control and Prevention estimates that approximately 20,000 individuals suffer symptoms of *E. coli* O157:H7 poisoning each year from undercooked food. Ask students how they can avoid food poisoning. (Possible answers include cooking meat thoroughly, choosing clean restaurants, and making sure that foods are served and stored at their proper temperatures.)

Discussion
- Is *E. coli* a gram-positive or gram-negative bacterium? (gram-negative)
- How does *E. coli* reproduce? (binary fission) How fast can the cells divide? (as often as every 20 minutes)
- Describe the genetic material of an *E. coli* cell. (It has one circular DNA molecule containing about 5,000 genes.)
- What is the function of pili? (Pili serve to attach *E. coli* to surfaces and to join bacterial cells in conjugation.)

Escherichia coli

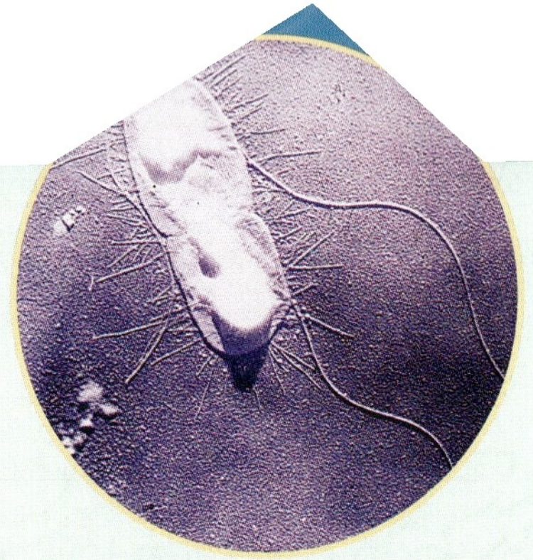

- **Scientific name:** *Escherichia coli*
- **Size:** Up to 1 μm
- **Habitat:** Inhabits the intestines of many mammals
- **Mode of nutrition:** Heterotrophic

Characteristics

Cell structure *E. coli* is a Gram-negative eubacterium. It has a rigid cell wall composed of peptidoglycan. Outside the cell wall lies the outer membrane, which is composed of lipids and polysaccharides.

Genetic material Like all bacteria, *E. coli* has a single DNA molecule in the form of a loop. *E. coli* has approximately 5,000 genes.

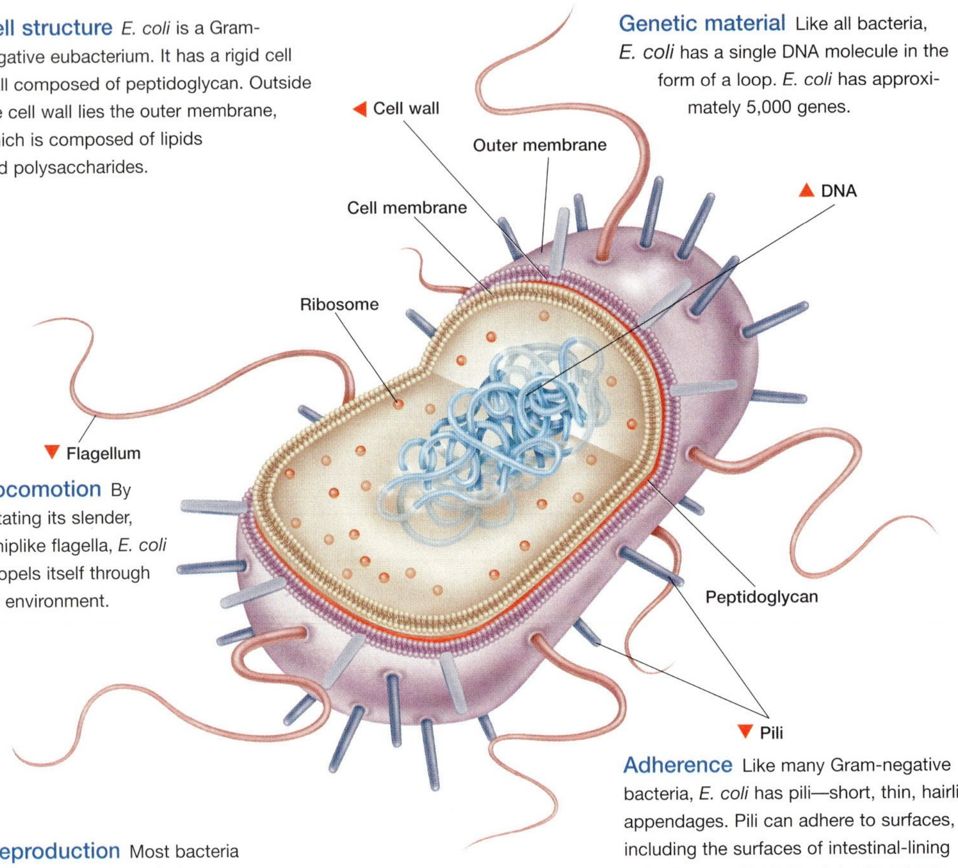

Locomotion By rotating its slender, whiplike flagella, *E. coli* propels itself through its environment.

Adherence Like many Gram-negative bacteria, *E. coli* has pili—short, thin, hairlike appendages. Pili can adhere to surfaces, including the surfaces of intestinal-lining cells. Pili also join bacterial cells prior to conjugation.

Reproduction Most bacteria reproduce by binary fission, a process by which a single cell divides into two identical new cells. *E. coli* can divide as often as every 20 minutes.

446

MISCONCEPTION ALERT

"Bad" Bacteria Students tend to think of bacteria as "bad" organisms because they are often described as sources of disease. Of the thousands of kinds of bacteria, however, only a few are harmful. Bacteria are nature's recyclers, and scientists are finding ways to use bacteria to produce desirable materials and degrade wastes. Biotechnologists insert genes into bacterial cells that allow them to make plastics, pharmaceuticals, pesticides, and foods. Bacteria can even be used to mine metals, such as copper and gold, and to clean up industrial wastes.

Pathogenic Bacteria

In order to understand infectious diseases, think of your body as a treasure chest full of resources. Your body has protein, minerals, fats, carbohydrates, and vitamins. You may want to keep and use these resources, but so do many other organisms, including the bacteria on and in your body. Bacteria have evolved various means of obtaining these resources from you. In some cases, the competition for the resources in your body can result in your becoming ill.

Bacteria Can Metabolize Their Host

Heterotrophic bacteria obtain nutrients by secreting enzymes that break down complex organic structures in their environment and then absorbing them. If that environment is your throat or lungs, this can cause serious problems.

For example, tuberculosis, a disease of the lungs, is caused by *Mycobacterium tuberculosis,* shown in **Figure 13.** Tuberculosis was once one of the most common causes of death. In most cases, infection occurs when tiny droplets of moisture that contain the bacteria are inhaled. Some bacteria settle in the lungs, where they grow using human tissue as their nutrients. The bacteria may also spread to other parts of the body. Symptoms may include coughing up sputum and blood, chest pain, fever, fatigue, weight loss, and loss of appetite. If left untreated, death may occur as quickly as within 18 months but more commonly within 5 years. Other important bacterial diseases are described in **Table 2.**

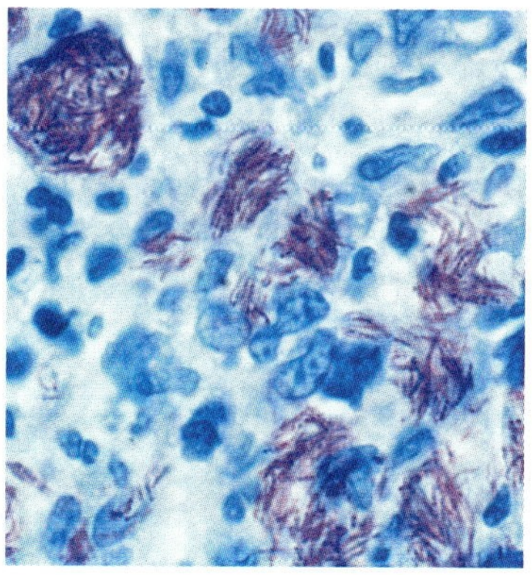

Figure 13 Tuberculosis in a lung. The red-stained structures in this light micrograph are *Mycobacterium tuberculosis,* which cause tuberculosis.

Table 2 Important Bacterial Diseases

Disease	Description of illness	Bacterium	How the disease is transmitted or caused
Anthrax (respiratory)	Fever, severe difficulty breathing	*Bacillus anthracis*	Inhalation of spores
Bubonic plague	Fever, bleeding, lymph nodes that form swellings called buboes; often fatal	*Yersinia pestis*	Bite of an infected flea
Cholera	Severe diarrhea and vomiting; often fatal if not treated	*Vibrio cholerae*	Drinking contaminated water
Dental cavities	Destruction of minerals in tooth	*Streptococcus mutans*	Dense collections of bacteria in mouth
Lyme disease	Rash, pain, swelling in joints	*Borrelia burgdorferi*	Bite of an infected tick
Tuberculosis	Fever, cough, difficulty breathing	*Mycobacterium tuberculosis*	Inhalation
Typhus	Headache, high fever	*Rickettsia prowazekii*	Bite of infected flea or louse

447

Using the Table — GENERAL

Have students examine **Table 2** and suggest ways that bacterial and viral infections can be prevented. Lead students to consider ways that pathogens can be spread such as through air, water, blood, food, pets, rodents, and so on. (Students may suggest that diseases can be prevented by sanitation, good hygiene, lifestyle changes, vaccinations, healthy immune systems, and proper storage and preparation of food.)

Demonstration — GENERAL

Ask students to bring in newspaper and magazine articles that discuss specific bacterial or viral infections. Make a list on the board or overhead of the diseases discussed in each article, and have students indicate whether each pathogen is a bacterium or virus. Have students discuss the methods of contracting each disease, the people commonly affected, and potential treatments. **LS Interpersonal**

HISTORY CONNECTION

The plague that spread through Europe in the fourteenth century killed more than 25 million people. Introduced by rats from the Black Sea region, it spread through trade routes to Europe. The disease is caused by the bacterium *Yersinia pestis* and is transmitted by fleas. It causes enlarged lymph nodes, called buboes—hence, the name bubonic plague. Fever, pain, and necrosis, or blackening of the skin, can also occur. This last symptom lead to the disease's other name: Black Death.

Teach, *continued*

Using the Figure

Have students examine **Figure 14.** Point out that the zone of clearance around the bacterial colony represents the destruction of red blood cells by *Streptococcus*. Ask students if they think *Streptococcus* species are the only bacteria that lyse red blood cells. (Hemolysins are also produced by *Staphylococcus* and *Clostridium* species.) English Language Learners

SKILL BUILDER —ADVANCED

Writing Skills Have students research and write a report about the guidelines for food handling and storage that are recommended by the Centers for Disease Control and Prevention. **LS Verbal**

SKILL BUILDER —ADVANCED

Graphing For decades the tuberculosis (TB) death rate in the United States was in decline. In the 1940s, health workers believed TB would be eradicated because the disease seemed to be under control. By 1985, the number of reported cases had plunged to only 22,201—one-fourth of the number of cases reported in 1953. However, the number of cases began to rise in 1986. Although TB is treatable and preventable, a cure takes about 6 months once a person's lungs are infected. Many patients stop taking their medication before they are completely cured. Have students research the number of TB cases from the 1940s to the present and make a bar graph showing the number of cases in each decade. **LS Logical**

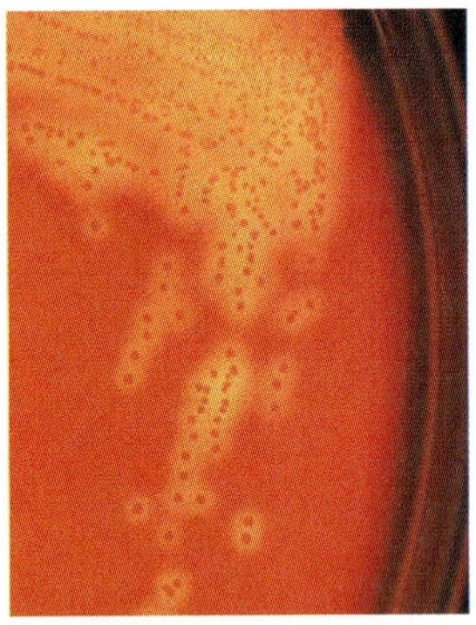

Figure 14 The effect of bacterial toxins. This species of *Streptococcus* secretes a toxin that destroys red blood cells. The agar contains red blood cells and clearly shows a zone around the bacteria where the toxin has destroyed the red blood cells.

Not all bacteria are lethal. For example, some bacteria cause everyday health problems, such as acne. Acne occurs in about 85 percent of teenagers. Bacteria, such as *Propionibacterium acnes*, normally grow in an oil gland of the skin. They metabolize a certain kind of oil produced by those glands. During puberty the oil glands increase the amount of oil produced, and the bacterial population on the skin increases greatly. The bacteria grow in the pores where the oil normally flows, forming pimples and blackheads.

Bacterial Toxins

The second way bacteria cause disease is by secreting chemical compounds into their environment. These chemicals, called **toxins,** are poisonous to eukaryotic cells, as shown in **Figure 14.** Toxins can be secreted into the body of an infected person or into a food in which bacteria are growing.

When bacteria grow in food and produce toxins, the toxins can cause illness in humans who eat those contaminated foods. This kind of illness is called an intoxication. For example, *Staphylococcus aureus* causes the most common type of food poisoning. The symptoms include nausea, vomiting, and diarrhea. This type of poisoning is painful but is seldom fatal.

Another type of intoxication that is fatal occurs when food is not canned properly. Sometimes canned food is not heated enough to kill endospore-forming bacteria, such as *Clostridium botulinum.* The bacteria can then grow and produce a deadly toxin that affects the nervous system. A person who eats food that contains this toxin then becomes ill with a disease called botulism, whose symptoms include double vision and paralysis. People with botulism may die because they are unable to breathe.

Some bacteria are responsible for other diseases reported in the news, such as *E. coli* O157:H7, the cause of several outbreaks of food poisoning in the United States. *E. coli* normally lives in our intestines. However, if it acquires DNA that codes for the toxin through conjugation, it can produce the toxin. *E. coli* poisoning is associated with raw or improperly cooked ground beef.

Most bacteria can be killed by boiling water or various chemicals. Using hot, soapy water to prevent contamination of our food utensils and food supply is one way of preventing disease. Many commercial antibacterial products can also be used to prevent bacterial contamination in the kitchen and in industrial food factories.

Biowarfare

Biowarfare is the deliberate exposure of people to biological toxins or pathogens such as bacteria or viruses. The United States government is justifiably concerned about the use of *bioweapons*—biological toxins or pathogens suitable for mass infection—against military personnel overseas and against civilians within the United States. Biologists are working on new approaches to recognize the onset of an attack with a bioweapon, to treat infected people, and to slow the spread of any outbreak of disease.

did you know?

Bacterial Toxins Some bacterial toxins can be used for medical purposes. Botulinum toxin is used to treat dystonia, an abnormal muscle rigidity that causes painful muscle spasms. The toxin inhibits the release of acetylcholine from the neurons in the neuromuscular junction, allowing the muscles to relax. Dermatologists are using the same toxin to erase "frown lines" and wrinkles on the face. Diluted toxin is injected under the skin to relax facial muscles and smooth out the wrinkles. The effect is temporary and must be repeated every 6 months.

Antibiotics

In 1928, the British bacteriologist Alexander Fleming noticed a fungus of the genus *Penicillium* growing on a culture of *S. aureus*. He saw that bacteria did not grow near the fungus. He concluded that the fungus was secreting a substance that killed the bacteria, as shown in **Figure 15.** Fleming isolated the substance and named it penicillin. In the early 1940s, scientists found that penicillin was effective in treating many bacterial diseases, such as pneumonia.

Different antibiotics interfere with different cellular processes. Because these processes do not occur in viruses, antibiotics are not effective against them. Other antibiotics, such as tetracycline and ampicillin, have been discovered in nature or imitated chemically.

Figure 15 Antibiotics are naturally produced. Alexander Fleming saw a plate of agar very similar to this one. Notice how the bacteria do not grow next to this fungus.

Antibiotic-Resistant Bacteria

In recent years, some bacteria have become resistant to antibiotics. Susceptible bacteria are eliminated from the population, and resistant bacteria survive and reproduce, thus passing on their resistance traits. Mutations for antibiotic resistance arise spontaneously in bacterial populations as errors in DNA replication. There are many individuals in a bacterial population, and bacteria multiply very rapidly (doubling their numbers in as few as 20 minutes). Therefore, a mutation that gives the bacteria a selective advantage can quickly spread throughout a population.

Antibiotic Misuse Mutations that confer resistance to antibiotics are strongly favored in bacterial populations being treated with an antibiotic. Usually, if the full course of the antibiotic is administered, all the targeted bacteria are killed and there is no chance for a resistant strain to develop. If antibiotic treatment ends prematurely, some of the bacteria may survive. Which ones? The ones most resistant to the antibiotic. A patient who does not take the full course of a prescribed antibiotic is setting the stage for the development of antibiotic-resistant bacteria.

Multiple-antibiotic Resistance A related problem can arise in a patient being treated with two or more antibiotics at the same time. This practice selects for bacteria that have acquired several antibiotic-resistance genes. A number of strains of *Staphylococcus aureus* associated with severe infections of hospital patients (so-called hospital staph) have appeared in recent years. These strains are resistant to penicillin and a wide variety of other antibiotics, so infections caused by these strains are very difficult to treat.

Recently, concern has arisen over the common use of antibacterial soaps. Antibacterial soaps are marketed as a means of protecting people from harmful bacteria. Their routine use, however, may favor bacteria resistant to the antibacterial agents in the soap. Ultimately, routine use of antibacterial soaps could reduce our ability to treat common bacterial infections.

Group Activity — Advanced

Writing **"Rational" Drug Design** One strategy that researchers are using to control resistant bacterial strains is called "rational" drug design. By looking at the genes and enzymes involved in drug resistance, scientists are trying to create drugs that bind to the active sites of the enzymes and render them inactive. Have student groups research this work and prepare presentations for class. Presentations might include models of enzyme action. Co-op Learning

Group Activity — General

Breakthroughs in Science Organize students into small groups to discuss the technological uses of bacteria. Have students propose future uses of bacteria and ask them to imagine the potential impact of their breakthroughs. Encourage students to consider potentially harmful, beneficial, and benign effects of their ideas on quality of life, economy, health, and the environment. Then have students produce a diagram illustrating the topic of their discussion to share with the class. Diagrams might feature the technological development in the center with its effects on society represented by spokes radiating outward. Co-op Learning English Language Learners

Trends in Genetic Engineering

Engineered Bacteria Show students pictures of an oil spill and a field with a growing crop. Ask students what useful roles bacteria might play in these pictures. Explain that bacteria are being genetically modified by biotechnologists to perform a variety of useful tasks, including cleaning up spilled petroleum products and removing PCBs from the environment. In agriculture, bacteria can take the place of some fertilizers and pesticides and can be used to prevent frost damage in strawberry plants. Human insulin, human growth hormone, and TPA (which dissolves blood clots) are now being made using engineered bacteria.

Close

Reteaching — BASIC

Have students research the proliferation of *Staphylococcus* infections that are contracted in hospitals. How are they spread? What strains are involved? Why are they particularly dangerous? How can they be prevented? Have students present their findings in a short written report.

Quiz — GENERAL

1. What are the three basic shapes of bacteria? (bacillus, coccus, spirillum)
2. How do heterotrophic bacteria obtain energy? (They obtain energy by feeding on organic material produced by other organisms.)
3. In what two ways do bacteria produce disease? (They metabolize their host's body and produce toxins.)

Alternative Assessment — GENERAL

Have students give a class presentation on composting to show that bacteria are an essential part of the decay process. Students should explain the type of metabolism used by the bacteria (heterotrophic) and the conditions under which the bacteria live in the compost pile. The presentation should also address the question of whether decay could occur in the absence of bacteria.

Figure 16 Swiss cheese. In making Swiss cheese, bacteria grow in the cheese and produce gas. As the cheese hardens these pockets of gas remain, giving the cheese its characteristic holes.

Importance of Bacteria

Despite the misery that some bacteria cause humans in the form of disease and food spoilage, much of what bacteria do is extremely important to our health and economic well-being.

Food and Chemical Production

Many of the foods that we eat are processed by specific kinds of bacteria. For example, many fermented foods are produced with the assistance of bacteria, as shown in **Figure 16.** These foods include pickles, buttermilk, cheese, sauerkraut, olives, vinegar, sourdough bread, and even some kinds of sausages.

Humans are able to use different bacteria to produce different kinds of chemicals for industrial uses, as shown in **Figure 17.** For example, different kinds of *Clostridium* species can make either acetone or butanol. These chemicals can be used to produce a large variety of other useful chemicals.

Genetic engineering companies use genetically engineered bacteria to produce their many products, such as drugs for medicine and complex chemicals for research.

Figure 17 Industrial fermenter. Bacteria can be used to produce useful chemicals such as in this fermenter.

Mining and Environmental Uses of Bacteria

Mining companies can use bacteria to concentrate desired elements from low-grade ore. Low-grade ore has a low percentage of the desired mineral, but it also has sulfur compounds. Chemoautotrophic bacteria can convert the sulfur into a soluble compound, leaving the desired mineral behind. The sulfur compound can be washed away with water, leaving only the desired mineral. This technique can be used to harvest copper or uranium.

Bacteria metabolize different organic chemicals and are therefore used to help clean up environmental disasters such as petroleum and chemical spills. Powders containing petroleum-metabolizing bacteria are used to help clean oil spills.

Section 2 Review

1. **Construct** a table that lists the seven ways bacteria differ from eukaryotic cells. SC.G.1.4.1
2. **List** the structures found in *E. coli.* SC.F.1.4.2
3. **Identify** the relationship between photosynthesis, heterotrophic metabolism, and chemoautotrophic metabolism. SC.F.1.4.3
4. **Describe** the relationship between metabolism, toxins, bacteria, and disease. SC.G.1.4.1
5. **List** three ways bacteria are helpful. SC.G.1.4.1
6. **Critical Thinking Defending a Theory** How does the growth of antibiotic resistance in bacteria support the theory of evolution by natural selection? SC.G.2.4.3
7. *FCAT Prep* Which disease is caused by inhaling a bacterium?
   A cholera
   B botulism
   C tuberculosis
   D *E. coli* food poisoning
   SC.F.1.4.8

Answers to Section Review

1. Differences include internal compartmentalization, cell size, multicellularity, chromosomes, reproduction, flagella, and metabolic diversity.
2. cell wall, pili, flagella, ribosomes, circular DNA
3. All are methods by which bacteria obtain energy. Photosynthesizers use light energy, chemoautotrophs use simple chemicals, and heterotrophs use complex organic material made by other organisms.
4. Bacteria may cause disease by metabolizing the host's cells or by secreting toxins.
5. Bacteria are used in the processing of many foods, the creation of medical and industrial chemicals, the fixation of nitrogen, mining, and environmental restoration.
6. Natural selection favors those organisms best suited to their environment. When an antibiotic is present, bacteria that are naturally resistant to it will survive over nonresistant bacteria. The resistant bacteria will eventually outnumber the less well-adapted bacteria.
7. **A.** Incorrect. Cholera is caused by drinking contaminated water. **B.** Incorrect. Botulism is caused by eating contaminated food. **C.** Correct. **D.** Incorrect. *E. coli* food poisoning is caused by eating contaminated food.

Section Review Correlations

SC.F.1.4.2: Item 2
SC.F.1.4.3: Item 3
SC.F.1.4.8: Item 7
SC.G.1.4.1: Items 1, 4, 5
SC.G.2.4.3: Item 6

Study ZONE CHAPTER HIGHLIGHTS

Key Concepts

1 Viruses

- Viruses consist of segments of a nucleic acid contained in a protein coat, and some have an envelope.
- Viruses do not have all of the characteristics of life and are therefore not considered to be alive.
- Viruses replicate inside living cells. They enter a cell by injecting their genetic material into the cell, by slipping through tears in the plant cell wall, or by binding to molecules on the cell surface.
- Some viruses replicate through a lytic cycle or a lysogenic cycle.
- HIV replicates inside immune system cells, eventually destroying them, leaving the host without adequate defense against disease.
- Emerging viruses are newly recognized viruses or viruses that have reappeared or spread to new areas.
- Viroids are infectious RNA molecules that cause disease in plants, and prions are infectious proteins that cause disease in certain animals.

2 Bacteria

- Bacteria differ from eukaryotes in their cellular organization, cell structures, and metabolic diversity.
- Bacteria can be classified into two groups according to their cell wall structure. Gram staining can be used to distinguish these two groups.
- Bacteria can transfer genes to one another by conjugation.
- Bacteria are grouped according to their ribosomal RNA sequences and the way they obtain energy.
- Bacteria cause disease by metabolizing nutrients in their host or by releasing toxins, which damage their host.
- Bacterial disease can usually be fought with soap, chemicals, and antibiotics.
- Bacteria are used to make foods, antibiotics, and chemicals; to fix nitrogen; to clean the environment; and to cycle important chemicals in the environment.

Key Terms

Section 1

virus (434)
pathogen (434)
capsid (435)
envelope (435)
glycoprotein (435)
bacteriophage (435)
lytic (436)
provirus (436)
lysogenic (437)
prion (441)
viroid (441)

Section 2

pilus (442)
bacillus (443)
coccus (443)
spirillum (443)
capsule (443)
antibiotic (443)
endospore (443)
conjugation (443)
anaerobic (444)
aerobic (445)
toxin (448)

451

Alternative Assessment — GENERAL

Have students choose two diseases, one caused by a bacterium and one caused by a virus. Students can use information from the textbook, do outside research, or create fictitious diseases. Ask students to draw plausible structures for the pathogens and to describe the method of infection, prevention, and treatment. Students' drawings must depict the characteristics of bacteria and viruses described in this chapter.

Chapter Resource File

- Science Skills Worksheet GENERAL
- Critical Thinking Worksheet ADVANCED
- Test Prep Pretest GENERAL
- Chapter Test GENERAL

Answer to Concept Map

The following is one possible answer to Performance Zone item 15 on the next page.

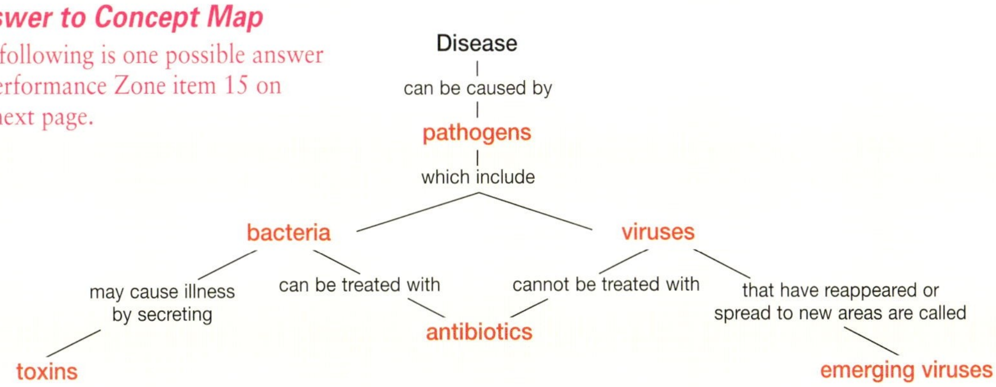

CHAPTER 20 **ZONE**

Understanding Key Ideas

1. d
2. a
3. c
4. c
5. d
6. c
7. b
8. The body's immune system acts as a selective force. Some of the viruses are eliminated. Those viruses that have features that enable them to escape destruction continue to multiply.
9. Scientists can compare anthrax genome markers in the unknown strain to genome markers of known strains in the "library" in an effort to find a match.
10. The answer to the concept map is found at the bottom of the Study Zone page.

Critical Thinking

11. Certain mutations enable HIV to resist antiviral drugs. As the virus is exposed to these drugs, those HIV that are resistant will survive and continue to multiply, becoming more common, while those HIV not resistant will be selected against and become less common.
12. The Native Americans had no natural immunity to the virus because they had never been exposed to it. Many Europeans were immune because they had been exposed to the virus, either through contact with it or by having and recovering from smallpox.

ZONE

Understanding Key Ideas

1. Unlike cells, viruses do not
a. grow.
b. have homeostasis.
c. metabolize.
d. All of the above
SC.G.1.4.1

2. What evidence led Stanley to conclude that TMV is not a living organism?
a. The extract of TMV crystallized.
b. TMV is made of RNA and protein.
c. TMV reproduces only in cells.
d. The virus poisons tobacco plants.
SC.H.1.4.1

3. HIV infects and destroys
a. skin cells. **c.** immune cells.
b. red blood cells. **d.** bacterial cells.
SC.F.1.4.8

4. Bacteria
a. always have flagella.
b. are smaller than viruses.
c. have aerobic or anaerobic metabolism.
d. have a nucleus.
SC.F.1.4.2

5. Bacteria that do not require sunlight and obtain energy by removing electrons from hydrogen-rich chemicals are called
a. heterotrophs.
b. photosynthetic bacteria.
c. cyanobacteria.
d. chemoautotrophs.
SC.F.1.4.3

6. Environmental spills of petroleum are sometimes cleaned up using
a. viroids. **c.** bacteria.
b. prions. **d.** bacteriophages.
SC.H.3.4.5

7. Identify the pilus in the photo below. SC.F.1.4.2

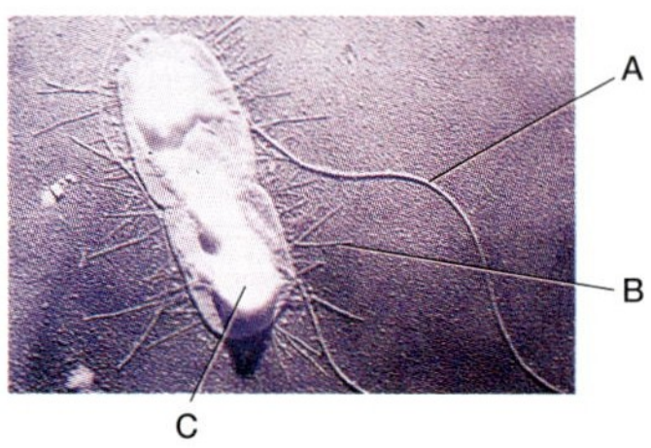

8. If cold viruses invade your body, your body's immune system may destroy most but not all of these viruses. How does your body's immune system affect the evolution of the cold viruses? SC.G.2.4.3

9. **BIOWatch** How can a "library" of anthrax genome markers help scientists identify unknown anthrax strains? SC.H.3.4.2

10. **Concept Mapping** Make a concept map describing the relationships of bacteria and viruses to diseases. Try to include the following terms in your map: *bacteria, viruses, pathogen, emerging viruses, antibiotics,* and *toxin.* SC.G.1.4.1

Critical Thinking

11. Applying Information How does the increase of resistance to antiviral drugs in HIV relate to the theory of evolution by natural selection? SC.G.2.4.3

12. Evaluating Results In the 1520s, the Spanish explorer Cortés and his armies introduced smallpox to the Americas. The death rate among the Native American people ranged from 50 to 90 percent compared with a death rate of about 10 percent among people in Europe. What accounts for the difference in death rates? SC.F.1.4.8

13. Career Connection **Virologist** Research the field of virology. Write a report on your findings that includes job description, training required, kinds of employers, growth prospects, and starting salary. SC.H.1.4.1

Alternative Assessment

14. Finding Information Research and write a report on a preventable viral disease, such as polio or smallpox. In your report, discuss the process scientists followed in identifying the cause of the disease, isolating the virus, formulating a vaccine, and testing the vaccine. SC.H.3.4.3

452

13. Virologists work with viruses and tissue cultures that are the hosts for viruses. Most virologists have a medical degree or a graduate degree in microbiology or virology. Employers may include pharmaceutical companies, university research labs, government research labs, public health labs, and hospitals. Growth potential is good. Starting salary will vary by region.

Alternative Assessment

14. Answers will vary depending on the viral disease chosen.

Chapter Review Correlations

SC.F.1.4.2: Items 4, 7
SC.F.1.4.3: Item 5
SC.F.1.4.8: Items 3, 12
SC.G.1.4.1: Items 1, 10
SC.G.2.4.3: Items 8, 11
SC.H.1.4.1: Items 2, 13
SC.H.3.4.2: Item 9
SC.H.3.4.3: Item 14
SC.H.3.4.5: Item 6

Assignment Guide

Section	Questions
1	1, 2, 3, 8, 10, 11, 12, 13, 14
2	1, 4, 5, 6, 7, 9, 10

FCAT Prep

For questions 1–6, write your answers on a separate sheet of paper.

1. Which of the following is a type of virus that infects bacteria? SC.G.1.4.1
   A. bacteriophage
   B. emerging virus
   C. glycoprotein
   D. viroid

2. The basic components of all viruses are a nucleic acid and what other structure? SC.F.1.4.5
   F. endospore
   G. glycoprotein
   H. icosahedron
   I. protein coat

3. What structure allows *E. coli* to move? SC.F.1.4.2
   A. flagellum
   B. nucleus
   C. peptidoglycan
   D. pili

4. When does a bacteriophage kill its host cell? SC.F.1.4.5
   F. during conjugation
   G. during the lytic cycle
   H. during the lysogenic cycle
   I. while assembling the capsid

5. READ INQUIRE EXPLAIN A person is infected by HIV, and the virus does not mutate while in the person's body.

   Why would that person not likely experience immune system failure? SC.F.1.4.8

6. READ INQUIRE EXPLAIN Many microbiologists oppose the use of antibiotics in patients with viral infections. They claim that antibiotics will not help the patient, because a virus does not carry out any of the life processes that antibiotics interrupt. Such overuse of antibiotics encourages the proliferation of resistant strains of bacteria.

   Part A Determine the life process that viruses need cells to carry out. SC.F.1.4.5

   Part B Infer how using an antibiotic promotes the proliferation of resistant strains of bacteria. SC.G.2.4.3

Test TIP

When faced with similar answers, define the answer choices and then use those definitions to narrow down the choices on a multiple-choice question.

FCAT Doctor

Question 4 Answer G is the correct choice. During the lytic cycle, viruses replicate using host cells. The new viruses burst out of the host, killing it in the process. Answer F is incorrect because conjugation is the process by which two bacteria exchange genetic material. It does not affect viral reproduction. Answer H is incorrect because, during the lysogenic cycle, the viral DNA remains embedded in the host genome. The host's activities are not affected, but it reproduces, creating infected daughter cells. Answer I is incorrect because assembling the capsid happens before the viruses burst out of the host cell and kill it.

Question 5 HIV replicates only in the macrophages until it mutates. When it mutates, HIV can then replicate in the T cells, destroying them, which destroys the immune system.

Question 6

Part A Viruses are not considered to be alive, because they cannot carry out the processes of life without the aid of cells from other organisms.

Part B Certain mutations enable bacteria to resist antibiotics. Bacteria that are resistant will survive and continue to multiply, becoming more common, while those that are not resistant will be selected against.

Answers

1. A
2. I
3. D
4. G
5. Only after HIV mutates does it infect T cells and destroy the immune system.

6A. Viruses use cells to reproduce. The cells are guided by information stored in the viral DNA.

6B. Mutated bacteria that are resistant to the antibiotic will be naturally selected.

Test Prep Correlations

SC.F.1.4.2: Item 3
SC.F.1.4.5: Items 2, 4, 6A
SC.F.1.4.8: Item 5
SC.G.1.4.1: Item 1
SC.G.2.4.3: Item 6B

Skills Practice

Staining and Observing Bacteria

Teacher's Notes

Time Required 45 minutes

Ratings

EASY → HARD

Teacher Preparation 3
Student Setup 3
Concept Level 2
Cleanup 2

Skills Acquired

- Classifying
- Collecting Data
- Experimenting
- Organizing and Analyzing Data

Scientific Methods

In this lab, students will:

- Make Observations
- Draw Conclusions
- Communicate the Results

Materials and Equipment

Materials for this lab can be ordered from WARD'S. See *Master Materials List* at the front of this book for catalog numbers. All cultures should be autoclaved before disposal. Have students place their used swabs in a centrally located container. The swabs may then be incinerated or autoclaved before disposal.

Safety Cautions

Make sure students wear safety goggles, lab aprons, and disposable gloves before obtaining their bacterial cultures. Remind students to use methylene blue carefully in step 7 and to keep it away from their faces, skin, and clothing. Be sure that students wipe their tables down with the rubbing alcohol before and after the lab. Be sure that open flames are isolated from the areas where alcohol is being used. Make sure that students dispose of their materials properly and wash their hands before leaving the lab.

Tips and Tricks

There should be sufficient wax pencils (or permanent marking pens) to supply three or four students with one pencil. You may purchase live cultures or lyophilized cultures from a biological supply house such as WARD'S. Purchase *Micrococcus luteus* (coccus), *Spirillum volutans* (spirillum), and *Bacillus megaterium* (bacillus). Standard isopropyl alcohol is sufficient for aseptic-technique washing down of the tables. Provide students with sufficient sterile cotton swabs that are wrapped individually or paired. Buy premixed nutrient broth from a biological supply house such as WARD'S, or prepare culture medium from dehydrated concentrate and sterilize in an autoclave at 120° C at 15 psi for 20 minutes. About one day before the lab, grow the bacteria in nutrient broth at 37° C. Prepare a sufficient amount of culture broth to provide

Skills Practice Lab

Staining and Observing Bacteria

SC.H.1.4.1

SKILLS

- Using aseptic techniques
- Using a microscope

OBJECTIVES

- **Prepare** and stain wet mounts of bacteria.
- **Identify** different types of bacteria by their shape.

MATERIALS

- wax pencil
- 3 microscope slides
- safety goggles
- lab apron
- disposable gloves
- rubbing alcohol
- paper towels
- 3 culture tubes of bacteria (A, B, and C)
- test-tube rack
- sterile cotton swabs
- Bunsen burner with striker
- microscope slide
- forceps or wooden alligator-type clothespin
- 150 mL beaker
- methylene blue stain in dropper bottle
- compound microscope

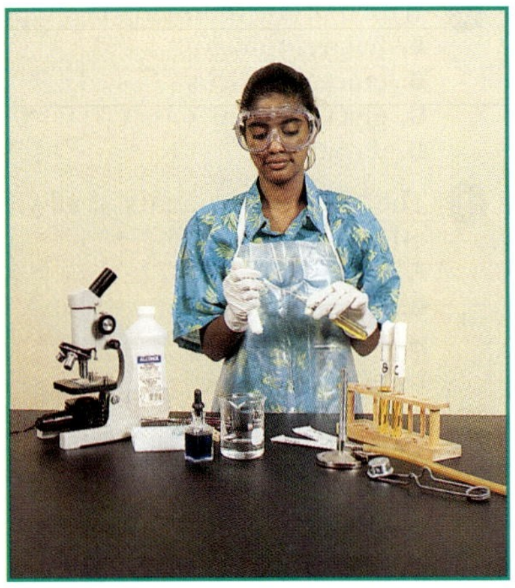

ChemSafety

CAUTION: Always wear safety goggles and a lab apron to protect your eyes and clothing.

CAUTION: Do not touch or taste any chemicals. Know the location of the emergency shower and eyewash station and how to use them. If you get a chemical on your skin or clothing, wash it off at the sink while calling to the teacher. Notify the teacher of a spill. Spills should be cleaned up promptly, according to your teacher's directions.

CAUTION: Glassware is fragile. Notify the teacher of broken glass or cuts. Do not clean up broken glass or spills with broken glass unless the teacher tells you to do so.

Before You Begin

Like all **prokaryotes**, bacteria are unicellular organisms that sometimes form filaments or loose clusters of cells. They are prepared for viewing by making a **smear**, a slide on which cells have been spread and dried. Treating the cells with a **stain** makes them more visible under magnification. In this lab, you will stain, identify, and compare and contrast different types of bacteria.

1. Write a definition for each boldface term in the paragraph above and for each of the following terms: strepto, staphylo, coccus, bacillus, spirillum.
2. Based on the objectives for this lab, write a question you would like to explore about different kinds of bacteria.

Procedure

PART A: Observing Live Bacteria

1. Put on safety goggles, a lab apron, and disposable gloves.
2. Use a wax pencil to label three microscope slides *A, B,* and *C*.
3. Use rubbing alcohol and paper towels to clean the surface of your lab table and gloves. Allow the table to air dry. **CAUTION: Alcohol is flammable. Do not use alcohol near an open flame.**
4. Light a Bunsen burner with a striker. **CAUTION: Keep combustibles away from flames. Do not light a Bunsen burner when others in the room are using alcohol.**

454

5. Beginning with culture A, make a smear of three different bacteria as follows. Remove the cap from a culture tube. *Note: Do not place the cap on the table.* Pass the opening of the tube through the flame of a Bunsen burner. Lightly touch the tip of a sterile swab to the bacterial culture. Pass the opening of the tube through the flame again, and replace the cap. Transfer a small amount of the culture to the appropriately labeled microscope slide by rubbing the swab on the slide, and spread out to cover about half of the total slide area. Dispose of the swab in a proper container. Repeat for cultures B and C.
6. Allow your smears to air dry. Using microscope slide forceps, pick up each slide one at a time and pass it over the flame several times. Let each slide cool.
7. Using microscope slide forceps, place one of your slides across the mouth of a 150 mL beaker half-filled with water. Place 2–3 drops of methylene blue stain on the dried bacteria. *Note: Do not allow the stain to spill into the beaker.* **CAUTION: Methylene blue will stain your skin and clothing.** Let the stain stay on the slide for 2 minutes. Then dip the slide into the water in the beaker several times to rinse it. Blot the slide dry with a paper towel. *Note: Do not rub the slide.*
8. Repeat step 7 for your other two slides.
9. Allow each slide to air dry, and then observe them with a microscope. Make a sketch of a few cells on each slide. Compare your sketches to Figure 10, and identify the type of bacteria on each slide.

PART B: Cleanup and Disposal

10. Dispose of slides, used swabs, solutions, and broken glass in the designated waste containers. Do not pour chemicals down the drain or put lab materials in the trash unless your teacher tells you to do so.
11. Clean up your work area and all lab equipment. *Clean the surface of your lab table with rubbing alcohol.* Return lab equipment to its proper place. Wash your hands thoroughly before you leave the lab and after you finish all work.

Analyze and Conclude

1. **Summarizing Results** Describe the shape and grouping of the cells of each type of bacteria you observed.
2. **Analyzing Methods** Why should the test tube caps from the culture tubes (in Part B) not be placed on the table?
3. **Evaluating Viewpoints** Evaluate the following advice: Always use caution when handling bacteria, even if the bacteria is known to be harmless.
4. **Drawing Conclusions** How did you classify the bacteria in cultures A, B, and C, as a coccus, a bacillus, or a spirillum?
5. **Further Inquiry** Write a new question about bacteria that could be explored with another investigation.

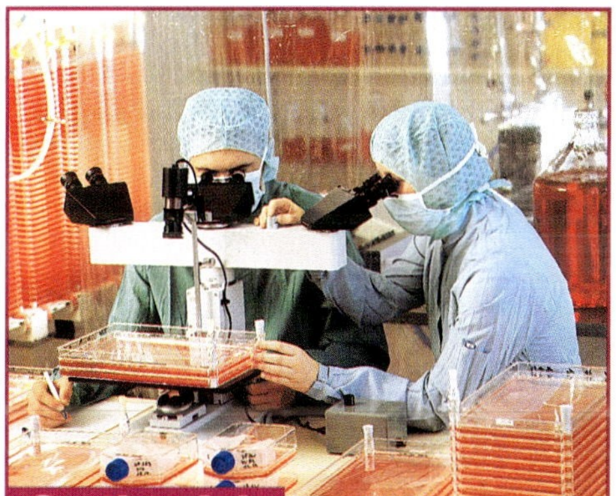

On the Job

Microbiologists are scientists who study organisms too small to be seen by the naked eye. Do research to discover how microbiologists better our lives. For more about careers, visit **go.hrw.com** and type in the keyword **HX4 Careers.**

groups of three or four students with about 5 mL of each culture. Transfer about 5 mL of culture to each tube with a sterile pipet. Do not allow students to access your stock cultures. If students' microscopes have an oil immersion lens, you may provide them with immersion oil so that they can view the bacteria at a higher magnification.

Each lab group should be given a set of bacterial cultures. If you allow groups to obtain samples directly from stock cultures, the stocks are likely to become contaminated. Use broth cultures, which will enable students to see chains of cocci.

Answers to Before You Begin

1. *prokaryote*—organism that has no nucleus or membrane-enclosed structures; *smear*—microscope slide that has bacterial cells smeared and dried on it; *stain*—a colored substance that makes cells more visible under magnification; *strepto*—prefix used to describe bacteria that form chains of cells; *staphylo*—prefix used to describe bacteria that form clusters of cells; *coccus*—spherical bacterium; *bacillus*—cylindrical- or rod-shaped bacterium; *spirillum*—corkscrew-shaped bacterium.
2. Answers will vary. For example: What physical differences can we observe in different types of bacteria when seen under a compound light microscope?

Answers to Analyze and Conclude

1. All three types of bacterial cells were observed. Cocci may form pairs, groups of four, chains, or clusters. Bacilli usually appear as single cells, although sometimes they are arranged in pairs or in chains. Spirilla almost always appear singly, and they differ in length and in the number and size of their spirals.
2. Culture tube caps should not be placed on lab tables because bacteria on the tables could touch the cap and contaminate the bacterial cultures.
3. Answers will vary. Students should mention that mistakes in identification, contamination, or mutations may result in the presence of harmful bacteria.
4. Answers will vary according to the type of bacteria placed into the marked tubes.
5. Answers will vary. For example: Are antiseptics equally effective against the three types of bacteria?

Answers to On the Job

Answers will vary. Microbiologists work closely with doctors and government health agencies to reduce the incidence of diseases caused by microorganisms.

Disease Detectives

Background

From the beginning of the twentieth century until the 1990s, deaths from infectious diseases declined in the United States. Death rates have been climbing recently for a number of reasons—AIDS, the evolution of drug-resistant strains of several pathogens, and the appearance of new, "emerging" diseases like hantavirus pulmonary syndrome. HPS is typical of emerging diseases in many ways. It is not a new disease (caused by a pathogen that has never infected humans) but rather a newly discovered one. It came to light only after an environmental disruption (record precipitation the winter before) triggered an unusually high number of cases.

The swift response to the HPS outbreak was a triumph for the CDC, especially the Special Pathogens Branch, which deals with emerging diseases. Several factors made the quick discovery of the new hantavirus possible. First, the CDC committed more than 100 people to pursuing the virus—some of them worked 16 hours a day, 7 days a week, for months. Second, PCR has made the analysis of genetic data much easier and faster. Third, through years of work on hantaviruses in other parts of the world, scientists had accumulated a large database to draw from.

A thorough knowledge of biology—coupled with insight and hard work—helped scientists track down a killer.

Disease Detectives

In May of 1993, a killer was on the loose in the Four Corners area—New Mexico, Arizona, Utah, and Colorado—in the southwestern United States. Over the previous month, three young, previously healthy people had died in a disturbingly similar way in northwestern New Mexico. All of the victims had come down with what seemed to be a mild case of the flu—with fever, muscle aches, weakness, and lung congestion. All three had been rushed to the hospital gasping for breath, but even emergency treatment could not save them.

Autopsies revealed that all three victims had drowned. Their lungs had filled with plasma—the liquid part of the blood—that had leaked from the tiny blood vessels within the lungs. What could cause such severe damage? Laboratory tests ruled out diseases with similar symptoms, including influenza, bubonic plague, and anthrax.

As doctors and researchers began to investigate further, additional deaths came to light. Fearing that these cases might signal the beginning of a disease outbreak, officials decided to ask for help from the federal Centers for Disease Control and Prevention (CDC) in Atlanta, Georgia.

Catching a Killer Virus

The CDC sent a team of disease investigators to the Four Corners area while researchers in Atlanta tested blood and tissue samples from the victims. One of the tests revealed that the victims had been exposed to a kind of virus known as a hantavirus.

This result was surprising. All known hantaviruses from North America were harmless to humans. Furthermore, the hantaviruses that cause disease in Asia and Europe attack the kidneys, not the lungs. But further testing confirmed the result. The killer was a hantavirus.

What is the virus like? Scientists used one of biology's most powerful tools to find out—the polymerase chain reaction, or PCR. PCR is a method of making many copies of a particular

Pinyon pine seeds

Source of the virus
Deer mice, which thrive on the seeds of pinyon pine, were found to carry hantavirus, a dangerous pathogen.

456

Teaching Strategies

- This story focuses on one geographical area. Have students locate the Four Corners area on a map. Explain how the area got its name—it is the only point where the borders of four states meet. Have them locate the Navajo Indian Reservation on the map. Explain that many of the first cases of HPS were found in Navajos.
- Ask students to describe the structure of a virus. Then have them explain how viruses reproduce. Ask students how viral reproduction differs from bacterial reproduction. Point out that most scientists do not consider viruses to be alive. Have students turn to the definition of *life* given in Chapter 1, and then have them identify the criteria for life that viruses do not meet.
- Have students look in a field guide for mammals and find the map showing the range of the deer mouse. Have them locate the range of the cotton rat, which also carries hantavirus.

Four Corners area
Researchers were surprised to find a deadly hantavirus here.

DNA strand. The copying process requires a primer, a short piece of DNA only a few nucleotides long that becomes the beginning of the new strand. To investigate the characteristics of the new hantavirus, CDC researchers made several primers that shared nucleotide sequences with known hantaviruses. By adding different primers to tissue samples from the victims, the scientists could "fish out" pieces of the virus's genes and then determine their nucleotide sequence.

Determining the Cause of the Outbreak

Within a month of being called in, CDC scientists had caught the killer and sequenced much of its genetic material. They called this new disease *hantavirus pulmonary syndrome,* or HPS.

But scientists still needed to learn how HPS is transmitted and why the outbreak occurred when it did. Elsewhere, researchers knew that hantaviruses are carried by rodents—rats, mice, and their relatives. To find out whether rodents also carried the new hantavirus, researchers trapped thousands of rodents in and around the homes of the victims. About 30 percent of the deer mice they caught tested positive for the virus. These small mice—about 15 cm long—live throughout most of the United States.

Why had the disease not appeared before in the Four Corners area? An unusually wet winter the year before led to a bumper crop of pinyon pine seeds, the deer mouse's favorite food. In turn, the deer mouse population boomed. The larger population probably meant that mice entered homes more often and people were more likely to contact them outdoors. ■

Infectious Disease Investigator

Disease investigator

Profile

Disease investigators never know what kind of phone calls they will get. Their work is often like detective work—trying to piece things together as they seek to find out who has been infected, where, and who else might have been exposed. The field of study that investigates the causes and controls of disease outbreaks is called *epidemiology*.

Job Description

Disease investigators are involved in preventing disease outbreaks and in stopping outbreaks that do occur from spreading. If there is an outbreak, these scientists have to quickly identify the disease, determine how people are being infected, and take action to prevent more people from getting sick. Most disease investigators work for local, state, or federal government agencies.

Science/Math Career Preparation

Microbiology	Mathematics
Zoology	Genetics
Botany	Biochemistry

Analyzing STS Issues

Science and Society

1. **What does the CDC do?** Researchers from the CDC played a key role in solving the hantavirus mystery. Use library resources or the Internet to find out more about the CDC. What is the CDC's mission? What kinds of people work there? Where does the CDC get its funding?
2. **How is the hantavirus controlled?** While researching the CDC, investigate further for information on the hantavirus. Hantavirus infections come from activities that involve humans and infected rodents coming into contact with one another. What measures can be taken to limit exposure to rodents?

Technology: PCR

3. **What are some other newly discovered diseases?** HPS is just one of many emerging, or newly discovered, diseases. Use library resources or the Internet to research three other emerging diseases that have come to light in the last 30 years. Find out where each disease was discovered, its cause, and its symptoms. How is PCR used in the study of emerging diseases?

Read Further

1. Centers for Disease Control and Prevention, www.cdc.gov
2. American Lung Association, www.lungusa.org
3. World Health Organization, www.who.int/en/

Analyzing STS Issues Answers

1. The CDC is an agency of the Health and Human Services Administration of the federal government. The mission of the CDC is to reduce death and injury from all causes. The CDC employs epidemiologists, statisticians, pathologists, virologists, bacteriologists, public health experts, occupational injury experts, public relations specialists, and others. The CDC receives its funding from federal taxes.
2. Risks greatly increase when humans enter or clean a rodent infested area. Occupying previously vacant cabins, barns, or dwelling that are actively infected with rodents should be minimized. Rodents should always be handled with gloves and wild rodents should not be kept as pets or research subjects. Also, avoid exposure to rodent droppings.
3. Answers will vary. Possible answers include Lyme disease, Legionnaires' disease, Ebola fever, Lassa fever, and AIDS. PCR can be used to produce large quantities of DNA from pathogenic viruses and bacteria of unknown identity. The DNA is then used to identify the pathogenic viruses and bacteria.

Chapter Opener		**SE** **Quick Review,** p. 459 ■ **SE** **Reading Activity,** p. 459 ■ **TE** **Using the Figure,** p. 458 GENERAL
45 min pp. 460–463 **Section 1** Characteristics of Protists	**OSP** **Lesson Plan*** **TR** **Bellringer*** **TR** **Section Outline*** **TR** **F14 Reproduction of *Chlamydomonas**** **TR** **F17 Reproduction of *Ulva****	**TE** **Demonstration,** p. 461 ◆ GENERAL **TE** **Using the Figure,** p. 462 **CRF** **Inquiry Lab** Slowing the Movement of Protozoans* GENERAL
& 4 • 90 min pp. 464–471 **Section 2** Protist Diversity	**OSP** **Lesson Plan*** **TR** **Bellringer*** **TR** **Section Outline*** **TR** **F18 Structure of *Euglena**** **TR** **F16 *Paramecium**** **TR** **F20 Life Cycle of *Plasmodium****	**TE** **Activity,** Amoebas, p. 464 ◆ GENERAL **TE** **Using the Figure,** p. 464 **TE** **Group Activity** Space Algae, p. 465 ADVANCED **SE** **Quick Lab** Observing Characteristics of Diatoms, p. 466 ■ **CRF** **Datasheets for In-Text Labs** Observing Characteristics of Diatoms* **TE** **Using the Figure,** p. 467 **TE** **Activity,** Test the Eyespot, p. 467 GENERAL **SE** **Data Lab** Interpreting Competition Among Protists, p. 468 ■ **CRF** **Datasheets for In-Text Labs** Interpreting Competition Among Protists* **TE** **Using the Figure,** p. 470 **TE** **Activity,** Grow Your Own Slime Mold, p. 470 ◆ GENERAL **CRF** **Exploration Lab** Protists—A Comparison* ◆ GENERAL
& 6 • 90 min pp. 472–474 **Section 3** Protists and Health	**OSP** **Lesson Plan*** **TR** **Bellringer*** **TR** **Section Outline*** **TR** **F19 Diseases Caused by Protists***	**TE** **Activity,** Wanted: Helpful Protist, p. 472 GENERAL **TE** **Demonstration,** p. 473 BASIC **SE** **Exploration Lab** Observing Protistan Responses to Light, pp. 478–479 ◆ ■ **CRF** **Datasheets for In-Text Labs** Observing Protistan Responses to Light*

& 8 • 90 min

Chapter Review and Assessment Resources

- **SE** **Study Zone,** p. 475 ■
- **SE** **Performance Zone,** p. 476 ■
- **SE** **FCAT Test Prep,** p. 477 ■
- **TR** **Concept Mapping***
- **CRF** **Vocabulary Review Worksheet*** ■ GENERAL
- **CRF** **Test Prep Pretest*** ■ GENERAL
- **CRF** **Chapter Test*** ■ GENERAL
- **CRF** **Chapter Test*** ADVANCED
- **OSP** **Test Generator***
- **OSP** **Test Item Listing***
- **OSP** **Modified Chapter Test*** BASIC

Online and Technology Resources

go.hrw.com — Holt Online Learning

Visit **go.hrw.com** for access to Holt Online Learning or enter the keyword **HX6 Home** for a variety of free online resources.

One-Stop Planner® CD-ROM

This CD-ROM package includes
- Lab Materials QuickList Software
- Holt Calendar Planner
- Customizable Lesson Plans
- Printable Worksheets
- ExamView® Test Generator
- Interactive Teacher Edition
- Holt PuzzlePro® Resources
- Holt PowerPoint® Resources

Guided Reading Audio CDs

These CDs are designed to help auditory learners and reluctant readers. (Audio CDs are also available in Spanish.)

Compression guide:
To shorten your instruction because of time limitations, omit blocks 1, 6, and 7.

KEY

- **TE** Teacher Edition
- **SE** Student Edition
- **OSP** One-Stop Planner
- **CRF** Chapter Resource File
- **TR** Teaching Transparency
- * Also on One-Stop Planner
- ■ Also Available in Spanish
- ◆ Requires Advance Prep

SKILLS DEVELOPMENT AND PRACTICE	SECTION REVIEW AND ASSESSMENT	CORRELATIONS
		Florida Sunshine Standards/FCAT
TE **Reading Skill Builder** Paired Reading, p. 462 BASIC **CRF** **Directed Reading Worksheet*** BASIC **CRF** **Active Reading Worksheet*** GENERAL	**SE** **Section 1 Review** ■ **TE** **Reteaching,** p. 463 BASIC **TE** **Quiz,** p. 463 GENERAL **TE** **Alternative Assessment,** p. 463 GENERAL **CRF** **Quiz*** ■ BASIC	SC.F.1.4.7, SC.F.2.4.1, SC.G.1.4.1
TE **Skill Builder** Writing Skills, p. 467 GENERAL **TE** **Reading Skill Builder** Reading Organizer, p. 470 BASIC **CRF** **Directed Reading Worksheet*** BASIC **CRF** **Active Reading Worksheet*** GENERAL	**SE** **Section 2 Review** ■ **TE** **Reteaching,** p. 471 BASIC **TE** **Quiz,** p. 471 GENERAL **TE** **Alternative Assessment,** p. 471 GENERAL **CRF** **Quiz*** ■ BASIC	SC.F.1.4.2, SC.G.1.4.1
TE **Skill Builder** Writing Skills, p. 473 GENERAL **CRF** **Directed Reading Worksheet*** BASIC **CRF** **Active Reading Worksheet*** GENERAL **CRF** **Critical Thinking Worksheet*** ADVANCED **CRF** **Science Skills Worksheet*** GENERAL	**SE** **Section 3 Review** ■ **TE** **Reteaching,** p. 474 BASIC **TE** **Quiz,** p. 474 BASIC **TE** **Alternative Assessment,** p. 475 ADVANCED **CRF** **Quiz*** ■ BASIC **Modified Worksheet*** BASIC	SC.H.3.4.3, SC.H.3.4.5

SCILINKS. NSTA
www.scilinks.org
Maintained by the **National Science Teachers Association**

Topic: Protista
SciLinks Code: HX4152

Topic: Reproduction in Protista
SciLinks Code: HX4158

Topic: Malaria Vaccine
SciLinks Code: HX4117

Topic: Protozoans
SciLinks Code: HX4153

CHAPTER 21

Overview

Tell students that they will learn about the amazingly diverse members of the Kingdom Protista. Many of the characteristics and reproductive strategies of these organisms are not seen in other kingdoms. The various phyla of protists are discussed and compared. Finally, beneficial protists and those that cause disease are introduced. Students will learn about the complex life cycle and multiple hosts of the protist that causes malaria, an important disease of humans.

Using the Figure

The Kingdom Protista contains the largest variety of organisms among the six kingdoms. As you can see in the photo at right, microscopic protists come in all sizes and shapes. *Spirogyra,* the long filamentous alga that contains a spiral-shaped ribbon of chloroplasts, is less than 100 μm wide. Not all protists are microscopic, however. Brown algae, known as kelp, are among the largest organisms on Earth.

Assessing Prior Knowledge

Before beginning this chapter, be sure your students can answer the questions in the ***Quick Review*** on the facing page. If they cannot, encourage them to reread the sections indicated.

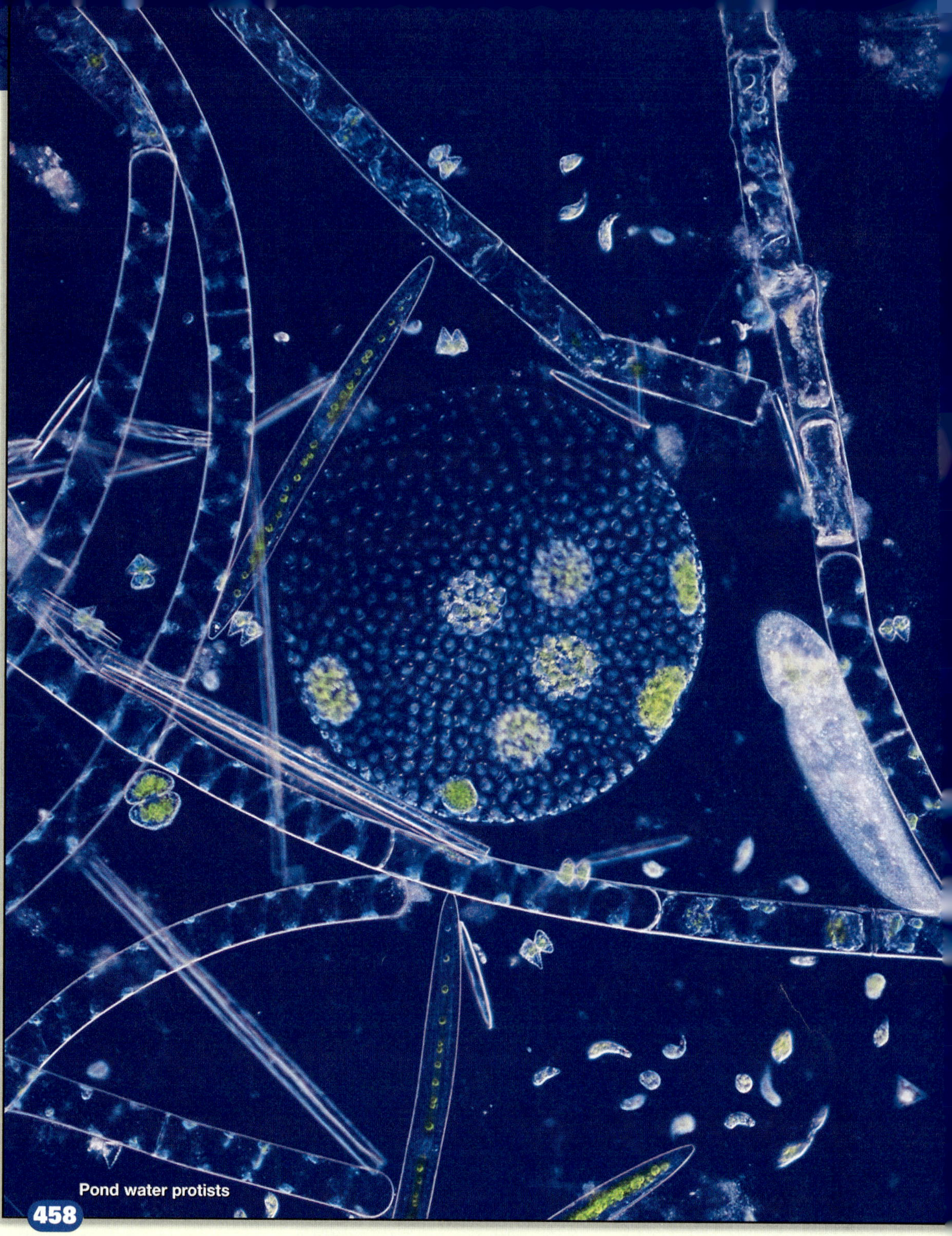

Pond water protists

458

Chapter Correlations | **Florida Sunshine Standards/FCAT**

SC.F.1.4.2 Body structures are uniquely designed and adapted for their function.

SC.F.2.4.1 There are mechanisms of asexual and sexual reproduction and different genetic advantages and disadvantages of asexual and sexual reproduction.

SC.G.1.4.1 There is great diversity and interdependence among living things.

SC.H.3.4.3 Scientists can bring information, insights, and analytical skills to matters of public concern and help people understand the possible causes and effects of events.

SC.H.3.4.5 The value of a technology may differ for different people and at different times.

CHAPTER

21 Protists

Florida Sunshine Standards/FCAT

SC.F.1.4.2 Body structures are uniquely designed and adapted for their function.

SC.F.2.4.1 There are mechanisms of asexual and sexual reproduction and different genetic advantages and disadvantages of asexual and sexual reproduction.

SC.G.1.4.1 There is great diversity and interdependence among living things.

SC.H.3.4.3 Scientists can bring information, insights, and analytical skills to matters of public concern and help people understand the possible causes and effects of events.

SC.H.3.4.5 The value of a technology may differ for different people and at different times.

Quick Review

Answer the following without referring to earlier sections of your book.

1. **Label** the structures of a eukaryotic cell. *(Chapter 3, Section 3)*
2. **Describe** the different kinds of photosynthetic pigments that make photsynthesis possible. *(Chapter 5, Section 2)*
3. **Summarize** the importance of sexual reproduction. *(Chapter 7, Section 2)*
4. **Define** the term *plankton*. *(Chapter 16, Section 3)*
5. **Summarize** the advantage of multicellularity. *(Chapter 18, Section 2)*

Did you have difficulty? *For help, review the sections indicated.*

Reading Activity

List on a separate piece of paper all the words, phrases, and ideas that you associate with the word protist. Share one or two of your ideas with your classmates. Inquire of each other or the teacher to clarify each of the ideas listed. After all discussion is completed, write notes on your paper about what you remember. Look over your notes to see what you know about protists based on your experience and discussion.

An amazing world exists in a drop of pond water. There you will find a whole landscape of single cells that are independent organisms—the protists—the most diverse of all the kingdoms.

Looking Ahead

Section 1
Characteristics of Protists
Diversity
Reproduction

Section 2
Protist Diversity
Ameboid Movement
Algae
Diatoms
Flagellates
Protistan Molds
Sporozoans

Section 3
Protists and Health
Protists and Humans
Malaria

internet connect

www.scilinks.org
National Science Teachers Association *sci*LINKS Internet resources are located throughout this chapter.

SciLINKS. Maintained by the National Science Teachers Association

459

Identifying Misconceptions

Students may assume that protists such as those pictured are animals simply because they do not appear to be plants or fungi. The protists that may be visible in a drop of pondwater, for example, show similarities to animals in that they are able to swim through water, have cilia and flagella, and may appear to have distinct body parts. Remind students that the protists are in fact not animals and that the kingdom Protista includes all eukaryotic organisms that cannot be classified within any other kingdom.

Quick Review Answers

1. Labeled structures should include the nucleus, nucleolus, cell membrane, mitochondria, ribosomes, rough and smooth ER, Golgi apparatus, and vacuoles. Plant cells should also show chloroplasts and cell walls.
2. Chlorophyll absorbs blue and red light and reflects green and yellow. Carotenoids absorb wavelengths of light that are different from those absorbed by chlorophyll and appear yellow and orange.
3. Sexual reproduction is important because it creates different combinations of genes, increasing genetic variability.
4. Plankton are a diverse community of small bacteria, algae, fish larvae, and small animals that drift in the ocean. Plankton provide food for aquatic animals and produce large amounts of oxygen through photosynthesis.
5. Multicellularity enables cells to specialize in different functions, allowing an organism to become more complex.

Chapter Resource File

- Vocabulary Worksheets
- Concept Mapping

Reading Activity Answers

Student lists should reflect the diversity of protists by saying that protists are both unicellular and multicellular; autotrophic, heterotrophic, and parasitic; large as well as microscopic; and reproduce both sexually and asexually.

Section 1

Focus

Overview

Before beginning this section review with your students the objectives listed in the Student Edition. Students will learn that protists are a varied and ancient group of organisms. The distinguishing characteristics of protists and their methods of reproduction are discussed.

Bellringer

Provide small groups of students with folders that contain pictures of protists. (Try to include some representatives of each group shown in **Table 1** on page 461.) Ask students to organize the protists into subgroups. LS Visual

Motivate

Discussion/ Question — GENERAL

Ask students what characteristics of an organism can help ensure the survival of its genes in the next generation. Encourage students to propose strategies that could help organisms perpetuate their genes. How might protists ensure that their genes are passed on? (Suggestions might include improved ways of storing and protecting genetic information or alternative forms of reproduction under different environmental conditions.) Tell students that evolutionary pressures have led to the diversity of reproductive strategies in protists.

Section Correlations

SC.F.2.4.1
SC.G.1.4.1

Section 1 Characteristics of Protists

Florida Sunshine Standards/FCAT

SC.F.2.4.1 There are mechanisms of asexual and sexual reproduction and different genetic advantages and disadvantages of asexual and sexual reproduction.

SC.G.1.4.1 There is great diversity and interdependence among living things.

Objectives

- **List** the characteristics of protists. SC.G.1.4.1
- **List** three environments where protists can be found. SC.G.1.4.1
- **Identify** the unifying features of protists. SC.G.1.4.1
- **Distinguish** asexual and sexual reproduction of *Chlamydomonas*. SC.F.2.4.1
- **Differentiate** two ways multicellular protists reproduce sexually. SC.F.2.4.1

Key Terms

protozoan
alga
zygospore
alternation of generations
sporangium

Diversity

The most diverse of all organisms, protists are mostly unicellular, microscopic organisms, such as the green scum you might find growing on a rock in a pond. But a few protists are as complex and multicellular as the massive house-high seaweed called kelp, which is found in the oceans.

Characteristics

The kingdom Protista *(proh TEES tuh)* consists of an unusually diverse assortment of eukaryotes that exhibit a broad array of characteristics, as summarized in **Figure 1.** For example, some protists are photosynthetic (like plants), some ingest food (like animals), and some absorb their food (like fungi).

Some protists have flagella or cilia, which they can use for locomotion or to collect nutrients. Other protists use other means of locomotion. Protists are found almost everywhere there is water. Many live in lakes and oceans, floating as plankton or anchored to rocks. They are also common inhabitants of damp soil and sand, and they thrive in moist environments, such as leaf litter. Some species are parasites.

Many protists have mechanisms for monitoring and responding to stimuli in their environment. For example, some protists have eyespots, small organelles containing light-sensitive pigments that detect changes in the quality and intensity of light.

The First Eukaryotes

Protists—the first eukaryotes—are thought to have evolved about 1.5 billion years ago through the process of endosymbiosis. The kingdom Protista contains life-forms similar to those that gave rise to the three kingdoms of multicellular organisms—fungi, plants, and animals.

Figure 1 Protist characteristics. These characteristics, which are found in many eukaryotes, first evolved in protists.

Protist Characteristic

- Sexual reproduction
- Multicellularity
- Mitosis and meiosis
- Complex flagella and cilia

Magnification: 1,174×

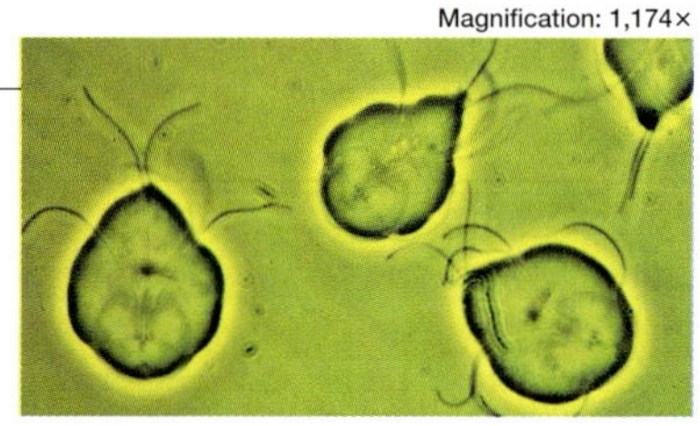

Chapter Resource File

- Directed Reading BASIC
- Active Reading GENERAL

Transparencies

TR Bellringer
TR F14 Reproduction of *Chlamydomonas*
TR F17 Reproduction of *Ulva*

One-Stop Planner CD-ROM

- Reading Organizers BASIC
- Reading Strategies BASIC

Two important eukaryotic features that evolved among the protists are sexual reproduction and multicellularity. Many protists reproduce only asexually, by mitosis; some use meiosis and sexual reproduction only in times of environmental stress; and others reproduce sexually most of the time.

Multicellularity involves a significant amount of coordination among specialized cells. This trait evolved independently in different groups of protists at different times. Early during the evolution of protists, complex flagella and cilia also appeared.

What Unites Protists?

The kingdom Protista contains all eukaryotes that cannot be classified as animals, plants, or fungi. Protists lack the specialized features that characterize the three other multicellular kingdoms. Unlike plants and animals, protists do not reproduce by forming embryos. Protists do not develop complex multicellular reproductive structures. The major phyla of protists are very different from one another and, with a few exceptions, are only distantly related.

Historically, scientists have referred to heterotrophic protists as **protozoa** and to photosynthetic protists as **algae**. These commonly used terms are not formal classification categories. The major phyla of protists are classified according to some of their distinguishing features in **Table 1.** This classification, however, does not reflect the evolutionary relationships of protists.

Table 1 Some Important Protistan Phyla

Phylum	Distinguishing features	Mode of nutrition
Rhizopoda (amoebas) Foraminifera (forams)	Move using pseudopodia	Heterotrophic
Chlorophyta (green algae) Rhodophyta (red algae) Phaeophyta (brown algae)	Typically multicellular	Photosynthetic 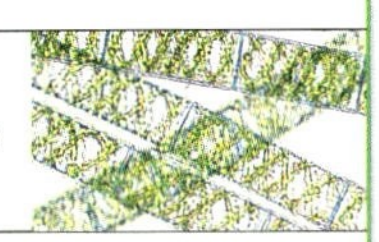
Bacillariophyta (diatoms) Dinoflagellata Euglenophyta	Typically unicellular 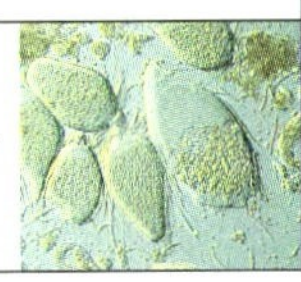	Typically photosynthetic
Kinetoplastida Ciliophora (ciliates)	Move using flagella/cilia	Heterotrophic
Acrasiomycota (cellular slime molds) Myxomycota (plasmodial slime molds) Oomycota	Funguslike	Heterotrophic 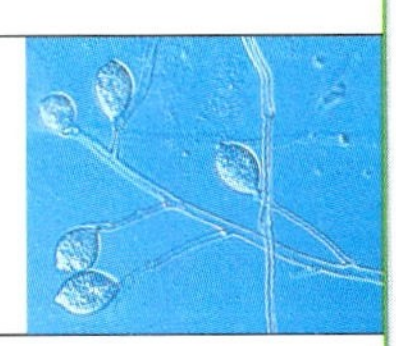
Apicomplexa (sporozoans)	Form spores	Heterotrophic

Teach

Teaching Tip — GENERAL

Eyespots Ask students to define the function of an eyespot. (Eyespots detect the quality and intensity of light.) Ask why it might be important for a protist to monitor the characteristics of light. Lead students to predict that some protists are photosynthetic and that they use information about light to maintain a homeostatic metabolism. This discussion will be continued in Section 2.

Teaching Tip — GENERAL

Pond Scum Pollution of fresh water can lead to a proliferation of green algae known as pond scum. When a small body of water is polluted with nitrates and phosphates from sewage, detergents, or runoff from fertilizers, green algae reproduce rapidly. When the algae die, bacteria decompose them, depleting the oxygen in the water. Ask students: What is the overall effect of explosive algae growth on other organisms living in the pond? (As the algae decompose, oxygen-dependent organisms, such as fish, may die.)

Demonstration — GENERAL

Fill several containers with water from a fish tank or a nearby pond and set them in the sun. Add small amounts of a nitrogen- and/or phosphorus-based fertilizer to some of the containers. Have students compare the resulting algal growth in each of the containers. Use tap water as a control. Ask students how this experiment simulates real-life conditions. (The containers represent bodies of water. The fertilizer represents agricultural runoff, lawn care products, and detergent-loaded wastewater.) Note that some of the observed "algal" growth may actually be photosynthetic prokaryotes. **LS Visual**

did you know?

Van Leeuwenhoek The first person to describe microorganisms was a Dutch cloth merchant named Anton van Leeuwenhoek. Van Leeuwenhoek is sometimes described as an amateur scientist. He is credited with the first descriptions of microorganisms such as bacteria and protists. Van Leeuwenhoek ground his own lenses and developed some of the first powerful microscopes (270×) to inspect cloth. He called the tiny organisms he viewed through his microscopes "animalcules" because he assumed that they were tiny animals.

Reproduction

internet connect
www.scilinks.org
Topic: Reproduction in Protists
Keyword: HX4158
SCILINKS. Maintained by the National Science Teachers Association

Reproduction in the unicellular green alga *Chlamydomonas* is typical of unicellular protists. Chlamydomonas is unusual in that some species can form colonies. Colonial *Chlamydomonas* lose their flagella and secrete a sticky fluid that binds each cell to its neighbors. *Chlamydomonas* species reproduce sexually and asexually.

As a mature organism, the single-celled protist is haploid. When it reproduces asexually, *Chlamydomonas* first absorbs its tail and divides by mitosis, producing two to eight haploid cells called zoospores, which remain within the wall of the parent cell. Mature zoospores break out of the parent cell and grow to become mature haploid cells.

Sexual Reproduction in Unicellular Protists

During environmental stress, such as a shortage of nutrients, *Chlamydomonas* species reproduce sexually. The haploid cell divides first by mitosis to produce haploid gametes. After they are released, a pair of gametes from different *Chlamydomonas* individuals fuse to form a pair. This pair of gametes then shed their cell walls and fuse into a diploid zygote with a thick protective wall called a **zygospore** *(ZIE goh spohr)*.

A zygospore can withstand unfavorable environmental conditions for long periods of time. When environmental conditions become favorable again, meiosis within the zygospore produces haploid cells, which break out of the zygospore wall. These haploid cells grow into mature cells, completing the sexual life cycle, as shown in **Figure 2.**

Figure 2 ***Chlamydomonas* reproduction.** The unicellular green algae in the genus *Chlamydomonas* reproduce sexually and asexually.

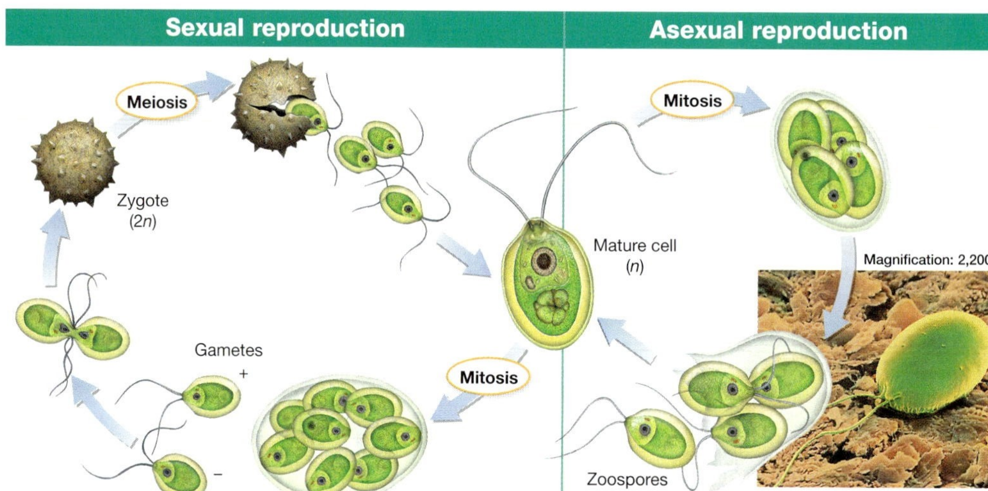

Teach, *continued*

READING SKILL BUILDER — **BASIC**

Paired Reading Assign students to cooperative pairs. Have each student read pages 462–463 silently. As they read, have students indicate passages or parts of figures they understand with a check mark and passages or parts of figures they do not understand with a question mark. Then, have the partners discuss what they did or did not understand.

Teaching Tip — GENERAL

Reproduction in Multicellular Protists Have students draw a Graphic Organizer similar to the one at the bottom of the page to summarize the two modes of reproduction found among multicellular green algae.

Using the Figure

Use **Figure 2** to trace sexual and asexual reproduction in *Chlamydomonas.* Which form of reproduction does it use most frequently? (asexual reproduction) Which does it use under environmental stress? (sexual reproduction) **LS Visual**

English Language Learners

Graphic Organizer

Use this graphic organizer with **Teaching Tip** on this page.

Conjugation → 2 haploid filaments → 1 diploid zygote → resting spore → haploid filament

Alteration of generations → haploid gametophyte → –gamete +gamete → diploid zygote → diploid sporophyte

Alteration of generations → sporophyte → sporangia → haploid zoospores → haploid gametophytes

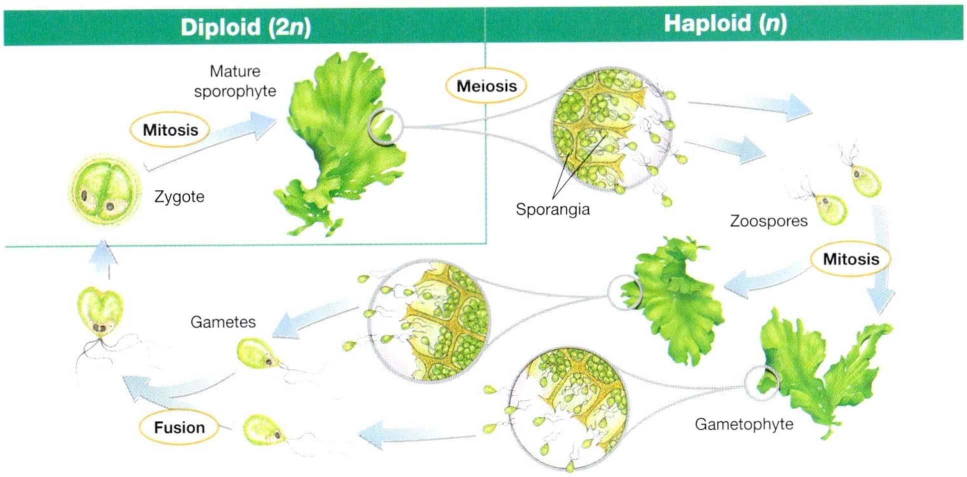

Figure 3 **Life cycle of *Ulva*.** The multicellular green alga *Ulva,* or sea lettuce, has a life cycle in which haploid and diploid individuals alternate.

Sexual Reproduction in Multicellular Protists

Sexual reproduction among multicellular protists occurs in many different ways. Two of these are alternation of generations and conjugation.

Alternation of generations *Ulva* is a common genus of marine green alga. **Figure 3** shows that the reproductive cycle of *Ulva,* called **alternation of generations,** is characterized by two distinct multicellular phases. The diploid, spore-producing phase is called the sporophyte generation. The haploid, gamete-producing phase is called the gametophyte generation. The adult sporophyte alga has reproductive cells called **sporangia** *(spoh RAN jee uh),* which produce haploid spores by meiosis. The spores grow into multicellular haploid gametophytes. The mature gametophyte produces haploid gametes that fuse and complete the life cycle by dividing through mitosis to form a new diploid sporophyte.

Conjugation *Spirogyra,* a filamentous green alga species, reproduces sexually by conjugation. Conjugation is the temporary union of two protists to exchange nuclear material. The process begins when two filaments align side by side. Part of the cell walls between adjacent algae form a bridge between the cells. The nucleus of one cell then passes through the tube into the adjacent cell. The two nuclei eventually form a resting spore, which produces a new haploid filament.

Section 1 Review

1. **Critical Thinking Relating Concepts** Describe two ways sexual reproduction can occur in multicellular protists. SC.F.2.4.1

2. **Summarize** why protists are not classified with the other three eukaryotic kingdoms. SC.G.1.4.1

3. **List** three characteristics of protists. SC.G.1.4.1

4. **FCAT Prep** Which of the following characteristics describes all organisms in the kingdom Protista? SC.G.1.4.1
   A eukaryotic
   B photosynthetic
   C parasitic
   D unicellular

463

Answers to Section Review

1. Multicellular protists may reproduce sexually when two haploid gametes fuse during the gametophyte phase of an alternation of generations cycle. Another method of sexual reproduction in protists is conjugation—the exchange of nuclear material between two individuals.
2. Protists lack the specialized features of plants, animals, and fungi.
3. Answers may include that protists may reproduce sexually or asexually; be unicellular or multicellular; divide by mitosis and meiosis; be heterotrophic, autotrophic, or both; have complex flagella and cilia; and may have light-sensing organelles.
4. **A.** Correct. All protists are eukaryotes.
   B. Incorrect. Not all protists are photosynthetic.
   C. Incorrect. Not all protists are parasitic.
   D. Incorrect. Protists can be either unicellular or multicellular.

Close

Reteaching — Basic

Concept Map Have students work individually or in small groups to make a concept map of the following terms: zoospores, haploid, diploid, unicellular, multicellular, zygospore, conjugation, gametes, alternation of generations, *Chlamydomonas, Spirogyra, Ulva,* zygote, and sporangia.

Quiz — General

1. What two eukaryotic features evolved among the protists? (sexual reproduction and multicellularity)
2. Differentiate between a zoospore and a zygospore in *Chlamydomonas.* (A zoospore is a haploid product of mitosis produced during asexual reproduction. A zygospore is a diploid zygote produced as a product of sexual reproduction.)
3. What is alternation of generations? (the reproductive cycle that alternates a diploid, spore-producing phase with a haploid, gamete-producing phase)

Alternative Assessment — General

Have students write a paragraph or make a model summarizing alternation of generation. The gametophyte and sporophyte generations should be clearly differentiated.

Section Review Correlations

SC.F.2.4.1: Item 1
SC.G.1.4.1: Items 2, 3, 4

Section 2

Focus

Overview

Before beginning this section review with your students the objectives listed in the Student Edition. This section introduces students to the characteristics of the various phyla of protists. Movement, structure, reproduction, and mode of nutrition are discussed for each phylum.

Bellringer

Collect live specimens of algae or obtain preserved specimens from a biological supply house. Have students examine the algae and compare their characteristics. LS Visual

Motivate

Activity — GENERAL

Amoebas Obtain a culture of live amoebas. Using a microprojector or video-microscope, project an image of an amoeba for students to view. Discuss the general characteristics of the amoeba, pointing out the nucleus and means of locomotion (pseudopodia). If time permits, repeat with *Paramecium* and *Euglena*. LS Visual

Using the Figure

Have students use **Figure 4** to create a step-by-step drawing or comic strip depicting the movement of a pseudopodium and the flow of cytoplasm as the amoeba engulfs its prey. Ask students to add a brief paragraph to explain their drawings. LS Visual

Section Correlations

SC.F.1.4.2
SC.G.1.4.1

Section 2 Protist Diversity

Florida Sunshine Standards/FCAT

SC.F.1.4.2 Body structures are uniquely designed and adapted for their function.

SC.G.1.4.1 There is great diversity and interdependence among living things.

Objectives

- **Identify** how amoebas and forams move. SC.F.1.4.2
- **Describe** the structure of diatoms. SC.F.1.4.2
- **Contrast** three kinds of algae. SC.G.1.4.1
- **Differentiate** three different kinds of flagellates. SC.G.1.4.1
- **Summarize** the general characteristics of a *Paramecium*. SC.G.1.4.1

Key Terms

amoeba
pseudopodium
diatom
euglenoid
kinetoplastid
cilium
plasmodium

Ameboid Movement

One of the most easily recognized groups of protists consists of amoebas *(uh MEE buhs)* and forams. Amoebas and forams are unicellular heterotrophs that have a unique form of locomotion called ameboid movement.

Amoebas

Amoebas, members of the phylum Rhizopoda *(RIE zoh POH duh)*, are protists that move by using flexible, cytoplasmic extensions. These extensions are called **pseudopodia** *(soo doh POH dee uh)*, from the Greek words *pseudo*, meaning "false," and *podium*, meaning "foot." Because an amoeba has no cell walls or flagella, it is extremely flexible. During ameboid movement, a pseudopodium bulges from the cell surface, stretches outward, and anchors itself to a nearby surface. The cytoplasm from the rest of the amoeba then flows into the pseudopodium. Pseudopodia can surround and engulf food particles, as shown in **Figure 4.** Amoebas live in both fresh water and salt water and are especially abundant in soil. Meiosis and sexual reproduction do not occur in most amoebas. They reproduce by fission, dividing into two new cells. The majority of amoebas are free-living, but some species are parasites, such as *Entamoeba histolytica*, the protist that causes amebic dysentery in humans. These organisms are transmitted in contaminated food or water.

Figure 4 Pseudopodia. This amoeba is using its pseudopodia to engulf a smaller organism (in yellow circle).

Magnification: 205×

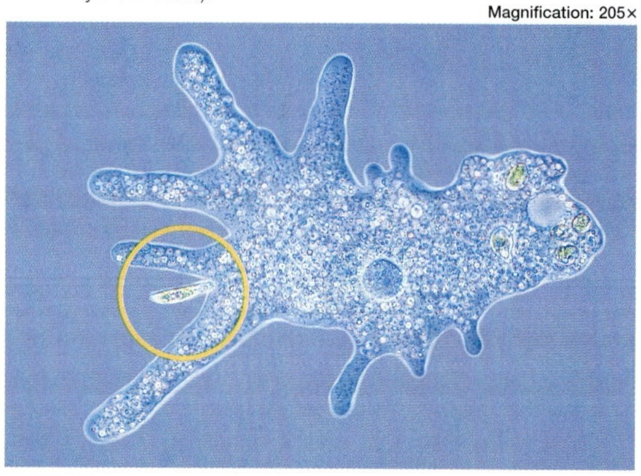

Foraminifera

Members of the phylum Foraminifera *(foh ram ih NIHF ur uh)*, or forams, are marine protists that typically live in sand or attach themselves to other organisms or rocks. Forams are characterized by their porous shells, called tests. Tests usually have many chambers arranged in a spiral shape. They resemble a tiny snail and consist of organic material that contains grains of calcium carbonate. Long, thin projections of cytoplasm extend through the pores in the tests to aid in swimming and in catching prey. Some forams also obtain nourishment from algae that live symbiotically under their tests.

The tests of dead forams have accumulated on ocean floors over millions of years. Their calcium carbonate shells helped form the limestone deposits that are important components of many land formations.

464

Chapter Resource File

- Directed Reading BASIC
- Active Reading GENERAL
- Data Sheet for Quick Lab GENERAL
- Data Sheet for Data Lab GENERAL

Transparencies

TR Bellringer

One-Stop Planner CD-ROM

- Reading Organizers BASIC
- Reading Strategies BASIC

Algae

Algae are protists that are strict phototrophs. Some are unicellular, others are multicellular. Algae are distinguished by the type of photosynthetic pigment they contain and by their cell or body shape.

Green Algae

Most green algae (phylum Chlorophyta) are freshwater unicellular organisms, but some are large, multicellular marine organisms, as shown in **Figure 5.** They also exist as a major part of microscopic marine plankton, inhabit damp soil, or even thrive within the cells of other organisms as symbionts. Green algae contain the same pigments found in the chloroplasts of plants. Most green algae have sexual and asexual reproductive stages.

Red Algae

Red algae (phylum Rhodophyta) are mostly multicellular organisms found in warm ocean waters. Their red pigments are efficient at absorbing the light that penetrates deep waters. Some red algae, as shown in Figure 5, have calcium carbonate in their cell walls. Others are used to make agar and carrageenan *(kayr uh GEE nuhn)*. Red algae have a complex life cycle, usually involving alternation of generations.

Brown Algae

Brown algae (phylum Phaeophyta) are multicellular and are found mostly in marine environments. The larger brown algae known as kelp grow along coasts and provide food and shelter for many different kinds of organisms. They are among the largest organisms on Earth. Most brown algae reproduce by alternation of generations.

Real Life

People eat and use algae.

Many red algae are used in making food and food additives. Agar is used to prevent baked goods from drying, to grow bacteria in Petri dishes, and to make gelatin capsules. Carrageenan is used in paints, dairy products, and cosmetics to help stabilize their different ingredients.

Finding Information

Read the labels on packages and cans at home, to find out if they contain any agar or carrageenan.

Figure 5 **Three phyla of algae.** *Ulva* is a species of green algae composed of a sheet of cells that is only two cells thick. Coralline alga is a red alga that contributes to the great coral reefs. Some species of the brown algae *Macrocystis* can grow to a length of more than 60 m.

Teach

Real Life

Answer

Student answers will vary. Ready-mix cakes, instant puddings, pie fillings, dairy products, and canned pet foods all contain agar or carrageenan.

Teaching Tip — GENERAL

Have You Had Your Algae Today? Have students keep a tally of the algae-derived products that they eat or use in one week. Encourage them to read food product labels. Some key terms that identify algae extracts are *lecithin, carrageenan, algin, emulsifier,* and *thickening agent.* Some of the many products that contain algae include pudding, jelly, jelly beans, ice cream, chocolate milk, marshmallow, processed cheese, and salad dressing. Have students write an essay about the algae in their life.

Group Activity — ADVANCED

Space Algae Have student groups research and give a class presentation on how algae might be useful in space. Green algae have been tested in the space program as a means of eliminating some of the cargo needed for long missions. The algae would feed shrimp, a food source for the astronauts, and would also recycle carbon dioxide and release oxygen. Because they regenerate continuously, the algae would add less weight and volume to the spacecraft than more traditional food and air sources.

Co-op Learning

MISCONCEPTION ALERT

- **Hidden Green Pigments** Both brown and red algae contain chlorophyll in addition to the pigments that make them look brown and red. A common misconception is that they lack chlorophyll; however, the green pigment is simply hidden by the other colors.

Integrating Physics and Chemistry

Brown Algae Stalks The brown algae *Macrocystis* can grow very long stalks because it is less dense than surrounding seawater. The *Macrocystis* stalks are supported by the upward force of water pressure, which is called buoyancy. Remind students that buoyancy increases as the density difference between two substances increases. Ask students what makes *Macrocystis* less dense than the surrounding water. (The base of each leaf blade contains gas-filled sacs.)

Teach, *continued*

Observing Characteristics of Diatoms

QUICK LAB

Skills Acquired
Observing, interpreting observations

Teacher's Notes
Have students experiment with different light angles using the flashlight. If a polarized light source is available (try using a polarizing filter from a camera), compare the appearance of the diatoms with the nonpolarized light source.

Answers to Analysis

1. In radial symmetry, a plane passing through the center of the diatom divides the organism into two halves. In bilateral symmetry, there is only one plane that divides the diatom into two mirror image halves.
2. Students may use shape, symmetry, and texture as useful characteristics.
3. Changing the angle of the light will change the observed reflectance and alter the appearance of surface features.

Diatoms

WORD Origins

The word *amoeba* is from the Greek *amoibe,* meaning "change." Knowing this makes it easier to remember that amoebas are able to change their shape.

Diatoms *(DIE uh tahmz),* members of the phylum Bacillariophyta, are photosynthetic, unicellular protists with unique double shells. The shells are made of silica and often have unique markings. Their shells are like small boxes with lids, one half fitting inside the other. Abundant in oceans and lakes, diatoms are important producers in the food chain. Diatoms can have one of two types of symmetry: radial (like a wheel) or bilateral (two-sided). The empty shells of diatoms form thick deposits that are mined commercially as diatomaceous earth. Diatomaceous earth is used as an abrasive or to add the sparkling quality to paint used on roads. Diatomaceous earth is also sold as a natural control for pests, such as slugs, and some insects, such as fleas. The sharp edges of the diatom shells cut into the body of the pest, leading to its death.

Diatoms secrete chemicals through holes in their shells, enabling them to move by gliding. Individuals are diploid and usually reproduce asexually. The two halves of the shell separate, and each half regenerates another matching half. As a consequence of this model of reproduction, diatoms tend to get smaller and smaller with each generation. When an individual gets too small because of repeated division, it slips out of its shell, grows to full size, and regenerates a new shell. Sexual reproduction in diatoms is rare.

QUICK LAB

Observing Characteristics of Diatoms SC.G.1.4.1

Try this activity to find out why diatomaceous earth is used to make abrasives, fine filters, and reflective paints.

Materials

pipet, water, microscope slide, toothpick, diatomaceous earth, coverslip, compound microscope

Magnification: 240×

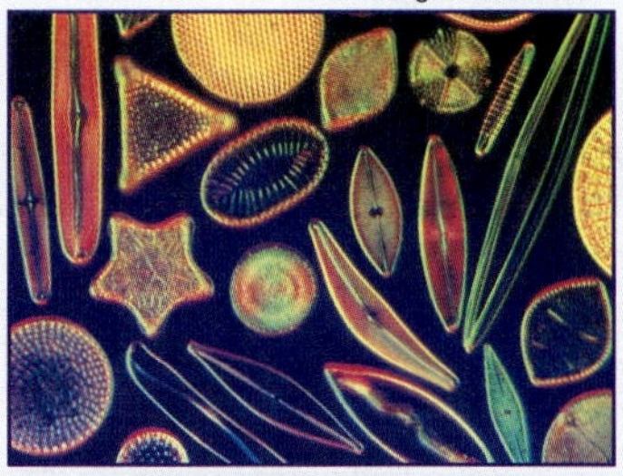

Procedure

1. Using a pipet, place a drop of water in the center of a clean microscope slide.
2. Use a toothpick to scoop up a small amount of diatomaceous earth and mix it with the water drop. Add a coverslip.
3. Observe your wet mount under both low and high power of a compound microscope.
4. Draw some of the diatoms you see.
5. Observe the wetmount under low power as your partner shines a flashlight (at a 45° angle) on the slide. Turn off the microscope's light source so that only the flashlight is lighting the slide. Record your observations.

Analysis

1. **Label** your drawings as radial or bilateral. Find out the meanings of these terms and how they apply to your diatom drawings.
2. **Select** some characteristics you observed that are useful in classifying particular species of diatoms.
3. **Interpret** what you observed when the flashlight was shone on the slide.

466

GEOGRAPHY CONNECTION GENERAL

The Sargasso Sea is named for its floating masses of *Sargassum natans.* This brown alga reproduces by fragmenting into small pieces, and the fragments cover millions of square miles between the Azores and the Bahamas. Nautical lore tells of ship graveyards trapped in the thick mat of algae. In truth, the area was avoided because it lies within the wind-sparse "horse" latitudes. Have students locate this area on a map or globe.

Integrating Physics and Chemistry

Silica Have students look up the chemical formula for silica and determine whether it is a mixture, element, or a compound. (compound) Ask students if diatomaceous earth is a mixture, element, or compound. (mixture) Have them defend their answers.

Flagellates

Flagellates are protists that move using flagella. The three major phyla of flagellates are the dinoflagellates, the zoomastigotes, and the euglenoids.

Dinoflagellates

Dinoflagellates *(DIE noh FLAJ uh layts)*, members of the phylum Dinoflagellata are unicellular and most have two flagella. A few kinds of dinoflagellates are found in fresh water, but most are marine and make up part of the plankton. Most dinoflagellates have a protective coat made of cellulose that is often encrusted with silica, giving them unusual shapes, as shown in **Figure 6.** Their flagella beat in two grooves—one encircling the body like a belt, the other perpendicular to it. As a result, dinoflagellates spin through the water like a top. A few dinoflagellates produce powerful toxins. The poisonous "red tides" that occur frequently in coastal areas are often associated with population explosions of dinoflagellates. Dinoflagellates usually reproduce asexually by mitosis, and they can be photosynthetic, heterotrophic, or both.

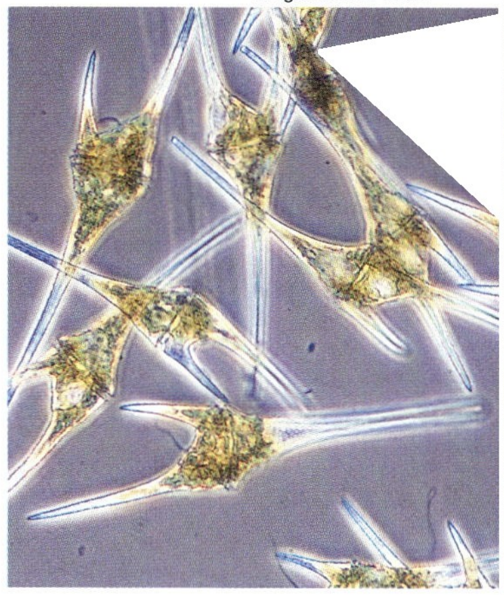

Figure 6 Dinoflagellates. This dinoflagellate, *Gonyaulax tamarensis*, moves using flagella.

Euglenoids

Euglenoids *(yoo GLEE noyds)*, members of the phylum Euglenophyta, are freshwater protists with two flagella. This group clearly shows the difficulty of classifying protists as animals or plants. About one-third of the 1,000 known species of euglenoids have chloroplasts and are photosynthetic; other species lack chloroplasts, ingest their food, and are heterotrophic. A member of *Euglena*, shown in **Figure 7,** has a protein scaffold called a pellicle *(PEHL ih kuhl)* inside the cell membrane. Since the pellicle is flexible, the euglenoid can change shape. A light-sensitive organ called the eyespot helps orient the movements of these organisms toward light. Reproduction in this phylum occurs by mitosis.

Kinetoplastids

Kinetoplastids, members of the phylum Kinetoplastida, are unicellular heterotrophs that have at least one flagellum, and some species have thousands. While most reproduce only asexually, some are known to produce gametes and reproduce sexually. Kinetoplastids are clearly related to euglenoids, and many taxonomists merge the two phyla together. Some kinetoplastids, such as the trypanosomes, cause diseases such as African sleeping sickness in humans and domestic animals.

Figure 7 *Euglena.* *Euglena* is a versatile protist. It contains chloroplasts and is photosynthetic, but it is also heterotrophic and can live without light.

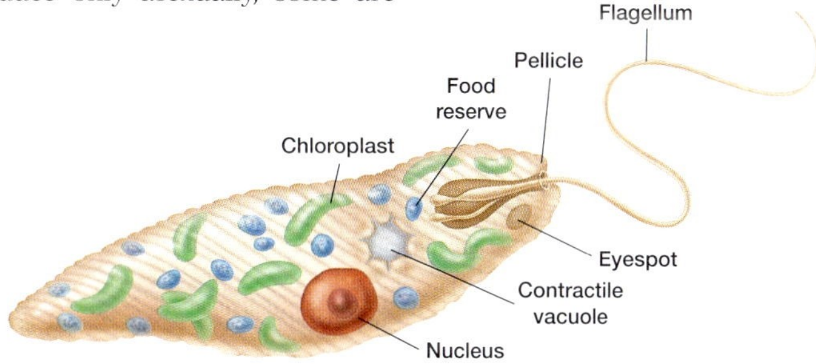

467

Teaching Tip — GENERAL

Red Tides The toxic substance produced by the dinoflagellate responsible for "red tides" can kill fish and invertebrates that consume the algae. Red tides can also be a human health risk. Ask students how red tides might result in human poisoning. (People who ingest mussels that feed on the toxic dinoflagellates can become ill and may even die.)

SKILL BUILDER — GENERAL

Writing Skills The dinoflagellate *Noctiluca* glows because of a chemical reaction that produces light. The waters that contain these organisms create luminescent, or "sparkling," waves. Encourage students to find out more about luminescence and *Noctiluca* and write a short report about the subject.

Using the Figure

Use **Figure 7** to point out the parts of *Euglena*. Make sure students understand that the pellicle is inside the cell membrane. Ask students how protists in the Euglenophyta phylum represent a classification challenge. (Some are photosynthetic, some are heterotrophic, and some are both.)

Activity — GENERAL

Test the Eyespot Have students design an experiment to test a euglenoid's response to light. Ask students to predict whether euglenoids will move toward or away from light. Guide them to conclude that euglenoids will move toward a light source because they use light to make food. **LS Interpersonal**

Transparencies

TR F18 Structure of *Euglena*
TR F16 *Paramecium*

Attention Grabber

Dynamite When nitroglycerin is absorbed by diatomaceous earth, it is stable, safe to handle, and can be transported. This compound is more commonly known as dynamite. In the absence of diatomaceous earth, the nitroglycerin is highly unstable and explodes with great force when jolted or heated.

Teach, continued

Interpreting Competition Among Protists

DATA LAB

Skills Acquired

Analyzing data, applying information, evaluating information

Teacher's Notes

Based on the graph, ask students to predict what might happen to the *P. caudatum* population after another few days. (In the presence of *P. aurelia*, it may die off completely.) Ask students to suggest conditions that might allow *P. caudatum* to thrive even in the presence of *P. aurelia*. (In the presence of a predator or if the *P. aurelia* were significantly older, they might die off and allow *P. caudatum* to survive.)

Answers to Analysis

1. *P. aurelia*
2. The variables were number of live organisms (dependent variable), time (independent variable), and whether the paramecia were grown alone or with another species.
3. Answers may vary. Sample answer: *P. aurelia* was the superior competitor, obtaining the necessary nutrients at the expense of *P. caudatum*.
4. Accept all well-supported answers. Students should realize that the results are valid only under these controlled conditions. Adding variables to simulate their natural habitat may yield results more closely resembling the natural world.

Magnification: 148×

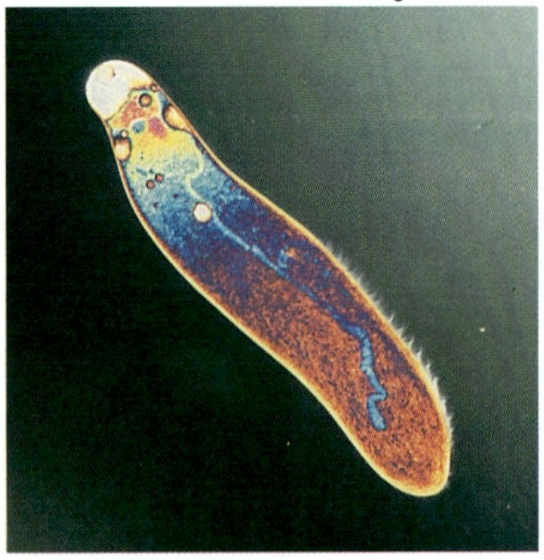

Figure 8 **Ciliate.** The cilia that surround this ciliate enable it to move and feed.

Ciliates

Ciliates are the most complex and unusual of the protists. All members of the phylum Ciliophora *(sih lee AWF oh ruh)* have large numbers of **cilia,** tightly packed rows of short flagella used for movement, as shown in **Figure 8.** Ciliates are complex unicellular heterotrophs. The body wall of ciliates is a tough but flexible outer pellicle that enables the organism to squeeze through or move around many obstacles. The pellicle consists of an outer membrane with numerous fluid-filled cavities beneath it. Ciliates, such as *Paramecium,* shown in *Up Close: Paramecium,* on the following page, form vacuoles for ingesting nutrients and regulating their water balance.

Most ciliates have two types of nuclei within their cells: micronuclei and larger macronuclei. The micronuclei contain normal chromosomes that divide by mitosis. Macronuclei contain small pieces of DNA derived from the micronuclei.

Reproduction in ciliates is usually by mitosis, with the body splitting in half. Cells divide asexually for about 700 generations and then die if sexual reproduction has not occurred. Most ciliates can engage in sexual reproduction through conjugation, a process in which two cells unite and exchange genetic material.

DATA LAB

Interpreting Competition Among Protists SC.G.1.4.1

Background

Protists, like all organisms, must compete with one another for nutrients. To examine the effects of competition between two species of *Paramecium,* equal numbers of the paramecia were grown together (dashed lines) and separately (solid lines). Study the graph at right, and answer the following questions.

Competition in Paramecia

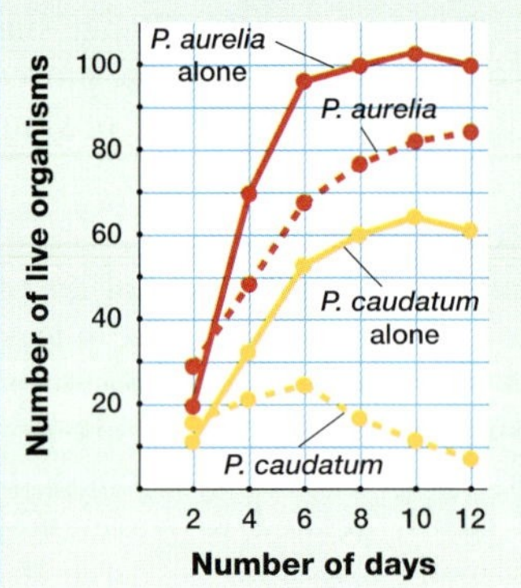

Analysis

1. **Identify** the *Paramecium* that grew best alone.
2. **Identify** the variables in this experiment.
3. **Critical Thinking** **Analyzing Data** How would you explain the difference in the growth curves in the group that had both organisms?
4. **Critical Thinking** **Predicting Outcomes** In a natural setting there would be more than two organisms present. Predict the effect that the presence of other organisms would have on the growth of the two species of *Paramecium.*

468

Career

Limnologist Limnologists are freshwater ecologists employed by universities and research and consulting firms, as well as local, state, and federal governments. Limnologists work with others to study and improve water quality and the health of aquatic ecosystems. Protists and other microorganisms are often at the center of their research. For example, Dr. Ruth Patrick, a limnologist, has invented the "diatometer"—a device used to identify diatoms. Encourage students to learn more about the work of limnologists.

Paramecium

- **Scientific name:** *Paramecium caudatum*
- **Size:** Microscopic; up to 1 mm long
- **Habitat:** Lives in freshwater streams and ponds
- **Diet:** Bacteria, small protists, organic debris

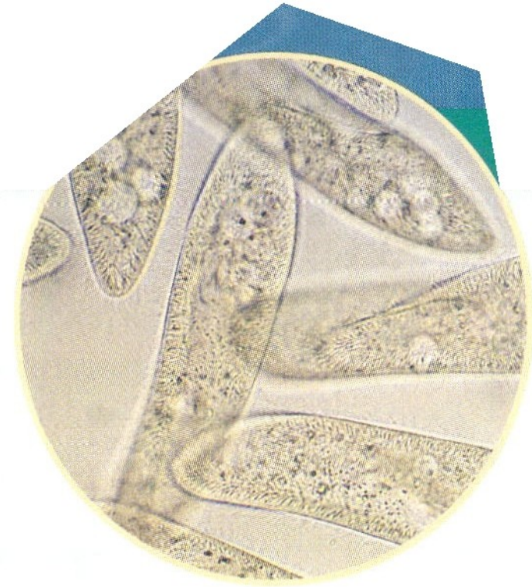

Magnification: 1,200×

Characteristics

Nuclei Members of the genus *Paramecium* have two nuclei. The macronucleus contains fragmented chromosomes used in routine cellular functions, and it divides by pinching in two. The micronucleus contains the cell's chromosomes and divides by mitosis.

Surface *P. caudatum,* a ciliate, is covered with thousands of cilia arranged in rows along the cell. Cilia beat in waves that move diagonally across the cell, causing the protist to spin through the water. *P. caudatum* is surrounded by a rigid protein covering called a pellicle.

Maintaining water concentration Like other freshwater protists, *P. caudatum* constantly absorbs water by osmosis. *P. caudatum* gets rid of excess water using its contractile vacuoles, saclike organelles that expand, collecting excess water, and then contract, squeezing water out of the cell.

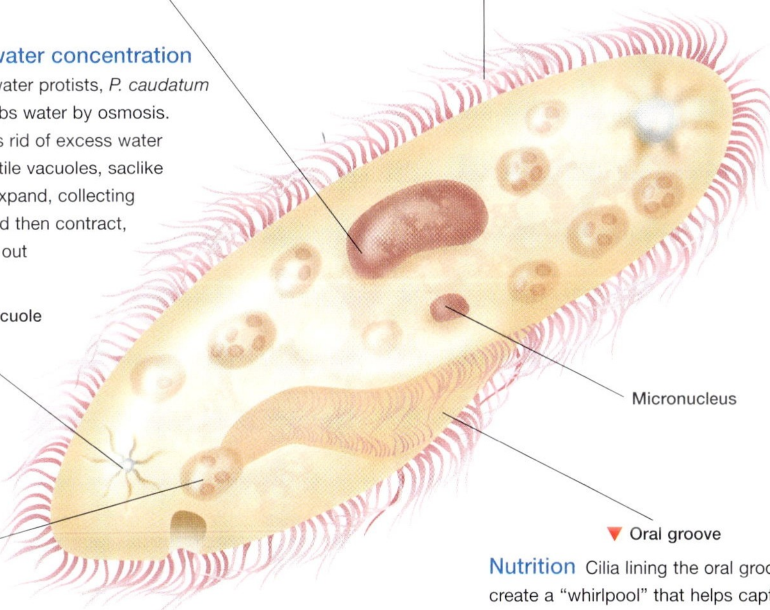

Nutrition Cilia lining the oral groove create a "whirlpool" that helps capture small bits of food. Food moves down the funnel-shaped groove and is engulfed in a food vacuole combining with digestive enzymes as it moves through the cell. Undigested food is released from the cell by exocytosis.

Genetic variation *P. caudatum* generally reproduces asexually by binary fission. Genes are shuffled during a sexual process called conjugation.

Up Close

Paramecium

Teaching Strategies

- Ask students to draw a flowchart to trace the movement of food through a paramecium. (Charts should begin with food entering the paramecium via the oral groove. At the mouth, food is engulfed by a food vacuole containing digestive enzymes. Undigested food is excreted by exocytosis.)
- Ask students to compare the methods used by amoebas and paramecia for capturing food. (Amoebas use their pseudopodia to surround and engulf food by endocytosis. Paramecia use cilia to create a whirlpool to pull food into the oral groove, where the food is engulfed by a food vacuole.)

Discussion

- How would a paramecium living in a clean environment differ from one living in a polluted environment? (Paramecia in polluted water may be less efficient at maintaining water balance, acquiring nutrients, and moving due to the pollutants and subsequent damage to cilia.)
- Ask students to describe an ideal environment for a paramecium. (Answers will vary but should include a food source and a freshwater environment that will allow the organism to maintain homeostasis.)

REAL WORLD CONNECTION

Plasmodial slime molds often grow in wet bark mulch that is used for landscaping. Although the slime mold does not harm plants, it is unsightly. The most common slime mold, *Fuligo septica*, is known as dog-vomit fungus. The plasmodium is usually yellow, orange, tan, or white and may be more than two feet in diameter. It will disappear on its own if the area is raked to promote air drying.

Integrating Physics and Chemistry

Review viscosity, which is the resistance of a fluid to flow, and remind students that paramecia move rather quickly relative to other protists. Therefore, when you want to study paramecia under the microscope, it is helpful to slow them down. One way to slow them is by adding a viscous material such as Detain™ to the slide. Because Detain™ has a higher viscosity than water, it impedes the movement of paramecia.

Teach, *continued*

Using the Figure

Ask students what would cause *D. discoideum* to change its form, as shown in **Figure 9.** (environmental stress such as food or water deprivation)

READING SKILL BUILDER — BASIC

Reading Organizer Ask students to make a table that summarizes the four protistan molds discussed in this section. Allow students to choose their own headings so that they can organize the information in their own terms. Possible headings include Phylum, Habitat, Appearance, and Type of Mold.

Activity — GENERAL

Grow Your Own Slime Mold

Let students grow their own slime molds in Petri dishes filled with dampened oatmeal. If a source of protistan molds is not readily available, check with a local university. English Language Learners

Figure 9 Cellular slime mold. The individual amoebas of *Dictyostelium discoideum* aggregate, move in a mass called a slug, and form a stalked structure that contains spores.

Protistan Molds

Protistan molds are heterotrophs with some mobility. They were once thought to be related to fungi because of a similar appearance and reproductive structures. However, the cell walls of protistan molds contain different carbohydrates than fungal cell walls. Also, protists carry out normal mitosis, whereas mitosis in fungi is unusual, as you will learn later in this book.

Cellular Slime Molds

Cellular slime molds, members of the phylum Acrasiomycota *(uh KRAZ ee oh mie koh tuh),* resemble amoebas but have distinct features. The individual organisms behave as separate amoebas, moving through the soil and ingesting bacteria. During environmental stress, the individual amoebas gather together and move toward a fixed center. There they form multicellular colonies called slugs. Each slug develops a base, a stalk, and a swollen tip that develops spores, as shown in **Figure 9.** Each of these spores, when released, becomes a new amoeba, which then begins to feed and repeat the life cycle.

Plasmodial Slime Molds

Plasmodial slime molds, members of the phylum Myxomycota *(MIHKS oh MIE koh tuh),* are a group of organisms that stream along as a **plasmodium,** a mass of cytoplasm that looks like an oozing slime. As they move, they engulf bacteria and other organic material, as shown in **Figure 10.** A plasmodial slime mold contains many nuclei, but they are not separated by cell walls. If the plasmodium begins to dry out or starve, it divides into many small mounds. Each mound produces a stalk tipped with a capsule in which haploid spores develop. The spores are highly resistant to hostile environmental conditions. When conditions are favorable, the spores germinate and become haploid cells that are either amoeboid or flagellated. These haploid cells are able to fuse into diploid zygotes, which undergo mitosis and form a new plasmodium.

Figure 10 Plasmodial slime mold. A plasmodial slime mold is a mass of cytoplasm containing many nuclei.

Other Molds

Oomycetes *(oh oh MIE seets),* members of the phylum Oomycota, are the water molds, white rusts, and downy mildews that often grow on dead algae and dead animals in fresh water. All members of the group either are parasites or feed on dead organic matter. Oomycetes are unusual in that their spores have two flagella: one pointed forward, the other pointed backward. Many oomycetes are plant pathogens, including *Phytophthora infestans,* which causes late blight in potatoes. This protist caused the failure of potato crops in Ireland between 1845–1850, which led to a famine that killed about 1 million people.

470

Trends in Agriculture

Potato Threat A new, more virulent form of the potato blight that was responsible for the Irish potato famine of 1845–1847 threatens current potato crops. The new blight, a type of water mold, is resistant to metalaxyl, the fungicide traditionally used against potato blight in the United States. The disease threatens a major global food source, as potatoes are grown and consumed around the world. Currently scientists are working with wild potatoes to produce new disease-resistant varieties.

Transparencies

TR F20 Life Cycle of *Plasmodium*

Sporozoans

Parasitic protists that form spores during their reproduction cycle are called sporozoans. They are members of the phylum Apicomplexa and are nonmotile, unicellular parasites. All sporozoans are parasitic and cause many serious diseases. Malaria, a sporozoan disease, kills more people than any other infectious disease. Sporozoans infect animals and are transmitted from host to host. There are about 4,500 named species in the phylum Apicomplexa.

Sporozoans have complex life cycles that involve both asexual and sexual reproduction. Sexual reproduction involves the fertilization of a large female gamete by a small, flagellated male gamete. The resulting zygote forms a thick-walled structure that makes the zygote resistant to drought and other unfavorable environmental conditions. A few sporozoans are listed in **Table 2.**

Many sporozoans are transmitted from one host to another by blood-feeding insects, such as mosquitoes, black flies, and midges. Other sporozoans are transmitted in the feces of an infected animal, such as the one shown in **Figure 11.** The parasites infect a new individual when that individual eats or drinks food or water that has been contaminated with infected feces. Some sporozoans are even transmitted through predation, such as when a cat becomes infected by eating an infected mouse.

Figure 11 ***Cryptosporidium.*** *Cryptosporidia* (stained purple) are found in the feces of infected animals. In the 1980s and 1990s, water supplies in several cities in the United States became infected with these sporozoans.

Table 2 Some Sporozoans and Their Hosts

Sporozoans	Disease	Hosts
Plasmodium sp.	Malaria	Humans and other vertebrates
Toxoplasma sp.	Toxoplasmosis	Humans and cats
Babesia sp.	Cattle tick fever	Cattle, mice, humans, deer, dogs
Cryptosporidium sp.	Cryptosporidiosis (diarrhea)	Cattle, humans, birds, deer, dogs, cats

Section 2 Review

1. **Describe** a pseudopodium. SC.F.1.4.2
2. **Identify** the role of diatoms in the aquatic food chain. SC.G.1.4.2
3. **Construct** a table that compares three kinds of algae. SC.G.1.4.1
4. **Explain** how protistan molds differ from fungi. SC.G.1.4.1
5. **Critical Thinking** **Evaluating Viewpoints** "*Euglena* is a protozoan, not an alga." Is this an accurate statement? Why or why not? SC.G.1.4.1
6. *FCAT Prep* What term describes the resistant stages formed by a parasitic protist that causes the disease *cryptosporidiosis*?
   A kinetoplastids
   B pseudopodia
   C slime molds
   D spores
   SC.F.1.4.2

Answers to Section Review

1. A pseudopodium, or "false foot," is an extension of the amoeba's cell membrane that allows it to move and engulf prey.
2. Diatoms are important producers in the aquatic food chain.
3. Tables will vary but should reflect the descriptions of green, red, and brown algae on page 465.
4. Protistan molds carry out normal mitosis; fungi do not. The cell walls of protistan molds contain a different carbohydrate from the carbohydrate in fungal cell walls.
5. Yes, euglenoids cannot be classified as algae because they are not strictly autotrophic.
6. **A.** Incorrect. *Cryptosporidium* is not a kinetoplast. **B.** Incorrect. *Cryptosporidium* does not have pseudopodia. **C.** Incorrect. *Cryptosporidium* does not form slime molds. **D.** Correct. *Cryptosporidium* forms spores.

Close

Reteaching — Basic

Ask students to pretend that they have been invited to give a presentation at a club of protist enthusiasts. Each student should focus on a particular protist discussed in Section 2. Presentations should include a drawing of the protist and a description of its characteristics, including reasons for classifying the organism in the Kingdom Protista. Encourage student creativity, allowing them to name their "pet" and to make up anecdotes to describe their protist's special characteristics. Have students share their creations and post their work around the room.

Quiz — General

1. How do amoebas reproduce? (Amoebas reproduce by fission; they have no method of sexual reproduction.)
2. What is the function of the contractile vacuole in *Paramecium?* (The contractile vacuole collects excess water from inside the cell and removes it.)
3. Describe the plasmodium of a plasmodial slime mold. (The plasmodium is a mass of cytoplasm containing many nuclei that are not separated by cell walls.)

Alternative Assessment — General

Have student pairs create a game in which one partner describes characteristics of a particular protist discussed in Section 2 and the other partner guesses the organism's name.

Section Review Correlations

SC.F.1.4.2: Items 1, 6
SC.G.1.4.1: Items 3, 4, 5
SC.G.1.4.2: Item 2

Section 3

Focus

Overview

Before beginning this section review with your students the objectives listed in the Student Edition. This section discusses how protists affect humans in beneficial and harmful ways. Malaria is used as an example of a disease-causing protist, and the complex life cycle of *Plasmodium* is explained.

Bellringer

Have students make a list of some characteristics of an ideal habitat for a microbe. (Students may list a moderate or warm climate in which the microbe can remain active throughout the year or poor sanitary conditions, which allow the microbe to spread.) Ask students what conclusions about human infection they can come to based on their list. (Human populations that face such conditions suffer the most from microbial diseases.)

Motivate

Activity — GENERAL

Wanted: Helpful Protist Have students write a want ad advertising for a beneficial protist. They should list requirements for the "job," for example, that it must be able to photosynthesize, digest cellulose, or decompose organic material.

Section Correlations

SC.H.3.4.3

SC.H.3.4.5

Section 3 Protists and Health

Florida Sunshine Standards/FCAT

SC.H.3.4.3 Scientists can bring information, insights, and analytical skills to matters of public concern and help people understand the possible causes and effects of events.

SC.H.3.4.5 The value of a technology may differ for different people and at different times.

Objectives

- **Identify** two ways that protists affect human health. SC.H.3.4.3
- **Name** three human diseases, other than malaria, caused by protists. SC.H.3.4.3
- **Summarize** how malaria is transmitted. SC.H.3.4.3
- **Evaluate** the methods used to control malaria. SC.H.3.4.5

Key Terms

sporozoite
merozoite

Protists and Humans

One of the greatest effects protists have on humans is that they cause disease. This effect can be measured in pain, death, and the medical costs of preventing and treating diseases. Some diseases caused by protists are listed in **Table 3.**

Protists also affect humans through the diseases they cause in livestock. The added cost of treating diseased livestock is passed on to consumers in the form of higher meat prices. This increased cost of living hinders progress in the developing world.

Beneficial Protists

There are many commensal protists that live in the digestive tracts of humans and in the digestive tracts of the animals that humans eat. Cattle could not digest the cellulose in the hay and grass they eat without the aid of commensal protists in their digestive tract.

Protists, which make up much of the plankton in the ocean, help to support food chains. Protists are also the single largest group of photosynthesizers on the planet. Because we all breathe oxygen, we all benefit from the gas that protists produce. Many protists are also detritivores, so they help recycle important chemicals, such as nitrogen, carbon, and phosphorus, in the environment.

Table 3 Diseases Caused by Protists

Disease	Description of illness	Protist	How the disease is transmitted
African sleeping sickness	Fever, weakness, lethargy	*Trypanosoma gambiense, Trypanosoma rhodesiense*	Bite from infected tsetse fly
Amebic dysentery	Bloody diarrhea, vomiting, extremely strong stomach cramps, fever	*Entamoeba histolytica*	Contaminated food or water
Giardiasis	Cramps, nausea, diarrhea, vomiting	*Giardia lamblia*	Contaminated food or water
Malaria	Fever, chills, sweats	*Plasmodium* sp.	Bite from infected mosquito
Toxoplasmosis	Primary danger is fetal infection; can cause convulsions, brain damage, blindness, and death in fetuses	*Toxoplasma gondii*	Contact with infected cats or improperly cooked meat

Chapter Resource File
- Directed Reading BASIC
- Active Reading GENERAL

Transparencies

TR Bellringer

TR F19 Diseases Caused by Protists

One-Stop Planner CD-ROM
- Reading Organizers BASIC
- Reading Strategies BASIC
- Occupational Applications Worksheet Nurse Practitioner GENERAL
- Supplemental Reading Guide *The Andromeda Strain* ADVANCED

Malaria

Malaria is one of the most deadly diseases in humans. Over 100 million people have malaria at any one time, and up to 3 million people, mostly children, die from it every year.

The symptoms include severe chills, fever, sweating, confusion, and great thirst. Victims die of anemia, kidney failure, or brain damage unless the disease is treated.

Malaria Life Cycle

Malaria is caused by several species of *Plasmodium* and is spread by the bite of certain mosquitoes. There are three stages in the *Plasmodium* life cycle, as shown in **Figure 12.** When an infected mosquito bites a human to obtain blood, it injects saliva that contains a chemical that prevents the blood from clotting. If the mosquito is infected with *Plasmodium,* it will also inject about 1,000 protists with its saliva. This infective stage of *Plasmodium* is called the **sporozoite.** Sporozoites infect the liver, where they rapidly divide and produce millions of cells of the second stage of the life cycle, called the **merozoite.** Merozoites infect red blood cells and divide rapidly. In about 48 hours, the blood cells rupture, releasing more merozoites and toxic substances. This begins a cycle of fever and chills that characterizes malaria. The cycle repeats every 48–72 hours (depending on which species is causing the infection) as more blood cells are infected and destroyed.

In the third stage, some of the merozoites in the blood develop into gametes. After these gametes are eaten by a mosquito, male and female gametes fuse to form zygotes. The zygotes divide many times to form many sporozoites, which migrate to the salivary glands of the mosquito. Unlike humans, the mosquito cannot regurgitate food or mix saliva and food in its mouth; therefore, the malaria parasite must mature in the mosquito before it can infect another human.

Figure 12 Life cycle of *Plasmodium.* *Plasmodium* has a complex life cycle that involves a mosquito, human blood, and liver cells.

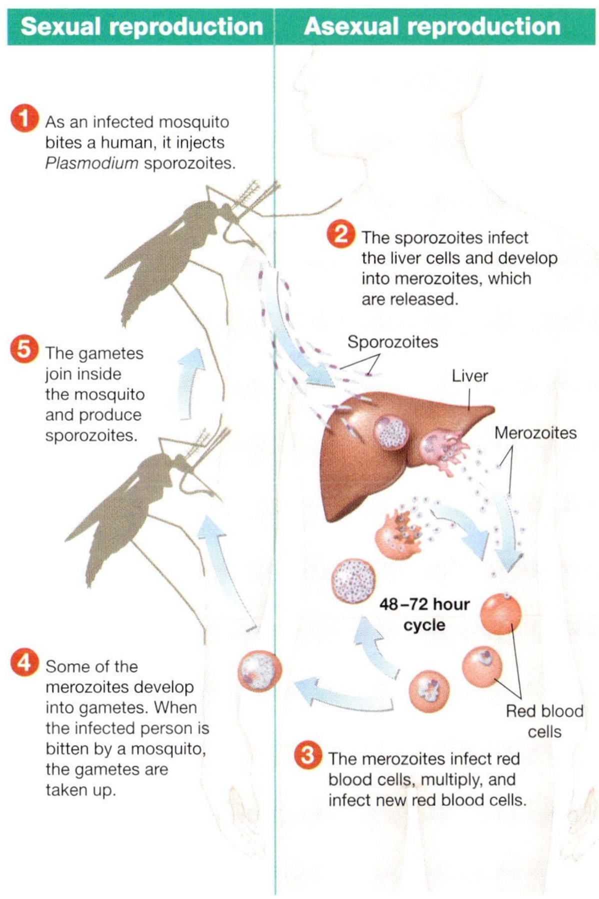

Teach

SKILL BUILDER — GENERAL

Writing Skills Before reading this section, ask students to write a short paragraph about how they would protect themselves from diseases caused by protists. After reading the section, have students edit their paragraphs as needed. (Precautions might include getting the appropriate vaccinations or medications, avoiding infested areas, watching for possible food or water contamination, and protecting themselves from the insects or other animals that can carry the pathogens.) **LS Verbal**

Teaching Tip — GENERAL

Food Chains and DDT Remind students that the use of DDT has been banned in the United States because it accumulates in the tissues of organisms as it passes through the food chain. The chemical is still in use in some malaria-infested countries to control mosquito populations. Review how some organisms, such as the bald eagle, were affected before and after the banning of DDT. (Before the ban, the shells of the eggs were so thin that many broke, and the unhatched chicks died. The eagle is now making a comeback.)

Demonstration — BASIC

Project **Figure 13** on an overhead projector. Point out the merozoites and the sporozoites. Be sure students can differentiate these two stages. **LS Visual** English Language Learners

Cultural Awareness

Controlling Malaria in Africa Ten percent of the world's population lives in tropical Africa, where 90 percent of the world's malarial infections occur. Three species of African mosquitoes carry and spread the pathogen to humans while feeding on their blood. In tropical Africa, one person can receive more than 300 bites by infected mosquitoes per year. International researchers hope to introduce a pathogen-inhibiting gene into the genomes of these mosquitoes. Releasing transgenic mosquitoes to spread the new gene throughout the *Plasmodium*-carrying mosquito populations could spare other continents, such as South America, from future malaria epidemics.

Teach, *continued*

BIOWatch

Making a Malaria Vaccine

Teaching Strategies
Why is it important for scientists to develop more than one protein to use in a potential vaccine? (Plasmodium goes through several stages in the body and each stage has different surface proteins. Scientists want the human body to recognize and attack any of the different stages.)

Discussion
Could these vaccines cause malaria? (No, the vaccine consists only of surface proteins and not disease-causing organisms.)

Close

Reteaching — BASIC

Have students work in cooperative groups to research and report on the life cycles of several disease-causing protists. Reports should include the method of infection, the symptoms, and where the disease is found. LS Verbal Co-op Learning

Quiz — BASIC

1. Explain why cattle could not live without protists. (Protists digest the cellulose in grass and hay. This makes it easier for the cattle to digest their food and obtain energy.)
2. What causes the cycle of fever and chills in malaria? (the rupturing of blood cells followed by the release of merozoites and toxic substances)

Section Review Correlations

SC.H.3.4.3: Items 1-4

Treating and Preventing Malaria

In the middle of the seventeenth century, quinine *(KWIE nien)*, a chemical derived from the bark of the cinchona tree *(Cinchona officialis,* found in South America) was discovered to be a remedy for malaria. Derivatives of quinine, such as chloroquine and primaquine, are now used to treat malaria.

Malaria can also be controlled by reducing mosquito populations. This is done by spraying insecticides, reducing mosquito breeding places, and introducing animals that will eat mosquito larvae, such as mosquito fish.

BIOWatch

Making a Malaria Vaccine

With their bodies racked by fever as high as 40°C (104°F) and chills so intense that their teeth chatter, up to 3 million people needlessly die of malaria every year. The worldwide epidemic of malaria is becoming more dangerous as malaria parasites become resistant to the standard drug treatments. To control the epidemic, scientists are trying to make a malaria vaccine.

Making the Malaria Vaccine Work

For a vaccine to work against a parasite, it must stimulate the body's immune cells to recognize and attack specific molecules on the surface of the parasite. But *Plasmodium* goes through several stages inside the human body, and each stage can have many different surface molecules. Therefore, a vaccine may cause the immune system to attack only one stage. If a few parasites survive, they could produce millions of new parasites. A second problem is that *Plasmodium* parasites spend much of their time inside human liver and red blood cells, where the immune system cannot find them.

Current Approaches

Scientists have identified surface proteins of *Plasmodium* sporozoites and merozoites and can make large amounts of these proteins. They hope that an injection of these proteins into human volunteers will give the volunteers' immune cells a chance to recognize the proteins and prepare for infection. If these people are later infected by a mosquito, their cells may be able to attack *Plasmodium* before the parasite can enter their liver cells.

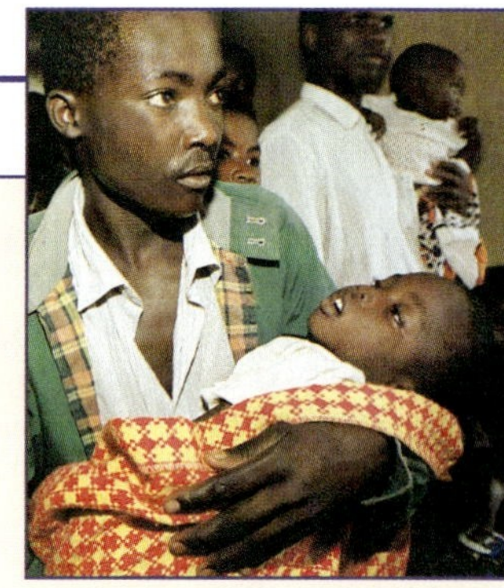

World-wide efforts to develop an effective malaria vaccine have produced many failures and a few partial successes. Scientists are still working on ways to make the immune system's response to the vaccines stronger and longer lasting.

internet connect
www.scilinks.org
Topic: Malaria Vaccine
Keyword: HX4117
SCILINKS. Maintained by the National Science Teachers Association

Section 3 Review

1. **Summarize** two different ways protists affect human health. SC.H.3.4.3
2. **Describe** three human diseases caused by protists. SC.H.3.4.3
3. **Describe** how malaria is transmitted. SC.H.3.4.3
4. FCAT Prep In the life cycle of *Plasmodium*, the parasite that causes malaria, where do the parasite's gametes combine to form a zygote?
   A in human blood cells
   B in the human liver
   C inside a mosquito
   D inside a tsetse fly
   SC.H.3.4.3

474

Answers to Section Review

1. Some protists cause disease, but others are beneficial inhabitants of the human gut. Many others contribute oxygen to the atmosphere and recycle carbon and nutrients in the environment.
2. Malaria is caused by *Plasmodium* sporozoans. The disease causes severe fever and can lead to anemia, kidney failure, and brain damage. See Table 3 on page 472 for a list and description of other diseases caused by protists.
3. Malaria is transmitted when a mosquito extracts *Plasmodium* gametes from an infected individual. The gametes mature into sporozoites in the mosquito and are injected into a new individual when the mosquito bites it, resulting in a new infection.
4. **A.** Incorrect. Gametes cannot fuse in blood cells. **B.** Incorrect. There are no gametes in the human liver. **C.** Correct. The gametes fuse in the mosquito. **D.** Incorrect. Tsetse flies are not part of the *Plasmodium* life cycle.

Study ZONE CHAPTER HIGHLIGHTS

Key Concepts

1 Characteristics of Protists

- The kingdom Protista contains the most diverse groups of eukaryotic organisms of any kingdom.
- Protists live in moist environments and can be either free-living or parasitic.
- Some protists are able to reproduce sexually in times of stressful environmental conditions.

2 Protist Diversity

- Amoebas and forams are protists that move using cytoplasmic extensions called pseudopodia.
- Diatoms are unicellular protists with glasslike double shells.
- Algae are strictly photosynthetic protists that can be multicellular or unicellular. Algae are classified according to the type of photosynthetic pigment they contain.
- Flagellates move with the use of flagella.
- Ciliates are unicellular protists that use cilia to move.
- Protistan molds resemble fungi, but they are considered protists.
- Cellular slime molds normally live as individual cells and aggregate during times of stress.
- Plasmodial slime molds live as colonial organisms and form sexual reproductive structures that form and spread spores.
- Oomycetes are molds that live as saprophytes or parasites.
- All sporozoans are parasitic and have complex life cycles.

3 Protists and Health

- Protists negatively affect human health by causing diseases in humans and their food sources.
- Protists positively affect human health through their participation in food webs, through commensal relationships with humans and their food sources, and by recycling vital resources.
- Protists cause diseases such as dysentery, giardiasis, malaria, and toxoplasmosis.
- Malaria is the most serious protist disease that affects humans. Drugs and mosquito control can be used to control malaria.

Key Terms

Section 1

protozoan (461)
alga (461)
zygospore (462)
alternation of generations (463)
sporangium (463)

Section 2

amoeba (464)
pseudopodium (464)
diatom (466)
euglenoid (467)
kinetoplastid (467)
cilium (468)
plasmodium (470)

Section 3

sporozoite (473)
merozoite (473)

Study ZONE

Alternative Assessment — ADVANCED

Ask students to write a short essay exploring the rise of drug-resistant protistan pathogens. Have students consider the phrase "survival of the fittest." Does evolution occur faster in microbes than in other organisms? (Yes, they multiply more quickly and thus have more opportunity for mutation and recombination. By killing the nonresistant microbes with drugs, more resources are available for the resistant strains to multiply.)

Chapter Resource File

- Science Skills Worksheet GENERAL
- Critical Thinking Worksheet ADVANCED
- Test Prep Pretest GENERAL
- Chapter Test GENERAL

Answer to Concept Map

The following is one of several answers to Performance Zone item 15.

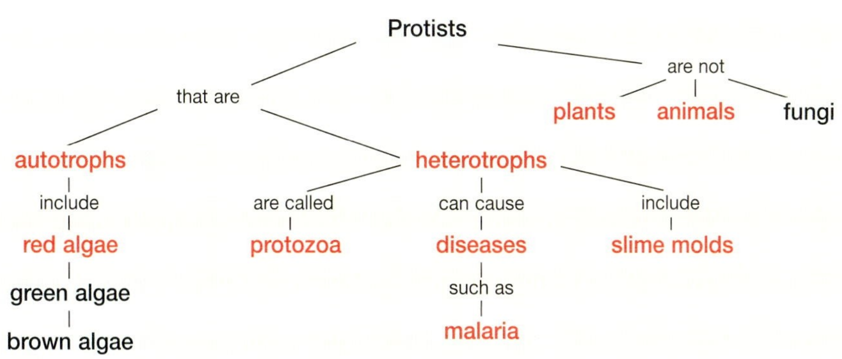

Performance ZONE

ANSWERS

Understanding Key Ideas

1. c
2. c
3. c
4. c
5. a
6. d
7. a
8. Creating a vaccine for malaria is difficult because the protists that cause the disease undergo several stages, each having different surface proteins. The protists also "hide" inside the human liver and blood cells.
9. The answer to the concept map is found at the bottom of page 475.

Critical Thinking

10. Uncontrolled dumping of sulfur could harm or kill other organisms in the river, upsetting the ecological balance. Sulfur would probably make the river water unusable by people who depend on it. Dumping sulfur into the river could cause the formation of large amounts of sulfuric acid in the river.
11. Her conclusions may be correct, as both heterotrophic and autotrophic euglenas exist. No chlorophyll is visible in specimen A, so it most likely is heterotrophic.
12. No, because the sporozoites need to mature in the salivary glands of a mosquito before they can infect another human.

Understanding Key Ideas

1. Two important eukaryotic features that evolved among protists are
a. photosynthesis and silica shells.
b. forams and pseudopodia.
c. sexual reproduction and multicellularity.
d. spores and microtubules.
SC.G.1.4.1

2. Which habitat is *least* likely to harbor any species of protist?
a. ocean waters
b. the human liver
c. a desert
d. leaf litter
SC.G.2.4.4

3. Which pair is *not* a correct match between a protist and its manner of reproduction?
a. Spirogyra: conjugation
b. Ulva: alternation of generations
c. Paramecium: alternation of generations
d. Chlamydomonas: gametes
SC.G.1.4.1

4. Pseudopodia are used by members of the phylum
a. Bacillariophyta.
b. Euglenophyta.
c. Rhizopoda.
d. Chlorophyta.
SC.G.1.4.1

5. Red algae differ from green in that red algae
a. always inhabit marine environments.
b. lack pigment.
c. are always multicellular.
d. display alternation of generations.
SC.G.1.4.1

6. A photosynthetic single-celled protist that moves using flagella would likely be classified as a member of which phylum?
a. Apicomplexa
b. Oomycota
c. Bacillariophyta
d. Euglenophyta
SC.G.1.4.1

7. How is amebic dysentery spread?
a. by drinking contaminated water
b. from the bite of a mosquito
c. from the bite of the tsetse fly
d. by eating overcooked meat
SC.G.1.4.1

8. BIOWatch Why has a successful malaria vaccine been so difficult to make? SC.H.3.4.3

9. **Concept Mapping** Make a concept map describing the protists. Try to include the following terms in your map: *red algae, malaria, protozoa, slime mold, autotrophs, plants, diseases, heterotrophs,* and *animals*. Include additional terms as needed. SC.G.1.4.1

Critical Thinking

10. **Evaluating Methods** People in a small village were being infected with a parasitic protist from water in a nearby river. Scientists decided to dump a large amount of sulfur into the river to kill the protists. What possible errors are there in using this method to prevent sickness? SC.H.3.4.3

11. **Justifying Conclusions** A scientist found two euglenas, below, and concluded that specimen A is heterotrophic, but specimen B is not Justify the scientist's conclusions.

Magnification: 206×

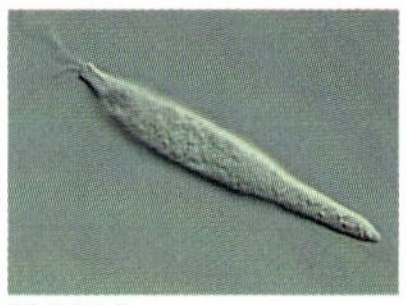
A

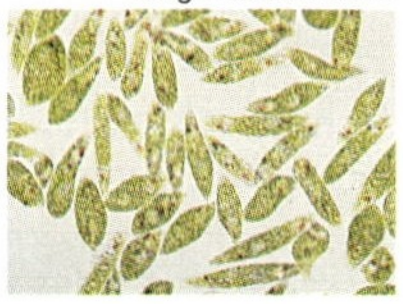
B

SC.H.2.4.2

12. **Applying Information** If while drawing blood from a patient with malaria, some of the blood splattered on your bare arm, could you catch this disease? SC.H.3.4.3

Alternative Assessment

13. **Finding Information** Use the media center or Internet resources to learn more about the history of plant diseases caused by protistan molds. Present an oral report that describes how scientists controlled these diseases in the past and how the information gained through research has created new strategies for disease prevention. Use photos, illustrations, and graphics to enhance your presentation. SC.H.3.4.5

476

Alternative Assessment

13. Answers will vary. An example is the discovery of potato blight, which is caused by the fungus *Phyophthora infestans*. Once researchers determined the life cycle of the mold and that the spores require a moist environment, they were able to recommend that the farmers keep the field around the infected plants dry. Today, farmers use fungicides and are trying to genetically modify the potato to make it more resistant to blight.

Assignment Guide

Section	Questions
1	1, 2, 3
2	4, 5, 6, 10, 11, 13
3	7, 9

Chapter Review Correlations

SC.G.1.4.1: Items 1, 3-7, 9
SC.G.2.4.4: Item 2
SC.H.3.4.3: Items 8, 10, 12
SC.H.2.4.2: Item 11
SC.H.3.4.5: Item 13

FCAT Prep

For questions 1–6, write your answers on a separate sheet of paper.

1 Which of the following are photosynthetic protists with boxlike double shells? SC.G.1.4.1

A. diatoms
B. euglenoids
C. kinetoplastids
D. plankton

2 What do slime molds form in order to reproduce? SC.F.2.4.1

F. plasmodia
G. pseudopodia
H. slugs
I. spores

3 Malaria-transmitting mosquitoes prefer a warm climate. What might happen to the incidence of malaria in the world if global warming continues?

A. The geographical range of the disease might expand, possibly increasing the incidence of malaria.
B. Fewer mosquitoes would carry the disease, so the incidence of malaria would possibly decrease.
C. Fewer people would be infected by the disease, because the increased temperatures would kill the protist.
D. Malaria-transmitting mosquitoes would not be able to breed, possibly decreasing the incidence of malaria. SC.G.2.4.6

4 In what stage is *Plasmodium* infective?

F. larva
G. merozoite
H. mosquito
I. sporozoite

SC.F.1.4.5

Base your answers to questions 5 and 6 on the graph below.

DAILY TEMPERATURE OF A MALARIA PATIENT

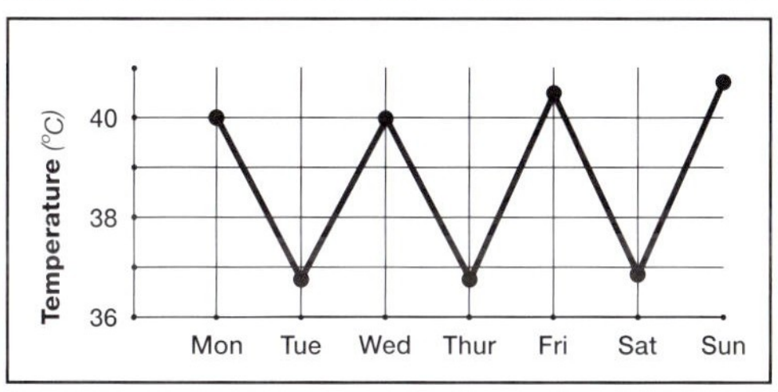

5 If the pattern continues, what will be the approximate temperature of the patient on Monday?

A. 37°C
B. 39°C
C. 41°C
D. 45°C

SC.G.1.4.1

6 What is the approximate difference between the high and low temperatures, to the nearest 1°C? SC.G.1.4.1

Test TIP

If you are unsure of the answer to a particular question, put a question mark beside it and go on to the next question. If you have time, go back and reconsider any question that you skipped. (Do not write in this book.)

477

FCAT Doctor

Question 3 Answer A is the correct choice. If global temperatures warmed, more areas might have a climate that encouraged the reproduction of malaria-transmitting mosquitoes. Thus, there would be more opportunity for the mosquitoes to infect people. Answer B is incorrect because global warming would not likely decrease the number of mosquitoes carrying the disease. Answer C is incorrect because even if the protist were harmed by increased temperatures, global warming would not make the entire planet inhospitable to the protist. Answer D is incorrect because even if some places became inhospitable to the mosquitoes, there would still be parts of the world where mosquitoes could breed.

Question 5 Answer A is the correct choice. The temperature cycle will be at 37°C on Monday. Answer B is incorrect because this temperature is more moderate than the high and low points in the cycle. Answer C is incorrect because the temperature will be at a low point on Monday. Answer D is incorrect because this temperature is significantly higher than the high points in the cycle.

Question 6 The correct answer is 3°C. The high temperatures are approximately 40°C. The low temperatures are approximately 37°C. The difference is 40°C–37°C = 3°C.

Answers

1. A
2. I
3. A
4. I
5. A
6. 3

Test Prep Correlations

SC.F.1.4.5: Item 4
SC.F.2.4.1: Item 2
SC.G.1.4.1: Items 1, 5, 6
SC.G.2.4.6: Item 3

Exploration Lab

Observing Protistan Responses to Light

Teacher's Notes

Time Required
45 to 75 minutes

Ratings

EASY → HARD

TEACHER PREPARATION 2

STUDENT SETUP 2

CONCEPT LEVEL 2

CLEANUP 2

Skills Acquired
- Collecting Data
- Designing Experiments
- Experimenting
- Organizing and Analyzing Data

Scientific Methods
In this lab, students will:
- Make Observations
- Ask Questions
- Test the Hypothesis
- Analyze the Results

Materials and Equipment
Materials for this lab can be ordered from WARD'S. See the *Master Materials List* at the front of this book for catalog numbers.

Safety Cautions
Discuss all safety symbols and caution statements with students. Protists and Detain™ may be washed down the drain.

Tips and Tricks
This investigation may take more than one class period. To save time, you may wish to make the "sun shades" in advance. (Students should consider that some protists benefit from light and others do not; this might affect the organisms' responses to light.)

Exploration Lab

Observing Protistan Responses to Light

SC.F.1.4.7

SKILLS
- Using scientific methods
- Using a microscope

OBJECTIVES
- **Identify** several different types of protists.
- **Compare** the structures, methods of locomotion and feeding, and behaviors of several different protists.
- **Relate** a protist's response to light to its method of feeding.

MATERIALS
- Detain™ (protist-slowing agent)
- microscope slides
- plastic pipets with bulbs
- mixed culture of protists
- toothpicks
- coverslips
- compound microscope
- protist references
- black construction paper
- scissors
- paper punch
- white paper
- sunlit window sill or lamp
- forceps

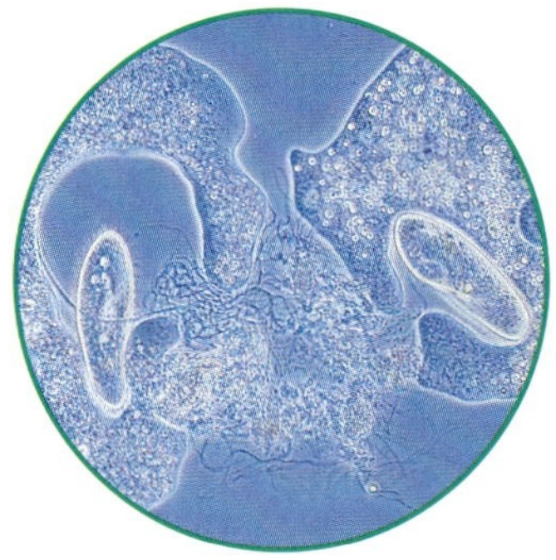

Before You Begin

Protists belong to the kingdom **Protista**, which is a diverse group of **eukaryotes** that cannot be classified as animals, plants, or fungi. Many protists are unicellular. Among the protists, there are **producers**, **consumers**, and **decomposers**. In this lab, you will observe live protists and compare their structures, methods of locomotion and feeding, and behaviors.

1. Write a definition for each boldface term in the paragraph above and for each of the following terms: cilia, flagellum, pseudopod.
2. Make a data table similar to the one below.
3. Based on the objectives for this lab, write a question you would like to explore about protists.

Procedure

PART A: Make Observations

1. **Caution: Do not touch your face while handling microorganisms.** Place a drop of Detain™ on a microscope slide. Add a drop of liquid from the bottom of a mixed culture of protists. Mix the drops with a toothpick. Add a coverslip. View the slide under low power of a microscope. Switch to high power.
2. Use references to identify the protists. Record data for each type of protist.
3. Repeat step 1 *without* using Detain™.
4. Punch a hole in a 40 × 20 mm piece of black construction paper that has a slight curl, as shown in the photo.

DATA TABLE

Protist	Color	Method of locomotion	Method of feeding	Other observations

478

Answers to Before You Begin

1. *protists*—member of Kingdom Protista

   Kingdom Protista—a diverse group of eukaryotic organisms that cannot be classified as animals, plants, or fungi

   eukaryotes—organisms that have complex cells with a nucleus enclosed by a membrane

   producer—an organism that makes its own food from light energy and inorganic molecules in its environment

   consumer—an organism that must consume other organisms to obtain energy

   decomposer—an organism that breaks down organic matter

   cilia—tightly packed rows of short hairlike structures used for movement

   flagellum—a whiplike structure used for movement

   pseudopod—a flexible cytoplasmic extension used for movement and obtaining food
2. Tables should be similar to the one on page 478.
3. Answers will vary. For example: How do *Euglena* and *Volvox* respond to light?

5. Place a wet mount of protists on a piece of white paper. Then put the paper and slide on a sunlit window sill or under a table lamp. Position the sun shade on top of the slide so that the hole is in the center of the coverslip.
6. To examine a slide, first view the area in the center of the hole under low power. *Note: Do not disturb the sun shade. Do not switch to high power.* Then have a partner carefully remove the sun shade with forceps while you observe the slide.

PART B: Design an Experiment

7. Work with members of your lab group to explore one of the questions written for step 3 of **Before You Begin.** To explore the question, design an experiment that uses the materials listed for this lab.

> **You Choose**
>
> **As you design your experiment, decide the following:**
>
> a. what question you will explore
> b. what hypothesis you will test
> c. how long you will expose protists to light
> d. how many times you will repeat your experiment
> e. what your control will be
> f. what data you will record and how you will make your data table

8. Write a procedure for your experiment. Make a list of all the safety precautions you will take. Have your teacher approve your procedure and safety precautions before you begin the experiment.
9. Set up and carry out your experiment.

PART C: Cleanup and Disposal

10. Dispose of lab materials and broken glass in the designated waste containers. Put protists in the designated containers. Do not put lab materials in the trash unless your teacher tells you to do so.
11. Clean up your work area and all lab equipment. Return lab equipment to its proper place. Wash your hands thoroughly before you leave the lab and after you finish all work.

Analyze and Conclude

1. **Summarizing Results** Describe the different types of locomotion you observed in protists, and give examples of each.
2. **Analyzing Results** Identify which protists were affected by light, and describe how they were affected.
3. **Inferring Conclusions** What is the relationship between a protist's response to light and its method of feeding?
4. **Further Inquiry** Write a new question about protists that could be explored with another investigation.

Do You Know?

Do research in the library or media center to answer these questions:

1. What livestock diseases caused by parasitic protists are most commonly found in the United States?
2. How do backpackers avoid getting diseases caused by protists that are transmitted in water?

Use the following Internet resources to explore your own questions about protists.

> **internet connect**
> **www.scilinks.org**
> **Topic:** Protists
> **Keyword:** HX4153
> SCILINKS. Maintained by the National Science Teachers Association

Answers to Procedure

Sample Procedure

1. Prepare five wet mounts of *Euglena* and five wet mounts of *Volvox*, all without any protist-slowing agent. View under the low power of a compound microscope, and record the observations.
2. Place the wet mounts on top of a piece of white paper on a sunlit window sill or under a table lamp. Position the "sun shade" so that the hole is in the center of the coverslip for each slide.
3. After 10 minutes, place each slide on the stage of a compound microscope and view the slide using the low-power objective. Record your observations.
4. Have one person remove the "sun shade" on each slide (using forceps) while another person continues to view the protists. Record your observations.

Answers to Analyze and Conclude

1. Protists move by using cilia, flagella, or pseudopodia. For example, *Amoeba* moves by flowing cytoplasm into a pseudopodium, *Paramecium* uses cilia beating in waves to propel the cell, *Euglena* uses a whiplike flagellum, and *Volvox* uses many flagella.
2. *Euglena* congregated within the area exposed to light. Once the shade was removed, they scattered. Euglenoids have an eyespot used to detect the light needed for photosynthesis.
3. Protists that depend on light for food were attracted to the light.
4. Answers will vary. For example: Do samples from the middle and upper areas of the culture jar have different protist populations from those taken at the bottom?

Answers to Do You Know?

1. Answers will vary. Students should report on livestock diseases such as blackhead, coccidiosis, and giardia.
2. Answers will vary. Students should report on safe water treatments that are practical for backpackers, such as portable water-filtering systems.

Sample data table:

Protist	Color	Method of Locomotion	Method of Feeding	Other Observations
Amoeba	Whitish	Pseudopodia	Consumer	Changes shape
Blepharisma	Pink	Cilia	Consumer	
P. caudatum	Clear	Cilia	Consumer	
Euglena	Green	Flagellum	Producer	Has "eye-spot"
Stentor	Bluish	Cilia	Consumer	Trumpet-shaped
Volvox	Green	Flagellum	Producer	Colonial

Chapter Opener		SE **Quick Review,** p. 481 ■ SE **Reading Activity,** p. 481 ■ TE **Using the Figure,** p. 480 GENERAL TE **Opening Activity,** p. 481 GENERAL
45 min pp. 482–484 **Section 1** Characteristics of Fungi	OSP **Lesson Plan*** TR **Bellringer*** TR **Section Outline*** TR **F21 *Penicillium* Mold***	TE **Activity** Mold Growth, p. 482 BASIC TE **Using the Figure,** p. 483 SE **Quick Lab** Observing Bread Mold, p. 483 ■ CRF **Datasheets for In-Text Labs** Observing Bread Mold*
& 4 • 90 min pp. 485–488 **Section 2** Fungal Diversity	OSP **Lesson Plan*** TR **Bellringer*** TR **Section Outline*** TR **F24 Structure of a Mushroom*** TR **F23 Life Cycle of Zygomycetes*** TR **F26 Life Cycle of Ascomycetes*** TR **F25 Life Cycle of Basidiomycetes***	TE **Using the Figure,** p. 486 GENERAL TE **Demonstration,** p. 486 ADVANCED TE **Group Activity** Fungal Industries, p. 487 TE **Activity** Spore Prints, p. 488 BASIC CRF **Exploration Lab** Fungi Phyla* GENERAL CRF **Skills Practice Lab** Determining the Best Food for Yeast* ◆ GENERAL
& 6 • 90 min pp. 490–492 **Section 3** Fungal Partnerships	OSP **Lesson Plan*** TR **Bellringer*** TR **Section Outline***	TE **Demonstration,** p. 490 BASIC TE **Using the Figure,** p. 491 SE **Data Lab** Analyzing the Effect of Mycorrhizae, p. 491 ■ CRF **Datasheets for In-Text Labs** Analyzing the Effect of Mycorrhizae* SE **Skills Practice Lab** Observing Yeast and Fermentation, pp. 496–497 ■ CRF **Datasheets for In-Text Labs** Observing Yeast and Fermentation*

& 8 • 90 min

Chapter Review and Assessment Resources

- SE **Study Zone,** p. 493 ■
- SE **Performance Zone,** p. 494 ■
- SE **FCAT Test Prep,** p. 495 ■
- TR **Concept Mapping***
- CRF **Vocabulary Review Worksheet*** ■ GENERAL
- CRF **Test Prep Pretest*** ■ GENERAL
- CRF **Chapter Test*** ■ GENERAL
- CRF **Chapter Test*** ADVANCED
- OSP **Test Generator***
- OSP **Test Item Listing***
- OSP **Modified Chapter Test*** BASIC

go.hrw.com Holt Online Learning

Visit go.hrw.com for access to Holt Online Learning or enter the keyword **HX6 Home** for a variety of free online resources.

One-Stop Planner® CD-ROM

This CD-ROM package includes
- Lab Materials QuickList Software
- Holt Calendar Planner
- Customizable Lesson Plans
- Printable Worksheets
- ExamView® Test Generator
- Interactive Teacher Edition
- Holt PuzzlePro® Resources
- Holt PowerPoint® Resources

Guided Reading Audio CDs

These CDs are designed to help auditory learners and reluctant readers. (Audio CDs are also available in Spanish.)

Compression guide:
To shorten your instruction because of time limitations, omit block 1.

KEY

- **TE** Teacher Edition
- **SE** Student Edition
- **OSP** One-Stop Planner
- **CRF** Chapter Resource File
- **TR** Teaching Transparency
- * Also on One-Stop Planner
- ■ Also Available in Spanish
- ◆ Requires Advance Prep

SKILLS DEVELOPMENT AND PRACTICE	SECTION REVIEW AND ASSESSMENT	CORRELATIONS
		Florida Sunshine Standards/FCAT
CRF Directed Reading Worksheet* BASIC **CRF** Active Reading Worksheet* GENERAL	**SE** Section 1 Review ■ **TE** Reteaching, p. 484 BASIC **TE** Quiz, p. 484 GENERAL **TE** Alternative Assessment, p. 484 GENERAL **CRF** Quiz* ■ BASIC	SC.F.1.4.2, SC.F.2.4.1, SC.G.1.4.1, SC.G.1.4.2, SC.H.1.4.1
TE Skill Builder Writing Skills, p. 487 ADVANCED **TE** Reading Skill Builder, p. 487 GENERAL **CRF** Directed Reading Worksheet* BASIC **CRF** Active Reading Worksheet* GENERAL	**SE** Section 2 Review ■ **TE** Reteaching, p. 488 BASIC **TE** Quiz, p. 488 GENERAL **TE** Alternative Assessment, p. 488 GENERAL **CRF** Quiz* ■ BASIC	SC.F.2.4.1, SC.G.1.4.1, SC.H.3.4.4
CRF Directed Reading Worksheet* BASIC **CRF** Active Reading Worksheet* GENERAL **CRF** Critical Thinking Worksheet* ADVANCED **CRF** Science Skills Worksheet* GENERAL	**SE** Section 3 Review ■ **TE** Reteaching, p. 492 BASIC **TE** Quiz, p. 492 GENERAL **TE** Alternative Assessment, p. 493 GENERAL **CRF** Quiz* ■ BASIC Modified Worksheet* BASIC	SC.F.1.4.1, SC.G.1.4.1

SCILINKS. NSTA
www.scilinks.org
Maintained by the **National Science Teachers Association**

Topic: Characteristics of Fungi
SciLinks Code: HX4037

Topic: Symbiosis of Fungi
SciLinks Code: HX4172

Topic: Lichens
SciLinks Code: HX4114

Overview

Tell students that the purpose of this chapter is to learn how the kingdom Fungi is characterized, how fungi obtain nutrients, and how fungi reproduce. Students will learn that the three groups of fungi differ primarily in their means of sexual reproduction. Finally, students will learn how fungi play a key role in the environment as partners with plants and algae.

Using the Figure

Tell students that the scarlet waxy cap mushroom (*Hygrophorus* sp.) shown here typically grows on mosses. Its scarlet-red color soon fades to orange or pale yellow. Mushrooms in this genus are waxy in appearance and can often be identified by their texture, habitat, and color. The scarlet waxy cap mushroom is listed as being nonpoisonous, but it may make people sick. Remind students that they should never eat a wild mushroom unless an expert has identified it as being safe.

Assessing Prior Knowledge

Before beginning this chapter, be sure your students can answer the questions in the ***Quick Review*** on the facing page. If they cannot, encourage them to reread the sections indicated.

Chapter Resource File

- Vocabulary Worksheets
- Concept Mapping

Scarlet waxy cap mushrooms

480

Chapter Correlations — **Florida Sunshine Standards/FCAT**

SC.F.1.4.2 Body structures are uniquely designed and adapted for their function.

SC.F.2.4.1 There are mechanisms of asexual and sexual reproduction and different genetic advantages and disadvantages of asexual and sexual reproduction.

SC.G.1.4.1 There is great diversity and interdependence among living things.

SC.G.1.4.2 The flow of energy through an ecosystem made up of producers, consumers, and decomposers carries out the processes of life, and some energy dissipates as heat and is not recycled.

SC.H.3.4.4 Funds for science research come from federal government agencies, industry, and private foundations, and this funding often influences the areas of discovery.

CHAPTER

22 Fungi

Florida Sunshine Standards/FCAT

SC.F.1.4.2 Body structures are uniquely designed and adapted for their function.

SC.F.2.4.1 There are mechanisms of asexual and sexual reproduction and different genetic advantages and disadvantages of asexual and sexual reproduction.

SC.G.1.4.1 There is great diversity and interdependence among living things.

SC.G.1.4.2 The flow of energy through an ecosystem made up of producers, consumers, and decomposers carries out the processes of life, and some energy dissipates as heat and is not recycled.

SC.H.3.4.4 Funds for science research come from federal government agencies, industry, and private foundations, and this funding often influences the areas of discovery.

Quick Review

Answer the following without referring to earlier sections of your book.

1. **Distinguish** meiosis from mitosis. *(Chapter 7, Section 1)*
2. **Summarize** the importance of mycorrhizae to plants. *(Chapter 12, Section 3)*
3. **Describe** the process of succession. *(Chapter 16, Section 1)*
4. **Describe** the meaning of symbiosis. *(Chapter 17, Section 1)*
5. **Define** the term *mutualism*. *(Chapter 17, Section 1)*

Did you have difficulty? *For help, review the sections indicated.*

Reading Activity

Before you begin this chapter, write the following statements on a separate piece of paper:

1. *Fungi are very similar to plants.*
2. *Fungi are economically valuable.*
3. *Many fungi are beneficial to other organisms.*

Take a minute or two to consider the statements and decide whether you agree or disagree with each. Share your opinions with the class. Read the chapter, and then reconsider the statements to see if your opinions are confirmed or changed.

Many fungi are edible, but some people are made sick by this mushroom. Never eat mushrooms growing in the wild. Some are fatally poisonous.

Looking Ahead

internet connect

www.scilinks.org
National Science Teachers Association *sci*LINKS Internet resources are located throughout this chapter.

SCILINKS. Maintained by the National Science Teachers Association

Opening Activity

Identifying Fungi Prior to class, put a mushroom in a paper bag. Have students determine the contents of the bag by asking "yes" or "no" questions. (This strategy will also help you determine student misconceptions regarding fungi.) Once the class has identified the contents of the bag, analyze students' responses and point out any misconceptions.

Quick Review Answers

1. Mitosis is cell division in which two new cells are formed, each having the same number of chromosomes as the original cell. In meiosis, two cell divisions usually form four new cells, each with half the number of chromosomes as the original.
2. The fungi in mycorrhizae are important to plants because they supply minerals to the plant roots.
3. In succession, fast-growing pioneer species begin to grow first. They make the environment more hospitable for other species, which then are able to grow. Each succeeding wave of new species changes the environment so that other species can grow.
4. In symbiosis, two or more species live together in a close relationship that may be beneficial or harmful to one or all of the species.
5. Mutualism is a symbiotic relationship in which both species benefit.

Reading Activity Answers

Students may think that fungi are similar to plants because they both grow in soil. After reading the chapter, they should recognize how different the two kingdoms are. Students may not realize that fungi are both economically valuable and beneficial to other organisms.

Section 1

Focus

Overview

Before beginning this section review with your students the objectives listed in the Student Edition. This section introduces structures unique to fungi that facilitate their survival. Fungal methods of obtaining nutrients and reproducing are introduced.

Bellringer

Ask students to write a short description of places where they have seen fungi growing and what these sites have in common. Have students describe the ideal environment in which fungi would thrive. (Fungi will survive almost anywhere with sufficient moisture and a carbon-rich food source.)

Motivate

Activity — BASIC

Have student groups place a piece of bakery bread in a labeled, zippered, plastic sandwich bag. (Note: Do not use commercial packaged bread; it contains preservatives.) Have students sprinkle a second piece of bread with a small amount of water and place it in a second labeled plastic bag. The bags should be sealed with tape and kept in a warm, dark place for a day or two. Students should observe mold growth on the moist bread but not on the dry bread. Ask what conditions promote mold growth. CAUTION: Some bread molds are health hazards. Do not allow students to open the bags containing the bread.

English Language Learners

Transparencies

TR Bellringer

TR F21 *Penicillium* Mold

Section Correlations

SC.F.1.4.2

SC.F.2.4.1

SC.G.1.4.1

SC.G.1.4.2

Section 1 Characteristics of Fungi

Florida Sunshine Standards/FCAT

SC.F.1.4.2 Body structures are uniquely designed and adapted for their function.

SC.G.1.4.1 There is great diversity and interdependence among living things.

SC.G.1.4.2 The flow of energy through an ecosystem made up of producers, consumers, and decomposers carries out the processes of life, and some energy dissipates as heat and is not recycled.

Objectives

- **List** the characteristics of the kingdom Fungi. SC.G.1.4.1
- **Describe** the structure of a typical fungus body. SC.F.1.4.2
- **Identify** how fungi obtain nutrients. SC.G.1.4.2
- **Relate** the way fungi obtain nutrients to their role in ecosystems. SC.G.1.4.2
- **Distinguish** the ways that fungi reproduce. SC.F.2.4.1

Key Terms

chitin
hypha
mycelium

Study TIP

Organizing Information

The numbered list on this page gives four important characteristics of fungi. Use this information to start a concept map that summarizes the characteristics of fungi.

Kingdom Fungi

Some of the most unusual organisms that exist today are members of the kingdom Fungi. Mushrooms and molds are common fungi that grow so rapidly they sometimes appear overnight.

The first fungi were probably unicellular eukaryotes. Fungi first appeared on Earth about 430 million years ago. In the past, biologists grouped fungi with plants because fungi are immobile, have a cell wall, and appear "rooted" in the soil, as the mushrooms do in **Figure 1.** However, the unique features of fungi indicate that they should be classified as a separate kingdom.

1. **Fungi are heterotrophic.** The stalk and cap of the mushroom are not green like the stem and leaves of a plant. Plants appear green because they contain chlorophyll, the pigment that aids in photosynthesis. Fungi do not contain chlorophyll. Rather, they obtain energy by breaking down organic material that they absorb from their environment.
2. **Fungi have filamentous bodies.** Plants are made of many cell and tissue types, but fungi are made of long, slender filaments. These filaments weave tightly together to form the fungus body and reproductive structures, such as the mushroom.
3. **Fungal cells contain chitin.** The cells of the mushroom, like the cells of all fungi, have walls made of **chitin** *(KIE tihn),* the tough polysaccharide found in the hard outer covering of insects. Plant cells have walls made of cellulose, a different polysaccharide.
4. **Fungi exhibit nuclear mitosis.** Mitosis in fungi is different from mitosis in plants and in most other eukaryotes. In most eukaryotes, the nuclear envelope disintegrates in prophase and re-forms in telophase. In dividing mushroom cells, however, the nuclear envelope remains intact from prophase to anaphase. As a result, spindle fibers form within the nucleus. The spindle fibers then drag chromosomes to opposite poles of the nucleus, rather than to opposite poles of the cell. Mitosis is complete when the nuclear envelope pinches in two.

Figure 1 Mushrooms

These mushrooms are actually the reproductive structures of a large network of filaments that makes up the body of a fungus.

482

Chapter Resource File

- Directed Reading BASIC
- Active Reading GENERAL
- Data Sheet for Quick Lab GENERAL

INCLUSION Strategies — GENERAL

- *Learning Disability*
- *Attention Deficit Disorder*

While some fungi are known to be harmful, some fungi are known to have beneficial uses. Using a word processor or pencil and paper, ask students to make one list of some harmful effects of fungi and another list of the beneficial uses of fungi. Students may record their findings as a public service warning about harmful fungi or beneficial uses of fungi.

Structures and Nutrients

Structures

In **Figure 2,** the fungus *Penicillium* is shown growing on an orange. The green and white fuzz you recognize as mold is actually the reproductive structures of the fungus. The body of the fungus lies within the tissues of the orange. All fungi except yeasts have bodies composed of slender filaments called **hyphae** *(HIE fee).*

When hyphae grow, they branch and form a tangled mass called a **mycelium** *(mie SEE lee uhm),* shown in Figure 2. A mycelium can be made of many meters of individual hyphae. This body organization provides a high surface-area-to-volume ratio, which makes a fungus well suited for absorbing nutrients from its environment.

Each hypha is a long string of cells divided by partial walls. Some species do not have walls between cells. Cytoplasm flows freely throughout the hypha.

Figure 2 ***Penicillium* mold**

This orange is covered with a fungus from the genus *Penicillium*.

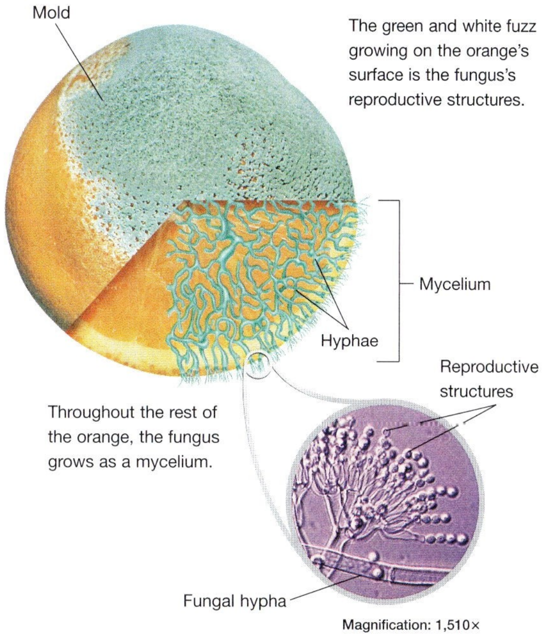

Nutrients

All fungi obtain nutrients by secreting digestive enzymes that break down organic matter in their environment. Fungi then absorb the decomposed molecules. Many fungi decompose nonliving organic matter, such as leaves, branches, dead animals, and waste. So, fungi are resource recyclers. Other fungi, such as the fungus that causes ringworm, are parasites that absorb nutrients from living hosts.

Parasitic fungi compete for nutrients with their hosts. In humans, they sometimes cause infectious diseases, such as athlete's foot and yeast infections. **Figure 3** warns of the fungus *Histoplasma capsulatum,* which invades the body's organs and causes life-threatening infections.

Fungi often grow on human foods, such as bread and fruit, making the food undesirable. Fungi are also known to attack nonfood materials, such as paper, cardboard, cloth, paint, and leather.

Some fungi are commercially valuable. Unicellular fungi called yeasts are useful in baking, brewing, and wine-making. Other fungi provide the flavor and aroma of certain cheeses. Many kinds of antibiotics, such as penicillin, are produced by fungi.

Figure 3 **Dangerous fungi.** *Histoplasma capsulatum* grows in the feces of bats and birds and infects humans when the dried spores are inhaled. Symptoms of infection include fever, cough, chills, chest pain, and muscle aches.

Graphic Organizer

Use this Graphic Organizer with **Teaching Tip** *on this page.*

Hyphae

- Form a mass called a mycelium
- Secrete enzymes to aid in absorbing food
- Grow quickly by passing nutrients through incomplete septa

Teach

Group Activity — GENERAL

Comparing Fungi and Plants

Have groups of students write the following words on index cards or strips of paper, with one word per card: chitin, cellulose, cell wall, roots, hyphae, spore, seed, autotroph, heterotroph. Have students sort the terms into a "plant" group and a "fungus" group. Ask students to discuss the differences and similarities between both groups. (Both plants and fungi have cell walls, and some plants make spores. Words associated only with plants are cellulose, roots, seed, and autotroph. Words in the fungus-only group are chitin, hyphae, and heterotroph.) **LS Interpersonal**

Co-op Learning

Teaching Tip

Observing Fungal Structure

Bring an orange covered with mold to class. Cut the orange in half and have students compare it to the orange shown in **Figure 2.** Have students locate the mycelium and the reproductive structures of the fungus in the figure and in the sample you brought. **CAUTION:** Students who are allergic to mold should not handle fungi and should wear a face mask. Then ask students to make a Graphic Organizer similar to the one at the bottom of this page to describe the structure and function of hyphae.

Integrating Physics and Chemistry

Have students decide if fungi are dependent on physical changes, chemical changes, or both when obtaining nutrients from their environments. Alternatively, ask students to find out what chemical reactions or changes occur when yeast is used in baking, brewing, and wine-making.

Teach, *continued*

Observing Bread Mold

QUICK LAB

Skills Acquired
Observing, interpreting results

Teacher's Notes
CAUTION: Use *only* prepared slides of bread mold. Some bread molds may contain cancer-causing substances. Do not handle moldy bread for this lab.

Answers to Analysis

1. Labels include hyphae, mycelium, and reproductive structures.
2. The mycelium made of hyphae is throughout the bread. The reproductive structures are on the surface.
3. The hyphae absorb nutrients from the bread. The mycelium provides a high surface-to-volume ratio for nutrition. Reproductive structures release spores.

Close

Reteaching — BASIC

Group students into teams. Tell students to complete a passage that you read aloud from the text. The first team to complete the passage wins. Play until you have reviewed all of the objectives for this section.

Quiz — GENERAL

True or False:

1. Fungi are autotrophs. (False—they are heterotrophs.)
2. Fungal spores are haploid. (True)

Section Review Correlations

SC.F.1.4.2: Item 2
SC.F.2.4.1: Item 5
SC.G.1.4.1: Items 1, 6, 7
SC.G.1.4.2: Items 3, 4

Figure 4 Puffball. *Lycoperdon perlatum,* a puffball, releases hundreds of thousands of spores through a small opening.

Reproduction

Fungi reproduce by releasing spores formed sexually or asexually in reproductive structures at the tips of hyphae. Reproductive structures grow high above the food source. This adaptation allows air currents to carry the spores to a new habitat. As you can see in **Figure 4,** fungal spores are so small and light that they remain suspended in the air for long periods of time; the wind can carry them great distances.

Fungal spores are haploid. Most spores are formed by mitosis during asexual reproduction. In sexual reproduction, hyphae from two mating types fuse. The fused hyphae form a sexual reproductive structure. On this structure, the fungus forms spores through fusion of the two genetically different nuclei.

QUICK LAB

Observing Bread Mold SC.H.1.4.1

You can use a microscope to see the individual threads of cells that make up the body of a fungus.

Materials

prepared slide of *Rhizopus*—black bread mold, compound microscope

Procedure

1. Examine a slide of black bread mold under low power of a microscope.
2. Move the slide to an area where you can clearly see threadlike structures.
3. Draw what you see in your lab notebook. Be sure to use at least one third of the page.
4. Move the slide to examine an area where you can clearly see the round bulblike structures.
5. Draw what you see in your notebook.

Analysis

1. **Label** the drawings you made, using **Figure 2** as a guide.
2. **Explain** where you would find each structure on the loaf of bread shown above.
3. **Critical Thinking** **Recognizing Relationships** Relate the structures you drew to their functions described in the text.

Section 1 Review

1. **Distinguish** the characteristics of fungi from those of plants. SC.G.1.4.1
2. **Compare** the characteristics of the mycelium with those of the reproductive structures. SC.F.1.4.2
3. **Summarize** the way fungi obtain nutrients. SC.G.1.4.2
4. **Describe** the role fungi play in the environment. SC.G.1.4.2
5. **Summarize** the different ways that fungi reproduce. SC.F.2.4.1
6. **Critical Thinking** **Evaluating Conclusions** Two of your fellow students insist that yeasts should be classified as protists because they are eukaryotic unicellular organisms. Evaluate their claim. SC.G.1.4.1
7. *FCAT Prep* Which of the following sentences describes an organism that cannot be a fungus?
   A It is unicellular.
   B It has cell walls.
   C It releases spores.
   D It is photosynthetic.
   SC.G.1.4.1

Answers to Section Review

1. Fungi are heterotrophic, filamentous, and have chitinous cell walls and nuclear mitosis. Plants are autotrophic and have tissues, cellulose cell walls, and no nuclear mitosis.
2. The mycelium contains hyphae, which are thin filaments of individual cells divided by partial walls. The mycelium grows in the food source. Reproductive structures grow high above the food source where the spores they contain can be carried by wind to a new location.
3. The fungal hyphae secrete digestive enzymes that break down the organic material in which they grow. The nutrients are then absorbed through the cell walls of the hyphae.
4. Fungi are decomposers, breaking down organic matter in the environment.
5. Fungi reproduce asexually by producing spores, which develop into new individuals. Fungi reproduce sexually when hyphae from compatible mating types fuse, forming a sexual reproductive structure and spores.
6. Yeasts have chitin in their cell walls, as do all fungi. Protists do not have chitin.
7. **A.** Incorrect. Yeasts are unicellular fungi. **B.** Incorrect. Fungi have cell walls. **C.** Incorrect. Fungi release spores. **D.** Correct. Fungi have no chlorophyll.

Fungal Diversity

Florida Sunshine Standards/FCAT

SC.F.2.4.1 There are mechanisms of asexual and sexual reproduction and different genetic advantages and disadvantages of asexual and sexual reproduction.

SC.G.1.4.1 There is great diversity and interdependence among living things.

SC.H.3.4.4 Funds for science research come from federal government agencies, industry, and private foundations, and this funding often influences the areas of discovery.

Section 2

Objectives

- **Describe** the characteristics used to classify fungi. SC.G.1.4.1
- **List** two commercial uses for fungi. SC.H.3.4.4
- **Describe** three phyla of fungi. SC.G.1.4.1
- **Distinguish** between the life cycles of zygomycetes, ascomycetes, and basidiomycetes. SC.F.2.4.1
- **Describe** the mushroom *Amanita muscaria.* SC.G.1.4.1

Key Terms

zygosporangium
stolon
rhizoid
ascus
yeast
budding
basidium

Reproductive Structures

You can see how diverse fungi are if you examine the types of reproductive structures that they have. Based on the types of structures produced during sexual reproduction, fungi can be classified in three phyla. **Table 1** below lists some of these characteristics.

Asexual Reproduction

A fourth group, the deuteromycetes, is composed of fungi in which no sexual stage has been seen. Traditionally, this group has been called a phylum. Through the use of molecular techniques, scientists have reclassified most of these asexually reproducing organisms into the phylum Ascomycota.

There are about 17,000 species without a sexual stage. Many of these fungi are economically important. For example, some species of *Penicillium* produce the antibiotic penicillin. Other species produce the unique flavors of some cheeses. Species of *Aspergillus* are used for fermenting soy sauce and producing citric acid. Most of the fungi that cause skin diseases, such as athlete's foot and ringworm, are also deuteromycetes.

Table 1 Three Sexually Reproducing Phyla of Fungi

Phylum	Distinctive characteristics	Examples
Zygomycota	Sexual spores are formed in zygosporangia; hyphae have no walls 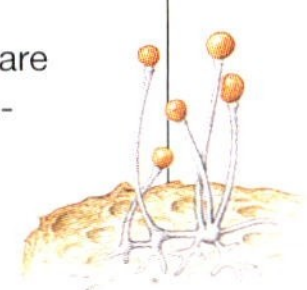	black bread molds
Ascomycota	Sexual spores are formed in asci; hyphae are divided by walls 	morels, truffles, yeasts, cup fungi
Basidiomycota	Sexual spores are formed in basidia; hyphae are divided by walls 	mushrooms, puffballs, rusts, smuts

internet connect

www.scilinks.org
Topic: Characteristics of Fungi
Keyword: HX4037

SCiLINKS. Maintained by the National Science Teachers Association

Section 2

Focus

Overview

Before beginning this section review with your students the objectives listed in the Student Edition. This section introduces the life cycles and reproductive structures that are used to classify fungi. The three phyla of sexually-reproducing fungi are described. A fourth group, those fungi that have no sexual stage, are discussed.

Bellringer

Ask students to make a list of characteristics that could be used to identify and classify an organism. Once completed, check to see if anyone's list includes reproductive structures. Introduce the lesson by explaining how fungal phyla are classified according to their methods of sexual reproduction.

Motivate

Discussion — Basic

Ask students if they have ever used soy sauce. Tell them that soy sauce is made from soy beans with the help of a fungus, *Aspergillus oryzae*. Ask students if they know of other foods that are made with the help of fungi. (The production of miso, a soybean paste used in Japanese cooking; tofu, a cheeselike soybean product used in different types of Asian cooking; wine; beer; and bread all rely on fungi.)

Chapter Resource File

- Directed Reading Basic
- Active Reading General

Transparencies

TR Bellringer
TR F24 Structure of a Mushroom

One-Stop Planner CD-ROM

- Reading Organizers Basic
- Reading Strategies Basic

Section Correlations

SC.F.2.4.1
SC.G.1.4.1
SC.H.3.4.4

Zygomycetes

If you place an uncovered loaf of bread near a windowsill, after a few days a cottony mold, shown in **Figure 5,** will cover its surface. Common black bread mold, *Rhizopus stolonifer,* is a member of the phylum Zygomycota *(zie goh my COHT uh).* Members of the phylum Zygomycota are named for the thick-walled sexual structures called **zygosporangia** *(zie goh spohr AN jee uh)* that characterize these members.

Species of *Rhizopus* and other zygomycetes usually live in the soil and feed on decaying plant and animal matter. The mycelia that grow along the surface of the bread are called **stolons** *(STOH lahnz).* The hyphae that anchor the fungus in the bread are called **rhizoids** *(RIE zoydz).* The hyphae of zygomycetes usually do not have walls.

Asexual reproduction in zygomycetes is much more common than sexual reproduction. During asexual reproduction, haploid spores are produced in the tips of specialized hyphae. It is the spores in fungi that cause fungal allergies in people. When they mature, these spores are shed and carried by the wind to new locations, where they germinate and start new mycelia. Reproduction in *Rhizopus* is shown in **Figure 6.**

Figure 5 Bread mold. *Rhizopus stolonifer* is often found growing on bread.

Figure 6 Life cycle of zygomycetes. Zygomycetes may reproduce sexually or asexually.

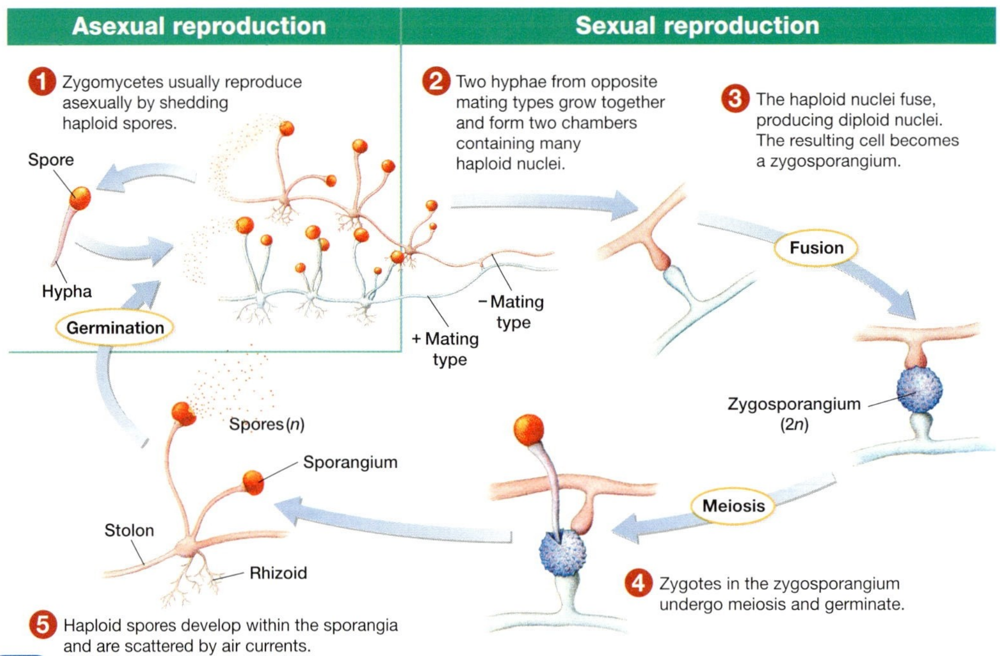

486

Teach

Using the Figure

Use **Figure 6** to help students understand the complex life cycle of a zygomycete. Have students trace the life cycle of the mold, and ask them for an explanation at each illustrated stage. **LS Visual**

Demonstration — ADVANCED

Display a series of scanning electron micrographs showing zygospore formation in *Rhizopus.* Ask students why zygospores are advantageous for fungi. (The protective zygospore can survive until environmental conditions are favorable for germination.) **LS Visual**

READING SKILL BUILDER — GENERAL

Reading Organizer Have students make a table on a piece of paper. Down the left side, they should write these headings: Zygomycota, Ascomycota, and Basidiomycota. Across the top, they should write these headings: Method of Reproduction, Structures, Habitat, and Example. As students read the section, they should complete the table. **LS Verbal**

Teaching Tip — GENERAL

Hyphae Have students make a drawing of *Rhizopus.* Stolons and rhizoids should be shown and labeled. Be sure students are using these two terms correctly. English Language Learners

PHYSICS CONNECTION

Tell students that the zygomycete *Pilobolus* grows in animal dung and produces a reproductive structure about 5 to 10 mm high. A clear structure below the spore acts as a lens, focusing sunlight on a photoreceptor inside the fungus. Explain that the reproductive structure bends towards the light and eventually ejects the spores to a distance of up to 2 m at a velocity near 50 km/h. Ask students to calculate the time it would take a spore to travel 2 m, based on the velocity of 50 km/h. (v = d/t where v is the velocity—50 km/h, d is the distance—2 m, and t is the time.) (The time it takes to travel 2 m is 0.14 s.)

Ascomycetes

The chestnut tree, *Castanea dentata,* was once a common tree in the eastern United States. Around 1890, a disease called chestnut blight wiped out virtually all the chestnut trees within a few years. Chestnut blight is caused by *Endothia parasitica,* a member of the phylum Ascomycota *(AS koh mie koh tuh).* Other ascomycetes include flavorful morels and truffles prized by gourmet chefs.

The ascomycetes are named for their characteristic sexual reproductive structure. The microscopic **ascus** *(AS kuhs)* is a saclike structure in which haploid spores are formed. Asci usually form within the interwoven hyphae of a cup-shaped fruiting body. Reproduction in a typical ascomycete is shown in **Figure 7.**

Ascomycetes usually reproduce asexually. Asexual spores form at the tips of the hyphae. The spores are not contained in any sac or structure. When the spores are released, air currents carry them to other places, where they may germinate and form new mycelia.

Yeast is the common name given to unicellular ascomycetes. There are about 350 species of yeasts. *Saccharomyces cerevisiae,* or baker's yeast, has been used for thousands of years to make bread and alcoholic beverages, such as beer. Other yeasts, such as *Candida albicans,* cause human disease. *C. albicans* causes thrush, a disease in which milk-white lesions form on the mouth, lips, and throat.

Most yeasts reproduce asexually by fission or budding. In **budding,** a small cell forms from a large cell and pinches itself off from the large cell.

Real Life

Many species of ascomycetes cause powdery mildew.
Powdery mildew is a fungal infection of plants. Grapes, wheat, and apples are examples of plants affected by powdery mildew.

Finding Information
Research ways that farmers protect their crops from powdery mildew.

Figure 7 Life cycle of ascomycetes. Ascomycetes can reproduce sexually or asexually.

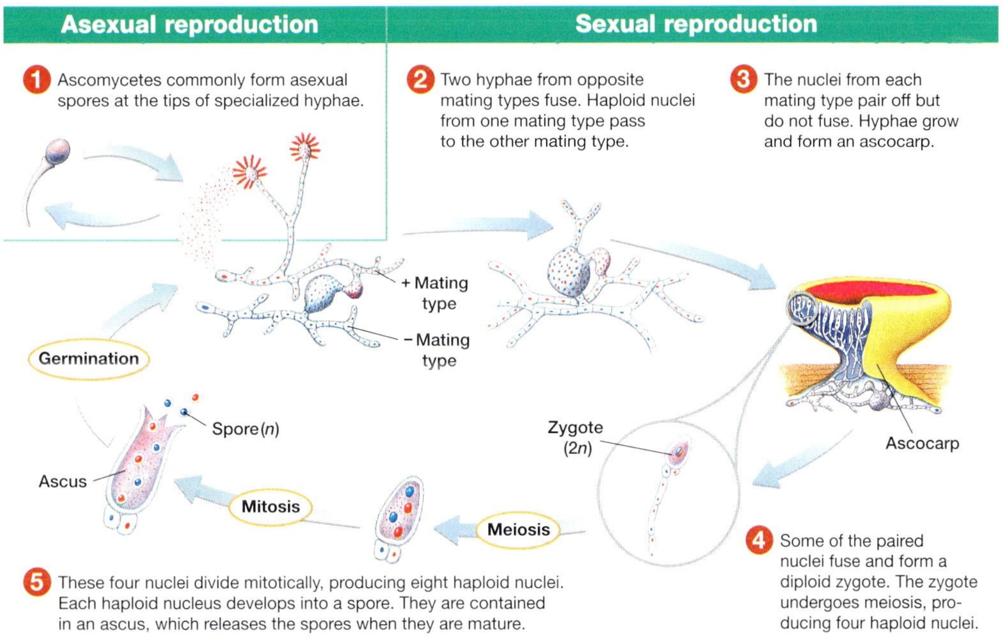

Real Life
Answer
Farmers can protect their crops from powdery mildew by planting only after the weather warms up, rotating crops, using fungicide-treated seeds, planting mildew-resistant cultivars, and applying fungicides to the plants.

Teaching Tip — ADVANCED

Comparing Fungi Ask students to compare and contrast the three sexually-reproducing phyla of fungi.

SKILL BUILDER — ADVANCED

Writing Skills Have students write an account of how the chestnut blight wiped out all the chestnut trees in the United States. Students should note that resistant cultivars of chestnut trees have been developed and are now being planted.

Group Activity

Fungal Industries Invite a speaker from a company that uses fungi in its production process (for example, a representative from a bakery, pharmaceutical company, winery, food-processing plant, or a cheese-making plant). Ask your guest to discuss the importance of fungi in his or her industry and how the fungi are cultivated for industrial use. Have students prepare questions for the speaker.

READING SKILL BUILDER — BASIC

Assign Chapter 22 of the *Holt Biology* Guided Reading Audio CD Program to help students achieve greater success in reading the chapter. English Language Learners

Transparencies

TR F23 Life Cycle of Zygomycetes
TR F26 Life Cycle of Ascomycetes
TR F25 Life Cycle of Basidiomycetes

Cultural Awareness

Bread-Mold Sandwich An Indonesian food called tempeh is made of soybeans that are boiled, skinned, and inoculated with various species of *Rhizopus* (bread mold). The tempeh is then fried, roasted, or diced in preparation for a meal. Today tempeh can be found in health food and natural food stores, as well as in some large supermarkets.

Teach, *continued*

Activity — BASIC

Spore Prints Have students place mushroom caps on paper, with the gill side down. Cover the mushrooms to keep them from drying and to keep the air still. After one or two days, prints will be made from the falling spores. Tell students that spore prints are used by mycologists to help identify and classify basidiomycetes. **LS Visual**

Close

Reteaching — BASIC

Collect or have students collect mushrooms growing nearby (or use prepared specimens). CAUTION: Warn students not to eat any of the samples and to wash their hands after they touch the mushrooms. Using field guides, have students identify which phylum each sample represents. **LS Visual**

Quiz — GENERAL

1. Is the asexual stage of fungi haploid or diploid? (haploid)
2. Where are spores produced in mushrooms? (in basidia, which line the gills of the mushroom)

Alternative Assessment — GENERAL

Have students use 3 × 5 in. cards to make flashcards. Cards should be labeled on one side with the name of a fungal structure. The other side of the card should be labeled with the name of the associated phylum. Have students use their cards to quiz one another. **LS Interpersonal**

Figure 8 Rust on wheat. Rust is a basidiomycete that attacks cereal crops, making them unfit for humans to eat.

Basidiomycetes

The kind of fungi with which you are probably most familiar—mushrooms—are members of the phylum Basidiomycota *(buh SIHD ee oh mie koh tuh).* Other basidiomycetes include toadstools, puffballs, jelly fungi, and shelf fungi. The **basidium** *(buh SIHD ee uhm)* is the club-shaped sexual reproductive structure for which the basidiomycetes are named. Spores are produced on this structure. You can see these spores in the *Up Close: Mushroom* feature later in this chapter. Asexual reproduction is rare among the basidiomycetes, except in some rusts and smuts. These two important groups of plant pathogens affect many crop plants, as shown in **Figure 8.** Sexual reproduction of a typical basidiomycete is illustrated in **Figure 9.** Many mushrooms are harmless, but many are also deadly, such as the *Amanita muscaria* (fly agaric), shown in the *Up Close* feature. Other *Amanita* species have names, such as death angel and destroying angel, that reflect the danger of their toxins for humans.

Figure 9 Life cycle of basidiomycetes

Basidiomycetes usually reproduce sexually by means of a fruiting body, also called a mushroom.

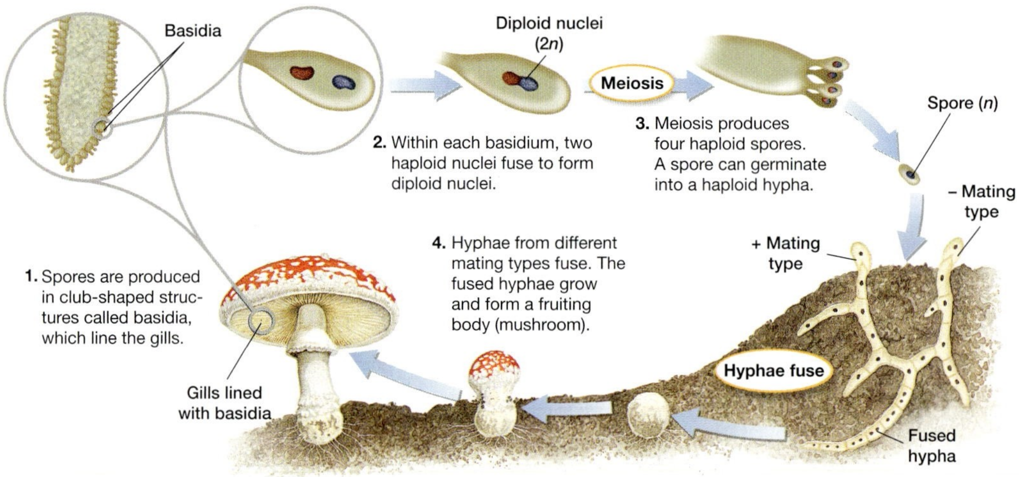

Section 2 Review

1. **Summarize** how fungi are classified. SC.G.1.4.1
2. **Describe** the distinctive characteristics of the three phyla of fungi, including their life cycles. SC.F.2.4.1
3. **Describe** the structure of the mushroom. SC.G.1.4.1
4. **Critical Thinking Justifying Conclusions** The fruiting body of some ascomycetes, such as truffles, is found below ground and gives off a delicious scent. Scientists hypothesize that this scent might contribute to spreading the spores. How might this scent help the spread of the fungus's spores? SC.F.1.4.2
5. *FCAT Prep* Which of the following roles makes rusts of interest to humans? SC.H.3.4.3
   - **A** to make bread
   - **B** to attack crop plants
   - **C** to produce penicillin
   - **D** to cause athlete's foot

488

Answers to Section Review

1. Fungi are classified into phyla based on their sexual reproductive structures.
2. See Figures on previous pages.
3. A mushroom consists of a stalk (stem) and a flattened cap composed of interwoven hyphae. The underside of the cap is lined with gills, which house the spore-producing basidia.
4. If the spores could be spread by an animal, the scent would alert and attract the animal to the underground fruiting body.
5. **A.** Incorrect. Yeast is used to make bread. **B.** Correct. Rusts are plant pathogens. **C.** Incorrect. Penicillin is not made by rusts. **D.** Incorrect. Athlete's foot is not caused by a rust.

Section Review Correlations

SC.F.1.4.2: Item 4
SC.F.2.4.1: Item 2
SC.G.1.4.1: Items 1, 3
SC.H.3.4.3: Item 5

Mushroom

- **Scientific name:** *Amanita muscaria*
- **Size:** 10–15 cm
- **Habitat:** Moist organic soils
- **Nutrition:** Heterotrophic

Characteristics

Cell structure *A. muscaria* and other fungi have cell walls made of chitin, a complex polysaccharide also found in the external skeleton of insects. In some fungi, hyphae are not divided into separate cells but have many nuclei in the same cytoplasm. In other fungi, hyphae are divided into cells by perforated walls called septa.

Reproduction Under proper conditions, underground hyphae grow upward and weave together to produce a mushroom. Mushrooms are the reproductive structures of fungi such as *A. muscaria.* A mushroom has a flattened cap attached to a stem called a stalk. The underside of the cap is lined with rows of gills. Thousands of club-shaped reproductive cells called basidia form on the gills. Through fusion and meiosis, each basidium produces spores that are released and form new hyphae.

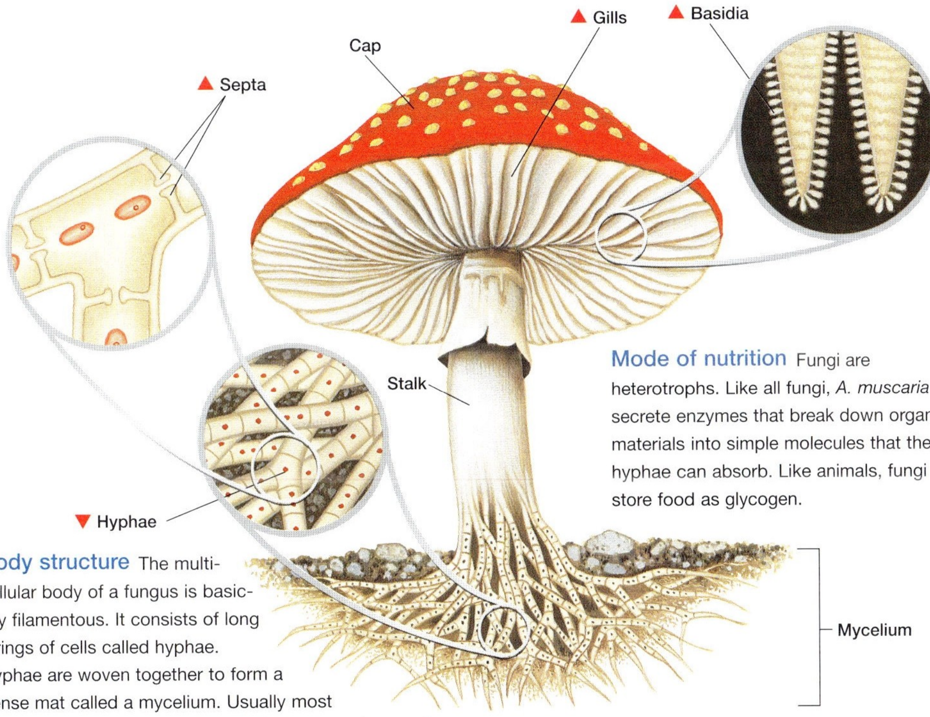

Mode of nutrition Fungi are heterotrophs. Like all fungi, *A. muscaria* secrete enzymes that break down organic materials into simple molecules that the hyphae can absorb. Like animals, fungi store food as glycogen.

Body structure The multicellular body of a fungus is basically filamentous. It consists of long strings of cells called hyphae. Hyphae are woven together to form a dense mat called a mycelium. Usually most of a mycelium is hidden within a substrate, such as soil.

489

INCLUSION Strategies

- *Learning Disability*
- *English as a Second Language*

Have students visit local restaurants in person or on the Internet to find different dishes that have mushrooms in the recipe. Have students interview a cook to determine what kinds of mushrooms are used in their recipes, how the mushrooms are prepared, and how popular the dishes made with mushrooms are. Students may be able to share a family recipe that includes mushrooms with the class.

MISCONCEPTION ALERT

Eating Mushrooms Contrary to popular belief, poisonous mushrooms cannot be discerned by a quick test such as smearing liquid from a mushroom on white paper and looking for a color change or looking for a color change in a broken stem. Only a trained mycologist should attempt to identify edible mushrooms.

Up Close

Mushroom

Teaching Strategies

- Bring a field guide of mushrooms and other fungi to class and have students look through it to gain an appreciation for fungal diversity. Have each student select a particular mushroom or fungus and draw a large, colored picture to display in the classroom.
- Have students write and illustrate a brief report about mushrooms. Illustrations might include a spore print, a labeled drawing, or a recipe.
- Give pairs of students a mushroom. Have them cut the cap crosswise and the stalk lengthwise and examine the structures with a hand lens. Provide slides, coverslips, microscopes, and thin tissue slices. Have students make wet mounts and observe the cell structure of the hyphae. Then ask students to describe the structures and their functions.

CAUTION: Students who are allergic to mold should not handle fungi and should wear a face mask.

Discussion

- Why is it incorrect to classify a mushroom as a plant? (Fungi are classified in their own kingdom because they are heterotrophic, have filamentous bodies, have cells that contain chitin, and exhibit nuclear mitosis. Plants lack these characteristics.)
- How are septa different from the cell walls of plants? (Septa have gaps that allow materials, even nuclei, to flow from one "cell" to another.)
- Although gills and hyphae serve different purposes, what common characteristic do these structures share? (Both gills and hyphae are structures with a large surface area, which allows the cells to carry out their role more efficiently.)

Focus

Overview

Before beginning this section review with your students the objectives listed in the Student Edition. This section discusses the two symbiotic relationships between fungi and photosynthesizers—mycorrhizae and lichens—and their importance to the environment.

Bellringer

Have students look at photos of colorful lichens or bring in live lichens growing on a rock for them to examine. Ask students to write a short paragraph explaining what kind of organism they think a lichen is and in which kingdom it is classified. (A lichen is not an organism but a partnership of organisms; it is classified in kingdom Fungi.) LS Visual

Motivate

Demonstration — Basic

Provide students with pictures or live specimens of the three common forms of lichen: crustose, foliose, and fruticose. Ask students to describe the appearance of each lichen and compare it to familiar objects. (crustose—crusty; foliose—leafy; fruticose—upright, stalky) English Language Learners

Section Correlations

SC.G.1.4.1

Section 3 Fungal Partnerships

Florida Sunshine Standards/FCAT

SC.G.1.4.1 There is great diversity and interdependence among living things.

Objectives

- **Distinguish** two symbiotic relationships that involve fungi. SC.G.1.4.1
- **Summarize** the ecological importance of mycorrhizae. SC.G.1.4.1
- **Describe** lichens. SC.G.1.4.1

Key Terms

mycorrhiza
lichen

internet connect
www.scilinks.org
Topic: Symbiosis of Fungi
Keyword: HX4172
SCILINKS. Maintained by the National Science Teachers Association

Symbiotic Relationships

Fungi are involved in many kinds of symbiotic associations with algae and plants. These mutualistic relationships play important roles in ecology. You may recall that mutualism is a type of symbiosis in which each partner benefits. In these symbiotic associations, the fungus (a heterotroph) provides minerals and other nutrients that it absorbs from the environment. The algae or plant (a photosynthesizer) provides the ability to use sunlight to power the building of carbohydrates.

Mycorrhizae

A **mycorrhiza** *(MIE koh RIE zuh)* is a type of mutualistic relationship formed between fungi and vascular plant roots. The hyphae help transfer phosphorus and other minerals from the soil to the roots of the plant, while the plant supplies carbohydrates to the fungus.

In the mycorrhizae of most species of plants, the hyphae penetrate the outer cells of the root. The fungus is usually a zygomycete. In **Figure 10,** you can see the hyphae that grow in the roots. Fossils show that the rootlike structures of the earliest plants often had mycorrhizae, which may have played an important role in the invasion of land by plants. Scientists think that when plants invaded the land the soil of that time completely lacked organic matter. However, plants with mycorrhizae can grow successfully in infertile soil. Some vascular plants survive today by continuing this partnership as mycorrhizae.

In many plants, the mycorrhizae do not physically penetrate the plant root but instead wrap around it. These nonpenetrating mycorrhizae represent relationships in which a particular species of plant has become associated with a particular fungus, usually a basidiomycete. These kinds of mycorrhizae are important because they aid the growth of many commercially significant trees, such as pines, oaks, beeches, and willows.

Some mycorrhizae produce economically important edible mushrooms, and the mycorrhizae of some ascomycete species produce an edible fruiting body called a truffle.

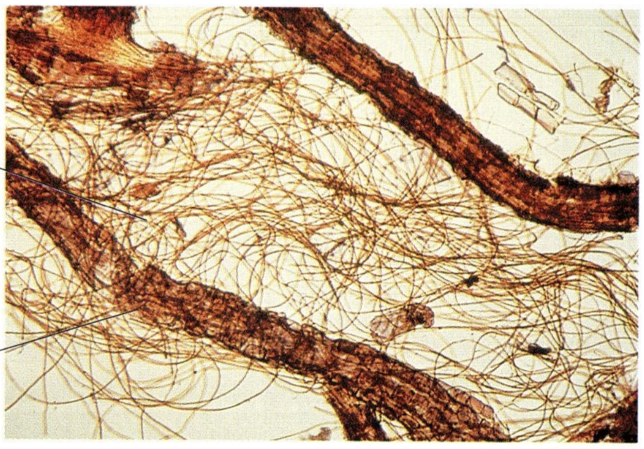

Figure 10 Mycorrhizae. The hyphae of the fungus in mycorrhizae sometimes appear as a tangled mass around the root of the plant.

490

Chapter Resource File

- Directed Reading Basic
- Active Reading General
- Data Sheet for Data Lab General

Transparencies

TR Bellringer

One-Stop Planner CD-ROM

- Reading Organizers Basic
- Reading Strategies Basic
- Supplemental Reading Guide *The Andromeda Strain* Advanced

Lichens

A **lichen** *(LIE kuhn)* is a symbiosis between a fungus and a photosynthetic partner such as a green alga, a cyanobacterium, or both. The photosynthetic partner provides carbohydrates. It is protected from the environment by the fungal partner, which helps it absorb mineral nutrients. In most lichens, the fungus is an ascomycete. When you look at a lichen, such as the ones in **Figure 11,** you are seeing the fungus. The photosynthetic partner is hidden between the layers of hyphae. Sunlight penetrates the layers of hyphae and enables the photosynthetic partner to carry out photosynthesis.

The tough construction of the fungus combined with the photosynthetic abilities of the alga, or cyanobacterium, has enabled lichens to colonize harsh habitats. Lichens have been found in arid desert regions and in the Arctic; they grow on bare soil, on tree trunks, and on sunbaked rocks. Recall that during succession, lichens are often the first colonists. They break down rocks and prepare the environment for other organisms. Lichens are a key component of primary succession because they are able to grow on rock and help break the rock down into soil. Lichens containing cyanobacteria carry out nitrogen fixation and introduce useful forms of nitrogen into the soil.

Figure 11 **Lichens.** The algae, shown as green cells in the micrograph, are the photosynthetic partners of the fungus growing in this British soldier lichen.

DATA LAB: *Analyzing the Effect of Mycorrhizae* SC.G.1.4.1

Background

Two groups of plants were planted in similar soils under similar conditions, but group A was grown in sterilized soil and group B was grown in nonsterilized soil. After 18 weeks of growth, a photograph was taken of the plants. Examine the photographs, and answer the following questions:

Analysis

1. **Compare** the growth of the two groups. Which grew faster?
2. **Explain** why one group grew better than the other group.
3. **Critical Thinking** **Inferring Relationships** Suggest a possible cause of slower growth in the smaller plants.
4. **Recommend** a course of action to restore growth in the stunted plants.

Career

Plant Pathologist Fungi are vitally important to the health of crops and forests as beneficial partners and as potential pathogens. Plant pathologists often wrestle with fungal diseases, including rusts, which have complex life cycles involving multiple hosts. Some researchers are investigating the use of fungi to control weeds.

did you know?

Lichens and Fallout Lichens in Finland, Norway, and Sweden absorbed radioactive fallout from the 1986 Chernobyl nuclear power disaster. Reindeer, a major food source for people living in the area, ate the radioactive lichens. More than 70,000 reindeer had to be destroyed because of their exposure to excessive radiation.

Teach

Using the Figure

Use **Figure 10** to explain how mycorrhizae greatly increase the absorbing capacity of plant roots. Whether the fungus enters the root or wraps around the outside of the root, the results are the same: the fungus gains carbohydrates and the plant gains needed minerals. LS **Visual**

DATA LAB: *Analyzing the Effect of Mycorrhizae*

Skills Acquired
Interpreting results, applying information, predicting outcomes

Teacher's Notes
In addition to increasing the uptake of minerals, mycorrhizae also increase water absorption and protect tiny rootlets from infection by soil pathogens. Despite their fertility, some prairie soils have trouble supporting pines and other trees because they lack the necessary mycorrhizal fungi.

Answers to Analysis

1. Group B
2. Sterilizing the soil kills all of the mycorrhizal fungi, which provide mineral nutrients to plants.
3. Without the symbiotic fungi, the plants did not have as many nutrients available to them.
4. Inoculate the soil of the stunted plants with mycorrhizal fungi.

Teach, continued

Exploring Further

Lichens as Environmental Watchdogs

Teaching Strategies
Organisms used to measure environmental health are indicator species. Lichens are indicators of air quality as aquatic insects are of water quality. Encourage students to find out more about indicator species.

Discussion
Why don't scientists test rain directly for pollutants? (Doing so would give a picture of the pollution that day in one area, but studying lichens can indicate how conditions have changed over time or over a wide area.)

Close

Reteaching — BASIC

Have students prepare a two-column table comparing lichens with mycorrhizae.

Quiz — GENERAL

1. What is a truffle? (A truffle is the edible fruiting body of certain ascomycete mycorrhizae.)
2. What does the photosynthetic part of a lichen contribute to the fungus? (It provides carbohydrates to the fungus.)

Section Review Correlations

SC.G.1.4.1: Items 1, 2, 3, 5

SC.G.2.4.2: Item 4

Lichens are able to survive drought and freezing by becoming dormant. When moisture and warmth return, lichens resume normal activities. In harsh environments, lichens may grow slowly. Some lichens that grow in the mountains appear to be thousands of years old and cover an area no larger than a fist. These lichens are among the oldest living organisms on Earth. Although lichens are known to survive extremes of temperatures, they are susceptible to chemical changes in their environment and so have become a living indicator of the amount of pollution in the environment in which they live.

Exploring Further

Lichens as Environmental Watchdogs

Since the 1950s scientists have found that most lichens require clean air to thrive. In the Los Angeles Basin, for example, rising smog levels have been linked to the disappearance of lichens. In the Pacific Northwest, lichens are most abundant in old-growth forests with good air quality. For these reasons, scientists have been using lichens to monitor air pollution.

Lichens as Indicators of Air Pollution

Lichens have no roots, so the nutrients they take up must come from the air. Rain, fog, and dew wet the surface of a lichen. When they are wet, lichens absorb nutrients and any pollutants that are in the air.

Lichens can live for centuries, making them well suited for studies of air-pollution changes that occur over a long time. Many lichen species also have large geographical ranges. Thus, a single species can indicate air quality at different distances from a source of pollution, such as a factory or power plant.

How Are Lichens Used?

To monitor air quality with lichens, scientists often map the distribution of lichens in an area. They count the number of lichen species, note how often each species occurs, and measure the total area covered by each species. Mapping studies done over many years can reveal long-term changes in lichen survival.

Pollutants from the stove pipe keep lichens from growing on part of this roof.

Scientists can obtain more detailed data on the effects of air pollution by determining the concentration of metals and other pollutants in lichen samples. They also assess the health of a lichen by measuring its chlorophyll content and its rate of photosynthesis.

To test air quality in places where lichens do not exist, scientists sometimes transplant healthy lichens from areas where they occur naturally. They then analyze the transplanted lichens for pollutants and look for any changes in the health of the lichens that may be caused by the move.

internet connect
www.scilinks.org
Topic: Lichens
Keyword: HX4114
SCILINKS. Maintained by the National Science Teachers Association

Section 3 Review

1. **Describe** two types of symbioses that involve fungi. SC.G.1.4.1
2. **Explain** how mycorrhizae are thought to have helped plants to colonize land. SC.G.1.4.1
3. **Identify** the organisms found in lichens. SC.G.1.4.1
4. **Summarize** how lichens promote the process of biological succession. SC.G.2.4.2
5. FCAT Prep What kind of relationship exists between a fungus and an alga in a lichen?
   A commensalism **C** parasitism
   B mutualism **D** predation
   SC.G.1.4.1

Answers to Section Review

1. Fungal symbioses include mycorrhizae and lichens.
2. Mycorrhizae may have provided plants with nutrients so they could grow in infertile soil.
3. A lichen is composed of a fungus and a photosynthesizer such as an alga or a cyanobacterium.
4. Lichens are often the first organisms to colonize an area, and they pave the way for growth of other organisms by breaking down rocks to form soil. Some lichens can also fix nitrogen, adding this nutrient to the soil in usable form.
5. **A.** Incorrect. In a lichen, neither partner remains unaffected. **B.** Correct. Both partners in a lichen benefit. **C.** Incorrect. In a lichen, neither partner is harmed. **D.** Incorrect. In a lichen, neither partner feeds on the other.

Study ZONE

Key Concepts

1 Characteristics of Fungi

- Fungi are eukaryotic heterotrophs. Their bodies are made up of slender woven filaments. Fungal cells contain chitin and go through nuclear mitosis.
- Fungi obtain nutrients by secreting digestive enzymes and absorbing the decomposed nutrients from their environment.
- Fungi decompose dead organic matter; they are an important resource recycler.
- Most fungi reproduce by releasing spores that are produced asexually and sexually.

2 Fungal Diversity

- Fungi are classified by their sexual reproductive structures.
- Fungi in which sexual reproduction has not been observed are referred to as deuteromycetes.
- Fungi in the phylum Zygomycota produce spores in thick-walled sexual structures called zygosporangia.
- Fungi in the phylum Ascomycota produce spores in a saclike structure called an ascus.
- Yeasts are unicellular ascomycetes that reproduce by budding.
- Fungi in the phylum Basidiomycota produce spores in a club-shaped structure called a basidium.

3 Fungal Partnerships

- Fungi can be involved in two types of symbioses, mycorrhizae or lichens.
- Mycorrhizae are symbiotic associations in which a fungus transfers minerals to a plant's roots, which in turn supply carbohydrates to the fungus.
- The fungal partner in a lichen protects the photosynthetic partner and provides the lichen with minerals. The photosynthetic partner provides the fungus with carbohydrates.

Key Terms

Section 1

chitin (482)
hypha (483)
mycelium (483)

Section 2

zygosporangium (486)
stolon (486)
rhizoid (486)
ascus (487)
yeast (487)
budding (487)
basidium (488)

Section 3

mycorrhiza (490)
lichen (491)

Alternative Assessment — GENERAL

Have students design a graphic organizer with three tiers. On the top tier, place the word Fungi. The second tier should contain the following terms: *medicine*, *agriculture*, *industry*, *ecology*, and *economics*. The third tier should be developed by the students and should contain examples to illustrate the terms in the second tier.

Chapter Resource File

- Science Skills Worksheet GENERAL
- Critical Thinking Worksheet ADVANCED
- Test Prep Pretest GENERAL
- Chapter Test GENERAL

Answer to Concept Map

The following is one possible answer to Performance Zone item 15 on the next page.

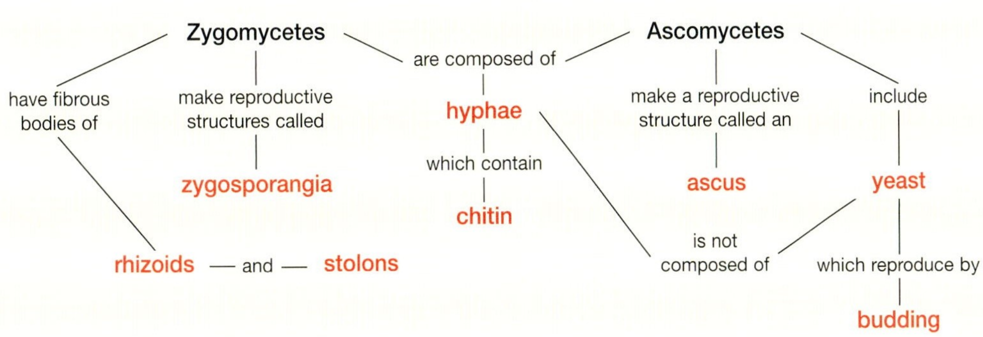

ANSWERS

Understanding Key Ideas

1. d
2. a
3. d
4. d
5. b
6. c
7. b
8. d
9. By mapping the location, population density, and composition of lichens, scientists can monitor air quality. Extremely sensitive to pollutants, lichens cannot survive in areas of high pollution.
10. The answer to the concept map is found at the bottom of the Study Zone page.

Critical Thinking

11. Many antibiotics, foods, and beverages could not be made. Human and agricultural diseases caused by fungi would decrease. After a period of time, dead material would begin to accumulate because fungi would not be available to decompose it. Plant growth would slow because of less available fixed nitrogen and lack of mycorrhizae. Eventually, succession might cease because there would be fewer pioneer organisms.
12. Plants already established in soil are better because they are more likely to have mycorrhizae than would bare-rooted plants.

Performance ZONE

Understanding Key Ideas

1. Fungi differ from plants in that fungi
   a. are multicellular.
   b. are immobile.
   c. have cell walls.
   d. are heterotrophic.
   SC.G.1.4.1
2. Which of the following characteristics is shared by all fungi and helps them obtain their nutrients?
   a. external digestion
   b. phagocytosis
   c. feeding on nonliving matter
   d. catching prey
   SC.G.1.4.2
3. Deuteromycetes are more difficult to classify than fungi of other phyla because
   a. they develop from zygosporangia.
   b. they are parasitic.
   c. they undergo meiosis.
   d. they do not reproduce sexually.
   SC.G.1.4.1
4. Some fungal associations no larger than a fist appear to be thousands of years old. These have been found
   a. in temperate forests.
   b. on well-irrigated alluvial plains.
   c. in fields of corn.
   d. in harsh environments, high in the mountains.
   SC.G.1.4.1
5. One might expect that plants without mycorrhizae are
   a. more likely to get fungal diseases.
   b. unsuccessful in the transfer of minerals from the soil to the roots.
   c. best suited to poor soil conditions.
   d. primitive and might soon become extinct.
   SC.G.1.4.1
6. In basidiomycetes, most of the mycelium grows
   a. above the ground as a mushroom.
   b. above the ground as an ascocarp.
   c. as a network of hyphae in the soil.
   d. as a network of hyphae within bread.
   SC.F.1.4.2
7. In many fungi, both asexual and sexual reproduction involve
   a. the fusion of opposite mating types.
   b. the shedding of haploid spores.
   c. the formation of basidia.
   d. budding.
   SC.F.2.4.1
8. Fungi differ from most eukaryotes in that they
   a. are heterotrophic.
   b. have a cell wall.
   c. reproduce sexually.
   d. exhibit nuclear mitosis.
   SC.G.1.4.1
9. **Exploring Further** Explain how scientists use lichens to monitor air quality and why lichens were chosen by scientists to monitor air quality. SC.H.3.4.2
10. **Concept Mapping** Construct a concept map that describes the structure and reproductive methods of different fungi. Use the following terms in your map: *chitin, hyphae, zygosporangia, stolon, rhizoid, ascus, yeast,* and *budding*. Use additional terms as necessary. SC.F.1.4.2

Critical Thinking

11. **Predicting Outcomes** If all fungi suddenly disappeared from Earth, what types of changes would you notice immediately? after a period of time? SC.D.2.4.1
12. **Forming Reasoned Opinions** When purchasing garden plants, is it better to buy bare root plants or plants that are established in a pot of soil? Explain your reasoning from an ecological point of view. SC.G.1.4.1

Alternative Assessment

13. **Summarizing Information** Use the media center or Internet resources to learn about the economic impact of fungi. Research fungi used as foods and the importance of certain fungi in the preparation or manufacture of food. Create a display to summarize your findings. SC.H.3.4.3

Alternative Assessment

13. Answers will vary. Students' reports might also address the historical significance of fungus-related foods, including the early, large-scale consumption of these foods in Europe and Asia and their rise in popularity in the United States.

Assignment Guide

Section	Questions
1	1, 2, 7, 8, 10, 13
2	3, 6, 10, 13
3	4, 5, 9, 11, 12, 13

Chapter Review Correlations

SC.D.2.4.1: Item 11
SC.F.1.4.2: Items 6, 10
SC.F.2.4.1: Item 7
SC.G.1.4.1: Items 1, 3, 4, 5, 8, 12
SC.G.1.4.2: Item 2
SC.H.3.4.2: Item 9
SC.H.3.4.3: Item 13

FCAT Prep

For questions 1–6, write your answers on a separate sheet of paper.

1. What carbohydrate makes up the cell walls of fungi?
   A. cellulose **C.** fructose
   B. chitin **D.** glucose
   SC.G.1.4.3
2. What kind of structure has a symbiotic relationship with fungi in mycorrhizae?
   F. algae **H.** lichens
   G. chloroplasts **I.** roots
   SC.G.1.4.1
3. How dose the zygosporangium of *Rhizopus stolonifer* function to ensure the survival of the species?
   A. It anchors the fungus to the food source.
   B. It causes allergies in humans who want to eat it.
   C. It dissolves organic matter for the fungus to absorb.
   D. It remains dormant during unsuitable conditions.
   SC.F.1.4.2
4. A scientist concludes that several cultures of cup-shaped ascocarps are a species of ascomycetes that reproduce only asexually. Why is the scientist's conclusion incorrect?
   F. Ascomycetes reproduce only sexually.
   G. Ascocarps are, in fact, produced by basidiomycetes.
   H. Unicellular ascomycetes reproduce sexually by budding.
   I. An ascocarp is a structure that is formed by sexual reproduction.
   SC.F.2.4.1

Base your answers to questions 5 and 6 on the graph below.

CHANGES IN THE TRUFFLES HARVESTED AND OAK FOREST AREA OVER 40 YEARS

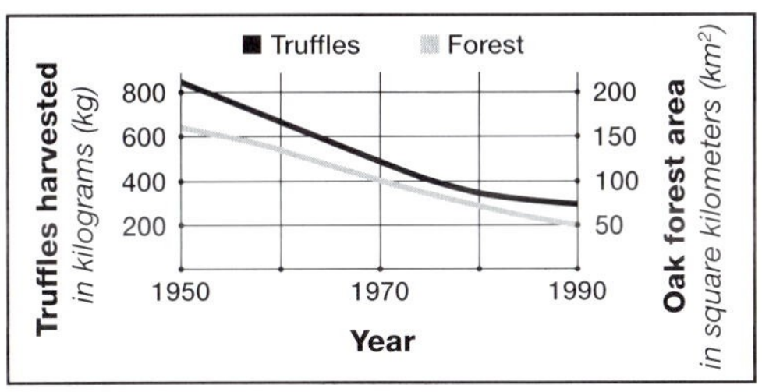

5. Which of the statements is supported by the data in the graph?
   A. The oak forest covered 125 km^2 in 1970.
   B. The truffle harvest decreased constantly between 1950 and 1975.
   C. The truffle harvest decreased most rapidly between 1980 and 1990.
   D. The oak forest area decreased by 50 percent between 1955 and 1970.
   SC.H.2.4.2
6. If the truffles harvested in 2000 were half of that in 1975, how many kilograms of truffles (to the nearest 25 kg) were harvested in 2000?
   SC.H.2.4.2

Test TIP

When using a graph to answer a question, study the data plotted on the graph to identify any trends in the data before you answer the question.

495

FCAT Doctor

Question 3 Answer D is the correct choice. The zygosporangium is a thick-walled structure that allows the fungus to survive until environmental conditions are more suitable for meiosis and germination. Answer A is incorrect because the fungus is anchored to the food source by hyphae called rhizoids. Answer B is incorrect because, while spores do cause allergies in humans, causing allergies does not ensure the survival of the species. Answer C is incorrect because this is not the specific function of the zygosporangium.

Question 4 Answer I is the correct choice. A species would not be classified as an ascomycete if it reproduced only asexually. Answer F is incorrect because ascomycetes reproduce both sexually and asexually. Answer G is incorrect because ascomycetes do produce ascocarps. Basidiomycetes produce basidia. Answer H is incorrect because unicellular ascomycetes, called yeast, reproduce asexually by budding.

Question 5 Answer B is the correct choice. The graph is a straight line during that period. Answer A is incorrect because the oak forest covered 100 km^2 in 1970. Answer C is incorrect because the graph leveled off between 1980 and 1990. Answer D is incorrect because the decrease from 150 km^2 to 100 km^2 is 33 percent, not 50 percent.

Question 6 The correct answer is 200. In 1975, 400 kg were harvested. Half of that is 200 kg.

Answers

1. B
2. I
3. D
4. I
5. B
6. 200

Test Prep Correlations

SC.F.1.4.2: Item 3
SC.F.2.4.1: Item 4
SC.G.1.4.1: Item 2
SC.G.1.4.3: Item 1
SC.H.2.4.2: Items 5, 6

Skills Practice Lab

Observing Yeast and Fermentation

Teacher's Notes

Time Required
One 50-minute class period

Ratings
EASY → HARD

Teacher Preparation: 3
Student Setup: 2
Concept Level: 3
Cleanup: 2

Skills Required
- Observing
- Measuring
- Collecting Data
- Analyzing Data

Scientific Methods
In this lab, students will:
- Make Observations
- Analyze the Results
- Draw Conclusions
- Communicate the Results

Materials and Equipment
Materials for this lab can be ordered from WARD'S. See *Master Materials List* at the front of this book for catalog numbers.

Safety Caution
Review all safety symbols with students before beginning the lab.

Skills Practice Lab

Observing Yeast and Fermentation SC.F.1.4.1

SKILLS
- Observing
- Measuring
- Collecting data
- Analyzing data

OBJECTIVE
- **Observe** the release of energy by yeast during fermentation.

MATERIALS
- 500 mL vacuum bottle
- 10 cm glass tubing
- 2-hole rubber stopper
- 250 mL beaker
- 75 g sucrose
- one package dry yeast
- thermometer
- 50 cm rubber tubing
- 150 mL limewater

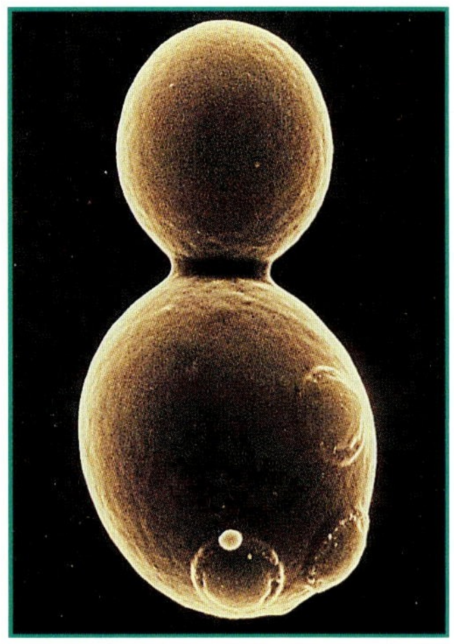
Yeast cells

ChemSafety

CAUTION: Always wear safety goggles and a lab apron to protect your eyes and clothing.

CAUTION: Glassware is fragile. Notify the teacher of broken glass or cuts. Do not clean up broken glass or spills with broken glass unless the teacher tells you to do so.

Before You Begin

Sucrose is a disaccharide—a carbohydrate made from two monosaccharides. It is one chemical made by plants to store the sun's energy. Yeast release the energy stored in sucrose in a process called **fermentation**. In this investigation you will have a chance to observe and measure the products of fermentation.

1. Write a definition for the boldface term in the paragraph above.
2. Make a data table like the one below.

DATA TABLE

Fermentation by Yeast

	Time	Date	Temperature		Time	Date	Temperature		Time	Date	Temperature
1.				8.				15.			
2.				9.				16.			
3.				10.				17.			
4.				11.				18.			
5.				12.				19.			
6.				13.				20.			
7.				14.				21.			

Tips and Tricks
Review the fundamentals of graphing prior to this investigation. Prepare a saturated limewater solution by combining approximately 700 mg of calcium carbonate per 500 ml of distilled water. Allow the solution to sit overnight and filter before using. The limewater should become cloudy as the carbon dioxide gas produced by the yeast flows into the flask. An indicator solution using bromthymol blue instead of limewater may also be also be used to detect the presence of carbon dioxide gas in the flask. This solution can be prepared by dissolving 0.25 g of bromthymol blue powder per 50 ml of distilled water. The solution should be green or blue. If it is yellow, the solution should be slowly titrated with a 0.1% sodium hydroxide solution only until it turns blue or dark green. The solution should turn yellow as the concentration of carbon dioxide gas present in the flask increases.

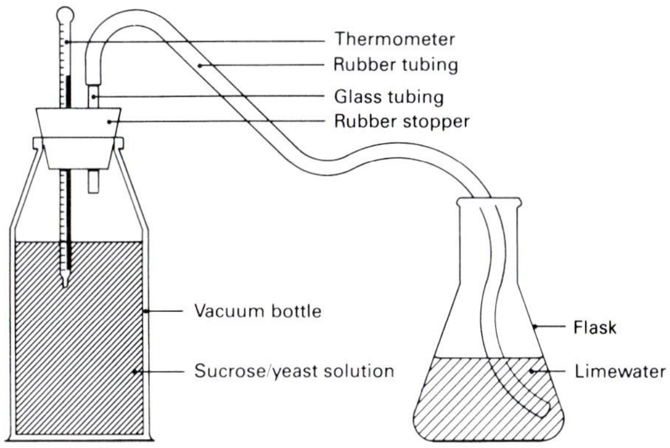

Procedure

1. Set up your vacuum bottle according to the diagram above.
2. Mix 75g of sucrose in 400 mL of water.
3. When the sucrose has dissolved, add one-half package of fresh yeast and stir.
4. Pour the sucrose-yeast solution into a vacuum bottle until it is approximately three-quarters full.
5. Adjust the thermometer so that it extends down into the sugar-yeast solution.
6. Record the temperature of the solution on the observation chart as soon as possible. Continue to record the temperature as often as possible during the next two days.

Analyze and Conclude

1. **Summarizing Results** Prepare a graph of your data, illustrating the temperature over time. Complete the graph by drawing a curve through the plotted points.
2. **Analyzing Data** What does the curved line plotted on the graph indicate?
3. **Drawing Conclusions** What can you conclude about the energy contained in sucrose?
4. **Predicting Patterns** What do you think would happen if there were only one hole in the stopper for the thermometer?
5. **Further Inquiry** If you know that fermentation liberates energy and gives off carbon dioxide and alcohol as waste products, how would you prove that fermentation is really taking place in the sugar-yeast solution?

Do You Know?

Do research in the library or media center to answer these questions:

1. How do mammals get energy from glucose when they cannot use respiration?
2. How do "naturally carbonated" sodas generate bubbles?

Use the following Internet resources to explore your own questions about fermentation.

internet connect
www.scilinks.org
Topic: Fermentation
Keyword: HX4080
SCLINKS. Maintained by the National Science Teachers Association

Answers to Before You Begin

1. Fermentation is the process by which cells release the energy in sugar without using oxygen.

Answers to Do You Know

1. Answers will vary but should include a discussion of anaerobic glucose metabolism, or glycolysis.
2. Answers will vary. "Naturally carbonated" beverages generate bubbles as a product of yeast fermentation. They are usually made using a sugar-yeast mixture as the main ingredient.

Answers to Analyze and Conclude

1. Students' graphs will vary. They should show that as fermentation progresses the temperature in the sucrose-yeast solution increases.
2. Answers will vary. Students should interpret the curved line as an indication that fermentation activity increases over time.
3. Answers will vary. Since the temperature of the sucrose-yeast solution increases over time, the students should conclude that as fermentation occurs large amounts of energy are being released. This would indicate that the sucrose molecules contain a large amount of energy in their molecular bonds, which is released as the yeast metabolizes the sucrose.
4. Answers will vary. The carbon dioxide gas produced by the yeast would increase the pressure in the bottle and the stopper would eventually come off.
5. Answers will vary. Students could test the sucrose-yeast solution at the beginning of the experiment for the presence of carbon dioxide by adding a small amount to an indicator solution.

UNIT 6 Exploring Plants

Chapters

More than 20 billion bushels of corn are harvested worldwide every year.

Corn: America's Crop

Yesterday...

Used for food by Native Americans beginning 10,000 years ago, corn was first cultivated in what is today called Mexico in 6000 to 5000 B.C. The Aztec Indians, the Maya, and the Inca all grew corn and featured it prominently in their traditions and ceremonies. Corn is an important food crop. **Where is the Corn Belt?**

Incas harvesting the corn crop

Today...

Cooked, steamed, roasted, ground into cornmeal, or popped, corn has long been a mainstay of the American diet. Corn is used to make cornstarch, corn syrup, breakfast cereals, cornmeal, salad dressing, corn oil, margarine, coloring agents, stabilizing agents, and countless other products. **Read to find out what corn and wheat have in common.**

Agricultural scientist

Tomorrow...

Scientists continue to develop new uses for corn and its products. For example, ethanol, or ethyl alcohol distilled from corn, can be used as a lead-free octane booster and as a replacement for regular leaded gasoline. **Discover how corn is able to conduct photosynthesis efficiently in intense heat.**

Automobile that runs on ethanol made from corn

internet connect

www.scilinks.org
Topic: Corn
Keyword: HX4052

SCILINKS. Maintained by the National Science Teachers Association

Chapter Opener		SE **Quick Review,** p. 501 ■ SE **Reading Activity,** p. 501 ■ TE **Using the Figure,** p. 500 TE **Opening Activity,** p. 501
& 3 • 90 min pp. 502–507 **Section 1** Adaptations of Plants	OSP **Lesson Plan*** TR **Bellringer*** TR **Section Outline*** TR **G31 Stomata and Guard Cells*** TR **G13 Vascular Tissue*** TR **G11 Structure of a Vascular Plant*** TR **G43 Structure and Function of Seeds***	TE **Using the Figure,** p. 503 SE **Quick Lab** Observing the Behavior of Stomata, p. 503 ■ CRF **Datasheets for In-Text Labs** Observing the Behavior of Stomata* TE **Activity** Seed Dispersal, p. 504 GENERAL TE **Group Activity** Observing Pollen Grain, p. 505 ADVANCED TE **Using the Figure,** p. 506 GENERAL CRF **Quick Lab** Inferring Leaf Function from Structure* BASIC
& 5 • 90 min pp. 508–515 **Section 2** Kinds of Plants	OSP **Lesson Plan*** TR **Bellringer*** TR **Section Outline*** TR **G39 Life Cycle of an Angiosperm*** TR **G8 Familiar Families of Angiosperms***	TE **Demonstration,** p. 508 BASIC TE **Using the Figure,** p. 510 TE **Demonstration,** p. 511 BASIC TE **Activity** Growing Fern Gametophytes, p. 511 BASIC SE **Data Lab** Analyzing the Effect of Climate on Plants, p. 512 ■ CRF **Datasheets for In-Text Labs** Analyzing the Effect of Climate on Plants* TE **Activity** Conifer ID, p. 512 GENERAL TE **Demonstration,** p. 514 GENERAL TE **Activity** Classifying Plants as Monocots or Dicots, p. 515 ADVANCED CRF **Skills Practice Lab** Mapping Biotic Factors* GENERAL
& 7 • 90 min pp. 516–522 **Section 3** Plants in Our Lives	OSP **Lesson Plan*** TR **Bellringer*** TR **Section Outline*** TR **G3 Medicines Originally Derived from Plants***	SE **Quick Lab** Distinguishing Fruits and Vegetables, p. 517 ■ CRF **Datasheets for In-Text Labs** Distinguishing Fruits and Vegetables* TE **Group Activity** Local Plant Processing, p. 519 ADVANCED TE **Demonstration,** p. 520 GENERAL TE **Activity** Paper Types, p. 520 GENERAL SE **Skills Practice Lab** Surveying Plant Diversity, pp. 526–527 ■ CRF **Datasheets for In-Text Labs** Surveying Plant Diversity*

& 9 • 90 min

Chapter Review and Assessment Resources

SE **Study Zone,** p. 523 ■
SE **Performance Zone,** p. 524 ■
SE **FCAT Test Prep,** p. 525 ■
TR **Concept Mapping***
CRF **Vocabulary Review Worksheet*** ■ GENERAL
CRF **Test Prep Pretest*** ■ GENERAL
CRF **Chapter Test*** ■ GENERAL
CRF **Chapter Test*** ADVANCED
OSP **Test Generator***
OSP **Test Item Listing***
OSP **Modified Chapter Test*** BASIC

Online and Technology Resources

Holt Online Learning

Visit go.hrw.com for access to Holt Online Learning or enter the keyword **HX6 Home** for a variety of free online resources.

One-Stop Planner® CD-ROM

This CD-ROM package includes
- Lab Materials QuickList Software
- Holt Calendar Planner
- Customizable Lesson Plans
- Printable Worksheets
- ExamView® Test Generator
- Interactive Teacher Edition
- Holt PuzzlePro® Resources
- Holt PowerPoint® Resources

Guided Reading Audio CDs

These CDs are designed to help auditory learners and reluctant readers. (Audio CDs are also available in Spanish.)

Compression guide:
To shorten your instruction because of time limitations, omit blocks 1, 6, and 7.

KEY

TE	Teacher Edition	CRF	Chapter Resource File	*	Also on One-Stop Planner
SE	Student Edition	TR	Teaching Transparency	■	Also Available in Spanish
OSP	One-Stop Planner			◆	Requires Advance Prep

SKILLS DEVELOPMENT AND PRACTICE	SECTION REVIEW AND ASSESSMENT	CORRELATIONS
		Florida Sunshine Standards/FCAT
TE **Skill Builder** Vocabulary, p. 503 BASIC **CRF** **Directed Reading Worksheet*** BASIC **CRF** **Active Reading Worksheet*** GENERAL	**SE** **Section 1 Review** ■ **TE** **Reteaching,** p. 507 BASIC **TE** **Quiz,** p. 507 GENERAL **TE** **Alternative Assessment,** p. 507 GENERAL **CRF** **Quiz*** ■ BASIC	SC.F.1.4.2, SC.F.2.4.1, SC.G.1.4.1
TE **Skill Builder** Math Skills, p. 509 ADVANCED **TE** **Reading Skill Builder** Reading Organizer, p. 511 BASIC **TE** **Reading Skill Builder** Discussion, p. 513 BASIC **CRF** **Directed Reading Worksheet*** BASIC **CRF** **Active Reading Worksheet*** GENERAL	**SE** **Section 2 Review** ■ **TE** **Reteaching,** p. 515 BASIC **TE** **Quiz,** p. 515 GENERAL **TE** **Alternative Assessment,** p. 515 ADVANCED **CRF** **Quiz*** ■ BASIC	SC.D.2.4.1, SC.G.1.4.1
TE **Reading Skill Builder** Anticipation Guide, p. 517 BASIC **TE** **Skill Builder** Math Skills, p. 518 GENERAL **TE** **Reading Skill Builder** Reading Organizer, p. 521 BASIC **CRF** **Directed Reading Worksheet*** BASIC **CRF** **Active Reading Worksheet*** GENERAL **CRF** **Critical Thinking Worksheet*** ADVANCED **CRF** **Science Skills Worksheet*** GENERAL	**SE** **Section 3 Review** ■ **TE** **Reteaching,** p. 522 BASIC **TE** **Quiz,** p. 522 GENERAL **TE** **Alternative Assessment,** p. 522 GENERAL **CRF** **Quiz*** ■ BASIC **Modified Worksheet*** BASIC	SC.D.2.4.1, SC.F.1.4.2, SC.G.1.4.1, SC.H.3.4.5

SCLINKS. NSTA
www.scilinks.org
Maintained by the **National Science Teachers Association**

Topic: Vascular Plants
SciLinks Code: HX4182

Topic: Structure and Function of Seeds
SciLinks Code: HX4169

Topic: Nonvascular Plants
SciLinks Code: HX4130

Topic: Seedless Vascular Plants
SciLinks Code: HX4161

Topic: Angiosperms
SciLinks Code: HX4007

Topic: Vegetarian Diets
SciLinks Code: HX4184

Topic: Medicines from Plants
SciLinks Code: HX4119

CHAPTER 23

Overview
Tell the students the purpose of this chapter is to recognize the importance and diversity of the plant kingdom. In the many plant species alive on Earth today, students can find evidence of the adaptations that allowed plants to first colonize land, then diversify and flourish. Students will learn the common characteristics of all plants, as well as the unique characteristics of different groups of plants. Some of these characteristics make plants extremely important to humans.

Using the Figure
Direct students' attention to the chapter opening figure. The blue and white bluebonnet that is seen throughout the picture is the state flower of Texas. It is a lupine species that is endemic, or native only to Texas. Throughout various regions, native plants provide ecological benefits such as sustaining ecosystem stability. They also create economic benefits through reduced maintenance and energy costs, and they enhance biological richness with beauty and regional natural heritage.

Assessing Prior Knowledge
Before beginning this chapter, be sure your students can answer the questions in the ***Quick Review*** on the facing page. If they cannot, encourage them to reread the sections indicated.

Chapter Resource File

- **Vocabulary Worksheets**
- **Concept Mapping**

Texas wildflowers

500

Chapter Correlations **Florida Sunshine Standards/FCAT**

SC.D.2.4.1 The systems on Earth and the quality of life are interconnected.

SC.F.1.4.2 Body structures are uniquely designed and adapted for their function.

SC.F.2.4.1 There are mechanisms of asexual and sexual reproduction and different genetic advantages and disadvantages of asexual and sexual reproduction.

SC.G.1.4.1 There is great diversity and interdependence among living things.

SC.H.3.4.5 The value of a technology may differ for different people and at different times.

CHAPTER

23 Introduction to Plants

Florida Sunshine Standards/FCAT

SC.D.2.4.1 The systems on Earth and the quality of life are interconnected.
SC.F.1.4.2 Body structures are uniquely designed and adapted for their function.
SC.F.2.4.1 There are mechanisms of asexual and sexual reproduction and different genetic advantages and disadvantages of asexual and sexual reproduction.
SC.G.1.4.1 There is great diversity and interdependence among living things.
SC.H.3.4.5 The value of a technology may differ for different people and at different times.

Quick Review

Answer the following without referring to earlier sections of your book.

1. **Describe** the process of mitosis. *(Chapter 6, Section 3)*
2. **Describe** the process of meiosis. *(Chapter 7, Section 1)*
3. **Identify** life cycles that have a gametophyte and life cycles that have a sporophyte. *(Chapter 7, Section 2)*
4. **Describe** the role of mycorrhizae. *(Chapter 12, Section 3 and Chapter 22, Section 3.)*
5. **List** the characteristics of the kingdom Plantae. *(Chapter 19, Section 3)*

Did you have difficulty? *For help, review the sections indicated.*

Reading Activity

Before you begin to read this chapter, survey each section and identify any subtitles, headings, and captions that signal the topic of discussion. As you read, locate other words in the body of the text that signal the sequential pattern.

Looking Ahead

Section 1
Adaptations of Plants
Establishment of Plants on Land
Vascular Tissue, Seeds, and Flowers
Plant Life Cycles

Section 2
Kinds of Plants
Nonvascular Plants
Seedless Vascular Plants
Gymnosperms
Angiosperms

Section 3
Plants in Our Lives
Plants as Food
Cereals
Nonfood Uses of Plants

Plants provide us with the food and oxygen that make life possible. They enrich our lives with beauty and sweet scents. Plants also provide buildings, paper, furniture, clothing, and medicines.

internet connect

www.scilinks.org
National Science Teachers Association *sci*LINKS Internet resources are located throughout this chapter.

SCILINKS. Maintained by the National Science Teachers Association

501

Opening Activity

Show students pictures or live specimens of several plants from different phyla. Ask students to point out the similarities and differences between the groups of plants. On the board or overhead projector, develop a chart organizing the characteristics that students point out.

Quick Review Answers

1. Mitosis is the process in which chromosomes are equally distributed between two new cells during cell division. During mitosis, chromosomes line up along the equator of the dividing cell and the chromatids are pulled apart to opposite ends of a cell.
2. In meiosis, a single cell undergoes two divisions to produce four gamete cells, each with half the number of chromosomes. During the first division of meiosis, homologous chromosome pairs line up side-by-side at the equator. These pairs are separated in the first division. The second division is like mitosis.
3. Plants, algae, and some protists have life cycles with a diploid phase called the sporophyte and a haploid phase called the gametophyte.
4. Mycorrhizae are symbiotic associations in which a fungus transfers minerals to a plant's roots, which supply sugars to the fungus.
5. Plants are photosynthetic eukaryotes. Plant cells have cell walls containing cellulose. Individual plants cannot move from place to place. Plants are the primary producers in terrestrial ecosystems.

Reading Activity Answers

Answers will vary but should include general characteristics shared by all plants, the unique features of different groups of plants, and the many ways that humans use plants.

Section 1

Focus

Overview

Before beginning this section review with your students the objectives listed in the Student Edition. This section introduces the adaptations that plants have made to life on land. Students will learn how groups of plants become increasingly well suited for dryer conditions. They will learn the common characteristics of all plants, as well as the characteristics that differentiate vascular from nonvascular plants.

Bellringer

Ask students to write three things that are important for the survival of an organism on land. (Students should recognize the need for obtaining water and nutrients, the need for maintaining homeostasis in a dry, sunny environment, and the need for reproduction out of water.)

Section 1 Adaptations of Plants

Florida Sunshine Standards/FCAT

SC.F.1.4.2 Body structures are uniquely designed and adapted for their function.

SC.F.2.4.1 There are mechanisms of asexual and sexual reproduction and different genetic advantages and disadvantages of asexual and sexual reproduction.

SC.G.1.4.1 There is great diversity and interdependence among living things.

Objectives

- **Summarize** how plants are adapted to living on land. SC.F.1.4.2
- **Distinguish** nonvascular plants from vascular plants. SC.G.1.4.1
- **Relate** the success of plants on land to seeds and flowers. SC.F.2.4.1
- **Describe** the basic structure of a vascular plant sporophyte. SC.F.1.4.2

Key Terms

cuticle
stoma
guard cell
vascular system
nonvascular plant
vascular plant
seed
embryo
seed plant
flower
phloem
xylem
shoot
root
meristem

Establishment of Plants on Land

Plants are the dominant group of organisms on land, based on weight. The kingdom Plantae is a very diverse group. Individuals range from less than 2 mm across to more than 100 m tall. Most plants are photosynthetic; they produce organic materials from inorganic materials by photosynthesis. A few plant species, like the one shown in **Figure 1,** live as parasites. Many parasitic plants cannot photosynthesize.

Plants probably evolved from multicellular aquatic green algae that could not survive on land. Multicellularity enabled plants to develop features that helped them live more successfully on land. Before plants could thrive on land, they had to be able to do three things: absorb nutrients from their surroundings, prevent their bodies from drying out, and reproduce without water to transmit sperm.

Absorbing Nutrients

Aquatic algae and plants take nutrients from the water around them. On land, most plants take nutrients from the soil with their roots. Although the first plants had no roots, fossils show that fungi lived on or within the underground parts of many early plants. So botanists think that fungi may have helped early land plants to get nutrients from Earth's rocky surface. Symbiotic relationships between fungi and the roots of plants are called *mycorrhizae*. Today, about 80 percent of all plant species form mycorrhizae.

Preventing Water Loss

The first plants lived at the edges of bodies of water, where drying out was not a problem. A watertight covering, which reduces water loss, made it possible for plants to live in drier habitats. This covering, called the **cuticle,** is a waxy layer that covers the nonwoody aboveground parts of most plants. But like the wax on a shiny car, the cuticle does not let oxygen or carbon dioxide pass through it. Pores called **stomata** *(STOH muh tuh)* (singular, stoma) permit plants to exchange oxygen and carbon dioxide. Stomata, which extend through the cuticle and the outer layer of cells, are found on at least some parts of most

Figure 1 **Rafflesia.** The flowers of *Rafflesia keithii,* of Malaysia, measure almost 1 m across and weigh up to 11 kg (24 lb). The plant, which has no stems or leaves, is parasitic on the roots of grape vines.

502

Chapter Resource File

- Directed Reading BASIC
- Active Reading GENERAL
- Datasheet for Quick Lab GENERAL

One-Stop Planner CD-ROM

- Reading Organizers BASIC
- Reading Strategies BASIC
- Occupational Application Worksheet Botanist GENERAL

Transparencies

TR Bellringer
TR G31 Stomata and Guard Cells

Section Correlations

SC.F.1.4.2
SC.F.2.4.1
SC.G.1.4.1

Figure 2 Stomata and guard cells

The surface of a leaf has numerous stomata, each of which is surrounded by a pair of guard cells.

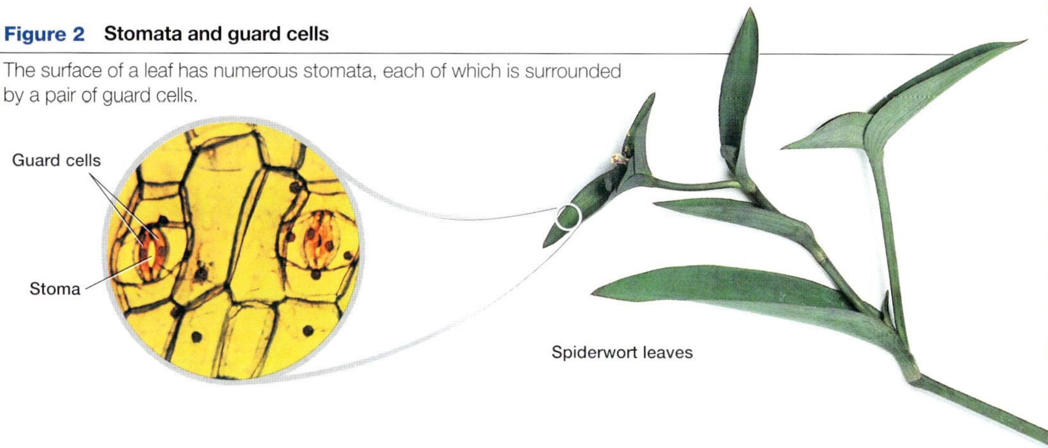

plants. A pair of specialized cells called **guard cells** border each stoma, as seen in **Figure 2.** Stomata open and close as the guard cells change shape.

Reproducing on Land

Aquatic algae reproduce sexually when sperm swim through the water and fertilize eggs. The sperm of most plants, however, must be able to move without water. In most plants, sperm are enclosed in a structure that keeps them from drying out. The structures that contain sperm make up pollen. Pollen permits the sperm of most plants to be carried by wind or animals rather than by water.

internet connect
www.scilinks.org
Topic: Plants of Florida
Keyword: HX6F16
SciLinks Maintained by the National Science Teachers Association

QUICK LAB: *Observing the Behavior of Stomata* SC.F.1.4.2

You can use nail polish to see that a leaf has many stomata.

Materials

clear nail polish, plant kept in light, plant kept in darkness, two 4–5 cm strips of clear tape, 2 microscope slides, compound microscope

Procedure

1. Paint a thin layer of clear nail polish on a 1 × 1 cm area of a leaf on a plant kept in light. Do the same using a plant kept in darkness. Let the nail polish dry for 5 minutes.
2. Place a 4–5 cm strip of clear tape over the nail polish on each leaf. Press the tape firmly to the nail polish.
3. Carefully pull the tape off each leaf. Stick each piece of tape to a microscope slide. Label it appropriately.
4. View each slide with a microscope, first under low power and then under high power.
5. Draw and label what you see on each slide.

Analysis

1. **Describe** any differences in the stomata of the two plants.
2. **Critical Thinking** **Drawing Conclusions** Which plant will lose water more quickly? Explain.

503

Teach

Using the Figure — GENERAL

Draw students' attention to **Figure 2.** Point out the guard cells and the stoma of the leaf. Be sure students understand that the guard cells form the opening that is called a stoma. Tell students that in many species, the majority of stomata are on the lower leaf surfaces. Ask students how this benefits plants. (The lower surface is usually not exposed to direct sunlight. This helps reduce water loss.)

SKILL BUILDER — BASIC

Vocabulary Have students use a dictionary to find the origin of the term *stoma.* (*Stoma* comes from the Greek word meaning "mouth.") English Language Learners

QUICK LAB: *Observing the Behavior of Stomata*

Skills Acquired
Analyzing results, making comparisons, drawing conclusions

Teacher's Notes
Instruct students to put the nail polish on the lower leaf surface because there will likely be more stomata there. Tell students that the nail polish is used to make an impression of the leaf surface and may also remove some of the cells on the leaf's surface.

Answers to Analysis

1. The students should note that the plant kept in the light has more open stomata.
2. The plant kept in the light will lose water more quickly because it has more open stomata. While this seems inefficient for preventing water loss, the stomata must be open during the day to allow photosynthesis to occur.

did you know?

Pine Stomata Pine trees often have sunken stomata. This adaptation helps them conserve water, an important ability for plants growing in cold climates. Many pine trees grow where the soil freezes and the winds are dry.

Teach

Teaching Tip — GENERAL

Tall Tendencies Ask students to brainstorm and suggest characteristics that would enable plants to grow tall. (a vascular system to carry water and nutrients; rigid cell walls to provide support; thickened stems) Ask students to speculate on the advantages of increased plant height. (Tall plants are able to obtain more sunlight due to reduced competition and shade. Tall plants are also able to send their pollen and fruit/seeds over greater distances than shorter plants.) **LS Intrapersonal** Co-op Learning

Activity — GENERAL

Seed Dispersal Bring a variety of fruits and cones containing seeds to class. Include some large colorful fruits (apples, oranges, grapes); some small fruits with wings (elm, maple) or umbrella-like appendages (dandelion); some cones with scales (pine, spruce, fir); and a coconut. Ask students to examine the fruits and cones and identify which of the fruits are dispersed by the wind. (winged fruits or small fruits with appendages, conifer seeds) Ask students which seeds are likely dispersed by animals. (showy, fleshy fruits) Ask students how the coconut is dispersed. (By water, since the coconut can float along coasts and from island to island.) **LS Visual**

Transparencies

TR G13 Vascular Tissue
TR G11 Structure of a Vascular Plant
TR G43 Structure and Function of Seeds

internet connect
www.scilinks.org
Topic: Vascular Plants
Keyword: HX4182
SciLinks. Maintained by the National Science Teachers Association

Vascular Tissue, Seeds, and Flowers

As plants adapted to land, they developed many features that helped in their success. There was no basic difference in structure between the aboveground and underground parts of the earliest plants. Later plants, however, had roots, stems, and leaves. One of the most important changes in plants was the development of *conducting tissues* that move water and other materials through the plant body.

Advantages of Conducting Tissue

The first plants were small. Materials were transported within their bodies by osmosis and diffusion. Today, most plants have strands of specialized cells that transport materials. These specialized cells are connected end to end like the sections in a pipeline, as shown in **Figure 3.** Some strands carry water and mineral nutrients from the roots to the leaves. Other strands carry organic nutrients from the leaves to wherever they are needed.

Specialized cells that transport water and other materials within a plant are found in vascular tissues. The existence of vascular tissue allowed for larger and more-complex plants. The larger, more-complex plants have a **vascular system,** a system of well-developed vascular tissues that distribute materials more efficiently. Three groups of plants alive today lack a vascular system. These relatively small plants that have no vascular system are called **nonvascular plants.** Plants that have a vascular system are called **vascular plants.**

Advantages of Seeds

After vascular tissue, the seed was the next important adaptation to appear in plants. A **seed** is a structure that contains the embryo of a plant. An **embryo** is an early stage in the development of plants and animals. Most plants living today are **seed plants**—vascular plants that produce seeds. The first seed plants appeared about 380 million years ago. Seeds offer a plant's offspring several survival advantages, which are summarized in **Figure 4.**

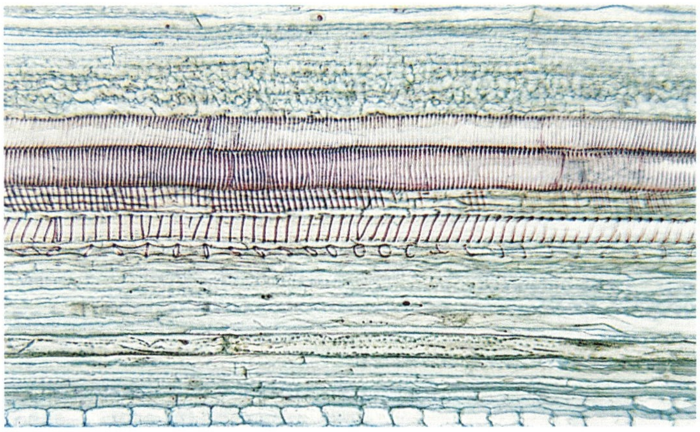

Figure 3 Vascular tissue. Thick-walled, tubular cells like these carry water from the tips of roots to the tips of leaves. Stacked end to end, these cells form tiny pipes called vessels.

504

Attention Grabber

Students may be surprised to learn that the tallest trees on Earth are in California. They are the coast redwoods *(Sequoia sempevirens)*, which can grow more than 110 m (360 ft). The tallest flowering plant is the Australian eucalyptus *(Eucalyptus regnans)*, which has been measured at 84 m (275 ft). The next tallest tree is the giant sequoia *(Sequiadendron giganteum)* of California, which can reach 76 m (250 ft).

EARTH SCIENCE CONNECTION

The seedless vascular plants were the most common plants on Earth about 300 million years ago. Many of these were very large trees, including a number of different tree ferns. When these plants died, their bodies became buried and only partly decomposed. The remains of their tissues formed underground coal deposits. People have dug up many of these carbon-rich deposits and burned the coal as a fossil fuel. The period of time in Earth's history when these plants were most abundant is called the Carboniferous period.

Figure 4 **Structure and function of seeds**

The structure of a seed helps it to perform its functions.

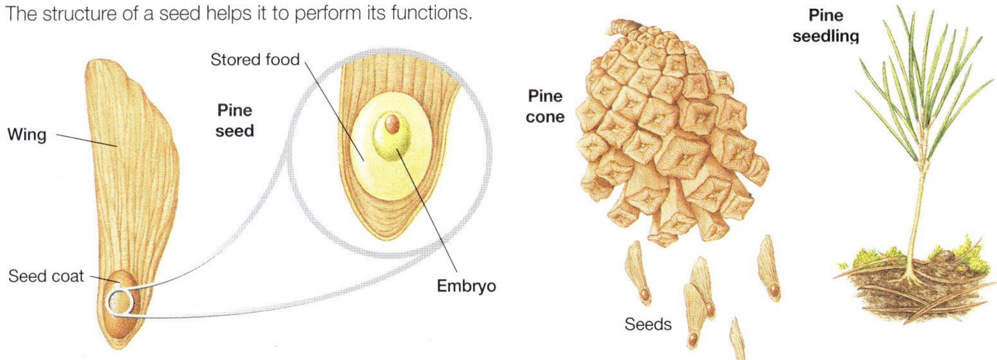

The seed coat of a pine seed covers and protects the embryo.

The stored food supply will nourish the embryo as it starts to grow.

A wing helps pine seeds disperse.

Pine seeds may not germinate for several years.

1. **Protection.** Seeds are surrounded by a protective cover called the seed coat. The seed coat protects the embryo from drying out and from mechanical injury and disease.
2. **Nourishment.** Most kinds of seeds have a supply of nutrients stored in them. These nutrients are a ready source of nourishment for a plant embryo as it starts to grow.
3. **Plant dispersal.** Seeds disperse (spread) the offspring of seed plants. Many seeds have structures that help wind, water, or animals carry them away from their parent plant. Dispersal prevents competition for water, nutrients, light, and living space between parents and offspring.
4. **Delayed growth.** The embryo in a seed is in a state of suspended animation. Most seeds will not sprout until conditions are favorable, such as when moisture is present and the weather is warm. Thus, seeds make it possible for plant embryos to survive through unfavorable periods such as droughts or cold winters.

internet connect
www.scilinks.org
Topic: Structure and Function of Seeds
Keyword: HX4169
SCiLINKS. Maintained by the National Science Teachers Association

Advantages of Flowers

The last important adaptation to appear as plants evolved was the **flower**, a reproductive structure that produces pollen and seeds. Flowers make plant reproduction more efficient. The pollen of the first seed plants was carried by wind. Large amounts of pollen are needed to ensure cross-pollination by wind—an inefficient system. Most plants living today are flowering plants—seed plants that produce flowers. The first flowering plants appeared more than 130 million years ago. Many flowers attract animals, such as insects, bats, and birds. As **Figure 5** shows, tiny pollen grains stick to animals, which carry pollen directly from one flower to another. Flowering plants that are pollinated by animals produce less pollen, and cross-pollination can occur between individuals that live far apart.

Figure 5 **Pollination.** This honeybee is covered with pollen grains containing the sperm of the plant it has just visited. The bee will transfer some of the pollen to the next flower it visits.

Teaching Tip — GENERAL

Durability of Seeds Tell students that seeds are adapted for survival in a wide variety of climates. Emphasize that seeds can survive in a dormant state if conditions are unfavorable for growth. Have students speculate about why so many seedlings appear after forest fires. (Fire clears the accumulated leaf litter on the forest floor. Many seeds need contact with bare mineral soil to germinate. Fire also opens some cones to allow seeds to fall out and germinate. Fire may also eliminate shade, allowing light to reach the forest floor and stimulate germination.)

Group Activity — ADVANCED

Observing Pollen Grain Have students work in groups of two or three. Have students make wet-mount slides of pollen from pine trees or observe prepared slides of pine pollen. Then have them draw and label the pollen grains and estimate their size based on the field of view. If possible, have students compare the pollen from several different species. Then have students examine the pollen from angiosperms, and compare the pollen structure of different groups. If pollen is not in season, use prepared slides or pollen from nursery grown flowers. Have groups prepare posters of the different pollen structures.
LS Visual

did you know?

Attractive Orchids Orchids have highly elaborate ways of attracting pollinators. One orchid species has a flower that resembles a female wasp. Male wasps pick up some pollen while trying to mate with the modified petals. In another species, the flower holds a small pool of water. When a bee tries to collect nectar, it falls into the water and can escape only through a narrow passageway that forces the bee to brush against the pollen-bearing anthers.

Teach, *continued*

Teaching Tip — BASIC

Life Cycle Review Ask students the following questions: How do scientists represent the chromosome number of a gametophyte? (n, or haploid) How do scientists denote the chromosome number of the sporophyte? (2n, or diploid) What is the name of the first sporophyte cell? (zygote) What process produces a zygote? (fertilization) What is the name of the first gametophyte cell? (spore) What process produces spores? (meiosis)

Using the Figure — GENERAL

Draw students' attention to **Figure 7.** Have students compare the relative sizes of sporophytes and gametophytes in the moss and the flowering plant. (Students should notice that the moss sporophyte is smaller than the gametophyte, while the flowering-plant gametophyte is considerably smaller than the flowering-plant sporophyte.)
LS Visual

Plant Life Cycles

In many algae, the zygote is the only diploid ($2n$) cell. It undergoes meiosis right after fertilization. So the bodies of these algae consist of haploid cells. In the ancestors of plants, however, meiosis was delayed. The zygote divided by mitosis and grew into a multicelled sporophyte that was diploid and produced haploid (n) spores by meiosis. The spores grew into multicelled gametophytes that were haploid and produced gametes by mitosis. As a result, plants have life cycles in which haploid plants that make gametes (gametophytes) alternate with diploid plants that make spores (sporophytes). A life cycle in which a gametophyte alternates with a sporophyte is called alternation of generations. The basic plant life cycle is shown in **Figure 6.**

Figure 6 Alternation of generations. In the life cycle of a plant, a diploid sporophyte generation alternates with a haploid gametophyte generation.

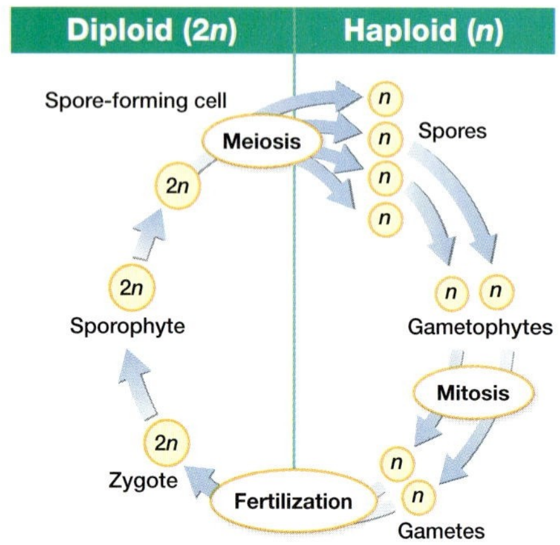

Unlike the green algae with alternation of generations, plants have gametophytes and sporophytes that look very different. In addition, the relative sizes of gametophytes and sporophytes changed as plants evolved, as **Figure 7** shows. In nonvascular plants, such as mosses, the gametophyte generation is dominant (most noticeable). In vascular plants, such as the flowering plants, the sporophyte generation is dominant. Like the presence of a vascular system, the relative sizes of gametophytes and sporophytes is a fundamental difference between the nonvascular plants and the vascular plants.

Figure 7 Nonvascular plants versus vascular plants

Gametophytes and sporophytes vary greatly in size.

REAL WORLD CONNECTION

Pollen allergies are commonly known as "hay fever." Such allergies are actually reactions to proteins found on the outside of the pollen grains. Different pollens cause allergies in different people because each type of pollen has a different protein in the coats of its pollen grains.

The Vascular-Plant Sporophyte

As the size of plant sporophytes increased, so did the complexity of their structure. An increase in size enables cell specialization and, therefore, the development of complex tissues and specialized structures. The following features characterize the sporophytes of most vascular plants.

Vascular System Larger bodies require an efficient vascular system for transporting materials internally. The sporophytes of vascular plants have a vascular system with two types of vascular tissue. Each type of vascular tissue contains strands of long, tubelike cells that are lined up end to end like sections of pipe. These strands of cells transport water and nutrients within a plant's body. Relatively soft-walled cells transport organic nutrients in a kind of tissue called **phloem** *(FLOH uhm)*. Hard-walled cells transport water and mineral nutrients in a kind of tissue called **xylem** *(ZIE luhm)*. The walls of the water-conducting cells in xylem are thickened, which helps support the plant body. This makes it possible for vascular plants to grow to great heights.

Distinctive Body Form Nearly all plants have a body that consists of a vertical shaft from which specialized structures branch, as shown in **Figure 8.** The part of a plant's body that grows mostly upward is called the **shoot.** In most plants, the part of the body that grows downward is called the **root.** Zones of actively dividing plant cells, called **meristems** *(MEHR uh stehmz)*, produce plant growth. The vertical body form results as new cells are made at the tips of the plant body. As vascular plants became better adapted to life on land, most developed the familiar plant structures—roots, stems, and leaves—which are complex structures made of several different types of specialized tissues.

Figure 8 Vascular plant sporophyte. The sporophytes of the vast majority of vascular plants have an aboveground shoot with stems and leaves and an underground root. Growth occurs in regions called *meristems*.

Section 1 Review

1. **Summarize** how plants are adapted to living successfully on land. SC.F.1.4.2
2. **Describe** two basic differences between nonvascular plants and vascular plants. SC.G.1.4.1
3. **Critical Thinking Relating Concepts** How have seeds and flowers made plants more successful on land? SC.F.2.4.1
4. **Critical Thinking Evaluating Conclusions** Why do you think vascular plants are more successful as land plants than are nonvascular plants? SC.F.1.4.2
5. FCAT Prep Water loss from most plants is reduced by what waxy layer?
   A cuticle
   B phloem
   C stomata
   D xylem
   SC.F.1.4.2

507

Close

Reteaching — BASIC

Have students work in pairs to review the key terms for this section. Encourage each pair of students to take turns asking and answering questions about the definitions of the terms. LS Verbal

Co-op Learning | English Language Learners

Quiz — GENERAL

1. True or false: nonvascular plants contain a well-developed system of tissues that transports water and nutrients. (False. Vascular plants contain transporting tissues.)
2. What is the primary role of flowers? (to promote pollination and thus increasing efficiency of reproduction)

Alternative Assessment — GENERAL

Students should work in pairs to design and create a single poster that illustrates the advantages of vascular seedless plants over nonvascular plants; the advantages of seeds over spores; and the advantages of flowering plants over non-flowering seed plants. LS Visual Co-op Learning

Answers to Section Review

1. formation of a waxy cuticle, vascular tissues, drought-resistant reproductive structures, and seed development
2. Vascular plants have a vascular system that transports water, minerals, and sugars throughout the plant; nonvascular plants do not. Nonvascular plant life cycles are dominated by the gametophyte; vascular plants have a dominant sporophyte.
3. Seeds and flowers have enable, land plants to distribute sperm, reproduce, and disperse offspring without water. Flowers have enabled land plants to produce offspring from fewer numbers of gametes because animal pollinators often aid in fertilization.
4. Vascular plants are more successful on land because adaptations have allowed them to reduce their dependence on water. These include cuticle protected leaves, vascular tissue, roots, tall stems, and drought resistant spores or seeds.
5. **A.** Correct. The waxy cuticle limits transpiration.**B.** Incorrect. The phloem carries sugars and organic compounds in vascular plants. **C.** Incorrect. Stomata are small holes in the leaves of a plant that open to allow gas exchange and close to limit water loss. **D.** Incorrect. The xylem carries water and minerals throughout vascular plants.

Section Review Correlations

SC.F.1.4.2: Items 1, 4, 5
SC.F.2.4.1: Item 3
SC.G.1.4.1: Item 2

Section 2

Focus

Overview

Before beginning this section review with your students the objectives listed in the Student Edition. This section introduces the defining characteristics of the different groups of plants. Students will also identify several examples of each group. These groups include nonvascular plants, vascular seedless plants, gymnosperms, and angiosperms.

Bellringer

On the board or overhead, write the following headings: *Nonvascular, Seedless Vascular, Gymnosperm, Monocot Angiosperm,* and *Dicot Angiosperm.* Ask students to write the headings in their notebook, and place at least one example under each heading.

Motivate

Demonstration — BASIC

Bring in living specimens or parts of specimens of a variety of plants. The collection should include as many of the following as possible: moss, liverwort, fern, horsetail, conifer, cycad, ginkgo, a flowering monocot (corn, iris, grass), and a flowering dicot (rose, daisy, tomato, etc.). Ask students to describe and compare the characteristics of each, noting both similarities and differences. (Differences may include size, presence/absence of spores, flowers, fruits, leaf shape, vein patterns, and stiffness.)

Section 2 Kinds of Plants

Florida Sunshine Standards/FCAT

SC.G.1.4.1 There is great diversity and interdependence among living things.

Objectives

- **Describe** the key features of the four major groups of plants. SC.G.1.4.1
- **Classify** plants into one of the 12 phyla of living plants. SC.G.1.4.1

Key Terms

rhizoid
rhizome
frond
cone
gymnosperm
angiosperm
fruit
endosperm
monocot
dicot

Nonvascular Plants

The brilliant green carpet of mosses you see in **Figure 9** is made up of thousands of individual plants. Living carpets of mosses are often found near streams, coastlines, and other moist places. But these tough little plants also live in some surprising places, such as cracks in city sidewalks and rocky mountaintops—any place where a little moisture can collect. The moisture makes it possible for these nonvascular plants to survive.

Nonvascular plants do not have a vascular system for transporting water and other nutrients within their bodies. This means that all nonvascular plants lack true roots, stems, and leaves, although most have structures that resemble them. True roots, stems, and leaves are complex structures that contain vascular tissues.

Key Features of Nonvascular Plants

In addition to the lack of true roots, stems, and leaves, nonvascular plants share several other features. These features are key adaptations that have enabled them to survive on land.

Figure 9 A carpet of mosses. Mosses grow in tightly packed mats that may contain dozens of plants per square inch.

Small Size All nonvascular plants are small and relatively simple. Water and other nutrients are transported within their bodies mostly by osmosis and diffusion, which move materials short distances. This greatly limits the size of a nonvascular plant's body.

Larger Gametophyte The gametophytes of nonvascular plants are larger and more noticeable than the sporophytes. Hairlike projections called **rhizoids** anchor the gametophytes to the surfaces on which they grow. The smaller, usually nongreen sporophytes grow on the gametophytes and depend on them for nutrients.

Require Water for Sexual Reproduction Nonvascular plants must be covered by a film of water in order for fertilization to occur. Eggs and sperm form in separate structures, which are often on separate plants. The gametophytes grow in mats of tightly packed individuals. When these mats are covered by a film of water, the sperm can easily swim to neighboring individuals and fertilize their eggs.

508

Chapter Resource File

- Directed Reading BASIC
- Active Reading GENERAL
- Data Sheet for Data Lab GENERAL

Transparencies

TR Bellringer

One-Stop Planner CD-ROM

- Reading Organizers BASIC
- Reading Strategies BASIC
- Portfolio Project Plant Focus GENERAL

Section Correlations

SC.G.1.4.1

Kinds of Nonvascular Plants

The nonvascular plants include the mosses and the two simplest groups of plants—liverworts and hornworts. Examples of these plants are shown in **Figure 10.**

Mosses The mosses (phylum Bryophyta) are the most familiar nonvascular plants. The "leafy" green plants that you recognize as mosses are gametophytes. Moss sporophytes, which are not green, grow from the tip of a gametophyte. Each sporophyte consists of a bare stalk topped by a spore capsule. Most mosses have a cuticle, stomata, and some simple conducting cells. The walls of the water-conducting cells in mosses are not thickened, as they are in a vascular plant. Mosses never get very large because their water-conducting cells carry water only short distances.

Liverworts Like the mosses, liverworts (phylum Hepatophyta) grow in mats of many individuals. Liverworts have no conducting cells, no cuticle, and no stomata. Their gametophytes are green. In some species, such as the common liverwort shown in Figure 10, the gametophytes of liverworts are flattened and have lobes. Structures that resemble stems and leaves make up the gametophytes of most liverworts, like those of the mosses. The sporophytes of liverworts are very small and consist of a short stalk topped by a spore capsule.

Hornworts The hornworts (phylum Anthocerophyta) are a small group of nonvascular plants that, like the liverworts, completely lack conducting cells. The sporophyte of a hornwort has both stomata and a cuticle. The gametophyte of a hornwort is green and flattened. Green hornlike sporophytes grow upward from the gametophytes.

internet connect
www.scilinks.org
Topic: Nonvascular Plants
Keyword: HX4130
SCILINKS. Maintained by the National Science Teachers Association

WORD Origins

The word *liverwort* combines the familiar word *liver* and the Old English word *wort,* meaning "herb." The word *liverwort* dates back to the Middle Ages, when it was thought that plants resembling certain body parts could cure diseases of those body parts.

Figure 10 Nonvascular plants

There are three phyla of nonvascular plants.

Polytrichum, a moss
(Phylum Bryophyta)

Marchantia, a liverwort
(Phylum Hepatophyta)

Anthoceros, a hornwort
(Phylum Anthocerophyta)

509

Teach

Teaching Tip — GENERAL

Mosses Have students observe different types of mosses using a stereomicroscope. They should find both gametophyte and sporophyte stages. Point out that the structures resembling leaves on the moss gametophyte are not considered to be true leaves. Tell students that a leaf is a complex structure that consists of several specialized tissues. Ask them whether they think the sporophyte and gametophyte are photosynthetic. (Only the gametophyte is photosynthetic.) Tell students that the sporophyte is dependent upon the gametophyte for survival. **LS Visual**

SKILL BUILDER — ADVANCED

Math Skills Have students determine the mass of a small sample of dry sphagnum (peat) moss, which is available at garden centers. Have students record the mass of the dry moss. Next, have them place this sample in a large beaker of water for 10–15 minutes. Tell students that 1kg of dry moss can soak up approximately 25 kg of water. Have students predict the mass of their moss after soaking in water. Have students remove the wet moss from the beaker and determine its mass. Tell students to compare their findings with their predictions. (The mass of the wet moss should be about 25 times greater than the mass of the dry moss.)

MEDICINE CONNECTION

Tell students that moss has been used as an antiseptic for centuries. Even in World War I, moss was used to treat soldier's wounds in emergency situations. The antiseptic qualities result from moss' natural absorbency and acidity.

MISCONCEPTION ALERT

Mistaken Identity A variety of plants and plantlike organisms are mistakenly called mosses. Examples include Irish moss, Spanish moss, and reindeer moss. Show photos of these examples and reinforce the concept that common names can be misleading. For example, Irish moss is a red alga, Spanish moss is a flowering vascular plant, and reindeer moss is a lichen.

Teach, *continued*

Teaching Tip — GENERAL

Local Habitats If your school grounds contain a number of different habitats, take the students on a tour of the campus. Otherwise, schedule a trip to a nearby park. Ask students to identify the habitats in which nonvascular plants would most likely be found. (moist habitats) Ask where vascular plants would most likely be found. (any habitat) Ask students to try to identify one nonvascular plant and one vascular plant. Have them explain what characteristics they used to determine vascularity. (nonvascular: moist habitat, small size, no flowers or seeds; vascular: any habitat, large or small, leaves with veins)

Real Life

Answers

Answers may vary. Lycopodium powder is used by the pharmaceutical industry as a coating to prevent the pills from sticking together and to disguise the taste of the medicine.

Using the Figure — GENERAL

Draw attention to **Figure 11.** Ask students how Cooksonia resembles plants that are alive today. (The upright growth habit of these plants and the location of the reproductive structures resemble modern plants.) Ask students how they differ from the plants they are most familiar with. (These primitive plants have no leaves, roots, flowers, seeds, or fruits.)

internet connect

www.scilinks.org
Topic: Seedless Vascular Plants
Keyword: HX4161

SCILINKS. Maintained by the National Science Teachers Association

Seedless Vascular Plants

Vascular plants that do not produce seeds are called seedless vascular plants. The earliest known seedless vascular plant, *Cooksonia,* is illustrated in **Figure 11.** The sporophytes of these ancient plants had branched, leafless stems that were only a few centimeters long. Spore-forming sporangia were located at the tips of the stems. *Rhynia,* another early seedless vascular plant, also had horizontal underground stems, or **rhizomes.**

Real Life

The spores of a common club moss, *Lycopodium,* form a powder that has several uses.

Herbalists use the spores to make a powder for treating skin disorders. The spores are also used to make photographic flash powder.

Finding Information
Find out how **Lycopodium** *powder is used by the pharmaceutical industry.*

Key Features of Seedless Vascular Plants

Seedless vascular plants are much larger and more complex than the nonvascular plants. Other key features enabled them to spread and adapt to drier habitats on land.

Vascular System Seedless vascular plants have a vascular system with both xylem and phloem. The water-conducting cells in the xylem are reinforced with *lignin,* a major part of wood. Because of their vascular system, seedless vascular plants grow much larger than nonvascular plants and also develop true roots, stems, and leaves.

Larger Sporophyte The sporophytes of seedless vascular plants are larger than the gametophytes. Their larger size makes it easier for the wind to carry away spores, which makes dispersal more efficient. The much smaller gametophytes of most seedless vascular plants develop on or below the surface of soil. As in the nonvascular plants, water is needed for fertilization. When there is enough water on or in the soil, the sperm swim to eggs and fertilize them.

Drought-Resistant Spores The spores of the seedless vascular plants have thickened walls that are resistant to drying. Such spores make it possible for a plant to live in drier habitats.

Figure 11 The earliest known vascular plant

Cooksonia, the oldest known vascular plant, lived about 410 million years ago.

Model of *Cooksonia*

Artist's rendition of *Cooksonia*

510

Cultural Awareness

Edible Seedless Vascular Plants The fiddleheads of ferns found in the forests of the northeastern United States during the spring were considered a delicacy by early Native American groups. Edible fiddleheads are actually the tightly coiled, emerging fronds of the fern, *Matteuccia struthiopteris.* Today, they are still used in American cuisine. You may want to have interested students research recipes for fiddleheads or prepare some dishes for the class.

Kinds of Seedless Vascular Plants

The seedless vascular plants include ferns and three other groups of plants known as fern allies—whisk ferns, club mosses, and horsetails. Like the ferns, the fern allies usually grow in moist places.

Ferns The ferns (phylum Pterophyta) are the most common and most familiar seedless vascular plants. Ferns grow throughout the world, but they are most abundant in the tropics. The plants you recognize as ferns are sporophytes. Most fern sporophytes have a rhizome that is anchored by roots and leaves called **fronds.** The coiled young leaves of a fern, shown in **Figure 12,** are called fiddleheads. Spores are produced in sporangia that grow in clumps on the lower side of fronds. The gametophytes of ferns are flattened, heart-shaped green plants that are usually less than 1 cm (0.4 in.) across.

Figure 12 **A fern.** This sword fern sporophyte has many fronds and fiddleheads. The inset shows a gametophyte at twice its actual size.

Club Mosses Unlike true mosses, the club mosses (phylum Lycophyta), have roots, stems, and leaves. Their leafy green stems branch from an underground rhizome. Spores develop in sporangia that form on specialized leaves. In some species, such as the one seen in **Figure 13,** clusters of nongreen spore-bearing leaves form a structure called a **cone.**

Horsetails The horsetails (phylum Sphenophyta) also have roots, stems, and leaves. The vertical stems of horsetails, which grow from a rhizome, are hollow and have joints. Whorls of scalelike leaves grow at the joints. Spores form in cones located at the tips of stems.

Whisk Ferns The whisk ferns (phylum Psilotophyta) probably most closely resemble the earliest vascular plants. Whisk ferns have highly branched stems and no leaves or roots. They produce spores in sporangia that form at the tips of short branches.

Figure 13 Fern allies

In addition to ferns, there are three other living phyla of seedless vascular plants that are known as fern allies.

Lycopodium, a club moss (Phylum Lycophyta)

Equisetum, a horsetail (Phylum Sphenophyta)

Psilotum, a whisk fern (Phylum Psilotophyta)

Cultural Awareness

Introduced Species Can Replace Native Species Cultures that spread to new regions of the Earth often bring familiar plant and animal species with them, sometimes with disastrous results. Settlers introduced pigs on the island of Hawaii as a source of food, but these animals wiped out native vegetation. Additionally, the settlers brought their own plants, which eventually crowded out native species. The people of Hawaii are now trying to reclaim their native plants, including tree ferns. Have students research introduced plants in their own area that are threatening native plant communities.

Demonstration — Basic

Bring examples of several kinds of ferns to class, and have students examine them. Also bring some pictures of tree ferns, which are native to many tropical regions. Have students note the shapes of the leaves and any fertile fronds or sori present on the lower surface of the fronds. Point out that there are more than 12,000 species of ferns on Earth today. Among these species we find a tremendous diversity of leaf form (from simple, rounded leaves to lacy fronds with hundreds of leaflets); plant shape (moss-like to tree form); and plant size (2.5 cm to 20 m tall).
LS Visual

Activity — Basic

Use a fern leaf from the demonstration above. Show students the clusters of sporangia (sori) on the lower surface of a frond. Let them remove a sorus and crush it on white paper using the eraser end of a pencil. Next, have them sprinkle some spores on a moist peat pellet and place the peat pellet in a sealed jar. The first structures they will observe will be the little green heart-shaped gametophytes.
LS Kinesthetic

Reading Skill Builder — Basic

Reading Organizer Encourage students having difficulty to use the information on this page to make a chart that outlines the differences between ferns, club mosses, horsetails, and whisk ferns. English language learner students may want to fill in the chart in their native language. English Language Learners

Teach, *continued*

Activity — GENERAL

Conifer ID Obtain cuttings of several conifers. Then have students construct their own keys, which they can use to identify the species. Students may look at cones, needle size, needle shape, needle organization, and needle color. Then supply students with keys or field guides and ask them to correctly identify their samples. LS Kinesthetic

DATA LAB *Analyzing the Effect of Climate on Plants*

Skills Acquired
Analyzing data, interpreting graphs, drawing conclusions

Teacher's Notes
Point out that the bar graph represents precipitation and that the line graph represents temperature.

Answers to Analysis

1. Answers may vary. Anchorage temperatures average at or below freezing for six months of the year. Precipitation is highest in summer and fall, and relatively low in other seasons.
2. Answers should predict that because the taiga shares a similar plant community, the taiga outside of Anchorage should have a fairly similar climate.
3. Yes. Explanations should rely on the map for reasoning. The taiga species are not found further north or south. Soil type or human activities are also valid explanations.

Figure 14 Juniper pollen. This juniper, a type of gymnosperm, releases clouds of pollen in late fall or early winter.

Gymnosperms

Gymnosperms *(JIHM noh spurmz)* are seed plants whose seeds do not develop within a sealed container (a fruit). The word *gymnosperm* comes from the Greek words *gymnos,* meaning "naked," and *sperma,* meaning "seed."

Key Features of Gymnosperms

Gymnosperms are among the most successful groups of plants. The following key features have made them successful on land.

Seeds All gymnosperms produce seeds. Seeds protect plant embryos, provide them with nutrients, and permit them to survive long periods of unfavorable conditions. In some plants, seeds also disperse new plants far from their parents.

Greatly Reduced Gametophytes All seed plants produce very tiny gametophytes of two types—male and female. The gametophytes form within the tissues of the sporophytes. Grains of pollen are male gametophytes. Female gametophytes form within structures that become seeds. In all but one species of gymnosperm, male and female gametophytes develop in male and female cones, respectively.

Wind Pollination The sperm of gymnosperms do not swim through water to reach and fertilize eggs. Instead, the sperm are carried to the structures that contain eggs by pollen, which can drift on the wind, as seen in **Figure14.** Wind pollination makes sexual reproduction possible even when conditions are very dry.

DATA LAB *Analyzing the Effect of Climate on Plants* SC.D.2.4.1

Background

The map at right shows the taiga of North America. The taiga is a vast forest of conifers, a type of gymnosperm. The graph shows average annual temperature and precipitation data for Anchorage, Alaska, which is located at the western edge of the taiga. Use the map and graph to answer the following questions.

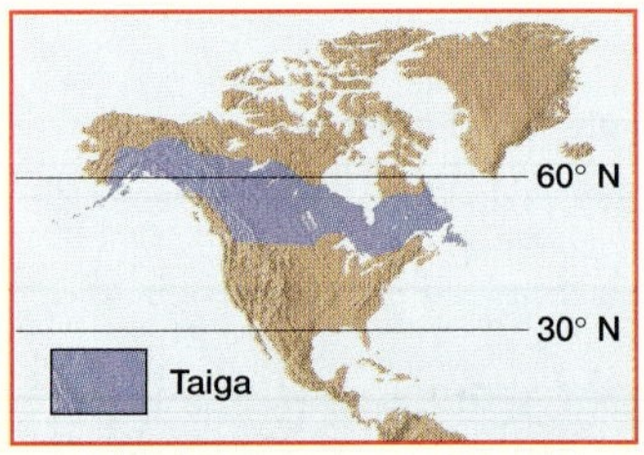

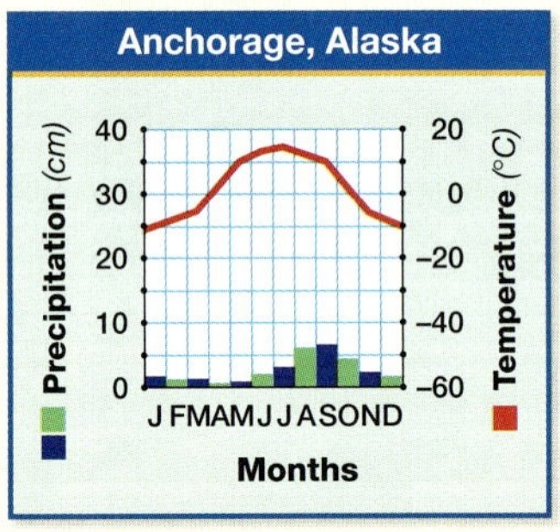

Analysis

1. **Describe** the climate of Anchorage, Alaska.
2. **Critical Thinking Predicting Patterns** What type of climate would you expect to find in other parts of the taiga?
3. **Critical Thinking Drawing Conclusions** Does climate appear to be an important factor in where the conifers of the taiga grow? Explain.

Kinds of Gymnosperms

Four groups of living seed plants are referred to as gymnosperms—conifers, cycads, ginkgo, and gnetophytes. Examples of each of these four groups are shown in **Figure 15** and **Figure 16.**

Conifers The conifers (phylum Coniferophyta) are the most familiar, and most successful, gymnosperms. Conifers have leaves that are either needle-like or reduced to tiny scales, as Figure 15 shows. These leaves are an adaptation for limiting water loss. Some of the tallest living plants, the redwoods of coastal California and Oregon, are conifers. The oldest trees in the world are thought to be bristlecone pines, another species of conifer that grows in the Rocky Mountains and Great Basin. Some bristlecone pines are about 5,000 years old. Vast forests of conifers grow in cool, dry regions of the world.

Figure 15 Juniper leaves and cones. The tiny scalelike leaves of junipers—a type of conifer—are an adaptation that limits water loss. The blue, berrylike structures are the female cones of this juniper.

Cycads The cycads (phylum Cycadophyta) have short stems and palmlike leaves. Cones that produce pollen and those that produce seeds develop on different plants. Cycads are widespread throughout the tropics.

Ginkgo The only living species of ginkgo (phylum Ginkgophyta), or maidenhair tree, has fan-shaped leaves that resemble the leaves of the maidenhair fern. The male and female gametophytes of ginkgo develop on separate trees. Ginkgo seeds do not develop within a cone.

Gnetophytes The gnetophytes (phylum Gnetophyta) are a diverse group of trees, shrubs, and vines that produce pollen and seeds in cones that resemble flowers. One type of gnetophyte, *Ephedra,* is common in the western United States.

Figure 16 Other gymnosperms

In addition to conifers, there are three other living phyla of gymnosperms.

Encephalartos, a cycad (Phylum Cycadophyta)

Leaves and seeds of *Ginkgo* (Phylum Ginkgophyta)

Ephedra (Mormon tea), a gnetophyte (Phylum Gnetophyta)

513

READING SKILL BUILDER — BASIC

Discussion Lead the class in brainstorming the characteristics they have observed in pine trees. (They stay green all year; have thin, needle-like leaves; have pinecones.) Stress that while this section provides additional information about gymnosperms, students already know a great deal about them. Encourage English language learners to translate the list into their native language. English Language Learners

Career

Botanist A botanist studies plants and their environments. Most botanists specialize in a particular area. For example, plant physiologists study the structural parts of plants and how they function; plant breeders study the ways in which plants reproduce and how their traits are inherited; horticulturists work with crops such as flowers, fruits, and vegetables; agronomists work with crops such as grains and dry beans; and foresters study all aspects of tree growth. Have students research a field of botany that interests them and prepare a written report on career opportunities in that field.

did you know?

Living Fossils The ginkgo, or maidenhair tree, is considered a "living fossil" because it was once thought to be extinct. The ginkgo is the only broadleaf gymnosperm still surviving, and it closely resembles fossil specimens. In 1994, another gymnosperm thought to be extinct for 100 million years was discovered in a rain forest near Sydney, Australia. A small population of Wollemi pines, a prehistoric type of conifer belonging to a family that was thought to have become extinct, was found by a park ranger. This ancient family of trees once spread into the Northern Hemisphere.

Teach, *continued*

Demonstration — GENERAL

Bring several examples of cones (pine, spruce, fir, yew) and flowers (iris, daisy, gladiolus, etc) to class, and have students examine and compare each of these types of reproductive structures. (Both produce seeds; flowers have petals, stamens, and pistils, and cones do not; cones have scales.) **LS Visual**

Teaching Tip — GENERAL

Charting Information Have students make a Graphic Organizer similar to the one at the bottom of this page to compare and contrast the characteristics of the major plant groups. Ask students to illustrate their graphic organizer with a drawing of a plant in each of the major groups. **LS Kinetic**

Transparencies

TR G39 Life Cycle of an Angiosperms
TR G8 Familiar Families of Angiosperms

internet connect
www.scilinks.org
Topic: Angiosperms
Keyword: HX4007
SCILINKS. Maintained by the National Science Teachers Association

Angiosperms

Most seed plants are flowering plants, or **angiosperms** *(AN jee oh spurmz)*. Angiosperms produce seeds that develop enclosed within a specialized structure called a **fruit,** as seen in **Figure 17.** The word *angiosperm* comes from the Greek words *angeion,* meaning "case," and *sperma,* meaning "seed."

Key Features of Angiosperms

Angiosperms are the most recent group of plants to evolve. The following key features made them the most successful group of plants.

Flowers The male and female gametophytes of angiosperms develop within flowers, which promote pollination and fertilization more efficiently than do cones. Some flowers, such as roses, are brightly colored or have strong scents. This attracts insects and other animals that carry pollen and increases the likelihood of cross-pollination. Other flowers, such as garden peas, are adapted for self-pollination, which often occurs before the flowers open. The flowers of many angiosperms, such as oaks and grasses, have small greenish flowers that are adapted for wind pollination. The female reproductive part of a flower also provides a pathway that enables sperm to reach and fertilize eggs without swimming through water.

Fruits Although fruits provide some protection for developing seeds, their primary function is to promote seed dispersal. The angiosperms produce many different types of fruits, which develop from parts of flowers. Many fruits are eaten by animals. The seeds are dispersed as they pass undigested from the animals' bodies. Other fruits have structures that help them float on wind or water. Some fruits even forcefully eject their seeds, flinging them away from the parent plant.

Endosperm The seeds of angiosperms have a supply of stored food called **endosperm** at some time during their development. In many angiosperms, the endosperm is absorbed by the embryo before the seeds mature.

Figure 17 Seeds in a fruit. These melons, which contain seeds, are the fruits of an angiosperm.

Graphic Organizer

Use this graphic organizer with **Teaching Tip** on this page.

Phylum	Vascular system	Roots	Seeds	Flowers	Dominant gametophyte	Dominant sporophyte
Liverworts					✔	
Mosses					✔	
Ferns	✔	✔				✔
Conifers	✔	✔	✔			✔
Flowering plants	✔	✔	✔	✔		✔

Kinds of Angiosperms

Botanists divide the angiosperms into two subgroups—monocots and dicots as **Figure 18** shows. The **monocots** are flowering plants that produce seeds with one seed leaf (cotyledon). Most monocots also produce flowers with parts that are in multiples of three and have long, narrow leaves with parallel veins. The **dicots** are flowering plants that produce seeds with two seed leaves. Most dicots also produce flowers with parts in multiples of two, four, or five and have leaves with branching veins. **Table 1** lists examples of some of the most familiar families of angiosperms.

Figure 18 A monocot and a dicot

Monocots and dicots differ in several ways.

Daylilies are monocots.

Roses are dicots.

Table 1 Familiar Families of Angiosperms

Subgroup	Family	Examples
 Monocots (class Monocotyledonae)	Iridaceae (iris)	Irises, gladiolus, crocus, blue-eyed grass
	Liliaceae (lily)	Daylilies, tulips, asparagus, aloe vera
	Poaceae (grass)	Wheat, corn, rice, lawn grasses
Dicots (class Dicotyledonae) 	Asteraceae (aster)	Daisies, sunflowers, lettuce, ragweed
	Brassicaceae (mustard)	Broccoli, cauliflower, turnips, cabbage
	Fabaceae (legume)	Beans, clovers, peas, peanuts, soybeans
	Rosaceae (rose)	Roses, apples, peaches, pears, plums
	Solanaceae (nightshade)	Potatoes, tomatoes, peppers, petunias

Section 2 Review

1. **Identify** three key features of each of the four major groups of plants. SC.G.1.4.1
2. **Classify** each of the following plants as one of the four major groups of plants: pine trees, carnations, sphagnum moss, and wood fern. SC.G.1.4.1
3. **Critical Thinking Evaluating Conclusions** Why are angiosperms said to be the most successful group of plants? SC.F.2.4.1
4. **Critical Thinking Recognizing Patterns** How are spores and pollen grains adapted for their functions, and how do their numbers impact their environment? SC.F.1.4.2
5. *FCAT Prep* What structure do ferns produce in order to reproduce?
   A cones
   B flowers
   C seeds
   D spores
   SC.F.1.4.2

515

Answers to Section Review

1. Nonvascular plants lack a vascular system, roots and true leaves; have a larger gametophyte generation; and produce spores. Seedless vascular plants have a vascular system, roots, and leaves; release spores; and have a larger sporophyte generation. Gymnosperms have a vascular system, roots, and leaves; produce spores; and have a large sporophyte. Angiosperms have a vascular system, roots, leaves, and flowers; produce fruits and seeds; and have a large sporophyte.
2. Pine trees are gymnosperms; carnations are angiosperms; sphagnum moss is a nonvascular plant; and wood fern is a seedless vascular plant.
3. Angiosperms are successful partly because they produce fruits that provide a mechanism for seed dispersal by wind, water, or animals.
4. Spores and pollen grains have coats that are resistant to harsh environmental conditions. They may be transported by the wind or by animals. The large amount of spores or pollen produced helps ensure reproduction and colonization of new environments.
5. **A.** Incorrect. Gymnosperms reproduce by producing cones. **B.** Incorrect. Angiosperms reproduce by producing flowers. **C.** Incorrect. Ferns do not produce seeds **D.** Correct.

Activity — ADVANCED

Classifying Plants as Monocots or Dicots Provide examples and pictures of a variety of plants, including both dicots and monocots. Let students determine whether the examples are dicots or monocots. Point out that some plants are not easily placed into one of the groups, such as pothos ivy. Botanists usually rely on flower structure in such cases. **LS Visual**

Close

Reteaching — BASIC

Have students write down the key characteristics of each major group of plants discussed in this section. For each major group, have students provide one example. Then choose a student to read a description aloud, and allow fellow students to interrupt as soon as they recognize the group being described. Continue until all groups have been described at least twice.

Quiz — GENERAL

1. Mosses, liverworts, and hornworts are examples of _______ plants. (nonvascular)
2. Gymnosperms reproduce by seeds that are produced in _______. (cones)
3. Plants that produce flowers with parts in groups of three are classified as _______. (monocots)

Alternative Assessment — ADVANCED

Have students create a guide to the plants in the area. Students should create a booklet with the following sections: *Nonvascular Plants, Seedless Vascular Plants, Gymnosperms, Monocots,* and *Dicots.* Students should identify and record at least two species for each section of their booklet.

Section Review Correlations

SC.F.1.4.2: Items 4, 5
SC.F.2.4.1: Item 3
SC.G.1.4.1: Items 1, 2

Section 3

Focus

Overview

Before beginning this section review with your students the objectives listed in the Student Edition. This section examines some of the many ways that humans use plants. Students will learn about food plants, including cereal grasses and other crops. This section will also discuss the many ways that trees can be used, including the chemicals and fibers obtained from trees and other plants.

Bellringer

Ask students to make a list of the different plant parts that can be eaten as food. (Students should list leaves, roots, stems, fruits, seeds, flowers, buds, bulbs, tubers, and shoots.) Ask students to name any other products that plants provide people. (Answers will vary, but may include wood, fibers for paper and textiles, medicines, chemicals, dyes, and oxygen.)

Motivate

Discussion — GENERAL

Ask students to consider all of the ways plants contribute to our survival and way of life. Allow students to brainstorm and develop a list, then open the question for discussion. Ask students how much of their food, clothing, books, furniture, pencils, paper, medicine, and other resources comes from plants. Ask them to consider life without plants. What would people use instead of plants? (Answers will vary.) LS Interpersonal

Section Correlations

SC.D.2.4.1
SC.G.1.4.1
SC.H.3.4.5

Section 3 Plants in Our Lives

Florida Sunshine Standards/FCAT

SC.D.2.4.1 The systems on Earth and the quality of life are interconnected.

SC.G.1.4.1 There is great diversity and interdependence among living things.

SC.H.3.4.5 The value of a technology may differ for different people and at different times.

Objectives

- **Identify** foods that come from plants and their dietary importance. SC.G.1.4.1
- **Describe** several ways that wood is used. SC.D.2.4.1
- **Explain** how plants are used to treat human ailments. SC.H.3.4.5
- **Identify** plants that are used to make paper and cloth. SC.H.3.4.5

Key Terms

vegetative part
cereal
grain

Plants as Food

Humans depend on plants in many ways. For one thing, plants store the extra nutrients they make or absorb in their bodies. Thus, plant parts contain organic nutrients (carbohydrates, fats, and proteins) and minerals (calcium, magnesium, and iron). All types of plant parts—roots, stems, leaves, flowers, fruits, and seeds—are eaten as food.

Fruits and Vegetables

The United States government identifies each of the foods that comes from a plant as an agricultural commodity. Each type of food is classified by a term—such as *fruit* or *vegetable*—that is registered in Washington, D.C. But, these terms have different meanings in botany. For example, to a botanist, a fruit is the part of a plant that contains seeds, and a **vegetative part** is any nonreproductive part of a plant. The foods that you think of as fruits—such as apples, bananas, and melons—are also fruits in the botanical sense. Vegetables, on the other hand, may be any botanical part of a plant, as you can see in **Figure 19.** Fruits and vegetables provide dietary fiber and are important sources of essential vitamins and minerals.

Figure 19 **Plant parts eaten as food**

The foods you eat come from different parts of plants.

516

REAL WORLD CONNECTION

Leafy vegetables are an important source of nutrition. Most greens can be eaten cooked or raw. Many varieties are rich in beta-carotene, vitamin C, folic acid, calcium, and iron. In general, darker leaves are more nutritious than lighter-colored leaves. Have students check references on nutrition (such as the USDA's *Composition of Foods*) for vitamin, mineral, and fiber content of a variety of leafy green vegetables, as well as other fruits and vegetables and commonly eaten snack foods. Allow students to compare their results with each other.

Chapter Resource File

- Directed Reading BASIC
- Active Reading GENERAL
- Datasheet for Quick Lab GENERAL

Transparencies

TR Bellringer

Root Crops

Potatoes are an important food staple in many parts of the world. Rich in calories and easy to grow, potatoes are an ideal crop for a small farm. Potatoes are classified as a root crop because they grow underground. But, potatoes are actually tubers, modified underground stems that store starch. Yams, an essential food crop in many tropical parts of the world, are roots. Sweet potatoes, carrots, radishes, turnips, beets, and cassava *(kuh SAH vuh)* are also important root crops. These vegetables are enlarged roots that store starch. Cassava, seen in **Figure 20,** is the staple food of more than 500 million people. Also known as manioc *(MAN ee awk),* cassava supplies more than one-third of the calories eaten in Africa.

Figure 20 Cassava. Cassava develops thick starch-filled roots up to 120 cm (4 ft) long. The roots are eaten like potatoes. Tapioca, shown above in the bowl, is made from cassava.

Legumes

Many members of the pea family, which are called *legumes,* produce protein-rich seeds in long pods. For example, about 45 percent of a soybean, the most important legume grown for food, is protein. Soybeans are often cooked and pressed into cakes called tofu *(TOH foo).* Peas, peanuts, and the many different types of beans are the seeds of legumes. Alfalfa, which is fed to livestock, is another important legume. Like many legumes, alfalfa has nitrogen-fixing bacteria, which add nitrogen compounds to the soil, in its roots. Therefore, alfalfa is also grown to enrich the soil.

QUICK LAB

Distinguishing Between Fruits and Vegetables SC.F.1.4.2

You can find out if a plant product is a fruit by cutting it open and examining its internal structure.

Materials

apple, banana, green bean, potato, squash, tomato, plastic knife

Procedure

1. Look at several familiar fruits and vegetables. Classify each one as either a fruit or a vegetable in the familiar sense.
2. **CAUTION: Sharp objects can cause injury. Handle knives carefully.** Use a plastic knife to cut open each fruit and vegetable.
3. Look at the fruits and vegetables again. Classify each by its botanical function—either a fruit or a vegetative part.

Analysis

1. **Compare** the familiar and botanical classifications you gave each fruit and vegetable.
2. **Critical Thinking Analyzing Data** Which fruits and vegetables did you classify differently?
3. **Critical Thinking Analyzing Results** Defend the classifications you made for item 2.
4. **Critical Thinking Drawing Conclusions** Based on your data, when is a vegetable a fruit?

Teach

READING SKILL BUILDER — BASIC

Anticipation Guide Before students begin this section, write the following statements on the board or overhead projector:

1. Wood is the most valuable resource that people get from plants.
2. Most of the food that people eat comes from fruits.
3. Plants produce many chemicals that are useful as medicines, but none that are poisonous.

Ask students to decide whether they agree or disagree with each statement. Ask volunteers to share the reasons for their opinions. Have students read the chapter to see if their opinions are confirmed or changed.

QUICK LAB

Distinguishing Between Fruits and Vegetables

Skills Acquired
Analyzing data, analyzing results, drawing conclusions

Teacher's Notes
Point out that some true fruits might not have obvious seeds. Commercially produced bananas are sterile triploids (3n). The fruit is large but the seeds are difficult to see and not viable.

Answers to Analysis

1. Apples and bananas are true fruits. Green beans, squash, and tomatoes are true fruits that most people would identify as vegetables. Potatoes are vegetative structures that are familiarly called vegetables.
2. Students will probably need to reclassify green beans, squashes, and tomatoes as fruits.
3. True fruits contain seeds. Many of the "vegetables" in this lab contain seeds which makes them true fruits.
4. A "vegetable" is a fruit if it is produced from a flower, contains parts of the flower (ovary), and contains seeds.

Cultural Awareness

Cassava The staple crop of Brazil and many other tropical and subtropical countries is a perennial shrub root called cassava, or manioc. It is asexually propagated from tuberous roots. These starchy roots contain compounds that form cyanide, which must be removed through processing before eating. The roots are usually ground into flour and used in cooking. Processing methods are complex, labor intensive, and pose a risk of cyanide poisoning to processors. In Brazil, toasted cassava meal, or "farofa," is served at most meals. Cassava is also used to make tapioca and beer in some cultures. There are bitter and sweet varieties of cassava, with varying levels of toxicity. Worldwide, cassava helps to feed more than 500 million people.

Teach, *continued*

Teaching Tip — GENERAL

Small Selection of Crops

Tell students that approximately 80,000 plant species produce edible parts, but only 50 of these are cultivated as major food sources. Seven of these species—rice, wheat, corn, barley, sorghum, potatoes, and cassava—make up 75 percent of the food consumed around the world.

SKILL BUILDER — GENERAL

Math Skills Tell students that of the 560 million metric tons of wheat produced around the world each year, 20 percent is used to feed livestock, and 80 percent to feed humans directly. Of the 520 million metric tons of rice produced, almost 100 percent feeds humans. Finally, of the 530 million metric tons of corn produced, 65 percent is used to feed livestock, the rest going directly to humans. Have students calculate the amount of each crop eaten by humans (in metric tons), then prepare bar graphs to display this information. Have students describe this information in writing, and rank each crop in terms of the amount eaten by humans. (Rice is the most important, at nearly 520 million metric tons; wheat is second, at 448; and corn is third, at 185.5 million metric tons.) LS **Logical**

Cereals

Most of the foods that people eat come directly or indirectly from the fruits of cereals. **Cereals** are grasses that are grown as food for humans and livestock. Cereal grasses produce large numbers of a type of edible, dry fruit called a **grain.** A grain contains a single seed with a large supply of endosperm. Each grain develops from a flower. The flowers of cereal grasses form in tightly packed clusters of many individual flowers. A grain is covered by a dry, papery husk called the *bran,* which includes the wall of the ovary and the seed coat. Cereal grains are rich in carbohydrates and also contain protein, vitamins, and dietary fiber. More than 70 percent of the world's cultivated farmland is used for growing cereal grains. In fact, more than half of the calories that humans consume come from just three cereal grasses: wheat, corn, and rice.

Figure 21 Wheat. Modern bread wheat is a hybrid of three wild species. The ripenend heads of bread wheat turn golden brown.

Wheat

For more than one-third of the world's population, wheat, seen in **Figure 21,** is the primary source of food. The endosperm of wheat grains, which is high in carbohydrates, is commonly ground into white flour and used to make breads and pasta. Vitamin-rich wheat germ consists of the embryos of wheat grains. Whole-wheat flour consists of the endosperm plus the germ and bran layers. Wheat grains are not always ground into flour. In the Middle East, wheat grains are often boiled or soaked, dried, and then pounded until they crack. The cracked grains, called *bulgur (BUL guhr),* are used in dishes such as tabbouleh *(tuh BOO lee)* and pilaf *(pih LAHF).* Most wheat is grown in temperate regions that have fertile soil and moderate rainfall. One of the world's best wheat-growing areas is the Great Plains region of the United States and Canada—a temperate grassland biome.

Figure 22 Corn. Each ear of corn developed from a flower spike that consisted of more than 500 flowers.

Corn

Corn, seen in **Figure 22,** is the most widely cultivated crop in the United States. American colonists of the 1600s and 1700s first learned how to grow corn from Native Americans. In the southeastern United States, corn was more widely grown than wheat, which does not grow as well in hot climates. Thus, foods that are made from corn—corn bread, corn pone, hominy, and grits—are a traditional component of the southeastern American diet. Corn is also one of the world's chief foods for farm animals. About 70 percent of the corn crop harvested in the United States is consumed by livestock. Other uses for corn include the production of corn syrup, margarine, corn oil, cornstarch, and fuel-grade ethanol. Most of the corn grown in the United States today comes from a region known as the Corn Belt, which includes Iowa, Nebraska, Minnesota, Illinois, and Indiana.

518

did you know?

Wheat Strains Help Mexico In 1944 new strains of wheat helped end food shortages in Mexico. Norman Borlaug, who helped launch the Green Revolution, was awarded the Nobel Prize in 1970 for developing new strains of wheat in Mexico. His research on wheat began in 1944. Within 20 years, Mexico had enough wheat not only to feed its own people but also to export to other countries. India and Pakistan experienced similar successes. Borlaug's high-yielding strains were produced from crosses with a dwarf variety.

One-Stop Planner CD-ROM

- **Reading Organizers** BASIC
- **Reading Strategies** BASIC
- **Portfolio Project** Plant Research Project GENERAL

Rice

For more than half of the people in the world, rice is the main part of every meal. Although it is low in protein, rice is an excellent source of energy-rich carbohydrates. While brown rice still has its vitamin-rich bran layers, white rice has been processed to remove the bran layers. This processing helps to prevent spoilage in stored rice. In societies where people eat mainly rice, vitamin-rich sauces such as soy sauce are often added to white rice to make meals more nutritious. The white rice you buy at a grocery store is enriched with added vitamins. Rice is often added to processed foods such as breakfast cereal, soup, baby food, and flour. In the United States, rice is grown in central California, in the Southeast, and along the Gulf Coast in fields such as the one shown in **Figure 23.**

Figure 23 Gulf coast rice field. Rice plants are grown in standing water.

FLORIDA BIOWatch

Searching for the First Flower

How did flowering plants evolve? The question fascinated Charles Darwin, who described the evolution of flowering plants as "an abominable mystery." More than one hundred years later, the mystery still intrigues paleobotanists, scientists who study prehistoric plants.

Little Evidence, Few Clues

One reason that scientists remained puzzled is because angiosperm fossils are rare. Fossils of any kind are formed by occasional, chance circumstances. Most fossils are formed when hard tissues, such as bones or wood, become buried and preserved in sediment. However, most flowering plants lack hard tissues and so are likely to decay quickly after death. Thus, few traces remain of Earth's earliest angiosperms.

A New Fossil

In 2002, David Dilcher, a paleobotanist at the University of Florida and the Florida Museum of Natural History, announced an exciting discovery. Dilcher and his colleagues at Jilin University in Changchun, China had found fossil evidence of an ancient, flowering, aquatic plant. Discovered in northeast China, the fossils are the earliest known evidence of flowering plants.

Flowers, Plain and Simple

Dilcher and his colleagues named the plant *Archaefructus sinensis,* which means "ancient fruit from China." The plant lived about 125 million years ago. Its simple flowers lacked petals but contained seeds surrounded by a fruit. Leaves near the flowers had swollen leaf bases, helping to lift the flowers to the water's surface to aid in pollination and seed dispersal. Previously, other fossils suggested that the first angiosperm was a woody plant. But *Archaefructus* was a small, herblike plant that lived in shallow lakebeds.

Angiosperms include economically important plants. Discovering the evolutionary relationships among the flowering plants may help scientists improve crop plants or develop valuable new plant varieties.

internet connect

www.scilinks.org

Topic: Science and Technology of Florida

Keyword: HX6F14

SCILINKS. Maintained by the National Science Teachers Association

519

Group Activity — Advanced

Local Plant Processing Have students work in groups of three or four. Ask them to prepare a report describing 10 plant-based products made within a 150km radius of the school. Some possible products include cereals and other crops, food processing, forestry, furniture, wood products, nurseries and greenhouses, textiles, plant-based cosmetics, plant-based medicines and supplements, and paper production or paper recycling. As part of the project, have students create a map with the school at its center and the various plant-based industries located on the map.

Co-op Learning

BIOWatch — General

Searching for the First Flower

Teaching Strategies

Tell students that many other fossils of early plants consisted solely of leaves and/or pollen. Ask students: How did the fossils of *Archaefructus sinensis* differ from other fossil finds? (The fossils of *Archaefructus* included flowers and fruits.)

Discussion

Ask students: What ancient events might have caused *Archaefructus sinensis* to become extinct? (Answers will vary. Possible answers include climate changes that caused lakebeds to dry up, or competition from other species of plants.

Cultural Awareness

Rice and Culture Rice plays an important part in the customs and economies of many cultures where it is a major food source. For example, Japanese place a shiminawa, or rice-straw rope, over an entrance to ward off evil. The largest rope in the world is a shiminawa that hangs over the entrance to the Izumo Shinto Shrine and weighs 6 metric tons. Rice is the symbol of the goddess of wealth and prosperity and is often worshiped as a goddess by the Hindu. The act of pouring rice over someone's head signifies prosperity and happiness. Students may find this amusing. Remind them of the familiar western custom of throwing rice at a wedding.

Teach, *continued*

Demonstration — GENERAL

Bring the following to class: some biodegradable cornstarch-based packing pellets, some polystyrene peanut pellets, 2 beakers of water, and iodine in a dropper bottle. To show the advantages of biodegradable products, drop a few of the biodegradable packing pellets into a beaker of water. Stir until dissolved. Add several drops of iodine to the beaker, and note any color change. Ask students to identify the composition of the pellets. (Purple indicates starch, from any corn in this case.) Repeat the procedure with the polystyrene peanuts. Point out that both products are reusable, but only the cornstarch-based pellets will easily biodegrade.

Activity — GENERAL

Paper Types Have students make wet-mount slides of various types of paper and then observe the slides under a microscope. Staining the paper with methylene blue will make the fibers more visible. Have students draw what they see and relate the fibers to the characteristics of the papers. (Students should notice that longer fibers are present in higher quality papers, while short fibers are present in soft papers such as tissue.) **LS Visual**

Teaching Tip — GENERAL

Products That Come From Trees
Have students make a Graphic Organizer similar to the one at the bottom of this page to illustrate the importance of trees. Ask them to use the vocabulary words applicable for this section, as well as any other words that they wish to include.

Figure 24 Latex. The milky sap of certain plants is called latex. This man is collecting latex from a rubber tree in Java, Indonesia.

Nonfood Uses of Plants

Plants are used by people for many purposes other than food. For example, rubber was first made from latex, the milky white sap of tropical trees of the genus *Hevea*. Latex is extracted from rubber trees by the method seen in **Figure 24.** Guayule *(gwah YOO lee)*, a member of the sunflower family that is native to the southwestern United States, is another source of natural rubber. (Most of today's rubber, however, is synthesized from petroleum, a nonrenewable resource.) The most important nonfood products obtained from plants are wood and fibers.

Wood

After food, wood is the single most valuable resource obtained from plants. Countless products, such as those shown in **Figure 25,** are made from wood. The wood from trees that have been cut down and sawed into boards is called lumber. Nearly 75 percent of the lumber cut in the United States is used for building construction. The rest is used to make products that contain wood, or it is ground and moistened to make wood pulp. Wood pulp is made into paper, rayon, and many other products. Finally, for more than a quarter of the world's people, wood is still the main source of fuel for heating and cooking.

Figure 25 Items made with wood

Furniture, buildings, boats, cabinets, and violins are made from wood.

520

Graphic Organizer

Use this graphic organizer with **Teaching Tip** on this page.

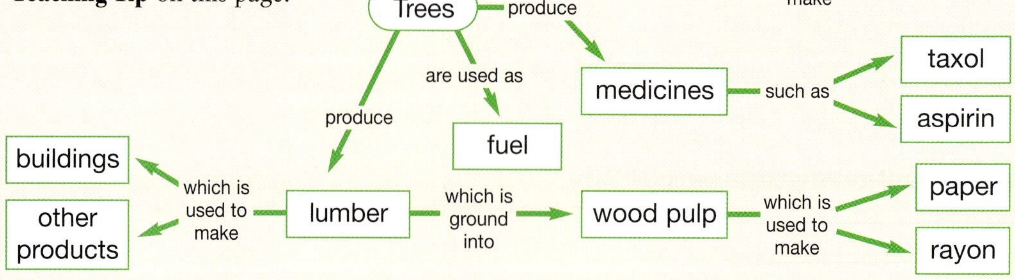

Figure 26 Sources of medicines

These two common garden plants are the sources of important medicines.

Rosy periwinkle, the original source of two cancer-fighting drugs

Foxglove, the source of a drug used to treat cardiac disorders

Medicines

People have always used substances obtained from plants to treat a variety of ailments. By studying the plants traditionally used to treat human ailments, researchers have developed many "modern" medicines. For example, solutions made by soaking the bark of willow trees, *Salix,* were a traditional cure for aches and pains. The pain-relieving chemical found in willows is called salicin *(SAL uh sihn).* Acetylsalicylic *(uh SEET l sal uh SIHL ihk)* acid, a derivative of salicin, was first sold in 1899 under the name "aspirin." Today, aspirin is the most widely used pain-relieving drug in the world.

Two familiar garden plants, seen in **Figure 26,** are important sources of life-saving medicines. The extremely poisonous leaves of the foxglove, *Digitalis purpurea,* yield digitalis *(dihj ih TAL ihs),* a drug that is used to stabilize irregular heartbeats and to treat cardiac disorders. The rosy periwinkle, *Catharanthus roseus,* is the source of two cancer-treatment drugs—vinblastine *(vihn BLAS teen)* and vincristine *(vihn KRIHS teen).* Vinblastine is often used to treat Hodgkin's disease, a type of cancer that affects the lymph nodes. Vincristine is used to treat childhood leukemia and other types of cancer. **Table 2** contains other examples of medicines that originally derived from plants.

Table 2 Some Drugs Originally Derived from Plants

Name	Source	Action
Caffeine	Tea leaves	Acts as a stimulant
Codeine	Poppy fruits	Relieves pain
Cortisone	Yam tubers	Relieves symptoms of allergies
Ephedrine	Ephedra stems	Acts as a decongestant
Taxol	Yew tree bark	Reduces the size of cancerous tumors

internet connect

www.scilinks.org

Topic: Medicines from Plants

Keyword: HX4119

SCILINKS. Maintained by the National Science Teachers Association

Teach, continued

READING SKILL BUILDER — BASIC

Reading Organizer Have students keep a list of plants and their uses as they read this section. When they have finished reading, have them make a table of the items in their list.

MEDICINE CONNECTION

The potato plant is a member of the nightshade family (Solanaceae). Many plants in this family produce alkaloids that are poisonous to people and animals. Some examples are tobacco, which produces the alkaloid nicotine; jimson weed, which produces the alkaloids atropine and scopolamine, and belladonna, which also produces atropine. Atropine and scopolamine are used to treat muscle spasms. The belladonna's name ("beautiful woman") refers to the old practice of placing drops of the plant extract into women's eyes, which resulted in dilated pupils. Potato plants are normally not poisonous, but if the tubers are exposed to light, they may turn green. These green portions contain alkaloids that are harmful in large quantities.

did you know?

Biotech Plants May Help Fight Cancer Genetic engineering offers hope of introducing the anticancer proteins found in foods, such as broccoli, into other fruits and vegetables. If genetic engineering is successful, people would have a wider variety of plant foods to choose from to obtain these proteins. Those who do not like to eat broccoli and related plants could get these proteins from other sources.

Trends in Textile Fabrication

Friendlier Rayon Rayon is a synthetic fabric made from chemically altered cellulose fibers, usually obtained from wood. Clothing, curtains, upholstery, and industrial products are made from rayon. New research is aimed at making rayon using a more environmentally friendly process.

Transparencies

TR G3 Medicines Originally Derived from Plants

Close

Reteaching — Basic

Have students work in groups of three or four. Have each group answer the following questions with as many examples as they can think of: What are the most important cereal grains? (wheat, rice, and corn) What are some other plants used as food sources? (lentils, soybeans, sugar cane, cassava, etc.) What are some of the important products of the timber industry? (lumber, paper, fuel, turpentine) What are some medicines derived from plants? (taxol, codeine, digitalis, ephedrine, etc.) When groups are finished, have students share their answers.

Quiz — General

1. List six plant parts that can be used as food. (Roots, stems, flowers, fruits, seeds, and leaves are all used as food.)
2. True or false: scientists have identified all of the possible medicines that can be made from plants. (False. Not all plants have been discovered, and there are chemicals produced by plants that have never been tested.)

Alternative Assessment — General

Assign students to cooperative groups to prepare a lunch menu that consists only of plant-derived foods, utensils, and furniture. Students should take care to design a fully nutritional meal. Provide a library cart of reserved books selected for the project.

Co-op Learning

Section Review Correlations

SC.D.2.4.1: Item 1

SC.G.1.4.1: Items 4, 6

SC.H.3.4.5: Items 2, 3, 5

Figure 27 Cotton

Cotton is the plant fiber that is most widely used to make cloth.

Cotton bolls that have split open, revealing cotton fibers

Indian woman spinning cotton into thread that will be woven into cloth

Fibers

If you were to look at this sheet of paper very closely through a magnifying glass, you would see that it is made of many interlocking fibers. These fibers are strands of cellulose, which is a component of the cell walls of plants. In plants, fibers help provide support for the plant body. The strength and flexibility of plant fibers make them ideal materials for making paper, cloth, and rope. Most of the fibers used to make paper come from wood. Paper-making fibers are also obtained from many other plants, including cotton, flax, rice, bamboo, and papyrus *(puh PIE ruhs)*.

For centuries, people have made clothing with cloth made of cotton, the world's most important plant fiber. As **Figure 27** shows, white fibers fill up the inside of a cotton boll *(bohl)*, the fruit of the cotton plant. Cotton thread is spun from the fine white fibers, which grow on cotton seeds. The stems of flax yield softer, more durable fibers that are used to make linen. More than 30 percent of the world's clothing is now made of synthetic fibers, but natural plant fibers are still prized for their durability and comfort. Sturdy fibers of hemp and sisal *(SIE suhl)* plants are used to make rope.

Study TIP

Reviewing Information Prepare flashcards for each of the Key Terms in this chapter. On each card, write the term on one side and its definition on the other side. Use the cards to review meanings of the Key Terms.

Section 3 Review

1. **Describe** several ways in which wood is used. SC.D.2.4.1
2. **List** five medicines that are derived from plants, and state how each is used. SC.H.3.4.5
3. **Name** two types of plants that provide fiber used in clothing. SC.H.3.4.5
4. **Critical Thinking Predicting Results** Name the three most important cereal grains and predict the results if one of them ceased to exist. SC.G.1.4.1
5. **Critical Thinking Evaluating Viewpoints** Justify the viewpoint that wood is the most important nonfood plant product. SC.H.3.4.5
6. **FCAT Prep** Which plant has nitrogen-fixing bacteria in its roots? SC.G.1.4.1
   A alfalfa
   B corn
   C potato
   D wheat

522

Answers to Section Review

1. Wood is used for fuel, lumber, furniture, tools and utensils, paper, rayon and turpentine.
2. Aspirin is used as a pain reliever and anti-inflammatory; digitalis is used to treat heart disorders; vinblastine is used to treat cancer; codeine is used as a pain reliever; cortisone is used to treat allergies; ephedrine is used as a decongestant; and taxol is used to treat cancer.
3. cotton, flax, hemp
4. Wheat, rice, and corn. Students should recognize that alternative food sources would need to be quickly utilized or massive starvation could result.
5. Answers will vary, but students should recognize that wood is used in many ways.
6. **A.** Correct. Alfalfa and members of the pea family are some of the plants that contain bacteria in root nodules that fix nitrogen into a form usable by plants. **B.** Incorrect. Corn, like all plants, needs nitrogen for healthy growth, but corn roots do not contain nitrogen-fixing bacteria. **C.** Incorrect. Potatoes have underground, starchy tubers, but do not contain nitrogen-fixing bacteria. **D.** Incorrect. Wheat needs nitrogen for healthy growth, but it does not contain nitrogen-fixing bacteria.

Study ZONE

Key Concepts

1 Adaptations of Plants

- To survive on land, plants must absorb mineral nutrients, prevent their bodies from drying out, and reproduce without water to transmit male gametes.
- Vascular plants have a system of well-developed tissues that transport water within a plant. The nonvascular plants lack a vascular system.
- Seeds protect and nourish a plant's embryo, disperse the offspring, and delay the growth of the embryo until conditions are favorable. Flowers make reproduction more efficient by promoting pollination.
- The sporophytes of vascular plants have a vascular system. Their bodies consist of an aboveground shoot and an underground root.

2 Kinds of Plants

- Nonvascular plants are small and lack vascular tissue. Mosses, liverworts, and hornworts are nonvascular plants.
- Seedless vascular plants produce spores with thickened walls that prevent them from drying out. Ferns, club mosses, horsetails, and whisk ferns are seedless vascular plants.
- Gymnosperms are seed plants that produce cones. Conifers, cycads, ginkgoes, and gnetophytes are gymnosperms.
- Angiosperms are seed plants that produce flowers and fruits. The angiosperms are classified as either monocots or dicots.

3 Plants In Our Lives

- All types of plant parts—roots, stems, leaves, flowers, fruits, and seeds—provide food for humans. Rice, corn, and wheat are cereal grasses and are our most important sources of food.
- Wood is a source of wood pulp used for making paper, lumber used for building materials, and fuel.
- Many important medicines are currently made from plants or were originally derived from plants.
- Plant fibers are used to make paper, cloth, and rope. The most important sources of plant fibers are wood and cotton.

Key Terms

Section 1

cuticle (502)
stoma (502)
guard cell (503)
vascular system (504)
nonvascular plant (504)
vascular plant (504)
seed (504)
embryo (504)
seed plant (504)
flower (505)
phloem (507)
xylem (507)
shoot (507)
root (507)
meristem (507)

Section 2

rhizoid (508)
rhizome (510)
frond (511)
cone (511)
gymnosperm (512)
angiosperm (514)
fruit (514)
endosperm (514)
monocot (515)
dicot (515)

Section 3

vegetative part (516)
cereal (518)
grain (518)

523

Alternative Assessment — GENERAL

Have students use the information in Chapter 23 to design a procedure for a question-and-answer game. Students may choose to model their game after a television game show or a trivia-type board game. Encourage students to write questions that vary in difficulty.

Answer to Concept Map

The following is one possible answer to Performance Zone item 15.

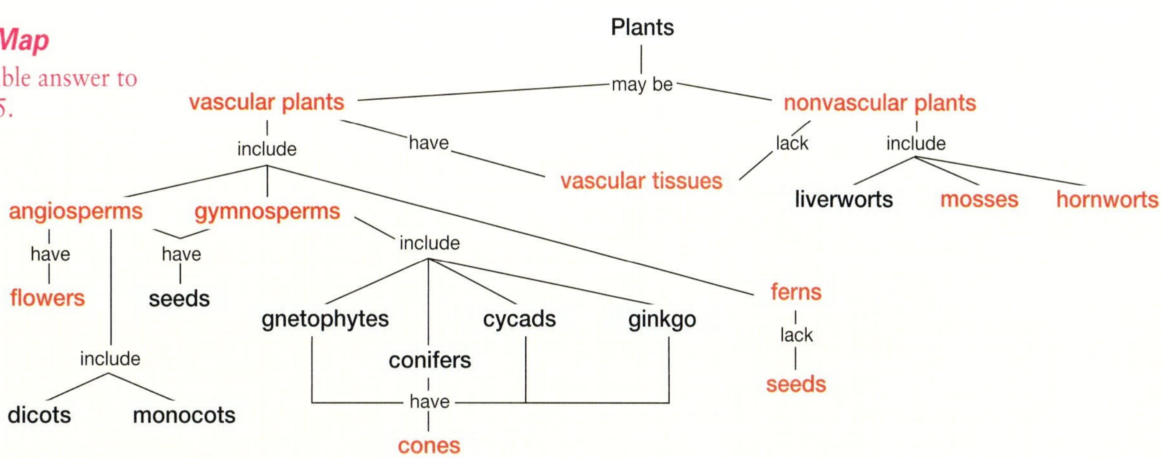

CHAPTER 23 ZONE

ANSWERS

Understanding Key Ideas

1. a
2. c
3. d
4. a
5. c
6. b
7. d
8. The fossils of *Archaefructus* included flowers and fruits.
9. Meiosis involves two divisions of the nucleus, which reduces the chromosome number in half in a cell and produces four haploid cells; in plants the cells that result directly from meiotic division are called spores.
10. The seeds would likely be dispersed by becoming attached to the fur of a mammal and being transported as the mammal moves around.
11. One possible answer to the concept map is found at the bottom of the Study Zone page.

Critical Thinking

12. The sporophytes are large and have many leaves that can intercept light from the sun. A vascular system enables the sporophytes to grow very large by distributing water and nutrients efficiently. The reproductive cells (eggs and sperm) are produced near the tops of the sporophyte, where they can readily be dispersed by wind or animals.
13. Scientists believe there are many undiscovered plants in the tropical rain forests, and that some of these plants have possible medicinal value.

Alternative Assessments

14. Answers will vary, but should include a microscope and microscope slides.

Chapter Review Correlations

SC.D.1.4.3: Item 8
SC.D.2.4.1: Item 13
SC.F.1.4.2: Items 1, 2, 9-12
SC.G.1.4.1: Items 3, 4, 5
SC.H.1.4.1: Item 14
SC.H.3.4.5: Items 6, 7

Assignment Guide

Section	Questions
1	1, 9, 12
2	2, 3, 4, 10, 11
3	5, 6, 8, 13, 14

Performance ZONE

Understanding Key Ideas

1. Seeds helped plants adapt to life on land by
   a. providing nourishment for embryos.
   b. protecting embryos from air pollution.
   c. sprouting during unfavorable weather.
   d. limiting the dispersal of plant offspring. SC.F.1.4.2
2. Which of the following is *not* a characteristic of vascular plants?
   a. xylem and phloem
   b. stems and leaves
   c. a dominant gametophyte
   d. a diploid sporophyte SC.F.1.4.2
3. Unlike angiosperms, gymnosperms
   a. are pollinated by wind.
   b. do not have seeds.
   c. have a diploid sporophyte generation.
   d. do not bear fruit. SC.G.1.4.1
4. Which of the following are *not* dicots?
   a. grass family
   b. rose family
   c. mustard family
   d. legume family SC.G.1.4.1
5. The most important sources of food are
   a. legumes.
   b. root crops.
   c. cereal grains.
   d. vegetables. SC.G.1.4.1
6. Drugs derived from the rosy periwinkle are used in the treatment of
   a. heart disease.
   b. leukemia.
   c. allergies.
   d. headaches. SC.H.3.4.5
7. Which of the following is *not* a source of fibers for both paper and cloth?
   a. flax c. wood
   b. cotton d. sisal SC.H.3.4.5
8. **BIOWatch** Most fossils of early plants consist of leaves or pollen. How did the fossils of *Archaefructus sinensis* differ from other fossil finds? SC.D.1.4.3
9. How does meiosis result in the production of haploid spores? (**Hint:** See Chapter 7, Section 1.) SC.F.1.4.2
10. Look at the cocklebur in the photograph below. It is a fruit that contains the seeds of a cocklebur plant. Suggest how this plant's seeds might be dispersed. SC.F.1.4.2

11. **Concept Mapping** Make a concept map that shows how plants are classified. Include the following terms in your map: *vascular plants, nonvascular plants, ferns, angiosperms, gymnosperms, mosses, cones, vascular tissue, seeds,* and *flowers*. SC.F.1.4.2

Critical Thinking

12. **Inferring Causes** Most plants have a vascular system and a sporophyte that is much larger that the gametophyte. How have these features contributed to the success of plants as they evolved on land? SC.F.1.4.2
13. **Forming Reasoned Opinions** Why is the loss of tropical rain forests and other types of forests of concern to medical science? SC.D.2.4.1

Alternative Assessment

14. **Selecting Technology** You are asked to compare the length and thickness of the fibers in several types of paper. Make a list of the laboratory equipment you would need for this task, and explain why you would need each piece. SC.H.1.4.1

FCAT Prep

For questions 1–6, write your answers on a separate sheet of paper.

1 What structure made it possible for plants to prevent water loss and to spread onto land?
A. cuticle
B. mycorrhizae
C. pollen
D. seed
SC.F.1.4.2

2 Which of the following is a diploid individual in plant life cycles?
F. gametophyte
G. epiphyte
H. sporophyte
I. zygospore
SC.F.1.4.2

3 Why might it be a reproductive advantage for a woody plant to be tall?
A. Seeds on tall plants are less likely to be eaten by insects.
B. Seeds on tall plants are more likely to be dispersed by grazing animals.
C. Seeds can disperse farther in the wind when located farther from the ground.
D. Seeds grown up high are more likely to get the sunlight they need to germinate.
SC.G.2.4.3

4 Which of the following is true of angiosperms?
F. They produce fruits.
G. They reproduce by releasing spores.
H. Their gametophytes develop inside cones.
I. Their sperm must swim through water to fertilize eggs.
SC.F.2.4.1

5 READ INQUIRE EXPLAIN Many people consider corn to be a cereal crop instead of a vegetable.

Differentiate between corn as an agricultural cereal crop and corn as a botanical fruit. SC.G.1.4.1

6 READ INQUIRE EXPLAIN People use many paper products every day. These include items such as notebook paper, paper towels, newspaper, and packaging. Some paper products, such as notebook paper and newspaper, are recyclable. When recycling paper products isn't possible, people can reduce their impact on the environment by using reusable instead of disposable products, buying products made with recycled paper, and buying products that have minimal paper packaging.

Part A What biological molecule gives paper its strength? SC.F.1.4.2

Part B Assess how recycling paper products reduces one's impact on the environment. SC.H.3.4.5

Test TIP

Carefully read the instructions, the question, and the answer options before choosing an answer.

FCAT Doctor

Question 3 Answer C is the correct choice. Seeds produced near the top of a tall plant can be dispersed farther in the wind than if they were located closer to the ground. Answer A is incorrect because insects eat the seeds of both tall and short plants. Answer B is incorrect because grazing animals would not disperse seeds that are located far above them. Answer D is incorrect because seeds of tall and short plants generally germinate in the ground, so there is no light advantage for the seeds of tall plants.

Question 5 Grains are the edible dried fruits of cereal crops. Fruits are structures that enclose the seeds of a plant.

Question 6

Part A Paper comes from plant material. Plant cell walls contain the carbohydrate cellulose, which gives plant cells (and paper) their structure.

Part B Making products from recycled materials usually saves energy, water, and other natural resources. How resources are used, or not used, has a direct impact on the environment.

Answers

1. A
2. H
3. C
4. F
5. Corn is a cereal crop because the part of the plant that is eaten consists of grains. All cereal grains are actually fruits, because they contain a seed that formed inside an ovary of a flower.

6A. Cellulose gives paper its strength.

6B. Paper products are made from plants such as trees. Recycling used paper instead of making new paper products from natural resources can reduce the number of trees removed from the environment.

Test Prep Correlations

SC.F.1.4.2: Items 1, 2, 6A
SC.F.2.4.1: Item 4
SC.G.1.4.1: Item 5
SC.G.2.4.3: Item 3
SC.H.3.4.5: Item 6B

Skills Practice Lab

Surveying Plant Diversity

Teacher's Notes

Time Required 45 minutes

Ratings

EASY → HARD

Teacher Preparation 1

Student Setup 1

Concept Level 2

Cleanup 1

Skills Acquired

- Classifying
- Collecting Data
- Identifying/Recognizing Patterns
- Inferring

Scientific Methods

In this lab, students will:

- Make Observations
- Ask Questions
- Test the Hypothesis
- Analyze the Results
- Draw Conclusions
- Communicate the Results

Materials and Equipment

Materials for this lab can be ordered from WARD'S. See *Master Materials List* at the front of this book for catalog numbers.

Safety Cautions

Review all safety symbols and caution statements with students. Remind students to be careful with glass containers and prepared slides. Review procedure for cleaning and disposing of broken glass.

Tips and Tricks

Review the four major plant groups before conducting lab. Set up at least two stations for each group of plants. Provide a stereomicroscope or hand lens for stations with mosses and ferns. Provide a compound microscope for stations with prepared slides.

Skills Practice Lab

Surveying Plant Diversity SC.G.1.4.1

SKILLS

- Observing
- Comparing

OBJECTIVES

- **Identify** similarities and differences among four phyla of living plants.
- **Relate** structural adaptations of plants to their success on land.

MATERIALS

- live or preserved specimens of mosses, ferns, conifers, and flowering plants
- stereomicroscope or hand lens
- compound microscope
- prepared slides of fern gametophytes

Before You Begin

Most plants are complex photosynthetic organisms that live on land. The ancestors of plants lived in water. As plants evolved on land, however, they developed adaptations that made it possible for them to be successful in dry conditions. All plant life cycles are characterized by **alternation of generations,** in which a haploid **gametophyte** stage alternates with a diploid **sporophyte** stage. Distinct differences in the relative sizes and structures of gametophytes and sporophytes are seen among the 12 phyla of living plants. In this lab, you will examine representatives of the four most familiar plant phyla.

1. Write a definition for each boldface term in the paragraph above and for the following terms: sporangium, spore, frond, cone, flower, fruit.
2. Make a data table similar to the one below.
3. Based on the objectives for this lab, write a question you would like to explore about the characteristics of plants.

Procedure

Part A: Conducting a Survey

1. Visit the station for each of the plants listed below, and examine the specimens there. Answer the questions, and record observations in your data table.
2. **Mosses** Examine a clump of moss with a stereomicroscope or hand lens. Make a sketch of what you see.
3. **Mosses** Examine a moss gametophyte with a sporophyte attached to it. Draw what you see, and label the parts you recognize. Label each part as haploid or diploid.
   a. Which stage of a moss has rootlike structures?
   b. Where are the spores of a moss produced?

DATA TABLE

Phylum name	Dominant generation	Major characteristics	Examples

526

Answers to Before You Begin

1. *alternation of generations*—a lifecycle that alternates between a haploid gametophyte stage and a diploid sporophyte stage; *gametophyte*—the gamete-producing haploid stage of a plant's life cycle; *sporophyte*— the spore-producing diploid stage of a plant's life cycle; *sporangium*—structure in which spores are produced; *spore*—haploid asexual reproductive structure that is resistant to environmental stress; *frond*—leaf of a fern; *cone*—reproductive structure of the sporophyte found in certain plants; *flower*—reproductive structure of angiosperm in which the gametophytes develop; *fruit*—mature plant ovary in angiosperms; adapted for dispersal
3. Answers will vary. For example: What characteristics are shared by all plants?

Answers to Procedure

Sample Procedure

1. Answers will vary depending on plant species used.
2. Drawings will vary depending on plant species used.
3. Drawings will vary depending on plant species used.
   a. gametophyte
   b. sporophyte

4. **Ferns** Examine the sporophyte of a fern, and look for evidence of reproductive structures on the underside of the fronds. Draw what you see. Label a leaf (frond), stem, root, and reproductive structure.
   a. How does water travel through a fern? List observations supporting your answer.
   b. What kind of reproductive cells are produced by fern fronds?
5. **Ferns** Examine a slide of a fern gametophyte with a compound microscope. Draw what you see, and label any structures you recognize.
6. **Conifers** Draw a part of a branch of one of the conifers at this station. Label a leaf, stem, and cone (if present).
   a. Is a branch of a pine tree part of a gametophyte or part of a sporophyte?
   b. In what part of a conifer would you look to find its reproductive structures?
7. **Conifers** Examine a prepared slide of pine pollen. Draw a few of the grains.
   a. What reproductive structure is found within a pollen grain?
   b. How does the structure of pine pollen aid in its dispersal by wind?
8. **Angiosperms** Draw one of the representative angiosperms at this station. Label a leaf, stem, root, and flower (if present). Indicate the sporophyte and location of gametophytes.
   a. Where do angiosperms produce sperm and eggs?
   b. How do the seeds of angiosperms differ from those of gymnosperms?
9. **Angiosperms** Examine several fruits. Draw and label the parts of one fruit.

PART B: Cleanup and Disposal

10. Dispose of broken glass in the designated waste containers. Do not put lab materials in the trash unless your teacher tells you to do so.
11. Wash your hands thoroughly before you leave the lab and after you finish all work.

Analyze and Conclude

1. **Analyzing Information** How are bryophytes different from the other major groups of plants?
2. **Recognizing Patterns** How do the gametophytes of gymnosperms and angiosperms differ from the gametophytes of bryophytes and ferns?
3. **Drawing Conclusions** What structures are present in both gymnosperms and angiosperms but absent in both bryophytes and ferns?
4. **Evaluating Hypotheses** Dispersal is the main function of fruits in angiosperms. Defend or refute this hypothesis. List observations you made during this lab to support your position.
5. **Inferring Conclusions** Based on their characteristics, which phylum of plants appears to be the most successful? Justify your conclusion.
6. **Further Inquiry** Write a new question about plant diversity that could be explored with another investigation.

On the Job

Drawing accurate diagrams of organisms is an important part of a plant taxonomist's research. Do research to discover the names of some famous plant taxonomists and scientific artists and to learn about some of the techniques used in their work. For more about careers, visit **go.hrw.com** and type in the keyword **HX4 Careers.**

4. Drawings will vary depending on plant species used.
   a. water travels through the cells of the vascular system. Veins are visible in the leaves.
   b. spores
5. Drawings will vary depending on plant species used. In general, fern gametophyte should be relatively heart-shaped.
6. Drawings will vary depending on plant species used.
   a. sporophyte
   b. cones
7. Drawings will vary depending on plant species used.
   a. sperm
   b. Appendages help the pollen grains to sail in the wind.
8. Drawings will vary depending on plant species used.
   a. in flowers
   b. Angiosperm seeds are produced in fruits; gymnosperm seeds are produced in cones.
9. Drawings will vary depending on plant species used.

Answers to Analyze and Conclude

1. The gametophyte is the most noticeable stage of the moss life cycle; the sporophyte is the most noticeable stage in other groups of plants.
2. Gymnosperm and angiosperm gametophytes develop within tissues of the sporophyte; bryophyte and fern gametophytes are distinct plants.
3. pollen and seeds
4. Answers will vary. Students should recognize that fruits often have structures that promote seed dispersal by wind, water, animals, or gravity. Many others have fleshy, edible parts that attract animal dispersal agents.
5. Answers will vary. Students should recognize that angiosperms are the most successful phylum of plants. The presence of flowers to attract pollinators and fruits to disperse seeds could be offered as reasons for the success of angiosperms. The great diversity and number of angiosperms could be offered as evidence of their success.
6. Answers will vary.

On the Job

Reports will vary depending on which taxonomist each student selects.

Section	Teaching Resources	Labs and Activities
Chapter Opener		SE **Quick Review,** p. 529 ■ SE **Reading Activity,** p. 529 ■ TE **Opening Activity,** p. 529 TE **Using the Figure,** p. 528
& 3 • 90 min pp. 530–533 **Section 1** Sexual Reproduction in Seedless Plants	OSP **Lesson Plan*** TR **Bellringer*** TR **Section Outline*** TR **G33 Life Cycle of a Moss*** TR **G34 Life Cycle of a Fern***	TE **Using the Figure,** p. 531 SE **Quick Lab** Observing a Fern Gametophyte, p. 532 ■ CRF **Datasheets for In-Text Labs** Observing a Fern Gametophyte*
& 5 • 90 min pp. 534–540 **Section 2** Sexual Reproduction in Seed Plants	OSP **Lesson Plan*** TR **Bellringer*** TR **Section Outline*** TR **G42 Seed Structure*** TR **G35 Life Cycle of a Conifer*** TR **G36 Floral Structure***	TE **Activity** Gymnosperm Flash Cards, p. 534 BASIC TE **Group Activity** Dissecting Seeds, p. 535 BASIC TE **Activity** Plant Propagator p. 536 ADVANCED SE **Quick Lab** Observing the Gametophytes of Pines, p. 536 ■ CRF **Datasheets for In-Text Labs** Observing the Gametophytes of Pines* TE **Using the Figure,** p. 537 TE **Group Activity** Diversity of Pollen Grains, p. 537 BASIC SE **Quick Lab** Observing the Arrangement of Parts of a Flower, p. 538 ■ CRF **Datasheets for In-Text Labs** Observing the Arrangement of Parts of a Flower* CRF **Exploration Lab** Fruits and Seeds* GENERAL CRF **Skills Practice Lab** Dissecting Flowers* GENERAL
& 7 • 90 min pp. 541–544 **Section 3** Asexual Reproduction	OSP **Lesson Plan*** TR **Bellringer*** TR **Section Outline*** TR **G45 Stems Modified for Vegetative Reproduction*** TR **G46 External Structures of *Kalanchoë**** TR **G47 Internal Structures of *Kalanchoë**** TR **G48 Methods of Vegetative Plant Propagation***	TE **Demonstration,** p. 541 BASIC SE **Exploration Lab** Observing the Effects of Nutrients on Vegetative Reproduction, pp. 548–549 ■ CRF **Datasheets for In-Text Labs** Observing the Effects of Nutrients on Vegetative Reproduction*

& 9 • 90 min

Chapter Review and Assessment Resources

- SE **Study Zone,** p. 545 ■
- SE **Performance Zone,** p. 546 ■
- SE **FCAT Test Prep,** p. 547 ■
- TR **Concept Mapping***
- CRF **Vocabulary Review Worksheet*** ■ GENERAL
- CRF **Test Prep Pretest*** ■ GENERAL
- CRF **Chapter Test*** ■ GENERAL
- CRF **Chapter Test*** ADVANCED
- OSP **Test Generator***
- OSP **Test Item Listing***
- OSP **Modified Chapter Test*** BASIC

Online and Technology Resources

Holt Online Learning

Visit go.hrw.com for access to Holt Online Learning or enter the keyword **HX6 Home** for a variety of free online resources.

One-Stop Planner® CD-ROM

This CD-ROM package includes
- Lab Materials QuickList Software
- Holt Calendar Planner
- Customizable Lesson Plans
- Printable Worksheets
- ExamView® Test Generator
- Interactive Teacher Edition
- Holt PuzzlePro® Resources
- Holt PowerPoint® Resources

Guided Reading Audio CDs

These CDs are designed to help auditory learners and reluctant readers. (Audio CDs are also available in Spanish.)

Compression guide:
To shorten your instruction because of time limitations, omit block 1.

KEY

TE	Teacher Edition	CRF	Chapter Resource File	*	Also on One-Stop Planner
SE	Student Edition	TR	Teaching Transparency	■	Also Available in Spanish
OSP	One-Stop Planner			◆	Requires Advance Prep

SKILLS DEVELOPMENT AND PRACTICE	SECTION REVIEW AND ASSESSMENT	CORRELATIONS
		Florida Sunshine Standards/FCAT
TE Reading Skill Builder Brainstorming, p. 531 GENERAL **TE Skill Builder** Math Skills, p. 532 GENERAL **CRF Directed Reading Worksheet*** BASIC **CRF Active Reading Worksheet*** GENERAL	**SE Section 1 Review** ■ **TE Reteaching,** p. 533 BASIC **TE Quiz,** p. 533 GENERAL **TE Alternative Assessment,** p. 533 GENERAL **CRF Quiz*** ■ BASIC	SC.F.2.4.1, SC.G.1.4.1, SC.H.1.4.1
TE Reading Skill Builder Reading Organizer, p. 535 GENERAL **TE Reading Skill Builder** Interactive Reading, p. 535 GENERAL **CRF Directed Reading Worksheet*** BASIC **CRF Active Reading Worksheet*** GENERAL	**SE Section 2 Review** ■ **TE Reteaching,** p. 540 BASIC **TE Quiz,** p. 540 GENERAL **TE Alternative Assessment,** p. 540 ADVANCED **CRF Quiz*** ■ BASIC	SC.F.1.4.2, SC.F.2.4.1, SC.H.1.4.1
CRF Directed Reading Worksheet* BASIC **CRF Active Reading Worksheet*** GENERAL **CRF Critical Thinking Worksheet*** ADVANCED **CRF Science Skills Worksheet*** GENERAL	**SE Section 3 Review** ■ **TE Reteaching,** p. 544 BASIC **TE Quiz,** p. 544 GENERAL **TE Alternative Assessment,** p. 544 GENERAL **CRF Quiz*** ■ BASIC **Modified Worksheet*** BASIC	SC.F.2.4.1, SC.G.1.4.3

SCILINKS. NSTA
www.scilinks.org
Maintained by the **National Science Teachers Association**

Topic: Crop Pollination
SciLinks Code: HX4053

Topic: Duckweed
SciLinks Code: HX4062

Overview

Tell students that the purpose of this chapter is to teach them about reproduction in plants. They will learn that sexual reproduction in plants involves alternating haploid and diploid phases in plant life cycles and about the sexual reproductive structures and life cycles of seed and seedless plants. Finally, students will learn about how plants reproduce asexually as well as sexually, and how people propagate many plants vegetatively.

Using the Figure

Ask students to identify the structures shown in the figure. (They are pollen grains.) Ask them how large these are, relatively. (They are microscopic.) Ask students to identify the function of the structures. (They are used for reproduction.) Explain that pollen contains cells that are precursors of sperm cells. Sperm cells are transported by pollen grains to an egg cell. Tell students that the wide variety of pollen forms is related to method of dispersal. Pollen that is transported by air currents often has external structures that help keep it aloft. Pollen that is transported by animals, such as insects or mammals, may have tiny barbs or be sticky.

Assessing Prior Knowledge

Before beginning this chapter, be sure your students can answer the questions in the ***Quick Review*** on the facing page. If they cannot, encourage them to reread the sections indicated.

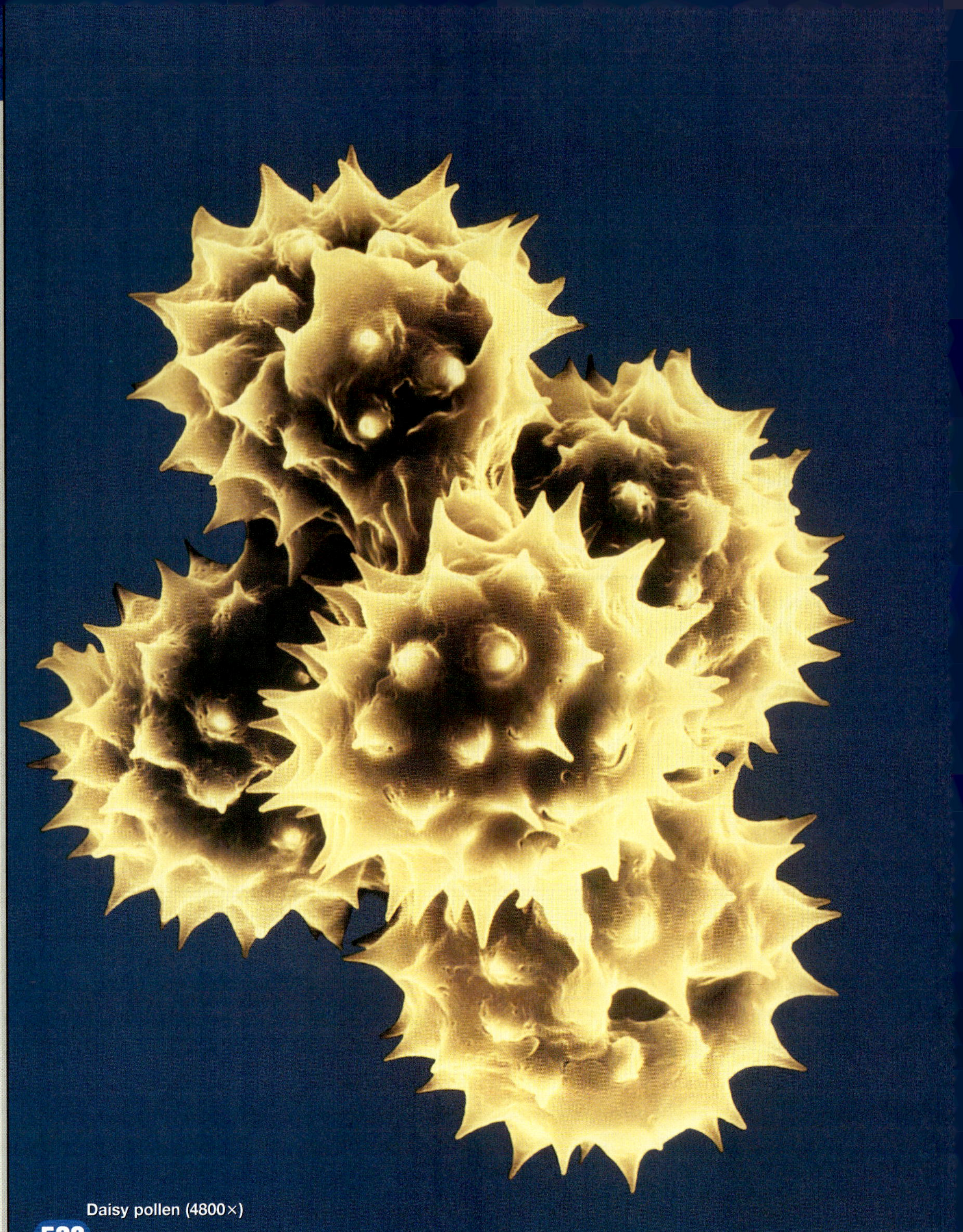

Daisy pollen (4800×)

528

Chapter Resource File

- Vocabulary Worksheet
- Concept Mapping

Chapter Correlations | **Florida Sunshine Standards/FCAT**

SC.F.1.4.2 Body structures are uniquely designed and adapted for their function.

SC.F.2.4.1 There are mechanisms of asexual and sexual reproduction and different genetic advantages and disadvantages of asexual and sexual reproduction.

SC.G.1.4.1 There is great diversity and interdependence among living things.

CHAPTER

24 Plant Reproduction

Florida Sunshine Standards/FCAT

SC.F.1.4.2 Body structures are uniquely designed and adapted for their function.

SC.F.2.4.1 There are mechanisms of asexual and sexual reproduction and different genetic advantages and disadvantages of asexual and sexual reproduction.

SC.G.1.4.1 There is great diversity and interdependence among living things.

Quick Review

Answer the following without referring to earlier sections of your book.

1. **Compare** the processes of mitosis and meiosis. *(Chapter 6, Section 3 and Chapter 7, Section 1)*
2. **Distinguish** sexual reproduction from asexual reproduction. *(Chapter 7, Section 2)*
3. **Differentiate** between plant gametophytes and sporophytes. *(Chapter 23, Section 1.)*
4. **Define** the terms *cone* and *fruit*. *(Chapter 23, Section 2)*

Did you have difficulty? *For help, review the sections indicated.*

Reading Activity

Before you begin to read this chapter, write down all of the key words for each of the three sections of the chapter. Then write a definition next to each word that you have heard of. As you read the chapter, write definitions next to the words that you did not previously know, and modify as needed any definitions you have written.

The characteristic shape of pollen grains differs from one plant species to the next. The spiky, outer layer is composed chiefly of a polymer of carotinoids—the pigments that give fall leaves their characteristic colors.

Looking Ahead

internet connect

www.scilinks.org
National Science Teachers Association *sci*LINKS Internet resources are located throughout this chapter.

SCI LINKS. Maintained by the National Science Teachers Association

Opening Activity

Plant Display Bring some plants with sexual and asexual reproductive structures to class and display them. To show the sexual reproductive structures of seedless plants, display fern fronds with sori and moss or liverwort reproductive structures. To show the sexual reproductive structures of seed plants, display cones from a conifer (pine, juniper, spruce, fir) and flowers from a flowering plant (rose, hibiscus, iris). To show asexual reproductive structures, display bulbs or corms (onions, tulip, gladiolus); rhizomes (iris, fern); plantlets on runners (strawberry, spider plant); and plantlets on leaves (kalanchoë). Discuss with students the form of reproduction each structure represents. Return to the display as you move through the chapter, and ask students to identify the structures and type of reproduction involved. **LS Visual**

Quick Review Answers

1. Mitosis results in the production of cells identical to the original dividing cell. Meiosis results in the production of cells that have half the number of chromosomes as the original dividing cell.
2. Sexual reproduction involves the union of parental gametes, each with half the number of chromosomes. Asexual reproduction involves a single parent producing offspring that are genetically identical to the parent.
3. Gametophytes contain haploid cells and produce gametes by mitosis. Sporophytes contain diploid cells and produce spores by meiosis.
4. A cone is the reproductive structure of a gymnosperm. A fruit is a reproductive structure of angiosperms.

Reading Activity Answers

Students may be able to define terms such as pollen, seed coat, and petal. After students read the chapter, their written definitions should reflect an increased understanding of the terms.

Section 1

Focus

Overview

Before beginning this section review with your students the objectives listed in the Student Edition. This lesson begins with a discussion of alternation of generations, which occurs in all plants. It continues with the life cycle of a moss to introduce the reproductive structures and life cycles of nonvascular plants. A fern is used to illustrate the reproductive structures and life cycles of seedless vascular plants.

Bellringer

Place a moss and a fern on your desk. Instruct students to write a short paragraph that describes a moss or fern plant. Tell them to be as descriptive as possible about the plant's general appearance and where it lives.

Motivate

Discussion/Question

Tell students that the sperm of a nonvascular plant must swim to an egg cell for fertilization to occur. Ask students to describe environmental conditions that would enable nonvascular plants to reproduce sexually. (Nonvascular plants are able to reproduce sexually wherever moisture is abundant. Shady areas near streams are ideal habitats. In drier habitats, nonvascular plants can reproduce sexually after a rainfall or heavy dew.) **LS Interpersonal**

Section 1 Sexual Reproduction in Seedless Plants

Florida Sunshine Standards/FCAT

SC.F.2.4.1 There are mechanisms of asexual and sexual reproduction and different genetic advantages and disadvantages of asexual and sexual reproduction.

SC.G.1.4.1 There is great diversity and interdependence among living things.

Objectives

- **Summarize** the life cycle of a moss. SC.F.2.4.1
- **Summarize** the life cycle of a fern. SC.F.2.4.1
- **Compare** and **Contrast** the life cycle of a moss with the life cycle of a fern. SC.G.1.4.1

Key Terms

archegonium
antheridium
sorus

Reproduction in Nonvascular Plants

The carpet of green you often see near streams and in moist, shady places is usually made up of mosses or liverworts. As you learned in the previous chapter, these small, relatively simple plants are nonvascular plants. They do not have a vascular system for distributing water and nutrients. Mosses and liverworts do not usually thrive outside moist places because they must be covered by a film of water to reproduce sexually.

Like all plants, nonvascular plants have a life cycle called alternation of generations. In this type of life cycle, a gamete-producing stage, or gametophyte, alternates with a spore-producing stage, or sporophyte. Gametophytes produce gametes (eggs and sperm) in separate multicellular structures. The structure that produces eggs is called an **archegonium** *(ark uh GOHN ee uhm)*. The structure that produces sperm is called an **antheridium** *(an thuhr IHD ee uhm)*. Sporophytes produce spores in a sporangium. The gametophytes of nonvascular plants are larger and more noticeable than are the sporophytes. This difference in size is very pronounced in the liverworts, as you can see in **Figure 1.**

Figure 1 Reproductive structures of a liverwort

The gametophytes of Marchantia, a common liverwort, produce male and female gametes on separate stalks.

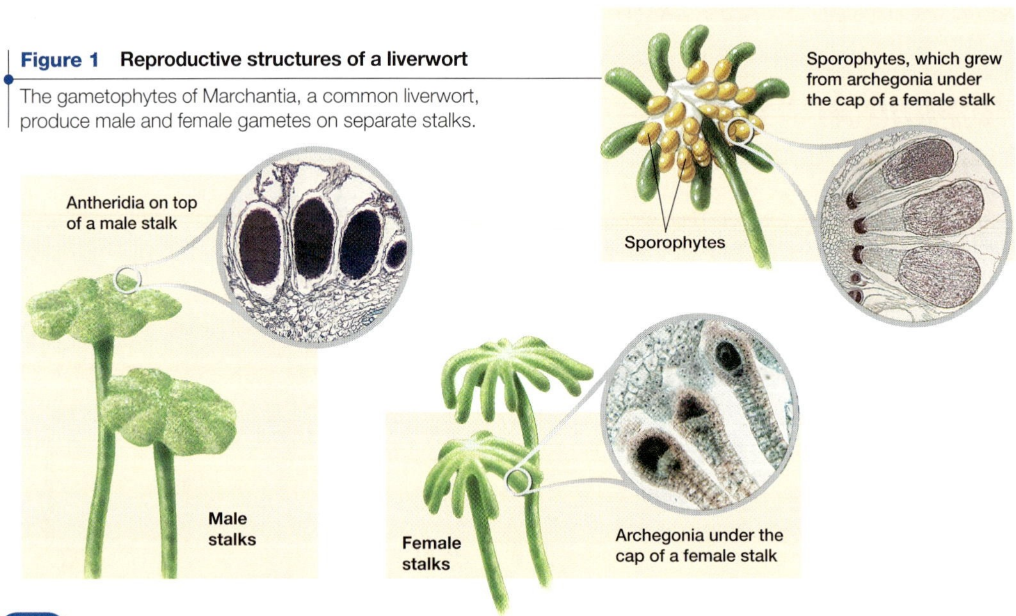

Transparencies

- TR G33 Life Cycle of a Moss
- TR G34 Life Cycle of a Fern

Section Correlations

SC.F.2.4.1
SC.G.1.4.1

Chapter Resource File

- Directed Reading BASIC
- Active Reading GENERAL
- Data Sheet for Quick Lab GENERAL

One-Stop Planner CD-ROM

- Reading Organizers BASIC
- Reading Strategies BASIC

Life Cycle of a Moss

The life cycle of a moss is summarized in **Figure 2.** Sexual reproduction results in a fertilized egg, or zygote. The diploid zygote grows into a new diploid sporophyte. As you can see, a moss sporophyte grows from a gametophyte and remains attached to it. The sporophyte consists of a bare stalk with a spore capsule (sporangium) at its tip. Spores form by meiosis inside the spore capsule. Therefore, as in all plants, the spores are haploid. The spore capsule opens when the spores are mature, and the spores are carried away by wind or water. When a moss spore settles to the ground, it germinates and grows into a "leafy" green gametophyte. Archegonia and antheridia form at the tips of the haploid gametophytes. Eggs and sperm form by mitosis inside the archegonia and antheridia. Remember, moss gametophytes grow in tightly packed clumps of many individuals. When water covers a clump of mosses, sperm can swim to nearby archegonia and fertilize the eggs inside them.

WORD Origins

The word *archegonium* comes from the Greek words *archegonos,* meaning "first of a race." Knowing this makes it easier to remember that a new and genetically different individual grows from an archegonium when its egg is fertilized.

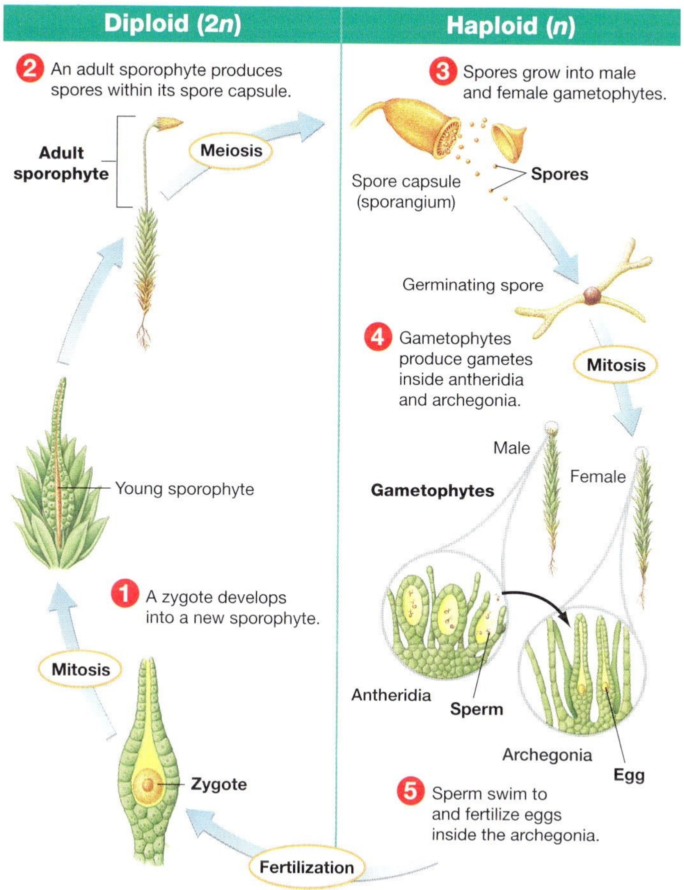

Figure 2 Moss life cycle. In mosses, a sporophyte that consists of a spore capsule on a bare stalk alternates with a "leafy," green gametophyte.

Teach

Using the Figure

Walk students through the moss life cycle shown in **Figure 2,** starting with the zygote. Point out that the zygote divides mitotically. Repeated mitotic cell divisions result in a multicellular, diploid organism called the sporophyte. In many mosses, the sporophytes are not green, and they consist of a bare stalk topped with a spore capsule. Reproductive cells in the spore capsule undergo meiosis, which reduces the chromosome number from diploid to haploid. Point out that the spores grow into either male or female gametophytes, which are haploid. In mosses, the gametophytes are green and have leaflike structures. Finally, point out that when fertilization (the union of an egg and a sperm) occurs and a new zygote is formed, the diploid chromosome number is restored, and a new sporophyte generation begins. **LS Visual**

READING SKILL BUILDER — GENERAL

Brainstorming Ask students to identify the haploid structures in a plant's life cycle (spores, male and female gametophytes, eggs, and sperm). Ask students which structures are unicellular (spores, sperm, and eggs) and which are multicellular (male and female gametophytes). Ask students to identify the haploid structures in an animal's life cycle (egg and sperm). Ask students if these structures are unicellular or multicellular (unicellular). Ask students to identify the diploid structures in a plant's life cycle (sporophyte, zygote). Ask which is unicellular (zygote) and which is multicellular (sporophyte). Ask students to identify the diploid structures in an animal's life cycle (all stages of development from the zygote to the adult). Ask which of these structures is unicellular (zygote) and which are multicellular (all other stages of development). **LS Interpersonal**

MISCONCEPTION ALERT

Spores The term *spore* does not have a single, concise definition as students might think. Two things are generally true of spores. First, they are unicellular; second, they have a protective outer covering. In some bacteria, spores form from regular cells when environmental conditions are harsh. These cells lose most of their water and develop a protective outer covering. Many of these bacteria also produce toxins, which allow the spores to remain viable in harsh environments. The spores of fungi and plants are haploid, drought-resistant, and easily transported by wind or water. The spores of fungi and plants develop into multicellular individuals (by mitotic cell division) without fusing with another cell.

Teach, continued

SKILL BUILDER — GENERAL

Math Skills Ask students to use the information in **Figure 3** to estimate the number of spores that one fern plant might produce. To help them with their calculations, have students assume that only half of the frond is visible in the photo, each sporangium produces 100 spores, and one plant produces 12 fronds. (A sample calculation would be as follows: 260 sori per frond × 30 sporangia per sorus × 100 spores per sporangium × 12 fronds per plant = 9,360,000 spores per plant.)
LS **Logical**

QUICK LAB — *Observing a Fern Gametophyte*

Skills Acquired
Observing, making comparisons, making conclusions

Teacher's Notes
Tell students that fern gametophytes develop on moist soil and are very small.

Answers to Analysis

1. Answers will vary but should be consistent with the diagrams in Figure 4.
2. A new sporophyte begins its growth in an archegonium. This process occurs because the egg remains in the archegonium, and a sperm cell swims from an antheridium to the archegonium, where it fertilizes the egg.

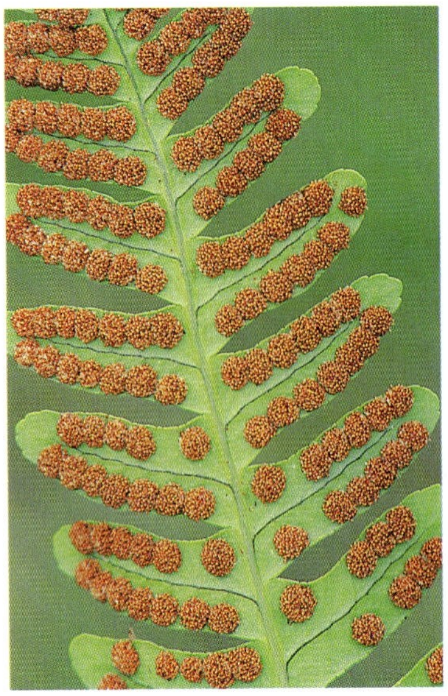

Figure 3 Sori on a fern frond. Many sori are visible on this portion of a frond from a polypody fern. Each sorus consists of about 20–30 sporangia.

Reproduction in Seedless Vascular Plants

You may recall that the seedless vascular plants include the whisk ferns, horsetails, club mosses, and ferns. The seedless vascular plants differ from the nonvascular plants because they have efficient water- and food-conducting systems of vascular tissue. Like the nonvascular plants, the seedless vascular plants thrive in moist, shady places. They can reproduce sexually only when a film of water covers the gametophyte. Eggs form in archegonia, and sperm form in antheridia. The archegonia and antheridia develop on the lower surfaces of the gametophytes. In many species of seedless vascular plants, both eggs and sperm are produced by the same individual. In some species, however, eggs and sperm are produced by separate gametophytes.

Unlike nonvascular plants, seedless vascular plants have sporophytes that are much larger than their gametophytes. Some ferns, for example, have sporophytes that are as large as trees. On the other hand, the gametophytes of ferns are thin, green, heart-shaped plants that are less than 1 cm (0.4 in.) across. The sporophytes produce spores in sporangia. In horsetails and club mosses, sporangia develop in conelike structures. In ferns, clusters of sporangia form on the lower surfaces of fronds, as shown in **Figure 3.** A cluster of sporangia on a fern frond is called a **sorus.** The word *sorus* comes from the Greek word *soros,* meaning "a heap."

QUICK LAB — *Observing a Fern Gametophyte* SC.H.1.4.1

You can observe the archegonia and antheridia of a fern gametophyte with a microscope.

Materials

prepared slide of a fern gametophyte with archegonia and antheridia, compound microscope

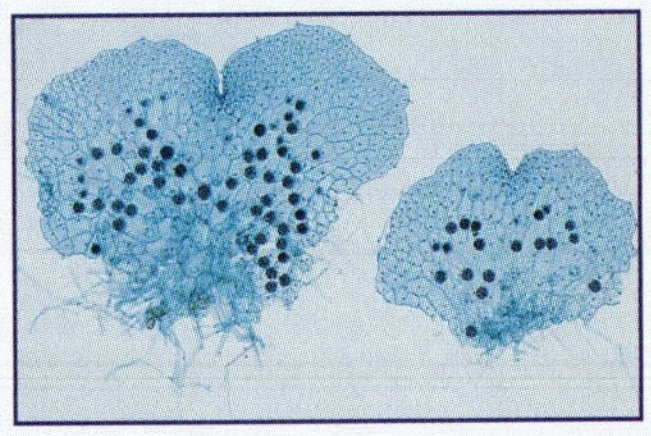

Fern Gametophytes (56×)

Procedure

1. Examine a slide of a fern gametophyte under low power of a microscope. Move the slide until you can see a cluster of archegonia. Now, switch to high power, and focus on one archegonium. Draw and label what you see.
2. Switch back to low power, and move the slide until you can see several egg-shaped structures. These are antheridia. Now, switch to high power, and focus on one antheridium. Draw and label what you see.

Analysis

1. **Describe** the appearance of an archegonium and an antheridium.
2. **Critical Thinking Drawing Conclusions** In which structure, an archegonium or antheridium, does the growth of a new sporophyte begin? Explain.

532

REAL WORLD CONNECTION

Garden centers and florists often receive calls about insects on ferns and flower arrangements that include fern leaves. When the fern leaves have sporangia on them, people often mistake the sori or sporangia for insects. A close look at these spore-producing structures reveals that they are lined up in rows across from each other. Insect pests, such as scale and mealy bugs, rarely line up in such regular patterns.

Life Cycle of a Fern

The life cycle of a fern is summarized in **Figure 4.** A fertilized egg, or zygote, grows into a new sporophyte. The diploid sporophyte produces spores by meiosis. The haploid spores fall to the ground and grow into haploid gametophytes. Fern gametophytes produce gametes by mitosis—eggs in archegonia and sperm in antheridia. Sperm swim to archegonia and fertilize the eggs inside them.

Figure 4 Fern life cycle. In ferns, a large sporophyte with leaves called fronds alternates with a small, green, heart-shaped gametophyte.

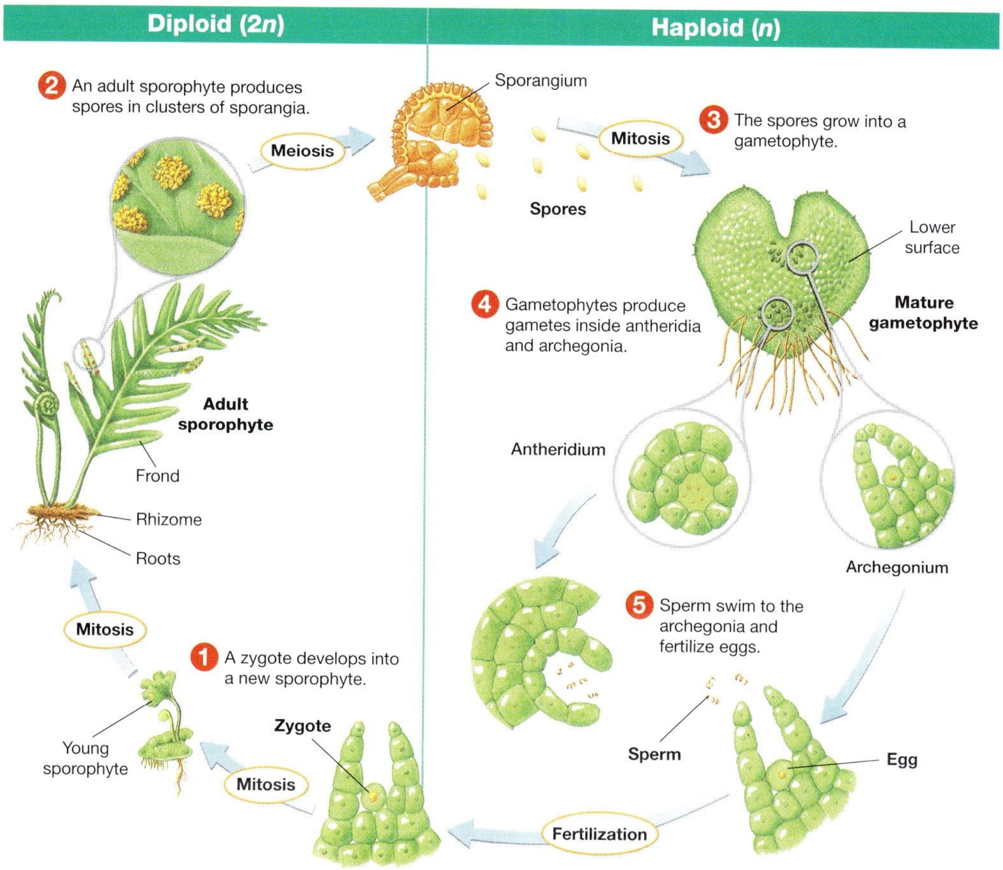

Section 1 Review

1. **List** five major steps in the life cycle of a moss. SC.F.2.4.1
2. **List** five major steps in the life cycle of a fern. SC.F.2.4.1
3. **Critical Thinking Forming Reasoned Opinions** Which reproductive structures, gametes or spores, are responsible for the dispersal (spread) of seedless plants? Justify your answer. SC.F.1.4.2
4. **Critical Thinking Analyzing Information** What are the major differences between the moss life cycle and the fern life cycle? SC.G.1.4.1
5. *FCAT Prep* What is the function of an archegonium?
   A to conduct water
   B to produce eggs
   C to produce sperm
   D to produce spores
   SC.F.1.4.2

533

Answers to Section Review

1. First, a zygote grows into a sporophyte by mitotic cell division. Next, the sporophyte forms spores following meiotic cell division. Then, the spores grow into gametophytes by mitotic cell division. After the gametophytes form eggs and sperm by mitotic cell division, eggs are fertilized by sperm and a new zygote forms.
2. First, a zygote grows into a sporophyte by mitotic cell division. Next, the sporophyte forms spores following meiotic cell division. Then, the spores grow into gametophytes by mitotic cell division. After the gametophytes form eggs and sperm by mitotic cell division, eggs are fertilized by sperm and a new zygote forms.
3. Spores are responsible for the dispersal of seedless plants. Spores are drought-resistant and can be transported in air or in water.
4. In mosses, the gametophyte is dominant; in ferns, the sporophyte is dominant. Mosses form separate male and female gametophytes, while many ferns form a gametophyte that produces both eggs and sperm.
5. **A.** Incorrect. There are no specialized structures for conducting water in the moss plant. **B.** Correct. **C.** Incorrect. The antheridium produces sperm. **D.** Incorrect. The sporangium produces spores.

Close

Reteaching — BASIC

Summarize the life cycles of mosses and ferns. Emphasize that all plants alternate between a haploid phase and a diploid phase in their life cycles. Tell students that in mosses and ferns, the multicellular forms in both phases are visible with the unaided eye. Point out that the gametophyte phase in mosses is dominant and that the sporophytes are not photosynthetic. Also point out that the sporophyte phase is dominant in ferns but that the young sporophytes are dependent on the gametophytes. LS Verbal

Quiz — GENERAL

1. What is the name of the single-celled, haploid reproductive structure of plants? (spore)
2. What is the meaning of the phrase *alternation of generations?* (It means that plants have a life cycle that alternates between haploid and diploid phases.)
3. What is the name of the structure that makes up the first stage of a new plant's life cycle? (zygote)

Alternative Assessment — GENERAL

Poster Have student groups make posters that depict the life cycles of mosses and ferns. Suggest that they use color-coding to identify the structures common to the life cycles of these two groups. (Both have a zygote, a sporophyte, spores, a gametophyte, antheridia, archegonia, eggs, and sperm.) Students also can use color-coding to highlight the differences between the life cycles of mosses and ferns. (Mosses have separate male and female gametophytes, while many ferns have both antheridia and archegonia on each gametophyte.) LS Verbal Co-op Learning

Section Review Correlations

SC.F.1.4.2: Items 3, 5
SC.F.2.4.1: Items 1, 2
SC.G.1.4.1: Item 4

Section 2

Focus

Overview

Before beginning this section review with your students the objectives listed in the Student Edition. This lesson introduces reproduction in seed plants, including seed structure and formation. Discussion of the reproductive structures and life cycle of gymnosperms features a coniferous plant. The lesson on reproductive structures and the life cycle of angiosperms includes a discussion of insects in the fertilization process and the formation of endosperm by double fertilization.

Bellringer

Give each student five different kinds of seeds and have them list similarities and differences. Examples include various beans, peas, maple seeds, pine seeds, coconuts, and poppy. Remind them of the discussion about pollen dispersal at the beginning of the chapter and point out that seeds are also adapted to different means of dispersal, including wind, water, and ingestion and elimination by animals. This exercise can lead into a discussion of seed structure and function.

Motivate

Activity — BASIC

Gymnosperm Flash Cards Have students write the key terms for this section on separate note cards. Then, have them write a definition on the back of each card as they read the section. Have students divide their cards into three groups: characteristics of gymnosperms only (none found only in gymnosperms); characteristics of angiosperms only (sepal, petal, stamen, anther, pistil, ovary, double fertilization); and characteristics of both gymnosperms and angiosperms (pollen grain, ovule, pollination, pollen tube, seed coat, cotyledon). Have students use their cards for the Reteaching exercise.
LS Verbal

Section Correlations

SC.F.1.4.2
SC.F.2.4.1

Section 2

Sexual Reproduction in Seed Plants

Florida Sunshine Standards/FCAT

SC.F.1.4.2 Body structures are uniquely designed and adapted for their function.

SC.F.2.4.1 There are mechanisms of asexual and sexual reproduction and different genetic advantages and disadvantages of asexual and sexual reproduction.

Objectives

- **Distinguish** the male and female gametophytes of seed plants. SC.F.1.4.2
- **Describe** the function of each part of a seed. SC.F.1.4.2
- **Summarize** the life cycle of a conifer. SC.F.2.4.1
- **Relate** the parts of a flower to their functions. SC.F.1.4.2
- **Summarize** the life cycle of an angiosperm. SC.F.2.4.1

Key Terms

pollen grain
ovule
pollination
pollen tube
seed coat
cotyledon
sepal
petal
stamen
anther
pistil
ovary
double fertilization

Reproductive Structures of Seed Plants

Reproduction in seed plants (gymnosperms and angiosperms) is quite different from reproduction in seedless plants. For one thing, you need a microscope to see the gametophytes of seed plants, as **Figure 5** shows. Also, spores are not released from seed plants. The spores remain within the tissue of a sporophyte and develop into two kinds of gametophytes—male gametophytes, which produce sperm, and female gametophytes, which produce eggs. The tiny gametophytes of seed plants consist of only a few cells. An immature male gametophyte of a seed plant is a **pollen grain,** which has a thick protective wall. A female gametophyte of a seed plant develops inside an **ovule** *(AHV yool),* which is a multicellular structure that is part of the sporophyte. Following fertilization, the ovule and its contents develop into a seed.

Because the gametophytes of seed plants are very small, seed plants are able to reproduce sexually without water. Wind and animals transport pollen grains to the structures that contain ovules. The transfer of pollen grains from the male reproductive structures of a plant to the female reproductive structures of a plant is called **pollination.** When a pollen grain reaches a compatible female reproductive structure, a tube emerges from the pollen grain. This tube, called a **pollen tube,** grows from a pollen grain to an ovule and enables a sperm to pass directly to an egg.

Figure 5 Seed plant gametophytes

The tiny gametophytes of angiosperms develop within specialized structures that form in the reproductive parts of a flower.

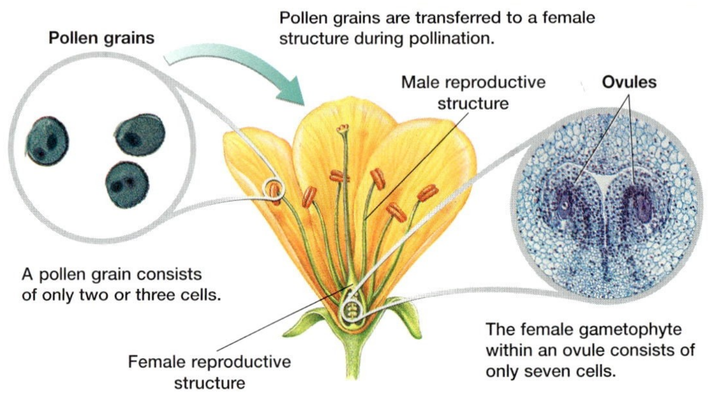

534

Chapter Resource File

- Directed Reading BASIC
- Active Reading GENERAL
- Data Sheet for Quick Lab GENERAL

Transparencies

TR Bellringer
TR G42 Seed Structure

Seeds

As you learned in the previous chapter, seeds contain the embryos of seed plants. A plant embryo is a new sporophyte. A seed forms from an ovule after the egg within it has been fertilized. The outer cell layers of an ovule harden to form the **seed coat** as a seed matures. The tough seed coat protects the embryo in a seed from mechanical injury and from a harsh environment. Different kinds of seeds germinate under different conditions. Some seeds must dry out before they germinate. Other seeds must be exposed to light. In still other seeds, chemicals that inhibit germination must be washed away by rain. Often, a seed must be exposed to cold temperatures, or the seed coat must be damaged, before the seed can germinate. Thus, seeds enable the embryos of seed plants to survive conditions that are unfavorable for plant growth for long periods of time.

Seeds also contain tissue that provides nutrients to plant embryos. In gymnosperms, this nutritious tissue is part of the female gametophyte. The seeds of angiosperms, however, develop a nutritious tissue called *endosperm.* Endosperm originates at the same time an egg is fertilized. In some angiosperms, such as corn and wheat, endosperm is still present in mature seeds. In other angiosperms, such as beans and peas, the nutrients in the endosperm have already been transferred to the embryo by the time a seed is mature.

Leaflike structures called **cotyledons** *(kah tuh LEE duhnz),* or seed leaves, are a part of a plant embryo. Cotyledons function in the transfer of nutrients to the embryo. The embryos of gymnosperms have two or more cotyledons. For example, pine embryos have eight cotyledons. In the flowering plants, the embryos of monocots have one cotyledon, and the embryos of dicots have two cotyledons. The structure of three types of seeds is shown in **Figure 6.**

Study TIP

Interpreting Graphics After reading the chapter, trace or make a sketch of Figure 6 without the labels. On separate pieces of paper, write down the labels. Without referring to your book, match the labels with the correct parts of your sketch.

Figure 6 Seed structure

Seeds have many similarities and differences in structure.

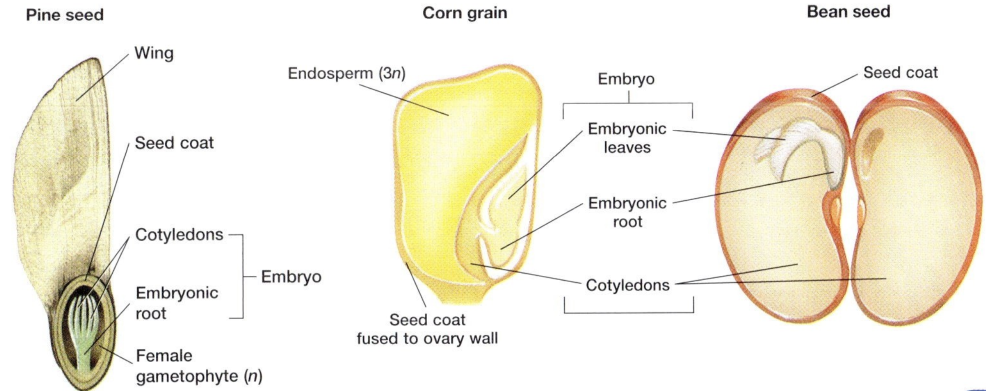

Teach

Teaching Tip

Origins of Embryos Tell students that in some plants, such as citrus, the embryo contained within the seed doesn't always develop from a fertilized egg. Instead, it develops from other cells in the ovule. Plants that grow from such embryos are identical to their female parent.

READING SKILL BUILDER — GENERAL

Reading Organizer As students read about the plant life cycles in this chapter, have them develop life cycle concept maps. Maps should include the main steps of the cycle, connecting words, and arrows showing the progression of steps.
LS Verbal

Group Activity — BASIC

Dissecting Seeds Bring a variety of seeds to class, including at least one type of gymnosperm seed (pine, spruce, fir), one type of dicot seed (bean, pea, melon, squash), and one type of monocot seed (corn, wheat, iris). Make a longitudinal cut through at least one seed of each type. Have students examine the whole and opened seeds using their unaided eyes and a dissecting microscope. Ask students to compare the characteristics of the three different types of seeds.
LS Visual Co-op Learning

One-Stop Planner CD-ROM

- Reading Organizers BASIC
- Reading Strategies BASIC

did you know?

A Sunflower Seed is Not Just a Seed A sunflower seed is a dry fruit (called an achene) with a seed inside. The edible part is the actual seed, which consists mainly of cotyledons. A corn kernel is also not just a seed but a dry fruit (called a caryopsis) with a seed inside. The outer covering of a corn kernel is not easily removed, as is the outer covering of a sunflower seed. The edible part of the corn fruit is the whole kernel.

Integrating Physics and Chemistry

There is a limit on how much protection a tough seed coat provides for a seed embryo against mechanical injury. A mechanical injury to a seed occurs when the kinetic energy of an object that strikes a seed exceeds the intrinsic strength of the seed coat. Remind students that kinetic energy—that is, energy of motion—depends on an object's mass and its speed. Consequently, more massive objects require less speed in order to damage a seed coat.

Teach, *continued*

Activity — ADVANCED

Plant Propagator Plant propagators work for wholesale and retail nurseries where they grow plants for sale. A plant propagator grows plants from seeds and by a variety of vegetative methods. Have students use library references or online databases to find out about plant propagation in the nursery and floral industries. Have students write reports summarizing their findings. **LS Verbal**

QUICK LAB *Observing the Gametophytes of Pines*

Skills Acquired
Observing, making comparisons, drawing conclusions

Teacher's Notes
Ask students to look for similarities between the two types of cones when they are examined under a microscope.

Answers to Analysis

1. Both have whorls of scales. But male cones are smaller than female cones and contain pollen, while female cones contain eggs and eventually seeds.
2. The growth rate of pollen tubes in pines is extremely slow.

Transparencies

TR G35 Life Cycle of a Conifer

Cones

Seed plants are the most successful of all plants. The success of the seed plants is due in part to the specialized structures in which seeds develop. In angiosperms, the ovules (immature seeds) are completely enclosed by sporophyte tissue at the time of pollination. In gymnosperms, the ovules are not completely enclosed by sporophyte tissue until after pollination.

The gametophytes of gymnosperms develop in cones, which consist of whorls (circles) of modified leaves called scales. Gymnosperms produce two types of cones. Male cones, or pollen cones, produce pollen grains within sacs that develop on the surface of their scales. Female cones, or seed cones, produce ovules on the surface of their scales. Many gymnosperms produce both male and female cones on the same plant. As shown in **Figure 7,** the numerous small pollen cones lie to the left of the large seed cone. In some gymnosperms, male and female cones form on separate plants.

Pollen cones produce large quantities of pollen grains that are carried by wind to female cones. At the time of pollination, the scales of a female cone are open, exposing the ovules. When a pollen grain lands near an ovule, a slender pollen tube grows out of the pollen grain and into the ovule. The sperm moves through the pollen tube and enters the ovule. Thus, the pollen tube delivers a sperm to the egg inside the ovule. Seed cones close up after pollination and remain closed until the seeds within them are mature. This process can take up to two years.

Figure 7 Male and female pine cones. This branch of an Austrian pine has an immature seed cone and many pollen cones.

QUICK LAB *Observing the Gametophytes of Pines* SC.H.1.4.1

You can observe the gametophytes of a pine with a microscope.

Materials

prepared slides of the following: male pine cone, female pine cone, pine ovule; hand lens; compound microscope

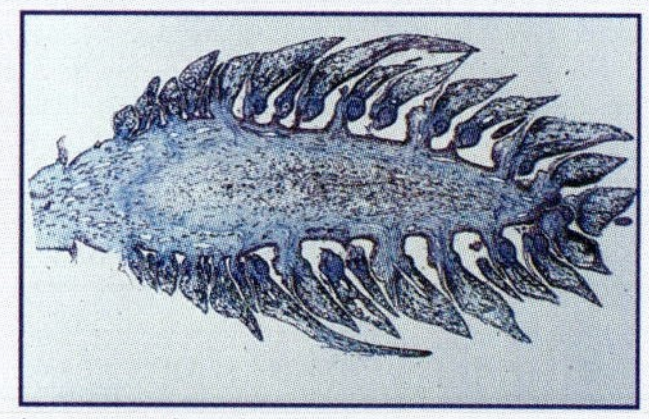

Immature female pine cone

Procedure

1. Examine prepared slides of male and female pine cones first with a hand lens and then under the low power of a microscope.
2. Make a sketch of each type of pine cone, and label the structures that you recognize.
3. Examine a prepared slide of a pine ovule under the low power of a compound microscope. Compare what you see with the photo above.
4. Draw a pine ovule, and label the following structures: scale, ovule, egg, pollen tube (if visible).

Analysis

1. **Compare** and **Contrast** the structure and contents of male and female pine cones.
2. **Critical Thinking Applying Information** It takes 15 months for a pine pollen tube to grow through the wall of a pine ovule. How would you describe the rate of pollen-tube growth in pines?

536

Cultural Awareness

George Washington Carver George Washington Carver (1864–1943) was a horticulturist whose agricultural research revolutionized the economy of the southern United States by encouraging farmers to grow more than just cotton. He urged farmers to grow sweet potatoes, peanuts, pecans, and soybeans. In 1896, the same year he received his master's degree at Iowa State College of Agriculture and Mechanic Arts, Carver was asked to head the Department of Agriculture at Tuskegee Institute, in Alabama. There he helped to develop more than 300 industrial products from peanuts, 100 from sweet potatoes, and 75 from pecans.

Life Cycle of a Conifer

Most gymnosperms are conifers, a group that includes pines. You can trace the stages in the life cycle of a pine in **Figure 8.** In pines, as in all plants, a diploid zygote results from sexual reproduction. The zygote develops into an embryo, which then becomes dormant (inactive). The embryo and the surrounding tissues form a seed. When their seeds are mature, seed cones open, and the seeds fall out. A seed of most pines has a wing that causes it to spin like the blade of a helicopter. Thus, pine seeds often travel some distance from their parent tree.

When conditions are favorable for growth, the embryos grow into new sporophytes. An adult pine tree produces both male and female cones. Spores form by meiosis, which occurs inside immature cones. The spores grow into gametophytes, which produce eggs and sperm by mitosis. After pollination, a pollen tube begins to grow from each pollen grain toward the eggs inside an ovule. Fertilization occurs as a sperm fuses with an egg, forming a zygote that will grow into a new sporophyte.

Figure 8 Conifer life cycle. In conifers, a very large sporophyte that produces cones alternates with tiny gametophytes that form on the scales of cones.

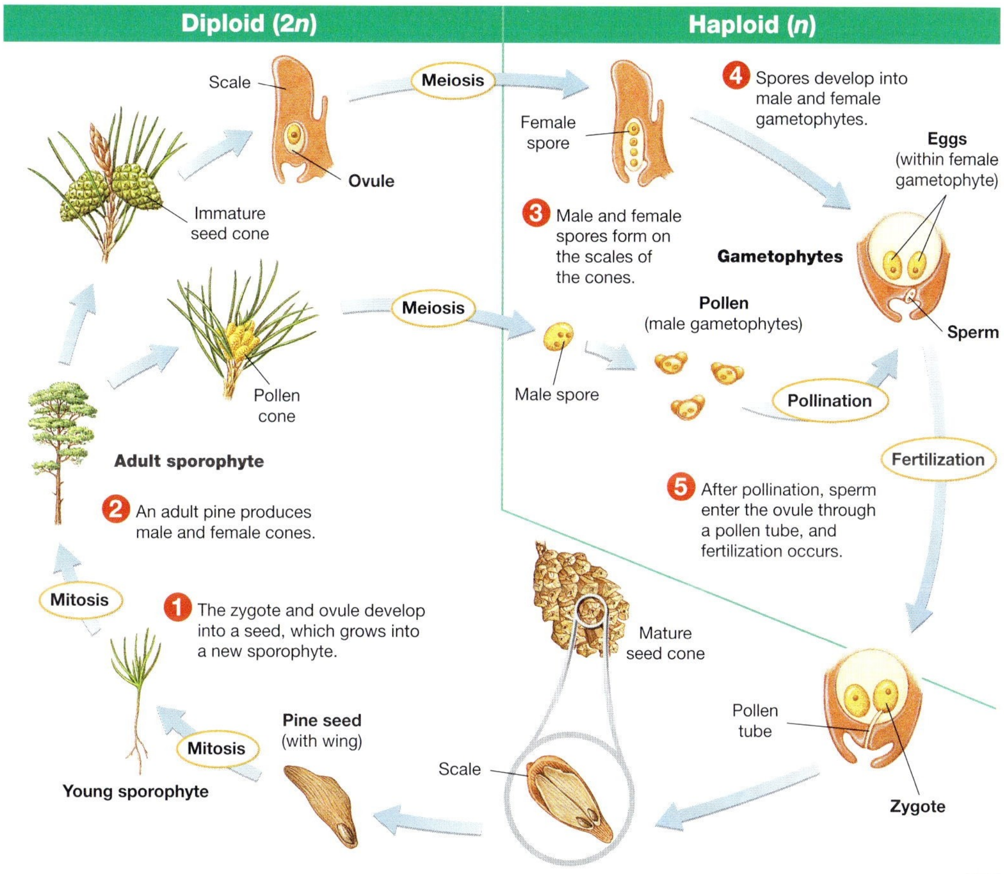

537

Using the Figure

Walk students through the gymnosperm life cycle shown in **Figure 8,** starting with the zygote. Point out that, as in animals and other multicellular eukaryotic organisms, repeated mitotic cell divisions of the zygote result in a multicellular diploid organism. In plants, the multicellular diploid organism is called the sporophyte. Reproductive cells in the mature sporophyte undergo meiosis. During this process, the chromosome number is reduced from diploid to haploid. Point out that there are two kinds of spores produced—male spores (microspores) and female spores (megaspores)—and that these spores grow into two different types of gametophytes—pollen grains, or male gametophytes (microgametophytes), that produce sperm cells and female gametophytes (megagametophytes) that produce egg cells. Both kinds of gametophyte are haploid. Finally, point out that when fertilization (the uniting of an egg and a sperm) occurs and a new zygote is formed, the diploid chromosome number is restored, and a new sporophyte generation begins.
LS Visual

Group Activity — BASIC

Diversity of Pollen Grains

Provide students with a variety of flowers with pollen, or give them pollen alone (for example, pollen from a pine tree). Have students make wet-mount slides of pollen samples to observe the intricately detailed coats of pollen grains. Have students draw what they observe. You could also provide them with pictures (for example, from a college biology textbook) showing electron micrographs of some pollen grains.
LS Kinesthetic English Language Learners

did you know?

What is a "Jumping Bean"? Jumping beans, which are grown in Central and South America, are the seeds of plants in the Euphorbiaceae family. A jumping bean contains a moth larva. The jumping movements occur when the moth larva inside changes position with a jerk. The jumping bean is also known as the "bronco bean" or "leaper."

Integrating Physics and Chemistry

When a pine seed falls to the ground, the force of gravity causes it to accelerate toward the earth's surface at approximately 9.8 m/s^2. However, a pine seed's wing reduces the rate of acceleration by giving the seed some upward lift as it spins in the air. Provide the students with some pine seeds and have them measure the mass of each seed. Next, have students drop each one from various heights and time how long it takes each seed to reach the ground. Then, have students calculate each pine seed's average acceleration using Newton's second law, $a = F/m$, and compare this number with the expected acceleration.

Teach, *continued*

QUICK LAB *Observing the Arrangement of Parts of a Flower*

Skills Acquired
Observing, making comparisons, drawing conclusions

Teacher's Notes
Provide a variety of monocot and dicot flowers for the class and have each student examine one of each type.

Ask students to think about how the flowers they examine are pollinated. (Flowers that are open and have stamens and stigmas exposed are likely cross-pollinated. Flowers that are closed and have stamens and stigmas that are not exposed are likely self-pollinated.)

Answers to Analysis

1. Answers may vary. Students should comment on color, size, and numbers of sepals and petals.
2. Answers will vary. Students should suggest that large, brightly colored petals may attract animal pollinators (such as insects) while flowers with small, inconspicuous petals are probably wind-pollinated.
3. Answers may vary. Students should note the following: monocots have flower parts in multiples of three; dicots have flower parts in multiples of two, four, or five.

Flowers

In angiosperms, gametophytes develop within flowers. The basic structure of a flower is shown in **Figure 9.** Flower parts are arranged in four concentric whorls. The outermost whorl consists of one or more **sepals** *(SEE puhlz),* which protect a flower from damage while it is a bud. The second whorl consists of one or more **petals,** which attract pollinators. The third whorl consists of one or more **stamens** *(STAY muhnz),* which produce pollen. Each stamen is made of a threadlike filament that is topped by a pollen-producing sac called an **anther.** The fourth and innermost whorl of a flower consists of one or more **pistils,** which produce ovules. Ovules develop in a pistil's swollen lower portion, which is called the ovary. Usually, a stalk, called the style, rises from the **ovary.** Pollen lands on and sticks to the stigma—the swollen, sticky tip of the style.

Flowers may or may not have all four of the basic flower parts. A flower that has all four parts is called a complete flower. Flowers that lack any one of the four types of parts are called incomplete flowers. If a flower has both stamens and pistils, it is called a perfect flower. Flowers that lack either stamens or pistils are called imperfect flowers.

Figure 9 Basic flower structure. The four basic parts of a flower—sepals, petals, stamens, and pistils—are arranged in concentric whorls.

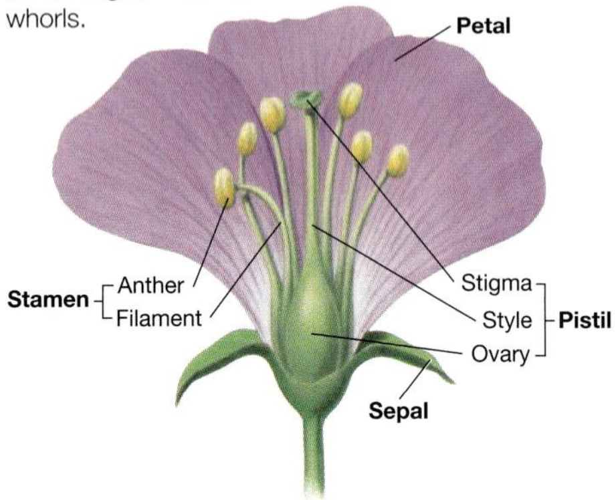

QUICK LAB *Observing the Arrangement of Parts of a Flower* SC.H.1.4.1

You can see how the parts of flowers are arranged by dissecting flowers.

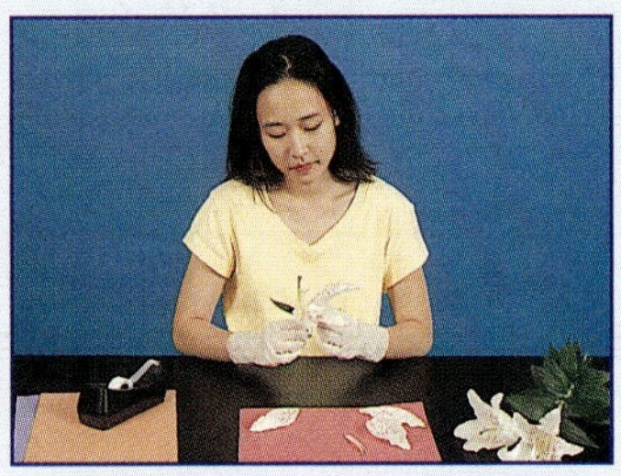

Materials
gloves, monocot flower, dicot flower, paper, tape

Procedure

1. Put on gloves. Examine a monocot flower and a dicot flower. Locate the sepals, petals, stamens, and pistil of each flower.
2. Separate the parts of each flower, and tape them to a piece of paper. Label each set of parts.
3. Count the number of petals, sepals, and stamens in each flower. Record this information below each flower.

Analysis

1. **Compare** and **Contrast** the appearance of the sepals and petals of each flower.
2. **Critical Thinking** **Forming a Hypothesis** For each flower, suggest a function for the petals based on their appearance.
3. **Critical Thinking** **Justifying Conclusions** Explain why each flower is from either a monocot or a dicot.

538

INCLUSION Strategies

- *Learning Disability*
- *Attention Deficit Disorder*

Using **Figure 9** as a guide, ask students to make a clay model of the basic flower structure. The parts of the flower can be labeled with straight pins and strips of paper stuck into the model. The model should be made to lie flat on a piece of cardboard for easy display. Students can show their understanding of the parts of the flower by describing them in a short presentation of tape recording.

Transparencies
TR G36 Floral Structure

Flowers and Their Pollinators

Many flowers have brightly colored petals, sugary nectar, strong odors, and shapes that attract animal pollinators. Flowers are a source of food for pollinators such as insects, birds, and bats. For example, bees eat nectar and collect pollen, which is a rich source of protein they feed to their larvae. A bee gets coated with pollen as it visits a flower and then carries that pollen to other flowers.

Bees locate flowers by scent first and then by color and shape. Bee-pollinated flowers are usually blue or yellow and often have markings that show the location of nectar. Moths, which feed at night, tend to visit heavily scented white flowers, which are easy to find in dim light. Flies may pollinate flowers that smell like rotten meat.

Many flowers are not pollinated primarily by insects. Red flowers, for instance, may be pollinated by hummingbirds. Some large white flowers that open at night are pollinated by nighttime visitors—bats, as seen in **Figure 10.** Many flowers, such as those of grasses and oaks, are pollinated by wind. Wind-pollinated flowers are usually small and lack bright colors, strong odors, and nectar.

Figure 10 **Bat pollination.** This lesser long-nosed bat pollinates an organ pipe cactus as it feeds on the pollen of the plant's flowers.

FORENSICS BIOWatch

Telltale Pollen

Few living things are as short-lived as a flower. Delicate and lovely, its petals wither and die in just a few days. Yet flowers contain some of the toughest structures found in nature: pollen grains.

A thick, decay-resistant wall surrounds each tiny pollen grain. Remarkably, each flowering plant species has its own unique pollen. As shown in the scanning electron micrograph above right, under the microscope, pollen grains may have unusual shapes or may be covered with distinctive ridges, knobs, or spikes. Because no two plant species have identical pollen, investigators often can use pollen to link an object or a suspect to a scene of a crime.

Floral Footprints

Different areas have different kinds of pollen-producing plants. Dirt, mud, or dust collected from a person's clothing, hair, or shoes may contain pollen that can reveal a suspect's past whereabouts or link him or her to a specific location. Pollen recovered from the air filters of a vehicle can show where the vehicle might have traveled. In imported goods, the presence of certain pollens can verify the country of origin.

Drug Detection

Pollen from marijuana (*Cannibis* spp.) found trapped in dust filters, air ducts, or cracks in the floors can confirm that marijuana was grown or processed in a certain location.

The mixture of different pollen types found in a sample can also be highly specific, much like a fingerprint. Law enforcement agencies can use this evidence to link plant-based drugs found on different people or at different places. Investigators can also use pollen evidence to pinpoint the region where a drug crop was grown.

Graphic Organizer

Use this graphic organizer with **Teaching Tip** on this page.

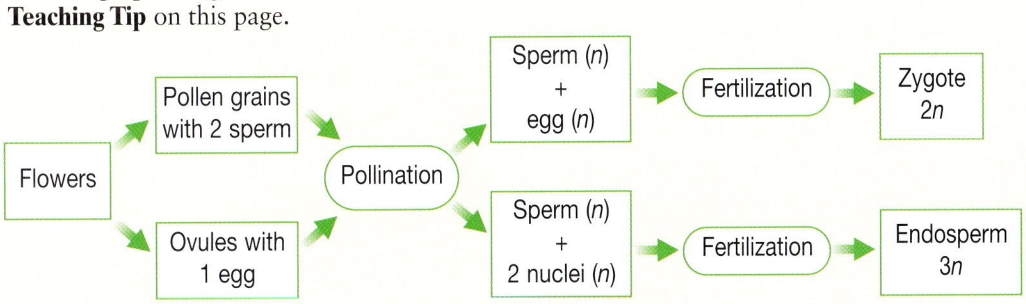

Teaching Tip — GENERAL

Double Fertilization in Angiosperms Have students make a Graphic Organizer similar to the one at the bottom of this page to illustrate the process of double fertilization in angiosperms, discussed on the following page. Ask students to use the applicable vocabulary words for this section, plus any other words that they wish to include. LS **Visual**

BIOWatch

Telltale Pollen

Teaching Strategies
Explain to students that scientists also use pollen grains to learn about climate change. Pollen trapped in soil sediments can become preserved as microfossils. These pollen grains reveal the past plant life of a geographic area, helping scientists learn how climates have changed throughout the last several thousand years. For example, sediments containing pine pollen suggest that an area might once have been covered with conifer forests during a colder, drier time.

Discussion
Law enforcement officials are preparing to search a building. How could forensic botany help them establish that marijuana had been present? (Law enforcement officials could look for marijuana pollen in the dust filters, air ducts, or cracks in the floors and walls.)

Forensics Labs Appendix
Turn to pp. 1048–1071 for forensics lab activities and more student information on forensics techniques.

Close

Reteaching — Basic

Have students construct a crossword puzzle using the terms presented in this section and the definitions they wrote as directed in the lesson opener Activity. LS Verbal

Quiz — General

1. What do male and female gametophytes produce, respectively? (sperm and eggs)
2. What are the three main components of a seed? (embryo, endosperm, seed coat)
3. Which is larger in a conifer, the sporophyte or the gametophyte? (sporophyte)
4. Which structure in a flower is the male equivalent of the pistil? (the stamen)

Alternative Assessment — Advanced

Provide students with a variety of cones and fruits. Try to include pine, spruce, and fir cones. Try to include peaches, tomatoes, grapes or blueberries, apples, beans, nuts, strawberries, pineapple or figs, and elm or maple fruits. Provide a botany textbook or field guide that students can use to classify the fruits. LS Visual

Section Review Correlations

SC.F.1.4.2: Items 1, 2, 5, 6
SC.F.2.4.1: Items 3, 4

Life Cycle of an Angiosperm

Figure 11 summarizes the life cycle of an angiosperm. Following fertilization, the zygote and the tissues of the ovule develop into a seed, which grows into a new sporophyte. The adult sporophytes of angiosperms produce spores by meiosis. These spores grow into gametophytes. The female gametophytes grow inside the ovules, which develop within the ovary of a pistil. The male gametophytes, or pollen grains, are produced in the anther of a stamen. A pollen grain contains two sperm cells. One sperm fuses with the egg, forming the zygote. The other sperm fuses with the haploid nuclei of two other cells produced by meiosis. The fusing of three haploid (n) cells forms a triploid ($3n$) cell that develops into endosperm. This is a process called **double fertilization.**

Figure 11 Angiosperm life cycle. In angiosperms, a large sporophyte alternates with a tiny gametophyte.

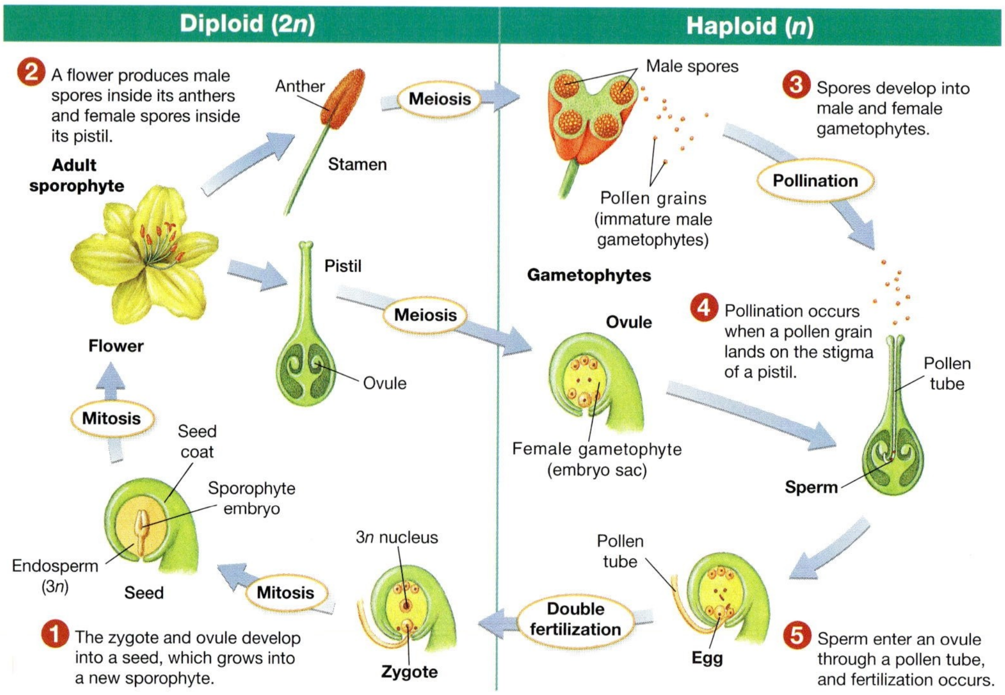

Section 2 Review

1. **Distinguish** pollen grains from ovules. SC.F.1.4.2
2. **Describe** the function of each part of a seed. SC.F.1.4.2
3. **Summarize** the life cycle of a conifer. SC.F.2.4.1
4. **Describe** the main events in the life cycle of an angiosperm? SC.F.2.4.1
5. **Critical Thinking Relating Concepts** How is each part of a flower suited to its function? SC.F.1.4.2
6. FCAT Prep What term is used to describe the sacs that produce pollen in angiosperms?
   A anthers **C** pistils
   B ovules **D** sepals
   SC.F.1.4.2

540

Answers to Section Review

1. A pollen grain is an immature male gametophyte of a seed plant. An ovule is a sporophyte structure that contains a female gametophyte in a seed plant.
2. The embryo is the young sporophyte, the food reserves (endosperm) nourish the young sporophyte when the seed germinates, and the seed coat protects the embryo from harsh environmental conditions.
3. First, a zygote grows into a sporophyte by mitotic cell division. Next, the sporophyte forms male and female spores in cones following meiotic cell division. Then, the spores grow into male and female gametophytes by mitotic cell division. After the gametophytes form eggs and sperm by mitotic cell division, eggs are fertilized by sperm and a new zygote forms. The zygote and surrounding tissue develop into a seed.
4. Answers may vary.
5. Answers should summarize the information in Figure 11.
6. **A.** Correct. **B.** Incorrect. The ovules produce female gametophytes. **C.** Incorrect. The pistils produce ovules. **D.** Incorrect. The sepals protect a flower from damage while it is a bud.

Asexual Reproduction

Section 3

Florida Sunshine Standards/FCAT — SC.F.2.4.1 There are mechanisms of asexual and sexual reproduction and different genetic advantages and disadvantages of asexual and sexual reproduction.

Vegetative Reproduction

Most plants are able to reproduce asexually. The new individuals that result from asexual reproduction are genetically the same as the parent plant. Plants reproduce asexually in a variety of ways that involve nonreproductive parts, such as stems, roots, and leaves. The reproduction of plants from these parts is called **vegetative reproduction.** Many of the structures by which plants reproduce vegetatively are modified stems, such as runners, bulbs, corms, rhizomes, and tubers. **Table 1** describes these structures.

Vegetative reproduction is faster than sexual reproduction in most plants. A single plant can spread rapidly in a habitat that is ideal for its growth by reproducing vegetatively. Therefore, a mass of hundreds or even thousands of individuals, such as a stand of grasses or ferns, may have come from one individual. To learn about one unique method of vegetative reproduction in one plant, look at *Up Close: Kalanchoë,* on the next two pages.

Table 1 Stems Modified for Vegetative Reproduction

Name	Description	Examples
Runner	Horizontal, above-ground stem	Airplane plant, Bermuda grass 
Bulb	Very short, stem with thick, fleshy leaves; only in monocots	 Onion, daffodil, tulip
Corm	Very short, thickened, underground stem with thin, scaly leaves	Gladiolus, crocus 
Rhizome	Horizontal underground stem	 Iris, fern, sugar cane
Tuber	Swollen, fleshy, underground stem	Potato, caladium 

Objectives

- **Describe** several types of vegetative reproduction in plants. SC.F.2.4.1
- **Distinguish** sexual reproduction in kalanchoës from asexual reproduction in kalanchoës. SC.F.2.4.1
- **Recommend** several ways to propagate plants. SC.F.2.4.1

Key Terms

vegetative reproduction
plant propagation
tissue culture

Section 3

Focus

Overview

Before beginning this section review with your students the objectives listed in the Student Edition. The purpose of this lesson is to introduce the different ways that plants reproduce without forming seeds. The lesson describes asexual reproduction in one plant species in detail. It concludes with a discussion of how plants can be propagated vegetatively.

Bellringer

Ask students to think of examples of how flowering plants reproduce without making seeds. Help them by asking about gardening, hiking, or other nature experiences they may have had. Students may have seen clumps of plants of the same type, strawberry runners, or clusters of bulbs. Ask students why all the plants in a lawn might look the same. (They are all clones from one plant.) **LS Interpersonal**

Motivate

Demonstration — BASIC

Show students example of bulbs (lily, onion, tulip); corms (crocus, gladiolus); rhizomes (iris); runners (airplane plant, strawberry); and tubers (potato). Point out that all of these are modified stems and therefore have buds that can grow into new plants. Demonstrate vegetative propagation by rooting an airplane plant or a potato. Be sure to include a bud in the section you set up for rooting. **LS Visual** English Language Learners

Chapter Resource File

- Directed Reading BASIC
- Active Reading GENERAL
- Data Sheet for Exploration Lab GENERAL

Transparencies

TR Bellringer

TR G45 Stems Modified for Vegetative Reproduction

One-Stop Planner CD-ROM

- Reading Organizers BASIC
- Reading Strategies BASIC
- Supplemental Reading Guide *A Feeling for the Organism: The Life and Work of Barbara McClintock* ADVANCED

Section Correlations

SC.F.2.4.1

Kalanchoë

Teaching Strategies Bring specimens of *Kalanchoë daigremontiana* to class. Let students propagate the plant by planting stem cuttings, leaf cuttings, or plantlets.

- Point out to students that the formation of plantlets along the leaf margin that occurs in kalanchoës is a very unusual form of asexual reproduction that occurs in only a few species. Tell students that these plants are often grown commercially because of their unusual appearance and method of asexual reproduction. Also tell students that because kalanchoës are easy to grow, they often become serious greenhouse weeds.
- For interest, take students to a greenhouse where succulents are grown, or have a horticulturist talk to students about the characteristics and propagation of kalanchoës and other succulents.
- CAM photosynthesis may be difficult for students to comprehend. Help students understand CAM photosynthesis by having them illustrate the process.

LS Visual

Transparencies

TR G46 External Structure of Kalanchoë

TR G47 Internal Structures of Kalanchoë

Kalanchoë

- **Scientific name:** *Kalanchoë daigremontiana*
- **Size:** Grows from 30 cm (about 1 ft) to 1 m (3.3 ft) tall
- **Range:** Native to southwestern Madagascar; cultivated worldwide
- **Habitat:** Semiarid tropical grassland with moist summers and well-drained, fertile soil
- **Importance:** Kalanchoës *(kal an KOH eez)* are grown as indoor potted plants and as outdoor perennials in warm climates.

External Structures

Leaves The fleshy leaves are bluish green, with purple markings and saw-toothed margins. Leaf blades range from 12 to 25 cm (about 5 to 10 in.) long. Leaves are arranged in pairs that are opposite one another.

Flowers A cluster of flowers forms on a flowering stalk that grows from the end of a stem. The flowers are bell-shaped and about 2.5 cm (1 in.) long. Flower parts occur in fours. Each flower produces many tiny seeds.

Plantlets Tiny new plants develop along leaf margins. These plantlets are a means of vegetative reproduction. When a plantlet falls to the ground, it grows into a new plant.

Stem and leaf cuttings Kalanchoës are often propagated vegetatively by planting stem and leaf cuttings.

Air roots The roots that grow from the stems and from plantlets originate from stem tissue.

542

Attention Grabber

Fruit Tree Grafting Many commercial fruit trees such as those that produce apples and citrus fruits, are composed of one variety of a species grafted onto another variety. The root stock is often that of a hardy variety that can withstand harsh environmental conditions and is resistant to common diseases. The scion (SEYE on), or the part that is grafted onto the root stock, is a variety that has desirable characteristics (such as flavor, color, early fruit set), but it is not always well adapted to the environmental conditions under which it is grown.

INCLUSION Strategies

• *Developmental Delay*

Bring a variety of vegetation, such as those listed in **Table 1,** for students to observe, touch, and sketch during class. Placing them in dirt, water, or in a cool dark place over time will allow roots and stems to sprout and begin to grow. The changes observed each day should be discussed, recorded, and drawn. Student understanding of the different stems used for vegetative reproduction would be greater for having observed the process in person.

Internal Structures

Leaf structure Kalanchoës are succulents, which means they have fleshy leaves and stems that store water. A kalanchoë leaf shows how some succulents are adapted for conserving water. A thick cuticle covers the leaf, and the epidermis (outer layer of cells) consists of several layers of cells. Relatively few, very small stomata dot the leaf surfaces.

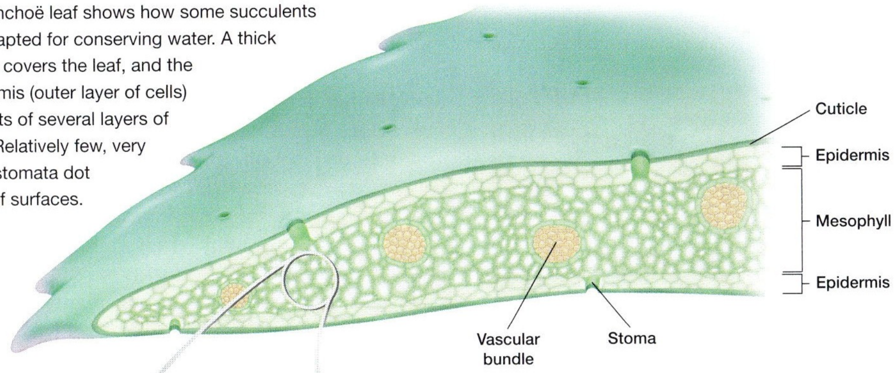

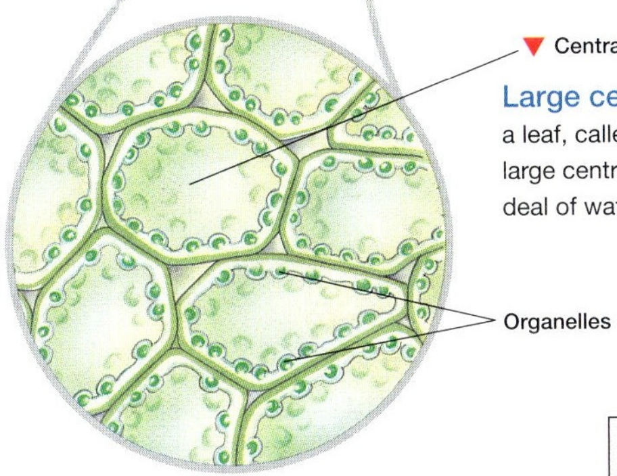

Mesophyll cells

Large central vacuole The cells inside a leaf, called the mesophyll cells, have a large central vacuole that can hold a great deal of water.

Mesophyll cell

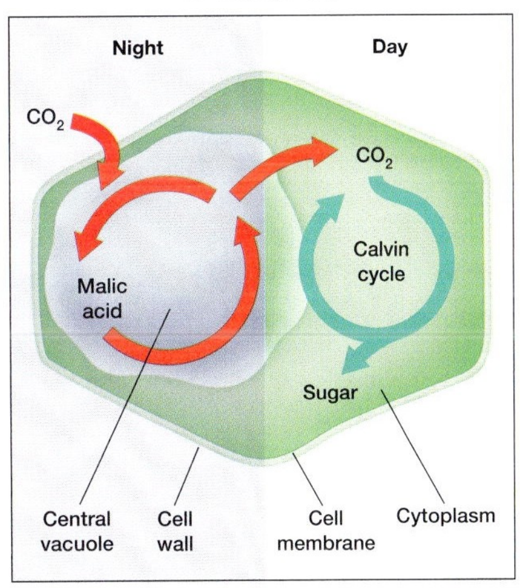

CAM photosynthesis Kalanchoës belong to the Crassulaceae family, a group of succulent plants that are adapted to hot climates. Photosynthesis in kalanchoës involves a process called crassulacean acid metabolism (CAM).

The stomata of CAM plants open only at night, unlike those of other plants. At night, the plants fix carbon dioxide by using it to make malic acid. The malic acid is stored in the large central vacuoles of the mesophyll cells. In daytime, the stomata remain closed, which prevents water loss. Carbon dioxide is released from malic acid during the day and used by the Calvin cycle to make sugar.

543

Up Close

Kalanchoë

Discussion

- Ask students to describe sexual and asexual reproduction in kalanchoës. (Kalanchoës are flowering plants that reproduce sexually by forming seeds following pollination and fertilization. They reproduce asexually by forming plantlets containing roots, stems, and leaves on the margins of the mother plant's leaves. They can also produce new plants from pieces of stems and leaves.)
- Ask students how the genetic composition of a plantlet compares with that of the mother plant. (The two are identical because the plantlets are clones of the mother plant.)
- Ask students to name some adaptations found in kalanchoës that enable them to survive in dry environments. (Kalanchoës have thick, fleshy leaves and stems that store water; the leaves have a thick cuticle, several layers of epidermal cells, and few stomata; the plants photosynthesize using the CAM process, which enables them to have their stomata open at night.) **LS Interpersonal**

SOCIAL STUDIES CONNECTION

Like *Kalanchoë daigremontiana*, many other horticulturally important plants are natives of Madagascar, which is an island off the southeast coast of Africa. One such plant is the rosy periwinkle, *Catharanthus roseus*, which is a popular bedding plant in the United States and the original source of two cancer treatment drugs. Have students use the Internet to research other economically important plants to find out where they are grown and how much income they bring to the countries where they are grown.

did you know?

Asexual reproduction is energy efficient. Asexual reproduction enables plants to produce many offspring by using less energy than is required for sexual reproduction. However, because offspring produced asexually are genetically identical to their parent, they are all equally susceptible to any environmental condition or disease to which the parent was susceptible. Sexual reproduction results in genetic variation among the offspring, which is particularly advantageous when the environment changes often, or when new disease-causing agents enter the environment.

Close

Reteaching — Basic

Ask students to describe characteristics of plants that would indicate that they have been asexually propagated. (uniformity of appearance, as in flower color, fruit color and size, plant size, etc.)

Quiz — General

1. What is the most common part of a plant that is involved in asexual reproduction? (stems)
2. Can plants that reproduce asexually also reproduce sexually? (yes)
3. How do CAM plants compare with most other plants in the way in which they undergo photosynthesis? (CAM plants have their stomata open at night, when the stomata of most other plants are closed. Carbon dioxide is fixed during the night, stored, and then used in the Calvin cycle to make sugar during the day.) **LS Verbal**

Alternative Assessment — General

Garden Design Have students use poster board to design a garden that includes both sexually and asexually reproducing plants. Students may wish to use magazines or plant catalogs to cut out pictures of plants used in their design, or they may draw simple schematics, similar to landscaping plans. Have students include the name of each plant and how it reproduces in table format. English Language Learners

Transparencies

TR G48 Methods of Vegetative Plant Propagation

Section Review Correlations

SC.F.2.4.1: Items 1-5

Plant Propagation

People grow plants for many purposes, such as for food, to beautify homes, or to sell. Most field crops, such as cereal grains, vegetables, and cotton, are grown from seed. Many other plants are grown from vegetative parts. Growing new plants from seed or from vegetative parts is called **plant propagation.**

Plants are often propagated using the structures the plants produce for vegetative reproduction. Bulbs and corms divide as they grow, forming many pieces that can each grow into a new plant. Rhizomes, roots, and tubers can be cut or broken into pieces with one or more buds that can grow into new shoots. But people also grow plants from vegetative parts that are not specialized for vegetative reproduction. For example, pieces of plants, such as the stems of ivys and the leaves of African violets, are cut from the parent plant. The cuttings are then used to grow new plants. **Figure 12** shows a method of propagating trees called budding. In another technique called **tissue culture,** pieces of plant tissue are placed on a sterile medium and used to grow new plants. **Table 2** summarizes some of the methods of vegetative plant propagation that are widely used to grow plants.

Figure 12 Budding pears. A bud from a desirable variety of pears is attached to a stem of another pear species. The bud will grow into a branch that produces the desirable variety of pears.

Table 2 Methods of Vegetative Plant Propagation

Method	Description	Examples
Budding and grafting	Small stems from one plant are attached to larger stems or roots of another plant.	Grape vines, hybrid roses, fruit and nut trees
Taking cuttings	Leaves or pieces of stems or roots are cut from one plant and used to grow new individuals.	African violets, ornamental trees and shrubs, figs
Tissue culture	Pieces of tissue from one plant are placed on a sterile medium and used to grow new individuals.	Orchids, potatoes, many houseplants

Section 3 Review

1. **Describe** four types of vegetative reproduction in plants, and give an example of each. SC.F.2.4.1
2. **Classify** methods of reproduction in kalanchoës as sexual or asexual. SC.F.2.4.1
3. **Recommend** five ways to propagate plants. SC.F.2.4.1
4. **Critical Thinking Justifying Conclusions** Why would someone choose to propagate a particular plant for commercial purposes by using vegetative structures instead of seed? SC.F.2.4.1
5. **FCAT Prep** What term is used to describe the aboveground stems that allow Bermuda grass to reproduce asexually? SC.F.2.4.1
   A corms
   B rhizomes
   C runners
   D tubers

Answers to Section Review

1. Bulbs are short stems with thick, fleshy leaves (daffodil, onion, tulip); corms are short, thickened underground stems with thin, scaly leaves (crocus, gladiolus); rhizomes are horizontal underground stems (fern, iris, sugar cane); tubers are swollen, fleshy, underground stems (potato, caladium); runners are aerial stems that form roots (airplane plant, kalanchoë); plantlets form on leaves (kalanchoë)
2. Formation of seeds is sexual reproduction. Formation of plantlets on the leaf margins or growth of new plants from pieces of stems or leaves is asexual reproduction.
3. planting seeds; planting vegetative reproductive structures such as bulbs, corms, tubers, rhizomes, or plantlets; budding or grafting; taking stem or leaf cuttings; tissue culture
4. Someone might choose to propagate a plant commercially using vegetative structures to produce a new crop rapidly and/or to produce a crop of identical plants.
5. **A.** Incorrect. Corms are short, underground stems. **B.** Incorrect. Rhizomes are horizontal, underground stems. **C.** Correct. **D.** Incorrect. Tubers are swollen, fleshy, underground stems.

Study ZONE

Key Concepts

1 Sexual Reproduction in Seedless Plants

- In mosses, the "leafy" green gametophytes are larger than the sporophytes, which consist of a bare stalk and a spore capsule.
- In the life cycle of a fern, the sporophytes are much larger than the gametophytes. The thin, green, heart-shaped gametophytes produce both sperm and eggs.
- Nonvascular plants and seedless vascular plants need water for fertilization because sperm must swim to eggs.

2 Sexual Reproduction in Seed Plants

- The tiny gametophytes of seed plants develop from spores that remain within sporophyte tissues. Male gametophytes are pollen grains. Female gametophytes develop inside ovules.
- A seed contains an embryo, which is a new sporophyte, and a supply of nutrients for the embryo. The cotyledons of an embryo help transfer nutrients to the embryo. A seed coat covers and protects a seed.
- In gymnosperms, male and female gametophytes develop in separate cones on the sporophytes. After fertilization, ovules develop into seeds, which grow into new sporophytes.
- Flowers have four types of parts—petals, sepals, stamens, and pistils. Petals attract pollinators. Sepals protect buds and may also attract pollinators. Pollen forms in the anthers of stamens. Seeds develop in the ovary of a pistil.
- In angiosperms, male and female gametophytes develop in the flowers of the sporophytes. After fertilization, ovules develop into seeds, which grow into new sporophytes.

3 Asexual Reproduction

- Vegetative reproduction is the growth of new plants from nonreproductive plant parts, such as stems, roots, and leaves.
- Kalanchoës are succulents that are often grown as potted plants and readily reproduce either vegetatively or by seeds.
- People often grow plants from their vegetative structures. This is called vegetative propagation.

Key Terms

Section 1

archegonium (530)
antheridium (530)
sorus (532)

Section 2

pollen grain (534)
ovule (534)
pollination (534)
pollen tube (534)
seed coat (535)
cotyledon (535)
sepal (538)
petal (538)
stamen (538)
anther (538)
pistil (538)
ovary (538)
double fertilization (540)

Section 3

vegetative reproduction (541)
plant propagation (544)
tissue culture (544)

Alternative Assessment — GENERAL

Assign students to four cooperative groups. Have each group draw and label a life cycle diagram of a different type of plant (moss, fern, gymnosperm, angiosperm). Collect the diagrams, and then redistribute them to the groups for evaluation, making sure that none of the groups gets its own diagram back. Discuss each life cycle diagram, making corrections as necessary.

Chapter Resource File

- Science Skills Worksheet GENERAL
- Critical Thinking Worksheet ADVANCED
- Test Prep Pretest GENERAL
- Chapter Test GENERAL

Answer to Concept Map

The following is one of several possible answers to Performance Zone item 15.

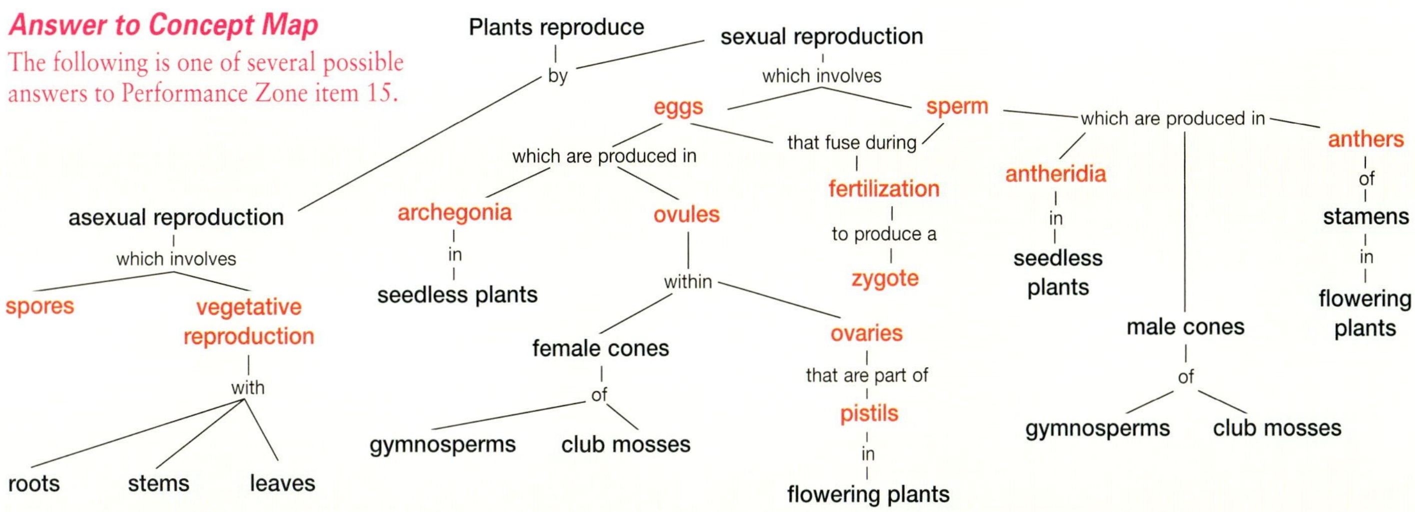

Performance ZONE

Understanding Key Ideas

1. Mosses and liverworts thrive in a moist environment because they need _____ for reproduction. SC.F.2.4.1
a. bees
b. birds
c. water
d. wind

2. The life cycle of a moss differs from the life cycle of a fern in that
a. the gametophyte is absent in ferns.
b. the sporophyte is absent in mosses.
c. moss spores do not form on leaves.
d. the gametophytes of mosses are green. SC.F.2.4.1

3. In angiosperms, the zygote and the first cell of the endosperm form by SC.F.1.4.5
a. mitosis.
b. meiosis.
c. pollination.
d. double fertilization.

4. Vegetative reproduction has *not* occurred when a new plant grows from a SC.F.2.4.1
a. leaf.
b. root.
c. stem.
d. seed.

5. Which of the following structures do kalanchoës produce for vegetative reproduction? SC.F.2.4.1
a. seeds
b. plantlets
c. flowers
d. bulbs

6. **BIOWatch** How can forensic botany help link a suspect to a location? SC.H.3.4.3

7. What is the function of the fruits in which seeds mature? (**Hint:** See Chapter 23, Section 1.) SC.F.1.4.2

8. Which of the following structures is *not* used to propagate dicots vegetatively?
a. tubers
b. rhizomes
c. seeds
d. stem cuttings SC.F.1.4.2

9. Look at the flower in the photograph below. It is the flower of the unicorn plant. How is this flower probably pollinated? Justify your answer. SC.G.1.4.1

10. **Concept Mapping** Make a concept map that explains how plants reproduce. Try to include the following terms in your map: *archegonium, antheridium, egg, sperm, ovule, zygote, stamen, anther, pistil, ovary, fertilization, spore,* and *vegetative reproduction.* SC.F.1.4.2

Critical Thinking

11. **Evaluating Conclusions** All nonvascular plants require a film of water for sperm to swim through and fertilize eggs. Therefore, many people conclude that nonvascular plants are not able to survive in very dry climates, such as deserts. Is this a valid conclusion? Justify your answer. SC.F.1.4.2

12. **Justifying Conclusions** A classmate has found a plant whose flowers lack petals and have many stamens. Your classmate tells you that the plant is wind-pollinated. Justify this conclusion. SC.F.1.4.2

13. **Evaluating Methods** You are asked to grow a large number of identical potted plants for a florist. The plants can be grown from either seeds or cuttings. Which method of plant propagation would you use? Justify your choice. SC.F.2.4.1

Alternative Assessment

14. **Career Connection Plant Breeder** Use the media center or Internet to find out about the field of plant breeding. Write a report on your findings. SC.F.2.4.1

546

ANSWERS

Understanding Key Ideas

1. c
2. c
3. d
4. d
5. b
6. Investigators can collect pollen from objects or from the suspect, then match the kinds of pollen to the pollen found in a potential location.
7. The dispersal of seeds is the main function of fruits.
8. c
9. Students should recognize that this flower is probably bee pollinated because it is yellow, has a landing platform, and has pollen guides showing the location of nectar.
10. One possible answer to Performance Zone item 15 is found on the bottom of the Study Zone page.

Critical Thinking

11. Answers will vary. Students should offer evidence to support their conclusion. Some students may recognize that this conclusion is not valid because nonvascular plants are found in dry climates. These plants survive in shady spots under trees and rocks, where moisture accumulates occasionally, and near springs, creeks, and rivers.
12. The primary role of petals is to attract animal pollinators. Therefore, the absence of petals is a strong indication that the plant is wind-pollinated.
13. Answers will vary, but students should recommend a type of vegetative propagation, such as cuttings or tissue culture, since these methods produce plants that are genetically identical to each other and to their parent plant.

Alternative Assessment

14. Plant breeders are scientists who work to improve plants that have commercial value. Plant breeders make crosses between plants that have desirable characteristics and select plants with the most desirable combination of traits. Some plant breeders use genetic engineering in their work. Plant breeders work for universities, federal agencies, and private industries. To become a plant breeder, you must obtain a master's degree or a Ph. D. in agronomy, horticulture, or plant breeding. Starting salary will vary by region.

Assignment Guide

Section	Questions
1	1, 2, 10, 11
2	3, 6, 7, 9, 10, 12, 13, 14
3	4, 5, 8, 13

Chapter Review Correlations

SC.F.1.4.2: Items 7, 8, 10, 11, 12
SC.F.1.4.5: Item 3
SC.F.2.4.1: Items 1, 2, 4, 5, 13, 14
SC.G.1.4.1: Item 9
SC.H.3.4.3: Item 6

FCAT Prep

For questions 1–7, write your answers on a separate sheet of paper.

1. What is the purpose of the archegonium in mosses and ferns? SC.F.1.4.2
   A. produce eggs
   B. produce sperm
   C. develop into a cone
   D. develop into a seed

2. In seed plants, what structure transfers sperm from a pollen grain directly to an egg in an ovule? SC.F.1.4.2
   F. endosperm **H.** pollinator
   G. pollen tube **I.** seed coat

3. What type of plant produces a seed with a single cotyledon? SC.G.1.4.1
   A. dicot
   B. gymnosperm
   C. monocot
   D. nonvascular

4. Which part of a flower produces eggs? SC.F.1.4.2
   F. petal **H.** sepal
   G. pistil **I.** stamen

5. What conclusion can be drawn about a plant whose flowers are large and colorful? SC.F.1.4.2
   A. The plant must be self-pollinated.
   B. The plant must be a gymnosperm.
   C. The plant must be wind-pollinated.
   D. The plant must be pollinated by insects.

6. READ INQUIRE EXPLAIN Plant breeders are scientists who work to improve plants that have commercial value.

   Estimate how plant breeders could breed plants with the most desirable flower color. SC.H.3.4.2

7. READ INQUIRE EXPLAIN Use of pesticides could reduce the number of plants in a geographic area. Pesticides can kill insect pollinators such as bees and flies. Reducing the number of insect pollinators could result in a reduced number of flowering plants.

   Part A Why would a reduction in the number of insect pollinators cause a reduction in the number of flowering plants? SC.F.2.4.1

   Part B Construct other ways to reproduce flowering plants when there are no insects. SC.G.1.4.1

Test TIP

For questions involving life cycles, draw as much of the life cycle as you can remember. Looking at such a model may help you understand the question better and help you determine the correct answer.

547

FCAT Doctor

Question 5 Answer D is the correct choice. Large, colorful flowers attract insects and other pollinators. Answer A is incorrect because there is no reason for a self-pollinating plant to have structures that attract pollinators. Answer B is incorrect because flowering plants are angiosperms. Answer C is incorrect because wind-pollinated plants do not need petals at all.

Question 6 By selecting plants with the desired flower color and crossing them, plant breeders can produce plants with the desired color of flower.

Question 7 Part A Most seed plants are flowering plants, or angiosperms. The flowers promote pollination and fertilization with their bright colors or strong scents.

Part B Birds, bats, and the wind could disperse the seeds, acting as pollinators for the flowering plants.

Answers

1. A
2. G
3. C
4. G
5. D
6. Plant breeders could make crosses between plants that have the desired color.

7A. Flowering plants are angiosperms and depend on pollination to reproduce.

7B. The plants could reproduce by having birds, mammals, or wind pollinate their flowers.

Test Prep Correlations

SC.F.1.4.2: Items 1, 2, 4, 5
SC.F.2.4.1: Item 7A
SC.G.1.4.1: Items 3, 7B
SC.H.3.4.2: Item 6

Exploration Lab

Observing the Effects of Nutrients on Vegetative Reproduction

Teacher's Notes

Time Required
Day 1, 45 minutes; then 10 minutes each day for 2 weeks

Ratings
EASY → HARD

TEACHER PREPARATION: 3

STUDENT SETUP: 2

CONCEPT LEVEL: 2

CLEANUP: 2

Skills Acquired
- Collecting Data
- Designing Experiments
- Experimenting
- Identifying/Recognizing Patterns
- Organizing and Analyzing Data
- Predicting

Scientific Methods
In this lab, students will:
- Make Observations
- Ask Questions
- Test the Hypothesis
- Analyze the Results
- Draw Conclusions

Materials and Equipment
Materials for this lab can be ordered from WARD's. See *Master Materials List* at the front of this book for catalog numbers.

Safety Cautions
- Be sure that students have read and understand all of the safety rules for working in the lab.
- Remind students to read the ChemSafety section before beginning the lab. Discuss all safety symbols and caution statements with students.
- Make sure students wear safety goggles and a lab apron.
- Make sure students wash their hands thoroughly before leaving the laboratory.

Exploration Lab

Observing the Effects of Nutrients on Vegetative Reproduction

SC.G.1.4.3

SKILLS
- Using scientific processes
- Observing
- Graphing and analyzing data

OBJECTIVES
- **Identify** the structures of duckweed.
- **Compare** vegetative reproduction of duckweed in different nutrient solutions.

MATERIALS
- safety goggles
- lab apron
- duckweed culture
- 5 Petri dishes
- stereomicroscope or hand lens
- glass-marking pen
- beakers
- pond water
- Knop's solution
- 0.1% fertilizer solution
- distilled water

ChemSafety

CAUTION: Always wear safety goggles and a lab apron to protect your eyes and clothing.

CAUTION: Do not touch or taste any chemicals. Know the location of the emergency shower and eyewash station and how to use them. If you get a chemical on your skin or clothing, wash it off at the sink while calling to the teacher. Notify the teacher of a spill. Spills should be cleaned up promptly, according to your teacher's directions.

CAUTION: Glassware is fragile. Notify the teacher of broken glass or cuts. Do not clean up broken glass or spills with broken glass unless the teacher tells you to do so.

Before You Begin

Duckweed is a common aquatic plant. Like many flowering plants, duckweed reproduces readily by **vegetative reproduction**, which is a type of **asexual reproduction**. As individual plants grow, they divide into smaller individuals. Several individuals may remain joined together, forming a mat. All plants require certain **mineral nutrients**, such as nitrogen, phosphorus, and potassium, for the growth of vegetative parts. In this lab, you will investigate the effect of nutrients on the vegetative reproduction of duckweed.

1. Write a definition for each boldface term in the paragraph above.
2. Based on the objectives for this lab, write a question you would like to explore about vegetative reproduction in duckweed.

Procedure

PART A: Make Observations

1. Place a duckweed plant in a Petri dish. Then place a few drops of water on the plant.
2. Observe the duckweed plant with a stereomicroscope or a hand lens. Sketch what you see. Label the structures that you recognize.

PART B: Design an Experiment

3. Work with members of your lab group to explore one of the questions written for step 2 of **Before You Begin.** To explore the question, design an experiment that uses the materials listed for this lab.

548

Tips and Tricks
Obtain duckweed from a pond or a quiet stream. You can also purchase duckweed from a garden center or from a biological supply company. To make a 0.1 percent fertilizer solution, place 1 g of a commercial fertilizer (with an analysis of at least 23-19-17) in a large container, and add enough distilled water to make 1 L of solution. Stir until the fertilizer is dissolved. Place enough materials for a class on a supply table. Set up labeled containers for the disposal of solutions, broken glass, and duckweed.

Before students begin the lab, conduct a class discussion in which you ask the following questions:

1. What is vegetative reproduction? (the asexual process of making new individuals from nonreproductive parts of an organism)
2. How do mineral nutrients affect the growth of vegetative structures in plants? (They provide additional elements needed for making organic macromolecules and for the growth of the vegetative parts.)

Require students to present a written procedure for their experiment and a list of all safety precautions before allowing them to gather materials for Part C of the lab.

You Choose

As you design your experiment, decide the following:

a. what question you will explore
b. what your hypothesis will be
c. what solutions to test
d. how much of each solution to use
e. how many individuals to use for each test
f. what your control will be
g. how you will judge which solution is the best
h. what data to record in your data table

4. Write a procedure for your experiment. Make a list of all the safety precautions you will take. Have your teacher approve your procedure and safety precautions before you begin the experiment.

PART C: Conduct Your Experiment

5. Put on safety goggles and a lab apron.

6. Set up your experiment. **CAUTION: Nutrient solutions are mild eye irritants. Avoid contact with your skin and eyes.** Complete step 8.

7. Conduct your experiment and collect data for two weeks.

PART D: Cleanup and Disposal

8. Dispose of solutions, broken glass, and duckweed in the designated waste containers. Do not pour chemicals down the drain or put lab materials in the trash unless your teacher tells you to do so.

9. Clean up your work area and all lab equipment. Return lab equipment to its proper place. Wash your hands thoroughly before you leave the lab and after you finish all work.

Analyze and Conclude

1. Summarizing Results Compare the appearance of plants growing in each nutrient solution with that of the plants in distilled water. Explain your observations.

2. Analyzing Data In which Petri dish did the greatest amount of growth (increase in numbers) take place?

3. Analyzing Results In which Petri dish did the least amount of growth take place?

4. Evaluating Hypotheses Did the results you observed agree with your hypothesis? If not, how are they different?

5. Recognizing Patterns As the number of new duckweed plants in a particular group increased, what happened to the group of plants?

6. Graphing Data Make a graph of your data. Label the *y*-axis "Number of plants," and the *x*-axis "Days." Use a different color to represent each solution you tested.

7. Drawing Conclusions What factors regulate the rate of vegetative reproduction in duckweed?

8. Evaluating Methods Why are the new duckweed plants produced by vegetative reproduction genetically the same as the parent plant?

9. Further Inquiry Write a new question about vegetative reproduction in duckweed that could be explored with another investigation.

Do You Know?

Do research in the library or media center to answer these questions:

1. What do the flowers of duckweeds look like?

2. What is the smallest species of flowering plants, and how small are they?

Use the following Internet resources to explore your own questions about duckweed.

internet connect
www.scilinks.org
Topic: Duckweed
Keyword: HX4062
SCILINKS. Maintained by the National Science Teachers Association

549

Students should see abundant vegetative reproduction in the 2-week period.

Dilute nutrient solutions with water, and pour them down the sink.

Wrap duckweed in newspaper and place into a trash can.

Answers to Before You Begin

1. *vegetative reproduction*—a type of asexual reproduction that results in the growth of new individuals from nonreproductive parts, such as leaves, stems (rhizomes, tubers, bulbs, corms), and roots; *asexual reproduction*—reproduction that occurs without the union of gametes that involve a single parent producing offspring that are genetically identical to the parent; *mineral nutrients*—elements and inorganic compounds that are needed for plant growth and development
2. For example: Which nutrients are needed for vegetative reproduction in duckweed?

Answers to Do You Know?

1. The flowers of duckweeds consist of only one or two stamens and a pistil, which are located in a pocket on the plant.
2. Two species of the genus *Wolfia* are less than 1 mm in diameter.

Sample Procedure

1. Label four Petri dishes to indicate one of each of these test solutions: distilled water, pond water, Knop's solution, 0.1% fertilizer.
2. Fill each Petri dish three-quarters full with the solution indicated on the dish.
3. Place six duckweed plants in each Petri dish. Cover the dishes, and place them in a well lighted area.
4. Observe the dishes every day for 2 weeks. Record data on number of plants in a dish, number of plants in each cluster, and general appearance of plants.

Answers to Analyze and Conclude

1. Plants grown in Knop's solution should be the largest and darkest green. Plants grown in pond water should also look healthy and green. Plants grown in dilute fertilizer solution or distilled water should be smaller and lighter green.
2. The greatest amount of growth should occur in the dish with Knop's solution.
3. The least amount of growth should occur in distilled water.
4. Answers will vary. Students should state how their results differ from their hypotheses.
5. The clusters divide as they get larger.
6. Answer will vary.
7. Students should recognize that more mineral nutrients will produce more growth.
8. Vegetative reproduction involves mitotic cell division, and mitosis results in the production of genetically identical cells
9. Answers will vary.

Chapter Opener

SE **Quick Review,** p. 551 ■
SE **Reading Activity,** p. 551 ■
TE **Using the Figure,** p. 550
TE **Opening Activity,** p. 551

---

3 • 90 min pp. 552–559

Section 1
The Vascular Plant Body

OSP **Lesson Plan***
TR **Bellringer***
TR **Section Outline***
TR **G14 Structure of Xylem***
TR **G15 Structure of Phloem***
TR **G17 Radish Root Structure***
TR **G23 Dicot and Monocot Stem Structures***
TR **G22 Structure of Stems***
TR **G28 Simple and Compound Leaves***
TR **G30 Structure of a Leaf***

TE **Activity** Wetting Agents, p. 552 GENERAL
TE **Group Activity** Tissue Types, p. 553 GENERAL
TE **Using the Figure,** p. 554 BASIC
TE **Demonstration,** p. 555 GENERAL
TE **Activity** Vascular Bundles, p. 556 GENERAL
SE **Quick Lab** Comparing the Structures of Roots and Stems, p. 557 ■
CRF **Datasheets for In-Text Labs** Comparing the Structures of Roots and Stems*
TE **Using the Figure,** p. 558 BASIC
SE **Quick Lab** Observing the Structures Inside a Leaf, p. 558 ■
CRF **Datasheets for In-Text Labs** Observing the Structures Inside a Leaf*
CRF **Quick Lab** Observing Stems* BASIC

---

5 • 90 min pp. 560–564

Section 2
Transport in Plants

OSP **Lesson Plan***
TR **Bellringer***
TR **Section Outline***
TR **G32 Control of Stomatal Opening***
TR **G26 Water Movement in Plants***
TR **G25 The Pressure-Flow Model***
TR **G9 Sugar Maple***
TR **G10 Sugar Maple, cont.***

TE **Activity** Surface Tension, p. 560 GENERAL
SE **Data Lab** Inferring the Rate of Transpiration, p. 561 ■
CRF **Datasheets for In-Text Labs** Inferring the Rate of Transpiration*
CRF **Skills Practice Lab** Transpiration* ◆ GENERAL
SE **Skills Practice Lab** Separating Plant Pigments, pp. 568–569 ■
CRF **Datasheets for In-Text Labs** Separating Plant Pigments*

---

7 • 90 min

Chapter Review and Assessment Resources

SE **Study Zone,** p. 565 ■
SE **Performance Zone,** p. 566 ■
SE **FCAT Test Prep,** p. 567 ■
TR **Concept Mapping***
CRF **Vocabulary Review Worksheet*** ■ GENERAL
CRF **Test Prep Pretest*** ■ GENERAL
CRF **Chapter Test*** ■ GENERAL
CRF **Chapter Test*** ADVANCED
OSP **Test Generator***
OSP **Test Item Listing***
OSP **Modified Chapter Test*** BASIC

Online and Technology Resources

go.hrw.com Holt Online Learning

Visit **go.hrw.com** for access to Holt Online Learning or enter the keyword **HX6 Home** for a variety of free online resources.

One-Stop Planner® CD-ROM

This CD-ROM package includes
- Lab Materials QuickList Software
- Holt Calendar Planner
- Customizable Lesson Plans
- Printable Worksheets
- ExamView® Test Generator
- Interactive Teacher Edition
- Holt PuzzlePro® Resources
- Holt PowerPoint® Resources

Guided Reading Audio CDs

These CDs are designed to help auditory learners and reluctant readers. (Audio CDs are also available in Spanish.)

Compression guide:
To shorten your instruction because of time limitations, omit block 1.

KEY

TE	Teacher Edition	CRF	Chapter Resource File	*	Also on One-Stop Planner
SE	Student Edition	TR	Teaching Transparency	■	Also Available in Spanish
OSP	One-Stop Planner			◆	Requires Advance Prep

SKILLS DEVELOPMENT AND PRACTICE	SECTION REVIEW AND ASSESSMENT	CORRELATIONS
		Florida Sunshine Standards/FCAT
TE **Reading Skill Builder** Reading Organizer, p. 553 BASIC TE **Skill Builder** Interpreting Graphics, p. 555 BASIC TE **Reading Skill Builder** Interactive Reading, p. 555 BASIC CRF **Directed Reading Worksheet*** BASIC CRF **Active Reading Worksheet*** GENERAL	SE **Section 1 Review** ■ TE **Reteaching,** p. 559 BASIC TE **Quiz,** p. 559 GENERAL CRF **Quiz*** ■ BASIC	SC.F.1.4.2
CRF **Directed Reading Worksheet*** BASIC CRF **Active Reading Worksheet*** GENERAL CRF **Critical Thinking Worksheet*** ADVANCED CRF **Science Skills Worksheet*** GENERAL	SE **Section 2 Review** ■ TE **Reteaching,** p. 564 BASIC TE **Quiz,** p. 564 GENERAL TE **Alternative Assessment,** p. 564 GENERAL CRF **Quiz*** ■ BASIC **Modified Worksheet*** BASIC	SC.F.1.4.1, SC.F.1.4.2, SC.H.3.4.6

SCILINKS. NSTA
www.scilinks.org
Maintained by the **National Science Teachers Association**

Topic: Vascular Plants
SciLinks Code: HX4182

Topic: Transpiration
SciLinks Code: HX4178

CHAPTER 25

Overview

Tell students the purpose of this chapter is to help them understand that plants are composed of structures that are modified to help them survive in their environments. Specialized cells in leaves, for example, are adapted for capturing sunlight and carrying out photosynthesis. Stems and roots are also adapted to perform their jobs, especially in transporting materials throughout the plant.

Using the Figure

Have students study the picture of the kokerboom. Ask them what adaptations it has or may have to allow it to survive the punishing environment of the Namibian desert. (Ideas might include thick bark; thick, fleshy leaves to hold water; and long, penetrating roots.)

Assessing Prior Knowledge

Before beginning this chapter, be sure your students can answer the questions in the ***Quick Review*** on the facing page. If they cannot, encourage them to re-read the section indicated.

A kokerboom, or quiver tree, *Aloe dichotoma*

550

Chapter Correlations — **Florida Sunshine Standards/FCAT**

SC.F.1.4.1 Body processes involve specific biochemical reactions governed by biochemical principles.

SC.F.1.4.2 Body structures are uniquely designed and adapted for their function.

SC.H.3.4.6 Scientific knowledge is used by those who engage in design and technology to solve practical problems, taking human values and limitations into account.

CHAPTER

25 Plant Structure and Function

Florida Sunshine Standards/FCAT

SC.F.1.4.1 Body processes involve specific biochemical reactions governed by biochemical principles.

SC.F.1.4.2 Body structures are uniquely designed and adapted for their function.

SC.H.3.4.6 Scientific knowledge is used by those who engage in design and technology to solve practical problems, taking human values and limitations into account.

Quick Review

Answer the following without referring to earlier sections of your book.

1. **Explain** what causes the adhesion and cohesion of water. *(Chapter 2, Section 2)*
2. **Describe** the organization of a plant cell. *(Chapter 3, Section 3)*
3. **Summarize** the steps in photosynthesis. *(Chapter 5, Section 2)*
4. **State** the relationship between stomata and guard cells. *(Chapter 23, Section 1)*
5. **Define** the terms *cuticle, stoma, xylem, phloem,* and *meristem. (Chapter 23, Section 1)*

Did you have difficulty? *For help, review the sections indicated.*

Reading Activity

Before you read this chapter, survey the subtitles, headings, captions, and words in boldface type. Try to identify the purpose of this chapter. As you read this chapter, create a descriptive reading organizer.

Looking Ahead

The structure of the quiver tree enables it to survive in the harsh southern African climate. For generations, the native San people have used the tree's bark and branches to make arrow quivers.

internet connect

www.scilinks.org

National Science Teachers Association *sci*LINKS Internet resources are located throughout this chapter.

SciLINKS. Maintained by the National Science Teachers Association

Opening Activity

Plant Similarities and Differences Show students three very different plants, such as a cactus, a tomato, and a Venus' flytrap. Have the class compose a list of things that all three plants need to survive. (All the plants need water, carbon dioxide, light energy, and minerals for making organic molecules.) Ask students why these three plants look so different. (They have specialized tissues to accommodate their needs.)

Quick Review Answers

1. Water is a polar molecule, with regions of + and − charge. These regions are attracted to charged regions on other polar molecules (adhesion) as well as the oppositely charged region on other water molecules (cohesion).
2. Plant cells have rigid cell walls surrounding a plasma membrane that contains the cytoplasm and cellular organelles. Many plant cells also contain chloroplasts and large central vacuoles.
3. Light energy is captured and converted to chemical energy within the chloroplasts. With carbon dioxide, chemical energy is used to form organic molecules.
4. Guard cells surround and control the opening and closing of the stomata.
5. *Cuticle*—the waxy coating on a leaf; *stoma*—a tiny opening on a leaf that allows gas exchange; *xylem*—vascular tissue that transports water; *phloem*—vascular tissue that transports sugar produced during photosynthesis; *meristem*—a tissue that divides rapidly, increasing the height and diameter of a plant.

Reading Activity Answers

The purpose of this chapter is to study the parts of vascular plants and how they function.

Chapter Resource File

- Vocabulary Worksheets
- Concept Mapping

Section 1

Focus

Overview

Before beginning this section review with your students the objectives listed in the Student Edition. Plants have leaves, stems, and roots that are modified to meet the plant's needs. Each structure contains cells that perform particular functions that allow the plant to survive in its environment.

Bellringer

Have students think of outdoor plants and how they think the leaves, stems, and roots are adapted to the plant's environment. (Example: cactus, reduced leaves to prevent water loss; thick, fleshy stems to store water; and extensive roots to maximize water absorption.)

Motivate

Activity — GENERAL

Wetting Agents Provide students with wax paper, water, liquid detergent, and two beakers. Have students place two drops of water on wax paper and ask them what happens when the paper is tilted. (The drops of water roll off the paper.) Then have students place two drops of soapy water on the wax paper and describe what they see. (The soapy water spreads out over the wax paper.) Tell students that the detergent acts as a "wetting agent" by interfering with the attraction water molecules have for each other. Wetting agents are often mixed with insecticides so the drops will spread out and provide better coverage on the leaves.
LS Kinesthetic

Section 1 The Vascular Plant Body

Florida Sunshine Standards/FCAT

SC.F.1.4.2 Body structures are uniquely designed and adapted for their function.

Objectives

- **Identify** the three kinds of tissues in a vascular plant's body, and state the function of each. SC.F.1.4.2
- **Compare** the structures of different types of roots, stems, and leaves. SC.F.1.4.2
- **Relate** the structures of roots, stems, and leaves to their functions. SC.F.1.4.2

Key Terms

dermal tissue
ground tissue
epidermis
cork
vessel
sieve tube
cortex
root hair
root cap
herbaceous plant
vascular bundle
pith
heartwood
sapwood
petiole
mesophyll

Tissues

Like your body, a plant's body is made of tissues that form organs. In vascular plants, there are three types of tissue systems—the dermal tissue system, ground tissue system, and vascular tissue system. As you have read, vascular tissue forms strands that conduct water, minerals, and organic compounds throughout a vascular plant. **Dermal tissue** forms the protective outer layer of a plant. **Ground tissue** makes up much of the inside of the nonwoody parts of a plant, including roots, stems, and leaves. **Figure 1** shows how these three tissues are arranged in a nonwoody dicot. The tissues are organized a little differently in other types of vascular plants.

Each type of tissue contains one or more kinds of cells that are specialized to perform particular functions. As you read further in this chapter, you will learn that some specialized plant cells lack organelles found in other plant cells. In fact, some plant cells cannot perform their functions until they have lost most of their organelles.

Figure 1 Plant tissues

The leaves, stems, and roots of a vascular plant contain all three kinds of plant tissues.

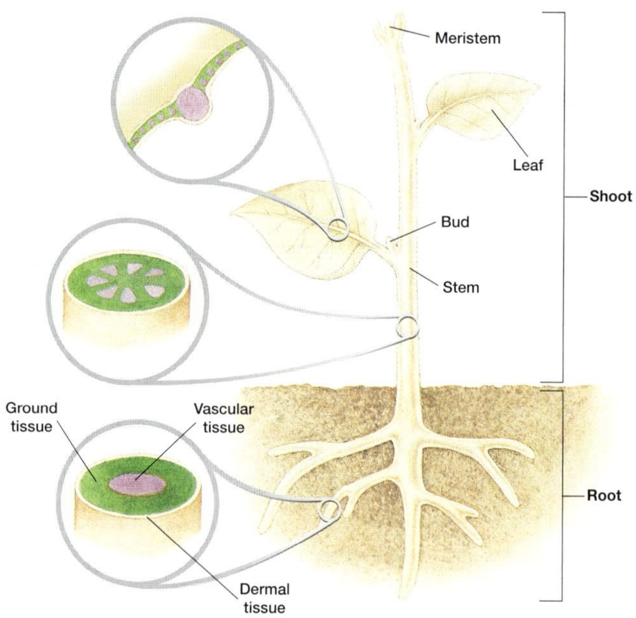

internet connect
www.scilinks.org
Topic: Vascular Plants
Keyword: HX4182
SciLINKS Maintained by the National Science Teachers Association

552

Chapter Resource File

- Directed Reading BASIC
- Active Reading GENERAL
- Data Sheet for Quick Lab GENERAL

Transparencies

TR Bellringer
TR G14 Structure of Xylem
TR G15 Structure of Phloem

Section Correlations

SC.F.1.4.2

Figure 2 **Dermal tissue.** Epidermis is one type of dermal tissue. Hairlike outgrowths of the epidermis, like the ones on this African violet leaf, help reduce water loss from the leaves of some plants.

Dermal Tissue System

Dermal tissue covers the outside of a plant's body. In the nonwoody parts of a plant, dermal tissue forms a "skin" called the **epidermis.** The word *epidermis* comes from the Greek words *epi,* meaning "upon," and *derma,* meaning "skin." The epidermis of most plants is made up of a single layer of flat cells. A waxy cuticle, which prevents water loss, coats the epidermis of the stems and leaves. Often, the cells of the epidermis have hairlike extensions or other structures, as **Figure 2** shows. Extensions of the epidermal cells on leaves and stems often help to slow water loss. Extensions of the epidermal cells on root tips help increase water absorption. The dermal tissue on woody stems and roots consists of several layers of dead cells that are referred to as **cork.** Cork cells contain a waterproof chemical and are not covered by a waxy cuticle. In addition to protection, dermal tissue also functions in gas exchange and in the absorption of mineral nutrients.

Ground Tissue System

Ground tissue makes up much of the inside of most plants. Most ground tissue consists of thin-walled cells that remain alive and keep their nucleus after they mature. In addition, ground tissue contains some thick-walled cells. Look for these different types of cells in the ground tissue shown in **Figure 3.** Ground tissue has different functions, depending on where it is located in a plant. The ground tissue in leaves, which is packed with chloroplasts, is specialized for photosynthesis. The ground tissue in stems and roots functions mainly in the storage of water, sugar, and starch. Throughout the body of a plant, ground tissue also surrounds and supports the third kind of plant tissue—vascular tissue.

Figure 3 **Ground tissue.** A variety of cells make up the ground tissue visible in this cross section of a wheat stem. Thin-walled cells make up most of the ground tissue. Thick-walled cells strengthen the stem.

Magnification: 122×

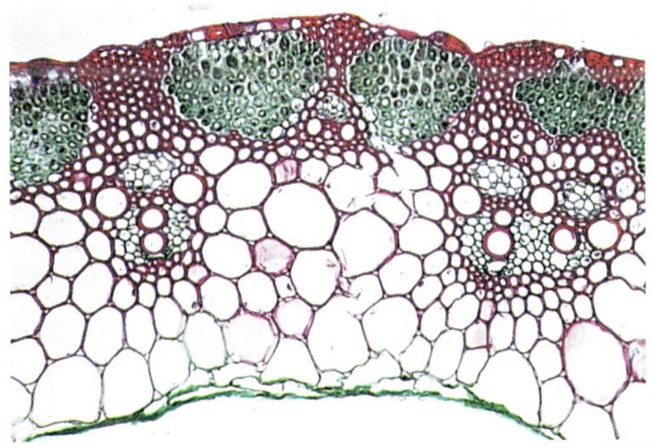

553

Teach

Group Activity — GENERAL

Tissue Types Provide each group of students with a microscope, three microscope slides, and three coverslips. Provide the class with a red onion, a potato, and a stalk of celery. Have students prepare wet-mount slides of the following: red-onion skin (dermal tissue), potato (ground tissue), and celery (dermal, ground, and vascular tissue). Have students draw what they observe in the field of view under the microscope and label the types of tissues they find in each example. Remind them that the field of view should be drawn as a circle.

Co-op Learning | English Language Learners

READING SKILL BUILDER — BASIC

Reading Organizer Have students write down each of the key terms for this section on separate note cards. Have them also make a note card with the words "vascular tissue" on it. As the students read the section, have them write a brief definition of each term on the back of its card. Have students use their cards in the Reteaching exercise at the end of the section. English Language Learners

Teaching Tip — BASIC

Plant Cells and Tissues

Have students construct a table summarizing the types of plant tissues and the cells they contain. In the table, students should classify each type of cell encountered as part of the dermal tissue, ground tissue, or vascular tissue. A completed table is found in the Graphic Organizer at the bottom of this page. LS Logical

One-Stop Planner CD-ROM

- Reading Organizers BASIC
- Reading Strategies BASIC
- Occupational Applications Botanist GENERAL

Graphic Organizer

Use this graphic organizer with **Teaching Tip** on this page.

Plant tissue	Cell types
Dermal tissue	Epidermal cells, guard cells, cork cells
Ground tissue	Mesophyll cells, cortex cells, pith cells
Vascular tissue	Vessel cells, tracheids, sieve-tube cells, companion cells

Teach, *continued*

Using the Figure — Basic

Have students examine **Figure 4.** Ask what similarities they notice between the conducting cells of the xylem and those of the phloem. (The cells are elongated and connected end to end; they have openings in their cell walls at the ends of the cells and also some in the sides.) Ask what these similarities indicate about the similarities in function of xylem and phloem. (The two types of tissues are adapted for transporting substances over long distances, as well as for allowing lateral transport.) Ask what differences they notice between the cells of the xylem and those of the phloem. (There are two types of conducting cells in xylem but only one type in phloem. Also, sieve-tube cells in phloem have companion cells, but the xylem cells do not.) **LS Visual**

Teaching Tip

Interconnected Vascular Tissues
Point out to students that the conducting cells of the xylem in a plant are all interconnected. The strands of vascular tissue in the leaves connect to the vascular tissues in the stem, which in turn connect to the vascular tissue in the roots. Water moves from the roots to the stem to the leaves through this vascular tissue. The same is true for phloem. Root cells underground need a sugar supply from photosynthetic leaves to survive, and the phloem cells can transport this sugar a great distance. Some animals, like porcupines, can eat a complete ring of bark around a tree. This severs the phloem cells, preventing sugar from reaching the roots, which causes the tree to die. This process is called girdling and can be a problem for orchardists.

Vascular Tissue Systems

Plants have two kinds of vascular tissue—xylem and phloem. Both xylem and phloem contain strands of cells that are stacked end to end and act like tiny pipes, as you can see in **Figure 4.** These strands of cells act as a plumbing system, carrying fluids and dissolved substances throughout a plant's body.

Xylem

Xylem has thick-walled cells that conduct water and mineral nutrients from a plant's roots through its stems to its leaves. The conducting cells in xylem must lose their cell membrane, nucleus, and cytoplasm before they can conduct water. At maturity, all that is left of these cells is their cell walls. One type of xylem cell found in all vascular plants is called a tracheid *(TRAY kee ihd)*. Tracheids are narrow, elongated, and tapered at each end. Water flows from one tracheid to the next through pits, which are thin areas in the cell walls. Gnetophytes and flowering plants also have a second type of xylem cell, which makes up conducting strands called **vessels.** The vessel cells are wider than tracheids and have large perforations in their ends. The perforations allow water to flow more quickly between vessel cells.

Figure 4 Xylem and phloem

Both xylem and phloem contain strands of tubular conducting cells that are stacked end to end like sections of pipe.

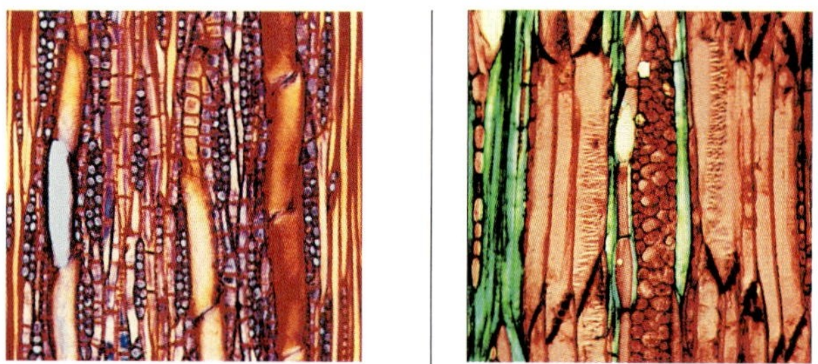

The conducting cells of xylem are tracheids and vessel cells. Tracheids are slender and have tapered ends. Vessel cells are larger and form vessels.

The conducting cells of phloem are sieve-tube cells, which form sieve tubes. Companion cells lie next to the sieve-tube cells.

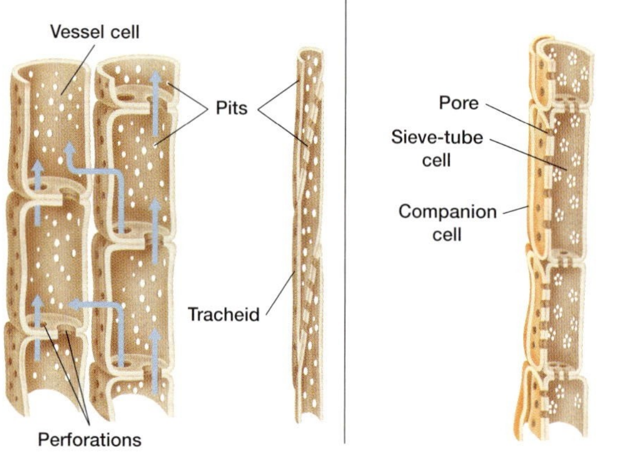

Phloem

Phloem contains cells that conduct sugars and other nutrients throughout a plant's body. The conducting cells of phloem have a cell wall, a cell membrane, and cytoplasm. These cells either lack organelles or have modified organelles. The conducting strands in phloem are called **sieve tubes.** Pores in the walls between neighboring sieve-tube cells connect the cytoplasms and allow substances to pass freely from cell to cell. Beside the sieve tubes are rows of companion cells, which contain organelles. Companion cells carry out cellular respiration, protein synthesis, and other metabolic functions for the sieve-tube cells.

did you know?

Aphids and Phloem. Scientists who were studying sugar transport in plants noticed that many aphids that feed on plants release a substance called "honeydew" from their anuses. Other insects commonly feed on this honeydew. The scientists analyzed the honeydew and found that it was high in sugars. They concluded that when aphids feed on plants, they must insert their mouthparts, called stylets, directly into the sugar-transporting cells. The scientists then conducted some studies in which they quickly cut the aphids off from their stylets while they were feeding and then made cross-sections of the plant tissues. In these sections, they found that the tips of the aphids' stylets ended in sieve-tube cells. This led the scientists to conclude that phloem is the sugar-transporting tissue in plants.

Roots

Most plants are anchored to the spot where they grow by roots, which also absorb water and mineral nutrients. In many plants, roots also function in the storage of organic nutrients, such as sugar and starch. Many dicots, such as carrots and radishes, have a large central root from which much smaller roots branch. This type of root system is called a *taproot* system. In contrast, most monocots, such as grasses, have a highly branched, fibrous root system, shown in **Figure 5.** Some plants have roots that grow from aboveground stems or leaves. Such roots are called adventitious *(ad ven TIH shuhs)* roots. The prop roots of corn and the aerial roots of orchids are examples of adventitious roots.

As **Figure 6** shows, a root has a central core of vascular tissue that is surrounded by ground tissue. The ground tissue surrounding the vascular tissue is called the **cortex.** Roots are covered by dermal tissue. An epidermis covers all of a root except for the root tip. The epidermal cells just behind a root tip often produce **root hairs,** which are slender projections of the cell membrane. Root hairs greatly increase the surface area of a root and its ability to absorb water and mineral nutrients. A mass of cells called the **root cap** covers and protects the actively growing root tip. A layer of cork replaces the epidermis in the older sections of a root. Many plants have roots that become woody as they get older. Layers of xylem replace the ground tissue in woody roots.

Figure 5 Fibrous root system. A fibrous roots system is made up of many roots that are about the same size.

Figure 6 Root structure. Roots have characteristic external and internal structures.

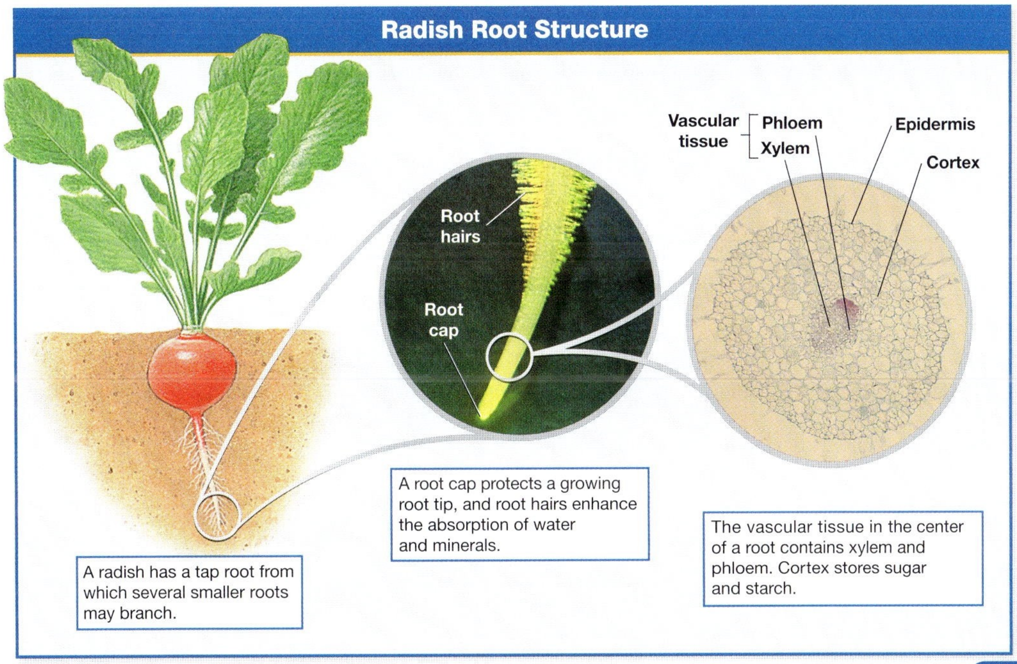

555

MISCONCEPTION ALERT

Plants Need Oxygen It is a common misconception that plants do not need oxygen because they produce it in photosynthesis. However, almost all plant cells have mitochondria, which break down carbon-containing molecules in the presence of oxygen and release energy for cellular functions. In leaves and green stems, oxygen is produced in photosynthesis. However, in roots, no oxygen is produced because there are no photosynthetic cells. Therefore, roots must obtain oxygen from the soil. This is why houseplants that are overwatered and crops that are flooded for extended periods of time die. This is also why gardeners try to keep their soil aerated, using organic matter such as mulch and compost, and earthworms that dig holes in the soil so water will drain. Trees such as mangroves and bald cypresses that live in swamps have roots that emerge above the water (pneumatophores) and obtain oxygen from the air.

Demonstration — GENERAL

Bring modified plant organs or pictures of these plant organs to class for the students to observe. Roots may include fleshy taproots adapted for sugar and starch storage (carrot, beet, and sweet potato) and aerial roots that extract water from the air (orchids and bromeliads). Stems may include bulbs (onion and lily), corms (gladiolus and crocus), rhizomes (iris), and tubers (Irish potato), adapted for sugar and starch storage; runners (spider plant and strawberry) and stolons (most lawn grasses), which grow new plants at nodes; and stems adapted for water storage (cactus). Leaves may include those adapted for moist, understory environments (philodendron); needles adapted for cold, windy environments (pine); those adapted for sugar and starch storage (cabbage); fleshy petioles adapted for sugar, starch, and/or water storage (celery and rhubarb); bracts (flowering dogwood and poinsettia), which are modified leaves adjacent to flowers and are sometimes showy; and leaves adapted for asexual reproduction (kalanchoë). Ask the students to compare the functions of these modified organs with their usual functions.

SKILL BUILDER — BASIC

Interpreting Graphics Have students look at **Figure 6.** Ask them to explain what the small and large circles in the figure represent. (The small circles show sections of a root that is to be examined. The large circles are the enlarged view of the same area. Note that the second large circle on the right is an enlarged cross section of the root.) **LS Visual** English Language Learners

READING SKILL BUILDER — BASIC

Interactive Reading Assign Chapter 25 of the *Holt Biology* Guided Reading Audio CD Program to help students achieve greater success in reading the chapter. English Language Learners

Teach, *continued*

Real Life

Answers

Answers will vary. Many different types of trees may be cut to produce either straight-grained lumber or an unusual pattern known as figured grain. Straight-grained lumber is used in making boards and planks for construction, decking, handles, and baseball bats. Plywood, veneers for furniture and paneling, musical instruments, and art objects are often made from figured grain.

Activity — GENERAL

Vascular Bundles Have students look at microscope slides of cross sections of various monocot (corn, iris, tulip, etc.) stems and dicot stems such as buttercup or geranium. Have them write the name of each plant cross section on a piece of paper and determine whether it is a monocot or dicot. They should refer to **Figure 7** for help. Have them pass the slides around the room until all students have seen each different species. Ask students to hypothesize about which type of plant stem, monocot or dicot, they think would have the capability to grow largest in diameter, and why. Discuss each answer volunteered. (Dicot stems often grow larger in diameter because their circular arrangement of vascular bundles (and accompanying vascular cambium) allows for organized, outward growth. Monocots typically do not grow as large as some dicots, such as an apple tree or lilac tree, because monocots have scattered vascular bundles.) LS **Visual**

Real Life

Xylem forms wood's characteristic markings, or grain.

Wood is usually cut parallel to the axis of a tree trunk, producing straight-grained lumber. In some trees, however, xylem vessels form interesting patterns.

Recognizing Patterns
Examine several wooden objects. Decide whether they have straight grain or another type of grain.

Stems

The shoots of most plants consist of stems and leaves. Stems support the leaves and house the vascular tissue, which transports substances between the roots and the leaves. Many plants have stems that are specialized for other functions. For example, the stems of cactuses store water. Potatoes are stems that are specialized for nutrient storage and for asexual reproduction.

Leaves are attached to a stem at points called nodes. The space between two nodes is called an internode. Buds that can grow into new branches are also located at the nodes on a stem. Look for these structures in **Figures 7** and **8.** Other features of a stem depend on whether the stem is woody or nonwoody.

Nonwoody Stems

A plant with stems that are flexible and usually green is called a **herbaceous** *(huhr BAY shuhs)* **plant.** Herbaceous plants include violets, clovers, and grasses. As Figure 7 shows, the stems of herbaceous plants contain bundles of xylem and phloem called **vascular bundles.** The vascular bundles are surrounded by ground tissue. In monocot stems, such as that of corn, illustrated in Figure 7, the vascular bundles are scattered in the ground tissue. In dicot stems, however, the vascular bundles are arranged in a ring. The ground tissue outside the ring of vascular bundles is called the cortex. The ground tissue inside the ring is called the **pith.** Herbaceous stems are covered by an epidermis. Stomata in the epidermis enable the stems to exchange gases with the outside air.

Figure 7 Herbaceous stems. The vascular bundles are arranged differently in dicots and monocots.

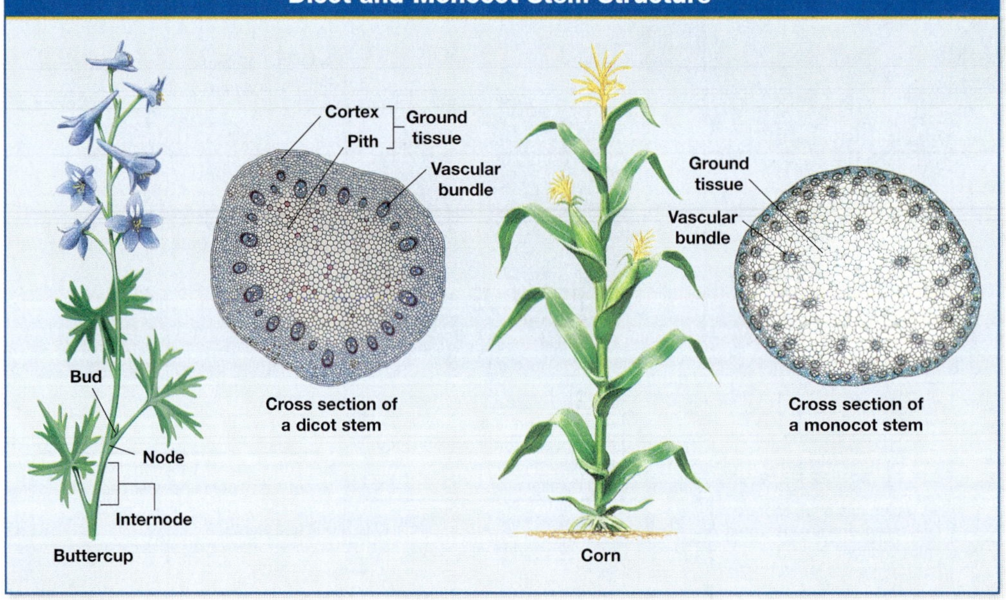

556

Career

Horticulturist Many colleges offer degrees in horticulture, a broad and diverse field involved with the production and utilization of a multitude of specialized plants. Horticulturists are involved in the production of fruits and vegetables; they also improve the human environment through beautification, conservation, and by providing the ornamental plants used in landscaping. Degrees can be obtained in fields such as horticulture crops, horticulture education, landscape contracting, horticulture science, and horticulture therapy. While the field of horticulture is based on a biological and agricultural science foundation, there is opportunity for individuals oriented toward business, social science, art, education, and international development. Salary and job opportunities vary with geographic region.

Woody Stems

Trees and shrubs, such as pines, oaks, roses, and hollies, have woody stems. As Figure 8 shows, woody stems are stiff and nongreen. Buds, which produce new growth, are found at the tips and at the nodes of woody stems. They exchange gases through pores in their bark. A young woody stem has a central core of pith and a ring of vascular bundles, which fuse into solid cylinders as the stem matures. Layers of xylem form the innermost cylinder and are the major component of wood. A cylinder of phloem lies outside the cylinder of xylem. Woody stems are covered by cork, which protects them from physical damage and helps prevent water loss. Together, the layers of cork and phloem make up the bark of a woody stem. A mature woody stem contains many layers of wood and is covered by a thick layer of bark. The wood in the center of a mature stem or tree trunk is called **heartwood.** The xylem in heartwood, which can no longer conduct water, provides support. **Sapwood,** which lies outside the heartwood, contains vessel cells that can conduct water.

Figure 8 Woody stems

Woody stems are typically stiff and nongreen.

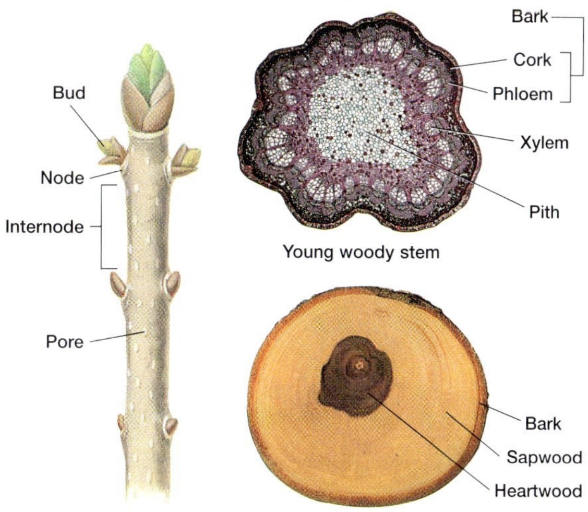

Quick Lab: *Comparing the Structures of Roots and Stems* SC.F.1.4.2

You can use a microscope to see differences in the internal structure of roots and stems.

Materials

compound microscope, prepared slide of a cross section of the following: dicot root, monocot root, dicot stem, monocot stem

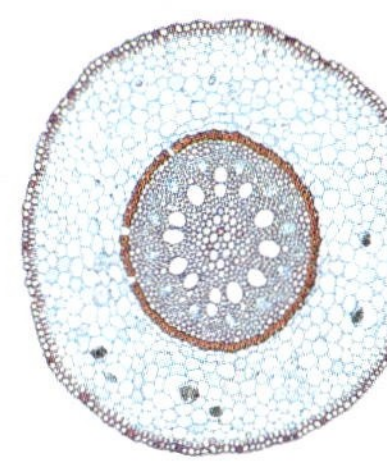

Cross section of a corn root

Procedure

1. View cross sections of dicot and monocot roots with a compound microscope. For each, draw and label what you see under low power. Then look at the vascular tissue in each root under high power. Draw what you see in each root, and label the xylem and phloem.
2. View cross sections of dicot and monocot stems with a compound microscope. For each, draw and label what you see under low power. Then look at a vascular bundle in each stem under high power. Draw each vascular bundle, and label the xylem and phloem.

Analysis

1. **Compare and contrast** the location of xylem and phloem in roots and stems.
2. **Compare and contrast** the arrangement and structure of the vascular bundles in monocot and dicot stems.
3. **Describe** the relationship between the structure and function of vascular tissue.

557

Teaching Tip — Advanced

Annual Rings in Woody Plants

Bring to class a cross section of a tree trunk with irregular growth rings. Ask students to relate the rings to the tree's history. (Close rings usually indicate slow growth; rings that are closer together on one side indicate growth against an obstacle.) Count the rings to estimate the age of the tree. If the cross section is of a recently cut tree from the local area, have students identify the ring that represents the most rapid growth. Then have them count the number of years this rapid growth occurred before the tree was cut down. Have students research to determine if the precipitation on or near that year was higher than normal. LS Logical

Quick Lab: *Comparing the Structures of Roots and Stems*

Skills Acquired

Observing, comparing

Teacher's Notes

Ask students to name some differences in the functions of roots and stems. Ask students to think about the differences between monocots and dicots as they observe the different kinds of stems.

Answers to Analysis

1. Xylem and phloem in the roots are located in a cylinder in the center of the root. Xylem and phloem in the stem are located in discrete vascular bundles in the cortex.
2. The vascular bundles in monocot stems are scattered throughout the cortex; the vascular bundles of dicot stems are arranged in a ring near the outside of the stem.
3. The structure of vascular tissue is specialized to conduct water and nutrients. For example, in order to conduct water, mature xylem cells lack cytoplasm and a nucleus.

Transparencies

- **TR G17** Radish Root Structure
- **TR G23** Dicot and Monocot Stem Structures
- **TR G22** Structure of Stems

Misconception Alert

Roses Do Not Have Thorns The structures that protect the stems of many members of the rose family are not really thorns. Botanically, thorns are modified stems that contain xylem. The "thorns" found on a rose are actually called prickles, which are derived only from dermal tissue. How can you tell them apart without using a microscope? Prickles are easily removed; thorns are not. To anyone who has ever cut roses for a flower arrangement, however, "A thorn by any other name is just as sharp."

Teach, *continued*

Using the Figure — BASIC

Have students examine the leaves in **Figure 9** and note the differences. One way to tell a leaf from a leaflet is to look for the bud in the axil of the leaf. Show students examples of simple and compound leaves. Ask students if there are buds in the axils of simple leaves. (yes) Ask students if there are buds in the axils of leaflets. (no) **LS Visual**

QUICK LAB *Observing the Structures Inside a Leaf*

Skills Acquired
Observing, drawing conclusions

Teacher's Notes
Ask students to note which leaf surface has the most stomata. (lower)

Answers to Analysis

1. A stoma is a slitlike opening between two guard cells. It allows carbon dioxide to enter the leaf. Water vapor and oxygen exit the leaf through the stomata.
2. Veins are distributed throughout the mesophyll. They contain xylem and phloem.
3. The palisade cells are located where the most light will hit the leaf. The spongy cells form a loosely packed layer (or layers) of irregularly shaped cells. The large open areas around these cells allow carbon dioxide to move freely from stomata on the lower leaf surface up to the palisade layer.

Leaves

Leaves are the primary photosynthetic organs of plants. Most leaves have a flattened portion, called the blade, that is often attached to a stem by a stalk called the **petiole** *(PEHT ee ohl)*. A leaf blade may be divided into two or more sections called leaflets, as shown in **Figure 9.** Leaves with an undivided blade are called simple leaves. Leaves with two or more leaflets are called compound leaves. Leaflets reduce the surface area of a leaf blade. Many plants have highly modified leaves that are specialized for particular purposes. For example, the spines of a cactus and the tendrils of a garden pea are modified leaves. Cactus spines are specialized for protection and water conservation, while garden-pea tendrils are specialized for climbing.

Figure 9 Simple and compound leaves

Most leaves consist of a flattened blade and a petiole that attaches to a stem.

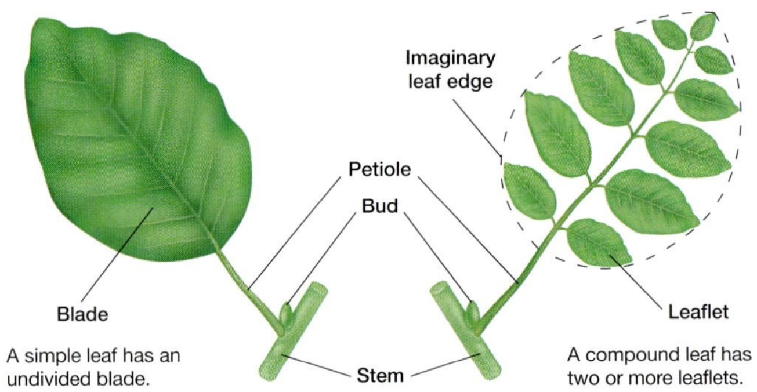

QUICK LAB *Observing the Structures Inside a Leaf* SC.F.1.4.2

With a microscope, you can see how a leaf is put together.

Materials

prepared slide of a leaf cross section, compound microscope

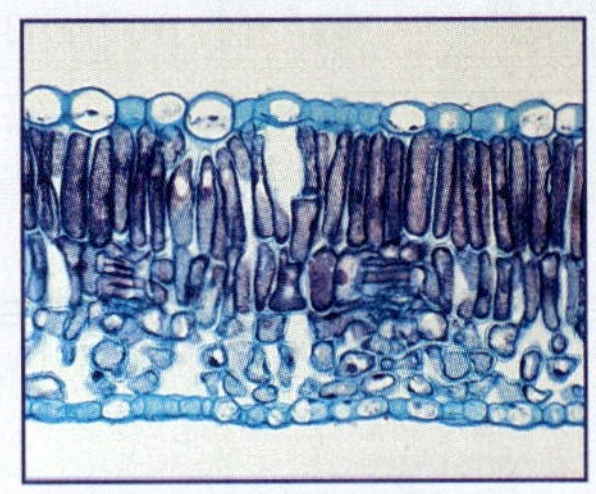

Cross section of a lilac leaf (530×)

Procedure

1. View a cross section of a leaf under low power of a compound microscope. Then switch to high power.
2. Use Figure 10 in this chapter to identify the following structures: stoma, guard cells, upper and lower epidermis, palisade layer, spongy layer, and vein.

Analysis

1. **Describe** a stoma, and relate the function of a stoma to your description.
2. **Describe** the location and contents of the veins.
3. **Critical Thinking**
Relating Concepts How do the location and structure of the palisade and spongy layers help a leaf perform photosynthesis?

558

Cultural Awareness

Edible and Medicinal Plants Historically, plants have been an important part of the diet of Native Americans of the Pacific Northwest. The root of the camas, a flower in the lily family, was the most important vegetable food for northwest Indians. Camas roots were about the size of onions and could be kept over winter to supply food during times of need. Digging for camas roots was an important activity for many northwest tribes. In addition to camas root, the Indians harvested many other plant species. Cattail shoots made delicious vegetables. Black moss could be harvested at any time of year and would be steamed and pressed into hard squares and dried. It could be served for up to three years. Other plants were used for medicinal or other purposes. Cascara bark is a potent laxative. Oregon grape was used as a tonic for infections and balsam oil was used for blisters and as a cough syrup. Wild parsnip was used as a repellent to keep flies and other insects away from drying salmon and meat.

Section Review Correlations

SC.F.1.4.2: Items 1-4

A leaf is a mass of ground tissue and vascular tissue covered by epidermis, as **Figure 10** shows. A cuticle coats the upper and lower epidermis. Both xylem and phloem are found in the veins of a leaf. Veins are extensions of vascular bundles that run from the tips of roots to the edges of leaves. In leaves, the ground tissue is called **mesophyll** *(MEHS oh fihl)*. Mesophyll cells are packed with chloroplasts, where photosynthesis occurs. The chlorophyll in chloroplasts makes leaves look green.

Most plants have leaves with two layers of mesophyll. One or more rows of closely packed, columnar cells make up the palisade layer, which lies just beneath the upper epidermis. A layer of loosely packed, spherical cells, called the spongy layer, lies between the palisade layer and the lower epidermis. The spongy layer has many air spaces through which gases can travel. Stomata, the tiny holes in the epidermis, connect the air spaces to the outside air.

Figure 10 Structure of a leaf

The internal structure of a leaf is highly organized.

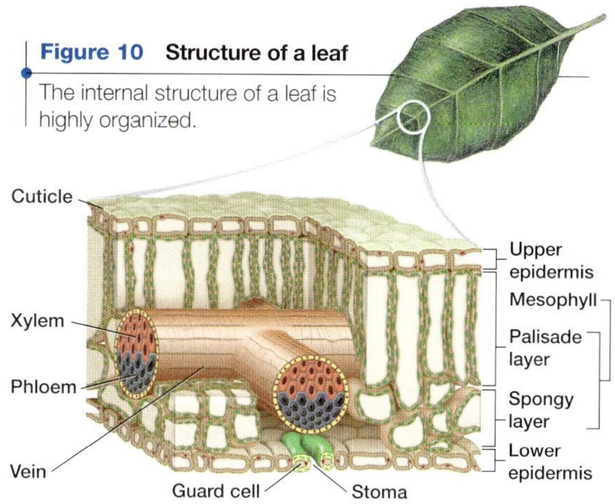

Exploring Further

C_3 Versus C_4 Leaves

Internal leaf structure varies among plants, depending on how they carry out photosynthesis. In photosynthesis, carbon dioxide from the air is added to organic molecules. This process, called carbon fixation, occurs during the Calvin cycle. Because the product of carbon fixation by the Calvin cycle is a three-carbon compound, plants that use *only* the Calvin cycle to fix carbon are called C_3 plants. The leaf structure shown in Figure 10 is typical of a C_3 plant. More than 90 percent of all plants are C_3 plants.

Cross section of a corn leaf (584×)

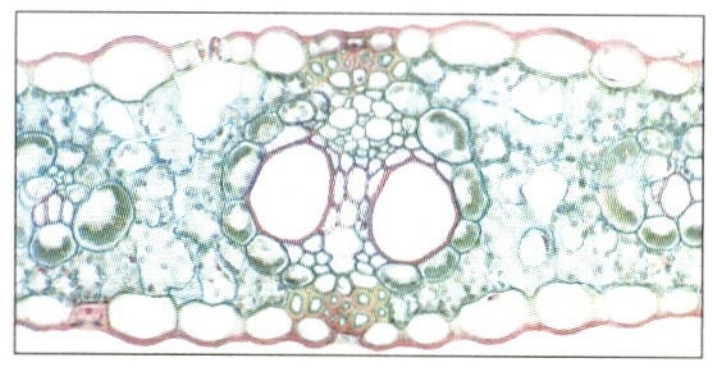

Plants such as corn and sugar cane use another chemical process to fix carbon before the Calvin cycle. The first products of this process are four-carbon compounds, so these plants are called C_4 plants. The leaves of C_4 plants have a different internal structure, as shown in the photo of a corn leaf above. C_4 plants are plentiful in the tropics because they conduct photosynthesis efficiently in high heat and intense light.

Section 1 Review

1. **Name** the three kinds of tissue that make up a plant, and list two functions for each. SC.F.1.4.2
2. **Summarize** the differences in the structure of the stems of monocots and dicots. SC.F.1.4.2
3. **Critical Thinking** **Forming Hypotheses** How do the structures of roots, stems, and leaves help them to carry out their functions? SC.F.1.4.2
4. *FCAT Prep* In which part of a plant would you find closely packed, columnar cells that contain many chloroplasts?
   A leaf
   B root
   C woody stem
   D nonwoody stem
   SC.F.1.4.2

559

Exploring Further — ADVANCED

C_3 Versus C_4 Leaves

Teaching Strategies
Have students examine the photo of the cross section of a corn leaf and compare its structure with the structure of the leaf shown in **Figure 10.**

Discussion
How does the structure of the corn leaf shown differ from the structure of a typical C_3 plant leaf shown in **Figure 10?** (The corn leaf has a ring of cells surrounding the veins, and it does not have palisade and spongy layers in the mesophyll.)

Close

Reteaching — BASIC

Have students take out the note cards they prepared in the Reading Skill Builder at the beginning of the section. Ask them to arrange the cards in five sets: terms relating to roots; terms relating to stems; terms relating to leaves; terms relating to roots and stems; and terms relating to roots, stems, and leaves. Have students review Section 1 to check the arrangement of their cards. English Language Learners

Quiz — GENERAL

True or False:

1. Vascular tissues in a plant may be considered to be analogous to arteries and veins in humans. (true)
2. The tissue that transports water from the roots up through a plant is phloem. (false, it is xylem)
3. Root hairs help to anchor a plant and increase the absorptive surface of the root. (true)

Transparencies

TR G28 Simple and Compound Leaves
TR G30 Structure of a Leaf

Answers to Section Review

1. Dermal tissue protects the plant and absorbs water and minerals. Ground tissue stores water, sugar, and starch and conducts photosynthesis. Vascular tissue transports water, minerals, and sugars throughout the plant.
2. Vascular bundles of a dicot stem are arranged in a ring; vascular bundles of a monocot stem are scattered throughout the ground tissue.
3. Roots have branches and root hairs, which help to anchor the plant and create a large surface area for absorption of water and nutrients. Stems have cells with rigid walls that keep the stem upright. These cells transport water, minerals, and sugars. Leaves are thin and flat, with a large surface area for gathering light. They have stomata, which enable gas exchange to occur.
4. **A.** Correct. Leaves have columnar cells with many chloroplasts. **B.** Incorrect. Roots do not have chlorophyll. **C.** Incorrect. Woody stems do not have chlorophyll. **D.** Incorrect. Nonwoody stems have vascular bundles.

Section 2

Focus

Overview

Before beginning this section review with your students the objectives listed in the Student Edition. All plant cells must have a supply of food and water, whether the cells are hundreds of feet above ground or underground. For sugar made in leaves to reach root cells and for underground water to reach leaves, plants must have a transport system often capable of moving substances great distances.

Bellringer

Ask students to write down three common substances that all cells need to survive. (Answers may vary, but likely will include food, water, and oxygen.) Then, have them write down how they think most plant cells obtain these substances. (*Food*—through photosynthesis; *water*—from the ground; *oxygen*—passes through tiny openings into the leaf) **LS Verbal**

Motivate

Activity — GENERAL

Surface Tension Work in teams of two. Give each team a penny, a medicine dropper, and a small glass of water. Have each group guess the number of drops of water they can put on the penny before the water spills over. Discuss with them why so many drops could be placed on the penny. Review the concept of cohesion and how it causes surface tension, and explain that cohesion is an important force in the transport of water in plants.
LS Interpersonal Co-op Learning

Section Correlations

SC.F.1.4.1
SC.F.1.4.2
SC.H.3.4.6

Section 2 Transport in Plants

Florida Sunshine Standards/FCAT
SC.F.1.4.1 Body processes involve specific biochemical reactions governed by biochemical principles.
SC.F.1.4.2 Body structures are uniquely designed and adapted for their function.
SC.H.3.4.6 Scientific knowledge is used by those who engage in design and technology to solve practical problems, taking human values and limitations into account.

Objectives

- **Relate** transpiration to the movement of water up a plant. SC.F.1.4.2
- **Describe** how guard cells regulate the rate of transpiration. SC.F.1.4.2
- **Recognize** several distinguishing features of sugar maple trees. SC.H.3.4.6
- **Describe** the process of translocation in a plant. SC.F.1.4.1

Key Terms

transpiration
source
sink
translocation

Movement of Water

You know that water and mineral nutrients move up from a plant's roots to its leaves through xylem. However, some trees have leaves that are more than 100 m (328 ft) above the ground. How do plants manage to get water so high? Simply put, water is pulled up through a plant as it evaporates from the plant's leaves, as **Figure 11** shows.

Step 1 Recall that the surfaces of leaves are covered with many tiny pores, the stomata. When the stomata are open, water vapor diffuses out of a leaf. This loss of water vapor from a plant is called **transpiration.** In most plants, more than 90 percent of the water taken in by the roots is ultimately lost through transpiration.

Step 2 The xylem contains a column of water that extends from the leaves to the roots. The cohesion of water molecules causes water molecules that are being lost by a plant to pull on the water molecules still in the xylem. This pull extends through the water in the xylem. Water is drawn upward in the same way liquid is drawn through a siphon. As long as the column of water in the xylem does not break, water will keep moving upward as transpiration occurs.

Step 3 Roots take in water from the soil by osmosis. This water enters the xylem and replaces the water lost through transpiration.

Figure 11

BIO graphic — Water Movement in Plants

Transpiration drives the movement of water through a plant.

1 Water vapor exits the leaves through stomata by transpiration.

2 The loss of water creates a pull that draws water up through the xylem.

3 Water drawn into the roots from the soil by osmosis moves up the stem.

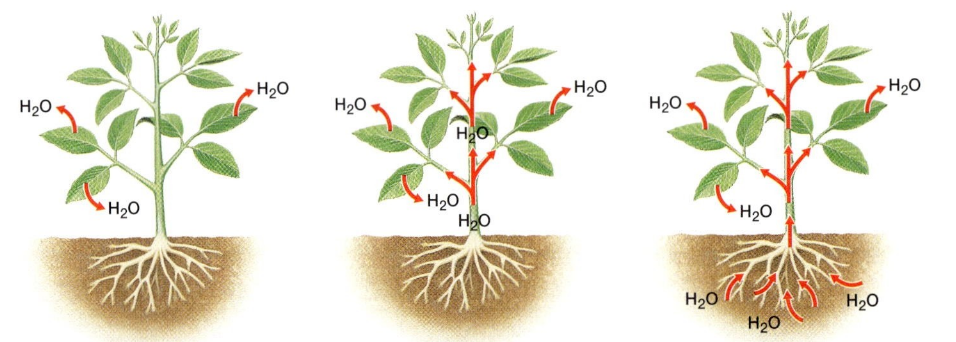

560

Chapter Resource File

- Directed Reading BASIC
- Active Reading GENERAL
- Data Sheet for Data Lab GENERAL
- Data Sheet for Skills Practice Lab GENERAL

Transparencies

TR Bellringer
TR G32 Control of Stomatal Opening
TR G26 Water Movement in Plants

Figure 12 **Control of stomatal opening**

Changes in the shape of guard cells cause stomata to open or close.

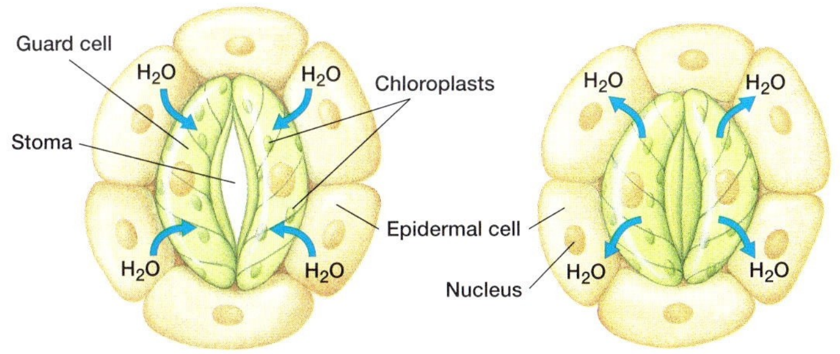

1. A stoma opens as the guard cells take in water, lengthen, and bow apart.
2. A stoma closes as the guard cells lose water, shorten, and come together.

Guard Cells and Transpiration

A stoma is surrounded by a pair of guard cells that are shaped like two cupped hands. Changes in water pressure within the guard cells cause the stoma to open or close, as shown in **Figure 12.** When the guard cells take in water, they swell. However, extra cellulose strands in their cell walls permit the cells to increase in length but not in diameter. As a result, guard cells that take in water bend away from each other, opening the stoma and allowing transpiration to proceed. When water leaves the guard cells, they shorten and move closer to each other, closing the stoma and stopping transpiration. Thus, the loss of water from guard cells for any reason causes stomata to close, stopping further water loss. This is an example of homeostasis in action.

internet connect

www.scilinks.org
Topic: Transpiration
Keyword: HX4178

SCiLINKS. Maintained by the National Science Teachers Association

DATA LAB *Inferring the Rate of Transpiration* SC.F.1.4.1

Background

The graph shows the rate of water movement in a plant during high humidity and during low humidity. The rate of water movement indicates the rate of transpiration. Use the graph to answer the questions below.

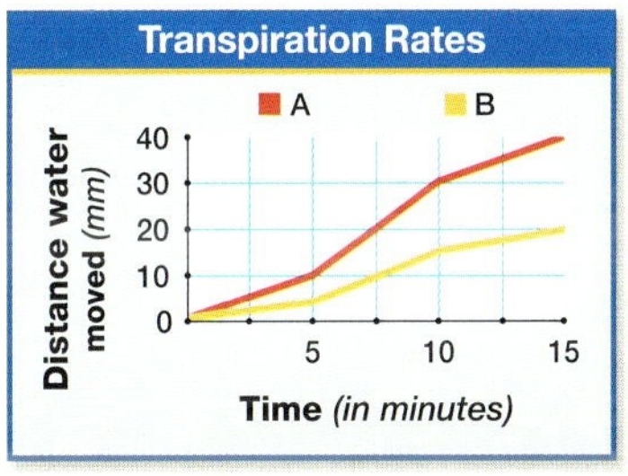

Analysis

1. **Determine** how far water had moved after 10 minutes under the condition represented by curve A.
2. **Critical Thinking** **Analyzing Results** After 15 minutes, how much farther had water moved under condition A than under condition B?
3. **Critical Thinking** **Recognizing Relationships** Which curve indicates a lower transpiration rate? Explain your reasoning.
4. **Critical Thinking** **Drawing Conclusions** Which curve shows the transpiration rate during low humidity? Justify your answer.

Teach

Teaching Tip

Guttation Tell students that sometimes in the early morning, droplets of water appear on the edges of the leaves of some plants. Show photos of leaves with these droplets, which look somewhat like dew except that they form at the tips of the veins. These droplets of water are forced out of the plant by root pressure, which develops as water moves into roots by osmosis. This phenomenon, called guttation, occurs only when the air is very moist and the stomata are closed.

DATA LAB *Inferring the Rate of Transpiration*

Skills Acquired
Interpreting, analyzing, drawing conclusions

Teacher's Notes
Point out that a properly prepared graph includes a short descriptive title, axes labeled with appropriate variables and units, numbers on both axes covering the entire range of the data including zero, and data points and/or lines clearly indicated and differentiated.

Answers to Analysis

1. 30 mm
2. about 20 mm farther
3. Curve B indicates a lower transpiration rate because it shows water moving a shorter distance per unit of time than curve A shows.
4. Curve A shows the transpiration rate during low humidity because it shows water moving a greater distance up a plant than curve B shows.

One-Stop Planner CD-ROM

- **Reading Organizers** BASIC
- **Reading Strategies** BASIC
- **Supplemental Reading** *A Feeling for the Organism: The Life and Work of Barbara McClintock* ADVANCED

did you know?

Dutch Elm Disease Dutch elm disease is a fungal disease that has killed thousands of American Elm *(Ulmus americana)* trees in this country, particularly in eastern areas. The fungal spores are transported from tree to tree by an elm bark beetle that burrows into the bark and eats it. The fungus can also be transported from tree to tree through root grafts that form when trees of the same type are planted close together and their roots grow together. The fungus grows into the xylem. The tree isolates the area where the fungus is living by growing new cells around it. This cuts off water flow through the affected xylem cells. The first visible symptom of Dutch elm disease in a tree is flagging, which occurs when all the leaves on a branch high in the tree die. If enough of the xylem of a tree becomes affected, the tree will die.

Up Close

Sugar Maple

Teaching Strategies

- Have students do library research on the sugar maple. Then have each student write an interesting fact about sugar maples on a piece of paper cut in the shape of a maple leaf. Draw the outline of a maple tree on a large sheet of butcher paper, and have students attach their leaves to the branches of the tree.
- Have students bring leaves and samaras from local maple trees and compare them with the leaves and samaras shown in this feature.
- Display samples of maple wood and photographs of various types of maple trees at different times of the year.
- Provide examples of foods that contain maple flavoring or maple sugar.

Teaching Tip

Fall Color Ask students why leaves are green. (They contain chlorophyll, which reflects green light.) There are also other pigments in leaves, such as carotenoids, which reflect yellow and orange light; xanthophylls (yellow pigments), which reflect orange light; and anthocyanins, which reflect red light. Usually we do not see these pigments because chlorophyll is more abundant and it masks the other colors. In the fall, chlorophyll is broken down and the other pigments can be seen.

Sugar Maple

- **Scientific name:** *Acer saccharum*
- **Size:** 12 to 37 m (40 to 121 ft) tall, canopy up to 14 m (46 ft) wide, trunk up to 1 m (3 ft) in diameter
- **Range:** Northeastern United States and adjacent regions of southeastern Canada
- **Habitat:** Northern temperate forests
- **Importance:** The wood of sugar maples is used to make furniture, musical instruments, and flooring. Their sap is made into maple syrup and maple sugar.

External Structures

Flowers Maple flowers appear before the leaves in early spring. The flowers lack petals and thus are incomplete. Most are imperfect (either male or female) as well.

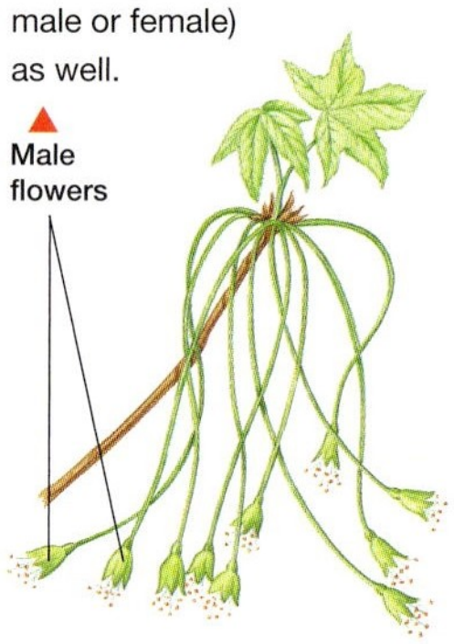

Seeds and fruits Seeds mature inside winged fruits called samaras. The wings enable the dispersal of seeds by wind.

Twigs and buds The twigs are green at first, and then they change to a glossy reddish brown. The reddish brown buds are conical, pointed, and about 4 mm (0.2 in.) long.

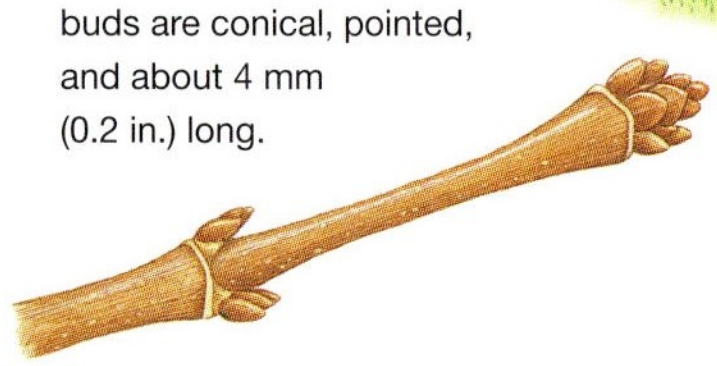

Bark Young sugar maples have smooth, light gray bark. The bark of older trees becomes dark gray or brown, rough, and deeply furrowed.

Leaves The leaves range from 7.5 to 15 cm (3 to 6 in.) across. Most leaves have five sharply toothed lobes. The main veins branch from a single point, like the fingers on a hand. In the fall, the leaves change to bright yellow, orange, or red, and fall from the tree.

562

HISTORY CONNECTION

In 1932 the infant son of Charles Lindbergh, the first person to fly solo across the Atlantic Ocean, was kidnapped and killed in Hopewell, New Jersey. The most damaging piece of evidence that led to the conviction of the suspect was a ladder used in the kidnapping. A wood technologist established by microscopic evidence that four types of wood were used in making the ladder. A detailed analysis of the grain and wood from one of the rungs matched the floorboards from the suspect's attic.

Internal Structures

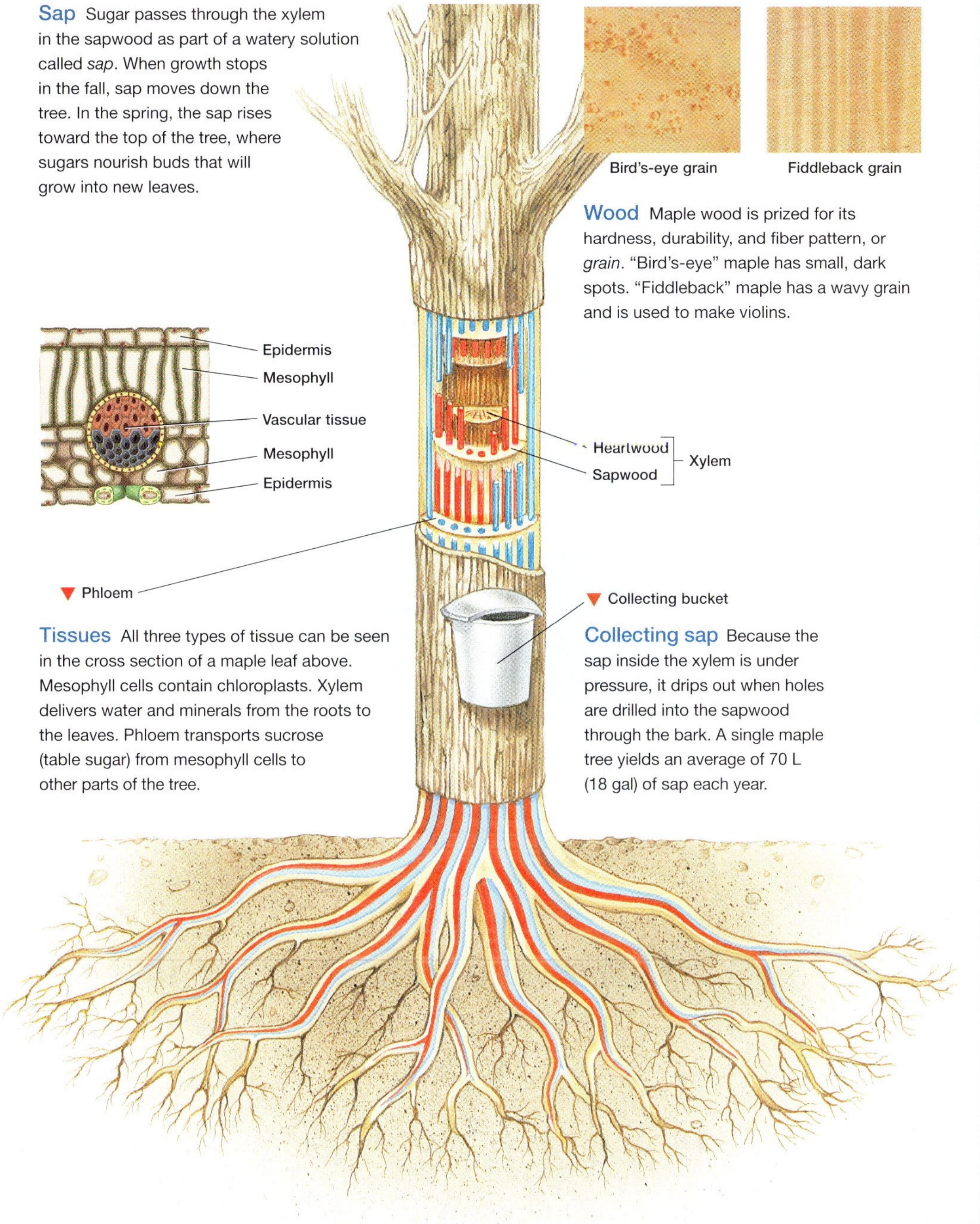

Sap Sugar passes through the xylem in the sapwood as part of a watery solution called *sap*. When growth stops in the fall, sap moves down the tree. In the spring, the sap rises toward the top of the tree, where sugars nourish buds that will grow into new leaves.

Bird's-eye grain   Fiddleback grain

Wood Maple wood is prized for its hardness, durability, and fiber pattern, or *grain*. "Bird's-eye" maple has small, dark spots. "Fiddleback" maple has a wavy grain and is used to make violins.

Tissues All three types of tissue can be seen in the cross section of a maple leaf above. Mesophyll cells contain chloroplasts. Xylem delivers water and minerals from the roots to the leaves. Phloem transports sucrose (table sugar) from mesophyll cells to other parts of the tree.

Collecting sap Because the sap inside the xylem is under pressure, it drips out when holes are drilled into the sapwood through the bark. A single maple tree yields an average of 70 L (18 gal) of sap each year.

563

Up Close

Sugar Maple

Discussion

- What is the economic importance of sugar maple trees in the United States? (They provide raw materials for making important products such as maple syrup, maple sugar, furniture, musical instruments, and wood flooring and furniture.)
- How might a sugar maple tree's production of maple sap be affected if it were growing in an environment where the temperature stayed the same year round, such as the tropics? (In an environment with constant temperature, there might not be as much sugar converted to starch for storage. As a result, there would not be a large amount of sap rising at any one time, and collecting the sap would be difficult.)
- Why is maple a more desirable wood for making furniture than pine? (Maple is a hardwood, which produces more durable wood than a softwood such as pine.)

Teaching Tip

Annual Rings Tell students that annual rings are not always "annual." While this is usually the case, an unusual rainy period during the normally dry summer months could cause a new spurt of growth and hence, another growth ring where there would normally not be one.

Graphic Organizer

Use this graphic organizer with **Teaching Tip** on the next page.

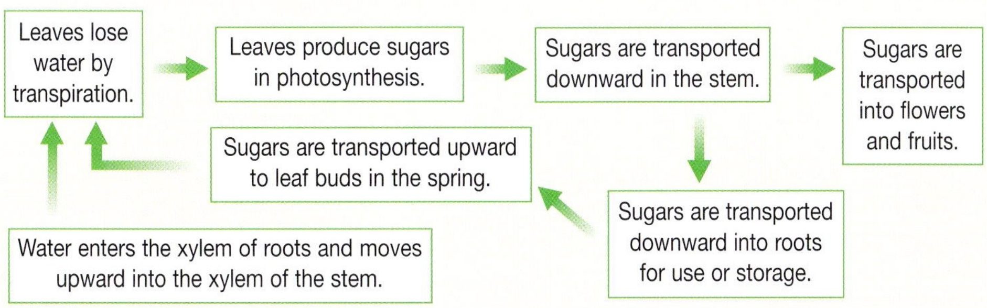

Transparencies

TR G25 The Pressure-Flow Model
TR G9 Sugar Maple
TR G10 Sugar Maple (cont.)

Teach, *continued*

Teaching Tip

Transpiration and Translocation

Have students prepare a Graphic Organizer similar to the one at the bottom of the previous page that summarizes the processes of transpiration and translocation. LS Logical

Close

Reteaching — BASIC

Have students write down the names of all of the cells involved in transpiration and translocation and then identify the role that each cell type plays in these processes.

Quiz — GENERAL

True or False:

1. In most plants more than 90 percent of the water taken in through the roots is lost through transpiration. (true)
2. When water is scarce, guard cells lose their pressure and sag open. (false; they sag and close)
3. Active transport is an important element of the pressure-flow model of sugar movement in plants. (true)

Alternative Assessment — GENERAL

Have students imagine they are a water molecule in the soil. They should explain everything that would happen from the time that they enter a root hair until the time they are transpired from a leaf. They should include the structures they pass through and the forces that act upon them. LS Verbal

Section Review Correlations

SC.F.1.4.1: Items 1, 4

SC.F.1.4.2: Items 2, 5, 6

SC.H.3.4.6: Item 3

Movement of Organic Compounds

Study TIP

Reviewing Information Work with a partner to review the pressure-flow model of translocation, covered on this page. Review the steps of model by quizzing each other on the events of each step.

Organic compounds move throughout a plant within the phloem. Botanists use the term **source** to refer to a part of a plant that provides organic compounds for other parts of the plant. For example, a leaf is a source because it makes starch during photosynthesis. A root that stores sugar is also a source. Botanists use the term **sink** to refer to a part of a plant that organic compounds are delivered to. Actively growing parts, such as root tips and developing fruits, are examples of sinks. The movement of organic compounds within a plant from a source to a sink is called **translocation.**

The movement of organic compounds in a plant is more complex than the movement of water for three reasons. First, water flows freely through empty xylem cells, but organic compounds must pass through the cytoplasm of living phloem cells. Second, water only moves up in xylem, while organic compounds move in all directions in phloem. Third, water can diffuse through cell membranes but organic compounds cannot. The German botanist Ernst Münch proposed a model of translocation in 1924. Münch's model, which is often called the pressure-flow model, is shown in **Figure 13.**

Step 1 Sugar from a source enters phloem cells by active transport.

Step 2 When the sugar concentration in the phloem increases, water enters the sieve tubes in phloem from xylem by osmosis.

Step 3 Pressure builds up inside the sieve-tube cells and pushes sugar through the sieve tubes. The sugar moves at a rate as high as 100 cm/h.

Step 4 Sugar moves from phloem cells into a sink by active transport.

Figure 13 BIO graphic

The Pressure-Flow Model

Translocation is described by this model.

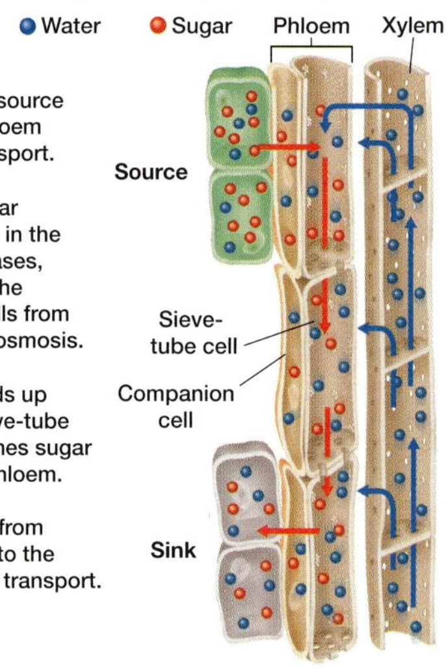

1. Sugar from a source enters the phloem by active transport.
2. When the sugar concentration in the phloem increases, water enters the sieve-tube cells from the xylem by osmosis.
3. Pressure builds up inside the sieve-tube cells and pushes sugar through the phloem.
4. Sugar moves from the phloem into the sink by active transport.

Section 2 Review

1. **Relate** the process of transpiration to the movement of water through a plant. SC.F.1.4.1
2. **Describe** how guard cells regulate transpiration. SC.F.1.4.2
3. **List** the features that make sugar maples economically important. SC.H.3.4.6
4. **Describe** how translocation occurs in plants. SC.F.1.4.1
5. **Critical Thinking** **Predicting Outcomes** When the soil is dry and it is very hot, how can a plant reduce its water loss? Explain your answer. SC.F.1.4.2
6. *FCAT Prep* In plants, what is active transport used to move?
   - **A** sugar into the phloem
   - **B** sugar into the xylem
   - **C** water into the phloem
   - **D** water into the xylem

   SC.F.1.4.2

564

Answers to Section Review

1. Transpiration is the loss of water from leaves. Water moves through a leaf to replace the lost water molecules. This draws water into the leaf from the stem, up the stem, from the roots into the stem, and from the soil into the roots.
2. When guard cells swell with water, they bend to form an opening, or stoma, through which water can exit a leaf in the process of transpiration. When guard cells lose water they sag, closing the stoma.
3. Maple trees provide raw materials for making maple syrup, maple sugar, musical instruments, wood flooring and furniture.
4. Sugar from a source is actively transported into a sieve-tube cell. Water then enters the sieve-tube cell by osmosis, and pressure builds up in the sieve-tube cell and pushes the solution through the sieve tube. Sugar is then actively transported out into a sink cell.
5. As a plant loses water, its guard cells lose pressure and sag, closing its stomata. This decreases water loss from transpiration and conserves water in the plant.
6. **A.** Correct. Active transport moves sugar into the phloem. **B.** Incorrect. Xylem moves water, not sugar. **C.** Incorrect. Water enters sieve tube cells by osmosis. **D.** Incorrect. Water moves freely through xylem.

Study ZONE CHAPTER HIGHLIGHTS

Key Concepts

1 The Vascular Plant Body

- A vascular plant's body contains three kinds of tissue systems—dermal tissue, ground tissue, and vascular tissue.
- Dermal tissue covers a plant. A thin layer of epidermis covers nonwoody parts. Several layers of cork cover woody parts.
- Ground tissue is specialized for photosynthesis in leaves and for storage and support in stems and roots.
- Vascular tissue conducts water, minerals, and organic compounds throughout the plant.
- Xylem contains vessels, which are made up of cells that conduct water only after they lose their cytoplasm. Water flows between cells through pits and perforations in their cell walls.
- Phloem contains sieve tubes, which are made up of cells that are still living. Substances pass between the cells through pores.
- Roots have a central core of vascular tissue that is surrounded by ground tissue and epidermal tissue. Root hairs on root tips increase the surface area available for absorption.
- Nonwoody stems contain bundles of xylem and phloem embedded in ground tissue. Woody stems have an inner core of xylem surrounded by a cylinder of phloem.
- Leaves are a mass of ground tissue and vascular tissue covered by epidermis. The ground tissue cells conduct photosynthesis. Gases are exchanged through the stomata in the epidermis.

2 Transport in Plants

- Transpiration, the loss of water from a plant's leaves, creates a pull that draws water up through xylem from roots to leaves.
- Guard cells control water loss by closing a plant's stomata when water is scarce. Thus, they also regulate the rate of transpiration.
- The sugar maple is a commercially valuable tree. Its hard wood is used to make furniture, musical instruments, and other products. Its sap is made into maple syrup and maple sugar.
- Organic compounds are pushed through the phloem from a source to a sink in a process called translocation.

Key Terms

Section 1

dermal tissue (552)
ground tissue (552)
epidermis (553)
cork (553)
vessel (554)
sieve tube (554)
cortex (555)
root hair (555)
root cap (555)
herbaceous plant (556)
vascular bundle (556)
pith (556)
heartwood (557)
sapwood (557)
petiole (558)
mesophyll (559)

Section 2

transpiration (560)
source (564)
sink (564)
translocation (564)

Alternative Assessment — GENERAL

Have students make a drawing of a plant, including its roots, stems, and leaves. (The drawings do not have to look very realistic, but they should clearly distinguish the structures shown.) Have students indicate the tissues and types of cells present in the roots, stems, and leaves. Have them also place labeled arrows indicating the movement of water and sugars in the processes of transpiration and translocation. LS Visual

Chapter Resource File

- Science Skills Worksheet GENERAL
- Critical Thinking Worksheet ADVANCED
- Test Prep Pretest GENERAL
- Chapter Test GENERAL

Answer to Concept Map

The following is one possible answer to Performance Zone item 15 on the next page.

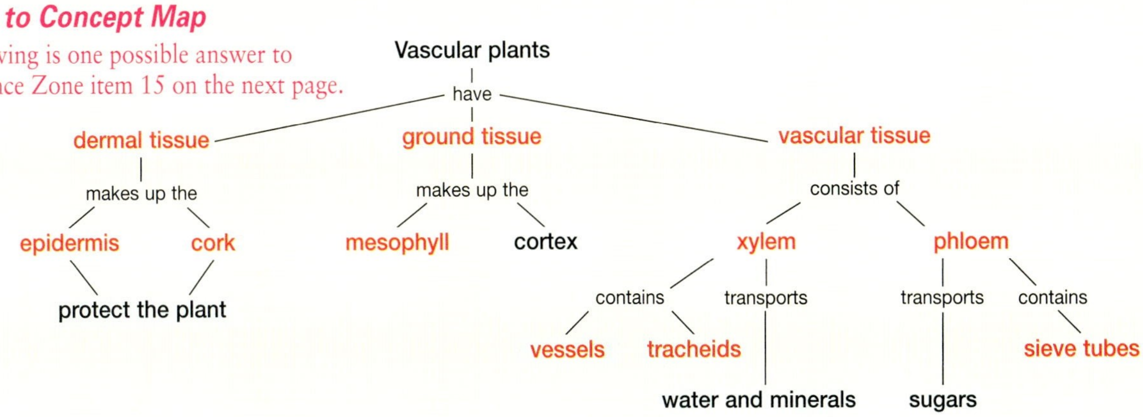

ANSWERS

Understanding Key Ideas

1. d
2. d
3. a
4. a
5. c
6. C_3 leaves have distinct palisade and spongy mesophyll layers; C_4 leaves lack distinct palisade and spongy mesophyll layers and have a layer of cells surrounding their vascular bundles.
7. The sugar maple's range is the northeastern United States and southeastern Canada. Its habitat is northern temperate forests.
8. d
9. Water enters into sieve tubes by osmosis when the cells become loaded with sugars. The water pressure that builds up as a result of the influx of water causes the sap in the sieve tube to move into an adjacent sieve tube. Movement of sugars from sieve tube to sieve tube in this manner results in translocation.
10. One possible answer to the concept map is found on the bottom of the Study Zone page.

Critical Thinking

11. Wilting helps a plant prevent water loss and thus, wilting helps maintain homeostasis.
12. Such a chemical would dissolve the waxy cuticle and allow the herbicide to reach epidermal cells and enter leaves, increasing its efficiency in killing the plant.

Understanding Key Ideas

1. Which of the following is *not* a function of ground tissue?
   a. storage
   b. support
   c. photosynthesis
   d. protection SC.F.1.4.2
2. Which of the following phrases describes the structure of a monocot stem?
   a. contains a ring of vascular bundles that surrounds a core of ground tissue
   b. contains a core of vascular tissue that is surrounded by a ring of ground tissue
   c. contains several layers of xylem that are surrounded by a ring of phloem
   d. contains vascular bundles that are scattered throughout the ground tissue SC.F.1.4.2
3. The stem in the photograph below is probably adapted for SC.F.1.4.2
   a. climbing. **c.** nutrient storage.
   b. water storage. **d.** conserving water.

4. The column of water in a plant's xylem can remain unbroken because of the
   a. cohesion of water molecules.
   b. repulsion between water molecules.
   c. strong walls in the xylem.
   d. stiff fibers in the bark. SC.F.1.4.1
5. Guard cells swell and become longer when
   a. carbon dioxide moves out of the cells.
   b. water moves out of the cells.
   c. water moves into the cells.
   d. oxygen moves into the cells. SC.F.1.4.8
6. **Exploring Further** How are C_3 leaves different from C_4 leaves? SC.G.1.4.1
7. **Up Close** What is the range and habitat of the sugar maple? SC.G.1.4.1
8. Organic compounds move through phloem
   a. by diffusion from a sink to a source.
   b. by diffusion from the leaves to the roots.
   c. by active transport within a sieve tube.
   d. because of the pressure created by the movement of water into the sieve tubes. SC.F.1.4.1
9. How is osmosis involved in translocation in a vascular plant? (**Hint:** See Chapter 4, Section 1.) SC.F.1.4.1
10. **Concept Mapping** Make a concept map that describes the organization of the vascular plant body. Try to include the following terms in your map: *cork, dermal tissue, epidermis, ground tissue, mesophyll, phloem, sieve tubes, tracheids, vascular tissue, vessels,* and *xylem.* SC.F.1.4.2

Critical Thinking

11. **Inferring Function** When a plant wilts, its stomata close. How does wilting help a plant maintain homeostasis? SC.F.1.4.2
12. **Evaluating Results** Some herbicides (weed killers) contain a chemical that breaks down waxy substances. Explain why such a chemical might be useful in a herbicide. SC.H.3.4.5
13. **Analyzing Information** Trace the path of a water molecule through a vascular plant, from the water molecule's entry into a root hair to its exit between two guard cells. Identify and describe the function of each structure the water molecule would encounter on its journey. SC.F.1.4.2

Alternative Assessment

14. **Organizing Information** Construct clay models of cross sections through a C_3 leaf and a C_4 leaf. Include the epidermis, mesophyll, veins, and guard cells in the model of each leaf. Use the models to explain to your class each leaf's structure and the function of its parts. SC.H.3.4.1

13. The molecule would cross the epidermis of the root hair, enter the ground tissue, which functions in storage, and then enter the xylem in the center of the root. It would enter a vessel or a tracheid and then be pulled upward through the stem and into the leaves. In the leaves, the water molecule would be pulled out of the xylem and into the spongy mesophyll. The water molecule would then evaporate and diffuse out of the leaf through a stoma, the opening between two guard cells.

Alternative Assessment

14. Student models will vary but should resemble the leaves shown in **Figure 10** and **Exploring Further.**

Assignment Guide

Section	Questions
1	1, 2, 3, 6, 7, 12, 14
2	4, 5, 8, 9, 10, 11, 13

Chapter Review Correlations

SC.F.1.4.1: Items 4, 8, 9
SC.F.1.4.2: Items 1, 2, 3, 10, 11, 13
SC.F.1.4.8: Item 5
SC.G.1.4.1: Items 6, 7
SC.H.3.4.1: Item 14
SC.H.3.4.5: Item 12

FCAT Prep

For questions 1–6, write your answers on a separate sheet of paper.

1 Where would xylem cells that no longer conduct water be found in a plant?

A. cortex **C.** pith
B. heartwood **D.** sapwood

SC.F.1.4.2

2 What is the function of guard cells?

F. They open and close stomata, regulating the rate of transpiration.
G. They cover the actively growing root tip and protect it from damage.
H. They increase the surface area of a root and its ability to absorb water.
I. They contain dozens of chloroplasts, the main site of photosynthesis in leaves.

SC.F.1.4.2

3 How does the change from open stomata to closed stomata affect a plant's ability to carry out photosynthesis? SC.F.1.4.2

A. It prevents the escape of glucose.
B. It has no effect on photosynthesis.
C. It blocks the entry of carbon dioxide.
D. It exposes more chlorophyll to sunlight.

4 Why would plants that have a fibrous root system rather than a taproot system be more likely to prevent soil erosion on a steep hill?

F. Their roots tend to store more nutrients and water.
G. Their roots penetrate deeper into lower layers of soil.
H. Their roots cover a larger area in the upper layers of the soil.
I. Their roots tend to grow from aboveground stems or leaves.

SC.G.1.4.1

Base your answers to questions 5 and 6 on the diagram below.

LEAF CELLS UNDER TWO CONDITIONS

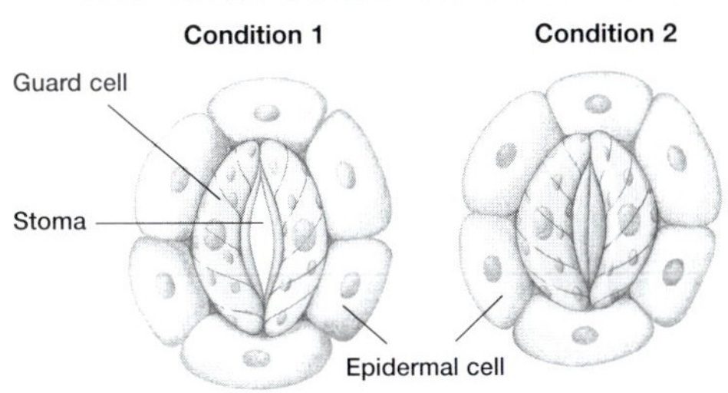

5 What caused the change from condition 1 to condition 2?

A. Sugar moved into the stoma.
B. Sugar moved into the guard cells.
C. Water moved out of the guard cells.
D. Water moved out of the epidermal cells.

SC.F.1.4.2

6 How many nuclei are shown in the group of cells under condition 2? SC.F.1.4.2

Test TIP

When using a diagram to answer a question, look in the image for evidence that supports your potential answer.

567

FCAT Doctor

Question 3 Answer C is the correct choice. Gas exchange takes place through the stomata only when they are open. Answer A is incorrect because glucose does not pass through stomata. Answer B is incorrect because most stomata are located in the leaves of a plant, the major photosynthetic organ. Answer D is incorrect because most stomata are located on the undersides of leaves where sunlight does not strike.

Question 4 Answer H is the correct choice. The upper layers of the soil are lost first in erosion. Answer F is incorrect because fibrous root systems are not best adapted for storage. Answer G is incorrect because this is true of taproots. Answer I is incorrect because this is true of adventitious roots.

Question 5 Answer C is the correct choice. Water moving out of the guard cells closes the stoma. Answer A is incorrect because sugar does not diffuse unaided through the cell membrane. Answer C is incorrect because if sugar moved into the guard cells, water would follow, and the cells would swell, causing them to bend open the stoma. Answer D is incorrect because water moving out of the epidermal cells would have little effect on the guard cells.

Question 6 The correct answer is 8. There are 6 epidermal cells and 2 guard cells, and each cell has a nucleus.

Answers

1. B
2. F
3. C
4. H
5. C
6. 8

Test Prep Correlations

SC.F.1.4.2: Items 1, 2, 3, 5, 6
SC.G.1.4.1: Item 4

Skills Practice Lab

SEPARATING PLANT PIGMENTS

Teacher's Notes

Time Required 30 minutes

Ratings

EASY → HARD

TEACHER PREPARATION 3
STUDENT SETUP 2
CONCEPT LEVEL 2
CLEANUP 2

Skills Acquired

- Collecting Data
- Constructing Models
- Experimenting
- Measuring

Scientific Methods

In this lab, students will:

- Make Observations
- Analyze the Results
- Draw Conclusions

Materials and Equipment

Materials for this lab can be ordered from WARD'S. See *Master Materials List* at the front of this book for catalog numbers. For every 30 students, order one WARD'S "Chromatography of Simulated Plant Pigments" kit. WARD'S refill kit contains all the consumable items. Set up labeled containers for the disposal of solutions, broken glass, and chromatograms.

Safety Cautions

Review all safety symbols with students before beginning the lab. Make sure students wear safety goggles and a lab apron. Caution students to use care when working with the stock solutions so that they do not spill. Caution students to take care when using the capillary tubes because they are breakable. Make sure students wash their hands before leaving the lab.

Skills Practice Lab

Separating Plant Pigments SC.F.1.4.1

SKILLS

- Performing paper chromatography
- Calculating

OBJECTIVES

- **Separate** the pigments that give a leaf its color.
- **Calculate** the R_f value for each pigment.
- **Describe** how paper chromatography can be used to study plant pigments.

MATERIALS

- safety goggles
- lab apron
- strip of chromatography paper
- scissors
- metric ruler
- pencil
- capillary tube
- drop of simulated plant pigments extract
- 10 mL graduated cylinder
- 5 mL of chromatography solvent
- chromatography chamber

ChemSafety

CAUTION: Always wear safety goggles and a lab apron to protect your eyes and clothing.

CAUTION: Do not touch or taste any chemicals. Know the location of the emergency shower and eyewash station and how to use them. If you get a chemical on your skin or clothing, wash it off at the sink while calling to the teacher. Notify the teacher of a spill. Spills should be cleaned up promptly, according to your teacher's directions.

CAUTION: Glassware is fragile. Notify the teacher of broken glass or cuts. Do not clean up broken glass or spills with broken glass unless the teacher tells you to do so.

Before You Begin

Pigments produce colors by reflecting some colors of light and absorbing or transmitting others. Pigments can be removed from plant tissues using **solvents,** chemicals that dissolve other chemicals. The pigments can then be separated from the solvent and from each other by using **paper chromatography.** The word *chromatography* comes from the Greek words *chromat,* which means "color," and *graphon,* which means "to write." The $\boldsymbol{R_f}$ is the ratio of the distance that a pigment moves relative to the distance that a solvent moves. Since the R_f for a compound is constant, scientists can use it to identify compounds. In this lab, you will learn how to use paper chromatography to separate a mixture of pigments.

1. Write a definition for each boldface term in the previous paragraph and for each of the following terms: chlorophyll a, chlorophyll b, carotene, xanthophyll.
2. Make a data table similar to the one below.

DATA TABLE

Band no.	Color	Pigment	Migration (in mm)	R_f value
1 (top)				
2				
3				
4				
Solvent				

568

Tips and Tricks

Review the procedure for creating a chromatogram, on which pigments show up as colored streaks. Students'chromatograms will be almost identical to those of real plant pigments. Have students practice using a capillary tube to ensure getting a tiny drop of extract on the chromatography paper. Emphasize the importance of using a pencil when marking a chromatogram. Emphasize the importance of placing only the tip of the chromatography paper in the solvent. To prevent contamination of the solvent, the pigment should never make direct contact with the solvent. Tell students to mark the distances traveled by the pigments with a pencil as soon as the chromatogram is removed from the solvent. The simulated plant pigments fade over time. Allow students to use calculators to calculate Rf values.

3. Based on the objectives for this lab, write a question you would like to explore about plant pigments or paper chromatography.

Procedure

Part A: Making a Chromatogram

1. Put on safety goggles and a lab apron. Use scissors to cut the bottom end of a strip of chromatography paper to a tapered end. **CAUTION: Sharp or pointed objects may cause injury. Handle scissors carefully.**
2. Draw a faint pencil line 1 cm above the pointed end of the paper strip, as shown in the photo on the facing page. Use a capillary tube to apply a tiny drop of the simulated plant pigments extract on the center of the line.
3. Pour 5 mL of chromatography solvent into a chromatography chamber. Pull the chromatography paper through the opening of the cap, and adjust the length of the strip so that a small portion of the tip end is immersed in the solvent. DO NOT immerse the pigment in the solvent.
4. Place the cap over the chromatography chamber. Carefully bend the end of the strip of chromatography paper over the cap, as shown in the photograph on the facing page. Be sure that the strip does not touch the walls of the chamber.
5. Remove the strip from the chromatography chamber when the solvent nears the top of the chamber (within 5–7 minutes).
6. With a pencil, mark the position of the uppermost end of the solvent and the farthest distance each pigment moved. Measure the distance that the solvent and each pigment moved. Record your observations and measurements in your data table. Tape or glue your chromatogram to your lab report. Label the pigment colors.
7. Use the formula below to calculate and record the R_f for each pigment.

$$R_f = \frac{\textit{Distance substance (pigment) traveled}}{\textit{distance solvent traveled}}$$

Part B: Cleanup and Disposal

8. Dispose of chromatography paper, solutions, and broken glass in the designated waste containers. Do not pour chemicals down the drain or put lab materials in the trash unless your teacher tells you to do so.
9. Clean up your work area and all lab equipment. Return lab equipment to its proper place. Wash your hands thoroughly before you leave the lab and after you finish all work.

Analyze and Conclude

1. **Summarizing Results** Describe what happened to the simulated plant pigments during the lab.
2. **Analyzing Data** How do your R_f values compare with those of your classmates?
3. **Inferring Conclusions** What is a chromatogram?
4. **Further Inquiry** Write a new question about plant pigments that could be explored with another investigation.

On the Job

Paper chromatography is used to separate a variety of chemicals from living tissue extracts. Do research to discover how paper chromatography is used in research laboratories. For more about careers, visit **go.hrw.com** and type in the keyword **HX4 Careers.**

Answers to Before You Begin

pigments—molecules that absorb some colors of light and reflect others; *solvents*—chemicals that can dissolve other chemicals; *paper chromatography*—technique of separating a mixture of chemicals by making them pass through a strip of paper in which the chemicals move at different speeds; *Rf*—the ratio of the distance a pigment moves relative to the distance a solvent moves; *chlorophyll a*—green photosynthetic pigment found in plants; *chlorophyll b*—green photosynthetic pigment found in plants; *carotene*—orange pigment found in plants; *xanthophyll*—yellow pigment found in plants

3. Answers will vary. For example: How many different pigments are found in the mixture?

Answers to Analyze and Conclude

1. The greenish mixture separated into four different colors: dark green, light green, yellow, and orange. The pigments moved different distances along the strip of chromatography paper. The orange moved the farthest distance, and the dark green moved the shortest distance.
2. Answers will vary. Students should conclude that the R_f values are the same for each pigment separated.
3. A chromatogram is a strip of chromatography paper that has pigments separated from a mixture along its length.
4. Answers will vary. For example: Can pigments in other substances such as fruit juices and ink be separated with the same solvent?

Answers to Procedure

Sample Data Table

Band No.	Color	Pigment	Migration distance (in mm)	R_f value
1 (top)	orange	carotene	37	0.92
2	yellow	xanthophyll	29	0.72
3	light green	chlorophyll a	17	0.42
4	dark green	chlorophyll b	8	0.20
Solvent	—	—	40	—

Chapter Opener

- **SE** **Quick Review,** p. 571 ■
- **SE** **Reading Activity,** p. 571 ■
- **TE** **Using the Figure,** p. 570 GENERAL
- **TE** **Opening Activity,** p. 571 BASIC

& 3 • 90 min pp. 572–578

Section 1

How Plants Grow and Develop

- **OSP** **Lesson Plan***
- **TR** **Bellringer***
- **TR** **Section Outline***
- **TR** **G64 Seed Germination***
- **TR** **G18 Apical Meristems***
- **TR** **G24 Development of a Woody Stem***

- **TE** **Activity** Viewing Seeds, p. 572 BASIC
- **TE** **Demonstration,** p. 575 GENERAL
- **TE** **Group Activity** Design a Flower Bed, p. 573 GENERAL
- **TE** **Activity** Apical Meristems, p. 574 GENERAL
- **TE** **Demonstration,** p. 575 GENERAL
- **TE** **Activity** The Debate about Food Irradiation, p. 576 ADVANCED
- **CRF** **Exploration Lab** Seeds and Respiration* GENERAL
- **CRF** **Inquiry Lab** What Direction Will it Grow?* GENERAL

& 5 • 90 min pp. 579–584

Section 2

Regulating Growth and Development

- **OSP** **Lesson Plan***
- **TR** **Bellringer***
- **TR** **Section Outline***
- **TR** **G20 Major Mineral Nutrients Required by Plants***
- **TR** **G40 Bread Wheat***
- **TR** **G49 Went's Experiment***
- **TR** **G51 Flowering and Photoperiodism***

- **TE** **Using the Figure,** p. 580
- **SE** **Quick Lab** Investigating the Effects of Ethylene on a Plant, p. 581 ■
- **CRF** **Datasheets for In-Text Labs** Investigating the Effects of Ethylene on a Plant*
- **SE** **Data Lab** Interpreting Annual Rings, p. 583 ■
- **CRF** **Datasheets for In-Text Labs** Interpreting Annual Rings*
- **TE** **Using the Figure,** p. 583 GENERAL
- **SE** **Math Lab** Analyzing the Effect of Cold on Seed Germination, p. 584 ■
- **CRF** **Datasheets for In-Text Labs** Analyzing the Effect of Cold on Seed Germination*
- **SE** **Skills Practice Lab** Comparing Bean and Corn Seedlings, pp. 588–589 ◆ ■
- **CRF** **Datasheets for In-Text Labs** Comparing Bean and Corn Seedlings*
- **Forensics and Applied Science Experiments** Effects of Acid Rain on Plants* GENERAL

& 7 • 90 min

Chapter Review and Assessment Resources

- **SE** **Study Zone,** p. 585 ■
- **SE** **Performance Zone,** p. 586 ■
- **SE** **FCAT Test Prep,** p. 587 ■
- **TR** **Concept Mapping***
- **CRF** **Vocabulary Review Worksheet*** ■ GENERAL
- **CRF** **Test Prep Pretest*** ■ GENERAL
- **CRF** **Chapter Test*** ■ GENERAL
- **CRF** **Chapter Test*** ADVANCED
- **OSP** **Test Generator***
- **OSP** **Test Item Listing***
- **OSP** **Modified Chapter Test*** BASIC

Holt Online Learning

Visit go.hrw.com for access to Holt Online Learning or enter the keyword **HX6 Home** for a variety of free online resources.

One-Stop Planner® CD-ROM

This CD-ROM package includes
- Lab Materials QuickList Software
- Holt Calendar Planner
- Customizable Lesson Plans
- Printable Worksheets
- ExamView® Test Generator
- Interactive Teacher Edition
- Holt PuzzlePro® Resources
- Holt PowerPoint® Resources

Guided Reading Audio CDs

These CDs are designed to help auditory learners and reluctant readers. (Audio CDs are also available in Spanish.)

Compression guide:
To shorten your instruction because of time limitations, omit blocks 1 and 6.

KEY

- **TE** Teacher Edition
- **SE** Student Edition
- **OSP** One-Stop Planner
- **CRF** Chapter Resource File
- **TR** Teaching Transparency
- * Also on One-Stop Planner
- ■ Also Available in Spanish
- ◆ Requires Advance Prep

SKILLS DEVELOPMENT AND PRACTICE	SECTION REVIEW AND ASSESSMENT	CORRELATIONS
		Florida Sunshine Standards/FCAT
TE Reading Skill Builder Interactive Reading, p. 573 BASIC **TE Skill Builder** Interpreting Visuals, p. 573 GENERAL **TE Reading Skill Builder** Paired Summarizing, p. 577 BASIC **CRF Directed Reading Worksheet*** BASIC **CRF Active Reading Worksheet*** GENERAL	**SE Section 1 Review** ■ **TE Reteaching,** p. 578 BASIC **TE Quiz,** p. 578 GENERAL **TE Alternative Assessment,** p. 578 GENERAL **CRF Quiz*** ■ BASIC	SC.F.1.4.2, SC.G.1.4.1, SC.H.1.4.1, SC.H.3.4.6
TE Reading Skill Builder Reading Hint, p. 580 BASIC **TE Reading Skill Builder** Reading Organizer, p. 582 BASIC **CRF Directed Reading Worksheet*** BASIC **CRF Active Reading Worksheet*** GENERAL **CRF Critical Thinking Worksheet*** ADVANCED **CRF Science Skills Worksheet*** GENERAL	**SE Section 2 Review** ■ **TE Reteaching,** p. 584 BASIC **TE Quiz,** p. 585 GENERAL **TE Alternative Assessment,** p. 584 GENERAL **CRF Quiz*** ■ BASIC **Modified Worksheet*** BASIC	SC.F.1.4.7, SC.F.1.4.8, SC.G.1.4.1, SC.G.1.4.3

Holt Biology Interactive Tutor CD-ROM

This CD-ROM consists of interactive activities that give students a fun way to extend their knowledge of biology concepts.

Unit 2 Photosynthesis

SCILINKS. NSTA

www.scilinks.org

Maintained by the **National Science Teachers Association**

Topic: Primary Growth in Plants
SciLinks Code: HX4148

Topic: Secondary Growth in Plants
SciLinks Code: HX4160

Topic: Genetic Engineering
SciLinks Code: HX4092

Topic: Plant Hormones
SciLinks Code: HX4140

Overview

Tell students that the purpose of this chapter is to demonstrate that as plants grow, they develop specialized cells, tissues, and structures in a coordinated manner. Plant growth and development require a balance of substances obtained from the environment and a balance of growth-regulating hormones produced by the plant.

Using the Figure

The oldest organisms on Earth, bristlecone pines are named for prickles on the scales of the female cones. The average age of a bristlecone pine is about 1,000 years, but a few are older than 4,000 years. A bristlecone pine named Methuselah has been determined to be over 4,700 years old. It is the oldest known living organism and was a seedling before the Great Pyramid of Giza was built. Growing at the tree line in the mountains of only six Western states, bristlecone pines are exposed to cold temperatures and dryness which results in slow growth.

Assessing Prior Knowledge

Before beginning this chapter, be sure your students can answer the questions in the ***Quick Review*** on the facing page. If they cannot, encourage them to reread the sections indicated.

Chapter Resource File

- **Vocabulary Worksheets**
- **Concept Mapping**

Bristlecone pine "flagged" by the wind

570

Chapter Correlations — **Florida Sunshine Standards/FCAT**

SC.F.1.4.2 Body structures are uniquely designed and adapted for their function.

SC.F.1.4.7 Organisms respond to internal and external stimuli.

SC.F.1.4.8 Cell behavior can be affected by molecules from other parts of the organism or even from other organisms.

SC.G.1.4.1 There is great diversity and interdependence among living things.

SC.G.1.4.3 The chemical elements that make up the molecules of living things are combined and recombined in different ways.

SC.H.3.4.6 Scientific knowledge is used by those who engage in design and technology to solve practical problems, taking human values and limitations into account.

CHAPTER

26 Plant Growth and Development

Florida Sunshine Standards/FCAT

SC.F.1.4.7 Organisms respond to internal and external stimuli.

SC.F.1.4.8 Cell behavior can be affected by molecules from other parts of the organism or even from other organisms.

SC.G.1.4.1 There is great diversity and interdependence among living things.

SC.G.1.4.3 The chemical elements that make up the molecules of living things are combined and recombined in different ways.

SC.H.3.4.6 Scientific knowledge is used by those who engage in design and technology to solve practical problems, taking human values and limitations into account.

Quick Review

Answer the following without referring to earlier sections of your book.

1. **Identify** the requirements for photosynthesis and cellular respiration. *(Chapter 5, Sections 2 and 3)*
2. **Define** *meristem, vascular tissue, xylem,* and *phloem*. *(Chapter 23, Section 1)*
3. **Identify** the parts of a seed. *(Chapter 23, Section 1)*
4. **Describe** the structures of stems and roots. *(Chapter 25, Section 1)*

Did you have difficulty? *For help, review the sections indicated.*

Reading Activity

Before you read this chapter, write a short list of all the things you already know about plant growth and development. Then list things you want to know about plant growth and development.

Looking Ahead

Section 1

How Plants Grow and Develop

Seeds Sprout
Meristems
Plant Development

Section 2

Regulating Growth and Development

Nutrients
Hormonal Control of Growth
Environmental Influences on Growth

internet connect

www.scilinks.org

National Science Teachers Association *sci*LINKS Internet resources are located throughout this chapter.

SCILINKS. Maintained by the National Science Teachers Association

Exposed to extreme cold and deprived of water, trees such as this one that grow near the tree line in the mountains are stunted. Like flags, they "stream" away from the prevailing wind.

Reading Activity Answer

Students may be quite familiar with seed germination but are probably "rusty" in their knowledge of plant tissues. Additionally, students may be unfamiliar with plant hormones, and may think that only animals have hormones.

Identifying Misconceptions

Students may think that the giant sequoias in California are the oldest known organisms in the world. Although these trees are very old—many ranging in age from 2,000 to 3,000 years—older bristlecone pines exist.

Opening Activity — BASIC

Germination Show students that plant germination can push caps off bottles. This demonstration will take a few days. Fill a container with beans. Fill another container with beans and water. Snap the caps onto the bottles. (Don't use child-proof bottles.) In a few days, the germinating beans will knock the caps off or even split the bottle apart. Ask students why the beans with water are more powerful than the beans without. (The beans with water are expanding as they germinate.)

Quick Review Answers

1. The requirements for photosynthesis are carbon dioxide, water, and sunlight. For cellular respiration, oxygen and glucose are required. Both sets of reactions take place in organelles of living cells, using enzymes.
2. Meristems are zones of actively dividing plant cells. Vascular tissue, comprised of xylem and phloem, transports water and other materials within plants. Xylem consists of hard-walled cells that transport water and mineral nutrients, and help support the plant body. Phloem consists of soft-walled cells that transport organic nutrients.
3. Seeds consist of a plant embryo, stored food, and a protective seed coat.
4. Herbaceous stems contain vascular bundles surrounded by ground tissue. Woody stems have a central core of pith and a ring of vascular bundles, which fuse into solid cylinders as the stem matures.

Section 1

Section 1 How Plants Grow and Develop

Florida Sunshine Standards/FCAT

SC.G.1.4.1 There is great diversity and interdependence among living things.
SC.H.3.4.6 Scientific knowledge is used by those who engage in design and technology to solve practical problems, taking human values and limitations into account.

Objectives

- **Compare** seed germination in beans and corn. SC.G.1.4.1
- **Contrast** annuals, biennials, and perennials. SC.G.1.4.1
- **Explain** how primary and secondary growth are produced. SC.F.1.4.2
- **Describe** several traits of bread wheat. SC.H.3.4.6
- **Contrast** development in plants and animals. SC.G.1.4.1

Key Terms

germination
perennial
annual
biennial
primary growth
secondary growth
apical meristem
cork cambium
vascular cambium
annual ring

Seeds Sprout

A seed contains a plant embryo that is in a state of suspended animation. Some embryos can remain in suspended animation inside a seed for thousands of years. Seeds sprout with a burst of growth in response to certain changes in the environment. These changes, such as rising temperature and increasing soil moisture, usually signal the start of favorable growing conditions.

Many seeds must be exposed to cold or to light before they can sprout. The seed coats of other seeds must be damaged before they can sprout. Exposure to fire, passing through the digestive system of an animal, and falling on rocks are several natural ways that seed coats are damaged. A seed cannot sprout until water and oxygen penetrate the seed coat. When water enters a seed, the tissues in the seed swell, and the seed coat breaks. If enough water and oxygen are available after the seed coat breaks, the young plant, or seedling, begins to grow.

Germination

A plant embryo resumes its growth in a process called **germination.** The first sign of germination is the emergence of the embryo's root. What happens next varies somewhat from one type of plant to another, as you can see in **Figure 1.** The young shoots of some plants, such as beans, form a hook. The hook protects the tip of the shoot from injury as it grows through the soil.

Figure 1 Seed germination

Beans and corn show two characteristic patterns of seed germination.

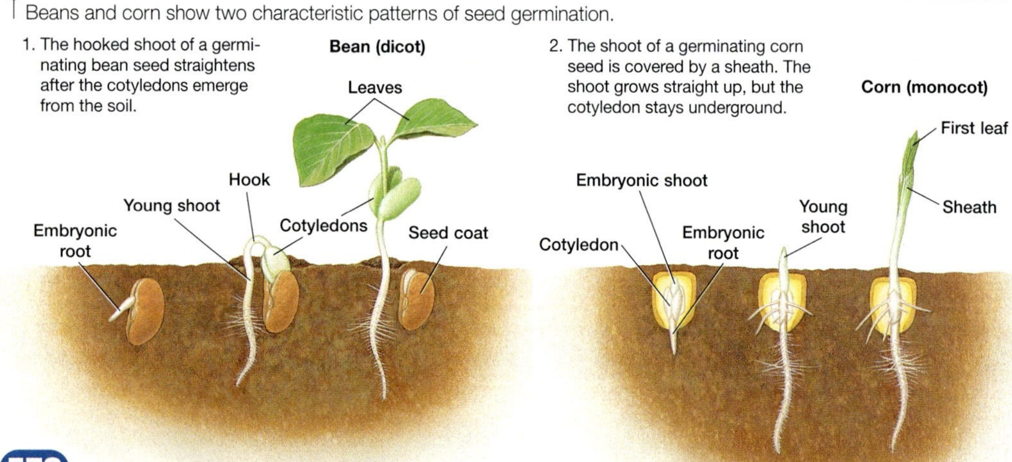

1. The hooked shoot of a germinating bean seed straightens after the cotyledons emerge from the soil.
2. The shoot of a germinating corn seed is covered by a sheath. The shoot grows straight up, but the cotyledon stays underground.

572

Focus

Overview

Before beginning this section review with your students the objectives listed in the Student Edition. This section begins by describing seed germination. Then, after defining perennials, annuals, and biennials in the context of plant life spans, the section explains meristems and their role in primary and secondary plant growth.

Bellringer

Ask students to draw the structure of a seed. (The drawings should show an embryo, a stored food supply, and a protective outer covering.) Ask students to share their drawings so that you can determine their prior knowledge before beginning this section. Allow ELL students to label their drawings in their native language and English. LS **Visual**

Motivate

Activity — BASIC

Provide students with bean seeds and corn seeds that you have germinated. (Place the seeds between sheets of wet paper towels, and put the wet towels with seeds in a closed black garbage bag. Place the bag in a warm spot for 3–4 days. Check seeds at days 2 and 3.) Have students slice the germinating seeds longitudinally, or provide pre-cut germinating seeds. Ask students to view the seeds under stereomicroscopes, and identify the structures shown in **Figure 1.**

Section Correlations

SC.F.1.4.2
SC.G.1.4.1
SC.H.3.4.6

Chapter Resource File

- Directed Reading BASIC
- Active Reading GENERAL
- Data Sheet for Skills Practice Lab GENERAL

Transparencies

TR G44 Seed Germination
TR Bellringer

One-Stop Planner CD-ROM

- Reading Organizers BASIC
- Reading Strategies BASIC
- Portfolio Project Plant Focus Worksheet GENERAL

The young shoots of other plants, such as corn, have a protective sheath around their shoots. In some plants, such as beans and marigolds, the cotyledons emerge from the soil and unfold aboveground. In other plants, such as corn and peas, the cotyledons remain underground. After the shoot of a seedling emerges, its roots and shoots continue to grow throughout its life.

Plant Life Spans

As you have read in a previous chapter, bristlecone pines are the oldest known trees. They are estimated to be about 5,000 years old. In contrast, some plants live for only a few weeks. Depending on how long it lives, a plant can be classified as one of three basic types: perennial, annual, or biennial.

Perennials Many herbaceous plants and all woody plants are perennials. A **perennial** is a plant that lives for several years. Most perennials reproduce many times during their life span. Others, like the herbaceous perennial shown in **Figure 2,** reproduce only once before they die. Chrysanthemums, daffodils, and irises are familiar herbaceous perennials. These plants store nutrients for the next season's growth in fleshy roots or underground stems. The aboveground shoots of herbaceous perennials often die after each season of growth. Trees, shrubs, and many vines are woody perennials. Some woody perennials drop their leaves each year. Plants that drop all of their leaves each year, such as elms, maples, and grapevines, are known as deciduous *(dee SIHJ oo uhs)* plants. Those that drop a few leaves at a time throughout the year, such as firs, pines, and junipers, are called evergreens.

Figure 2 A herbaceous perennial. Century plants live for many years but reproduce only once. Like the dried-up plant on the left, this flowering century plant will die when its seeds are mature.

Annuals Sunflowers, beans, corn, and many weeds are annuals. An **annual** is a plant that completes its life cycle (grows, flowers, and produces fruits and seeds) and then dies within one growing season. Virtually all annuals are herbaceous plants. Most annuals grow rapidly when conditions are favorable. Individual plants can become quite large if they get enough water and nutrients.

Biennials Carrots, parsley, and onions are biennials. A **biennial** is a flowering plant that takes two growing seasons to complete its life cycle. During the first growing season, biennials produce roots and shoots. The shoots consist of a short stem and a rosette (circular cluster) of leaves. The roots store nutrients. In the second growing season, a biennial plant uses the stored nutrients to produce a flowering stalk. The plant dies after flowering and producing fruits and seeds.

Teach

READING SKILL BUILDER — BASIC

Interactive Reading Assign Chapter 26 of the *Holt Biology* Guided Reading Audio CD Program to help students achieve greater success in reading the chapter.

SKILL BUILDER — GENERAL

Interpreting Visuals Bring students' attention to **Figure 1.** Have students compare the structures of the germinating corn seed and bean seed. Ask students how monocot and dicot seedlings differ in the way they emerge from the soil. (Dicot seedlings have a stem that hooks as it emerges to protect the delicate tip, whereas monocots grow straight out of the soil. Monocot seedlings have a protective sheath that guards the tip.) Ask students what other differences they notice. (The cotyledon and seed coat of the corn seed remain underground, while the cotyledons of the bean emerge from the soil.) English Language Learners

Group Activity — GENERAL

Design a Flower Bed Have students work in groups of four to design a flower garden that will provide color in every season (or in three seasons, if you live in an area with a harsh winter climate). Let students use nursery catalogs to find examples of annuals, biennials, and perennials, as well as to determine where each plant will grow and when it will bloom. Assign each group a set of conditions under which its garden must grow (e.g., size, distribution of sun and shade, soil type). Finally, tell them to consider aesthetic factors (e.g., color combinations, taller plants in back, shorter plants in front). Have each student design a graphic that shows the group's garden from the top during one of the seasons and identifies their plants. LS Visual

REAL WORLD CONNECTION

Take students on a walk around the school grounds or a park, and have them keep a log of the plants they see, using general terms such as flower, tree, grass, or shrub. Have students write a paragraph that answers the following questions based on what they observe. What types of plants did you see? Did you see more of one type of plant than another? If so, what might account for the difference? (More annuals tend to be present in spring and summer; only woody perennials may be present in winter.)
LS Interpersonal

did you know?

Plant Collections The Royal Botanic Gardens, Kew, located about 10 km southwest of London, England, is recognized as a global center of excellence in the study of plant diversity. The world's largest documented living and preserved plant collections are located there.

Teach, *continued*

Activity — GENERAL

Apical Meristems Let students observe prepared slides of the apical meristems of the shoots and roots of several plants. Have them draw what they see and label the parts identified in **Figure 3.** Have students write a brief paragraph describing how the cells in the apical meristems differ from those in the surrounding tissues. (The cells of the apical meristems are small and unspecialized. The surrounding cells tend to be larger and longer and have thicker walls, and some have specialized structures such as chloroplasts or root hairs.) **LS Visual**

Teaching Tip — GENERAL

Removal of the Apical Meristem Ask students what they think would happen if the apical meristem of a stem or root were removed. (Removing or damaging the apical meristems of a plant would stop or limit the vertical growth of the stem or root.) This frequently occurs in nature, as tips of stems are broken off by wind or animals, killed by disease, or eaten by insects or other animals. Lateral buds often begin to grow after the apical meristem is removed. If the plant is a tree, one of these lateral buds will take over as the main trunk, and the tree will take on its normal form again in a few years. If the plant is a shrub or an herbaceous plant, the growth of many lateral buds will make the plant bushier, and it will not grow as tall.

Transparencies

TR G18 Apical Meristems

Study TIP

Compare and Contrast
To compare and contrast *primary growth* and *secondary growth,* make a two-column list. In one column write the ways in which the two are alike. In the other column, write the ways in which they are different.

Meristems

Plants grow by producing new cells in regions of active cell division called meristems. Almost all plants grow in length by adding new cells at the tips of their stems and roots. Growth that increases the length or height of a plant is called **primary growth.** Many plants also become wider as they grow taller. Growth that increases the width of stems and roots is called **secondary growth.** After new cells are formed by cell division, they grow and undergo differentiation. Recall from your reading that differentiation is the process by which cells become specialized in form and function.

Primary Growth

Apical *(AP ih kuhl)* **meristems,** which are located at the tips of stems and roots, produce primary growth through cell division. As shown in **Figure 3,** apical meristems are regions of small, undifferentiated cells. To better understand how primary growth occurs in most plants, imagine a stack of dishes. As you add more dishes to the top, the stack grows taller but not wider. Similarly, the cells in the apical meristems of most plants add more cells to the tips of a plant's body. New cells are added through cell division. The cells then lengthen. Thus, primary growth makes a plant's stems and roots get longer without becoming wider. To learn about primary growth in a monocot, look at *Up Close: Bread Wheat* later in this section.

The tissues that result from primary growth are called *primary tissues.* The new cells produced by apical meristems differentiate into the primary dermal, ground, and vascular tissues of roots, stems, and leaves. Some of the cells produced by the root apical meristem also become part of the root cap. These cells replace cells that are worn away as the root pushes through the soil.

internet connect
www.scilinks.org
Topic: Primary Growth in Plants
Keyword: HX4148
SCILINKS. Maintained by the National Science Teachers Association

Figure 3 Apical meristems

Both shoot tips and root tips contain apical meristems, where cell division occurs.

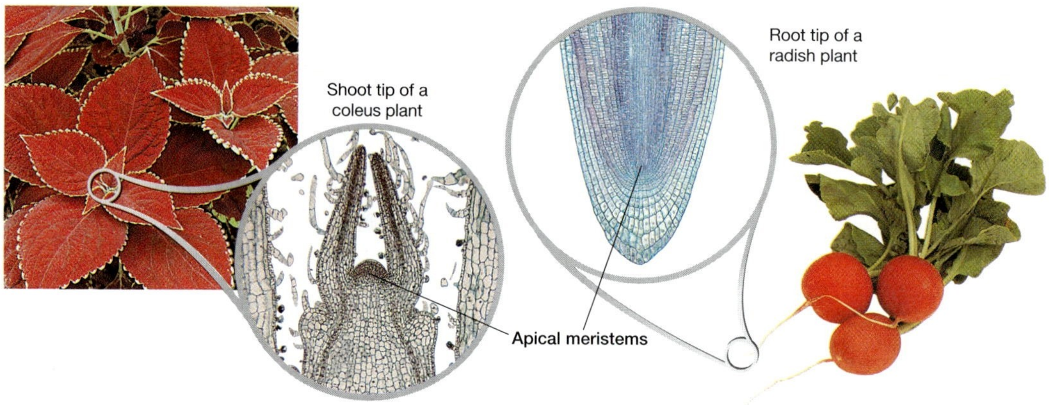

574

MEDICINE CONNECTION

Flowering plants and ferns are the original sources of approximately 25% of all drugs prescribed each year in North America. In fact, botany was considered a branch of medicine until the early to mid 1800s.

Cultural Awareness

The Asian Art of Bonsai Bonsai is the cultivation of miniature specimens of plants that would normally be much larger. The effect is obtained by frequent, selective pruning of both the shoot and the roots. Bonsai plants are typically grown in small containers. The word bonsai comes from the Japanese characters *bon,* meaning "tray," and *sai,* meaning "plant." The Chinese practiced bonsai in approximately the third century B.C., but it was the Japanese Buddhist monks who perfected the art.

Secondary Growth

Secondary growth occurs in parts of many herbaceous plants, such as in carrot roots. However, it is most dramatic in woody plants. Secondary growth is produced by cell division in two meristems, which form thin cylinders near the outside of woody stems and roots. One meristem, called the **cork cambium** *(KAM bee uhm)*, lies within the bark and produces cork cells. The other meristem, called the **vascular cambium,** lies just under the bark and produces vascular tissues. The tissues that result from secondary growth are called secondary tissues. **Figure 4** shows how woody stems develop.

Step 1 A young woody stem has a ring of vascular bundles between the cortex and the pith. Each vascular bundle contains primary xylem and primary phloem.

Step 2 Vascular cambium develops between the primary xylem and the primary phloem in each vascular bundle. Secondary phloem is produced toward the outside of the stem. Secondary xylem is produced toward the inside of the stem. The cork cambium forms when the epidermis is stretched and broken as the stem grows in diameter.

Step 3 Eventually, the vascular bundles merge into solid cylinders. No cortex or primary phloem remains. The cork, cork cambium, and secondary phloem make up the bark. The vascular cambium and secondary xylem lie inside the bark. Thick layers of secondary xylem, or wood, often form rings. Since one new ring is usually formed each year, the rings are called **annual rings.**

internet connect

www.scilinks.org
Topic: Secondary Growth in Plants
Keyword: HX4160

SCI LINKS. Maintained by the National Science Teachers Association

BIO graphic

Figure 4

Development of a Woody Stem

The wood in a woody stem results from secondary growth.

1. Initially, the stem is covered by epidermis and contains cortex, pith, and a ring of vascular bundles with primary xylem and phloem.
2. A vascular cambium forms between the xylem and phloem in each vascular bundle. Cork cambium forms under the epidermis.
3. In a mature stem, the vascular cambium adds new layers of secondary xylem and phloem each year.

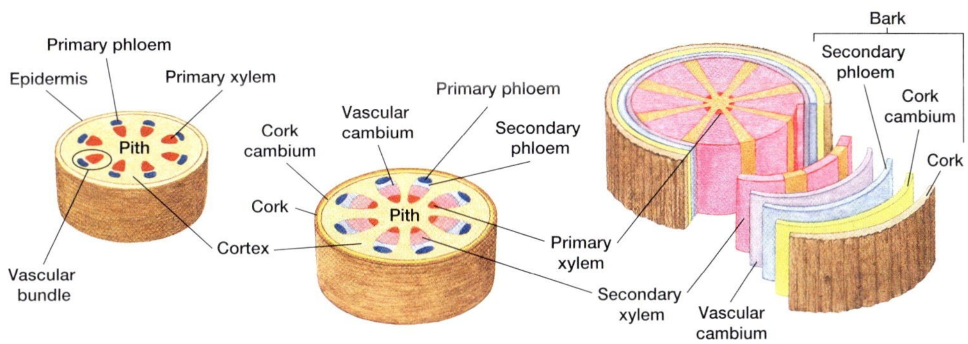

Demonstration — GENERAL

Provide students with prepared slides of a cross section of a woody stem. Have them examine the slides under a compound microscope. Point out that vascular cambium is only one layer of cells. It is found just under the bark of a woody stem and can be located in microscope slides of woody stem cross sections by finding the phloem cells and the large, thickened xylem cells. Also point out that both xylem and phloem develop from the vascular cambium cells, so they develop in rows on either side of this "meristematic" tissue. Have students locate the layer between the two types of cells in a woody stem cross section. This is the row of vascular cambium cells. Have students describe how xylem and phloem originate from vascular cambium. (Xylem and phloem originate through cell division and cell differentiation.) **LS Visual**

Teaching Tip — GENERAL

Summarizing Plant Growth

Have students prepare a sequential diagram that summarizes how herbaceous plants grow by primary growth and how woody plants grow by both primary and secondary growth. An example of a completed diagram appears in the Graphic Organizer on this page. **LS Visual**

Transparencies

TR G24 Development of a Wood Stem

Graphic Organizer

Use this graphic organizer with **Teaching Tip** on this page.

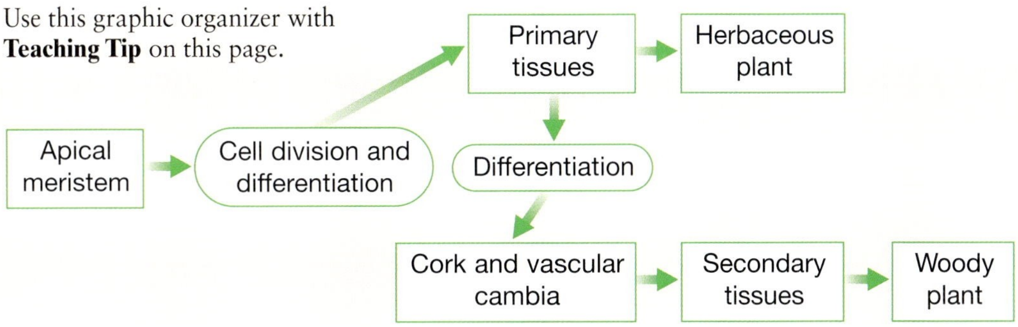

Up Close

Bread Wheat

Teaching Strategies

- Show students specimens of living or preserved wheat plants.
- Let students dissect grains, or kernels, of wheat and locate the endosperm, bran, and germ of the kernels. Explain that the endosperm serves as a food supply for the embryo when it begins to grow. White wheat flour contains only endosperm and whole wheat flour contains both endosperm and bran.
- For interest, have students make bread so that they notice the elasticity of the dough, which is caused by gluten.

Activity — ADVANCED

The Debate about Food Irradiation Food irradiation is the process of exposing food to radiation—the energy emitted by the unstable nuclei of certain atoms in the form of waves or rays. As it travels through the food, radiation kills organisms such as bacteria and insects. Have students research the history of and the current debate over food irradiation in preparation for a class debate. (Pro: Food irradiation makes the food supply safer, provides a better quality of food, and extends the shelf life of food. Con: irradiation induces chemical changes in food, which may affect the color, odor, and texture of food and lower the nutritional value of food.)

Bread Wheat

- **Scientific name:** *Triticum aestivum*
- **Size:** 0.3 to 0.8 m (1 to 2.5 ft) tall
- **Range:** Agricultural regions worldwide
- **Habitat:** Cultivated fields in temperate and subtropical grasslands
- **Importance:** Wheat is the principal staple food in temperate regions of the world. The grains of *Triticum aestivum* are usually ground into flour that is used to make bread.

External Structures

Leaves Wheat, a member of the grass family, is a monocot. Its leaves are long but only about 2 cm (0.8 in.) wide, with parallel veins. The leaves are attached to the stem by a sheath, which wraps around the stem.

Stems Wheat plants have hollow, jointed stems called culms. Given plenty of space, mature plants may develop as many as 100 culms. Each culm in a plant may have from three to six leaves.

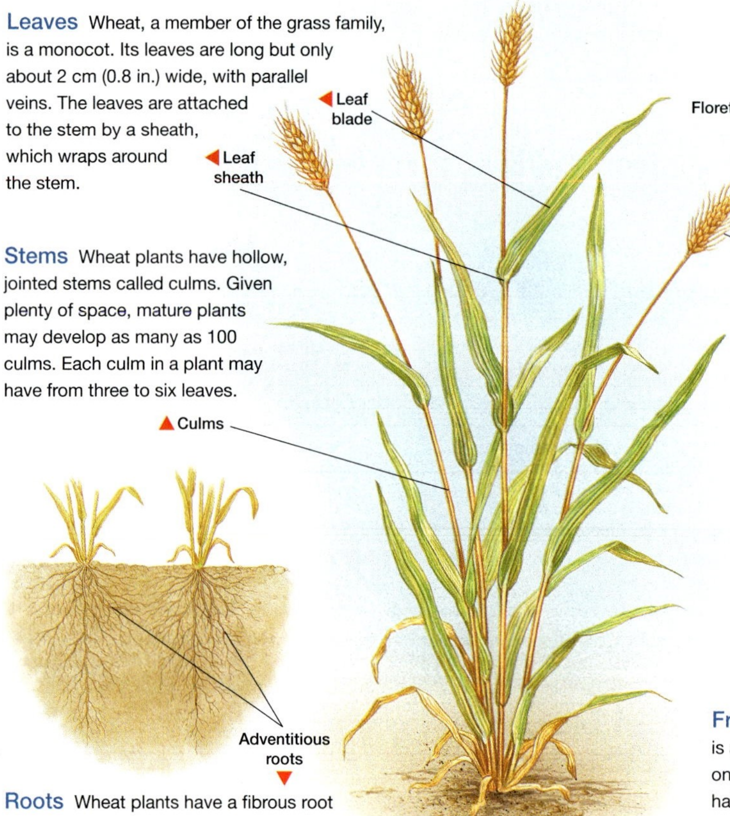

Roots Wheat plants have a fibrous root system composed of adventitious roots, as do most grasses. The extensive root system may be more than 2.2 m (7 ft) deep.

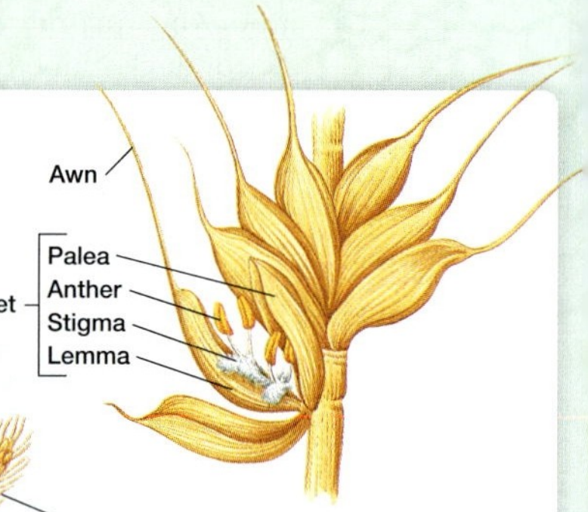

Flowers The flowers, which occur in dense clusters called spikes, develop at the top of each culm. Spikes range from 5 to 13 cm (2 to 5 in.) in length. Like all grass flowers, wheat flowers lack petals and sepals. Instead, two modified leaves called the palea and the lemma enclose the stamens and pistil of each tiny flower, or floret. The lemmas of some bread-wheat varieties have a long bristle called an awn.

Fruit A kernel, or grain, of wheat is a one-seeded fruit with a crease on one side and a brush of tiny hairs at one end. The grains are high in gluten, a sticky mixture of proteins that make dough elastic.

576

Attention Grabber

Wheat Varieties Students may be surprised to learn that more than 200 varieties of wheat are grown in the United States. Hard red wheats, the varieties grown for making bread flour, belong to the species *Triticum aestivum.* Durum wheats, or macaroni wheats, which are grown for making pasta because they contain less gluten, belong to the species *Triticum dicoccum.* Wheat may be planted in either the fall or the spring. The term winter wheat refers to a crop that is planted in the fall, overwinters as seedlings, and is harvested in late spring or summer. The term spring wheat refers to a crop that is planted in the spring for harvest in the fall.

Internal Structures

Fruit structure A wheat kernel is about 85 percent starchy endosperm. The kernel's outer layers, called the bran, make up about 12 percent of the kernel. The bran consists of the ovary wall, seed coat, and aleurone layer, which contains protein and oils. The embryo, or wheat germ, makes up less than 3 percent of the kernel.

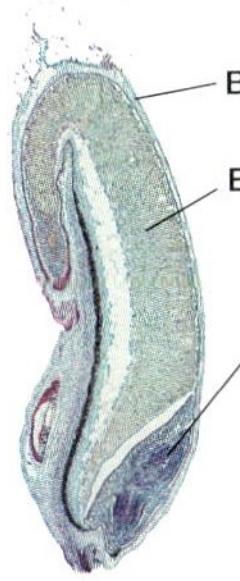

Kernel (longitudinal section)

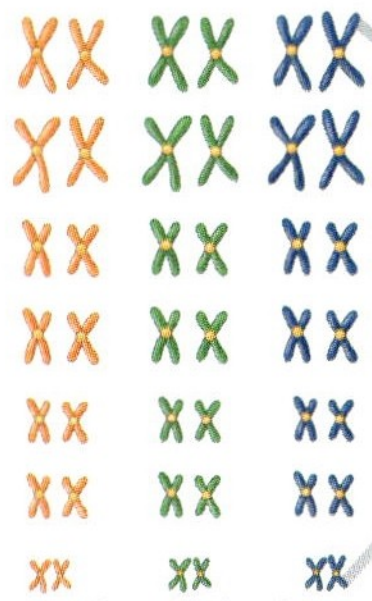

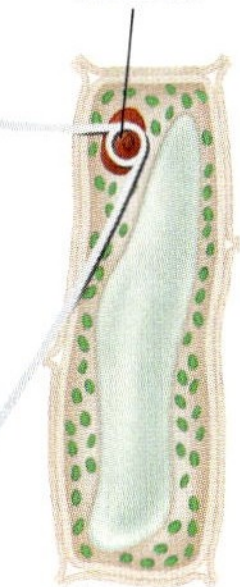

Chromosomes Bread-wheat cells have 42 chromosomes. A karyotype reveals that a bread-wheat cell has three sets of 14 chromosomes (7 pairs), shown here in different colors. Thus, bread wheat is not diploid ($2n$). Instead, it is hexaploid ($6n$)—a polyploid with six of each kind of chromosome. Polyploidy, having many sets of chromosomes, is common among cultivated plants. Among plants, polyploidy enables hybrids to result from crosses among different species. Such is the case with bread wheat, which is a natural hybrid of three closely related species of wheat.

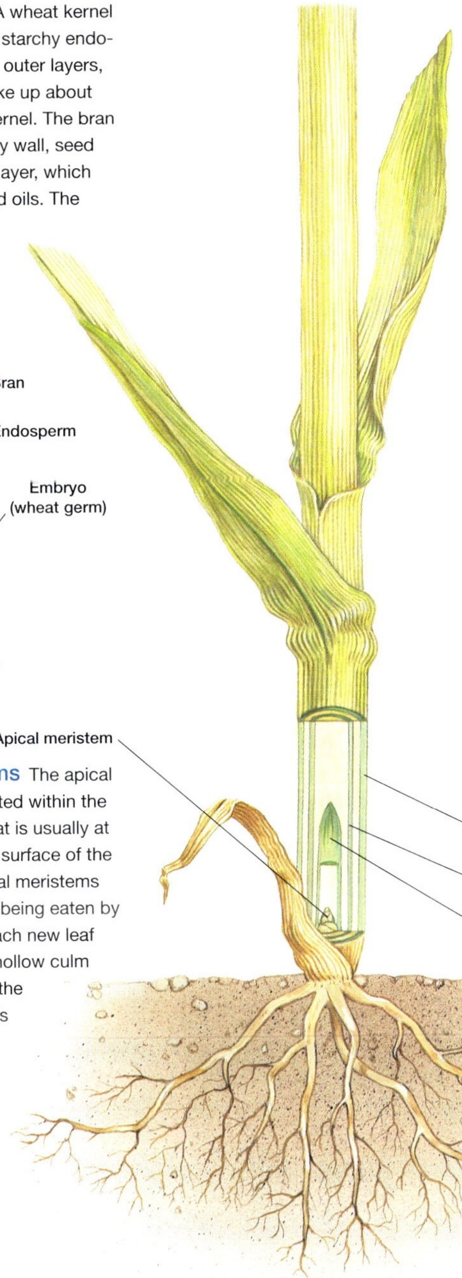

Apical meristems The apical meristems are located within the culms at a point that is usually at or just beneath the surface of the soil. Thus, the apical meristems are protected from being eaten by grazing animals. Each new leaf grows up within a hollow culm and emerges from the sheath of the culm's uppermost leaf.

577

Up Close

Bread Wheat

Discussion

- The chemical 2, 4-D is used to kill dicots. Will it kill wheat plants? Why or why not? (No, wheat is a monocot as it has the typical characteristics of monocots such as parallel leaf veins, fibrous roots, and seeds with a single cotyledon.)
- How did our ancestors help produce the bread wheat varieties that exist today? (From wild hybrids, they selected plants with desirable characteristics such as higher yields, larger seeds, and more gluten.)
- How would a human karyotype differ from the wheat karyotype? (A human karyotype normally has 23 pairs of chromosomes, while a wheat karyotype has seven sets of six similar chromosomes, in which there are three homologous pairs.)
- Is polyploidy beneficial to all living organisms? Why or why not? (No, polyploidy may be beneficial in plants because it causes plants to be larger, have showier flowers, and be more vigorous. In most animals, including humans, polyploidy is usually detrimental.) LS Interpersonal

Trends in Agriculture

Municipal Sewage In recent years, a great deal of research has been conducted to investigate the use of municipal sewage as a fertilizer. Such wastes must receive treatments that remove or kill organisms that might pose a health hazard to humans. One serious concern about these materials is the tendency they have to contain excesses of heavy metals such as lead and cadmium, especially if certain types of industries contribute to the waste. These metals can be harmful to humans even at low levels. Municipal wastes are not recommended for use in the fertilization of food crops for this reason. LS Intrapersonal

Reading Skill Builder — Basic

Paired Summarizing Pair students. Have one student read silently about the external structures of bread wheat and the other about the internal structures. Then have one student summarize his or her section aloud without referring to the text. The partner should listen without interrupting and be prepared to point out any inaccuracies or omissions in the summary. At this point students can refer to the text. Have students switch roles. Co-op Learning

BIOWatch

Saving Florida's Orchids

Teaching Strategies
Tell students that many of Florida's native species are endangered for similar reasons: poaching, habitat loss, and drainage of wetlands. Some orchid species are not reproducing due to the lack of pollinators such as ground-nesting bees.

Debate
Tell students that many orchid species worldwide have become endangered due to being dug up and sold as houseplants. Have students debate whether the sale of orchids should be regulated.

Close

Reteaching — BASIC

Have students work in cooperative groups to summarize the growth of a plant, from a seed to a mature adult. Co-op Learning

Quiz — GENERAL

1. When you buy seeds in a packet, why have they not germinated? (Water and oxygen must penetrate the seed coat for the seed to sprout.)
2. When you plant perennials, what would you expect to happen after one season of growth? (The plant would live for more than two years.)

Alternative Assessment — GENERAL

Provide students with prepared microscope slides of plant tissues, and ask them to identify meristematic and vascular tissues by drawing what they see. LS Visual

Section Review Correlations

SC.F.1.4.2: Items 3, 5

SC.G.1.4.1: Items 1, 2, 4

Plant Development

internet connect
www.scilinks.org
Topic: Plant Life Spans
Keyword: HX4141
SCILINKS. Maintained by the National Science Teachers Association

Genes guide the development of both plants and animals, but their patterns of development are very different. As an animal develops, sets of genes that control development are inactivated and may not be used again. Most animals stop developing when they become adults. Plants, in contrast, continuously make new cells in meristems. These cells differentiate and replace or add to existing tissues. Thus, a plant continues to develop throughout its life.

Many cells in a mature plant can activate all of their genes. Such cells can divide and form masses of undifferentiated cells. In a sense, they can reverse their development. These cells can undergo differentiation and develop into a mature plant. A technique called *tissue culture* is used to grow new plants from tissue that can reverse its development. A tissue culture is prepared by placing tissue on a sterile nutrient medium. Masses of undifferentiated cells that form grow into plants that are genetically identical to the parent plant.

FLORIDA BIOWatch: Saving Florida's Orchids

Southwest Florida is home to more species of native orchids than any other region of the United States. Many of these species are rare or endangered. And in the Florida Panther National Wildlife Refuge, the survival of the orchids is intertwined with that of other organisms in the swamp ecosystem—including panthers and certain types of fungi.

When wildlife biologist Larry Richardson took a job at the refuge, he was supposed to preserve habitat for the Florida panther. But he soon realized that native orchids were a part of that habitat, and were also endangered. So Richardson studied ways to germinate and replant the orchids within the refuge. He found that the native orchids depend on symbiotic fungi in the soil to help them grow after they germinate. He developed ways to propagate orchids along with their specific fungal partners. This process has resulted in faster and more successful rates of propagation than was previously possible.

internet connect
www.scilinks.org
Topic: Plants of Florida
Keyword: HX6F16
SCILINKS. Maintained by the National Science Teachers Association

Section 1 Review

1. **Compare** and contrast the germination of bean seeds and corn seeds. SC.G.1.4.1
2. **Summarize** the basic differences between annuals, biennials, and perennials. SC.G.1.4.1
3. **Explain** how primary growth and then secondary growth produce a woody stem. SC.F.1.4.2
4. **Critical Thinking Analyzing Methods** In what ways does plant development differ from animal development? SC.G.1.4.1
5. **FCAT Prep** What is growth that increases the width of a plant's stems and roots called?
   A germination
   B differentiation
   C primary growth
   D secondary growth
   SC.F.1.4.2

578

Answers to Section Review

1. Corn seedlings grow straight up, and the cotyledon stays in the soil. Bean seedlings form a hook as they grow, and the two cotyledons emerge.
2. Annuals live for one growing season, reproduce once, and then die. Biennials live for two growing seasons, reproduce once during the second season, and then die. Perennials live for more than two years and may reproduce many times.
3. Primary growth lengthens the stem and forms new primary dermal, ground, and vascular tissues. Secondary growth increases the stem's diameter by producing layers of cork and secondary xylem and phloem.
4. Plant development is continuous and "reversible," while animal development is not.
5. **A.** Incorrect. Germination is the growth of a plant embryo. **B.** Incorrect. Differentiation is the process by which cells become specialized in form and function. **C.** Incorrect. Primary growth increases the length or height of a plant. **D.** Correct. Secondary growth increases the width of stems and roots.

Regulating Growth and Development

Section 2

Florida Sunshine Standards/FCAT

SC.F.1.4.7 Organisms respond to internal and external stimuli.

SC.F.1.4.8 Cell behavior can be affected by molecules from other parts of the organism or even from other organisms.

SC.G.1.4.3 The chemical elements that make up the molecules of living things are combined and recombined in different ways.

Objectives

- **Identify** the major nutrients plants need to grow. SC.G.1.4.3
- **Describe** how plant hormones control plant growth. SC.F.1.4.8
- **Relate** environmental factors to plant growth. SC.F.1.4.7

Key Terms

mineral nutrient
auxin
hormone
apical dominance
tropism
photoperiodism
dormancy

Nutrients

Like all multicellular organisms, plants grow by adding new cells through cell division. Plants must have a steady supply of the raw materials they use to build new cells in order to grow. Plants need only two raw materials—carbon dioxide and water—to make all the carbohydrates in their tissues. As you learned earlier in this book, carbon dioxide and water are needed for photosynthesis.

Like animals, plants also need oxygen for cellular respiration. Although the green parts of a plant produce oxygen during photosynthesis, most of the oxygen used by leaves and stems comes from the air. Roots, which usually do not carry out photosynthesis, get oxygen from the air spaces between soil particles. If the soil around a plant's roots becomes compacted or saturated with water, it may not provide enough oxygen for the roots, and the plant could die.

However, carbon dioxide, water, and oxygen do not satisfy all of a plant's needs for raw materials. Plants also require small amounts of at least 14 **mineral nutrients,** which are elements absorbed mainly as inorganic ions. **Table 1** lists the six mineral nutrients needed in the greatest amounts for healthy plant growth and describes the importance of each nutrient. Commercial fertilizers may contain most of these mineral nutrients.

Table 1 Major Mineral Nutrients Required by Plants

Nutrient	Importance
Nitrogen	Part of proteins, nucleic acids, chlorophylls, ATP, and coenzymes; promotes growth of green parts
Phosphorus	Part of ATP, ADP, nucleic acids, phospholipids of cell membranes, and some coenzymes
Potassium	Needed for active transport, enzyme activation, osmotic balance, and stomatal opening
Calcium	Part of cell walls; needed for enzyme activity and membrane function
Magnesium	Part of chlorophyll; needed for photosynthesis and activation of enzymes
Sulfur	Part of some proteins and coenzyme A; needed for cellular respiration

579

Section 2

Focus

Overview

Before beginning this section review with your students the objectives listed in the Student Edition. This section focuses on factors that influence plant growth and development, including nutrients, hormones, and environmental influences such as sunlight, temperature, and the seasons.

Bellringer

Ask students to identify the substances plants need for survival. After taking attendance, write their responses on the board. (Initially, they will probably identify light, water, and carbon dioxide.) Prompt them to identify other needs by asking why people fertilize their yards. Ask students how they think plants might regulate their growth and development without having a nervous system as animals do. (They produce hormones.) LS Interpersonal

Motivate

Discussion/Question — ADVANCED

Ask students what growth factors soil provides plants, and whether plants can grow without soil. (Usually, soil provides minerals that plants need for growth. Using a practice known as hydroponics, plants are grown with their roots immersed in water through which oxygen is bubbled and to which the minerals needed for plant growth are added.) Consider the possibility of developing a small hydroponic garden in the classroom. LS Verbal

Chapter Resource File

- Chapter Resource File
- Directed Reading BASIC
- Active Reading GENERAL
- Data Sheet for Quick Lab GENERAL
- Data Sheet for Data Lab GENERAL
- Data Sheet for Math Lab GENERAL

One-Stop Planner CD-ROM

- Reading Organizers BASIC
- Reading Strategies BASIC
- Supplemental Reading Guide *A Feeling for the Organism The Life and Work of Barbara McClintock* ADVANCED

Section Correlations

SC.F.1.4.7
SC.F.1.4.8
SC.G.1.4.3

Teach

Real Life

Answer

The chemicals found in Agent Orange are 2,4-D (2,4-dichlorophenoxyacetic acid), 2,4,5-T (2,4,5-trichloro-phenoxyacetic acid), and trace amounts of dioxin. Exposure to Agent Orange has been blamed for miscarriages, skin diseases, cancers, and birth defects.

READING SKILL BUILDER — BASIC

Reading Hint Tell students that this textbook uses typographic aids, such as boldfaced words and in-text pronunciation guides, to help them learn new terms. Tell students that before they read a section of the book, to find the boldfaced terms and determine their meanings by reading the sentence in which each term appears.

Using the Figure

Walk students through **Figure 5,** which summarizes the steps of Frits Went's classic experiment on the effects of auxin. Help students to understand that when the agar is purple, it contains auxin. Ask students to state the significance of step 4. (It is a control designed to show that agar is not the cause of the effect.) Point out to students that Frits Went's experiments only worked for grasses. Studies in other plants such as sunflowers and radishes have not supported Went's hypotheses.

Transparencies

TR G20 Major Mineral Nutrients Required by Plants

TR G40 Bread Wheat

TR G49 Went's Experiment

Real Life

The herbicide Agent Orange contains two synthetic auxins.

This herbicide was used to defoliate plants in the jungles of Vietnam during the Vietnam War. One ingredient of Agent Orange is still commonly used as a herbicide.

Finding Information

Find out the names of the chemicals in Agent Orange, and identify the health problems to which it has been linked.

Hormonal Control of Growth

For centuries, people have known that plants bend strongly toward a light source as their shoots elongate. In the 1920s, the Dutch biologist Frits Went hypothesized that a chemical produced in the shoot tip of a grass causes this bending response. Went named the growth-promoting chemical that causes stems to bend **auxin** *(AWK sin)*. The steps in Went's experiment are summarized in **Figure 5.**

Step 1 Went removed the tip of an oat shoot and placed the tip on an agar block. Auxin diffused from the tip into the block.

Step 2 Went then transferred the agar block to the cut end of a shoot, which was followed by growth of the shoot.

Step 3 When Went placed an agar block with auxin on either side of cut shoots, the shoots grew in the opposite direction.

Step 4 As a control, Went placed an agar block without auxin on the cut end of other shoots. These shoots did not grow.

Auxin

Auxin is one of many plant hormones. The word *hormone* comes from the Greek word *horman,* meaning "to set in motion." A **hormone** is a chemical that is produced in one part of an organism and transported to another part, where it causes a response. Auxin causes plant cell walls to become more acidic, which allows the cells

Figure 5

BIO graphic — The Steps in Went's Experiment

Auxin apparently caused oat seedlings to elongate and bend toward light.

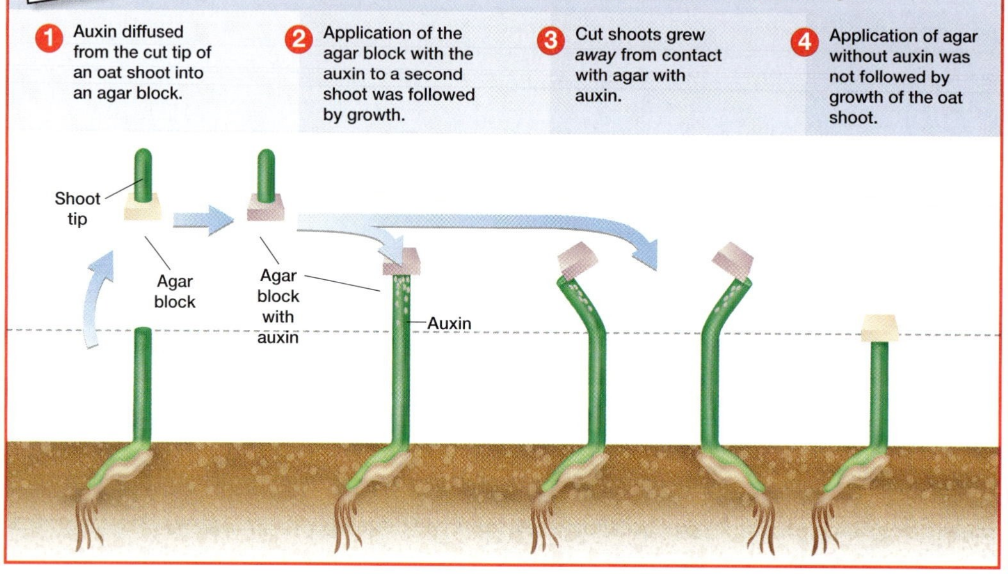

580

did you know?

Darwin's Plant Research Charles Darwin did many experiments with plants. In the mid-1800s, Darwin and his son Francis did the earliest experiments on phototropism. They discovered that a grass seedling responds to light coming from one direction with growth on the opposite (shaded) side, resulting in the seedling bending toward the light source.

Integrating Physics and Chemistry

Have students review Went's experiment on the hormone auxin. Ask them to determine if his experimental results demonstrate a physical or chemical change in the plant. Have them defend their answers. In addition, instruct students to analyze and critique Went's experiment as to its strengths and weaknesses using the information provided.

to elongate as they grow. Auxin accumulates on the dark side of a stem. As a result, the cells on the dark side of a stem elongate more than the cells on the light side. The difference in elongation causes the stem to grow toward the light. Auxin also inhibits the growth of the buds along a stem. This inhibition is called **apical dominance.** Cutting off the tip of a stem removes the source of auxin and enables the other buds to grow. That is why pruning the stems of a plant makes the plant become bushier.

Hormones in Agriculture

More than a century ago, citrus farmers discovered that they could cause citrus fruits to ripen by storing them in a room heated by a kerosene stove. The ripening was caused by ethylene, which is a gaseous organic compound produced when kerosene is incompletely burned. Most plant tissues produce ethylene. Today, ethylene is used to promote the ripening of tomatoes, bananas, and other fruits that are harvested before they ripen. Ethylene also loosens the fruit of cherries, blackberries, and blueberries, making it easier to harvest these crops mechanically.

Gibberellins *(jihb uhr EHL ihnz)* are produced in developing shoots and seeds. They stimulate stem elongation, fruit development, and seed germination. Gibberellins are used to enlarge Thompson seedless grapes. Other seedless fruits treated with gibberellins include apples, cucumbers, mandarin oranges, and peaches.

Cytokinins *(sie toh KIE nihnz),* which are produced in root tips, stimulate cell division and slow the aging of some plant organs. Cytokinins are sprayed on cut flowers to keep them fresh and on fruits and vegetables to extend their shelf life. Cytokinins are added to tissue-culture media to cause undifferentiated cells to form shoots.

internet connect
www.scilinks.org
Topic: Plant Hormones
Keyword: HX4140
SCILINKS. Maintained by the National Science Teachers Association

QUICK LAB: *Investigating the Effects of Ethylene on a Plant* SC.F.1.4.8

You can use a ripe apple to see one of the effects of ethylene on plants.

Materials

4 L glass jars with lids (2), 2 plants in 5 cm pots, small ripe apple

Procedure

1. **Place** a plant inside one of the jars. Tightly secure the lid.
2. **Place** the other plant and the apple inside the other jar. Tightly secure the lid.
3. **Observe** both jars for several days. Record what you see.

Analysis

1. **Describe** any changes in the plant in each jar.
2. **Critical Thinking** **Drawing Conclusions** A ripe apple gives off ethylene gas. Based on your observations, how does ethylene affect a plant?

QUICK LAB: *Investigating the Effects of Ethylene on a Plant*

Skills Acquired
Observing, making comparisons, drawing conclusions

Teacher's Notes
Tell students that ethylene affects plants in many ways. For example, when some plants are exposed to ethylene gas, they do not elongate normally. Ethylene also promotes the production of male flowers in plants such as squash and cucumbers. Ethylene is used by citrus-fruit growers to ripen fruits that are picked while they are still green.

Answers to Analysis

1. Answers will vary. Plants in jars containing an apple should begin to lose their leaves after a few days.
2. Answers will vary. From the results of this experiment, students should conclude that ethylene promotes plant aging and leaf drop.

MISCONCEPTION ALERT

Anthropomorphizing Plants It is common for students to attribute human characteristics to nonhuman organisms, such as plants and animals. Thus, students might think that a plant's response to light, for example, is somehow conscious, or that plants bend toward the light because they see it. Emphasize that a plant's response to light, gravity, or other environmental stimuli is simply a hormonal stimulus-response. Plants do not perceive environmental stimuli in the same way that humans and animals do.

Attention Grabber

Rotten Apples The expression "One rotten apple spoils the whole barrel" is commonly used to refer to people whose unacceptable behavior may influence others to behave unacceptably. This expression comes from the observation by food growers and handlers that a rotten apple in a barrel of apples results in all the apples rotting. The explanation is that rotting fruit give off ethylene, which promotes the ripening of other nearby fruits. The rotting fruit is infected with a fungus, which easily infects other fruit after they ripen.

Teach, *continued*

READING SKILL BUILDER — BASIC

Reading Organizer Before students finish reading this section, have them make a list of the following terms: auxin, tropism, phototropism, gravitropism, and thigmotropism. As they learn about the terms, have them write a brief definition of each term. Have them save their list of terms and definitions for use later in the Reteaching exercise at the end of this section. LS **Verbal**

Teaching Tip — GENERAL

Tissue Culture Tell students that the growth of whole plants from pieces of plants in tissue culture requires the addition of plant hormones in order for them to differentiate properly. The relative amounts of the hormones affect the process of differentiation. Typically, auxins and cytokinins are first added to the culture medium to promote the formation of a mass of undifferentiated cells. Next a high level of auxins and a low level of cytokinins are added to the culture medium. This promotes root formation. Then a low level of auxins and a high level of cytokinins are added to the culture medium. This promotes shoot formation.

WORD Origins

The word *thigmotropism* comes from the Greek words *thigma*, meaning "touch," and *tropos*, meaning "turn."

Environmental Influences on Growth

Because most plants are anchored in one spot, they cannot move from an unfavorable environment to a more favorable one as animals do. Instead, plants respond to their environment by adjusting the rate and pattern of their growth. For example, a plant that receives plenty of water and mineral nutrients may grow much faster and larger than it would if it received very little water and mineral nutrients. Also, a plant grown in full sun may grow much faster and larger than it would if it were grown in the shade or indoors. So the availability of light and nutrients affect the rate of plant growth. Many of a plant's responses to environmental stimuli, however, are triggered by the hormones that regulate plant growth.

Tropisms

A **tropism** *(TROH piz uhm)* is a response in which a plant grows either toward or away from a stimulus. Auxin is responsible for producing tropisms. **Figure 6** shows examples of three common types of tropisms. *Phototropisms* are responses to light. Responses to gravity are called *gravitropisms*. A *thigmotropism* is a response to touch. If a plant grows toward a stimulus, the response is called a positive tropism. If a plant grows away from the direction of the stimulus, the response is called a negative tropism. Thus, a shoot that grows up out of the ground shows both positive phototropism (growing toward the light) and negative gravitropism (growing away from the pull of gravity).

Figure 6 **Tropisms**

Tropisms are growth responses that occur either toward or away from a stimulus.

The bending of an amaryllis toward the light is a positive phototropism.

The upward growth of shoots is a negative gravitropism; the downward growth of roots is a positive gravitropism.

The coiling of grapevine tendrils around a wire is a thigmotropism.

Career

Horticulturist Horticulturists study fruits, vegetables, and ornamental plants. These plant scientists are interested in producing better fruit, vegetable, and ornamental crops with less pesticides, fertilizers, and water. Have students work in teams to use library references or on-line databases to find information about the role of horticulturists in food production. Have students investigate current research projects in horticulture, and have them report their findings to their classmates.

Trends in Plant Genetics

Genetic Diversity The National Plant Germplasm System (NPGS) is a cooperative effort to preserve the genetic diversity of plants. When plants lose their genetic diversity, which happens when only a few varieties of a species are cultivated, these species can be wiped out easily by a single disease or environmental condition to which they are sensitive. However, when many varieties of a species are cultivated, some varieties will have genes that confer resistance to various pests and diseases, or allow the plant to survive certain adverse conditions.

Interpreting Annual Rings SC.H.1.4.1

DATA LAB

Background

The annual rings of a woody stem provide important clues to annual variations in the rainfall an area receives over time. Thick rings form in years with heavy rainfall. Relatively thin rings form in dry years. Use the photo at right to answer the following questions.

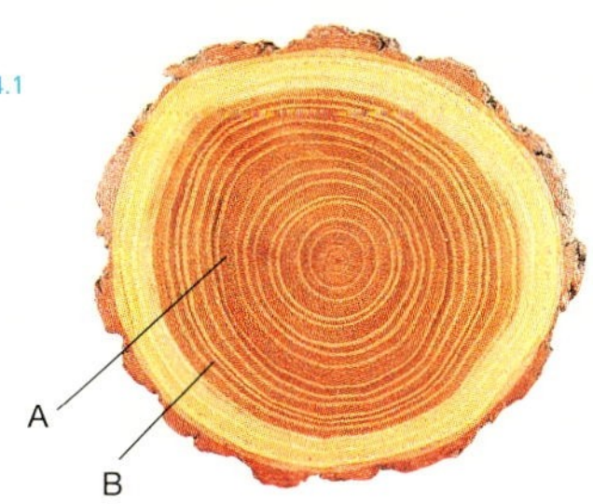

Analysis

1. **Critical Thinking** **Interpreting Data** What do the annual rings indicate about the climate where this plant grew?
2. **Critical Thinking** **Drawing Conclusions** Which ring, A or B, indicates a year when this plant received more rainfall?
3. **Critical Thinking** **Making Predictions** How will the annual rings of a nearby tree of the same age and species compare with those of this tree?

Photoperiodism

Certain plants bloom in the spring and others bloom in the summer or fall. Some plants bloom as soon as they reach a mature size. In many plants, seasonal patterns of flowering and other aspects of growth and development are caused by changes in the length of days and nights. The response of a plant to the length of days and nights is called **photoperiodism.**

Most plants can be categorized as one of three types in reference to photoperiodism. A plant that responds when days become shorter than a certain number of hours is said to be a short-day plant. A plant that responds when days become longer than a certain number of hours is called a long-day plant. Plants whose growth and development are not affected by day length are known as day-neutral plants. It is really the length of the nights, however, rather than the length of the days that controls photoperiodism, as Figure 7 shows. Knowledge of photoperiodism is very important to the nursery and floral industries. The length of days and nights is controlled artificially in greenhouses where plants such as poinsettias and chrysanthemums are grown. Thus, commercial growers force the plants to produce flowers at times of the year when they ordinarily would not. This makes it possible for poinsettias to be available for Christmas and chrysanthemums to be available year round.

Figure 7 **Flowering and photoperiodism.** Long-day plants flower when nights are short. Short-day plants flower when nights are long. If a flash of light interrupts a long night, long-day plants flower and short-day plants do not.

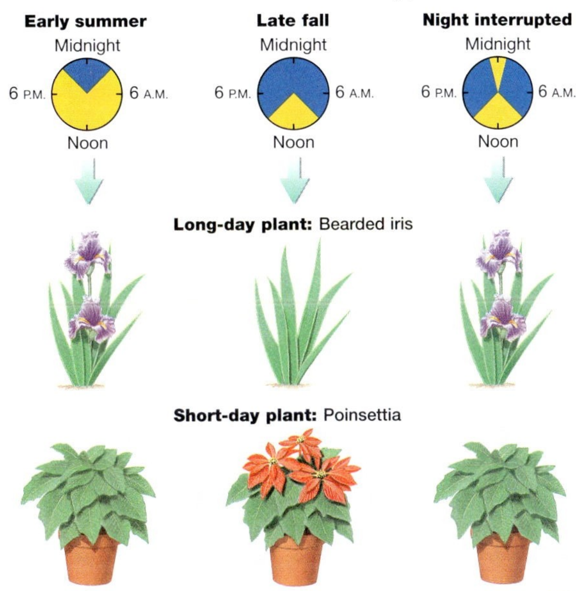

Interpreting Annual Rings

DATA LAB

Skills Acquired

Making observations, analyzing data, making conclusions

Teacher's Notes

Point out to students that it is the number of rings in a tree trunk that indicates its age, not the diameter of the trunk. A number of environmental factors can influence the width of an annual ring.

Answers to Analysis

1. Answers will vary. The rings indicate that the tree grew in a temperate climate and that the amount of rainfall varied from year to year.
2. A
3. They should be very similar.

Using the Figure — GENERAL

Walk students through **Figure 7** by first following what happens to long-day (short-night) and short-day (long-night) plants when the days are long and the nights are short. Then follow what happens to long-day (short-night) and short-day (long-night) plants when the days are short and the nights are long. Finally, follow what happens to long-day (short-night) and short-day (long-night) plants when they are grown under long nights that are interrupted by a flash of light in the middle of the night. Tell students that poinsettias are short-day (long-night) plants that are desirable ornamentals for the Christmas holidays. Ask students if these plants would normally produce flowers in late December. (yes) Ask students what would happen if lights were turned on during the night in a greenhouse where poinsettias are grown? (The plants would not bloom.)

MISCONCEPTION ALERT GENERAL

Plant "Food" Fertilizers are often referred to as plant "food." However, the term food is usually reserved for organic substances that provide a source of energy and organic molecules for living things when they are broken down. Because minerals do not contain usable energy and are not broken down in plants, they should not be referred to as food. Plants do not eat food; rather, they produce their own organic molecules during photosynthesis.

Transparencies

TR G51 Flowering and Photoperiodism

Figure 8 **Bud dormancy.** Thick scales cover the dormant buds on this twig from an apple tree.

Responses to Temperature

Temperature affects growth and development in many plants. For example, most tomato plants will not produce fruit if the nighttime temperatures are too high. Many plants that flower in early spring will not produce flowers until exposed to cold temperatures for a certain number of hours. Most deciduous woody plants drop their leaves in the fall in response to cooler temperatures and shorter periods of daylight. Thick, protective scales develop around their buds, as **Figure 8** shows. After a period of low temperatures, the buds begin growing into new leaves or sections of woody stem.

Dormancy is the condition in which a plant or a seed remains inactive, even when conditions are suitable for growth. Many plants and seeds remain dormant until they have been exposed to low temperatures for several weeks. Dormancy helps plants to survive by keeping buds from growing and seeds from germinating during warm spells before winter has ended.

MATH LAB: *Analyzing the Effect of Cold on Seed Germination* SC.F.1.4.7

Background

In some plants, a period of low temperatures is needed to break seed dormancy. The graph at right shows how being stored at a low temperature (4°C) affected the ability of apple seeds to germinate. Use the graph to answer the following questions.

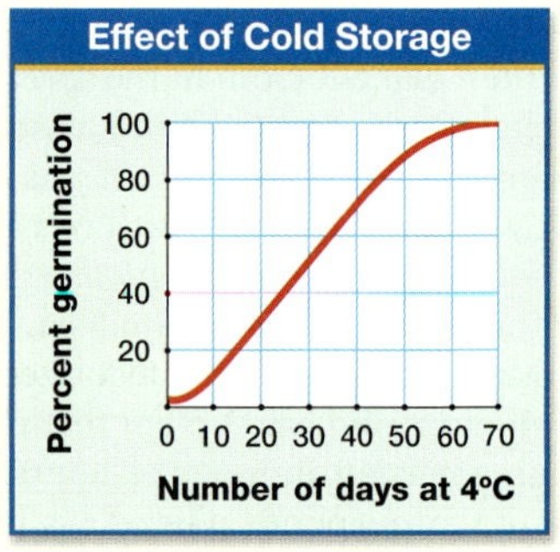

Analysis

1. **Summarize** the overall effect of cold temperatures on the germination of apple seeds.
2. **Calculate** the number of weeks that apple seeds must be stored at 4°C for at least 80 percent of the seeds to germinate.
3. **Critical Thinking** **Interpreting Graphs** What percentage of apple seeds germinate after storage at 4°C for 20 days?
4. **Critical Thinking** **Predicting Patterns** What percentage of apple seeds will germinate after being stored at 4°C for 80 days?

Section 2 Review

1. **List** the six mineral nutrients that plants require in the greatest amounts. SC.G.1.4.3
2. **Explain** how auxin causes a stem to grow toward a light source. SC.F.1.4.7
3. **Describe** an example of a negative gravitropism. SC.F.1.4.7
4. **Critical Thinking** **Predicting Outcomes** Why is it an advantage for plant growth and development to be regulated by environmental stimuli? SC.F.1.4.7
5. FCAT Prep When a vine responds to the touch of a fence wire by growing a tendril around the wire, the vine is exhibiting what tropism? SC.F.1.4.7
   A gravitropism **C** phototropism
   B thigmotropism **D** photoperiodism

584

MATH LAB: *Analyzing the Effect of Cold on Seed Germination*

Skills Acquired
Interpreting graphs, analyzing data

Teacher's Notes
Seeds with true dormancy cannot germinate until the plant hormone abscisic acid is either washed away or broken down.

Answers to Analysis
1. Cold temperatures promote the germination of apple seeds.
2. 45 days ÷ 7 days/week = 6.43 weeks
3. approximately 30 percent
4. 100 percent

Close

Reteaching — BASIC
Have students list the boldfaced terms in this section with their definitions. Have them create a crossword puzzle using the definitions as clues and the terms as words in the puzzle. LS Logical

Quiz — GENERAL
True or False:
1. Minerals and water satisfy all of a plant's needs for raw materials. (false)
2. Auxin is a type of plant hormone. (true)

Alternative Assessment — GENERAL
Ask students to summarize the roles of the plant hormones discussed in this section.

Answers to Section Review
1. nitrogen, phosphorus, potassium, calcium, magnesium, sulfur
2. Auxin accumulates on the dark side of a stem. As a result, the cells on the dark side of a stem elongate more than the cells on the light side. This results in the stem bending toward the light source in some plants.
3. An example of a negative gravitropism is a shoot growing upward from the soil after a seed germinates.
4. It is an advantage for plant growth to be regulated by environmental stimuli because plants are unable to move to a new environment when conditions become unfavorable.
5. **A.** Incorrect. Gravitropism is the response of a plant to gravity. **B.** Correct. Thigmotropism is the response of a plant to touch. **C.** Incorrect. Phototropism is the response of a plant to light. **D.** Incorrect. Photoperiodism is the response of a plant to the length of days and nights.

Section Review Correlations
SC.F.1.4.7: Items 2-5
SC.G.1.4.3: Item 1

ZONE

Key Concepts

1 How Plants Grow and Develop

- Germination is the resumption of growth by the embryo in a seed. Water and oxygen must penetrate the seed coat before germination can occur.
- Annuals complete their life cycle in one growing season. Biennials complete their life cycle in two growing seasons. Perennials live several years and may reproduce many times.
- Apical meristems located at the tips of shoots and roots produce primary growth. The tissues that result from primary growth are known as primary tissues.
- Secondary growth increases a plant's stem and root width. In woody stems, secondary growth is produced by the cork cambium and vascular cambium, two meristems near the outside of the stem.
- Bread wheat is a cereal grass with long leaves, hollow stems, and clusters of tiny flowers. The fruits, or grains, are usually ground into flour that is used to make bread.
- Plants develop throughout their lives. Plant development is reversible. Many mature plant cells can divide to form masses of undifferentiated cells, which can develop into a new plant.

2 Regulating Plant Growth and Development

- Plants need at least 14 mineral nutrients for growth. They also need carbon dioxide and water for photosynthesis and oxygen for cellular respiration.
- Hormones regulate plant growth and development. Auxin is a hormone that causes shoots to elongate and inhibits the growth of lateral buds.
- Plants modify their growth in response to the direction of light, gravity, and touch. Such growth responses are called tropisms.
- Seeds and many mature plants survive periods of unfavorable environmental conditions by becoming dormant.
- In many plants, seasonal patterns of flowering and other aspects of growth and development are caused by changes in the length of days and nights.

Key Terms

Section 1

germination (572)
perennial (573)
annual (573)
biennial (573)
primary growth (574)
secondary growth (574)
apical meristem (574)
cork cambium (575)
vascular cambium (575)
annual ring (575)

Section 2

mineral nutrient (579)
auxin (580)
hormone (580)
apical dominance (581)
tropism (582)
photoperiodism (583)
dormancy (584)

Alternative Assessment

Organize students into groups. Have each group use the information in Chapter 26 to prepare a multimedia presentation on plant growth and development that they could give to a middle-school science class or to the general public.

Chapter Resource File

- Science Skills Worksheet GENERAL
- Critical Thinking Worksheet ADVANCED
- Test Prep Pretest GENERAL
- Chapter Test GENERAL

Answer to Concept Map

The following is one of several possible answers to Performance Zone item 15.

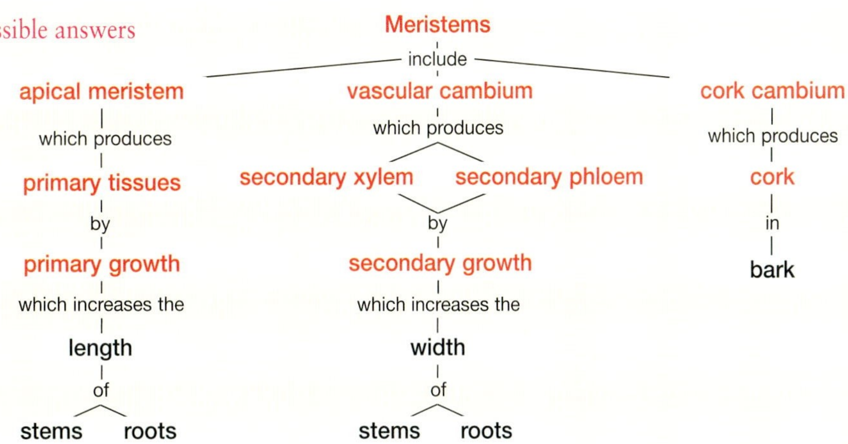

CHAPTER 26 **ZONE**

ANSWERS

1. a
2. d
3. b
4. d
5. b
6. b
7. Cytokinins stimulate cell division and slow aging. They are used in agriculture to prolong freshness of cut flowers, fruits, and vegetables and to stimulate shoot growth in tissue cultures. Gibberellins stimulate stem elongation, fruit development, and seed germination. They are used in agriculture to induce fruit formation without seeds and to enlarge seedless fruits.
8. Wheat cells have six of each kind of chromosome rather than two of each kind as in diploid organisms.
9. These organisms all share the same habitat, and certain species of orchids and fungi are symbiotic.
10. One possible answer to the concept map is found at the bottom of the Study Zone page.

Critical Thinking

11. The energy for a carrot plant's second year of growth comes from compounds that are stored in the root.
12. The cells in these plant parts are able to reverse their development. These cells can produce cells that are undifferentiated, which can then undergo differentiation to form all of the tissues necessary to grow an entire new plant.

Chapter Review Correlations

SC.F.1.4.2: Items 10, 11
SC.F.1.4.7: Item 5
SC.F.1.4.8: Items 3, 4, 7
SC.F.2.4.2: Item 8
SC.G.1.4.1: Items 1, 9
SC.G.1.4.3: Items 2, 6
SC.H.3.4.3: Items 12, 14
SC.H.3.4.5: Item 13

Performance ZONE

Understanding Key Ideas

1. Plant and animal development differ in that plant development
   a. is continuous and reversible.
   b. stops after a plant reaches maturity.
   c. is not affected by environment.
   d. is controlled by genes that cannot be reactivated.
   SC.G.1.4.1
2. Which of the following is *not* a raw material needed for plant growth?
   a. carbon dioxide
   b. water
   c. oxygen
   d. vitamins
   SC.G.1.4.3
3. Auxin causes cells to
   a. have less flexible cell walls.
   b. elongate more as they grow.
   c. bend toward light.
   d. develop lateral buds.
   SC.F.1.4.8
4. Which of the following caused the different growth patterns of the two pots of mums shown below?
   a. apical meristem
   b. auxin
   c. apical dominance
   d. all of the above
   SC.F.1.4.8

5. The response of a plant to the length of days and nights is
   a. gravitropism.
   b. photoperiodism.
   c. phototropism.
   d. thigmotropism.
   SC.F.1.4.7
6. Major mineral nutrients required by plants include all of the following except
   a. nitrogen. c. phosphorus.
   b. lead. d. potassium.
   SC.G.1.4.3
7. How do cytokinins and gibberellins affect plant growth, and how are these hormones used in agriculture? SC.F.1.4.8
8. **Up Close** How does the chromosome number of bread wheat differ from that of diploid organisms? SC.F.2.4.2
9. **BIOWatch** What is the connection between Florida's native orchids, panthers, and fungi? SC.G.1.4.1
10. **Concept Mapping** Construct a concept map that describes growth in vascular plants. Try to include the following terms in your map: *meristems, apical meristem, primary growth, primary tissues, cork cambium, secondary growth, cork, secondary phloem, secondary xylem,* and *vascular cambium.* SC.F.1.4.2

Critical Thinking

11. **Applying Information** Carrots are biennials. What role does the root of a carrot plant play in the plant's second year of growth? SC.F.1.4.2
12. **Forming Reasoned Opinions** Why is it possible for people to grow new plants from pieces of leaf, stem, or root tissue in which the cells have already undergone differentiation? SC.H.3.4.3

Alternative Assessment

13. **Finding and Communicating Information** Use the media center or Internet resources to learn how commercial growers produce plants such as poinsettias and chrysanthemums that flower at times when they would not flower in nature. Summarize your findings in a written report. SC.H.3.4.5
14. **Career Connection** Agronomist Agronomists study soil management and crop production. Write a report that includes a job description, training required, kinds of employers, growth prospects, and starting salary. SC.H.3.4.3

586

Alternative Assessment

13. Answers will vary.
14. Agronomists are plant scientists who traditionally study crops and soils. Many agronomists work in the field, in government and at universities. Others work for private industry or state agricultural extension services. Positions are available for individuals with bachelor's degrees, master's degrees, and Ph. D.'s in agronomy or soil science. Starting salary will vary by region.

Assignment Guide

Section	Questions
1	8, 9, 10, 12
2	1, 2, 3, 4, 5, 6, 7, 11, 13

FCAT Prep

For questions 1–6, write your answers on a separate sheet of paper.

1 The emergence of what part of the embryo is the first sign of germination in a bean?
A. cotyledon
B. hooked shoot
C. root
D. sheathed shoot
SC.F.1.4.2

2 What term describes plants that live for several years or more?
F. annuals
G. biennials
H. perennials
I. terminals
SC.G.1.4.1

3 Some seeds are soaked in acid before they are packaged and sold to farmers. What is the purpose of this treatment?
A. Acid treatment is required for germination of both wild and cultivated seeds.
B. Acid prevents the seed from drying out, because a dried seed will not germinate even when moistened.
C. Acid deactivates the seed so that it will not germinate until the farmer soaks it in a solution containing a base.
D. Acid will probably damage the seed coat so that water can more easily penetrate the seed coat, promoting germination. SC.H.3.4.5

4 In which tissues does cell division that increases the diameter of a woody stem occur?
F. apical meristems
G. pith
H. primary tissues
I. vascular cambium
SC.F.1.4.2

5 READ INQUIRE EXPLAIN Dormancy is the condition in which seeds or buds do not grow even when environmental conditions are favorable for growth to occur.

Assess how dormancy contributes positively to a plant's survival. SC.F.1.4.7

6 READ INQUIRE EXPLAIN In an experiment, a student placed a green banana in each of ten plastic bags. The student also placed a ripe pear in five of the bags and then sealed all of the bags. The bananas in the bags without pears took longer to ripen than the bananas in the bags with pears.

Part A Evaluate the results of this experiment. SC.F.1.4.8

Part B How might a banana farmer use the results of this experiment to increase production on a farm? SC.H.3.4.5

Test TIP

If you are not sure about the spelling of certain words when answering a short-response question, look at the question itself to see if the word appears in the question.

FCAT Doctor

Question 3 Answer D is the correct choice. Seeds treated with acid germinate more quickly and more uniformly than do untreated seeds. Answer A is incorrect because many wild and cultivated seeds can germinate without being treated. Answer B is incorrect because treating a seed with acid will not prevent the seed from losing water. Answer C is incorrect because the seed can still germinate, even after being treated with acid and without being treated with a base.

Question 5 It is not always beneficial for a seed to germinate the moment conditions are right for growth. For example, winter may still not be over, even though the temperature rises for one day.

Question 6

Part A A chemical (ethylene) within the pears promoted ripening in the bananas.

Part B Student answers may vary, but should be reasonable and well thought out. Answer should include the use of ripe pears to speed up the ripening process of the bananas.

Answers

1. C
2. H
3. D
4. I
5. Dormancy helps plants to survive by keeping buds from growing and seeds from germinating during warm spells before winter has ended.

6A. Ripe pears give off ethylene, which promotes ripening in the bananas.

6B. Sample answer: The farmer might pick the bananas before they are ripe to prevent damaging them and then use pieces of ripe pears to speed up the ripening process for selling the bananas to customers.

Test Prep Correlations

SC.F.1.4.2: Items 1, 4
SC.F.1.4.7: Item 5
SC.F.1.4.8: Item 6A
SC.G.1.4.1: Item 2
SC.H.3.4.5: Items 3, 6B

Skills Practice Lab

Comparing Bean and Corn Seedlings

Teacher's Notes

Time Required Three 30-minute lab periods

Ratings

EASY → HARD

TEACHER PREPARATION: 2
STUDENT SETUP: 2
CONCEPT LEVEL: 2
CLEANUP: 1

Skills Acquired
- Collecting Data
- Inferring
- Interpreting
- Measuring

Scientific Methods

In this lab, students will:
- Make Observations
- Ask Questions
- Analyze the Results

Materials and Equipment

Materials for this lab can be ordered from WARD'S. See *Master Materials List* at the front of this book for catalog numbers.

Safety Cautions

Be sure that students have read and understand all of the safety rules for working in the lab. Caution students to use scalpels carefully to avoid injuring themselves and others. Make sure students wash their hands before leaving the lab.

Skills Practice Lab

Comparing Bean and Corn Seedlings

SC.G.1.4.1

SKILLS
- Comparing
- Drawing
- Relating

OBJECTIVES
- **Observe** the structures of bean seeds and corn kernels.
- **Compare** and **Contrast** the development of bean embryos and corn embryos as they grow into seedlings.

MATERIALS
- 6 bean seeds soaked overnight
- stereomicroscope
- 6 corn kernels soaked overnight
- scalpel
- paper towels
- 2 rubber bands
- 150 mL beakers (2)
- glass-marking pen
- metric ruler

Before You Begin

A **seed** contains an inactive plant **embryo.** A plant embryo consists of one or more **cotyledons,** an embryonic shoot, and an embryonic **root.** Seeds also contain a supply of nutrients. In **monocots,** the nutrients are contained in the **endosperm.** In **dicots,** the nutrients are transferred to the cotyledons as seeds mature. A seed **germinates** when the embryo begins to grow and breaks through the protective **seed coat.** The embryo then develops into a young plant, or **seedling.** In this lab, you will examine bean seeds and corn kernels and then germinate them to observe the development of their seedlings.

1. Write a definition for each boldface term in the paragraph above.
2. Based on the objectives for this lab, write a question you would like to explore about seedling development.

Procedure

PART A: Observing Seed Structure

1. Remove the seed coat of a bean seed, and separate the two fleshy halves of the seed.
2. Locate the embryo on one of the halves of the seed. Examine the bean embryo with a stereomicroscope. Draw the embryo, and label the parts you can identify.
3. Examine a corn kernel, and locate a small light-colored oval area. **CAUTION: Sharp or pointed objects may cause injury. Handle scalpels carefully.** Use a scalpel to cut the kernel in half along the length of this area.
4. Locate the corn embryo, and examine it with a stereomicroscope. Draw the embryo, and label the parts you can identify.

PART B: Observing Seedling Development

5. Fold a paper towel in half as shown in the photo at the top of the next page. Set five corn kernels on the paper towel. Roll up the paper towel, and put a rubber band around the roll. Stand the roll in a beaker with 1 cm of water in the bottom. Add water to the beaker as needed to keep the paper towels wet, but do not allow the corn kernels to be covered by water.
6. Repeat step 5 with five bean seeds.

Tips and Tricks

Bean seeds and corn kernels may be purchased at a nursery, hardware store, or feed store, or they may be ordered from a biological supply company. Soak bean seeds and corn kernels prior to this investigation. Set up labeled containers for the disposal of solutions, broken glass, and seedlings. This investigation must be done in three 30-minute periods over 5 days. The seed coat of the bean should be easy to remove if the seeds have soaked long enough.

Answers to Before You Begin

1. *seed*—reproductive structure of a plant that consists of an embryo, a stored food supply, and a seed coat

   embryo—an early stage in the development of an organism

   cotyledons—structures in a seed that are part of the embryo and function in its nourishment

   root—portion of a plant that anchors the plant, absorbs water and mineral nutrients, and usually grows belowground

   monocot—plant characterized by seeds with one cotyledon

7. After three days, unroll the paper towels and examine the corn and bean seedlings. Use a glass-marking pen to mark the roots and shoots of the developing seedlings. Starting at the seed, make a mark every 0.5 cm along the root of each seedling. And again starting at the seed, make a mark every 0.5 cm along the stem of each seedling.
8. Draw a corn seedling and a bean seedling in your lab report. Label the parts of each seedling. Also show the marks you made on each seedling, and indicate the distance between the marks.
9. Using a fresh paper towel, roll up the seeds, place the rolls in the beakers, and add fresh water to the beakers.
10. After two more days reexamine the seedlings. Measure the distance between the marks. Repeat step 8.

PART C: Cleanup and Disposal

11. Dispose of seeds, broken glass, and paper towels in the designated waste containers. Do not put lab materials in the trash unless your teacher tells you to do so.
12. Clean up your work area and all lab equipment. Return lab equipment to its proper place. Wash your hands thoroughly before you leave the lab and after you finish all work.

Analyze and Conclude

1. **Relating Concepts** Corn and beans are often cited as representative examples of monocots and dicots, respectively. Relate the seed structure of each to the terms *monocotyledon* and *dicotyledon*.
2. **Summarizing Results** What parts of a plant embryo were observed in all seedlings on the third day?
3. **Drawing Conclusions** In which part or parts of bean seedlings and corn seedlings do the seedlings grow in length? Explain.
4. **Forming Hypotheses** How are the tender young shoots of bean seedlings and corn seedlings protected as the seedlings grow through the soil?
5. **Evaluating Viewpoints** Defend the following statement: There are both similarities and differences in seed structure and seedling development in beans and corn.
6. **Further Inquiry** Write a new question about seedling development that could be explored with another investigation.

On the Job

Plant physiology is the study of the processes that occur in plants. Do research to discover where plant physiologists work and what types of research are currently being conducted in the field of plant physiology. For more about careers, visit **go.hrw.com** and type in the keyword **HX4 Careers.**

Answers to Procedure

2.

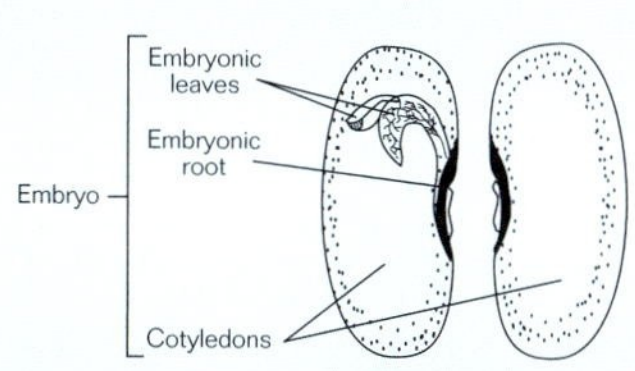

Bean seed

4.

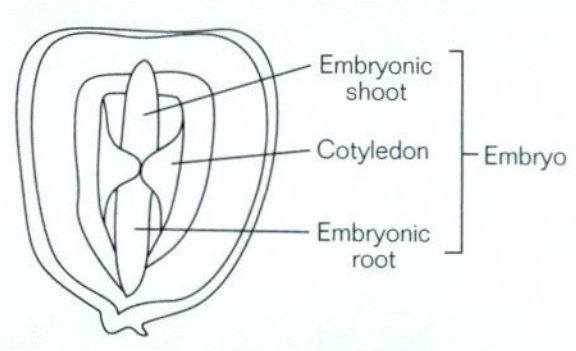

Corn grain

8.

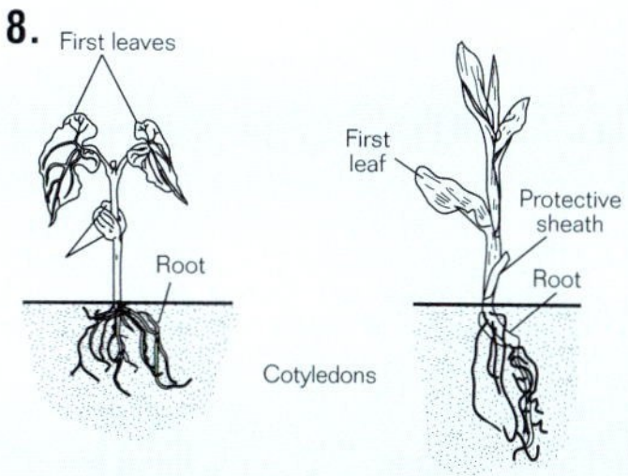

Bean seedling Corn seedling

endosperm—triploid (3n) nutritive tissue that develops from the union of a sperm with two nuclei in an ovule

dicot—plant characterized by seeds with two cotyledons

germinate—to begin growth, as in a seed

seed coat—protective outer covering of a seed

seedling—a young plant that has developed from a germinating seed

2. Answers will vary. For example: What part of a seedling emerges first?

Answers to Analyze and Conclude

1. Corn seeds have only one cotyledon and thus are monocotyledons; bean seeds have two cotyledons and thus are dicotyledons.
2. Embryonic leaves and roots are observed in all seedlings on the third day.
3. Bean seedlings grow at the tips of their roots and stems. The distance between the marks has changed at the tips of the stems and the tips of the roots. Corn seedlings grow in length at the tips of their roots. They do not have a visible stem. The distance between the marks changed only on the roots.
4. Bean shoots have a hook in their embryonic stems. The hook pushes through the soil before the cotyledons, which enclose the embryonic shoot. Corn shoots are surrounded and protected by a sheath as they push through the soil.
5. In both beans and corn, the embryo consists of an embryonic shoot, an embryonic root, and cotyledons. As bean embryos and corn embryos develop, their shoots grow up, become green, and form leaves, while their roots grow down and do not become green. Bean seeds have two cotyledons; corn kernels have only one. Corn seeds have endosperm; mature bean seeds do not. The shoots of beans hook as they germinate; a corn shoot grows straight up.
6. Answers will vary.

Answer to On the Job

Answers will vary.

UNIT 7 Exploring Invertebrates

Chapters

Beekeeper examining a honeycomb frame full of honey

Honeybees

Yesterday...

Humans have raised bees for their honey and their beeswax for thousands of years. American colonists brought bees from England and early pioneers took them along on the journey west. **What are the characteristic features of insects?**

French beekeeper in the early 1700s

Today...

Honeybees are valued today not only as a source of honey but also for their role as pollinators of crops such as almonds, cherries, apples, and pears. Migratory beekeepers carry bees from state to state, renting their bees to farmers in time to ensure a successful harvest. **Discover how a beehive is a highly organized social system.**

Honeybee

Tomorrow...

In the United States, bee-dependent crops are worth \$10 billion a year. Scientists continue to study different varieties of bees in an effort to find those that are the most efficient pollinators.

Honeybee research

internet connect

www.scilinks.org
Topic: Honeybees
Keyword: HX4099

SCILINKS. Maintained by the National Science Teachers Association

Chapter Opener

- SE **Quick Review,** p. 593 ■
- SE **Reading Activity,** p. 593 ■
- TE **Using the Figure,** p. 592 GENERAL
- TE **Opening Activity,** p. 593 GENERAL

& 3 • 90 min pp. 594–603

Section 1

Characteristics of Animals

- OSP **Lesson Plan***
- TR **Bellringer***
- TR **Section Outline***
- TR **H3 Origins of Animals Tissues and Organs***
- TR **H4 Radial and Bilateral Symmetries***
- TR **H9 Three Body Plans of Symmetrical Animals***

- TE **Demonstration,** p. 594 GENERAL
- TE **Using the Table,** p. 596
- TE **Activity** Following the Directions, p. 598 BASIC
- TE **Demonstration,** p. 598 GENERAL
- TE **Demonstration,** p. 599 GENERAL
- SE **Quick Lab** Recognizing Symmetry, p. 599 ■
- CRF **Datasheets for In-Text Labs** Recognizing Symmetry*
- TE **Using the Figure,** p. 600
- TE **Using the Figure,** p. 601 BASIC
- TE **Activity** Other Phyla, p. 602 ADVANCED
- TE **Group Activity** Evolutionary Milestones, p. 602 GENERAL
- SE **Data Lab** Exploring the Animal Kingdom, p. 603 ■
- CRF **Datasheets for In-Text Labs** Exploring the Animal Kingdom*
- CRF **Quick Lab** Recognizing Patterns of Symmetry* BASIC

& 5 • 90 min pp. 604–610

Section 2

Animal Body Systems

- OSP **Lesson Plan***
- TR **Bellringer***
- TR **Section Outline***
- TR **H5 Open and Closed Circulatory Systems***

- TE **Demonstration,** p. 604
- TE **Demonstration,** p. 605
- TE **Activity** Open Vs. Closed Circulatory Systems, p. 606 BASIC
- TE **Using the Figure,** p. 607
- SE **Math Lab** Calculating Filtration Rate in the Human Kidney, p. 608 ■
- CRF **Datasheets for In-Text Labs** Calculating Filtration Rate in the Human Kidney*
- TE **Using the Figure,** p. 609
- CRF **Skills Practice Lab** Embryonic Development* GENERAL
- SE **Skills Practice Lab** Surveying Invertebrate Diversity, pp. 614–615 ◆ ■
- CRF **Datasheets for In-Text Labs** Surveying Invertebrate Diversity*

& 7 • 90 min Chapter Review and Assessment Resources

- SE **Study Zone,** p. 611 ■
- SE **Performance Zone,** p. 612 ■
- SE **FCAT Test Prep,** p. 613 ■
- TR **Concept Mapping***
- CRF **Vocabulary Review Worksheet*** ■ GENERAL
- CRF **Test Prep Pretest*** ■ GENERAL
- CRF **Chapter Test*** ■ GENERAL
- CRF **Chapter Test*** ADVANCED
- OSP **Test Generator***
- OSP **Test Item Listing***
- OSP **Modified Chapter Test*** BASIC

go.hrw.com Holt Online Learning

Visit go.hrw.com for access to Holt Online Learning or enter the keyword **HX6 Home** for a variety of free online resources.

One-Stop Planner® CD-ROM

This CD-ROM package includes
- Lab Materials QuickList Software
- Holt Calendar Planner
- Customizable Lesson Plans
- Printable Worksheets
- ExamView® Test Generator
- Interactive Teacher Edition
- Holt PuzzlePro® Resources
- Holt PowerPoint® Resources

Guided Reading Audio CDs

These CDs are designed to help auditory learners and reluctant readers. (Audio CDs are also available in Spanish.)

Compression guide:
To shorten your instruction because of time limitations, omit block 1.

KEY

TE	Teacher Edition	CRF	Chapter Resource File	*	Also on One-Stop Planner
SE	Student Edition	TR	Teaching Transparency	■	Also Available in Spanish
OSP	One-Stop Planner			◆	Requires Advance Prep

SKILLS DEVELOPMENT AND PRACTICE	SECTION REVIEW AND ASSESSMENT	CORRELATIONS
		Florida Sunshine Standards/FCAT
TE Reading Skill Builder Interactive Reading, p. 595 BASIC **TE Skill Builder** Graphing, p. 595 ADVANCED **TE Reading Skill Builder** Reading Organizer, p. 596 BASIC **TE Skill Builder** Math, p. 597 GENERAL **TE Skill Builder** Writing, p. 598 ADVANCED **TE Reading Skill Builder** Discussion, p. 601 BASIC **CRF Directed Reading Worksheet*** BASIC **CRF Active Reading Worksheet*** GENERAL	**SE Section 1 Review** ■ **TE Reteaching,** p. 603 BASIC **TE Quiz,** p. 603 GENERAL **TE Alternative Assessment,** p. 603 GENERAL **CRF Quiz*** ■ BASIC	SC.D.1.4.3, SC.F.1.4.2, SC.G.1.4.1, SC.H.3.4.1
TE Skill Builder Writing Skills, p. 609 ADVANCED **CRF Directed Reading Worksheet*** BASIC **CRF Active Reading Worksheet*** GENERAL **CRF Critical Thinking Worksheet*** ADVANCED **CRF Science Skills Worksheet*** GENERAL	**SE Section 2 Review** ■ **TE Reteaching,** p. 610 BASIC **TE Quiz,** p. 610 GENERAL **TE Alternative Assessment,** p. 610 GENERAL **CRF Quiz*** ■ BASIC **Modified Chapter Test*** BASIC	SC.F.1.4.1, SC.F.1.4.2, SC.F.2.4.1, SC.G.1.4.1

SCILINKS. NSTA
www.scilinks.org
Maintained by the **National Science Teachers Association**

Topic: Origin of Tissues
SciLinks Code: HX4132

Topic: Body Symmetry
SciLinks Code: HX4128

Topic: Body Cavity
SciLinks Code: HX4027

Topic: Animal Groups
SciLinks Code: HX4010

Topic: Animal Reproduction
SciLinks Code: HX4011

Overview

Tell students that the purpose of this chapter is to teach them about the structural, developmental, and reproductive strategies that all animals share, and some of the variations in these characteristics that exist among animals. They will learn about the basic characteristics of the major groups of animals. Finally, they will learn about the structure and function of some of the animal body systems, as well as some of the variations in these characteristics that exist among animals.

Using the Figure

Direct students' attention to the photograph on this page. Ask them to identify reasons why the organisms shown are classified as animals. (They can move and they do not make their own food.) Ask them to identify how they are different from mammals, such as themselves. (Answers will vary but students should note that they do not have a head or a bony skeleton; some students may note that they don't have very complex body systems compared to those of mammals.) Tell them also that jellyfish are among the most primitive organisms and lack many of the characteristics found in mammals and other more advanced organisms. For example, jellyfish exhibit little specialization in body structures that carry out specific tasks. Animals that live in water ingest food from the water around them, eliminate their wastes into the water, and shed reproductive cells into the water. Point out that the earliest animals on Earth are believed to have lived in water.

Assessing Prior Knowledge

Before beginning this chapter, be sure your students can answer the questions in the ***Quick Review*** on the facing page. If they cannot, encourage them to reread the sections indicated.

Moon jellyfish

592

Chapter Correlations — Florida Sunshine Standards/FCAT

SC.D.1.4.3 Changes in Earth's climate, geological activity, and life forms may be traced and compared.

SC.F.1.4.2 Body structures are uniquely designed and adapted for their function.

SC.F.2.4.1 There are mechanisms of asexual and sexual reproduction and different genetic advantages and disadvantages of asexual and sexual reproduction.

SC.G.1.4.1 There is great diversity and interdependence among living things.

CHAPTER

27 Introduction to Animals

Florida Sunshine Standards/FCAT

SC.D.1.4.3 Changes in Earth's climate, geological activity, and life forms may be traced and compared.

SC.F.1.4.2 Body structures are uniquely designed and adapted for their function.

SC.F.2.4.1 There are mechanisms of asexual and sexual reproduction and different genetic advantages and disadvantages of asexual and sexual reproduction.

SC.G.1.4.1 There is great diversity and interdependence among living things.

Quick Review

Answer the following without referring to earlier sections of your book.

1. **Define** the term *metabolism*. *(Chapter 1, Section 2)*
2. **Describe** the process of diffusion. *(Chapter 4, Section 1)*
3. **Summarize** how gametes are formed. *(Chapter 6, Section 1)*
4. **State** the relationship of a phylum to a kingdom. *(Chapter 14, Sections 1 and 2.)*
5. **Define** the terms *tissue* and *organ*. *(Chapter 19, Section 3)*

Did you have difficulty? *For help, review the sections indicated.*

Reading Activity

Before you begin to read, survey the chapter, noting the red headings at the tops of pages and the blue subheadings. Use these heads to make an outline of the chapter, and leave space after each heading. Fill in important facts as you read.

Looking Ahead

internet connect

www.scilinks.org
National Science Teachers Association *sci*LINKS Internet resources are located throughout this chapter.

SCILINKS. Maintained by the National Science Teachers Association

The beautiful moon jellyfish, *Aurelia arita,* is found in the Atlantic and Pacific Oceans. People used to call jellyfishes *sea lungs* because their pulsating movement seems similar to breathing.

Opening Activity — GENERAL

Place a potted plant and a natural sponge on your desk. Ask students what the plant and the sponge have in common. (Both organisms are stationary; the plant is alive and the sponge was once alive.) Ask students how the two are different. (The plant can make its own food. Some students will say that the sponge is an animal.)

Quick Review Answers

1. The sum of all of the chemical reactions that occur in an organism, including those involving the breakdown of substances taken into the body and the formation of specific substances needed by the body.
2. Diffusion is the process in which materials move from an area where they are more concentrated to an area where they are less concentrated.
3. Gametes are formed when precursor cells in the body divide by meiosis to produce haploid cells that mature into gametes capable of fertilization.
4. A kingdom is made up of a number of related phyla of organisms.
5. A tissue is composed of a collection of cell types that work together to carry out a related function; an organ is a collection of tissues that work together to carry out a related major body function.

Reading Activity Answers

Answers will vary. Ask students to exchange outlines with a neighbor and evaluate each other's outlines for accuracy.

Chapter Resource File

- Vocabulary Worksheets
- Concept Mapping

Focus

Overview

Before beginning this section review with your students the objectives listed in the Student Edition. The purpose of this lesson is to describe to students the features that all animals have in common. They will learn about some of the major developments that occurred in animals as they evolved and how the major groups of animals differ from each other.

Bellringer

Ask students to write down the smallest animal and the largest one they can think of. Ask them to describe in a few sentences the major characteristics of both.

Motivate

Demonstration — GENERAL

Have students compare a projected image of *Daphnia* with that of any large, familiar mammal. Have students list the similarities and differences between the two animals. (Similarities: both move and neither can make its own food. Differences: size, shape, body covering, complexity, and methods of reproduction.) **LS Visual**

Section Correlations

SC.D.1.4.3
SC.F.1.4.2
SC.G.1.4.1

Section 1

Characteristics of Animals

Florida Sunshine Standards/FCAT

SC.D.1.4.3 Changes in Earth's climate, geological activity, and life forms may be traced and compared.
SC.F.1.4.2 Body structures are uniquely designed and adapted for their function.
SC.G.1.4.1 There is great diversity and interdependence among living things.

Objectives

- **Identify** the features that animals have in common. SC.G.1.4.1
- **Distinguish** radial symmetry from bilateral symmetry. SC.F.1.4.2
- **Summarize** the importance of a body cavity. SC.F.1.4.2
- **Identify** how scientists determine evolutionary relationships among animals. SC.D.1.4.3

Key Terms

blastula
ectoderm
endoderm
mesoderm
body plan
asymmetrical
radial symmetry
bilateral symmetry
cephalization
coelom
acoelomate
pseudocoelomate
coelomate
phylogenetic tree

General Features of Animals

Humans have long marveled at and depended on animals. When Linnaeus first classified animals in the mid-1700s, he counted 4,236 kinds. Since then, many new animal species have been identified, and the count is now over a million. We share the planet with a fantastic array of animal forms, and our lives intersect in many ways. For example, if it were not for insects, such as the honeybee, many of our crops would not be pollinated. The prices of insect-pollinated fruits and vegetables would increase dramatically. And items such as apples and cucumbers might not be available at your local supermarket.

Think about a snail, a lizard, a hawk, an elephant—all are quite different. But the features these four animals share are just as important as their differences in size, shape, and behavior. Like all animals, they exhibit the following features—heterotrophy, mobility, multicellularity, diploidy, and sexual reproduction. They also exhibit blastula formation, cells organized into tissues, and the absence of a cell wall. These features are a legacy inherited from their common ancestor.

Heterotrophy

Animals are heterotrophs—that is, they cannot make their own food. Some animals that live in the ocean remain in one place and consume tiny particles of food that they filter from passing sea water. But most animals, such as those shown in **Figure 1,** move from place to place searching for food. Once food is located, it is eaten and then digested in a cavity inside the animal's body.

Figure 1 Heterotrophy. Unlike plants, animals must seek out food sources in their environment.

594

Chapter Resource File

- Directed Reading BASIC
- Active Reading GENERAL
- Data Sheet for Quick Lab GENERAL
- Data Sheet for Data Lab GENERAL

One-Stop Planner CD-ROM

- Reading Organizers BASIC
- Reading Strategies BASIC

Transparencies

TR Bellringer
TR H3 Origins of Animal Tissues and Organs
TR H4 Radial and Bilateral Symmetries
TR H9 Three Body Plans of Symmetrical Animals

Figure 2 Mobility. A tiger finds food by roaming over its territory, which usually ranges from 10 to 30 mi^2.

Mobility

Animals are unique among living things in being able to perform rapid, complex movements. Animals move by means of muscle cells, specialized cells that are able to contract with considerable force. Animals can swim, crawl, walk, run, and even fly. In fact, flight has evolved four times among animals, in insects, pterosaurs (extinct reptiles from the time of the dinosaurs), birds, and bats. The tiger shown in **Figure 2** moves through its environment searching for food. The size of the tiger's home range depends on the availability of prey species.

Multicellularity

All animals are multicellular. Some are too small to be seen clearly with the naked eye, like *Daphnia,* shown in **Figure 3.** Others, such as some enormous whales, are larger than a city bus. In spite of their differences in body size, there is little difference in the size of most of the cells that make up these animals. The cells on the skin of your hand are roughly the same size as the cells in the heart of a whale or in the wing muscle of a hummingbird.

Figure 3 Life in a drop of water. *Daphnia,* commonly called water fleas, belong to a group of animals known as crustaceans.

Diploidy

With few exceptions, animals are diploid *(DIP loyd),* meaning adults have two copies of each chromosome, one inherited from their father and one from their mother. Only their gametes (egg and sperm) are haploid. In contrast, many plants have four or more copies of each chromosome, while fungi often have only one. A great advantage of diploidy is that it permits an animal to exchange genes between the two copies of a set of chromosomes, creating new combinations of genes.

595

Teach

Teaching Tip — GENERAL

Animal Features Point out to students that terms and phrases such as heterotrophy, multicellularity, diploidy, absence of a cell wall, and reproduce sexually are used to describe features of animals. Help students connect these terms and phrases to their own concrete experiences by asking them to choose any animal they are familiar with. Have them write the name of the animal and then present evidence that their animal is heterotrophic, multicellular, lacks cell walls, and reproduces sexually. Have a few students explain to the class how they know that their animals have the features common to all animals. LS Verbal

READING SKILL BUILDER — BASIC

Interactive Reading Assign Chapter 27 of the *Holt Biology* Guided Audio CD Program to help students achieve greater success in reading the chapter. LS Auditory

SKILL BUILDER — ADVANCED

Graphing Have students work in groups of two or three to gather information on the speed that different animals can travel. (They could use library or Internet resources to collect the information.) Have them graph, illustrate, and display their data. Tell students to be certain to include both aquatic and terrestrial animals in their research. LS Interpersonal

did you know?

Some fish are triploid. Through gene technology, an egg can be made triploid. The eggs of some commercially raised salmon are exposed to a stimulus, such as a sudden temperature change, that interferes with the eggs' chromosomes as they separate during cell division. The result is triploid salmon that do not mature sexually. Since these salmon do not invest any energy in reproduction, they feed longer and grow larger, which increases the yield when the salmon are harvested.

Integrating Physics and Chemistry

After reminding students that momentum equals mass × velocity, have them find the tiger's momentum in **Figure 2** if one assumes that the tiger has a mass of 150 kg and is moving at a velocity of 2 m/s. In order to inquire further about their understanding of velocity, ask the students what two properties are associated with velocity. (speed and direction)

Teach, *continued*

Using the Table

Have students study **Table 1.** Then name a body structure and ask students which embryonic tissue layer gave rise to the structure. (Structures of ectodermal origin include skin, hair, fingernails, brain, spinal cord, and nerves. Structures of mesodermal origin include the heart, arteries, veins, muscles, kidneys, bladder, uterus, testes, and thyroid gland. Structures of endodermal origin include the lungs, stomach, small intestine, gallbladder, pancreas, esophagus, and colon.) **LS Visual**

READING SKILL BUILDER — **BASIC**

Reading Organizer Have students start a list of words that can be explained by their etymologies. Ask them to split each word into its parts and to define each part. Tell them that knowing the meanings of word roots can help them understand other words. For example, the root word *derm* means "skin." Placing prefixes in front of *-derm* creates words that name the three primary tissue layers and indicate their origin. *Ecto-* means "outside," *endo-* means "within," and *meso-* means "middle." **LS Verbal**

Magnification: 620×

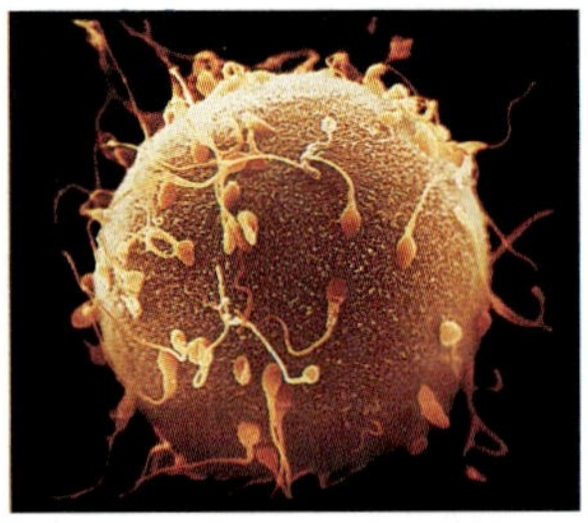

Figure 4 Egg and sperm. This human egg is surrounded by sperm, only one of which will fertilize the egg.

Sexual Reproduction

Almost all animals reproduce sexually by producing gametes, as do many plants, fungi, and protists. As shown in **Figure 4,** the females' egg cells are much larger than the males' sperm cells. Unlike the egg cells, the sperm cells of animals have a flagella and are highly mobile.

Absence of a Cell Wall

Among the cells of multicellular organisms, only animal cells lack rigid cell walls. The absence of a cell wall has allowed animals mobility that other multicellular organisms do not have. You may not realize this, but there are cells moving about in your body all the time. Cells called *macrophages*, for example, act as mobile garbage collectors, crawling over tissues and removing debris. A cell with a rigid cell wall could never perform such a task.

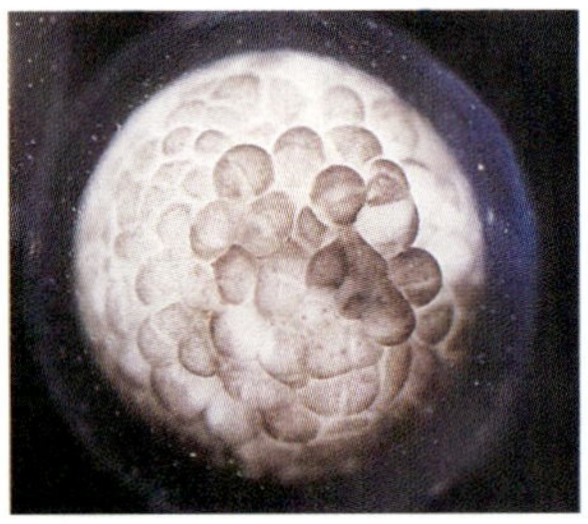

Figure 5 Blastula. The individual cells of a blastula form a sphere.

Blastula Formation

In all animals except sponges, the zygote (fertilized egg cell) undergoes cell divisions that form a hollow ball of cells called a **blastula** *(BLAS tyoo luh)*, shown in **Figure 5.** Cells within the blastula eventually develop into three distinct layers of cells—**ectoderm, endoderm,** and **mesoderm.** These layers are called the primary tissue layers because they give rise to all of the tissues and organs of the adult body. (A few simple invertebrates, such as *Hydra* and their kin, develop only two tissue layers, endoderm and ectoderm.) **Table 1** lists the three primary tissue layers and summarizes the body tissues and organs to which they give rise. Note that the table includes some organs, such as the urinary bladder, found only in vertebrates. The organs of vertebrates are complex structures containing cells that arise from more than one primary tissue layer. For example, the digestive system is formed primarily from endoderm and mesoderm.

Table 1 Origin of Animal Tissues and Organs

Primary tissue layer	Gives rise to
Ectoderm	Outer layer of skin; nervous system; sense organs, such as eyes
Endoderm	Lining of digestive tract; respiratory system; urinary bladder; digestive organs; liver; many glands
Mesoderm	Most of the skeleton; muscles; circulatory system; reproductive organs; excretory organs

596

Attention Grabber

Animal Adaptations Making generalizations about organisms is very useful for helping us understand how animals are adapted to the environment in which they live. However, there are numerous exceptions to these generalizations. For example, while animals are noted for their ability to move around and obtain food or a mate and avoid danger, some adult animals such as coral and sea anemones are not motile. We can modify the generalization about motility in animals to state correctly that all animals are motile during at least some part of their life cycle. Likewise, while animals are noted for being diploid organisms, some adult male animals such as ants are actually haploid. Animals are noted for reproducing sexually. However, sea anemones reproduce by splitting in two, two flatworms can develop from one that is cut in two, and aphids can reproduce by the development of unfertilized eggs. Finally, while animals are noted for being multicellular, some jellyfish are actually colonies of mainly unicellular organisms.

Tissues

The cells of all animals except sponges are organized into structural and functional units called tissues. Recall that tissues are groups of cells with a common structure that work together to perform a specific function. For example, the cells of adipose *(AD uh pohs)* tissue, shown in **Figure 6,** are specialized for storing fat. The cells of muscle tissue are specialized to contract, producing movement. The cells of nerve tissue are specialized to conduct signals.

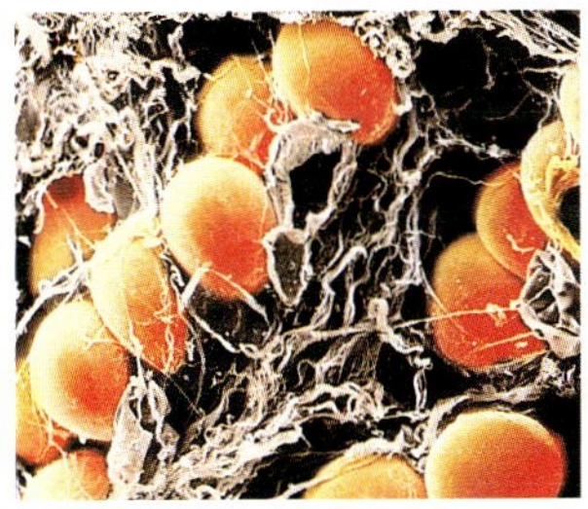

Figure 6 Adipose tissue. These adipose cells are specialized for storing fat, and each one contains a fat droplet. Together the cells make up adipose tissue.

Exploring Further

From Zygote to Gastrula

Just as you did, this sea urchin began life as a zygote but quickly developed into an embryo made up of layers of tissue—endoderm, ectoderm, and mesoderm. A variation of this pattern of development is found in all animals except sponges and is evidence that animal life arose from a common ancestor. The illustrations at right show the sea urchin's pattern of development in its first hours of life.

Cell numbers increase

During a process called *cleavage,* the cells of the zygote divide, doubling the number of cells with each division. However, the mass of the developing embryo does not increase, and the cells formed are progressively smaller. After about 3 hours, the zygote has become a solid ball of cells. Cell division continues until a blastula is formed.

Cells change locations

During a second process, called *gastrulation,* the blastula begins to collapse inward. At the same time, its cells move about, changing their location within the blastula. The cells begin to vary in size and form three primary tissue (germ) layers. Gastrulation is complete, and the zygote is now an embryo.

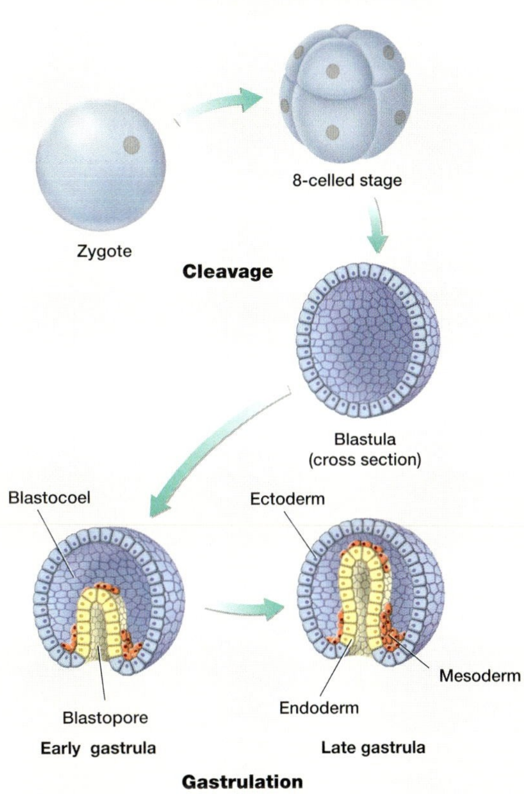

internet connect
www.scilinks.org
Topic: Origin of Tissues
Keyword: HX4132
SCILINKS. Maintained by the National Science Teachers Association

597

did you know?

Twins form in one of two ways. Dizygotic (fraternal) twins result when ovulation produces two eggs that are fertilized by different sperm. Because dizygotic twins develop from different sperm and eggs, they can be as genetically dissimilar as any two siblings. Monozygotic (identical) twins result when an embryo splits in two early in cleavage, giving rise to two genetically identical embryos.

The incidence of twin pregnancies is about 11 per 1,000 pregnancies (or 1 in every 90 pregnancies). The number of dizygotic twins is 4–50 per 1,000 pregnancies, and the number of monozygotic twins is 1–5 per 1,000 pregnancies.

SKILL BUILDER — GENERAL

Math Skills Have students calculate how many cells will be present in a developing zygote after 3, 5, and 10 divisions. (8, 32, 1024) Have students write a general formula for calculating the number of cells after n divisions. (number of cells = 2^n) Ask students what assumption must be made for this formula to be valid. (Each cell must divide during each cell division.) **LS Logical**

Exploring Further

From Zygote to Gastrula

Teaching Strategies

Tell students that although all animals except sponges develop a blastula, the way the blastula develops can vary. The number of cells in a blastula and the length of time it takes for the blastula and the gastrula to form depend on the species.

Tell students that not all blastulas look like that of a sea urchin. Show students photos or drawings of a frog's blastula. Ask them why these two blastulas might have different shapes. (A frog's blastula has much more yolk.)

Discussion

Have students look at the five drawings. What do the colors represent on the bottom two drawings? (blue, ectoderm; yellow, endoderm; red, mesoderm)

What is a zygote, and how many cells does it have? (a fertilized egg; one cell)

What happens during cleavage? (The cells divide. Each division doubles the number of cells in the embryo.)

What happens during gastrulation? (The cells of the blastula begin to move about and vary in size. As they move, they form three distinct layers—ectoderm, endoderm, and mesoderm.)

Teach, *continued*

SKILL BUILDER — ADVANCED

Writing Skills Have students research symmetry using the Web site in the Internet Connect box on the following page. Have students write a report summarizing their findings. Students should include illustrations or models with their report. **LS Verbal**

Activity — BASIC

Following the Directions After introducing the terms anterior, posterior, ventral, and dorsal to your students, reinforce this anatomical terminology by having students sketch the outline of a worm. Then direct them to draw different structures on their worm. For example, you might ask them to shade the dorsal surface black; and draw eyes on the anterior end. Finish by drawing the completed worm on the board. **LS Visual**

Demonstration — GENERAL

Use some common classroom objects to illustrate different types of symmetry. For instance, a crumpled piece of paper is asymmetrical, a roll of tape is radially symmetrical, and a stapler is bilaterally symmetrical. Point out that the symmetry does not have to be perfect in every detail to be considered a type of symmetry. For example, a sea anemone is radially symmetrical even though all its tentacles might not be exactly the same size. Then give students a series of food items to classify as asymmetrical (scrambled eggs, popcorn), radially symmetrical (pizza, bagel), or bilaterally symmetrical (taco, hot dog, loaf of bread). **LS Visual**

Body Symmetry

Figure 7 Asymmetry. Animals that grow in an irregular pattern, such as this sponge, show asymmetry.

All animals have their own particular **body plan,** a term used to describe an animal's shape, symmetry, and internal organization. An animal's body plan results from a pattern of development programmed into the animal's genes by natural selection. Sponges, such as the one shown in **Figure 7,** have the simplest body plan of all animals. Sponges are **asymmetrical,** or irregular in shape, and sometimes their shape depends on where they are growing. The body plans of virtually all other animals show a definite body shape and symmetry.

Radial Symmetry

Some of the first animals to evolve in the ancient oceans had radial symmetry. Animals with **radial symmetry** have body parts arranged around a central axis, somewhat like the spokes around a bicycle wheel. A plane passing through the central axis divides the organism into roughly equal halves, as shown in **Figure 8.** Today's radially symmetrical animals are aquatic. Most move slowly or drift in ocean currents.

Bilateral Symmetry

The bodies of all other animals show **bilateral symmetry,** a body design in which there are distinct right and left halves. A plane passing through the animal's midline divides the animal into mirror image halves, as shown in **Figure 9.** There is a dorsal (top) and a ventral (bottom) surface plus an anterior (front) end and a posterior (back) end.

Figure 8 Radial symmetry. Each of the planes passing through a sea anemone's central axis divides it into roughly identical halves.

598

Cultural Awareness

Totem Poles Many Native American tribes have family totems, which are symbols for their family, clan, or tribe. A specific animal is associated with the totem, and some groups consider the animal totem to be the ancestor of the family. Often there is a prohibition against killing or eating that particular animal. In addition to human figures, animals seen on totem poles include minks, orcas, frogs, seals, hawks, eagles, ravens, and small marine animals.

Bilateral symmetry was a major evolutionary change in animals because it enabled different parts of the body to become specialized in different ways. For example, most bilaterally symmetrical animals have evolved an anterior concentration of sensory structures and nerves, a process called **cephalization** *(SEF uhl lih ZAY shuhn)*. Animals with cephalic ends, or heads, are often active and mobile. With sensory organs concentrated in the front, such animals can more easily sense food and danger.

internet connect
www.scilinks.org
Topic: Body Symmetry
Keyword: HX4028
SciLinks. Maintained by the National Science Teachers Association

Figure 9 Bilateral symmetry. The body parts of this gray squirrel are arranged so that it has a right and a left half that are mirror images of each other.

QUICK LAB

Recognizing Symmetry SC.H.3.4.1

You can use the letters of the alphabet to better understand the nature of symmetry.

Materials

envelope containing letters of the alphabet

Procedure

1. Spread the letters on the table in front of you so you can see all of them.
2. Sort the letters into groups based on their symmetry, using the terms *asymmetry, radial symmetry,* and *bilateral symmetry.* For example, the letters *A* and *T* show bilateral symmetry. The letter *J* is asymmetrical.

Analysis

1. **Propose** a definition for each kind of symmetry you found in the letters.
2. **List** any letters you found difficult to classify, explaining why it was difficult to classify these letters.
3. **Identify** the letters that show the same kind of symmetry as sponges.
4. **Identify** two or three animals that you might be familiar with that have the same kind of symmetry as the letter *M.*
5. **Critical Thinking Evaluating Methods** What are some strengths and weaknesses of using symmetry as a way of classifying or describing organisms?

599

Demonstration — GENERAL

Provide students with a poster, overhead transparency, or model of the internal anatomy of the human body. Point out to students that, overall, the body is bilaterally symmetrical (one arm on each side, one eye on each side, one lung on each side, etc.). Next place a square or rectangular mirror on the midline of the illustration or model. Ask students to identify ways in which the internal anatomy of the body is asymmetrical. (The spleen is on the left, the liver is on the right, the heart is slightly left of center, the right lung has three lobes and the left lung has two, the stomach is on the left, the appendix is on the right, the arteries branching off the aorta are asymmetrical, etc.)

QUICK LAB

Recognizing Symmetry

Skills Acquired
Comparing, categorizing, applying information

Teacher's Notes
Tell students that they may identify some letters that do not fit neatly into a category. They should categorize these letters as they think best.

Answers to Analysis

1. Definitions will vary but could be similar to those in the text.
2. Answers will vary. Students may have difficulty with letters such as O. Some may say it shows radial symmetry, while others may classify it as having bilateral symmetry.
3. Answers will vary, but students may include F, G, J, L, N, P, Q, R.
4. Answers will vary; accept any bilaterally symmetrical animal.
5. Answers will vary. Students may think that considering symmetry is helpful when the type of symmetry is clear-cut but less helpful when there is a question of what type of symmetry is displayed.

Career

Aquatic Biologist An aquatic biologist is a scientist who specializes in the biology of organisms that live in water, including oceans, rivers, ponds, and lakes. Many aquatic biologists study the complex ecosystems of these environments. They investigate the physical and chemical features of the water in addition to the biological factors. For example, they might gauge the effects of pollutants on the organisms that live in a lake.

Teach, *continued*

Teaching Tip — BASIC

Urban Animals Provide students with the information contained in the Real World Connection at the bottom of this page. Work with students to create a list of animals that are adapted to life in your community. **LS Intrapersonal**

Using the Figure

Have students compare the cross section of the acoelomate flatworm in **Figure 10** with the cross section of the pseudocoelomate roundworm. Point out the solid body construction of the flatworm versus the hollow body construction of the roundworm. Then have students compare the body plans of the flatworm and roundworm to that of the coelomate earthworm. Draw students' attention to the placement of the three primary tissues in each figure. Tell students that while pseudocoelomates have muscles lining their body wall, coelomates have an advantage because they have an additional layer of muscles surrounding their gut. Ask students to name some human structures that contain a muscle layer derived from mesoderm. (blood vessels, digestive tract, uterus) **LS Visual**

WORD Origins

The term *coelomate* is from the Greek *koilia*, meaning "body cavity." Placing prefixes in front of this word changes its meaning. For example, *a-* means "without" and *pseudo-* means "false."

Internal Body Cavity

Bilaterally symmetrical animals have one of three basic kinds of internal body plans, each illustrated in **Figure 10.** The body plan may include a body cavity, or **coelom** *(SEE luhm)*, a fluid-filled space found between the body wall and the digestive tract (gut). This space is lined with cells that come from mesoderm. Animals with no body cavity are called **acoelomates** *(ay SEEL oh mayts)*. The space between an acoelomate's body wall and gut is completely filled with tissues. Other animals, called **pseudocoelomates** *(SOO doh seel oh mayts)*, have a body cavity located between the mesoderm and endoderm. Their body cavity is called a pseudocoelom (false coelom).

Coelomates have a true coelom, a body cavity located entirely within the mesoderm. Because the mesodermal layer lines the body wall and wraps around the gut, the gut and other internal organs of coelomates are suspended within the coelom.

A true coelom provides an internal space where mesoderm and endoderm can be in contact with each other during embryonic development. This aided the evolution of complex organs made of more than one type of tissue. Because these internal organs are suspended in a fluid-filled coelom, they are protected from the movement of surrounding muscles. Thus, an animal can move about without damaging the organs or interfering with their function.

internet connect
www.scilinks.org
Topic: Body Cavity
Keyword: HX4027
SCILINKS. Maintained by the National Science Teachers Association

Figure 10 **Three body plans**

Bilaterally symmetrical animals have one of three basic body plans.

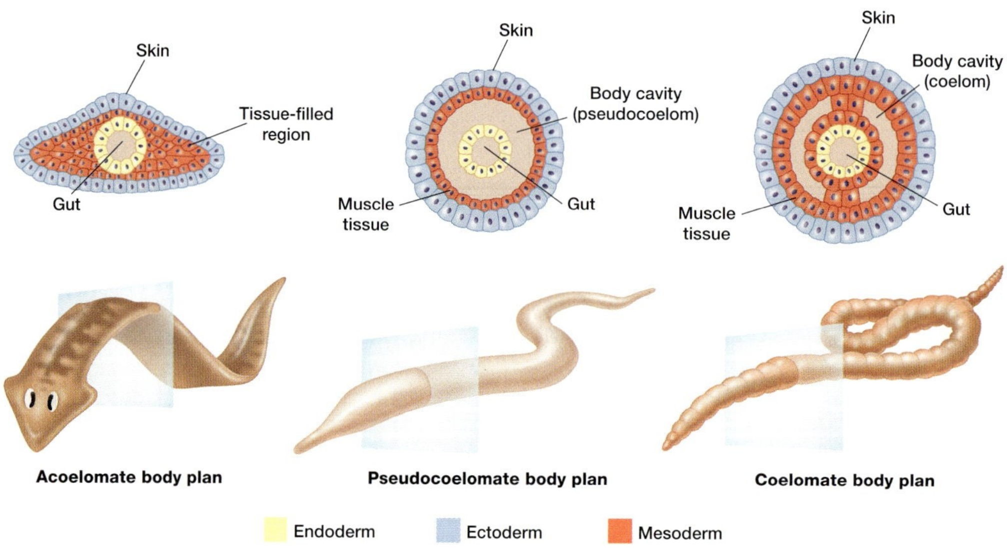

600

REAL WORLD CONNECTION

A variety of animals have adapted to living with humans. These animals are attracted to food, water, nesting sites, and the protection from predators they find within the human community. In the United States, examples of animals that have successfully adapted to urban or suburban life include gulls, peregrine falcons, owls, doves, geese, coyotes, rabbits, raccoons, squirrels, mice, chipmunks, opossums, turtles, lizards, snakes, cockroaches, ants, and wasps.

Body Segmentation

Segmented animals are composed of a series of repeating, similar units called *segments*. Segmentation underlies the organization of all "advanced" animals and is easy to observe in some animals, such as earthworms. Crustaceans (lobsters and their kin), spiders, and insects also show some degree of body segmentation, but it may be difficult to observe during their adult stage. For example, segments are not as apparent in the adult moth in **Figure 11,** but they are clearly visible in its immature caterpillar stage.

In vertebrates, segments are not visible externally, but there is evidence of segmentation in a vertebrate embryo. Vertebrate muscles develop from repeated blocks of tissue called *somites (SOH mietz),* and the vertebral column (backbone) consists of a stack of very similar vertebrae.

In segmented worms, each segment can move independently, permitting great flexibility and mobility. Therefore, a segmented worm's long body can move in ways that are often quite complex. For example, consider an earthworm as it crawls along a flat surface. It lengthens some parts of its body while shortening others. Many scientists think segmentation evolved as an adaptation that permitted more-efficient burrowing.

In highly segmented animals, such as earthworms, each segment repeats many of the organs in the adjacent segment. As a result, an injured animal can still perform vital life functions. Segments are not totally independent of each other, however. Materials pass from one segment to another through a circulatory system that connects them. Nerves also connect each segment to a brain that coordinates the body's movements. Segmentation also offers evolutionary flexibility. A small change in an existing segment can produce a new type of segment with a different function. For example, some segments are modified for feeding, while others are modified for reproduction.

Figure 11 Segmentation. This caterpillar shows clearly defined segments. In its adult form—a cecropia moth—the segments are more difficult to observe.

601

Using the Figure — Basic

Direct students' attention to **Figure 11.** Have students count the number of segments they can see in the caterpillar. Tell students that most of the segments are fused in the adult form. Then have students try to count the number of fused segments on the moth's abdomen. Ask: What clue is present that can help you distinguish the segments? (color pattern) **LS Visual**

Reading Skill Builder — Basic

Discussion Ask students to point out examples of segmentation in the external structure of the human body. Students will find this difficult, so tell them that although vertebrates are considered segmented animals, vertebrate segmentation is strictly internal. The backbone and rib cage are composed of repeating units, and the muscles are made up of multiple blocks of tissue. Segmentation is also evident in the development of the brain.

Technology Connection

The Polymerase Chain Reaction (PCR) is a process in which numerous copies of DNA molecules can be produced from incredibly small amounts of biological material. PCR allows a scientist to mimic the cell's own natural DNA replication process in a test tube. Using this technique, scientists can easily characterize and compare the genetic material from different organisms of the same or different species. The process is used in forensics to test for matches between the blood of crime suspects and blood or tissue found on victims. It has been used in the sequencing of the human genome in the Human Genome Project. It is also used to evaluate evolutionary relationships between organisms. Kary Mullis won the Nobel Prize for Chemistry in 1993 for his discovery of the PCR process.

Teach, *continued*

Activity — Advanced

Other Phyla Have students research one phylum that is not shown on the phylogenetic tree in **Figure 12.** (Phyla that are not represented in Figure 12 include Ctenophora, comb jellies; Nemertea, ribbon worms; Bryozoa, bryozoans; Brachiopoda, brachiopods; and Phoronida, phoronids.) Ask them to write a report about the phylum and tell where it would be located if it were shown on this tree.
LS Verbal

Group Activity — General

Evolutionary Milestones Draw a Graphic Organizer on the board similar to the one shown at the bottom of the page, but do not include the names of the phyla. Scramble the order of the phyla and list them separately. Beginning with Multicellularity, have a student come to the board and write the correct phylum next to this evolutionary milestone. Then ask the student to explain what the description means. When you get to Deuterostomes and Notochord, tell students that they will learn these terms in Chapter 31.
LS Visual Co-op Learning

internet connect
www.scilinks.org
Topic: Animal Groups
Keyword: HX4010
SCILINKS. Maintained by the National Science Teachers Association

Kinds of Animals

Kingdom Animalia contains about 35 major divisions called *phyla* (singular, phylum), depending on how certain organisms are classified. The animals in the various phyla show an extraordinary range of body forms, internal body systems, and behaviors.

To visually represent the relationships among various groups of animals, scientists often use a type of branching diagram called a phylogenetic tree. A **phylogenetic tree,** such as the one in **Figure 12,** shows how animals are related through evolution. Clues to animal relationships can be found in the fossil record and by comparing the anatomy and physiology of living animals. Clues are also found by comparing patterns of development in animal embryos. The most direct evidence of evolutionary relationships, however, comes from comparing the DNA in the genes of various animal species. For example, scientists long debated whether giant pandas are related more closely to bears or to raccoons. Comparing their DNA showed that pandas are more closely related to bears than to raccoons.

Figure 12 Evolutionary milestones. This phylogenetic tree shows one hypothesis of the evolutionary relationships among nine of the major animal phyla. The circled numbers indicate important milestones in the evolution of the animal body, as listed in the table below.

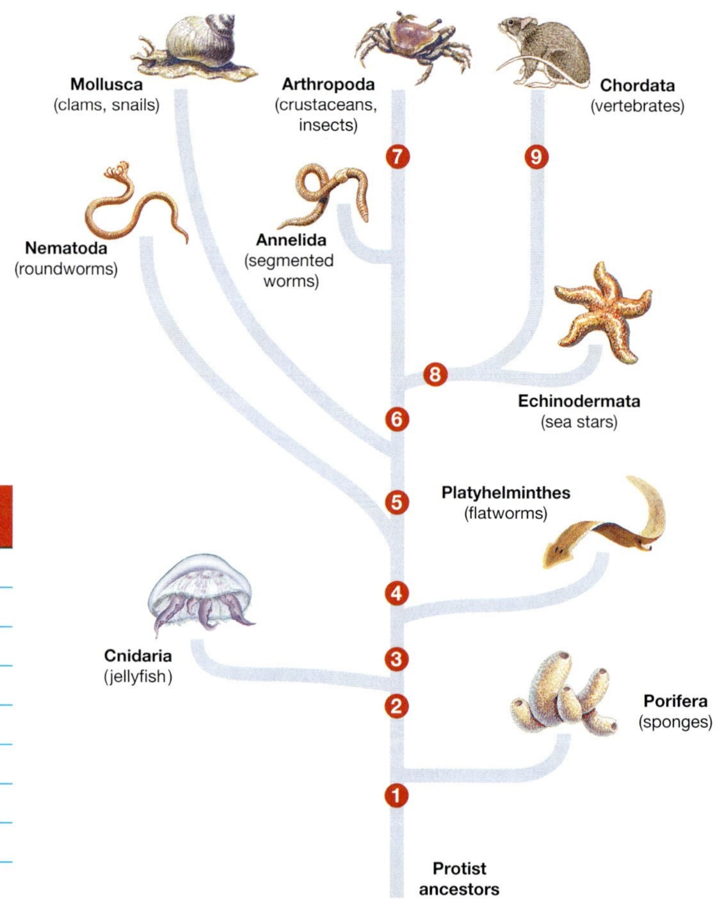

	Evolutionary milestones
1	Multicellularity
2	Tissues
3	Bilateral symmetry
4	Body cavity
5	Coelom
6	Segmentation
7	Jointed appendages
8	Deuterostomes
9	Notochord

Graphic Organizer

Use this graphic organizer with the **Group Activity** on this page.

Phylum	Evolutionary milestone
Chordata	Notochord
Echinodermata	Deuterostomes
Arthropoda	Jointed appendages
Annelida	Segmentation
Mollusca	Coelom
Nematoda	Pseudocoelom
Platyhelminthes	Bilateral symmetry
Cnidaria	Tissues
Porifera	Multicellularity

Section Review Correlations

SC.D.1.4.3: Item 4
SC.F.1.4.2: Items 2, 3
SC.G.1.4.1: Items 1, 5

The animal kingdom is often divided into two groups: invertebrates (animals without a backbone) and vertebrates (animals with a backbone). As you study these two groups, you will discover how a series of key evolutionary innovations in the body plan of animals has led to today's animals. The first animals to evolve did not have body cavities. After body cavities appeared, segmentation evolved. Later, jointed appendages (such as legs) evolved. Backbones evolved even later. Scientists consider these innovations (shown in Figure 12 as numbered milestones) when they attempt to determine evolutionary relationships among different groups of animals. Watch for these evolutionary stages and milestones as you continue studying invertebrates in later chapters in this book.

DATA LAB

Exploring the Animal Kingdom SC.G.1.4.1

Background

You can find out more about the animal phyla by referring to the section "A Six-Kingdom System" in the Appendix. Turn to this section and locate the information for kingdom Animalia. Follow the procedure below to evaluate this information.

Procedure

1. Read the introductory paragraph for kingdom Animalia. Then quickly skim over the information presented. Do not read it word for word, but observe how the information is divided into sections.
2. Choose one phylum, and read all of the information about it.
3. As you read, notice how color is used and what types of information are given, for example, number of species found and habitat.

Analysis

1. **List** at least three types of information you found for the phylum you read about.
2. **Analyze** how color is used to distinguish between the different entries on a page.
3. **Propose** a way that you might use this information when studying about a particular animal phylum.

Section 1 Review

1. **Describe** each of the eight features animals have in common. SC.G.1.4.1
2. **Summarize** the difference between radial symmetry and bilateral symmetry. SC.F.1.4.2
3. **Compare** the body plans of acoelomates, pseudocoelomates, and coelomates. SC.F.1.4.2
4. **Critical Thinking** **Relating Concepts** In 1994, Western scientists first observed the Vietnamese saola, a hoofed mammal. The saola was proved to be related to wild cattle and buffalo. How do you think scientists identified the saola's closest relatives? SC.D.1.4.3
5. *FCAT Prep* Which characteristic is found in all animals?
   A tissues
   B body cavity
   C multicellularity
   D body segmentation
   SC.G.1.4.1

603

Answers to Section Review

1. Student descriptions will vary but should mention heterotrophy, mobility, multicellularity, diploidy, sexual reproduction, absence of a cell wall, blastula formation, and tissues.
2. Animals with radial symmetry can be divided into equal parts by a plane drawn through the center of the animal. They have no right or left side. Animals with bilateral symmetry can be divided into identical right and left halves by a plane that passes down their midline. They have dorsal and ventral sides and anterior and posterior ends.
3. Acoelomates have no body cavity. Pseudocoelomates have a body cavity located between the mesoderm and the endoderm. Coelomates have a body cavity that is completely surrounded by mesoderm.
4. Scientists compared the saola's DNA with that of animals they thought the saola was related to. The DNA evidence pointed to wild cattle and buffalo.
5. **A.** Incorrect. The bodies of sponges, which are animals, do not have their cells organized into tissues. **B.** Incorrect. Flatworms, jellyfish, and sponges do not have body cavities. **C.** Correct. **D.** Incorrect. Mollusks, roundworms, flatworms, jellyfish, and sponges do not have segmentation.

DATA LAB

Exploring the Animal Kingdom

Skills Acquired
Observing, comparing, summarizing

Answers to Analysis

1. Acceptable answers include number of organisms in the phylum, general characteristics of the phylum, classes that belong to the phylum, and examples of organisms in the phylum.
2. The kingdoms are in bold red print; the phyla and subphyla are in bold blue print; the classes are in bold green print; and the orders are in bold black print.
3. Answers will vary.

Close

Reteaching — BASIC

Have students work in pairs to make two drawings: one showing the three types of symmetry and one showing the three animal body plans. LS Interpersonal

Quiz — GENERAL

1. Which kind of symmetry is exhibited by most animals? (bilateral symmetry)
2. What is a coelom? (It is a body cavity that is located entirely within mesodermal tissue.)
3. What is a blastula? (A hollow ball of cells formed by early cell divisions of an animal zygote.)

Alternative Assessment — GENERAL

Form teams of three or four students. Assign each team one of the characteristics of animals. Have each team write a brief description of the characteristic and develop an accompanying visual. Teams should then present their information to the class. LS Interpersonal Co-op Learning

Focus

Overview

Before beginning this section review with your students the objectives listed in the Student Edition. The purpose of this lesson is to teach students about particular functions in animals that are carried out by specialized cells and the tissues and organs they form.

Bellringer

Ask students to make a list of the substances they need to take into their bodies in order to survive and then describe how each substance enters the body. (oxygen—taken in by breathing; nutrients, energy, water—taken in by eating and drinking.)

Motivate

Demonstration

Make a two-column table on the board. Label one column "Think about" and the other "Don't think about." Tell students that people engage in many activities that they make conscious decisions about, but they also engage in activities that they make no decisions about. Place an entry in each column. For example: Think about—how my hair looks; Don't think about—digesting my breakfast. Ask students to suggest other entries.

Section 2 Animal Body Systems

Florida Sunshine Standards/FCAT

SC.F.1.4.2 Body structures are uniquely designed and adapted for their function.

SC.F.2.4.1 There are mechanisms of asexual and sexual reproduction and different genetic advantages and disadvantages of asexual and sexual reproduction.

Objectives

- **Summarize** the functions of the digestive, respiratory, circulatory, nervous, skeletal, and excretory systems. SC.F.1.4.2
- **Compare** a gastrovascular cavity with a one-way digestive system. SC.F.1.4.2
- **Differentiate** open from closed circulatory systems. SC.F.1.4.2
- **Distinguish** asexual from sexual reproduction. SC.F.2.4.1

Key Terms

gastrovascular cavity
respiration
gill
open circulatory system
closed circulatory system
hydrostatic skeleton
exoskeleton
endoskeleton
hermaphrodite
external fertilization
internal fertilization

Tissues and Organs

As you go about your day, you make decisions that involve thinking—what to wear to school, how to solve a problem, where to sit during lunch. But a lot that happens during your day does not require your thought. You digest your food, you balance yourself as you walk, and your heart beats. These and thousands of other functions are carried out by your body. But how does your body, or the body of any animal, carry out these tasks?

Simple animals like sponges carry out the many tasks of living with little specialization in the cells of their body. More-complex animals have evolved tissues and organs that are specialized to perform specific functions. Six important functions of these tissues and organs are digestion, respiration, circulation, conduction of nerve impulses, support, and excretion.

Digestion

Single-celled organisms and sponges digest their food within their body cells. This means that their food source cannot be larger than their individual cells. All other animals digest their food extracellularly (outside of their body cells) within a digestive cavity, as shown in **Figure 13.** Digestive enzymes that are released into the cavity begin the breakdown of food, permitting the animal to prey on organisms larger than its body cells. Simple animals, such as the

Figure 13 Extracellular digestion

A hydra has a gastrovascular cavity, while a roundworm has a digestive tract in which food travels in one direction only.

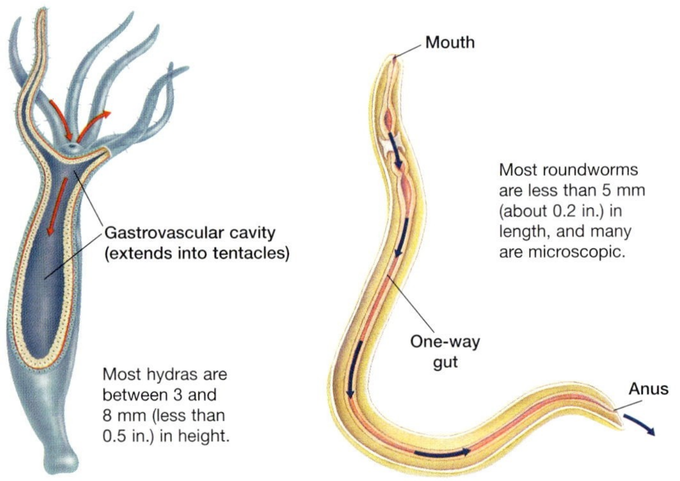

internet connect
www.scilinks.org
Topic: Body Systems
Keyword: HX4029
SciLinks Maintained by the National Science Teachers Association

Integrating Physics and Chemistry

Ask students to determine whether each of the six important functions of tissues and organs are more the result of physical or chemical changes or both. For example, whereas digestion is the result of chemical changes in food matter, circulation would seem to be more of a physical change—that is, movement—of blood cells and plasma.

Chapter Resource File

- Directed Reading BASIC
- Active Reading GENERAL
- Data Sheet for Math Lab GENERAL

One-Stop Planner CD-ROM

- Reading Organizers BASIC
- Reading Strategies BASIC
- Supplemental Reading Guide Journey to the Ants ADVANCED

Section Correlations

SC.F.1.4.2
SC.F.2.4.1

hydra and flatworms, have a **gastrovascular cavity,** a digestive cavity with only one opening. There can be no specialization within a gastrovascular cavity because every cell is exposed to all stages of food digestion.

Other animals have a digestive tract (gut) with two openings, a mouth and an anus. The anus is the opening through which undigested food leaves the body. In a digestive tract, food moves in one direction, from mouth to anus. Unlike a gastrovascular cavity, a one-way digestive system allows for specialization. For example, there may be a section of gut for food storage, a section for breaking down food into small pieces, and a section for the chemical digestion of food. Eventually the food is broken down into molecules small enough to pass through the lining of the gut and into the bloodstream.

Respiration

In simple animals, such as jellyfish, oxygen gas and carbon dioxide gas are exchanged directly with the environment by diffusion. The uptake of oxygen and the release of carbon dioxide, called **respiration,** can take place only across a moist surface, such as the damp skin of an earthworm. In larger, more complex animals, simple diffusion cannot provide for adequate gas exchange. Most large animals have specialized respiratory structures.

Some aquatic (and a few terrestrial) animals respire with **gills,** very thin projections of tissue that are rich in blood vessels. Gills provide a large surface for gas exchange. **Figure 14** shows the gills of a mud puppy. Gills are not suitable for most terrestrial animals because gills do not function unless they are kept moist. A variety of respiratory organs, such as lungs, have evolved in many terrestrial animals that allow them to respire on dry land.

Figure 14 **Gills.** The feathery gills of this mud puppy are supported by its watery environment.

Real Life

Lung tissue is delicate.

Many products, such as spray paints and some glues, contain chemicals that can damage or destroy lung tissue. This reduces the surface area available for gas exchange, reducing oxygen intake. When using chemical products, always follow the instructions on the warning label. If necessary, wear a protective mask.

Finding Information

Research the human disease emphysema and find out what a person can do to prevent emphysema.

605

Teach

Demonstration

To demonstrate the difference between a gastrovascular cavity and a one-way gut, show students a cup and a length of tubing. Demonstrate how water can enter and leave the cup through only one opening. Swirl the water around a bit to show how it is exposed to all areas of the inner surface of the cup. Then fill the tube with water. Use your fingers to push the water through the tube, duplicating the action of muscles. Point out how the water moves in only one direction through the tube. Once the water passes through a section of the tube, it will not return to that section again. **LS Visual**

Real Life

Answer

Emphysema causes shortness of breath, due to the loss of lung elasticity. It is usually caused by smoking or exposure to other types of air pollution.

Transparencies

TR Bellringer

TR H5 Open and Closed Circulatory Systems

did you know?

Feeding in Animals Animals are heterotrophs, and so they must absorb nutrients or ingest food sources. Ingestive eaters, the majority of animals, use a mouth to ingest food. Absorptive feeders, such as tapeworms, live in a digestive system of another animal and absorb nutrients directly through the body wall. Filter feeders, such as oysters, collect small organisms from the surrounding water. Substrate feeders, such as earthworms and termites, eat the material (soil or wood) they burrow through. Fluid feeders, such as aphids, pierce the body of a plant or animal and withdraw fluids.

Teach, *continued*

Teaching Tip — BASIC

Transport in Plants and Animals Ask students to think about how nutrients are transported in plants. (Mineral nutrients are transported in the xylem by water that rises from the roots to the leaves, drawn by transpiration of water from leaves and loss of the water vapor through stomata. Sugars are transported in the phloem.) Have students contrast this with how nutrients are transported in animals. (In most animals, nutrients enter the circulatory fluid from the gut. A heart pumps the fluid through the body, delivering nutrients to the body's cells.)

Activity — BASIC

Open Vs. Closed Circulatory Systems Ask students to make a two-column list that will allow them to compare and contrast open and closed circulatory systems. In one column, ask them to write the ways in which the two systems are alike. In the other column, ask them to write the ways in which the two systems are different.

Figure 15 Circulatory systems

In an open circulatory system, fluid leaves the circulatory vessels, but in a closed circulatory system, blood remains in the blood vessels.

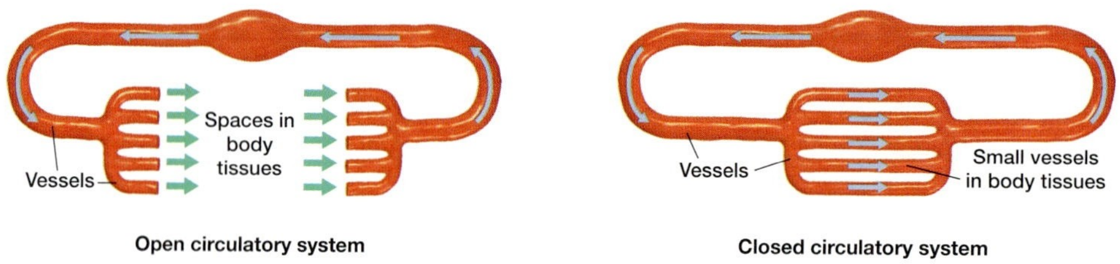

Circulation

In simple animals, body cells are exposed to either the external environment or the gastrovascular cavity. None of the body cells are far away from sources of oxygen or nourishment. More-complex animals, however, have tissues that are several cell layers thick. Many of the cells in this tissue are not close enough to the surface of the cell layer to exchange materials directly with the environment. Oxygen and nutrients must be transported to these body cells by a circulatory system.

Two types of circulatory systems are shown in **Figure 15.** In an **open circulatory system,** a heart pumps fluid containing oxygen and nutrients through a series of vessels out into the body cavity. There the fluid washes across the body's tissues, supplying them with oxygen and nutrients. The fluid collects in open spaces in the animal's body and flows back to the heart. In a **closed circulatory system,** a heart pumps blood through a system of blood vessels. These blood vessels form a network that permits blood flow from the heart to all of the body's cells and back again. The blood remains in the vessels and does not come in direct contact with the body's tissues. Instead, materials pass into and out of the blood by diffusing through the walls of the blood vessels.

Conduction of Nerve Impulses

Nerve cells (neurons) are specialized for carrying messages in the form of electrical impulses (conduction). These cells coordinate the activities in an animal's body, enabling the animal to sense and respond to its environment. Members of all of the major animal phyla except sponges have nerve cells. **Figure 16** shows the arrangement of nerve cells in three animals—a hydra, a flatworm, and a grasshopper. The simplest arrangements of nerves are found in animals like hydras and jellyfishes. All of their nerve cells are similar and are linked to one another in a web called a nerve net. There is little coordination among the nerve cells in a nerve net.

Bilaterally symmetric animals have clusters of neurons called *ganglia*. The ganglia at the anterior end of the animal body became

Study TIP

Reviewing Information

Work with a partner to review the body functions in this section. Review each function by asking each other questions that include the boldface terms. For example, ask, "What is the difference between a gastrovascular cavity and a one-way gut?"

MISCONCEPTION ALERT

Brains Students may have heard the expression "bird brain" used to indicate that someone is not smart. Birds actually have very complex brains and many bird species are capable of learning complex tasks. Even though most animals may not appear smart to us, each species (except sponges) has evolved a nervous system that enables it to be successful in its environment.

Integrating Physics and Chemistry

Have students compare and contrast how nerve cells carry electrical impulses with the movement of electrons in an electrical circuit. How are they alike and unalike? Is the arrangement of the nerve cells within a body more like a series or parallel circuit? Before answering this question think about the effects of a brain versus spinal cord injury.

Figure 16 **Nervous systems**

The hydra has a simple nerve net, while the flatworm and the grasshopper have more-complex nervous systems.

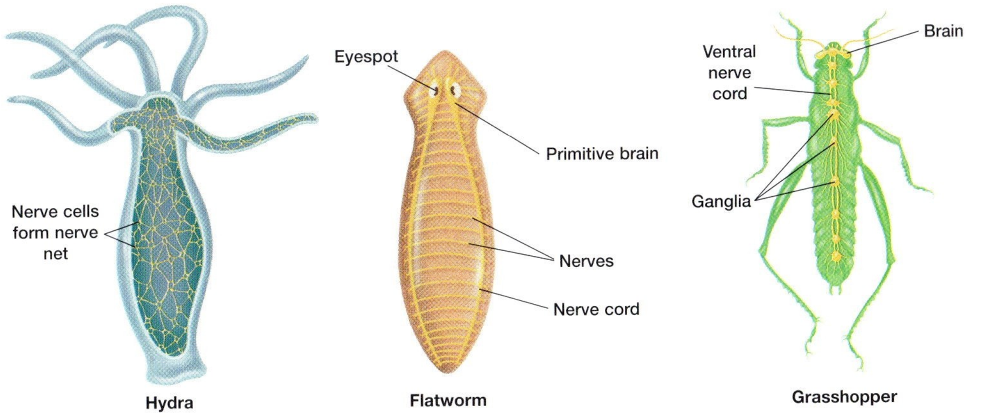

larger and more complex, forming a brain-like structure, as seen in the flatworm. More-complex invertebrates, such as the grasshopper, have brains with sensory structures, such as eyes, associated with them. These cephalized animals—that is, animals with heads—could interact with their environments in more-complex ways than could other animals.

Support

An animal's skeleton provides a framework that supports its body. It is also vital to animal movement. All animals move using the same force: the contraction (shortening) of muscle tissue against a framework such as the skeleton provides.

Hydrostatic skeleton Many soft-bodied invertebrates have a hydrostatic skeleton. A **hydrostatic skeleton** consists of water that is contained under pressure in a closed cavity, such as a gastrovascular cavity or a coelom. Imagine a balloon filled with water. The water presses against the balloon, supporting it. If pressure is applied to the balloon in any place, the water must shift, altering the shape of the balloon. The hydrostatic skeleton of a hydra, shown in **Figure 17,** is formed by its gastrovascular cavity. Other soft-bodied invertebrates, such as earthworms, have a fluid-filled coelom that serves as a hydrostatic skeleton. In both cases, muscle forces exerted against the hydrostatic skeleton aid movement.

Exoskeleton Other invertebrates, such as insects, clams, and crabs, have a type of skeleton known as an exoskeleton. An **exoskeleton** is a rigid external skeleton that encases the body of an animal. An exoskeleton supports movement in a different manner than a hydrostatic skeleton does. The muscles of animals with exoskeletons are attached to the inside of the skeleton, which provides a surface for them to pull against. Exoskeletons also protect an organism's soft internal parts.

Magnification: 70×

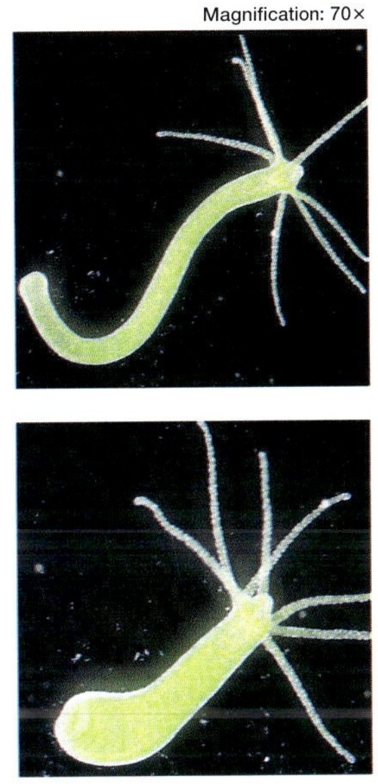

Figure 17 **Hydrostatic skeleton.** When the hydra closes the opening to its gastrovascular cavity and contracts muscles in its body wall, its body elongates. When water is released from the cavity and other muscles contract, the hydra's body shortens.

607

OCEANOGRAPHY CONNECTION

Over 200 different species of jellyfish are found throughout the world's oceans and seas. These invertebrates, animals lacking backbones, belong to the phylum Cnidaria. They occur in many shapes and sizes, ranging from the tiny, spherical thimble jellyfish of the Caribbean to the largest jellyfish of them all, the Arctic lion's mane, whose tentacles may stretch over 100 feet in length. These creatures are neither jelly nor fish.

Jellyfish can be found in all the world's oceans, and a few even inhabit freshwater. Over 95% water themselves, they have no heart, blood, brain, or gills. The bell, or body of a jellyfish, has one or more mouths on its central oral surface. Some jellyfish have frilly oral arms around the mouth. Jellyfish feed on small, drifting animals called zooplankton—which includes other jellyfish, juvenile fish, and larval crustaceans. Tentacles, long string-like structures that surround the bell rim, trap food. These tentacles can be longer or shorter than the diameter of the jellyfish's bell.

Teaching Tip — GENERAL

Hydrostatic Skeleton Tell students that hydrostatic skeletons are found in some invertebrates with a coelom, such as an earthworm. Explain that the force of muscles against the fluid-filled coelom produces movement. Ask students why the coelom needs to be filled with fluid for this to happen. (Fluid cannot be compressed when a force is applied to it; a gas can be compressed.)

Teaching Tip

Amazing Jellyfish Provide students with the information contained in the Oceanography Connection at the bottom of this page. Tell them that jellyfish and hydra are closely related and belong to the same phylum, Cnidaria. Jellyfish have a stage in their life cycle when they have a body form that is very similar to the form of hydra. This form is essentially an upside-down jellyfish, with its tentacles waving upward in the water.

Using the Figure

Have students look at the three nervous systems in **Figure 16.** Tell them that a hydra's nerve cells are found throughout its thin body wall. Each cell is linked to other nerve cells. In contrast to a flatworm, a hydra has no central region with a greater concentration of nerve cells. Point out that as animals became more complex, a more complex system of nerves was needed to coordinate activities and to handle sensory stimuli. A cluster of neurons at the anterior end (primitive brain) became increasingly complex until a true brain evolved. **LS Visual**

Integrating Physics and Chemistry

Review Newton's third law of motion with the students: *for every action force, there is an equal and opposite reaction force.* Then ask students what is the action force and reaction force when muscle tissue contracts against a body's framework such as a skeleton so as to produce an animal's movement.

Teach, *continued*

MATH LAB *Calculating Filtration Rate in the Human Kidney*

Skills Acquired
Calculating, applying information

Teacher's Notes
Tell students that they will calculate how much fluid (filtrate) the kidneys filter from the blood each day. Remind them that most of the filtrate is reabsorbed by the kidneys; only a small percentage is excreted as urine.

When students finish, have them look at the answer to analysis item 3. Tell them to visualize 180 1 L bottles of soda. As a class, calculate the percentage of fluid that is reabsorbed if a person excretes 2.0 L of urine a day. (about 99%; (2.0 L / 180) × (100%) = 1.1% filtrate excreted; 100%–about 1% = 99%)

Answers to Analysis

1. 7,500 mL per hour (125 mL/min × 60 min/h)
2. 180,000 mL per day (7,500 mL/h × 24 h/d)
3. 180 L per day (180,000 mL/d ÷ 1,000 mL/L)
4. If the kidneys could not return water to the body, a person would become dehydrated very quickly.

Endoskeleton An **endoskeleton** is composed of a hard material, such as bone, embedded within an animal. The most familiar type of endoskeleton is that of humans and other vertebrates. As with exoskeletons, muscles attached to the endoskeleton alternately contract and relax, enabling the animal to move.

Excretion

The term *excretion* refers to the removal of wastes produced by cellular metabolism. These waste products leave the cell by crossing the cell membrane and are then removed from the body. Some waste products are highly toxic and will poison an organism if not removed. The most important of these toxic wastes is ammonia. As ammonia forms, it dissolves in body fluids, becoming more dilute and thus less toxic. Simple aquatic invertebrates and some fishes excrete ammonia into the water through their skin or gills by diffusion. This is effective, but results in the loss of a lot of water.

Other animals, especially terrestrial animals, need to minimize water loss. A variety of ways have evolved in these animals by which wastes are removed from the body. One way is by converting ammonia to nontoxic chemicals, like urea. As the excretory system eliminates these wastes, water and other useful substances are returned to the body. Thus, eliminating metabolic wastes is linked to maintaining the body's water balance. For example, a mammal's kidneys filter fluid from the blood. The kidneys then concentrate the metabolic wastes filtered from the fluid and excrete them as concentrated urine. Simultaneously, the kidneys regulate the water content of the body by making the urine more or less dilute as necessary.

MATH LAB *Calculating Filtration Rate in the Human Kidney* SC.F.1.4.1

Background

The human kidney filters fluid from the blood at the rate of approximately 125 mL per minute. However, only a small percentage of this fluid is excreted as urine—adult humans normally excrete between 1.5 and 2.3 L of urine a day.

Analysis

1. **Calculate** how many milliliters of fluid the human kidneys filter each hour.
2. **Calculate** how many milliliters of fluid the kidneys filter each day.
3. **Critical Thinking** **Analyzing Data** Convert your answer in item 2 from milliliters to liters. For help, see "Math and Problem-Solving Skills: SI Measurement" in the Appendix. To better visualize the quantity of fluid represented by your answer, think about the volume of fluid contained in a 1 L bottle of soda.
4. **Critical Thinking** **Predicting Outcomes** What would happen if the kidneys could not return water to the body?

MISCONCEPTION ALERT

Egestion The process of excretion should not be confused with the elimination of undigested food from the body. Food is eaten, or ingested. However, while undigested food travels through the digestive system, it never enters the body's cells. Thus, un-digested food is *egested*, or passed out, rather than excreted.

Integrating Physics and Chemistry

Point out to students that when the toxic waste ammonia (NH_3) dissolves in the body's water it forms ammonium ions (NH^{4+}) and hydroxide ions (OH^-). Since the number of hydroxide ions increases in this chemical reaction, ammonia is classified as a base. For additional practice, have students look up the chemical formula for urea and write the chemical equation for the conversion of ammonia to urea.

Reproductive Strategies

While the reproductive system of an individual animal is not essential to its survival, reproduction is necessary if the species is to survive. There are two types of reproduction in animals, asexual and sexual.

Asexual Reproduction

Reproduction that does not involve the fusion of two gametes is called asexual reproduction. A sponge, for example, can reproduce by fragmenting its body. Each fragment grows into a new sponge. Some species of sea anemone reproduce by pulling themselves in half, forming two new adult anemones, as shown in **Figure 18.**

An unusual method of asexual reproduction is parthenogenesis *(pahr thuh noh JEN uh sis),* in which a new individual develops from an unfertilized egg. Parthenogenesis is common among insects. In honeybees, for example, a queen bee mates only once and stores the sperm. Although she has mated, the queen bee has the ability to lay unfertilized eggs that develop by parthenogenesis into male bees, called drones. Female bees develop by sexual reproduction, when the queen releases stored sperm to fertilize her eggs. A few species of fishes, amphibians, and lizards reproduce by parthenogenesis. Animals that reproduce asexually are usually able to also reproduce sexually.

Sexual Reproduction

In sexual reproduction, a new individual is formed by the union of a male and a female gamete. Gametes are produced in the sex organs. The testes *(TEHS teez)* produce the male gametes (sperm), and the ovaries produce the female gametes (eggs). Some species of animals, called **hermaphrodites** *(huhr MAF roh dietz),* have both testes and ovaries. Each individual functions as both a male and a female. But a hermaphrodite's sperm and eggs are usually produced at different times, so self-fertilization does not occur.

Real Life

Pregnant sea horse fathers give birth.

A female sea horse inserts her eggs into a pouch on the male's abdomen, where they are fertilized. The male incubates the eggs until they are fully developed and then gives birth to tiny sea horses.

Finding Information

Read about the role of the male bluegill sunfish in nesting.

internet connect

www.scilinks.org

Topic: Animal Reproduction

Keyword: HX4011

SCILINKS. Maintained by the National Science Teachers Association

Figure 18 Asexual reproduction. This pink-tipped surf anemone is in the process of pulling into two halves, each of which will be an adult sea anemone.

Real Life

Answer

Research on the nesting behavior of the bluegill sunfish has shown that the male of the species defends the nest, guards the eggs, and helps bring up the hatchlings.

Teaching Tip — Basic

Egg Shells and Internal Fertilization Ask students to hypothesize how a bird egg inside its hard shell can be fertilized. (Answers may vary.) Explain that internal fertilization occurs before the shell is formed. After fertilization, as the egg moves through the oviduct, albumin (egg white) is deposited around the egg. Next, a fibrous shell membrane covers the albumin, and finally a hard shell of calcium salts covers the egg. It takes about 24 hours after the egg is released into the oviduct, where it may or may not be fertilized, for the shell to be completed and the egg to be ready for laying.

Using the Figure — General

Use the photograph in **Figure 18** to help explain reproduction by fission (splitting). Ask students why dividing into two halves is considered asexual reproduction. (Only one parent is involved.) Ask why it is beneficial for a sea anemone to reproduce asexually. (If the parent is well adapted to its environment, the offspring will also be well adapted.) **LS Visual**

SKILL BUILDER — Advanced

Writing Skills Have students research animal reproduction using the Web site in the Internet Connect box on this page. Have students write a report summarizing their findings. **LS Verbal**

Attention Grabber

Hermaphroditism Some animals have ovaries and testes at different times during their lives. Some fish, for example, all begin life as females. Only the largest individuals develop into males. If the male in a group dies, the largest surviving female in the group will then become a male. There are other animals that change from male to female during their lives. These sex changes are called sequential hermaphroditism.

did you know?

Fertilization Animals that reproduce by external fertilization usually produce and release a large number of eggs. Sea urchins and sea stars release millions of eggs into the ocean at one time. These eggs are vulnerable to predators, and only a small fraction of the offspring will survive. In contrast, animals that reproduce by internal fertilization usually produce far fewer eggs. Human females usually only release one egg each month. Internal development of the embryo ensures a higher survival rate, so fewer eggs are needed.

Close

Reteaching — Basic

Have students work in pairs to make a set of flash cards for the boldface words in the chapter. One side of each card should have a word, and the reverse side should have its definition. Ask students to test each other using their flash cards. **LS Interpersonal**

Quiz — General

1. What is extracellular digestion? (digestion that occurs outside of cells; usually it occurs inside the body, as in animals)
2. Why must the surface area of gills and lungs be kept moist? (because diffusion can only occur across a moist surface)
3. Why is an endoskeleton critical for large terrestrial organisms? (because they must support their bodies and move around without the buoyancy provided by water)

Alternative Assessment — Advanced

Write the terms gastrovascular cavity, digestive tract, nerve net, hydrostatic skeleton, exoskeleton, and endoskeleton on the board. Ask students to define each term and to give examples of animals that have these features. **LS Verbal**

Section Review Correlations

SC.F.1.4.2: Items 1, 2, 3, 6

SC.F.2.4.1: Items 4, 5

Figure 19 External fertilization. Coral on the Great Barrier Reef, off the coast of Australia, release billions of gametes in a mass spawning that occurs once a year.

Figure 20 Sea turtle. This green sea turtle digs a nest in the sand, where she deposits her eggs.

Among animals that rarely meet members of their own species, hermaphrodites can have a reproductive advantage. A hermaphrodite's chances of reproductive success are increased because it can either fertilize eggs or have its own eggs fertilized. Many simple invertebrates, including slugs and earthworms, and some fishes are hermaphrodites.

When sperm are released during sexual reproduction, their moving flagella propel them toward the egg. Most aquatic animals simply release the male and female gametes near one another in the water, where fertilization occurs. This method is called **external fertilization** because the egg is fertilized outside of the female's body. Often large numbers of gametes are released during external fertilization, but only a small percentage of the resulting fertilized eggs will survive to develop into adults. **Figure 19** shows the mass release of gametes by coral polyps on a coral reef.

External fertilization is not practical for animals that live on land because gametes dry out quickly when exposed to air. Most terrestrial animals reproduce sexually by means of internal fertilization. In **internal fertilization,** the union of the sperm and egg occurs within the female's body. The male places semen *(SEE muhn)*, a fluid containing sperm and fluid secretions, directly into the female's body. In this way, fertilization takes place in a moist environment, and the gametes are protected from drying out.

Once fertilization occurs, developing eggs must be kept moist. This is not a problem for aquatic animals whose eggs are covered by a jellylike coat. The eggs of terrestrial animals need more protection. Their eggs have a shell that protects them from drying out and provides a degree of protection from physical damage. Some animals, such as the sea turtle shown in **Figure 20,** place their fertilized eggs in a safe place and leave them. Other animals remain with their eggs to protect them. In most mammals and a few other species, the eggs develop internally, and living young emerge from their mother's body.

Section 2 Review

1. **Summarize** the functions of the six body systems discussed. SC.F.1.4.2
2. **Describe** how a gastrovascular cavity differs from a one-way gut. SC.F.1.4.2
3. **Compare** open and closed circulatory systems. SC.F.1.4.2
4. **Describe** how asexual reproduction differs from sexual reproduction. SC.F.2.4.1
5. **Critical Thinking Justifying Conclusions** Which method of fertilization, external or internal, is more practical for most terrestrial animals? Justify your answer. SC.F.2.4.1
6. *FCAT Prep* Which two body systems in most animals are involved in taking up oxygen from the environment and transporting oxygen to body cells?
   A digestive and respiratory
   B respiratory and circulatory
   C circulatory and nervous
   D nervous and excretory
   SC.F.1.4.2

610

Answers to Section Review

1. The digestive system breaks down food; the respiratory system exchanges oxygen and carbon dioxide; the circulatory system transports nutrients and oxygen and removes wastes; the nervous system carries nerve impulses; the skeletal system provides a framework to support the body and enable movement; the excretory system removes cellular waste products.
2. A gastrovascular cavity has one opening to the exterior environment; a one-way gut has two.
3. Fluid leaves an open circulatory system and washes across the body cells. Fluid (blood) does not leave a closed circulatory system.
4. Asexual reproduction does not involve the fusion of two gametes, and the organism produced is an exact genetic copy of its single parent. Sexual reproduction involves the union of genetically different gametes.
5. For most terrestrial animals, internal fertilization is more practical. The sperm, egg, and zygote must remain moist to survive.
6. **A.** Incorrect. The digestive system is not involved. **B.** Correct. **C.** Incorrect. The nervous system is not involved. **D.** Incorrect. Neither system is involved in oxygen uptake or transport.

Study ZONE

Key Concepts

1 Characteristics of Animals

- All animals share these general features: heterotrophy, mobility, multicellularity, sexual reproduction, diploidy, the absence of a cell wall, cells organized as tissues, and blastula formation.
- Animals with radial symmetry have body parts arranged around a central axis. Animals with bilateral symmetry have a distinct right and left half, and most display cephalization.
- Animals have one of three basic body plans: acoelomate, pseudocoelomate, and coelomate.
- Segmentation in body structure underlies the organization of all advanced animals.
- There are about 35 animal phyla, which contain an extraordinary range of body forms and body systems.
- Scientists classify animals using several different types of data, which include comparing anatomy and physiology, patterns of development, and DNA.
- The animal kingdom is divided in two groups: vertebrates and invertebrates.

2 Animal Body Systems

- Body systems are specialized to carry out different tasks.
- Simple animals have a gastrovascular cavity with only one opening, while more-complex animals have a one-way gut.
- Simple animals exchange gases directly through their skin. More-complex aquatic animals use gills, while terrestrial animals use a variety of respiratory organs, such as lungs.
- In an open circulatory system, circulatory fluid leaves the vessels and enters the body cavity. In a closed circulatory system, blood remains in the vessels.
- While simple animals have little coordination among their nerve cells, complex animals have nerve cords and a brain with associated sensory structures.
- For most animals, eliminating wastes is linked to maintaining the correct water balance in their body.
- Asexual reproductive methods include fragmentation, splitting in two, and parthenogenesis. In sexual reproduction, male and female gametes combine to form a new individual.

Key Terms

Section 1

blastula (596)
ectoderm (596)
endoderm (596)
mesoderm (596)
body plan (598)
asymmetrical (598)
radial symmetry (598)
bilateral symmetry (598)
cephalization (599)
coelom (600)
acoelomate (600)
pseudocoelomate (600)
coelomate (600)
phylogenetic tree (602)

Section 2

gastrovascular cavity (605)
respiration (605)
gill (605)
open circulatory system (606)
closed circulatory system (606)
hydrostatic skeleton (607)
exoskeleton (607)
endoskeleton (608)
hermaphrodite (609)
external fertilization (610)
internal fertilization (610)

611

Alternative Assessment

Assign students to cooperative work groups. Have each group write questions and answers for certain pages of the book. Collect the questions and answers and use them to quiz the groups. You can keep students in the same groups or form new ones. Co-op Learning

Chapter Resource File

- Science Skills Worksheet GENERAL
- Critical Thinking Worksheet ADVANCED
- Test Prep Pretest GENERAL
- Chapter Test GENERAL

Answer to Concept Map

The following is one possible answer to Performance Zone item 15.

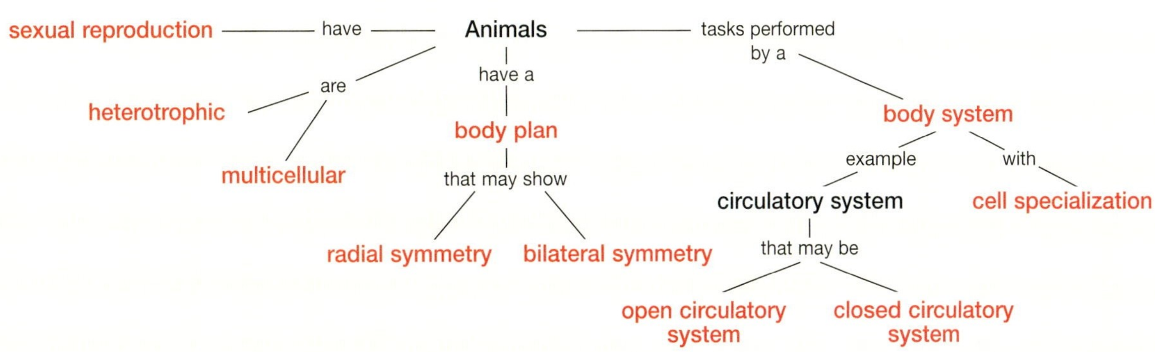

CHAPTER 27 ZONE

ANSWERS

Understanding Key Ideas

1. c
2. c
3. b
4. c
5. b
6. a
7. The kidneys filter waste from body fluids so that the waste can be excreted without excessive fluid loss.
8. Gastrulation is the process in which the blastula folds inward and cells move or change in size. It leads to the formation of three primary tissue (germ) layers. Each layer gives rise to cells that become a part of specific tissues and organs.
9. All of the cells in the body of most animals are diploid (have two sets of chromosomes) except for the sex cells (gametes), which are haploid (have one set of chromosomes).
10. A possible answer to the concept map is found at the bottom of the Study Zone page.

Critical Thinking

11. Radial symmetry is beneficial for sessile animals or animals that float freely in water because food may come from any direction. Bilateral symmetry is better suited to terrestrial animals that must forage or hunt for food.
12. Cephalization is the concentration of sensory organs in the front end of an animal. Terrestrial animals tend to move through their environment head first. Having more sensory organs in the front end permits them to seek food and sense that they may be approaching danger.

Chapter Review Correlations

SC.F.1.4.2: Items 3, 5-8, 10, 11, 12
SC.F.2.4.2: Item 9
SC.G.1.4.1: Items 1, 2, 4
SC.G.2.4.3: Item 13
SC.H.1.4.7: Item 14
SC.H.3.4.3: Item 15

Performance ZONE

Understanding Key Ideas

1. Which of the following organisms do *not* have cells organized into tissues?
   a. hydras
   b. sea urchins
   c. sponges
   d. sea anemones
   SC.G.1.4.1
2. An animal's body plan includes all of the following except
   a. internal organization.
   b. shape.
   c. size.
   d. symmetry.
   SC.G.1.4.1
3. The presence of a true body cavity (coelom) allows
   a. direct exchange of oxygen and carbon dioxide with the environment.
   b. specialization of the gut.
   c. cephalization.
   d. bilateral symmetry.
   SC.F.1.4.2
4. Which of the following does *not* have nerve cells?
   a. hydra
   b. jellyfish
   c. sponge
   d. flatworm
   SC.G.1.4.1
5. Skeletal systems provide all of the following except
   a. protection for an animal's soft parts.
   b. absorption of nutrients from food.
   c. a framework for supporting the body.
   d. a framework for muscles to pull against.
   SC.F.1.4.2
6. Which of the following is *not* true of kidneys?
   a. They remove oxygen from the body.
   b. They remove wastes from the bloodstream.
   c. They produce concentrated urine.
   d. They help balance the body's water content.
   SC.F.1.4.2
7. How do kidneys function to reduce water loss in terrestrial animals? SC.F.1.4.2
8. **Exploring Further** Summarize the process of gastrulation, relating it to the formation of tissue layers and the development of tissues and organs. SC.F.1.4.2
9. Explain how the terms *diploid* and *haploid* apply to what you have learned in this chapter. (**Hint:** See Chapter 6, Section 1.) SC.F.2.4.2
10. **Concept Mapping** Construct a concept map that describes the characteristics of animals. Try to include the following terms in your concept map: *heterotrophic, multicellular, cell specialization, body plan, radial symmetry, bilateral symmetry, body system, open circulatory system, closed circulatory system,* and *sexual reproduction.* SC.F.1.4.2

Critical Thinking

11. **Forming Reasoned Opinions** Radially symmetric animals, such as hydras, are not found on land. However, bilaterally symmetrical animals live on land and in water. Propose a hypothesis to explain why radially symmetrical animals are best suited to aquatic life. SC.F.1.4.2
12. **Evaluating Hypotheses** Defend the position that cephalization gives terrestrial animals an advantage as they seek food. SC.F.1.4.2
13. **Justifying Conclusions** Support the argument that segmentation makes an animal more evolutionarily flexible. SC.G.2.4.3

Alternative Assessment

14. **Career Connection** Zookeeper Zookeepers are responsible for the day-to-day care of animals in their charge. Research a zookeeper's responsibilities, and write a report that includes a job description, training required, kinds of employers, growth prospects, and starting salary. SC.H.1.4.7
15. **Finding and Communicating Information** To learn what role public aquariums play in educating the public about aquatic life, develop a presentation designed to attract additional donations. SC.H.3.4.3

612

13. Segmentation is the division of the body into repeating units. Segmentation offers evolutionary flexibility because a small change in one segment can produce a new type of segment.

Alternative Assessment

14. Zookeepers feed animals, clean their cages, and are responsible for bathing and exercising some animals. Often a zookeeper will specialize in caring for certain types of animals, such as big cats or hoofed mammals. The zookeeper understands the behavior of the animals and is usually the first to notice any sign of sickness. Training includes a four-year college degree with training in animal science, zoology, marine biology, conservation biology, wildlife management, or animal behavior preferred. Employers include zoos and aquariums. Growth prospects are fair. Starting salary will vary by region.
15. Answers will vary. Accept any answers that show a thoughtful consideration of the use of funds.

Assignment Guide

Section	Questions
1	2, 3, 8, 9, 11, 12, 13
2	1, 4, 5, 6, 7, 9, 10, 14, 15

FCAT Prep

For questions 1–6, write your answers on a separate sheet of paper.

1. What is a defining characteristic of heterotrophic organisms?
   - **A.** They are capable of flight.
   - **B.** They obtain food from their environment.
   - **C.** They remain in one place throughout their lifetime.
   - **D.** They make their own food from inorganic materials.

   SC.G.1.4.1

2. Which of the following cells in animals has two copies of each chromosome? SC.F.2.4.2
   - **F.** a gamete
   - **G.** a germ cell
   - **H.** a diploid cell
   - **I.** a haploid cell

3. How does the excretory system of an animal help maintain homeostasis?
   - **A.** It provides a framework of support and is vital to movement.
   - **B.** It helps break down food that the animal eats for energy.
   - **C.** It helps maintain the balance of salts and water in the body.
   - **D.** It helps maintain the balance of oxygen and carbon dioxide gases in the blood.

   SC.F.1.4.2

4. Where does gas exchange take place in many aquatic phyla?
   - **F.** gills
   - **G.** gut
   - **H.** kidneys
   - **I.** lungs

   SC.F.1.4.2

Base your answers to questions 5 and 6 on the diagram below. The letters *A–D* represent milestones in the evolution of animals.

PHYLOGENETIC TREE OF ANIMAL PHYLA

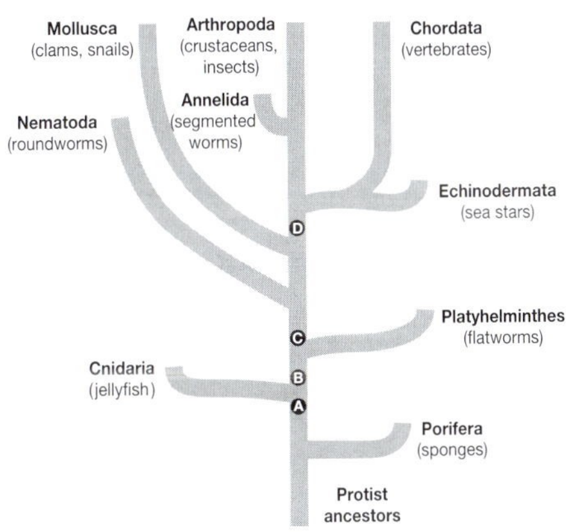

5. Which letter represents the presence of tissues?
   - **A.** *A*
   - **B.** *B*
   - **C.** *C*
   - **D.** *D*

   SC.D.1.4.3

6. How many phyla branch off after the point that represents the development of a body cavity? SC.D.1.4.3

Test TIP

When using an illustration that has labels to answer a question, read the labels carefully, and then check that the answer you choose matches your interpretation of the labels.

613

FCAT Doctor

Question 3 Answer C is the correct choice. Homeostasis is the maintenance of a stable internal environment. The excretory system keeps the concentration of salts in the body stable. Answer A is incorrect because this is the job of the skeletal system. Answer B is incorrect because this is the job of the digestive system. Answer D is incorrect because this is the job of the respiratory system.

Question 5 Answer A is the correct choice. Protists and sponges, which are located before the development of tissues, do not have tissues with specialized cells. Answer B is incorrect because step *B* represents the development of bilateral symmetry. Answer C is incorrect because step *C* represents the development of a body cavity. Answer *D* is incorrect because step D represents the development of segmentation.

Question 6 The correct answer is 6. The development of a body cavity is represented by point C on the phylogenetic tree. There are 6 phyla that branch off after that point: Nematoda, Mollusca, Echinodermata, Chordata, Annelida, and Arthropoda.

Answers

1. B
2. H
3. C
4. F
5. A
6. 6

Test Prep Correlations

SC.D.1.4.3: Items 5, 6
SC.F.1.4.2: Items 3, 4
SC.F.2.4.2: Item 2
SC.G.1.4.1: Item 1

Skills Practice Lab

Surveying Invertebrate Diversity

Teacher's Notes

Time Required
Two 45-minute periods

Ratings
EASY → HARD

Teacher Preparation 1
Student Setup 1
Concept Level 2
Cleanup 1

Skills Acquired
- Classifying
- Communicating
- Identifying/Recognizing Patterns
- Inferring
- Interpreting

Scientific Methods
In this lab, students will:
- Make Observations
- Ask Questions
- Draw Conclusions
- Communicate the Results

Materials and Equipment
Materials for this lab can be ordered from WARD'S. See *Master Materials List* at the front of this book for catalog numbers.

Safety Cautions
Caution students to use extreme care when observing the preserved specimens. The preservative can leak if the jars are tilted. Since some animals may be preserved in glass jars, students should be careful not to break the jars. Tell students to wash their hands if they come into contact with the preservative. Also remind students to wash their hands before leaving the laboratory.

Skills Practice Lab

Surveying Invertebrate Diversity

SC.G.1.4.1

SKILLS
- Observing
- Comparing

OBJECTIVES
- **Observe** the similarities and differences among groups of invertebrates.
- **Relate** the structural adaptations of invertebrates to their evolution.

MATERIALS
- safety goggles
- lab apron
- preserved or living specimens of invertebrates
- prepared slides of sponges, hydras, planarians, and nematodes
- compound microscopes
- hand lenses or stereomicroscopes
- probes

ChemSafety

CAUTION: Always wear safety goggles and a lab apron to protect your eyes and clothing.

CAUTION: Do not touch or taste any chemicals. Know the location of the emergency shower and eyewash station and how to use them. If you get a chemical on your skin or clothing, wash it off at the sink while calling to the teacher. Notify the teacher of a spill. Spills should be cleaned up promptly, according to your teacher's directions.

CAUTION: Glassware is fragile. Notify the teacher of broken glass or cuts. Do not clean up broken glass or spills with broken glass unless the teacher tells you to do so.

Before You Begin

Invertebrates include all animals except those with backbones. Every phylum of the kingdom Animalia except the phylum Chordata consists only of invertebrates. In this lab, you will examine representatives of eight phyla of animals. You will see many similarities and differences in **body plan**—shape, symmetry, and internal organization.

1. Write a definition for each boldface term in the paragraph above and for the following terms: radial symmetry, bilateral symmetry, dorsal, ventral, anterior, posterior, cephalization, segmentation.
2. Describe the three basic body plans found in animals.
3. Make a data table similar to the one on the right to record observations about invertebrates.
4. Based on the objectives for this lab, write a question you would like to explore about the characteristics of invertebrates.

DATA TABLE

Phylum	Symmetry	Body plan	Other	Examples

Procedure

PART A: Conducting a Survey

1. Put on safety goggles and a lab apron.
2. Visit each invertebrate station, and examine the specimens there. Answer the questions, and record observations in your data table.

614

Tips and Tricks
Have students read the lab before they come to class. Before class, have students make their data table and an answer sheet for the questions at each station. This will make it easier for students to answer the questions, because not all students will start with Station 1. Have materials on the lab tables when students arrive. Students should work in groups of two to four and move clockwise from station to station.

Answers to Before You Begin
1. *invertebrates*—animals without backbones; *body plan*—the shape, symmetry, and internal organization of the body; *radial symmetry*—symmetry in all directions about a central axis; *bilateral symmetry*—two-sided symmetry; *dorsal*—the top side; *ventral*—the bottom side; *anterior*—the front end; *posterior*—the rear end; *cephalization*—the anterior concentration of sensory structures and nerves; *segmentation*—the division of the body into repeating, similar units
2. An acoelomate has no body cavity. A pseudocoelomate has a body cavity between the mesoderm and the endoderm. A coelomate has a body cavity in the mesoderm.
4. Students' questions will vary. For example: What characteristics are shared by all invertebrates?

3. **Sponges** Examine each specimen.

a. Describe the shape of a sponge.
b. What do you think is the role of the many holes, or pores, in a sponge?
c. Examine a prepared slide of a sponge with a compound microscope. What do you notice about the organization of the cells in sponges?

4. **Cnidarians** Examine each specimen.

a. Divide the cnidarian specimens into two groups. What feature did you use to make your division?
b. How many body openings does a cnidarian have?
c. Examine a prepared slide of a hydra. What do you notice about the organization of the cells in cnidarians?

5. **Flatworms and Roundworms** Examine each specimen.

a. How does a flatworm differ from a roundworm in external appearance?
b. Do any of the worms appear to be segmented? Explain.
c. Examine prepared slides of planarians and nematodes. How many body openings does each have?

6. **Mollusks** Examine each specimen.

a. In what ways do the mollusks differ in external appearance?
b. Which group of mollusks has the most noticeable "feet"?

7. **Annelids** Examine each specimen.

a. How are an earthworm and a leech similar? How are they different?
b. Describe any differences you see in the segments of the annelid worm.

8. **Arthropods** Examine each specimen.

a. What characteristic do you observe in all arthropod appendages?
b. How does the number of walking legs differ among these arthropods?

9. **Echinoderms** Examine each specimen.

a. The word *echinoderm* means "spiny skin." Why is this name appropriate?
b. What does an echinoderm's ventral surface look like?

PART B: Cleanup and Disposal

10. Dispose of broken glass in the designated waste containers. Do not put lab materials in the trash unless your teacher tells you to do so.

11. Wash your hands thoroughly before you leave the lab and after you finish all work.

Analyze and Conclude

1. **Summarizing Data** Which animal phyla show cephalization, and which do not?
2. **Recognizing Patterns** What type of symmetry is found with cephalization?
3. **Recognizing Patterns** What characteristics do annelids and arthropods share?
4. **Analyzing Methods** Were you able to identify the type of body plan found in all of the specimens? Explain.
5. **Further Inquiry** Write a new question about invertebrates that could be explored with another investigation.

On the Job

Parasitology is the study of parasites. Do research to discover how parasitologists help protect you and your pets from diseases caused by parasites. For more about careers, visit **go.hrw.com** and type in the keyword **HX4 Careers.**

Answers to Procedure

3. **A.** Sponges have an irregular shape; they are asymmetrical. **B.** The pores allow water, which carries in food particles and oxygen and carries out wastes, to flow through the sponge. **C.** The cells in a sponge are of several types that are loosely arranged in several layers but are not organized into tissues.
4. **A.** body shape; There are polyps and medusas. **B.** one **C.** The cells are of several types and are in two tissue layers.
5. **A.** Flatworms are flat and vary in length. Roundworms have long, tubular bodies. **B.** Students may respond "yes" if they view a tapeworm because its body looks somewhat segmented. Explain that only coelomates show true segmentation. Tapeworms do have a series of similar sections, but this is different from true segmentation. **C.** planarians have one; nematodes have two.
6. **A.** Mollusks may have a two-part shell, single shell, or no shell. **B.** Gastropods—the foot is a large muscular structure.
7. **A.** Both have bilateral symmetry and a body divided into many segments. Earthworms are tubular and much longer than they are wide; leeches are flattened and not nearly as long. **B.** Segments at the anterior and posterior ends are smaller than those in the middle.
8. **A.** They are jointed. **B.** Millipedes have two pairs of legs per body segment, centipedes have one pair of legs per body segment, insects have a total of three pairs, spiders have four, and decapod crustaceans have five.
9. **A.** Most echinoderms are covered by numerous spines that grow out of the skin. **B.** The ventral surface has several grooves in which there are numerous tube feet.

Answers to Analyze and Conclude

1. All but the sponges, cnidarians, and echinoderms exhibit cephalization.
2. bilateral symmetry
3. segmentation, bilateral symmetry, and cephalization
4. No, body plans cannot be determined by examining an animal's exterior.
5. Answers will vary. For example: How does an arthropod's outer covering differ from a worm's?

Answer to On the Job

Answers will vary. Parisitologists study the life cycles of parasites and search for medicines to treat diseases caused by parasites.

Chapter Opener

- **SE** **Quick Review,** p. 617 ■
- **SE** **Reading Activity,** p. 617 ■
- **TE** **Using the Figure,** p. 616 GENERAL
- **TE** **Opening Activity,** p. 617 GENERAL

& 3 • 90 min pp. 618–621

Section 1
Sponges

- **OSP** **Lesson Plan***
- **TR** **Bellringer***
- **TR** **Section Outline***
- **TR** **H10 Structure of a Sponge***
- **TR** **H11 Sexual Reproduction in Sponges***

- **TE** **Demonstration,** p. 618 BASIC
- **TE** **Using the Figure,** p. 619 GENERAL
- **TE** **Group Activity** Gemmule Survival Pods, p. 621 ADVANCED

& 5 • 90 min pp. 622–628

Section 2
Cnidarians

- **OSP** **Lesson Plan***
- **TR** **Bellringer***
- **TR** **Section Outline***
- **TR** **H12 Two Cnidarian Body Forms***
- **TR** **H13 Cnidarian Body Plan***
- **TR** **H15 Sexual Reproduction in *Obelia****

- **TE** **Using the Figure,** p. 623 ADVANCED
- **SE** **Math Lab** Estimating Size Using a Microscope, p. 623 ■
- **CRF** **Datasheets for In-Text Labs** Estimating Size Using a Microscope*
- **TE** **Using the Figure,** p. 624 BASIC
- **TE** **Using the Figure,** p. 625 GENERAL
- **TE** **Using the Figure,** p. 626 GENERAL
- **TE** **Demonstration,** p. 627 ◆ BASIC
- **TE** **Using the Figure,** p. 627 GENERAL

& 7 • 90 min pp. 629–634

Section 3
Flatworms and Roundworms

- **OSP** **Lesson Plan***
- **TR** **Bellringer***
- **TR** **Section Outline***
- **TR** **H19 Planarian***

- **TE** **Activity** Life of a Planarian p. 630 GENERAL
- **SE** **Quick Lab** Observing Planarian Behavior p. 631 ■
- **CRF** **Datasheets for In-Text Labs** Observing Planarian Behavior*
- **TE** **Using the Figure,** p. 631 GENERAL
- **TE** **Activity** Flukes p. 632 GENERAL
- **TE** **Using the Figure,** p. 632 ADVANCED
- **TE** **Group Activity** Create a Life Cycle p. 633 GENERAL
- **SE** **Data Lab** Identifying Parasites p. 634 ■
- **CRF** **Datasheets for In-Text Labs** Identifying Parasites*
- **CRF** **Quick Lab** Preventing Parasites* BASIC
- **CRF** **Skills Practice** What Does a Planarian Prefer?* GENERAL
- **SE** **Exploration Lab** Observing Hydra Behavior, pp. 638–639 ■
- **CRF** **Datasheets for In-Text Labs** Observing Hydra Behavior*

& 9 • 90 min

Chapter Review and Assessment Resources

- **SE** **Study Zone,** p. 635 ■
- **SE** **Performance Zone,** p. 636 ■
- **SE** **FCAT Test Prep,** p. 637 ■
- **TR** **Concept Mapping***
- **CRF** **Vocabulary Review Worksheet*** ■ GENERAL
- **CRF** **Test Prep Pretest*** ■ GENERAL
- **CRF** **Chapter Test*** ■ GENERAL
- **CRF** **Chapter Test*** ADVANCED
- **OSP** **Test Generator***
- **OSP** **Test Item Listing***
- **OSP** **Modified Chapter Test*** BASIC

Online and Technology Resources

go.hrw.com Holt Online Learning

Visit go.hrw.com for access to Holt Online Learning or enter the keyword **HX6 Home** for a variety of free online resources.

One-Stop Planner® CD-ROM

This CD-ROM package includes
- Lab Materials QuickList Software
- Holt Calendar Planner
- Customizable Lesson Plans
- Printable Worksheets
- ExamView® Test Generator
- Interactive Teacher Edition
- Holt PuzzlePro® Resources
- Holt PowerPoint® Resources

Guided Reading Audio CDs

These CDs are designed to help auditory learners and reluctant readers. (Audio CDs are also available in Spanish.)

Compression guide:
To shorten your instruction because of time limitations, omit blocks 1 and 8.

KEY

TE Teacher Edition
SE Student Edition
OSP One-Stop Planner
CRF Chapter Resource File
TR Teaching Transparency
* Also on One-Stop Planner
■ Also Available in Spanish
◆ Requires Advance Prep

SKILLS DEVELOPMENT AND PRACTICE	SECTION REVIEW AND ASSESSMENT	CORRELATIONS
		Florida Sunshine Standards/FCAT
TE **Skill Builder** Vocabulary p. 619 BASIC TE **Reading Skill Builder** Paired Summarizing, p. 620 BASIC CRF **Directed Reading Worksheet*** BASIC CRF **Active Reading Worksheet*** GENERAL	SE **Section 1 Review** ■ TE **Reteaching,** p. 621 BASIC TE **Quiz,** p. 621 GENERAL TE **Alternative Assessment,** p. 621 GENERAL CRF **Quiz*** ■ BASIC	SC.F.1.4.2, SC.F.2.4.1, SC.G.1.4.1
TE **Reading Skill Builder** Interactive Reading, p. 623 BASIC TE **Reading Skill Builder** Discussion, p. 626 GENERAL CRF **Directed Reading Worksheet*** BASIC CRF **Active Reading Worksheet*** GENERAL	SE **Section 2 Review** ■ TE **Reteaching,** p. 628 BASIC TE **Quiz,** p. 628 GENERAL CRF **Quiz*** ■ BASIC	SC.F.1.4.2, SC.F.2.4.1, SC.G.1.4.1, SC.H.1.4.1
CRF **Directed Reading Worksheet*** BASIC CRF **Active Reading Worksheet*** GENERAL CRF **Critical Thinking Worksheet*** ADVANCED CRF **Science Skills Worksheet*** GENERAL	SE **Section 3 Review** ■ TE **Reteaching,** p. 634 BASIC TE **Quiz,** p. 634 GENERAL CRF **Quiz*** ■ BASIC **Modified Worksheet*** BASIC	SC.F.1.4.1, SC.F.1.4.2, SC.F.1.4.7, SC.F.2.4.1, SC.G.1.4.1

SCiLINKS NSTA
www.scilinks.org
Maintained by the **National Science Teachers Association**

Topic: Cnidarians
SciLinks Code: HX4048

Topic: Coral Reefs
SciLinks Code: HX4051

Topic: Flukes
SciLinks Code: HX4084

Topic: Roundworms
SciLinks Code: HX4159

Topic: Hydra
SciLinks Code: HX4102

CHAPTER 28

Overview

Tell students the purpose of this chapter is to introduce them to invertebrate animals. Although the simple invertebrates introduced in this chapter are not as complex as many other animals, they are more complex than their name implies. Simple invertebrates have evolved to occupy a broad array of lifestyles, and their body plans and behaviors reflect this evolution.

Using the Figure

Ask students to look at the photograph of the azure vase sponge pictured. Ask them to think of various uses for sponges. (cleaning, mopping, washing dishes) Tell them that ancient Greeks used sponges in their helmets for padding, and the Romans used them for paintbrushes and mops. While people today often used manufactured, or synthetic sponges, natural sponges are still the best choice for many tasks including applying make-up and painting.

Assessing Prior Knowledge

Before beginning this chapter, be sure your students can answer the questions in the ***Quick Review*** on the facing page. If they cannot, encourage them to reread the sections indicated.

Chapter Resource File

- Vocabulary Worksheets
- Concept Mapping

Azure vase sponge

616

Chapter Correlations | **Florida Sunshine Standards/FCAT**

SC.F.1.4.2 Body structures are uniquely designed and adapted for their function.

SC.F.2.4.1 There are mechanisms of asexual and sexual reproduction and different genetic advantages and disadvantages of asexual and sexual reproduction.

SC.G.1.4.1 There is great diversity and interdependence among living things.

CHAPTER

28 Simple Invertebrates

Florida Sunshine Standards/FCAT

SC.F.1.4.2 Body structures are uniquely designed and adapted for their function.

SC.F.2.4.1 There are mechanisms of asexual and sexual reproduction and different genetic advantages and disadvantages of asexual and sexual reproduction.

SC.G.1.4.1 There is great diversity and interdependence among living things.

Quick Review

Answer the following without referring to earlier sections of your book.

1. **Describe** the process of osmosis. *(Chapter 4, Section 1)*
2. **Identify** the organisms collectively called plankton. *(Chapter 17, Section 3 and Chapter 21, Section 2)*
3. **Define** the term *colonial organization*. *(Chapter 19, Section 1)*
4. **Distinguish** between ectoderm, mesoderm, and endoderm. *(Chapter 27, Section 1)*
5. **Distinguish** between acoelomates, pseudocoelomates, and coelomates. *(Chapter 27, Section 1)*

Did you have difficulty? *For help, review the sections indicated.*

Reading Activity

Before you read this chapter, write a short list of all the things you know about the members of each of the following phyla: sponges, cnidarians, flatworms, and roundworms. Then write a list of the things that you want to know about these simple invertebrates. Save your list, and to assess what you have learned, see how many questions you can answer after reading this chapter.

Since ancient times, people have used sponges for a variety of functions. The ancient Greeks placed sponges in their helmets for padding, and the Romans used them for paintbrushes and mops. The beautiful azure vase sponge ranges in color from intense blue to pinkish-purple.

Looking Ahead

internet connect

www.scilinks.org

National Science Teachers Association *sci*LINKS Internet resources are located throughout this chapter.

SCILINKS. Maintained by the National Science Teachers Association

617

Opening Activity — GENERAL

Sponges Have several natural sponge skeletons and several synthetic sponges available for students to examine. Emphasize that observation is an important part of scientific inquiry. Ask students to make a list of how the natural and synthetic sponges are alike and how they are different. (For example, natural sponges are asymmetrical, while synthetic sponges are symmetrical.) **LS Visual**

Quick Review Answers

1. Osmosis is the movement of water through a membrane from an area of high concentration to an area of lower concentration.
2. Plankton is a biological community drifting freely in the upper waters of the ocean and consisting mostly of microscopic organisms.
3. *Colonial organization* describes a collection of cells that are permanently associated but in which little or no integration of cell activities occurs.
4. In animals, ectoderm is the outer layer of embryonic tissue from which the skin and nervous system develop. The mesoderm is the middle layer from which the skeleton and muscles develop, and endoderm is the inner layer from which the digestive organs develop.
5. Acoelomates are animals that lack a coelom (body cavity), pseudocoelomates are animals that have a body cavity located between the endoderm and the mesoderm, and coelomates are animals that have a body cavity located entirely within the mesoderm.

Reading Activity Answers

Students may know that many of these organisms are aquatic and some are parasites. They may want to know how humans may be affected by parasitic worms or about the stinging mechanism in jellyfish.

Section 1

Focus

Overview

Before beginning this section review with your students the objectives listed in the Student Edition. This section explains the characteristics and life history of sponges. Students will learn how sponges receive nutrients, reproduce, and how their body is structurally supported.

Bellringer

Challenge students to think about different structures for different lifestyles. Ask them to describe on paper how an animal that cannot move is able to feed and protect itself. Have them imagine that they are stuck on the floor and cannot ever move again. How will they get food? How will they protect themselves from large mobile predators? (Answers may vary. Some students might say that they would just wait until food came close enough, and then they would reach out and grab it. For protection they might curl up into a ball if a predator came too close.)

Motivate

Demonstration — BASIC

Pour a mixture of pebbles and water through a spaghetti strainer, trapping the pebbles inside the strainer. Show the pebbles to your students, and tell them to imagine that the pebbles are food particles. Explain that the collars of the choanocytes that line the inside of the sponge trap food particles in water in a similar manner.

Section Correlations

SC.F.1.4.2
SC.F.2.4.1
SC.G.1.4.1

Section 1 Sponges

Florida Sunshine Standards/FCAT

SC.F.1.4.2 Body structures are uniquely designed and adapted for their function.

SC.F.2.4.1 There are mechanisms of asexual and sexual reproduction and different genetic advantages and disadvantages of asexual and sexual reproduction.

SC.G.1.4.1 There is great diversity and interdependence among living things.

Objectives

- **Summarize** the general features of sponges. SC.G.1.4.1
- **Describe** how sponge cells receive nutrients. SC.F.1.4.2
- **Describe** how a sponge's body is structurally supported. SC.F.1.4.2
- **Distinguish** between sexual and asexual reproduction in sponges. SC.F.2.4.1

Key Terms

ostia
oscula
sessile
choanocyte
amoebocyte
spongin
spicule
gemmule

The Simplest Animals

Sponges are so unlike other animals that early naturalists classified them as plants. It wasn't until the late 1700s that scientists using improved microscope technology began studying sponges closely. Scientists then realized that sponges are animals. The bodies of most sponges completely lack symmetry and consist of little more than masses of specialized cells embedded in a gel-like substance called mesohyl *(MEHZ oh hil)*. You could say that a sponge's body is somewhat like chopped fruit in gelatin. The chopped fruit represents the specialized cells, and the gelatin represents the mesohyl.

Sponge cells are not organized into tissues and organs. However, they do have a key property of all animal cells—cell recognition. A simple lab experiment can demonstrate that sponge cells can recognize other sponge cells. A living sponge can be passed through a fine silk mesh, causing the individual cells to separate. On the other side of the mesh, the individual sponge cells will recombine to form a new sponge.

Sponges have a body wall penetrated by tiny openings, or pores, called **ostia** *(AHS tee uh)*, through which water enters. The name of the phylum, Porifera, refers to this system of pores. Sponges also have larger openings, or **oscula,** through which water exits. You can see the many oscula of the sponge shown in **Figure 1.** Sponges are also **sessile** *(SEHS uhl)*. Early in their lives, sponges attach themselves firmly to the sea bottom or some other submerged surface, like a rock or coral reef. They remain there for life. Sponges can have a diameter as small as 1 cm (0.4 in.) or as large as 2 m (6.6 ft).

Figure 1 Sponge. The small openings in this sponge's body are ostia. The larger openings are oscula.

Most sponges are bag-shaped and have a large internal cavity. One or more oscula (singular, osculum) are located in the top of

Evolutionary Milestone

1 Multicellularity

The bodies of all animals, including sponges (phylum Porifera), are multicellular—made of many cells. Although the sponge is composed of several different cell types, these cells show only a small degree of coordination with each other.

Chapter Resource File

- Directed Reading BASIC
- Active Reading GENERAL

Transparencies

TR Bellringer

One-Stop Planner CD-ROM

- Reading Organizers BASIC
- Reading Strategies BASIC

Figure 2 Sponge interior

Water enters the sponge through many small pores (ostia) in its body wall and exits through the osculum, an opening at the top of the sponge.

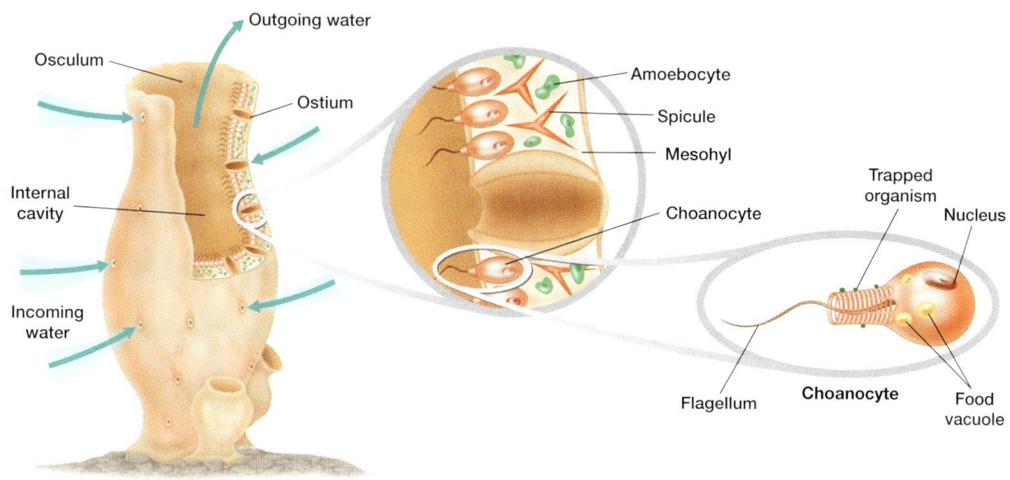

the body wall, as shown in **Figure 2.** Lining the internal cavity of a sponge is a layer of flagellated cells called **choanocytes** *(koh AN oh siets),* or collar cells. The flagella of these cells extend into the body cavity. As the flagella beat, water is drawn in through the pores in the body wall. The water is driven through the body cavity before it exits through the osculum.

As sea water passes through the sponge's body cavity, the collar cells function as sieves. These cells trap plankton and other tiny organisms in the small hairlike projections on the collar. The trapped organisms are then pulled into the interior of the collar cells, where they are digested intracellularly (within the cell). As sea water leaves the sponge, wastes are carried away in it.

How do the other sponge cells, such as those in the body wall, survive if the collar cells take in all of the food? The collar cells release nutrients into the mesohyl where other specialized cells, called *amoebocytes (uh MEE boh siets),* pick up the nutrients. **Amoebocytes** are sponge cells that have irregular amoeba-like shapes. They move about the mesohyl, supplying the rest of the sponge's cells with nutrients and carrying away their wastes.

Protistan Ancestors

The choanocytes of sponges very closely resemble a kind of protist called a choanoflagellate, shown in **Figure 3.** Ancient choanoflagellates are thought by many scientists to be the ancestors of sponges. Other free-swimming colonial flagellates closely resemble sponge larvae, however, and some scientists believe organisms similar to these other flagellates were the true ancestors of sponges.

Figure 3 **Choanoflagellate.** Ancient choanoflagellates similar to the one shown above may be the ancestors of sponges.

619

Teach

SKILL BUILDER — Basic

Vocabulary Tell students that the term Porifera literally means "pore bearing". Ask students to find the derivatives of other vocabulary words in this section. English Language Learners **LS Verbal**

Using the Figure — General

Work with students to help them understand the relationship of the two close-ups to the whole sponge shown in **Figure 2.** Point out that in the close-up on the left, the flagella of the collar cells are extending into the internal cavity of the sponge. The close-up on the right shows one collar cell. **LS Visual**

INCLUSION Strategies — General

- *Attention Deficit Disorder*
- *Gifted and Talented*

Have students write log entries from an "Internet Deep-sea Dive." Entries in their log should include the identification of different examples of invertebrates seen on their dives, the classes and scientific names of each of the examples, and the characteristics that make each of the examples invertebrates. Additionally, students may include locales of the dives and equipment required to make these deep-sea dives.

did you know?

Carnivorous Sponges In 1994, scientists discovered a carnivorous sponge living in a marine cave near France. The previously unknown species captures small crustaceans and other animals. It traps its prey using filaments covered with spikes of silica. Small animals become trapped in the filaments. Then certain cells of the sponge migrate to the prey, envelope it, and slowly digest it.

MATH CONNECTION

Tell students that a sponge that is 10 cm tall and 1 cm in diameter pumps 22.5 L of water through its body every day. To give students an idea of how much water this is, show them a 3 L bottle. Ask students to calculate the number of bottles it would take to hold the amount of water circulated by a sponge in one day. (22.5 L ÷ 3 L/bottle = 7.5 bottles) Then ask students to calculate the number of bottles it would take to hold the amount of water that a sponge circulates in a year. (22.5 L × 365 days/year = 8,212.5 L per year; 8,212.5 L ÷ 3 L/bottle = 2,737.5 bottles) **LS Logical**

Transparencies

TR H10 Structure of a Sponge

Teach, *continued*

Real Life

Answer

Luffa sponges have a more symmetrical structure, exhibiting an almost radial symmetry, and are cylindrical. True sponges, however, are usually asymmetrical and globular.

READING SKILL BUILDER — BASIC

Paired Summarizing Have students read the information on sponge skeletons to themselves. Assign partners, and have one student summarize aloud what has been read while the other student listens. The listener should point out inaccuracies and add ideas that the other student left out. Have students alternate roles and repeat this process. Students should work together during a final clarification process. You may want to pair ELL students with native English speakers. **LS Verbal** English Language Learners

Teaching Tip — GENERAL

Charting Invertebrates Have students make a Graphic Organizer similar to the one at the bottom of this page. Tell them to only fill in the empty cells in the row labeled *Sponges*. Tell students that they will fill in the other rows as they learn about the simple invertebrates in Sections 2 and 3. **LS Visual**

Section Review Correlations

SC.F.1.4.2: Items 1, 2, 5, 6, 7

SC.F.2.4.1: Item 4

SC.G.1.4.1: Item 3

Real Life

What is a luffa sponge?
A luffa sponge really isn't a sponge at all but a gourd. When dried, the fibrous material found in the gourd forms a "skeleton" similar to that of some sponges, and it can be used for many of the same purposes.

Comparing Structures
Obtain a natural sponge and a luffa sponge, and compare the nature of their "skeletons."

Sponge Diversity

As any snorkeler can tell you, brilliantly colored sponges abound in warm, shallow sea waters. Other marine sponges live at great depths, and a few species even live in fresh water. Rather than being a simple baglike shape, the body wall of some sponges, such as the azure vase sponge on the first page of this chapter, may contain hundreds of folds that are sometimes visible as fingerlike projections. These folds increase a sponge's size and surface area.

Sponge Skeletons

To prevent the sponge from collapsing in on itself, the sponge body is supported by a skeleton. A sponge's skeleton, however, does not have a fixed framework like a human skeleton does. Instead, the skeletons of most sponges are composed of spicules. A **spicule** is a tiny needle composed of silica or calcium carbonate. A few sponges have skeletons composed of a resilient, flexible protein fiber called **spongin.** Some sponges contain both spongin and spicules. These supporting structures are found throughout the mesohyl.

Taxonomists group sponges into three types based on the composition of their skeletons. Calcareous sponges have spicules composed of calcium carbonate. Glass sponges and demosponges have spicules made of silica. Some demosponges also contain spongin. In some species the spongin is reinforced with spicules of silica. The three classes of sponges are shown in **Figure 4.**

Figure 4 Three types of sponges

Sponges have skeletons made of spicules, spongin, or both.

620

Graphic Organizer

Use this graphic organizer with **Teaching Tip** on this page.

Animal group	Lifestyle	Structures/functions
Sponges	Sessile filter feeders	Amoebocytes / carry food and wastes Spicules / support and protection Spongin / support
Cnidarians		
Flatworms		
Roundworms		

Reproduction

Sponges can reproduce asexually. A remarkable property of sponges is that they regenerate when they are cut into pieces. Each bit of sponge, however small, will grow into a complete new sponge. As you might suspect, sponges frequently reproduce by shedding fragments, each of which develops into a new individual. Sponges also reproduce by budding. A third form of asexual reproduction occurs in some freshwater sponges. When living conditions become harsh (cold or very dry), some freshwater sponges form **gemmules** *(JEHM yools)*, clusters of amoebocytes encased in protective coats. Sealed in with ample food, the cells survive even if the rest of the sponge dies. When conditions improve, the cells grow into a new sponge.

Sexual reproduction is also common among sponges. Most sponges are hermaphrodites, meaning they produce both eggs and sperm. Since eggs and sperm are produced at different times, self-fertilization is avoided. In most species of sponges, sperm cells from one sponge enter another sponge through its pores, as shown in **Figure 5.** Collar cells on the receiving sponge's interior pass the sperm into the mesohyl, where the egg cells reside, and fertilization occurs. The fertilized eggs develop into larvae and leave the sponge. After a brief free-swimming stage, the larvae attach themselves to an object and develop into new sponges.

Figure 5 **Sexual reproduction in sponges**

In most species of sponges, sperm from one sponge fertilize eggs from another sponge.

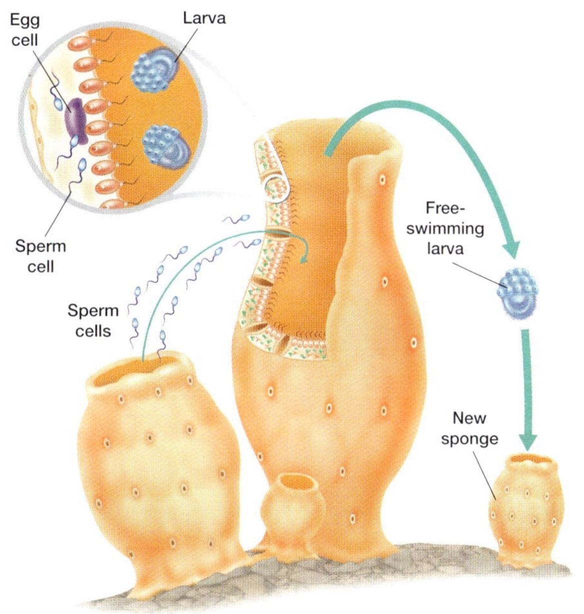

Section 1 Review

1. **Draw** a simple sketch of a sponge body plan, and label all the parts you include. SC.F.1.4.2
2. **Summarize** how a sponge feeds and distributes nutrients. SC.F.1.4.2
3. **Describe** the three types of sponge skeletons. SC.G.1.4.1
4. **Compare** asexual and sexual reproduction in sponges. SC.F.2.4.1
5. **Critical Thinking Forming Hypotheses** What advantage might there be to a free-swimming larval stage in sponges? SC.F.1.4.2
6. **Critical Thinking Determining Factual Accuracy** Evaluate this statement: Sponges have two cell layers, mesohyl and collar cells. SC.F.1.4.2
7. **FCAT Prep** What is one function of choanocytes in a sponge? SC.F.1.4.2
   A circulating water **C** fertilizing eggs
   B distributing nutrients **D** supporting the body

621

Answers to Section Review

1. Sketches will vary but should include the labels in Figure 2.
2. A sponge's collar cells trap small organisms. The cells digest the organisms and release nutrients into the mesohyl, where amoebocytes distribute the nutrients to other cells.
3. Calcareous sponges have skeletons with calcium carbonate spicules. Glass sponges and demosponges have silica spicules. Some demosponges also have spongin.
4. Sponges reproduce asexually by budding. Some freshwater sponges reproduce by forming gemmules. Sponges reproduce sexually by fertilizing other sponges. Sperm enters a sponge through the pores and passes into the mesohyl, where fertilization of the eggs occurs.
5. Free-swimming larvae can colonize new areas.
6. This statement is false. The mesohyl is not a cell layer; it is a gel-like substance that has cells embedded in it.
7. **A.** Correct. The flagella of choanocytes circulate water. **B.** Incorrect. Amoebocytes distribute nutrients. **C.** Incorrect. Fertilized eggs grow into sponges. **D.** Incorrect. Spicules and sponging fibers support the body.

Group Activity — Advanced

Gemmule Survival Pods

Survival pods have been used for submarines. Ask students what a properly equipped survival pod should contain. (water, food, oxygen, the means to get rid of waste, protection from the outside environment) List suggestions on the board. Ask students to work in pairs to take each item and suggest how a gemmule meets that need. (water stored or gained by osmosis, stored food, oxygen gained by diffusion, the means to get rid of waste by the amoebocytes, protection by the protective coat) **LS Verbal**

Transparencies

TR H11 Sexual Reproduction in Sponges

Close

Reteaching — Basic

Show students a diagram of a sponge. Provide them with labels bearing the names of sponge cells and structures. Have students place the labels on the diagram. **LS Visual**

Quiz — General

1. Name the two sponge cell types discussed in the text. (choanocytes and amoebocytes)
2. Describe how water is circulated in the sponge and why this is important. (Choanocytes circulate the water using their flagella, which brings in water carrying food and oxygen and eliminates water carrying waste products.)

Alternative Assessment — General

Have students draw a diagram of a sponge and label its different cells and structures. Tell students to be certain to include egg and sperm cells in their diagram and indicate how sperm enter the sponge. Students should add arrows to show the path water takes as it flows through the sponge. **LS Visual**

Section 2

Focus

Overview

Before beginning this section review with your students the objectives listed in the Student Edition. This section discusses the characteristics of cnidarians, as well as examples of cnidarians and their basic body forms. Students will learn about the three different classes of cnidarians and how cnidarians reproduce.

Bellringer

Encourage students to examine **Figure 6,** which shows the difference between medusas and polyps. Have them write a short song, poem, or limerick to help them remember how these two forms are different. Have volunteers recite their work to the class.

Motivate

Identifying Preconceptions — GENERAL

Show students photographs of jellyfish, hydra, sea anemones, and corals. Ask students about their own experiences with these cnidarians. Have they ever seen a jellyfish, a Portuguese man-of-war, a coral, or a sea anemone? Have they ever touched a cnidarian? What did it feel like? Where have they seen cnidarians-in fresh water or salt water? Take this opportunity to explain that although most cnidarians are marine, some, such as hydras, live in fresh water.
LS Intrapersonal

Section 2 Cnidarians

Florida Sunshine Standards/FCAT

SC.F.1.4.2 Body structures are uniquely designed and adapted for their function.

SC.F.2.4.1 There are mechanisms of asexual and sexual reproduction and different genetic advantages and disadvantages of asexual and sexual reproduction.

SC.G.1.4.1 There is great diversity and interdependence among living things.

Objectives

- **Describe** the two cnidarian body forms. SC.G.1.4.1
- **Summarize** how cnidocytes function. SC.F.1.4.2
- **Summarize** the life cycle of *Obelia.* SC.F.2.4.1
- **Compare** three classes of cnidarians. SC.G.1.4.1
- **Compare** asexual and sexual reproduction in cnidarians. SC.F.2.4.1

Key Terms

medusa
polyp
cnidocyte
nematocyst
basal disk
planula

Two Body Forms

As the fragile bell of a jellyfish moves rhythmically through the water or the flowerlike sea anemone sways gently in the ocean currents, it's easy to be caught up in the mystery and beauty of these animals. But don't be deceived by their allure, for jellyfish and sea anemones are carnivores that can inflict a vicious sting. Along with hydras and corals, these animals belong to the phylum Cnidaria *(nih DAIR ee uh).* Cnidarians have two basic body forms, as shown in **Figure 6,** and both show radial symmetry. **Medusa** *(muh DOO suh)* forms are free-floating, jellylike, and often umbrella-shaped. **Polyp** *(PAHL ihp)* forms are tubelike and are usually attached to a rock or some other object. A fringe of tentacles surrounds the mouth, located at the free end of the body. Many cnidarians exist only as medusas, while others exist only as polyps. Still others alternate between these two phases during the course of their life cycle.

The cnidarian body has two layers of cells, as illustrated by the hydra in **Figure 7.** The outer layer derives from ectoderm, and the inner layer derives from endoderm. As in the sponge, there is a middle layer of mesoglea. But cnidarians differ from sponges in that cnidarians' cells are arranged into tissues.

internet connect
www.scilinks.org
Topic: Cnidarians
Keyword: HX4048
SciLinks Maintained by the National Science Teachers Association

Figure 6 Cnidarian body forms

The two body forms of cnidarians—medusa and polyp—consist of the same body parts arranged differently.

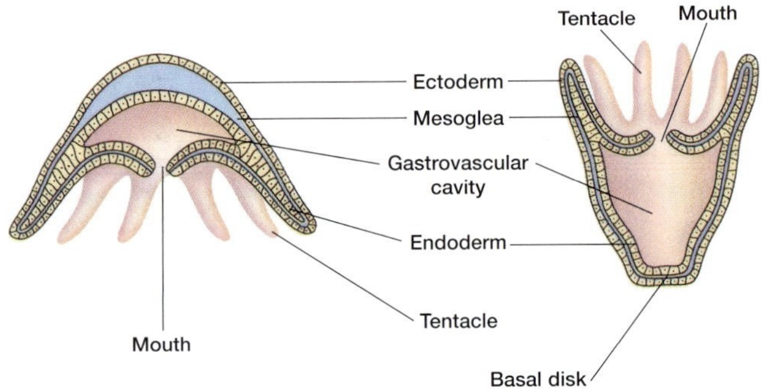

Evolutionary Milestone

2 Tissues

The cnidarian body plan is more complex than that of a sponge—it contains specialized tissues that carry out particular functions. The tissues, however, are not organized into organs.

622

Chapter Resource File
- Directed Reading BASIC
- Active Reading GENERAL
- Data Sheet for Math Lab GENERAL

Transparencies
TR Bellringer
TR H12 Two Cnidarian Body Forms
TR H13 Cnidarian Body Plan

One-Stop Planner CD-ROM
- Reading Organizers BASIC
- Reading Strategies BASIC
- Basic Skills Worksheet Microscope Magnification BASIC

Section Correlations
SC.F.1.4.2
SC.F.2.4.1
SC.G.1.4.1

Cnidocytes

Flexible fingerlike tentacles surround the opening to the gastrovascular cavity of cnidarians. Located on the tentacles are stinging cells called **cnidocytes** *(NIH doh siets)*, also shown in Figure 7. Cnidocytes are the distinguishing characteristic of the animals in the phylum Cnidaria. Within each cnidocyte is a small barbed harpoon called a **nematocyst** *(nehm AAH toh sihst)*. Nematocysts are used for defense and to spear prey. Some nematocysts contain deadly toxins, while others contain chemicals that stun but do not kill. When triggered, the nematocyst explodes forcefully and sinks into the cnidarian's prey. The captured prey is then pushed into the cnidarian's gastrovascular cavity by the tentacles.

Extracellular Digestion

In cnidarians, digestion begins extracellularly (outside the cell), in the gastrovascular cavity. Enzymes break food down into small fragments. Then cells lining the cavity engulf the fragments, and digestion is completed intracellularly. This allows cnidarians to feed on organisms larger than their own individual cells.

Figure 7 Cnidarian body plan

Like all cnidarians, this hydra is composed of tissues derived from endoderm and ectoderm.

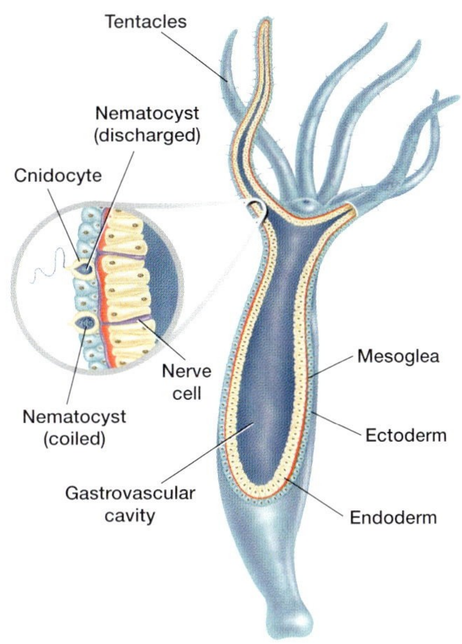

MATH LAB

Estimating Size Using a Microscope SC.H.1.4.1

You can use the microscope to estimate the size of cnidarians that are too small to measure directly.

Materials

transparent millimeter ruler, compound microscope with low-power objective or a dissecting microscope, prepared slide of a medusa or polyp

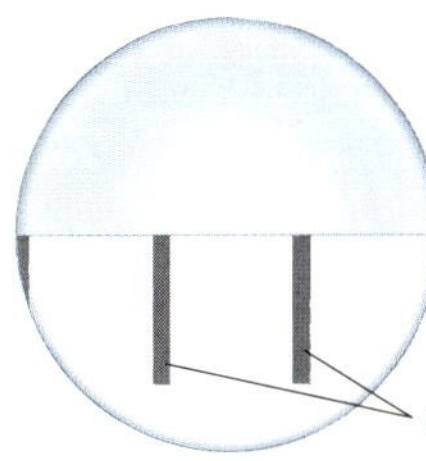

Procedure

1. Identify the millimeter marks along the edge of the ruler.
2. With the microscope on low power (4× or lower), place the ruler on the stage and focus on the millimeter marks.
3. Adjust the ruler so that one edge lies across the diameter of the field, as shown above. Then measure the diameter of the field of view in millimeters.
4. Remove the ruler, and place the prepared slide on the stage. Identify the tentacles, gastrovascular cavity, and mouth.
5. Estimate the length and width of your organism as a ratio of the width of the field of view. For example, the length of your organism may appear to cover about two-thirds of the field of view.

Analysis

1. **Calculate** the size of your organism in millimeters by multiplying the ratio you found in step 5 by the width of the field of view you found in step 3.
2. **Describe** the body plan of the organism you viewed using terms from step 4.

Graphic Organizer

Use this graphic organizer with **Teaching Tip** *on this page.*

Animal groups	Lifestyles	Structures/functions
Cnidarians	Free-floating or attached	Gastrovascular cavity/digestion Tentacles/food getting Cnidocytes/protection and food getting

Teach

READING SKILL BUILDER — BASIC

Interactive Reading Assign Chapter 28 of the *Holt Biology* Guided Audio CD Program to help students achieve greater success in reading the chapter. **LS Auditory**

Using the Figure — ADVANCED

Ask students to locate the gastrovascular cavity of the cnidarian in **Figure 7.** Ask them how cnidarian digestion compares with that of the sponge. (The sponge does not digest extracellularly in the cavity, but within individual choanocytes.) Point out the tissue layers. Ask students how a tissue in the cnidarians differs from the cells in a sponge. (A tissue is a group of cells with a common structure and function. Sponges do not have tissues.) **LS Visual**

Teaching Tip — GENERAL

Graphic Organizer Refer students to their graphic organizer from Section 1. Ask the class to now fill out the cnidarians' row. **LS Visual**

MATH LAB

Estimating Size Using a Microscope

Skills Acquired

Collecting data, measuring

Teacher's Notes

Review with students the proper procedures for using a compound microscope.

Answers to Analysis

1. Answers will vary, but should be calculated by multiplying the fraction of the field the organism took up by the size in mm measured across the field of view.
2. Answers will vary. Refer to Figure 6 for an illustration.

Hydrozoans

The most primitive cnidarians are members of class Hydrozoa. Most species of hydrozoans are colonial marine organisms whose life cycle includes both polyp and medusa stages. Freshwater hydrozoans are less common, but are familiar to many people because they are often studied in school laboratories.

Figure 8 Freshwater hydra

This tiny hydra is attached to the leaf of a small aquatic plant. One way a hydra can move is by tumbling.

Magnification: 34×

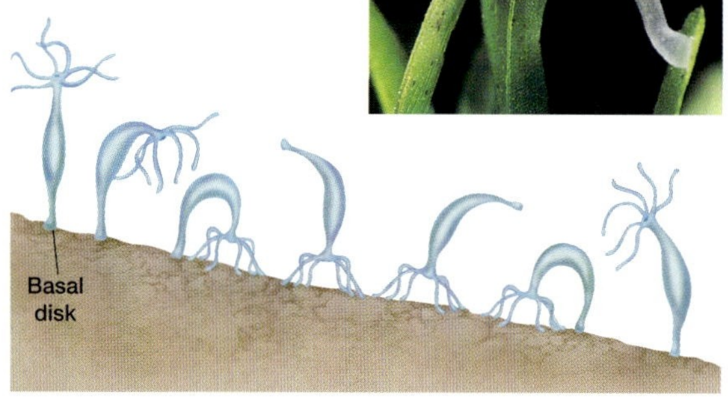

Freshwater Hydrozoa

The abundant freshwater genus *Hydra* is unique among hydrozoans because it has no medusa stage and exists only as a solitary polyp. Hydras live in quiet ponds, lakes, and streams. They attach to rocks or water plants by means of a sticky secretion they produce in an area of their body called the **basal disk**. Hydras can glide around by decreasing the stickiness of the material secreted by their basal disk. Sometimes hydras move by tumbling, as shown in **Figure 8.** To tumble, the hydra bends its body over and touches the surface it is attached to with its tentacles. Then it pulls its basal disk free, flipping it over to the other side of its tentacles. The basal disk then reattaches, and the hydra returns to an upright position. Most hydras are brown or white, like the one in Figure 8. Others appear green because of the algae living within their cells.

Marine Hydrozoa

Marine hydrozoans are typically far more complex than freshwater hydrozoans. Often many individuals live together, forming colonies. The cells of the colony lack the interdependence that characterizes the cells of multicellular organisms. However, they often exhibit considerable specialization. For example, the colonial Portuguese man-of-war (genus *Physalia*) incorporates both medusas and polyps. A gas-filled float (probably a highly modified polyp) allows *Physalia* to float on the surface of the water. Dangling below the float are tentacles that can reach 60 m (about 197 ft). These tentacles are used to stun and entangle prey. Their nematocysts are tipped with powerful neurotoxins (nerve poisons) that are dangerous and may be fatal to humans. *Physalia*, shown in **Figure 9,** has other specialized polyps and medusas, each carrying out a different function, such as feeding or sexual reproduction.

Teach, *continued*

Using the Figure — Basic

Refer students to the illustration of the tumbling hydra in **Figure 8.** Tumbling may look like a difficult way to move, so compare it to the movement of a flexible toy spring that can "walk" down a set of steps. A hydra is similar in that once it connects one end of its body to a solid surface, the rest of its body will flip over due to elasticity and momentum. LS Visual

did you know?

Stolen Nematocysts Some sea slugs (nudibranchs) feed on cnidarians and save the untriggered nematocysts within their own bodies. When other animals try to eat the sea slug, the stored-up nematocysts are triggered. As a result of this painful encounter, many predators learn to avoid sea slugs.

Cultural Awareness

A Mythological Hydra The name *hydra* is a reference to the nine-headed serpent that was slain by Hercules, a figure in Greek mythology. According to Greek myth, the slaying of the Hydra was one of the twelve labors that Hercules was assigned to perform. Killing the Hydra was not an easy task for Hercules because when one head of the Hydra was cut off, two others would grow in its place.

Reproduction in Hydrozoans

Most hydrozoans are colonial organisms whose polyps reproduce asexually by forming small buds on the body wall. The buds develop into polyps that eventually separate from the colony and begin living independently. Many hydrozoans are also capable of sexual reproduction. Some species of *Hydra* are hermaphrodites, but in most species the sexes are separate.

The genus *Obelia* is typical of many marine colonial hydrozoans. *Obelia* lives in colonies that form when one polyp asexually produces buds that do not separate from it. Eventually, there are numerous polyps attached to one stem, forming the colony. The *Obelia* colony shown in **Figure 10** is branched like deer antlers, with various polyps attached to the branched stalks. The reproductive polyps give rise asexually to male and female medusas. These medusas leave the polyps and grow to maturity in the ocean waters.

During sexual reproduction, the medusas release sperm or eggs into the water. The gametes fuse and produce zygotes that develop into free-swimming, ciliated larvae called **planulae** *(PLAN yoo lee)*. The planulae eventually settle on the ocean bottom and develop into new polyps. Each polyp gives rise to a new colony by asexual budding, and the life cycle is repeated.

Figure 9 ***Physalia.*** A single Portuguese man-of-war colony can contain 1,000 individual medusas and polyps.

Figure 10 Reproduction in *Obelia*

In *Obelia's* life cycle, the medusa stage (sexual) and the polyp stage (asexual) alternate.

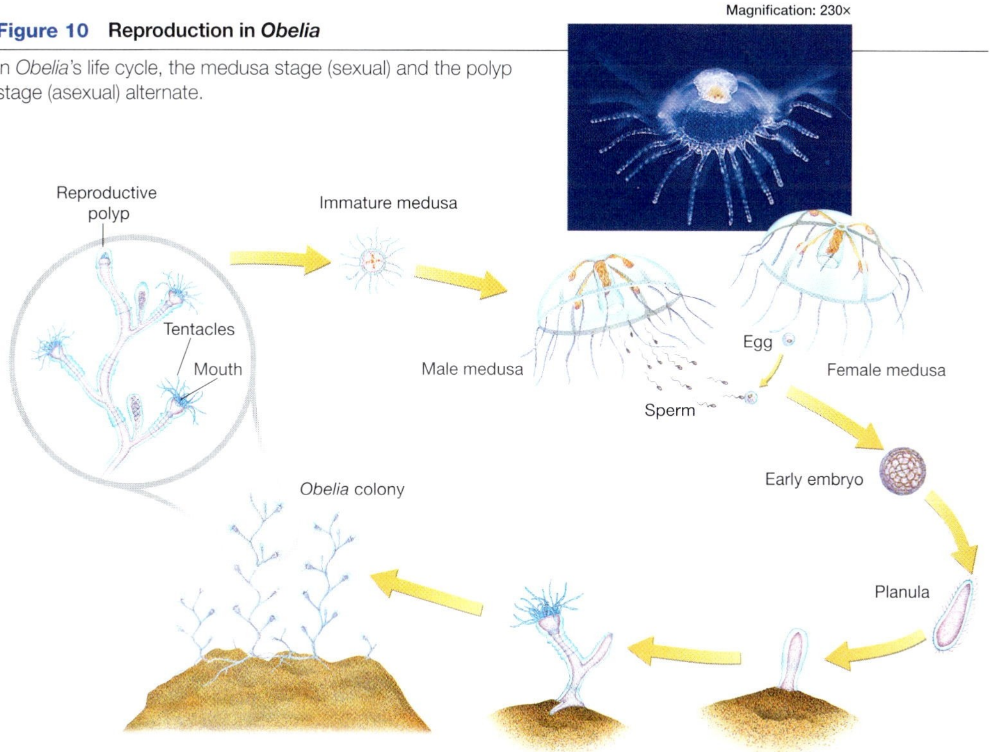

Teaching Tip — GENERAL

Planulae There are both negative and positive aspects to the existence of the free-swimming planulae of *Obelia*. Ask students about the dangers to free-swimming larvae in the ocean, and why it is an advantage for larvae to be able to swim far from their parents. (Larvae may starve, be eaten, wash up on shore, or not find a suitable place to settle down. Motile larvae allow *Obelia* to spread to new places.) **LS Verbal**

Using the Figure — GENERAL

Check for student understanding of the life cycle of *Obelia*, shown in **Figure 10.** Ask students the following questions: Which stages are free-swimming and which are sessile? (medusa, polyp) Which stage reproduces asexually? (polyp stage) In which stage are there males and females that reproduce sexually? (medusa stage) **LS Visual**

Transparencies

TR H15 Reproduction in *Obelia*

MISCONCEPTION ALERT

The Portuguese Man-of-War Many people think that the Portuguese man-of-war *(Physalia)* is a jellyfish because it is shaped somewhat like one. However, the term jellyfish is used for scyphozoans, a class of cnidarians in which the medusa form dominates the life cycle. *Physalia* is actually a hydrozoan and exists as a colony of medusas and polyps.

Teach, *continued*

READING SKILL BUILDER — GENERAL

Discussion Jellyfish have toxins in their nematocysts that paralyze and kill fish. Ask students why it is important for a simple floating animal like a jellyfish to immobilize its prey. (A jellyfish cannot stalk or chase prey, and it has no jaws or claws to capture prey.) **LS Logical**

Using the Figure — GENERAL

Have students study **Figure 11** to compare and contrast the medusa and polyp stages of a jellyfish. Ask students to name other animals that have different bodies during different stages of their life. (hydrozoans, insects, frogs, and so on) Then ask students to compare *Aurelia* medusae and polyps to those of *Obelia*. (They are almost identical.) Ask students to identify the big difference in the life cycles of hydrozoans and scyphozoans. (Hydrozoans spend most of their life as a polyp; scyphozoans spend most of their life as a medusa.) **LS Visual**

Polyps

Medusa

Figure 11 Marine jellyfish *Aurelia*. *Aurelia* polyps are about the size of hydras. The free-swimming medusas range from 10 cm (3.9 in.) to 25 cm (9.8 in.) in diameter.

WORD Origins

The name Scyphozoa is from the Greek *skyphos*, meaning "cup," and *zoia*, meaning "animal." The name refers to the fact that members of this class spend most of their lives as medusas, which have the shape of an inverted cup.

Scyphozoans

Cnidarians belonging to the class Scyphozoa *(sie fuh ZOH uh)* are the organisms usually referred to as true jellyfish. Scyphozoans are active predators that ensnare and sting prey with their tentacles. The toxins contained within the nematocysts of some species are extremely potent. Scyphozoans range in size from as small as a thimble to as large as a queen-size mattress.

The jellyfish seen in the ocean are medusas, which reproduce sexually. However, most species of jellyfish also go through an inconspicuous polyp stage at some point in their life cycle. The stinging nettle, *Aurelia,* shown in **Figure 11,** is one of the most familiar jellyfishes. *Aurelia*'s tiny polyps hang from rocky surfaces. Periodically the polyps release young medusas into the water. The *Aurelia* life cycle is similar to that of *Obelia,* pictured on the previous page. The major difference is that *Aurelia* spends most of its life as a medusa, while *Obelia* spends most of its life as a polyp.

Jellyfish Relatives

Related to the jellyfish are the cubozoans, or box jellies. As their name implies, cubozoans have a cube-shaped medusa. Their polyp stage is inconspicuous, and in some species, it has never been observed. Most box jellies are only a few centimeters in height, although some are 25 cm (10 in.) tall. A tentacle or group of tentacles is found at each corner of the "box." Stings of some species, such as the sea wasp, can inflict severe pain and even death among humans. The sea wasp lives in the ocean along the tropical northern coast of Australia.

Other relatives are members of the phylum Ctenophora *(tehn AW for uh),* which includes the comb jellies. Comb jellies differ from true jellyfish in two major ways—they have only a medusa stage and they have no cnidocytes. Their tentacles are covered with a sticky substance that traps plankton, the comb jelly's main prey. Although a comb jelly is only about 2.5 cm (1 in.) in diameter, its tentacles can be 10 times as long.

626

Cultural Awareness

Edible Jellyfish In many Asian countries, some species of jellyfish are considered delicacies. Their taste has been compared with that of pickles. Although jellyfish have stinging cells, the salting process used to prepare them as food breaks down the toxins in these cells.

Anthozoans

The largest class of cnidarians is class Anthozoa. Anthozoans exist only as polyps. The most familiar anthozoans are the brightly colored sea anemones and corals. Other members of this class are known by such fanciful names as sea pansies, sea fans, and sea whips.

Anthozoans, such as the sea anemone shown in **Figure 12,** typically have a thick, stalklike body topped by a crown of tentacles that usually occur in groups of six. Nearly all of the shallow-water species contain symbiotic algae, such as dinoflagellates. The anthozoans provide a place for the dinoflagellates to live in exchange for some of the food that the dinoflagellates produce. The brilliant color of most anthozoans is actually that of dinoflagellates living within it. Some anthozoans reproduce asexually by forming buds, but they also reproduce sexually by releasing eggs and sperm into the ocean, where fertilization occurs. The fertilized eggs develop into planulae that settle and develop into polyps.

Figure 12 Sea anemone. When threatened, the sea anemone quickly retracts its tentacles and compresses its body.

Sea Anemones

Sea anemones are a large group of soft-bodied polyps found in coastal areas all over the world. Many species are quite colorful, and most do not grow very large, only from 5 mm (0.2 in.) to 100 mm (4.0 in.) in diameter. Sea anemones feed on fish and other marine life that happen to swim within reach of their tentacles.

Sea anemones are highly muscular and relatively complex animals. When touched, most sea anemones retract their tentacles into their body cavity and contract into a tight ball. Sea anemones often reproduce asexually by slowly pulling themselves into two halves. This method of reproduction often results in large populations of genetically identical sea anemones.

Corals

Most coral polyps live in colonies called reefs, such as the one shown in **Figure 13.** Each polyp secretes a tough, stonelike outer skeleton of calcium carbonate that is cemented to the skeletons of its neighbors. (Some corals called soft corals do not secrete hard exoskeletons.) Only the top layer of a coral reef contains living coral polyps. When coral polyps die, their skeletons remain and provide a

Figure 13 Coral. This coral reef is made up of hundreds of thousands of individual coral polyps. When the polyps feed (inset), they extend their tentacles from the protection of their stony skeleton.

SKILL BUILDER — ADVANCED

Writing Skills Have students research the relationship between some species of sea anemones and certain types of fishes, such as the damselfish and the clownfish. The clownfish gets protection by hiding in the tentacles, and the sea anemone gets cleaned as the clownfish eats undigested debris and parasites. Students should investigate how these fishes are able to avoid discharging the anemone's nematocysts. Ask students to write a brief report. **LS Verbal**

Demonstration — BASIC

Collect photographs or specimens of coral for students to observe. Try to have some photos showing coral polyps. Some students find it difficult to believe that cnidarians, such as sea anemones and corals, are animals. Ask students to defend the classification of anthozoans in the animal kingdom. **LS Visual**

Using the Figure — GENERAL

Have students look closely at the insert of the coral polyp shown in **Figure 13.** Ask how the polyp resembles the polyp form of other cnidarians. (It is cup-shaped and contains a ring of upward-facing tentacles surrounding the opening to its exterior.) **LS Visual**

Integrating Physics and Chemistry

Have students identify the elements found in the compound calcium carbonate. Then, ask them to explain the bonding relationships between the elements based on their respective locations on the periodic table.

Teach, *continued*

BIOWatch

Exploring Florida's Coral Reefs

Discussion

Ask students: Why were researchers surprised to find a living coral reef at Pulley Ridge? (Most coral reefs are found in shallow waters, but Pulley Ridge is much deeper than most reefs.)

Close

Reteaching — BASIC

Have students make a table that shows the similarities and differences between the three classes of cnidarians: Hydrozoa, Scyphozoa, and Anthozoa. (Similarities: radial symmetry, aquatic lifestyle, presence of cnidocytes. Differences: predominance of a polyp or medusa body form, the maintenance of a solitary versus colonial lifestyle.) LS Verbal

Quiz — GENERAL

1. List the names and locations of the two layers of tissues found in the cnidarians. (ectoderm on the outer surface, endoderm on the inner surface)
2. Describe the digestive process of the cnidarians. (Digestion begins extracellularly inside the gastrovascular cavity. Digestion is completed intracellularly.)

foundation for new coral polyps. Over thousands of years, these formations build up into coral reefs where hundreds of thousands of polyps live together on top of old skeletons. Coral reefs are found primarily in tropical regions of the world, where the ocean water is warm and clear, an environment that is ideal for the corals and the dinoflagellates that live inside them.

FLORIDA BIOWatch

Exploring Florida's Coral Reefs

South of Miami, some 6,000 coral reefs form a graceful arc extending from Key Biscayne through the Gulf of Mexico. Reefs provide shelter and breeding sites for many plants and animals and attract millions of Florida visitors annually.

A Deep-water Surprise

Marine scientists from the University of South Florida recently discovered the deepest living coral reef ever found in the United States. The reef lies at a depth of 70 meters on Pulley Ridge, a submerged barrier island area west of the Dry Tortugas. Working with scientists from the U.S. Geological Survey (USGS), the Florida researchers used mini-submarines and submersible cameras to explore Pulley Ridge and gather data. The scientists were surprised to find a remarkable variety of corals, algae, and fishes there.

A Unique Kind of Reef

Unlike Pulley Ridge, most coral reefs are found in shallow water. Organisms in coral reefs depend on light filtering down from the surface. The coral polyps that form reefs live symbiotically with zooxanthellae, photosynthetic dinoflagellates that require sunlight. Most reefs grow vertically, but the Pulley Ridge coral grows flat, an adaptation to the low light.

Researchers have found that the corals on Pulley Ridge are much healthier than those from most shallow-water reefs. To remain healthy, coral reefs need warm, clear water that is low in phosphate and nitrate nutrients. Officials who oversee the Gulf of Mexico must now determine how to protect the reef to ensure its long-term survival.

internet connect

www.scilinks.org
Topic: Coral Reefs
Keyword: HX4051

SCILINKS. Maintained by the National Science Teachers Association

Section 2 Review

1. **Compare** the two body forms of cnidarians. SC.G.1.4.1
2. **Relate** cnidocytes and nematocysts to food gathering. SC.F.1.4.2
3. **Draw** and label the life cycle of *Obelia*. SC.F.2.4.1
4. **Summarize** the similarities and differences in the three classes of cnidarians described. SC.G.1.4.1
5. **Distinguish** between the two types of asexual reproduction found in cnidarians. SC.F.2.4.1
6. **Critical Thinking Forming Hypotheses** Some cnidarians are unique in exhibiting polyp and medusa forms. How might their two body forms give them an advantage over species that have only one body form? SC.F.1.4.2
7. *FCAT Prep* Where does digestion take place in cnidarians?
   A only extracellularly
   B only intracellularly
   C in a digestive tract
   D in a gastrovascular cavity and intracellularly
   SC.F.1.4.2

628

Answers to Section Review

1. A medusa is often umbrella shaped. A polyp is tube-like and closed at one end.
2. Cnidocytes are stinging cells used in feeding. They contain a nematocyst, which when triggered, is shot into the cnidarian's prey.
3. Drawing should be similar to Figure 10.
4. All cnidarians have stinging cells on their tentacles and two cell layers—a middle layer of mesoglea, and a gastrovascular cavity. Hydrozoans are primarily colonial and spend most of their life as a polyp. Scyphozoans exist primarily as a medusa. Anthozoans are only polyps and may be solitary or colonial.
5. A colonial cnidarian may reproduce by budding. Some may reproduce asexually by slowly pulling themselves in half.
6. The two body forms exploit different environments, so these cnidarians have access to a broader array of resources while spreading their risks between two lifestyles.
7. **A.** Incorrect. Digestion begins extracellularly. **B.** Incorrect. Digestion occurs intracellularly after the cells of the cavity engulf nutrients. **C.** Incorrect. Cnidarians do not have a digestive tract. **D.** Correct. Digestion occurs in a gastrovascular cavity and intracellularly.

Section Review Correlations

SC.F.1.4.2: Items 2, 6, 7
SC.F.2.4.1: Items 3, 5
SC.G.1.4.1: Items 1, 4

Flatworms and Roundworms

Section 3

Florida Sunshine Standards/FCAT

SC.F.1.4.2 Body structures are uniquely designed and adapted for their function.
SC.F.2.4.1 There are mechanisms of asexual and sexual reproduction and different genetic advantages and disadvantages of asexual and sexual reproduction.
SC.G.1.4.1 There is great diversity and interdependence among living things.

Flatworms

When you think of a worm, you probably visualize a creature with a long, tubular body, such as an earthworm. You might be less familiar with flatworms and roundworms. The flatworms are the largest group of acoelomate worms. Although the flatworm body plan is relatively simple, it is a great deal more complex than that of a sponge or cnidarian. Flatworms have a middle tissue layer, the mesoderm. And unlike sponges and cnidarians, the flatworm has tissues that are organized into organs.

The flatworm's body is bilaterally symmetrical and flat, like a piece of tape or ribbon. As a result, each cell in the animal's body lies very close to the exterior environment. This permits dissolved substances, such as oxygen and carbon dioxide, to pass efficiently through the flatworm's solid body by diffusion. In addition, portions of the flatworm's highly-branched gastrovascular cavity run close to practically all of its tissues. This gives each cell ready access to food molecules. Most flatworms have no respiratory or circulatory system.

Flatworms belong to phylum Platyhelminthes, which contains three major classes: Turbellaria, Cestoda, and Trematoda. They range in size from free-living forms less than 1 mm (0.04 in.) in length to parasitic intestinal tapeworms several meters long.

Objectives

- **Compare** the three classes of flatworms. SC.G.1.4.1
- **Summarize** the life cycle of a blood fluke. SC.F.2.4.1
- **Describe** the body plan of a roundworm. SC.F.1.4.2
- **Summarize** the life cycle of the roundworm *Ascaris.* SC.F.2.4.1

Key Terms

proglottid
fluke
tegument

Turbellaria

Almost all members of class Turbellaria are free-living marine flatworms, such as the one shown in **Figure 14.** However, marine flatworms are rarely studied by students because they are difficult to raise in captivity. Instead, students usually study a freshwater turbellarian such as *Dugesia,* one of a group of flatworms commonly called planarians. *Dugesia* is shown in *Up Close: Planarian,* on the following page.

Figure 14 Marine flatworm. Most free-living flatworms are marine species that swim with graceful wavelike movements.

Evolutionary Milestone

3 Bilateral Symmetry

Flatworms were likely the first bilaterally symmetrical animals, with left and right halves that mirror each other. Like all bilaterally symmetrical animals, flatworms have a distinct anterior (cephalic) end.

Section 3

Focus

Overview

Before beginning this section review with your students the objectives listed in the Student Edition. This section explains the characteristics and life histories of flatworms and roundworms. Students will explore the three different classes of flatworms, as well as the anatomy of both flatworms and roundworms.

Bellringer

The word *worm* can refer to a wide variety of organisms. Ask students to think about what a worm is, and ask them to list on paper the names or describe the types of worms they can remember. Ask students where these worms live and what they eat. Have volunteers share their descriptions with the rest of the class.
LS Verbal

Motivate

Discussion/Question — Basic

Ask students what they might do if they were in a rowboat that had a small leak. (Students may answer that they would use a small container to bail out the water as it seeps in.) Explain to students that all organisms have mechanisms that help them maintain proper water balance. Tell them that water seeps into *Dugesia*, the flatworm they will study on the next page, like it seeps into a leaky boat. Ask them to suggest ways in which *Dugesia* might maintain water balance. (The excretory system expels water so that the animal does not swell up.)
LS Logical

Chapter Resource File

- Directed Reading Basic
- Active Reading General
- Data Sheet for Quick Lab General

Transparencies

TR Bellringer
TR H19 Planarian

One-Stop Planner CD-ROM

- Reading Organizers Basic
- Reading Strategies Basic
- Supplemental Reading Guide *Journey to the Ants: A Story of Scientific Exploration* Advanced

Section Correlations

SC.F.1.4.2
SC.F.2.4.1
SC.G.1.4.1

Up Close

Discussion Guide the discussion by posing the following questions:

- Instead of a circulatory system that delivers nutrients to tissues, what does a planarian have? (Branches of the digestive tract reach the tissues directly.)
- How do planarians reproduce asexually? (They tear themselves in two, and each half regenerates to form a complete worm.)
- How do planarians reproduce sexually? (They are hermaphrodites that fertilize each other's eggs. Protective capsules surround groups of fertilized eggs, which hatch in 2–3 weeks.)

Teach

Activity — GENERAL

Writing Life of a Planarian Have your students imagine a day in the life of a planarian and write a story about the planarian's adventures. Their story should include information about the planarian's nervous system, water balance, reproduction, feeding, digestion, and excretion. Caution them to avoid anthropomorphic descriptions of the planarian, though a little creative license should be allowed. Have students volunteer to read their stories to the class. **LS Verbal**

Planarian

- **Scientific name:** *Dugesia* sp.
- **Size:** Average length of 3–15 mm (0.1–0.6 in.)
- **Range:** Worldwide
- **Habitat:** Cool, clear, permanent lakes and streams
- **Diet:** Protozoans and dead and dying animals

Dugesia feeding

Characteristics

Nervous System Sensory information gathered by the brain is sent to the muscles by two main nerve cords that are connected by cross branches. Light-sensitive structures called eyespots are connected to the brain. The eyespots are close to each other, giving *Dugesia* a cross-eyed appearance.

Feeding *Dugesia,* a free-living flatworm, must extend its muscular pharynx out of its centrally located mouth in order to feed.

Reproduction *Dugesia* reproduces asexually in the summer by attaching its posterior end to a stationary object and stretching until it breaks in two, each of which will become a complete animal. Sexual reproduction also occurs. Individuals are hermaphrodites, and two individuals simultaneously transfer sperm to each other. Eggs of both individuals are fertilized and are released in clusters enclosed in a protective capsule. Several capsules are laid at a time, and the eggs inside hatch in 2 to 3 weeks.

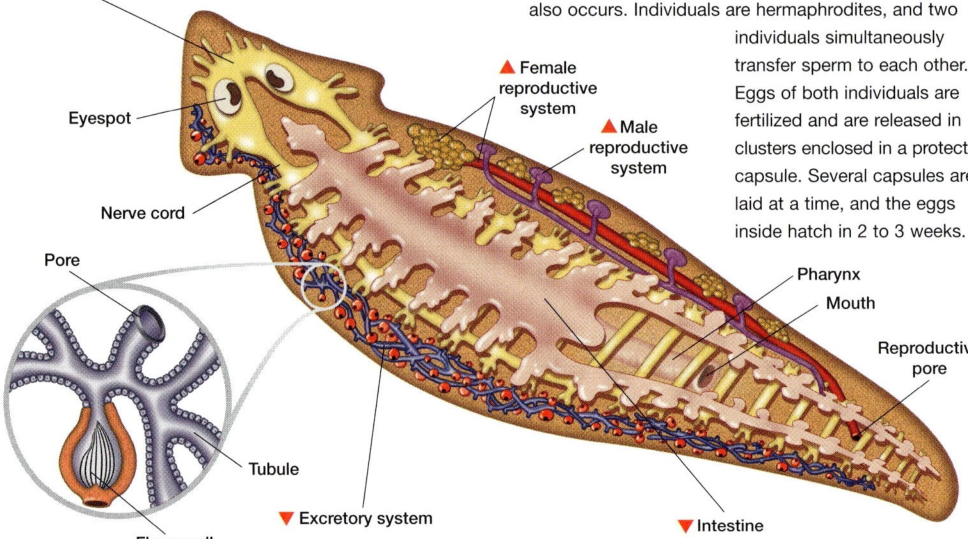

Water Balance Because *Dugesia*'s body cells contain more solutes than fresh water does, water continually enters its body by osmosis. Excess water moves into a network of tiny tubules that run the length of *Dugesia*'s body. Side branches are lined with many flame cells, specialized cells with beating tufts of cilia that resemble a candle flame. The beating cilia draw water through pores to the outside of the worm's body.

Digestion The highly branched intestine enables nutrients to pass close to all of the flatworm's tissues. Nutrients are absorbed through the intestinal wall. Undigested food is expelled through the mouth.

Trends in Neurology

Impulsive Invertebrates Scientists have used many types of invertebrates, including planarians and other flatworms, to try and understand more about nerve cells and the nervous system. More than 50 years ago, biologists used giant axons from squids to learn how nerve impulses travel along nerve-cell processes. More recent experiments with *Aplysia,* a large sea slug, have helped scientists better understand the neurological bases of learning. Also, researchers in Germany have recently made a computer chip that can send signals to a single nerve cell in a living leech. The leech's nerve cell can return the signal back to the computer chip. The researchers are hoping that this technology will help them develop computers that can communicate with nerve cells in a human body.

Observing Planarian Behavior

SC.H.1.4.1

Most bilaterally symmetrical organisms have sense organs concentrated in one end of the animal. You can observe how this arrangement affects the way they explore their environment.

Materials

eyedropper, live culture of planaria, small culture dish with pond water, hand lens or dissecting microscope, forceps, and small piece of raw liver (3–7 cm)

Procedure

1. Using the tip of the eyedropper, place a planarian in the culture dish with pond water.
2. Using the hand lens or dissecting microscope, observe the planarian as it adjusts to its environment. Determine which end of the planarian contains sensory apparatus for exploring the environment.
3. Using forceps, place the liver in the pond water about 1 cm behind the planarian.
4. Observe the planarian's response. If the planarian approaches the liver, move the liver to a different position.
5. Continue observing the planarian for 5 minutes, moving the liver frequently.

Analysis

1. **Describe** the planarian's means of locomotion.
2. **Describe** how the planarian responded to the liver.
3. **Contrast** the feeding behavior of planarians with that of hydras, described earlier in this chapter.
4. **Critical Thinking** **Evaluating an Argument** Evaluate this statement: Bilateral symmetry gives planaria an advantage when feeding because sensory organs are concentrated in one end. Support your opinion with the observations you made on planaria.

Cestoda

Class Cestoda is made up of a group of parasitic flatworms commonly called tapeworms. Tapeworms use their suckers and a few hooklike structures, shown in **Figure 15,** to permanently attach themselves to the inner wall of their host's intestines. Food is then absorbed from the host's intestine directly through the tapeworm's skin. Tapeworms grow by producing a string of rectangular body sections called **proglottids** *(proh GLAHT ihds)* immediately behind their head. (Each proglottid is a complete reproductive unit, a fact that makes tapeworms easy to spread.) These sections are added continually during the life of the tapeworm. The long, ribbonlike body of a tapeworm may grow up to 12 m (40 ft) long.

Figure 15 Tapeworm

A tapeworm's body consists of a head and a series of proglottids.

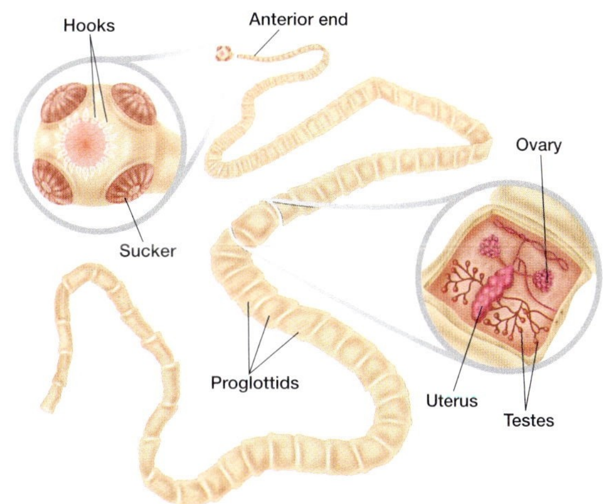

631

Observing Planarian Behavior

Skill Acquired
Observing, comparing, evaluating conclusions

Teacher's Notes

Before students begin, demonstrate how to place planarian in a culture. Planarians are photonegative and should be kept in the dark as much as possible. Use opaque pans, and keep the water temperature close to 18°C. Prior to the lab, avoid feeding the planarians for a few days. This will increase their feeding response during the lab.

Answers to Analysis

1. Planarians move by contracting and expanding their body as they grip the surface. They often turn their head from side to side as they move.
2. Planarians likely will turn their heads in the direction of the liver before moving toward it.
3. Hydras are relatively stationary feeders and use cnidocytes to sting and kill prey that is moving towards them. Planarians are bilaterally symmetrical and have sensory organs located at one end of their body. They are able to detect food sources and move toward them.
4. Answers will vary. Students should support their arguments with their observations.

Cultural Awareness

Tapeworm Infections Students may presume that tapeworm infections are relatively rare among people living in developed countries. Even in the United States, however, tapeworms from pigs and cattle can infect humans. Researchers estimate that about 1 percent of American cattle are infected with beef tapeworms. A significant fraction of all beef consumed in the United States is not federally inspected, and lightly infected beef can be missed during inspections. As a result, if a person eats undercooked roast beef, hamburgers, or steaks from an infected animal, the chance of becoming infected with beef tapeworm is significant.

Using the Figure — GENERAL

Have students study the different parts of the tapeworm shown in **Figure 15.** Ask them to hypothesize the advantages of the hooks on the anterior end and the redundancy of the body sections. (The hook holds on to the host's intestinal wall. The redundant segments aid in the proliferation of the tapeworm.) LS **Visual**

Teach, *continued*

Activity — GENERAL

Writing **Flukes** Have students research flukes using the Internet. Have students write a report summarizing their findings. Ask students to illustrate their reports and share them with the class. LS **Verbal**

Using the Figure — ADVANCED

Have students analyze **Figure 16** from the perspective of a public health investigator. Divide the class into small groups of two or three students, and have them devise a plan to combat blood fluke infections. LS **Logical** Co-op Learning

Teaching Tip — GENERAL

Graphic Organizer Refer students to their graphic organizer from Section 1. Divide the class into small groups and ask them to fill in the cells for flatworm and roundworms. Ask the class to list at least three major structures and their functions. LS **Visual**

internet connect
www.scilinks.org
Topic: Flukes
Keyword: HX4084
SCILINKS. Maintained by the National Science Teachers Association

Most tapeworm infections occur in vertebrates, and about a dozen different kinds of tapeworms commonly infect humans. One of the tapeworms that infects humans is the beef tapeworm, *Taenia saginata*. Beef tapeworm larvae live in the muscle tissue of infected cattle, where they form enclosed fluid-filled sacs called cysts. Humans become infected when they eat infected beef that has not been cooked to a temperature high enough to kill the larvae.

Trematoda

The largest flatworm class, Trematoda, consists of parasitic worms called **flukes.** Some flukes are endoparasites, or parasites that live *inside* their hosts. Endoparasites have a thick protective covering of cells called a **tegument** that prevents them from being digested by their host. Other flukes are ectoparasites, or parasites that live on the *outside* of their hosts.

Flukes have very simple bodies with few organs. Flukes do not have well-developed digestive systems. Rather, they take their nourishment directly from their hosts. Flukes have one or more suckers that they use to attach themselves to their host. They use their muscular pharynx to suck in nourishment from the host's body fluids.

Most flukes have complex life cycles involving more than one host, one of which may be a human. Blood flukes of the genus *Schistosoma* are responsible for the disease schistosomiasis *(shihs tuh soh MIE uh sihs)*, a major public health problem in the tropics. Infection occurs when people use or wade in water contaminated with *Schistosoma* larvae. The larval parasites bore through a person's skin and make their way to blood vessels in the intestinal wall. They block blood vessels, resulting in bleeding of the intestinal wall and damage to the liver. As shown in **Figure 16,** the life cycle of blood flukes includes a particular species of snail as an intermediate host.

Figure 16 *Schistosoma* life cycle

In the life cycle of blood flukes, snails are intermediate hosts and humans are final hosts.

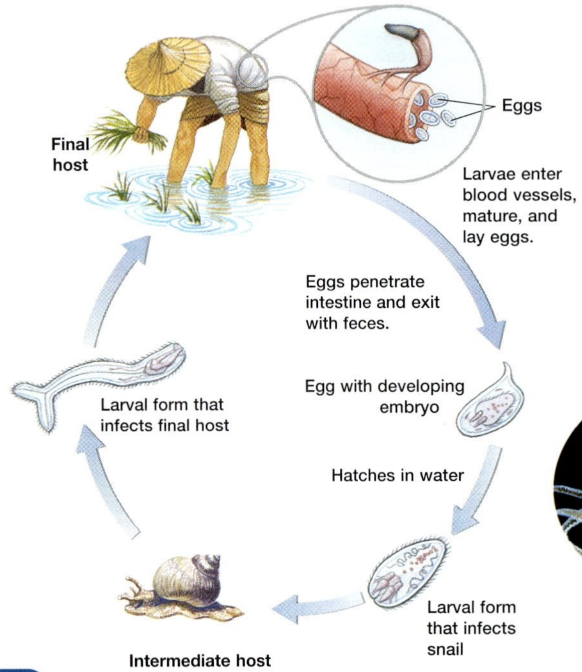

Adult male blood flukes are thick-bodied, while adult females are threadlike.

632

Graphic Organizer

Use this graphic organizer with **Teaching Tip** on this page.

Animal groups	Lifestyles	Structures/functions
Flatworms	Free-living or parasitic	Egg and sperm cells/ reproduction Flame cells/excretion Pharynx/feeding

Roundworms

If you have a dog, you may be familiar with roundworms, some of which are canine parasites. Treatment for roundworms is a common reason for a trip to the vet, as shown in **Figure 17.** Roundworms (nematodes) are members of the phylum Nematoda and are characterized by the presence of a body cavity called a *pseudocoelom*. Movement of the fluid within the roundworm's pseudocoelom serves as a simple circulatory and gas exchange system. Oxygen and carbon dioxide move by diffusion into and out of the fluid. Nutrients from the digestive system also diffuse into the fluid and are distributed to the body cells.

Roundworms have long, cylindrical bodies and are the simplest animals to have a one-way digestive system. A flexible, thick layer of epidermis and cuticle form a protective cover and give the roundworm's body its shape. Beneath this cover, a layer of muscle extends along the length of the worm. These long muscles pull against the cuticle and the pseudocoelom (fluid-filled body cavity), whipping the worm's body from side to side. While some roundworms grow to be a foot or more in length, most are microscopic or only a few millimeters long. The vast majority of roundworms are free-living, active hunters.

Roundworm Infections

About 50 roundworm species are plant or animal parasites that cause considerable economic damage to crops and inflict terrible human suffering. Plant roundworms may attack any part of the plant—leaves, stem, roots—depending on the species. They feed on the living plant cells, causing wilting and withering of the plant. At least 14 species of roundworms infect humans. Three sources of human infection are *Ascaris lumbricoides*, *Trichinella spiralis*, and members of the genus *Necator*, commonly called hookworms.

The eggs of *Ascaris* are carried through human waste to the soil, where they can live for years. If ingested, the eggs enter the intestine, where they develop into larvae. The larvae bore through the blood vessels in the intestine and enter the bloodstream, which carries them to the lungs, causing respiratory distress. Some larvae may wander into the ducts of the pancreas or gallbladder, causing a blockage. Eventually, the larvae return

Magnification: 120×

Figure 17
Roundworms in pets. When a dog or cat has to be wormed, it is usually due to a roundworm infection caused by *Toxocara canis* or *Toxocara cati.*

Evolutionary Milestone

4 **Body Cavity**

Roundworms have a pseudocoelom, a body cavity that forms between the gut and the body wall. All pseudocoelomates have a one-way gut in which food passes into the mouth and out of the anus.

Integrating Physics and Chemistry

Since oxygen and carbon dioxide can only move by diffusion into and out of the roundworm's pseudocoelom, increasing the surface area through which solutes can pass will increase the rate of diffusion. Have students compare and contrast flatworms and roundworms and then discuss what structural features the respective worms have so as to increase surface area and thus increase the rate of diffusion.

Career

Parasitologist Parasitologists study the life cycles of parasites, animals, or plants that live in or on other organisms and take their nourishment from them. Some parasitologists specialize in studying animal parasites. Some medical doctors specialize in the diagnosis and treatment of people with parasitic infections. Other parasitologists specialize in studying plant parasites and their effects on agriculture. Group students into teams, and have them use library and on-line resources to research parasitology careers.

Teaching Tip GENERAL

Roundworms and Plants
Nematodes are organisms that gardeners have to deal with because they can destroy the health of a plant. When a gardener buys plants for a vegetable garden, the plants may be marked with the letter *N*. The *N* stands for "nematode resistant." Because roundworms that infect plants are so common in some areas of the country, it is often necessary to breed plants that are resistant to them. But some free-living nematodes—called beneficial nematodes—are good for plants. Beneficial nematodes kill many organisms in the soil that harm plants. Have interested students find out more about nematodes and plants.
LS Verbal

Group Activity GENERAL

Create a Life Cycle Divide the class into small groups. Have each group pick either *Ascaris* or *Necator*. Ask them to research their worm in a zoology textbook. Tell them to diagram a life cycle for their worm similar to the life cycles diagrammed earlier in the chapter. Display some of the life cycles in the classroom. LS Visual

Co-op Learning

INCLUSION Strategies GENERAL

• *Gifted and Talented*

Have the student visit or interview a veterinarian on the telephone to learn how roundworms affect cats and dogs. The student should find out if there are more than one type of roundworm that infects cats and dogs, what symptoms a pet infected with roundworms would have, and the treatment and care required for a pet infected with roundworms. The student should be able to report on the life cycle of the roundworm with a pet as a host when they present their interview findings to the class.

Teach, *continued*

DATA LAB

Identifying Parasites

Skills Acquired
Interpreting graphs

Teacher's Notes
Have students review the information about roundworms.

Answers to Analysis

1. Drug 2 slowed egg release in A and Drug 1 in B.
2. *Schistosoma:* intestinal blood and liver; *Ascaris:* intestines
3. A—*Schistosoma,* B—*Ascaris*
4. Drug 1 targets the intestine, which *Ascaris* infects. Drug 2 targets blood vessels, which *Schistosoma* infects.
5. Dams provide habitat for snails harboring *Schistosoma.*

Close

Reteaching — BASIC

Have students create fact sheets explaining the differences between flatworms and roundworms.
LS Verbal

Quiz — GENERAL

1. Compare a free-living worm with a parasitic worm. (A host supplies parasitic worms most of their needs; free-living worms do not depend on their host.)
2. How could a blood fluke infection be prevented? (better sanitation, controlling snail population)

Figure 18 Roundworm ***Ascaris.*** These adult *Ascaris* are in the stomach of a brown pelican.

to the intestine, where they mature and mate. Adult *Ascaris* may grow up to 0.3 m (1 ft) in length. **Figure 18** shows an *Ascaris* species whose final host is a bird.

Like *Ascaris, Trichinella* and *Necator* have complex life cycles that can involve a human host. *Trichinella* infects pigs and causes a serious disease called trichinosis *(trihk ih NOH sihs)* in humans. Infection with *Trichinella* can be avoided by not eating undercooked pork. Members of the genus *Necator* live mostly in the warm, moist soils of the tropics. Infection can occur when people step barefooted on soil containing hookworm larvae, which can enter the blood vessels if they penetrate the soles of the feet.

DATA LAB

Identifying Parasites

SC.G.1.4.1

Background

This graph shows how two drugs affect the release of eggs in a human infested with two parasites—*Schistosoma* and *Ascaris.* Drug 1 works by killing adult parasites in the intestines. Drug 2 works by killing adult parasites in the blood vessels. Use the graph and your knowledge of parasitic infections to answer the analysis questions.

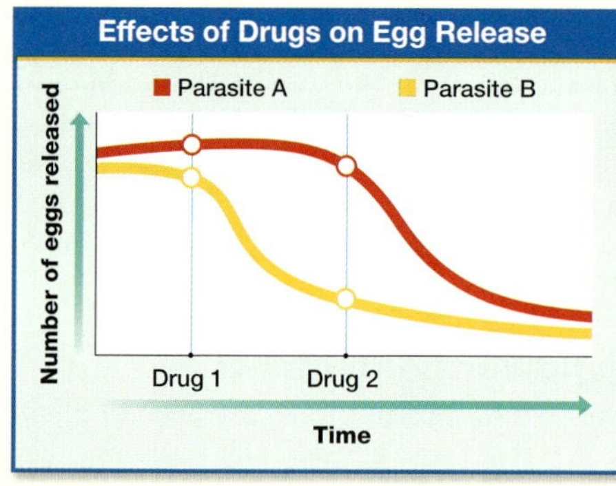

Analysis

1. **Describe** the response of the parasites to the two different drug treatments.
2. **Identify** the main human organs and tissues infected by the adult stages of *Schistosoma* and *Ascaris*. Use your textbook if necessary.
3. **Identify** which curve on the graph shows *Schistosoma* egg production and which shows *Ascaris* egg production.
4. **Critical Thinking Justifying Conclusions** Explain why you made the identifications you did in item 3.
5. **Critical Thinking Forming Hypotheses** *Schistosoma* spends part of its life cycle as a parasite of snails. Hypothesize a reason for an increase in the number of cases of schistosomiasis in villages near where hydroelectric dams have been built.

internet connect
www.scilinks.org
Topic: **Roundworms**
Keyword: **HX4159**
SCILINKS. Maintained by the National Science Teachers Association

Section 3 Review

1. **Compare** the internal and external anatomy of a planarian with that of a parasitic flatworm. SC.G.1.4.1
2. **Summarize** in words or with a diagram the life cycle of a blood fluke. SC.F.2.4.1
3. **Summarize** the life cycle of *Ascaris.* SC.F.2.4.1
4. **Describe** a major innovation in body plan that first occurred in roundworms. SC.F.1.4.2
5. FCAT Prep Which organ system is missing in a planarian?
   A digestive
   B nervous
   C reproductive
   D respiratory
   SC.F.1.4.2

Answers to Section Review

1. Parasitic flatworms have suckers and hooks on their anterior end, while planarians do not. Planarians have eyespots, while parasitic flatworms do not. Planarians have a highly branched intestine, while parasitic flatworms do not have a digestive system.
2. Answers will vary but should include the basic steps shown in Figure 16.
3. The eggs are carried through human waste and if ingested, enter the intestine and develop into larvae. The larvae bore into blood vessels and are carried to the lungs and pancreatic or bile ducts and eventually return to the intestine to reproduce.
4. Roundworms have a pseudocoelom.
5. **A.** Incorrect. The planarian has highly branched intestines. **B.** Incorrect. The planarian has a brain and nerve cords. **C.** Incorrect. The planarian is hermaphroditic. **D.** Correct. Oxygen and carbon dioxide are exchanged directly by body cells by diffusion.

Section Review Correlations

SC.F.1.4.2: Items 4, 5
SC.F.2.4.1: Items 2, 3
SC.G.1.4.1: Item 1

ZONE

Key Concepts

1 Sponges

- Sponges lack symmetry and tissues.
- Sponges are sessile filter feeders that draw sea water through pores into an internal cavity, trapping tiny aquatic organisms.
- The sponge's supportive skeleton is composed of soft spongin fibers, hard spicules, or a combination of both.
- Sponges that reproduce sexually are usually hermaphrodites. Sponges also reproduce asexually.

2 Cnidarians

- Cnidarians are radially symmetrical, with bodies made up of tissue. Their body form may be a medusa or a polyp.
- Cnidocytes are stinging cells found in the tentacles of cnidarians. Harpoon-like nematocysts are located within the cnidocytes.
- Most hydrozoans are colonial organisms that reproduce asexually, though many forms can also reproduce sexually.
- Jellyfish are active predators, and some have extremely potent toxins within their nematocysts.
- Jellyfish spend most of their lives as medusas and usually reproduce sexually.
- Sea anemones and corals have thick, stalklike polyp bodies. Their life cycle includes no medusa form.

3 Flatworms and Roundworms

- Flatworms have flattened bodies that lack a body cavity. Most flatworms, such as planarians and marine flatworms, are free-living, but others, such as flukes and tapeworms, are parasites.
- Tapeworms are intestinal parasites that absorb food directly through their skin.
- Flukes are endoparasitic flatworms. They have a protective covering called a tegument that keeps them from being digested by their host.
- Roundworms have a pseudocoelom and a one-way gut. Most are free-living, but some are animal parasites.

Key Terms

Section 1

ostia (618)
oscula (618)
sessile (618)
choanocyte (619)
amoebocyte (619)
spongin (620)
spicule (620)
gemmule (621)

Section 2

medusa (622)
polyp (622)
cnidocyte (623)
nematocyst (623)
basal disk (624)
planula (625)

Section 3

proglottid (631)
fluke (632)
tegument (632)

Alternative Assessment — GENERAL

Have students compare and contrast the simple invertebrates discussed in this chapter. Students may wish to create a chart to accompany their narrative that lists characteristics that all simple invertebrates have in common and those that are unique to each group. Students may also wish to use illustrations to explain certain points. LS Visual

Chapter Resource File

- Science Skills Worksheet GENERAL
- Critical Thinking Worksheet ADVANCED
- Test Prep Pretest GENERAL
- Chapter Test GENERAL

Answer to Concept Map

The following is one of several possible answers to Performance Zone item 15.

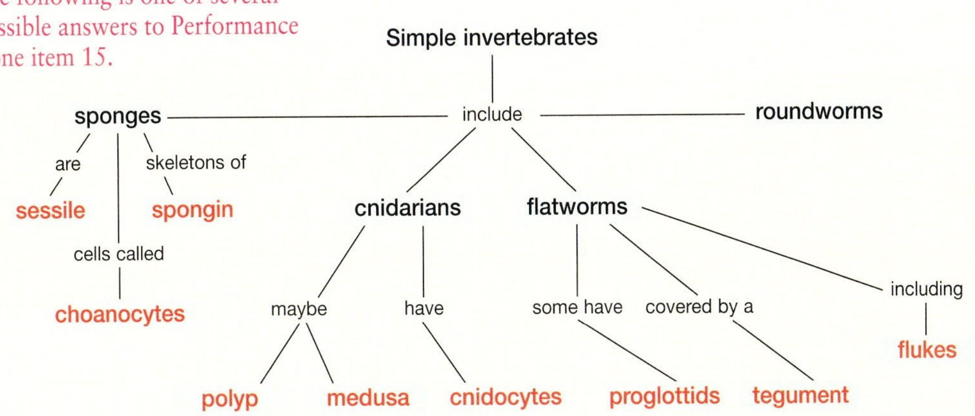

Understanding Key Ideas

1. b
2. a
3. c
4. a
5. b
6. Answers will vary. Reasons why no animal groups evolved from sponges might include that they are well adapted to their environments and have not been subject to strong selective pressure.
7. Parasitic flukes take their nourishment from their hosts by absorbing nutrients through their skin from the host's body fluids.
8. *acoelomate*—hydrozoans, cnidarians, anthozoans, flatworms; *pseudocoelomates*—roundworms; *coelomates*—none.
9. The Pulley Ridge coral reef is much deeper than most reefs, and its corals grow flat rather than vertically.
10. One possible answer to the concept map is found at the bottom of the Study Zone page.

Performance ZONE

Understanding Key Ideas

1. Which of the following is *not* a characteristic of sponges?
   a. body wall penetrated by many pores
   b. cells organized into tissues
   c. collar cells that trap food particles
   d. amoebocytes that transport food
   SC.G.1.4.1
2. What prevents self-fertilization among sponges?
   a. Gametes are released at different times.
   b. Few male sponges exist.
   c. Sponges are hermaphrodites.
   d. Encounters between members of the same species are rare.
   SC.F.2.4.1
3. A Portuguese man-of-war and a hydra are similar in that both SC.G.1.4.1
   a. are colonial.
   b. contain medusas and polyps.
   c. are hydrozoans.
   d. produce planulae.
4. Which sequence reflects the life cycle of *Obelia*? SC.F.2.4.1
   a. polyp ⟶ medusa ⟶ planula
   b. medusa ⟶ polyp ⟶ planula
   c. planula ⟶ medusa ⟶ polyp
   d. polyp ⟶ planula ⟶ medusa
5. Identify the function of the structure shown below.
   a. respiration
   b. water removal
   c. feeding
   d. digestion
   SC.F.1.4.2

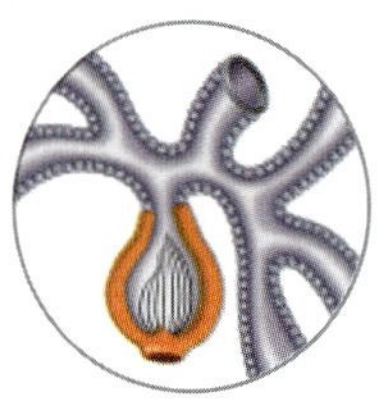

6. Porifera has been called a dead-end phylum. List some possible reasons why no animal group evolved from the sponges. SC.D.1.4.4
7. How are parasitic flukes able to live when they no longer possess a well-developed digestive system? SC.F.1.4.2
8. Classify all of the organisms covered in this chapter as either acoelomate, pseudocoelomate, or coelomate. (**Hint:** See Chapter 27, Section 1.) SC.G.1.4.1
9. **BIOWatch** What is the difference between the Pulley Ridge coral reef and most other reefs? SC.G.1.4.1
10. **Concept Mapping** Make a concept map that shows the major characteristics of sponges, cnidarians, flatworms, and roundworms. Include the following terms in your map: *sessile, choanocyte, spongin, medusa, polyp, cnidocyte, fluke, tegument,* and *proglottid.* SC.G.1.4.1

Critical Thinking

11. **Applying Information** Which animal would tend to have more water enter its body—a marine flatworm or a freshwater flatworm? Explain your answer. SC.G.1.4.1
12. **Evaluating Conclusions** A student concludes that infection with *Schistosoma* is more difficult to prevent than is infection with *Trichinella*. Evaluate this conclusion. SC.G.1.4.1

Alternative Assessment

13. **Identifying Structures** Make an anatomical drawing of the interior of a sponge, cnidarian, flatworm, or roundworm. Identify the species, and label at least 10 structures. Distribute copies of your drawing to your classmates. SC.F.1.4.2
14. **Forming a Model** In groups of three, research how one of the three different types of coral reefs—fringing, barrier, and atoll—is formed. Then build a model of your reef type or make a map showing where such reefs are located. Set up an exhibit of your work, and use a tape recording to create a "tour." SC.H.3.4.1

Critical Thinking

11. A freshwater flatworm would tend to have more water enter its body because the concentration of water is higher on the outside and will go into the body through the process of osmosis from an area of highest concentration to one of lower concentration.
12. Answers will vary. Many students may agree because *Trichinella* can be avoided simply by eating well-cooked pork or by not eating pork at all. *Schistosoma* is contracted by contact with contaminated water. It may be difficult for a person to avoid a contaminated water source, especially if it is the only water source available.

Alternative Assessment

13. Drawings will vary.
14. Models and exhibits will vary.

Assignment Guide

Section	Questions
1	1, 2, 6, 8, 10, 13
2	3, 4, 8, 9, 10, 13, 14
3	5, 7, 8, 10, 11, 12, 13

Chapter Review Correlations

SC.D.1.4.4: Item 6
SC.F.1.4.2: Items 5, 7, 13
SC.F.2.4.1: Items 2, 4
SC.G.1.4.1: Items 1, 3, 8-12
SC.H.3.4.1: Item 14

FCAT Prep

For questions 1–6, write your answers on a separate sheet of paper.

1. What is a sponge's protein skeleton composed of? SC.F.1.4.2
   A. amoebocytes
   B. mesoglea
   C. spicules
   D. spongin

2. What are specialized stinging cells found in cnidarians called? SC.F.1.4.2
   F. choanocytes
   G. cnidocytes
   H. medusas
   I. polyps

3. Which of the following is an anthozoan? SC.G.1.4.1
   A. hydra
   B. jellyfish
   C. box jelly
   D. sea anemone

4. What term describes the covering that protects endoparasites from the actions of digestive enzymes? SC.F.1.4.2
   F. basal disk
   G. osculum
   H. proglittid
   I. tegument

5. READ INQUIRE EXPLAIN Individuals of a single species of sponge may vary in appearance depending on their environment. Factors that affect sponge shape include differences in the material on which they grow, availability of space, and the speed and temperature of water currents.

   Analyze how these factors might make the classification of sponges confusing. SC.G.1.4.1

6. READ INQUIRE EXPLAIN When locating sunken ships, some treasure seekers use dynamite to blast away portions of the ocean floor. Blasting disturbs the ocean floor, where many invertebrates live. It also can cause harm to invertebrates that are filter feeders.

   Part A Describe the process used by filter feeders to obtain food. SC.F.1.4.2
   Part B Explain the harm that blasting causes to filter feeders. SC.G.2.4.6

Test TIP

If you find a particular question difficult, put a light pencil mark beside it and keep working. (Do not write in this book). As you answer other questions, you may find information that helps you answer the difficult question.

FCAT Doctor

Question 4 Answer I is the correct choice. Endoparasites, such as flukes, live in the intestines of their hosts. Tegument protects them from being digested like food. Answer F is incorrect because a basal disk is the area on the body of a hydra that allows it to attach to rocks or plants. Answer G is incorrect because an osculum is an opening in a sponge through which water exits. Answer H is incorrect because a proglittid is a rectangular body section of a tapeworm.

Question 5 A biologist attempting to classify sponges would need to be very familiar with the environment in which they live in order to avoid misclassification.

Question 6

Part A Filter feeders feed by allowing water to flow through sieve-like filtering structures that filter out nutrients.

Part B Tiny particles stirred up in the water by blasting could block the filtering structures in filter feeders.

Answers

1. D
2. G
3. D
4. I
5. If the variables that determine the shape of a sponge were not considered, two similar-looking sponges might be mistakenly identified as the same sponge, or vice versa.
6A. As water passes through the filter feeder's body, specialized structures or cells trap plankton and other tiny organisms, which are later digested.
6B. Particles stirred up by the blast fill the water, which interferes with filtering.

Test Prep Correlations

SC.F.1.4.2: Items 1, 2, 4, 6A
SC.G.1.4.1: Items 3, 5
SC.G.2.4.6: Item 6B

Exploration Lab

Observing Hydra Behavior

Teacher's Notes

Time Required 45 minutes

Ratings

EASY → HARD

Teacher Preparation 2
Student Setup 2
Concept Level 3
Cleanup 2

Skills Acquired
- Collecting Data
- Communicating
- Designing Experiments

Scientific Methods
In this lab, students will:
- Make Observations
- Ask Questions
- Analyze the Results
- Draw Conclusions

Materials and Equipment
Materials for this lab can be ordered from WARD'S. See *Master Materials List* at the front of this book for catalog numbers.

Safety Cautions
Caution students to handle glassware carefully to avoid injuring themselves or others.

Exploration Lab

Observing Hydra Behavior SC.F.1.4.7

SKILLS
- Using scientific processes
- Observing

OBJECTIVES
- **Observe** a hydra finding and capturing prey.
- **Determine** how a hydra responds to stimuli.

MATERIALS
- silicone culture gum
- microscope slide
- 2 medicine droppers
- *Hydra* culture
- *Daphnia* culture
- concentrated beef broth
- filter paper cut into pennant shapes
- forceps
- stereomicroscope

Magnification: 260×

Hydra feeding on *Daphnia*

Before You Begin

Cnidarians are carnivorous animals. A common cnidarian is **Hydra**, a freshwater organism that feeds on smaller freshwater animals, such as water fleas *(Daphnia)*. Hydras find food by responding to stimuli, such as chemicals and touch. The way an animal responds to stimuli is called **behavior**. The tentacles of a cnidarian are armed with **nematocysts**, shown below, which are used in defense and in capturing prey. When a hydra receives stimuli from potential prey, its nematocysts spring out and harpoon or entangle the prey. In this lab, you will observe the feeding behavior of hydras to determine how they find and capture prey.

1. Write a definition for each boldface term in the paragraph above.
2. Based on the objectives for this lab, write a question you would like to explore about the feeding behavior of hydras.

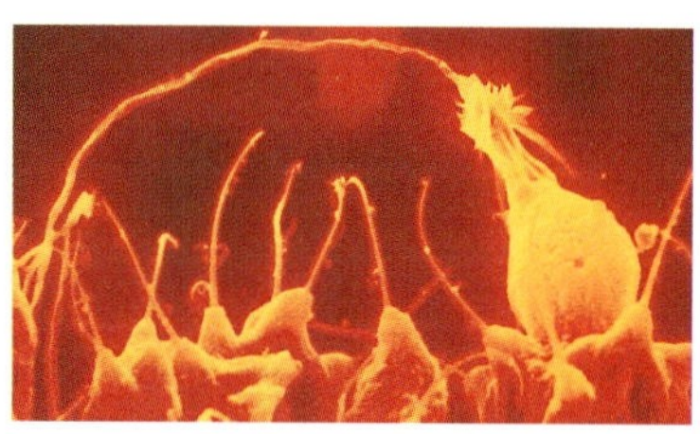

Nematocyst discharged

Procedure

PART A: Make Observations

1. To make an experimental pond for observing hydras, squeeze out a long piece of silicone culture gum. Arrange it to form a circular well on a microscope slide, as shown in the photograph below. **CAUTION: Glassware is fragile. Notify the teacher promptly of any broken glass or cuts.**

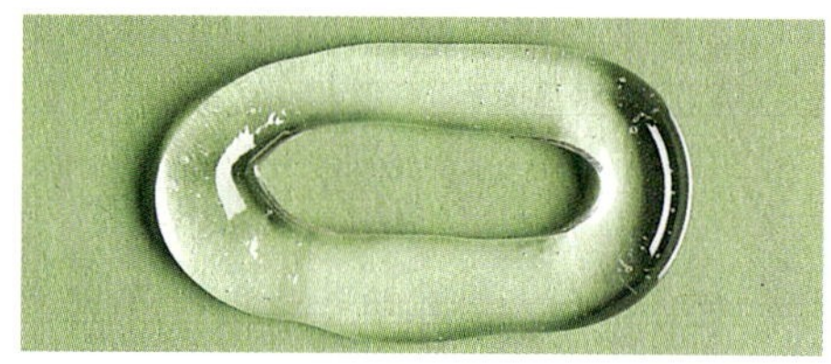

2. With a medicine dropper, gently transfer a hydra from its culture dish to the well on the slide, making sure the water covers the animal. **CAUTION: Handle hydras gently to avoid injuring them.** Allow the hydra to settle, then examine it under the high power of a stereomicroscope. Draw a hydra and label the body stalk, mouth, and tentacles.

Tips and Tricks
Demonstrate how to pick up hydras and *Daphnia* with a medicine dropper. If the tip of the dropper is too small, the dropper can be inverted. Instead of *Daphnia*, you may also use brine shrimp or a thread soaked in anything that has glutathione (e.g., liver) to elicit capture behavior in the hydra. Discuss some of the basic principles of sensory cells, neurons, and effector cells, relating the information to the way hydras capture prey. To observe whether a hydra responds to a chemical stimulus (nutrient), the students can hold a pennant-shaped piece of filter paper with forceps and move the long tip of the pennant near, but not touching, the hydra's tentacles (as a control). After observing and recording the hydra's response to the filter paper, they can dip the same piece of filter paper in beef broth and repeat the procedure. To observe how a hydra responds to touch, students can use the long tip of a clean pennant-shaped piece of filter paper and touch the hydra's tentacles, mouth, disk, and stalk. To observe feeding behavior, students can use a medicine dropper to transfer live *Daphnia* to the well with the hydra on the microscope slide.

PART B: Design an Experiment

3. Work with the members of your lab group to explore one of the questions written for step 2 of **Before You Begin.** To explore the question, design an experiment that uses the materials listed for this lab.
4. Write a procedure for your experiment. Make a list of all the safety precautions you will take. Have your teacher approve your procedure and safety precautions before you begin the experiment.

You Choose

As you design your experiment, decide the following:

- **a.** what question you will explore
- **b.** what hypothesis you will test
- **c.** how to observe a hydra's feeding behavior
- **d.** how to test a hydra's response to a stimulus, such as a chemical or a touch
- **e.** what your test groups and controls will be
- **f.** what to record in your data table

PART C: Conduct Your Experiment

5. Set up and carry out your experiment. **CAUTION: Handle hydras gently to avoid injuring them.**
6. Allow hydras to settle before exposing them to a test condition. If your hydra does not respond after a few minutes, obtain another hydra from the culture dish. Repeat your procedure.

PART D: Cleanup and Disposal

7. Dispose of lab materials and broken glass in the designated waste containers. Put hydras and daphnias in the designated containers. Do not put lab materials in the trash unless your teacher tells you to do so.
8. Clean up your work area and all lab equipment. Return lab equipment to its proper place. Wash your hands thoroughly before you leave the lab and after you finish all work.

Analyze and Conclude

1. **Analyzing Results** Describe a hydra's response to chemicals (beef broth).
2. **Analyzing Results** Describe a hydra's response to touch.
3. **Drawing Conclusions** How does a hydra detect its prey?
4. **Justifying Conclusions** Give evidence to support your conclusion about how hydras detect prey.
5. **Inferring Conclusions** Based on your observations, how do you think a hydra behaves when it detects a threat in its natural habitat?
6. **Inferring Conclusions** What happens to food that has not been digested by a hydra?
7. **Inferring Conclusions** How is a hydra adapted to a sedentary lifestyle?
8. **Further Inquiry** Write a new question about the behavior of hydras that could be explored with another investigation.

Do You Know?

Do research in the library or media center to answer these questions:

1. What different kinds of food does a hydra eat?
2. How is the feeding method of a hydra different from that of a sponge?

Use the following Internet resources to explore your own questions about sponges and cnidarians.

internet connect
www.scilinks.org
Topic: Hydra
Keyword: HX4102
SCILINKS. Maintained by the National Science Teachers Association

Sample Procedure

Use a medicine dropper to transfer live Daphnia to the well with the hydra on the microscope slide. Observe the hydra carefully under the stereomicroscope. Record observations in the data table. If the hydra does not respond, repeat the experiment with another hydra.

Answers to Before You Begin

1. *cnidarian*—an organism of the phylum Cnidaria that has a hollow gut with a single opening and flexible, fingerlike tentacles
   hydra—common freshwater cnidarian with a body stalk and tentacles surrounding the mouth
   behavior—how an animal responds to a stimulus
   nematocysts—cells found on a cnidarian that are used in defense and in capturing prey
2. Answers will vary. For example: How does a hydra respond to a *Daphnia* that swims close to it?

Answers to Analyze and Conclude

1. The hydra should show a feeding response to the beef broth, which includes expansion of its mouth, movement of its tentacles, and elongation of its body.
2. The hydra contracts its body.
3. A hydra responds to chemicals in the water to detect the presence of prey.
4. The hydra exhibited predatory behavior in response to the beef broth. Yet the hydra displayed nonpredatory behavior when touched.
5. Based on the hydra's response to touch, students should conclude that a hydra contracts its body when threatened.
6. Undigested food is released from the hydra's mouth.
7. Tentacles and nematocysts enable a hydra to capture prey that drift past it in slow-moving water. Its ability to contract when touched protects the hydra from predators.
8. Answers will vary. For example: How does a hydra respond to light?

Sample Data Table

Observations of Hydra

Response to filter paper	No response
Response to beef broth	Moves toward filter paper
Response to touch	Contracts
Feeding behavior	Tentacles pull the *Daphnia* to the hydra's mouth.

Answers to Do You Know?

1. Hydra will capture anything that is of the appropriate size, but their most common foods include small crustaceans, worms, and insect larvae.
2. Both sponges and hydras are sessile. However, sponges obtain food passively by filter feeding. They have no specialized structures for capturing food. Hydras are active feeders that have tentacles and nematocysts for capturing food.

Chapter Opener

SE **Quick Review,** p. 641 ■
SE **Reading Activity,** p. 641 ■
TE **Using the Figure,** p. 640 GENERAL
TE **Opening Activity,** p. 641 GENERAL

& 3 • 90 min pp. 642–650
Section 1
Mollusks

OSP **Lesson Plan***
TR **Bellringer***
TR **Section Outline***
TR **H24 Mollusk Body Plan***

TE **Activity** Mollusks on the Menu, p. 643 GENERAL
TE **Using the Figure,** p. 644 GENERAL
SE **Quick Lab** Modeling an Open Circulatory System, p. 645 ■
CRF **Datasheets for In-Text Labs** Modeling an Open Circulatory System*
TE **Activity** Snail Behavior, p. 646 BASIC
TE **Demonstration,** p. 647 GENERAL
TE **Activity** Identifying Mollusks, p. 647 BASIC
TE **Activity** Internet Research p. 648 ADVANCED
TE **Activity** Major Characteristics of Mollusks, p. 649 GENERAL
TE **Demonstration,** p. 649 BASIC
SE **Data Lab** Analyzing the Molluscan Body Plan, p. 650 ■
CRF **Datasheets for In-Text Labs** Analyzing the Molluscan Body Plan*
CRF **Quick Lab** Snails* BASIC

& 5 • 90 min pp. 651–656
Section 2
Annelids

OSP **Lesson Plan***
TR **Bellringer***
TR **Section Outline***
TR **H25 Anatomy of the Earthworm***

TE **Demonstration,** p. 651
SE **Quick Lab** Modeling a Closed Circulatory System, p. 652 ■
CRF **Datasheets for In-Text Labs** Modeling a Closed Circulatory System*
TE **Using the Figure,** p. 653 GENERAL
TE **Group Activity** Comparing Mollusks and Annelids, p. 654 GENERAL
TE **Group Activity** Worm Observations, p. 654 GENERAL
CRF **Skills Practice Lab** Earthworm Dissection* GENERAL
SE **Skills Practice Lab** Observing Characteristics of Clams, pp. 660–661 ■
CRF **Datasheets for In-Text Labs** Observing Characteristics of Clams*

& 7 • 90 min
Chapter Review and Assessment Resources

SE **Study Zone,** p. 657 ■
SE **Performance Zone,** p. 658 ■
SE **FCAT Test Prep,** p. 659 ■
TR **Concept Mapping***
CRF **Vocabulary Review Worksheet*** ■ GENERAL
CRF **Test Prep Pretest*** ■ GENERAL
CRF **Chapter Test*** ■ GENERAL
CRF **Chapter Test*** ADVANCED
OSP **Test Generator***
OSP **Test Item Listing***
OSP **Modified Chapter Test*** BASIC

go.hrw.com Holt Online Learning

Visit go.hrw.com for access to Holt Online Learning or enter the keyword **HX6 Home** for a variety of free online resources.

One-Stop Planner® CD-ROM

This CD-ROM package includes
- Lab Materials QuickList Software
- Holt Calendar Planner
- Customizable Lesson Plans
- Printable Worksheets
- ExamView® Test Generator
- Interactive Teacher Edition
- Holt PuzzlePro® Resources
- Holt PowerPoint® Resources

Guided Reading Audio CDs

These CDs are designed to help auditory learners and reluctant readers. (Audio CDs are also available in Spanish.)

Compression guide:
To shorten your instruction because of time limitations, omit blocks 1 and 6.

KEY

TE	Teacher Edition	CRF	Chapter Resource File	*	Also on One-Stop Planner
SE	Student Edition	TR	Teaching Transparency	■	Also Available in Spanish
OSP	One-Stop Planner			◆	Requires Advance Prep

SKILLS DEVELOPMENT AND PRACTICE	SECTION REVIEW AND ASSESSMENT	CORRELATIONS
		Florida Sunshine Standards/FCAT
TE Reading Skill Builder Interactive Reading, p. 643 BASIC **TE** Skill Builder Vocabulary p. 644 **TE** Reading Skill Builder Paired Summarizing, p. 646 BASIC **CRF** Directed Reading Worksheet* BASIC **CRF** Active Reading Worksheet* GENERAL	**SE** Section 1 Review ■ **TE** Reteaching, p. 650 BASIC **TE** Quiz, p. 650 GENERAL **TE** Alternative Assessment, p. 650 GENERAL **CRF** Quiz* ■ BASIC	SC.D.1.4.4, SC.F.1.4.2, SC.G.1.4.1, SC.H.3.4.1
CRF Directed Reading Worksheet* BASIC **CRF** Active Reading Worksheet* GENERAL **CRF** Critical Thinking Worksheet* ADVANCED **CRF** Science Skills Worksheet* GENERAL	**SE** Section 2 Review ■ **TE** Reteaching, p. 656 BASIC **TE** Quiz, p. 656 GENERAL **TE** Alternative Assessment, p. 656 GENERAL **CRF** Quiz* ■ BASIC Modified Worksheet* BASIC	SC.F.1.4.2, SC.G.1.4.1, SC.H.1.4.1, SC.H.3.4.1

SCILINKS. NSTA
www.scilinks.org
Maintained by the **National Science Teachers Association**

Topic: Mollusks
SciLinks Code: HX4124

Topic: Annelids
SciLinks Code: HX4012

Topic: Leeches
SciLinks Code: HX4113

Overview

Tell students that the purpose of this lesson is to teach them about the similarities and differences between the three major classes of mollusks – the bivalves, the gastropods, and the cephalopods – and about the physical characteristics and various adaptations of segmented worms.

Using the Figure

Direct students' attention to the figure opening this chapter. Tell them that an octopus in the open sea might seem easy pickings for predators such as moray eels, sea lions and bigger octopuses, but the octopus can marshal a dazzling array of defenses. Like a squid, it can disorient a pursuer with a burst of purplish-black ink. Should it lose an arm to the jaws of a predator, it can grow a new one. More impressive still, an octopus can change color in less than a second, to disappear into its surroundings.

Assessing Prior Knowledge

Before beginning this chapter, be sure your students can answer the questions in the ***Quick Review*** on the facing page. If they cannot, encourage them to reread the sections indicated.

Common octopus

640

Chapter Correlations | **Florida Sunshine Standards/FCAT**

SC.D.1.4.4 Earth's systems and organisms are the result of a long, continuous change over time.

SC.F.1.4.2 Body structures are uniquely designed and adapted for their function.

SC.G.1.4.1 There is great diversity and interdependence among living things.

CHAPTER

29 Mollusks and Annelids

Florida Sunshine Standards/FCAT
SC.D.1.4.4 Earth's systems and organisms are the result of a long, continuous change over time.
SC.F.1.4.2 Body structures are uniquely designed and adapted for their function.
SC.G.1.4.1 There is great diversity and interdependence among living things.

Quick Review

Answer the following without referring to earlier sections of your book.

1. **Describe** the process of diffusion. *(Chapter 4, Section 1)*
2. **Define** the term *plankton*. *(Chapter 17, Section 3)*
3. **Define** the terms *gill* and *hydrostatic skeleton*. *(Chapter 27, Section 2)*
4. **Distinguish** between an open and closed circulatory system. *(Chapter 27, Section 2)*

Did you have difficulty? *For help, review the sections indicated.*

Reading Activity

As you read this chapter, draw and label a diagram of an idealized mollusk. Your mollusk should have all of the anatomical features described in the chapter (although not all mollusks possess all features).

Looking Ahead

Section 1
Mollusks
A True Coelom
Key Characteristics of Mollusks
Body Plans of Mollusks

Section 2
Annelids
The First Segmented Animals
Annelid Groups

The octopus is one of the most intelligent of the invertebrates. It can perform complex behaviors, including building its own house out of debris found on the ocean floor.

internet connect

www.scilinks.org
National Science Teachers Association *sci*LINKS Internet resources are located throughout this chapter.

SCiLINKS. Maintained by the National Science Teachers Association

Chapter Resource File

- Vocabulary Worksheets
- Concept Mapping

Opening Activity

Mollusk Shells Invite students to bring to class any shell collections they might have. Display the specimens with mollusks or mollusk shells. (Examples include clams, mussels, scallops, oysters, and snails.) Ask students to look for similarities among the mollusk shells and the other shells.

Quick Review Answers

1. In the process of diffusion, a substance moves from an area where it is more concentrated to an area where it is less concentrated.
2. Plankton are microscopic organisms belonging to the Kingdom Protista or Kingdom Animalia that float on the surface of Earth's oceans.
3. Gills are membranous structures through which oxygen is taken up from the environment, which is usually a watery one. A hydrostatic skeleton is a fluid-filled body cavity that provides structural support and a framework against which muscles can push.
4. In a closed circulatory system, blood is always kept inside blood vessels or a heart. In an open circulatory system, blood can move out of the blood vessels into cavities of the body.

Reading Activity Answers

Students' drawings should resemble **Figure 2,** with all of the structures labeled.

Section 1

Focus

Overview

Before beginning this section review with your students the objectives listed in the Student Edition. Tell students that the purpose of this lesson is to teach them about the general structural and physiological characteristics of mollusks, including their body plan and organ systems. They will also learn about distinguishing characteristics of the three major groups of mollusks: the gastropods, bivalves, and cephalopods.

Bellringer

Ask students to think about a snail, a clam, and an octopus. Tell students that these are all mollusks. Ask them to write down some characteristics that these animals have in common. (Students will most likely note the prominent foot, or its modification.)

Motivate

Identifying Preconceptions — BASIC

Ask students to hypothesize why some mollusks secrete a shell and others do not. Guide their thinking by asking how mollusks without shells might protect themselves. (Students should recall that squids and octopuses spray ink and can propel themselves away from danger. Slugs secrete noxious substances that discourage predators.)

Section 1 Mollusks

Florida Sunshine Standards/FCAT

SC.D.1.4.4 Earth's systems and organisms are the result of a long, continuous change over time.

SC.F.1.4.2 Body structures are uniquely designed and adapted for their function.

SC.G.1.4.1 There is great diversity and interdependence among living things.

Objectives

- **Summarize** the evolutionary relationship between mollusks and annelids. SC.D.1.4.4
- **Describe** the key characteristics of mollusks. SC.G.1.4.1
- **Describe** excretion, circulation, respiration, and reproduction in mollusks. SC.F.1.4.2
- **Compare** the body plans and feeding adaptations of gastropods, bivalves, and cephalopods. SC.F.1.4.2

Key Terms

trochophore
visceral mass
mantle
foot
radula
nephridium
adductor muscle
siphon

A True Coelom

While most of the simple invertebrates you read about in the last chapter may be unfamiliar to you, chances are good that you have seen many mollusks and annelids. Snails, slugs, oysters, clams, scallops, octopuses, and squids are all mollusks. If you have seen an earthworm, then you know what an annelid is. While a snail may not seem to have much in common with an earthworm, these two very different-looking animals are related.

Mollusks and annelids were probably the first major groups of organisms to develop a true coelom. (Recall that in animals that have a true coelom, or body cavity, the gut and other internal organs are suspended from the body wall and cushioned by the fluid within the coelom.) Another feature shared by mollusks and annelids is a larval stage called a **trochophore** *(TRAHK oh fawr)*, which develops from the fertilized egg. In some species, the trochophore, shown in **Figure 1,** is free-swimming and propels itself through the water by movement of cilia on its surface. The presence of a trochophore larva in mollusks and annelids suggests that they share a common ancestor.

Members of the phylum Mollusca make up the second largest animal phyla, exceeded only by phylum Arthropoda. Mollusks are abundant in almost all marine, freshwater, and terrestrial habitats. There are more species of terrestrial mollusks than there are of terrestrial vertebrates. These mollusks often go unnoticed because people are not accustomed to looking for them. Seven classes of mollusks make up the phylum Mollusca. The three major classes are Gastropoda (snails and slugs), Bivalvia (clams, oysters, and scallops), and Cephalopoda (octopuses and squids).

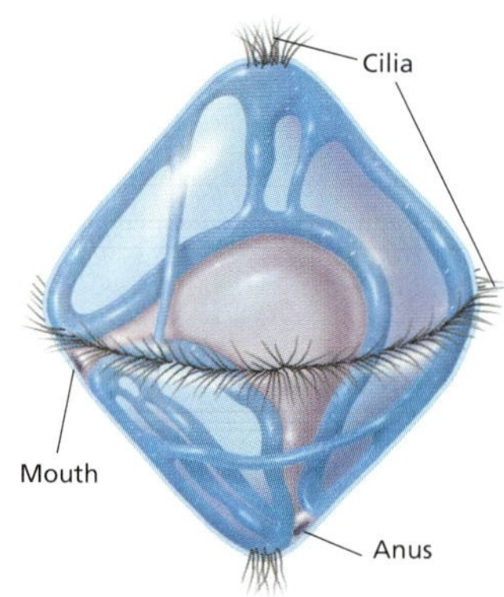

Figure 1 Trochophore larva

The microscopic trochophore larva has a belt of cilia that circles its body. The beating of the cilia propels the trochophore through the water.

Evolutionary Milestone

5 Coelom

A true coelom develops entirely within the mesoderm. Contact between the mesoderm and endoderm during the development of the embryo leads to the development of complex organs.

Chapter Resource File

- Directed Reading BASIC
- Active Reading GENERAL
- Data Sheet for Quick Lab GENERAL
- Data Sheet for Data Lab GENERAL

One-Stop Planner CD-ROM

- Reading Organizers BASIC
- Reading Strategies BASIC

Transparencies

TR H24 Mollusk Body Plan

Section Correlations

SC.D.1.4.4
SC.F.1.4.2
SC.G.1.4.1

Key Characteristics of Mollusks

Despite their varied appearance, the members of the different groups of mollusks share a number of key characteristics.

1. **Body cavity.** The body cavity in mollusks is a true coelom, although in most species it is reduced to a small area immediately surrounding the heart.
2. **Symmetry.** Most mollusks exhibit bilateral symmetry.
3. **Three-part body plan.** The body of every mollusk has three distinct parts: the visceral mass, the mantle, and the muscular foot, as shown in **Figure 2.** The **visceral** *(VIS uhr uhl)* **mass** is a central section that contains the mollusk's organs. The mantle is wrapped around the visceral mass like a cape. The **mantle** is a heavy fold of tissue that forms the outer layer of the body. Finally, every mollusk has a muscular region called a **foot,** which is used primarily for locomotion.
4. **Organ systems.** Mollusks have organ systems for excretion, circulation, respiration, digestion, and reproduction.
5. **Shell.** Many mollusks have either one or two shells that serve as an exoskeleton, protecting their soft body. The shell is composed of protein that is strengthened by calcium carbonate, an extremely hard mineral.
6. **Radula.** All mollusks except bivalves have a **radula** *(RAJ uh luh)*, a tongue-like organ located in their mouth. The radula, shown in Figure 2, has thousands of pointed, backward-curving teeth arranged in rows. When a mollusk feeds, it pushes its radula out of its mouth, and the teeth scrape fragments of food off rocks or plant matter. Mollusks that are predators use their radula for attacking their prey.

internet connect
www.scilinks.org
Topic: Mollusks
Keyword: HX4124
SciLinks. Maintained by the National Science Teachers Association

internet connect
www.scilinks.org
Topic: Aquariums and Museums of Florida
Keyword: HX6F01
SciLinks. Maintained by the National Science Teachers Association

Figure 2 Three-part body plan

All mollusk bodies are composed of a visceral mass, a mantle, and a foot. Most mollusks also have a radula.

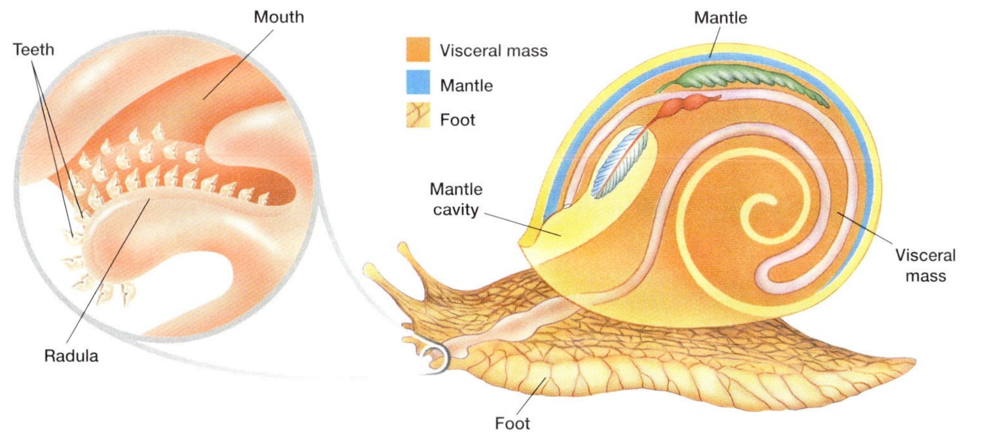

Teach

Activity — GENERAL

Mollusks on the Menu Provide students with menus from several seafood restaurants and seafood cookbooks from the library. Ask students to identify all the mollusks they can find. (Students should be able to identify oysters, clams, and scallops. Depending on the menu or cookbook, there may also be snails [escargot], squids [calamari], octopuses, abalone, and mussels.) Ask students what part of the body is eaten for each mollusk they identify. (oysters, clams, snails, abalone, mussels – the whole body, except the shell; squids and octopuses – the foot, which is modified into tentacles and the mantle; scallops – the adductor muscle)

READING SKILL BUILDER — BASIC

Interactive Reading Assign Chapter 29 of the *Holt Biology* Guided Audio CD Program to help students achieve greater success in reading the chapter.

did you know?

Banana Slugs Banana slugs live on damp foggy forest floors on the west coast of North America only. They like foggy summers and rainy winters, and hide in damp places. The yellow banana slug is only found in the small region between the Salinas Valley and the San Francisco peninsula of California. It is the mascot for the University of California at Santa Cruz.

MISCONCEPTION ALERT

Mantle versus Shell Students often confuse the mantle with the shell or assume that the mantle composes part of the shell. Be sure students understand the distinction between the two: the mantle, which is composed of tissue, secretes the shell, which is composed of protein and calcium carbonate.

Teaching Tip

Excretory Strategies Relate the excretion of wastes in nephridia to cleaning up a table after a meal. Nephridia get rid of wastes as well as substances that the mollusk needs; thus, some substances must be reabsorbed before the wastes are discharged. The process is similar to throwing away everything on the table after the meal, and then going through the trash to recover the plates, silverware, and leftovers.

Using the Figure —GENERAL

Use **Figure 3** to point out that mollusks have advanced organ systems. Ask students to categorize each of the organs as part of one of the following systems: digestive, circulatory, respiratory, or excretory. (*digestive system*—mouth, radula, intestine, stomach; *circulatory system*—heart; *respiratory system*—gill; *excretory system*— nephridium) **LS Visual**

SKILL BUILDER

Vocabulary *Nephros* is the Greek word for "kidney." Nephritis is inflammation of the kidneys, a nephrectomy is the removal of a kidney, a nephron is the structure in a kidney that filters blood, and nephrology is the field of medicine dedicated to the study and treatment of the kidneys.

Figure 3 Mollusk body plan

Although mollusks vary greatly in body form, they all have complex organ systems.

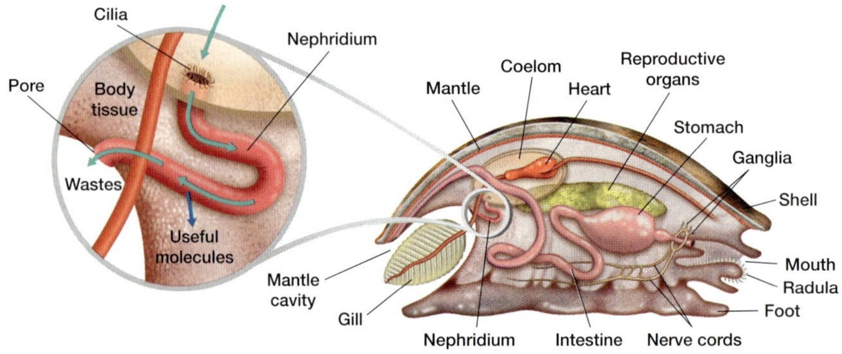

Organ Systems

Mollusks are the only coelomates without segmented bodies. Like the roundworms, mollusks have a one-way digestive system. Mollusks, however, are more complex than roundworms. As a group, mollusks are quite diverse, and no one mollusk can represent the phylum as a whole. **Figure 3** shows the basic mollusk body with organs that are characteristic of the phylum.

Excretion Mollusks are one of the earliest evolutionary lines to have developed an efficient excretory system. A mollusk's coelom is a collecting place for waste-laden body fluids. The beating of cilia pulls the fluid from the coelom into tiny tubular structures called **nephridia** *(nee FRIHD ee uh),* also shown in Figure 3. The nephridia recover useful molecules (sugars, salts, and water) from the coelomic fluid. The recovered molecules are reabsorbed into the mollusk's body tissues. The remaining fluid waste leaves the mollusk's body through a pore that opens into the mantle cavity. Nephridia are found in all coelomate animals except arthropods and chordates.

Circulation The digestive tube of mollusks and other coelomates is surrounded by mesoderm, which acts as a barrier to the diffusion of nutrients into the cells of the body. Mollusks have a circulatory system. Recall that in a circulatory system, blood carries nutrients and oxygen to tissues and removes waste and carbon dioxide. Most mollusks have a three-chambered heart and an open circulatory system. Octopuses and squids are exceptions because they each have a closed circulatory system.

Respiration Respiration among mollusks is carried out in a variety of ways. Most mollusks respire with gills, which are located in the mantle cavity. Mollusk gills extract 50 percent or more of the dissolved oxygen from the water that passes over them.

In freshwater snails, ciliated gills beat on the inner surface of the mantle. This causes a continuous stream of water to pass over the

644

Trends in Marine Biology

Mollusk Classes The living mollusks are divided into the following seven classes:

Aplacophora contains about 250 species of marine, wormlike animals that have no shell. Monoplacophora has about a dozen living species; they are very small and covered with a shell. They live at great ocean depths. Polyplacophora has about 800 species of animals known as chitons. Chitons have a shell of eight plates that usually overlap. Scaphopods have elongate, tubular shells that are open at both ends. All 350 or so species are marine.

The cephalopods include the octopus, squid, cuttlefish, and nautilus. The foot of cephalopods is modified into tentacles, and only a few have a shell. All 600 to 650 species are marine predators or scavengers. Gastropods usually have a single, coiled shell and a well-developed foot. They are found in marine, freshwater, and terrestrial habitats. There are about 60,000 species of gastropods. Bivalves have two valves connected by a flexible ligament. Most are filter feeders. There are about 10,000 living species of bivalves.

gills. Most terrestrial snails have no gills. Instead, the thin membrane that lines their empty mantle cavity functions like a primitive lung. This membrane must be kept moist for oxygen to diffuse across it. Therefore, terrestrial snails, shown in **Figure 4,** are most active at night or after it rains when air has a high moisture content. During dry weather, a terrestrial snail pulls back into its shell and plugs the opening with a wad of mucus to keep water in. Sea snails also lack gills, and gas exchange takes place directly through their skin.

Reproduction Most species of mollusks have distinct male and female individuals, although some snails and slugs are hermaphrodites. Certain species of oysters and sea slugs are able to change from one sex to the other and back again. Many marine mollusks are moved from place to place as their trochophore larvae drift in the ocean currents. Octopuses, squids, freshwater snails, and some freshwater mussels have no free-swimming larvae. In these mollusks, the larval stage occurs within the egg, and a juvenile-stage mollusk hatches from the egg.

Figure 4 **Terrestrial snails.** Terrestrial snails are most active when the air is moist.

QUICK LAB

Modeling an Open Circulatory System SC.H.3.4.1

You can model an open circulatory system using simple items to represent the heart, blood vessels, blood, and body tissues of a living organism.

Materials

surgical tubing, 15 cm (about 6 in.) piece; clear plastic tubing, 15 cm (about 6 in.) and 7.5 cm (about 3 in.) pieces; shallow pan filled with water; eyedropper; food coloring

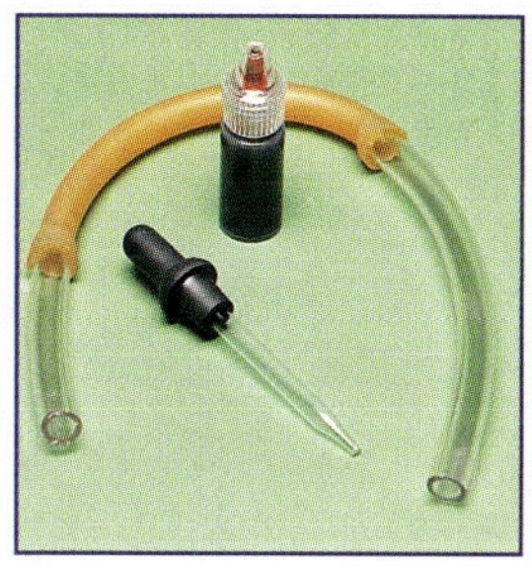

Procedure

1. Connect the surgical tubing to the two pieces of clear plastic tubing, as shown above.
2. Place the tubing into the tray filled with water. Allow the tubing to fill with water and rest on the bottom.
3. With the tubing still submerged, use an eyedropper to place two drops of food coloring into the short piece of clear plastic tubing.
4. With your thumb and index finger, squeeze along the piece of surgical tubing to pump the food coloring through the system.
5. As you continue to pump, observe the movement of food coloring.

Analysis

1. **Describe** what happened when you squeezed along the tubing.
2. **Identify** the structures represented by the pan of water, the surgical tubing, and the clear plastic tubing.
3. **Critical Thinking** **Evaluating Results** Evaluate your model's efficiency at pumping blood through the system.
4. **Critical Thinking** **Analyzing Methods** How does this model differ from a real circulatory system?
5. **Critical Thinking** **Analyzing Methods** How could you modify the model to make it more accurate?

645

Teaching Tip — GENERAL

Oxygen Delivery Systems Point out to students that it takes both respiratory and circulatory systems to deliver oxygen to the tissues of mollusks. If either system stops functioning, the mollusk will die. Ask students to consider the various ways a mollusk could suffocate. (oxygen-deficient water, nonfunctioning cilia, nonfunctioning heart, loss of blood, dried out gills or lungs)

QUICK LAB

Modeling an Open Circulatory System

Skills Acquired
Applying information, modeling, observing, evaluating

Teacher's Notes
Make sure that students remove any air from the tubing once it is submerged. Use a dark dye. If the dye diffuses too quickly in the water, try using an oil-based dye or small, lightweight, colored beads.

Answers to Analysis

1. The dye moves away from the pumping action but is also drawn back into the open tubing.
2. The pan of water represents the body and tissues of the mollusk, surgical tubing represents the heart, and the clear plastic tubing represents the blood vessels.
3. Answers will vary.
4. In a real system, there would be many more vessels carrying the blood out to the tissues and back toward the heart. The oxygenated blood would also be absorbed by the various tissues, and the used blood would be reoxygenated by the gills.
5. Answers will vary, but may include adding models for individual organs, adding more "blood," and adding more blood vessels.

did you know?

Reproduction in Oysters Oysters are broadcast spawners, ejecting sperm and eggs into the water, where fertilization occurs. Oysters may spawn multiple times during the season, with each female releasing millions of eggs during each spawning event. Oysters are "protandric" because they develop first as males and then become females due to regression of the testes and subsequent development of the ovaries.

Fertilized eggs develop in a free-floating environment, and a trochophore larva hatches from the egg 6-9 hours after fertilization. After about three weeks, the larvae progress through a number of larval stages. The last larval stage has a muscular foot that the larva uses to crawl on the bottom, searching for a suitable place to attach. Attachment sites include almost any hard surface such as other living oysters, oyster shell, rocks, docks, pilings, and glass bottles. Larvae often settle on top of other oyster shells, forming large beds or oyster reefs. Once attached, the newly settled oysters stay in the same location throughout their lives.

Teach, *continued*

Teaching Tip

Mollusk Shells The beauty, diversity, and hardness of mollusk shells accounts for their wide use as decorations. Even the inside linings of shells are used. Shells of freshwater mussels are used to make pearl buttons, and mother-of-pearl from oysters is used to decorate ornamental boxes, jewelry, and musical instruments.

Activity — Basic

Snail Behavior Allow students to explore snail behavior. Obtain several land snails from gardens, wooded areas, or a biological supply house. Have students devise ways to test the snails' responses to touch, light, moisture, and gravity. Caution students to use extreme care in handling snails, making sure not to let them dry out. When not in use, keep the snails in a cool, moist terrarium with pieces of lettuce. LS Kinesthetic

Reading Skill Builder — Basic

Paired Summarizing Encourage students to pause after each paragraph they read to summarize the main points to a partner. You might model the technique by having a student read a paragraph out loud and summarize it. Then have another student read the next paragraph and summarize it. LS Interpersonal

English Language Learners

Study TIP

Organizing Information
Make a table to organize information about mollusks. Across the top, write the headings *Gastropods, Bivalves,* and *Cephalopods.* Along the side, write *Foot, Shell, Feeding, Reproduction.* Add information (and perhaps more terms along the side) to the table as you read.

Body Plans of Mollusks

The basic mollusk three-part body plan differs in each class of mollusks. As you read about the different classes of mollusks, you will see how the mollusk shell and foot are adapted for many different living conditions.

Gastropods

Gastropods—snails and slugs—are primarily a marine group that has successfully invaded freshwater and terrestrial habitats. They range in size from microscopic forms to the sea hare *Aplysia,* which reaches 1 m (almost 40 in.) in length. Most gastropods have a single shell. During the evolution of slugs and nudibranchs *(NOO dih branks),* the shell was lost completely. **Figure 5** shows three terrestrial tree snails and a nudibranch (sea slug). The foot of gastropods is adapted for locomotion. Terrestrial species secrete mucus from the base of their foot, forming a slimy path that they can glide along. Most gastropods have a pair of tentacles on their head with eyes often located at the tips.

Gastropods display varied feeding habits. Many are herbivores that scrape algae off rocks using their radula. Some terrestrial snails can be serious garden and agricultural pests, using their radula to saw off leaves. Sea slugs and many other gastropods are active predators. Whelks and oyster drills, for example, use their radula to bore holes in the shells of other mollusks. Then they suck out the soft tissues of their prey. In cone shells, such as the one shown in **Figure 6,** the radula is modified into a kind of poison-tipped harpoon that is shot into prey. The poison paralyzes the prey, which is then swallowed whole.

Throughout human history, snails have been a source of food for humans. Land snails belonging to the genus *Helix* are raised on snail farms and are consumed in great quantities. While freshwater snails are rarely eaten, a few marine species, such as conchs *(KAHNGKS),* are considered delicacies.

Figure 5 Snails and slugs. These Florida tree snails live where the air is moist enough to keep them from drying out. Sea slugs are often brilliantly colored, and most are under 15 cm (6 in.) in length.

646

Cultural Awareness

Uses of Shells Mollusk shells have served many different functions in various cultures, especially in making jewelry and crafts. In addition, shells have been used for several practical purposes. Phoenicians and Romans used Murex sea snails to make a purple dye for coloring fabric. Native North Americans carved large clam shells into wampum beads and used other shells as money. Filipinos cut thin Placuna oyster shells to fit into wooden frames as windowpanes.

Figure 6 Cone shell. This cone shell searches the ocean bottom for prey. Once located, the prey is secured by the cone shell's radula and swallowed whole.

Bivalves

Most bivalves are marine, but some also live in fresh water. Many species of freshwater mussels are found throughout the rivers and lakes of North America and are important links in aquatic food chains. Oysters and mussels are important sources of food for humans. All bivalves have a two-part hinged shell. The valves, or shells, of a bivalve are secreted by the mantle. Two thick muscles, the **adductor muscles,** connect the valves. When these muscles are contracted, they cause the valves to close tightly. While most species of mollusks are sessile, some can move from place to place quite fast if necessary. For example, a swimming scallop opens and closes its valves rapidly. This pushes it along with the jets of water released when its valves snap shut.

Bivalves are unique among the mollusks because they do not have a distinct head region or a radula. A nerve ganglion above their foot serves as a simple brain. Most bivalves have some type of simple sense organs. For example, some bivalves have sensory cells located along the edge of their mantle that respond to light and touch.

Most bivalves are either male or female, but a few species are hermaphroditic. Bivalves reproduce sexually by releasing sperm and eggs into the water, where fertilization occurs. The fertilized eggs develop into free-swimming trochophore larvae. The larvae of a few freshwater mussels are brooded in a pouch within the mollusk's gills. The larvae are then released into the water, and they complete their larval stage as parasites on fish. This is a very unusual life cycle for a mollusk.

Most bivalves are filter feeders. Many, such as the clam illustrated in **Figure 7,** use their muscular foot to dig down into the sand. Once there, the cilia on their gills draw in sea water through hollow tubes called **siphons** *(SIE fuhns).* The water moves down one siphon tube, over the gills, and out the other siphon tube. The gills are used for feeding as well as respiration. A sticky mucus covers the gills, and as water moves over the gills, small marine organisms and organic material become trapped in the mucus. The cilia then direct the food-laden mucus to the bivalve's mouth.

WORD Origins

The class name *Bivalvia* comes from the Latin *bi,* meaning "two," and *valva,* meaning "part of a door."

Figure 7 Clam. Many bivalves, like this clam, burrow into sand or mud and feed by drawing sea water in one siphon and expelling it out the other.

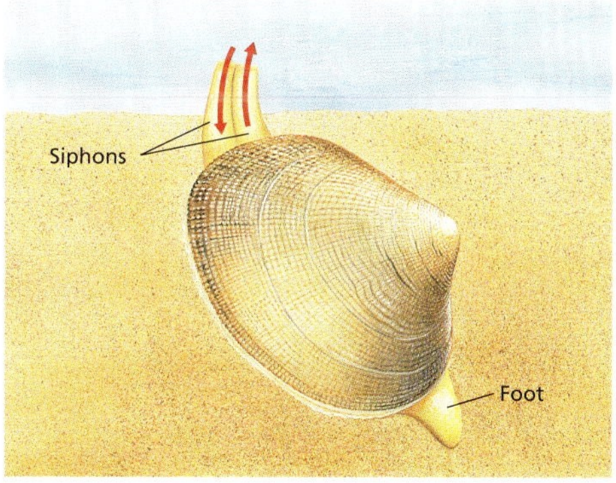

647

Demonstration — GENERAL

Show students a dissected fresh or preserved mollusk such as a clam or oyster. Ask them to sketch the mollusk and label the following parts: shell, mantle, gonads, foot, gut, heart, and gills. LS Visual English Language Learners

Activity — BASIC

Identifying Mollusks Obtain several fossil mollusks, and set them out for students to view. Fossil mollusks are some of the most common and inexpensive fossils you can obtain, and they are easily identifiable as bivalves, gastropods, or shelled cephalopods. Have students match modern varieties of mollusks to their fossils. LS Visual

GEOGRAPHY CONNECTION

The zebra mussel is a small, freshwater bivalve native to Europe. At some point, zebra mussels hitchhiked across the Atlantic Ocean in the ballast water of ships. Although they were not here in detectable numbers before 1985, they have infested waters in the Great Lakes, where they have no natural predators. Although small, the mussels attach themselves to any hard surface, including ship hulls, docks, and even other mussels.

Teach, *continued*

Teaching Tip

A Smashing Success Tell students that some bivalves have evolved to survive the crushing pressure of waves. Muscles and other bivalves that often cling to exposed rocks must be able to withstand daily battering by waves. They survive the poundings by having shells that are very hard and are shaped so as to deflect onrushing water.

Activity — ADVANCED

Internet Research Have students research mollusk consumption using the Web site in the Internet Connect box on page 666. Have students write a report summarizing their findings.

BIOWatch

Fighting Bacteria in Raw Oysters

Teaching Strategies

- Tell students that a single oyster can pump and filter more than 100 liters of water in 24 hours. Ask students: Why is it advantageous for oysters to filter large amounts of water? (Oysters must filter large amounts of water in order to obtain sufficient amounts of food.)

Discussion

Ask students: What is the best way to prepare oysters? (Answers will vary, but all methods should ensure that oysters are thoroughly cooked before serving.)

Figure 8 **Pearl oyster.** The best pearls come from oysters belonging to the genus *Pinctada*.

Like clams, oysters and scallops use their gills to filter food from the water. Oysters are permanently attached to rocks in the open water, where they feed. Scallops swim, and water passes over their gills as they move.

Many species of bivalves, such as the oyster in **Figure 8,** produce pearls. Pearls form when a tiny foreign object, such as a grain of sand, becomes lodged between the mollusk's mantle and shell. Bivalves respond to these irritants by coating them with thin sheets of nacre *(NAY kuhr),* also called mother-of-pearl. Nacre is the same hard, shiny substance that composes the inner shell surface. Successive layers of nacre are added until the foreign body is completely enclosed in the newly formed pearl.

While many bivalves form pearls, only a few species produce the beautifully colored nacre essential for gem-quality pearls. In fine pearls, the nacre contains tiny, overlapping mineral crystals. These crystals act like prisms, breaking up any light that falls on them into rainbows of color. This is what gives these pearls their iridescence. The mineral crystals found in the nacre of ordinary pearls are larger and do not reflect light as beautifully.

FLORIDA BIOWatch

Fighting Bacteria in Raw Oysters

Oysters flourish in Florida's estuaries, where nutrient-rich fresh water meets the salt water. A popular seafood item, oysters can be fried, baked, steamed, or broiled. Some people, however, choose to eat them raw.

A Dangerous Bacterium

Oysters are filter feeders. As they feed, they strain single-celled algae from the ocean water. Oysters sometimes also take in an unwelcome guest: the bacterium *Vibrio vulnificus.* A relative of the organism that causes cholera, *Vibrio* is naturally present in marine environments.

Thoroughly cooking oysters and other seafood will kill the *Vibrio* bacteria. But persons who eat raw or undercooked oysters containing *Vibrio* may become ill. In healthy people, *Vibrio* can cause diarrhea, vomiting, and abdominal pain. In persons with weakened immune systems or chronic liver, stomach, or blood disease, *Vibrio* can infect the bloodstream. Although rare, approximately 50 percent of severe *Vibrio* bloodstream infections are fatal. If diagnosed early enough, the disease caused by *Vibrio* can be treated with antibiotics. However, since the symptoms resemble those of the flu, diagnosis often comes too late for antibiotics to be effective.

A New Weapon

University of Florida scientists are using bacteria-attacking viruses, or *phages,* to combat *Vibrio*. The researchers collected naturally occurring phages that prey solely on *Vibrio* bacteria. Then they grew the phages in the

lab. When concentrated solutions of the phages were injected into *Vibrio*-infected mice, the *Vibrio* bacteria were destroyed and the mice recovered. Unlike antibiotics, which leave the body, the phages continued to grow and multiply until they had destroyed all of the *Vibrio* bacteria.

648

OCEANOGRAPHY CONNECTION

Between 150 to 200 species of octopus inhabit the world's oceans. Coastal octopus species thrive in warm rocky shallows off the coasts of the southeastern United States, western Central America, and Japan, as well as in the Mediterranean and the Caribbean. These can weigh up to 50 pounds and have a 10-foot arm span. The species that comes closest in size to the monsters of science fiction is the giant Pacific octopus, found off the western coast of North America and across the northern Pacific to Japan. The biggest ever captured weighed more than 600 pounds and measured 31 feet from arm tip to arm tip. Despite their impressive growth, Pacific giants rarely live longer than three years. At the other extreme, adults of some pygmy species weigh less than a penny, grow to an arm span of less than 2 inches, and are lucky to live six months.

Cephalopods

Squids, octopuses, cuttlefish, and nautiluses are all cephalopods. Most of their body is made up of a large head attached to tentacles (a foot divided into numerous parts), as shown in **Figure 9.** The tentacles are equipped with either suction cups or hooks for seizing prey. Squids have 10 tentacles, while octopuses have eight. The nautilus has 80–90 tentacles, although they are not nearly as long as those of the other cephalopods. Although cephalopods evolved from shelled ancestors, most modern cephalopods lack an external shell. The nautilus is the only living cephalopod species that still has an outer shell. Squids as well as the cuttlefish have a small internal shell. Cuttlefish "bones" are often attached to bird cages to provide calcium for canaries and other pet birds.

Cephalopods are the most intelligent of all invertebrates. They have a complex nervous system that includes a well-developed brain. Cephalopods are capable of exhibiting complex behaviors. Octopuses can easily be trained to distinguish between classes of objects, such as between a square and a cross, and they are the only invertebrates with this ability.

The structure of a cephalopod eye is similar in many ways to that of a vertebrate eye, and some species have color vision. The eyes of a squid can be very large. A giant squid that washed up on a beach in New Zealand in 1933 had eyes that were 40 cm (about 15.75 in.) across. At over 20 m (65 ft) in length, the giant squid is the largest of all invertebrates and has the largest eyes known in any animal.

Like most aquatic mollusks, cephalopods draw water into their mantle cavity and expel it through a siphon. In squids and octopuses, this system functions as a means of jet propulsion. When threatened, they quickly close their mantle cavity, causing water to shoot forcefully out of the siphon. Squids and octopuses can also

WORD Origins

The name *cephalopod* comes from the Greek *kephalicos,* meaning "head," and *pous,* meaning "foot."

Figure 9 Cephalopods. Like all cephalopods, the squid is an active predator. Cuttlefish are agile swimmers that hunt at night, seeking small fishes and crustaceans. The nautilus swims with its coiled shell positioned over its head.

Activity — GENERAL

Major Characteristics of Mollusks Ask students to develop a table similar to the Graphic Organizer at the bottom of this page, to compare the major characteristics of bivalves, gastropods, and cephalopods. LS Logical

Demonstration — BASIC

Have students test the suction of dry and wet suction cups. Ask them whether suction cups work better when dry or wet. (They work better when wet.) Cephalopods have suction cups on their tentacles that have tremendous gripping power. Even sperm whales can receive suction disk scars during fights with giant squids. The suction disks have tiny hooks that can damage the flesh. LS Kinesthetic

DATA LAB: *Analyzing the Molluscan Body Plan*

Skills Acquired
Observing, applying information, predicting

Teacher's Notes
If possible, display examples of shells from a cephalopod, gastropod, and bivalve.

Answers to Analysis

1. A. cephalopod; B. gastropod; C. bivalve
2. The shells of the different classes have evolved to support different ways of life.
3. In cephalopods, the foot is modified into tentacles and aids in movement and in capturing food. The flat-bottomed foot of gastropods secretes a substance that allows the animal to glide over surfaces. The foot of a bivalve allows it to bury itself in the sand.
4. Slugs would be more abundant in areas with acidic soils because the environment cannot support many shell-building organisms.

Graphic Organizer

Use this graphic organizer with the **Activity** on this page.

Characteristics	Bivalves	Gastropods	Cephalopods
Have head, foot, and visceral mass	Yes; head undeveloped	Yes; torsion of visceral mass	Yes; foot divided into tentacles
Have true coelom	Yes	Yes	Yes
Have shells	Two shells	Most have one shell	No, except chambered nautilus
Use locomotion	Most do not	Yes	Yes

Close

Reteaching — BASIC

Ask students to list the different functions of the mantle with examples of mollusks that exhibit such functions. (primitive lung – terrestrial snails; shell formation – bivalves; jet propulsion – squids and octopuses)

Quiz — GENERAL

1. What is the most prominent characteristic of a mollusk? (its foot)
2. What body systems do mollusks share with mammals? (excretory, circulatory, respiratory, reproductive)
3. How are cephalopods different from gastropods and bivalves? (Cephalopods have a foot that is modified into tentacles, and they usually lack a shell.)

Alternative Assessment — GENERAL

Have students choose a mollusk that affects humans. Then have them write a report on how the mollusk has been affected by humans and how humans have been affected by the mollusk.
LS Verbal

Section Review Correlations

SC.F.1.4.2: Items 3, 6
SC.G.1.4.1: Items 1, 2, 4
SC.H.3.4.3: Item 5

release a dark fluid that clouds the water and conceals the direction of their escape. The ink of the cuttlefish contains a reddish brown pigment called sepia. For centuries this ink was used by artists as a pigment and is found in many famous paintings.

All cephalopods are active marine predators. They feed on fish, mollusks, crustaceans, and worms. Once the prey has been snared by the tentacles, it is pulled to the mouth, where it is torn apart by strong, beaklike jaws. The cephalopod's radula then pulls the pieces into the mouth.

DATA LAB

Analyzing the Molluscan Body Plan SC.F.1.4.2

Background

Mollusks share many common characteristics, yet there is great variety among the classes. The drawings on the right show how the shell (brown) and foot (green) vary in three classes of mollusks. Use the drawings to answer the analysis questions.

Analysis

1. **Determine** the class of mollusk *A*, mollusk *B*, and mollusk *C*.
2. **Compare** the shell modifications. Why might a shell suited to one mollusk be inappropriate for another?
3. **Critical Thinking Identifying Functions** For each class shown, explain how the foot is useful for the animal's environment or kind of movement.
4. **Critical Thinking Predicting Outcomes** Terrestrial snails and slugs are nearly identical except that slugs do not have a shell. Acidic forest soils are often poor in minerals, including calcium. Alkaline or neutral soils are rich in minerals. In which kind of soil would you be more likely to find a slug? Explain your answer.

Section 1 Review

1. **Identify** two characteristics that mollusks and annelids have in common. SC.G.1.4.1
2. **Summarize** six characteristics common to most groups of mollusks. SC.G.1.4.1
3. **Describe** how a nephridium functions in waste removal. SC.F.1.4.2
4. **Compare** the distinguishing features of each of the three major classes of mollusks. SC.G.1.4.1
5. **Critical Thinking Forming Hypotheses** A chemical pollutant accidentally spills into a bay. One of the effects of this chemical is that it paralyzes cilia. The next day almost all of the oysters in the bay are dead. Develop a hypothesis that explains why the oysters died. SC.H.3.4.3
6. *FCAT Prep* A gastropod's radula is part of what organ system?
   A circulatory
   B digestive
   C excretory
   D respiratory
   SC.F.1.4.2

650

Answers to Section Review

1. Mollusks and annelids both exhibit true coeloms, and most members of both groups develop from larvae called trochophores.
2. Most mollusks share the following key characteristics: coelom, bilateral symmetry, a three-part body plan (visceral mass, mantle, and foot), organ systems, a shell, and a radula.
3. A nephridium funnels coelomic fluid, which includes wastes, into the mantle cavity, where it is excreted to the outside environment.
4. Most gastropods have a single shell, a single large foot, and, on their head, tentacles tipped with eyes. Bivalves have a two-part hinged shell and lack a head and radula. Most bivalves possess one or more siphons used for filter feeding. Cephalopods have large eyes on a large head attached to numerous tentacles.
5. The beating of cilia moves water over the gills. Therefore, the oysters suffocated when the chemical paralyzed their cilia.
6. **A.** Incorrect. The heart is part of the circulatory system. **B.** Correct. The radula is part of the digestive system. **C.** Incorrect. The nephridium is part of the excretory system. **D.** Incorrect. The gill is part of the respiratory system.

Annelids

Section 2

Florida Sunshine Standards/FCAT

SC.F.1.4.2 Body structures are uniquely designed and adapted for their function.

SC.G.1.4.1 There is great diversity and interdependence among living things.

The First Segmented Animals

You have probably heard the expression "a can of worms," which calls up an image of a lot of wiggly, wriggly creatures. An earthworm may come to mind, but there are many different species of segmented worms. Worms might not look like much, but this group of coelomates belongs to an ancient group, phylum Annelida. Annelid fossils can be found in rock that is 530 million years old. Scientists think that annelids evolved in the sea, where two-thirds of today's annelid species live. Most other annelid species are terrestrial earthworms. Annelids range in size from less than 1 mm (0.04 in.) long to more than 3 m (10 ft) long.

Annelids, such as the earthworm and fireworm shown in **Figure 10,** are easily recognized by their segments, which are visible as a series of ringlike structures along the length of their body. Each segment contains digestive, excretory, circulatory, and locomotor (movement) organs. Some of the segments are modified for specific functions, such as reproduction, feeding, or sensation. A well-developed **cerebral ganglion,** or primitive brain, is located in one anterior segment. The brain is connected to a nerve cord that runs along the underside of the worm's body.

Objectives

- **Identify** the major change in body plan that distinguishes annelids from mollusks. SC.G.1.4.1
- **Describe** the basic annelid body plan. SC.F.1.4.2
- **Describe** the annelid digestive system. SC.F.1.4.2
- **Compare** the three classes of annelids. SC.G.1.4.1

Key Terms

cerebral ganglion
septa
setae
parapodium

Figure 10 Annelids

The ringlike segments of this earthworm and marine fireworm identify them as annelids.

Earthworm

Fireworm

Evolutionary Milestone

6 **Segmentation**

Annelids were the first organisms to have a body plan based on repeated body segments. Segmentation underlies the body organization of all coelomate animals except mollusks.

651

Section 2

Focus

Overview

Before beginning this section review with your students the objectives listed in the Student Edition. Tell students that the purpose of this lesson is to teach them about the distinguishing structural and functional characteristics of the annelids, the segmented worms. They will also learn about the three major groups of annelids: the marine or Polychaete worms, the Oligochaetes, and the leeches or Hirudinea.

Bellringer

Ask students to write a short description of the evolutionary advantages of body segmentation. (It allowed specialization of the body segments.)

Motivate

Demonstration

Obtain earthworms from garden soil or a bait shop, and have students observe their behavior. Ask students to describe how they can tell which end of the earthworm is the head. If students say that the head is the leading end, remind them that squid sometimes move tail first. Direct their attention to the earthworm's mouth. Caution students to keep their worms moist.

Chapter Resource File

- Directed Reading BASIC
- Active Reading GENERAL
- Data Sheet for Quick Lab GENERAL

Transparencies

TR Bellringer
TR H25 Anatomy of the Earthworm

One-Stop Planner CD-ROM

- Reading Organizers BASIC
- Reading Strategies BASIC

Section Correlations

SC.F.1.4.2
SC.G.1.4.1

Teach

Modeling a Closed Circulatory System

QUICK LAB

Skills Acquired
Observing, modeling, interpreting results, applying information, evaluating

Teacher's Notes
Have students conduct Quick Lab: Modeling an Open Circulatory System before conducting this lab. Ask students how oxygen and nutrients get to the tissues that need them if the blood never touches the body tissues directly. (The oxygen and nutrients pass through the vessel walls.) Use a dark dye. If the dye diffuses too quickly in the water, try using an oil-based dye or small, lightweight colored beads.

Answers to Analysis

1. The "blood" moved through the blood vessels.
2. The pan of water represents the body and tissues, surgical tubing represents the heart, and clear plastic tubing represents the blood vessels.
3. Answers will vary.
4. Answers will vary, but may include adding model tissues and increasing the number of model blood vessels.
5. Students should note that the closed system exerted a greater pressure on the fluid than the open system.

internet connect
www.scilinks.org
Topic: Annelids
Keyword: HX4012
SCILINKS. Maintained by the National Science Teachers Association

Internal body walls, called **septa,** separate the segments of most annelids. Nutrients and other materials pass between the segments through the circulatory system. Sensory information is delivered by a nerve cord that connects nerve centers in the segments to the brain.

Characteristics of Annelids

In addition to segmentation, annelids share a number of other characteristics.

1. **Coelom.** The fluid-filled coelom is large and is located entirely within the mesoderm.
2. **Organ systems.** The organ systems of annelids show a high degree of specialization and include a closed circulatory system and excretory structures called nephridia. The gut has different regions that perform different functions in digestion.
3. **Bristles.** Most annelids have external bristles called **setae** *(SEET ee).* The paired setae located on each segment provide traction as the annelid crawls along. Some annelids, like the fireworm shown in Figure 10 on the previous page, also have fleshy appendages called **parapodia** *(par uh POH dee uh).*

QUICK LAB

Modeling a Closed Circulatory System

SC.H.3.4.1

You can model a closed circulatory system using simple items to represent the heart, blood vessels, blood, and body tissues of a living organism.

Materials

clear plastic tubing, 30 cm (about 12 in.) piece; surgical tubing, 15 cm (about 6 in.) piece; shallow pan filled with water; eyedropper, food coloring

Procedure

1. Connect one end of the clear tubing to the surgical tubing. Submerge in the pan.
2. Use the eyedropper to insert several drops of food coloring into the surgical tubing. Close the tubing by attaching the two free ends together, as shown above.
3. With your thumb and index finger, squeeze along the piece of surgical tubing to pump the food coloring through the system.
4. Observe the food coloring as it moves through the tubing.

Analysis

1. **Describe** what happened when you pumped the food coloring through the system.
2. **Identify** what structures the pan of water, the surgical tubing, and the clear plastic tubing represent.
3. **Evaluate** your model's efficiency at pumping blood through the system.
4. **Critical Thinking**
Analyzing Methods How could you modify the model to make it more accurate?
5. **Critical Thinking**
Inferring Relationships If you did the Quick Lab in the previous section that models an open circulatory system, recall what happened in that system. Which model do you think exerted a greater pressure on the fluid in the tube?

MISCONCEPTION ALERT

Insect Larvae Many people mistake grubs, maggots, and caterpillars for worms. These animals are arthropods, not annelids, and are the larval form of certain insects (beetles, flies, and moths or butterflies, respectively). Have students list the annelid characteristics that are shared by these unrelated insects. (Answers may vary, but arguments could be made for each of the four characteristics of annelids.)

Annelid Groups

Annelids differ in the number of setae (bristles) they have on each segment, and not all annelids have parapodia. These two external characteristics are used to classify annelids.

Marine Worms

Marine segmented worms are members of class Polychaeta *(PAHL ih keet uh)*, the largest group of annelids. Polychaetes live in virtually all ocean habitats. They are often beautiful, showing unusual forms and iridescent colors. A distinctive characteristic of polychaetes is the pair of fleshy, paddle-like parapodia that occur on most of their segments. The parapodia, which usually have setae, are used to swim, burrow, or crawl. Parapodia also greatly increase the surface area of the polychaete's body, making gas exchange between the animal and the water more efficient.

Many polychaetes are burrowing species, but others live in protective tubes formed by the hardened secretions of glands located on their segments. Grains of sand or other foreign material may be cemented into the tube. Such tubeworms, like the feather duster shown in **Figure 11,** live with only their head stuck out of the tube. Featherlike head structures trap food particles from the water that passes over them. Other species of polychaetes feed by pumping water through their body. Free-swimming polychaetes, such as *Nereis* shown in **Figure 12,** are predators that use their strong jaws to feed on small animals.

Figure 11 **Feather duster.** Feather dusters filter-feed by trapping food particles in their featherlike head structures.

Figure 12 ***Nereis***

Nereis, a polychaete worm, grasps its prey in its jaws, which open when it thrusts out its pharynx.

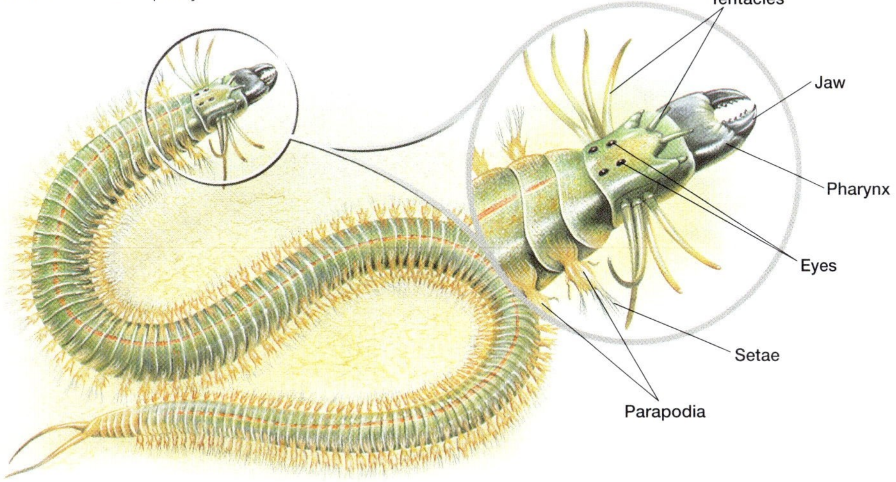

Teaching Tip — GENERAL

The Ultimate Worm Ask students to list the anatomical features that make earthworms more advanced than flatworms or roundworms. Help students recall that earthworms have a segmented body, a true coelom, a highly specialized gut, a closed circulatory system, and external bristles.

Using the Figure — GENERAL

Ask a volunteer to examine **Figure 11** and offer an explanation for how feather duster worms acquire food. (Bristle-like appendages filter small organisms and organic matter out of water that flows over them.) Have the class name an advantage and a disadvantage of this feeding strategy. (Advantage: Allows for a sessile lifestyle. Disadvantage: Potential food must pass close by to be captured.) **LS Visual**

did you know?

Importance of Polychaetes Polychaetes are some of the most important animals in marine benthic (bottom-dwelling) communities. Though few people notice polychaetes or even know of their existence, these worms exert a strong influence in marine benthic food webs. Some of the 10,000 species so far described eat animals, others eat plants, and many more provide food for commercially important fish.

Career

Oceanographer Oceanography is a diverse field with branches in the biological, chemical, physical, geological, and engineering fields. A graduate degree is usually required, along with an undergraduate degree in biology, chemistry, geology, or engineering. Biological oceanographers study the interrelationships of marine life, energy transfer, and the impact of human activities on marine communities. Encourage students to learn more about this field at the library or on the Internet.

Teach, *continued*

Group Activity — GENERAL

Comparing Mollusks and Annelids Have students make a table similar to the one in the Graphic Organizer on the bottom of this page. Students' tables should compare body cavity, body plan, organ systems, and larvae. Ask students to write a paragraph explaining the differences.
LS Logical

Group Activity — GENERAL

Worm Observations Obtain samples of the three different classes of segmented worms. Try to obtain a predatory marine worm (such as *Nereis*), an earthworm, and a leech. Have students observe the worms and provide answers to the following questions: How is each worm's method of movement and the medium in which it lives reflected in its body structures? (All types of segmented worms have flexible, muscular bodies; *Nereis* and the earthworm have body appendages or extensions that enable them to move across a surface while the leech has a smooth body with a sucker that it can use for attachment.) How is each worm's method of obtaining food reflected in its body structures? (*Nereis* has jaws used for preying on other organisms; the leech has piercing structures; the earthworm does not have a well-developed head region and ingests soil particles as it moves along.) How is each worm's method of respiration reflected in its body structures? (All of these worms respire by diffusing oxygen across their body surfaces; their bodies are long and narrow, which results in a high surface-to-volume area for diffusion.)
Co-op Learning

Real Life

"The early bird catches the worm."

Because earthworms respire through their skin, they must stay moist. As a result, earthworms come to the surface of the soil only at night, when the air is more moist. The "early bird" catches worms that have not yet burrowed in for the day.

Earthworms

Earthworms and some related freshwater worms are members of the class Oligochaeta *(AHL ih goh KEET uh)*. Oligochaetes have no parapodia and only a few setae on each segment. Earthworms lack the distinctive head region of polychaetes and have no eyes. They do, however, have light-sensitive and touch-sensitive organs located at each end of their body. They have other sensory cells that detect moisture.

Earthworms, such as the ones in **Figure 13,** are highly specialized scavengers. They literally eat their way through the soil, consuming their own weight in soil every day. As they tunnel, earthworms take in organic matter and other materials using their muscular pharynx. The ingested soil moves through their one-way gut, down the esophagus and into a storage chamber called the *crop*. From here, the soil moves to an area called the *gizzard*. The grinding action of the gizzard crushes the soil particles together, breaking them down. The crushed material moves to the intestine, which extends to the posterior end of the earthworm's body. Digested food molecules are absorbed into the intestinal wall, and the remaining material passes out through the anus in a form called castings. The tunneling activity of earthworms allows air to penetrate the soil, and their castings fertilize it. Rich, organic soil may contain thousands of earthworms per acre.

Hydrostatic Skeleton

The fluid within the coelom of each body segment creates a hydrostatic skeleton that supports the segment. Each segment contains muscles that pull against this hydrostatic skeleton. Circular muscles wrap around the segment, while longitudinal muscles span its length. As shown in the *Up Close: Earthworm,* on the next page, when the circular muscles contract, the segment becomes longer. When the longitudinal muscles contract, the segment bunches up, increasing in diameter. An earthworm crawls by alternately contracting the two sets of muscles in its segments. The brain coordinates the muscular activity of each body segment, thus controlling movement.

Figure 13 Earthworms burrowing. Earthworms come to the surface only at night or during heavy rains. During dry or cold weather, they burrow deep into the soil and become inactive.

654

Graphic Organizer

Use this graphic organizer with the **Group Activity: Comparing Mollusks and Annelids** on this page.

	Body cavity	Body plan	Organ systems	Larvae
Mollusks	True coelom	Muscular foot, head, visceral mass	Circulatory (usually open system), respiratory, digestive, excretory	Trochophore
Annelids	True coelom	Segmented with bristles	Circulatory (closed), respiratory, digestive, excretory	Trochophore

Earthworm

- **Scientific name:** *Lumbricus terrestris*
- **Size:** Grows up to 30 cm (12 in.) long
- **Range:** Europe; eastern and northwestern North America
- **Habitat:** Damp soil
- **Diet:** Organic matter contained in soil

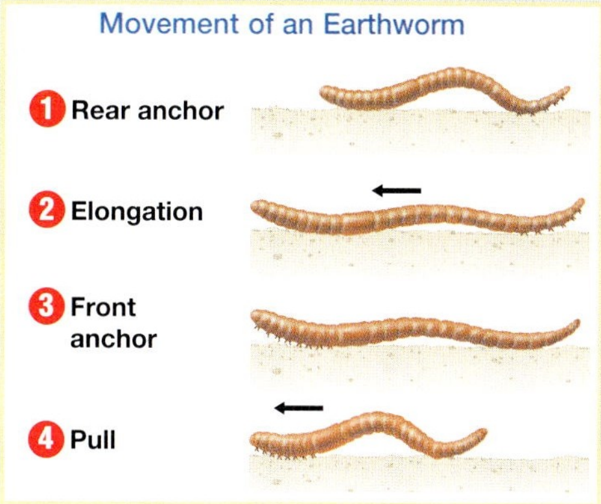

Characteristics

Respiration Oxygen and carbon dioxide diffuse through the earthworm's skin. This exchange can take place only if the worm's skin is kept moist.

Movement As shown in the diagram, (1) first the earthworm anchors several of its rear segments by sinking their setae into the ground. (2) The worm then contracts the circular muscles in front of the anchored segments. This causes the anterior segments to elongate. (3) Then the setae in front of the stretched region are anchored and the rear setae are released. (4) The circular muscles relax and the longitudinal muscles contract, pulling the rear segments forward.

Digestion Earthworms "eat" soil, which is ground up in a thick, muscular gizzard. Food molecules pass across the walls of the intestine and are absorbed into the bloodstream.

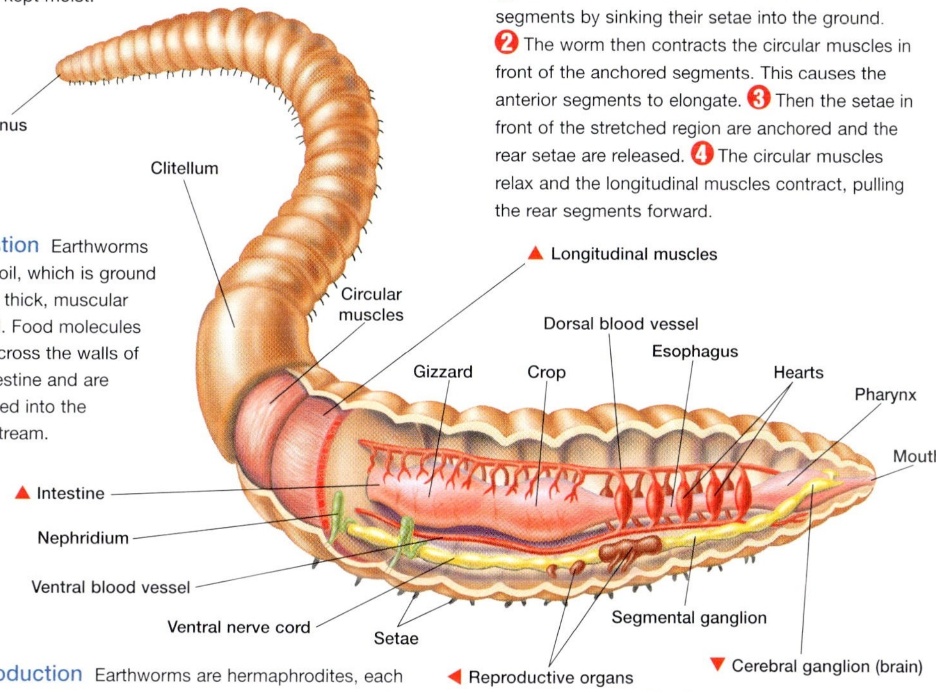

Reproduction Earthworms are hermaphrodites, each individual containing both sexes. Mating occurs when two earthworms join ventrally head to tail, exchanging sperm. During egg laying, the clitellum (a thickened, glandular ring of cells) of each worm secretes a mucous cocoon that encloses the fertilized eggs. Young worms emerge from the cocoon several weeks later.

Brain The brain coordinates the muscular activity of each body segment. It also processes sensory information from light-sensitive and touch-sensitive organs located at each end of the body.

655

Up Close

Earthworm

Teaching Strategies
Earthworms, which have no teeth, eat their way through soil and grind it up for nutrients. Ask students what structure earthworms use to grind up soil. (gizzard) Ask them which other toothless animals use gizzards to grind food. (birds) Finally, ask why food needs to be ground up. (Smaller particles are digested more easily.)

Discussion
- How do earthworms loosen soil and help recycle soil nutrients? (Earthworms make holes in the soil by moving through it, digesting soil matter into simpler components.)
- What is the role of the clitellum? (It secretes a protective layer of chitin for the fertilization and incubation of eggs.)

BIOWatch

Leeches Make a Comeback

Teaching Strategies
Have students research past and present uses of leeches in medicine and write a brief (one-page) paper that describes their findings.

Discussion
Ask if any students have ever had a leech attach while swimming or wading in a pond or stream. Did it hurt as it attached? (Usually there is no sensation.) Did you bleed after the leech was removed? (If the leech was fully attached and feeding, students should answer yes because leeches use anticoagulants to facilitate feeding. The bleeding stops soon after the leech is removed.)

HISTORY CONNECTION

Charles Darwin (1809–1882), the British naturalist famous for his theory of evolution by natural selection, was fascinated with the ways that earthworms aerate and enrich soil. In his last botanical book, *The Formation of Vegetable Mould Through the Action of Worms,* Darwin noted the service of earthworms in recycling organic matter in soil. He also observed that an earthworm could ingest its weight in soil each day and that in one year, the earthworms inhabiting one hectare could digest 22 to 40 metric tons of soil. The work was published only 6 months prior to Darwin's death and was called a pioneering study in quantifying ecology.

Figure 14 Leech. Most species of leeches are small, 2.5 to 5.0 cm (1 to 2 in.) long.

Leeches

When people hear the word *leech,* they usually associate it with bloodsucking, and for a good reason. A leech, shown in **Figure 14,** has suckers at both ends of its body. Most species are predators or scavengers, but some are parasites of vertebrates and crustaceans.

Leeches are the only members of class Hirudinea *(hihr yoo DIHN ee uh).* Leeches lack both setae and parapodia. The body of a leech is flattened, and unlike other annelids, its segments are not separated internally.

BIOWatch

Leeches Make a Comeback

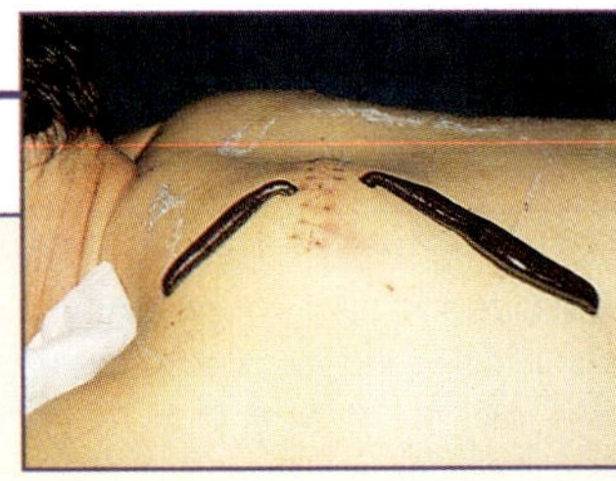

For many centuries, it was commonly believed that an excess amount of blood was the cause of a wide range of illnesses, from a fever or headache to severe heart disease. A standard treatment for these conditions was bloodletting using leeches. Physicians applied leeches to the patient's body, allowing the leeches to suck out the patient's "bad blood."

Use in Microsurgery

Although doctors no longer believe in "bad-blood," leeches are making a comeback in the field of health care. One use of leeches is during surgery to reattach severed limbs, fingers, or toes. In this type of operation, called microsurgery, the surgeon uses tiny instruments and a microscope to reconnect tendons, blood vessels, and nerves. It is not possible to reconnect the smallest of the blood vessels, so circulation in the reattached part is usually poor. Often tissues in the region die, and the reattached part cannot heal. By applying leeches to suck out the accumulated blood, tissues remain healthy until new blood vessels grow and circulation is restored to normal. As a result, the success rate of surgery for reattachments has increased.

Other Applications

Leeches possess other useful qualities. Their saliva contains *anticoagulants*—substances that prevent blood from clotting—and enzymes that can break up blood clots. It is not necessary to apply leeches to a patient to take advantage of these chemicals. Today these substances are produced through genetic engineering and have proven useful in the treatment of some heart patients.

internet connect
www.scilinks.org
Topic: Leeches
Keyword: HX4113
SCILINKS. Maintained by the National Science Teachers Association

Section 2 Review

1. **Summarize** how you can tell if a wormlike organism is an annelid worm. SC.G.1.4.1
2. **Relate** an annelid's septa to its overall body plan. SC.F.1.4.2
3. **Describe** the major features of an earthworm's digestive system. SC.F.1.4.2
4. **Compare** the external appearance of marine annelids, earthworms, and leeches. SC.G.1.4.1
5. *FCAT Prep* What happens to a segment in an earthworm when the circular muscles in that segment contract?
   - **A** It shortens.
   - **B** It elongates.
   - **C** It bends to one side.
   - **D** It increases in diameter.

   SC.F.1.4.2

656

Close

Reteaching — Basic

Have students pick an annelid in this chapter and pretend that the animal is a pet. Ask them to provide a picture of the pet and prepare a report about it. In their report, students should identify the kind of habitat necessary to keep the pet, describe how and what the animal is fed, and provide any necessary warnings for handling.

Quiz — General

1. Why is "poly-" an appropriate prefix for polychaete worms? (They have many body segments.)
2. Why are earthworms likely to be found in richly organic soil but not in a very sandy soil? (Because they use the energy they obtain from organic matter in the soil as their source of energy for survival.)
3. Why is it advantageous for parasitic leeches to be able to distend their bodies by ingesting a huge meal? (Because they have to search for food, consuming a large meal at one time would last them until they can find another source of food.)

Alternative Assessment — General

Have students combine their new understanding of annelids with their artistic skills and draw a "new" species of annelid. Tell them to label their annelids and to make sure that the creatures include both the fundamental annelid traits as well as any reasonable adaptations to their designated environment.
LS Visual

Section Review Correlations

SC.F.1.4.2: Items 2, 3, 5
SC.G.1.4.1: Items 1, 4

Answers to Section Review

1. Annelid worms have segments, and most have external bristles.
2. Septa are internal body walls that separate the segments of most annelids. Blood vessels and a nerve cord connect the segments to one another, and to the annelid's brain.
3. The earthworm's digestive system is a highly modified gut consisting of the following regions: mouth, pharynx, esophagus, crop, gizzard, intestine, and anus.
4. Marine annelids are often very colorful and have a head region with eyes; their segments usually have a pair of fleshy extensions called parapodia. Earthworms lack parapodia and have no head region. Usually there are several bristly setae on each segment of an earthworm. Leeches lack setae and parapodia, and their bodies are generally flat.
5. **A.** Incorrect. Contraction of the longitudinal muscles results in shortening. **B.** Correct. **C.** Incorrect. Contraction of the circular muscles does not cause the worm to bend to one side. **D.** Incorrect. Contraction of the circular muscles does not increase the diameter of the worm.

Study ZONE

CHAPTER HIGHLIGHTS

Key Concepts

1 Mollusks

- Many mollusks and annelids have a larval form called a trochophore.
- The mollusk body has three distinct parts: a visceral mass, a mantle, and foot.
- All mollusks except bivalves have a rasping tonguelike radula.
- Mollusks have a true coelom and well-developed organs.
- Most mollusks respire with gills but some respire with a primitive lung.
- Nephridia enable mollusks to recover the useful substances from their bodily wastes.
- Gastropods (snails and slugs) live in oceans, in fresh water, and on land.
- Bivalves (clams, oysters, and their kin) are aquatic and have hard shells called valves that protect their soft bodies.
- Gastropods and bivalves have an open circulatory system.
- Cephalopods (octopuses, squids, and their kin) have a well-developed head region, many tentacles, and a closed circulatory system. Most cephalopods have no external shell.

2 Annelids

- Annelids are coelomate worms that have segmented bodies and complex nervous systems.
- Annelids are classified according to the presence or absence of setae and parapodia.
- Annelids respire through their skin, and they have a closed circulatory system.
- Earthworms burrow through the soil, ingesting it as they crawl.
- Marine polychaetes have parapodia and setae. Some are active predators and others are filter feeders.
- Leeches lack parapodia and setae, and their segments are not separated internally. They may be aquatic or terrestrial, and some are parasites.

Key Terms

Section 1

trochophore (642)
visceral mass (643)
mantle (643)
foot (643)
radula (643)
nephridium (644)
adductor muscle (647)
siphon (647)

Section 2

cerebral ganglion (651)
septa (652)
seta (652)
parapodium (652)

Study ZONE

Alternative Assessment

Have students design a new species of mollusk. Students might draw and describe their animal or create a model. They should give their creation a scientific name based on its features and describe its habitat, reproductive habits, and feeding habits. Designs must include the key characteristics of a mollusk.

Chapter Resource File

- Science Skills Worksheet GENERAL
- Critical Thinking Worksheet ADVANCED
- Test Prep Pretest GENERAL
- Chapter Test GENERAL

Answer to Concept Map

The following is one possible answer to Performance Zone item 15.

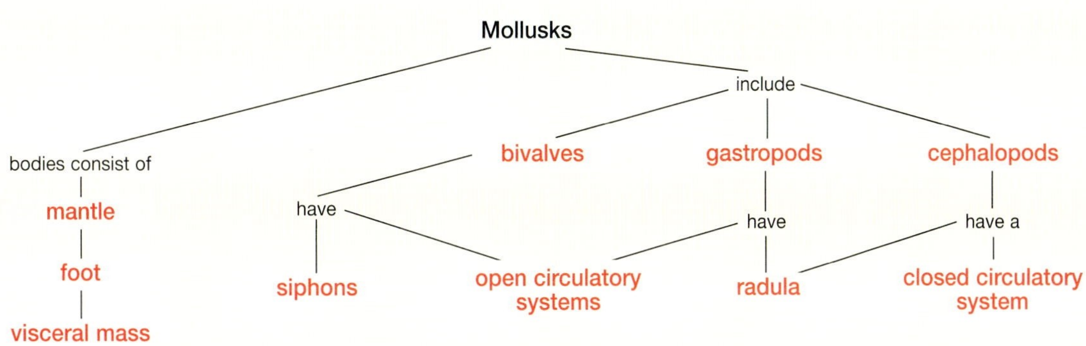

ANSWERS

Understanding Key Ideas

1. a
2. b
3. a
4. c
5. d
6. d
7. b
8. Persons with weakened immune systems or chronic liver, stomach, or blood disease should ensure that oysters have been fully cooked before eating.
9. Leeches remove blood from damaged tissue, which helps keep the tissues healthy until new blood vessels can form.
10. A possible answer to the concept map is found at the bottom of the Study Zone page.

Critical Thinking

11. The food of aquatic annelids is not as coarse and does not require as much grinding for digestion as the food of terrestrial annelids.
12. The coelom allowed for the evolution of complex organs composed of more than one tissue type. A coelom also allows the internal organs to be suspended from the body wall and cushioned by fluid. This allows mollusks to move around without damaging their organs or interfering with their function.

Alternative Assessment

13. Answers will vary. Students should list a variety of mollusks that are used for human consumption. They should also describe areas where each mollusk can be found.
14. Answers will vary. Worm farmers have varying degrees of education, from high school through graduate school. They all share a working knowledge of growing, breeding, and harvesting worms on a large scale. Worm farmers usually own their own businesses or work as part of a larger agricultural products business. They may also be employed by waste management agencies. The growth potential for this field is fair. Starting salary will vary by region.

Chapter Review Correlations

SC.D.1.4.3: Item 11
SC.D.1.4.4: Item 12
SC.F.1.4.2: Items 2, 6, 7, 10
SC.G.1.4.1: Items 1, 3, 4, 5, 13
SC.H.3.4.2: Item 14
SC.H.3.4.3: Item 8
SC.H.3.4.5: Item 9

Understanding Key Ideas

1. Both mollusks and annelids
   - **a.** are coelomates.
   - **b.** have at least a remnant of a shell.
   - **c.** have no larval form.
   - **d.** have a visceral mass.

   SC.G.1.4.1
2. Terrestrial snails respire
   - **a.** with gills.
   - **b.** with a primitive lung.
   - **c.** through their skin.
   - **d.** through their siphon.

   SC.F.1.4.2
3. Which of the following is *not* true of bivalves?
   - **a.** They have a distinctive head region.
   - **b.** They have sense organs.
   - **c.** They have an open circulatory system.
   - **d.** Most are filter feeders.

   SC.G.1.4.1
4. Cephalopods have all of the following characteristics *except*
   - **a.** bilateral symmetry.
   - **b.** a three-part body plan.
   - **c.** an open circulatory system.
   - **d.** a true coelom.

   SC.G.1.4.1
5. Annelids are divided into three classes. This classification is based on the number of setae and the presence or absence of
   - **a.** segments.
   - **b.** hearts.
   - **c.** a gizzard.
   - **d.** parapodia. SC.G.1.4.1
6. Blood in the circulatory system of an annelid
   - **a.** flows into its body cavity.
   - **b.** delivers carbon dioxide to its tissues.
   - **c.** passes through gills.
   - **d.** stays within its circulatory system.

   SC.F.1.4.2
7. Earthworm movement requires all of the following *except*
   - **a.** circular muscles.
   - **b.** secretion of mucus.
   - **c.** muscle contractions.
   - **d.** traction provided by setae.

   SC.F.1.4.2
8. **BIOWatch** People who have what conditions should be especially careful about consuming oysters? What should all people be sure to check about the oysters that they eat? SC.F.1.4.7
9. **BIOWatch** How can leeches be beneficial following microsurgery? SC.H.3.4.5
10. **Concept Mapping** Make a concept map that shows the characteristics and diversity of the mollusks. Include the following words in your map: *foot, visceral mass, mantle, radula, siphons, gastropod, bivalve, cephalopod, open circulatory system,* and *closed circulatory system.* SC.F.1.4.2

Critical Thinking

11. **Recognizing Patterns** The gizzards of annelids that have returned to an aquatic environment over evolutionary time are smaller and less muscular than those of terrestrial annelids, such as earthworms. How do the differences between the two types of gizzards represent adaptations for aquatic and terrestrial feeding? SC.D.1.4.3
12. **Relating Concepts** Explain the significance of a coelom in the evolution of mollusks. SC.D.1.4.4

Alternative Assessment

13. **Finding and Communicating Information** Use the library or Internet resources to compile a list of mollusks used by humans as food sources. For each mollusk listed, tell where it is harvested and which part of the mollusk is eaten. Prepare a brochure to summarize your findings. Include a visual that informs the reader of the geographic location of the food source. SC.G.1.4.1
14. **Career Connection** Worm Farmer Research the field of growing segmented worms for use in research, as fishing bait, and for soil improvement. Your report should include a job description, training required, kinds of employers, growth prospects, and starting salary. SC.H.3.4.2

Assignment Guide

Section	Questions
1	3, 4, 8, 12, 13
2	1, 2, 5, 6, 7, 9, 11, 14

FCAT Prep

For questions 1–6, write your answers on a separate sheet of paper.

1. What structure do mollusks use to extract useful molecules from coelomic fluid?
   A. ganglia
   B. nephridia
   C. parapodia
   D. radula

SC.F.1.4.2

2. Which of the following is a part of the mollusk's three-part body plan?
   F. antenna
   G. mantle
   H. parapodia
   I. radula

SC.F.1.4.2

3. What structures connect the valves of bivalves?
   A. adductor muscles
   B. parapodia
   C. siphons
   D. visceral mass

SC.F.1.4.2

4. What trait easily identifies an annelid?
   F. body cavity
   G. cephalization
   H. nephridia
   I. segmentation

SC.G.1.4.1

5. READ INQUIRE EXPLAIN Classification of organisms has traditionally been based on physical similarities. However, gastropods, bivalves, and cephalopods are physically quite different.

   Assess why gastropods, bivalves, and cephalopods are all grouped into the phylum Mollusca, despite their physical differences.

SC.D.1.4.3

6. READ INQUIRE EXPLAIN A particular bivalve mollusk called a shipworm can do extensive damage to ocean pier pilings (supports) by burrowing into them with its radula. As a science project, a student decides to determine if a new paint can reduce shipworm damage more than other paints do.

   Part A What is the function of a radula? SC.F.1.4.2

   Part B Evaluate the variables that the student must control in order for the experimental results to be considered valid. SC.H.1.4.1

Test TIP

Whenever possible, highlight or underline numbers or words that are critical to the correct understanding of a question.

659

FCAT Doctor

Question 1 Answer B is the correct choice. Mollusks have nephridia, which are tubular structures that recover useful molecules, such as sugars, salts, and water, from coelomic fluid. Answer A is incorrect because ganglia are clusters of neurons that interpret sensory information in many kinds of simple invertebrates. Answer C is incorrect because parapodia are fleshy appendages on polychaete worms and are used for locomotion. Answer D is incorrect because a radula is used to scrape fragments of food off rocks or plant matter. The radula is a tongue-like organ located in the mouth of all mollusks except bivalves.

Question 5 All mollusks have a coelom, a three-part body plan – the visceral mass, the mantle, and the muscular foot – a trocophore larval stage, and a common evolutionary ancestor.

Question 6

Part A A radula is a tongue-like organ located in a mollusk's mouth. The radula has backward-curving teeth arranged in rows.

Part B For an experiment to be valid, only one variable can be tested. Therefore, every variable other than the type of paint used must be identical in order to achieve valid results.

Answers

1. B
2. G
3. A
4. I
5. Gastropods, bivalves, and cephalopods all have a coelom, a three-part body plan, and a trocophore larval stage.

6A. A radula scrapes fragments of food off rocks or plant matter.

6B. All variables except for the type of paint must be identical.

Test Prep Correlations

SC.D.1.4.3: Item 5
SC.F.1.4.2: Items 1, 2, 3, 6A
SC.G.1.4.1: Item 4
SC.H.1.4.1: Item 6B

Skills Practice Lab

Observing Characteristics of Clams

Teacher's Notes

Time Required 45 minutes

Ratings

EASY → HARD

Teacher Preparation 2

Student Setup 2

Concept Level 2

Cleanup 2

Skills Acquired

- Inferring
- Interpreting
- Measuring
- Predicting

Scientific Methods

In this lab, students will:

- Make Observations
- Ask Questions
- Draw Conclusions

Materials and Equipment

Materials for this lab can be ordered from WARD'S. See *Master Materials List* at the front of this book for catalog numbers.

Safety Cautions

Make sure students wear safety goggles, lab aprons, and disposable gloves throughout the investigation.

Before students begin step 8, remind them to keep hydrochloric acid (HCl) away from skin and clothing. If acid is spilled, thoroughly flush all affected areas with running water and mop up the spill with cloths designed for chemical spill cleanup. Avoid having students handle acid spills.

Skills Practice Lab

Observing Characteristics of Clams SC.H.1.4.1

SKILLS

- Observing
- Testing for the presence of a chemical

OBJECTIVES

- Observe the behavior of a live clam.
- Examine the structure and composition of a clam shell.

MATERIALS

- safety goggles
- lab apron
- live clam
- small beaker or dish
- eyedropper
- food coloring
- glass stirring rod
- clam shell
- Petri dish
- scalpel
- stereomicroscope
- 0.1 M HCl

ChemSafety

CAUTION: Always wear safety goggles and a lab apron to protect your eyes and clothing.

CAUTION: Do not touch or taste any chemicals. Know the location of the emergency shower and eyewash station and how to use them. If you get a chemical on your skin or clothing, wash it off at the sink while calling to the teacher. Notify the teacher of a spill. Spills should be cleaned up promptly, according to your teacher's directions.

CAUTION: Glassware is fragile. Notify the teacher of broken glass or cuts. Do not clean up broken glass or spills with broken glass unless the teacher tells you to do so.

Before You Begin

Clams are **mollusks**, and they have a two-part shell. The body of a clam consists of a visceral mass and a muscular **foot**. There is no definite head. Two tubes, an **incurrent siphon** and an **excurrent siphon**, extend from the body on the side opposite the foot. Like all mollusks, clams have a shell composed of **calcium carbonate**. A membrane called the **mantle** lines the shell and forms successive rings of shell as a clam grows. The **umbo** is the oldest part of a clam shell. In this lab, you will examine live clams and clam shells.

1. Write a definition for each boldface term in the paragraph.
2. Based on your objectives, write a question you would like to explore about clams.

Procedure

PART A: Observe a Live Clam

1. Put on safety goggles and a lab apron.
2. Place a live clam in a small beaker or shallow dish of water. Using an eyedropper, apply two drops of food coloring near the clam, as shown above.
3. Observe and record what happens to the food coloring.
4. Using a stirring rod, touch the clam's mantle. **CAUTION: Touch the clam gently to avoid injuring it.**
5. Observe and record the clam's response to touch.

660

Tips and Tricks

Keep the clams in a freshwater aquarium with green algae and a layer of sand until lab time. Review the mollusk body plan before the lab. Before students begin step 7, tell them to use extreme care when chipping clam shells with their scalpel. Have students use a scalpel that has a permanent blade. Broken glass and pieces of clam shell can be thrown in the trash. Unused 0.1 M HCl solution can be neutralized by carefully adding 0.1 M NaOH until the pH is between 6 and 8. The resulting solution can then be poured down the drain.

Answers to Before You Begin

1. *mollusks*—animals with a soft body that secrete a shell containing calcium carbonate; *foot*—part of the mollusk body that extends from the shell and aids in locomotion; *incurrent siphon*—tube that takes water into a bivalve's shell; *excurrent siphon*—tube that expels water from a bivalve's shell; *calcium carbonate*—mineral found in a mollusk's shell; *mantle*—tissue that lines a mollusk's shell and secretes the shell; *umbo*—the oldest part of a bivalve's shell
2. Answers will vary. For example: How does a clam respond to touch?

PART B: Observe a Clam Shell

6. Examine the concentric growth rings on the shell. Locate the knob-shaped umbo on the shell. Count and record the number of growth rings on the clam shell.

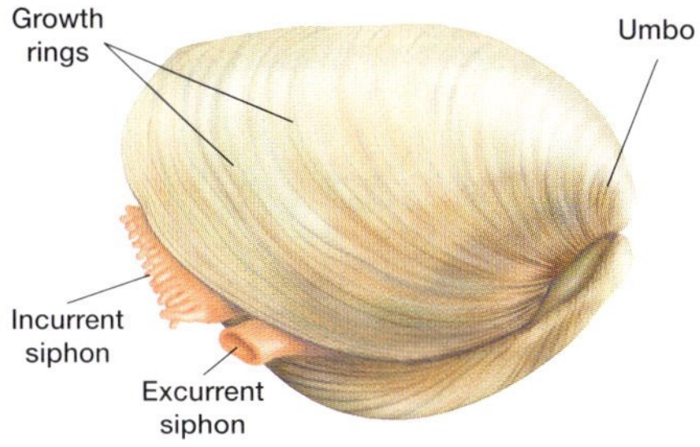

7. Place the clam shell in a Petri dish. Use a scalpel to chip away part of the shell to expose its three layers. **CAUTION: Sharp or pointed objects may cause injury. Handle scalpels carefully.** View the shell's layers with a stereomicroscope. The outermost layer protects the clam from acids in the water. The innermost layer is mother-of-pearl, the material that forms pearls.
8. The middle layer of the shell contains crystals of calcium carbonate. To test for the presence of this compound, place one drop of 0.1 M HCl on the middle layer of the shell. **CAUTION: Hydrochloric acid is corrosive. Avoid contact with skin, eyes, and clothing. Avoid breathing vapors.** If calcium carbonate is present, bubbles of carbon dioxide will form in the drop. Record your observations.

PART C: Cleanup and Disposal

9. Dispose of solutions, broken glass, and pieces of clam shell in the waste containers designated by your teacher. Do not pour chemicals down the drain or put lab materials in the trash unless your teacher tells you to do so.
10. Clean up your work area and all lab equipment. Return live clams to the stock container. Return lab equipment to its proper place. Wash your hands thoroughly before you leave the lab and after you finish all work.

Analyze and Conclude

1. **Analyzing Results** Find the incurrent and excurrent siphons of the clam in the illustration on this page. Using this information, explain your observations in step 3.
2. **Drawing Conclusions** What is the purpose of a clam's shell?
3. **Making Predictions** Based on your observations, how do you think clams respond when they are touched or threatened in their natural habitat?
4. **Forming a Hypothesis** What does a clam take in from water that passes through its body?
5. **Inferring Relationships** Water that enters a clam's incurrent siphon passes over the clam's gills. How does this help the clam respire?
6. **Further Inquiry** Write a new question about clams that could be explored with another investigation.

On the Job

Geologists use hydrochloric acid to test rocks for the presence of calcium carbonate. Do research to discover the role that mollusks play in the formation of rocks. For more about careers, visit **go.hrw.com** and type in the keyword **HX4 Careers.**

Answers to Procedure

3. The food coloring will be taken in by the incurrent siphon. Students will observe colored water being expelled near the excurrent siphon.
5. Answers will vary. Students will probably observe that the clam draws in its siphons and foot and closes its shell tightly when touched.
6. Answers will vary. The number of growth rings will depend on the clam's age.
8. Students should observe bubble formation, indicating the presence of calcium carbonate in the middle layer of the shell.

Answers to Analyze and Conclude

1. The food coloring was drawn into the clam through the incurrent siphon and released through the excurrent siphon.
2. The shell protects the clam from injury and predators.
3. Clams draw in their siphons and foot and close their shells tightly when threatened or touched.
4. Food and oxygen are taken in from the water.
5. A mollusk's gills can extract dissolved oxygen from water that passes over them.
6. Answers will vary. For example: What types of food do clams prefer?

Answers to On the Job

Bivalve shells are hard and fossilize well. Also, they live in aquatic environments where they are readily deposited with other sediments to form sedimentary rock, which contains almost all known fossils.

Chapter Opener

- SE **Quick Review,** p. 663 ■
- SE **Reading Activity,** p. 663 ■
- TE **Using the Figure,** p. 680 GENERAL
- TE **Opening Activity,** p. 681 GENERAL

& 3 • 90 min pp. 664–669
Section 1
Features of Arthropods

- OSP **Lesson Plan***
- TR **Bellringer***
- TR **Section Outline***
- TR **H26 Characteristics of Arthropods***
- TR **H31 Exploration of an Insect***

- TE **Demonstration,** p. 664 ◆ BASIC
- TE **Using the Figure,** p. 666 BASIC
- SE **Quick Lab** Evaluating Jointed Appendages, p. 667 ■
- CRF **Datasheets for In-Text Labs** Evaluating Jointed Appendages*
- CRF **Quick Lab** Comparing Arthropods* BASIC

pp. 670–672
on 2
rs and Other chnids

- OSP **Lesson Plan***
- TR **Bellringer***
- TR **Section Outline***
- TR **H30 Anatomy of a Brown Recluse Spider***

- TE **Activity** Poisonous Spiders p. 670 ADVANCED

& 7 • 90 min pp. 673–679
Section 3
Insects and Their Relatives

- OSP **Lesson Plan***
- TR **Bellringer***
- TR **Section Outline***
- TR **H36 Insect Mouthparts***
- TR **H41 Complete Metamorphosis***
- TR **H40 Incomplete Metamorphosis***
- TR **H35,H39 Anatomy of a Grasshopper***

- TE **Demonstration,** p. 674 BASIC
- TE **Using the Figure,** p. 674
- SE **Data Lab** Analyzing the Effects of Pesticide Use, p. 678 ■
- CRF **Datasheets for In-Text Labs** Analyzing the Effects of Pesticide Use*

& 9 • 90 min pp. 680–682
Section 4
Crustaceans

- OSP **Lesson Plan***
- TR **Bellringer***
- TR **Section Outline***
- TR **H32 Comparison of Crustaceans and Insects***

- TE **Demonstration,** p. 680 ◆ GENERAL
- SE **Data Lab** Relating Molting to Mortality Rates, p. 682 ■
- CRF **Datasheets for In-Text Labs** Relating Molting to Mortality Rates*
- CRF **Skills Practice** Crayfish Dissection* ◆ GENERAL
- SE **Exploration Lab** Observing Pill Bug Behavior, pp. 686–687 ■
- CRF **Datasheets for In-Text Labs** Observing Pill Bug Behavior*

0 & 11 • 90 min
Chapter Review and Assessment Resources

- SE **Study Zone,** p. 683 ■
- SE **Performance Zone,** p. 684 ■
- SE **FCAT Test Prep,** p. 685 ■
- TR **Concept Mapping***
- CRF **Vocabulary Review Worksheet*** ■ GENERAL
- CRF **Test Prep Pretest*** ■ GENERAL
- CRF **Chapter Test*** ■ GENERAL
- CRF **Chapter Test*** ADVANCED
- OSP **Test Generator***
- OSP **Test Item Listing***
- OSP **Modified Chapter Test*** BASIC

Holt Online Learning

Visit go.hrw.com for access to Holt Online Learning or enter the keyword **HX6 Home** for a variety of free online resources.

One-Stop Planner® CD-ROM

This CD-ROM package includes
- Lab Materials QuickList Software
- Holt Calendar Planner
- Customizable Lesson Plans
- Printable Worksheets
- ExamView® Test Generator
- Interactive Teacher Edition
- Holt PuzzlePro® Resources
- Holt PowerPoint® Resources

Guided Reading Audio CDs

These CDs are designed to help auditory learners and reluctant readers. (Audio CDs are also available in Spanish.)

Compression guide:
To shorten your instruction because of time limitations, omit blocks 1 and 10.

KEY

TE	Teacher Edition	CRF	Chapter Resource File	*	Also on One-Stop Planner
SE	Student Edition	TR	Teaching Transparency	■	Also Available in Spanish
OSP	One-Stop Planner			◆	Requires Advance Prep

SKILLS DEVELOPMENT AND PRACTICE	SECTION REVIEW AND ASSESSMENT	CORRELATIONS
		Florida Sunshine Standards/FCAT
TE Reading Skill Builder Reading Organizer, p. 665 BASIC TE Reading Skill Builder Interpreting Visuals, p. 665 GENERAL TE Reading Skill Builder Interactive Reading, p. 666 GENERAL CRF Directed Reading Worksheet* BASIC CRF Active Reading Worksheet* GENERAL	SE Section 1 Review ■ TE Reteaching, p. 669 BASIC TE Quiz, p. 669 GENERAL TE Alternative Assessment, p. 669 GENERAL CRF Quiz* ■ BASIC	SC.D.1.4.4, SC.F.1.4.2, SC.G.1.4.1
TE Skill Builder Vocabulary, p. 672 GENERAL CRF Directed Reading Worksheet* BASIC CRF Active Reading Worksheet* GENERAL	SE Section 2 Review ■ TE Reteaching, p. 672 BASIC TE Quiz, p. 672 GENERAL TE Alternative Assessment, p. 672 GENERAL CRF Quiz* ■ BASIC	SC.F.1.4.2, SC.G.1.4.1
TE Skill Builder Interpreting Visuals, p. 675 GENERAL CRF Directed Reading Worksheet* BASIC CRF Active Reading Worksheet* GENERAL	SE Section 3 Review ■ TE Reteaching, p. 679 BASIC TE Quiz, p. 679 GENERAL TE Alternative Assessment, p. 679 GENERAL CRF Quiz* ■ BASIC	SC.F.1.4.2, SC.G.1.4.1, SC.G.2.4.6
CRF Directed Reading Worksheet* BASIC CRF Active Reading Worksheet* GENERAL CRF Critical Thinking Worksheet* ADVANCED CRF Science Skills Worksheet* GENERAL	SE Section 4 Review ■ TE Reteaching, p. 682 BASIC TE Quiz, p. 682 GENERAL CRF Quiz* ■ BASIC Modified Worksheet* BASIC	SC.F.1.4.2, SC.G.1.4.1, SC.H.1.4.1

SCILINKS. NSTA
www.scilinks.org
Maintained by the **National Science Teachers Association**

Topic: Arthropods
SciLinks Code: HX4014

Topic: Biological Pest Control
SciLinks Code: HX4021

Topic: Mites and Disease
SciLinks Code: HX4123

Topic: Crustaceans
SciLinks Code: HX4054

Overview

Tell students that the purpose of this chapter is to describe the arthropods, a diverse animal phylum. In spite of their diversity, arthropods share many characteristics, which include jointed appendages, an exoskeleton, and segmentation. The first section describes arthropods in general, while the remaining 3 sections describe various groups of arthropods in particular: the spiders, insects, and crustaceans.

Using the Figure — GENERAL

The butterfly pictured here is a European Swallowtail, *Papilio machaon.* The arthropod characteristics visible are wings and a distinct head. Tell students that butterflies are good examples of arthropods that undergo stages of change in their life cycles, from egg to larva to pupa to the adult butterfly. Some larvae (caterpillars) of the European Swallowtail are camouflaged by looking like bird droppings on leaves. Others have large false eyes to ward off enemies.

Assessing Prior Knowledge

Before beginning this chapter, be sure your students can answer the questions in the ***Quick Review*** on the facing page. If they cannot, encourage them to reread the sections indicated.

European swallowtail

Chapter Correlations | **Florida Sunshine Standards/FCAT**

SC.D.1.4.4 Earth's systems and organisms are the result of a long, continuous change over time.

SC.F.1.4.2 Body structures are uniquely designed and adapted for their function.

SC.G.1.4.1 There is great diversity and interdependence among living things.

CHAPTER

30 Arthropods

Florida Sunshine Standards/FCAT

SC.D.1.4.4 Earth's systems and organisms are the result of a long, continuous change over time.

SC.F.1.4.2 Body structures are uniquely designed and adapted for their function.

SC.G.1.4.1 There is great diversity and interdependence among living things.

Quick Review

Answer the following without referring to earlier sections of your book.

1. **Summarize** the evolutionary movement of arthropods from the sea onto land. *(Chapter 12, Section 3)*
2. **Define** the term *chitin*. *(Chapter 22, Section 1)*
3. **Describe** the circulation of blood in an open circulatory system. *(Chapter 27, Section 2)*
4. **Define** the term *sessile*. *(Chapter 28, Section 1)*

Did you have difficulty? *For help, review the sections indicated.*

Reading Activity

As you read through this chapter, make an outline of the material presented. You may need to first review outlining methods, such as using Roman numerals, letters, and Arabic numerals. Review your outline to be sure that the main ideas are included.

The beautiful European swallowtail butterfly, *Papilio machaon*, produces one of two colors of pupa. Brown pupae are found on brown leaves or stems. Green pupae are found on green leaves or stems.

Looking Ahead

internet connect

www.scilinks.org

National Science Teachers Association *sci*LINKS Internet resources are located throughout this chapter.

SCiLINKS. Maintained by the National Science Teachers Association

Opening Activity — GENERAL

Arthropods I Know Place students in groups of three. Ask each group to draw on their prior knowledge and the list of arthropod characteristics in Section 1 to develop a list of 20 arthropods (such as ants, spiders, grasshoppers, etc.) Ask them to find as many images of arthropods on their list as they can, collecting images from the Internet, magazines, or other resources. Then ask students to point out and label arthropod characteristics on the hard copies of the images. English Language Learners

Quick Review Answers

1. The first animals to successfully invade land from the sea were arthropods, a kind of animal with a hard outer skeleton and jointed limbs. Biologists think a type of scorpion was the first arthropod to live on land.
2. Chitin is a tough polysaccharide found in the hard outer covering of arthropods.
3. In an open circulatory system, a heart pumps fluid through a series of vessels out into the body cavity. There the fluid washes across the body's tissues, collects in open spaces, and then flows back to the heart.
4. Sessile means fixed to one spot.

Reading Activity Answers

Students should use Roman numerals for the first level of the outline, capital letters for the second level, and Arabic numerals for the third, each followed by a period. The second level should be indented with respect to the first, and the third indented with respect to the second. Continuing that pattern of indenting, students should use a lower case italicized letter followed by a parenthesis for the fourth level, and Arabic numerals in parentheses for the fifth. Encourage students to use the chapter headings in their outlines.

Chapter Resource File

- Vocabulary Worksheets
- Concept Mapping

Section 1

Focus

Overview

Before beginning this section review with your students the objectives listed in the Student Edition. Section 1 describes the common characteristics of arthropods. The arthropod family tree shows that this phylum of organisms is divided into two groups: those with jaws and those with fangs or pincers.

Bellringer

Tell students that one of the characteristics of arthropods is an exoskeleton, or "outer skeleton," constructed of the strong, chemically-resistant material called chitin. Ask students to brainstorm a list of advantages and disadvantages of an exoskeleton. (Advantages: protection, muscle attachment, and reduction of water loss. Disadvantages: Does not "stretch" for growth, can be heavy, and can be vulnerable to breaking or cracking.) LS Visual

Motivate

Demonstration — Basic

Bring in the exoskeleton of a lobster, crab, tarantula, or other arthropod. Allow students to examine it with a hand lens and try to find the areas where muscles were attached to the interior surface. English Language Learners

Section 1 Features of Arthropods

Florida Sunshine Standards/FCAT

SC.D.1.4.4 Earth's systems and organisms are the result of a long, continuous change over time.

SC.F.1.4.2 Body structures are uniquely designed and adapted for their function.

SC.G.1.4.1 There is great diversity and interdependence among living things.

Objectives

- **Summarize** the evolutionary relationship of arthropods and annelids. SC.D.1.4.4
- **Identify** the three subphyla of arthropods. SC.G.1.4.1
- **Describe** the characteristics of arthropods. SC.F.1.4.2
- **Describe** how growth occurs in arthropods. SC.F.1.4.2

Key Terms

appendage
thorax
cephalothorax
compound eye
molting
trachea
spiracle
Malpighian tubule

Jointed Appendages

Whether you are looking at a scorpion or a leaf-footed bug, as shown in **Figure 1,** when you see an arthropod, you will probably notice its appendages. An **appendage** is a structure that extends from the arthropod's body wall. Unlike the parapodia and setae of annelids, arthropod appendages have joints that bend. The phylum name, Arthropoda, literally means "joint foot." A variety of jointed appendages are found in arthropods, including legs for walking, antennae for sensing the environment, and mouthparts for sucking, ripping, and chewing food.

Arthropods almost certainly share a distant common ancestor with the annelid worms. Like annelids, arthropods have a coelom and a segmented body. Arthropod fossils, some as much as 600 million years old, are among the oldest, best-preserved fossils of multicellular animals. Among the most numerous of the early arthropods were the now-extinct trilobites, which lived in the sea. Like modern arthropods, trilobites had segmented bodies and jointed appendages, and they were the first animals to have eyes capable of forming images. Trilobites became extinct about 250 million years ago. The first terrestrial arthropods were probably scorpions similar to the modern scorpion shown in Figure 1.

Figure 1 Arthropods

This leaf-footed bug and hairy desert scorpion belong to phylum Arthropoda.

Leaf-footed bug

Scorpion

Evolutionary Milestone

7 Jointed appendages

Arthropods were the first animals to have jointed appendages. Joints permit powerful movement, aiding locomotion. Jointed appendages became specialized in many different ways, helping to create the vast diversity seen among the arthropods.

664

Chapter Resource File

- Directed Reading Basic
- Active Reading General
- Data Sheet for Quick Lab General

Transparencies

TR Bellringer
TR H26 Characteristics of Arthropods
TR H31 Exploration of an Insect

One-Stop Planner CD-ROM

- Reading Organizers Basic
- Reading Strategies Basic
- Occupational Application Biology Teacher General

Section Correlations

SC.D.1.4.4
SC.F.1.4.2
SC.G.1.4.1

Arthropod Diversity

If a prize were given for sheer numbers, it would go to the arthropods. The total number of arthropod species exceeds that of all other kinds of animals combined. There may be 5,000,000 or more species of arthropods. There are more species of beetles alone than there are of vertebrates. Scientists estimate that 10^{18} arthropods are alive at any one moment! The great majority of arthropods are small, about 1 mm (0.04 in.) in length. Some parasitic mites are only 80 µm (0.003 in.) long. The largest arthropods are gigantic crabs 3.6 m (12 ft) across, found in the sea near Japan.

Living arthropods are traditionally divided into two groups, arthropods with jaws and arthropods with fangs or pincers. As shown in **Figure 2,** arthropods with jaws belong to either subphylum Uniramia *(yoo nuh RAY mee uh)* or to subphylum Crustacea *(kruhs TAY shuh)*. Arthropods with fangs or pincers belong to subphylum Chelicerata *(chuh LIS uh rahd uh)*. Each of these three subphyla represents a distinct evolutionary line.

internet connect
www.scilinks.org
Topic: Arthropods
Keyword: HX4014
SCILINKS. Maintained by the National Science Teachers Association

Figure 2 **Phylogenetic tree**

This phylogenetic tree shows the relationships among the three arthropod subphyla which form two groups: those with jaws and those with fangs or pincers.

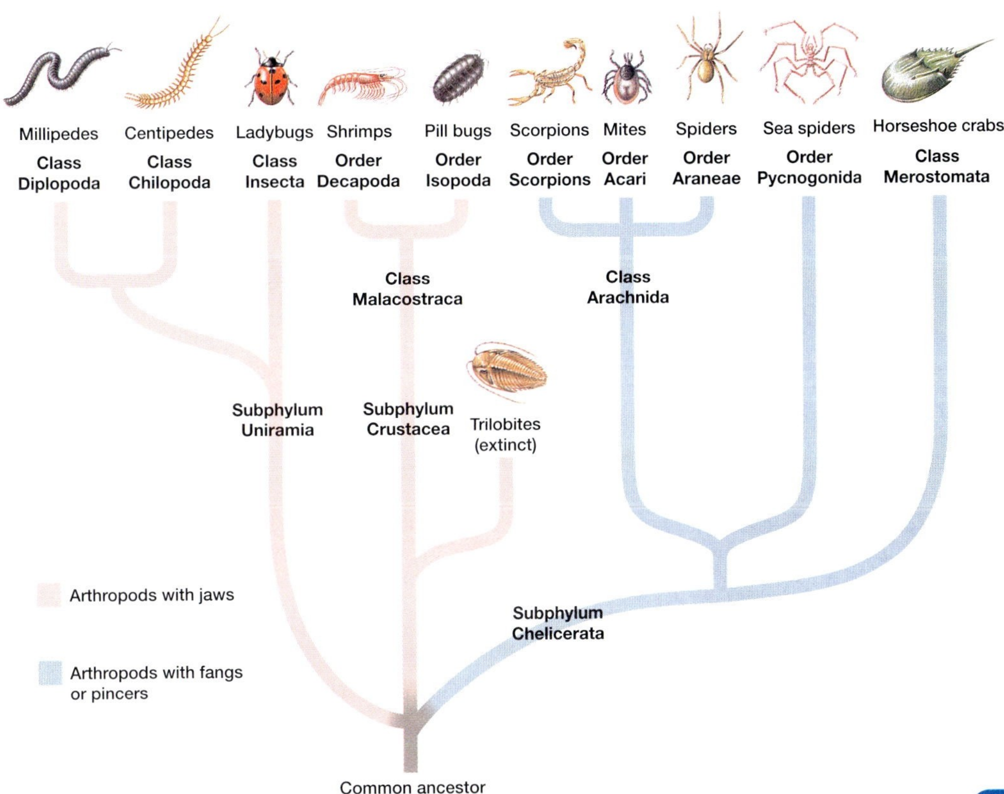

665

Teach

READING SKILL BUILDER — **BASIC**

Reading Organizer As students read this section, have them put the various classes of arthropods and their descriptions on index cards. The class name should be placed on one side and the description on the other. In addition, students could make cards with the class name on one side and examples on the other. Have students use the cards as flash cards to review the section. **LS Verbal**

READING SKILL BUILDER — **GENERAL**

Interpreting Visuals Explain to students that the phylogenetic tree in **Figure 2** "reads" like a graph up the Y axis, but not from left to right across the X axis. Explaining further: Time progresses up the Y axis. That is, points on the tree closer to the bottom of the page denote time longer ago, and points on the tree closer to the top of the figure denote more recent time. However, organismal change does not progress from left to right along the X axis. Organismal change is shown whether moving to the left or right.

Teaching Tip — **GENERAL**

Jointed Appendages Have students read the Evolutionary Milestone on page 664. Remind them that jointed appendages were a major evolutionary development. Have students point out the joints in the appendages shown in **Figure 1.** **LS Visual**

did you know?

"Jointed Foot" The prefix "arthro" comes from a Greek word meaning "joint," and the prefix "pod" comes from another Greek word meaning "foot." The word *arthritis* refers to an inflammation of the joints. A *podiatrist* is a doctor who treats conditions of the feet and a *pseudopod* is also known as a "false foot."

Teach, *continued*

READING SKILL BUILDER — GENERAL

Interactive Reading Assign Chapter 30 of the *Holt Biology* Guided Audio CD Program to help students achieve greater success in reading the chapter. LS Auditory

Teaching Tip — GENERAL

Arthropod Segmentation Have students look through this chapter and find examples of trends in arthropod body plans. Ask: Which are more segmented, larvae or adults? (larvae) Which types of arthropods show more segmentation? (Answers will vary. Millipedes and centipedes are highly segmented.)

Using the Figure — BASIC

Have students discuss which characteristics from the list can be seen in the two arthropods pictured in **Figure 3.** (Students can see the external structures on the list, but not the internal structures.) LS Visual

Arthropod Body Plan

While arthropods may be quite different in appearance, they share a number of internal and external features, which are summarized in **Figure 3.** There is great variation in appearance among arthropod species, and not every species has every feature listed. However, these features are characteristic of the phylum as a whole.

Segmentation

In arthropods, individual body segments often exist only during the larval stage. For example, when you look at a butterfly larva (a caterpillar), you can easily see that it has many segments. However, if you look closely at an adult butterfly, you will see only three body regions. In most arthropods the many body segments fuse during development to form three distinct regions—the head, the **thorax** (midbody region), and the abdomen. In some arthropods, such as the crab shown at the top in Figure 3, the head is fused with the thorax to form a body region called the **cephalothorax.**

Figure 3 Arthropod characteristics. These eight characteristics are typical of arthropods, although not all arthropods show each characteristic.

Characteristics of Arthropods

- Jointed appendages
- Segmentation
- Distinct head, often with compound eyes
- Exoskeleton
- Respiration by gills, tracheae, or book lungs
- Open circulatory system
- Excretion through Malpighian tubules
- Wings on many arthropods

Figure 4 Compound eyes. The compound eye of this house fly is made of 800 or more individual units.

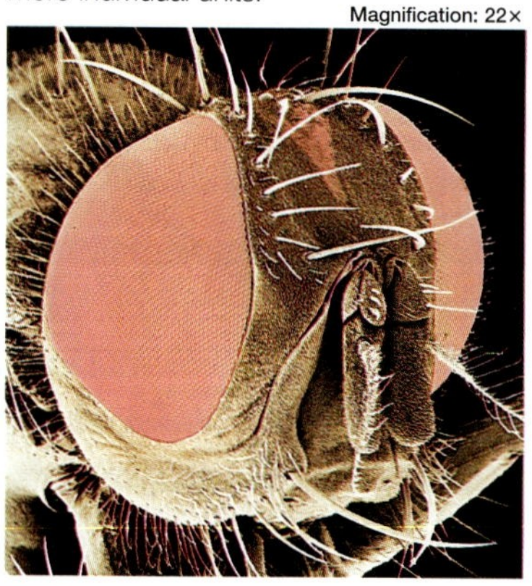

Compound Eyes

Many arthropods have compound eyes, shown in **Figure 4.** A **compound eye** is an eye composed of multiple individual visual units, each with its own lens and retina. The brain receives input from each of the units, and then composes an image of an object. While the image formed is not as clear as what you see, arthropods see motion much more quickly. This is why it is so difficult to sneak up on a fly. Some arthropods also have simple, single-lens eyes that do not form images, but simply distinguish light from dark.

Most insects have both compound and simple eyes. In dragonflies and locusts, these simple eyes function as horizon detectors. The ability to see the horizon helps the insect stabilize its position during flight.

666

MISCONCEPTION ALERT

Compound Eyes The individual units in a compound eye do not all see the same image. To study the imaging capabilities of ommatidia, the segments that make up compound eyes, photographers isolated eyes from insects and shot photos through them. These pictures did not show thousands of identical images as one might expect and as is often portrayed in science fiction movies. Instead, the photographers found the visual effect to be similar to that of a mosaic.

Exoskeleton

Figure 5 Jointed appendages. The joints in the legs of this praying mantis permit it to perform many complex movements, such as manipulating prey.

The outer layer of the arthropod body is a rigid exoskeleton (often called a shell) composed primarily of chitin. The exoskeleton is thin and flexible where the joints of the appendages are located. Muscles attached to the interior surfaces of the exoskeleton can pull against it, causing the animal's joints to bend. As shown in **Figure 5,** many arthropods can use their jointed appendages to perform complex movements. While chitin is tough, it is brittle and breaks easily. As arthropods increase in size, their exoskeletons must become thicker to withstand the pull of larger muscles without breaking. However, an increase in thickness of the exoskeleton adds weight, restricting the size arthropods can reach.

The exoskeleton of the different arthropod groups varies greatly in thickness. If you have ever attempted to swat a large insect, you know that its exoskeleton can be difficult to crush. Crustaceans, for example, have a thick, relatively inflexible exoskeleton. In comparison, the exoskeleton of other insects and some arachnids is fairly soft and flexible. Regardless of the nature of an arthropod's exoskeleton, it provides protection from injury and helps to prevent water loss.

QUICK LAB

Evaluating Jointed Appendages SC.F.1.4.2

To understand the importance of jointed appendages, test your range of movement without and with bending your joints.

Materials

meterstick, paper, and pencil

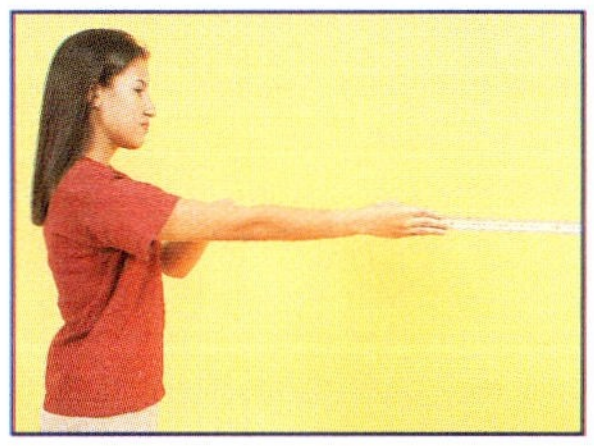

Procedure

1. Work in pairs, and assign one person to be the test subject and one person to record the data.
2. The test subject extends one arm straight out in front of the body. The subject then places a meterstick along the inside of the arm, as shown in the illustration. The elbow should not be bent.
3. The recorder measures and records the distance along the meterstick that the test subject can reach with extended (not bent) fingers.
4. The test subject now tries to increase the range of movement by bending the fingers only. The recorder measures and records the closest and farthest distance along the meterstick that can be reached.
5. The test subject now tries to increase the range of motion by bending the elbow. The recorder measures and records the closest and farthest distance along the meterstick that can be reached.

Analysis

1. **Describe** how eating breakfast might be different if you did not have joints on your fingers and at your elbows.
2. **Predict** the advantages an animal with jointed appendages has over an animal without jointed appendages when capturing and consuming food.
3. **Predict** the advantages for an arthropod that has sense organs (eyes and odor detectors) on the ends of jointed appendages.

Teaching Tip — GENERAL

Armor Relate the weight of an exoskeleton to that of the body armor worn by the knights of the Middle Ages. The knights had to be assisted onto their horses, and the weight of the armor was very tiring. Even today, law enforcement officers sometimes dislike wearing bulletproof vests because they are too heavy and hot to wear.

QUICK LAB

Evaluating Jointed Appendages

Skills Acquired

Measuring, predicting, drawing conclusions

Teacher's Notes

Bring in a crab exoskeleton to demonstrate the range of motion in one jointed appendage.

Answers to Analysis

1. A person's arms would be useless for bringing food to one's own mouth.
2. An animal with jointed appendages would have a much greater range of motion, which would be better for manipulating food.
3. Answers will vary. One advantage is that the animal could move its sense organs by moving particular appendages instead of its whole body.

Integrating Physics and Chemistry

In order to distinguish mass and weight, point out to students that when the thickness of an exoskeleton increases, more mass (i.e., matter) is added. The increased mass then results in a proportional increase in the pull of gravity on the arthropod which thereby increases its weight, which is a measure of the force of gravity. Finally, ask students how an exoskeleton limits the size and movement of arthropods.

Cultural Awareness

Crickets To the Japanese, crickets are considered pets and a delightful source of musical and artistic pleasure. Favorite pet crickets are often given elaborate cages and beautiful porcelain water dishes. Japanese cricket owners even "tickle" their crickets with delicate hand-carved brushes to encourage them to sing. To the British, a cricket on the hearth is a sign of good luck.

Figure 6 Molting. This green cicada is emerging from its old exoskeleton, which it leaves behind as a colorless ghost of itself.

Molting

A tough exoskeleton protects an arthropod from predators and helps prevent water loss. But an exoskeleton cannot grow larger, so an arthropod cannot simply grow bigger, as many other animals do. Imagine blowing up a balloon inside a soft drink can—after a certain point, the balloon cannot get any bigger. Arthropods have the same problem. In a process called **molting,** or ecdysis *(EHK duh sihs),* they shed and discard their exoskeletons periodically. Molting is triggered by the release of certain hormones. Just before molting, a new exoskeleton forms beneath the old one. When the new exoskeleton is fully formed, the old one breaks open. The arthropod emerges in its new, still-soft exoskeleton, as shown in **Figure 6.** The new exoskeleton hardens within a few hours or a few days, depending on the species.

Respiration

The majority of terrestrial arthropods respire through a network of fine tubes called **tracheae** *(TRAY kee ee),* as shown in **Figure 7.** Air enters the arthropod's body through structures called **spiracles** and passes into the tracheae, delivering oxygen throughout the body. Valves that control the flow of air through the spiracles and prevent water loss were a key adaptation for the first arthropods that invaded land more than 400 million years ago.

Real Life

Head lice often infect entire schools.

Head lice are notoriously difficult to control. Having clean hair will not prevent a head lice infection, and anti-lice shampoos often fail to kill immature head lice.

Finding Information

Research the latest remedies for treating head lice, and find out how to avoid infection.

Figure 7 Tracheal system of a beetle

A complex series of hollow tubes called tracheae run through the bodies of most terrestrial arthropods.

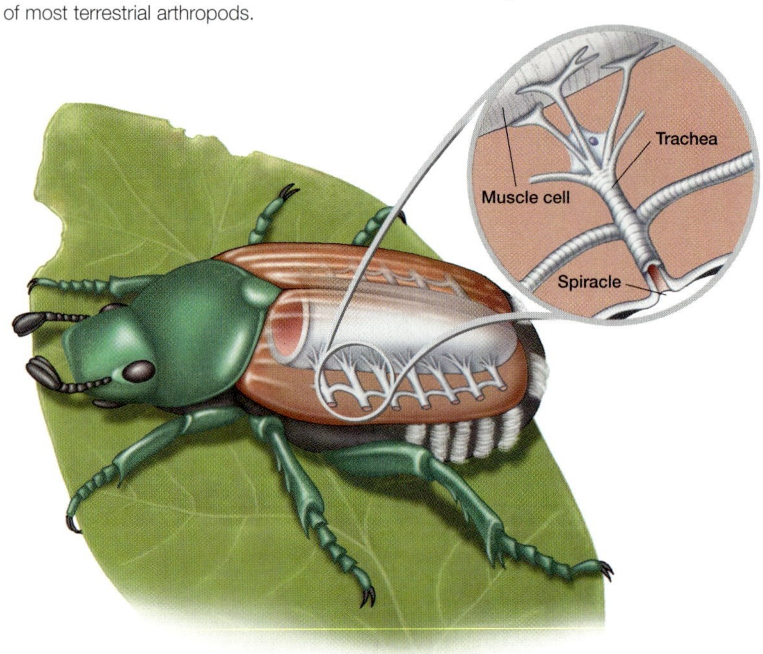

Teach, *continued*

Teaching Tip — BASIC

Molting Ask students to brainstorm some of the problems inherent in molting. (Emerging from an exoskeleton can be hazardous to an arthropod. For some time after emerging from an outgrown exoskeleton, an arthropod is vulnerable to predators and the environment because its new exoskeleton has not yet hardened.) Ask students how arthropods can protect themselves during molting. (Arthropods usually hide until their new exoskeletons have hardened.)

Teaching Tip — GENERAL

Arthropod Respiratory Organs
Insects have hollow branching tubes called tracheae that carry air into the body's tissues. Spiders have leaflike plates called book lungs that have a large surface area to exchange oxygen and carbon dioxide. Lobsters have gills that also have a large surface area for gas exchange.

Real Life

Answer

Topical products for head lice contain the insecticide permethrin. Most products are shampoos or creams used to treat the scalp. After such treatments, health care practitioners suggest combing the hair with a fine-toothed comb to remove the eggs. Creams are also available to treat the skin all over the body. The spread of lice can usually be prevented if people avoid sharing combs, brushes, hats, towels, and other items that may transfer lice or eggs from one person to another.

MISCONCEPTION ALERT

Albino Roaches? People often see what they think are "albino" insects, such as roaches. These white insects are not really albinos. They have just shed their exoskeletons recently and full coloration has not yet returned.

Section Review Correlations

SC.D.1.4.4: Item 1
SC.F.1.4.2: Items 3, 4, 6
SC.G.1.4.1: Items 2, 5

Excretion

Terrestrial arthropods have a unique excretory system that efficiently conserves water and eliminates metabolic wastes. This system is composed of excretory units called Malpighian *(mal PIHG ee uhn)* tubules. **Malpighian tubules** are slender, fingerlike extensions from the arthropod's gut that are bathed by blood. Water and small dissolved particles in the blood move through the tubules and into the arthropod's gut. As this fluid moves through the gut, most of the water, valuable ions, and metabolites from the fluid are reabsorbed into the arthropod's body tissues. Metabolic wastes remain in the gut and eventually leave the body through the anus. You can see the Malpighian tubules on the grasshopper in *Up Close: Grasshopper,* in Section 3 of this chapter.

FORENSICS BIOWatch

Crawly Clues

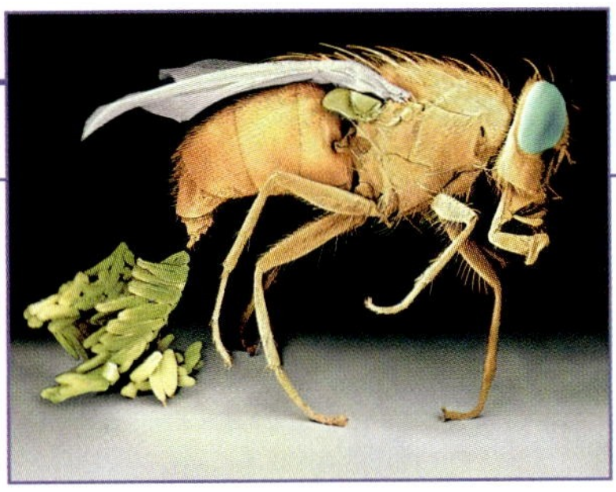

Entomologists are scientists who study the taxonomy, structure, and behavior of insects. Forensic entomologists apply their knowledge of insects to help solve crimes or resolve legal issues.

Postmortem Parade

Forensic medical entomologists are often asked to determine the time of a crime victim's death. To many insects, a victim's body is a banquet table. Blowflies (family Calliphoridae) and flesh flies (family Sarcophagidae) feed upon decaying human flesh. After the moment of death, certain species will appear at predictable intervals in a kind of grim postmortem parade. Blowflies, for example, typically infest a corpse beginning about two days after death. Other insect may come later.

Each species passes through each stage of metamorphosis at a precise rate. Insects develop most quickly in warmer weather. An experienced entomologist can sometimes identify the insect species based on the larval and pupal stages collected from the corpse. In some cases, the entomologist will raise the eggs, larvae, and pupae to adulthood to determine the species.

Location

Sometimes the kinds of insects found on a corpse can help determine the location of death. For example, blowflies are usually found in urban areas. A blowfly-infested corpse found along a rural road might have been killed in the city, and then moved to the countryside.

Section 1 Review

1. **State** the evolutionary relationship of the arthropods and the annelids. SC.D.1.4.4
2. **Describe** the three externally visible characteristics common to all arthropods. SC.G.1.4.1
3. **Summarize** how compound eyes function. SC.F.1.4.2
4. **Describe** how arthropods grow. SC.F.1.4.2
5. **Critical Thinking** **Relating Concepts** Draw a concept map of the three subphyla of arthropods, with two examples of each subphylum. SC.G.1.4.1
6. **FCAT Prep** Oxygen is delivered throughout an arthropod's body by what feature?
   A blood
   B tracheae
   C exoskeleton
   D Malpighian tubules
   SC.F.1.4.2

Answers to Section Review

1. Annelids and arthropods almost certainly share a common ancestor.
2. Jointed appendages, an exoskeleton, and segmentation.
3. Compound eyes receive images from each lens and send them to the brain, where one image is formed.
4. Arthropods cannot grow without shedding their exoskeleton. A new exoskeleton forms, and the old one splits and is molted. The new exoskeleton hardens, and the cycle begins again.
5. Concept maps will vary but should include the subphyla Uniramia, Crustacea, and Chelicerata, and two examples in each subphylum.
6. **A.** Incorrect. Although the blood does carry some oxygen, the body's primary source of oxygen is its network of tracheae. **B.** Correct. The tracheae are a network of fine tubes that deliver oxygen throughout the arthropod body. **C.** Incorrect. The exoskeleton covers the body, protecting the arthropod. **D.** Incorrect. Malphighian tubules function in excretion.

BIOWatch

Crawly Clues

Teaching Strategies
Explain to students that when visiting a crime scene, a forensic entomologist makes careful observations and measurements, including the temperature of the area around and under the victim.

Discussion
In January, investigators find a blowfly-infested body in a rural Montana field. The victim had been reported missing two days earlier. What conclusions might a forensic entomologist suggest? (It is highly unusual to find a large number of blowflies in a corpse that cannot be more than two days old, especially in cold temperatures. Also, blowflies are primarily found in urban areas; their presence in a corpse found in rural Montana might suggest that the murder occurred elsewhere.)

Close

Reteaching — BASIC

Have students write the characteristics of arthropods on index cards and quiz one another with them.
LS Interpersonal

Quiz — GENERAL

True or False:

1. One characteristic of arthropods is a closed circulatory system. (False. Arthropods have open circulatory systems.)
2. When an arthropod molts, it sheds its exoskeleton. (True)

Alternative Assessment — GENERAL

Have students write review questions and answers for a trivia-type game. Pair each student with a partner, and have the students in each pair take turns asking their partner the questions.
LS Interpersonal

Section 2 Spiders and Other Arachnids

Florida Sunshine Standards/FCAT

SC.F.1.4.2 Body structures are uniquely designed and adapted for their function.

SC.G.1.4.1 There is great diversity and interdependence among living things.

Objectives

- **Summarize** the characteristics of arachnids. SC.G.1.4.1
- **Identify** the internal and external characteristics of brown recluse spiders. SC.F.1.4.2
- **Compare** spiders, ticks, and mites. SC.G.1.4.1
- **Identify** the health threats posed by some arachnids. SC.G.1.4.1

Key Terms

chelicera
pedipalp
spinneret

Arachnid Modifications

Perhaps no other group of animals is more disliked and feared by humans than the arachnids—spiders, scorpions, ticks, mites, and daddy longlegs. While it is true that some spiders and scorpions are highly venomous, in general these creatures do more good than harm. For example, many spiders are major predators of insect pests, and gardeners usually welcome them. Arachnids *(uh RAK nihdz)* form the largest class in subphylum Chelicerata. Two minor classes, marine horseshoe crabs and sea spiders, also belong to this subphylum. The members of subphylum Chelicerata have mouthparts called **chelicerae** *(kuh LIS uh ree)* that are modified into pincers or fangs, as shown in **Figure 8.**

The arachnid body is made up of a cephalothorax and an abdomen. There are no antennae, and the first pair of appendages are chelicerae. The second pair of appendages are **pedipalps,** which are modified to catch and handle prey. (The pedipalps are sometimes specialized for sensory or even reproductive functions.) Following the pedipalps are four pairs of appendages called walking legs.

All arachnids except some mites are carnivores, and most are terrestrial. Since arachnids do not have jaws, they are able to consume only liquid food. To do so, the arachnid first injects its prey with powerful enzymes that cause the prey's tissues to liquefy. Then the arachnid sucks the liquid food into its stomach.

Figure 8 Chelicerae. The baboon spider's pointed black chelicerae (fangs) and its two pair of pedipalps are clearly seen in this close-up of its head region.

Spiders

The chelicerae of spiders are modified into fangs. Poison glands located in the spider's anterior end secrete a toxin through these fangs. The toxin kills or paralyzes the prey. The spider then injects enzymes into the prey that digest its tissues, and the spider sucks up the liquid food. Spiders are important predators of insects in almost every terrestrial ecosystem. Only two species of spiders living in the United States, the black widow and brown recluse, are dangerous to humans. Not all spiders build beautiful webs as the orb-builders do. Most spiders can secrete sticky strands of silk from appendages called **spinnerets** located at the end of the abdomen. Tubes located on some spinnerets do not produce silk. Instead, they excrete a sticky substance that the spider can use to make some silk strands adhesive.

Section 2

Focus

Overview

Before beginning this section review with your students the objectives listed in the Student Edition. Section 2 describes the characteristics of arachnids: spiders, scorpions, ticks, mites, and daddy longlegs.

Bellringer

Ask students to list what comes to mind when they hear the word "spider." For homework, have students determine whether their preconceptions are accurate. For example, if students list "poisonous" have them research which spiders are poisonous. Are there poisonous spiders in your state? If students list words like "scary," have them clarify what characteristics of spiders make them scary and have them determine the accuracy of these characteristics.

Motivate

Activity — Advanced

Poisonous Spiders Have students search the library or an on-line database for information about poisonous spiders around the world, such as the Australian funnel-web spider, whose bite can be lethal if left untreated. Have students make a pamphlet describing the spider and its habitat, the symptoms caused by the spider's bite, and the measures a person should take if bitten. LS Verbal

Section Correlations

SC.F.1.4.2
SC.G.1.4.1

Chapter Resource File

- Directed Reading BASIC
- Active Reading GENERAL

One-Stop Planner CD-ROM

- Reading Organizers BASIC
- Reading Strategies BASIC

Transparencies

TR Bellringer
TR H30 Anatomy of a Brown Recluse Spider

Brown Recluse Spider (Violin Spider)

- **Scientific name:** *Loxosceles reclusa*
- **Size:** Length of females, up to 10 mm (0.5 in.); males are smaller
- **Range:** South-central United States, from central Texas to Alabama, north to southern Ohio
- **Habitat:** Dark, dry sheltered sites outdoors or indoors
- **Diet:** Small insects

Characteristics

Cephalothorax Six eyes, in pairs, form a semicircle around the front of the cephalothorax. Two chelicerae and two pedipalps are located next to the mouth. Four pairs of walking legs attach to the cephalothorax, which is marked on top with a distinctive violin shape.

Abdomen The abdomen contains most of the spider's organs. Spinnerets located here are used to spin small, irregular webs.

Reproduction During mating, the male uses its pedipalps modified into sperm storage organs to insert sperm into the female's body. The female lays an average of 20–50 eggs inside a silk cocoon that she spins and hangs in her web.

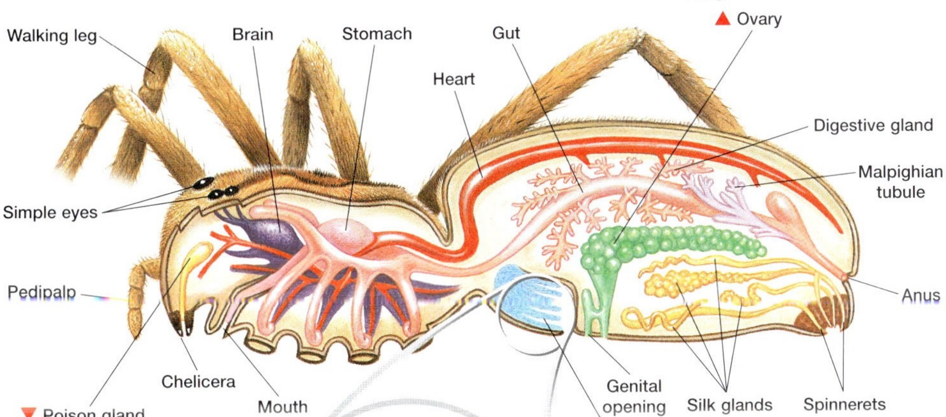

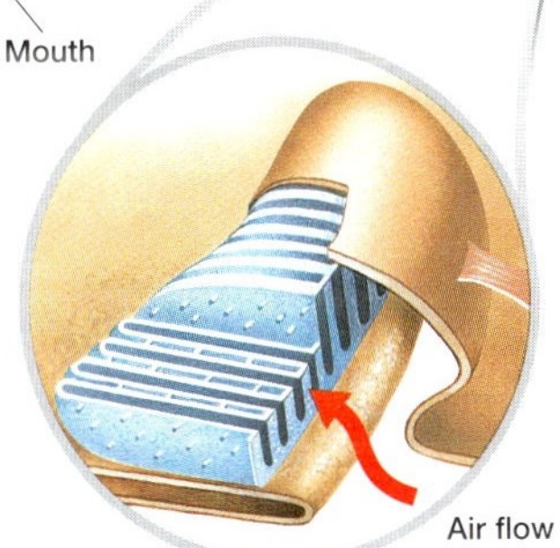

Poison gland Poison glands are connected to the fanged chelicera. The venom of the brown recluse is harmful to many humans. The severity of a person's reaction depends on the individual's sensitivity and how much venom is injected.

Respiration Book lungs contain stacks of thin blood-filled plates of tissue. Air enters the book lungs through slits on the underside of the spider's abdomen and passes over them. Blood flowing through the plates picks up oxygen by diffusion.

did you know?

Spider Silk All spiders have silk glands and organs called spinnerets that spin the silk. Some spiders spin webs that are used to catch prey. Others may use webs to protect their eggs and young. Spiders do not discard their old webs but eat them to recycle the protein in them. Spiders also use silk to wrap up prey, or they spin sticky balls to throw at prey. Burrow-dwellers line their underground homes with silk.

Up Close

Brown Recluse Spider

Teaching Strategies Have students look up the word recluse in a dictionary. (A recluse is one who lives in seclusion or isolation.) Explain that during the day, brown recluses hide in corners or other dark areas, such as old boxes that are seldom opened. They are active at night. Caution students that if they live in an area where brown recluses live, they should wear gloves when they clean out an area that has been left undisturbed for some time.

Explain that many spiders spin at least two varieties of silk threads. Dry threads are used by orb-weaving spiders to form the web's radial framework; sticky threads are arranged in a spiral around the radial threads. Ask students what advantage a spider gains by spinning both dry and sticky threads. (Spiders avoid becoming trapped in their own webs by walking on the dry radial threads.)

Discussion

- Ask students what the adaptive advantage of venom production might be in bees, ants, scorpions, spiders, and centipedes. (Bees and ants use venom for defense. Scorpions, spiders, and centipedes use venom to capture prey.)
- Look at the art showing the interior of the spider. What structures are present that are characteristic of arthropods in general? (jointed appendages, open circulatory system, Malpighian tubules)
- What characteristics of arachnids are present? (pedipalps, chelicerae)

Teach, *continued*

SKILL BUILDER — GENERAL

Vocabulary Arachnids are named for Arachne, a weaver in Greek mythology. Arachne boastfully challenged the goddess Athena to a weaving contest. Arachne produced a beautiful tapestry. Athena was so enraged by Arachne's skill that she changed Arachne into a spider.
LS **Verbal**

Close

Reteaching — BASIC

Have students create crossword puzzles using the vocabulary terms in this section. Then have them trade puzzles and try to solve them.
LS **Verbal**

Quiz — GENERAL

1. Define the term "chelicerae." (Chelicerae are arachnid mouthparts that are modified into fangs or pincers.)
2. Explain the meaning of the following statement: All arachnids except some mites are carnivores. (The statement means that most arachnids feed on animals, not on plants.)

Alternative Assessment — ADVANCED

Have students write an essay on arachnids, highlighting the organism they think is the most fascinating and explaining why. LS **Verbal**

internet connect
www.scilinks.org
Topic: Mites and Disease
Keyword: HX4123
SciLinks. Maintained by the National Science Teachers Association

Scorpions and Mites

Two other familiar groups of arachnids are scorpions and mites. Like spiders, they have chelicerae and pedipalps, but these structures are modified differently.

Scorpions

Scorpions have long, slender, segmented abdomens that end in a venomous stinger used to stun their prey. The stinger-tipped abdomen is usually folded forward over the rest of the scorpion's body, a trait that makes scorpions instantly recognizable. The pedipalps of scorpions are large, grasping pincers, which are used not for defense but for seizing food and during sexual reproduction.

Mites

Mites are by far the largest group of arachnids. Some mites, including chiggers and ticks, are well known to humans because of their irritating bites. They are easily recognizable because their head, thorax, and abdomen are fused into a single, unsegmented body. Most adult mites, such as the one shown in **Figure 9,** are quite small, typically less than 1 mm (0.04 in.) long, but ticks grow larger. Many aquatic mites are herbivores, while terrestrial mites are usually predators.

Most mites are not harmful, but some are plant and animal pests. While feeding, plant mites may pass viral and fungal infections to the plant. Blood-sucking ticks attach themselves to a host, often a human. Lyme disease is spread by bites from infected deer ticks, like the one shown in Figure 9.

Dust mite

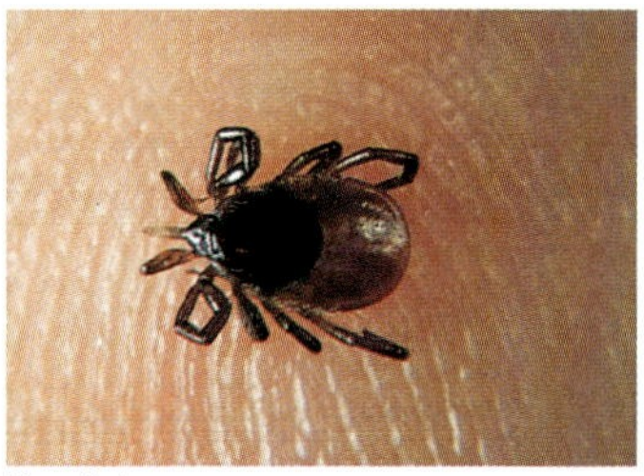
Deer tick

Figure 9 Mites. The house dust mite is a major cause of allergies in humans. The bite of an infected deer tick can cause Lyme disease.

Section 2 Review

1. **Compare** the body plan of spiders, scorpions, and mites, including differences in appendages. SC.F.1.4.2
2. **Describe** the body and coloring of a brown recluse spider. SC.F.1.4.2
3. **Critical Thinking Summarizing Information** Explain why a tick bite is more a cause for concern than the bite of most spiders. SC.G.1.4.1
4. **Critical Thinking Evaluating Conclusions** Is an animal that has grasping pincers, a segmented body, and two antennae correctly identified as a scorpion? Why or why not? SC.F.1.4.2
5. FCAT Prep What is the relationship between a tick and a dog an example of?
   - **A** commensalism
   - **B** homeostasis
   - **C** mutualism
   - **D** parasitism

   SC.G.1.4.1

Answers to Section Review

1. *Spider*—fangs modified from chelicerae; pedipalps; two segments (cephalothorax and abdomen). *Scorpion*—stinger; pedipalps modified into pincers; long, slender, segmented body. *Mite*—chelicerae and pedipalps; head, thorax, and abdomen fused into one.
2. They are up to 10 mm in length, dark brown, with a violin shape on their cephalothorax.
3. Only two spiders in the United States are poisonous to humans. In contrast, many ticks transmit diseases, such as Lyme disease.
4. The identification is not correct. Scorpions have no antennae.
5. **A.** Incorrect. Commensalism is a relationship in which one organism benefits and the other is neither harmed nor helped. **B.** Incorrect. Homeostasis is the maintenance of stable internal conditions in spite of changes in the external environment. **C.** Incorrect. Mutualism is a relationship in which both organisms benefit. **D.** Correct. Parasitism is a relationship in which one organism benefits and the other is harmed but usually not killed.

Section Review Correlations

SC.F.1.4.2: Items 1, 2, 4
SC.G.1.4.1: Items 3, 5

Insects and Their Relatives

Section 3

Florida Sunshine Standards/FCAT

SC.F.1.4.2 Body structures are uniquely designed and adapted for their function.

SC.G.1.4.1 There is great diversity and interdependence among living things.

Insect Diversity

Anyone who has ever been on a picnic in a wooded area does not have to be told that insects are numerous. Ants, mosquitoes, gnats, flies, bees, crickets—they all want to join in while the cicadas sing in the background. These animals all belong to the arthropod subphylum Uniramia, an enormous group of mostly terrestrial arthropods that have chewing mouthparts called **mandibles** (jaws). Uniramians consist of three classes: Insecta (insects), Diplopoda (millipedes), and Chilopoda (centipedes).

The insects are by far the largest group of organisms on Earth, with more than 700,000 named species. Most scientists agree that there may be several million insect species in existence, with most of the undiscovered species living in the tropics. As shown in **Figure 10,** more than 50 percent of all named animal species are insects. More than 90 percent of these species belong to one of the four orders shown in **Table 1.** To read about other orders of insects, see "A Six-Kingdom System of Classification" in the Appendix of this book.

Objectives

- **Describe** the characteristics of insects. SC.G.1.4.1
- **Compare** complete and incomplete metamorphosis. SC.F.1.4.2
- **Identify** the external and internal structures of the Eastern Lubber grasshopper. SC.F.1.4.2
- **Compare** millipedes and centipedes with insects. SC.G.1.4.1

Key Terms

mandible
metamorphosis
chrysalis
pupa
nymph
caste

Figure 10 Species of insects

The predominance of insects, especially beetles (Coleoptera), in the living world is illustrated by the blue section of this pie chart.

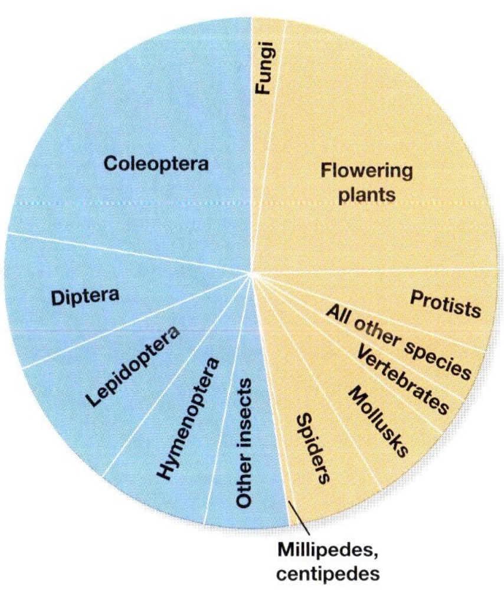

Table 1 Four Orders of Insects

Order	Examples		Number of species
Coleoptera "shield winged"	Beetles, weevils		350,000
Diptera "two winged"	Flies, mosquitoes		120,000
Lepidoptera "scale winged"	Butterflies, moths		120,000
Hymenoptera "membrane winged"	Ants, wasps, bees		100,000

673

Transparencies

TR Bellringer
TR H36 Insect Mouthparts
TR H41 Complete Metamorphosis
TR H40 Incomplete Metamorphosis
TR H35, H39 Anatomy of a Grasshopper

Chapter Resource File

- Directed Reading BASIC
- Active Reading GENERAL
- Data Sheet for Data Lab GENERAL

One-Stop Planner CD-ROM

- Reading Organizers BASIC
- Reading Strategies BASIC
- Supplemental Reading *Journey to the Ants* ADVANCED

Section 3

Focus

Overview

Before beginning this section review with your students the objectives listed in the Student Edition. Section 3 describes the general body plan and life cycles of insects. Then, after discussing insect flight and the social insects, Section 3 includes the centipedes and millipedes, members (with the insects) of the Subphylum Uniramia.

Bellringer

Prepare a handout for students with the following facts: *There are more arthropod species than all other animal species combined. Eighty-five percent of all animals are arthropods. Arthropods are our main competitors for food. If left unchecked (by nature), they could take over the world. There are 200 million insects for every human on Earth. Insects destroy 10–15 percent of the world's food supply each year. Insects have been on Earth for approximately 300 million years. Bush crickets have ears on their knees.* Have the students read the facts while you take attendance, then ask students which fact was the most interesting to them and why.

Motivate

Identifying Preconceptions—GENERAL

People use the terms "insects" and "bugs" freely to refer to many organisms, some of which are not insects. Ask students to list 5 organisms that they think are insects. Then have students compare the characteristics of the organisms they listed with those of insects. Have students determine whether or not the organisms on their lists are truly insects.

Section Correlations

SC.F.1.4.2
SC.G.1.4.1

Teach

Demonstration — BASIC

Bring examples of live or preserved insects to class for display, or bring detailed photographs of insects from a nature magazine. Allow students to compare the animals. Ask them to point out examples of the arthropod characteristics. (jointed appendages, segmentation, exoskeleton, wings, spiracles, etc.) LS Visual

Using the Figure

Use **Figure 11** to point out how the mouthparts of different insects are adapted for different functions. Grasshoppers cut and chew their food, mosquitoes puncture skin to draw out blood, and flies soak up fluids with their mouthparts. LS Visual

Teaching Tip

Jumping Fleas Dr. Miriam Rothschild is a world-class entomologist and the world's expert on fleas. In 1975 she published a paper giving the first description of the flea's jumping mechanism. Using high-speed photography that records 10,000 frames per second, Dr. Rothschild concluded that a flea's ability to leap is equivalent to a human jumping over a tower 30,000 times before exhaustion sets in. She also found that a jumping rabbit flea has 230 times more acceleration than a spacecraft reentering the Earth's atmosphere after a trip to the moon. Dr. Rothschild catalogued the Rothschild Collection of Fleas, housed in the British Museum (Natural History), in six illustrated volumes.

Study TIP

Organizing Information
The numbered list on this page lists three important characteristics of insects. Use this information to draw a concept map that summarizes insect characteristics.

Insect Body Plan

Insects are primarily a terrestrial group, and aquatic insects probably had terrestrial ancestors. Although the great majority of insects are small (some are only a few centimeters in length), others are much larger. The African Goliath beetle, for example, exceeds 10 cm (4 in.) in length. Generally, the larger insects live in tropical areas. Despite great variation in their size, all insects share the same general body plan, made up of three body sections.

1. **Head.** Located on an insect's head are mandibles, specialized mouthparts, and one pair of antennae. The mandibles and mouthparts of different insect species are adapted for eating different foods, as shown in **Figure 11.** In addition, an insect's head usually has a relatively large pair of compound eyes and a pair of antennae. Like the mouthparts, antennae vary greatly in size and shape.
2. **Thorax.** The thorax is composed of three fused segments. Attached to the thorax are three pairs of jointed walking legs. Some insects, such as fleas, lice, and silverfish, lack wings, but other adult insects have one or two pairs attached to the thorax.
3. **Abdomen.** The abdomen is composed of 9 to 11 segments. In adult insects, there are no wings or legs attached to the abdomen.

Turn the facing page to learn more about one particular insect, the grasshopper, in *Up Close: Eastern Lubber Grasshopper.*

Figure 11 Insect mouthparts

The mouthparts of the different insect species are adapted for different functions.

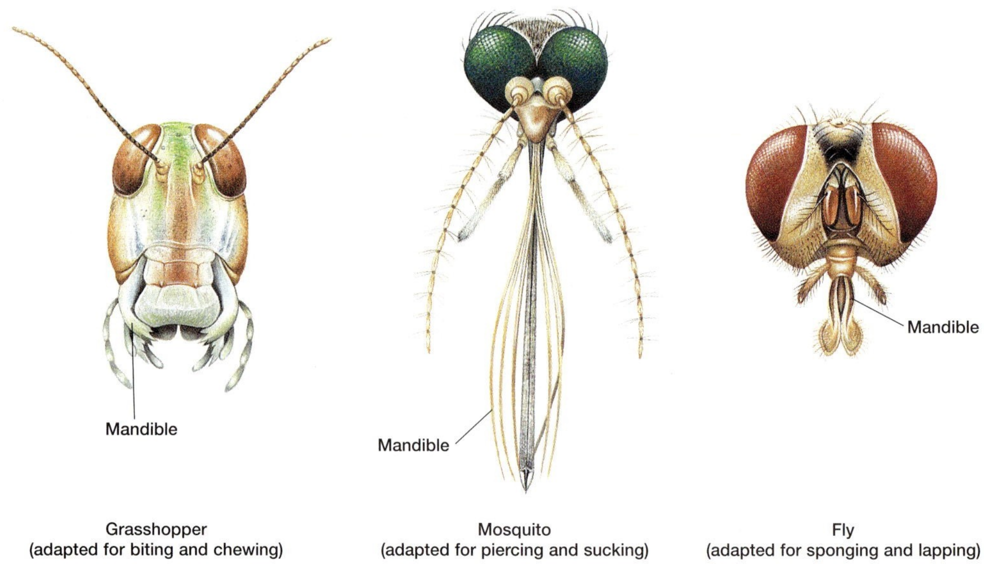

674

Career

Forensic Entomologist Maggots crawling out of eye sockets are usually the stuff of horror films. But to forensic entomologists, insects and cadavers are part of their work. Forensic entomology is the science of using insect evidence to determine information related to a crime. By using insect evidence gathered from and around a corpse, a forensic entomologist can determine approximate time and location of death and may help provide information in investigations of sudden death, traffic accidents with no obvious cause, and other types of crime. Most forensic entomologists hold a Ph.D. degree. However, a student with a BS degree and a major in entomology would be qualified for a position as a research assistant, technician, or staff member in a research laboratory.

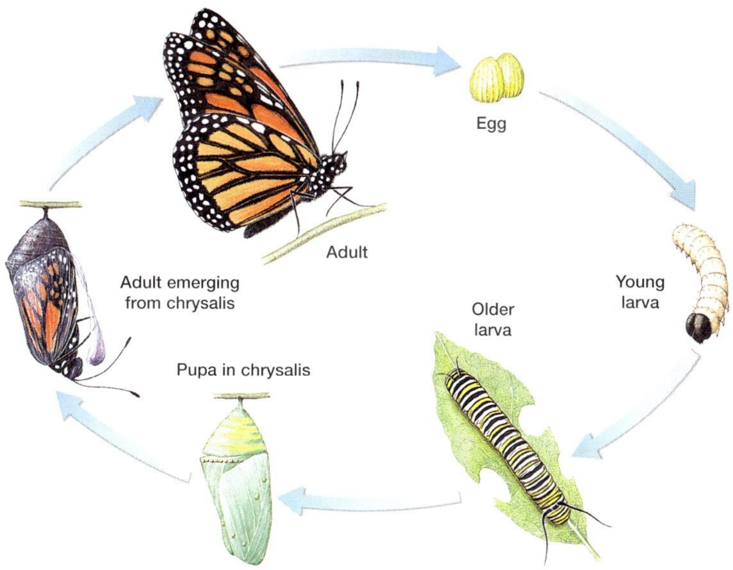

Figure 12 Complete metamorphosis. A complete metamorphosis includes a pupa stage, which is passed inside a chrysalis.

Insect Life Cycle

The life cycles of most insects are complex, and often several molts are required before the adult stage is reached. During the last molt, the young insect undergoes a dramatic physical change called **metamorphosis.**

Complete Metamorphosis Almost all insect species undergo "complete" metamorphosis, as shown in **Figure 12.** In complete metamorphosis, the wingless, wormlike larva encloses itself within a protective capsule called a **chrysalis** *(KRIHS uh lihs)*. Here, it passes through a **pupa** stage, in which it changes into an adult.

A complete metamorphosis is a complex life cycle. The larvae can, however, exploit different habitats and food sources than adults. For example, the larvae of nectar-drinking butterflies are caterpillars that eat leaves! This ecological separation of young from adults eliminates competition. This increases the chance of survival for each phase of the life cycle.

Incomplete Metamorphosis A smaller number of species develop into adults in a much less dramatic incomplete metamorphosis, as shown in **Figure 13.** In these species, the egg hatches into a juvenile, or **nymph** *(NIHMF)*, that looks like a small, wingless adult. After several molts, the nymph develops into an adult.

Figure 13 Incomplete metamorphosis. The nymph passes through several molts before it becomes an adult.

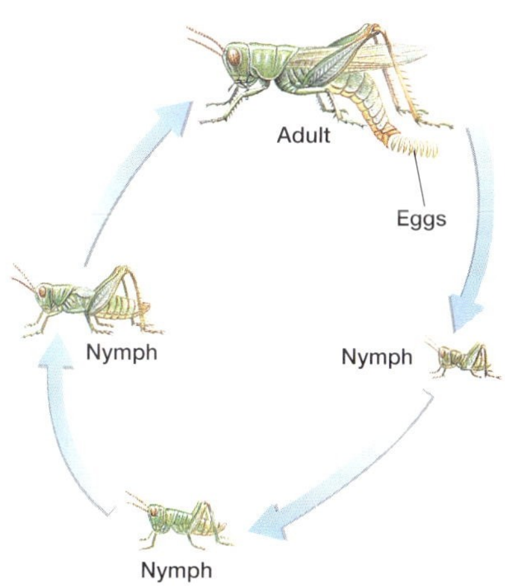

675

Teaching Tip — ADVANCED

Silk Encourage students to research the history of the discovery of silk. The Chinese were the first to produce silk fabrics, and for about 3,000 years, they were the only people who knew how to cultivate silkworms. Chinese legends colorfully tell of the discovery of silk by Emperor Huangdi's wife in about 2700 B.C. Have students create a poster or diorama of the history of silk.

Teaching Tip — BASIC

Metamorphosis Ask students to compare complete and incomplete metamorphosis in a Graphic Organizer similar to the one at the bottom of this page. Students may use either generalizations or specific insects in their comparisons. LS Visual

SKILL BUILDER — GENERAL

Interpreting Visuals Tell students that each visual in **Figures 13 and 14** depicts a cycle. Since there is no beginning or end in a cycle, the student may choose any point to begin analyzing the cycle. However, when comparing two cycles, tell students to choose an equivalent point in each to begin, such as the adult, to make the comparison easier.

Graphic Organizer

Use this graphic organizer with **Teaching Tip: Metamorphosis** on this page.

Complete metamorphosis: Egg → Larva → Pupa → Adult

Incomplete metamorphosis: Egg → Nymph → Adult

Up Close

Eastern Lubber Grasshopper

Teaching Strategies Have students look up grasshoppers in field guides that are used for insect identification. Ask them to compare the external structures of the eastern lubber grasshopper with those of other grasshoppers. (The eastern lubber grasshopper is a short-horned grasshopper. Long-horned grasshoppers, such as katydids and Mormon crickets, have long, threadlike antennae.)

Ask students to list the parts of the digestive system that help to physically break down food as well as the parts that chemically break down food. (Physical digestion is performed by the labrum, labium, mandible, maxilla, and gizzard. Chemical digestion is performed by the midgut and coelom.)

Eastern Lubber Grasshopper

- **Scientific name:** *Romalea microptera*
- **Size:** 5 cm (2 in.) to 6.5 cm (2.6 in.) in length
- **Range:** Eastern United States
- **Habitat:** Fields and meadows
- **Diet:** Grasses and other leafy vegetation

External Structures

Thorax The thorax is composed of three fused segments, each with a pair of legs. The front two pairs are walking legs. The rear pair is larger jumping legs.

During mating season, males "sing" to potential mates by rubbing a row of pegs on a jumping leg against ridges on a forewing.

Head Two antennae contain sense organs for touch and smell. On each side of the head is a very large compound eye. Located high on the forehead are three light-detecting ocelli.

Wings Grasshoppers have a pair of leathery forewings that protect the more delicate flying wings (This grasshopper is flightless).

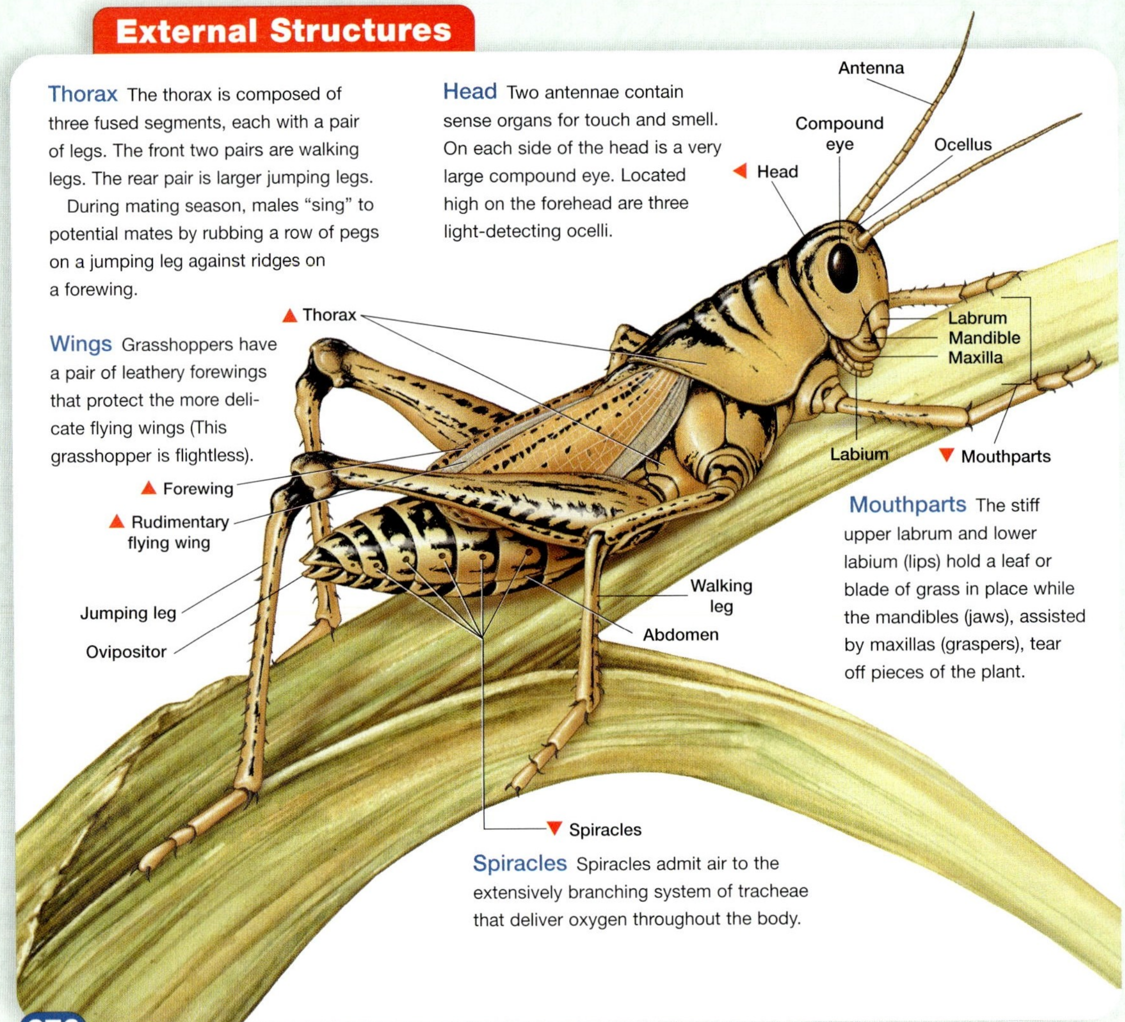

Mouthparts The stiff upper labrum and lower labium (lips) hold a leaf or blade of grass in place while the mandibles (jaws), assisted by maxillas (graspers), tear off pieces of the plant.

Spiracles Spiracles admit air to the extensively branching system of tracheae that deliver oxygen throughout the body.

676

Attention Grabber

Romance and the Grasshopper Almost everyone knows that grasshoppers produce sounds that resemble chirping "songs." But most people don't know that it is generally male grasshoppers that play these songs as part of their courtship behaviors. They generally have a repertoire of three. The male grasshopper's initial song attracts the female, and when a female comes close, the male plays a more "intimate" melody. However, when another male is close by and competing for the female, the male plays the rivalry song. Some females play no songs in response, while the females of some species respond with short bursts of song.

Internal Structures

Reproductive system The female collects the male's sperm in a storage pouch called a seminal receptacle. Later, the female digs a hole using two pairs of pointed ovipositors. As she releases the eggs into the hole, they are fertilized by the stored sperm.

Circulatory system A long blood vessel with a series of muscular "hearts" runs along the grasshopper's back. Blood is pumped out of the open system and bathes the body tissues directly before returning to the heart.

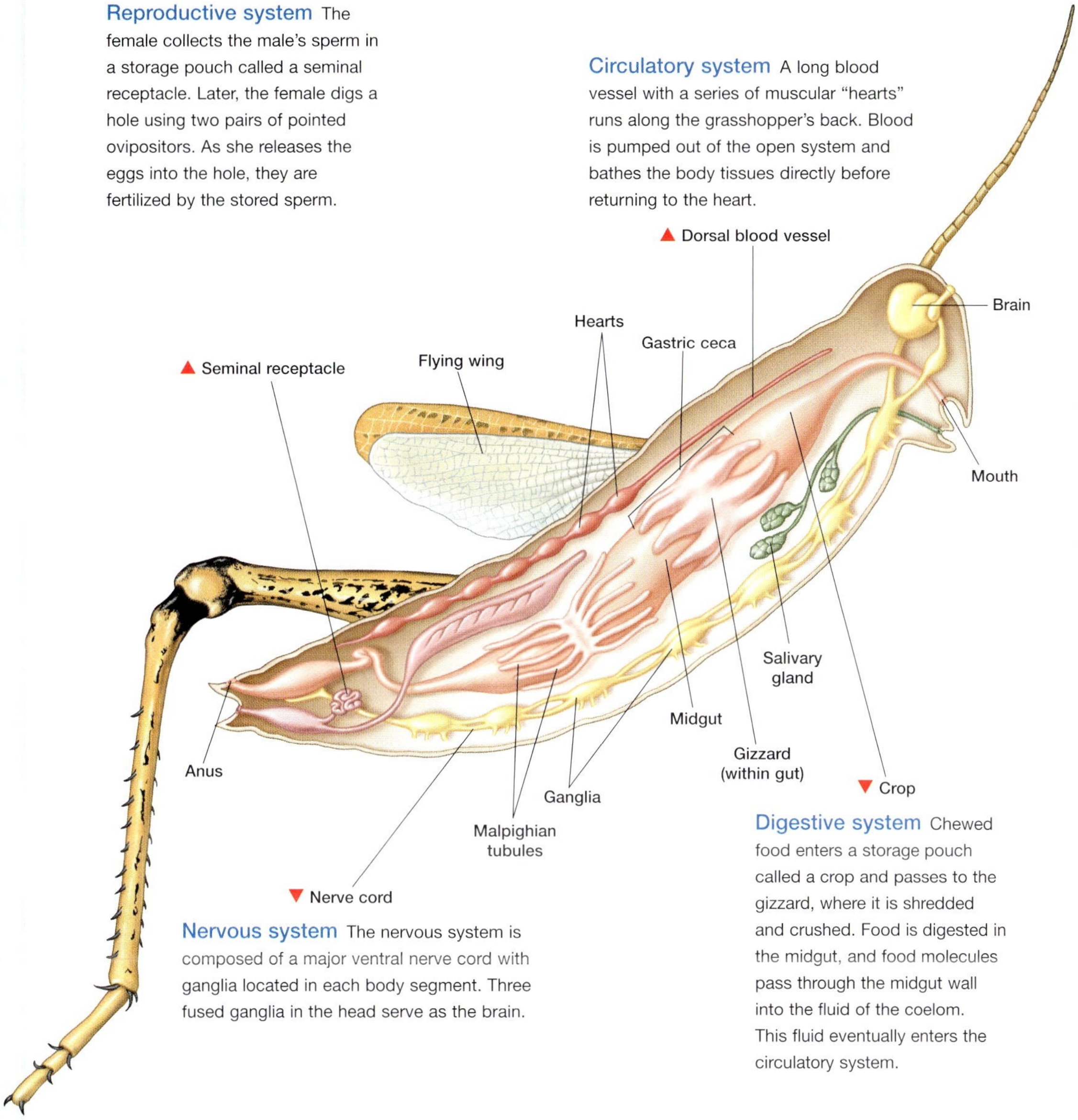

Nervous system The nervous system is composed of a major ventral nerve cord with ganglia located in each body segment. Three fused ganglia in the head serve as the brain.

Digestive system Chewed food enters a storage pouch called a crop and passes to the gizzard, where it is shredded and crushed. Food is digested in the midgut, and food molecules pass through the midgut wall into the fluid of the coelom. This fluid eventually enters the circulatory system.

Up Close

Eastern Lubber Grasshopper

Discussion

- How do grasshopper eyes make it almost impossible to sneak up on this insect? (Grasshoppers have compound eyes on each side of their head and three light-detecting ocelli on their forehead.)
- How can you tell that this grasshopper is a female from its external structures? (This grasshopper has ovipositors, which are used to deposit fertilized eggs.)
- What structures do grasshoppers use to move slowly? How do they make quick escapes? (Grasshoppers use their walking legs to move slowly. When they need to escape quickly, they use their jumping legs or wings.)
- What are the three main sections of a grasshopper's body, and what is a major body function that occurs in each section? (The grasshopper has a head, a thorax, and an abdomen. Touching, smelling, seeing, and chewing occur in the head; walking and jumping occur in the thorax; and respiration, reproduction, and excretion occur in the abdomen.)
- What would happen if every grasshopper were to freeze during a cold winter? (The species would continue to survive because grasshopper eggs stay dormant in winter and hatch in the spring.)

did you know?

Silk is an agricultural product. Today, silk is made with cultivated silkworms, usually caterpillars of the species *Bombyx mori.* The silk-making process begins as the female lays 300 to 500 eggs on special paper provided by silk farmers. The eggs develop into larvae. When the silkworm larvae are ready to begin metamorphosis, they spin a cocoon made of one continuous silk thread, in which they will develop into a pupa and from which they will emerge as a moth. Silk farmers do not allow most pupas to become moths, because a moth will shatter the silk cocoon as it leaves. Instead, farmers kill most pupae and process the silk cocoons. First the cocoon is unwound. Then several processes strengthen the threads. Finally, the thread is dyed and woven into fabric.

Teach, *continued*

Analyzing the Effects of Pesticide Use

DATA LAB

Skills Acquired
Reading graphs, interpreting information

Teacher's Notes
Ask students to name all the pesticides they can. Discuss the problems associated with their use.

Answers to Analysis

1. Populations were stable during the first 4 years. Both populations cycled between the same high and low values each year.
2. As the population of the beneficial species increases, pest numbers decrease, and vice versa.
3. The number of beneficial species and insect pests decreased. The pest population eventually recovered and further increased. The beneficial species population did not recover.
4. Before the application of pesticide, the pest population reached an annual maximum of about 6,000 per acre. After the pesticide application, the pest population rose to 10,000 per acre.
5. Answers will vary. Example: If the beneficial species is more susceptible to the pesticide than the pest species, then the pest population will recover and ultimately increase while the beneficial population will not recover.

Figure 14 Insect flight. This stop-action series shows how this insect's wings move during flight.

Flight

Insects were the first animals to have wings. For more than 100 million years, until flying reptiles appeared, insects were the only flying organisms. Flight, illustrated in **Figure 14,** was a great evolutionary innovation. Flying insects were able to reach previously inaccessible food sources and to escape quickly from danger.

An insect's wings develop from saclike outgrowths of the body wall of the thorax. The wings of adult insects are composed entirely of chitin, strengthened by a network of tubes called veins (which carry air and a bloodlike substance). In most insects, the power stroke of the wing during flight is downward, and it is produced by strong flight muscles. When at rest, most insects fold their wings over their abdomen, but a few insects are unable to do this. Dragonflies, for example, keep their wings outstretched when they rest beside a pond. Most insects have two pairs of wings. A few groups of insects, such as fleas and lice, are wingless.

In most insects only one pair of wings is functional for flight. In some species, the second pair of wings serves another purpose. For example, in grasshoppers and beetles, the forewings act as protective wing covers. In flies, the hindwings are modified into knoblike structures that help control stability during flight.

Analyzing the Effects of Pesticide Use SC.G.2.4.6

DATA LAB

Background

In nature, insect pests are usually kept in balance by the presence of predators, including other insects. The use of some pesticides can upset this balance, as shown in the graph below. Examine the graph, and answer the analysis questions.

Analysis

1. **Identify** the years during which the two insect populations appear to maintain stability in relation to each other. Justify your answer with data from the graph.
2. **Describe** the relationship between the two insect species before year 4.
3. **Describe** the changes in the two populations after the use of a pesticide.
4. **Compare** the annual changes in population size of the pest species before and after the use of a pesticide.
5. **Critical Thinking Developing Hypotheses** Propose a hypothesis that might explain the dramatic changes that occur in the insect populations after the use of pesticides.

Changes in Two Insect Populations

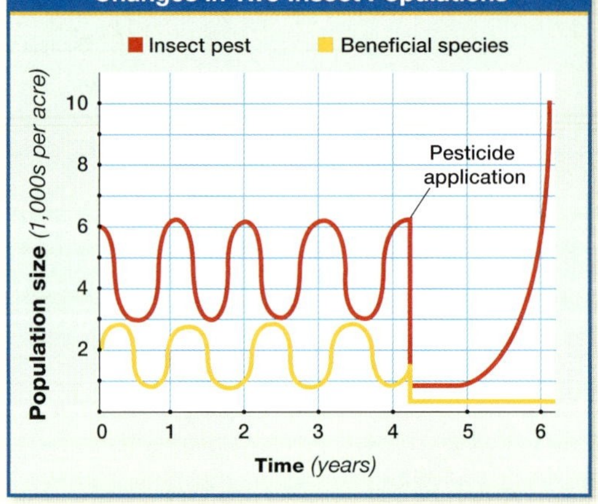

REAL WORLD CONNECTION GENERAL

Pest control Invite a pest-control authority to class to discuss the problems and risks associated with termites and other common household pests, such as ants and cockroaches. Ask the visitor to explain when such insects are considered a hazard. Have students ask questions about new methods used to control pests and about how to avoid infestations.

Integrating Physics and Chemistry

Review the concepts of work (*work* = *force* × *distance*) and power (*power* = *work/time*) by having students calculate and compare the work and power done in flight by two imaginary lacewings. For example, assume that during its flight lacewing A exerted a force of 2 N over a distance of 10 m in 4 s while lacewing B exerted a force of 3 N over a distance of 15 m in 10 s. Remind students that the joule (N-m) is the work unit and watt (J/s) is the power unit. Ask students which lacewing did the most work and which one had the highest power output.

Section Review Correlations

SC.F.1.4.2: Item 5
SC.F.2.4.1: Item 2
SC.G.1.4.1: Items 1, 3, 4

Social Insects

Figure 15 **Termite colony.** Most of the members of this termite colony are unable to reproduce. The queen, with her enormous abdomen, is the egg-laying machine of the colony.

Two orders of insects, Hymenoptera (ants, bees, and wasps) and Isoptera (termites), have elaborate social systems. These insects often live in highly organized societies of genetically related individuals. Within these insect societies, there is a marked division of labor, with different kinds of individuals performing specific functions. The role played by an individual in a colony is called its **caste.** Caste is determined by a combination of heredity; diet, especially as a larva; hormones; and *pheromones,* chemical substances used for communication. In the termite colony shown in **Figure 15,** for example, small, active members called workers gather the food, raise the young, and excavate tunnels. Other, larger termites, called soldiers, defend the colony with their immense jaws. Both workers and soldiers are sterile. Reproduction is a function of only the queen and king.

Insect Relatives

Centipedes and millipedes, shown in **Figure 16,** have similar bodies. Each has a head region followed by numerous similar segments. Each segment bears one or two pairs of legs. Centipedes have one pair of legs per segment and can have up to 173 segments. Modern millipedes have from 11 to 100 or more body segments, and most millipede segments have two pairs of legs. While centipedes are carnivores, most millipedes are herbivores.

Centipede

Millipede

Figure 16 **Centipedes and millipedes.** Centipedes are carnivorous predators, while millipedes are herbivores that feed on decayed vegetation.

Section 3 Review

1. **Relate** the Eastern Lubber grasshopper's body plan to that of a typical insect. SC.G.1.4.1
2. **Compare** the life cycles of grasshoppers and butterflies. SC.F.2.4.1
3. **Identify** the distinguishing characteristics of insects, millipedes, and centipedes. SC.G.1.4.1
4. **Critical Thinking Forming Hypotheses** Based on the information given in Table 1, what characteristic is key to determining an insect's classification? Support your answer. SC.G.1.4.1
5. *FCAT Prep* What are the sense organs in the grasshopper's antennae for?
   - **A** touch and smell
   - **B** smell and hearing
   - **C** hearing and vision
   - **D** vision and touch

   SC.F.1.4.2

679

Answers to Section Review

1. The Eastern Lubber grasshopper has a body plan like that of a typical insect. It has a head, a thorax, and an abdomen; three pairs of jointed legs; antennae; and an exoskeleton.
2. Grasshoppers go through several nymph stages during incomplete metamorphosis. Butterflies go through complete metamorphosis, with egg, larva, pupa, and adult stages.
3. Insects have three body sections, three pairs of legs, and antennae; millipedes have many body segments with two pairs of legs per segment, and are herbivores; centipedes have many body segments with one pair of legs per segment, and are carnivores.
4. Wings are the key characteristic in classification. They are described for each class.
5. **A.** Correct. Two antennae contain sense organs for touch and smell. **B.** Incorrect. Grasshoppers' antennae contain sense organs for smell but not hearing. **C.** Incorrect. Grasshoppers see with compound eyes, not their antennae. **D.** Incorrect. Grasshoppers' antennae contain sense organs for touch, but they see with their compound eyes.

Teaching Tip — GENERAL

Bees in Agriculture Tell students that the widespread use of pesticide sprays has resulted in the loss of beehives. Have students discuss the effects the loss of these hives might have on consumers. (Because bees are needed to pollinate many crops, reduced agricultural production and higher prices would result.) Tell students that some farmers have to rent mobile beehives to pollinate their crops.

Close

Reteaching — BASIC

Divide the class into four groups. Write the name of each of the four insect orders listed in **Table 1** on slips of paper. Have a representative from each group pick one of the folded slips out of a bowl or beaker, making sure they cannot see the writing on the slips. Then have each group ask questions of the other groups about insects in their order. The questions should be phrased so that they can be answered "yes" or "no."
LS Interpersonal

Quiz — GENERAL

1. List the structures usually found on the insect head. (mandibles, specialized mouth parts, and antennae)
2. What is the relationship between insects and centipedes? (Insects and centipedes are both arthropods in the Subphylum Uniramia. However, they belong to different Classes.)

Alternative Assessment — GENERAL

Collect a variety of images of insects, centipedes, and millipedes from nature magazines, the Internet, or other resources you may have. Have students work in groups of three and distribute the images equally among the groups. Have students develop twenty questions (and answers) about their images. Using their images and the questions they developed, have groups quiz each other.

Focus

Overview

Before beginning this section review with your students the objectives listed in the Student Edition. Section 4 focuses on crustaceans. This section first describes terrestrial crustaceans, such as pill bugs, sow bugs, sand fleas, and land crabs. Next, Section 4 discusses aquatic crustaceans, such as crabs, lobsters, shrimps, and crayfish. The section ends with the story of sessile crustaceans, the barnacles.

Bellringer

Invite students to name as many crustaceans as they can. If they have difficulty, suggest that they think about crustaceans they may have seen in seafood restaurants and aquaria.

Motivate

Demonstration — GENERAL

Bring in a few crayfish in a bucket for students to observe. Have students predict how the crayfish will move if a stick is suddenly placed in front of them. Students may be surprised when the crayfish shoot off backwards. Have students observe the crayfish's movements for several minutes. LS Visual

Section Correlations

SC.F.1.4.2
SC.G.1.4.1

Section 4 Crustaceans

Florida Sunshine Standards/FCAT

SC.F.1.4.2 Body structures are uniquely designed and adapted for their function.

SC.G.1.4.1 There is great diversity and interdependence among living things.

Objectives

- **Summarize** how crustaceans and insects are similar and dissimilar. SC.G.1.4.1
- **Describe** the body plan of decapods. SC.F.1.4.2

Key Terms

nauplius
krill

Crustacean Habitats

Just as insect species dominate on land, crustaceans abound in the world's oceans. Their great numbers have earned them the nickname "the insects of the sea." Many are microscopic creatures that drift as plankton in the ocean currents. While primarily marine, members of subphylum Crustacea are also found in fresh water and in a few terrestrial habitats. Crustaceans include crabs, lobsters, crayfish, shrimps, barnacles, water fleas *(Daphnia)*, and pill bugs.

Many crustaceans have a distinctive larval form called a **nauplius** *(NAW plee uhs)*. The nauplius, shown in **Figure 17,** has three pairs of branched appendages. Like insects, the nauplius undergoes a series of molts before it takes on its adult form.

Adult crustaceans also have mandibles, as insects do. But crustaceans differ from insects in a number of important respects, as summarized in **Table 2.**

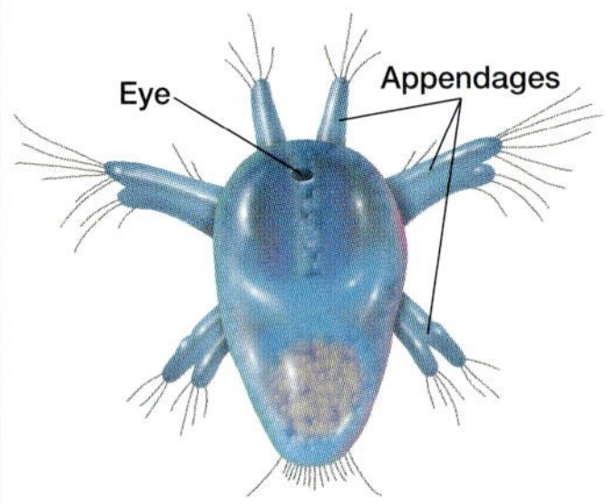

Figure 17 Nauplius. A microscopic, free-swimming nauplius larva is a developmental stage of almost all crustaceans.

Terrestrial Crustaceans

Only a few crustacean groups have successfully invaded terrestrial habitats. The most widespread group of terrestrial crustaceans is composed of the pill bugs and sow bugs. They live among leafy ground litter found in gardens and woods. Pill bugs and sow bugs belong to a group called isopods and are the only crustaceans that are truly terrestrial. Another group, the sand fleas, includes several thousand species typically found along beaches. In addition, a few species of land crabs live in damp areas. Land crabs are only partly adapted to terrestrial living. They are active primarily at night, when the air is more moist. Their life cycle is tied to the ocean, where the larvae live until maturity.

Table 2 Comparison of Crustaceans and Insects

Characteristic	Crustaceans	Insects
Nature of appendages	Most are branched at the end	Unbranched at the end
Antennae	Two pairs	One pair
Chewing appendages	Usually three pairs	One pair
Location of appendages	Cephalothorax and abdomen	Head and thorax
Respiration	Gills	Tracheal system

internet connect
www.scilinks.org
Topic: Crustaceans
Keyword: HX4054
SciLinks Maintained by the National Science Teachers Association

680

Chapter Resource File

- Directed Reading BASIC
- Active Reading GENERAL
- Data Sheet for Data Lab GENERAL
- Data Sheet for Data Lab GENERAL

Transparencies

TR Bellringer
TR H32 Comparing Crustaceans and Insects

One-Stop Planner CD-ROM

- Reading Organizers BASIC
- Reading Strategies BASIC

Aquatic Crustaceans

Crustaceans are a major food source for humans and some animals. The members of some orders of crustaceans are quite small. Common are fairy shrimps, water fleas, ostracods, and tiny copepods *(KOH puh pahds)*. Copepods are among the most abundant multicellular organisms on Earth and are a key food source in the marine food chain. Another small marine crustacean, *Euphausia superba,* swarms in huge groups and is known by its common name, **krill.** Krill, shown in **Figure 18,** are the chief food for many marine species.

Figure 18 **Krill.** Found in icy Antarctic waters, krill are the favorite food of many marine animals.

Decapods

Large marine crustaceans such as shrimps, lobsters, and crabs, along with the freshwater crayfish shown in **Figure 19,** have five pairs of legs and are often referred to as decapods. Almost one quarter of all crustaceans are decapods. The head and thorax of decapods are fused into a single cephalothorax, which is covered on top by a protective shield called a carapace.

In crayfish and lobsters, the anterior pair of legs are modified into large pincers called chelipeds *(KEE luh pehdz)*. Appendages called swimmerets are attached to the underside of the abdomen and are used in swimming and in reproduction. Flattened, paddle-like appendages called uropods are at the end of the abdomen. Many decapods have a telson, or tail spine. Decapods can propel themselves through the water by forcefully flexing their abdomen.

Figure 19 **Crayfish**

Like all decapods, the crayfish has five pairs of legs.

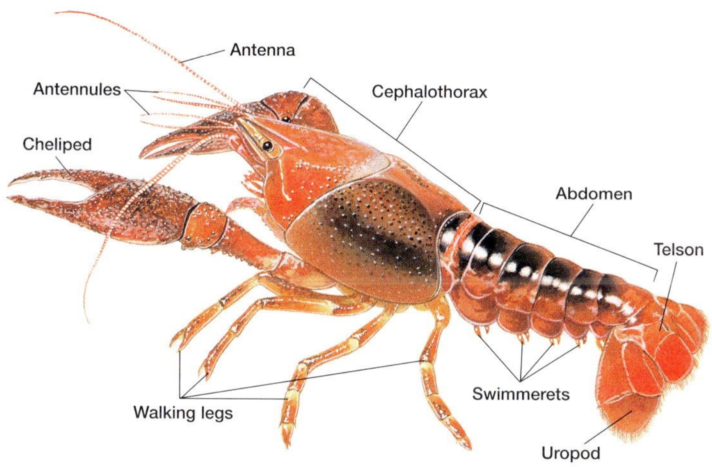

Teach

Teaching Tip — GENERAL

Health Some crabs have been found to carry human parasites. In Asia, freshwater crabs can carry lung flukes, which can then be transmitted to humans who eat the crabs. In Africa, onchocerciasis, an eye disease that is transmitted by flies, is also carried on the backs of river crabs, where the fly larvae live. However, most decapods are not harmful to people when they are cooked properly. Students in the United States are often unaware of diseases that afflict people in countries that do not have effective public health systems. Encourage students to find out more about these parasitic diseases.

did you know?

How the fiddler crab got its name. The male fiddler crab has one large cheliped (pincer) and one small one. The male waves its large cheliped back and forth, much like playing a fiddle, when it is in the presence of a female.

MISCONCEPTION ALERT

Pill Bugs Although their name implies otherwise, pill bugs or potato bugs are actually terrestrial crustaceans, not insects. Pill bugs belong to the Order Isopoda (the isopods). Whereas insects have three pairs of legs and breathe with a system of tracheae, pill bugs have seven pairs of legs and breathe with gills.

Sessile Crustaceans

Barnacles are a group of crustaceans that are sessile as adults. Free-swimming larvae attach themselves to a rock, post, or some other submerged object, where they remain. Hard plates that can open and close protect the barnacle's body. When feeding, barnacles extend their jointed feeding appendages (legs) through the open plates. Their feathery legs stir food from the water into the barnacles' mouth. Unlike most crustaceans, barnacles are hermaphrodites. However, they do not usually fertilize their own eggs.

DATA LAB: Relating Molting to Mortality Rates SC.H.1.4.1

Background

During the soft-shell stage that follows molting, many crustaceans die of disease or are eaten by predators. The bar graph below shows the percent mortality for crabs over a 9-month period. Study the data, and answer the analysis questions.

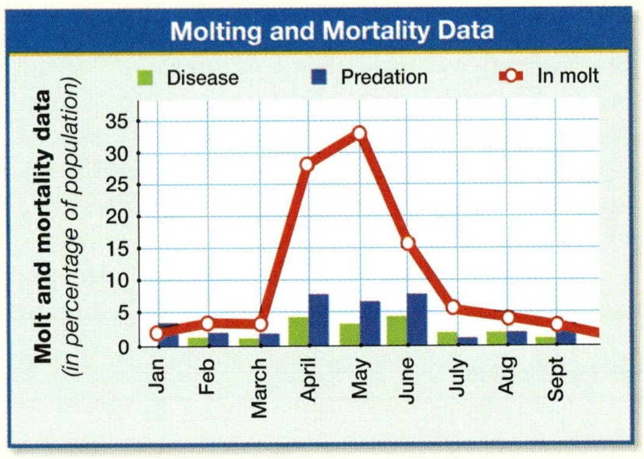

Analysis

1. **Summarize** what the data in the graph tell you about crab mortality.
2. **Summarize** what the graph shows about molting in crabs.
3. **Describe** the relationship between the mortality rates and molting periods of crabs.
4. **Critical Thinking** **Developing Hypotheses** Propose a hypothesis that explains the relationship between the percent of crabs molting and mortality rates.
5. **Critical Thinking** **Making Predictions** Most states have laws that require crab fishers to return molting crabs to the water. How might the length of time a molting crab is exposed to air or how roughly a crab is handled affect whether the crab survives being caught and released?

Section 4 Review

1. **Compare** the body plan of a crustacean with that of an insect. SC.G.1.4.1
2. **Describe** the external body plan of a decapod. SC.F.1.4.2
3. **Critical Thinking** **Making Predictions** Pill bugs respire with gills. How might this affect the distribution of pill bugs in an ecosystem? SC.F.1.4.2
4. **FCAT Prep** Which structures are adaptations of crustaceans to life in aquatic biomes?
   A antennae
   B gills
   C mandibles
   D walking legs
   SC.F.1.4.2

682

Teach, *continued*

DATA LAB: *Relating Molting to Mortality Rates*

Skills Acquired
Reading graphs, applying information

Teacher's Notes
Have students infer what soft-shelled crabs are and when they are most readily available in restaurants.

Answers to Analysis

1. Most crabs die in April, May, and June.
2. Most crabs molt in April, May, and June.
3. During their molting season, many crabs fall prey to predators and disease.
4. When the crabs molt, their soft shells make them more vulnerable.
5. Answers will vary. Crabs are fragile after molting and are more susceptible to damage from rough handling and to disease.

Close

Reteaching — BASIC
Show photographs of a lobster and a water flea, and ask students to identify the similarities and differences. LS Visual

Quiz — GENERAL
True or False:
1. Pill bugs and sow bugs are terrestrial crustaceans, not insects. (True)
2. Lobsters and crabs are called decapods because they have five pairs of legs. (True)

Answers to Section Review

1. Answers should summarize the information given in Table 2 of this section.
2. Decapods have antennae, pincers, a cephalothorax with walking legs, and a segmented abdomen.
3. You would most likely find pill bugs only in moist environments.
4. **A.** Incorrect. Antennae work well in terrestrial environments, sensing touch and smell. **B.** Correct. Gills allow organisms to breathe in water. **C.** Incorrect. Mandibles work equally well in terrestrial and aquatic environments. **D.** Incorrect. Walking legs work well in terrestrial environments and are not very useful in aquatic environments.

Section Review Correlations

SC.F.1.4.2: Items 2, 3, 4
SC.G.1.4.1: Item 1

Study ZONE

Key Concepts

1 Features of Arthropods

- All arthropods have a coelom, a segmented body, and jointed appendages that are modified to perform different functions.
- Arthropods have an exoskeleton made of chitin, which they discard periodically in a process called molting.
- Arthropods are grouped into three subphyla: Chelicerata, Uniramia, and Crustacea.

2 Spiders and Other Arachnids

- Members of subphylum Chelicerata have mouthparts, called chelicerae, that are modified into fangs or pincers.
- Spiders have a head and a cephalothorax, no antennae, six or eight pairs of simple eyes, a pair of fangs and pedipalps, and four pairs of walking legs.
- Scorpions have pedipalps modified into large, grasping pincers. A stinger is located at the end of their abdomen.
- Mites have body parts that are fused to form an unsegmented body. Many mites transmit diseases.

3 Insects and Their Relatives

- Insects make up more than half of all named animal species.
- All insects have a body plan with three body sections (head, thorax, and abdomen), three pairs of legs (all attached to the thorax), and one pair of antennae.
- The life cycles of insects are complex and involve a process of change called metamorphosis, during which larvae change into the adult insects.

4 Crustaceans

- Crustaceans have a distinctive larval form called a nauplius.
- Copepods and krill, which are tiny marine crustaceans, are the chief food of many marine species.
- Most crustaceans have branched appendages, two pairs of antennae, three chewing appendages, walking legs attached to the thorax, and gills. Like insects, crustaceans have jaws called mandibles.

Key Terms

Section 1

appendage (664)
thorax (666)
cephalothorax (666)
compound eye (666)
molting (668)
trachea (668)
spiracle (668)
Malpighian tubule (669)

Section 2

chelicera (670)
pedipalp (670)
spinneret (670)

Section 3

mandible (673)
metamorphosis (675)
chrysalis (675)
pupa (675)
nymph (675)
caste (679)

Section 4

nauplius (680)
krill (681)

683

Study ZONE

Alternative Assessment — GENERAL

Have students construct concept maps that show the similarities and differences among the following arthropods: arachnids, insects, millipedes and centipedes, and crustaceans.

Chapter Resource File

- Science Skills Worksheet GENERAL
- Critical Thinking Worksheet ADVANCED
- Test Prep Pretest GENERAL
- Chapter Test GENERAL

Answer to Concept Map

The following is one possible answer to Performance Zone item 15.

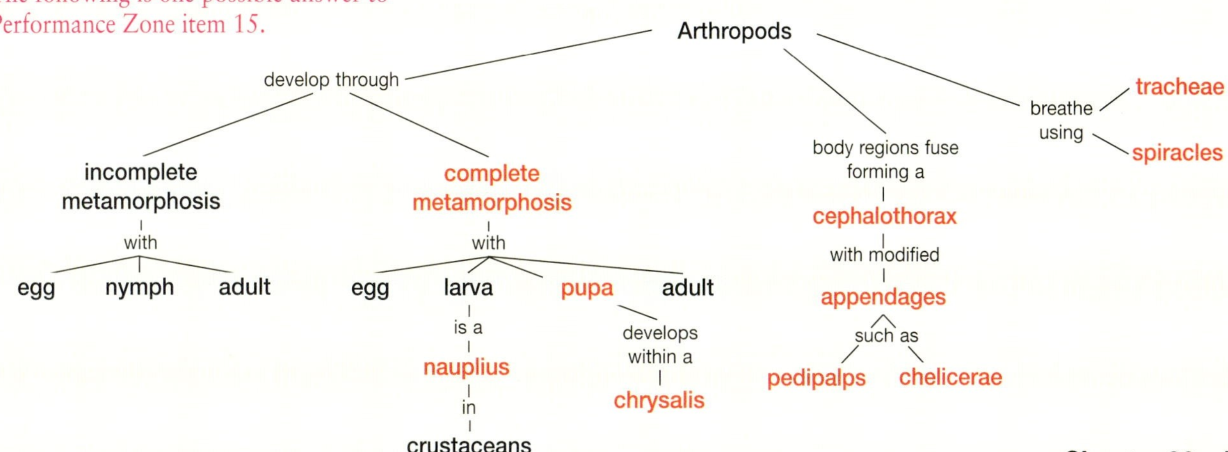

ANSWERS

Understanding Key Ideas

1. c
2. c
3. d
4. c
5. a
6. d
7. b
8. a
9. A forensic entomologist could analyze the insect parts to determine their species. This could reveal information about where the suspect's car had been driven.
10. The answer to the concept map is at the bottom of the Study Zone page.

Critical Thinking

11. Spiders eat insects. When spiders are removed, insects have fewer predators. Killing the spiders probably led to an increase in the insect population.

Alternative Assessment

12. Models will vary but should show the external structures illustrated in the drawings of the spider or grasshopper depicted in the Up Close features in this chapter.
13. Answers will vary, but brochures should be well organized and present the group's findings accurately.

Chapter Review Correlations

SC.D.1.4.3: Item 1
SC.F.1.4.2: Items 2-5, 10
SC.G.1.4.1: Items 6, 7, 13
SC.G.1.4.2: Item 8
SC.G.2.4.2: Item 11
SC.H.3.4.1: Item 12
SC.H.3.4.3: Item 9

Performance ZONE CHAPTER REVIEW

Understanding Key Ideas

1. What evidence suggests that arthropods are closely related to annelids?
   a. Arthropods and annelids have gills.
   b. Both groups have marine species.
   c. Segmentation is present in both groups.
   d. Arthropods have vestigial parapodia.
   SC.D.1.4.3
2. Which is a feature of the arthropod body plan?
   a. a hydrostatic support system
   b. pharyngeal slits
   c. an exoskeleton
   d. a nonsegmented body
   SC.F.1.4.2
3. Arthropods molt because
   a. their body grows faster than their shell.
   b. of damage to their exoskeleton.
   c. their exoskeleton cracks and lets in water.
   d. their hard exoskeleton cannot grow larger.
   SC.F.1.4.2
4. In adult insects
   a. the abdomen has wings.
   b. there are two pairs of antennae.
   c. the legs are attached to the thorax.
   d. the first appendages are chelicerae.
   SC.F.1.4.2
5. Which of the following sequences shows a complete metamorphosis?
   a. egg → larva → pupa → adult
   b. egg → larva → adult
   c. egg → young juvenile → older juvenile
   d. egg → pupa → winged juvenile → adult
   SC.F.1.4.2
6. Millipedes and centipedes differ in that millipedes
   a. are terrestrial and segmented.
   b. have one pair of legs on each segment.
   c. have poisonous fangs.
   d. are herbivores.
   SC.G.1.4.1
7. Which of the following is *not* a crustacean?
   a. lobster c. copepod
   b. scorpion d. water flea
   SC.G.1.4.1
8. Copepods are said to be the most important animals on Earth because they are
   a. a critical link in the marine food chain.
   b. found in both the ocean and fresh water.
   c. accomplished predators.
   d. easier to collect and study than other arthropods. SC.G.1.4.2
9. **BIOWatch** Why might a forensic entomologist need to raise adult insects from eggs or immature forms? SC.H.3.4.3
10. **Concept Mapping** Construct a concept map that outlines the three major groups of arthropods and that gives the characteristics for each group. Try to include the following terms in your concept map: *appendages, cephalothorax, tracheae, spiracles, chelicerae, pedipalps, complete metamorphosis, chrysalis, pupa,* and *nauplius.* SC.F.1.4.2

Critical Thinking

11. **Recognizing Logical Inconsistencies** A neighbor commented that there was an increased number of insects around her house and that she was killing every spider she saw. How might these actions affect the number of insects around a house? SC.G.2.4.2

Alternative Assessment

12. **Forming a Model** Using papier mâché or some other material, make a model of a grasshopper or a spider. Decide in advance how many features of insects you must show to make an adequate model. Present your model to the class, and describe the structures that you modeled. SC.H.3.4.1
13. **Communicating Information** Use the media center or Internet sources to learn more about diseases transmitted by arthropods. Develop a brochure that presents your findings. The brochure should discuss the species that transmit the disease, as well as the symptoms and treatment of the disease. SC.G.1.4.1

684

Assignment Guide

Section	Questions
1	1, 2, 3, 13
2	2, 10, 11, 12
3	4, 5, 6, 9, 10, 11, 12
4	7, 8, 10, 13

FCAT Prep

For questions 1–6, write your answers on a separate sheet of paper.

1. An unknown arthropod has three body segments, one pair of antennae, and three pairs of jointed legs. What kind of arthropod is it?
   A. insect
   B. arachnid
   C. aquatic crustacean
   D. terrestrial crustacean
   SC.G.1.4.1

2. Through what structures do insects respire? SC.F.1.4.2
   F. lungs
   G. ganglia
   H. gills
   I. spiracles

3. The wings of the atlas moth, *Attacus atlas,* look much like the head of a snake. Why might it benefit a flying insect to look like a snake?
   A. Its prey might think the insect is a snake and run away from it.
   B. Snakes might come near the insect and protect it from predators.
   C. Predators might think the insect is a snake and therefore not attack it.
   D. The markings act like camouflage and allow it to hide on the head of a snake.
   SC.G.2.4.3

4. On what body part are a spider's spinnerets located?
   F. abdomen
   G. cephalothorax
   H. pedipalps
   I. thorax
   SC.F.1.4.2

Base your answers to questions 5 and 6 on the diagram below.

GRASSHOPPER EXTERNAL STRUCTURE

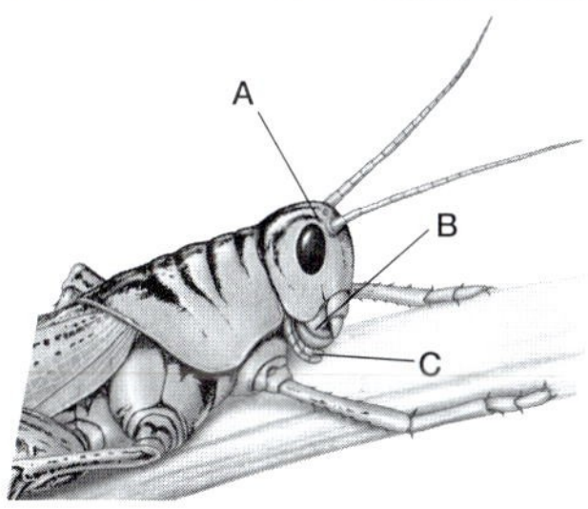

5. What are structures *B* and *C* used for? SC.F.1.4.2
   A. inhaling and exhaling air
   B. biting and chewing leaves
   C. catching and handling prey
   D. sponging and lapping liquids

6. How many legs does a grasshopper have? SC.F.1.4.2

Test TIP

When using a diagram to answer questions, carefully study each part of the figure as well as any lines or labels used to indicate parts of the figure.

685

FCAT Doctor

Question 1 Answer A is the correct choice. These are characteristics of Class Insecta. Answer B is incorrect because arachnids have two body segments, no antennae, a pair of chelicerae, a pair of pedipalps, and four pairs of jointed legs. Answers C and D are incorrect because aquatic and terrestrial crustaceans have two pairs of antennae instead of one and can have more than three pairs of legs.

Question 3 Answer C is the correct choice. Looking like an animal that a predator won't eat is an advantage for the insect. Answer A is incorrect because scaring away prey is not an advantage for any animal. Answer B is incorrect because it is not likely that snakes would think that the insect is another snake and then protect it. Answer D is incorrect because a snake's head is not a safe place to hide.

Question 5 Answer B is the correct choice. The mandibles and the maxillas are used to bite and chew leaves. Answer A is incorrect because air enters and leaves the grasshopper through the spiracles. Answer C is incorrect because grasshoppers are herbivores and do not catch and handle prey. Answer D is incorrect because the grasshopper has no structures for sponging and lapping liquids.

Question 6 The correct answer is 6. All insects have 3 pairs of jointed legs that are attached to the thorax.

Answers

1. A
2. I
3. C
4. F
5. B
6. 6

Test Prep Correlations

SC.F.1.4.2: Items 2, 4, 5, 6
SC.G.1.4.1: Item 1
SC.G.2.4.3: Item 3

Exploration Lab

Observing Pill Bug Behavior

Teacher's Notes

Time Required 40–50 minutes

Ratings

EASY → HARD

Teacher Preparation 2

Student Setup 2

Concept Level 3

Cleanup 2

Skills Acquired

- Designing Experiments
- Experimenting
- Collecting Data
- Organizing and Analyzing Data

Scientific Methods

In this lab, students will:

- Make Observations
- Test their Hypotheses
- Analyze the Results
- Draw Conclusions

Materials and Equipment

Materials for this lab can be ordered from WARD'S. See *Master Materials List* at the front of this book for catalog numbers.

Safety Cautions

Discuss all safety symbols and caution statements with students. Remind students to be careful when using scissors.

Tips and Tricks

Encourage students to read the lab and complete the Before You Begin section before they come to class on the day of the lab. Remind students that although pill bugs have an exoskeleton, they are delicate creatures and should be probed very gently. Before students begin the lab, conduct a class discussion in which you ask the following questions: What are the characteristics of a crustacean? What is the ideal habitat for land-dwelling isopods?

Exploration Lab

Observing Pill Bug Behavior SC.H.1.4.1

SKILLS

- Using scientific methods
- Observing

OBJECTIVES

- **Identify** arthropod characteristics in a pill bug.
- **Observe** the behavior of pill bugs on surfaces with different textures.
- **Infer** the adaptive advantages of pill bug behaviors.

MATERIALS

- 4 adult pill bugs
- 2 Petri dishes
- stereomicroscope or hand lens
- blunt probe
- fabrics with different textures
- scissors
- transparent tape
- clock or watch with second hand

ChemSafety

CAUTION: Always wear safety goggles and a lab apron to protect your eyes and clothing.

CAUTION: Do not touch or taste any chemicals. Know the location of the emergency shower and eyewash station and how to use them. If you get a chemical on your skin or clothing, wash it off at the sink while calling to the teacher. Notify the teacher of a spill. Spills should be cleaned up promptly, according to your teacher's directions.

CAUTION: Glassware is fragile. Notify the teacher of broken glass or cuts. Do not clean up broken glass or spills with broken glass unless the teacher tells you to do so.

Before You Begin

Pill bugs live in moist terrestrial environments, such as under rocks and logs. Like other **crustaceans,** pill bugs respire with gills and have hard outer shells and jointed appendages. They respond to a **stimulus,** such as light, moisture, or touch, by moving toward or away from the stimulus or by curling into a ball. In this lab, you will look for arthropod characteristics in pill bugs and observe the behavior of pill bugs on surfaces with different textures.

1. Write a definition for each boldface term in the paragraph above.
2. Based on the objectives for this lab, write a question you would like to explore about pill bug characteristics and behavior.

Procedure

PART A: Make Observations

1. Place a pill bug in a Petri dish, and observe it with a stereomicroscope or hand lens. Observe it from a dorsal viewpoint as well as from the side. List the characteristics that tell you the pill bug is an arthropod.
2. Touch the pill bug with a blunt probe. **CAUTION: Touch pill bugs gently to avoid injuring them.** Record your observations.

PART B: Design an Experiment

3. Work with the members of your lab group to explore one of the questions written for step 2 of **Before You Begin.** To explore the question, design an experiment that uses the materials listed for this lab.

Distribute the pill bugs to students in small paper cups. Students should perform their experiments using at least three pill bugs.

Answers to Before You Begin

1. *pill bugs*—small crustaceans in the order Isopoda that live in moist terrestrial environments

   crustaceans—arthropods with mandibles, two pairs of antennae, and branched appendages.

   stimulus—any action or agent that causes or changes an activity in an organism
2. Answers will vary. For example: Do pill bugs prefer rough or smooth surfaces?

Answers to Procedure

1. Arthropod characteristics of pill bugs include an exoskeleton, jointed appendages, and a segmented body.
2. When students touch a pill bug with a blunt probe, the pill bug should curl into a ball.

You Choose

As you design your experiment, decide the following:

- **a.** what question you will explore
- **b.** what hypothesis you will test
- **c.** which four different fabrics to use
- **d.** how many times to test each fabric
- **e.** the length of each test
- **f.** what your control will be
- **g.** what data to record in your data table

4. Write a procedure for your experiment. Make a list of all the safety precautions you will take. Have your teacher approve your procedure and safety precautions before you begin the experiment.

5. Set up and carry out your experiment. **CAUTION: Sharp or pointed objects can cause injury. Handle scissors carefully.**

PART C: Cleanup and Disposal

6. Dispose of fabric scraps and broken glass in the designated waste containers. Put pill bugs in the designated container. Do not put lab materials in the trash unless your teacher tells you to do so.

7. Clean up your work area and all lab equipment. Return lab equipment to its proper place. Wash your hands thoroughly before you leave the lab and after you finish all work.

Analyze and Conclude

1. **Analyzing Methods** Why did you test several pill bugs in this investigation instead of just one pill bug?
2. **Analyzing Results** Did all of your pill bugs show a similar pattern of movement? Explain.
3. **Graphing Results** Make a graph of your data. Plot the average time spent on the material on the *y*-axis and the type of material on the *x*-axis.
4. **Analyzing Results** Rank the fabrics according to the total amount of time spent on them by the pill bugs.
5. **Drawing Conclusions** Which fabric texture do pill bugs seem to prefer?
6. **Inferring Conclusions** How is a pill bug's response to disturbances an advantage?
7. **Inferring Conclusions** How is being able to detect surface texture helpful to pill bugs in their natural habitat?
8. **Further Inquiry** Write a new question about pill bugs that could be explored with another investigation.

Do You Know?

Do research in the library or media center to answer these questions:

1. What are the advantages and disadvantages of biological pest control when compared with the use of chemical pesticides?
2. What role do crustaceans play in the lives of humans?

Use the following Internet resources to explore your own questions about arthropods.

internet connect
www.scilinks.org
Topic: Biological Pest Control
Keyword: HX4021
SCILINKS. Maintained by the National Science Teachers Association

4. One way of conducting the experiment would be as follows: Trace the outline of the bottom of the Petri dish on one of the fabrics. Cut out the circle from the fabric and fold it in half; then cut along the fold to produce two half-circles. Repeat this procedure for all fabrics used. Tape together two half-circles, each of a different fabric, and place them on the bottom of a Petri dish, tape side down. To study behavior, place a pill bug in the center of the circle and record the time it spends on each fabric and the path it follows. Repeat for each fabric used. To extend the investigation, students can tape together the four fabrics in all possible combinations, which would result in six different kinds of two-fabric circles. Then have students calculate the total time pill bugs spend on each type of fabric.

Answers to Analyze and Conclude

1. Several pill bugs were tested to improve the reliability of the data.
2. Answers may vary. The pill bugs should show a similar pattern of movement.
3. Graphs will vary but should accurately reflect the data. Students should make bar graphs, with each type of material represented by a bar.

Sample Graph

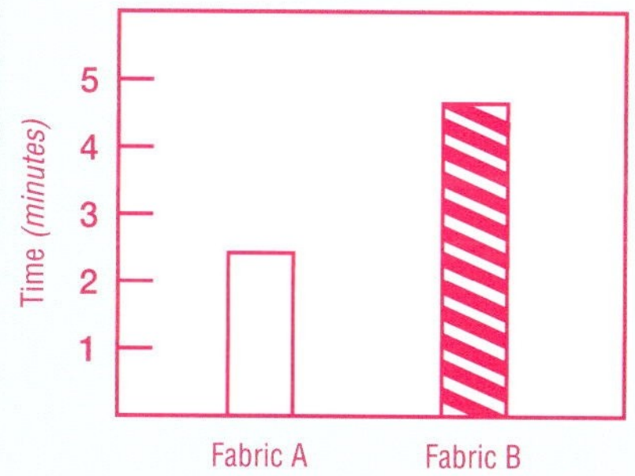

4. Answers will depend on the types of fabrics used. The highest ranking should go to the fabric on which the pill bugs spent the most time.
5. Answers will depend on the types of fabrics used. Pill bugs should spend more time on rough-textured fabrics, such as wool, than on smooth-textured fabrics, such as silk.
6. The pill bug's response to disturbances allows it to protect its gills and underside by rolling into a ball.
7. Being able to distinguish between rough textures and smooth textures may help pill bugs find soft substrates and avoid crawling over surfaces that might expose them to predators.
8. Answers will vary. For example: Do pill bugs prefer moist or dry environments? Do pill bugs prefer light or dark environments?

Answers to Do You Know?

1. Answers will vary. Insecticides have higher research, development, and application costs; cause more environmental pollution; upset the natural balance of organisms; and require annual applications. Biological control has lower research, development, and application costs; does not cause environmental pollution; and does not upset the natural balance of organisms. Both types of control are restricted in their general applicability.
2. Answers will vary. Crustaceans are a source of food and can also be crushed and used as fertilizer. Crustaceans can be destructive, damaging rice paddies and ship hulls. Some are used as bait for fishing.

Insect-Resistant Crops

Background

People have been fighting pests with chemicals since at least 1000 B.C. But synthetic chemical pesticides did not come into general use until after World War II, when DDT and related petroleum-based insecticides became widely available. At first, the new pesticides were so effective that some scientists predicted that insect pests could be wiped out. But the first DDT-resistant insects were discovered in 1946 in Sweden, and resistance quickly spread to other countries and species.

As resistance became more severe, scientists began to notice another problem with pesticides: they were killing large number of other organisms—birds, fish, helpful insects, spiders, and others. In 1962, Rachel Carson's book *Silent Spring* brought this problem to the attention of the public and helped start the environmental movement. In 1972, the EPA banned DDT for all but a few uses, and in 1976, dieldrin and aldrin—two widely used pesticides—were also banned. Recognition of the destructive effects of chemical pesticides stimulated the search for more environmentally friendly alternatives. *Bt* toxin is one example of such an alternative.

Should the genes for insect resistance be transferred into crops?

Insect-Resistant Crops

Each year, Americans use hundreds of millions of tons of pesticides to protect their homes, crops, forests, livestock, and pets from insects and other pests. A drawback of pesticides is that they remain effective only for a short time because pests acquire resistance to the effects of pesticides. Pest control has become a race between the scientists who are trying to invent new pesticides and the ever-evolving pests. And because of pests' adaptability—some insects have evolved resistance to a new insecticide within a single growing season—pests seem to have the edge.

Pesticides from Bacteria

One pesticide with many advantages is the *Bt* toxin, which is derived from the common soil bacterium *Bacillus thuringiensis*. This bacterium produces a toxin that binds to a receptor protein in the digestive system of some insects, killing the insect. But only a few groups of insects are susceptible, including beetles, butterflies, flies, and ants. *Bt* toxin does not harm most beneficial insects, spiders and other arthropod predators of insects, or vertebrates, including humans.

Bt toxin has two other things going for it. First, different strains of the bacterium make different toxins that are effective against different pests. This gives farmers a choice of weapons to use, depending on which insects are attacking their crops. It also reduces the likelihood that resistance will evolve, because the odds that an insect will be resistant to more than one toxin are small. Second, *Bt* toxin breaks down into harmless byproducts within 2 or 3 days of being sprayed on a field. As a result, pests are exposed

Boll weevil
Boll weevils cause extensive damage to United States cotton crops each year.

Teaching Strategies

- The approval of plants engineered to express *Bt* toxin has stirred up passions among scientists and non-scientists alike. Have students analyze newspaper and magazine articles expressing different views on this subject. Have students summarize each article's position and then list the arguments each article cites to support its position. Ask students whether each article resorts to emotional appeal or rhetoric instead of logic. Ask students to identify which article they find most convincing and to explain their choice.
- Point out that the Bt toxin is one example of the importance of biodiversity. Before this valuable toxin was discovered, *Bacillus thuringiensis* was, as one writer put it, "just a bacterium found in ordinary dirt."

Harvesting cotton
Seed pods (called bolls) contain fibers that are used to make fabric.

to less of the toxin, and the probability of their evolving resistance is reduced.

Genetic Engineering of Insect Resistance

Through genetic engineering, the genes for *Bt* toxins have been inserted into several crop plants, including corn, cotton, and potatoes. These plants now produce their own insecticides. During the 1990s, the Environmental Protection Agency (EPA), the federal agency that regulates pesticides, approved the sale of crops genetically engineered to produce *Bt* toxin. Since then, the area of farmland planted with the crops has grown rapidly.

Not everyone agrees with the EPA's decision that the crops are safe. Critics also note that the question is not whether resistance will become widespread but how fast that will happen. They point out that the genetically engineered plants produce the toxin continuously, which increases evolutionary pressure on the pests to adapt.

Supporters of the EPA's decision note that the farmers who buy the seeds are required to plant part of their land with nonengineered crops. This land will serve as a "refuge" on which susceptible insects can live. Supporters reason that if any *Bt*-resistant insects appear, their resistance traits will be swamped when they interbreed with the nonresistant individuals from the refuge. Finally, because some pests attack both corn and cotton, the EPA prohibited the planting of genetically engineered corn and cotton in the same state. This should reduce the pests' exposure to *Bt* toxin, supporters say. ■

Agricultural Scientist

Soil scientist

Profile

Agricultural scientists study farm crops and animals, seeking ways to increase crop yield and quality, control pests and weeds more effectively, and conserve soil and water.

Job Description

Agricultural scientists apply their knowledge of biology to agricultural issues, such as improving the nutritional value of crops. Crop scientists study the breeding and management of crops and sometimes use genetic engineering to develop crops resistant to pests and drought.

Science/Math Career Preparation

Botany	Biochemistry
Zoology	Technical writing
Microbiology	Mathematics
Chemistry	Economics

Analyzing STS Issues

Science and Society

1. **What are alternatives to synthetic chemical pesticides?** Use library resources or the Internet to research integrated pest management. How does it differ from traditional methods of controlling pests? What are some of its shortcomings?
2. **Should genetically engineered crops be labeled?** Current regulations do not require labeling of genetically engineered crops that are sold in stores. Many consumer advocates contend that all such crops should be labeled so that consumers know what they are eating. Research both sides of this issue, and then write a logical, persuasive essay arguing either for or against labeling.
3. **What is biological control?** Biological control is an alternative to chemical pesticides. Using library resources or the Internet, research the topic of biological control. What is biological control? Describe one example of biological control.

Technology: Genetic Engineering

4. **How do scientists genetically engineer plants?** Use library resources or the Internet to find the two main methods for inserting genes into plant cells. Why do scientists want to add genes for herbicide resistance to plants? What are the benefits to plants such as cotton?

689

Read Further

1. U.S. Department of Agriculture, www.usda.gov
2. Council for Agricultural Science and Technology, www.cast-science.org
3. National Council for Agricultural Education, www.agedhq.org

Answers to Analyzing STS Issues

1. Integrated pest management uses pesticides only as a last resort. It emphasizes other, more environmentally friendly pest-control techniques such as natural predators or diseases of the pests, crop rotation to prevent growth of large pest populations, and planting when insect populations are low. It has been criticized for relying too heavily on pesticides.
2. Answers will vary. Students' essays should be logical.
3. Biological control involves using the natural enemies or diseases of the pest to manage its population. Examples include Australian ladybird beetles used to control scale insects and parasitic wasps used to control caterpillars.
4. Scientists use two main techniques for inserting genes into plants: infecting the plant with genetically engineered bacterium that carries the target gene or firing tiny metal pellets covered with DNA into the cell. Scientists want to put herbicide-resistant genes into plants in order to reduce the amount of herbicide used to control weeds. Effective weed management is critical to cotton production. Cotton seedlings grow slowly when they are young, and are not as competitive as the weeds. Weeds can cause serious reductions in the amount of cotton harvested.

Chapter Opener

SE **Quick Review,** p. 691 ■
SE **Reading Activity,** p. 691 ■
TE **Using the Figure,** p. 690 GENERAL
TE **Opening Activity,** p. 691 GENERAL

& 3 • 90 min pp. 692–699

Section 1
Echinoderms

OSP **Lesson Plan***
TR **Bellringer***
TR **Section Outline***
TR **I46 Patterns of Embryonic Development***
TR **H44 Anatomy of a Sea Star***

TE **Demonstration,** p. 692 GENERAL
TE **Group Activity** Symmetry, p. 693 GENERAL
TE **Demonstration,** p. 693 BASIC
TE **Demonstration,** p. 695 GENERAL
SE **Data Lab** Determining How Predators Affect Prey, p. 695 ◆ ■
CRF **Datasheets for In-Text Labs** Determining How Predators Affect Prey*
TE **Demonstration,** p. 697 GENERAL
TE **Group Activity** Tide Pool Diorama, p. 698 ADVANCED
CRF **Skills Practice Lab** Sea Star Dissection* ◆ GENERAL
CRF **Quick Lab** Modeling Tube Feet* ◆ BASIC

& 5 • 90 min pp. 700–702

Section 2
Invertebrate Chordates

OSP **Lesson Plan***
TR **Bellringer***
TR **Section Outline***
TR **H45 Lancelet Interior***
TR **H47 Adult Tunicate***

TE **Demonstration,** p. 700 BASIC
TE **Using the Figure,** p. 701 GENERAL
SE **Quick Lab** Comparing the Structures of the Notochord and Nerve Cord, p. 702 ◆ ■
CRF **Datasheets for In-Text Labs** Comparing the Structures of the Notochord and Nerve Cord*
SE **Skills Practice Lab** Analyzing Sea Star Anatomy, pp. 706–707 ◆ ■
CRF **Datasheets for In-Text Labs** Analyzing Sea Star Anatomy*

& 7 • 90 min

Chapter Review and Assessment Resources

SE **Study Zone,** p. 703 ■
SE **Performance Zone,** p. 704 ■
SE **FCAT Test Prep,** p. 705 ■
TR **Concept Mapping***
CRF **Vocabulary Review Worksheet*** ■ GENERAL
CRF **Test Prep Pretest*** ■ GENERAL
CRF **Chapter Test*** ■ GENERAL
CRF **Chapter Test*** ADVANCED
OSP **Test Generator***
OSP **Test Item Listing***
OSP **Modified Chapter Test*** BASIC

---

go.hrw.com Holt Online Learning

Visit go.hrw.com for access to Holt Online Learning or enter the keyword **HX6 Home** for a variety of free online resources.

One-Stop Planner® CD-ROM

This CD-ROM package includes
- Lab Materials QuickList Software
- Holt Calendar Planner
- Customizable Lesson Plans
- Printable Worksheets
- ExamView® Test Generator
- Interactive Teacher Edition
- Holt PuzzlePro® Resources
- Holt PowerPoint® Resources

Guided Reading Audio CDs

These CDs are designed to help auditory learners and reluctant readers. (Audio CDs are also available in Spanish.)

Compression guide:
To shorten your instruction because of time limitations, omit block 1.

KEY

TE	Teacher Edition	CRF	Chapter Resource File	*	Also on One-Stop Planner
SE	Student Edition	TR	Teaching Transparency	■	Also Available in Spanish
OSP	One-Stop Planner			◆	Requires Advance Prep

SKILLS DEVELOPMENT AND PRACTICE	SECTION REVIEW AND ASSESSMENT	CORRELATIONS
		Florida Sunshine Standards/FCAT
TE **Skill Builder** Interpreting Visuals, p. 693 BASIC TE **Reading Skill Builder** Interactive Reading, p. 694 BASIC TE **Integrating Physics and Chemistry,** p. 694 TE **Skill Builder** Interpreting Visuals, p. 696 BASIC TE **Reading Skill Builder** Paired Summarizing, p. 697 BASIC CRF **Directed Reading Worksheet*** BASIC CRF **Active Reading Worksheet*** GENERAL	SE **Section 1 Review** ■ TE **Reteaching,** p. 699 BASIC TE **Quiz,** p. 699 GENERAL TE **Alternative Assessment,** p. 699 GENERAL CRF **Quiz*** ■ BASIC	SC.F.1.4.2, SC.G.1.4.1, SC.G.2.4.2
CRF **Directed Reading Worksheet*** BASIC CRF **Active Reading Worksheet*** GENERAL CRF **Critical Thinking Worksheet*** ADVANCED CRF **Science Skills Worksheet*** GENERAL	SE **Section 2 Review** ■ TE **Reteaching,** p. 702 BASIC TE **Quiz,** p. 702 GENERAL TE **Alternative Assessment,** p. 702 GENERAL CRF **Quiz*** ■ BASIC **Modified Worksheet*** BASIC	SC.G.1.4.1

SCILINKS NSTA
www.scilinks.org
Maintained by the **National Science Teachers Association**

Topic: Echinoderms
SciLinks Code: HX4065

Topic: Bioassay
SciLinks Code: HX4019

Topic: Invertebrate Chordates
SciLinks Code: HX4105

CHAPTER 31

Overview

In this chapter, students will study how echinoderms develop, what characteristics all echinoderms display, and how the echinoderm classes differ. Students will also describe the characteristics common to all chordates and be able to distinguish the two invertebrate chordate phyla.

Using the Figure

Direct students' attention to the photo of the ruby brittle star. Tell them that ruby brittle stars are echinoderms. Fossil and embryological evidence shows that echinoderms are relatively close to chordates in evolutionary development.

Assessing Prior Knowledge

Before beginning this chapter, be sure your students can answer the questions in the ***Quick Review*** on the facing page. If they cannot, encourage them to reread the sections indicated.

Chapter Resource File

- Vocabulary Worksheets
- Concept Mapping

Ruby brittle star

690

Chapter Correlations — **Florida Sunshine Standards/FCAT**

SC.F.1.4.2 Body structures are uniquely designed and adapted for their function.

SC.G.1.4.1 There is great diversity and interdependence among living things.

CHAPTER

31 Echinoderms and Invertebrate Chordates

Florida Sunshine Standards/FCAT

SC.F.1.4.2 Body structures are uniquely designed and adapted for their function.

SC.G.1.4.1 There is great diversity and interdependence among living things.

Quick Review

Answer the following without referring to earlier sections of your book.

1. **Distinguish** invertebrates from vertebrates. *(Chapter 19, Section 3)*
2. **Define** the terms *gastrula* and *blastula*. *(Chapter 27, Section 1)*
3. **Distinguish** between radial symmetry and bilateral symmetry. *(Chapter 27, Section 1)*

Did you have difficulty? *For help, review the sections indicated.*

Reading Activity

Before you begin to read this chapter, write down all of the key words for both sections of the chapter. Then, write a definition next to each word that you have heard of. As you read the chapter, write definitions next to the words that you did not previously know, and modify as needed your original definitions of words familiar to you.

Looking Ahead

Ruby brittle stars are echinoderms that are commonly found among corals and sponges in deeper parts of coral reefs. Brittle stars have radial symmetry, five flexible arms and a hard internal skeleton. They are the fastest moving echinoderms.

internet connect

www.scilinks.org

National Science Teachers Association *sci*LINKS Internet resources are located throughout this chapter.

SCILINKS. Maintained by the National Science Teachers Association

691

Opening Activity—GENERAL

Diversity in Invertebrates

Display a phylogenetic diagram of the major animal phyla. Review the invertebrate animal phyla that have been discussed so far: Sponge, Cnidaria, Platyhelminthes, Nematoda, Mollusca, Annelida, and Arthropoda. Point out Echinodermata and Chordata on the phylogenetic diagram. Ask students to name some animals from each phylum. (Students may be familiar with sea stars, sand dollars, and sea urchins. Less familiar to them will be sea cucumbers, lancelets, and tunicates.) **LS Visual**

Quick Review Answers

1. Vertebrates have a hollow dorsal nerve cord and a notochord that typically is replaced with a set of bones called the vertebral column. Invertebrates do not have a dorsal nerve cord and lack both a notochord and a vertebral column.
2. A blastula is a hollow ball of cells that develops from a zygote and has cells arranged in layers that will develop into the primary germ layers of ectoderm, mesoderm, and endoderm. A gastrula is a further-developed embryo that contains the three primary germ layers.
3. An object or organism that is radially symmetrical can be divided into two symmetrical halves by any plane that crosses its central axis. An object or organism that is bilaterally symmetrical can be divided into two symmetrical halves only by a plane that crosses the midline of the body.

Reading Activity Answers

Students may not know many of the key words for this chapter before they read it. Encourage them to write whatever they can think of and then carefully and completely write each definition as they read it. Suggest that they also write down the roots for each word and their meanings.

Focus

Overview

Before beginning this section review with your students the objectives listed in the Student Edition. This lesson describes the similarities and differences between protostome and deuterostome development and introduces students to the groups of organisms that show each of these developmental patterns. Students will study echinoderm structure and function and review examples of several different groups of echinoderms.

Bellringer

Ask students to list the names of as many non-fish marine organisms as they can think of. (Most will be either mollusks or echinoderms.)

Motivate

Demonstration — GENERAL

Bring in several preserved or dried specimens of echinoderms such as sea cucumbers, sea stars, and sand dollars. You can supplement these specimens with pictures in books. Allow students to inspect them and come up with as many common characteristics as they can. List those characteristics on the board or overhead, and refer to the list as you teach the specific characteristics. LS Visual

Section 1 Echinoderms

Florida Sunshine Standards/FCAT

SC.F.1.4.2 Body structures are uniquely designed and adapted for their function.

SC.G.1.4.1 There is great diversity and interdependence among living things.

Objectives

- **Compare** the developmental pattern found in protostomes with that found in deuterostomes. SC.G.1.4.1
- **Describe** the major characteristics of echinoderms. SC.G.1.4.1
- **Summarize** how the sea star's water vascular system functions. SC.F.1.4.2

Key Terms

blastopore
protostome
deuterostome
ossicle
water-vascular system
skin gill

Animal Development

If you have been to a saltwater aquarium, you're sure to have seen echinoderms, which are spiny invertebrates that live on the ocean bottom. How could echinoderms like the brittle star shown on the first page of this chapter be related to animals such as chordates, which are primarily vertebrates? The answer lies in their early development. As an embryo develops, it goes through a gastrula stage. As shown in **Figure 1,** a gastrula has an opening to the outside called the **blastopore.** In acoelomate animals, the mouth develops from or near the blastopore. This pattern of development also occurs in some coelomate animals, such as annelids, mollusks, and arthropods. Animals with mouths that develop from or near the blastopore are called **protostomes.**

Some animals follow a different pattern of development. In phylums Echinodermata and Chordata, the anus—not the mouth—develops from or near the blastopore. (The mouth forms later, on another part of the embryo.) Animals with this pattern of development are called **deuterostomes,** also shown in Figure 1. If you know the origin of these two terms, it's easy to remember the differences between the two developmental patterns. The term *protostome* is from the Greek *protos,* meaning "first," and *stoma,* meaning "mouth." The prefix *deutero-* is from the Greek *deuteros,* meaning "second." In deuterostomes, the anus develops first and the mouth develops second.

Figure 1 Embryonic development

The development of an animal embryo follows one of two patterns.

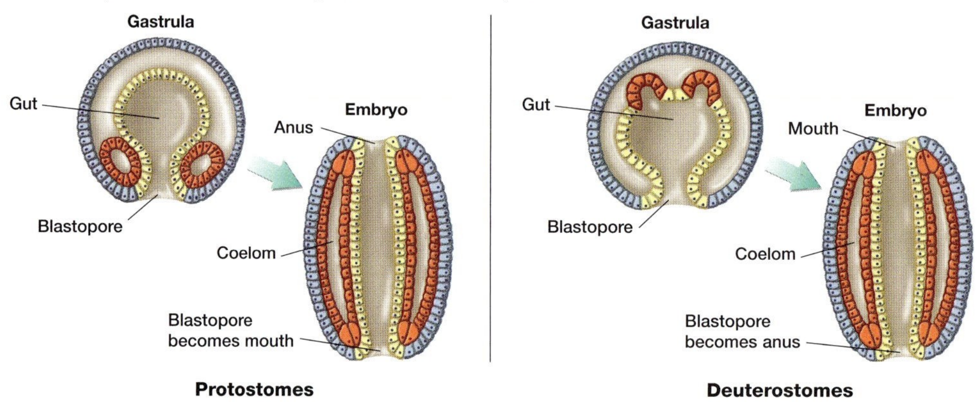

Chapter Resource File

- Directed Reading BASIC
- Active Reading GENERAL
- Data Sheet for Data Lab GENERAL

Transparencies

TR Bellringer
TR I46 Patterns of Embryonic Development

One-Stop Planner CD-ROM

- Reading Organizers BASIC
- Reading Strategies BASIC

Section Correlations

SC.F.1.4.2
SC.G.1.4.1

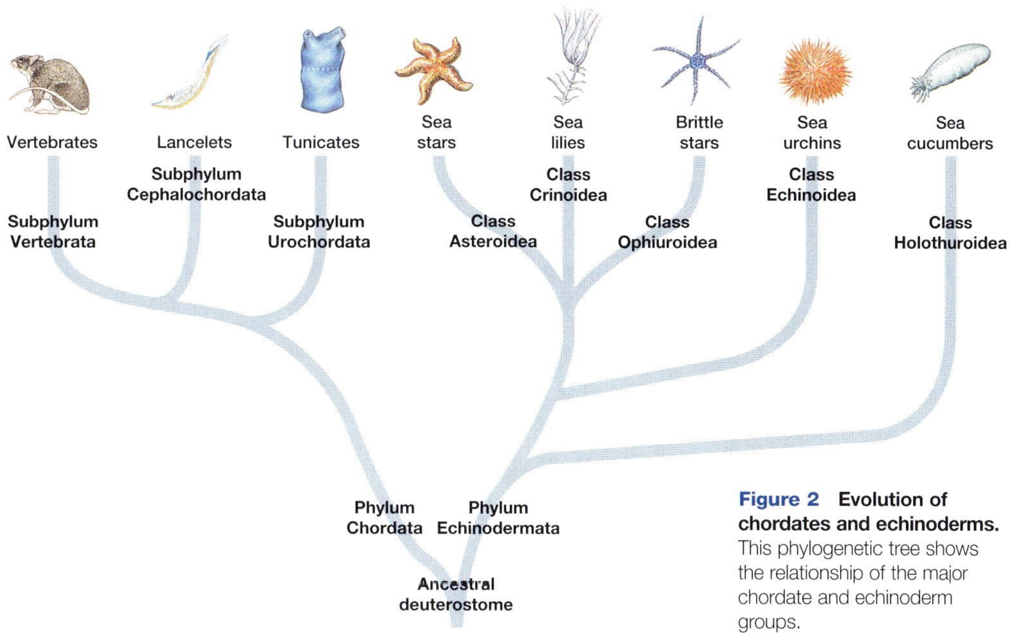

Figure 2 Evolution of chordates and echinoderms. This phylogenetic tree shows the relationship of the major chordate and echinoderm groups.

The first deuterostomes were marine echinoderms that evolved more than 650 million years ago. They were also the first animals to develop an endoskeleton. Today, most people are familiar with echinoderms known as "starfish," which are not really fish and are more properly called sea stars. In addition to sea stars, many other animals commonly seen along the sea shore—sea urchins, sand dollars, and sea cucumbers—are echinoderms. All are marine, and all are radially symmetrical as adults.

Chordates, as well as a few other small phyla, are also deuterostomes. (Humans and all other vertebrates are chordates.) Like the echinoderms, chordates have an internal skeleton. This developmental similarity unites these seemingly dissimilar animal phyla. It also leads scientists to believe that chordates and echinoderms derived from a common ancestor, as shown in the phylogenetic tree in **Figure 2.** The identity of the ancestral deuterostome is not known. The fossil record indicates that echinoderms, such as the sea lily in **Figure 3,** were abundant in the ancient seas.

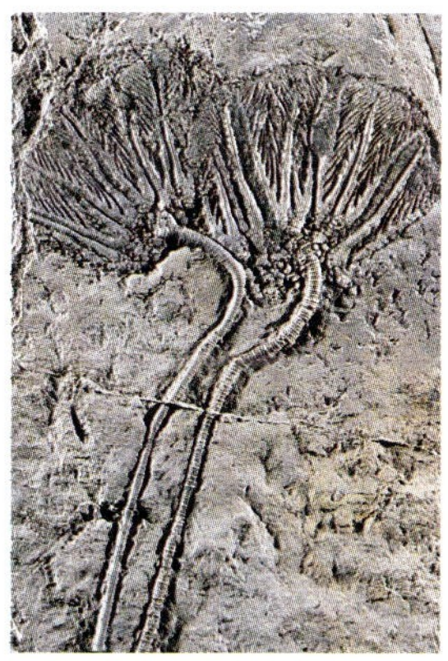

Figure 3 Fossil sea lily. Sea lilies such as the one preserved by this fossil were plentiful in the ancient oceans.

Evolutionary Milestone

8 Deuterostomes

Echinoderms are coelomates that have a deuterostome pattern of embryo development. The same pattern of development occurs in the chordates.

693

Integrating Physics and Chemistry

Calcium is a component of the echinoderm skeleton. Have students locate the element calcium on the periodic table and state the number of its outer or valence electrons, bonding pattern, physical properties, and chemical behavior based on its location on the periodic table.

Trends in Medicine

Body Part Regeneration The importance of chemical signals in the development of deuterostomes has significant medical implications for the regeneration of lost body parts. Scientists are working to discover factors that cause cells to change and grow so that someday it might be possible to re-grow a severed spinal cord or an amputated limb.

Teach

SKILL BUILDER — BASIC

Interpreting Visuals Have students examine **Figure 2.** Ask them to identify the first group of echinoderms that evolved (sea cucumbers) and the most recently evolved group(s) of echinoderms. (sea stars, sea lilies, brittle stars) Ask them to identify the evolutionary relationship between echinoderms and chordates—did one evolve from the other? (Echinoderms and chordates evolved from a common ancestor that was neither an echinoderm nor a chordate.) **LS Visual**

Group Activity — GENERAL

Symmetry All adult echinoderms display pentaradial symmetry. *Penta* means "five." Therefore, echinoderms have five or approximately multiples of five arms. Have groups of students create displays that define and model various forms of symmetry. Encourage them to compare objects made by humans with those found in nature. They can use pictures and clay models to make their comparisons.
LS Kinesthetic Co-op Learning

Demonstration — BASIC

Display any echinoderms you or your students may have collected such as sea stars or sand dollars. Emphasize the five-part radial symmetry of the specimens. Point out that during drying, the skin of sand dollars disintegrates, and only the hard internal skeleton remains. Ask students how echinoderms differ from arthropods. (Echinoderms have endoskeletons, no body segments, and radial symmetry; arthropods have exoskeletons, body segments, and bilateral symmetry.) **LS Visual**

Teach, *continued*

Teaching Tip

Five-Part System The water vascular system of a typical sea star has five main parts. Water flows through this system to circulate oxygen and assist movement. Have students construct a Graphic Organizer similar to the one on the bottom of this page to illustrate the path of water through a sea star.

Teaching Tip

Sea Urchin Embryology Tell students that sea urchins hatch from eggs as microscopic floating larvae. These delicate, elegant-looking creatures have two mirror-image halves, just like beetles, birds and bats. They float for weeks near the ocean's surface fanning even tinier organisms into their mouths. Then, a change occurs. A small bundle of cells inside the organism begin to grow. All of the other cells die, and the tiny bundle settles to the ocean floor to grow into something that looks like a calcified ball several inches in diameter. Instead of two mirror-image halves (bilateral symmetry) seen in the larva, the adult urchin is organized more like five equal pie slices (pentaradial symmetry), a nearly spherical version of its cousin the starfish. Adult urchins creep slowly around the ocean bottom using a five-toothed structure to scrape food from rocks. Studies of both body shape and particularly molecular composition indicate that their ancestors were bilateral.

Real Life

Do sand dollars resemble the sun?

To the Chumash Indians of southern California, the lines radiating from a sand dollar's mouth resemble the sun's rays. Sand dollars, once plentiful in the Pacific Ocean, became the Chumash's symbol for the "newborn" sun of the winter solstice.

Modern Echinoderms

Many of the most familiar animals seen along the seashore—sea stars, sea urchins, sand dollars—are echinoderms. Echinoderms are also common in the deep ocean. While all echinoderms are marine, the different classes of echinoderms vary considerably in the details of their body design. Despite their apparent differences, all echinoderms share four fundamental characteristics.

1. **Endoskeleton.** Echinoderms have a calcium-rich endoskeleton composed of individual plates called **ossicles.** When ossicles first form in young echinoderms, they are enclosed in living tissue, so they are a true endoskeleton. Even though the ossicles of adult echinoderms appear to be external, they are covered by a thin layer of skin (although sometimes the skin is worn away). In adult sea stars and in many other echinoderms, a large number of these plates are fused together. The fused plates function much like an arthropod exoskeleton. They provide sites for muscle attachment and shell-like protection. In most echinoderms, the plates of the endoskeleton bear spines that project upward through their skin.
2. **Five-part radial symmetry.** All echinoderms are bilaterally symmetrical as larvae. During their development into adults, the larvae's body plan becomes radially symmetrical. Most adult echinoderms, such as the one shown on the left in **Figure 4,** have a five-part body plan with arms that radiate from a central point. However, the number of arms can vary. Echinoderms have no head or brain. Instead, the nervous system consists of a central ring of nerves with branches extending into each of the arms. Although echinoderms are capable of complex response patterns, each arm acts more or less independently. Many species, including sea stars, can regenerate a new arm if a portion of an arm is lost. In some species of sea stars, a complete animal can regenerate from an arm connected to a portion of the central disk. However, a complete sea star cannot regenerate from an arm alone.

WORD *Origins*

The term *echinoderm* is from the Greek *echinos,* meaning "spiny," and *derma,* meaning "skin."

Figure 4 Five-part body plan. The echinoderm five-part body plan is easily seen in this colorful African species. Other sea stars, such as the sunstar, have more than five arms.

694

Graphic Organizer

Use this graphic organizer with **Teaching Tip** *on this page.*

Madeporite or Sieve Plate → Stone canal → Ring canal → Radial canal → Ampulla of Tube Feet

3. **Water-vascular system.** Echinoderms have a water-filled system of interconnected canals and thousands of tiny hollow tube feet called a **water-vascular system.** In some echinoderms, such as the sea star, the tube feet extend outward through openings in the ossicles. In some species, each tube foot has a sucker at its tip. Many echinoderms use their tube feet to crawl across the sea floor. The water-vascular system also functions in feeding and gas exchange. A sea star can use the hundreds of tube feet on its arms to pull the valves of a bivalve open. Some gas exchange and waste excretion takes place through the thin walls of the tube feet.

4. **Coelomic circulation and respiration.** The echinoderm body cavity functions as a simple circulatory and respiratory system. Particles, including respiratory gases, move freely throughout the large, fluid-filled coelom. Many echinoderms have skin gills that aid respiration and waste removal. **Skin gills,** shown in **Figure 5,** are small, fingerlike projections that grow among the echinoderm's spines. These projections create an increased surface area through which respiratory gases can be exchanged. Skin gills also function as excretory structures, and wastes that accumulate in them are released into the surrounding water.

You can learn more about the structure of one particular echinoderm, the sea star, in *Up Close: Sea Star,* on the following page.

Figure 5 **Skin gills.** An echinoderm's skin gills function as both respiratory and excretory organs.

internet connect
www.scilinks.org
Topic: Echinoderms
Keyword: HX4065
SCILINKS. Maintained by the National Science Teachers Association

DATA LAB

Determining How Predators Affect Prey

SC.G.2.4.2

Background

Sea stars can be very effective predators, and they frequently eat mollusks. The chart at right shows the relative number of two species of mollusks before and after the introduction of a predatory sea star. Study the chart, and answer the Analysis questions.

Sea Star Predation of Mollusks

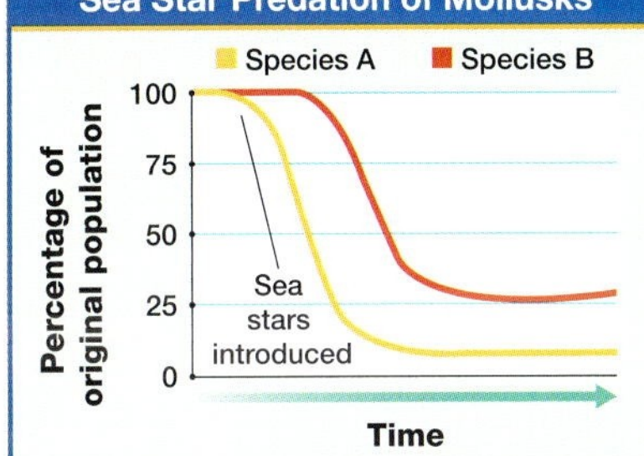

Analysis

1. **Compare** the relative sizes of the two mollusk populations before the introduction of the sea star.
2. **Identify** the preferred prey of the sea star, and use the data presented in the graph to support your answer.
3. **Critical Thinking** **Analyzing Data** When the sea star began preying on the nonpreferred species, the preferred species had dropped to what percent of its original population?
4. **Critical Thinking** **Inferring Relationships** What factors might cause the sea star to begin consuming a nonpreferred species, even when its preferred prey is still present?
5. **Critical Thinking** **Predicting Outcomes** Predict the relative abundance of the two species of mollusks if the sea star remains in the area indefinitely.

Teaching Tip

Strength of Tube Feet Tell students that the legs of sea stars can be very powerful when working together. An individual foot can be pulled off a surface easily. However, when hundreds of tube feet work together, a sea star can exert tremendous strength—enough to pull apart the valves of a clam or hold the animal to a rock when a wave pounds on it.

Demonstration —— GENERAL

You can use an ordinary dropper to represent tube feet. Show students that when an ampulla is filling with water, suction is created at the open end. When the ampulla is expelling water, the suction is gone, and water leaves the open end. This may be easier for students to see if you add a small amount of glitter to the water. LS Visual

DATA LAB

Determining How Predators Affect Prey

Skills Acquired
Analyzing data, predicting results

Teacher's Notes
Show students a picture of a sea star eating a mollusk.

Answers to Analysis

1. The populations were approximately the same size.
2. Species A; species A's population was the first to decline and declined more rapidly than species B's population.
3. about 50 percent
4. Accept any reasonable answer. The most obvious factor is relative abundance—non-preferred prey will be easier to find as preferred prey becomes scarce.
5. There will be about 10 percent of the original population of species A and 25 percent of the original population of species B as long as both populations continue to reproduce and replace lost individuals.

OCEANOGRAPHY CONNECTION

Along the coast in parts of California and Alaska, sea otters and kelp forests flourish together. What's the connection? Sea urchins. Sea urchins eat giant kelp, and when there are enough sea urchins, they can roam the forest floor and graze entire forests. Where urchins abound, it's hard for new giant kelp plants to survive and grow. But, otters eat urchins. And where otters are present, urchins have to hole up in deep crevices to survive. By controlling sea urchins, sea otters help kelp forests thrive. A single otter consumes more than 5,000 pounds per year of crab, snails, sea stars, urchins, and other invertebrates. When 17th- and 18th-century hunters slaughtered otters for their fur, the kelp forests were decimated. Kelp populations in many areas have rebounded since, and once again provide a home for a wide variety of invertebrates and vertebrates, including many fish species.

Up Close

Sea Star

Teaching Strategies
Ask students to explain the evolutionary link between sea stars and chordates. Point out that although the relationship between adult echinoderms and chordates may not be apparent, the embryos of both groups have certain developmental stages in common. Echinoderms not only share our deuterostome pattern of embryonic development, but also have bilateral symmetry as larvae and endoskeletons as adults. **LS Logical**

SKILL BUILDER — BASIC

Interpreting Visuals Pose the following questions as you discuss the illustration. **LS Visual**

1. What makes a tube foot extend? (contraction of the water-filled ampulla)
2. What makes it withdraw? (The ampulla relaxes and refills with water.)
3. Why can't sea stars live in terrestrial environments? (They wouldn't be able to move about because they need water to operate their water vascular system.)

Sea Star

- **Scientific name:** *Asterias vulgaris*
- **Size:** Typically from 15 to 30 cm (6 to 12 in.) in diameter
- **Range:** East coast of North America
- **Habitat:** Intertidal; often on hard, rocky surfaces
- **Diet:** Slow-moving or sessile species, including mollusks, crustaceans, polychaetes, and corals

Characteristics

Water-vascular system Water enters and leaves the system through pores in the madreporite *(ma druh PAWR it),* a sievelike structure that filters out large particles. Water then moves into the ring canal and passes into the radial canals.

Tube feet Hundreds of tube feet extend from the bottom of the radial canals, and each foot is connected to a waterfilled sac called an ampulla *(am PUHL uh).* When water is pumped from the sacs into the tube feet, they expand outward. Suckers on the ends of the tube feet attach firmly to solid surfaces. When muscles force water back into the ampulla, the tube feet shorten, pulling *Asterias* forward.

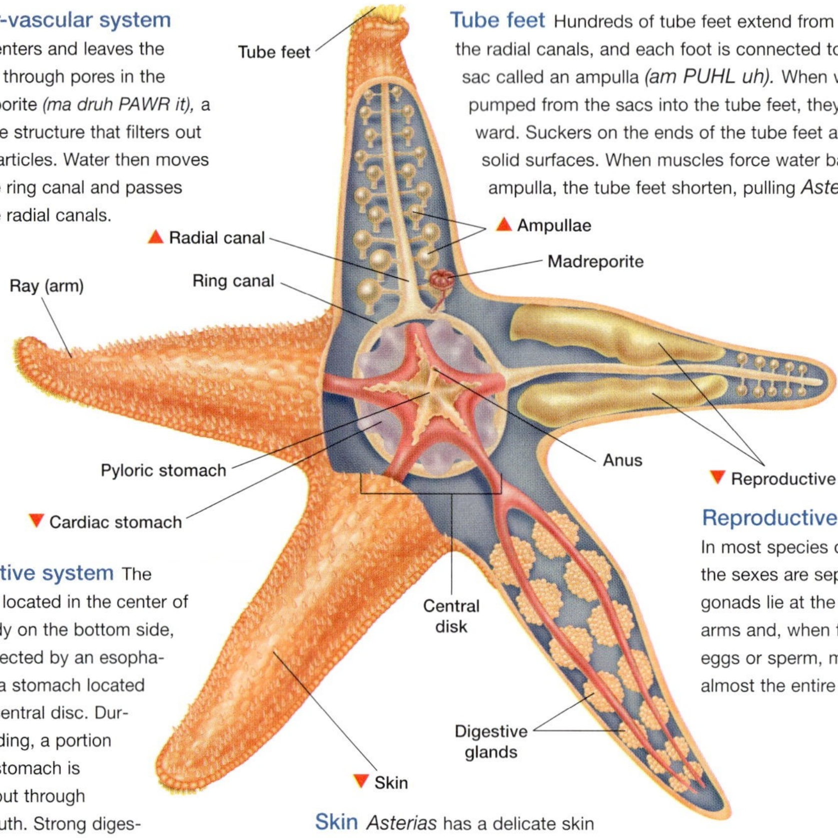

Reproductive system In most species of *Asterias* the sexes are separate. The gonads lie at the base of the arms and, when filled with eggs or sperm, may occupy almost the entire arm.

Digestive system The mouth, located in the center of the body on the bottom side, is connected by an esophagus to a stomach located in the central disc. During feeding, a portion of the stomach is thrust out through the mouth. Strong digestive juices liquefy the prey, which is then ingested.

Skin *Asterias* has a delicate skin stretched over an endoskeleton of spiny plates.

696

Attention Grabber

Sea Star Reproduction In addition to asexual reproduction (any arm can regenerate a new body as long as part of the central disc is present), starfish normally undergo sexual reproduction. A female sea star can deposit as many as 2.5 million eggs at one time. The eggs are deposited in the water, where they are fertilized by the sperm of a male sea star.

Transparencies

TR H44 Anatomy of a Sea Star

Echinoderm Diversity

Echinoderms are one of the most numerous of all marine phyla. In the past, they were even more plentiful than they are now. There are more than 20 extinct classes of echinoderms and an additional six classes of living members. As you saw on the phylogenetic tree that appeared earlier in this section, the living classes include sea stars, sea lilies, brittle stars, sea urchins, and sea cucumbers. The recently discovered sea daisy does not appear on the phylogenetic tree because its relationship to the other echinoderms is not fully understood.

Sea Stars

Sea stars are the echinoderms most familiar to people. Almost all species of sea stars are carnivores, and they are among the most important predators in many marine ecosystems. For example, the crown-of-thorns sea star eats coral polyps. In 1 year, a single crown-of-thorns can consume up to 6 m^2 of a reef. Over time, this sea star can destroy an entire coral reef ecosystem. Other sea stars prey on bivalve mollusks, whose shells they pull open with their powerful tube feet, as shown in **Figure 6.**

The ossicles of many species of sea stars produce pincerlike structures called pedicellaria *(ped uh suh LAH ree uh)*. Pedicellaria contain their own muscles and nerves, and they snap at anything that touches them. This action prevents small organisms from attaching themselves to the surface of the sea star.

Figure 6 **Sea star.** This sea star is using its tube feet to pry open the shell of a clam. Then it will feed on the clam's soft tissues.

internet connect
www.scilinks.org
Topic: Aquariums and Museums of Florida
Keyword: HX6F01
SciLinks. Maintained by the National Science Teachers Association

Brittle Stars

The sea star's relatives, the brittle stars and sea baskets, make up the largest class of echinoderms. Brittle stars have slender branched arms that they move in pairs to row along the ocean floor. Their arms break off easily, a fact that gives brittle stars their name. Brittle stars and sea baskets live primarily on the ocean bottom, and they usually hide under rocks or within crevices in coral reefs. Although a few species are predators, most brittle stars are filter feeders or feed on food in the ocean sediment.

Sea Lilies and Feather Stars

The sea lilies and feather stars are the most ancient and primitive living echinoderms. They differ from all other living echinoderms because their mouth is located on their upper, rather than lower, surface. Sea lilies are sessile and are attached to the ocean floor by a stalk that is about 60 cm (23 in.) long. Feather stars, shown in **Figure 7,** use hooklike projections to attach themselves directly to the ocean bottom or a coral reef. They sometimes crawl or swim for short distances.

Figure 7 **Feather star.** The feathery arms of these feather stars are adapted for filter feeding.

697

Teaching Tip

Moving Materials Into and Out of Cells Tell students that the exchange of gases occurs on the surface of the sea star because particles tend to move from an area of greater concentration to an area of lower concentration. The uptake of oxygen molecules by skin gills occurs because ocean water contains a greater concentration of oxygen molecules than the skin gills contain. Therefore, oxygen enters the skin gills by diffusion.

Demonstration — GENERAL

Invite a marine zoologist or aquarium store employee to discuss echinoderms. If possible, have your guest bring in live specimens or videotapes to show how echinoderms move and feed. **LS Auditory**

READING SKILL BUILDER — BASIC

Paired Summarizing Encourage students to pause after each paragraph they read to summarize the main points to a partner. You might model the technique by having a student read a paragraph out loud and summarize it. Then have another student read the next paragraph and summarize it. You may want to pair ELL students wth native English speakers. **LS Verbal** English Language Learners

INCLUSION Strategies

• Attention Deficit Disorder • Learning Disability

Have the students download pictures of sea stars, brittle stars, sea lilies, sea urchins, sand dollars, and sea cucumbers. Ask the students to create a class poster, echinoderm notebook, or bulletin board with these pictures. On index cards, have the students identify each of the echinoderms with their scientific name and a brief description as to what characteristics they have that determines they are echinoderms.

REAL WORLD CONNECTION

Clam farmers see sea stars as a nuisance because sea stars devour clams. In the past, the farmers cut the sea stars into several pieces and flung them back into the sea. They did not know that sea stars could regenerate from a severed arm, as long as one piece of the central disc was present. The clam farmers were actually making the problem worse! Today, clam farmers spread lime on the clam beds. Lime kills sea stars but leaves clams unharmed.

Teach, *continued*

Group Activity — ADVANCED

Tide Pool Diorama Have students work in small groups to conduct library and/or Internet research on organisms found in tide pools. They can choose whichever ocean coast they prefer for their tide pool study. Instruct them to find out about the most common kinds of animals as well as plants that are found in the type of tide pool they choose. Ask them to search for pictures of these organisms in books or on the Internet. Then have each group construct a diorama depicting the organisms found in the type of tide pool they have selected. (Students should find at least several kinds of echinoderms, such as sea urchins and sea stars; several kinds of mollusks, such as chitons, mussels, whelks, and nudibranchs; several kinds of arthropods, including crabs and barnacles; cnidarians, such as sea anemones; tunicates, such as sea squirts; annelids, such as tube worms; sponges; fish; water plants; and plankton.) Display the tide pool dioramas for the whole class to observe.

LS Kinesthetic Co-op Learning

Sand dollar

Sea urchin

Figure 8 Sea urchin and sand dollar. Sea urchins usually live on rocky ocean bottoms, while sand dollars live on sandy ocean bottoms.

Sea Urchins and Sand Dollars

The sea urchins and sand dollars, shown in **Figure 8,** lack distinct arms but have the basic five-part body plan seen in other echinoderms. Both sea urchins and sand dollars have a hard, somewhat flattened endoskeleton of fused plates covered with spines protruding from it. The spines provide protection and, in some species of sea urchins, contain a venom that causes a severe burning sensation. In some other species of sea urchin, a specialized type of pedicellarium contains a toxin used to paralyze prey. Sea urchins are found on the ocean bottoms while sand dollars live in sandy areas along the sea coast.

Sea Cucumbers

Sea cucumbers are soft-bodied, sluglike animals without arms. They differ from other echinoderms in that their ossicles are small and are not fused together. Because of this, the sea cucumber's long, cylindrical body is soft. Often the body has a tough, leathery exterior. The sexes of most sea cucumbers are separate, but some species are hermaphrodites.

Sea cucumbers feed by trapping tiny organisms present in the sea water. Their mouth, located at one end of the body, is surrounded by several dozen tube feet modified into tentacles. The tentacles are covered with a sticky mucus that entraps plankton. Periodically, the sea cucumber draws its tentacles into its mouth and cleans off the plankton and mucus. The tentacles are then coated with a fresh supply of mucus. When threatened, a sea cumber has an unusual means of defending itself. As shown in **Figure 9,** the sea cucumber can release a number of sticky threads from its anus to entrap its attacker.

Figure 9 Sea cucumber. When threatened, a sea cucumber releases sticky threads that entrap its attacker.

698

Cultural Awareness

California's Sea Urchin Fishery A growing demand for traditional Japanese cuisine in the United States has created a growing market for California sea urchins. Although the sea urchins were once considered pests by kelp harvesters and sports divers, they are now commercially harvested. Most of California's sea urchin catch is exported to Japan.

Section Review Correlations

SC.F.1.4.2: Items 3, 4

SC.G.1.4.1: Items 1, 2

Sea Daisies

In 1986, a new class of echinoderm was discovered: strange disk-shaped little animals called sea daisies. Less than 1 cm (0.39 in.) in diameter, these creatures were first found in deep waters off the coast of New Zealand. Only a few species are known. Sea daisies have five-part radial symmetry but no arms. Their tube feet are located around the edges of the disk rather than along the radial lines, like they are in other echinoderms.

Marine Organisms Help Keep Coastal Water Clean

If you were swimming or fishing in Florida's coastal waters, you probably would not be able to detect the presence of toxic chemicals in the water, the sediments, or the sea life. To help protect humans and marine organisms, scientists have developed several tests to monitor marine environments for potential health hazards. A *bioassay* is the use of a living organism or cell culture to test for the presence of a substance. The *sea urchin fertilization bioassay* makes use of the fact that the gametes of sea urchins are very sensitive to pollutants.

Using Sea Urchins

For this bioassay, samples of ocean water, sediment, and industrial wastes that are discharged into the ocean are collected periodically from different sites. Then, under controlled conditions in a laboratory, sperm and eggs from sea urchins are added to the samples. (In most echinoderms, gametes are normally fertilized in open water.) After a short waiting period, scientists compare the fertilization success rate in the collected water samples with the fertilization success rate found in control water samples. If the test samples show a lower fertilization rate, the samples may contain toxic pollutants.

Taking Action

If the bioassay indicates the presence of pollutants, more specific tests may be needed to identify which pollutants are present. If the pollutants can be traced to a factory or sewage treatment plant, the facility may be required to remove or neutralize chemicals and sediments from its discharge.

Recording water temperature in a bioassay tank

In the future, it may be possible to clean up some pollutants by using aquatic plants that have the ability to remove toxic chemicals from water. This process, known as *phytoremediation,* is a promising new area of research for marine biology.

internet connect
www.scilinks.org
Topic: Bioassay
Keyword: HX4019
SCI LINKS. Maintained by the National Science Teachers Association

Section 1 Review

1. **Summarize** why echinoderms are considered to be more closely related to tunicates, lancelets, and vertebrates than to other animals. SC.G.1.4.1
2. **Summarize** the four major echinoderm characteristics. SC.G.1.4.1
3. **Describe** how the sea stars use their water-vascular system to move along the sea floor. SC.F.1.4.2
4. *FCAT Prep* In an echinoderm, what structure carries out the functions of a circulatory system?
   A coelom
   B madreporite
   C ossicles
   D pedicellariae
   SC.F.1.4.2

699

Answers to Section Review

1. Echinoderms and chordates share a common pattern of embryonic development (both are deuterostomes), and both have an internal skeleton. These similarities suggest that the groups share a common ancestor.
2. All echinoderms have an endoskeleton composed of ossicles, five-part radial symmetry, a water-filled vascular system, and coelomic circulation and respiration.
3. Water enters the madreporite and travels into the ring canal and radial canals to the ampulla of the tube feet. When water enters the tube feet, they expand, and suckers attach to the surface. When water reenters the ampulla, the tube feet shorten and pull the sea star forward.
4. **A.** Correct. The coelom acts as a circulatory system. **B.** Incorrect. The madreporites function in the water vascular system but are not involved in gas exchange. **C.** Incorrect. The ossicles are part of the echinoderm's skeleton and are not involved in circulation. **D.** Incorrect. Pedicellariae are part of the skeletal system and are not involved in circulation.

BIOWatch

Marine Organisms Help Keep Coastal Water Clean

Teaching Strategies
Use articles from newspapers and magazines to point out recent problems and solutions concerning Florida's water quality. If possible, visit a water treatment plant or have someone from such a facility come to school to answer students' questions.

Discussion
- What is a bioassay? (using living organisms or cell cultures to test for the presence of a substance)
- How do scientists use the sea urchin gametes to tell if water samples are contaminated? (Fewer fertilizations will occur in contaminated water.)

Close

Reteaching — Basic

Make an illustrated chart of echinoderm classes and the characteristics of each class. Leave some of the characteristics blank and have students fill them in.

Quiz — General

1. The part of the digestive tract that develops at the ________ of an animal determines whether it is a protostome or a deuterostome. (blastopore)
2. Is the skeleton of an echinoderm more like that of an arthropod or a vertebrate? Explain your answer. (a vertebrate because it is an endoskeleton)

Alternative Assessment — General

Have students write review questions for a trivia-type game about the groups of echinoderms.

Section 2

Focus

Overview

Before beginning this section review with your students the objectives listed in the Student Edition. This lesson introduces the four structural and functional characteristics of the chordate phylum. It concludes with the characteristics of two invertebrate chordate phyla, the tunicates and the lancelets.

Bellringer

Put a labeled diagram of a lancelet on the overhead. Have students list the characteristics not found on other animals they have studied. (Labeled structure should include those in **Figure 10**. Student lists should include notochord, dorsal nerve cord, postanal tail, and pharyngeal pouches.)

Motivate

Demonstration —— BASIC

Show students pictures of the embryonic stages of several vertebrates, such as humans, sheep, fish, and chickens. Have them compare the animals and point out which structures they have in common. (pharyngeal pouches, notochord, dorsal nerve cord, tail) Tell students that it is possible for an animal to be a chordate but not a vertebrate. **LS Visual**

Section 2 Invertebrate Chordates

Florida Sunshine Standards/FCAT

SC.G.1.4.1 There is great diversity and interdependence among living things.

Objectives

- **Describe** the characteristics of chordates. SC.G.1.4.1
- **Define** the term *invertebrate chordate.* SC.G.1.4.1
- **Compare** tunicates and lancelets. SC.G.1.4.1

Key Terms

chordate
notochord
pharyngeal pouch
invertebrate chordate

The Chordate Skeleton

The second major group of deuterostomes are the **chordates.** Chordates have a very different kind of endoskeleton from that of echinoderms. The chordate endoskeleton is completely internal. During the development of the chordate embryo, a stiff rod called the **notochord** develops along the back of the embryo. Using muscles attached to this rod, early chordates could swing their backs from side to side, enabling them to swim through the water. The development of an internal skeleton was an important step that led to the evolution of vertebrates. The endoskeleton, which muscles attach to, made it possible for animals to grow large and to move quickly.

Other Chordate Characteristics

Chordates also share three other characteristics. They have a single, hollow, dorsal nerve cord with nerves attached to it that travel to different parts of the body. Chordates also have a series of **pharyngeal pouches.** Pharyngeal pouches of aquatic chordates develop in the wall of the pharynx and develop into the gill structure later in the animals' development. In terrestrial chordates, the pharyngeal pouches develop into different structures, such as the parathyroid gland and the inner ear. Another chordate characteristic is a postanal tail, which is a tail that extends beyond the anus. All chordates have all four of these characteristics at some time in their life, even if it is only briefly as embryos. **Figure 10** shows these chordate characteristics as seen in the body of an adult lancelet.

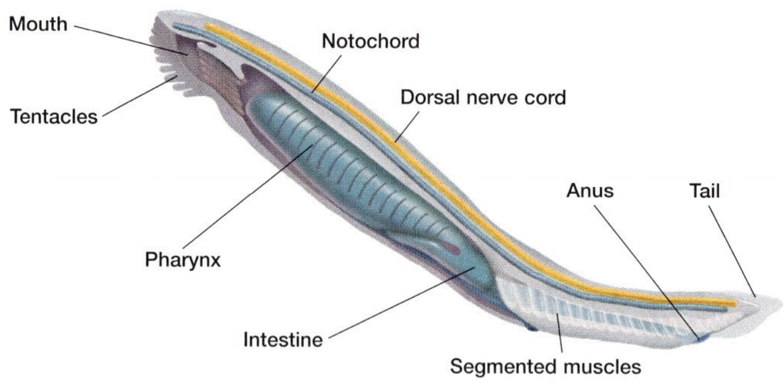

Figure 10 Lancelet interior. Adult lancelets possess all of the characteristics of chordates.

Evolutionary Milestone

9 Notochord

Tunicates, lancelets, and all the vertebrates belong to phylum Chordata (chordates). Chordates are coelomate animals that have a flexible, dorsal rod called a **notochord.** In vertebrate chordates, the notochord is replaced during embryonic development by a vertebral column (backbone).

700

Chapter Resource File

- Directed Reading BASIC
- Active Reading GENERAL
- Data Sheet for Quick Lab GENERAL

One-Stop Planner CD-ROM

- Reading Organizers BASIC
- Reading Strategies BASIC
- Supplemental Reading Guide *Journey to the Ants* ADVANCED

Transparencies

TR Bellringer
TR H45 Lancelet Interior
TR H47 Adult Tunicate

Section Correlations

SC.G.1.4.1

Invertebrate Chordates

Phylum Chordata is divided into three subphyla. The vast majority of chordate species belong to subphylum Vertebrata, which you will study in the next unit. Two other subphyla, Urochordata (the tunicates) and Cephalochordata (the lancelets), contain a small number of species. Because members of these two subphyla are chordates that do not have backbones, they are called **invertebrate chordates.**

Study TIP

Organizing Information
Make a table to organize information about invertebrate chordates. Across the top, write the headings *Adult characteristics* and *Larval characteristics.* Along the sides, write *Lancelet* and *Tunicate.* Add information to the table as you read Section 2.

Tunicates

Only the free-swimming tunicate larvae have a nerve cord, notochord, and postanal tail. These features are lost during the larvae's transformation into adulthood. However, adult tunicates, shown in **Figure 11,** retain their pharyngeal slits. Most adult tunicates are sessile, filter-feeding marine animals. A tough sac, called a tunic, develops around the adult's body and gives tunicates their name. Cilia beating within the tunicate cause water to enter the incurrent siphon. The water circulates through the tunicate's body, passes though its pharyngeal slits, and leaves the body through the excurrent siphon. As water passes through the slits in the pharynx, food is filtered from it and passed into the stomach. Undigested food passes to the anus, which empties into the excurrent siphon.

All tunicates are hermaphrodites, and some are also able to reproduce asexually by budding. While some tunicates are solitary, budding can result in colonies of identical tunicates.

Figure 11 Adult tunicate

Pharyngeal slits are the only chordate characteristic retained by adult tunicates.

Adult tunicate, dorsal view

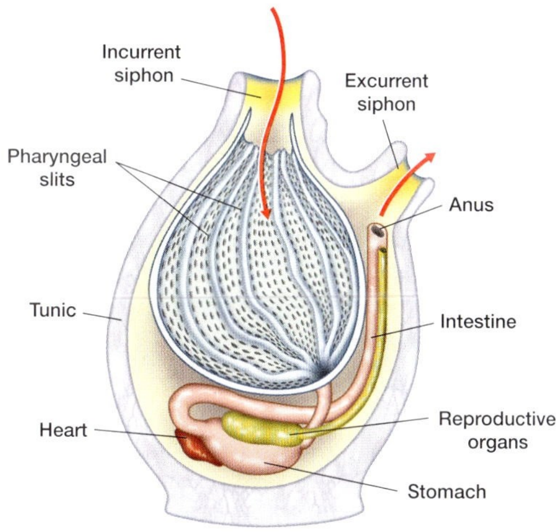

Tunicate interior

701

Teach

Teaching Tip

Lancelets Have Transparent Skin
Tell students that, unlike the skin of other vertebrates, the skin of a lancelet is transparent. This feature allows us to see the rows of ossicles along the side of the notochord.

Using the Figure — GENERAL

Point out the segmented muscles in the lancelet shown in **Figure 10.** Tell students that these muscle blocks are called myomeres and that they are actually modified body segments present in chordate muscle tissues. The "flaky" texture of many fish filets is an example of these stacked muscle layers. Then, draw student's attention to **Figure 11.** Point out the arrows that indicate the direction of water flow in the tunicate interior. Have students trace the water's path with their finger. **LS Visual**

did you know?

Other Deuterostome Phyla
In addition to echinoderms and chordates, there are two other deuterostome phyla. Both groups live in the sea. Arrow worms belong to the phylum Chaetognatha. Their size ranges from 2.5 to 10 cm in length, and they have straight bodies that resemble darts or torpedoes. Acorn worms, phylum Hemichordata, are characterized by their long proboscis. These bottom-dwellers grow much longer than arrow worms, ranging from 20 mm to 2.5 m in length.

Career

Marine Biologist Have interested students research the career of a marine biologist and write a report that includes the type of degree needed, classes to take in high school and college, summer experiences, helpful volunteer experiences, working environment, starting salary, and growth potential.

INCLUSION Strategies

- *Attention Deficit Disorder*
- *Learning Disability*
- *Developmental Delay*

Using an index card or construction paper, have the students write a postcard to the teacher describing the lancelets seen on a recent imaginary vacation. The students' description should include the characteristics that make the lancelet an invertebrate chordate and where the lancelet was found. Student may also draw a picture of the lancelet for the front of the postcard.

Comparing the Structures of the Notochord and Nerve Cord

Skills Acquired
Describing and identifying structures

Teacher's Notes
Ask students why they should use a cross section, not longitudinal sections, for this lab.

Answers to Analysis

1. The nerve cord is located dorsally and has a small diameter. The wider notochord is just below the nerve cord.
2. bilateral symmetry
3. Lancelets: bilateral symmetry; Adult echinoderms: pentaradial symmetry
4. protection for the nerve cord, points for muscle attachment, and flexibility

Close

Reteaching — BASIC

Have students compare lancelets and tunicates.

Quiz — GENERAL

1. Which chordate characteristic do adult tunicates retain? (pharyngeal slits)
2. How do lancelets obtain food? (by filtering water over their pharyngeal pouches)

Alternative Assessment — GENERAL

Have students build a chordate model that includes the four chordate characteristics.

Figure 12 **Lancelet.** Lancelets bury their tails in sediment.

Lancelets

Lancelets, shown in **Figure 12,** receive their name from their bladelike shape. Although lancelets may resemble fish, they are not fish. Lancelet fossils have been found in rocks over 550 million years—lancelets are much older than any fish species is. Lancelets are scaleless chordates only a few centimeters long. The lancelets' V-shaped bundles of muscles are arranged in a series of repeating segments. Lancelets are found worldwide in shallow ocean water. They spend most of their time with their mouths protruding from mud or sand. The beating of cilia that line the front end of their digestive tract draws water through the mouth and pharynx and out the pharyngeal slits. Lancelets feed on microscopic protists that they filter out of the water. Unlike tunicates, the sexes are separate in lancelets.

QUICK LAB: *Comparing the Structures of the Notochord and Nerve Cord* SC.F.1.4.2

The notochord and hollow nerve cord are two important characteristics of all chordates. While both are located on an animal's dorsal side, they differ in size, structure, and location. You can compare the two when viewing a cross section of an adult lancelet.

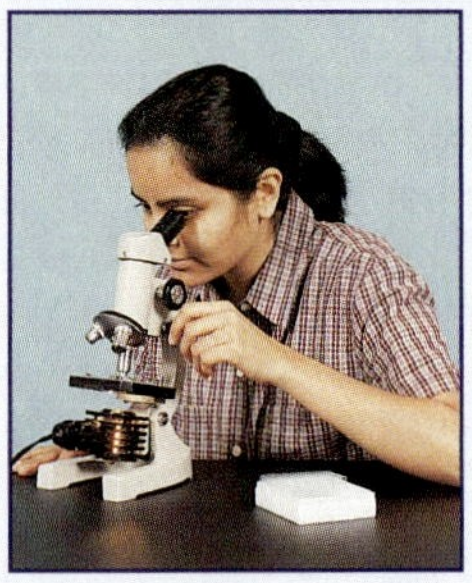

Materials

compound microscope, prepared slide of the cross section of an adult lancelet

Procedure

1. Place a prepared slide of a cross section of an adult lancelet under the microscope.
2. Locate the dorsal side of the specimen, and turn the slide so the dorsal side is on top.
3. Locate the notochord and hollow nerve cord. If visible, locate the intestine.
4. Sketch the specimen, and label the dorsal and ventral sides, the notochord, the nerve cord, and the intestines.

Analysis

1. **Describe** the structure, location, and size of the nerve cord and the notochord.
2. **Identify** the kind of symmetry observed in the adult lancelet.
3. **Compare** the lancelet's symmetry with the symmetry of adult echinoderms.
4. **Critical Thinking** **Forming Hypotheses** In vertebrate chordates, the notochord becomes a backbone that encases the nerve cord. Why might this arrangement be an advantage to an animal?

Section 2 Review

1. **Describe** the characteristics common to all chordates. SC.G.1.4.1
2. **Summarize** why tunicates and lancelets are classified as invertebrate chordates. SC.G.1.4.1
3. **Compare** the chordate characteristics found in adult tunicates with those found in adult lancelets. SC.G.1.4.1
4. *FCAT Prep* The pharyngeal slits of a tunicate play a role in what system?
   A digestive
   B nervous
   C reproductive
   D skeletal
   SC.F.1.4.2

702

Answers to Section Review

1. At some point during their development, all chordates have a notochord; a hollow, dorsal nerve cord; pharyngeal pouches; and a postanal tail.
2. Tunicates and lancelets possess all four basic chordate characteristics, but they lack a backbone.
3. Adult tunicates retain their pharyngeal pouches. The nerve cord, notochord, and tail are lost after the larval stage. Adult lancelets have a notochord, a dorsal nerve cord, a postanal tail, and pharyngeal slits.
4. **A.** Correct. The pharygeal slits work with the digestive system. **B.** Incorrect. The nerve cord is lost in adult tunicates. **C.** Incorrect. Reproductive organs are located next to the pharynx. Tunicates are hermaphroditic, and some reproduce by budding. **D.** Incorrect. The tunic provides support and protection.

Section Review Correlations

SC.F.1.4.2: Item 4
SC.G.1.4.1: Items 1, 2, 3

Key Concepts

1 Echinoderms

- During embryonic development, a protostome's mouth develops from the blastopore. In a deuterostome, the anus forms from the blastopore and the mouth forms later from a different opening.
- Because echinoderms and chordates are both deuterostomes, scientists believe that both groups were derived from a common ancestor.
- Echinoderms lack a head or brain. Their nervous system consists of a central ring of nerves with branches extending into each of the five parts of its body plan.
- Echinoderms share four characteristics: an endoskeleton composed of ossicles; five-part radial symmetry; a water-vascular system; and coelomic circulation and respiration.
- In many echinoderm species, respiration and waste removal are performed by skin gills.
- Echinoderms are a diverse group consisting of seven classes: sea stars, brittle stars, sea lilies, feather stars, sea urchins and sand dollars, sea cucumbers, and sea daisies.

2 Invertebrate Chordates

- At some point in their lives, all chordates have a notochord, a dorsal nerve chord, pharyngeal pouches, and a postanal tail.
- Phylum Chordata includes invertebrate and vertebrate chordates.
- Invertebrate chordates do not have a backbone (vertebral column). Two invertebrate subphyla are Urochordata (tunicates) and Cephalochordata (lancelets).
- Tunicate larvae have a nerve cord, notochord, pharyngeal pouches, and a postanal tail. As adults, they lose most of these characteristics.
- Lancelets retain their notochord, dorsal nerve cord, pharyngeal pouches, and postanal tail into adulthood.

Key Terms

Section 1

blastopore (692)
protostome (692)
deuterostome (692)
ossicle (694)
water-vascular system (695)
skin gill (695)

Section 2

chordate (700)
notochord (700)
pharyngeal pouch (700)
invertebrate chordate (701)

703

Alternative Assessment — GENERAL

Have students write questions to be used in the exam for this chapter. They should write several types of questions: multiple choice, true/false, fill-in-the-blank, essay. Be sure students provide the correct answers as well. Use the best questions in the test for this chapter.

Chapter Resource File

- Science Skills Worksheet GENERAL
- Critical Thinking Worksheet ADVANCED
- Test Prep Pretest GENERAL
- Chapter Test GENERAL

Answer to Concept Map

The following is one of several possible answers to Performance Zone item 15.

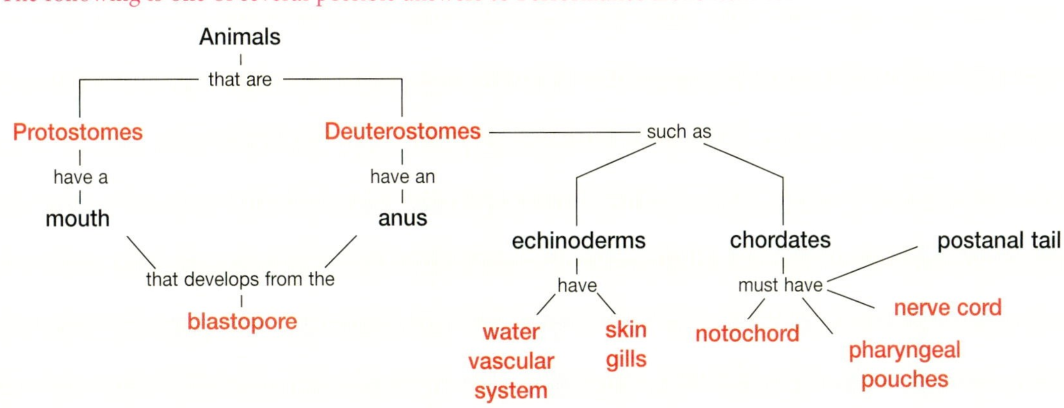

ANSWERS

Understanding Key Ideas

1. c
2. d
3. a
4. c
5. b
6. b
7. The notochord represented the beginnings of an internal skeleton. An internal skeleton provided points for muscle attachment and allowed vertebrates to grow large and move quickly.
8. Scientists can run specific tests to determine what contaminants are present. If the toxicity can be traced to a specific source, waste can be treated before it is discharged. Contaminated sediments can be removed or decontaminated. In the future, plants may be used to remove toxic chemicals from contaminated water.
9. In most crustaceans, gas exchange occurs across gills. Insects have a system of tracheae and spiracles for gas exchange. In echinoderms, gas exchange occurs as dissolved oxygen moves through the water-filled coelomic cavity. Many echinoderms also have skin gills that provide extra surface area for gas exchange.
10. One possible answer to the concept map is found at the bottom of the Study Zone page.

Chapter Review Correlations

SC.F.1.4.2: Items 7, 9, 10, 11
SC.G.1.4.1: Items 1-6, 13
SC.G.2.4.2: Item 14
SC.G.2.4.6: Item 8
SC.H.1.4.1: Item 12

Performance ZONE

Understanding Key Ideas

1. The phylum characterized by radial symmetry and a water-vascular system is
   a. Chordata.
   b. Arthropoda.
   c. Echinodermata.
   d. Cnidaria.
   SC.G.1.4.1
2. The presence of which of the following characteristics is typical among echinoderms?
   a. nauplius larvae
   b. a notochord
   c. an exoskeleton
   d. a water-vascular system
   SC.G.1.4.1
3. Which of the following pairs is a correct match between an adult echinoderm and one of its characteristics?
   a. sea cucumber: leathery epidermis
   b. sand dollar: five arms
   c. sea lily: free-swimming
   d. sea star: sessile
   SC.G.1.4.1
4. Which echinoderm group includes species that specialize in hunting bivalves?
   a. sea cucumbers
   b. sand dollars
   c. sea stars
   d. sea lilies
   SC.G.1.4.1
5. Animals with a notochord, a dorsal nerve cord, and pharyngeal slits are members of the phylum
   a. Echinodermata.
   b. Chordata.
   c. Annelida.
   d. Cnidaria.
   SC.G.1.4.1
6. Which of the following do adult chordates and adult echinoderms have in common?
   a. a nonsegmented body
   b. an internal skeleton
   c. a water-vascular system
   d. bilateral symmetry
   SC.G.1.4.1
7. In what way was the notochord an important development in the evolution of vertebrates? SC.F.1.4.2
8. **BIOWatch** Once preliminary tests determine that coastal waters are polluted, what kinds of actions might be taken to reduce the pollution? SC.G.2.4.6
9. Gas exchange is essential to life. Compare how gas exchange occurs in most crustaceans, insects, and echinoderms. (**Hint:** See Chapter 30, Sections 1 and 4.) SC.F.1.4.2
10. **Concept Mapping** Construct a concept map that outlines the development and body plan of echinoderms and invertebrate chordates. Try to include the following terms in your map: *blastopore, protostome, deuterostome, water-vascular system, skin gills, notochord, nerve cord,* and *pharyngeal slits*. SC.F.1.4.2

Critical Thinking

11. **Evaluating Conclusions** Sea cucumbers and sea lilies are relatively sessile. Their larvae, however, are capable of swimming. Explain how swimming larvae provide an advantage for these echinoderms. SC.F.1.4.2
12. **Forming Reasoned Opinions** A scientist collects several specimens of an unidentified animal. After conducting an in-depth study of the mysterious species, the scientist observes that they have tube feet, an endoskeleton, and a protostome pattern of embryonic development. Why is the classification of these organisms difficult? SC.H.1.4.1
13. **Inferring Relationships** Explain why the larval form of an organism can be valuable in determining relationships among species. SC.G.1.4.1

Alternative Assessment

14. **Finding and Communicating Information** Use the media center or Internet resources to learn more about the crown-of-thorns sea star, or *Acanthaster planci*, which is abundant on the Great Barrier Reef. Find out why this echinoderm poses such a threat to its environment, and prepare a report on your findings. Include steps being taken by the Australian government to deal with the problem. SC.G.2.4.2

Critical Thinking

11. Answers will vary. Swimming enables these echinoderm larvae to find underexploited living spaces and food sources.
12. Answers will vary. The species has characteristics of echinoderms, including tube feet and an endoskeleton. However, echinoderms are deuterostomes. Thus, classification into an existing phylum is difficult.
13. In many animal species, the larval form may have characteristics not present in the adult form and that are similar to characteristics of other, possibly related species.

Alternative Assessment

14. *Acanthaster planci* is dangerous to the Great Barrier Reef because it is a voracious eater. *A. planci* often eats in large groups, secreting a chemical that attracts more sea stars to the area to feed. Marine biologists in Australia have tried injecting sea stars with poisons and asking amateur divers to remove them by hand.

Assignment Guide

Section	Questions
1	1, 2, 3, 4, 6, 8, 9, 10, 11, 13, 14
2	5, 7, 10, 12, 13

FCAT Prep

For questions 1–6, write your answers on a separate sheet of paper.

1. Which of the following patterns of development is characteristic of echinoderms and lancelets?
   A. acoelomate **C.** protostome
   B. deuterostome **D.** pseudocoelomate
   SC.G.1.4.1
2. A scientist comes across a colony of tunicates in which all of the individuals are identical to one another. What conclusion might the scientist draw about the colony?
   F. The tunicates are mobile and feed by trapping prey as a group.
   G. The colony arose when two different species of tunicates mated.
   H. The tunicates in the colony arose from asexual reproduction by budding.
   I. These tunicates are hermaphrodites and reproduced sexually to form the colony.
   SC.G.1.4.1
3. What is the difference between a protostome and a deuterostome?
   A. Echinoderms are protostomes, while vertebrates are deuterostomes.
   B. Echinoderms are deuterostomes, while invertebrate chordates are protostomes.
   C. Protostomes were the first to develop an endoskeleton and include all chordates.
   D. The blastopore becomes a mouth in protostomes and an anus in deuterostomes.
   SC.G.1.4.1
4. What structure coordinates the movements of a sea star?
   F. mantle **H.** radula
   G. nerve ring **I.** tube foot
   SC.F.1.4.2

Base your answers to questions 5 and 6 on the diagram below.

ADULT TUNICATE

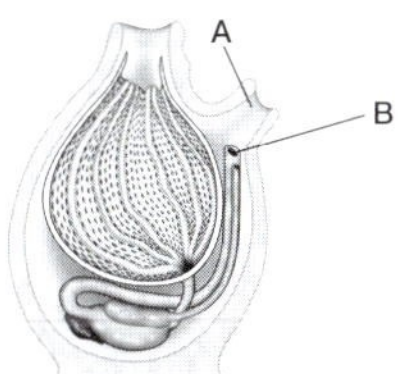

5. What is the relationship between structures *A* and *B*?
   A. *B* collects food from the water that enters the body through *A*.
   B. *B* extracts oxygen from the water that enters the body through *A*.
   C. *B* releases gametes into the water that leaves the body through *A*.
   D. *B* releases undigested food into the water that leaves the body through *A*.
   SC.F.1.4.2
6. If a tunicate filters 12 liters of seawater per hour, how many hours would it take for it to filter 420 liters of seawater? SC.F.1.4.2

Test TIP

If you are unsure of the correct answer to a multiple-choice question, start by crossing out answers that you know are wrong. Reducing the number of answer choices in this way may help you choose the correct answer.

705

FCAT Doctor

Question 2 Answer H is the correct choice. Asexual reproduction results in identical offspring. Answer F is incorrect because most tunicates are sessile filter feeders. Answer G is incorrect because two different species cannot mate and produce fertile offspring. Answer I is incorrect because the offspring of sexual reproduction are not as likely to be identical.

Question 3 Answer D is the correct choice. The two differ in the body structures that form during embryonic development. Answer A is incorrect because both echinoderms and vertebrates are deuterostomes. Answer B is incorrect because both echinoderms and chordates are deuterostomes. Answer C is incorrect because deuterostomes, not protostomes, were the first to develop an endoskeleton and include all chordates.

Question 5 Answer D is the correct choice. Remains of food that are not filtered out by the digestive system are released from the anus (B) and exit the excurrent siphon (A). Answer A is incorrect because food does not enter through the excurrent siphon. Answer B is incorrect because oxygen does not enter through the excurrent siphon and the anus. Answer C is incorrect because gametes do not leave the body through the anus.

Question 6 The correct answer is 35. At 12L/hour, it would take 420L/12L, or 35 hours.

Answers

1. B
2. H
3. D
4. G
5. D
6. 35

Test Prep Correlations

SC.F.1.4.2: Items 4, 5, 6
SC.G.1.4.1: Items 1, 2, 3

Skills Practice Lab

Analyzing Sea Star Anatomy

Teacher's Notes

Time Required 50 minutes

Ratings

EASY → HARD

Teacher Preparation 3
Student Setup 2
Concept Level 2
Cleanup 3

Skills Acquired
- Identifying/Recognizing Patterns
- Collecting Data
- Interpreting
- Communicating

Scientific Methods

In this lab, students will:
- Make Observations
- Analyze the Results
- Draw Conclusions
- Communicate the Results

Materials and Equipment

Materials for this lab can be ordered from WARD'S. See *Master Materials List* at the front of this book for catalog numbers. You will need one preserved sea star, dissection tray, set of dissecting instruments, and hand lens or dissecting microscope for each student or lab group. Set up labeled containers for the disposal of sea star body parts, dissecting pins, and gloves.

Safety Cautions

Review all safety symbols with students before beginning the lab. Make sure students wear safety goggles, gloves and a lab apron, and wear this protective gear yourself when handling preserved specimens. Caution students to use care when working with sharp instruments. Caution students to keep their hands away from their faces during the lab. Make sure students wash their hands before leaving the lab.

Skills Practice Lab

Analyzing Sea Star Anatomy SC.F.1.4.2

SKILLS
- Observing
- Collecting data
- Inferring

OBJECTIVES
- **Observe** anatomical structures of an echinoderm.
- **Infer** function of body parts from structure.

MATERIALS
- disposable gloves
- preserved sea star
- dissection tray
- dissection scissors
- hand lens
- dissecting microscope
- forceps
- blunt probe
- sharp probe
- dissection pins

ChemSafety

CAUTION: Always wear safety goggles and a lab apron to protect your eyes and clothing.

CAUTION: Do not touch or taste any chemicals. Know the location of the emergency shower and eyewash station and how to use them. If you get a chemical on your skin or clothing, wash it off at the sink while calling to the teacher. Notify the teacher of a spill. Spills should be cleaned up promptly, according to your teacher's directions.

CAUTION: Glassware is fragile. Notify the teacher of broken glass or cuts. Do not clean up broken glass or spills with broken glass unless the teacher tells you to do so.

Before You Begin

Sea stars are members of the phylum Echinodermata, a group of invertebrates that also includes sand dollars, sea urchins, and sea cucumbers. **Echinoderms** share four main characteristics: an **endoskeleton**, **five-part radial symmetry**, a **water-vascular system**, and circulation and respiration through their **coelom.**

1. Write a definition for each boldface term above.
2. Create a data table like the one shown to the right.
3. Based on the objectives for this lab, write a question you would like to explore about sea star anatomy.

DATA TABLE

Function of Sea Star Structures

Structure	Observations	Inferred function
Madreporite		
Spine		
Skin gill		

Procedure

PART A: External Anatomy

1. **CAUTION: Put on safety goggles, a lab apron, and protective gloves.** As you observe the sea star body structures, record your observations and your inference of each structure's function in the table. On a separate sheet of paper or in your lab notebook, draw and label the sea star and the structures that you observe.

706

Tips and Tricks

Leave sea stars in a flowing water bath for at least 4 hours to remove preservative before the lab begins. Demonstrate for students safe techniques for using dissection instruments and dissecting pins. Encourage students to observe other students' specimens and to note external and internal differences between individuals of a single species.

2. Using forceps, hold a preserved sea star under running water to gently but thoroughly remove excess preservative. Then place the sea star in a dissecting tray.
3. Refer to the diagram of a sea star in **Up Close** in this chapter to locate the madreporite on the upper surface of the sea star.
4. Use a hand lens to observe the sea star's spines. Are they distributed in any recognizable pattern? Are they exposed or covered by tissue? Are they movable or fixed?
5. Use the dissecting microscope to look for small skin gills, If any are present, describe their location and structure.
6. Examine the sea star's lower surface. Find the mouth, and use forceps or a probe to gently move aside any soft tissues. What structures are found around the mouth?
7. Locate the tube feet. Describe their distribution. Using a dissecting microscope, observe and then draw a single tube foot.

PART B: Internal Anatomy

8. **CAUTION: Scissors, probes, and pins are sharp. Use care not to puncture your gloves or injure yourself or others.** Using scissors and forceps, carefully the cut the body wall away from the upper surface of one of the sea star's arms. Start near the end of the arm and work toward the center.
9. Find the digestive glands in the arm you have opened. Then, locate the short branched tube that connects the digestive glands to the pyloric stomach.
10. Cut the tube that connects the digestive glands to the stomach, and move the digestive glands out of the arm. Look for the reproductive organs.
11. Locate the two rows of ampullae that run the length of the arm.
12. Carefully remove the body wall from the upper surface of the central region of the sea star. Locate the pyloric stomach and the cardiac stomach.
13. Remove the stomachs and find the ring canal and the radial canals. In which direction does water move through these canals?
14. Dispose of sea stars and sea star body parts in the waste container designated by your teacher. Do not put lab materials in the trash unless your teacher tells you to do so.
15. Clean up your work area and all lab equipment. Return lab equipment to its proper place. Wash your hands thoroughly before you leave the lab and after you finish all work.

Analyze and Conclude

1. **Analyzing Results** What type of symmetry is found in the sea star?
2. **Inferring Relationships** What is the relationship between the ampullae and the tube feet?
3. **Making Predictions** How does a sea star use its stomach during feeding?
4. **Making Predictions** If the ring canals and radial canals did not function properly, how would this affect the sea star's ability to move and feed?
5. **Further Inquiry** Write a new question about echinoderms that could be explored with another investigation.

Use the following Internet resources to explore your own questions about echinoderms.

internet connect
www.scilinks.org
Topic: Echinoderms
Keyword: HX4065
SCILINKS. Maintained by the National Science Teachers Association

Answers to Before You Begin

1. *echinoderm*—a radially symmetrical marine invertebrate that has an endoskeleton, such as a sea star, a sea urchin, or a sea cucumber

   endoskeleton—an internal skeleton made of bone and cartilage

   five-part radial symmetry—a body plan in which five similar parts of an animal's body are organized in a circle around a central axis

   water-vascular system—a system of canals filled with a watery fluid that circulates throughout the body of an echinoderm

   coelom—a body cavity that contains the internal organs
2. Tables should resemble Data Table on student's page.
3. Answers will vary. For example: Which of the sea star's body structures are modified for feeding?

Answers to Procedure

1. Students' drawings should be clear and labeled.
4. Spines are usually short, scattered over the surface, exposed, and fixed.
5. Skin gills are distributed between the spines but may not be visible in preserved specimens.
6. Five pairs of movable spines surround the mouth.
7. Tube feet are arranged in two rows along the length of each arm's oral surface.
13. Water moves from the ring canal to the radial canals.

Answers to Analyze and Conclude

1. Sea stars have five-part radial symmetry.
2. The ampullae are hollow and are connected to the tube feet. The ampullae pump water into the tube feet, causing the tube feet to expand outward.
3. Part of the stomach is thrust through the mouth, and digestive juices liquefy the prey, which then is ingested.
4. Because the sea star's water vascular system is essential to locomotion and feeding, the sea star would not be able to move about or feed normally if the ring canals and radial canals malfunctioned.
5. Answers will vary. For example: How do the bodies of sea urchins differ from those of sea stars?

Chapters

Only a few wild tigers remain in the world.

Yesterday...

Native to Asia, tigers once lived throughout much of the southern half of the continent. Tigers were killed in large numbers as hunters killed them for their skins and humans cleared the forests in which they lived. **What characteristics of tigers are representative of other mammals?**

Prince of Wales, Edward VII, on a hunting expedition in India

Today...

Tigers everywhere are at critically low numbers. Poaching, the illegal killing of tigers for their bones, skin, and other body parts, continues in many countries. In India, Nepal, and Russia, protected areas have been set aside to help preserve tiger habitat. **Discover how a mammal's teeth provide clues about its diet.**

Tiger skin confiscated by the police in Delhi, India

Tomorrow...

The long-term survival of tigers may depend on breeding programs in zoos and nature reserves. In some areas, more rigorous enforcement of anti-poaching laws is helping to protect wild populations. **Find out to which order of mammals tigers belong.**

Bengal tiger cubs

internet connect

www.scilinks.org
Topic: Tigers
Keyword: HX4176

SCILINKS. Maintained by the National Science Teachers Association

Chapter Opener		SE **Quick Review,** p. 711 ■ SE **Reading Activity,** p. 711 ■ TE **Using the Figure,** p. 710 GENERAL TE **Opening Activity,** p. 711 GENERAL
3 & 4 • 135 min pp. 712–720 **Section 1** Vertebrates in the Sea and on Land	OSP **Lesson Plan*** TR **Bellringer*** TR **Section Outline*** TR **H6 Vertebrate Skeleton (Cat)*** TR **I47 Major Vertebrate Organ Systems*** TR **I2 Evolution of Jaws*** TR **I9 From Fin to Limb***	TE **Activity** Animal List, p. 713 GENERAL TE **Group Activity** Learning the Time Periods, p. 714 GENERAL TE **Demonstration,** p. 714 GENERAL TE **Group Activity** Shark!, p. 715 ADVANCED SE **Data Lab** Using Timelines and Phylogenic Trees, p. 716 ■ CRF **Datasheets for In-Text Labs** Using Timelines and Phylogenic Trees* TE **Demonstration,** p. 717 SE **Data Lab** Comparing the Surface Areas of Gills, p. 719 ■ CRF **Datasheets for In-Text Labs** Comparing the Surface Areas of Gills*
& 6 • 90 min pp. 721–730 **Section 2** Terrestrial Vertebrates	OSP **Lesson Plan*** TR **Bellringer*** TR **Section Outline*** TR **I16 Breakup of Pangaea***	TE **Activity** Trumping With Dinosaurs, p. 723 ADVANCED TE **Group Activity** Dinosaur Adaptations, p. 723 ◆ GENERAL TE **Demonstration,** p. 726 ◆ TE **Demonstration,** p. 727 ◆ TE **Demonstration,** p. 728 ◆ GENERAL TE **Activity** Mammal Comparisons, p. 729 GENERAL
& 8 • 90 min pp. 731–734 **Section 3** Evolution of Primates	OSP **Lesson Plan*** TR **Bellringer*** TR **Section Outline*** TR **I41 Comparison of Gorilla and Australopithecine Skeletons*** TR **I42 Hominids in the Fossil Record***	TE **Activity** Primate Graphic Organizer, p. 733 GENERAL
& 10 • 90 min pp. 735–738 **Section 4** The Genus *Homo*	OSP **Lesson Plan*** TR **Bellringer*** TR **Section Outline*** TR **I43 Theory of Hominid Migration***	TE **Activity** Hominid Sizes, p. 735 ◆ GENERAL TE **Demonstration,** p. 736 ◆ TE **Group Activity** Extranuclear and Y Chromosome DNA, p. 737 ADVANCED SE **Exploration Lab** Comparing Hominid Skulls, pp. 742–743 ■ GENERAL CRF **Datasheets for In-Text Labs** Comparing Hominid Skulls* ◆ GENERAL

1 & 12 • 90 min **Chapter Review and Assessment Resources**

- SE **Study Zone,** p. 739 ■
- SE **Performance Zone,** p. 740 ■
- SE **FCAT Test Prep,** p. 741 ■
- TR **Concept Mapping***
- CRF **Vocabulary Review Worksheet*** ■ GENERAL
- CRF **Test Prep Pretest*** ■ GENERAL
- CRF **Chapter Test*** ■ GENERAL
- CRF **Chapter Test*** ADVANCED
- OSP **Test Generator***
- OSP **Test Item Listing***
- OSP **Modified Chapter Test*** BASIC

Holt Online Learning

Visit go.hrw.com for access to Holt Online Learning or enter the keyword **HX6 Home** for a variety of free online resources.

One-Stop Planner® CD-ROM

This CD-ROM package includes
- Lab Materials QuickList Software
- Holt Calendar Planner
- Customizable Lesson Plans
- Printable Worksheets
- ExamView® Test Generator
- Interactive Teacher Edition
- Holt PuzzlePro® Resources
- Holt PowerPoint® Resources

Guided Reading Audio CDs

These CDs are designed to help auditory learners and reluctant readers. (Audio CDs are also available in Spanish.)

Compression guide:
To shorten your instruction because of time limitations, omit blocks 1, 3, 6, and 8.

KEY

TE	Teacher Edition	**CRF**	Chapter Resource File	*	Also on One-Stop Planner
SE	Student Edition	**TR**	Teaching Transparency	■	Also Available in Spanish
OSP	One-Stop Planner			◆	Requires Advance Prep

SKILLS DEVELOPMENT AND PRACTICE	SECTION REVIEW AND ASSESSMENT	CORRELATIONS
		Florida Sunshine Standards/FCAT
TE Skill Builder Distinguishing Vertebrate, Vertebrae, and Vertebra, p. 713 BASIC **TE Skill Builder** Interpreting Visuals, p. 715 GENERAL **TE Skill Builder** Observing Fish, p. 717 **CRF Directed Reading Worksheet*** BASIC **CRF Active Reading Worksheet*** GENERAL	**SE Section 1 Review** ■ **TE Reteaching,** p. 720 BASIC **TE Quiz,** p. 720 GENERAL **TE Alternative Assessment,** p. 720 GENERAL **CRF Quiz*** ■ BASIC	SC.D.1.4.3, SC.F.1.4.2, SC.G.1.4.1
TE Reading Skill Builder Reading Hint, p. 726 BASIC **TE Reading Skill Builder** Paired Summarizing, p. 727 BASIC **CRF Directed Reading Worksheet*** BASIC **CRF Active Reading Worksheet*** GENERAL	**SE Section 2 Review** ■ **TE Reteaching,** p. 730 BASIC **TE Quiz,** p. 730 GENERAL **TE Alternative Assessment,** p. 730 GENERAL **CRF Quiz*** ■ BASIC	SC.D.1.4.3, SC.D.1.4.4, SC.G.1.4.1
TE Skill Builder Graphing, p. 733 GENERAL **CRF Directed Reading Worksheet*** BASIC **CRF Active Reading Worksheet*** GENERAL	**SE Section 3 Review** ■ **TE Reteaching,** p. 734 BASIC **TE Quiz,** p. 734 GENERAL **TE Alternative Assessment,** p. 734 GENERAL **CRF Quiz*** ■ BASIC	SC.D.1.4.3, SC.D.1.4.4, SC.F.1.4.2, SC.G.1.4.1
TE Reading Skill Builder Reading Organizer, p. 737 BASIC **CRF Directed Reading Worksheet*** BASIC **CRF Active Reading Worksheet*** GENERAL **CRF Critical Thinking Worksheet*** ADVANCED **CRF Science Skills Worksheet*** GENERAL	**SE Section 4 Review** ■ **TE Reteaching,** p. 738 BASIC **TE Quiz,** p. 738 GENERAL **TE Alternative Assessment,** p. 738 GENERAL **CRF Quiz*** ■ BASIC **Modified Worksheet*** BASIC	SC.D.1.4.3, SC.G.1.4.1

SCILINKS. NSTA
www.scilinks.org
Maintained by the **National Science Teachers Association**

Topic: Evolution of Jaws
SciLinks Code: HX4075

Topic: Evolution of Limbs
SciLinks Code: HX4076

Topic: Pangaea
SciLinks Code: HX4135

Topic: Dinosaur Extinction
SciLinks Code: HX4057

Topic: Birds
SciLinks Code: HX4024

Topic: Early Mammals
SciLinks Code: HX4063

Overview

Tell students the purpose of this chapter is to illustrate the major characteristics seen in vertebrates. The chapter will introduce key adaptations that led to the appearance of fishes and amphibians. Students will also learn why dinosaurs became the dominant land vertebrate, and possible reasons for their extinction. The chapter will explore the rise of mammals, and the relationship between reptiles and birds. Finally, students will learn about early primates, and the evolution of hominid primates and humans.

Using the Figure

Tell students that environmental changes over millions of years have resulted in changes in vertebrate structure and function through the process of evolution by natural selection. These changes have enabled vertebrates to exploit new habitats and adopt new ways of life. This chapter summarizes many of these changes and their possible origins.

Assessing Prior Knowledge

Before beginning this chapter, be sure your students can answer the questions in the ***Quick Review*** on the facing page. If they cannot, encourage them to reread the sections indicated.

Chapter Resource File

- Vocabulary Worksheets
- Concept Mapping

ımmoth skeleton

Chapter Correlations — **Florida Sunshine Standards/FCAT**

SC.D.1.4.3 Changes in Earth's climate, geological activity, and life forms may be traced and compared.

SC.D.1.4.4 Earth's systems and organisms are the result of a long, continuous change over time.

SC.F.1.4.2 Body structures are uniquely designed and adapted for their function.

SC.G.1.4.1 There is great diversity and interdependence among living things.

CHAPTER

32 Introduction to Vertebrates

Florida Sunshine Standards/FCAT

SC.D.1.4.3 Changes in Earth's climate, geological activity, and life forms may be traced and compared.

SC.D.1.4.4 Earth's systems and organisms are the result of a long, continuous change over time.

SC.F.1.4.2 Body structures are uniquely designed and adapted for their function.

SC.G.1.4.1 There is great diversity and interdependence among living things.

Quick Review

Answer the following without referring to earlier sections of your book.

1. **Summarize** how mass extinctions have affected life on Earth. *(Chapter 12, Section 2)*
2. **Define** the terms *vertebrate* and *continental drift*. *(Chapter 12, Section 3)*
3. **Distinguish** between a carnivore and a herbivore. *(Chapter 16, Section 2)*
4. **Define** the terms *bilateral symmetry*, *cephalization*, and *coelom*. *(Chapter 27, Section 1)*
5. **Summarize** the function of genes. *(Chapter 10, Section 1)*
6. **Explain** the importance of phylogenetic relationships. *(Chapter 13, Section 2)*

Did you have difficulty? *For help, review the sections indicated.*

Reading Activity

As you read the first section of this chapter, complete a reading organizer to describe the series of adaptations in body structure for agnathans, ostracoderms, acanthodians, and placoderms.

Mammoths were large mammals with tusks and a trunk similar to those of modern elephants. Now extinct, mammoths lived throughout North America starting about 1.7 million years ago.

Looking Ahead

Section 1
Vertebrates in the Sea and on Land

Adaptations of Vertebrates
The First Vertebrates
Evolution of Fishes
Modern Fishes
Evolution of Amphibians
Modern Amphibians

Section 2
Terrestrial Vertebrates

Early Reptiles
Evolution of Dinosaurs
Evolution of Birds
Modern Birds
Modern Reptiles
Evolution of Mammals
Modern Mammals

Section 3
Evolution of Primates

Characteristics of Primates
Nonhuman Primates
Early Hominids

Section 4
The Genus *Homo*

Homo habilis
Homo erectus
Homo sapiens

Opening Activity

Ask students how the internal, bony skeleton of vertebrates is an advantage over the external skeleton of many invertebrates. (An internal skeleton [paired with muscles, tendons, and ligaments] provides for more mobility by producing a versatile system of levers and pulleys. Calcium-containing bone is also harder and stronger than invertebrate exoskeletons.)

Identifying Preconceptions

Students do not often recognize that all vertebrates evolved from aquatic ancestors. Jawless fishes were the first organisms to have a backbone. This adaptation eventually led to the development of amphibians, reptiles, dinosaurs, birds, and mammals. Have students examine **Figures 4 and 11,** to understand the evolutionary relationship between each class of vertebrates.

Reading Activity Answers

Students' reading organizers should include information that agnathans are the most primitive of these fish, and that they lacked jaws. They should note that ostracoderms also lacked jaws. The acanthodians had jaws and spines on their bodies. Placoderms had jaws and bony plates on their heads.

Quick Review Answers

1. Mass extinctions have resulted in losses of up to 96 percent of all species of animals.
2. A vertebrate is an animal with a backbone. Continental drift is the movement of Earth's landmasses over geologic time.
3. A carnivore is an animal that eats other animals. An herbivore is an animal that eats plants and other primary producers.
4. Bilateral symmetry is a body plan in which the right and left halves look the same. Cephalization is an anterior concentration of sensory structures and nerves. A coelom is a fluid-filled space between the body wall and the gut.
5. The function of genes is to carry the instructions for making a protein.
6. Phylogenetic relationships provide very strong evidence for evolution, because they reflect relationships indicated by the fossil record.

Section 1 Vertebrates in the Sea and on Land

Focus

Overview

Before beginning this section review with your students the objectives listed in the Student Edition. This section introduces students to the characteristics of vertebrates. They will learn about the first vertebrates, which were jawless, other fishes that evolved bony jaws and fins, and the cartilaginous fishes. Finally, they will learn about the characteristics of amphibians.

Bellringer

Ask students to make a pair of lists, one with characteristics of fishes and the other with characteristics of amphibians. Ask them to circle any traits that are not common to both of these groups of animals.
LS Verbal

Chapter Resource File

- Directed Reading BASIC
- Active Reading GENERAL
- Data Sheet for Data Lab GENERAL
- Data Sheet for Data Lab GENERAL

One-Stop Planner CD-ROM

- Reading Organizers BASIC
- Reading Strategies BASIC

Florida Sunshine Standards/FCAT

SC.F.1.4.2 Body structures are uniquely designed and adapted for their function.

SC.G.1.4.1 There is great diversity and interdependence among living things.

Objectives

- **Identify** the key characteristics of vertebrates. SC.G.1.4.1
- **Describe** two adaptations found in early fishes. SC.F.1.4.2
- **Identify** the relationship of fishes to amphibians. SC.G.1.4.1
- **Summarize** the key adaptations of amphibians for life on land. SC.F.1.4.2

Key Terms

vertebra
agnathan
acanthodian
cartilage

Adaptations of Vertebrates

If you go to a zoo, many of the animals you see—the lions, elephants, snakes, turtles, and birds—are vertebrates, just as people are. Vertebrates are chordates with a backbone. Even though these vertebrates appear different, they are very much alike internally, indicating that in the distant past they had a common ancestor. Present-day differences in vertebrates reflect their different evolutionary paths.

What are the internal similarities shared by vertebrates? First, vertebrates are chordates with a backbone. They take their name from the individual segments, called **vertebrae** *(VUR tuh bree)* (singular, *vertebra*), that make up the backbone. **Figure 1** shows a typical vertebrate and its endoskeleton. In most vertebrates, the backbone completely replaces the notochord found in invertebrate chordates.

The backbone provides support for and protects a dorsal nerve cord. It also provides a site for muscle attachment. These functions paved the way for the development of an internal skeleton that allowed vertebrates to grow larger than their invertebrate ancestors. In addition to a backbone, vertebrates have a bony skull that encases and protects their brain.

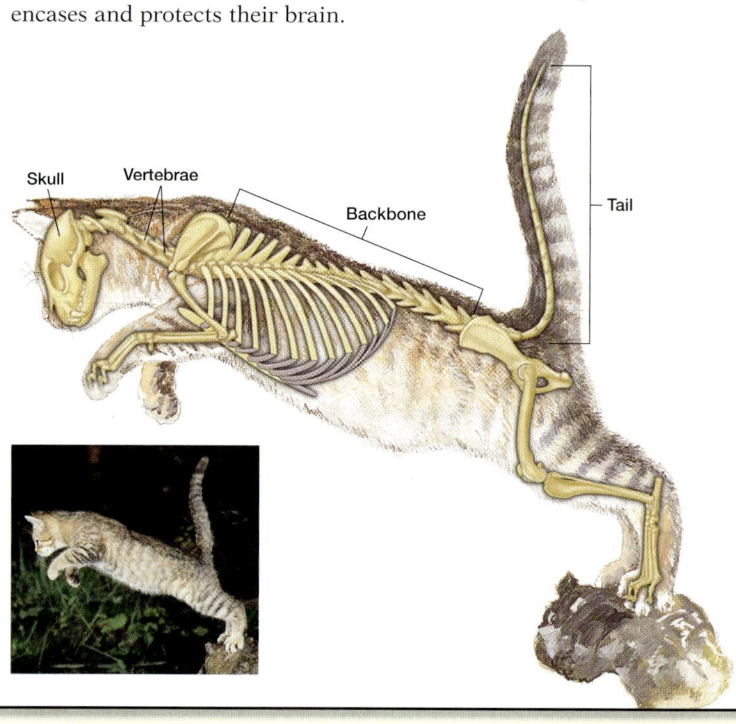

Figure 1 Vertebrate. This cat's skeleton includes a skull and a backbone composed of vertebrae.

712

Integrating Physics and Chemistry

Calcium is an important component of the endoskeleton of vertebrates. Have students locate the element calcium on the periodic table and state the number of its outer or valence electrons, bonding pattern, physical properties, and chemical behavior based on its location on the periodic table.

Transparencies

TR Bellringer
TR H46 Vertebrate Skeleton (Cat)
TR I47 Major Vertebrate Organ System

Section Correlations

SC.F.1.4.2
SC.G.1.4.1

Vertebrates share a number of other characteristics, including segmentation, bilateral symmetry, and two pairs of jointed appendages, such as limbs or fins. They exhibit cephalization and have complex brains and sense organs. All vertebrates have a true coelom and a closed circulatory system with a chambered heart.

The tissues of vertebrates compose organs. In turn, the organs compose organ systems. Vertebrate organ systems tend to be more complex than the organ systems found in invertebrates. For example, recall the grasshopper's digestive system, seen in *Up Close* in an earlier chapter. Its digestive tract is divided into several specialized regions. However, the only associated structures are salivary glands. In contrast, the cat's digestive system, shown in **Figure 2,** consists of a digestive tract, together with two accessory organs, the liver and the pancreas. These complex organs secrete enzymes that aid digestion. Figure 2 shows the major organ systems of a typical vertebrate.

Real Life

How many vertebrae do giraffes have in their neck?

Nearly all vertebrates, including giraffes, have seven neck vertebrae. At almost 6 m (20 ft) tall, the giraffe is the tallest of all vertebrates, and its seven cervical (neck) vertebrae are greatly elongated.

Figure 2 Major organ systems

A vertebrate's organ systems perform a variety of functions.

Vertebrate Organ Systems

1 Nervous system: monitors the environment; controls and coordinates body functions.

2 Circulatory system: carries blood and substances dissolved in it around the body.

3 Digestive system: prepares food for use by animal's cells; removes solid wastes from body.

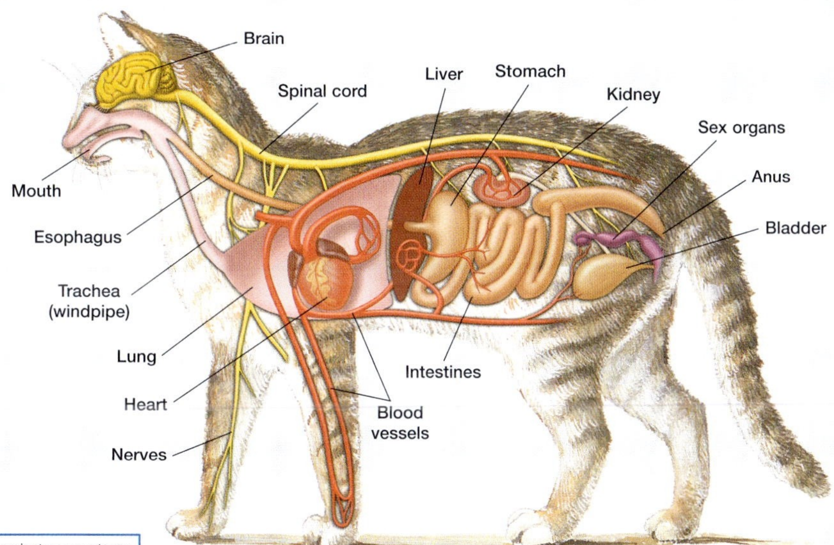

4 Respiratory system: exchanges gases (O_2 and CO_2) between blood and the animal's environment.

5 Reproductive system: produces and carries eggs or sperm; usually allows for internal fertilization of egg and internal development of offspring.

6 Excretory system: removes wastes from the body.

713

Motivate

Activity — GENERAL

Animal List Instruct each student to make a list of 10 animals. Give no further instructions. Next have students put an asterisk by each animal that has a backbone. (A high percentage of the animals listed will likely be vertebrates.) Point out that while invertebrates, such as insects, mollusks, and echinoderms, are animals, most people think only of vertebrates when they think of animals.

Teach

SKILL BUILDER — BASIC

Distinguishing Vertebrate, Vertebrae, and Vertebra Write the words vertebrate, vertebrae, and vertebra on the board or overhead projector. Have students feel the large bump on the back of their neck. Point out that this bump is part of a single vertebra, the entire column is made up of many vertebrae, and animals that have this column of bones are called vertebrates.

Teaching Tip — ADVANCED

Paleontologist Provide your students with the information in the Career feature at the bottom of this page of the Teacher's edition. Tell students that paleontologists study fossilized ancient life forms. Using their observations, these scientists infer the relationships between extinct and living organisms. They also use evidence they find to make inferences about the ecology of previous periods in Earth's history. Paleontologists use a variety of methods to analyze the remains they examine. Have students work in teams to use library or Internet references to find more information about paleontology.
LS Interpersonal

Career

Paleontologist Paleontology is the study of the history of life on Earth, as reflected in the fossil record. There are many subdivisions in the field of paleontology, including vertebrate paleontology, invertebrate paleontology, micropaleontology, paleobotany, taphonomy, biostratigraphy, and paleoecology. Paleontologists also are frequently involved in studies of evolutionary biology. A strong college background in the sciences is absolutely essential for an aspiring paleontologist, with concentrations in both biology and geology. A doctoral degree (Ph.D.) is almost always necessary for any serious professional career in paleontology. Most professional paleontologists are college and university professors that work in departments of geology. A smaller number of professional paleontologists work in museums, generally conducting research. Some work for government surveys, usually in geological mapping or other applied geological problem solving. Paleontologists can also work for major oil companies, helping to search for petroleum.

Teach, continued

Group Activity — GENERAL

Learning the Time Periods

Learning the sequence of the periods in Earth's history is difficult. A shortcut is to have students make up a creative mnemonic device. Divide the class into small groups, explain what a mnemonic device is, and have each group invent one for the Earth's time periods. Do not give any hints. Write each group's results on the board and discuss the merits of each one. (Example: Queen Tracy's Crown Jewels Taught Paleontologists Carbon-containing Diamonds Stand Old Catastrophes.) Then, if students learn that the Jurassic period began about 200 million years ago (mya), the Devonian about 400 mya, and the Cambrian about 600 mya, they can get a sense of how the other time periods fit in. LS Verbal Co-op Learning

Demonstration — GENERAL

With all textbooks closed, display in random order seven large pictures—one each of a lamprey, a shark, a bony fish, an amphibian, a reptile, a mammal, and a bird. In game-show style, have a volunteer come forward and arrange the pictures in the order in which each class of vertebrate evolved. If the student is incorrect, have the student return to his or her seat. Tell the next volunteer how many pictures are out of order, but not which ones. Continue until the correct order is reached, at which point you can ring a bell to signify success. LS Visual

The First Vertebrates

The first chordates evolved about 550 million years ago. At that time, many different groups of organisms appeared in the shallow seas that covered a large portion of Earth's continents. According to the fossil record, the first vertebrates appeared about 50 million years later. The first vertebrates were fishes similar to the one shown in **Figure 3.** Unlike most of the fishes you are familiar with, the earliest fishes, called **agnathans,** had neither jaws nor paired fins. But agnathans did have a backbone, which provided a central axis for muscle attachment. As their muscles pulled against the backbone, the agnathans propelled themselves along the ocean bottom.

Figure 3 Ancient fish. This fossil fish is typical of the early jawless fishes.

Within another 50 million years, jawless fishes had diversified into a great variety of species. The major group was the ostracoderms *(ahs TRAK uh durms),* which had primitive fins and massive plates of bony tissue on their body. Jawless fishes dominated the oceans for about 100 million years, until they were replaced by new kinds of fishes that were hunters. **Figure 4** shows one hypothesis about evolutionary relationships among fishes.

Figure 4 Evolution of fishes

This simplified phylogenetic tree shows one hypothesis about the relationship between the early fishes and the fishes and amphibians that evolved later.

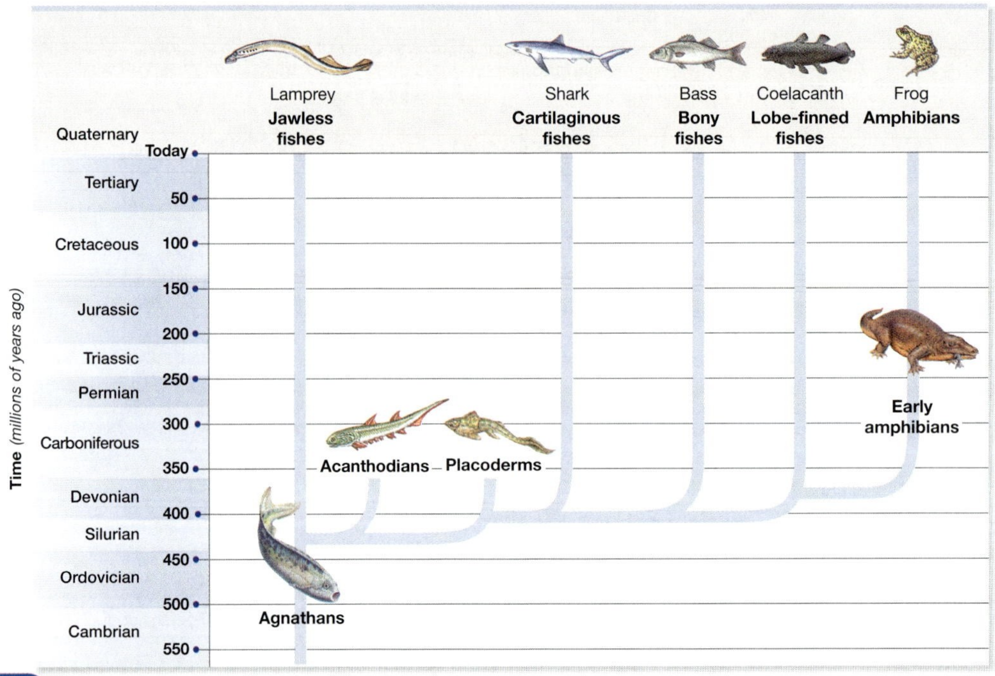

REAL WORLD CONNECTION

As of 2002, aquaculture provided more than 30 percent of the world's seafood supply. Aquaculture production is growing, while wild fisheries are declining. In 2000, Americans consumed an average of 15.6 pounds of fishery products (edible meat) per person. To meet this increasing demand, the U.S. imported more than $10.2 billion of edible seafood in 2001 and at least 40 percent of this was cultured. Dependence on imported seafood is a major contributor to the overall U.S. trade deficit. In 2001, the trade deficit in edible seafood exceeded $7 billion, double the amount from 1997. Farmed shrimp and salmon are a significant portion of imported seafood, but the U.S. also imports farmed catfish, tilapia, oysters, mussels, scallops and other familiar seafood. Most of the seafood sold in restaurants and in grocery stores consists of farmed products. Ornamental fish are also farmed and imported from around the world. The U.S. is now the largest market for these fish that grace many public and private aquariums.

Evolution of Fishes

To survive as predators in the water, fishes must meet two important challenges. The first challenge is that of pursuing prey through the water. Fishes are fast and agile swimmers. The bodies of fishes are streamlined and many are flattened sideways, which makes it easy for them to move through the water. Fishes also have paired fins supported by spines, which provide fine control of movement.

The second challenge is that of grasping prey once it is within reach. About 430 million years ago, the **acanthodians** *(uh kan THOH dee uhns)*, or spiny fishes, appeared. Acanthodians had strong jaws with jagged, bony edges that served as teeth, enabling them to hold onto prey. The development of jaws in fishes was a key evolutionary innovation. As shown in **Figure 5,** jaws are thought to have evolved from gill arch supports made of **cartilage** *(KAHRT lihj)*, a lightweight, strong, flexible tissue.

The spiny fishes had internal skeletons of cartilage, although some fossils indicate that their skeletons also contained some bone. Their scales also contained small plates of bone. The presence of bone in the spiny fishes foreshadowed the much larger role that bone would play in their descendants.

About 20 million years after the acanthodians appeared, the placoderms evolved. Placoderms *(PLAK uh durms)* were jawed fishes with massive heads armored with bony plates. By the end of the Devonian period, almost all of the early fishes, including the placoderms, had disappeared. After dominating the seas for almost 50 million years, they were replaced by swifter swimmers—the sharks and bony fishes.

internet connect
www.scilinks.org
Topic: Evolution of Jaws
Keyword: HX4075
SciLinks. Maintained by the National Science Teachers Association

Figure 5 Evolution of jaws

The development of jaws was a key evolutionary innovation.

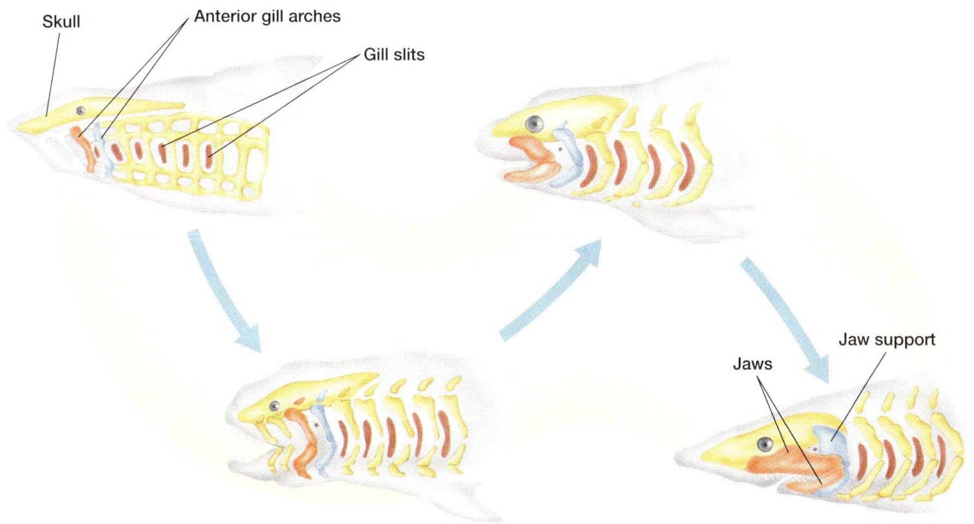

SKILL BUILDER — GENERAL

Interpreting Visuals Direct students' attention to **Figure 5** on this page. Identify the structures in color in the figure. (yellow—bones; orange—anterior gill arch or jaw bones; blue—second anterior gill arch or jaw bone; magenta—gill slits) Trace the changes in the size, shape, and position of the two anterior gill arches as they evolved into jawbones that are hinged together. Ask students: what advantage was provided by the development of jaws? (The animal with the jaws could use a lot of leverage to bite and chew with its teeth.) **LS Visual**

Group Activity — ADVANCED

Shark! Have students work in small groups to conduct library or Internet research on sharks. Each group should choose or be assigned one group of sharks to investigate. Examples they could research are the Great White Shark, the Whale Shark, the Tiger Shark, the Leopard Shark, and the Blue Shark. Instruct each group to collect information on their shark's size, its other physical characteristics, its habitat, its primary and secondary food sources, whether or not it has been known to bite humans, and any other interesting facts. Ask each group to prepare an informational booklet about the shark they studied. Their booklets should include a picture or drawing of their shark.
LS Interpersonal Co-op Learning

Transparencies

TR I2 Evolution of Jaws

REAL WORLD CONNECTION

In 2001, the Food and Drug Administration (FDA) issued an advisory to pregnant women and women of childbearing age who may become pregnant to avoid eating shark, swordfish, king mackerel, or tilefish. As a matter of prudent public health advice, the FDA recommended that nursing mothers and young children not eat these fish as well. Some fish such as shark, swordfish, king mackerel, and tilefish contain high levels of an organic form of mercury, called methylmercury. Methylmercury can cause severe neurological damage, and is especially damaging to an unborn baby's developing nervous system. These long-lived, larger fish accumulate the highest levels of methylmercury, therefore posing the greatest health risk. A tragic example of the effects of bioaccumulation of methylmercury took place in Minimata Bay in Japan during the 1950s. Mercury contamination of bay waters by a local chemical plant lead to mass poisoning of the local population of fishermen and their families. Many children that were exposed to methylmercury *in utero* were born with severe cerebral palsy.

Teach, *continued*

Using Timelines and Phylogenetic Trees

DATA LAB

Skills Acquired
Analyzing, interpreting

Teacher's Notes
Have students work in pairs, with one student analyzing **Figure 4** and the other analyzing the timeline. After answering questions 1–3 using one of the graphics, students should double-check their answers against the other graphic. They should then compare their answers.

Analysis Answers:

1. When the jawed fishes appeared, plants, arthropods, and fungi were invading the land.
2. The Ordovician period began about 500 mya and ended about 440 mya. The Silurian period began about 440 mya and ended about 410 mya.
3. Of the fishes living during the Devonian period, only the jawless, cartilaginous, bony, and lobe-finned fishes have descendants living today.
4. Answers will vary. Some students may say a timeline would be more useful, because it would show when various musical groups appeared on the music scene. Others may say a tree would be because it would show when certain related events occurred, such as when the lead singer of a band went solo, but the rest of the band continued on its own.

Sharks and Bony Fishes

By the end of the Devonian period, almost all of the early fishes had disappeared. At about the same time (400 million years ago), sharks and bony fishes appeared. Sharks and bony fishes, which are superior swimmers, thrived in the Devonian seas. Most have streamlined bodies that are well adapted for rapid movement through the water. Both sharks and bony fishes also have an assortment of movable fins that greatly aid their swimming. However, their evolutionary pathways differ. Sharks, such as the extinct *Hybodus*, shown in **Figure 6,** have a skeleton composed of cartilage. Calcium carbonate (the mineral oyster shells are made of) is deposited in the outer layers of the cartilage. A thin layer of bone covers this reinforced cartilage. The result is a very light but strong skeleton. Bony fishes, on the other hand, have a skeleton made of bone, which is heavier and less flexible than cartilage. Bony fishes have a swim bladder, which compensates for this increased weight. This gas-filled sac buoys them in the water, just as an air-filled balloon buoys a swimmer.

Figure 6 ***Hybodus.*** Early sharks, such as *Hybodus*, were among the first vertebrates to have jaws.

DATA LAB *Using Timelines and Phylogenetic Trees* SC.D.1.4.3

Background

Use the timeline in Section 2 of the chapter titled "History of Life on Earth" and the phylogenetic tree in Figure 4 in this chapter to answer the analysis questions. Tell which graphic you used to answer each question.

Analysis

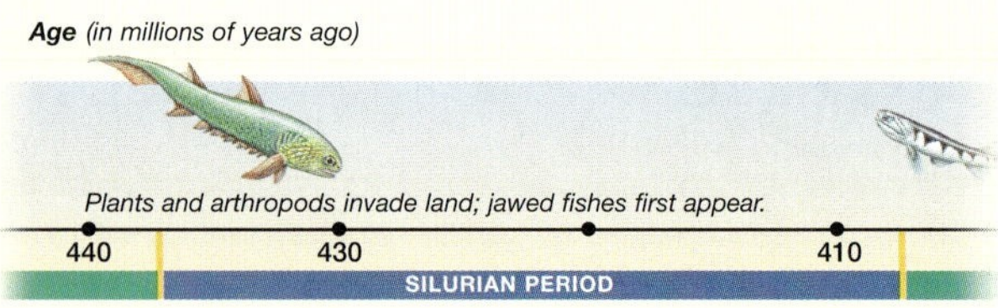

1. **Critical Thinking Interpreting Graphics** When the jawed fishes appeared, what was occurring on land?
2. **Critical Thinking Interpreting Graphics** What are the approximate beginning and ending dates of the Ordovician period and the Silurian period?
3. **Critical Thinking Interpreting Graphics** Of the fishes living during the Devonian period, which groups have descendants living today?
4. **Critical Thinking Analyzing Methods** Imagine that you are giving a presentation on the history of a particular type of music, such as rock and roll. Which format, a timeline or a tree, would better suit your presentation? Explain your answer.

Cultural Awareness

Japanese Seafood Eating seafood is a part of everyday life for the island-dwelling inhabitants of Japan. The Japanese are well known for their consumption of raw seafood. The two forms in which raw seafood is eaten are *sashimi* and *sushi*. Sashimi consists of thinly sliced raw fish or mollusks. Sushi is rice flavored with wine vinegar (*shari*), that is raw or cooked fish, or formed into a roll with fish, shellfish, or vegetables wrapped in seaweed (*nori*). Both dishes are typically dipped in *shoyu* (soy sauce), which is spiced up a by a dash or two of *wasabi* (horseradish).

Yakimono, or *agemono*, are catch-all terms for any number of grilled foods. *Okonomiyaki* is another Japanese favorite; the name means "cook what you like." This consists of a flat, pancake-like dish topped with cabbage, shrimp, squid, octopus, or diced vegetables. *Unagi*, or eel, is usually cooked over hot coals and brushed with a soy and sake (rice wine) sauce.

Modern Fishes

Today there are more fishes than any other group of vertebrates, both in terms of numbers of individuals and numbers of species. If you consider that water covers three-fourths of Earth's surface and that fishes live in virtually every aquatic habitat, this fact is not surprising. Today's fishes belong to one of three major groups: the agnathans, the cartilaginous *(kart'l AJ uh nuhs)* fishes, and the bony fishes. The agnathans (hagfishes and the lampreys) resemble the early jawless fishes.

The first cartilaginous fishes (sharks) and the bony fishes evolved at about the same time, 400 million years ago. These two groups of fishes likely evolved from the same early, jawless fishes that gave rise to the acanthodians and the placoderms. The shark's relatives, the skates and rays, evolved later.

Bony fishes are the fishes most familiar to us. Bony fishes make up about 95 percent of modern fish species. Because bony fishes are adapted to many different environments, they vary greatly in size, color, and shape. **Figure 7** shows three examples of bony fishes.

Figure 7 **Bony fishes**

Bony fishes live in a wide variety of habitats, and there is great variation in their appearance.

Teaching Tip — GENERAL

Sharks Are Not Boneheads

Remind students about the characteristics of cartilage and bone by having them find examples of each kind of material on their own head. Ask: What is the advantage of having cartilage at the end of the nose instead of bone? (It could break easily if it were made of bone.) What is the advantage of having a skull made of bone? (It protects the delicate brain inside.) How does the cartilaginous skeleton of sharks give them an advantage of maneuverability? (It is more flexible and lighter than a bony skeleton.) **LS Logical**

SKILL BUILDER

Writing **Observing Fish** Give students an opportunity to observe fish, either on video or in a classroom aquarium. Ask them to record their observations in a notebook. These observations should include different types of fins and how each is used, different types of mouths adapted for different diets, and different body shapes and movements of fish through water. Have students write a short essay, summarizing their observations. **LS Intrapersonal**

Demonstration

Fill a 1-L bottle with water and set it beside another 1-L bottle full of air. Ask: Which bottle contains more oxygen? (The bottle of air contains more oxygen.) Explain that air contains about 21 percent oxygen, whereas water contains less than 1 percent. Stagnant, warm water contains even less oxygen. If fishes obtain oxygen only from the water that passes through their gills, they can suffocate in stagnant water. If a goldfish is kept in a fish bowl without an air pump, it will come to the surface to gulp air because the warm, still water in the bowl is low in oxygen.

MEDICINE CONNECTION

The therapeutic potential of cartilage has been investigated for more than 30 years. Cartilage products have been tested as treatments for cancer, psoriasis, and arthritis, as well as a number of other conditions.

The first report that shark cartilage contains at least one angiogenesis (new blood vessel growth) inhibitor was published in the early 1980s, and the only published report to date of a clinical trial of shark cartilage as a treatment for cancer appeared in the late 1990s. The body of a shark is about 6 percent cartilage. At least three different mechanisms of action have been proposed to explain the anticancer potential of cartilage: 1) it is toxic to cancer cells; 2) it stimulates the immune system; and 3) it inhibits angiogenesis. There is only limited evidence to support the first two mechanisms of action; however, the evidence in favor of the third mechanism is more substantial. None of the reported data have been from phase III clinical trials, and the data that have been reported are inconclusive. Additional clinical studies are currently under way.

Teach, *continued*

Teaching Tip — GENERAL

Ocean Emigration Split your class into small groups of two or three students and have them discuss their answers to the following questions: What conditions might have caused vertebrates to leave the ocean and invade the land? What was unfavorable about an aquatic environment or favorable about a terrestrial one? Then have everyone in the class list the advantages of a terrestrial environment. (Advantages include new plant and animal food sources, less competition, fewer predators, and more room.) **LS Verbal**

Using the Figure — GENERAL

Direct students' attention to **Figure 8.** Have students examine the colored bones in the drawing of the lobe-finned fish. Tell them the name of each bone: femur—red; tibia—blue; fibula—purple; humerus—yellow; ulna—orange; radius—green. Have students find the corresponding bones in the drawing of the early amphibian. Then ask them to locate these bones on their own bodies. (*femur*—thigh; *tibia and fibula*—lower leg; *humerus*—upper arm; *ulna and radius*—forearm) **LS Visual**

Transparencies

TR I9 From Fin to Limb

Evolution of Amphibians

The first group of vertebrates to live on land were the amphibians, which appeared about 370 million years ago. Amphibians probably share a common ancestor with modern lungfishes and other lobe-finned fishes. As shown in **Figure 8,** the pattern of bones in an amphibian's limbs bears a strong resemblance to that of a lobe-finned fish. While several species of lobe-finned fishes exist today, those species thought to be ancestral to amphibians are extinct.

The "Age of Amphibians"

Although amphibians first appeared in the Devonian period, they increased greatly in numbers during the Carboniferous period. During this time, which began what biologists call the "age of amphibians," the number of amphibian families increased from 14 to about 34. (Recall that a family is a taxonomic category composed of one or more genera.) By the late Carboniferous period, much of what was to become North America was covered by low-lying tropical swamplands. Amphibians thrived in this moist environment, sharing it with early reptiles.

In the Permian period that followed, amphibians reached their greatest diversity, increasing to 40 families. In the early Permian period, a remarkable change occurred among amphibians—many of them began to leave the marshes for dry uplands. By the middle Permian, 60 percent of all amphibian species were living in dry environments. *Eryops,* illustrated in **Figure 9,** was typical of amphibians of this period and was well adapted for life on land.

Figure 8 From fin to limb

As shown by the colors below, the pattern of limb bones in an early amphibian bears a distinct resemblance to the pattern of fin bones in a lobe-finned fish.

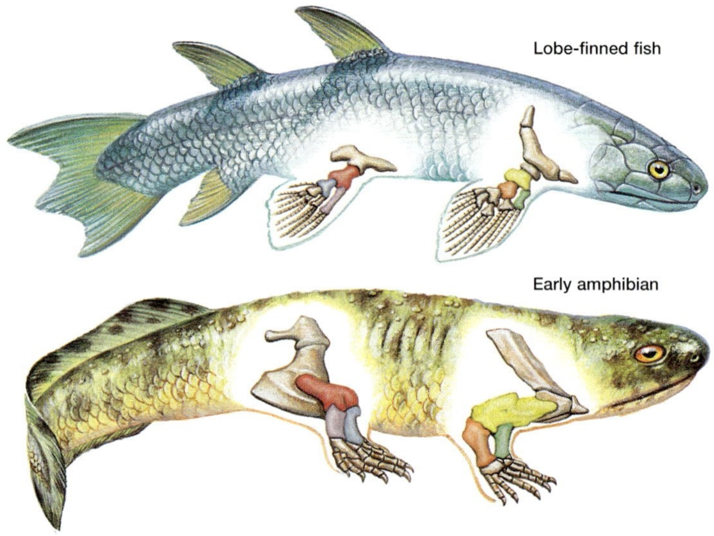

ENVIRONMENTAL SCIENCE CONNECTION

Class Amphibia, which includes frogs, toads, newts, salamanders and caecilians, first appeared on Earth during the Devonian period (360 to 408 million years ago). While many amphibian species have evolved and gone extinct over the millennia, more than 5,000 species exist today. Amphibians have adapted over this expanse of time to occupy different habitats including wetlands, forests, deserts, prairies, and savannas. Amphibian populations are in decline in many areas of the world. Countless areas that previously hosted a range of healthy amphibian populations now have fewer or no frogs, toads, and salamanders. A few species, including Costa Rica's Monteverde golden toad and Australia's Gastric brooding frog, are now believed to be extinct. Scientists conducting field research have produced compelling evidence that habitat loss, climate change, UV radiation, contaminants and pollutants, disease, and predation by invasive species are possible causes for the decline of amphibians.

Adaptations for Life on Land

Life on land is quite different from life in the water. Thus, a number of major adaptations allowed some species to successfully invade land.

1. **Legs.** Legs support the body's weight as well as allow movement from place to place.
2. **Lungs.** The delicate structure of a fish's gills depends on water for support. On land, lungs carry out gas exchange.
3. **Heart.** Walking on land requires a greater expenditure of energy than swimming. Land animals tend to have higher metabolic rates than aquatic animals. This, in turn, requires greater amounts of oxygen. The structure of the vertebrate heart allows oxygen to be delivered to the body efficiently.

Modern amphibians have such adaptations, but many groups are not well-adapted to dry land. For example, amphibian eggs are not watertight, and modern amphibians must seek out water or damp areas in which to reproduce.

Figure 9 ***Eryops.*** *Eryops* grew to 1.8 m (6 ft) long and probably lived much like a modern alligator, moving in and out of freshwater habitats.

DATA LAB: *Comparing the Surface Areas of Gills*

SC.F.1.4.2

Background

Air contains more oxygen than water, so why do fish die when removed from water? To understand what happens to gills when they are removed from water, follow the procedure below.

Materials

cellophane wrap, scissors, toothpick, ruler, container of water

Procedure

1. Cut an 8 × 6 cm piece of cellophane wrap. Use the piece of wrap and a toothpick to make a model of fish gills, as shown in the drawing.
2. Measure the approximate length and width of the model. Calculate the area of the model by using the formula $a = l \times w$.
3. Submerge the model in water and allow it to float at the top. Notice any change in shape of the model.
4. Grasp the model by the toothpick and pull it gently through the water. Then remove the model and place it on the table.
5. Without rearranging the model in any way, measure its approximate length and width. Calculate the area of the model again.

Analysis

1. **Summarize** any difference you observed between the gill model in the water and the wet model out of the water.
2. **Compare** the area obtained in step 2 with that obtained in step 5. If the areas are different, identify which was larger.
3. **Critical Thinking** **Analyzing Data** Consider what you know about the requirements for gas exchange across a gill's surface. Do the data you obtained suggest a reason why fishes cannot live out of water? Explain your answer.
4. **Critical Thinking** **Comparing Structures** Your model is two-dimensional. To calculate the surface area of an actual gill, you would need to know another measurement. What is that measurement, and why is it important?

Teaching Tip — GENERAL

Classification Copy the Graphic Organizer at the bottom of the page onto the board, but do not copy the words in parentheses. Ask a volunteer to fill in one blank and then give an example of an animal that would fit in the same row. Continue until all blanks are filled in. Characteristics and examples may vary, as there are many correct responses. **LS Intrapersonal**

DATA LAB: *Comparing the Surface Areas of Gills*

Skills Acquired

Comparing, calculating, interpreting results

Teacher's Notes

Before students begin, ask them how they would measure the classroom floor to order new floor covering. (Multiply the length by the width.) Ask students what units they would use in placing their order. (square feet or square yards) Ask students to think about how they would measure the room to fill it with water, but do not solicit answers.

Analysis Answers:

1. The gill model in the water will appear larger and flatter.
2. The area calculated in step 2 will be larger than the area calculated in step 5, because the "gills" will have collapsed on themselves.
3. The data suggest that when a fish is out of water, the surface area of its gills decreases significantly. Because the surface area decreases, there is insufficient gas exchange, and the fish will die unless returned to the water.
4. One would need to know the depth of the gills to determine the total surface area. This is important because this third dimension increases the surface area of the gills.

Graphic Organizer

Use this graphic organizer with the **Teaching Tip** on this page.

Name of Group	Characteristic	Example
Agnathans	(No jaws)	(Lamprey)
Cartilaginous fishes	(Cartilage skeleton)	(Shark)
(Bony fishes)	Bony skeleton	(Trout)
(Amphibians)	Soft, moist skin	(Frog)

Close

Reteaching — BASIC

Have students review the six major concept headings in Section 1. Then ask them to write a question for each heading that can be answered with the heading itself. (For example: What is a vertebrate? A vertebrate is a chordate with a backbone.)

Quiz — GENERAL

1. What are the three groups of fishes that have lived on Earth? (agnathans [jawless fishes], acanthodians [spiny fishes], and placoderms [cartilaginous and bony fishes])
2. What was the first group of vertebrates that appeared on Earth? (jawless fishes)
3. What was the first group of vertebrates that lived on land? (amphibians)

Alternative Assessment — GENERAL

Have students make a two-column chart with the headings Fishes and Amphibians. For each column, have students enter information about the history of that group, starting with the earliest data. For example, the first entry under Fishes might be: "Jawless fishes appeared about 500 mya."

Modern Amphibians

The middle Permian period marked the peak of amphibian success. By the end of this period a new kind of vertebrate, a reptile called a *therapsid,* had become common and began to replace the amphibians. By the end of the Triassic period that followed, there were only 15 families of amphibians, including the first species of frogs. And by the Jurassic period, only three groups of amphibians remained—frogs, salamanders, and caecilians. The age of amphibians was over.

All of today's amphibians are descendants of the amphibians that survived into the Jurassic period. They are found in aquatic and moist habitats throughout the temperate and tropical regions of the world. Frogs and toads make up the largest, and probably the most familiar, group of modern amphibians. Salamanders and newts are far less numerous. Caecilians account for less than 1 percent of today's amphibian species. **Figure 10** shows a representative sample of today's amphibians.

Figure 10 Amphibians

Today's amphibians are descendants of the three amphibian groups that survived into the Jurassic period.

Section 1 Review

1. **Describe** two characteristics that distinguish vertebrates from invertebrates. SC.G.1.4.1
2. **Discuss** two adaptations that enabled early fishes to dominate the oceans. SC.F.1.4.2
3. **Relate** the structure of the limbs of a lobe-finned fish to the evolution of amphibians. SC.G.1.4.1
4. **Critical Thinking** **Evaluating a Hypothesis** Evaluate this statement: Amphibians are not fully adapted for life on land. SC.F.1.4.2
5. FCAT Prep According to fossil evidence, what feature did the first fishes lack?
   A backbone
   B fins
   C jaws
   D muscles
   SC.D.1.4.3

Answers to Section Review

1. Vertebrates have a backbone made of individual vertebrae and a dorsal nerve cord. They also have a bony skull that encases and protects the brain. Vertebrates have a true coelom and a closed circulatory system with a chambered heart.
2. Early fishes developed streamlined bodies that were flattened sideways, to move easily through the water. They also developed paired fins supported by spines, which provided fine movement control. These adaptations enabled early fishes to move more efficiently when catching prey and escaping predators.
3. The pattern of bones in an early amphibian's limbs bears a distinct resemblance to that of the fin bones of a lobe-finned fish.
4. True. All amphibians are still dependent on water for reproduction. Because their eggs are not waterproof, amphibians must rely on an external water source to keep the eggs moist. If an egg dries out, the embryo within it will die.
5. **A.** Incorrect. The first fishes had a backbone. **B.** Incorrect. The first fishes had fins. **C.** Correct. The first fishes were jawless. **D.** Incorrect. The first fishes had muscles.

Section Review Correlations

SC.D.1.4.3: Item 5
SC.F.1.4.2: Items 2, 4
SC.G.1.4.1: Items 1, 3

Terrestrial Vertebrates

Section 2

Florida Sunshine Standards/FCAT

SC.D.1.4.3 Changes in Earth's climate, geological activity, and life forms may be traced and compared.

SC.D.1.4.4 Earth's systems and organisms are the result of a long, continuous change over time.

SC.G.1.4.1 There is great diversity and interdependence among living things.

Early Reptiles

To understand the problem that water loss can be, all you need to do is stay outside on a hot day. It won't be long before you are sweating and thinking about a cool drink. If you don't increase your fluid intake and cool off, you may start to feel dizzy and nauseated. Fluid loss is a problem for all **terrestrial** animals, that is, animals that live on land. The adaptations that permitted amphibians to live on land further developed in reptiles. Two very important adaptations for terrestrial life evolved in reptiles. Reptiles were the first animals to have skin and eggs that are both almost watertight, and they differ from amphibians in this respect.

When reptiles first evolved, about 320 million years ago, Earth was entering a long, dry period. Early reptiles were better suited to these conditions than amphibians were, and the reptiles quickly diversified. Within 50 million years, reptiles had replaced amphibians as the dominant terrestrial vertebrates. **Figure 11** shows one hypothesized evolutionary relationship of terrestrial vertebrates.

Objectives

- **Summarize** why dinosaurs became the dominant land vertebrates. SC.D.1.4.4
- **Contrast** ectotherms with endotherms. SC.G.1.4.1
- **Identify** the dinosaurlike and the birdlike features of *Archaeopteryx.* SC.D.1.4.3
- **Summarize** why mammals replaced dinosaurs. SC.D.1.4.4

Key Terms

terrestrial
thecodont
Pangaea
ectothermic
endothermic
therapsid

Figure 11 Evolution of terrestrial vertebrates

This simplified phylogenetic tree shows one hypothesis of the relationship between early reptiles and the vertebrates that evolved from them.

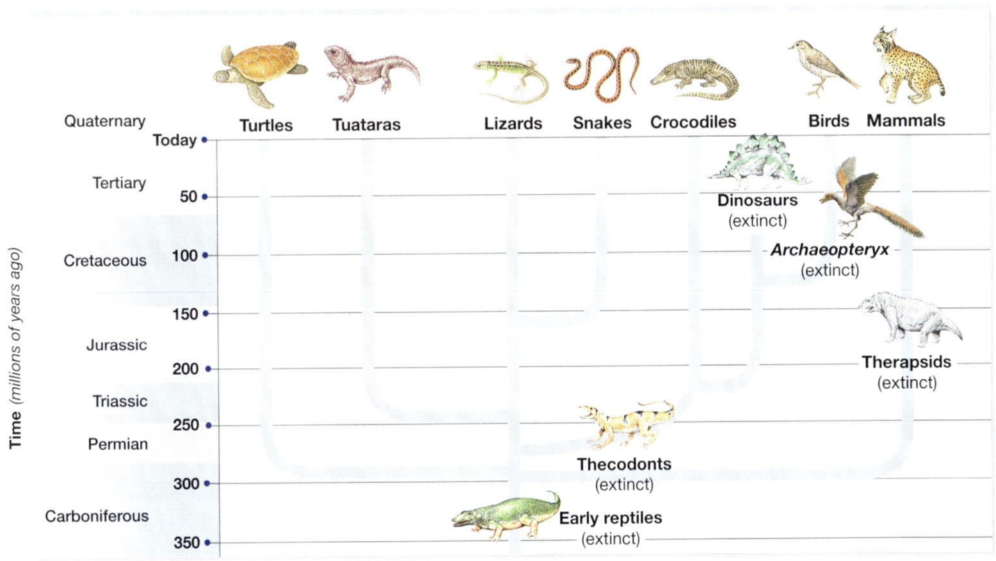

Section 2

Focus

Overview

Before beginning this section review with your students the objectives listed in the Student Edition. This section introduces students to the characteristics of reptiles that enabled them to be so successful. Students will learn about the major groups of reptiles that have lived on Earth, including the dinosaurs. They will learn about the evolution of birds, apparently from a reptilian ancestor, and their adaptations to living on land. Finally, this section will discuss the proliferation of mammals after the dinosaur extinctions, and the characteristics of living mammals.

Bellringer

Ask students to draw a picture of an imaginary, four-legged animal that spent its days on land and its nights in the water. Ask them to draw a picture of an imaginary four-legged animal that lived exclusively on land. Finally, ask them to draw a picture of an imaginary land animal that lacks legs, one that flies, and one that walks on just two legs. Have them discuss why they drew certain characteristics for each animal with other students. LS Interpersonal

Chapter Resource File

- Directed Reading BASIC
- Active Reading GENERAL

Transparencies

TR Bellringer

One-Stop Planner CD-ROM

- Reading Organizers BASIC
- Reading Strategies BASIC
- Supplemental Reading Guide *Through a Window: My Thirty Years with the Chimpanzees of Gombe* ADVANCED

Section Correlations

SC.D.1.4.3
SC.D.1.4.4
SC.G.1.4.1

Motivate

Discussion/Question — GENERAL

Ask the students who have pet reptiles to describe the pros and cons of owning their particular reptile. Ask these students what advice they would have for other students who might be interested in owning a reptile. LS Verbal

Teach

Group Activity — GENERAL

Visiting Herpetologist Invite a local herpetologist or reptile enthusiast to your class to show and discuss both exotic and local reptiles. The day before your guest visits, have each student make a list of five questions they would like to ask about reptiles. Ask your guest to bring turtles, lizards, and nonvenomous snakes, if possible. Encourage students to handle docile snakes to see that they are not "slimy." Remind students that reptiles can carry diseases, such as salmonellosis. Instruct all students who have handled the visitor's reptiles to wash their hands with soap and water. LS Kinesthetic

Transparencies

TR I16 Breakup of Pangea

Figure 12 Breakup of Pangaea. The supercontinent Pangaea began to break up during the Jurassic period. By the late Cretaceous period today's continents began to be recognizable.

Evolution of Dinosaurs

Beginning about 235 million years ago, dinosaurs dominated life on land for roughly 150 million years. They evolved from the **thecodonts,** an extinct group of crocodile-like reptiles. During their long history, dinosaurs changed a great deal, in part because the world they inhabited changed. Thus, dinosaurs represent a long parade of change and adaptation—animals that lived at different times and were adapted to very different environments.

One factor that affected dinosaur evolution was the movement of the continents, which radically altered Earth's climate. When the dinosaurs first appeared, all of Earth's landmasses were joined in a single supercontinent called **Pangaea** *(pan GEE uh),* shown in **Figure 12.** There were few mountain ranges over this enormous stretch of land, and the interior was dry. Coastal climates were much the same all over the world—quite warm, with a dry season followed by a very wet rainy season. As Pangaea broke apart, the climates of the various landmasses varied. Some species of dinosaurs could not adapt and became extinct, while new kinds flourished.

internet connect

www.scilinks.org
Topic: Pangaea
Keyword: HX4135

SCILINKS. Maintained by the National Science Teachers Association

Triassic Dinosaurs

The oldest known dinosaur fossils are in rocks from the early Triassic period, about 235 million years ago. One of the first known dinosaurs, *Eoraptor,* illustrated in **Figure 13,** was a 30 cm (1 ft) long bipedal (two-footed) carnivore. By the end of the Triassic period some 22 million years later, small, carnivorous dinosaurs were very common and had largely replaced the thecodonts.

There are at least three reasons why dinosaurs were so successful.

1. **Leg structure.** Legs positioned directly under the body provided good support for the dinosaur's body weight, enabling dinosaurs to be faster and more agile runners than the thecodonts. Dinosaurs were better able to catch prey and escape from predators.
2. **Drought resistance.** Dinosaurs were well adapted to the dry conditions found in Pangaea during the late Triassic period.

Figure 13 ***Eoraptor.*** *Eoraptor* was only 30 cm (1 ft) long—about the size of a chicken.

722

did you know?

Dinosaur Sizes One of the largest dinosaurs was *Brachiosaurus* ("arm lizard"); it reached 23 m in length and 12 m in height (about the length of two large school buses and the height of a four-story building). Fragmentary leg bones and vertebrae of even larger dinosaur species are known, but these skeletal remains are too incomplete to determine their exact size. Several of these (*Argentinasaurus* and *Amphicoelias*) might have been one and a half to two times larger than *Brachiosaurus*. The smallest dinosaurs were just slightly larger than a chicken. A larger dinosaur, *Compsognathus* ("pretty jaw"), was 1 m (3.4 ft) long and probably weighed about 2.5 kg (about 5.5 lb). These three dinosaur types all lived during the Jurassic period. *Mussaurus* ("mouse lizard") was claimed as the smallest dinosaur, but it is now known to be the hatchling of a dinosaur type that was much larger than *Compsognathus* when fully grown.

Figure 14 Sauropod

The sauropod *Brachiosaurus* stood 12.5 m (41 ft) tall and reached up to 23 m (75 ft) in length. It weighed 81,000 kg (89 tons), more than 14 African elephants or 1,500 students.

3. **Extinction of other animal species.** At the end of the Triassic period a large meteorite struck northeastern Canada. (Its impact site, the Manicouagan Crater, is still visible today.) This event might have been responsible for the great loss of animal diversity that occurred at the end of the Triassic period. Thecodonts and many other species became extinct, but the dinosaurs survived.

internet connect

www.scilinks.org
Topic: Comparing Dinosaurs
Keyword: HX4049

SciLinks. Maintained by the National Science Teachers Association

The Jurassic Period— The Golden Age of Dinosaurs

The Jurassic period is considered the golden age of dinosaurs because of the variety and abundance of dinosaurs that lived during this time. They included the largest land animals of all time, the sauropods *(SAWR oh pawdz)*. As **Figure 14** shows, sauropods had enormous, barrel-shaped bodies, heavy columnlike legs, and very long necks and tails. Sauropods were the dominant herbivores of the Jurassic period.

By the late Jurassic period, the carnivorous theropods *(THEHR uh pawdz)* were common. Theropods stood on two powerful legs and had short arms. Their large heads were equipped with sharp teeth, and each foot had sickle-shaped claws used for ripping open prey. This anatomy was well suited for rapid running and quick, slashing attacks. Theropods preyed on the large herbivorous dinosaurs and were the dominant terrestrial predators until the end of the Cretaceous period. **Figure 15** shows a representative theropod.

Figure 15 **Theropod.** The theropod *Velociraptor* was about 1.8 m (6 ft) long, about the height of a large human.

MISCONCEPTION ALERT

Humans and Dinosaurs A surprising number of people think that people and dinosaurs lived at the same time. This is not the case. After the dinosaurs died out, nearly 65 million years passed before people appeared on Earth. However, small mammals (including shrew-sized primates) *were* alive at the time of the dinosaurs. Many scientists who study dinosaurs (vertebrate paleontologists) now think that birds are direct descendants of one line of carnivorous dinosaurs, and some consider that they in fact represent modern living dinosaurs. This theory remains under discussion and shows that there is still much we don't know about dinosaurs.

Activity — ADVANCED

Trumping With Dinosaurs Using library or Internet resources, have students create dinosaur fact cards, each containing a dinosaur's name, where it was found, its height and weight, the geologic period when it lived, its method of locomotion, and the type of food it ate. Instruct them to play a trumping card game in which the stack of cards is dealt equally to all players (four to seven cards per player). The first player should look at his or her first card and choose which category they think will beat the other cards. For example, they may think they have the tallest dinosaur, or the heaviest, or the oldest. They lay this card in the center, state the challenge category, and then the other players each lay their first card down. The player whose card beats the others in that category keeps all the cards and starts the next round. The winner is the one with the most cards at the end. For categories such as "geologic period" and "type of food," the players will need to agree beforehand which characteristic will trump the others. **LS Interpersonal**

Group Activity — GENERAL

Dinosaur Adaptations Obtain a set of high quality plastic dinosaurs, such as those available through some museum shops. Divide the class into groups of three or four students, and give each group two or three dinosaur models, along with the names of the dinosaurs. Each group should appoint a recorder. The recorder should write the name of each dinosaur model at the top of a column on a sheet of paper. Each group should examine each of its models and determine how the dinosaur it represents was adapted for survival in its environment. (For example: *Brachiosaurus*—a long neck that enabled it to reach tall vegetation to eat, strong legs that enabled it to move quickly away from enemies and to obtain food.) After 10–15 minutes, have each group present their models and speculate about each dinosaur's adaptations. Allow time for other students to ask questions. **LS Logical**

Teach, *continued*

Teaching Tip — BASIC

Thermoregulation Tell students that because reptiles are ectothermic they must avoid temperature extremes. Ask students how reptiles cope with hot climates and cold winters. (Some reptiles that live in hot climates remain in the shade or come out only at night; some have a higher heat tolerance than mammals; and some that live in regions with harsh winters hibernate.) **LS Logical**

Teaching Tip

Dinosaurs Were Successful Tell students that dinosaurs have been portrayed as large, stupid, sluggish creatures that were doomed to extinction. Have students evaluate popular images of dinosaurs versus what scientists now think to be true about them. Remind students that during the time when dinosaurs and mammals coexisted, dinosaurs were able to diversify and dominate the Earth, whereas mammals remained small and mainly nocturnal. It was not until the extinction of dinosaurs that mammals began to flourish. Also remind students that dinosaurs dominated the Earth for about 150 million years. Point out that extinction has been the rule rather than the exception through history.

Real Life

Answer

Answers may vary and may include that the prospective pet owners should be aware of a reptile's needs for food, water, heat, shelter, and space.

Real Life

Reptiles make fascinating pets!

But there are drawbacks to their increasing popularity. Some reptiles are dangerous and difficult to care for, and many carry harmful bacteria in their intestinal tracts. Often reptiles offered for sale have been removed from their natural environment, causing a decline in their native populations.

Forming Reasoned Opinions

Evaluate the pros and cons of owning a reptile. What advice would you give to someone who wants a pet reptile?

Were Dinosaurs Warmblooded?

Ectothermic animals, such as today's reptiles, are animals whose metabolism is too slow to produce enough heat to warm their bodies. Such animals must absorb heat from their environment. Thus, their body temperature changes as the temperature of their environment changes. Ectotherms are sometimes referred to as "coldblooded." Mammals and birds, in contrast, are **endothermic** animals. They maintain a high, constant body temperature by producing heat internally. Endotherms are sometimes referred to as "warmblooded."

Were dinosaurs ectotherms or endotherms? For a long time it was assumed that dinosaurs were ectotherms, as modern reptiles are. However, new evidence indicates that at least some dinosaurs were endotherms. The microscopic structure of the bones of some dinosaurs resembles that of modern endotherms, both in growth rate and growth pattern. In addition, chemical analysis indicates that the bones of some Cretaceous dinosaurs probably formed under endothermic conditions. A lively debate on this question will probably continue for some time.

The Cretaceous Extinction

Toward the end of the Cretaceous period, sea levels began to fall and the climate began to cool. Many kinds of dinosaurs began to decrease in number. Then, 65 million years ago, all dinosaurs abruptly disappeared from the fossil record. No one knows for certain why this happened. Most scientists now agree that the major contributing cause was the impact on Earth of one or more meteorites or asteroids, the largest of which may have been 8 to 16 km (5 to 10 mi) in diameter. A crater 320 km (200 mi) wide off the coast of Mexico's Yucatán peninsula is very likely the site where this meteorite or asteroid collided with Earth approximately 65 million years ago, at the end of the Cretaceous period. A thin line of sediment marks the end of the Cretaceous period in rocks throughout the world. This sediment is rich in iridium, a mineral rare in Earth's crust but common in meteorites.

Scientists think that such an impact would have thrown large amounts of material from Earth's surface into the atmosphere. This material would have blocked out sunlight for a considerable period of time and created a prolonged, worldwide cold period. The endothermic birds and mammals, which were relatively small and insulated with feathers and fur, survived. The smaller ectothermic reptiles and amphibians also survived. But the dinosaurs did not survive.

Scientists are not sure why the dinosaurs did not survive this period of intense cold. Disease or competition from mammals might have led to their extinction. However, if most Cretaceous dinosaurs were endotherms, as some scientists now think they were, the dinosaurs would not have been able to survive the cold. They lacked the insulation of birds and mammals and could not lower their activity level as smaller reptiles and amphibians could.

internet connect

www.scilinks.org
Topic: Dinosaur Extinction
Keyword: HX4057
SCILINKS. Maintained by the National Science Teachers Association

Integrating Physics and Chemistry

Divide students into teams and have them analyze, review, and critique the question about whether dinosaurs were ectotherms or endotherms. Ask students to identify the strengths and weaknesses of each position using the evidence and information furnished in the textbook. If time permits, have students defend their positions in a debate format. For an alternative exercise, have students analyze, review, and critique the scientific hypothesis regarding the extinction of dinosaurs at the end of the Cretaceous period. Ask students to identify strengths and weaknesses of the meteorite hypothesis using the evidence and information furnished in the textbook.

Evolution of Birds

The earliest known bird is *Archaeopteryx* (meaning "ancient wing"), illustrated in **Figure 16.** *Archaeopteryx* was about the size of a crow and shared many features with small theropods. For example, it had a long reptilian tail. This is a feature of a dinosaur, not of a bird. *Archaeopteryx* had no breastbone to anchor its flight muscles, as do modern birds. And unlike the hollow bones of modern birds, *Archaeopteryx's* bones were solid. It also had forelimbs similar to those of a dinosaur. Because of these dinosaurlike features, several *Archaeopteryx* fossils were originally classified as dinosaurs.

The impressions of feathers on some *Archaeopteryx* fossils raised questions about their classification as dinosaurs. *Archaeopteryx* fossils appear avian (birdlike) due to the feathers on their wings and tails. *Archaeopteryx* also had other avian features, notably a fused collar bone—the wishbone—which dinosaurs did not have. Today, most biologists agree that *Archaeopteryx* is very closely related to the small dinosaur *Compsognathus*. Some biologists go so far as to classify *Archaeopteryx* as a "feathered dinosaur." Recent fossil discoveries in China, such as the fossil shown in **Figure 17,** support this opinion. However, many biologists continue to classify birds in a separate class, Aves, because of their distinct features.

The fossil record now reveals a diverse collection of toothed birds with hollow bones and breastbones adapted for flight. By the early Cretaceous period, only 15 million years after *Archaeopteryx* lived, a variety of birds with many of the features of modern birds had evolved.

Figure 16 ***Archaeopteryx.*** The *Archaeopteryx* fossil below shows clear impressions of feathers surrounding the wings. The artist's interpretation of *Archaeopteryx* shows both its dinosaurian and avian features.

Figure 17 A feathered dinosaur? This dinosaur fossil, *Protoarchaeopteryx*, found in China in 1996, shows impressions of feathers along its back.

Teaching Tip — GENERAL

Fossilized Birds Ask students for possible reasons why relatively few bird fossils have been found. (The thin, hollow bones break easily and must have near-perfect conditions to fossilize.) Show a skeleton of a bird and point out the fused pelvic bones and fused collarbones (wishbone, or furcula). Have students note the long neck and very short tail on the skeleton. Ask students to compare the bird tail with the tail of *Archaeopteryx*. Break or cut a chicken leg bone to show that the bone is hollow. **LS Logical**

Teaching Tip

Evolution of Flight Flight is an amazing accomplishment. It evolved in only three groups in the 500 million years of vertebrate history. True flight involves the flapping of wings, rather than simply soaring, parachuting, or gliding. The three groups of vertebrates that exhibit or exhibited true flight are the pterosaurs (which were reptiles), the birds, and the bats (which are mammals).

Attention Grabber

Largest Bird *Aepyornis maximus*, the elephant bird, is an extinct member of the Ratitae, a class of flightless birds that includes the ostrich, emu, kiwi, cassowary, and rhea. The elephant bird once lived in Madagascar but became extinct approximately 400 years ago. At half a ton, *Aepyornis* is believed to have been the heaviest bird that ever lived. Its eggs, with a capacity of about 7.5 L (2 gal), are the largest known—even larger than the biggest dinosaur eggs. The elephant bird was made famous when the great science fiction writer H. G. Wells published a short story entitled "*Aepyornis* Island" in 1894. While a number of *Aepyornis* eggs have been recovered over time, only very rarely are they intact. CT scanning of an unbroken egg has enabled scientists to generate images of the embryonic bones without damaging the egg. The resulting developmental data will prove invaluable in ascertaining this bird's evolutionary relationships.

Teach, continued

READING SKILL BUILDER — Basic

Reading Hint Have the class read this page and the previous page that describes the evolution of birds. Ask them to spot large words that they cannot pronounce, such as *Archaeopteryx*, Cretaceous, and *Compsognathus*. Make a list of these words on the board or overhead projector and pronounce them correctly for the class. Have the entire class repeat each word after you say it. English Language Learners
LS Auditory

Demonstration

Provide students with pictures of some flightless birds, such as the penguin, the kiwi, the cassowary, and the ostrich. All of these birds evolved from ancestors that could fly. Tell students that there are only about 40 species of flightless birds alive today, out of the more than 8,000 species of birds. Flightlessness is an adaptation that usually evolves on islands where predators are absent. However, not all flightless birds are found on islands. Ostriches live on the African savannas. Ostriches can't fly away and escape predators, but they have evolved another defense. They can use the claws on their feet as dangerous weapons when threatened. The wing bones of flightless birds are relatively small. Tell students that flying birds do have a keel on their breastbone. The keel anchors the strong muscles that move a bird's wings during flight. Flightless birds have either a flat breastbone or a greatly reduced keel.

internet connect
www.scilinks.org
Topic: Birds
Keyword: HX4024
SciLinks Maintained by the National Science Teachers Association

Modern Birds

There are more species of birds than of any other terrestrial vertebrate. All but a few of the modern orders of birds are thought to have arisen after the Cretaceous extinction. Since the impressions of feathers are rarely fossilized and since modern birds have hollow delicate bones, the fossil record of birds is incomplete. Relationships among the families of modern birds are mostly inferred from studies of the degree of DNA similarity among living birds. These studies suggest that the ostrich and its relatives belong to the oldest group of living birds. Ducks, geese, and waterfowl likely arose next and were followed by a diverse group of woodpeckers, parrots, swifts, and owls. Next came the songbirds, which include 60 percent of today's bird species. The most recent birds to appear were the more specialized orders, including many birds of prey, flamingos, and penguins. **Figure 18** shows representative modern birds.

Birds live in a wide variety of environments and differ greatly in appearance. While most birds have wings and can fly, the ostrich has wings and does not fly. The penguin's wings have been modified to be used as flippers that propel it swiftly through ocean waters. Striking colors often found in male birds help them attract a mate. All birds eat frequently because they have a high metabolic rate and typically store little body fat. The beaks of birds are adapted to the type of food the bird eats. Most birds consume small, energy-rich meals of fruits, seeds, worms, or insects. Hawks and owls are predators; they eat rabbits, rodents, and other small mammals, and sometimes snakes and lizards. Gulls and pelicans eat fish. Vultures feed on dead animals.

internet connect
www.scilinks.org
Topic: Animals of Florida
Keyword: HX6F17
SciLinks Maintained by the National Science Teachers Association

Figure 18 Modern birds

Although birds vary greatly in appearance, they all have feathers, a key defining characteristic.

did you know?

Birders The American Birding Association conducted a survey of 2,000 birders in 1996. Their findings on the characteristics of birders included the following: 66% were male and 34% were female; 98% were white; 76% were between the ages of 31 and 65 years; 31% stated that they had started birding before the age of 21 years; 34% stated their highest level of education completed as college and 42% stated the highest level as graduate school. The birders' talents included the following: 34% could identify more than 400 birds by sight without the use of a field guide; and 21% could identify more than 200 birds by their song without the use of a field guide.

Modern Reptiles

Of the 16 orders of reptiles known to have existed, only four remain today. Representatives of these orders are shown in **Figure 19.** The turtles have changed very little in structure since the time of the dinosaurs. The vast majority of living reptiles belong to the second group to evolve—snakes and lizards. Tuataras belong to the third group of surviving reptiles to evolve.

The fourth line of living reptiles, the crocodilians—crocodiles and their relatives, including the familiar alligators—appeared on Earth much later than the first three groups. Crocodilians have changed very little in more than 200 million years. Like dinosaurs, crocodilians are descendants of the thecodonts.

In some ways, such as the structure of their heart, crocodilians resemble birds far more than they resemble other living reptiles. And crocodilians are the only living reptiles that care for their young. What does this mean in terms of their relationships to other vertebrate species? Today, many biologists think that birds are direct descendants of the dinosaurs. If this is true, then crocodilians and birds are more closely related to dinosaurs, and to each other, than they are to other living reptiles. This close evolutionary relationship may account for some of their similarities.

Study TIP

Reading Effectively

Locate the thecodonts in Figure 11. Notice how the lines leading to the crocodilians and birds stem from the thecodonts. This suggests that crocodiles and birds come from a different lineage than the other reptile groups.

Figure 19 Reptiles

Of the four living orders of reptiles, the order that includes crocodiles and alligators is the most recent. Crocodiles and their close relatives differ in several ways from other reptiles.

727

Teaching Tip — BASIC

Reptile Zoogeography Tell students that no alligators or crocodiles are native to the Pacific Northwest or northeastern United States. Ask them why this might be true. (The climate in the winter is too cold for these large ectotherms to survive in these locations.) Ask students how they think the number of reptile species found in the United States compares with the number found in Canada, and why. (The United States has many more reptile species than Canada, most likely because it has a much warmer climate.) **LS Logical**

READING SKILL BUILDER — BASIC

Paired Summarizing Have students read the pages on the history of reptiles silently. Then assign partners and have one student summarize orally what they have read while the other student listens. The listener should point out inaccuracies and add ideas that were left out. Have students alternate roles and repeat this process for each of the pages. Students should work together during a final clarification process, referring to the text as needed. **LS Auditory**

Demonstration

Show students a photograph of a female alligator or crocodile near her nest. Tell students that these reptiles lay 20 to 90 eggs and then bury them in sand, mud, or plant debris. Explain that female alligators and the females of some species of crocodiles remain near the nest and protect the eggs while the embryos inside are developing. After the young hatch, the mother may help them to escape from the nest and may even carry them in her mouth to water. Some mother alligators protect their young until they are over a year old.

Attention Grabber

The Cobra One of the deadliest reptiles, the Indian King Cobra, is also one of the most fascinating to the people of India. In a tradition that goes back at least 2,000 years, Indians compare areas of energy located throughout the human body, collectively termed *kundalini* ("coiled one"), to a cobra. They believe that this energy, like its cobra counterpart, must be treated with great respect. A very popular yoga exercise position is called The Cobra.

Teach, *continued*

Demonstration — GENERAL

Mammalian milk is a rich source of food, but it also contains other substances that affect the life of young mammals. Milk has factors important to the immunity of young mammals. Bring empty cartons of milk to class, and have students read the list of nutrients on the carton labels. Finish by asking what kinds of foreign substances may be present in milk and how they might get there. (Answers include pollutants from the air and water, such as pesticides, heavy metals and fertilizers; substances that the female may have ingested, such as medications or hormones; pathogens the mother might have in her blood, such as HIV; and helpful antibodies the mother produced as a result of previous exposure to pathogens. The vitamin D in commercial cow's milk is added during the processing of the milk.) Point out to students that commercial milk is pasteurized, so it would not contain pathogens. Human milk is known to be not only nutritious for babies, but also protective against diseases because of the mother's antibodies that are present in the milk. **LS Logical**

Evolution of Mammals

Figure 20 ***Eozostrodon.*** Only 12 cm (5 in.) in total length, *Eozostrodon* is typical of the small, early nocturnal mammals.

The first mammals appeared about 220 million years ago, just as the dinosaurs were evolving from thecodonts. It is most likely that mammals were descendants of the **therapsids,** an extinct order of animals that were probably endotherms. Mammals are the only vertebrates that have fur and mammary glands. Early mammals, such as *Eozostrodon,* illustrated in **Figure 20,** were small—about the size of mice. They were insect-eating tree dwellers that were active at night.

For 155 million years, while the dinosaurs flourished, mammals were a minor group that changed little. Only five orders of mammals arose in that time, and their fossils are scarce. In the Cretaceous extinction, most animal species larger than a small dog disappeared. The smaller reptiles, including lizards, turtles, crocodilians, and snakes, survived. Mammals and birds also survived.

Diversification of Mammals

Just as after earlier mass extinctions, the stage was again set for a great evolutionary play. But the world's climate was no longer dry, as it was after the Triassic extinction. The adaptations that served reptiles well in dry climates were no longer so important. Mammals and birds became the dominant vertebrates on land. In the Tertiary period, mammals rapidly diversified, taking over many of the ecological roles once dominated by dinosaurs. Mammals reached their greatest diversity in the late Tertiary period, about 15 million years ago. At that time, tropical conditions existed over much of Earth. During the last 15 million years, world climates have changed, and the area covered by tropical habitats has decreased. As a result, the number of mammalian species has declined.

internet connect
www.scilinks.org
Topic: Discovering Florida's Past
Keyword: HX6F02
SCILINKS. Maintained by the National Science Teachers Association

Ice Age Mammals Today, all very large land animals are mammals. However, many large land mammals existed during the last ice age (about 2 million to 10,000 years ago). At that time, many species of enormous mammals such as the Irish elk, illustrated in **Figure 21,** roamed Earth.

Figure 21 **Irish elk.** The Irish elk was actually a species of large deer that lived throughout Europe, northern Africa, and northern Asia.

728

did you know?

The Amazing Platypus The platypus was thought to be a hoax when its discovery was first announced. When the first platypus skin was sent back to London from Australia in 1799, zoologists thought that this creature with a duck-like bill and beaver-like tail must surely be a joke. After careful investigation, the zoologists were amazed to discover that they were examining an extraordinary animal, one like no other on Earth.

Giant ground sloths lived then and weighed three tons, as large as a modern elephant. Enormous vegetarian cave bears populated Europe during the ice ages. Large lionlike saber-toothed cats hunted with jaws that opened an incredible 120 degrees to allow the cat to drive its huge upper pair of saber teeth into prey. The shoulders of Irish elk, shown in Figure 21, were over 2.1 m (7 ft) high, and the span of the antlers could be up to 3.65 m (12 ft). All of these giant mammals are now extinct. Today only elephants approach the size of the enormous mammoths and giant camels that roamed Earth then.

Later Mammals Today, there are more than 4,500 species of mammals, and they inhabit virtually every habitat on Earth. They are found in jungles, deserts, and on polar ice. They live in fresh-water streams, and deep in the ocean. They even populate the air—bats are among the most successful of all mammal groups. Mammals range in size from the blue whale, which may weigh as much as 136,000 kg (150 tons), to the tiny Kitti's hognose bat, weighing about 1.5 grams (less than 0.1 oz). Most mammals are covered with a dense coat of fur, but some have little fur. Whales have only a few hairs.

internet connect

www.scilinks.org
Topic: Early Mammals
Keyword: HX4063

SCiLINKS. Maintained by the National Science Teachers Association

FORENSICS BIOWatch

If Bones Could Talk

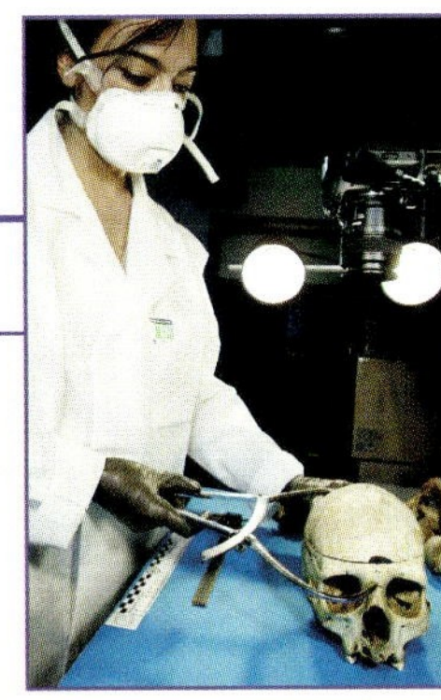

When a murder victim has been dead for weeks, months, or years, clues can be few and far between. In these cases, homicide investigators often seek the aid of forensic physical anthropologists. Physical anthropology is the scientific study of human variation, classification, and evolution. Forensic anthropologists use their knowledge of skeletal anatomy and biology to help solve crimes.

Silent Witness

Long after soft body tissue has decayed, bones remain, bearing silent witness to their owner's fate. In the hands of an experienced forensic anthropologist, telltale evidence will often emerge. If a victim suffered a violent death (for example, a hammer blow to the head), an anthropologist can analyze signs of physical injury. Using a microscope, the anthropologist can see precise details of tool marks, which may suggest a likely murder weapon. Such details can also help investigators distinguish saw marks, for example, from postmortem damage caused by a gnawing rodent. Based on the condition of the body, forensic anthropologists can also estimate the time that has passed since death.

Who Is It?

Many times a criminal investigation will seek to determine the victim's identity. Skeletal remains can reveal the victim's age, sex, and height. Based on patterns of muscle attachment, a forensic anthropologist can even determine whether the victim was right- or left-handed. Computer-assisted modeling techniques let anthropologists create a possible "face" that fits the underlying bone structure of the skull. A law enforcement agency or a member of the public may recognize this image. The anthropologist can also adjust the image to "age" the subject, a technique often used when investigating child kidnappings.

729

Trends in Marine Biology

Marine Mammals When it comes to diving deeply, marine mammals as different as seals and blue whales employ the same physiological adaptations to travel the maximum distance with minimum effort. Using miniature video system recorders carried by the animals themselves, scientists observed that diving mammals begin their descents into water with a few powerful strokes. They then continue down mostly in a glide, which reduces their demands for oxygen.

These mammals also share an anatomical feature that makes a gliding descent possible and protects them from getting "the bends." In humans and other land animals, air gets trapped as the lungs are compressed in a dive, forcing nitrogen into the bloodstream. This can produce a painful and often life-threatening condition known as "the bends." The lungs of diving mammals collapse progressively as water pressure increases, so that air is forced out and into the upper part of the respiratory system. As the increasing pressure compresses the air in the animals' respiratory system into a smaller and smaller volume, they become less buoyant.

Activity — GENERAL

Mammal Comparisons Have students make a chart comparing monotremes, marsupials, and placental mammals. The chart headings should be Type of Mammal, Examples, Where Embryo Develops, and Food Source for the Young. (Sample answer: Monotreme, duck-billed platypus, egg outside mother's body, milk; Marsupial, kangaroo, pouch on mother's body, milk; Placental, human, within mother's body, milk.) LS **Visual**

BIOWatch — GENERAL

If Bones Could Talk

Teaching Strategies

Explain to students that forensic anthropology plays an important role in investigations of international human rights violations. Examination of mass graves can reveal irrefutable scientific evidence of torture and assassinations. Forensic anthropologists have helped establish the cause of death for the thousands of citizens who disappeared during the last military dictatorship in Argentina, as well as those who died in mass killings in Rwanda and the former Yugoslavia.

Discussion

Explain why a forensic anthropologist is an important member of a criminal investigative team? (A forensic anthropologist can determine the age, sex, and height of the victim. If a victim was murdered, a forensic anthropologist may be able to find evidence suggesting a particular murder weapon. In some cases, the anthropologist can also construct a hypothetical "face" based on underlying bone structure.)

Forensics Labs Appendix

Turn to pp. 1048–1071 for forensics lab activities and more student information on forensics techniques.

Modern Mammals

The first group of mammals to evolve laid eggs, as did their early ancestors. The only surviving direct descendants of these early mammals are two species of echidna *(ee KIHD nuh)* and the duck-bill platypus, shown in **Figure 22.** These egg-laying mammals are in a group of their own called the monotremes.

Mammals other than monotremes show one of two patterns of development. Both of these groups give birth to live young, but their embryos develop in different ways.

1. **Marsupials.** Marsupial *(mahr SOO pee uhl)* mammals include kangaroos, also shown in Figure 22, opossums, koalas, and wombats. Their young are born at a very immature stage and complete their development in their mother's pouch (called a marsupium).
2. **Placentals.** Placental mammals include dogs, cats, horses, sheep, gorillas, humans, and most of the other mammals you are familiar with. Placental mammals, such as the goat seen in Figure 22, develop within their mother's body and are nourished by an organ called the placenta. In a later chapter, you will learn more about the mammals.

Figure 22 Mammals. There are three types of mammals—placentals, marsupials, and monotremes—each represented below.

Section 2 Review

1. **Describe** two key adaptations that allowed reptiles to live and reproduce on land. SC.D.1.4.4
2. **Summarize** the factors that contributed to the dinosaurs' success on land. SC.D.1.4.4
3. **Compare** ectothermic and endothermic animals. SC.G.1.4.1
4. **Relate** changes in the world's climate to the increase in numbers of mammals. SC.D.1.4.3
5. **Critical Thinking Forming Reasoned Opinions** Evaluate the change in classification of *Archaeopteryx* from dinosaur to bird. SC.H.1.4.3
6. **FCAT Prep** At the end of the Cretaceous period 65 million years ago, natural selection resulted in the diversification of what group?

   A mammals **C** thecodonts
   B sauropods **D** theropods
   SC.D.1.4.4

730

Close

Reteaching — BASIC

Tell students to make two columns on a sheet of paper. Have them label the first column Reptiles and the second column Birds. Then ask them to list as many key reptilian and avian facts or characteristics as possible under the appropriate headings. Tell students to write on only one side of the paper. Students should then compare their lists with those of a partner and add to their lists any important characteristics they missed. Ask students to cut out each of the facts, leaving the column headings intact. Then have them turn over and shuffle the pieces of paper. Each student should turn his or her partner's facts face-up, one-by-one, and place the fact under the appropriate column heading. When finished, partners should check each other's work.

Quiz — GENERAL

1. From what other group of vertebrates are birds thought to have evolved? (Birds are thought to have evolved from dinosaurs.)
2. What size were most of the animals that survived extinction of the dinosaurs? (They were relatively small.)

Alternative Assessment — GENERAL

Have students look at **Figure 11.** Have them trace the line that leads from early reptiles to today's birds and mammals. Ask students to explain in their own words the events that led from the early reptiles to modern birds and mammals.

Section Review Correlations

SC.D.1.4.3: Item 4
SC.D.1.4.4: Items 1, 2, 6
SC.G.1.4.1: Item 3
SC.H.1.4.3: Item 5

Answers to Section Review

1. The watertight egg permitted eggs to be laid on land without the embryos drying out. Watertight skin prevented the loss of body fluids.
2. Legs positioned directly under the body permitted dinosaurs to be faster and more agile. Dinosaurs were well adapted to living in the dry conditions. Many species became extinct by the end of the Triassic period, which allowed the dinosaurs to take over many ecological niches.
3. Ectotherms must obtain heat from the environment because their metabolism is too slow to generate sufficient internal heat. Endotherms have a rapid metabolism, which produces enough heat to keep their bodies warm.
4. At the end of the Cretaceous period, the world began to cool. Mammals, with their high metabolic rate and insulating fur, could flourish. Dinosaurs could not lower their activity level. Their lack of insulation and high metabolic rate made it difficult to survive.
5. Answers will vary, but should indicate that this was due to new fossils with feathers.
6. **A.** Correct. **B.** Incorrect. Sauropods became extinct. **C.** Incorrect. Thecodonts were extinct. **D.** Incorrect. Theropods became extinct.

Evolution of Primates

Section 3

Florida Sunshine Standards/FCAT

SC.D.1.4.3 Changes in Earth's climate, geological activity, and life forms may be traced and compared.
SC.D.1.4.4 Earth's systems and organisms are the result of a long, continuous change over time.
SC.F.1.4.2 Body structures are uniquely designed and adapted for their function.
SC.G.1.4.1 There is great diversity and interdependence among living things.

Characteristics of Primates

Fossil evidence indicates that small, insect-eating mammals with large eyes and small, sharp teeth lived about 80 million years ago, during the age of dinosaurs. These ancient mammals were the ancestors of the first primates. A **primate** is a member of the mammalian order Primates, which includes prosimians, monkeys, apes, and humans. The first primates evolved about 50 million years ago. These animals had two features that enabled them to stalk and capture prey in the branches of trees.

1. **Grasping hands and feet** The grasping hands and feet of primates enable them to cling to their mothers when they are young, grip limbs, hang from branches, and seize food.
2. **Forward orientation of the eyes** Unlike the eyes of their ancestors, which were located on the sides of the head, the eyes of primates are positioned at the front of the face. This forward placement of the eyes produces overlapping binocular vision that enables the primate brain to judge distance more precisely (depth perception). The ability to judge depth is very important for an animal that leaps from branch to branch high above the ground. Some other mammals have binocular vision, but only primates have both binocular vision and grasping hands.

Prosimians

According to the fossil record, the modern primates that most closely resemble early primates are the prosimians. A **prosimian** is a member of a group of mostly night-active primates that live in trees. Modern prosimians include lorises, lemurs, and tarsiers, such as the one shown in **Figure 23.** Fossil evidence indicates that prosimians were common about 38 million years ago in North America, Europe, Asia, and Africa.

Figure 23 Prosimian. The tarsier, *left*, has bendable, clawed fingers and toes and forward-facing eyes, which are key adaptations for life in the trees.

Objectives

- **Name** two unique features of primates. SC.F.1.4.2
- **Contrast** prosimians with monkeys. SC.G.1.4.1
- **Distinguish** monkeys from apes. SC.G.1.4.1
- **Describe** the evolutionary relationship between humans and apes. SC.D.1.4.3
- **Identify** the evidence that indicates human ancestors walked upright before their brains enlarged. SC.D.1.4.4

Key Terms

primate
prosimian
diurnal
hominid

Section 3

Focus

Overview

Before beginning this section review with your students the objectives listed in the Student Edition. This section introduces the characteristics of the earliest primates, how upright-walking primates evolved later, and the variety of hominid types that have been found on Earth. Students will also learn about the characteristics that distinguish primates from other vertebrates, and those that distinguish hominids from other primates.

Bellringer

Ask students to make a sketch of an animal that has the characteristics that the first primate would have had. Ask them to describe or sketch what an early human would have looked like.

Motivate

Discussion — GENERAL

Before students begin reading, explain that the focus of this lesson is the evolution of primates. Work with the class to create a list of those characteristics that primates share with other living mammals, and those characteristics unique to primates. LS Logical

Chapter Resource File

- Directed Reading BASIC
- Active Reading GENERAL

Career

Anthropologist Physical anthropologists, also called biological anthropologists, are concerned with humans and other primates as physical organisms, and with their places in the scheme of biological evolution. These scientists deal with the classification of early forms of humans, the physical differences between species, human genetics, modes of physiological adaptation, and reaction to different physical environments. Physical anthropologists may work for international or private organizations, medical or research establishments. Most work for universities and museums.

Transparencies

TR Bellringer

One-Stop Planner CD-ROM

- Reading Organizers BASIC
- Reading Strategies BASIC

Section Correlations

SC.D.1.4.3
SC.D.1.4.4
SC.F.1.4.2
SC.G.1.4.1

Teach

Teaching Tip — BASIC

Human Hands Instruct students to look at their own hands and observe how their fingers bend and how their nails are attached to their fingers. Ask students to name some uses for hands and some uses for paws. (Uses for hands include: grasping, climbing, pinching, holding, pushing, pointing, waving, slapping, using tools. Uses for paws include: running, hopping, digging, climbing, pushing.) If students have mammalian pets at home, have them compare their hands to their dog or cat's paws. Discuss the differences in class. LS **Visual**

Using the Figure

Direct students' attention to **Figure 24.** Tell them that many people have the misconception that, according to evolutionary theory, humans descended from gorillas or chimpanzees. Remind students that branch points on an evolutionary tree indicate where a common ancestor of two groups diverged. Show students on the phylogenetic tree where the branch leading to modern gorillas split off from the branch leading to modern humans. This split occurred close to 7 million years ago. Show students on the tree where the branch leading to modern chimpanzees split off from the branch leading to modern humans. This split occurred close to 4 million years ago. Tell students that the most recent common ancestor in both of these cases didn't look like either one of its modern day descendants.

Nonhuman Primates

WORD Origins
The word *diurnal* is from the Latin *diurnis*, meaning "daily." Knowing this makes it easier to remember that diurnal primates are active during the day.

About 36 million years ago, a revolution occurred in how primates live. Many primate species became diurnal. **Diurnal** *(die UR nuhl)* animals are active during the day, and they sleep at night. The evolution of a diurnal pattern gave primates more opportunities to feed and enabled them to better detect predators. Modern day-active primates include monkeys and apes.

Monkeys

Feeding mainly on fruits and leaves, monkeys were among the first primates to have opposable thumbs. An opposable thumb—such as your own—stands out at an angle from the other fingers and can be bent inward toward them to hold an object. This gives the hand a greatly increased level of ability to manipulate objects.

Apes

Apes, which share a common ancestor with monkeys, first appeared about 30 million years ago. Modern apes include gibbons, orangutans, gorillas, and chimpanzees. Apes have larger brains with respect to their body size than monkeys, and none of the apes have tails.

The phylogenetic tree shown in **Figure 24** represents the theoretical evolutionary relationships between modern apes and humans. DNA analysis has shown that the genes of humans and chimpanzees are remarkably similar. Indeed, human and chimpanzee DNA nucleotide sequences may differ by as little as 5 percent. For example, the 287 amino acids that make up two kinds of chains of protein in human hemoglobin are identical to the amino acids in chimpanzee hemoglobin.

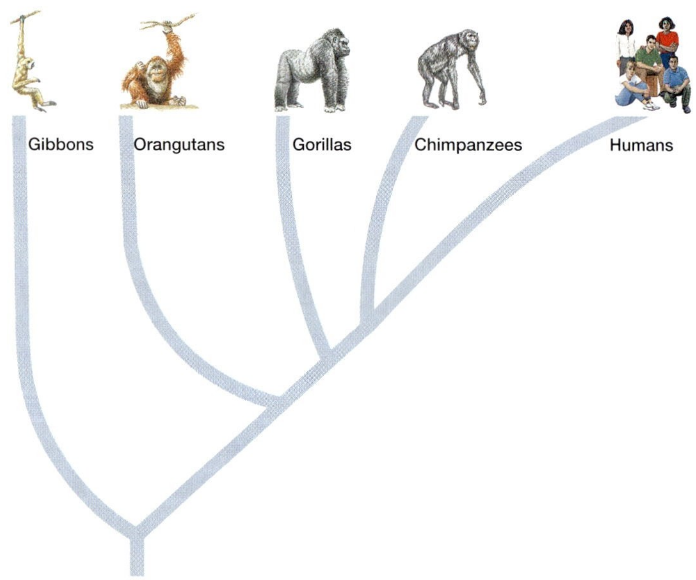

Figure 24 Phylogenetic tree of apes and humans. The evolutionary group that led to modern gibbons diverged earlier than other primate groups.

HISTORY CONNECTION

Louis Leakey (1903–1972) was a tireless promoter of the study of human origins, and exerted tremendous impact on the prevailing conception of early humans. Mary Leakey (1913–1996) was one of the world's most renowned hunters of early human fossils, credited with many discoveries that have changed the way scientists conceive human evolution. For more than 30 years, Richard Leakey (b. 1944) has made international headlines for his anthropological, political, and environmental work in Kenya. Heir to the legacy of his parents, Richard Leakey has been credited with some of this century's most successful paleoanthropologic finds. Maeve Leakey (b. 1942) impressed the world with her 1999 discovery of a 3.5 million-year-old skull and partial jaw believed to belong to new branch of hominids. Louise Leakey (b. 1972) represents the third generation of paleontologists in her renowned family. Her most significant contribution thus far has been her participation in the 1999 discovery of *Kenyanthropus platyops*.

Early Hominids

Hominids *(HAHM ih nihds)* are primates that walk upright on two legs. Hominids are members of the group that led to the evolution of humans. According to the fossil record, hominids first appeared on Earth about 5 million to 7 million years ago. All early primates walked on all four limbs. What, then, caused hominids to stand up and walk on two legs? Hypotheses about this aspect of human evolution differ. Fifteen million years ago, the world's climate began to cool, and the great forests of Africa were largely replaced by savannas (treeless plains). Natural selection may have favored primate forms adapted to living on the ground.

Australopithecines

The early hominids best represented by fossil finds belong to the group known as australopithecines *(aw stray loh PIHTH uh seenz)*. Australopithecines belong to the genus *Australopithecus*. **Figure 25** shows Lucy, the fossil remains of an australopithecine that lived more than 3 million years ago. Their fossil remains indicate that australopithecines walked upright on two legs; thus they are classified as hominids. As **Table 1** shows, the skeletons of australopithecines differ from those of modern apes in several ways.

The brains of australopithecines had a slightly greater volume, relative to body weight, than the brain of an ape. Some australopithecine species weighed about 18 kg (40 lb) and were approximately 1.1 m (3.5 ft) tall, about the size of a small chimpanzee. Other australopithecine species were larger, weighing more than 45 kg (100 lb) and standing more than 1.5 m (5 ft) tall. Their brains, 400 cm^3 in volume, were generally as large as those of modern chimpanzees. Australopithecine brains were much smaller, however, than the brains of modern humans, which are about 1,350 cm^3 (83 in.^{3}).

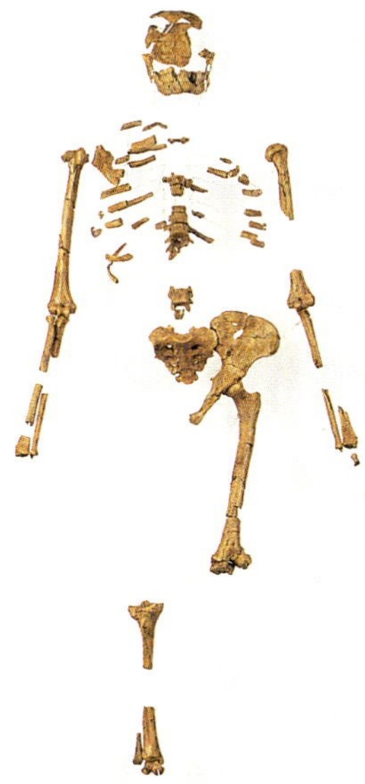

Figure 25 Lucy. Lucy's leg bones indicate that she walked upright. She stood about 1 m (3 ft) tall.

Table 1 Comparison of Gorilla and Australopithecine Skeletons

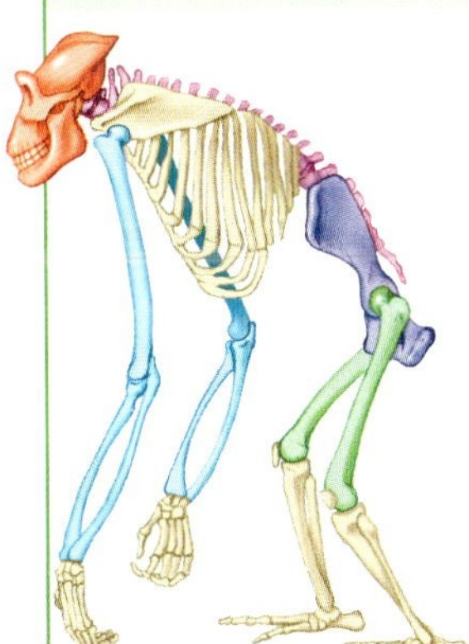

Gorilla	Australopithecine
Skull atop C-shaped spine; spinal cord exits near rear of skull	Skull atop S-shaped spine; spinal cord exits at bottom of skull
Arms longer than legs; arms and legs used for walking	Arms shorter than legs; only legs used for walking
Tall and narrow pelvis, allowing the body weight to shift forward	Bowl-shaped pelvis, centering the body weight over the legs
Femurs (thighbones) angled away from pelvis when walking upright	Femurs angled inward, directly below body to carry its weight

Skull   Spine   Arms   Pelvis   Femurs

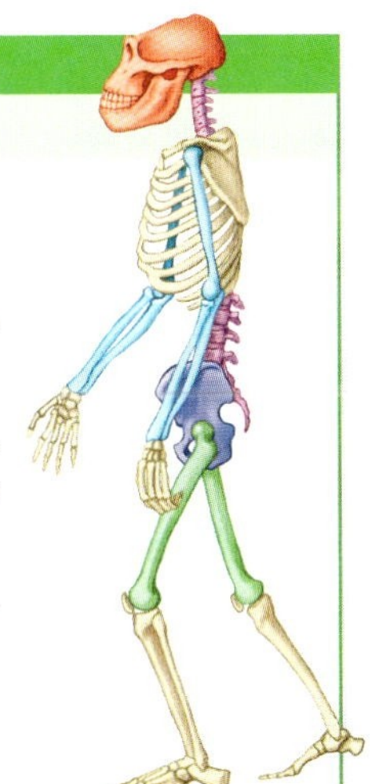

Teaching Tip — GENERAL

Which Hemisphere? Ask students why it is unlikely that the first hominids evolved in North or South America. (Students should remember that evidence indicates that humans have evolved from the line that gave rise to apes. Because apes never lived in North and South America, it is unlikely that the hominids originated there.) **LS Logical**

SKILL BUILDER — GENERAL

Graphing Have students draw a phylogenetic tree of primates using the information they used for the Activity: Primate Graphic Organizer below. Their trees should show a primate common ancestor at the bottom of the tree and a modern prosimian at the end of a long main branch. The primate common ancestor should be labeled with the traits "grasping hands and feet" and "binocular vision." A line for monkeys, apes, and all hominids should branch off the main branch and should be labeled with the trait "opposable thumb." A line for all hominids should branch off the secondary branch and be labeled with the traits "upright walking" and "large brain." Finally, a line for the genus *Homo* should branch off the tertiary branch and be labeled with the trait "tool user." **LS Visual**

Activity — GENERAL

Primate Graphic Organizer Have students prepare a Graphic Organizer in the format of a table, using the organisms and traits shown below. After reading about each type of primate, have students check any and all appropriate boxes in the table to indicate the presence of a specific characteristic. **LS Verbal**

Transparencies

TR I41 Comparison of Gorilla and Australopithecine Skeletons

TR I42 Hominids in the Fossil Record

Graphic Organizer

Use this graphic organizer with **Activity** on this page.

Group	Grasping	Binocular	Opposable	Upright	Large	Tool
Prosimian	✓	✓				
Monkey	✓	✓	✓			
Ape	✓	✓	✓		✓	
Early hominid	✓	✓	✓	✓	✓	
Homo species	✓	✓	✓	✓	✓	✓

Close

Reteaching — BASIC

Use a model or picture of a human skeleton to point out the parts of a skeleton that differ between gorillas and hominids. Use **Table 1,** on the previous page, and have students determine whether each description applies to gorillas or hominids. LS **Visual**

Quiz — GENERAL

1. What advantage do forward-facing eyes confer to those animals that possess this trait? (Forward-facing eyes make it possible for animal to have binocular vision, which results in the formation of a three-dimensional image and better depth perception.)
2. How are apes different from monkeys? (Apes have larger, more developed brains than monkeys, and none of the apes have tails.)

Alternative Assessment — GENERAL

Read the following passage from Darwin's *Descent of Man.* "Thus we have given to man a pedigree of prodigious length... The world... appears as if it had long been preparing for the advent of man: and this, in one sense, is strictly true, for he owes his birth to a long line of progenitors. If any single link in this chain had never existed, man would not have been exactly what he is now." Ask students to interpret the passage. (The evolution of humans can be traced through a long chain to the origin of the first living things. If any ancestors in this chain had not evolved, modern humans would not be the same as they are.) LS **Logical**

Section Review Correlations

SC.D.1.4.3: Item 4

SC.F.1.4.2: Items 1, 5

SC.G.1.4.1: Items 2, 3

WORD Origins

The word *ardipithecus* is from the Afar (the people who live in the area where this fossil was found) word *ardi*, meaning "ground" or "floor," and from the Greek word *pithecus*, meaning "ape." Knowing this makes it easier to remember that *Ardipithecus* is the genus name of one of the oldest known hominids.

Other Early Hominids

In the past few years, scientists have discovered fossils of hominids that lived at the same time or earlier than *Australopithecus*. The relationship between early hominid species, shown in **Figure 26,** is uncertain. These early fossils indicate that the hominid line is an old one, extending back past 6 million years. The discovery in 2002 of a 6 million to 7 million year old fossil in central Africa, is creating lively debate among scientists about the origins of the hominids. This new species, *Sahelanthropus tchadensis*, appears at the same time more primitive and more modern than the australopithecines. Some scientists think that *Sahelanthropus* represents a species at the root of the family tree of hominids. The modern features of this very old fossil also suggest to some scientists that australopithecines may not have been ancestral humans, but rather a group of species that left no descendants.

Until more early hominid fossils are discovered, the historical picture remains too unclear to draw any firm conclusion about true lines of descent. What is clear is that over the last 6 million years, a variety of different hominids have existed, with more than one species living at one time. As new fossils are found, our picture of the root of the human phylogenetic tree may come into sharper focus.

Figure 26 Hominid evolution. Several species of hominids arose and then died out.

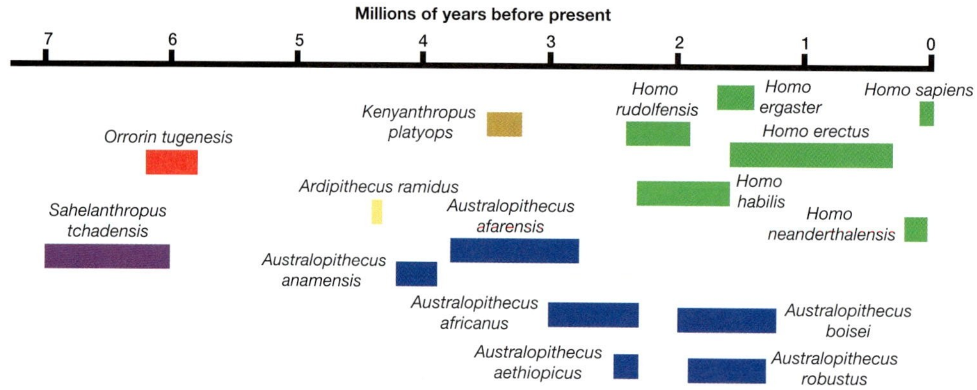

Section 3 Review

1. **Describe** two unique features of primates. SC.F.1.4.2
2. **Describe** one characteristic that distinguishes prosimians from monkeys and one characteristic that distinguishes monkeys from apes. SC.G.1.4.1
3. **Identify** and explain the evidence that closely links humans to chimpanzees. SC.G.1.4.1
4. **Critical Thinking Comparing Structures** Summarize the differences between the skeleton of an ape and the skeleton of an australopithecine. SC.D.1.4.3
5. **FCAT Prep** What evidence would help scientists establish whether or not a primate was bipedal?
   A skull size
   B pelvis shape
   C femur weight
   D number of ribs
   SC.F.1.4.2

734

Answers to Section Review

1. Primates have grasping hands and feet, and forward orientation of the eyes.
2. Monkeys have opposable thumbs, while prosimians do not. Prosimians are mainly nocturnal, while monkeys are mainly diurnal. Apes do not have tails and they have larger brains than monkeys have.
3. DNA nucleotide sequences in chimpanzees and humans may differ by only about 5%. 287 amino acids in human hemoglobin are identical to those amino acids in chimpanzee hemoglobin.
4. The ape has a C-shaped spine, and the spinal cord exits near the rear of the skull. The australopithecine spine is S-shaped, and the spinal cord exits at the bottom of the skull. Ape arms are longer than legs, and are used for walking. Australopithecine arms are shorter than legs, and they were bipedal.
5. **A.** Incorrect. Skull size would not provide evidence of bipedalism. **B.** Correct. **C.** Incorrect. Femur weight would not be evidence. **D.** Incorrect. Number of ribs would not be evidence.

The Genus *Homo*

Section 4

Florida Sunshine Standards/FCAT

SC.D.1.4.3 Changes in Earth's climate, geological activity, and life forms may be traced and compared.

SC.G.1.4.1 There is great diversity and interdependence among living things.

Homo habilis

Our genus, *Homo*, is composed of at least three species. Of these three, modern humans, *Homo sapiens sapiens*, is the only surviving species. The first members of the genus *Homo* appeared on Earth more than 2 million years ago. In the early 1960s, hominid bones were discovered near the site where *Australopithecus* fossils had been found. Stone tools were scattered among the bones. Scientists disagreed about whether this fossil was that of an early human or an australopithecine. Reconstruction of the bones of the skull indicated that the volume of the brain was about 640 cm^3 (39 in.^{3}). This is larger than the australopithecine brain volume of 400 to 550 cm^3 (24 to 34 in.^{3}).

Because of its association with tools, this hominid was named *Homo habilis*. The Latin word *homo* means "man," and the Latin word *habilis* means "handy." Bones discovered in 1987 indicate that *Homo habilis* stood about 1.2 m (4 ft) tall. Fossils indicated that *Homo habilis* lived in Africa for about 500,000 years and then became extinct.

Objectives

- **Compare** *H. habilis* with australopithecines. SC.G.1.4.1
- **Describe** the characteristics of *Homo erectus*. SC.G.1.4.1
- **Describe** the evidence that suggests that *H. sapiens* evolved in Africa. SC.D.1.4.3
- **Compare** Neanderthals with modern humans. SC.G.1.4.1

Key Terms

None

Figure 27 Richard Leakey and *Homo habilis* skull

Homo habilis lived in eastern Africa.

Richard Leakey (seated, far right) and his co-workers found one of the most complete *Homo habilis* skulls to date.

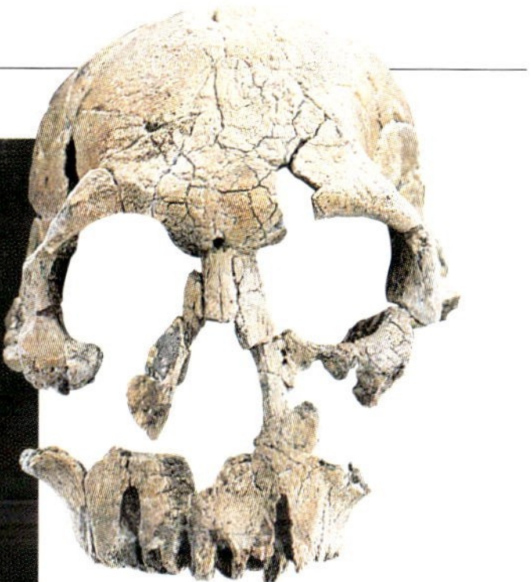

Homo habilis had a brain volume of about 700 cm^3 on the average and many of the characteristics of modern human skulls.

Section 4

Focus

Overview

Before beginning this section review with your students the objectives listed in the Student Edition. This section introduces students to the characteristics of the earliest known species of humans, *Homo habilis*. Students will also learn about *H. erectus*, and where it lived.

Bellringer

Ask students to sketch a picture of a setting where early humans might have found themselves. They can include drawings of figures if they choose. (Their settings should be rather primitive, but should indicate strong human tendencies for creating settlements.)

Motivate

Activity — GENERAL

Hominid Sizes Split the class into five groups to make life-size hominid outlines on butcher paper. Tape several large sheets of butcher paper on the wall. Have one group make a grid that is marked off every 10 cm from the floor to 200 cm up the wall. The other four groups should pick *Australopithecus afarensis* (1.1 m tall), *Homo habilis* (1.2 m tall), *H. erectus* (1.5 m tall), or *H. sapiens* (1.7 m tall) and create a life-size paper outline. Have students label each species and attach it to the grid. LS **Visual**

Trends in Linguistics

Sign Language Primates are largely visual animals, possibly as a result of a common adaptation to visual predation. Vision is also much more highly developed than any other sense, including hearing, in humans. Nonhuman primates have much better control over hand movements than over vocalization, which consists mainly of emotional expressions. This perhaps explains why attempts to teach chimpanzees a version of sign language have been much more successful than attempts to teach them human vocal language.

Chapter Resource File

- Directed Reading BASIC
- Active Reading GENERAL

One-Stop Planner CD-ROM

- Reading Organizers BASIC
- Reading Strategies BASIC
- Supplemental Reading Guide *Origin of Species* ADVANCED

Section Correlations

SC.D.1.4.3
SC.G.1.4.1

Teaching Tip

Tool Use Inform students that the tools made by *Homo habilis* were crude by modern standards and included only a few varieties. Scientists have found about a dozen types used by *H. erectus*. Many of these tools had characteristic shapes. For example, teardrop-shaped hand axes are often found at sites where *H. erectus* lived.

Demonstration

Increasing brain size is an important aspect of human evolution. To help students appreciate how much hominid brain size has increased, get seven large, clear jars or bottles, label each container for a hominid species, and fill each with the volume of water equal to the hominid's estimated brain size. Brain sizes, measured as cranial capacities, are given in the following table:

Cranial Capacity of Hominids

Species Name	Cranial Capacity
Australopithecus afarensis	375-400 cm^3 (22.5-24 $in.^3$)
Australopithecus africanus	400-500 cm^3 (24-30 $in.^3$)
Australopithecus boisei	650 cm^3 (39 $in.^3$)
Australopithecus robustus	530 cm^3 (32 $in.^3$)
Homo habilis	700 cm^3 (42 $in.^3$)
Homo erectus	1000 cm^3 (60 $in.^3$)
Homo neanderthalensis	1500 cm^3 (90 $in.^3$)
Homo sapiens (modern)	1350 cm^3 (81 $in.^3$)

LS Visual

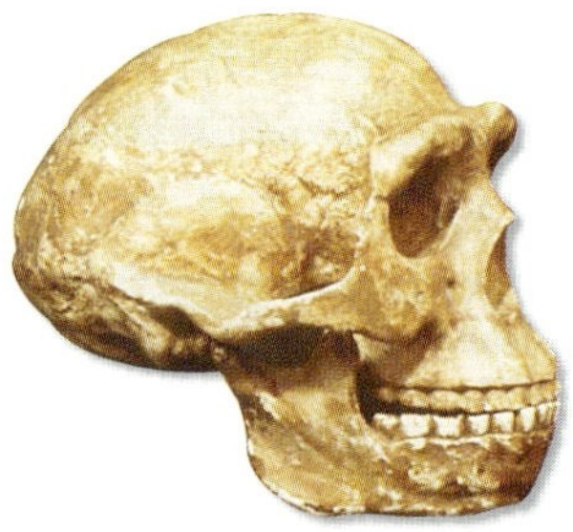

Figure 28 ***Homo erectus.*** *Homo erectus* had large brow ridges and smaller teeth than *Homo habilis*.

Homo erectus

Our understanding of *Homo habilis* is limited because it is based on only a few fossil specimens. In contrast, scientists have found many fossils of the species that replaced *Homo habilis*—the species *Homo erectus*.

The first *Homo erectus* specimen found, called Java Man, was discovered in 1891 on the island of Java, in southeast Asia. *Homo erectus* was larger than *Homo habilis*—about 1.5 m (5 ft) tall. *Homo erectus* also had a large brain of about 1,000 cm^3 (60 $in.^3$). Java Man may be more than about 700,000 years old. Similar fossils found in China are thought to be as old as 500,000 years. In 1976, a *Homo erectus* skull that may be 1.5 million years old was discovered in eastern Africa. **Figure 28** shows a fossil *Homo erectus* skull.

Homo erectus evolved in Africa and migrated into Asia and Europe, as shown in **Figure 29.** *Homo erectus* probably lived in groups of 20 to 50 individuals. They hunted large animals, used fire, and made both stone and bone tools. *Homo erectus* survived for more than 1 million years. The species disappeared about 200,000 years ago, as early modern humans emerged. Most scientists think that *Homo erectus* was the direct ancestor of our species, *Homo sapiens*.

Figure 29 **Hominid migration**

Homo erectus evolved in Africa and migrated to Asia and Europe. Later, *Homo sapiens* appeared in Africa and migrated to Europe and Asia.

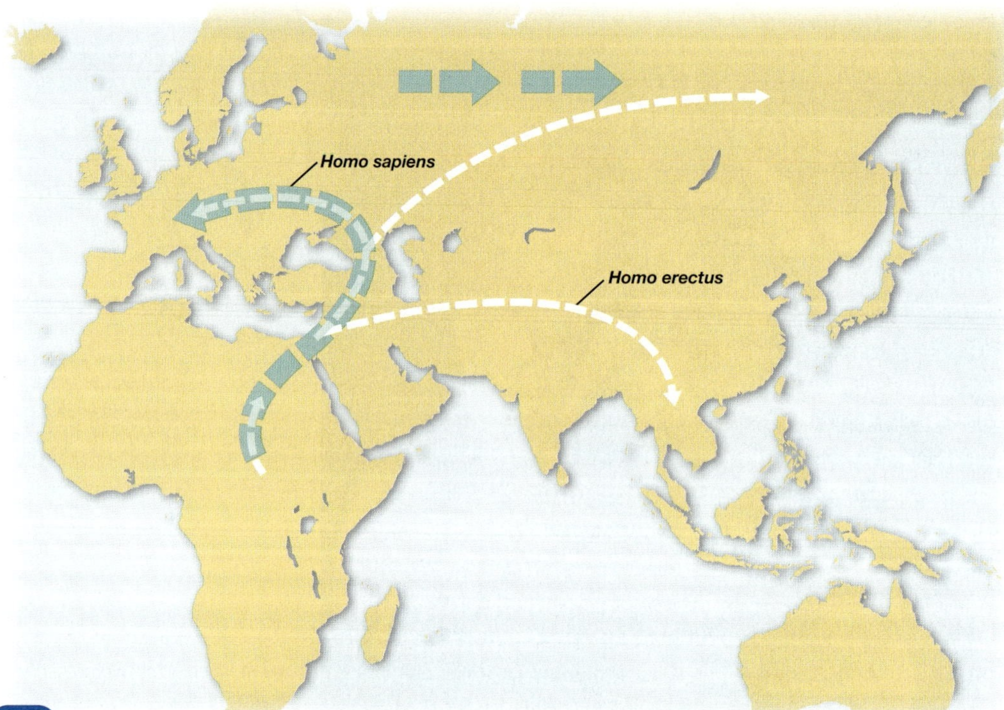

736

Transparencies

TR Bellringer

TR I43 Theory of Hominid Migration

Homo sapiens

Homo sapiens is the only surviving species of the genus *Homo*. The name *Homo sapiens* is from the Latin *homo*, meaning "man," and *sapiens*, meaning "wise." *Homo sapiens* is a newcomer to the hominid family. *Homo sapiens* has not existed as long as *Homo erectus* did. Early *Homo sapiens* left behind many fossils and artifacts, including the first known paintings, such as the one shown in **Figure 30.**

Figure 30 Cave painting. This thundering herd of horses was painted on the wall of a cave in Lascaux, France. The cave paintings may be about 20,000 years old.

African Origins?

In the past, some scientists proposed that independent *Homo erectus* groups living in Africa, Europe, and Asia interbred and that *Homo sapiens* thus arose as a new species at the same time all over Earth. Most scientists, however, argue that *Homo sapiens* appeared in one place (Africa) and then migrated to Europe and Asia, where the species gradually replaced *Homo erectus*.

Exploring Further

DNA Dating

To answer the question of when and where *Homo sapiens* evolved, researchers are studying the DNA of modern humans. Recall that DNA tends to accumulate more and more mutations as time passes. Therefore, individuals that share a recent common ancestor have relatively similar DNA. In contrast to this, individuals who share only a *distant* common ancestor have less similar DNA. Thus individuals in older human populations should be more genetically different from each other than individuals in newer populations are.

Mitochondrial DNA

Human DNA has been looked at in two ways that shed light on the issue. The first approach focuses on mitochondrial DNA (mDNA). Humans inherit mDNA from only their mother—it is found in the egg cell. mDNA is thus passed down unchanged from generation to generation. Researchers began by comparing the mDNA of individuals from different ethnic backgrounds. Based on their data, the researchers hypothesize that all modern humans share a common ancestor that lived about 170,000 years ago. According to fossil evidence, this was about the time *Homo sapiens* appeared in Africa.

The researchers used the mDNA data to generate a human "family tree." When analyzed in this way, the data show a distinct branch that arose 52,000 years ago between Africans and non-Africans. This supports the hypothesis that some modern *Homo sapiens* left Africa and spread to Europe and Asia. They thus retraced the path taken by *Homo erectus* many years before.

Y Chromosomal DNA

Researchers have also analyzed DNA on the Y chromosome. Like mDNA, DNA on the Y chromosome does not undergo recombination during meiosis. Thus DNA on the Y chromosomes is passed down unchanged in males from one generation to the next. Researchers analyzed DNA variations in more than 1,000 European males. They found that 80 percent of the males studied share a single pattern of DNA. This suggests that modern Europeans share a relatively recent common ancestor. The data indicate the shared pattern arose about 40,000 to 50,000 years ago. This piece of evidence suggests that modern humans came to Europe at that time.

Taken together, the analysis of both mDNA and Y chromosome DNA indicate that modern humans arose only once, in Africa.

737

HISTORY CONNECTION

In 1987, scientists at the University of California at Berkeley created a human family tree using mitochondrial DNA data. The scientists calculated the length of time it would take to accumulate the pattern of mutations that are seen in modern human populations. They concluded that the ancestor of all modern humans arose in Africa about 150,000 years ago. This common ancestor was widely reported as the "mitochondrial Eve." This designation was given because mitochondrial DNA is passed in the egg from the maternal parent to offspring. The confusing oversimplification appeared to leave out Adam. In the past decade, Adam has been added to the evolutionary analysis through examination of Y chromosome mutations over time. In 1997, scientists further revolutionized human molecular anthropology when they isolated nuclear DNA from a 40,000-year-old specimen of a Neanderthal. More recent studies have concluded that modern humans share a common ancestor who lived about 170,000 years ago (± 50,000 years).

Teaching Tip

Neanderthal Health Tell students that scientists know that Neanderthals cared for their sick and injured. Neanderthal skeletons have been found with healed broken bones. The fractures probably incapacitated the individuals so others must have cared for them until their bones healed. Furthermore, scientists found an adult male's skeleton with severe deformities. The individual would probably have been dependent on others his entire life.

READING SKILL BUILDER — Basic

Reading Organizer Students may need an organizational framework to help them remember the hominids discussed in Sections 3 and 4. Have students make a chart with these column headings: Hominid, Discoverer, Place Discovered, and Date Discovered. Tell students to complete as many columns as possible for each of the hominid species mentioned in these sections. **LS Visual**

Group Activity — Advanced

Extranuclear and Y Chromosome DNA Have students work in two large groups. Assign one to work on mitochondrial DNA and the other to work on Y chromosome DNA. Have each group designate members who will conduct library or Internet research to find the most current information available on the genes located in these segments of DNA. Have each group also designate members to conduct library or Internet research on the functions of the genes identified. Finally, have each group designate members to construct a poster depicting the DNA segment they have researched, the genes it codes for, and the functions of these genes. Have each group prepare and give an oral presentation on their findings to the whole class. **LS Interpersonal** Co-op Learning

Close

Reteaching — BASIC

Assume that a skull has just been found. Ask students what characteristics would indicate that the fossil was of a Neanderthal skull. (The skull would need a large cranial capacity, protruding mouth bones, and heavy, bony brow ridges.)

Quiz — GENERAL

1. For what characteristic was *Homo habilis* named? (*Habilis* is Latin for "handy," so this human ancestor was named for its ability to use tools.)
2. What species is most likely to have been the direct ancestor of our own species, *Homo sapiens*? (*Homo erectus*.)
3. What molecular evidence suggests that *Homo sapiens* originated in Africa and then radiated out to other places? (Variations in mitochondrial and Y chromosome DNA support this hypothesis.)

Alternative Assessment — GENERAL

Pair each student with a partner. Have one student in each group summarize orally the information presented in this section without referring to the text. The other student should listen and ask clarifying questions to ensure that the final account is accurate and complete.

Section Review Correlations

SC.D.1.4.3: Items 2, 3, 5

SC.G.1.4.1: Items 1, 4

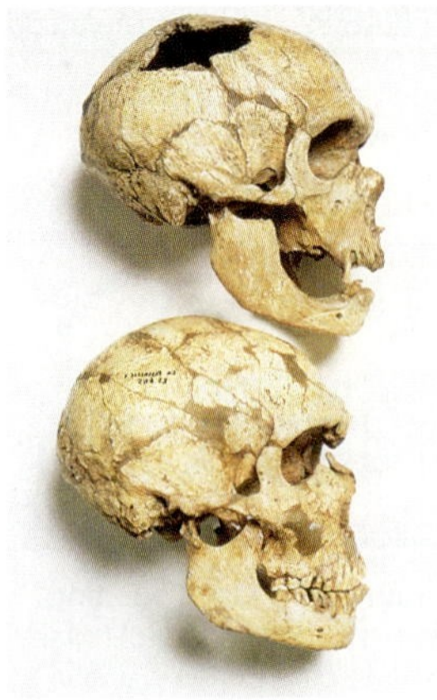

Figure 31 Neanderthals. Neanderthals were distinguished by massive ridges over the eyes.

Homo sapiens in Europe

Members of the species *Homo sapiens* first appeared in Europe about 130,000 years ago. The fossils of a group of *Homo sapiens* called Neanderthals (*nee AN dur TALZ*—the *h* is silent) were found in 1856 in the Neander Valley of Germany. The Neanderthals were short and powerfully built. Their skulls were massive, as shown in **Figure 31,** with protruding faces and heavy, bony brow ridges. The average Neanderthal brain was slightly larger than that of modern humans, although there is much overlap in size. Many scientists that study human evolution now classify Neanderthals as a separate human species, *Homo neanderthalensis*.

Neanderthals became more and more abundant in Europe and Asia, and by 70,000 years ago, they had become fairly common. Neanderthals cared for their injured and sick. They also commonly buried their dead and often placed food, weapons, and even flowers with the bodies. Such attention to the dead suggests that they may have believed in a life after death. Neanderthals were the first hominids to show evidence of abstract thinking.

Modern *Homo sapiens*

About 34,000 years ago, the European Neanderthals were abruptly replaced by *Homo sapiens* of modern appearance. These early humans are thought to have evolved first in Africa and then migrated to Europe and Asia. Early modern humans lived by hunting. They had complex patterns of social organization, and they probably used language. They spread across Siberia and reached North America at least 12,000 years ago, when a land bridge connected Siberia and Alaska.

The ability of humans to make and use tools effectively has been important to the success of humans. Humans use symbolic language and can shape concepts out of experience. Written language has enabled humans to transmit concepts from one generation to another. Humans can also reason abstractly, applying existing knowledge to new situations. Humans also have cultural evolution. Through culture, we have found ways to change our environment to suit our needs, rather than changing ourselves in response to the environment. This is both an exciting potential and an enormous responsibility.

Section 4 Review

1. **Describe** how *Homo habilis* differs from australopithecines. SC.G.1.4.1
2. **Describe** the evidence that identifies *Homo erectus* as the first human species to have left Africa. SC.D.1.4.3
3. **Analyze** the two hypotheses of the origin of *Homo sapiens.* SC.D.1.4.3
4. **Contrast** Neanderthals with modern humans. SC.G.1.4.1
5. FCAT Prep What evidence have scientists used to estimate when and where modern humans evolved? SC.D.1.4.3
   A mDNA
   B cave paintings
   C analysis of language
   D articles buried with the dead

738

Answers to Section Review

1. *H. habilis* had a larger brain and made and used tools.
2. *H. erectus* skeletons have been found in Asia and Europe in addition to Africa, whereas *H. habilis* fossils have only been found in Africa.
3. One hypothesis holds that *H. sapiens* arose independently and simultaneously in many locations in Africa, Europe, and Asia from interbreeding groups of *H. erectus.* A second hypothesis holds that *H. sapiens* arose in only one place, Africa, and then migrated to Europe and Asia, replacing *H. erectus* as they migrated.
4. Neanderthals had slightly larger brains than modern humans. They were also shorter and stockier and had massively boned faces.
5. **A.** Correct. **B.** Incorrect. Cave paintings were also made by Neanderthals. **C.** Incorrect. Language is not preserved in the fossil record. **D.** Incorrect. Neanderthals also buried articles with their dead.

Study ZONE

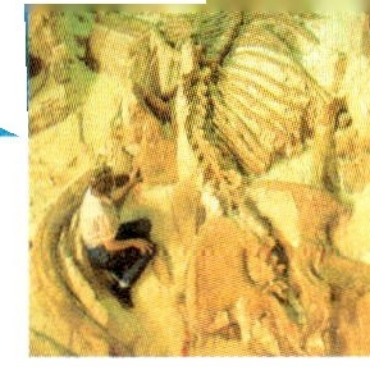

Key Concepts

1 Vertebrates in the Sea and on Land

- Vertebrates have an endoskeleton that includes a backbone, bilateral symmetry, cephalization, a coelom, a closed circulatory system, and a chambered heart.
- The first vertebrates, jawless fishes, evolved about 500 million years ago.
- Spiny fishes with jaws appeared 430 million years ago.
- Cartilaginous sharks and bony fishes evolved about 400 million years ago. Bony fishes account for more than 90 percent of today's fishes.
- Amphibians, which appeared about 370 million years ago, were the first group of vertebrates to live on land.

2 Terrestrial Vertebrates

- Reptiles evolved about 320 million years ago and have watertight skin and a watertight egg.
- Dinosaurs dominated the land for 150 million years.
- Modern birds may be the descendants of a group of small theropods, a type of dinosaur.
- The first mammals appeared 220 million years ago and took over many ecological roles once dominated by dinosaurs.

3 Evolution of Primates

- The first primates evolved 50 million to 60 million years ago.
- Primates have grasping hands and binocular vision.
- DNA evidence indicates that humans are more closely related to chimpanzees than to any other primate species.
- Early hominids walked upright before their brains became significantly larger.

4 The Genus *Homo*

- *Homo habilis* had a brain much larger than that of australopithecines and used tools.
- *Homo erectus* evolved in Africa about 1.5 million years ago and may have been the direct ancestor of *Homo sapiens*.
- *Homo sapiens* probably evolved in Africa and migrated to the rest of the world.

Key Terms

Section 1

vertebra (712)
agnathan (714)
acanthodian (715)
cartilage (715)

Section 2

terrestrial (721)
thecodont (722)
Pangaea (722)
ectothermic (724)
endothermic (724)
therapsid (728)

Section 3

primate (731)
prosimian (731)
diurnal (732)
hominid (733)

Section 4

None

Alternative Assessment — GENERAL

Assign each student to make up 12 questions, one from each of the red concept heads in Chapter 32. Have them provide two true/false, two short answer, four multiple choice, and four essay questions with answers. Collect the questions and use the best ones to conduct an oral quiz, reading them as students write the answers. Number each question you use so you can go back and have students check their answers. Discuss correct responses to items students have questions about. LS Intrapersonal

Chapter Resource File

- Science Skills Worksheet GENERAL
- Critical Thinking Worksheet ADVANCED
- Test Prep Pretest GENERAL
- Chapter Test GENERAL

Answer to Concept Map

The following is one of several possible answers to Performance Zone item 15.

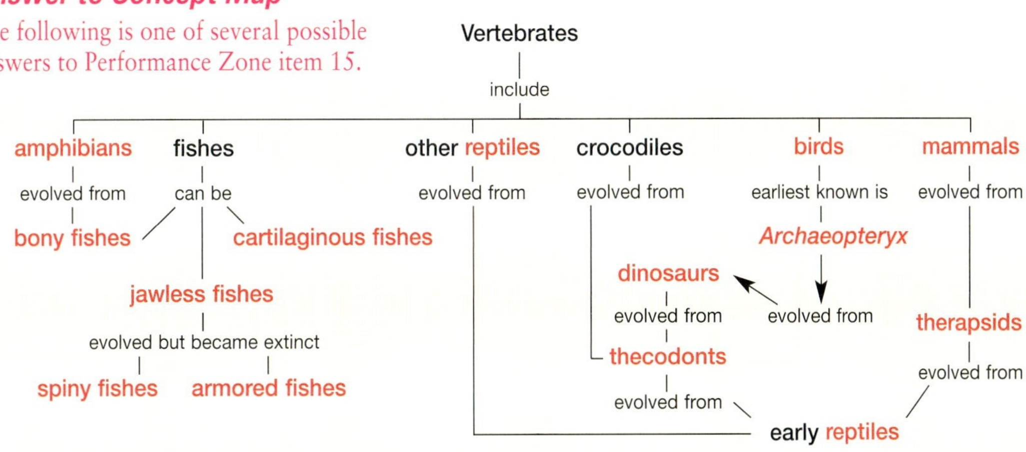

ANSWERS

Understanding Key Ideas

1. c
2. c
3. b
4. d
5. c
6. b
7. d
8. Crocodiles and birds have similar hearts, and both care for their young.
9. Answers will vary, but could include a blow from a blunt instrument (such as a hammer), a gunshot wound, or a stabbing.
10. The answer to the concept map is found at the bottom of the previous page.

Critical Thinking

11. The eyes of the first primates faced forward. The resulting binocular vision gave primates excellent depth perception, which enabled them to judge distances accurately when moving through trees.
12. Although stone tools have been associated only with the genus *Homo*, there were australopithecians alive 2 million years ago. It is therefore risky, in the absence of other data, to conclude that only the genus *Homo* could be responsible for the tools.

Alternative Assessment

13. Guides will vary but should include the following information: backbone made of individual vertebrae (a spinal column), dorsal nerve cord protected by the backbone, bony skull, paired appendages, bilateral symmetry, cephalization with complex brain and sense organs, and complex organ systems.
14. Models and guides will vary but should show information similar to that presented in Figure 5 or Figure 8.

Chapter Review Correlations

SC.D.1.4.3: Items 4, 5, 8
SC.D.1.4.4: Item 6
SC.F.1.4.2: Items 2, 11
SC.G.1.4.1: Items 1, 3, 7, 10, 12, 13
SC.H.3.4.1: Item 14
SC.H.3.4.3: Item 9

Performance ZONE CHAPTER REVIEW

Understanding Key Ideas

1. To survive as predators, fish had to adapt to which of the following two challenges?
   a. low oxygen conditions, shallow seas
   b. avoiding predators, grasping prey
   c. pursuing prey in water, grasping prey
   d. filtering invertebrates from the water, avoiding predators
   SC.G.1.4.1
2. Which of the following is *not* an amphibian adaptation for life on land?
   a. legs
   b. lungs
   c. watertight skin
   d. more efficient heart
   SC.F.1.4.2
3. The first fully terrestrial vertebrates were
   a. frogs.
   b. reptiles.
   c. dinosaurs.
   d. mammals.
   SC.G.1.4.1
4. As Pangaea broke apart,
   a. dinosaurs became more alike.
   b. mammals became dominant.
   c. dinosaur species flourished.
   d. world climates changed.
   SC.D.1.4.3
5. Mammals are thought to have survived the climatic changes that caused the extinction of dinosaurs because the bodies of mammals
   a. have little insulation.
   b. are ectothermic.
   c. were relatively small.
   d. lacked insulation.
   SC.D.1.4.3
6. Which sequence reflects the order in which the major groups of vertebrates are thought to have evolved?
   a. bony fishes → reptiles → amphibians
   b. bony fishes → amphibians → reptiles
   c. amphibians → mammals → reptiles
   d. birds → mammals → reptiles
   SC.D.1.4.4
7. What evidence convinced scientists that *Archaeopteryx* should be classified as a bird rather than a small dinosaur?
   a. teeth
   b. hollow bones
   c. hard-shelled eggs
   d. fused collarbone
   SC.G.1.4.1
8. What evidence is there that crocodiles and birds are likely to be more closely related to each other than to other living reptiles? SC.D.1.4.3
9. **BIOWatch** List three causes of death that could be discovered through the work of forensic anthropologist. SC.H.3.4.3
10. **Concept Mapping** Construct a concept map that shows the relationships among the various living and nonliving vertebrates. Include the following terms in your concept map: *jawless fishes, spiny fishes, armored fishes, cartilaginous fishes, bony fishes, amphibians, thecodonts, dinosaurs, reptiles, therapsids, mammals, Archaeopteryx,* and *birds.* SC.G.1.4.1

Critical Thinking

11. **Recognizing Relationships** Describe the positioning of the eyes in the first primates. How did this positioning of the eyes improve the primates' ability to navigate their environment? SC.F.1.4.2
12. **Evaluating Hypotheses** Some stone tools are more than 2 million years old. Evaluate the hypothesis that they must have been made by a species from the genus *Homo*. SC.G.1.4.1

Alternative Assessment

13. **Recognizing Relationships** Vertebrates share several characteristics. Develop an illustrated guide that informs readers about these characteristics. SC.G.1.4.1
14. **Forming a Model** Make a series of models or create a mural that shows either the evolution of jaws or the evolution of limbs. Create a written guide or an audiotape that guides the viewer through your work. SC.H.3.4.1

Assignment Guide

Section	Questions
1	1, 2, 6, 10, 13, 14
2	3, 4, 5, 6, 7, 8, 9, 10
3	11
4	12

FCAT Prep

For questions 1–6, write your answers on a separate sheet of paper.

1 Which of the following describes reptiles and birds?
A. They are cold-blooded.
B. They are endotherms.
C. They have feathers.
D. They have watertight eggs.
SC.G.1.4.1

2 What were the earliest fishes called?
F. acanthodians
G. agnathans
H. thecodonts
I. therapsids
SC.G.1.4.1

3 Which of the following are night-active primates?
A. ectoderms
B. endoderms
C. prosimians
D. thecodonts
SC.G.1.4.1

4 Apes have larger brains than monkeys. What do apes lack that monkeys have? SC.G.1.4.1
F. a tail
G. bipedalism
H. depth perception
I. opposable thumbs

5 READ INQUIRE EXPLAIN A key difference between aquatic (water) and terrestrial (land) ecosystems is the amount of available water.

Compare the adaptations of reptiles to land to the adaptations of plants to land. SC.G.1.4.1

6 READ INQUIRE EXPLAIN Fossil evidence collected in the 1980s in Arctic Alaska suggests that some dinosaurs were year-round residents of areas that reached freezing temperatures and were in total darkness during the winter months. This evidence of "Arctic dinosaurs" contradicts the hypothesis that the extinction of dinosaurs was due to a climate change produced by a meteor collision that sent debris into the Earth's atmosphere.

Part A Infer how the climate could be changed by a meteor colliding with Earth. SC.D.1.4.3

Part B Assess the evidence of "Arctic dinosaurs" and its contradiction of the meteor hypothesis. SC.H.1.4.3

Test TIP

Take time to read each question completely on a standardized test, including all of the answer choices. Consider each answer choice before determining which one is correct.

FCAT Doctor

Question 1 Answer D is the correct choice. Both birds and reptiles have watertight eggs. Answer A is incorrect because reptiles are cold-blooded, and birds are warm-blooded. Answer B is incorrect because birds are endotherms, and reptiles are ectotherms. Answer C is incorrect because birds have feathers, but reptiles do not.

Question 5 Reptiles developed waterproof skin and watertight eggs. Plants developed vascular systems and stomata.

Question 6

Part A A reduction in the amount of sunlight that reaches Earth would cause a dip in average global temperatures.

Part B The hypothesis that cold killed the dinosaurs relies on the theory that dinosaurs had inadequate insulation and could not slow down their metabolism to survive the cold. New hypotheses would have to be developed to account for the "Arctic dinosaurs."

Answers

1. D
2. G
3. C
4. F
5. Both groups of organisms had to adapt ways of retaining water.

6A. Debris from the collision sent into the atmosphere could block the sun and cool the Earth.

6B. Dinosaurs that could survive the Arctic climate would not go extinct when debris blocked the sun and cooled the Earth.

Test Prep Correlations

SC.D.1.4.3: Item 6A
SC.G.1.4.1: Items 1-5
SC.H.1.4.3: Item 6B

Exploration Lab

Comparing Hominid Skulls

Teacher's Notes

Time Required

One 40-minute class period

Ratings

EASY → HARD

Teacher Preparation: 2
Student Setup: 1
Concept Level: 2
Cleanup: 1

Skills Acquired

- Collecting Data
- Identifying/Recognizing Patterns
- Inferring
- Interpreting
- Organizing and Analyzing Data

Scientific Methods

In this lab, students will:

- Make Observations
- Test the Hypothesis
- Analyze the Results
- Draw Conclusions

Materials and Equipment

Materials for this lab can be ordered from WARD'S. See *Master Materials List* at the front of this book for catalog numbers.

Tips and Tricks

If life-size casts of skulls are available, substitute these for the diagrams provided, or use them to supplement the investigation.

The checklist of features has incorporated conversion factors of 40 for calculating square area and 1,000 for calculating cubic volume of life-size skulls. If life-size casts of skulls are available for students to measure, eliminate the factor of 40 in calculating square area. Substitute 4.2, or $4/3 \times \pi$, for the factor of 1,000 in calculating cubic volume.

Exploration Lab

Comparing Hominid Skulls

SC.D.1.4.3

SKILLS

- Measuring
- Comparing anatomical features

OBJECTIVES

- **Identify** differences and similarities between the skulls of apes and the skulls of humans.
- **Identify** differences and similarities between the fossilized skulls of hominids.
- **Classify** the features of hominid skulls as apelike, humanlike, or intermediate.

MATERIALS

- metric ruler
- protractor

Ape

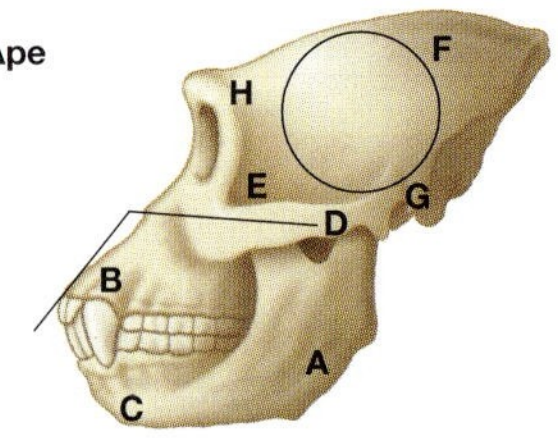

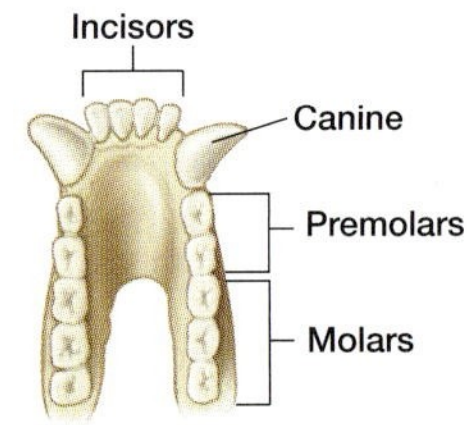

Human

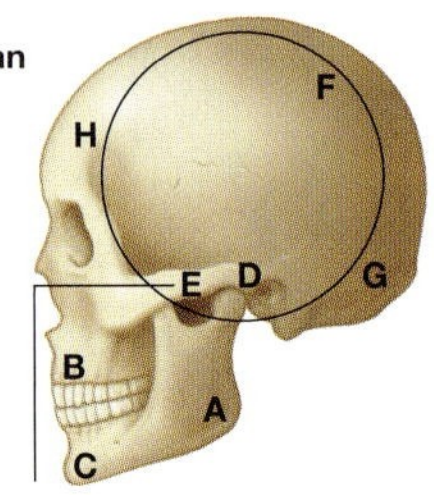

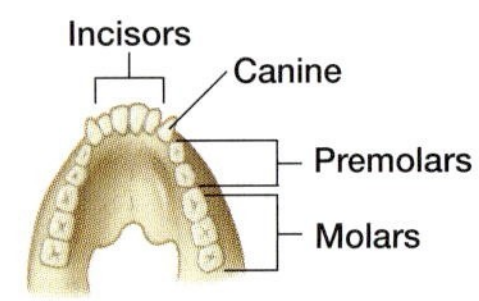

Before You Begin

Modern **apes** and humans share a **common ancestor.** Much of our understanding of human evolution is based on the study of the fossilized remains of **hominids.** By studying fossilized bones and identifying similar and dissimilar structures, scientists can infer the **anatomy,** or body structure, of a species. In this lab, you will identify differences and similarities between the skulls of apes, early hominids, and humans.

1. Write a definition for each boldface term in the paragraph above.
2. Make a data table similar to each one shown at right.
3. Based on the objectives for this lab, write a question you would like to explore about human evolution.

DATA TABLE 1

Name	Cranial capacity (cm^3)	Lower face area (cm^2)	Brain area (cm^2)	Jaw angle (degrees)	Brow ridge	Teeth
Ape						
Human						

DATA TABLE 2

Name				
Australopithecus robustus				
Australopithecus africanus				
Homo erectus				
Neanderthal				

742

Answers to Before You Begin

1. *apes*—tailless primates with long arms, a broad chest, and a larger, more developed brain than monkeys; *common ancestor*—one species from which two or more species evolved; *hominids*—members of the family Hominidae of the order Primates, and are characterized by opposable thumbs, no tail, longer lower limbs, and erect bipedalism; *anatomy*—the body structure of an organism.
2. Have students copy the data tables provided.
3. Answers will vary. For example: How have facial features changed as hominids have evolved?

How to Interpret the Features of a Skull

Cranial capacity: Use the circles drawn on the skulls to estimate brain volume, or cranial capacity. Measure the radius of each circle in centimeters. Then cube this number, and multiply the result by 1,000 to calculate the approximate life-size cranial capacity in cubic centimeters.

Lower face area: Measure *A* to *B* and *C* to *D* in centimeters for each skull. Multiply these two numbers together, and multiply the product by 40 to approximate the life-size lower face area in square centimeters.

Brain area: Measure *E* to *F* and *G* to *H* in centimeters for each skull. Multiply these two numbers and multiply the product by 40 to approximate the life-size brain area in square centimeters.

Jaw angle: Note the two lines that come together near the nose of each skull. Use a protractor to measure the inside angle made by the lines and to determine how far outward the jaw projects.

Brow ridge: Note the presence or absence of a bony ridge above the eye sockets.

Teeth: Count the number of each kind of teeth in the lower jaw.

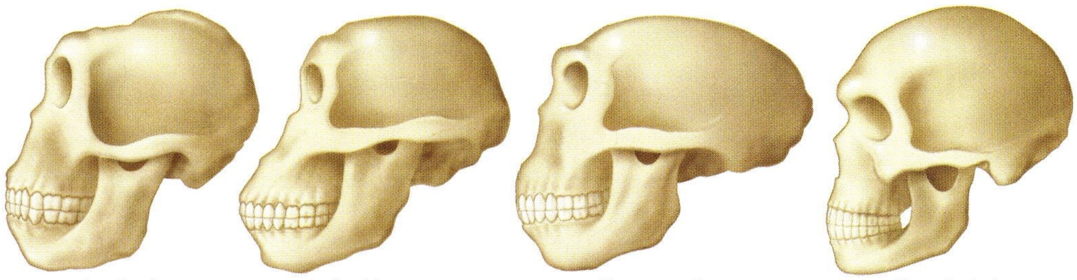

A. robustus | *A. africanus* | *Homo erectus* | Neanderthal

Procedure

PART A: Ape Skulls and Human Skulls

1. Examine the diagrams of the skull and jaw of an ape and a human. Look for similarities and differences between the features listed in the chart "How to Interpret the Features of a Skull." Record your observations and measurements for each feature listed in Data Table 1.

PART B: Fossil Hominids

2. Examine the four fossil hominid skulls. On the hominid skulls, observe and measure four features that are listed in the chart "How to Interpret the Features of a Skull." Use the human skull as a model for taking measurements. Record your observations and measurements in Data Table 2.
3. Compare your data for the hominids with your data for the modern ape and human. Classify each feature of the hominid skulls as being apelike, humanlike, or intermediate by writing an *A*, *H*, or *I* next to your observation or measurement for that feature.
4. Using your data, predict the order in which the hominids shown here evolved.

Analyze and Conclude

1. **Summarizing Results** How did skull structure change as hominids evolved?
2. **Drawing Conclusions** Which fossil skull is most apelike? most humanlike?
3. **Further Inquiry** Write a new question about human evolution that could be explored with another investigation.

On the Job

Anthropology is the scientific study of humans. Do research to learn about a famous anthropologist, such as Louis or Mary Leakey. For more about careers, visit **go.hrw.com** and type in the keyword **HX4 Careers.**

Answers to On the Job

Answers will vary. Sample answer: Louis and Mary Leakey were a husband-and-wife team. Louis Leakey discovered fossil hominids in Tanzania that showed that human evolution was centered in Africa, not Asia. Mary Leakey described the sequence in which early stone tools came into use. She also discovered one of the first humanlike skulls that indicated that humanlike beings lived in prehistoric eastern Africa. She led a team that discovered a trail of footprints preserved in volcanic ash thought to be from humanlike creatures that lived approximately 3,700,000 years ago.

Answers to Procedure

1. See Sample Data Table 1.
2. See Sample Data Table 2.
3. See Sample Data Table 2.
4. Students should predict that *A africanus* evolved first, followed by *A. robustus, H. erectus,* and Neanderthal.

Answers to Analyze and Conclude

1. As hominids evolved, the brow ridge was minimized, the brain area increased, and the jaw became less prominent.
2. Students should recognize that the Neanderthal skull is most humanlike because of its large brain area and nonprotruding jaw. Students may disagree about which hominid skull is most apelike, *A. robustus* or *A. africanus.* Both skulls have a small brain area, a brow ridge, and a protruding jaw.
3. Answers will vary. For example: What was the probable diet of fossil hominids based on their tooth and jaw structure?

Sample Data Table 1:

Name	Cranial capacity (cm^3)	Lower face area (cm^2)	Brain area (cm^2)	Jaw angle (degrees)	Brow ridge	Teeth
Ape	422	420	230	130	Yes	6 molars, 4 premolars, 2 canines, 4 incisors
Human	1953	161	264	92	No	Same as above

Sample Data Table 2:

Name	Lower face area (cm^2)	Brain area (cm^2)	Jaw angle (degrees)	Brow ridge
A. robustus	300 (I)	260 (I)	115 (I)	Yes
A. africanus	252 (I)	190 (A)	125 (I)	Yes
H. erectus	312 (I)	252 (I)	111 (I)	Yes
Neanderthal	300 (I)	308 (H)	115 (I)	Yes

Chapter Opener

- **TE** **Using the Figure,** p. 744 GENERAL
- **SE** **Quick Review,** p. 745 ■
- **SE** **Reading Activity,** p. 745 ■
- **TE** **Opening Activity,** p. 745 GENERAL

& 3 • 90 min pp. 746–750

Section 1
The Fish Body

- **OSP** **Lesson Plan***
- **TR** **Bellringer***
- **TR** **Section Outline***
- **TR** **I8 Respiration in Fishes***
- **TR** **I7 Fish Heart Structure***

- **TE** **Using the Figure,** p. 747 GENERAL
- **TE** **Demonstration,** p. 748 ◆
- **TE** **Demonstration,** p. 749 ◆ GENERAL
- **SE** **Data Lab** Analyzing Ion Excretion in Fish, p. 749 ■
- **CRF** **Datasheets for In-Text Labs** Analyzing Ion Excretion in Fish*

& 5 • 90 min pp. 751–757

Section 2
Today's Fishes

- **OSP** **Lesson Plan***
- **TR** **Bellringer***
- **TR** **Section Outline***
- **TR** **I3 Lateral Line in Fishes***
- **TR** **I5 External Structures of Fish–Yellow Perch***
- **TR** **I6 Internal Structures of Fish–Yellow Perch***

- **TE** **Using the Figure,** p. 752
- **TE** **Demonstration,** p. 753 ◆ GENERAL
- **SE** **Quick Lab** Modeling the Action of a Swim Bladder, p. 756 ◆ ■
- **CRF** **Datasheets for In-Text Labs** Modeling the Action of a Swim Bladder*
- **CRF** **Exploration Lab** Schooling Behavior in Fishes* ◆ GENERAL

& 7 • 90 min pp. 758–764

Section 3
Amphibians

- **OSP** **Lesson Plan***
- **TR** **Bellringer***
- **TR** **Section Outline***
- **TR** **I11 Amphibian Heart Structure***
- **TR** **I13 Life Cycle of a Frog***
- **TR** **I14 External Structure of a Frog–Leopard Frog***
- **TR** **I15 Internal Structure of a Frog–Leopard Frog***

- **TE** **Activity** Naming Amphibians, p. 758 GENERAL
- **TE** **Using the Figure,** p. 759 ADVANCED
- **TE** **Demonstration,** p. 760 ◆ GENERAL
- **TE** **Using the Figure,** p. 760 GENERAL
- **TE** **Using the Figure,** p. 761 GENERAL
- **CRF** **Skills Practice Lab** Frog Dissection* ◆ GENERAL
- **SE** **Exploration Lab** Observing a Live Frog, pp. 768–769 ◆ ■
- **CRF** **Datasheets for In-Text Labs** Observing a Live Frog* ◆

& 9 • 90 min

Chapter Review and Assessment Resources

- **SE** **Study Zone,** p. 765 ■
- **SE** **Performance Zone,** p. 766 ■
- **SE** **FCAT Test Prep,** p. 767 ■
- **TR** **Concept Mapping***
- **CRF** **Vocabulary Review Worksheet*** ■ GENERAL
- **CRF** **Test Prep Pretest*** ■ GENERAL
- **CRF** **Chapter Test*** ■ GENERAL
- **CRF** **Chapter Test*** ADVANCED
- **OSP** **Test Generator***
- **OSP** **Test Item Listing***
- **OSP** **Modified Chapter Test*** BASIC

Holt Online Learning

Visit go.hrw.com for access to Holt Online Learning or enter the keyword **HX6 Home** for a variety of free online resources.

One-Stop Planner® CD-ROM

This CD-ROM package includes
- Lab Materials QuickList Software
- Holt Calendar Planner
- Customizable Lesson Plans
- Printable Worksheets
- ExamView® Test Generator
- Interactive Teacher Edition
- Holt PuzzlePro® Resources
- Holt PowerPoint® Resources

Guided Reading Audio CDs

These CDs are designed to help auditory learners and reluctant readers. (Audio CDs are also available in Spanish.)

Compression guide:
To shorten your instruction because of time limitations, omit blocks 1, 3, 6, and 8.

KEY

TE	Teacher Edition	CRF	Chapter Resource File	*	Also on One-Stop Planner
SE	Student Edition	TR	Teaching Transparency	■	Also Available in Spanish
OSP	One-Stop Planner			◆	Requires Advance Prep

SKILLS DEVELOPMENT AND PRACTICE	SECTION REVIEW AND ASSESSMENT	CORRELATIONS
		Florida Sunshine Standards/FCAT
CRF Directed Reading Worksheet* BASIC CRF Active Reading Worksheet* GENERAL	SE Section 1 Review ■ TE Reteaching, p. 750 BASIC TE Quiz, p. 750 GENERAL TE Alternative Assessment, p. 750 GENERAL CRF Quiz* ■ BASIC	SC.F.1.4.1, SC.F.1.4.2, SC.F.2.4.1, SC.G.1.4.1
CRF Directed Reading Worksheet* BASIC CRF Active Reading Worksheet* GENERAL	SE Section 2 Review ■ TE Reteaching, p. 757 BASIC TE Quiz, p. 757 GENERAL TE Alternative Assessment, p. 757 GENERAL CRF Quiz* ■ BASIC	SC.F.1.4.2, SC.G.1.4.1
TE Reading Skill Builder Reading Organizer, p. 759 GENERAL CRF Directed Reading Worksheet* BASIC CRF Active Reading Worksheet* GENERAL CRF Critical Thinking Worksheet* ADVANCED CRF Science Skills Worksheet* GENERAL	SE Section 3 Review ■ TE Reteaching, p. 764 BASIC TE Quiz, p. 764 GENERAL TE Alternative Assessment, p. 764 GENERAL CRF Quiz* ■ BASIC Modified Worksheet* BASIC	SC.F.1.4.2, SC.G.1.4.1, SC.H.1.4.1

SCILINKS. NSTA
www.scilinks.org
Maintained by the **National Science Teachers Association**

Topic: Kidneys
SciLinks Code: HX4110

Topic: Sharks
SciLinks Code: HX4163

Topic: Circulation in Amphibians
SciLinks Code: HX4043

Overview

Tell students the purpose of this chapter is to describe the characteristics of fishes and amphibians and to demonstrate how each group is uniquely adapted to their environments. Modifications of their skeletons and body systems gave fish an advantage in water while other modifications enabled amphibians to make a successful transition between life in water and life on land.

Using the Figure

Tell students to look at the photograph of the leopard frog on this page. Ask them if the frog is a fish or an amphibian. (amphibian) Then ask them to explain why it is an amphibian and why it is not a fish. Ask them to identify traits that frogs and other amphibians have in common with fish. Then ask them to identify from the picture (or previous knowledge) some special adaptations frogs have for their environment and life cycle. (Allow students to freely share their ideas without confirming or correcting their ideas.)

Assessing Prior Knowledge

Before beginning this chapter, be sure your students can answer the questions in the ***Quick Review*** on the facing page. If they cannot, encourage them to reread the sections indicated.

Leopard frog

744

Chapter Correlations — Florida Sunshine Standards/FCAT

SC.F.1.4.1 Body processes involve specific biochemical reactions governed by biochemical principles.

SC.F.1.4.2 Body structures are uniquely designed and adapted for their function.

SC.F.2.4.1 There are mechanisms of asexual and sexual reproduction and different genetic advantages and disadvantages of asexual and sexual reproduction.

SC.G.1.4.1 There is great diversity and interdependence among living things.

CHAPTER

33 Fishes and Amphibians

Florida Sunshine Standards/FCAT

SC.F.1.4.1 Body processes involve specific biochemical reactions governed by biochemical principles.

SC.F.1.4.2 Body structures are uniquely designed and adapted for their function.

SC.F.2.4.1 There are mechanisms of asexual and sexual reproduction and different genetic advantages and disadvantages of asexual and sexual reproduction.

SC.G.1.4.1 There is great diversity and interdependence among living things.

Quick Review

Answer the following without referring to earlier sections of your book.

1. **Describe** the process of osmosis. *(Chapter 4, Section 1)*
2. **Define** the term *gills*. *(Chapter 27, Section 2)*
3. **State** the function of the excretory systems. *(Chapter 27, Section 2)*
4. **Distinguish** between external and internal fertilization. *(Chapter 27, Section 2)*
5. **Define** the term *metamorphosis*. *(Chapter 30, Section 3)*

Did you have difficulty? *For help, review the sections indicated.*

Reading Activity

Take a break after reading each section of this chapter, and closely study the figures in the section. Reread the figure captions, and for each one write out a question that can be answered by referring to the figure and its caption. Refer to your list of figures and questions as you review the concepts addressed in the chapter before you complete the Performance Zone chapter review.

The Northern Leopard frog, *Rana pipiens*, has become scarce in areas where the populations had once been dense. This amphibian is now being bred in protected habitats in those areas to increase its numbers.

Looking Ahead

internet connect

www.scilinks.org
National Science Teachers Association *sci*LINKS Internet resources are located throughout this chapter.

SCILINKS. Maintained by the National Science Teachers Association

Opening Activity — Basic

Jaw Function Bring several sizes and types of pliers to class. Have students pick up different objects using the tools. Ask students to decide which tools exert the most force and which are best for picking up small objects. Relate the function of the tools to jaw function. Ask students to name the vertebrate group in which jaws first evolved. (fishes) **LS Kinesthetic**

Quick Review Answers

1. Osmosis is the diffusion of water molecules across a selectively permeable membrane (from an area of higher concentration to an area of lower concentration).
2. Gills are extremely thin projections of blood vessel-rich tissue that are the sites of gas exchange in aquatic (and a few terrestrial) animals.
3. The function of the excretory system is to remove wastes produced by cellular metabolism. This waste elimination is also linked to maintaining proper water balance.
4. External fertilization occurs when male and female gametes join outside either parent's body (usually in water) while internal fertilization is when the union of sperm and egg occurs inside the female's body.
5. Metamorphosis is a process of dramatic physical change through which an immature organism passes as it grows to its adult stage.

Chapter Resource File

- Vocabulary Worksheets
- Concept Mapping

Reading Activity Answers

Students should have one question for each of the 18 figures in the chapter—four in Section 1, six in Section 2, and eight in Section 3. Encourage students to ask meaningful questions about the figures, not questions that focus on an unimportant word in the caption.

Section 1

Focus

Overview

Before beginning this section review with your students the objectives listed in the Student Edition. This section introduces students to the characteristics of fishes and their special adaptations to life in water.

Bellringer

K-W-L Have students make short individual lists of all the things they already **K**now about fish. Then have students list things they **W**ant to know about fish. Have students save their lists for later use in the Reteaching at the end of Section 1.

Motivate

Discussion — GENERAL

Write the following words on the board: starfish, sea horse, sea cow, lamprey, eel, and stingray. Ask students to identify which organisms they think are fish. Ask students to write down whether it is a fish or not and why. (The starfish is not a fish—it is an echinoderm; and the sea cow, or manatee, is also not a fish—it is a mammal. The sea horse, lamprey, eel, and stingray are all fish. This demonstrates that fishes are a very diverse group of organisms that all share important characteristics the students will learn about in Section 1.) **LS Verbal**

Section Correlations

SC.F.1.4.1
SC.F.1.4.2
SC.F.2.4.1
SC.G.1.4.1

Section 1 The Fish Body

Florida Sunshine Standards/FCAT

SC.F.1.4.1 Body processes involve specific biochemical reactions governed by biochemical principles.

SC.F.1.4.2 Body structures are uniquely designed and adapted for their function.

SC.G.1.4.1 There is great diversity and interdependence among living things.

Objectives

- **Describe** the characteristics of modern fishes. SC.G.1.4.1
- **Summarize** how fish obtain oxygen. SC.F.1.4.2
- **Summarize** how blood circulates through a fish. SC.F.1.4.2
- **Contrast** how marine and freshwater fishes balance their salt and water content. SC.F.1.4.1
- **Describe** two methods of reproduction in fishes. SC.F.2.4.1

Key Terms

gill filament
gill slit
countercurrent flow
nephron

Key Characteristics of Modern Fishes

What makes a goldfish instantly recognizable as a fish? You might name characteristics such as its fins, gills, scales, and typical fish shape as traits that contribute to the goldfish's "fishiness." But some fishes don't look quite so fishy. This is because the term *fish* refers to any member of one of three general categories of vertebrates: Agnatha (jawless fishes), Chondrichthyes (cartilaginous fishes), and Osteichthyes (bony fishes). The great diversity of fishes found today reflects various adaptations that enable fishes to live in the oceans and fresh waters around the world. Fishes vary in size from whale sharks longer than a moving van to gobies no larger than your fingernail. Despite the variation seen among fishes, shown in **Figure 1,** all share certain key characteristics.

1. **Gills.** Fishes normally obtain oxygen from the oxygen gas dissolved in the water around them. They do this by pumping a great deal of water through their mouths and over their gills.
2. **Single-loop blood circulation.** Blood is pumped from the heart to the capillaries in the gills. From the gills, blood passes to the rest of the body and then returns to the heart. (Lungfishes, which have a double-loop circulation, are an exception.)
3. **Vertebral column (backbone).** All fishes have an internal skeleton made of either cartilage or bone, with a vertebral column surrounding the spinal cord. The brain is fully encased within a protective covering called the *skull* or *cranium*.

To learn about one common fish, read *Up Close: Yellow Perch* later in this chapter.

Figure 1 Fish diversity. While these three fish appear quite different externally, they share a number of characteristics.

Stingray

Sea horse

Trout

746

Chapter Resource File

- Directed Reading BASIC
- Active Reading GENERAL
- Data Sheet for Data Lab GENERAL

Transparencies

TR Bellringer
TR I8 Respiration in Fishes
TR I7 Fish Heart Structure

One-Stop Planner CD-ROM

- Reading Organizers BASIC
- Reading Strategies BASIC
- Problem Solving Worksheet Solution Concentration GENERAL

Gills

One challenge faced by all animals is the need to get enough oxygen for cellular respiration. Sponges, cnidarians, many flatworms and roundworms, and some annelids obtain oxygen by diffusion through the body surface. Other marine invertebrates, such as mollusks, arthropods, and echinoderms have gills, which are specialized respiratory organs. Fishes also respire with gills.

If you look closely at the face of a swimming fish, you will notice that as it swims, the fish continuously opens and closes its mouth, as if it were trying to eat the water. What looks like eating is actually breathing. The major respiratory organ of a fish is the gill, shown in **Figure 2.** Gills are made up of rows of **gill filaments**—fingerlike projections through which gases enter and leave the blood. The gill filaments hang like curtains between a fish's mouth and cheeks. At the rear of the cheek cavity is an opening called a **gill slit.** When a fish "swallows," water is forced over the gills and out through the gill slits.

This swallowing procedure is the core of a great change in gill design shown by fishes—countercurrent flow, also shown in Figure 2. In **countercurrent flow,** water passes over the gills in one direction as blood flows in the opposite direction through capillaries in the gills. Countercurrent flow ensures that oxygen diffuses into the blood over the entire length of the capillaries in the gills. Due to this arrangement, the gills of bony fishes are extremely efficient respiratory organs. Fish gills can extract up to 85 percent of the oxygen in the water passing over them.

Figure 2 Fish gill structure. Countercurrent flow increases the gill's efficiency.

Respiration in Fishes

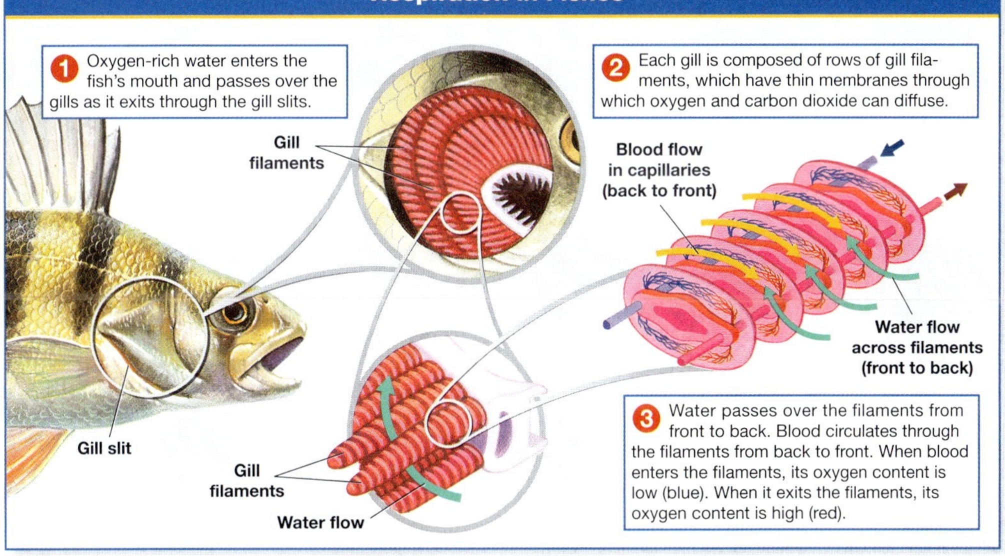

Integrating Physics and Chemistry

Several factors affect the amount of oxygen gas dissolved in the water around fish, including temperature and pressure. Remind students that as temperature increases, a solvent holds less dissolved gas. In addition, as pressure increases, a solvent holds more dissolved gas. Ask student where a higher concentration of dissolved oxygen gas is likely to be found, near the surface or deeper in the water? (Higher concentrations of dissolved oxygen will be found in deep water because the temperature decreases and pressure increases as one moves into deeper water.)

INCLUSION Strategies

• *Learning Disability*

Have students research the fish known as the sea horse. Students should give an oral report to the class that includes the characteristics of this fish, descriptions or pictures of different species of sea horses, and a description of the life cycle of the sea horse. In the course of their research, students may encounter other fishes that do not necessarily look like a "typical" fish. Encourage students to report on these fish as well.

Teach

Using the Figure — GENERAL

Have students locate the portion of **Figure 2** that shows countercurrent flow. Point out that water flows over the fish's gill from front to back. At the same time, the blood in the capillaries in the gills flows from back to front. Explain that this arrangement permits gas exchange to occur all along the capillaries in the gills. If the water and blood flowed in the same direction, the oxygen content of the two equilibrate at about 50 percent saturation, far lower than in countercurrent flow. **LS Visual**

Teaching Tip — ADVANCED

Understanding Countercurrent Flow Have students make a chart, dividing their paper into two columns: *same direction flow* and *opposite direction flow*. Each column should be further subdivided into O_2 *in water* and O_2 *in fish*. Have students record *80%* O_2 saturation in water and *20%* O_2 saturation in fish gill capillaries for both flow directions. Assume that 10% of the O_2 can be exchanged by diffusion for each subsequent chart row (representing a certain length of gill capillary). Under *same direction flow*, the next row would be 70% in the water and 30% in the fish. What is the maximum O_2 content that can be achieved in this fish if it had a same direction flow? (50%; both fluids continue their respective increase and decrease by 10% intervals until there is no further concentration difference at 50%.) To complete the other part of the chart, students need to realize that the O_2 content of water remains at 80% since new water is continually passing over the gills. The second row under *opposite direction flow* would be 80% and 30%. What is the maximum O_2 content that can be achieved if the fish had an opposite direction flow? (80%; O_2 content in the fish increases by 10% intervals until there is no further difference in O_2 content at 80%.) **LS Logical**

Teach, *continued*

Demonstration

To demonstrate the simple tubular "heart" of ancestral chordates, use an aquarium siphon (available at pet stores) that has a bulb attached to it. Insert a small piece of rubber or glass tubing into the siphon to keep the valve open. Place the end of the siphon in a jar of colored water and fill the rest of the tube with water. Place the other end of the siphon in another jar containing an equal volume of clear water. Squeeze the bulb, representing the primitive "heart," to show how blood is inefficiently moved in both directions. Water will flow out in both directions, and when the bulb is released, water will be sucked in from both ends. The level of each jar will go down very slightly when the bulb is squeezed and rise an equal amount when it is released. You can also demonstrate the value of valves by showing the efficiency of the pump when the valve is operating properly. With the valve enabled, water will move from one jar to the other. LS Visual

WORD Origins

The term *sinus venosus* is Latin, as are many anatomical terms. The Latin word *sinus* means "bend," and the Latin word *venosus* refers to veins. The sinus venosus is a bent collecting chamber that conducts blood into the heart.

Circulation of Blood

Chordates that were ancestral to the vertebrates had a simple tubular "heart." This structure was little more than a specialized zone of one artery that had more muscle tissue than the other arteries had. When a tubular heart contracts, blood is pushed in both directions, and circulation is not very efficient.

For an organism with gills, such as a fish, a simple tubular heart is not an adequate pump. The tiny capillaries in a fish's gills create resistance to the flow of blood, so a stronger pump is needed to overcome this resistance. In fishes, the tubular heart of early chordates has been replaced with a simple chamber-pump heart, shown in **Figure 3.** The chamber-pump heart can be thought of as a tube with four chambers in a row.

1. **Sinus venosus** *(SIE nuhs vuh NOH suhs).* This collection chamber acts to reduce the resistance of blood flow into the heart.
2. **Atrium.** Blood from the sinus venosus fills this large chamber, which has thin, muscular walls.
3. **Ventricle.** The third chamber is a thick-walled pump with enough muscle tissue to contract strongly, forcing blood to flow through the gills and eventually to the rest of the body.
4. **Conus arteriosus** *(KOH nuhs ahr TIHR ee oh suhs).* This chamber is a second pump that smoothes the pulsations and adds still more force.

The fish heart represents one of the great evolutionary changes found in vertebrates—a heart that pumps fully oxygenated blood through a single circulatory loop to the body's tissues.

Figure 3 Fish chamber-pump heart. These four steps show how blood flows through the heart of a fish.

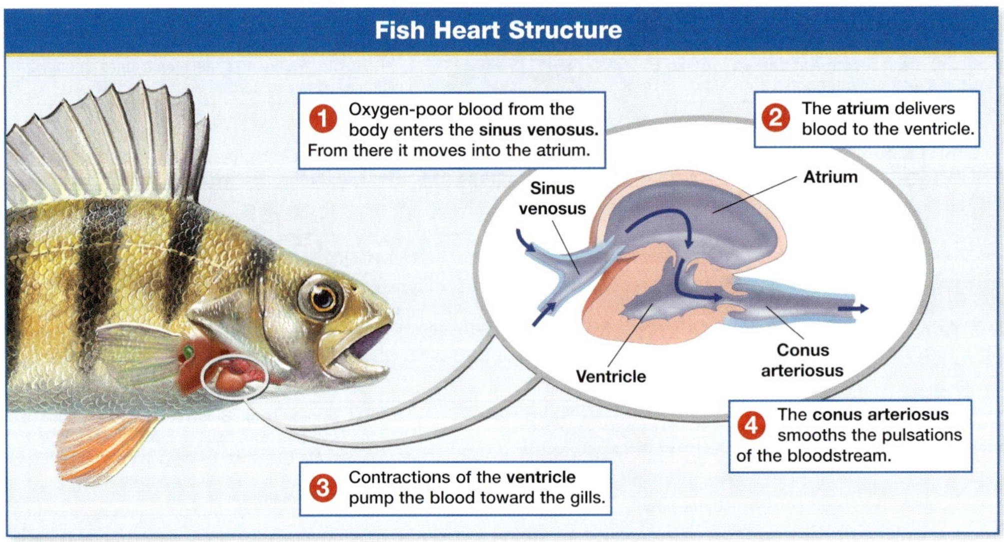

748

CHEMISTRY CONNECTION

Invite a chemistry teacher to your class to discuss the amazing chemical properties of water. Topics could include cohesion, surface tension, heat of vaporization, heat of fusion, polar covalent and ionic bonds, and the relationship between the temperature and density of water.

Career

Aquaculturist Aquaculture, or fish farming, is one of the fastest-growing sectors of agriculture. Both freshwater and saltwater species can be raised commercially. Operating a fish farm requires a knowledge of how to keep the fish's environment clean, proper feeding techniques, and how to control fish infections, such as those caused by bacteria and parasites.

Kidneys

Vertebrates have a body that is about two-thirds water, and most will die if the amount of water in their body falls much lower than this. Therefore, minimizing dehydration (water loss) has been a key evolutionary challenge facing all vertebrates. Even some fishes must cope with the problem of water loss. If this seems strange to you, remember that osmosis causes a net movement of water through membranes toward regions of higher ion concentration.

The ion (salt) concentration of sea water is three times that of the tissues of a marine bony fish. As a result, these fishes lose water to the environment through osmosis. To make up for the water they lose, marine bony fishes drink sea water and actively pump excess ions out of their body. Freshwater fishes have the opposite problem. Because their bodies contain more ions than the surrounding water, they tend to take in water through osmosis. The additional water dilutes their body salts, so freshwater fishes regain salts by actively taking them in from their environment.

Although the gills play a major role in maintaining a fish's salt and water balance, another key element is a pair of kidneys. Kidneys are organs made up of thousands of nephrons. **Nephrons** are tubelike units that regulate the body's salt and water balance and remove metabolic wastes from the blood. Excess water and bodily wastes leave the kidneys in the form of a fluid called *urine*. Marine fishes excrete small amounts of urine and rid their bodies of ammonia largely through their gills. Freshwater fishes excrete large amounts of dilute urine.

internet connect
www.scilinks.org
Topic: Kidneys
Keyword: HX4110
SCiLINKS. Maintained by the National Science Teachers Association

DATA LAB *Analyzing Ion Excretion in Fish* SC.F.1.4.1

Background

A few species of fish, such as adult salmon, are able to move between saltwater and freshwater environments. The graph at right shows the excreted ion concentration of a fish as it travels from one body of water to another. Examine the graph, and answer the analysis questions.

Ion Excretion in Fish

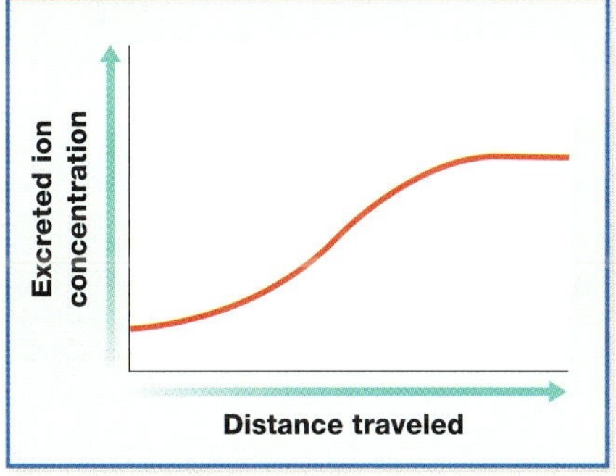

Analysis

1. **Determine** if the fish is losing or gaining ions by excretion as it travels.
2. **Critical Thinking Inferring Conclusions** Is the fish traveling from fresh to salt water or from salt to fresh water?
3. **Summarize** the reasoning you used to answer item 2.

Demonstration — GENERAL

The day before this lesson, fill two beakers with distilled water. Add one tablespoon of salt to one beaker and mix well. Cut two equal lengths of potato. Put one potato piece in each beaker, and let it sit overnight. The next day, tell students that each potato piece represents a freshwater fish. Tell them one fish has been placed in fresh water, and one has been placed in the ocean. Remove the potato pieces, and let students observe each. Ask students which "fish" was in the ocean. (the limp one) Ask them how they arrived at their conclusions. (In the ocean, the water concentration was greater in the potato than in the beaker, so water diffused out of it; the opposite is true for the freshwater potato.) Ask students why fishes have difficulty living in water environments other than their own. (They would need some mechanism to regulate their water and salt concentrations.) LS Visual

DATA LAB *Analyzing Ion Excretion in Fish*

Skills Acquired
Analyzing, inferring

Teacher's Notes

After finishing the three analysis questions, have students speculate on what mechanisms allow fish to move successfully between fresh water and salt water. (They should come up with ideas about osmosis through the kidneys and gills.)

Answers to Analysis

1. losing ions by excretion
2. from fresh water to salt water
3. The graph shows that the farther the fish travels, the greater its excreted ion concentration. A fish's tissues have a lower concentration of ions than is found in salt water, and they lose water through osmosis. One would expect the fish to excrete more ions as the water becomes saltier.

MATH CONNECTION

Salmon return to spawn in the stream where they were hatched. During spawning, females lay from 2,000 to 10,000 eggs. The young salmon hatch and make their way downstream to the sea; most perish during this journey. A Canadian study reported that fewer than 1 percent, or approximately 20 to 100 hatchlings, of the young from each female survive to spawn. ($2,000 \times 0.01 = 20$, and $10,000 \times 0.01 = 100$) LS Logical

Real Life

How many eggs do female fish lay during spawning season?

The number of eggs laid during spawning varies from species to species.

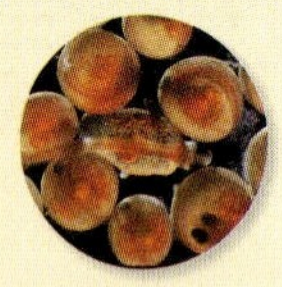

Some fishes, such as sea horses, lay between a dozen and several hundred eggs. These eggs are well protected, and most develop into young sea horses.

Finding Information

Find out how many eggs are laid by cod.

Reproduction

The sexes are separate in most fishes, and generally fertilization takes place externally. In a process called *spawning*, male and female gametes are released near one another in the water, as shown in **Figure 4.** A yolk sac within each egg contains nutrients the developing embryo will need for growth. The yolk sac remains attached to the hatchling fish but is quickly used up. Then, the growing fish must seek its own food. More likely than not, it will become food for some larger animal. Many species of fishes release large numbers of eggs that are fertilized in a single spawning season. This practice helps ensure that some individuals will survive to maturity.

The eggs of sharks, skates, and rays are fertilized inside the female's body. During mating, the male uses two organs called *claspers* to insert sperm into the female. In most species, the eggs develop inside the female and the young are born live. A few species of sharks lay eggs.

Figure 4 Fish spawning. These salmon spawn in shallow river waters. Thousands of eggs are released in a single mating, but only a small percentage of hatchlings live until adulthood.

Section 1 Review

1. **Discuss** the key characteristics found in all fishes. SC.G.1.4.1
2. **Describe** how countercurrent flow aids a fish in obtaining oxygen. SC.F.1.4.2
3. **Summarize** why the fish heart and circulatory system are considered important evolutionary changes. SC.D.1.4.4
4. **Contrast** reproduction in sharks with that of most other fishes. SC.F.2.4.1
5. **Critical Thinking Forming a Hypothesis** A student removes Fish A from a saltwater aquarium and Fish B from a freshwater aquarium. By mistake, the student returns each fish to the wrong aquarium. The next day, both fish are dead. Form a hypothesis that explains why. SC.F.1.4.1
6. *FCAT Prep* Which chamber of a fish's heart generates most of the force that pumps blood through the body?

A atrium
B ventricle
C sinus venosus
D conus arteriosus

SC.F.1.4.2

750

Answers to Section Review

1. Gills allow fish to extract oxygen from the water. Single-loop blood circulation allows blood to be pumped through the gills and back to the heart. A bony vertebral column surrounds and protects the spinal cord.
2. Countercurrent flow ensures that oxygen diffuses into the blood over the entire length of the gill capillaries.
3. The fish heart pumps blood through a single-loop circulatory pathway. The blood goes to the gills and then to the body tissues. As a result, fully oxygenated blood is delivered to the tissues.
4. Fertilization in sharks is internal; most other fishes have external fertilization.
5. Fish A was unable to prevent the inflow of water by osmosis when placed in fresh water. Fish B was unable to prevent the outward flow of water by osmosis when placed in salt water. In each case, the resulting imbalance in the fish's salt concentration caused the fish's death.
6. **A.** Incorrect. The atrium is large but its muscular walls are thin. **B.** Correct. The ventricle has thick, muscular walls capable of pumping blood with strong contractions. **C.** Incorrect. The sinus venosus is a collection chamber for oxygen-poor blood from the body. **D.** Incorrect. The conus arteriosus is a small chamber that smoothes the blood flow and provides only slightly more force.

Real Life

Answer

Cod can lay up to 500,000 eggs per kilogram of body weight.

Close

Reteaching — BASIC

K-W-L Tell students to return to their lists of things they want to know about fish, which they created for the Bellringer. Have them place check marks next to the questions that they are now able to answer. Students should finish by making a list of what they have **L**earned. Ask students:

- Which questions are still unanswered?
- What new questions do you have?

Quiz — GENERAL

1. What is the major respiratory organ of fishes? (gill)
2. What organ eliminates waste and helps maintain water and salt balance in fishes? (kidney)

Alternative Assessment — GENERAL

Have students make two columns on a sheet of paper. Have them label one column *External characteristics* and the other column *Internal characteristics*. Have them review the chapter and list the major internal and external characteristics of fishes.

Section Review Correlations

SC.D.1.4.4: Item 3
SC.F.1.4.1: Item 5
SC.F.1.4.2: Items 2, 6
SC.F.2.4.1: Item 4
SC.G.1.4.1: Item 1

Today's Fishes

Section 2

Florida Sunshine Standards/FCAT

SC.F.1.4.2 Body structures are uniquely designed and adapted for their function.

SC.G.1.4.1 There is great diversity and interdependence among living things.

Jawless Fishes

Perhaps the most unusual fishes found today are the surviving jawless fishes, the lampreys and hagfishes. These primitive creatures have changed little over the past 330 million years. Little is known about hagfishes, which is not surprising when you consider where they live—on the ocean floor at depths as great as 1,700 m (about 1 mi). Lampreys are better understood and are found in both salt and fresh water. Interestingly, all species of lampreys must return to fresh water to reproduce, suggesting that their ancestors lived in fresh water.

Lampreys and hagfishes have scaleless, eel-like bodies with multiple gill slits and unpaired fins. Their skeletons are made of cartilage, a strong fibrous connective tissue, and both kinds of fishes retain their notochord into adulthood. Hagfishes, such as the one shown in **Figure 5,** are scavengers of dead and dying animals on the ocean bottom. Because of this behavior, they are sometimes called the "vultures of the sea." When threatened, a hagfish can produce huge quantities of slime from its roughly 200 slime glands. Recently, biologists have discovered that hagfishes are far more numerous than once thought and play a vital role in the ecology of the oceans.

Most lampreys, such as the one shown in Figure 5, are parasitic on other living fishes. A lamprey has a suction-cup-like structure around its mouth that it uses to attach itself to its host. After attachment, the lamprey gouges out a wound with its rough tongue, feeding on blood and bits of flesh from the wound.

Objectives

- **Distinguish** between the three general categories of modern fishes. SC.G.1.4.1
- **Describe** the major external and internal characteristics of the yellow perch. SC.F.1.4.2
- **Summarize** features of bony fishes. SC.G.1.4.1

Key Terms

lateral line
operculum
swim bladder
teleost

internet connect

www.scilinks.org
Topic: Aquariums and Museums of Florida
Keyword: HX6F01

SCILINKS. Maintained by the National Science Teachers Association

Figure 5 **Hagfish and lamprey**

These two modern jawless fishes have changed little over the past 330 million years.

Hagfish

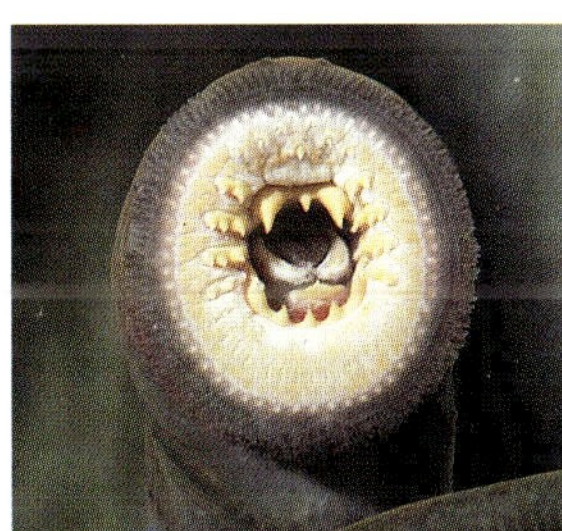

Lamprey

Focus

Overview

Before beginning this section review with your students the objectives listed in the Student Edition. This section introduces students to the three groups of fishes that exist today, and explains how they differ from each other.

Bellringer

Have students draw a quick picture of the strangest fish they have ever seen (and name it if possible). Tell students to list characteristics that it has in common with all fishes. Then ask them to list characteristics that they think are unique to this fish.

Motivate

Discussion — BASIC

Provide students with pictures of the following fishes: lamprey, eel, hagfish, hammerhead shark, dogfish shark, stingray, salmon, trout, and bass. Display the pictures at the front of the room, and explain that they are classified into three distinct groups. Have students try to determine which fish belong in the same groups and why. List the characteristics students mention as similar or dissimilar for each fish. (Students may note jaws or lack of jaws, paired or unpaired fins, scales or no scales, gill slits or no gill slits, etc.) Discuss how structural characteristics are often important in classifying organisms, and how the fishes are good examples of this.
LS Visual

Chapter Resource File

- Directed Reading BASIC
- Active Reading GENERAL
- Data Sheet for Quick Lab GENERAL

Transparencies

TR Bellringer
TR I3 Lateral Line in Fishes
TR I5 External Structures of Fish–Yellow Perch
TR I6 Internal Structures of Fish–Yellow Perch

One-Stop Planner CD-ROM

- Reading Organizers BASIC
- Reading Strategies BASIC
- Basic Skills Worksheet Mass and Density BASIC

Section Correlations

SC.F.1.4.2
SC.G.1.4.1

Teach

Teaching Tip — BASIC

Natural Selection of Body Shape To look for similarities in body shape, have students diagram the profile of several living and nonliving things that move quickly through water. Examples you might provide include a shark, a tuna, a bottlenose dolphin, a seal, a submarine, and a torpedo. All have the basic shape shown below. Ask students why this shape occurs in things that need to move rapidly through water. (Direct students to think about water resistance and the need to direct movement through water using fins.) **LS Visual**

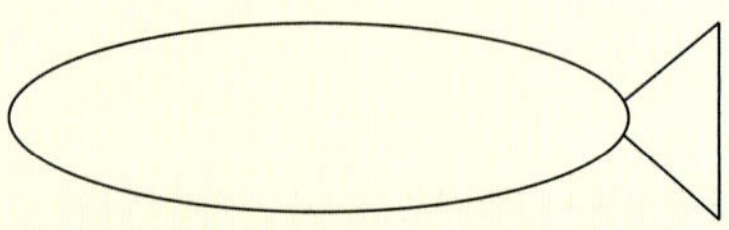

Using the Figure — GENERAL

In **Figure 6** point out how the shark's teeth are arranged in rows on its jaw, with the newer teeth behind the old ones. Explain that a shark's teeth and scales are very similar in structure and that the teeth probably evolved from scales. Both have an inner pulp cavity that is covered by a layer of dentine and then by enamel. The shark's teeth are not embedded in its jaw as human teeth are. Rather, they sit on top of the jaw and therefore break off easily. **LS Visual**

internet connect
www.scilinks.org
Topic: Sharks
Keyword: HX4163
SciLinks. Maintained by the National Science Teachers Association

Cartilaginous Fishes

Sharks, skates, and rays are cartilaginous *(KAHRT'l AJ uh nuhs)* fishes. Their skeletons are made of cartilage strengthened by the mineral calcium carbonate (the material oyster shells are made of). Calcium carbonate is deposited in the outer layers of cartilage and forms a thin layer that reinforces the cartilage. The result is a very light but strong skeleton.

The shark's light, streamlined body allows it to move quickly through the water in search of prey. Its skin contains cone-shaped placoid scales, which give the skin a rough texture. As you can see in **Figure 6,** the shark's scales and teeth are quite similar in structure. This is because the teeth are actually modified scales. The teeth are arranged in 6 to 10 rows along the shark's jaw. The teeth in front are pointed and sharp and are used for biting and cutting. Behind the front teeth, rows of immature teeth are growing. When a front tooth breaks or is worn down, a replacement tooth moves forward. One shark may use more than 20,000 teeth during its lifetime. This system of tooth replacement guarantees that the teeth being used are always new and sharp.

Two smaller groups of cartilaginous fishes, the skates and rays, have flattened bodies that are well adapted to life on the sea floor. Rays are usually less than 1 m (3.3 ft) long, while skates are typically smaller. However, the giant manta ray may be up to 7 m (23 ft) wide. Most species of skates and rays have flattened teeth that are used to crush their prey, mainly small fishes and invertebrates.

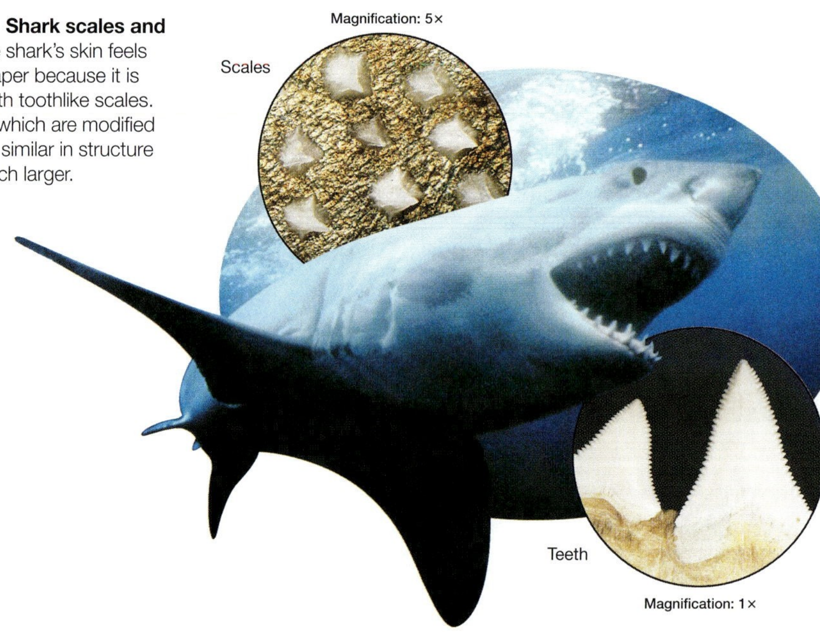

Figure 6 Shark scales and teeth. The shark's skin feels like sandpaper because it is covered with toothlike scales. The teeth, which are modified scales, are similar in structure but are much larger.

MISCONCEPTION ALERT

Fear of Sharks Shark attacks are often highly publicized. This has created a fear of shark attacks that is often greater than the actual probability of an attack. Bees, wasps, and snakes are responsible for far more fatalities each year. In the United States, the annual risk of death from being struck by lightning is 30 times greater than that from a shark attack. An attack by a shark is also less common than such beach-related injuries as spinal damage, dehydration, jellyfish and stingray stings, and sunburn.

Bony Fishes

Jawless and cartilaginous fishes are not as diverse as bony fishes, which are the most numerous of all the fishes. In addition to a strong, internal skeleton made completely of bone, bony fishes have a series of unique structural adaptations that contribute greatly to their success.

1. **Lateral line system.** Bony fishes have a fully developed lateral line system. The **lateral line,** shown in **Figure 7,** is a sensory system that extends along each side of a bony fish's body. As moving water presses against the fish's sides, nerve impulses from ciliated sensory cells within the lateral line permit the fish to perceive its position and rate of movement. For example, a trout moving upstream to spawn uses its lateral line system to obtain the sensory information it needs to orient its head upstream.

The lateral line system also enables a fish to detect a motionless object by the movement of water deflected by that object. The way that a fish detects an object with its lateral line and the way that you hear music with your inner ear are quite similar. Both processes share the same basic mechanism—sensory cells with cilia detect vibrations and send this information to the brain.

Study TIP

Organizing Information
Use the information on this page and the next four pages to draw a concept map that summarizes the characteristics of bony fishes. On your concept map, include information from *Up Close: Yellow Perch.*

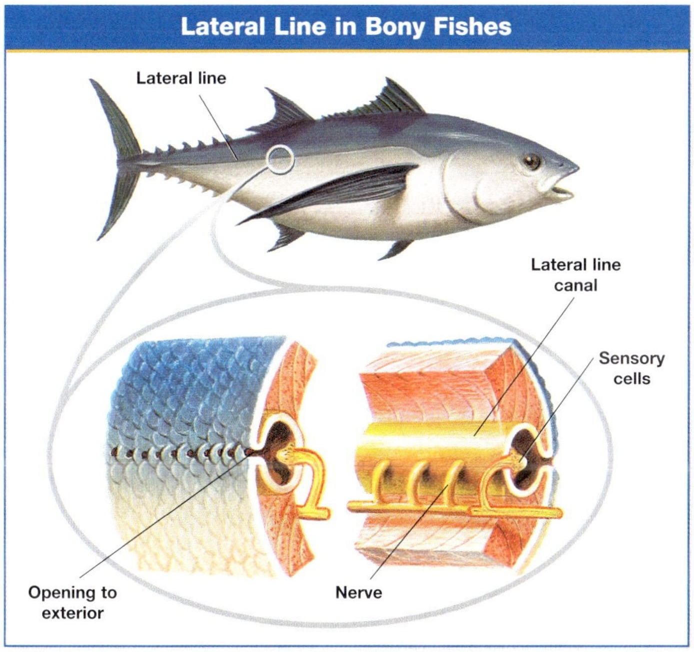

Figure 7 Lateral line. The lateral line contains sensory cells that help a fish perceive its position in the water.

Demonstration — GENERAL

Obtain a transparent tropical fish called a glass catfish, commonly available in pet stores. This fish gives a real-life view of the internal organs of a bony fish. Have students view the glass catfish and see what internal organs they can identify. Have them study the drawings of the internal structure of the yellow perch for help in identifying the organs. LS Visual English Language Learners

Teaching Tip

Train Your Fish Fish have the ability to detect pressure changes, such as sound waves, through their lateral line system, so it is possible to train a fish to come to the surface in response to a sound. Tell students who have fish to tap the side of their aquarium very gently and in the same manner each time they feed the fish. After a couple of weeks, the fish will come to the top to feed whenever the aquarium is tapped in this manner.

Activity — GENERAL

Fish and Diet It is a good idea to include some fish in one's diet. Have students go to local grocery stores and record the different types of fish offered. Remind students to look for canned and frozen fish as well as fresh fish. Some students may want to survey local restaurants to determine what kind of fish they offer. Have students research these fish to determine where they were caught or raised; their habitats; their nutrient, vitamin, and mineral contents; and any other interesting characteristics of the fish, such as characteristic behaviors. LS Kinesthetic

Integrating Physics and Chemistry

A fish's lateral line senses other objects in the water by detecting reflected water movements from nearby objects. These water movements or waves are an example of a mechanical wave—a wave that requires a medium. The energy of the vibration that sets a wave into motion moves through the medium as the wave passes through. Vibrating particles pass energy down the line to neighboring particles. Ask students to identify the medium that perpetuates the waves from an object to the lateral line of a fish.

did you know?

Salmon return to their hatching stream. Pacific salmon find their way to the stream where they hatched primarily by the sense of smell. Vegetation and minerals in the water give each stream a unique "flavor." Young salmon imprint on this flavor and can remember it 4 or 5 years later when they return to spawn. All Pacific salmon die shortly after spawning. Interestingly, when young salmon (smote stage) migrate to the sea, they go downstream backward.

Yellow Perch

Teacher's Notes

Have students relate the perch's external structures to its lifestyle and habitat. Have students pay special attention to the specialized senses and swimming adaptations of this successful fish. Then ask students to classify each labeled internal structure of the perch according to the following body systems: nervous system, circulatory system, reproductive system, digestive system, respiratory system, excretory (urinary) system, muscular system, and skeletal system.

Discussion

1. How does the yellow perch control roll, pitch, and yaw when it swims? (The dorsal fins prevent roll, the caudal fin controls pitch, and the anal fin controls yaw.)

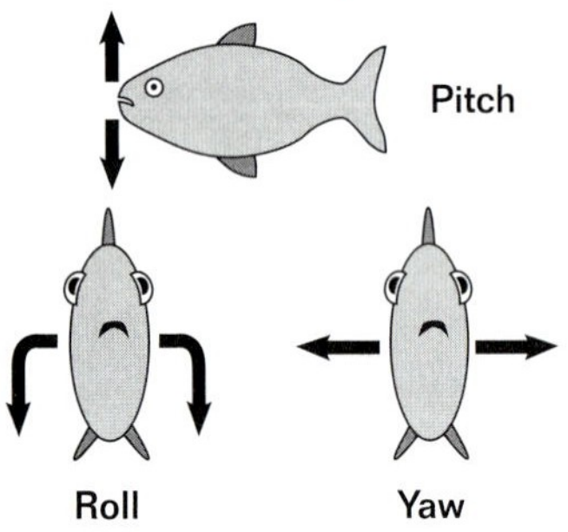

2. How would a perch find food if it were kept in the dark? (It could smell the food and sense the food's movement with its lateral line.)

Yellow Perch

- **Scientific name:** *Perca flavescens*
- **Size:** About 0.3 m (1 ft) long and up to 2.3 kg (5 lb)
- **Range:** Found in lakes and rivers from the Great Lakes to the Atlantic coast and as far south as South Carolina
- **Habitat:** Lives concealed among vegetation or submerged tree roots
- **Diet:** Feeds on insect larvae, crustaceans, and other fishes

External Structures

Lateral line The lateral line is a sense organ that detects vibrations in water that are caused by currents or pressure waves. The perch uses this sensory information to direct its movement as it swims and to detect objects in its environment, including predators and prey.

Opercula Movements of the opercula draw water into the perch's mouth. The water then moves over the gills, where oxygen and carbon dioxide are exchanged. Then the water is forced out through the gill slits.

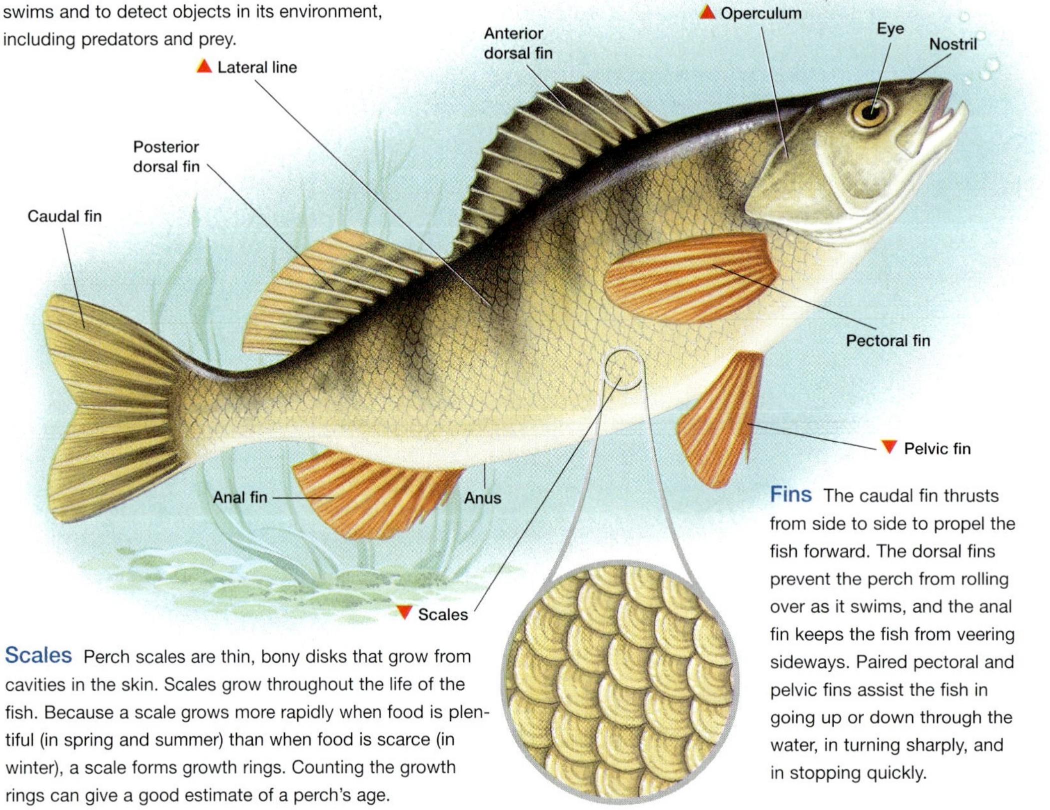

Scales Perch scales are thin, bony disks that grow from cavities in the skin. Scales grow throughout the life of the fish. Because a scale grows more rapidly when food is plentiful (in spring and summer) than when food is scarce (in winter), a scale forms growth rings. Counting the growth rings can give a good estimate of a perch's age.

Fins The caudal fin thrusts from side to side to propel the fish forward. The dorsal fins prevent the perch from rolling over as it swims, and the anal fin keeps the fish from veering sideways. Paired pectoral and pelvic fins assist the fish in going up or down through the water, in turning sharply, and in stopping quickly.

754

Reproductive organs Yellow perch produce gametes during their breeding season in the spring. The female lays strings of eggs that are fertilized externally. In warm water, the young hatch within days and grow quickly. In cold water, the development of the eggs may take much longer.

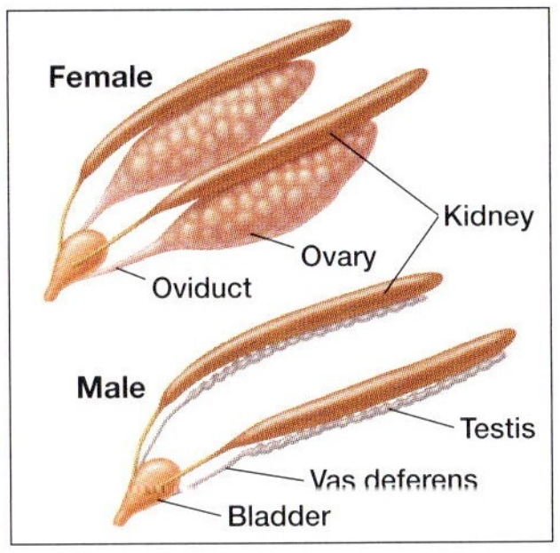

▲ Reproductive organs

Brain The optic lobe receives information from the eyes, and the olfactory bulbs receive information from chemical-sensing cells. The cerebrum processes this and other information. The cerebellum coordinates muscle activity, and the medulla oblongata controls the function of many internal organs.

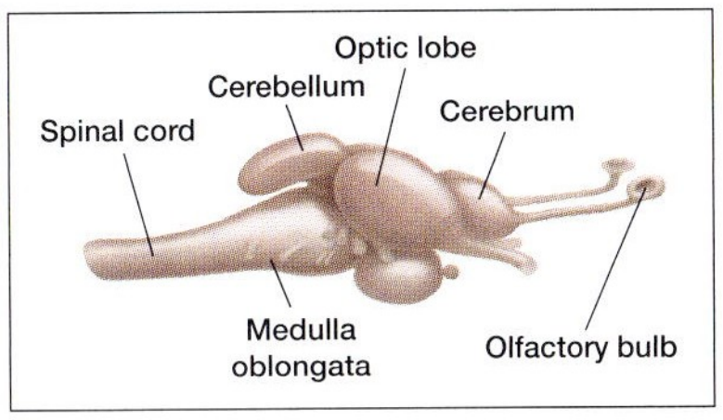

▲ Brain

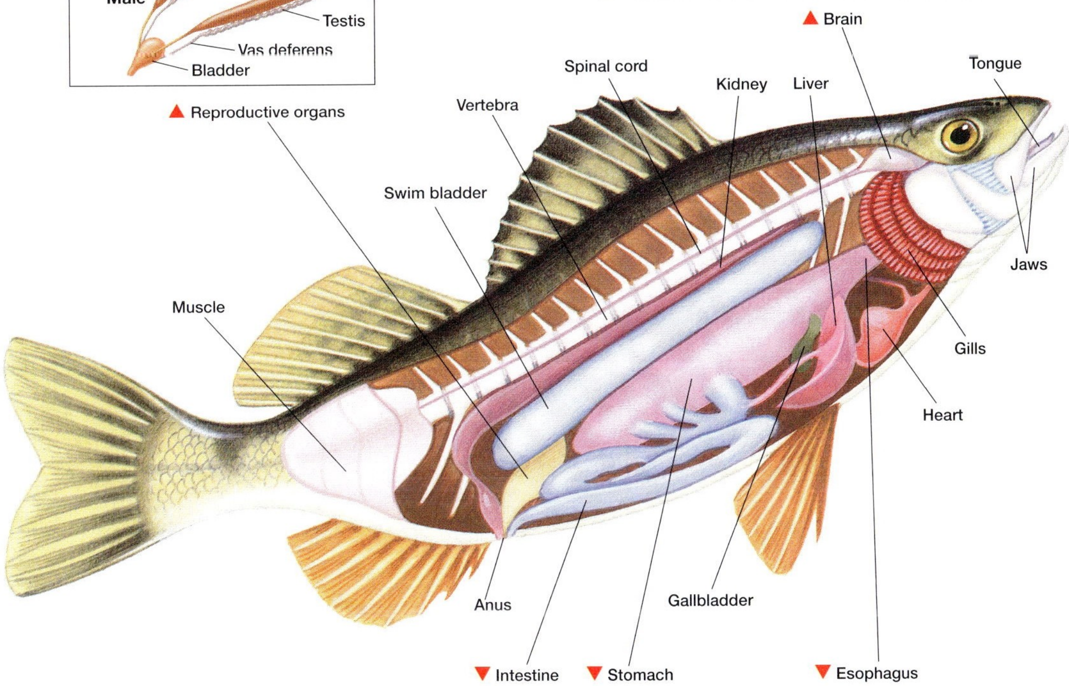

▼ Intestine ▼ Stomach ▼ Esophagus

Digestive system The digestive system reflects a basic arrangement of structures found in all vertebrates. Food enters the mouth and passes from the esophagus to the stomach. The liver secretes bile, and the pancreas secretes enzymes into a short intestine. The bile and enzymes help break down the food. Absorption of digested food occurs through the inner lining of the intestine. Undigested material exits through the anus.

755

Up Close

Yellow Perch

3. Why are the mouth and opercula needed for a perch to respire? (Movements of the opercula draw water in through the open mouth and force the water over the gills, where oxygen is removed from the water.)

4. What would happen to a yellow perch that was missing each body system? (nervous system—it could not move or sense its environment; circulatory system—its cells could not get oxygen and nutrients or get rid of wastes; reproductive system—it could not produce offspring; respiratory system—it could not obtain oxygen or eliminate carbon dioxide; digestive system—it could not break down food; excretory (urinary) system—it could not get rid of nitrogen wastes; muscular system—it could not move; skeletal system—its body would collapse)

5. What would happen if the swim bladder of a perch was suddenly unable to absorb more gas or get rid of gas? (The perch would be unable to achieve neutral buoyancy at different levels in the water. It would have to use more energy to move up or down in the water.)

6. Why is the number of sperm cells produced by the perch so much greater than the number of egg cells? (Because of external fertilization, most sperm cells will not reach the egg cells. A great number is needed to ensure that each egg is fertilized.)

Teach, *continued*

QUICK LAB *Modeling the Action of a Swim Bladder*

Skills Acquired
Observing, analyzing, applying knowledge

Teacher's Notes
Have all students drop their raisins into the soft drink at the same time. For a more dramatic effect, have students use a graduated cylinder instead of a beaker.

Answers to Analysis

1. The raisins sank to the bottom. As bubbles collected on them, the raisins rose to the top. When they reached the top, they sank again.
2. The bubbles clung to the raisins and imparted buoyancy, causing the raisins to rise. When the raisins reached the top, the bubbles burst, and the raisins sank again.
3. The bubbles formed on the outside of the raisin. A swim bladder is internal. Also, the amount of gas in the swim bladder can be increased or decreased, unlike the bubbles on the raisins.
4. It would take a great deal of energy to stay in one position under water with two lungs full of air. A swim bladder offers a fish tremendous savings of energy because the fish controls the amount of gas in the swim bladder. This makes more energy available for daily activities, such as catching prey or escaping predators.

Figure 8 Operculum. When a fish's mouth opens, water enters and the opercula close over the gills. When the mouth closes, the opercula open and water moves across the gills and out of the fish.

2. **Gill cover.** Most bony fishes have a hard plate, an **operculum,** that covers the gills on each side of the head. Movements of certain muscles and of the opercula, shown in **Figure 8,** permit a bony fish to draw water over the gills, which enables the fish to take in oxygen. By using this mechanism, most bony fishes can move water over their gills while remaining stationary in the water. A bony fish doesn't have to swim forward with its mouth open to move water over its gills. This ability to respire without swimming enables a bony fish to conserve energy that can be spent chasing after prey and escaping from predators.
3. **Swim bladder.** The density of the fish body is slightly greater than that of sea water. How then do bony fishes keep from sinking? Bony fishes contain a special gas sac called a **swim bladder.** By adjusting the gas content of the swim bladder, bony fishes can regulate their buoyancy. As the swim bladder fills, the fish rises, and as it empties, the fish sinks. The swim bladder of early bony fishes was connected to their throat, and they gulped air to fill it. The swim bladder of modern bony fishes, shown on the previous page, does not have a direct passage to the mouth. Instead, gas is exchanged between the bloodstream and the swim bladder. This permits the fish to maintain or change its depth in the water.

There are two groups of bony fishes, the ray-finned fishes and the lobe-finned fishes. The yellow perch described in *Up Close: Yellow Perch* on the previous two pages is a common type of ray-finned fish.

QUICK LAB *Modeling the Action of a Swim Bladder* SC.F.1.4.2

Most fish use a swim bladder to regulate their depth in water. As gas enters the swim bladder, the fish rises in the water. As gas is expelled, the fish sinks to a lower depth.

Materials

100 mL beaker or small glass; cold, clear, carbonated soft drink; 2 very dry raisins

Procedure

1. Fill a 100 mL beaker with a cold, carbonated soft drink.
2. Drop two raisins into the beaker, and observe what happens over the next 5 minutes.

Analysis

1. **Describe** what happened after you dropped the raisins into the soft drink.
2. **Forming Hypotheses** Develop a hypothesis to explain your observations.
3. **Critical Thinking Analyzing Results** How does the lifting of the raisins differ from the use of a swim bladder to control buoyancy?
4. **Critical Thinking Forming Reasoned Opinions** Think about the energy you would have to expend to keep yourself in one position under water. What advantage might a swim bladder provide to a fish?

did you know?

Swim bladders can be damaged. Swim bladders on some fish have been known to leak or rupture, releasing the gas into the body cavity. In some instances, the fish floats to the top of the water, causing it to be easy prey for hungry predators. Only by frenzied swimming can the fish manage to stay under water, and it will soon tire and return to float on the top. Ask students to imagine trying to dive under water while holding a large balloon.
LS Intrapersonal

Integrating Physics and Chemistry

Remind students that buoyancy increases as the density of an object decreases. Ask them what happens to a fish's average density and therefore its buoyancy when the swim bladder increases or decreases in size. Also, ask students whether a fish's swim bladder should increase of decrease in size if a fish wants to swim in deeper water.

Ray-Finned Bony Fishes

Ray-finned bony fishes, such as the ones shown in **Figure 9,** comprise the vast majority of living fishes. Their fins are supported by bony structures called *rays*. **Teleosts** *(TEL ee ahsts),* such as the yellow perch you saw in *Up Close,* are the most advanced of the ray-finned bony fishes. Teleosts have highly mobile fins, very thin scales, and completely symmetrical tails. About 95 percent of all living fish species are teleosts.

Figure 9 **Pacific bluefin tuna**

At sexual maturity, Pacific bluefin tuna can weigh about 136 kg (300 lb), although some grow larger.

Lobe-Finned Bony Fishes

Only seven species of lobe-finned fishes survive today. One species is the coelacanth *(SEE luh kanth),* shown in **Figure 10,** and the other six species are all lungfishes. The lobe-finned fishes have paired fins that are structurally very different from the fins of ray-finned fishes. In many lobe-finned fishes, each fin consists of a long, fleshy, muscular lobe that is supported by a central core of bones. These bones are connected by joints, like the joints between the bones in your hand. Bony rays are found only at the tips of each lobed fin. Muscles within each lobe can move the bony rays independently of each other.

Scientists have debated whether the direct ancestor of amphibians was a coelacanth or a lungfish. However, recent evidence has led biologists to believe that it was neither. The ancestor of the amphibians most likely was a third type of lobe-finned fish that is now extinct.

Figure 10 **Coelacanth.** Coelacanths were thought to have been extinct for millions of years, until one was caught off the coast of Africa in 1938. Coelacanths can reach up to nearly 3 m (9.8 ft) in length.

Section 2 Review

1. **Compare** the three categories of modern fishes. SC.G.1.4.1
2. **Summarize** the role of the operculum in fish respiration. SC.F.1.4.2
3. **Summarize** how the swim bladder can be viewed as an energy-saving mechanism. SC.F.1.4.2
4. **Describe** the digestive process in a yellow perch. SC.F.1.4.2
5. **Relate** a yellow perch's lateral line system to the human ear. SC.F.1.4.2
6. **Critical Thinking** **Evaluating Conclusions** An unidentified species of fish has rough skin, several rows of teeth, and no opercula. Based on these characteristics, a student infers that the fish has a swim bladder. Explain why you agree or disagree with this conclusion. SC.F.1.4.2
7. *FCAT Prep* What is the mouth of the lamprey specialized for?
   A chewing seaweed
   B straining plankton
   C parasitizing other fish
   D scavenging dead animals
   SC.F.1.4.2

757

Close

Reteaching — Basic

Have students work in small groups to make a table that compares each type of fish they have studied in this section. Across the top of their table they should write *Jawless Fish*, *Cartilaginous Fish*, and *Bony Fish*. The titles of the rows down the side of the table will vary from group to group. Have groups fill in the table with information they have learned. Each group can then present its table to the class. Have individual students copy his or her group's table and place it in his or her portfolio.

Quiz — General

1. How are the skeletons of sharks, skates, and rays similar to those of jawless fishes? (Both groups of fishes have skeletons made of cartilage.)
2. What are teleosts and how common are they? (Teleosts are ray-finned bony fishes with highly mobile fins, thin scales, and symmetrical tails. They represent 95% of all living fish species.)

Alternative Assessment — General

After reviewing the section, have each student choose a favorite fish. Have them give a short oral report, explaining to the class how the particular characteristics of their chosen fish enable it to survive and give it an advantage over other types of fish. Have each student write the name and category of their fish on the board prior to their presentation and provide a picture or illustration of the fish for class viewing.

Answers to Section Review

1. Agnathans: jawless, scaleless, have unpaired fins. Cartilaginous fishes: cartilaginous skeletons, streamlined bodies, placoid scales. Bony fishes: bony skeletons, opercula, swim bladders, well-developed lateral line system.
2. The operculum pumps water through the mouth and over the gills for oxygen extraction.
3. A fish can maintain a particular depth in the water by adjusting the amount of gas in the bladder rather than by using muscular force.
4. Food enters the mouth and moves to the stomach. Substances in the intestine help break down food. Nutrients are absorbed in the intestine and undigested food exits the body through the anus.
5. Both have sensory cells that sense vibrations and send nerve impulses to the brain.
6. Disagree; these are characteristics of cartilaginous fishes, which have no swim bladder.
7. **A.** Incorrect. See answer C. **B.** Incorrect. Whales have comb-like teeth for straining plankton. **C.** Correct. The lamprey's mouth is like a suction-cup to attach itself to its host. **D.** Incorrect. Hagfishes have mouths adapted to eating dead animals.

Section Review Correlations

SC.F.1.4.2: Items 2-7
SC.G.1.4.1: Item 1

Section 3

Focus

Overview

Before beginning this section review with your students the objectives listed in the Student Edition. This section introduces students to how amphibians differ from fish and discusses the adaptations amphibians have for life in and out of water.

Bellringer

Tell students to think about how big the largest frog species gets. Then ask them to list some things that a frog that size would eat. (The largest frog species is *Gigantorana goliath* of West Africa. It can be more than 30 cm (1 ft) long and can weigh over 3.3 kg (7.5 lb). It eats other animals as large as rats and ducks.

Motivate

Activity — GENERAL

Explain that a caecilian is a legless, wormlike animal. Ask students to close their books, and write the words Apoda, Anura, and Urodela at the top of three columns on the board. Tell students that these order names come from the Greek language. To the side write *a*—"without," *oura*—"tail," *pous*—"foot," and *delos*—"visible." Ask students to place the following amphibians into the proper columns by thinking of each animal's characteristics and what the three column headings mean: toad, salamander, bullfrog, caecilian, newt, and tree frog. (Apoda—caecilian; Anura—toad, bullfrog, and tree frog; Urodela—salamander and newt) LS Verbal

Section Correlations

SC.F.1.4.2
SC.G.1.4.1

Section 3 Amphibians

Florida Sunshine Standards/FCAT

SC.F.1.4.2 Body structures are uniquely designed and adapted for their function.

SC.G.1.4.1 There is great diversity and interdependence among living things.

Objectives

- **Summarize** the characteristics of modern amphibians. SC.G.1.4.1
- **Compare** the three orders of living amphibians. SC.G.1.4.1
- **Describe** the major external and internal characteristics of the leopard frog. SC.F.1.4.2

Key Terms

lung
pulmonary vein
septum

Key Characteristics of Modern Amphibians

The next time you see a frog, consider that it is a surviving member of an ancient amphibian group, the first vertebrates to walk on land. The croaking and peeping of frogs, such as the one shown in **Figure 11,** make it difficult not to notice them, but their smaller, quieter relatives live nearby, hidden in damp habitats. Class Amphibia contains three orders of living amphibians: order Anura (frogs and toads), order Urodela (salamanders and newts), and order Apoda (caecilians). Most amphibians share five key characteristics.

1. **Legs.** The evolution of legs was an important adaptation for living on land. Frogs, toads, salamanders, and newts have four legs. Caecilians lost their legs during the evolutionary course of adapting to a burrowing existence.
2. **Lungs.** Although larval amphibians have gills, most adult amphibians breathe with a pair of lungs. Lungless salamanders are an exception.
3. **Double-loop circulation.** Two large veins called *pulmonary veins* return oxygen-rich blood from the lungs to the heart. The blood is then pumped to the tissues at a much higher pressure than in the fish heart.
4. **Partially divided heart.** The atrium of the amphibian heart is divided into left and right sides, but the ventricle is not. A mixture of oxygen-rich and oxygen-poor blood is delivered to the amphibian's body tissues.
5. **Cutaneous respiration.** Most amphibians supplement their oxygen intake by respiring directly through their moist skin. Cutaneous respiration ("skin breathing") limits the maximum body size of amphibians because it is efficient only when there is a high ratio of skin surface area to body volume.

Figure 11 **Spring peeper.** In some areas, the call of the spring peeper is one of the first signs of spring.

Lungs

Although air contains about 20 times as much oxygen as sea water does, gills cannot function as respiratory organs when out of water. Thus, one of the major challenges that faced the first land vertebrates was that of obtaining oxygen from air. The evolutionary solution to this challenge was the lung.

A **lung** is an internal, baglike respiratory organ that allows oxygen and carbon dioxide to be exchanged between the air and the bloodstream. The amount of oxygen a lung can absorb depends on its internal surface area. The greater the surface area, the greater the amount of oxygen that can be absorbed. In amphibians, the

758

Chapter Resource File

- Directed Reading BASIC
- Active Reading GENERAL

One-Stop Planner CD-ROM

- Reading Organizers BASIC
- Reading Strategies BASIC
- Supplemental Reading Guide Through a Window ADVANCED

Transparencies

TR Bellringer
TR I11 Amphibian Heart Structure
TR I13 Life Cycle of a Frog
TR I14 External Structure of a Frog–Leopard Frog
TR I15 Internal Structure of a Frog–Leopard Frog

lungs are hardly more than sacs with folds on their inner membrane that increase their surface area, as shown in **Figure 12.** With each breath, fresh oxygen-rich air is drawn into the lungs. There it mixes with a small volume of air that has already given up most of its oxygen. Because of this mixing, the respiratory efficiency of lungs is much less than that of gills. Because there is much more oxygen in air than there is in water, however, lungs do not have to be as efficient as gills. Many amphibians also obtain oxygen through their thin, moist skin.

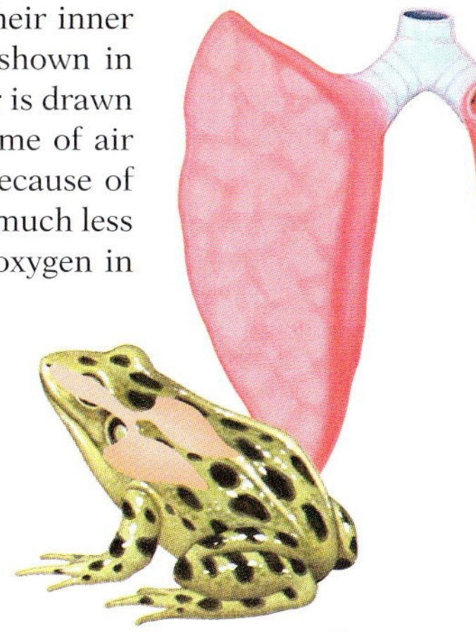

Figure 12 Amphibian lungs. The lungs of an amphibian are sacs with a folded internal membrane that provides a large surface for gas exchange.

Double-Loop Circulation

As amphibians evolved and became active on land, their circulatory system changed, resulting in a second circulatory loop. This change allowed more oxygen to be delivered to their muscles. **Figure 13** compares the single-loop circulation of most fishes with the double-loop circulation of amphibians (also found in lungfishes). Notice that amphibians have a pair of blood vessels not found in fishes, the pulmonary veins. The **pulmonary veins** carry oxygen-rich blood from the amphibian's lungs to its heart. The heart pumps the oxygen-rich blood to the rest of the body. The advantage of this arrangement is that oxygen-rich blood can be pumped to the amphibian's tissues at a much higher pressure than it can in fishes. (Recall that in fish, blood is pumped through the gills before reaching the body organs. As a result, much of the force of the heartbeat is lost.)

Figure 13 Circulatory loops

Circulation in fishes involves a single loop. Amphibians have a second loop that goes from the heart to the lungs and back to the heart.

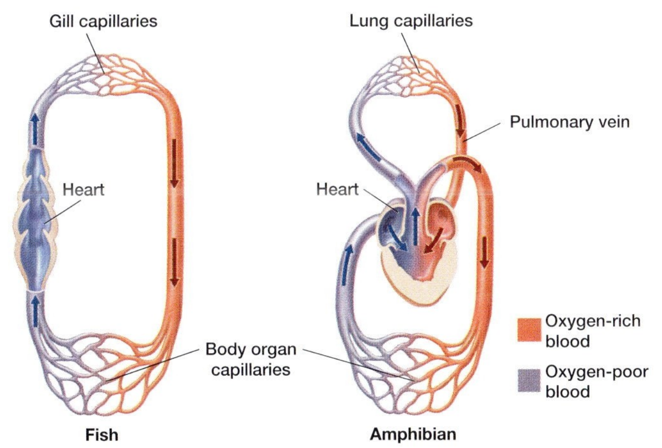

Teach

SKILL BUILDER — GENERAL

Reading Organizer Have students create a Venn diagram similar to the one shown in the Graphic Organizer at the bottom of this page. As students read Section 3, have them list, in the appropriate regions of the diagram, the characteristics that are exclusive to fish, exclusive to amphibians, and common to both groups. After the diagram is complete, have each student use the information in the diagram to write a paragraph that either supports or rejects the placement of fish and amphibians into two separate groups. **LS Logical**

Teaching Tip — BASIC

Gas Exchange Tell students that although most amphibians have lungs, some species of salamanders exchange oxygen and carbon dioxide through gills, even as adults. Ask students where they would expect to find these amphibians and why. (in an aquatic environment; because gills collapse in air) Tell students that many species of salamanders have neither lungs nor gills and respire entirely through their skin. **LS Verbal**

Using the Figure — ADVANCED

Use **Figure 13** to compare the circulatory pathways of fish and amphibians. Ask: Which system is a single loop and which is a double loop? (The fish system is a single loop, and the amphibian system is a double loop.) What is the function of the second loop in the amphibian system? (The second loop creates a circulatory pathway to and from the lungs.) What advantage does the double-loop system provide? (The amphibian's heart pumps oxygen-rich blood to the amphibian's tissues at much higher pressures than can happen in the fish's single-loop system.) **LS Visual**

Graphic Organizer

Use this graphic organizer with the **Skill Builder** on this page.

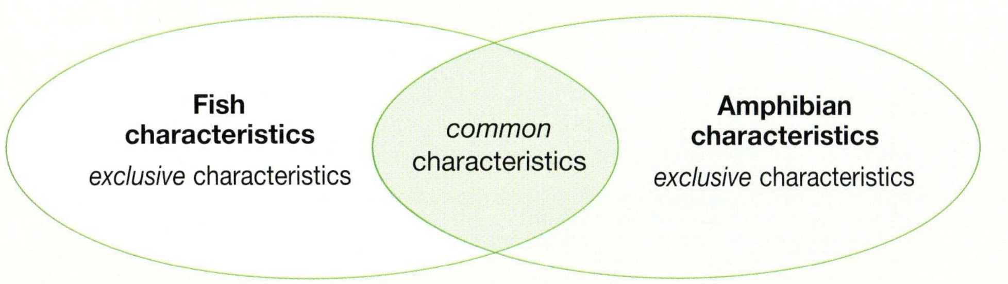

Teach, *continued*

Teaching Tip — GENERAL

Comparing Heart Anatomy

Have students compare the heart of the amphibian in **Figure 14** with that of the fish in **Figure 3.** Ask them to list the similarities and differences between the two. Then have each student compare his or her list with that of a partner. English Language Learners **LS Visual**

Demonstration — GENERAL

To illustrate that oxygen-rich and oxygen-poor blood tends to stay separate in the amphibian's ventricle, show how two liquids with different densities tend to separate. Fill a graduated cylinder about half full with corn syrup. Ask students what they think will happen if you add colored water to the cylinder. (The clear corn syrup will remain at the bottom.) Then fill the rest of the cylinder with colored water. In a second cylinder, reverse the order in which you add the liquids. Ask students to predict what will happen when you add the corn syrup to the colored water. (The syrup will sink to the bottom.) **LS Visual**

Using the Figure — GENERAL

Have students closely examine **Figure 14.** Tell them that the heart shown in the frog is drawn in dorsal view, with the blue sinus venosus on top. Explain that the heart shown in the enlargement is drawn in ventral view. The sinus venosus is not visible in the enlargement, but the three blue veins that carry blood to the sinus venosus are visible. **LS Visual**

internet connect

www.scilinks.org

Topic: Circulation in Amphibians

Keyword: HX4043

SCILINKS. Maintained by the National Science Teachers Association

Circulation of Blood

Not only did the path of circulation in amphibians change, but several important changes occurred in the heart. As you read about these changes, use **Figure 14** to trace the flow of blood through the amphibian heart.

The sinus venosus continues to deliver oxygen-poor blood from the body to the right side of the heart, as shown in step 1. (You can see the sinus venosus on the left side of Figure 14.) In addition, oxygen-rich blood from the lungs enters the left side of the heart directly, as shown in step 2.

A dividing wall known as the **septum** separates the amphibian atrium into right and left halves. You cannot see the septum in Figure 14 as it is beneath the conus arteriosus. The septum prevents the complete mixing of oxygen-rich and oxygen-poor blood as each enters the heart. As shown in step 3, both types of blood empty into a single ventricle, where some mixing of oxygen-rich and oxygen-poor blood occurs. However, due to the anatomy of the ventricle, the two streams of blood remain somewhat separate, as shown in step 4. Oxygen-rich blood tends to stay on the side that directs blood toward the body. Oxygen-poor blood tends to stay on the side that directs blood toward the lungs.

A number of amphibians have a spiral valve that divides the conus arteriosus. The spiral valve also helps to keep the two streams of blood separate as they leave the heart. Even so, some oxygen-poor blood is delivered to the body's tissues. Recall, however, that amphibians also obtain oxygen through their skin. This additional oxygen partly offsets the limitations of their circulatory system.

Figure 14 Amphibian heart. These four steps show how blood flows through the heart of an amphibian.

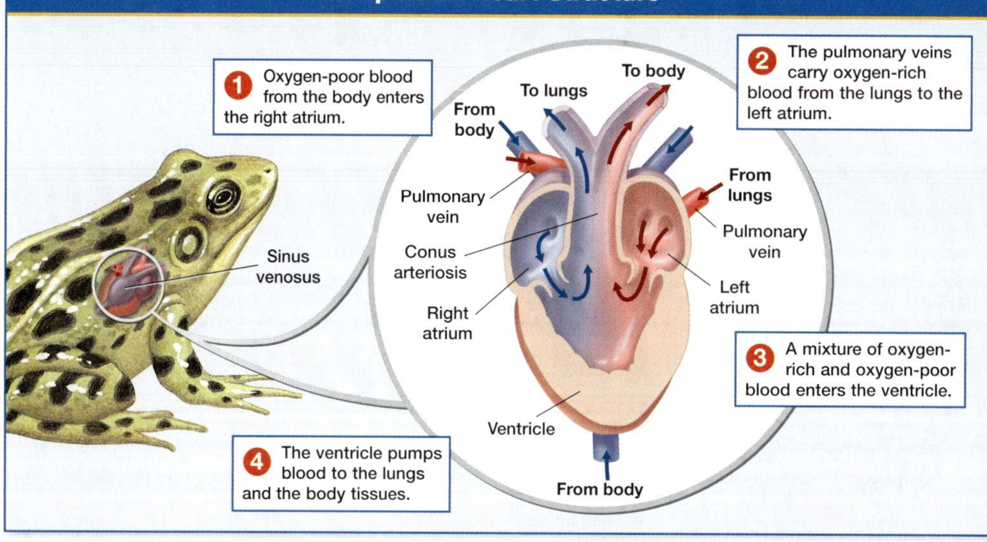

760

MISCONCEPTION ALERT

Do toads cause warts? Some students may have the false idea that they can get warts from toads. Explain that this is not true. Warts are caused by viruses that are not carried on toads. The rough, "warty" appearance of the toad's skin is an adaptation for living successfully in dry habitats and is not due to skin problems.

Integrating Physics and Chemistry

Fish gills and amphibian lungs have large surface areas for gas exchange which increases the amount of oxygen absorbed. Other factors, such as stirring or shaking and usually heating, also increase the rate at which a solute dissolves in a solvent. However, remind students that even though the rate at which a solute dissolves varies, a solute's solubility is constant at a given temperature.

Frogs and Toads

The order Anura is made up of frogs and toads that live in environments ranging from deserts to rain forests, valleys to mountains, and ponds to puddles. Adult anurans are carnivorous, eating a wide variety of small prey. Some species have a sticky tongue that they extend rapidly to catch their prey. The frog body, particularly its skeleton, is adapted for jumping, and its long muscular legs provide the power. To learn about the leopard frog, see *Up Close: Leopard Frog* on the next two pages. Toads, such as the one shown in **Figure 15,** are very similar to frogs but have squat bodies and shorter legs. Their skin is not smooth like that of a frog but is covered with bumps.

Figure 15 **Toad.** Toads like this common Asian toad have dry, bumpy skin and relatively short legs.

Reproduction in Frogs

Like most living amphibians, frogs depend on the presence of water to complete their life cycle. The female releases her eggs into the water and a male's sperm fertilize them externally. After a few days, the fertilized eggs hatch into swimming, fishlike larval forms called *tadpoles*. Tadpoles breathe with gills and feed mostly on algae. After a period of growth, the body of the tadpole changes into that of an adult frog. The rate at which tadpoles develop depends on the species and the availability of food. This process of dramatic physical change, called *metamorphosis*, is shown in **Figure 16.**

Figure 16 **Frog life cycle**

The transition of a larval frog (tadpole) to an adult involves a complex series of external and internal body changes.

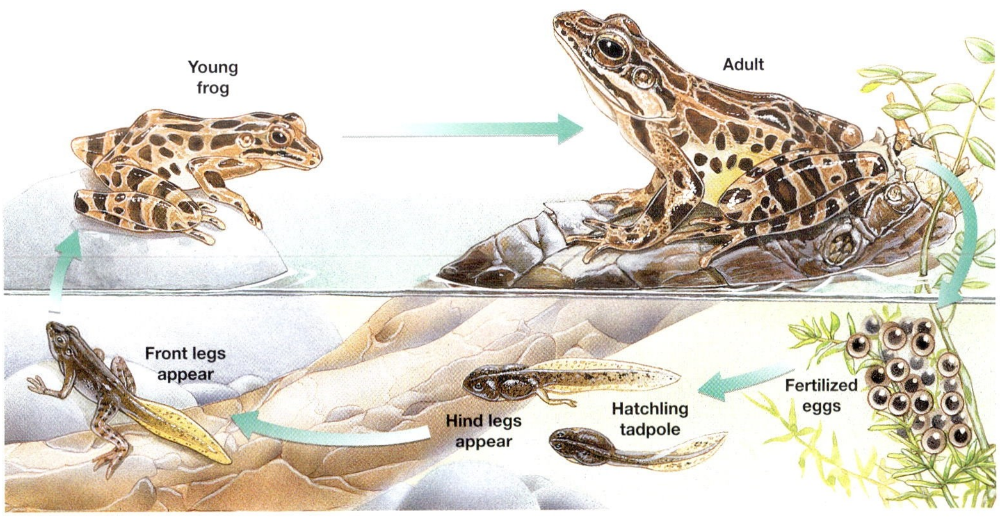

761

Teaching Tip — BASIC

Tadpole Metamorphosis

Purchase some "hind limb bud-stage" tadpoles from a biological supply company. Keep the tadpoles in a small, aerated aquarium, and have students make written or drawn observations every few days. Students will note that the tail gets shorter, the body gets squarer, and the hind limb buds develop into hind legs. Some tadpoles also may begin to develop front legs, but the mortality rate at this stage is high. Point out that while these dramatic external changes are taking place, important internal changes are occurring as well. In most species, the tadpole's digestive system changes to accommodate the switch from an herbivorous to a carnivorous diet. The respiratory system changes from gills to lung-and-skin respiration in the adult frog. If neither plant nor animal food is available to the metamorphosing tadpole at the precise time, it can starve. Likewise, if the tadpole does not have a place to rest its head above water, it can drown as its respiratory system changes.

LS Visual — English Language Learners

Using the Figure — GENERAL

Have students examine **Figure 16.** Have them note the external changes that are taking place in the frog. Then ask students to hypothesize about what internal changes are also occurring. Have them think about the different lifestyles of a frog and a tadpole. (digestive system—herbivore to carnivore; respiratory system—gills to lungs)

LS Visual

INCLUSION Strategies

- Developmental Delay
- Learning Disability
- Attention Deficit Disorder

Have students create a poster showing the differences between frogs and toads. The poster should compare and contrast the characteristics of these amphibians, describing the individual characteristics of the frog and the toad, as well as the similarities or differences of how and where these two amphibians live. The poster should have drawings or pictures of different frogs and toads.

did you know?

Some frogs have developed amazing ways to prevent their eggs from drying out. The female Surinam toad (actually a frog) of South America carries her eggs in pockets of skin on her back. As many as 60 young pass through the tadpole stage while imbedded in her back and then emerge as small frogs. Another species of frog in Australia swallows the eggs and keeps them in the stomach until the young emerge out of the adult's mouth.

Up Close

Leopard Frog

Teacher's Notes

- Have pairs of students make two columns on a sheet of paper. Ask them to label one *Water* and the other *Land*. As students read about the leopard frog, have them put the characteristics of the frog into one or both of the columns, depending on the nature of each characteristic. For example, *Tympanic membrane* would go in both columns, while *Webbed toes* would go under *Water*. Lead a discussion in which conflicting results are analyzed.
- Have students compare the organ systems of the frog with those of the yellow perch. Have them describe the similarities and differences.

Discussion

Guide discussion by posing the following questions:

1. How do you think the leopard frog got its name? (its spotted appearance)
2. How is the ear of the frog similar to the lateral line of the yellow perch? (Both structures contain ciliated sensory cells.)
3. Why do frogs produce so many eggs and sperm? (Their eggs and larvae are eaten by many predators.)
4. How do leopard frogs breathe? (with lungs and through their skin)

Leopard Frog

- **Scientific name:** *Rana pipiens*
- **Size:** Body length (legs excluded) of 5–9 cm (2–3.5 in.)
- **Range:** From northern Canada to southern New Mexico and from eastern California to the Atlantic coast
- **Habitat:** Lives in the short grass of meadows and around ponds
- **Diet:** Feeds on crickets, mosquitoes, and other small animals

External Structures

Tympanic membrane When sound causes the tympanic membrane (eardrum) to vibrate, a tiny bone transmits the vibrations to the inner ear. There, ciliated sensory cells (similar to those found in the lateral line of a fish) detect the vibrations and help the frog maintain balance. Leopard frogs hear well in both water and air.

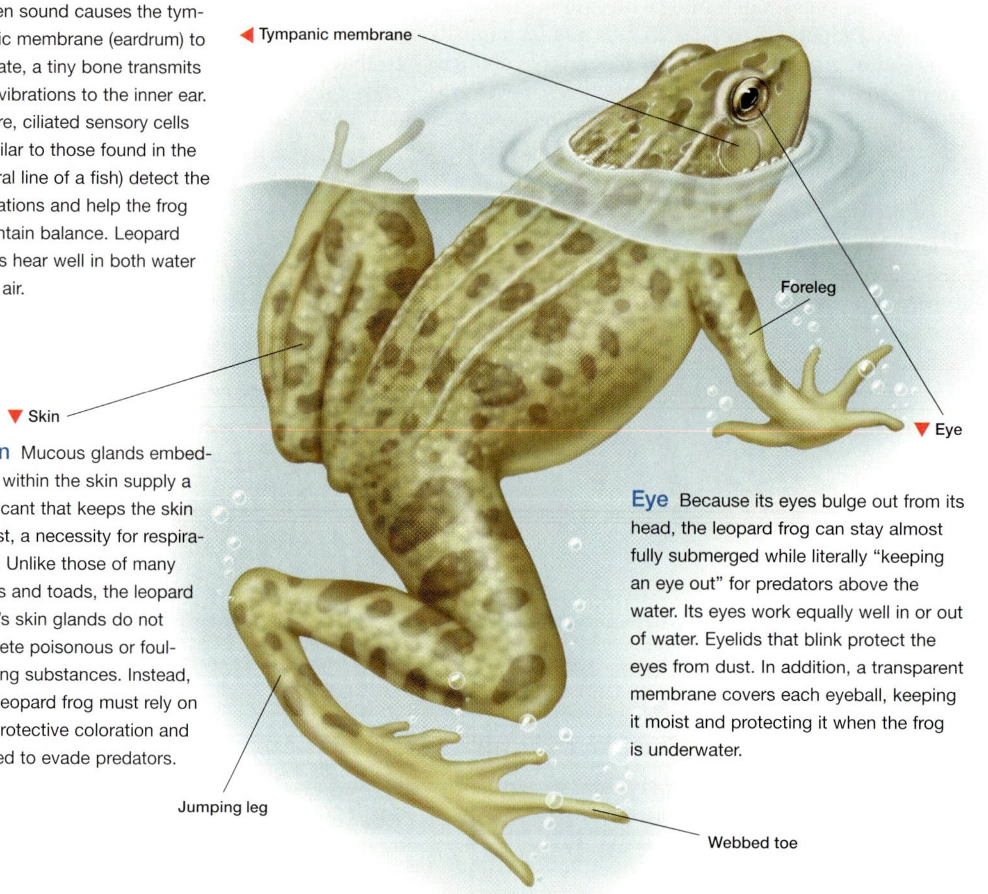

Skin Mucous glands embedded within the skin supply a lubricant that keeps the skin moist, a necessity for respiration. Unlike those of many frogs and toads, the leopard frog's skin glands do not secrete poisonous or foul-tasting substances. Instead, the leopard frog must rely on its protective coloration and speed to evade predators.

Eye Because its eyes bulge out from its head, the leopard frog can stay almost fully submerged while literally "keeping an eye out" for predators above the water. Its eyes work equally well in or out of water. Eyelids that blink protect the eyes from dust. In addition, a transparent membrane covers each eyeball, keeping it moist and protecting it when the frog is underwater.

762

did you know?

A frog's ears are external. Most frogs and toads have a tympanic membrane on the surface of both sides of their head. The tympanic membranes of a human are called eardrums, and each is protected inside the ear canal.

Brain The frog's brain differs from the fish's brain in that its components are more complex. For example, the larger, more complex cerebrum of a frog is able to process a wider assortment of sensory information than the cerebrum of a fish can.

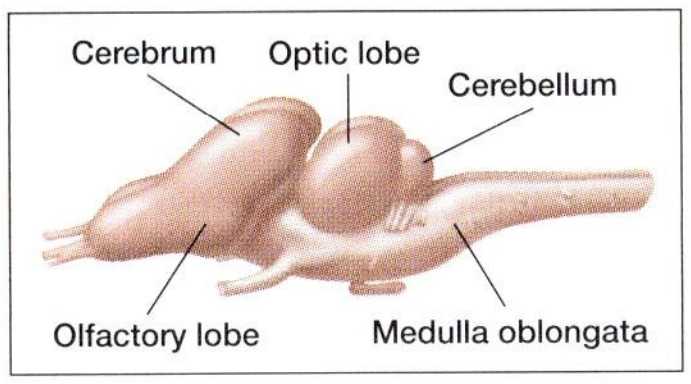

Tongue and jaw The tongue flicks out with great speed, curls around the prey, and returns to the mouth. Two large teeth that project from the roof of the mouth impale struggling prey. In addition the upper jaw is lined with small, sharp teeth that prevent the prey from escaping. Food is not chewed but swallowed whole.

Skeleton The skeletal system of the leopard frog (and all other modern frogs) has only nine vertebrae and no ribs. The rest of the vertebrae are fused into a single bone (urostyle). When a frog is sitting, the urostyle points upward, which gives the frog its characteristic humped back. When a frog jumps the long hind legs extend, which produces a powerful thrust forward.

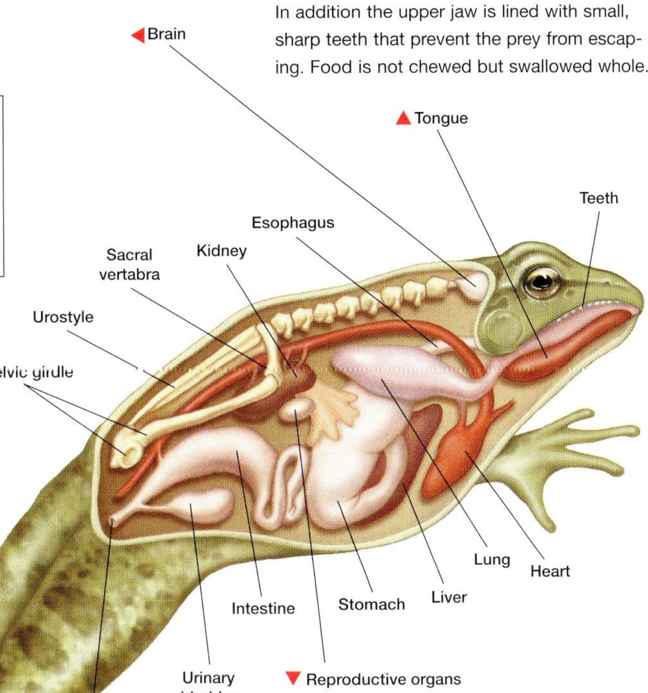

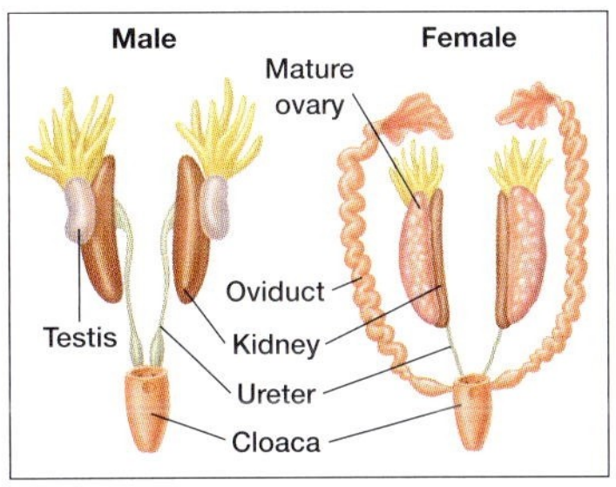

Cloaca Undigested material passes into the cloaca, a chamber that opens to the outside of the body. Urine from the kidneys travels to the bladder and then passes into the cloaca, as do gametes from the reproductive organs. All of these materials exit the body through the cloacal opening.

Reproductive organs Prior to breeding, the reproductive organs of male and female leopard frogs produce enormous numbers of gametes. The female releases a large cluster of eggs into the water. The male then discharges his sperm over them, fertilizing them externally.

Up Close

Leopard Frog

5. Why is a fish more agile in the water than a frog? (A more-developed cerebellum in the fish enables it to have better muscle coordination. Also, a fish's streamlined body and fins are modifications for movement through water.)
6. Why will a fish suffocate if it is taken out of water? (Its gills will collapse, dry out, and no longer absorb oxygen.)
7. How does a frog change its depth in water? (It swims using its webbed feet.)
8. Why is it vital for male frogs to call to females? (Frogs are camouflaged and may not find each other by sight alone. Each species of frog has its own mating call, which prevents females from mating with the wrong species of male.)
9. What is the advantage of a frog of having its eyes on the top of its head? (In water, the frog can keep most of its body hidden from predators while still being able to watch for them or for prey.)
10. How can a frog stay underwater for so long with such small lungs? (It can also absorb oxygen through its skin.)

Salamanders and Caecilians

Salamanders have elongated bodies, long tails, and smooth, moist skin. There are about 369 species of salamanders, all belonging to the order Urodela. They typically range from 10 cm to 0.3 m (4 in. to 1 ft) in length. However, giant Asiatic salamanders of the genus *Andrias* grow as long as 1.5 m (5 ft) and weigh up to 41 kg (90 lb). Because salamanders need to keep their skin moist, most are unable to remain away from water for long periods, although some salamander species manage to live in dry areas by remaining inactive during the day.

Figure 17 Salamander. This four-toed salamander has deposited its eggs in a damp, mossy environment.

Salamanders lay their eggs in water or in moist places, as shown in **Figure 17.** Fertilization is usually external. A few species of salamanders practice a type of internal fertilization in which the female picks up a sperm packet that has been deposited by the male and places it in her cloaca. Unlike frog and toad larvae, salamander larvae do not undergo a dramatic metamorphosis. The young that hatch from salamander eggs are carnivorous and resemble small versions of the adults, except that the young usually have gills. A few species of salamanders, such as the North American mudpuppy and the Texas spring salamander, never lose their larval characteristics. They retain their external gills as adults.

Figure 18 Caecilian. Like most caecilians, this one from Colombia, South America, burrows beneath the soil and is rarely seen.

Caecilians

Caecilians (order Apoda) are a highly specialized group of tropical, burrowing amphibians with small, bony scales embedded in their skin. They feed on small invertebrates found in soil. These legless, wormlike animals, shown in **Figure 18,** grow to about 0.3 m (1 ft) long, although some species can be up to 1.2 m (4 ft) long. During breeding, the male deposits sperm directly into the female. Depending on the species, the female may bear live young or lay eggs that develop externally. Caecilians are rarely seen, and scientists do not know a lot about their behavior.

Section 3 Review

1. **Summarize** how amphibians take in oxygen. SC.F.1.4.2
2. **Contrast** the single-loop circulation of fish with the double-loop circulation of amphibians. SC.G.1.4.1
3. **Compare** the external characteristics of each order of amphibians. SC.G.1.4.1
4. **Compare** reproduction and development in frogs and salamanders. SC.F.2.4.1
5. **Relate** the tongue of the leopard frog to its feeding habits. SC.F.1.4.2
6. **Explain** why it is difficult to "sneak up" on a frog. SC.F.1.4.2
7. FCAT Prep In a frog's heart, what is true of the blood that enters the left atrium?
   A comes from the body **C** goes to the body
   B comes from the lungs **D** goes to the lungs
   SC.F.1.4.2

Close

Reteaching — Basic

Have students make a list of characteristics that distinguish amphibians from fish. (Answers might include that amphibians have legs and lungs, can live on land, exchange gases through their skin, and have a double-loop circulatory system.)

Quiz — General

1. What is the main respiratory organ of most adult amphibians? (lungs)
2. What are two important ways that most amphibians depend on the presence of water? (for reproduction and for cutaneous respiration)
3. How are toads different from frogs? (Toads have squat bodies with shorter legs and their skin is dry and bumpy to prevent water loss in drier environments; frogs have moist skin and longer bodies and legs.)

Alternative Assessment — General

Have students work in groups to prepare a poster showing the frog's life cycle. The poster should include drawings of each of the four stages of a frog's life. Next to each drawing, students should add a paragraph that explains what is illustrated by the drawing and what they have learned about it. For example, next to the drawing of frog eggs, the paragraph could explain how frog eggs are fertilized and where they are laid.

Section Review Correlations

SC.F.1.4.2: Items 1, 5, 6, 7

SC.F.2.4.1: Item 4

SC.G.1.4.1: Items 2, 3

Answers to Section Review

1. through gills, lungs, or skin
2. In single-loop circulation, blood is pumped through gills and then travels through the body. In double-loop circulation, blood is pumped to the lungs and then travels back to the heart, which pumps it to the rest of the body. Blood flows faster and more efficiently in a double-loop system.
3. *Apoda*—small, bony scales embedded in the skin and legless; *Urodela*—elongated bodies with tails; *Anura*—elongated hind legs and no tail
4. Except for a few species of salamanders, both have external fertilization. Tadpoles undergo a radical metamorphosis to become frogs; larval salamanders resemble adults with gills.
5. The tongue can be flicked out at great speed to capture prey.
6. Bulging eyes give a wide field of vision. Large tympanic membranes readily detect sounds.
7. **A.** Incorrect. Oxygen-poor blood from the body is received in the right atrium. **B.** Correct. The left atrium receives oxygen-rich blood from the lungs. **C.** Incorrect. From the left atrium, blood goes to the ventricle. **D.** Incorrect. Blood goes to the ventricle from the left atrium.

Study ZONE

CHAPTER HIGHLIGHTS

Key Concepts

1 The Fish Body

- All fishes have gills and a backbone, and they circulate oxygen-rich blood from their gills directly to body tissues.
- Countercurrent flow maximizes the amount of oxygen that can be extracted from water.
- The four-chambered fish heart collects oxygen-poor blood from the body and pumps it through the gills where it receives oxygen. Oxygen-rich blood then circulates to the rest of the body.
- A fish relies on its gills and a pair of kidneys to regulate its salt and water balance.
- Most fishes fertilize their eggs externally as males and females release their gametes near one another in the water.

2 Today's Fishes

- Hagfishes and lampreys are the only surviving jawless fishes.
- Sharks have light, highly streamlined bodies well suited for rapid swimming, which makes them swift and efficient predators.
- Bony fishes are the most diverse and abundant group of fishes.
- Bony fishes have an internal skeleton made completely of bone, a swim bladder, a lateral line sensory system, and a set of gill covers called *opercula*.

3 Amphibians

- Most amphibians have legs, breathe with lungs and through their skin, and have two circulatory loops.
- Most amphibians supplement their oxygen intake through cutaneous respiration—respiration through their moist skin.
- An amphibian lung is basically an air sac with a large surface area for gas exchange.
- The amphibian heart pumps oxygen-poor blood to the lungs and receives oxygen-rich blood from the lungs. The oxygen-rich blood is then pumped to the body.
- Salamanders are semiaquatic predators with tails, and caecilians are legless amphibians specialized for burrowing.

Key Terms

Section 1

gill filament (747)
gill slit (747)
countercurrent flow (747)
nephron (749)

Section 2

lateral line (753)
operculum (756)
swim bladder (756)
teleost (757)

Section 3

lung (758)
pulmonary vein (759)
septum (760)

765

Alternative Assessment — ADVANCED

Have students make a timeline of the evolution of vertebrates from the first jawless fishes through amphibians. Next to each of the organisms listed on the timeline, students should describe the new characteristics that arose in the organism through natural selection.

Chapter Resource File

- Science Skills Worksheet GENERAL
- Critical Thinking Worksheet ADVANCED
- Test Prep Pretest GENERAL
- Chapter Test GENERAL

Answer to Concept Map

The following is one possible answer to Performance Zone item 15.

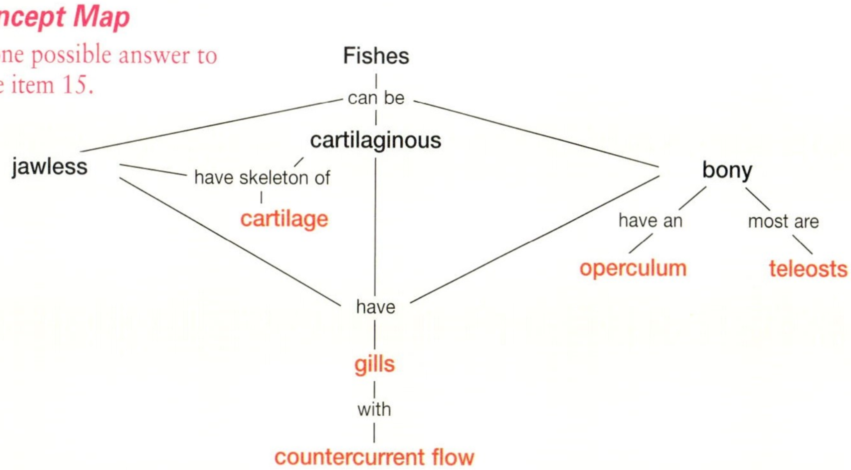

CHAPTER 33 ZONE

ANSWERS

Understanding Key Ideas

1. d
2. b
3. d
4. a
5. d
6. b
7. a
8. a
9. Metamorphosis in amphibians takes place in gradual stages—from egg to tadpole to frog. In many insects, after the egg and larval stages, a pupa is formed. After a period of time, the adult form emerges. The adult looks quite different from the larva.
10. One possible answer to the concept map is found at the bottom of the Study Zone page.

Critical Thinking

11. Marine fish lose water to their environment through osmosis, so they drink a lot of sea water. Because sea water contains a high concentration of salts, marine fish must actively pump excess salt out of their body. Freshwater fish actively take in salts from their environment.
12. Osmotic balance refers to the concentration of salts in the body fluids. The operculum is the cover over the gills. Cutaneous respiration is the process of taking in oxygen through the skin.

Performance ZONE

CHAPTER REVIEW

Understanding Key Ideas

1. Which of the following is *not* a key characteristic of fishes?
a. vertebral column
b. gills
c. single-loop circulation
d. tympanic membrane SC.G.1.4.1

2. Which of the following shark characteristics is *not* an adaptation for predation?
a. streamlined body
b. internal fertilization of eggs
c. sharp, replaceable teeth
d. lightweight skeleton SC.F.1.4.2

3. Yellow perch and sharks share all of the following characteristics *except*
a. gills.
b. an internal skeleton.
c. a single-loop circulatory system.
d. a swim bladder.
SC.G.1.4.1

4. A shark's skeleton is
a. composed of cartilage.
b. composed of bone.
c. very dense.
d. quite rigid. SC.F.1.4.2

5. Most adult amphibians respire
a. through their skin.
b. through their skin and gills.
c. through their lungs.
d. through their skin and lungs.
SC.G.1.4.1

6. Which of the following is *not* a characteristic of amphibians?
a. lungs
b. heart with two ventricles
c. cutaneous respiration
d. double-loop circulation
SC.G.1.4.1

7. Which of the following characteristics of leopard frogs is *not* an adaptation for avoiding predators?
a. fast, flicking tongue
b. skeleton adapted for jumping
c. spotted, greenish-brown skin
d. position of the eyes
SC.G.2.4.3

8. How do tadpoles differ from frogs?
a. Tadpoles have gills; frogs do not.
b. Tadpoles are carnivorous; frogs are herbivorous.
c. Frogs show body symmetry; tadpoles do not.
d. Frogs live in water; tadpoles live in damp vegetation.
SC.G.1.4.1

9. Compare amphibian metamorphosis with insect metamorphosis. (**Hint:** See Chapter 30, Section 3.) SC.G.1.4.1

10. **Concept Mapping** Construct a concept map describing the characteristics of jawless, cartilaginous, and bony fishes. Try to include the following terms in your concept map: *gills, countercurrent flow, cartilage, operculum,* and *teleosts.* SC.G.1.4.1

Critical Thinking

11. **Inferring Relationships** Explain how marine and freshwater fishes differ in the way they maintain their salt and water balance.
SC.F.1.4.1

12. **Distinguishing Relevant Information** A student is writing a paper on the evolution of the heart. Which of the following terms do not pertain to her topic? Explain. *Sinus venosus, pulmonary veins, septum, osmotic balance, atrium, operculum, conus arteriosus,* and *cutaneous respiration.* SC.F.1.4.2

Alternative Assessment

13. **Forming a Model** Construct a model that shows how water passes over the gills of a bony fish. Then explain in writing why countercurrent flow increases respiratory efficiency. SC.H.3.4.1

14. **Finding and Communicating Information** Use the media center or Internet resources to learn more about amphibians that live in your area. Create an illustrated reference table that includes their scientific and common names and information about size, habitat, diet, and population size.
SC.G.1.4.1

766

Alternative Assessment

13. Models should resemble the drawings in Figure 2. Counter-current flow ensures that oxygen diffuses into the blood over the entire length of the capillaries in the gills.
14. Tables will vary. The table should list frogs and toads separately from salamanders and newts. Presentations will vary. They should explain all features of the references table.

Assignment Guide

Section	Questions
1	1, 10, 11, 12, 13
2	2, 3, 4, 10, 12
3	5, 6, 7, 8, 9, 10, 12, 14

Chapter Review Correlations

SC.F.1.4.1: Item 11
SC.F.1.4.2: Items 2, 4, 12
SC.G.1.4.1: Items 1, 3, 5, 6, 8, 9, 10, 14
SC.G.2.4.3: Item 7
SC.H.3.4.1: Item 13

FCAT Prep

For questions 1–6, write your answers on a separate sheet of paper.

1 What organ in bony fishes senses pressure changes in water?
A. gill
B. lateral line
C. operculum
D. septum
SC.F.1.4.2

2 What organ in bony fishes regulates buoyancy?
F. atrium
G. conus arteriosus
H. lateral line
I. swim bladder
SC.F.1.4.2

3 A newspaper article reports that some carp in a local pond are approximately 50 years old. How can this claim be verified?
A. by counting the number of growth rings on a scale from one of the fish
B. by comparing the size of the fish with younger fish from the same pond
C. by comparing the color of the fish with younger fish from the same pond
D. by counting the layers of dark and light layers in a cross section of one fish
SC.H.3.4.3

4 What group of amphibians is legless?
F. caecelians
G. lampreys
H. toads
I. skates
SC.G.1.4.1

5 READ INQUIRE EXPLAIN Tadpoles live in water, breathe with gills, and feed mostly on algae.

Differentiate between the digestive system function of a tadpole and that of an adult frog. SC.F.1.4.2

6 READ INQUIRE EXPLAIN A scientist writes the following description of a vertebrate circulatory system: Oxygen-rich blood is pumped from the capillaries to a small chamber in the heart. From the small chamber, the blood is pumped to a larger chamber in the heart where it mixes with some oxygen-poor blood. The blood is then pumped to the rest of the body. When it returns from the body, the oxygen-poor blood enters another small chamber in the heart. From that chamber the blood is pumped to the larger chamber and then on to the capillaries.

Part A Which type of animal has this kind of circulatory system?
SC.F.1.4.2
Part B Predict the type of environment in which you would most likely find an animal with this kind of circulatory system. SC.G.1.4.1

Test TIP

When a question uses a word you don't know, try to determine the meaning of the word by breaking it down into smaller parts and inferring the meaning of each part.

767

FCAT Doctor

Question 3 Answer A is the correct choice. Counting the growth rings on a fish's scale can give a good idea of its age. Answer B is incorrect because size does not reliably tell a fish's age, as its size is affected by environmental conditions. Answer C is incorrect because color does not give information about the age of carp. Answer D is incorrect because trees, not fish, have cross sections that show growth rings.

Question 5 The proteins and other materials that a carnivore must digest require different enzymes from those of herbivores.

Question 6

Part A Amphibians have a double-loop circulatory system in which oxygenated blood mixes with deoxygenated blood in the ventricle.

Part B Amphibians with this kind of circulatory system would live in moist habitats on land.

Answers

1. B
2. I
3. A
4. F
5. Most tadpoles are herbivores, while adult frogs are carnivores.

6A. Amphibians have this type of circulatory system.

6B. You would find this animal in a moist habitat on land.

Test Prep Correlations

SC.F.1.4.2: Items 1, 2, 5, 6A
SC.G.1.4.1: Items 4, 6B
SC.H.3.4.3: Item 3

Exploration Lab

Observing a Live Frog

Teacher's Notes

Time Required
One 45-minute class period

Ratings
EASY → HARD

TEACHER PREPARATION: 2
STUDENT SETUP: 1
CONCEPT LEVEL: 1
CLEANUP: 1

Skills Acquired
- Collecting Data
- Communicating
- Inferring
- Organizing and Analyzing Data

Scientific Methods
In this lab, students will:
- Make Observations
- Ask Questions
- Analyze the Results
- Draw Conclusions

Materials and Equipment
Materials for this lab can be ordered from WARD'S. See *Master Materials List* at the front of this book for catalog numbers. If you use frogs that are not native to your area, do not release them into the local environment when the investigation is completed. Indigenous frogs may be released into an appropriate habitat. Students may take indigenous frogs home to keep as pets if they have an appropriate habitat for them.

Safety Cautions
Have students follow the guidelines from the National Association of Biology Teachers for handling vertebrate animals.

Exploration Lab

Observing a Live Frog SC.H.1.4.1

SKILLS
- Observing
- Relating

OBJECTIVES
- **Examine** the external features of a frog.
- **Observe** the behavior of a frog.
- **Explain** how a frog is adapted to life on land and in water.

MATERIALS
- live frog in a terrarium
- live insects (crickets or mealworms)
- 600 mL beaker
- aquarium half-filled with dechlorinated water

Before You Begin

Frogs, which are **amphibians**, are adapted for living on land and in water. For example, a frog's eyes have an extra eyelid called the **nictitating membrane**. This eyelid protects the eye when the frog is underwater and keeps the eye moist when the frog is on land. The smooth skin of a frog acts as a respiratory organ by exchanging oxygen and carbon dioxide with the air or water. The limbs of a frog enable it to move both on land and in water. In this lab, you will examine a live frog in both a terrestrial environment and an aquatic environment.

1. Write a definition for each boldface term in the paragraph above and for the following term: tympanic membrane.
2. Make a data table similar to the one below.

DATA TABLE

Behavior/structure	Observations
Breathing	
Eyes	
Legs	
Response to food	
Response to noise	
Skin	
Swimming behavior	

3. Based on the objectives for this lab, write a question you would like to explore about frogs.

Procedure

PART A: Observing a Frog

1. Observe a live frog in a terrarium. Closely examine the external features of the frog. Make a drawing of the frog. Label the eyes, nostrils, tympanic membranes, front legs, and hind legs.
2. Watch the frog's movements as it breathes air into and out of its lungs. Record your observations.
3. Look closely at the frog's eyes, and note their location. Examine the upper and lower eyelids as well as a third transparent eyelid called a *nictitating membrane*. Describe how the eyelids move.

768

Tips and Tricks
Provide each group with a frog in a small terrarium. Have groups take turns releasing their frog into the freshwater aquarium. After students complete their observations, remove each frog from the aquarium yourself to avoid injury to the frogs.

Answers to Before You Begin
1. *amphibians*—animals that are adapted for living both on land and in water, usually referring to vertebrates in the class Amphibia; *nictitating membrane*—the extra eyelid of frogs' eyes; *tympanic membrane*—membrane located on a frog's head that receives vibrations and transmits them to the inner ear.
2. Students' data tables should resemble the one on this page.
3. Answers will vary. For example: How are frogs adapted for living on land and in water?

4. Study the frog's legs, and note the difference between the front and hind legs.

5. Place a live insect, such as a cricket or a mealworm, into the terrarium. Observe how the frog reacts.

6. Tap the side of the terrarium farthest from the frog, and observe the frog's response.

7. Place a 600 mL beaker in the terrarium. **CAUTION: Handle live frogs gently. Frogs are slippery! Do not allow a frog to injure itself by jumping from a lab table to the floor.** Carefully pick up the frog, and examine its skin. How does it feel? Now place the frog in the beaker. Cover the beaker with your hand, and carry it to a freshwater aquarium. Tilt the beaker, and gently lower it into the water until the frog swims out.

8. Watch the frog float and swim. Notice how the frog uses its legs to swim. Also notice the position of the frog's head. As the frog swims, bend down to view the underside of the frog. Then look down on the frog from above. Compare the color on the dorsal and ventral sides of the frog.

PART B: Cleanup and Disposal

9. Dispose of broken glass in the designated waste containers. Put live animals in the designated containers. Do not pour chemicals down the drain or put lab materials in the trash unless your teacher tells you to do so.

10. Clean up your work area and all lab equipment. Return lab equipment to its proper place. Wash your hands thoroughly before you leave the lab and after you finish all work.

Analyze and Conclude

1. **Summarizing Information** How does a frog use its hind legs for moving on land and in water?

2. **Recognizing Relationships** How does the position of a frog's eyes benefit the frog while it is swimming?

3. **Analyzing Data** What features of an adult frog provide evidence that it has an aquatic life and a terrestrial life?

4. **Analyzing Methods** Were you able to determine in this lab how a frog hears? Explain.

5. **Inferring Conclusions** What can you infer about a frog's field of vision from the position of its eyes?

6. **Forming Hypotheses** How is the coloration on the dorsal and ventral sides of a frog an adaptive advantage?

7. **Further Inquiry** Write a new question about frogs that could be explored with another investigation.

On the Job

Herpetology is the study of reptiles and amphibians. Do research to discover how herpetologists are working with the Declining Amphibian Task Force (FROGLOG) to solve the mystery of the worldwide decline in amphibian populations. For more about careers, visit **go.hrw.com** and type in the keyword **HX4 Careers**.

Answers to Procedure

1. Students' drawing should resemble the illustrations in the Up Close feature of Section 3.
2. The floor of the frog's mouth moves up and down as it breathes.
3. The eyes are on top of the head. The upper and lower eyelid are fixed. The third eyelid moves down to cover the eye.
4. The front legs are short. The hind legs are long and heavily muscled; they fold alongside the frog's body.
5. The frog may strike at the insect or worm, catching it with its tongue.
6. Answers will vary. Tapping might startle the frog.
7. The frog probably has smooth, moist skin.
8. The frog uses its hind legs to move through the water. The frog's head stays partially above water. The ventral side of the frog is light in color; the dorsal side is dark.

Answers to Analyze and Conclude

1. A frog uses its long hind legs for jumping on land and for swimming in water.
2. Having eyes on the top of its head enables a frog to see above the waterline while the rest of the frog is submerged.
3. *aquatic life*—webbed feet, eyes on top of head, respires through skin; *terrestrial life*—jumping legs, breathes with lungs
4. Answers will vary. Students should recognize that although they observed the tympanic membrane, they cannot determine how a frog hears by observing its actions.
5. A frog can see in almost all directions.
6. As viewed from above, the dark dorsal skin blends in with the color of the water, making the frog less visible to predators from above. Viewed from below, the light ventral skin blends in with the color of the sky, making the frog less visible to aquatic predators.
7. Answers will vary. For example: How do two frogs interact on land and in water?

Answers to On the Job

Herpetologists study the anatomy, physiology, and ecology of reptiles and amphibians. Student reports should provide supporting data for all statements regarding the worldwide decline in amphibian populations. Scientists are researching several possible reasons for the decline, including pollution, acid precipitation, habitat loss, climate change, and disease.

Chapter Opener		TE **Using the Figure,** p. 770 SE **Quick Review,** p. 771 ■ SE **Reading Activity,** p. 771 ■
& 3 • 90 min pp. 772–777 **Section 1** The Reptilian Body	OSP **Lesson Plan*** TR **Bellringer*** TR **Section Outline*** TR **I18 Key Features of Reptiles*** TR **I20 Changes in Lizard Body Temperature*** TR **I19 Reptilian Heart Structure***	TE **Activity** Favorite Reptile, p. 772 BASIC TE **Using the Figure,** p. 773 BASIC SE **Data Lab** Identifying Ectotherms, p. 773 ■ CRF **Datasheets for In-Text Labs** Identifying Ectotherms* SE **Quick Lab** Modeling Watertight Skin, p. 774 ◆ ■ CRF **Datasheets for In-Text Labs** Modeling Watertight Skin* TE **Activity** Comparing Eggs, p. 775 ADVANCED TE **Group Activity** Local Snakes, p. 776 ADVANCED
& 5 • 90 min pp. 778–783 **Section 2** Today's Reptiles	OSP **Lesson Plan*** TR **Bellringer*** TR **Section Outline*** TR **I21 External Structure of a Timber Rattlesnake*** TR **I22 Internal Structure of a Timber Rattlesnake***	TE **Demonstration,** p. 778 ◆ GENERAL TE **Using the Figure,** p. 782 GENERAL CRF **Exploration Lab** Using a Dichotomous Key to Identify Lizards* ◆ GENERAL
& 7 • 90 min pp. 784–792 **Section 3** Characteristics and Diversity of Birds	OSP **Lesson Plan*** TR **Bellringer*** TR **Section Outline*** TR **I23 Characteristics of Birds*** TR **I27 Contour Feather Structure*** TR **I48 Avian Skeleton*** TR **I29 Avian Heart Structure*** TR **I30 Avian Lung Structure*** TR **I25 External Structures of Birds–Bald Eagle*** TR **I26 Internal Structures of Birds–Bald Eagle*** TR **I31 Avian Adaptations***	TE **Demonstration,** p. 784 ◆ BASIC TE **Demonstration,** p. 785 ◆ GENERAL TE **Activity** Origin of Birds, p. 787 ADVANCED TE **Using the Table,** p. 791 BASIC TE **Demonstration,** p. 791 ◆ BASIC TE **Group Activity** Avian Adaptations, p. 791 ADVANCED SE **Math Lab** Calculating Average Bone Density, p. 792 ■ CRF **Datasheets for In-Text Labs** Calculating Average Bone Density* CRF **Skills Practice** Conducting a Bird Survey* GENERAL SE **Exploration Lab** Observing Color Change in Anoles, pp. 796–797 ■ GENERAL CRF **Datasheets for In-Text Labs** Observing Color Change in Anoles* ◆

& 9 • 90 min Chapter Review and Assessment Resources

- SE **Study Zone,** p. 793 ■
- SE **Performance Zone,** p. 794 ■
- SE **FCAT Test Prep,** p. 795 ■
- TR **Concept Mapping***
- CRF **Vocabulary Review Worksheet*** ■ GENERAL
- CRF **Test Prep Pretest*** ■ GENERAL
- CRF **Chapter Test*** ■ GENERAL
- CRF **Chapter Test*** ADVANCED
- OSP **Test Generator***
- OSP **Test Item Listing***
- OSP **Modified Chapter Test*** BASIC

Online and Technology R...

go.hrw.com Holt Online Learning

Visit go.hrw.com for access to Holt Online Learning or enter the keyword **HX6 Home** for a variety of free online resources.

One-Stop Planner® CD-ROM

This CD-ROM package includes
- Lab Materials QuickList Software
- Holt Calendar Planner
- Customizable Lesson Plans
- Printable Worksheets
- ExamView® Test Generator
- Interactive Teacher Edition
- Holt PuzzlePro® Resources
- Holt PowerPoint® Resources

Guided Reading Audio CDs

These CDs are designed to help auditory learners and reluctant readers. (Audio CDs are also available in Spanish.)

Compression guide:
To shorten your instruction because of time limitations, omit block 1.

KEY

TE	Teacher Edition	CRF	Chapter Resource File	*	Also on One-Stop Planner
SE	Student Edition	TR	Teaching Transparency	■	Also Available in Spanish
OSP	One-Stop Planner			◆	Requires Advance Prep

SKILLS DEVELOPMENT AND PRACTICE	SECTION REVIEW AND ASSESSMENT	CORRELATIONS
		Florida Sunshine Standards/FCAT
TE **Reading Skill Builder** Interactive Reading, p. 773 GENERAL TE **Skill Builder** Math Skills, p. 776 GENERAL CRF **Directed Reading Worksheet*** BASIC CRF **Active Reading Worksheet*** GENERAL	SE **Section 1 Review** ■ TE **Reteaching,** p. 777 BASIC TE **Quiz,** p. 777 GENERAL TE **Alternative Assessment,** p. 777 GENERAL CRF **Quiz*** ■ BASIC	SC.F.1.4.1, SC.F.1.4.2, SC.F.1.4.7, SC.G.1.4.1, SC.H.3.4.1
TE **Skill Builder** Writing Skills, p. 782 BASIC CRF **Directed Reading Worksheet*** BASIC CRF **Active Reading Worksheet*** GENERAL	SE **Section 2 Review** ■ TE **Reteaching,** p. 783 BASIC TE **Quiz,** p. 783 GENERAL TE **Alternative Assessment,** p. 783 ADVANCED CRF **Quiz*** ■	SC.F.1.4.2, SC.G.1.4.1
TE **Reading Skill Builder** Paired Summarizing, p. 785 BASIC CRF **Directed Reading Worksheet*** BASIC CRF **Active Reading Worksheet*** GENERAL CRF **Critical Thinking Worksheet*** ADVANCED CRF **Science Skills Worksheet*** GENERAL	SE **Section 3 Review** ■ TE **Reteaching,** p. 792 BASIC TE **Quiz,** p. 792 GENERAL TE **Alternative Assessment,** p. 792 GENERAL CRF **Quiz*** ■ BASIC **Modified Worksheet*** BASIC	SC.F.1.4.2, SC.F.1.4.7, SC.G.1.4.1, SC.H.1.4.1

SCILINKS. NSTA

www.scilinks.org

Maintained by the **National Science Teachers Association**

Topic: Characteristics of Reptiles
SciLinks Code: HX4038

Topic: Amniotic Egg
SciLinks Code: HX4005

Topic: Characteristics of Birds
SciLinks Code: HX4036

Topic: Adaptations of Reptiles
SciLinks Code: HX4003

Overview

Tell the students that this chapter will provide an overview of the characteristics of reptiles and birds. Students will learn the distinguishing features of snakes, lizards, turtles, tortoises, crocodilians, and tuataras. They will study the unique adaptations of birds, and note how each bird is uniquely adapted to its way of life.

Using the Figure

Have students look at the picture of the young reptile emerging from its egg on this page. Ask students to look for adaptations that make this animal better suited to life on land than a salamander. (tough, watertight egg; claws; hatches with strong limbs; scales for protection and to prevent drying out)

Assessing Prior Knowledge

Before beginning this chapter, be sure your students can answer the questions in the ***Quick Review*** on the facing page. If they cannot, encourage them to reread the sections indicated.

770 A reptile emerging from an amniotic egg

Chapter Correlations — Florida Sunshine Standards/FCAT

SC.F.1.4.2 Body structures are uniquely designed and adapted for their function.

SC.F.1.4.7 Organisms respond to internal and external stimuli.

SC.G.1.4.1 There is great diversity and interdependence among living things.

CHAPTER

34 Reptiles and Birds

Florida Sunshine Standards/FCAT

SC.F.1.4.2 Body structures are uniquely designed and adapted for their function.

SC.F.1.4.7 Organisms respond to internal and external stimuli.

SC.G.1.4.1 There is great diversity and interdependence among living things.

Quick Review

Answer the following without referring to earlier sections of your book.

1. **Describe** the process of molting in arthropods. *(Chapter 30, Section 1)*
2. **Distinguish** between ectotherms and endotherms. *(Chapter 32, Section 2)*
3. **Summarize** the evolutionary relationships between reptiles and birds. *(Chapter 32, Section 2)*
4. **Relate** countercurrent flow to the efficiency of the fish gill. *(Chapter 33, Section 1)*

Did you have difficulty? *For help, review the sections indicated.*

Reading Activity

Write down the title of this chapter and the titles of its three sections on a piece of paper or in your notebook. Leave a few blank lines after each section title. Then, write down what you think you will learn in each section. Save your list, and after you finish reading this chapter, check off everything that you learned that was on your list.

Ancient amphibians gave rise to a new group of animals that were able to lay eggs on dry land—the reptiles. The amniotic egg is a significant evolutionary milestone for land-dwelling animals, such as reptiles and birds.

Looking Ahead

Section 1
The Reptilian Body
- *Key Characteristics of Reptiles*
- *Water Retention*
- *Respiration*
- *Reproduction*

Section 2
Today's Reptiles
- *Lizards and Snakes*
- *Other Orders of Reptiles*

Section 3
Characteristics and Diversity of Birds
- *Key Characteristics of Birds*
- *Adaptations of Birds*

internet connect

www.scilinks.org

National Science Teachers Association *sci*LINKS Internet resources are located throughout this chapter.

SCILINKS. Maintained by the National Science Teachers Association

771

Identifying Misconceptions

Students may have the misconception that the class Reptilia is unsuccessful because of the extinction of the dinosaurs. While reptiles may be smaller in size than in past eras, they remain diverse and numerous, inhabiting all of the continents except Antarctica. With about 7000 species, reptiles are among the most diverse groups of terrestrial vertebrates, second only to birds.

Quick Review Answers

1. Prior to molting, arthropods grow a new exoskeleton just beneath the old one. The old exoskeleton splits and the arthropod emerges. In a short while, the new exoskeleton expands and hardens. This process allows arthropods to grow.
2. Ectotherms have a relatively slow metabolism that does not produce enough heat to maintain a constant temperature. Endotherms have a faster metabolism that uses the energy from food to maintain a constant warm temperature.
3. Fossil evidence suggests that birds evolved from reptiles. *Archaeopteryx*, the earliest known bird, had several reptile-like features including teeth and a long tail.
4. In countercurrent flow, water passes over the gills in the opposite direction of blood flowing through the gills. This allows a very efficient absorption of oxygen—up to 85%.

Chapter Resource File

- Vocabulary Worksheets
- Concept Mapping

Reading Activity Answers

Answers will vary. Ask students to read any entries from their lists that were not answered. Discuss these unanswered questions as a class, and invite students to do additional research if their questions remain unanswered.

Section 1

Focus

Overview

Before beginning this section review with your students the objectives listed in the Student Edition. In this section, students will learn how reptiles are well-adapted to life on land. Terrestrial adaptations include a strong skeleton, watertight skin covered with scales, watertight eggs, larger lungs with greater surface area, internal fertilization, and a partially divided heart.

Bellringer

Ask students to design a survival capsule for themselves. The capsule can only be 1.5 m high by 1 m wide. The capsule will not be opened for three weeks. Nothing, other than gasses, can pass through the wall of the capsule. What should the capsule contain? (Answers will vary. Students should include an ample water supply and food. Students should also include an area to store waste and a means to ensure fresh air.) Ask students to compare their capsule with the amniotic egg described in this section.

Motivate

Activity — Basic

Favorite Reptile Have each student draw a picture of his or her favorite reptile. Have volunteers share their drawings with the class and tell why they chose their particular reptile. Lead a discussion of how reptiles are viewed in popular culture. LS Visual

Section Correlations

SC.F.1.4.2
SC.F.1.4.7
SC.G.1.4.1

Section 1 The Reptilian Body

Florida Sunshine Standards/FCAT

SC.F.1.4.2 Body structures are uniquely designed and adapted for their function.
SC.F.1.4.7 Organisms respond to internal and external stimuli.
SC.G.1.4.1 There is great diversity and interdependence among living things.

Objectives

- **Describe** the key characteristics of reptiles. SC.G.1.4.1
- **Relate** a reptile's ectothermic metabolism to its activity level. SC.F.1.4.7
- **Summarize** the adaptations that enable reptiles to live on land. SC.F.1.4.2

Key Terms

amniotic egg
oviparous
ovoviviparous

internet connect
www.scilinks.org
Topic: Characteristics of Reptiles
Keyword: HX4038
SciLinks. Maintained by the National Science Teachers Association

Key Characteristics of Reptiles

Many people react with fear or repulsion when they see a snake slither across a yard or field. But snakes and their reptile relatives are important members of most ecosystems, and they kill large numbers of insect pests and small rodents. It's true that some reptiles—venomous snakes and crocodilians (crocodiles and alligators)—are dangerous. Most reptiles, however, live quietly and go about their business, preferring to avoid humans.

Members of class Reptilia live throughout the world in a wide variety of habitats, except in the coldest regions, where it is impossible for ectotherms to survive. Reptiles share certain fundamental characteristics, features they retain from the time when reptiles replaced amphibians as the dominant terrestrial vertebrates. **Figure 1** summarizes these key features.

Reptiles have a strong, bony skeleton, and most have two pairs of limbs, although snakes and some lizards are legless. The legs of reptiles are positioned more directly under their body than are the limbs of amphibians. Thus, reptiles can move more easily on land than amphibians can. Unlike amphibians, reptiles have toes with claws, which are used for climbing and digging. Claws also enable reptiles to get a good grip on the ground, allowing many reptiles to run quickly for short distances.

The nervous system of a reptile is very similar to that of an amphibian. Like their dinosaur ancestors, modern reptiles have a brain that is small in relation to their body. For example, an alligator about 2.5 m (8 ft) long has a brain that is about the size of a walnut. Despite this small brain size, reptiles are capable of complex behaviors, including elaborate courtship.

Figure 1 Characteristics of living reptiles. This male anole is extending his dewlap, a display used during courtship and when defending territory.

Key Features of Reptiles

- Strong, bony skeleton and toes with claws
- Ectothermic metabolism
- Dry, scaly skin, almost watertight
- Amniotic eggs, almost watertight
- Respiration through well-developed lungs
- Ventricle of heart partly divided by a septum
- Internal fertilization

Chapter Resource File

- Directed Reading — Basic
- Active Reading — General
- Data Sheet for Data Lab — General
- Data Sheet for Quick Lab — General

One-Stop Planner CD-ROM

- Reading Organizers — Basic
- Reading Strategies — Basic
- Basic Skills Worksheets
  Reading a Thermometer
  Temperature Conversions — Basic

Ectothermic Metabolism

Reptiles' ectothermic metabolism is too slow to generate enough heat to warm their bodies, so they must absorb heat from their surroundings. As a result, a reptile's body temperature is largely determined by the temperature of its environment. Many reptiles regulate their temperature behaviorally, by basking in the sun to warm up or seeking shade to cool down. **Figure 2** shows that a lizard can maintain a relatively constant body temperature throughout the day by moving between sunlight and shade. At very low temperatures, most reptiles become sluggish and unable to function. Intolerance of cold generally limits their geographical range and, in temperate climates, forces them to remain inactive through the winter.

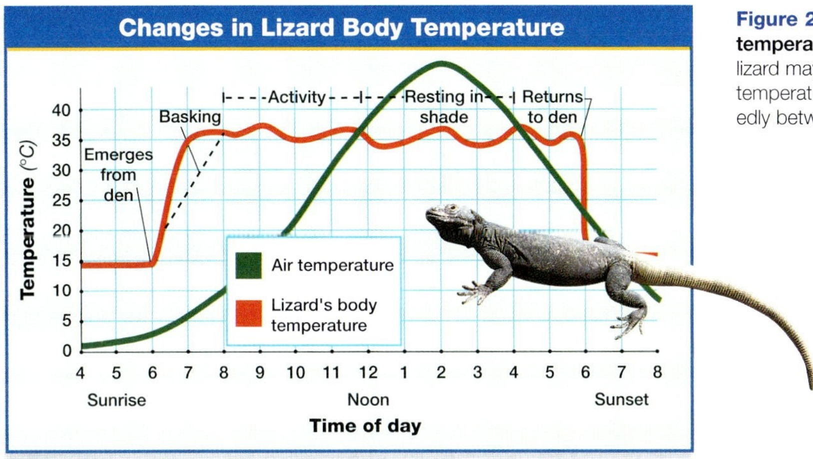

Figure 2 Body temperature in a lizard. A lizard may regulate its body temperature by moving repeatedly between sun and shade.

DATA LAB: *Identifying Ectotherms*

SC.F.1.4.1

Background

The body temperature of all animals changes during the course of a day. How it changes can help you identify an animal as an ectotherm or an endotherm.

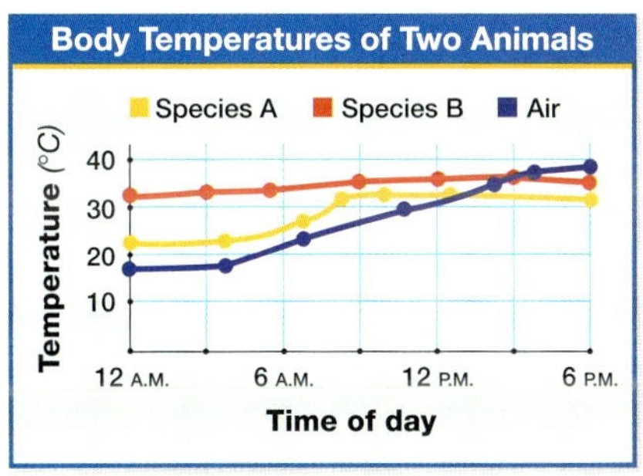

Analysis

1. **Analyze** the data and determine which animal species, A or B, is most likely an ectotherm. Explain your reasoning.
2. **Identify** the time of day the animal you identified as an ectotherm reaches its lowest body temperature.
3. **Identify** the time of day the animal you identified as an ectotherm reaches its highest body temperature.
4. **Propose** a reason why the ectotherm's body temperature is highest at this time.
5. **Predict** what the endotherm's graph line would look like if it were extended to show body temperature between 6 P.M. and midnight.

773

Teach

READING SKILL BUILDER — GENERAL

Interactive Reading Assign Chapter 34 of the *Holt Biology* Guided Reading Audio CD Program to help students achieve greater success in reading the chapter.

Using the Figure — BASIC

Have students draw a graph that depicts the changes in human body temperature for a 24-hour period. (The graph should show that the human body remains at about 37°C for the entire day.) Then ask them to compare their graphs to the one for the lizard's body temperature in **Figure 2.** Ask: Why are the graphs different? (The lizard depends on external sources for heat, so its body temperature tracks the temperature of the environment. Humans produce their own heat, so body temperature is not dependent on environmental temperature changes.) LS **Visual**

DATA LAB: *Identifying Ectotherms*

Skills Acquired

Analyzing data, interpreting graphs, predicting patterns

Teacher's Notes

Remind students that *ecto-* refers to "outside," *endo-* refers to "within," and *therm* refers to "heat." Hence, *ectotherm* means "outside heat" and *endotherm* means "heat within."

Answers to Analysis

1. Species A (yellow curve) probably is an ectotherm. Its temperature increases during the day, when the air temperature increases.
2. around 12 A.M.
3. around 12 P.M.
4. It has been exposed to sunlight-warmed air since early in the morning.
5. It would decline steadily from its value at 6 P.M. to its value at 12 A.M.

Transparencies

- **TR** Bellringer
- **TR I18** Key Features of Reptiles
- **TR I20** Changes in Lizard Body Temperature
- **TR I19** Reptilian Heart Structure

MISCONCEPTION ALERT

Cold-blooded Emphasize that the terms *ectothermic* and *coldblooded* are not synonyms, although they are often mistakenly used that way. Many reptiles experience body temperatures as high as or higher than human body temperature. For example, the body temperature of the desert iguana in western North America regularly reaches 42°C (108°F), which would be fatal for a human.

Teach, *continued*

QUICK LAB

Modeling Watertight Skin

Skills Acquired
Calculating, analyzing data, inferring conclusions

Teacher's Notes
Be sure to have students wipe the pan of the scale clean between weighings. Place an incandescent lamp on each table to simulate a desert environment.

Answers to Analysis

1. The mass of the skinless grape should decrease. The mass of the intact grape should stay the same.
2. The skin prevents water from evaporating from an intact grape, so the grape's weight does not change. Without its skin, a grape loses water and becomes lighter.
3. The skinless grape represents an amphibian's skin. The intact grape represents a reptile's skin.
4. Watertight skin, as in the intact grape, prevents water loss. This allows an animal to survive in dry environments. Most amphibians, like the skinless grape, would dry up if out of water or moist environment for an extended period of time.

Study TIP

Reviewing Information
Reread the bulleted list of key features of reptiles in Figure 1. Then write them down on a separate piece of paper, leaving room to write notes about each characteristic. As you read, summarize how the information relates to a particular characteristic.

Water Retention

Amphibians such as frogs cannot be considered fully terrestrial because they lose too much water through their skin. Amphibians must stay moist to avoid dehydration, and their method of reproduction requires a moist environment. Reptiles have evolutionary adaptations that free them from the water requirements of amphibians.

Watertight Skin

Terrestrial animals face a serious problem of water loss as water evaporates through their skin. Modern reptiles have evolved a skin made of light, flexible scales. These scales overlap and form a protective, almost watertight skin that minimizes water loss, as shown in **Figure 3.**

Figure 3 Reptilian scales. The scales of a reptile's skin form a tight seal that retains moisture within the reptile's body.

QUICK LAB

Modeling Watertight Skin

SC.H.3.4.1

Scales make a reptile's skin almost watertight. This is one of reptiles' adaptations to terrestrial life. You can use grapes to model and compare water loss in different types of skin.

Materials

forceps, 2 grapes, balance, Petri dish

Procedure

1. Find the mass of one grape, and record it in a data table. Then place the grape in an open Petri dish.
2. Using forceps, peel the skin from the second grape. Find and record the mass of the peeled grape. Then place it in the same Petri dish, but do not let the two grapes touch.
3. Wait 15 minutes, and then find and record the mass of each grape again.

Analysis

1. **Calculate** the difference between the original and final masses of each grape.
2. **Propose** an explanation for any changes in mass you observed.
3. **Determine** which grape represents an amphibian's skin and which represents a reptile's skin.
4. **Describe** how a watertight skin is an adaptation to terrestrial life. Include information you have learned in this lab in your explanation.

774

did you know?

Introduced Species Cause Problems In the western United States, the tansy ragwort plant is poisonous to cattle. The cinnabar moth, whose larvae eat the tansy ragwort, was introduced into the region to control this plant. However, the larvae accumulate toxins that kill the northern alligator lizards that prey on the larvae. Populations of these lizards may be at risk in certain areas of the west.

Watertight Eggs

For a reptile living on dry land, reproduction presents another serious water-loss problem. Without a watery environment, both sperm and eggs will dry out. A reptile's fertilized eggs need a moist environment in which to develop. As you will read later in this chapter, the first problem is overcome by internal fertilization.

The nature of a reptile's amniotic *(am nee AHT ic)* egg solves the second problem. An **amniotic egg** contains both a water supply and a food supply and is key to a reptile's success as a terrestrial animal. Because the egg's tough shell makes it essentially watertight, it does not dry out, even in very dry habitats. Most reptiles, all birds, and three species of mammals reproduce by means of amniotic eggs with shells. (Other mammals produce amniotic eggs, but the embryo develops within the female's uterus rather than within a shell. You will learn about the development of these eggs in a later chapter.) The formation of amniotic eggs with shells suggests that these three groups of animals evolved from a common ancestor.

BIOWatch: The Amniotic Egg

Both reptiles and birds have amniotic eggs, which are very much alike internally. Although a reptile's eggshell is leathery and a bird's is hard, both are almost watertight. However, the shells are porous enough to allow oxygen to enter the egg and carbon dioxide to leave. The shell and the albumen (egg white) lying beneath it protect and cushion the developing embryo. The albumen is also a source of protein and water for the embryo.

Within the egg, four specialized membranes—the amnion, the yolk sac, the allantois, and the chorion—play important roles in maintaining a stable environment in which the embryo can develop.

The amnion *(AM nee awn)* encloses the embryo within a watery environment. In a sense, this membrane creates a little pond that substitutes for the water in which amphibians lay their eggs. This watery enclosure also protects the embryo by cushioning it.

The yolk sac contains the yolk, the developing embryo's food supply. The embryo absorbs nourishment from the yolk through blood vessels connecting its gut and the yolk sac.

The allantois *(uh LAHN toh is)* is a sac that stores waste products from the embryo. It also serves as the embryo's organ for gas exchange. Blood vessels in its walls carry oxygen to and carbon dioxide from the embryo.

Surrounding the amnion, yolk sac, and allantois is a membrane called the chorion *(KAWR ee awn)*. The chorion allows oxygen to enter the egg and carbon dioxide to leave.

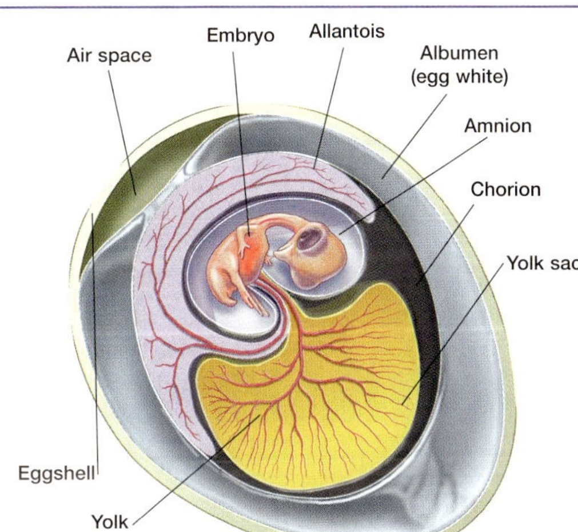

internet connect
www.scilinks.org
Topic: Amniotic Egg
Keyword: HX4005
SCILINKS. Maintained by the National Science Teachers Association

775

BIOWatch

The Amniotic Egg

Teaching Strategies
Give each lab group a hard-boiled egg. Have students crack it open and gently peel away the shell. Ask them to identify the chorion, yolk, and albumen in the egg.

Discussion

- What is the advantage of having an egg with a "pond" in it rather than just laying eggs in a pond? (Eggs can be laid out of the water, underground, in a tree, etc. Organisms can live in a wider variety of environments.)
- How is the amniotic egg similar to the high-tech fabrics used in outdoor clothing? (It is impervious to water but pervious to gases such as carbon dioxide and oxygen.)
- What is the advantage of having a storage sac for waste products (the allantois)? (Without it, waste products could build up in the blood and become toxic to the embryo.)
- Why does oxygen enter and carbon dioxide leave the egg? (As the embryo gives off carbon dioxide, its concentration inside the egg increases, which causes it to diffuse out of the egg. Oxygen is more concentrated outside the egg, so it diffuses inward.)

Attention Grabber

Green Turtles Tour the Atlantic Green turtles migrate more than 2,000 km (1,250 mi) from the waters off Brazil to the beaches of Ascension Island in the central Atlantic Ocean. A female green turtle will lay her eggs on the same beach where she hatched. Scientists do not yet know for certain how she finds her birthplace.

Activity — ADVANCED

Comparing Eggs Using field guides, books, and Internet resources, have students research the eggs of birds, reptiles, and amphibians. Ask each student to make life-size, full-color drawings of several types of eggs on small pieces of paper. When students are finished, assemble the drawings into three large posters for comparison: *The Eggs of Amphibians, The Eggs of Reptiles, and The Eggs of Birds.* **LS Interpersonal**

Teach, *continued*

Teaching Tip — BASIC

Surface Area Ask students which melts faster—multiple cubes of ice or a single block of ice of the same weight? (multiple cubes of ice) Lead students into a discussion of the importance of surface area in living things. Relate the cubes of ice to alveoli in the lungs. **LS Verbal**

SKILL BUILDER — GENERAL

Math Skills Ask students to find the surface area of an 8 mL cube and the combined surface area of eight 1 mL cubes. Remind students that the surface area of a cube is found by $l \times w \times 6$ *(where l = length, w = width, and 6 = the number of sides)*. Point out that an 8-mL cube has two-centimeter sides and a 1-mL cube has one-centimeter sides. (Although the overall volume is the same, the 8-mL cube has a surface area of 24 cm^2 and eight 1-mL cubes have a combined surface area of 48 cm^2.) **LS Logical**

Group Activity — ADVANCED

Local Snakes Use a field guide to determine what kinds of snakes are found in your area. Give students a list of these snakes and their scientific names. Have groups research these snakes and report to the class on the snakes' habitat, range, reproduction, behavior, feeding habits, and any other relevant information. **LS Verbal** Co-op Learning

Respiration

Because most reptiles are far more active than amphibians, they have greater metabolic requirements for oxygen. Their bodies meet this demand in several ways.

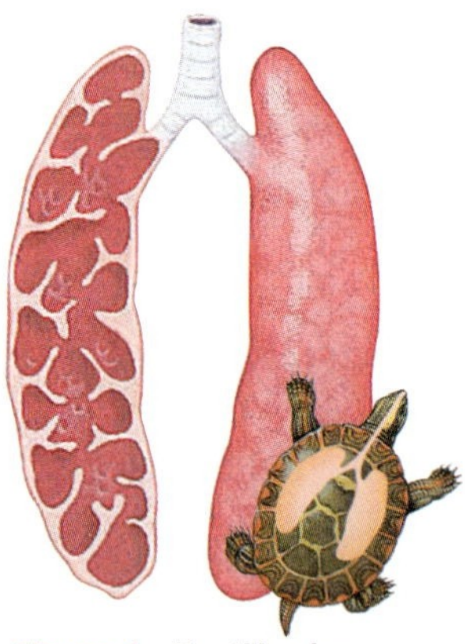

Figure 4 Reptilian lungs. The lungs of reptiles contain numerous internal folds.

Lungs

A reptile's scaly skin does not permit gas exchange, so reptiles cannot use their skin as an additional respiratory surface, as many amphibians can. However, the lungs of most reptiles have many internal folds, as shown in **Figure 4.** These folds greatly increase the respiratory surface area of a reptile's lungs. In addition, reptiles have strong muscles attached to their rib cage. The action of these muscles helps to move air into and out of the lungs, increasing the lungs' efficiency.

Heart

Recall that the ventricle of the amphibian heart is not divided by a septum. Oxygen-poor blood and oxygen-rich blood mix somewhat in the amphibian's ventricle. In most reptiles, however, the septum extends into the ventricle, partly dividing it into right and left halves, as shown in **Figure 5.** The septum enables a much better, but still incomplete, separation of oxygen-rich and oxygen-poor blood. As a result, oxygen is delivered to the body cells more efficiently than in amphibians.

Unlike most reptiles, crocodilians have a heart with a completely divided ventricle that consists of two pumping chambers. This arrangement fully separates the lung circulation from the body circulation. Thus, the delivery of oxygen throughout the body is further improved in these animals.

Figure 5 Reptilian heart. In most reptiles, the ventricle of the heart is partly divided by a septum.

Reptilian Heart Structure

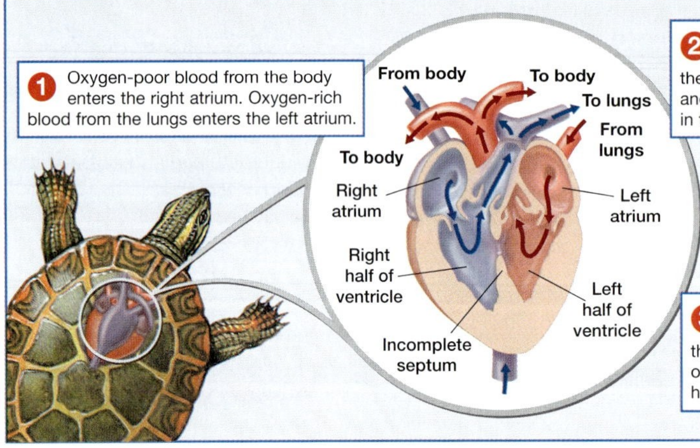

1 Oxygen-poor blood from the body enters the right atrium. Oxygen-rich blood from the lungs enters the left atrium.

2 An incomplete septum partly divides the ventricle. Thus, there is less mixing of oxygen-rich and oxygen-poor blood than there is in the amphibian heart.

3 Arteries carry oxygen-rich blood from the left half of the ventricle to the body and oxygen-poor blood from the right half of the ventricle to the lungs.

776

SOCIAL STUDIES CONNECTION

Snakes in Literature Throughout history, many cultures have used snakes in literary, cultural, and religious settings, such as *The Bible*, and in stories such as the Greek myth of Medusa. Even children's stories like *Aladdin* and *The Jungle Book* include snakes. Snakes are often portrayed as evil, cold-blooded animals. Have students research this phenomenon in literature and write an essay on whether the reputation of the snake is deserved.

Section Review Correlations

SC.F.1.4.2: Items 3, 5
SC.F.1.4.7: Items 2, 6
SC.G.1.4.1: Item 1
SC.H.1.4.4: Item 4

Reproduction

Unlike the eggs of most amphibians, reptilian eggs are fertilized within the female, a process called internal fertilization. The male reptile introduces his semen directly into the female's body. The semen contains sperm and fluid secretions. Internal fertilization protects the gametes from drying out, even though the adult animals are fully terrestrial.

Many reptiles are **oviparous** *(oh VIHP urh uhs)*, meaning the young hatch from eggs, as shown in **Figure 6.** In most cases, the eggs are not protected by the parents. Most snakes and lizards, all turtles and tortoises, and all crocodilians are oviparous. All birds and three species of mammals are also oviparous.

Some species of snakes and lizards are **ovoviviparous,** which means the female retains the eggs within her body until shortly before hatching, or the eggs may hatch within the female's body. Although the embryos receive water and oxygen from the female, their nourishment comes from the yolk sac. The offspring of a snake, shown in Figure 6, are born able to fend for themselves. In ovoviviparous reptiles, the eggs are less vulnerable to predators.

WORD Origins

The term *ovoviviparous* comes from three different Latin words: *ovum,* meaning "egg," *vivus,* meaning "alive," and *parere,* meaning "to bring forth or bear."

Figure 6 Reproduction. Sea turtles hatch from eggs buried on sandy beaches. The eggs of some species of snakes are incubated within the female's body, and the young are born live.

Sea turtles

Hatchling snakes

Section 1 Review

1. **Identify** seven characteristics of reptiles. SC.G.1.4.1
2. **Describe** how the ectothermic nature of reptiles influences their physical activity and feeding habits. SC.F.1.4.7
3. **Summarize** the skin and egg adaptations that allow reptiles to live on land. SC.F.1.4.2
4. **Critical Thinking Forming Reasoned Opinions** Data show that an animal's temperature changes over the course of a day. A student asserts that this proves the animal is an ectotherm. What must the student consider before making such a claim? SC.H.1.4.4
5. **Describe** how reptiles meet their need for more oxygen than amphibians require. SC.F.1.4.2
6. *FCAT Prep* If a lizard's internal temperature sensors detect a decrease in body temperature, the lizard can maintain homeostasis by what method?
   A resting in the shade
   B slowing its metabolism
   C basking in the sunshine
   D speeding up its metabolism
   SC.F.1.4.7

Answers to Section Review

1. strong, bony skeletons, and toes with claws; ectothermic metabolism; dry scaly skin, almost watertight; amniotic eggs; well-developed lungs; partly or completely divided ventricle; internal fertilization
2. Because they are ectotherms, reptiles must thermoregulate behaviorally. They can be physically active and hunt for food only when their body temperature is within a certain critical range.
3. The almost watertight skin and amniotic eggs of reptiles minimize water loss on land.
4. The student would need to compare the animal's temperature changes to changes in the environmental temperature.
5. In contrast to amphibians, reptiles have well-developed lungs with numerous folds, as well as strong rib muscles.
6. **A.** Incorrect. Resting in the shade would not increase body temperature. **B.** Incorrect. A slower metabolism would not increase body temperature. **C.** Correct. Ectotherms maintain body temperature by basking. **D.** Incorrect. Ectotherms cannot alter their metabolism to maintain homeostasis.

Close

Reteaching — BASIC

Have students pair up and list each of the seven key features of reptiles, which are given in **Figure 1,** on individual index cards. On the back of each card, have them write a one-word clue to identify the characteristic on the front of the card. Ask students to shuffle the cards and turn them so only the clue words are showing. Have one student choose a card at random and read the clue word. His or her partner should respond with the key characteristic. If the student answers correctly, his or her partner should set the card aside. Students should continue the game until all of the cards are set aside. Students should then switch roles and repeat the process.

LS Interpersonal Co-op Learning

Quiz — GENERAL

True or False:

1. As in most amphibians, fertilization in reptiles is external. (False. Fertilization is internal, an adaptation to life on land.)
2. The heart of most reptiles only partially separates blood from the lungs and blood from the body. (True. Except in crocodilians, freshly oxygenated blood mixes in the heart with oxygen-depleted blood.)

Alternative Assessment — GENERAL

Have teams of students write brief descriptions of what they think are the five most important things they have learned about reptiles. Have teams present their descriptions to the class, with each team member participating in the presentation.

LS Verbal Co-op Learning

Focus

Overview

Before beginning this section review with your students the objectives listed in the Student Edition. In this section, students are introduced to the orders of the reptiles, one of the most diverse classes of terrestrial vertebrates. Students will also learn the distinguishing characteristics of all the orders.

Bellringer

Before reading this section, ask students to list as many reptiles as they can. When they are done, have them indicate which reptiles on their list are closely related. After reading the section, have students make corrections and add any missing groups of reptiles to their lists. (Students should indicate that snakes and lizards are related, turtles and tortoises are related, and that crocodiles and alligators are related. Students are probably unfamiliar with the tuataras.)

Motivate

Demonstration — GENERAL

Obtain a picture of a skink, and cover its body so that only its head shows. Ask students if the animal is a lizard or a snake. Then reveal the rest of the animal, pointing out how similar some lizards are to snakes. Ask students how to tell a lizard from a snake. (Most lizards have legs and external ears, a pectoral girdle, and many have movable eyelids. Snakes lack these features.) LS Visual

Section 2 Today's Reptiles

Florida Sunshine Standards/FCAT

SC.F.1.4.2 Body structures are uniquely designed and adapted for their function.

SC.G.1.4.1 There is great diversity and interdependence among living things.

Objectives

- **Compare** the four living orders of reptiles. SC.G.1.4.1
- **Describe** the timber rattlesnake's adaptations for locating and capturing prey. SC.F.1.4.2
- **Compare** the parental care of crocodilians with that of other reptiles. SC.G.1.4.1

Key Terms

carapace
plastron

Lizards and Snakes

You've probably walked by a snake or lizard without even knowing it was there. Most are quiet, and their coloration often conceals them from view. Even if you visited the jungles of South America, you might not notice an anaconda unless it moved. What's an anaconda? It's the world's largest snake, frequently reaching 5 m (about 16 ft) in length. The largest anaconda ever found was twice that long. Very large anacondas have been known to prey on jaguars. After such a meal, the anaconda may not eat again for up to a year.

Snakes and lizards belong to order Squamata. A distinguishing characteristic of this order is a lower jaw that is only loosely connected to the skull. This allows the mouth to open wide enough to accommodate large prey and explains how an anaconda can swallow a jaguar. This ability is a contributing factor to the success of snakes and most lizards as predators.

Lizards

Common lizards include iguanas, chameleons, geckos, anoles, and horned lizards (often mistakenly called "horny toads"). A few species of lizards are herbivores, but most are carnivores. Most lizards are small, measuring less than 30 cm (1 ft) in length, but lizards that belong to the monitor family can be quite large. The Komodo dragon of Indonesia, shown in **Figure 7,** is the largest monitor lizard. It can be up to 3 m (10 ft) in length and weigh up to 125 kg (275 lb). The tail of some species of lizards, such as the gecko shown in Figure 7, breaks off easily when seized by a predator, allowing the lizard to escape. Lizards can regenerate a new tail, but it does not have any vertebrae in it.

Gecko

Komodo dragon

Figure 7 Lizards. Geckos are small reptiles, rarely exceeding 24 cm (10 in.) in length. The Komodo dragon is the world's largest lizard.

778

Chapter Resource File

- Directed Reading BASIC
- Active Reading GENERAL

Transparencies

TR Bellringer
TR I21 External Structures of a Timber Rattlesnake
TR I22 Internal Structures of a Timber Rattlesnake

One-Stop Planner CD-ROM

- Reading Organizers BASIC
- Reading Strategies BASIC
- Occupational Application Worksheet: Emergency Medical Technician GENERAL

Section Correlations

SC.F.1.4.2
SC.G.1.4.1

Snakes

Snakes probably evolved from lizards during the Cretaceous period. The close relationship between lizards and snakes is reflected in their many similarities. In fact, it is often difficult to distinguish the legless species of lizards from snakes. Snakes lack movable eyelids and external ears, as do several species of lizards. Also, both snakes and lizards molt periodically, shedding their outer layers of skin.

Body Structure The skeleton of snakes is unique. Most snakes have no trace of a pectoral girdle (the supporting bones for the bones of the forelimbs), which is found even in legless lizards. The snake's jaw is very flexible because it has five points of movement. (Your jaw, in contrast, has only one movement point.) One of these points is the chin, where the halves of the lower jaw are connected by an elastic ligament. This ligament permits the lower jaw to spread apart when a large meal is being swallowed. The African egg-eating snake, shown in **Figure 8,** can swallow eggs that are much larger than its head in a process that can take an hour or more.

Feeding While many snakes simply seize their prey and swallow it whole, some snakes use other methods to subdue their prey. Many very large snakes, such as anacondas, boas, and pythons, are constrictors, as are some smaller species, such as king snakes. Constrictors wrap their body around their prey, gradually squeezing tighter and tighter until the prey suffocates. The snakes then swallow their prey whole, even if the prey is very large. Like all snakes, constrictors have no teeth that are suited for cutting and chewing.

Some snakes kill their prey with venom (poison). Of the 13 or more commonly recognized families of snakes, only four are venomous: (1) cobras, kraits, and coral snakes; (2) sea snakes; (3) adders and vipers; and (4) rattlesnakes, water moccasins, and copperheads. In most venomous snakes, modified salivary glands produce a venom that is injected into the victim through grooved or hollow teeth. The African boomslang and twig snakes produce venom but do not inject it. Instead, they bite their prey with fangs located at the back of their mouth. Grooved teeth direct the venom into their victim's wound. You can read more about the biology of snakes in *Up Close: Timber Rattlesnake,* on the following pages.

Real Life

Need a lizard?

Instead of calling an exterminator to rid their homes of unwanted pests, some adventuresome homeowners keep a Tokay gecko. This gecko preys voraciously on mice and insects and is very efficient at ridding a home of these pests.

Finding Information

Investigate the pros and cons of using geckos for pest control. Who should try this method and who should avoid it?

internet connect

www.scilinks.org
Topic: Animals of Florida
Keyword: HX6F17

SCILINKS Maintained by the National Science Teachers Association

Figure 8 **Snake feeding.** Snakes have flexible jaws that allow them to swallow prey much larger than their head.

Teach

Real Life

Answer

People who have other pets that might harass a gecko, such as a cat or a dog, should not try to use geckos for pest control.

Teaching Tip — BASIC

Dangerous Snakes Ask students what venomous snake is the most dangerous in the world. Record their responses. (examples: king cobra, sea snake, coral snake, and Australian tiger snake) Ask why a particular species is considered more dangerous than others. Tell them that many different venomous snakes are touted as the "most dangerous snake alive," or the "most poisonous snake on Earth." Point out that several factors determine how "dangerous" a snake is. Among these factors are the toxicity of its venom, the type of venom (hemolytic or neurotoxic), and the amount of venom a bite victim receives. In India, there are about 900,000 snakebites each year, resulting in about 9,000 deaths. Many of these bites are delivered by the aggressive king cobra, which is frequently encountered by India's large human population. LS Verbal

MEDICINE CONNECTION

The American Red Cross recommends the following first-aid treatment for snakebite:

- Wash the bite with soap and water.
- Immobilize the area and keep it lower than the heart.
- Get medical help.

Medical professionals are nearly unanimous in their views of what **not** to do for snakebite:

- No ice or any other type of cooling on the bite
- No tourniquets
- No electric shock
- No incisions in the wound

Up Close

Timber Rattlesnake

Teaching Strategies
Help students understand the rattlesnake's ability to sense its prey in the dark by using a warm object such as a heating pad. Place the heating pad on a table and allow it to warm the surface. Remove the heating pad and have students move their hands above the tabletop without touching the table itself. They should be able to feel the heat radiating from the warmed spot. Have students determine how close their hands must be to detect the heat. Inform them that a rattlesnake can locate warm prey from a distance of 1 m (39 in.). These snakes, however, sense heat with their pit organs, not through their skin.

On a sheet of paper, have students list each major heading in the Up Close feature. After they read each section, have them hypothesize how that particular feature of the timber rattlesnake is an adaptation for survival. Then have students exchange papers with a partner and read their partner's paper, noting how their partner may have identified a different survival value for some of the features.

Timber Rattlesnake

- **Scientific name:** *Crotalus horridus*
- **Size:** Typically 90–150 cm (36–60 in.) long; maximum 189 cm (74 in.)
- **Range:** Eastern and central United States, from northern New York to northern Florida and west, to central Texas
- **Habitat:** Prefers thick brush, dense woodland, or swamp
- **Diet:** Primarily small mammals

External Structures

Rattle The rattle typically consists of 5 to 7 interlocking rings made of keratin, a protein. When shaken, it produces a rattling sound that serves as a warning. Contrary to popular belief, the snake does not add a rattle each year. Instead, each time the snake sheds its skin during molting, a new ring is added to the base of the rattle. The more rapidly the snake grows, the more rattles it accumulates during a given time. This is why the number of rattles a snake has increases with the size of the snake.

Pit organ Between each eye and nostril of the rattlesnake is an organ that can detect infrared radiation. The snake can locate a warm-bodied animal in a cool, nighttime environment by detecting the difference in infrared radiation emitted by the animal and the cooler background. Thus, a rattlesnake can hunt in total darkness.

780

MISCONCEPTION ALERT GENERAL

Slimy Snakes Contact a local nature center, zoo, or herpetological society to arrange for a live, non-poisonous snake and handler to visit the class. Have students make a list of some of their preconceptions about snakes. For instance, many students believe that a snake's skin is slimy, like that of a frog. Another misconception is that the tongue of a snake is poisonous, and that snakes will attack people for no reason. Show the class a live snake, and allow students to touch and examine the snake. Help to dispel misconceptions by discussing possible origins for such fears. Remain sensitive to the fact that many people will continue to fear snakes.

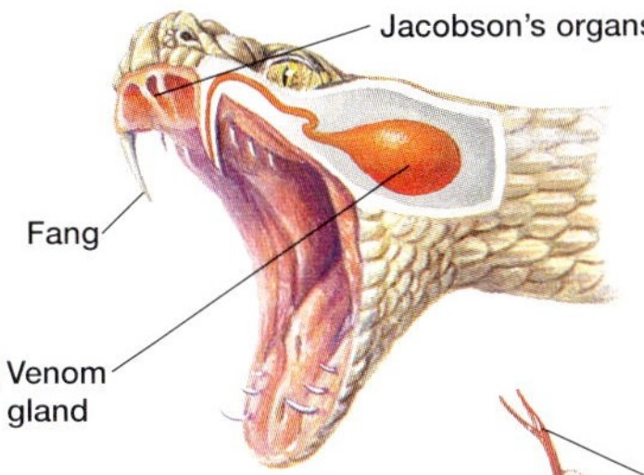

Venom glands The timber rattlesnake has hollow upper front teeth, or fangs. When the rattlesnake strikes, these hinged fangs swing forward from the roof of the mouth and inject venom deep into the prey. The venom contains hemotoxins, proteins that attack the circulatory system, destroying red blood cells and causing internal hemorrhaging. Modified salivary glands in the upper jaw produce the venom.

Jacobson's organs Flicking its forked tongue into the air, the rattlesnake takes in chemical samples from the environment. These chemicals are transferred to two depressions in the roof of the mouth called Jacobson's organs, which detect the odor of the chemicals. The snake uses these organs to follow the scent trail of prey.

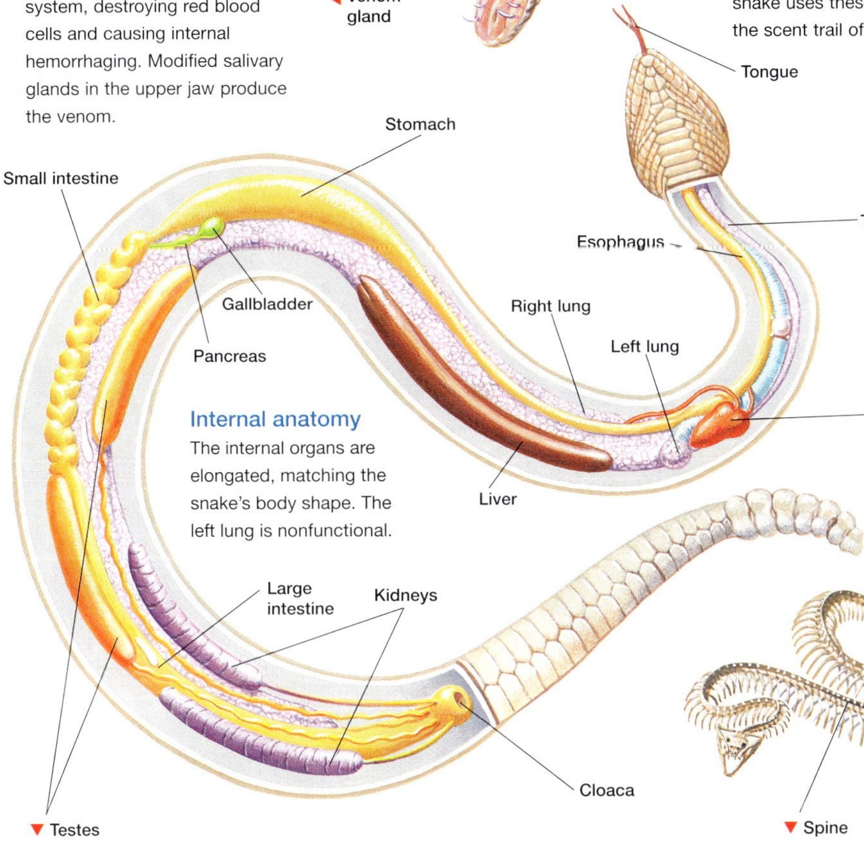

Internal anatomy The internal organs are elongated, matching the snake's body shape. The left lung is nonfunctional.

Reproductive structures This male rattlesnake produces sperm in his testes. Female timber rattlesnakes are ovoviviparous. A female carries her fertilized eggs in her body while they develop. Each egg has a thin membrane through which water and oxygen pass from the mother to the embryo. All nourishment is provided by the egg's yolk. After the eggs hatch in the mother's body, the live young are ejected and must fend for themselves.

Spine The rattlesnake's spine is made up of several hundred vertebrae, each with its own pair of attached ribs. It provides the framework for thousands of muscles that manipulate not only the skeleton but also the snake's skin, causing the overlapping scales to extend or lie flat.

Up Close

Timber Rattlesnake

Discussion

- Explain why it would be easier for a timber rattlesnake to kill a mouse than a lizard at night. (The mouse is an endotherm and the pit viper's heat-sensitive pits can detect its body heat. The lizard is an ectotherm and would not necessarily be warmer than its surroundings.)
- What is the advantage to a rattlesnake of announcing its presence to other animals by rattling? (Answers may vary but may include that by rattling the snake keeps large mammals, such as cattle, from stepping on it.)
- How is an ovoviviparous snake, such as the timber rattlesnake, different from a viviparous snake in the way developing embryos are sustained? (Ovoviviparous mothers supply no nutrients to the developing young, while viviparous mothers do.)

did you know?

Triple-jointed Jaws The tiny snakes called threadsnakes have triple-jointed jaws. When a threadsnake burrows into the nests of ants or other social insects, parts of its lower jaw rotate like a pair of swinging doors. It is thought that this action helps the snake to eat quickly and escape injury or death from ant stings.

Attention Grabber

Venomous Snakes About 99 percent of venomous snakebites in the United States are from pit vipers. About 8,000 snakebites occur in the United States each year, and around 9 to 15 victims die. Every state except Maine, Alaska, and Hawaii is home to at least one poisonous snake species.

Other Orders of Reptiles

The remaining orders of living reptiles contain far fewer species than the order Squamata does. There are about 250 or more species of turtles (which generally live in water) and tortoises (which live on land), all classified in the order Chelonia. The order Crocodilia is composed of 25 species of large, aquatic reptiles. The order Rhynchocephalia *(RING koh seh FAY lee uh)* contains only two species of tuataras.

Turtles and Tortoises

Turtles and tortoises, shown in **Figure 9,** differ from other reptiles in that their bodies are encased within a hard, bony, protective shell. Many of them can pull their head and legs into the shell for effective protection from predators. While most tortoises have a dome-shaped shell, water-dwelling turtles have a streamlined, disk-shaped shell that permits rapid maneuvering in water. Turtles and tortoises lack teeth but have jaws covered by sharp plates, which form powerful beaks. Many are herbivores but some, such as the snapping turtle, are aggressive carnivores.

Today's turtles and tortoises differ little from the earliest known turtle fossils, which date to more than 200 million years ago. This evolutionary stability may reflect the adaptive aspects of their basic shell-covered body structure. The shell is made of fused plates of bone covered with horny shields or tough, leathery skin. In either case, the shell consists of two basic parts. The **carapace** is the dorsal (top) part of the shell, and the **plastron** is the ventral (bottom) portion. The vertebrae and ribs of most species are fused to the inside of the carapace, as shown in **Figure 10.** The shell provides the support for all muscle attachments in the torso.

Figure 9 Turtle and tortoise. Like other sea turtles, this green sea turtle (top) spends virtually its entire life in the sea. The Galápagos tortoise (bottom) spends its life on land.

Figure 10 Turtle interior. In this ventral view, a turtle's plastron has been removed to show the relationship of the vertebral column, ribs, pelvis, and pectoral girdle to the carapace.

Crocodiles and Alligators

Of all the living reptiles, the crocodilians are most closely related to the dinosaurs. In addition to crocodiles and alligators, shown in **Figure 11,** the order Crocodilia includes the alligator-like caimans and the long-snouted gavial. Crocodilians are aggressive carnivores. Some are quite large. American alligators can reach 5.5 m (18 ft) in length, and Nile crocodiles can reach 6 m (20 ft) in length and weigh 750 kg (1,650 lb). Crocodilians generally capture prey by stealth, often floating just beneath the water's surface near the shore. When an animal comes to the water to drink, the crocodilian explodes out of the water and seizes its prey. The crocodilian then hauls the prey back into the water to be drowned and eaten. The bodies of crocodilians are well adapted for this form of hunting. Their eyes are high on the sides of the head, and their nostrils are on top of the snout. As a result, they can see and breathe while lying nearly submerged in the water. Crocodilians have a very strong neck and an enormous mouth studded

782

Teach, *continued*

Using the Figure — General

Point out in **Figure 10** that a turtle's shoulders lie within its rib cage. Students should recognize that their own shoulders are outside of their ribs. Tell students that a turtle is attached to its shell and cannot crawl out of it, as cartoon turtles often do. **LS Visual**

SKILL BUILDER — Basic

Writing Skills Have students imagine that they are the first person from their country to see a turtle. Ask students to write a description of a turtle for an audience that has never seen a turtle before. Encourage them to be both descriptive and accurate. **LS Verbal**

Teaching Tip

Endangered Crocodilians Tell students that although crocodilians are fierce predators, many of the 25 species of crocodilians are endangered or threatened. Overhunting of crocodilians for their hides, which are used to make leather goods, is the primary cause of their decline.

Section Review Correlations

SC.F.1.4.2: Items 2, 5, 6

SC.G.1.4.1: Items 1, 3, 4

Cultural Awareness

Turtle Myths The turtle appears as a central character in mythologies from around the world. An ancient Chinese myth describes how the turtle Kwei created the universe. The Chinese Book of Rites names the turtle as one of the four benevolent spirit animals. Hindu myths from India describe the world as being supported by four elephants standing on a turtle's back. And in North America, several native cultures imagined the world as an island on the back of a great turtle.

did you know?

Alligators' Sounds Alligators are unusual among reptiles in being able to make definite vocalizations. The male alligator bellows loudly during mating season. Vocal sacs on each side of his throat inflate when he calls. When alligators hatch from their buried eggs, they make a sound that is almost like a bark. The sound signals their mother that it is time to open the nest. Hatchlings also have a distress call they use to alert their mother if they feel threatened.

American alligator

Australian crocodiles

Figure 11 **Crocodilians.** In general, the snouts of alligators are shorter and broader than those of crocodiles.

with sharp teeth. A valve in the back of the mouth prevents water from entering the lungs when crocodilians feed underwater.

Unlike other living reptiles, crocodilians care for their young after hatching. For instance, a female American alligator builds a nest of rotting vegetation for her eggs. After the eggs hatch, the mother may tear open the nest to free the hatchlings. The young alligators remain under her protection for up to a year.

Tuataras

The two living species of tuataras are members of the genus *Sphenodon* and are native to New Zealand. *Sphenodon punctatus*, the more common species, is shown in **Figure 12.** Tuataras are lizardlike reptiles up to 70 cm (2 ft) long. Unlike most reptiles, tuataras are most active at low temperatures. They burrow or bask in the sun during the day and feed on insects, worms, and other small animals at night. Tuataras are sometimes called living fossils because they have survived almost unchanged for 150 million years. Since the arrival of humans in New Zealand about 1,000 years ago, the tuatara's range has diminished, and their numbers are declining.

Figure 12 **Tuatara.** Tuataras live on only a few small islands in New Zealand.

Section 2 Review

1. **Describe** the characteristics shared by lizards and snakes. SC.G.1.4.1
2. **Describe** the function of two different organs that help snakes locate their prey. SC.F.1.4.2
3. **Summarize** the ways turtles and tortoises differ from other reptiles. SC.G.1.4.1
4. **Compare** the parental care shown by alligators with that shown by most other reptiles. SC.G.1.4.1
5. **Critical Thinking** **Recognizing Relationships** How does the position of a crocodile's nostrils and eyes relate to its method of hunting? SC.F.1.4.2
6. *FCAT Prep* What is the pit organ of a rattlesnake sensitive to?
   A airborne chemicals
   B faint sounds
   C ground vibrations
   D infrared radiation
   SC.F.1.4.2

783

Answers to Section Review

1. Snakes and lizards both have scaly skin and molt periodically. Their lower jaw is loosely connected to the skull.
2. Jacobson's organs, located in the roof of the mouth, detect microscopic airborne particles, which the brain interprets as scent. Pit vipers use their pit organs to detect heat.
3. Their bodies are encased in a protective shell, and they lack teeth.
4. Alligators and other crocodilians are the only reptiles to care for their young. Some alligators build nests for their young and care for them for up to a year after they hatch.
5. The crocodile is able to keep most of its body submerged and hidden while still being able to breathe and look for prey.
6. **A.** Incorrect. The Jacobson's organs in the roof of the mouth are sensitive to airborne chemicals. **B.** Incorrect. Snakes lack ears and do not hear faint sounds. **C.** Incorrect. Snakes can feel ground vibrations through their bodies, not the pit organs. **D.** Correct. Using pit organs, warm-blooded animals can be found even in the dark.

Close

Reteaching — BASIC

Write the following words or phrases on the board: lizards and snakes, turtles and tortoises, crocodiles and alligators, and tuataras. Have students copy them, leaving space for additional information. Have students write distinguishing traits of each group in the appropriate spaces without using their textbooks. (lizards and snakes-lower jaw has loose connection to skull, no pectoral girdle in snakes, molt periodically, may lack eyelids; turtles and tortoises-hard shell, lack teeth, jaws have a powerful beak; crocodiles and alligators-long snout, eyes high on head, nostrils on top of head, large mouth with sharp teeth; tuataras-lizardlike, active at night) When the students have finished their work, have them share their results with a partner, recording any characteristics they omitted but their partner identified. **LS Verbal** Co-op Learning

Quiz — GENERAL

1. The reptiles that share the most characteristics with snakes are the _______ (lizards)
2. Unlike other reptiles, _______ care for their young after they hatch. (crocodilians)
3. The shells of turtles and tortoises consist of a plastron on the underside and a _______, which fuses with the rib cage. (carapace)

Alternative Assessment — ADVANCED

Assign students to work in groups of three. Each group will research the reptiles of an assigned continent. For each reptile, include the following: order, appearance (drawing), size, diet, habitat, range, and status (common, rare, threatened, or endangered). When groups have finished their research and drawings, the entire collection can be assembled on a large mural of the continents. Ask a local elementary school to display the mural. **LS Interpersonal** Co-op Learning

Focus

Overview

Before beginning this section review with your students the objectives listed in the Student Edition. In this section, students will summarize the key features of birds. These include several reptile-like characteristics, such as amniotic eggs and scale-covered feet and legs. Unique adaptations include feathers for flight and insulation, a strong yet very lightweight skeleton, a highly efficient respiratory system and completely divided heart. These features allow flight for nearly all birds. Students will also examine the highly variable beaks and feet of birds, which allow different species to make use of a range of diets and habitats.

Bellringer

Ask students to list the unique adaptations of birds that allow them to fly. (Students may list feathers, lightweight or hollow bones, wings, rapid metabolism, and an efficient heart and respiratory system.)

Motivate

Demonstration — BASIC

Bring a feather to class and ask students what kind of animal it is from. When students answer "a bird," ask them to think of any bird that does not have feathers. (There are none.) Ask them if they can think of any living animals that have feathers but are not considered birds. (no) Emphasize that having feathers is a unique classification characteristic for the class Aves. LS Visual

Section Correlations

SC.F.1.4.2
SC.G.1.4.1

Section 3 Characteristics and Diversity of Birds

Florida Sunshine Standards/FCAT

SC.F.1.4.2 Body structures are uniquely designed and adapted for their function.

SC.G.1.4.1 There is great diversity and interdependence among living things.

Objectives

- **Summarize** the key characteristics of birds. SC.G.1.4.1
- **Describe** how a bird's feathers and bone structure aid flight. SC.F.1.4.2
- **Summarize** how a bird's lungs and heart are adapted for high efficiency. SC.F.1.4.2
- **Relate** the structure of a bird's feet and beak to its habits and diet. SC.F.1.4.2

Key Terms

contour feather
preen gland
down feather

internet connect
www.scilinks.org
Topic: Characteristics of Birds
Keyword: HX4036
SCILINKS. Maintained by the National Science Teachers Association

Key Characteristics of Birds

Why do people use the expression "free as a bird"? Most likely it comes from a bird's ability to fly seemingly wherever it wishes. Through human history, the gift of flight has been celebrated in stories, poetry, and songs. But there is more to birds than flight; in fact, some species of birds can't fly.

The birds you see today are the modern members of class Aves. Unlike their reptilian relatives, birds usually lack teeth and have a tail that is greatly reduced in length. But they do retain some reptilian characteristics. For instance, birds lay amniotic eggs that are very similar to those of reptiles, and the feet and legs of birds are covered with scales. Other characteristics unique to birds distinguish them from all other animals. The most obvious is the presence of feathers and the modification of the forelimbs into wings. **Figure 13** lists some distinguishing features of birds. To learn more about the anatomy and habits of one bird, see *Up Close: Bald Eagle* later in this section.

Feathers

Feathers are modified reptilian scales that develop from tiny pits, called follicles, in the skin. Just as snakes and lizards replace their skin by molting, birds molt and replace their feathers. However, few birds shed all of their feathers at one time.

Birds have two main types of feathers: contour feathers and down feathers. **Contour feathers** cover the bird's body and give adult birds their shape. Specialized contour feathers, called *flight feathers*, are found on a bird's wings and tail. These feathers help provide lift for flight. As shown in **Figure 14,** a contour feather has many branches called barbs. Each barb has many projections, called

Figure 13 Characteristics of birds. Like most birds, this tern is well adapted to flight.

Characteristics of Birds

- Forelimbs modified into wings
- Body covered with feathers
- Lightweight bones
- Endothermic metabolism
- Super-efficient respiratory system
- Heart with completely divided ventricle

784

Chapter Resource File

- Directed Reading BASIC
- Active Reading GENERAL
- Data Sheet for Math Lab GENERAL

One-Stop Planner CD-ROM

- Reading Organizers BASIC
- Reading Strategies BASIC
- Supplemental Reading Guide *Through a Window* ADVANCED

Transparencies

TR Bellringer
TR I23 Characteristics of Birds
TR I27 Contour Feature Structure
TR I48 Avian Skeleton
TR I29 Avian Heart Structure
TR I30 Avian Lung Structure

Figure 14 **Contour feather**

The structure of a contour feather helps create a smooth, aerodynamic surface, aiding flight.

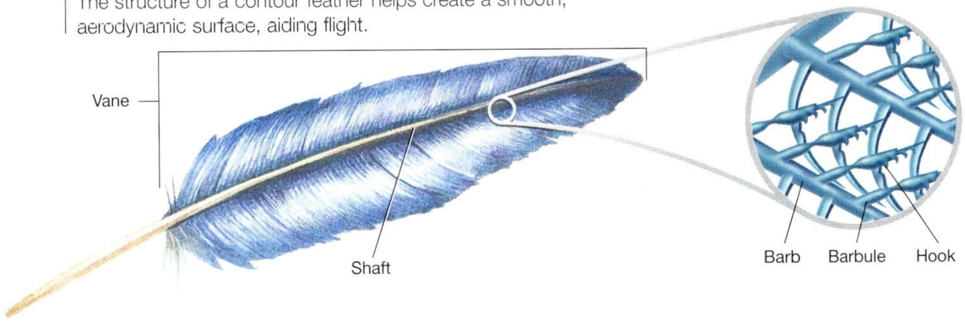

barbules, that are equipped with microscopic hooks. These hooks link the barbs to one another, giving the feather a continuous surface and a sturdy but flexible shape. With use, the connections become undone. When you see a bird pulling its feathers through its beak, it is relinking these connections. This process is called *preening*. Preening also serves another function. Most birds have a gland called a **preen gland** which secretes oil. When a bird preens, it spreads the oil over its feathers, cleaning and waterproofing them.

Down feathers cover the body of young birds and are found beneath the contour feathers of adults. Their soft, fluffy structure provides good insulation for the bird, helping the bird conserve body heat.

Feathers are important for other reasons too. Their coloration may be protective (as camouflage) or may be important in the selection of a mate. For example, the feathers of some birds allow them to blend in with their surroundings. In other species, the males develop special plumage during the breeding season.

Strong, Lightweight Skeleton

If you have ever picked up a bird, such as a parakeet, you may have been surprised at how light it was compared to a mammal of a similar size. This is because the bones of birds are thin and hollow. Many of the bones are fused, making a bird's skeleton more rigid than a reptile's. The fused sections form a sturdy frame that anchors muscles during flight. The power for flight (or for swimming underwater in the case of some birds, like penguins) comes from large breast muscles that can make up 30 percent of a bird's body weight. These muscles stretch from the wing to the breastbone. The breastbone is greatly enlarged and bears a prominent keel for muscle attachment, as illustrated in **Figure 15.** Muscles also attach to the fused collarbones (wishbone). No other living vertebrates have a keeled breastbone or fused collarbones.

Figure 15 **Avian skeleton.** A bird's large, keeled attachment point for flight muscles, while its fused collarbones (wishbone) help absorb the stresses of flight.

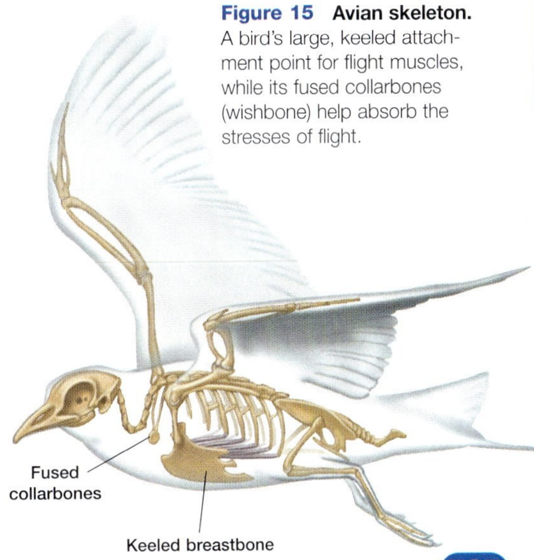

Teach

READING SKILL BUILDER — **BASIC**

Paired Summarizing After students have read silently about birds, have pairs of students summarize what they have read without looking at the textbook. One student should listen without interrupting but should be prepared to point out any inaccuracies in the summary, and to add any ideas that were left out. Students may refer to the textbook during this clarification process. You may want to pair ELL students with native English speakers. English Language Learners

Teaching Tip — GENERAL

Preening Birds often sit and preen their feathers. Students have probably observed this behavior. Have students hypothesize about the function of preening and the reason that birds spend so much time doing it. After a period of discussion, point out that when birds preen, they smooth out and clean their feathers, making them more aerodynamic. Also note that birds spread oil from a gland near the base of their tail to waterproof the feathers and make them more resistant to breakage, much the way hair conditioner works. LS Logical

Demonstration — GENERAL

Bring cooked, cleaned chicken bones to class for students to examine. Cut a few bones in half so students can see the hollow structure. If you have a wishbone, allow students to pull it gently to note its flexibility. The cartilage keel on the breast of a chicken is also interesting to examine because it provides an attachment surface for flight muscles. LS Visual

Career

Veterinarian When most people think of a veterinarian, they think of a "dog or cat doctor." Many students will know that veterinarians also treat livestock. However, most people probably do not know that veterinarians also treat birds and reptiles. Have students investigate the training required to become a veterinarian, and have them find out where vet schools are located in your region of the country.

MISCONCEPTION ALERT

Bird Nests Children who find bird nests are often warned not to touch them. The children are told, "the mother won't return to the nest if you touch it." In reality, most birds have a very poor sense of smell and probably would not know if their nest had been touched by a human (exceptions are flightless birds, ducks, and vultures). To compensate, many birds have keen vision and hearing. LS Interpersonal

Teach, *continued*

Teaching Tip

Heat Source The high temperature of birds is a by-product of their rapid metabolism. You may wish to use the analogy of a car engine for metabolism. If the engine is running, heat is produced. The faster the engine is running, the more heat is produced. Ectothermic reptiles are like cars whose engines run very slowly all the time, producing little heat. In contrast, birds (and mammals) keep their engines running at high speed all of the time, producing much heat.

Study TIP

Interpreting Graphics
After studying Figure 18, use your own words to summarize how one breath of air circulates through a bird's lungs. Remember to include both the inhalation and the exhalation cycle.

Endothermic Metabolism

Birds are endotherms; that is, they generate enough heat through metabolism to maintain a high body temperature. Birds maintain body temperatures ranging from 40°C to 42°C (104°F to 108°F), which is significantly higher than the body temperature of most mammals. For comparison, your body temperature is 37°C (98°F). These high temperatures are due to a high rate of metabolism, which satisfies the increased energy requirements of flight.

Completely Divided Ventricle

As in crocodilians, the ventricle of birds is completely divided by a septum, as shown in **Figure 16.** Oxygen-rich and oxygen-poor blood are kept separate, meaning that oxygen is delivered to the body cells more efficiently. The sinus venosus, which is a prominent part of the fish heart, is not a separate chamber of the heart in birds (or mammals). However, a small amount of tissue from it remains in the wall of the right atrium. This tissue is the point of origin of the heartbeat and is known as the heart's pacemaker.

Highly Efficient Lungs

Birds such as the geese shown in **Figure 17** use a considerable amount of energy when they fly. Since birds often fly for long periods of time, their cellular demand for energy exceeds that of even the most active mammals. How do birds get the energy they need?

Recall that reptiles meet their increased need for oxygen with lungs that have a larger surface area than the lungs of amphibians. But there is a limit to how much the efficiency of a lung can be improved just by increasing its surface area. Another way to

Figure 16 Avian heart. A bird's heart has a complete septum.

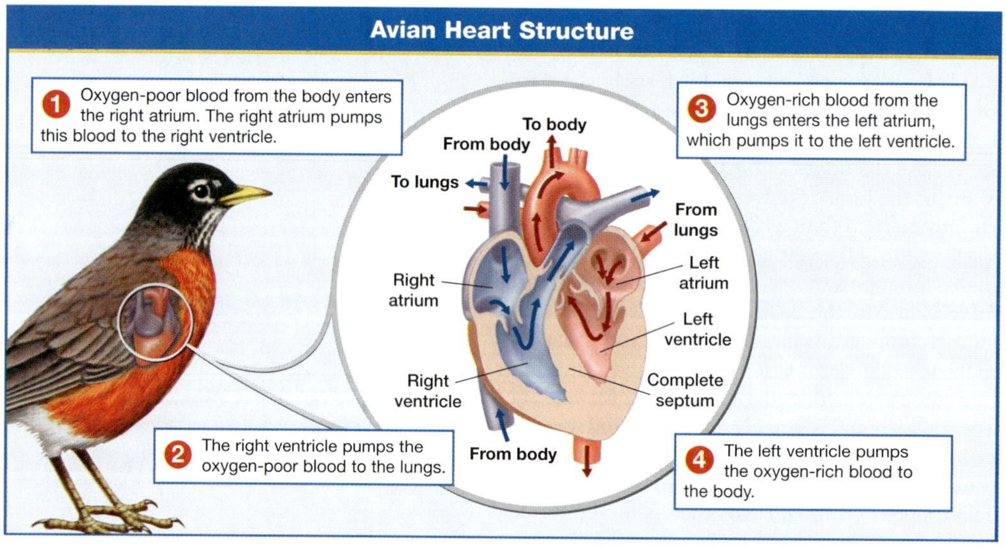

Cultural Awareness

Eagle Feathers The only people in North America who can legally own eagle feathers are Native Americans. The eagle is so highly valued by tribes throughout the United States that its feathers must be earned through personal sacrifice, and then they may be used only in special ceremonies. For instance, if a Winnebago pow-wow dancer accidentally drops an eagle feather during a performance, the dance is stopped until the feather is purified by an elder and then reclaimed by the dancer, who is not allowed to dance again for a year.

increase the efficiency of a lung is to have air pass over its respiratory surface in one direction only, just as water flows over a fish's gills in one direction. This is what happens in birds. One-way air flow is possible in birds because they have air sacs connected to their lungs, as shown in **Figure 18.** There is no gas exchange in the air sacs. They simply act as holding tanks.

There are two important advantages to one-way air flow. First, the lungs are exposed only to air that is almost fully oxygenated, increasing the amount of oxygen transported to the body cells. Second, the flow of blood in the lungs runs in a different direction than the flow of air does. Unlike the flow of water and blood in fish gills, the flow of air and blood in bird lungs are not completely opposite (countercurrent). Nevertheless, the difference in direction does increase oxygen absorption.

These three characteristics—endothermic metabolism, a completely divided ventricle, and highly efficient lungs—provide the energy a bird needs for takeoff and sustained flight. They enable a hummingbird to flap its wings rapidly (20–80 beats per second) as it hovers by a flower. They also permit migrating birds to fly thousands of kilometers without stopping. One species of shorebirds called the lesser yellowlegs flies across the open ocean from Massachusetts to Martinique in the West Indies. Incredibly, some of these birds cover this distance of 3,220 km (about 2,000 mi) in less than 6 days. Note, however, that many birds, such as gulls and vultures, remain aloft for long periods of time using little energy. These birds take advantage of upward air movements that lift them.

Figure 17 Flight. These barnacle geese expend an enormous amount of energy during take off and flight.

Figure 18 Avian respiration. A single breath of air stays in a bird's respiratory system for two cycles of inhalation and exhalation.

Avian Lung Structure

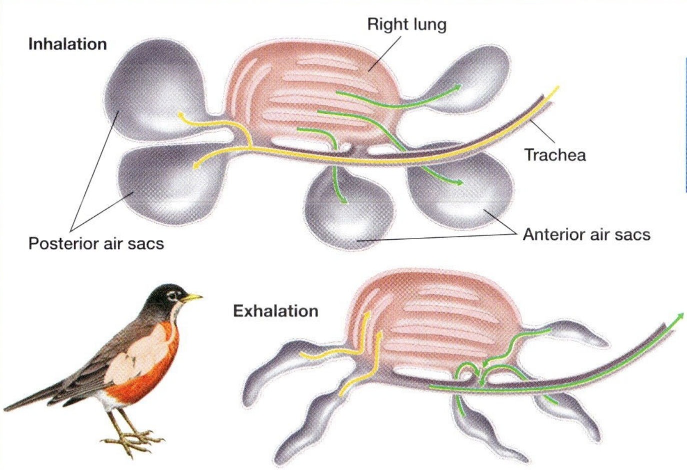

1 During inhalation, most of the fresh air (yellow) inhaled is pulled into the posterior air sacs. At the same time, stale air (green) from the previous inhalation is pulled into the anterior air sacs.

2 During exhalation, fresh air (yellow) from the posterior air sacs enters the lungs. At the same time, stale air (green) from the previous inhalation moves from the anterior air sacs out of the body.

787

Teaching Tip — GENERAL

White or Dark Meat? The dark color of some chicken meat is partly due to the presence of myoglobin, a hemoglobin-related molecule that helps provide oxygen to muscles that must contract vigorously and repeatedly. Ask students why a duck has dark breast meat but a chicken has white breast meat. (The duck flies long distances, but the chicken does not.) Ask what chickens usually use for locomotion and what color that meat is. (They usually use their legs to walk, and the legs contain dark meat.)

Activity — ADVANCED

Origin of Birds Ask students to identify the living reptiles that are most closely related to birds. (crocodilians) Have them name two characteristics that crocodilians and birds share. (a heart with completely divided ventricle, parental care of young, amniotic eggs, and vocalization) Have students research recent fossil discoveries that have changed views on the origin of birds. **LS Verbal**

did you know?

Brood Parasites The brown-headed cowbird is a brood parasite. A female brown-headed cowbird will fly to a nest full of eggs, roll one or two eggs out, and lay the same number as the number she has displaced. When the owner of the nest returns, it will incubate all the eggs, including the cowbird's. The cowbird eggs often have a shorter incubation time and hatch before the eggs of the incubating species. Therefore, the cowbird nestlings are fed first and get a head start in their development over the other chicks in the nest. Cowbird nestlings may even eject the nestlings of the incubating species, thereby eliminating competition for food.

Up Close

Bald Eagle

Teaching Strategies

- The largest bald eagle nests in North America were found in Vermillion, Ohio—3.7 m (12 ft) deep, 2.6 m (8.5 ft) in diameter, and weighing 1,800 kg (2 tons)-and in Florida—6.1 m (20 ft) deep and 2.9 m (9.5 ft) across. An eagle nest is called an aerie.
- After World War II, the numbers of bald eagles in the lower 48 states fell drastically because of the widespread use of the pesticide DDT. Because DDT breaks down very slowly, it accumulates in the food chain and causes the eggshells of bald eagles and many other birds to become thin and fragile. Affected eggs break when a parent attempts to incubate them. In 1972, the use of DDT was banned in the United States. By 1978, fewer than 500 breeding pairs of bald eagles remained in the United States outside Alaska. A slow recovery of the eagles began, and by the summer of 1999, six states had removed bald eagles from the endangered species list.
- Show students a picture of a large but immature bald eagle and ask them what kind of bird it is. Point out that the bald eagle does not get its white head and tail until its fourth or fifth year of life.

Bald Eagle

- **Scientific name:** *Haliaeetus leucocephalus*
- **Size:** Wingspan is typically over 2 m (6.5 ft), and body weight often exceeds 7 kg (15 lb)
- **Range:** Nearly all of North America, from Florida to northern Alaska
- **Habitat:** Forested areas near water that have tall trees for perching and nesting
- **Diet:** Fish, small mammals, birds, carrion

External Structures

Eyes Vision is a bald eagle's most important sense. The bird's keen eyesight allows it to see prey at great distances. Its visual acuity is 3–4 times higher than ours.

Feathers The body of the bald eagle is covered with feathers everywhere except the feet and the beak, which are bare. Both sexes develop the characteristic white head and neck at maturity.

Grasping feet The bald eagle has large feet and talons—the hind talon may be 5 cm (2 in.) long. The talons are used to snatch fish from the water while the eagle is flying. When the muscles of the legs contract, the tendons in the lower legs are pulled, and the talons lock together around the fish.

Beak The beak is massive, with an elongated, sharp, downward-curving tip. Because they have no teeth, bald eagles do not chew their food. Instead, they use their beak to tear their prey into portions that they swallow whole.

788

Transparencies

TR I25 External Structures of Birds–Bald Eagle

TR I26 Internal Structures of Birds–Bald Eagle

TR I31 Avian Adaptations

Internal Structures

Brain In the ratio of brain size to body size, birds rank second among vertebrates, behind only mammals. The large cerebellum receives and integrates information from the muscles, eyes, and inner ears. This makes possible the precise control of movement and balance necessary for flight. The optic lobe is large because it processes input from the eagle's most important sense organs—the eyes. The cerebrum performs many functions, including evaluation of sensory information, control of behavior, and learning.

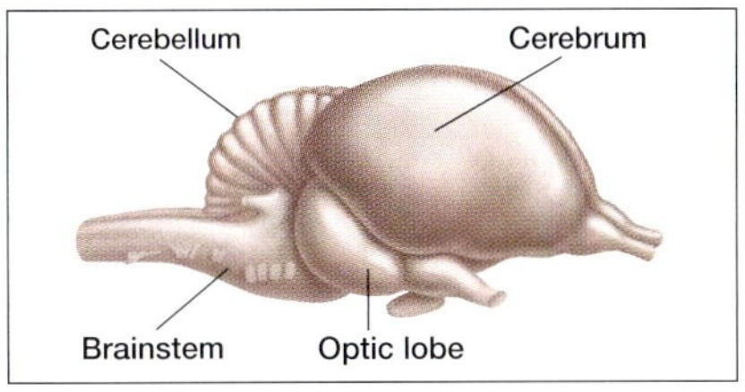

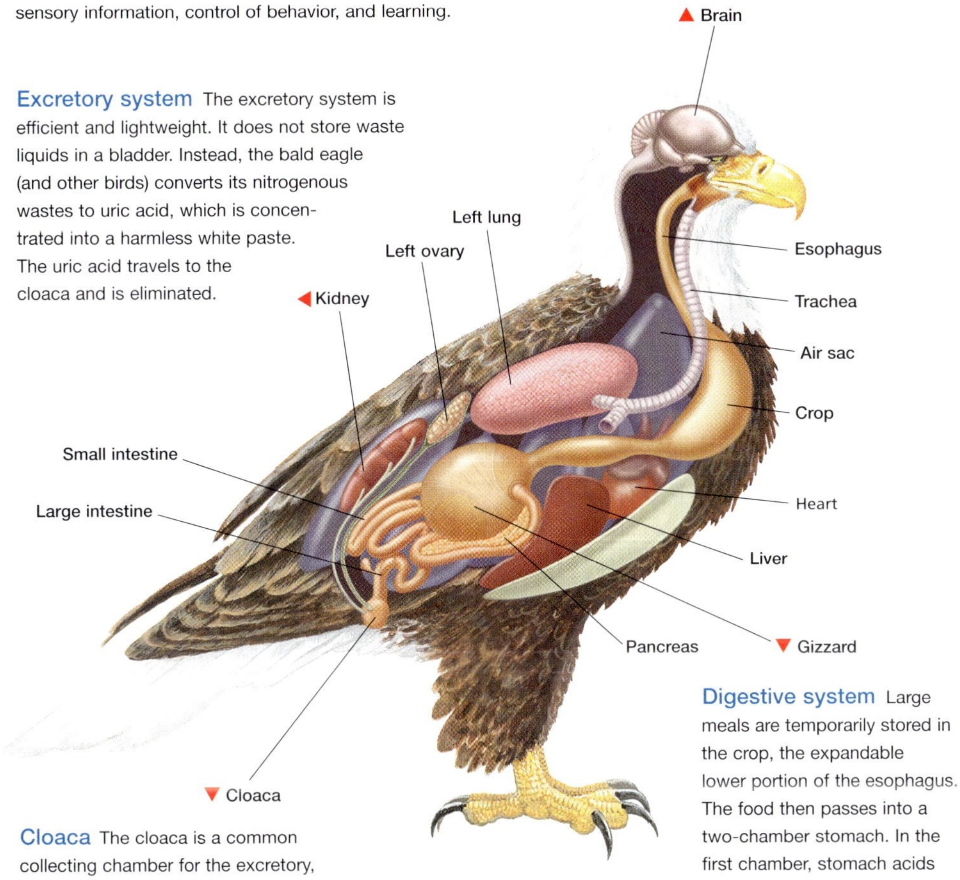

Excretory system The excretory system is efficient and lightweight. It does not store waste liquids in a bladder. Instead, the bald eagle (and other birds) converts its nitrogenous wastes to uric acid, which is concentrated into a harmless white paste. The uric acid travels to the cloaca and is eliminated.

Cloaca The cloaca is a common collecting chamber for the excretory, digestive, and reproductive systems. As fertilized eggs travel down the female eagle's oviduct, egg white, membranes, and the shell are added. The completed egg then passes into the cloaca and out of the female's body.

Digestive system Large meals are temporarily stored in the crop, the expandable lower portion of the esophagus. The food then passes into a two-chamber stomach. In the first chamber, stomach acids begin breaking down the food. The partially digested food is then passed to the second chamber, the gizzard, where it is ground and crushed. Undigested material is eliminated through the cloaca.

789

Up Close

Bald Eagle

Discussion

- Identify three differences between the internal structures of the timber rattlesnake and the bald eagle. (The rattlesnake has one functional lung, lacks air sacs, has neither a crop nor a gizzard, has venom-producing glands, and does not have hollow bones.)
- Would the heat-sensitive pits of the timber rattlesnake be an effective way of sensing prey for a bald eagle? (No, the rattlesnake's pits are sensitive only to nearby heat sources, and the bald eagle spots its prey from great distances.)
- Why might it be a disadvantage for an eagle to have a urinary bladder? (Urine contains a greater amount of water, which is heavy. The extra weight would be a disadvantage in flight.)
- Why might teeth be a disadvantage to a bird? (Teeth are heavy and require heavy jaws.) **LS Visual**

Activity — General

Eagle Eyes Eagles have much better vision than humans do. To simulate how much better, tape a worksheet on the wall and have students stand about 3 m away and try to read it. Then have them move 1 m away and read the worksheet. Tell them that an eagle could see the paper at 3 m as well as humans can at 1 m.

Teaching Tip — GENERAL

Reptiles and Birds Have students make a three-column chart like the one in the Graphic Organizer shown at the bottom of this page. They should use the information on birds in Section 3 and the material on reptiles in Sections 1 and 2. Students should use the following characteristics for both reptiles and birds: type of body covering, type of heart, mode of reproduction, type of metabolism, and parental care of offspring. LS **Visual**

Teaching Tip

Introduced Pests House sparrows and European starlings are two common birds that may be familiar to students. Inform students that these two species were introduced into the United States from Europe. Both species have become serious pests. These exotic species out compete native songbirds for food and nesting sites. In fact, they are thought to be at least partly responsible for the decline in the numbers of bluebirds.

Figure 19 Penguin. The penguin's wings are adapted for swimming rather than flying.

Adaptations of Birds

While there is great diversity among the 28 orders of birds, 60 percent of all bird species belong to order Passeriformes. These birds, also known as the songbirds, number approximately 5,300 species and are by far the largest group of terrestrial vertebrates. Birds are adapted for different ways of life, and you can tell a great deal about the habits and diet of a bird by examining its beak (bill), legs, and feet. Carnivorous birds such as hawks have curved talons for seizing prey and a sharp beak for tearing apart their meal. The beaks of ducks are flat for shoveling through water or mud, and their webbed feet enable them to swim. Finches are seed eaters, and their short, thick beak is adapted for crushing seeds while their curved toes enable them to cling to branches. Other birds, such as the penguins shown in **Figure 19,** are flightless, and their wings and feet are modified for swimming.

During the evolutionary history of birds, their beaks, legs, and feet have been adapted to the particular environment the birds live in, as shown in **Table 1.** Some birds are more highly specialized than others, and many birds are highly flexible in their eating habits. The song sparrow, for example, has a strong bill that it uses in winter to crack hard seeds. In summer, the sparrow uses its bill to catch soft-bodied insects.

Table 1 Avian Adaptations

Type of bird	Beak adaptations	Foot adaptations
Songbirds (e.g., cardinal, robin) 	**Seed-cracking:** Short, thick, strong beak   **Insect-catching:** Long, slender beak for probing	**Perching:** Toes can cling to branches; one toe points backward 
Hummingbirds 	**Probing:** Thin, slightly curved beak for inserting into flowers to sip nectar 	**Hovering:** Legs so small the bird cannot walk on the ground; tiny feet 

Graphic Organizer

Use this graphic organizer with **Teaching Tip: Reptiles and Birds.**

Characteristic	Reptiles	Birds
Body covering	Scaly skin	Feathers
Type of heart	Partially or completely divided ventricle	Completely divided ventricle
Reproduction	Oviparous or ovoviviparous	Oviparous
Metabolism	Ectothermic	Endothermic
Parental care	Little or none	Until offspring can fly

Type of bird	Beak adaptations	Foot adaptations
Woodpeckers 	**Drilling:** Strong, chisel-like beak 	**Grasping:** Feet with two toes pointing forward and two pointing backward 
Parrots 	**Cracking, tearing:** Short, stout, hooked beak used to crack seeds and nuts and to tear vegetation 	**Climbing/grasping:** Strong toes, two pointing forward, two pointing backward; adapted for perching, climbing, and holding food 
Birds of prey 	**Tearing:** Curved, pointed beak for pulling apart prey 	**Grasping:** Powerful, curved talons for seizing and gripping prey 
Ducks 	**Sieving:** Long, flattened, rounded bill 	**Swimming:** Three toes linked by webs for improved swimming 
Long-legged waders 	**Fishing:** Long, slender, spear-shaped beak for fishing 	 **Wading:** Long legs; toes spread out over a large area to support bird on soft surfaces

Using the Table — BASIC

Have students pair up. Instruct the students to cover the two columns on the right side of **Table 1** with a sheet of paper. Have one student uncover the Beak adaptations column, choose one, and read the description. The other student should try to identify the type of bird whose beak is described. This process should continue until the second student has correctly identified each bird. The students should then switch roles and repeat the process using the Foot adaptations column. **LS Verbal**

Demonstration — BASIC

Show students pictures of several different types of birds. Be sure to include some with distinctly different beaks and feet. A typical assortment would be: woodpecker, heron, pelican, grosbeak, chickadee, goose, and hummingbird. Ask students to examine the different types of beaks and feet and to hypothesize about each bird's habitat and diet. **LS Visual**

Group Activity — ADVANCED

Avian Adaptations Divide the class into small groups, and assign each group a particular feeding strategy and bird habitat. Have each group use a field guide to identify five species of birds that share the assigned feeding strategy and habitat. Students should then use drawings or written descriptions to demonstrate the similarities in beaks, feet, and behavior among birds with similar lifestyles. Have each group share its findings with the entire class. **LS Visual** Co-op Learning

Trends in Ornithology

Researchers around the country have been monitoring bird populations each year to determine long-term trends. One apparent trend: of the 42 bird species that are listed as threatened or endangered, 22 are associated with wetlands. The trend holds true for other animals as well. Have students research the threatened and endangered birds in your area. In their reports, students should discuss the probable causes of decline, as well as plans for species recovery.

REAL WORLD CONNECTION

Birding (once called birdwatching) has been described as the second-favorite outdoor activity in the United States, after gardening. Invite a local naturalist or member of a nearby chapter of the Audubon Society to visit your classroom and discuss tips on how to attract and identify backyard birds. Make a list of local birds and show slides of each bird to your class. Slides can be purchased from many sources, including the Cornell University Laboratory of Ornithology.

Other Adaptations

There are many groups of birds, each of which is adapted to its particular living conditions. For example, gulls and terns have streamlined bodies that are adapted for flying over the water in search of fish. Owls' excellent low-light vision enables them to survive as nocturnal hunters. For a list of the orders of birds, see "Classification in Kingdoms and Domains" in the Appendix.

MATH LAB: *Calculating Average Bone Density* SC.H.1.4.1

Background

Density is the ratio of the mass of an object to its volume. Several teams of students determined the density of bones from two different animals. You can use their data to practice calculating average bone density.

DATA TABLE

Bone type	Team 1	Team 2	Team 3	Team 4
Animal 1	1.6 g/cm^3	1.0 g/cm^3	1.2 g/cm^3	1.4 g/cm^3
Animal 2	2.3 g/cm^3	1.8 g/cm^3	1.8 g/cm^3	2.1 g/cm^3

1. **Add the densities of one bone type.** For example, if three bone samples have densities of 3.0, 3.1, and 2.9 g/cm^3, their sum would be 9.0 g/cm^3.
2. **Divide the sum of the densities by the number of samples.**

$$\text{Average density} = \frac{\text{sum of the densities}}{\text{number of samples}} = \frac{9.0\text{ g/cm}^3}{3} = 3.0\text{ g/cm}^3$$

Analysis

1. **Calculate** the average bone density for each of the two animals in the data table. Express your answer in grams per cubic centimeter.
2. **Critical Thinking Evaluating Methods** Why is it important to analyze several samples and obtain the average of your data?
3. **Critical Thinking Drawing Conclusions** Based on your answer to item 1, which of the two animals is more likely to be a bird?

Section 3 Review

1. **Identify** the adaptations of birds for flight. SC.F.1.4.2
2. **Summarize** how birds obtain the energy necessary for flight. SC.F.1.4.2
3. **Relate** the bald eagle's methods of hunting and feeding to its external body features. SC.F.1.4.2
4. **Critical Thinking Evaluating Hypotheses** A student examines a bird that has delicate, perching feet with long, slender toes. Its beak is small but slightly long and pointed. The student concludes that the bird is a seed-eating songbird. Do you agree? Explain your reasoning. SC.F.1.4.2
5. **FCAT Prep** Which structure is part of the excretory, digestive, and reproductive systems of a bird?
   A cloaca
   B gizzard
   C kidney
   D ovary
   SC.F.1.4.2

792

MATH LAB: *Calculating Average Bone Density*

Skills Acquired

Analyzing data, calculating

Teacher's Notes

After students read the explanatory paragraph, have them glance at the data and predict which animal has the higher average bone density.

Answers to Analysis

1. Animal 1: 1.3 g/cm^3; Animal 2: 2.0 g/cm^3
2. A certain amount of variation is normal in biological systems.
3. Animal 1

Close

Reteaching — BASIC

Have each student develop ten questions for this section, and use the questions in a review game. LS Interpersonal

Quiz — GENERAL

True or False:

1. Birds are ectotherms. (False. Birds are endotherms.)
2. Birds have a two-way airflow during respiration. (False. Birds have a one-way airflow.)

Alternative Assessment — GENERAL

Have student groups choose one bird adaptation. Groups should write and illustrate a summary of the adaptation. Post the summaries together. LS Verbal

Section Review Correlations

SC.F.1.4.2: Items 1-5

Answers to Section Review

1. forelimbs modified as wings; feathers; streamlined shape; lightweight bones, endothermic metabolism; highly efficient lungs; heart with completely divided ventricle
2. An endothermic body and a rapid metabolism provide the energy necessary for flight.
3. The bald eagle has strong feet with sharp talons for grabbing prey, and a massive, hooked beak for tearing prey into pieces it can eat.
4. The small, pointed beak indicates that it is not a seedeater because seedeaters have thick beaks for crushing seeds. The foot structure, however, is consistent with the songbird anatomy.
5. **A.** Incorrect. The kidneys play an important excretory role in birds, but are not closely associated with the other systems. **B.** Correct. The cloaca is an opening that allows wastes to leave the body. In females, it also allows sperm to enter the body and fertilized eggs to leave the body. **C.** Incorrect. The gizzard is the second part of the stomach that helps to grind food; it is not part of the excretory or reproductive systems. **D.** Incorrect. The ovary is a reproductive organ; it is not part of the digestive or excretory systems.

Study ZONE

Key Concepts

1 The Reptilian Body

- Reptiles have a strong, bony skeleton.
- Reptiles are ectothermic.
- Reptiles have nearly watertight skin and eggs, both of which enable them to be terrestrial animals.
- Reptiles have paired lungs that have a greater surface area for gas exchange than the lungs of amphibians.
- Reptiles have a double-loop circulatory system. Most have a ventricle that is partly divided into right and left halves, resulting in incomplete separation of oxygen-rich and oxygen-poor blood.
- Reptilian fertilization is internal.

2 Today's Reptiles

- Snakes and lizards (order Squamata) share many characteristics, such as periodic molting, but snakes have no legs.
- The shells of turtles and tortoises (order Chelonia) are made of fused plates of bone covered with horny shields or leathery skin.
- Unlike other reptiles, crocodilians (order Crocodilia), care for their young after hatching. They also have a completely divided ventricle.
- There are only two species of tuataras (order Rhynchocephalia).

3 Characteristics and Diversity of Birds

- Birds are endotherms. Their high rate of metabolism helps them meet the large energy requirements for flight.
- A bird's contour feathers give the bird its shape and aid flight. Its down feathers provide insulation.
- The bones of birds are thin and hollow, and many of them are fused; all are adaptations for flight.
- One-way airflow through the lungs provides the large amounts of oxygen birds need for flight.
- The ventricle of the bird heart is completely divided by a septum.

Key Terms

Section 1

amniotic egg (775)
oviparous (777)
ovoviviparous (777)

Section 2

carapace (782)
plastron (782)

Section 3

contour feather (784)
preen gland (785)
down feather (785)

Alternative Assessment — GENERAL

Have each student use the red and blue subheadings in this chapter to create questions based on each subheading, and then correctly answer it. For example, "Reptiles Share Several Key Characteristics," could become "What key characteristics do reptiles share?" LS Verbal

Chapter Resource File

- Science Skills Worksheet GENERAL
- Critical Thinking Worksheet ADVANCED
- Test Prep Pretest GENERAL
- Chapter Test GENERAL

Answer to Concept Map

The following is one possible answer to Performance Zone item 15 on the following page.

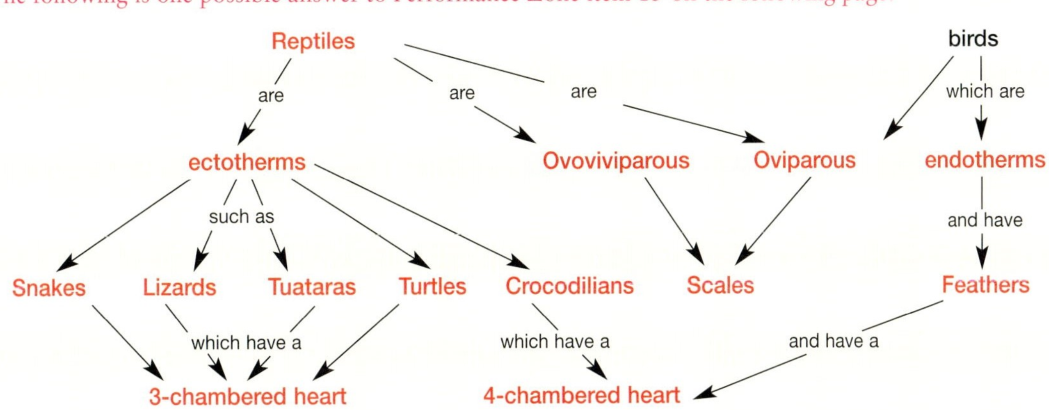

CHAPTER 34 ZONE

ANSWERS

Understanding Key Ideas

1. b
2. b
3. a
4. c
5. a
6. d
7. d
8. d
9. The amnion membrane stores water; the yolk sac stores food; the allantois stores wastes; and the chorion allows oxygen to enter the egg and carbon dioxide to exit.
10. One possible answer to the concept map is found at the bottom of the Study Zone page.

Critical Thinking

11. Because it mixes oxygen-rich and oxygen-poor blood, a three-chambered heart would not distribute enough oxygen to the hummingbird's muscles to sustain flight.
12. Animals that invest a long period of parental care typically have fewer offspring than animals with little or no parental care, but the offspring that are cared for are more likely to live to adulthood. In both cases, only a small number of offspring survive to breeding age.

Alternative Assessment

13. Answers will vary depending on birds chosen. Students should acknowledge their sources and include an illustration of each bird.
14. Students should research what type of food and habitat is appropriate for the lizard they choose and set up their terrarium accordingly.

Performance ZONE CHAPTER REVIEW

Understanding Key Ideas

1. Which is *not* an adaptation of reptiles for life on land?
   a. watertight skin
   b. external fertilization
   c. amniotic egg
   d. kidneys SC.G.1.4.1

2. The heart of most reptiles has
   a. no septum.
   b. a partly divided ventricle.
   c. a fully divided ventricle.
   d. two pumping chambers. SC.F.1.4.2

3. In reptiles, fertilization
   a. is internal.
   b. is external.
   c. always occurs in water.
   d. does not occur.
   SC.F.2.4.1

4. Snakes differ from lizards in that snakes do *not* have
   a. lungs. **c.** a pectoral girdle.
   b. kidneys. **d.** a flexible jaw.
   SC.G.1.4.1

5. Jacobson's organs are involved in the sense of
   a. smell. **c.** sight.
   b. hearing. **d.** touch.
   SC.F.1.4.2

6. The feathers of most birds are well adapted for
   a. swimming and repelling water.
   b. expelling heat and feeding.
   c. flying and conducting heat.
   d. flying and insulating. SC.F.1.4.2

7. The foot illustrated below is most likely of a bird adapted for
   a. wading. **c.** perching.
   b. grasping. **d.** swimming.
   SC.F.1.4.2

8. Which group of living reptiles is most closely related to birds?
   a. snakes **c.** rhynchocephalians
   b. turtles **d.** crocodilians SC.G.1.4.1

9. **BIOWatch** Name the four membranes contained in an amniotic egg, and describe how they make the egg an independent life-support system. SC.F.1.4.2

10. **Concept Mapping** Construct a concept map that describes the characteristics of both reptiles and birds. Include the following terms in your map: *ectotherm, endotherm, oviparous, ovoviviparous, scales, feathers, reptiles, snakes, lizards, tuataras, turtles, crocodilians, three-chambered heart,* and *four-chambered heart.* SC.G.1.4.1

Critical Thinking

11. **Predicting Outcomes** How might having a three-chambered heart, like that of most reptiles, affect a hummingbird in flight? SC.F.1.4.2

12. **Recognizing Logical Connections** How might a long period of parental care be related to the number of offspring an animal produces? SC.G.2.4.3

Alternative Assessment

13. **Being a Team Member and Communicating** Work with two or three of your classmates to find out what kinds of birds are common in your area. Select at least six birds to explore in depth and research the following: its habitat, its food, its beak and foot adaptations, and where it winters. Present the information you gathered in an illustrated guide. Make copies of your guide available for interested students. SC.F.1.4.2

14. **Organizing Information** Create a habitat in a terrarium for a small lizard, and observe the lizard's behavior. Make a labeled drawing of the environment you create, and keep a journal of your observations. SC.H.2.4.2

794

Assignment Guide

Section	Questions
1	1, 2, 3, 9, 14
2	4, 5, 8, 9, 10
3	6, 7, 8, 9, 10, 11, 12, 13

Chapter Review Correlations

SC.F.1.4.2: Items 2, 5, 6, 7, 9, 11, 13
SC.F.2.4.1: Item 3
SC.G.1.4.1: Items 1, 4, 8, 10
SC.G.2.4.3: Item 12
SC.H.2.4.2: Item 14

FCAT Prep

For questions 1–6, write your answers on a separate sheet of paper.

1 What term describes the eggs of reptiles and birds? SC.F.1.4.2
A. amniotic **C.** ovoviviparous
B. oviparous **D.** externally fertilized

2 What is the function of the preen gland? SC.F.1.4.2
F. secreting oil **H.** stimulating egg production
G. sensing chemicals **I.** changing body temperature

3 When a piece of a lizard's tail breaks off, the separated portion may wiggle about forcefully. How is this an advantage for the lizard? SC.G.2.4.3
A. The movement of the tail frightens the predator away.
B. Eating the tail provides the lizard with energy needed to escape.
C. Predators are distracted, giving the lizard time to escape.
D. The wiggling tail can injure a predator, preventing it from pursuing the lizard.

4 Many viviparous snakes and lizards live in cold climates. Why might viviparity be advantageous in such environments? SC.G.2.4.2
F. The embryos are able to avoid freezing by maintaining their own body temperature.
G. The young stay warm inside the mother, who can maintain a constant body temperature.
H. Maintaining a constant body temperature allows the mother to warm the eggs by sitting on them.
I. The young carried inside the mother receive some heat from her body due to muscle movements and basking.

Base your answers to questions 5 and 6 on the chart below.

NORMAL RANGES OF BODY TEMPERATURE

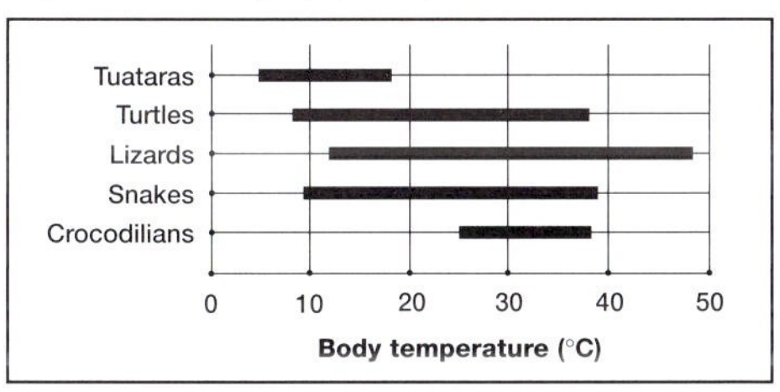

5 Which of these statements is supported by the data in the chart?
A. Turtles and snakes have similar body temperature ranges.
B. Lizards always have a higher body temperature than tuataras.
C. Crocodilians have a greater body temperature range than tuataras.
D. Some tuataras can have a higher body temperature than some crocodilians.
SC.G.1.4.1

6 How broad is the range of body temperatures for snakes, to the nearest 10°C? SC.G.1.4.1

Test TIP

Scan the answer choices for words such as *never* and *always*. Such words often are used in incorrect statements because they are too broad.

FCAT Doctor

Question 4 Answer I is the correct choice. Viviparous snakes retain their eggs in their bodies until they hatch. They are ectothermic, so any warmth the mother can provide comes from the environment. Answer F is incorrect because the embryos are not endothermic. Answer G is incorrect because the mother is not endothermic. Answer H is incorrect because the mother is not endothermic, and the eggs remain in her body until they hatch.

Question 5 Answer A is the correct choice. According to this data, turtles and snakes have a body temperature range of about 9°C–38°C. Answer B is incorrect because lizards often have a higher body temperature than tuataras, but their temperature ranges overlap. Answer C is incorrect because, while crocodilians have a greater body temperature than tuataras, the ranges of body temperatures are about the same for both. Answer D is incorrect because the lowest body temperature for a crocodilian is higher than the highest body temperature for a tuatara.

Question 6 The correct answer is 30. Range is the difference between the highest temperature, 40°C, and the lowest temperature, 10°C. For snakes the range is 30°C.

Answers

1. A
2. F
3. C
4. I
5. A
6. 30

Test Prep Correlations

SC.F.1.4.2: Items 1, 2
SC.G.1.4.1: Items 5, 6
SC.G.2.4.3: Items 3, 4

Exploration Lab

Observing Color Change in Anoles

Teacher's Notes

Time Required

Two 45-minute lab periods

Ratings

EASY → HARD

TEACHER PREPARATION (2)
STUDENT SETUP (1)
CONCEPT LEVEL (2)
CLEANUP (1)

Skills Acquired

- Observing
- Designing experiments
- Collecting data
- Communicating

Scientific Methods

In this lab, students will:

- Make Observations
- Ask Questions
- Test the Hypothesis
- Analyze the Results
- Draw Conclusions

Materials and Equipment

Materials for this lab can be ordered from WARD'S. See *Master Materials List* at the front of this book for catalog numbers.

Safety Cautions

Remind students to handle anoles with care. These lizards are easily frightened and can run quickly. If an anole is picked up by its tail, the tail may break off.

Exploration Lab

Observing Color Change in Anoles

SC.F.1.4.7

SKILLS

- Using scientific methods
- Observing

OBJECTIVES

- **Observe** live anoles.
- **Relate** the color of an anole to the color of its surroundings.

MATERIALS

- glass-marking pencil
- 2 large, clear jars with wide mouths and lids with air holes
- 2 live anoles
- 6 shades each of brown and green construction paper, ranging from light to dark (2 swatches of each shade)

Before You Begin

Lizards are a group of **reptiles.** There are 250–300 species of anoles, lizards in the genus *Anolis*. Like chameleons, anoles can change color, ranging from brown to green. Anoles live in shrubs, grasses, and trees. Light level, temperature, and other factors, such as whether the animal is frightened or has eaten recently, can all affect the color of an anole. When anoles are frightened, they usually turn dark gray or brown and are unlikely to respond to other **stimuli.** Anoles generally change color within a few minutes. In this lab, you will observe the ability of anoles to change color when they are placed on different background colors. You will also determine how this ability might be an advantage to anoles.

1. Write a definition for each boldface term in the paragraph above.
2. Make a data table similar to the one below.
3. Based on the objectives for this lab, write a question you would like to explore about the color-changing behavior of anoles.

DATA TABLE

Anole	Color 1		Color 2	
	Change	Time	Change	Time
1				
2				

Procedure

PART A: Make Observations

1. Observe live anoles in a terrarium. Make a list of characteristics that indicate that anoles are reptiles.
2. Work with a partner to place anoles to be studied in separate glass jars. **CAUTION: Handle anoles gently, and follow instructions carefully. Anoles run fast and are easily frightened. Plan your actions before you start.** By working efficiently, you can keep your anole from becoming overly frightened. Carefully pick up one anole by grasping it firmly but gently around the shoulders. Do not pick up anoles by their tail. Place the anole in a glass jar. Quickly and carefully place a lid with air holes on the jar.
3. When anoles become overly frightened, they remain dark. While you are designing your experiment, do not disturb your anoles, and let them recover from your handling.

PART B: Design an Experiment

4. Work with members of your lab group to explore one of the questions written for step 3 of **Before You Begin.** To explore the question, design an experiment that uses the materials listed for this lab.

796

Tips and Tricks

At least one hour before the lab exercise begins, place half the anoles in a terrarium lined with black construction paper. These anoles should turn brown. Place the remaining anoles in a terrarium lined with white construction paper. These anoles should turn green. Each terrarium should be illuminated by a fluorescent light and should contain a shallow dish of water.

Because anoles move quickly, you may want to place them in the jars for students in advance. Cover jars with lids that have air holes. Keep all jars in the same environment. Color change may be subtle and may be affected by feeding patterns or epinephrine (adrenaline) if the lizard is frightened. Very cool temperatures can induce a change to brown. Clean terrariums and jars after you have removed the anoles. Responsible and interested students may wish to adopt the anoles that are no longer needed in the classroom. Show students how to properly set up a vivarium to care for the anoles.

You Choose

As you design your experiment, decide the following:

a. what question you will explore
b. what hypothesis you will test
c. how many anoles you will need
d. what background colors you will use
e. how many times you will test each background with an anole
f. how long you will observe each test and how you will keep track of time
g. what your control will be
h. what data to record in your data table

5. Write a procedure for your experiment. Make a list of all the safety precautions you will take. Have your teacher approve your procedure and safety precautions before you begin the experiment.

6. Set up and carry out your experiment.

PART C: Cleanup and Disposal

7. Dispose of construction paper and broken glass in the designated waste containers. Put anoles in the designated container. Do not put lab materials in the trash unless your teacher tells you to do so.

8. Clean up your work area and all lab equipment. Return lab equipment to its proper place. Wash your hands thoroughly before you leave the lab.

Analyze and Conclude

1. Summarizing Results Briefly state how the variable you tested influenced the color-changing behavior of anoles.

2. Evaluating Results Did any unplanned variables influence your data? (For example, was there a loud noise, or was a jar suddenly moved?)

3. Analyzing Methods How could your experiment be modified to improve the certainty of your results?

4. Analyzing Data Were there any inconsistencies in your data? (For example, two anoles reacted in different ways.) If so, offer an explanation for them.

5. Drawing Conclusions After considering your data, make a statement about color-changing behavior in anoles.

6. Further Inquiry Write a new question about anoles that could be explored with another investigation.

Do You Know?

Do research in the library or media center to answer these questions:

1. What other behaviors are characteristic of anoles?

2. How is a chameleon different from an anole?

Use the following Internet resources to explore your own questions about lizards that change color.

internet connect
www.scilinks.org
Topic: Adaptations of Reptiles
Keyword: HX4003
SCILINKS. Maintained by the National Science Teachers Association

Answers to Before You Begin

1. *reptiles*—ectothermic vertebrates with scaly, watertight skin; lungs; and a heart with partially or completely divided ventricle.
   stimuli—environmental factors that influence the behavior of an organism
2. Student data tables should resemble the example on this page.
3. Answers will vary. One example is: How quickly do anoles change color after they move to a new background color?

Answers to Procedure

1. Easily observed characteristics include scaly skin and toes with claws.
4. Experimental designs will vary. One example: Place each jar containing an anole on top of a sheet of paper that is a different color than the anole. Note whether each anole changes color and, if it does, the color it becomes and the time required for the color change.
5. Procedures will vary, based on the questions that students choose to answer. Be sure that safety precautions are listed.

Sample Data Table

Anole	Green paper		Brown paper	
	Change	Time	Change	Time
Brown at start	green	3 s	stayed brown	
Green at start	stayed green		brown	3 s

Answers to Analyze and Conclude

1. Students' answers should clearly describe how each test was conducted and should state their results in terms of color changes in the anoles. See Sample Data Table.
2. Anoles may react to stimuli other than the independent variable selected. Students may list several uncontrolled variables, such as temperature or environmental stressors, which could have affected their results.
3. Answers will depend on the groups' experimental designs. Students should suggest improving their methods by eliminating or controlling as many uncontrolled variables as possible.
4. Answers will vary. Students should include a possible explanation for any inconsistency observed.
5. Answers will vary. For example: Anoles change to a color that most closely matches the color of their environment.
6. Answers will vary. One possibility: Would anoles react differently if there were two in each jar?

Answers to Do You Know?

1. Male anoles extend the dewlap under their chin as a display during courtship and when defending their territory.
2. Unlike anoles, chameleons have opposable toes and a strong, curled tail—two features that help them grasp tree branches. Chameleons also have large eyes that can move independently of each other.

Chapter Opener		**SE** **Quick Review,** p. 799 ■ **SE** **Reading Activity,** p. 799 ■ **TE** **Using the Figure,** p. 798 GENERAL **TE** **Opening Activity,** p. 799 GENERAL
& 3 • 90 min pp. 800–806 **Section 1** The Mammalian Body	**OSP** **Lesson Plan*** **TR** **Bellringer*** **TR** **Section Outline*** **TR** **I32 Characteristics of Mammals*** **TR** **I36 Mammalian Lungs*** **TR** **I38 External Structure of a Grizzly Bear*** **TR** **I39 Internal Structure of a Grizzly Bear***	**SE** **Quick Lab** Evaluating the Insulation Value of Hair, p. 801 ◆ ■ **CRF** **Datasheets for In-Text Labs** Evaluating the Insulation Value of Hair* **TE** **Using the Figure,** p. 802 BASIC **TE** **Demonstration,** p. 802 ◆ GENERAL **CRF** **Quick Lab** Comparing Winter Dormancy in Mammals* BASIC **CRF** **Skills Practice Lab** Fetal Pig Dissection* ◆ GENERAL
& 5 • 90 min pp. 807–816 **Section 2** Today's Mammals	**OSP** **Lesson Plan*** **TR** **Bellringer*** **TR** **Section Outline*** **TR** **I40 Comparison of Bats and Whales***	**TE** **Activity** Greatest Mammal, p. 807 GENERAL **TE** **Activity** Placental Mammals, p. 810 GENERAL **TE** **Demonstration,** p. 811 ◆ BASIC **SE** **Data Lab** Comparing Gestation Periods, p. 811 ■ **CRF** **Datasheets for In-Text Labs** Comparing Gestation Periods* **TE** **Using the Figure,** p. 812 BASIC **TE** **Group Activity** Orders of Mammals, p. 812 ADVANCED **TE** **Demonstration,** p. 813 BASIC **TE** **Group Activity** Mammal Orders, p. 813 ADVANCED **TE** **Demonstration,** p. 814 GENERAL **TE** **Using the Figure,** p. 815 BASIC **TE** **Group Activity** Mammal Helpers, p. 815 GENERAL **TE** **Group Activity** Local Zoo or Wildlife Park, p. 815 ◆ ADVANCED **SE** **Exploration Lab** Exploring Mammalian Characteristics, pp. 820–821 ■ **CRF** **Datasheets for In-Text Labs** Exploring Mammalian Characteristics* ◆

& 7 • 90 min

Chapter Review and Assessment Resources

- **SE** **Study Zone,** p. 817 ■
- **SE** **Performance Zone,** p. 818 ■
- **SE** **FCAT Test Prep,** p. 819 ■
- **TR** **Concept Mapping***
- **CRF** **Vocabulary Review Worksheet*** ■ GENERAL
- **CRF** **Test Prep Pretest*** ■ GENERAL
- **CRF** **Chapter Test*** ■ GENERAL
- **CRF** **Chapter Test*** ADVANCED
- **OSP** **Test Generator***
- **OSP** **Test Item Listing***
- **OSP** **Modified Chapter Test*** BASIC

go.hrw.com Holt Online Learning

Visit go.hrw.com for access to Holt Online Learning or enter the keyword **HX6 Home** for a variety of free online resources.

One-Stop Planner® CD-ROM

This CD-ROM package includes

- Lab Materials QuickList Software
- Holt Calendar Planner
- Customizable Lesson Plans
- Printable Worksheets
- ExamView® Test Generator
- Interactive Teacher Edition
- Holt PuzzlePro® Resources
- Holt PowerPoint® Resources

Guided Reading Audio CDs

These CDs are designed to help auditory learners and reluctant readers. (Audio CDs are also available in Spanish.)

Compression guide:
To shorten your instruction because of time limitations, omit block 1.

KEY

TE	Teacher Edition	CRF	Chapter Resource File	*	Also on One-Stop Planner
SE	Student Edition	TR	Teaching Transparency	■	Also Available in Spanish
OSP	One-Stop Planner			◆	Requires Advance Prep

SKILLS DEVELOPMENT AND PRACTICE	SECTION REVIEW AND ASSESSMENT	CORRELATIONS
		Florida Sunshine Standards/FCAT
TE Reading Skill Builder K-W-L, p. 801 BASIC TE Skill Builder Writing Skills, p. 801 GENERAL TE Reading Skill Builder Interactive Reading, p. 803 BASIC CRF Directed Reading Worksheet* BASIC CRF Active Reading Worksheet* GENERAL	SE Section 1 Review ■ TE Reteaching, p. 806 BASIC TE Quiz, p. 806 GENERAL TE Alternative Assessment, p. 806 GENERAL CRF Quiz* ■ BASIC	SC.F.1.4.1, SC.F.1.4.2, SC.G.1.4.1
TE Reading Skill Builder Writing Skills, p. 808 ADVANCED TE Reading Skill Builder Paired Reading, p. 810 BASIC CRF Directed Reading Worksheet* BASIC CRF Active Reading Worksheet* ◆ GENERAL CRF Critical Thinking Worksheet* ADVANCED CRF Science Skills Worksheet* ◆ GENERAL	SE Section 2 Review ■ TE Reteaching, p. 816 BASIC TE Quiz, p. 816 GENERAL TE Alternative Assessment, p. 816 ◆ GENERAL CRF Quiz* ■ BASIC Modified Worksheet* BASIC	SC.D.1.4.3, SC.F.1.4.2, SC.F.2.4.1, SC.G.1.4.1, SC.G.2.4.3

SCILINKS. NSTA
www.scilinks.org
Maintained by the **National Science Teachers Association**

Topic: Mammals
SciLinks Code: HX4118

Topic: Dairy Products
SciLinks Code: HX4056

Topic: Placental Mammals
SciLinks Code: HX4138

Overview

Tell students the purpose of this chapter is to recognize the incredible diversity of mammals in both anatomy and physiology. Students will understand how this diversity allows mammals to occupy a tremendous variety of habitats.

Using the Figure

Have students study the photograph of the mountain lion. Ask them to list all the features the mountain lion has that help it to survive. Have students share their lists with the class and make a master list on the board or overhead. Discuss how each adaptation helps this mammal to survive. (Examples: Powerful legs to help chase and catch prey; two eyes facing forward for binocular vision to spot prey; whiskers to sense its surroundings; claws for grabbing and holding prey; large ears that can move to detect and pinpoint sounds; fur for insulation, etc.)

Assessing Prior Knowledge

Before beginning this chapter, be sure your students can answer the questions in the ***Quick Review*** on the facing page. If they cannot, encourage them to reread the sections indicated.

Mountain lion

798

Chapter Correlations — Florida Sunshine Standards/FCAT

SC.D.1.4.3 Changes in Earth's climate, geological activity, and life forms may be traced and compared.

SC.F.1.4.2 Body structures are uniquely designed and adapted for their function.

SC.F.2.4.1 There are mechanisms of asexual and sexual reproduction and different genetic advantages and disadvantages of asexual and sexual reproduction.

SC.G.1.4.1 There is great diversity and interdependence among living things.

SC.G.2.4.3 Genetic variation of offspring contributes to population control in an environment, and natural selection ensures that those who are best adapted to their surroundings survive to reproduce.

CHAPTER

35 Mammals

Florida Sunshine Standards/FCAT

SC.D.1.4.3 Changes in Earth's climate, geological activity, and life forms may be traced and compared.

SC.F.1.4.2 Body structures are uniquely designed and adapted for their function.

SC.F.2.4.1 There are mechanisms of asexual and sexual reproduction and different genetic advantages and disadvantages of asexual and sexual reproduction.

SC.G.1.4.1 There is great diversity and interdependence among living things.

SC.G.2.4.3 Genetic variation of offspring contributes to population control in an environment, and natural selection ensures that those who are best adapted to their surroundings survive to reproduce.

Quick Review

Answer the following without referring to earlier sections of your book.

1. **Define** the term *metabolism*. *(Chapter 1, Section 2)*
2. **Compare** the diets of carnivores and herbivores. *(Chapter 16, Section 2)*
3. **Summarize** the difference between endothermic and ectothermic animals. *(Chapter 32, Section 2)*
4. **Describe** the breakup of Pangaea. *(Chapter 32, Section 2)*

Did you have difficulty? *For help, review the sections indicated.*

Reading Activity

After reading each section of this chapter, closely study the figures in the section. Reread the figure captions, and for each one, write a question that can be answered by referring to the figure and its caption. As you review the chapter, but before you complete the Performance Zone chapter review, refer to your list of figures and questions.

The mountain lion, *Puma concolor*, has the widest distribution of any species of native mammals in the Western Hemisphere—from Canada throughout South America. It is now threatened or endangered in much of its range.

Looking Ahead

internet connect

www.scilinks.org

National Science Teachers Association *sci*LINKS Internet resources are located throughout this chapter.

SCILINKS. Maintained by the National Science Teachers Association

799

Identifying Misconceptions

Students do not often recognize that the word "animal" refers to a wide array of organisms. When most people hear the word "animal," they think of a mammal. Point out to students that mammals are just a very small part of the animal kingdom. Vertebrate orders include mammals, birds, reptiles, amphibians, and fish—all animals. However, surprising to many, invertebrates, things like sponges, worms, clams, insects, spiders, and starfish, are also animals.

Quick Review Answers

1. Metabolism is the sum of all the chemical processes occurring in an organism.
2. Carnivores primarily eat other animals; herbivores eat plants.
3. Endotherms produce their own body heat through metabolism; ectotherms get their body heat from the environment.
4. Pangaea was a single supercontinent. It began to break apart about 200 million years ago to form the current continents. This breakup caused the new landmasses to have different climates. It is hypothesized that some animals could not adapt to the changes and became extinct.

Reading Activity Answers

As an example, students might view the duckbill platypus photo on the second page of this section and write, "How can these animals find food with such tiny eyes?" *(Answer from the caption, "Using its soft beak as a scanner, the platypus swims along underwater, locating worms and other small creatures by tracking the faint electric signals emitted by their beating hearts.")*

Chapter Resource File

- Vocabulary Worksheets
- Concept Mapping

Section 1

Focus

Overview

Before beginning this section review with your students the objectives listed in the Student Edition. This section provides an in-depth look at the key characteristics that all mammals share.

Bellringer

Ask students to make two lists. On one list they should write the names of as many large terrestrial herbivores as they can think of. On the other list they should write the names of as many large terrestrial carnivores as possible. Have students determine what percentage of the animals on their lists are mammals. (Answers will vary.) Point out that today these two ecological roles are dominated by mammals. During the age of dinosaurs, however, reptiles dominated these roles.

Motivate

Discussion/Question — BASIC

Ask how many students have pets. Then find out how many of those pets are mammals. Ask students why they think most people prefer mammals to fish, amphibians, reptiles, or birds. For fun, have students share their favorite pet tricks or stories. LS Intrapersonal

Section Correlations

SC.F.1.4.2
SC.G.1.4.1

Section 1 The Mammalian Body

Florida Sunshine Standards/FCAT

SC.F.1.4.2 Body structures are uniquely designed and adapted for their function.

SC.G.1.4.1 There is great diversity and interdependence among living things.

Objectives

- **Describe** three functions of hair. SC.F.1.4.2
- **Relate** a mammal's teeth to its diet. SC.F.1.4.2
- **Summarize** how mammals maintain a high body temperature. SC.F.1.4.2
- **Relate** the characteristics of mammals to one mammal, the grizzly bear. SC.G.1.4.1
- **Describe** parental care in mammals. SC.G.1.4.1

Key Terms

hair
alveolus
mammary gland
weaning

Key Characteristics of Mammals

If you were to look out over an African landscape, you would notice the big mammals—lions, elephants, antelopes, and zebras. Your eye would not as readily pick out the many birds, snakes, lizards, and frogs that also live there. The fact that almost all of today's large, land-dwelling vertebrates are mammals makes them much more easily noticed. While most mammals are terrestrial, some, like whales, swim in the sea, and others—the bats—fly through the air. In spite of their differences, mammals share key characteristics that are summarized in **Figure 1.**

Mammals are well adapted for terrestrial living and are able to retain water more efficiently than reptiles. That's because the mammalian kidney has an exceptional ability to concentrate waste products in a small volume of urine. Later in this book, you will find out how the mammalian kidney functions.

Hair

Of all animal species, only mammals have hair. Even whales and dolphins, which appear to be hairless, have a few sensitive bristles on their snout. A **hair** is a filament composed mainly of dead cells filled with the protein keratin. The evolutionary origin of hair is unknown, but hair probably did not derive from reptilian scales.

The primary function of hair is insulation. Mammals, such as the polar bear shown in **Figure 1,** typically maintain body temperatures higher than the temperature of their surroundings. As a result, they tend to lose body heat. However, the dense coat of hair that covers most mammals holds heat in. Like other mammals that live in cold environments, the polar bear has a layer of fat under its skin that provides additional insulation from the cold. Humans have a sparse

Figure 1 Characteristics of mammals. Despite their vast external differences, mammals share a number of distinctive features.

Characteristics of Mammals

- Hair
- Diverse and specialized teeth
- Endothermic metabolism
- Mammary glands that produce milk

Chapter Resource File

- Directed Reading BASIC
- Active Reading GENERAL
- Data Sheet for Quick Lab GENERAL

Transparencies

TR Bellringer
TR I32 Characteristics of Mammals

covering of hair and a limited amount of body fat; in most climates we need clothes to provide adequate insulation.

Hair has functions other than insulation. The coloration and pattern of a mammal's coat often enable the animal to blend in with its surroundings. A small brown mouse is almost invisible against the dark forest floor. The orange and black stripes of a Bengal tiger, shown in **Figure 2,** conceal it in the tall, orange-brown grass in which it hunts. Some animals, such as the arctic fox, show a seasonal change in the color of their coat from white in winter to brown in summer that provides protective coloration year round. The color of a mammal's coat may also be a clear signal; the black and white fur of a skunk, for instance, warns would-be predators to stay away.

In some animals, specialized hairs serve a sensory function. The whiskers of cats and dogs are stiff hairs that are very sensitive to touch. Mammals that are active at night or that live underground often rely on their whiskers for information about the environment. Other specialized hairs can be used as a defensive weapon. For example, when threatened, porcupines defend themselves by raising their sharp, barbed quills, also shown in Figure 2.

Figure 2 Functions of hair. This tiger's stripes and reddish fur help the tiger blend in with the surrounding grasses. Porcupines often use their quills for defense.

QUICK LAB

Evaluating the Insulation Value of Hair SC.F.1.4.1

When you are getting dressed on a cold day, why are you are more likely to choose a wool sweater than a cotton one? In this lab you will compare the insulating abilities of the animal fiber wool and the vegetable fiber cotton.

Materials

MBL or CBL system with appropriate software, temperature probe, 1 wool sock, beaker of ice, graph paper (optional), 1 cotton sock

Procedure

1. Set up an MBL/CBL system to collect data from a temperature probe at 6-second intervals for 100 data points.
2. Find and record the room temperature.
3. Insert the end of the probe into one thickness of wool sock. Then place the sock-covered probe into a beaker of ice. Collect temperature data for 10 minutes.
4. View the graph of your data. If possible, print out the graph. Otherwise, plot the graph on graph paper.
5. Repeat steps 1–4 using a cotton sock and fresh ice.

Analysis

1. **Analyze** your data and determine which sock was the better insulator.
2. **Summarize** why these results are of importance to mammals.

Teach

READING SKILL BUILDER — BASIC

K-W-L Before students read this chapter, have them write short individual lists of all the things they already **K**now (or think they know) about mammals. Ask them to contribute their entries to a master list on the board or overhead. Then have students list things they **W**ant to know about mammals. Have students save their lists for use later in the **Reteaching** on the last lesson page of this chapter. **LS Intrapersonal**

SKILL BUILDER — GENERAL

Writing Skills Draw students' attention to the bottom picture in **Figure 2.** Instruct each student to write a persuasive paragraph or two on one of these topics: "Porcupines have a better defensive system than lions" or "Lions have a better defensive system than porcupines." Have volunteers from each viewpoint read their finished work to the class. Discuss their ideas in class. **LS Verbal**

QUICK LAB *Evaluating the Insulation Value of Hair*

Skills Acquired
Observing, analyzing data, inferring

Teacher's Notes
Be sure students let the ice sit in the water for a sufficient time so the temperature can stabilize. Remind students to keep the probe in the tip of the sock for both trials. The probe should be at the same depth in the beaker for both trials.

Answers to Analysis

1. Answers will vary. If done correctly, the wool sock should be the better insulator.
2. Mammals' lives often depend on the conservation of their body warmth by hair or fur.

One-Stop Planner CD-ROM

- **Reading Organizers** BASIC
- **Reading Strategies** BASIC

did you know?

Hairs Are Dead Cells Sprays, conditioners, mousses, dyes, gels, and all kinds of special hair treatments cost Americans billions of dollars each year. Yet the very hair on which we spend so much money to "bring to life" is actually nothing more than dead cells. As hair cells are produced in the scalp, they die and are pushed outward. These long strands of dead cells are combed, brushed, dried, sprayed, and even ironed and "mayonnaised" by some people in the quest for the perfect "do."

Teach, *continued*

Demonstration — GENERAL

Show your students a dog or cat skull. Point to the different types of teeth and ask students what each tooth is specialized for. (Incisors—biting, cutting, and holding; canines—puncturing; premolars and molars—grinding and crushing.) Tell students that skulls are often identified by a dental formula, written like this: 2/2, 1/1, 2/2, 2/3. The numbers represent the number of teeth in one-half of a jaw. The first number is for the upper jaw; the second number is for the lower jaw. The first pair of numbers represents incisors, followed by numbers for canines, premolars, and molars. The mammal with this dental formula would have, on one side of its mouth, two upper and two lower incisors, one upper and one lower canine, two upper and two lower premolars, and two upper and three lower molars. What would the total number of teeth be in this mammal? (15 × 2 sides = 30)

Using the Figure — BASIC

Have students examine **Figure 3.** Point out that the coyote's teeth function like a pair of scissors. Students should see that the coyote's teeth are unsuited for grinding plant material. The deer's molars, on the other hand, are well-suited for this task because they function like a grinding mill. Ask students to identify what type of tooth is reduced in size in deer and give a reason why. (Deer, which are herbivores, have small canines.) **LS Visual**

Mammalian Teeth

Unlike some other vertebrates whose teeth are constantly being lost and replaced, mammals usually have only two sets of teeth. The first set, commonly called baby teeth or milk teeth, is replaced by permanent teeth, which are not replaced if lost or damaged. Animals use their teeth in a variety of ways—to secure and chew food, for protection or as a threat signal, and to perform tasks, as when a beaver cuts down trees to make a dam.

In most mammals, four types of teeth can be recognized: incisors, canines, premolars, and molars. Each type of tooth performs a different function in eating. Incisors, the front teeth, are for biting and cutting. Behind them are canines used for stabbing and holding. Lining the jaw are the premolars and molars. As a mammal chews, its upper and lower molars fit together, crushing and grinding the food more thoroughly than a reptile's teeth can. The resulting smaller bits of food can be quickly digested, which permits a mammal to eat enough food to fuel its endothermic metabolism.

A mammal's teeth are specialized for the food it eats, and it is usually possible to determine a mammal's diet by examining its teeth. **Figure 3** shows the differences between the teeth of a coyote (a carnivore) and those of a deer (a herbivore). The coyote has long canine teeth that are suited for grasping prey, and its sharp molars can cut off pieces of flesh. In contrast, the deer's canines are small, and it uses its incisors to nip off selected pieces of plant material. The deer's premolars and molars are flat and covered with ridges that form a surface on which plant material can be ground.

internet connect
www.scilinks.org
Topic: Mammals
Keyword: HX4118
SCILINKS. Maintained by the National Science Teachers Association

Figure 3 Specialized teeth

A mammal's teeth provide clues about its diet.

802

REAL WORLD CONNECTION — BASIC

Adult humans usually have 12 molars—6 in each side of their mouth. The back molars, the so-called wisdom teeth, are the last to erupt, typically emerging between the ages of 17 and 25. Not everyone has four wisdom teeth; some people have two or none. If the wisdom teeth grow in at an angle, they can force other teeth out of alignment and may have to be removed. Ask how many students have had their wisdom teeth removed and have them share their experiences if they wish. **LS Intrapersonal**

Transparencies

TR I36 Mammalian Lungs

Body Temperature

Like birds, mammals are endotherms, generating heat internally through the rapid metabolism of food. Because a mammal's body temperature remains relatively constant regardless of the temperature of its surroundings, mammals can be active at any time of day or night. They also can live in very cold climates, where most ectothermic amphibians and reptiles cannot. In addition, endothermic metabolism permits mammals to sustain activities that require high levels of energy, such as running or flying long distances.

To maintain the high metabolic rate required by an endotherm, a mammal must eat about 10 times as much food as an ectotherm of similar size. Metabolizing this food requires a considerable amount of oxygen. Mammals, like birds, have respiratory and circulatory systems that are very efficient at acquiring and distributing oxygen.

Respiratory System

Mammalian lungs, shown in **Figure 4,** have a large internal surface area that aids the exchange of oxygen and carbon dioxide. Thus, mammalian lungs are much more efficient at obtaining oxygen from the air than are reptilian and amphibian lungs. Respiration in mammals is aided by the diaphragm, a sheet of muscle that separates the chest cavity from the abdominal cavity. When the diaphragm contracts, the chest cavity enlarges, drawing air into the lungs.

The lungs of mammals contain small, grape-shaped chambers called **alveoli** *(al VEE uh lie),* (singular, *alveolus*), also shown in Figure 4. Alveoli provide a very large respiratory surface area. In more active mammals, the alveoli are smaller and more numerous, further increasing the surface area for diffusion.

Heart and Circulatory System

Like crocodiles and birds, mammals have a four-chambered heart with a septum that completely divides the ventricle. The division of the ventricle creates two pumping chambers, one for each loop of the mammal's double-loop circulatory system. One chamber pumps oxygen-rich blood to the body, while the other pumps oxygen-poor blood to the lungs. Because the two do not mix, only oxygen-rich blood is delivered to the tissues, a condition vital for meeting the oxygen needs of endotherms.

You will learn more about the mammalian respiratory and circulatory systems later in this book. You can read more about the anatomy of one mammal in *Up Close: Grizzly Bear* on the following two pages.

Real Life

Some wild mammals love city life!

Rats and mice have adapted quite successfully to urban environments. Some large cities contain as many rodents as people. But when rodents decide to move into our living space, they cause trouble. In addition to being destructive, some harbor diseases.

Finding Information

Find out what people can do to discourage wild mammals from invading their dwellings.

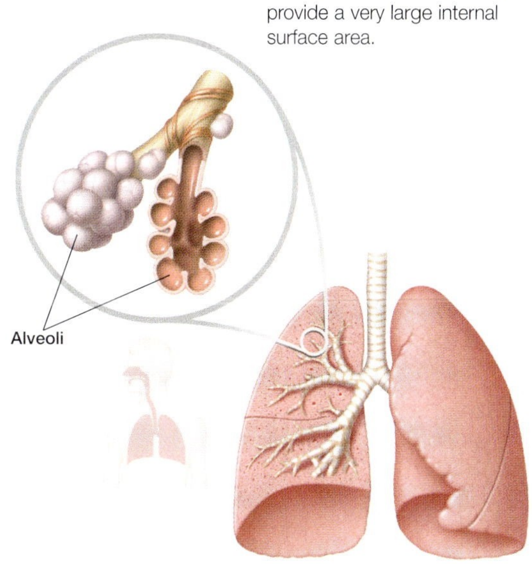

Figure 4 Mammalian lungs. The lungs of mammals contain clusters of alveoli that provide a very large internal surface area.

803

Real Life

Answer

Invasion of a home by wild mammals can be discouraged by removing food and water sources and by modifying exterior openings such as dryer and attic vents.

Teaching Tip — BASIC

Metabolic Rate and Body Size Explain that the rate at which mammals metabolize their food is often related to their body size. Body size determines an organism's surface area-to-volume ratio. As a result, shrews (small insectivores) lose heat so rapidly that they use 80 to 90 percent of their energy just trying to stay warm. Large mammals, such as elephants, have trouble dissipating excess body heat because their surface area–to-volume ratio is low. Ask students which animal, the shrew or the elephant, they would expect to eat more with respect to body mass. (shrew) Because the shrew loses so much heat to the environment, it must eat an amount of food equal to its own body mass each day. **LS Logical**

READING SKILL BUILDER — BASIC

Interactive Reading Assign Chapter 35 of the *Holt Biology* Guided Reading Audio CD Program to help students achieve greater success in reading the chapter. **LS Auditory**

Teaching Tip — GENERAL

Hibernation Ask students how some small mammals that have a high metabolic rate can survive cold winters when there is little food available. (They hibernate.) Ask students to speculate on how hibernation helps them survive. (During hibernation the animal's body temperature drops. As the animal's body temperature decreases, its metabolism slows and the animal's need for food decreases.) **LS Verbal**

MISCONCEPTION ALERT

Hibernation Contrary to popular belief, bears do not undergo true hibernation. In true hibernation, an animal's body temperature drops to nearly that of the environment. Bears of the northern forests den for several months but do not hibernate in the true sense. Their heart rate may decrease from 40 to 70 beats per minute to about 8 to 12 beats per minute. However, their body temperature only drops 3 to 7° C (5 to 9° F). Other mammals, such as ground squirrels, do undergo true hibernation. During a period of hibernation, a ground squirrel's respiratory rate decreases from 200 breaths per minute to 4 or 5 per minute, and its heart rate decreases from 150 to 5 beats per minute.

Up Close

Grizzly Bear

Teacher's Notes Tell students that the grizzly bear is classified as a threatened species in the contiguous United States, a vulnerable species in Canada, and an at-risk species in the Canadian province of Alberta. Several factors are involved in the decline of grizzly bear populations. Human-caused mortality is the major factor. Bear deaths may be caused by hunting or through loss and fragmentation of habitats. Where habitats are fragmented, bear populations frequently occur at low densities. Grizzly bears have very few offspring during their lifetime, so their populations cannot recover quickly from human-caused losses.

Grizzly Bear

- **Scientific name:** *Ursus arctos horribilis*
- **Size:** Males average 160 kg (350 lb) and can reach 1.2 m (4 ft) at the shoulder; females are smaller
- **Range:** Alaska and western Canada, with small populations in Washington, Idaho, Montana, and Wyoming
- **Habitat:** Tundra and mountainous forests and meadows
- **Diet:** Omnivorous; primary diet is vegetation; hunts insects, small mammals, and fish; eats carrion

Female with cub

External Structures

Fur Thick fur—ranging from yellowish-brown to black—covers the body. The name "grizzly" comes from silver-tipped hairs that are often sprinkled over the bear's head and back.

Senses Grizzlies have good hearing but relatively poor eyesight. They rely primarily on their excellent sense of smell to follow an odor trail or catch the scent of distant food.

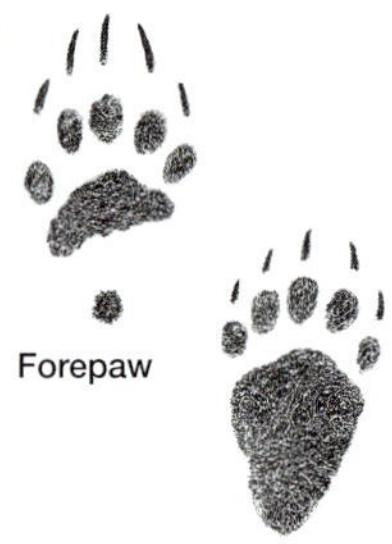

Strong limbs Grizzlies are extremely strong and have great endurance. The hump atop the bear's back is a knot of strong muscles that power the forelimbs. The paws are tipped with curved claws up to 10 cm (4 in.) long. One swat of the forepaw can kill an adult moose or elk. Unlike a cat's, the bear's claws do not retract, and they are not adapted for climbing.

HISTORY CONNECTION

During the Lewis and Clark expedition of 1804–1806, members of the Corps of Discovery had heard of the strength and ferocity of grizzly bears and were curious and anxious to see one. In his journal, William Clark wrote, "20th. of October Satturday 1804—Great numbers of Buffalow Elk & Deer, Goats. our hunters killed 10 Deer & a Goat to day and wounded a white Bear, I saw several fresh tracks of those animals which is 3 times as large as a man track." On May 5, 1805 Meriwether Lewis wrote of one encounter: "a most tremendious looking anamal, and extreemly hard to kill notwithstanding he had five balls through his lungs and five others in various parts he swam more than half the distance across the river to a sandbar & it was at least twenty minutes before he died; [he] made the most tremendous roaring from the moment he was shot." A week later Lewis concluded, "I find that the curiossity of our party is pretty well satisfyed with rispect to this anamal."

Brain Like most mammals, the bear has a well-developed cerebrum, the portion of the brain where higher mental functioning occurs. The cerebellum, a center for motor coordination, is also large and connects directly to the portions of the cerebrum that govern motor activity.

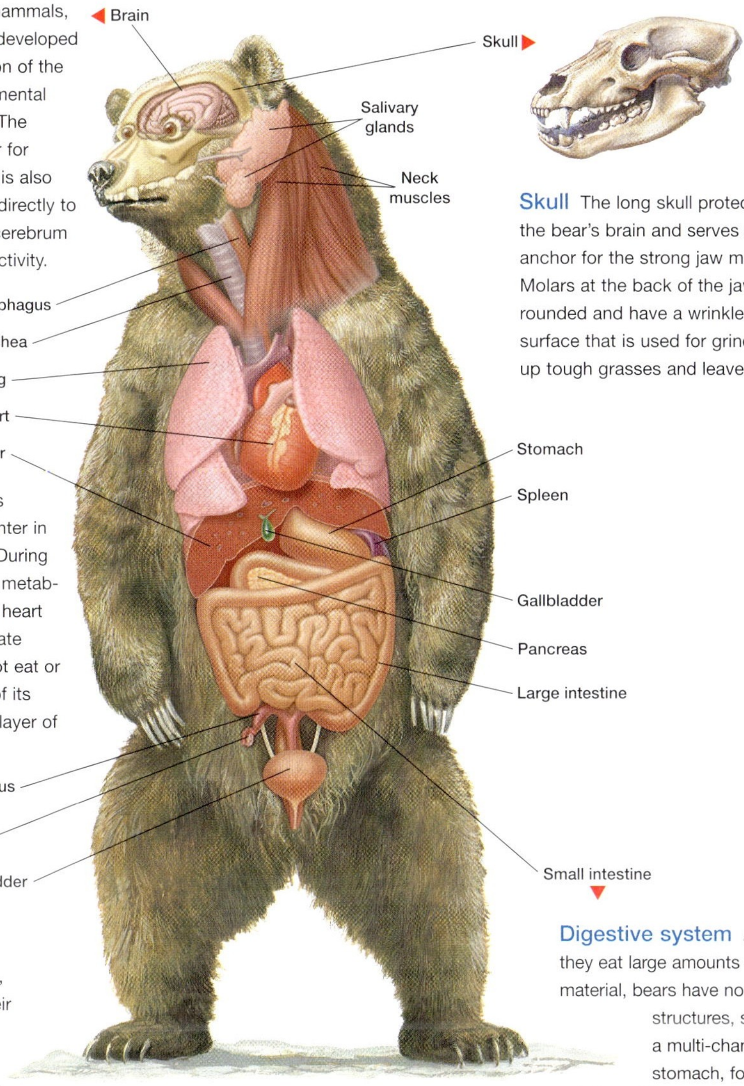

Skull The long skull protects the bear's brain and serves as an anchor for the strong jaw muscles. Molars at the back of the jaw are rounded and have a wrinkled surface that is used for grinding up tough grasses and leaves.

Fat layer Grizzlies snooze away the winter in underground dens. During this time, the bear's metabolism slows, and its heart rate and breathing rate decrease. It does not eat or drink, obtaining all of its energy from a thick layer of stored fat.

Reproductive system Like all placental mammals, grizzlies nourish their embryos through a placenta. Mating occurs from May to June, but the fertilized eggs are not implanted in the uterus until late fall. Females reproduce every 2–4 years. One to four cubs about the size of a rat are born in late winter. The cubs suckle their mother's rich milk, and by the time they emerge from the den in spring, their weight may have increased 20-fold. Cubs usually remain with their mother at least 2 years.

Digestive system Although they eat large amounts of plant material, bears have no specialized structures, such as a multi-chambered stomach, for digesting cellulose. However, a bear's small and large intestines are relatively long, which helps break down hard-to-digest plant material. Bacteria in the large intestine also contribute to the digestion of plants.

805

Up Close

Grizzly Bear

Discussion

- Identify several differences between the internal organs of the grizzly bear and the bald eagle, shown in the chapter Reptiles and Birds. (The bald eagle has a crop for storing food, a two-part stomach, air sacs connected to the lungs, and a cloaca.)
- How is the period of dormant inactivity during the winter advantageous to the grizzly bear? (By lowering its metabolic rate, the bear can survive the winter when food is scarce.)
- Most experts agree that if a person were to come upon a grizzly bear unexpectedly in the wild, one of the worst things to do is to turn and run. Why? (Grizzly bears can easily outrun the swiftest runner. Furthermore, running stimulates the bear to treat a person as prey, making an attack much more likely.)
- Fertilized eggs do not implant in the female's uterus until late fall. What is the adaptive value of this timing? (It ensures that the young will be born in springtime, when there will be ample food for the mother to produce milk.)

Transparencies

TR 138 External Structures of a Grizzly Bear

TR 139 Internal Structures of a Grizzly Bear

MISCONCEPTION ALERT

Digestive Systems Students may think that the digestive systems of all vertebrate species are roughly the same. In birds, the crop, an enlarged segment of the esophagus, can store food temporarily. The bear has no such feature. In the gizzard of birds, partially digested food from the stomach is mechanically digested. The digestive system of the bear, however, lacks these specializations and is very similar to other mammalian, including human, digestive systems.

Close

Reteaching — BASIC

Write the following characteristics on the board: *Hair, Endothermy, Mammary glands, Specialized teeth,* and *Placental nourishment.* Ask students to explain the adaptive value of each characteristic and state specific examples. LS Verbal

Quiz — GENERAL

True or False:

1. The primary function of hair is camouflage. (false, the primary function is insulation)
2. Mammals have diverse and specialized teeth. (true)
3. Mammals maintain a constant body temperature. (true)

Alternative Assessment — GENERAL

Writing Tell students to suppose that they are writing a guide to mammalian characteristics for a natural history museum. They should develop a simple, illustrated guide in a brochure format.
LS Verbal

Study TIP

Reviewing Information
Work with a partner to review the four characteristics of mammals. For each characteristic, both partners should write a question. Exchange questions, and find the answers to your partner's questions.

Parental Care

Mammals are unique among the vertebrates in the way that they nourish their young after birth. **Mammary glands** located on the female's chest or abdomen produce a nutrient-rich energy source called milk and give this class its name, Mammalia. Milk is rich in protein, carbohydrates (chiefly the sugar lactose), and fat. It also contains water, which prevents dehydration, and minerals, such as calcium, that are critical to early growth. Young mammals are nourished on milk from birth until **weaning,** the time when their mother stops nursing them.

Unlike other vertebrates, young mammals are dependent on their mother for a relatively long period, receiving milk and other food, protection, and shelter from her. For most animals, once the young can fend for themselves, the mother leaves them. The participation of the father in raising the young varies from species to species. Young mammals, such as the sea otter shown in **Figure 5,** often learn necessary survival skills during the time they are dependent on their mother. For primates, such as the chimpanzees in Figure 5, this early learning is especially important, and primates have the longest period of parental dependency of all mammals.

Figure 5 Parental care. This young otter will spend 6 to 8 months with its mother. In contrast, this baby chimp will remain close to its mother for several years.

Section 1 Review

1. **Describe** three functions of hair. SC.F.1.4.2
2. **Compare** the functions of the different types of mammalian teeth. SC.F.1.4.2
3. **Relate** the mammal's heart and respiratory systems to its endothermic metabolism. SC.F.1.4.2
4. **Summarize** the ways in which mammals provide parental care. SC.G.1.4.1
5. **Summarize** how the grizzly bear exhibits characteristics typical of mammals. SC.G.1.4.1
6. **Critical Thinking** **Justifying Conclusions** You and your lab partner examine a mammalian skull and jaw that contains only incisor teeth. Your partner concludes that you do not have enough information to identify the specimen as a herbivore or carnivore. Evaluate this conclusion. SC.F.1.4.2
7. *FCAT Prep* If two species of mammals are the same size but one is more active than the other, the more active mammal is likely to have what feature?
   A fewer alveoli
   B smaller alveoli
   C smaller diaphragm
   D smaller lungs
   SC.F.1.4.2

806

Answers to Section Review

1. insulation, protection, and sensory function
2. Incisors are for biting/cutting, canines are for stabbing/holding, and premolars/molars are for crushing/grinding.
3. The four-chambered heart keeps oxygen-rich and oxygen-poor blood separate. A diaphragm aids respiration, and lung alveoli provide a large surface for gas exchange. These features help oxygen reach cells quickly to meet the high energy demand of endotherms.
4. Females provide milk. Adults provide protection and teach survival skills to young.
5. A grizzly bear exhibits fur, specialized teeth, a placenta, and mammary glands.
6. Correct. If the missing teeth were molars, the animal would be a herbivore. If the missing teeth were canines, the animal would be a carnivore.
7. **A.** Incorrect. This would reduce gas exchange and limit activity. **B.** Correct. Smaller, more numerous alveoli increase the surface area for respiratory exchange and allow for more activity. **C.** Incorrect. This would decrease breathing efficiency. **D.** Incorrect. This would reduce gas exchange.

Section Review Correlations

SC.F.1.4.2: Items 1, 2, 3, 6, 7
SC.G.1.4.1: Items 4, 5

Today's Mammals

Section 2

Florida Sunshine Standards/FCAT

SC.D.1.4.3 Changes in Earth's climate, geological activity, and life forms may be traced and compared.

SC.G.2.4.3 Genetic variation of offspring contributes to population control in an environment, and natural selection ensures that those who are best adapted to their surroundings survive to reproduce.

Mammalian Diversity

Cave paintings, some perhaps 40,000 years old, show that human life has long been intertwined with the lives of animals. The vast majority of the animals shown in ancient cave paintings are mammals. While some of the mammals represented are extinct, the descendents of many (horses, hyenas, rhinoceroses, panthers, bison, and lions) still live. They are a part of the astounding array of mammals that share our planet with us. In their diversity of size, anatomy, and habitat, mammals surpass all other vertebrate groups. Mammals range in size from tiny shrews that weigh about 1.5 g (less than 0.1 oz) to gigantic blue whales that can weigh up to 136,000 kg (150 tons).

Consider also the differences between a bat and a whale, compared in **Figure 6.** These two mammals are adapted to live in very different environments. Bats fly and are active primarily at night, while whales are permanently aquatic. However, both groups face a similar challenge—how to navigate in an environment where visibility is often limited. Bats and some whales have a similar solution to this problem: they use echolocation, which works something like the sonar of a ship. In echolocation, animals emit high-frequency sound waves. As the waves travel, they strike objects in the environment, and a portion of each wave is reflected back to the animal. The brain interprets the reflected wave, or echo, revealing the object's size and location.

Objectives

- **Recognize** how mammals are adapted to different environments. SC.G.2.4.3
- **Compare** reproductive patterns in monotremes, marsupials, and placental mammals. SC.F.2.4.1
- **Relate** the distribution of monotremes and marsupials to the breakup of Pangaea. SC.D.1.4.3

Key Terms

placenta
gestation period
ungulate
cud

Comparison of Bats and Whales

Bats
- Forelimbs modified into wings and covered with leathery skin
- Body covered with hair
- Active at night; use echolocation to navigate

Whales
- Forelimbs flattened and paddle-shaped; no hind limbs
- Nearly hairless, streamlined body
- Communicate with sound; some use echolocation

Figure 6 Comparison of bats and whales. Although both are mammals, bats and whales are adapted to live in different environments.

807

Chapter Resource File
- Directed Reading BASIC
- Active Reading GENERAL
- Data Sheet for Data Lab GENERAL

Transparencies

TR Bellringer

TR I40 Comparison of Bats and Whales

One-Stop Planner CD-ROM
- Reading Organizers BASIC
- Reading Strategies BASIC
- Occupational Application Worksheet Wildlife Biologist GENERAL
- Supplemental Reading Guide: *Through a Window: My Thirty Years with the Chimpanzees of Gombe* ADVANCED

Section 2

Focus

Overview

Before beginning this section review with your students the objectives listed in the Student Edition. This section describes the incredible diversity of mammal forms and their methods of reproduction. The section concludes with an overview of the orders of placental mammals.

Bellringer

Write the following categories on the board or overhead: Terrestrial, Marine, Carnivorous, and Herbivorous. Have students copy the categories on a sheet of paper and make a list of mammals that fit into each category. When they are finished, make a master list from the student lists. Discuss the differences in the mammals listed. If a duckbill platypus is not listed, ask students if they know how it is different from the other mammals listed. Tell students that in this section they will learn about how the orders of mammals differ. English Language Learners

Motivate

Activity — GENERAL

Greatest Mammal Working in groups, have students come up with the "All-Time Greatest Mammal." Tell students they must make a list of the characteristics of their chosen mammal that makes it so great. Have one member of each group announce its mammal and describe its great traits. Evaluate each trait by class discussion and determine whether it is a true mammalian trait. In the end, have the class vote for the All-Time Greatest Mammal as described by each group.
LS Interpersonal Co-op Learning

Section Correlations

SC.D.1.4.3
SC.F.2.4.1
SC.G.2.4.3

Teach

READING SKILL BUILDER — ADVANCED

Writing Skills When the first specimens of the platypus arrived in Europe in the 1790s, the scientists who examined them suspected a hoax. The platypus seemed to be composed of the body parts of two different animals. Like mammals, it had hair, a single jawbone, and three middle-ear bones. But it also had a duck's bill. When scientists could find no evidence of stitching or tampering, they concluded that the platypus, however odd, is indeed a mammal. Have each student write a letter as if he or she was the person who sent the first platypus to Europe from Australia. In the letter they should describe the mammalian characteristics of the platypus and attempt to convince the platypus recipients that the animal they are receiving is not a hoax. **LS Verbal**

Teaching Tip

Monotreme Adaptations Female monotremes have no nipples, thus the young do not suckle. Instead, milk oozes out from pores in the skin, much like sweat, and the young lap it up. Male platypuses have a poisonous spur on their hind legs, which they use in combat with other males over control of breeding sites. Most students will be unfamiliar with platypuses. If possible, show an educational video on these amazing mammals.

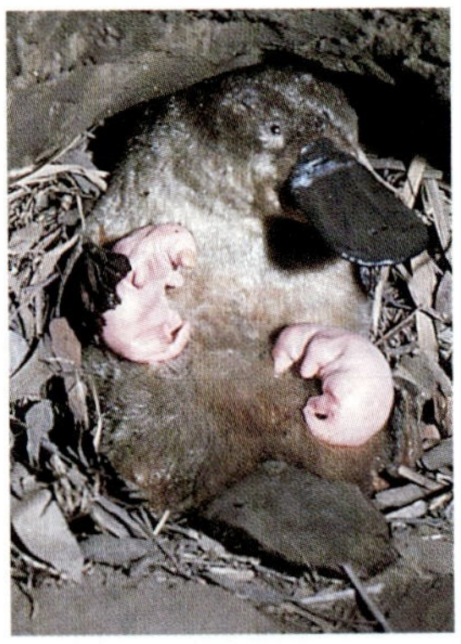

Figure 7 Platypus. Although these platypus young hatched from eggs, they drink milk produced by their mother's mammary glands.

Reproduction

All mammals reproduce by internal fertilization: The male releases sperm into the female's reproductive tract where one or more eggs are fertilized. But there is some variation among mammals in how and where the fertilized egg develops. Present-day mammals are divided into three groups—monotremes, marsupials, and placental mammals—based on their pattern of development.

Monotremes

The monotremes, the most primitive of the mammals, are represented by only one order, Monotremata. All modern monotremes live in Australia, New Guinea, and South America. This distribution may be the result of the movement of these landmasses by continental drift.

The three living monotreme species—the duckbill platypus and two species of echidnas *(ee KIHD nuhz)*, or spiny anteaters—show a curious mix of characteristics. They have shoulders and forelimbs that are quite reptilian in appearance. Among living mammals, only monotremes reproduce by laying eggs. Their eggs, like reptile eggs, have leathery shells. Like birds, the female monotreme incubates her eggs with her body heat, and at hatching, the newborns are only partially developed. Also like birds and reptiles, monotremes have a cloaca, a common passageway for the digestive, reproductive, and urinary systems. No other mammals have a cloaca. Unlike other mammals, adult monotremes do not have true teeth. Because of these characteristics, monotremes may more closely resemble early mammals than do many other living mammal species.

Two features monotremes share with other mammals are hair and milk production. Unlike other mammals, however, a female monotreme does not have nipples and young monotremes do not nurse. Instead, the young lap up milk that oozes from glands located on their mother's belly.

The platypus, shown in **Figure 7,** inhabits lakes and streams in eastern and southern Australia. Its broad, flat tail and webbed front feet make it an excellent swimmer. The platypus uses its flat bill to probe for crustaceans, worms, and snails. Echidnas, shown in **Figure 8,** are found in parts of Australia and in New Guinea. They are terrestrial and have very strong, sharp claws and an elongated, beaklike snout which they use for burrowing and for digging out insects and other invertebrates.

Figure 8 Echidna. Echidnas are covered with sharp, barbless quills. They use their powerful claws and long, beaklike snout to dig for their main food, termites and ants.

808

Cultural Awareness

Makah Whale Hunt In 1855 the Makah Indian Tribe signed a treaty with the United States in which the Makahs were promised the right to engage in whaling, a tribal tradition for more than 2,000 years. Whales are central to the Makah culture. They are represented in their songs, dances, designs, and basketry. The conduct of a whale hunt requires rituals and ceremonies that are deeply spiritual. In the 1920s, due to the scarcity of gray whales, whaling was halted. In 1994 the gray whale was taken off the endangered species list, and in 1996 the gray whale population was estimated at 22,263. Amid much controversy, protest, and constant media watch, the Makahs resumed the hunt for whales in the spring of 1999. Offshore the state of Washington, they were successful in harvesting a gray whale. Its meat was distributed to tribal members.

Marsupials

Order Marsupialia includes not just the well-known kangaroos, but wombats, wallaroos, koalas, and the opossums shown in **Figure 9.** In marsupial mammals the young are born only days or weeks after fertilization—tiny and incompletely developed except for their front limbs. Without any parental help, the newborns crawl to their mother's nipples, usually located in a pouch (the marsupium) on her abdomen. There each newborn attaches itself to a nipple and continues its growth and development for several months. When the young marsupials are able to function on their own, they emerge from the pouch, although they continue to return there to nurse.

Today, many marsupial species are found in the Australian region—Australia, New Guinea, and a few nearby islands—where about half of mammalian species are marsupials. This limited distribution is the result of the breakup of Pangaea. About 70 million years ago, the Australian region separated from the continents of Antarctica and South America. As placental mammals had not yet reached the Australian region, the marsupials there developed in isolation.

Figure 9 Virginia opossum. At birth, these four young opossums looked like the pink newborn opossum, which is smaller than your thumb.

FLORIDA BIOWatch: Florida's Endangered Manatees

Early Florida explorers thought the mysterious creatures were monsters—or mermaids! Manatees are marine mammals. An average adult Florida manatee is about 3 m (10 feet) long and weighs close to 540 kg (1,200 pounds). Manatees are slow, shallow swimmers. In Florida, they live in rivers, springs, and coastal waters, where they feed on aquatic plants. Female manatees reproduce slowly, producing one calf every two or three years.

Manatees in Decline

Today, manatees are a symbol of the struggle to conserve Florida's natural resources. Coastal development and pollution have reduced their numbers. Manatees are also vulnerable to collisions with powerboats, which cause one-quarter of all manatee deaths. Florida's manatees are federally protected under both the Endangered Species Act and the Marine Mammal Protection Act. Yet their reproductive rate cannot keep up with deaths due to collisions and habitat loss.

Monitoring Manatees

Biologists at the Florida Fish and Wildlife Research Institute maintain a manatee identification catalog. The catalog contains photos of hundreds of individual manatees, noting distinctive permanent scars from boat propellers or fungal infections. Tracking individuals over time helps researchers learn survival rates, reproductive rates, habitat use, and movement patterns.

internet connect
www.scilinks.org
Topic: Aquariums and Museums of Florida
Keyword: HX6F01
SCILINKS. Maintained by the National Science Teachers Association

Teaching Tip — BASIC

Natural Selection at Birth

Marsupial newborns are so underdeveloped that their hind limbs have not yet formed. They must creep to their mother's pouch using their forelimbs, guided by their sense of smell. Each newborn attaches to a nipple and stays attached to it while it completes its development. But a female marsupial may give birth to more babies than she has nipples, and any newborn that does not get a nipple will not survive. Ask students to describe the features of newborn marsupials that would enhance their chances of survival. (Their sense of smell must be keen enough to find the pouch, they must have strong forelimbs to crawl to the pouch, and they must be able to crawl quickly. In addition, the order in which the young are born may affect their survival.) **LS Verbal**

BIOWatch: Florida's Endangered Manatees

Teaching Strategies

Point out that manatees cannot survive extended exposure to cold water. During the winter, manatees tend to migrate southward. They often gather at natural warm water springs and areas of warm water discharge from power plants.

Discussion

- Ask students: What is the impact of the growing human population on the Florida manatee? (Development and pollution have caused habitat loss for the manatee.)
- Ask students: Why are manatees especially vulnerable to injuries caused by powerboats? (Manatees are slow, shallow swimmers. They can easily be run over by fast-moving power boats.)

did you know?

Calcium from Bones In many Asian countries, lactose-intolerant individuals (those who cannot digest milk), get their calcium in an interesting way. Many food dishes, such as sweet and sour pork, contain vinegar (an acid) and small bones. The vinegar dissolves the calcium out of the bones and when a person sucks on the bones, he or she ingests calcium.

Teach, *continued*

READING SKILL BUILDER — BASIC

Paired Reading Assign students to cooperative pairs and have each student copy on paper the red, blue, and green headings in Section 2. Then have the students read the pages silently, writing key words that will refer them to sentences they do not understand. When finished, have the partners share and discuss the segments of the section they did not understand. **LS Verbal**

Co-op Learning

Teaching Tip — GENERAL

Mammalian Evolution Have students create a Graphic Organizer that features a cladogram similar to the one at the bottom of this page. Use the graphic organizer to show the relationships among monotremes, marsupials, and the placental mammals. Then have students locate the positions on the cladogram where the following characters should be written in: hair, milk production, live birth, and placenta. **LS Visual**

Activity — GENERAL

Placental Mammals Have each student make three columns on a sheet of paper. Instruct them to label the columns *Monotremes*, *Marsupials*, and *Placentals*. Show the students photos of 10 to 15 different mammals. Have students put the animals into one of the three categories. Then, tell the students which category they belong in. Have students correct their papers. **LS Visual**

internet connect
www.scilinks.org
Topic: Placental Mammals
Keyword: HX4138
SCILINKS. Maintained by the National Science Teachers Association

Placental Mammals

The young of placental mammals develop within the female's uterus, where they are nourished by nutrients from her blood. An organ called the **placenta** allows the diffusion of nutrients and oxygen from the mother's blood, across placental membranes, and into the blood of the developing fetus. Waste materials from the fetus diffuse in the opposite direction and are eliminated by the mother's excretory system.

The period of time between fertilization and birth is called the **gestation** *(jeh STAY shuhn)* **period.** Most placental mammals have a longer gestation period than marsupial mammals do, and their young are more completely developed at birth. Some placental mammals, like the foal in **Figure 10,** can stand and walk within a few hours of birth. Others, like the rabbits in Figure 10, are born blind, deaf, and helpless.

Placental mammals vary greatly in size, shape, diet, and habits. They live in a variety of habitats, from hot, moist rainforests to the frigid tundra. Although mammals share many similarities, species have different characteristics. For example, some placental mammals are adapted for running, leaping, swimming, or flying.

Alone among vertebrates, some placental mammals have hooves, horns, or antlers. Hooves, such as the zebra hoof shown in **Figure 11,** are specialized pads that cover the toes of many running mammals. Hooves are made of keratin, a versatile protein that is a component of many mammalian structures. The horns of sheep, cattle, and antelopes are composed of a core of bone surrounded by a sheath of keratin. This bony core is firmly attached to the skull, and the horn is never shed. (The horn of a rhinoceros is composed not of bone but of hairlike fibers of keratin that form a hard structure.)

internet connect
www.scilinks.org
Topic: Animals of Florida
Keyword: HX6F17
SCILINKS. Maintained by the National Science Teachers Association

Other placental mammals, such as deer and elk, grow and shed a set of antlers each year. Antlers, grown only by the male, are composed of bone but are not covered by a keratin sheath. While they

Figure 10 Young mammals

Newborn foals are on their feet nursing within a couple of hours, while young rabbits do not even open their eyes until they are about 10 days old.

Mare and foal

Baby rabbits

810

Graphic Organizer

Use this graphic organizer with the **Teaching Tip** found on this page.

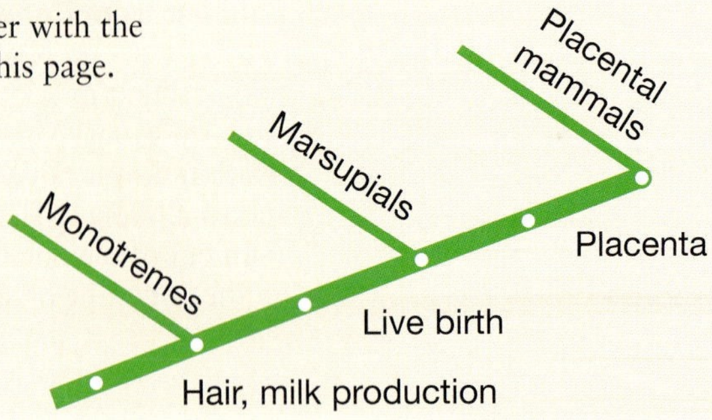

are growing, the antlers are covered by a thin layer of soft skin called velvet. When the antlers are fully grown, the velvet dries up and comes off. The male uses his antlers during breeding season to attract females and to combat other males. After the breeding season is over, the antlers are shed, and a new pair grows the next year. Because the male grows a larger pair of antlers each year, antler size gives an indication of a male deer's age.

Domestic Animals Domestic animals are animals that have been kept and bred by people for special purposes. Most domestic animals are placental mammals whose association with humans dates back at least 2,000 years. These animals include dogs, cats, cattle, horses, donkeys, mules, rabbits, sheep, goats, pigs, camels, llamas, and alpacas.

Different breeds of domestic mammals have been developed through selective breeding. For example, some breeds of goats produce more milk than others. Their milk is used to produce a variety of dairy products. Other goats, such as angora goats, are bred for their fine hair, which is spun into yarn. Some domestic mammals are hybrids of two different species. Mules, for example, are the offspring of a female horse (mare) and a male donkey. Like most hybrids, mules are sterile. The mule's strength, endurance, and surefootedness make it valuable as a pack animal and for chores such as plowing.

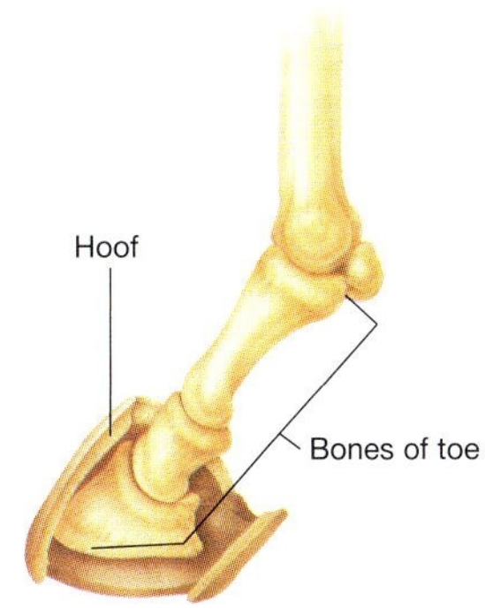

Figure 11 **Hoof.** A zebra's foot is modified so that the zebra walks on a single toe that is covered with a hoof made of keratin.

DATA LAB: *Comparing Gestation Periods* SC.F.2.4.1

Background

If you have ever raised gerbils or hamsters, you know that they can produce several litters of young each year. That's because they have a very short gestation period compared to other mammals. Use the table below to find out more about gestation periods.

Analysis

1. **Critical Thinking** **Recognizing Patterns** State a generalization about the relationship between the length of the gestation period and the number of offspring per pregnancy.
2. **Critical Thinking** **Forming Hypotheses** Propose a hypothesis to explain this relationship.
3. **Recommend** a way that the information in the table could be rearranged to show this relationship more clearly.

Gestation Periods in Mammals

Mammal	Gestation period	Offspring per pregnancy
Vampire bat	210 days	1
Gerbil *	19–21 days	4–7
Human	about 265 days	1
Quarter horse	332–342 days	1
Black spider monkey	226–232 days	1
Grey squirrel	44 days usually	3
Rabbit *	about 31 days	3–6
Sperm whale	420–430 days	1
Arctic wolf	63 days	4–5

*More than two litters per year

Demonstration — BASIC

Allow students to handle both horns and antlers. Students should be able to feel that antlers are made of bone, whereas horns are not. (State wildlife agencies or local hunters are possible sources of both antlers and horns.) Point out that the keratin in horns is the same material comprising fingernails and hair, and is nonliving tissue. Bone is a living tissue, complete with an internal blood supply, which is not visible in the dead bone being demonstrated in class. **LS Kinesthetic** English Language Learners

Teaching Tip — GENERAL

Antler Function Ask students how antlers benefit a male deer. (Students say that antlers function as weapons. Antlers may also function as "ornaments" in attracting females.) Ask students if antlers can be detrimental to a deer. (Yes; sometimes they get locked up with other male deer during battles for females, causing both deer to starve to death.) **LS Interpersonal**

DATA LAB: *Comparing Gestation Periods*

Skills Acquired

Recognizing relationships, making hypotheses, organizing information

Answers to Analysis

1. In general, the longer the gestation period is, the fewer offspring are born.
2. Student hypotheses will vary.
3. The table could be rearranged to show the gestation period or number of offspring per pregnancy in decreasing or increasing numbers.

MATH CONNECTION — GENERAL

Tell students that the average life span of a mouse is 2 years and that each mouse needs 1 square meter of grass-covered land to survive. Suppose that every 3 months each female mouse gives birth to a litter of 8, half of which are female.

A pair of adult mice—a male and a pregnant female—arrives, floating on a log, at a grassy island. The island is the size of a football field (91 m × 49 m). The day they arrive, the female gives birth. There are no predators on the island, and the mice and their descendants survive until the island can no longer provide enough space and food. Have students calculate the following:

1. How many mice will there be after 3 months? after 6 months? after 9 months? after 1 year? (after 3 months, 56; after 6 months, 250; after 9 months, 1,250; after 1 year, 6,250)
2. When can the island no longer support the entire population of mice? (The island has an area of 4,459 square meters and can no longer support the mice before the end of the first year.) **LS Logical**

Teach, *continued*

Using the Figure — BASIC

Divide the class into 12 groups. Assign each group one of the photographs shown in **Figures 12** through **23.** Have each group examine their assigned photograph and make a list of adaptations that particular mammal has for survival. Have each group share their list of adaptations with the class and encourage class discussion about each mammal order.
LS Visual Co-op Learning

Teaching Tip — GENERAL

Bat Navigation In the late 1700s, the Italian scientist Lazzaro Spallanzani found that blinded bats could still fly and capture prey. However, deafened bats could not avoid objects in their path and they could not capture prey. He concluded that bats depend on sound to navigate. Have students discuss how Spallanzani may have conducted this experiment. LS Logical

Group Activity — ADVANCED

Orders of Mammals Assign students to cooperative groups of two to four. Assign each group one of the mammalian orders to research. For example, groups could research migratory habits, if any; general physical and behavioral characteristics; feeding habits; and significance of the order to humans or to the biosphere. Group members should compile their findings in a report and present it to the class.
LS Verbal Co-op Learning

Figure 12 Mouse

Figure 13 Bat

Figure 14 Shrew

Modern Placental Mammals

There are 19 orders of placental mammals, which include more than 90 percent of all mammal species. Terrestrial placental mammals inhabit all continents except Antarctica, and aquatic placental mammals inhabit all oceans. The following 12 orders contain the vast majority of mammal species.

Order 1 Order Rodentia

Over 40 percent of all placental mammals are gnawing mammals called *rodents*. Rodents are distinguished from most other mammals by their teeth, which are specialized for gnawing. All rodents have two pairs of large, curving incisor teeth that grow continuously. As the rodent gnaws, the back side of the incisors wears away faster than the front, creating a sharp chisel edge on the teeth. The success of this group of herbivores can be attributed to their intelligence, small size, and rapid rate of reproduction. Most rodents, such as the mouse shown in **Figure 12,** are small, but a few are larger. For example, beavers can weigh between 18 and 43 kg (between 40 and 95 lb).

Order 2 Order Chiroptera

This order is composed of bats, the only mammals capable of true flight. A bat's front limbs are modified into wings. The thumb, which is not attached to the wing, has a curved claw that can be used for clinging or grasping. Bats generally live in groups and are active only at night. During the day, bats hang upside down in caves or some other protected place, as shown in **Figure 13.** Most bats are carnivorous, using echolocation to find insects, which they catch while in flight. Other bats eat fruit or nectar from night-blooming flowers.

Order 3 Order Insectivora

Insectivores are the mammals most similar to the ancestors of the placental mammals. As their name implies, these small mammals eat mainly insects, but their diet may also include fruit, small birds, and snakes, as well as other insectivores. Insectivores have an enormous appetite, and some, such as the shrew shown in **Figure 14,** must eat more than two times their body weight daily to fuel their extremely high metabolic rate.

MISCONCEPTION ALERT

The Truth About Bats Apart from snakes, bats have been the subject of more misconceptions, superstitions, and old wives' tales than any other group of animals. For example, some people think that all bats suck blood. The truth is, many bats are fruit eaters, and even vampire bats don't suck blood—they nick the skin with their sharp teeth and then lap up the blood like a dog drinking from its water dish. Others believe that most bats carry rabies, however, the percentage of rabid bats is actually very low, less than one percent.

Order 4 Order Carnivora

Some of the best-known animals are the flesh-eating hunters called *carnivores*. Scientists generally divide this order into two subgroups: the cat family and the dog family, represented by the fox shown in **Figure 15.** Carnivores are strong and extremely intelligent, and they have keen senses of smell, vision, and hearing—characteristics that have enabled them to become successful hunters. Their long canine teeth are specialized for capturing prey and tearing flesh. Some members of this order are no longer carnivorous: raccoons and bears are omnivores, while pandas are herbivores.

Figure 15 Fox

Order 5 Order Pinnipedia

This order of marine carnivores includes seals and sea lions that feed at sea but return to land to mate, rear their young, and rest. All four limbs are modified as flippers for swimming, and their bodies are streamlined for rapid movement through the water. An insulating layer of blubber protects them from the cold ocean waters. Most species live in large colonies called rookeries which are headed by a large male, such as the walrus shown in **Figure 16.**

Figure 16 Walrus

Order 6 Order Primates

Humans belong to the order Primates. Other members of this order include the prosimians, which are active at night, and monkeys and apes, which are active during the day. Most nonhuman primates, like the monkey in **Figure 17,** are tree dwellers, and many of their characteristic features are adaptations for living in trees. Flexible, grasping hands and feet aid tree-dwelling primates in climbing. Monkeys and prosimians have a tail that is used for balance. The excellent depth perception of the primates is critical for those that live in the trees. Primates are extremely curious and their ability to learn is exceptional.

Figure 17 Monkey

Demonstration — BASIC

Allow students to observe real hooves. Pet stores often sell goat or cow hooves for dogs to chew. Provide students with a hand lens to examine the hooves. Remind students that hooves are composed of keratin. Have students compare the animal hoof material with their own fingernails. **LS Visual**

Teaching Tip — GENERAL

Whales with Pinnipeds Whales and pinnipeds are adapted for life in the water. Have students construct a table that lists similarities and differences between these aquatic mammals. (Similarities: modified flipper-forelimbs, feed in the water, they breathe air, and have streamlined bodies; Differences: whales lack hind limbs—pinnipeds have paddlelike hind limbs; whales cannot leave the water—pinnipeds mate and sleep on land.) **LS Verbal**

Group Activity — ADVANCED

Mammal Orders Have students work in small groups. Write the name of each of the first 12 orders of mammals listed in this section on separate slips of paper and put the 12 slips of paper in a box. Have one member of each group come forward and draw two slips of paper from the box, reading the names of the two orders they have chosen. If there are more than six groups, have the students return the paper slips to the box and have the remaining groups randomly choose two slips of paper also. Each group should make a chart with *Similarities* and *Differences* listed at the top. They should then research their two orders and fill in the chart with as much information as they can find. Group spokespersons should share their lists with the class. **LS Verbal** Co-op Learning

Cultural Awareness — GENERAL

Folk Medicine Threatens the Tiger's Survival Tigers are disappearing at a rapid rate because humans are encroaching on their habitat and overhunting them. Poachers kill tigers for their pelts, which can be worth as much as $15,000. Tigers are also slaughtered to feed the large market for tiger-based folk medicines, which are used mainly in Asia. Folk remedies made from the bones, eyes, and other parts of the tiger purportedly lengthen life, instill vigor, and cure diseases and impotence. There is no medical evidence, however, that these medications are effective.

Teach, *continued*

Demonstration — GENERAL

Show a picture or slide of a bandicoot. Ask students if they think it is a monotreme, marsupial, or placental mammal, and if there is any way to tell from the picture. (usually not, unless the animal is large and a pouch is visible) Explain that marsupials have a yolk-sac placenta but aren't considered placental mammals. These "placental" marsupials are the bandicoots, insectivorous inhabitants of New Guinea and Australia. The size of bandicoots varies from that of a rat to that of a small dog. Tell students the bandicoot's placenta is not as efficient as that of a placental mammal, nor is it as intimately associated with the mother's tissues. In bandicoots the fertilized egg adheres to the lining of the uterus. Blood vessels in the lining of the uterus exchange materials with blood vessels in the fertilized egg. In contrast, the fertilized eggs of some placental mammals embed themselves in the lining of the uterus. Fingerlike outgrowths penetrate into the uterine lining and greatly increase the surface area through which nutrients can be absorbed. The gestation period in bandicoots, like other marsupials, is short. LS Visual

Teaching Tip — ADVANCED

Writing **Little-known Mammals** Have students research a mammal that they know little about, such as a tapir, baleen whale, pika, dugong, or manatee, and write a short, one-paragraph description to share with the class. Encourage English language learners to research an animal from their native country. LS Interpersonal English Language Learners

Figure 18 Sheep

Order 7 Order Artiodactyla

Mammals belonging to this order and the following order, Perissodactyla, are classified as **ungulates** *(UNG gyoo lits)*, mammals with hoofs. Ungulates walk, not on their entire foot as most vertebrates do, but with their weight supported by their hoof-covered toes. Most ungulates are herbivores that live together in herds. The young are well developed at birth and can move along with the herd within a day or two. An ungulate's main means of defense are the security of the herd and the ability to run very fast when danger approaches.

Artiodactyls have an *even* number of toes within their hooves. Many of these mammals have a stomach with a storage chamber called a *rumen*. Microbes in the rumen break down the cellulose in the plant material artiodactyls eat. Mammals with a rumen regurgitate partly digested food, called **cud,** rechew it, and swallow it again for further digestion. This order includes pigs, hippopotamuses, camels, deer, antelope, cattle, goats, and giraffes as well as sheep, such as the one shown in **Figure 18.**

Figure 19 Rhinoceroses

Order 8 Order Perissodactyla

Ungulates with an *odd* number of toes within their hooves are classified as perissodactyls. This order includes horses, zebras, tapirs, and rhinoceroses, such as the two shown in **Figure 19.**

Perissodactyls do not chew their cud. Instead of a rumen, they have a cecum, a pouch branching from their large intestine. The cecum contains microbes that digest the cellulose in their diet. Perissodactyls are far less numerous than artiodactyls.

Figure 20 Orca

Order 9 Order Cetacea

Cetaceans are divided into two groups: the predatory toothed whales, dolphins, and porpoises, and the filter-feeding baleen whales. Whales, such as the orca shown in **Figure 20,** are probably descendants of land mammals that returned to the sea about 50 million years ago and have adapted to a fully aquatic life. Their streamlined bodies have front limbs modified into flippers, no hind limbs, and a broad, flat tail for swimming. A nostril called a blowhole is located on top of their head. Cetaceans are very intelligent animals that communicate with each other by making sounds that we hear as a series of clicks.

did you know?

The Odd Opossum The opossum (*Didelphis marsupialis*) is North America's only marsupial. This uniqueness was not lost on explorers of the New World. Captain John Smith, an early explorer of the Chesapeake Bay, described the opossum as a creature with "an head like a swine, and a taile like a rat, and is of the bignes of a cat." Regarding the female's pouch, he wrote, "Under her belly she hath a bagge, wherein she lodgeth, carrieth, and sucketh her young." European naturalists were at a loss as to how to classify this new creature. They described it as being "half fox, half monkey" and even gave it the scientific name *Simia vulpa* (literally, "monkey-fox").

The widely known act of "playing possum" seems to be an involuntary threat response in which an opossum "plays dead." Some predators will not eat an already-dead animal and will abandon the "dead" opossum in search of live prey. The opossum then "wakes up" and is on its way.

Order 10 Order Lagomorpha

This order is composed of rabbits and hares. Like rodents, they have one pair of long, continually growing incisors, but they also have an additional pair of peg-like incisors that grow just behind the front pair. Rabbits and hares have long hind legs and are specialized for hopping, as shown in **Figure 21.** Rabbits build nests that the female lines with fur. The young are born furless and their eyes are closed. Hares do not construct nests. Their young are born with fur and their eyes are open.

Figure 21 Rabbit

Order 11 Order Sirenia

These somewhat barrel-shaped marine animals include the dugongs and manatees, shown in **Figure 22.** Like whales, they have front limbs modified as flippers and no hind limbs. A flattened tail is used for propulsion through the tropical oceans, estuaries, and rivers where sirenians live, grazing on aquatic plants. Despite their appearance and habitat, sirenians are closely related to elephants and are often called *sea cows*.

Figure 22 Manatees

Order 12 Order Proboscidea

The two living species of this order, the African elephant and the Indian elephant shown in **Figure 23,** are the largest land animals alive today. Their long, boneless trunk is really an elongated nose and upper lip and is used for a variety of tasks. An elephant's upper incisor teeth are modified into long ivory tusks. Elephants live in herds made up of a dominant male, a number of females, and young of varying ages.

Figure 23 Indian elephant

Using the Figure — Basic

Have students examine the picture of the rabbit running in **Figure 21.** Ask students to point out two specialized adaptations that help the rabbit survive. (The rabbit has large hind legs for swift running, large ears for acute hearing, and coloration that provides camouflage.) Ask students if they can think of a function of the large ears that is not associated with hearing. (They act as radiators to dissipate heat and cool the rabbit.) **LS Visual**

Group Activity — General

Mammal Helpers Have students work in groups. Each group should brainstorm to come up with a list of ways that mammals help humans other than being food for them. You might assign each group a particular order or let each group identify mammals from any order. (Examples: Camels, elephants, and donkeys are often used for packing and travel; horses are used for working with livestock; dogs are used as guide animals for the blind; and dogs, cats, rats, gerbils, hamsters, mice, rabbits, etc. are common pets.) **LS Interpersonal**

Group Activity — Advanced

Local Zoo or Wildlife Park If possible, organize a trip to a local zoo or wildlife park. Preview the area before the trip. Make a list of all the mammals the students will see. Have students make a brochure with drawings and information about each mammal. Reproduce the brochure and distribute copies to each student for the trip. **LS Visual**

did you know?

Armadillos The nine-banded armadillo (*Dasypus novemcinctus*) is one of the few mammalian species whose range is expanding. In the late 1800s armadillos lived only as far north as central Texas. Now "dilloes," as some locals call them, inhabit most of Texas and all of Louisiana. Armadillos also have spread to parts of Kansas, Oklahoma, Arkansas, Mississippi, New Mexico, Georgia, Florida, and Alabama.

Close

Reteaching — BASIC

K-W-L Tell students to return to the lists of things they **W**ant to know about mammals, which they created using the Reading Skill Builder exercise on the fifth page of Section 2. Have them place check marks next to the questions that they are now able to answer. Students should finish by making a list of what they have **L**earned. Ask students the following:

- Which questions are still unanswered?
- What new questions do you have?

You may wish to assign projects that require students to research their unanswered questions.
LS Intrapersonal

Quiz — GENERAL

True or False:

1. Mammals have the least diverse body forms of all vertebrates. (false)
2. Mammal embryos develop in one of three major ways. (true)
3. Most of today's mammals are marsupials. (false; most are placentals)

Alternative Assessment — GENERAL

Have students design a brochure for their ideal zoo. Using old magazines or newspapers, have students include pictures of placental, marsupial, and monotreme mammals. Include a few nonmammals, such as fish, birds, reptiles, and amphibians. Ask students to identify each animal and classify each mammal into the proper order.
LS Visual.

Section Review Correlations

SC.D.1.4.3: Item 2
SC.F.1.4.2: Items 4, 5
SC.F.2.4.1: Item 1
SC.G.2.4.3: Item 3

The seven remaining orders of placental mammals are summarized in **Table 1.** These orders contain few species, but some are quite interesting. The hyrax, for example, looks as if it could be kin to rabbits or rodents. But traits such as its hooflike nails lead some biologists to think the hyrax is most closely related to elephants or perissodactyls.

Table 1 Orders of Placental Mammals

Order		Description
	Edentata anteaters, armadillos	Toothless or with poorly developed teeth that lack enamel; found only in the Western Hemisphere
	Macroscelidea elephant shrews	Ground-dwelling insect eaters; long, flexible snout; hop about somewhat like small kangaroos
	Scandentia tree shrews	Omnivorous; small, squirrel-like mammals; long snout, sharp teeth; live mainly on the ground, despite their name
	Pholidota pangolins (scaly anteaters)	Body covered with overlapping scales; no teeth; very long tongue for capturing ants
	Hyracoidea hyraxes	Rabbitlike body; short ears; four hoofed toes on front feet; three hoofed toes on back feet
	Dermoptera flying lemurs	Squirrel-like; glide on a sheet of skin stretching between their forelegs and hind legs
	Tubulidentata aardvarks	Nocturnal; piglike body; big ears; long snout used to feed on ants and termites

Section 2 Review

1. **Compare** the reproductive patterns of monotremes, marsupials, and placental mammals. SC.F.2.4.1
2. **Relate** the location of modern-day marsupials to the breakup of Pangaea. SC.D.1.4.3
3. **Describe** how artiodactyls and perissodactyls are adapted for digesting plant material. SC.G.2.4.3
4. **Summarize** the ways in which aquatic mammals are adapted to life in the water. SC.F.1.4.2
5. *FCAT Prep* In which biome would you most likely find a mammal that uses baleen for filter feeding?
   A tundra
   B savanna
   C deep ocean
   D temperate grassland
   SC.F.1.4.2

816

Answers to Section Review

1. Monotremes lay eggs; marsupials are born underdeveloped and attach to a nipple in the mother's pouch to complete development; placentals develop in the mother's uterus, getting food through the placenta.
2. About 70 million years ago, the southern region of Pangaea became separated from the rest of Pangaea as it broke apart.
3. Many artiodactyls are cud chewers. Perissodactyls do not chew cud. A special intestinal pouch, the cecum, contains microbes that digest the cellulose in their food.
4. They have streamlined bodies, insulating blubber, front limbs modified as flippers, and flat tails for swimming. In pinnipeds, all four limbs are modified for swimming.
5. **A.** Incorrect. The tundra is a terrestrial biome unsuitable for baleen whales. **B.** Incorrect. A tropical grassland such as the savanna is unsuitable for baleen whales. **C.** Correct. Baleen is found in whales in the deep ocean. **D.** Incorrect. Grasslands would be unsuitable for baleen whales.

Study ZONE CHAPTER HIGH

Key Concepts

1 The Mammalian Body

- Mammals are the only animals with hair. The primary function of hair is to insulate a mammal's body, though it can also provide camouflage or a clear signal through coloration, serve a sensory function, or be a defensive weapon.
- Mammals usually have two sets of teeth in their lifetime. Teeth in the second set are not replaced, even if lost or damaged.
- The four types of mammalian teeth are highly specialized: incisors are for biting and cutting; canines are for stabbing and holding; and premolars and molars crush and grind the food.
- Mammals are endotherms, generating heat internally through the rapid metabolism of food. Endothermy is made possible by highly efficient respiratory and circulatory systems.
- Mammals nurse their young with milk from the mammary glands of the female.

2 Today's Mammals

- In terms of anatomy and habitat, mammals are the most diverse of all vertebrate groups.
- All mammals reproduce by internal fertilization.
- Monotremes consist of three species that have a cloaca and lay eggs.
- Marsupials give birth to incompletely developed young that complete their development in the mother's pouch.
- Placental mammals nourish their unborn young in the uterus through the placenta.
- The different mammalian species have evolved a variety of adaptations that permit them to live in a wide range of habitats.
- Many aquatic mammals have a layer of blubber that insulates them from the cold. Their forelimbs are modified into flippers, and they have no hind limbs. A flattened tail aids in swimming.
- Ungulates have digestive systems modified for digesting cellulose. Even-toed ungulates chew their cud.

Key Terms

Section 1

hair (800)
alveolus (803)
mammary gland (806)
weaning (806)

Section 2

placenta (810)
gestation period (810)
ungulate (814)
cud (814)

Alternative Assessment — GENERAL

Divide students into groups. Have each group develop five sets of questions and answers for assigned pages of the chapter. Have each group pass its questions (but not answers) to another group for it to answer. Continue in this manner until every group has answered every set of questions. LS Verbal

Co-op Learning

Chapter Resource File

- Science Skills Worksheet GENERAL
- Critical Thinking Worksheet ADVANCED
- Test Prep Pretest GENERAL
- Chapter Test GENERAL

Answer to Concept Map

The following is one possible answer to Performance Zone item 15.

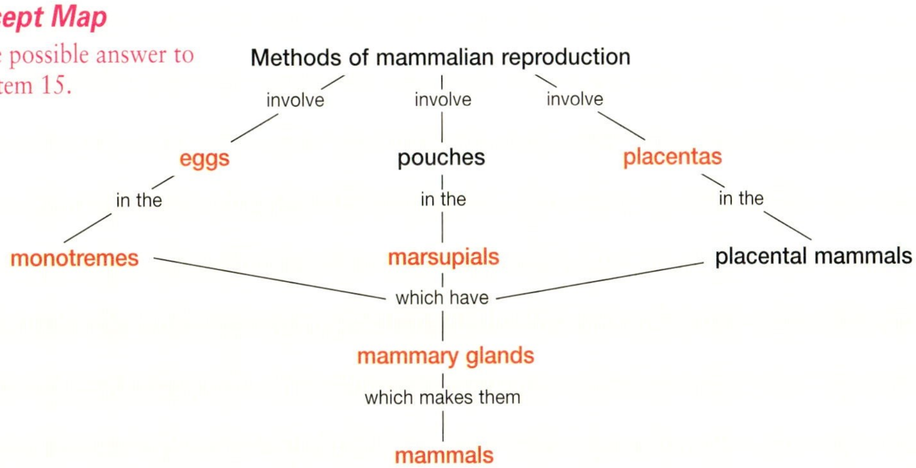

CHAPTER 35 ZONE

ANSWERS

Understanding Key Ideas

1. b
2. b
3. d
4. c
5. a
6. d
7. d
8. c
9. coastal development, pollution, and collisions with powerboats
10. One possible answer to the concept map is found at the bottom of the Study Zone page.

Critical Thinking

11. The high metabolic rate produces much heat. Hair helps prevent heat loss.
12. A mammal requires large amounts of food to fuel its metabolism. It takes in large amounts of oxygen to "burn" (metabolize) the food.
13. Answers may vary. Both hair and down feathers are light and trap air for insulation.
14. Answers may vary. Students may conclude that vertebrates that mature quickly do not need long periods of parental care. Some students may point out that the process of natural selection shapes each species' behavior in a way that is optimal for survival and reproduction.

Alternative Assessment

15. Answers will vary but should include a more detailed account than is found on page 809.

Performance ZONE CHAPTER REVIEW

Understanding Key Ideas

1. The primary function served by mammalian hair is
a. camouflage.
b. insulation.
c. defense.
d. sensory.
SC.F.1.4.2

2. Which of the following is *not* a type of mammal tooth?
a. canine
b. baleen
c. molar
d. incisor
SC.F.1.4.2

3. Endothermic metabolism permits mammals to do all of the following *except*
a. live in very cold climates.
b. generate heat internally.
c. run or fly for long distances.
d. eat less than ectotherms.
SC.F.1.4.2

4. Care given by mammals to their young
a. begins after the young are weaned.
b. is similar to that given by reptiles.
c. involves nursing and teaching survival skills.
d. ends soon after the young are born.
SC.G.1.4.1

5. Monotremes differ from marsupials in that monotremes
a. lay eggs.
b. are viviparous.
c. nourish unborn via the placenta.
d. do not have mammary glands.
SC.G.1.4.1

6. Artiodactyls differ from perissodactyls in that artiodactyls
a. have a cecum.
b. do not have a rumen.
c. seldom live in herds.
d. have an even number of toes.
SC.F.1.4.2

7. Which of the following is *not* a mammalian adaptation for aquatic living?
a. limbs modified as flippers
b. layer of blubber
c. streamlined body
d. keen sense of smell
SC.F.1.4.2

8. Which of the following grizzly bear features is *not* typical of most mammals?
a. thick hair
b. specialized teeth
c. powerful claws
d. placental reproduction SC.F.1.4.2

9. **BIOWatch** What are the main threats to the survival of Florida's endangered manatees? SC.G.2.4.6

10. **Concept Mapping** Make a concept map that describes the different methods of mammalian reproduction. Include the following terms in your map: *monotreme, marsupial, mammal, mammary gland, egg,* and *placenta.* SC.G.1.4.1

Critical Thinking

11. Drawing Conclusions How do hair and a high rate of metabolism help a mammal maintain homeostasis? SC.F.1.4.2

12. Applying Information A mammal must eat about 10 times as much food as an ectotherm of similar size. What role might the respiratory system play in this need? SC.F.1.4.2

13. Comparing Structures In what way are a mammal's hair and a bird's down feathers alike? How are they different? SC.F.1.4.2

14. Evaluating Conclusions Some mammal species must care for their young for many years before they reach maturity and can survive on their own. Can you conclude that all vertebrate young would benefit from this type of parental care? Explain why or why not. SC.G.1.4.1

Alternative Assessment

15. Recognizing Patterns Find out more about how plate tectonics can explain the pattern of distribution of mammals on Earth. Analyze and critique the theory, and write a report on your findings. Include a description of how the breakup of Pangaea led to the predominance of placental mammals on all the continents except Australia. SC.D.1.4.3

818

Assignment Guide

Section	Questions
1	1, 2, 3, 4, 8, 9, 11, 12, 13, 14
2	5, 6, 7, 10, 15

Chapter Review Correlations

SC.D.1.4.3: Item 15
SC.F.1.4.2: Items 1, 2, 3, 6, 7, 8, 11, 12, 13
SC.G.1.4.1: Items 4, 5, 10, 14
SC.G.2.4.6: Item 9

FCAT Prep

For questions 1–6, write your answers on a separate sheet of paper.

1. Clusters of what structures greatly increase the respiratory surface of the lungs? SC.F.1.4.2
   A. alveoli
   B. mammary glands
   C. placentas
   D. ungulates

2. After what stage does a young animal no longer nurse? SC.G.1.4.1
   F. gestation
   G. hibernating
   H. insulating
   I. weaning

3. What term describes an animal with hooves? SC.F.1.4.2
   A. cetacean
   B. marsupial
   C. monotreme
   D. ungulate

4. Which statement best describes the function of the mammalian heart? SC.F.1.4.2
   F. Two pumping chambers supply oxygen-rich blood to the lungs.
   G. Four pumping chambers supply oxygen-poor blood to the lungs.
   H. One pumping chamber supplies blood to the lungs and body organs.
   I. One pumping chamber supplies oxygen-poor blood to the lungs, and one pumping chamber supplies oxygen-rich blood to the body organs.

5. READ INQUIRE EXPLAIN Placental mammals include zebra, horses, goats, dogs, cats, rabbits, and humans.

   Assess the function that the placenta serves in placental mammals. SC.F.1.4.2

6. READ INQUIRE EXPLAIN Mammals are the only animals with hair, but not all mammalian hair serves the same function. Hair serves some unusual functions in some mammals. For example, some mammals have specialized hairs that act as defensive weapons. Porcupine quills are stiff, sharp hair structures that can injure an attacker. Rhinoceros horns are formed from hair. Though biologists are not certain of the horn's function, mother rhinoceroses have been known to use their horns to protect their young from predators.

   Part A What is the primary function of hair? SC.F.1.4.2

   Part B Is this a function of human hair? Explain your answer. SC.F.1.4.2

Test TIP

Before looking at the answer choices for a question, try to answer the question yourself.

FCAT Doctor

Question 4 Answer I is the correct choice. The right ventricle supplies oxygen-poor blood to the lungs, and the left ventricle supplies oxygen-rich blood to the body tissues. Answer F is incorrect because only one chamber pumps blood to the lungs, and the blood is oxygen-poor. Answer G is incorrect because only one chamber, the right ventricle, pumps oxygen-poor blood to the lungs. Answer H is incorrect because, in the mammalian four-chambered heart, oxygen-rich blood and oxygen-poor blood are not mixed.

Question 5 Placental mammals nourish their unborn young in the uterus through the placenta.

Question 6

Part A Many mammals may use hair to attract a mate. Others have hairs that have sensory functions, such as the whiskers of a cat. Still others have hairs that serve as defensive weapons, such as the spines of a porcupine. However, these are not hair's primary function. Mammalian hair keeps the body warm, helping it to maintain a constant body temperature—hair's primary function.

Part B Humans have a sparse covering of hair and a limited amount of body fat. Most humans need clothes to provide adequate insulation and help conserve body warmth.

Answers

1. A
2. I
3. D
4. I
5. It diffuses nutrients and oxygen to a fetus.
6A. The primary function of hair is to insulate the body.
6B. To some degree, human hair does insulate the body. When you get cold, your body hairs rise up and keep a thin layer of warm air near your body. Hair on your head keeps your head from getting cold.

Test Prep Correlations

1. **SC.F.1.4.2:** Items 1, 3-6B
2. **SC.G.1.4.1:** Item 2

Exploration Lab

Exploring Mammalian Characteristics

Teacher's Notes

Time Required 45 Minutes

Lab Rating

EASY → HARD

Teacher Preparation 2
Student Setup 1
Concept Level 1
Cleanup 1

Skills Acquired

- Classifying
- Collecting Data
- Inferring
- Organizing and Analyzing Data

Scientific Methods

In this lab, students will:
- Make Observations
- Analyze the Results
- Draw Conclusions

Materials and Equipment

Materials for this lab can be ordered from WARD'S. See *Master Materials List* at the front of this book for catalog numbers. Provide specimens, photos, or diagrams of a variety of mammalian and nonmammalian skulls. For hand-washing, antibacterial soap is recommended.

Exploration Lab

Exploring Mammalian Characteristics

SC.F.1.4.2

SKILLS
- Observing
- Drawing
- Inferring

OBJECTIVES
- **Examine** distinguishing characteristics of mammals.
- **Infer** the functions of mammalian structures.

MATERIALS
- hand lens or stereomicroscope
- prepared slide of mammalian skin
- compound microscope
- mirror
- specimens or pictures of vertebrate skulls (some mammalian, some nonmammalian)

Magnification: 28×

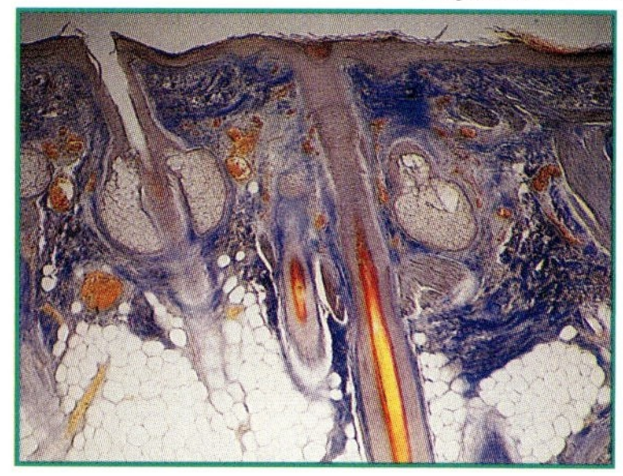

Mammalian skin (human scalp)

Before You Begin

Mammals are vertebrates with **hair**, **mammary glands**, a single lower jawbone, and specialized teeth. Other characteristics of mammals include **endothermy** and a four-chambered heart. Mammals also have **oil (sebaceous) glands** in their skin, and most have **sweat glands.** In this lab, you will examine some of the characteristics of mammals that distinguish them from other vertebrates.

1. Write a definition for each boldface term in the paragraph above.
2. Make a data table similar to the one below.

DATA TABLE

Mammal	Incisors	Canines	Premolars	Molars
Human				

3. Based on the objectives for this lab, write a question you would like to explore about the characteristics of mammals.

Procedure

PART A: Examining Mammalian Skin

1. Use a hand lens to look at several areas of your skin, including areas that appear to be hairless. Record your observations.
2. Look at a prepared slide of mammalian skin under low power of a compound microscope. Notice the glands in the skin. Look for the oil (sebaceous) glands and the sweat glands. Draw and label an example of each type of gland.

PART B: Examining Mammalian Teeth and Skulls

3. Wash your hands thoroughly with soap and water. Use a mirror to look in your mouth. Identify the four kinds of mammalian teeth you see.
4. Count each kind of tooth on one side of your lower jaw. Multiply the number of each kind of tooth by four, and record these numbers in the appropriate columns of your data table. Wash your hands again before continuing.
5. Look at the skulls of several mammals. Identify the kinds of teeth in each skull. For each skull, find the number of each kind of tooth as you did in step 4.
6. Look at the skulls of several nonmammalian vertebrates, and compare nonmammalian teeth to mammalian teeth.
7. Compare the jaws of mammalian skulls to those of nonmammalian vertebrates. As you look at each skull, notice the structure

820

Safety Cautions

Review all safety symbols with students before beginning the lab. Advise students to handle specimens and laboratory equipment with care. Warn students not to touch their faces or eyes after handling samples.

Tips and Tricks

Point out features of skulls that vary among animals, such as the position of the foramen magnum, the opening where the brain stem exits the skull. In bipedal animals, it is at the bottom of the skull, but in quadrupeds, it is at the back of the skull.

of the lower jawbone and how the upper jaw-bone and the lower jawbone connect.

PART C: Cleanup and Disposal

8. Dispose of broken glass in the waste container designated by your teacher.

9. Clean up your work area and all lab equipment. Return lab equipment to its proper place. Wash your hands thoroughly before you leave the lab and after you finish all work.

Analyze and Conclude

1. **Summarizing Information** List the characteristics that distinguish mammals from other vertebrates.
2. **Interpreting Graphics** Compare the amount of hair on humans to that on the mammals shown in the photographs below.
3. **Inferring Relationships** What role, if any, might hair or fur play in enabling mammals to be endotherms?
4. **Forming Hypotheses** Besides the role of hair you identified in item 3 above, what other roles do you think hair might play in mammals?
5. **Recognizing Patterns** Where are the oil (sebaceous) glands located in the skin of mammals?
6. **Forming Hypotheses** Do you think the mammals in the photos on the left below have more sweat glands or fewer sweat glands than humans? Explain.
7. **Comparing Structures** How is the mammalian jaw different from nonmammalian jaws?
8. **Inferring Conclusions** Based on the shape of your teeth, would you classify humans as carnivores (meat eaters), herbivores (plant eaters), or omnivores (meat and plant eaters)? Explain.
9. **Evaluating Conclusions** Justify the following conclusion: The kinds and shapes of a mammal's teeth can be used to determine its diet.
10. **Further Inquiry** Write a new question about the characteristics of mammals that could be explored with a new investigation.

On the Job

Comparative anatomy is the study of the anatomical similarities and differences between organisms. Do research to discover how comparative anatomy is used to hypothesize the relationships among the animal phyla. For more about careers, visit **go.hrw.com** and type in the keyword **HX4 Careers.**

Answers to Before You Begin

1. *mammal*—vertebrate that has hair, mammary glands, and specialized teeth
   hair—strands of protein that grow from the skin of mammals
   mammary glands—milk-producing glands of female mammals
   endothermy—the generation of one's own body heat by metabolism
   oil (sebaceous) glands—oil-producing glands located in the hair follicles of mammalian skin
   sweat glands—water-secreting glands in the skin of some mammals
2. Data table should be similar to that shown in student text.
3. Answers will vary. For example: How do the teeth of humans compare to those of other mammals?

Answers to Procedure

1. Students should note that they can see pores and hairs (except on the fingers, palms, toes, and soles of the feet).
2. Students' drawings should show oil and sweat glands.
3. Students should identify incisors, canines, premolars, and molars.
4. A typical answer is two incisors, one canine, two premolars, and three molars.
5. Answers will vary. The number and types of teeth will vary according to the animal.
6. Answers will vary. Nonmammalian teeth will differ significantly from mammalian teeth.

Answers to Analyze and Conclude

1. Hair, single jawbone, specialized teeth, and mammary glands
2. Humans have far less hair than skunks and more hair than dolphins.
3. Hair serves as insulation from extreme external temperatures and helps maintain a constant internal temperature.
4. Answers will vary. Hair enables a mammal to gather sensory information from its surroundings. It also helps to protect a mammal's skin from water loss, sunburn, and injury.
5. Sebaceous glands are located inside the hair follicles
6. Answers will vary. Students should recognize that skunks, as well as dolphins, are not likely to have as many sweat glands as humans. Humans have very little hair for insulation. Sweat glands help to cool the body. Skunks have sufficient hair for efficient insulation. Dolphins live in water and have no need for sweat glands.
7. Mammals have a single jawbone that connects to the skull at a joint.
8. Answers may vary. Students should recognize that humans are omnivores. Humans have sharp incisors and canines for tearing meat and flat molars and premolars for grinding plant material.
9. Answers will vary. Carnivores have long, sharp canines for grabbing prey and tearing meat. Herbivores have flattened teeth for grinding and crushing plant material.
10. Answers will vary. For example: How is a mammal's muzzle shape related to its diet?

Chapter Opener

TE **Using the Figure,** p. 822
SE **Quick Review,** p. 823 ■
SE **Reading Activity,** p. 823 ■
TE **Opening Activity,** p. 823 BASIC

& 3 • 90 min pp. 824–829

Section 1
Evolution of Behavior

OSP **Lesson Plan***
TR **Bellringer***
TR **Section Outline***
TR **144 Nest Building Behavior***

TE **Group Activity** Animal Trainer, p. 825 GENERAL
TE **Using the Figure,** p. 826 BASIC
TE **Activity,** p. 826 ■ ADVANCED
TE **Group Activity** Learning in Pets, p. 827 GENERAL
SE **Quick Lab** Recognizing Learned Behavior, p. 828 ■
CRF **Datasheets for In-Text Labs** Recognizing Learned Behavior*
CRF **Exploration Lab** Responses in the Fruit Fly* GENERAL
CRF **Inquiry Lab** Observing Animal Behavior* GENERAL

& 5 • 90 min pp. 830–836

Section 2
Types of Behavior

OSP **Lesson Plan***
TR **Bellringer***
TR **Section Outline***
TR **145 Animal Behavior***

TE **Group Activity** Animal Behavior, p. 831 GENERAL
TE **Demonstration,** p. 833 GENERAL
TE **Group Activity** Tracking Whales, p. 833 ADVANCED
SE **Quick Lab** Observing Territorial Behavior in Crickets, p. 834 ■
CRF **Datasheets for In-Text Labs** Observing Territorial Behavior in Crickets *
SE **Exploration Lab** Studying Nonverbal Communication, pp. 840–841 ■
CRF **Datasheets for In-Text Labs** Studying Nonverbal Communication*

& 7 • 90 min

Chapter Review and Assessment Resources

SE **Study Zone,** p. 837 ■
SE **Performance Zone,** p. 838 ■
SE **FCAT Test Prep,** p. 839 ■
TR **Concept Mapping***
CRF **Vocabulary Review Worksheet*** ■ GENERAL
CRF **Test Prep Pretest*** ■ GENERAL
CRF **Chapter Test*** ■ GENERAL
CRF **Chapter Test*** ADVANCED
OSP **Test Generator***
OSP **Test Item Listing***
OSP **Modified Chapter Test*** BASIC

Holt Online Learning

Visit go.hrw.com for access to Holt Online Learning or enter the keyword **HX6 Home** for a variety of free online resources.

One-Stop Planner® CD-ROM

This CD-ROM package includes
- Lab Materials QuickList Software
- Holt Calendar Planner
- Customizable Lesson Plans
- Printable Worksheets
- ExamView® Test Generator
- Interactive Teacher Edition
- Holt PuzzlePro® Resources
- Holt PowerPoint® Resources

Guided Reading Audio CDs

These CDs are designed to help auditory learners and reluctant readers. (Audio CDs are also available in Spanish.)

Compression guide:
To shorten your instruction because of time limitations, omit blocks 1, 2, and 3.

KEY

- **TE** Teacher Edition
- **SE** Student Edition
- **OSP** One-Stop Planner
- **CRF** Chapter Resource File
- **TR** Teaching Transparency
- * Also on One-Stop Planner
- ■ Also Available in Spanish
- ◆ Requires Advance Prep

SKILLS DEVELOPMENT AND PRACTICE	SECTION REVIEW AND ASSESSMENT	CORRELATIONS
		Florida Sunshine Standards/FCAT
TE Reading Skill Builder Interactive Reading, p. 825 BASIC **CRF Directed Reading Worksheet*** BASIC **CRF Active Reading Worksheet*** GENERAL	**SE Section 1 Review** ■ **TE Reteaching,** p. 829 BASIC **TE Quiz,** p. 829 GENERAL **TE Alternative Assessment,** p. 829 GENERAL **CRF Quiz*** ■ BASIC	SC.G.2.4.3, SC.H.2.4.2
TE Reading Skill Builder Reading Organizer, p. 831 BASIC **TE Skill Builder** Vocabulary, p. 835 **TE Reading Skill Builder** Discussion, p. 835 BASIC **CRF Directed Reading Worksheet*** BASIC **CRF Active Reading Worksheet*** GENERAL **CRF Critical Thinking Worksheet*** ADVANCED **CRF Science Skills Worksheet*** GENERAL	**SE Section 2 Review** ■ **TE Reteaching,** p. 836 BASIC **TE Quiz,** p. 836 GENERAL **TE Alternative Assessment,** p. 836 GENERAL **CRF Quiz*** ■ BASIC **Modified Worksheet*** BASIC	SC.F.1.4.7, SC.G.2.4.3, SC.H.1.4.1, SC.H.2.4.2

Holt Biology Interactive Tutor CD-ROM

This CD-ROM consists of interactive activities that give students a fun way to extend their knowledge of biology concepts.

Unit 7 Ecosystem Dynamics

SCILINKS NSTA

www.scilinks.org

Maintained by the **National Science Teachers Association**

Topic: Animal Behavior
SciLinks Code:HX4008

Topic: Conditioning
SciLinks Code: HX4050

Topic: Imprinting
SciLinks Code: HX4104

Topic: Animal Communication
SciLinks Code: HX4009

Topic: Tracking Animals by Satellite
SciLinks Code: HX4177

Overview

Tell students that this chapter focuses on what influences behavior in animals. Students will learn to distinguish between innate and learned behaviors. Finally, students will categorize animal behaviors, describe how animals communicate, and explain how they choose a mate.

Using the Figure

Direct students' attention to the photograph on this page. Tell students that scientists know that some behavior is influenced by genes and therefore can be acted upon by natural selection. Also tell them that the different kinds of behavior an animal exhibits are adaptive. However, behavior is not determined entirely by genes. In this chapter, students will examine how genes and learning interact to shape behavior.

Assessing Prior Knowledge

Before beginning this chapter, be sure your students can answer the questions in the ***Quick Review*** on the facing page. If they cannot, encourage them to reread the sections indicated.

Chapter Correlations — Florida Sunshine Standards/FCAT

SC.F.1.4.7 Organisms respond to internal and external stimuli.

SC.G.2.4.3 Genetic variation of offspring contributes to population control in an environment, and natural selection ensures that those who are best adapted to their surroundings survive to reproduce.

SC.H.1.4.1 Investigations are conducted to explore new phenomena, to check on previous results, to test how well a theory predicts, and to compare different theories.

SC.H.2.4.2 Scientists control conditions in order to obtain evidence, but when that is not possible for practical or ethical reasons, they try to observe a wide range of natural occurrences to discern patterns.

CHAPTER

36 Animal Behavior

Florida Sunshine Standards/FCAT

SC.F.1.4.7 Organisms respond to internal and external stimuli.

SC.G.2.4.3 Genetic variation of offspring contributes to population control in an environment, and natural selection ensures that those who are best adapted to their surroundings survive to reproduce.

SC.H.1.4.1 Investigations are conducted to explore new phenomena, to check on previous results, to test how well a theory predicts, and to compare different theories.

SC.H.2.4.2 Scientists control conditions in order to obtain evidence, but when that is not possible for practical or ethical reasons, they try to observe a wide range of natural occurrences to discern patterns.

Quick Review

Answer the following without referring to earlier sections of your book.

1. **Describe** the relationship of genes to inherited traits. *(Chapter 30, Section 1)*
2. **Relate** natural selection to adaptation. *(Chapter 32, Section 2)*

Did you have difficulty? *For help, review the sections indicated.*

Reading Activity

Before you read this chapter, write a short list of all of the things you know about animal behavior. Then write a list of the things that you want to know about animal behavior. Save your list, and to assess what you have learned, see how many of your own questions you can answer after reading this chapter.

The great egret, *Casmerodius albus,* lays from three to five eggs. Young birds may stay with their parents for more than six weeks.

Looking Ahead

internet connect

www.scilinks.org

National Science Teachers Association *sci*LINKS Internet resources are located throughout this chapter.

SCILINKS. Maintained by the National Science Teachers Association

823

Opening Activity — BASIC

Learned or Innate? Ask students who own pets to describe one behavior of their pet. (example: a dog "shaking hands") Then ask if a particular behavior is learned or not. (The example behavior is learned.) Ask students what criteria they used to determine whether or not a behavior is learned. (Answers will vary. Students must realize that learned behavior involves prior exposure to the behavior.) If all students describe learned behaviors, ask them to think of some behaviors that do not have to be learned. (example: eating)

Quick Review Answers

1. Genes are segments of DNA that act as a code for different traits. When genes are passed down from one generation to another, the traits that they code for are inherited by the new generation.
2. Natural selection occurs when organisms with certain inherited characteristics reproduce while organisms with other inherited characteristics do not reproduce or do so in smaller numbers. Organisms that are able to reproduce and pass along their traits are considered adapted to their environment, and the traits are considered adaptations because they enhance survival.

Reading Activity Answers

Students will likely have a lot of ideas about animal behavior. They might not realize that many behaviors are inherited. Also, they might think that behaviors are only those activities that are conscious ones.

Chapter Resource File

- Vocabulary Worksheets
- Concept Mapping

Focus

Overview

Before beginning this section review with your students the objectives listed in the Student Edition. In this lesson, students will identify animal behaviors. They also will describe the roles that natural selection and genes play in determining behavior. Finally, they will be able to recognize that behavior is influenced by the environment and learning as well as genetics.

Bellringer

Ask students to write down an example of an animal behavior. Ask them to write about an adaptive advantage that the behavior might confer. (For example, a kitten's meow is adaptive because it results in the kitten's mother focusing her attention on the kitten and providing him food or meeting some other need.)

Motivate

Discussion/ Question — BASIC

Ask student volunteers to describe at least one thing they have taught their pets. For example, did students house-train their dogs by using a punishment (such as a swat with a newspaper) after an "accident," or did they use a reward (such as a dog treat or praise) if the dog urinated while outdoors? Ask students which method (positive or negative) they believe is more effective, and why.
LS Intrapersonal

Section Correlations

SC.G.2.4.3
SC.H.2.4.2

Section 1 Evolution of Behavior

Florida Sunshine Standards/FCAT

SC.G.2.4.3 Genetic variation of offspring contributes to population control in an environment, and natural selection ensures that those who are best adapted to their surroundings survive to reproduce.

SC.H.2.4.2 Scientists control conditions in order to obtain evidence, but when that is not possible for practical or ethical reasons, they try to observe a wide range of natural occurrences to discern patterns.

Objectives

- **Distinguish** between "how" and "why" questions about behavior. SC.H.2.4.2
- **Describe** how natural selection shapes behavior. SC.G.2.4.3
- **Compare** innate and learned behaviors. SC.G.2.4.3
- **Summarize** how behavior is influenced by both heredity and learning. SC.G.2.4.3

Key Terms

behavior
innate behavior
fixed action pattern behavior
learning
conditioning
reasoning
imprinting

What Is Behavior?

A squirrel buries a nut; a hungry baby cries. A frog jumps into a pond to avoid a predator, and a driver applies the brakes when approaching a red traffic light. These are all examples of behavior. A **behavior** is an action or series of actions performed by an animal in response to a stimulus. The stimulus might be something in the environment, such as a sound, a smell, a color, or another individual. The stimulus can also be related to the internal state of the animal, such as being hungry or cold. For example, when under threat, the lizard shown in **Figure 1** flares out the folds of skin around its head. This gives it a threatening appearance, which tells potential enemies to stay away.

Scientists studying behavior investigate two kinds of questions—"how" questions and "why" questions. "How" questions are about how a behavior is triggered, controlled, and performed. For instance, consider the squirrel, also shown in Figure 1. "How" questions about squirrel behavior might include "How does a squirrel select which nuts to bury?" "How does it choose where to bury the nut?" and "How does it remember where the nut is?"

However, answering "how" questions provides only a partial understanding of a behavior. Scientists also try to answer "why" questions, such as "Why do squirrels bury nuts?" "Why" questions concern the reasons a behavior exists and are really questions about the evolution of behavior. A study of animal behavior may therefore seek to identify the benefits of a particular behavior.

Figure 1 Animal behavior. This Australian frilled lizard and European red squirrel are engaging in behaviors typical of their species.

Chapter Resource File
- Directed Reading BASIC
- Active Reading GENERAL
- Data Sheet for Quick Lab GENERAL

Transparencies
TR Bellringer

One-Stop Planner CD-ROM
- Reading Organizers BASIC
- Reading Strategies BASIC

Natural Selection and Behavior

Recall that evolution by natural selection is a process by which populations change in response to their environment. Natural selection favors traits that improve the likelihood that an individual will survive to reproduce. Over time, traits that provide a reproductive advantage become more common. Traits that do not provide an advantage become less common and may disappear.

An understanding of natural selection can help answer a "why" question. A good example of this is seen in East African lions, which live in small groups called *prides*. Each pride contains several adult females, several young lions (called cubs), and one or more adult males. The adult males father all the cubs and defend the pride against other males. But a male or group of males usually can control a pride for only a couple of years. Then they are forced out by younger males who take over the pride. When this happens, the new males often kill all the young cubs in the pride, as shown in **Figure 2.** In contrast, male lions are usually quite tolerant of their own offspring, also shown in Figure 2.

Figure 2 Behavior in male lions. After taking over a pride, a male lion often kills the young cubs in the pride. However, the males are usually tolerant of their own cubs.

To understand this behavior, we need to understand how the new males would benefit from it. The new males will control the pride for only a few years, so they have a very short time in which to reproduce. But female lions with cubs will not breed until their cubs are grown, which may take more than two years. If a female's cubs die, however, she will mate again almost immediately.

Why do the new males kill the cubs? One hypothesis suggests that the new males will father more cubs as a result of this behavior. Note, however, that this does not suggest that male lions are aware that they are killing the offspring of other males, or that they understand how they will benefit from this behavior.

Individual Selection

You may have heard it said that a trait or behavior ensures the survival of the species. This once popular belief is now considered false. Most scientists now agree that natural selection favors traits that contribute to the survival and reproduction of *individuals*, not species. The actions of male lions support this idea. Cub-killing increases the already high death rate among cubs and actually reduces the likelihood that the species will survive. Because natural selection favors traits that benefit individuals, the male lions usually will behave in ways that are favorable for them, not for the pride as a whole.

internet connect
www.scilinks.org
Topic: Animal Behavior
Keyword: HX4008
SCILINKS. Maintained by the National Science Teachers Association

825

Teach

Teaching Tip — GENERAL

Recognizing Relationships Point out to students that scientists seek to answer "how" and "why" questions in all areas of biology. For example, geneticists have learned much about how meiosis occurs—how the chromosomes line up and separate and how cytokinesis occurs. But they are also interested in why meiosis occurs and what benefits it brings to the organism. Ask students to come up with some "how" and "why" questions about subjects they studied in previous chapters. Then ask students to answer their questions.

READING SKILL BUILDER — BASIC

Interactive Reading Assign Chapter 36 of the *Holt Biology* Guided Audio CD Program to help students achieve greater success in reading the chapter. English Language Learners

Group Activity — GENERAL

Animal Trainer Invite a dog obedience trainer to visit your class and to discuss the most effective methods he or she uses to train dogs. Ask the trainer to share his or her favorite stories of incidents that have occurred during training. Beforehand, have each student formulate five questions to ask the trainer. Be sure to ask for the trainer's opinion on negative reinforcement versus positive reinforcement. If school rules allow, have the trainer bring a dog and demonstrate training skills in the classroom. LS Interpersonal

Cultural Awareness

Australian Aborigines To the early European colonists of Australia, the absence of aboriginal technology indicated that they were a very primitive people. The Aborigines could produce only a limited variety of tools because they did not know how to mine, smelt, or work metal. They had to rely on hunting and gathering because they did not practice agriculture. Most groups did not wear clothes, even in winter. The largest "structures" they created were piles of discarded shells. However, as the Europeans soon learned, the Aborigines compensated for their technological "deficiencies" with a deep and detailed knowledge of their environment, which allowed them to survive in areas where European explorers and settlers perished. In particular, the Aborigines were experts on the behavior of animals, and they used this knowledge to capture game and to find water in dry areas.

Teach, *continued*

Teaching Tip — GENERAL

Innate Behaviors Ask students to name several innate behaviors. (for example, suckling, eating, and drinking) Ask them why they think it is important for these behaviors to be instinctive rather than learned. Next ask them to come up with a hypothetical behavior that would be a disadvantage to an animal. (for example, a bird refusing to fly when it is capable of doing so) Ask: What would most likely happen to this bird? Would its genes be passed on to the next generation?

Using the Figure

Have students examine the hybrid lovebird in **Figure 4.** Ask: What nesting behaviors does this bird exhibit that are similar to those of the Fischer's lovebird? (It carries nesting materials in its beak.) What nesting behaviors are like those of the peach-faced lovebird? (It places nesting materials in its feathers.) What problem does it encounter? (It fails to let go of the nesting material in its mouth, so the material does not stay tucked into its feathers.) **LS Visual**

Activity — ADVANCED

Ask students how biologists know that the orb spider does not learn to build its web. Ask each student to design an experiment that would demonstrate that web building by the orb spider is innate. (Designs should include isolation of the eggs so that the young spiders have no contact with the parent. The design of the isolated spiders' webs should be compared against that of the parents' web.) **LS Logical**

Transparencies

TR I44 Nest Building Behavior

Figure 3 Fixed action pattern behavior. Like all web-building spiders, the orb spider is genetically predisposed to build her web the same way each time.

Genetically Influenced Behavior

From years of observation and experimentation, biologists have learned that many kinds of animal behaviors are influenced by genes. Genetically programmed behavior is often called **innate behavior,** or more commonly, instinct. The orb spider, shown in **Figure 3,** builds her web the same way every time. There is little or no variation in what she does, and her female offspring will build their webs in the same manner without being taught. This type of innate behavior is called **fixed action pattern behavior** because the action always occurs the same way.

Demonstrating the Genetic Basis of Behavior

Nest building is an innate behavior exhibited by most birds, including African lovebirds, shown in **Figure 4.** These small parrots construct their nests from materials that they collect and carry back to the nest site. One species of lovebird, Fischer's lovebird, carries a single long strip of nesting material in its beak. A second, closely related species, the peach-faced lovebird, carries several short strips of nesting material tucked into the feathers near its tail.

Evidence that these behaviors have a genetic basis comes from studies in which the two types of lovebirds were interbred. As shown in Figure 4, the resulting hybrid birds showed nesting behaviors that resembled those of both parents. They chose medium-length strips of nesting material and tried to place the strips in the feathers near their tail. But the hybrid birds were rarely successful because they did not let go of the strips after placing them in their feathers. Eventually, some of the hybrid birds learned to carry the nesting material in their beak.

Figure 4 Nest building behavior

The hybrid offspring of two lovebird species show nest-building behaviors similar to those of both parents.

826

Attention Grabber

Bower Birds The male bower bird has a fascinating and complex courting behavior. Instead of relying on its plumage or song to attract a mate, the bower bird constructs an elaborate structure called a bower using leaves, twigs, and other items it finds on the ground. The bower bird goes one step further and decorates the bower with colorful items ranging from feathers, pebbles, berries and shells to coins, spoons, and even pieces of aluminum foil. However, the bowers aren't nests for raising chicks; they are strictly for attracting females for mating. When a female arrives and inspects the bower, the male struts in front of the bower and sings. Mating takes place after the female deems the bower acceptable and enters it. The female then leaves the bower and builds a nest nearby.

There are 17 species of bower birds in Australia and New Guinea, each with its own style of bower. For example, the satin bower bird paints the walls of its bower with chewed berries or charcoal. Others use moss to create a lawn.

Learning and Behavior

Behaviors are influenced by genes, but to what degree can behaviors be modified by experience? The development of behaviors through experience is called **learning.** Learning can influence the expression of innate behavior and the expression of behaviors that are not innate. One simple kind of learning is habituation. In habituation, an animal learns to ignore a frequent, harmless stimulus. For example, birds may at first stay away from a garden that has a new scarecrow. But if the position of the scarecrow is not changed on a regular basis, the birds learn to ignore it and go into the garden unafraid.

Study TIP

Recognizing Differences
To understand the difference between classical and operant conditioning, think about how the behavior is learned. In classical conditioning, the stimulus has no relationship to an activity. In operant conditioning, there is a direct relationship.

Classical Conditioning

A more complex type of learning is **conditioning,** or learning by association. One of the most famous studies of conditioning was Russian psychologist Ivan Pavlov's work with dogs, carried out in the late 1890s and early 1900s. When Pavlov presented meat powder (a stimulus) to a hungry dog, the dog salivated—an innate response to food. At the same time the dog received the meat powder, Pavlov also presented the dog with a second, unrelated stimulus—a ringing bell. After repeated trials, the dog learned to associate the ringing bell with the meat powder and would salivate in response to the bell alone. The dog became conditioned to associate the ringing of the bell with a reward (meat powder). This type of conditioning, in which an animal comes to associate an unrelated response with a stimulus, is called *classical conditioning*.

internet connect
www.scilinks.org
Topic: Conditioning
Keyword: HX4050
SCILINKS. Maintained by the National Science Teachers Association

Trial-and-Error Learning

Animals also learn by trial-and-error that performing a certain action will result in a reward or a punishment. For example, a dog may learn to avoid a particular cat after being scratched on the nose once or twice. When trial-and-error learning occurs under highly controlled conditions, it is called *operant conditioning*.

Operant conditioning was demonstrated in another famous set of experiments conducted by the American psychologist B. F. Skinner. Skinner studied learning in rats by placing them in a "Skinner box," illustrated in **Figure 5.** Once inside, the rat would explore the box. Occasionally, it would accidentally press a lever, and a pellet of food would appear. At first, the rat would ignore the lever and continue to move about, but it soon learned to press the lever to obtain food. This sort of trial-and-error learning is of major importance to most vertebrates, and it influences many behaviors essential to survival, such as searching for food.

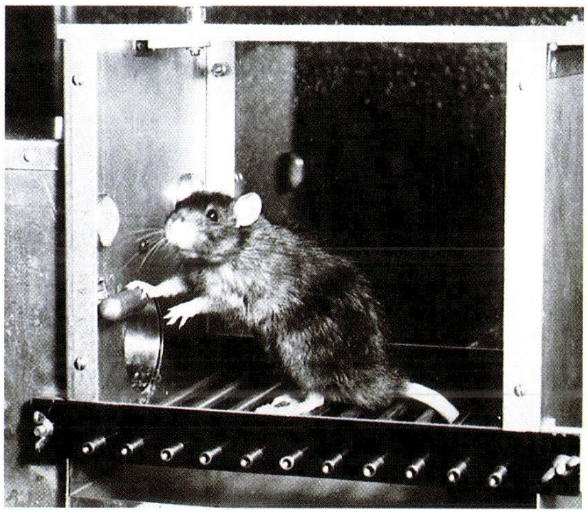

Figure 5 Skinner box. When placed in a Skinner box, this rat learned by trial-and-error to push a lever to receive a reward of food.

827

Teaching Tip — GENERAL

Pavlov's Experiments Pavlov's dogs initially would salivate only in response to the meat powder stimulus. Salivation is an unconditioned response, meaning it occurs automatically and does not have to be learned. Remind students that they show a similar behavior when they sense food. Ask students to speculate about what the function of this response might be. (Salivation prepares the mouth to receive food by lubricating the mouth and increasing the concentration of a digestive enzyme found in saliva.) LS Intrapersonal

Group Activity — GENERAL

Learning in Pets Divide the class into two-person teams, and have each team design an experiment to test the ability of a pet to learn a new behavior. Suggest some possibilities, such as teaching a dog a new behavior or having a mouse perform some sort of activity to obtain food. Encourage students to test different conditions on their own time to determine if one method is more successful. For example, a dog that performs a certain behavioral response can be rewarded with food. In a second conditioning experiment, the same dog can be rewarded with petting when it exhibits the desired response. Students can then correlate the learning time with the type of reward. If students plan to carry out their experiments, be sure they treat all animals in a humane manner. Guidelines for working with vertebrates are available from the National Association of Biology Teachers. Have students write a report describing their procedure and the results of their experiments. LS Interpersonal Co-op Learning

Trends in Psychology

Think Positively In a clinical study, researchers found that attitude and mindset really can affect a person's overall health. Researchers Martin Seligman and Gregory Buchanan, both of the University of Pennsylvania, studied the behavior of university freshman, including coping skills and levels of depression experienced by the students.

Incoming freshman first took a survey to identify their general attitudes and ability to cope. Students who were identified as the most pessimistic by this survey participated in the study. The study included a workshop group and control group. In the workshop group, students learned how to dispute their own negative thinking as well as social skills for averting depression.

Results of the study show that after an 18-month follow-up, 22 percent of the students who participated in the workshop experienced moderate or severe depression, versus 32 percent from the control group.

Teach, *continued*

Recognizing Learned Behavior

QUICK LAB

Skills Acquired
Observing, collecting data, analyzing data, evaluating

Teacher's Notes
Have students work in small groups. Before they begin, have each group formulate a hypothesis about what the sow bugs will do.

Answers to Analysis

1. Answers will vary depending on sow bug behavior.
2. Answers will vary depending on sow bug behavior.
3. Answers will vary, but make sure that students' answers are supported by evidence.
4. If the sow bugs show a preference for the wet paper towel and if no other factors influence the direction they turn, the bugs should exhibit random turning behavior if dry paper toweling is used.

Reasoning

The ability to analyze a problem and think of a possible solution is called **reasoning.** Reasoning involves using experience to develop an insight into how to solve a new problem. The dog shown in **Figure 6** cannot think of a solution to the situation it is in. Humans and some other primates show the ability to reason. For example, in one experiment a chimpanzee was placed in a room with some boxes and a banana hung high overhead. Although it had not been in a similar situation, the chimpanzee stacked up the boxes to reach the banana, a behavior that required reasoning.

Figure 6 An inability to reason. Although the solution seems obvious to humans, this dog is unable to figure out how to reach the food.

There has been much research and debate over the ability of other animals, such as dolphins, some parrots, and even octopuses, to reason. So far, there is no clear evidence that other animals can reason. Many animals can, however, learn complex tasks. Some birds living in cities have learned to remove the foil covering from bottles of nonhomogenized milk to reach the cream at the top. Japanese snow monkeys have learned to float grain on water to separate the grain from sand.

QUICK LAB — *Recognizing Learned Behavior*

SC.H.2.4.2

Sow bugs must keep moist to survive. Follow the procedure below to see if sow bugs can learn to find moisture.

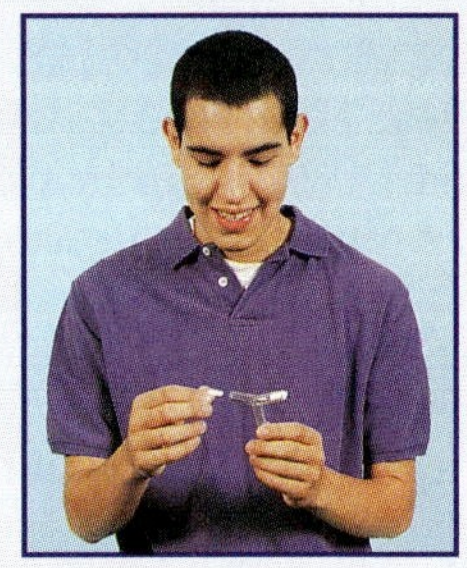

Materials

small wads of paper towel (one moist and one dry), T-maze made of two 6 cm (about 2.25 in.) pieces of 1.25 cm (0.5 in.) clear vinyl tubing, sow bug, blunt probe

Procedure

1. Place the moist paper wad in the open end of the left side of the T, and place the dry paper wad on the right side.
2. Place the sow bug at the bottom of the T. If it does not start to crawl, gently prod it with a blunt probe.
3. Observe what the sow bug does when it reaches the T section. Retrieve the sow bug and perform as many trials as time allows.
4. Keep a record of the results of each trial.
5. Using the same sow bug, repeat this procedure for three days.

Analysis

1. **Summarize** your sow bug's behavior, in writing or on a graph.
2. **Describe** any trend in behavior that you observed.
3. **Determine** if your sow bug modified its behavior through learning, using evidence to support your answer.
4. **Evaluate** the value of performing a final trial in which the T-maze contains two dry paper wads.

SOCIOLOGY CONNECTION

The National Household Education Survey found the following for 3- to 5-year-olds not yet enrolled in kindergarten in 1999:

- 61% whose mother's highest education was less than high school were read to three or more times in the past week by a family member, compared with
- 76% whose mother had a high school diploma or equivalent,
- 85% whose mother had completed vocational education or some college,
- 91% whose mother had a college degree, and
- 93% whose mother had a graduate/professional degree or training.

Genetic and Learned Aspects of Behavior

Do genes determine most behaviors, or do animals usually learn how to behave from experience? Over the last century, this topic has been debated. Some scientists argued that most behaviors are genetically programmed because different individuals in the same species act in the same ways. Other scientists claimed that behaviors are shaped by an animal's experiences. Most biologists who study animal behavior have come to think that animal behavior, particularly the complex behavior of vertebrates, has both genetic and learned components.

internet connect
www.scilinks.org
Topic: Imprinting
Keyword: HX4104
SCINKS. Maintained by the National Science Teachers Association

Imprinting

Learning that can occur only during a specific period early in the life of an animal and cannot be changed once it occurs is called **imprinting.** Imprinting is easily observed in young geese and ducks, which have no innate recognition of their mother. Instead, these birds are genetically programmed to follow the first moving object they see during a short period immediately after they hatch.

There is great survival value to this behavior, as the young must follow their mother as she leads them to water, helps them find food, and keeps them out of danger. However, the young will follow any object they see during this period just as they would their mother—including toy wagons, boxes, and balloons. Once the young birds imprint on an object, they prefer to follow it, even when given the opportunity to follow a member of their own species.

Konrad Lorenz, a Nobel Prize-winning pioneer in the study of animal behavior, observed imprinting when he raised a group of newly hatched goslings (young geese) by hand and found that they imprinted on him. **Figure 7** shows Konrad Lorenz leading his "family" of goslings. The goslings ability to imprint on an object during a sensitive period is not a learned behavior; it is programmed into their genes. However, the process of imprinting is a form of learned behavior. Thus learning determines the final shape of this genetically based behavior.

Figure 7 **Imprinting.** These goslings imprinted on Konrad Lorenz and followed him around just as if he were their mother.

Section 1 Review

1. **Describe** the difference between "how" and "why" questions in regard to animal behavior. SC.H.2.4.2
2. **Summarize** how cub killing by male lions supports the hypothesis that natural selection shapes behavior. SC.G.2.4.3
3. **Distinguish** between and give an example of innate and learned behavior. SC.G.2.4.3
4. **Analyze** the behaviors involved in imprinting. SC.G.2.4.3
5. **Critical Thinking** **Forming Reasoned Opinions** A friend is teaching his dog a new trick in which it is rewarded each time the trick is performed correctly. The friend says his method is called *classical conditioning*. Evaluate your friend's use of this term. SC.H.3.4.3
6. *FCAT Prep* When an Australian frilled lizard flares the folds of skin around its head in response to a threat, the lizard exhibits what characteristic? SC.G.2.4.3
   A imprinting **C** classical conditioning
   B innate behavior **D** trial-and-error learning

829

Close

Reteaching — Basic

Have students select any behavior that is described in this section. Ask them to write a list of "why" and "how" questions that a scientist could ask about that particular behavior. (For example: Why does a gosling follow any object shortly after hatching? How does this behavior change, if at all, over time?)

Quiz — General

1. A behavior occurs in response to a ________ . (stimulus)
2. What does the experiment with hybrid lovebirds demonstrate? (It demonstrates that some behaviors are determined entirely by genes.)
3. When a young child learns how to build a tower of blocks by repeating her efforts until she succeeds, what kind of learning is the child demonstrating? (trial-and-error learning)

Alternative Assessment — General

Have students observe the behavior of a pet or other animal and write a report describing their observations. In their reports, students should describe the behaviors of their subject and the context for each behavior. Students should also explain the possible adaptive value of each behavior they observe.

Answers to Section Review

1. "How" questions are about how a behavior is triggered, controlled, and performed. "Why" questions concern the adaptive value of a behavior.
2. By killing the cubs, the male lion increases his chances of fathering his own cubs, thereby increasing the likelihood that his "cub-killing" genes will be passed on.
3. Innate behavior, such as eating, requires no experience. Learned behavior, such as a dog salivating in response to a ringing bell, is modified by experience.
4. In imprinting, an animal is genetically programmed to learn something during a critical period early in its life. Once learned, the imprinted information cannot be changed.
5. The friend's usage is incorrect. In classical conditioning, the stimulus has no relationship to the activity. This is a case of trial-and-error learning.
6. **A.** Incorrect. Imprinting involves learning during a specific period early in life. **B.** Correct. The lizard does not need to be taught the behavior to exhibit it. **C.** Incorrect. Classical conditioning involves learning to associate a stimulus with an unrelated response. **D.** Incorrect. Trial-and-error learning is a learned behavior that involves repeated attempts to achieve a correct behavior.

Section Review Correlations

SC.G.2.4.3: Items 2, 3, 4, 6
SC.H.2.4.2: Item 1
SC.H.3.4.3: Item 5

Section 2

Focus

Overview

Before beginning this section review with your students the objectives listed in the Student Edition. Students will be able to describe six different categories of animal behavior and some of the influences that shape these behaviors, both genetic and learned. Students will develop an understanding of the nature of communication between animals and what influences communication. Finally, students will explain mate selection in animals.

Bellringer

Ask students to write a sentence describing the eating behavior of a dog and another sentence describing the eating behavior of a cat. Ask them finally to write a sentence that characterizes the differences in eating behaviors of dogs and cats. (Dogs are generalists and eat most any food, while cats are specialists and eat primarily meat.)

Motivate

Discussion/Question — GENERAL

Have the class write the six categories of behavior from **Figure 9** on the board. Then, have students write an example for each type of behavior below each of the six categories. Allow healthy disagreement among students as you lead a discussion of the characteristics of each category.

Section 2 Types of Behavior

Florida Sunshine Standards/FCAT

SC.F.1.4.7 Organisms respond to internal and external stimuli.
SC.H.1.4.1 Investigations are conducted to explore new phenomena, to check on previous results, to test how well a theory predicts, and to compare different theories.

Objectives

- **Discuss** six types of animal behavior. SC.H.1.4.1
- **Discuss** how animals use signals. SC.F.1.4.7
- **Summarize** how sexual selection can influence evolution. SC.G.2.4.3

Key Terms

sexual selection

Categories of Animal Behavior

As you sit on a park bench, a pigeon approaches, and you soon realize that it expects food. You toss out a bit of your sandwich and the pigeon eats it, immediately looking up for more. This urban pigeon has been conditioned to seek food from people. Has its human supplier been conditioned too?

Behavior is an animal's most immediate way of dealing with its environment. Because the environment is complex and can change rapidly, most animals have many different kinds of behavior, each suited to a particular situation. For instance, a squirrel may perform one kind of behavior when it finds a nut on the ground—it digs a hole. It performs a completely different behavior when a snake approaches—it runs for shelter—because digging a hole would not help it escape from the snake.

Like the squirrel, the musk oxen of the Arctic display many different types of behavior, some of them cooperative. When predators, usually wolves, appear, the adult musk oxen form a defensive circle around their young, as shown in **Figure 8.** The tight circle and the danger of injury from the adult's horns and hooves usually prevent a successful attack. While musk oxen can run, running would not protect the herd in the way that their group defense does.

Biologists have classified the behaviors animals perform into several broad categories. **Figure 9** shows some of these categories and gives an example of each.

Figure 8 Musk oxen. When threatened, adult musk oxen form a defensive circle around their young.

Chapter Resource File

- Directed Reading BASIC
- Active Reading GENERAL
- Data Sheet for Quick Lab GENERAL

Transparencies

TR Bellringer
TR 145 Animal Behavior

One-Stop Planner CD-ROM

- Reading Organizers BASIC
- Reading Strategies BASIC
- Supplemental Reading Guide *Journey to the Ants* ADVANCED
- Supplemental Reading Guide *Through a Window* ADVANCED

Section Correlations

SC.F.1.4.7
SC.G.2.4.3
SC.H.1.4.1

Influences on Behavior

While many of the behaviors illustrated in Figure 9 may seem different from one another, they all tend to favor survival and reproductive success. To gain a better idea of the nature of animal behavior, let's examine one of these behaviors—foraging behavior—in detail. *Foraging* is finding and getting food.

Animals can be divided into two broad groups based on the range of food items each group consumes. Specialists feed primarily or exclusively on one kind of food. Some species of ants, for example, eat only spider eggs. Generalists, in contrast, consume many different kinds of food. For example, some insects eat the leaves of a wide variety of plants. Generalists are typically less efficient than specialists at

Figure 9 Animal behavior. Although their methods differ, all animals engage in at least some of the behaviors shown below.

Foraging behavior

Locate, obtain, and consume food

A raccoon searches along streams and ponds for fish, frogs, crayfish, and small rodents. It also hunts for insects and fruit in woodlands.

Migratory behavior

Move to a more suitable environment as seasons change

Monarch butterflies migrate thousands of kilometers, from the United States to central Mexico.

Defensive behavior

Protect from predators

When threatened, a hognose snake turns onto its back and plays dead.

Territorial behavior

Protect a resource for exclusive use

Like many wild cats, this young cheetah claws on trees, leaving a scent that marks its territory.

Courtship behavior

Attract a mate

During its breeding season, the male stickleback fish develops a bright coloring and builds an elaborate nest to attract a female.

Parental care

Ensure survival of young

This robin is feeding an insect to its offspring.

Teach

Group Activity — GENERAL

Animal Behavior Copy the Graphic Organizer at the bottom of this page onto the board or an overhead, but do not copy the entries in parentheses. Have volunteers come up one at a time and fill in the blanks. Students' answers may vary from the ones given. LS Logical Co-op Learning

Teaching Tip

Migratory Behavior Tell students that migration may bring a bird to a much more favorable environment where resources, such as food and nesting sites, are available, where predators are rare, or where the climate is moderate. Costs of migration include the energy required to fly long distances and the possibility of dying during the long migration.

READING SKILL BUILDER — BASIC

Reading Organizer Have students make a table with the following headings: Parental Care, Courtship Behavior, Defensive Behavior, Foraging Behavior, Migratory Behavior, and Territorial Behavior. As students read this section, they should assign each behavior mentioned in the text to one of the categories in the table. Inform students that some behaviors will not fit into any of the categories, so it will not be possible to classify all the behaviors described in the section. LS Verbal

Graphic Organizer

Use this graphic organizer with **Group Activity** on this page.

Type of behavior	Function of behavior	Example of behavior
(Parental care)	Ensuring survival of young	(Bird feeds worm to nestling.)
(Courtship behavior)	(Attracting a mate)	Male frogs croak to attract a mate.
(Defensive behavior)	Protection from predators	(Hognose snake plays dead.)
(Foraging behavior)	(Locating, obtaining, and consuming food)	A raccoon searches along streams and ponds for rodents.
(Migratory behavior)	(Moving to a more suitable environment)	Birds fly south for the winter.
Territorial behavior	(Protecting a resource for exclusive use)	(An aquarium fish forces all other fish to one end of the tank.)

Teach, *continued*

BIOWatch

Tracking Sea Turtles by Satellite

Teaching Strategy
Ask students to use Internet and library resources to research animal tracking. Each student should identify an animal that has been tracked by electronic means (not necessarily by satellite—some animals are tracked by other means). Have each student write a brief report that tells why the animal is tracked, what data are collected, and how scientists use this information.

Discussion
Ask students: By using the loggerhead tracking data on the Internet, how could students help wildlife researchers conserve the threatened turtles? (Accept any reasonable response. Sample answer: A group of students could "adopt" a specific turtle. The students could track their turtle's movements and periodically send reports to scientists. This might allow researchers to track more turtles than they could otherwise.)

internet connect
www.scilinks.org
Topic: Animal Communication
Keyword: HX4009
SCILINKS. Maintained by the National Science Teachers Association

feeding on any one type of food. However, generalists have the advantage of being able to collect more than one kind of food. Which approach to foraging is better? When one kind of food source is plentiful, specialists forage more successfully. But when food sources are diverse and no particular one of them is more common, generalists find more to eat.

For predators, food typically comes in a variety of sizes. Larger food items contain more energy. But larger items are harder to capture, and they are usually less abundant.

Foraging thus involves a trade-off between a food's energy content and its availability. Animals tend to feed on prey that maximize their energy intake per unit of foraging time. This approach is called optimal foraging. Natural selection has favored the evolution of foraging behaviors of this sort. Sometimes, however, animals will consume foods that are low sources of energy. Often this is because those foods supply an important nutrient. The location of the food source may also allow the consumer to avoid being captured by some other predator.

FLORIDA BIOWatch

Tracking Sea Turtles by Satellite

The long-distance movements of sea turtles, whales, birds, butterflies, bats, and other animals are among the most fascinating examples of animal behavior. But in some cases, little is known about these movements. To learn more, biologists have made use of satellite technology.

How Satellite Tracking Works

In March 2003, Florida researchers captured eight loggerhead sea turtles in Florida Bay and fitted them with satellite transmitters, as shown in the photo at right. The transmitter consists of electronic circuitry, a battery, and an antenna. For use in small animals, these parts must weigh as little as possible. Satellites orbiting above Earth pick up the radio signals produced by the transmitter and relay the signals to a central computer on land. The computer analyzes the information and calculates the animal's location. From all over the world, biologists can connect to this computer through the Internet. They can access data on the movements of specific animals, and plot those movements on a map.

New Kinds of Data

Satellite tracking allows biologists to track animals that would otherwise be hard to follow or observe. For example, most of what was previously known about sea turtles was based on observing their nesting behavior on beaches. But loggerheads spend 90 percent of their time at sea. Using satellite tracking, biologists have learned that some loggerheads look for food thousands of miles from the beaches where they nest! The new information will help researchers better understand and protect loggerheads and other animals near Florida Bay.

internet connect
www.scilinks.org
Topic: Tracking Animals by Satellite
Keyword: HX4177
SCILINKS. Maintained by the National Science Teachers Association

832

Integrating Physics and Chemistry

Remind students that satellites orbiting Earth are not in a gravity-free environment even though they are in outer space. Actually, they are free-falling around Earth due to the constant force-to-mass ratio, which accelerates them toward Earth at 9.8 m/s^2. However, when their forward movement or velocity is combined with their free-fall acceleration toward Earth, a constant orbital trajectory is produced.

Figure 10 Animal communication. A dog's play bow means "I'm available to play," while a snarl is a signal of aggression and sometimes fear.

Communication

You approach an unfamiliar dog and it begins to bark. You know that if you go closer, the dog might bite you, so you stop and talk to it. The dog continues to bark but not so aggressively, and it begins to wag its tail. You and the dog have each responded to a signal given by the other.

A signal can be a sound, posture, movement, color, scent, or facial expression. These signals are sent and received through all of the senses familiar to us—sight, hearing, smell, touch, and taste. Animals use signals to influence the behavior of other animals. Because they face a variety of social situations in which communication is needed, animals usually have several different signals, each suited to a different situation, as shown in **Figure 10.**

Natural selection has shaped animal signals so that they reach the intended receiver efficiently and stimulate a response. To be transmitted efficiently, a signal must be able to travel through the environment from sender to receiver. A signal must also be recognizable to the receiver, or it won't have any effect on behavior. Consider the loud mating call emitted by the male túngara frog. The call carries a long distance, reaching even far-off females. At night, when túngara frogs are active, a loud call is the best way to communicate. Visual signals, such as colors and movements, would be noticeable from only a short distance away and would not be nearly as effective at attracting a mate.

Figure 11 Primate communication. Vervet monkeys have distinct calls that identify different kinds of predators.

Primate Communication

Among animals, vocal communication may be most developed in the primates. Some primates, such as those shown in **Figure 11,** have a "vocabulary" of calls that allows individuals to communicate the identity of specific predators, such as eagles, leopards, and snakes. Chimpanzees and gorillas can be taught by humans to recognize and use a large number of symbols to communicate abstract concepts. Chimpanzees and gorillas cannot talk, however, because they are physically unable to produce the

REAL WORLD CONNECTION

Isolated in worlds of their own, people with autism appear indifferent and remote and are unable to form emotional bonds with others. Although people with this baffling brain disorder can display a wide range of symptoms and disabilities, many are incapable of understanding other people's thoughts, feelings, and needs. Often, language and intelligence fail to develop fully, making communication and social relationships difficult. Many people with autism engage in repetitive activities, like rocking or banging their heads, or rigidly following familiar patterns in their everyday routines. Some are painfully sensitive to sound, touch, sight, or smell. Children with autism do not follow the typical patterns of child development. In some children, hints of future problems may be apparent from birth. In most cases, the problems become more noticeable as the child slips farther behind other children the same age. With appropriate support, many people with autism can be trained to do meaningful work and participate in the life of the community.

Demonstration — GENERAL

Animal behaviorists (ethologists) are scientists who study how animals behave and the adaptive significance of behavior. These scientists consider all the possible influences on animal behavior, such as heredity, physiology, environment, food, weather, and the behavior of other animals. Invite an animal behaviorist to the class to speak. Ask the speaker to include the following in his or her talk: recommended high school courses, college degree needed, working conditions, employment opportunities, and salary ranges. Have each student prepare several questions to ask the speaker. Have students write a short paper describing what they have learned.
LS Verbal

Group Activity — ADVANCED

Tracking Whales Teachers and students can track animal movements throughout the world on the Internet using WhaleNet. This educational Web site provides updated satellite data on blue whales, harbor porpoises, and loggerhead turtles, as well as archived satellite data on right whales, sperm whales, elephant seals, gray seals, harbor seals, and hooded seals. Another Web site is maintained by the Sea Turtle Survival League. Both sites show the dates and tracking locations of animals on a world or regional map, which is updated periodically. Have students make maps and plot the movements of individual animals. Keep the maps on the wall for other students to observe. Some of the transmitter batteries function for only 2 to 3 months, so it is a good idea for your students to choose several animals to track.
LS Visual Co-op Learning

Teach, *continued*

Real Life GENERAL

Answer

Answers will vary depending on the foreign language of the interviewee. **LS Interpersonal**

QUICK LAB *Observing Territorial Behavior in Crickets*

Skills Acquired
Observing, collecting data, organizing data, analyzing

Teacher's Notes
Tell students to remain quiet and as still as possible so the crickets are not affected by any noise or movements. Have students place the objects around the crickets so the crickets are about the same distance from each object.

Answers to Analysis

1. Answers will vary depending on the crickets' behavior.
2. Answers will vary depending on the crickets' behavior.
3. Answers will vary. Students should not attribute a sense of purpose to the crickets' actions.
4. Answers will vary. Hypotheses should describe the adaptive value of each behavior.

Real Life

Just say /r/.
If you are having trouble mastering the throaty sound of a French /r/, just remember that a French student struggles with the English /th/ sound, and will probably pronounce *the* as *zee.*

Finding Information
Interview a native speaker of a foreign language. Ask him or her to identify three sounds in his or her language that are difficult for English speakers to master.

sounds of speech. Some researchers believe that chimpanzees can combine symbols they have learned in meaningful ways. But chimpanzees cannot rearrange symbols to form a new sentence with a different meaning. That is a very complex task, which only humans can perform.

In humans, language develops at a very early age. Infants begin to learn language by trial-and-error during the "babbling baby" phase of childhood, at about six months of age. At first, infants all over the world babble the same consonant sounds, including sounds they have never heard. Soon, however, the infants pick out the sounds used by the people around them and repeat only those sounds. The other sounds begin to drop away and are forgotten. Children quickly and effortlessly learn a vocabulary of thousands of words, a feat no chimpanzee can do. This ability to learn language rapidly seems to be genetically influenced. But language is not the only form of human communication. Evidence suggests that odors and nonverbal signals (body language) may also be important.

QUICK LAB *Observing Territorial Behavior in Crickets*

SC.H.2.4.2

The chirp of a male cricket attracts females and warns other males to stay away from his territory. You can study chirping behavior by observing crickets in an aquarium.

Materials

5 male crickets, each marked with a different color; 5 unmarked female crickets; covered aquarium; slice of apple and of potato; small plastic jar; 5 cm (2 in.) square of cardboard

DATA TABLE

Cricket Behavior

Cricket	Apple	Potato	Jar	Tent	Female
Blue					
Yellow					
Red					
Green					
White					

Procedure

1. Place the crickets and food in an aquarium. Make two shelters by turning the plastic jar on its side and by folding the cardboard in half to form a tent-like structure.
2. Make a chart like the one above to record the behavior of the male crickets.
3. Observe the crickets for 10 minutes. Among the males, look for territorial (aggressive) behaviors—chirping, stroking others with antennae, pushing others away, etc.
4. For each observation of aggressive behavior, record the color of the aggressive male and where the behavior occurred—for example, next to the jar or the tent.
5. For each cricket, tally the number of times aggressive behavior was observed. Make a list that ranks each cricket, placing the cricket with the highest tally on top.
6. Then tally the numbers for where the behaviors occurred. Rank the locations.

Analysis

1. **Critical Thinking Analyzing Data** Were any crickets more aggressive than the others? Give evidence to support your answer.
2. **Describe** the circumstances in which most aggressive behavior occurred.
3. **Propose** a reason to explain your answer to item 2.
4. **Critical Thinking Forming Hypotheses** For each aggressive behavior you observed, form a hypothesis that explains its function.

MISCONCEPTION ALERT

Ability to Reason Students often believe that an animal understands why it behaves in a certain way. We don't know what animals are thinking when they engage in their behaviors. The scientific evidence that has been collected, however, indicates that nonhuman animals don't have the same kind of reasoning abilities that humans have.

INCLUSION Strategies

- *Gifted and Talented*
- *Attention Deficit Disorder*

Students may wish to study an animal that lives in or around their home or school. Have students create a list of how and why questions about the animal's behavior that they would like to understand. Have students write their hypothetical answers and then research to find scientific answers to their questions. The findings may be reported to the class or written as a report.

Reproductive Behavior

When ready to mate, animals produce signals to communicate with potential partners. Each species usually produces a unique courtship signal that ensures that individuals do not mate with individuals of another species. For example, the flash patterns of fireflies differ between species. A female firefly recognizes a male of her own species by the pattern of his flashes, and she will ignore any male that has a different flash pattern. The chemical produced by a female silk moth attracts only males of her own species. Many species of insects, amphibians, and birds produce unique sounds or songs to attract mates. A white-crowned sparrow will respond to the song of another white-crowned sparrow, but it will totally ignore the song of an English sparrow.

During the breeding season, animals make several important "decisions" concerning mating and parenting. These include how many mates to have and how much time and energy to devote to rearing offspring. These decisions are all aspects of an animal's reproductive strategy, a set of behaviors that have evolved and that maximize reproductive success.

Mate Choice

Males and females usually differ in their reproductive strategies. In many animals, females do not mate with the first male they encounter. Instead, the female seems to evaluate the male before she decides whether to mate. This behavior, called *mate choice*, has been observed in many invertebrate and vertebrate species. Female túngara frogs, for instance, have been observed "shopping around" among calling males. A female will sit near a male and listen to his call for several minutes, then move on to another male and listen to his call. She may evaluate several males before choosing one with which to mate.

Sexual Selection What characteristics do animals use in choosing a mate? When Charles Darwin considered this question more than a century ago, he made an important discovery about evolution. Darwin noticed that males often have extreme characteristics that they use in their courtship displays. Take, for example, widowbirds, shown in **Figure 12.** During the breeding season, the male widowbird grows an extremely long tail, up to five times longer than the female's. How did such differences between the sexes evolve? The long tail of the male widowbird cannot be essential for survival, since the female bird survives quite well with a much shorter tail.

Figure 12 Widowbirds. During breeding season, the tail of the male widowbird, shown above, grows to more than three times the length of his body. At other times of the year, it is similar in length to that of the female, which is the bird on the left in the inset photograph.

835

Teaching Tip

Anthropomorphism Tell students that attributing a human trait to a nonhuman is known as anthropomorphism. People often attribute human emotions to animals. For example, someone might say that a mother bird is sad because one of her chicks fell out of the nest. While the female bird may engage in behavior that shows it understands and is concerned about the situation, we can't say that the bird feels sadness as humans do.

SKILL BUILDER

Vocabulary The study of behavior is known as ethology, from the Greek word *ethos*, meaning "character" or "custom."

READING SKILL BUILDER — Basic

Discussion Tell students that two mechanisms are involved in sexual selection. The first is competition between individuals of the same sex, usually males, for access to females. This form of selection has resulted in the evolution of weaponry such as antlers and horns, and increases in male size that provide some individuals with an advantage over others for access to females. The second mechanism is mate choice by individuals of the selecting sex, usually females. This form of selection has resulted in the evolution of traits such as the showy tail of the peacock, elaborate coloration in birds and fishes and many kinds of vocalizations. As with many other animals, sexual size dimorphism has evolved in humans, probably due to male-male competition. Males generally have more developed musculature than females due to the effects of testosterone at puberty. Lead the class in a discussion of examples of both forms of sexual selection in animals, including humans.

Attention Grabber

Depression In the United States, the greatest health problem that affects behavior is depression. It is known as "the common cold of mental health problems." Students should be aware that it is not unusual for a person to be saddened by divorce, the death of a loved one, or other major losses. But when the sadness does not go away, it becomes what authorities call "clinical depression." More than 15 million Americans suffer clinical depression at any given moment. The dark cloud of depression has an even darker lining — suicide. Suicide is now a leading cause of death among teens and young adults. Fortunately, depression is highly treatable when diagnosed accurately and attacked with some combination of antidepressant medication and the many non-drug therapies. John McIntyre, M.D., a past president of the American Psychiatric Association, says, "There is virtually no one who cannot be helped." **LS Intrapersonal**

Close

Reteaching — Basic

Have students go through this section and turn each subhead into a question on a sheet of paper. Then ask them to write an answer to each question. Ask volunteers to share their questions and answers with the class.

Quiz — General

1. What is the selective advantage of territorial behavior? (It preserves natural resources for use by an individual or his group.)
2. Do animals that don't vocalize engage in communication? How? (Animals that don't communicate using vocalizations may communicate by posture, movement, color, scent, or facial expression.)
3. In most animal species, which sex chooses the mate? (female)

Alternative Assessment — General

Have students review the types of behavior shown in **Figure 9.** Then have them locate every example of animal behavior mentioned in the chapter. If a behavior fits in one of the categories listed in the figure, have them place it there as an example. (For example: defensive behavior—musk oxen form a defensive circle around their young.) If students identify a behavior that does not fit into a category, have them make a new category for it. Students should justify the need for any category they create.

Figure 13 Gorillas. On average, male gorillas are 50 percent larger than females. Full grown males can weigh as much as 200 kg (440 lb).

Darwin recognized that extreme traits, such as the male widowbird's tail, could have evolved if they helped males attract or acquire mates. He proposed the mechanism of sexual selection to account for such traits. **Sexual selection** is selection in which a mate is chosen based on a certain trait or set of traits. Thus, traits that increase the ability of individuals to attract or acquire mates appear with increased frequency. Even traits that have a negative effect on survival can evolve in this way, provided that their benefits to reproduction are high enough.

It is usually females that select mates based on their physical traits. One explanation for this tendency is that reproduction has a greater metabolic cost and requires a greater investment from females. For example, eggs are less numerous and much larger than sperm. (Humans' eggs are about 195,000 times larger!) In mammals, females are also responsible for gestation and milk production. Males may show mate choice as well if their parental involvement is high. This happens in crickets, where the sperm packet a male deposits makes up 30 percent of his body weight, contributes nutrition for the female, and helps her develop eggs.

Competition by Males What kinds of traits help males acquire mates? Because male animals usually compete among themselves for the chance to mate with a female, selection has often favored traits that make males more intimidating or better at combat. For example, in many animal species, such as the gorillas shown in **Figure 13,** the male is much larger than the female, and the largest male will have the most opportunities to mate. Other examples of traits related to male competition are antlers in deer and moose, horns in bighorn sheep, and manes in lions. The extreme traits of some male animals, such as large size and tusks in walrus, often permit males to assess each other. Males that are not physically well matched rarely engage in serious fights. Thus, some extreme traits may reduce conflict among males rather than increase it.

Competition among males can also take subtle forms. In some species, a male can gain a reproductive advantage over other males by interfering with their reproduction. For instance, in some species of worms, butterflies, and snakes, the male seals the female's reproductive tract after mating, thus preventing other males from mating with her.

Section 2 Review

1. **Describe** the function of six different animal behaviors. SC.H.1.4.1
2. **Summarize** in words and with examples the ways in which animals use signals. SC.F.1.4.7
3. **Discuss** how selection can account for the extreme traits found in the males of some species. SC.G.2.4.3
4. *FCAT Prep* What specific signal produced by the male firefly causes a female to recognize males of her own species?
   A chemical
   B flash pattern
   C sound
   D type of nest
   SC.F.1.4.7

836

Answers to Section Review

1. *parental care*—ensures the survival of offspring; *courtship*—attracts mate; *defensive*—predator protection; *foraging*—food location; *migratory*—movement to better environment; *territorial*—protects a resource
2. For example: *sound*—dog barking indicating alarm; *posture*—person with hands on hips indicating irritation; *movement*—threat display indicating impending attack; *color*—bright plumage indicating quality of genes; *scent*—dog marking territory; *facial expression*—frowning indicating displeasure
3. The most intimidating male (for example, the deer with the largest antlers) has a better chance of becoming the dominant male. He will mate with more females and pass on more of his genes than subordinate males will.
4. **A.** Incorrect. The signal the male produces is visual, not olfactory. **B.** Correct. Male fireflies have species-specific flash patterns that the females recognize. **C.** Incorrect. The signal the male produces is visual, not auditory. **D.** Incorrect. The signal the male produces involves flashes of light, not production of a nest.

Section Review Correlations

SC.F.1.4.7: Items 2, 4
SC.G.2.4.3: Item 3
SC.H.1.4.1: Item 1

Study ZONE

Key Concepts

1 Evolution of Behavior

- A behavior is an action or a series of actions performed in response to a stimulus.
- Natural selection favors behavioral traits that increase the likelihood of an individual's survival and reproduction.
- Genetically programmed behaviors are called innate behaviors, instincts, or fixed action pattern behaviors, and there is little or no variation in how they are performed.
- Learning is the modification of behavior by experience. Learning may occur by association with an unrelated stimulus (classical conditioning) or by trial-and-error (one type of which is operant conditioning).
- Reasoning is the ability to think of a possible solution to a problem.
- Many behaviors, especially complex behaviors, have both genetic and learned aspects.
- Learning determines the final shape of many genetically based behaviors, such as imprinting.

2 Types of Behavior

- Animal behaviors fall into several broad categories, which include parental care, courtship behavior, defensive behavior, foraging behavior, migratory behavior, and territorial behavior.
- Most animals use signals, often vocal or visual, to communicate with one another.
- Primates are unique among animals in using symbols to communicate.
- The human ability to learn language rapidly during childhood seems to be genetically influenced.
- By the mechanism of sexual selection, traits that increase the ability of an individual to attract a mate appear with increased frequency.

Key Terms

Section 1

behavior (824)
innate behavior (826)
fixed action pattern behavior (826)
learning (827)
conditioning (827)
reasoning (828)
imprinting (829)

Section 2

sexual selection (836)

Alternative Assessment — ADVANCED

Insect societies, like those of humans, exhibit complex behavioral patterns. Have students use library and online resources to obtain information about a behavioral pattern exhibited by a particular social insect. Have them prepare a brief written report and share their findings with the class.

Chapter Resource File

- Science Skills Worksheet GENERAL
- Critical Thinking Worksheet ADVANCED
- Test Prep Pretest GENERAL
- Chapter Test GENERAL

Answer to Concept Map

The following is one of several possible answers to Performance Zone item 15.

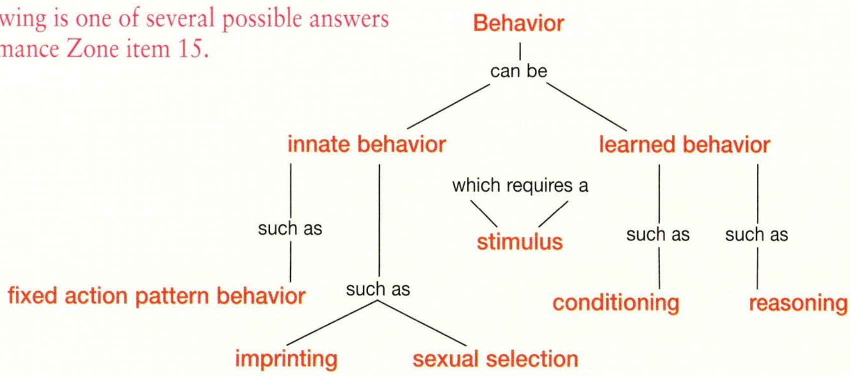

CHAPTER 36 ZONE

ANSWERS

Understanding Key Ideas

1. c
2. c
3. c
4. a
5. d
6. d
7. c
8. Sample answer: Electronic tracking could supply continuous and specific information on an animal's movements; trapping only pinpoints one location to which the animal has traveled. However, both methods might pose risks to the animal.
9. In natural selection, traits that help an animal survive to reproductive age tend to be passed on to the next generation. Harmful traits will be selected against. Like physical traits, many behaviors are important for survival and reproduction. For example, it is just as important to survival for an animal to locate, obtain, and consume food (foraging behavior) as it is for it to be able to run swiftly from predators. Animals that are poor at foraging may not live to reproduce.
10. One possible answer to the concept map is found at the bottom of the Study Zone page.

Critical Thinking

11. Answers may vary. The flaw in this argument is that the variable of learning ("nurture") has not been eliminated.
12. Answers may vary. In some bird species, females prefer males that are more brightly colored or extravagantly plumed. Since the female chooses her mate, males with these traits are more likely to be chosen. As a result, these traits appear with greater frequency in the male population.

Alternative Assessment

13. Answers will vary. All students in the group should make a significant contribution to the presentation.
14. Answers will vary. Several ideas may explain the "how" of this phenomenon. For example, the whales may follow prey into shallow waters. One possible "why" answer is that whales that do this (without becoming stranded) are more likely to find enough food to survive and reproduce.

Chapter Review Correlations

SC.F.1.4.7: Items 3, 10, 13
SC.F.2.4.1: Item 7
SC.G.1.4.2: Item 6
SC.G.2.4.3: Items 1, 2, 4, 5, 9, 12
SC.H.1.4.4: Item 11
SC.H.3.4.2: Item 8
SC.H.3.4.3: Item 14

Assignment Guide

Section	Questions
1	1, 2, 3, 9, 10, 11, 13, 14
2	4, 5, 6, 7, 8, 10, 12

CHAPTER REVIEW ZONE

Understanding Key Ideas

1. A male lion that takes over a pride may kill all the young cubs in order to ensure the survival and reproduction of
   a. the females.
   b. the pride.
   c. himself.
   d. his siblings.
   SC.G.2.4.3
2. The nest-building behavior of Fischer's lovebirds is SC.G.2.4.3
   a. learned.
   b. a result of operant conditioning.
   c. innate.
   d. gradually learned.
3. Which of the following best represents classical conditioning?
   a. the feeding behavior shown by rats in Skinner boxes
   b. a male túngara frog calling to a female
   c. a dog salivating at the sound of a bell
   d. a primate giving a warning signal to members of its troop
   SC.F.1.4.7
4. Parental care is performed in order to
   a. ensure survival of young.
   b. protect parents from predators.
   c. protect a resource for exclusive use.
   d. locate, obtain, and consume food.
   SC.G.2.4.3
5. The purpose of foraging behavior is to
   a. ensure survival of young.
   b. protect individuals from predators.
   c. protect a resource for exclusive use.
   d. locate, obtain, and consume food.
   SC.G.2.4.3
6. Which of the following behaviors is least likely to be associated with food resources?
   a. territorial c. foraging
   b. migratory d. defensive SC.G.1.4.2
7. Which of the following best demonstrates the mechanism of sexual selection?
   a. the bright colors of a poisonous frog
   b. the ability of a lion to run fast
   c. the long tail of a male widowbird
   d. the web-spinning behavior of a spider
   SC.F.2.4.1
8. **BIOWatch** What might be the advantages of tracking animals electronically versus getting data by trapping animals? SC.H.3.4.2
9. Summarize the mechanism of natural selection, and explain why behaviors are just as important to survival and reproduction as physical features are. (**Hint:** See Chapter 13, Section 1.) SC.G.2.4.3
10. **Concept Mapping** Construct a concept map describing animal behavior. Use the following terms in your map: *behavior, stimulus, innate behavior, fixed action pattern behavior, learned behavior, conditioning, reasoning, imprinting,* and *sexual selection.* SC.F.1.4.7

Critical Thinking Skills

11. **Evaluating an Argument** "A child's behavior closely resembles the behavior of its parents. Therefore, most human behaviors are genetically controlled." Explain the logical flaw in this argument. SC.H.1.4.4
12. **Recognizing Relationships** In many bird species, the male is more brightly colored and heavily plumed than the female. Explain this in terms of sexual selection and evolution. SC.G.2.4.3

Alternative Assessment

13. **Finding and Communicating Information** Use the media center or Internet resources to learn more about imprinting. Work with a small group to find examples of imprinting, and develop a class presentation on the subject. SC.F.1.4.7
14. **Summarizing Information** At times, some species of whales seem to deliberately swim into waters that are shallow. Often they end up stranded on the beach, where they cannot survive. Using the Internet or other resources, search for information you can use to answer the "how" and "why" questions about this phenomenon. Present your findings in a report to your class. SC.H.3.4.3

838

FCAT Prep

For questions 1–6, write your answers on a separate sheet of paper.

1. What is learning by association? SC.F.1.4.7
   A. assuming
   B. conditioning
   C. imprinting
   D. reasoning

2. The orb spider builds her web in exactly the same way every time. What is this an example of? SC.G.1.4.1
   F. learned behavior
   G. random behavior
   H. inherited behavior
   I. fixed action pattern behavior

3. What is the term for the ability to analyze a problem and think of a possible solution? SC.F.1.4.7
   A. assuming
   B. conditioning
   C. imprinting
   D. reasoning

4. How can a behavior that results in the death of one offspring evolve by natural selection? SC.G.2.4.3
   F. The behavior has no adaptive value.
   G. The behavior protects all the young in a group.
   H. The behavior leads to the death of all the young in a group.
   I. The behavior helps more of one individual's offspring to survive.

5. READ INQUIRE EXPLAIN Captured male widowbirds whose tails have been cut mate with half as many females as wild males with longer tails.

   Design an experiment to determine whether female widowbirds prefer males with longer tails. SC.H.2.4.2

6. READ INQUIRE EXPLAIN Imprinting is learning that can occur only during a specific period early in the life of an animal and cannot be changed once it occurs. Many birds, such as ducks and geese, imprint on their mother shortly after they hatch. Goslings can even imprint on humans. When they do, they will follow the human around in single file just as if he or she were their mother.

   Part A Analyze how imprinting behavior includes both genetics and learned factors. SC.F.1.4.7

   Part B Why might natural selection have favored the evolution of imprinting behavior? SC.G.2.4.3

Test TIP

When using experimental data to answer a question, determine the constants, variables, and control before answering the question.

839

FCAT Doctor

Question 4 Answer I is the correct choice. A behavior that involves killing the offspring of another individual can help more of another individual's offspring survive. An example can be found in male lions, which kill the offspring of other males. Answer F is incorrect because natural selection favors behaviors with adaptive value. Answer G is incorrect because all the young in a group may not be protected by the death of one. Answer H is incorrect because natural selection does not favor the death of all young, as the species would become extinct.

Question 5 Capturing and handling may have affected the behavior of the widowbirds. Including a control group of captured males who have normal tails will help to validate the data.

Question 6

Part A The ability to imprint during a sensitive period early in the life of an animal is programmed into the animal's genes. However, the process of imprinting is a learned behavior.

Part B The imprinting behavior is how goslings learn to recognize and follow their mother.

Answers

1. B
2. I
3. D
4. I
5. Answers may vary, but the experiment should include a control group of captured males whose tails are of normal length.

6A. The animal that imprints is genetically preprogrammed to imprint on the first moving object it sees. The individual experience that leads to this imprinting causes the specific learned behavior of following that particular moving object.

6B. Goslings that follow their mother are more likely to find food and avoid danger.

Test Prep Correlations

SC.F.1.4.7: Items 1, 3, 6A
SC.G.1.4.1: Item 2
SC.G.2.4.3: Items 4, 6B
SC.H.2.4.2: Item 5

Exploration Lab

Studying Nonverbal Communication

Teacher's Notes

Time Required 45 minutes of classroom time; 30 minutes of time outside of class

Ratings

EASY ——→ HARD

Teacher Preparation 1
Student Setup 1
Concept Level 2
Cleanup 1

Skills Acquired
- Identifying/Recognizing Patterns
- Collecting Data
- Organizing and Analyzing Data
- Interpreting

Scientific Methods
In this lab, students will:
- Make Observations
- Analyze the Results
- Draw Conclusions

Materials and Equipment
Students will need a stopwatch or a clock with a second hand, paper, and a pencil.

Safety Cautions
Caution students to observe people in a safe setting and in an unobtrusive way that cannot be construed as offensive. Behavior changes quickly.

Tips and Tricks
Form groups of two or three students. It is important that members of each group have time together outside of class to conduct their observations. Each observation will require the presence of at least two members of the group. Suggest that at least one student of the pair or group observe the behavior while another student records observations.

Exploration Lab

Studying Nonverbal Communication SC.H.1.4.1

SKILLS
- Observing
- Analyzing
- Graphing

OBJECTIVES
- **Recognize** that posture is a type of nonverbal communication.
- **Observe** how human posture changes during a conversation.
- **Determine** the relationship of gender to the postural changes that occur during a conversation.

MATERIALS
- stopwatch or clock with a second hand
- paper
- pencil

Before You Begin

People communicate nonverbally with their **posture**, or body position. The position of the body while standing is called the **stance**. In an **equal stance**, the body weight is supported equally by both legs. In an **unequal stance**, more weight is supported by one leg than by the other. In this lab, you will observe and analyze how stance changes during conversations between pairs of people who are standing.

1. Write a definition for each boldface term in the paragraph above.
2. Make a data table similar to the one below. The sample data entered in row 1 show how to enter data. Do not copy these data.

DATA TABLE

Pairs	Gender		15-s intervals		
	Involved	Observed	15 s	30 s	45 s
1	F, M	M	U, W	E	E
2					
3					

3. Based on the objectives for this lab, write a question you would like to explore about nonverbal communication.

Procedure

PART A: Observing Behavior

1. Work in a group of two or three to observe conversations between pairs of people. Each conversation must last between 45 seconds and 5 minutes. One person in your group should be the timekeeper and the other group members should record data. **Note:** Be sure that your subjects are unaware they are being observed.
2. Observe at least three conversations. Record the genders of the two participants in each conversation and the gender of the one person whose posture you observe. **Note:** Be sure that the timekeeper accurately clocks the passage of each 15-second interval.
3. For each 15-second interval, record all of the changes in stance by the person you are observing. For example, note every time your subject shifts from an equal stance to an unequal stance, or vice versa. To record the stance simply, you may write *E* to identify an equal stance and *U* to identify an unequal stance.

Answers to Before You Begin

1. *posture*—body position; *stance*—position of the body while standing; *equal stance*—stance in which the body weight is supported equally by both legs; *unequal stance*—stance in which more of the body weight is supported by one leg than by the other
2. Data tables should be similar to the sample on page 840.
3. Answers will vary. For example: What effects do conflicting postures have on an observer?

4. If the subject assumes an unequal stance, also record the number of weight shifts from one foot to the other. Indicate a weight shift simply by writing *W*.

5. When a conversation ends, write down whether the pair departed together or separately. To record this, write *T* to indicate departing together or *S* to indicate departing separately.

6. After you have completed each observation, tally the total number of weight shifts within each 15-second block. **IMPORTANT!** Retain data only for conversations that last at least 45 seconds. If a conversation ends before you have collected data for 45 seconds, observe another conversation.

PART B: Analyzing Data

7. After all observations have been completed, combine the data from all of the groups in your class. Analyze the data, without regard to gender.
   a. Determine the most common stance during the first 15 seconds of a conversation, the middle 15 seconds, and the last 15 seconds. Make a bar graph to summarize the class data.
   b. Find the average number of weight shifts in the beginning, middle, and end intervals. Make a bar graph to summarize the class data.

8. Repeat step 7, but analyze the data according to gender this time.

9. Compile the data and make bar graphs for each of the following: males talking with a male, males talking with a female, females talking with a male, and females talking with a female. Compare these graphs with the ones you made in Step 7.

Analyze and Conclude

1. **Analyzing Results** Which stance was used most often during a conversation?
2. **Recognizing Relationships** Which behavior most often signals that a conversation is about to end: stance change or weight shift?
3. **Drawing Conclusions** Do males and females differ in their departure signals? Justify your conclusion.
4. **Forming Hypotheses** What do you think might be an adaptive significance of a departure signal?
5. **Forming Reasoned Opinions** What other behaviors you observed were forms of nonverbal communication? Justify your answer.
6. **Further Inquiry** Write a new question about animal behavior that could be explored with a new investigation.

On the Job

Psychology is the study of human and animal behavior. Do research to discover how psychology is used to treat behavioral disorders in humans or in pets. For more about careers, visit **go.hrw.com** and type in the keyword **HX4 Careers.**

Answers to Procedure

Observations in steps 2–6 will vary. Check that students have correctly entered all observations in their data table.

7. a. Answers will vary. Students may find that the most common stance is with equal weight on each leg. Check that the bar graphs reflect the combined class data.
   b. Answers will vary. Students may find that the number of weight shifts increased toward the end of intervals. Check that the bar graphs reflect the combined class data.
8. Answers will vary. Check that students have correctly analyzed the combined class data by gender. The bar graphs should reflect the analysis by gender.
9. Answers will vary. Check that students have correctly compiled the data. The bar graphs should reflect the compiled data.

Answers to Analyze and Conclude

1. Answers will vary. Students will most likely find that the most common stance is with equal weight on each leg.
2. Answers will vary. Students should support their answers by citing their data.
3. While researchers have not found significant differences between the sexes in these signals, your students may see some trends due to the small sample size. This could lead to an interesting discussion about statistics.
4. Answers will vary, but students may suggest that nonverbal cues may make the departure expected and less abrupt. This may ensure that the person leaving does not insult the other person.
5. Answers will vary, but students may mention facial expressions and head, hand, arm, or eye movements.
6. Answers will vary. For example: What postural changes occur when two dogs greet each other? How does nonverbal communication differ among different cultures?

Answers to On the Job

Answers will vary. Behavior modification techniques commonly used by psychologists include systematic desensitization, in which an individual is trained to relax in the presence of fear-producing stimuli; aversion therapy, in which an aversive stimulus is paired with an undesirable behavior; and biofeedback, in which individuals are trained to monitor their own physiological processes in cases in which a behavioral disorder may have a physiological basis.

Chapters

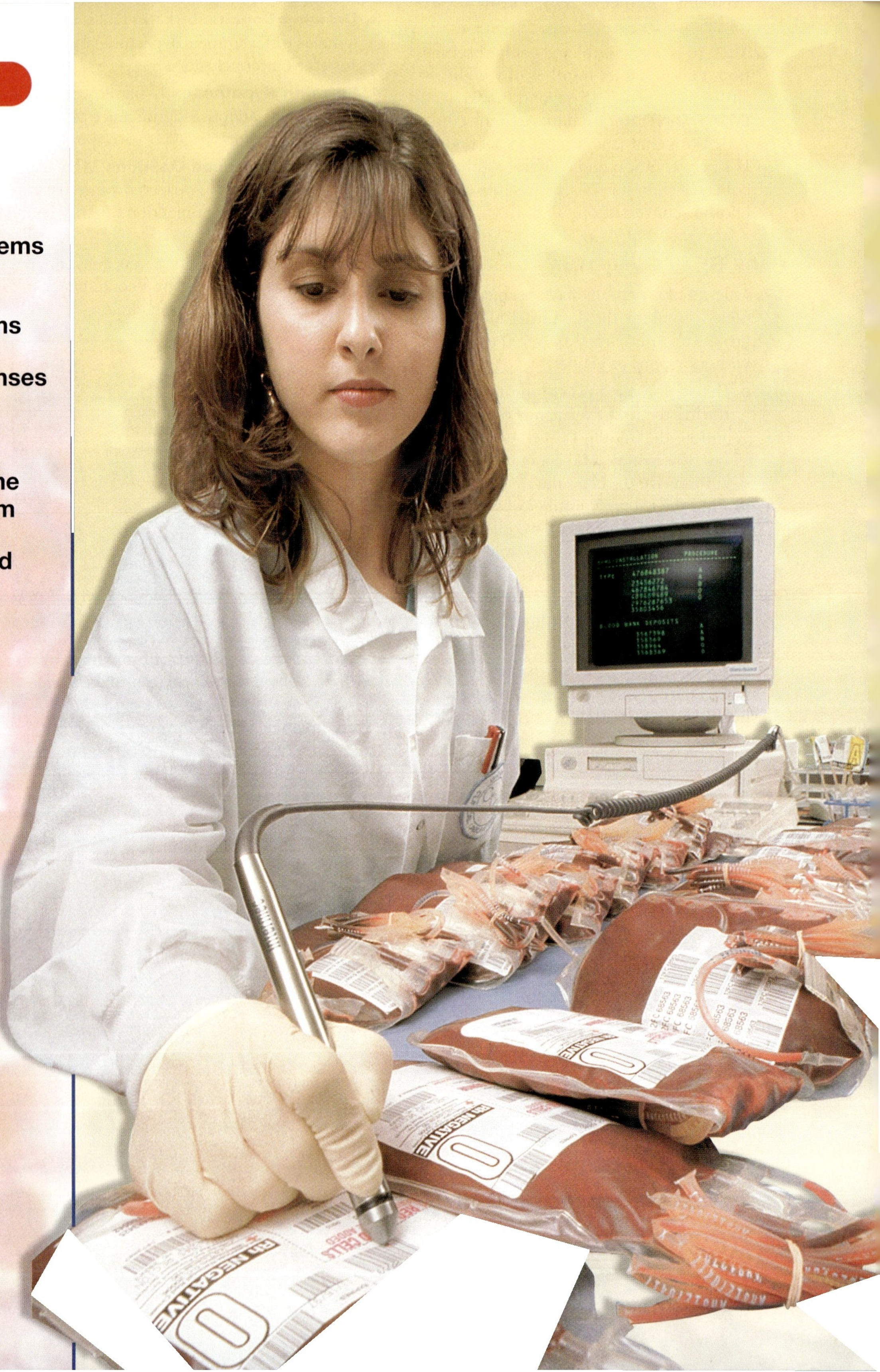

Donated blood saves thousands of lives every year.

Circulation

Yesterday...

Colleagues of William Harvey, a 17th century English anatomist, considered his conclusion that blood flows one way in a continuous system of vessels to be a strange idea. **Read to learn how valves prevent blood from flowing backwards.** In the 1940s, Dr. Charles Drew, the first African-American to earn an M.D. at Columbia University, developed commercial procedures for the safe transfusion of plasma, the liquid component of blood.

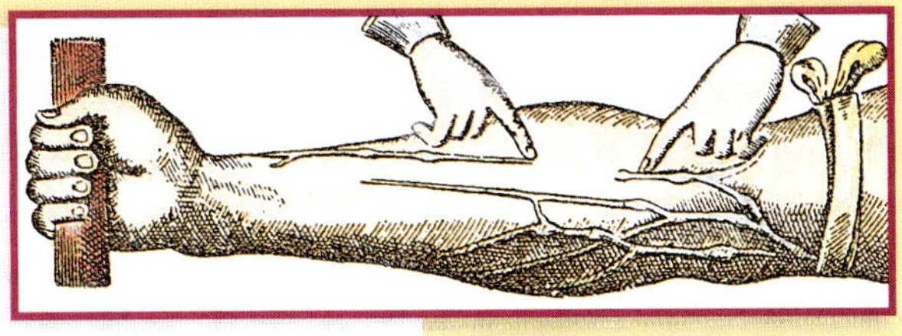

William Harvey's theory of blood circulation

Dr. Charles Drew

Today...

Scientists know that lifestyle choices—for example, what we eat and how much we exercise—affect the health of our circulatory system. Eating a healthful diet and exercising regularly will reduce the likelihood of hypertension (commonly called high blood pressure) and lower the risk of heart attack. **What are the dangers of hypertension?**

Moderate exercise

Tomorrow...

Blood is the river of life. Researchers are currently exploring artificial substitutes for red blood cells. **What are the components of blood?**

Artificial blood research

internet connect

www.scilinks.org
Topic: Blood
Keyword: HX4025

SCILINKS. Maintained by the National Science Teachers Association

Chapter Opener

- TE **Using the Figure,** p. 844
- SE **Quick Review,** p. 845 ■
- SE **Reading Activity,** p. 845 ■
- TE **Opening Activity,** p. 845 Basic

& 3 • 90 min pp. 846–849

Section 1
Body Organization

- OSP Lesson Plan*
- TR Bellringer*
- TR J1 Human Body Tissues*
- TR J3 Major Organ Systems of the Human Body*
- TR J4 Cavities of the Human Body*
- TR J5 Inside the Human Coelom*

- TE **Activity** Organizational Hierarchy, p. 848 Basic

& 5 • 90 min pp. 850–855

Section 2
Skeletal System

- OSP Lesson Plan*
- TR Bellringer*
- TR J7 Structure of Bone*
- TR J8 Compact Bone*
- TR J9 Effects of Osteoporosis*
- TR J12 The Human Knee*

- TE **Using the Figure,** p. 851 Basic
- TE **Group Activity** Spinal Flexibility, p. 853 ◆ Basic
- TE **Demonstration,** p. 853 ◆ Basic
- TE **Group Activity** Orthopedic Sports Injuries, p. 854 Advanced
- CRF **Quick Lab** Comparing Skeletal Joints* ◆ Basic

& 7 • 90 min pp. 856–860

Section 3
Muscular System

- OSP Lesson Plan*
- TR Bellringer*
- TR J17 Opposing Muscles in the Arm*
- TR J14 Skeletal Muscle*
- TR J15 Contraction of a Muscle*
- TR J16 Interaction of Muscle Protein Filaments*

- TE **Demonstration,** p. 856 Basic
- TE **Using the Figure,** p. 857 General
- TE **Demonstration,** p. 857 ◆ General
- TE **Using the Figure,** p. 858 Basic
- TE **Activity** Tendons in Fingers, p. 858 Basic
- TE **Using the Figure,** p. 859 Basic

& 9 • 90 min pp. 861–864

Section 4
Skin, Hair, and Nails

- OSP Lesson Plan*
- TR Bellringer*
- TR Section Outline*
- TR J18 Structure of Skin*

- CRF **Datasheets for In-Text Labs** Life Span and Lifestyle* ◆
- CRF **Inquiry Lab** How Far Apart Are Your Touch Receptors?* General
- SE **Exploration Lab** Analyzing the Work of Muscles, pp. 868–869 ■
- CRF **Datasheets for In-Text Labs** Analyzing the Work of Muscles*

---

0 & 11 • 90 min **Chapter Review and Assessment Resources**

- SE **Study Zone,** p. 865 ■
- SE **Performance Zone,** p. 866 ■
- SE **FCAT Test Prep,** p. 867 ■
- TR Concept Mapping*
- CRF Vocabulary Review Worksheet* ■ General
- CRF Test Prep Pretest* ■ General
- CRF Chapter Test* ■ General
- CRF Chapter Test* Advanced
- OSP Test Generator*
- OSP Test Item Listing*
- OSP Modified Chapter Test* Basic

Online and Technology Resources

Holt Online Learning

Visit go.hrw.com for access to Holt Online Learning or enter the keyword **HX6 Home** for a variety of free online resources.

One-Stop Planner® CD-ROM

This CD-ROM package includes

- Lab Materials QuickList Software
- Holt Calendar Planner
- Customizable Lesson Plans
- Printable Worksheets
- ExamView® Test Generator
- Interactive Teacher Edition
- Holt PuzzlePro® Resources
- Holt PowerPoint® Resources

Guided Reading Audio CDs

These CDs are designed to help auditory learners and reluctant readers. (Audio CDs are also available in Spanish.)

Compression guide:
To shorten your instruction because of time limitations, omit blocks 1, 2, and 3.

KEY

TE	Teacher Edition	CRF	Chapter Resource File	*	Also on One-Stop Planner
SE	Student Edition	TR	Teaching Transparency	■	Also Available in Spanish
OSP	One-Stop Planner			◆	Requires Advance Prep

SKILLS DEVELOPMENT AND PRACTICE	SECTION REVIEW AND ASSESSMENT	CORRELATIONS
		Florida Sunshine Standards/FCAT
TE Reading Skill Builder Interactive Reading, p. 847 BASIC CRF Directed Reading Worksheet* BASIC CRF Active Reading Worksheet* GENERAL	SE Section 1 Review ■ TE Reteaching, p. 849 BASIC TE Quiz, p. 849 GENERAL CRF Quiz* ■ BASIC	SC.F.1.4.2
TE Reading Skill Builder Paired Reading, p. 852 BASIC CRF Directed Reading Worksheet* BASIC CRF Active Reading Worksheet* GENERAL	SE Section 2 Review ■ TE Reteaching, p. 855 BASIC TE Quiz, p. 855 GENERAL TE Alternative Assessment, p. 855 GENERAL CRF Quiz* ■ BASIC	SC.F.1.4.2, SC.H.3.4.3
CRF Directed Reading Worksheet* BASIC CRF Active Reading Worksheet* GENERAL	SE Section 3 Review ■ TE Reteaching, p. 860 BASIC TE Quiz, p. 860 GENERAL TE Alternative Assessment, p. 860 GENERAL CRF Quiz* ■ BASIC	SC.F.1.4.2, SC.F.1.4.3
TE Skill Builder Writing Skills, p. 863 ADVANCED CRF Directed Reading Worksheet* BASIC CRF Active Reading Worksheet* GENERAL CRF Critical Thinking Worksheet* ADVANCED CRF Science Skills Worksheet* GENERAL	SE Section 4 Review ■ TE Reteaching, p. 864 BASIC TE Quiz, p. 864 GENERAL TE Alternative Assessment, p. 864 ADVANCED CRF Quiz* ■ BASIC Modified Worksheet* BASIC	SC.F.1.4.2, SC.H.1.4.1, SC.H.3.4.3

SCILINKS. NSTA
www.scilinks.org
Maintained by the **National Science Teachers Association**

Topic: Organ Systems
SciLinks Code: HX4131

Topic: Joint Disorders
SciLinks Code: HX4107

Topic: Muscle Structure
SciLinks Code: HX4126

Topic: Muscle Contraction
SciLinks Code: HX4125

Topic: Forensic Analysis
SciLinks Code: HX4087

Topic: Skin Cancer
SciLinks Code: HX4165

CHAPTER 37

Overview

Tell students that in this chapter they will learn about the internal organization of the body from cells to organ systems. Students will be able to explain how bones grow and work with each other to move the body, and about the mechanics of muscle movement. Finally, they will describe the structure of the skin and common skin disorders.

Using the Figure

Direct students' attention to the chapter opening figure. Tell them that the human body is composed over 100 trillion cells. These cells do not function independently. Rather, they require interactions with the other cells. The organization of the body allows for an efficient interdependence of all cells. This interdependence extends to the tissues, organs, and organ systems, which together make up the human body.

Assessing Prior Knowledge

Before beginning this chapter, be sure your students can answer the questions in the ***Quick Review*** on the facing page. If they cannot, encourage them to reread the sections indicated.

Hurdler

844

Chapter Correlations — Florida Sunshine Standards/FCAT

SC.F.1.4.2 Body structures are uniquely designed and adapted for their function.

SC.F.1.4.3 Membranes are sites for chemical synthesis and essential energy conversions.

SC.H.3.4.3 Scientists can bring information, insights, and analytical skills to matters of public concern and help people understand the possible causes and effects of events.

CHAPTER

37 Introduction to Body Structure

Florida Sunshine Standards/FCAT

SC.F.1.4.2 Body structures are uniquely designed and adapted for their function.

SC.F.1.4.3 Membranes are sites for chemical synthesis and essential energy conversions.

SC.H.3.4.3 Scientists can bring information, insights, and analytical skills to matters of public concern and help people understand the possible causes and effects of events.

Quick Review

Answer the following without referring to earlier sections of your book.

1. **Describe** the role of ATP in metabolism. *(Chapter 5, Section 1)*
2. **Summarize** the two stages of cellular respiration. *(Chapter 5, Section 1)*
3. **Distinguish** between tissues, organs, and organ systems. *(Chapter 20, Section 3)*

Did you have difficulty? *For help, review the sections indicated.*

Reading Activity

Before you read this chapter, survey the subtitles, headings, captions, and words in boldface type. Try to identify the purpose of this chapter. As you read each section, create a descriptive reading organizer.

Looking Ahead

internet connect

www.scilinks.org

National Science Teachers Association *sci*LINKS Internet resources are located throughout this chapter.

SCILINKS. Maintained by the National Science Teachers Association

An activity as complicated as jumping hurdles while running full speed requires a tremendous amount of energy and the coordination of many body systems.

845

Opening Activity — BASIC

Doing Simple Tasks Have a volunteer perform a simple task, such as opening a jar, tuning a radio, or eating a cracker. Ask what systems are being used to accomplish each task. (nervous system, skeletal system, muscular system) **LS Kinesthetic**

Quick Review Answers

1. ATP is the primary source of energy for the chemical reactions of metabolism that require energy.
2. In the first stage of cellular respiration, the carbon-carbon bonds of a glucose molecule are broken and the energy contained in them is stored primarily as NADH. In the second stage of cellular respiration, hydrogen atoms in NADH are removed and transported through a membrane transport protein that is also an enzyme; this enzyme catalyzes the formation of ATP from ADP.
3. Tissues are body structures that are made up of cells that together perform a related body function. Organs are made up of two or more tissues that together perform a related body function. Organ systems are made up of two or more organs that together perform a major body function.

Reading Activity Answers

Answers will vary but should contain an accurate summary of the chapter and its purpose.

Chapter Resource File

- Vocabulary Worksheets
- Concept Mapping

Section 1

Focus

Overview

Before beginning this section review with your students the objectives listed in the Student Edition. Students will distinguish among the structure and function of the four tissue types found in the body. They will recognize how tissues are organized into organs and then further into organ systems, the structure and function of the body's cavities, and how endothermy helps maintain homeostasis.

Bellringer

Ask students to write down what causes a black eye. Instruct them to read the Real Life feature on the next page when they have finished and compare their answer to the one given there.

Motivate

Discussion/ Question — GENERAL

Lead a class discussion about all of the activities that occur in the stomach. Ask students if they have muscle tissue in their stomach and, if so, what it does. (Yes; muscle tissue in the stomach mixes food and stomach secretions and moves the food into the small intestines.) Ask them if they have nervous tissue in their stomach and, if so, what it does. (Yes; nervous tissue in the stomach stimulates muscles to contract and glands to secrete, and sends signals such as hunger and satiety to the brain.)

Section Correlations

SC.F.1.4.2

Section 1 Body Organization

Florida Sunshine Standards/FCAT

SC.F.1.4.2 Body structures are uniquely designed and adapted for their function.

Objectives

- **Identify** four levels of structural organization within the human body. SC.F.1.4.2
- **Analyze** the four kinds of body tissues. SC.F.1.4.2
- **List** the body's major organ systems. SC.F.1.4.2
- **Evaluate** the importance of endothermy in maintaining homeostasis. SC.F.1.4.2

Key Terms

epithelial tissue
nervous tissue
connective tissue
muscle tissue
body cavity

Levels of Structural Organization

The human body contains more than 100 trillion cells and more than 100 kinds of cells. How do these cells work together? The body is structurally organized into four levels: cells, tissues, organs, and organ systems. Recall that a tissue is a group of similar cells that work together to perform a common function. The cell types of the body are grouped by function into four basic kinds of tissues: epithelial, nervous, connective, and muscle tissues. These tissues, shown in **Figure 1,** are the building blocks of the human body.

Four Kinds of Tissues

Epithelial tissue There are many different kinds of epithelial *(ehp ih THEE lee uhl)* tissue. **Epithelial tissue** lines most body surfaces, and it protects other tissues from dehydration and physical damage. An epithelial layer is usually no more than a few cells thick. These cells are typically flat and thin, and they contain only a small amount of cytoplasm. Epithelial tissue is constantly being replaced as cells die.

Figure 1 Body tissues

Cells of the body are grouped into different kinds of tissues.

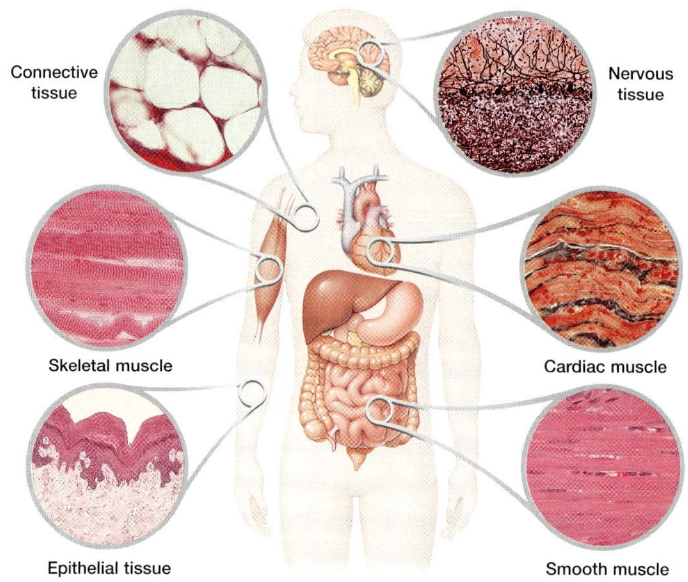

846

Chapter Resource File

- Directed Reading BASIC
- Active Reading GENERAL

Transparencies

TR Bellringer
TR J1 Human Body Tissues
TR J3 Major Organ Systems of the Human Body

One-Stop Planner CD-ROM

- Reading Organizers BASIC
- Reading Strategies BASIC

Nervous Tissue The nervous system is made of **nervous tissue.** Nervous tissue consists of nerve cells and their supporting cells. Nerve cells carry information throughout the body. You will learn more about the nervous system in Chapter 41.

Connective Tissue Various types of **connective tissue** support, protect, and insulate the body. Connective tissue includes fat, cartilage, bone, tendons, and blood. Some connective tissue cells, such as those in bone, are densely packed. Others, such as those found in blood, are farther apart from each other.

Muscle Tissue Three kinds of **muscle tissue** enable the movement of body structures by muscle contraction. The three kinds of muscle tissues are skeletal muscle, smooth muscle, and cardiac muscle.

1. **Skeletal muscle.** Skeletal muscle is called voluntary muscle because you can consciously control its contractions. Skeletal muscles move bones in the trunk and limbs.
2. **Smooth muscle.** Smooth muscle is called involuntary muscle because you cannot consciously control its slow, long-lasting contractions. Smooth muscles line the walls of blood vessels and hollow organs. Some contract only when stimulated by signal molecules; others contract spontaneously.
3. **Cardiac muscle.** Cardiac muscle is involuntary and is found in the heart. The powerful, rhythmic contractions of cardiac muscle pump blood to all body tissues. Groups of cardiac cells contract all at once, stimulating adjacent groups of cells to contract.

Real Life

A black eye is the common name for bruised tissues around the eye that may also be swollen.

The dark color of the bruise is caused by blood oozing into spaces between the tissues from tiny blood vessels that were broken when the eye was injured.

Stem Cells

Every human starts life as a single fertilized egg, which rapidly divides into a small cluster of cells. After about 5 days, a small ball of a few hundred cells is formed, which encloses a mass of embryonic stem cells. These early, undifferentiated cells will give rise to all of the types of cells of the developing body. Embryonic stem cells are immortal—that is, they divide indefinitely. And embryonic stem cells are not yet specialized. Indeed any embryonic stem cell is capable of becoming any type of tissue found in the adult body.

Because they can develop into any tissue, embryonic stem cells offer the possibility of repairing damaged tissues. Stem cell therapy in mice has been shown to repair heart muscle and to produce functional nerve cells in the brain. The use of human embryonic stem cells is very controversial. Because obtaining embryonic stem cells destroys an early embryo, therapeutic use of embryonic stem cells raises serious ethical issues.

Adults also have stem cells. Stem cells in bone marrow produce different types of blood cells. The adult brain contains stem cells that develop into new nerve cells. Adult stem cells are not as versatile as embryonic stem cells, and they are not immortal. Most stop reproducing after fewer than 100 cell divisions. Scientists are now at work on several therapeutic applications of adult stem cells.

internet connect

www.scilinks.org
Topic: Medicine in Florida
Keyword: HX6F10

SCILINKS. Maintained by the National Science Teachers Association

847

Teach

Teaching Tip

Fast-Twitch and Slow-Twitch

There are two types of skeletal muscle fibers, fast-twitch and slow-twitch. Fast-twitch muscle fibers work quickly and powerfully for performing tasks such as weightlifting. Slow-twitch muscle fibers contract slowly, use less energy, and are used for longer activities, such as distance running. Both types of fibers exist in each of the body's skeletal muscles in varying proportions depending on the muscle type.

READING SKILL BUILDER — BASIC

Interactive Reading Assign Chapter 37 of the *Holt Biology* Guided Audio CD Program to help students achieve greater success in reading the chapter.

INCLUSION Strategies

• *Attention Deficit Disorder*

Have students design a class chart of the four kinds of tissue found in the body. The chart should include the kind of tissue, where in the body the tissue is found, the body organs made-up of this tissue, and the function of each type of tissue. Students may also create a chart to identify the different types of muscle and where in the body the different muscle tissues are found.

Cultural Awareness

Qi and The Body In some Asian countries, many people believe that the body has an energy field, call qi (chee), which moves through the body through channels called meridians. The function of qi is thought to be similar to the functions of the nervous and circulatory systems—moving information and energy within the body. Acupuncture, an ancient treatment for pain and many other ailments, is based on the concept of qi. Many scientists now believe that acupuncture is effective for pain because it activates the body's endogenous opiates. LS Verbal

Teach, *continued*

Teaching Tip — GENERAL

Tissue and Organ Transplants Ask students to think of body parts that have been transplanted in humans. Have them identify whether the part is a tissue or an organ. (Their list may include organs such as the heart, kidney, liver, lung, skin, and pancreas, and tissues such as blood, bone marrow, cornea, and bone.) **LS Visual**

Activity — BASIC

Organizational Hierarchy Have students make a Graphic Organizer similar to the one at the bottom of this page to demonstrate that simpler structures in the body are components of more complex structures. Ask students to use the terms cell, tissue, organ, organ system, and body. **LS Visual**

Teaching Tip — ADVANCED

Homeostasis Point out to students that the maintenance of homeostasis in a multicellular organism is a difficult and complex process. Communication must be maintained among organ systems, organs, tissues, and even individual cells. Multicellular organisms have elaborate signaling systems that are only now being discovered. Many diseases result from the disruption of homeostasis. Have students select an endocrine or metabolic disorder and write a short report about it. Diseases to suggest include diabetes mellitus, goiter, hypo- or hyperthyroidism, Cushing's syndrome, and porphyria. **LS Verbal**

internet connect
www.scilinks.org
Topic: Organ Systems
Keyword: HX4131
SCILINKS. Maintained by the National Science Teachers Association

Organ Systems

Body organs are made of combinations of two or more types of tissues working together to perform a specific function. The heart, for example, contains cardiac muscle tissue and connective tissue, and the heart is stimulated by nervous tissue. Each organ belongs to at least one organ system, which is a group of organs that work together to carry out major activities or processes. The different organs in an organ system interact to perform a certain function, such as digestion. The digestive system is composed of the mouth, throat, esophagus, stomach, intestines, liver, gallbladder, and pancreas. Some organs function in more than one organ system. The pancreas, for example, functions in both the digestive system and the endocrine system. **Table 1** lists the body's major organ systems.

Table 1 Major Organ Systems of the Body

System	Major structures	Functions
Circulatory	Heart, blood vessels, blood (cardiovascular) lymph nodes and vessels, lymph (lymphatic)	Transports nutrients, wastes, hormones, and gases
Digestive	Mouth, throat, esophagus, stomach, liver, pancreas, small and large intestines 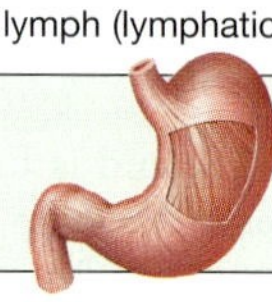	Extracts and absorbs nutrients from food; removes wastes; maintains water and chemical balances
Endocrine	Hypothalamus, pituitary, pancreas and many other endocrine glands	Regulates body temperature, metabolism, development, and reproduction; maintains homeostasis; regulates other organ systems
Excretory	Kidneys, urinary bladder, ureters, urethra, skin, lungs	Removes wastes from blood; regulates concentration of body fluids
Immune	White blood cells, lymph nodes and vessels, skin	Defends against pathogens and disease
Integumentary	Skin, nails, hair	Protects against injury, infection, and fluid loss; helps regulate body temperature
Muscular	Skeletal, smooth, and cardiac muscle tissues	Moves limbs and trunk; moves substances through body; provides structure and support
Nervous	Brain, spinal cord, nerves, sense organs 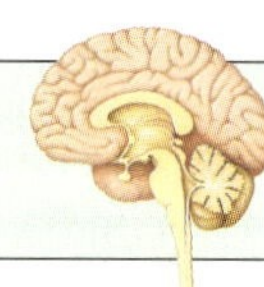	Regulates behavior; maintains homeostasis; regulates other organ systems; controls sensory and motor functions
Reproductive	Testes, penis (in males); ovaries, uterus, breasts (in females)	Produces gametes and offspring
Respiratory	Lungs, nose, mouth, trachea	Moves air into and out of lungs; controls gas exchange between blood and lungs
Skeletal	Bones and joints 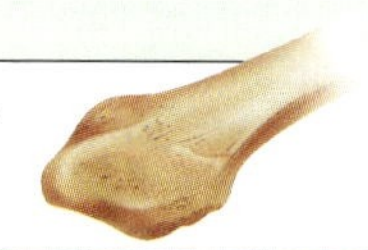	Protects and supports the body and organs; interacts with skeletal muscles, produces red blood cells, white blood cells, and platelets

Graphic Organizer

Use this graphic organizer with **Activity** on this page.

Body → (is composed of) → Organ systems → (are composed of) → Organs → (are composed of) → Tissues → (are composed of) → Cells

Body Cavities

The body contains four large fluid-filled spaces, or **body cavities,** that house and protect the major internal organs. Within the body cavities, shown in **Figure 2,** organs are suspended in fluid that supports their weight and prevents them from being deformed by body movements. These organs are also protected by bones and muscles. For example, your heart and lungs are protected by the rib cage and the sternum inside the thoracic *(thoh RAS ik)* cavity. Your brain, encased within the cranial *(KRAY nee uhl)* cavity, is protected by the skull. Your digestive organs, located in the abdominal cavity, are protected by the pelvis and abdominal muscles. Your spinal cord is protected by the vertebrae that surround the spinal cavity.

Endothermy

Like all mammals, humans are endotherms. Humans maintain a fairly constant internal temperature of about 37°C (98.6°F). Your body uses a great deal of energy to maintain a stable internal condition. For example, a large percentage of the energy you consume in food is devoted to maintaining your body temperature. You would not survive very long if your temperature fell much below the normal range. Very high temperatures, such as occur with fever, are also dangerous because they can inactivate critical enzymes. Your body maintains a constant temperature due to the flow of blood through blood vessels just under the skin. To release heat to the air, blood flow is increased to these vessels. To retain heat, blood is shunted away from the skin. As an endotherm, you can remain active at external temperatures that would slow the activity of ectotherms. Endothermy enables you to sustain strenuous activity, such as exercise, for a long time.

To maintain homeostasis, the body's organ systems must function smoothly together. The nervous system and the endocrine system operate on negative feedback with other organ systems. This promotes stability throughout the body. In addition to temperature regulation, homeostasis involves adjusting metabolism, detecting and responding to environmental stimuli, and maintaining water and mineral balances.

Figure 2 Body cavities. Many organs and organ systems are encased in protective body cavities.

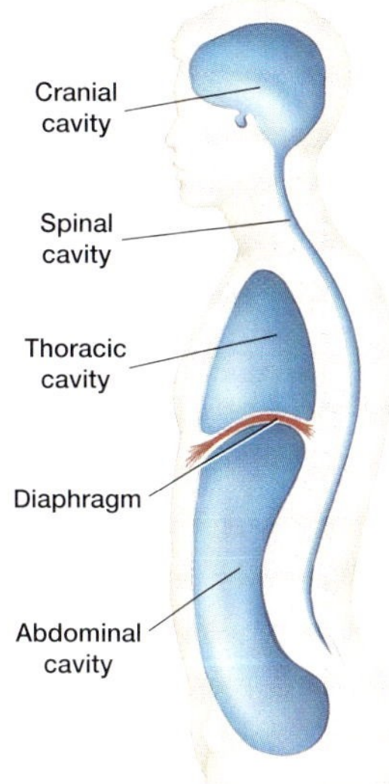

Section 1 Review

1. **Summarize** the four levels of structural organization in the body. SC.F.1.4.2
2. **List** the four different kinds of body tissues, and give an example of each kind. SC.F.1.4.2
3. **Describe** the relationship between organs and organ systems. SC.F.1.4.2
4. **Relating Concepts** How is endothermy advantageous to humans? SC.F.1.4.2
5. **Critical Thinking Inferring** Why should fever be controlled during an illness? SC.H.3.4.3
6. *FCAT Prep* In which part of the body would you most likely find flat, thin cells that contain only a small amount of cytoplasm?
   A bone
   B cardiac muscle
   C digestive tract lining
   D skeletal muscle
   SC.F.1.4.2

Answers to Section Review

1. *cell*—individual unit; *tissue*—group of similar cells that work together to perform a common function; *organ*—combination of two or more types of tissues that work together; *organ system*—group of organs that work together to perform a specific function
2. *epithelial*—lines body surfaces; *nervous*—nerve cells; *connective*—fat, cartilage, bone, tendon, blood; *muscle*—skeletal, smooth, cardiac
3. An organ system is composed of two or more organs that work together.
4. Endothermy allows humans to function in many environments and to engage in strenuous activities for long periods.
5. It is important to control a fever because high temperatures can inactivate essential enzymes.
6. **A.** Incorrect. Bone tissue is composed of densely packed cells. **B.** Incorrect. Cardiac muscle tissue is composed of interconnected cells. **C.** Correct. The lining of the digestive tract is epithelial tissue, which is composed of thin, flat cells. **D.** Incorrect. Skeletal muscle is not composed of flat, thin cells.

Close

Reteaching — BASIC

Use a relationship with which students are more familiar to help them learn the levels of organization in the body. For example, a house is part of a city, which is part of a county, which is part of a state, which is part of the United States.

Quiz — GENERAL

1. What is the main function of epithelial tissue? (to protect other tissues from dehydration and physical damage.)
2. What type of tissue is blood? (connective)
3. Which organ system regulates the concentration of body fluids? (excretory)

Integrating Physics and Chemistry

Point out to students that just as bones and muscle protect organs, seat belts also provide vital protection, since according to Newton's first law of motion, *an object in motion maintains its velocity unless it experiences an unbalanced force.* The tendency to continue moving is called inertia and explains the need for seat belts because they provide the unbalanced force necessary to stop *you* from moving once the car has stopped. Without seat belts disastrous consequences—such as being thrown from a car—could result.

Transparencies

TR J4 Cavities of the Human Body
TR J5 Inside the Human Coelom

Section Review Correlations

SC.F.1.4.2: Items 1-4, 6
SC.H.3.4.3: Item 5

Section 2

Focus

Overview

Before beginning this section review with your students the objectives listed in the Student Edition. Students will describe the components of the skeletal system and the structure of bone, how bones grow and how they lose density, and the structure and function of the body's joints and how they can become damaged.

Bellringer

Have students try to perform several activities without bending their fingers, such as picking up a pencil, taking off or putting on a jacket, or turning a page in a book. Then, have them write two or three sentences about joints and their importance.

Motivate

Discussion/Question — Basic

Ask students what they already know about different types of joints. Tell them the names of the types of joints. Ask them to provide an example of each type and to describe the movement for which each type is adapted.

Teaching Tip

Immature humans have 33 vertebrae. During development, vertebrae of the sacrum and coccyx fuse so that there are 26 separate bones.

Section Correlations

SC.F.1.4.2
SC.H.3.4.3

Section 2 Skeletal System

Florida Sunshine Standards/FCAT

SC.F.1.4.2 Body structures are uniquely designed and adapted for their function.

SC.H.3.4.3 Scientists can bring information, insights, and analytical skills to matters of public concern and help people understand the possible causes and effects of events.

Objectives

- **Distinguish** between the axial skeleton and the appendicular skeleton. SC.F.1.4.2
- **Analyze** the structure of bone. SC.F.1.4.2
- **Summarize** the process of bone development. SC.F.1.4.2
- **List** two ways to prevent osteoporosis. SC.H.3.4.3
- **Identify** the three main classes of joints. SC.F.1.4.2

Key Terms

axial skeleton
appendicular skeleton
bone marrow
periosteum
Haversian canal
osteocyte
osteoporosis
joint
ligament

The Skeleton

What keeps your body from collapsing like a limp noodle? An internal skeleton of bones shapes and supports your body. Your skeleton provides protection for internal organs and, along with muscles, enables movement with a versatile system of levers and joints. Muscles pull against bones at joints, moving the limbs and the trunk. Your skeleton is made mostly of bone, a type of hard connective tissue that is constantly being formed and replaced. The human skeleton, shown in **Figure 3,** contains 206 individual bones. Of these, 80 bones form the **axial skeleton,** which includes bones of the skull, spine, ribs, and sternum. The other 126 bones, including those of the arms, legs, pelvis, and shoulder, form the **appendicular** *(ap uhn DIHK yoo luhr)* **skeleton.**

Skeleton

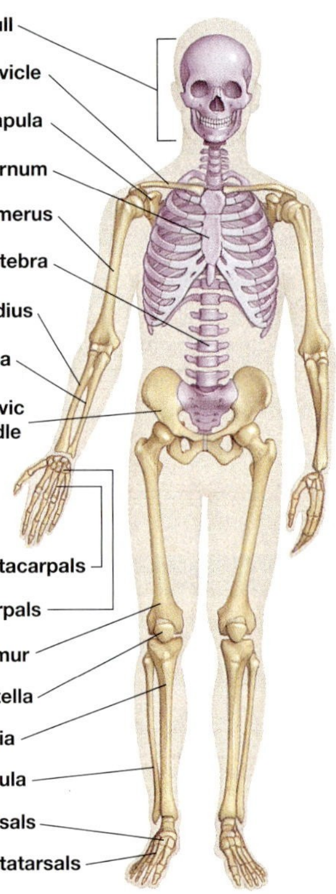

Figure 3 Skeleton. Bones of the appendicular skeleton "hang" from bones of the axial skeleton (purple). Immature humans have 33 vertebrae. During development, vertebrae of the sacrum and coccyx fuse so that there are 26 separate bones.

Axial Skeleton

The most complex part of the axial skeleton is the skull. Of the 29 bones in the skull, 8 bones form the cranium, which encases the brain. The skull also contains 14 facial bones, 6 middle-ear bones, and a single bone that supports the base of the tongue. The skull is attached to the top of the spine, or backbone, which is a flexible, curving column of 26 vertebrae that supports the center of the body. Curving forward from the middle vertebrae are 12 pairs of ribs, which form a protective rib cage around the heart and lungs.

Appendicular Skeleton

The appendicular skeleton forms the appendages or limbs—the shoulders, arms, hips, and legs. The arms and legs are attached to the axial skeleton at the shoulders and hips, respectively. The shoulder attachment, called the pectoral girdle, contains two large, flat shoulder blades, or scapulas, and two slender, curved

Chapter Resource File
- Directed Reading — Basic
- Active Reading — General

Transparencies
- TR Bellringer
- TR J7 Structure of Bone
- TR J8 Compact Bone

One-Stop Planner CD-ROM
- Reading Organizers — Basic
- Reading Strategies — Basic
- Occupational Application Worksheet Physical Therapist — General

collarbones, or clavicles. The clavicles connect the scapulas to the upper region of the sternum and hold the shoulders apart. This arrangement enables full rotation of the arms about the shoulder. The hip attachment, called the pelvic girdle, contains two large pelvic bones. The pelvic bones distribute the weight of the upper body evenly down the legs.

WORD Origins

The word *periosteum* is from the Greek *peri*, meaning "around," and *osteon*, meaning "bone."

Structure of Bone

As shown in **Figure 4,** bones are made of a hard outer covering of compact bone surrounding a porous inner core of spongy bone. Compact bone is a dense connective tissue that provides a great deal of support. Spongy bone is a loosely structured network of separated connective tissue. Some cavities in spongy bone are filled with a soft tissue called **bone marrow.** Red bone marrow begins the production of all blood cells and platelets. The hollow interior of long bones is filled with yellow bone marrow. Yellow bone marrow consists mostly of fat, which stores energy. Bones are surrounded and protected by a tough exterior membrane called the **periosteum** *(pair ee AHS tee uhm)*. The periosteum contains many blood vessels that supply nutrients to bones.

Figure 4 **Structure of bone**

Many bones contain bone marrow, blood vessels, and both compact and spongy bone tissue.

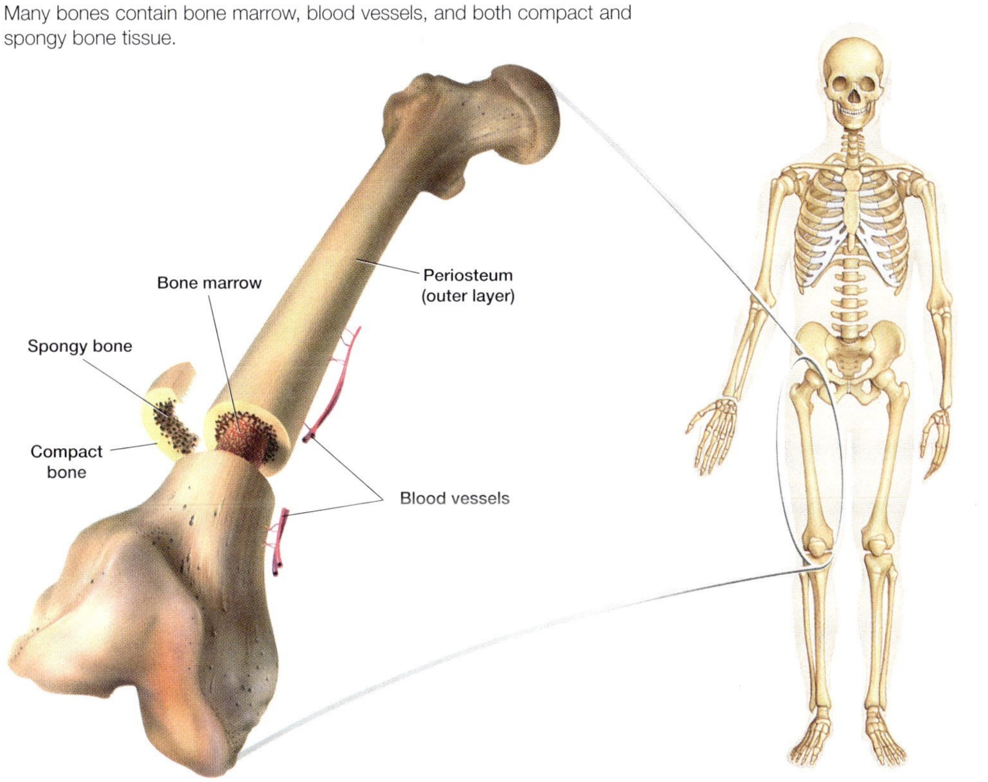

Teach

Teaching Tip

Tissues in Bone Point out to students that, while bone is composed primarily of connective tissue, it also contains the other three types of body tissues. Bone has nerve cells that can send pain signals to the brain if a bone is broken. Bone also contains smooth muscle tissue in the walls of the blood vessels that carry materials into and out of bone, as well as epithelial tissue in the walls of these blood vessels. Bone also has a wide variety of connective tissues, including blood, fat, mineralized connective tissue, as well as ligaments and tendons attached to the bones.

Teaching Tip

Mnemonics for Bone Names
Tell students some simple ways to remember the names of bones. For example, the scapula is large and flat like a spatula; the ulna ends at the elbow (both first letters are vowels); the radius radiates around the ulna; the metacarpals are where fingers meet the carpals; the sternum looks like a tie (stern people wear ties); the fibula is the little bone in the lower leg (a little lie is a fib); tarsals are bones in your feet (you get tar on the soles of your feet) **LS Verbal**

Using the Figure — Basic

Direct students' attention to **Figure 3.** Point out the bones of the axial skeleton. (bones of the skull, spine, and rib cage) Ask students to identify the functions of these bones. (to enclose and protect internal organs and provide vertical support) Point out the bones of the appendicular skeleton. (the bones of the pectoral and pelvic girdles, arms, and legs) Ask students to identify the functions of these bones. (to provide attachment of the limb bones to the axial skeleton and to enable gross motor movement) **LS Visual**

INCLUSION Strategies

• ***Gifted and Talented***

Using reference materials or the Internet, have students research the condition known as osteoporosis. The students should learn about the condition and its cause, why it affects women more than men, and what types of action can be taken to prevent the onset of osteoporosis. Students can then develop a taped or live public service announcement or brochure to share the information they have learned with others.

did you know?

Hyoid Bone The hyoid bone is the origin for the tongue muscles. You can find it by feeling the neck, lateral to the Adam's apple, directly beneath the carotid arteries. If you push gently on one side of the throat, you can feel the bone easily on the opposite side. This bone can be broken by choking, making it very painful to speak or swallow. **LS Kinesthetic**

Teach, *continued*

Real Life GENERAL

Answer

Answers will vary. Encourage students to use the media center to find nonfictional accounts of crime solving. LS **Verbal**

Teaching Tip

Living Bones Students may think that bone has few of the traits of living tissue. Emphasize that the bony endoskeleton of vertebrates grows as the animal matures. This is unlike the hard exoskeleton of invertebrates, which must be shed for the animal to grow. The non-living components of bone—crystalline material that surrounds the living bone cells—give bone its strength and hardness.

READING SKILL BUILDER BASIC

Paired Reading Pair each student with a partner. Have each student read silently about bone growth and osteoporosis. After students finish reading, ask one student from each pair to summarize what he or she understood, referring to the text when needed. The second student should add anything omitted. Both students should then describe what they found confusing and should work together to understand all the passages. Have them prepare a list of questions to pose to the class about passages that are still unclear after their collaboration. LS **Interpersonal**

Real Life

Years after an unsolved murder is committed, the victim's bones may hold clues to the crime. Forensic anthropologists have solved many cases by analyzing bones and other human remains found at crime scenes.

Finding Information
Read some accounts of crimes solved by forensic anthropologists to learn how these scientists analyze bones.

Growth of Bones

In early development, the skeleton is made mostly of cartilage, a type of connective tissue that serves as a template for bone formation. During development, most cartilage is gradually replaced by bone as minerals are deposited. Deposits of calcium and other minerals harden bones, enabling them to withstand stress and provide support. In compact bone, new bone cells are added in layers around narrow, hollow channels called **Haversian canals.** Haversian canals extend down the length of a bone, and they contain blood vessels that enter the bone through the periosteum.

As shown in **Figure 5,** layers of new bone cells form several concentric rings around Haversian canals. These rings form columns that enable the bone to withstand tremendous amounts of stress. Eventually, bone cells called **osteocytes** *(AHS tee oh siets)* become embedded within the bone tissue. Osteocytes maintain the mineral content of bone. The blood vessels that run through each Haversian canal supply the osteocytes with nutrients needed to maintain bone cells. Bones continue to thicken and elongate through adolescence as bone cells replace cartilage. Bone elongation occurs at the ends of long bones. Cartilage degenerates as new bone cells are added, causing bones to lengthen.

Figure 5 **Compact bone**

In compact bone, concentric rings of bone surround Haversian canals.

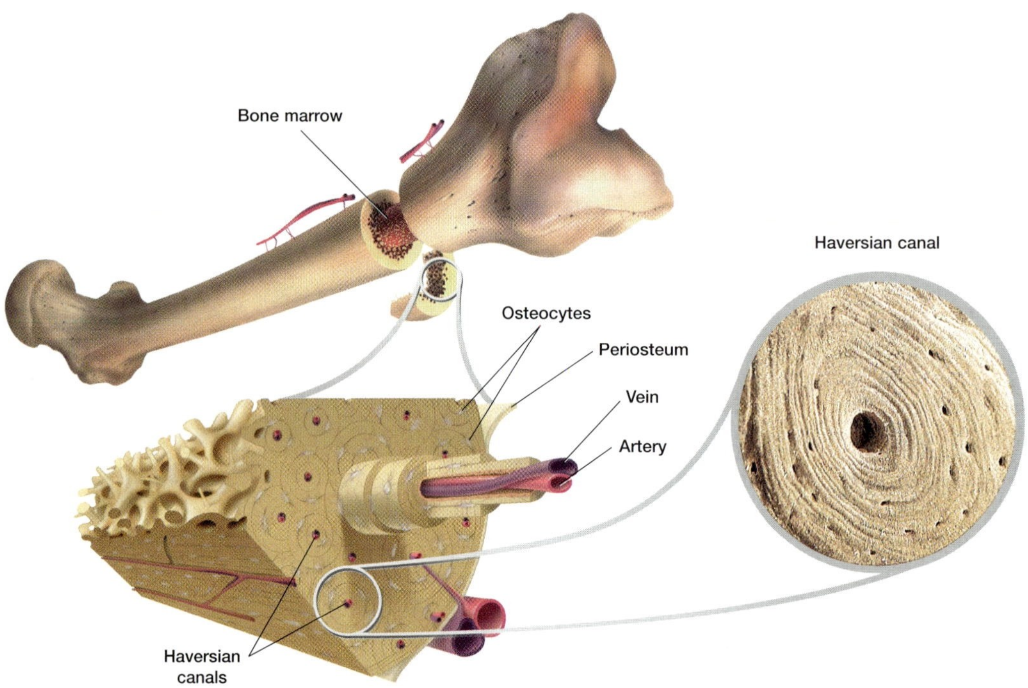

MISCONCEPTION ALERT

Calcium in Bones Most people know that calcium is needed for healthy bones. However, most people don't know that bone tissue serves as a storage site for calcium. If calcium is needed elsewhere in the body (such as for nerve or muscle function), it may be released from the bones in a process called reabsorption. Conversely, if calcium is abundant in the body, it may be added to the bones. Parathyroid hormone, produced in the parathyroid glands, stimulates the release of calcium from bone cells into the blood. This hormone is released in response to low blood calcium levels detected by the parathyroid glands. Calcitonin, produced in the thyroid gland, stimulates the absorption of calcium into bone cells from the blood. This hormone is released in response to high blood calcium levels.

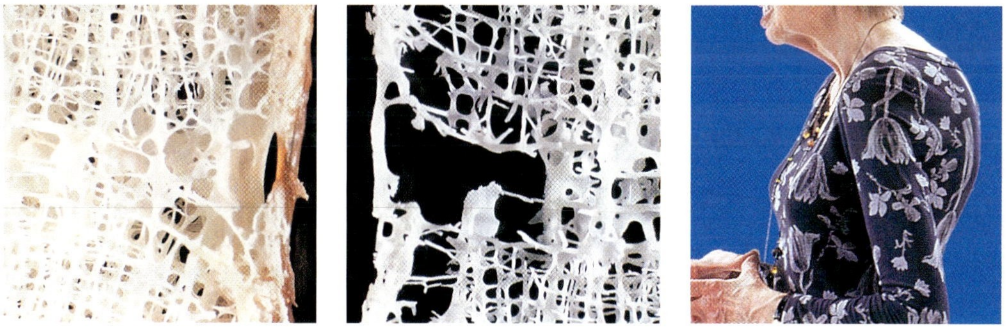

Normal bone   Bone in osteoporosis   Dowager's hump

Figure 6 Effects of osteoporosis. Compare the density of a normal bone with that of a bone weakened by osteoporosis. One sign of osteoporosis is the familiar "dowager's hump" of the back, caused by curvature of the spine.

Osteoporosis

In young adults, the density of bone usually remains constant. However, around the age of 35, bone replacement gradually becomes less efficient and some bone tissue is lost. Severe bone loss, as shown in **Figure 6,** can lead to a condition called **osteoporosis** *(ahst ee oh puh ROH sihs),* which means "porous bone." Bones affected by osteoporosis become brittle and are easily fractured.

Although both women and men lose bone tissue as they age, more women than men are affected by osteoporosis. Because women's bones are usually smaller, women cannot afford to lose as much bone tissue as men. In addition, the production of sex hormones decreases after menopause. This decrease in hormone production has been linked to an increased rate of bone loss in women following menopause.

You can take action now to prevent future osteoporosis. Building strong bones now will make you less likely to be affected by osteoporosis later in life. A mineral-rich diet that includes dairy products, green leafy vegetables, whole grains, and legumes such as those shown in **Figure 7,** together with regular exercise throughout your life will help maintain bone density.

Figure 7 Preventing osteoporosis. A proper diet and regular exercise help reduce the risk of osteoporosis.

853

Teaching Tip

Strong Bones Tell students that, although bones are lightweight, they are very strong. A human long bone has the tensile strength of cast iron, but it weighs only one-third as much. A bone can resist 25,000 lb/in.^{2} (1,800 kg/cm^2) of compression and 15,000 lb/in.^{2} (1,100 kg/cm^2) of tension. **LS Logical**

Group Activity — BASIC

Spinal Flexibility Have students work in groups to create models of the vertebral column. Provide each group with empty or partly filled sewing thread spools and a very long shoelace or cording. (about 2 feet long) Instruct students to thread the lacing or cording through successive spools, tying a knot between each spool. (It might be necessary to tie one knot on top of another in order to keep the knot from slipping through the spools.) Have each group create a string of spools on their lacing or cording. Next ask students to make observations on the flexibility of each individual "joint" in the string of spools and on the whole length of spools, and to compare this flexibility to that of the human vertebral column. Point out that the small range of motion of the joints between each vertebra has a compounding effect. Although each joint allows only a small amount of movement between two individual bones, when added across all 26 vertebrae, the human spine is very flexible.

Demonstration

Soak a cooked, clean chicken bone in vinegar overnight to remove much of the calcium from the bone. Compare the flexibility of the soaked bone with that of a cooked, clean bone that has not been soaked. Relate these ideas to what happens to bone due to osteoporosis. **LS Visual**

REAL WORLD CONNECTION

Exercise is as important for building and maintaining strong bones as it is for building strong muscles. Exercise becomes especially important as people grow older, when bone mass begins to decrease due to hormonal and other physiological changes.

Two types of exercises are important for maintaining healthy bones: weight-bearing and resistance exercises. Weight-bearing exercises are exercises in which your bones and muscles bear your body weight and include activities such as jogging, walking, soccer, tennis, stair climbing, and dancing. Swimming and bicycling are not weight-bearing exercises. Resistance exercises involve using the bones and muscles to work against a force or weight and include activities such as free weights and weight machines found at gyms and health clubs.

Teach, *continued*

Teaching Tip

The Soft Spot Many students who have younger siblings know that a baby has a soft spot on the top of its head. Explain that the soft spot is an area where sections of the skull have not grown together and fused. This area will become an immovable joint as the baby matures. Tell students that several layers of tough membranes protect the brain, but care must be taken not to injure the baby's vulnerable head.

Group Activity —— ADVANCED

Orthopedic Sports Injuries Have students work in small groups to research sports injuries to the skeletal or muscular systems. Ask each group to choose a specific type of injury (e.g., jumper's knee, shin splints, runner's knee, tennis elbow), or a category of injuries (e.g., injuries to a specific body part, injuries commonly associated with a specific sport, or injuries of a particular tissue such as bones, muscles, tendons, or ligaments). Ask each group to prepare a poster or brochure that describes the injury type they have researched and includes labeled illustrations. Have students present their findings to the class or display their projects so that other class members can examine them all.

Joints

A **joint** is where two bones meet. Pads of cartilage cushion the ends of the bones of a joint, enabling the joint to withstand great pressure and stress. The bones of a joint are held together by strong bands of connective tissue called **ligaments.** Ligaments not only help stabilize joints but also prevent joints from moving too far in any one direction. Many sports-related injuries to ligaments are caused by an impact that forces a joint to overextend. Injury occurs because the impact exceeds the tension that the ligaments can withstand.

Three Main Types of Joints

The skeletal system contains three main types of joints that enable varying degrees of movement: immovable joints, slightly movable joints, and freely movable joints. Examples of the three types of joints are shown in **Figure 8.**

Figure 8 Types of joints. The body contains immovable, slightly movable, and freely movable joints.

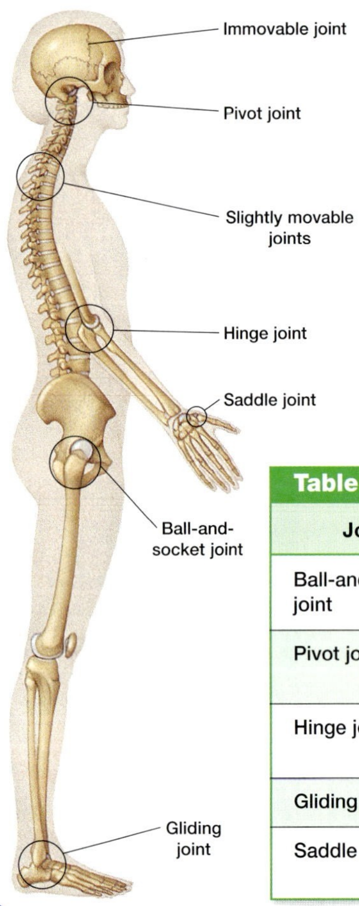

Immovable Joints Tight joints that permit little or no movement of the bones they join are called immovable joints. The cranial bones of the skull are joined by sutures, a type of immovable joint in which the bones are separated by only a thin layer of connective tissue.

Slightly Movable Joints Joints that permit limited movement of the bones they join are called slightly movable joints. For example, the vertebrae of the spine are joined by cartilaginous joints, which are a kind of slightly movable joint in which a bridge of cartilage connects the bones. Slightly movable joints are also located between bones of the rib cage.

Freely Movable Joints In joints that permit movement, the direction of bone movement is determined by the structure of the joint. Joints that permit the most movement are called freely movable joints. Some kinds

Table 2 Movable Joints

Joint	Type of movement	Examples
Ball-and-socket joint	All types	Shoulders and hips
Pivot joint	Rotation	Top of spine (turning of head)
Hinge joint	Bending and straightening	Elbows, knuckles of fingers and toes
Gliding joint	Sliding motion	Wrists and ankles
Saddle joint	Rotation, bending, and straightening	Base of thumbs

did you know?

Artificial Hip Joint In 1963, a British orthopedic surgeon revolutionized arthritic hip therapy when he created an artificial hip joint. He created a metal (vitalium) ball and plastic (polyethylene) socket. The plastic socket is cemented to the pelvis. This invention has helped many people, including children who are born with a hip socket that is too shallow and bones that are not aligned properly as well as arthritis patients who suffer disabling pain in the hip joint.

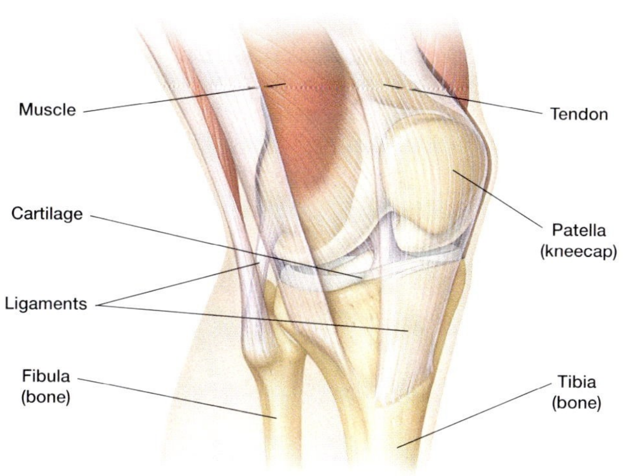

Figure 9 Knee. The knee is an example of a freely movable joint.

of freely movable joints are listed in **Table 2.** The structure of one freely movable joint, the knee, is shown in **Figure 9.**

internet connect
www.scilinks.org
Topic: Joint Disorders
Keyword: HX4107
SCILINKS. Maintained by the National Science Teachers Association

Disorders of Joints

Recall that ligaments hold the bones of a joint together. A lining of tissue that surrounds the joint secretes a lubricating fluid that reduces friction at the ends of the bones. When a disease afflicts the bones, connective tissue, or lubricating tissues in a freely movable joint, the joint's ability to move may be greatly impaired.

Rheumatoid arthritis is a painful inflammation of freely movable joints. This condition occurs when cells of the immune system attack the tissues around joints, severely damaging the joints. Symptoms of rheumatoid arthritis include stiffening and swelling of the joints. Osteoarthritis is a similar disorder that causes the degeneration of cartilage that covers the surfaces of bones. As the cartilage wears away, the bones rub together, causing pain.

Section 2 Review

1. **Distinguish** between the axial skeleton and the appendicular skeleton. SC.F.1.4.2
2. **Differentiate** between compact bone and spongy bone. SC.F.1.4.2
3. **Describe** how bones elongate in development. SC.F.1.4.2
4. **List** the three main types of joints, and give an example of each type. SC.F.1.4.2
5. **Analyzing Information** Why are women more likely than men to develop osteoporosis? SC.H.3.4.3
6. **Critical Thinking Relating Concepts** The bones of a newborn baby are made mostly of cartilage. Why is that an advantage during childbirth? SC.F.1.4.2
7. *FCAT Prep* What effect does regular exercise have on the skeletal system?
   A reduces bone mass
   B leads to osteoporosis
   C maintains bone density
   D makes bones more porous
   SC.F.1.4.7

855

Answers to Section Review

1. The axial skeleton makes up the axis or center of the skeleton and is composed of the skull, spine, ribs, and sternum. The appendicular skeleton is composed of the remainder of the bones.
2. Compact bone is dense; spongy bone is loosely connected and porous.
3. Bone cells gradually replace cartilage at the ends of the bones.
4. *immovable*—cranium; *slightly movable*—vertebrae; *freely movable*—knee, elbow, hip, and shoulder
5. Women usually have less bone tissue to lose, and after menopause their production of sex hormones decreases.
6. Cartilage is more flexible than bone, making passage through the birth canal easier.
7. **A.** Incorrect. Regular exercise increases bone mass. **B.** Incorrect. Regular exercise helps protect against osteoporosis. **C.** Correct. Regular exercise throughout life helps to maintain bone density. **D.** Incorrect. Regular exercise makes bones less porous.

Close

Reteaching — BASIC

Have students examine prepared slides of cartilage and bone. Ask them to explain how the structure of cartilage and bone is related to the function each performs in the body. (Students' responses will vary but should describe the hardness of bone, the role of bone in support and protection, the flexibility of cartilage, and the role of cartilage as a precursor of bones.) **LS Visual**

Quiz — GENERAL

1. Name one major function of the axial skeleton. (vertical support, protection of internal organs)
2. What is the relationship between cartilage and bone? (In vertebrates, the skeleton of the embryo is composed of cartilage. As the young animal grows, most of the cartilage is generally replaced by bone tissue.)
3. Which kind of joint has the greatest range of motion? (freely movable joints)

Alternative Assessment — GENERAL

Have students make models of joints and bone cross sections using door hinges, old antennas, poster board, plastic foam, marking pens, and other readily available materials. On cross sections, students should label the parts of the bone. Place the models into their portfolios.

Transparencies

TR J9 Effects of Osteoporosis
TR J12 The Human Knee

Section Review Correlations

SC.F.1.4.2: Items 1-4, 6
SC.F.1.4.7: Item 7
SC.H.3.4.3: Item 5

Section 3

Focus

Overview

Before beginning this section review with your students the objectives listed in the Student Edition. Students will understand how the skeleton is moved by muscles, how muscle tissue is constructed, and how muscle tissue contracts. They will describe aerobic and anaerobic muscle contraction and develop an appreciation for how exercise affects muscle health.

Bellringer

Have students alternately bend and straighten their arms or legs while feeling the muscles in the front and the back. Tell them to write a description of what they feel when the limb is straightened versus bent. This activity can lead to a discussion of how muscles work in opposition to bend and straighten the limbs at the joints.

Motivate

Demonstration — Basic

Have a volunteer hold his or her right arm straight out, palm up, at shoulder level. Ask him or her to place the left thumb and index finger lightly on the right biceps about two inches apart and raise the right palm to the shoulder. Have the student explain what happens to the fingers. (The fingers should move closer together, demonstrating that the muscle is actually shortening.) The demonstration works best with the sleeve rolled up. LS Kinesthetic

Section Correlations

SC.F.1.4.2
SC.F.1.4.3

Section 3 Muscular System

Florida Sunshine Standards/FCAT

SC.F.1.4.2 Body structures are uniquely designed and adapted for their function.

SC.F.1.4.3 Membranes are sites for chemical synthesis and essential energy conversions.

Objectives

- **Describe** the action of muscle pairs in moving the body. SC.F.1.4.2
- **Relate** the structure of a skeletal muscle to the muscle's ability to contract. SC.F.1.4.2
- **Describe** how energy is supplied to muscles for contraction. SC.F.1.4.3

Key Terms

tendon
flexor
extensor
actin
myosin
myofibril
sarcomere

Muscles and Movement

Every time you move, you use your muscles. Walking and running, for example, require precisely timed and controlled contractions of many skeletal muscles. When you lift a heavy object, the total force produced by muscle contractions in your arm must overcome the weight of the object. Muscles in your jaw contract and enable you to chew food with your teeth. Even when you are idle, many skeletal muscles, including those in your back and neck, remain partially contracted to maintain balance and posture.

Movement of the Skeleton

Muscles can move body parts because muscles are attached to bones of the skeleton. As shown in **Figure 10,** most skeletal muscles are attached to bones by strips of dense connective tissue called **tendons.** One attachment of the muscle, the origin, is a bone that remains stationary during a muscle contraction. The muscle pulls against the origin. The other attachment, the insertion, is the bone that moves when the muscle contracts. Movement occurs when a muscle contraction pulls the muscle's insertion toward its origin.

Skeletal muscles are generally attached to the skeleton in opposing pairs. One muscle in a pair pulls a bone in one direction, and the other muscle pulls the bone in the opposite direction. In the limbs, each opposing pair of muscles includes a flexor muscle and an extensor muscle, as shown in Figure 10. A **flexor** muscle causes a joint to bend. An **extensor** muscle causes a joint to straighten.

Figure 10 Muscle pair

Pairs of opposing muscles work together to move bones at joints.

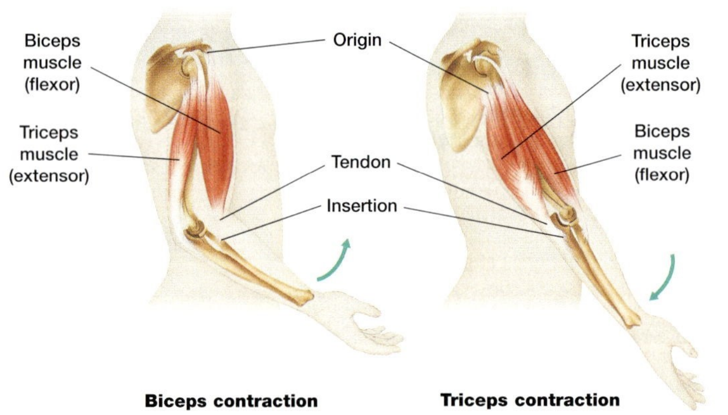

856

Chapter Resource File

- Directed Reading Basic
- Active Reading General

Transparencies

TR Bellringer
TR J17 Opposing Muscles in the Arm
TR J14 Skeletal Muscle
TR J15 Contraction of a Muscle

One-Stop Planner CD-ROM

- Reading Organizers Basic
- Reading Strategies Basic

Figure 11 Skeletal muscle

In skeletal muscle, contraction occurs within the sarcomeres of muscle fibers.

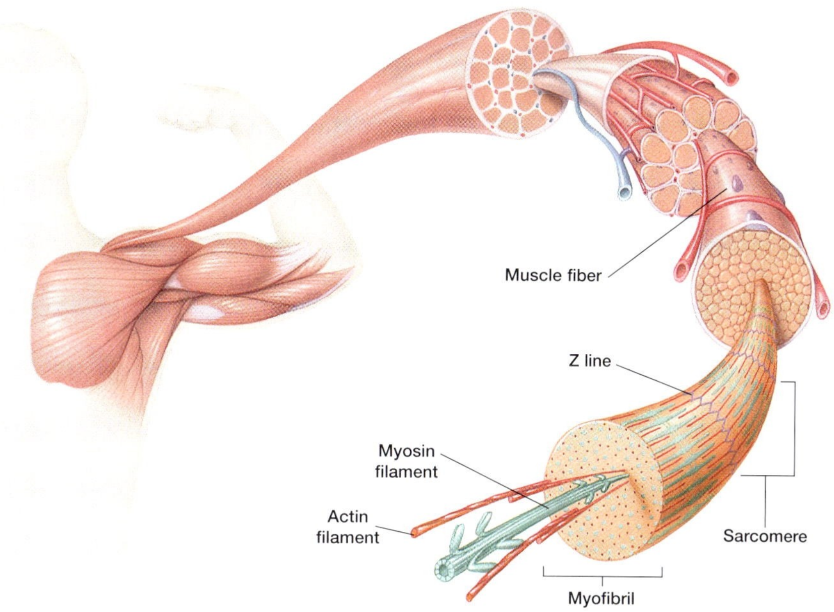

Muscle Structure

Muscles contain some connective tissue, which holds muscle cells together and provides elasticity. Muscle tissue also contains large amounts of contractile protein filaments. These protein filaments, called **actin** and **myosin** *(MIE oh sihn)*, enable muscles to contract. Actin and myosin are usually found in the cytoskeleton of eukaryotic cells, but they are far more abundant in muscle cells. Other characteristics of muscle tissue include the ability to stretch or expand and the ability to respond to stimuli, such as signal molecules released by nerve cells.

Skeletal muscle tissue consists of many parallel elongated cells called muscle fibers. As shown in **Figure 11,** each muscle fiber contains small cylindrical structures called **myofibrils** *(mie oh FIE bruhlz)*. Myofibrils have alternating light and dark bands that produce a characteristic striated, or striped, appearance when viewed under a microscope. In the center of each light band is a structure called a Z line, which anchors actin filaments. The area between two Z lines is called a **sarcomere** *(SAHR koh mihr)*. Thus, a myofibril is a grouping of sarcomeres linked end to end. Each sarcomere contains overlapping thin and thick protein filaments that move and interact with each other. The thin filaments are actin, and the thick filaments are myosin. The filaments run parallel to one another along the length of the sarcomere. The dark bands that occur in the middle of the sarcomere are regions where the thick filaments and the thin filaments overlap.

internet connect
www.scilinks.org
Topic: Muscle Structure
Keyword: HX4126
SciLinks. Maintained by the National Science Teachers Association

857

did you know?

Achilles Tendon The Achilles tendon is the thickest and strongest tendon in the human body. According to ancient Greek mythology, Achilles' mother dipped him in the magical River Styx to make him an invulnerable warrior. As she immersed him, she held him by one heel and forgot to dip him a second time so the heel she held could become wet too. Therefore, the place where she held him remained untouched by the magic water of the Styx and that part stayed mortal or vulnerable. He was killed when an arrow pierced his one weak point, the place on his heel where his mother held on to him. To this day, any weak point is called an "Achilles' heel". The Achilles tendon is named for this mythical hero.

Teach

Using the Figure

Direct students' attention to **Figure 11.** Point out the muscle fibers, myofibrils, sarcomeres, Z lines, actin, and myosin. Make sure that students understand the separate levels of organization shown in the figure. Help students understand that muscles comprise many bundles of muscle fibers, that a muscle fiber contains many myofibrils, and that a myofibril consists of actin and myosin. Remind students that muscle contractions begin at the microscopic level—with actin and myosin.

Demonstration — GENERAL

Provide students with diagrams of part or all of a few insects' skeletal and muscular systems, including at least one flying insect. Provide them also with diagrams of parts or all of a few vertebrate's skeletal and muscular systems, including at least one bird. Ask students to compare the attachments of muscles to skeleton in the insects and the vertebrates. (The insects' muscles are attached to their exoskeletons and the vertebrates' muscles are attached to their endoskeletons.) Ask students to compare skeletal movements in the insects and the vertebrates. (Both types of animals move their skeletons by contracting muscles; the insects' muscles are inside of their exoskeletons and the vertebrates' muscles are external to their endoskeletons.) Ask students to compare the muscle and skeletal structures and functions involved in flight in an insect and a bird. (The insect does not have muscles in its wing as the bird does. The insect moves it wings by the actions of muscles attached to different points on the exoskeleton. These actions cause the wings to move back and forth. The bird moves it wings by the actions of muscles in the chest area of the body and in the wings.)

Teach, *continued*

Teaching Tip — GENERAL

Sarcomeres Ask students to research muscle contraction using the Web site in the Internet Connect box on the next page. Have them draw a diagram of a sarcomere before and during a contraction, and ask them to annotate their drawings with explanations of the steps of muscle contraction. **LS Visual**

Using the Figure — BASIC

Direct students' attention to **Figure 12.** Work with students to help them understand the events of muscle contraction. Interlace the fingers of your hands to show how actin and myosin slide past each other in the sarcomeres. Have students contract a muscle (such as the gastrocnemius of the calf), and point out that they should feel the muscle become shorter and thicker. Explain that this is the result of the contraction of many sarcomeres within the muscle. **LS Kinesthetic**

Activity — BASIC

Have students curl their middle finger under and place their fingertips down on a table. Have them try to raise their fingers one by one. They should be able to lift all fingers except the ring finger. Explain that each finger has a tendon, and those tendons attach to a muscle in the arm. The tendons in the middle and ring finger are linked together in the hand, connecting their movement. **LS Kinesthetic**

Muscle Contraction

How does a muscle contract? Muscle contraction occurs in the sarcomeres of myofibrils. The overlapping arrangement of the thick and thin protein filaments in a sarcomere enables muscle contraction. The events of a muscle contraction are summarized in **Figure 12.**

Step 1 Before a muscle is stimulated, the sarcomere is relaxed. Myosin and actin filaments partially overlap one another.

Step 2 A muscle contraction usually begins when a muscle fiber is stimulated by signal molecules released by a nerve cell. This causes myosin and actin filaments to "slide" along one another so that they overlap even more. The sarcomere becomes shorter as the Z lines are pulled closer together.

Step 3 The sarcomere is fully contracted, and myosin and actin completely overlap one another. This shortening of sarcomeres occurs down the entire length of the muscle fiber.

What determines the force of contraction? A muscle exerts the greatest force when all of its fibers are contracted. When a fiber is stimulated, its sarcomeres contract. The total amount of force a muscle exerts depends on how often muscle fibers are stimulated and how many muscle fibers contract.

How is the force of muscular contraction controlled? As different numbers of fibers in a muscle contract at one time, the total force generated by contraction varies. For example, the total amount of force needed to lift a pencil is much less than the force needed to lift a brick. Thus, fewer muscle fibers in your arm contract when you lift a pencil than when you lift a brick.

The set of muscle fibers activated by a nerve cell is called a motor unit. Every time a nerve cell activates its motor unit, all the fibers in that unit contract. Muscles that require a finer degree of control, such as muscles that move the fingers, have only a few muscle fibers in each motor unit. Large muscles, such as muscles in the leg, have several hundred muscle fibers in each motor unit.

Figure 12

BIO graphic **Muscle Contraction**

During a muscle contraction, sarcomeres in myofibrils shorten.

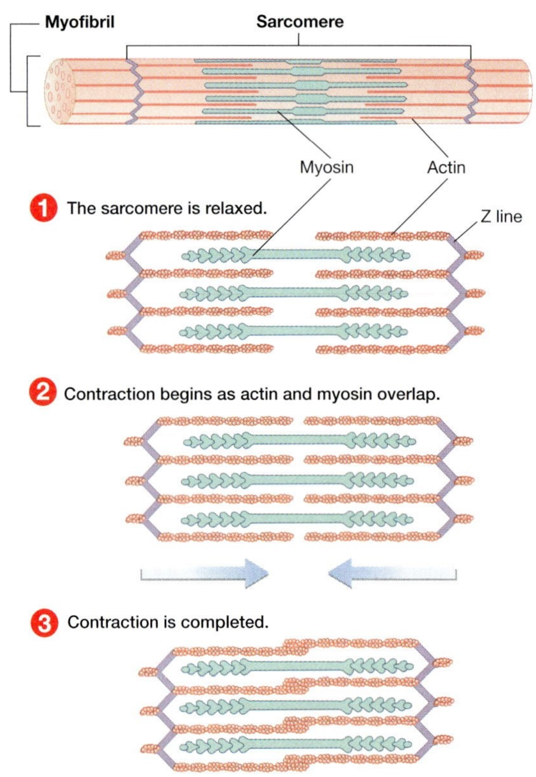

858

MEDICINE CONNECTION

Because movement requires energy, students might assume that ATP is consumed at the moment muscles contract. A key discovery made by the British physiologist and biophysicist Archibald V. Hill, however, helped biologists determine the sequence of events. Hill discovered that muscle cells use ATP after, not during, contraction. ATP does not cause contraction, rather ATP prepares the muscle for the next contraction. In 1922 Hill received a Nobel Prize in medicine and physiology for his work on muscles. His work was continued by Hugh Huxley, who formulated a theory that explained the role of ATP in cocking myosin heads prior to the contraction of the sarcomere.

Interaction of Myosin and Actin

How do actin and myosin cause sarcomeres to shorten during a muscle contraction? Myosin filaments have long, finger-like projections with an enlarged "head" at one end. Actin filaments contain many sites to which myosin can bind during a muscle contraction. Stimulation of a muscle fiber leads to the exposure of these binding sites on actin filaments. As shown in **Figure 13,** the myosin heads attach to the binding sites on actin filaments and then rotate, causing myosin to move relative to actin.

internet connect
www.scilinks.org
Topic: Muscle Contraction
Keyword: HX4125
SCILINKS. Maintained by the National Science Teachers Association

Step 1 Muscle contraction begins as a myosin head attaches to an exposed binding site on an actin filament.

Step 2 The myosin head rotates, causing the actin filament to "slide" against the myosin filament. This sliding causes the filaments to overlap one another.

Step 3 ATP is used as the myosin head detaches and snaps back into its original position. The myosin head reattaches to actin at a binding site farther along the actin filament. When the myosin heads cannot move farther, they release momentarily and reposition themselves to grab the actin and pull again. Thus, the myosin heads "walk" along actin filaments, essentially "stepping" at each available binding site. This grabbing and pulling action is repeated, causing the sarcomere to shorten as the Z lines are pulled closer together.

A lot of energy is needed to power a muscle contraction. ATP is used each time a myosin head moves from one binding site on an actin filament to another. Without ATP, myosin heads would remain attached to actin filaments, keeping the muscle contracted. ATP is also used to regulate calcium ions in muscle cells. Calcium ions are needed for binding sites to be exposed on actin filaments. Without calcium ions and ATP, a muscle could not contract.

BIO graphic

Figure 13

Interaction of Muscle Protein Filaments

During a muscle contraction, myosin and actin move against one another.

1 Muscle contraction begins as a myosin head attaches to a binding site on an actin filament.

2 The myosin head rotates, causing the actin filament to "slide" against the myosin filament.

3 ATP is used as the myosin head detaches from the binding site and snaps back into its original position.

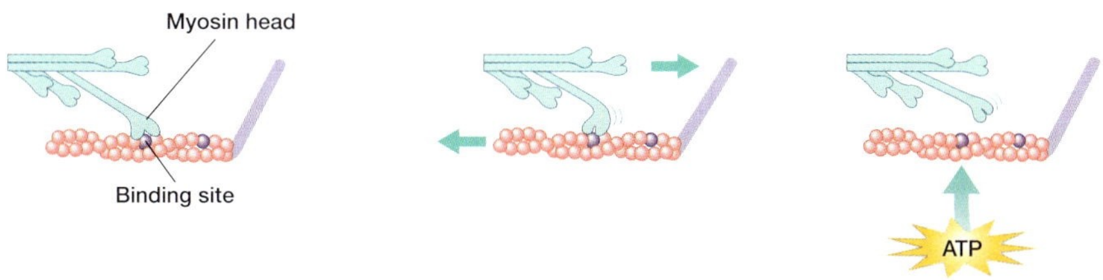

859

Teaching Tip

Rigor Mortis After death, ATP production stops. Without ATP, myosin heads attach to actin filaments but cannot detach. As a result, all the muscles are locked in their contracted position, making the body rigid. This condition is known as rigor mortis ("stiffness of death"), and it persists for up to 25 hours after death.

Using the Figure — Basic

Have four students stand in a straight line, one behind the other, with their hands on their hips. Have a fifth student face the leader of the line and slowly pull herself along the line by grasping the elbow of each member of the line in succession. Ask students to relate **Figure 13** to the demonstration. (The fifth student represents myosin; the others are actin. The fifth student's hand is the myosin head.) **LS Kinesthetic**

Integrating Physics and Chemistry

Explain to students that resistance exercises such as weight lifting often involve a simple machine known as a human forearm, which is an example of a third-class lever. Your biceps muscle attaches near the elbow, which is the fulcrum. It contracts a short distance to move the forearm a long distance. Have students identify other third-class levers of the human body. Ask them if any human body joints are like other simple machines such as the inclined plane or pulley.

Transparencies

TR J16 Interaction of Muscle Protein Filaments

Attention Grabber

Face Muscles The human face has 44 separate muscles, four of which are devoted to chewing and 40 to facial expression. No other species has such capacity for making faces.

It is now clear that extraordinary amounts of information can be gleaned by systematically measuring facial expression. Researchers have recently found a way to measure changes in facial muscle tension that are too small to create visible movements. They have discovered that when normal people merely think about certain mildly emotional topics, such as recalling happy or sad experiences, tension increases slightly in some facial muscles. Sensitive facial expression measurement techniques such as this aid in determining subtleties of feelings and attitudes, especially when verbal reports are untrustworthy or unavailable.

Close

Reteaching — BASIC

Have students create a flip book of 20 pages showing the mechanics of muscle contraction. The first and last pages should have the following labels: actin, myosin, myosin head, and sarcomere.

Quiz — GENERAL

1. What are tendons? (They are strips of connective tissue that attach skeletal muscles to bones.)
2. What is a sarcomere? (A sarcomere is a very small subunit of a muscle cell.)
3. How is calcium involved in muscle function? (Calcium enables myosin heads to bind to actin filaments in the initial stage of muscle contraction.)

Alternative Assessment — GENERAL

Ask students to draw a diagram to illustrate the role of ATP in muscle contraction. Their diagrams should show each step in the muscle contraction process and indicate where ATP is involved.

Integrating Physics and Chemistry

Describe how energy and work are related and distinguish both terms from power. Whereas energy or work is the ability to change or move matter (i.e., *work* = *force* × *distance*), power is a measure of how much work gets done in a certain amount of time. Power is therefore derived from the ratio of energy or work to time (i.e., *power* = *work/time*).

Real Life

What is the best way to increase muscle size, strength, and endurance?

If you want to have large muscles and be strong enough to lift large loads, you should train with heavy weights. On the other hand, repeatedly lifting lighter weights builds muscle endurance.

Figure 14 Resistance exercise. Weight lifting and other resistance exercises primarily involve anaerobic energy pathways in muscles.

Aerobic and Anaerobic Energy Pathways

The ATP used to power contractions is usually supplied by aerobic respiration. During prolonged exercises, such as a long-distance walk, oxygen is consumed at a sustainable, steady rate, and aerobic respiration yields most of the ATP. However, during brief, intense activities, such as sprinting and weight lifting, anaerobic processes take over. Most of the ATP used in such activities comes from glycolysis, as the oxygen available to muscle cells rapidly decreases.

When both anaerobic and aerobic energy pathways become insufficient for muscle contraction, muscles can use only glycogen as an energy source. As glycogen is used up, the body begins to use fat as an energy source. When ATP consumption exceeds ATP production, muscle fatigue and soreness may result, leaving muscle fibers unable to recover from contraction.

Exercise and Fitness

Consistent aerobic exercise makes the heart pump more efficiently and thus increases the energy available to muscles as a result of improved blood circulation. More oxygen is extracted by the body with each breath, increasing the oxygen supply to muscles. More ATP is available for muscle contractions, thereby reducing muscle fatigue. The increase in muscle efficiency results in greater endurance, or the ability to continue exercising.

Resistance exercises, such as weight lifting, shown in **Figure 14,** can increase muscle size and strength. Resistance exercises are mostly anaerobic, so they do not usually improve the uptake of oxygen to muscles. Muscle mass is increased by resistance training. The amount of tension and the rate of exercise are both important factors. However, the short-term demands of such strength training do not cause the circulatory changes that increase endurance.

Excessive exercise or failure to warm up properly can lead to muscle injury. Muscles can tear if they are stretched too far during strenuous exercise. If excessive stress causes tendons to become inflamed, a painful condition called tendinitis results.

Section 3 Review

1. **Describe** how muscle pairs work together to move body parts. SC.F.1.4.2
2. **Compare** the roles of thick and thin filaments in muscle contraction. SC.F.1.4.2
3. **Identify** the energy pathway that is primarily involved with exercises that increase muscle size and strength. SC.F.1.4.3
4. **Critical Thinking Applying Information** What causes muscle cramping after rigorous exercise or a repeated movement? SC.H.3.4.3
5. **FCAT Prep** Which main organ systems are involved when you flex your arm at the elbow?
   A muscular, skeletal, immune
   B nervous, muscular, skeletal
   C skeletal, excretory, nervous
   D endocrine, muscular, immune
   SC.F.1.4.2

Answers to Section Review

1. Many skeletal muscles exist in opposing pairs. Flexors bend joints and extensors straighten them.
2. Thick (myosin) filaments have heads that attach to thin (actin) filaments. The heads rotate, causing actin to slide against myosin. This process repeats many times during a single muscle contraction.
3. the anaerobic pathway
4. Answers may include: muscle fatigue, dehydration, and salt imbalance.
5. **A.** Incorrect. The immune system is not involved in muscle contraction. **B.** Correct. All three of these systems are involved. **C.** Incorrect. The excretory system is not involved in muscle contraction. **D.** Incorrect. The endocrine and immune systems are not involved in muscle contraction.

Section Review Correlations

SC.F.1.4.2: Items 1, 2, 5
SC.F.1.4.3: Item 3
SC.H.3.4.3: Item 4

Skin, Hair, and Nails

Section 4

Florida Sunshine Standards/FCAT

SC.F.1.4.2 Body structures are uniquely designed and adapted for their function.

SC.H.3.4.3 Scientists can bring information, insights, and analytical skills to matters of public concern and help people understand the possible causes and effects of events.

Skin

The skin, which makes up about 15 percent of your total body weight, is the largest organ of the body. Many specialized structures are found in the skin, which along with the hair and nails, forms the integumentary system. The skin protects the body from injury, provides the first line of defense against disease, helps regulate body temperature, and prevents the body from drying out through evaporation. As shown in **Figure 15,** the skin is made mostly of connective tissue and layers of epithelial tissue. The two primary layers of skin are the epidermis and the dermis.

Epidermis

The **epidermis** is the outermost layer of the skin. About as thick as this page, the epidermis is made of several layers of epithelial cells. The part of the epidermis you see when you look in a mirror is a thin layer of flattened, dead cells that contain keratin. **Keratin** is a protein that makes skin tough and waterproof. The cells of the epidermis are continuously damaged by the environment. They are scraped, ripped, worn away by friction, and dried out because of moisture loss. Your body deals with this damage not by repairing the cells, but by replacing them.

Objectives

- **Analyze** the structure and function of the epidermis. SC.F.1.4.2
- **Describe** how the dermis helps the body maintain homeostasis. SC.F.1.4.2
- **Summarize** how hair and nails are formed. SC.F.1.4.2
- **Identify** various types of skin disorders. SC.H.3.4.3

Key Terms

epidermis
keratin
melanin
dermis
hair follicle
subcutaneous tissue
sebum

Figure 15 Structure of skin

Skin has two distinct layers that contain many blood vessels, nerve cells, muscles, hairs, and glands.

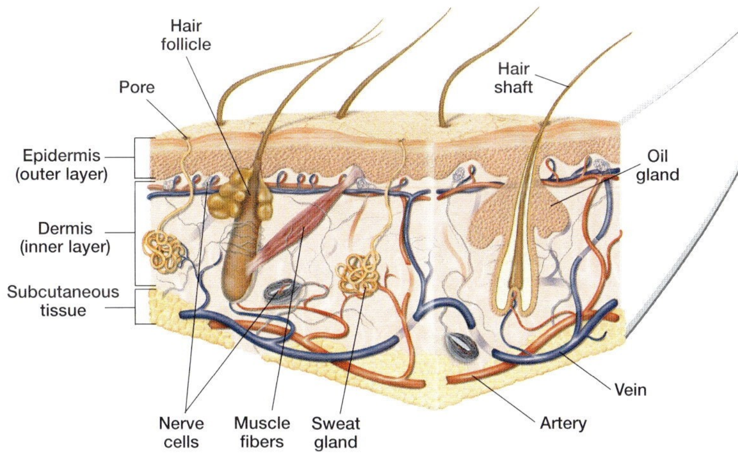

861

Chapter Resource File

- Directed Reading BASIC
- Active Reading GENERAL

Transparencies

TR Bellringer
TR J18 Structure of Skin

One-Stop Planner CD-ROM

- Reading Organizers BASIC
- Reading Strategies BASIC
- Supplemental Reading Guide *Microbe Hunters* ADVANCED

Section 4

Focus

Overview

Before beginning this section review with your students the objectives listed in the Student Edition. Students will examine the composition of the different layers of the skin as well of the functions of the skin. They will describe the structure of hair and nails. Finally, they will develop an understanding of skin disorders including their causes, symptoms, and treatments.

Bellringer

Ask students to write down all of the functions they can think of that involve the skin. (containing other organs, reduces water loss, regulates body temperature, removes toxins, protects against injury, protects against infection)

Motivate

Discussion

Ask students for a show of hands to the following questions: Who likes to go to the beach? Who tans? Who burns? Explain that a suntan is actually a defense mechanism of the skin. Melanin is a dark pigment that is present in varying degrees in different individuals. Additional melanin is produced by the skin when it is exposed to ultraviolet radiation. We call this additional melanin a tan; the darkened skin protects the dermis somewhat from damaging radiation. Emphasize the importance of using SPF 15 or higher sunscreen to prevent sunburn. Burns early in life increase the chances of skin cancer later. LS Intrapersonal

Section Correlations

SC.F.1.4.2
SC.H.3.4.3

Teach

Teaching Tip — GENERAL

Skin Layers Have students make a Graphic Organizer similar to the one at the bottom of this page to demonstrate the relationship between the skin's two primary layers and the subcutaneous tissue. Below the name of each of the three layers, have students describe its components. LS Visual

BIOWatch: Courtroom Science

Teaching Strategies
Have students collect newspaper, magazine, or Internet articles about the uses of forensic technologies in the courtroom.

Discussion
Should these relatively new technologies be used in a courtroom? (Answers will vary but may include that although they are scientifically valid, they have not been tested sufficiently.)

What are some of the drawbacks of using hair for drug testing? (Answers will vary but may include that some individuals may cut their hair to avoid detection of their drug use.)

Should employers be able to do random drug testing on hair? (Answers will vary but may include that it is an accurate indicator of drug use and provides a long-term record, but it may violate an individual's rights in certain cases.)

Forensics Labs Appendix
Turn to pp. 1048–1071 for forensics lab activities and more student information on forensics techniques.

The outermost cells of the skin are continually shed and replaced by a layer of actively dividing cells at the base of the epidermis. As new skin cells form, they migrate upward and produce large amounts of keratin. These cells are shed about a month after they reach the surface. The inner layer of the epidermis also contains cells that produce the pigment melanin. **Melanin** *(MEHL uh nihn)* ranges in color from yellow to reddish brown to black, and it helps determine skin color. People with more melanin tend to have darker skin, and people with less melanin usually have lighter skin. Melanin also absorbs ultraviolet (UV) radiation, protecting the skin from exposure to sunlight. Exposure to UV radiation increases the production of melanin. This is why some people become "tan" after exposure to sunlight. However, UV radiation has been shown to cause skin cancer, especially in people with light skin. Thus, you should avoid excessive exposure to sunlight and wear sunscreen when outdoors.

Dermis

The **dermis** is the functional layer of skin that lies just beneath the epidermis. Connective tissue in the dermis makes the skin tough and elastic. The dermis contains many nerve cells, blood vessels, **hair follicles,** and specialized skin cells. Sensations of touch,

FORENSICS BIOWatch: Courtroom Science

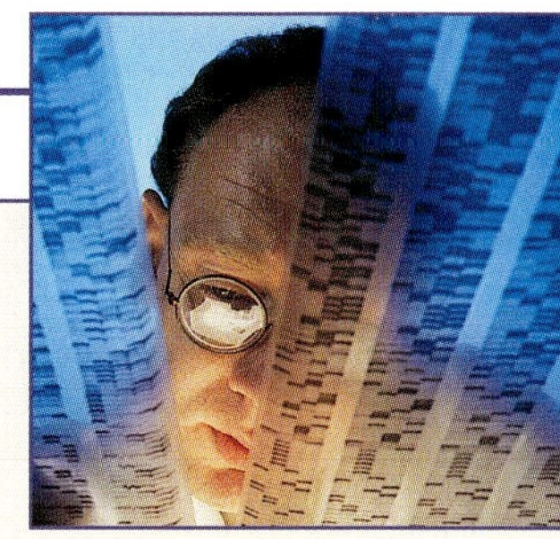

Forensic scientist

In the United States, medical evidence is often used in criminal trials. Hairs left at a crime scene can provide enough information to make or break a case. To the unaided eye, the differences between hairs amount to no more than color, coarseness, and whether the hairs are curly or straight. Using a microscope, however, scientists can distinguish more than two dozen characteristics in a single hair. Experts compare hairs found at a crime scene with hairs of suspects. These comparisons can rule out a suspect when the hairs are different, but they usually are not enough to convict a suspect without other evidence.

DNA Fingerprinting

Hair consists of cells that contain DNA. Thus, scientists can get additional evidence from hair through DNA fingerprinting. First used as forensic evidence in the 1980s, DNA fingerprinting is now fairly common in many kinds of criminal trials. By 1997, DNA fingerprinting had been used in more than 50,000 cases in the United States.

Drug Testing

Hair can also provide evidence of drug use. Hair collects drugs that are delivered to hair follicles by blood. Hair provides a longer record of drug exposure than blood or urine does. Head hair grows an average of about 1 cm (0.4 in.) per month; a 10 cm (4 in.) length of hair gives evidence of a person's drug use during the last 10 months. Such evidence has been introduced in cases involving illegal drug use.

Graphic Organizer

Use this graphic organizer with **Teaching Tip** on this page.

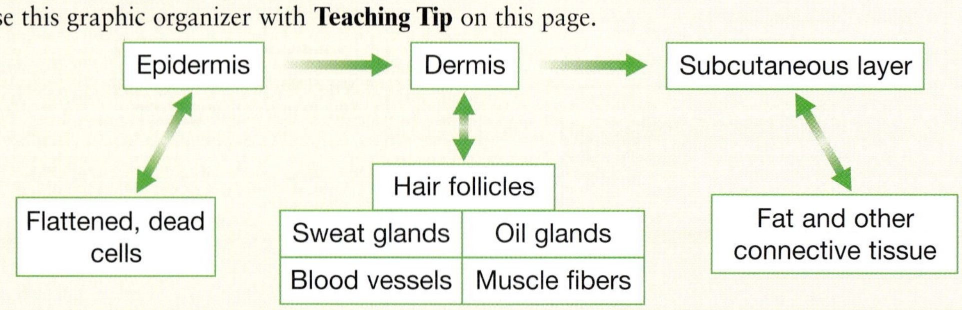

temperature, and pain originate in nerve cells within the dermis. The dermis also contains tiny muscles that are attached to hair follicles in your skin. When you get cold, these muscles contract and pull the hair shafts upright, helping to insulate the body. These muscles also cause goose bumps on the skin's surface.

Temperature Regulation A network of blood vessels in the dermis provides nourishment to the living cells of the skin. These blood vessels also help regulate body temperature by either radiating heat into the air or conserving heat. If your body gets too hot, blood vessels just under the skin dilate so that blood flows near the skin's surface, releasing heat from the body. This is why people with light complexions turn slightly red during strenuous exercise, as shown in **Figure 16.** If your body gets too cold, the blood vessels constrict, reducing heat loss.

Sweat glands in the dermis also help remove excess body heat. The evaporation of sweat from the skin's surface removes heat more efficiently than the dilation of blood vessels. Most sweat is about 99 percent water and 1 percent dissolved salts and acids. Certain sweat glands located in body areas with dense hair, such as the armpits, also secrete proteins and fatty acids. Because these substances provide a rich food source for bacteria, stale sweat often releases the offensive odor of bacterial waste products.

Figure 16 **Cooling mechanism.** Flushed (reddened) skin is a sign that the body is overheated and is releasing heat from the skin.

Subcutaneous Tissue

Subcutaneous tissue, located beneath the skin just under the dermis, is a layer of connective tissue made mostly of fat. Subcutaneous tissue acts as a shock absorber, provides additional insulation to help conserve body heat, and stores energy. Subcutaneous tissue also anchors the skin to underlying organs. The thickness of subcutaneous tissue varies in different parts of the body. For example, the eyelids have very little, while the buttocks and thighs may have a lot. The pads of subcutaneous tissue in the soles of your feet may be more than 6 mm (0.25 in.) thick.

Hair and Nails

Hair and nails are derived from the epidermis. Hair follicles produce individual hairs, which help protect and insulate the body. Shown in **Figure 17,** hair is made mostly of dead, keratin-filled cells. A shaft of hair grows up from the hair follicle and through the skin's surface. Each hair on your head grows for several years. Then the follicle enters a resting phase for several months, and the hair is eventually shed. Hair color is primarily determined by the presence of the pigment melanin. Blonde hair and red hair typically contain less melanin than brown hair and black hair.

Nails are produced by specialized epidermal cells located in the light, semicircular area at the base of each nail. These cells become filled with keratin as they are pushed outward by new cells. Nails protect the tips of the fingers and toes and continue to grow throughout life.

Figure 17 **Hair.** An average human head contains about 100,000 hairs.

Teaching Tip

Hair Growth Tell students that hair grows at different rates in different regions of the body. The growth rate also differs with age and sex. Hair follicles go through cycles that average 4 years of growth followed by several months of inactivity. With the beginning of a new active growth cycle, the old hair is pushed out (if it has not fallen out already). Out of more than 100,000 hairs, only about 50 hairs a day are lost. Several factors can increase hair loss. These include dietary factors, high fever, stress, drugs, radiation, and some hormonal factors related to pregnancy.

Teaching Tip — BASIC

Homeothermy Tell students that when a person's body temperature rises above 41°C (106°F), enzymes in the brain begin to denature. Have students describe the appearance of a raw egg white. (clear, runny, slimy) Then have them describe the appearance of a cooked egg white. (white, firm) Explain that the egg white changes when it is cooked because its proteins become misshapen, or denatured. Enzymes also are proteins, and at high temperatures they no longer function normally.

SKILL BUILDER — ADVANCED

Writing Skills Ask students to research skin cancer using the Web site in the Internet Connect box on the next page. Have them write a paragraph on the risk factors and symptoms of two relatively common carcinomas: basal cell carcinoma and squamous cell carcinoma. Ask students to identify the primary risk factor for nearly all types of skin cancer. (exposure of the skin to the sun) **LS Verbal**

Integrating Physics and Chemistry

Remind students that UV radiation is an electromagnetic wave or disturbance that carries energy through space and therefore does not need a medium in order to travel. In addition, mention that the vibrations in electromagnetic waves are caused by fluctuations in electric and magnetic fields, which are perpendicular to the direction the wave travels and thus behave like transverse waves.

Trends in Neurobiology

New Technique Medical researchers have developed a technique for severing nerves within the chest cavity that appear to be connected with excessive sweating (especially of the palms) and flushing of the upper body (especially of the face). Surgeons use an endoscope to view the inside of the thoracic cavity, and as a result the procedure involves minimal invasion to surrounding tissue. The procedure has a success rate of about 95 percent.

Close

Reteaching — BASIC

Use rapid-fire questions to review the layers of the skin and the important structures in each layer. Some key items: artery (dermis), melanin (epidermis), dead cells (epidermis), fat (subcutaneous tissue), middle layer (dermis).

Quiz — GENERAL

1. Identify one function of skin. (protects the body from injury, protects the body from disease, helps regulate body temperature, prevents the body from drying out)
2. What is melanin and what is its function? (Melanin is a skin pigment that absorbs ultraviolet radiation, thus protecting the skin against this radiation.)
3. What kind of tissue is hair made of? (epithelial)
4. What kinds of tissues are found in the skin? (nervous, muscular, epithelial, connective)

Alternative Assessment — ADVANCED

Have students make a three-dimensional model of the skin using construction paper, tissue paper, yarn, glue, or other art materials. Ask them to show and explain their models to the class when they finish.

Section Review Correlations

SC.F.1.4.2: Items 1, 2, 3, 6
SC.F.2.4.3: Item 4
SC.H.3.4.3: Item 5

internet connect
www.scilinks.org
Topic: Skin Cancer
Keyword: HX4165
SCLINKS. Maintained by the National Science Teachers Association

Skin Disorders

The skin is continuously exposed to damaging factors such as insect bites, microorganisms, and ultraviolet radiation. Injuries such as scrapes and blisters are often minor and usually heal rapidly without permanent scarring. Burns, however, can be very serious and can result in permanent scarring or even death. Some skin disorders are the result of changes that occur within the body over time.

Acne

The most common skin problem for teenagers is acne *(AK nee)*, a chronic inflammatory condition that involves the skin's oil-producing glands. Oil glands in the dermis release **sebum** *(SEE buhm)*, an oily secretion that lubricates the skin. Sebum is released through ducts, or pores, into nearby hair follicles. These oil glands are especially active during adolescence. Acne is caused by excessive secretion of sebum, which blocks pores with oil, dirt, and bacteria. Makeup and other cosmetic products can contribute to clogging. As a result, the surrounding tissue becomes infected and inflamed, and the pores accumulate pus, producing pimples. Serious acne may need to be treated using antibiotics. Although acne cannot be prevented, it can usually be managed with proper skin care.

Figure 18 Skin cancer. In its early stages, a carcinoma may look like a wart. A malignant melanoma often looks like a mole that changes in size, shape, or color.

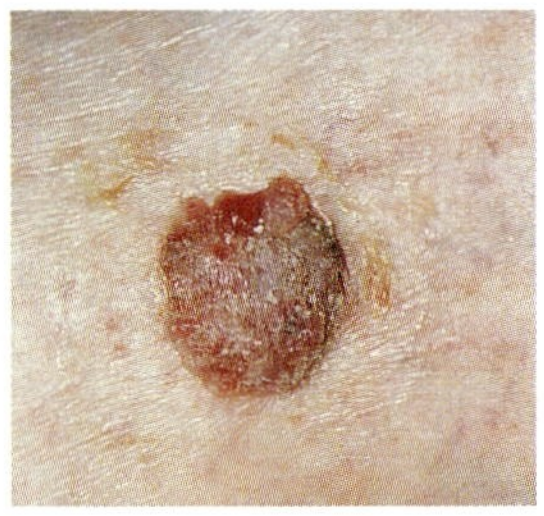

Carcinoma

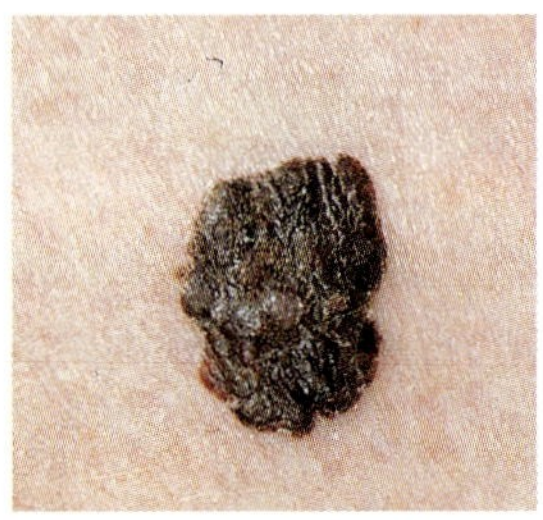

Malignant melanoma

Skin Cancer

Skin cancer can result from genetic mutations caused by overexposure to UV radiation. The most common types of skin cancer are carcinomas *(kahr sih NOH mahz)*, which originate in skin cells that do not produce pigments. If they are detected early, carcinomas can be treated. A small percentage of skin cancers are caused by mutations that occur in pigment-producing skin cells. These cancers, called malignant melanomas *(mehl uh NOH mahz)*, grow very quickly and spread easily to other parts of the body. About 8 out of 10 skin cancer deaths are from malignant melanomas. A carcinoma and a malignant melanoma are shown in **Figure 18.** You can reduce the risk of skin cancer by avoiding overexposure to either natural or artificial UV radiation and by using protective sunscreens.

Section 4 Review

1. **Describe** the structure of the epidermis. SC.F.1.4.2
2. **List** two ways that the dermis helps regulate body temperature. SC.F.1.4.2
3. **Summarize** how nails are formed. SC.F.1.4.2
4. **Identify** the most common cause of skin cancer and how it can be avoided. SC.F.2.4.3
5. **Critical Thinking Recognizing Relationships** Why is a third-degree burn, which destroys the epidermis and dermis of the skin, such a serious injury? SC.H.3.4.3
6. *FCAT Prep* When your body temperature becomes too low, what happens to the blood vessels just under your skin?
   A They dilate.
   B They constrict.
   C They release sweat.
   D They raise hair shafts.
   SC.F.1.4.2

Answers to Section Review

1. The epidermis is composed of a thin layer of dead, flattened, keratin-containing cells.
2. The dermis helps regulate body temperature by controlling the diameter of blood vessels. A large diameter vessel dissipates heat; a small diameter vessel conserves heat. Also, the dermis releases sweat to cool the body.
3. At the base of each nail are specialized epidermal cells. These cells fill with keratin, which forms the nail, as they are pushed outward by new cells.
4. Overexposure to UV radiation is the most common cause of skin cancer. Using sunscreen and limiting sun exposure can help reduce the risk.
5. The epidermis and dermis perform several vital functions, including waterproofing the body, keeping pathogens out, and regulating body temperature. Loss of the dermis and epidermis removes the barrier between the interior tissues and the external environment.
6. **A.** Incorrect. These blood vessels dilate when the body temperature becomes too high. **B.** Correct. Constriction helps to reduce heat loss. **C.** Incorrect. Blood vessels do not release sweat; sweat is released when the body temperature becomes too high. **D.** Incorrect. Hair shafts are not raised by blood vessels.

Study ZONE

Key Concepts

1 Body Organization

- Cells are grouped into four types of body tissues: epithelial tissue, nervous tissue, connective tissue, and muscle tissue.
- Body organs contain several types of body tissues.
- Organs are grouped into organ systems in which organs interact to perform a certain function, such as digestion.
- Endothermy enables the body to maintain homeostasis at all times, regardless of the temperature outside the body.

2 Skeletal System

- The skeleton supports the body, provides protection for internal organs, and enables movement.
- The 206 bones of the skeleton are divided into the axial skeleton and the appendicular skeleton.
- Bones are made of hard compact bone surrounding porous spongy bone.
- Early in development, the skeleton is mostly cartilage. Bones harden as calcium and other mineral deposits build up.
- Bones thicken and elongate as development continues.
- Three kinds of joints fasten bones together: immovable joints, slightly movable joints, and freely movable joints.

3 Muscular System

- Muscles are attached to bones by tendons.
- Muscle pairs move parts of the body by pulling on bones.
- Sarcomeres shorten during muscle contraction.
- During muscle contraction, actin and myosin interact.
- Energy is required for muscles to contract.

4 Skin, Hair, and Nails

- The skin consists of two layers: the epidermis and the dermis.
- Subcutaneous tissue anchors skin to underlying organs.
- Hair and nails are derived from the epidermis.
- Most skin disorders are caused by damage to the epidermis.

Key Terms

Section 1

epithelial tissue (846)
nervous tissue (847)
connective tissue (847)
muscle tissue (847)
body cavity (849)

Section 2

axial skeleton (850)
appendicular skeleton (850)
bone marrow (851)
periosteum (851)
Haversian canal (852)
osteocyte (852)
osteoporosis (853)
joint (854)
ligament (854)

Section 3

tendon (856)
flexor (856)
extensor (856)
actin (857)
myosin (857)
myofibril (857)
sarcomere (857)

Section 4

epidermis (861)
keratin (861)
melanin (862)
dermis (862)
hair follicle (862)
subcutaneous tissue (863)
sebum (864)

Alternative Assessment — GENERAL

Have students write trivia questions for a review game. These questions can be based on the review questions found at the end of each section.

Chapter Resource File

- Science Skills GENERAL
- Critical Thinking ADVANCED
- Test Prep Pretest GENERAL
- Chapter Test GENERAL

Answer to Concept Map

The following is one of several possible answers to Performance Zone item 15.

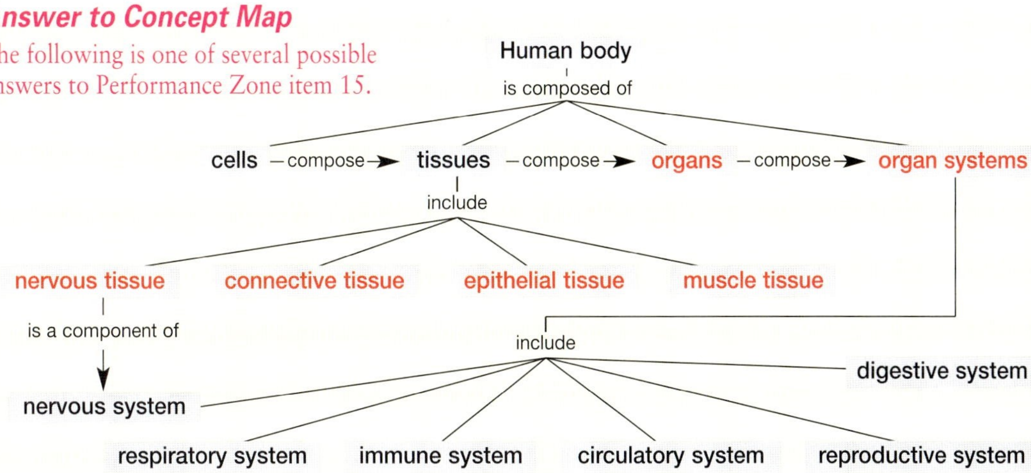

ANSWERS

Understanding Key Ideas

1. c
2. d
3. c
4. a
5. a
6. d
7. increased secretion of oil, clogged pores, dirt, makeup, and bacteria
8. *ball and socket*—shoulder, pivoting; *hinge*—knee, bending and straightening
9. DNA fingerprinting and analysis of hair, blood, and urine samples. Answers will vary, but identification by DNA sampling is the most specific.
10. The answer to the concept map is found on the previous page.

Critical Thinking

11. Muscle contraction requires ATP, which is produced when food is broken down during cellular respiration.
12. The bones thicken and become stronger with age as bone replaces cartilage.
13. Washing skin too often can deplete the natural bacteria-killing substance, resulting in an overgrowth of bacteria.

Alternative Assessment

14. Answers will vary.

Understanding Key Ideas

1. The thoracic cavity contains SC.F.1.4.2
   a. the spinal cord.
   b. the brain.
   c. organs of the respiratory system.
   d. organs of the reproductive system.
2. Which of the following is a function of the skeletal system? SC.F.1.4.2
   a. support
   b. protection
   c. movement
   d. All of the above
3. Resistance exercises
   a. decrease muscle endurance.
   b. decrease muscle strength.
   c. increase muscle size.
   d. increase the number of muscle cells. SC.F.1.4.2
4. Shortening of sarcomeres causes
   a. muscles to contract.
   b. Z lines to move apart.
   c. muscles to relax.
   d. None of the above SC.F.1.4.2
5. The dermis helps regulate body temperature by producing
   a. sweat.
   b. acne.
   c. oil.
   d. sebum. SC.F.1.4.2
6. The risk of developing skin cancer is increased by SC.F.2.4.3
   a. eating oily foods.
   b. using sunscreen.
   c. exercising.
   d. sunbathing.
7. What factors contribute to acne, shown in the photograph below? SC.F.1.4.8

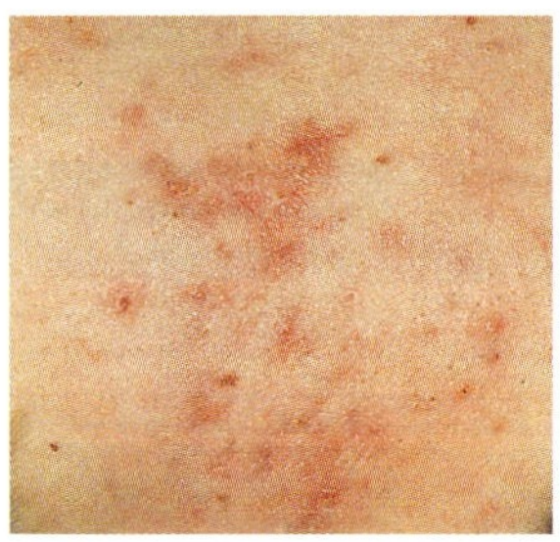

8. Give an example of a ball-and-socket joint and a hinge joint. What type of movement is permitted by each of these joints? SC.F.1.4.2
9. BIOWatch What are some ways in which forensic science can help in identifying a criminal? Which method produces the most compelling evidence? SC.H.3.4.3
10. **Concept Mapping** Make a concept map that illustrates the body's four levels of structural organization. Try to include the following terms in your map: *muscle tissue, connective tissue, epithelial tissue, nervous tissue, organ,* and *organ system.* SC.F.1.4.2

Critical Thinking

11. **Inferring Relationships** Relate the events of muscle contraction to cellular respiration. SC.F.1.4.2
12. **Inferring Relationships** Young thoroughbred horses that are raced too early in life have an increased risk of breaking the bones in their legs. Using this information, what can you infer about bone development in horses? SC.H.3.4.3
13. **Predicting Outcomes** Oil glands in the skin secrete a substance that helps kill bacteria. What might happen if you washed your skin too often? SC.F.1.4.2

Alternative Assessment

14. **Relating Structure to Function** Use a compound light microscope to compare prepared slides of bone, muscle, and epithelial tissue. Write a short report that describes the structural differences between cells. Relate the structure of cells to their functions. SC.F.1.4.2
15. **Communicating** Find out about the causes, symptoms, and treatment of muscular disorders such as muscular dystrophy and Lou Gehrig's disease. Summarize your findings in an oral report. SC.H.3.4.3

15. Neuromuscular disorders are usually genetic in nature, although some muscular dystrophies may be autoimmune in origin. Symptoms of Lou Gehrig's disease (amyotrophic lateral sclerosis) include muscle tissue degeneration and paralysis, followed by death. In the muscular dystrophies, muscle tissue gradually is replaced by fat, leading to immobility. Muscle degeneration often affects the ability to breathe, leading to respiratory infections.

Assignment Guide

Section	Questions
1	1, 10
2	2, 8, 12, 14
3	3, 4, 11, 14, 15
4	5, 6, 7, 9, 13, 14

Chapter Review Correlations

SC.F.1.4.2: Items 1-5, 8, 10, 11, 13, 14
SC.F.1.4.8: Item 7
SC.F.2.4.3: Item 6
SC.H.3.4.3: Items 9, 12, 15

FCAT Prep

For questions 1–6, write your answers on a separate sheet of paper.

1. What do the bones of the arms and legs form?
   A. the vertebrae **C.** the pectoral girdle
   B. the axial skeleton **D.** the appendicular skeleton
   SC.F.1.4.2
2. What is the sequence of the levels of organization found in the body?
   F. muscle → muscle cell → muscle tissue → muscular system
   G. muscle cell → muscle tissue → muscle → muscular system
   H. muscle cell → muscular system → muscle → muscle tissue
   I. muscular system → muscle tissue → muscle cell → muscle
   SC.F.1.4.2
3. Red blood cells are produced in red bone marrow. How are red blood cells transported to the rest of the body?
   A. They are taken up by the intestine, which is part of the digestive system.
   B. They are taken up by the lungs, which are part of the respiratory system.
   C. They enter lymph vessels in bone and are carried by the lymphatic system.
   D. They enter blood vessels in bone and are carried by the circulatory system.
   SC.F.1.4.2
4. Which of the following is made mostly of connective tissue?
   F. dermis **H.** hair
   G. epidermis **I.** keratin
   SC.F.1.4.2

Base your answers to questions 5 and 6 on the diagram below.

SARCOMERE UNDER TWO CONDITIONS

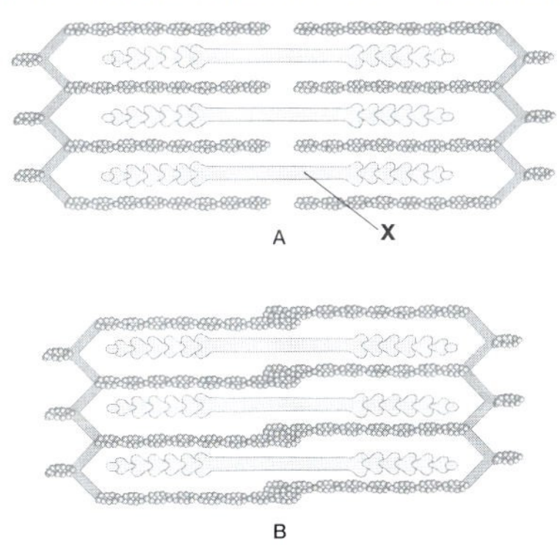

5. When a sarcomere changes from condition *A* to condition *B*, what happens to the muscle cell that contains it?
   A. It contracts. **C.** It rotates.
   B. It relaxes. **D.** It slides.
   SC.F.1.4.2
6. How many myosin filaments are represented in the sarcomere under condition *A*? SC.F.1.4.2

Test TIP

When using a diagram to answer a question, look in the image for evidence that supports your potential answer.

867

FCAT Doctor

Question 3 Answer D is the correct choice. Small blood vessels in the bone are part of the circulatory system, which delivers many needed materials through the body. Answer A is incorrect because this is not a function of the digestive system. The intestine takes up nutrients from food. Answer B is incorrect because this is not a function of the respiratory system. The lungs take up oxygen from the air. Answer C is incorrect because this is not a function of the lymphatic system. Lymph vessels carry lymph, which transports fluid throughout the body.

Question 5 Answer A is the correct choice. The diagram represents the shortening of a sarcomere, which causes muscle to contract. Answer B is incorrect because shortening of the sarcomere contracts muscle rather than relaxing it. Answer C is incorrect because muscle fiber cells do not rotate. Answer D is incorrect because a muscle fiber cell does not slide; it contracts.

Question 6 The correct answer is 3. The contractile protein filaments of muscle tissue consist of actin and myosin filaments. The thin filaments are actin, and the thick filaments are myosin. The diagram under condition *A* shows 8 complete actin filaments and 3 myosin filaments.

Answers

1. D
2. G
3. D
4. F
5. A
6. 3

Test Prep Correlations

SC.F.1.4.2: Items 1-6

Exploration Lab

Analyzing the Work of Muscles

Teacher's Notes

Time Required
One 50-minute class period.

Ratings
EASY → HARD

Teacher Preparation 1
Student Setup 1
Concept Level 1
Cleanup 1

Skills Required
- Using Scientific Methods
- Data Collection
- Data Interpretation

Scientific Methods
In this lab, students will:
- Make Observations
- Analyze the Results
- Draw Conclusions
- Communicate the Results

Materials and Equipment
Materials for this lab can be ordered from WARD'S. See *Master Materials List* at the front of this book for catalog numbers.

Safety Caution
Any student feeling pain from the exercise should stop exercising immediately.

Exploration Lab

Analyzing the Work of Muscles SC.H.1.4.1

SKILLS
- Using scientific methods
- Data collection
- Data interpretation

OBJECTIVE
- **Relate** muscles to the work they do.
- **Observe** the effects of fatigue

MATERIALS
- watch with second hand
- graph paper
- spring hand grips

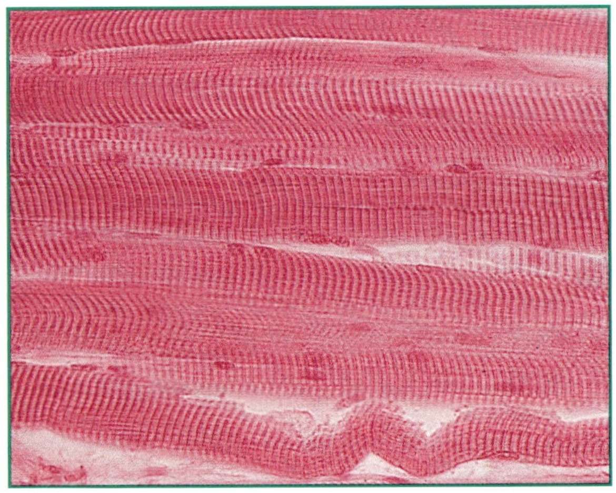

Striated muscle

Before You Begin

Muscles are attached to bones. As muscles contract, they move the bones to which they are attached. This is a basic type of work accomplished by the human body. As muscles are used, lactic acid builds up, resulting in **fatigue**. In this lab you will investigate how fatigue affects the amount of work that muscles can do.

1. Write a definition for the bold face term in the preceding paragraph.
2. Create a data table like the one below.

DATA TABLE
for Muscle Contractions

Time	Number of Muscle Contractions in 10-Second Intervals													
	1st 10 sec.	2nd 10 sec.	3rd 10 sec.	4th 10 sec.	5th 10 sec.	6th 10 sec.	7th 10 sec.	8th 10 sec.	9th 10 sec.	10th 10 sec.	11th 10 sec.	12th 10 sec.	13th 10 sec.	14th 10 sec.
Trial 1														
Trial 2														
Trial 3														
Trial 4														

868

Tips and Tricks
Some students' hands may not be large enough to grip the hand grips. Pair students who have small hands and give them a tennis ball to squeeze.

Procedure

1. Perform this investigation with a partner. Designate one laboratory partner to observe and record while the other performs the experiment.
2. Hold the spring hand grips in your left hand if you are righthanded, or in your right hand if you are lefthanded. Squeeze the grips rapidly and as hard as possible at a steady pace, until complete fatigue is experienced in the muscles of your hand and forearm.
3. The recorder should count and record the number of squeezes for every 10 seconds.
4. Allow the experimenter to rest for one minute and repeat the procedure for two more trials.
5. Record the data in your table. Some spaces may be left blank.
6. Switch roles with your partner and repeat the procedure.

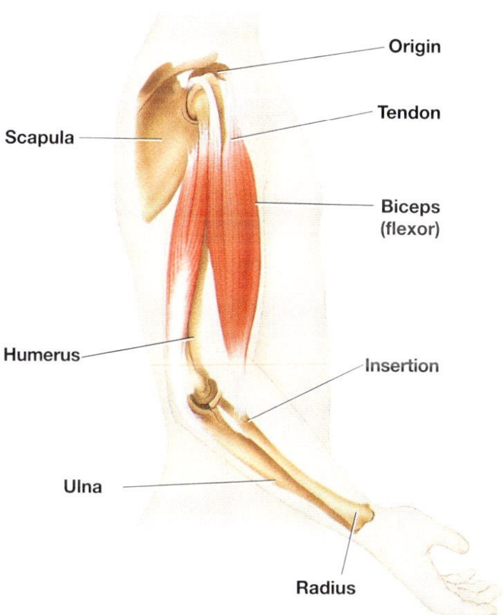

Biceps Muscle

Analyze and Conclude

1. **Summarizing Results** Plot the results of the three trials on a graph. The *X* axis should be used for time in seconds, and the *Y* axis for the number of muscle contractions.
2. **Analyzing Data** Account for the differences in the amount of work done by the muscles during the three trials.
3. **Drawing Conclusions** What is the relationship between the work muscles can do and fatigue?
4. **Predicting Patterns** How does the work done in the muscles of your hands and arms relate to the work done by the muscle of your heart?
5. **Further Inquiry** Compare the charts and the graphs of the athletes and nonathletes.

Do You Know?

Do research in the library or media center to answer these questions:

1. Some bacteria produce lactic acid. Explain how lactic acid from bacteria can be used as a food preservative.
2. Some sports drink labels claim to include ingredients that neutralize lactic acid. Why would they make this claim? Can a sports drink eliminate lactic acid from muscles?

Use the following Internet resources to explore your own questions about muscle contraction.

internet connect

www.scilinks.org
Topic: Muscle Contraction
Keyword: HX4125

SCILINKS. Maintained by the National Science Teachers Association

Answers to Before You Begin

1. Fatigue means "to become tired."

Answers to Analyze and Conclude

1. Answers may vary, but the number of muscle contractions probably decreases over time.
2. The muscles have more energy, in the form of ATP, during the first trial, but the supply is quickly used and lactic acid is formed as a waste product. Time is needed for ATP molecules to re-form and lactic acid to dissipate. Therefore, not as much work is done in the second and third trials.
3. When muscles are fatigued, they can do less work than when they are not fatigued.
4. The strenuous work done by the muscles in the hand and arm increases the heart rate, demonstrating that work done by the cardiac muscle increases also. The cardiovascular system helps to support the work done by muscle activity.
5. Answers may vary, but athletes may have muscles that show fatigue less quickly than nonatheletes.

Answers to Do You Know?

1. Answers may vary. Lactic acid produced by bacteria has been shown to inhibit the growth of pathogenic microorganisms in some foods. The lactic acid destroys or permeates the surrounding membrane of the microorganism, causing its anaerobic life processes to cease.
2. Answers may vary. Some sports drinks contain bicarbonate, which the manufacturers claim, can neutralize lactic acid in the muscles. Others contain carbohydrates which, when broken down, are thought to supply oxygen to muscle cells so that aerobic processes can continue.

Chapter Opener

- **SE** **Quick Review,** p. 871
- **SE** **Reading Activity,** p. 871
- **TE** **Using the Figure,** p. 870 GENERAL
- **TE** **Opening Activity,** p. 871 BASIC

& 3 • 90 min pp. 872–879

Section 1
The Circulatory System

- **OSP** **Lesson Plan***
- **TR** **Bellringer***
- **TR** **Section Outline***
- **TR** **J22 Cardiovascular System***
- **TR** **J21 Blood Vessels***
- **TR** **J29 Blood Clotting Cascade***
- **TR** **J25 Systemic Circulation***

- **TE** **Activity** Circulatory Function, p. 872 GENERAL
- **TE** **Using the Figure,** p. 873
- **TE** **Demonstration,** p. 874 ◆ BASIC
- **TE** **Using the Figure,** p. 875 GENERAL
- **SE** **Quick Lab** Mapping the Valves in Veins, p. 875
- **CRF** **Datasheets for In-Text Labs** Mapping the Valves in Veins*
- **TE** **Demonstration,** p. 876 ◆ BASIC
- **TE** **Using the Figure,** p. 877 GENERAL
- **TE** **Activity** Blood Composition, p. 877 GENERAL
- **TE** **Group Activity** Blood Diseases, p. 877 ADVANCED

& 5 • 90 min pp. 880–884

Section 2
The Heart

- **OSP** **Lesson Plan***
- **TR** **Bellringer***
- **TR** **Section Outline***
- **TR** **J26 Path of Bloodflow Through the Heart***
- **TR** **J24 Circulatory Loops In the Human Body***
- **TR** **J19 Electrical Regulation of the Heart***
- **TR** **J20 Electrocardiogram***

- **TE** **Demonstration,** p. 880 ◆ BASIC
- **TE** **Using the Figure,** p. 881 BASIC
- **TE** **Demonstration,** p. 881 GENERAL
- **TE** **Demonstration,** p. 882 ◆
- **TE** **Group Activity** Measuring Blood Pressure, p. 883 ◆ GENERAL
- **CRF** **Skills Practice** Modeling Capillaries* GENERAL
- **CRF** **Quick Lab** Relating Cell Structure to Function* BASIC
- **SE** **Forensics Labs Appendix,** Blood Typing, p. 1053
- **Forensics and Applied Science Experiments** Blood Typing* GENERAL

& 7 • 90 min pp. 885–890

Section 3
The Respiratory System

- **OSP** **Lesson Plan***
- **TR** **Bellringer***
- **TR** **Section Outline***
- **TR** **J31 The Human Respiratory System***
- **TR** **J33 Inhalation and Exhalation***
- **TR** **J32 O_2 and CO_2 in the Blood***

- **SE** **Math Lab** Calculating the Amount of Air Respired, p. 886
- **CRF** **Datasheets for In-Text Labs** Calculating the Amount of Air Respired*
- **TE** **Demonstration,** p. 887 ◆ BASIC
- **TE** **Demonstration,** p. 888 BASIC
- **SE** **Quick Lab** Modeling the Role of Bicarbonate in Homeostasis, p. 889
- **CRF** **Datasheets for In-Text Labs** Modeling the Role of Bicarbonate in Homeostasis*
- **SE** **Exploration Lab** Determining Lung Capacity, pp. 894–895
- **CRF** **Datasheets for In-Text Labs** Determining Lung Capacity*

& 9 • 90 min

Chapter Review and Assessment Resources

- **SE** **Study Zone,** p. 891
- **SE** **Performance Zone,** p. 892
- **SE** **FCAT Test Prep,** p. 893
- **TR** **Concept Mapping***
- **CRF** **Vocabulary Review Worksheet*** GENERAL
- **CRF** **Test Prep Pretest*** GENERAL
- **CRF** **Chapter Test*** GENERAL
- **CRF** **Chapter Test*** ADVANCED
- **OSP** **Test Generator***
- **OSP** **Test Item Listing***
- **OSP** **Modified Chapter Test*** BASIC

Online and Technology Resources

Holt Online Learning

Visit go.hrw.com for access to Holt Online Learning or enter the keyword **HX6 Home** for a variety of free online resources.

One-Stop Planner® CD-ROM

This CD-ROM package includes

- Lab Materials QuickList Software
- Holt Calendar Planner
- Customizable Lesson Plans
- Printable Worksheets
- ExamView® Test Generator
- Interactive Teacher Edition
- Holt PuzzlePro® Resources
- Holt PowerPoint® Resources

Guided Reading Audio CDs

These CDs are designed to help auditory learners and reluctant readers. (Audio CDs are also available in Spanish.)

Compression guide:
To shorten your instruction because of time limitations, omit block 1.

TE Teacher Edition	**CRF** Chapter Resource File	* Also on One-Stop Planner
SE Student Edition	**TR** Teaching Transparency	■ Also Available in Spanish
OSP One-Stop Planner		◆ Requires Advance Prep

		Florida Sunshine Standards/FCAT
TE Reading Skill Builder Interactive Reading, p. 873 BASIC **TE Skill Builder** Math Skills, p. 876 BASIC **CRF Directed Reading Worksheet*** BASIC **CRF Active Reading Worksheet*** GENERAL	**SE Section 1 Review** ■ **TE Reteaching,** p. 879 BASIC **TE Quiz,** p. 879 GENERAL **TE Alternative Assessment,** p. 879 GENERAL **CRF Quiz*** ■ BASIC	SC.F.1.4.2, SC.F.1.4.5, SC.H.3.4.3
CRF Directed Reading Worksheet* BASIC **CRF Active Reading Worksheet*** GENERAL	**SE Section 2 Review** ■ **TE Reteaching,** p. 884 BASIC **TE Quiz,** p. 884 GENERAL **TE Alternative Assessment,** p. 884 GENERAL **CRF Quiz*** ■ BASIC	SC.F.1.4.2, SC.H.3.4.3, SC.H.3.4.5
TE Reading Skill Builder Brainstorming, p. 887 BASIC **TE Reading Skill Builder** Reading Organizer, p. 888 BASIC **CRF Directed Reading Worksheet*** BASIC **CRF Active Reading Worksheet*** GENERAL **CRF Critical Thinking Worksheet*** ADVANCED **CRF Science Skills Worksheet*** GENERAL	**SE Section 3 Review** ■ **TE Reteaching,** p. 890 BASIC **TE Quiz,** p. 890 GENERAL **TE Alternative Assessment,** p. 890 GENERAL **CRF Quiz*** ■ BASIC **Modified Worksheet*** BASIC	SC.F.1.4.2, SC.F.1.4.5, SC.H.1.4.1, SC.H.3.4.1, SC.H.3.4.3

SciLinks NSTA
www.scilinks.org
Maintained by the **National Science Teachers Association**

Topic: White Blood Cells
SciLinks Code: HX4190

Chapter Opener		TE **Using the Figure,** p. 898 GENERAL SE **Quick Review,** p. 899 ■ SE **Reading Activity,** p. 899 ■ TE **Opening Activity,** p. 899 GENERAL
& 3 • 90 min pp. 900–905 **Section 1** Your Body's Need for Food	OSP **Lesson Plan*** TR **Bellringer*** TR **Section Outline*** TR **J45 Nutrients in Food*** TR **J48 USDA Food Pyramid*** TR **J47 Trace Elements*** TR **J46 Vitamins***	TE **Using the Figure,** p. 901 GENERAL TE **Demonstration,** p. 903 ◆ BASIC TE **Activity** Analyzing Diet, p. 904 GENERAL TE **Using the Figure,** p. 904 TE **Group Activity** Vitamin Nutrition, p. 904 ◆ BASIC CRF **Quick Lab** Reading Labels: Nutritional Information* BASIC **Forensics and Applied Science Experiments** Identifying Organic Compounds in Foods* GENERAL **Forensics and Applied Science Experiments** Energy in Foods* GENERAL
& 5 • 90 min pp. 906–911 **Section 2** Digestion	OSP **Lesson Plan*** TR **Bellringer*** TR **Section Outline*** TR **J49 Digestive System in the Human Body*** TR **J51 Teeth*** TR **J52 The Stomach and Accessory Digestive Organs*** TR **J54 Villi in the Small Intestine*** TR **J55 Cross Section of the Small Intestine***	TE **Demonstration,** p. 906 ◆ BASIC TE **Activity** Salivation, p. 907 ◆ GENERAL TE **Demonstration,** p. 907 ◆ BASIC TE **Demonstration,** p. 909 ◆ GENERAL TE **Using the Figure,** p. 909 ◆ TE **Activity** Viewing Digestive Organs, p. 909 ◆ ADVANCED SE **Quick Lab** Modeling the Function of Bile, p. 911 ■ CRF **Datasheets for In-Text Labs** Modeling the Function of Bile*
& 7 • 90 min pp. 912–916 **Section 3** Excretion	OSP **Lesson Plan*** TR **Bellringer*** TR **Section Outline*** TR **J59 Organs of Excretion*** TR **J56 Human Kidney Structure*** TR **J58 How a Kidney Machine Works***	TE **Activity** Observing Sweat, p. 912 ◆ GENERAL TE **Using the Figure,** p. 913 TE **Demonstration,** p. 914 ◆ GENERAL TE **Activity** Interview, p. 915 GENERAL CRF **Inquiry Lab** What Secrets Does Urine Hold?* GENERAL SE **Exploration Lab** Lactose Digestion, pp. 920–921 ◆ ■ CRF **Datasheets for In-Text Labs** Lactose Digestion*

& 9 • 90 min

Chapter Review and Assessment Resources

- SE **Study Zone,** p. 917 ■
- SE **Performance Zone,** p. 918 ■
- SE **FCAT Test Prep,** p. 919 ■
- TR **Concept Mapping***
- CRF **Vocabulary Review Worksheet*** ■ GENERAL
- CRF **Test Prep Pretest*** ■ GENERAL
- CRF **Chapter Test*** ■ GENERAL
- CRF **Chapter Test*** ADVANCED
- OSP **Test Generator***
- OSP **Test Item Listing***
- OSP **Modified Chapter Test*** BASIC

go.hrw.com Holt Online Learning

Visit go.hrw.com for access to Holt Online Learning or enter the keyword **HX6 Home** for a variety of free online resources.

One-Stop Planner® CD-ROM

This CD-ROM package includes
- Lab Materials QuickList Software
- Holt Calendar Planner
- Customizable Lesson Plans
- Printable Worksheets
- ExamView® Test Generator
- Interactive Teacher Edition
- Holt PuzzlePro® Resources
- Holt PowerPoint® Resources

Guided Reading Audio CDs

These CDs are designed to help auditory learners and reluctant readers. (Audio CDs are also available in Spanish.)

Compression guide:
To shorten your instruction because of time limitations, omit block 1.

KEY

TE	Teacher Edition	CRF	Chapter Resource File	*	Also on One-Stop Planner
SE	Student Edition	TR	Teaching Transparency	■	Also Available in Spanish
OSP	One-Stop Planner			◆	Requires Advance Prep

SKILLS DEVELOPMENT AND PRACTICE	SECTION REVIEW AND ASSESSMENT	CORRELATIONS
		Florida Sunshine Standards/FCAT
TE Reading Skill Builder Interactive Reading, p. 901 BASIC TE Skill Builder Interpreting Visuals, p. 902 ADVANCED CRF Directed Reading Worksheet* BASIC CRF Active Reading Worksheet* GENERAL	SE Section 1 Review ■ TE Reteaching, p. 905 BASIC TE Quiz, p. 905 GENERAL TE Alternative Assessment, p. 905 GENERAL CRF Quiz* ■ BASIC	SC.G.1.4.3, SC.H.3.4.3, SC.H.3.4.4
TE Reading Skill Builder Reading Organizer, p. 907 BASIC TE Skill Builder Graphing, p. 910 BASIC CRF Directed Reading Worksheet* BASIC CRF Active Reading Worksheet* GENERAL	SE Section 2 Review ■ TE Reteaching, p. 911 BASIC TE Quiz, p. 911 GENERAL TE Alternative Assessment, p. 911 GENERAL CRF Quiz* ■ BASIC	SC.F.1.4.2, SC.H.3.4.1
TE Skill Builder Math Skills, p. 913 GENERAL CRF Directed Reading Worksheet* BASIC CRF Active Reading Worksheet* GENERAL CRF Critical Thinking Worksheet* ADVANCED CRF Science Skills Worksheet* GENERAL	SE Section 3 Review ■ TE Reteaching, p. 916 BASIC TE Quiz, p. 916 GENERAL TE Alternative Assessment, p. 916 GENERAL CRF Quiz* ■ BASIC Modified Worksheet* BASIC	SC.F.1.4.2, SC.G.1.4.3, SC.H.3.4.3

SCILINKS. NSTA
www.scilinks.org
Maintained by the **National Science Teachers Association**

Topic: Poisons
SciLinks Code: HX4142

Topic: Kidney Diseases
SciLinks Code: HX4109

Topic: Kidney Dialysis
SciLinks Code: HX4108

Topic: Lactose Intolerance
SciLinks Code: HX4111

CHAPTER 39

Overview

Tell students the purpose of this chapter is to understand the importance of the various nutrients to their bodies, how their bodies are specialized to digest those nutrients, and how their bodies eliminate metabolic wastes. The metabolic processes that each system performs require materials and energy ultimately obtained from food and the activity of the digestive system. Likewise, the same metabolic processes generate toxic waste products. Eventually, these materials are removed from the body by the process of excretion.

Using the Figure

Tell students to look at the photograph of the human intestine villi on this page. Ask students if they see a doorway in the picture. Assure them that there is one. Ask students to explain how food they eat gets into their bodies and bloodstream. Explain that most nutrients are absorbed into the bloodstream through villi in the small intestine just like those shown in the photograph. Tell the students that if they were a nutrient inside the small intestine, these villi would be their "doorway" into the body and bloodstream.

Assessing Prior Knowledge

Before beginning this chapter, be sure your students can answer the questions in the ***Quick Review*** on the facing page. If they cannot, encourage them to reread the sections indicated.

Villi in human intestine (1,280×)

898

Chapter Correlations | **Florida Sunshine Standards/FCAT**

SC.F.1.4.2 Body structures are uniquely designed and adapted for their function.

SC.G.1.4.3 The chemical elements that make up the molecules of living things are combined and recombined in different ways.

SC.H.3.4.3 Scientists can bring information, insights, and analytical skills to matters of public concern and help people understand the possible causes and effects of events.

SC.H.3.4.4 Funds for science research come from federal government agencies, industry, and private foundations, and this funding often influences the areas of discovery.

CHAPTER

39 Digestive and Excretory Systems

Florida Sunshine Standards/FCAT

SC.F.1.4.2 Body structures are uniquely designed and adapted for their function.
SC.G.1.4.3 The chemical elements that make up the molecules of living things are combined and recombined in different ways.
SC.H.3.4.3 Scientists can bring information, insights, and analytical skills to matters of public concern and help people understand the possible causes and effects of events.
SC.H.3.4.4 Funds for science research come from federal government agencies, industry, and private foundations, and this funding often influences the areas of discovery.

Quick Review

Answer the following without referring to earlier sections of your book.

1. **Differentiate** between carbohydrates, fats, and proteins. *(Chapter 2, Section 3)*
2. **Compare** saturated and unsaturated fatty acids. *(Chapter 2, Section 3)*
3. **Describe** the role of enzymes in chemical reactions. *(Chapter 2, Section 4)*
4. **Summarize** the function of cellular respiration. *(Chapter 5, Section 3)*
5. **Discuss** the balance of water and salt in vertebrates. *(Chapter 33, Section 1)*

Did you have difficulty? *For help, review the sections indicated.*

Reading Activity

Before you begin to read, survey the chapter, noting the red headings at the tops of pages and the blue subheadings. Use these heads to make an outline of the chapter, leaving space after each heading. Fill important facts in as you read.

The multiple, protruding villi of the small intenstine greatly increase the surface area through which nutrients are absorbed.

Looking Ahead

Section 1
Your Body's Need for Food
Food
Energy and Building Materials
Vitamins, Minerals, and Water

Section 2
Digestion
Breaking Down Food
The Small Intestine
The Large Intestine

Section 3
Excretion
Water and Metabolic Wastes
The Kidneys

internet connect

www.scilinks.org
National Science Teachers Association *sci*LINKS Internet resources are located throughout this chapter.

SciLINKS. Maintained by the National Science Teachers Association

899

Chapter Resource File
- Vocabulary Worksheets
- Concept Mapping

Reading Activity Answers

Students should have 20 headings as a part of their outlines with at least one main idea for each heading. (There are eight red main headings with twelve blue subheadings—or seven total headings in Section 1, eight in Section 2, and five in Section 3.)

Opening Activity — GENERAL

Anticipation Guide Before students begin this chapter, write the following statements on the board or overhead:

- Fifty percent of your body's fuel should come from carbohydrates.
- The majority of nutrients in food are released and absorbed in the stomach.
- Both the kidneys and the liver cleanse the blood of toxic substances.

Give students time to form opinions about each statement. Ask volunteers with differing views about each statement to share the reasons for their opinions with the class. Then have students read the chapter to see if their opinions are confirmed or changed in any way. Have students reevaluate their opinions in Reteaching at the end of this chapter. LS Verbal

Quick Review Answers

1. Carbohydrates are organic compounds made of carbon, hydrogen, and oxygen. Lipids (fats) are non-polar organic compounds that store energy and form an important part of cell membranes. Proteins are organic compounds made of amino acids that form a variety of molecules specialized for certain functions.
2. In saturated fatty acids, all the carbon atoms in the hydrocarbon chain have single covalent bonds. In unsaturated fatty acids, some of the carbon atoms are linked by double covalent bonds.
3. Enzymes are substances that increase the speed of chemical reactions by reducing the activation energy of the reactions.
4. The function of cell respiration is to convert the stored chemical energy in glucose to usable chemical energy in adenosine triphosphate (ATP).
5. In vertebrates, salt and water balance is largely regulated by the action of the kidneys.

Section 1

Focus

Overview

Before beginning this section review with your students the objectives listed in the Student Edition. This section introduces students to the role each nutrient plays in maintaining a healthy body. Students will also learn about health disorders related to malnutrition.

Bellringer

Have students list the nutrients they think they get from eating a small bag of chips and a bottle of soda pop. Then have them estimate the number of calories this "snack" provides. If possible, bring in samples of each—empty bags and bottles are okay. Students may be amazed at how many servings are in even these "individual sizes" of food and drink and how that multiplies the calories—and the relatively poor nutrition. LS Logical

Motivate

Identifying Preconceptions — GENERAL

Have students list the nutrients necessary for proper body functioning and food sources where each nutrient can be found. (For example: Calcium is necessary for healthy teeth and bones; calcium can be found in milk.) Have students revisit the list after reading Section 1. LS Verbal

Section 1 Your Body's Need for Food

Florida Sunshine Standards/FCAT

SC.G.1.4.3 The chemical elements that make up the molecules of living things are combined and recombined in different ways.

SC.H.3.4.4 Funds for science research come from federal government agencies, industry, and private foundations, and this funding often influences the areas of discovery.

Objectives

- **Identify** five nutrients found in foods. SC.G.1.4.3
- **Relate** the role of carbohydrates, proteins, lipids, vitamins, minerals, and water in maintaining a healthy body. SC.G.1.4.3
- **Describe** each of the parts of the USDA food guide pyramid. SC.H.3.4.4
- **Name** one health disorder associated with high levels of saturated fats in the diet. SC.H.3.4.3

Key Terms

nutrient
digestion
calorie
vitamin
mineral

Food

Your body uses energy to move, to grow, and even to lie still and sleep. The amount of energy you need depends on many factors, including your age, your sex, your rate of growth, and your level of physical activity. Different activities use different amounts of energy, as shown in **Figure 1.**

You obtain energy from the nutrients in the foods and beverages you consume. A **nutrient** is a substance required by the body for energy, growth, repair, and maintenance. Nutrients in food and beverages include carbohydrates, lipids, proteins, vitamins, and minerals. Each nutrient plays a different role in keeping your body healthy. Water is essential for life and for maintaining health.

The large molecules in food must be broken down in order to be absorbed into the blood and carried to cells throughout the body. The process of breaking down food into molecules the body can use is called **digestion.** Your cells then break the chemical bonds of the digested food molecules and use the energy that is released to make ATP during the process of cellular respiration.

The energy available in food is measured by using a unit called a calorie. A **calorie** is the amount of heat energy required to raise the temperature of 1 g of water 1°C (1.8°F). The greater the number of calories in a quantity of food, the more energy the food contains. Because a calorie represents a very small amount of energy, nutritionists use a unit called the Calorie (with a capital C), which is equal to 1,000 calories. On food labels and throughout this book, the word *calories* represents Calories (1,000 calories).

Figure 1 Energy required for common activities. Quiet activities require just a little more energy than what it takes to keep you alive. Strenuous activities require more energy.

900

Chapter Resource File

- Directed Reading BASIC
- Active Reading GENERAL

Transparencies

TR Bellringer
TR J45 Nutrients in Food

Section Correlations

SC.G.1.4.3
SC.H.3.4.3
SC.H.3.4.4

Energy and Building Materials

Each nutrient plays a different role in maintaining a healthy body. Carbohydrates, proteins, and lipids are involved in providing both energy and building materials to the cells.

Carbohydrates

Carbohydrates that exist as single sugar molecules are called monosaccharides or simple carbohydrates. Carbohydrates made of two or many sugar molecules linked together by chemical bonds are called complex carbohydrates. Complex carbohydrates must be digested (broken down) into simple sugars before cells can use their energy.

Many foods contain carbohydrates, as shown in **Figure 2.** Glucose, fructose, and other simple sugars are found in fruits, honey, and onions. Glucose, a simple sugar, is used by cells for energy, and it can be directly absorbed into your bloodstream. Table sugar contains sucrose, two simple sugars linked together. Starches are long chains of sugars found in cereal grains and in vegetables such as potatoes, beans, and corn. Cellulose is a major component of plant cell walls and is found in all foods that come from plants. Cellulose, which is a major part of fiber, does not provide energy because we do not have the enzymes to digest it. However, cellulose aids in human digestion by stimulating the walls of the digestive tract to secrete mucus, which helps pass food through the digestive tract.

If excess carbohydrates are consumed, they are stored as the carbohydrate glycogen in the liver and in some muscle tissue. Glycogen can later be broken back down into glucose when the body needs energy. The remainder of the excess glucose is converted to fat and stored in fatty tissue.

Study TIP

Organizing Information Make a table to organize information about food nutrients and water. Across the top, write the headings *Carbohydrates, Proteins, Lipids, Vitamins, Minerals,* and *Water.* Along the left side, write *Functions, Food sources,* and *Additional comments*. Add information to the table as you read Section 1.

Figure 2 Nutrients in food

Although most foods contain a mix of nutrients, some foods are richer than others in a specific nutrient.

Carbohydrate-rich foods	Protein-rich foods	Fat-rich foods
(Carbohydrates contain 4 calories per gram.) **Breads, pasta, grains, cereals, potatoes, fruits**	(Proteins contain 4 calories per gram.) **Fish, eggs, poultry, beef pork, nuts, legumes, milk, cheese, tofu**	(Fats contain 9 calories per gram.) **Milk, cheese, meats, butter, olives, avocados, fried foods, oils, chips**

Teach

READING SKILL BUILDER — BASIC

Interactive Reading Assign Chapter 39 of the *Holt Biology* Guided Audio CD Program to help students achieve greater success in reading the chapter. **LS Auditory** English Language Learners

Using the Figure — GENERAL

Have students study **Figure 2.** Ask students why it is important for weight-conscious people to get a balance of nutrients from carbohydrate- and protein-rich foods while limiting fat intake. (Carbohydrates take less time to digest, providing quick energy and satisfying hunger quickly. Proteins take longer to digest providing energy and satisfying hunger over a longer period of time. Both release only 4 Calories per gram. Although fats provide energy and satisfy hunger over a longer period of time than carbohydrates or proteins do, they contain more than twice the number of Calories per gram.) Tell students that insufficient protein in the diet can cause a person to eat more because hunger and energy requirements are not met for as long as they need to be. However, some protein-rich foods contain fats and should be eaten in small amounts. **Figure 2** shows proteins contain hidden fats and should be consumed in lesser quantities than carbohydrates. **LS Visual**

One-Stop Planner CD-ROM

- **Reading Organizers** BASIC
- **Reading Strategies** BASIC
- **Problem Solving** Using Food Labels to Calculate Percentage of Nutrients and Calories GENERAL

Integrating Physics and Chemistry

Have students identify some of the various chemical bonds that are broken in the mouth, stomach, and small intestine by digestive enzymes. For example, the mouth's saliva contains *amylase* which breaks the chemical bonds of long starch molecules, while *pepsin,* which is added in the stomach, begins the chemical breakdown of protein molecules.

HISTORY CONNECTION

ATP is the energy currency of all living cells. Fritz Lipmann isolated ATP from muscle tissue in 1929. In research he began in 1941, he discovered that ATP and similar compounds shuttle energy around the cell, once the energy is captured from carbohydrates, fats, and proteins during cellular respiration. In 1953 Lipmann shared the Nobel Prize in physiology or medicine for his discoveries.

Proteins

The digestive products of proteins—amino acids—are normally used by the body for making other protein molecules, such as enzymes and antibodies. When more protein is eaten than is needed by the cells, the amino acids are used for energy or converted to fat. The body requires 20 different amino acids to function. A child's or teen's body can make 10 of the amino acids from other amino acids. The other 10, called essential amino acids, must be obtained directly from food. Most animal products, such as eggs, milk, fish, poultry, and beef, contain all of the essential amino acids. No single plant food contains all of the essential amino acids. But eating certain combinations of two or more plant products can supply all the essential amino acids. Adults must get eight essential amino acids from food.

The guidelines for healthy eating are summarized in the USDA (U.S. Department of Agriculture) food guide pyramid, as shown in **Figure 3.** The pyramid lists the daily number of servings needed from each food group to obtain a variety of nutrients in your diet.

Lipids

Lipids, organic compounds that are insoluble in water, are used to make steroid hormones and cell membranes and to store energy. Fats are lipids that store energy in plants and animals. Fats are

Figure 3 The USDA food guide pyramid

Food groups at the bottom of the pyramid should be eaten in greater amounts than those at the top.

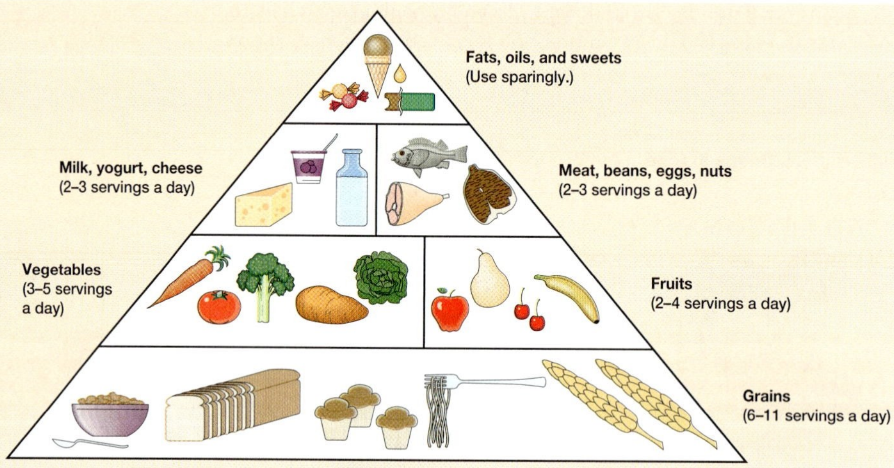

Serving Sizes

Milk, yogurt, cheese
- 1 cup of milk
- 1 cup of yogurt
- 1.5 oz of cheese
- 2 oz of processed cheese

Meat, beans, eggs, nuts
- 1 egg
- 2 tbsp of peanut butter
- 2–3 oz of meat, chicken, or fish
- $\frac{1}{2}$ cup of cooked beans

Vegetables
- $\frac{1}{2}$ cup of vegetables
- $\frac{3}{4}$ cup of vegetable juice
- 1 cup of raw, leafy vegetables

Fruits
- 1 apple, banana, or orange
- $\frac{1}{2}$ cup of fruit
- $\frac{3}{4}$ cup of fruit juice

Grains
- 1 slice of bread
- $\frac{1}{2}$ cup of rice or pasta
- $\frac{1}{2}$ cup of hot cereal
- 1 oz cold cereal
- 1 tortilla

Teach, *continued*

Teaching Tip — GENERAL

Analyzing Vegetarian Diets
Have students make a list of protein food sources and note whether the food contains all of the essential amino acids or only some of them. Then discuss the different types of vegetarian diets such as those that omit only red meat, omit all meat but fish, omit all meat but include animal products such as milk and eggs, and omit all animal products. Have students create a meal for each of these diet types and decide how it is difficult to obtain a nutritionally complete meal for each diet. LS **Logical**

SKILL BUILDER — ADVANCED

Interpreting Visuals Have students look at **Figure 3** and **Figure 4.** Ask students to write a brief essay explaining how the percentage of calories from each dietary nutrient in Figure 4 relates to the number of servings of each food group in Figure 3. (Carbohydrates are mainly found in the lower tiers of the pyramid, which corresponds to the high percentage of calories from carbohydrates. Proteins are mainly in the second level and lipids are in the first and second levels of the pyramid, which corresponds to lower percentages. The percentage of calories from proteins may seem small compared to the percentage from fats since their number of recommended servings are similar, but fats carry twice the calories that proteins do.) LS **Verbal**

Transparencies

TR J48 USDA Food Pyramid

Trends in Nutrition

Food Guides Around the World Countries around the world use different graphics to illustrate their dietary guidelines. Canada uses a four-banded rainbow, with each color representing one of its food groups and the width of each band representing the amount recommended. Israel uses a chalice with water at the top, largest section and grains at the wide base of the chalice. The Philippines uses a six-pointed star. The United Kingdom, Germany, and Norway use a wheel or dinner plate divided into sections that represent the food groups. The size of each section indicates its relative importance to the total diet.

stored around organs and act as padding and insulation. Fats also act as solvents for fat-soluble vitamins.

Although lipids are essential nutrients, too much fat in the diet is known to harm several body systems. For example, a diet high in saturated fats is linked to high blood-cholesterol levels, which in turn may be connected to cardiovascular diseases. It is recommended that a person limits his or her consumption of saturated fats and that most of the fats in the diet be unsaturated.

Energy-Giving Nutrient	Percentage of Daily Calories
Carbohydrates	45 to 65
Protein	10 to 15
Fats	25 to 35

Figure 4 **A balanced meal.** The percentages of the day's total calories that should come from each nutrient are shown. About half of your day's calories should come from foods high in complex carbohydrates.

Balancing Nutrients and Energy

Regardless of their source, the excess calories you eat will be stored as either glycogen or body fat, and you will gain weight. If you use more calories than you take in, additional energy will be obtained from your body's energy stores, and you will lose weight. Your diet and overall activity level determine in part whether you store excess calories as glycogen or as fat. **Figure 4** summarizes what percentage of the day's total calories should come from each nutrient.

Obesity is described as being more than 20 percent heavier than your ideal body weight. Obesity significantly increases an individual's risk of diabetes, heart disease, osteoarthritis, and many other disorders. Regular physical activity is important in maintaining energy balance.

BIOWatch

Obesity and Health

The National Center for Health Statistics reported that by the year 2000, 31 percent of adults in the United States were obese. Medically, obesity is defined as having a body mass index (BMI) of 30 or higher. To calculate your BMI, first multiply your weight in pounds by 703, and then divide by your height in inches squared.

$$\text{BMI} = \frac{\text{weight in pounds} \times 703}{(\text{height in inches})^2}$$

For example, someone who is 6'0" (72 inches) tall and who weighs 189 pounds has a body mass index of (189 × 703) ÷ (72 × 72) = 25.6.

Unfortunately, being obese often has a serious consequence—diabetes. Late-onset diabetes affected 14 million Americans in 2000, and 80 percent of those so affected were obese. Like the rate of obesity, the rate of late-onset diabetes is soaring, up 40 percent in the last 10 years.

Late-onset diabetes is a disorder in which the hormone insulin is unable to induce the body's cells to take up glucose from the blood. Glucose-starved tissues consume their own protein, and waste away. Diabetes is the leading cause of kidney failure, blindness, and amputation in adults.

What is the link between obesity and late-onset diabetes? Researchers have recently found that fat cells produce a hormone called *resistin* that prompts cells to resist insulin. Mice given resistin by researchers lose much of the ability to respond to insulin and fail to take up blood sugar. Drugs that inhibit resistin restore insulin's effectiveness. Researchers hypothesize that resistin blocks the same glucose-transporting molecules in the plasma membrane that insulin activates.

Demonstration — BASIC

Bring some French fries and a plain baked potato to class, or show students a picture of these foods. Have students examine the baked potato and the French fries carefully, and ask them to brainstorm differences between the two foods with regard to cooking, taste, and nutritional value. (Unlike the baked potato, French fries are peeled, sliced, fried in oil or fat, and salted.) Ask students how the body is affected by a diet high in oils, fats, and salt. (Oils and fats increase caloric intake. Saturated fats may increase cholesterol levels. A diet high in salt can cause fluid to be retained in the body. Over a long period of time, high salt intake can contribute to hypertension.) **LS Visual**

BIOWatch

Obesity and Health

Teacher's Notes

Be sensitive to those students in your class that may be overweight. The Centers for Disease Control and Prevention publishes the *Morbidity and Mortality Weekly Report*. If possible, bring some of these to class. Have students conduct research about these topics and present their research in oral reports and/or fact sheets.

Discussion

Ask students: Is a person who is 5'8" and 180 pounds obese? (No, the person's BMI is 27.4.) What hormone does not work appropriately in diabetes? (insulin) What hormone causes cells not to respond to insulin? (resistin) What type of cell produces resistin? (fat cells) How does that fact link obesity with diabetes? (Obese people have more fat cells, which produce more resistin, which causes more insulin to be blocked, which can lead to diabetes.)

Cultural Awareness

Mediterranean Diets and Heart Disease

The French, Greeks, and other southern Europeans suffer from less heart disease than either their northern European neighbors or Americans. The Mediterranean diet is thought to play a role. The Mediterranean diet is rich in canola oil, fish, fruits, cereals, and beans. Even when researchers account for age, sex, alcohol intake, and lifestyle differences, eating a Mediterranean diet appears to reduce the probability of heart problems.

MISCONCEPTION ALERT

Athletes, Proteins, and Carbohydrates

A common misconception is that athletes need to eat more protein to improve their athletic performance and maintain greater muscle mass. Actually, a diet rich in complex carbohydrates is much more effective in sustaining intense muscle activity than a diet high in protein.

Teach, *continued*

Activity — GENERAL

Analyzing Diet Have students list the kinds and amounts of each food they eat in a day. The next day, have students compare their number of servings for each food group to the recommended amounts given in the food pyramid shown in **Figure 3**, as well as the amount of vitamins and minerals they are receiving. Ask students to plan how they can reasonably change their diet to make it more balanced. **LS Intrapersonal**

Using the Figure

Diagnosing Deficiencies Tell students to imagine that they are physicians. Ask them to use information in **Table 1** to answer the following questions: If a child suffers from frequent internal infections and has difficulty seeing in dim light, what problem would you suspect? (The child likely is suffering from a vitamin A deficiency.) What foods would you prescribe to someone who has sore gums and tends to bruise easily? (fruit, especially citrus fruits) **LS Interpersonal**

Group Activity — BASIC

Vitamin Nutrition Organize students into groups. Give each group a few boxes of cereal. Ask the groups to identify the vitamins that are listed on the cereal labels. Have students use **Table 1** to list the role of each vitamin and the effects of a deficiency in that vitamin. **LS Visual**

Co-op Learning

Vitamins, Minerals, and Water

Vitamins, minerals, and water are required in our diets. They do not provide energy, but they contribute to many different functions, including regulating the reactions that release energy.

Vitamins Many different **vitamins,** organic substances that occur in many foods in small amounts, are necessary in trace amounts for the normal metabolic functioning of the body.

Vitamins dissolve in either water or fat, as summarized in **Table 1.** Fat-soluble vitamins—vitamins A, D, E, and K—can be stored in body fat. Excessive amounts of vitamins A and D can be toxic. Excess water-soluble vitamins—vitamin C and the B vitamins—are excreted in urine and must be replenished by the diet.

Table 1 Vitamins

Vitamin	Food sources	Role	Effects of deficiency
Water-soluble			
Vitamin B_1 (thiamin)	Most vegetables, nuts, whole grains, organ meats	Assists in carbohydrate metabolism, helps nerves and heart to function properly	Digestive disturbances, impaired senses
Vitamin B_2 (riboflavin)	Fish, poultry, cheese, yeast, green vegetables 	Needed for healthy skin and tissue repair, assists in carbohydrate metabolism	Blurred vision, cataracts, cracking of skin, lesions of intestinal lining
Vitamin B_3 (niacin)	Whole grains, fish, poultry, tomatoes, legumes, potatoes	Keeps skin healthy, assists in carbohydrate metabolism	Mental disorders, diarrhea, inflamed skin
Vitamin B_{12} (cobalamin)	Meat, poultry, milk, dairy products	Needed for formation of red blood cells	Reduced number of red blood cells
Vitamin C (ascorbic acid)	Citrus fruits, strawberries, potatoes 	Needed for wound healing, healthy gums and teeth	Swollen and bleeding gums, loose teeth, slow-healing wounds
Fat-soluble			
Vitamin A (retinol)	Butter, eggs, liver, carrots, green leafy vegetables, sweet potatoes	Keeps eyes and skin healthy, needed for strong bones and teeth	Infections of urinary and digestive systems, night blindness
Vitamin D (cholecalciferol)	Salmon, tuna, fish liver oils, fortified milk, cheese 	Assists in calcium uptake by the gut, needed for strong bones and teeth	Bone deformities in children, loss of muscle tone
Vitamin E (tocopherol)	Many foods, especially wheat and other vegetable oils, olives, whole grains	Protects cell membranes from damage by reactive oxygen compounds (free radicals)	Reduced number of red blood cells, nerve tissue damage in infants
Vitamin K (menadione)	Leafy green vegetables, liver, cauliflower	Needed for normal blood clotting	Bleeding caused by prolonged clotting time

Attention Grabber

Vitamin Overdoses Some people have the idea that if a little is good, then a lot must be better. Unfortunately, this is not the case with certain vitamins. For example, overdoses of vitamin A can cause enlargement of the liver and spleen. Overdoses of vitamin D can cause kidney damage and calcification of soft tissues. Although vitamins A and D are both fat-soluble, making them more difficult for the body to eliminate, even the water-soluble vitamins have some negative effects if overused. Massive doses of vitamin C can cause kidney stones and enhance blood coagulation. Megadoses of niacin, one of the B vitamins, can cause liver damage and gout.

Section Review Correlations

SC.F.1.4.5: Item 6
SC.G.1.4.3: Items 1, 2, 4
SC.H.3.4.3: Item 5
SC.H.3.4.4: Item 3

Minerals Different minerals are required to maintain a healthy body. **Minerals** are naturally occurring inorganic substances that are used to make certain body structures and substances, for normal nerve and muscle function, and to maintain osmotic balance. Some minerals are essential for enzyme function. Minerals are not produced by living organisms. Minerals must be replaced on a daily basis because they are soluble in water. Teeth and bones require the minerals calcium and phosphorus. Iron is required for transporting oxygen. Magnesium, calcium, sodium, potassium, and zinc help regulate function of the nerves and muscles.

Trace elements, such as those listed in **Table 2,** are minerals present in the body in small amounts. Humans usually obtain adequate amounts of the required trace elements directly from the plants they eat or indirectly from animals that have eaten plants.

Water You can survive only a few days without water, though you can live several weeks without food. Water is used by the body as a medium to transport gases, nutrients, and waste products. Water also plays a role in regulating body temperature. Two-thirds of the body's weight is water.

Table 2 Trace Elements

Trace element	Food sources	Role
Iodine	Seafood, plants grown in iodine-rich soil, iodized table salt	Synthesis of thyroid hormones
Cobalt	Leafy vegetables, liver, kidney	Synthesis of vitamin B_{12}
Zinc	Meat, shellfish, dairy products	Synthesis of digestive enzymes, proper immune function
Molybdenum	Legumes, cereals, milk	Protein synthesis
Manganese	Whole grains, nuts, legumes	Hemoglobin synthesis, urea formation
Selenium	Meat, seafood, cereal grains	Preventing chromosome breakage

Section 1 Review

1. **Predict** four nutrients that would be found in a serving of green beans. SC.G.1.4.3
2. **Compare** the functions of carbohydrates and proteins in maintaining a healthy body. SC.G.1.4.3
3. **Describe** the type of information the USDA food guide pyramid provides. SC.H.3.4.4
4. **Evaluate** the roles vitamins, minerals, and water play in maintaining a healthy body. SC.G.1.4.3
5. **Critical Thinking Applying Information** Your friend wants to feed her elderly grandmother more food in order to keep her healthy. Is this a good idea? Explain. SC.H.3.4.3
6. **FCAT Prep** Which of the following is a function of lipids in the body?
   - **A** make proteins
   - **B** make glycogen
   - **C** make steroid hormones.
   - **D** enhance enzyme activity

   SC.F.1.4.5

Answers to Section Review

1. carbohydrates, vitamins, minerals, protein
2. Both can be used as a source of energy, but the body usually uses amino acids from proteins as building blocks for other proteins.
3. The food guide recommends the number of servings of each different type of food group a person should eat daily to obtain a variety of nutrients in the diet.
4. Vitamins enhance enzyme activity. Minerals are necessary for making certain body structures and substances, for normal nerve and muscle functions, for maintaining osmotic balance, and for enzyme function. Water transports gases, nutrients, and waste and regulates body temperature.
5. Answers will vary. Students who disagree might argue that elderly people have a lower metabolic rate, and thus need fewer calories in a balanced diet. Students who agree might point out that elderly people tend to eat less and may get sick if they become malnourished.
6. **A.** Incorrect. Proteins are made of amino acids. **B.** Incorrect. Glycogen is made of carbohydrates. **C.** Correct. Lipids make steroid hormones. **D.** Incorrect. Enzyme activity is enhanced by vitamins and minerals.

Teaching Tip — GENERAL

Calcium and Vitamin D Ask students to identify the predominant mineral in milk. (calcium) Ask students to identify what nutrient is added to milk, (vitamin D) and then ask students why. (Accept any answers that tie vitamin D and calcium together.) Tell students that vitamin D is needed for the absorption of calcium from the intestines. Explain that rickets, a disease that affects growing bones and is caused by a vitamin D deficiency, used to be common in the United States. In the 1930s the federal government mandated the enrichment of milk with vitamin D. Today rickets is virtually unknown in the United States. LS Verbal

Close

Reteaching — BASIC

Bring in a label from a cereal box. List the nutrients in a table on the board or overhead. Have students write the function of each nutrient. Also, students can review their lists of nutrients and sources from the Motivate: Identifying Preconceptions activity at the start of this section.

Quiz — GENERAL

1. Which nutrient type should make up a majority of your diet? (carbohydrates)
2. Name two nutrients that do not provide energy but are needed for proper enzyme function. (vitamins and minerals)
3. What nutrient could you live only a few days without? (water)

Alternative Assessment — GENERAL

Have students describe any fad diets they may have read about or tried. Have them analyze these diets in terms of how they may fail to provide certain nutrients.

Transparencies

TR J47 Trace Elements
TR J46 Vitamins

Focus

Overview

Before beginning this section review with your students the objectives listed in the Student Edition. This section introduces students to the functions of the digestive organs and how food is broken down and absorbed.

Bellringer

Tell students to write down in correct sequence all the organs (at least 5) through which their food passes as it travels along the digestive tract. Then have them try to list any glands or organs (3) that are found along the digestive tract that help digestion but through which food does not pass.

Motivate

Demonstration — Basic

Hang an 8 m (26 ft) piece of clothesline or rope at the front of the room. Explain to students that it represents the approximate length of the digestive system. Use clothespins and markers to label the clothesline as follows:

PART	LENGTH	AVERAGE TIME FOOD SPENDS HERE
Mouth	8 cm (3 in.)	5–30 s
Pharynx/ esophagus	25 cm (10 in.)	5–10 s
Stomach	15 cm (6 in.)	2–6 h
Small intestine	4–6 m (13–20 ft)	3–6 h
Large intestine	1.2 m (4 ft)	up to 2 d

LS Visual

Section 2 Digestion

Florida Sunshine Standards/FCAT

SC.F.1.4.2 Body structures are uniquely designed and adapted for their function.

Objectives

- **Relate** the four major functions of the digestive system to the processing of food. SC.F.1.4.2
- **Summarize** the path of food through the digestive system and the major digestive processes that occur in the mouth, stomach, small intestine, and large intestine. SC.F.1.4.2
- **Describe** how nutrients are absorbed from the digestive system into the bloodstream or lymphatic system. SC.F.1.4.2
- **Identify** the role of the pancreas and liver in digestion. SC.F.1.4.2

Key Terms

amylase
esophagus
pepsin
lipase
villus
colon

Breaking Down Food

Imagine you just ate your favorite meal. What happens to that food? Before your body can use the nutrients in the food you eat, the large food molecules must be broken down physically and chemically. The process of breaking down food into molecules the body can use is called digestion. The digestive system is the body system that is involved in the taking in and processing of food for use by your body cells. The digestive system takes in food, breaks it down into molecules small enough for the body to absorb, and gets rid of undigested molecules and waste.

As shown in **Figure 5,** the digestive system is made up of a long, winding tube, the digestive tract, that begins at the mouth and winds through the body to the anus. Food travels more than 8 m (26 ft) through your digestive tract. The digestive tract includes the mouth, pharynx, esophagus, stomach, small intestine, large intestine, and rectum. Although the liver and pancreas *(PAN kree uhs)* are not part of the digestive tract, they deliver secretions into the digestive tract through ducts (tubes).

Digestive System

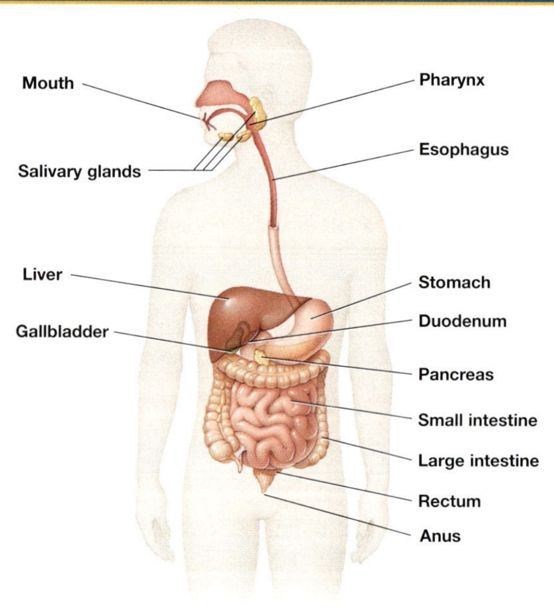

Figure 5 **Processing food.** The digestive system breaks down food into individual nutrient molecules that can be absorbed into the bloodstream.

906

Chapter Resource File

- Directed Reading Basic
- Active Reading General
- Data Sheet for Quick Lab General

Transparencies

TR Bellringer
TR J49 Digestive System in the Human Body

One-Stop Planner CD-ROM

- Reading Organizers Basic
- Reading Strategies Basic

Section Correlations

SC.F.1.4.2

Starting Digestion

The digestion of food begins as soon as the food is ingested. The teeth rip and chew food into shreds, and the tongue mixes the pieces with a watery solution called saliva. Taste buds on the tongue are sensitive to certain chemicals in the food. Saliva is secreted into the mouth by three pairs of salivary glands, shown in Figure 5. Saliva moistens and lubricates the food so that it can be swallowed more easily.

Saliva also contains **amylases** *(AM uh lay sehs)*, enzymes that begin the breakdown of carbohydrates such as starch, into monosaccharides (single sugars). The mechanical action of chewing and the chemical action of amylase are both part of the digestion of carbohydrates in the mouth.

Notice in **Figure 6** that the structure of our teeth helps in the breakdown of food. The two front teeth, the incisors, cut food. The cuspids, or canines, shred food. The back teeth, the molars, crush and grind food.

After passing through the region in the back of the throat called the pharynx *(FAIR ihnks)*, the food triggers a swallowing response. The action of swallowing moves the epiglottis (a flap of cartilage) over the opening of the trachea—the tube that leads to the lungs. This action prevents food from entering the trachea and eventually the lungs. Instead, food enters the esophagus *(ih SAHF uh guhs)*.

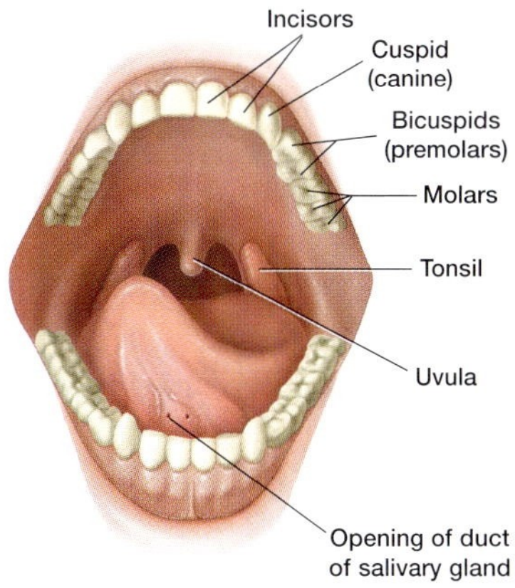

Figure 6 Teeth and digestion. Canine and incisor teeth are used for cutting and tearing food. The molars are used to grind food.

The Esophagus

The **esophagus** is a long tube that connects the mouth to the stomach. No digestion takes place in the esophagus. Its role is to act as a kind of descending elevator, moving food down to the stomach.

The esophagus is about 25 cm (10 in.) long. The lower two-thirds of the esophagus is wrapped in sheets of smooth muscle. Food does not simply fall into the stomach; it is pushed down, as shown in **Figure 7.** Successive rhythmic waves of smooth muscle contraction in the esophagus, called peristaltic *(pehr uh STAHL tihk)* contractions, or peristalsis, move the food toward the stomach. Peristalsis can be thought of as waves moving through the muscle with the area where the wave is passing causing the muscle to narrow. It takes about 5 to 10 seconds for food to pass down the esophagus and into the stomach.

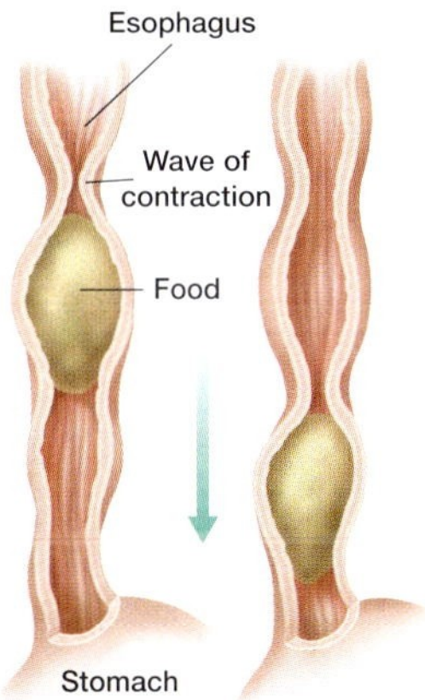

Figure 7 Peristalsis moves food. Food is pushed down the esophagus and toward the stomach by waves of smooth muscle contractions in the wall of the esophagus.

The Stomach

Food exits the esophagus and enters the stomach through a muscular valve called a sphincter *(SFIHNGK tuhr)*. The sphincter prevents acid-soaked food in the stomach from making its way back into the esophagus. The stomach is a saclike organ located just beneath the diaphragm. Besides temporarily storing food, the stomach, shown in **Figure 8** on the next page, also mechanically breaks down food and chemically unravels and breaks down proteins.

Teach

READING SKILL BUILDER — BASIC

Reading Organizer Have students create a reading organizer that lists the organs of the digestive system, their functions, and associated enzymes. If students are confused about any of these, encourage them to use outside sources of information such as an encyclopedia.
LS Verbal

Activity — GENERAL

Salivation Organize the class into groups. Give each group two test tubes with stoppers, four saltine crackers, water, a straw or stirring rod, and an iodine solution. In one test tube stir two crushed crackers with water, then add a few drops of iodine and invert to mix. The solution in the test tube will turn blue-black, indicating the presence of starch. Have a volunteer chew up the other two crackers without swallowing, and spit them into the second test tube. Add some water, stir, and then add a few drops of iodine, and invert to mix. The solution starts out blue-black, but as salivary amylase enzymes start to break the starch down into sugars, the color will fade to purple and eventually disappear. LS Kinesthetic Co-op Learning

Demonstration — BASIC

Cut the legs off of an old pair of pantyhose. Place a tennis ball in the toe and tie off the other end. You can show the process of peristalsis by rhythmically squeezing the tennis ball from one end to the other. Remind students that this happens automatically after swallowing, and several "swallows" (boli) can travel down the esophagus at one time. English Language Learners
LS Visual

INCLUSION Strategies

• Learning Disability • Attention Deficit Disorder

Using **Figure 5,** have students write each body part responsible for digestion on a note card. Students will then write the function of each body part in the digestive process. The cards should be numbered on the back to indicate the order the food is processed through the body. The students may study the digestive process by attempting to order the note cards by body part and function. Students may also use the note cards to present the digestive process to the class, small groups, or pairs.

REAL WORLD CONNECTION

Halitosis (bad breath) results when decomposing food particles accumulate in the mouth, allowing bacteria to flourish. Saliva aids in washing away the food particles. Therefore, people who suffer from any disease that inhibits saliva secretion usually have problems with dental caries and bad breath, as well as difficulty talking, swallowing, and eating.

Transparencies

TR J51 Teeth

Teach, continued

BIOWatch

Dental Records

Teaching Strategies
Forensic odontologists also give expert testimony in civil cases involving dental issues such as professional malpractice, insurance disputes, or the presence of dental injuries and/or physical neglect in children, adults, or the elderly.

Discussion
Explain why forensic odontology might provide the only way to identify victims of disasters such as airplane crashes or fires. (The bodies of victims of disasters such as airplane crashes or fires are often badly damaged, leaving few remains for analysis except teeth.)

Forensics Labs Appendix
Turn to pp. 1048–1071 for forensics lab activities and more student information on forensics techniques.

Transparencies

TR J52 The Stomach and Accessory Digestive Organs

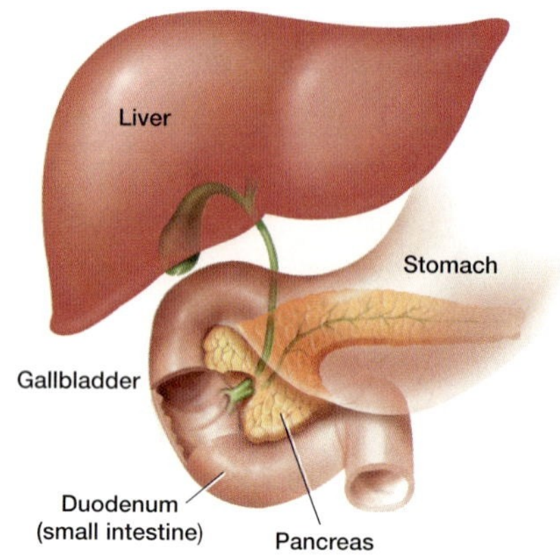

Figure 8 The stomach and accessory digestive organs. Many organs are involved in the complete breakdown of nutrients.

When food enters the stomach, gastric juice is secreted by the cells that line the inside of the stomach. Gastric juice is a combination of hydrochloric *(HIE droh klawr ihk)* acid (HCl) and pepsin. The acid breaks the bonds in proteins and unfolds large protein chains into single protein strands. **Pepsin,** a digestive enzyme secreted by the stomach, cuts the single protein strands into smaller chains of amino acids. Pepsin is effective only in an acidic environment.

The stomach mixes its contents by using peristaltic waves. Swallowed food can spend from 2 to 6 hours in the stomach. Your stomach secretes about 2 L (2.11 qt) of HCl every day, which creates a solution about 3 million times more acidic than your bloodstream. The hormone gastrin regulates the synthesis of HCl. Thus, HCl is made only when the pH in the stomach is higher than about 1.5.

A coating of mucus protects the lining of the stomach from gastric acid. Bicarbonate in the stomach helps neutralize digestive fluids. Blood circulation in the stomach lining also helps protect stomach tissues.

FORENSICS BIOWatch — Dental Records

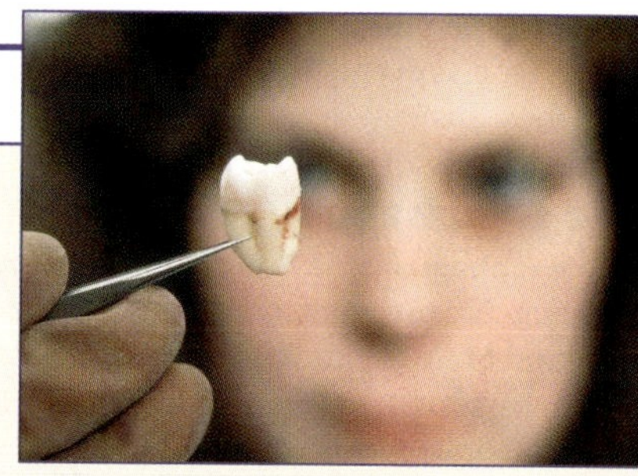

The next time you eat an apple, pause between bites to look at the apple. You've left bite marks, patterns that match your teeth—and no one else's. If the apple were found at a crime scene, a forensic odontologist could use the bite mark evidence to search for a suspect.

Forensic odontologists use dental evidence to identify human remains of missing or unknown persons. Forensic odontologists also use dental features to link suspects to crimes. Odontology is the scientific study of the teeth, their anatomy, development, and diseases. Forensic odontologists compare crime scene evidence with the dental records of a possible suspect or victim.

Identifying the Dead

When an unidentified body is found, forensic odontologists examine the teeth and the jaw. Tooth enamel is the hardest biological substance known. Teeth are a reliable source of evidence, lasting long after most of the body has decayed. The size of the teeth and the amount of wear help odontologists estimate a victim's age.

Forensic odontologists play an important role in the identification of victims of catastrophic events, such as airplane crashes, fires, or terrorist attacks. Disasters such as these result in many casualties and the bodies of victims are often badly damaged. After odontologists have completed their work, law enforcement officials can match forensic data with known dental records, X rays, or photographs.

Linking a Suspect to a Crime

Sometimes human bite marks are found on the body of the victim of a crime. Forensic odontologists analyze bite marks in cases of assault or homicide. Computer modeling can enable odontologists to match a bite mark to a dental mold of a suspect's mouth. The odontologist may also collect saliva from the bite wound for later DNA testing.

Career

Counselor Counselors give clients moral support and an opportunity to discuss personal or work-related problems. Counselors may be professional social workers, psychologists, or community workers interested in helping people. Employers include treatment centers, hospital-based programs, or government agencies. Each state has its own requirements for counselors. In some states counselors must earn a certificate. Have students interview a counselor that works with persons with eating disorders such as anorexia or bulimia. Ask each student to compile a list of 10 questions and answers from the interview.

The Small Intestine

Food passes from the stomach into the small intestine when a sphincter between the two organs opens. The small intestine is a coiled tubular organ about 6 m (19.8 ft) long that is continuous with the stomach and that functions mainly in the digestion and absorption of nutrients. The word *small* refers to the small diameter of the small intestine as compared with the diameter of the large intestine—not to its length. Peristalsis mixes the food, which remains in the small intestine for about 3 to 6 hours.

The first part of the small intestine, the duodenum *(doo oh DEE nuhm),* receives secretions from the pancreas, liver, and gallbladder, as shown in Figure 8. Cells that line the small intestine and the pancreas secrete digestive enzymes involved in completing the digestion of carbohydrates into monosaccharides, proteins into amino acids, and lipids into fatty acids and glycerol.

Before fats can be digested by pancreatic enzymes called **lipases** *(LIE pays uhs),* the fats must first be treated with bile, a greenish fluid produced by the liver. Bile breaks up fat globules into tiny fat droplets, a process called emulsification *(ee MUHL suh fih kay shuhn).* The gallbladder, a green muscular sac attached to the liver, concentrates and stores bile until it is needed in the small intestine.

Most absorption (passage of nutrients to the blood or lymph) occurs in the small intestine. The lining of the small intestine is covered with fine fingerlike projections called **villi** (singular form, villus). Villi, shown in **Figure 9,** are too small to see with the naked eye. In turn, the cells covering each villus have projections on their outer surface called microvilli. The villi and microvilli greatly increase the area available for absorption of nutrients. Sugars and amino acids enter capillaries in the villi and are carried in the blood to the liver for further metabolism. Fatty acids and glycerol enter lymphatic vessels in the villi and eventually enter the bloodstream.

Figure 9 Villi in the small intestine

Inside each villus are capillaries and lymphatic vessels where nutrients enter the bloodstream.

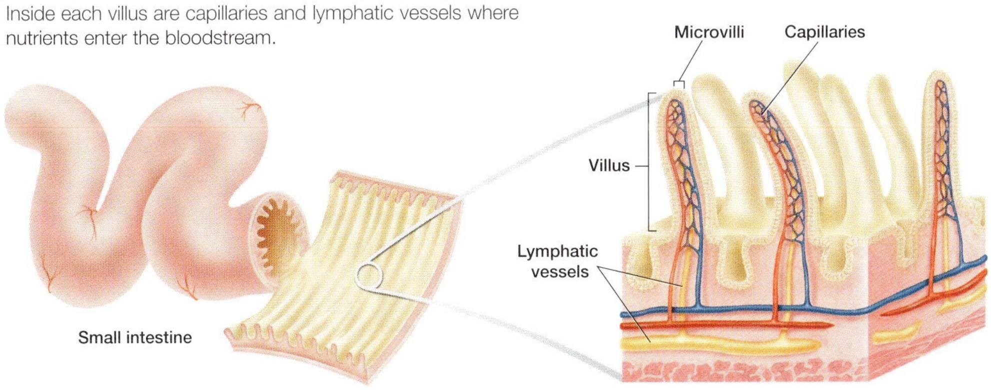

Demonstration — GENERAL

You will need two beakers of tap water, a sugar cube, a crushed sugar cube, and two stirring rods. At the same time, drop the sugar cube and crushed sugar into two separate beakers. Stir rapidly and observe which sugar dissolves first. Ask students for an explanation. (The crushed sugar cube has more surface area, so it dissolves faster.) Relate this principle to the increase of surface area in the intestine brought about by the villi and microvilli. LS Verbal

Using the Figure

Have students look at the structure of the small intestine in **Figure 9.** Use two pieces of yarn to show how surface area increases with the number of villi. (This can be demonstrated on the overhead.) Make a circle with a short piece of yarn. Use a much longer piece of yarn to simulate villi inside of the shorter piece. When you compare the lengths of the yarn, it becomes clear that villi (and microvilli) significantly increase surface area. LS Visual

Activity — ADVANCED

Viewing Digestive Organs

Provide students with a compound light microscope and prepared slides of several different digestive organs. You might want to include cross sections of tissue from the stomach, the small intestine, the liver, and the pancreas. Have students observe the specimens, compare the different cross sections, and make sketches of them. Encourage students to link the structures they observe with the functions of each organ. For example, students may relate the structure of villi to the function of nutrient absorption of the small intestine. English Language Learners LS Visual

Integrating Physics and Chemistry

Explain to students that acids are substances that react with water to form hydrogen ions, H^+. Since gastric juice contains hydrochloric acid, a strong acid, its pH is significantly below 7, it would have a sour taste, and it is corrosive.

MISCONCEPTION ALERT

Digestion and Absorption Although some digestion occurs in the mouth and stomach, most chemical digestion occurs in the small intestine. All nutrient absorption into the bloodstream occurs in the small intestine, not the stomach. However, both alcohol and aspirin are absorbed from the stomach directly into the bloodstream, accounting for their fast-acting effects on the body.

Transparencies

TR J54 Villi in the Small Intestine

TR J55 Cross Section of the Small Intestine

Focus

Teaching Tip — General

Enzymes and Nutrients Help students remember enzymes and the nutrients they break down straight by making a Graphic Organizer similar to the one at the bottom of this page. Remind students that the suffix *-ase* means "to break apart." LS Visual

SKILL BUILDER — Basic

Graphing Provide students with the following average lengths of the digestive organs: esophagus—25 cm, stomach—25 cm, small intestine—6 m, large intestine—1.5 m. Then have students prepare a bar graph or another graphic to compare the average lengths of each digestive organ listed. You might also want to have students measure and label the lengths using adding-machine tape or a skein of yarn and compare them to the illustration of the digestive system in this chapter. English Language Learners LS Logical

Section Review Correlations

SC.F.1.4.2: Items 1-5

SC.F.1.4.5: Item 6

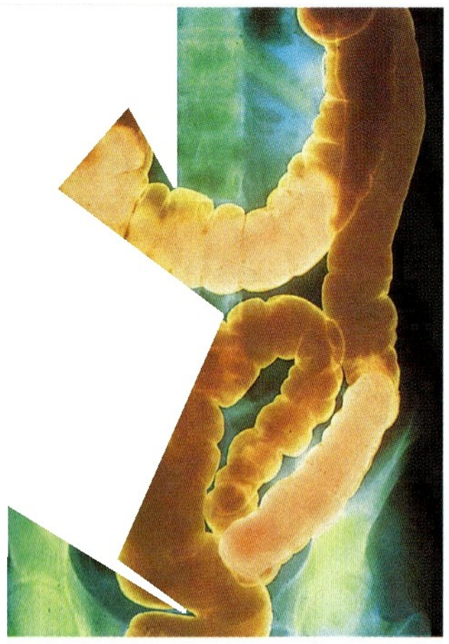

Figure 10 X ray of the large intestine (colon). Stretched out, the large intestine, which appears orange, is about 1 m (3.3 ft) long.

The Large Intestine

All components of food that are not for energy production (for example, cellulose) are considered wastes. The wastes move into the large intestine. The large intestine, or **colon** *(KOH luhn),* shown in **Figure 10,** is much shorter than the small intestine. However, the diameter of the large intestine is about three times larger than the diameter of the small intestine. The colon is not coiled like the small intestine. Instead, it is composed of three relatively straight segments. No digestion takes place in the colon.

The volume of material that flows through the digestive system each day is large—about 10 L of food, drink, and secretions enter the small intestine. But the amount of material that leaves the body as waste is small. This is because almost all of the fluids and solids (about 90–95 percent) are absorbed during their passage through the small intestine. Mostly mineral ions and water are absorbed through the wall of the large intestine.

Most of the colon's contents are dead cells, mucus, digestive secretions, bacteria, and yeast. A thriving colony of bacteria live in the human colon. These microbes synthesize many compounds that your body needs and cannot get easily from the food you eat, including vitamin K and several B vitamins. In addition, bacteria aid in transforming and compacting the undigested materials into the final waste product, feces.

The final segment of the large intestine is the rectum. Solids in the colon pass into the rectum as a result of peristalsis in the large intestine. From the rectum, the solid feces are eliminated from the body through the anus. Undigested material passes through the large intestine and is expelled through the anus in 12 to 24 hours.

Balancing water absorption in the intestine is important. Wastes rushed through the large intestine before the remaining water is absorbed result in diarrhea (watery feces). When food remains in the colon for long periods of time, which causes much water to be absorbed, constipation (hard feces) results. Hard feces are difficult to pass.

The Liver's Role in Digestion

The human liver, shown in Figure 8, is about the size of a football and weighs more than 1.4 kg (3.1 lb). It presses upward against the diaphragm and occupies the upper right side of the abdominal cavity.

The liver plays several roles in digestion. The liver secretes bile, which aids in the emulsification of fats. Bile also promotes the absorption of fatty acids and the fat-soluble vitamins A, D, E, and K. Bile pigments (the products of hemoglobin breakdown) give bile a yellowish green color. Jaundice, a condition in which the eyes, skin, and urine become abnormally yellow, is a result of increased amounts of bile pigments in the blood. Jaundice often occurs as a result of hepatitis, an inflammation of the liver.

Graphic Organizer

Use this graphic organizer with **Teaching Tip** on this page.

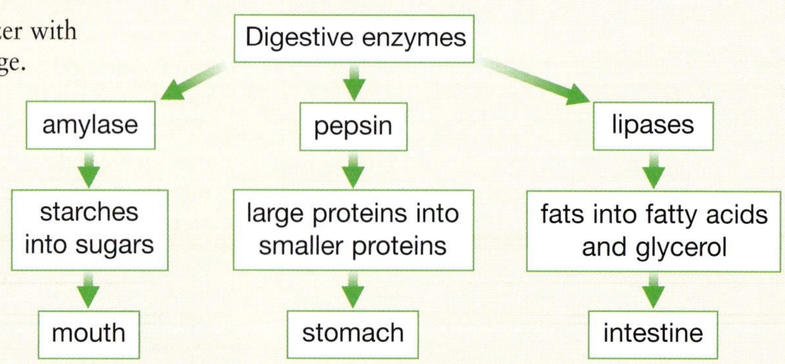

The Liver's Role in Metabolism

Digested food molecules in the bloodstream are transported to the liver. The liver stabilizes blood sugar by converting extra sugar to glycogen for storage. The liver then breaks down the glycogen when it is needed for energy. The liver also modifies amino acids. Fat-soluble vitamins and iron are stored in the liver. The liver monitors the production of cholesterol and detoxifies poisons. If the liver is unable to change a substance's harmful form, it stores it. In this way, toxins, including heavy metals and pesticides, accumulate in the liver.

The liver can also be damaged by viral infections, chronic drug or alcohol use, and traumatic injury. As a result of any of these, healthy liver cells are destroyed and replaced by scar tissue. The scarring of the liver is called cirrhosis *(suh ROH sihs).*

internet connect
www.scilinks.org
Topic: Poisons
Keyword: HX4142
SciLinks Maintained by the National Science Teachers Association

QUICK LAB

Modeling the Function of Bile

SC.H.3.4.1

You can use a detergent and cooking oil to simulate the effect bile has on breaking up (emulsifying) fats as part of digestion.

Materials

two 250 mL beakers, water, cooking oil, dish detergent, stirring rod, graduated cylinder

Procedure

1. Label one beaker *A* and one beaker *B*. Fill each beaker halfway with water.
2. Add 10 mL of the cooking oil to each beaker.
3. While stirring, slowly add 10 drops of dish detergent to beaker *B* only.

Analysis

1. **Describe** how oil reacts with the water.
2. **Describe** what happened to the oil when the dish detergent was added.
3. **Compare** the effect of dish detergent on oil with the effect of bile on fats.
4. **Critical Thinking** **Inferring Conclusions** Do the detergents and bile increase or decrease the surface area of oil? In the case of bile, how does this help the digestive process?

Section 2 Review

1. **Summarize** the path a piece of cheese pizza would follow through the digestive system. SC.F.1.4.2
2. **Relate** the role of the mouth, stomach, small intestine, and large intestine in the digestion of a piece of cheese pizza. SC.F.1.4.2
3. **Locate** the area of the digestive system where nutrients are absorbed into the bloodstream. SC.F.1.4.2
4. **State** how the liver and pancreas are involved in digestion. SC.F.1.4.2
5. **Critical Thinking** **Applying Information** A person has a small intestine that has villi but a reduced number of microvilli. Would you expect this person to be underweight or overweight? Explain. SC.F.1.4.2
6. FCAT Prep The enzyme pepsin is involved in the digestion of which kind of molecule?
   A fats
   B proteins
   C starches
   D monosaccharides
   SC.F.1.4.5

911

Answers to Section Review

1. mouth, pharynx, esophagus, stomach, small intestine, large intestine, rectum
2. *mouth*—mechanical and chemical digestion of carbohydrates in the pizza; *stomach*—mechanical digestion and chemical digestion of the proteins in the pizza; *small intestine*—chemical digestion of the carbohydrates, proteins, and lipids in the pizza, and absorption; *large intestine*—reabsorption of water and compaction of waste
3. small intestine
4. Both the pancreas and the liver provide digestive enzymes. The liver also provides bile to the small intestine, maintains blood sugar levels, modifies amino acids, stores vitamins and minerals, regulates cholesterol production, and detoxifies poisons.
5. The person would be underweight—more food will pass out as waste. The person will probably also be malnourished.
6. **A.** Incorrect. Fats are digested by lipases. **B.** Correct. Proteins are digested by the enzyme pepsin. **C.** Incorrect. Starches are digested by amylase enzymes. **D.** Incorrect. Monosaccharides are already broken down into absorbable units.

QUICK LAB

The Function of Bile

Skills Acquired
Relating concepts, inferring conclusions, comparing

Teacher's Notes
Detergents are made of molecules with hydrophobic and hydrophilic ends. The hydrophobic ends surround the grease and pull it toward the water, breaking up the grease.

Answers to Analysis

1. The oil spreads away from the water and the oil droplets associate.
2. The oil was broken up.
3. Detergent breaks up oil like bile breaks up fat.
4. They both increase surface area. For the bile this speeds up the breakdown of foods.

Close

Reteaching — BASIC

Have students make flashcards with a digestive organ named on one side and its function on the other. Have students arrange them in order from mouth to anus and identify each as part of the digestive tract or as accessory organs.

Quiz — GENERAL

1. What muscular movements cause food to be passed through the digestive tract? (peristalsis)
2. What two organs are important in digestion but are not a part of the digestive tract? (liver and pancreas)

Alternative Assessment — GENERAL

Refer to the clothesline from the Demonstration on the first page of this section. Remove the labels, and have students identify the parts of the digestive system and what happens to food in each of those parts.

Section 3

Teach, *continued*

Overview

Before beginning this section review with your students the objectives listed in the Student Edition. This section introduces students to how metabolic wastes are removed from the blood and excreted from the body. Students will also learn about the causes and consequences of kidney damage.

Bellringer

Tell students that the blood must be cleaned regularly. Then ask students to speculate, without looking in the textbook, how the body cleans blood. Once students have written a description of the cleaning process, allow them to check their answers against their textbook. LS Verbal

Motivate

Activity — GENERAL

Observing Sweat Have a volunteer put his or her hand in a clear plastic bag and tape it loosely to the wrist. Study the bag after 10 or 15 minutes. What do students observe? (tiny droplets of moisture) Explain that sweat is constantly evaporating from the skin. If time permits, have the student rub some of their own sweat onto individual slides and observe salt crystals. Red dye will make the crystals easier to see. LS Kinesthetic

Section Correlations

SC.F.1.4.2
SC.H.3.4.3

Section 3 Excretion

Florida Sunshine Standards/FCAT

SC.F.1.4. Body structures are uniquely designed and adapted for their function.

SC.H.3.4.3 Scientists can bring information, insights, and analytical skills to matters of public concern and help people understand the possible causes and effects of events.

Objectives

- **Identify** major wastes produced by humans and the organ or tissue where they are eliminated from the body. SC.F.1.4.2
- **Relate** the role of nephrons to the filtering of blood in the kidneys. SC.F.1.4.2
- **Summarize** how nephrons form urine. SC.F.1.4.2
- **Describe** the path of urine through the human urinary system. SC.F.1.4.2
- **Predict** how kidney damage might affect homeostasis and threaten life. SC.H.3.4.3

Key Terms

excretion
urea
nephron
urine
ureter
urinary bladder
urethra

Water and Metabolic Wastes

Cleaning up, though not always a pleasant chore, must be done to maintain a healthy living environment. In the same way, our bodies must get rid of wastes to maintain health. Food residues are eliminated from the body in the form of feces. Other wastes produced as a result of metabolic reactions that occur in the body must also be eliminated. For example, water and carbon dioxide are produced during cellular respiration. During the metabolism of proteins and nucleic acids, a toxic nitrogen-containing waste, ammonia, is formed.

The body must remove wastes. It must also maintain osmotic balance and stable pH by either excreting or conserving salts and water. **Excretion** is the process that rids the body of toxic chemicals, excess water, salts, and carbon dioxide while maintaining osmotic and pH balance.

The organs involved in excretion are shown in **Figure 11.** Carbon dioxide (and some water vapor) is transported to your lungs by the circulatory system and excreted every time you exhale. Excess water is excreted through the skin in sweat and through the kidneys in urine. In the liver, ammonia is converted to a much less toxic nitrogen waste called **urea** *(yoo REE uh)*, which is then carried by the bloodstream to the kidneys, where it is removed from the blood.

Organs of Excretion

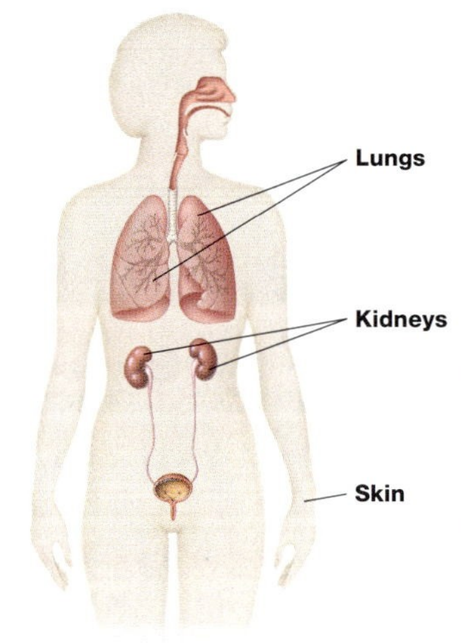

1. The lungs excrete carbon dioxide and water vapor in exhaled air.
2. The kidneys excrete nitrogen wastes, salts, water, and other substances in urine.
3. The skin excretes water, salts, small amounts of nitrogen wastes, and other substances in sweat.

Figure 11 **Organs of excretion.** The lungs, the kidneys, and the skin all function as excretory organs. The main excretory products are carbon dioxide, water, and nitrogen wastes (urea).

912

Chapter Resource File

- Directed Reading BASIC
- Active Reading GENERAL

Transparencies

TR Bellringer
TR J59 Organs of Excretion

One-Stop Planner CD-ROM

- Reading Organizers BASIC
- Reading Strategies BASIC
- Supplemental Reading Guide *Microbe Hunters* ADVANCED

The Kidneys

The kidneys are extremely important organs because of their role in regulating the amount of water and salts contained in blood plasma. The kidneys are a pair of bean-shaped, reddish brown organs located in the lower back. Each kidney is the size of a small fist. The body has to maintain a certain level of salts in the blood plasma and in the fluid surrounding cells, or serious harm to the body's cells and organ systems can result.

WORD Origins

The word *glomerulus* is from the French *gloméruli*, meaning "a compact cluster." Knowing this makes it easier to remember that the glomerulus is a compact cluster of capillaries within a nephron.

Blood Filters

Each kidney is a complex organ composed of roughly 1 million microscopic blood-filtering units called nephrons *(NEHF rahns)*, as shown in **Figure 12.** **Nephrons** are tiny tubes in the kidneys. One end of a nephron is a cup-shaped capsule surrounded by a tight ball of capillaries that filters wastes from the blood, retains useful molecules, and produces urine. Three different phases occur as the blood flows through a nephron: filtration, reabsorption and secretion.

Filtration Filtration begins at the cup-shaped capsule called Bowman's capsule. Within each Bowman's capsule an arteriole enters and splits into a fine network of capillaries called a glomerulus *(gloh MEHR yoo luhs)*. The glomerulus acts as a filtration device. The blood pressure inside the capillaries forces a fluid composed of water, salt, glucose, amino acids, and urea into the hollow interior of the Bowman's capsule. This fluid is called filtrate. Blood cells, proteins, and other molecules too large to cross the membrane remain in the blood.

Figure 12 Kidneys and nephrons

The kidneys filter out toxins, urea, water, and mineral salts from the blood as fluid passes through the microscopic filtering units called nephrons.

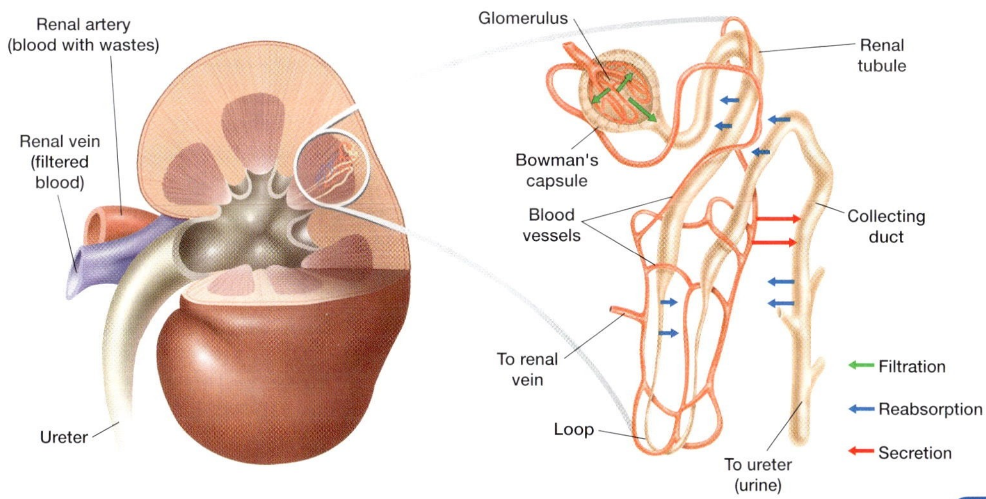

REAL WORLD CONNECTION

Exercise and sweating deplete the body of many necessary materials. Sports drinks are beverages that consist of water with about a 6 percent concentration of carbohydrates and electrolytes (salts, particularly sodium). Sports drinks are thought to improve performance in athletes who are exercising for more than 90 minutes because the carbohydrates maintain blood sugar (glucose) levels during exercise when muscle glycogen (stored glucose) levels start to run critically low. The electrolytes in the sports drink are supposed to allow the carbohydrates and water in the drink to be absorbed into the body faster.

Teach

Teaching Tip — BASIC

Kidneys Ask students to point to the location of their kidneys. They are located in the middle of the back, near the spinal cord, just under the last ribs. English Language Learners
LS Kinesthetic

Using the Figure

Review how passive transport processes (blood pressure) account for the filtration of blood in the nephron. Have students locate where filtration occurs in the nephron shown in **Figure 12.** (in Bowman's capsule) Review how active transport processes account for the reabsorption of materials as blood passes through the nephron and both passive and active processes are involved in secretion. Have students locate where reabsorption and secretion occur in the nephron shown in the figure. (the renal tubules) Point out that because of active transport, the kidney has an extremely high energy requirement, even higher than that of the heart, when comparing their relative weights.
LS Visual

SKILL BUILDER — GENERAL

Math Skills Tell students that the heart pumps about 5 L (1.3 gal) of blood per minute. Keeping in mind that one-fourth of this volume travels to the kidneys, have students calculate the volume of blood that is filtered each day. (5 L/min × 60 min/hr × 24 h × 0.25 = 1,800 L per day) Ask students to calculate how much of the blood is actually filtered if about 55–60 percent of the 5 L of blood is plasma, and only 18–20 percent of the plasma, water, and small solutes are actually filtered out of the blood and into the nephron capsule. (1,800 L × 0.55 × 0.18 = 178.2 L are actually filtered per day.) **LS Logical**

Transparencies

TR J56 Human Kidney Structure

Teach, *continued*

Real Life

Answers

Answers will vary. Researchers have found nanobacteria in kidney stones. It is thought that the bacteria cause kidney stones in genetically susceptible humans. The mineral shells that cause calcium deposits in the body form around the bacteria. Nanobacterial infections can be treated with antibiotics.

Teaching Tip — GENERAL

Water Treatment Plant Ask the manager from a water treatment plant to come and discuss the processes the plant uses to remove substances from water. Have students write a short report that relates this information to how the kidney removes substances from the blood. **LS Auditory**

Demonstration — GENERAL

Obtain a pig or sheep kidney from a local butcher. Cut the kidney in half lengthwise. Have students identify the obvious structures. Ask students if they can identify the smell. (ammonia) Ask students how kidneys are prepared before cooking. (You must soak the kidney in salt water to release the ammonia before cooking it.) **LS Visual**

Real Life

Kidney stones are painful.

Kidney stones form from crystallized mineral and urea salts. The crystals form "stones" that block the passage of urine and result in extreme pain. Ultrasound waves can be used to crush kidney stones until they are small enough to pass with the urine.

Finding Information

Research the role of nanobacteria in the formation of kidney stones.

Reabsorption and Secretion Reabsorption begins when the filtrate passes from the Bowman's capsules into the renal tubules—long, narrow tubes connected to Bowman's capsules. Renal tubules bend at their center, which forms a loop. As the filtrate passes through the renal tubules, the tubules extract from the filtrate a variety of useful molecules, including glucose, ions, and some water. These substances reenter the bloodstream through capillaries that wrap around the tubule. This arrangement prevents these molecules from being eliminated from the body in the urine. Some substances can pass from the blood into the filtrate in a process called secretion.

Urine Formation The **urine** that is excreted from the body is formed from the water, urea, and various salts that are left after the absorption and secretion processes. Collecting ducts receive fluid from several nephron segments and empty the urine into areas of the kidneys that lead to the ureters. **Ureters** *(yoo REET uhrs)* are tubes that carry the urine from the kidney to the urinary bladder. The collecting duct removes much of the water from the filtrate that passes through it. As a result, human urine can be very concentrated. In fact, it can be as much as four times more concentrated than blood plasma is.

Elimination of Urine

The ureters, shown in **Figure 13,** have smooth muscle in their walls. The slow, rhythmic contractions of this muscle move the urine through the ureters. The ureters direct the urine into the **urinary bladder,** a hollow, muscular sac that stores urine. The urinary bladder gradually expands as it fills. The average urinary bladder can hold up to about 0.6 L (0.63 qt) of urine. The urinary bladders of males tend to be larger than those of females.

Figure 13 **The organs of urinary excretion.** Urine exits the kidneys by way of two ureters that empty into a storage organ called the urinary bladder. Urine exits the body through the urethra.

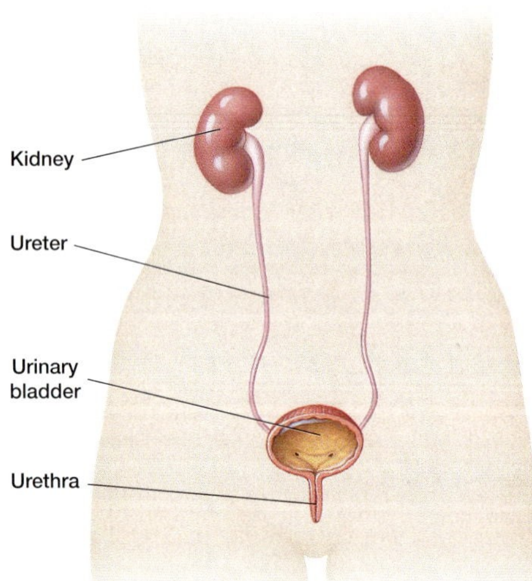

Muscular contractions of the bladder force urine out of the body. Urine leaves the bladder and exits the body through a tube called the **urethra** *(yoo REE thruh)*. A healthy adult eliminates from about 1.5 L (1.6 qt) to 2.3 L (2.4 qt) of urine a day, depending on the volume of fluid he or she consumes.

In females the urethra lies in front of the vagina and is only about 2.5 cm (1 in.) long. Such a short length makes it easy for bacteria and other pathogens to invade the female urinary system, which explains why females are more prone to urinary infections than males are. There is no connection between the urethra and the genital (reproductive) system in females.

In males the urethra passes through the penis. In males, both sperm and urine exit the body through the urethra. The tube that carries sperm from the testes eventually merges with the urethra.

Attention Grabber

Daily Fluid Intake It is ironic that as food conscious as Americans are, they often pay little attention to the other side of the coin—their fluid intake. In fact, people can survive without food much longer than they can survive without water. On average, an adult needs 1.5–3 L of water per day, some of it being supplied by foods with high water content such as lettuce, watermelon, and citrus fruits. Any combination of foods and beverages that provides the needed amount of water is acceptable, but most nutritionists recommend 6–8 glasses of liquid a day over and above the amount contained in food.

The elimination of urine from the body through the urethra is called urination. When the bladder fills with urine, stretch receptors in the bladder's wall send nerve impulses to the spinal cord. In response, the spinal cord returns impulses to the bladder and urethra simultaneously. These impulses cause contraction of the bladder's muscular walls and relaxation of the rings of muscle closing off the urethra. The bladder then empties its contents through the urethra. In older children and adults, the brain overrides this urination reflex, which delays the release of urine until a convenient time.

internet connect
www.scilinks.org
Topic: Kidney Diseases
Keyword: HX4109
SCILINKS. Maintained by the National Science Teachers Association

Damage to the Kidneys

Because of the vital role played by the kidneys in maintaining homeostasis, diseases affecting these organs may eventually threaten life. If one kidney is lost in an accident or by disease, the other may enlarge and do the work of both. Nephrotic syndrome refers to a number of signs and symptoms that result from damage to the glomeruli, which leads over time to kidney failure. The most common causes of kidney failure are infection, diabetes, high blood pressure, and damage to the kidneys by the body's own immune system.

Because of their function in excretion, kidneys often are exposed to hazardous chemicals that have entered the body through the lungs, skin, or gastrointestinal tract. Household substances that, in concentration, can damage kidneys include paint, varnishes, furniture oils, glues, aerosol sprays, air fresheners, and lead. When kidneys fail, toxic wastes, such as urea, accumulate in the plasma, and blood-plasma ion levels increase to dangerous levels. If both kidneys fail, there are only two treatment options.

Kidney Dialysis Kidney dialysis, also called hemodialysis *(HEE moh die AL uh sihs)*, is a procedure for filtering the blood by using a dialysis machine, as shown in **Figure 14.** A dialysis machine, just as the nephrons in the kidney, sorts small molecules in the blood, keeping some and discarding others. Dialysis machines are sometimes used for kidneys that are damaged but either will eventually heal or be replaced by a kidney transplant.

Kidney Transplants A more permanent solution to kidney failure is transplantation of a kidney from a healthy donor. A major problem with kidney transplants is common to all organ transplants—rejection of the transplanted organ by the recipient's immune system. Recall that the cells of your body have "self-markers," or antigens, on their surfaces that identify the cells to your immune system so it will not attack them. The combination of these antigens displayed on your body's cells is as unique as your fingerprints.

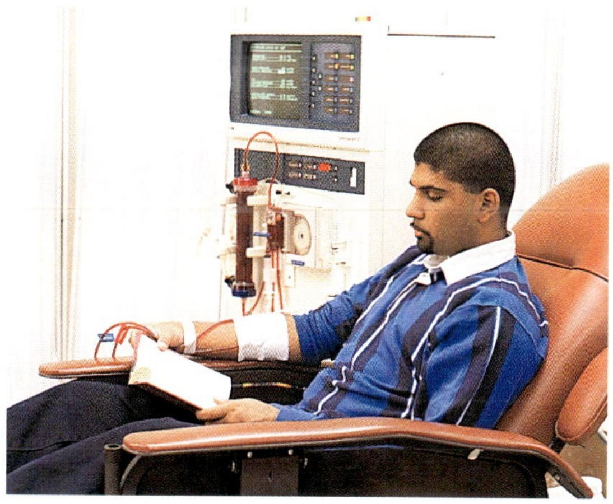

Figure 14 Hemodialysis. Hemodialysis has prolonged the lives of many people with damaged or diseased kidneys. The dialysis machine functions like a kidney in that it filters urea and excess ions from the blood.

Activity — GENERAL

Interview Have students interview a person who has had kidney stones or talk to a physician about them. Students should ask about the causes of kidney stones and the treatment. If possible, allow students to bring in a kidney-stone specimen for the class to observe. LS Auditory

Teaching Tip — BASIC

Concept Map Have students define the new terms in this section and use the terms to construct a concept map. Make sure the maps contain appropriate words and phrases linking the terms. LS Verbal

Transparencies

TR J58 How a Kidney Machine Works

did you know?

Urine Output High blood pressure, drinking lots of fluids, and cold weather promote increased urine output. High blood pressure increases the filtration pressure in the kidneys. Increased fluid consumption raises the blood pressure and decreases the output of antidiuretic hormone (ADH). Cold weather increases the metabolic rate and thus increases the rate of filtration. Low blood pressure, reduced intake of fluids, and hot weather decrease urine output for the converse reasons.

Career

Dialysis Technician Dialysis technicians operate dialysis machines and monitor and record a patient's weight and vital signs before, during, and after dialysis. Have students research the career of a dialysis technician and prepare a short report. The report should include a job description, training required, employer, growth potential, starting salary, and job satisfaction.

Only identical twins have the same set of antigens. The more closely related two individuals are, the more likely they are to have common antigens. This is why tissue transplants are more likely to succeed if the donor and recipient are closely related. But even in close matches, there is some chance of transplant rejection. To reduce chances of rejection, the recipient is treated with drugs that suppress the activity of the immune system.

Exploring Further

Kidney Dialysis

People whose kidneys are damaged cannot filter their blood. Kidney dialysis is one option for artificially filtering the blood. In kidney dialysis, tubes called catheters are surgically inserted into an artery and a vein, usually on a forearm. The catheters are equipped with valves. Every few days the catheters are connected to a dialysis machine, as shown to the right.

Blood Is Filtered

Blood passes from the patient's artery into the dialysis machine. Inside the machine, the blood travels through many hollow tubes, each of which is surrounded by a thin, permeable membrane. Waste materials and ions that have accumulated in the person's blood diffuse through the membrane into a fluid that has the same makeup as normal blood plasma and is free of wastes. The filtered blood is then returned to the person's vein.

Dialysis is not a permanent solution to kidney failure. A single healthy kidney can meet all of the homeostatic needs of the body, but no dialysis machine can. Dialysis patients must carefully manage their salt, protein, and water intake because the dialysis machine cannot regulate these blood components as well as the kidney can.

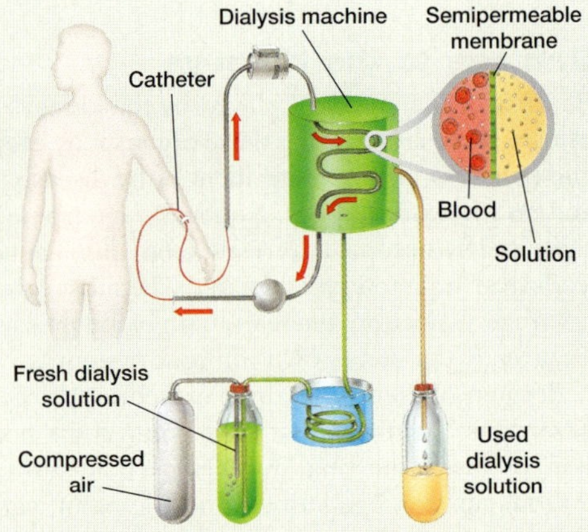

internet connect
www.scilinks.org
Topic: Kidney Dialysis
Keyword: HX4108
SCILINKS. Maintained by the National Science Teachers Association

Section 3 Review

1. **Identify** how the carbon dioxide in your body is produced and excreted. SC.G.1.4.3
2. **Relate** the following terms to the formation of urine: filtration, reabsorption, and secretion. SC.F.1.4.2
3. **Name** the liquid stored inside the collecting duct of a nephron. SC.G.1.4.3
4. **Summarize** how urine is stored and eliminated from the body. SC.F.1.4.2
5. **Critical Thinking** **Applying Information** A doctor has just informed a patient that his urine contains a high sugar concentration. Explain why this may indicate damaged kidneys. SC.G.1.4.3
6. **FCAT Prep** Which system carries the substances that are removed from the body by the excretory system to the kidneys?
   A circulatory system
   B digestive system
   C nervous system
   D respiratory system
   SC.F.1.4.2

916

Exploring Further

Kidney Dialysis

To demonstrate osmosis, bring in some dialysis tubing and show students how permeable it is. Pour a starch solution into the tubing, which is immersed in an iodine and water solution. The iodine will diffuse into the tubing, turning the starch solution blue-black.

Discussion

Ask students why dialysis machines are less effective than kidneys. (Machines do not dialyze constantly as the body does. Levels of salt, protein and water cannot be closely regulated with a machine.)

Close

Reteaching — BASIC

Have students return to their statements in the Opening Activity for this chapter. Ask students if their opinions have been confirmed or changed in any way.

Quiz — GENERAL

1. What are the three organs of excretion? (lungs, kidneys, skin)
2. What tubelike structures are the main blood-filtering units of the kidneys? (nephrons)

Alternative Assessment — GENERAL

Ask students to summarize the function of the urinary system, using as many science terms as possible.

Answers to Section Review

1. Carbon dioxide is produced during cellular respiration and excreted through the lungs.
2. *filtration*—kidneys filter blood, removing water, salt, glucose, amino acids, and urea; *reabsorption*—useful molecules and water are reabsorbed by the bloodstream; *secretion*—wastes and toxic materials pass from blood into filtrate
3. urine
4. Urine is stored in the bladder. When stretch receptors in the bladder send nerve impulses to the spinal cord, the bladder muscles relax and urination occurs.
5. Sugar should be reabsorbed by the bloodstream in the kidney.
6. **A.** Correct. Wastes to be removed from the body are carried to the kidney by the circulatory system. **B.** Incorrect. The digestive system does not carry waste products to the kidney for removal. **C.** Incorrect. The nervous system does not carry waste products to the kidney. **D.** Incorrect. The respiratory system can carry metabolic waste (carbon dioxide) but carries it to the lungs.

Section Review Correlations

SC.F.1.4.2: Items 2, 4, 6
SC.G.1.4.3: Items 1, 3, 5

Key Concepts

1 Your Body's Need for Food

- Food and beverages provide the nutrients and water required by the body for growth, energy, repair, and maintenance.
- Carbohydrates and lipids provide most of the body's energy. Proteins are normally used for making other proteins.
- The USDA food guide pyramid graphically summarizes the daily recommended servings from each food group.
- Vitamins enhance the activity of enzymes and regulate the release of energy. Minerals are used to make certain body structures and substances, for normal nerve and muscle function, to maintain osmotic balance, and for enzyme function.
- Water acts as a lubricant, solvent, and coolant, and as a support medium for cells and tissues.

2 Digestion

- Teeth break down food into smaller pieces. Amylase begins the breakdown of starch to sugars. The stomach stores and mechanically breaks down food. Stomach acid and pepsin chemically break down proteins.
- Most chemical digestion occurs in the small intestine with the help of secretions from the pancreas, liver, and gallbladder.
- Usable compounds are absorbed into capillaries or lymphatic vessels in villi. Compounds not absorbed are eventually excreted as feces.
- The liver releases bile, helps to maintain blood sugar levels, and detoxifies poisons.

3 Excretion

- The skin, lungs, and kidneys are specialized to excrete wastes.
- Nephrons in the kidneys filter wastes from the blood. Most of the water, some of the salts, and all of the sugar and amino acids in the filtrate are reabsorbed into the bloodstream. The water, urea, and salts that remain in the nephron are eliminated as urine.
- Kidney dialysis and organ transplants are treatment options when both kidneys fail.

Key Terms

Section 1

nutrient (900)
digestion (900)
calorie (900)
vitamin (904)
mineral (905)

Section 2

amylase (907)
esophagus (907)
pepsin (908)
lipase (909)
villus (909)
colon (910)

Section 3

excretion (912)
urea (912)
nephron (913)
urine (914)
ureter (914)
urinary bladder (914)
urethra (914)

917

Study ZONE

Alternative Assessment — GENERAL

Commercial diet programs are big business. Have students research commercial diet programs and report how each claims to bring about weight loss. Encourage students to look for information on how these programs might affect either the digestive or excretory systems. For example, a high-protein diet places added stress on the liver and kidneys, because the body must produce and then rid itself of more urea, secondary to higher nitrogen intake. Have students present their research in a multimedia format. **LS Verbal**

Chapter Resource File

- Science Skills Worksheet GENERAL
- Critical Thinking Worksheet ADVANCED
- Test Prep Pretest GENERAL
- Chapter Test GENERAL

Answer to Concept Map

The following is one possible answer to Performance Zone item 15 on the next page.

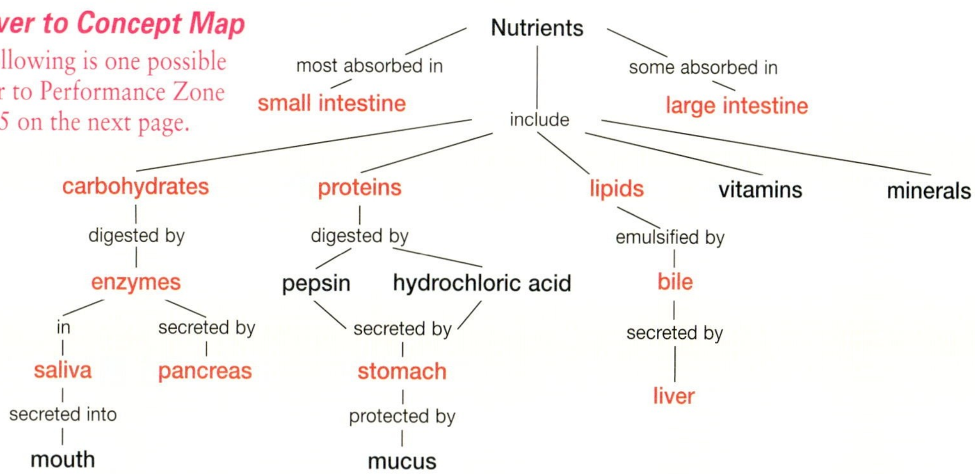

ANSWERS

Understanding Key Ideas

1. c
2. a
3. b
4. d
5. c
6. A. grains, 6–11 servings a day;
   B. fruits, 2–4 servings a day
7. d
8. A forensic odontologist would examine the half-eaten sandwich and make careful records of the bite marks. These records could be matched with known dental records, X rays, or photographs from possible suspects.
9. Blood moves through tubes composed of semipermeable membranous material, which are bathed in fluid with the same composition as clean plasma. Wastes diffuse out of the tubes and into the "plasma." The filtered blood circulates into the patient's body.
10. One possible answer to the concept map is found at the bottom of the Study Zone page.

Critical Thinking

11. Answers will vary. Students who agree might argue that water is more readily available so that their need to conserve water is not as critical as in humans.
12. acidic foods such as tomatoes, oranges, and other citrus fruits
13. Bacteria live in the human colon where they synthesize vitamin K and several B vitamins, and aid in the elimination of feces.

Performance ZONE

CHAPTER REVIEW

Understanding Key Ideas

1. The primary function of carbohydrates is to
   a. break down molecules.
   b. aid in digestion.
   c. supply the body with energy.
   d. regulate the flow of acid.
   SC.G.1.4.3
2. Food from the _______ food group should be eaten in the greatest abundance.
   a. grains
   b. fats, oils, and sweets
   c. vegetables
   d. milk, yogurt, and cheese
   SC.G.1.4.3
3. The body needs vitamins because they
   a. supply energy.
   b. activate enzymes.
   c. function as enzymes.
   d. act as hormones.
   SC.G.1.4.3
4. The _______ are involved in excretion.
   a. kidneys and stomach
   b. liver and pancreas
   c. pancreas and kidneys
   d. kidneys and lungs
   SC.F.1.4.2
5. During secretion in the kidney, substances move from
   a. the filtrate to the blood.
   b. the urethra to the bladder.
   c. the blood to the filtrate.
   d. the bladder to the urethra.
   SC.F.1.4.2
6. **Identifying Information** Identify the food groups represented by *A* and *B*. Indicate the number of servings that should be eaten daily for *A* and *B*. SC.G.1.4.3

A B

7. Which substance is *not* a waste eliminated from the body through the kidneys? SC.F.1.4.2
   a. urea
   b. water
   c. salts
   d. oxygen
8. **BIOWatch** A suspect leaves a half-eaten sandwich at a crime scene. How would the sandwich be useful to a forensic odontologist? SC.H.3.4.3
9. **Exploring Further** Summarize how a kidney dialysis machine filters blood. SC.H.3.4.5
10. **Concept Mapping** Make a concept map that shows how nutrients are digested. Try to include the following words in your map: *carbohydrates, proteins, lipids, enzymes, saliva, pancreas, stomach, bile, liver, small intestine,* and *large intestine.* SC.F.1.4.2

Critical Thinking

11. **Evaluating Conclusions** The length of the looped tubule in a nephron varies among mammal species. A friend believes the looped tubules of mammals that live in the water would be shorter than those found in humans. Do you agree or disagree? Explain. SC.F.1.4.2
12. **Recommending Information** Calcium needs an acidic environment for optimal absorption. What kinds of foods would you recommend be combined with calcium-rich foods in order to maximize absorption? SC.G.1.4.3
13. **Relating Concepts** Describe the symbiotic role of bacteria in the human intestine. SC.G.1.4.1

Alternative Assessment

14. **Communicating** Write an article that discusses diuretics (substances that increase urine excretion) for your school or local newspaper. Emphasize diuretics that most people have heard of, such as the caffeine in coffee and soft drinks. SC.F.1.4.8

918

Alternative Assessment

14. Answers will vary. Students' lists might include caffeine, parsley, juniper, buchu leaves, uva ursi, and other herbs, as well as prescription drugs.

Assignment Guide

Section	Questions
1	1, 2, 3, 6, 8, 10
2	10
3	4, 5, 7, 9

Chapter Review Correlations

SC.F.1.4.2: Items 4, 5, 7, 10, 11
SC.F.1.4.8: Item 14
SC.G.1.4.1: Item 13
SC.G.1.4.3: Items 1, 2, 3, 6, 12
SC.H.3.4.3: Item 8
SC.H.3.4.5: Item 9

FCAT Prep

For questions 1–6, write your answers on a separate sheet of paper.

1. What biological molecules make up proteins?
   A. amino acids
   B. fatty acids
   C. glycerol
   D. monosaccharides
   SC.G.1.4.3
2. Which of the following correctly pairs the enzyme with the food molecule it digests?
   F. lipase, fat **H.** lipase, starch
   G. amylase, fat **I.** pepsin, starch
   SC.F.1.4.5
3. A friend believes that a vegetarian diet would decrease his intake of saturated fat and cholesterol. Which statement supports this idea?
   A. Many people who eat nonvegetarian diets get more iron and protein.
   B. Some vegetarians have fewer health problems than non-vegetarians.
   C. Some vegetarian diets include foods that are high in fats and cholesterol.
   D. In most nonvegetarian diets, fats and cholesterol come from animal products.
   SC.H.3.4.3
4. What is the main function of dietary fiber?
   F. to provide energy
   G. to maintain osmotic balance
   H. to provide materials for making enzymes
   I. to help food pass through the digestive tract
   SC.F.1.4.8

Base your answers to questions 5 and 6 on the table below.

FOOD LABEL

Nutrition Facts	Amount/serving	% DV*	Amount/serving	% DV*
	Total fat 1 g	1%	Total carb. 43 g	14%
Serv. size 2 oz (56 g/$\frac{1}{8}$ box)	Sat. fat 0 g	0%	Dietary fiber 2 g	8%
Servings per container 8	Cholesterol 0 mg	0%	Sugars 3 g	
Calories 210 Fat Cal. 10	Sodium 0 mg	0%	Protein 6 g	
*Percent Daily Values (DV) are based on a 2,000 Calorie diet.	Vitamin A 0% • Vitamin C 0% • Calcium 2% • Iron 10% Thiamin 30% • Riboflavin 10% • Niacin 15%			

5. Approximately what percentage of the calories in this food come from fats?
   A. 1 percent **C.** 10 percent
   B. 5 percent **D.** 14 percent
   SC.G.1.4.3
6. How many grams of carbohydrates are in 4 servings of this food?
   SC.G.1.4.3

Test TIP

Sometimes only a portion of a graph or table is needed to answer a question. Focus only on the necessary information to avoid confusion.

FCAT Doctor

Question 3 Answer D is the correct choice. This statement indicates that avoiding animal products could reduce your intake of fats and cholesterol. Answers A and B are incorrect because these statements do not address cholesterol and fats. Answer C is incorrect because this statement indicates that even with a vegetarian diet, your friend is still at risk of having a diet that is high in saturated fat and cholesterol.

Question 5 Answer B is the correct choice. To determine the percentage of calories from fat, the number of fat calories (10) should be divided by the total calories (210), which gives about 5%. Answer A is incorrect because the percentage of fat that this food contributes to the Daily Values is 1%, not the percentage of its total calories that are from fats. Answer C is incorrect because the fat calories per serving is 10, which is not the percentage of its total calories that are from fats. Answer D is incorrect because the percentage of carbohydrates that a serving contributes to the Daily Values is 14%, not the percentage of its total calories that are from fats.

Question 6 The correct answer is 172. One serving of this food has 43 g of carbohydrates. Four servings would have 4 × 43, or 172 g.

Answers

1. A
2. F
3. D
4. I
5. B
6. 172

Test Prep Correlations

SC.F.1.4.5: Item 2
SC.F.1.4.8: Item 4
SC.G.1.4.3: Items 1, 5, 6
SC.H.3.4.3: Item 3

Exploration Lab

Lactose Digestion

Teacher's Notes

Time Required

Two 45-minute class periods

Ratings

EASY → HARD

- Teacher Preparation
- Student Setup
- Concept Level
- Cleanup

Skills Acquired

- Collecting Data
- Designing Experiments
- Experimenting
- Inferring
- Organizing and Analyzing Data

Scientific Methods

In this lab, students will:

- Make Observations
- Ask Questions
- Test the Hypothesis
- Analyze the Results
- Draw Conclusions

Materials and Equipment

Materials for this lab can be ordered from WARD'S. See *Master Materials List* at the front of this book for catalog numbers. All milk products can be poured down the drain. Glucose test strips can be placed in the trash.

Safety Cautions

Discuss all safety symbols and caution statements with students.

Exploration Lab

Demonstrating Lactose Digestion SC.G.1.4.3

SKILLS

- Using scientific methods
- Observing
- Comparing

OBJECTIVES

- **Describe** the relationship between enzymes and the digestion of food molecules.
- **Evaluate** the ability of a milk-treatment product to promote lactose digestion.
- **Infer** the presence of lactose in milk and foods that contain milk.

MATERIALS

- milk-treatment product (liquid)
- toothpicks
- depression slides
- droppers
- whole milk
- glucose solution
- glucose test strips

ChemSafety

CAUTION: Always wear safety goggles and a lab apron to protect your eyes and clothing.

CAUTION: Do not touch or taste any chemicals. Know the location of the emergency shower and eyewash station and how to use them. If you get a chemical on your skin or clothing, wash it off at the sink while calling to the teacher. Notify the teacher of a spill. Spills should be cleaned up promptly, according to your teacher's directions.

CAUTION: Glassware is fragile. Notify the teacher of broken glass or cuts. Do not clean up broken glass or spills with broken glass unless the teacher tells you to do so.

Before You Begin

People with a condition known as **lactose intolerance** often experience stomach and intestinal pain, bloating, and diarrhea when they eat foods that contain milk. These symptoms result from an inability to digest lactose, a sugar found in milk. **Lactose** is a disaccharide made of one glucose unit and one galactose unit. Lactose molecules are broken down into glucose and galactose molecules during **digestion.** People who cannot digest lactose do not produce **lactase,** the digestive enzyme that aids the breakdown of lactose. In this lab, you will investigate a milk-treatment product that is designed to aid lactose digestion.

1. Write a definition for each boldface term in the paragraph above.
2. List at least 10 foods that contain milk.
3. Make a data table similar to the one below.
4. Based on the objectives for this lab, write a question you would like to explore about enzymes and digestion.

DATA TABLE

Solution	Result (+ or –)	Interpretation

Procedure

PART A: Design an Experiment

1. Read the information sheet that comes with the milk-treatment product. Discuss with your lab group what the product is and what it does. Write a summary of your discussion for your lab report.

920

Tips and Tricks

Glucose test strips and commercial milk-treatment products that contain yeast-derived lactase are available from most pharmacies. Buy several boxes of the glucose test strips so the lab groups each have a color guide. Avoid buying the yeast-derived lactase in solid form. Do not use milk-treatment product tablets. The tablets will give positive glucose results. Each lab group will need copies of the information sheet that accompanies the product. When you read the students' protocols for approval, direct students to use only drops of milk and milk-treatment product instead of full quarts of milk. Glucose test strips are more convenient than Benedict's test because they do not require a hot water bath and they give faster results.

Before students begin the lab ask how you could tell if a milk-treatment product digests lactose. The presence of glucose and/or galactose in the milk after addition of the treatment product would indicate that lactose has been broken down. Some students will be unsure about what controls to use. Instruct students to test the glucose test strip for a positive glucose test in a known glucose solution.

2. Work with the members of your lab group to explore one of the questions written for step 4 of **Before You Begin.** To explore the question, design an experiment that uses the materials listed for this lab.

You Choose

As you design your experiment, decide the following:

a. what question you will explore
b. what hypothesis you will test
c. what your controls will be
d. how much milk and milk-treatment product to use for each test
e. how to determine whether lactose was broken down
f. what data to record in your data table

3. Write the procedure for your group's experiment. Make a list of all the safety precautions you will take. Have your teacher approve your procedure and safety precautions before you begin the experiment.

4. Set up your group's experiment, and collect data.

PART B: Cleanup and Disposal

5. Dispose of solutions, broken glass, and glucose test strips in the designated waste containers. Do not pour chemicals down the drain or put lab materials in the trash unless your teacher tells you to do so.

6. Clean up your work area and all lab equipment. Return lab equipment to its proper place. Wash your hands throughly before you leave the lab and after you finish all work.

Analyze and Conclude

1. **Summarizing Information** What are the milk-treatment product's ingredients?
2. **Recognizing Relationships** What is the relationship between lactose and lactase?
3. **Analyzing Methods** What role did the glucose solution play in your experiment?
4. **Drawing Conclusions** What does the milk-treatment product do to milk?
5. **Analyzing Conclusions** How do your results justify your conclusion?
6. **Evaluating Methods** Why should you test the milk-treatment product with glucose test strips?
7. **Analyzing Results** What do you infer from the results of this lab about treatments for other medical problems resulting from enzyme deficiencies?
8. **Forming Reasoned Opinions** As a person grows older, will he or she be more likely or less likely to develop lactose intolerance? Explain your answer.
9. **Predicting Patterns** Do you think lactose intolerance might be inherited? Explain your answer.
10. **Further Inquiry** Write a new question about enzymes and digestion that could be explored with another investigation.

Do You Know?

Do research in the library or media center to answer these questions:

1. What are some other food-treatment products that contain digestive enzymes?
2. Why does the improper breakdown of certain food molecules cause symptoms such as stomach pain, gas, and diarrhea?

Use the following Internet resources to explore your own questions about lactose intolerance.

internet connect
www.scilinks.org
Topic: Lactose Intolerance
Keyword: HX4111
SCLINKS. Maintained by the National Science Teachers Association

921

Answers to Before You Begin

1. *lactose intolerance*—an inability to digest the lactose in milk or milk-containing products due to the lack of the enzyme lactase
   lactose—a disaccharide found in milk and composed of a glucose unit and galactose unit
   digestion—the breakdown of food molecules into simpler molecules that can be taken up by cells
   lactase—the enzyme that catalyzes the breakdown of lactose
2. Answers will vary. Examples may include milk, cheese, butter, ice cream, and cottage cheese.
3. Students should have a data table similar to the one shown.
4. Answers will vary. For example: How effective is a milk-treatment product in breaking down lactose?

Answers to Do You Know?

1. Answers will vary. Students could list products such as bean-treatment products.
2. Stomach pain results from a buildup of gas, which is produced by bacteria as a waste product when breaking down food.

Answers to Procedure

1. Use a glucose strip to test water treated with 3 drops of milk-treatment product (as a negative control).
2. Use a glucose strip to test an untreated glucose solution (as a positive control).
3. Use a glucose strip to test the milk-treatment product itself (to verify the product does not contain glucose).
4. Use a glucose strip to test untreated milk and other untreated foods.
5. Use a glucose strip to test milk and other foods treated with the milk-treatment product.

Answers to Analyze and Conclude

1. yeast-derived lactase
2. Lactose is a disaccharide found in milk. Lactase is an enzyme that catalyzes the breakdown of lactose into glucose and galactose. Lactose is the substrate for the enzyme lactase.
3. The glucose solution is a positive control for the experiment.
4. It breaks down the lactose in the milk into glucose and galactose.
5. Answers may vary. Students indirectly tested for the breakdown of lactose by testing for the presence of glucose. Milk that was treated with the milk-treatment product tested positive for glucose, while milk that was not treated was negative for glucose.
6. to eliminate the milk-treatment product as a possible source of glucose
7. Other disorders caused by enzyme deficiencies might be treated with products that contain the missing enzymes.
8. Answers may vary. The risk of lactose intolerance increases with age because lactase is required most during infancy and early childhood when milk is a major part of the diet.
9. Answers may vary. As an enzyme, lactase is a protein and proteins are encoded in DNA, the molecule responsible for inheritance. Certain populations have a higher incidence of lactose intolerance than others.
10. Answers will vary. For example: How would blood glucose levels compare in people eating treated dairy products with those eating untreated dairy products?

Chapter Opener		SE **Quick Review,** p. 923 ■ SE **Reading Activity,** p. 923 ■ TE **Using the Figure,** p. 922 GENERAL	
& 3 • 90 min pp. 924–926 **Section 1** Two Lines of Nonspecific Defenses	OSP **Lesson Plan*** TR **Bellringer*** TR **Section Outline*** TR **J36 Inflammatory Response***	TE **Group Activity** Sequencing Lines of Defense, p. 925 ◆ GENERAL CRF **Quick Lab** Calculating Contagiousness* BASIC CRF **Skills Practice** Testing for Vitamin C* GENERAL	
& 5 • 90 min pp. 927–929 **Section 2** Immune Response	OSP **Lesson Plan*** TR **Bellringer*** TR **Section Outline*** TR **J102 Immune Response***	TE **Demonstration,** p. 927 ◆ GENERAL TE **Using the Figure,** p. 928 BASIC	
& 7 • 90 min pp. 930–932 **Section 3** Disease Transmission and Prevention	OSP **Lesson Plan*** TR **Bellringer*** TR **Section Outline***	TE **Activity** Mutating Viruses, p. 931 ADVANCED SE **Quick Lab** Simulating Antigen Activity, p. 932 ■ CRF **Datasheets for In-Text Labs** Simulating Antigen Activity*	
& 9 • 90 min pp. 933–936 **Section 4** Disorders of the Immune System	OSP **Lesson Plan*** TR **Bellringer*** TR **Section Outline*** TR **J41 Number of Helper T Cells Following HIV Infection***	TE **Activity** Immune System Disorders, p. 933 GENERAL TE **Using the Figure,** p. 934 GENERAL TE **Demonstration,** p. 935 ◆ BASIC SE **Data Lab** Tracking the Spread of AIDS, p. 935 ■ CRF **Datasheets for In-Text Labs** Tracking the Spread of AIDS* SE **Skills Practice Lab** Simulating Disease Transmission, pp. 940–941 ■ CRF **Datasheets for In-Text Labs** Simulating Disease Transmission*	

) & 11 • 90 min

Chapter Review and Assessment Resources

- SE **Study Zone,** p. 937 ■
- SE **Performance Zone,** p. 938 ■
- SE **FCAT Test Prep,** p. 939 ■
- TR **Concept Mapping***
- CRF **Vocabulary Review Worksheet*** ■ GENERAL
- CRF **Test Prep Pretest*** ■ GENERAL
- CRF **Chapter Test*** ■ GENERAL
- CRF **Chapter Test*** ADVANCED
- OSP **Test Generator***
- OSP **Test Item Listing***
- OSP **Modified Chapter Test*** BASIC

Online and Technology Resources

go.hrw.com Holt Online Learning

Visit go.hrw.com for access to Holt Online Learning or enter the keyword **HX6 Home** for a variety of free online resources.

One-Stop Planner® CD-ROM

This CD-ROM package includes
- Lab Materials QuickList Software
- Holt Calendar Planner
- Customizable Lesson Plans
- Printable Worksheets
- ExamView® Test Generator
- Interactive Teacher Edition
- Holt PuzzlePro® Resources
- Holt PowerPoint® Resources

Guided Reading Audio CDs

These CDs are designed to help auditory learners and reluctant readers. (Audio CDs are also available in Spanish.)

Compression guide:
To shorten your instruction because of time limitations, omit block 1.

KEY

TE	Teacher Edition	CRF	Chapter Resource File	*	Also on One-Stop Planner
SE	Student Edition	TR	Teaching Transparency	■	Also Available in Spanish
OSP	One-Stop Planner			◆	Requires Advance Prep

SKILLS DEVELOPMENT AND PRACTICE	SECTION REVIEW AND ASSESSMENT	CORRELATIONS
		Florida Sunshine Standards/FCAT
TE Reading Skill Builder Interactive Reading, p. 925 BASIC CRF Directed Reading Worksheet* BASIC CRF Active Reading Worksheet* GENERAL	SE Section 1 Review ■ TE Reteaching, p. 926 BASIC TE Quiz, p. 926 GENERAL TE Alternative Assessment, p. 926 GENERAL CRF Quiz* ■ BASIC	SC.F.1.4.2, SC.F.1.4.5, SC.F.1.4.8
CRF Directed Reading Worksheet* BASIC CRF Active Reading Worksheet* GENERAL	SE Section 2 Review ■ TE Reteaching, p. 929 BASIC TE Quiz, p. 929 GENERAL TE Alternative Assessment, p. 929 GENERAL CRF Quiz* ■ BASIC	SC.F.1.4.2, SC.F.1.4.8
TE Reading Skill Builder Paired Summarizing, p. 931 BASIC TE Skill Builder Interpreting Visuals, p. 931 GENERAL CRF Directed Reading Worksheet* BASIC CRF Active Reading Worksheet* GENERAL	SE Section 3 Review ■ TE Reteaching, p. 932 BASIC TE Quiz, p. 932 GENERAL CRF Quiz* ■ BASIC	SC.F.1.4.5, SC.F.1.4.8, SC.H.3.4.5
TE Skill Builder Math Skills, p. 935 GENERAL CRF Directed Reading Worksheet* BASIC CRF Active Reading Worksheet* GENERAL CRF Critical Thinking Worksheet* ADVANCED CRF Science Skills Worksheet* GENERAL	SE Section 4 Review ■ TE Reteaching, p. 936 BASIC TE Quiz, p. 936 GENERAL TE Alternative Assessment, p. 936 GENERAL CRF Quiz* ■ BASIC Modified Worksheet* BASIC	SC.F.1.4.5, SC.F.1.4.8, SC.H.2.4.2, SC.H.3.4.1

SCLINKS. NSTA
www.scilinks.org
Maintained by the **National Science Teachers Association**

Topic: Immune Systems
SciLinks Code: HX4103

Topic: HIV Transmission
SciLinks Code: HX4098

Topic: Asthma
SciLinks Code: HX4015

CHAPTER 40

Overview

Tell students that the purpose of the chapter is to explore the way the body protects itself against disease, how diseases are spread, and how they may be prevented. Students will learn that the immune system sometimes fails, often with serious consequences.

Using the Figure

Imagine a science fiction movie in which a massive blob of protoplasm streams across the countryside "gobbling up" every living thing in its way. It sounds far-fetched, but it happens every day in our bodies, as can be seen in the photograph. Ask students if they have ever been to the doctor, had a blood test, and were found to have a "high white count?" Ask students what this may mean. (It means a "high white blood cell count," and it tells the physician that the body is making lots of extra white blood cells, most likely in response to an infection.)

Assessing Prior Knowledge

Before beginning this chapter, be sure your students can answer the questions in the ***Quick Review*** on the facing page. If they cannot, encourage them to reread the sections indicated.

White blood cell attacking bacteria (6,480×)

922

Chapter Correlations — Florida Sunshine Standards/FCAT

SC.F.1.4.2 Body structures are uniquely designed and adapted for their function.

SC.F.1.4.5 Complex interactions among the different kinds of molecules in the cell cause distinct cycles of activity governed by proteins.

SC.F.1.4.8 Cell behavior can be affected by molecules from other parts of the organism or even from other organisms.

SC.H.3.4.5 The value of a technology may differ for different people and at different times.

CHAPTER

40 The Body's Defenses

Florida Sunshine Standards/FCAT

SC.F.1.4.2 Body structures are uniquely designed and adapted for their function.

SC.F.1.4.5 Complex interactions among the different kinds of molecules in the cell cause distinct cycles of activity governed by proteins.

SC.F.1.4.8 Cell behavior can be affected by molecules from other parts of the organism or even from other organisms.

SC.H.3.4.5 The value of a technology may differ for different people and at different times.

Quick Review

Answer the following without referring to earlier sections of your book.

1. **Identify** the role of receptor proteins in cellular communication. *(Chapter 4, Section 2)*
2. **Explain** the relationship between HIV and AIDS. *(Chapter 21, Section 1)*
3. **Differentiate** between antibodies and antigens. *(Chapter 38, Section 1)*

Reading Activity

Write down the title of this chapter and the titles of its four sections on a piece of paper or in your notebook. Leave a few blank lines after each section title. Then write down what you think you will learn in each section. Save your list, and after you finish reading this chapter, check off everything that you learned that was on your list.

A white blood cell (macrophage) ingests bacteria as part of the immune system's response to infection. The macrophage degrades bacterial proteins into peptides, which then form antigens on the cell's surface.

Looking Ahead

internet connect

www.scilinks.org

National Science Teachers Association *sci*LINKS Internet resources are located throughout this chapter.

SCILINKS. Maintained by the National Science Teachers Association

Chapter Resource File

- Vocabulary Worksheets
- Concept Mapping

Identifying Misconceptions

Students may think that the immune system only defends the body against pathogens such as bacteria and viruses. Make sure students are aware that the immune system also defends the body against defective body cells as well, such as cancer cells.

Quick Review Answers

1. Cells must communicate with each other to coordinate growth, metabolism, and other activities. Cells that do not lie next to each other cannot communicate directly with other cells. Instead, some cells release signal molecules that carry information to nearby cells and throughout the body. Receptor proteins located in the cell membrane bind to the specific signal molecules, enabling the cell to respond to the signal molecule.
2. AIDS (Acquired immunodeficiency syndrome) is a disease in which an individual is unable to defend his or her own body against infections that do not normally occur in healthy individuals. HIV, the human immunodeficiency virus, causes AIDS.
3. Antigens are substances, often foreign proteins, which produce an immune response in the body. Antibodies are defensive proteins produced by the immune system in response to a specific antigen.

Reading Activity Answers

Items not checked off after reading the chapter might be good topics for independent research.

Focus

Overview

Before beginning this section review with your students the objectives listed in the Student Edition. This section introduces students to the nonspecific defenses of the human body, including skin and mucous membranes. Students will then learn that pathogens that make it past the first line of defense face the inflammatory and temperature responses that can be very effective in killing the invader.

Bellringer

Ask students to make a list of the many ways the body prevents infectious organisms from entering. (skin, mucous membranes, nose hairs, etc.) When they are finished, have them share their lists. Ask them if the barriers on their lists can distinguish between the types of infections and can prevent specific ones. (no)

Motivate

Discussion — Basic

Ask students what happens when they get a wood splinter in their finger and they don't get it out within the first day or so. (The area turns red, becomes sore, and may swell.) What happens after several more days, if the splinter isn't removed? (Pus may form around the splinter; years ago this was called "festering.") Ask them why they think the pus is whitish in color. (White blood cells are converging on the area of the wood splinter and it is being digested.) LS Verbal

Section Correlations

SC.F.1.4.2

SC.F.1.4.5

SC.F.1.4.8

1 Nonspecific Defenses

Florida Sunshine Standards/FCAT

SC.F.1.4.2 Body structures are uniquely designed and adapted for their function.

SC.F.1.4.5 Complex interactions among the different kinds of molecules in the cell cause distinct cycles of activity governed by proteins.

SC.F.1.4.8 Cell behavior can be affected by molecules from other parts of the organism or even from other organisms.

Objectives

- **Describe** how skin and mucous membranes defend the body. SC.F.1.4.2
- **Compare** the inflammatory response with the temperature response. SC.F.1.4.8
- **Identify** proteins that kill or inhibit pathogens. SC.F.1.4.5
- **Analyze** the roles of white blood cells in combating pathogens. SC.F.1.4.8

Key Terms

pathogen
mucous membrane
inflammatory response
histamine
complement system
interferon
neutrophil
macrophage
natural killer cell

Two Lines of Nonspecific Defenses

Some animals, including turtles, clams, and armadillos, defend themselves with their hard armor shells. However, even armor will not protect against the most dangerous enemies that they or the human body faces—harmful bacteria, viruses, fungi, and protists. You, as well as most animals, survive because your body's immune system defends against these pathogens. A **pathogen** is a disease-causing agent. The immune system consists of cells and tissues found throughout the body. The body uses both nonspecific and specific defense mechanisms to detect and destroy pathogens, thereby preventing or reducing the severity of infection.

First Line of Nonspecific Defenses

The body's surface defenses are nonspecific, meaning they do not target specific pathogens. Your skin is the first of your immune system's nonspecific defenses against pathogens. Skin acts as a nearly impenetrable barrier to invading pathogens, keeping them outside the body. This barrier is reinforced with chemical weapons. Oil and sweat make the skin's surface acidic, inhibiting the growth of many pathogens. Sweat also contains the enzyme lysozyme, which digests bacterial cell walls.

Mucous membranes cover some body surfaces that come into contact with pathogens. **Mucous** *(MYOO kuhs)* **membranes** are layers of epithelial tissue that produce a sticky, viscous fluid called mucus. Mucous membranes line the digestive system, nasal passages, lungs, respiratory passages, and the reproductive tract. Like the skin, mucous membranes serve as a barrier to pathogens and produce chemical defenses. Cells lining the bronchi and bronchioles in the respiratory tract secrete a layer of mucus that traps pathogens before they can reach the warm, moist lungs, which are an ideal breeding ground for microorganisms. Cilia on cells of the respiratory tract continually sweep mucus toward the opening of the esophagus. Mucus then can be swallowed, sending pathogens to the stomach, where they are digested by acids and enzymes.

Skin and mucous membranes work to prevent any pathogens from entering the body. Occasionally these defenses are penetrated. You take pathogens into your body when you breathe, because many microbes and microbial spores are suspended in the air. Other pathogens may be present in the food you eat. Pathogens can also enter through wounds or open sores. When invaders reach deeper tissue, a second line of nonspecific defenses takes over.

924

Chapter Resource File

- Directed Reading BASIC
- Active Reading GENERAL

One-Stop Planner CD-ROM

- Reading Organizers BASIC
- Reading Strategies BASIC
- Occupational Application Worksheet Sanitarian GENERAL

Attention Grabber

SCID To illustrate our dependence on a functional immune system, show students a picture of David, the "boy in the bubble." David was born with severe combined immune deficiency (SCID), a disease characterized by the inability to produce B cells and T cells, which play key roles in the immune system. SCID patients must live in a protected environment because the slightest infection can lead to death. David died at the age of 12 after an unsuccessful bone marrow transplant.

Second Line of Nonspecific Defenses

What happens when pathogens break through your body's first line of defense? When the body is invaded, four important nonspecific defenses take action: the inflammatory response; the temperature response; special proteins that kill or inhibit pathogens; and white blood cells, which attack and kill pathogens.

Inflammatory response Injury or local infection, such as a cut or a scrape, causes an inflammatory response. An **inflammatory response** is a series of events that suppress infection and speed recovery. Imagine that a splinter has punctured your finger, creating an entrance for pathogens, as shown in **Figure 1.** Infected or injured cells in your finger release chemicals, including histamine. **Histamine** *(HIHST uh meen)* causes local blood vessels to dilate, increasing blood flow to the area. Increased blood flow brings white blood cells to the infection site, where they can attack pathogens. This also causes swelling and redness in the infected area. The whitish liquid, or pus, associated with some infections contains white blood cells, dead cells, and dead pathogens.

Temperature response When the body begins its fight against pathogens, body temperature increases several degrees above the normal value of about 37°C (98.6°F). This higher temperature is called a fever, and it is a common symptom of illness that shows the body is responding to an infection. Fever is helpful because many disease-causing bacteria do not grow well at high temperatures. Although fever may slow the growth of bacteria, very high fever is dangerous because extreme heat can destroy important cellular proteins. Temperatures greater than 39°C (103°F) are considered dangerous, and those greater than 41°C (105°F) can be fatal.

internet connect
www.scilinks.org
Topic: Immune Systems
Keyword: HX4103
SciLinks. Maintained by the National Science Teachers Association

Figure 1 Inflammatory response

When pathogens penetrate your body, an inflammatory response is triggered.

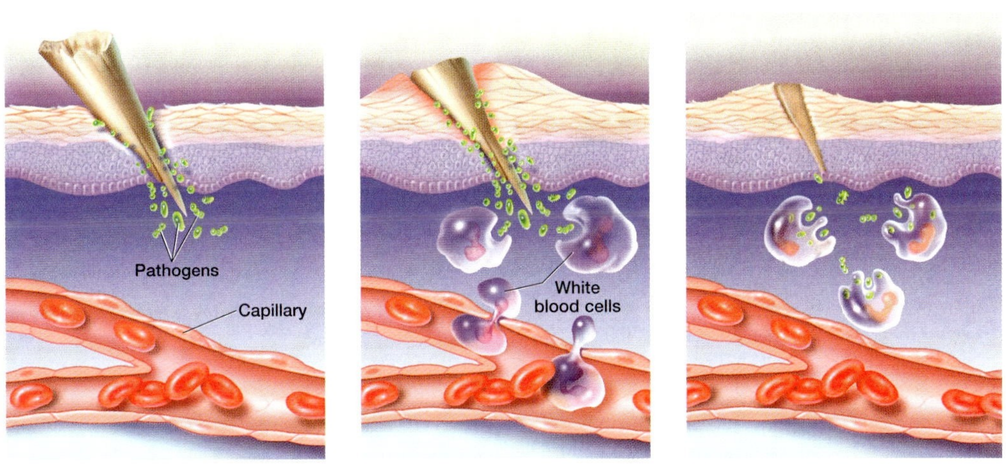

1. When the skin is punctured, pathogens enter the body.
2. Blood flow to the area increases, causing swelling and redness.
3. White blood cells attack and destroy the pathogens.

925

Teach

READING SKILL BUILDER — BASIC

Interactive Reading Assign Chapter 40 of the *Holt Biology* Guided Audio CD Program to help students achieve greater success in reading the chapter. LS Auditory

Teaching Tip — ADVANCED

Mosquito Bites Ask students why they think mosquito bites turn red, swell, and itch if a mosquito is simply sucking up blood? (The mosquito injects a small amount of saliva, which contains an anticoagulant. The anticoagulant initiates the inflammatory response, making the bite site swell, turn red, and itch.) LS Logical

Group Activity — GENERAL

Sequencing Lines of Defense Work in groups of three or four. On index cards, have each group write the name of each defensive protector the body has which is considered a first line of defense. Do the same with the second line of defense. On the back of each card write the function that each particular defense has in the body. Have the groups shuffle the cards and then try to organize them into first and second lines of defense. They should check their answers in this section, then reshuffle the cards and try again until they get them right several times in a row. Next, they should turn the cards over and try to do the same thing with the functions of each defense. Finally, they should use the cards as flash cards to quiz each other on the various components of the first and second lines of defense and how they function. LS Kinesthetic

Co-op Learning

did you know?

Symbiotic Bacteria Some of the body's protection from pathogens comes from symbiotic bacteria. For example, in the vagina, beneficial bacteria inhibit the growth of harmful bacteria and fungi. Use of antibiotics to kill pathogens often kills the body's beneficial bacteria as well, which can lead to vaginal and urinary tract infections.

Transparencies

TR Bellringer

TR J36 Inflammatory Response

Close

Reteaching — Basic

Assign students to cooperative groups of three. Have each group write a comic sketch or draw cartoons to show how the first line of defense operates in protecting the body against pathogens. (Sketches and drawings may depict the skin keeping out invaders, pathogens sticking to mucous membranes, phagocytes engulfing bacteria, and aspects of the inflammatory response.) English Language Learners

Quiz — General

True or False:

1. The inflammatory response targets specific pathogens. (false; it is nonspecific)
2. Fever can be helpful because many disease-causing bacteria do not grow well at high temperatures. (true)
3. The first lines of nonspecific defense help to keep pathogens out of the body. (true)

Alternative Assessment — General

Have students make up one test question for each subsection in this lesson. Have students supply the answer for their questions. Then collect all the papers with the questions and read them to students, skipping over redundant questions. Number the questions as you go through them. Have each student answer the questions on a piece of paper as you read them. When finished, go through each question and discuss the answers.

Section Review Correlations

SC.F.1.4.5: Item 3

SC.F.1.4.8: Items 1, 2, 4

Magnification: 2,280×

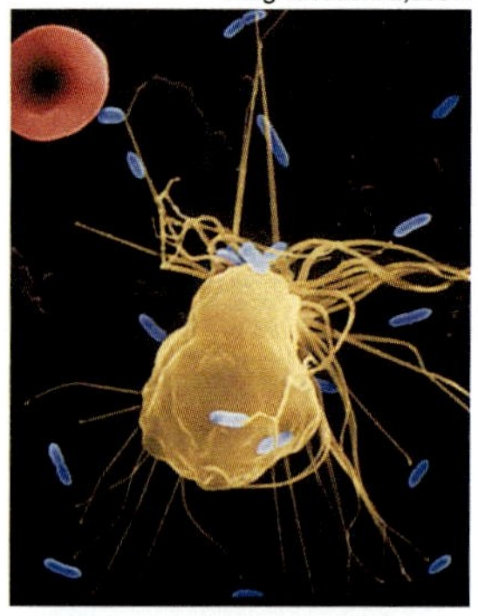

Figure 2 Macrophage. Cytoplasmic extensions of this macrophage (yellow) are capturing bacteria (blue).

Magnification: 14,250×

Figure 3 Natural killer cell. This natural killer cell (yellow) is attacking a cancer cell (pink).

Proteins Various proteins also provide nonspecific defenses. One defense mechanism, called the **complement system,** consists of about 20 different proteins. Complement proteins circulate in the blood and become active when they encounter certain pathogens. Then some of these proteins form a membrane attack complex (MAC), a ring-shaped structure. The MAC punches a hole in the cell membrane, causing the cell to leak and die. Another nonspecific defense is **interferon** *(in tuhr FEER ahn),* a protein released by cells infected with viruses. Interferon causes nearby cells to produce an enzyme that prevents viruses from making proteins and RNA.

White blood cells The most important counterattacks in the second line of nonspecific defenses are carried out by three kinds of white blood cells: neutrophils, macrophages, and natural killer cells. These cells patrol the bloodstream, wait within the tissues for pathogens, and then attack the pathogens. Each kind of cell uses a different mechanism to kill pathogens.

1. **Neutrophils.** A **neutrophil** *(NOO truh fihl)* is a white blood cell that engulfs and destroys pathogens. The most abundant type of white blood cell, neutrophils engulf bacteria and then release chemicals that kill the bacteria—and themselves. Neutrophils can also squeeze between cells in the walls of capillaries to attack pathogens at the site of an infection.
2. **Macrophages.** White blood cells called **macrophages** *(MA kroh fay jez),* shown in **Figure 2,** ingest and kill pathogens they encounter. They also clear dead cells and other debris from the body. Most macrophages travel through the body in blood, lymph, and fluid between cells. Macrophages are concentrated in particular organs, especially the spleen and lungs.
3. **Natural killer cells.** A **natural killer cell** is a large white blood cell that attacks cells infected with pathogens. Natural killer cells destroy an infected cell by puncturing its cell membrane. Water then rushes into the infected cell, causing the cell to swell and burst. One of the body's best defenses against cancer, natural killer cells can detect and kill cancer cells, as shown in **Figure 3,** before a tumor can develop.

Section 1 Review

1. **Describe** how the inflammatory and temperature responses help defend against infection. SC.F.1.4.8
2. **Identify** the role of white blood cells in the second line of nonspecific defenses. SC.F.1.4.8
3. **Critical Thinking Relating Concepts** Explain why taking a drug that reduces fever might delay rather than speed up your recovery from an infection. SC.F.1.4.5
4. **FCAT Prep** In the inflammatory response, what substance do infected or injured cells release, causing local blood vessels to dilate?
   A mucus
   B histamine
   C interferon
   D complement proteins
   SC.F.1.4.8

926

Answers to Section Review

1. The inflammatory response increases blood flow; this brings white blood cells to an injury site. The temperature defense raises tissue temperature; this kills or inhibits some pathogens.
2. Neutrophils engulf pathogens. Macrophages ingest and kill them and rid the body of dead cells. Natural killer cells puncture membranes of infected cells.
3. The drug inhibits the temperature response. Reducing body temperature may allow pathogens to thrive.
4. **A.** Incorrect. Mucus does not cause blood vessels to dilate. **B.** Correct. Histamine increases blood flow; this brings white blood cells to the infected area, where they can attack pathogens. **C.** Incorrect. Interferon causes cells to make an enzyme that prevents viruses from making proteins and RNA. **D.** Incorrect. Complement proteins move in the blood and are activated by pathogens.

Immune Response

Section 2

Florida Sunshine Standards/FCAT

SC.F.1.4.2 Body structures are uniquely designed and adapted for their function.

SC.F.1.4.8 Cell behavior can be affected by molecules from other parts of the organism or even from other organisms.

Specific Defenses

What happens when pathogens occasionally overwhelm your body's nonspecific defenses? Pathogens that have survived the first and second lines of nonspecific defenses still face a third line of specific defenses—the immune response. The immune response consists of an army of individual cells that rush throughout the body to combat specific invading pathogens. The immune response is not localized in the body, nor is it controlled by a single organ. It is more difficult to evade than the nonspecific defenses.

Objectives

- **List** four kinds of immune-system cells, and describe their functions. SC.F.1.4.2
- **Describe** how white blood cells recognize pathogens. SC.F.1.4.8
- **Identify** the role of helper T cells in the immune response. SC.F.1.4.8
- **Compare** the role of T cells with that of B cells in the immune response. SC.F.1.4.8

Key Terms

cytotoxic T cell
B cell
helper T cell
antigen
plasma cell
antibody

Cells Involved in the Immune Response

White blood cells are produced in bone marrow and circulate in blood and lymph. Of the 100 trillion or so cells in your body, about 2 trillion are white blood cells. Four main kinds of white blood cells participate in the immune response: macrophages, cytotoxic T cells, B cells, and helper T cells. Each kind of cell has a different function. Macrophages consume pathogens and infected cells. **Cytotoxic** *(sie toh TAHKS ihk)* **T cells** attack and kill infected cells. **B cells** label invaders for later destruction by macrophages. **Helper T cells** activate both cytotoxic T cells and B cells. Macrophages can attack any pathogen. B cells and T cells, however, respond only to pathogens for which they have a genetically programmed match. These four kinds of white blood cells interact to remove pathogens from the body.

Recognizing Invaders

To understand how the third line of defenses works, imagine that you have just come down with influenza—the flu. You have inhaled influenza virus particles, but they were not all trapped by mucus in the respiratory tract. The virus has begun to infect and kill your cells. At this point, macrophages begin to engulf and destroy the virus.

An infected body cell will display antigens of an invader on its surface. An **antigen** *(AN tih jihn)* is a substance that triggers an immune response. Antigens typically include proteins and other parts of viruses or pathogen cells. Antigens are present on the surface of the infected body cell. White blood cells of the immune system are covered with receptor proteins that respond to infection by binding to specific antigens on the surfaces of the infecting microbes. These receptors recognize and bind to antigens that match their particular shape, as shown in **Figure 4.**

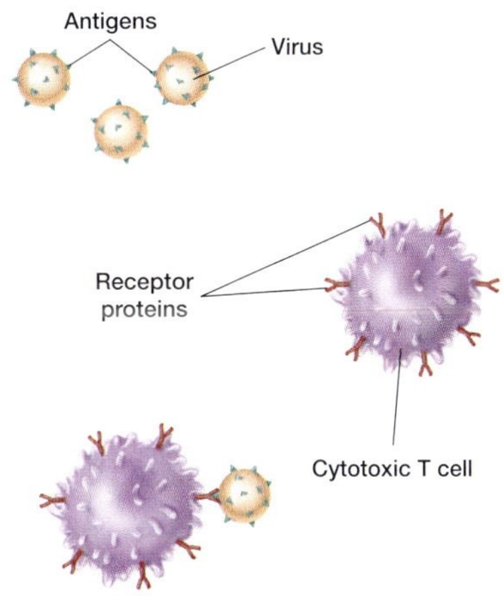

Figure 4 Antigens. Some cells of the immune system have receptor proteins that bind to specific antigens.

Section 2

Focus

Overview

Before beginning this section review with your students the objectives listed in the Student Edition. This section introduces students to the body's third line of defense, the immune response. Students will learn that in order to destroy pathogens, the third line of defense must be able to recognize an invader as foreign.

Bellringer

Show the class an artist's rendering of viruses bursting out of an infected cell. Ask students what they think is happening in the picture. (Ideas will vary, but some students should identify the cell and the many particles spilling out of it.) Discuss student ideas, and then point out the infected cell and viruses. Explain that viruses must infect other cells to reproduce. **LS Verbal**

Motivate

Demonstration — GENERAL

To show that microorganisms are ubiquitous, gather five sterile Petri dishes filled with sterile nutrient agar. Collect microorganisms from five different areas in the room by sweeping surfaces with sterile swabs moistened with sterile distilled water. Inoculate one dish with one swab and tape each dish closed with masking tape. Label each dish and place all five upside down in a warm, dark place or in an incubator kept at room temperature for several days. At the end of this period, show students the colonies of microorganisms that have grown on the agar. Explain that most of the microorganisms in the dishes probably are harmless but that some may be pathogens. Dispose of swabs and Petri dishes after sterilizing them in an autoclave or pressure cooker at 121°C for 15 minutes or soaking them for 30 minutes in chlorine bleach. **LS Visual**

Chapter Resource File

- Directed Reading BASIC
- Active Reading GENERAL

One-Stop Planner CD-ROM

- Reading Organizers BASIC
- Reading Strategies BASIC

Section Correlations

SC.F.1.4.2
SC.F.1.4.8

INCLUSION Strategies

- *Attention Deficit Disorder*
- *Gifted and Talented*

White blood cells attack particles that invade the body that can cause diseases or infections. Have students diagram the way a white blood cell would eliminate pathogens that invade the body. Also, have the students call a local clinic to determine how white blood cell counts are used to identify different kinds of infections. Students may present their findings to the class in an oral or a tape recorded report.

Teach

Teaching Tip — GENERAL

Immune Cells Have students create a Graphic Organizer similar to the one at the bottom of this page. Along the top of the table they should list the following categories: Type of cell, Function, and Location in the body. In the first column, they should list all the types of cells that play a role in the body's defenses. (macrophage, neutrophil, natural killer cell, helper T cell, cytotoxic T cell, B cell, plasma cell, and memory cell) Then ask students to complete the information in the second and third columns.
LS Visual

Using the Figure — BASIC

Once you have discussed **Figure 5,** have students answer the following questions: What types of white blood cells are involved in the immune response? (helper T, cytotoxic T, and B cells) What type of cell is involved in a passive, humoral defense? (B cell) What type of cell is active and destroys pathogens? (cytotoxic T cell) What is the role of helper T cells? (They regulate the T- and B-cell actions.)
LS Verbal

Integrating Physics and Chemistry

Explain to students that when antibodies bind to antigens, a chemical change occurs due to a chemical reaction between the two biological molecules. As a result of this chemical change, the invading antigen is altered so as to incapacitate it or ultimately make it more susceptible to destruction or elimination by the body's immune system.

Transparencies

TR Bellringer
TR J102 Immune Response

Section Review Correlations

SC.F.1.4.2: Item 1
SC.F.1.4.5: Item 5
SC.F.1.4.8: Items 2, 3, 4, 6

The Immune Response Has Two Main Parts

Two distinct processes work together in an immune response. One is the B cell response, a defense that aids the removal of extracellular pathogens from the body. The other is the T cell response, a defense that involves the destruction of intracellular pathogens by cytotoxic T cells. Both the T cell response and the B cell response are regulated by helper T cells. Both responses, which happen simultaneously, are summarized in **Figure 5.**

BIO graphic

Figure 5

Immune Response

The immune response involves several kinds of white blood cells.

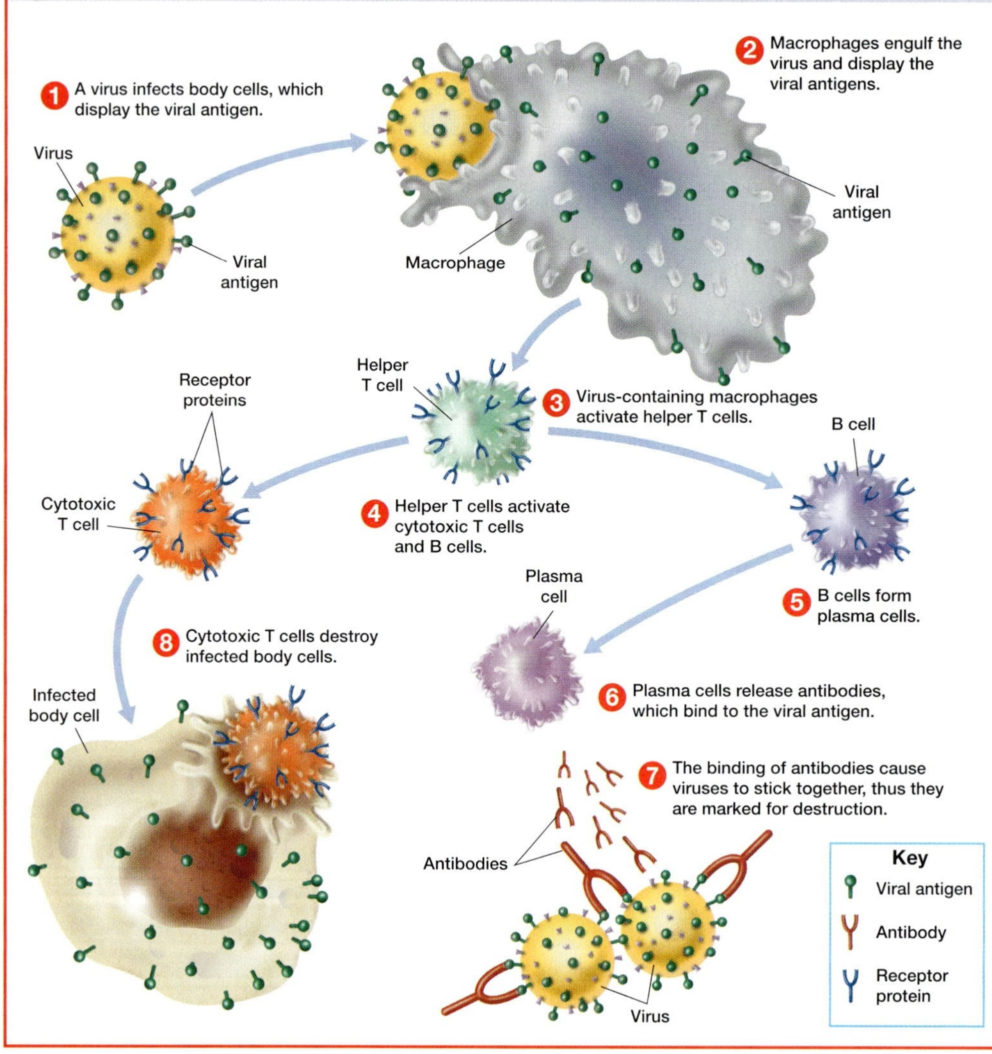

Graphic Organizer

Use this graphic organizer with **Teaching Tip** on this page.

Type of cell	Function	Location in the body
Macrophage	Ingests and kills pathogens	Spleen, lungs, blood, lymph, interstitial fluids
Neutrophil	Engulfs and destroys pathogens	Bloodstream, infection sites
Natural killer cell	Punctures infected cells	Infected cells
Helper T cell	Activates cytotoxic T and B cells	Bloodstream, infection sites
Cytotoxic T cell	Punctures infected cells	Infection sites
B cell	Labels invaders for destruction by macrophages	Infection sites
Plasma cell	Releases antibodies	Bloodstream
Memory cell	Protects against defeated pathogens	Bloodstream

Step 1 When a virus infects body cells, the infected cells display the viral antigen on their surfaces.

Step 2 Macrophages engulf the virus and display the viral antigens on their surfaces.

Step 3 Receptor proteins on helper T cells bind to the viral antigen displayed by the macrophages. The macrophages release a protein called interleukin-1 *(ihn tuhr LOO kihn)*.

Step 4 Interleukin-1 activates helper T cells, but helper T cells do not attack pathogens directly. Instead, helper T cells activate cytotoxic T cells and B cells. Stimulation by interleukin-1 causes helper T cells to release interleukin-2. Interleukin-2 stimulates further division of helper T cells and cytotoxic T cells, amplifying the body's response to the infection.

Step 5 Interleukin-2 released by helper T cells also activates B cells. Activated B cells divide and develop into plasma cells. **Plasma cells** are cells that release Y-shaped antibodies into the blood. An **antibody** is a defensive protein produced upon exposure to a specific antigen, which can bind to that antigen.

Step 6 Plasma cells divide repeatedly and make large numbers of antibodies. Plasma cells release antibodies into the bloodstream where they attach to the viruses. Antibodies bind to the viral antigen and mark the virus for destruction.

Step 7 The binding of antibodies cause viruses and antigens to stick together, forming clumps that can be easily identified and destroyed by macrophages.

Step 8 Activated cytotoxic T cells destroy infected cells by puncturing their cell membranes. Your body makes millions of different T cells, each with receptor proteins that bind to a specific antigen. Receptor proteins on cytotoxic T cells bind to the viral antigen displayed by infected cells. For example, any of your body's cells that bear traces of an influenza virus will be destroyed by cytotoxic T cells with receptor proteins that bind to the antigen of that virus.

Real Life

How often do you wash your hands?

Hand washing is an effective way to prevent the spread of disease. A recent study found that only 68 percent of adults wash their hands after using public restrooms.

Inferring Relationships
Why is it crucial that food preparers wash their hands after using a restroom?

Study TIP

Reading Effectively
Antigens trigger an immune response. Remember that an antigen is an *anti*body-*gen*erating substance.

Section 2 Review

1. **List** the different kinds of white blood cells involved in the immune response. SC.F.1.4.2
2. **Describe** how white blood cells recognize and bind to pathogens. SC.F.1.4.8
3. **Compare** the roles of B cells and T cells in the immune response. SC.F.1.4.8
4. **Recognizing Relationships** Explain the role of helper T cells in the immune response. SC.F.1.4.8
5. **Critical Thinking Predicting Outcomes** How would an enzyme that destroys interleukins affect the immune response? SC.F.1.4.5
6. *FCAT Prep* Which cells produce antibodies and release them into the blood?
   A cytotoxic T cells
   B helper T cells
   C macrophages
   D plasma cells
   SC.F.1.4.8

Real Life Answer

Food that is touched can be contaminated with microbes that cause intestinal diseases. The disease-causing microbes can then be passed on to a person that eats the contaminated food.

Teaching Tip — ADVANCED

Negative Feedback Interleukin-2 released by helper T cells stimulates the division of another type of T cell, the suppressor T cell. These cells divide slowly and become common a week or more after infection. They gradually shut down the immune response by inhibiting the production of B cells and cytotoxic T cells. Ask students what kind of feedback system this would be. (a negative feedback system) **LS Logical**

Close

Reteaching — BASIC

Have students review the function of T cells in the immune response. Have them make index cards with sequential words and drawings depicting T-cell function. Have students pair off. One student should shuffle the cards, and the other student should put them in the correct order. Students should then switch roles. **LS Interpersonal** English Language Learners

Quiz — GENERAL

True or False:

1. The immune response of the body is a defense that is nonspecific. (false; it is specific)
2. Antigens trigger the immune response. (true)
3. Plasma cells make antibodies. (true)

Alternative Assessment — GENERAL

Make copies of **Figure 5,** but without the explanations at each step. Have students, without using their book, fill in the explanations of what is happening at each of the eight steps. **LS Visual**

Answers to Section Review

1. macrophages, cytotoxic T cells, B cells, helper T cells
2. White blood cells recognize invaders because infected cells display surface antigens. White blood cells have receptors that bind to these antigens.
3. B cells become plasma cells that release antibodies that bind to antigens to mark them for destruction by macrophages. T cells are stimulated by interleukin-1 to produce interleukin-2, which activates helper T cells and cytotoxic T cells that puncture cell membranes.
4. Helper T cells bind to viral antigens on macrophage surfaces, which stimulates the release of interleukin-1 by macrophages. Helper T cells also activate B cells and cytotoxic T cells.
5. Destroying interleukins would short-circuit the immune response.
6. **A.** Incorrect. Cytotoxic T cells are white blood cells that attack and kill infected cells. **B.** Incorrect. Helper T cells function to activate both cytotoxic T cells and B cells. **C.** Incorrect. Macrophages ingest and kill pathogens. **D.** Correct. Plasma cells release specialized, Y-shaped proteins called antibodies into the blood.

Section 3

Focus

Overview

Before beginning this section review with your students the objectives listed in the Student Edition. This section introduces students to the modes in which diseases are transmitted and how transmission can be prevented. Students will also delve into the history of pathology and immunology.

Bellringer

Ask students how many of them have had chicken pox. For those who have had chicken pox, ask them if they have had it more than once. (Except in rare situations, most will answer "no.") Tell students they can develop immunity to certain diseases and this will be discussed in this section.

Motivate

Identifying Preconceptions —GENERAL

Make photocopies or a transparency of a vaccination record. Give students the copies of the vaccination record. Ask students why it is important to maintain such records. (If vaccinations are not kept up-to-date, a person is at risk of contracting a disease.) Ask students if they know about other less common vaccinations such as those that protect against typhoid fever, yellow fever, and so on. (Answers will vary.) LS Intrapersonal

Section Correlations

SC.F.1.4.5
SC.F.1.4.8
SC.H.3.4.5

Section 3

Disease Transmission and Prevention

Florida Sunshine Standards/FCAT

SC.F.1.4.5 Complex interactions among the different kinds of molecules in the cell cause distinct cycles of activity governed by proteins.
SC.F.1.4.8 Cell behavior can be affected by molecules from other parts of the organism or even from other organisms.
SC.H.3.4.5 The value of a technology may differ for different people and at different times.

Objectives

- **List** five ways diseases can be transmitted to humans. SC.F.1.4.8
- **Summarize** Koch's postulates for identifying pathogens. SC.F.1.4.8
- **Analyze** how the body produces immunity to pathogens. SC.F.1.4.5
- **Describe** how vaccines produce immunity to pathogens. SC.H.3.4.5

Key Terms

Koch's postulates
immunity
vaccination
vaccine
antigen shifting

Disease Transmission

In general, you can get infectious diseases in any of five different ways: through person-to-person contact, air, food, water, and animal bites. Diseases transferred from person to person are considered contagious, or communicable. For example, when a person sneezes, droplets of saliva and mucus carrying pathogens are expelled from the mouth and nose, as shown in **Figure 6.** If another person breathes these droplets, the pathogens can infect that person. People directly transmit some diseases by kissing, shaking hands, touching sores, or having sexual contact. People can also transmit diseases indirectly through objects contaminated with pathogens, such as drinking glasses, toys, plumbing, and needles used to inject drugs or in tatooing.

By minimizing exposure to pathogens, you can decrease your chances of becoming ill. For example, to prevent illnesses caused by bacteria found in foods that contain animal products, these foods should always be cooked thoroughly. Utensils and other surfaces that foods touch should be sanitized.

Figure 6 Disease transmission. When a person sneezes, pathogens are expelled from the mouth and nose.

Detecting Disease

The German physician Robert Koch (1843–1910) established a procedure for diagnosing causes of infection. Koch determined that bacteria cause anthrax, a disease that afflicts cattle, sheep, goats, and humans. Anthrax is a serious disease although it is not passed from person to person. In an experiment, Koch isolated bacteria from a cow with anthrax and then infected a healthy cow with the bacteria. The healthy cow developed anthrax and had the same bacteria that the first cow had. In his research, Koch developed the following four-step procedure, known as **Koch's postulates**, as a guide for identifying specific pathogens.

1. The pathogen must be found in an animal with the disease and not in a healthy animal.
2. The pathogen must be isolated from the sick animal and grown in a laboratory culture.
3. When the isolated pathogen is injected into a healthy animal, the animal must develop the disease.
4. The pathogen should be taken from the second animal and grown in a laboratory culture. The cultured pathogen should be the same as the original pathogen.

930

Chapter Resource File

- Directed Reading BASIC
- Active Reading GENERAL
- Data Sheet for Quick Lab GENERAL

One-Stop Planner CD-ROM

- Reading Organizers BASIC
- Reading Strategies BASIC

Transparencies

TR Bellringer

Long-Term Protection

The specific immune response is very powerful, and it can be a long-lasting defense. After an immune response, some B cells and T cells become memory cells that continue to patrol your body's tissues. Some memory cells provide lifelong protection against previously encountered pathogens. If a pathogen ever appears again, memory cells activate antibody production against that pathogen. As shown in **Figure 7,** a second exposure to the same pathogen causes a sharp increase in antibody concentration. This enables macrophages to destroy the pathogen before you become ill. You are said to be "immune," or resistant, to the disease caused by that pathogen.

Primary and Secondary Immune Responses

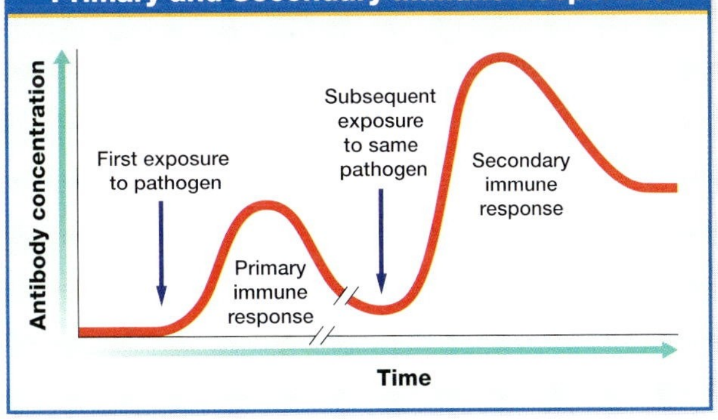

Figure 7 Immune responses. The first time you are exposed to a pathogen, your immune system responds normally. If you become exposed to the same pathogen again, antibody production increases quickly.

Resistance to Disease

Resistance to a particular disease is called **immunity.** It has long been observed that individuals who recover from an infectious disease develop an immunity to that disease. This knowledge preceded the development of immunology, a branch of science that deals with antigens, antibodies, and immunity. Immunologists study the body's defenses and ways to help protect against disease.

In 1796, an English doctor named Edward Jenner performed an experiment that marks the beginning of immunology. Smallpox, which is caused by a virus, was a common and deadly disease then. Jenner observed that milkmaids who had contracted cowpox, a mild form of smallpox, rarely became infected with smallpox. Jenner hypothesized that cowpox produced protection against smallpox. To test his hypothesis, Jenner infected healthy people with cowpox. As Jenner had predicted, many of the people he infected never developed smallpox, even though they had been exposed to the virus. We now know that smallpox and cowpox are caused by two similar viruses. The cowpox infection caused an immune response that later prevented smallpox infection in Jenner's patients.

Vaccination Jenner's procedure of injecting the cowpox virus to produce resistance to smallpox is called vaccination. **Vaccination** *(vak sih NAY shuhn)* is a medical procedure used to produce immunity. You have probably been to the doctor for vaccination to guard against various diseases. Modern vaccination usually involves an injection, or "shot," of a vaccine under the skin. A **vaccine** *(vak SEEN)* is a solution that contains a dead or weakened pathogen or genetic material from a pathogen.

A vaccine triggers an immune response against the pathogen without symptoms of infection. For several days after you are vaccinated, your immune system develops antibodies and memory

931

Teach

READING SKILL BUILDER — BASIC

Paired Summarizing Pair each student with a partner. Have each student read Section 3 silently while making notes on a piece of paper. Have students write down and put a question mark next to anything he or she does not clearly understand. After they finish reading, ask one student to summarize and the second to add anything omitted. Both readers should then help each other with any passages that either (or both) did not understand. Have them create a list of questions to ask the class. Discuss each group's questions as a class. English Language Learners
LS Verbal

SKILL BUILDER — GENERAL

Interpreting Visuals Have students examine **Figure 7.** Ask them why the second peak in the line is so much higher than the first peak. (Memory cells in the body recognize the pathogen when it invades a second time, and they immediately produce a massive amount of antibodies to destroy the pathogen.) A person is not even aware of this tremendous immune response in the body. **LS Visual**

Activity — ADVANCED

Mutating Viruses Changes in the surface proteins of many viruses, such as cold and influenza, can allow them to re-infect individuals many times. Ask students how this can be so. (The mutations prevent the immune system from recognizing the virus as one that is has encountered before, so they are not able to immediately destroy the virus.) Have students create an illustration, model, or computer animation illustrating their answer. Students may want to also include information about why it may be difficult to create a vaccine for a mutating virus. **LS Kinesthetic**

Attention Grabber

Disease Warfare In September 2001, spores of the bacteria *Bacillus anthracis* was sent by mail to several U.S. locations. Twenty-two confirmed or suspected cases of anthrax infection resulted. Five persons died. The *B. anthracis* spores were sent in at least 5 letters to Florida, New York City, and Washington, D.C. Because of these deadly attacks, the U.S. government stepped up its efforts to produce new anthrax vaccine. One of the first cases of the modern use of biological weapons in the United States occurred in 1984. About 3,000 followers of an Indian guru named Bhagwan Shri Rashneesh settled near the north central Oregon town of The Dalles. Attempting to take over the county government, the Bhagwan's followers cultured salmonella bacteria and secretly sprinkled it on restaurant salad bars, doorknobs, and other common locations, hoping to sicken enough people so they could not vote in the local election. In all, 751 people became ill, but the election was not compromised. Some of the Bhagwan's followers were prosecuted and jailed.

QUICK LAB

Simulating Antigen Activity

Skills Acquired
Inferring, predicting

Teacher's Notes
Before the lab, review antigens and antibodies.

Answers to Analysis

1. AB
2. Antibodies are bound to antigens.
3. Type A would clot only with anti-A serum. Type B would clot only with anti-B serum.

Close

Reteaching — BASIC

Invite a health professional to speak to the class about disease transmission and prevention. Have students write questions to ask the speaker. **LS Auditory**

Quiz — GENERAL

True or False:

1. Hand-washing after using the restroom is an important way to stop the spread of disease. (true)
2. A vaccination is a medical procedure that "tricks" the body into producing an immune response. (true)

Alternative Assessment — GENERAL

Have each student research one common disease. Have students make an outline and report orally to the class on their chosen disease. **LS Verbal**

Real Life

Flu can be deadly.
In 1918, an influenza (flu) epidemic killed more than 20 million people. To prevent this from happening again, scientists track the antigen shifting of flu viruses and target the new viral antigens for vaccines.

cells against the pathogen. You develop a long-lasting immunity to the disease. In 1977, smallpox became the first infectious disease to be eradicated from the public by vaccination. Vaccination has also reduced the incidence of many other diseases, including measles, polio, tetanus, and diphtheria.

Antigen shifting You can get the flu even if you have already been infected or vaccinated. Influenza viruses constantly mutate over time. The viruses produce new antigens that your immune system does not recognize, a process known as **antigen shifting.** With subsequent exposure to the virus, your body must make new antibodies.

QUICK LAB

Simulating Antigen Activity

SC.F.1.4.8

Using simulated blood, you can see what happens when antigens encounter specific antibodies.

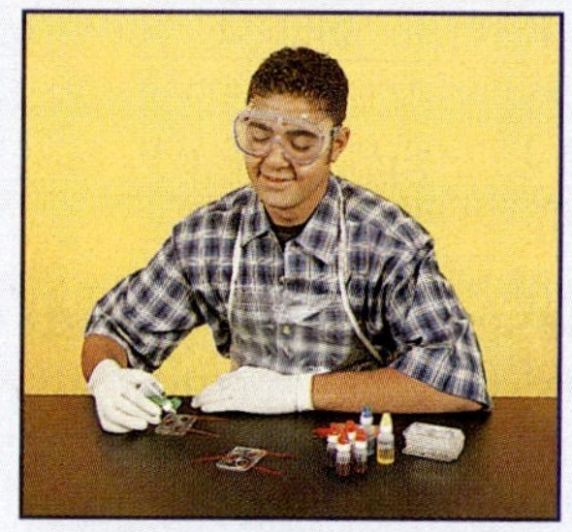

Materials

safety goggles, disposable gloves, lab apron, 2 blood-typing trays, simulated blood (types AB and O), simulated anti-A and anti-B blood-typing serums, 4 toothpicks

Procedure

1. Put on safety goggles, disposable gloves, and a lab apron.
2. Place 3–4 drops of type AB simulated blood into each well in a clean blood-typing tray. **CAUTION: Use only simulated blood provided by your teacher.**
3. Add 3–4 drops of anti-A blood-typing serum to one well. Stir the mixture for 30 seconds using a toothpick. Add 3–4 drops of anti-B blood-typing serum to the other well. Use a new toothpick to stir the mixture. Look for clumps separating from the mixtures.
4. Repeat steps 2 and 3 using simulated type O blood.
5. Dispose of your materials according to your teacher's directions. Clean up your work area and wash your hands.

Analysis

1. **Determine** which blood type has antigens that are recognized by the blood-typing sera.
2. **Evaluating Results** What does clumping of the blood mixtures indicate?
3. **Predicting Outcomes** What would happen if you did the same experiment using type A blood and type B blood?

Section 3 Review

1. **List** two ways that diseases can be transmitted between people. SC.F.1.4.8
2. **Summarize** Koch's postulates for identifying specific pathogens. SC.F.1.4.8
3. **Describe** how vaccination produces immunity. SC.H.3.4.5
4. **Critical Thinking** **Relating Concepts** Explain why you cannot get many diseases more than once. SC.F.1.4.5
5. **FCAT Prep** What causes smallpox?
   A bacterium
   B fungus
   C protist
   D virus
   SC.F.1.4.8

932

Answers to Section Review

1. Answers might include kissing, shaking hands, sharing eating utensils, engaging in sexual activity, and so on.
2. The pathogen is found in sick, but not in healthy, animals. The pathogen can be isolated and grown in the laboratory. Injection of the pathogen into a healthy animal causes the disease. The pathogen isolated from the second animal is the same as the one isolated from the first animal.
3. Vaccination induces the immune system to develop antibodies and memory cells for a specific pathogen.
4. Memory cells immediately recognize the pathogen, and the immune system mounts a response, quickly eliminating the pathogen.
5. **A.** Incorrect. Bacteria, unlike smallpox, are cellular organisms. **B.** Incorrect. A fungus is a cellular organism. Smallpox, a virus, is not made of cells. **C.** Incorrect. Protists are single-celled eukaryotes. Viruses are not made of cells. **D.** Correct. The once-deadly disease is caused by a virus.

Section Review Correlations

SC.F.1.4.5: Item 4
SC.F.1.4.8: Items 1, 2, 5
SC.H.3.4.5: Item 3

Disorders of the Immune System

Section 4

Florida Sunshine Standards/FCAT

SC.F.1.4.5 Complex interactions among the different kinds of molecules in the cell cause distinct cycles of activity governed by proteins.

SC.F.1.4.8 Cell behavior can be affected by molecules from other parts of the organism or even from other organisms.

Objectives

- **Describe** several autoimmune diseases. SC.F.1.4.5
- **Summarize** how HIV disables the immune system. SC.F.1.4.5
- **List** five ways HIV is transmitted. SC.F.1.4.8
- **Identify** causes of an allergic reaction. SC.F.1.4.8

Key Terms

autoimmune disease
AIDS
HIV
CD4
allergy

Autoimmune Diseases

The ability of your immune system to distinguish cells and antigens of your body from foreign cells and antigens is crucial to the fight against pathogens. In some people, the immune system cannot distinguish between the body's antigens and foreign antigens, causing an autoimmune disease. In an **autoimmune disease,** the body launches an immune response against its own cells, attacking body cells as if they were pathogens. The immune system cannot distinguish between antigens of "self" and "nonself." This effect may be caused by the inappropriate production of antibodies specific to the antigens of body cells.

Autoimmune diseases affect organs and tissues in various areas of the body. For example, multiple sclerosis *(skleh ROH sihs)* usually strikes people between the ages of 20 and 40. Multiple sclerosis (MS) is generally thought to be an auto-immune disease. In people with multiple sclerosis, the immune system attacks and gradually destroys insulating material surrounding nerve cells in the brain, in the spinal cord, and in the nerves leading from the eyes to the brain. This impairs and may eventually stop the functioning of these nerve cells. Multiple sclerosis causes problems with vision, speech, and coordination. **Table 1** lists and describes several autoimmune diseases.

Table 1 Autoimmune Diseases

Disease	Areas affected	Symptoms
Graves' disease	Thyroid gland	Weakness, irritability, heat intolerance, increased sweating, weight loss, insomnia
Multiple sclerosis (MS)	Nervous system	Weakness, loss of coordination, problems with vision and speech
Rheumatoid arthritis	Joints	Severe pain, fatigue, disabling inflammation of joints
Systemic lupus erythematosus (SLE)	Connective tissue, joints, kidneys	Facial skin rash, painful joints, fever, fatigue, kidney problems, weight loss
Type I diabetes	Insulin-producing cells in pancreas	Increased blood glucose level, excessive urine production, problems with vision, weight loss, fatigue, irritability

933

Section 4

Focus

Overview

Before beginning this section review with your students the objectives listed in the Student Edition. This section introduces students to disorders of the immune system. It also describes situations in which the body's immune system launches an attack on its own cells, including how the human immunodeficiency virus (HIV) affects the immune system, and what causes allergic reactions.

Bellringer

Have students make a list of all the organs they can think of that have been transplanted from one person to another. (Answers may include such organs as skin, heart, kidneys, and lungs.) Ask them why some transplants fail even though the surgery was technically a success. (Some students should mention organ rejection, which can lead to a discussion of immunosuppressant drugs and their effectiveness.)

Motivate

Activity — GENERAL

Immune System Disorders Have students work in groups to list as many allergies and disorders of the immune system as they can think of. Students should rank them on a scale of 1–10, from most critical to the least critical. (Example, hay fever might be ranked as an 8, whereas a food allergy may be a 1 or 2.) Have each group share its list and discuss why different groups have different rankings of different allergies. LS Interpersonal

Co-op Learning

Chapter Resource File

- Directed Reading BASIC
- Active Reading GENERAL
- Data Sheet for Data Lab GENERAL

One-Stop Planner CD-ROM

- Reading Organizers BASIC
- Reading Strategies BASIC
- Supplemental Reading Guide
  Microbe Hunters ADVANCED

Transparencies

TR Bellringer

TR J41 Number of Helper T Cells Following HIV Infection

Section Correlations

SC.F.1.4.5
SC.F.1.4.8

Teach

Teaching Tip — ADVANCED

Organ Rejection Drugs such as cyclosporine and corticosteroids, which are immunosuppressants, are given to recipients of organ transplants. Have students hypothesize as to why this is necessary and what precautions transplant recipients must take. (Immunosuppressive therapy reduces the recipient's immune response to antigens found in the transplanted organ. Precautions against infection are necessary because the patient's immune system is weakened.) Have students explain why transplant patients are more likely than others to develop cancer. (One of the functions of the immune system is to destroy cancer cells before they spread.) LS **Verbal**

Using the Figure — GENERAL

Have students study the graph in **Figure 9.** Ask them what happens to the number of helper T cells per milliliter of blood over time following an HIV infection. (The number of helper T cells decreases.) Ask students why this happens. (HIV infects helper T cells and kills them.) Ask students how the number of helper T cells per milliliter of blood relates to AIDS. (When the number falls below 200 helper T cells per milliliter of blood, AIDS is diagnosed.) LS **Visual**

HIV Infection

Before 1981, **AIDS,** or acquired immunodeficiency syndrome, was unknown. Between 1981 and 2000, more than 448,000 Americans died of AIDS. Since then, the total number of people living with HIV in the United States has increased to more than 850,000. AIDS is a disease caused by **HIV,** or the human immunodeficiency virus.

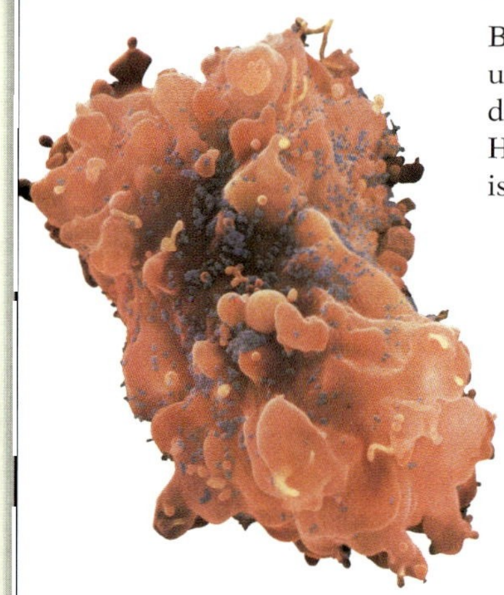

Figure 8 HIV. Small HIV particles (purple) surround a helper T cell (orange).

Many scientists think HIV evolved from a virus similar to one that infects nonhuman primates in Africa. A mutation enables HIV to recognize a receptor protein called **CD4** on some human cells. HIV, shown in **Figure 8,** enters white blood cells by binding to CD4. HIV usually invades helper T cells, which begin to produce HIV soon after infection. As helper T cells die, the immune system gradually weakens and becomes overwhelmed by pathogens that it would normally detect and destroy. The body becomes susceptible to other diseases, called opportunistic infections, that generally cause illness only in people with weakened immune systems.

Testing for HIV

Antibodies to HIV can be detected in blood. Someone whose blood contains antibodies to HIV is said to be HIV positive. A diagnosis of AIDS may be made based on several criteria, including a helper T cell count less than 200 cells/mL of blood. **Figure 9** shows how the number of helper T cells may decline over time in an HIV-positive person.

The time between HIV infection and the onset of AIDS can exceed 10 years, and this time period is increasing as new treatments for HIV infection are developed. A person with HIV may feel and appear healthy but can infect other people. In the United States, the number of deaths caused by AIDS has dropped from more than 38,000 in 1996 to about 22,000 in 1997, and to about 15,000 in 2000. This decrease does not reflect a decline in HIV infection, but rather more effective drug therapies, which postpone onset of the disease.

Figure 9 Onset of AIDS. The graph at right shows the decline over time in the number of helper T cells in a person infected with HIV.

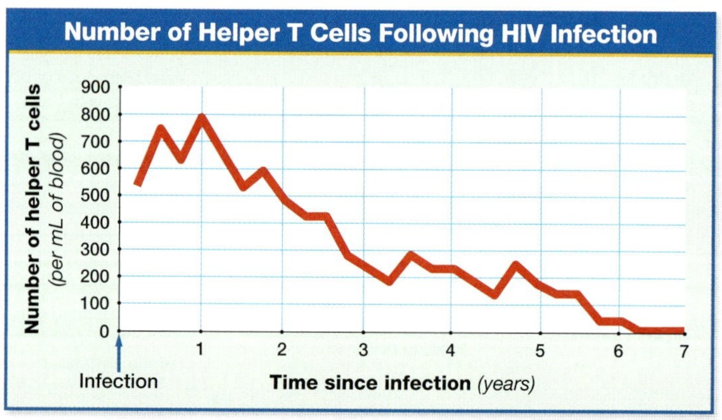

Cultural Awareness

Sir Frank Macfarlane Burnet Sir Frank Macfarlane Burnet was an Australian scientist who discovered and described acquired immunological tolerance to transplanted tissues. He also developed techniques of virus culture in chick embryos, which have become standard laboratory practice. Sir Macfarlane Burnet was elected to the Royal Society in 1942, received the Order of Merit in 1958, and was awarded the Nobel Prize for medicine in 1960. His book, *Natural History of Infectious Disease*, which was co-authored with David O. White, was published by Cambridge University Press in 1972.

Transmission of HIV

You can become infected with HIV if you come in contact with body fluids—including the blood—of an infected person. The most common method of HIV transmission is through sexual contact. Use of a latex condom during sexual contact reduces but does not eliminate the risk of getting or spreading HIV. Many people infected with HIV do not know they are infected. The only sure way to prevent HIV infection by sexual contact is through abstinence (the conscious decision to refrain from sexual activity).

HIV can be passed between drug users who share a hypodermic needle because HIV-infected blood often remains in the needle or syringe. Several years ago, many people became infected with HIV after receiving transfusions of HIV-contaminated blood. This is very unlikely now because blood made available for transfusion is tested for HIV. In addition, pregnant or nursing mothers can pass HIV to their infants through blood and breast milk.

HIV is not transmitted through the air, on toilet seats, by kissing or handshaking, or by any other medium where HIV-infected white blood cells could not survive. Although HIV has been found in saliva, tears, and urine, these body fluids usually contain too few HIV particles to cause an infection. Mosquitoes and ticks do not transmit HIV because they do not carry HIV-infected white blood cells.

internet connect

www.scilinks.org
Topic: HIV Transmission
Keyword: HX4098

SCILINKS Maintained by the National Science Teachers Association

DATA LAB: *Tracking the Spread of AIDS* SC.H.2.4.2

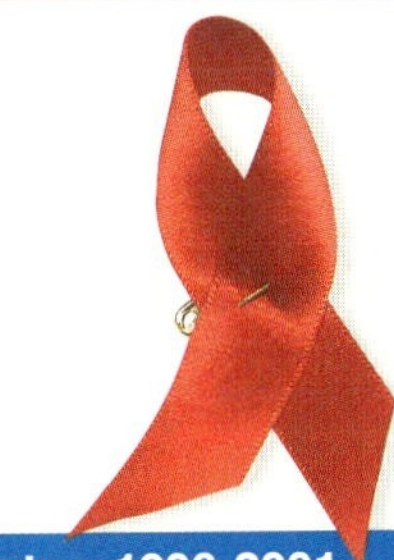

Background

The graph below shows the total AIDS cases reported in the United States between 1996 and 2001. Use the graph to answer the following questions:

Analysis

1. **Describe** how the number of people with AIDS has changed since 1996.
2. **Inferring Relationships** Is the number of Americans infected with HIV most likely greater than or less than the number of people with AIDS? Explain why.
3. **Evaluating Data** The graph indicates that the number of new AIDS cases reported each year has decreased since 1996. Suggest a possible reason for this decline.

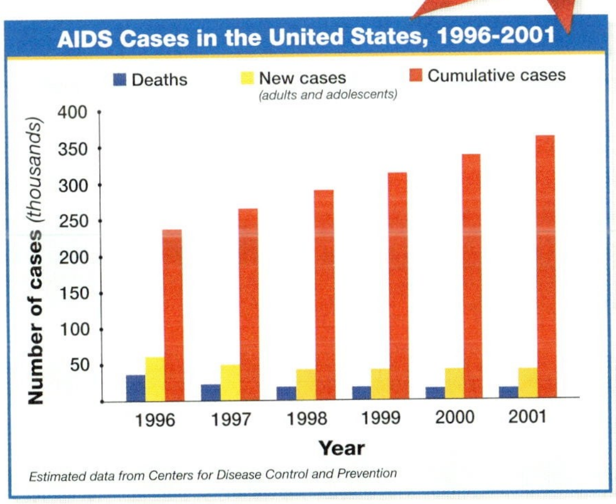

935

Demonstration — BASIC

Invite a healthcare professional to give a presentation to the class about how HIV is transmitted and the best ways to protect yourself from transmission. Encourage students to take notes throughout the presentation. Afterwards, have students make an outline of the main points made by the speaker. LS Auditory

SKILL BUILDER — GENERAL

Math Skills Ask students why HIV keeps spreading in the population even though there is much information about how to prevent it from spreading. (Many people are infected with HIV and do not know it, so they continue to spread the disease.) If one person infected 5 other people in Year One and then died, and each of those infected 5 people in one year, and then died (and so on), how many people would be infected after just 10 years? (Year One: 5 infected; Year 2: 25; Year 3: 125; Year 4: 625; Year 5: 3,125; Year 6: 15,625; Year 7: 78,125; Year 8: 390,625; Year 9: 1,953,125; Year 10: 9,765,625 infected!) LS Logical

DATA LAB: *Tracking the Spread of AIDS*

Skills Acquired
Interpreting, analyzing, inferring

Teacher's Notes
Review with students how to read a bar graph.

Answers to Analysis

1. The cumulative number of AIDS cases has increased since 1996. However, the number of new AIDS cases has declined since 1996.
2. Greater; everyone who has AIDS is infected with HIV, but not everyone who is infected with HIV has developed AIDS.
3. New drugs postpone the onset of AIDS in HIV-infected individuals.

MISCONCEPTION ALERT

AIDS-related Infections Some students may think that HIV itself causes the death of AIDS patients. Actually HIV results in cancers and infections by opportunistic pathogens that are unusual in individuals with healthy immune systems. Infections and cancers associated with AIDS include Kaposi's sarcoma, *Pneumocystis carinii* pneumonia, candidiasis, and others. Kaposi's sarcoma is a cancer of the skin and mucous membranes that often spreads through other organs. *Pneumocystis carinii* pneumonia is the most common of these AIDS-related diseases. Candidiasis, commonly known as thrush, is a fungal infection producing a thick, white coating of the mouth, tongue, and other components of the digestive system. Other opportunistic pathogens include cytomegalovirus, which produces inflammation of the retina, colon, and adrenal glands; and tuberculosis and *Salmonella* infection, both caused by bacteria.

BIOWatch

Asthma

Invite a person who has asthma to class to share his or her experiences with this disease.

Discussion

Ask students: Do you agree or disagree with the reasons given for high inner-city asthma rates? (Answers will vary.)

Close

Reteaching — BASIC

Have students write and act out a one-minute commercial about allergy relief. They should address what causes allergies, what the symptoms are, and how allergy medication works.

Quiz — GENERAL

True or False:

1. HIV attacks the very immune system designed to prevent infections. (true)
2. HIV and AIDS are the same thing. (false; HIV is the virus that can lead to the condition called AIDS)

Alternative Assessment — GENERAL

Have each student conduct research about one of the diseases discussed in this section. Students should write a report including the disease name, its scientific name, where it is commonly found, how it is transmitted, what happens to it when it infects a new host, and how it can be prevented or treated. **LS Verbal**

Section Review Correlations

SC.F.1.4.5: Items 1, 4

SC.F.1.4.8: Items 2, 3, 5

Allergic Reactions

Many health problems are caused by inappropriate responses of the immune system. One example is an allergic reaction. An **allergy** is the body's inappropriate response to a normally harmless antigen. Allergy-causing antigens include pollen, the feces of dust mites, fungal spores, and substances found in some foods and drugs. Most allergic reactions are merely uncomfortable. Cells exposed to allergy-causing antigens release histamine. Histamine causes swelling, redness, increased mucus production, runny nose, itchy eyes, and nasal congestion. Most allergy medicines contain antihistamines, which are drugs that prevent the action of histamine. Severe allergic reactions, such as asthma, can be life threatening if they are not treated immediately.

BIOWatch: Asthma

Asthma is an inflammation of the respiratory tract often caused by an allergic reaction to substances in the air. Asthma affects about 15 million Americans and causes more than 5,000 deaths each year. Inner-city residents get asthma three times as often as people who live outside cities. In some cities, the death rate from asthma is eight times the national average. Some scientists think increased asthma rates in inner-city residents is related to pollution, emotional stress, and limited access to health care. One study suggests that cockroach feces may cause asthma in many inner-city children.

Asthma Attack

During an asthma attack, the respiratory passages become inflamed and swollen. Then mucus collects in the lungs, restricting airflow. Finally, muscles that surround the bronchial tubes tighten, causing shortness of breath.

Treating Asthma

Asthma sufferers can take medicines that increase airflow by relaxing bronchial-tube muscles, but their effects wear off after a few hours. Other medicines provide long-lasting relief by preventing or reducing inflammation.

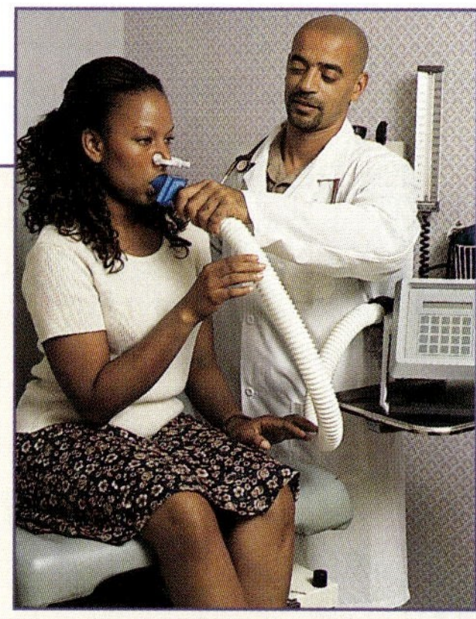

Measuring lung capacity

internet connect

www.scilinks.org

Topic: Asthma

Keyword: HX4015

SCILINKS. Maintained by the National Science Teachers Association

Section 4 Review

1. **Describe** the cause of autoimmune diseases. SC.F.1.4.5
2. **List** two ways that HIV can be transmitted and two ways that it cannot. SC.F.1.4.8
3. **Critical Thinking Recognizing Relationships** Explain why it might take several weeks after exposure to HIV for a person's HIV antibody test to be positive. SC.F.1.4.8
4. **Distinguish** between HIV infection and AIDS. SC.F.1.4.5
5. *FCAT Prep* What is a common symptom of an allergic reaction to airborne antigens?
   A itchy eyes
   B opening nasal passages
   C reduced mucus production
   D weakened immune response
   SC.F.1.4.8

936

Answers to Section Review

1. Autoimmune diseases occur because the body lacks the ability to distinguish between self and non-self cells.
2. HIV can be transmitted by sexual contact, shared hypodermic needles, breast milk, and from a mother to a fetus. HIV cannot be transmitted by shaking hands, toilet seats, insect bites, or through the air.
3. The immune system might take several weeks to build up enough HIV antibodies to be detected.
4. HIV infection is diagnosed by the presence of HIV antibodies in the blood. AIDS is diagnosed if there are less than 200 helper T cells/mL blood.
5. **A.** Correct. A common symptom of an allergic reaction to airborne antigens is itchy eyes, caused by the release of histamine from cells exposed to the antigen. **B.** Incorrect. Airborne antigens would likely cause swelling of nasal passages. **C.** Incorrect. Airborne antigens usually increase mucus production. **D.** Incorrect. An airborne antigen, like any antigen, would stimulate a strong immune response.

Study ZONE CHAPTER HIGH...

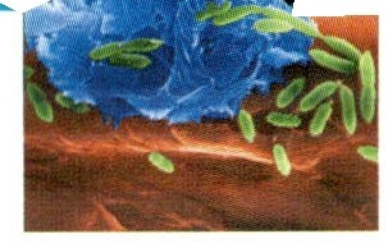

Key Concepts

1 Nonspecific Defenses

- Skin and mucous membranes act as barriers to pathogens.
- The inflammatory response increases blood flow to an infected area, while the temperature response inhibits bacterial growth.
- Complement proteins form a membrane attack complex (MAC). Interferon stimulates cells and inhibits viruses.
- Neutrophils, macrophages, and natural killer cells use different methods to attack and destroy invading pathogens.

2 Immune Response

- Receptors on white blood cells bind to specific antigens.
- The T cell response is a defense in which cytotoxic T cells destroy pathogens.
- The B cell response is a defense in which antibodies mark pathogens for destruction by white blood cells.

3 Disease Transmission and Prevention

- Diseases are transmitted to humans through person-to-person contact, air, food, water, and animal bites.
- Biologists use Koch's postulates to identify pathogens.
- Memory cells can produce long-term immunity to pathogens.
- Vaccination produces long-term immunity to pathogens.
- Antigen shifting makes the immune response of memory cells ineffective.

4 Disorders of the Immune System

- In an autoimmune disease, the immune system attacks body cells as if they were pathogens.
- HIV, the virus that causes AIDS, invades helper T cells, causing them to produce more HIV particles and eventually die.
- HIV is transmitted by HIV-infected white blood cells in body fluids, through sexual contact or by the sharing of a hypodermic needle with an infected person.
- An allergic reaction is an inappropriate response to normally harmless antigens.

Key Terms

Section 1

pathogen (924)
mucous membrane (924)
inflammatory response (925)
histamine (925)
complement system (926)
interferon (926)
neutrophil (926)
macrophage (926)
natural killer cell (926)

Section 2

cytotoxic T cell (927)
B cell (927)
helper T cell (927)
antigen (927)
plasma cell (929)
antibody (929)

Section 3

Koch's postulates (930)
immunity (931)
vaccination (931)
vaccine (931)
antigen shifting (932)

Section 4

autoimmune disease (933)
AIDS (934)
HIV (934)
CD4 (934)
allergy (936)

937

Alternative Assessment — GENERAL

Have students work in cooperative groups. Ask each group to develop a question to send to a newspaper health advice columnist dealing with any topic covered in this chapter. Collect the questions and redistribute them to other groups. Each group should then write an appropriate response to the question. LS Interpersonal

Chapter Resource File

- Science Skills Worksheet GENERAL
- Critical Thinking Worksheet ADVANCED
- Test Prep Pretest GENERAL
- Chapter Test GENERAL

Answer to Concept Map

The following is one possible answer to the Performance Zone item 15.

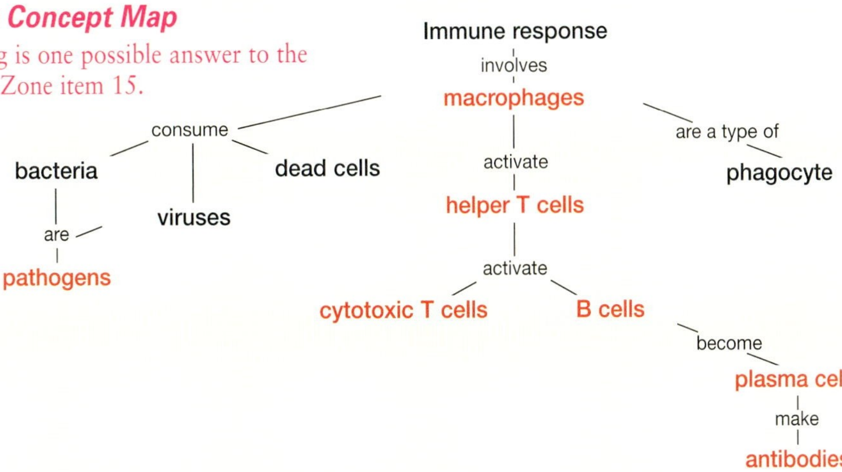

ANSWERS

Understanding Key Ideas

1. b
2. a
3. a
4. b
5. b
6. **a.** Macrophages ingest and destroy pathogens and clear debris. Neutrophils engulf and poison pathogens. **b.** Helper T cells activate cytotoxic T cells, which seek and destroy pathogens. **c.** Immunity is the body's resistance to pathogens that have been previously encountered. A vaccine is a dose of dead or deactivated pathogens given to stimulate immunity. **d.** An allergy is the body's overreaction to a normally harmless antigen. Histamine causes an allergic response that includes swelling, redness, increased mucus production, runny nose, itchy eyes, and nasal congestion.
7. Macrophages consume pathogens and infected cells; cytotoxic T cells kill infected cells; B cells label invaders for destruction; helper T cells activate cytotoxic T cells and B cells.
8. Receptors on the T cells recognize and bind to antigens that match their particular shape.
9. The respiratory passages become inflamed and swollen, mucus collects in the lungs, and muscles around the bronchial tubes tighten, all leading to reduced airflow.
10. One possible answer to the concept map is found at the bottom of the Study Zone page.
11. between 49 and 54 months

Chapter Review Correlations

SC.F.1.4.2: Items 7, 12
SC.F.1.4.8: Items 3-6, 8-11, 14
SC.F.2.4.3: Item 2
SC.H.1.4.1: Item 15
SC.H.1.4.2: Item 1
SC.H.3.4.3: Item 13

Performance ZONE

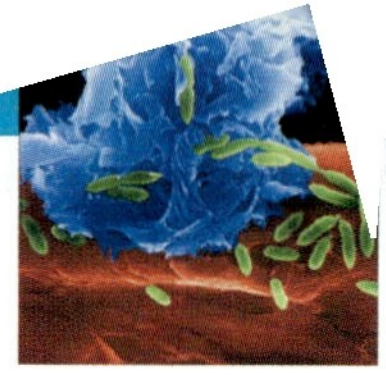

Understanding Key Ideas

1. Robert Koch
a. treated smallpox patients.
b. established a four-step procedure for identifying pathogens.
c. perfected vaccination.
d. identified complement proteins.
SC.H.1.4.2

2. Flu vaccinations are given each year because
a. influenza viruses mutate often.
b. influenza is caused by bacteria.
c. very few memory cells are produced.
d. macrophages cannot engulf flu viruses.
SC.F.2.4.3

3. HIV can be transmitted by
a. sexual contact.
b. mosquito bites.
c. shaking hands.
d. vaccination only.
SC.F.1.4.8

4. Rheumatoid arthritis is an example of
a. an allergic reaction.
b. an autoimmune disease.
c. an AIDS-related infection.
d. a bacterial infection.
SC.F.1.4.8

5. HIV disables the immune system by
a. blocking the action of macrophages.
b. destroying helper T cells.
c. activating production of B cells.
d. All of the above
SC.F.1.4.8

6. For each pair of terms, explain the differences in their meanings. SC.F.1.4.8
a. macrophage, neutrophil
b. immunity, vaccine
c. allergy, histamine

7. Name three types of white blood cells, and explain their roles in the immune system. SC.F.1.4.2

8. How do cytotoxic T cells recognize antigens? SC.F.1.4.8

9. **BIOWatch** What symptoms are usually associated with an asthma attack? SC.F.1.4.8

10. **Concept Mapping** Make a concept map that describes the immune response. Include the following terms in your map: *pathogen, macrophage, helper T cell, cytotoxic T cell, B cell, plasma cell,* and *antibody.* SC.F.1.4.8

11. The graph below shows the decrease in the number of helper T cells in a person with AIDS. How many months after infection did the onset of AIDS occur? SC.F.1.4.8

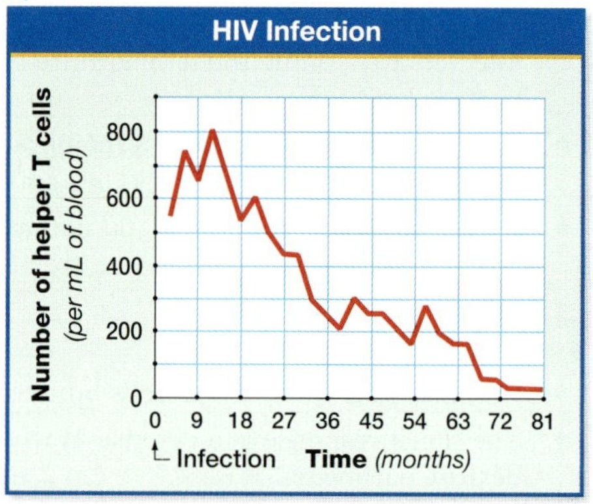

Critical Thinking

12. Analyzing Information Plasma cells contain a large Golgi apparatus and large amounts of rough endoplasmic reticulum. How is the presence of these organelles related to the function of plasma cells? SC.F.1.4.2

13. Inferring Relationships People who are severely burned often die from infection. Given what you know about disease transmission, explain why this is common. SC.H.3.4.3

Alternative Assessment

14. Summarizing Information Use the media center or the Internet to research three different vaccines. Make a large chart or table on poster board listing the pathogens they protect against, their effectiveness, side effects, and boosters required, if any. Present your chart to your class. SC.F.1.4.8

15. Career Focus Immunologist Research the field of immunology, and write a report on your findings. Your report should include a job description, education and training required, kinds of employers, growth prospects, and starting salary. SC.H.1.4.1

938

Critical Thinking

12. The Golgi apparatus and endoplasmic reticulum are responsible for protein mobilization. Because plasma cells produce antibodies, high protein synthesis activity would be expected.
13. The barrier of the skin prevents many infections. Severe burns result in skin loss and subsequent infection.

Alternative Assessment

14. Answers will vary.
15. Immunologists are scientists who diagnose, treat, and research immunological diseases and disorders. Immunologists attend medical school after college. Immunologists work in private practice, in hospitals, and for pharmaceutical companies. The growth potential of the field is excellent. Starting salary will vary by region.

Assignment Guide

Section	Questions
1	6a
2	7, 8, 10, 12
3	1, 2, 6b, 14
4	3, 4, 5, 6c, 9, 11, 15

FCAT Prep

For questions 1–6, write your answers on a separate sheet of paper.

1. Which of the following is a nonspecific defense?
   A. antibodies
   B. the B cell response
   C. the T cell response
   D. the inflammatory response
   SC.F.1.4.8

2. What role do mucous membranes play in the immune system?
   F. causing blood clots
   G. producing antibodies
   H. activating helper T cells
   I. secreting mucous, which traps pathogens
   SC.F.1.4.2

3. B cells and cytotoxic T cells are stimulated by interleukin-2. What kind of cell releases interleukin-2?
   A. helper T cells
   B. macrophages
   C. natural killer cells
   D. neutrophils
   SC.F.1.4.6

4. Which of the following statements is true of plasma cells?
   F. They engulf pathogens.
   G. They produce antibodies.
   H. They result from cytotoxic T cells.
   I. They are directly stimulated by interleukin-1.
   SC.F.1.4.2

5. READ INQUIRE EXPLAIN A person can become infected with HIV if he or she comes into contact with the body fluids of an infected person.

   Under what circumstances can a child be born with HIV? SC.F.1.4.8

6. READ INQUIRE EXPLAIN Scientific research into treatments and a possible cure for AIDS is an ongoing process. One important development is a clinical trial of a potential HIV vaccine conducted in 15 U.S. medical centers. Also, a new medication has been developed which combines two proven medications (3Tc and AZT) into one tablet. This medication may help people with HIV live longer, healthier lives. Another medication, alitretinoin, treats lesions of Kaposi's sarcoma.

   Part A Explain the relationship between HIV and AIDS. SC.F.1.4.8

   Part B How could an HIV vaccine prevent people from developing AIDS? SC.F.1.4.8

Test TIP

To help you learn the stages of the immune response, make a note card describing each stage, mix the cards up, and practice reordering the stages.

FCAT Doctor

Question 2 Answer I is the correct choice. Mucous membranes are layers of tissue that produce a sticky, viscous fluid that traps pathogens. Answer F is incorrect because platelets, not mucous membranes, cause blood to clot. Answer G is incorrect because plasma cells, not mucous membranes, produce antibodies. Answer H is incorrect because virus-containing macrophages, not mucous membranes, activate helper T cells.

Question 5 An unborn baby can get the HIV virus only from its mother.

Question 6

Part A HIV invades helper T cells. As these cells die, the immune system weakens and becomes overwhelmed by pathogens that it would normally detect and destroy. The body then becomes susceptible to cancers and infections – the acquired immunodeficiency syndrome.

Part B Vaccines contain a dead or weakened pathogen that triggers an immune response against the pathogen without symptoms of infection.

Answers

1. D
2. I
3. A
4. G
5. When the mother is infected with HIV (even if the father is HIV negative), the baby can be born with HIV.
6A. Acquired immunodeficiency syndrome, AIDS, is the disease caused by the human immunodeficiency virus, HIV.
6B. The vaccine would cause the body to produce HIV-fighting cells before a person is infected with the virus.

Test Prep Correlations

SC.F.1.4.2: Items 2, 4
SC.F.1.4.6: Item 3
SC.F.1.4.8: Items 1, 5, 6A, 6B

Skills Practice Lab

SIMULATING DISEASE TRANSMISSION

Teacher's Notes

Time Required 30 minutes

Ratings

EASY ⟶ HARD

TEACHER PREPARATION 3

STUDENT SETUP 3

CONCEPT LEVEL 2

CLEANUP 3

Skills Acquired

- Collecting Data
- Communicating
- Identifying/Recognizing Patterns
- Interpreting
- Organizing and Analyzing Data

Scientific Methods

In this lab, students will:

- Make Observations
- Analyze the Results
- Draw Conclusions
- Communicate the Results

Materials and Equipment

Materials for this lab can be ordered from WARD'S. See *Master Materials List* at the front of this book for catalog numbers.

Safety Cautions

Review all safety symbols with students before beginning the lab.

Skills Practice Lab

Simulating Disease Transmission

SC.H.3.4.1

SKILLS

- Modeling
- Organizing and analyzing data

OBJECTIVES

- **Simulate** the transmission of a disease.
- **Determine** the original carrier of the disease.

MATERIALS

- safety goggles
- lab apron
- disposable gloves
- dropper bottle of unknown solution
- large test tube
- indophenol indicator

ChemSafety

CAUTION: Always wear safety goggles and a lab apron to protect your eyes and clothing.

CAUTION: Do not touch or taste any chemicals. Know the location of the emergency shower and eyewash station and how to use them. If you get a chemical on your skin or clothing, wash it off at the sink while calling to the teacher. Notify the teacher of a spill. Spills should be cleaned up promptly, according to your teacher's directions.

CAUTION: Glassware is fragile. Notify the teacher of broken glass or cuts. Do not clean up broken glass or spills with broken glass unless the teacher tells you to do so.

Before You Begin

Communicable diseases are caused by **pathogens** and can be transmitted from one person to another. You can become infected by a pathogen in several ways, including by drinking contaminated water, eating contaminated foods, receiving contaminated blood, and inhaling infectious **aerosols** (droplets from coughs or sneezes). In this lab, you will simulate the transmission of a communicable disease. After the simulation, you will try to identify the original infected person in the closed class population.

1. Write a definition for each boldface term in the paragraph above.
2. Make data tables similar to the ones shown at right.
3. Based on the objectives for this lab, write a question you would like to explore about disease transmission.

DATA TABLE 1

Round number	Partner's name

DATA TABLE 2

Name of infected person	Names of infected person's partners		
	Round 1	Round 2	Round 3

Procedure

PART A: Simulate Disease Transmission

1. Put on safety goggles, a lab apron, and gloves.
2. You will be given a dropper bottle of an unknown solution. When your teacher says to begin, transfer 3 dropperfuls of your solution to a clean test tube.

940

Tips and Tricks

Before students begin, remind them of safety practices, and emphasize that they should not allow solutions to touch their skin or clothes. Use stock dropper bottles of distilled water for half the students and dropper bottles containing 10 percent ascorbic acid for the rest of the students. Both solutions will be clear. A 10 percent solution of ascorbic acid can be prepared by dissolving 10 g of ascorbic acid in 50 mL of water and diluting it to 100mL with water. Each student will need a dropper bottle of unknown solution, a large test tube, and indophenol indicator solution. If the procedure is followed precisely, the route of transmission can be traced easily. Be sure to warn students not to select a new partner until each round is over and all students have exchanged liquids for that round. If one student makes a mistake, it can make it very difficult to trace the route of transmission.

3. Select a partner for Round 1. Record the name of this partner in Data Table 1.
4. Pour the contents of one of your test tubes into the other test tube. Then pour half the solution back into the first test tube. You and your partner now share any pathogens either of you might have.
5. On your teacher's signal, select a new partner for Round 2. Record this partner's name in Data Table 1. Repeat step 4.
6. On your teacher's signal, select another new partner for Round 3. Record this partner's name. Repeat step 4.
7. Add one dropperful of indophenol indicator to your test tube. "Infected" solutions will stay colorless or turn light pink. "Uninfected" solutions will turn blue. Record the results of your test.

PART B: Trace the Disease Source

8. If you are infected, write your name and the name of your partner in each round on the board or on an overhead projector. Mark your infected partners. Record all the data for your class in Data Table 2.
9. To trace the source of the infection, cross out the names of the uninfected partners in Round 1. There should be only two names left. One is the name of the original disease carrier. To find the original disease carrier, place a sample from his or her dropper bottle in a clean test tube, and test it with indophenol indicator.
10. To show the disease transmission route, make a diagram similar to the one below. Show the original disease carrier and the people each disease carrier infected.

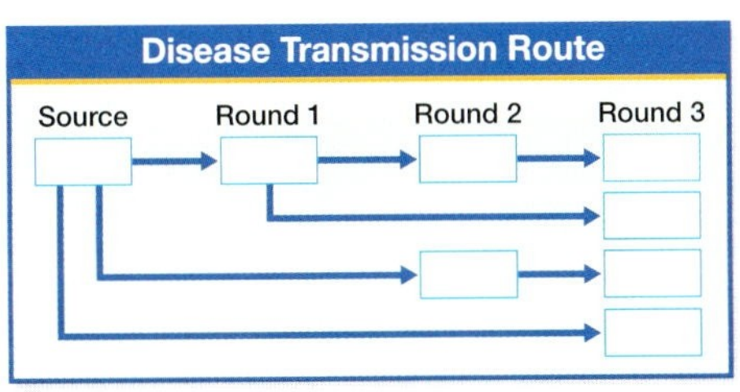

PART C: Cleanup and Disposal

11. Dispose of solutions and broken glass in the designated waste containers. Do not pour chemicals down the drain unless your teacher tells you to do so.
12. Clean up your work area and all lab equipment. Return lab equipment to its proper place. Wash your hands thoroughly before you leave the lab and after you finish all work.

Analyze and Conclude

1. **Interpreting Data** After Round 3, how many people were "infected"? Express this number as a percentage of your class.
2. **Relating Concepts** What do you think the clear fluids each student started with represent? Explain why.
3. **Drawing Conclusions** Can someone who does not show any symptoms of a disease transmit that disease? Explain.
4. **Further Inquiry** Write a new question about disease transmission that could be explored with another investigation.

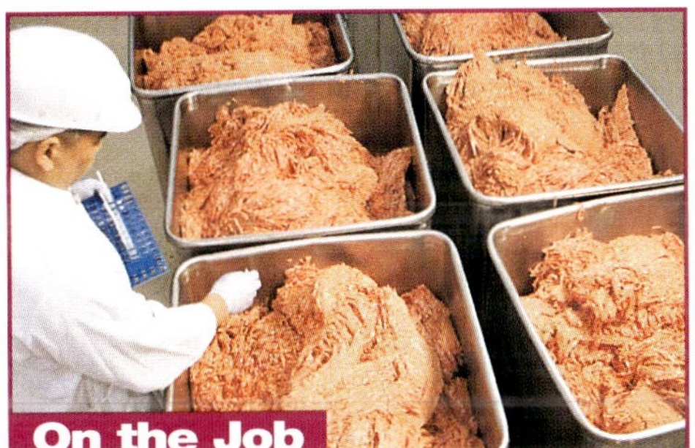

On the Job

Public health officials, such as food inspectors, study and work to prevent the spread of diseases in human populations. Do research to find out how public health officials trace the origin of communicable diseases. For more about careers, visit **go.hrw.com** and type in the keyword **HX4 Careers.**

Answers to Before You Begin

1. *communicable disease*—a disease that can be transmitted from one person to another
   pathogen—a disease-causing agent such as a bacterium or a virus
   aerosol—a mist of tiny droplets of a liquid
2. Tables should resemble Data Tables 1 and 2.
3. Answers will vary. Sample answer: Can you determine whether a disease has been caused by the passing of pathogens from person to person or by environmental conditions?

Answers to Procedure

2. One half of the students will have an acidic solution.
3. Students should record their partner's name in Data Table 1.
4. After they share solutions, most students will have acidic mixtures ("infected" + "non-infected" or "infected" + "infected" solutions). A few students will have non-acidic solutions ("non-infected" + "non-infected" solutions).
5. Students should record the name of their second partner in Data Table 1.
6. Each repetition of the infection activity will increase the number of "infected" individuals.
7. Most solutions will remain colorless or turn pink.
8. All students will need to announce whether or not they are infected for this step.
9. If the procedure is followed precisely, the route of transmission can be traced easily.
10. A student volunteer should construct a diagram of the route of infection on the board or overhead.

Answers to Analyze and Conclude

1. Answers will vary depending on class size.
2. The clear fluids represent uncontaminated aerosols, food, water, or blood (distilled water) and contaminated aerosols, food, water, or blood (ascorbic acid).
3. Yes; some pathogens must multiply in the body before symptoms appear; during this period of time they can be transmitted to another person.
4. Answers will vary. For example: Would dividing the class into two or three quarantined groups before repeating the experiment affect disease transmission rate?

Answers to On the Job

Answers will vary. Students should include information on what products and places public health officials inspect and what an inspection entails.

Chapter Opener

- **SE** **Quick Review,** p. 943 ■
- **SE** **Reading Activity,** p. 943 ■
- **TE** **Using the Figure,** p. 942 GENERAL
- **TE** **Opening Activity,** p. 943 GENERAL

& 3 • 90 min pp. 944–949

Section 1

Neurons and Nerve Impulses

- **OSP** Lesson Plan*
- **TR** Bellringer*
- **TR** Section Outline*
- **TR** J60 Structure of a Neuron*
- **TR** J61 Conduction of a Nerve Impulse*
- **TR** J62 Synaptic Transmission*

- **TE** **Using the Figure,** p. 945
- **TE** **Activity** Information Transfer, p. 945 BASIC
- **SE** **Data Lab** Analyzing Changes During a Nerve Impulse, p. 946 ■
- **CRF** **Datasheets for In-Text Labs** Analyzing Changes During a Nerve Impulse*
- **TE** **Using the Figure,** p. 947
- **TE** **Using the Figure,** p. 948

& 5 • 90 min pp. 950–955

Section 2

Structures of the Nervous System

- **OSP** Lesson Plan*
- **TR** Bellringer*
- **TR** Section Outline*
- **TR** J64 Human Brain*
- **TR** J65 Human Spinal Cord*
- **TR** J67 Knee-Jerk Response*
- **TR** J68 Autonomic Nervous System*

- **TE** **Demonstration,** p. 950 BASIC
- **TE** **Using the Figure,** p. 951
- **TE** **Demonstration,** p. 952 GENERAL
- **CRF** **Inquiry Lab** What Can A Survey Tell You* GENERAL

& 7 • 90 min pp. 956–960

Section 3

Sensory Systems

- **OSP** Lesson Plan*
- **TR** Bellringer*
- **TR** Section Outline*
- **TR** J69 Types of Sensory Receptors*
- **TR** J73 Structure of the Eye*
- **TR** J70 Structure of the Ear*

- **TE** **Activity** Phosphenes, p. 956 GENERAL
- **TE** **Activity** Visual Feedback and Touch, p. 957 GENERAL
- **TE** **Demonstration,** p. 958 ◆ GENERAL
- **SE** **Quick Lab** Demonstrating Your Blind Spot, p. 958 ■
- **CRF** **Datasheets for In-Text Labs** Demonstrating Your Blind Spot*
- **TE** **Demonstration,** p. 959
- **CRF** **Exploration Lab** Vision* GENERAL

& 9 • 90 min pp. 961–966

Section 4

Drugs and the Nervous System

- **OSP** Lesson Plan*
- **TR** Bellringer*
- **TR** Section Outline*
- **TR** J77 Psychoactive Drugs of Abuse*
- **TR** J75, J76 Action of Cocaine*
- **TR** J78 Effects of Blood Alcohol Concentration*

- **TE** **Demonstration,** p. 961 ◆ GENERAL
- **TE** **Using the Figure,** p. 962
- **TE** **Group Activity** Drugs and Their Effects, p. 963 GENERAL
- **TE** **Demonstration,** p. 964 GENERAL
- **TE** **Activity** BAC, p. 964 BASIC
- **TE** **Group Activity** Social Drug Use, p. 965 GENERAL
- **SE** **Exploration Lab** Calculating Reaction Times, pp. 970–971 ■ GENERAL
- **CRF** **Datasheets for In-Text Labs** Calculating Reaction Time*

0 & 11 • 90 min

Chapter Review and Assessment Resources

- **SE** **Study Zone,** p. 967 ■
- **SE** **Performance Zone,** p. 968 ■
- **SE** **FCAT Test Prep,** p. 969 ■
- **TR** Concept Mapping*
- **CRF** Vocabulary Review Worksheet* ■ GENERAL
- **CRF** Test Prep Pretest* ■ GENERAL
- **CRF** Chapter Test* ■ GENERAL
- **CRF** Chapter Test* ADVANCED
- **OSP** Test Generator*
- **OSP** Test Item Listing*
- **OSP** Modified Chapter Test* BASIC

Online and Technology Resources

go.hrw.com Holt Online Learning

Visit go.hrw.com for access to Holt Online Learning or enter the keyword **HX6 Home** for a variety of free online resources.

One-Stop Planner® CD-ROM

This CD-ROM package includes
- Lab Materials QuickList Software
- Holt Calendar Planner
- Customizable Lesson Plans
- Printable Worksheets
- ExamView® Test Generator
- Interactive Teacher Edition
- Holt PuzzlePro® Resources
- Holt PowerPoint® Resources

Guided Reading Audio CDs

These CDs are designed to help auditory learners and reluctant readers. (Audio CDs are also available in Spanish.)

Compression guide:
To shorten your instruction because of time limitations, omit block 1.

KEY

TE	Teacher Edition	CRF	Chapter Resource File	*	Also on One-Stop Planner
SE	Student Edition	TR	Teaching Transparency	■	Also Available in Spanish
OSP	One-Stop Planner			◆	Requires Advance Prep

SKILLS DEVELOPMENT AND PRACTICE	SECTION REVIEW AND ASSESSMENT	CORRELATIONS
		Florida Sunshine Standards/FCAT
TE **Reading Skill Builder** Paired Summarizing, p. 945 BASIC TE **Skill Builder** Vocabulary, p. 946 BASIC TE **Reading Skill Builder** Interactive Reading, p. 947 BASIC TE **Skill Builder** Math Skills, p. 947 GENERAL CRF **Directed Reading Worksheet*** BASIC CRF **Active Reading Worksheet*** GENERAL	SE **Section 1 Review** ■ TE **Reteaching,** p. 949 BASIC TE **Quiz,** p. 949 GENERAL TE **Alternative Assessment,** p. 949 GENERAL CRF **Quiz*** ■ BASIC	SC.F.1.4.2, SC.F.1.4.5, SC.F.1.4.6
TE **Reading Skill Builder** Discussion, p. 952 GENERAL TE **Reading Skill Builder** Paired Summarizing, p. 954 BASIC CRF **Directed Reading Worksheet*** BASIC CRF **Active Reading Worksheet*** GENERAL	SE **Section 2 Review** ■ TE **Reteaching,** p. 955 BASIC TE **Quiz,** p. 955 GENERAL TE **Alternative Assessment,** p. 955 GENERAL CRF **Quiz*** ■ BASIC	SC.F.1.4.2
CRF **Directed Reading Worksheet*** BASIC CRF **Active Reading Worksheet*** GENERAL	SE **Section 3 Review** ■ TE **Reteaching,** p. 960 BASIC TE **Quiz,** p. 960 GENERAL TE **Alternative Assessment,** p. 960 GENERAL CRF **Quiz*** ■ BASIC	SC.F.1.4.2, SC.F.1.4.7
SE **Forensics Labs Appendix** The Counterfeit Drugs, p. 1062 CRF **Forensics and Applied Science Experiments** The Counterfeit Drugs* GENERAL	TE **Skill Builder** Writing Skills, p. 962 ADVANCED TE **Reading Skill Builder** Anticipation Guide, p. 966 BASIC CRF **Directed Reading Worksheet*** BASIC CRF **Active Reading Worksheet*** GENERAL CRF **Critical Thinking Worksheet*** ADVANCED	SC.F.1.4.1, SC.F.1.4.8

Holt Biology Interactive Tutor CD-ROM

This CD-ROM consists of interactive activities that give students a fun way to extend their knowledge of biology concepts.

Unit 1 Cell Transport and Homeostasis

SCILINKS. NSTA
www.scilinks.org
Maintained by the **National Science Teachers Association**

Topic: Spinal Cord
SciLinks Code: HX4168

Topic: Autonomic Nervous System
SciLinks Code: HX4017

Topic: Sensory Receptors
SciLinks Code: HX4162

Topic: Chemical Senses
SciLinks Code: HX4041

Topic: Blood Alcohol Concentration
SciLinks Code: HX4026

CHAPTER 41

Overview

Tell students the purpose of this chapter is to explain how the nervous system allows the human body to respond to stimuli. The nervous system transmits information from muscles, organs, and sensory receptors to the brain. In the brain, the information is integrated, and signals are returned to muscles and glands. The nervous system thus helps maintain homeostasis. The actions of the nervous system are rapid but of a shorter duration than chemical messages such as hormones. Through the transmission of nerve impulses, an organism can respond quickly to internal or external stimuli.

Using the Figure

Ask students to look at the nerve-cell network illustrated on this page. Ask them to suggest the function of this network. The neuron is the most basic unit of the network. The junction between a nerve cell and another nerve cell or muscle cell is the synapse. Networks of neurons constantly gather, integrate, interpret, and respond to information about the body's internal condition and about external, environmental conditions.

Assessing Prior Knowledge

Before beginning this chapter, be sure your students can answer the questions in the ***Quick Review*** on the facing page. If they cannot, encourage them to reread the sections indicated.

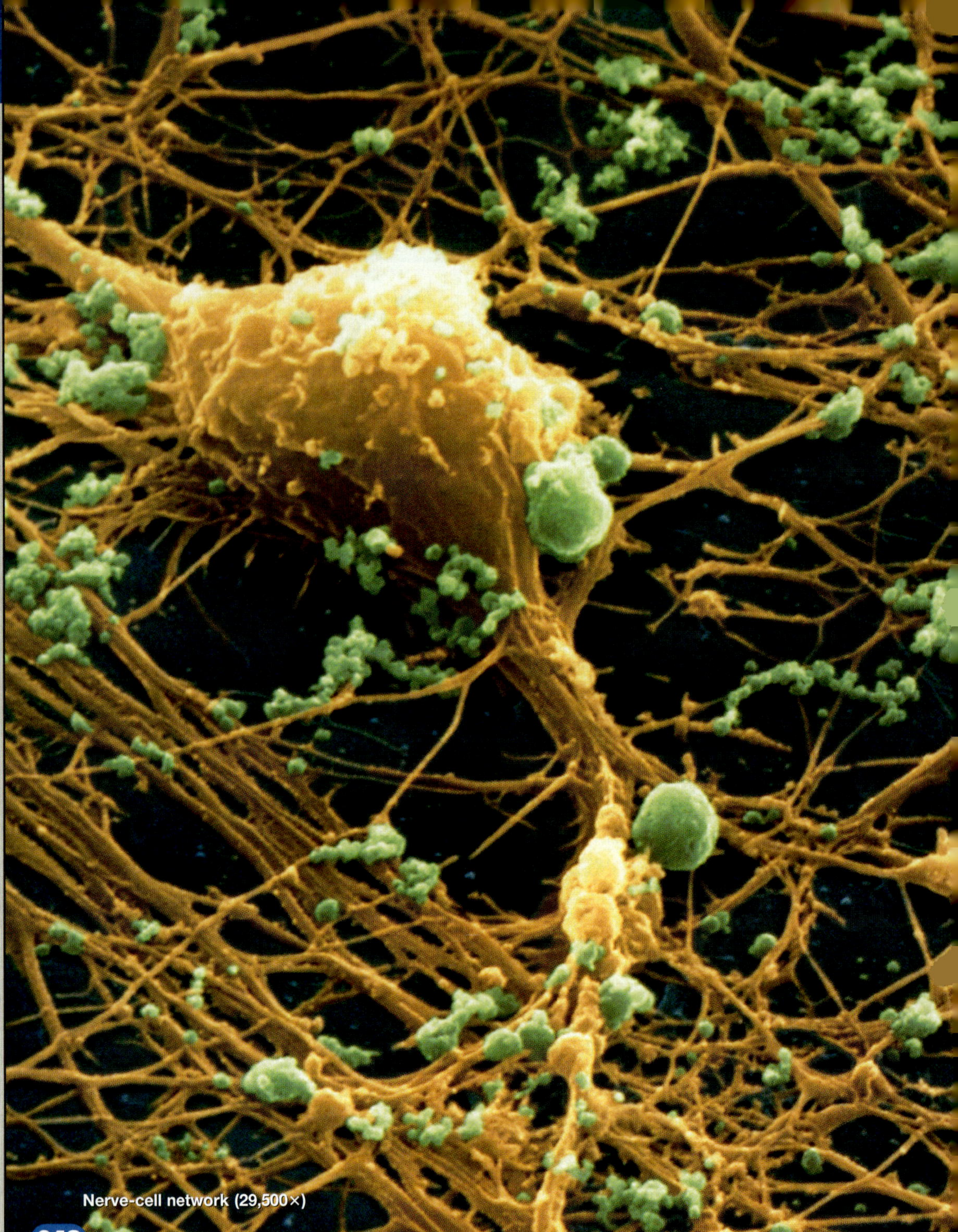

Nerve-cell network (29,500×)

942

Chapter Correlations — Florida Sunshine Standards/FCAT

SC.F.1.4.2 Body structures are uniquely designed and adapted for their function.

SC.F.1.4.5 Complex interactions among the different kinds of molecules in the cell cause distinct cycles of activity governed by proteins.

SC.F.1.4.6 Separate parts of the body communicate with each other using electrical and/or chemical signals.

SC.F.1.4.7 Organisms respond to internal and external stimuli.

SC.F.1.4.8 Cell behavior can be affected by molecules from other parts of the organism or even from other organisms.

CHAPTER

41 Nervous System

Florida Sunshine Standards/FCAT

SC.F.1.4.5 Complex interactions among the different kinds of molecules in the cell cause distinct cycles of activity governed by proteins.

SC.F.1.4.6 Separate parts of the body communicate with each other using electrical and/or chemical signals.

SC.F.1.4.7 Organisms respond to internal and external stimuli.

SC.F.1.4.8 Cell behavior can be affected by molecules from other parts of the organism or even from other organisms.

Quick Review

Answer the following without referring to earlier sections of your book.

1. **Describe** the importance of ion channels in cell transport. *(Chapter 4, Section 1)*
2. **Identify** the role of sodium-potassium pumps in cells. *(Chapter 4, Section 2)*
3. **Distinguish** between endocytosis and exocytosis. *(Chapter 4, Section 2)*
4. **List** three functions of receptor proteins. *(Chapter 4, Section 2)*

Did you have difficulty? *For help, review the sections indicated.*

Reading Activity

Copy the following statements in your notebook:

1. *Addiction is a purely psychological response to drug use.*
2. *Reflexes occur before the brain is aware of danger.*
3. *Prolonged exposure to loud noise can cause permanent hearing loss.*

Before you read this chapter, write down whether you agree or disagree with each statement. After you have finished reading the chapter, decide whether or not you still agree with your first response.

Nerve-cell networks like this one transmit thought, emotions, and sensations by conducting electrochemical signals from cell to cell.

Looking Ahead

internet connect

www.scilinks.org

National Science Teachers Association *sci*LINKS Internet resources are located throughout this chapter.

SCILINKS. Maintained by the National Science Teachers Association

Opening Activity — GENERAL

Nervous System Function Ask students to brainstorm about the functions of the nervous system. Write their responses on the chalkboard. Make three columns on the chalkboard with the following headings: *Monitoring internal and external environments, Coordinating bodily activities,* and *Functioning of intellect and memory.* Ask students to place their responses into the appropriate categories. Then ask them to explain why they grouped their responses in this way. **LS Verbal**

Quick Review Answers

1. An ion channel is a transport protein in a cell membrane through which ions can pass.
2. A sodium-potassium pump is a carrier protein that transports sodium ions out of, and potassium ions into, a cell.
3. Endocytosis is the movement of a substance by a vesicle to the inside of a cell. Exocytosis is the movement of a substance by a vesicle to the outside of a cell.
4. Receptor proteins bind to specific signal molecules, causing the cell to respond, forming either open or closed ion channels.

Reading Activity Answers

Students may agree or disagree with the statements. They may reassess their lists after studying the chapter.

Chapter Resource File

- Vocabulary Worksheets
- Concept Mapping

Section 1

Focus

Overview

Before beginning this section review with your students the objectives listed in the Student Edition. This section explains the structure and function of neurons, the events that occur in synaptic transmission, and the steps of a nerve impulse.

Bellringer

Ask students to write definitions for electrons and ions and to suggest a function of ions in nerve cells. Ask them to check their definitions by looking them up in the glossary. (Electrons are negatively charged particles that move around the nucleus of an atom; ions are atoms or molecules that have either a positive charge [resulting from the loss of one or more electrons] or a negative charge [resulting from the gain of one or more electrons]; nerve cells have electrical charges that are different from the electrical charges in the fluid around them.)

Interactive Biology Tutor

- **Unit 1—Cell Transport and Homeostasis** This engaging tutorial reinforces cell membrane structure and function.

Section Correlations

SC.F.1.4.2
SC.F.1.4.5
SC.F.1.4.6

Section 1

Neurons and Nerve Impulses

Florida Sunshine Standards/FCAT

SC.F.1.4.2 Body structures are uniquely designed and adapted for their function.
SC.F.1.4.5 Complex interactions among the different kinds of molecules in the cell cause distinct cycles of activity governed by proteins.
SC.F.1.4.6 Separate parts of the body communicate with each other using electrical and/or chemical signals.

Objectives

- **Analyze** the structure and function of neurons. SC.F.1.4.2
- **Describe** how the resting potential is established in a neuron. SC.F.1.4.5
- **Sequence** the steps of a nerve impulse. SC.F.1.4.6
- **List** the events that occur in synaptic transmission of a nerve impulse. SC.F.1.4.6

Key Terms

neuron
dendrite
axon
nerve
membrane potential
resting potential
action potential
synapse
neurotransmitter

Neurons

If your body used only chemical signals to send messages, your interaction with the environment would be slow. A quicker means of communication is needed, especially if your brain has an urgent message for the muscles in your legs, such as "Contract quickly! A speeding car is headed this way!" In addition to chemical signals, your nervous system uses electrical signals to send messages rapidly throughout your body.

The nervous system contains a complex network of nerve cells, or **neurons** *(NOO rahns)*. Neurons, such as those shown in **Figure 1,** are specialized cells that transmit information throughout the body. Neurons enable many important functions, such as movement, perception, thought, emotion, and learning.

Structure of Neurons

A neuron's unique structure enables it to conduct electrical signals called nerve impulses. Neurons communicate by transmitting nerve impulses to body tissues and organs, including muscles, glands, and other neurons. Neurons vary greatly in form, but a typical neuron is similar to the one shown in **Figure 2.** **Dendrites** *(DEHN driets)*, which extend from the cell body of the neuron, are the "antennae" of the neuron. Dendrites receive information from other cells. The neuron's cell body collects information from dendrites, relays this information to other parts of the neuron, and maintains the general functioning of the neuron. An **axon** is a long membrane-covered extension of the cytoplasm that conducts nerve impulses. The ends of an axon are called axon terminals. When a neuron communicates with other cells, it does so at its axon terminals.

Nervous tissue consists mostly of neurons and their supporting cells. Bundles of axons are called **nerves.** The arrangement of axons in a nerve is similar to a telephone cable with many different communication channels, each carried by a separate wire. Nerves appear as fine, white threads when viewed with the unaided eye.

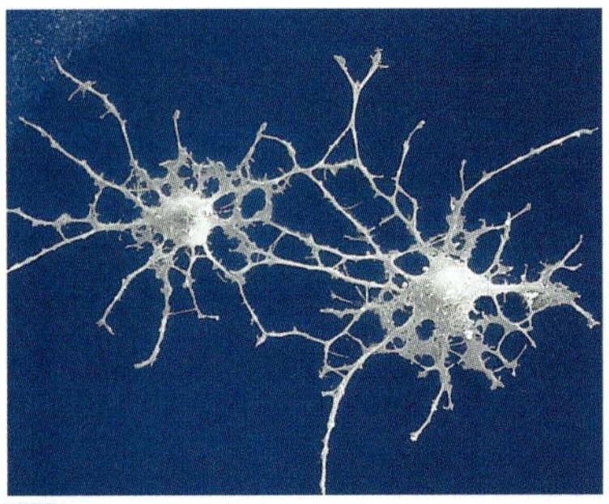

Figure 1 Two neurons. An average adult human brain contains about 100 billion neurons.

Insulated Neurons

Many neurons have a layer of insulation on their axon called a myelin *(MIE uh lihn)* sheath, as shown in Figure 2. Myelin is produced by supporting cells that surround the axon. The presence of myelin causes nerve impulses to move faster down the axon. The myelin sheath is interrupted at intervals, leaving gaps called nodes of Ranvier

944

Chapter Resource File

- Directed Reading BASIC
- Active Reading GENERAL
- Data Sheet for Data Lab GENERAL

Transparencies

TR Bellringer
TR J60 Structure of a Neuron

Figure 2 **Myelinated neuron.** A myelin sheath covers the axons of many neurons. Myelin increases the speed of nerve impulses.

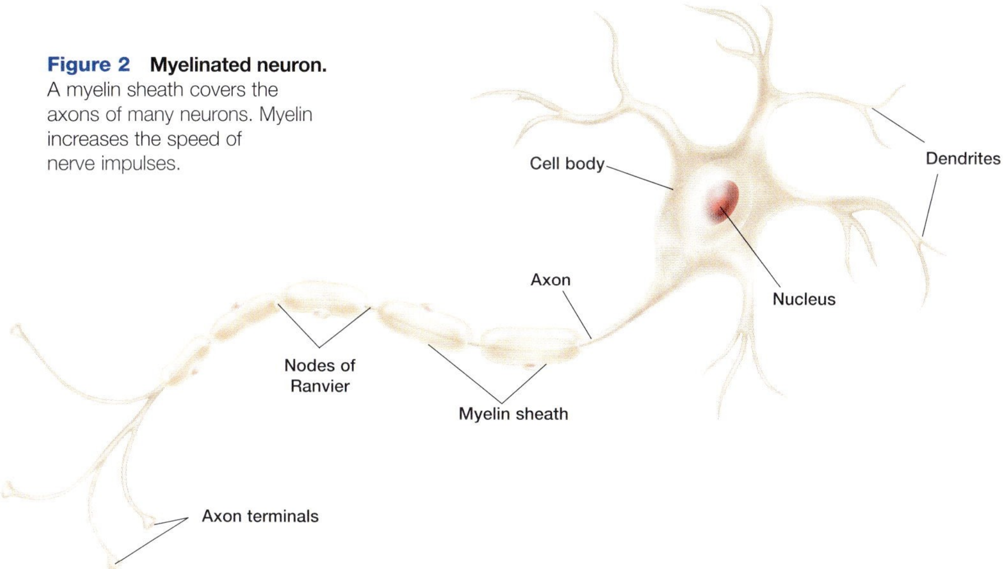

(RAHN vee ay), where the axon membrane is exposed to the surrounding fluid. Conduction of nerve impulses is faster in myelinated axons because nerve impulses "jump" from node to node as they move down the axon. Myelin is especially beneficial in neurons that must transmit information very rapidly, such as those involved with quick movement.

The speed of impulse conduction is also related to axon diameter. A large-diameter axon conducts impulses faster than a small-diameter axon, assuming both axons are either myelinated or unmyelinated.

internet connect
www.scilinks.org
Topic: Neurons
Keyword: HX4129
SCILINKS. Maintained by the National Science Teachers Association

Neuron Function

All cells have an electrical charge on the inner surface of the cell membrane that is different from the electrical charge of the fluid outside the cell. The difference in electrical charge across the cell membrane, called the **membrane potential,** results from the movement of ions into and out of the cell. This movement depends on the relative concentration of ions inside and outside the cell, the ability of the ions to diffuse across the cell membrane, and the electrical charge of the ions. The membrane potential is expressed as voltage, like that of a battery.

Ions diffuse across a neuron's cell membrane by passing through proteins that act as ion channels. Each type of channel allows only specific ions to pass. Certain channels are voltage-gated—that is, whether they are open or closed depends on the membrane potential. Even a small change in the membrane potential can affect the permeability of the cell membrane to certain ions. As these ions diffuse into or out of the neuron, they in turn affect the membrane potential.

945

did you know?

Phenylketonuria Phenylketonuria, or PKU, is a disease caused by an enzyme deficiency that leads to a buildup of phenylpyruvic acid in the body. This results in the destruction of the myelin sheaths on neurons. If undetected, PKU can lead to severe mental retardation and even death. PKU is easy to detect, so all newborns in the United States are tested for it. If found, a low phenylalanine diet is prescribed. Food products such as diet sodas carry a message to phenylketonurotics warning them that the product contains phenylalanine.

One-Stop Planner CD-ROM

- Reading Organizers BASIC
- Reading Strategies BASIC

Motivate

Using the Figure

Have students observe prepared slides or photographs of neurons with simple morphology (such as stellate neurons). Remind students to look for cell bodies, axons, and dendrites. Have students draw and label what they see and compare their drawings with the neuron shown in **Figure 2.** LS Visual

Activity — BASIC

Information Transfer Ask students to recall the game "Telephone" in which one person whispers a statement to his or her neighbor, who whispers the statement to his or her neighbor, and so on until the last person in the room has heard the statement. Discuss the similarities and differences between this game and nerve impulses as they pass through the body. (Similarities: a nerve impulse, like a whispered statement, passes from one cell [person] to the next in line. Differences: the whispered statement usually is garbled at the end of the line [the end of the game], but nerve impulses retain their integrity.) LS Interpersonal

Teach

READING SKILL BUILDER — BASIC

Paired Summarizing Pair students with a partner. Pair English language learners with proficient English speakers. Have each student read silently about resting potential and action potential on the following page. Then have one of the students summarize aloud what has been read without referring to the text. The partner should listen without interrupting and be prepared to point out any inaccuracies or omissions in the summary. At this point students can refer to the text. Have students switch roles and repeat the activity. LS Interpersonal

English Language Learners

Teach, *continued*

SKILL BUILDER — BASIC

Vocabulary The term *resting potential* is misleading because there is constant activity in a neuron even when the neuron is not conducting a nerve impulse. Ions are continuously being transported and diffusing across the membrane.

DATA LAB *Analyzing Changes During a Nerve Impulse*

Skills Acquired
Analyzing, interpreting, inferring, drawing conclusions

Teacher's Notes
Encourage students to study **Figure 3** on page 947 if they need help answering these questions.

Answers to Analysis

1. The action potential lasts about 1.5 milliseconds.
2. Voltage gated sodium channels are open at point A.
3. Voltage gated potassium channels are open at point B.
4. an influx of positively charged sodium ions
5. an outflow of positively charged potassium ions

Study TIP

Reading Effectively
The membrane potential of a cell is expressed in millivolts (mV). A millivolt is equal to one-thousandth of a volt (V). As shown in the graph in the Data Lab below, a neuron's membrane potential can be positive or negative. The resting potential of an average neuron is about –70 mV. During an action potential, the membrane potential of the neuron reaches about +40 mV.

Resting Potential

When a neuron is not conducting a nerve impulse, the neuron is said to be at rest. The membrane potential of a neuron at rest is called the **resting potential.** In a typical neuron, the resting potential is negative, about –70 millivolts (mV). At the resting potential, the inside of the cell is negatively charged with respect to the outside of the cell. Why is the resting potential negative? Recall that sodium-potassium pumps actively transport sodium ions, Na^+, out of a cell and potassium ions, K^+, into the cell. This results in a greater concentration of sodium ions outside the cell than inside the cell, and a greater concentration of potassium ions inside the cell than outside the cell. In a neuron, voltage-gated sodium channels are closed at the resting potential. Thus, very few sodium ions can diffuse into the cell, despite their strong concentration gradient. Some voltage-gated potassium channels are open at the resting potential. Potassium ions can therefore diffuse out of the cell down their concentration gradient, carrying their positive charge with them. Neurons also contain negatively charged proteins that are too large to exit the cell.

Action Potential

When a neuron is conducting a nerve impulse, changes occur in the cell membrane of the neuron. A nerve impulse is also called an action potential. An **action potential** is a local reversal of polarity—from a negative charge to a positive charge—inside the neuron. An action potential moves down an axon like a flame burning down a fuse. The events of an action potential are summarized in **Figure 3.**

DATA LAB *Analyzing Changes During a Nerve Impulse* SC.F.1.4.6

Background

The graph below illustrates changes that occur in the membrane potential of a neuron during an action potential. Use the graph to answer the following questions. Refer to Figure 3 as needed.

Analysis

1. **Determine** about how long an action potential lasts.
2. **State** whether voltage-gated sodium channels are open or closed at point *A*.
3. **State** whether voltage-gated potassium channels are open or closed at point *B*.
4. **Critical Thinking Recognizing Relationships** What causes the membrane potential to become less negative at point *A*?
5. **Critical Thinking Recognizing Relationships** What causes the membrane potential to become more negative at point *B*?

Action Potential

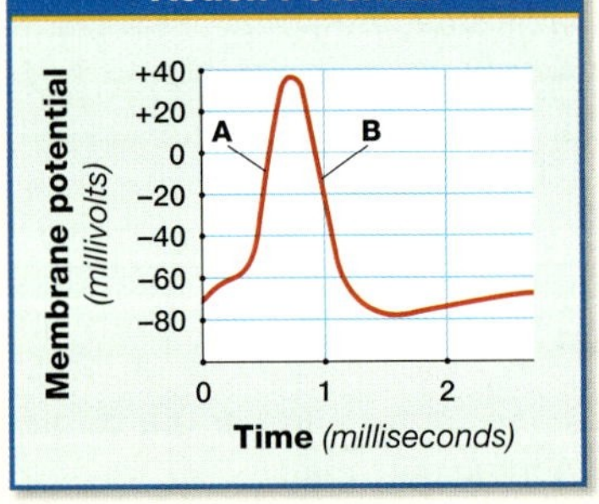

REAL WORLD CONNECTION

The cause of multiple sclerosis (MS) is unknown, but many scientists think that it results from the actions of the immune system against the myelin sheaths of nerves of the central nervous system. The myelin sheaths in the brain and spinal cord become sclerotic, or hard, causing poor impulse conductivity. Sometimes the symptoms of MS disappear altogether, but they may return after periods of remission. Current therapies for MS include corticosteroids and beta interferon.

Transparencies

TR J61 Conduction of a Nerve Impulse
TR J62 Synaptic Transmission

One-Stop Planner CD-ROM

- Reading Organizers BASIC
- Reading Strategies BASIC

Figure 3

BIO graphic: Conduction of a Nerve Impulse

An action potential moves rapidly down an axon.

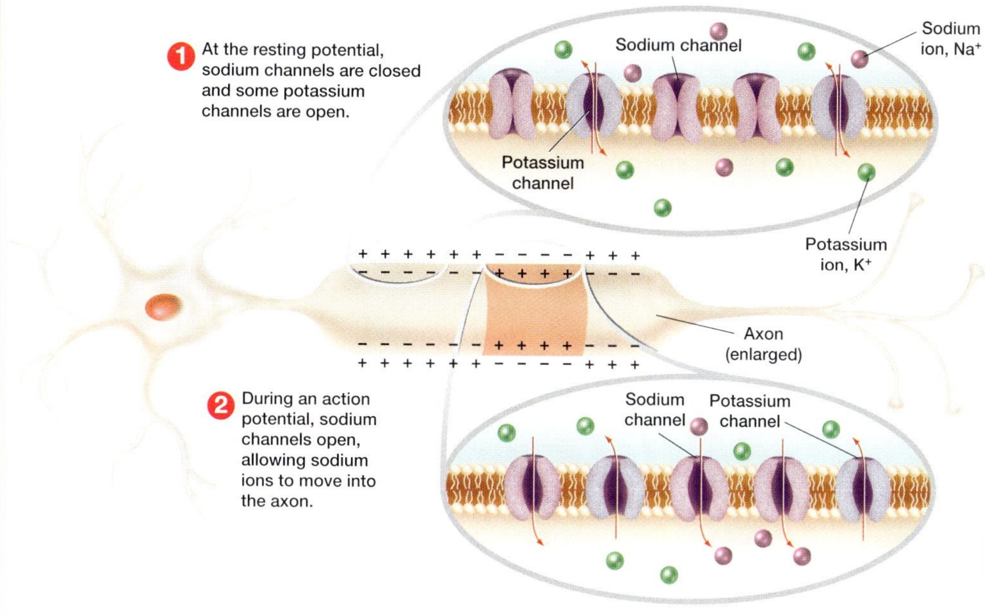

Step 1 At the resting potential, the inside of the neuron is negatively charged with respect to the outside of the neuron. The neuron is ready to conduct an action potential.

Step 2 An action potential begins when a stimulus, such as a signal molecule, causes a local change in the membrane potential to a more positive value. This change causes voltage-gated sodium channels to open, and sodium ions rapidly flow into the axon. For a brief moment, the membrane potential approaches about +40 mV as the inside of the axon becomes positively charged. This sudden local reversal of polarity begins a chain reaction that causes voltage-gated sodium channels to open down the entire length of the axon. As each sodium channel opens, sodium ions flow into the axon. The action potential conducts rapidly down the axon toward the axon terminals.

Voltage-gated sodium channels close immediately after the action potential has passed. Then additional voltage-gated potassium channels open, allowing potassium ions to flow out of the axon. As a result, the membrane potential becomes negative again immediately after the action potential. The resting potential is fully restored as sodium-potassium pumps reestablish the original concentrations of sodium ions and potassium ions inside and outside the axon. The neuron cannot conduct another action potential until that time.

947

Using the Figure

Figure 3 illustrates action potential conduction along an axon. The axon of this neuron has been enlarged to show that the membrane potential becomes positive during an action potential. Tell students that the reversal of polarity during an action potential is caused by the rapid influx of sodium ions through voltage-gated sodium channels. After an action potential has passed, voltage-gated potassium channels remain open, causing potassium ions to diffuse out of the cell. Thus, the membrane repolarizes shortly after the action potential passes. **LS Visual**

READING SKILL BUILDER — Basic

Interactive Reading Assign Chapter 41 of the *Holt Biology* Guided Audio CD Program to help students achieve greater success in reading the chapter.

SKILL BUILDER — General

Math Skills A single neuron may have synapses onto as many as 1,000 other neurons. Have students calculate how many neurons could be stimulated if each of these 1,000 neurons forms synapses with another 1,000 neurons. (one million neurons) Students should readily see how quickly and widely nerve impulses can travel. In reality, some of these synapses are excitatory, while others are inhibitory. Each neuron "evaluates" the impulses it receives, averages the input, and responds accordingly. For example, if the overall total input is excitatory, the receiving neuron initiates an impulse in a neighboring cell. **LS Logical**

MEDICINE CONNECTION

Dr. Emmeline Edwards, an African-American scientist who grew up in Haiti, studies the role that neurochemicals play in coping with stress. Information gained from her research can be applied toward the development of drugs for treating human depression. Dr. Edwards works at the Department of Psychiatry at the State University of New York, Stony Brook.

Attention Grabber

Squid Neurons The squid (*Loligo* sp.) was one of the first organisms used to measure the resting potential of a neuron. Because of its large, long axons that innervate muscles responsible for expelling water from the body cavity, the squid makes an ideal specimen for studies in cellular neurophysiology. Some of these axons are as large as 1 mm in diameter, making them easy to work with.

Communication Between Neurons

A junction at which a neuron meets another cell is called a **synapse** *(SIHN aps)*, shown in **Figure 4.** At synapses, neurons usually do not touch the cells they communicate with. Between an axon terminal and a receiving cell is a tiny gap called a synaptic cleft. At a synapse, the transmitting neuron is called a presynaptic neuron, and the receiving cell is called a postsynaptic cell.

When a nerve impulse arrives at an axon terminal of a presynaptic neuron, the impulse cannot cross the synaptic cleft. Instead, the impulse triggers the release of signal molecules called **neurotransmitters** into the synaptic cleft. Neurotransmitter molecules are produced by neurons and are stored inside vesicles. There are many different neurotransmitters and several mechanisms of neurotransmitter action. For example, in human muscles the principal neurotransmitter is a chemical called acetylcholine *(as ee tihl KOH leen)*. The brain utilizes several neurotransmitters such as glutamate *(GLOO tuh mayt)* and dopamine.

Figure 4 Synapse. A synapse is a junction at which signals are transmitted between a neuron and another cell.

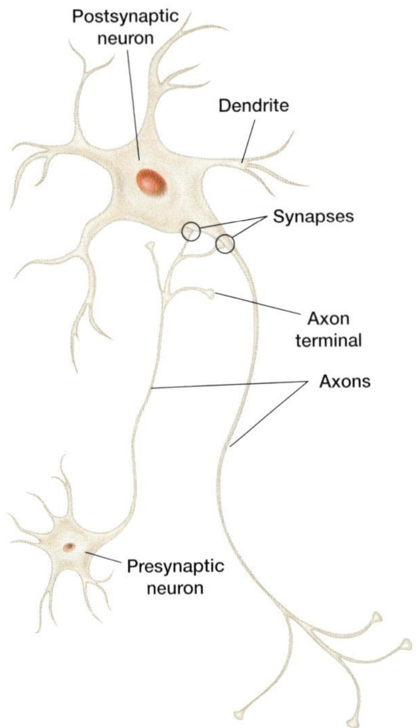

Release of Neurotransmitter

A nerve impulse causes a presynaptic neuron to release neurotransmitter molecules into the synaptic cleft. When an action potential reaches an axon terminal of the presynaptic neuron, vesicles that contain neurotransmitter molecules fuse with the cell membrane. This releases neurotransmitter molecules into the synaptic cleft by exocytosis. Neurotransmitter molecules diffuse across the synaptic cleft and interact with the postsynaptic cell. As shown in **Figure 5,** neurotransmitter molecules bind to receptor proteins on the postsynaptic cell. In some cells, ion channels open when a neurotransmitter binds to these receptor proteins. Such channels are called chemical-gated ion channels; whether these channels are open or closed depends on the binding of a chemical—in this case a neurotransmitter molecule.

A neurotransmitter may either excite or inhibit the activity of the postsynaptic cell it binds to. For example, when the neurotransmitter opens chemical-gated ion channels, ions move across the cell membrane of the postsynaptic cell. This causes the membrane potential of the postsynaptic cell to change depending on the charge of the ions that move into or out of the cell. If positively charged ions enter a postsynaptic neuron, an action potential may be produced (excitation). On the other hand, if positively charged ions flow out of the postsynaptic neuron, or if negatively charged ions enter the neuron, an action potential may be suppressed (inhibition).

948

Teach, *continued*

Using the Figure

Have students identify the dendrites and the cell body, which receive inputs, in **Figure 4.** Then have students identify the axon terminals, which allow the signal to pass to the next neuron in a circuit. Tell students that neurotransmitter molecules are released at the axon terminals, and they then travel to adjacent dendrites and cell bodies. An impulse can travel in only one direction—away from the cell body of the neuron and toward its axon terminals. **LS Visual** English Language Learners

Teaching Tip — GENERAL

Summarizing an Action Potential Have students summarize in writing the events involved in the conversion of an action potential into a chemical signal at the synapse. (Answers may vary but should include depolarization, movement of the action potential down the axon, release of neurotransmitter substance from the axon terminal, diffusion across the synaptic cleft, and binding to the post-synaptic neuron's membrane.) **LS Verbal**

Using the Figure

Figure 5 illustrates chemical synaptic transmission. Neurotransmitter molecules released from a presynaptic neuron bind to receptor proteins on a postsynaptic cell, opening ion channels in the postsynaptic cell. Emphasize to students that neurotransmitters can be either excitatory or inhibitory depending on the nature of the neuron they stimulate and the mechanism of their action. **LS Visual**

did you know?

Neurotransmitters and Emotional Health Extremely small amounts of neurotransmitters underlie all of our moods, from ecstasy to depression. The neurotransmitters norepinephrine and dopamine are involved in feelings of pleasure. If the levels of these neurotransmitters are too low in the brain, we might feel depressed. Endorphins and enkephalins are substances that block pain signals. Their production increases during physical stress, and runners who experience the increase refer to the sensation as a "runner's high." Another compound, substance P, is released in response to certain painful stimuli, such as ingestion of capsaicin, a chemical irritant in hot peppers. Our neurotransmitter levels react to emotional and physical stressors, such as our thoughts and environment. You literally can have an "attitude" that can affect your health.

Neurotransmitter molecules do not remain in the synaptic cleft indefinitely. Instead, most neurotransmitter molecules are cleared from the synaptic cleft very shortly after they are released. Many presynaptic neurons reabsorb neurotransmitter molecules and use them again. At other synapses, neurotransmitter molecules are broken down by enzymes or other chemicals. This happens, for example, at the synapses between neurons and skeletal muscle cells. The reuptake or breakdown of the neurotransmitter molecules ensures that their effect on postsynaptic cells is not prolonged.

Figure 5 **Synaptic transmission**

Neurotransmitter molecules are released from a presynaptic neuron, diffuse across the synaptic cleft, and interact with a postsynaptic cell.

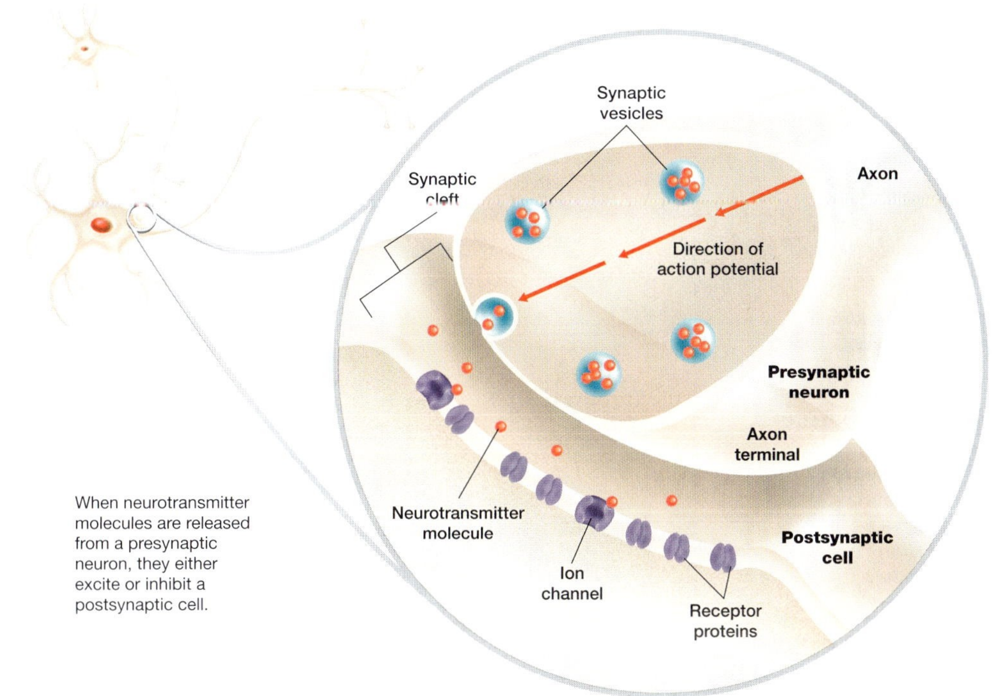

Section 1 Review

1. **Describe** the structure of a typical neuron. SC.F.1.4.2
2. **Describe** how the movement of ions across the cell membrane determines the membrane potential. SC.F.1.4.1
3. **Summarize** the events involved in the synaptic transmission of a nerve impulse. SC.F.1.4.6
4. **Critical Thinking** **Inferring Relationships** How does the membrane potential affect the permeability of a neuron's cell membrane? SC.F.1.4.3
5. **FCAT Prep** What is the junction at which a neuron communicates with another neuron or a muscle cell called?
   A nerve
   B synapse
   C myelin sheath
   D neurotransmitter
   SC.F.1.4.6

949

Close

Reteaching — BASIC

Have students use clay or another moldable material to build a model of a neuron and a synapse. Ask them to label the model, showing at what points the message is electrical and at what points the message is chemical. Have the students use arrows to show the direction of the impulse. **LS Visual**

Quiz — GENERAL

1. Compare action potential and resting potential. (At resting potential, sodium channels are closed; during action potential, sodium channels are open.)
2. What is membrane potential? (the difference in electrical charge across the cell membrane)
3. What must happen to the cell membrane of a neuron for an action potential to occur? (Its potential must become less negative.)

Alternative Assessment — GENERAL

Have each student write a question on one side of a 3×5 card for a trivia-type review game of this section. Have them write the answer to the question on the reverse side of the card. Collect the cards and divide the class into two teams. Ask the questions and award points to each team for correct responses.

Answers to Section Review

1. A typical neuron has a bulky cell body studded with dendrites, which are membrane-covered processes. One axon, which is a long, membrane-covered process, extends from the cell body to adjacent cells.
2. When ions move across the cell membrane, the relative concentration of ions changes inside and outside the cell. This results in a change of charge. The difference in electrical charge across a membrane is the membrane potential.
3. At the synaptic cleft, the presynaptic neuron releases a neurotransmitter substance. The neurotransmitter diffuses across the synaptic cleft and binds to the postsynaptic cell, which begins a depolarization wave in the next cell.
4. The membrane potential determines whether voltage-gated ion channels are opened or closed.
5. **A.** Incorrect. A nerve is a bundle of axons. **B.** Correct. The synapse is the junction at which two neurons communicate. **C.** Incorrect. The myelin sheath is the fatty outer layer found on many neurons. **D.** Incorrect. A neurotransmitter is a signal molecule that transmits nerve impulses across a synapse.

Section Review Correlations

SC.F.1.4.1: Item 2
SC.F.1.4.2: Item 1
SC.F.1.4.3: Item 4
SC.F.1.4.6: Items 3, 5

Focus

Overview

Before beginning this section review with your students the objectives listed in the Student Edition. This section explores the central nervous system, the peripheral nervous system, the autonomic nervous system, and the somatic nervous system. The major parts and functions of the brain are discussed, as well as the spinal cord and spinal reflex.

Bellringer

Ask students to make a list of reflex behaviors. (Examples include blinking, swallowing, kneejerk, recoiling from a hot surface.) Have students indicate which, if any, of these behaviors require conscious thought.

Motivate

Demonstration — Basic

Have a volunteer stand at the front of the room. While talking to the class, suddenly snap your fingers directly in front of the volunteer's face. He or she will blink reflexively when you snap your fingers. Ask students if a person can overcome a reflex such as blinking when something comes toward their eyes unexpectedly. (Answers will vary.) Try snapping again a few more times. Ask the student volunteer if he or she was able to control the blinking reflex. (Yes, if the student is expecting the noise, he or she will probably be able to consciously control their blinking response over time.) LS Kinesthetic

Section 2 Structures of the Nervous System

Florida Sunshine Standards/FCAT

SC.F.1.4.2 Body structures are uniquely designed and adapted for their function.

Objectives

- **Distinguish** between the central nervous system and the peripheral nervous system. SC.F.1.4.2
- **Identify** the major parts of the brain and their functions. SC.F.1.4.2
- **Describe** the structure of the spinal cord. SC.F.1.4.2
- **Sequence** the events of a spinal reflex. SC.F.1.4.2
- **Compare** the somatic nervous system with the autonomic nervous system. SC.F.1.4.2

Key Terms

central nervous system
peripheral nervous system
sensory neuron
motor neuron
brain
cerebrum
cerebellum
brain stem
thalamus
hypothalamus
spinal cord
reflex
interneuron

Central Nervous System

Neurons are the most important cells of the nervous system. The functions of the nervous system depend on the complex interaction between billions of neurons. Networks of neurons constantly gather, integrate, interpret, and respond to information about the body's internal state and environmental conditions. How are neurons organized in the nervous system? As shown in **Figure 6,** there are two main divisions of the nervous system—the central nervous system, shown in orange, and the peripheral nervous system, shown in purple. The **central nervous system** (CNS) consists of the brain and the spinal cord. The CNS is the control center of the body. The CNS interprets and responds to information from the environment and from within the body. The **peripheral nervous system** (PNS) contains sensory neurons and motor neurons. **Sensory neurons** send information from sense organs, such as the skin, to the CNS. **Motor neurons** send commands from the CNS to muscles and other organs.

Nervous System

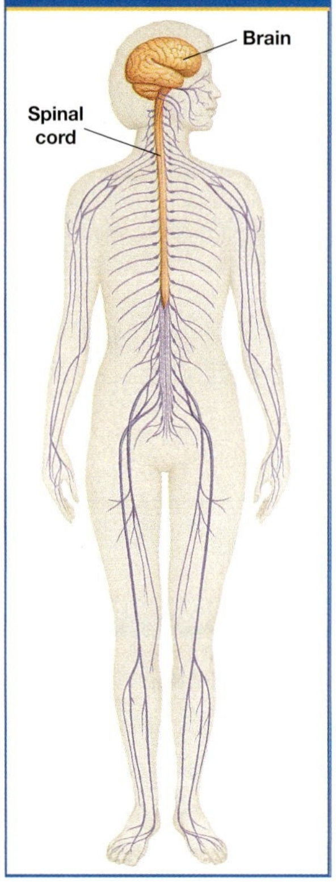

Figure 6 Nervous system. The central nervous system (orange) consists of the brain and the spinal cord. The peripheral nervous system (purple) branches throughout the body.

Brain

The **brain** is the body's main processing center. Encased entirely within the skull, the brain contains about 100 billion neurons. An average adult brain weighs about 1.5 kg (3 lb). Thoughts, feelings, emotions, behavior, perception, and memories are controlled by your brain. Your brain also enables you to learn and process information, such as the text in this book. Scientists have determined the location of various functions in the brain. The brain consists of three major parts, shown in **Figure 7**—the cerebrum, the cerebellum, and the brain stem.

950

Chapter Resource File

- Directed Reading BASIC
- Active Reading GENERAL

Transparencies

TR Bellringer

TR J64 Structure of the Human Brain

Career

Medical Imaging Technician Have students prepare a report on the job of a medical imaging technician. They should include descriptions of the different medical procedures these workers perform, such as computerized axial tomography (CAT), magnetic resonance imaging (MRI), and positron emission tomography (PET). They should also include a description of training necessary and starting salary. Have them write a brief report on what they found. Ask for volunteers to read their reports to the class.

Section Correlations

SC.F.1.4.2

Cerebrum The **cerebrum** *(seh REE bruhm)* is the largest part of the brain. The capacity for learning, memory, perception, and intellectual function resides in the cerebrum. The cerebrum has a folded outer layer with many bumps and grooves. A long, deep groove down the center divides the cerebrum into right and left halves, or hemispheres. The cerebral hemispheres communicate through a connecting band of axons called the corpus callosum *(KOR puhs kuh LOH suhm)*. In general, the left cerebral hemisphere receives sensations from and controls movements of the right side of the body. The right cerebral hemisphere receives sensations from and controls movements of the left side of the body.

Most sensory and motor processing occurs in the cerebral cortex *(KOHR teks)*, the folded, thin (2–4 mm) outer layer of the cerebrum. The cerebral cortex contains about 10 percent of the brain's neurons. The folded outer surface of the cerebrum is the cerebral cortex, which has a large surface area. The cerebral cortex is primarily involved with the functioning of sensory systems.

Cerebellum The **cerebellum** *(ser uh BEL uhm)*, which is located at the posterior base of the brain, regulates balance, posture, and movement. The cerebellum smooths and coordinates ongoing movements, such as walking, by timing the contraction of skeletal muscles. The cerebellum integrates and responds to information about body position from the cerebrum and the spinal cord to control balance and posture.

Brain Stem At the base of the brain is the stalklike **brain stem**. The brain stem is a collection of structures leading down to the spinal cord and connecting the cerebral hemispheres with the cerebellum. The lower brain stem consists of the midbrain, the pons, and the

Real Life

Each year, nearly 250 bicyclists die because of brain injuries.

Wearing a bicycle helmet reduces the risk of head trauma by more than 70 percent.

Evaluating Viewpoints
Should there be helmet laws for bicyclists just as there are seatbelt laws for automobile drivers?

Figure 7 **Brain.** The cerebrum is divided into two hemispheres. This view shows the right hemisphere.

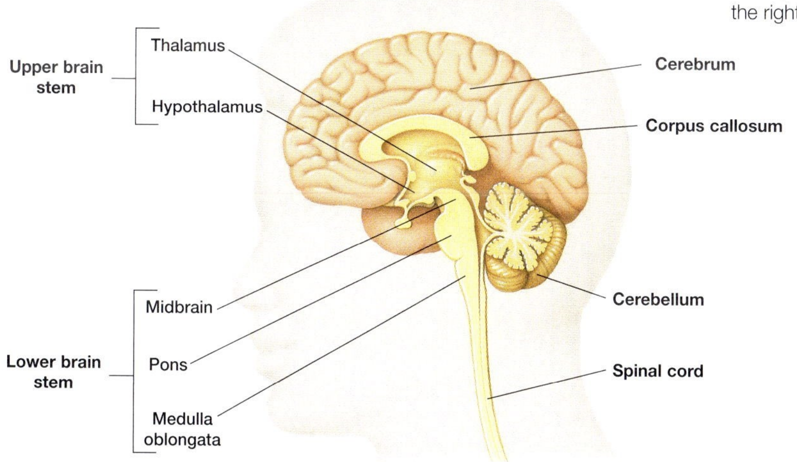

Graphic Organizer

Use this graphic organizer with **Teaching Tip** *on this page.*

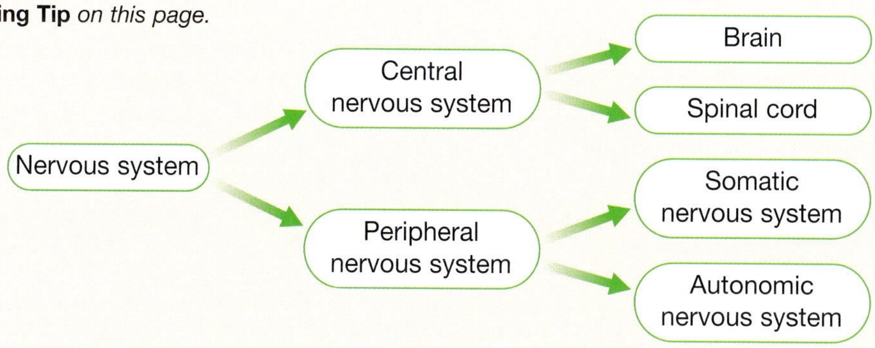

Teach

Teaching Tip

Nervous System As students read about the nervous system, have them construct a Graphic Organizer similar to the one at the bottom of this page that describes the main structures of the nervous system. Have them use the following terms: autonomic nervous system, brain, central nervous system, nervous system, peripheral nervous system, spinal cord, and somatic nervous system.

Real Life

Answer

Answers will vary. Proponents of helmet laws may cite the large cost, usually borne by the public, of caring for head-injury victims.

Teaching Tip

Meningitis Remind students that the brain is encased in the skull. In addition, tough protective membranes known as meninges cover the brain. A disease known as meningitis results when bacteria or viruses cause inflammation of the meninges. Meningitis can be fatal if not treated quickly.

Using the Figure

Have students review the parts of the brain as labeled in **Figure 7.** Point out that the brain is divided into a right and a left half, which are connected by the corpus callosum. Ask students if they think they are right-brained or left-brained. Although in most people, both parts of the brain work together, some people rely more on one side or the other. People who rely more on their right brain tend to be more imaginative and intuitive, whereas people who rely more on their left brain tend to be more logical and analytical. **LS Visual** English Language Learners

Teach, *continued*

READING SKILL BUILDER — BASIC

Discussion Ask students if they think there are structural or physical differences in the brains of people who are considered to be geniuses. Ask students if they think Albert Einstein's brain, for example, might show differences that indicate his genius. Dr. Einstein was a brilliant scientist who won a Nobel Prize in physics in 1921. He permitted his brain to be preserved after his death for study. Scientists at McMaster University, Ontario, Canada compared the size and shape of his brain to those of 91 men and women of average intelligence and found that Einstein's brain was 15% wider in one region. Other studies have been done to determine if his brain was structurally different from the brains of average human brains. To date, no conclusive findings have been made to relate any differences in brain structure to intelligence.

Demonstration — GENERAL

Have students work in pairs to demonstrate reaction time. Have one student of each pair hold a ruler vertically. Have the other student position his or her fingers 1 inch below the ruler, in a pinching mode. (This works best if the forearm rests on a table.) Have the first student drop the ruler without warning. The other student should try to catch it between the thumb and index finger. The shorter the segment allowed to fall through the fingers, the quicker the reaction time. **LS Kinesthetic**

Study TIP

Interpreting Graphics
As you look at Figure 8, think of the spinal cord as a busy two-way highway with sensory traffic going north and motor traffic going south.

medulla oblongata *(mi DUHL uh ahb lahn GAHT uh)*. These structures relay information throughout the CNS and play an important role in homeostasis by regulating vital functions, such as heart rate, breathing rate, body temperature, and sleep.

The upper brain stem contains important relay centers that direct information to and from different parts of the brain. The **thalamus** *(THAL uh muhs)* is a critical site for sensory processing. Sensory information from all parts of the body converges on the thalamus, which relays the information to appropriate areas of the cerebral cortex. Below the thalamus, at the base of the brain, is the hypothalamus. The **hypothalamus,** along with the medulla oblongata, helps regulate many vital homeostatic functions, such as breathing and heart rate. The hypothalamus is responsible for feelings of hunger and thirst. It also regulates many functions of the endocrine system by controlling the secretion of many hormones.

The thalamus and hypothalamus are linked to some areas of the cerebral cortex by an extensive network of neurons called the limbic system. The limbic system includes structures of both the brain stem and the cerebrum. The limbic system has an important role in memory, learning, and various emotions, such as pleasure and anger.

Spinal Cord

The **spinal cord,** shown in **Figure 8,** is a dense cable of nervous tissue that runs through the vertebral column. The spinal cord extends from the medulla oblongata through the vertebrae to a level just below the ribs. The spinal cord links the brain to the PNS. The brain receives information that travels upward through the spinal cord. Through the spinal cord, the brain also sends commands that control the rest of the body. In addition to relaying messages, the spinal cord functions in reflexes. A **reflex** is a sudden, involuntary contraction of muscles in response to a stimulus.

Figure 8 Spinal cord
Spinal nerves have a dorsal root and a ventral root that diverge as they enter the spinal cord.

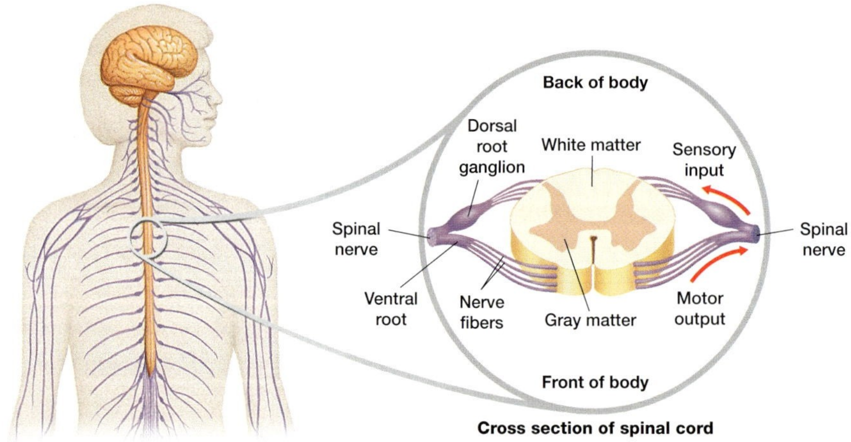

952

REAL WORLD CONNECTION

Medical scientists think that some psychosomatic illnesses may be mediated by the hypothalamus through its connections to the autonomic nervous system and the endocrine system. Many stress-related illnesses, such as ulcers, asthma, and high blood pressure, are clearly connected to hypothalamic functions.

MISCONCEPTION ALERT

Brain Size Students may believe that modern *Homo sapiens* have the largest brains of any hominid in history. Remind students that Neanderthals had larger brains than those of some modern humans. The brains of modern humans are much more convoluted than those of Neanderthals. This suggests that intelligence may be related more to the extent of the convolutions (which is related to the area of the cerebral cortex) than to overall brain size.

The spinal cord is linked to the PNS through 31 pairs of spinal nerves. The spinal nerves, which branch from the spinal cord, carry information to and from the CNS. Spinal nerves in the upper part of the spinal cord branch into the arms and upper body, and spinal nerves in the lower part of the spinal cord branch into the legs and lower body. Each spinal nerve has a dorsal root and a ventral root. Dorsal roots contain sensory neurons, which carry information from areas of sensory input to the CNS. Ventral roots contain motor neurons, which carry motor responses from the CNS to muscles, glands, and other organs. As shown in Figure 8, dorsal and ventral roots come together to form the spinal nerves near the spinal cord.

The spinal cord contains a core of gray matter covered by a sheath of white matter, as shown in Figure 8. Gray matter contains the cell bodies of neurons, whereas white matter contains the axons of neurons. Included in the gray matter are **interneurons,** which link neurons to each other.

BIOWatch

Spinal Cord Injury

Unlike most other parts of the body, the spinal cord does not heal after an injury. Damaged neurons stop conducting nerve impulses at the site of injury, permanently paralyzing the legs or, in some cases, all four limbs.

Every year, spinal cord injuries—whether incurred in athletics or automobile accidents—leave nearly 15,000 Americans partially or totally paralyzed. In 1995, actor Christopher Reeve, shown in the photo at right, injured his spinal cord after falling headfirst from a horse. The fall broke vertebrae in Reeve's neck, paralyzing him from the neck down.

A treatment currently available for people with spinal cord injuries is an anti-inflammatory drug called methylprednisolone. If given within 8 hours after the spinal cord is injured, the drug can improve chances of recovery. Even with this drug, however, recovery is usually far from complete.

Stopping Cell Death

Cells continue to die near the site of a spinal cord injury for several weeks after injury occurs. Myelin-producing cells die, leaving neurons in the spinal cord unable to function. Some scientists think that stopping the death of these cells could help avoid paralysis. In experiments on rats, researchers have found that a cell-death inhibitor improves the rats' ability to use their hind legs after a spinal cord injury. Researchers are investigating other cell-death inhibitors that could be used on humans.

Bridging the Gap

After the spinal cord is injured, damaged axons begin to regrow. However, their growth is inhibited by substances in the spinal cord. Peripheral nerves lack these substances, so the axons in these nerves can regrow quite well. To stimulate the growth of axons in the injured spinal cord, researchers have grafted pieces of peripheral nerves into the spinal cord. The nerve grafts provide tunnels for regrowing axons. Rats with such nerve grafts begin to show signs of recovery within 3 weeks. Within a year, they can support their own weight. Similar grafts have not yet been tried on humans.

internet connect
www.scilinks.org
Topic: Spinal Cord
Keyword: HX4168
SCLINKS. Maintained by the National Science Teachers Association

Teaching Tip — ADVANCED

Encourage interested students to use the Internet Connect box on this page to find out more about current research into the cure for spinal cord injury. Have students give a brief oral report to the class presenting what they find.
LS Verbal

BIOWatch

Spinal Cord Injury

Teaching Strategies

Bring in pictures of Christopher Reeve and other people who have paralysis as a result of spinal cord injuries. If the information is available, talk about how they were injured. Were any of the injuries preventable?

Discussion

- Why does paralysis occur? (Nerve impulses cannot cross the site of injury in the spinal cord.)
- Is it possible for a person to have arms but not legs paralyzed as the result of a spinal cord injury? Explain your answer. (No, because everything below the spinal cord injury is affected. If the injury is above the spinal cord area that controls the arms, all four limbs are affected.)
- Would you volunteer to be a test case for the methods described if you had a spinal cord injury? Why or why not? (Answers will vary.)

Transparencies

TR J65 Structure of the Human Spinal Cord
TR J67 Knee-Jerk Response
TR J68 Physiological Effects of the Autonomic Nervous System

MISCONCEPTION ALERT

Brain Tumors Because most neurons do not undergo mitosis in adult humans, students may think that a brain cannot develop a tumor. Brain tumors, however, are not formed from neurons. They arise from the supportive glial cells in the nervous system, which do undergo mitosis.

Teach, continued

READING SKILL BUILDER — BASIC

Paired Summarizing Have students choose a partner. Have both partners read silently about the peripheral nervous system on this page and the facing page. Then have one student summarize aloud what has been read without referring to the textbook. The partner should listen without interrupting and be prepared to point out any inaccuracies or omissions in the summary. Students may refer to the text to verify the facts. Then have students switch roles.
LS Interpersonal

INCLUSION Strategies

- *Attention Deficit Disorder*
- *Learning Disability*

Have students work with a partner to make observations of how the body keeps a sense of balance. With the partner standing perfectly still, feet together, and his or her eyes closed for one minute, have the student observe from the side to see how much the subject wavers back and forth. Next, observe the subject from behind to see if there is any wavering from side to side. Write these observations down. Repeat the test after switching roles. Students may report to the class their findings.

Study TIP

Interpreting Graphics
As you look at Figure 9, notice that in a spinal reflex, motor neurons stimulate muscles in the same region in which the stimulus that caused the reflex originated.

Peripheral Nervous System

The peripheral nervous system connects the brain and the spinal cord to the rest of the body. In addition to the 31 pairs of spinal nerves, 12 pairs of cranial nerves connect the brain with areas in the head and neck. The PNS contains two principal divisions—the sensory division and the motor division. The sensory division directs sensory information to the central nervous system. The motor division carries out responses to sensory information. The motor division of the PNS consists of two independent systems—the somatic nervous system and the autonomic nervous system.

Somatic Nervous System

Most motor neurons that stimulate skeletal muscles are under our conscious control. These neurons are part of the somatic nervous system. Some activity in the somatic nervous system, such as spinal reflexes, is involuntary. A spinal reflex is a self-protective motor response. Spinal reflexes are extremely rapid because they usually involve the spinal cord but do not involve the brain.

The knee-jerk reflex, shown in **Figure 9,** is an example of a spinal reflex. When the ligament below your kneecap is tapped, your lower leg suddenly kicks forward. Tapping the ligament stimulates a sensory neuron, shown in red. The sensory neuron sends a nerve impulse to the spinal cord and excites a motor neuron, shown in green, which causes the quadriceps to contract. This causes the leg to extend rapidly. The sensory neuron also stimulates an interneuron, shown in blue. The interneuron inhibits a motor neuron that would normally cause the hamstrings to contract, allowing the hamstrings to relax.

Figure 9 Knee-jerk reflex

When the ligament below the patella is tapped, the quadriceps contracts, the hamstrings relax, and the leg rapidly extends.

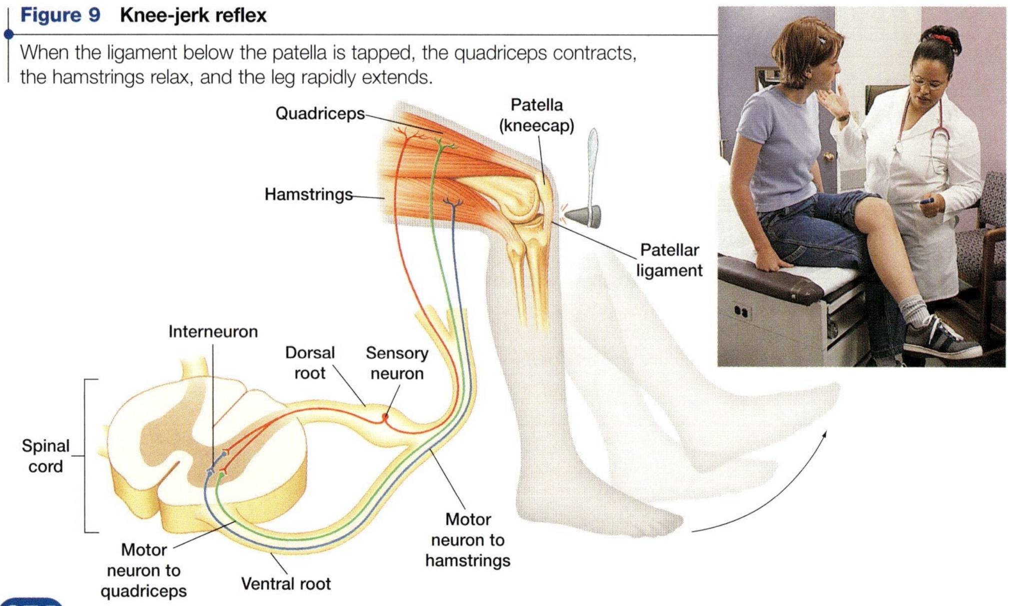

954

Attention Grabber

Referred Pain Referred pain is pain that originates in a location different from where it is felt. Referred pain is important in clinical diagnoses. For example, inadequate oxygenation of the heart muscle often results in pain being referred to the chest wall and shoulder. Inflammation of the appendix, in the lower right quadrant of the abdomen, is sometimes felt as pain in the navel area.

Integrating Physics and Chemistry

Relate membrane, resting, and action potentials to **potential difference.** Potential difference is the change in the electrical potential energy of a charged particle divided by its mass and is measured in volts (V). In addition, remind students that electrical potential energy is increased if a force pushes electrical charges in the direction opposite the electric force. Check their understanding by asking what biological mechanism pushes electrical charges opposite the electric force. (the sodium-potassium pumps.)

Autonomic Nervous System

Peripheral motor neurons that regulate smooth muscles do not require our conscious control. These neurons are part of the autonomic nervous system, which regulates heart rate and blood flow by controlling contractions of cardiac muscle in the heart and smooth muscle lining the walls of blood vessels. It also controls muscles in the digestive, urinary, respiratory, and reproductive systems, as well as the secretions of many glands.

Two divisions of the autonomic nervous system—the parasympathetic division and the sympathetic division—maintain stability in the body by counterbalancing each other's effects. The parasympathetic division is most active under normal conditions. It keeps your body functioning even when you are not active. For example, you continue to breathe when you fall asleep.

The sympathetic division dominates in times of physical or emotional stress. It controls the "fight-or-flight" response that you experience during a stressful situation, such as "nervousness" when taking a pop quiz. The sympathetic division increases blood pressure, heart rate, and breathing rate. It also directs blood flow toward your heart and skeletal muscles. Effects of the autonomic nervous system are summarized in **Table 1.**

internet connect
www.scilinks.org
Topic: Autonomic Nervous System
Keyword: HX4017
SCILINKS. Maintained by the National Science Teachers Association

Table 1 Physiological Effects of the Autonomic Nervous System

Organ	Effect of sympathetic division	Effect of parasympathetic division
Eyes	Pupils dilate	Pupils constrict
Heart	Heart rate increases	Heart rate decreases
Lungs	Bronchioles dilate	Bronchioles constrict
Intestines	Gastric secretions decrease	Gastric secretions increase
Blood vessels	Blood vessels dilate	Little or none

Section 2 Review

1. **Name** the two main divisions of the nervous system, and state their general functions. SC.F.1.4.2
2. **Compare** the functions of the cerebellum and the brain stem. SC.F.1.4.2
3. **Distinguish** between dorsal roots and ventral roots of the spinal cord. SC.F.1.4.2
4. **Name** the division of the autonomic nervous system that is more active under normal conditions. SC.F.1.4.2
5. **Critical Thinking** **Comparing Functions** Why is a spinal reflex more rapid than a voluntary movement? SC.F.1.4.6
6. *FCAT Prep* When a sudden stretch of the quadriceps muscle triggers the knee-jerk reflex, how is homeostasis maintained?
   A The patella elongates.
   B The hamstrings contract.
   C The quadriceps elongate.
   D The quadriceps contract.
   SC.F.1.4.7

Close

Reteaching — Basic

On the chalkboard or overhead projector, list the major concepts presented in this section. Have the class review the text to identify examples that can be used to illustrate each of these concepts.

Quiz — General

1. What might be the result from damage to the cerebellum? (loss of balance)
2. What controls the "fight or flight" response? (sympathetic division of the autonomic nervous system)
3. What do the hypothalamus and the medulla oblongata regulate? (vital homeostatic functions)

Alternative Assessment — General

Have pairs of students take turns lightly tapping each other's leg slightly below the knee. If possible, have them use a rubber-headed hammer borrowed from the school nurse. Ask students to each write a brief summary of their observations and the cause of the effect produced. Tell them that their summary must include the following terms: reflex, axon, synapse, motor neuron, sensory neuron, spinal cord.

Answers to Section Review

1. The two main divisions of the nervous system are the central nervous system and the peripheral nervous system. The central nervous system, consisting of the brain and spinal cord, is the control center of the body. The peripheral nervous system connects the brain and spinal cord to the rest of the body.
2. The cerebellum regulates balance and the timing of movement, while the brain stem regulates vital functions such as breathing, heart rate, body temperature, and sleep.
3. Dorsal roots contain sensory neurons; ventral roots contain motor neurons.
4. The parasympathetic division is more active.
5. A spinal reflex travels only as far as the spinal cord, whereas a voluntary movement involves the brain and therefore takes longer than a spinal reflex.
6. **A.** Incorrect. The patella does not change. **B.** Incorrect. The hamstrings will relax if the quadriceps contract. **C.** Incorrect. The quadriceps will elongate if the hamstrings contract. **D.** Correct. For the knee to jerk, the quadriceps must contract.

Section Review Correlations

SC.F.1.4.2: Items 1-4
SC.F.1.4.6: Item 5
SC.F.1.4.7: Item 6

Focus

Overview

Before beginning this section review with your students the objectives listed in the Student Edition. This section explores the senses. The structures and mechanisms of touch, vision, hearing, taste, and smell are described.

Bellringer

Ask students each to select an object in the room. Ask them to point to a specific place on the object with their index finger. Direct them to focus on that place, first with both eyes open, then with just the left eye closed, and finally with just the right eye closed. Have students write down what happens to the image and offer an explanation. (Each eye registers an image that is slightly different from the other. When both eyes are used, the brain combines the images to create a new image.)

Motivate

Activity — GENERAL

Phosphenes Have students close their eyes and press lightly on their eyelids. Ask what they see. (patterns of colors) These patterns are called phosphenes. The nerves in the eyes send messages to the brain, which interprets these messages as colors. Then ask what this phenomenon tells them about the sense of sight. (The eyes respond to visual stimuli. The brain is interpreting pressure stimuli as visual stimuli.) **LS Visual**

Section Correlations

SC.F.1.4.2
SC.F.1.4.7

Section 3 Sensory Systems

Florida Sunshine Standards/FCAT

SC.F.1.4.2 Body structures are uniquely designed and adapted for their function.

SC.F.1.4.7 Organisms respond to internal and external stimuli.

Objectives

- **List** five types of sensory receptors and the stimuli to which they respond. SC.F.1.4.7
- **Identify** sites of sensory processing in the brain. SC.F.1.4.2
- **Analyze** the structure of the eye and its role in the visual system. SC.F.1.4.2
- **Describe** how the ear detects sound and helps maintain balance. SC.F.1.4.7
- **Compare** the senses of taste and smell. SC.F.1.4.7

Key Terms

sensory receptor
retina
rod
cone
optic nerve
cochlea
semicircular canal

Perception of Stimuli

The perception of everything you respond to in the environment, such as the horn of a passing car or cold rain on your face, is made possible by sensory systems. Sensory systems are essential to survival, and they enable us to experience both pleasurable and painful stimuli. Sensory systems help maintain homeostasis by constantly adjusting body conditions to respond to changes in the environment. This requires the integration of the peripheral nervous system and the central nervous system. The sensory division of the PNS collects information about sensory stimuli in and around the body. The sensory information is sent to the brain, which processes the information and, if necessary, generates a motor response to the stimuli.

How does the nervous system detect sensory stimuli? Specialized neurons called **sensory receptors** detect sensory stimuli and then convert the stimuli to electrical signals, in the form of nerve impulses, that can be interpreted by the brain. Although sensory receptors are located throughout the body, they are most concentrated in the sense organs—the eyes, ears, nose, mouth, and skin. **Table 2** lists several types of sensory receptors and some of their locations.

Sensory Receptors

Mechanoreceptors throughout the body respond to physical stimuli—such as pressure and tension—that cause distortion or bending of tissue. These stimuli alter the electrical activity of mechanoreceptors. Many mechanoreceptors are found in the skin, and they are concentrated in very sensitive areas, including the face, hands, fingertips, and neck. Pain receptors, which respond to potentially harmful stimuli—such as intense heat or cold and tissue damage—are responsible for painful sensations. Pain is a very important sensation

internet connect
www.scilinks.org
Topic: Sensory Receptors
Keyword: HX4162
SCILINKS. Maintained by the National Science Teachers Association

Table 2 Types of Sensory Receptors

Receptor type	Stimuli	Locations
Thermoreceptors	Temperature change	Skin, hypothalamus
Pain receptors	Tissue damage	All tissues and organs except the brain
Mechanoreceptors	Movement, pressure, tension	Skin, ears, muscles
Photoreceptors	Light	Eyes
Chemoreceptors	Chemical	Tongue, nose

Chapter Resource File

- Directed Reading BASIC
- Active Reading GENERAL
- Data Sheet for Quick Lab GENERAL

Transparencies

TR Bellringer
TR J69 Types of Sensory Receptors
TR J63 Cerebral Cortex

Integrating Physics and Chemistry

Remind students that electric circuits—like nerve pathways—usually have a switch, which opens and closes the path in order to stop or permit the flow of charges through the electric current. Whereas the switch is often a mechanical device in an electric circuit, it is the synapse for nerve pathways.

Figure 10 **Processing sites and lobes of the cerebral cortex**

Specific areas of the cerebral cortex control different functions of the body.

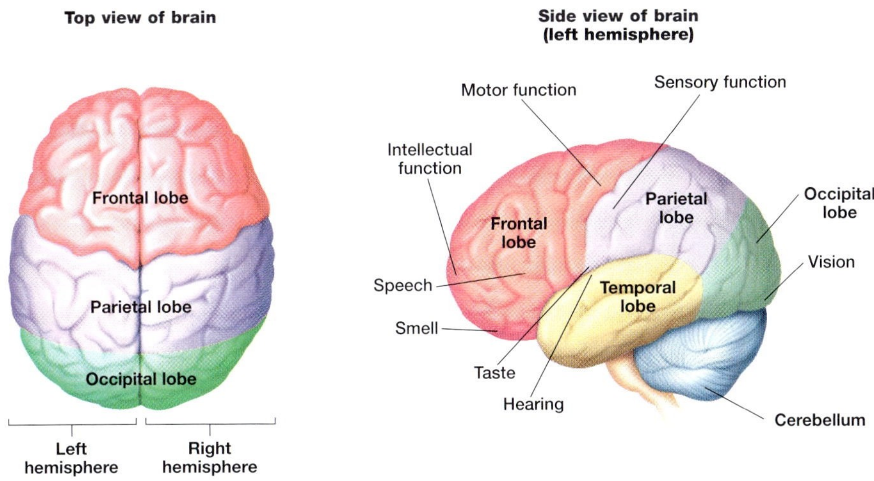

because it informs you that something is wrong in your body. Many self-protective responses, such as reflexes, are initiated by pain receptors. *Thermoreceptors*, located in the skin and hypothalamus, detect changes in temperature. Thermoreceptors play an important role in homeostasis, helping to keep the body temperature within its normal range.

Sensory receptors are located throughout the body, and sensory input from these receptors enters the central nervous system in an organized fashion. Sensory stimuli that originate in the lower body enter the lower part of the spinal cord. Sensory stimuli that originate in the upper body enter the upper part of the spinal cord and the brain.

Processing of Sensory Information

Recall that the cerebral cortex contains a large percentage of the brain's neurons. Many of the neurons in the cerebral cortex are responsible for processing incoming sensory information from the sense organs. The thalamus relays information from the sense organs to certain regions of the cerebral cortex. As shown in **Figure 10,** deep grooves divide the cerebral hemispheres into four general areas, or lobes: the occipital *(ahk SIP ih tuhl)* lobe, the parietal *(puh RIE uh tuhl)* lobe, the temporal lobe, and the frontal lobe. Sensory neurons from the different sense organs come together at certain regions in the cerebral cortex. For example, most visual processing takes place in the occipital lobe, located at the back of the head. Similarly, processing of sound is carried out within the temporal lobe.

957

Attention Grabber

Phineas Gage On September 13, 1848, an accidental explosion on a Vermont railroad caused an extremely unusual injury. A railway foreman named Phineas Gage was setting explosive charges with a tamping iron, a heavy iron rod. One of the charges exploded, sending the tamping iron through the left side of his skull. It entered under his left cheekbone, exited through the top of his head, and landed 25 yards away. Most of the left frontal lobe of his brain was destroyed, but it is said that Gage never even lost consciousness. After seven months of recuperation, Gage was well enough to return to work. His physical and intellectual functions were unchanged, but his personality was completely different. Once polite and patient, he became profane, impatient, and rude, and was unable to resume his work as a foreman. His horrific injury, however, greatly influenced physicians and scientists who were attempting to determine how function is localized in the brain.

Teach

Teaching Tip — Basic

Touch Receptors Tell students that although touch receptors are spread throughout the skin, some areas have a larger concentration of touch receptors than others. For example, lips and hands have large numbers of touch receptors situated close together. It is easy to distinguish two points at very close distances on skin covering the lips and hands. However, on the back there are areas where touch receptors are several millimeters apart. Tell students that they may not be able to distinguish two different points that are touched on their backs because of this.

Activity — General

Visual Feedback and Touch

Have students work in pairs. One student should hold his or her arms out and crossed so that the palms meet, and the fingers are interlocked. Have students bend their elbows and move the hands down and in toward the body. Keep the elbows bent, and bring the hands close to the body, twisting the hands upward. Have the partner point to a finger without touching it. The student should try to move that finger quickly. Ask students to share what happens. (The student will have difficulty identifying and moving the finger indicated by his or her partner. This is because the eyes confirm body positions that are monitored by positional receptors, called proprioceptors. If the hands are in an unusual position, proprioceptors and the eyes send conflicting messages to the brain.) Ask students if it is easier to identify the finger if the partner first touches that finger. (Yes, because touch receptors give additional information to the brain.) **LS Kinesthetic**

One-Stop Planner CD-ROM

- Reading Organizers — Basic
- Reading Strategies — Basic

Teach, *continued*

Demonstration — GENERAL

You will need a cardboard tube from a roll of paper towels for this demonstration. Tell a student volunteer to keep both eyes open, and to place the tube over his or her left eye. Have the student hold his or her right hand next to the end of the tube, and move it slowly along the tube toward his or her face. Ask the student what he or she observes. (It should appear as though the tube passes through the palm of the hand. The brain automatically combines the images seen with both eyes into one picture.) **LS Visual**

QUICK LAB *Demonstrating Your Blind Spot*

Skills Acquired
Summarizing, drawing conclusions, applying information

Teacher's Notes
Make sure students are closing their right eyes and are looking at the O. It should disappear about 1 ft from the end of the nose. Try the lab with the X and the left eye as well.

Answers to Analysis

1. rods and cones
2. There are no photoreceptors (rods and cones) at that spot.
3. The X disappears when it is in the part of the visual field mediated by the part of the retina where there are no photoreceptors.

Eyes

Humans have very good eyesight. Our eyes enable us to see in color and to distinguish fine details and movement. The structure of the eye is shown in **Figure 11.** Light enters the eye through the pupil. Light then passes through the lens, a thick, transparent disk that focuses light on the retina. The **retina** is a lining on the back inner surface of the eye that consists of photoreceptors and neurons. The retina contains two types of photoreceptors—rods and cones—which convert light energy to electrical signals that can be interpreted by the brain. **Rods** respond best to dim light. **Cones** respond best to bright light and enable color vision. The retina also contains many other neurons that process visual information. The axons of some of these neurons make up the **optic nerve.** The optic nerve exits through the back of the eye and runs along the base of the brain to the thalamus. The thalamus then relays visual information to the occipital lobe of the cerebral cortex, where the information is processed.

Figure 11 Structure of eye. Light enters the eye through the pupil and is focused on the retina, which contains photoreceptors.

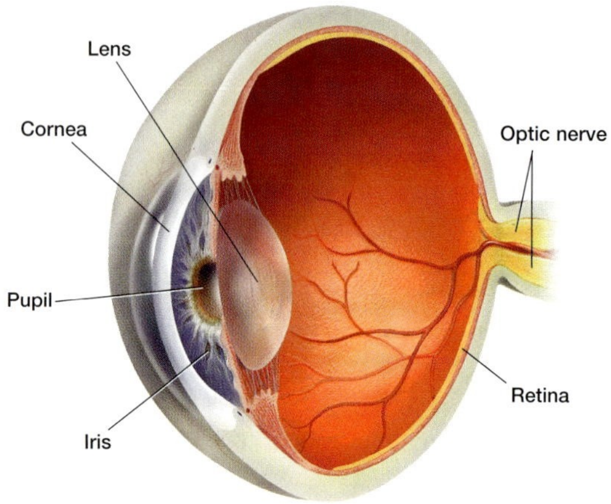

QUICK LAB *Demonstrating Your Blind Spot* SC.F.1.4.7

The blind spot in your visual field corresponds to the site where the optic nerve exits the back of the eye. There are no photoreceptors at the site where the optic nerve exits. Use the procedure below to demonstrate your blind spot.

Materials

unlined 3 × 5 index card, pencil

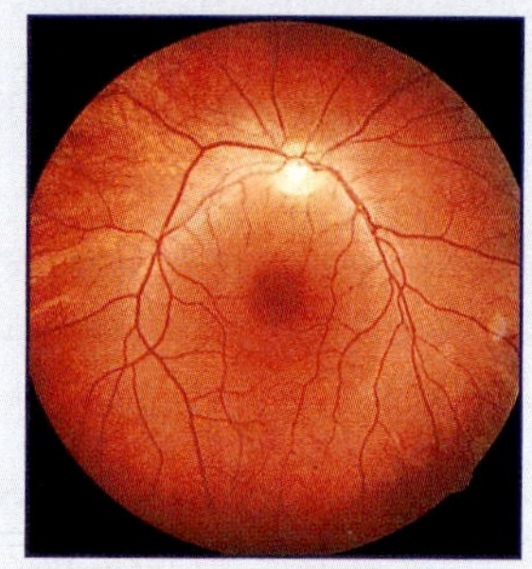

Retina

Procedure

1. On the index card, draw an *X* about 1 in. from the left side of the card. Draw an *O* about the same size 3 in. to the right of the *X*.
2. Hold your index card in front of you at arm's length. Close your right eye and stare at the *O* with your left eye. Slowly move the card toward you while continuing to stare at the *O* until the *X* disappears from view.

Analysis

1. **Name** the two kinds of photoreceptors found in the retina.
2. **Propose** why you cannot see images that fall on the site where the optic nerve exits the eye.
3. **Critical Thinking Relating Concepts** What is the relationship between the structure of the retina and the disappearance of the *X* on the index card?

958

did you know?

Glaucoma Glaucoma is an eye disease in which increased pressure inside the eye causes damage to the retina and the optic nerve. The cause of the increased pressure is a buildup of the aqueous humor, the fluid interior of the eye. Glaucoma is the second most common cause of blindness.

Transparencies

TR J73 Structure of the Eye
TR J70 Structure of the Ear

Attention Grabber

Tell students that if they stare at a picture long enough and then look at a white wall, they will see the same image shapes, but with different colors. Give students pieces of black construction paper with large blue spots in the middle. After they have stared at the black paper for a minute, have them look at a white wall. Ask students what color spot they see on the white wall. (Students should see a red spot.)

Ears

How do your ears enable you to hear? Your ears convert the energy in sound waves to electrical signals that can be interpreted by your brain. **Figure 12** shows the structure of the ear. Sound waves enter the ear through the ear canal and strike the tympanic *(tim PAN ik)* membrane, or eardrum, causing it to vibrate. Behind the eardrum, three small bones of the middle ear—the hammer, anvil, and stirrup—transfer the vibrations to a fluid-filled chamber within the inner ear. This chamber, called the **cochlea** *(KAHK lee uh),* is coiled like a snail's shell, and it contains mechanoreceptors called hair cells. Hair cells rest on a membrane that vibrates when waves enter the cochlea. Waves of different frequencies cause different parts of the membrane to vibrate and thus stimulate different hair cells. When hair cells are stimulated, they generate nerve impulses in the auditory nerve. The impulses travel to the brain stem through the auditory nerve. The thalamus then relays the information to the temporal lobe of the cerebral cortex, where the auditory information is processed.

WORD Origins

The word *cochlea* is from the Greek *kochlias,* meaning "snail shell."

Keeping Your Balance

The ears not only enable you to hear but also help you maintain equilibrium. The **semicircular canals** are fluid-filled chambers in the inner ear that contain hair cells. Clusters of these hair cells respond to changes in head position with respect to gravity. When your head moves, the hair cells are stimulated according to the magnitude and direction of the fluid's movement, and they send electrical signals to the brain. Signals generated by the hair cells enable the brain to determine the orientation and position of the head.

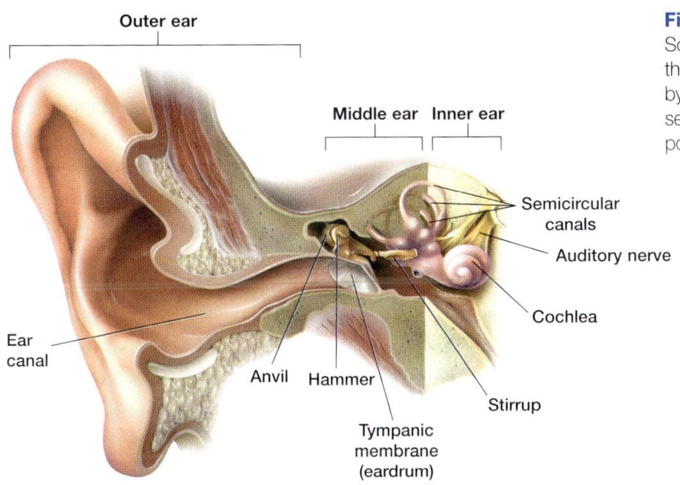

Figure 12 Structure of ear. Sound waves are transmitted to the inner ear and are detected by mechanoreceptors. The semicircular canals detect the position of the head.

Demonstration

Show the class a chart used to test people for colorblindness. Determine if any students are colorblind. Point out that colorblindness is caused by a chemical disorder in the cones of the eye. Complete colorblindness is extremely rare. A deficiency in the red or green cones is most common, affecting about 5 percent of the population. **LS Visual**

Teaching Tip

Measuring Sound Play recorded music at a volume high enough to be disconcerting to an average listener. Tell students that the loudness of sound is measured in decibels, ranging from a sound of 0 decibels (dB), which is just audible, to a sound of 140 dB, which is painful. A jet taking off can produce a sound of 150 dB and cause hearing loss. This is why people who work on the tarmac at airports wear protective earplugs and headsets.

INCLUSION Strategies

- *Attention Deficit Disorder*
- *Learning Disability*

Working with a partner, have students make observations on sound detection. With the subject seated and eyes closed, have the other student use a clicker for the sound test. The student should "click" the clicker at different points around the student and record where the subject detected the sound origination. After recording observations and switching roles, students should compare the observation data. They may report their findings to the class.

Integrating Physics and Chemistry

When light passes through the lens of an eye, light waves are bent and focused on the retina. Review the following properties of light: polarization, reflection, and refraction. Ask students to identify and describe which property of light occurs as light passes through the lens.

did you know?

The "Five" Senses Students have heard of the five senses: hearing, smell, taste, sight, and touch. But these are misleading categories. Touch actually consists of four distinct types of sensory receptors: hot, cold, pressure, and pain. Another sense, equilibrium or balance, is equally important but not usually discussed in terms of the senses. Equilibrium is maintained through fluids in the semicircular canals of the ear. Complex sensory organs are responsible for some of the senses, such as hearing and sight. Other senses are the result of small clusters of receptors, such as taste buds.

Close

Reteaching — BASIC

Assign students to cooperative groups of four. Have each group choose the sensory system they would least like to lose. Instruct each group to defend its choice.

Quiz — GENERAL

1. List the four lobes of the brain. (occipital, parietal, temporal, frontal)
2. What structures in the ear are responsible for equilbrium? (semicircular canals)
3. What is a sensory receptor? (a neuron that sends information from the sense organ to the central nervous system)

Alternative Assessment — GENERAL

Ask students to prepare a diagram that traces a sound from its origin to the brain. Make sure that students use the following key terms in their diagram: source, tympanic membrane, auditory nerve, anvil, stirrup, cochlea, ear canal, temporal region of the cerebral cortex, mechanoreceptors, and hammer.

Chemical Senses

internet connect
www.scilinks.org
Topic: Chemical Senses
Keyword: HX4041
SCILINKS. Maintained by the National Science Teachers Association

Embedded within the surface of the tongue are 2,000–5,000 taste buds. Most taste buds are located within small projections on the surface of the tongue. A taste bud, shown in **Figure 13,** is a cluster of 50–100 taste cells. Taste cells are chemoreceptors that detect at least four basic chemical substances: sugars (sweet), acids (sour), alkaloids (bitter), and salts (salty). Each taste cell is generally sensitive to all tastes but is most sensitive to only one of them. A taste bud is stimulated when food molecules dissolved in saliva bind to taste cells. Taste cells generate electrical signals that can be interpreted by the brain.

Chemoreceptors that detect odors, called olfactory *(ahl FAK tuh ree)* receptors, are located in the roof of the nasal passage. Chemicals in the air stimulate olfactory receptors, which generate electrical signals that can be interpreted by the brain. Your sense of smell affects your enjoyment of food. When you have a bad cold and your nose is stuffed up, your food may seem to have little taste.

Figure 13 Location and structure of taste buds

A taste bud is a cluster of taste cells surrounding a taste pore.

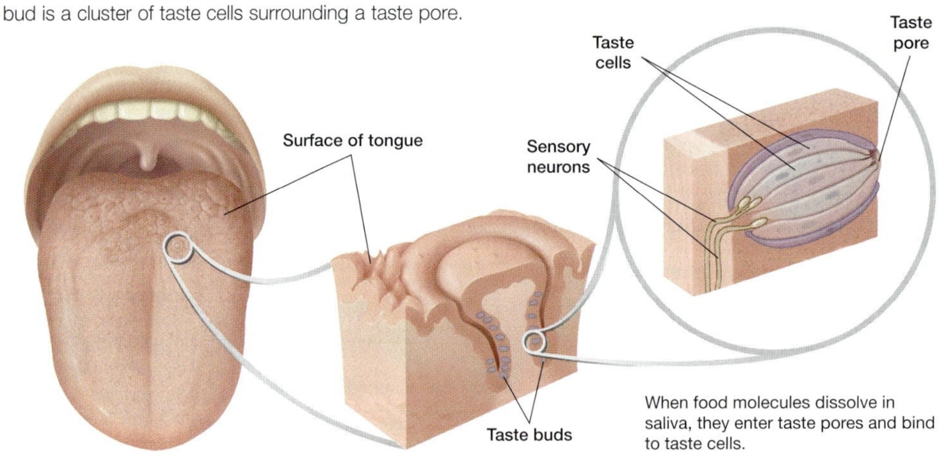

Section 3 Review

1. **List** two different types of sensory receptors and the kinds of stimuli to which they respond. SC.F.1.4.7
2. **Sequence** the events that occur when light enters the eye. SC.F.1.4.7
3. **Describe** how sound waves are transmitted through the ear. SC.F.1.4.7
4. **Critical Thinking Comparing Structures** Distinguish between taste cells and olfactory receptors. SC.F.1.4.7
5. FCAT Prep What would a person who had defects in both cochleas likely be unable to detect?
   A colors
   B odors
   C sounds
   D tastes
   SC.F.1.4.2

960

Answers to Section Review

1. Answers may vary. Sample answers: pain receptors respond to potentially harmful stimuli; thermo-receptors respond to temperature changes; and mechano-receptors respond to pressure and tension.
2. Light enters the eye through the pupil. The lens focuses light on the retina. Rods and cones on the retina convert light into electrical signals sent along the axons of the optic nerve, which runs to the thalamus, and finally to the occipital lobe of the cerebrum where the information is processed.
3. Sound waves enter the ear and strike the tympanic membrane, making it vibrate. The vibrations pass through the hammer, anvil, and stirrup to the cochlea where mechanoreceptors send impulses via the auditory nerve to the temporal lobe of the cerebrum.
4. Chemicals in food are detected by one or more of four kinds of taste receptors on the tongue. Olfactory receptors bind chemicals in the air.
5. **A.** Incorrect. The cone cells detect colors. **B.** Incorrect. Odors are detected by the olfactory receptors. **C.** Correct. The hair cells in the cochlea transfer sound waves into nerve impulses. **D.** Incorrect. Tastes are detected by the taste buds.

Section Review Correlations

SC.F.1.4.2: Item 5
SC.F.1.4.7: Items 1-4

Drugs and the Nervous System

Section 4

Florida Sunshine Standards/FCAT SC.F.1.4.8 Cell behavior can be affected by molecules from other parts of the organism or even from other organisms.

Objectives

- **Identify** types of psychoactive drugs, and describe their effects. SC.F.1.4.8
- **Describe** how drug addiction develops. SC.F.1.4.8
- **Describe** effects of commonly abused drugs on the nervous system. SC.F.1.4.8

Key Terms

psychoactive drug
addiction
tolerance
withdrawal
stimulant
depressant

Psychoactive Drugs

Many different kinds of drugs are available to the public. Advertisements tell you about pain relievers, antacids, cough syrups, and other medications that can help you feel better. Drugs can prevent, treat, or cure many different illnesses. However, drugs, whether legal or illegal, can also be misused or abused.

In the broadest sense, a drug is a chemical that alters body structures or biological functions. Drugs that alter the functioning of the central nervous system are known as **psychoactive drugs.** Many medications, such as those prescribed by doctors to treat mental disorders, contain psychoactive drugs. Caffeine, found in coffee and soft drinks, is also a psychoactive drug. Alcohol, marijuana, and cocaine are examples of commonly abused psychoactive drugs. Psychoactive drugs also include many other substances, such as inhalants. Many psychoactive drugs produce physiological dependence and addiction. Abuse of psychoactive drugs can damage the body, and in some cases, can result in death. **Table 3** lists several classes of commonly abused psychoactive drugs.

Table 3 Psychoactive Drugs of Abuse

Drug	Examples	Psychoactive effects	Risks associated with use
Depressants	Barbiturates (sedatives), tranquilizers, alcohol	Decreased activity of the central nervous system	Drowsiness, depression, brain or nerve damage, coma, respiratory failure
Stimulants	Cocaine, crack, nicotine, amphetamines	Increased activity of the central nervous system	Aggressive behavior, paranoia, cardiac arrest, high blood pressure, brain damage
Inhalants	Nitrous oxide, ether, paint thinner, glue, cleaning fluid, aerosols	Disorientation, confusion, memory loss	Brain damage, kidney and liver damage, respiratory failure
Hallucinogens	LSD, PCP, MDMA (ecstasy), peyote (mescaline), psilocybe mushroom	Sensory distortion, anxiety, hallucinations, numbness	Depression, paranoia, aggressive behavior
THC	Marijuana, hashish	Short-term memory loss, impaired judgment	Lung damage, loss of motivation
Narcotics	Heroin, morphine, codeine, opium	Feeling of well-being, sedation, impaired sensory perception, impaired reflexes	Coma, respiratory failure

Section 4

Focus

Overview

Before beginning this section review with your students the objectives listed in the Student Edition. This section explores examples of psychoactive drugs, addiction, and some commonly abused drugs.

Bellringer

Ask students to list on paper as many physical and social effects of smoking cigarettes as they can. Have them compare their list to the text on page 965, Effects of Tobacco.

Motivate

Demonstration — GENERAL

Demonstrate the effect of secondhand smoke. Put several smoking cigarettes in an ashtray. Caution: Burn cigarettes only under a vent hood. Place a piece of glass approximately 5 cm (2 in.) away from the cigarettes. After the cigarettes have burned completely, wipe the glass with a white cloth and show the cloth to the class. LS Visual

Chapter Resource File

- Directed Reading BASIC
- Active Reading GENERAL

Transparencies

TR Bellringer

One-Stop Planner CD-ROM

- Reading Organizers BASIC
- Reading Strategies BASIC
- Occupational Applications Worksheet
  Pharmacist
  Forensic Toxicologist GENERAL

Section Correlations

SC.F.1.4.8

Teach

Using the Figure

Emphasize that addiction is psychological as well as physiological. Ask students to study **Figure 15.** Have them explain the relationship between the craving for an addictive drug and the changes in the number of receptor proteins on postsynaptic cells. **LS Visual**

SKILL BUILDER —Advanced

Writing Skills A number of potentially addictive substances, including opiates and cocaine, have now been linked to specific receptors in the brain. Have students research how the work of neurochemist Solomon H. Snyder led to a greater understanding of how potentially addictive substances affect the operation of receptors in the brain. Have students write reports that describe the evidence that led Snyder to infer the existence of endorphins, the naturally occurring opiate-like substances in brain tissue. **LS Verbal**

Figure 14 Coca plant. Cocaine is derived from the coca plant, *Erythroxylon coca.*

Drug Addiction and Neuron Function

Addiction is a physiological response caused by use of a drug that alters the normal functioning of neurons and synapses. Once a neuron or synapse has been altered by a drug, it cannot function normally unless the drug is present. With repeated exposure to a drug, a person addicted to the drug develops tolerance to the drug. **Tolerance** is a characteristic of drug addiction in which increasing amounts of the drug are needed to achieve the desired sensation. **Withdrawal** is a set of emotional and physical symptoms caused by removal of the drug from the body. The severity of drug addiction is evident in recovering addicts who experience withdrawal when they stop taking an addictive drug. Symptoms of withdrawal may include vomiting, headache, depression, and seizures. Withdrawal from barbituates, and withdrawal in cases of severe alcohol addiction, can cause death and should be supervised by a doctor.

A Model of Drug Addiction

Cocaine is a highly addictive stimulant found in the leaves of the coca plant, *Erythroxylon coca,* shown in **Figure 14.** A **stimulant** is a drug that generally increases the activity of the central nervous system. Despite being illegal, cocaine is still used by many people.

Recall that in synaptic transmission, neurotransmitter molecules are released from a presynaptic neuron and bind to receptor proteins on a postsynaptic cell. Some neurotransmitter molecules are

Figure 15

BIO graphic

Action of Cocaine

Cocaine alters the function of dopamine-producing neurons in the limbic system.

1 Normal synapse
Dopamine is reabsorbed by the presynaptic neuron.

2 Synapse with cocaine
Cocaine blocks the reabsorption of dopamine.

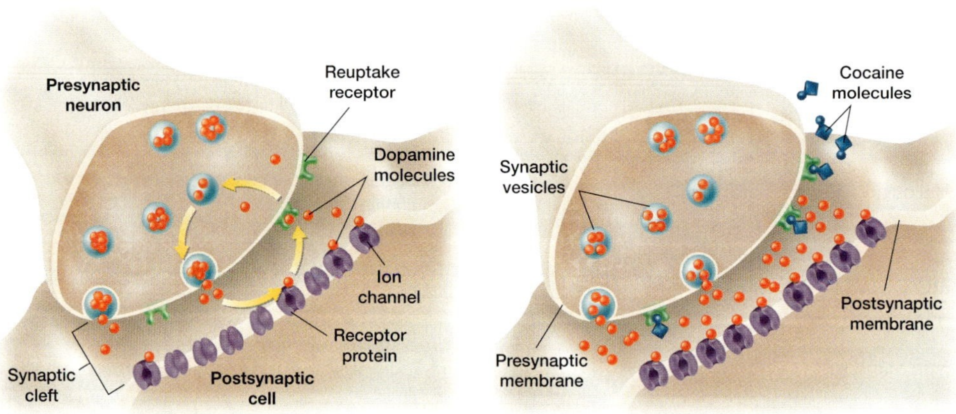

962

did you know?

Dangerous Street Drugs A street drug called "croak" results from the mixture of methamphetamines and crack cocaine. Methamphetamines stimulate the release of norepinephrine and dopamine. Crack cocaine prevents the reabsorption of dopamine. Extremely high heart rates and blood pressure can result when large amounts of dopamine enter and remain in synapses. Drug dealers often mix deadly combinations of drugs to make more money, to keep the users "hooked," and in some cases to fool the users into thinking they are using a "pure" drug.

reabsorbed by presynaptic neurons after they have been released into the synaptic cleft. Cocaine is an example of a drug that interferes with a presynaptic neuron's ability to reabsorb, or reuptake, neurotransmitter molecules. Cocaine affects dopamine *(DOH pah meen)* neurons in the limbic system, which plays an important role in the sensation of pleasure. The mechanism of cocaine action is summarized in **Figure 15.**

Step 1 At a normal synapse, reuptake receptors move molecules of dopamine in the synaptic cleft back into the presynaptic neuron.

Step 2 Cocaine blocks the reuptake of dopamine molecules by interfering with these reuptake receptors.

Step 3 As a result, excess dopamine remains in the synaptic cleft, overstimulating the postsynaptic cell. Overstimulation produces an intense feeling of exhilaration and well-being. Because the post synaptic cell has been overstimulated, the number of dopamine receptors will decrease over time.

Step 4 If cocaine is removed from the synaptic cleft, the number of dopamine molecules returns to normal. This level is now too low to adequately stimulate the postsynaptic cell because it has fewer receptor proteins. Addiction occurs because more cocaine must be taken to maintain adequate stimulation of the postsynaptic cell.

Real Life

Caffeine is a stimulant found in many foods and beverages.

Chocolate, coffee, tea, and some soft drinks often contain caffeine.

Calculating

Make a list of everything you eat and drink (including medicines) for 24 hours; then use a reference table to determine your total caffeine intake.

3 Overstimulated postsynaptic cell
The number of receptor proteins on the postsynaptic cell decreases.

4 Cocaine removed from synapse
Dopamine release returns to normal, but the postsynaptic cell is understimulated.

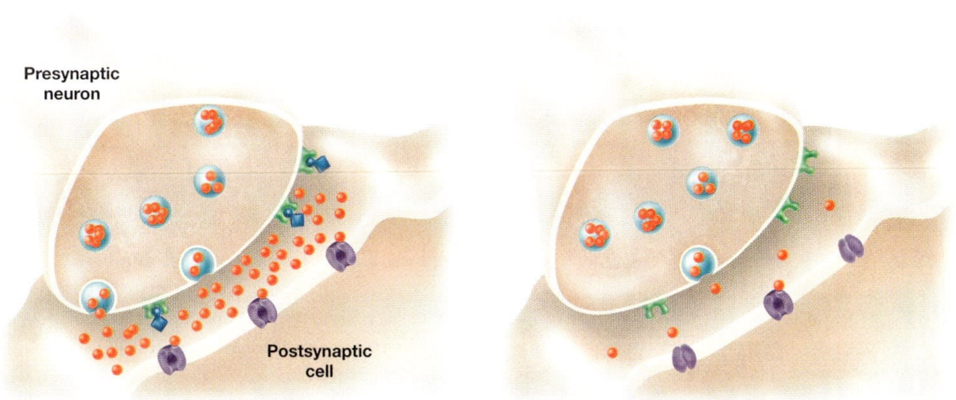

Teaching Tip — GENERAL

Effects of Drugs Have students choose a class of frequently abused drugs (such as opiates, depressants, or stimulants) or a specific drug (such as cocaine, marijuana, or alcohol). Ask them to make a pamphlet about the effects of the drug(s), the dangers of using the drug(s), and where to seek help for drug addiction. **LS Verbal**

Teaching Tip

Natural Diuretics Alcohol and caffeine are diuretics, which cause the excretion of excessive amounts of urine. Alcohol inhibits the release of antidiuretic hormone. This causes more water to be excreted in the urine. Caffeine circulates in the bloodstream to the kidneys where it inhibits the reabsorption of solutes and water. Thus, caffeine also increases urine production.

Group Activity — GENERAL

Drugs and Their Effects Assign students to groups of four. Have each group make a table listing different kinds of drugs, their desired effects, and their possible side effects. For each drug listed, have students categorize the drugs as medicinal or social, legal or illegal, and prescription or over the counter. Their tables might include stimulants, depressants, narcotics, hallucinogens, inhalants, hormone replacements, analgesics, antibiotics, antihistamines, and appetite suppressants. Students should find that many drugs have more side effects than desired effects.
LS Verbal Co-op Learning

Transparencies

- **TR J77** Psychoactive Drugs of Abuse
- **TR J75** Action of Cocaine
- **TR J76** Action of Cocaine, cont.
- **TR J78** Effects of Blood Alcohol Concentration

Figure 16 Breath test. Law enforcement officials use a device that detects the level of alcohol vapors in the breath to estimate the BAC of drunk-driving suspects.

Alcohol

Of all the psychoactive drugs, alcohol (ethanol) is one of the most widely used and abused. Alcohol, found in wine, beer, and liquor, is a depressant that produces a sense of well-being when taken in small amounts. A **depressant** is a drug that generally decreases the activity of the central nervous system. As more alcohol is consumed, reaction time increases, and coordination, judgment, and speech become impaired. This produces a state of intoxication known as being "drunk." Drunkenness results as the blood-alcohol concentration (BAC) increases. BAC can be measured by a breath test, illustrated in **Figure 16,** that detects the level of alcohol vapors in the breath. **Table 4** shows the effects of alcohol at various concentrations in the blood.

Table 4 Effects of Blood Alcohol Concentration

BAC*	Condition
0.02–0.04	Slight impairment and sedation
0.05–0.06	Slight impairment of coordination; increased reaction time
0.07–0.09	Slurred speech; blurred vision; intoxication
0.10–0.15	Severe intoxication; impaired coordination, vision, and balance
0.15–0.30	Dizziness; confusion; inability to walk; extremely severe intoxication
0.30–0.50	Unconsciousness
0.50–0.60	Coma or death

*in mg of alcohol per mL of blood

Alcohol is absorbed into the blood through the stomach and small intestine. Alcohol affects neurons throughout the nervous system, changing the shape of receptor proteins. Such widespread changes in receptor proteins have various effects on normal brain functioning.

Addiction to alcohol, or alcoholism, is the most prevalent drug-abuse problem in the United States. People who drink excessive amounts of alcohol over long periods of time develop serious health problems. For example, many alcoholics do not eat properly when drinking heavily. This can lead to malnutrition, abnormalities in the circulatory system, and inflammation of the stomach lining. In addition, the liver begins to use alcohol as an energy source. After exposure to alcohol over time, the liver accumulates fat deposits. If drinking of alcohol continues, a potentially fatal liver condition called cirrhosis *(sih ROH sis)* may develop. In a cirrhotic liver, cells are replaced with scar tissue, and liver functioning is impaired.

internet connect
www.scilinks.org
Topic: Blood Alcohol Concentration
Keyword: HX4026
SciLinks. Maintained by the National Science Teachers Association

Teach, *continued*

Demonstration — GENERAL

Show the class a videotape, in slow motion if possible, of someone performing a complicated high dive or balance-beam routine. Have students relate this type of performance to the function of the cerebellum. Alcohol has a significant effect on the synapses in the cerebellum. Have students discuss alcohol's effects on balance and coordination. LS Interpersonal

Activity — BASIC

BAC Have students research laws for blood alcohol concentration using the web site in the Internet Connect box on this page. Have students write a report summarizing their findings. LS Verbal

Demonstration — GENERAL

Invite a local law enforcement officer to come to the class to discuss DUI (Driving Under the Influence) and DWI (Driving While Intoxicated), incidence of DUI/DWI, and the dangers of drinking and driving. Have the officer tell students what blood alcohol concentration is considered "legally drunk" in your state. Have students find this BAC level in **Table 4.**

Attention Grabber

Alcohol Concentration Tell students that beer, wine, and liquor differ in their ethanol concentrations. However, a 12 oz. can of beer, a 4 oz. glass of wine, and a 1 oz. shot of liquor all contain about the same amount of absolute ethanol. Tell students that "proof" reflects the alcohol concentration in a beverage. Alcohol concentration is determined by dividing the proof number in half—80 proof liquor has a 40 percent alcohol content.

Nicotine

About 50 million Americans smoke cigarettes despite convincing evidence that smoking causes mouth cancer, heart disease, lung cancer, and emphysema. So why do people continue to smoke? Many smokers say they would like to stop smoking but find the habit too difficult to overcome. They are addicted to nicotine, a drug in cigarette smoke.

Effects of Nicotine

Nicotine is the highly addictive stimulant found in the leaves of the tobacco plant, *Nicotiana tabacum*, shown in **Figure 17.** Nicotine is extremely toxic; a dose of only 60 mg is lethal in humans. Tobacco leaves are dried or crushed and are then smoked in cigarettes, cigars, and pipes. Tobacco is also chewed and snuffed.

Nicotine quickly enters the bloodstream and circulates through the body. In the brain, nicotine mimics the action of the neurotransmitter acetylcholine. Scientists have extensively studied the behavior of the brain when exposed to nicotine. Nicotine binds to brain cells at specific sites usually reserved for acetylcholine. These sites are the central controls of the brain—mechanisms the brain uses to adjust levels of many of its activities. Like twisting the dial on a central control, the binding of nicotine to these sites produces many changes. After a while, the smoker's body makes adjustments, and systems almost return to normal—as long as the smoker keeps smoking. Take away the nicotine, however, and all those adjustments throw everything out of balance all at once. The only way to keep things "normal" is to keep smoking. The smoker is addicted.

Effects of Tobacco

Smokers get more than nicotine from cigarette smoke. Inhaled smoke contains hundreds of toxic and mutagenic chemicals that pass through the mouth, air passages, and lungs. These chemicals, also called tars, are produced by burning tobacco. Because tars and other chemicals in tobacco smoke are powerful mutagens, smoking causes lung cancer. Almost all cases of lung cancer, a major cause of death in the United States, are attributed to smoking.

In the United States, smoking-related illnesses cause more than 400,000 deaths each year. Smoking is associated with cancer of the mouth and larynx, and smoking may increase the risk of cancer of the pancreas and bladder. Smoking is also a major contributor to often-fatal respiratory disorders, such as emphysema. The tars in smoke irritate mucous membranes in the mouth, nose, and throat. They accumulate in the lungs and paralyze cilia that move debris from the lungs. Tars also blacken lung tissue and decrease breathing capacity. People who are exposed to secondhand smoke are at risk for the same diseases as people who smoke. Women who smoke during pregnancy are more likely to have miscarriages or to give birth to stillborn babies.

Figure 17 Tobacco plant. Tobacco leaves are dried and crushed and are then smoked in cigarettes, cigars, and pipes. Tobacco is also chewed and snuffed.

Real Life

Is smokeless tobacco harmful?

The use of smokeless tobacco, such as chewing tobacco, causes cancers of the lips, mouth, and gums. When chewing tobacco is placed between the cheek and gum, nicotine and other chemicals are absorbed into the bloodstream.

Finding Information

Find out about mouth cancers caused by tobacco.

965

Group Activity — GENERAL

Social Drug Use Have students bring in newspaper and magazine articles that deal with smoking cigarettes and drinking alcohol. Have some extra articles on hand to use if needed. Assign students to groups of four and ask students to consider the image of social drug use that is portrayed in the media. Have each group discuss its article and report back to the class with a brief summary. Display the articles on a bulletin board, and encourage the class to read them as they study this section.

LS Interpersonal Co-op Learning

Real Life — ADVANCED

Answer

About 90 percent of people who develop cancer of the mouth, pharynx, and larynx use tobacco in some form. Smoking cigarettes, cigars, or pipes is associated with cancers in the mouth, pharynx, larynx, esophagus, and other organs. Using smokeless tobacco is associate with cancers of the cheek, gums, and lips.

Source: American Cancer Society

REAL WORLD CONNECTION

Nicotine is extremely toxic. In fact, a sulfate derivative of nicotine has been used as a commercial insecticide for years. The lower doses of nicotine experienced by smokers are not immediately noticeable as toxic, however. Initially, nicotine elevates blood pressure by constricting blood vessels and increasing the heart rate. It stimulates the release of free fatty acids into the bloodstream. Smoking also adds carbon monoxide to the blood. This added carbon monoxide causes fatty acids to be deposited on the inside of the arteries and thus is a major contributor to cardiovascular disease. The combined effect of the strain of cardiovascular disease and the lack of oxygen contributes heavily to the damage seen in smokers' cardiovascular systems.

READING SKILL BUILDER — BASIC

Anticipation Guide Ask volunteers to discuss how their opinions formed in the Reading Activity on the opening page of this chapter changed or stayed the same as a result of their reading. Students should point to specific passages in the text that support their reasoning.
LS Verbal

Close

Reteaching — BASIC

Have students work in pairs. Ask each student to write definitions for the key terms listed on the first page in this section. Ask student pairs to quiz each other on the term, and then describe the effects of psychoactive drugs on the central nervous system.

Quiz — GENERAL

1. Why are pain receptors important? (They notify you that body tissues have been injured.)
2. What is the function of neurotransmitters? (They transmit nerve impulses across the synapse.)
3. Compare stimulants with depressants. (A stimulant generally increases the activity of the central nervous system; a depressant generally decreases the activity.)

Alternative Assessment — BASIC

Have students design an informative poster to educate people about the dangers of smoking.

Section Review Correlations

SC.F.1.4.8: Items 1-6

Figure 18 Opium poppy. Opium is a narcotic derived from the poppy plant, *Papaver somniferum*.

Figure 19 Hemp. Marijuana is produced from the hemp plant, *Cannabis*.

Drugs of Abuse

Narcotics are extremely addictive psychoactive drugs that relieve pain and induce sleep. Some of the most potent narcotics are derived from the poppy plant, *Papaver somniferum,* shown in **Figure 18.** The sap that oozes from the cut seed pod forms a thick, gummy substance called opium. Drugs derived from opium, called opiates or narcotics, include codeine *(KOH deen)*, morphine, and heroin, a more potent form of morphine. Codeine is widely prescribed by doctors for pain relief. Morphine is one of the most effective pain-relieving drugs used today. Heroin addiction and abuse are among the most serious illegal-drug problems in society.

Recall that pain receptors throughout the body detect painful stimuli. As uncomfortable as it may feel, pain plays a very important role in the body. Pain notifies you that body tissues have been injured or damaged. Imagine how your body would look and function today if you did not have the ability to sense pain. Pain begins as a signal at damaged nerve endings. Nerve impulses generated by pain receptors travel to the spinal cord toward the brain. After reaching the spinal cord, a pain signal is suppressed by a class of neurotransmitters called enkephalins *(ihn KEHF uh lihnz)*. When enkephalins bind to neurons in the spinal cord, they prevent pain signals from reaching the brain.

Narcotics mimic the action of enkephalins by binding to the same receptor proteins in the spinal cord. These receptor proteins are called opiate receptors because scientists observed opiates binding to them before enkephalins were ever discovered. Narcotics also affect the limbic system, producing a feeling of well-being.

Marijuana

In addition to alcohol and tobacco, marijuana, though illegal, is a widely consumed drug. Marijuana comes from various species of the hemp plant, *Cannabis,* shown in **Figure 19.** Hashish also comes from the hemp plant. The active ingredient in marijuana and hashish is commonly known as THC. When marijuana is smoked, it may cause disorientation, impaired judgment, short-term memory loss, and general loss of motivation. Scientists continue to research the effects of THC on the nervous system.

Section 4 Review

1. **Describe** how tolerance to a drug develops. SC.F.1.4.8
2. **Summarize** how cocaine produces addiction. SC.F.1.4.8
3. **Distinguish** between stimulants and depressants. Give an example of each. SC.F.1.4.8
4. **Critical Thinking Recognizing Relationships** What do all psychoactive drugs have in common? SC.F.1.4.8
5. **Critical Thinking Applying Information** Why is drug addiction considered a physiological condition? SC.F.1.4.8
6. **FCAT Prep** Cocaine interferes with the normal functions of the limbic system by blocking what function?
   A sensory perception
   B reuptake of dopamine
   C synaptic transmission
   D release of enkephalins
   SC.F.1.4.8

966

Answers to Section Review

1. Repeated exposure to a drug leads to tolerance, a condition in which increased amounts of the drug are needed to achieve the desired effect.
2. Cocaine overstimulates postsynaptic neurons by blocking dopamine reuptake, ultimately decreasing the number of dopamine receptors. When cocaine is removed, insufficient dopamine is available to stimulate postsynaptic cells at a normal level.
3. Stimulants, such as nicotine, increase central nervous system activity. Depressants, such as alcohol, decrease central nervous system activity.
4. All psychoactive drugs alter the function of the central nervous system.
5. Addictive drugs alter the normal functioning of neurons and synapses.
6. **A.** Incorrect. Sensory perception is everything your senses respond to in the environment. **B.** Correct. It interferes with the presynaptic neuron's ability to reabsorb the neurotransmitter. **C.** Incorrect. Synaptic transmission is the transmission of signals between cells. **D.** Incorrect. This prevents pain signals from reaching the brain.

Study ZONE CHAPTER HIGHLIGHTS

Key Concepts

1 Neurons and Nerve Impulses

- Neurons are specialized cells that rapidly transmit information as electrical signals throughout the body.
- At the resting potential, the inside of a neuron is negatively charged with respect to the outside of the neuron.
- An action potential moves rapidly down an axon.
- Synaptic transmission involves the release of neurotransmitters at synapses.

2 Structures of the Nervous System

- The central nervous system consists of the brain and spinal cord.
- The brain contains three major parts: the cerebrum, the cerebellum, and the brain stem.
- The spinal cord links the brain to the peripheral nervous system, which branches throughout the body.

3 Sensory Systems

- Sensory receptors detect various sensory stimuli.
- Photoreceptors in the eyes convert light into electrical signals that are interpreted by the brain.
- The ear converts sound into electrical signals that are interpreted by the brain.
- The semicircular canals monitor the position of the head.
- Taste and smell are related chemical senses.

4 Drugs and the Nervous System

- Psychoactive drugs affect the central nervous system.
- Drug addiction involves physiological changes in neurons.
- Alcohol is an addictive depressant that widely affects the central nervous system.
- Nicotine is an addictive stimulant found in tobacco products.

Key Terms

Section 1

neuron (944)
dendrite (944)
axon (944)
nerve (944)
membrane potential (945)
resting potential (946)
action potential (946)
synapse (948)
neurotransmitter (948)

Section 2

central nervous system (950)
peripheral nervous system (950)
sensory neuron (950)
motor neuron (950)
brain (950)
cerebrum (951)
cerebellum (951)
brain stem (951)
thalamus (952)
hypothalamus (952)
spinal cord (952)
reflex (952)
interneuron (953)

Section 3

sensory receptor (956)
retina (958)
rod (958)
cone (958)
optic nerve (958)
cochlea (959)
semicircular canal (959)

Section 4

psychoactive drug (961)
addiction (962)
tolerance (962)
withdrawal (962)
stimulant (962)
depressant (964)

Alternative Assessment — GENERAL

Divide the class into four groups of students. Assign one section of this chapter to each group. Tell student groups that they are to write a creative poem, short play, or very short story (three paragraphs) that includes all of the key terms found in their assigned section. Tell students that the definitions of each term should appear in their writing. When students have completed their creative writing, ask for volunteers from each group to present their product to the rest of the class.

Chapter Resource File

- Science Skills Worksheet GENERAL
- Critical Thinking Worksheet ADVANCED
- Test Prep Pretest GENERAL
- Chapter Test GENERAL

Answer to Concept Map

The following is one of several possible answers to Performance Zone item 15.

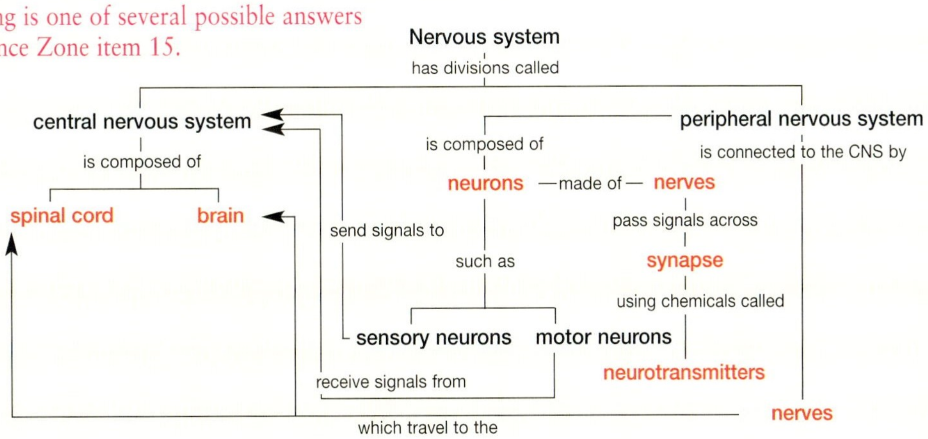

Performance ZONE

CHAPTER REVIEW

Understanding Key Ideas

1. A myelin sheath on the axon of a neuron
   a. covers the axon completely.
   b. decreases the rate of impulse conduction.
   c. increases the rate of impulse conduction.
   d. has no effect on impulse conduction.
   SC.F.1.4.2
2. When a neuron is at the resting potential,
   a. the inside is positively charged.
   b. the outside is negatively charged.
   c. the inside is negatively charged.
   d. None of the above
   SC.F.1.4.6
3. During an action potential,
   a. sodium ions flow into a neuron.
   b. sodium ions flow out of a neuron.
   c. potassium ions flow into a neuron.
   d. there is no movement of ions.
   SC.F.1.4.6
4. In a spinal reflex, the signal travels
   a. immediately to the brain.
   b. to the spinal cord and out to a muscle.
   c. only through sensory neurons.
   d. only through motor neurons.
   SC.F.1.4.6
5. Drug addiction is considered a physiological condition because addictive drugs
   a. can be purchased illegally.
   b. must be injected.
   c. alter the functioning of neurons.
   d. are used in social settings.
   SC.F.1.4.8
6. The diagram below shows the brain of a fish. How is the cerebrum of the fish brain different from that of a human brain? What do the large olfactory bulbs of the fish brain indicate about the relative importance of the sense of smell to the fish? SC.F.1.4.2

7. Action potentials travel in only one direction along a neuron—toward the axon terminals and away from the cell body. What structures of the neuron ensure that this pattern is always followed? SC.F.1.4.6
8. List three ways that the binding of a neurotransmitter to a receptor protein on a postsynaptic cell could cause changes in the cell. (**Hint:** See Chapter 4, Section 2.) SC.F.1.4.6
9. **BIOWatch** Although neurons in the spinal cord do not grow and regenerate when they have been injured, scientists have developed ways to help prevent paralysis. Describe two of these ways. SC.H.3.4.5
10. **Concept Mapping** Make a concept map that describes the structures and functions of the nervous system. Try to include the following terms: *spinal cord, brain, neuron, nerve, synapse,* and *neurotransmitter*. SC.F.1.4.2

Critical Thinking

11. **Inferring Relationships** People who suffer from vertigo feel dizzy and disoriented in certain situations. What is the relationship between vertigo and the semicircular canals? SC.H.3.4.3
12. **Recognizing Relationships** Suggest a possible defect of the retina or of retinal cells that would cause colorblindness, a condition in which a person cannot distinguish between certain colors. SC.H.3.4.3

Alternative Assessment

13. **Summarizing Information** Research the causes and symptoms of various disorders of the nervous system caused by the degeneration of neurons. Some disorders include Alzheimer's disease, Parkinson's disease, and multiple sclerosis. Find out how these disorders affect the nervous system. List the types of drugs or other methods used to treat these disorders. SC.H.3.4.3

968

ANSWERS

Understanding Key Ideas

1. c
2. c
3. a
4. b
5. c
6. The fish cerebrum is much smaller than a human cerebrum (relative to other parts of the brain). The sense of smell probably is very important to the fish.
7. Voltage-gate sodium channels, which close shortly after an action potential, cannot open again for a brief moment after the action potential. This prevents the nerve impulse from moving in the opposite direction.
8. When binding to receptor proteins on the postsynaptic cell, some neurotransmitters cause ion channels to open, some stimulate the formation of second messengers in the cell, and others stimulate enzyme activity in the cell.
9. Cell-death inhibitors are used for myelin-forming cells. Peripheral nerves are grafted to stimulate axon growth.
10. One possible answer to the concept map is found at the bottom of the previous page.

Critical Thinking

11. The semicircular canals detect acceleration and orientation of the body. Eyes detect movement, sending that message to the brain. If the hair cells in the semicircular canals do not, however, detect the same degree of movement, they send a conflicting signal to the brain. These conflicting signals can cause dizziness and vertigo.
12. Cone cells enable color vision. If damage or a chemical disorder occurs in these cells, color blindness could be the result.

Alternative Assessment

13. Sample answer: Alzheimer's disease: cause: unknown—origin of plaque buildups in the brain may be neurochemical, environmental, infectious, or genetic; symptoms: impaired memory, loss of intellectual function, atrophy of brain tissue, degradation of language ability; part of the nervous system: brain; available treatment: none available—symptoms are treated with medication, physical therapy, and occupational therapy.

Chapter Review Correlations

SC.F.1.4.2: Items 1, 6, 10
SC.F.1.4.6: Items 2, 3, 4, 7, 8
SC.F.1.4.8: Item 5
SC.H.3.4.3: Items 11, 12, 13
SC.H.3.4.5: Item 9

Assignment Guide

Section	Questions
1	1, 2, 3, 7, 8
2	4, 6, 9, 13
3	11, 12
4	5

FCAT Prep

For questions 1–6, write your answers on a separate sheet of paper.

1 Which of the following is a sensory receptor that is stimulated by light?

A. cochlea **C.** interneuron
B. cone **D.** optic nerve

SC.F.1.4.7

2 If the sodium–potassium pump of a neuron failed, what effect would this likely have on the neuron's function?

F. The concentrations of positive and negative ions would cause the neuron to be negative inside.
G. Voltage-gated potassium channels and voltage-gated sodium channels would no longer function.
H. The neuron could not conduct another action potential until the resting potential was fully restored.
I. The concentration of sodium ions would be higher outside the cell and that of potassium ions would be higher inside.

SC.F.1.4.5

3 Which part of the ear contains mechanoreceptors?

A. cochlea **C.** hammer
B. eardrum **D.** stirrup

SC.F.1.4.2

4 Which structure increases the speed at which the axon conducts action potentials?

F. axon terminal
G. dendrite
H. myelin sheath
I. nucleus

SC.F.1.4.2

5 READ INQUIRE EXPLAIN During an epileptic seizure, many neurons in the brain produce large bursts of action potentials, causing the body to become rigid and to jerk or convulse.

Using what you know about the brain's control of muscles and posture, what could be deduced from these symptoms? SC.H.3.4.3

6 READ INQUIRE EXPLAIN Depression affects several million Americans. Symptoms of depression include sadness, anger, poor communication, withdrawal, and indifference to surroundings. Depression may be triggered by the loss of a friend or relative, a major disappointment at work, prescription drugs, prolonged illness, alcohol or drug withdrawal, or hormones. Treatments include counseling, several types of drugs, and exercise.

Part A Assess which division of the nervous system is most affected by depression. SC.F.1.4.2

Part B Where would drugs that treat depression by simulating neurotransmitters be effective in neural tissue? SC.F.1.4.8

Test TIP

Allow a few minutes at the end of the test-taking period to check for mistakes made in marking answers.

969

FCAT Doctor

Question 2 Answer H is the correct choice. The sodium–potassium pump helps the neuron achieve the resting potential by pumping sodium ions out of the cell and potassium ions into the cell. Answer F is incorrect because the normal resting potential when the pump is working is negative inside. Answer G is incorrect because the functioning of voltage-gated ion channels is not directly related to the functioning of the sodium–potassium pump. Answer I is incorrect because the normal resting potential of the cell has these concentrations, which are maintained by the sodium–potassium pump.

Question 5 The muscles that are rigid throughout the body are all being stimulated by nerve impulses.

Question 6

Part A The brain in the central nervous system is the body's main processing center. It controls thoughts, feelings, emotions, and memories, and would be most involved in symptoms of depression.

Part B Drugs may target neurotransmitters in order to regulate sensations and moods.

Answers

1. B
2. H
3. A
4. H
5. Many muscles are being stimulated at once.

6A. The central nervous system interprets and responds to information from the environment and would therefore be most involved in symptoms such as sadness, anger, poor communication, withdrawal, and indifference to surroundings.

6B. Drugs that simulate neurotransmitters would be effective in the synaptic cleft between a presynaptic neuron and a postsynaptic cell.

Test Prep Correlations

SC.F.1.4.2: Items 3, 4, 6A
SC.F.1.4.5: Item 2
SC.F.1.4.7: Item 1
SC.F.1.4.8: Item 6B
SC.H.3.4.3: Item 5

Exploration Lab

Calculating Reaction Times

Teacher's Notes

Time Required
One 50-minute class period

Ratings
EASY → HARD

TEACHER PREPARATION 1
STUDENT SETUP 1
CONCEPT LEVEL 1
CLEANUP 1

Skills Acquired
- Collecting Data
- Constructing Materials
- Experimenting
- Organizing and Analyzing Data

Scientific Methods
In this lab, students will:
- Make Observations
- Ask Questions
- Test the Hypothesis
- Draw Conclusions

Materials and Equipment
Materials for this lab can be ordered from WARD'S. See *Master Materials List* at the front of this book for catalog numbers.

Safety Cautions
Remind students to handle the meter stick carefully and to keep it away from their face and far away from the face and eyes of their classmates. Allow students to have a practice trial. Instruct students to look only at the ruler and not at their partner. Point out that looking at their partner could distort the results of the investigation because the partner might give a clue about when he or she will drop the meterstick.

Exploration Lab

Calculating Reaction Times SC.F.1.4.1

SKILLS
- Measuring
- Calculating

OBJECTIVES
- **Determine** human reaction times.
- **Design** an experiment that measures changes in reaction times.

MATERIALS
- meterstick

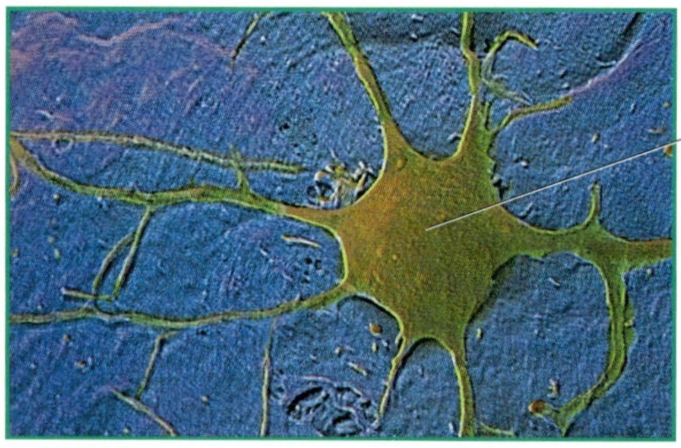

Neuron

Before You Begin
When you want to move your hand, your brain must send a message all the way to the muscles in your arms. How long does that take? In this exercise, you will work with a partner to see how quickly you can react. In this lab, you will investigate reaction times and design an experiment to investigate influences on reaction times.

1. Create a Data Table to record reaction times.
2. Write a hypothesis about an influence on reaction times. (For example: "People who have eaten breakfast have faster reaction times than people who have not eaten breakfast.")

Procedure
PART A: Calculating Reaction Times

1. Sit in a chair and have a partner stand facing you while holding a meterstick in a vertical position.

DATA TABLE

Hand: trial number	Subject 1 reaction time (s)	Subject 2 reaction time (s)
Left: 1		
Left: 2		
Left: 3		
Left: average		
Right: 1		
Right: 2		
Right: 3		
Right: average		

970

Answers to Before You Begin
1. Data table should be similar to the sample on this page.
2. Hypotheses may vary, but should be reasonable and testable. Sample answer: the dominant hand will have a faster reaction time than the non-dominant hand.

Answers to Procedure
Observations in steps 2–6 will vary. Check that students have correctly entered all observations in their data table.

10. Answers may vary. Students may choose any method for testing reaction time as long as the results are measurable.
11. Answers will vary. Students should write a logical and sequential set of procedures, with any precautions that apply.
12. Answers will vary. Students' data should be presented in a manner so that it is clear what the measurements represent.

2. Hold your thumb about 3 cm from your fingers near the bottom end of the stick. The meterstick should be positioned to fall between your thumb and fingers.
3. Tell your partner to let go of the meterstick without warning.
4. When your partner releases the meter stick, catch the stick by pressing your thumb and fingers together. Your partner should be ready to catch the top of the meterstick if it begins to tip over.
5. Record the number of centimeters the stick dropped before you caught it. The distance that the meterstick falls before you catch it can be used to evaluate your reaction time.
6. Repeat the procedure several times, and calculate the average number of centimeters.
7. Try this procedure with your other hand.
8. Close your eyes and have your partner say "now," when the stick is released.
9. Exchange places with your partner, and repeat the procedure.

PART B: Designing Your Own Experiment

10. Work with the members of your lab group to explore one of the hypotheses written in the **Before You Begin** section of this lab.

You Choose

As you design your experiment, decide the following:

a. what hypothesis you will explore
c. how you will test the hypothesis
d. what the controls will be
e. how many trials to perform
f. what data to record in your data table.

11. Write a procedure for your experiment. Make a list of all the safety precautions you will take. Have your teacher approve your procedure and safety precautions before you begin the experiment.
12. Set up your group's experiment and collect data.

Analyze and Conclude

1. **Summarizing Results** What was your fastest reaction time?
2. **Analyzing Data** How does your reaction time when using your dominant hand compare with your reaction time when using your other hand?
3. **Drawing Conclusions** Why may each hand have a different reaction time? Why may each person have a different reaction time? Compared to earlier trials, was the reaction time in step 8 faster, slower or the same? If the time was faster or slower hypothesize a reason for the difference.
4. **Predicting Patterns** Compile the data gathered by each pair in your class. Can you identify any trends in the data? (For example, do males and females have the same average reaction times?)
5. **Further Inquiry** Write a new question about reaction times that could be explored in another investigation.

Do You Know?

Do research in the library or media center to answer these questions:

1. How can athletes improve their reaction times?
2. What factors can influence how fast a person reacts to a stimulus?

Use the following Internet resources to explore your own questions about neurons.

internet connect
www.scilinks.org
Topic: Neurons
Keyword: HX4129
SCILINKS. Maintained by the National Science Teachers Association

971

Answers to Analyze and Conclude

1. Answers may vary.
2. Answers may vary, but reaction times are often faster in dominant hands.
3. Answers may vary. Reaction times calculated when the subject has closed eyes will probably be slower. Communication is an additional step, and sending and receiving the information that the stick is dropping takes time.
4. Answers may vary.
5. Answers may vary. Sample answer: Would training be a method of improving reaction time? Do students who eat breakfast have faster reaction times than students who skip breakfast? Can caffeine increase reaction time?

Answers to Do You Know?

1. Answers may vary. Training is an example of a method for athletes to improve reaction time.
2. Answers may vary. Examples of things which can influence reaction time are nutrition, sleep deprivation, and drug use.

Chapter Opener		TE **Using the Figure,** p. 972 GENERAL SE **Quick Review,** p. 973 ■ SE **Reading Activity,** p. 973 ■ TE **Opening Activity,** p. 973 GENERAL
& 3 • 90 min pp. 974–976 **Section 1** Hormones	OSP Lesson Plan* TR Bellringer* TR Section Outline* TR J79 The Endocrine System* TR J83 Major Endocrine Glands* TR J88 Maintaining a Normal Blood Glucose Level*	TE **Group Activity** Exocrine or Endocrine? p. 975 GENERAL CRF **Skills Practice Lab** Modeling the Endocrine System* GENERAL CRF **Quick Lab** Graphing Growth Rate Data* GENERAL
& 5 • 90 min pp. 977–981 **Section 2** How Hormones Work	OSP Lesson Plan* TR Bellringer* TR Section Outline* TR J80 Hormones Acting on Target Cells* TR J81 How Amino Acid-Based Hormones Work* TR J82 How Steroid Hormones Work* TR J90 Negative Feedback*	TE **Demonstration,** p. 977 ◆ GENERAL TE **Using the Figure,** p. 978 GENERAL TE **Demonstration,** p. 979 ◆ GENERAL TE **Using the Figure,** p. 979 GENERAL
& 7 • 90 min pp. 982–988 **Section 3** The Major Endocrine Glands	OSP Lesson Plan* TR Bellringer* TR Section Outline* TR J84 The Hypothalamus and Pituitary Gland* TR J85 Hormones Secreted by the Pituitary Gland* TR J86 The Adrenal Gland*	TE **Demonstration,** p. 983 ◆ GENERAL TE **Group Activity** Riddles, p. 985 GENERAL TE **Group Activity** Pituitary Hormones, p. 985 ◆ GENERAL SE **Math Lab** Analyzing Blood-Glucose Regulation, p. 986 ■ CRF **Datasheets for In-Text Labs** Analyzing Blood-Glucose Regulation* SE **Exploration Lab** The Effect of Epinephrine on Heart Rate, pp. 992–993 ◆ ■ CRF **Datasheets for In-Text Labs** The Effect of Epinephrine on Heart Rate*

& 9 • 90 min **Chapter Review and Assessment Resources**

- SE **Study Zone,** p. 989 ■
- SE **Performance Zone,** p. 990 ■
- SE **FCAT Test Prep,** p. 991 ■
- TR Concept Mapping*
- CRF Vocabulary Review Worksheet* ■ GENERAL
- CRF Test Prep Pretest* ■ GENERAL
- CRF Chapter Test* ■ GENERAL
- CRF Chapter Test* ADVANCED
- OSP Test Generator*
- OSP Test Item Listing*
- OSP Modified Chapter Test* BASIC

go.hrw.com Holt Online Learning

Visit go.hrw.com for access to Holt Online Learning or enter the keyword **HX6 Home** for a variety of free online resources.

One-Stop Planner® CD-ROM

This CD-ROM package includes
- Lab Materials QuickList Software
- Holt Calendar Planner
- Customizable Lesson Plans
- Printable Worksheets
- ExamView® Test Generator
- Interactive Teacher Edition
- Holt PuzzlePro® Resources
- Holt PowerPoint® Resources

Guided Reading Audio CDs

These CDs are designed to help auditory learners and reluctant readers. (Audio CDs are also available in Spanish.)

Compression guide:
To shorten your instruction because of time limitations, omit block 1.

KEY

TE	Teacher Edition	CRF	Chapter Resource File	*	Also on One-Stop Planner
SE	Student Edition	TR	Teaching Transparency	■	Also Available in Spanish
OSP	One-Stop Planner			◆	Requires Advance Prep

SKILLS DEVELOPMENT AND PRACTICE	SECTION REVIEW AND ASSESSMENT	CORRELATIONS
		Florida Sunshine Standards/FCAT
TE Reading Skill Builder Paired Summary, p. 975 BASIC CRF Directed Reading Worksheet* BASIC CRF Active Reading Worksheet* GENERAL	SE Section 1 Review ■ TE Reteaching, p. 976 BASIC TE Quiz, p. 976 GENERAL TE Alternative Assessment, p. 976 GENERAL CRF Quiz* ■ BASIC	SC.F.1.4.2, SC.F.1.4.6
TE Skill Builder Interpreting Graphics, p. 980 GENERAL CRF Directed Reading Worksheet* BASIC CRF Active Reading Worksheet* GENERAL	SE Section 2 Review ■ TE Reteaching, p. 981 BASIC TE Quiz, p. 981 GENERAL TE Alternative Assessment, p. 981 GENERAL CRF Quiz* ■ BASIC	SC.F.1.4.6
TE Reading Skill Builder Paired Summary, p. 984 BASIC TE Skill Builder Vocabulary, p. 985 GENERAL CRF Directed Reading Worksheet* BASIC CRF Active Reading Worksheet* GENERAL CRF Critical Thinking Worksheet* ADVANCED CRF Science Skills Worksheet* GENERAL	SE Section 3 Review ■ TE Reteaching, p. 988 BASIC TE Quiz, p. 988 GENERAL TE Alternative Assessment, p. 988 GENERAL CRF Quiz* ■ BASIC Modified Worksheet* BASIC	SC.F.1.4.2, SC.F.1.4.6

SCILINKS NSTA
www.scilinks.org
Maintained by the **National Science Teachers Association**

Topic: Anabolic Steroids
SciLinks Code: HX4006

Topic: Hormones and Body Fat
SciLinks Code: HX4101

Topic: Melatonin
SciLinks Code: HX4121

Topic: Hormones
SciLinks Code: HX4100

CHAPTER 42

Overview

Tell students the purpose of this chapter is to show that chemical messengers called hormones have a tremendous impact on cellular, and consequently, bodily function. Feedback systems and other complex interactions of hormones and target tissues can have a dramatic effect on a person's well-being. Disruption of these interactions can lead to serious illness.

Using the Figure

Ask students to look at the photo of Gail Devers. Ask them what kind of feelings she would be having after winning a gold medal at the Olympics. (Answers may include happiness, elation, and joy.) Have each student make a list of all the emotions they can think of. Ask students if they think hormones can affect these emotions. (Yes.) Have students also note Gail Devers' large bicep muscle. Ask students why the muscles in males will get much larger than in females, even though they may go through the same fitness routine. (Male hormones have a greater affect on muscle size than female hormones.)

Assessing Prior Knowledge

Before beginning this chapter, be sure your students can answer the questions in ***Quick Review*** on the facing page. If they cannot, encourage them to reread the sections indicated.

Gail Devers

972

Chapter Correlations — Florida Sunshine Standards/FCAT

SC.F.1.4.2 Body structures are uniquely designed and adapted for their function.

SC.F.1.4.6 Separate parts of the body communicate with each other using electrical and/or chemical signals.

CHAPTER

42 Hormones and the Endocrine System

Florida Sunshine Standards/FCAT

SC.F.1.4.2 Body structures are uniquely designed and adapted for their function.

SC.F.1.4.6 Separate parts of the body communicate with each other using electrical and/or chemical signals.

Quick Review

Answer the following without referring to earlier sections of your book.

1. **Describe** the function and structure of glycogen. *(Chapter 2, Section 3)*
2. **Describe** the action of enzymes. *(Chapter 2, Section 4)*
3. **Summarize** the location and function of receptors. *(Chapter 3, Section 2)*
4. **Summarize** the role of DNA and mRNA in protein synthesis. *(Chapter 10, Section 1)*
5. **Describe** the role of the sympathetic nervous system. *(Chapter 41, Section 2)*

Did you have difficulty? *For help, review the sections indicated.*

Reading Activity

Before you read this chapter, write a short list of all of the things you know about glands and hormones. Then write a list of the things that you want to know about glands and hormones. Save your lists, and to assess what you have learned, see how many of your own questions you can answer after reading this chapter.

Severe weight loss, shaking fits, and loss of vision in one eye were symptoms experienced in 1988 by American sprinter Gail Devers. She was later diagnosed with Grave's disease, a condition in which excessive amounts of hormones are made by the thyroid gland. Devers went on to win the gold medal in the 100 m dash in two Summer Olympics.

Looking Ahead

Section 1

Hormones

Coordination of Activities
Endocrine Glands and Tissues

Section 2

How Hormones Work

Target Cells
Receptors
Feedback Mechanisms

Section 3

The Major Endocrine Glands

Endocrine Control
The Thyroid and Parathyroid Glands
The Adrenal Glands
The Pancreas and Other Organs

internet connect

www.scilinks.org

National Science Teachers Association *sci*LINKS Internet resources are located throughout this chapter.

SciLINKS. Maintained by the National Science Teachers Association

973

Opening Activity — GENERAL

Hormonal Imbalance Show the class photographs of people who obviously have a permanent or temporary hormonal imbalance (for example, someone who is extremely short or tall, or a bearded female.) Ask students if they know the type of molecule that is responsible (hormone). Tell students that hormones are very important to maintaining homeostasis. They are secreted in tiny concentrations in the blood or extracellular fluid. Either too much or too little of a hormone can have serious consequences. LS **Visual**

Identifying Misconceptions

Students do not often recognize that hormones act on cells in a very specific manner. Tell students that, in this chapter, they will learn how hormones transmit messages that change cell activity in a predictable way.

Quick Review Answers

1. Glycogen is a polysaccharide made of chains of glucose and is used for energy storage.
2. Enzymes lower the activation energy needed to start a chemical reaction. Enzymes are specific in the reactions they catalyze.
3. Most receptor proteins are embedded in the lipid bilayer. They function by recognizing and binding to signal molecules outside the cell.
4. DNA stores the genetic information for the cell. mRNA carries the DNA code to ribosomes in the cytoplasm to be translated into proteins.
5. The sympathetic nervous system is important during times of physical or emotional stress. It increases blood pressure, heart rate, breathing rate, and directs blood flow toward heart and skeletal muscles in the "fight-or-flight" response.

Chapter Resource File

- Vocabulary Worksheets
- Concept Mapping

Reading Activity Answers

Students may know that glands produce hormones, and that hormones are a major factor in bodily changes at puberty. They may be familiar with some of the more common hormones such as testosterone, estrogen, and insulin.

Section 1

Focus

Overview

Before beginning this section review with your students the objectives listed in the Student Edition. This section introduces students to the chemicals called hormones. It describes what hormones are, what they do, and why it is important that hormones are carefully regulated in the body.

Bellringer

At the beginning of a class, while students are settling down and paying little attention to you, make a loud, startling noise. Immediately afterward, ask students to make a list of their feelings in response to the noise. List responses on the board or overhead. At intervals, ask if anyone still feels jittery. Tell students that the effects they felt were the result of a hormone.

Motivate

Identifying Preconceptions —GENERAL

Ask students what the difference is between the nervous system and the endocrine system. (In general, the nervous system sends electrical signals, while the endocrine system sends chemical signals.) After discussion, point out to students that, although it is convenient to distinguish between the endocrine and nervous systems, the lines between these two systems are blurred. There is considerable overlap and coordination between them, and this helps the body maintain homeostasis.

Section Correlations

SC.F.1.4.2
SC.F.1.4.6

Section 1 Hormones

Florida Sunshine Standards/FCAT

SC.F.1.4.2 Body structures are uniquely designed and adapted for their function.

SC.F.1.4.6 Separate parts of the body communicate with each other using electrical and/or chemical signals.

Objectives

- **Identify** four major functions of hormones. SC.F.1.4.6
- **Differentiate** between endocrine and exocrine glands. SC.F.1.4.2
- **Compare** the endocrine and nervous systems. SC.F.1.4.2
- **Distinguish** the functions of endorphins from the functions of prostaglandins. SC.F.1.4.6

Key Terms

hormone
endocrine gland

Coordination of Activities

Reacting to fear, growing taller, and developing male or female characteristics are all activities in the body that are partially regulated by hormones *(HAWR mohnz)*. **Hormones** are substances secreted (released) by cells that act to regulate the activity of other cells in the body.

You—and the cyclist shown in **Figure 1**—make hormones that keep your body functioning properly. The functions of hormones include the following:

1. Regulating growth, development, behavior, and reproduction
2. Coordinating the production, use, and storage of energy
3. Maintaining homeostasis (temperature regulation, metabolism, excretion, and water and salt balance)
4. Responding to stimuli from outside the body

Hormones act as chemical messengers, carrying instructions that cause cells to change their activities. For example, some hormones cause the cells of the heart to increase the rate at which the heart is beating. In the past, it was thought that hormones, once secreted from a cell, had to be transported through the bloodstream to reach the cells they were to act on. Today, we know that some hormones act directly on adjacent cells without traveling through the blood.

The instructions a hormone carries are determined by both the hormone itself and the cell it affects. For example, a hormone may instruct a cell to make a specific protein or to activate a specific enzyme. The same hormone can also instruct a different cell to alter the permeability of its cell membrane or even to release another hormone. Hormones can instruct muscle cells to relax and nerve cells to fire action potentials.

Each hormone is very specific about which types of cells can receive its instructions. Each hormone acts like a key that opens a lock on or inside the cell. A hormone will act only on cells with the right lock. The locks, as discussed later in this chapter, are receptors on or inside the cell.

Figure 1 Hormones and homeostasis. Combining activities, such as water balance and temperature regulation, requires coordination. Such coordination is maintained by hormones.

974

Chapter Resource File

- Directed Reading BASIC
- Active Reading GENERAL

One-Stop Planner CD-ROM

- Reading Organizers BASIC
- Reading Strategies BASIC

INCLUSION Strategies

- *Attention Deficit Disorder*
- *Learning Disability*

Ask students to use a word processor or pencil and paper to create a list of the four major functions of hormones. Using the list, have students create and record a public service announcement to inform others of the functions of hormones. In the announcement, have students give specific details about how hormones accomplish their functions as described in the text. Have volunteers present a live version of their announcement to the class.

Endocrine Glands and Tissues

A gland is an organ whose primary function is to secrete materials into other regions of the body. **Endocrine** *(EN doh krihn)* **glands** are ductless organs that secrete hormones directly into either the bloodstream or the fluid around cells (extracellular fluid). In addition to endocrine glands, several other organs contain cells that secrete hormones. These organs include the brain, stomach, small intestine, kidney, liver, and heart.

All of the endocrine glands and hormone-secreting tissues collectively make up the endocrine system, shown in **Figure 2.** The endocrine system coordinates all of the body's sources of hormones.

Some organs, such as the pancreas, are both endocrine and exocrine *(EHKS oh krihn)* glands. Exocrine glands deliver substances through ducts (tubelike structures). The ducts transport the substances to specific locations inside and outside the body. Sweat glands, mucous glands, salivary glands, and other digestive glands are examples of exocrine glands. The exocrine part of the pancreas produces digestive enzymes and delivers them to the small intestine through ducts. The endocrine part of the pancreas secretes two hormones into the bloodstream that regulate blood glucose levels.

The Endocrine System

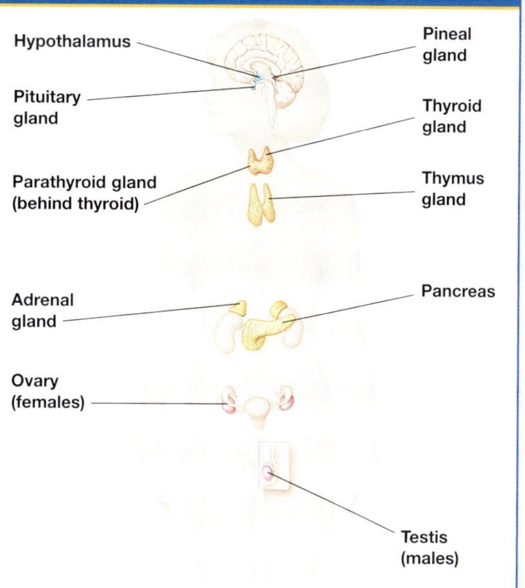

Figure 2 Coordinating the body's activities. Endocrine glands are located throughout the human body. In addition to the organs shown above, many other organs secrete hormones.

Hormones and Neurotransmitters Are Chemical Messengers

As you learned in the previous chapter, the nervous system is also involved in coordinating the body's activities. The endocrine system and nervous system interact in this shared role. However, with some exceptions, each system acts through different chemical messengers and in different ways.

The chemical messengers of the nervous system are known as neurotransmitters, while the chemical messengers of the endocrine system are called hormones. Some nerve cells, however, are capable of secreting hormones, and several chemicals serve as both hormones in the endocrine system and neurotransmitters in the nervous system. For example, epinephrine is both a neurotransmitter and a hormone. When secreted from a nerve cell, epinephrine conveys messages to other neurons. When secreted from an endocrine cell in the adrenal gland, epinephrine acts as a "fight-or-flight" hormone.

Teach

Group Activity — GENERAL

Exocrine or Endocrine? Write the names of these glands on separate index cards: adrenal gland, ovary, thyroid gland, pancreas, testes, pituitary gland, parathyroid gland, thymus gland, hypothalamus, sweat glands, mucous glands, salivary glands, intestinal glands, and another "pancreas" card. Pass the cards out to students. Make two columns on the board. Label one "endocrine" and the other "exocrine." Have students come up and, using tape, stick their card in the column where they believe it belongs. When finished, ask students what the difference is between exocrine and endocrine glands. (Exocrine glands secrete substances through ducts, and endocrine glands do not.) After discussion, have other students go to the board and rearrange the cards. Have the remaining students check the text to make final corrections. Discuss why each gland fits in the category it is in, and ask why there are two cards for the pancreas. (It is part of both exocrine *and* endocrine systems.) **LS Logical** Co-op Learning

READING SKILL BUILDER — BASIC

Paired Summarizing Have students read silently the first page of Section 1. Then assign partners and have one student, the "reteller," summarize orally what has been read while the other student, the "listener," listens to the retelling. The listener should point out inaccuracies and add ideas that were left out. Have students alternate roles and repeat this process for each page of the section. Students should work together during a final clarification process, referring to the text as needed. **LS Auditory**

Transparencies

TR Bellringer
TR J79 The Endocrine System
TR J83 Major Endocrine Glands
TR J88 Maintaining a Normal Blood Glucose Level

REAL WORLD CONNECTION

Although prostaglandins are known for increasing pain, they play a vital role in labor and delivery. Prostaglandins not only stimulate uterine muscle contractions, but also promote blood clotting. Blood clotting is important during childbirth because it minimizes hemorrhaging during and after delivery. Prostaglandin-inhibiting drugs, such as aspirin and ibuprofen, may delay labor, or they may inhibit blood clotting and promote excessive bleeding, which could result in severe blood loss or even death.

Career — GENERAL

Endocrinologist Invite a medical professional who specializes in hormone therapy to discuss hormones, hormone disorders, and treatments available. Ask the guest to bring photographs of people with endocrine disorders and micrographs of endocrine tissues, if possible. In addition, ask the guest speaker to discuss her or his education, and job opportunities in this field of work. In advance, have the class prepare a list of questions they would like to have answered. Also have them jot down and ask any questions they might think of during the presentation. **LS Intrapersonal**

Teach, *continued*

Real Life

Answers

Drugs that inhibit prostaglandin production work by inhibiting the body's production of cyclooxygenase, an enzyme involved in prostaglandin synthesis. Long-term use of drugs that inhibit prostaglandin production can lead to stomach ulcers or kidney damage.

Close

Reteaching — BASIC

Have each student write three quiz questions related to the material in this section. Then divide the class into small groups, and have the students quiz each other using their own questions. Co-op Learning

Quiz — GENERAL

True or False:

1. Hormones regulate many processes, including growth, development, and behavior. (True.)
2. Most of the endocrine glands are located in the brain. (False. Most are found in the body.)

Alternative Assessment — GENERAL

On an outline of a human body (similar to **Figure 2**), have students draw in the glands of the endocrine system and correctly label each one. Below each label they should indicate the hormone(s) produced by that gland and the hormone's function.

Real Life

Prostaglandins can cause pain, fever, and inflammation.

These symptoms, however, can be treated. Many nonprescription drugs, such as aspirin, ibuprofen, naproxen sodium, and ketoprofen, work by inhibiting prostaglandin production.

Finding Information

What symptoms can occur with long-term use of these drugs?

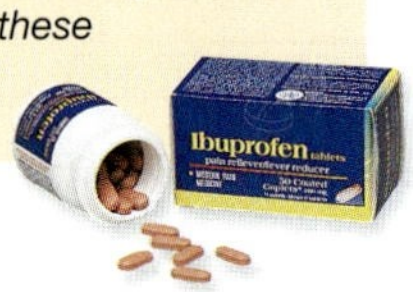

Another difference between the endocrine and nervous systems is that neurotransmitters are fast-acting and usually short-lived messengers. Hormones are usually slower-acting and longer-lived. The effects of some can last for days, weeks, or even years.

Finally, neurotransmitters are released from nerve cells directly to adjacent target cells. Endocrine cells can release hormones either into the bloodstream, where they travel to the cell they are to act on, or into the extracellular fluid to act on nearby cells.

Hormonelike Substances

The human body has many substances that regulate cellular activities much as hormones do. These substances were not initially considered hormones because they are not secreted into the bloodstream. However, most scientists today classify them as hormones. These substances include a large number of chemicals called neuropeptides, which are secreted by the nervous system, as well as chemicals called prostaglandins *(prahs tuh GLAN dihnz)*, which are secreted by most cells.

There are several different groups of neuropeptides. Enkephalins *(ehn KEHF uh lihnz)*, which were discussed in the previous chapter, are a group of neuropeptides that inhibit pain messages traveling toward the brain. Endorphins *(ehn DOHR fihnz)*, which are thought to regulate emotions, influence pain, and affect reproduction, are another important group of neuropeptides. Unlike neurotransmitters, enkephalins and endorphins tend to affect many cells near the nerve cells that release them.

Prostaglandins are modified fatty acids that have a variety of functions. They tend to accumulate in areas where tissues are disturbed or injured. There are dozens of different prostaglandins, and they produce a variety of effects. For example, some prostaglandins cause the constriction of blood vessels. The constricted blood vessels in turn affect blood pressure and body temperature. Other prostaglandins cause blood vessels to dilate, producing inflammation. A headache may result when blood vessels swell and their walls press against nerves in the brain.

Section 1 Review

1. **Describe** four ways in which hormones coordinate the activities of the body. SC.F.1.4.6
2. **Name** the type of gland that secretes substances into the bloodstream or extracellular fluid. SC.F.1.4.2
3. **Differentiate** the actions and chemical messengers in the endocrine system from those in the nervous system. SC.F.1.4.6
4. **Recognizing Relationships** Compare endorphins to neurotransmitters. SC.F.1.4.6
5. *FCAT Prep* Antidiuretic hormone is secreted when the body becomes dehydrated. How might this hormone help maintain water balance?
   A by causing profuse sweating
   B by stimulating an appetite for salt
   C by causing the release of dilute urine
   D by promoting water reuptake from urine
   SC.F.1.4.6

976

Answers to Section Review

1. Answers will vary, but might include regulating growth, development, behavior, and reproduction; producing, using, and storing energy; maintaining homeostasis; and reacting to external stimuli.
2. Endocrine glands secrete substances into the blood stream or extracellular fluid.
3. The endocrine system relies on relatively slow-acting hormones, with longer-lived responses. The nervous system relies on faster-acting neurotransmitters with shorter-lived responses.
4. Both chemicals are secreted by neurons. Unlike neurotransmitters, which are released directly to adjacent nerve cells, endorphins affect many cells near the target cells that produce them.
5. **A.** Incorrect. Profuse sweating would cause more water loss and greater dehydration. **B.** Incorrect. Ingesting more salt would cause more dehydration. **C.** Incorrect. Release of dilute urine would cause more water loss and greater dehydration. **D.** Correct. Reabsorption of water by the kidney is increased by antidiuretic hormone.

Section Review Correlations

SC.F.1.4.2: Item 2

SC.F.1.4.6: Items 1, 3, 4, 5

How Hormones Work

Section 2

Florida Sunshine Standards/FCAT — SC.F.1.4.6 Separate parts of the body communicate with each other using electrical and/or chemical signals.

Target Cells

After hormones are released from the cell in which they are made, they bind to and act only on target cells. A **target cell** is a specific cell that a hormone binds to and acts on (carries the message to). Imagine what would happen if hormones were not specific. All the cells in the body would respond to the hormone, resulting in uncoordinated activities, such as activation of many enzymes.

A hormone recognizes a target cell because the target cell has specific receptors. The receptors are located either on the surface of the target cell (on the cell membrane) or inside the cell (in the cytoplasm or nucleus). Recall that a receptor is a protein to which a molecule binds. A hormone's shape matches that of a particular receptor protein much like a key fits into a lock, as shown in **Figure 3.** Thus, a hormone binds only to cells that have a particular receptor protein, ignoring all other cells.

Types of Hormones

Most hormones are classified as either amino-acid-based or steroid hormones. **Amino-acid-based hormones** are hormones made of amino acids (either a single modified amino acid or a protein composed of 3 to 200 amino acids). Most amino-acid-based hormones are water soluble. **Steroid hormones** are lipid hormones that the body makes from cholesterol. Steroid hormones are fat soluble.

Objectives

- **Relate** how hormones act only on specific cells. SC.F.1.4.6
- **Summarize** how amino-acid-based hormones function. SC.F.1.4.6
- **Summarize** how steroid and thyroid hormones function. SC.F.1.4.6
- **Relate** how negative feedback is used to regulate hormone levels. SC.F.1.4.6

Key Terms

target cell
amino-acid-based hormone
steroid hormone
second messenger
negative feedback

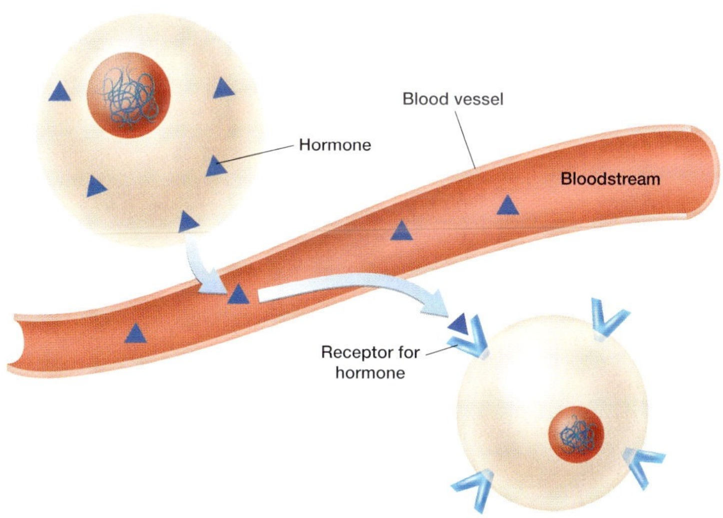

Figure 3 Hormones act on target cells. Hormones travel in the blood or in the fluid around cells to reach their target cells. The binding of the hormone with its receptor signals the target cell to change its activity.

977

Section 2

Focus

Overview

Before beginning this section review with your students the objectives listed in the Student Edition. Hormones act only on specific cells called target cells. The way a hormone acts on a cell is dependent on the type of hormone. This section introduces students to different types of hormones and how they work. An imbalance in hormone levels in the body can have serious results, so they must be closely regulated. Most hormones in the human body are regulated by negative feedback systems.

Bellringer

Have students write a short, nontechnical explanation of how a thermostat on a furnace helps to keep the room temperature at a comfortable level. When finished, ask for volunteers to read their explanations. They should note that when the temperature rises *above* a certain set-point, the thermostat turns the furnace off; when the room cools *below* a certain set-point, the thermostat turns the furnace back on. Explain that this is a negative feedback system and that many hormones are controlled in a similar way. **LS Logical**

Motivate

Demonstration — GENERAL

Obtain several different locks, each opened by a different key. On each lock, tape a slip of paper with a different cellular activity written on it (for example, "activate an enzyme," "increase blood flow," etc.). Tell students that the keys represent hormones. Give one key to a student and ask the student to try to open all of the locks. The student will be able to open only one lock. Have the student read out the cellular activity. Tell the students that in the same way, only the cell with the right keyhole will respond to the hormone. The hormone's message is then read by a second messenger, which results in a change in activity. **LS Kinesthetic**

Chapter Resource File
- Directed Reading BASIC
- Active Reading GENERAL

One-Stop Planner CD-ROM
- Reading Organizers BASIC
- Reading Strategies BASIC
- Occupational Applications Pharmacist GENERAL

Transparencies
- TR Bellringer
- TR J80 Hormones Acting on Target Cells
- TR J81 How Amino-Acid-Based Hormones Work
- TR J82 How Steroid Hormones Work
- TR J90 Negative Feedback

Teach, *continued*

Teaching Tip — GENERAL

Types of Hormones Have students organize information about the two main classes of hormones in a Graphic Organizer similar to the one at the bottom of this page. **LS Logical**

Teaching Tip — GENERAL

Receptor Sites Ask students if they can think of any other molecule that has a highly specific place (an active site) to which a protein or other substrate must bind in order to complete its function. (Enzyme-substrate interactions, like hormone-receptor protein interactions, are highly specific.) **LS Logical**

Using the Figure — GENERAL

BIO graphic Have students examine **Figure 4.** Point out that the hormone is the "first messenger" because it is the first molecule to bring the message to the cell telling the cell what it must do. In Step 1, the hormone binds to a receptor located on the cell membrane. When the hormone binds to the receptor, the receptor changes its shape. This shape change eventually results in the conversion of ATP to cyclic AMP, as shown in Step 2. Cyclic AMP, then acts as the "second messenger" by initiating a series of enzyme activations (or deactivations, depending on the hormone) as shown in Step 3. In the example in the figure, Step 4 illustrates how cyclic AMP starts the activation of enzymes that catalyze the breakdown of glycogen to individual glucose molecules. **LS Visual**

Real Life

How Hormones Work
Alfred Gilman and Martin Rodbell were two of the first scientists to determine how one of the amino-acid-based hormones works. They received the Nobel Prize for medicine in 1994 for their discoveries.

Receptors

When a hormone binds to a specific receptor on a target cell, the hormone brings the target cell a message. What happens after the hormone binds, however, depends on the type of hormone.

Amino-Acid-Based Hormones

Amino-acid-based hormones are not fat soluble, and most bind to cell membrane receptors, as shown in **Figure 4.**

Step 1 When an amino-acid-based hormone binds to a receptor, the shape of the receptor changes.

Step 2 This change in shape eventually activates a **second messenger,** a molecule that passes the message from the first messenger (the hormone) to the cell. For example, when glucagon, a hormone secreted by the pancreas, binds to a receptor, an enzyme is activated that converts ATP to a second messenger called cyclic AMP (cAMP).

Step 3 The second messenger then activates or deactivates certain enzymes in a cascade fashion. That is, one enzyme activates another enzyme, which activates yet another, and so on. In the case of glucagon, the second messenger cAMP activates a series of enzymes that breaks down glycogen into glucose.

Step 4 Eventually the activity of the target cell is changed by the final enzyme in the cascade—even though the hormone never enters the cell!

BIO graphic

Figure 4

How Amino-Acid-Based Hormones Work

Most amino-acid-based hormones, such as glucagon, bind to cell-membrane receptors, which activate second messengers that relay the hormone's message.

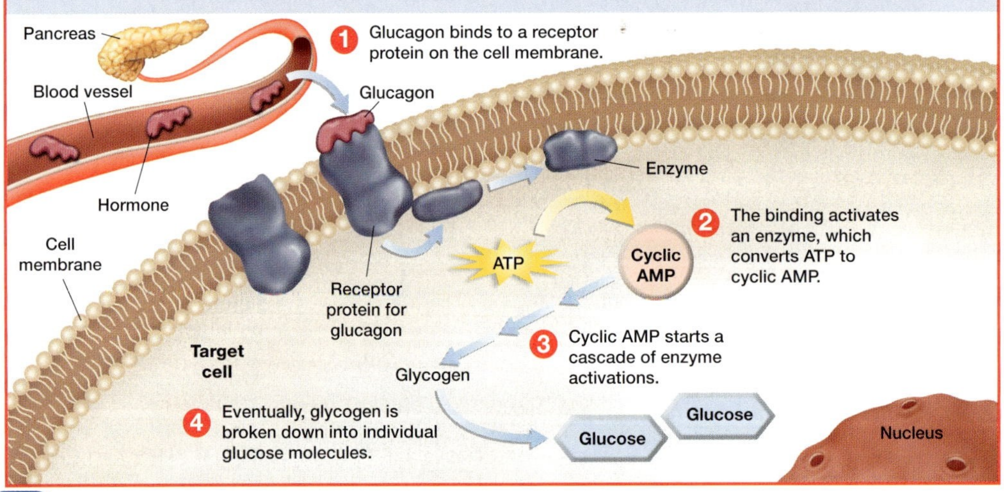

978

Graphic Organizer

Use this graphic organizer with **Teaching Tip: Types of Hormones** on this page.

Amino-acid-based hormones	Steroid and thyroid hormones
Water soluble	Fat soluble
Bind to receptors on the surface of the cell membrane	Bind to receptors in the cytoplasm or nucleus
Example: glucagon	Example: cortisol
Activate or deactivate enzymes	Activate or deactivate gene expression

Section Correlations

SC.F.1.4.6

Steroid and Thyroid Hormones

Because steroid and thyroid hormones are fat soluble, they readily pass through the cell membranes of their target cells. Steroid hormones bind to receptors located in a target cell's cytoplasm or its nucleus; thyroid hormones bind to receptors in a target cell's nucleus.

Cortisol is a steroid hormone made in the adrenal glands and released in response to stressful situations, such as the one shown in **Figure 5.** How steroid hormones such as cortisol work is summarized in **Figure 6.**

Step 1 The hormone diffuses through the cell membrane and binds to its receptor. The hormone and receptor form a hormone-receptor complex in the cytoplasm.

Step 2 The hormone-receptor complex enters the nucleus of the cell and binds to DNA.

Step 3 Depending on the hormone and the target cell, the binding either activates or inactivates a gene. That is, either the gene is transcribed and the resulting mRNA is translated into protein, or transcription and translation are inhibited.

Step 4 The target cell's activities are altered. For example, cortisol stimulates the making of enzymes that break down proteins and fats into glucose.

If the receptor for a steroid or thyroid hormone is located in the nucleus, the hormone enters the nucleus and binds to the receptor there. The hormone-receptor complex then binds to and affects the DNA in the same manner as it does with receptors in the cytoplasm.

Figure 5 Steroid hormones and stress. Stressful situations, such as taking an exam, result in the release of the steroid hormone cortisol.

Figure 6

BIO graphic — How Steroid Hormones Work

The steroid hormone–receptor complex binds to DNA in the nucleus and activates or inactivates transcription and translation of a gene.

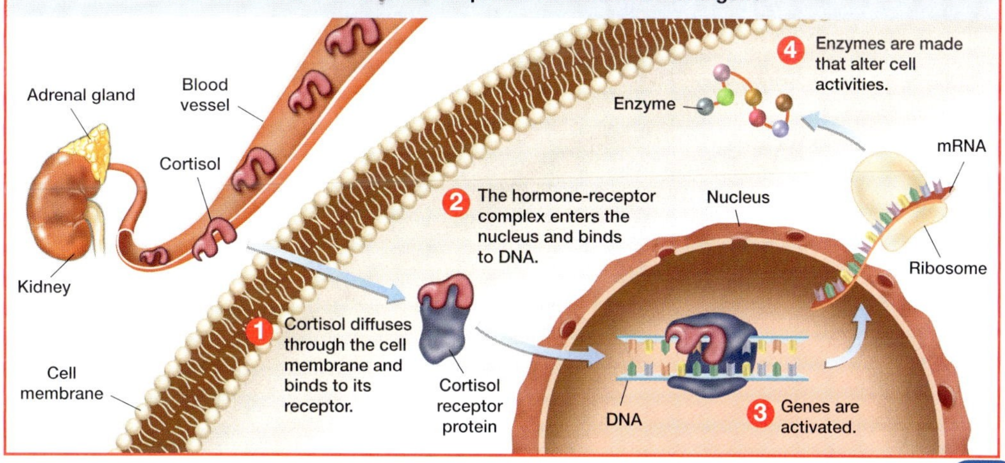

Demonstration — GENERAL

Obtain 4 small glass beakers or flasks, water, plain gelatin (a protein), light-colored cooking oil, and vitamin E capsules (oil-filled). To demonstrate that lipids are fat soluble and many proteins are not, place a small amount of gelatin in each of two flasks, one half-filled with water and one half-filled with cooking oil. Repeat the procedure, using separate flasks, with the vitamin E capsules. Swirl the liquids in the flasks until each substance has dissolved in one of the liquids. Ask students which substance is fat-soluble. (The vitamin E.) Which is not? (The protein—it is water-soluble.) Which type of substance can pass through a plasma membrane by diffusion? (A fat-soluble substance.) Why? (The plasma membrane has a double layer of phospholipids.) Could the gelatin enter a cell? (No.) Tell students that like gelatin, protein-based peptide hormones cannot enter cells, while fat-soluble steroid hormones can. **LS Kinesthetic**

Using the Figure — GENERAL

Have students compare **Figures 4 and 6.** Point out that because steroid hormones are fat-soluble, they penetrate the cell membrane to bind to a receptor either in the cytoplasm (as shown in Step 1) or in the nucleus (not shown). Once the hormone binds to the receptor, the hormone-receptor complex binds to DNA and activates a gene. Depending on the hormone, a gene can also be deactivated. In the figure, the gene is activated, mRNA is transcribed, and a new protein is synthesized (Step 4). **LS Visual**

did you know?

Steroid Hormones and Transplant Patients

The class of steroid hormones called corticosteroids inhibits the immune system when the hormones are present at high levels. Patients who undergo a transplant operation might receive a corticosteroid, such as cortisol, to reduce the immune defenses of the body. This allows the patient's body to accept, rather than reject, the transplanted organ.

Integrating Physics and Chemistry

Check students' understanding of the nature of solutes and solvents and the ability to form a solution by asking them to explain why amino-acid hormones, which are generally water-soluble, bind to the cell membrane of target cells, while steroid hormones, which are fat soluble, pass through the cell membrane of target cells. If they need assistance, review cell membrane structure, polar/non-polar concepts, and the "like dissolves like" rule.

Teach, *continued*

Teaching Tip — GENERAL

Feedback Loops Have students describe what would happen if part of the endocrine system were damaged and began to function as a positive feedback system, wherein a change in one direction promoted a continued change in the same direction. (There would be a constant oversecretion of those hormones.) LS Logical

SKILL BUILDER — GENERAL

Interpreting Graphics Have students examine **Figure 7.** Ask them what the lines would most likely look like if the graph were extended for double the current length. (The red line would most likely drop because of the low level of stimulating substance, and the yellow line would begin to rise because of the falling level of secondary substance.) LS Logical

Teaching Tip

Sex Hormones Tell students that the gonads produce three main types of sex hormones: androgens (testosterone), estrogens (estradiol), and progestins (progesterone). These sex hormones are found in both males and females, but in different proportions. Females have a high ratio of estrogens to androgens, and males have a high ratio of androgens to estrogens. Ask students what might happen if the hormone ratio were disrupted. (The male body could be more feminized, showing breast development, higher voice, and smaller muscle development; the female body could be masculinized, with such traits as beard growth and deeper voice.) LS Logical

Section Review Correlations

SC.F.1.4.2: Item 1

SC.F.1.4.6: Items 2-5

WORD Origins

The word *oxytocin* is from the Greek *oxys,* meaning "quick," and *tokos,* meaning "childbirth."

Feedback Mechanisms

The human body makes more than 40 hormones, and the body must regulate the release of the hormones. Nerve impulses alone can increase or decrease secretion of some hormones. For example, a baby nursing on a mother's breast stimulates the release of the hormone oxytocin, which in turn stimulates the release of milk from the mother's mammary glands.

Recall that homeostasis is essential in all living things. The endocrine system plays an important role in homeostasis. Different hormones moving through the bloodstream affect specific target tissues, and the amounts of various hormones must be maintained within a very narrow range.

In many cases the level of a hormone in the blood turns production of the hormone off and on through feedback mechanisms. Feedback mechanisms detect the amount of hormones in circulation or the amount of other chemicals produced because of hormone action. The endocrine system then adjusts the amount of hormones being made or released.

If high levels of a hormone stimulate the output of even *more* hormone, the regulation is called positive feedback. For example, the hormone that stimulates egg release also regulates the female hormone estrogen. A rise in estrogen levels, however, will stimulate the release of *more* of the regulatory hormone.

In humans, the release of most hormones is regulated through negative feedback, as shown in **Figure 7.** In **negative feedback,** a change in one direction stimulates the control mechanism to counteract further change in the same direction. For example, high levels of a hormone inhibit the production of more hormone, whereas low levels of a hormone stimulate the production of more hormone. The liver also plays a role in negative feedback by removing hormones from the blood and breaking them down. Negative feedback works like a person trying to maintain a certain speed in a car by pressing or releasing the gas pedal (accelerator).

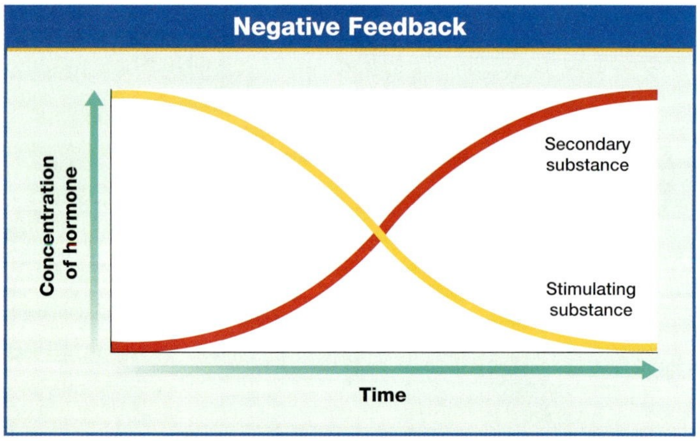

Figure 7 Negative feedback. In negative feedback, a secondary substance inhibits production of its initial stimulating substance.

980

did you know?

Heart Hormones The first heart hormone was discovered in the early 1990s. This hormone, atrial natriuretic factor, results in lower blood volume and pressure. In 1999, three more heart hormones were discovered: vessel dilator, which removes sodium and lowers blood pressure; long-acting natriuretic peptide, which removes salt from the blood; and kaliuretic peptide, which removes potassium from the blood. Researchers are currently testing whether one or more of these heart hormones can be used to treat congestive heart failure.

INCLUSION Strategies

- *Learning Disability*
- *Attention Deficit Disorder*

Have students use note cards to record the four steps of how amino-acid based hormones work and the four steps of how steroid and thyroid hormones work. Students should use one card per step, writing the number of the step on the back. Students may study each process by attempting to order the cards in each process and check their accuracy by the numbers on the back.

BIOWatch

Anabolic Steroids Are Dangerous

Many athletes use anabolic (protein-building) steroids and other hormone therapies to increase the size of their muscles and improve their performance. The unnatural use of steroids disrupts the feedback mechanisms that regulate hormone concentrations in the body.

Which Steroid Hormones Are Used?

The steroids used by athletes include synthetic hormones that mimic the male sex hormone testosterone. Many precursors to testosterone (such as androstenedione) are also used. Testosterone is secreted during puberty, when it stimulates many of the characteristics associated with being a man. For example, hair grows on the face, the underarms, and the pubic area; the voice deepens; and bigger muscles develop in the arms, legs, shoulders, and elsewhere.

Do Steroids Really Improve an Athlete's Performance?

When athletes inject steroids, they are trying to stimulate the production of proteins in the muscle cells as a way of increasing muscle mass and strength. In large doses, steroids can promote increases in mass, strength, and endurance.

Many Side Effects Are Associated with the Use of Steroids.

There are many side effects that accompany steroid use. When steroids are used before the skeleton matures completely, they stop the bones from growing. The body never reaches adult height and may look distorted. Liver cancer and other liver disorders may also result from steroid use. Some males who use anabolic steroids develop enlarged breasts and shriveled testes. Females who use these chemicals may develop facial hair, deepening of the voice, and male-pattern baldness. Finally, the virus that causes AIDS can be transmitted if shared needles are used to inject the steroids. *The long-term risks to health are often greater than any benefits from the use of steroids.*

internet connect
www.scilinks.org
Topic: Anabolic Steroids
Keyword: HX4006
SCILINKS. Maintained by the National Science Teachers Association

Section 2 Review

1. **Name** the structures found on or inside cells that allow hormones to recognize their target cells. SC.F.1.4.2
2. **Relate** how an amino-acid-based hormone changes a cell's activity. SC.F.1.4.6
3. **Relate** how a steroid or thyroid hormone changes a cell's activity. SC.F.1.4.6
4. **Analyzing Graphics** Use **Figure 7** to describe how hormone levels are regulated by negative feedback. SC.F.1.4.6
5. *FCAT Prep* *X* and *Y* are hormones. *X* stimulates the secretion of *Y*, which exerts negative feedback on the cells that secrete *X*. What happens when the blood level of *Y* decreases?
   A Less *X* is secreted.
   B More *X* is secreted.
   C Secretion of *X* stops.
   D Less *Y* is secreted.
   SC.F.1.4.6

Answers to Section Review

1. Structures that allow hormones to recognize target cells are called receptors.
2. When an amino-acid-based hormone binds to a receptor protein, that receptor protein changes shape. This causes the activation of a second messenger, which then activates or deactivates enzymes in a cascading fashion.
3. Steroid or thyroid hormones enter the cell and bind to receptors in the cell. The hormone-receptor complexes then bind to DNA, initiating or inhibiting gene transcription.
4. In a negative feedback system, high hormone levels cause hormone production to be reduced, and low hormone levels cause production to increase.
5. **A.** Incorrect. See answer B. **B.** Correct. As Y decreases, the production of X is no longer inhibited. **C.** Incorrect. See answer B. **D.** Incorrect. As Y decreases, the production of X is no longer inhibited. The resulting increase in X will stimulate production of Y.

BIOWatch

Anabolic Steroids are Dangerous

Teaching Strategies

Have students bring in articles about athletes who have been disqualified from competitions because of steroid use.

Discussion

- Why do you think anabolic steroids are banned from athletic events? (They are dangerous, and they give an athlete an unfair advantage.)
- What side effects are associated with steroid use? (Side effects include liver cancer, enlarged breasts, and shrunken testes in men; facial hair and deeper voices in women.)
- Should there be more drug testing of athletes? (Answers will vary.) LS **Verbal**

Close

Reteaching — BASIC

Have students describe to a partner how protein hormones and neurotransmitters are similar in the way they affect their target cells.
LS **Interpersonal**

Quiz — GENERAL

True or False:

1. Any particular hormone will work on a target cell. (False. Hormones are very specific.)
2. In negative feedback hormone systems, a secondary substance inhibits production of its initial stimulating substance. (True.)

Alternative Assessment — GENERAL

Have students use a large piece of paper to make a diagram of a negative feedback loop. Diagrams should include drawings of the glands and/or organs involved, and clear explanations about what is happening next to each step of the loop. LS **Visual**

Chapter Opener		TE **Using the Figure,** p. 994 SE **Quick Review,** p. 995 ■ SE **Reading Activity,** p. 995 ■ TE **Opening Activity,** p. 995
& 3 • 90 min pp. 996–998 **Section 1** Male Reproductive System	OSP **Lesson Plan*** TR **Bellringer*** TR **Section Outline*** TR **J91 Structure of the Testes*** TR **J94 Male Reproductive System*** TR **J92 Sperm Cell*** TR **J93 Male Hormones and Reproduction***	TE **Using the Figure,** p. 997 Basic
& 5 • 90 min pp. 999–1003 **Section 2** Female Reproductive System	OSP **Lesson Plan*** TR **Bellringer*** TR **Section Outline*** TR **J96 Female Reproductive System*** TR **J97 Ovarian and Menstrual Cycles***	TE **Using the Figure,** p. 1000 TE **Group Activity** Cervical Cancer, p. 1001 Advanced SE **Data Lab** Analyzing Hormone Secretion, p. 1003 ■ CRF **Datasheets for In-Text Labs** Analyzing Hormone Secretion* CRF **Quicklab** Blood Typing, Preg. & Hemolytic Disease* Basic
& 7 • 90 min pp. 1004–1007 **Section 3** Development	OSP **Lesson Plan*** TR **Bellringer*** TR **Section Outline*** TR **J98 Early Zygote Development*** TR **J101 Structure of the Placenta***	TE **Demonstration,** p. 1004 General TE **Group Activity** Fetal Alcohol Syndrome, p. 1006 General CRF **Inquiry Lab** Will a Hormone Affect Frog Metamorphosis* General
& 9 • 90 min pp. 1008–1010 **Section 4** Sexually Transmitted Diseases	OSP **Lesson Plan*** TR **Bellringer*** TR **Section Outline*** TR **J35 Sexually Transmitted Diseases*** TR **J43 Known Routes of HIV Transmission***	TE **Activity** STDs, p. 1008 Basic TE **Activity** Incidence of STDs, p. 1009 General SE **Skills Practice Lab** Observing Embryonic Development, p. 1014 ■ CRF **Datasheets for In-Text Labs** Observing Embryonic Development*

0 & 11 • 90 min

Chapter Review and Assessment Resources

- SE **Study Zone,** p. 1011 ■
- SE **Performance Zone,** p. 1012 ■
- SE **FCAT Test Prep,** p. 1013 ■
- TR **Concept Mapping***
- CRF **Vocabulary Review Worksheet*** ■ General
- CRF **Test Prep Pretest*** ■ General
- CRF **Chapter Test*** ■ General
- CRF **Chapter Test*** Advanced
- OSP **Test Generator***
- OSP **Test Item Listing***
- OSP **Modified Chapter Test*** Basic

Online and Technology Resources

Holt Online Learning

Visit go.hrw.com for access to Holt Online Learning or enter the keyword **HX6 Home** for a variety of free online resources.

One-Stop Planner® CD-ROM

This CD-ROM package includes
- Lab Materials QuickList Software
- Holt Calendar Planner
- Customizable Lesson Plans
- Printable Worksheets
- ExamView® Test Generator
- Interactive Teacher Edition
- Holt PuzzlePro® Resources
- Holt PowerPoint® Resources

Guided Reading Audio CDs

These CDs are designed to help auditory learners and reluctant readers. (Audio CDs are also available in Spanish.)

Compression guide:
To shorten your instruction because of time limitations, omit block 1.

KEY

- **TE** Teacher Edition
- **SE** Student Edition
- **OSP** One-Stop Planner
- **CRF** Chapter Resource File
- **TR** Teaching Transparency
- * Also on One-Stop Planner
- ■ Also Available in Spanish
- ◆ Requires Advance Prep

SKILLS DEVELOPMENT AND PRACTICE	SECTION REVIEW AND ASSESSMENT	CORRELATIONS
		Florida Sunshine Standards/FCAT
TE Reading Skill Builder K-W-L, p. 997 BASIC **CRF** Directed Reading Worksheet* BASIC **CRF** Active Reading Worksheet* GENERAL	**SE** Section 1 Review ■ **TE** Reteaching, p. 998 BASIC **TE** Quiz, p. 998 GENERAL **TE** Alternative Assessment, p. 998 GENERAL **CRF** Quiz* ■ BASIC	SC.F.1.4.2
TE Skill Builder Vocabulary, p. 1000 GENERAL **TE** Skill Builder Math Skills, p. 1001 GENERAL **TE** Skill Builder Interpreting Graphics, p. 1002 GENERAL **TE** Reading Skill Builder Interactive Reading, p. 1002 BASIC **CRF** Directed Reading Worksheet* BASIC **CRF** Active Reading Worksheet* GENERAL	**SE** Section 2 Review ■ **TE** Reteaching, p. 1003 BASIC **TE** Quiz, p. 1003 GENERAL **TE** Alternative Assessment, p. 1003 GENERAL **CRF** Quiz* ■ BASIC	SC.F.1.4.2, SC.F.1.4.6
TE Skill Builder Interpreting Visuals, p. 1005 GENERAL **TE** Reading Skill Builder Reading Hint, p. 1006 BASIC **CRF** Directed Reading Worksheet* BASIC **CRF** Active Reading Worksheet* GENERAL	**SE** Section 3 Review ■ **TE** Reteaching, p. 1007 BASIC **TE** Quiz, p. 1007 GENERAL **TE** Alternative Assessment, p. 1007 GENERAL **CRF** Quiz* ■ BASIC	SC.F.1.4.2, SC.F.1.4.8
TE Skill Builder Math Skills, p. 1009 GENERAL **CRF** Directed Reading Worksheet* BASIC **CRF** Active Reading Worksheet* GENERAL **CRF** Critical Thinking Worksheet* ADVANCED **CRF** Science Skills Worksheet* GENERAL	**SE** Section 4 Review ■ **TE** Reteaching, p. 1010 BASIC **TE** Quiz, p. 1010 GENERAL **TE** Alternative Assessment, p. 1010 GENERAL **CRF** Quiz* ■ BASIC Modified Worksheet* BASIC	SC.F.1.4.2, SC.F.1.4.8, SC.G.1.4.1

Holt Biology Interactive Tutor CD-ROM

This CD-ROM consists of interactive activities that give students a fun way to extend their knowledge of biology concepts.

Unit 4 Cell Reproduction

SCILINKS. NSTA
www.scilinks.org

Maintained by the **National Science Teachers Association**

Topic: Transportation of Sperm
SciLinks Code: HX4179

Topic: Ovarian Cycle
SciLinks Code: HX4133

Topic: Ultrasound
SciLinks Code: HX4180

Overview

Tell students that the purpose of this chapter is to explain human reproduction and development. The chapter first describes the anatomy and physiology of the male and female reproductive systems. It then describes fertilization, pregnancy, and development of the embryo/fetus. Lastly, the chapter explains how sexually transmitted diseases (STDs) can be passed from person to person, focusing on bacterial and viral STDs.

Using the Figure

The photo is of a human fetus at 6 months. The white hair covering the fetus is called lanugo (lah-NEW-go). It is usually shed during the ninth month of development before birth, but it is present on premature infants. With the exception of the nervous system, most of the development of fetal body systems has taken place by this time, but the fetus could survive on its own only with medical intervention. During the seventh through ninth months, fetal systems will mature, and the fetus will grow and double its weight several times. Development of the nervous system, including development of the brain, will continue past birth.

Assessing Prior Knowledge

Before beginning this chapter, be sure your students can answer the questions in the ***Quick Review*** on the facing page. If they cannot, encourage them to reread the sections indicated.

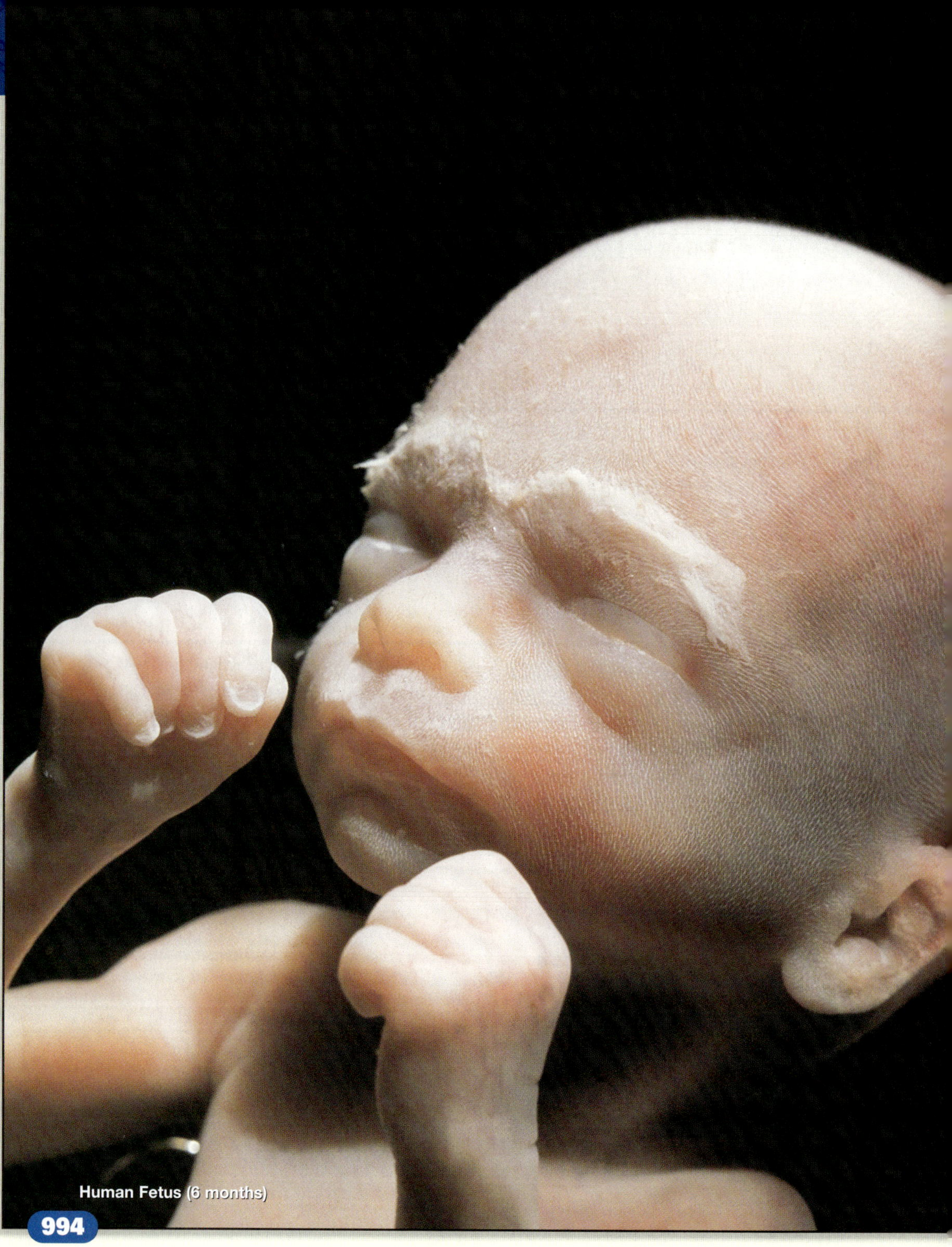

Human Fetus (6 months)

994

Chapter Correlations | **Florida Sunshine Standards/FCAT**

SC.F.1.4.2 Body structures are uniquely designed and adapted for their function.

SC.F.1.4.6 Separate parts of the body communicate with each other using electrical and/or chemical signals.

SC.F.1.4.8 Cell behavior can be affected by molecules from other parts of the organism or even from other organisms.

SC.G.1.4.1 There is great diversity and interdependence among living things.

CHAPTER

43 Reproduction and Development

Florida Sunshine Standards/FCAT

SC.F.1.4.2 Body structures are uniquely designed and adapted for their function.

SC.F.1.4.6 Separate parts of the body communicate with each other using electrical and/or chemical signals.

SC.F.1.4.8 Cell behavior can be affected by molecules from other parts of the organism or even from other organisms.

SC.G.1.4.1 There is great diversity and interdependence among living things.

Quick Review

Answer the following without referring to earlier sections of your book.

1. **Sequence** the phases of meiosis. *(Chapter 7, Section 1)*
2. **List** the stages of animal development in the zygote. *(Chapter 27, Section 1)*
3. **Describe** the functions of sex hormones. *(Chapter 42, Section 3)*
4. **Summarize** the role of the hypothalamus in the endocrine system. *(Chapter 42, Section 3)*

Did you have difficulty? *For help, review the sections indicated.*

Reading Activity

Take a break after reading each section of this chapter, and closely study the figures in the section. Reread the figure captions, and for each one, write out a question that can be answered by referring to the figure and its caption. Refer to your list of figures and questions as you review the concepts addressed in the chapter before you complete the Performance Zone chapter review.

At six months, the human fetus is about 35.6 cm (14 in.) long and has a mass of about 908 g (weighs about 2 lb). The fetus has some brain wave activity, and its eyes will open soon.

Looking Ahead

Section 1
Male Reproductive System
The Testes

Section 2
Female Reproductive System
The Ovaries
The Ovum

Section 3
Development
Fertilization
Pregnancy

Section 4
Sexually Transmitted Diseases
STDs

internet connect

www.scilinks.org
National Science Teachers Association *sci*LINKS Internet resources are located throughout this chapter.

SCILINKS. Maintained by the National Science Teachers Association

995

Opening Activity

Rates of Development Draw a small dot on the board. Ask students how the size of this dot compares with a newly fertilized egg. (It is larger than a fertilized egg.) Tell students that a fertilized egg is about 0.14 mm (0.005 in.) in diameter. Then show students a doll that is about the size of a newborn baby. Tell students that the average length of a newborn baby is approximately 51 cm (20 in.). Have them calculate a fetus's average rate of growth per month (5.7 cm [2.2 in.] per month) and their own average rate of growth per month since birth (for a 166-cm tall [65 in.] 16-year-old student, about 0.6 cm [0.23 in.] per month). How does the increase in size during prenatal development compare with that of postnatal growth? (It is much greater during prenatal development.) **LS Logical**

Quick Review Answers

1. The phases of meiosis are: prophase I, metaphase I, anaphase I, telophase I, cytokinesis; prophase II, metaphase II, anaphase II, telophase II, cytokinesis.
2. The stages of animal development are the zygote, eight-celled stage, blastula, and gastrula.
3. The sex hormones affect the formation of gametes and control sexual behavior and cycles. They also stimulate the development of secondary sex characteristics, such as breast size, hair growth, and muscle development.
4. The hypothalamus coordinates the activities of the nervous and endocrine systems and controls many body functions, including body temperature, blood pressure, and emotions.

Chapter Resource File

- Vocabulary Worksheets
- Concept Mapping

Reading Activity Answers

Students should be certain to ask questions that can be answered by the figure and its caption. For example, an appropriate question for Figure 1 is "What is the function of the testes?"

Focus

Overview

Before beginning this section review with your students the objectives listed in the Student Edition. Section 1 describes the male reproductive system, focusing on the production, maturation, storage, and structure of sperm. The section ends with a discussion of how sperm are transported within the male body and delivered to the female reproductive tract.

Bellringer

Have students list reproductive hormones and their functions. (Examples: FSH—regulates development of male and female gametes; LH—stimulates egg release from the ovary; estrogen, progesterone, and testosterone—regulate gamete formation and sexual behavior and cycles) After taking attendance, list these hormones and functions on the board or overhead, adding to the list as you work through the chapter.

Motivate

Identifying Preconceptions —GENERAL

Ask students if they believe that they are at risk for developing cancer. High school students often think that cancers are diseases of old people. However, testicular cancer is a disease that generally strikes young men between the ages of 15 and 35 years. Discuss American Lance Armstrong, the multiple-time Tour de France winner, and his battle to overcome testicular cancer. LS Interpersonal

Section Correlations

SC.F.1.4.2

Section 1

Male Reproductive System

Florida Sunshine Standards/FCAT

SC.F.1.4.2 Body structures are uniquely designed and adapted for their function.

Objectives

- **Describe** how sperm are produced. SC.F.1.4.2
- **Identify** the major structures of the male reproductive system. SC.F.1.4.2
- **Relate** the structure of a sperm cell to its functions. SC.F.1.4.2
- **Sequence** the path taken by sperm as they leave the body. SC.F.1.4.2

Key Terms

testes
seminiferous tubules
epididymis
vas deferens
seminal vesicles
prostate gland
bulbourethral glands
semen
penis

The Testes

What are the roles of a human male in sexual reproduction? Recall that sexual reproduction involves the formation of a diploid zygote from two haploid sex cells, or gametes, through fertilization. The roles of a male in sexual reproduction are to produce sperm cells—the male gametes—and to deliver the sperm cells to the female reproductive system to fertilize an egg cell—the female gamete.

Where are sperm produced? Two egg-shaped **testes** *(TEHS teez),* or testicles, are the gamete-producing organs of the male reproductive system. The testes are located in the scrotum *(SKROHT uhm),* an external skin sac. The testes first form inside the abdominal cavity then move down into the scrotum either before or shortly after birth. The normal body temperature of 37°C (98°F) is too high for sperm to complete development. In the scrotum the temperature is about 3°C lower than it is in the rest of the body, making it an ideal location for sperm production.

Production of Sperm

The testes begin to produce sperm during the adolescent stage of development known as puberty *(PYOO buhr tee).* As shown in **Figure 1,** each testis contains hundreds of compartments packed with many tightly coiled tubules, called **seminiferous** *(sehm uh NIHF uhr uhs)* **tubules.** Sperm cells are produced through meiosis in the lining of the seminiferous tubules. Thus, human sperm cells contain only 23 chromosomes (the haploid number) instead of the usual 46 chromosomes (the diploid number) found in other body cells. Two hormones released by the anterior pituitary regulate the functioning of the testes. Luteinizing hormone (LH) stimulates secretion of the sex hormone testosterone. Follicle-stimulating hormone (FSH), along with testosterone, stimulates sperm production in the seminiferous tubules. Cells located between the seminiferous tubules secrete testosterone.

Figure 1 Testes

The testes produce sperm cells.

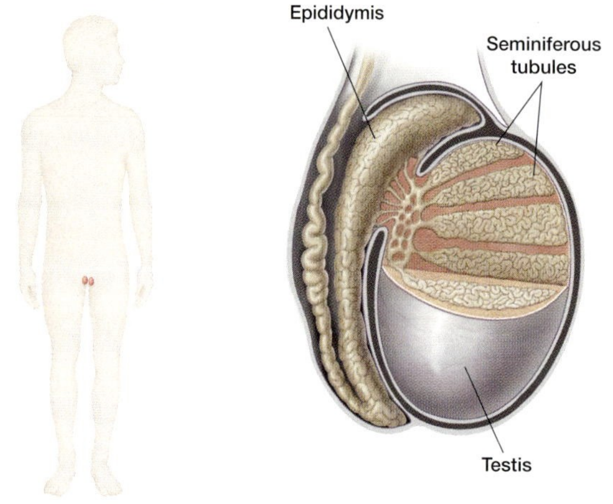

Chapter Resource File

- Directed Reading BASIC
- Active Reading GENERAL

Transparencies

TR Bellringer
TR J91 Structure of the Testes
TR J94 Male Reproductive System

One-Stop Planner CD-ROM

- Reading Organizers BASIC
- Reading Strategies BASIC

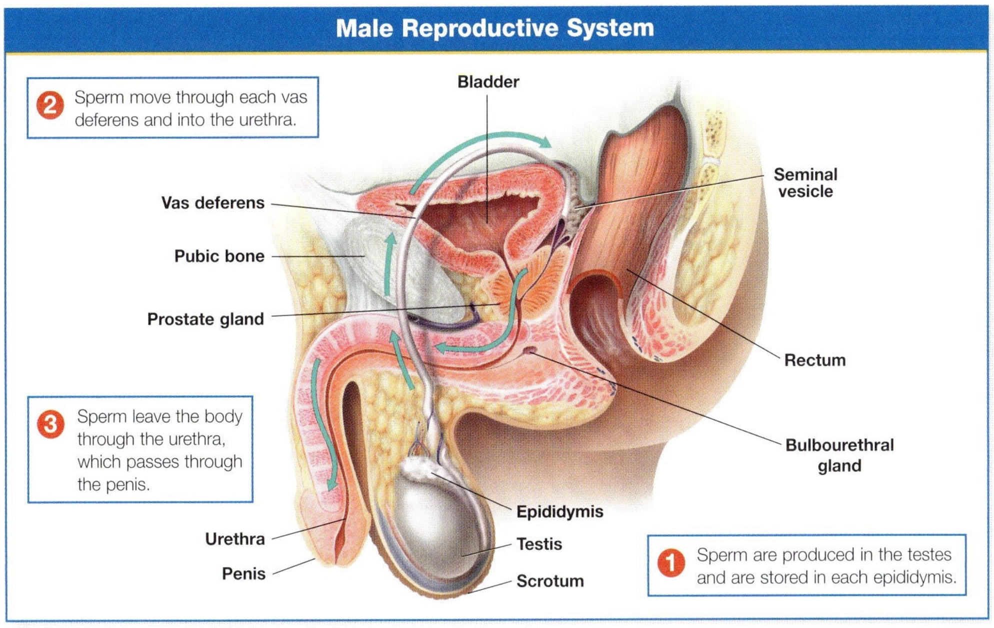

Figure 2 **Male reproductive system.** The arrows indicate the path taken by sperm cells from the testes as they exit the body.

Maturation and Storage of Sperm

A typical adult male produces several hundred million sperm cells each day. After being produced in the seminiferous tubules, the sperm travel through a series of long tubes though they are not yet capable of swimming. Sperm then enter a long coiled tube called the **epididymis** *(ehp uh DIHD ih mihs)*, shown in Figure 1. Within each epididymis, the sperm mature and become capable of moving.

The epididymis is also the site where most of the sperm are stored. From the epididymis, some sperm move to another long tube, the **vas deferens** *(vas DEHF uh rehnz)*. Sperm move through the vas deferens and into the urethra, as shown by the arrows in **Figure 2.** Sperm leave the body by passing through the urethra, the same duct through which urine exits the body.

Structure of Mature Sperm

As shown in **Figure 3,** a mature sperm cell consists of a head with very little cytoplasm, a midpiece, and a long tail. Enzymes at the tip of the head help the sperm cell penetrate an egg cell during fertilization. The midpiece contains many mitochondria that supply sperm with the energy needed to propel themselves through the female reproductive system. The tail of a sperm cell is a powerful flagellum that whips back and forth, enabling the sperm cell to move. ATP produced in the mitochondria power the whiplike movements of the tail. During fertilization, only the head of a sperm enters an egg, so a father's mitochondria are not passed to offspring.

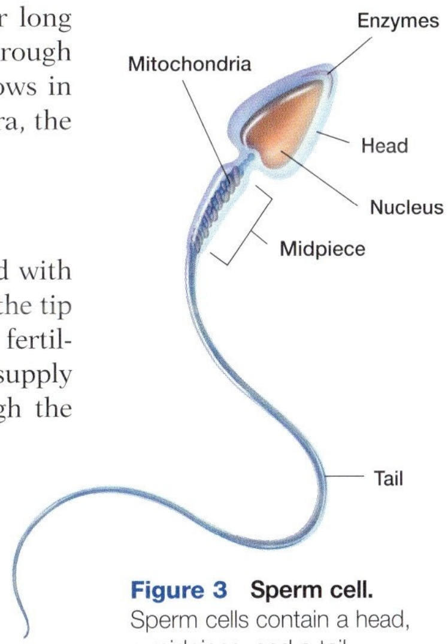

Figure 3 **Sperm cell.** Sperm cells contain a head, a midpiece, and a tail.

Teach

Teaching Tip — GENERAL

Vasectomy Explain to students that a vasectomy is a minor operation performed on adult males as a method of birth control. In a vasectomy, both vas deferens are cut to stop sperm from exiting the body. Ask students if they think a vasectomy would affect secondary sex characteristics. (No. Cutting the vas deferens does not affect the production of male hormones, only the release of sperm, therefore secondary sex characteristics will not be affected.)

Using the Figure — BASIC

Have students identify what part of the sperm shown in **Figure 3** must penetrate the egg for fertilization to take place. (The head region, which contains paternal DNA [genes], must be delivered into the egg.) What parts of a sperm cell help it move toward and fertilize an egg? (Sperm cells have a tail, which propels the sperm, and mitochondria, which provide energy for movement.) LS Visual

Teaching Tip — BASIC

Path of Sperm Have students make a Graphic Organizer similar to the one at the bottom of this page to trace the movement of sperm cells from the epididymis to the urethra. LS Visual

READING SKILL BUILDER — BASIC

K-W-L Before students continue further in this chapter, have them write short lists of all the things they **K**now about reproduction, development, and STDs. Write some of these ideas on the board or overhead. Then have students list things they **W**ant to know about reproduction, development, and STDs. Have students save their lists for use later at the end of Section 4. LS Verbal

Graphic Organizer

Use this graphic organizer with **Teaching Tip: Path of Sperm** on this page.

Seminiferous tubules → Epididymis → Vas deferens → Urethra

Close

Reteaching — BASIC

Show the class an overhead transparency of the male reproductive system. Have students identify the structures and functions of the various parts. Review quickly, calling on students to name the structure or function.

Quiz — GENERAL

1. Where are sperm produced? (Sperm are produced in the testes.)
2. What is the function of the prostate? (The prostate secretes an alkaline fluid, which becomes part of the semen.)
3. What is the physical mechanism by which the penis becomes erect? (When the three cylinders of spongy tissue within the penis become filled with blood, the penis becomes erect.)

Alternative Assessment — GENERAL

Provide students a graphic showing the male reproductive system. Ask students to label the structures in the figure and draw the path of sperm through the system using a colored pencil.

Transparencies

TR J92 Sperm Cell

TR J93 Male Hormones and Reproduction

internet connect

www.scilinks.org

Topic: Transportation of Sperm

Keyword: HX4179

SCILINKS. Maintained by the National Science Teachers Association

Semen

As sperm cells move into the urethra, they mix with fluids secreted by three exocrine glands: the seminal vesicles, the prostate gland, and the bulbourethral glands. These fluids nourish the sperm and aid their passage through the female reproductive system. The **seminal vesicles,** which lie between the bladder and the rectum, produce a fluid rich in sugars that sperm use for energy. The **prostate** *(PRAHS tayt)* **gland,** which is located just below the bladder, secretes an alkaline fluid that neutralizes the acids in the female reproductive system. Before semen leaves the body, the **bulbourethral** *(buhl boh yoo REE thruhl)* **glands** also secrete an alkaline fluid that neutralizes traces of acidic urine in the urethra. The mixture of these secretions with sperm is called **semen** *(SEE muhn).*

Delivery of Sperm

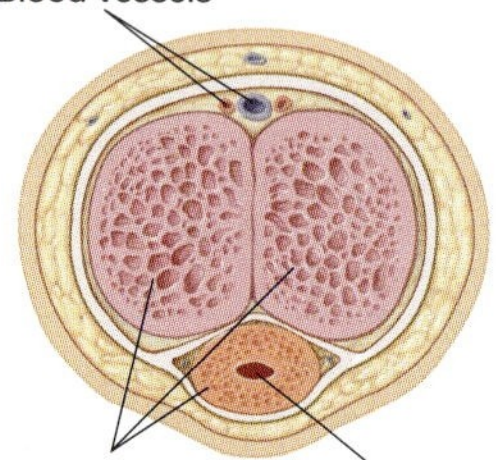

Figure 4 Cross section of penis. The penis contains three cylinders of spongy tissue. When the spaces in these cylinders fill with blood, the penis becomes erect.

The urethra passes through the **penis,** the male organ that deposits sperm in the female reproductive system during sexual intercourse. During sexual arousal, blood flow to the penis increases. As shown in **Figure 4,** the penis contains three cylinders of spongy tissue. Small spaces separate the cells of the spongy tissue. Blood collects within these spaces, causing the penis to become rigid and erect.

Sperm exit the penis through ejaculation *(ee jak yoo LAY shun),* the forceful expulsion of semen. During ejaculation, muscles around each vas deferens contract, moving sperm into the urethra. Muscles at the base of the penis force semen out of the urethra.

After the semen is deposited in the female reproductive system, sperm swim until they encounter an egg cell or until they die. If sperm are unable to reach an egg, fertilization does not occur. Covering the penis with a thin rubber sheath called a condom helps prevent fertilization during sexual intercourse. Abstaining from sexual intercourse is the surest way to prevent fertilization.

About 3.5 mL of semen, containing 300–400 million sperm, is expelled during ejaculation. Because most sperm die in the female reproductive system, fertilization usually requires a high sperm count. Males with fewer than 20 million sperm per mL of semen are generally considered sterile.

Section 1 Review

1. **Sequence** the path that mature sperm take from the testes to the outside of the body. SC.F.1.4.2
2. **Describe** the role of each part of a mature sperm cell. SC.F.1.4.2
3. **Recognizing Relationships** How do secretions by exocrine glands help the delivery of sperm to the female reproductive system? SC.F.1.4.6
4. **Describe** the functions of the testes. SC.F.1.4.2
5. **Inferring Relationships** If a male's left vas deferens is blocked, how is his sperm count affected? Explain your answer. SC.F.1.4.2
6. *FCAT Prep* What structure secretes a fluid containing sugars that sperm use for energy?
   A epididymis
   B prostate gland
   C seminal vesicles
   D vas deferens
   SC.F.1.4.2

Section Review Correlations

SC.F.1.4.2: Items 1, 2, 4, 5, 6

SC.F.1.4.6: Item 3

Answers to Section Review

1. Mature sperm travel from the epididymis, through the vas deferens, and out the urethra.
2. The head contains genetic information and enzymes that help the sperm cell penetrate the egg. The midpiece contains mitochondria, which supply energy. The tail is a flagellum, which enables the sperm cell to move.
3. Exocrine secretions provide alkalinity, which decreases the acidity of the male urethra and the female reproductive tract. Exocrine secretions also provide nourishment and a fluid environment in which the sperm can swim.
4. Testes produce gametes called sperm and the sex hormone testosterone.
5. The sperm count likely is reduced by half as mature sperm from the left epididymis will be blocked.
6. **A.** Incorrect. The epididymis does not secrete fluid; it is a coiled tube in which sperm mature. **B.** Incorrect. The prostate gland secretes an alkaline fluid that neutralizes the acids in the female reproductive system. **C.** Correct. The seminal vesicles provide a fluid rich in sugars, which is used for energy. **D.** Incorrect. The vas deferens does not secrete fluid; it is a tube through which sperm move.

Female Reproductive System

Section 2

Florida Sunshine Standards/FCAT

SC.F.1.4.2 Body structures are uniquely designed and adapted for their function.

SC.F.1.4.6 Separate parts of the body communicate with each other using electrical and/or chemical signals.

The Ovaries

Each month, the female reproductive system prepares for a possible pregnancy by producing a mature egg cell—the female gamete. After sperm have been deposited and fertilization has occurred, the role of the male in reproduction is complete. If pregnancy occurs, the female reproductive system will nourish and protect the fertilized egg through nine months of development.

Production of Eggs

Two egg-shaped ovaries, shown in **Figure 5,** are located within the abdominal cavity. The **ovaries** *(OH vuh reez)* are the gamete-producing organs of the female reproductive system. Females are born with all of the egg cells they will ever produce. At birth, the ovaries contain about 2 million immature egg cells that already have begun the first division of meiosis. Like sperm cells, egg cells contain 23 chromosomes (the haploid number) because eggs also are formed through meiosis.

After meiosis begins, egg cells become stalled in prophase of the first meiotic division. When a female reaches puberty, the increased production of sex hormones enables meiosis to resume. However,

Objectives

- **Describe** how eggs are produced. SC.F.1.4.2
- **Identify** the major structures of the female reproductive system. SC.F.1.4.2
- **Analyze** the events of the ovarian and menstrual cycles. SC.F.1.4.6

Key Terms

ovary
ovum
fallopian tube
uterus
vagina
ovarian cycle
ovulation
follicle
corpus luteum
menstrual cycle
menstruation

Figure 5 **Female reproductive organs**

The ovaries produce egg cells. The uterus nurtures the fetus during pregnancy.

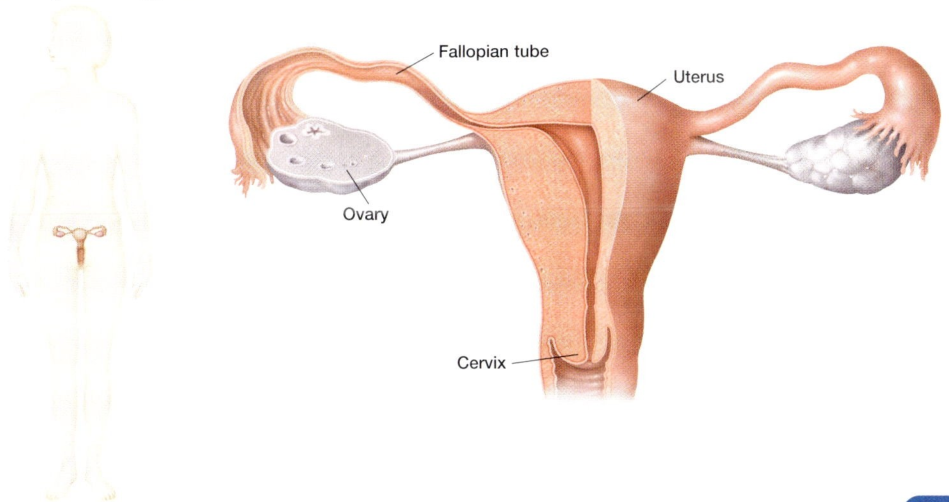

999

Section 2

Focus

Overview

Before beginning this section review with your students the objectives listed in the Student Edition. Section 2 describes the female reproductive system, focusing on the production and maturation of eggs, and ovulation. The section ends with a discussion of the ovarian and menstrual cycles.

Bellringer

Have students pull out the lists they started in the Bellringer activity in Section 1. Ask them to add any details learned about male reproductive hormones. (Examples: LH—released by anterior pituitary, stimulates secretion of testosterone; FSH—released by anterior pituitary, stimulates sperm production in seminiferous tubules) Tell them they will study similar details about female hormones in this section.

Motivate

Discussion/ Question — GENERAL

Ask students at what age women can no longer become pregnant and give birth. (A woman's reproductive life generally ends at menopause, which occurs at about age 50.) Ask students at what age men can no longer father children. (A man's reproductive life never ends, unless disability or disease renders him unable to father children.) Tell students that the probability of birth defects increases with maternal age, beginning at about age 35, but not with paternal age. Ask students to explain why these differences exist between male and female reproductive characteristics. (A woman has all the eggs she will ever have at birth. As a woman ages, her eggs age, increasing the probability of genetic damage. Eventually ovulation ceases, and a woman can no longer bear children without medical intervention. A man manufactures sperm from puberty through the rest of his life.) LS **Logical**

Transparencies

TR Bellringer

TR J96 Female Reproductive System

Section Correlations

SC.F.1.4.2

SC.F.1.4.6

Teach

Using the Figure

Have students locate the point at which sperm enter the female reproductive system illustrated in **Figure 6.** (the vagina) Have them trace the path the egg travels and locate where the sperm meets the egg. (in the upper third of the fallopian tubes) **LS Visual**

Teaching Tip

Sperm and Egg Size Tell students that eggs are much larger than sperm. Ask students what might account for this large difference in size. Explain to students that the difference in the sizes of sperm and eggs is due to the difference between the processes of sperm formation (spermatogenesis) and egg formation (oogenesis). During spermatogenesis, equal divisions of the cytoplasm follow meiosis I and meiosis II, resulting in four equal-sized sperm cells having little cytoplasm. During oogenesis, unequal divisions of the cytoplasm follow meiosis I and meiosis II, resulting in one large egg and three tiny cells that degenerate.

SKILL BUILDER — GENERAL

Vocabulary The word *cervix* means "neck" in Latin. Have students explain why this is an appropriate term for the ring of strong muscles located in the lower part of the uterus. (Answers will vary but could include the similarity of the cervix to the neck of a bottle.) English Language Learners **LS Verbal**

Female Reproductive System

1. An ovary releases an ovum into a fallopian tube about once every month.
2. The ovum moves through the fallopian tube toward the uterus.
3. When fertilization occurs, the fertilized ovum develops in the uterus.

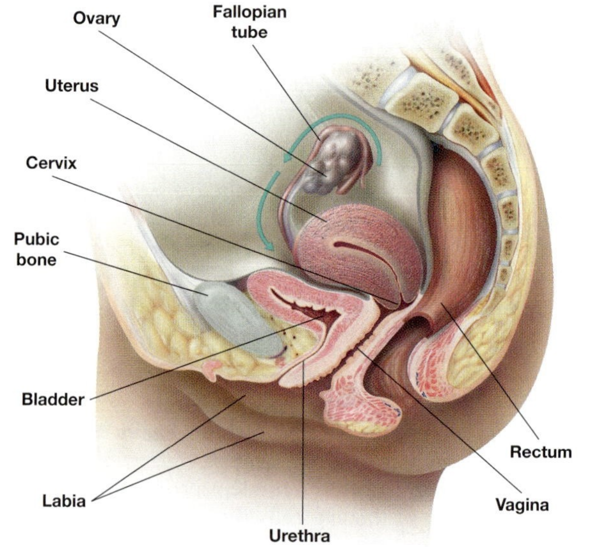

Figure 6 Female reproductive system. The arrows indicate the path taken by an ovum from an ovary to the uterus.

normally only one immature egg cell matures each month. In the lifetime of a female, only 300–400 egg cells will mature. When an egg cell matures, it is called an **ovum** *(OH vuhm).*

Structures of the Female Reproductive System

An ovum is released from an ovary about every 28 days. Cilia sweep the ovum into a fallopian tube. Each **fallopian** *(fuh LOH pee uhn)* **tube** is a passageway through which an ovum moves from an ovary toward the uterus. Smooth muscles lining the fallopian tubes contract rhythmically, moving the ovum down the tube and toward the uterus, as shown by the arrows in **Figure 6.** An ovum's journey through a fallopian tube usually takes three to four days to complete. If the ovum is not fertilized within 24–48 hours, it dies. An ovum, shown in **Figure 7,** is many times larger than a sperm cell and can be seen with the unaided eye.

Figure 7 Ovum. Notice the great difference in size between the sperm and the ovum.

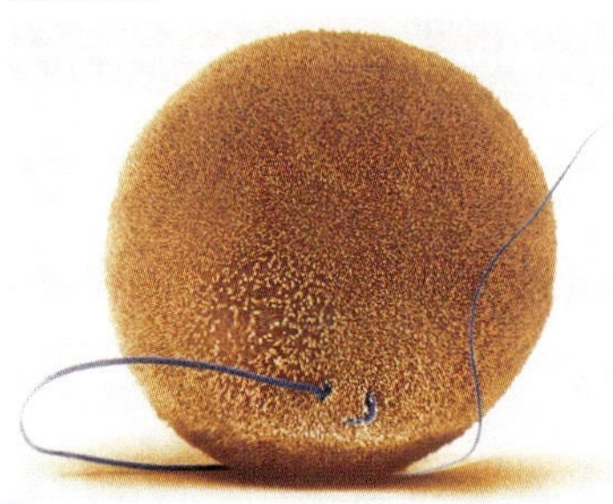

The **uterus** *(YOO tuh ruhs)* is a hollow, muscular organ about the size of a small fist. If fertilization occurs, development will take place in the uterus. During sexual intercourse, sperm are deposited inside the **vagina** *(vuh JIE nuh),* a muscular tube that leads from the outside of the female's body to the entrance to the uterus, called the *cervix (SUR vihks).* A soft rubber cap called a *diaphragm (DIE uh fram)* can be used to cover the cervix and help prevent fertilization by blocking the passage of sperm into the uterus. A diaphragm is more effective when used with a sperm-killing chemical, or spermicide. During childbirth, a baby passes through the cervix and leaves the mother's body through the vagina.

1000

Chapter Resource File
- Directed Reading BASIC
- Active Reading GENERAL

One-Stop Planner CD-ROM
- Reading Organizers BASIC
- Reading Strategies BASIC
- Data Sheet for Data Lab GENERAL

MISCONCEPTION ALERT

Fetal Development Many young children think that babies develop in the stomach. Older students are often confused about the location of the uterus. Point out that the uterus sits above the bladder, yet under the intestine. As the fetus grows, compression of the abdominal and thoracic organs can become a problem late in the pregnancy, causing an expectant mother to experience fatigue, backache, indigestion, circulatory problems, and/or reduced lung capacity.

The Ovum

The ovaries prepare and release an ovum in a series of events collectively called the **ovarian cycle.** The release of an ovum from an ovary is called **ovulation** *(ahv yoo LAY shuhn).* The ovum is then swept into the fallopian tube and begins to move toward the uterus, awaiting fertilization. Although the duration of the ovarian cycle varies from female to female, the cycle generally spans about 28 days.

internet connect
www.scilinks.org
Topic: Ovarian Cycle
Keyword: HX4133
SCILINKS. Maintained by the National Science Teachers Association

Phases of the Ovarian Cycle

Follicular phase The ovarian cycle has two distinct phases: the follicular phase *(fuh LIK yoo luhr)* and the luteal phase. These phases are regulated by hormones released by the hypothalamus and the anterior pituitary. The events of the ovarian cycle are summarized in **Figure 8.** In an ovary, egg cells mature within follicles. A **follicle** *(FAHL i kuhl)* is a cluster of cells that surrounds an immature egg cell and provides the egg with nutrients. During the follicular phase of the ovarian cycle, hormones regulate the completion of an egg cell's maturation. The follicular phase, which marks the beginning of the ovarian cycle, begins when the anterior pituitary releases follicle-stimulating hormone (FSH) and luteinizing hormone (LH) into the bloodstream. FSH causes the follicle to develop. The follicle produces estrogen, a sex hormone that aids in the growth of the follicle.

Figure 8 Ovarian cycle

During the ovarian cycle, ovulation occurs about every 28 days.

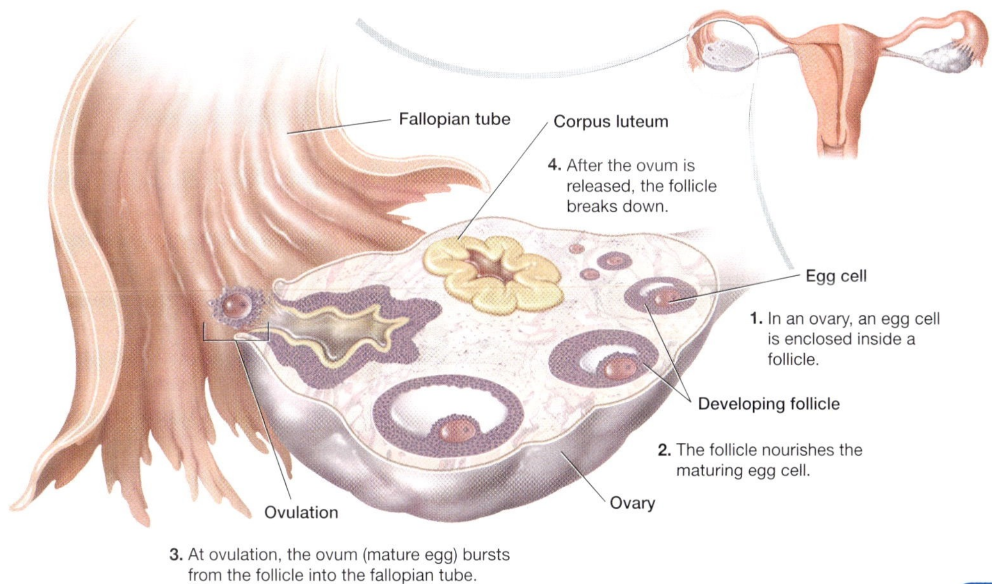

SKILL BUILDER — GENERAL

Math Skills At puberty, females have more than 400,000 immature egg cells in their ovaries. Have students calculate the percentage of these eggs that will ovulate, assuming that 13 ovarian cycles occur per year and that the reproductive life span of a woman is approximately 40 years. (40 years × 13 eggs/year = 520 eggs; 520 eggs/400,000 eggs = 0.13%) **LS Logical**

Teaching Tip — GENERAL

PMS Premenstrual syndrome (PMS) refers to symptoms such as mood changes, depression, anxiety, irritability, headaches, and bloating just before menstruation. Ask students what might cause these symptoms. Explain to students that hormonal changes due to the menstrual cycle have been blamed for the syndrome. Treatments for PMS include low-dose combined oral contraceptives, nutritional approaches, counseling, and exercise.

Group Activity — ADVANCED

Writing **Cervical Cancer** Cervical cancer, like testicular cancer, can develop in young people. Cervical cancer often strikes women between the ages of 20 and 40 years. Divide the class into groups and ask each group to research the link between this cancer and the sexually transmitted human papilloma virus (HPV) and identify any other risk factors for this type of cancer. Have students create a poster or information pamphlet describing the risk factors for cervical cancer and how these risks can be reduced. Co-op Learning

MISCONCEPTION ALERT

Fertility and Youth It is a common belief that a girl cannot become pregnant before she menstruates. However, ovulation precedes a menstrual period. Therefore, a young girl must ovulate before her first menstrual period (menarche). It is therefore possible, though unlikely, for a girl to become pregnant before menarche.

Transparencies

TR J97 Ovarian and Menstrual Cycles

Teach, *continued*

Teaching Tip — GENERAL

Stress Tests Ask students to explain why certain environmental factors, such as stress and prolonged, strenuous exercise, can affect the female reproductive cycle. (These can influence hormone levels, thus affecting the female reproductive cycle.)

SKILL BUILDER — GENERAL

Interpreting Graphics Ask students why the three graphics in **Figure 9** are stacked one atop the other. (All three use the same *x*-axis, which shows days of the menstrual cycle. By looking at all three, one can compare the menstrual cycle, the ovarian cycle, and hormone concentration as the cycle proceeds through the month.) Pick a day or portion of the 28-day cycle and ask students to explain the events of the menstrual cycle, the ovarian cycle, and hormone concentration at that time. **LS Visual**

READING SKILL BUILDER — BASIC

Interactive Reading Assign Chapter 43 of the *Holt Biology* Guided Audio CD Program to help students achieve greater success in reading the chapter.

Real Life

Answer

Menstrual cramps often are treated with drugs that inhibit prostaglandin formation, such as ibuprofen and naproxin sodium.

Ovulation At first the small increase in the level of estrogen prevents further release of FSH and LH from the anterior pituitary. This is caused by a negative feedback mechanism. But as the follicle approaches maturity, it begins to secrete large amounts of estrogen. The anterior pituitary responds to this high level of estrogen by greatly increasing secretion of LH. This increase in LH secretion is caused by a positive feedback mechanism. This surge of LH causes the egg cell to complete the first meiotic division, and it causes the follicle and the ovary to rupture. When the follicle bursts, ovulation occurs, as shown in Figure 8.

Luteal phase The luteal *(LOOT ee uhl)* phase of the ovarian cycle follows the follicular phase, as shown in **Figure 9.** After ovulation occurs, LH causes the cells of the ruptured follicle to grow, forming a corpus luteum. A **corpus luteum** *(KOHR puhs LOOT ee uhm)* is a yellowish mass of follicular cells that functions like an endocrine gland. LH causes the corpus luteum to secrete estrogen and progesterone, another sex hormone. Estrogen and progesterone inhibit the release of FSH and LH. This prevents the development of new follicles during the luteal phase.

Figure 9 Ovarian and menstrual cycles. The ovarian cycle is regulated by hormones produced by the hypothalamus and the pituitary gland. The menstrual cycle is regulated by hormones produced by the follicle and the corpus luteum.

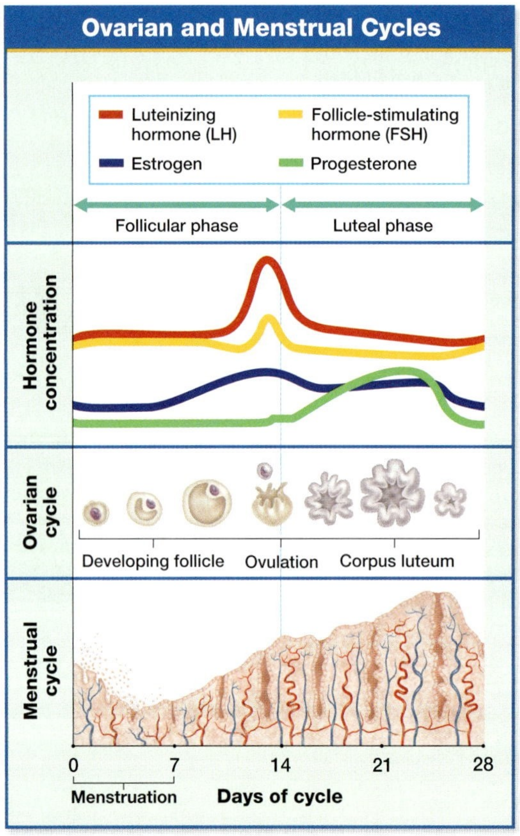

Preparation for Pregnancy

Progesterone signals the body to prepare for fertilization. If fertilization occurs, the corpus luteum continues to produce progesterone for several weeks. If fertilization does not occur, production of progesterone slows and eventually stops, marking the end of the ovarian cycle. Prescription drugs containing relatively large doses of synthetic estrogen and progesterone-like hormones have been designed to disrupt the ovarian cycle and prevent ovulation.

Menstrual cycle While changes occur in the ovaries during the ovarian cycle, changes also occur in the uterus, as shown in Figure 9. The series of changes that prepare the uterus for a possible pregnancy each month is called the **menstrual** *(MEN struhl)* **cycle.** The menstrual cycle lasts about 28 days.

The menstrual cycle is influenced by the changing levels of estrogen and progesterone during the ovarian cycle. Prior to ovulation, increasing levels of estrogen cause the lining of the uterus to thicken. After ovulation, high levels of both estrogen and progesterone cause further development and maintenance of the uterine lining. If pregnancy does not occur, the levels of estrogen and progesterone decrease. This

did you know?

Pregnancy Side Effects At least four major systems—the digestive, urinary, respiratory, and cardiovascular systems—change significantly during pregnancy. Until a woman can adjust to the elevated progesterone and estrogen levels in her body, she may experience excessive salivation and nausea, which is called "morning sickness." In addition, pressure in her abdomen may cause stomach acid to wash backward into her esophagus, causing chronic heartburn. Total blood volume rises, and both blood pressure and pulse rate increase. The pregnant woman's heart pumps 20 to 40 percent more blood, depending on the stage of pregnancy. Pressure by the uterus on pelvic blood vessels may cause varicose veins to result from the restricted blood flow back from the legs.

Section Review Correlations

SC.F.1.4.2: Items 1, 4, 5

SC.F.1.4.6: Items 2, 3

decrease causes the lining of the uterus to shed, marking the end of the menstrual cycle. The end of the menstrual cycle coincides with the end of the luteal phase of the ovarian cycle.

Menstruation

When the lining of the uterus is shed, blood vessels break and bleeding results. A mixture of blood and discarded tissue then leaves the body through the vagina. This process, called **menstruation** *(men STRAY shuhn)*, usually occurs about 14 days after ovulation. At the end of the ovarian and menstrual cycles, estrogen and progesterone levels are low. Negative feedback of estrogen and progesterone thus causes the pituitary to again begin to produce FSH and LH, starting the cycles again.

Women eventually stop menstruation, usually between the ages of 45 and 55. After this event, called *menopause*, a woman no longer ovulates and thus moves out of the childbearing phase of her life. During menopause, many women experience symptoms, such as hot flashes, caused by a decrease in estrogen production. Estrogen, which can be taken to relieve symptoms of menopause, is a widely used prescription drug in the United States.

Real Life

Many women have cramps during their menstrual periods.

The release of a prostaglandin stimulates contractions of muscles in the wall of the uterus, which causes cramps in the uterus.

Finding Information
Research ways that menstrual cramps can be treated.

DATA LAB — *Analyzing Hormone Secretions*

SC.F.1.4.6

Background

The ovarian and menstrual cycles are regulated by hormones secreted by the hypothalamus, the pituitary gland, and the ovaries. Feedback mechanisms play a major role in these cycles. Use Figure 9 and the explanation in the text to answer the following questions.

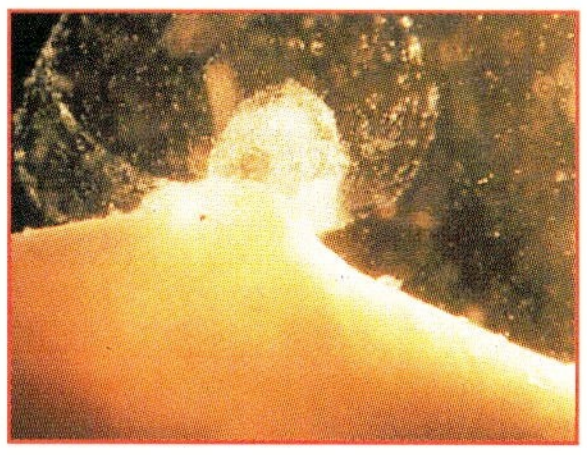

Ovulation

Analysis

1. **Identify** the hormones that are secreted in large amounts prior to ovulation.
2. **Describe** the effect of estrogen production on the secretion of LH.
3. **Critical Thinking** **Analyzing Concepts** What type of feedback mechanism causes a decrease in the secretion of LH and FSH during the luteal phase?
4. **Critical Thinking** **Analyzing Concepts** What type of feedback mechanism causes the surge of LH secretion during the follicular phase?

Section 2 Review

1. **Describe** the functions of ovarian follicles. SC.F.1.4.2
2. **Compare** the regulatory roles of LH and FSH. SC.F.1.4.6
3. **Recognizing Relationships** What causes the lining of the uterus to thicken and then to be shed during the menstrual cycle? SC.F.1.4.6
4. **Relating Concepts** How could the maturation of an egg cell be halted in the ovary? SC.F.1.4.2
5. ***FCAT Prep*** When do the egg cells of a human female begin meiosis?
   A before she is born
   B at the start of puberty
   C during ovulation
   D during menstruation
   SC.F.1.4.2

Answers to Section Review

1. The follicle encloses the immature egg and provides nutrients to it. The follicle also secretes estrogen, which encourages ripening of the follicle.
2. LH and FSH both induce the follicle to produce estrogen. Increased secretion of LH causes the follicle to rupture and ovulation to occur. Continued secretion of LH stimulates the growth of the corpus luteum. Estrogen and progesterone secreted by the corpus luteum suppress further release of LH and FSH.
3. Rising estrogen and progesterone levels cause the uterine lining to thicken in preparation for pregnancy. When estrogen and progesterone levels drop, the lining is shed.
4. Reduced levels of FSH and/or LH could halt egg cell development.
5. **A.** Correct. At birth, the ovaries contain about 2 million immature egg cells that already have begun the first division of meiosis. **B.** Incorrect. Meiosis resumes in individual eggs beginning at puberty. **C.** Incorrect. At ovulation, all but the final meiotic division of an egg has taken place. **D.** Incorrect. During menstruation, the unfertilized eggs, which have previously undergone meiotic divisions, leave the body with the menstrual flow.

DATA LAB — *Analyzing Hormone Secretion*

Skills Acquired
Interpreting graphs, analyzing data, inferring relationships

Teacher's Notes
Remind students to use the key to properly assign hormone names to the colored curves.

Answers to Analysis

1. Levels of LH, estrogens, and FSH rise substantially just prior to ovulation.
2. A peak in estrogen production precedes and causes a spike in LH production.
3. Negative feedback—as progesterone increases, LH and FSH decrease.
4. Positive feedback—as estrogen increases, LH also increases.

Close

Reteaching — BASIC

Use **Figure 9** to correlate hormone levels with uterine lining formation and menstruation.

Quiz — GENERAL

True or False:

1. Fertilization takes place in the uterus. (False. Fertilization takes place in the fallopian tube. The developing embryo then implants in the uterus.)
2. The menstrual cycle prepares the uterine lining for a possible pregnancy each month. (True)
3. The corpus luteum develops from a ruptured follicle after ovulation. (True)

Alternative Assessment — GENERAL

Have students create graphics of the female reproductive system that show the location of the egg during each phase of the ovarian cycle. (Graphics should be similar to Figure 8.)

Focus

Overview

Before beginning this section review with your students the objectives listed in the Student Edition. Section 3 describes development of the embryo and fetus, beginning with conception and ending with birth. The Biowatch feature describes the use of ultrasound, a fetal imaging technique that has been in use since the 1960s.

Bellringer

Ask students to make a list of environmental influences that could damage the fetus and create a composite list on the board. Tell students that these harmful environmental influences are called teratogens. Alcohol is a teratogen that can cause fetal alcohol syndrome. Other teratogens include German measles virus (rubella), cocaine, and X rays.

Motivate

Demonstration — GENERAL

Have students draw a 0.5-mm dot on a sheet of paper and label the dot "14 days." Have them draw a 1-mm line segment below the dot and label it "18 days." Tell them to continue, drawing line segments of 4 mm, 8 mm, 19 mm, 38 mm, and 63 mm and labeling them "4 weeks," "6.5 weeks," "9 weeks," "11 weeks," and "15 weeks," respectively. Tell students that these are the approximate lengths of a developing human through the 15th week of pregnancy. **LS Visual**

Section 3 Development

Florida Sunshine Standards/FCAT

SC.F.1.4.2 Body structures are uniquely designed and adapted for their function.

SC.F.1.4.8 Cell behavior can be affected by molecules from other parts of the organism or even from other organisms.

Objectives

- **Sequence** the events of fertilization, cleavage, and implantation. SC.F.1.4.2
- **Summarize** the three trimesters of pregnancy. SC.F.1.4.2
- **Describe** the effects of drug use on development. SC.F.1.4.8

Key Terms

cleavage
blastocyst
implantation
gestation
pregnancy
embryo
placenta
fetus

Fertilization

If sperm are present in the female reproductive system within a few days after ovulation, fertilization may occur. To fertilize an ovum, a sperm cell must swim to a fallopian tube, where fertilization usually occurs. During fertilization, a sperm cell penetrates an ovum by releasing the enzymes at the tip of its head. These enzymes break down the jellylike outer layers of the ovum. The head of the sperm enters the ovum, and the nuclei of the ovum and sperm fuse together. This produces a diploid cell called a *zygote*.

Cleavage and Implantation

In the first week after fertilization, the zygote undergoes a series of internal divisions known as **cleavage**, as shown in **Figure 10.** Cleavage produces many smaller cells—first two cells, then four, then eight, and so on—within the zygote. Cleavage continues as the zygote moves through the fallopian tube toward the uterus. By the time it reaches the uterus, the zygote is a hollow ball of cells called a **blastocyst** (*BLAS toh sist*). About six days after fertilization, the blastocyst burrows into the lining of the uterus in an event called **implantation**. There it will undergo development, eventually forming a living human.

Figure 10 Early zygote development

Fertilization, cleavage, and implantation occur after ovulation.

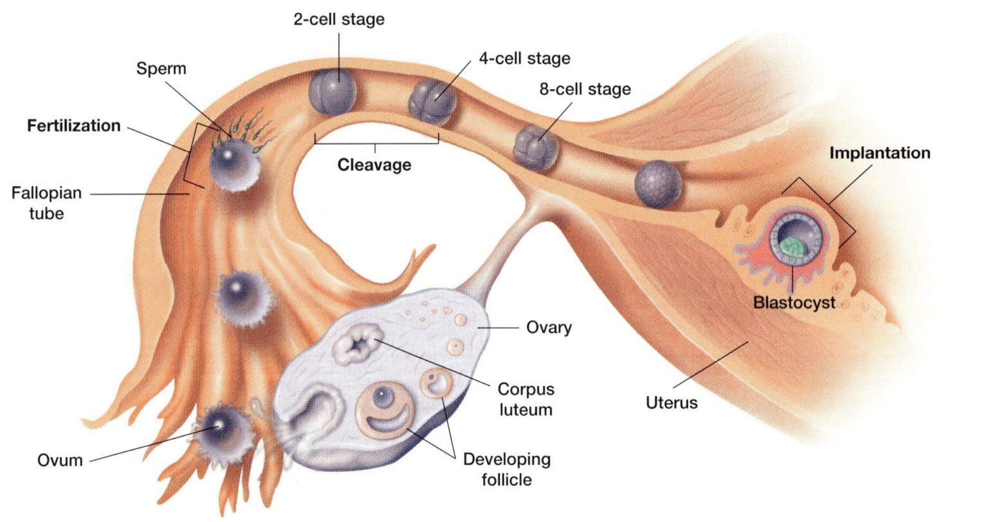

Chapter Resource File

- Directed Reading BASIC
- Active Reading GENERAL

Transparencies

TR Bellringer
TR J98 Early Zygote Development
TR J101 Structure of the Placenta

Section Correlations

SC.F.1.4.2
SC.F.1.4.8

Pregnancy

Development begins with a single diploid cell from which billions of other cells arise. The uterus provides protection and nourishment during development. Human development takes about 9 months—a period known as **gestation** *(jes TAY shuhn),* or pregnancy. The 9 months of **pregnancy** are often divided into three trimesters, or 3-month periods. For the first 8 weeks of pregnancy, the developing human is called an **embryo** *(EHM bree oh).*

First Trimester

Supportive membranes The most crucial events of development occur very early in the first trimester. In the second week after fertilization—shortly after implantation—the embryo grows rapidly. Membranes that will protect and nourish it also develop. One of these membranes, the amnion *(AM nee ahn),* encloses and protects the embryo. Another membrane, the chorion *(KOHR ee ahn),* interacts with the uterus to form the placenta.

The **placenta** *(plah SEHN tah)* is the structure through which the mother nourishes the embryo. As shown in **Figure 11,** the mother's blood normally never mixes with the blood of the embryo. Instead, nutrients in the mother's blood diffuse through the placenta and are carried to the embryo through blood vessels in the umbilical *(uhm BIL i kuhl)* cord.

The waste products of the embryo also pass through the placenta into the mother's blood. Most other substances, including drugs and pathogens, can also diffuse through the placenta. Thus, if the mother ingests any harmful substances, the embryo is also affected. For example, alcohol use by pregnant women, especially during early pregnancy, is a leading cause of birth defects. Fetal alcohol

Figure 11 Placenta

The developing human is nourished through the placenta.

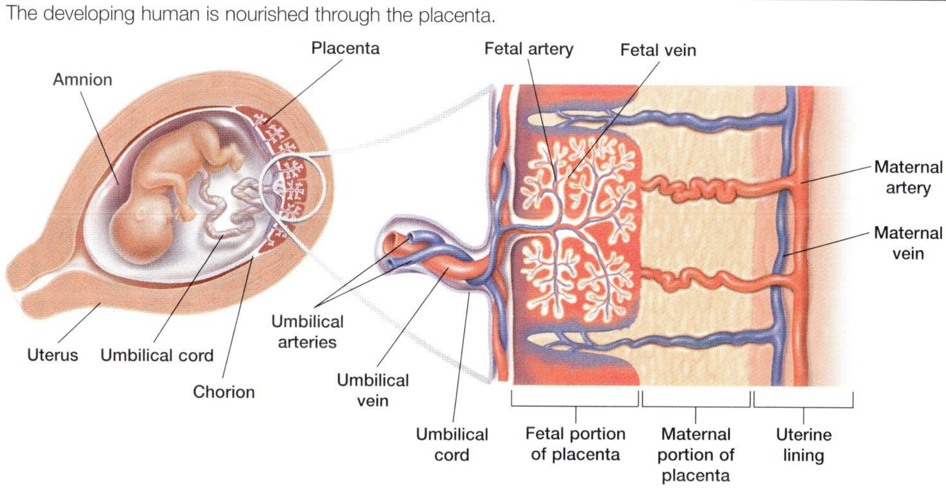

Teach

Teaching Tip — BASIC

Fetal Development Obtain a copy of *Life* magazine's August 1990 cover article titled "How Life Begins" or a similar article that is well illustrated with photographs. Have students read the article, observe the pictures, and relate the information to this chapter.

SKILL BUILDER — GENERAL

Interpreting Visuals Ask students to explain what relationship the right side of **Figure 11** has to the left side of **Figure 11.** (The figure to the right is an enlargement of the area encircled in the figure to the left.) Ask students to describe what is being shown in the figure to the right and why it was important to show an enlargement of this area. (The figure to the right shows an enlargement of a portion of the placenta, which has shared fetal and maternal tissues. An enlargement was necessary to show how the blood vessels of the fetus and the blood vessels of the mother come into close contact, how maternal blood bathes fetal vessels, and how fetal and maternal blood do not co-mingle as many people erroneously think.) LS Visual

One-Stop Planner CD-ROM

- Reading Organizers BASIC
- Reading Strategies BASIC
- Occupational Applications Medical Sonographer GENERAL

Cultural Awareness

Calculating Age Western cultures count age in terms of time that has passed since birth. Some Asian cultures, however, take the approximate time spent in the uterus into account; therefore, a newborn is considered to be 1 year old.

Career

Genetic Counselor Genetic counselors help couples assess their risk of having a child with an inherited disorder. Genetic counselors review evidence of genetic disorders in the family backgrounds of prospective parents and then make predictions. Many genetic counselors are physicians with intensive training in genetics. Have students research the career of a genetic counselor. They should prepare a pamphlet that includes a job description, education and training, starting salary, and career outlook.

syndrome is a collection of birth defects in which a baby may have a deformed face and often severe mental, behavioral, and physical retardation. Women should abstain from alcohol and avoid all unnecessary drugs throughout pregnancy.

Development of embryo As the placenta forms, the inner cells of the blastocyst form the three primary tissue layers—endoderm, mesoderm, and ectoderm. By the end of the third week, blood vessels and the gut begin to develop, and the embryo is about 2 mm (0.08 in.) long. In the fourth week, the arms and legs also begin to form, and the embryo more than doubles in length to about 5 mm (0.2 in.). By the end of the fourth week, all of the major organs begin to form, and the heart begins to beat.

During the second month, the final stage in embryonic development takes place. The arms and legs take shape. Within the body cavity, the major internal organs, including the liver and pancreas, are evident. By the end of the second month, the embryo is about 22 mm (0.9 in.) long and weighs about 1 g (0.036 oz).

Development of fetus From the eighth week of pregnancy until childbirth, the developing human is called a **fetus** *(FEET uhs)*. By the end of the first trimester, the sex of the fetus can be distinguished. A fetus has recognizable body features, and its organ systems have begun to form, as shown in **Figure 12.**

Figure 12 Development of fetus

A fetus has recognizable body characteristics.

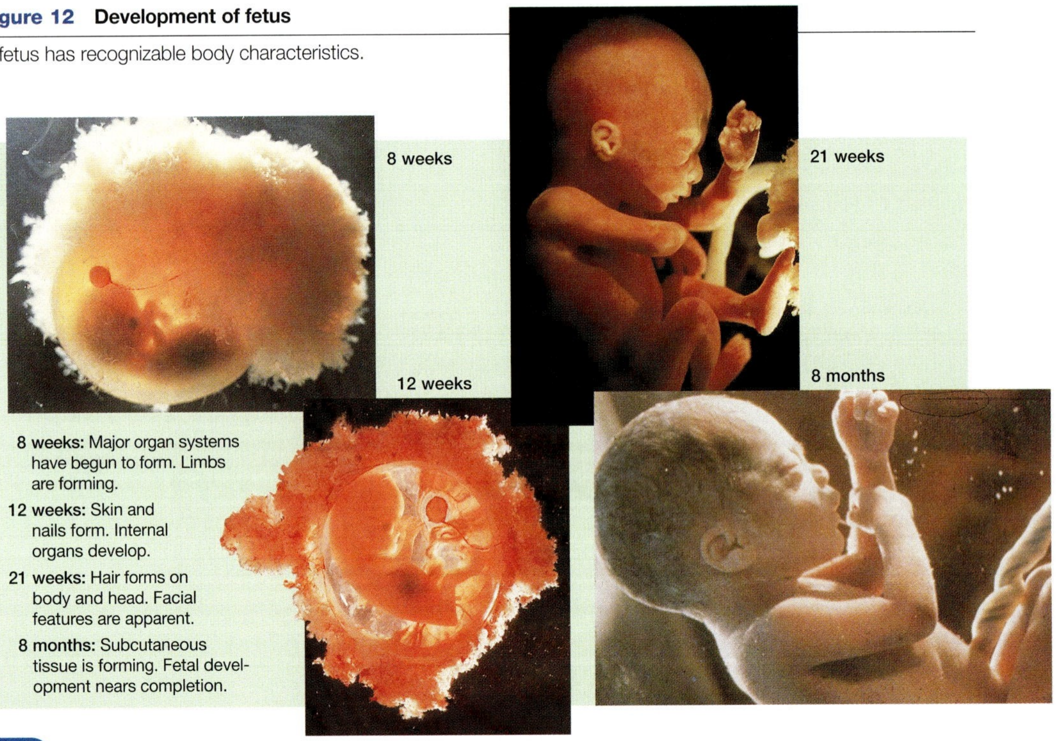

8 weeks: Major organ systems have begun to form. Limbs are forming.

12 weeks: Skin and nails form. Internal organs develop.

21 weeks: Hair forms on body and head. Facial features are apparent.

8 months: Subcutaneous tissue is forming. Fetal development nears completion.

Teaching Tip — GENERAL

Health A mother-to-be must have a sufficient supply of the B-vitamin folic acid in her body before pregnancy, or her fetus may be at risk for neural tube defects, which are birth defects of the brain or spine. Folic acid can be found in some enriched foods and vitamin pills. Pregnant women should follow their doctor's advice about all vitamin intake. Have students visit the folic acid home page of the Centers for Disease Control and Prevention for more information on folic acid and healthy pregnancies.

READING SKILL BUILDER — BASIC

Reading Hint Tell students to develop a time line of human development based on information in the section. Have them divide the time line into trimesters and the trimesters into weeks. From their reading, have students fill in their time lines with descriptions of development, noting when the terms "embryo" and "fetus" are used. **LS Visual**

Group Activity — GENERAL

Fetal Alcohol Syndrome Place students in groups of three. Have them develop designs for posters that point out the dangers of drinking alcohol during pregnancy. After receiving approval for their designs from the teacher, groups should construct their posters and present their designs and ideas to the class.
LS Interpersonal Co-op Learning

MEDICINE CONNECTION

Placenta previa is a condition in which the placenta develops in the lower part of the uterus and obstructs the cervical opening to the vagina (birth canal). This may cause hemorrhaging as the fetus grows, and a Cesarean section may be necessary to deliver the infant. Abruptio placentae is a condition in which the placenta begins tearing away from the uterine wall, causing hemorrhaging. Both of these conditions can lead to miscarriage and even death of the mother. However, treatments for both conditions are often successful.

Integrating Physics and Chemistry

Explain to students that ultrasound imaging works because the speed and reflective behavior of the high-frequency sound waves varies depending on the density of the material through which the sound waves travel. Since sound waves are longitudinal and mechanical waves, they travel faster through denser materials. By converting these observed density differences and time variations between return reflections, an image of the fetus or other internal structures is produced.

Section Review Correlations

SC.F.1.4.2: Items 1, 2, 5, 6

SC.F.1.4.8: Items 3, 4

Second and Third Trimesters

During the second and third trimesters, the fetus grows rapidly as its organs become functional. By the end of the third trimester, the fetus is able to exist outside the mother's body. After about 9 months of development, the fetus leaves the mother's body in a process called *labor*, which usually lasts several hours. During labor, the walls of the uterus contract, expelling the fetus from the uterus, as shown in **Figure 13.** The fetus leaves the mother's body through the vagina. The placenta and the umbilical cord are expelled after the baby is born. After birth, development is far from complete. Physical growth and neurological development continue for years after birth.

Figure 13 Childbirth

During childbirth, the fetus exits the mother's body through the vagina.

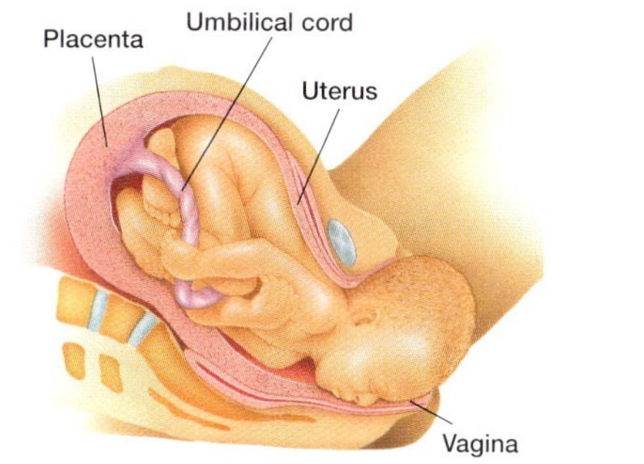

BIOWatch: Ultrasound Imaging

Since the 1960s, several methods have been developed for looking inside the body without surgery. One such method is ultrasound imaging. To make an ultrasound image, a physician holds a small probe against a patient's skin. The probe emits high-frequency sound waves, which produce echoes when they reflect off structures in the patient's body. The probe detects the echoes, which are converted to an image on a video screen.

Medical Uses of Ultrasound

Ultrasound measurements of the size of the embryo or fetus can indicate its age, while other signs indicate its health. For example, heart motions can usually be seen by 7 weeks. In addition, many fetal abnormalities can be diagnosed using ultrasound.

Advantages of Ultrasound

The biggest advantage of ultrasound imaging is its safety. Unlike X rays, ultrasound does not involve ionizing radiation, which can cause mutations. Ultrasound imaging has no known harmful effects.

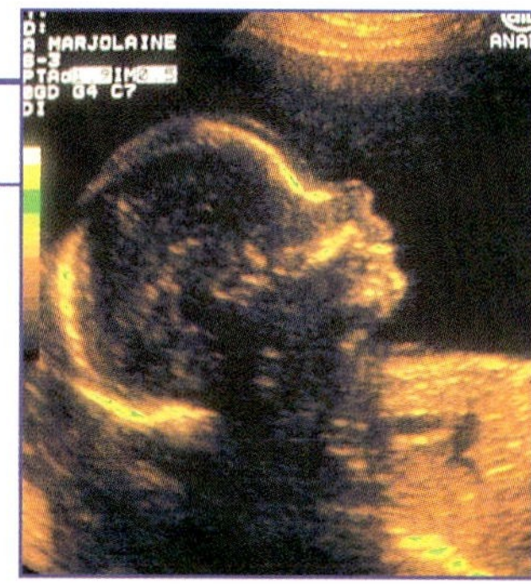

Ultrasound image of fetus

internet connect

www.scilinks.org
Topic: Ultrasound
Keyword: HX4180
SciLinks. Maintained by the National Science Teachers Association

Section 3 Review

1. **Summarize** the events in development that occur in the embryo's first month. SC.F.1.4.2
2. **Describe** the function of the placenta. SC.F.1.4.2
3. **Describe** fetal alcohol syndrome. SC.F.1.4.8
4. **Relating Concepts** Why are some drugs harmful when they are taken during pregnancy? SC.F.1.4.8
5. **Predicting Outcomes** What might happen if more than one egg were released from the ovaries prior to fertilization? SC.F.1.4.2
6. *FCAT Prep* Following fertilization, cleavage begins while the zygote is in what structure?
   A an ovary
   B the uterus
   C the placenta
   D a fallopian tube
   SC.F.1.4.2

Answers to Section Review

1. In the first month after fertilization, the zygote implants in the uterine wall; the amnion, chorion, and placenta form; and major organs begin to develop.
2. In the placenta, the blood systems of the mother and fetus come in close association (but do not mix). Here, oxygen and nutrients pass to the fetus, and fetal carbon dioxide and wastes are removed.
3. Fetal alcohol syndrome is caused by the consumption of alcohol by a pregnant woman. Her child is born with mental and physical defects, including an abnormal head and mental retardation.
4. Many drugs can cross the placenta from mother to fetus.
5. A multiple birth may occur if all eggs are fertilized.
6. **A.** Incorrect. When in the ovary, the egg has not yet ovulated. Fertilization and cleavage can take place only after ovulation. **B.** Incorrect. By the time the egg reaches the uterus, it already has undergone cleavage several times. **C.** Incorrect. The placenta is an organ that develops after fertilization takes place. **D.** Correct. Fertilization and then cleavage take place when the zygote is in the fallopian tube.

BIOWatch: Ultrasound Imaging

Teaching Strategies

Obtain a sonograph on videotape. Use the videotape to stimulate discussion about the activity.

Discussion

- What is ultrasound used for during pregnancy? (to detect pregnancy, determine the size and age of a fetus, check for fetal health and abnormalities)
- Other than pregnancy, can you think of other medical uses for ultrasound? (Answers will vary. Ultrasound also is used to examine internal organs such as the heart.)

Close

Reteaching — Basic

Show the class the opening segment of the movie *Look Who's Talking*. Have students describe the sequence of events that occurs from conception to birth. Point out that the effects of the painkiller administered to the mother also affected the baby.

Quiz — General

1. What occurs during implantation? (the blastocyst burrows into the lining of the uterus)
2. In which trimester does the embryo develop? (first)
3. How long is the period of gestation in humans? (9 months)

Alternative Assessment — General

Ask students to create a booklet that tells the story of fetal development from the point of view of the fetus. Tell students to be creative and include drawings or photos in their booklet.

Section 4

Focus

Overview

Before beginning this section review with your students the objectives listed in the Student Edition. Section 4 describes diseases spread by sexual contact and discusses bacterial and viral sexually transmitted diseases in detail.

Bellringer

Ask students to list risk factors for contracting sexually transmitted diseases. Explain to them that a risk factor is a characteristic or behavior that increases an individual's chances of developing a condition or disease. After taking attendance, make a composite list on the board. Compare that list with these risk factors: having sex with infected individuals, having sex with multiple partners, having sex with people you do not know well, having sex while under the influence of drugs, and having sex without using a condom.

Motivate

Activity — BASIC

STDs Have students generate a class list of STDs based on their prior knowledge. Develop a chart using this list, including columns for symptoms and treatments. Amend the information as students read the section.

Section Correlations

SC.F.1.4.8

SC.G.1.4.1

Section 4

Sexually Transmitted Diseases

Florida Sunshine Standards/FCAT

SC.F.1.4.8 Cell behavior can be affected by molecules from other parts of the organism or even from other organisms.

SC.G.1.4.1 There is great diversity and interdependence among living things.

Objectives

- **Identify** the causes and symptoms of several bacterial STDs. SC.G.1.4.1
- **Identify** the causes and symptoms of some viral STDs. SC.F.1.4.8
- **Compare** the treatment and cure rates of viral STDs with those of bacterial STDs. SC.F.1.4.8

Key Terms

gonorrhea
syphilis
chlamydia
pelvic inflammatory disease
genital herpes

STDs

Recall that disease-causing pathogens are transmitted in many ways. Pathogens present in body fluids, such as semen, can be passed from one person to another through sexual contact. Diseases spread by sexual contact are called *sexually transmitted diseases*, or *STDs*. Both viruses and bacteria can cause STDs. **Table 1** lists several types of STDs. Abstinence is the only sure way to protect yourself from contracting an STD.

Bacterial STDs

Most STDs that are caused by bacteria can be successfully treated and cured with antibiotics. Unfortunately, the early symptoms of most bacterial STDs are very mild and often are not detected. Early detection and treatment are necessary to prevent serious consequences that can result from infection. For example, untreated bacterial STDs can cause sterility in both men and women. Three major bacterial STDs are gonorrhea, syphilis, and chlamydia.

Gonorrhea *(gahn uh REE uh)* is a bacterial STD that causes painful urination and a discharge of pus from the penis in males. In females, gonorrhea sometimes causes a vaginal discharge but more

Table 1 Sexually Transmitted Diseases

Disease	Symptoms	Pathogen
AIDS	Immune-system failure and susceptibility to opportunistic infections	Human immunodeficiency virus (HIV)
Chlamydia	Painful urination and penile discharge in males; vaginal discharge and abdominal pain in females	*Chlamydia trachomatis* (bacterium)
Genital herpes	Painful blisters on genital region, thighs, or buttocks and flulike symptoms	Herpes simplex virus (HSV)
Genital warts	Warts on genital or anal region	Human papilloma virus (HPV)
Gonorrhea	Painful urination and penile discharge in males; vaginal discharge and abdominal pain in females	*Neisseria gonorrhoeae* (bacterium)
Hepatitis B	Flulike symptoms and yellowing of skin	Hepatitis B virus
Syphilis	Chancre on penis in males; chancre in vagina or on cervix in females; fever and rash	*Treponema pallidum* (bacterium)

Chapter Resource File

- Directed Reading BASIC
- Active Reading GENERAL

Transparencies

TR Bellringer

TR J35 Sexually Transmitted Diseases

TR J43 Known Routes of HIV Transmission

One-Stop Planner CD-ROM

- Reading Organizers BASIC
- Reading Strategies BASIC
- Occupational Applications Nurse Practitioner GENERAL

often has no symptoms. In males, untreated gonorrhea can spread to the vas deferens, epididymis, or testes. In females, it can spread to the fallopian tubes and cause pain and scarring that may lead to infertility. Some strains of gonorrhea are resistant to commonly used antibiotics, such as penicillin.

Syphilis *(SIHF uh lihs)* is a serious bacterial STD that usually begins with the appearance of a small, painless ulcer called a *chancre (SHAHN kuhr)* 2–3 weeks after infection. In males, the chancre usually appears on the penis. In females, the chancre may form inside the vagina or on the cervix. If syphilis is not treated, it may cause fever, swollen lymph glands, or a rash like the one shown in **Figure 14** a few weeks after infection. These symptoms disappear without treatment. Years later, however, syphilis may cause destructive lesions on the nervous system, blood vessels, bones, and skin. A pregnant woman infected with syphilis can also transmit the disease to the fetus. As a result, the fetus may be stillborn or suffer serious damage to organ systems.

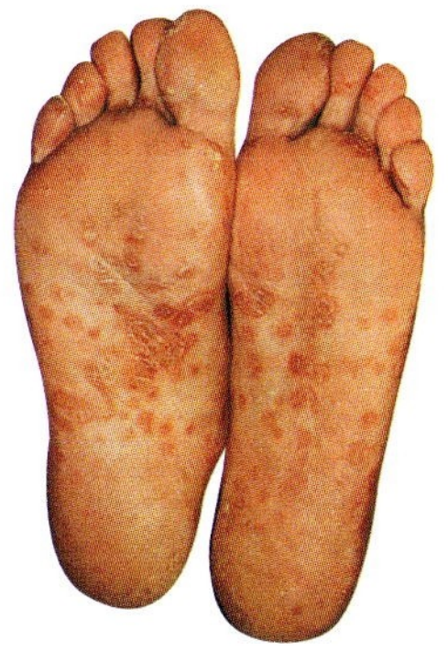

Figure 14 **Syphilis.** A rash such as this one is a symptom of the second stage of syphilis. Even at this stage, syphilis can be cured by treatment with antibiotics.

Chlamydia *(kluh MIHD ee ah)* is the most common bacterial STD in the United States. The symptoms of chlamydia are similar to those of a mild case of gonorrhea: painful urination in males and vaginal discharge in females. Like gonorrhea, chlamydia often is not detected. Chlamydia, even more than gonorrhea, is likely to cause scar tissue in infected fallopian tubes, leading to infertility, or the inability to become pregnant.

Pelvic inflammatory disease One of the most common causes of infertility in women is **pelvic inflammatory disease**, or **PID**. PID is a severe inflammation of the uterus, ovaries, fallopian tubes, or abdominal cavity that results from a bacterial STD that has gone untreated. **Figure 15** shows the damage that PID can cause in the fallopian tubes. Most cases of PID are the result of gonorrhea or chlamydia infections.

Figure 15 **Pelvic inflammatory disease**

Most cases of PID result from gonorrhea or chlamydia infections.

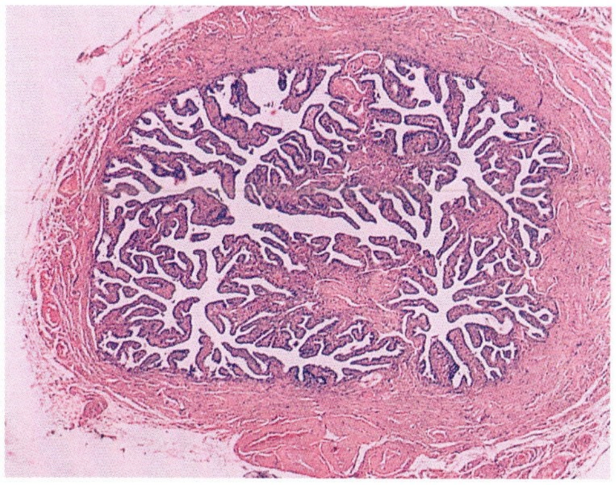

A normal fallopian tube has a highly folded lining and many spaces through which gametes can pass.

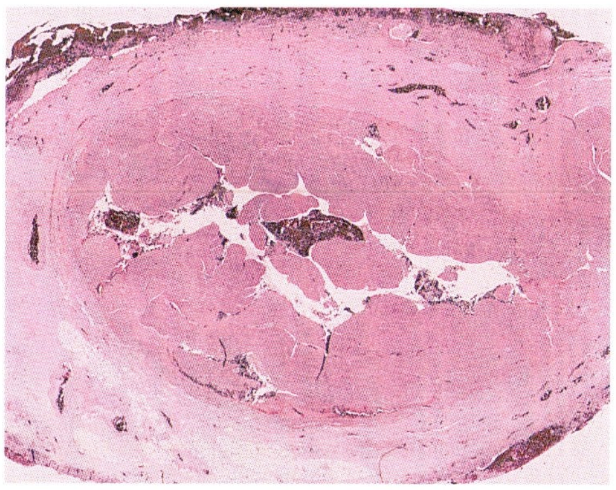

In a fallopian tube scarred by PID, many of these spaces have become blocked with tissue.

1009

Teach

SKILL BUILDER — GENERAL

Math Skills Tell students that they are epidemiologists for the Centers for Disease Control and Prevention (CDC). They are trying to determine who has been infected with a new sexually transmitted disease (STD) within the last two months. Only one person with the disease has come to the hospital for treatment. If this person had sex within the past two months with three "contacts," how many people will students have to interview? (4) If each of these "contacts" had sex with three additional "contacts" within the past two months, how many people will students have to interview? (13) What does this scenario tell about the transmission of STDs? (The transmission of STDs can spread to many people rapidly in a population where people are having sex with multiple partners.) **LS Logical**

Activity — GENERAL

Incidence of STDs Have students research how many people each year contract each of the seven STDs presented in **Table 1.** Create a bar graph based on their findings. Alternatively, divide the class into groups and have each group graph the incidence of one disease over time.

Teaching Tip — BASIC

STDs are "Blind" Remind students that acquiring sexually transmitted diseases results from sexual contact with infected individuals. It does not matter what kind of personal hygiene the person has, how much education the person has, or how much money the person makes. STDs infect people from all walks of life. **LS Interpersonal**

INCLUSION Strategies

- *Gifted and Talented*
- *Attention Deficit Disorder*

Using **Table 1,** have students create an extended chart of STDs. The chart should include the disease, symptoms, pathogen, and treatments. Students may use reference materials, the Internet, or interviews with public health agencies or clinics to collect information. An additional column may be added to the chart to report expected fetal outcomes if the disease is active during pregnancy and delivery.

Cultural Awareness

Diagnosing Syphilis Dr. William Hinton, a pathologist and renowned authority on sexually transmitted diseases, developed the Davies-Hinton Test for the detection of syphilis. Hinton was the first *African-American* professor to teach at Harvard Medical School.

Viral STDs

Because viruses are not affected by antibiotics, STDs that are caused by viruses cannot be treated and cured with antibiotics. AIDS and genital herpes are two common viral diseases that are transmitted through sexual contact. Other common viral STDs include genital warts and hepatitis B.

AIDS is a fatal disease caused by the human immunodeficiency virus (HIV). Transmission through sexual contact is the most common way that people become exposed to HIV. HIV destroys the immune system of infected individuals by attacking white blood cells. People with AIDS generally die from opportunistic infections that persist only in people with weakened immune systems.

The number of HIV infections among teenagers and young adults has increased dramatically over the last decade. AIDS is now the leading killer of African-American men between the ages of 25 and 44. More than 460,000 people in the United States have already died from AIDS. The number of young adults in the United States with AIDS has increased drastically over the last 15 years. While the number of new AIDS cases reported has decreased each year since 1993 due to improved drug treatments, new HIV infections have not decreased, and are most frequent among young adults—like the young woman shown in **Figure 16.** Researchers are trying to develop new treatments for AIDS.

Genital herpes is caused by herpes simplex virus (HSV). About 70 percent of genital herpes infections are caused by HSV-2. The rest are caused by HSV-1, which more commonly causes cold sores, or fever blisters, around and inside the mouth.

Symptoms of genital herpes include periodic outbreaks of painful blisters in the genital region, as shown in **Figure 17,** and flu-like aches and fever. Antiviral drugs can temporarily eliminate the blisters caused by genital herpes, but they cannot eliminate HSV from the body. Although genital herpes is not life threatening, it can have serious consequences. Women with genital herpes have a greater risk of developing cervical cancer. Like HIV, herpes simplex virus can be passed from mother to fetus during pregnancy or birth. Infants infected with HSV may suffer severe damage to their nervous system or even die as a result of the infection.

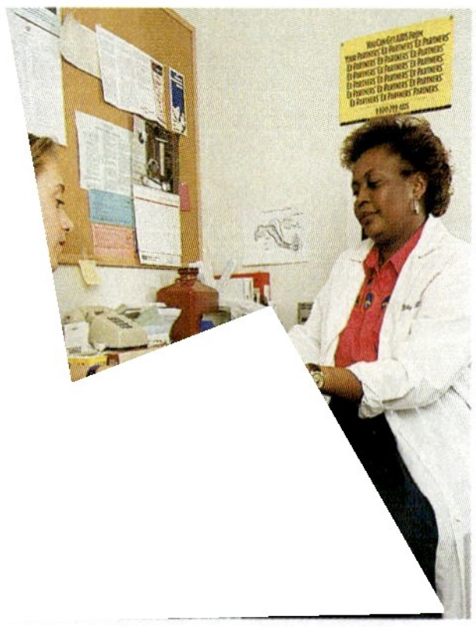

Figure 16 Counseling. Counseling is available to people with HIV.

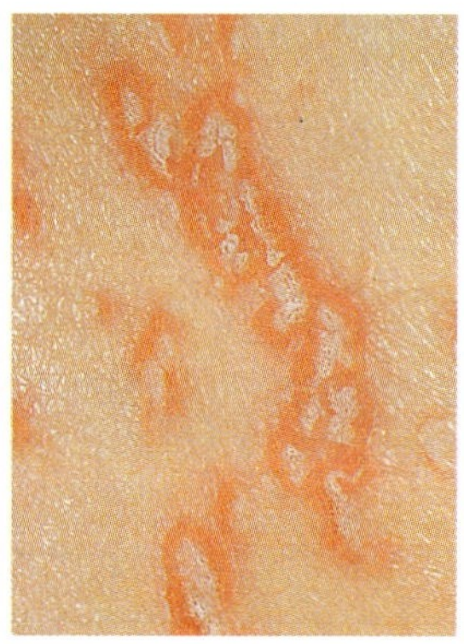

Figure 17 Genital herpes. Blisters caused by genital herpes may appear on or near the genitalia.

Section 4 Review

1. **Name** three common STDs caused by bacteria. Why is early detection of these diseases important? SC.G.1.4.1
2. **Describe** how HIV weakens the immune system of an infected individual. SC.F.1.4.8
3. **Recognizing Differences** What is the main difference between the treatment of viral STDs and the treatment of bacterial STDs? SC.F.1.4.8
4. **List** three symptoms of genital herpes. SC.F.1.4.8
5. **Applying Information** How can you best protect yourself from contracting a sexually transmitted disease? SC.F.1.4.8
6. *FCAT Prep* Which of the following is a sexually transmitted disease that can be treated with antibiotics? SC.F.1.4.8
   A AIDS
   B syphilis
   C hepatitis B
   D genital herpes

Answers to Section Review

1. Gonorrhea, syphilis, and chlamydia are bacterial STDs. Left untreated, they can cause sterility and other serious, permanent damage.
2. HIV attacks white blood cells, eventually destroying the immune system.
3. Bacterial STDs can be treated and usually cured with antibiotics. Drugs for viral STDs can provide relief from symptoms but cannot cure the disease.
4. Symptoms of genital herpes include painful genital blisters, flulike aches, and fever.
5. Answers will vary. Students may suggest abstinence, use of condoms, and monogamy with another monogamous person.
6. **A.** Incorrect. HIV is a virus, and viruses are not susceptible to antibiotics. **B.** Correct. Syphilis is a bacterial STD, which can be treated with antibiotics. **C.** Incorrect. Hepatitis B is a virus, and viruses are not susceptible to antibiotics. **D.** Incorrect. Genital herpes is a viral STD, and viruses are not susceptible to antibiotics.

Close

Reteaching — BASIC

Tell students to return to the lists of things they **W**ant to know that they created using the Reading Skill Builder that addresses reproduction, development, and STDs. Have them place check marks next to questions that they are now able to answer. Students should finish by making a list of what they have **L**earned. Conclude by asking students the following questions:

- Which questions are still unanswered?
- What new questions do you have?

Have students research any unanswered or new questions.

Quiz — GENERAL

1. Which type of STDs can be treated and cured with antibiotics? (Bacterial STDs)
2. What is the most common bacterial STD in the United States? (Chlamydia)
3. Which STD has the following symptoms: painful urination and a discharge of pus from the penis? (Gonorrhea)

Alternative Assessment — GENERAL

Copy **Table 1** and cut it into separate cells. Provide groups of three students with the cells of the complete table. Ask them to place the STDs in one column and then use the remaining cells to create an accurate table that describes the symptoms and pathogen for each STD.

Section Review Correlations

SC.F.1.4.8: Items 2-6
SC.G.1.4.1: Item 1

Study ZONE

CHAPTER HIGHLIGHTS

Key Concepts

1 Male Reproductive System

- Sperm cells are produced by meiosis in the testes.
- Sperm mature and are stored in each epididymis.
- A mature sperm cell consists of a head, a midpiece, and a long, powerful tail.
- Sperm move through the epididymis and the vas deferens and exit the body through the urethra.

2 Female Reproductive System

- Egg cells are produced by meiosis in the ovaries.
- An egg cell matures in a 28-day ovarian cycle.
- The menstrual cycle prepares the uterus for pregnancy.
- The menstrual and ovarian cycles are regulated by hormones.

3 Development

- After fertilization, cleavage and implantation occur.
- The human gestation period is about 9 months.
- The most crucial events of development occur during the first trimester of pregnancy.
- The mother nourishes the fetus through the placenta.
- Primary tissue layers develop into organs and tissues.
- The fetus leaves the mother's body during labor.

4 Sexually Transmitted Diseases

- Many STDs are caused by bacteria and viruses.
- Syphilis is a severe bacterial STD that can have destructive effects on the nervous system, bones, and skin if untreated.
- Gonorrhea and chlamydia are common bacterial STDs that can scar the fallopian tubes and lead to infertility.
- AIDS is a viral STD in which HIV destroys immune-system cells, leaving the body vulnerable to opportunistic infections.
- Genital herpes is a viral STD that causes blistering.

Key Terms

Section 1

testes (996)
seminiferous tubules (996)
epididymis (997)
vas deferens (997)
seminal vesicles (998)
prostate gland (998)
bulbourethral glands (998)
semen (998)
penis (998)

Section 2

ovary (999)
ovum (1000)
fallopian tube (1000)
uterus (1000)
vagina (1000)
ovarian cycle (1001)
ovulation (1001)
follicle (1001)
corpus luteum (1002)
menstrual cycle (1002)
menstruation (1003)

Section 3

cleavage (1004)
blastocyst (1004)
implantation (1004)
gestation (1005)
pregnancy (1005)
embryo (1005)
placenta (1005)
fetus (1006)

Section 4

gonorrhea (1008)
syphilis (1009)
chlamydia (1009)
pelvic inflammatory disease (1009)
genital herpes (1010)

Alternative Assessment — GENERAL

Have students work in cooperative groups. Each group should develop a question to send to a newspaper or magazine health-advice columnist dealing with any topic covered in this chapter. Collect the questions and redistribute them to other groups. Each group should then write an appropriate response to the question.

Chapter Resource File

- Science Skills Worksheet GENERAL
- Critical Thinking Worksheet ADVANCED
- Test Prep Pretest GENERAL
- Chapter Test GENERAL

Answer to Concept Map

The following is one possible answer to Performance Zone item 15.

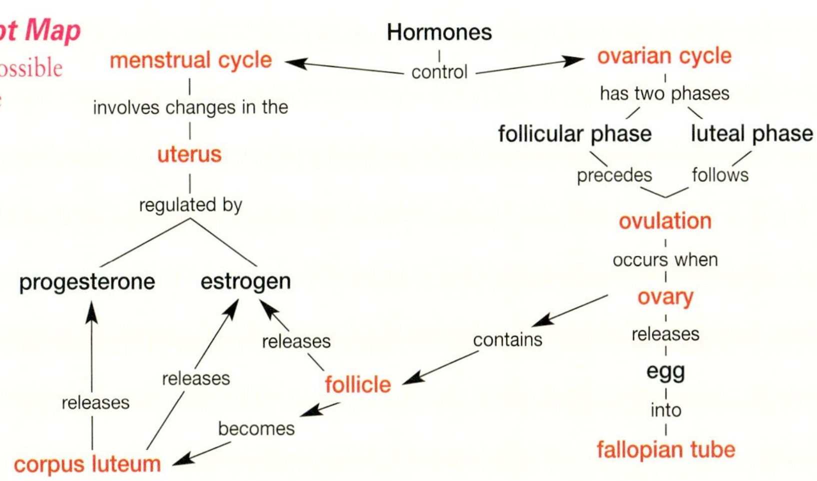

Understanding Key Ideas

1. c
2. d
3. a
4. c
5. d
6. The head contains genetic information and enzymes. The midpiece contains mitochondria, which supply energy. The tail is a flagellum, which allows the sperm to move.
7. FSH and LH are released causing the follicle to produce estrogen which encourages follicle maturation. A small increase in estrogen has a negative feedback effect on the production of FSH and LH. Later in the cycle, increasing estrogen stimulates LH release.
8. Birth control pills contain combinations of estrogen and progesterone type hormones which interrupt the normal hormone levels of the menstrual cycle and therefore maturation of eggs.
9. Ultrasound can be used to detect pregnancy, estimate fetal age, and detect abnormalities in the developing baby. It is also used to examine internal organs.
10. The answer to the concept map can be found at the bottom of the previous page.

Critical Thinking

11. The man has 60 million ÷ 3.5 mL = 17 million sperm per mL. Having fewer than 20 million sperm per mL of semen is normally an indicator of sterility.
12. The cell would not survive.
13. The fetus must have essential nutrients to develop properly.

Chapter Review Correlations

SC.F.1.4.2: Items 1, 2, 6, 10, 12
SC.F.1.4.6: Item 7
SC.F.1.4.8: Items 3, 4, 8, 13
SC.G.1.4.1: Item 5
SC.H.3.4.3: Item 11
SC.H.3.4.5: Items 9, 14

Performance ZONE — CHAPTER REVIEW

Understanding Key Ideas

1. The correct pathway of sperm is from the
   a. testes to vas deferens to epididymis.
   b. epididymis to urethra to vas deferens.
   c. testes to epididymis to vas deferens.
   d. urethra to vas deferens to testes.
   SC.F.1.4.2
2. Which of the following is *not* a function of the female reproductive system?
   a. production of gametes
   b. nourishment of the fetus
   c. maturation of eggs
   d. secretion of FSH
   SC.F.1.4.2
3. Which of the following sexually transmitted diseases *cannot* be treated with antibiotics? SC.F.1.4.8
   a. genital herpes
   b. syphilis
   c. gonorrhea
   d. chlamydia
4. Which of the following is *not* true for human development?
   a. Alcohol and drugs taken during pregnancy may harm the embryo or fetus.
   b. Crucial events occur during the first trimester of pregnancy.
   c. Drugs and alcohol taken during pregnancy cannot cause birth defects.
   d. Normal development may be affected by viral diseases.
   SC.F.1.4.8
5. A symptom associated with the earliest stage of syphilis is SC.G.1.4.1
   a. painful urination.
   b. blisters in the genital area.
   c. fever blisters and cold sores.
   d. a painless chancre.
6. The diagram below shows a mature human sperm cell. Explain the roles of the structures labeled A, B, and C in the sperm's ability to fertilize an egg. SC.F.1.4.2

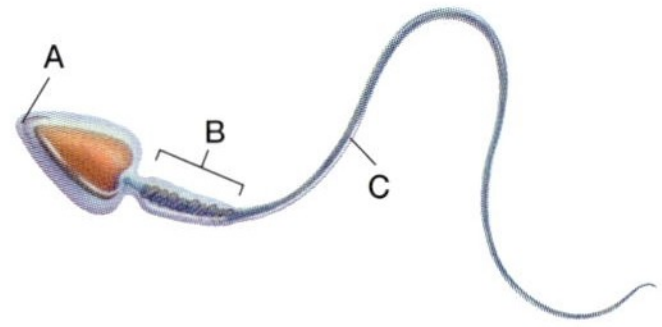

7. **Recognizing Patterns** Describe the role of feedback mechanisms in the maturation of an egg cell during the ovarian cycle of an adult female. SC.F.1.4.6
8. How do birth control pills, which contain synthetic estrogen and progesterone-like hormones, prevent pregnancy? SC.F.1.4.8
9. **BIOWatch** Describe two different medical uses for ultrasound imaging. What can be learned about a fetus by using this technology? Find out other ways ultrasound imaging is used in medicine. SC.H.3.4.5
10. **Concept Mapping** Make a concept map that describes the ovarian and menstrual cycles. Include the following terms in your map: *ovary, fallopian tube, uterus, ovarian cycle, follicle, ovulation, corpus luteum,* and *menstrual cycle.* Include additional terms in your map as needed. SC.F.1.4.2

Critical Thinking

11. **Making Inferences** A man interested in fathering children wants to know his sperm count. He finds out that he has a sperm count of fewer than 60 million sperm in a 3.5 mL sample of semen. If you were his physician, what would you tell him about the results of the test? SC.H.3.4.3
12. **Predicting Outcomes** What do you think would happen if more than one sperm were able to fertilize an egg? SC.F.1.4.2
13. **Inferring Relationships** Why should a pregnant woman eat many healthy foods? SC.F.1.4.8

Alternative Assessment

14. **Interpreting Information** Research some causes of and treatments for infertility for both men and women. What are the benefits and shortcomings of each type of treatment? Summarize your findings in a written report. SC.H.3.4.5

1012

Alternative Assessment

14. Many factors cause infertility, including hormone imbalance, decreased sperm count, ovarian or fallopian tube scarring from STDs, ovulation dysfunction, and chronic disease. Some treatments are *in vitro* fertilization (which is very expensive) and fertility drugs (which increase chances for multiple births).

Assignment Guide

Section	Questions
1	1, 6, 10, 11, 14
2	2, 7, 9
3	4, 8, 12, 13
4	3, 5

FCAT Prep

For questions 1–6, write your answers on a separate sheet of paper.

1 In the 1960s, many women who took a tranquilizer called *thalidomide* early in pregnancy gave birth to babies with serious limb defects. Women who took the drug later in pregnancy gave birth to normal babies. What does this tell you about fetal development?
A. Critical periods of limb formation occur early in pregnancy.
B. The arms and legs are the last structures in the fetus to form.
C. Taking *thalidomide* during pregnancy is essential to having a normal baby.
D. Babies cannot form normally if a mother takes *thalidomide* during pregnancy.
SC.H.3.4.3

2 Where does fertilization usually take place?
F. cervix **H.** fallopian tubes
G. epididymis **I.** uterus
SC.F.1.4.2

3 How are semen and sperm related?
A. Both are stored in the bladder.
B. Sperm and semen are both gametes.
C. Semen is made up of fluids and sperm.
D. Both are produced in the prostate gland.
SC.F.1.4.2

4 When does an embryo develop endoderm, mesoderm, and ectoderm?
F. during cleavage **H.** during gastrulation
G. during fertilization **I.** during implantation
SC.F.1.4.2

Base your answers to questions 5 and 6 on the graph below.

HORMONE CONCENTRATIONS DURING THE OVARIAN CYCLE

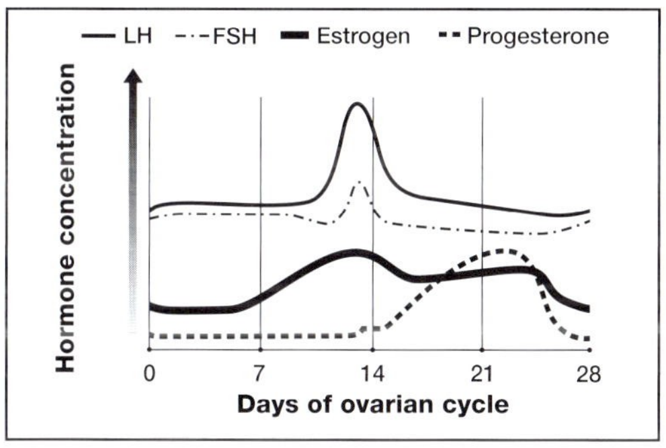

5 About when during the ovarian cycle is the concentration of progesterone highest?
A. day 7 **C.** day 23
B. day 13 **D.** day 28
SC.F.1.4.6

6 On approximately what day during the ovarian cycle does ovulation occur? SC.F.1.4.6

Test TIP

When analyzing a graph, pay attention to its title. The title should tell you what is plotted on the graph and provide some context for the data.

1013

FCAT Doctor

Question 2 Answer H is the correct choice. Normally, the egg is fertilized in the fallopian tubes and travels to the uterus for implantation. Answer F is incorrect because the cervix is the entrance to the uterus from the vagina. An egg could not implant in the uterus if it were fertilized here. Answer G is incorrect because fertilization does not occur in the male's body, where the epididymis is located. Answer I is incorrect because fertilization usually occurs in the fallopian tubes, before the egg reaches the uterus.

Question 5 Answer C is the correct choice. On day 23, the concentration of progesterone is highest, as indicated by a peak in the graph. Answer A is incorrect because on day 7, the concentration of progesterone is low. Answer B is incorrect because on day 13, the concentration of progesterone has risen only slightly from its low. Answer D is incorrect because on day 28, the concentration of progesterone has fallen to its low.

Question 6 The correct answer is 14. Ovulation occurs approximately on day 14 in women with a 28-day cycle, after a surge of LH.

Answers

1. A
2. H
3. C
4. H
5. C
6. 14

Test Prep Correlations

SC.F.1.4.2: Items 2, 3, 4
SC.F.1.4.6: Items 5, 6
SC.H.3.4.3: Item 1

Skills Practice Lab

OBSERVING EMBRYONIC DEVELOPMENT

Teacher's Notes

Time Required
One 50-minute class period

Ratings
EASY → HARD

TEACHER PREPARATION 1
STUDENT SETUP 1
CONCEPT LEVEL 3
CLEANUP 1

Skills Acquired
- Observing
- Comparing and Contrasting
- Making Drawings
- Drawing Conclusions

Scientific Methods
In this lab, students will:
- Make Observations
- Analyze the Results
- Draw Conclusions
- Communicate the Results

Materials and Equipment
Materials for this lab can be ordered from WARD'S. See *Master Materials List* at the front of this book for catalog numbers.

Safety Information
Review microscope and slide safety. Tell students not to scrape the microscope objectives on the slides. If a slide breaks, remind students to be careful of its sharp edges.

Skills Practice Lab

Observing Embryonic Development

SC.F.1.4.2

SKILLS
- Observing
- Comparing and contrasting
- Making drawings
- Drawing conclusions

OBJECTIVES
- **Identify** the stages of early animal development.
- **Describe** the changes that occur during early development.
- **Compare** the stages of human embryonic development with those of echinoderm embryonic development.

MATERIALS
- prepared slides of sea star development, including
- unfertilized egg
- zygote
- 2-cell stage
- 4-cell stage
- 8-cell stage
- 16-cell stage
- 32-cell stage
- 64-cell stage
- blastula
- early gastrula
- middle gastrula
- late gastrula
- compound light microscope
- paper and pencil

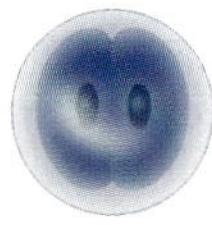
2-cell stage

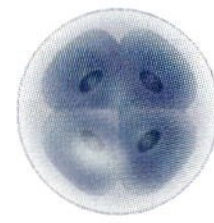
4-cell stage

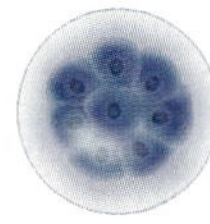
8-cell stage

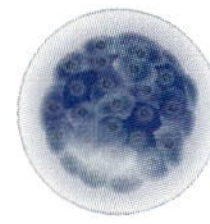
64-cell stage

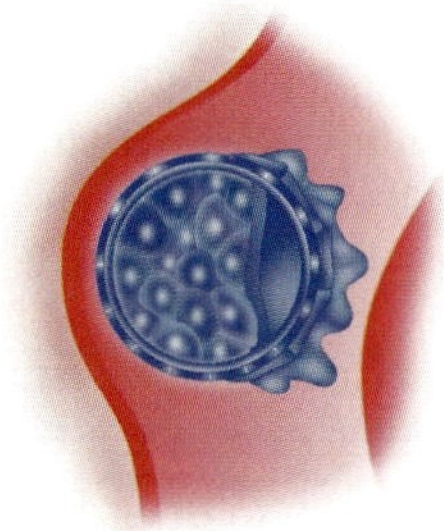
Blastocyst

ChemSafety

CAUTION: Always wear safety goggles and a lab apron to protect your eyes and clothing.

CAUTION: Glassware is fragile. Notify the teacher of broken glass or cuts. Do not clean up broken glass or spills with broken glass unless the teacher tells you to do so.

Before You Begin

Most members of the animal kingdom begin life as a single cell—the fertilized egg, or **zygote.** The early stages of development are quite similar in different species. Cleavage follows fertilization. During cleavage, the zygote divides many times without growing. The new cells migrate and form a hollow ball of cells called a **blastula.** The cells then begin to organize into the three primary germ layers: endoderm, mesoderm, and ectoderm. During this process, the developing organism is called a **gastrula.**

1. Write a definition for each boldface term in the preceding paragraph.
2. Based on the objectives for this lab, write a question you would like to explore about embryonic development.

Tips and Tricks
Remind students that their drawings should be informative rather than artistic. If possible, have students compare the stages of sea-star development to photographs of the mammalian stages.

Procedure

1. Obtain a set of prepared slides that show star eggs at different stages of development. Choose slides labeled unfertilized egg, zygote, 2-cell stage, 4-cell stage, 8-cell stage, 16-cell stage, 32-cell stage, 64-cell stage, blastula, early gastrula, middle gastrula, late gastrula, and young sea star larva. (Note: *Blastula* is the general term for the embryonic stage that results from cleavage. In mammals, a blastocyst is a modified form of the blastula.)
2. Examine each slide using a compound light microscope. Using the microscope's low-power objective first, focus on one good example of the developmental stage listed on the slide's label. Then switch to the high-power objective, and focus on the image with the fine adjustment.
3. In your lab report, draw a diagram of each developmental stage that you examine (in chronological order). Label each diagram with the name of the stage it represents and the magnification used. Record your observations as soon as they are made. Do not redraw your diagrams. Draw only what you see; lab drawings do not need to be artistic or elaborate. They should be well organized and include specific details.
4. Compare your diagrams with the diagrams of human embryonic stages shown at left.
5. Clean up your materials and wash your hands before leaving the lab.

Analyze and Conclude

1. **Summarizing Results** Compare the size of the sea star zygote with that of the blastula. At what stage does the embryo become larger than the zygote?
2. **Analyzing Data** What is the earliest stage in which all of the cells in the embryo no longer look exactly alike? How do cell shape and size change during successive stages of development?
3. **Drawing Conclusions** From your observations of changes in cellular organization, why do you think the blastocoel (the space in the center of the hollow sphere of cells of a blastula) is important during embryonic development?
4. **Predicting Patterns** How are the symmetries of a sea star embryo and a sea star larva different from the symmetry of an adult sea star? Would you expect to see a similar change in human development? What must happen to the sea star gastrula before it becomes a mature sea star?
5. **Further Inquiry** How do your drawings of sea star embryonic development compare with those of human embryonic development? Based on your observations, in what ways do you think sea star embryos could be used to study early human development?

Do You Know?

Do research in the library or media center to answer these questions:

1. How do identical twins develop?
2. How do fraternal twins develop?
3. What is *in vitro* fertilization?

Use the following Internet resources to explore your own questions about embryonic development and cloning.

internet connect

www.scilinks.org
Topic: Cloning
Keyword: HX4047

SCILINKS. Maintained by the National Science Teachers Association

Answers to Before You Begin

1. A zygote is a fertilized egg. A blastula is a hollow ball of cells that develops shortly after fertilization. A gastrula is a three-layered structure that forms from a developing blastula.
2. Answers may vary.

Answers to Analyze and Conclude

1. The zygote and the blastula are the same size. By the time the embryo develops into a gastrula, it is larger than the zygote.
2. When the embryo reaches the gastrula stage, some cells look different from others. Until the gastrula stage, all cells are roughly spherical. After that, some become flattened, elongated, or irregularly shaped. From the first cleavage until the gastrula stage, the cells get progressively smaller. After that, some cells begin to grow.
3. A blastocoel enables invagination (folding) to occur during gastrulation, when the three primary germ layers form.
4. Both sea-star embryos and larvae have bilateral symmetry, whereas the adult has radial symmetry. No; both human embryos and adults have bilateral symmetry. Before the gastrula becomes a mature sea star, it develops into a larva, which has bilateral symmetry. Because the adult has radial symmetry, a repositioning of body organs must occur.
5. Sample answer: Sea-star embryos can be used as a model of early human development because they exhibit similar stages of cell division and differentiation. Sea-star eggs are a good choice for the study of early embryonic development because they are large, numerous, and easy to culture, and because their development is similar to that of the mammals.

Becoming an Active Reader

What Is an Active Reader?

Active reading means thinking and interacting with the text before, during, and after reading it. You might be wondering, "Isn't everyone *thinking* while they are reading?" Not necessarily. Many students read without consciously directing their thinking. Active readers are different—they have developed habits of thinking to use *before, during,* and *after* they read. They use these habits to successfully reach their goal: to understand what they're reading about and remember it well enough to be able to pass the test.

Before You Read

Active readers begin by getting familiar with the reading assignment. This is called previewing. The word *preview* means "to look before." Looking over the chapter before you read helps you predict what you will be learning. It can also help you recall what you already know about a subject—this can help you learn more material and remember it longer.

Scan the Horizon

An active reader will turn to the first page of a chapter, ask questions, and use the information on the page to find answers. Here are some questions to ask.

1. **What do I know?** Questions at the beginning of each chapter in the **Quick Review** prompt you to remember vital information from previous chapters.
2. The **Reading Activity** will help you more fully engage the text. Many Reading Activities encourage you to survey your own knowledge of the chapter's topic. All Reading Activities will help you become more aware of the information that will be presented in the chapter.
3. **What's here?** If you're in an unfamiliar city, you might use a road map to find your way around. In this textbook, the section titled **Looking Ahead** is your map, preparing you for what's coming up. Take time to really look at the organization of the chapter. This organization tells you the focus and the purpose of the chapter.

Get to Know the Neighborhood

Now you're ready to begin the first section. Start by recognizing the elements that are there to help you:

- **Objectives** tell you what you should be able to do after reading the section. Read the objectives before you read the section.
- **Key Terms** are the lesson's key vocabulary words listed in order of appearance. They appear in **bold** black print, highlighted in yellow, within the lesson.

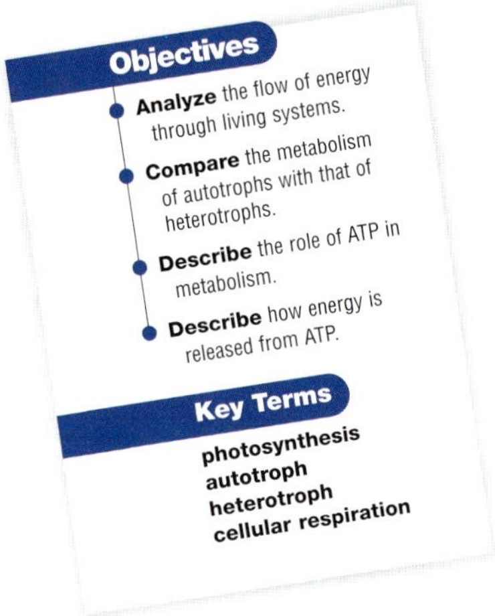

Find Your Way

As you look over the chapter, you will notice text headings in **bold** red print and text subheadings in **bold** blue print. Each red heading is a main idea for the lesson. Red headings are always located at the top of a page, making it easy for you to scan the chapter and find the main ideas. Smaller, blue headings are words or phrases. These subheadings identify important topics that are covered under each red heading. As you read, notice that bold blue print is used for references to nearby photos, illustrations, and tables (ex. **Figure 1, Table 2**).

As You Read

As they read, active readers remain self-aware. They ask themselves questions about how well they understand what they are reading. As you read the lesson, try asking yourself questions.

Monitor your understanding.
- What did I just learn from that paragraph?
- Do I need to read this paragraph again?
- What can I say about that illustration?
- Can I respond to the Objectives now?

Express what you know.
- What can I say about what I just learned?

Ask more questions.
- What is it that I want to know now?
- How does this affect my life?

Study Tips along the way will give you tips for improvement. Look for the following types of Study Tips as you read.

Reading Effectively strategies for finding meaning, understanding content, and identifying main ideas

Organizing Information tips on how to organize what you're learning

Comparing and Contrasting recognizing similarities and differences

Reviewing Information how to review—and remember—what you're reading

Interpreting Graphics suggestions for understanding graphs and illustrations

After You Read

When finished reading, active readers take a few moments to assess their overall level of understanding. Ask yourself:

- What can I say about each illustration or figure?
- Did I learn what I expected to learn?
- What do I know about each lesson title, Objective, heading, subheading, figure, or Key Term in the text?
- What can I tell others about what I know?

Check Your Comprehension

At the end of each section, **Section Review** questions will let you see how much you have learned. Each Section Review question measures how well you mastered an Objective at the beginning of the section. If you cannot answer a Section Review question, go back to the section of text that covers the Objective and re-read the information on that page.

Are You Ready for the Test?

When you have finished reading the chapter, turn to the **Study Zone** page at the end of the chapter to review the **Key Concepts** (main ideas) and the **Key Terms.** If you understand the Key Concepts and Key Terms, then you may answer the questions in the **Performance Zone** to see how well you've learned the material.

Study TIP
Reading Effectively
Use the objectives and the red subheadings at the top of the pages to help you identify the main ideas presented in Section 1.

Study TIP
Reading Effectively
As you read, notice that the names of most enzymes, such as amylase and catalase, end with *-ase*. This will help you identify other enzymes you will encounter in this book.

WORD Origins
The word *filial* is from the Latin *filialis*, meaning "of a son or daughter." Knowing this makes it easier to remember that the F generations refer to any generation following the parental (P) generation.

Concept Mapping

What's the Best Way to Study?

Although there are many ways to study, *concept mapping* may help you understand ideas by showing you their connections to other ideas. A concept map not only identifies the major concepts from a chapter or your class notes but also shows the relationships between the concepts, much as a road map illustrates how highways and other roads are linked to cities.

Identifying Concepts

Suppose you have just finished reading a section of a chapter and you would like to make a concept map as a study aid. How do you begin? First you need to identify the concepts in that section. For instance, examine the following words:

- pool
- grass
- tree
- water
- sky
- playing
- biking
- raining
- thinking

All of these words are concepts, which usually form a picture in your mind. Now read the following series of words:

- the
- to
- has
- when
- was
- be
- with
- can

Are they concepts? No. They do not form a picture in your mind. They are linking words. Linking words play an important role in concept mapping; you use them to connect concepts in your map. See if you can identify the concepts and linking words in the concept map shown in **Figure A** on the facing page. As you can see, the concepts are listed in boxes, and the linking words and lines are used to connect the concepts. Notice that one linking word can be used to link several concepts. You can also use arrows to help link concepts more clearly.

Study TIP

Organizing Information

You can use concept maps as a study guide to help you organize and review information.

Organizing Concepts

Some concepts are more general and include other concepts. These general concepts will be the main ideas in your map. Determine the main idea for the following list of concepts:

- Tokyo
- Mexico City
- Seoul
- New York
- Bombay

They are cities, of course. Now determine the main idea for the following list:

- car
- bus
- train
- bicycle
- truck

Each of these concepts is an example of a vehicle.

In a concept map, the main concept should go at the top. The more specific concepts and examples should go below. Capitalize the first letter of the main concept. Write the other concepts using all lowercase letters. For example, in the concept map shown in Figure A, the main concept is "Biology." The more specific concepts listed below it in the map, such as "taxonomy," describe biology. Study this concept map before you try one yourself. Then read the following paragraph and make a list of any important words you think should be defined.

What is life? What is the difference between living and nonliving things? If you were in a wilderness area, it would be easy for you to pick out the living and nonliving things. The animals and plants are the living organisms. Organisms are made of substances organized into living systems. The rocks, air, water, and soil you see are nonliving. They contribute substances to the living organisms.

Connecting Ideas

Two general ideas prevail in the preceding paragraph: living and nonliving. You could make two separate concept maps, the main concept of one being "Living" and the other being "Nonliving,"

Figure A

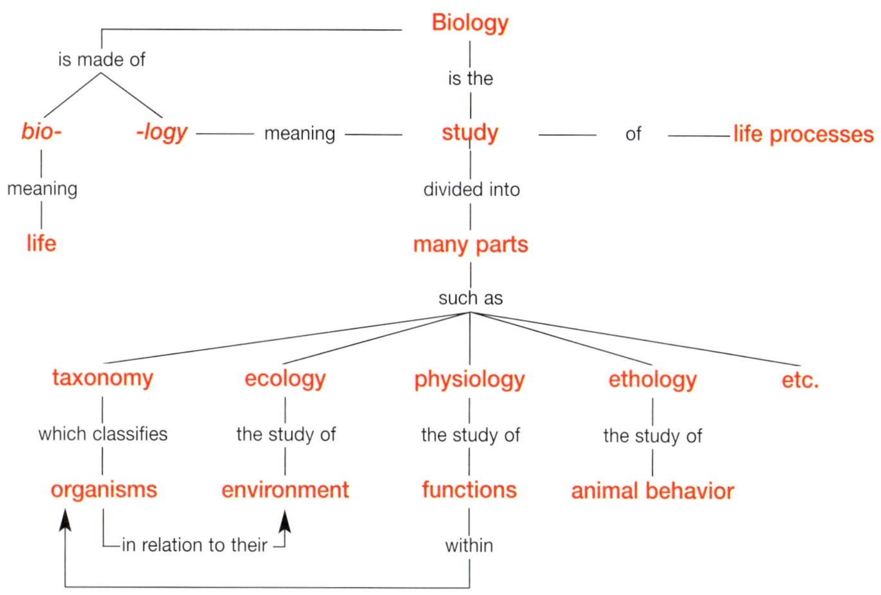

or you could make one map and use "Natural things" as the main concept. Natural things can be living or nonliving. Therefore, the concepts "living" and "nonliving" are parts of the main concept "Natural things" and should be placed below it in the map.

You have learned that living things are organisms and that they are made of substances that are organized into living systems. Living things are different from but related to nonliving things. So you should make the connection between living and nonliving things near the top of the concept map. The rest of the map should relate living and nonliving things. Nonliving things contribute substances that are used by living systems.

Listing Concepts

The next step is to put the concepts in order from the most general to the most specific. You can do this by writing the concepts on pieces of paper. Some concepts will share the same rank and be equally specific. Remember that the examples are the most specific and will be at the bottom of the map.

Now begin to rearrange the concepts you have written on the pieces of paper. Start with the most general; get the main idea. Then, if there are two or more concepts that are equally specific, place them on the same level. For example, "Natural things" is the main idea. Then "living" and "nonliving" are placed under "Natural things" on the same row or level.

Now continue to lay out all the other concepts under the subconcepts in the first row or level until you have used them all. You can rearrange your pieces of paper any time, so keep pushing them around as if they were pieces of a jigsaw puzzle until you have arranged them the way you think they belong.

Linking Concepts

Now make the connections between the concepts. Use lines to connect the concepts, and write linking words on the lines to show or tell why they are connected. Use linking words for all the lines connecting all the concepts. Glue or tape down your concept papers if you want to make the map permanent, or use a separate piece of paper to draw a sketch showing the way you have arranged the concepts.

Now you have the completed concept map shown in **Figure B** on the following page. If you had a choice between reading the paragraph or looking at this map, you would probably agree that the map shows the concepts more clearly. This map gives you the main idea more quickly, and it is easier to understand all the ideas because their relationships to other ideas are shown.

Features of Good Concept Maps

Remember, practice is the key to good concept mapping. You will get better as you go along. Here are some things to remember:

- A concept map does not have to be symmetrical. It can have more concepts on one side than on the other.
- There are no "perfect" concept maps, only maps that come closer to the meanings of the concepts. As the mapmaker, you must make your map work for you.
- Do not put more than three words in a concept box.
- Do not have more than four concept boxes in a row without branching out.
- Connect every pair of concepts with linking words. Use as few linking words as possible.

If the relationships you have made between any two concepts are wrong, your teacher will help you sort out your misconception. Even if your relationships are absolutely correct, maps made by your classmates may be different. These maps could be equally correct, even though they may look nothing like yours. Everyone thinks a little bit differently, and as a result, other people may see different relationships between certain concepts.

As you practice making concept maps, your teacher will examine your linking words more closely. Because the linking words and the lines between concept boxes relate concepts, your linking words will tell you if you really understand how concepts are connected.

A concept map should always have the following characteristics:

- It is two dimensional—not just a list of concepts connected by lines.
- It shows concepts in order of importance.
- It contains many branches with no more than four concept boxes in a row and no more than three words in each concept box.
- It contains only concepts in the boxes and only linking words on the lines.

Evaluating Your Skills

For the first map you make on your own, think about something you know very well. Do you play a team sport or an individual sport? Do you have a hobby? Do you enjoy a particular kind of music? Whatever topic you choose, use it as the main concept for your concept map. This will be more fun and easier because you know this topic so well.

Figure B

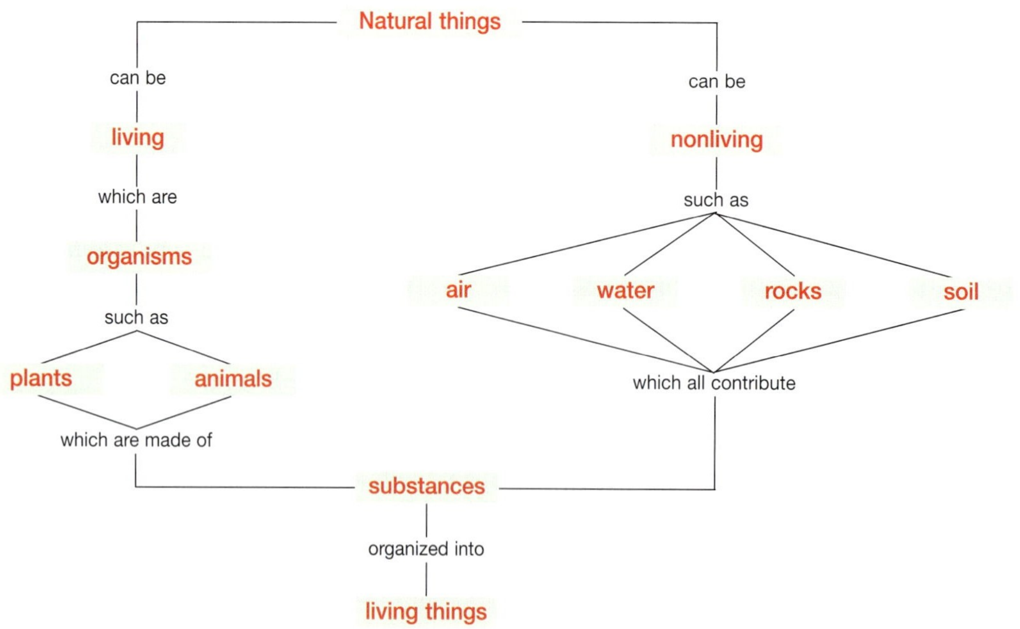

Analyzing Word Parts

Determining the Meanings of Words

The challenge of understanding a new word can often be simplified by carefully examining the word. Many words can be divided into three parts: a prefix, root, and suffix. The **prefix** consists of one or more syllables placed in front of a **root.** The root is the main part of the word. The **suffix** consists of one or more syllables at the end of a root. Prefixes and suffixes modify or add to the meaning of the root. A knowledge of common prefixes, roots, and suffixes can give you clues to the meaning of unfamiliar words and can help make learning new words easier. For example, each of the word parts in **Table 1** can be combined with the root *derm* to form a word.

Table 2 lists word prefixes and suffixes commonly used in biology. Each word part is followed by its usual meaning, an example of a word in which it is used, and a definition of that word. Examine the definition and the example. Decide whether each word part in the first column is a prefix or suffix, depending on how the word part is used in the example.

Table 1 Word Parts

Prefix	Root	Suffix
hypo-	derm	*-ic*
pachy-	derm	
	derm	*-atology*
	derm	*-atologist*
	derm	*-atitis*

Use Table 1 to form five words using the root *derm.* Then use the list of word parts and their definitions to write what you think is each word's meaning. An example is shown below.

Example:	**Dermatologist**
derm (root):	skin
-logy (suffix):	the study of
-ist (suffix):	someone who practices or deals with something
dermatologist:	someone who studies or deals with skin

Table 2 Word Prefixes and Suffixes

Prefix or suffix	Definition	Example
a-	not, without	asymmetrical: not symmetrical
ab-	away, apart	abduct: move away from the middle
-able	able	viable: able to live
ad-	to, toward	adduct: move toward the middle
amphi-	both	amphibian: type of vertebrate that lives both on land and in water
ante-	before	anterior: front of an organism
anti-	against	antibiotic: substance, such as penicillin, capable of killing bacteria
arche-	ancient	*Archaeopteryx:* a fossilized bird
arthro-	joint	arthropod: jointed-limbed organism belonging to the phylum Arthropoda
auto-	self, same	autotrophic: able to make its own food
bi-	two	bivalve: mollusk with two shells
bio-	life	biology: the study of life
blast-	embryo	blastula: hollow ball stage in the development of an embryo
carcin-	cancer	carcinogenic: cancer-causing
cereb-	brain	cerebrum: part of the vertebrate brain
chloro-	green	chlorophyll: green pigment in plants needed for photosynthesis

Table 2 Word Prefixes and Suffixes

Prefix or suffix	Definition	Example
chromo-	color	chromosome: structure found in eukaryotic cells that contains DNA
chondro-	cartilage	Chondrichthyes: cartilaginous fish
circ-	around	circulatory: system for moving fluids through the body
-cide	kill	insecticide: a substance that kills insects
co-, con-	with, together	conjoined twins: identical twins physically joined by a shared portion of anatomy at birth
-cycle	circle	pericycle: layer of plant cells
cyt-	cell	cytology: the study of cells
de-	remove	dehydration: removal of water
derm-	skin	dermatology: study of the skin
di-	two	diploid: full set of chromosomes
dia-	through	dialysis: separating molecules by passing them through a membrane
ecol-	dwelling, house	ecology: the study of living things and their environments
ecto-	outer, outside	ectoderm: outer germ layer of developing embryo
-ectomy	removal	appendectomy: removal of the appendix
endo-	inner, inside	endoplasm: cytoplasm within the cell membrane
epi-	upon, over	epiphyte: plant growing upon another plant
ex-, exo-	outside of	exobiology: the search for life elsewhere in the universe
gastro-	stomach	gastropod: type of mollusk
-gen	type	genotype: genes in an organism
-gram	write or record	climatogram: depicting the annual precipitation and temperature for an area
hemi-	half	hemisphere: half of a sphere
hetero-	different	heterozygous: different alleles inherited from parents
hist-	tissue	histology: the study of tissues
homeo-	the same	homeostasis: maintain a constant condition
hydro-	water	hydroponics: growing plants in water instead of soil
hyper-	above, over	hypertension: blood pressure higher than normal
hypo-	below, under	hypothalamus: part of the brain located below the thalamus
-ic	of or pertaining to	hypodermic: pertaining to under the skin
inter-	between, among	interbreed: breed within a family or strain
intra-	within	intracellular: inside a cell
iso-	equal	isogenic: having an identical set of genes
-ist	someone who practices or deals with something	biologist: someone who studies life
-logy	study of	biology: the study of life
macro-	large	macromolecule: large molecule, such as DNA or proteins

Table 2 Word Prefixes and Suffixes

Prefix or suffix	Definition	Example
mal-	bad	malnourishment: poor nutrition
mega-	large	megaspore: larger of two types of spores produced by some ferns and flowering plants
meso-	in the middle	mesoglea: jellylike material found between outer and inner layers of coelenterates
meta-	change	metamorphosis: change in form
micro-	small	microscopic: too small to be seen with unaided eye
mono-	one, single	monoploid: one set of alleles
morph-	form	morphology: study of the form of organisms
neo-	new	neonatal: newborn
nephr-	kidney	nephron: functional unit of the kidneys
neur-	neuron	neurotransmitter: chemical released by a neuron
oo-	egg	oogenesis: gamete formation in female diploid organisms
org-	living	organism: living thing
-oma	swelling	carcinoma: cancerous tumor
orth-	straight	orthodontics: the practice of straightening teeth
pachy-	thick	pachyderm: thick-skinned animal, such as an elephant
para-	near, on	parasite: organism that lives on and gets nutrients from another organism
path-	disease	pathogen: disease-causing agent
peri-	around	pericardium: membrane around the heart
photo-	light	phototropism: bending of plants toward light
phyto-	plants	phytoplankton: plankton that consists of plants
poly-	many	polypeptide: sequence of many amino acids joined together to form a protein
-pod	foot	pseudopod: false foot that projects from the main part of an amoeboid cell
pre-	before	prediction: a forecast of events before they take place
-scope	instrument used to see something	microscope: instrument used to see very small objects
semi-	partially	semipermeable: allowing some particles to move through
-some	body	chromosome: structure found in eukaryotic cells that contains DNA
sub-	under	substrate: molecule on which an enzyme acts
super-, supra-	above	superficial: on or near the surface of a tissue or organ
syn-	with	synapse: junction of a neuron with another cell
-tomy	to cut	appendectomy: operation in which the appendix is removed
trans-	across	transformation: the transfer of genetic material from one organism to another
ur-	referring to urine	urology: study of the urinary tract
visc-	organ	viscera: internal organs of the body

Safe Laboratory Practices

It is your responsibility to protect yourself and other students by conducting yourself in a safe manner while in the laboratory. You can avoid accidents in the laboratory by following directions, handling materials carefully, and taking your work seriously. Read the following general safety guidelines below before attempting to do work in the laboratory. Make sure you understand all safety guidelines before entering the laboratory. If necessary, ask your teacher for clarification of laboratory rules and procedures.

General Guidelines for Laboratory Safety

- **Only perform experiments specifically assigned by your teacher.** Do not attempt any laboratory procedure without your teacher's direction, and do not work alone in the laboratory.
- **Familiarize yourself with the investigation and all safety precautions before entering the lab.** Be aware of the potential hazards of the required materials and procedures. Before you begin, ask your teacher to explain any parts of an investigation that you do not understand.
- **Before beginning work, tie back long hair, roll up loose sleeves, and put on any required personal protective equipment as directed by your teacher.** Avoid or confine loose clothing that could knock things over, catch on fire, or absorb chemical solutions. Nylon and polyester fabrics burn and melt more readily than does cotton. Do not wear open-toed shoes, sandals, or canvas shoes in the laboratory.
- **Always wear a lab apron and safety goggles.** Wear this equipment even if you are not working on an experiment. Laboratories contain chemicals that can damage your clothing, skin, and eyes. If your safety goggles cloud up or are uncomfortable, ask your teacher for help. Lengthening the strap slightly, washing the goggles with soap and warm water, or using an anti-fog spray may help the problem.
- **No contact lenses are allowed in the lab.** Even if you are wearing safety goggles, chemicals could get between contact lenses and your eyes and cause irreparable eye damage. If your doctor requires that you wear contact lenses instead of glasses, then you should wear eye-cup safety goggles—similar to goggles worn for underwater swimming—in the lab. Ask your doctor or your teacher how to use eye-cup safety goggles to protect your eyes.
- **Know the location of all safety and emergency equipment used in the laboratory.** Ask your teacher where the nearest eyewash stations, safety blankets, safety shower, fire extinguisher, first-aid kit, and chemical spill kit are located.
- **Immediately report any accident, incident, or hazard—no matter how trivial—to your teacher.** Any incident involving bleeding, burns, fainting, chemical exposure, or ingestion should also be reported immediately to the school nurse or to a physician.
- **In case of fire, alert your teacher and leave the lab.** Standard fire-safety procedures should be followed.
- **Do not have or consume food or drink in the lab.** Do not store or eat food in the laboratory.

- **Do not fool around in the lab.** Take your lab work seriously and behave appropriately in the laboratory. Be aware of your classmates' safety as well as your own at all times.
- **Do not apply cosmetics in the lab.** Some hair-care products and nail polish are highly flammable.
- **Keep your work area neat and uncluttered.** Have only books and other materials that are needed to conduct the experiment in the laboratory.
- **Clean your work area at the conclusion of each lab period as directed by your teacher.** Broken glass, chemicals, and other laboratory waste products should be disposed of in separate special containers. Dispose of waste materials as directed by your teacher.
- **Wash your hands with soap and hot water after each lab period.** Wash your hands at the conclusion of each lab period and before leaving the laboratory to avoid contamination.

Key to Safety Symbols and Their Precautions

Before you begin working in the laboratory, familiarize yourself with the following safety symbols, which are used throughout this textbook, and guidelines that you should follow when you see these symbols.

Eye Safety

- **Wear approved safety goggles as directed.** Safety goggles should always be worn in the laboratory, especially when you are working with a chemical or solution, a heat source, or a mechanical device.
- **In case of eye contact, do the following:** Go to an eyewash station immediately and flush your eyes (including under the eyelids) with running water for at least 15 minutes. Hold your eyelids open with your thumb and fingers, and roll your eyeball around. While doing this, have another student notify your teacher.
- **Do not wear contact lenses in the lab.** Chemicals can be drawn up under a contact lens and into the eye. If you must wear contacts prescribed by a physician, tell your teacher. You must also wear approved eye-cup safety goggles to help protect your eyes.

- **Do not look directly at the sun through any optical device or lens system, and do not reflect direct sunlight to illuminate a microscope.** Such actions concentrate light rays to an intensity that can severely burn your retinas, possibly causing blindness.

Hand Safety

- **Do not cut objects while holding them in your hand.** Dissect specimens in a dissecting tray.
- **Wear protective gloves when working with an open flame, chemicals, solutions, or wild or unknown plants.**

Safety with Gases

- **Do not inhale any gas or vapor unless directed to do so by your teacher.** Never inhale pure gases.
- **Handle materials that emit vapors or gases in a well-ventilated area.** This work should be done in an approved chemical fume hood.

Sharp-Object Safety

- **Use extreme care when handling all sharp and pointed instruments, such as scalpels, sharp probes, and knives.**
- **Do not use double-edged razor blades in the laboratory.**
- **Do not cut objects while holding them in your hand.** Cut objects on a suitable work surface. Always cut in a direction away from your body.

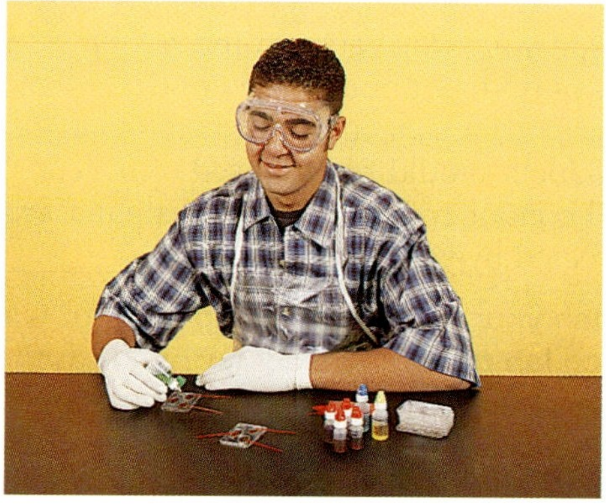

Animal Care and Safety

- **Do not approach or touch any wild animals.** When working outdoors, be aware of poisonous or dangerous animals in the area.
- **Always get your teacher's permission before bringing any animal (including pets) into the school building.**
- **Handle animals only as directed by your teacher.** Mishandling or abusing any animal will not be tolerated.

Heating Safety

- **Be aware of any source of flames, sparks, or heat (open flames, electric heating coils, hot plates, etc.) before working with flammable liquids or gases.**
- **When heating chemicals or solutions in a test tube, do not point the test tube toward anyone.**
- **Avoid using open flames.** If possible, work only with hot plates that have an "On-Off" switch and an indicator light. Do not leave hot plates unattended. Do not use alcohol lamps. Turn off hot plates and open flames when they are not being used.
- **Know the location of laboratory fire extinguishers and fire-safety blankets.**
- **Use tongs or appropriate insulated holders when handling heated objects.** Heated objects often do not appear to be hot. Do not pick up an object with your hand if it could be warm.
- **Keep flammable substances away from heat, flames, and other ignition sources.**

Hygienic Care

- **Keep your hands away from your face and mouth while working in the lab.**
- **Wash your hands thoroughly before leaving the laboratory.**
- **Remove contaminated clothing immediately.** If you spill caustic substances on your skin or clothing, use the safety shower or a faucet to rinse. Remove affected clothing while under the shower, and call to your teacher. (It may be temporarily embarrassing to remove clothing in front of your classmates, but failure to rinse a chemical off your skin could result in permanent damage.)
- **Launder contaminated clothing separately.**
- **Use the proper technique demonstrated by your teacher when handling bacteria or other microorganisms.** Treat all microorganisms as if they are pathogens. Do not open Petri dishes to observe or count bacterial colonies.
- **Return all stock and experimental cultures to your teacher for proper disposal.**

Glassware Safety

- **Inspect glassware before use; do not use chipped or cracked glassware.** Use heat-resistant glassware for heating materials or storing hot liquids.
- **Do not attempt to insert glass tubing into a rubber stopper without specific instruction from your teacher.**
- **Immediately notify your teacher if a piece of glassware breaks. Do not attempt to clean up broken glass.**

Proper Waste Disposal

- **Clean and sanitize all work surfaces and personal protective equipment after each lab period as directed by your teacher.**
- **Dispose of all sharp objects (such as broken glass) and other contaminated materials (biological or chemical) in special containers as directed by your teacher.**

Electrical Safety

- **Do not use equipment with frayed electrical cords or loose plugs.**
- **Fasten electrical cords to work surfaces using tape.** This will prevent tripping and will ensure that equipment cannot fall off the table.
- **Do not use electrical equipment near water or with wet hands or clothing.**
- **Hold the rubber cord when you plug in or unplug equipment.** Do not touch the metal prongs of the plug, and do not unplug equipment by pulling on the cord.

Clothing Protection

- **Wear an apron or laboratory coat at all times in the laboratory to prevent chemicals or chemical solutions from contacting skin or clothes.**

Plant Safety

- **Do not ingest any plant part used in the laboratory (especially commercially sold seeds).** Do not touch any sap or plant juice directly. Always wear gloves.
- **Wear disposable polyethylene gloves when handling any wild plant.**
- **Wash hands thoroughly after handling any plant or plant part (particularly seeds). Avoid touching your face and eyes.**
- **Do not inhale or expose yourself to the smoke of any burning plant.** Smoke contains irritants that can cause inflammation in the throat and lungs.
- **Do not pick wildflowers or other plants unless directed by your teacher.**

Chemical Safety

- **Always wear safety goggles, gloves, and a lab apron or coat when working with any chemical or chemical solution to protect your eyes and skin.**
- **Do not taste, touch, or smell any chemicals or bring them close to your eyes, unless specifically instructed by your teacher.** If you are directed by your teacher to note the odor of a substance, do so by waving the fumes toward you with your hand. Do not pipette any chemicals by mouth; use a suction bulb as directed by your teacher.
- **Know the location of the emergency lab shower and eyewash and how to use them.** If you get a chemical on your skin or clothing, wash it off at the sink while calling to your teacher.
- **Always handle chemicals or chemical solutions with care.** Check the labels on bottles, and observe safety procedures. Do not return unused chemicals or solutions to their original containers. Return unused reagent bottles or containers to your teacher.
- **Do not mix any chemicals unless specifically instructed by your teacher.** Otherwise harmless chemicals can be poisonous or explosive if combined.
- **Do not pour water into a strong acid or base.** The mixture can produce heat and splatter.
- **Report any spill immediately to your teacher.** Spills should be cleaned up promptly as directed by your teacher.

ChemSafety

CAUTION: Always wear safety goggles and a lab apron to protect your eyes and clothing.

CAUTION: Do not touch or taste any chemicals. Know the location of the emergency shower and eyewash station and how to use them. If you get a chemical on your skin or clothing, wash it off at the sink while calling to the teacher. Notify the teacher of a spill. Spills should be cleaned up promptly, according to your teacher's directions.

CAUTION: Glassware is fragile. Notify the teacher of broken glass or cuts. Do not clean up broken glass or spills with broken glass unless the teacher tells you to do so.

Using a Compound Light Microscope

Parts of the Compound Light Microscope

- The **eyepiece** magnifies the image, usually 10×.
- The **low-power objective** further magnifies the image, up to 4×.
- The **high-power objectives** further magnify the image, from 10× to 43×.
- The **nosepiece** holds the objectives and can be turned to change from one objective to another.
- The **body tube** maintains the correct distance between the eyepiece and the objectives. This is usually about 25 cm (10 in.), the normal distance for reading and viewing objects with the unaided eye.
- The **coarse adjustment** moves the stage up and down in large increments to allow gross positioning and focusing of the objective lens.
- The **fine adjustment** moves the stage slightly to bring the image into sharp focus.
- The **stage** supports a slide that contains the viewed specimen.
- The **stage clips** secure the slide in position for viewing.
- The **diaphragm** (not labeled), located under the stage, controls the amount of light allowed to pass through the object being viewed.
- The **light source** provides light for viewing the image. It can be either a light reflected with a mirror or an incandescent light from a small lamp. Never use reflected direct sunlight as a light source.
- The **arm** supports the body tube.
- The **base** supports the microscope.

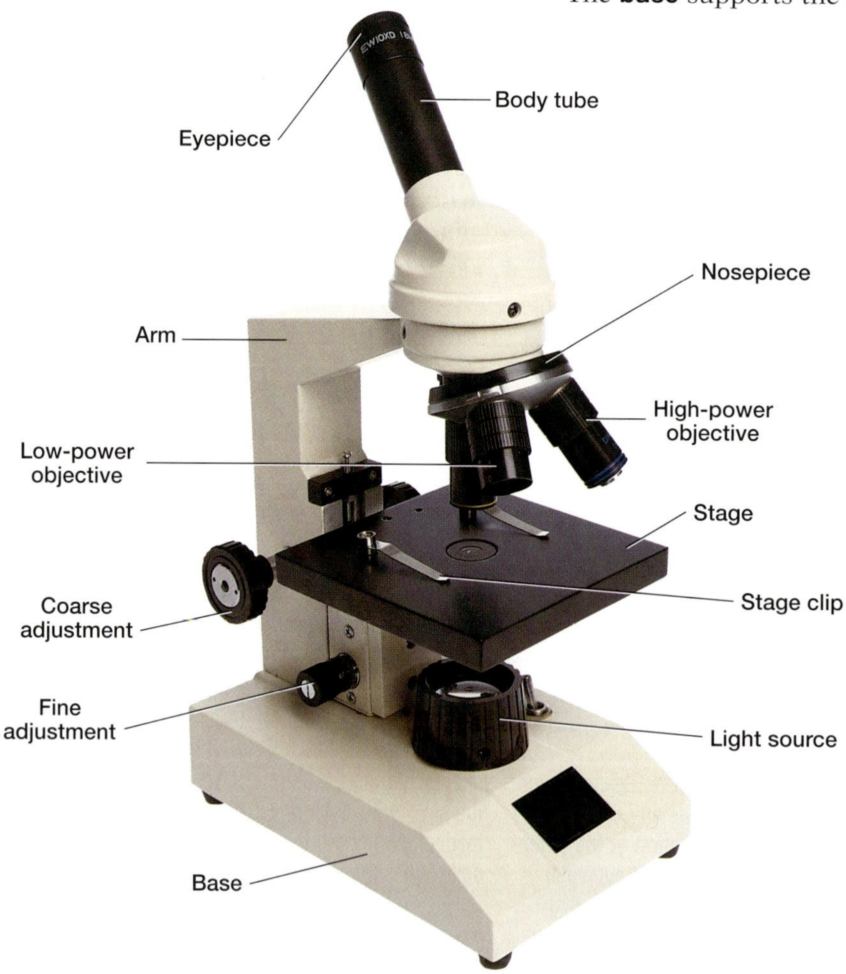

Proper Handling and Use of the Compound Light Microscope

1. Carry the microscope to your lab table using both hands, one supporting the base and the other holding the arm of the microscope. Hold the microscope close to your body.
2. Place the microscope on the lab table at least 5 cm (2 in.) from the edge of the table.
3. Check to see what type of light source the microscope has. If the microscope has a lamp, plug it in, making sure that the cord is out of the way. If the microscope has a mirror, adjust it to reflect light through the hole in the stage.

   CAUTION: If your microscope has a mirror, do not use direct sunlight as a light source. Using direct sunlight can damage your eyes.

4. Adjust the revolving nosepiece so that the low-power objective is aligned with the body tube.
5. Place a prepared slide over the hole in the stage, and secure the slide with the stage clips.
6. Look through the eyepiece, and move the diaphragm to adjust the amount of light that passes through the specimen.
7. Now look at the stage at eye level. Slowly turn the coarse adjustment to raise the stage until the objective almost touches the slide. Do not allow the objective to touch the slide.
8. While looking through the eyepiece, turn the coarse adjustment to lower the stage until the image is in focus. Never focus objectives downward. Use the fine adjustment to achieve a sharply focused image. Keep both eyes open while viewing a slide.
9. Make sure that the image is exactly in the center of your field of vision. Then switch to the high-power objective. Focus the image with the fine adjustment. Never use the coarse adjustment at high power.
10. When you are finished using the microscope, remove the slide. Clean the eyepiece and objectives with lens paper, and return the microscope to its storage area.

Procedure for Making a Wet Mount

1. Use lens paper to clean a glass slide and coverslip.

   CAUTION: Glass slides and coverslips break easily. Handle them carefully. Notify your teacher if you break a slide or coverslip.

2. Place the specimen that you wish to observe in the center of the slide.
3. Using a medicine dropper, place one drop of water on the specimen.
4. Position the coverslip so that it is at the edge of the drop of water and at a 45° angle to the slide. Make sure that the water runs along the edge of the coverslip.
5. Lower the coverslip slowly to avoid trapping air bubbles.
6. If a stain or solution will be added to a wet mount, place a drop of the staining solution on the microscope slide along one side of the coverslip. Place a small piece of paper towel on the opposite side of the coverslip.
7. As the water evaporates from the slide, add another drop of water by placing the tip of the medicine dropper next to the edge of the coverslip, just as you would if adding stains or solutions to a wet mount. If you have added too much water, remove the excess by using the corner of a paper towel as a blotter. Do not lift the coverslip to add or remove water.

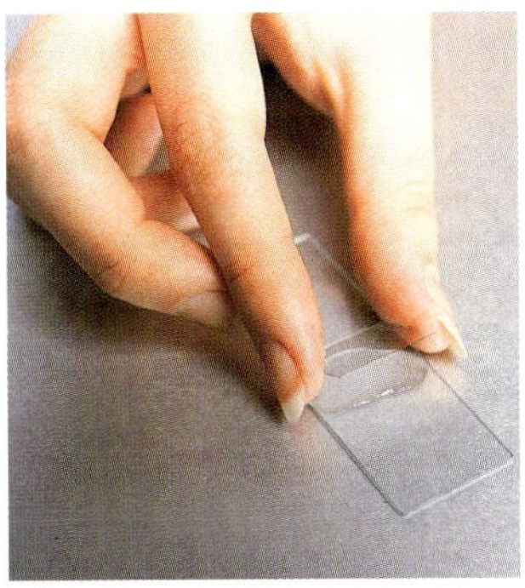

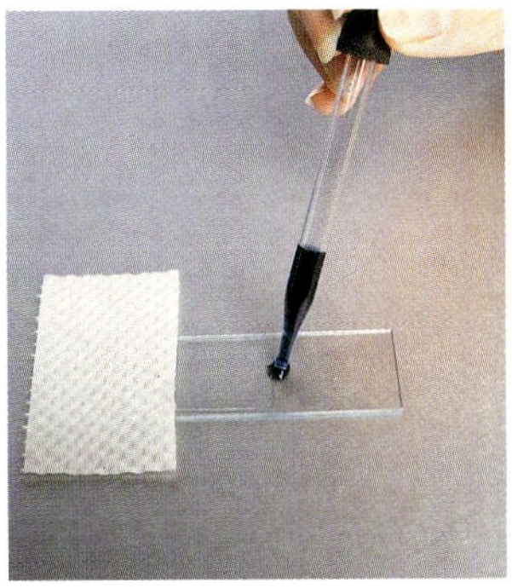

Determining Mass and Temperature

Reading a Balance

A single-pan balance, such as the one shown at right, has one pan and three or four beams. The scale of measure for each beam depends on the model of the balance. When an object is placed on the pan, the riders are moved along the beams until the mass on the beams equals the mass of the object in the pan.

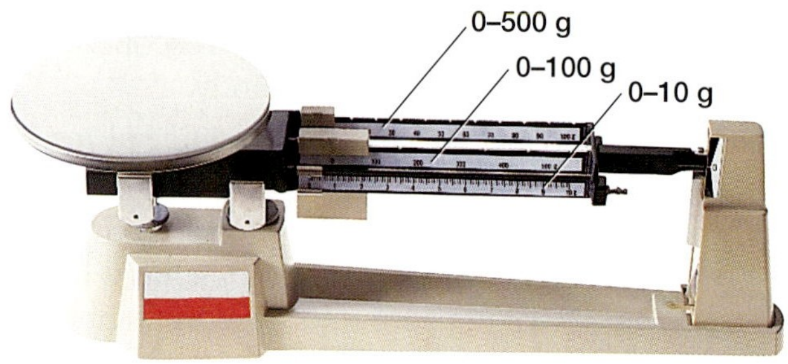

Measuring Mass

When determining the mass of a chemical or powder, use weighing or filter paper. Determine the mass of the paper, and subtract that mass from the total mass. Use the following procedure for determining the mass of objects:

1. Make sure the balance is on a level surface and the pan is allowed to move freely. Position all the riders at zero. If the pointer does not come to rest in the middle of the scale, calibrate the balance using the adjustment knob (usually located under and to the left of the pan).

CAUTION: Never place a hot object or chemical directly on a balance pan.

2. Place the object on the pan.

3. Move the largest rider along the beam to the right until it is at the last notch that does not move the pointer below the zero point in the middle of the scale.

4. Follow the same procedure with the next rider.

5. Move the smallest rider until the pointer rests at zero in the middle of the scale.

6. Add up the readings on all the beams to determine the mass of the object.

Practice Exercises

1. Determine the mass of each of the following items using a single-pan balance:

a. an empty 250 mL beaker
b. 250 mL beaker filled with 100 mL of water
c. 250 mL beaker filled with 100 mL of vegetable oil
d. a house key
e. a small book
f. a paper clip or small safety pin

2. Determine the mass of each object represented by the balance readings shown.

a.

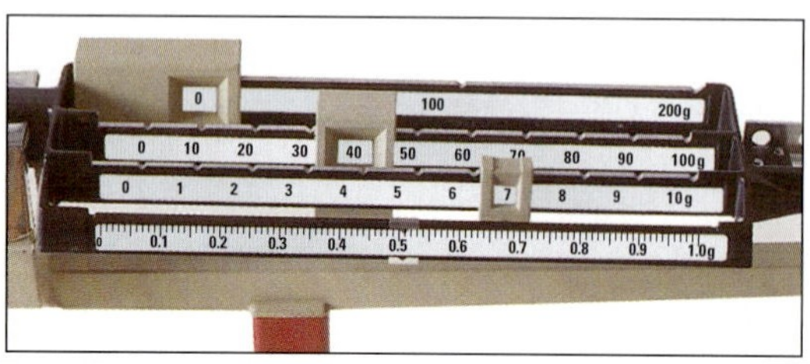

b.

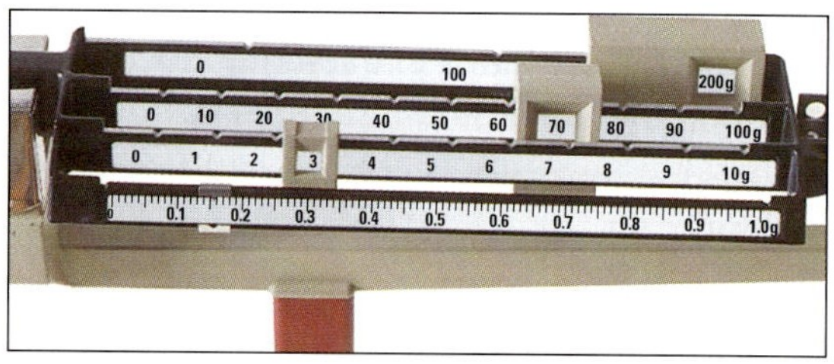

c.

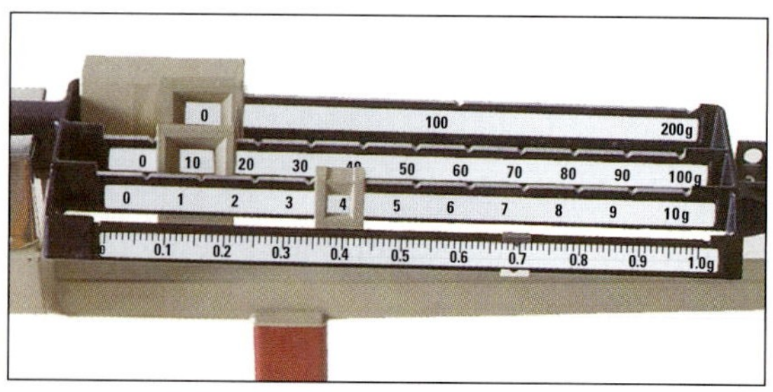

Reading a Thermometer

Many laboratory thermometers are the bulb-type shown below. The sensing bulb of the thermometer is filled with a colored liquid (alcohol) that expands when heated. When the liquid expands, it moves up the stem of the thermometer through the capillary tube. Thermometers usually measure temperature in degrees Celsius (°C).

Measuring Temperature

Use the following procedure when measuring the temperature of a substance.

1. Carefully lower the bulb of the thermometer into the substance. The stem of the thermometer may rest against the side of the container, but the bulb should never rest on the bottom where heat is being applied. If the thermometer has an adjustable clip for the side of the container, the thermometer can be suspended in the liquid.

CAUTION: Do not hold a thermometer in your hand while measuring the temperature of a heated substance.

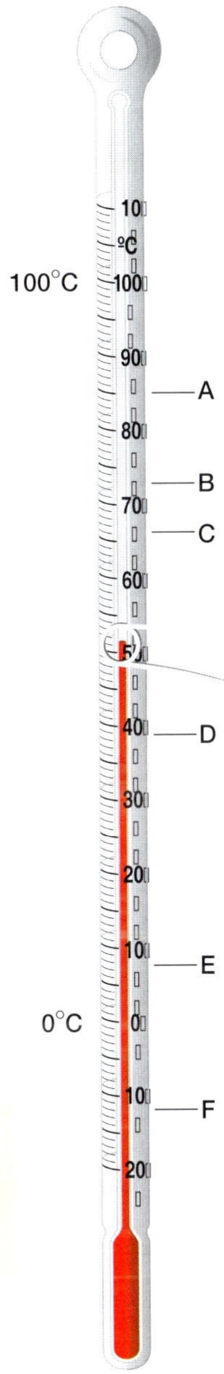

2. Gently rotate the thermometer in the clip. Watch the rising colored liquid in the capillary tube. When the liquid in the capillary tube stops rising, note the whole-degree increment nearest the top of the liquid column. If your thermometer is marked in tenths of a degree, count the increments up or down from the nearest whole degree to obtain your reading. For example, if the top of the colored liquid column is closest to the 51°C mark but somewhat above it, as shown below, what is the accurate temperature reading? Because the top of the column about one-half of a degree above the 51°C mark, the temperature is 51.5°C. Add one-half of a degree to 51°C to obtain your reading.

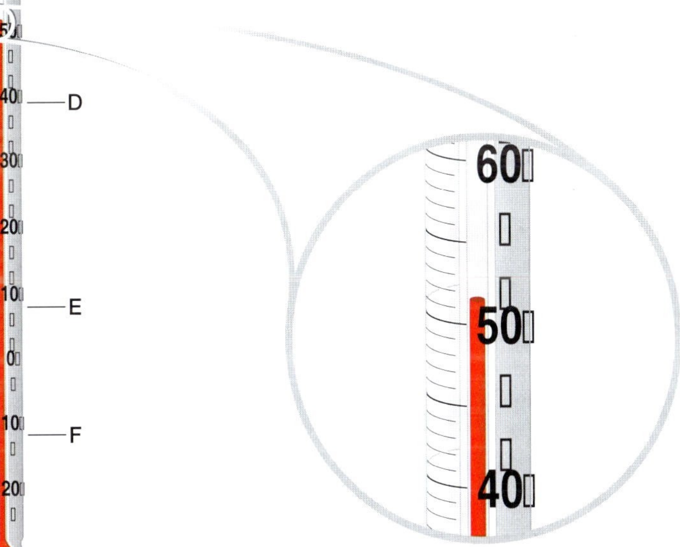

Practice Exercises

Use the thermometer shown above to answer the following questions:

1. Identify the scale used for this thermometer.

2. Determine whether this thermometer is marked only in whole degrees or in tenths of degrees.

3. Estimate the temperature reading on this thermometer.

4. SKILL Interpreting Variables What would be the temperature reading if the top of the column were resting at each of the following points?

a. *A*
b. *B*
c. *C*
d. *D*
e. *E*
f. *F*

SI Measurement

SI Units

Scientists throughout the world use the metric system. The official name of the metric system is the Système International d'Unités, or the International System of Measurements. It is usually referred to simply as SI. Most measurements in this book are expressed in metric units. You will always use metric units when you take measurements in the lab.

SI Prefixes

SI is a decimal system; that is, all relationships between SI units are based on powers of 10. Most units have a prefix that indicates the relationship of that unit to a base unit. For example, the SI base unit for length is the meter. A meter equals 100 centimeters (cm), or 1,000 millimeters (mm). A meter also equals 0.001 kilometer (km). **Table 1** summarizes the prefixes used in SI units.

Table 1 SI Prefixes

Prefix	Symbol	Factor of base unit
giga-	G	1,000,000,000
mega-	M	1,000,000
kilo-	k	1,000
hecto-	h	100
deka-	da	10
deci-	d	0.1
centi-	c	0.01
milli-	m	0.001
micro-	μ	0.000001
nano-	n	0.000000001
pico-	p	0.000000000001

Conversion Factors

Conversion between SI units requires a conversion factor. For example, to convert from meters to centimeters, you need to know the relationship between meters and centimeters.

$$1 \text{ cm} = 0.01 \text{ m} \quad \text{or} \quad 1 \text{ m} = 100 \text{ cm}$$

If you need to convert 15.5 centimeters to meters, you could do either of the following:

$$15.5 \text{ cm} \times \frac{1 \text{ m}}{100 \text{ cm}} = 0.155 \text{ m}$$

or

$$15.5 \text{ cm} \times \frac{0.01 \text{ m}}{1 \text{ cm}} = 0.155 \text{ m}$$

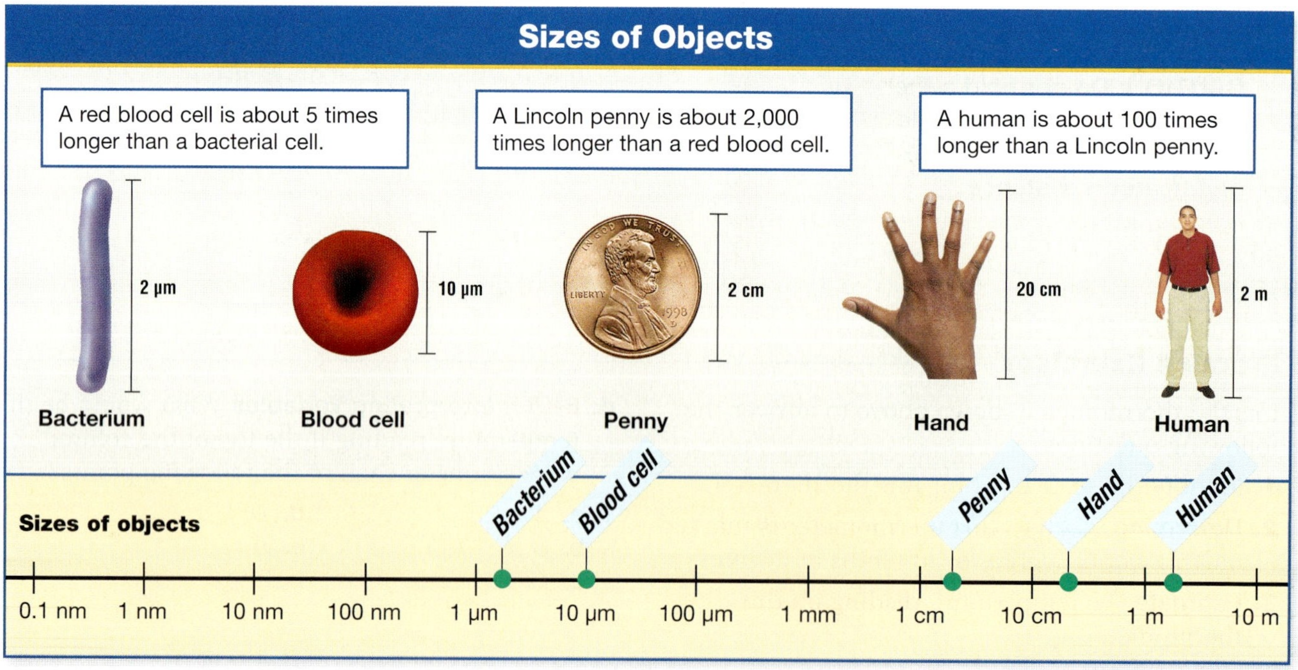

Base Units

In this book, you will see three fundamental quantities represented by base units in SI: mass, length, and time. The base units of these quantities are the kilogram (kg), the meter (m), and the second (s). These quantities, their abbreviations, and their equivalent measurements are listed in **Table 2.**

Derived Units

Other important quantities, such as area (m^2) and liquid volume (m^3), are expressed in derived units. A derived unit is a combination of one or more base units. Like base units, derived units can be expressed using SI prefixes. These quantities are listed in **Table 3.**

Table 2 Conversions for SI Base Units

Mass: unit = kilogram (kg)
1 kilogram (kg) = 1,000 g
1 gram (g) = 0.001 kg
1 milligram (mg) = 0.001 g
1 microgram (μg) = 0.000001 g
Length: unit = meter (m)
1 kilometer (km) = 1,000 m
1 meter (m) = 100 cm
1 centimeter (cm) = 0.01 m
1 millimeter (mm) = 0.001 m
1 micrometer (μm) = 0.000001 m
Time: unit = second (s)
1 minute (min) = 60 s
1 hour (h) = 3,600 s = 60 min
1 day (d) = 24 h

Table 3 Conversions for SI Derived Units

Area: unit = square meter (m^2)
1 square kilometer (km^2) = 100 ha
1 hectare (ha) = 10,000 m^2
1 square meter (m^2) = 10,000 cm^2
1 square centimeter (cm^2) = 100 mm^2
Liquid volume: unit = cubic meter (m^3)
1 cubic meter (m^3) = 1 kL
1 kiloliter (kL) = 1,000 L
1 liter (L) = 1,000 mL
1 milliliter (mL) = 0.001 L
1 cubic centimeter (cm^3) = 1 mL
Mass density: unit = kilograms per cubic meter (kg/m^3)
Temperature: unit = degrees Celsius (°C)
Velocity: unit = meters per second (m/s)

Temperature

In SI, the Celsius scale is used to express temperature. In the Celsius scale, 0°C (32°F) is the freezing point of water, and 100°C (212°F) is the boiling point of water. You can use the temperature scale shown below to convert between the Celsius scale and the Fahrenheit scale, which is commonly used in the United States. You can also use the following equation to convert between degrees Celsius (T_C) and degrees Fahrenheit (T_F):

$$T_F = \frac{9}{5} T_C + 32$$

For example, to convert 0°C to degrees Fahrenheit, do the following:

$$T_F = \frac{9}{5}(0°C) + 32°F = 0 + 32°F = 32°F$$

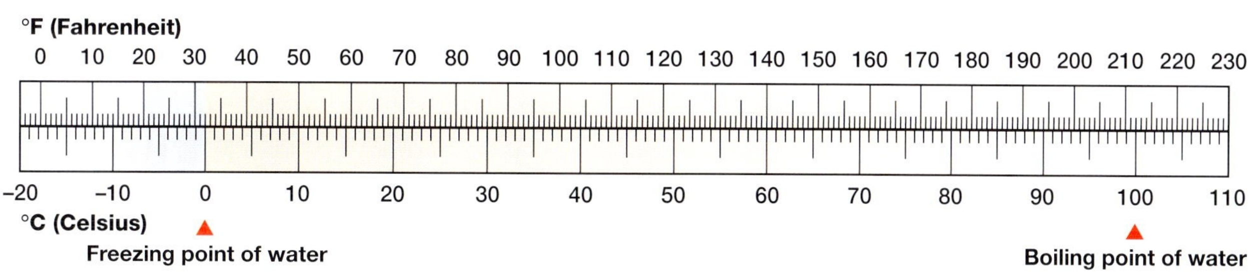

Graphing

Line Graphs

Line graphs, such as the one shown at right, are most often used to compare or relate one or more sets of data that show continuous change. In the graph shown at right, both "Daily salt intake" and the "Systolic pressure" are the **variables,** or sets of data that are being compared.

In graphs, values are assigned to the independent variable. In this case, "Daily salt intake" is the **independent variable.** "Systolic pressure" is called the **dependent variable** because blood pressure is affected by salt intake, according to the graph. Each set of data—the independent and dependent variables—is called a data pair.

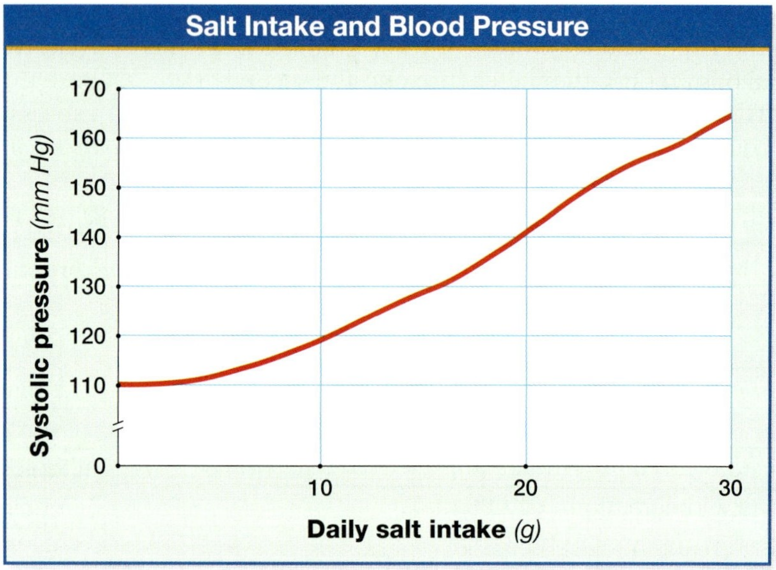

Another way to think about independent and dependent variables is to think about the amount of sleep you get. You know that how alert or tired you feel often depends on the number of hours of sleep that you had the night before. The amount of sleep is the independent variable; your alertness is the dependent variable. Studying biology, you will see many examples of dependent and independent variables represented in graphs.

When you are making a line graph using data pairs, first organize data pairs into a table. Then draw the horizontal and vertical axes of your graph. Be sure to label each axis, including units where appropriate. Refer to your data table to determine the scale and interval of each axis. Make sure that the scale and interval of each axis are consistent. Plot each data pair on the graph. Then connect the plotted data points to make a line, or curve. Finally, give the graph a title that clearly indicates the relationship between the data shown by the graph.

Characteristics of Line Graphs

Important characteristics of line graphs include the following:

- The independent variable is graphed on the horizontal *(x)* axis.
- The dependent variable is graphed on the vertical *(y)* axis.
- Both axes are labeled.
- An appropriate scale and interval are used on each axis. The same scale and interval must be used for the total length of the axis.
- Reasonable starting points are used for each axis.
- The data pairs are plotted as accurately as possible.
- The title of the graph accurately reflects the data presented.
- If more than one set of data is presented on a graph, a key must accompany the graph.
- The graph is easy to understand and interpret.

Bar Graphs

Sometimes it is not appropriate to use a line graph to represent data. A bar graph, such as the one shown at right, is appropriate for data that are not continuous. A bar graph is a good indicator of trends if the data are taken over a sufficiently long period of time. For example, studying color variations in moths requires that data be collected over a long period of time. Even after years of study, predictions can still be difficult to make with certainty. Notice that a bar graph is also useful in comparing multiple sets of data, such as those for the light and dark moths found in the woods near Birmingham and Dorset.

Color Variations in Moths

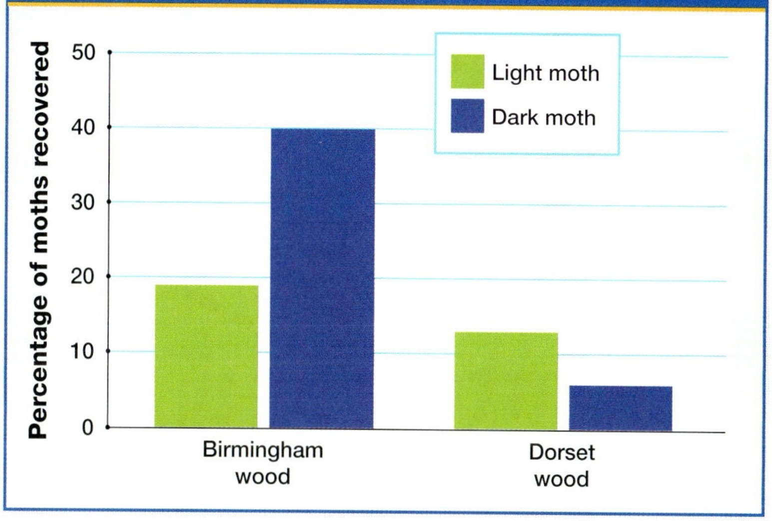

Characteristics of Bar Graphs

Important characteristics of bar graphs include the following:

- An appropriate scale is used on each axis.
- Reasonable starting points are used for each axis.
- A key accompanies the graph.
- The axes are labeled.
- Data are accurately plotted.
- The title of the graph accurately reflects the data presented.
- The graph is easy to understand and interpret.

Using Graphs to Make Predictions

Graphs show trends in data that may not be obvious from a data chart or table. Examine the graph of salt intake and blood pressure on the facing page. Do you notice any trends? Do you think you can conclude that salt intake affects blood pressure? The process of going beyond the data points in a graph to determine a relationship between the data is called **extrapolation.** The further we extrapolate, the less certain we can be of our predictions.

We can also use graphs to link information quickly between two sets of data pairs. For example, we can read the graph to determine that when the subject takes in 10 g of salt daily, the subject's systolic blood pressure is about 120 mm Hg. Linking information between two sets of data pairs in this way is called **interpolation.** Interpolation will help you identify relationships between sets of data in data pairs.

Study TIP

Interpreting Graphics
When reading graphs, identify the dependent and independent variables. Then see if you can determine a relationship between the two variables by interpolation.

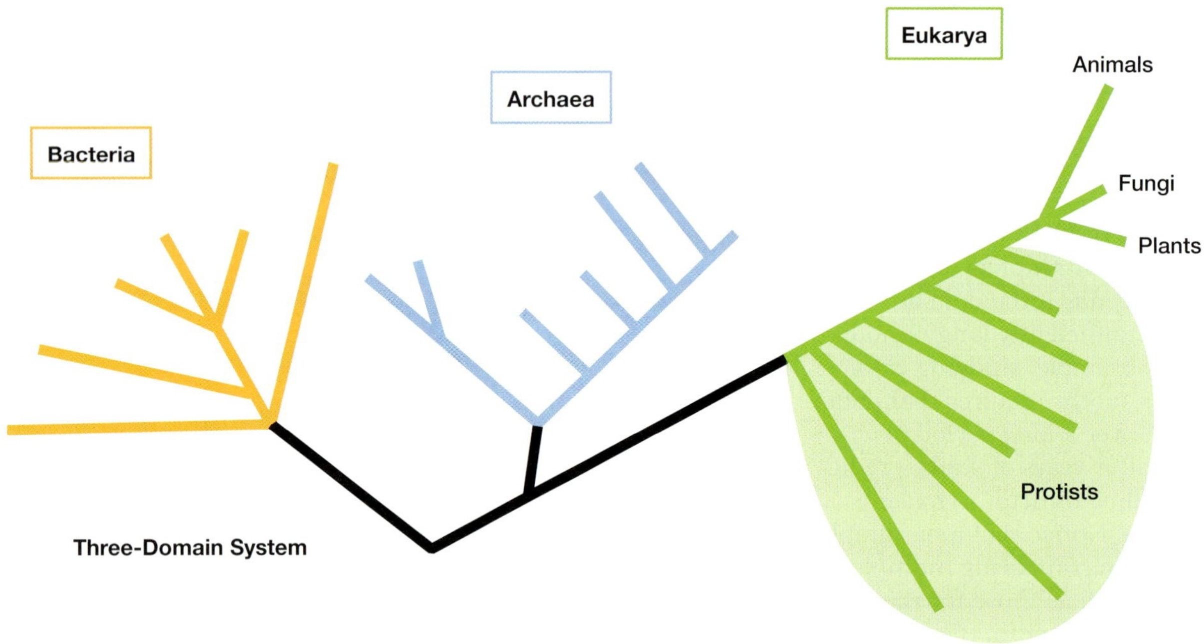

Three-Domain System

The classification system used in this book is based on both the commonly recognized six-kingdom system and the newer, three-domain system. A **kingdom** is a group of related phyla, and a **domain** is a group of related kingdoms. In the three-domain system illustrated above, all living things are grouped into three distinct domains based on similarities in their nucleic acid sequences. Two of the three domains consist of prokaryotes and one consists of eukaryotes.

The domain **Bacteria** is thought to be the oldest of the three domains. The domain Bacteria consists of the single kingdom Eubacteria—prokaryotic microbes commonly called *bacteria*. A second prokaryotic domain is **Archaea.** Archaea is composed of the single kingdom Archaebacteria—prokaryotic microbes called *archaebacteria*. Although both are prokaryotic microbes, archaebacteria and bacteria differ greatly.

The third domain, **Eukarya,** is composed of all of the eukaryotic organisms. In the four kingdoms within the domain Eukarya are found the animals, plants, fungi, and protists (the kingdoms Animalia, Plantae, Fungi, and Protista). The phylogenetic tree below shows the division of organisms into six kingdoms.

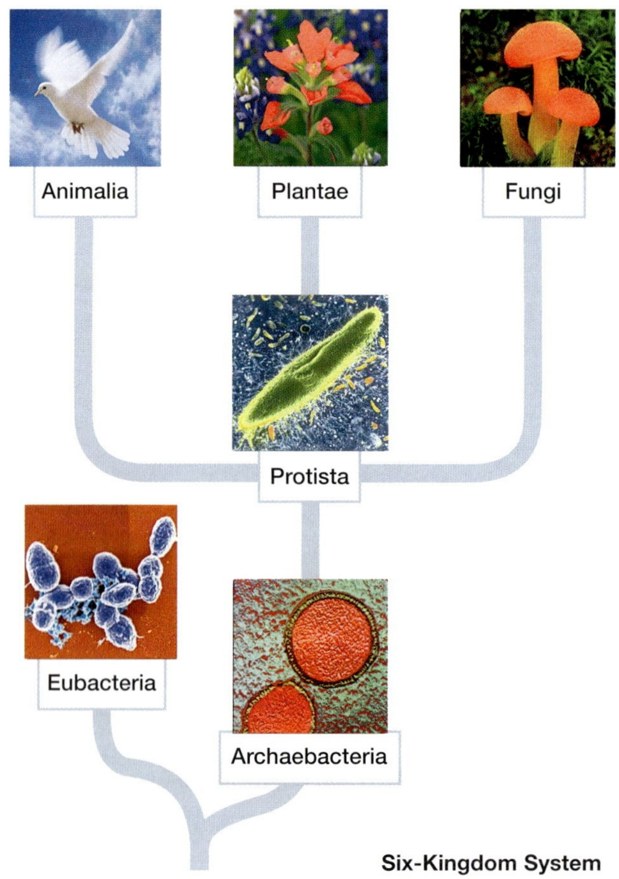

Six-Kingdom System

The information on the following pages is organized according to kingdoms. Not every phylum or group in each kingdom is discussed, however, and the classification of some groups is controversial. For example, biologists do not agree how the kingdoms Eubacteria, Archaebacteria, and Protista should be divided into phyla.

In this book, these kingdoms are divided into convenient and commonly recognized groups. Biologists also disagree about the number of species in various groups. Unless stated otherwise, the numbers given represent approximate numbers of living, named species.

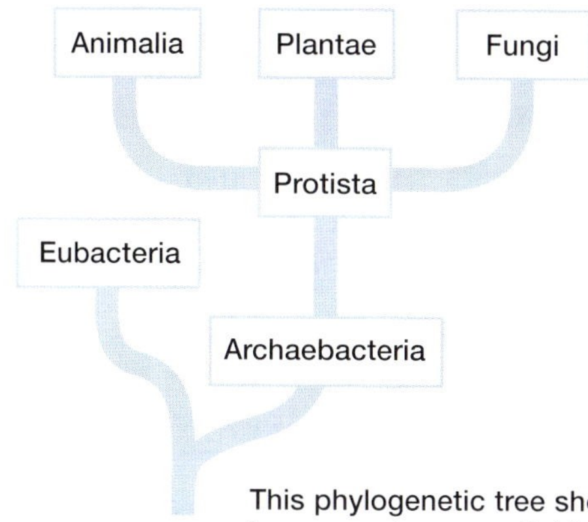

This phylogenetic tree shows how organisms are divided into six kingdoms.

Kingdom Eubacteria

More than 4,000 species
Typically unicellular; prokaryotic; without membrane-bound organelles; nutrition mainly heterotrophic (by absorption), but some are photosynthetic or chemosynthetic; reproduction usually by fission or budding.

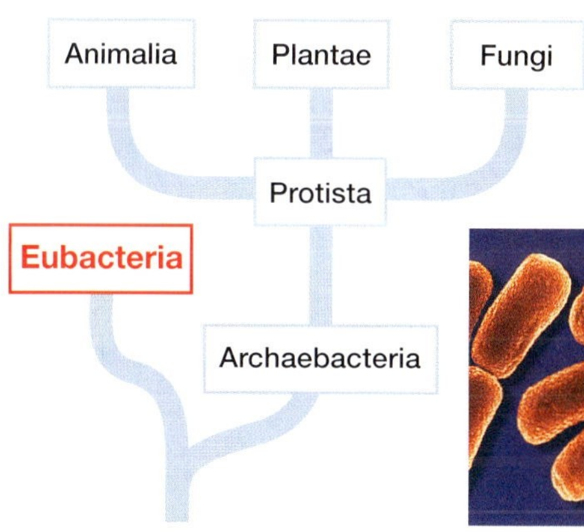

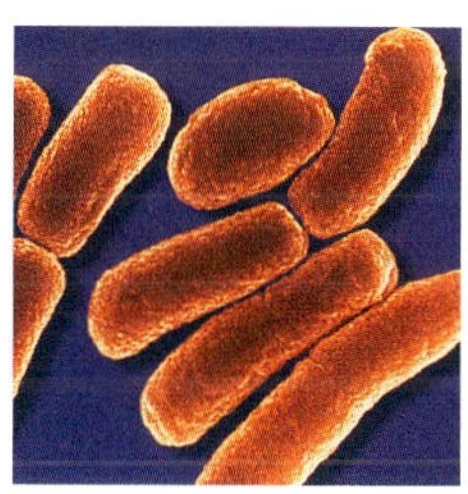

Escherichia coli

Cyanobacteria

Photosynthetic; surrounded by a pigmented covering; common on land and in the ocean; probably ancestors of chloroplasts in some protists.
Examples: *Anabaena, Oscillatoria, Spirulina*

Chemoautotrophs

Ancient bacteria that can grow without sunlight or other organisms; derive energy from reduced gases—ammonia (NH_3), methane (CH_4), hydrogen sulfide (H_2S); play critical roles in Earth's nitrogen cycles; includes nitrobacteria and sulfur bacteria.
Examples: *Nitrosomonas, Nitrobacter*

Enterobacteria

Typically rigid, rod-shaped, heterotrophic bacteria; can be aerobic or anaerobic; have flagella; responsible for many serious diseases of plants and humans.
Examples: *Escherichia coli, Salmonella typhimurium*

Pseudomonads

Straight or curved rods with flagella at one end; strict aerobes; common in soil; many are plant pathogens.
Example: *Pseudomonas aeruginosa*

Spirochaetes

Long, spiral cells; flagella originating at each end; responsible for several serious diseases.
Examples: *Treponema pallidum, Borrelia burgdorferi*

Actinomycetes

Filamentous bacteria that are often mistaken for fungi; spore-producing; sources of antibiotics including streptomycin, tetracycline, and chloramphenicol; cause diseases including dental plaque, leprosy, and tuberculosis.
Example: *Mycobacterium tuberculosis*

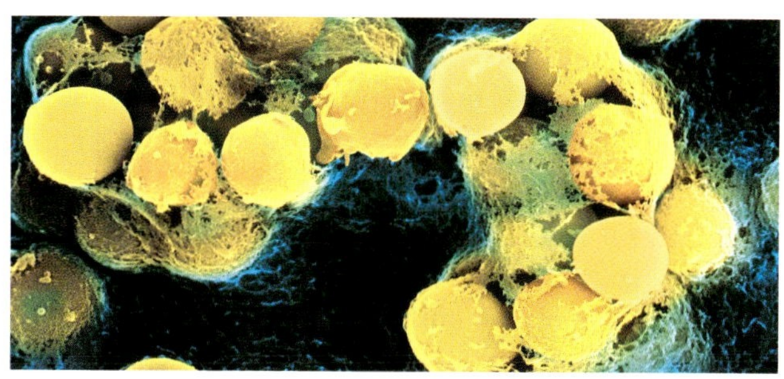

Streptococcus

Rickettsias

Parasitic bacteria found within the cells of vertebrates and arthropods; cause serious diseases.
Example: *Rickettsia rickettsii*

Gliding and budding bacteria

Rod-shaped cells; secrete slimy polysaccharides; often aggregate into gliding masses; live mainly in soil.
Example: *Myxobacteria*

Kingdom Archaebacteria

Fewer than 100 described species
Includes anaerobic and aerobic bacteria adapted to extreme environments; prokaryotic; differ from eubacteria in structure of cell wall and cell membrane; similarities to eukaryotes suggest that archaebacteria are more closely related to eukaryotes than to eubacteria; asexual reproduction only. There is evidence they also exist freely in oceans and soil but have not yet been cultured in labs.

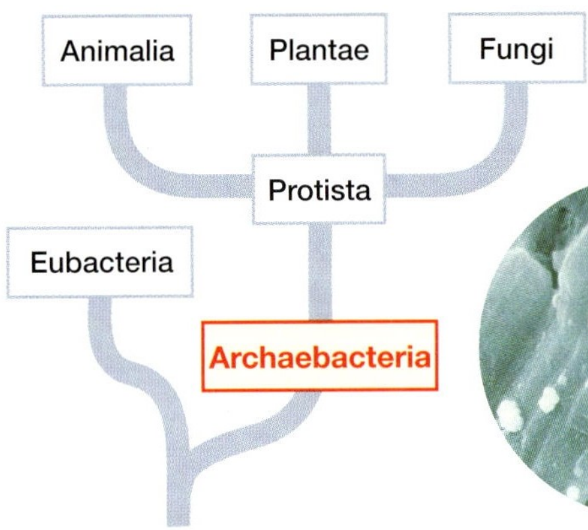

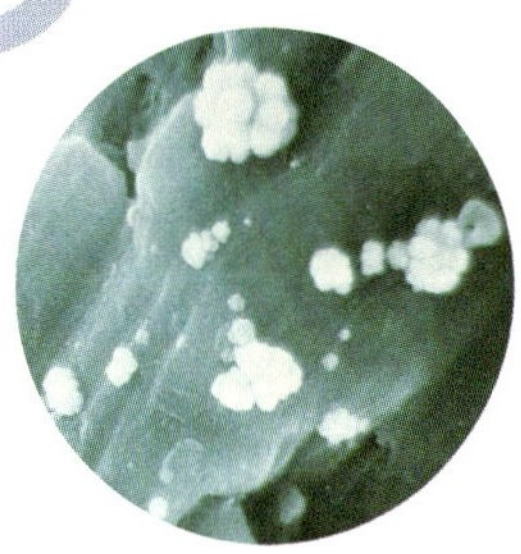

Acidianus brierleyi

Methanogens

Anaerobic methane producers; most species use carbon dioxide as a carbon source; inhabit soil, swamps, and the digestive tracts of animals, particularly grazing mammals such as cattle; produce nearly 2 trillion kilograms (2 billion tons) of methane gas annually.
Example: *Methanobrevibacter ruminatium*

Thermoacidophiles

Inhabit hot, acidic environments; can tolerate high temperatures; require sulfur; mostly anaerobic.
Example: *Sulfolobus solfataricus*

Extreme halophiles

Inhabit environments with very high salt content (salinity 15 to 20 percent), including the Dead Sea and the Great Salt Lake; many aerobic; gram-negative.
Example: *Halobacteroides holobius*

Kingdom Protista

About 43,000 species
Includes eukaryotes that are not plants, fungi, or animals; the most structurally diverse kingdom; both unicellular and multicellular; membrane-bound nucleus; nearly all have chromosomes, mitochondria, and internal compartments; many have chloroplasts; most have cell walls; reproduce sexually and asexually; aquatic or parasitic; many live in soil.

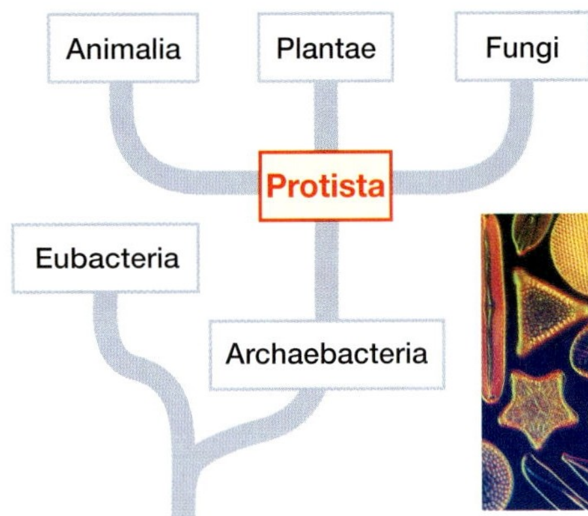

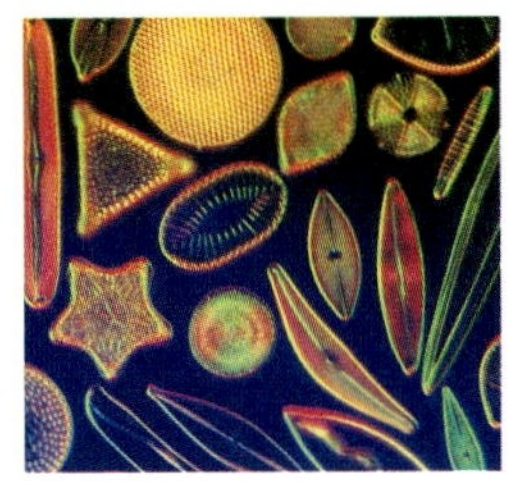

Diatoms

Phylum Rhizopoda

About 300 species
Unicellular and heterotrophic; amorphously shaped cells that move using cytoplasmic extensions called pseudopods; includes amoebas.

Phylum Foraminifera

About 300 species
Unicellular and heterotrophic; marine; have shells of organic material with pores through which many cytoplasmic threads project; includes forams.
Example: *Heterostegina depressa*

Phylum Chlorophyta

About 7,000 species
Unicellular, colonial, and multicellular; photosynthetic; contain chlorophylls *a* and *b;* contain chloroplasts similar to those of plants; scientists think plants descended from this group; includes green algae.
Examples: *Chlamydomonas, Chorella, Oedogonium, Spirogyra, Ulva, Volvox*

Phylum Rhodophyta

About 4,000 species
Almost all are multicellular; photosynthetic; most are marine; contain chlorophyll *a* and phycobilins; chloroplasts probably evolved from symbiotic cyanobacteria; includes red algae.
Example: *Porphyra*

Phylum Phaeophyta
About 1,500 species
Multicellular and photosynthetic; nearly all are marine; contain chlorophylls *a* and *c* and fucoxanthin, which is the source of their brownish color; includes brown algae.
Examples: *Fucus, Laminaria, Postelsia, Sargassum*

Phylum Bacillariophyta
More than 11,500 species
Unicellular and photosynthetic; secrete a unique shell made of opaline silica that resembles a box with a lid; chloroplasts resemble those of brown algae; contain chlorophylls *a* and *c* and fucoxanthin; includes diatoms.

Phylum Dinoflagellata
More than 2,100 species
Unicellular; heterotrophic and autotrophic species; mostly marine; body enclosed within two cellulose plates; contain chlorophylls *a* and *c* and carotenoids; includes dinoflagellates.
Examples: *Gonyaulax, Noctiluca*

Phylum Euglenophyta
About 1,000 species
Unicellular; both photosynthetic and heterotrophic species; asexual; most live in fresh water; chloroplasts are similar to those of green algae and are thought to have evolved from the same symbiotic bacteria; includes euglenoids.
Example: *Euglena*

Phylum Kinetoplastida
About 3,000 species
Mostly unicellular; heterotrophic; all have at least one flagellum; includes zoomastigotes.
Examples: *Giardia, Leishmania, Trypanosoma*

Phylum Ciliophora
About 8,000 species
Very complex single cells; heterotrophic; have rows of cilia and two types of cell nuclei; includes ciliates.
Examples: *Didinium, Paramecium, Stentor, Vorticella*

Phylum Acrasiomycota
About 70 species
Heterotrophic; amoeba-shaped cells that aggregate into a moving mass called a slug when they are deprived of food; cells within the slug retain their membranes and do not fuse; a slug produces spores that form new amoebas elsewhere; includes cellular slime molds.
Example: *Dictyostelium*

Phylum Myxomycota
About 500 species
Heterotrophic; individuals stream along as a multinucleate mass of cytoplasm; can give rise to spores that start a new individual in a more favorable environment; includes plasmodial slime molds.
Example: *Physarum*

Phylum Oomycota
About 580 species
Heterotrophic; unicellular parasites or decomposers; cell walls composed of cellulose, not chitin as in fungi; includes water molds, white rusts, and downy mildews.
Example: *Phytophthora*

Phylum Apicomplexa
About 3,900 species
Unicellular; heterotrophic; nonmotile; spore-forming parasites of animals; have complex life cycles; asexual and sexual reproduction; includes sporozoans.
Examples: *Plasmodium, Toxoplasma*

Kingdom Fungi

About 77,000 species
Eukaryotic, terrestrial heterotrophs with nutrition by absorption; all but yeasts are multicellular; body is typically composed of filaments (called hyphae) and is multinucleate, with incomplete divisions (called septae) between cells; cell walls made of chitin; about 17,000 species (known as deuteromycetes) without a sexual stage.

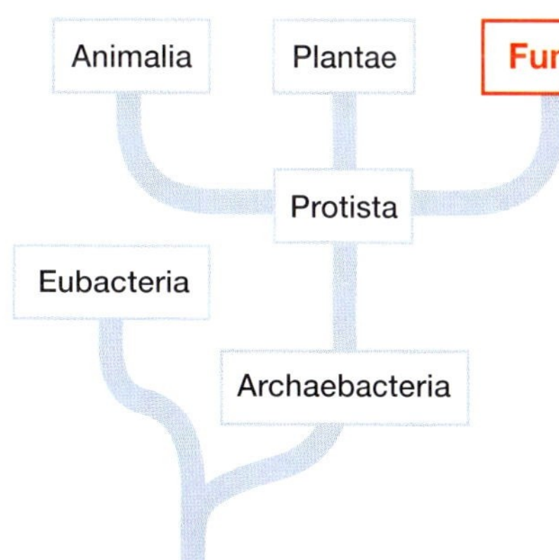

Scarlet mushroom

Phylum Zygomycota
About 665 species
Usually lack septae; fusion of hyphae leads to formation of zygote, which divides by meiosis when it germinates; terrestrial or parasitic; includes bread molds.
Examples: *Pilobolus, Rhizopus*

Phylum Ascomycota

About 30,000 species

Hyphae usually have perforated septae; fusion of hyphae leads to formation of densely interwoven mass that contains characteristic microscopic reproductive structures called asci (singular, ascus); many fungi formally grouped as Fungi Imperfecti are now grouped here; terrestrial, marine, and freshwater species; includes brewer's and baker's yeasts, molds, morels, and truffles.

Examples: *Neurospora, Saccharomyces*

Phylum Basidiomycota

About 16,000 species

Hyphae usually have incomplete septae; reproduction is typically sexual; fusion of hyphae leads to the formation of densely interwoven reproductive structure (mushroom) with characteristic microscopic structures called basidia (singular, basidium); includes mushrooms, toadstools, shelf fungi, rusts, and smuts.

Fungal Associations

About 20,000 species

Fungi form symbiotic associations with plants, green algae, and cyanobacteria.

Lichen

Lichens

About 15,000 species

Mutualistic relationships between fungi (almost always ascomycetes) and cyanobacteria, green algae, or both; the photosynthetic partners actually live among the hyphae of the fungus; the fungus derives energy from its photosynthetic partners.

Mycorrhizae

About 5,000 species

Mutualistic relationships between fungi and the roots of plants; 80 percent of all plants have mycorrhizae associated with their roots; the plant provides sugars to the fungi; in return, the fungi serve as accessory roots, greatly increasing the surface area available for the absorption of nutrients.

Kingdom Plantae

About 280,000 species

Multicellular; eukaryotic; mostly autotrophic; mostly terrestrial organisms containing tissues and organs; cell walls with cellulose; contain chlorophylls *a* and *b* in plastids; life cycle is alternation of generations.

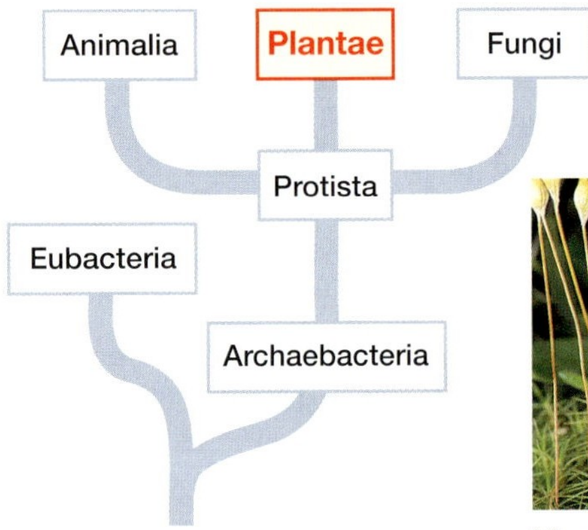

Moss

Phylum Bryophyta

About 10,000 species

Nonvascular plants; gametophytes are larger than sporophytes; sporophytes grow on gametophytes; have simple conducting tissue; lack roots, stems, and leaves; includes mosses.

Example: *Sphagnum*

Phylum Hepatophyta

About 6,000 species

Nonvascular plants; gametophytes are larger than sporophytes; sporophytes grow on gametophytes; lack stomata, roots, stems, and leaves; includes liverworts.

Example: *Marchantia*

Phylum Anthocerophyta

About 100 species

Nonvascular plants; gametophytes are larger than sporophytes; sporophytes grow on gametophytes; sporophytes have stomata; lack roots, stems, and leaves; includes hornworts.

Example: *Anthoceros*

Phylum Pterophyta

About 11,000 species

Seedless vascular plants; sporophytes are larger than gametophytes; sporophytes have roots, stems, and leaves that produce spores on their lower surfaces; gametophytes are small, flat, and independent; includes ferns.

Example: *Salvinia*

Phylum Lycophyta

About 1,000 species

Seedless vascular plants; sporophytes are larger than gametophytes; sporophytes produce spores in cones, resemble moss gametophytes, and have roots, stems, and leaves; gametophytes are small, flat, and independent; includes club mosses.

Examples: *Lycopodium, Selaginella*

Phylum Sphenophyta

15 species

Seedless vascular plants; sporophytes are larger than gametophytes; sporophytes produce spores in cones and have roots, leaves, and jointed stems; gametophytes are small, flat, and independent; includes horsetails.

Example: *Equisetum*

Phylum Psilotophyta

Several species

Seedless vascular plants; sporophytes are larger than gametophytes; sporophytes produce spores in sporangia at tips of stems and have roots and stems but no leaves; gametophytes are small, flat, and independent; includes whisk ferns.

Example: *Psilotum*

Phylum Coniferophyta

About 550 species

Gymnosperms, seed plants that produce naked seeds; sporophytes are mostly evergreen trees or shrubs with needlelike or scalelike leaves; male and female gametophytes are microscopic and develop from spores produced within cones on sporophytes; includes pines, spruces, firs, larches, and yews.

Examples: *Pinus, Taxus*

Phylum Cycadophyta

About 100 species

Gymnosperms, seed plants that produce naked seeds; sporophytes are evergreen trees and shrubs with palmlike leaves; male and female gametophytes are microscopic and develop from spores produced within cones on separate sporophytes; includes cycads.

Example: *Cycas*

Phylum Ginkgophyta

1 species

Gymnosperm, seed plant that produces naked seeds; sporophyte is a deciduous tree with fan-shaped leaves and fleshy seeds; male and female gametophytes are microscopic and develop from spores produced by separate sporophytes; includes *Ginkgo biloba.*

Phylum Gnetophyta

About 70 species

Gymnosperms, seed plants that produce naked seeds; sporophytes are shrubs or vines with some angiosperm characteristics; male and female gametophytes are microscopic and develop from spores produced within cones on sporophytes; includes gnetophytes.

Examples: *Ephedra, Welwitschia*

Phylum Anthophyta

About 250,000 species

Angiosperms, seed plants that produce seeds within a fruit; sporophytes are trees, shrubs, herbs, or vines that produce flowers; male and female gametophytes are microscopic and develop from spores produced within the reproductive structures of a flower; includes flowering plants.

Examples: *Aster, Prunus, Quercus, Zea*

Class Monocotyledones

About 70,000 species

Embryos have one cotyledon; flower parts in multiples of three; leaf veins parallel; vascular bundles scattered through stem tissue; includes grasses, sedges, lilies, irises, palms, and orchids.

Class Dicotyledones

About 180,000 species

Embryos have two cotyledons; flower parts in multiples of two, four, or five; leaves with netlike veins; vascular bundles in stems are arranged in rings; includes daisies, roses, maples, and elms.

Rose

Kingdom Animalia

More than 1 million species
Multicellular, eukaryotic, heterotrophic organisms; nutrition mainly by ingestion; most have specialized tissues, and many have complex organs and organ systems; no cell walls or chloroplasts; sexual reproduction predominates; both aquatic and terrestrial forms.

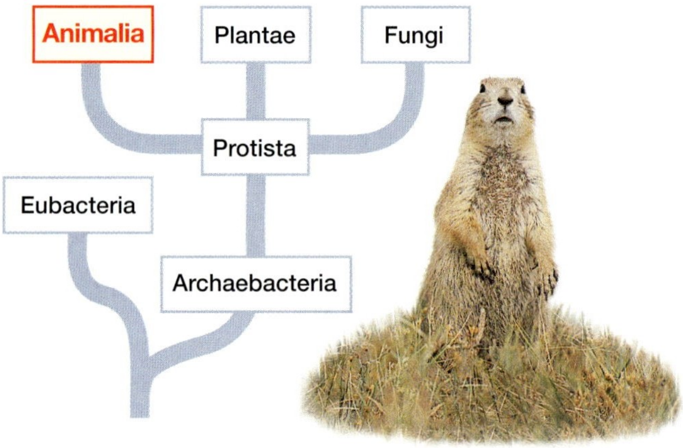

Prairie dog

Phylum Porifera

About 9,000 species
Asymmetrical; lack tissues and organs; body wall consists of two cell layers, penetrated by numerous pores; internal cavity is lined with unique food-filtering cells called choanocytes; sexual and asexual reproduction; mostly marine; includes sponges.

Phylum Cnidaria

About 10,000 species
Radially symmetrical; most have distinct tissues; baglike body of two cell layers; gelatinous; marine and freshwater species.

Class Hydrozoa

About 2,700 species
Most have both polyp and medusa stages in life cycle; includes hydras.
Examples: *Hydra, Obelia, Physalia*

Class Scyphozoa

About 200 species
Exclusively marine; medusa stage dominant; includes jellyfish.
Example: *Aurelia*

Class Anthozoa

About 6,200 species
Marine; solitary or colonial; medusa stage absent; includes sea anemones, corals, and sea fans.

Phylum Ctenophora

About 100 species
Radially symmetrical; transparent, gelatinous bodies resembling jellyfish; marine; includes comb jellies.

Phylum Platyhelminthes

About 20,000 species
Bilaterally symmetrical acoelomates; body flat and ribbonlike, without true segments; organs present; three germ layers; includes flatworms.

Class Turbellaria

More than 3,000 species
Mostly free-living aquatic or terrestrial forms; includes planarians.
Example: *Dugesia*

Class Cestoda

About 1,500 species
Specialized internal parasites; no digestive system; body sections called proglottids; hooked scolex for attaching to host; includes tapeworms.
Example: *Taenia saginata*

Class Trematoda

About 6,000 species
Internal parasites, with mouth at anterior end; often have complex life cycle with alternation of hosts; cause disease in humans and other animals; includes flukes.
Examples: *Schistosoma, Chlonorchis sinensis*

Phylum Nematoda

More than 12,000 species
Tiny, parasitic, unsegmented worms; long, slender body; pseudocoelomates; includes roundworms.
Examples: *Ascaris, Trichinella spiralis, Necator, Toxocara canis, Toxocara cati*

Phylum Mollusca

About 110,000 species
Soft-bodied animals with a true coelom; three-part body consisting of foot, visceral mass, and mantle; protostomes; most have a unique rasping tongue (radula); terrestrial, freshwater, and marine.

Cuttlefish

Class Polyplacophora

About 600 species

Elongated body and reduced head; similar to ancestral mollusk form; includes chitons.

Class Gastropoda

About 80,000 species

Visceral mass twisted during development; head, distinct eyes, and tentacles usually present; includes gastropods, such as snails, slugs, and whelks.

Class Bivalvia

10,000 species

Two shells connected by a hinge; no radula; large, wedge-shaped foot; includes bivalves, such as oysters, clams, and scallops.

Class Cephalopoda

More than 600 species

Foot modified into tentacles; includes cephlapods, such as squids, octopuses, nautilus, and cuttlefish.

Phylum Annelida

About 12,000 species

Serially segmented worms; bilaterally symmetrical; protostomes.

Class Polychaeta

About 8,000 species

Fleshy outgrowths called parapodia extend from segments; many bristles (setae); marine; includes feather dusters.

Example: *Nereis*

Class Oligochaeta

About 3,100 species

Head not well developed; no parapodia; few setae; terrestrial and freshwater forms; includes earthworms.

Class Hirudinea

About 600 species

Body flattened; no parapodia; usually suckers at both ends; many are external parasites; includes leeches.

Phylum Arthropoda

About 1 million species

Segmented bodies with paired, jointed appendages; bilaterally symmetrical; chitinous exoskeleton; protostomes; aerial, terrestrial, and aquatic forms.

Subphylum Chelicerata

Distinguished by absence of antennae and presence of chelicerae; all appendages unbranched; four pairs of walking legs; two body regions (cephalothorax and abdomen); predominantly terrestrial.

Class Arachnida

About 57,000 species

Terrestrial; use book lungs and tracheae for respiration; four pairs of legs; includes spiders, scorpions, ticks, and mites.

Class Merostomata

5 species

Cephalothorax covered by protective "shell"; sharp spike on tail; includes horseshoe crabs.

Class Pycnogonida

About 1,000 species

Small marine predators or parasites; usually four pairs of legs; includes sea spiders.

Subphylum Crustacea

About 35,000 species

Two pairs of antennae, mandibles, and appendages with two branches; predominantly aquatic.

Crab

Class Malacostraca

About 20,000 species

Typically five pairs of legs; two pairs of antennae; most are aquatic; includes crayfish, lobsters, crabs, shrimp, sow bugs, and krill.

Subphylum Uniramia

Antennae, mandibles, and unbranched appendages.

Class Insecta

About 750,000 species

Head, thorax, and abdomen; three pairs of legs, all attached to thorax; usually two pairs of wings.

Order Coleoptera: includes beetles, ladybugs, and weevils.

Order Diptera: includes flies, mosquitoes, gnats, and midges.

Order Lepidoptera: includes butterflies and moths.

Order Hymenoptera: includes bee ants, wasps, hornets, and ichneumon fly.

Beetle

Order Hemiptera: includes water striders, water boatmen, back swimmers, bedbugs, squash bugs, stink bugs, and assassin bugs.

Order Homoptera: includes cicadas, aphids, leaf hoppers, and scale insects.

Order Orthoptera: includes grasshoppers, cockroaches, walking sticks, praying mantises, and crickets.

Order Odonata: includes dragonflies and damsel flies.

Order Neuroptera: includes ant lions and lacewings.

Order Thysanura: includes silverfish, bristletails, and firebrats.
Order Anoplura: includes sucking lice.
Order Isoptera: includes termites.
Order Ephemeroptera: includes mayflies.
Order Siphonaptera: includes fleas.
Order Dermaptera: includes earwigs.

Class Chilopoda
About 2,500 species
Body flattened and consisting of 15–170 or more segments; one pair of legs attached to each segment; includes centipedes.

Class Diplopoda
About 10,000 species
Elongated body of 15–200 segments; two pairs of legs per segment; herbivorous; includes millipedes.

Phylum Echinodermata
About 6,000 species
Deuterostomes; adults radially symmetrical with five-part body plan; most forms have water vascular system with tube feet for locomotion; marine.

Class Asteroidea
About 1,500 species
Body usually with five arms and double rows of tube feet on each arm; mouth directed downward; includes sea stars.

Sea star

Class Ophiuroidea
About 2,000 species
Usually with five slender, delicate arms or rays; includes brittle stars and basket star.

Class Crinoidea
About 600 species
Mouth faces upward and is surrounded by many arms; includes sea lilies and feather stars.

Class Echinoidea
About 900 species
Body spherical, oval, or disk-shaped; arms lacking but five-part body plan still apparent; includes sea urchins and sand dollars.

Class Holothuroidea
About 1,500 species
Elongated, thickened body with tentacles around the mouth; includes sea cucumbers.

Phylum Chordata
About 42,500 species
Bilaterally symmetrical; deuterostomes; coelom present; have notochord, dorsal nerve cord, pharyngeal slits, and tail; aquatic and terrestrial.

Subphylum Urochordata
About 1,250 species
Saclike covering, or tunic, in adults; larvae are free-swimming and have nerve cord and notochord; marine; includes tunicates.

Subphylum Cephalochordata
23 species
Small and fishlike with a permanent notochord; filter feeders; includes lancelets.

Subphylum Vertebrata
About 40,000 species
Most of the notochord is replaced by a spinal column composed of vertebrae that protect the dorsal nerve cord; recognizable head containing a brain.

Jawless fishes
63 species
Freshwater or marine eel-like fishes without true jaws, scales, or paired fins; cartilaginous skeleton; includes lampreys and hagfish.

Cartilaginous fishes
About 850 species
Fishes with jaws and paired fins; gills present; no swim bladder; cartilaginous skeleton; includes sharks, rays, and skates.

Bony fishes
About 18,000 species
Freshwater and marine fishes with gills attached to gill arch; jaws and paired fins; bony skeleton; most have swim bladder; includes rayfins, such as tuna, sea horse, trout, perch, sturgeon, and angelfish; and lobefins, such as coelacanth and lungfish.

Class Amphibia
About 4,200 species
Freshwater or terrestrial; gills present at some stage; skin often slimy and lacking scales; eggs typically laid in water and fertilized externally.

Order Anura: includes frogs and toads.
Order Urodela: includes salamanders and newts.
Order Apoda: includes caecilians.

Class Reptilia
About 7,000 species
Terrestrial or semiaquatic vertebrates; breathe using lungs at all stages; body covered by scales; most species lay amniotic eggs covered with a protective shell; fertilization internal.

Order Squamata: includes lizards and snakes.
Order Chelonia: includes turtles and tortoises.

Order Crocodilia: includes alligators, crocodiles, gavials, and caimans.
Order Rhynchocephalia: includes tuataras.

Class Aves

About 9,000 species
Body covered with feathers; forelimbs modified into wings; four-chambered heart; endothermic; lay shelled, amniotic eggs.

Order Passeriformes: includes robins, bluebirds, sparrows, warblers, and thrushes.
Order Apodiformes: includes swifts and hummingbirds.
Order Piciformes: includes woodpeckers, sapsuckers, flickers, and toucans.
Order Psittaciformes: includes parrots, parakeets, macaws, and cockatoos.
Order Charadriiformes: includes snipes, sandpipers, plovers, gulls, terns, auks, puffins, and ibises.

Tern

Order Columbiformes: includes pigeons and doves.
Order Falconiformes: includes hawks, falcons, eagles, kites, and vultures.
Order Gaviiformes: includes loons.
Order Gruiformes: includes cranes, coots, gallinules, and rails.
Order Anseriformes: includes ducks, geese, and swans.
Order Strigiformes: includes owls.
Order Ciconiiformes: includes herons, bitterns, egrets, storks, spoonbills, and ibises.
Order Pelecaniformes: includes pelicans, cormorants, and gannets.
Order Galliformes: includes pheasants, turkeys, quails, partridges, and grouse.
Order Procellariiformes: includes albatrosses and petrels.
Order Cuculiformes: includes cuckoos and roadrunners.
Order Caprimulgiformes: includes goatsuckers, whippoorwills, and nighthawks.
Order Coraciiformes: includes kingfishers.
Order Sphenisciformes: includes penguins.
Order Struthioniformes: includes ostriches.
Order Apterygiformes: includes kiwis.

Class Mammalia

About 4,400 species
Hair on at least part of body; young nourished with milk secreted by mammary glands; endothermic; breathe with lungs.

Order Rodentia: includes squirrels, woodchucks, mice, rats, muskrats, and beavers.

Bat

Order Chiroptera: includes bats.
Order Insectivora: includes moles and shrews.
Order Marsupialia: includes opossums, kangaroos, koalas, and wallabies.
Order Carnivora: includes bears, weasels, mink, otters, skunks, lions, tigers, and wolves.
Order Primates: includes monkeys, lemurs, gibbons, orangutans, gorillas, chimpanzees, and humans.
Order Artiodactyla: includes hippopotamuses, camels, llamas, deer, giraffes, cattle, sheep, and goats.
Order Cetacea: includes whales, porpoises, and dolphins.
Order Lagomorpha: includes rabbits, hares, and pikas.
Order Pinnipedia: includes seals, sea lions, and walruses.
Order Edentata: includes armadillos, sloths, and anteaters.
Order Macroscelidea: includes elephant shrews.
Order Perissodactyla: includes tapirs, rhinoceroses, horses, and zebras.
Order Scandentia: includes tree shrews.
Order Hyracoidea: includes hyraxes.
Order Pholidota: includes pangolins.
Order Sirenia: includes sea cows, dugongs, and manatees.
Order Monotremata: includes duckbill platypus and spiny anteaters.
Order Dermoptera: includes flying lemurs.
Order Proboscidea: includes elephants.
Order Tubulidentata: includes aardvark.

Polar bear

Key:

Atomic number — 6
Symbol — C
Name — Carbon
Average atomic mass — 12.0107

Period	Group 1	Group 2	Group 3	Group 4	Group 5	Group 6	Group 7	Group 8	Group 9
1	1 **H** Hydrogen 1.007 94								
2	3 **Li** Lithium 6.941	4 **Be** Beryllium 9.012 182							
3	11 **Na** Sodium 22.989 770	12 **Mg** Magnesium 24.3050							
4	19 **K** Potassium 39.0983	20 **Ca** Calcium 40.078	21 **Sc** Scandium 44.955 910	22 **Ti** Titanium 47.867	23 **V** Vanadium 50.9415	24 **Cr** Chromium 51.9961	25 **Mn** Manganese 54.938 049	26 **Fe** Iron 55.845	27 **Co** Cobalt 58.933 200
5	37 **Rb** Rubidium 85.4678	38 **Sr** Strontium 87.62	39 **Y** Yttrium 88.905 85	40 **Zr** Zirconium 91.224	41 **Nb** Niobium 92.906 38	42 **Mo** Molybdenum 95.94	43 **Tc** Technetium (98)	44 **Ru** Ruthenium 101.07	45 **Rh** Rhodium 102.905 50
6	55 **Cs** Cesium 132.905 43	56 **Ba** Barium 137.327	57 **La** Lanthanum 138.9055	72 **Hf** Hafnium 178.49	73 **Ta** Tantalum 180.9479	74 **W** Tungsten 183.84	75 **Re** Rhenium 186.207	76 **Os** Osmium 190.23	77 **Ir** Iridium 192.217
7	87 **Fr** Francium (223)	88 **Ra** Radium (226)	89 **Ac** Actinium (227)	104 **Rf** Rutherfordium (261)	105 **Db** Dubnium (262)	106 **Sg** Seaborgium (266)	107 **Bh** Bohrium (264)	108 **Hs** Hassium (277)	109 **Mt** Meitnerium (268)

* The systematic names and symbols for elements greater than 110 will be used until the approval of trivial names by IUPAC.

58 **Ce** Cerium 140.116	59 **Pr** Praseodymium 140.907 65	60 **Nd** Neodymium 144.24	61 **Pm** Promethium (145)	62 **Sm** Samarium 150.36
90 **Th** Thorium 232.0381	91 **Pa** Protactinium 231.035 88	92 **U** Uranium 238.028 91	93 **Np** Neptunium (237)	94 **Pu** Plutonium (244)

internet connect

go.hrw.com

Topic: Periodic Table
Go To: go.hrw.com
Keyword: HOLT PERIODIC

Visit the HRW Web site for updates on the periodic table.

Hydrogen
Semiconductors
(also known as *metalloids*)

Metals
Alkali metals
Alkaline-earth metals
Transition metals
Other metals

Nonmetals
Halogens
Noble gases
Other nonmetals

Group 10	Group 11	Group 12	Group 13	Group 14	Group 15	Group 16	Group 17	Group 18
								2 **He** Helium 4.002 602
			5 **B** Boron 10.811	6 **C** Carbon 12.0107	7 **N** Nitrogen 14.0067	8 **O** Oxygen 15.9994	9 **F** Fluorine 18.998 4032	10 **Ne** Neon 20.1797
			13 **Al** Aluminum 26.981 538	14 **Si** Silicon 28.0855	15 **P** Phosphorus 30.973 761	16 **S** Sulfur 32.065	17 **Cl** Chlorine 35.453	18 **Ar** Argon 39.948
28 **Ni** Nickel 58.6934	29 **Cu** Copper 63.546	30 **Zn** Zinc 65.409	31 **Ga** Gallium 69.723	32 **Ge** Germanium 72.64	33 **As** Arsenic 74.921 60	34 **Se** Selenium 78.96	35 **Br** Bromine 79.904	36 **Kr** Krypton 83.798
46 **Pd** Palladium 106.42	47 **Ag** Silver 107.8682	48 **Cd** Cadmium 112.411	49 **In** Indium 114.818	50 **Sn** Tin 118.710	51 **Sb** Antimony 121.760	52 **Te** Tellurium 127.60	53 **I** Iodine 126.904 47	54 **Xe** Xenon 131.293
78 **Pt** Platinum 195.078	79 **Au** Gold 196.966 55	80 **Hg** Mercury 200.59	81 **Tl** Thallium 204.3833	82 **Pb** Lead 207.2	83 **Bi** Bismuth 208.980 38	84 **Po** Polonium (209)	85 **At** Astatine (210)	86 **Rn** Radon (222)
110 **Ds** Darmstadtium (281)	111 **Uuu*** Unununium (272)	112 **Uub*** Ununbium (285)	113 **Uut*** Ununtrium (284)	114 **Uuq*** Ununquadium (289)	115 **Uup*** Ununpentium (288)			

A team at Lawrence Berkeley National Laboratories reported the discovery of elements 116 and 118 in June 1999. The same team retracted the discovery in July 2001. The discovery of elements 113, 114, and 115 has been reported but not confirmed.

63 **Eu** Europium 151.964	64 **Gd** Gadolinium 157.25	65 **Tb** Terbium 158.925 34	66 **Dy** Dysprosium 162.500	67 **Ho** Holmium 164.930 32	68 **Er** Erbium 167.259	69 **Tm** Thulium 168.934 21	70 **Yb** Ytterbium 173.04	71 **Lu** Lutetium 174.967
95 **Am** Americium (243)	96 **Cm** Curium (247)	97 **Bk** Berkelium (247)	98 **Cf** Californium (251)	99 **Es** Einsteinium (252)	100 **Fm** Fermium (257)	101 **Md** Mendelevium (258)	102 **No** Nobelium (259)	103 **Lr** Lawrencium (262)

The atomic masses listed in this table reflect the precision of current measurements. (Values listed in parentheses are the mass numbers of those radioactive elements' most stable or most common isotopes.)

Forensics Lesson

Teaching Strategy

Students likely have been introduced to forensics by popular television, but they may not know that the science has a relatively brief history. Tell students that tests for the presence of blood were developed in the early 1800s, and in the early 1900s, a method for determining ABO blood type was discovered. While differences between human fingerprints have been know for centuries, a method of classifying fingerprints for forensic use was not developed until the end of the 1800s. Have students choose a forensic area of specialization, investigate its history, and report their findings to the class.

Forensics Lesson

An Introduction to Forensic Science

A police takes a plaster cast of a tire track from the mud of a crime scene. A botanist uses a microscope to identify the source of pollen from the clothing of a crime suspect. A rescue worker sifts through debris after a fire, looking for the bodies of victims. In each of these cases, a small amount of evidence will be used to piece together a story or an identity.

Background

When most people think about forensic science, they recall crime dramas from television, in which a crime scene investigator seizes the smallest bits of evidence and ties them together to reconstruct the crime, enabling the identification of even the most cunning criminal.

Many of the scientists who work in forensics are involved in law enforcement. But there's much more to forensics than that. Basically, forensics is the making of knowledge and information available in a public forum, such as a court of law. A forensic scientist is a person who applies scientific knowledge and techniques to the investigation of evidence for the purpose of identification of a person or to establish a sequence of events that took place in the past.

For example, the evidence associated with a crime may be samples of hair, fibers, paint, glass, soil, blood, or plant material. Forensics scientists from different areas of specialization analyze evidence using a variety of different techniques, depending on the type of evidence.

You are familiar with some forensic techniques. For example, fingerprint analysis and blood typing have been in use for many years. The use of DNA to identify a person is now commonplace in the courtroom. Other forensic areas of specialization may be less familiar.

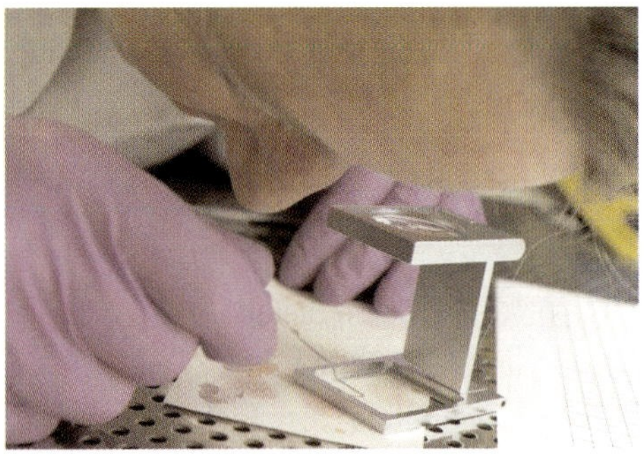

Figure 1 Document examination

Areas of Specialization in Forensics

Document Examination

Document examiners attempt to determine the authenticity of documents. Such determinations, such as the one shown in **Figure 1,** focus on handwriting or the machine (such as a typewriter, copier, or fax) used to produce the document, the inks used, and the material on which it is written.

Researchers, are now looking at ways to analyze the language patterns in a document, including particular words, sentence construction, and verb tenses, in order to help identify the author of the document.

Latent Fingerprints Research

Fingerprints at a crime scene, such as the one shown in **Figure 2,** are vital to many forensic investigations. The old method of dusting for fingerprints is very time consuming, however, and investigators may miss prints.

New methods are being developed that allow an entire room, or even an entire house, to be scanned for fingerprints in a matter of hours. One method uses fumes, produced when dried cyanoacrylate glue is

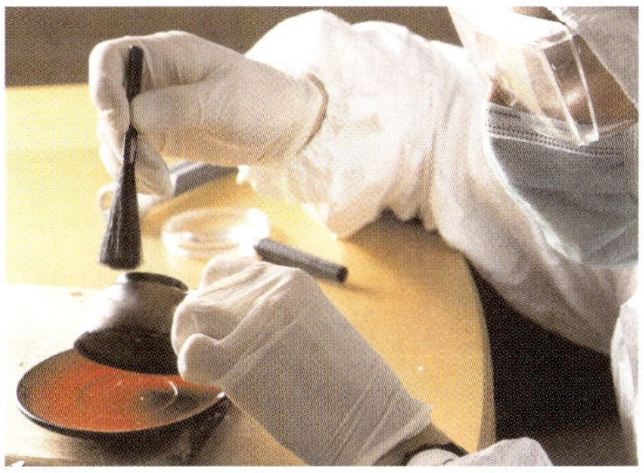

Figure 2 Fingerprint analysis

burned, to make latent, or hidden, fingerprints visible.

Trace Evidence Examination

Glass fragments, paint chips, and gunpowder residue are examples of trace evidence—physical evidence recovered from a crime scene or from the body of a victim or a suspect. Analyzing very small samples, such as the one shown in **Figure 3,** usually involves both physical and chemical tests, and may destroy the sample. New research is underway to find ways to analyze very small samples. Some methods being investigated are sophisticated types of chromatography, laser scanning, fluorescent imaging, and use of plasma devices. A trace evidence researcher, needs to have a broad background in chemistry, physics, and materials science.

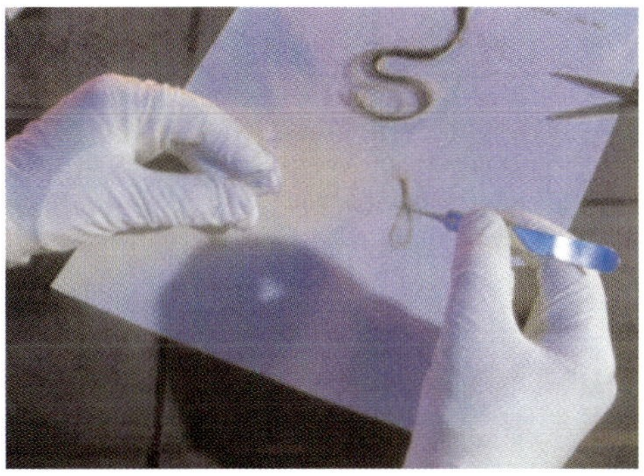

Figure 3 Trace evidence analysis

Forensic Dentistry

If a dead body has been badly burned or is decomposed, identification of the remains may be very difficult. In such cases, the only part of the body left intact may be the jaws and teeth. A forensic dentist can identify human remains based on the dental records, if they exist, of possible victims. The forensic dentist may perform a dental exam during an autopsy, which may include X rays and charts of the teeth and the skull.

In other cases, the forensic dentist may examine human bite marks on a victim for comparison with the tooth patterns of a suspect. The examination of dental injuries is another task that may be required of the forensic dentist.

Firearms Analysis

A gun leaves a set of scratches as unique as a fingerprint on each bullet fired from the gun. Because these scratches are unique to each gun, it is possible to match a particular gun to a fired bullet or to tell if two bullets were fired from the same gun. A firearms analyst, such as the one shown in **Figure 4,** compares scratch patterns in an attempt to match the bullet to a particular gun. Automated analysis using a computer database is the goal of research in this area. The scratch pattern from a bullet would be scanned into a computer for comparison with the database of known weapons.

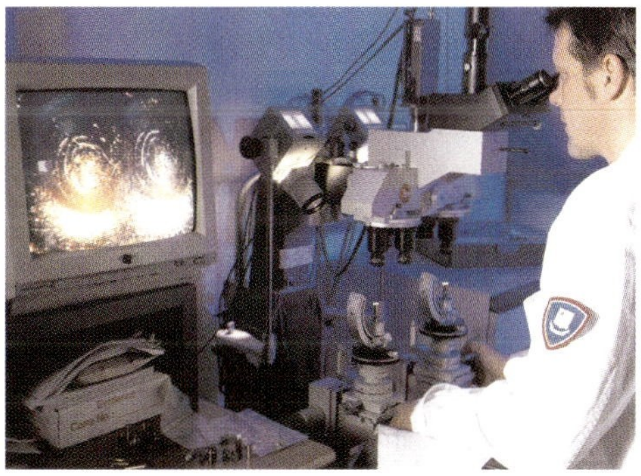

Figure 4 Firearms analysis

Hot Topic: Forensic mtDNA Analysis

One of the newest areas of forensic science involves the analysis of mitochondrial DNA (mtDNA), a type of DNA that is not found in the nuclei of cells. Rather, mtDNA occurs in mitochondria, small structures about the size of a bacterium that are found in the cell material surrounding the nucleus. Mitochondria, such as the ones shown in **Figure 5,** exist in large numbers in every cell, numbering from several to a thousand or more. Mitochondria have their own DNA, and each mitochondrion contains several identical copies of it. A person's mitochondria and thus their mitochondrial DNA are always inherited from the mother, whereas nuclear DNA is inherited from both parents.

One advantage of using mtDNA for forensic analysis comes from the number of mitochondria in each cell. This abundance allows a very small sample to provide a large mtDNA sample. Another advantage is that mtDNA exists in all types of cells and therefore any type of body tissue can be used, including bone.

One disadvantage is that, although the total size of mtDNA is small compared to nuclear DNA, an entire mitochondrion must be analyzed. This must be done because the number of variations in mtDNA is not large and the variations are very subtle.

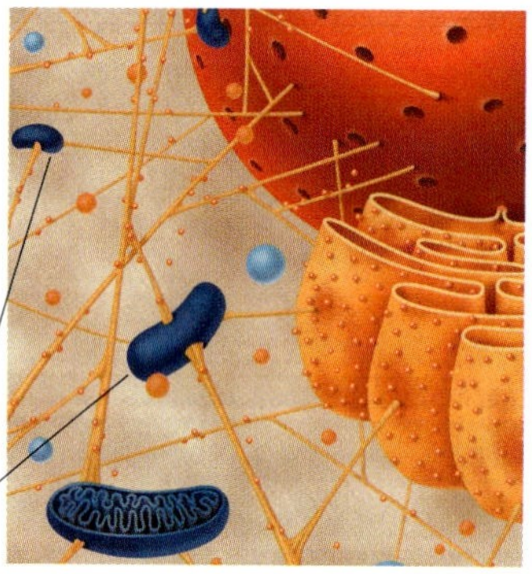

Figure 5 Mitochondria

Figure 6 The Russian royal family

Forensic Genealogy

Forensic investigators can use mtDNA analysis to aid in the identification of human remains. A person may be identified if their mtDNA can be compared to that from another person descended along maternal lines. One such investigation used the mtDNA obtained from the female human remains in a grave in Russia thought to be that of the last Russian czar, his wife, the czarina, and one of their daughters, shown in **Figure 6.** The mtDNA from Prince Phillip of the British royal family was used for the comparison because Prince Phillip is a direct descendant of the czarina's mother.

The mtDNA samples matched and thus the members of the Russian royal family were identified more than seventy years after they were killed during the Russian revolution. A related analysis proved that a woman claiming to be the missing princess Anastasia, another daughter of the czar, was not related.

Topic Questions

1. Describe in one sentence what a forensic scientist does.
2. What new method is being developed for determining the author of a document?
3. How can a firearms analyst tell what gun a certain bullet was fired from?
4. Where is mtDNA found, and how does it differ from the DNA in the nucleus?

1050

Answers to Topic Questions

1. Answers may vary. Sample answer: A forensic scientist uses scientific techniques to identify pieces of evidence and make the results known to the public.
2. Scientists analyze the various aspects—such as the words and sentence structure—of the language used to write the document
3. Each gun leaves a unique pattern of scratches on a bullet fired from that gun. Analysis of the scratch pattern on a bullet plus comparison with a database of known firearms can identify the weapon that fired the bullet.
4. mtDNA is found in mitochondria, which are small structures in the material surrounding the nucleus of a cell. Mitochondrial DNA has a very small number of base pairs compared with the DNA in the nucleus. mtDNA is inherited from only the mother, whereas nuclear DNA is inherited from both parents.

Forensics Lesson

Identification

Biological evidence is highly prized because it is not subject to the same sources of error as other types of evidence about identity. Eyewitnesses often get confused or make mistakes. Other types of physical evidence may show how the crime was committed, but may not offer much information about who committed it.

Analysis of blood or DNA evidence, on the other hand, can give investigators very specific information that can positively identify, or eliminate, a suspect as having been at the scene of the crime. Biological evidence such as this has its limitations, too. Contamination of samples can easily invalidate results, and laboratory specialists must be very careful in their analysis, as the samples they study are often very small. But since blood and DNA analysis methods were first developed, they have been important tools of identification for forensic scientists.

Blood Typing

Human blood can be one of four types—A, B, AB, or O—depending on which kinds of molecules, called antigens, are attached to the outer cell membrane of the red blood cells. These blood antigen molecules come in two forms: A antigens and B antigens.

A person's blood is type A if only A antigens are present on the red blood cells. The blood is type B if it has only B antigens. If both kinds of antigens are present, the blood type is AB. If neither kind of antigen is present, the blood is type O.

Determining a person's blood type is important not only for identification, but also to ensure that the person receives the correct blood type if a blood transfusion is needed. Giving a person the wrong blood type in transfusion can be fatal.

How Identification by Blood Type Works

Identification of blood type is very simple because red cells clump together when incompatible blood types are mixed. This clumping, called *agglutination*, is shown in **Figure 1.** *Serum* is the watery component of blood. The serum of type-A blood contains antibodies against the antigens of the cells in type-B blood. These antibodies, called *agglutinins*, attack the type-B red cells and cause them to clump together. Conversely, the serum from type-B blood contains antibodies that cause type-A red cells to clump. So, if A-serum is added to a blood sample and clumping occurs, the sample contains B antigens. However, because the sample could be type-B or type-AB, a test with B-serum is needed. If a second sample of the same blood forms clumps when combined with B-serum, A antigens are also present, and the blood is type-AB. The red blood cells in type-O blood (which carry neither A nor B antigens) will not form clumps when serum from type-A blood, -B, or -AB blood is added. Type O blood or serum (which contains antibodies to both A and B antigens) will, however, cause clumping of red blood cells in type-A, -B, or -AB whole blood.

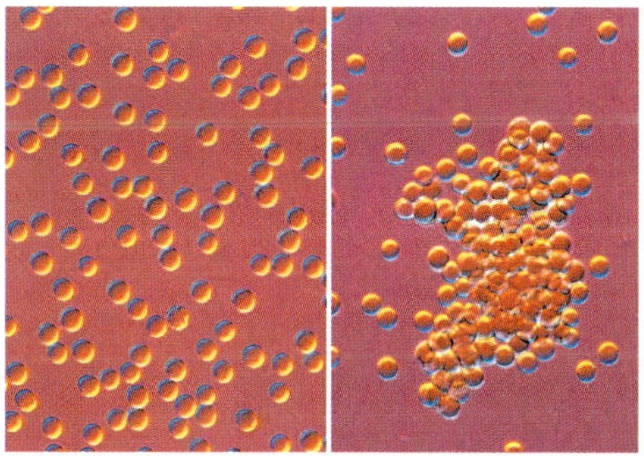

Figure 1 Normal blood cells (left); agglutinated blood cells (right)

Forensics Lesson

Teaching Strategy

Explain to students that blood typing was very important in forensics as well as health care throughout the 1900s, but for most modern forensic investigations, blood typing has been surpassed by DNA analysis. Blood typing may still be used, for example, to exclude a person from a pool of suspects. Have students research the methods of DNA analysis currently in use in major forensic laboratories. Suggest that students begin their searches with the Websites of large, urban police departments.

DNA Fingerprinting

With the exception of identical twins, no two people have the same DNA sequence. Scientists estimate that the genes of a human consist of over three billion nucleotide pairs (also called "base pairs"). Each nucleotide is made up of a phosphate group bonded to a deoxyribose sugar molecule, which is bonded to one of only four nitrogenous bases—adenine, guanine, cytosine, and thymine. A person's entire genetic code consists of a very long sequence of these base pairs in a certain arrangement, which varies from person to person.

Given the extreme length of a DNA molecule, you may think that comparing DNA from one person to DNA of another person is very difficult. In fact, the process is fairly simple. Scientists have discovered certain enzymes in bacteria, called restriction enzymes, that digest—that is, break—DNA chains in specific places. A specific restriction enzyme digests a DNA strand only where certain nucleotide sequences occur. These nucleotide sequences are found in all human DNA, but their placement along the strand varies from person to person.

How Identification by DNA Works

When a restriction enzyme is added to a DNA sample, the products will be DNA fragments of various sizes, depending on where the target sequence of nucleotides occurred. These fragments—called *restriction fragments*—can be separated according to size by a process called *electrophoresis,* shown in **Figure 2.** During electrophoresis, DNA fragments migrate through a gel toward an electrode at various speeds according to their electrical charges (which are, in turn, determined by the size of the fragments). The result is a pattern, called a *DNA fingerprint,* which is unique to one person's DNA. You can see in **Figure 2** that the patterns for samples II and III match, while the pattern for sample I is different.

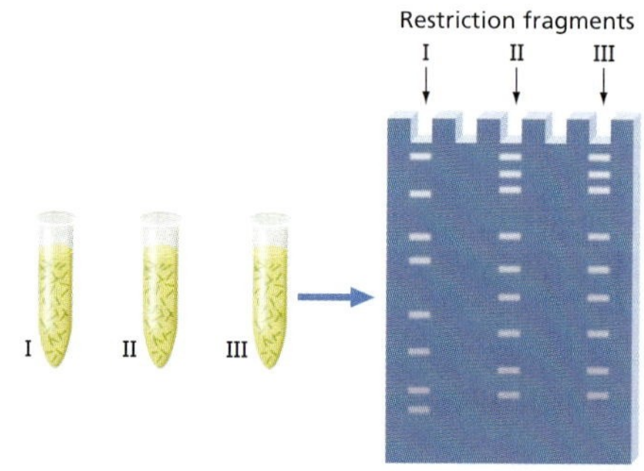

Figure 2 DNA-restriction enzyme mixtures (left); restriction fragments in gel (right)

DNA from a different person will produce a different DNA fingerprint. This is because some of the restriction fragments of the second person's DNA differ in size from those of the first person's DNA. An unknown DNA sample can be matched with a known sample of a person's DNA to an accuracy of 1 in 10 billion people. Due to its accuracy, DNA fingerprinting is often used to determine identity in criminal cases and to establish hereditary relationships, such as paternity.

Topic Questions

1. Explain why the blood cells of type B blood clump together when it is mixed with type A blood.
2. A student carrying out blood typing finds that the blood sample forms clumps when type A serum is added. The sample forms no clumps when type B serum is added. What is the type of the blood sample?
3. How does DNA determine most of the unique characteristics of a single person? Can a person have characteristics that are not determined by DNA? Explain and give two examples.
4. What characteristics of DNA fragments determine their separation during electrophoresis?

Answers to Topic Questions

1. Anti-B agglutins found in type A blood attack red blood cells in type B blood.
2. The sample is type B or 0.
3. DNA forms the code for many of the molecules synthesized during life processes. These molecules determine many of an individual's characteristics. People have many characteristics unrelated to DNA. These characteristics are acquired from the effects of the environment in which the person lives. Examples could include scars, growth and development differences due to diet or other environmental factors, diseases, differences in education, behavioral differences, and life choices. Students may mention the differences that allow them to distinguish between identical twins.
4. Fragments have characteristic lengths based on their nucleotide sequences. The fragments separate according to size (length of DNA chain).

Forensics Skills Lab

Blood Typing

SKILLS
- Experimenting
- Measuring
- Predicting
- Organizing and Analyzing Data

OBJECTIVES
- **Determine** the ABO and Rh blood types of unknown simulated blood samples.

MATERIALS
- blood samples, of unknown type from four subjects, simulated
- serum, anti-A, simulated (blood-typing)
- serum, anti-B, simulated (blood-typing)
- serum, anti-Rh, simulated (blood-typing)
- blood typing trays (4)
- toothpicks (12)
- wax pencil

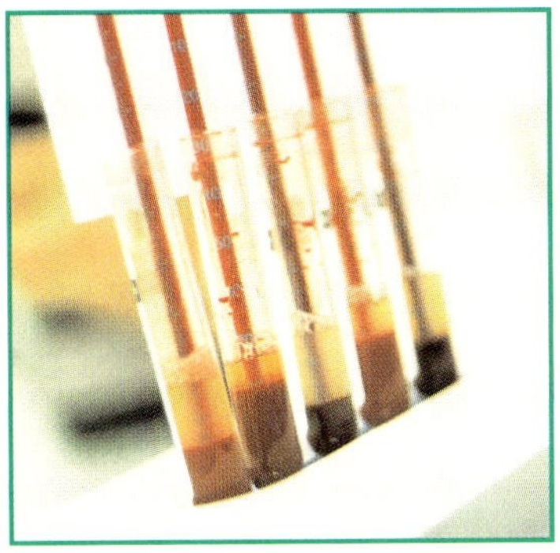

ChemSafety

CAUTION: Always wear safety goggles and a lab apron to protect your eyes and clothing.

CAUTION: Do not touch or taste any chemicals. Know the location of the emergency shower and eyewash station and how to use them. If you get a chemical on your skin or clothing, wash it off at the sink while calling to the teacher. Notify the teacher of a spill. Spills should be cleaned up promptly, according to your teacher's directions.

CAUTION: Glassware is fragile. Notify the teacher of broken glass or cuts. Do not clean up broken glass or spills with broken glass unless the teacher tells you to do so.

Background

Blood is one of the most common kinds of physical evidence at a crime scene where someone was injured. One of the first things a lab technician in a forensics lab must learn is how to determine ABO blood type from a blood sample. In this lab, you will determine the blood type of samples of simulated human blood from four different people.

Blood typing is performed using *antiserum*, blood serum that contains specific antibodies. For ABO blood typing, antibodies against the A and B antigens are used. These antibodies are called *anti-A* and *anti-B agglutinins*. If clumping—called *agglutination*—occurs in the test blood only when it is exposed to anti-A serum, the blood contains the A antigen and is said to be type A. If clumping occurs in the test blood only when it is exposed to anti-B serum, the blood contains the B antigen and is said to be type B. If agglutination occurs with anti-A and separately, anti-B sera, the blood is type AB, which has both A and B antigens. If no agglutination occurs with either serum type, the blood type is O. This information is summarized in **Table 1.**

Another type of marker protein on the surface of red blood cells is the Rh factor, so named because it was originally identified in rhesus monkeys. People whose blood contains the Rh factor are said to be Rh positive (Rh+). People whose blood does not contain the Rh factor are Rh negative (Rh–).

Forensics Lab

Blood Typing

Teacher's Notes

Time Required
One 50-minute class period

Ratings
EASY → HARD

Teacher Preparation 4
Student Setup 2
Concept Level 3
Cleanup 2

Skills Acquired
- Experimenting
- Measuring
- Predicting
- Organizing and Analyzing Data

Scientific Methods
In this lab, students will
- Make Observations
- Analyze the Results
- Draw Conclusions

Materials and Equipment
Materials for this lab can be ordered from WARD'S. See *Master Materials List* at the front of this book for catalog numbers.

Visit go.hrw.com and type in the keyword **Holt Forensics** to use the online Lab Materials Quicklist for the Forensics Labs.

Safety Cautions
Discuss all safety symbols and caution statements with students. Have students wear lab aprons. DO NOT allow students to test any blood other than the simulated blood you provide. Tell students that gloves provide protection from blood-borne pathogens and are necessary when working with body fluids.

Tips and Tricks
Remind students to use a new stirrer (toothpick) for each test and to break used stirrers in two before placing them in the designated container.

TABLE 1: Agglutination Reaction of ABO Blood-typing Sera

Blood Type	Serum antibodies	Agglutination reaction of red blood cells below with blood serum in "Blood Type" column			
		A	*B*	*AB*	*O*
A	Anti-B	–	+	+	–
B	Anti-A	+	–	+	–
AB	None	–	–	–	–
O	Anti-A Anti-B	+	+	+	–

A person with Rh– blood has no antibodies to Rh+ blood unless the person was exposed to Rh+ blood at an earlier age. No agglutination occurs the first time an Rh– person receives a blood transfusion from an Rh+ person. Agglutination can occur, however, the second time the Rh– person receives Rh+ blood. In addition to testing for ABO blood type, it is also important to test blood for transfusion for its Rh factor.

PART A: Typing Blood

Procedure

1. Put on safety goggles and a lab apron. With a wax pencil, label each of four blood-typing trays as follows:

   Tray 1—Mr. Thomas, Tray 2—Ms. Chen, Tray 3—Mr. Juarez, Tray 4—Ms. Brown.

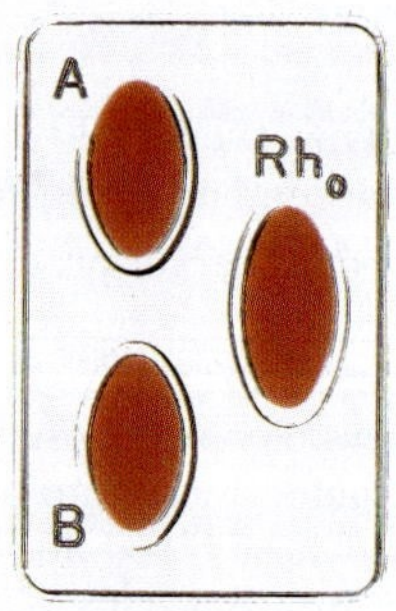

Figure 1 Blood-typing tray

2. **CAUTION: Use ONLY the simulated blood provided by your teacher. Do not use real blood.** Place 3 to 4 drops of Mr. Thomas's simulated blood in each of the A, B, and Rh wells of Tray 1 as shown in **Figure 1.**
3. Place 3 to 4 drops of Ms. Chen's simulated blood in each of the A, B, and Rh wells of Tray 2.
4. Place 3 to 4 drops of Mr. Juarez's simulated blood in each of the A, B, and Rh wells of Tray 3.
5. Place 3 to 4 drops of Ms. Brown's simulated blood in each of the A, B, and Rh wells of Tray 4.
6. Add 3 to 4 drops of the simulated anti-A serum to each A well on the four trays.
7. Add 3 to 4 drops of the simulated anti-B serum to each B well on the four trays.
8. Add 3 to 4 drops of the simulated anti-Rh serum to each Rh well on the four trays.
9. Use separate toothpicks to stir each sample of serum and blood. Record your observations in your lab report in a table like **Data Table A.** Indicate an agglutination reaction with a + and no reaction with a –. Also record your observations of each test. *Note: A positive test is indicated by obvious clumping of the red blood cells. Rh+ blood samples will undergo an agglutination reaction when exposed to anti-Rh serum.*

DATA TABLE A

	Agglutination reaction of blood sample with blood serum				
	Anti-A serum	Anti-B serum	Anti-Rh serum	Blood type	Observations
Tray 1: Mr. Thomas	+	–	+	A+	Answers may vary.
Tray 2: Ms. Chen	–	+	–	B–	Answers may vary.
Tray 3: Mr. Juarez	+	+	+	AB+	Answers may vary.
Tray 4: Ms. Brown	–	–	–	O–	Answers may vary.

DATA TABLE A

Agglutination reaction of blood samples with blood serum					
	Anti-A serum	Anti-B serum	Anti-Rh Serum	Blood type	Observations
Tray 1: Mr. Thomas					
Tray 2: Ms. Chen					
Tray 3: Mr. Juarez					
Tray 4: Ms. Brown					

PART B: Cleanup and Disposal

10. Dispose of paper and broken glass in the designated waste containers. Do not put lab materials in the trash unless your teacher tells you to do so.

11. Clean up your work area and all lab equipment. Return lab equipment to its proper place. Wash your hands thoroughly before you leave the lab.

Analyze and Conclude

1. **Applying Concepts** What factors determine the ABO blood types?
2. **Applying Concepts** What is the difference between an antigen and agglutinin?
3. **Applying Concepts** If Ms. Brown were serving as a blood donor, what ABO blood type(s) could receive her blood safely?
4. **Applying Concepts** Which person among the four represented by the simulated blood samples can receive donated blood from Ms. Chen? Explain your answer.
5. **Inferring Conclusions** People with type O blood who are also Rh– are commonly called universal donors. Explain why.
6. **Inferring Conclusions** A person with what blood type would be considered a universal recipient? Explain your answer.

Extension

1. **Further Inquiry**
The first baby with Rh+ blood born to a woman with Rh– blood usually has no health problems. The second Rh+ child, however, can be seriously threatened before birth if the mother produces antibodies against the Rh antigens of her baby. Find out why this happens and what treatment is given to babies in this situation to save their lives.
2. **Further Inquiry**
Find out what an emergency medical technician gives to a patient when the technician administers an emergency transfusion in the field. Why do technicians use this substance instead of blood for transfusing their patients?

internet connect
www.scilinks.org
Topic: Forensics in Florida
Keyword: HX6F09
SCILINKS. Maintained by the National Science Teachers Association

1055

Answers to Analyze and Conclude

1. The ABO blood typing is based on the presence or absence of antigens A and B, which are encoded by genes.
2. Antigens are substances found on the surface of red blood cells; agglutinins are antibodies found in plasma.
3. Ms. Brown has type O– blood. Therefore, she can give blood to groups O–, A–, B–, and AB.
4. Mr. Juarez can receive blood from Ms. Chen. Ms. Chen has blood type B–, which can be received by Mr. Juarez, who has blood type AB+. A person with type AB+ blood has A, B, and Rh antigens and therefore can receive blood from a person with type A, B, AB, or O (whether Rh positive or negative).
5. Type O blood carries no blood cell antigens to react with the recipient's agglutinins. Type O blood can, therefore, be received by people with any blood type.
6. A person with type AB+ blood would be considered a universal recipient because he or she can receive types A, B, AB, and O blood and either Rh+ or Rh– blood.

Forensics Lab

DNA Fingerprinting

Teacher's Notes

Time Required

Three 50-minute class periods and 15 minutes another day

Ratings

EASY → HARD

TEACHER PREPARATION 1
STUDENT SETUP 2
CONCEPT LEVEL 2
CLEANUP 2

Skills Acquired

- Collecting data
- Experimenting
- Identifying patterns
- Recognizing patterns
- Inferring
- Interpreting
- Analyzing data
- Predicting

Scientific Methods

In this lab, students will

- Make Observations
- Form a Hypothesis
- Analyze the Results
- Draw Conclusions

Materials and Equipment

Materials for this lab can be ordered from WARD'S. See *Master Materials List* at the front of this book for catalog numbers.

Visit go.hrw.com and type in the keyword **Holt Forensics** to use the online Lab Materials Quicklist for the Forensics Labs.

To cast 0.8% agarose gels yourself, refer to product literature supplied with prepared agarose. Prepare 1× TBE running buffer according to manufacturer's instructions. A 37° C incubator may be used in place of the 37° C water bath.

Forensics Lab

DNA Fingerprinting

SKILLS

- Collecting data
- Identifying patterns
- Inferring
- Interpreting
- Analyzing data
- Predicting

OBJECTIVES

- **Perform** a restriction digestion of DNA samples.
- **Evaluate** the results of simulated DNA fingerprints.
- **Identify** a hypothetical burglar by analyzing simulated DNA fingerprints.

MATERIALS

- safety goggles
- lab apron
- gloves
- microcentrifuge tubes (6)
- microcentrifuge tube rack/float
- micropipet, 10 or 20 μL
- micropipet tips (30)
- DNA samples (5)
- *Pvu II* restriction enzyme
- *Pvu II* reaction buffer, 10×
- water bath, 37° C
- loading dye
- water bath, 65° C
- agarose gel on gel tray (0.8 percent)
- electrophoresis system, battery powered
- beaker, 500 mL
- graduated cylinder, 250 mL
- TBE running buffer, 1×, 350 mL
- gel staining tray
- DNA stain
- Lambda DNA/*Hind III* marker
- resealable plastic bag

ChemSafety

CAUTION: Always wear safety goggles and a lab apron to protect your eyes and clothing.

CAUTION: Do not touch or taste any chemicals. Know the location of the emergency shower and eyewash station and how to use them. If you get a chemical on your skin or clothing, wash it off at the sink while calling to the teacher. Notify the teacher of a spill. Spills should be cleaned up promptly, according to your teacher's directions.

CAUTION: Glassware is fragile. Notify the teacher of broken glass or cuts. Do not clean up broken glass or spills with broken glass unless the teacher tells you to do so.

Background

The police are investigating the burglary of a residence. As the police arrived, the masked burglar smashed a glass door with a chair and quickly escaped. Police found small bits of bloodstained fabric on some of the pieces of the glass door. Four suspects have been brought in, and DNA samples have been collected from each. The police lab thus has the DNA from the blood samples found at the crime scene and DNA samples taken from each of the four suspects.

In this lab, you will perform some of the experimental procedures involved in DNA fingerprinting and use your results to identify the burglar.

Procedure

PART A: Restriction Enzyme Digestion of DNA

In the first step of DNA fingerprinting, samples of known and unknown origin are obtained and then *digested,* or cut into small

Safety Cautions

Students should be supervised at all times during this investigation. Discuss all safety symbols and caution statements with students. Be sure that students know the correct procedure for making electrical connections. Do not come in personal contact with or allow metal or any conductive material to come in contact with the reservoir buffer or the electrophorectic cell while the battery/power supply is on. If using a power supply, be sure to follow manufacturer's directions and precautions.

Disposal

DNA stain, DNA samples and 1× TBE buffer can be flushed down the drain with copious amounts of water. Stained agarose gels and other materials used in this lab can be put in the trash.

fragments, by a restriction enzyme. In this case, the samples would include DNA from the blood sample found at the scene (the burgler's DNA) and DNA from each of the suspects.

1. Put on safety goggles and a lab apron.
2. Obtain five microcentrifuge tubes.
3. Label tubes as follows: Crime Scene, Suspect #1, Suspect #2, Suspect #3, Suspect #4.
4. Set the micropipet to 2 µL and place a clean tip on the pipet.
5. Add 2 µL of *Pvu II* 10× reaction buffer to each tube. *Note: It may be helpful to place the pipet tip against the side of the microcentrifuge tube when dispensing small volumes.*
6. Set the micropipet to 10 µL.
7. Using a fresh pipet tip for each sample, add 10 µL of each DNA sample to the appropriate tube.
8. Set the micropipet to 2 µL.
9. Add 2 µL of *Pvu II* restriction enzyme to each tube. Again, be sure to use a fresh micropipet tip for each sample.
10. Close each tube and snap the tubes in a downward motion with your wrist to force all of the reagents to the bottom of the tubes.
11. Incubate all five tubes at 37° C for 45–60 minutes.
12. Add 2 µL of loading dye to each tube.
13. Incubate all five tubes for 5 minutes at 65° C to stop the restriction enzyme activity.
14. Place the samples in the freezer if you will not be using them immediately. Make sure all samples for your group are stored together and labeled with your group name.

PART B: Gel Electrophoresis of DNA Samples

The next step of DNA fingerprinting is to separate the restriction fragments by gel electrophoresis. In gel electrophoresis, the digested DNA samples are loaded into wells on a jellylike slab called a *gel.* The gel is then exposed to an electrical current. DNA has a negative electrical charge, so the restriction fragments are attracted to the positive pole when an electric current is applied. Smaller fragments travel farther and faster through the gel than do longer ones. Recall that restriction fragments harvested from one person's DNA differ somewhat from the restriction fragments of another person. So the pattern made by restriction fragments migrating through the gel also differs from person to person. A special dye, called a *loading dye,* is added to the DNA samples for two reasons. The loading dye is heavy, which helps the samples to stay in the wells. The loading dye also runs slightly faster than the DNA samples and indicates when the gel has finished running.

15. Place the tray with the gel on the lab bench.
16. Set the pipet to 10 µL and place a new tip on the end of your micropipet.
17. Open the microtube containing the DNA marker and use the pipet to load 10 µL of the marker DNA into the well in Lane 1 of an agarose gel. To do this, place both elbows on the lab table, lean over the gel, and slowly lower the micropipet tip into the opening of the well before depressing the plunger. *Note: Do not jab the micropipet tip through the bottom of the well.*
18. Set the pipet to 15 µL and place a new tip on the end of your micropipet.
19. Open the microtube containing the crime scene sample and use the pipet to load 15 µL of the sample into the well in Lane 2 of an agarose gel.

Tips and Tricks

This lab works best in groups of three to six students.

You may pour agarose gels in advance and store them in the refrigerator for up to one week prior to the electrophoresis portion of the lab. Place the gel, on a casting tray, in a resealable bag with 1–2 mL 1× TBE buffer.

Remove DNA samples and enzymes from the freezer and allow them to thaw 15 minutes prior to performing the lab. After samples have thawed, tap them on the countertop to make sure all of the liquid is on the bottom of the tube and not in the cap of the vial. Place and keep samples on ice.

Demonstrate how to use wrist action to snap the tube so that the liquid is forced to the bottom of the tube.

To visualize some of the DNA as the gel is being run, prestain the gel and buffer by adding 1 µL of DNA stain concentrate per mL of 1× TBE running buffer and by adding 1 µL of DNA stain concentrate per mL of liquefied agarose.

Staining and destaining following electrophoresis will still be necessary.

Gels may also be loaded *after* the 1× TBE running buffer has been added.

Loading the gel "wet" decreases the likelihood that the dye samples will be washed out of the wells, but makes it harder to see the wells and to avoid puncturing the bottom of the wells when loading. To "wet" load the gel, place a strip of black or dark colored paper under the electrophoresis chamber so that it is under the wells. Add 1× TBE running buffer until it covers the gel by 2–3 mm. Use a micropipet to load each well. Do not overload the wells.

Demonstrate how to use a micropipet to load samples into gel wells. Caution students to be very careful not to puncture the gel with the micropipet tips when loading samples.

(continued on p. 1058)

1× TBE buffer may be stored at room temperature for up to 18 months. If the concentrated buffer contains a white precipitate, it should be discarded.

When using a battery operated electrophoresis chamber, it will take up to two hours for the DNA to separate. If you use three batteries instead of five, the DNA can be separated overnight. For shorter running times (under an hour), use a power supply set between 75 V and 125 V.

Prepare stain immediately before staining by adding stain concentrate to warm (50–55° C) water. Warm stain results in enhanced banding of the DNA fragments.

Dilute stain may be saved and reused several times. For best results, reheat the stain before reusing.

For long-term storage of gels (up to a month), add several drops of dilute stain to the 1× TBE in the bag to prevent DNA bands from fading. Gels can be restained if necessary.

20. Using a new micropipet tip for each tube, repeat step 19 for each of the remaining samples. Place each sample in a well as follows:

Lane 1: DNA marker

Lane 2: Crime scene DNA

Lane 3: Suspect #1 DNA

Lane 4: Suspect #2 DNA

Lane 5: Suspect #3 DNA

Lane 6: Suspect #4 DNA

21. Follow your teacher's instructions to carefully place the agarose gel (still in a gel-casting tray) in the chamber of an electrophoresis apparatus, such as the one shown in **Figure 1,** so that the wells are closest to the negative electrode.

22. Pour approximately 350 mL of 1× TBE running buffer into a beaker.

23. Gently and slowly pour the running buffer from the beaker into one side of the electrophoresis chamber until the gel is completely covered (approximately 1 to 2 mm above the top surface of the gel). *Note: Pouring too fast will rinse your DNA sample out of the wells. Be careful not to overfill the chamber with buffer.*

24. Place the cover on the electrophoresis chamber. Wipe off any spills around the electrophoretic apparatus before doing the next step.

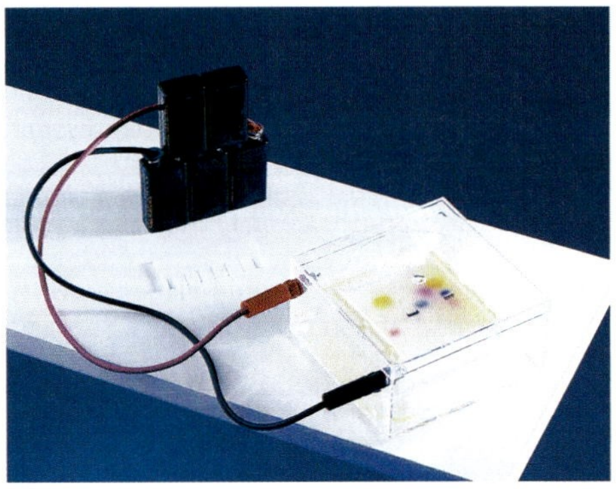

Figure 1 Electrophoresis set-up

25. Connect five 9 V alkaline batteries as instructed by your teacher. **CAUTION: Do not touch both ends of the patch cords or both terminals on the battery pack at the same time.**

26. Connect the red (positive) patch cord to the red terminal on the chamber and the red terminal on the battery pack. Follow the same procedure with the black (negative) patch cord and the black terminals.

27. Observe the migration of the loading dye along the gel toward the red (positive) electrode.

28. Disconnect the battery pack when the loading dye band has run halfway off the end of the gel. *Note: If gels have been run overnight, this step will take place on the following day.*

29. Remove the cover from the electrophoresis chamber.

30. Carefully lift the gel tray (containing the gel) from the chamber onto a piece of paper towel. Notch one side of the gel so that you can identify the lanes. *Note: Agarose gels are very fragile, so use extreme care when handling the gel.*

31. Use a metric ruler to measure the distance of the dye bands in Lane 6 (in mm) from each of the six sample wells. *Note: Be sure to measure from the center of the well to the center of the band.*

32. Place the gel in a resealable bag and add 1–2 mL of 1× TBE buffer and refrigerate the gel until the next lab period. If there is enough time (at least 45 minutes) left in the class period, proceed to step 33.

PART C: Staining the Gel

After the DNA fragments have been separated, the gel can be stained to visualize the banding pattern of each DNA fingerprint.

33. Gently place the gel in the staining tray.

34. Put on safety goggles and a lab apron. Pour approximately 100 mL of warm dilute stain into the staining tray so that it covers the gel.

35. Cover the tray and let the gel stain for approximately 30 minutes.

36. Carefully decant the used stain. *Note: Make sure the gel remains flat and does not move up against the corner. Decant the stain directly to a sink drain and flush with water.*

37. Add distilled water or tap water to the staining tray. Do not pour water directly onto the gel.

38. Allow the gel to de-stain either overnight or for 20 minutes. To de-stain in 20 minutes, gently rock the tray, and change the water several times. Overnight de-staining does not require a change of water.

39. View the gel against a white sheet of paper. Sketch the bands you see on the blank gel, which will look similar to the image in **Figure 2.**

40. Store the gel in a resealable plastic bag with 1–2 mL of TBE buffer.

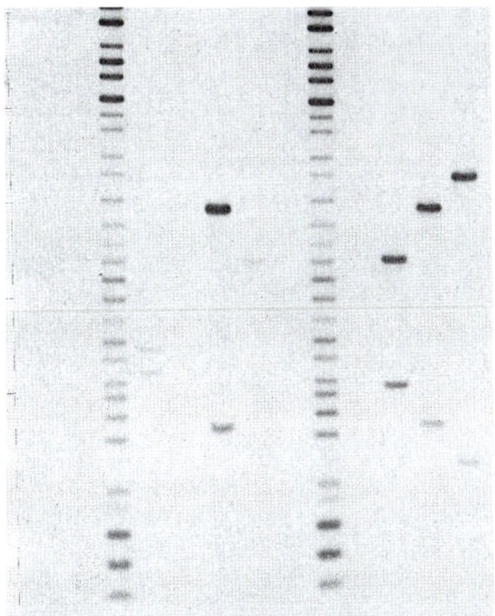

Figure 2 DNA fingerprint

PART D: Cleanup and Disposal

41. Dispose of paper and broken glass in the designated waste containers. Do not put lab materials in the trash unless your teacher tells you to do so.

42. Clean up your work area and all lab equipment. Return lab equipment to its proper place. Wash your hands thoroughly before you leave the lab.

Analyze and Conclude

1. **Explaining Events** Where are the smallest DNA fragments located on the gel? the largest? Explain why.
2. **Explaining Events** Why was loading dye added to each sample prior to loading the samples on the gel.
3. **Examining Data** Are any of the DNA fingerprints identical? If so, which ones?
4. **Drawing Conclusions** Based on your results, which suspect appears to be the burglar? Explain your answer.
5. **Interpreting Information** Explain why each suspect has a different banding pattern (or DNA fingerprint).
6. **Making Predictions** What do you think would happen if you placed your gel in the electrophoresis chamber with the wells containing DNA next to the red electrode instead of the black electrode?
7. **Making Inferences** From what types of crime scene evidence (other than blood) might police obtain DNA?

Extension

1. **Research and Communications** Research another way that DNA fingerprinting is used in society and present your results to your classmates in a short oral report.
2. **Research and Communications** Research the procedures involved in collecting evidence at a crime scene.

1059

Answers to Analyze and Conclude

1. The smallest DNA fragments are the bands located near the bottom of the gel. The largest DNA fragments are the bands located near the wells. All of the DNA has a negative charge and is attracted to the positive red anode. Because the smaller fragments move more quickly and easily through the agarose than the larger fragments, they are able to move farther down the gel.
2. Loading dye runs slightly faster than the DNA samples and indicates when the DNA fragment have separated and the gel is done running. DNA loading dye is also heavier than the electrophoresis buffer, which helps the DNA samples to stay in the wells.
3. The DNA fingerprint from the crime scene DNA should match the DNA fingerprint of Suspect #3.
4. Suspect #3, because the DNA fingerprint from the crime scene DNA matches the DNA fingerprint of Suspect #3.
5. The slight differences (or RFLPs) in each person's DNA sequence ensure that each person (except identical twins) has a unique set of sites where restriction enzymes can bind and cut the DNA. Because RFLPs, the exact number and size of fragments produced by digestion with a specific restriction enzyme varies from person to person. Separation of the fragments by gel electrophoresis results in a unique banding pattern.
6. The DNA would be attracted to the red anode and would quickly run off the top of the gel and into the buffer.
7. Answers will vary. DNA could be extracted from skin, hair, semen, and saliva.

Forensics Lesson

Teaching Strategy

Explain to students that chromatography is a general term for a range of methods for separating out and identifying the components of a chemical mixture. Often, solving a crime may involve identifying an unknown substance or finding traces of drugs or toxins in a person's body. When a toxin is suspected as a cause of death, the forensic laboratory will perform *toxicological analysis*. For such analyses, a forensics laboratory may be equipped with a liquid or gas chromatograph with integrated mass spectrometry capabilities. Prior to chromatography and mass spectrometry, it was difficult for forensic scientists to produce definitive evidence of, for example, poisoning as a cause of death. Have students research the kinds of crimes other than murder for which chromatography and mass spectrometry are used.

Forensics Lesson

Chromatography

Green paint has been thrown all over several walls of a local business. The police have several suspects and have found some green paint at each of the suspect's homes. How can they determine if any of the suspect samples match the paint thrown on the walls?

Background

Chromatography is a technique used to separate a mixture of different substances. It was originally developed in 1903 by a Russian botanist, Mikhail Tswett, who used this approach to separate colored plant pigments. *Chromatography* means "to write with colors" and comes from the Greek words *chroma*, color, and *graphein*, to write.

There are several types of chromatography, each one depending on the nature of the substance in a mixture. In most types of chromatography, polarity is the basic principle at work. A polar molecule has an uneven distribution of electric charge. Water is the most familiar example of a polar substance. Its uneven distribution of charge means that one side of a water molecule strongly attracts the opposite side of the water molecule next to it, as shown in **Figure 1.** This attraction between opposite charges means that polar substances—even different polar substances—have strong attraction to each other. On the other hand, nonpolar substances, such as oils, have a much stronger attraction for each other than for polar substances.

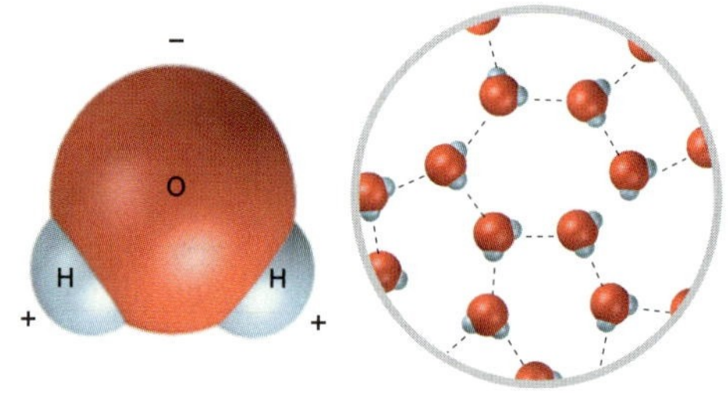

Figure 1 Water molecule (left); attraction between polar molecules (right)

How Chromatography Works

All kinds of chromatography involve two systems, or *phases*, in contact with each other. The mobile phase is a solvent that dissolves some or all of the substances to be separated in the mixture. The stationary phase on which the mixture is placed is often some type of solid material such as chromatography paper or a thin layer of a gel coated onto a glass or plastic plate. **Table 1,** shows some examples of different kinds of chromatography.

TABLE 1: SOME TYPES OF CHROMATOGRAPHY

Category	Mobile phase	Stationary phase	Separating principle
Paper chromatography	some solvent; can be water, methanol, etc.	paper	polarity of mixture components relative to mobile phase
Thin-layer chromatography	some solvent; can be water, methanol, etc.	silica gel plate	polarity of mixture components relative to mobile phase
Column chromatography	some solvent; can water, methanol, etc.	powdered adsorbent packed in a glass column	polarity of mixture components relative to mobile phase

As the mobile phase moves through or across the stationary phase, the stationary phase separates the components of the mixture. The result of the process is called a *chromatogram,* shown in **Figure 2,** so named for the colored bands produced when the separation involves a mixture of colored substances.

A paper chromatogram is produced by dabbing a sample of the mixture to be separated near one end of a piece of chromatography paper (the stationary phase). Another part of the chromatography paper is then wetted with the solvent. While the paper is in contact with the solvent, the solvent is wicked through the paper by capillary action.

The substances that make up the sample mixture are dissolved by the solvent and move along the paper with the solvent. As the solvent moves through the paper, the molecules of the various dissolved substances are attracted to both the solvent and the paper. The strengths of these attractions are determined by how polar the substance is. Because these attractions are different for each substance in a mixture, each substance moves through the paper at a different rate. The substances with a greater attraction to the solvent move faster and farther along the paper. Those with a greater attraction to the paper move more slowly, and consequently not as far as other substances.

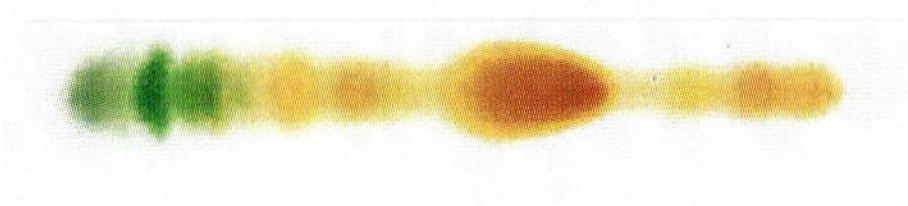

Figure 2 Chromatogram of pigments in a species of algae

Polar solvents, such as water, methanol, and acetic acid, tend to work best with substances that are also polar, such as water-based inks and dyes. A less polar solvent such as petroleum ether, cyclohexane, or methylene chloride will tend to work best with nonpolar or weakly polar substances, such as oil-based inks and dyes.

In addition, large molecules tend to move more slowly than small molecules because the large molecules cannot easily pass through the stationary phase material. The result of the differences in attraction and size cause the various substances to separate into distinct areas on the paper. The paper is then removed from the solvent and allowed to dry, producing a permanent record that can be analyzed or processed further.

How to Use Chromatography

Chromatography is often used as a tool for comparing the makeup of two or more mixtures. If the chromatogram is the same for each mixture, then the mixtures are probably the same. If the chromatograms are different, then the mixtures are probably different. Often, different samples to be tested are placed on the same piece of chromatography paper. This way, the conditions that create the chromatogram are the same for all the mixtures.

Topic Questions

1. In chromatography, what are the functions of the mobile and stationary phases?
2. Red dye can be made from a mixture of yellow dye and magenta (a purplish blue) dye. How could you determine if a particular red dye is made from a single dye or from a mixture of yellow and magenta dyes?
3. Two mixtures are compared using chromatography. Why is it desirable to place samples of both mixtures on the same paper when making the chromatogram?

Answers to Topic Questions

1. The mobile phase carries the mixture to be separated by chromatography. The stationary phase attracts some of the components of the mixture.
2. Produce a paper chromatogram of the dye to see if the dye separates into yellow and magenta. If it separates, then the dye is a mixture of yellow and magenta dyes. If the chromatogram produces only a red spot, then the dye is a single red dye.
3. Placing both mixtures on the same paper ensures that both mixtures are separated under the same set of conditions. Small differences in the solvent or paper and the length of time the paper is in the solvent are all factors that can affect the results of the separation.

Forensics Lab

The Counterfeit Drugs

Teacher's Notes

Time Required
One 50-minute class period

Ratings
EASY → HARD

TEACHER PREPARATION 1
STUDENT SETUP 2
CONCEPT LEVEL 2
CLEANUP 2

Skills Acquired
- Experimenting
- Inferring
- Interpreting
- Communicating

Scientific Methods
In this lab, students will
- Make Observations
- Analyze the Results
- Draw Conclusions
- Communicate the Results

Materials and Equipment
Materials for this lab can be ordered from WARD'S. See *Master Materials List* at the front of this book for catalog numbers. Additional items not listed in the student materials list are
- candies, common small chocolate tablets with hard colored shell (brown-colored ones only; 20 per class)
- food coloring, brown

Visit go.hrw.com and type in the keyword **Holt Forensics** to use the online Lab Materials Quicklist for the Forensics Labs.

Forensics Lab

The Counterfeit Drugs

SKILLS
- Experimenting
- Inferring
- Interpreting
- Communicating

OBJECTIVE
- **Determine** by strip paper chromatography if a suspect dye contains a particular pigment.

MATERIALS
- distilled water
- pigment solutions to be tested
- strips of chromatography paper (2)
- tape
- beakers, small (2)
- eyedropper
- paper clips, small
- pencil
- scissors

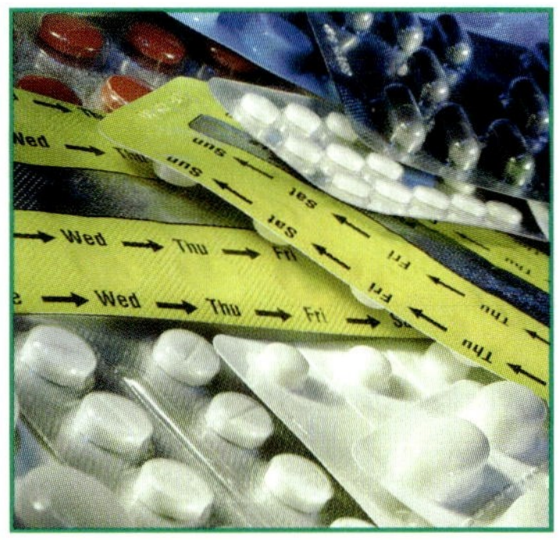

ChemSafety

CAUTION: Always wear safety goggles and a lab apron to protect your eyes and clothing.

CAUTION: Do not touch or taste any chemicals. Know the location of the emergency shower and eyewash station and how to use them. If you get a chemical on your skin or clothing, wash it off at the sink while calling to the teacher. Notify the teacher of a spill. Spills should be cleaned up promptly, according to your teacher's directions.

CAUTION: Glassware is fragile. Notify the teacher of broken glass or cuts. Do not clean up broken glass or spills with broken glass unless the teacher tells you to do so.

Background

A small pharmaceutical company has been accused of illegally manufacturing and exporting a counterfeit version of a popular blood pressure medication. The Food and Drug Administration (FDA) suspects that the company manufactures a generic version of the patented drug and labels it identically to the real drug.

The FDA, along with the Drug Enforcement Administration (DEA), has confiscated a sample of the drug as it was being loaded onto a ship. Officials need to determine whether or not their sample is the brand-name drug or the identical-looking, counterfeit version of the medication.

While the active chemical compound of the brand name drug and the counterfeit drug is the same, officials know that there is a difference between the genuine and the counterfeit pills. While the pills appear to be identical, the pigments used in the brown color coating of the pills differ. The brown coating used in the counterfeit pills contains a mixture of water-soluble blue and orange pigments. The coating used in the genuine version contains a homogeneous brown dye, also water-soluble. In this lab, you will use chromatography to test a sample of brown pigment to determine if the pill it came from is genuine or counterfeit.

Procedure

PART A: Performing Chromatographic Analysis

1. Put on safety goggles and a lab apron. Cut a length of chromatography paper equal to the depth

Safety Cautions
Discuss all safety symbols and caution statements with students. Since glassware and scissors are manipulated, there is the potential for glass cuts and punctures. Have students wear protective goggles and lab aprons.

Disposal
All liquids can be safely disposed of down the drain. Notify janitorial staff when disposing of broken glass.

Tips and Tricks
Chromatography paper is very inexpensive and may be purchased in either pre-cut strips, sheets, or in rolls. The size of the beakers that are used will determine the chromatogram lengths.

To prepare a dye sample for the "counterfeit drug," place about 20 brown-colored chocolate candies into a 50 mL beaker, and add a teaspoonful of water. The candy consists of an outer dye coating, a white sugar coating, and a chocolate interior. Stir the candy until the brown coating is dissolved in the water, and then pour this liquid into another small beaker. The dye coating of the brown candies contains blue and orange, which should separate out of the beaker you are using. You may check to make sure the length is correct by taping the top of the paper to a pencil and lowering the paper into the beaker, allowing the pencil to rest across the top of the beaker.

2. Once the paper is cut to the correct length, remove it from the beaker. Clip a paper clip to the bottom of the paper to keep the paper hanging straight down while it is in the beaker. (Keep the pencil attached to the top of the paper.) Using an eyedropper, dab some of the suspect dye onto the chromatography paper about 2 cm above where the water level will be in the bottom of the beaker.
3. Allow the dye to dry for a few minutes, and then repeat this spotting process several times (at the same place on the paper each time) to build up a concentration of pigment that will yield good results.
4. Now pour distilled water into the beaker to a depth of about 2 cm. It is important to have the test spot be 2 cm above the water level, and not immersed. Make sure that the dye to be tested does not come into contact with the water. Refer to **Figure 1** which shows the correct relationship between the dye spot and the water level.

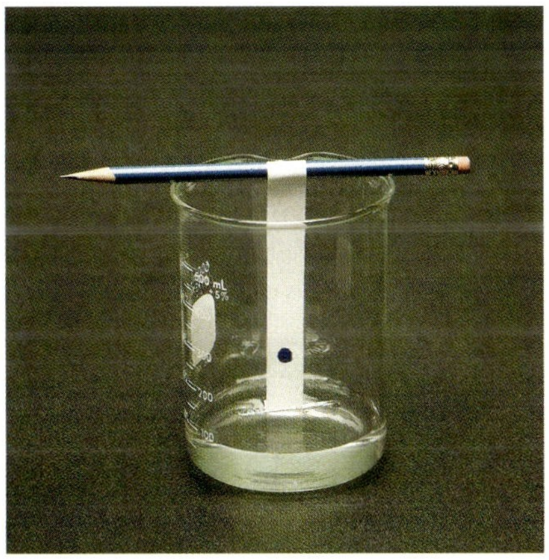

Figure 1 Paper chromatography setup

5. To conduct a chromatographic run, set up the beaker, paper, pencil, and water as shown in **Figure 1.** Allow the water to move up the paper by capillary action and separate the pigments in the dye. This process may take several minutes, so plan accordingly.
6. When the water migrates to within a few millimeters from the top of the paper, stop the run by removing the pencil and attached paper from the beaker. Empty the water from the beaker, place the pencil back across the top of the beaker, and allow the chromatogram to dry overnight.

PART B: Cleanup and Disposal

7. Dispose of paper and broken glass in the designated waste containers. Do not put lab materials in the trash unless your teacher tells you to do so.
8. Clean up your work area and all lab equipment. Return lab equipment to its proper place. Wash your hands thoroughly before you leave the lab.

Analyze and Conclude

1. **Explaining Events** Why do the pigments of a dye separate out on the chromatography paper?
2. **Analyzing Methods** Why is it important that tests in cases like this one be carried out such that the analysts do not know what products they are testing?
3. **Drawing Conclusions** Is the sample you tested from the genuine drug or the counterfeit drug? How can you tell?

internet connect
www.scilinks.org
Topic: Forensics in Florida
Keyword: HX6F09
SciLinks. Maintained by the National Science Teachers Association

upon chromatographic testing and enable students to identify the "drug" as the "counterfeit" version, as described in the case description.

If you choose to provide some lab groups with a dye that will give a negative result (the "genuine drug"), you might want to use a brown food coloring. Test the food coloring first to ensure that it does not give a similar result as the dyes from the brown-colored chocolate candies.

It is important that the dye be spotted repeatedly on the chromatography paper to obtain enough pigment to test. A common student error is to let the runs proceed too long, in which case all of the pigments will be carried to the very top of the paper. So it is important to instruct students to remove the paper strips from the beakers when the solvent migrates to within a few centimeters of the pencil and while the pigments are still spread out on the chromatography paper.

Answers to Analyze and Conclude

1. Molecules of pigments in the dye, paper, and water are all polar, so the pigments will travel along the paper with the water. The different dyes have different attractions to the water and paper, however, so they travel up the paper to different heights.
2. If the analysts do not know which samples are being tested, the results can be considered to be free of bias; that is, the outcomes were determined only by the natures of the samples, not by what the analysts were expecting to find.
3. Results will depend on the dye solutions you provide to the lab groups. Dye solutions that yield the blue and orange components should be considered from the counterfeit version of the drug. Those solutions that did not produce the blue and orange components (for instance, if you used a food coloring dye as the genuine version) should be considered as not proven to have been illegally imported.

Forensics Lab

THE QUESTIONABLE AUTOGRAPH

Teacher's Notes

Time Required
Two 50-minute class periods

Ratings
EASY ——— HARD

TEACHER PREPARATION 2
STUDENT SETUP 3
CONCEPT LEVEL 3
CLEANUP 1

Skills Acquired
- Designing experiments
- Experimenting
- Inferring
- Interpreting
- Communicating

Scientific Methods
In this lab, students will
- Make Observations
- Analyze the Results
- Draw Conclusions
- Communicate the Results

Materials and Equipment
Materials for this lab can be ordered from WARD'S. See *Master Materials List* at the front of this book for catalog numbers. Additional items not listed in the student materials list are
- beaker, 1000 mL

Visit go.hrw.com and type in the keyword **Holt Forensics** to use the online Lab Materials Quicklist for the Forensics Labs.

Forensics Lab

The Questionable Autograph

SKILLS
- Experimenting
- Inferring
- Interpreting
- Communicating

OBJECTIVES
- **Determine** by examination of pigments through thin layer chromatography if an autograph was written with a particular type of pen.

MATERIALS
- thin layer chromatography plates (2)
- teasing needle or pushpin
- single-hole puncher
- suspect autographs, and ink samples from an authentic pen
- spot plates (4)
- dropper bottles of distilled water and methanol
- open-ended capillary tubes (2)
- ruler
- graduated cylinder, 10 mL
- 2-oz. glass bottles with caps (2)

ChemSafety

CAUTION: Always wear safety goggles and a lab apron to protect your eyes and clothing.

CAUTION: Methanol is poisonous. Wear protective gloves. Avoid prolonged exposure to vapors and use in the hood or a well-ventilated area, as directed by the teacher. Keep methanol away from heat and flames, as it is flammable.

CAUTION: Do not touch or taste any chemicals. Know the location of the emergency shower and eyewash station and how to use them. If you get a chemical on your skin or clothing, wash it off at the sink while calling to the teacher. Notify the teacher of a spill. Spills should be cleaned up promptly, according to your teacher's directions.

CAUTION: Glassware is fragile. Notify the teacher of broken glass or cuts. Do not clean up broken glass or spills with broken glass unless the teacher tells you to do so.

Background

The United States Postal Service has been called on to investigate a report of mail fraud. Through an online auction, a sports fan has purchased a ticket stub autographed by his favorite ballplayer. Among the ballplayer's many commercial endorsements is one for a certain brand of pen—he agreed to use only that brand of pen to sign autographs. The suspicious sports fan thinks the signature on the ticket stub looks like it was made using an ordinary ballpoint pen.

The online auction company is about to ban the seller because of other complaints, and an investigation of mail fraud is underway. Postal inspectors need to determine whether or not the autograph was signed with the right kind of pen.

In this lab, you will use thin layer chromatography (TLC) to compare the ink from the signed ticket stub with ink from the type of pen the athlete would have used.

Because some inks are water soluble while others dissolve in methanol, you will need to conduct four thin layer chromatography (TLC) runs: two (one for the suspect ink and one for the authentic ink) using water as the solvent and another two of the same samples using methanol as the solvent.

Safety Cautions
Methanol (methyl alcohol) can be absorbed through the skin during prolonged exposure. Wear protective gloves and goggles when handling this chemical and remember to check students for allergy to latex if latex gloves are to be used. This alcohol is flammable and is a potential vapor hazard. To avoid inhalation, handle only in a well-ventilated area or hood.

Since glassware and teasing needles are manipulated, there is the potential for glass cuts and punctures. Handle sharp objects with care. As usual, appropriate protective goggles and aprons should be worn in the event an accident may produce flying projectiles. Notify janitorial staff when disposing of broken glass.

In the event a large quantity of methanol is spilled, students should be evacuated from the room until cleanup is complete.

Disposal
Carefully consult the supplier's MSDS (Material Safety Data Sheet) for specific warnings, emergency procedures, handling instructions, and directions for proper disposal of methanol.

Procedure

PART A: Performing Thin Layer Chromatography

1. Put on safety goggles, protective gloves, and a lab apron. Score two TLC plates, such as the one shown in **Figure 1,** in half lengthwise by scratching a line in the TLC gel coating with the teasing needle or pushpin, to allow two runs per plate.
2. To extract ink from an autograph, use a hole punch to punch about 25 holes in areas containing ink. Place half of these small disks into a spot plate well and then add a few drops of water to dissolve and remove the ink from the paper.
3. Place the rest of the paper disks containing autograph ink into another spot plate well, and add a few drops of methanol to dissolve and remove the ink from the paper. **CAUTION: Methanol is poisonous and flammable.**
4. Repeat steps 2 and 3 with ink samples from the authentic pen.
5. You will use one of the TLC plates to run the suspect and authentic inks side by side with water as the solvent, and the other TLC plate to run the suspect and authentic inks side by side with methanol as the solvent. Use a capillary tube to spot some of the extracted ink onto the TLC plates about 2 cm from one end. Allow the initial spot to dry, then repeat the process over the same location a few times to build up a concentration of pigment that will yield good results. Use a different capillary tube for each of the two ink samples.
6. Pour water to a depth of 1 cm in one of the beakers. Pour methanol to a depth of 1 cm in the other beaker. It is important that the test spots are 1 cm above the solvent level and not immersed or even touching the solvent initially.
7. Place the TLC plates, as upright as possible, into the beakers. The solvents will move up the plates and separate the pigments in the ink. Observe the process carefully, as it may take just a few minutes, or up to 30 minutes, for the solvent to move to the top of the plate. When the solvent level has risen to within 1 cm of the top of the plate, remove the plate from the bottle. Allow the TLC plates to dry overnight before examining them.

Figure 1 Thin-layer chromatography plate

PART B: Cleanup and Disposal

8. Dispose of paper and broken glass in the designated waste containers. Do not put lab materials in the trash unless your teacher tells you to do so.
9. Clean up your work area and all lab equipment. Return lab equipment to its proper place. Wash your hands thoroughly before you leave the lab.

Analyze and Conclude

1. Are the pigments contained in the suspect and authentic inks similar?
2. If the two inks tested are identical, does this prove with certainty that the autograph is authentic?
3. What factors could affect the accuracy of your results? How could your procedure have been improved in order to control for these factors?
4. What additional tests could be conducted in order to make the results of the case more certain?

Tips and Tricks

It will probably take one complete class period for the student groups to prepare for the runs and another period to actually conduct them. Black, thicker tip, water-soluble markers contain several pigments, as do permanent markers. It is not obvious by simple examination whether an autograph has been written with a water-soluble or a permanent marker, so you can have different autographs for classes or lab groups.

It is important that the ink extracted from the autographs be layered to obtain enough pigment to test. In addition, students should let the first spot dry before applying another layer, exactly on top of the first spot. Make sure the students have the test spots situated well above the solvent in the bottles. A common error is for students to allow the runs to proceed too long, in which case all of the pigments will be carried to the very top of the plate and be smeared together. So it is important to instruct students to remove the plates when the solvent migrates to within a centimeter of the top of the plate.

Answers to Analyze and Conclude

1. Students should note a difference between the pigments of the two inks.
2. No; the result demonstrates that the autograph is written by a pen consistent with the pen the ballplayer endorses.
3. Student answers may vary. If sufficient ink is not spotted on the chromatograpy plate to show a good separation, results may be inconclusive.
4. Student answers may vary. Students may suggest that a handwriting expert could analyze the ballplayer's handwriting in comparison to the questionable autograph.

Forensics Lesson

Teaching Strategy

Tell students that broken glass is a common type of evidence at crime scenes, as are other fragments of materials that may link criminals or victims to another part of a crime scene. Tiny fragments may be recovered from clothing, hair, and other places around a crime scene. Density is a simple physical property that can be compared to help determine whether a sample of material, such as a shard of glass, has the same origin as some other sample of material. Another property that is used to distinguish among samples of glass is the *refractive index*, a measure of the degree to which light is bent by the glass sample. Have students brainstorm other ways that a sample of material might be analyzed to determine its origin.

Forensics Lesson

Determination of Density

You find a shard of glass and wonder if it is from a window or eyeglasses. A large, old necklace may be made of gold or of a metal far less valuable. One of the easiest to analyze properties of an object is its density, which may be a clue to its identity or source.

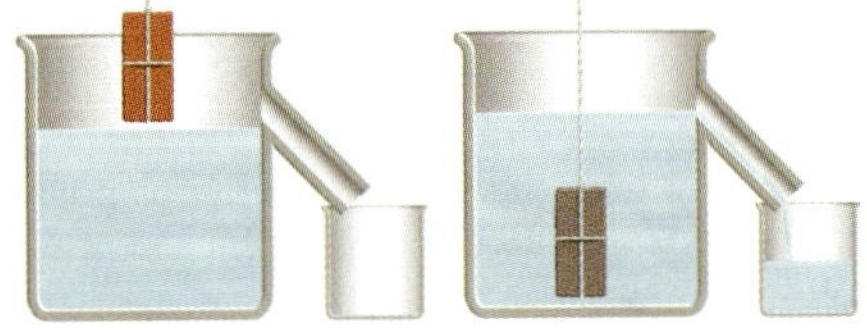

Figure 1 Object displacing its own volume of a fluid

Background

Density is the ratio of the mass of an object to its volume. This ratio can be expressed as an equation, as shown below.

density = mass/volume

also written $d = m/V$

An object has a density characteristic of the material of which it is made. At standard temperature and pressure, for example, the density of water is 1.0. When a solid object is placed in fluid, the difference between its density and the fluid's density determines whether the object will sink or float.

- If the density of the object is *greater* than the density of the fluid, the object will sink. The volume of fluid displaced will be the same as the volume of the object, as shown in **Figure 1.** (The fluid displaced by the brick in **Figure 1** is shown as overflow in the smaller beaker. This overflow fluid volume is equal to the volume of the brick.)
- If the density of the object is *the same* as the density of the fluid, the object will neither sink nor float: it will remain suspended in the fluid. The volume of fluid displaced will be equal to the volume of the object.
- If the density of the object is *less* than the density of the fluid, the object will float. The volume of fluid displaced will have a weight equal to the weight of the object.

Archimedes' Principle

Archimedes, a Greek who lived in the third century BCE, discovered an important fact about buoyant force, the upward force exerted on an object by the fluid in which it is immersed. Archimedes' principle states that for an object in a fluid, a buoyant force is exerted by the fluid on the object that is equal to the weight of the fluid displaced by the object. Archimedes' principle can be stated in a very simple equation, shown below.

buoyant force = weight of displaced fluid

Because of buoyant force, a submerged object will have an apparent weight that is less than its weight in air. (You experience buoyant force in a swimming pool and even in a bathtub.)

A floating object, on the other hand, appears to be weightless. This is because a floating object is less dense than the fluid, so only part of its volume displaces the fluid, and the weight of the displaced fluid is the same as the weight of the entire object.

Archimedes' principle can be used to determine the density of an object submerged in a fluid.

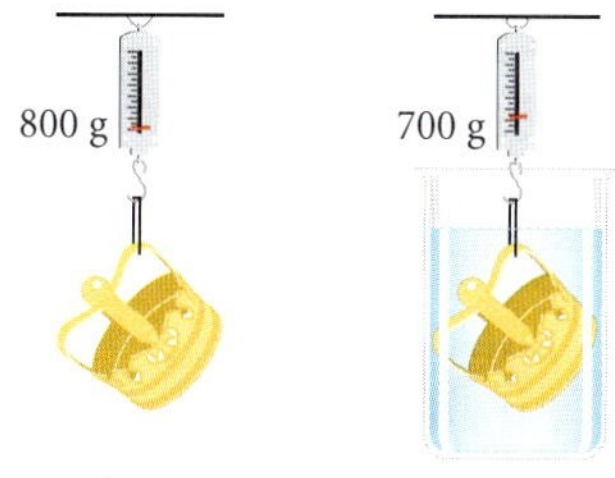

Figure 2 Weight of crown in air (left) and in water (right)

Density Determination of an Irregularly-Shaped Object

The density of a relatively large but irregularly shaped object, such as the crown shown in **Figure 2,** is most easily determined by using Archimedes' principle. In this method, the weight of an object is compared to its apparent weight when it is submerged in a fluid.

Density Determination of a Very Small Object

To determine the density of very small objects such as a tiny piece of glass or plastic, it is often more convenient and accurate to place the object in a fluid of known density to determine whether the object sinks or floats. The density of the fluid can then be adjusted by adding a second fluid that mixes with the first fluid but has a different density. The second fluid is added, dropwise, until the object being tested is suspended—neither floating nor sinking. At this point, the density of the fluid and the density of the object are the same.

Accurate results require careful recording of the number of drops of fluid added. Knowing the exact number of drops of each fluid added, and the density of each fluid, you can calculate the final volume of the mixtures using weighted averages. For this, a "weight" is given to the density of each fluid based on how much of each was present in the mixture at the point where it achieved the same density as the glass sample.

First, multiply the density of each fluid by the number of drops that were present. To compute the final density of the mixture, add the two weighted densities together and divide that by the total number of drops of fluid in the mixture.

For example, say you have glass sample that is suspended in a mixture. To make the mixture, you had added 30 drops of a fluid with a density of 2.50 g/mL, plus 5 drops of water, which has a density of about 0.998 g/mL. First, you weight the density of each fluid by multiplying it by the number of drops in your solution, and add these values together:

$$(30 \text{ drops} \times 2.50 \text{ g/mL})$$
$$+(5 \text{ drops} \times 0.998 \text{ g/mL})$$
$$= 80.0 \text{ drops} \bullet \text{g/mL}$$

To compute the density of the fluid mixture, divide this weighted total by the total number of drops:

$$\text{density} = \frac{80.0 \text{ drops} \bullet \text{g/mL}}{35 \text{ drops}} = 2.29 \text{ g/mL}$$

Because 2.29 g/mL is the final density of the mixture when the glass sample was suspended, we know 2.29 g/mL is also the density of the glass sample.

Topic Questions

1. What is the mathematical definition of density?

2. State Archimedes' principle.

3. An object placed in a fluid sinks and displaces fluid. What is the relationship between the volume of fluid displaced and the volume of the object?

4. If an object is submerged in a fluid, how is the weight of the object affected? Why?

5. Suppose that to determine the density of an object, you use the weighted-difference technique. What information is necessary for the determination to be successful? Explain.

Answers to Topic Questions

1. Mass divided by volume.

2. The buoyant force on an object in a fluid is equal to the weight of the fluid displaced by the object.

3. The volume of fluid displaced is the same as the volume of the object.

4. The weight of the object decreases. The fluid exerts a force on the object equal to the weight of the fluid displaced by the object.

5. You would need to know the density of the fluid used. The density of the fluid is used in the calculation to determine the volume of the object. The volume is required to determine the density because density is mass per unit volume.

Forensics Lab

THE PARKING LOT COLLISION

Teacher's Notes

Time Required
One 50-minute class period

Ratings
EASY → HARD

TEACHER PREPARATION 2
STUDENT SETUP 2
CONCEPT LEVEL 3
CLEANUP 2

Skills Acquired
- Designing experiments
- Experimenting
- Collecting data
- Inferring
- Interpreting
- Measuring
- Organizing and analyzing data
- Communicating

Scientific Methods
In this lab, students will
- Make Observations
- Analyze the Results
- Draw Conclusions
- Communicate the Results

Materials and Equipment
Materials for this lab can be ordered from WARD'S. See *Master Materials List* at the front of this book for catalog numbers. Additional items not listed in the student materials list are
- beaker, 1000 mL

Visit go.hrw.com and type in the keyword **Holt Forensics** to use the online Lab Materials Quicklist for the Forensics Labs.

Forensics Lab

The Parking Lot Collision

SKILLS
- Designing experiments
- Experimenting
- Collecting data
- Inferring
- Interpreting
- Measuring

OBJECTIVES
- **Determine** the density of a small glass chip by the method of suspension.

MATERIALS
- safety goggles
- lab apron
- gloves
- dropper bottles of distilled water and saturated ZnI_2 solution
- glass sample
- test tube, 2.5 mL with cap
- amber bottle for liquid disposal and recovery

ChemSafety

CAUTION: Always wear safety goggles and a lab apron to protect your eyes and clothing.

CAUTION: Do not touch or taste any chemicals. Know the location of the emergency shower and eyewash station and how to use them. If you get a chemical on your skin or clothing, wash it off at the sink while calling to the teacher. Notify the teacher of a spill. Spills should be cleaned up promptly, according to your teacher's directions.

CAUTION: Glassware is fragile. Notify the teacher of broken glass or cuts. Do not clean up broken glass or spills with broken glass unless the teacher tells you to do so.

Background

The high school campus police are following up on the report of a fender bender. A student said she was exiting the school when she saw another student back into the front of her car with a large truck, and then leave the scene. The police soon found the alleged perpetrator, who denied any involvement in the collision that left the front of the small car crushed, grill and headlights broken.

While talking to the driver of the truck, the officer noticed a shard of glass stuck in the soft rubber covering of the truck's bumper. When questioned, the driver of the truck claimed not to know the source of the glass, but announced confidently that she had not run into anyone's car. In this lab, you will use the suspension method to determine the density of the piece of glass retrieved from the bumper of the suspect's vehicle.

Procedure

PART A: Design an Experiment

1. Work with members of your lab group to design a procedure to test the density of the glass fragment in order to solve the case. If the glass is the same density as headlight glass, the incident and the suspect will require further investigation. You will use the method of density determination by suspension and with the lab materials provided. See "Forensics Lesson: Density" that precedes this lab, for hints.
2. Write a procedure for your experiment. Create a data table that clearly displays your calculations and results. Make a list of all the safety precautions you will take. Have your teacher approve your procedure and safety precautions before you begin the experiment.

Safety Cautions
Students should be supervised at all times during this investigation. Discuss all safety symbols and caution statements with students. Have students wear protective goggles, gloves, and aprons. Because glass fragments are manipulated, there is the potential for glass cuts and punctures. Students should wear appropriate protective gloves and work with caution when handling glass fragments.

Warn students to use appropriate protective gear in order not to get any ZnI_2 solution on their skin or in their eyes since it is corrosive. Accidents should immediately be reported to the instructor. Caution students to avoid inhaling vapors and use the solution only in well ventilated areas or in the hood.

Zinc iodide is a corrosive material which presents a dust hazard. Since inhalation should be avoided, preparation in the hood is recommended. Use appropriate gloves, aprons, and goggles when preparing this chemical.

Disposal
Notify janitorial staff when disposing of broken glass. Carefully consult the supplier's MSDS (Material Safety Data Sheet) for specific warnings, emergency procedures, handling instructions, and directions for proper disposal. While the initial cost of ZnI_2 might be of con-

Part B: Conduct Your Experiment

3. Put on safety goggles, protective gloves, and a lab apron.

4. **CAUTION; ZnI_2 is corrosive and is a poison. Do not taste or touch.** You will use two clear liquids of different density, saturated zinc iodide solution and distilled water. Glass densities are closer to the density of the saturated zinc iodide solution, so you will want to start by counting 25 drops of the zinc iodide solution into the test tube with the glass sample in it. Then add water dropwise, keeping track of the number of drops until suspension is achieved. Refer to **Figure 1** as a guide to achieving suspension of the glass fragment in the mixture.

5. **CAUTION: Glass samples have sharp edges. Handle them with care.** When the glass fragment is suspended, you will be able to determine the density of the mixture by calculating a weighted average from the densities of the two liquids (based on the number of drops added). If your technique and calculations are accurate, this calculated density of the mixture when the glass piece is suspended will be the same as the density of the glass piece.

Figure 1 Glass denser than solution (left); glass same density as solution (right)

TABLE 1: DENSITY VALUES OF TESTED SUBSTANCES

Substance	Characteristic density range
Water at 20°C	0.998 g/mL
Window-pane glass	2.47–2.56 g/mL
Headlight glass	2.47–2.63 g/mL
Ophthalmic glass	2.65–2.81 g/mL
Saturated ZnI_2 solution	2.73 g/mL

Density values you will need are listed in **Table 1.** If the suspect glass fragment in this case is determined to be headlight glass, investigation into the incident will continue.

Part C: Cleanup and Disposal

6. Dispose of paper and broken glass in the designated waste containers. Do not put lab materials in the trash unless your teacher tells you to do so.

7. Clean up your work area and all lab equipment. Return lab equipment to its proper place. Wash your hands thoroughly before you leave the lab.

Analyze and Conclude

1. What is the density of the glass sample? Show how you calculated your result.

2. What statement can you make based on the density of the glass sample?

3. What factors could affect the accuracy of your results? How could your procedure have been improved in order to control for these factors?

4. How would you interpret a value of 2.50 g/mL for the suspect glass chip?

1069

cern, each lab group will use at most a few milliliters of saturated stock solution, and it is possible to reuse the chemical simply by allowing the water to evaporate and then re-dissolving the recovered solid into a saturated stock solution. This should be done in a hood using appropriate chemical safety techniques. If disposal becomes necessary at a later time, follow MSDS disposal procedures.

Although the glass will not float on the liquid's surface, it can easily be seen to migrate to the top of the liquid. If headlight glass is used as the suspect sample, the glass density (approximately 2.60 g/mL) will be less than the density of the ZnI_2 solution (2.73 g/mL). Therefore, the glass will move upward in the solution. At that point, add only one drop of distilled water to the test tube, place the cap on the test tube, thoroughly mix the liquids, and again observe the movement of the glass chip in the mixture. The goal is to have the glass chip neither float on the top of the liquid nor sink to the bottom, but rather remain suspended in the middle of the liquid.

When the chip is suspended, the density values of the chip and liquid mixture are equal. If the glass moves toward the solution's surface, it is less dense than the solution, in which case the liquid must be made less dense by adding a

(continued on p. 1147)

Tips and Tricks

The usual method of determining density by dividing the mass of an object by its volume is not possible in this instance because of the very small mass and also the irregular shape of the glass chip to be tested. The method of suspension, sometimes termed the method of flotation, is a practical alternative. The density of the chip matches the density of the solution at the point where a glass chip neither sinks nor floats in the solution in which it is immersed. It is then possible to find the density of the solution using weighted averages as shown in the sample data on page 1047.

While this technique of determining density is the one most frequently used in crime labs, it can also be the most expensive. Bromoform (d = 2.89 g/mL) and bromobenzene (d = 1.49 g/mL) are both clear liquids and are mutually miscible. These properties make them the preferred choice for forensic density analysis, but they are hazardous, costly, and difficult to dispose of. The materials used in this activity are a much safer alternative.

To prepare 100 mL of ZnI_2 stock solution, dissolve 221 g of ZnI_2 in 50 mL of distilled and deionized water. Then add a small piece of mossy zinc to the solution to prevent oxidation of the iodide to iodine. Before using the solution, check the value of the stock solution density by dividing its mass by its volume. Because the solution will be slightly yellow, the analysis is best performed in direct sunlight so that the glass chip may be seen more easily. Upon completion of the lab, place all the liquids into a large beaker and simply allow the water to evaporate until the stock solution is again saturated. Check the density of the stock solution before reusing. The ZnI_2 solution is light sensitive, so store it in an amber bottle.

Forensics Lab

The Sports Shop Theft

Teacher's Notes

Time Required
One 50-minute class period

Ratings
EASY → HARD

Teacher Preparation 2
Student Setup 3
Concept Level 3
Cleanup 2

Skills Acquired
- Designing experiments
- Experimenting
- Collecting data
- Inferring
- Interpreting
- Measuring
- Organizing and analyzing data
- Communicating

Scientific Methods
In this lab, students will
- Make Observations
- Analyze the Results
- Draw Conclusions
- Communicate the Results

Materials and Equipment
Materials for this lab can be ordered from WARD'S. See *Master Materials List* at the front of this book for catalog numbers.

Visit go.hrw.com and type in the keyword **Holt Forensics** to use the online Lab Materials Quicklist for the Forensics Labs.

Forensics Lab

The Sports Shop Theft

SKILLS
- Designing experiments
- Experimenting
- Collecting data
- Inferring
- Interpreting
- Measuring
- Organizing and analyzing data
- Communicating

OBJECTIVES
- **Determine** the comparative densities of two small glass chips by the method of suspension in a gradient column.

MATERIALS
- safety goggles
- lab apron
- gloves
- dropper bottles of distilled water (d = 0.998 g/mL) and saturated ZnI2 solution (d = 2.73 g/ml)
- size 0 cork stoppers
- markers
- crime scene and suspect glass samples
- glass tubing, about 25 cm in length, 1/4 in. inside diameter
- ring stand
- clamp

- graduated cylinder, 10 mL
- pipette
- small bottle to support the glass tubing
- amber bottle, for liquid disposal and recovery

ChemSafety

CAUTION: Always wear safety goggles and a lab apron to protect your eyes and clothing.

CAUTION: Do not touch or taste any chemicals. Know the location of the emergency shower and eyewash station and how to use them. If you get a chemical on your skin or clothing, wash it off at the sink while calling to the teacher. Notify the teacher of a spill. Spills should be cleaned up promptly, according to your teacher's directions.

CAUTION: Glassware is fragile. Notify the teacher of broken glass or cuts. Do not clean up broken glass or spills with broken glass unless the teacher tells you to do so.

Background
The police are investigating the burglary of a sporting goods store. The thief broke a pane of glass in the door to gain access and headed straight for the store's valuable collection of jerseys and collectors-edition posters autographed by popular ball players. The investigating officer suspected that the job was planned—that the thief knew the layout of the store and where to find the most valuable items. The owner told the police that he had had to fire a young man the previous week. When the police paid a visit to the former employee, they asked to inspect the shoes in his closet. The employee denied that he had stolen any shoes, but that was not what interested the investigating officer. She found several shards of glass in one pair of sports shoes, and she asked where it came from. The former employee—now a suspect—said that he had accidentally stepped on his own eyeglasses, breaking the lenses.

In this lab, you will use the density column gradient method to compare the density of the glass from the shoes with the density of both ophthalmic glass (used for lenses) and of windowpane glass.

1070

Safety Cautions
Students should be supervised at all times during this investigation. Discuss all safety symbols and caution statements with students. Have students wear protective goggles, gloves, and aprons. Because glass fragments are manipulated, there is the potential for glass cuts and punctures. Students should wear appropriate protective gloves and work with caution when handling glass fragments.

Warn students to use appropriate protective gear in order not to get any ZnI_2 solution on their skin or in their eyes since it is corrosive. Accidents should immediately be reported to the instructor. Caution students to avoid inhaling vapors and use the solution only in well ventilated areas or in the hood.

Zinc iodide is a corrosive material which presents a dust hazard. Since inhalation should be avoided, preparation in the hood is recommended. Use appropriate gloves, aprons, and goggles when preparing this chemical.

Disposal
Notify janitorial staff when disposing of broken glass. Carefully consult the supplier's MSDS (Material Safety Data Sheet) for specific warnings, emergency procedures, handling instructions, and directions for proper disposal.

Procedure

PART A: Design an Experiment

1. Work with members of your lab group to design a procedure to test the comparative densities of the glass fragments in order to solve the case, using the method of density determination using a gradient column and with the lab materials provided. See "Forensics Lesson: Density" earlier in this lab appendix, for hints.
2. Write a procedure for your experiment. Create a data table that clearly displays your calculations and results. Make a list of all the safety precautions you will take. Have your teacher approve your procedure and safety precautions before you begin the experiment.

PART B: Conduct Your Experiment

3. Put on safety goggles, protective gloves, and a lab apron.
4. **CAUTION: Glass samples have sharp edges.** Handle them with care. You will use the density gradient column method for this analysis. This method uses glass tubing as a column, which will contain several layers of liquid, each of a different density. The densest layer will be found at the bottom and the least dense layer at the column's top. When an object is placed into a liquid and the substance neither sinks nor floats, but instead remains suspended, the density of the object and the liquid are the same. When dropped into the column, the suspect and crime scene glass samples will be suspended in the layers of the column that match their densities. If the two samples stop in the same density layer, that would indicate that they have the same density.
5. Your teacher will discuss with you the nature of the liquids to be used and how many layers your column should contain. See **Figure 1** for a diagram of the suspension column setup.

Figure 1 Suspension column

6. **CAUTION; ZnI_2 is corrosive and is a poison. Do not taste or touch.** Different concentrations of ZnI_2 are commonly used in the column. After you prepare each layer in a graduated cylinder, pour the layer carefully down the side of the column so that the different layers are not mixed. Mark the top of each layer with a marker. Density values you will need are listed in **Table 1** in the lab "The Parking Lot Collision" that precedes this one.

PART C: Cleanup and Disposal

7. Dispose of paper and broken glass in the designated waste containers. Do not put lab materials in the trash unless your teacher tells you to do so.
8. Clean up your work area and all lab equipment. Return lab equipment to its proper place. Wash your hands thoroughly before you leave the lab.

Analyze and Conclude

1. What is the density of the glass sample? the crime scene sample?
2. What can you conclude about the case based on the density of the suspect glass sample as compared with the density of the crime scene sample?
3. What factors could affect the accuracy of your results? How could you have controlled for these factors?

While the initial cost of ZnI_2 might be of concern, each lab group will use at most a few milliliters of saturated stock solution, and it is possible to reuse the chemical simply by allowing the water to evaporate and then re-dissolving the recovered solid into a saturated stock solution. This should be done in a hood using appropriate chemical safety techniques. If disposal becomes necessary at a later time, follow MSDS disposal procedures.

The use of a calibrated pipette can greatly increase accuracy because of the small volumes of water required to make layers of specific density. If using burettes instead of glass tubing, larger layer volumes will be required.

Simply use multiples of the "recipe" for the 1.5 mL layers in the table immediately before this paragraph. If a greater number of layers is desired, amounts of each liquid required may be determined by substituting desired d values into the equation for x in the Sample Data on p. 1147.

An interesting calibration of the layers may be accomplished by simply dropping crystals of known density into the column. They should migrate to the layer matching their density. Materials appropriate for this approach are as follows (many of these chemicals are hazardous, so MSDS (Materials Safety Data

(continued on p. 1147)

Tips and Tricks

To prepare 100 mL of ZnI_2 stock solution, dissolve 221 g of ZnI_2 in 50 mL of distilled and deionized water. Then add a small piece of mossy zinc to the solution to prevent oxidation of the iodide to iodine. Before using the solution, check the value of the stock solution density by dividing its mass by its volume. Because the solution will be slightly yellow, the analysis is best performed in direct sunlight so that the glass chip may be seen more easily. Upon completion of the lab, place all the liquids into a large beaker and simply allow the water to evaporate until the stock solution is again saturated. Check the density of the stock solution before reusing. The ZnI_2 solution is light sensitive, so store it in an amber bottle.

Students will most likely require assistance in deriving an equation for determining the density of each layer in the column. It is easy to change the outcome of the case by varying the type of suspect and crime scene glass chips given to each lab group or class. Windowpane, headlight, ophthalmic lens, and other types of glass are easily obtained, stored, and reused. The drop size obtained from regular droppers is relatively consistent, but calibration and use of a pipette would increase accuracy in the preparation of the layers in the column. The 25 cm lengths of glass tubing can accommodate five different 1.5 mL layers, so the 1.5 mL layer volume should be suggested to the students. Layer densities and liquid volumes would be as follows:

Density (g/mL)	mL ZnI_2	mL H_2O
2.50	1.30	0.20
2.55	1.34	0.16
2.60	1.39	0.11
2.65	1.43	0.07
2.70	1.47	0.03

Pronunciation Key

Sound	As In	Phonetic Respelling
ahy	bat	(BAT)
ay	face	(FAYS)
ah	lock argue	(LAHK) (AHR gyoo)
ow	out	(OWT)
ch	chapel	(CHAP uhl)
eh	test	(TEHST)
ai	rare	(RAIR)
ee	eat feet ski	(EET) (FEET) (SKEE)
ih	bit	(BIHT)
ie	idea	(ie DEE uh)
y	ripe	(RYP)

Sound	As In	Phonetic Respelling
ihng	going	(GOH ihng)
k	card kite	(KAHRD) (KEYET)
ng	anger	(ANG guhr)
oh	over	(OH vuhr)
aw	dog horn	(DAWG) (HAWRN)
oy	foil	(FOYL)
u	pull	(PUL)
oo	pool	(POOL)
s	cell sit	(SEHL) (SIHT)
sh	sheep	(SHEEP)

Sound	As In	Phonetic Respelling
th	that thin	(THAT) (THIHN)
uh	cut	(CUHT)
ur	fern	(FURN)
y	yes	(YEHS)
yoo	globule	(GLAHB yool)
yu	cure	(KYUR)
z	bags	(BAGZ)
zh	treasure	(TREHZH uhr)
uh	medal pencil onion	(MEHD uhl) (PEHN suhl) (UHN yuhn)
uhr	paper	(PAY puhr)

A

abiotic factor an environmental factor that is not associated with the activities of living organisms (340)

ABO blood group system a system used to classify human blood by antigens found on the surface of red blood cells (878)

acanthodian an early fish; the earliest known vertebrate to have jaws (715)

acid any compound that increases the number of hydronium ions when dissolved in water; acids turn blue litmus paper red and react with bases and some metals to form salts (33)

acid rain precipitation that has a pH below normal and has an unusually high concentration of sulfuric or nitric acids, often as a result of chemical pollution of the air from sources such as automobile exhausts and the burning of fossil fuels (386)

acoelomate an animal that lacks a coelom, or body cavity (600)

actin a protein responsible for the contraction and relaxation of muscle (857)

action potential a sudden change in the polarity of the membrane of a neuron, gland cell, or muscle fiber that facilitates the transmission of electrical impulses (946)

activation energy the minimum amount of energy required to start a chemical reaction (39)

active site the site on an enzyme that attaches to a substrate (41)

active transport the movement of chemical substances, usually across the cell membrane, against a concentration gradient; requires cells to use energy (81)

adaptation the process of becoming adapted to an environment; an anatomical, physiological, or behavioral change that improves a population's ability to survive (279)

addiction a physiological or psychological dependence on a substance, such as alcohol or drugs (962)

adductor muscle the thick muscle that joins the two valves in mollusks and that causes the shell to open (647)

adhesion the attractive force between two bodies of different substances that are in contact with each other (31)

adrenal gland one of the two endocrine glands located above each kidney (985)

aerobic describes a process that requires oxygen (104)

aggregation a grouping of cells or other organisms (418)

agnathan a member of a class of primitive, jawless fishes (714)

AIDS acquired immune deficiency syndrome, a disease caused by HIV, an infection that results in an ineffective immune system (934)

allele one of the alternative forms of a gene that governs a characteristic, such as hair color (167)

allergy a physical response to an antigen, which can be a common substance that produces little or no response in the general population (936)

alternation of generations the alternation of sexual reproduction and asexual reproduction in certain plants and animals (463)

alveolus any of the tiny air cells of the lungs where oxygen and carbon dioxide are exchanged (886)

amino acid any one of 20 different organic molecules that contain a carboxyl and an amino group and that combine to form proteins (36)

amino acid–based hormone a hormone that is made up of simple amino acids, peptides, or proteins (977)

amniotic egg a type of egg that is produced by reptiles, birds, and egg-laying mammals and that contains a large amount of yolk; usually surrounded by a leathery or hard shell within which the embryo and its embryonic membranes develop (775)

amylase an enzyme that breaks down starches into sugars (907)

anaerobic describes a process that does not require oxygen (104)

anemia a condition in which the oxygen-carrying ability of red blood cells is reduced and the production of red blood cells decreases (877)

angiosperm a flowering plant that produces seeds within a fruit (514)

annual a plant that completes its life cycle, reproduces, and dies within one growing season (573)

annual ring in secondary xylem (wood), the growth ring formed in one season (575)

anther in flowering plants, the tip of a stamen, which contains the pollen sacs where grains form (538)

antheridium a reproductive structure that produces male sex cells in flowerless and seedless plants (530)

antibiotic a substance that can inhibit the growth of or kill some microorganisms (443)

antibody a protein that reacts to a specific antigen or that inactivates or destroys toxins (929)

anticodon a region of tRNA that consists of three bases complementary to the codon of mRNA (212)

antigen a substance that stimulates an immune response (927)

antigen shifting the production of new antigens by a virus as it mutates over time (932)

aorta the main artery in the body; it carries blood from the left ventricle to systemic circulation (882)

apical dominance the inhibition of lateral bud growth on the stem of a plant by auxin produced in the terminal bud (581)

apical meristem the growing region at the tips of stems and roots in plants (574)

appendage a structure that extends from the main body, such as a limb, tentacle, fin, or wing (664)

appendicular skeleton the bones of the arms and legs (850)

aquifer a porous rock that stores and allows the flow of groundwater (393)

Archaebacteria a classification kingdom made up of bacteria that live in extreme environments; differentiated from other prokaryotes by various important chemical differences (258)

archegonium a female reproductive structure of small, nonvascular plants that produces a single egg and in which fertilization and development take place (530)

artery a blood vessel that carries blood away from the heart to the body's organs (873)

arthropod a member of the phylum Arthropoda, which includes invertebrate animals such as insects, crustaceans, and arachnids; characterized by having segmented bodies and paired appendages (266)

ascus the spore sac where ascomycetes produce ascospores (487)

asexual reproduction reproduction that does not involve the union of gametes and in which a single parent produces offspring that are genetically identical to the parent (150)

asymmetrical irregular in shape; without symmetry (598)

atom the smallest unit of an element that maintains the properties of that element (28)

ATP adenosine triphosphate, an organic molecule that acts as the main energy source for cell processes; composed of a nitrogenous base, a sugar, and three phosphate groups (37)

atrium a chamber that receives blood that is returning to the heart (881)

autoimmune disease a disease in which the immune system attacks the organism's own cells (933)

autosome any chromosome that is not a sex chromosome (122)

autotroph an organism that produces its own nutrients from inorganic substances or from the environment instead of consuming other organisms (94)

auxin a plant hormone that regulates cell elongation (580)

axial skeleton the bones of the skull and vertebral column (850)

axon an elongated extension of a neuron that carries impulses away from the cell body (944)

B

B cell a white blood cell that matures in bones and makes antibodies (927)

bacillus a rod-shaped bacterium (443)

bacteriophage a virus that infects bacteria (192)

basal disk an area on cnidarians, such as hydras, jellyfish, and corals, that enables them to adhere to surfaces (624)

base any compound that increases the number of hydroxide ions when dissolved in water; bases turn red litmus paper blue and react with acids to form salts (33)

base pairing rules the rules stating that cytosine pairs with guanine and adenine pairs with thymine in DNA, and that adenine pairs with uracil in RNA (197)

basidium a structure that produces asexual spores in basidiomycetes (488)

behavior an action that an individual carries out in response to a stimulus or to the environment (824)

biennial a plant that has a two-year life cycle (573)

bilateral symmetry a condition in which two equal halves of a body mirror each other (598)

binary fission a form of asexual reproduction in single-celled organisms by which one cell divides into two cells of the same size (119)

binomial nomenclature a system for giving each organism a two-word scientific name that consists of the genus name followed by the species name (300)

biodiversity the number and variety of organisms in a given area during a specific period of time (341)

biogeochemical cycle the circulation of substances through living organisms from or to the environment (350)

biological magnification the accumulation of increasingly large amounts of toxic substances within each successive link of the food chain (391)

biological species a group of organisms than can reproduce only among themselves and that are usually contained in a geographic region (305)

biology the scientific study of living organisms and their interactions with the environment (6)

biomass organic matter that can be a source of energy; the total mass of the organisms in a given area (349)

biome a large region characterized by a specific type of climate and certain types of plant and animal communities (372)

biotic factor an environmental factor that is associated with or results from the activities of living organisms (340)

blastocyst the modified blastula stage of mammalian embryos (1004)

blastopore an opening that develops in the blastula (692)

blastula the stage of an embryo before gastrulation (596)

blood pressure the force that blood exerts on the walls of the arteries (882)

body cavity any cavity that houses organs, such as the thoracic, abdominal, or pelvic cavity (849)

body plan an animal's shape, symmetry, and internal organization (598)

bone marrow soft tissue inside bones where red and white blood cells are produced (851)

brain the mass of nerve tissue that is the main control center of the nervous system (950)

brain stem the stemlike portion of the brain that connects the cerebral hemispheres with the spinal cord and that maintains the necessary functions of the body, such as breathing and circulation (951)

bronchus one of the two tubes that connect the lungs with the trachea (886)

budding asexual reproduction in which a part of the parent organism pinches off and forms a new organism (487)

bulbourethral gland one of the two glands in the male reproductive system that add fluid to the semen during ejaculation (998)

C

calorie the amount of energy needed to raise the temperature of 1 g of water 1°C; the Calorie used to indicate the energy content of food is a kilocalorie (900)

Calvin cycle a biochemical pathway of photosynthesis in which carbon dioxide is converted into glucose using ATP (102)

cancer a tumor in which the cells begin dividing at an uncontrolled rate and become invasive (12)

capillary a tiny blood vessel that allows an exchange between blood and cells in tissue (873)

capsid a protein sheath that surrounds the nucleic acid core in a virus (435)

capsule in mosses, the part that contains spores; in bacteria, a protective layer of polysaccharides around the cell wall (443)

carapace a shieldlike plate that covers the cephalothorax of some crustaceans (782)

carbohydrate any organic compound that is made of carbon, hydrogen, and oxygen and that provides nutrients to the cells of living things (34)

carbon fixation the synthesis of organic compounds from carbon dioxide, such as in photosynthesis (102)

carnivore an animal that eats other animals (346)

carotenoid a class of pigments that are present mostly in plants and that aid in photosynthesis (98)

carrier protein a protein that transports substances across a cell membrane (80)

carrying capacity the largest population that an environment can support at any given time (322)

cartilage a flexible and strong connective tissue (715)

caste a group of insects in a colony that have a specific function (679)

cell in biology, the smallest unit that can perform all life processes; cells are covered by a membrane and have a nucleus and cytoplasm (7)

cell cycle the life cycle of a cell; in eukaryotes, it consists of a cell-growth period in which DNA is synthesized and a cell-division period in which mitosis takes place (125)

cell membrane a phospholipid layer that covers a cell's surface and acts as a barrier between the inside of a cell and the cell's environment (56)

cell theory the theory that states that all living things are made up of cells, that cells are the basic units of organisms, that each cell in a multicellular organism has a specific job, and that cells come only from existing cells (55)

cell wall a rigid structure that surrounds the cell membrane and provides support to the cell (57)

cellular respiration the process by which cells produce energy from carbohydrates; atmospheric oxygen combines with glucose to form water and carbon dioxide (95)

central nervous system the brain and the spinal cord; its main function is to control the flow of information in the body (950)

central vacuole a large cavity or sac that is found in plant cells or protozoans and that contains air or partially digested food (66)

centromere the region of the chromosome that holds the two sister chromatids together during mitosis (119)

cephalization the concentration of nerve tissue and sensory organs at the anterior end of an organism (599)

cephalothorax in arachnids and some crustaceans, the body part made up of the head and the thorax (666)

cereal any grass that produces grains that can be used for food, such as rice, wheat, corn, oats, or barley (518)

cerebellum a posterior portion of the brain that coordinates muscle movement and controls subconscious activities and some balance functions (951)

cerebral ganglion one of a pair of nerve-cell clusters that serve as a primitive brain at the anterior end of some invertebrates, such as annelids (651)

cerebrum the upper part of the brain that receives sensation and controls movement (951)

chelicera in arachnids, either of a pair of appendages used to attack prey (670)

chitin a carbohydrate that forms part of the exoskeleton of arthropods and other organisms, such as insects, crustaceans, fungi, and some algae (482)

chlamydia a bacterial sexually transmitted disease marked by painful urination and vaginal discharge (1009)

chlorofluorocarbons hydrocarbons in which some or all of the hydrogen atoms are replaced by chlorine and fluorine; used in coolants for refrigerators and air conditioners and in cleaning solvents; their use is restricted because they destroy the ozone (abbreviation, CFC) (387)

chlorophyll a green pigment that is present in most plant cells, that gives plants their characteristic green color, and that reacts with sunlight, carbon dioxide, and water to form carbohydrates (98)

chloroplast an organelle found in plant and algae cells where photosynthesis occurs (66)

choanocyte any of the flagellate cells that line the cavities of a sponge (619)

chordate an animal that at some stage in its life cycle has a dorsal nerve, a notochord, and pharyngeal pouches; examples include mammals, birds, reptiles, amphibians, fish, and some marine lower forms (700)

chromatid one of the two strands of a chromosome that become visible during meiosis or mitosis (119)

chromosome in a eukaryotic cell, one of the structures in the nucleus that are made up of DNA and protein; in a prokaryotic cell, the main ring of DNA (119)

chrysalis the hard-shelled pupa of certain insects, such as butterflies (675)

cilium a hairlike structure arranged in tightly packed rows that projects from the surface of some cells (58)

cladistics a phylogenetic classification system that uses shared derived characters and ancestry as the sole criterion for grouping taxa (327)

cladogram a diagram that is based on patterns of shared, derived traits and that shows the evolutionary relationships between groups of organisms (309)

class a taxonomic category containing orders with common characteristics (302)

cleavage in biological development, a series of cell divisions that occur immediately after an egg is fertilized (1004)

climate the average weather conditions in an area over a long period of time (371)

clone an organism that is produced by asexual reproduction and that is genetically identical to its parent; to make a genetic duplicate (150)

closed circulatory system a circulatory system in which the heart circulates blood through a network of vessels that form a closed loop; the blood does not leave the blood vessels, and materials diffuse across the walls of the vessels (606)

cnidocyte a stinging cell of a cnidarian (623)

coccus a sphere-shaped bacterium (443)

cochlea a coiled tube that is found in the inner ear and that is essential to hearing (959)

codominance a condition in which both alleles for a gene are fully expressed (178)

codon in DNA, a three-nucleotide sequence that encodes an amino acid or signifies a start signal or a stop signal (211)

coelom a body cavity that contains the internal organs (600)

coelomate an animal that has a body cavity in which the internal organs are located (600)

coevolution the process in which long-term, interdependent changes take place in two species as a result of their interactions (362)

cohesion the force that holds molecules of a single material together (31)

colon a section of the large intestine (910)

colonial organism a collection of genetically identical cells that are permanently associated but in which little or no integration of cell activities occurs (418)

commensalism a relationship between two organisms in which one organism benefits and the other is unaffected (364)

community a group of species that live in the same habitat and interact with each other (340)

competition the relationship between species that attempt to use the same limited resource (365)

competitive exclusion the exclusion of one species by another due to competition (369)

complement system a system of proteins that circulate in the bloodstream and that combine with antibodies to protect against antigens (926)

complementary base pairing a characteristic of nucleic acids in which the sequence of bases on one strand is paired to the sequence of bases on the other (197)

compound a substance made up of atoms of two or more different elements joined by chemical bonds (29)

compound eye an eye composed of many light detectors separated by pigment cells (666)

concentration gradient a difference in the concentration of a substance across a distance (74)

conditioning the process of learning by association (827)

cone in animals, a photoreceptor within the retina that can distinguish colors and is very sensitive to bright light (958)

cone in plants, a seed-bearing structure (511)

connective tissue a tissue that has a lot of intracellular substance and that connects and supports other tissues (847)

consumer an organism that eats other organisms or organic matter instead of producing its own nutrients or obtaining nutrients from inorganic sources (345)

continental drift the hypothesis that states that the continents once formed a single landmass, broke up, and drifted to their present locations (268)

contour feather one of the most external feathers that cover a bird and that help determine its shape (784)

control group in an experiment, a group that serves as a standard of comparison with another group to which the control group is identical except for one factor (17)

convergent evolution the process by which unrelated species become more similar as they adapt to the same kind of environment (307)

cork the outer layer of bark of any woody plant (553)

cork cambium a layer of tissue under the cork layer where cork cells are produced (575)

coronary artery one of the two arteries that supply blood directly to the heart (882)

corpus luteum the structure that forms from the ruptured follicle in the ovary after ovulation; it releases hormones (1002)

cortex in plants, the primary tissue located in the epidermis; in animals, the outermost portion of an organ (555)

cotyledon the embryonic leaf of a seed (535)

countercurrent flow in fish gills, an arrangement whereby water flows away from the head and blood flows toward the head (747)

crossing-over the exchange of genetic material between homologous chromosomes during meiosis; can result in genetic recombination (144)

cud partly digested food that is regurgitated, rechewed, and reswallowed for further digestion by mammals that have a rumen (814)

cuticle a waxy or fatty and watertight layer on the external wall of epidermal cells (502)

cyanobacteria a bacterium that can carry out photosynthesis, such as a blue-green alga (258)

cystic fibrosis a fatal genetic disorder in which excessive amounts of mucus are secreted, blocking intestinal and bronchial ducts and causing difficulty in breathing (12)

cytokinesis the division of the cytoplasma of a cell; cytokinesis follows the division of the cell's nucleus by mitosis or meiosis (125)

cytoplasm the region of the cell within the membrane that includes the fluid, the cytoskeleton, and all of the organelles except the nucleus (56)

cytoskeleton the cytoplasmic network of protein filaments that plays an essential role in cell movement, shape, and division (56)

cytotoxic T cell a type of T cell that recognizes and destroys cells infected by virus (927)

D

decomposer an organism that feeds by breaking down organic matter from dead organisms; examples include bacteria and fungi (347)

dendrite a cytoplasmic extension of a neuron that receives stimuli (944)

density-dependent factor a variable affected by the number of organisms present in a given area (322)

density-independent factor a variable that affects a population regardless of the population density, such as climate (324)

deoxyribose a five-carbon sugar that is a component of DNA nucleotides (194)

dependent variable in an experiment, the variable that is changed or determined by manipulation of one or more factors (the independent variables) (17)

depressant a drug that reduces functional activity and produces muscular relaxation (964)

derived character a unique characteristic of a particular group of organisms (307)

dermal tissue the outer covering of a plant (552)

dermis the layer of skin below the epidermis (862)

detritivore a consumer that feeds on dead plants and animals (346)

deuterostome an animal whose mouth does not derive from the blastopore and whose embryo has indeterminate cleavage (692)

diabetes mellitus a serious disorder in which cells are unable to obtain glucose from the blood; caused by a deficiency of insulin or lack of response to insulin (987)

diaphragm a dome-shaped muscle that is attached to the lower ribs and that functions as the main muscle in respiration (886)

diatom a unicellular alga that has a double shell that contains silica (466)

dicot a dicotyledonous plant; an angiosperm that has two cotyledons, net venation, and flower parts in groups of four or five (515)

differentiation the process in which the structure and function of the parts of an organism change to enable specialization of those parts (419)

diffusion the movement of particles from regions of higher density to regions of lower density (75)

digestion the breaking down of food into chemical substances that can be used for energy (900)

diploid a cell that contains two haploid sets of chromosomes (121)

directional selection a natural selection process in which one genetic variation is selected and that causes a change in the overall genetic composition of the population (332)

dispersion in optics, the process of separating a wave (such as white light) of different frequencies into its individual component waves (the different colors) (321)

diurnal describes animals that are active during the day and sleep at night (732)

DNA deoxyribonucleic acid, the material that contains the information that determines inherited characteristics (37)

DNA fingerprint the pattern of bands that results when an individual's DNA fragments are separated (237)

DNA helicase an enzyme that unwinds the DNA double helix during DNA replication (198)

DNA polymerase an enzyme that catalyzes the formation of the DNA molecule (199)

DNA replication the process of making a copy of DNA (198)

dormancy a state in which seeds, spores, bulbs, and other reproductive organs stop growth and development and reduce their metabolism, especially respiration (584)

double fertilization the process by which one of the two sperm nuclei fuses with the egg nucleus to produce a diploid zygote and the other fuses with the polar nuclei to produce a triploid endosperm (540)

double helix the spiral-staircase structure characteristic of the DNA molecule (194)

down feather a soft feather that covers the body of young birds and provides insulation to adult birds (785)

E

ecology the study of the interactions of living organisms with one another and with their environment (9)

ecosystem a community of organisms and their abiotic environment (340)

ectoderm the outermost of the three germ layers of an embryo that develops into the epidermis and epidermal tissues, the nervous system, external sense organs, and the mucous membranes lining the mouth and anus (596)

ectothermic describes the ability of an organism to maintain its body temperature by gaining heat from the environment (724)

electron microscope a microscope that focuses a beam of electrons to magnify objects (51)

electron transport chain a series of molecules, found in the inner membranes of mitochondria and chloroplasts, through which electrons pass in a process that causes protons to build up on one side of the membrane (100)

electrophoresis the process by which electrically charged particles suspended in a liquid move through the liquid because of the influence of an electric field (231)

element a substance that cannot be separated or broken down into simpler substances by chemical means; all atoms of an element have the same atomic number (28)

embryo an organism in an early stage of development of plants and animals; in humans, a developing individual is referred to as an embryo from the second through the eighth week of pregnancy (504)

endocrine gland a ductless gland that secretes hormones into the blood (975)

endocytosis the process by which a cell membrane surrounds a particle and encloses the particle in a vesicle to bring the particle into the cell (83)

endoderm the innermost germ layer of the animal embryo; develops into the epithelium of the pharynx, respiratory tract, digestive tract, bladder, and urethra (596)

endoplasmic reticulum a system of membranes that is found in a cell's cytoplasm and that assists in the production, processing, and transport of proteins and in the production of lipids (63)

endosperm a triploid (3n) tissue that develops in the seeds of angiosperms and that provides food for a developing embryo (514)

endospore a thick-walled protective spore that forms inside a bacterial cell and resists harsh conditions (443)

endosymbiosis a mutually beneficial relationship in which one organism lives within another (259)

endothermic describes the ability of an organism to maintain body temperature by producing heat internally (724)

energy the capacity to do work (38)

energy pyramid a triangular diagram that shows an ecosystem's loss of energy, which results as energy passes through the ecosystem's food chain; each row in the pyramid represents a trophic (feeding) level in an ecosystem, and the area of a row represents food chain the pathway of energy transfer through various stages as a result of the feeding patterns of a series of organisms (346)

envelope a membranelike layer that covers the capsids of some viruses (435)

enzyme a type of protein that speeds up metabolic reactions in plant and animals without being permanently changed or destroyed (40)

epidermis the outer surface layer of cells of a plant or animal (553)

epididymis the long, coiled tube that is on the surface of a testis and in which sperm mature (997)

epinephrine a hormone that is released by the adrenal medulla and that rapidly stimulates the metabolism in emergencies, decreases insulin secretion, and stimulates pulse and blood pressure; also called adrenaline (985)

epithelial tissue a tissue composed of cells that form a barrier between an organism and its external environment (846)

equilibrium in chemistry, the state in which a chemical reaction and the reverse chemical reaction occur at the same rate such that the concentrations of reactants and products do not change (74)

esophagus a long, straight tube that connects the pharynx to the stomach (907)

Eubacteria a classification kingdom that contains all prokaryotes except archaebacteria (258)

eukaryote an organism made up of cells that have a nucleus enclosed by a membrane, multiple chromosomes, and a mitotic cycle; eukaryotes include animals, plants, and fungi but not bacteria or cyanobacteria (58)

evolution a change in the characteristics of a population from one generation to the next; the gradual development of organisms from other organisms since the beginnings of life (9)

excretion the process of eliminating metabolic wastes (912)

exocytosis the process by which a substance is released from the cell through a vesicle that transports the substance to the cell surface and then fuses with the membrane to let the substance out (83)

exon the portion of the DNA sequence in a gene that contains the sequence of amino acids in a chain and the beginning and the end of a coding sequence (218)

exoskeleton a hard, external, supporting structure that develops from the ectoderm (607)

experiment a procedure that is carried out under controlled conditions to discover, demonstrate, or test a fact, theory, or general truth (17)

extensor a muscle that extends a joint (856)

external fertilization the union of gametes outside the bodies of the parents, as in many fishes and amphibians (610)

extinct describes a species of organisms that has died out completely (282)

F

F_1 generation the first generation of offspring obtained from an experimental cross of two organisms (164)

F_2 generation the second generation of offspring, obtained from an experimental cross of two organisms; the offspring of the F_1 generation (164)

facilitated diffusion the transport of substances through a cell membrane along a concentration gradient with the aid of carrier proteins (80)

fallopian tube a tube through which eggs move from the ovary to the uterus (1000)

family the taxonomic category below the order and above the genus (302)

fermentation the breakdown of carbohydrates by enzymes, bacteria, yeasts, or mold in the absence of oxygen (108)

fertilization the union of a male and female gamete to form a zygote (153)

fetus a developing human from seven or eight weeks after fertilization until birth (1006)

fixed action pattern behavior an innate behavior that is characteristic of certain species (826)

flagellum a long, hairlike structure that grows out of a cell and enables the cell to move (57)

flexor a muscle that bends a limb or other body part (856)

flower the reproductive structure of a flowering plant that usually consists of a pistil, stamens, petals, and sepals (505)

fluke a parasitic flatworm of the class Trematoda (632)

follicle a small, narrow cavity or sac in an organ or tissue, such as the ones on the skin that contain hair roots or the ones in the ovaries that contain the developing eggs (1001)

food chain the pathway of energy transfer through various stages as a result of the feeding patterns of a series of organisms (346)

food web a diagram that shows the feeding relationships between organisms in an ecosystem (347)

foot an appendage that some invertebrates use to move; the lower part of a vertebrate's leg (431)

fossil the trace or remains of an organism that lived long ago, preserved in sedimentary rock (258)

frond the leaf of a fern or palm (511)

fruit a mature plant ovary; the plant organ in which the seeds are enclosed (514)

fundamental niche the largest ecological niche where an organism or species can live without competition (365)

G

gamete a haploid reproductive cell that unites with another haploid reproductive cell to form a zygote (118)

gametophyte in alternation of generations, the phase in which gametes are formed; a haploid individual that produces gametes (154)

gastrovascular cavity a cavity that serves both digestive and circulatory purposes in some cnidarians (605)

gemmule an asexual reproductive structure produced by some freshwater sponges (621)

gene a segment of DNA that is located in a chromosome and that codes for a specific hereditary trait (8)

gene cloning the process of isolating a gene sequence in the genome of an organism and inserting the gene sequence into a plasmid vector for production in large numbers (229)

gene expression the manifestation of the genetic material of an organism in the form of specific traits (208)

gene flow the movement of genes into or out of a population due to interbreeding (328)

genetic code the rule that describes how a sequence of nucleotides, read in groups of three consecutive nucleotides (triplets) that correspond to specific amino acids, specifies the amino acid sequence of a protein (211)

genetic drift the random change in allele frequency in a population (328)

genetic engineering a technology in which the genome of a living cell is modified for medical or industrial use (228)

genetics the science of heredity and of the mechanisms by which traits are passed from parents to offspring (162)

genital herpes a sexually transmitted disease that is caused by a herpes simplex virus (1010)

genotype the entire genetic makeup of an organism; also the combination of genes for one or more specific traits (168)

genus the level of classification that comes after family and that contains similar species (301)

germination the beginning of growth or development in a seed, spore, or zygote, especially after a period of inactivity (572)

gestation in mammals, the process of carrying young from fertilization to birth (1005)

gestation period in mammals, the length of time between fertilization and birth (810)

gill in mushrooms, a structure that is located on the underside of the cap and bears the spores; in aquatic animals, a respiratory structure that consists of many blood vessels surrounded by a membrane that allows for gas exchange (605)

gill slit a perforation between two gill arches through which water taken in through the mouth of a fish passes over the gills and out of the fish's body (747)

glucagon a hormone that is produced in the pancreas and that raises the blood glucose level (986)

glycolysis the anaerobic breakdown of glucose pyruvic acid, which makes a small amount of energy available to cells in the form of ATP (105)

glycoprotein a protein to which carbohydrate molecules are attached (435)

Golgi apparatus cell organelle that helps make and package materials to be transported out of the cell (64)

gonorrhea a sexually transmitted disease that is caused by bacteria and that results in inflammation of the mucous membranes in the urinary and reproductive tracts (1008)

gradualism a model of evolution in which gradual change over a long period of time leads to biological diversity (282)

grain the edible seed or seedlike fruit of a cereal grass (518)

greenhouse effect the warming of the surface of Earth and the lower atmosphere as a result of carbon dioxide and water vapor, which absorb and reradiate infrared radiation (388)

ground tissue a type of plant tissue other than vascular tissue that makes up much of the inside of a plant (552)

groundwater the water that is beneath the Earth's surface (351)

guard cell one of a pair of specialized cells that border a stoma and regulate gas exchange (503)

gymnosperm a woody vascular seed plant whose seeds are not enclosed by an ovary or fruit (512)

H

habitat the place where an organism usually lives (340)

hair in mammals, one of the many long and thin structures that grow out from the skin (800)

hair follicle a depression in the skin that encloses a hair and its root (862)

half-life the time required for half of a sample of a radioactive substance to disintegrate by radioactive decay or by natural processes (252)

haploid describes a cell, nucleus, or organism that has only one set of unpaired chromosomes (121)

Hardy-Weinberg principle the principle that states that the frequency of alleles in a population does not change unless evolutionary forces act on the population (326)

Haversian canal a channel containing blood vessels in compact bone tissue (852)

heart attack the death of heart tissues due to a blockage of their blood supply (884)

heartwood the nonconducting older wood in the center of a tree trunk (557)

helper T cell a white blood cell necessary for B cells to develop normal levels of antibodies (927)

herbaceous plant a plant that is soft and green instead of woody (556)

herbivore an organism that eats only plants (346)

heredity the passing of genetic traits from parent to offspring (8)

hermaphrodite an organism that has both male and female reproductive organs (609)

heterotroph an organism that obtains organic food molecules by eating other organisms or their by products and that cannot synthesize organic compounds from inorganic materials (95)

heterozygous describes an individual that has two different alleles for a trait (167)

histamine a chemical that stimulates the autonomous nervous system, secretion of gastric juices, and dilation of capillaries (925)

HIV human immunodeficiency virus, the virus that causes AIDS (12)

homeostasis the maintenance of a constant internal state in a changing environment; a constant internal state that is maintained in a changing environment by continually making adjustments to the internal and external environment (8)

hominid a member of the family Hominidae of the order Primates; characterized by opposable thumbs, no tail, relatively long lower limbs, and bipedalism; examples include modern humans and their ancestors (733)

homologous chromosomes chromosomes that have the same sequence of genes, that have the same structure, and that pair during meiosis (120)

homologous structures anatomical structures that share a common ancestry (286)

homozygous describes an individual that has identical alleles for a trait on both homologous chromosomes (167)

hormone a substance that is made in one cell or tissue and that causes a change in another cell or tissue located in a different part of the body (580)

Human Genome Project a research effort to sequence and locate the entire collection of genes in human cells (233)

hydrostatic skeleton in many invertebrates, the cavity that is filled with water and that has a support function (607)

hypertonic describes a solution whose solute concentration is higher than the solute concentration inside a cell (77)

hypha a nonreproductive filament of a fungus (421)

hypothalamus the region of the brain that coordinates the activities of the nervous and endocrine systems and that controls many body activities related to homeostasis (952)

hypothesis a theory or explanation that is based on observations and that can be tested (16)

hypotonic describes a solution whose solute concentration is lower than the solute concentration inside a cell (77)

I

immunity the ability to resist or to recover from an infection or disease (931)

implantation the process by which the newly fertilized egg in the blastocyst stage embeds itself in the lining of the uterus (1004)

imprinting learning that occurs early and quickly in a young animal's life and that cannot be changed once learned (829)

incomplete dominance a condition in which a trait in an individual is intermediate between the phenotype of the individual's two parents because the dominant allele is unable to express itself fully (177)

independent assortment the random distribution of the pairs of genes on different chromosomes to the gametes (146)

independent variable the factor that is deliberately manipulated in an experiment (17)

inflammatory response a protective response of tissues affected by disease or injury, characterized by redness, swelling, and pain (925)

innate behavior an inherited behavior that does not depend on the environment or experience (826)

insulin a hormone that is produced by a group of specialized cells in the pancreas and that lowers blood glucose levels (986)

interferon a protein that is produced by cells infected by a virus and that can protect uninfected cells from reproduction of the virus (926)

internal fertilization fertilization of an egg by sperm that occurs inside the body of a female (610)

interneuron a neuron located between the afferent neuron and the final neuron in a neural chain (953)

interphase a period between two mitotic or meiotic divisions during which the cell grows, copies its DNA, and synthesizes proteins (125)

intron a section of DNA that does not code for an amino acid and that is transcribed into RNA but is removed before it is translated (218)

invertebrate an animal that does not have a backbone (424)

nematocyst in cnidarians, a stinging cell that is used to inject a toxin into prey (623)

nephridium a tubule through which some invertebrates eliminate wastes (644)

nephron the functional unit of the kidney (913)

nerve a collection of nerve fibers through which impulses travel between the central nervous system and other parts of the body (944)

nervous tissue the tissue of the nervous system, which consists of neurons, their supporting cells, and connective tissue (847)

neuron a nerve cell that is specialized to receive and conduct electrical impulses (944)

neurotransmitter a chemical substance that transmits nerve impulses across a synapse (948)

neutrophil a large leukocyte that contains a lobed nucleus and many cytoplasmic granules (926)

niche the position (way of life) of a species in an ecosystem in terms of the physical characteristics (such as size, location, temperature, and pH) of the area where the species lives and the function of the species in the biological community (365)

nitrogen fixation the process by which gaseous nitrogen is converted into ammonia, a compound that organisms can use to make amino acids and other nitrogen-containing organic molecules (363)

nonvascular plant the three groups of plants (liverworts, hornworts, and mosses) that lack specialized conducting tissues and true roots, stems, and leaves (504)

norepinephrine a chemical that is both a neurotransmitter produced by the sympathetic nerve endings in the autonomic nervous system and a hormone secreted by the adrenal medulla to stimulate the functions of the circulatory and respiratory systems (abbreviation, NE) (985)

normal distribution a distribution of numerical data whose graph forms a bell-shaped curve that is symmetrical about the mean (331)

notochord the rod-shaped supporting axis found in the dorsal part of the embryos of all chordates, including vertebrates (700)

nucleic acid an organic compound, either RNA or DNA, whose molecules are made up of one or two chains of nucleotides and carry genetic information (37)

nucleotide in a nucleic-acid chain, a subunit that consists of a sugar, a phosphate, and a nitrogenous base (37)

nucleus in a eukaryotic cell, biology, a membrane-bound organelle that contains the cell's DNA and that has a role in processes such as growth, metabolism, and reproduction (58)

nutrient a substance or compound that provides nourishment (or food) or raw materials needed for life processes (900)

nymph an immature stage of some insects that is similar in function and structure to the adult (675)

O

observation the process of obtaining information by using the senses; the information obtained by using the senses (14)

omnivore an organism that eats both plants and animals (346)

oogenesis the production, growth, and maturation of an egg, or ovum (148)

open circulatory system a type of circulatory system in which the circulatory fluid is not contained entirely within vessels; a heart pumps fluid through vessels that empty into spaces called sinuses (606)

operator a short sequence of viral or bacterial DNA to which a repressor binds to prevent transcription (mRNA synthesis) of the adjacent gene in an operon (216)

operculum in fish, a hard plate that is attached to each side of the head, that covers gills, and that is open at the rear (756)

operon a unit of gene regulation and transcription in bacterial DNA that consists of a promoter, an operator, and one or more structural genes (216)

optic nerve the nerve that connects the retina of the eye to the brain and that transmits impulses that contribute to the sense of sight (958)

order the taxonomic category below the class and above the family (302)

organ a collection of tissues that carry out a specialized function of the body (419)

organ system a group of organs that work together to perform body functions (419)

organelle one of the small bodies that are found in the cytoplasm of a cell and that are specialized to perform a specific function (58)

osmosis the diffusion of water or another solvent from a more dilute solution (of a solute) to a more concentrated solution (of the solute) through a membrane that is permeable to the solvent (76)

ossicle one of the small, calcium carbonate plates that make up the endoskeleton of an echinoderm (694)

osteocyte a bone cell (852)

ovarian cycle a series of hormone-induced changes in which the ovaries prepare and release a mature ovum each month (1001)

ovary in the female reproductive system of animals, an organ that produces eggs; in flowering plants, the lower part of a pistil that produces eggs in ovules (538)

oviparous describes organisms that produce eggs that develop and hatch outside the body of the mother (777)

ovoviviparous describes organisms that produce eggs that develop and hatch inside the body of the mother (777)

ovulation the release of an ovum from a follicle of the ovary (1001)

ovule a structure in the ovary of a seed plant that contains an embryo sac and that develops into a seed after fertilization (534)

ovum a mature egg cell (149)

P

P generation parental generation, the first two individuals that mate in a genetic cross (164)

paleontologist a scientist who studies fossils (285)

Pangaea a single landmass that existed for about 40 million years before it began to break apart and form the continents that we know today (722)

parapodium in polychaetes, one of the two appendages that are used for locomotion or gas exchange (652)

parasitism a relationship between two species in which one species, the parasite, benefits from the other species, the host, and usually harms the host (362)

passive transport the movement of substances across a cell membrane without the use of energy by the cell (74)

pathogen a virus, microorganism, or other substance that causes disease; an infectious agent (454)

pedigree a diagram that shows the occurrence of a genetic trait in several generations of a family (175)

pedipalp one of the second pair of appendages that are beside the mouth of an arachnid and that are used for chewing and handling prey (670)

pelvic inflammatory disease a pelvic infection of the upper female reproductive system, including the uterus, ovaries, fallopian tubes, and other structures; it is a sexually transmitted disease (1009)

penis the male organ that transfers sperm to a female and that carries urine out of the body (998)

pepsin an enzyme that is found in gastric juices and that helps break down proteins into smaller molecules (908)

perennial a plant whose underground vegetative parts live for more than two years and whose upper parts die and regrow seasonally or annually (573)

periosteum the fibrous tissue that covers bones (851)

peripheral nervous system all of the parts of the nervous system except for the brain and the spinal cord (the central nervous system); includes the cranial nerves and nerves of the neck, chest, lower back, and pelvis (950)

petal one of the ring or rings of the usually brightly colored, leaf-shaped parts of a flower (538)

petiole the stalk that attaches a leaf to the stem of a plant (558)

pH a value used to express the acidity or alkalinity of a solution; it is defined as the logarithm of the reciprocal of the concentration of hydronium ions; a pH of 7 is neutral, a pH of less than 7 is acidic, and a pH of greater than 7 is basic (16)

pharynx in flatworms, the muscular tube that leads from the mouth to the gastrovascular cavity; in animals with a digestive tract, the passage from the mouth to the larynx and esophagus (886)

phenotype an organism's appearance or other detectable characteristic that results from the organism's genotype and the environment (166)

phloem the tissue that conducts food (sugars, amino acids, and mineral nutrients) in vascular plants (507)

phospholipid a lipid that contains phosphorus and that is a structural component in cell membranes (60)

photoperiodism the response of plants to seasonal changes in the relative length of nights and days (583)

photosynthesis the process by which plants, algae, and some bacteria use sunlight, carbon dioxide, and water to produce carbohydrates and oxygen (94)

phylogenetic tree a branching diagram that shows how organisms are related through evolution (602)

phylogeny the evolutionary history of a species or taxonomic group (307)

phylum the taxonomic group below kingdom and above class (302)

pigment a substance that gives another substance or a mixture its color (98)

pilus a short, thick appendage that allows a bacterium to attach to another bacterium (442)

pioneer species a species that colonizes an uninhabited area and that starts an ecological cycle in which many other species become established (343)

pistil the female reproductive part of a flower that produces seeds and consists of an ovary, style, and stigma (538)

pith the tissue that is located in the center of the stem of most vascular plants and that is used for storage (556)

pituitary gland an endocrine gland that is located at the base of the brain, stores and releases hormones produced by the hypothalamus, and secretes hormones under the control of the hypothalamus (983)

placenta the structure that attaches a developing fetus to the uterus and that enables the exchange of nutrients, wastes, and gases between the mother and the fetus (1005)

plankton the mass of mostly microscopic organisms that float or drift freely in the waters of aquatic (freshwater and marine) environments (378)

plant propagation the practice of reproducing plants from seeds or from vegetative parts (544)

planula the free-swimming, ciliated larva of a cnidarian (625)

plasma in biology, the liquid component of blood (876)

plasma cell a type of white blood cell that produces antibodies (929)

plasmid a circular DNA molecule that is usually found in bacteria and that can replicate independent of the main chromosome (229)

plasmodium the multinucleate cytoplasm of a slime mold that is surrounded by a membrane and that moves as a mass (470)

plastron the bottom, or ventral, portion of a turtle's shell (782)

platelet a fragment of a cell that is needed to form blood clots (877)

point mutation a mutation in which only one nucleotide or nitrogenous base in a gene is changed (219)

pollen grain the structure that contains the male gametophyte of seed plants (534)

pollen tube a tubular structure that grows from a pollen grain, enters the embryo sac, and allows the male reproductive cells to move to the ovule (534)

pollination the transfer of pollen from the male reproductive structures (the anthers) to the tip of a female reproductive structure (the pistil) of a flower in angiosperms or to the ovule in gymnosperms (534)

polygenic trait a characteristic of an organism that is determined by many genes (177)

polyp a form of a cnidarian that has a cylindrical, hollow body and that is usually attached to a rock or to another object (622)

population a group of organisms of the same species that live in a specific geographical area and interbreed (278)

population density the number of individuals of the same species that live in a given unit of area (321)

predation an interaction between two species in which one species, the predator, feeds on the other species, the prey (362)

prediction a statement made in advance that expresses the results that will be obtained from testing a hypothesis if the hypothesis is supported; the expected outcome if a hypothesis is accurate (16)

preen gland in birds, a special gland that secretes oil that a bird spreads over its feathers to clean and waterproof them (785)

pregnancy the period of time between conception and birth (1005)

primary growth the growth that occurs as a result of cell division at the tips of stems and roots and that gives rise to primary tissue (574)

primary productivity the total amount of organic material that the autotrophic organisms of an ecosystem produce (345)

primary succession succession that begins in an area that previously did not support life (343)

primate a member of the order primates, the group of mammals that includes humans, apes, and monkeys; typically distinguished by highly developed brains, forward-directed eyes, use of the hands, and varied locomotion (731)

prion an infectious particle that consists only of a protein and that does not contain DNA or RNA (441)

probability the likelihood that a possible future event will occur in any given instance of the event; the mathematical ratio of the number of times one outcome of any event is likely to occur to the number of possible outcomes of the event (173)

probe a strand of RNA or single-stranded DNA that has been labeled with a radioactive element or fluorescent dye and that is used to bind with and identify a specific gene in genetic engineering (231)

producer an organism that can make organic molecules from inorganic molecules; a photosynthetic or chemosynthetic autotroph that serves as the basic food source in an ecosystem (345)

profundal zone the zone in a freshwater habitat to which little sunlight penetrates (376)

proglottid one of the many body sections of a tapeworm; contains reproductive organs (631)

prokaryote an organism that consists of a single cell that does not have a nucleus or cell organelles; an example is a bacterium (57)

prosimian a member of a suborder of primates that are primarily arboreal and nocturnal, such as a lemur, loris, or tarsier (731)

prostate gland a gland in males that contributes to the seminal fluid (998)

protein an organic compound that is made of one or more chains of amino acids and that is a principal component of all cells (36)

protist an organism that belongs to the kingdom Protista (261)

protostome an organism whose embryonic blastopore develops into the mouth, whose coelom arises by schizocoely, and whose embryo has determinate cleavage (692)

protozoan a single-celled protist that can be aquatic or parasitic, that has organelles enclosed by a membrane, and that can move independently; examples include amebas and paramecia (461)

provirus viral DNA that has attached to a host cell's chromosome and that is replicated with the chromosome's DNA (436)

pseudocoelomate an animal that has a pseudocoelom, or false body cavity (600)

pseudopodium a retractable, temporary cytoplasmic extension that functions in food ingestion and movement in certain ameboid cells (464)

psychoactive drug a substance that has a significant effect on the mind or on behavior (961)

pulmonary vein the vein that carries oxygenated blood from the lungs to the heart (759)

pulse the rhythmic pressure of the blood against the walls of a vessel, particularly an artery (883)

punctuated equilibrium a model of evolution in which short periods of drastic change in species, including mass extinctions and rapid speciation, are separated by long periods of little or no change (282)

Punnett square a graphic used to predict the results of a genetic cross (170)

pupa the immobile, nonfeeding stage between the larva and the adult of insects that have complete metamorphosis; as a pupa, the organism is usually enclosed in a cocoon or chrysalis and undergoes important anatomical changes (675)

R

radial symmetry a body plan in which the parts of an animal's body are organized in a circle around a central axis (598)

radiometric dating a method of determining the age of an object by estimating the relative percentages of a radioactive (parent) isotope and a stable (daughter) isotope (252)

radula a rasping, tonguelike organ that is covered with chitinous teeth and that is used for feeding by many mollusks (643)

realized niche the range of resources that a species uses, the conditions that the species can tolerate, and the functional roles that the species plays as a result of competition in the species' fundamental niche (367)

reasoning the act of drawing a conclusion from facts or assumption (828)

receptor protein a protein that binds specific signal molecules, which causes the cell to respond (84)

recessive describes a trait or an allele that is expressed only when two recessive alleles for the same characteristic are inherited (167)

recombinant DNA DNA molecules that are artificially created by combining DNA from different sources (228)

red blood cell a disc-shaped cell that has no nucleus, that contains hemoglobin, and that transports oxygen in the circulatory system (876)

reflex an involuntary and almost immediate movement in response to a stimulus (952)

replication fork a Y-shaped point that results when the two strands of a DNA double helix separate so that the DNA molecule can be replicated (199)

repressor a regulatory protein that binds to an operator and blocks transcription of the genes of an operon (216)

reproduction the process of producing offspring (7)

reproductive isolation the inability of members of a population to successfully interbreed with members of another population of the same or a related species (281)

resolution in microscopes, the ability to form images with fine detail (51)

respiration the exchange of oxygen and carbon dioxide between living cells and their environment; includes breathing and cellular respiration (605)

resting potential the electric potential across the cell membrane of a nerve cell or muscle cell when the cell is not active (946)

restriction enzyme an enzyme that destroys foreign DNA molecules by cutting them at specific sites (229)

retina the light-sensitive inner layer of the eye, which receives images formed by the lens and transmits them through the optic nerve to the brain (958)

Rh factor one of several blood-group antigens carried on the surface of red blood cells (879)

rhizoid a rootlike structure in nonvascular plants, such as mosses or liverworts, that holds the plants in place and aids in absorption (486)

rhizome a horizontal, underground stem that provides a mechanism for asexual reproduction (510)

ribosome a cell organelle composed of RNA and protein; the site of protein synthesis (56)

RNA ribonucleic acid, a natural polymer that is present in all living cells and that plays a role in protein synthesis (37)

RNA polymerase an enzyme that starts (catalyzes) the formation of RNA by using a strand of a DNA molecule as a template (209)

rod one of the two types of light-detecting cells in the eye; rods can detect dim light and play a major role in noncolor and night vision (958)

root the mainly underground organ of vascular plants that holds plants in place and absorbs and stores water and minerals from the soil (507)

root cap the protective layer of cells that covers the tip of a root (555)

root hair an extension of the epidermis of a root that increases the root's surface area for absorption (555)

rRNA ribosomal RNA, an organelle that contains most of the RNA in the cell and that is responsible for ribosome function (212)

r-strategist a species that is adapted for living in an environment where changes are rapid and unpredictable; characterized by rapid growth, high fertility, short life span, small body size, and exponential population growth (324)

S

sapwood the tissue of the secondary xylem that is distributed around the outside of a tree trunk and is active in transporting sap (557)

sarcomere the basic unit of contraction in skeletal and cardiac muscle (857)

scanning electron microscope a microscope that produces an enlarged, three-dimensional image of an object by using a beam of electrons rather than light (54)

sebum the oily secretion of the sebaceous glands (864)

second messenger a molecule that is generated when a specific substance attaches to a receptor on the outside of a cell membrane, which produces a change in cellular function (85)

secondary growth plant growth that results from cell division in the cambia, or lateral meristems, and that causes the stems and roots to thicken (574)

secondary succession the process by which one community replaces another community that has been partially or totally destroyed (343)

seed a plant embryo that is enclosed in a protective coat (504)

seed coat the protective, outer covering of a seed (535)

seed plant a plant that produces seeds (504)

semen the fluid that contains sperm and various secretions produced by the male reproductive organs (998)

semicircular canal the fluid-filled canal in the inner ear that helps maintain balance and coordinate movements (959)

seminal vesicle one of two glandular structures in male vertebrates that hold and secrete seminal fluid (998)

seminiferous tubule one of the many tubules in the testis where sperm are produced (996)

sensory neuron a neuron that carries stimuli from a sense organ to the central nervous system (950)

sensory receptor a specialized structure that contains the ends of sensory neurons and that responds to specific types of stimuli (956)

sepal in a flower, one of the outermost rings of modified leaves that protect the flower bud (538)

septum a dividing wall, or partition, such as the wall between adjacent cells in a fungal hypha, the internal wall between adjacent segments of an annelid, and the thick wall between the right and left chambers of the heart (760)

sessile describes an organism that remains attached to a surface for its entire life and does not move (618)

seta one of the external bristles or spines that project from the body of an annelid (652)

sex chromosome one of the pair of chromosomes that determine the sex of an individual (122)

sex-linked trait a trait that is determined by a gene found on one of the sex chromosomes, such as the X chromosome or the Y chromosome in humans (175)

sexual reproduction reproduction in which gametes from two parents unite (150)

sexual selection an evolutionary mechanism by which traits that increase the ability of individuals to attract or acquire mates appear with increasing frequency in a population; selection in which a mate is chosen on the basis of a particular trait (836)

shoot the portion of a plant that grows mostly above the ground; includes the stems and leaves (507)

sieve tube in the phloem of a flowering plant, a conducting tube that is made up of a series of sieve-tube members stacked end to end (554)

sink any place where a plant stores or uses organic nutrients, such as sugar or starches (564)

sinoatrial node a mass of cardiac muscle cells that lies at the junction of the superior vena cava with the right atrium and that initiates and regulates contraction of the heart (abbreviation, SA node) (882)

siphon a hollow tube of bivalves used for sucking in and expelling sea water (647)

skin gill a transparent structure that projects from the surface of a sea star and that enables respiration (695)

sodium-potassium pump a carrier protein that uses ATP to actively transport sodium ions out of a cell and potassium ions into the cell (81)

solution a homogeneous mixture of two or more substances uniformly dispersed throughout a single phase (32)

sorus a cluster of spores or sporangia (532)

source a part of a plant that makes sugars and other organic compounds and from which these compounds are transported to other parts of the plant (564)

speciation the formation of new species as a result of evolution by natural selection (291)

species a group of organisms that are closely related and naturally mate to produce fertile offspring; also the level of classification below genus and above subspecies (9)

sperm the male gamete (sex cell) (148)

spermatogenesis the process by which male gametes form (148)

spicule a needle of silica or calcium carbonate in the skeleton of some sponges (620)

spinal cord a column of nerve tissue running from the base of the brain through the vertebral column (952)

spindle a network of microtubules that forms during mitosis and moves chromatids to the poles (128)

spinneret an organ that spiders and certain insect larvae use to produce silky threads for webs and cocoons (670)

spiracle an external opening in an insect or arthropod, used in respiration (668)

spirillum a spiral-shaped bacterium (443)

spongin a fibrous protein that contains sulfur and composes the fibers of the skeleton of some sponges (620)

sporangium a specialized sac, case, capsule, or other structure that produces spores (463)

spore a reproductive cell or multicellular structure that is resistant to environmental conditions and that can develop into an adult without fusion with another cell (154)

sporophyte in plants and algae that have alternation of generations, the diploid individual or generation that produces haploid spores (154)

sporozoite a sporozoan that has been released from the oocyst and is ready to penetrate a new host cell (473)

stabilizing selection a type of natural selection in which the average form of a trait is favored and becomes more common (332)

stamen the male reproductive structure of a flower that produces pollen and consists of an anther at the tip of a filament (538)

steroid a type of lipid that consists of four carbon rings to which various functional groups are attached and that usually has a physiological action (977)

stimulant a drug that increases the activity of the body or the activity of some part of the body (962)

stolon in plants, a creeping stem that can develop roots and shoots at its nodes or at its tip to form new individuals; the creeping hypha of some fungi that gives rise to new individuals (486)

stoma one of many openings in a leaf or a stem of a plant that enable gas exchange to occur (plural, stomata) (502)

stroke a sudden loss of consciousness or paralysis that occurs when the blood flow to the brain is interrupted (884)

subcutaneous tissue the layer of cells that lies beneath the skin (863)

substrate a part, substance, or element that lies beneath and supports another part, substance, or element; the reactant in reactions catalyzed by enzymes (41)

succession the replacement of one type of community by another at a single location over a period of time (343)

swim bladder in bony fishes, a gas-filled sac that is used to control buoyancy (756)

symbiosis a relationship in which two different organisms live in close association with each other (364)

synapse the junction at which the end of the axon of a neuron meets the end of a dendrite or the cell body of another neuron or meets another cell (948)

syphilis a sexually transmitted disease caused by the bacterium Treponema pallidum (1009)

T

target cell a specific cell to which a hormone is directed to produce a specific effect (977)

taxonomy the science of describing, naming, and classifying organisms (300)

teleost a group of ray-finned fishes that have a caudal fin, scales, and a swim bladder; the largest group of bony fishes (757)

tendon a tough connective tissue that attaches a muscle to a bone or to another body part (856)

terrestrial describes an organism that lives on land (721)

test cross the crossing of an individual of unknown genotype with a homozygous recessive individual to determine the unknown genotype (172)

testes the primary male reproductive organs, which produce sperm cells and testosterone (singular, testis) (996)

thalamus the part of the brain that directs incoming sensory and motor signals to the proper region (952)

thecondont the extinct reptile from which dinosaurs evolved (722)

theory an explanation for some phenomenon that is based on observation, experimentation, and reasoning (19)

therapsid the extinct order of mammal-like reptiles that likely gave rise to mammals (728)

thorax in higher vertebrates, the part of the body between the neck and the abdomen; in other animals, the body region behind the head; in arthropods, the mid-body region (666)

thylakoid a membrane system found within chloroplasts that contains the components for photosynthesis (99)

tissue a group of similar cells that perform a common function (419)

tissue culture the technique for growing living cells in an artificial medium (544)

tolerance the condition of drug addiction in which greater amounts of a drug are needed to achieve the desired effect (962)

toxin a substance that is produced by one organism and that is poisonous to other organisms (449)

trachea in insects, myriapods, and spiders, one of a network of air tubes; in vertebrates, the tube that connects the pharynx to the lungs (668)

transcription the process of forming a nucleic acid by using another molecule as a template; particularly the process of synthesizing RNA by using one strand of a DNA molecule as a template (208)

transfer RNA an RNA molecule that transfers amino acids to the growing end of a polypeptide chain during translation (212)

transformation the transfer of genetic material in the form of DNA fragments from one cell to another or from one organism to another (191)

transgenic animal an animal into which cloned genetic material has been transferred (241)

translation the portion of protein synthesis that takes place at ribosomes and that uses the codons in mRNA molecules to specify the sequence of amino acids in polypeptide chains (208)

translocation the movement of a segment of DNA from one chromosome to another, which results in a change in the position of the segment; also the movement of soluble nutrients from one part of a plant to another (564)

transpiration the process by which plants release water vapor into the air through stomata; also the release of water vapor into the air by other organisms (351)

trochophore a free-swimming, ciliated larva of many worms and some mollusks (642)

trophic level one of the steps in a food chain or food pyramid; examples include producers and primary, secondary, and tertiary consumers (345)

tropism the movement of all or part of an organism in response to an external stimulus, such as light or heat; movement is either toward or away from the stimulus (582)

true-breeding describes organisms or genotypes that are homozygous for a specific trait and thus always produce offspring that have the same phenotype for that trait (164)

U

ungulate a hoofed mammal (814)

uracil one of the four bases that combine with sugar and phosphate to form a nucleotide subunit of RNA; uracil pairs with adenine (208)

urea the principal nitrogenous product of the metabolism of proteins that forms in the liver from amino acids and from compounds of ammonia and that is found in urine and other body fluids (912)

ureter one of the two narrow tubes that carry urine from the kidneys to the urinary bladder (914)

urethra the tube that carries urine from the urinary bladder to the outside of the body (914)

urinary bladder a hollow, muscular organ that stores urine (914)

urine the liquid excreted by the kidneys, stored in the bladder, and passed through the urethra to the outside of the body (914)

uterus in female mammals, the hollow, muscular organ in which a fertilized egg is embedded and in which the embryo and fetus develop (1000)

V

vaccination the administration of treated microorganisms into humans or animals to induce an immune response (931)

vaccine a substance prepared from killed or weakened pathogens and introduced into a body to produce immunity (190)

vagina the canal in the female that extends from the vulva to the cervix and that receives the penis during sexual intercourse (1000)

valve a fold of membranes that controls the flow of a fluid (871)

vas deferens a duct through which sperm move from the epididymis to the ejaculatory duct at the base of the penis (997)

vascular bundle in a plant, a strand of conducting tissue that contains both xylem and phloem (556)

vascular cambium in a plant, the lateral meristem that produces secondary xylem and phloem (575)

vascular plant a plant that has a vascular system composed of xylem and phloem, specialized tissues that conduct materials from one part of the plant to another (504)

vascular system a conducting system of tissues that transport water and other materials in plants or in animals (504)

vascular tissue the specialized conducting tissue that is found in higher plants and that is made up mostly of xylem and phloem (422)

vector in biology, any agent, such as a plasmid or a virus, that can incorporate foreign DNA and transfer that DNA from one organism to another; an intermediate host that transfers a pathogen or a parasite to another organism (229)

vegetative part any nonreproductive part of a plant (516)

vegetative reproduction a type of asexual reproduction in which new plants grow from nonreproductive plant parts (541)

vein in biology, a vessel that carries blood to the heart (873)

vena cava one of the two large veins that carry blood from the body tissues to the heart (881)

ventricle one of the two large muscular chambers that pump blood out of the heart (881)

vertebra one of the 33 bones in the spinal column (backbone) (712)

vertebrate an animal that has a backbone; includes mammals, birds, reptiles, amphibians, and fish (267)

vesicle a small cavity or sac that contains materials in a eukaryotic cell; forms when part of the cell membrane surrounds the materials to be taken into the cell or transported within the cell (63)

vessel in plants, a tubelike structure in the xylem that is composed of connected cells that conduct water and mineral nutrients; in animals, a tube or duct that carries blood or another bodily fluid (554)

vestigial structure a structure in an organism that is reduced in size and function and that may have been complete and functional in the organism's ancestors (286)

villus one of the many tiny projections from the cells in the lining of the small intestine; increases the surface area of the lining for absorption (909)

viroid an infectious agent that consists of a small strand of RNA and that causes disease in plants (441)

virulent describes a microorganism or virus that causes disease and that is highly infectious (190)

virus a nonliving, infectious particle composed of a nucleic acid and a protein coat; it can invade and destroy a cell (434)

visceral mass the central section of a mollusk's body that contains the mollusk's organs (643)

vitamin an organic compound that participates in biochemical reactions and that builds various molecules in the body; some vitamins are called coenzymes and activate specific enzymes (904)

W

water vascular system a system of canals filled with a watery fluid that circulates throughout the body of an echinoderm (695)

weaning the time when an animal's dependence on its mother for food (milk) and protection comes to an end (806)

white blood cell a type of cell in the blood that destroys bacteria, viruses, and toxic proteins and helps the body develop immunities (877)

withdrawal the set of symptoms associated with the removal of an addictive drug from the body (962)

Y

yeast a very small, unicellular fungus that ferments carbohydrates into alcohol and carbon dioxide; used to ferment beer and to leaven bread and used as a source of vitamins and proteins (487)

Z

zygosporangium in members of the phylum Zygomycota, a sexual structure that is formed by the fusion of two gametangia and that contains one or more zygotes that resulted from the fusion of gametes produced by the gametangia (486)

zygospore in some algae, a thick-walled protective structure that contains a zygote that resulted from the fusion of two gametes (462)

zygote the cell that results from the fusion of gametes; a fertilized egg (121)

A

abiotic factor / factor abiótic un factor ambiental que no está asociado con las actividades de los seres vivos (340)

ABO blood group system / sistema de grupo sanguíneo ABO un sistema que se usa para clasificar la sangre humana en función de los antígenos que se encuentran en la superficie de los glóbulos rojos (878)

acanthodian / acantodio un pez antiguo; el primer vertebrado con mandíbulas que se conoce (715)

acid / ácido cualquier compuesto que aumenta el número de iones de hidrógeno cuando se disuelve en agua; los ácidos cambian el color del papel tornasol a rojo y forman sales al reaccionar con bases y con algunos metales (33)

acid rain / lluvia ácida precipitación con un pH inferior al normal, que tiene una concentración inusualmente alta de ácido sulfúrico y ácido nítrico como resultado de la contaminación química del aire por fuentes tales como los escapes de los automóviles y la quema de combustibles fósiles (386)

acoelomate / acelomado un animal que no tiene celoma, o cavidad en el cuerpo (600)

actin / actina una proteína responsable de la contracción y relajación de los músculos (857)

action potential / potencial de acción un cambio súbito en la polaridad de la membrana de una neurona, célula glandular o fibra muscular, el cual facilita la transmisión de impulsos eléctricos (946)

activation energy / energía de activación la cantidad mínima de energía que se requiere para iniciar una reacción química (39)

active site / sitio activo el sitio en una enzima que se une al sustrato (41)

active transport / transporte activo el movimiento de substancias químicas, normalmente a través de la membrana celular, en contra de un gradiente de concentración; requiere que la célula gaste energía (81)

adaptation / adaptación el proceso de adaptarse a un ambiente; un cambio anatómico, fisiológico o en la conducta que mejora la capacidad de supervivencia de una población (279)

addiction / adicción una dependencia fisiológica o psicológica de una substancia, tal como el alcohol o las drogas (962)

adductor muscle / músculo aductor el músculo grueso que une las dos válvulas en los moluscos y hace que la concha se abra (647)

adhesion / adhesión la fuerza de atracción entre dos cuerpos de diferentes substancias que están en contacto (31)

adrenal gland / glándula suprarrenal una de las dos glándulas endocrinas ubicadas arriba de cada riñón (985)

aerobic / aeróbico término que describe un proceso que requiere oxígeno (104)

aggregation / conglomerado un grupo de células u otros organismos (418)

agnathan / agnato un miembro de una clase de peces primitivos, sin mandíbulas (714)

AIDS / SIDA síndrome de inmunodeficiencia adquirida, enfermedad causada por una infección de VIH, la cual resulta en un sistema inmunológico ineficiente (934)

allele / alelo una de las formas alternativas de un gene que rige un carácter, como por ejemplo, el color del cabello (167)

allergy / alergia una reacción física a un antígeno, el cual puede ser una substancia común que produce una reacción ligera o que no produce ninguna reacción en la población general (936)

alternation of generations / alternacia de generaciones la alternación entre reproducción sexual y asexual que se da en ciertas plantas y animales (463)

alveoli / alveolo cualquiera de las diminutas células de aire de los pulmones, en donde ocurre el intercambio de oxígeno y dióxido de carbono (886)

amino acid / aminoácido cualquiera de las 20 distintas moléculas orgánicas que contienen un grupo carboxilo y un grupo amino y que se combinan para formar proteínas (36)

amino acid–based hormone / hormona derivada de aminoácidos una hormona que está formada por aminoácidos simples, péptidos o proteínas (977)

amniotic egg / huevo amniótico un tipo de huevo que es producido por los reptiles, las aves y los mamíferos que ponen huevos y que contiene una gran cantidad de yema; normalmente está rodeado por una cáscara áspera y dura, dentro de la cual se desarrollan el embrión y sus membranes embrionarias (775)

amylase / amilasa enzima que descompone los almidones en azúcares (907)

anaerobic / anaeróbico término que describe un proceso que no requiere oxígeno (104)

anemia / anemia condición en la que se reduce la capacidad de los glóbulos rojos de transportar oxígeno y la producción de glóbulos rojos disminuye (877)

angiosperm / angiosperma una planta que da flores y que produce semillas dentro de la fruta (514)

annual / anual planta que completa su ciclo de vida, se reproduce y muere en una estación de cultivo (573)

annual ring / anillo anual en el xilema secundario (madera), el anillo de crecimiento que se forma en una estación (575)

anther / antera en las plantas que dan flores, la punta del estambre, que contiene los sacos de polen donde se forman los granos (538)

antheridium / anteridio una estructura reproductiva que produce células sexuales masculinas en las plantas que no dan flores ni producen semillas (530)

antibiotic / antibiótico una substancia que inhibe el crecimiento de algunos microorganismos o los mata (443)

antibody / anticuerpo una proteína que reacciona ante un antígeno específico o que inactiva o destruye toxinas (929)

anticodon / anticodón una región del ARNt formada por tres bases que complementan el codón del ARNm (212)

antigen / antígeno una substancia que estimula una respuesta inmunológica (927)

antigen shifting / cambio antigénico la producción de antígenos nuevos por un virus cuando éste muta con el paso del tiempo (932)

aorta / aorta la arteria principal del cuerpo; transporta sangre del ventrículo izquierdo a la circulación sistémica (882)

apical dominance / dominancia apical inhibición del crecimiento lateral de un brote en el tallo de una planta debido a la producción de auxina en el brote terminal (581)

apical meristem / meristemo apical la región de crecimiento en la punta de los tallos y raíces de las plantas (574)

appendage / apéndice una estructura que se extiende del cuerpo principal, como por ejemplo, una extremidad, un tentáculo, una aleta o un ala (664)

appendicular skeleton / esqueleto apendicular los huesos de los brazos y piernas (850)

aquifer / acuífero un cuerpo rocoso o sedimento que almacena agua subterránea y permite que fluya (393)

archaebacteria / arqueobacteria un organismo procariótico que se diferencia de otros procariotes por la composición de su membrana y pared celular (258)

archegonium / arquegonio una estructura reproductiva femenina de ciertas plantas pequeñas y no vasculares, que produce un solo óvulo y en el cual ocurren la fertilización y el desarrollo (530)

artery / arteria un vaso sanguíneo que transporta sangre del corazón a los órganos del cuerpo (873)

arthropod / artrópodo miembro del phylum Arthropoda, el cual incluye a animales invertebrados tales como insectos, crustáceos y arácnidos, caracterizados por tener un cuerpo segmentado y un par de apéndices (266)

ascus / asca el saco de esporas donde los ascomicetos producen acosporas (487)

asexual reproduction / reproducción asexual reproducción que no involucra la unión de gametos, en la que un solo progenitor produce descendencia que es genéticamente igual al progenitor (150)

asymmetrical / asimétrico de forma irregular; sin simetría (598)

atom / átomo la unidad más pequeña de un elemento que conserva las propiedades de ese elemento (28)

ATP / ATP adenosín trifosfato; molécula orgánica que funciona como la fuente principal de energía para los procesos celulares; formada por una base nitrogenada, un azúcar y tres grupos fosfato (37)

atrium / aurícula una cámara que recibe la sangre que regresa al corazón (881)

autoimmune disease / enfermedad autoinmune una enfermedad en la que el sistema inmunológico ataca las células del propio organismo (933)

autosome / autosoma cualquier cromosoma que no es un cromosoma sexual (122)

autotroph / autótrofo un organismo que produce sus propios nutrientes a partir de substancias inorgánicas o del ambiente, en lugar de consumir otros organismos (94)

auxin / auxina una hormona vegetal que regula el alargamiento de las células (580)

axial skeleton / esqueleto axial los huesos del cráneo y la columna vertebral (850)

axon / axón una extensión alargada de una neurona que transporta impulsos hacia fuera del cuerpo de la célula (944)

B

B cell / célula B un glóbulo blanco de la sangre que madura en los huesos y fabrica anticuerpos (927)

bacillus / bacilo una bacteria que tiene forma de bastón (443)

bacteriophage / bacteriófago un virus que infecta a las bacterias (192)

basal disk / disco basal un área de los celenterados, como las hidras, medusas y corales, que les permite adherirse a las superficies (624)

base / base cualquier compuesto que aumenta el número de iones de hidróxido cuando se disuelve en agua; las bases cambian el color del papel tornasol a azul y forman sales al reaccionar con ácidos (33)

base pairing rules / regla de apareamiento de las bases las reglas que establecen que en el ADN, la citosina se une con la guanina y la adenina se une a la timina, y que en el ARN, la adenina se une con el uracilo (197)

basidium / basidio una estructura que produce esporas asexuales en los basidiomicetos (488)

behavior / conducta una acción que un individuo realiza en respuesta a un estímulo o a su ambiente (824)

biennial / bienal una planta que tiene un ciclo de vida de dos años (573)

bilateral symmetry / simetría bilateral una condición en la que dos mitades iguales de un cuerpo son imágenes de espejo una de otra (598)

binary fission / fisión binaria una forma de reproducción asexual de los organismos unicelulares, por medio de la cual la célula se divide en dos células del mismo tamaño (119)

binomial nomenclature / nomenclatura binomial un sistema para darle a cada organismo un nombre científico de dos palabras, el cual está formando por el género seguido de la especie (300)

biodiversity / biodiversidad la variedad de organismos que se encuentran en un área determinada, la variación genética dentro de una población, la variedad de especies en una comunidad o la variedad de comunidades en un ecosistema (341)

biogeochemical cycle / ciclo biogeoquímico la circulación de substancias del ambiente a los seres vivos y de los seres vivos al ambiente (350)

biological magnification / magnificación biológica la acumulación de cantidades cada vez mayores de substancias tóxicas en cada eslabón sucesivo de la cadena alimenticia (391)

biological species / especie biológica un grupo de organismos que sólo se reproducen entre ellos mismos y que normalmente están limitados a una región geográfica (305)

biology / biología el estudio científico de los seres vivos y sus interacciones con el medio ambiente (6)

biomass / biomasa materia orgánica que puede ser una fuente de energía; la masa total de los organismos en un área determinada (349)

biome / bioma una región extensa caracterizada por un tipo de clima específico y ciertos tipos de comunidades de plantas y animales (372)

biotic factor / factor biótico un factor ambiental que está asociado con las actividades de los seres vivos o que resulta de ellas (340)

blastocyst / blastocisto la etapa de bástula modificada de los embriones de los mamíferos (1004)

blastopore / blastoporo una abertura que se desarrolla en la blástula (692)

blastula / blástula la etapa en la que se encuentra un embrión antes de la gastrulación (596)

blood pressure / presión sanguínea la fuerza que la sangre ejerce en las paredes de las arterias (882)

body cavity / cavidad corporal cualquier cavidad que aloja órganos, tales como la cavidad torácica, abdominal o pélvica (849)

body plan / plan corporal la forma, simetría y organización interna de un animal (598)

bone marrow / médula ósea tejido blando que se encuentra en el interior de los huesos, donde se producen los glóbulos blancos y los glóbulos rojos (851)

brain / encéfalo la masa de tejido nervioso que es el centro principal de control del sistema nervioso (950)

brain stem / tronco encefálico la porción del cerebro que tiene forma de tronco, la cual conecta los hemisferios cerebrales con la médula espinal y mantiene las funciones necesarias del cuerpo, tales como la respiración y la circulación (951)

bronchus / bronquio uno de los dos tubos que conectan los pulmones con la tráquea (886)

budding / gemación reproducción asexual en la que una parte del organismo progenitor se separa y forma un nuevo organismo (487)

bulbourethral gland / glándula bulbouretral una de las dos glándulas del aparato reproductor masculino que añaden líquido al semen durante la eyaculación (998)

C

calorie / caloría la cantidad de energía que se requiere para aumentar la temperatura de 1 g de agua en 1°C; la Caloría que se usa para indicar el contenido energético de los alimentos es la kilocaloría (900)

Calvin cycle / ciclo de Calvin una vía bioquímica de la fotosíntesis en la que el dióxido de carbono se convierte en glucosa usando ATP (102)

cancer / cáncer un tumor en el cual las células comienzan a dividirse a una tasa incontrolable y se vuelven invasivas (12)

capillary / capilar diminuto vaso sanguíneo que permite el intercambio entre la sangre y las células de los tejidos (873)

capsid / cápside una cubierta de proteína que rodea el centro de ácido nucleico de un virus (435)

capsule / cápsula en los musgos, la parte que contiene las esporas; en las bacterias, una capa protectora de polisacáridos que se encuentra alrededor de la pared celular (443)

carapace / caparazón una placa parecida a un escudo que cubre el cefalotórax de algunos crustáceos (782)

carbohydrate / carbohidrato cualquier compuesto orgánico que está hecho de carbono, hidrógeno y oxígeno y que proporciona nutrientes a las células de los seres vivos (34)

carbon fixation / fijación del carbono la síntesis de compuestos orgánicos a partir del dióxido de carbono, como ocurre durante la fotosíntesis (102)

carnivore / carnívoro un animal que se alimenta de otros animales (346)

carotenoid / carotenoide una clase de pigmentos que se encuentran presentes principalmente en las plantas y que ayudan en la fotosíntesis (98)

carrier protein / proteína transportadora una proteína que transporta substancias a través de la membrana celular (80)

carrying capacity / capacidad de carga la población más grande que un ambiente puede sostener en cualquier momento dado (322)

cartilage / cartílago un tejido conectivo flexible y fuerte (715)

caste / casta un grupo de insectos en una colonia que tienen una función específica (679)

cell / célula en biología, la unidad más pequeña que puede realizar todos los procesos vitales; las células están cubiertas por una membrana y tienen un núcleo y citoplasma (7)

cell cycle / ciclo celular el ciclo de vida de una célula; en los eucariotes, consiste de un período de crecimiento celular en el que el ADN se sintetiza, y un período de división celular en el que ocurre la mitosis (125)

cell membrane / membrana celular una capa de fosfolípidos que cubre la superficie de la célula y funciona como una barrera entre el interior de la célula y el ambiente de la célula (56)

cell theory / teoría celular la teoría que establece que todos los seres vivos están formados por células, que las células son las unidades fundamentales de los organismos y que las células provienen únicamente de células existentes (55)

cell wall / pared celular una estructura rígida que rodea la membrana celular y le brinda soporte a la célula (57)

cellular respiration / respiración celular el proceso por medio del cual las células producen energía a partir de los carbohidratos; el oxígeno atmosférico se combina con la glucosa para formar agua y dióxido de carbono (95)

central nervous system / sistema nervioso central el cerebro y la médula espinal; su principal función es controlar el flujo de información en el cuerpo (950)

central vacuole / vacuola central una cavidad o bolsa grande que se encuentra en las células vegetales o en los protozoarios y que contiene aire o alimentos parcialmente digeridos (66)

centromere / centrómero la región de un cromosoma que mantiene unidas las dos cromátidas hermanas durante la mitosis (119)

cephalization / cefalización la concentración de tejido nervioso y órganos sensoriales en la parte anterior de un organismo (599)

cephalothorax / cefalotórax en los arácnidos y algunos crustáceos, la parte del cuerpo constituida por la cabeza y el tórax (666)

cereal / cereal cualquier hierba que produce granos que pueden ser usados como alimento, tales como el arroz, trigo, maíz, avena o centeno (518)

cerebellum / cerebelo una porción posterior del cerebro que coordina el movimiento de los músculos y controla las actividades subconscientes y algunas funciones de equilibrio (951)

cerebral ganglion / ganglio cerebral uno de un par de conjuntos de células nerviosas que funcionan como si fueran un cerebro primitivo en la parte anterior de algunos invertebrados, tales como los anélidos (651)

cerebrum / cerebro la parte superior del encéfalo que recibe las sensaciones y controla el movimiento (951)

chelicera / quelíceros en los arácnidos, uno de los dos apéndices usados para atacar a las presas (670)

chitin / quitina un carbohidrato que forma parte del exoesqueleto de los artrópodos y de otros organismos, como por ejemplo, insectos, crustáceos, hongos y algunas algas (482)

chlamydia / clamidia una enfermedad bacteriana transmitida sexualmente caracterizada por dolor al orinar y descargas vaginales (1009)

chlorofluorocarbons / clorofluorocarbonos hidrocarburos en los que algunos o todos los átomos de hidrógeno son reemplazados por cloro y flúor; se usan en líquidos refrigerantes para refrigeradores y aires acondicionados y en solventes para limpieza; su uso está restringido porque destruyen las moléculas de ozono de la estratosfera (abreviatura: CFC) (387)

chlorophyll / clorofila un pigmento verde presente en la mayoría de las células vegetales que les da a las plantas su color verde característico y que reacciona con la luz del Sol, el dióxido de carbono y el agua para formar carbohidratos (98)

chloroplast / cloroplasto un organelo que se encuentra en las células vegetales y en las células de las algas, en el cual se lleva a cabo la fotosíntesis (66)

choanocyte / coanocito cualquiera de las células flageladas que cubren las cavidades de una esponja (619)

chordate / cordado un animal que, en alguna etapa de su ciclo de vida, tiene un nervio dorsal, un notocrodio y bolsas faríngeas; entre los ejemplos se encuentran los mamíferos, aves, reptiles, anfibios, peces y algunas formas marinas inferiores (700)

chromatid / cromátida una de las dos hebras de un cromosoma que se vuelve visible durante la meiosis o mitosis (119)

chromosome / cromosoma en una célula eucariótica, una de las estructuras del núcleo que está hecha de ADN y proteína; en una célula procariótica, el anillo principal de ADN (119)

chrysalis / crisálida la pupa de cubierta dura de ciertos insectos, como las mariposas (675)

cilium / cilio una estructura parecida a un pelo ordenada en hileras muy comprimidas, que se proyecta a partir de la superficie de algunas células (58)

cladistics / cladística un sistema de clasificación filogénica en el que los únicos criterios de agrupación de los taxa son los caracteres comunes derivados y la ascendencia (327)

cladogram / cladograma un diagrama basado en modelos de caracteres comunes derivados, que muestra las relaciones evolutivas entre grupos de organismos (309)

class / clase una categoría taxonómica que contiene órdenes con características comunes (302)

cleavage / segmentación en el desarrollo biológico, una serie de divisiones celulares que ocurren inmediatamente después de que un óvulo es fecundado (1004)

climate / clima las condiciones promedio del tiempo en un área durante un largo período de tiempo (371)

clone / clon un organismo producido por reproducción asexual que es genéticamente idéntico a su progenitor; clonar significa hacer un duplicado genético (150)

closed circulatory system / aparato circulatorio cerrado un aparato circulatorio en el que el corazón hace que la sangre circule a través de una red de vasos que forman un circuito cerrado; la sangre no sale de los vasos sanguíneos y los materiales pasan a través de las paredes de los vasos por difusión (606)

cnidocyte / cnidocito una célula urticante de los cnidarios (623)

coccus / coco una bacteria que tiene forma de esfera (443)

cochlea / cóclea un tubo enrollado que se encuentra en el oído interno y es esencial para poder oír (959)

codominance / codominancia una condición en la que los dos alelos de un gene están totalmente expresados (178)

codon / codón en el ADN, una secuencia de tres nucleótidos que codifica un aminoácido o indica una señal de inicio o una señal de terminación (211)

coelom / celoma una cavidad del cuerpo que contiene los órganos internos (600)

coelomate / celomado un animal que tiene una cavidad en el cuerpo donde se encuentran los órganos internos (600)

coevolution / coevolución la evolución de dos o más especies que se debe a su influencia mutua, a menudo de un modo que hace que la relación sea más mutuamente beneficiosa (362)

cohesion / cohesión la fuerza que mantiene unidas a las moléculas de un solo material (31)

colon / colon una sección del intestino grueso (910)

colonial organism / organismo colonial un conjunto de células genéticamente idénticas que están asociadas permanentemente, pero en el que no se da una gran integración de las actividades celulares (418)

commensalism / comensalismo una relación entre dos organismos en la que uno se beneficia y el otro no es afectado (364)

community / comunidad un grupo de varias especies que viven en el mismo hábitat e interactúan unas con otras (340)

competition / competencia la relación entre dos especies (o individuos) en la que ambas especies (o individuos) intentan usar el mismo recurso limitado, de modo que ambas resultan afectadas negativamente por la relación (365)

competitive exclusion / exclusión competitiva la exclusión de una especie por otra debido a la competencia (369)

complement system / sistema complementario un sistema de proteínas que circulan en el torrente sanguíneo y que se combinan con anticuerpos para proteger en contra de los antígenos (926)

complementary base pairing / complementariedad de las bases nitrogenadas una característica de los ácidos nucleicos en que la secuencia de bases de una hebra está acoplada con la secuencia de bases de la otra (197)

compound / compuesto una substancia formada por átomos de dos o más elementos diferentes unidos por enlaces químicos (29)

compound eye / ojo compuesto un ojo compuesto por muchos detectores de luz separados por células de pigmentos (666)

concentration gradient/gradiente de concentración una diferencia en la concentración de una substancia a través de una distancia (74)

conditioning / condicionamiento aprendizaje por asociación (827)

cone / cono en los animales, un fotorreceptor de la retina que distingue colores y es muy sensible a la luz brillante (958)

cone / cono en las plantas, una estructura portadora de semillas (511)

connective tissue / tejido conectivo un tejido que tiene mucha substancia intracelular, y que conecta y sostiene otros tejidos (847)

consumer / consumidor un organismo que se alimenta de otros organismos o de materia orgánica, en lugar de producir sus propios nutrientes o de obtenerlos de fuentes inorgánicas (345)

continental drift / deriva continental la hipótesis que establece que alguna vez los continentes formaron una sola masa de tierra, se dividieron y se fueron a la deriva hasta terminar en sus ubicaciones actuales (268)

contour feather / pluma de contorno una las plumas más externas que cubren a un ave y que sirven para determinar su forma (784)

control group / grupo de control en un experimento, un grupo que sirve como estándar de comparación con otro grupo, al cual el grupo de control es idéntico excepto por un factor (17)

convergent evolution / evolución convergente el proceso por medio del cual especies no relacionadas se vuelven más parecidas a medida que se adaptan al mismo tipo de ambiente (307)

cork / corcho la capa externa de corteza de cualquier planta leñosa (553)

cork cambium / cámbium de corcho una capa de tejido que se encuentra debajo de la capa de corcho, en la cual se producen las células de corcho (575)

coronary artery / arteria coronaria una de las dos arterias que suministran sangre directamente al corazón (882)

corpus luteum / cuerpo lúteo la estructura que se forma a partir de los folículos rotos del ovario después de la ovulación; libera hormonas (1002)

cortex / corteza en las plantas, el tejido primario ubicado en la epidermis; en los animales, la porción externa de un órgano (555)

cotyledon / cotiledón la hoja embrionaria de una semilla (535)

countercurrent flow / flujo a contracorriente en las branquias de los peces, un arreglo por el cual el agua fluye alejándose de la cabeza y la sangre fluye hacia la cabeza (747)

crossing-over / entrecruzamien el intercambio de material genético entre cromosomas homólogos durante la meiosis; puede resultar en la recombinación genética (144)

cud / bolo alimenticio de los rumiante comida parcialmente digerida que es regurgitada, masticada y tragada nuevamente con el fin de digerirla más por los mamíferos que tienen rumen (814)

cuticle / cutícula una capa cerosa o grasosa e impermeable ubicada en la pared externa de las células de la epidermis (502)

cyanobacteria / cianobacteria una bacteria que efectúa la fotosíntesis, como por ejemplo, el alga verdiazul (258)

cystic fibrosis / fibrosis quística un trastorno genético mortal en el que se secretan cantidades excesivas de moco, lo cual bloquea los conductos intestinales y bronquiales y causa dificultad al respirar (12)

cytokinesis / citocinesis la división del citoplasma de una célula; la citocinesis ocurre después de que el núcleo de la célula se divide por mitosis o meiosis (125)

cytoplasm / citoplasma la región de la célula dentro de la membrana, que incluye el líquido, el citoesqueleto y los organelos, pero no el núcleo (56)

cytoskeleton / citoesqueleto red citoplásmica de filamentos de proteínas que juega un papel esencial en el movimiento, forma y división de la célula (56)

cytotoxic T cell / célula T citotóxica un tipo de célula T que reconoce y destruye las células infectadas por un virus (927)

D

decomposer / descomponedor un organismo que desintegra la materia orgánica de organismos muertos y se alimenta de ella; entre los ejemplos se encuentran las bacterias y los hongos (347)

dendrite / dendrita la extensión citoplásmica de una neurona que recibe estímulos (944)

density-dependent factor / factor dependiente de la densidad una variable afectada por el número de organismos presentes en un área determinada (322)

density-independent factor / factor independiente de la densidad una variable que afecta a una población independientemente de la densidad de la población, por ejemplo, el clima (324)

deoxyribose / desoxirribosa azúcar de cinco carbonos que es un componente de los nucleótidos de ADN (194)

dependent variable / variable dependiente en un experimento, la variable que se cambia o que se determina al manipular dos o más factores (las variables independientes) (17)

depressant / depresor un medicamento que reduce la actividad funcional y produce relajación muscular (964)

derived character / carácter derivado una característica especial de un grupo particular de organismos (307)

dermal tissue / tejido dérmico la cubierta exterior de una planta (552)

dermis / dermis la capa de piel que está debajo de la epidermis (862)

detritivore / detritívoro un consumidor que se alimenta de plantas y animales muertos (346)

deuterostome / deuteróstomo un animal cuya boca no se deriva del blastoporo y cuyo embrión presenta segmentación indeterminada (692)

diabetes mellitus / diabetes mellitus un serio trastorno en el que las células no pueden obtener glucosa de la sangre; es causado por una deficiencia de insulina o por una falta de respuesta a la insulina (987)

diaphragm / diafragma un músculo en forma de cúpula que está unido a las costillas inferiores y que es el músculo principal de la respiración (886)

diatom / diatomea un alga unicelular que tiene una concha doble la cual contiene sílice (466)

dicot / dicotiledónea una angiosperma con dos cotiledones, venación en forma de red y partes florales en grupos de cuatro o cinco (515)

differentiation / diferenciación el proceso por medio del cual la estructura y función de las partes de un organismo cambian para permitir la especialización de dichas partes (419)

diffusion / difusión el movimiento de partículas de regiones de mayor densidad a regiones de menor densidad (75)

digestion / digestión la descomposición de la comida en substancias químicas que se usan para generar energía (900)

diploid / diploide una célula que contiene dos juegos de cromosomas haploides (121)

directional selection / selección direccional un proceso de selección natural en el cual se selecciona una variación genética que origina un cambio en la composición genética global de la población (332)

dispersion / dispersión en óptica, el proceso de separar una onda que tiene diferentes frecuencias (por ejemplo, la luz blanca) de las ondas individuales que la componen (los distintos colores) (321)

diurnal / diurno término que describe animales que son activos durante el día y duermen en la noche (732)

DNA / ADN ácido desoxirribonucleico, el material que contiene la información que determina las características que se heredan (37)

DNA fingerprint / huella de ADN el patrón de bandas que se obtiene cuando los fragmentos de ADN de un individuo se separan (237)

DNA helicase / ADN helicasa una enzima que separa las hebras de la doble hélice del ADN durante la replicación del ADN (198)

DNA polymerase / ADN polimerasa una enzima que actúa como catalizadora en la formación de la molécula de ADN (199)

DNA replication / replicación del ADN el proceso de hacer una copia del ADN (198)

dormancy / letargo un estado en el que las semillas, esporas, bulbos y otros órganos reproductores dejan de crecer y desarrollarse y reducen su metabolismo, sobre todo la respiración (584)

double fertilization / fecundación doble el proceso por medio del cual uno de los dos núcleos de los espermatozoides se une con el núcleo del óvulo para producir un cigoto diploide, y el otro núcleo se une con el núcleo polar para producir un endosperma triploide (540)

double helix / doble hélice la estructura en forma de escalera en espiral característica de la molécula del ADN (194)

down feather / plumón una pluma suave que cubre el cuerpo de las crías de las aves y sirve como aislante en las aves adultas (785)

E

ecology / ecología el estudio de las interacciones de los seres vivos entre sí mismos y entre sí mismos y su ambiente (9)

ecosystem / ecosistema una comunidad de organismos y su ambiente abiótico (340)

ectoderm / ectodermo la capa más externa de las tres capas germinales de un embrión, que al desarrollarse se convierte en la epidermis y en los tejidos epidérmicos, el sistema nervioso, los órganos externos de los sentidos y las membranas mucosas que cubren la boca y el ano (596)

ectothermic / ectotérmico término que describe la capacidad de un organismo de mantener su temperatura corporal al obtener calor del ambiente (724)

electron microscope / microscopio electrónico microscopio que enfoca un haz de electrones para aumentar la imagen de los objetos (51)

electron transport chain / cadena de transporte de electrones una serie de moléculas que se encuentran en las membranas internas de las mitocondrias y cloroplastos y a través de las cuales pasan los electrones en un proceso que hace que los protones se acumulen en un lado de la membrana (100)

electrophoresis / electroforesis el proceso por medio del cual las partículas con carga eléctrica que están suspendidas en un líquido se mueven por todo el líquido debido a la influencia de un campo eléctrico (231)

element / elemento una substancia que no se puede separar o descomponer en substancias más simples por medio de métodos químicos; todos los átomos de un elemento tienen el mismo número atómico (28)

embryo / embrión un organismo en una de las primeras etapas del desarrollo de las plantas y animales; en los humanos, un individuo en desarrollo se denomina embrión de la segunda a la octava semana de embarazo (504)

endocrine gland / glándula endocrina una glándula sin conductos que secreta hormonas a la sangre (975)

endocytosis / endocitosis el proceso por medio del cual la membrana celular rodea una partícula y la encierra en una vesícula para llevarla al interior de la célula (83)

endoderm / endodermo la capa germinal interna del embrión de un animal; al desarrollarse se convierte en el epitelio de la faringe, tracto respiratorio, tracto digestivo, vejiga y uretra (596)

endoplasmic reticulum / retículo endoplásmico un sistema de membranas que se encuentra en el citoplasma de la célula y que tiene una función en la producción, procesamiento y transporte de proteínas y en la producción de lípidos (63)

endosperm / endosperma un tejido triploide (3n) que se desarrolla en las semillas de las angiospermas y que provee alimento para el embrión en desarrollo (514)

endospore / endospora una espora protectiva que tiene una pared gruesa, se forma dentro de una célula bacteriana y resiste condiciones adversas (443)

endosymbiosis / endosimbiosis una relación mutuamente beneficiosa en la que un organismo vive dentro de otro (259)

endothermic / endotérmico término que describe la capacidad de un organismo de mantener su temperatura corporal al producir calor internamente (724)

energy / energía la capacidad de realizar un trabajo (38)

energy pyramid / pirámide de energía un diagrama con forma de triángulo que muestra la pérdida de energía que ocurre en un ecosistema a medida que la energía pasa a través de la cadena alimenticia del ecosistema; cada hilera de la pirámide representa un nivel trófico (de alimentación) en el ecosistema, y el área de la hilera representa la energía almecenada en ese nivel trófico (348)

envelope / envoltura una capa similar a una membrana que cubre las cápsides de algunos virus (435)

enzyme / enzima un tipo de proteína que acelera las reacciones metabólicas en las plantas y animales, sin ser modificada permanentemente ni ser destruida (40)

epidermis / epidermis la superficie externa de las células de una planta o animal (553)

epididymis / epidídimo el conducto largo y enrollado que se encuentra en la superficie de los testículos, en el que los espermatozoides maduran (997)

epinephrine / epinefrina una hormona liberada por la médula suprarrenal que estimula el metabolismo rápidamente en casos de emergencia, disminuye la secreción de insulina y estimula el pulso y la presión sanguínea; también se llama adrenalina (985)

epithelial tissue / tejido epitelial un tejido compuesto por células que forman una barrera entre un organismo y su ambiente externo (846)

equilibrium / equilibrio en química, el estado en el que un proceso químico y el proceso químico inverso ocurren a la misma tasa, de modo que las concentraciones de los reactivos y los productos no cambian (74)

esophagus / esófago un conducto largo y recto que conecta la faringe con el estómago (907)

Eubacteria / Eubacteria un reino de clasificación que agrupa a todos los procariotes, excepto a las arqueobacterias (258)

eukaryote / eucariote un organismo cuyas células tienen un núcleo contenido en una membrana, múltiples cromosomas y un ciclo mitótico; los eucariotes incluyen animales, plantas y hongos, pero no bacterias ni algas (58)

evolution / evolución un cambio en las características de una población de una generación a la siguiente; el desarrollo gradual de organismos a partir de otros organismos desde los inicios de la vida (9)

excretion / excreción el proceso de eliminar desechos metabólicos (912)

exocytosis / exocitosis el proceso por medio del cual una substancia se libera de la célula a través de una vesícula que la transporta a la superficie de la célula en donde se fusiona con la membrana para dejar salir a la substancia (83)

exon / exón la porción de la secuencia del ADN de un gene que contiene la secuencia de aminoácidos en una cadena, y el inicio y el fin de una secuencia de codificación (218)

exoskeleton / exoesqueleto una estructura de soporte, dura y externa, que se desarrolla a partir del ectodermo (607)

experiment / experimento un procedimiento que se lleva a cabo bajo condiciones controladas para descubrir, demostrar o probar un hecho, teoría o verdad general (17)

extensor / extensor un músculo que extiende una articulación (856)

external fertilization / fecundación externa la unión de gametos afuera del cuerpo de los padres, como en el caso de muchos peces y anfibios (610)

extinct / extinto término que describe a una especie de un organismo que ha desaparecido completamente (282)

F

F_1 generation / generación F_1 la primera generación de descendencia que se obtiene de la cruza experimental de dos organismos (164)

F_2 generation / generación F_2 la segunda generación de descendencia que se obtiene de la cruza experimental de dos organismos de una generación F_1 (164)

facilitated diffusion / difusión facilitada el transporte de substancias a través de la membrana celular de una región de mayor concentración a una de menor concentración con la ayuda de proteínas transportadoras (80)

fallopian tube / trompa de Falopio un conducto a través del cual se mueven los óvulos del ovario al útero (1000)

family / familia la categoría taxonómica debajo del orden y arriba del género (302)

fermentation / fermentación la descomposición de carbohidratos por enzimas, bacterias, levaduras o mohos, en ausencia de oxígeno (108)

fertilization / fecundación la unión de un gameto masculino y femenino para formar un cigoto (153)

fetus / feto un ser humano en desarrollo de las semanas siete a ocho después de la fecundación hasta el nacimiento (1006)

fixed action pattern behavior / conducta de patrón fijo de acción una conducta innata que es característica de ciertas especies (826)

flagellum / flagelo una estructura larga parecida a una cola, que crece hacia el exterior de una célula y le permite moverse (57)

flexor / flexor un músculo que dobla una extremidad u otra parte del cuerpo (856)

flower / flor la estructura reproductiva de una planta que da flores, que normalmente consiste en un pistilo, estambres, pétalos y sépalos (505)

fluke / trematodo un gusano plano parasítico de la clase Trematoda (632)

follicle / folículo una bolsa o cavidad angosta y pequeña en un órgano o tejido, como las que se encuentran en la piel y contienen las raíces de los pelos, o las que se encuentran en los ovarios y contienen los óvulos en desarrollo (1001)

food chain / cadena alimenticia la vía de transferencia de energía través de varias etapas, que ocurre como resultado de los patrones de alimentación de una serie de organismos (346)

food web / red alimenticia un diagrama que muestra las relaciones de alimentación entre los organismos de un ecosistema (347)

foot / pie un apéndice que algunos invertebrados usan para moverse; la parte inferior de la pierna de un vertebrado (431)

fossil / fósil los rastros o restos de un organismo que vivió hace mucho tiempo, conservados en rocas sedimentarias (258)

frond / fronda la hoja de un helecho o palma (511)

fruit / fruto un ovario maduro de planta; el órgano de una planta donde se encuentran contenidas las semillas (514)

fundamental niche / nicho fundamental el nicho ecológico más grande en el que un organismo o especie vive sin experimentar competencia (365)

G

gamete / gameto una célula reproductiva haploide que se une con otra célula reproductiva haploide para formar un cigoto (118)

gametophyte / gametofito en generaciones alternadas, la fase en la que los gametos se forman; un individuo haploide que produce gametos (154)

gastrovascular cavity / cavidad gastrovascular una cavidad que tiene funciones digestivas y circulatorias en algunos cnidarios (605)

gemmule / gémula una estructura asexual reproductiva producida por algunas esponjas de agua dulce (621)

gene / gene un segmento de ADN ubicado en un cromosoma, que codifica para un carácter hereditario específico (8)

gene cloning / clonación de genes el proceso por medio del cual se aísla la secuencia de un gene del genoma de un organismo y esta secuencia se inserta en un vector plásmido para producir el gene en grandes cantidades (229)

gene expression / expresión de los genes la manifestación del material genético de un organismo en forma de caracteres específicos (208)

gene flow / flujo de genes el movimiento de genes a una población o fuera de ella debido al entrecruzamiento (328)

genetic code / código genético la regla que describe la forma en que una secuencia de nucleótidos, leídos en grupos de tres nucleótidos consecutivos (triplete) que corresponden a aminoácidos específicos, especifica la secuencia de aminoácidos de una proteín (211)

genetic drift / deriva genética el cambio aleatorio en la frecuencia de los alelos de una población (328)

genetic engineering / ingeniería genética una tecnología en la que el genoma de una célula viva se modifica con fines médicos o industriales (228)

genetics / genética la ciencia de la herencia y de los mecanismos por los cuales los caracteres son transmitidos de padres a hijos (162)

genital herpes / herpes genital una enfermedad transmitida sexualmente causada por el virus herpes simplex (1010)

genotype / genotipo la constitución genética completa de un organismo; también, la combinación de genes para uno o más caracteres específicos (168)

genus / género el nivel de clasificación que viene después de la familia y que contiene especies similares (301)

germination / germinación el comienzo del crecimiento o desarrollo de una semilla, espora o cigoto, sobre todo después de un período de inactividad (572)

gestation / gestación en los mamíferos, el proceso de llevar a las crías de la fecundación al nacimiento (1005)

gestation period / período de gestación en los mamíferos, el tiempo que transcurre entre la fecundación y el nacimiento (810)

gill / branquia en los animales acuáticos, una estructura respiratoria que está formada por muchos vasos sanguíneos rodeados por una membrana que permite el intercambio gaseoso (605)

gill / laminilla en los hongos, una estructura que se ubica en la parte inferior de sombrerete y que contiene las esporas (605)

gill slit / apertura branquial una perforación entre dos arcos branquiales a través de la cual el agua que un pez toma por la boca pasa sobre las branquias y hacia el exterior del cuerpo del pez (747)

glucagon / glucagón una hormona producida en el páncreas que aumenta el nivel de glucosa en la sangre (986)

glycolysis / glicólisis la descomposición anaeróbica de ácido pirúvico glucosa, la cual hace que una pequeña cantidad de energía en forma de ATP esté disponible para las células (105)

glycoprotein / glicoproteína una proteína que tienen unidas moléculas de carbohidratos (435)

Golgi apparatus / aparato de Golgi un organelo celular que ayuda a hacer y a empacar los materiales que serán transportados al exterior de la célula (64)

gonorrhea / gonorrea una enfermedad transmitida sexualmente producida por bacterias, que resulta en la inflamación de las membranas mucosas de los tractos urinario y reproductor (1008)

gradualism / gradualismo un modelo de evolución en el que un cambio gradual a través de un largo período de tiempo conlleva a la diversidad biológica (282)

grain / grano la semilla comestible, o fruta similar a una semilla, de un cereal (518)

greenhouse effect / efecto de invernadero el calentamiento de la superficie terrestre y de la parte más baja de la atmósfera, el cual se produce cuando el dióxido de carbono, el vapor de agua y otros gases del aire absorben radiación infrarroja y la vuelven a irradiar (388)

ground tissue / tejido basal un tipo de tejido vegetal diferente del tejido vascular y que constituye gran parte del interior de una planta (552)

groundwater / agua subterránea el agua que está debajo de la superficie de la Tierra (351)

guard cell / célula oclusiva una de las dos células especializadas que se encuentran al borde de un estoma y regulan el intercambio gaseoso (503)

gymnosperm / gimnosperma una planta leñosa y vascular, la cual produce semillas que no están contenidas en un ovario o fruto (512)

H

habitat / hábitat el lugar donde un organismo vive normalmente (340)

hair / pelo en los mamíferos, una de las muchas estructuras largas y delgadas que crecen a partir de la piel (800)

hair follicle / folículo piloso una depresión en la piel que contiene un pelo y su raíz (862)

half-life / vida media el tiempo que tarda la mitad de una muestra de una substancia radiactiva en desintegrarse por desintegración radiactiva o por procesos naturales (252)

haploid / haploide término que describe a una célula, núcleo u organismo que tiene sólo un juego de cromosomas que no están asociados en pares (121)

Hardy-Weinberg principle / principio de Hardy-Weinberg el principio que establece que la frecuencia de alelos en una población no cambia a menos que fuerzas evolutivas actúen en la población (326)

Haversian canal/canal de Havers un canal que contiene vasos sanguíneos en los huesos compactos (852)

heart attack / ataque cardíaco la muerte de los tejidos del corazón debido a una obstrucción de su suministro sanguíneo (884)

heartwood / duramen la madera más vieja y que no conduce la electricidad, que se encuentra en el centro de un tronco de árbol (557)

helper T cell / célula T auxiliar un glóbulo blanco de la sangre necesario para que las células B desarrollen niveles normales de un anticuerpo (927)

herbaceous plant / planta herbácea una planta que es suave y verde, en vez de leñosa (556)

herbivore / herbívoro un organismo que sólo come plantas (346)

heredity / herencia la transmisión de caracteres genéticos de padres a hijos (8)

hermaphrodite / hermafrodita un organismo que tiene órganos reproductores tanto masculinos como femeninos (609)

heterotroph / heterótrofo un organismo que obtiene moléculas de alimento al comer otros organismos o sus productos secundarios y que no puede sintetizar compuestos orgánicos a partir de materiales inorgánicos (95)

heterozygous / heterocigoto término que describe a un individuo que tiene dos alelos distintos para un mismo carácter (167)

histamine / histamina una substancia química que estimula el sistema nervioso autónomo, la secreción de jugos gástricos y la dilatación de capilares (925)

HIV / VIH virus de inmunodeficiencia humana; el virus que causa el SIDA (12)

homeostasis / homeostasis la capacidad de mantener un estado interno constante en un ambiente en cambio; un estado interno constante que se mantiene en un ambiente en cambio al hacer ajustes continuos al ambiente interno y externo (8)

hominid / homínido miembro de la familia Hominidae del orden de los primates; caracterizado por tener pulgares oponibles y extremidades inferiores relativamente largas, ser bípedo y no tener cola; incluye a los seres humanos modernos y a sus ancestros (733)

homologous chromosomes / cromosomas homólogos cromosomas con la misma secuencia de genes, que tienen la misma estructura y que se acoplan durante la meiosis (120)

homologous structures / estructuras homólogas estructuras que comparten un ancestro común (286)

homozygous / homocigoto término que describe a un individuo que tiene alelos idénticos para un carácter en los dos cromosomas homólogos (167)

hormone / hormona una substancia que es producida en una célula o tejido, la cual causa un cambio en otra célula o tejido ubicado en una parte diferente del cuerpo (580)

Human Genome Project / Proyecto del Genoma Humano un esfuerzo de investigación para determinar la secuencia y ubicación de todo el conjunto de genes de las células humanas (233)

hydrostatic skeleton / esqueleto hidrostático la cavidad llena de agua de muchos invertebrados que tiene una función de sostén (607)

hypertonic / hipertónico término que describe una solución cuya concentración de soluto es más alta que la concentración del soluto en el interior de la célula (77)

hypha / hifa un filamento no-reproductor de un hongo (421)

hypothalamus / hipotálamo la región del cerebro que coordina las actividades de los sistemas nervioso y endocrino y que controla muchas actividades del cuerpo relacionadas con la homeostasis (952)

hypothesis / hipótesis una teoría o explicación basada en observaciones y que se puede probar (16)

hypotonic / hipotónico término que describe una solución cuya concentración de soluto es más baja que la concentración del soluto en el interior de la célula (77)

I

immunity / inmunidad la capacidad de resistir una infección o enfermedad, o de recuperarse de ella (931)

implantation / implantación el proceso por medio del cual el óvulo fecundado en la etapa de blastocisto se adhiere a la cubierta interior del útero (1004)

imprinting / impresión aprendizaje que ocurre rápidamente al inicio de la vida de un animal joven y que una vez que se aprende no se puede cambiar (829)

incomplete dominance / dominancia incompleta una condición en la que un carácter de un individuo es intermedio entre el fenotipo de los dos padres del individuo porque el alelo dominante no puede expresarse por completo (177)

independent assortment / distribución independiente la distribución al azar de pares de genes de diferentes cromosomas a los gametos (146)

independent variable / variable independiente el factor que se manipula deliberadamente en un experimento (17)

inflammatory response / reacción inflamatoria una reacción de protección de los tejidos afectados por una enfermedad o lesión, caracterizada por enrojecimiento, inflamación y dolor (925)

innate behavior / conducta innata una conducta heredada que no depende del ambiente ni de la experiencia (826)

insulin / insulina una hormona que es producida por un grupo de células especializadas en el páncreas y que reduce los niveles de glucosa en la sangre (986)

interferon / interferón una proteína que producen las células infectadas por un virus y que puede proteger a las células que no han sido infectadas contra la reproducción del virus (926)

internal fertilization / fecundación interna fecundación de un óvulo por un espermatozoide, la cual ocurre dentro del cuerpo de la hembra (610)

interneuron / interneurona una neurona ubicada entre la neurona aferente y la neurona final en una cadena neural (953)

interphase / interfase un período entre dos divisiones mitóticas o meióticas durante las cuales la célula crece, copia su ADN y sintetiza proteínas (125)

intron / intrón una sección del ADN que no codifica para ningún aminoácido y que se transcribe al ARN pero se elimina antes de ser traducida (218)

invertebrate / invertebrado un animal que no tiene columna vertebral (424)

ion / ion un átomo, radical o molécula que ha ganado o perdido uno o más electrones y que tiene una carga negativa o positiva (30)

ion channel / canal iónico un poro en la membrana celular a través del cual pueden pasar los iones (78)

isotonic solution / solución isotónica una solución cuya concentración de soluto es igual a la concentración de soluto en el interior de la célula (77)

J

joint / articulación un lugar donde se unen dos o más huesos (854)

K

karyotype / cariotipo una distribución de cromosomas que se encuentra en las células de un individuo en la metafase o en la mitosis, los cuales están ordenados en pares homólogos y en orden de mayor a menor (122)

keratin / queratina una proteína dura que forma el cabello, las plumas de las aves, las uñas y los cuernos (861)

kingdom / reino la categoría taxonómica más alta, que contiene un grupo de phyla similares (302)

Koch's postulates / postulados de Koch un procedimiento de cuatro etapas que formuló Robert Koch para identificar patógenos específicos y para determinar la causa de una determinada enfermedad (930)

Krebs cycle / ciclo de Krebs una serie de reacciones bioquímicas que convierten el ácido pirúvico en dióxido de carbono y agua; es la vía principal de oxidación en las células animales, bacterianas y vegetales, y libera energía (106)

krill / krill un crustáceo marino pequeño que es la fuente principal de alimentos de la ballena barbada (681)

K-strategist / estratega K una especie caracterizada por maduración lenta, pocas crías, crecimiento lento de la población, reproducción en etapas tardías de la vida y una densidad de población cercana a la capacidad de carga del ambiente (325)

L

lac operon / operón lac un sistema de genes cuyo gene operador y sus tres genes estructurales controlan el metabolismo de la lactosa en E. Coli (216)

larynx / laringe el área de la garganta que contiene las cuerdas vocales y que produce sonidos vocales (886)

lateral line / línea lateral una línea apenas visible que se encuentra a ambos lados del cuerpo de un pez y que recorre la longitud del cuerpo, marcando la ubicación de los órganos de los sentidos que detectan vibraciones en el agua (753)

law of independent assortment / ley de la distribución independiente la ley que establece que los genes se separan de manera independiente durante la meiosis (169)

law of segregation / ley de la segregación la ley de Mendel que establece que pares de cromosomas homólogos se separan en la meiosis de modo que sólo un cromosoma de cada par esté presente en cada gameto (169)

learning / aprendizaje el desarrollo de conductas por medio de la experiencia o práctica (827)

lichen / liquen una masa de células de hongos y de algas que crecen juntas en una relación simbiótica y que normalmente se encuentran en rocas o árboles (491)

life cycle / ciclo de vida todos los sucesos en el crecimiento y desarrollo de un organismo hasta que el organismo llega a su madurez sexual (152)

ligament / ligamento un tipo de tejido que mantiene unidos los huesos en una articulación (854)

light microscope / microscopio óptico un microscopio que usa un rayo de luz visible que pasa a través de uno o más lentes para magnificar la imagen de un objeto (57)

limnetic zone / zona limnética el área de un hábitat de agua dulce que está lejos de la costa, pero que se encuentra cerca de la superficie (376)

lipase / lipasa una enzima que descompone moléculas de grasa en ácidos grasos y glicerol (909)

lipid / lípido un tipo de substancia bioquímica que no se disuelve en agua, como por ejemplo, las grasas y esteroides; los lípidos almacenan energía y forman las membranas celulares (35)

lipid bilayer / bicapa lipídica la estructura básica de la membrana biológica, formada por dos capas de fosfolípidos (60)

littoral zone / zona litoral una zona poco profunda del hábitat de agua dulce donde la luz llega al fondo y nutre a las plantas (376)

logistic model / modelo logístico un modelo del crecimiento de la población que supone que los niveles finitos de recursos limitan el crecimiento de la población (323)

lung / pulmón el órgano central del aparato respiratorio en el que el oxígeno del aire se intercambia con el dióxido de carbono de la sangre (758)

lymphatic system / sistema linfático un conjunto de órganos cuya función principal es recolectar el fluido extracelular y regresarlo a la sangre; los órganos de este sistema incluyen los nodos linfáticos y los vasos linfáticos (875)

lysosome / lisosoma un organelo celular que contiene enzimas digestivas (64)

M

macrophage / macrófago una célula del sistema inmunológico que envuelve a los patógenos y otros materiales (926)

magnification / magnificación el aumento del tamaño aparente de un objeto mediante el uso de lentes o espejos (51)

Malpighian tubule / tubo de Malpighi un tubo excretorio que se abre hacia la parte trasera del intestino de la mayoría de los insectos y ciertos artrópodos (669)

mammary gland / glándula mamaria una glándula que se encuentra en el pecho de los mamíferos hembra y que secreta leche (806)

mandible / mandíbula un tipo de parte de la boca que se encuentra en algunos artrópodos y que se usa para perforar y chupar la comida; la parte inferior de la quijada (673)

mantle / manto en biología, una capa de tejido que cubre el cuerpo de muchos invertebrados (643)

mass extinction / extinción masiva un episodio durante el cual grandes cantidades de especies se extinguen (263)

medusa / medusa una etapa sexual del ciclo de vida de un cnidario, que nada libremente, tiene la apariencia de un aguamala y la forma de un paraguas; también, un aguamala o hidra (622)

meiosis / meiosis un proceso de división celular durante el cual el número de cromosomas disminuye a la mitad del número original por medio de dos divisiones del núcleo, lo cual resulta en la producción de células sexuales (gametos o esporas) (144)

melanin / melanina un pigmento que ayuda a determinar el color de la piel (862)

membrane potential / potencial de membrana la diferencia en potencial eléctrico entre los dos lados de una membrana celular (945)

menstrual cycle / ciclo menstrual el ciclo reproductor femenino, caracterizado por un cambio mensual en el revestimiento del útero y una descarga de sangre (1002)

menstruation / menstruación la descarga de sangre y tejido de desecho del útero durante el ciclo menstrual (1003)

meristem / meristemo una región de células vegetales no diferenciadas que son capaces de dividirse y desarrollarse en tejidos vegetales especializados (507)

mesoderm / mesodermo en un embrión, la capa de células intermedia que da origen a los músculos, sangre y varios sistemas (596)

mesophyll / mesófilo en las hojas, el tejido que se encuentra entre capas de epidermis, donde ocurre la fotosíntesis (559)

metabolism / metabolismo la suma de todos los procesos químicos que ocurren en un organismo (7)

metamorphosis / metamorfosis una fase del ciclo de vida de muchos animales durante la cual ocurre un cambio rápido del organismo inmaduro al adulto; un ejemplo es el cambio de larva a adulto en los insectos (675)

mineral / mineral un sólido natural, normalmente inorgánico, que tiene una composición química característica, una estructura interna ordenada y propiedades físicas y químicas características (905)

mitochondrion / mitocondria en las células eucarióticas, el organelo celular rodeado por dos membranas que es el lugar donde se lleva a cabo la respiración celular, la cual produce ATP (65)

mitosis / mitosis en las células eucarióticas, un proceso de división celular que forma dos núcleos nuevos, cada uno de los cuales posee el mismo número de cromosomas (125)

molecule / molécula la unidad más pequeña de una substancia que conserva todas las propiedades físicas y químicas de esa substancia; puede estar formada por un átomo o por dos o más átomos enlazados uno con el otro (29)

molting / mudar la muda de un exoesqueleto, piel, plumas o cuernos, los cuales son reemplazados por partes nuevas (668)

monocot / monocotiledónea una planta que produce semillas que sólo tienen un cotiledón (515)

monohybrid cross / cruza monohíbrida una cruza entre individuos que involucra un par de caracteres contrastantes (164)

monosaccharide/monosacárido un azúcar simple que es una subunidad fundamental de los carbohidratos (34)

motor neuron / neurona motora una célula nerviosa que transmite impulsos nerviosos del sistema nervioso central a los músculos y a las glándulas (950)

mRNA / ARNm ARNm mensajero; una molécula de ARN de una sola hebra que codifica la información para hacer una proteína (211)

mucous membrane / membrana mucosa la capa de tejido epitelial que cubre las superficies internas del cuerpo y que secreta moco (924)

multiple alleles / alelos múltiples más de dos alelos (versiones del gene) para un carácter genético (178)

muscle tissue / tejido muscular el tejido formado por células que se contraen y relajan para producir movimiento (847)

mutation / mutación un cambio en la secuencia de la base de nucleótidos de un gene o de una molécula de ADN (8)

mutualism / mutualismo una relación entre dos especies en la que ambas se benefician (265)

mycelium / micelio una masa de filamentos de hongos, o hifas, que forma el cuerpo de un hongo (483)

mycorrhiza / micorriza una asociación simbiótica entre los hongos y las raíces de las plantas (265)

myofibril / miofibrilla una fibra que se encuentra en las células de los músculos estriados, la cual es responsable de la contracción muscular (857)

myosin / miosina la proteína más abundante en los tejidos musculares, la cual es el elemento constitutivo principal de los filamentos gruesos de las fibras musculares (857)

N

natural killer cell / célula asesina natural un glóbulo blanco presente en individuos que no han sido inmunizados, el cual destruye una variedad de células (926)

natural selection / selección natural el proceso por medio del cual los individuos que tienen condiciones favorables y que están mejor adaptados a su ambiente sobreviven y se reproducen con más éxito que los individuos que no están tan bien adaptados (9)

nauplius / naupilia la larva de la mayoría de los crustáceos, la cual nada libremente (680)

nematocyst / nematocisto en los cnidarios, una célula urticante que se usa para inyectar una toxina en una presa (623)

nephridium / nefridio túbulo a través del cual algunos invertebrados eliminan desechos (644)

nephron / nefrona la unidad funcional del riñón (913)

nerve / nervio un conjunto de fibras nerviosas a través de las cuales se desplazan los impulsos entre el sistema nervioso central y otras partes del cuerpo (944)

nervous tissue / tejido nervioso el tejido del sistema nervioso, formado por neuronas, sus células de apoyo y el tejido conectivo (847)

neuron / neurona una célula nerviosa que está especializada en recibir y transmitir impulsos eléctricos (944)

neurotransmitter / neurotransmisor una substancia química que transmite impulsos nerviosos por una sinapsis (948)

neutrophil / neutrófilo un leucocito grande que contiene un núcleo lobulado y muchos gránulos citoplásmicos (926)

niche / nicho la posición única que ocupa una especie, tanto en lo que se refiere al uso de su hábitat como en cuanto a su función dentro de una comunidad ecológica (365)

nitrogen fixation / fijación de nitrógeno el proceso por medio del cual el nitrógeno gaseoso se transforma en amoniaco, un compuesto que los organismos utilizan para elaborar aminoácidos y otras moléculas orgánicas que contienen nitrógeno (363)

nonvascular plant / planta no vascular los tres tipos de plantas (hepáticas, milhojas y musgos) que carecen de tejidos transportadores y de raíces, tallos y hojas verdaderas (504)

norepinephrine / norepinefrina una substancia química que es un neurotransmisor producido por las terminaciones nerviosas simpáticas en el sistema nervioso, y también una hormona secretada por la médula suprarrenal para estimular las funciones de los aparatos circulatorio y respiratorio (abreviatura: NE) (985)

normal distribution / distribución normal una distribución de datos numéricos cuya gráfica forma una curva en forma de campana que es simétrica respecto a la media (331)

notochord / notocordio el eje de soporte que tiene forma de bastoncillo y está ubicado en la parte dorsal de los embriones de todos los cordados, incluyendo los vertebrados (700)

nucleic acid / ácido nucleico un compuesto orgánico, ya sea ARN o ADN, cuyas moléculas están formadas por una o más cadenas de nucleótidos y que contiene información genética (37)

nucleotide / nucleótido en una cadena de ácidos nucleicos, una subunidad formada por un azúcar, un fosfato y una base nitrogenada (37)

nucleus / núcleo en una célula eucariótica, biología, un organelo cubierto por una membrana, el cual contiene el ADN de la célula y participa en procesos tales como el crecimiento, metabolismo y reproducción (58)

nutrient / nutriente una substancia o compuesto que proporciona nutrición (o alimento) o materias primas que se necesitan para llevar a cabo procesos vitales (900)

nymph / ninfa una etapa inmadura de algunos insectos que es similar en función y estructura al adulto (675)

O

observation / observación el proceso de obtener información por medio de los sentidos; la información que se obtiene al usar los sentidos (14)

omnivore / omnívoro un organismo que come tanto plantas como animales (346)

oogenesis / oogénesis la producción, crecimiento y maduración de un óvulo (148)

open circulatory system / aparato circulatorio abierto un tipo de aparato circulatorio en el que el fluido circulatorio no está totalmente contenido en los vasos sanguíneos; un corazón bombea fluido por los vasos sanguíneos, los cuales se vacían en espacios llamados senos (606)

operator / operador una secuencia corta de ADN viral o bacteriano a la que se une un represor para impedir la transcripción (síntesis de ARNm) del gene adyacente en un operón (216)

operculum / opérculo en los peces, una placa dura que se encuentra adherida a cada lado de la cabeza, cubre las branquias y está abierta en la parte trasera (756)

operon / operón una unidad de regulación y transcripción de los genes en el ADN bacteriano, formada por un promotor, un operador y uno o más genes estructurales (216)

optic nerve / nervio óptico el nervio que conecta la retina del ojo con el cerebro y que transmite impulsos que contribuyen al sentido de la vista (958)

order / orden la categoría taxonómica que se encuentra debajo de la clase y arriba de la familia (302)

organ / órgano un conjunto de tejidos que desempeñan una función especializada en el cuerpo (419)

organ system / aparato (o sistema) de órganos un grupo de órganos que trabajan en conjunto para desempeñar funciones corporales (419)

organelle / organelo uno de los cuerpos pequeños que se encuentran en el citoplasma de una célula y que están especializados para llevar a cabo una función específica (58)

osmosis / ósmosis la difusión de agua u otro solvente de una solución más diluida (de un soluto) a una solución más concentrada (del soluto) a través de una membrana que es permeable al solvente (76)

ossicle / osículo una de las pequeñas placas de carbonato de calcio que forman el endoesqueleto de un equinodermo (694)

osteocyte / osteocito una célula ósea (852)

ovarian cycle / ciclo ovárico una serie de cambios inducidos por hormonas en los cuales los ovarios preparan y liberan un óvulo maduro todos los meses (1001)

ovary / ovario en el aparato reproductor femenino de los animales, un órgano que produce óvulos; en las plantas con flores, la parte inferior del pistilo que produce óvulos (538)

oviparous / ovíparo término que describe organismos que producen huevos que se desarrollan fuera del cuerpo de la madre, y cuyas crías también salen del cascarón fuera del cuerpo de la madre (777)

ovoviviparous / ovovivíparo término que describe a organismos que producen huevos que se desarrollan dentro del cuerpo de la madre, y cuyas crías también salen del cascarón dentro del cuerpo de la madre (777)

ovulation / ovulación la liberación de un óvulo de un folículo del ovario (1001)

ovule / óvulo una estructura del ovario de una planta con semillas que contiene un saco embrionario y se desarrolla para convertirse en una semilla después de la fecundación (534)

ovum / óvulo una célula sexual madura (149)

P

P generation / generación P generación parental; los primeros dos individuos que se aparean en una cruza genética (164)

paleontologist / paleontólogo un científico que estudia los fósiles (285)

Pangaea / Pangea una sola masa de tierra que existió durante aproximadamente 40 millones de años y luego comenzó a separarse para formar los continentes, tal como los conocemos en la actualidad (722)

parapodium / parapodio en los poliquetos, uno de los dos apéndices que se usan para locomoción o para el intercambio de gases (652)

parasitism / parasitismo una relación entre dos especies en la que una, el parásito, se beneficia de la otra, el huésped, y normalmente lo daña (362)

passive transport / transporte pasivo el movimiento de substancias a través de una membrana celular sin que la célula tenga que usar energía (74)

pathogen / patógeno un virus, microorganismo u otra substancia que causa enfermedades; un agente infeccioso (454)

pedigree / pedigrí un diagrama que muestra la incidencia de un carácter genético en varias generaciones de una familia (175)

pedipalp / pedipalpo uno de los dos pares de apéndices que se encuentran junto a la boca de un arácnido y que sirven para masticar y manipular a las presas (670)

pelvic inflammatory disease / enfermedad pélvica inflamatoria una infección pélvica del aparato reproductor femenino superior, incluyendo al útero, ovarios, trompas de Falopio y otras estructuras; es una enfermedad que se transmite sexualmente (1009)

penis / pene el órgano masculino que transfiere espermatozoides a una hembra y que lleva la orina hacia el exterior del cuerpo (998)

pepsin / pepsina una enzima que se encuentra en los jugos gástricos y que sirve para descomponer proteínas en moléculas más pequeñas (908)

perennial / perenne una planta cuyas partes vegetativas subterráneas viven más de dos años y cuyas partes superiores mueren y vuelven a crecer estacional o anualmente (573)

periosteum / periostio el tejido fibroso que cubre los huesos (851)

peripheral nervous system / sistema nervioso periférico todas las partes del sistema nervioso, excepto el encéfalo y la médula espinal (el sistema nervioso central); incluye los nervios craneales y los nervios del cuello, pecho, espalda baja y pelvis (950)

petal / pétalo una de las partes de una flor que normalmente tienen colores brillantes y forma de hoja, las cuales forman uno de los anillos de una flor (538)

petiole / pecíolo el pedúnculo que une una hoja al tallo de una planta (558)

pH / pH un valor que expresa la acidez o la alcalinidad (basicidad) de un sistema; cada número entero de la escala indica un cambio de 10 veces en la acidez; un pH de 7 es neutro, un pH de menos de 7 es ácido y un pH de más de 7 es básico (16)

pharynx / faringe en los gusanos planos, el tubo muscular que va de la boca a la cavidad gastrovascular; en los animales que tienen tracto digestivo, el conducto que va de la boca a la laringe y al esófago (886)

phenotype / fenotipo la apariencia de un organismo u otra característica perceptible que resulta debido al genotipo del organismo y a su ambiente (166)

phloem / floema el tejido que transporta alimento (azúcares, aminoácidos y nutrientes minerales) en las plantas vasculares (507)

phospholipid / fosfolípido un lípido que contiene fósforo y que es un componente estructural de la membrana celular (60)

photoperiodism / fotoperiodicidad la respuesta de las plantas a los cambios de estación durante la duración relativa de las noches y de los días (583)

photosynthesis / fotosíntesis el proceso por medio del cual las plantas, algas y algunas bacterias utilizan la luz solar, dióxido de carbono y agua para producir carbohidratos y oxígeno (94)

phylogenetic tree / árbol filogenético un diagrama ramificado que muestra cómo se relacionan los organismos a través de la evolución (602)

phylogeny / filogenia la historia evolutiva de una especie o grupo taxonómico (307)

phylum / phylum el grupo taxonómico que se ubica debajo del reino y arriba de la clase (302)

pigment / pigmento una substancia que le da color a otra substancia o mezcla (98)

pilus / pilus un apéndice corto y grueso que le permite a una bacteria unirse a otra (442)

pioneer species / especie pionera una especie que coloniza un área deshabitada y empieza un ciclo ecológico en el cual se establecen muchas otras especies (343)

pistil / pistilo la parte reproductora femenina de una flor, la cual produce semillas y está formada por el ovario, estilo y estigma (538)

pith / médula el tejido que se ubica en el centro del tallo de la mayoría de las plantas vasculares y que se utiliza para almacenamiento (556)

pituitary gland / glándula pituitaria una glándula endocrina que se ubica en la base del encéfalo, almacena y libera hormonas producidas por el hipotálamo y secreta hormonas bajo el control del hipotálamo (983)

placenta / placenta la estructura que une al feto en desarrollo con el útero y que permite el intercambio de nutrientes, desechos y gases entre la madre y el feto (1005)

plankton / plancton la masa de organismos casi microscópicos que flotan o se encuentran a la deriva en aguas (dulces y marinas) de ambientes acuáticos (378)

plant propagation / propagación vegetal la práctica de reproducir plantas a partir de semillas o partes vegetativas (544)

planula / plánula la larva ciliada de un cnidario, la cual nada libremente (625)

plasma / plasma en biología, el componente líquido de la sangre (876)

plasma cell / célula plasmática un tipo de glóbulo blanco que produce anticuerpos (929)

plasmid / plásmido una molécula de ADN circular que se encuentra comúnmente en las bacterias y que puede duplicarse independientemente del cromosoma principal (229)

plasmodium / plasmodio el citoplasma plurinucleado de un moho de fango, el cual está rodeado por una membrana y se mueve como si fuera una masa (470)

plastron / plastrón la porción inferior, o ventral, del caparazón de una tortuga (782)

platelet / plaqueta el fragmento de una célula que se necesita para formar coágulos sanguíneos (877)

point mutation / mutación puntual una mutación en la que sólo cambia un nucleótido o una base nitrogenada en un gene (219)

pollen grain / grano de polen la estructura que contiene el gametofito masculino en las plantas con semilla (534)

pollen tube / tubo de polen una estructura tubular que crece a partir de un grano de polen, entra al saco embrionario y permite que las células reproductoras masculinas se muevan al óvulo (534)

pollination / polinización la transferencia de polen de las estructuras reproductoras masculinas (las anteras) de una flor a la punta de la estructura reproductora femenina (el pistilo) en las angiospermas o al óvulo en las gimnospermas (534)

polygenic trait / carácter poligénico una característica de un organismo que está determinada por muchos genes (177)

polyp / pólipo una forma de un cnidario que tiene un cuerpo hueco y cilíndrico y que normalmente está unido a una roca o a otro objeto (622)

population / población un grupo de organismos de la misma especie que viven en un área geográfica específica y se reproducen entre sí (278)

population density / densidad de población el número de individuos de la misma especie que viven en una unidad superficial determinada (321)

predation / depredación la interacción entre dos especies en la que una especie, el depredador, se alimenta de la otra especie, la presa (362)

prediction / predicción una afirmación que se hace por anticipado, la cual expresa los resultados que se obtendrán al poner a prueba una hipótesis si ésta es corroborada; el resultado esperado si la hipótesis es correcta (16)

preen gland / glándula uropígea en las aves, una glándula especial que secreta grasa que el ave esparce en sus plumas para limpiarlas e impermeabilizarlas (785)

pregnancy / embarazo el período de tiempo que transcurre entre la concepción y el nacimiento (1005)

primary growth / crecimiento primario el crecimiento que ocurre como resultado de la división celular en las puntas de los tallos y raíces y que da lugar al tejido primario (574)

primary productivity / productividad primaria la cantidad total de material orgánico que producen los organismos autótrofos de un ecosistema (345)

primary succession / sucesión primaria sucesión que comienza en un área donde previamente no podía existir la vida (343)

primate / primate un miembro del orden de los primates, el grupo de mamíferos entre los que se encuentran los seres humanos, simios y monos; normalmente se distinguen por tener cerebros muy desarrollados, ojos que miran hacia delante, uso de las manos y locomoción variada (731)

prion / prión una partícula infecciosa formada únicamente por una proteína y que no contiene ni ADN ni ARN (441)

probability / probabilidad término que describe qué tan probable es que ocurra un posible evento futuro en un caso dado del evento; la proporción matemática del número de veces que es posible que ocurra un resultado de cualquier evento respecto al número de resultados posibles del evento (173)

probe / sonda una hebra de ARN o una sola hebra de ADN que se ha marcado con un elemento radiactivo o con un color fluorescente, y que se usa en ingeniería genética para enlazarse con un gene específico e identificarlo (231)

producer / productor un organismo que elabora moléculas orgánicas a partir de moléculas inorgánicas; un autótrofo fotosintético o quimiosintético que funciona como la fuente fundamental de alimento en un ecosistema (345)

profundal zone / zona profunda la zona de un hábitat de agua dulce en la que entra poca luz solar (376)

proglottid / proglótido una de las muchas secciones corporales de una tenia; contiene los órganos reproductores (631)

prokaryote / procariote un organismo que está formado por una sola célula y que no tiene núcleo ni organelos celulares; un ejemplo es una bacteria (57)

prosimian / prosimio un miembro de un suborden de los primates que es principalmente arbóreo y nocturno, tal como el lémur, lorí o tarsius (731)

prostate gland / glándula próstata una glándula que contribuye al fluido seminal en los machos (998)

protein / proteína un compuesto orgánico que está hecho de una o más cadenas de aminoácidos y que es el principal componente de todas las células (36)

protist / protista un organismo que pertenece al reino Protista (261)

protostome / protóstomo un organismo cuyo blastoporo embriónico se desarrolla para convertirse en la boca, cuyo celoma surge por esquizocelia y cuyo embrión tiene segmentación determinada (692)

protozoan / protozoario un protista unicelular que puede ser acuático o parasítico, tiene organelos cubiertos por una membrana y se puede mover independientemente; entre los ejemplos se encuentran las amebas y los paramecios (461)

provirus / provirus ADN viral que se ha unido al cromosoma de una célula huésped y se ha duplicado con el ADN del cromosoma (436)

pseudocoelomate / pseudocelomado un animal que tiene un pseudoceloma, o cavidad falsa del cuerpo (600)

pseudopodium / pesudópodo una extensión citoplásmica retráctil y temporal que tiene una función en la ingestión de alimentos y en el movimiento de algunas células ameboides (464)

psychoactive drug / droga psicoactiva una substancia que tiene un efecto considerable en la mente o en el comportamiento (961)

pulmonary vein / vena pulmonar la vena que lleva sangre oxigenada de los pulmones al corazón (759)

pulse / pulso la presión rítmica de la sangre contra las paredes de un vaso sanguíneo, particularmente de una arteria (883)

punctuated equilibrium / equilibrio puntuado un modelo de evolución en el que períodos cortos en los que ocurren cambios drásticos en una especie (incluyendo extinciones masivas y especiación rápida) están separados por períodos largos en los que ocurren muy pocos cambios o en los que no ocurre ningún cambio (282)

Punnett square / cuadro de Punnett una gráfica que se usa para predecir los resultados de una cruza genética (170)

pupa / pupa la etapa inmóvil y que no se alimenta, entre la larva y el adulto de insectos que experimentan una metamorfosis completa; en la etapa de pupa, el organismo normalmente está encerrado en un capullo o crisálida y sufre importantes cambios anatómicos (675)

R

radial symmetry / simetría radial un plan corporal en el que las partes del cuerpo del animal están organizadas en un círculo alrededor de un eje central (598)

radiometric dating / datación radiométrica un método para determinar la edad de un objeto estimando los porcentajes relativos de un isótopo radiactivo (precursor) y un isótopo estable (hijo) (252)

radula / rádula un órgano áspero similar a una lengua, que está cubierto de dientes quitinosos y que muchos moluscos utilizan para alimentarse (643)

realized niche / nicho realizado la gama de recursos que una especie usa, las condiciones que la especie tolera y los papeles funcionales que la especie juega como resultado de la competencia en su nicho fundamental (367)

reasoning / razonamiento el acto de sacar una conclusión a partir de hechos o suposiciones (828)

receptor protein / proteína receptora una proteína que liga moléculas señal específicas, lo cual hace que la célula responda (84)

recessive / recesivo término que describe un carácter o un alelo que se expresa sólo cuando se heredan dos alelos recesivos de la misma característica (167)

recombinant DNA / ADN recombinante moléculas de ADN que son creadas artificialmente al combinar ADN de diferentes fuentes (228)

red blood cell / glóbulo rojo una célula que tiene forma de disco y que no tiene núcleo, contiene hemoglobina y transporta oxígeno en el aparato circulatorio (876)

reflex / reflejo un movimiento involuntario y prácticamente inmediato en respuesta a un estímulo (952)

replication fork / horquilla de replicación un punto que tiene forma de Y, el cual se produce cuando las dos hebras de una doble hélice de ADN se separan de modo que la molécula de ADN pueda duplicarse (199)

repressor / represor una proteína reguladora que se une a un operador y bloquea la transcripción de los genes de un operón (216)

reproduction / reproducción el proceso de producir descendencia (7)

reproductive isolation / aislamiento reproductivo la incapacidad de los miembros de una población de reproducirse exitosamente con miembros de otra población de la misma especie o de una especie relacionada (281)

resolution / resolución en los microscopios, la capacidad de formar imágenes con detalles precisos (51)

respiration / respiración en biología, el intercambio de oxígeno y dióxido de carbono entre células vivas y su ambiente; incluye la respiración y la respiración celular (605)

resting potential / potencial de reposo el potencial eléctrico que existe en la membrana celular de una célula nerviosa o de una célula muscular cuando la célula no está activa (946)

restriction enzyme / enzima de restricción una enzima que destruye moléculas de ADN extraño cortándolas en sitios específicos (229)

retina / retina la capa interna del ojo, sensible a la luz, que recibe imágenes formadas por el lente ocular y las transmite al cerebro por medio del nervio óptico (958)

Rh factor / factor Rh uno de varios antígenos de los grupos sanguíneos que son transportados en la superficie de los glóbulos rojos (879)

rhizoid / rizoide una estructura parecida a una raíz que se encuentra en las plantas no vasculares, tales como los musgos o las hepáticas, la cual mantiene a las plantas en su lugar y tienen una función en la absorción (486)

rhizome / rizoma un tallo horizontal subterráneo que proporciona un mecanismo de reproducción asexual (510)

ribosome / ribosoma un organelo celular compuesto de ARN y proteína; el sitio donde ocurre la síntesis de proteínas (56)

RNA / ARN ácido ribonucleico; un polímero natural que se encuentra en todas las células vivas y que juega un papel en la síntesis de proteínas (37)

RNA polymerase / ARN polimerasa una enzima que comienza (cataliza) la formación de ARN usando una hebra de una molécula de ADN como plantilla (209)

rod / bastoncillo uno de los dos tipos de células detectoras de luz que hay en el ojo; los bastoncillos pueden detectar luz tenue y juegan un papel importante en la visión nocturna y sin color (958)

root / raíz el órgano principalmente subterráneo de las plantas vasculares, el cual mantiene a las plantas en su lugar y absorbe y almacena agua y minerales del suelo (507)

root cap / cofia la capa protectora de células que cubre la punta de una raíz (555)

root hair / pelo radicular una extensión de la epidermis de una raíz, la cual aumenta el área superficial de la raíz para la absorción (555)

rRNA / ARNr ARN ribosomal; un organelo que contiene la mayor parte del ARN en la célula y que es responsable del funcionamiento de los ribosomas (212)

r-strategist / estratega r una especie que está adaptada para vivir en un ambiente donde los cambios son rápidos e impredecibles; se caracteriza por presentar crecimiento rápido, alta fertilidad, período de vida corto, cuerpo pequeño y crecimiento exponencial de la población (324)

S

sapwood / albura el tejido del xilema secundario que se distribuye en el exterior del tronco de un árbol y que tiene una función en el transporte de la savia (557)

sarcomere / sarcómero la unidad fundamental de contracción del músculo esquelético y cardíaco (857)

scanning electron microscope / microscopio electrónico de barrido un microscopio que produce una imagen agrandada tridimensional de un objeto usando un haz de electrones en lugar de luz (54)

sebum / sebo la secreción grasosa de las glándulas sebáceas (864)

second messenger / mensajero secundario una molécula que se genera cuando una substancia específica se une a un receptor en el exterior de la membrana celular, lo cual produce un cambio en la función celular (85)

secondary growth / crecimiento secundario crecimiento de las plantas que ocurre como resultado de la división celular en los cámbiums, o meristemos laterales, y que hace que se engruesen los tallos y las raíces (574)

secondary succession / sucesión secundaria el proceso por medio del cual una comunidad reemplaza a otra, la cual ha sido parcial o totalmente destruida (343)

seed / semilla el embrión de una planta que está encerrado en una cubierta protectora (504)

seed coat / cubierta seminal la cubierta exterior y protectora de una semilla (535)

seed plant / planta con semillas una planta que produce semillas (504)

semen / semen el fluido que contiene espermatozoides y varias secreciones producidas por los órganos reproductores masculinos (998)

semicircular canal / canal semicircular el canal lleno de fluido ubicado en el oído interno, el cual ayuda a mantener el equilibrio y a coordinar los movimientos (959)

seminal vesicle / vesícula seminal una de las dos estructuras glandulares en los vertebrados macho, las cuales acumulan y secretan fluido seminal (998)

seminiferous tubule / túbulo seminífero uno de los muchos túbulos que hay en los testículos, en donde se producen los espermatozoides (996)

sensory neuron / neurona sensorial una neurona que lleva estímulos de un órgano sensorial al sistema nervioso central (950)

sensory receptor / receptor sensorial una estructura especializada que contiene los extremos de las neuronas sensoriales y que responde a tipos específicos de estímulos (956)

sepal / sépalo en una flor, uno de los anillos más externos de hojas modificadas que protegen el capullo de la flor (538)

septum / septo una pared divisoria, o partición, tal como la pared que se encuentra entre células adyacentes en las hifas de los hongos, la pared interna que se encuentra entre segmentos adyacentes de un anélido y la pared gruesa que se encuentra entre las cámaras derecha e izquierda del corazón (760)

sessile / sésil término que describe a un organismo que permanece unido a una superficie durante toda su vida y no se mueve (618)

seta / seda una de las cerdas o espinas externas que se proyectan del cuerpo de un anélido (652)

sex chromosome / cromosoma sexual uno de los dos cromosomas que determinan el sexo de un individuo (122)

sex-linked trait / carácter ligado al sexo un carácter que es determinado por un gene que se encuentra en uno de los cromosomas sexuales, tal como el cromosoma X o el cromosoma Y en los seres humanos (175)

sexual reproduction / reproducción sexual reproducción en la que se unen los gametos de los dos padres (150)

sexual selection / selección sexual un mecanismo evolutivo por medio del cual los caracteres que aumentan la capacidad de los individuos de atraer o adquirir una pareja aparecen con más frecuencia en una población; selección en la que se elige una pareja con base en un carácter o caracteres particulares (6)

shoot / brote la porción de una planta que crece principalmente sobre el suelo; incluye los tallos y las hojas (507)

sieve tube / tubo criboso en el floema de una planta que da flores, un tubo de transporte, que está hecho de una serie de miembros de tubos cribosos apilados de un extremo al otro (554)

sink / sumidero cualquier lugar donde una planta almacena o usa nutrientes orgánicos, tales como azúcares o almidones (564)

sinoatrial node / nodo sinoauricular una masa de células de músculo cardíaco que se encuentra en la unión entre la vena cava superior y el atrio derecho, y que inicia y regula la contracción del corazón (abreviatura: nodo SA) (882)

siphon / sifón un conducto hueco que los bivalvos utilizan para sorber y expulsar el agua de mar (647)

skin gill / branquia dérmicas una estructura transparente que protege la superficie de las estrellas de mar y les permite respirar (695)

sodium-potassium pump / bomba de sodio-potasio una proteína transportadora que utiliza el ATP para efectuar el transporte activo de iones de sodio hacia el exterior de la célula y de iones de potasio hacia el interior de la célula (81)

solution / solución una mezcla homogénea de dos o más sustancias dispersas de manera uniforme en una sola fase (32)

sorus / soro un grupo de esporas o esporangios (532)

source / fuente la parte de una planta que elabora azúcares y otros compuestos orgánicos y a partir de la cual estos compuestos se transportan a otras partes de la planta (564)

speciation / especiación la formación de especies nuevas como resultado de la evolución por selección natural (291)

species / especie un grupo de organismos que tienen un parentesco cercano y que pueden aparearse de modo natural para producir descendencia fértil; tambien, el nivel de clasificación debajo de género y arriba de subespecie (9)

sperm / espermatozoide el gameto masculino (célula sexual) (148)

spermatogenesis / espermatogénesis el proceso por medio del cual se forman los gametos masculinos (148)

spicule / espícula una aguja de sílice o carbonato de calcio que se encuentra en el esqueleto de algunas esponjas (620)

spinal cord / médula espinal una columna de tejido nervioso que se origina en la base del cerebro y corre a lo largo de la columna vertebral (952)

spindle / huso mitótico una red de microtúbulos que se forma durante la mitosis y que mueve cromátidas a los polos (128)

spinneret / hilera un órgano que utilizan las arañas y algunas larvas de insectos para producir hilos sedosos con los que hacen redes y capullos (670)

spiracle / espiráculo una abertura externa de un insecto o artrópodo, que se usa en la respiración (668)

spirillum / espirilo una bacteria que tiene forma en espiral (443)

spongin / espongina una proteína fibrosa que contiene azufre y que forma las fibras del esqueleto de algunas esponjas (620)

sporangium / esporangio una bolsa, cubierta, cápsula u otra estructura especializada que produce esporas (463)

spore / espora una célula reproductora o estructura multicelular que resiste las condiciones ambientales y que se puede desarrollar para convertirse en un adulto sin necesidad de fusionarse con otra célula (154)

sporophyte / esporofito en las plantas y algas que tienen generaciones alternas, el individuo o generación diploide que produce esporas haploides (154)

sporozoite / esporozoito un esporozoario que ha sido liberado de un ooquiste y está listo para entrar a una nueva célula huésped (473)

stabilizing selection / selección de estabilización un tipo de selección natural en la que se favorece la forma promedio de un carácter, el cual se vuelve más común (332)

stamen / estambre la estructura reproductora masculina de una flor, que produce polen y está formada por una antera ubicada en la punta del filamento (538)

steroid / esteroide un tipo de lípido que está formado por cuatro anillos de carbono a los cuales se encuentran unidos diversos grupos funcionales, y que normalmente tiene acción fisiológica (977)

stimulant / estimulante una droga que aumenta la actividad del cuerpo o la actividad de alguna parte del cuerpo (962)

stolon / estolón en las plantas, un tallo que crece y que al desarrollarse da lugar a raíces y brotes en los nodos o en la punta para formar individuos nuevos; la hifa de algunos hongos, la cual crece y da lugar a individuos nuevos (486)

stoma / estoma una de las muchas aberturas de una hoja o de un tallo de una planta, la cual permite que se lleve a cabo el intercambio de gases (502)

stroke / ataque de apoplejía una pérdida súbita de la conciencia o parálisis que ocurre cuando se interrumpe el flujo sanguíneo al cerebro (884)

subcutaneous tissue / tejido subcutáneo la capa de células que se encuentra debajo de la piel (863)

substrate / sustrato una parte, sustancia o elemento que se encuentra debajo de otra parte, sustancia o elemento y lo sostiene; el reactivo en reacciones que son catalizadas por enzimas (41)

succession / sucesión el reemplazo de un tipo de comunidad por otro en un mismo lugar a lo largo de un período de tiempo (343)

swim bladder / vejiga natatoria en los peces óseos, una bolsa llena de gas que se usa para controlar la flotabilidad (756)

symbiosis / simbiosis una relación en la que dos organismos diferentes viven estrechamente asociados uno con el otro (364)

synapse / sinapsis el punto en el cual el extremo del axón de una neurona se une con el extremo de una dendrita o con el cuerpo de la célula de otra neurona, o bien, se encuentra con otra célula (948)

syphilis / sífilis una enfermedad transmitida sexualmente, producida por la bacteria Treponema pallidum (1009)

T

target cell / célula blanco una célula específica a la que se dirige una hormona para producir un efecto específico (977)

taxonomy / taxonomía la ciencia de describir, nombrar y clasificar organismos (300)

teleost / teleósteo un grupo de peces con aletas rayadas que tienen una aleta caudal, escamas y una vejiga natatoria; el grupo más grande de peces óseos (757)

tendon / tendón un tejido conectivo duro que une un músculo con un hueso o con otra parte del cuerpo (856)

terrestrial / terrestre término que describe a un organismo que vive en la tierra (721)

test cross / cruza de prueba el cruzamiento de un individuo cuyo genotipo se desconoce con un individuo homocigoto recesivo para determinar el genotipo desconocido (172)

testes / testículos los principales órganos reproductores masculinos, los cuales producen espermatozoides y testosterona (996)

thalamus / tálamo la parte del cerebro que dirige a la región apropiada las señales sensoriales y motoras que se reciben (952)

thecondont / tecodonte el reptil extinto a partir del cual evolucionaron los dinosaurios (722)

theory / teoría una explicación sobre algún fenómeno que está basada en la observación, experimentación y razonamiento (19)

therapsid / terápsido el orden extinto de reptiles parecidos a mamíferos que posiblemente dio origen a los mamíferos (728)

thorax / tórax en los vertebrados superiores, la parte del cuerpo que se encuentra entre el cuello y el abdomen; en otros animales, la región del cuerpo que se encuentra detrás de la cabeza; en los artrópodos, la región media del cuerpo (666)

thylakoid / tilacoide un sistema de membranas que se encuentra dentro de los cloroplastos y que contiene los componentes para que se lleve a cabo la fotosíntesis (99)

tissue / tejido un grupo de células similares que llevan a cabo una función común (419)

tissue culture / cultivo de tejidos la técnica de cultivar células vivas en un medio artificial (544)

tolerance / tolerancia el estado de adicción a una droga en el que se necesitan mayores cantidades de la droga para obtener el efecto deseado (962)

toxin / toxina una sustancia que un organismo produce y que es venenosa para otros organismos (449)

trachea / tráquea en los insectos, miriápodos y arañas, uno de una red de conductos de aire; en los vertebrados, el conducto que une la faringe a los pulmones (668)

transcription / transcripción el proceso de formar un ácido nucleico usando otra molécula como plantilla; en particular, el proceso de sintetizar ARN usando una de las hebras de la molécula de ADN como plantilla (208)

transfer RNA / ARN de transferencia una molécula de ARN que transfiere aminoácidos al extremo en crecimiento de una cadena de polipéptidos durante la traducción (212)

transformation / transformación la transferencia de material genético en forma de fragmentos de ADN de una célula a otra o de un organismo a otro (191)

transgenic animal / animal transgénico un animal al que se le ha transferido material genético clonado (241)

translation / traducción la porción de la síntesis de proteínas que tiene lugar en los ribosomas y que usa los codones de las moléculas de ARNm para especificar la secuencia de aminoácidos en las cadenas de polipéptidos (208)

translocation / translocación el movimiento de un segmento de ADN de un cromosoma a otro, lo cual resulta en un cambio en la posición del segmento; también, el movimiento de nutrientes solubles de una parte a otra de una planta (564)

transpiration / transpiración el proceso por medio del cual las plantas liberan vapor de agua al aire por medio de los estomas; también, la liberación de vapor de agua al aire por otros organismos (351)

trochophore / trocófora una larva ciliada de muchos gusanos y algunos moluscos, la cual nada libremente (642)

trophic level / nivel trófico uno de los pasos de la cadena alimenticia o de la pirámide alimenticia; entre los ejemplos se encuentran los productores y los consumidores primarios, secundarios y terciarios (345)

tropism / tropismo el movimiento de un organismo o de una parte de él en respuesta a un estímulo externo, como por ejemplo, la luz o el calor; el movimiento puede ser hacia el estímulo o en sentido opuesto a él (582)

true-breeding / variedad pura término que describe organismos o genotipos que son homocigotos para un carácter específico y, por lo tanto, producen descendencia que tiene el mismo fenotipo para ese carácter (164)

U

ungulate / ungulado un animal que tiene pezuñas (814)

uracil / uracilo una de las cuatro bases que se combinan con un azúcar y un fosfato para formar una subunidad de nucleótido de ADN; el uracilo se une a la adenina (208)

urea / urea el principal producto nitrogenado que se obtiene del metabolismo de las proteínas, se forma en el hígado a partir de aminoácidos y compuestos de amoníaco y se encuentra en la orina y otros fluidos del cuerpo (912)

ureter / uréter uno de los dos tubos angostos que llevan orina de los riñones a la vejiga urinaria (914)

urethra / uretra el tubo que lleva orina de la vejiga urinaria al exterior del cuerpo (914)

urinary bladder / vejiga urinaria un órgano hueco y muscular que almacena orina (914)

urine / orina el líquido que excretan los riñones, se almacena en la vejiga y pasa a través de la uretra hacia el exterior del cuerpo (914)

uterus / útero en los mamíferos hembras, el órgano hueco y muscular en el que se incrusta el óvulo fecundado y en el que se desarrollan el embrión y el feto (1000)

V

vaccination / vacunación la administración a seres humanos o animales de organismos que han sido tratados para inducir una respuesta inmunológica (931)

vaccine / vacuna una sustancia que se prepara a partir de organismos patógenos muertos o debilitados y se introduce al cuerpo para producir inmunidad (190)

vagina / vagina el canal de las hembras que se extiende de la vulva al cuello del útero y que recibe al pene durante el coito (1000)

valve / válvula un pliegue de membranas que controla el flujo de un fluido (874)

vas deferens / conducto deferente un conducto a través del cual los espermatozoides se mueven del epidídimo al conducto eyaculatorio que está en la base del pene (997)

vascular bundle / haz vascular en una planta, una hebra de tejido de transporte que contiene tanto xilema como floema (556)

vascular cambium / cámbium vascular en una planta, el meristemo lateral que produce xilema y floema secundarios (575)

vascular plant / planta vascular una planta que tiene un sistema vascular formado por xilema y floema, tejidos especializados que transportan materiales de una parte de la planta a otra (504)

vascular system / sistema vascular un sistema de transporte de los tejidos que lleva agua y otros materiales en las plantas o en los animales (504)

vascular tissue / tejido vascular el tejido especializado de transporte que se encuentra en las plantas superiores y que está formado principalmente por xilema y floema (422)

vector / vector en biología, cualquier agente, como por ejemplo un plásmido o un virus, que tiene la capacidad de incorporar ADN extraño y de transferir ese ADN de un organismo a otro; un huésped intermediario que transfiere un organismo patógeno o un parásito a otro organismo (229)

vegetative part / parte vegetativa cualquier parte no reproductiva de una planta (516)

vegetative reproduction / reproducción vegetativa un tipo de reproducción asexual en el que crecen plantas nuevas a partir de partes plantas que no se reproducen (541)

vein / vena en biología, un vaso que lleva sangre al corazón (873)

vena cava / vena cava una de las dos venas grandes que llevan sangre de los tejidos del cuerpo al corazón (881)

ventricle / ventrículo una de las dos cámaras musculares grandes que bombean sangre hacia el exterior del corazón (881)

vertebra / vértebra uno de los 33 huesos de la columna vertebral (espina dorsal) (712)

vertebrate / vertebrado un animal que tiene columna vertebral; incluye a los mamíferos, aves, reptiles, anfibios y peces (267)

vesicle / vesícula una cavidad o bolsa pequeña que contiene materiales en una célula eucariótica; se forma cuando parte de la membrana celular rodea los materiales que van a ser llevados al interior la célula o transportados dentro de ella (63)

vessel / vaso en las plantas, una estructura tubular que se encuentra en el xilema y que está formada por células conectadas que transportan agua y nutrientes minerales; en los animales, un tubo o conducto que lleva sangre y otros fluidos del cuerpo (554)

vestigial structure / estructura vestigial una estructura de un organismo, cuyo tamaño y función están reducidos, pero que es posible que haya estado completa y que haya sido funcional en los ancestros del organismo (286)

villus / vellosidad intestina una de las muchas proyecciones diminutas de las células que se encuentran en la pared interior del intestino delgado; aumenta el área superficial de la pared para absorción (909)

viroid / viroide un agente infeccioso que está constituido por una hebra pequeña de ARN y que produce enfermedades en las plantas (441)

virulent / virulento término que describe a un microorganismo o virus que causa enfermedades y que es altamente infeccioso (190)

virus / virus una partícula infecciosa sin vida formada por un ácido nucleico y una cubierta de proteína; puede invadir una célula y destruirla (434)

visceral mass / masa visceral la sección central del cuerpo de un molusco, la cual contiene sus órganos (643)

vitamin / vitamina un compuesto orgánico que participa en las reacciones bioquímicas y que forma varias moléculas en el cuerpo; algunas vitaminas se llaman coenzimas y activan enzimas específicas (904)

W

water vascular system / sistema vascular acuoso un sistema de canales que están llenos de un fluido acuoso que circula por todo el cuerpo de los equinodermos (695)

weaning / destete el momento en el que termina la dependencia que tiene un animal en su madre para obtener alimento (leche) y protección (806)

white blood cell / glóbulo blanco un tipo de célula de la sangre que destruye bacterias, virus y proteínas tóxicas, y que ayuda al cuerpo a desarrollar inmunidad (877)

withdrawal / abstinencia la serie de síntomas asociados con la remoción del cuerpo de una droga adictiva (962)

Y

yeast / levadura un hongo unicelular muy pequeño que fermenta a los carbohidratos y los convierte en alcohol y dióxido de carbono; se usa para fermentar cerveza, para hacer pan y como fuente de vitaminas y proteínas (487)

Z

zygosporangium / en los miembros del
phylum Zygomycota, una estructura sexual que se forma debido a la fusión de dos gametangios y que contiene uno o más cigotos que se formaron a partir de la fusión de gametos producidos por los gametangios (486)

zygospore / zigospora en algunas algas, una estructura protectora que tiene una pared gruesa y que contiene un cigoto que se formó a partir de la fusión de dos gametos (462)

zygote / cigoto la célula que resulta debido a la fusión de los gametos; el óvulo fecundado (121)

Note: Boldface page numbers refer to primary discussions. Page numbers followed by f *refer to figures. Page numbers followed by* t *refer to tables.*

A

C

D

E

H

I

J

K

Q

R

W

X

Y

Z

Academic Reviewers

(continued from page iv)

Chris Nice, Ph.D.
Assistant Professor
Department of Biology
Texas State University
San Marcos, Texas

Martin Nickels, Ph.D.
Professor
Department of Anthropology
Illinois State University
Normal, Illinois

Eva Oberdörster, Ph.D.
Lecturer
Department of Biological Sciences
Southern Methodist University
Dallas, Texas

Brian L. Pagenkopf, Ph.D.
Assistant Professor
Department of Chemistry
University of Texas
Austin, Texas

Barron Rector, Ph.D.
Assistant Professor and Extension Range Specialist
Texas Agricultural Extension Service
Texas A&M University
College Station, Texas

Michael Ryan, Ph.D.
Clark Hubbs Regents Professor in Zoology
The University of Texas
Austin, Texas

Dork Sahagian
Research Professor, Stratigraphy and Basin Analysis, Geodynamics
Global Analysis, Interpretation, and Modeling Program
University of New Hampshire
Durham, New Hampshire

Miles Silman, Ph.D.
Assistant Professor of Biology
Department of Biology
Wake Forest University
Winston-Salem, North Carolina

Richard Storey, Ph.D.
Dean of the Faculty and Professor of Biology
Colorado College
Colorado Springs, Colorado

Mary Wicksten, Ph.D.
Professor of Biology
Department of Biology
Texas A&M University
College Station, Texas

Teacher Reviewers

Robert Akeson
Science Teacher
Boston Latin School
Boston, Massachusetts

Robert Baronak
Biology Teacher
Donegal High School
Mount Joy, Pennsylvania

Betty Bates
Biology Teacher
DeKalb High School
DeKalb, Texas

David Blinn
Biology Teacher
Wrenshall High School
Wrenshall, Minnesota

Cindy Copolo, Ph.D.
Science Education Consultant
Summit Solutions
Bahama, North Carolina

Linda Culp
Science Chair and Science Teacher
Thorndale High School
Thorndale, Texas

Katherine Cummings
Science Teacher
Currituck County High School
Currituck, North Carolina

Alonda Droege
Biology Teacher
Highline High School
Burien, Washington

Benjamin Ebersole
Science Teacher
Donegal High School
Mount Joy, Pennsylvania

Richard Filson
Science Teacher
Edison High School
Stockton, California

Randa Flinn
Science Teacher
Northeast High School
Fort Lauderdale, Florida

Arthur Goldsmith
Biology and Earth Sciences Teacher
Hallandale High School
Hallandale, Florida

Marguerite A. Graham
Science Teacher
Gulliver Preparatory School
Miami, Florida

Teacher Reviewers *(continued from page 1136)*

Stacey Jeffress
Environmental Science Teacher
El Dorado High School
El Dorado, Arkansas

Bruce Katt
Biology Teacher
Georgetown High School
Georgetown, Texas

Debbie Keel
Science Teacher
Cypress Creek High School
Houston, Texas

Mike Lubich
Science Teacher
Mapletown High School
Greensboro, Pennsylvania

Thomas Manerchia
Environmental Science Teacher, Retired
Archmere Academy
Claymont, Delaware

Jason Marsh
Biology Teacher
Moose Lake Community School
Moose Lake, Minnesota

Ray McLarty
Science Teacher
Hononegah High School
Rockdon, IL

Betsy McGrew
Science Teacher
Star Charter School
Austin, Texas

Tammie Niffenegger
Science Chair and Science Teacher
Port Washington High School
Port Washington, Wisconsin

Donna Norwood
Science Teacher
Monroe High School
Charlotte, North Carolina

Denice Lee Sandefur
Science Chairperson
Nucla High School
Nucla, Colorado

Bert Sherwood
Secondary Science Instructional Specialist
Socorro Independent School District
El Paso, Texas

Joe Stanaland
Biology Teacher
Lake Travis High School
Austin, Texas

Tyson Yager
Science Teacher
Wichita High School East
Wichita, Kansas

Staff Credits

Executive Editor, High School Sciences and Health
Mark Grayson

Managing Editor
Debbie Starr

Editorial Development Team
Karen Ross, Senior Editor
Roxanne McCauley
Dyanne Semerjibashian

Copyeditors
Dawn Marie Spinozza, Copyediting Manager
Anne-Marie De Witt
Simon Key
Jane A. Kirschman
Kira J. Watkins

Editorial Support Staff
Mary Anderson
Soojinn Choi
Jeanne Graham
Shannon Oehler
Stephanie Sanchez
Tanu'e White

Editorial Interns
Kristina Bigelow
Erica Garza
Sarah Ray
Kenneth G. Raymond
Kyle Stock
Audra Teinert

Online Products
Bob Tucek, Executive Editor
Wesley M. Bain
Catherine Gallagher
Douglas P. Rutley

Book Design
Kay Selke, Director of Book Design

Media Design
Richard Metzger, Design Director
Chris Smith, Developmental Designer

Cover Design
Kay Selke, Director of Book Design

Image Acquisitions
Curtis Riker, Director
Jeanie Taylor, Photo Research Manager
Diana Goetting, Senior Photo Researcher

Publishing Services
Carol Martin, Director
Jeff Robinson, Manager, Ancillary Design

Production
Eddie Dawson, Senior Production Manager
Beth Sample, Project Manager
Teresa McQueen, Senior Production Coordinator

Technology Services
Laura Likon, Director
Juan Baquera, Manager
JoAnn Stringer, Manager

Senior Technology Services Analysts
Katrina Gnader
Lana Kaupp

Production Artists
Sara Buller
Margaret Sanchez
Patty Zepeda

eMedia
Melanie Baccus, eMedia Coordinator
Ed Blake, Design Director
Kimberly Cammerata, Design Manager
Lydia Doty, Senior Project Manager
Marsh Flournoy, Technology Project Manager
Dakota Smith, Quality Assurance Analyst
Cathy Kuhles, Technical Assistant
Tara F. Ross, Senior Project Manager
Ken Whiteside, Manager, Application Development

Photography Credits

Abbreviation Code
AA=Animals, Animals; BC=Bruce Coleman, Inc.; DRK=DRK Photo; ES=Earth Scenes; GH=Grant Heilman Photography; GI=Getty Images; HRW/SD=Holt, Rinehart and Winston/Sam Dudgeon; MP=Minden Pictures; PA=Peter Arnold, Inc.; PR=Photo Researchers; PT=Phototake; RLM=Robert & Linda Mitchell; SP/FOCA= Sergio Purtell/Foca; TF=Tim Fuller; VU=Visuals Unlimited

Abbreviations used: (t) top, (b) bottom, (c) center, (l) left, (r) right, (bkgd) background

Forensics BIOWatch icon (magnifying glass) © Artville/GI, (fingerprint) © Steve Allen/Brand X Pictures/GI; Forensics Labs icon (handprint) © Image 100/Royalty Free/CORBIS, (magnifying glass) © PhotoDisc/GI; Forensics Labs border © Digital Vision/GI; Florida BIOWatch icon © Sea World, Inc./CORBIS

COVER AND TITLE PAGE © Michael Aw/Photodisc/GI; **ii** Don Riepe/PA; **iii** UNEP (Brunner)/PA; **v** Art Wolfe/GI; **vii** (t) Norbert Wu/PA, (b) Arthur Tilley/GI; **viii** (t) E.R. Degginger/AA/ES, (b) Jerry L. Ferrara/PR; **ix** R.H. Virdee/GH; **xi** (t) William H. Mullins/NASC/PR, (b) D. Cavagnaro/VU; **xii** PhotoDisc; **xiii** (t) William H. Mullins/NASC/PR., (b) Dan Guravich/PR; **xiv** SP/FOCA; **xv** Tom McHugh/PR; **xvi** A. Witte & C. Mahaney/GI; **xvii** (t) Marty Cordano/DRK, (c) Jeff Rotman/PA, (b) Scott Bauer/Agricultural Research Service, USDA; **xviii** HRW/SD; **xix** Patti Murray/AA/ES; **xx** HRW/SD; **xxi** C. Milkins/OSF/AA/ES; **CHAPTER 1: 2-3** (bkgd) Dr. Dennis Kunkel/PT; **2** PhotoDisc; **3** (b) C. Meckes/Ottawa/PR, (c) Wood Ransaville Harlin, Inc., (t) Corbis; **4** TF; **6** UNEP/Brunner/PA; **7** (b) TF, (c) Zig Leszczynski/AA/ES, (t) M. Abbey/VU; **8** (b) Walter Hodges/GI, (tl) Dan Guravich/NASC/PR; **9** (b) Gerard Lacz/PA, (t) Jamie Harron-Papilio/Corbis; **11** Mauro Fermariello/SPL/PR; **12** Scott Camazine/PR; **13** Simon Fraser/RVI, Newcastle-Upon-Tyne/SPL/PR; **14** Zig Leszczynski/AA/ES; **15** (l) John Harte, (r) Jeff Smith/Fotosmith; **16** (b) RLM, (t) HRW/SD; **17** HRW/SD; **18** (b) J. Reid/Custom Medical Stock Photo, (t) Dan Nedrelo; **19** Will & Deni/PR; **20** Sinclair Stammers/SPL/PR; **21** TF; **22** TF; **24** HRW/SD; **CHAPTER 2: 26** Georg Gerster/PR; **29** HRW/SD; **30** (bc) Dennis Kunkel/PT, (bl) SP/FOCA, (tl) Paul Hermansen/GI; **31** Marc Epstein/DRK; **33** HRW/SD; **34** SP/FOCA; **35** HRW/SD; **36** TF; **38** (l) HRW/SD, (r) G.I. Bernard/AA/ES; **43** Georg Gerster/PR; **44** Georg Gerster/PR; **46** HRW/SD; **CHAPTER 3: 48** Robert Brons/BPS/GI; **51** (b, c) HRW/SD, (cl) Dr. Dennis Kunkel/PT, (cr) Victoria Smith/HRW, (tr) M. Abbey/PR; **52** (l) SP/FOCA/HRW, (r) Manfred Kage/PA; **53** (bl) Sinclair Stammers/SPL/PR, (br) Dr. Tony Brain/SPL/PR, (t) Victoria Smith/HRW; **54** (l) Philippe/Plailly/Eurelios/SPL/PR, (r) David M. Philips/VU; **56** Michael Abbey/PR; **57** (b) Chris Bjornberg/PR, (t) John Cardmore/BPS/GI; **58** Michael Gabridge/VU; **59** Volker Steger/Christian Bardele/SPL/PR; **60** Dr. Dennis Kunkel/PT; **62** Don Fawcett/VU; **63** R. Boldender/D. Fawcett/VU; **65** Don Fawcett/VU; **67** Robert Brons/BPS/GI; **68** (b) Michael Gabridge/VU, (t) Robert Brons/BPS/GI; **70** E.R. Degginger/Color-Pic, Inc.; **71** Barry Bomzer/GI; **CHAPTER 4: 72** Dr. Dennis Kunkel/PT; **75** HRW/SD; **76** HRW/SD; **78** Victor Scocozza/GI; **79** Dr. David Scott/PT; **81** Doug Wechsler; **84** TF; **85** HRW/SD; **86** Drs. Riqiang Fu and Timothy A. Cross, National High Magnetic Field Laboratory (NHMFL); **87** Dr. Dennis Kunkel/PT; **88** Dr. Dennis Kunkel/PT; **90** Ward's Natural Science; **CHAPTER 5: 92** Steve Gettle/ENP Images; **99** Dr. E. R. Degginger/Color-Pic; **101** C. Milkins/OSF/AA/ES; **103** Alan & Linda Detrick/PR; **108** John Cowell/GH; **109** AFP/Corbis; **111** Steve Gettle/ENP Images; **112** (b) BPA/SS/PR, (t) Steve Gettle/ENP Images; **114** HRW/SD; **115** HRW/SD; **CHAPTER 6: 116** Professors P.M. Motta & J. van Blerkom/SPL/PR; **118** (t) HRW/SD, (b) Stephen J. Krasemann/DRK; **119** Institut Pasteur/CNRI/PT; **120** (b) David M. Phillips/VU, (l, r) PhotoDisc; **121** (bc) Dennis

Kunkel/PT, (bl) Evelyn Gallaroo/PA, HRW/SD, (t) David M. Philips/VU; **122** (bl) Custom Medical Stock Photo, (br) TF, (tl) CNRI/SPL/PR; **123** ISM/PT; **127** Nancy Kedersha/PR; **129** Ariel Skelley/The Stock Market; **130** John D. Cunningham/VU; **131** (bl, c, br) John D. Cunningham/VU, (tr) David M. Philips/VU; **132** (cr) HRW/SD, (tl) R. Calentine/VU; **133** Professors P.M. Motta & J. van Blerkom/SPL/PR; **134** Professors P.M. Motta & J. van Blerkom/SPL/PR; **136** (l) HRW/SD, John D. Cunningham/VU; **137** HRW/SD; **138** (b) Dr. Tony Brian/SPL/PR, (t) Moredun Scientific/PR; **139** Jose Luis Pelaez, Inc./Corbis; **140-141** (bkgd) Dee Breger/PR; **140** Roger Tully/GI; **141** (b) GH, (c) Francoise Sauze/SPL/PR, (c) W & D Downey/Hulton-Getty/GI; **CHAPTER 7: 142** Yorgos Nikas/GI; **148** Norbert Wu/PA; **149** Professor P.M. Motta, G. Macchiarelli, S.A, Nottola/SPL/PR; **150** BPA/SS/PR; **151** (b) HRW/SD, (t), (l), (tr) PhotoDisc; **152** Robert Calentine/VU; **153** (b) L.J. Vitt, (c), (t) PhotoDisc; **154** Alan & Linda Detrick/PR; **155** Yorgos Nikas/GI; **156** Yorgos Nikas/GI; **157** (b) David M. Phillips/VU, (c) Jason Burns/Ace/PT, (cl) Petit Format/Nestle/PR, (cr) Yorgos Nikas/GI, (t) PhotoDisc; **158** (tc) HRW/SD, (tr) Paul A. Zahl/PR; **159** HRW/SD; **CHAPTER 8: 160** Gary J. Benson; **162** Culver Pictures, Inc.; **165** Nigel Cattlin/Holt Studios International/PR; **167** (c) Barry Runk/GH, (t) Christian Grzimek/Okapia/PR; **168** (b) Jeff Noble/International Stock Photography, (tl) Picture Press/Corbis, (tr) Mark S. Skalny/VU; **169** Angelina Lax/PR; **173** Image Shop/PT; **174** (cr) Lynn M. Stone/DRK, (cl) HRW/SD, (t) Ken Lax/HRW Photo; **175** (b) G.R. 'Dick' Roberts, (t) GI; **176** Ben Klaffke; **177** Arthur Tilley/GI; **178** Yoav Levy/PT; **179** (l) Dan Guravich/PR, (r) Tom Walker/GI, (t) Darrell Gulin/Corbis; **180** Meckes/Ottawa/PR; **181** Meckes/Ottawa/PR; **183** Gary J. Benson; **184** Gary J. Benson; **CHAPTER 9: 188** GI; **190** Alfred Pasieka/PA; **193** Phil Degginger/Color-Pic, Inc.; **195** Victoria Smith/HRW; **196** (b, t) SS/PR, (c) SPL/PR; **199** Dr. Gopal Murti/SPL/PR; **201** GI; 202 GI; **204** (c) SP/FOCA, (r) HRW/SD; **205** HRW/SD; **CHAPTER 10: 206** Runk Schoenberger/GH; **209** Jacana/PR; **210** (b) HRW/SD, (t) O.L. Miller, B .R. Beatty, D.W. Fawcett/VU; **214** John Langford/HRW; **216** Topham /The Image Works; **220** Photo Courtesy of the Gene Therapy Research Team and The University of Pennsylvania, The University of Florida Gainesville, and Cornell University; **221** Runk Schoenberger/GH; **222** Runk Schoenberger/GH; **224** HRW/SD; **225** HRW/SD; **CHAPTER 11: 226** Ted Horowitz/The Stock Market; **228** (c) Biophoto Associates/PR, (l) James Gerholdt/PA, (r) David M. Phillips/VU; **232** HRW/SD; **233** Charles Thatcher/GI; **234** Thomas Bishop/Custom Medical Stock Photo; **235** HRW Photo/Digital Vision Ltd.; **237** S. Fraser/PR, Inc.; **238** Agricultural Research Service, USDA; **241** Roslin Institute/PT; **243** Ted Horowitz/The Stock Market; **244** Ted Horowitz/The Stock Market; **246** Mark Joseph/GI; **248-249** (bkgd) Gregory G. Dimijian/PR; **248** RLM; **249** (b) Philip Nayson/PR, (c) Alejandro Balaguer/GI, (t) The Granger Collection, New York; **CHAPTER 12: 250** Steve Bloom/Masterfile; **253** HRW/SD; **254** Sojourner ®, Mars Rover ® and spacecraft design and images © copyright 1996-97 California Institute of Technology. All rights reserved. Further reproduction prohibited; **257** Sydney Fox/VU; **259** Newcomb & Wergin/BPS/GI; **261** (b) Lewis Trusty/AA/ES, (tl) A.M. Siegelman/VU, (tr) M. Abbey/VU; **262** James L. Amos/Corbis; **263** Martin Wendler/PA; **265** (l) Courtesy of Hans Steur, (r) Dana Richter/VU; **266** Runk/Schoenberger/GH; **267** Eric Kamp/PT; **268** Wolfgang Bayer/BC; **269** Steve Bloom/Masterfile; **270** Steve Bloom/Masterfile; **272** HRW/SD; **273** Rich Frishman/GI; **CHAPTER 13: 274** RLM; **276** Bridgeman Art Library; **279** SS/PR; **280** HRW/SD; **281** (c) Tom & Pat Leeson/DRK, (r) Tom & Pat Leeson/DRK; **283** (bc) E.R. Degginger/AA/ES, (l) RLM; **288** N. Pecnik/VU; **289** (b) Camr/A. B. Dowsett/PR, (t) S. Camazine/PR; **290** Vireo/Academy of Natural Sciences; **291** (b) Gary Mezaros/VU, (t) Zig Leszczynski/AA/ES; **293** RLM; **294** RLM; **296** TF; **297** Brian Milne/AA/ES; **CHAPTER 14: 298** Jack Dykinga; **300** Jerry L. Ferrara/PR; **301** (bc), (bl), (tc) SP/FOCA/HRW, (br) TF, (tl) Leonare Lee Rue III/NASC/PR, (tr) GH; **303** (b) Stephen Dalton/AA/ES, (t) Jack Kelly Clark/University of California; **304** HRW/SD; **305** (c) Jeff Lepore/NASC/PR, (l) Zig Leszczynski/AA/ES, (r) Ron Kimball Photography; **306** (c) Scala/Art Resource, (t) Tim Flannery; **307** (b) Gregory Ochocki/PR, (br) RLM; **311** Jack Dykinga; **312** (c) RLM, (t) Jack Dykinga; **314** (bc) BIOS (Klein/Hubert)/PA, (bl) RLM, (br) Barry Runk/GH, (tc) BIOS (Klein/Hubert)/PA, (tl) John Gilmore, (tr) Runk/Schoenberger/GH; **315** HRW/SD; **316** (c) Art Wolfe/GI; **316-317** (bkgd) Carr Clifton/MP; **317** (b) Stephen J. Krasemann/DRK, (c) Art Wolf/GI, (t) AA/ES; **CHAPTER 15: 318** Anup Shah/AA/ES; **320** (c) Peter Unmack, (l) Tom McHugh/Steinhare Aquarium/PR; **320** (r) Richard Laird/GI, (c) Manfred Danegger/PR; **321** (l) Manfred Mehlig/GI, (r) Jack Wilburn/AA/ES; **322** Corbis; **323** Shin Yoshino/MP; **324** (l) Tom McHugh/PR, (r) Johnny Johnson/DRK; **325** Mark Lotz/Florida Fish & Wildlife Conservation Commission.; **326** David Scharf/PA; **327** Michelle Bridwell/HRW; **328** (b) R.H. Virdee/GH, (t) Stan Osolinski; **329** HRW/SD; **330** UPI/Corbis-Bettmann; **331** HRW/SD; **333** Anup Shah/AA/ES; **334** Anup Shah/AA/ES; **336** HRW/SD; **337** HRW/SD; **CHAPTER 16: 338** Marc Chamberlain/GI; **340** Merlin D. Tuttle/Bat Conservation International/PR; **341** (l) Patti Murray/AA/ES, (r) Runk/Schoenberger/GH, (t) Jim Strawser/GH; **342** (b) Randal Alhadeff/HRW, (tl), (tr) Jim Tuten/AA/ES; **343** (b), (c) Richard Kolar/AA/ES, (t) Tom Bean/DRK; **344** HRW/SD; **346** Alfred Pasieka/SPL/PR, Inc; **350** Grant Heilman/GH; **354** © Peter Bisset/GI; **355** Marc Chamberlain/GI; **356** Marc Chamberlain/GI; **358** HRW/SD; **359** HRW/SD; **CHAPTER 17: 360** Jack Dykinga; **362** Anthony Mercieca/PR; **363** (b) Patti Murray/AA/ES, (t) Bill Beatty/VU; **364** (b) RLM, (t) Hans Pfletschinger/PA; **368** Anne Wertheim/AA/ES; **370** (b) Runk/Schoenberger/GH, (inset) RLM; **371** D. Cavagnaro/DRK; **373** (b) Art Wolfe/GI, (t) Tom Bean/GI; **374** (b) Victoria Smith/HRW, (c) Arthur C. Smith, III/GH, (t) Wolfgang Kaehler; **375** (tr) © Mark Newman/SuperStock, (cr) Doug Wechsler/AA/ES, (br) Biosphere Photographie/AA/ES; **377** (tr) © David Muench/CORBIS, (cr) © Randy Wells/GI; **378** (tl) © Jeff Hunter/Getty Images, (c) Manfred Kage/PA, (bl) ©2006 Norbert Wu; **379** Jack Dykinga; **380** Jack Dykinga; **382** (b) Victoria Smith/HRW, (t) Manfred Kage/PA; **CHAPTER 18: 384** Johnny Johnson/DRK; **386** Phil Degginger/Color-Pic, Inc.; **387** NOAA; **390** John S. Lough/VU; **392** D. Halleux/ PA; **394** Chad Ehlers/GI; **396** (l) Lester Lefkowitz/Corbis, (r) E. R. Degginger/Color-Pic, Inc.; **397** (t) Tony Freeman/PhotoEdit, (cr) © Dave G. Houser/CORBIS; **398** (b) Larry Kolvooro/HRW, (c) Courtesy of Chesapeake Bay Foundation, (t) SuperStock; **399** (b) Randy Wells, (t) Seth Resnick; **400** (c) HRW/TF, (r) HRW/SD, (l) Steve Smith/GI; **401** Johnny Johnson/DRK; **402** Johnny Johnson/DRK; **404** LSF/OSF/AA/ES; **405** Tom Tracy/GI; **CHAPTER 19: 406** (b) Enrico Ferorelli/Masterfile, (c) Stuart Westmorland/GI, (t) VCG/GI; **407** (bl) Michael P. Moore, (br) Scott Camazine/PR; **408** (c) Jeff Corwin/GI; **408-409** (bkgd) Ray Pfortner/PA; **409** (b) NIBSC/SPL/PR, (c) James King-Holmes/SPL/PR, (t) Corbis; **410** Larry Ulrich/DRK; **412** (bl) Gary D. Gaugler/PR, (br) Wolfgang Baumeister/PR, (c) Dr. Dennis Kunkel/PT, (tc) Adam Jones/PR, (tl) Myron Jay Dorf/Corbis, (tr) Rod Planck/PR; **414** (l) Andrew Syred/SPL/PR, (r) TF; **415** Corale L. Brierley/VU; **416** Micheal Fogdan/AA; **418** (l) Rakosky/Custom Medical Stock Photo, (r) Robert Calentine/VU; **419** N. Pecnik/VU; **420** (b) Gerald & Buff Corsi/VU, (t) D. Despommier/PR; **421** Ken Wagner/PT; **422** (l) Rod Planck Photography, (r) Tom Bean/DRK; **423** (b) Barbara Gerlach/DRK, (c) Jim Steinberg/PR, (tl) Wayne Lynch/DRK, (tr) John Gerlach/DRK; **424** (l) David Scharf, (r) Flip Nicklin/MP, (t) Patrick Ward/Masterfile; **425** Fred Bavendam/PA; **426** James McCullagh/VU; **427** Larry Ulrich/DRK; **428** (b) Hal Harrison/GH, (t) Larry Ulrich/DRK; **430** (c) Dr. E.R. Degginger/Color-Pic Inc., (l) Gerry Ellis/ENP Images, (r) Photodisc; **431** Fred Bruemmer/DRK; **CHAPTER 20: 432** S. Lowry/Univ. Ulster/GI; **434** Barry Dowsett/SPL/PR; **435** (r) E.O.S./Gelderglom/PR, (c) Dr. O. Bradfute/PA, (l) NIBSC/SPL/PR; **438** Oliver Meckes/Gelderblom/PR; **440** Runk/Schoenberger/GH; **441** Germain Rey/Liaison Agency; **442** Fred Hossler/VU; **443** (c) G. Shih and R. Kessel/VU, (l) Fran Heyl Associates, (r) John D. Cunningham/VU; **444** Rakosky/Custom Medical Stock Photo; **445** (br) Luis M. Alvarez/AP/Wide World Photos, (tl) C.P. Vance/VU, (tr) E. Weber/VU; **446** Fred Hossler/VU; **447** Cecil H. Fox/Science Source/PR; **448** Michael Abbey/PR; **449** Christine Case/VU; **450** (b) Richard T. Nowitz/PT, (t) TF; **451** S. Lowry/Univ. Ulster/GI; **452** (b) Fred Hossler/VU, (t) S. Lowry/Univ. Ulster/GI; **454** HRW/SD; **455** Michael Rosenfeld/GI; **456** (bc) Gilbert S. Grant/PR, (bl) Judd Cooney/PT, (br) Scott Camazine/Science Source/PR, (t) M. Weber/Index Stock Imagery, Inc.; **457** Maximilian Stock/SPL/PR; **CHAPTER 21: 458** RLM; **460** George J. Wilder/VU; **461** (b) Fran Heyl Associates, (c) M. Abbey/PR, (t) Runk Schoenberger/GH; **462** Microfield Scientific Ltd./SPL/PR; **464** Robert Brons/GI; **465** (c) Gary R. Robinson/VU, (l) Mark E. Gibson/VU, (r) Darryl Torckler/GI; **466** Kent Wood; **467** Roland Birke/PA; **468** Manfred Kage/PA; **469** RLM; **470** (b) Bill Beatty/AA/ES, (t) Cabisco/VU; **471** (b) Tetsu Yamazaki/International Stock Photography, (t) Centers for Disease Control; **474** Dufka/Archive Photos; **475** RLM; **476** RLM; **477** (l) Courtesy of Richard Triemer, Rutgers University, (r) M. Abbey/VU; **478** Robert Brons/BPS/GI; **479** HRW/SD; **CHAPTER 22: 480** Jeff Lepore/PR; **482** Michael Gadomski/AA/ES; **483** (b) Courtesy of Marshall Lyon/CDC, (c) Mike Eichelberger/VU, (inset) HRW/SD, (t) Noble Proctor/PR; **484** (b) Tom Bean/DRK, (t) Stephen Dalton/NASC/PR; **486** (b) E. R. Degginger/Color-Pic, Inc., (t) Barry Runk/GH; **488** (t),Runk/Schoenberger/GH, (b) Cabisco/VU; **489** Marty Cordano/DRK; **490** BPA/SS/PR; **491** (b) Runk/Schoenberger/GH, (tl) V. Ahmadjian/VU, (tr) Doug Wechsler/AA/ES; **492** D. Cavagnaro/DRK; **493** Jeff Lepore/PR; **494** Jeff Lepore/PR; **496** J. Rorsdyke/SPL/PR; **498-499** (bkgd) George Lepp/GI; **498** Corbis; **499** (b) Howard Buffett/GH, (c) Keith Weller/Agricultural Research Service, USDA, (t) The Granger Collection, New York; **CHAPTER 23: 500** Adam Jones/Pictor International ltd./PictureQuest; **502** Michael J. Doolittle/The Image Works; **503** (b) HRW/SD, (tl) L. Mellichamp/VU, (tr) HRW/SD; **504** RLM; **505** RLM; **506** (l) Michael P. Gadomski/PR, (r) Jane Grushow/GH; **508** Uniphoto Picture Agency; **509** (c) Runk/Schoenberger/GH, (l) Ed Reschke/PA, (r) Runk/Schoenberger/GH; **510** Heather Angel/Natural Visions; **511** (bc) Ed Reschke/PA, (bl) Runk/Schoenberger/GH, (br) RLM, (tl) Dwight Kuhn, (tr) Michael Ederegger/DRK; **512** RLM; **513** (bc) Runk/Schoenberger/GH, (bl) Patti Murray/AA/ES, (br) Gerald & Buff Corsi/VU, (t) HRW/SD; **514** Guy Marche/GI; **515** (l) Alan & Linda Detrick/PR, (r) William H. Mullins/NASC/PR; **517** (t, l, b) TF, (tr) Peter Holden/VU; **518** (b) Alan Pitcairn from GH, Inc., (t) Photodisc; **519** (b) TF/HRW. (t) Grant Heilman from GH, Inc.; **520** (bc) Bob Thomas/GI, (bl) Robert Kusel/GI, (br) Howard Kingsnorth/GI, (cl) Nick Gunderson/GI, (cr) James Rowan/GI, (t) Paolo Koch/PR; **521** (c) Patti Murray/AA/ES, (l) RLM, (r) Larry Lefever/GH; **522** (l) Inga Spence/VU, (r) Ric Ergenbright/Corbis; **523** Adam Jones/Pictor International, ltd./PictureQuest; **524** (b) David M. Schleser/PR, (t) Adam Jones/Pictor International ltd./PictureQuest; **526** HRW/SD; **527** Bill Bachman/PR; **CHAPTER 24: 528** S. Lowry/University of Ulster/GI; **530** (c) E. R. Degginger/Color-Pic, Inc., (l), (r) RLM; **532** (b) E. R. Degginger/Color-Pic, Inc., (t) Ed Reschke/PA; **534** RLM; **536** (b) Dr. E.R. Degginger/AA/ES, (t) Runk/Schoenberger/GH; **538** HRW/SD; **539** (b) Eye of Science/PR, (t) Merlin D. Tuttle/Bat Conservation International; **542** HRW/SD; **544** (l, r) Wiliam D. Adams; **545** S. Lowry/University of Ulster/GI; **546** (c) RLM, (t) S. Lowry/University of Ulster/GI; **548** RLM; **CHAPTER 25: 550** Art Wolfe/PR; **553** (b) Ed Reschke/PA, (tl) Lefever/Grushow/GH, (tr) E.R. Degginger/Color-Pic, Inc.; **554** George J. Wilder/VU; **555** (bc) Cabisco/VU, (br) Herb Charles Ohlmeyer/Fran Heyl Assocs., (t) R. Calentine/VU; **556** (bl, br) Runk/Schoenberger/GH; **557** (c) Runk/Schoenberger/GH, (t) John D. Cunningham/VU; **558** RLM; **559** Dwight Kuhn/DRK; **562** Walter H. Hodge/PA; **563** Dominic Oldershaw; **565** Art Wolfe/PR; **566** (b) Scott Camazine/PR, (t) Art Wolfe/PR; **568** HRW/SD; **569** Geoff Tompkinson/SPL/PR; **CHAPTER 26: 570** (bkgd) Arthur C. Smith III/GH; **573** D. Cavagnaro/VU; **574** (cl) RLM, (cr) R.F. Evert, (l) Robert P. Comport/AA/ES, (r) TF; **576** Darrell Gulin/DRK; **577** R.F. Evert; **578** © Joel Sartore/National Geographic/GI; **581** HRW/SD; **582** (c) R. Calentine/VU, (l) Barry L. Runk/GH, (r) Ed Reschke/PA; **583** Runk/Schoenberger/GH; **584** Runk/Schoenberger/GH; **585** Arthur C. Smith III/GH; **586** (b) Walter Chandoha, (t) Arthur C. Smith III/GH; **588** Michael P. Gadomski/PR; **589** (b) Michael Rosenfeld/GI, (t) HRW/SD; **590-591** (bkgd) Gerry Ellis/MP; **590** Maria Stenzel/National Geographic Society Image Collection; **591** (c) Christopher Burk/GI, (t) GI; **CHAPTER 27: 592** Bill Kamin/VU; **594** (l) PhotoDisc, (r) TF; **595** (bl) Manfred Kage/PA, (br) Jeff Smith/Fotosmith, (t) Tim Fitzharris/MP; **596** (b)

Carolina Biological Supply/PT, (t) David Philips/The Population Council/PR; **597** (b) Jeff Rotman/PA, (t) Quest/SPL/PR; **598** (b) Don Riepe/PA, (t) Mary Beth Angelo/PR; **599** (b) TF, (t) W.J. Weber/VU; **601** (l) Dr. E.R. Degginger/Color-Pic, Inc., (r) Don Riepe/PA; **603** Stephen Dalton/AA/ES; **605** (b) E.R. Degginger/Color-Pic, Inc., (t) Mitch Kezar/GI; **607** Runk/Schoenberger/GH; **608** SP/FOCA; **609** (b) Brandon D. Cole, (t) Dr. Paul Zahl/PR; **610** (b) MP, (t) Franklin J. Viola; **611** Bill Kamin/VU; **612** Bill Kamin/VU; **614** Jeff Smith/Fotosmith; **615** Agricultural Research Service, USDA; **CHAPTER 28: 616** J.W. Mowbray/NASC/PR; **618** Jeffrey L. Rotman/PA; **620** (bc) Manfred Kage/PA, (bl) Cabisco/VU, (br) RLM, (cr) Runk/Schoenberger/GH, (c) Neil G. McDaniel/NASC/PR, (cl) David J. Wrobel/Biological Photo Service, (t) TF; **624** Runk/Schoenberger/GH; **625** (b) RLM, (t) Runk/Schoenberger/GH; **626** (l) David J. Wrobel, (r) Bryan Mullennix/GI; **627** (bl) Fred Bavendam, (br) Herv Chaumeton/PR, (t) Fred Bavendam/MP; **628** Paul Osmond/Deep Sea Images; **629** Hal Beral/VU; **631** RLM; **632** Sinclair Stammers/SS/PR; **633** (b) TF, (t) SPL/PR; **634** RLM; **635** J.W. Mowbray/NASC/PR; **636** J.W. Mowbray/NASC/PR; **638** (b) G. Shih-R. Kessel/VU, (c) HRW/SD, (t) Carolina Biological Supply/PT; **CHAPTER 29: 640** John Seagrim/Masterfile; **645** (b) HRW/SD, (t) Robert Maier/AA/ES; **646** (l) J.H. Robinson/AA/ES, (r) MC Chamberlain/DRK; **648** (b) © CuboImages srl/Alamy Photos, (t) George H. Harrison/GH; **649** (c) Joyce & Frank Burek/AA/ES, (l) Runk/Schoenberger/GH, (r) Douglas Faulkner/NASC/PR; **651** (l) Ed Reschke/PA, (r) Hal Beral/VU; **652** HRW/SD; **653** John Anderson/AA/ES; **654** (b) J.P. Ferraro/Jacana/PR, (t) Barbara Gerlach/DRK; **656** (b) St. Barholomew's Hospital/SPL/PR, (t) RLM; **657** John Seagrim/Masterfile; **658** John Seagrim/Masterfile; **660** Jeff Smith/Fotosmith; **661** TF; **CHAPTER 30: 662** Mervyn Rees/GI; **664** (l) Raymond A. Mendez/AA/ES, (r) Joe McDonald/AA/ES; **666** (b) Dr. Jeremy Burgess/SPL/PR, (c) Art Wolfe/GI, (t) Dr. E.R. Degginger/Color-Pic, Inc.; **667** (b) HRW/SD, (t) Ed Reschke/PA; **668** RLM; **669** Volker Steger/PR; **670** RLM; **671** RLM; **672** (b) Scott Camazine/NASC/PR, (l) David Scharf/PA; **676** R. Al Simpson/VU; **678** Stephen Dalton/NASC/PR; **679** (bl) Marty Cordano/DRK, (br) Karl Weidmann/FOCI Image Library/PR, (t) RLM; **681** Flip Nicklin/MP; **683** Mervyn Rees/GI; **684** Mervyn Rees/GI; **686** RLM; **687** HRW/SD; **688** (b) GH, (c) Henry Ausloos/AA/ES, (t) Strawser/GH; **689** Scott Bauer/Agricultural Research Service, USDA; **CHAPTER 31: 690** Marilyn Kazmers/PA; **693** Phil Degginger/Geological Enterprises/Color-Pic, Inc.; **694** (bl) Dr. E. R. Degginger/Color-Pic, Inc., (br) Andrew J. Martinez/PR, (t) TF; **695** Joyce & Frank Burek/AA/ES; **696** Andrew J. Martinez/PR; **697** (b) Fred Bavendam, (t) Fred Winner/Jacana/PR; **698** (b) Kelvin Aitken/PA, (tl) D. Shale/OSF/AA/ES, (tr) Bruce Watkins/AA/ES; **699** David Thielk/HRW; **701** Gregory G. Dimijian, M.D./PR; **702** (b) HRW/SD, (t) Grant Heilman/GH; **703** Marilyn Kazmers/PA; **704** Marilyn Kazmers/PA; **706** Gregory Brown/AA/ES; **708-709** (bkgd) Nigel J.H. Smith/AA/ES; **708** Tim Davis/GI; **709** (b) K. Schafer and M. Hill/GI, (c) Valmik Thapar/PA, (t) Hulton/GI; **CHAPTER 32: 710** Tom Bean/DRK; **712** Robert Maier/AA/ES; **713** Jim Tuten/AA/ES; **714** John Cancalosi/Natural Selection; **717** (b) E.R. Degginger/AA/ES, (l) Tom McHugh/PR, (r) M. Gibbs/OSF/AA/ES; **720** (bl) William Leonard/DRK, (br) Zig Leszczynski/AA/ES, (tl) Carl R. Sams II/PA, (tr) James P. Rowan/DRK; **724** TF/HRW Photo; **725** (b) O. Louis Mazzatenta/National Geographic Society Image Collection, (t) S. Nielsen/DRK; **726** (bc) David Welling/AA/ES, (bl) E. R. Degginger/Color-Pic, Inc., (br) Stephen J. Krasemann/DRK, (t) Rob & Ann Simpson/VU; **727** (b) Tom J. Ulrich/VU, (c) Joe McDonald/VU, (tl) C.C. Lockwood/DRK, (tr) M. H. Sharp/PR; **728** Dr. E.R. Degginger/Color-Pic, Inc.; **729** Mauro Fermariello/PR; **730** (c) Tom McHugh/PR, (l) D. Cavagmaro/VU, (r) Tom McHugh/PR; **731** Frans Lanting/MP; **733** TF; **734** (l) Gordon Gahan/National Geographic Society Image Collection, (r) David L. Brill/Atlanta, artifact credit; National Museum of Kenya, Nairobi; **735** Educational Images/Custom Medical Stock Photo; **737** Douglas Mazonowicz/Art Resource, NY; **738** John Reader/SPL/PR; **739** Tom Bean/DRK; **740** Tom Bean/DRK; **743** Kenneth Garrett/National Geographic Society Image Collection; **CHAPTER 33: 744** Stephen Dalton/OSF/AA/ES; **746** (c) Fred Bavendam, (l) Bill Kamin/VU, (r) Kelvin Aitken/PA; **750** (b) Stuart Westmoreland/GI, (t) George Bernard/AA/ES; **751** (b) Hans Reinhard/Okapia/NASC/PR, (c), (l) Tom McHugh/NASC/PR; **752** (c) Kelvin Aitken/PA, (l inset) Doug Perrine/PA, (r inset) Wayne & Karen Brown; **754** John G. Shedd Aquarium; **756** HRW/SD; **757** (b) Tom McHugh/PR, (t) Richard Herrmann Photography; **758** R. Lindholm/VU; **761** RLM; **762** John Mitchell/PR; **764** (b) Juan M. Renjifo/AA/ES, (t) Zig Leszczynski/AA/ES; **765** Stephen Dalton/OSF/AA/ES; **766** Stephen Dalton/OSF/AA/ES; **768** (b) RLM, (t) HRW/SD; **769** T.A. Wiewandt/DRK; **CHAPTER 34: 770** Dr. E.R. Degginger/AA/ES; **772** Joe McDonald/AA/ES; **773** G.C. Kelley/PR; **774** (b) HRW/SD, (t), (l), (tr) Mark A. Chappell/AA/ES; **777** (l) Zig Leszczynski/AA/ES, (r) Michael Fogden/DRK; **778** (l) Tom McHugh/PR, (r) Roland Seitre/PA; **779** Michael Fogden/DRK; **780** Dr. E. R. Degginger/Color-Pic, Inc.; **782** (b) SP/FOCA, (c) Tom Walker/GI, (t) DR & TL Schrichte/GI; **783** (b) Zig Leszczynski/AA/ES, (tl) Fred Whitehead/AA/ES, (tr) David Austen/GI; **784** Richard During/GI; **787** Keith Scholey/Masterfile; **788** Alan & Sandy Carey/PR; **790** Tim Davis/GI; **793** Dr. E.R. Degginger/AA/ES; **794** Dr. E.R. Degginger/AA/ES; **796** Dr. E. R. Degginger/Color-Pic, Inc.; **797** Jeff Smith/Fotosmith; **CHAPTER 35**: **798** Darrell Bulin/GI; **800** Dan Guravich/PR; **801** (b) HRW/SD, (c) J. Sneesby/B. Wilkins/GI, (t) Stanley Breeden/DRK; **802** (b) SP/FOCA, (tl) Carl R. Sams II/PA, (tr) RLM; **804** Kevin Schafer; **806** (r) Tom McHugh/PR, (l), Jeff Foott; **807** (b) James D. Watt/Masterfile, (t) Joe McDonald/AA/ES; **808** (c) Darrell Gulin/GI, (b) Tom McHugh/PR, (t) Jean Phillippe Varin/NASC/PR; **809** (b) Douglas Faulkner/PR, (t) Dr. E.R. Degginger/AA/ES, (c) John Cancalosi/PA; **810** (l) Robert Maier/AA/ES, (r) Fred Habegger/GH; **812** (b) Sturgis McKeever/PR, (c) Art Wolfe/GI, (t) Rod Planck/PR; **813** (b) Norbert Wu/Masterfile, (c) Leo Keeler/AA/ES, (t) Jeremy Woodhouse/DRK; **814** (b) Fred Felleman/GI, (c) M. Harvey/DRK, (t) Stephen J. Krasemann/DRK; **815** (b) John Downer/Masterfile, (c) Douglas Faulkner/PR, (r) Thomas Kitchin/Natural Selection; **817** Darrell Gulin/GI; **818** Darrell Gulin/GI; **820** Ed Reschke/PA; **821** (bl) Flip Nicklin/MP, (br) Phil Schermeister/Corbis, (t) Arthur C. Smith III/GH; **CHAPTER 36: 822** J. Eastcott & Y. Momatiuk/Masterfile; **824** (l) Klaus Uhlenhut/AA/ES, (r) Neil McIntyre/Masterfile; **825** (b) Tim Davis/PR, (t) Robin Brandt/Natural Selection; **826** S. J. Krasemann/PR; **827** Omikron/Science Source/PR; **828** HRW/SD; **829** Thomas McAvoy/LIFE Magazine, © TIME Inc.; **830** Jeff Lepore/PR; **831** (bl) N. Nosing/AA/ES, (br) Gary W. Griffen/AA/ES, (tc) Patti Murray/AA/ES, (tl) Stephen J. Krasemann/PR, (tr) William H. Beatty/VU; **832** © David M. Barron/oxygengroup; **833** (b) Samuel R. Maglione/PR, (tl) Charles Palek/AA/ES, (tr) Francois Gohier/PR; **835** (l) Hans Reinhard/BC, (r) J.C. Carton/BC; **836** Martin Harvey/The Wildlife Collection; **837** J. Eastcott & Y. Momatiuk/Masterfile; **838** J. Eastcott & Y. Momatiuk/Masterfile; **839** (b) Robin Brandt/Natural Selection, Thomas/McAvoy/LIFE Magazine, © TIME Inc., (tl) Francois Gohier/PR; **840** HRW/SD; **841** TF; **842-843** (bkgd) GW Willis/BPS/GI; **842** Lester Lefkowitz/The Stock Market; **843** (b) Richard Nowitz/PT; **843** (c) Jim Cummings/GI, (t), (l), (tr) The Granger Collection, New York; **CHAPTER 37: 844** Telegraph Photo Library/GI; **846** (br) Carolina Biological Supply/PT, (bl) G. W. Willis/GI, (tr) John D. Cunningham/VU, (cr) Eric Grave/SS/PR, (tl) Ed Reschke/PA; **847** TF; **852** Andrew Syred/GI; **853** (bl) Coronado Frank Wing, (br) Caroline Wood/International Stock Photography, (tl, tc) Dr. M. Klein/PA, (tr) Yoav Levy/PT; **860** TF; **861** SP/FOCA; **862** Telegraph Colour Library/GI; **863** TF; **864** (b) James Stevenson/SS/PR, (t) Dr. P. Marazzi/SS/PR; **865** Telegraph Photo Library/GI; **866** (b) Dr. P. Marazzi/SPL/PR, (t) Telegraph Photo Library/GI; **868** TF; **CHAPTER 38: 870** Uniphoto Picture Agency; **873** Ed Reschke/PA; **874** John D. Cunningham/VU; **875** TF; **876** (b) Uniphoto Picture Agency, (t) Merlin D. Tuttle/Bat Conservation International; **877** (c) Don Fawcett/PR, (l) David M. Phillips/VU, (r) Robert Caughey/VU; **878** Veronica Burmeister/VU; **878** (b) Dr. Jurgen Scriba/PR; **882** TF; **883** John Feingersh/Tom Stack & Associates; **886** (b) Brian Seitz/PR, (t) Petit Format/Nestle/SS/PR; **889** HRW/SD; **890** (b) James Stevenson/SPL/PR, (t) A. Glauberman/SS/PR; **891** Uniphoto Picture Agency; **892** Uniphoto Picture Agency; **894** Yoav Levy/PT; **895** Geoff Tompkinson/SPL/PR; **896** (tc) VU, (tr) ©Martin M. Rotker/PR; **897** (tr) S. O'Brien/Custom Medical Stock Photo, (bl) ©Image 100/CORBIS; **CHAPTER 39: 898** Meckes/Ottawa/PR; **900** Michelle Bridwell/HRW; **903** HRW; **904** SP/FOCA; **905** (b) Davies & Starr, Inc./GI, (t) Russell Diane/HRW; **908** © Stephen Ferry/Liaison/Getty Images; **910** CNRI/SPL/PR; **911** HRW/SD; **914** Runk/Schoenberger/GH; **915** TF; **917** Meckes/Ottawa/PR; **918** Meckes/Ottawa/PR; **920** TF; **CHAPTER 40: 922** Dr. Dennis Kunkel/PT; **926** (b) Meckes/Ottawa/PR, (t) Dr. Dennis Kunkel/PT; **929** HRW/SD; **930** Mednet/PT; **932** HRW/SD; **934** Lennart Nilsson/Boehringer Ingelheim Int'l GmbH; **935** Paul Hardy/The Stock Market; **936** David M. Grossman/PR; **937** Dr. Dennis Kunkel/PT; **938** Dr. Dennis Kunkel/PT; **940** TF; **941** Mark Richards/PhotoEdit; **CHAPTER 41: 942** Oliver Meckes/Ottawa/PR; **944** Andrew Leonard/PR; **951** TF; **953** Damian Dovarganes/AP/Wide World Photos; **954** TF; **958** A. L. Blum/VU; **962** Gregory G. Dimijian/PR; **963** Michelle Bridwell/HRW Photo; **964** (b),Jim Strawser/GH, (t) Scott Camazine/PR; **966** Dr. E.R. Degginger/Color-Pic, Inc.; **967** Oliver Meckes/Ottawa/PR; **968** Oliver Meckes/Ottawa/PR; **970** Dr. David Scott/PT NYC; **CHAPTER 42: 972** Reuters/Wolfgang Rattay/GI; **974** David Young-Wolff/PhotoEdit; **976** HRW/SD; **979** The Image Works; **981** Terry Allen/International Stock Photography; **984** John Paul Kay/PA; **986** (b), (c) SP/FOCA, (t) Carolina Biological Supply/PT; **987** GI; **989** Reuters/Wolfgang Rattay/Archive Photos; **990** Reuters/Wolfgang Rattay/Archive Photos; **992** M.I. Walker/PR; **CHAPTER 43: 994** Lennart Nilsson/Albert Bonniers Förlag; **1000** David M. Philips/PR; **1003** C. Edelmann/La Villette/SS/PR; **1006** (bl, tl) Lennart Nilsson/Albert Bonniers Förlag, (tr) D. Bromhall/OSF/AA/ES, (br) Petit Format/Nestle/PR; **1007** Eurelios/PT; **1009** (bl) Cabisco/VU, (br) Frederick C. Skvara, M.D., (t) L. Moskowitz M.D./Medichrome; **1010** (b) SPL/Custom Medical Stock Photo, (t) Michelle Bridwell/HRW; **1011** Lennart Nilsson/Albert Bonniers Förlag; **1012** Lennart Nilsson/Albert Bonniers Förlag; **1016** HRW/Peter Van Steen; **1024** HRW/SD; **1025** SP/FOCA; **1026** HRW/SD; **1028** SP/FOCA; **1029** SP/FOCA; **1030** SP/FOCA; **1031** Dr. Dennis Kunkel/PT; **1036** (tl) Myron Jay Dorf/Corbis, (tc) Adam Jones/PR, (tr) Rod Planck/PR, (c) DR. Dennis Kunkel/PT, (bl) Gary D. Gaugler/PR, (br) Wolfgang Baumeister/PR; **1037** (l) Fran Heyl Associates (r) S. Lowry/Univ. Ulster/GI; **1038** (l) Corale L. Brierley/VU, (r) Kent Wood; **1039** Jeff Lepore/PR; **1040** (l) Doug Wechsler/AA/ES, (r) Ed Reschke/PA; **1041** William H. Mullins/NASC/PR; **1042** (l) Gerry Ellis/ENP Images, (r) Joyce and Frank Burek/AA; **1043** (t) E. R. Degginger, (b) Art Wolfe/ GI; **1044** RLM; **1045** (c) Richard During/GI, (t) Joe McDonald/AA, (b) Dan Guravich/PR; **1048** (cr) ©Medical-On-Line; **1049** (tl) Mauro Fermariello/Photo Researchers, Inc.; **1049** (cr) MAURO FERMARIELLO/Photo Researchers, Inc.; **1049** (bl) Dr. Jurgen Scriba/Photo Researchers, Inc.; **1050** (cr) © Underwood & Underwood/CORBIS; **1051** (c) ©ISM/J. C. Revy/Phototake; **1051** (cr) ©ISM/J. C. Revy/Phototake; **1053** (tr) © Royalty Free/CORBIS; **1054** (bl) © Ward's Natural Science; **1056** (tr) © Mark Costantini/San Francisco Chronicle/Corbis ; **1058** (bl) © Ward's Natural Science; **1059** (bl) Courtesy of Cellmark Diagnostics, Inc., Germantown, Maryland; **1061** (bl) © Ward's Natural Science; **1062** (tr) © Royalty Free/CORBIS; **1063** (bl) Sam Dudgeon/HRW; **1064** (tr) age fotostock/Tom Grill; **1068** (tr) © Christopher Griffin/Alamy Photos; **1069** (bl) Sam Dudgeon/HRW; **1069** (bc) Sam Dudgeon/HRW; **1070** (tr) © H. David Seawell/CORBIS; **1071** (cr) Denis Fagan/HRW

The Parking Lot Collision

Teacher Notes (continued from p. 1069)

drop of distilled water. If the chip sinks, it is more dense, and a drop of ZnI_2 solution should be added to increase the liquid's density. It is imperative that the liquids are thoroughly mixed after the addition of each drop and that the exact number of drops to achieve suspension are counted.

Because most types of glass are closer in density to that of the ZnI_2 solution, the analysis should begin by the student placing the glass chip into the test tube and then adding enough ZnI_2 drop by drop until the glass chip can clearly be seen moving to the top of the liquid level.

Students will most likely require assistance to derive the weighted-average equation for determining the density. It is easy to change the outcome of the case by varying the type of glass chips given to each lab group or class. Window-pane, headlight, ophthalmic lens, and other types of glass are easily obtained, stored, and reused. Because the individual glass chips used will vary in size, each lab group's results will differ. The drop size obtained from regular droppers is relatively consistent, but the calibration and use of a pipette would increase accuracy. Students should be instructed that if one drop of water in the ZnI_2 solution causes the chip to move one way, and a drop of the other liquid causes movement in the opposite direction, they are to disregard the very last drop and count the previous one as a half drop. Density values to within ±0.04 g/mL of the accepted values are readily achieved with this method.

Sample Data

For a small chip of headlight glass, around 25 drops of ZnI_2 solution and 2 drops of distilled water will be required to achieve suspension.

d = density

x = drops ZnI_2 (d = 2.73 g/mL)

y = drops distilled water (d = 0.998 g/mL)

$$d = [(x \times 2.73) + (y \times 0.998)]/(x + y)$$

$$d = [(25 \times 2.73) + (2 \times 0.998)]/27 = 2.60 \text{ g/mL}$$

Answers to Analyze and Conclude

1. Student answers may vary according to the sample they are given to analyze.
2. The glass sample may or may not be the same density as the glass in the broken headlight. Even if they are equal in density, one cannot conclude that they come from the same source.
3. Student answers may vary. If careful count of quantities (drops) is not kept, the results will be inaccurate.
4. The glass chip could be headlight glass or window glass.

The Sports Shop Theft

Teacher Notes (continued from p. 1071)

Sheets) should be consulted before use):

$NaClO_3$	d = 2.49 g/mL
$KClO_4$	d = 2.52 g/mL
NaF	d = 2.56 g/mL
$MgSO_4$	d = 2.66 g/mL
$KMnO_4$	d = 2.70 g/mL
$AgClO_4$	d = 2.80 g/mL

Sample Data

To compute the density for a layer, the following equation is used.

d = layer density

x = volume, in mL, of saturated ZnI_2 solution (d = 2.73 g/mL)

1.50 – x = volume, in mL, of distilled water (d = 0.998 g/mL)

$$d = (2.73x + (0.998)(1.50 - x))/1.50$$

Consequently, the expression for x, the volume of ZnI_2 to be used in a layer, becomes

$$x = (1.50d - 1.497)/1.732$$

Using this equation, students may determine the densities that they want for each layer, substitute each desired d value, and compute x (volume of ZnI_2 solution) and 1.50 – x (volume of water) for each layer.

Answers to Analyze and Conclude

1. Student answers may vary according to the sample they are given to analyze.
2. The glass sample may or may not be the same density as the glass from the crime scene. Even even if they are equal in density, one cannot conclude that they come from the same source.
3. Student answers may vary. If care is not taken to avoid mixing layers as the column is filled, the experiment will not produce valid results.